MOLECULAR BASIS OF NUTRITION AND AGING

A Volume in the Molecular Nutrition Series

MOLECULAR BASIS OF NUTRITION AND AGING

A Volume in the Molecular Nutrition Series

Edited by

Marco Malavolta
Translational Research Center on Nutrition and Ageing,
Scientific and Technologic Pole, INRCA, Ancona, Italy

Eugenio Mocchegiani
Translational Research Center on Nutrition and Ageing,
Scientific and Technologic Pole, INRCA, Ancona, Italy

AMSTERDAM • BOSTON • HEIDELBERG • LONDON
NEW YORK • OXFORD • PARIS • SAN DIEGO
SAN FRANCISCO • SINGAPORE • SYDNEY • TOKYO

Academic Press is an imprint of Elsevier

Academic Press is an imprint of Elsevier
125 London Wall, London EC2Y 5AS, UK
525 B Street, Suite 1800, San Diego, CA 92101-4495, USA
50 Hampshire Street, 5th Floor, Cambridge, MA 02139, USA
The Boulevard, Langford Lane, Kidlington, Oxford OX5 1GB, UK

Copyright © 2016 Elsevier Inc. All rights reserved.

No part of this publication may be reproduced or transmitted in any form or by any means, electronic or mechanical, including photocopying, recording, or any information storage and retrieval system, without permission in writing from the publisher. Details on how to seek permission, further information about the Publisher's permissions policies and our arrangements with organizations such as the Copyright Clearance Center and the Copyright Licensing Agency, can be found at our website: www.elsevier.com/permissions.

This book and the individual contributions contained in it are protected under copyright by the Publisher (other than as may be noted herein).

Notices

Knowledge and best practice in this field are constantly changing. As new research and experience broaden our understanding, changes in research methods, professional practices, or medical treatment may become necessary.

Practitioners and researchers must always rely on their own experience and knowledge in evaluating and using any information, methods, compounds, or experiments described herein. In using such information or methods they should be mindful of their own safety and the safety of others, including parties for whom they have a professional responsibility.

To the fullest extent of the law, neither the Publisher nor the authors, contributors, or editors, assume any liability for any injury and/or damage to persons or property as a matter of products liability, negligence or otherwise, or from any use or operation of any methods, products, instructions, or ideas contained in the material herein.

British Library Cataloguing-in-Publication Data
A catalogue record for this book is available from the British Library.

Library of Congress Cataloging-in-Publication Data
A catalog record for this book is available from the Library of Congress.

ISBN: 978-0-12-801816-3

For information on all Academic Press publications
visit our website at http://www.elsevier.com/

Typeset by MPS Limited, Chennai, India

Contents

List of Contributors   xiii
Introduction to the Molecular Basis of Nutrition and Aging   xix
Series Preface   xxiii
Acknowledgments   xxv

I

INTRODUCTORY ASPECTS ON AGING AND NUTRITION

1. Molecular and Cellular Basis of Aging   3
SURESH I.S. RATTAN

Introduction   3
Biological Principles of Aging   4
Occurrence, Accumulation, and Consequences of Molecular Damage   4
Homeodynamics and the Homeodynamic Space   5
Nutrition and Food for Aging Interventions   6
Nutritional Hormetins   6
Conclusions   7
Summary Points   7
References   7

2. Unraveling Stochastic Aging Processes in Mouse Liver: Dissecting Biological from Chronological Age   11
L.W.M. VAN KERKHOF, J.L.A. PENNINGS, T. GUICHELAAR, R.V. KUIPER, M.E.T. DOLLÉ AND H. VAN STEEG

Introduction   12
Pathological Parameters Are Only Partially Associated with Chronological Age   13
Intraorgan Specific Biological Phenotypes   15
Tissue-Specific Biological Phenotypes   16
Gene Expression Profiles Related to Pathological Aging Parameters   16
Gene Expression Profiles Correlating with Pathological Parameters Are Largely Specific to the Pathological Parameters   18
Future Perspectives   18

Conclusion   19
References   19

3. Nutrigenomics and Nutrigenetics: The Basis of Molecular Nutrition   21
JEAN-BENOIT CORCUFF AND AKSAM J. MERCHED

Introduction   21
Nutrigenetics of Omega-3 PUFA in CVD   22
Nutrigenetics of Omega-3 PUHA in Cancer   24
DNA Damage and Nutrients   25
Cellular Senescence and Nutrients   27
Epigenetics and Nutrients   27
Summary Points   28
References   28

4. Diet and Longevity Phenotype   31
FRANCESCO VILLA, CHIARA CARMELA SPINELLI AND ANNIBALE A. PUCA

Introduction   31
Main Text   32
Summary   37
References   38

5. Nutrition in the Elderly: General Aspects   41
EUGENIO MOCCHEGIANI

Introduction   43
Malnutrition in the Elderly: Definition and General Aspects   44
Causes of Malnutrition in the Elderly   45
Malnutrition: Possible Correction with Supplements in the Elderly   46
Nutrient-Sensing Pathways   47
Nutrient-Gene Interaction in the Elderly (Nutrigenomic Approach)   49
Conclusions and Perspectives   50
Summary Points   51
Acknowledgments   52
References   52

6. Nutrition in the Hospitalized Elderly   57
WAFAA MOSTAFA ABD-EL-GAWAD AND DOHA RASHEEDY

Introduction   57
Pathogenesis of Malnutrition in Hospitalized Elderly   58

Why Are Hospitalized Older Adults Nutritionally Vulnerable?	58
Clinical Consequences of Malnutrition in Hospital	60
Malnutrition Screening and Assessment in Hospitalized Elderly	61
Management of Hospitalization-Associated Malnutrition	64
Postdischarge Plan	67
Nutritional Issues in Special Groups of the Hospitalized Elderly Population	68
Recommendations	69
Conclusion	69
Summary	69
References	70

7. Drug–Nutrient Interactions in the Elderly 73
SULAIMAN SULTAN, ZOE HEIS AND ARSHAD JAHANGIR

Introduction	75
Types of Drug–Nutrition Interactions	75
Factors Affecting DNI in the Elderly	79
Effect of Drugs on Nutritional Status in Elderly	82
Drug Interactions with Vitamins Supplements in the Elderly	82
Drug–Herb Interactions in the Elderly	95
DNI in the Elderly: Challenges and Future Directions	95
Summary	103
References	104

8. Nutritional Biomarkers of Aging 109
ANNE SIEPELMEYER, ANTJE MICKA, ANDREAS SIMM AND JÜRGEN BERNHARDT

Introduction	109
The Antioxidant Network	110
Discussion	114
Nutritional Biomarkers of Aging	116
Conclusion	118
Summary	118
References	119

9. Food Preferences in the Elderly: Molecular Basis 121
LORENZO M. DONINI, ELEONORA POGGIOGALLE AND VALERIA DEL BALZO

Introduction: The Concept of Nutritional Frailty	121
Physiological Modifications Leading to Reduced Energy and/or Nutrient Intake	122
Modifications of Chemosensory Functions and Food Preferences in the Elderly	123
Modifications of Clinical and Nutritional Status Leading to Changes in Food Preferences	123
Other Determinants of Food Choice and Conclusions	124
Summary	124
References	125

II
MOLECULAR AND CELLULAR TARGETS

10. Telomeres, Aging, and Nutrition 129
VARINDERPAL DHILLON, CAROLINE BULL AND MICHAEL FENECH

Introduction	129
Telomeres and Aging	131
Nutrition and Telomeres	132
Knowledge Gaps	137
Conclusion	137
Summary	137
References	137

11. mTOR, Nutrition, and Aging 141
GIUSEPPE D'ANTONA

Mechanistic Target of Rapamycin	141
Nutrients and Energy Status as Upstream Regulators of mTORC1 and mTORC2	142
Downstream Targets of mTORC1 and mTORC2	147
mTOR in Senescence	149
Effects of mTORC1 in Age-Related Diseases	150
Summary	152
References	153

12. Lipid Peroxidation, Diet, and the Genotoxicology of Aging 155
PETR GRÚZ

Introduction	156
Flexible Fatty Acid Composition of Human Cells	156
Lipid Peroxidation and Its Consequences for Genome Stability	160
Protective Mechanisms Against Lipid Peroxidation	164
Resistance to Oxidative Damage in Long-Lived Species	166
Concluding Remarks	167
Summary	168
References	168

13. Accumulation of Damage Due to Lifelong Exposure to Environmental Pollution as Dietary Target in Aging 177
GABBIANELLI ROSITA, FEDELI DONATELLA AND NASUTI CINZIA

Introduction	177
Xenobiotics	178
Pyrethroids	179
Acute and Chronic Exposure	180
Early Life Exposure and Long-Term Effects	180
Biomarkers of Damage	183
Summary	185
References	185

14. Nutritional Impact on Anabolic and Catabolic Signaling   189
MIKLÓS SZÉKELY, SZILVIA SOÓS, ERIKA PÉTERVÁRI AND MÁRTA BALASKÓ

Introduction	189
Characteristic Age-Related and Diet-Induced Changes in Body Mass/Composition in Rats	191
Peripheral Signaling, Central Processing	191
Diet-Induced Changes of Body Mass/Composition	196
Conclusions	199
Acknowledgment	200
Summary	200
References	200

15. Aging and Its Dietary Modulation by FoxO1 Phosphorylation/Acetylation   205
DAE HYUN KIM, BYUNG P. YU AND HAE Y. CHUNG

Introduction	206
Altered FoxO1 Target Genes During Aging and Their Modulation by CR	207
Epigenetic Influences on FoxO1	208
Modulation of FoxO1 by Calorie Restriction	209
Conclusion	210
Summary	210
References	210

16. Epigenetic Responses to Diet in Aging   213
DIANNE FORD

Introduction	214
Epigenetic Modification of the Mammalian Genome	214
Epigenetic Alterations Associated with Aging: Possible Functional Consequences and Relationship to a Global Signature of Aging	215
The Role of Environmental Versus Genetic Factors in Shaping Aging-Related Epigenetic Drift	217
The Impact of Diet on Epigenetic Alterations Associated with Aging	217
Epigenetic Actions of Specific Dietary Components and Practices	218
Interrelationships Between Age-Associated and Diet-Induced Epigenetic Alterations	220
Potential Mechanisms Through Which Diet Could Induce Epigenetic Alterations That Protect Against Aging	221
The Functional Consequences of Stem Cell Aging and the Role of Epigenetic Alterations	222
Sirtuins as Mediators of Diet-Induced Epigenetic Alterations That Counteract Aging	222
The Utility of Epigenetic Alterations as a Nutritionally Modifiable Marker of the Aging Trajectory	223
Summary	223
References	223

17. The Controversy Around Sirtuins and Their Functions in Aging   227
YU SUN AND WEIWEI DANG

Introduction	228
Enzymatic Activities of Sirtuins	228
Sirtuins in Nonmammalian Model Organisms	228
Mammalian Sirtuins	230
Does Overexpression of Sirtuins Extend Lifespan?	232
Do Sirtuins Mediate the Longevity Effects of Calorie Restriction?	234
Is Resveratrol a Sirtuin Activator?	234
Are Sirtuins Proto-Oncogenes or Tumor Suppressors?	235
The Potential of Sirtuins as Therapeutic Targets	236
Conclusions	237
Summary	237
Acknowledgments	237
References	237

18. DNA Damage, DNA Repair, and Micronutrients in Aging   243
MARÍA MORENO-VILLANUEVA

Introduction	243
Conclusions	247
Summary	247
References	248

19. Neuroprotective Mechanisms of Dietary Phytochemicals: Implications for Successful Brain Aging   251
SERGIO DAVINELLI, GIOVANNI SCAPAGNINI, GUIDO KOVERECH, MARIA LUCA, CARMELA CALANDRA AND VITTORIO CALABRESE

Introduction	252
Oxidative Damage and Brain Aging	252
Brain Inflammation as a Form of Oxidative Stress	253
Neuroprotection of Phytochemicals in Brain Aging	254
Neuroprotective Phytochemicals and Modulation of Stress Signaling Pathways	256
Conclusions	258
Summary	259
References	259

20. Nutritional Modulation of Advanced Glycation End Products   263
MA. EUGENIA GARAY-SEVILLA, CLAUDIA LUEVANO-CONTRERAS AND KAREN CHAPMAN-NOVAKOFSKI

Introduction	263
AGEs Formation In Vivo	264
Formation of Dietary AGEs	265
Methods for Measuring AGEs	265
Absorption, Metabolism, and Elimination of dAGEs	266
Molecular Action	267
AGEs and Normal Aging	268
AGEs and Chronic Diseases Associated with Older Age	269

Dietary Interventions for dAGEs	271	Molecular Basis and Modifiable Risk Factors of Cardiovascular Disease	316
Summary	273	Nutrients, Foods and ASCVD Risk Factors	316
References	273	Functional Foods	322
		Dietary Patterns and Clinical Outcomes	324

21. miRNAs as Nutritional Targets in Aging   277

ROBIN A. McGREGOR AND DAE Y. SEO

Medical Nutrition Therapy for Management of Cardiovascular Risk Factors	326
Summary	326
References	327

Introduction	277
miRNA Biogenesis and Processing	278
Circulating miRNAs	278
Aging Associated miRNAs in Different Tissues	279
Dietary Modification of miRNAs	280
miRNAs Modulated by Protein	280
miRNAs Modulated by Carbohydrates	280
miRNAs Modulated by Dietary Fat and Fatty Acids	281
miRNA Modulation by Dietary Micronutrients	282
miRNA Modulation by Vitamin D	282
miRNA Modulation by Folate	283
miRNA Modulation by Vitamin A	283
miRNA Modulation by Vitamin C	284
miRNA Modulation by Vitamin E	284
miRNA Modulation by Dietary Minerals	284
miRNA Modulation by Flavonoids	285
miRNA Modulation by Polyphenols	285
miRNAs Modulated by Curcumin	286
miRNAs Modulated by Milk	286
Conclusion	287
Summary	287
Acknowledgments	287
References	287

24. The Influences of Dietary Sugar and Related Metabolic Disorders on Cognitive Aging and Dementia   331

SHYAM SEETHARAMAN

Introduction	331
The Influence of Dietary Sugar on Brain and Cognitive Impairment	332
Metabolic Disorders	334
Metabolic Disorders and Brain Aging	335
Dementia and Metabolic Disturbances	337
Diet and Brain Health	339
Conclusions	340
Final Remarks	340
Summary	340
References	341

25. Dietary Factors Affecting Osteoporosis and Bone Health in the Elderly   345

PAWEL GLIBOWSKI

22. Nutritional Modulators of Cellular Senescence In Vitro   293

MAURO PROVINCIALI, ELISA PIERPAOLI, FRANCESCO PIACENZA, ROBERTINA GIACCONI, LAURA COSTARELLI, ANDREA BASSO, RINA RECCHIONI, FIORELLA MARCHESELLI, DOROTHY BRAY, KHADIJA BENLHASSAN AND MARCO MALAVOLTA

Introduction	294
Cellular Senescence	294
Cellular Senescence in Aging and Age-Related Diseases	296
Strategies to Target Cellular Senescence with Therapeutical Perspective	297
Nutritional Factors and Cellular Senescence "In Vitro"	300
Comments and Conclusions	307
Summary	308
References	308

Introduction	345
Osteoporosis	345
Dietary Reference Values for Calcium	346
Food Sources of Calcium	347
Dietary Reference Values for Vitamin D	348
Food Sources of Vitamin D	349
The Effect of Dairy Products on Osteoporosis	350
Vitamin K and Bone Health	351
Seminutritional Factors Affecting Osteoporosis and Risk of Fractures	351
Protein Consumption and Bone Health	352
Future Trends	353
Summary	353
References	353

III
SYSTEM AND ORGAN TARGETS

26. The Aging Muscle and Sarcopenia: Interaction with Diet and Nutrition   355

EMANUELE MARZETTI, RICCARDO CALVANI, MATTEO TOSATO AND FRANCESCO LANDI

23. Nutrition, Diet Quality, and Cardiovascular Health   315

LI WANG, GEETA SIKAND AND NATHAN D. WONG

Introduction	315

Introduction	355
Age-Related Changes in Dietary Intake and Eating Habits	356
Nutritional Interventions Against Sarcopenia and Frailty: Preliminary Considerations	356
Protein Supplementation: A Matter of Quantity and Quality	357
Vitamin D	358

Creatine	358
Polyunsaturated Fatty Acids	358
Amino Acid Metabolites and Precursors: β-Hydroxy β-Methylbutyrate and Ornithine α-Ketoglutarate	359
Conclusion	359
Summary	359
References	360

27. Nutritional Status and Gastrointestinal Health in the Elderly   363

LUZIA VALENTINI, AMELIE KAHL AND ANN-CHRISTIN LINDENAU

Introduction	363
Enteric Nervous System	364
Taste and Smell (Chemosensation)	364
The Anorexia of Aging	364
Oropharyngeal Capacities and Swallowing	365
Esophagus, Achalasia, and Gastroesophageal Reflux Disease	365
Stomach, Gastric Emptying, Postprandial Hypotension, and Acid Secretion	366
Small Bowel, Nutrient Absorption, and Small Intestinal Permeability	366
Colon, Constipation, and Gut Microbiota	368
Conclusion	369
Summary	370
References	371

28. Can Nutritional Intervention Counteract Immunosenescence in the Elderly?   375

SARAH J. CLEMENTS AND SIMON R. CARDING

Introduction	375
Aging and the Innate Immune System	376
Aging and the Adaptive Immune System	377
Targets of Nutritional Intervention	379
Can Nutrition Impact on Immune Components Affected by Age?	380
Conclusions	386
Summary	386
References	386

29. Glucose Metabolism, Insulin, and Aging: Role of Nutrition   393

MASSIMO BOEMI, GIORGIO FURLAN AND MARIA P. LUCONI

Alpha and Beta Cell Mass	394
Insulin Secretion, Metabolism, and Clearance	394
Insulin Action	396
A Link Between Insulin and Central Nervous System Aging: Brain Insulin Resistance	398
Preventing Glucose Metabolism Alteration	400
Nutrition and Aging	402
Conclusions	403
Summary	403
References	404

30. Nutritional Status in Aging and Lung Disease   411

R. ANTONELLI INCALZI, N. SCICHILONE, S. FUSCO AND A. CORSONELLO

Introduction	412
Nutrition and Lung Disease: Pathological Pathways	412
Nutritional Assessment for Pulmonary Disease	414
Nutritional Intervention in Lung Disease	416
Conclusions	418
Summary Points	418
References	419

31. How Nutrition Affects Kidney Function in Aging   423

CHRISTINA CHRYSOHOOU, GEORGIOS A. GEORGIOPOULOS AND EKAVI N. GEORGOUSOPOULOU

Introduction	424
Renal Function in Aging	425
Polyunsaturated Fatty Acids	426
Mediterranean Type of Diet	427
Hyperuricemia and Diet	428
Polyphenols and Renal Function	429
Summary	430
References	430

32. The Role of Nutrition in Age-Related Eye Diseases   433

BAMINI GOPINATH

Age-Related Changes in the Eye	433
Degenerative Eye Diseases	434
The Relationship Between Nutrition and Healthy Ocular Structure and Function	436
Antioxidants and Age-Related Eye Diseases	436
Conclusions and Recommendations	442
Summary	442
References	443

IV

HEALTH EFFECTS OF DIETARY COMPOUNDS AND DIETARY INTERVENTIONS

33. Vitamin D Nutrient-Gene Interactions and Healthful Aging   449

MARK R. HAUSSLER, RIMPI K. SAINI, MARYA S. SABIR, CHRISTOPHER M. DUSSIK, ZAINAB KHAN, G. KERR WHITFIELD, KRISTIN P. GRIFFIN, ICHIRO KANEKO AND PETER W. JURUTKA

Vitamin D: From Nutrient to Tightly Regulated Hormone	449
The Kidney Is the Nexus of Healthful Aging	451

Mechanism of Gene Regulation by Liganded VDR	452
VDR-Mediated Control of Networks of Genes Vital for Healthful Aging	454
Conclusion and Perspectives	465
Summary	467
References	467

34. Carotenoid Supplements and Consumption: Implications for Healthy Aging — 473

KARIN LINNEWIEL-HERMONI, ESTHER PARAN AND TALYA WOLAK

Introduction	473
Structure and Function	474
Carotenoids and Oxidative Stress	474
Carotenoids and Vascular Health and Atherosclerosis	475
Effect of Carotenoids on Skin Aging	478
Carotenoids and Cancer Prevention	480
Conclusions	484
Summary	484
Acknowledgment	484
References	484

35. Mechanisms of Action of Curcumin on Aging: Nutritional and Pharmacological Applications — 491

ANA C. CARVALHO, ANDREIA C. GOMES, CRISTINA PEREIRA-WILSON AND CRISTOVAO F. LIMA

Introduction	492
Curcumin and Its Traditional Uses	493
Biochemical and Molecular Targets of Curcumin: An Overview	494
Curcumin and Aging	499
Nutritional and Pharmacological Applications of Curcumin in Aging	502
Concluding Remarks	506
Summary	506
Acknowledgments	506
References	506

36. One-Carbon Metabolism: An Unsung Hero for Healthy Aging — 513

EUNKYUNG SUH, SANG-WOON CHOI AND SIMONETTA FRISO

Introduction	513
One-Carbon Metabolism	514
Aging, Age-Associated Disease, and One-Carbon Metabolism	515
Inflammation and One-Carbon Metabolism	517
Interactions Between One-Carbon Metabolism Genes and Nutrients in Aging	518
Conclusion and Future Perspectives	520
Abbreviations	520
Summary	520
References	521

37. Iron Metabolism in Aging — 523

LAURA SILVESTRI

Introduction	524
Dietary Iron and Intestinal Iron Absorption	525
Regulation of Cellular Iron Homeostasis	525
Regulation of Systemic Iron Homeostasis	527
Deregulation of the Hepcidin-Ferroportin Axis	530
Iron Homeostasis and Hepcidin in the Elderly	532
Future Perspective	534
Summary	534
Acknowledgments	534
References	534

38. Dietary Mineral Intake (Magnesium, Calcium, and Potassium) and the Biological Processes of Aging — 537

NICOLAS CHERBUIN

Introduction	537
Dietary Intake and Biological Mechanisms Implicated in the Aging Process	538
Dietary Intake and Cardiovascular and Metabolic Effects	540
Dietary Intake and Brain Aging	543
Dietary Intake and Cognitive Aging	545
Discussion	547
Summary	547
Acknowledgments	547
References	547

39. Zinc: An Essential Trace Element for the Elderly — 551

PETER UCIECHOWSKI AND LOTHAR RINK

Introduction	551
Zinc Intake, Zinc Deficiency, and Zinc Status of the Elderly	552
Zinc and Immunosenescence	554
Zinc Supplementation in Elderly Subjects	558
Epigenetic Events, Polymorphism and Genetic Markers: Relationship to Zinc Deficiency in the Elderly	561
Conclusions	562
Summary Points	562
References	562

40. Testing the Ability of Selenium and Vitamin E to Prevent Prostate Cancer in a Large Randomized Phase III Clinical Trial: The Selenium and Vitamin E Cancer Prevention Trial — 567

BARBARA K. DUNN, ELLEN RICHMOND, DARRELL E. ANDERSON AND PETER GREENWALD

Introduction	568
Rationale for Select	571
SELECT	573
Implications and Future Directions	580
Summary	581
References	581

41. Iodine Intake and Healthy Aging 583
LEYDA CALLEJAS, SHWETHA MALLESARA AND PHILIP R. ORLANDER

Introduction	583
Sources of Iodine	584
Iodine and Its Role in the Human Body	585
Epidemiology	587
Iodine and the Human Lifespan	592
Summary	596
References	596

42. Vitamin B12 Requirements in Older Adults 599
ESMÉE L. DOETS AND LISETTE CPGM DE GROOT

Introduction	599
Vitamin B12	599
Approaches for Estimating Vitamin B12 Requirements	600
Vitamin B12 Requirements—A Factorial Approach	600
Vitamin B12 Requirements—Associations Between Vitamin B12 Intake, Health Outcomes, and Biomarkers	600
From Estimates of Requirements to Vitamin B12 Recommendations	604
Concluding Remarks	604
Summary	605
Acknowledgments	605
References	605

43. Vitamin C, Antioxidant Status, and Cardiovascular Aging 609
AMMAR W. ASHOR, MARIO SIERVO AND JOHN C. MATHERS

Introduction	610
Factors That Impair Endothelial Function with Aging	611
Antioxidant Vitamins as a Strategy to Delay Vascular Aging	613
Cardiovascular Effects of Vitamin C	614
Genetic Influences on Vitamin C Status and Impact on Cardiovascular Risk	615
Conclusions	616
Summary	617
References	617

44. Omega-3 Fatty Acids in Aging 621
NATALIA ÚBEDA, MARÍA ACHÓN AND GREGORIO VARELA-MOREIRAS

Introduction	622
Lipids, Aging, and Brain Function	623
Omega-3 Fatty Acids and Brain Health in Aging	624
Omega-3 Fatty Acids and Cardiovascular Disease	626
Effects of Omega-3 Fatty Acids on Immune Function in Normal Aging	627
Effects of Omega-3 Fatty Acids on Muscle Mass and Function in Normal Nonpathological Aging	628
Effects of Omega-3 Fatty Acids on Bone Health	628
Omega-3 and Cancer in Aging	629
Effects of Omega-3 Fatty Acids on Quality of Life and Mortality in Normal Aging	630

Final Conclusions	631
Summary	632
References	632

45. Vitamin E, Inflammatory/Immune Response, and the Elderly 637
EUGENIO MOCCHEGIANI AND MARCO MALAVOLTA

Introduction	638
Biology and Intake of Vitamin E	638
Vitamin E, Inflammatory/Immune Response, and Aging	639
Vitamin E—Gene Interactions	641
Conclusions and Perspectives	644
Summary Points	644
References	645

46. Polyphenols and Aging 649
E. PAUL CHERNIACK

Introduction	649
Overview of the Process of Aging	649
Plant Polyphenol Function and Xenobiotic Effects on the Aging Process	650
Intracellular Polyphenol Effects	651
Organ System Polyphenol Effects	652
Future Directions	653
Summary	654
References	654

47. Potential of Asian Natural Products for Health in Aging 659
BERNICE CHEUNG, MACY KWAN, RUTH CHAN, MANDY SEA AND JEAN WOO

Introduction	659
Asian Natural Products	660
Limitations of Research	672
Summary	673
References	673

48. Calorie Restriction in Humans: Impact on Human Health 677
ERIC RAVUSSIN, L. ANNE GILMORE AND LEANNE M. REDMAN

Why Calorie Restriction?	677
Mechanisms to Achieve Caloric Restriction	678
Evidence of the Benefits for Caloric Restriction in Humans	678
Potential Mechanisms for CR and More Healthful Aging	682
Psychological and Behavioral Effects of CR	686
CR and the Development of Chronic Disease	687
Could CR Increase Longevity in Humans?	688
Conclusion	689
Summary	690
Acknowledgments	690
References	690

49. Prebiotics and Probiotics in Aging Population: Effects on the Immune-Gut Microbiota Axis   693
THEA MAGRONE AND EMILIO JIRILLO

Introduction	693
The Relationship Between Gut Microbiota and Immune System in Elderly	694
Food Intake and Modulation of the Immune-Gut Microbiota Axis	698
Effects of Prebiotics and Probiotics on the Immune-Gut Microbiota Axis	698
Future Trends	700
Summary Points	701
Abbreviations	701
Acknowledgment	701
References	702

50. Vegetables and Fruit in the Prevention of Chronic Age-Related Diseases   707
KIRSTEN BRANDT

Effects of Intake of Vegetables and Fruit on Mortality and Morbidity	708
Constituents of Vegetables and Fruit That May Affect Age-Related Morbidity	711
Opportunities to Improve Intake and Effect	717
Summary	718
References	719

51. Current Nutritional Recommendations: Elderly Versus Earlier Stage of Life   723
CAROL WHAM AND MICHELLE MILLER

Introduction	724
Meeting the Nutritional Needs of Older Persons: Current Recommendations	725
Energy	725
Protein	726
Fat	729
Carbohydrate	729
Dietary Fiber	729
Water	730
Micronutrients	730
Summary	731
References	732

Index   **735**

List of Contributors

Wafaa Mostafa Abd-El-Gawad Geriatrics and Gerontology Department, Faculty of Medicine, Ain Shams University, Cairo, Egypt

María Achón Departamento de Ciencias Farmacéuticas y de la Salud, Facultad de Farmacia, Universidad CEU San Pablo, Boadilla del Monte, Madrid, Spain

Darrell E. Anderson Gray Sourcing, Inc., La Mesa, CA, USA

Ammar W. Ashor Human Nutrition Research Centre, Institute of Cellular Medicine, Newcastle University, Newcastle on Tyne, UK; College of Medicine, University of Al-Mustansiriyah, Baghdad, Iraq

Márta Balaskó Department of Pathophysiology and Gerontology, Medical School, University of Pécs, Pécs, Hungary

Andrea Basso Nutrition and Aging Centre, Scientific and Technological Pole, Italian National Institute of Health and Science on Aging (INRCA), Ancona, Italy

Khadija Benlhassan Immunoclin Corporation, Washington, DC, USA

Jürgen Bernhardt BioTeSys GmbH, Esslingen, Germany

Massimo Boemi UOC Malattie Metaboliche e Diabetologia, INRCA-IRCCS, Ancona, Italy

Kirsten Brandt Food Quality and Health Research Group, Human Nutrition Research Centre, School of Agriculture, Food and Rural Development, Newcastle University, Newcastle upon Tyne, UK

Dorothy Bray Immunoclin Corporation, Washington, DC, USA

Caroline Bull CSIRO Food and Nutrition, Adelaide, SA, Australia

Vittorio Calabrese Department of Biomedical and Biotechnological Sciences, University of Catania, Catania, Italy

Carmela Calandra Department of Medical and Surgical Sciences and Advanced Technologies, University of Catania, Catania, Italy

Leyda Callejas Division of Endocrinology, Diabetes and Metabolism, University of Texas Health Science Center, Houston, TX, USA

Riccardo Calvani Department of Geriatrics, Neurosciences and Orthopedics, Catholic University of the Sacred Heart School of Medicine, Rome, Italy

Simon R. Carding Gut Health & Food Safety Research Programme, Institute of Food Research, Norwich, UK; Norwich Medical School, University of East Anglia, Norwich, UK

Ana C. Carvalho Department of Biology, CITAB - Centre for the Research and Technology of Agro-Environmental and Biological Sciences, University of Minho, Braga, Portugal Department of Biology, CBMA - Centre of Molecular and Environmental Biology, University of Minho, Braga, Portugal

Ruth Chan Department of Medicine and Therapeutics, The Chinese University of Hong Kong, Shatin, Hong Kong Center for Nutritional Studies, The Chinese University of Hong Kong, Shatin, Hong Kong

Karen Chapman-Novakofski Department of Food Science and Human Nutrition, College of Agricultural, Consumer, and Environmental Sciences, University of Illinois, Urbana-Champaign, IL, USA

Nicolas Cherbuin Centre for Research on Ageing, Health and Wellbeing, The Australian National University, Canberra, ACT, Australia

E. Paul Cherniack Division of Geriatrics and Palliative Medicine, Miller School of Medicine, University of Miami, Bruce W. Carter Miami VA Medical Center, Miami, FL, USA

Bernice Cheung Department of Medicine and Therapeutics, The Chinese University of Hong Kong, Shatin, Hong Kong Center for Nutritional Studies, The Chinese University of Hong Kong, Shatin, Hong Kong

Sang-Woon Choi Chaum Life Center, School of Medicine, CHA University, Seoul, Korea

Christina Chrysohoou 1st Cardiology Clinic, University of Athens, Athens, Greece

Hae Y. Chung Molecular Inflammation Research Center for Aging Intervention (MRCA), College of Pharmacy, Pusan National University, Busan, Korea

Nasuti Cinzia Pharmacology Unit, School of Pharmacy, University of Camerino, Camerino, Italy

Sarah J. Clements Gut Health & Food Safety Research Programme, Institute of Food Research, Norwich, UK

Jean-Benoit Corcuff Department of Nuclear Medicine, CHU de Bordeaux, Pessac, France; Integrated Neurobiology and Nutrition, University of Bordeaux, Pessac, France

A. Corsonello Unit of Geriatric Pharmacoepidemiology, Italian National Research Center on Aging (INRCA), Cosenza, Italy

Laura Costarelli Nutrition and Aging Centre, Scientific and Technological Pole, Italian National Institute of Health and Science on Aging (INRCA), Ancona, Italy

Weiwei Dang Huffington Center on Aging, Baylor College of Medicine, Houston, TX, USA

Giuseppe D'Antona Department of Public Health, Experimental and Forensic Medicine, CRIAMS Sport Medicine Centre Voghera, University of Pavia, Pavia, Italy

Sergio Davinelli Department of Medicine and Health Sciences, University of Molise, Campobasso, Italy

Lisette CPGM de Groot Division of Human Nutrition, Wageningen University and Research Center, Wageningen, The Netherlands

Valeria del Balzo Food Science and Human Nutrition Research Unit, Medical Pathophysiology, Food Science and Endocrinology Section, Department of Experimental Medicine, Sapienza University of Rome, Rome, Italy

Varinderpal Dhillon CSIRO Food and Nutrition, Adelaide, SA, Australia

Esmée L. Doets FBR, Fresh Food and Chains, Wageningen University and Research Center, Wageningen, The Netherlands

M.E.T. Dollé Centre for Health Protection, National Institute for Public Health and the Environment (RIVM), Bilthoven, The Netherlands

Fedeli Donatella Molecular Biology Unit, School of Pharmacy, University of Camerino, Camerino, Italy.

Lorenzo M. Donini Food Science and Human Nutrition Research Unit, Medical Pathophysiology, Food Science and Endocrinology Section, Department of Experimental Medicine, Sapienza University of Rome, Rome, Italy

Barbara K. Dunn Chemopreventive Agent Development Research Group, Division of Cancer Prevention, National Cancer Institute/National Institutes of Health, Bethesda, MD, USA

Christopher M. Dussik School of Mathematical and Natural Sciences, Arizona State University, Glendale, AZ, USA

Michael Fenech CSIRO Food and Nutrition, Adelaide, SA, Australia

Dianne Ford Human Nutrition Research Centre and Institute for Cell and Molecular Biosciences, Newcastle University, Newcastle upon Tyne, UK

Simonetta Friso Department of Medicine, School of Medicine, University of Verona, Verona, Italy

Giorgio Furlan UOC Malattie Metaboliche e Diabetologia, INRCA-IRCCS, Ancona, Italy

S. Fusco Unit of Geriatric Pharmacoepidemiology, Italian National Research Center on Aging (INRCA), Cosenza, Italy

Ma. Eugenia Garay-Sevilla Department of Medical Science, University of Guanajuato, Leon, Mexico

Georgios A. Georgiopoulos 1st Cardiology Clinic, University of Athens, Athens, Greece

Ekavi N. Georgousopoulou Department of Nutrition-Dietetics, School of Health and Education, Harokopio University, Athens, Greece

Robertina Giacconi Nutrition and Aging Centre, Scientific and Technological Pole, Italian National Institute of Health and Science on Aging (INRCA), Ancona, Italy

L. Anne Gilmore Pennington Biomedical Research Center, Baton Rouge, LA, USA

Pawel Glibowski Department of Milk Technology and Hydrocolloids, University of Life Science in Lublin, Lublin, Poland

Andreia C. Gomes Department of Biology, CBMA - Centre of Molecular and Environmental Biology, University of Minho, Braga, Portugal

Bamini Gopinath Centre for Vision Research, The Westmead Institute, The University of Sydney, Sydney, NSW, Australia

Peter Greenwald Office of the Director, National Cancer Institute/National Institutes of Health, Bethesda, MD, USA

Kristin P. Griffin Department of Basic Medical Sciences, College of Medicine, University of Arizona, Phoenix, AZ, USA

Petr Grúz Division of Genetics and Mutagenesis, National Institute of Health Sciences, Tokyo, Japan

T. Guichelaar Centre for Infectious Disease Control, National Institute for Public Health and the Environment (RIVM), Bilthoven, The Netherlands

Mark R. Haussler Department of Basic Medical Sciences, College of Medicine, University of Arizona, Phoenix, AZ, USA

Zoe Heis Sheikh Khalifa bin Hamad Al Thani Center for Integrative Research on Cardiovascular Aging (CIRCA), Aurora University of Wisconsin Medical Group, Aurora Cardiovascular Services, Aurora Research Institute, Aurora Health Care, Milwaukee, WI, USA

R. Antonelli Incalzi Department of Geriatric Medicine, University Campus Bio-Medico, Rome, Italy

Arshad Jahangir Sheikh Khalifa bin Hamad Al Thani Center for Integrative Research on Cardiovascular Aging (CIRCA), Aurora University of Wisconsin Medical Group, Aurora Cardiovascular Services, Aurora Research Institute, Aurora Health Care, Milwaukee, WI, USA

LIST OF CONTRIBUTORS

Emilio Jirillo Department of Basic Medical Sciences Neuroscience and Sensory Organs, University of Bari, Bari, Italy

Peter W. Jurutka Department of Basic Medical Sciences, College of Medicine, University of Arizona, Phoenix, AZ, USA; School of Mathematical and Natural Sciences, Arizona State University, Glendale, AZ, USA

Amelie Kahl Department of Agriculture and Food Sciences, University of Applied Sciences Neubrandenburg, Neubrandenburg, Germany

Ichiro Kaneko Department of Basic Medical Sciences, College of Medicine, University of Arizona, Phoenix, AZ, USA; School of Mathematical and Natural Sciences, Arizona State University, Glendale, AZ, USA

Zainab Khan School of Mathematical and Natural Sciences, Arizona State University, Glendale, AZ, USA

Dae Hyun Kim Molecular Inflammation Research Center for Aging Intervention (MRCA), College of Pharmacy, Pusan National University, Busan, Korea

Guido Koverech Department of Biomedical and Biotechnological Sciences, University of Catania, Catania, Italy

R.V. Kuiper Department of Laboratory Medicine, Karolinska Institutet, Stockholm, Sweden

Macy Kwan Department of Medicine and Therapeutics, The Chinese University of Hong Kong, Shatin, Hong Kong

Francesco Landi Department of Geriatrics, Neurosciences and Orthopedics, Catholic University of the Sacred Heart School of Medicine, Rome, Italy

Cristovao F. Lima Department of Biology, CITAB - Centre for the Research and Technology of Agro-Environmental and Biological Sciences, University of Minho, Braga, Portugal

Ann-Christin Lindenau Department of Agriculture and Food Sciences, University of Applied Sciences Neubrandenburg, Neubrandenburg, Germany

Karin Linnewiel-Hermoni Clinical Biochemistry and Pharmacology, Ben-Gurion University of the Negev and Soroka Medical Center, Beer Sheva, Israel

Maria Luca Department of Medical and Surgical Sciences and Advanced Technologies, University of Catania, Catania, Italy

Maria P. Luconi UOC Malattie Metaboliche e Diabetologia, INRCA-IRCCS, Ancona, Italy

Claudia Luevano-Contreras Department of Medical Science, University of Guanajuato, Leon, Mexico

Thea Magrone Department of Basic Medical Sciences Neuroscience and Sensory Organs, University of Bari, Bari, Italy

Marco Malavolta Nutrition and Aging Centre, Scientific and Technological Pole, Italian National Institute of Health and Science on Aging (INRCA), Ancona, Italy; Translational Center Research on Nutrition and Ageing, Scientific and Technologic Pole, INRCA, Ancona, Italy

Shwetha Mallesara Division of Endocrinology, Diabetes and Metabolism, University of Texas Health Science Center, Houston, TX, USA

Fiorella Marcheselli Center of Clinical Pathology and Innovative Therapy, Italian National Research Center on Aging (INRCA-IRCCS), Ancona, Italy

Emanuele Marzetti Department of Geriatrics, Neurosciences and Orthopedics, Catholic University of the Sacred Heart School of Medicine, Rome, Italy

John C. Mathers Human Nutrition Research Centre, Institute of Cellular Medicine, Newcastle University, Newcastle on Tyne, UK

Robin A. McGregor Cardiovascular and Metabolic Disease Center, College of Medicine, Inje University, Busan, Republic of Korea

Aksam J. Merched UFR of Pharmaceutical Sciences; INSERM Bordeaux Research in Translational Oncology (BaRITOn), University of Bordeaux, Bordeaux, France

Antje Micka BioTeSys GmbH, Esslingen, Germany

Michelle Miller Nutrition and Dietetics, Flinders University, Adelaide, Australia

Eugenio Mocchegiani Translational Center Research on Nutrition and Aging, Scientific and Technologic Pole INRCA, Ancona, Italy

María Moreno-Villanueva Department of Biology, Molecular Toxicology Group, University of Konstanz, Konstanz, Germany

Philip R. Orlander Division of Endocrinology, Diabetes and Metabolism, University of Texas Health Science Center, Houston, TX, USA

Esther Paran Hypertension and Vascular Research Laboratory, Ben-Gurion University of the Negev and Soroka Medical Center, Beer Sheva, Israel

J.L.A. Pennings Centre for Health Protection, National Institute for Public Health and the Environment (RIVM), Bilthoven, The Netherlands

Cristina Pereira-Wilson Department of Biology, CITAB - Centre for the Research and Technology of Agro-Environmental and Biological Sciences, University of Minho, Braga, Portugal

Erika Pétervári Department of Pathophysiology and Gerontology, Medical School, University of Pécs, Pécs, Hungary

Francesco Piacenza Nutrition and Aging Centre, Scientific and Technological Pole, Italian National Institute of Health and Science on Aging (INRCA), Ancona, Italy

Elisa Pierpaoli Advanced Technology Center for Aging Research, Scientific Technological Area, Italian National Institute of Health and Science on Aging (INRCA), Ancona, Italy

Eleonora Poggiogalle Food Science and Human Nutrition Research Unit, Medical Pathophysiology, Food Science and Endocrinology Section, Department of Experimental Medicine, Sapienza University of Rome, Rome, Italy

Mauro Provinciali Advanced Technology Center for Aging Research, Scientific Technological Area, Italian National Institute of Health and Science on Aging (INRCA), Ancona, Italy

Annibale A. Puca Dipartimento di Medicina e Chirurgia, Università degli Studi di Salerno, Salerno, Italy; IRCCS Multimedica, Milan, Italy

Doha Rasheedy Geriatrics and Gerontology Department, Faculty of Medicine, Ain Shams University, Cairo, Egypt

Suresh I.S. Rattan Laboratory of Cellular Ageing, Department of Molecular Biology and Genetics, Aarhus University, Aarhus, Denmark

Eric Ravussin Pennington Biomedical Research Center, Baton Rouge, LA, USA

Rina Recchioni Center of Clinical Pathology and Innovative Therapy, Italian National Research Center on Aging (INRCA-IRCCS), Ancona, Italy

Leanne M. Redman Pennington Biomedical Research Center, Baton Rouge, LA, USA

Ellen Richmond Gastrointestinal and Other Cancers Research Group, Division of Cancer Prevention, National Cancer Institute/National Institutes of Health, Bethesda, MD, USA

Lothar Rink Institute of Immunology, Medical Faculty, RWTH Aachen University, Aachen, Germany

Gabbianelli Rosita Molecular Biology Unit, School of Pharmacy, University of Camerino, Camerino, Italy.

Marya S. Sabir School of Mathematical and Natural Sciences, Arizona State University, Glendale, AZ, USA

Rimpi K. Saini Department of Basic Medical Sciences, College of Medicine, University of Arizona, Phoenix, AZ, USA

Giovanni Scapagnini Department of Medicine and Health Sciences, University of Molise, Campobasso, Italy

N. Scichilone Biomedical Department of Internal and Specialist Medicine, Section of Pulmonology, University of Palermo, Palermo, Italy

Mandy Sea Department of Medicine and Therapeutics, The Chinese University of Hong Kong, Shatin, Hong Kong Center for Nutritional Studies, The Chinese University of Hong Kong, Shatin, Hong Kong

Shyam Seetharaman Department of Psychology, St. Ambrose University, Davenport, IA, USA

Dae Y. Seo Cardiovascular and Metabolic Disease Center, College of Medicine, Inje University, Busan, Republic of Korea

Anne Siepelmeyer BioTeSys GmbH, Esslingen, Germany

Mario Siervo Human Nutrition Research Centre, Institute of Cellular Medicine, Newcastle University, Newcastle on Tyne, UK

Geeta Sikand Division of Cardiology, Medical Sciences, University of California, Irvine, CA, USA

Laura Silvestri Division of Genetics and Cell Biology, IRCCS San Raffaele Scientific Institute, Milan, Italy; Vita Salute University, Milan, Italy

Andreas Simm Klinik für Herz- und Thoraxchirurgie, Universitätsklinikum Halle (Saale), Halle (Saale), Germany

Szilvia Soós Department of Pathophysiology and Gerontology, Medical School, University of Pécs, Pécs, Hungary

Chiara Carmela Spinelli Istituto di Tecnologie Biomediche, CNR Segrate, Milan, Italy

Eunkyung Suh Chaum Life Center, School of Medicine, CHA University, Seoul, Korea

Sulaiman Sultan Sheikh Khalifa bin Hamad Al Thani Center for Integrative Research on Cardiovascular Aging (CIRCA), Aurora University of Wisconsin Medical Group, Aurora Cardiovascular Services, Aurora Research Institute, Aurora Health Care, Milwaukee, WI, USA

Yu Sun Huffington Center on Aging, Baylor College of Medicine, Houston, TX, USA

Miklós Székely Department of Pathophysiology and Gerontology, Medical School, University of Pécs, Pécs, Hungary

Matteo Tosato Department of Geriatrics, Neurosciences and Orthopedics, Catholic University of the Sacred Heart School of Medicine, Rome, Italy

Natalia Úbeda Departamento de Ciencias Farmacéuticas y de la Salud, Facultad de Farmacia, Universidad CEU San Pablo, Boadilla del Monte, Madrid, Spain

Peter Uciechowski Institute of Immunology, Medical Faculty, RWTH Aachen University, Aachen, Germany

Luzia Valentini Department of Agriculture and Food Sciences, University of Applied Sciences Neubrandenburg, Neubrandenburg, Germany

L.W.M. van Kerkhof Centre for Health Protection, National Institute for Public Health and the Environment (RIVM), Bilthoven, The Netherlands

H. van Steeg Centre for Health Protection, National Institute for Public Health and the Environment (RIVM), Bilthoven, The Netherlands; Department of Human Genetics, Leiden University Medical Center, Leiden, The Netherlands

Gregorio Varela-Moreiras Departamento de Ciencias Farmacéuticas y de la Salud, Facultad de Farmacia, Universidad CEU San Pablo, Boadilla del Monte, Madrid, Spain

Francesco Villa Istituto di Tecnologie Biomediche, CNR Segrate, Milan, Italy

Li Wang Division of Cardiology, Medical Sciences, University of California, Irvine, CA, USA

Carol Wham School of Food and Nutrition, Massey University, Auckland, New Zealand

G. Kerr Whitfield Department of Basic Medical Sciences, College of Medicine, University of Arizona, Phoenix, AZ, USA

Talya Wolak Hypertension and Vascular Research Laboratory, Ben-Gurion University of the Negev and Soroka Medical Center, Beer Sheva, Israel; Hypertension Unit, Faculty of Health Sciences, Ben-Gurion University of the Negev and Soroka Medical Center, Beer Sheva, Israel

Nathan D. Wong Division of Cardiology, Medical Sciences, University of California, Irvine, CA, USA

Jean Woo Department of Medicine and Therapeutics, The Chinese University of Hong Kong, Shatin, Hong Kong Center for Nutritional Studies, The Chinese University of Hong Kong, Shatin, Hong Kong

Byung P. Yu Department of Physiology, The University of Texas Health Science Center at San Antonio, San Antonio, TX, USA

Introduction to the *Molecular Basis of Nutrition and Aging*

This book meets the goal of the Molecular Nutrition series by providing a unique collection of fascinating contributions on the fundamentals of the nutritional connection between aging and health. The great challenge of this book is to address a field that benefits from a marked popular interest with a critical scientific perspective.

Hence, the book is organized in one introductory section followed by three parts that each individually take the reader on a journey from molecules through cells and tissues up to the whole organism. This is likely the most comprehensive volume ever published on the impact of nutrition on aging as it encompasses a wide field of knowledge from the molecular basis to clinical interventions.

The book structures the scientific contributions around both well-established and the most innovative and novel potential targets for intervention to affect the biology of aging. The development of this structure and thus the content of the contributions are focused on those nutritional factors and interventions that are known to act on these targets, culminating in nutritional advice or identification of priorities for future research.

The book is particularly timely at this point of the revolution in biogerontology. Many scientists are currently considering the concept that age-related pathology is not separate and distinct from the physiological process of aging. The free radical theory of aging is under question and no longer appears to be the most plausible and promising explanation for the process of aging. In this context, belief in the likely effectiveness of antioxidants as antiaging treatments has been weakening while new discoveries on epigenetic, nutrigenomic, and nutrigenetic, as well as around cellular hallmarks of aging have led to the view that other mechanisms should be explored to develop nutritional interventions that can delay or even partially reverse aging.

This revolution is likely to have an impact on clinical nutrition, not only in the area of primary and secondary prevention but also in the management of the oldest old affected by age-related pathologies. This exceptional volume could be considered a key repository of information on well-established and novel mechanisms of action that may underlie observed effects of nutritional compounds on health during aging. The book also aims to give practical advice as well as to stimulate question and discussion around the most debated aspects of this field. The overarching goal of the editors is to provide fully referenced information to advanced students, scientists, and health professionals, to foster the development of new ideas, scientific projects, and clinical trials, which will be pivotal to improve the health of current and future older adults.

Marco Malavolta and Eugenio Mocchegiani, the editors, are internationally recognized biogerontologists with particular expertise in the area of nutritional intervention. Both editors put substantial effort into selecting the leading contributors from the scientific community to assure the quality and comprehensiveness of the book to generate a benchmark in the field. The book features the work of more than 85 scientists chosen from among the best-recognized and internationally distinguished researchers, clinicians, and epidemiologists in the fields of biogerontology, geriatric and nutritional science. Each contribution provides comprehensive information on a specific theme to guide the reader's understanding of the molecular mechanisms that lie beneath the impact of nutrients and dietary interventions on health during aging.

Common aspects of the 51 chapters are an explanation of key terms by a mini-dictionary, an abstract, and the presence of short informative summary points. The volume contains more than 100 tables and 80 figures, an extensive and detailed index and more than 3000 up-to-date references that provide the reader with an extraordinary source of information for further insight into specific themes.

Part I of the book contains nine chapters to introduce the reader to the basic concepts of nutrition and aging. Chapter 1 sets out the basis of biogerontology currently used to explain the development of the phenotype of aging. In brief, the molecular and cellular basis of aging lies in the occurrence and accumulation of damage that is accompanied by defects in maintenance and repair systems finally leading to dysregulated function, increased vulnerability to stress, and reduced ability to adapt and remodel. Importantly, age may be predicted or defined as biological age, in contrast to chronological age. This concept is considered in Chapter 8, which focuses on human biomarkers of aging, and in Chapter 2, which reports a meaningful example of a gene expression study performed in mouse liver tissues. Those interventions, including nutritional ones, that are able to minimize the occurrence and accumulation of molecular damage and that mimic the metabolism of individuals genetically predisposed to longevity (Chapter 4) have the potential to promote beneficial health effects. The development of appropriate tools for the diagnosis of nutritional status and individual response to nutrients that will be likely used in future for a personalized approach to nutrition is complementary to this knowledge. A rational approach to personalized nutrition in aging can be founded on diet—gene interactions (Chapters 2 and 3) and drug—nutrient interactions (Chapter 7), as well as in the changes in food preferences of the elderly (Chapter 9). The nutrition of hospitalized older people deserved particular attention, where intervention with nutritional drinks/oral supplements and, where applicable, supportive supplementary tube feeding must be considered (Chapter 5). Although still early for translation of the body of scientific knowledge presented here, practical application of this knowledge continues to grow and medical practitioners as well as dietitians will soon apply these new discoveries to tailor nutritional advice as a precise intervention to promote health and prevent disease in the older population.

Part II of the book is dedicated specifically to the molecular hallmarks of aging, which are presented and discussed from the perspective of them providing targets for nutritional intervention. Among the primary hallmarks of aging, likely to cause the accumulation of damage, telomere attrition (Chapter 10), epigenetic alterations (Chapters 16, 17, and 21), and genomic instability (Chapters 13 and 18) are discussed from a nutritional perspective. Specific key concepts include the possibility to manipulate with nutritional interventions the activity of the Polycomb-group proteins (key epigenetic regulators of stem cell self-renewal and cellular senescence), microRNAs (a class of small noncoding RNAs, which are powerful post-transcriptional regulators of gene expression), the cellular capacity to repair DNA damage, as well as the exposure to xenobiotics and other DNA damaging pollutants.

The key players in the response to age-related damage could be considered secondary hallmarks of aging. These molecular mediators are a focus of intense debate and contradictory findings generated in the main experimental models. Hence, major emphasis in this part of the book is given to the phenomenon of deregulated nutrient sensing including its primary targets, that is, the target of rapamycin (TOR) kinase (mTOR in the case of mammals) (Chapter 11) and the FOXO family of transcription factors (Chapter 15), the energy production problems and the related consequences of mitochondrial dysfunction and oxidative stress (Chapters 12 and 19), as well as to the hot topic of modulation of cellular senescence (Chapter 22). Features of the integrative hallmarks of aging are also discussed, with a focus on alteration of the molecular signals that regulate organismal energy balance (Chapter 14) and on the alteration of intercellular communication caused by exogenous and endogenous advanced glycated end products (Chapter 20).

The major challenges that emerge from this section are to disentangle the direct and indirect effects of nutrients and dietary interventions on these candidate hallmarks as well as to provide a quantitative estimation of their impacts on improving human health.

Part III of the book is dedicated to the aging of systems and organs with an emphasis on those changes that compromise physiological effectiveness in the organism. Systems and organs are thus described as the targets of dietary and nutritional interventions aimed to preserve their function during aging. Each chapter of this section is dedicated to a different system or organ and to the chronic noncommunicable diseases that originate from its dysfunction in aging. Although there is no specific focus on cancer (which can originate in different organs and tissues), all other global major causes of death and disability are considered. Thus, the impact of nutrition on cardiovascular aging and diseases (Chapter 23), brain aging and dementia (Chapter 24), disorders of carbohydrate metabolism and diabetes (Chapter 29), as well as lung aging and chronic respiratory diseases (Chapter 30), are all covered comprehensively.

Particular attention is also given to the impact of nutrition on changes and diseases that affect the musculoskeletal apparatus in the elderly, including bone aging and osteoporosis (Chapter 25), as well as frailty and sarcopenia (Chapter 26) and its consequences.

The role of nutritional factors implicated in the progression of renal failure or in preserving kidney function (Chapter 31) as well as in preventing or delaying degenerative diseases of the eye (Chapter 32) are also critically addressed. A specific chapter is dedicated to the impact of nutrition on the decline of the immune system in aging (Chapter 28) and another to the consequences of physiological and pathophysiological changes associated with the aging of the human gastrointestinal tract on nutritional status (Chapter 27).

This part of the book provides important information related to medical nutrition therapies with an impact on specific diseases of the elderly.

Part IV of the book is dedicated to dietary and nutritional interventions with a focus on the effects of single nutrients, dietary factors, supplements, or other nutritional strategies aimed to prevent or manage age-related diseases and preserve health and longevity. Particular attention is also given to the metabolism, absorption, bioavailability, distribution, and extraction of selected nutritional factors in the context of aging. The interaction of aging with a wide range of micronutrients is reviewed and discussed, including vitamins A (Chapter 34), B (Chapter 42), C (Chapter 43), D (Chapter 33), and E (Chapter 45), folic acid (Chapter 36), essential elements, such as zinc (Chapter 39), selenium (Chapter 40), iron (Chapter 37), and iodine (Chapter 41), as well as essential minerals (ie, magnesium, calcium, and potassium) (Chapter 38). Dietary interventions that impact on the macronutrient pool, such as caloric restriction in humans (Chapter 48) and omega-3 polyunsaturated fatty acids (Chapter 44), are also considered along with the impact on aging of bioactive nonnutritional factors, including polyphenols (Chapter 46) and the bioactive natural compounds prevalent in the Asian diet (Chapter 47), with a chapter dedicated to curcumin (Chapter 35). Beyond single micro- and macronutrients, this section also considers the potential of probiotics and prebiotics (Chapter 49) as well as the impact of dietary vegetables and fruit (Chapter 50) in preserving health in aging. A final chapter (Chapter 51) is dedicated to a revision of current nutritional recommendations with a specific focus on the needs of elderly people.

Series Preface

In this series on *Molecular Nutrition*, the editors of each book aim to disseminate important material pertaining to molecular nutrition in its broadest sense. The coverage ranges from molecular aspects to whole organs, and the impact of nutrition or malnutrition on individuals and whole communities. It includes concepts, policy, preclinical studies, and clinical investigations relating to molecular nutrition. The subject areas include molecular mechanisms, polymorphisms, SNPs, genomic-wide analysis, genotypes, gene expression, genetic modifications, and many other aspects. Information given in the Molecular Nutrition series relates to national, international, and global issues.

A major feature of the series that sets it apart from other texts is the initiative to bridge the transintellectual divide so that it is suitable for novices and experts alike. It embraces traditional and nontraditional formats of nutritional sciences in different ways. Each book in the series has both overviews and detailed and focused chapters.

Molecular Nutrition is designed for nutritionists, dieticians, educationalists, health experts, epidemiologists, and health-related professionals, such as chemists. It is also suitable for students, graduates, postgraduates, researchers, lecturers, teachers, and professors. Contributors are national or international experts, many of whom are from world-renowned institutions or universities. It is intended to be an authoritative text covering nutrition at the molecular level.

V.R. Preedy
Series Editor

Acknowledgments

The editors, Marco Malavolta and Eugenio Mocchegiani, would like first to thank all the authors and to recognize the excellent quality of their contributions, which together have produced a book of such high quality. All coauthors, young scientists, and contract researchers, who directly or indirectly have contributed with their enthusiasm and creativity, also deserve particular thanks. Marco Malavolta and Eugenio Mocchegiani also thank the Editorial Project Manager, Jeffrey Rossetti, for his encouragement to begin and continue this ongoing journey of discovery and learning in the field of molecular nutrition in aging. A special thanks is for Prof. Dianne Ford, for her precious assistance during the editing of the introduction to this book. Without the dedication and creative contributions of these scientists and authors this book would not have been possible.

PART I

INTRODUCTORY ASPECTS ON AGING AND NUTRITION

CHAPTER

1

Molecular and Cellular Basis of Aging

Suresh I.S. Rattan

Laboratory of Cellular Ageing, Department of Molecular Biology and Genetics, Aarhus University, Aarhus, Denmark

KEY FACTS

- Signs of biological aging appear progressively and exponentially during the period of survival beyond the ELS of a species.
- There are no gerontogenes evolved with a specific function of causing aging and eventual death.
- The role of genes in aging and longevity is mainly at the level of longevity-assurance in evolutionary terms.
- The phenotype of aging is highly differential and heterogeneous at all levels of biological organization.
- Aging is characterized by a stochastic occurrence, accumulation, and heterogeneity of damage in macromolecules.
- Mild stress-induced activation of defense and repair processes helps to maintain health and prolong longevity.

Dictionary of Terms

- *Essential lifespan (ELS):* optimal duration of life as required by the evolutionary life history of a species. ELS of a species is different from both the average lifespan (ALS) of a cohort of a population, and the maximum lifespan (MLS) recorded for an individual within a species. For example, ELS for *Homo sapiens* is considered to be about 45 years, whereas the present ALS for the populations of industrially developed countries is about 80 years, and the MLS recorded so far for human beings is 122 years.
- *Homeodynamics:* in contrast to the machine-based conceptual model homeostasis, which means the same state, the term homeodynamics incorporates the dynamic nature of the living systems, which is not static but constantly changing, remodeling, and adapting.
- *Homeodynamic space:* a conceptual term to describe the "survival ability" or the "buffering capacity" of a biological system; it is comprised of three main categories of biological processes—stress response, damage control, and continuous remodeling.
- *Hormesis:* biphasic dose response in which the negative or toxic consequences of exposure to high levels of a stressor are observed to be reversed (positive or beneficial) at low levels. Moderate physical exercise is the paradigm for physiological hormesis. The science and study of hormesis is known as Hormetics.
- *Hormetin:* a condition that induces hormesis; three main types of hormetins are: physical hormetins (temperature, irradiation, mechanical tension); nutritional hormetins (spices and other NNFC, calorie restriction, fasting); and mental hormetins (psychological challenge, meditation).

INTRODUCTION

Improving human health and longevity through nutrition is one of the longest running themes in history. While dreams of a perfect food for eternal youth and immortality may still occupy the minds of some, modern scientific knowledge has opened up novel approaches toward understanding and utilizing nutrition in a more realistic and rational way. However, in order to fully appreciate and evaluate the possible approaches toward modulating aging, it can be useful to have an overview and understanding of the current status of biogerontology—the study of the biological basis of aging.

This chapter aims to provide a general review of the molecular and cellular basis of aging, mechanistic theories of aging, homeodynamic mechanisms of survival, maintenance, and repair, followed by a discussion of nutrition-based aging interventions, especially the nutritional hormetins that bring about their health beneficial effects by stress-induced hormesis.

BIOLOGICAL PRINCIPLES OF AGING

Modern biogerontology can be considered to originate in the second half of the 20th century with the writings and experimental findings of Peter Medawar [1], Denham Harman [2], and Leonard Hayflick [3,4]. It can be safely said that the biological bases of aging are now well understood and a distinctive framework has been established [5–7]. This framework has been developed from numerous theoretical analyses, and hundreds of descriptive and interventional experimental studies performed on a wide variety of biological systems with a range of life histories and traits. Four main biological principles can be derived from these, which cover evolutionary, genetic, differential, and molecular aspects of aging and longevity (Table 1.1).

Thus, aging is an emergent and epigenetic metaphenomenon, which is neither determined by any gerontogenes, nor is it controlled by a single mechanism. Furthermore, individually no tissue, organ, or system becomes functionally exhausted even in very old organisms; and it is their interconnectedness, interaction, and interdependence that determine the survival of the whole. Various ideas have been put forward to explain the mechanistic basis of aging, and generally all of them incorporate, in one or the other way, molecular damage, molecular heterogeneity, and metabolic imbalance as the cause of aging. These ideas include virtual gerontogenes [12], system failure [13], unregulated growth-related quasiprograms [14], and metabolic instability [15,16]. Most importantly, almost all these views directly or indirectly reject the notion of the evolution of any specific and real genes for aging.

OCCURRENCE, ACCUMULATION, AND CONSEQUENCES OF MOLECULAR DAMAGE

As discussed in detail previously [9,17], molecular damages within a cell arise constantly mainly from the following sources: (i) reactive oxygen species (ROS) and other free radicals (FR) formed by the action of external inducers of damage (eg, UV-rays), and as a consequence of intrinsic cellular metabolism involving oxygen, metals, and other metabolites; (ii) nutritional glucose and its metabolites, and their biochemical interactions with ROS and FR; and (iii) spontaneous errors in biochemical processes, such as DNA duplication, transcription, posttranscriptional processing, translation, and posttranslational modifications. Occurrence of molecular damage has led to the formulation of at least two mechanistic theories of biological aging, which have been the basis of most of the experimental aging research during the last 50 years [9].

The first one of these is the so-called free radical theory of aging (FRTA), which arose from the premise that a single common biochemical process may be responsible for the aging and death of all living beings [18,19]. There is abundant evidence to show that a variety of ROS and other FR are indeed involved in the occurrence of molecular damage that can then lead to structural and functional disorders, diseases, and death. The chemistry and biochemistry of FR are very well worked out, and the cellular and organismic consequences are also well documented [20]. However, the main criticism raised against FRTA is with respect to its lack of incorporation of the essential and beneficial role of FR in the normal functioning and survival of biological systems [21,22]. Furthermore, FRTA presents FR as the universal cause of damage without taking into account the differences in the wide range of FR-counteracting mechanisms in different species, which effectively determine the extent of damage occurrence and accumulation. Additionally, a large body of data that shows the contrary and/or lack of predictable and expected beneficial results of antioxidant and FR-scavenging therapies has restricted the application of FRTA [22–25].

TABLE 1.1 Principles of Aging and Longevity

1. *Evolutionary life history principle*: Aging is an emergent phenomenon seen primarily in the period of survival beyond the natural lifespan of a species, termed "essential lifespan" (ELS) [8,9].
2. *Nongenetic principle*: There is no fixed and rigid genetic program that determines the exact duration of survival of an organism, and there are no real gerontogenes whose sole function is to cause aging [9].
3. *Differential principle*: The progression and rate of aging is different in different species, organisms within a species, organs and tissues within an organism, cell types within a tissue, subcellular compartments within a cell type, and macromolecules within a cell [9].
4. *Molecular mechanistic principle*: Aging is characterized by a stochastic occurrence, accumulation, and heterogeneity of damage in macromolecules, leading to the failure of maintenance and repair pathways [9–11].

The second major mechanistic theory that incorporates the crucial role of macromolecular damage is the so-called protein error theory of aging (PETA). The history of PETA, also known as the error catastrophe theory, is often marked with controversy [9,17,26]. Since the spontaneous error frequency in protein synthesis is generally several orders of magnitude higher than that in nucleic acid synthesis, the role of protein errors and their feedback in biochemical pathways has been considered to be a crucial one with respect to aging. Several attempts have been made to determine the accuracy of translation in cell-free extracts, and most of the studies show that there is an age-related increase in the misincorporation of nucleotides and amino acids [26–30]. It has also been shown that there is an age-related accumulation of aberrant DNA polymerases and other components of the transcriptional and translational machinery [31,32].

Further evidence in support of PETA comes from experiments which showed that an induction and increase in protein errors can accelerate aging in human cells and bacteria [26,33,34]. Similarly, an increase in the accuracy of protein synthesis can slow aging and increase the lifespan in fungi [35–37]. Therefore, it is not ruled out that several kinds of errors in various components of the protein synthetic machinery and in mitochondria do have long term effects on cellular stability and survival [29,30]. However, almost all these methods have relied on indirect in vitro assays, and so far direct, realistic, and accurate estimates of age-related changes in errors in cytoplasmic and mitochondrial proteins, and their biological relevance, have not been made. Similarly, applying methods such as two-dimensional gel electrophoresis, which can resolve only some kinds of misincorporations, have so far remained insensitive and inconclusive [26,38,39].

Both the FRTA and PETA provide molecular mechanisms for the occurrence of molecular damage. Additionally, nutritional components, especially the sugars and metal-based micronutrients, can induce, enhance, and amplify the molecular damage either independently or in combination with other inducers of damage. It is important to point out that although the action of the damaging agents is mainly stochastic, the result of whether a specific macromolecule will become damaged and whether damage can persist depends both on its structure, localization, and interactions with other macromolecules, and on the activity and efficiency of a complex series of maintenance and repair pathways, discussed below [9,17]. Understanding the quantitative and qualitative aspects of molecular damage in terms of their biological relevance is one of the most challenging aspects of the present biogerontological research.

Whatever the reason for the occurrence of molecular damage, accumulation of damage in DNA, RNA, proteins, and other macromolecules is a well-established molecular phenotype of aging. Since there is an extremely low probability that any two molecules become damaged in exactly the same way and to the same extent, an increase in molecular heterogeneity is inevitable. Increased molecular heterogeneity is the fundamental basis for the molecular, biochemical, cellular, and physiological changes happening during aging. Such age-related changes include genomic instability, mutations, dysregulated gene expression, cellular senescence, cell death, impaired intercellular communication, tissue disorganization, organ dysfunctions, increased vulnerability to stress, reduced ability to adapt and remodel, and increased chances of the emergence of age-related diseases [7,10,41].

HOMEODYNAMICS AND THE HOMEODYNAMIC SPACE

Another way to understand aging is by understanding the processes of life and their intrinsic limitations. Survival of an organism is a dynamic tug between the occurrence of damage and the processes of maintenance and repair systems (MARS). The main MARS that comprise the longevity-assurance processes are listed in Table 1.2.

Another way of conceptualizing MARS is the idea of "homeodynamic space," which may also be considered as the survival ability or the buffering capacity of a biological system [10]. The term "homeodynamics," meaning "the same dynamics," is distinct from the classical term homeostasis, that means "the same state," which ignores the reality of ever-dynamic, ever-changing and yet appearing to remain the same, dynamic living systems [42]. Biological systems—cells, tissues, organs, organisms, and populations—are never static, and therefore the most commonly used term homeostasis is wrong for living systems.

Three main characteristics of the homeodynamic space are the abilities to control the levels of molecular

TABLE 1.2 Main MARS in a Biological System

1. Nuclear and mitochondrial DNA repair
2. RNA and protein repair
3. Defenses against ROS and other FR
4. Removal of defective macromolecules and organelles by autophagy, lysosomes, and proteasomes
5. Detoxification of chemicals and nutritional metabolites
6. Sensing and responding to intra- and extracellular stress
7. Innate and adaptive immune responses and apoptosis
8. Wound healing, tissue regeneration, and other higher order processes, including thermal regulation, neuroendocrine balance, and daily rhythms

damage, to respond to external and internal stress, and to constantly remodel and adapt in dynamic interactions. A large number of molecular, cellular, and physiological pathways and their interconnected networks, including MARS listed in Table 1.2, determine the nature and extent of the homeodynamic space of an individual.

At the species level, biological evolutionary processes have assured the essential lifespan (ELS) of a species by optimizing for homeodynamic space through MARS, which are also the main target of evolutionary investment, stability, and selection [16,43–47]. However, the period of survival beyond ELS is characterized by the progressive shrinkage of the homeodynamic space characterized by reduced ability to tolerate stress, to control molecular damage, and to adapt and remodel. Shrinkage of the homeodynamic space leads to an increase in the zone of vulnerability, reduced buffering capacity, and increased probabilities for the onset and emergence of chronic diseases [10]. Major chronic conditions, for example, metabolic disorders, depression, dementia, malnutrition, and several types of age-related cancers, are mostly due to the generalized failure and dysregulation of processes of life and their interactive networks, and not due to any specific cause(s) [48–51]. Thus, aging in itself is not a disease, but is a condition that allows the emergence of one or more diseases in some, but not all, old people.

NUTRITION AND FOOD FOR AGING INTERVENTIONS

There is a lot of scientific and social interest in the real and potential power of food in improving health, preventing diseases, and extending the lifespan [52–55]. However, in scientific research and experimentation, often little or no distinction is made between nutrition and food, which is a gross omission in a social context. As discussed elsewhere [56], nutrition is the amalgamation of various components, such as proteins, carbohydrates, fats, and minerals, which are needed for the survival, growth, and development of a biological system. However, food is what, why, and how we eat for survival, health, and longevity. This distinction between nutrition and food is a very important variable for humans, and may be equally important for other animal models used in research, where the appearance, the smell, the texture, and the taste of the food matter. None of the nutritional components is by itself either good or bad, and none of the foods is either healthy or unhealthy. Nutrition can lead to either good effects or bad effects; and the food can have consequences making us either healthy or unhealthy. It is the quantity, quality, frequency, and emotional satisfaction that determine whether any particular food can help us achieve the aim of maintaining and improving health, and delaying, preventing, or treating a disease [56].

Some food components in the diet of human beings do not have any nutritional value in the normal sense of providing material for the structure, function, and energy requirements of the body [57]. Such nonnutritional food components (NNFC) usually come from spices, herbs, and the so-called vegetables and fruits, for example, onion, garlic, ginger, shallot, chive, and chilies [58]. Different combinations of NNFC are integral parts of different food cultures in different social setups, and carry a wide range of claims made for their health beneficial and longevity promoting effects. Not all such claims for NNFC have been scientifically tested and confirmed, and often very little is known about their biochemical mode of action. However, recent research in the field of hormesis is unraveling some of the mechanistic basis for the effects of NNFC [59].

Hormesis is the positive relationship between low-level stress and health [59–61]. Whereas uncontrolled, severe, and chronic stress is recognized as being harmful for health, single- or multiple-exposures to mild stress are generally health beneficial. Moderate exercise is the best example of such a phenomenon of mild stress-induced physiological hormesis. Exercise initially increases the production of FR, acids, and other potentially harmful biochemicals in the body, but the cellular responses to stress, in increasing defense and repair processes, protect and strengthen the body. Such conditions, which induce hormesis, are called hormetins, and are categorized as physical, mental, and nutritional hormetins [62,63].

NUTRITIONAL HORMETINS

Among different types of hormetins, nutritional hormetins, especially those derived from plant sources, have generated much scientific interest for their potential health beneficial effects. This is because of the realization that not all chemicals found in plants are beneficial for animals in a direct manner, but rather they cause molecular damage by virtue of their electrochemical properties [64]. Several NNFC and their constituent chemical entities, such as flavonoids or bioflavonoids, are nutritional hormetins. This is because they directly or indirectly induce one or more stress responses, such as Nrf2 activation, heat shock response (HSR), unfolded protein response, and sirtuin response [63,65]. After the initial recognition of disturbance or damage caused by a stressor, numerous downstream biochemical processes come into play, including the synthesis and activation of chaperones, stimulation of protein turnover, induction of autophagy, and an increase in antioxidant enzymes [65].

Several NNFC have been shown to achieve the antioxidant effects by the activation of Nrf2 transcription factor. This activation generally happens following the electrophilic modification/damage of its inhibitor protein Keap1, which then leads to the accumulation, heterodimerization, nuclear translocation, and DNA binding of Nrf2 at the antioxidant response element, resulting in the downstream expression of a large number of the so-called antioxidant genes, such as heme oxygenase HO-1, superoxide dismutase, glutathione, and catalase [64,66,67]. Some well-known phytochemicals and plant extracts which strongly induce Nrf2-mediated hormetic response include curcumin, quercetin, genistein, eugenol coffee, turmeric, rosemary, broccoli, thyme, clove, and oregano [64,68].

Another stress response pathway that has been studied in detail and can be the basis for identifying novel nutritional hormetins is the HSR. Induction of proteotoxic stress, such as protein misfolding and denaturation, initiates HSR by the intracellular release of the heat shock transcription factor from their captorproteins, followed by its nuclear translocation, trimerization, and DNA-binding for the expression of several heat shock proteins (HSP) [69,70]. A wide range of biological effects then occur which involve HSP, such as protein repair, refolding, and selective degradation of abnormal proteins leading to the cleaning up and an overall improvement in the structure and function of the cells. Various phytochemicals and nutritional components have been shown to induce HSR and have health beneficial effects including antiaging and longevity promoting effects. Some examples of nutritional hormetins involving HSR are phenolic acids, polyphenols, flavonoids, ferulic acid [71,72], geranylgeranyl, rosmarinic acid, kinetin, zinc [72–74], and the extracts of tea, dark chocolate, saffron, and spinach [75]. Further screening of animal and plant components for their ability to induce HSR can identify other potential nutritional hormetins.

Other pathways of stress response, which are involved in initiating hormetic effects of nutritional components are the NFkB, FOXO, sirtuins, DNA repair response, and autophagy pathways. Resveratrol and some other mimetics of calorie restriction work by the induction of one or more of these pathways [74,76]. Discovering novel nutritional hormetins by putting potential candidates through a screening process for their ability to induce one or more stress pathways in cells and organisms can be a promising strategy [63].

CONCLUSIONS

The molecular and cellular bases of aging lie in the progressive failure of MARS that leads to the emergence of the senescent phenotype. There are no gerontogenes with the specific evolutionary function to cause aging and death of an individual. The concept of homeodynamic space can be a useful one in order to identify a set of measurable, evidence-based, and demonstratable parameters of health, robustness, and resilience. Age-related health problems, for which there are no clear-cut causative agents, may be better tackled by focusing on health mechanisms and their maintenance, rather than disease management and treatment. Biogerontological and other research on life processes and lifestyle-related diseases have shown that the issues of aging, quality of life, and longevity need to be approached with health-oriented paradigms.

SUMMARY POINTS

- Molecular and cellular bases of aging lie in the occurrence and accumulation of damage.
- Imperfections of the MARS that comprise the homeodynamic space for survival lead to a progressive failure of homeodynamics.
- Impaired and dysregulated function, increased vulnerability to stress, and reduced ability to adapt and remodel are the major signs of aging.
- Aging is a continuum of life-history in which some changes can lead to the clinical diagnosis as emergence of one or more diseases.
- Approaches for intervention, prevention, and modulation of aging require means to minimize the occurrence and accumulation of molecular damage.
- Mild stress-induced hormesis caused by physical, biological, and nutritional hormetins is a promising holistic strategy for strengthening the homeodynamics.
- Some food components, which induce one or more pathways of stress response, are potential nutritional hormetins, and can have health- and longevity-promoting effects.

References

[1] Medawar PB. An unsolved problem in biology. London. H.K. Lewis; 1952.
[2] Harman D. Aging: a theory based on free radical and radiation chemistry. J Gerontol 1956;11(3):298–300.
[3] Hayflick L, Moorhead PS. The serial cultivation of human diploid strains. Exp Cell Res 1961;25:585–621.
[4] Hayflick L. The limited in vitro lifetime of human diploid cell strains. Exp Cell Res 1965;37:614–36.
[5] Holliday R. Aging is no longer an unsolved problem in biology. Ann N Y Acad Sci 2006;1067:1–9.

[6] Hayflick L. Biological aging is no longer an unsolved problem. Ann N Y Acad Sci 2007;1100:1–13.

[7] Rattan SIS. Biogerontology: from here to where? The Lord Cohen Medal Lecture-2011. Biogerontology 2012;13(1):83–91.

[8] Rattan SIS. Biogerontology: the next step. Ann N Y Acad Sci 2000;908:282–90.

[9] Rattan SIS. Theories of biological aging: genes, proteins and free radicals. Free Rad Res 2006;40:1230–8.

[10] Rattan SIS. Homeostasis, homeodynamics, and aging. In: Birren J, editor. Encyclopedia of gerontology. 2nd ed. London: Elsevier Inc; 2007. p. 696–9.

[11] Holliday R, Rattan SIS. Longevity mutants do not establish any "new science" of ageing. Biogerontology 2010;11(4):507–11.

[12] Rattan SIS. Gerontogenes: real or virtual? FASEB J 1995;9:284–6.

[13] Gavrilov LA, Gavrilova NS. The reliability theory of aging and longevity. J Theor Biol 2001;213:527–45.

[14] Blagosklonny MV. Cell cycle arrest is not yet senescence, which is not just cell cycle arrest: terminology for TOR-driven aging. Aging (Albany, NY) 2012;4(3):159–65.

[15] Demetrius L. Calorie restriction, metabolic rate and entropy. J Gerontol Biol Sci 2004;59A(9):902–15.

[16] Demetrius L. Boltzmann, Darwin and directionality theory. Phys Rep 2013;530:1–85.

[17] Rattan SIS. Increased molecular damage and heterogeneity as the basis of aging. Biol Chem 2008;389(3):267–72.

[18] Harman D. Free radical theory of aging: an update. Ann N Y Acad Sci 2006;1067:10–21.

[19] Harman D. Origin and evolution of the free radical theory of aging: a brief personal history, 1954–2009. Biogerontology 2009;10(6):773–81.

[20] Kalyanaraman B. Teaching the basics of redox biology to medical and graduate students: oxidants, antioxidants and disease mechanisms. Redox Biol 2013;1(1):244–57.

[21] Linnane AW, Eastwood H. Cellular redox regulation and prooxidant signaling systems. A new perspective on the free radical theory of aging. Ann N Y Acad Sci 2006;1067:47–55.

[22] Linnane AW, Kios M, Vitetta L. Healthy aging: regulation of the metabolome by cellular redox modulation and prooxidant signaling systems: the essential roles of superoxide anion and hydrogen peroxide. Biogerontology 2007;8(5):445–67.

[23] Le Bourg E, Fournier D. Is lifespan extension accompanied by improved antioxidant defences? A study of superoxide dismutase and catalase in Drosophila melanogaster flies that lived in hypergravity at young age. Biogerontology 2004;5(4):261–4.

[24] Le Bourg E. Antioxidants and aging in human beings. In: Rattan SIS, editor. Aging interventions and therapies. Singapore: World Scientific Publishers; 2005. p. 85–107.

[25] Howes RM. The free radical fantasy: a panoply of paradoxes. Ann N Y Acad Sci 2006;1067:22–6.

[26] Holliday R. The current status of the protein error theory of aging. Exp Gerontol 1996;31(4):449–52.

[27] Rattan SIS. Synthesis, modifications and turnover of proteins during aging. Exp Gerontol 1996;31(1–2):33–47.

[28] Rattan SIS. Transcriptional and translational dysregulation during aging. In: von Zglinicki T, editor. Aging at the molecular level. Dordrecht: Kluwer Academic Publishers; 2003. p. 179–91.

[29] Holliday R. Streptomycin, errors in mitochondria and ageing. Biogerontology 2005;6(6):431–2.

[30] Hipkiss A. Errors, mitochondrial dysfunction and ageing. Biogerontology 2003;4:397–400.

[31] Srivastava VK, Miller S, Schroeder M, Crouch E, Busbee D. Activity of DNA polymerase alpha in aging human fibroblasts. Biogerontology 2000;1(3):201–16.

[32] Srivastava VK, Busbee DL. Replicative enzymes and ageing: importance of DNA polymerase alpha function to the events of cellular ageing. Ageing Res Rev 2002;1:443–63.

[33] Nyström T. Translational fidelity, protein oxidation, and senescence: lessons from bacteria. Ageing Res Rev 2002;1:693–703.

[34] Nyström T. Aging in bacteria. Curr Opin Microbiol 2002;5:596–601.

[35] Silar P, Picard M. Increased longevity of EF-1a high-fidelity mutants in Podospora anserina. J Mol Biol 1994;235:231–6.

[36] Silar P, Rossignol M, Haedens V, Derhy Z, Mazabraud A. Deletion and dosage modulation of the eEF1A gene in Podospora anserina: effect on the life cycle. Biogerontology 2000;1(1):47–54.

[37] Holbrook MA, Menninger JR. Erythromycin slows aging of Saccharomyces cerevisiae. J Gerontol Biol Sci 2002;57A(1):B29–36.

[38] Hipkiss A. Accumulation of altered proteins and ageing: causes and effects. Exp Gerontol 2006;41:464–73.

[39] Kirkwood TBL. A systematic look at an old problem. Nature 2008;451(7 Feb):644–7.

[40] Rattan SIS. Aging is not a disease: implications for intervention. Aging Dis 2014;5(3):196–202.

[41] Lopez-Otin C, Blasco MA, Partridge L, Serrano M, Kroemer G. The hallmarks of aging. Cell 2013;153(6):1194–217.

[42] Yates FE. Order and complexity in dynamical systems: homeodynamics as a generalized mechanics for biology. Math Comput Model 1994;19:49–74.

[43] Kirkwood TBL. Time of our lives. London: Weidenfeld & Nicolson; 1999.

[44] Holliday R. Ageing: the paradox of life. Dordrecht: Springer; 2007.

[45] Martin GM. Modalities of gene action predicted by the classical evolutionary theory of aging. Ann N Y Acad Sci 2007;1100:14–20.

[46] Gladyshev VN. The origin of aging: imperfectness-driven nonrandom damage defines the aging process and control of lifespan. Trends Genet 2013;29(9):506–12.

[47] Wensink MJ, van Heemst D, Rozing MP, Westendorp RG. The maintenance gap: a new theoretical perspective on the evolution of aging. Biogerontology 2012;13(2):197–201.

[48] Tacutu R, Budovsky A, Yanai H, Fraifeld VE. Molecular links between cellular senescence, longevity and age-related diseases—a systems biology perspective. Aging (Albany NY) 2011;3(12):1178–91.

[49] Walter S, Atzmon G, Demerath EW, et al. A genome-wide association study of aging. Neurobiol Aging 2011;32(11):2109 e15–2109 e28

[50] O'Neill D. 2012—That was the year that was. Age Ageing 2013;42(2):140–4.

[51] Watson J. Oxidants, antioxidants and the current incurability of metastatic cancers. Open Biol 2013;3(1):120144.

[52] Ferrari CKB. Functional foods, herbs and neutraceuticals: towards biochemical mechanisms of healthy aging. Biogerontology 2004;5(5):275–89.

[53] Mutch DM, Wahli W, Williamson G. Nutrigenomics and nutrigenetics: the emerging faces of nutrition. FASEB J 2005;19:1602–16.

[54] Mattson MP. Dietary factors, hormesis and health. Ageing Res Rev 2008;7:43–8.

[55] Naik SR, Thakare VN, Joshi FP. Functional foods and herbs as potential immunoadjuvants and medicines in maintaining healthy immune system: a commentary. J Compl Integr Med 2010;7(1): Article 46

[56] Rattan SIS. Nutrition and food for health and longevity. Int J Nutr Pharm Neur Dis 2015;5(2):45.

[57] Gutteridge JM, Halliwell B. Antioxidants: molecules, medicines, and myths. Biochem Biophys Res Commun 2010;393(4):561–4.

References

[58] Aggarwal BB, Kunnumakkara AB, Harikumar KB, Tharakan ST, Sung B, Anand P. Potential of spice-derived phytochemicals for cancer prevention. Planta Med 2008;74(13):1560—9.

[59] Rattan SIS, Le Bourg E, editors. Hormesis in health and disease. Boca Raton, FL: CRC Press; 2014.

[60] Calabrese EJ, Bachmann KA, Bailer AJ, et al. Biological stress response terminology: integrating the concepts of adaptive response and preconditioning stress within a hormetic dose-response framework. Toxicol Appl Pharmacol 2007;222(1):122—8.

[61] Le Bourg E, Rattan SIS, editors. Mild stress and healthy aging: applying hormesis in aging research and interventions. Dordrecht: Springer; 2008.

[62] Rattan SIS. Hormesis in aging. Ageing Res Rev 2008;7:63—78.

[63] Rattan SIS. Rationale and methods of discovering hormetins as drugs for healthy ageing. Expert Opin Drug Discov 2012;7(5):439—48.

[64] Balstad TR, Carlsen H, Myhrstad MC, et al. Coffee, broccoli and spices are strong inducers of electrophile response element-dependent transcription in vitro and in vivo—studies in electrophile response element transgenic mice. Mol Nutr Food Res 2011;55(2):185—97.

[65] Demirovic D, Rattan SI. Establishing cellular stress response profiles as biomarkers of homeodynamics, health and hormesis. Exp Gerontol 2013;48(1):94—8.

[66] Calabrese V, Cornelius C, Dinkova-Kostova AT, Calabrese EJ, Mattson MP. Cellular stress responses, the hormesis paradigm, and vitagenes: novel targets for therapeutic intervention in neurodegenerative disorders. Antioxid Redox Signal 2010;13(11):1763—811.

[67] Calabrese V, Cornelius C, Mancuso C, et al. Cellular stress response: a novel target for chemoprevention and nutritional neuroprotection in aging, neurodegenerative disorders and longevity. Neurochem Res 2008;33(12):2444—71.

[68] Lima CF, Pereira-Wilson C, Rattan SI. Curcumin induces heme oxygenase-1 in normal human skin fibroblasts through redox signaling: relevance for anti-aging intervention. Mol Nutr Food Res 2011;55(3):430—42.

[69] Verbeke P, Fonager J, Clark BFC, Rattan SIS. Heat shock response and ageing: mechanisms and applications. Cell Biol Int 2001;25(9):845—57.

[70] Liberek K, Lewandowska A, Zietkiewicz S. Chaperones in control of protein disaggregation. EMBO J 2008;27:328—35.

[71] Barone E, Calabrese V, Mancuso C. Ferulic acid and its therapeutic potential as a hormetin for age-related diseases. Biogerontology 2009;10(2):97—108.

[72] Son TG, Camandola S, Mattson MP. Hormetic dietary phytochemicals. Neuromol Med 2008;10(4):236—46.

[73] Berge U, Kristensen P, Rattan SIS. Hormetic modulation of differentiation of normal human epidermal keratinocytes undergoing replicative senescence in vitro. Exp Gerontol 2008;43:658—62.

[74] Sonneborn JS. Mimetics of hormetic agents: stress-resistance triggers. Dose Response 2010;8(1):97—121.

[75] Wieten L, van der Zee R, Goedemans R, et al. Hsp70 expression and induction as a readout for detection of immune modulatory components in food. Cell Stress Chaperones 2010;15(1):25—37.

[76] Longo VD. Linking sirtuins, IGF-I signaling, and starvation. Exp Gerontol 2009;44(1—2):70—4.

CHAPTER 2

Unraveling Stochastic Aging Processes in Mouse Liver: Dissecting Biological from Chronological Age

L.W.M. van Kerkhof[1], J.L.A. Pennings[1], T. Guichelaar[2], R.V. Kuiper[3], M.E.T. Dollé[1] and H. van Steeg[1,4]

[1]Centre for Health Protection, National Institute for Public Health and the Environment (RIVM), Bilthoven, The Netherlands [2]Centre for Infectious Disease Control, National Institute for Public Health and the Environment (RIVM), Bilthoven, The Netherlands [3]Department of Laboratory Medicine, Karolinska Institutet, Stockholm, Sweden [4]Department of Human Genetics, Leiden University Medical Center, Leiden, The Netherlands

KEY FACTS

- Accumulation of cellular damage is a central feature of aging, resulting in functional decline and increased vulnerability to pathology and disease. This accumulation occurs over time, but is not exclusively time-dependent.
- Concerning biomedical issues, age may rather be predicted or defined as biological age, in contrast to chronological age.
- Examples of determinants of biological age are functional and physiological parameters, including pathological parameters.
- Age-related pathological phenotypes accumulate at different rates per individual and independent of other pathological endpoints within the same tissue, or similar endpoints across tissues.
- Within one individual, severe scorings of pathological parameters are rarely present for multiple parameters, that is, individuals score "old" on one parameter but not on the others, within and between organs.
- Using age-related phenotypes, gene-expression profiles and biological pathways are identified which are different from the profiles and pathways identified when using chronological age as the sole determiner of age.
- As all age-related phenotypes show some correlation with age, this new approach gives insight in phenotype-specific age markers that could be used to stratify individuals in their personal aging trajectories.
- Appropriate weighing factors between phenotype-specific markers may ultimately lead to the derivation of a biological age score for the entire individual organism.

Dictionary of Terms

- *Chronological age:* age expressed as the amount of time a person has lived in years, weeks, or days (eg, passport age).
- *Biological age:* age expressed as a measurement of biologically relevant parameters such as vulnerability to death and/or disease, risk of functional decline, or frailty index.
- *Gene expression profiles/whole transcriptome analysis:* profiles of the amount of mRNA expression in a

certain set of samples. Whole transcriptome analysis refers to the ability to analyze a large set of genes in a single sample.
- *Age-related phenotypes:* phenotype is a term used to parse a disease or set of symptoms in more stable components. In our case, phenotypes are used to parse symptoms related to aging into measurable parameters which are (partially) independent of each other.
- *Markers of frailty:* frailty refers to the age-related decline in physical, mental, and/or social functioning and as such is a concept of "unhealthy aging." Markers of frailty can be used to identify people at risk for accumulated aging phenotypes.
- *Lipofuscin:* pigment granules that are mainly composed of lipid-containing residues of lysosomal digestion. Accumulation of lipofuscin is associated with aging and can be observed in several organs (such as liver, nerve cells, kidney).
- *Karyomegaly:* refers to the presence of enlarged cell nuclei, and is associated with aging.
- *Liver vacuolization:* vacuolization potentially caused by increased glycogen storage, fat storage, or cellular swelling and may represent a (reversible) degenerative stage.

INTRODUCTION

Aging can be broadly defined as the functional decline occurring in organisms in a time-dependent manner. Aging is a complex process comprising a wide variety of interconnected features [1–7]. A central feature is the accumulation of cellular damage which results in functional decline and increased vulnerability to pathology and disease [1,2]. This accumulation occurs over time, but is not exclusively time-dependent. Human lifespan varies from less than 10 years for the severe progeria patients to over 100 years for centenarians [8–10]. This variation in human lifespan is partly related to genetic variation [8–10]. However, even in genetically identical animals and monozygotic twins lifespan, fitness, and biological functions of, for example, the immune system vary substantially [11–14], indicating that other factors than time and genetic variation are important determinants of biological aging as well. It is likely that the heterogeneity in lifespan observed within and between species is determined by a balance between damaging exposures and resilience to this damage [15]. It is important to note that biological aging is a stochastically driven process; there are components of coincidence involved in the net damage that occurs. Hence, variations in lifespan and fitness at older age arise from differences in exposure to damaging properties from the environment, differences in the characteristics of the damage done, and the body's innate ability to repair and compensate for this damage.

The functional decline associated with progressing chronological age has been proven difficult to mechanistically dissect or translate into consistent biomarkers of this decline. Likely, this is partly due to the complex interconnection of involved mechanisms that are difficult to disentangle and partly due to the use of chronological age (time) as the most important determinant in defining "young" and "old" animals or humans. Each individual organism follows its own path of aging, and, as a result, generalization of groups based on chronological age results in a heterogeneous set of aging processes, particularly in older groups. Consequently, this hampers investigations into the mechanisms underlying aging. Instead of using chronological age as the most important determinant, we propose to take biological age-ranks for a better assessment of the underlying mechanisms.

The idea of investigating biological age was proposed in 1969 by Alex Comfort [16] and has received attention ever since (eg, see Refs. [2,12,15,17–21]). Biological age refers to the age of an organism expressed in terms of biological fitness based on parameters that relate to the functional decline or vulnerability to death, in contrast to the chronological age, which depends exclusively on time. Hence, for a parameter to be informative of biological age it should associate with functional decline, vulnerability to death, or lifespan, and not necessarily with chronological age.

Several researchers have investigated whether physiological and/or functional parameters were associated with vulnerability to death or lifespan and, as such, whether these parameters were suitable markers for biological age. For example, Levine investigated several methods to use a set of physiological factors (eg, C-reactive protein, white blood cell count, serum urea nitrogen) to predict biological age [15]. The value of the predictions was tested by applying them to a large cohort of subjects (30–75 years) of which the physiological parameters and lifespan were known. He concluded that the Klemera and Doubal method, in which biological age is calculated based on a combined score of multiple aging dependent fitness parameters (including, eg, C-reactive protein, cholesterol levels, forced expiratory volume, and blood pressure) is better at predicting mortality than chronological age [15]. A slightly different approach was recently used by Tomas-Loba et al., who related metabolic profiles to chronological age and consecutively investigated if these could predict biological age in mice known to have a short or long lifespan [19]. The metabolic signature associated with chronological aging was able to predict aging produced by telomere-shortening.

However, in this study the researchers did not investigate if the metabolic profile was related to actual lifespan or functional decline.

A study by Holly et al. identified a set of six genes from human peripheral blood leukocytes that was predicative for chronological age [17]. This set was used in a second set of subjects to investigate if they could identify subjects with a younger predicted biological age (ie, healthier phenotype) compared to their chronological age. Subsequently, this group of "younger" subjects was compared to the remainder of the group based on several functional and physiological parameters associated with aging. They observed that the "predicted biologically younger" group scored better on the functional and physiological parameters (such as muscle strength, c-reactive protein, and several others), indicating that their biological age was lower than their chronological age.

In summary, these studies indicate that physiological and functional parameters can be used to predict or determine biological age. For example, it has been proposed that predicted biological age can be used to estimate an individual's chance of success when undergoing a complex surgery [22]. Another important functionality that using biological age enables is the investigation of mechanisms underlying aging and the search for pathology specific markers or frailty markers. In this chapter, we describe a method that takes into account a measure of biological aging, histopathological parameters in the liver, instead of chronological age as a mere determinant for age, with the aim of investigating mechanisms underlying aging.

PATHOLOGICAL PARAMETERS ARE ONLY PARTIALLY ASSOCIATED WITH CHRONOLOGICAL AGE

To investigate biological aging-related processes, different pathological parameters were scored at six chronological ages (13, 26, 52, 78, 104, 130 weeks) in female C57BL/6J mice [12,18]. Pathology scores were used to "rerank" the animals, independent of their chronological age, to derive a biological age ranking for each specific phenotype (Fig. 2.1). In Fig. 2.1, animals with the same rank are separated by chronological age for visualization purposes only, analyses were performed with original rankings.

We focused on three different liver pathologies (lipofuscin index, karyomegaly, and liver vacuolization) and to allow comparisons between organs we determined the severity of lipofuscin accumulation in the brain (brain lipofuscinosis). This focus on the liver was chosen considering the importance of metabolic processes of the liver in aging [2]. These three different pathological endpoints were selected from a set of seven initially analyzed endpoints [12], based on their scoring: higher dynamic scores are favorable in correlation analyses. Lipofuscin refers to the pigment granules that are mainly composed of lipid-containing residues of lysosomal digestion. Accumulation of lipofuscin is associated with chronological aging and can be observed in several tissues (such as liver, brain, nerve cells, kidney). Lipofuscin accumulation in the liver is expressed as the lipofuscin index (spot count × spot size × spot intensity, for all parameters the average of three fields is used). In the brain, lipofuscin accumulation is scored as stages of lipofuscinosis, which refers to the accumulation of lipofuscin in neurons. Karyomegaly refers to the presence of enlarged cell nuclei, associated with aging, and liver vacuolization is likely a consequence of increased glycogen storage, fat storage, or cellular swelling and may represent a (reversible) degenerative stage. The liver vacuolization was not further specified. For some animals certain pathology scores were not assessed. Therefore, the number of animals is unequal: liver lipofuscin 51 animals, liver karyomegaly 49 animals, liver vacuolization 49 animals, and brain lipofuscinosis 46 animals.

The pathology parameters partially correlate with chronological age (Table 2.1 and Fig. 2.1(A)). The strongest correlations are observed for lipofuscin accumulation in liver ($R = 0.85$) and brain ($R = 0.79$). The other two pathological parameters measured in liver correlate less strongly with chronological age: karyomegaly $R = 0.54$ and liver vacuolization $R = 0.52$ (Table 2.1). In Fig. 2.1(B), individual mice are ranked based on the separate pathology scores and colors indicate chronological age. Here the partial correlations observed between pathological parameters and chronological age are visualized, which exemplifies the concept of biological aging being partially distinct from chronological aging and emphasizes the importance of using biological endpoints. For example, based on lipofuscin scoring, a liver sample of a 2-year-old mouse could be considered "younger" than the liver of a 1-year-old mouse (mice are indicated with arrows in Fig. 2.1, red arrow = 130 weeks of age, green arrow = 52 weeks of age).

Interestingly, the variation observed within the age groups increases with chronological age for lipofuscin accumulation in liver and brain, with low levels of variation at a young age and increased variation at older ages (Fig. 2.1(A)). This is a phenomenon described for other factors as well: for example, variation of immune parameters increases with chronological age [23]. For liver karyomegaly

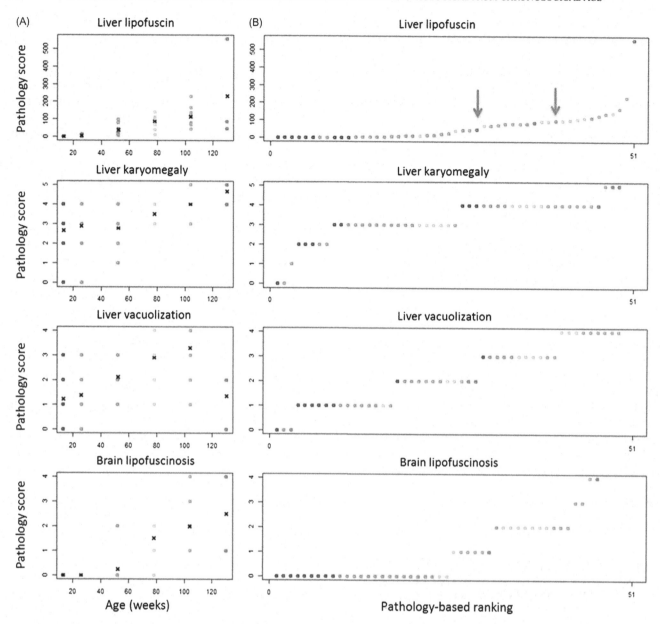

FIGURE 2.1 Ranking of mice based on chronological age (A) and pathological endpoint (B). This figure shows that pathology scores are partially associated with chronological age. Color indicates chronological age. Blue, 13 weeks; light blue, 26 weeks; green, 52 weeks; yellow, 78 weeks; orange, 104 weeks; red, 130 weeks. Black cross (panel A) indicates average values per group. Arrows indicate an example of a chronologically old mouse (130 weeks, *red arrow*) with a relatively low lipofuscin index (48.67, ranking 29/51) and a chronologically younger mouse (52 weeks, *green arrow*) with a relatively high lipofuscin index (97.3, ranking 40.5/51), see also section Intraorgan Specific Biological Phenotypes.

TABLE 2.1 Correlation Matrix of Chronological Age and Pathology Parameters: Values Represent the Spearman Rank Correlation

Correlation matrix of chronological age and pathology parameters					
	Age (weeks)	Liver lipofuscin	Liver karyomegaly	Liver vacuolization	Brain lipofuscinosis
Liver lipofuscin	0.85	1.00			
Liver karyomegaly	0.54	0.47	1.00		
Liver vacuolization	0.52	0.49	0.44	1.00	
Brain lipofuscinosis	0.79	0.61	0.55	0.45	1.00

van Kerkhof et al.: Expression profiles of aging-associated liver pathology phenotypes: unraveling stochastic aging processes.

and liver vacuolization, an increase in variation with age is not clearly observed (Fig. 2.1). This might be caused by the characteristics of these parameters. For example, for karyomegaly, large variation is already present at a young age. In addition, "ceiling effects" might occur, when pathology is abundantly present and is scored as the highest level, further increases in pathology do not result in a higher score. For liver vacuolization, scores might be underestimated when the total number of "young" small nuclei are decreased, since then size variation might be less evident and scoring might be affected. In addition, for liver vacuolization, a selection process might occur during aging in which animals that develop a more severe pathology score have a poorer survival, resulting in higher numbers of animals with a low pathology score in the older age groups (ie, the animals that do survive).

In summary, pathology parameters only partially correlate with chronological age, that is, within chronological age groups variation in pathology scores is clearly observed. This illustrates the concept of biological aging being partially distinct from chronological aging and emphasizes the importance of using biologically based endpoints.

INTRAORGAN SPECIFIC BIOLOGICAL PHENOTYPES

Fig. 2.2 visualizes in 3D the correlation observed between liver pathology parameters, of which the correlation coefficients are presented in Table 2.1. These results show that the severity of the lipofuscin index, karyomegaly, and hepatocellular vacuolization differs within one individual. The lipofuscin index correlates weakly with karyomegaly ($R = 0.47$) and with hepatocellular vacuolization ($R = 0.49$) (Table 2.1 and Fig. 2.2). Interestingly, only a few animals score "old" on all three parameters. For example, when the criterion for an "old" mouse is set to having for each parameter a score that is the "oldest" approximately 10%: lipofuscin index ≥ 130 (5 out of 51 mice), karyomegaly score ≥ 5 (3 out of 49 mice), and liver vacuolization score ≥ 4 (9 out of 49 mice), only one mouse meets this criterion (age = 104 weeks, indicated with orange arrow and number 1 in Fig. 2.2). When the criterion is set to having the approximately 40% "oldest" scores on liver pathology: lipofuscin index ≥ 90 (14 out of 51), karyomegaly score ≥ 4 (23 out of 49), and liver vacuolization score ≥ 3 (20 out of 49), only five mice meet these criteria. These results indicate that there is a low level of overlap in individual animals in intraorgan

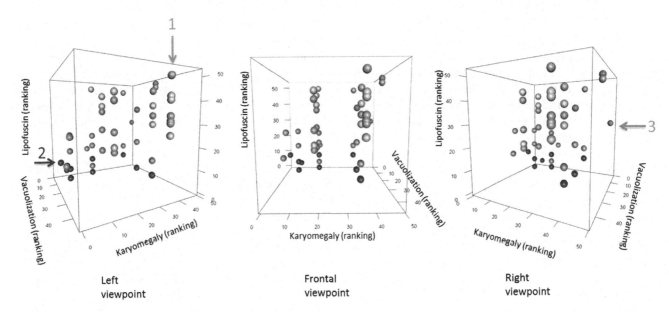

FIGURE 2.2 Pathology ranking for multiple endpoints in the liver. Mice are arranged by liver lipofuscin (y-axis), liver vacuolization (x-axis), and liver karyomegaly (z-axis). This figure visualizes the weak correlation between the different liver pathological parameters. Panels represent different viewpoints to the 3D matrix. Bead size reflects distance from sight to position of beads in the 3D matrix. Color indicates chronological age. Blue, 13 weeks; light blue, 26 weeks; green, 52 weeks; yellow, 78 weeks; orange, 104 weeks; red, 130 weeks. Arrows indicate examples of mice: *orange arrow* (1) indicates a mouse with a high score (top 10%, see section Tissue-Specific Biological Phenotypes) on all pathology parameters: lipofuscin index = 232, ranking 50/51, karyomegaly score = 4, ranking 36.4/48, vacuolization score = 4, ranking 45/45, age = 104 weeks; *blue arrow* (2) indicates a mouse with low pathology scores: lipofuscin index = 1, ranking 10/51, karyomegaly score = 0, ranking 1.5/48, vacuolization score = 1, ranking 10.5/45, age = 13 weeks; *red arrow* (3) indicates a mouse with high variation among pathology scores: lipofuscin index = 48.67, ranking 29/51, karyomegaly score = 5, ranking 48/48, vacuolization score = 0, ranking 2/45, age = 130 weeks.

pathology endpoints. The chronological age of the "40% oldest" five mice was 104 weeks for two mice and 78 weeks for three mice. Interestingly, none of the chronologically oldest mice (130 weeks) have an "old" pathology score on all three liver pathology parameters. An example of a 130-week-old mouse is indicated with the red arrow in Fig. 2.2, modest lipofuscin index (48.67, ranking 29/51), high karyomegaly score (5, ranking 48/48), and a low vacuolization score (0, ranking 2/45). Keeping in mind that the median survival of these mice is around 103 weeks of age [12], the mice in the oldest age group result from a strong selection bias for successful aging. Although it is not yet clear if the investigated pathological phenotypes directly relate to survival and fitness, these data suggest that having multiple severe pathology scores is related to poor survival. Hence, it appears that a selection process occurs during aging, resulting in the finding that some animals have a more beneficial or compensatory aging scenario and, therefore, survive better (eg, the 130-week-old animals which are not in the top 40% biologically "oldest" group). These findings strengthen the importance of using biological phenotypes and indicate that within an organ pathological processes occur largely independent of each other, indicating the need for intraorgan specific biological phenotypes.

TISSUE-SPECIFIC BIOLOGICAL PHENOTYPES

For the use of pathological endpoints as determinants of biological age, the question arises whether these processes occur in multiple organs in a similar fashion. Therefore, the severity of lipofuscin accumulation in the liver and brain were compared. Fig. 2.3 shows that there is only partial correlation between the severity of lipofuscin accumulation in the liver and brain ($R = 0.61$) (Table 2.1), although it is the highest correlation observed between the pathological parameters included. For example, the animal with the highest liver lipofuscin score has a relatively low brain lipofuscin score (Fig. 2.3, right most animal indicated with a black arrow, age 130 weeks). This indicates that within one individual tissue-specific aging processes occur, which should be considered as (partially) separate phenotypes, underlying the stochastic nature of aging processes.

GENE EXPRESSION PROFILES RELATED TO PATHOLOGICAL AGING PARAMETERS

The scoring of the different pathological parameters was used to investigate gene expression changes

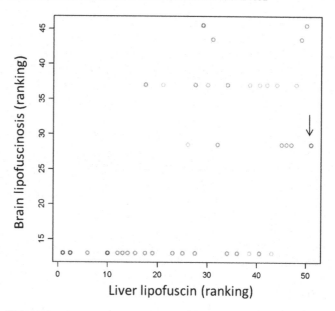

FIGURE 2.3 Pathology ranking for lipofuscin accumulation in liver and brain. Mice are arranged by liver lipofuscin ranking (x-axis) and brain lipofuscinosis ranking (y-axis) to allow comparison of both parameters. This figure visualizes the partial overlap between lipofuscin scores in liver and brain. Color indicates chronological age. Blue, 13 weeks; light blue, 26 weeks; green, 52 weeks; yellow, 78 weeks; orange, 104 weeks; red, 130 weeks. Arrow indicates an animal with high liver lipofuscin ranking, but only modest brain lipofuscinosis ranking (see also section Gene Expression Profiles Related to Pathological Aging Parameters).

associated with these parameters. Genes that alter expression in a manner that follows the kinetics of a certain pathological parameter might represent processes related to that biological endpoint's specific aging course. Gene expression is involved in and affected by most cellular processes, and therefore, whole transcriptome analysis allows investigation of several of the different aging processes simultaneously. For methodology of the whole transcriptome analysis, see Refs. [12,18]. For each of the endpoints, we identified genes that have an $R^2 > 0.5$ ($R > 0.707$ or $R < -0.707$; at least 50% of the variation in ranking is related to the pathological endpoint), using the Spearman rank correlation analysis. The number of genes with significant correlation to a pathological endpoint was highest for liver lipofuscin (287), followed by karyomegaly (225), brain lipofuscinosis (186), and liver vacuolization (111). Heat maps for the most significantly correlating genes ($|R| > 0.8$) are presented in Fig. 2.4(A–D). These expression patterns are not very consistent across the various pathological endpoints, which is in line with our previous findings that there are different genes associated with the different pathological endpoints [12,18].

Functional annotation analysis indicated that genes positively associated with liver lipofuscin are mainly associated with immunological processes, such as

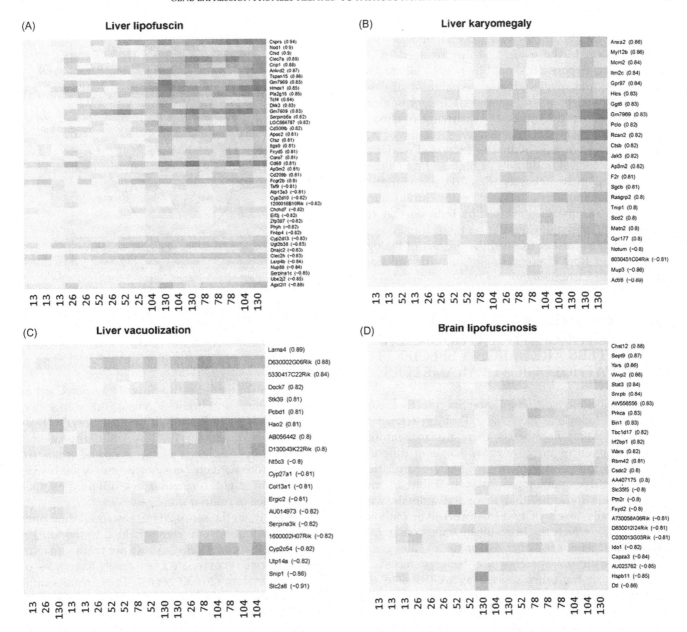

FIGURE 2.4 Heat maps of genes correlating with pathological parameters: (A) liver lipofuscin, (B) liver karyomegaly, (C) liver vacuolization, and (D) brain lipofuscinosis. The genes correlating most significantly |R| > 0.8 are shown. Gene names and correlation coefficient (in brackets) are shown on the y-axis. Age of the animals is shown on the x-axis in weeks.

immune response (20 upregulated genes), inflammatory response (9 genes), and oxidative stress response (3 genes), all indicating that stress at the cellular or tissue level occurs. Additionally, these genes were enriched for terms such as phagocytosis, vesicle mediated transport, and lysosome, all of which point to uptake and degradation of unwanted materials or waste, such as done by macrophages. Among the genes downregulated in relation to liver lipofuscin, a notable number of mitochondria-associated genes were detected. Mitochondrial dysfunction has previously been associated with chronological aging [2,24]. Genes associated with karyomegaly were also, albeit less clearly, enriched for genes involved in immune response- or lysosome-related processes as well as oxidative stress. In addition, there was enrichment for apoptosis-associated genes. In contrast, no significant functional enrichment was found among the genes associated with liver vacuolization, nor with brain lipofuscin. This, again, indicates that there are different functional processes involved in the various pathological endpoints. Interestingly, several of the pathways, such as the immune-related pathways, differ from the pathways that are detected when using chronological age as the sole determinant of age [12,18], indicating the value of using pathology-related endpoints.

With respect to the genes that are involved in the inflammatory response it should be noted that chronological aging and many aging-related chronic diseases are accompanied by alteration and decline of immune functions, such as the increased occurrence of chronic inflammation, a process known as "inflammaging" [25]. Moreover, in line with the increased variation found for biological endpoints at higher chronological age presented in this chapter (Fig. 2.1), also variation of immune parameters increase with chronological age [23]. In summary, these results indicate that the process of inflammaging should be taken into account in our search to further unravel the process of biological aging and how biological aging may be functionally related to immunology.

GENE EXPRESSION PROFILES CORRELATING WITH PATHOLOGICAL PARAMETERS ARE LARGELY SPECIFIC TO THE PATHOLOGICAL PARAMETERS

As described in sections Tissue-Specific Biological Phenotypes and Gene Expression Profiles Related to Pathological Aging Parameters, there are large intraindividual differences in the severity scores of different pathologies, however, there is some correlation between the liver pathologies as well (Table 2.1 and Fig. 2.2). This indicates that there are animals with similar severity levels of two (or three) liver pathologies (Fig. 2.2), which might result in some contamination of the gene sets, that is, some of the genes associated with one process are being detected in the analysis of the other process as well due to the ranking approach. To determine this possible bias in the gene sets, we investigated the overlap in these gene sets. For most endpoints very little overlap is observed (Fig. 2.5), only between liver lipofuscin and karyomegaly a substantial overlap (62 genes) was observed (overlap is 22% of lipofuscin correlating genes and 28% of karyomegaly correlating genes). Therefore, some bias might be present in these results.

For lipofuscin accumulation in liver and brain, some overlap in genes correlating with these endpoints might be expected, since these phenotypes have similarities. However, of the genes correlating with lipofuscin accumulation in liver and brain only four genes are overlapping (of 287 genes associated with liver lipofuscin and 186 genes associated with brain lipofuscinosis). To determine if the nonoverlapping genes were mainly tissue-specific genes, we used a mouse tissue data set [26] (www.biogps.org, top 1% tissue-specific genes). Interestingly, only low levels of tissue-specific genes

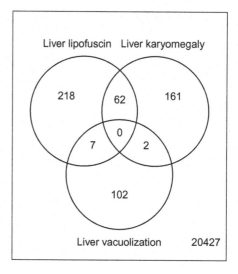

FIGURE 2.5 Venn diagram of genes correlating with the liver pathological endpoints ($R^2 > 0.5$, $R > 0.707$, or $R < -0.707$). Within the liver gene set, a significant correlation was observed for 287 genes, for 225 genes with liver karyomegaly, and for 111 with liver vacuolization. This figure shows the overlap among these genes. Of the gene set, 20,427 genes did not show a correlation with any pathological parameter.

were observed in the set of lipofuscin associated genes in liver (10 genes) and brain (3 genes). Taken together, these results might indicate that lipofuscinosis in liver and brain are different. Possibly, cellular processes that underlie lipofuscinosis in the liver and brain are different, or alternatively, but not mutually exclusive, liver and brain might to respond in a different manner to similar types of cellular damage.

Interestingly, in the set of 283 genes correlating only with liver lipofuscin, some genes belong to the top 1% of immune cell specific genes. These are mainly macrophages and cell types related to these phagocytes of the immune system (26–30 genes). These results indicate that macrophages, or macrophage-like cells, such as Kupffer cells of the liver, are likely involved with liver lipofuscin accumulation, which is in line with the results reported in the section Gene Expression Profiles Related to Pathological Aging Parameters.

FUTURE PERSPECTIVES

We have shown that ranking complex data sets (ie, gene expression data) according to lesion specific severity, results in the detection of unique markers sets for each of these endpoints. These markers are lost in the background noise when a chronological time ranking is used. As all markers show some correlation with age, this new approach gives insight in phenotype-specific age markers that may be used to stratify individuals during their respective aging trajectories. This type of approach can be applied in large scale molecular epidemiology studies since it can be performed with any phenotype related to biological aging, such as functional measurements (eg, grip strength) and physiological measurements (eg, blood pressure) or combinations of these (eg, a frailty index) and can be applied to a variety of complex data sets, such as metabolomics and epigenomics data sets. Application of this approach in multiple studies on different phenotypes and different data sets will greatly enhance our understanding of the processes underlying aging and will lead to the discovery of new biomarkers of unhealthy aging, that is, biomarkers of frailty. These biomarkers of frailty will aid in the early detection of unhealthy aging. In addition, using appropriate weighing factors between phenotype-specific markers may ultimately lead to the derivation of a biological age score for the entire individual organism and to the derivation of a personalized biological age score for individuals within the population comprising our aging society.

CONCLUSION

We have shown that pathological endpoints (as phenotypes of biological aging) accumulate at different rates per individual and independently of other pathological endpoints within the same tissue, or similar endpoints across tissues. Furthermore, by ranking complex data sets (ie, gene expression data) according to lesion specific severity, unique sets of markers are found for each of these endpoints. For example, reranking of mice according to their lipofuscin score revealed the involvement of immune processes, such as inflammation, indicating that the process of "inflammaging" should be taken into account in our search to further unravel the process of biological aging. These genes are lost in the background noise, that is, are not identified, when a chronological time ranking is used. As all phenotypes show some correlation with age, this new approach gives insight in phenotype-specific age markers that may be used to stratify individuals during their personal aging trajectories. This approach will lead to more insight into the mechanisms underlying aging. Appropriate weighting factors between phenotype-specific markers may ultimately lead to the derivation of a biological age score for the entire individual organism.

References

[1] Garinis GA, van der Horst GT, Vijg J, Hoeijmakers JH. DNA damage and ageing: new-age ideas for an age-old problem. Nat Cell Biol 2008;10(11):1241−7.
[2] Lopez-Otin C, Blasco MA, Partridge L, Serrano M, Kroemer G. The hallmarks of aging. Cell 2013;153(6):1194−217.
[3] Mocchegiani E, Malavolta M. Possible new antiaging strategies related to neuroendocrine-immune interactions. Neuroimmunomodulation 2008;15(4−6):344−50.
[4] Morimoto RI, Cuervo AM. Proteostasis and the aging proteome in health and disease. J Gerontol A Biol Sci Med Sci 2014;69 (Suppl. 1):S33−8.
[5] Oh J, Lee YD, Wagers AJ. Stem cell aging: mechanisms, regulators and therapeutic opportunities. Nat Med 2014;20(8):870−80.
[6] Xi H, Li C, Ren F, Zhang H, Zhang L. Telomere, aging and age-related diseases. Aging Clin Exp Res 2013;25(2):139−46.
[7] Zane L, Sharma V, Misteli T. Common features of chromatin in aging and cancer: cause or coincidence? Trends Cell Biol 2014.
[8] Cleaver JE, Lam ET, Revet I. Disorders of nucleotide excision repair: the genetic and molecular basis of heterogeneity. Nat Rev Genet 2009;10(11):756−68.
[9] Iannitti T, Palmieri B. Inflammation and genetics: an insight in the centenarian model. Hum Biol 2011;83(4):531−59.
[10] Kraemer KH, Patronas NJ, Schiffmann R, Brooks BP, Tamura D, DiGiovanna JJ. Xeroderma pigmentosum, trichothiodystrophy and Cockayne syndrome: a complex genotype-phenotype relationship. Neuroscience 2007;145(4):1388−96.
[11] Hollander CF, Solleveld HA, Zurcher C, Nooteboom AL, Van Zwieten MJ. Biological and clinical consequences of longitudinal studies in rodents: their possibilities and limitations. An overview. Mech Ageing Dev 1984;28(2−3):249−60.
[12] Jonker MJ, Melis JP, Kuiper RV, et al. Life spanning murine gene expression profiles in relation to chronological and pathological aging in multiple organs. Aging Cell 2013;12(5):901−9.
[13] Melis JP, Wijnhoven SW, Beems RB, et al. Mouse models for xeroderma pigmentosum group A and group C show divergent cancer phenotypes. Cancer Res 2008;68(5):1347−53.
[14] Brodin P, Jojic V, Gao T, et al. Variation in the human immune system is largely driven by non-heritable influences. Cell 2015;160(1−2):37−47.
[15] Levine ME. Modeling the rate of senescence: can estimated biological age predict mortality more accurately than chronological age? J Gerontol A Biol Sci Med Sci 2013;68(6):667−74.
[16] Comfort A. Test-battery to measure ageing-rate in man. Lancet 1969;2(7635):1411−14.
[17] Holly AC, Melzer D, Pilling LC, et al. Towards a gene expression biomarker set for human biological age. Aging Cell 2013;12(2):324−6.
[18] Melis JP, Jonker MJ, Vijg J, Hoeijmakers JH, Breit TM, van Steeg H. Aging on a different scale—chronological versus pathology-related aging. Aging 2013;5(10):782−8.
[19] Tomas-Loba A, Bernardes de Jesus B, Mato JM, Blasco MA. A metabolic signature predicts biological age in mice. Aging Cell 2013;12(1):93−101.
[20] DeCarlo CA, Tuokko HA, Williams D, Dixon RA, MacDonald SW. BioAge: Toward a multi-determined, mechanistic account of cognitive aging. Ageing Res Rev 2014;18c:95−105.
[21] Seroude L. Differential gene expression and aging. Scientific World J 2002;2:618−31.
[22] Jackson SH, Weale MR, Weale RA. Biological age—what is it and can it be measured? Arch Gerontol Geriatr 2003;36(2):103−15.
[23] Shen-Orr SS, Furman D. Variability in the immune system: of vaccine responses and immune states. Curr Opin Immunol 2013;25(4):542−7.
[24] Katic M, Kennedy AR, Leykin I, et al. Mitochondrial gene expression and increased oxidative metabolism: role in increased lifespan of fat-specific insulin receptor knock-out mice. Aging cell 2007;6(6):827−39.
[25] Franceschi C, Bonafe M, Valensin S, et al. Inflamm-aging. An evolutionary perspective on immunosenescence. Ann N Y Acad Sci 2000;908:244−54.
[26] Lattin JE, Schroder K, Su AI, et al. Expression analysis of G protein-coupled receptors in mouse macrophages. Immunome Res 2008;4:5.

CHAPTER

3

Nutrigenomics and Nutrigenetics: The Basis of Molecular Nutrition

Jean-Benoit Corcuff[1,2] *and Aksam J. Merched*[3,4]

[1]Department of Nuclear Medicine, CHU de Bordeaux, Pessac, France [2]Integrated Neurobiology and Nutrition, University of Bordeaux, Pessac, France [3]UFR of Pharmaceutical Sciences [4]INSERM Bordeaux Research in Translational Oncology (BaRITOn), University of Bordeaux, Bordeaux, France

KEY FACTS
- The integration of the human genome project with nutritional, genetic research, and health outcomes studies has led to the emergence of nutrigenetics and nutrigenomics.
- Nutrigenetics investigate in a systematic fashion the effect of genetic makeup on individual dietary response.
- Nutrigenomics focus in the role of dietary factors and bioactive nutrient products in gene expression and stability.
- Some nutrients contribute to limit DNA damages caused by aging.
- Epigenetic changes may influence expression of genes involved in nutrient metabolism and signaling. Conversely, nutrients may modulate processes that affect epigenetics.

INTRODUCTION

Many previous and ongoing large efforts aimed at gathering information to link genetic variations to disease risk factors are promoting the emergence of precision or personalized medicine. Moreover, based in their genetic variability, individuals differ in their abilities to metabolize nutrients and to respond to diet. Growing data on gene–disease and gene–nutrient interactions have increased the interest in individualized nutritional therapy with the general public and health practitioners.

The ingestion of nutrients and other so-called food bioactive components generates biological responses sustained by multiple networks of linked physiological processes. The latter include food absorption, molecular transformation, nutrient transport, molecular uptake and storage in cells, and cellular mechanisms of action, metabolism, and elimination. It was recognized decades ago that some individual genes were necessary for the metabolism of very specific nutrients. Indeed, some monogenic diseases have been linked to defects in metabolism pathways as all depend on the expression of a myriad of genes and subsequent protein translation. Thus, (i) genetic variants of these genes may exert diverse functional effects, and (ii) various nutrients may differently affect genes' expression. Applied to nutrition, great progress in genetics and genomics delineated two new fields: nutrigenetics and nutrigenomics. Nutrigenetics studies the effect of genetic variation on dietary response (Fig. 3.1), whereas nutrigenomics studies the effect of nutrients on DNA and gene expression; it often includes the effects on DNA structure such as the epigenome (Fig. 3.2). Nutrigenetics and nutrigenomics both aim to better understand nutrient–gene interactions. An underlying but not exclusive goal is to envision personalized nutrition strategies for optimal health and disease prevention. The latter are obviously goals to be reached for optimal and if possible delayed aging.

 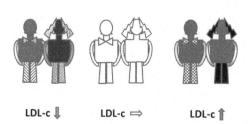

Nutritional recommendation for the general population: intake of omega 3 fatty acid improves your blood lipids

Interindividual variations in response to fish oil supplementation: variability in the concentration of serum LDL cholesterol

FIGURE 3.1 **Nutrigenetics and personalized nutrition; one size doesn't fit all.** The cartoon illustrates the fact that dietary recommendation, which are designed for the general population and based on different metabolic outcomes, may not be optimized for genetic subgroups. Here, heterogeneity in LDL-cholesterol (LDL-C) was observed in response to dietary intake, for example, 2.5 g EPA + DHA/day supplement for a 6-week period [1]. Among other factors, individual genetic variability, for example, in apoE, contributed to this heterogeneous response to changes in dietary fat.

 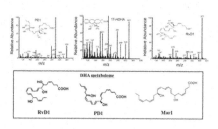  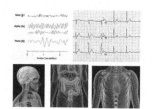

Diet or dietary supplements | Dietary bioactive byproducts | Differential regulation of genes or DNA integrity | Pleiotropic effects

FIGURE 3.2 **Nutrigenomic exploration of the role of dietary factors and bioactive nutrient products.** Fish and fish oil rich in omega-3 PUFA such as DHA and their bioactive end-products affect specific gene expression in several cellular targets leading to different health outcomes within the nervous, colorectal, and cardiovascular systems (see text for more information).

Aging is a process that combines different elements leading to two nonexclusive clinical situations. The first could be called healthy aging and the second is where some pathologies develop because of the extended life (eg, cancer, atherosclerosis, dementia, etc.). The distinction between these two processes may not be easy although it is important to mention that some elderly subjects have no clearly defined pathology, hence healthy aging. Subsequently, to describe nutrigenomic or nutrigenetic approaches of these situations is different. For instance, it is often more straightforward to imagine a link between nutrients, dyslipidemia, apolipoprotein polymorphisms and atherosclerosis, diabetes or cancer than to explore genes that are not expressed in healthy old subjects. Hence, many clinical or experimental nutrigenomic or nutrigenetic approaches target identified pathologies sustained by the idea that if a disease can be prevented or cured one can come back to a healthy aging situation. However, thanks to major advances in technologies investigating the whole genome, transcriptome, epigenome, or metabolome at affordable costs experimentally approaching healthy aging is now easier. A very didactic review enumerates the cellular hallmarks of aging [2].

The chapter will focus on nutrigenetics of cardiovascular diseases (CVD) and cancer. We will use n-3 polyunsaturated fatty acids (PUFA) intake as an example, taking into account gene variants encoding PUFA targets but also those involved in PUFA processing, as either genetic variability can affect response to PUFA intake. We will also discuss some aspects of nutrigenomics related to the effects of nutrients on cellular senescence, DNA damage, and epigenetics.

NUTRIGENETICS OF OMEGA-3 PUFA IN CVD

Inflammation and perturbed lipid metabolism are major players in atherosclerosis development. The relationship between elevated circulating lipids and atherosclerosis and cardiovascular diseases has been clearly established. Indeed, better nutrition remains the cornerstone to metabolic homeostasis and atherosclerosis prevention. Optimization of lipid metabolism depends on the interplay of complex biochemical pathways involving numerous enzymes, receptors, activators, and other factors. Genetic variability of these players is a key modulator of the final lipid phenotype. Unfortunately, much of the traditional nutritional research-based recommendations on reference values and guidelines to the general population have not paid sufficient attention to individual genetic variability; a much more nuanced approach towards such recommendations is warranted.

Dietary fatty acids such as omega-3 PUFAs are an excellent example of a widely studied nutrient that interacts with the genetic makeup in correlation with cardiovascular disease and many other human diseases [3]. A growing number of epidemiologic data attributed atheroprotective properties to fish and fish oil rich in omega-3 PUFA, in particular eicosapentaenoic acid (EPA) and docosahexaenoic acid (DHA) used in nutritional supplementation. However, many of these studies observed large variations in lipid phenotype, for example, levels of LDL-c, triglycerides, and HDL-c, which are probably related to the interactions between these nutrients and the lipid traits [1]. Here we summarize the interactions of individual gene variations in apolipoprotein (apo)AI, apoA5, apoE, TNFalpha, PPAR alpha, NOS3, lipoxygenases (ALOX) -5 and 12/15 genes [4] with omega-3PUFA intake in modulating lipid metabolism and cardiovascular outcome.

Studies of associations between plasma apoAI concentrations and circulating HDL-c levels have been inconsistent probably because of the complex genetic and dietary contribution to the control of HDL level. Indeed, the Framingham Study has reported significant interaction in women between a common genetic polymorphism in the promoter region of the APOA1 gene, the −75G/A SNP and PUFA intake in determining plasma HDL-c concentration [5]. Carriers of the A allele at the −75 G/A polymorphism show an increase in HDL-c concentrations with increased intakes of PUFA, whereas those of the more common G allele show the opposite effect by displaying lower HDL-c levels as the intake of PUFA increases.

Tumor necrosis factor-alpha (TNF-alpha) is a proinflammatory cytokine that can have an impact on lipid metabolism by modulating the expression of lipoprotein lipase (LPL); proliferator activated receptors (PPARs); apolipoprotein (apo) A-I, apo A-IV, and apo E; and lecithin:cholesterol acyltransferase. TNF-alpha itself is also known to be modulated by dietary PUFAs. Two common genetic polymorphisms in the promoter region of the TNF-alpha gene were known to alter the transcriptional activity of the cytokine gene in vitro. The potential role of these TNF-alpha genotypes on the association between dietary PUFA intake and serum lipid concentrations was investigated among individuals with type 2 diabetes in the Canadian trial of dietary Carbohydrate in Diabetes study [6]. PUFA intake was positively associated with serum HDL-c and apoA-I level in carriers of the −238A allele, but negatively associated in those with the −238GG genotype. PUFA intake was inversely associated with HDL-c level in carriers of the −308A allele, but not in those with the −308GG genotype. A stronger nutrigenetic effect was observed when the polymorphisms at the 2 positions (−238/−308) were combined. If validated these associations suggest that subjects with low levels of HDL-c, carriers of the A allele at the APOA1 −75 G/A and the alleles −238A, 308GG at the TNF-alpha −238G/A and −308G/A polymorphisms may benefit from diets containing higher percentages of PUFA.

Apolipoprotein A-V (ApoA-V) is a component of lipoproteins including HDL and chylomicrons and is involved in postprandial lipoprotein metabolism. By activating LPL, a key enzyme in the metabolism of TG, apoA-V is an important determinant of plasma TG concentrations. In the Framingham study, authors demonstrated a significant interaction between −1131T/C SNP, which is associated with greater TG concentrations in carriers of the C allele and PUFA intake [7]. The −1131C allele was associated with an increased fasting TG and remnant-like particle–TG concentrations only in subjects consuming 46% of energy from PUFA.

Nitric oxide synthase (NOS3) is responsible for the production of nitric oxide (NO), which is involved in the regulation of vascular function and blood pressure. SNPs in the NOS3 gene were found to be associated with a number of CVD risk markers including dyslipidemia and inflammation. The effects of NOS3 polymorphisms were investigated in a cohort of 450 patients with metabolic syndrome from the LIPGENE study, who participated in a 12-week dietary intervention to alter dietary fatty acid composition and amount [8]. The study demonstrated that carriers of the minor allele for rs1799983 SNP showed a negative correlation between plasma TG concentrations and plasma omega-3 PUFA status compared with subjects homozygous for the major allele. Following omega-3 PUFA supplementation, subjects with the minor alleles had a better response to changes in plasma omega-3 PUFA than major allele homozygous carriers. This highlights the potential benefit for individual carriers of the minor allele at rs1799983 in NOS3 in omega-3 PUFA supplementation to achieve reduction of plasma TG concentrations.

Peroxisome proliferator-activated receptor a (PPAR-a) is a ligand-dependent transcription factor that is a key regulator of lipid homeostasis. Because of its strong involvement in activating LPL, clearance of plasma triglycerides, and upregulation of HDL-c, PPARA has been a drug target for pharmaceutical companies. The most frequently studied polymorphism of the PPARA gene was the Leu162Val variant in which the minor allele was associated with lipid metabolism and atherosclerosis. However, there was controversy on the potential effect of this SNP on plasma TG and apoC-III concentrations. Subsequent studies showed that dietary PUFA intake can modulate the effects of this SNP on lipid metabolism [9]. The 162V allele was associated with higher TG and apoC-III levels only in subjects consuming a low-PUFA diet. Conversely, high consumption of PUFA diet in 162V subjects was related with the opposite effect on apoC-III [9].

ALOXs are involved in the biosynthesis of specialized bioactive lipid mediators from n-3 essential PUFAs, coined resolvins and protectins, which possess potent anti-inflammatory and proresolving actions that stimulate the resolution of acute inflammation [10]. Genetic studies on the role of 12/15-lipoxygenases in human atherosclerosis suggest that ALOX15 locus is more commonly associated with either a relatively neutral or an atheroprotective effect [11]. On the other hand, these human genetic studies did not consistently show an association of functional variants in ALOX15 with clinical endpoints of atherosclerosis. Given the potent actions of lipoxins, resolvins, protectins and maresins in models of human disease, deficiencies in resolution pathways may contribute to many diseases including atherosclerosis [12]. Lipidomic and metabolomic analysis of these bioactive lipid mediators provide a powerful approach to investigate the nutrigenetic and nutrigenomic effect of omega-3 PUFAs in health and disease.

NUTRIGENETICS OF OMEGA-3 PUHA IN CANCER

Cancer is a complex and heterogeneous disease with different types and forms also occurring at the interplay between genetic predisposition and environmental factors including nutrition. Many nutrients affect DNA integrity, cancer cell proliferation, and metabolic needs.

For these reason, many studies have investigated the interaction between intake of nutrients like n-3 fatty acids and the risk of different types of cancer such as breast, prostate, and colorectal cancers. As evidenced by experimental data, marine n-3 fatty acids may provide protection against many cancers, by neutralizing cytotoxic lipid peroxidation products, modulating inflammation, and preserving DNA integrity. In this section, we give examples of specific interactions of individual gene variations in glutathione S-transferases (GSTs), X-ray repair cross-complementing proteins (XRCC), Poly (ADP-ribose) polymerase (PARP), cyclooxygenase-2 (COX-2) or prostaglandin synthase (PTGS), lipoxygenases (5, 12, and 15 ALOX), with omega-3 PUFA intake in modulating these protective features and ultimately cancer outcome.

Experimental studies have demonstrated a direct role for the peroxidation products of marine n-3 fatty acids in breast cancer protection suggesting that the GSTs may be major catalysts in the elimination of these beneficial by-products [13]. GSTs are polymorphic with three well-characterized isozymes, GSTM1, T1, and P1, in which GSTM1 null, GSTT1 null, and GSTP1 AB/BB possess low GST activity. Genetic analysis on 258 breast cancer cases in the Singapore Chinese Health Study found a stronger protection of the n-3 fatty acids in the low than high activity genotype subgroups. Indeed, the risk of breast cancer in high versus low consumers of marine n-3 fatty acids, was reduced by half in women with the genotype GSTP1 AB/BB genotype, with similar borderline significance results for the other genotypes. The study concluded for the first time that the Chinese women carriers of the genetic polymorphisms encoding lower or no enzymatic activity of GSTM1, GSTT1, and/or GSTP1 experienced more breast cancer protection from marine n-3 fatty acids than those with high activity genotypes. Unfortunately, subsequent studies were not able to replicate these results.

XRCC1 (X-ray repair cross-complementing protein 1) and XRCC3 (X-ray repair complementing defective repair in Chinese hamster cells 3) are involved in the efficient repair of DNA and may play an important role in the repair or modulation of the effects of PUFA intake on cancer colorectal cancer risk. A large sigmoidoscopy-based case-control study (753 cases and 799 controls) in Los Angeles County, investigated possible associations between single-nucleotide polymorphisms in the XRCC1 (codons 194 Arg/Trp and codon 399 Arg/Gln) and XRCC3 (codon 241 Thr/Met) genes and colorectal adenoma risk and their possible role as modifiers of the effect of monounsaturated fatty acid, the ratio of omega-6/omega-3 polyunsaturated fatty acids, and antioxidant intake [14]). Although they did not find evidence of gene—dietary fat interactions for the XRCC3 codon 241 polymorphism, the study reported that high monounsaturated fatty acid intake was associated with adenoma risk only among subjects with the XRCC1 codon 194 Arg/Arg and codon 399 Gln/Gln combined genotypes. High omega-6/omega-3 polyunsaturated fatty acid ratios were associated with adenoma risk among subjects with the XRCC1 codon 194 Arg/Arg and codon 399 Gln/Gln or the codon 194 Arg/Trp or Trp/Trp and codon 399 Arg/Arg or Arg/Gln combined genotypes. These data suggest that the XRCC1 polymorphisms may modify the effect of unsaturated fatty acid intake on colorectal adenoma risk.

A nested case-control study of colorectal cancer (1181 controls and 311 cases) on Chinese participants in the Singapore Chinese Health Study [15] extended the previous study by simultaneously analyzing 7 SNP in various DNA repair genes: XRCC1 (Arg194Trp, Arg399Gln), OGG1 (Ser326Cys), PARP (Val762Ala, Lys940Arg), and XPD (Asp312Asn, Lys751Gln). The study observed that the PARP Val762Ala SNP modified the association between marine n-3 PUFA and rectal cancer risk, with no evidence of interaction among colon cancer. Notably, high intake of marine n-3 PUFA was associated with increased rectal cancer risk among carriers of at least one PARP codon 762 Ala allele. The latter is associated with reduced enzymatic activity. These data suggest that among carriers of a PARP protein with reduced enzymatic activity, diets high in marine n-3 PUFAs might be harmful for the rectum. The PARP protein plays an important role in maintaining genomic stability, apoptosis, and regulating transcription (as coactivator of the beta-catenin-TCF-4 complex). The fact that the authors have not observed similar findings for total n-3 PUFAs suggests that the association is driven by the long-chain EPA and DHA present mostly in fish.

COX-2 is a key enzyme in involved in the biosynthesis of eicosanoids derived from arachidonic acid and inflammation. Because prostate cancer tissues express high levels of COX-2, this enzyme may play a very important role in carcinogenesis. n-3 Fatty acids, such as EPA and DHA, induce metabolic shift from the biosynthesis of arachidonic acid-derived eicosanoids which can promote prostate cancer towards the less inflammatory E- and D-series resolvins [16]. Two independent studies demonstrated that the effect of long chain n-3 fatty acids on the risk of prostate cancer may be modified by COX-2 genetic variation. The first is the Cancer Prostate in Sweden Study, a population-based case-control study [17] that analyzed 5 SNPs in the COX-2 gene in 1378 prostate cancer cases and 782 controls. The authors reported that frequent consumption of fatty fish and marine fatty acids appears to reduce the risk of prostate cancer, and this association is modified by genetic variation in the COX-2 gene. Indeed, eating fatty fish (eg, salmon-type fish) once or more per week, compared to never, was associated with reduced risk of prostate cancer. They also found a significant interaction between salmon-type fish intake and a SNP in the COX-2 gene (rs5275: +6365 T/C), but not with the 4 other SNPs examined and strong inverse associations with increasing intake of salmon-type fish among carriers of the variant allele, but no association among carriers of the more common allele. The second is a case-control study of 466 men diagnosed with aggressive prostate cancer and 478 controls recruited by major medical institutions in Cleveland, Ohio [18]. After analyzing nine polymorphisms in the COX-2 gene, the study found that increasing intake of long chain (LC) n-3 was strongly associated with a decreased risk of aggressive prostate cancer. The association was stronger among men with SNP rs4648310 (+8897 A/G). Men with low LC n-3 intake and the variant rs4648310 SNP had an increased risk of disease which was reversed by increasing intake of LC n-3. The study did not find a similar pattern of interaction with rs5275 as reported in the previous one. The SNP rs4648310 and rs5275 are located 2.4 kb apart and exhibit weak linkage disequilibrium in their population. The functional effect of rs5275, an intronic variant, and rs4648310, flanking the 3′ end of the COX-2 gene, on COX-2 activity is not yet known. However, the combined findings of these two studies support the hypothesis that the effect of LC n-3 fatty acids on the risk of prostate cancer may be modified by COX-2 genetic variation and that LC n-3 modifies prostate inflammation through the COX-2 enzymatic pathway.

The interaction of n3-PUFAs intake with other gene variants affecting the enzymatic conversion of these PUFAs was investigated in colorectal cancer [19] bearing in mind that one of the proposed mechanisms by which n-3 fatty acids may inhibit promotion or progression of carcinogenesis is through their suppressive effect on the synthesis of AA-derived eicosanoids. A population-based case-control study of colon (case $n = 1574$), rectal cancer (case $n = 791$) and disease free controls ($n = 2969$) investigated interactions between dietary fatty acid intake and 107 candidate polymorphisms in PTGS1, PTGS2 (COX pathway), ALOX12, ALOX5, ALOX15, and FLAP (5-lipoxygenase-activating protein). The authors reported statistically significant increases in colon cancer risk for low DHA acid intake among those with the PTGS1 rs10306110 (-1053 A > G) variant genotypes and rectal cancer risk for low total fat intake among those with the variant PTGS1 rs10306122 (7135 A > G). The ALOX15 rs11568131 (10,339 C > T) wild-type in combination with a high inflammation score (low EPA intake, high AA intake, no regular NSAID use, high BMI, smoking) was associated with increased colon cancer risk. Rectal cancer risk was inversely associated with a low inflammation score among PTGS2 rs4648276 (3934 T > C] variant allele carriers. Overall, these data provide some evidence for interactions between dietary fat intake and genetic variation in genes involved in eicosanoid metabolism and colorectal cancer risk.

DNA DAMAGE AND NUTRIENTS

Nutrigenomics investigates the effects of dietary factors and bioactive nutrient products in gene expression and genome stability (Fig. 3.2); it often includes the effects on DNA structure such as the epigenome. This section focuses on the effects of nutrients on DNA damage mainly correlated with aging. Indeed, the accumulation of somatic damage is considered a major cause of aging. Reactive oxygen species (ROS), the natural by-products of oxidative energy metabolism, are often considered as major culprits of molecular damages [20]. There are arguments for a role of nutrients opposing the deleterious action of ROS and possibly other causes of DNA damage thus improving genomic stability.

Age-related alterations of DNA structure are partly reversed by enzymes devoted to DNA repair. Some of these enzymes are dependent on nutrients such as selenium. Selenium (Se), an essential micronutrient, is a component of the unusual amino acids selenocysteine (Se-Cys) and selenomethionine (Se-Met). It is also a component of some antioxidant enzymes (glutathione peroxidases and thioredoxin reductases). Various selenoproteins demonstrate antioxidant effects, reducing the biological impact of ROS on DNA amongst other biomolecules [21]. Se-related protection against DNA or chromosome damage has been shown in both in vivo and in vitro studies. Individuals with the lowest serum selenium levels had significantly higher levels of overall accumulated DNA damage [22]. A supplementation study on young healthy nonsmokers showed a reduction of DNA breakage after ingestion of a Se-containing antioxidant supplement [23]. Se may also protect mitochondrial DNA damages due to the local very high production of ROS [21].

Se is not the only nutrient to have been studied, vegetable-derived nutrients have also been investigated. For instance, DNA damage has been studied in women under a vegetarian diet [24]. In 60–70-year-old subjects, this diet was linked to significantly reduced values of DNA breaks with oxidized purines and DNA breaks with oxidized pyrimidines. In a randomized, controlled intervention study carotenoid supplementation decreased DNA damage [25]. More generally, a Mediterranean diet supplemented with coenzyme Q10 diet improved oxidative DNA damage in elderly subjects [26]. In vitro, potential antiaging molecules, such as rapamycin and metformin, but also nutrients (berberine, resveratrol, vitamin D3) protect cultured cells from DNA damage investigated with the expression of phosphorylated histone H2AX (γH2AX) [27, 28]. Thus, although as yet limited and fragmented, there are arguments for actions of some nutrients initiating on DNA repair or at least limiting damage. However, experimental conditions and control groups have to be carefully monitored. Indeed, for instance, the effects of resveratrol on DNA damage and redox status depend on the individual basal DNA status (athletes versus sedentary subjects) [29].

Some studies focused on a more localized DNA damage, telomeric damage. Telomeres, the ends of the DNA linear molecules, are subjected to a progressive shortening at each somatic cell division in cells devoid of significant telomerase activity. Telomeres are repeats of a species-specific hexamer sequence associated with the shelterin protein complex. The ends of linear chromosomes are unable to fully replicate during each cycle of cell division and thus telomeric DNA is shortened by 30-200 bp. Briefly, this DNA shortening is interpreted by the cells as DNA damage and drives the cells to a senescent state or to apoptosis. Besides cell division, various environmental, physiological, or psychological stressors may shorten telomeres. Targeting the latter may retard the signal for somatic cell senescence. Conversely, many tumoral cells exhibit telomerase activity or an alternative lengthening of telomeres that prevents telomere shortening, which is a means for the cells to escape telomere shortening-related apoptosis. Associative studies have shown that vitamins D and E, dietary fiber, and omega-3 fatty acid intakes in the diet correlated with longer telomeres. Conversely, processed meat, alcohol intake, and low fruit and vegetable intakes were negatively associated with telomere length (for review see [30]). Folates and vitamin B12 may for instance be indirectly implicated in telomere length. When a diet is deficient in these two nutrients, concentrations of homocystein rise. This has been correlated to a reduction of telomere length [31, 32]. Briefly, folate deficiency might induce telomere attrition and/or dysfunction by: excision of increased uracil in the telomeric hexamer repeats; aberrant epigenetic state of the subtelomeric DNA; and inefficient binding of the shelterin proteins [33]. This shows that there are arguments for a role for nutrients deficiencies in telomere length. The molecular processes are multiple and complex. For instance, Se can also increase telomerase activity and TERT expression [34]. However, many studies are observational studies and interventional studies are waited for. Obviously, interventional studies on telomere length will have to be conducted in animal models with shorter lifespan then human lifespan. This is a limitation as telomere may not display similar shortening nor have similar consequences in all species [35].

Food intake and its consequences on telomere length may not be related to a specific nutrient but more largely to metabolism, especially glucose metabolism [36]. The length of telomeres is associated with a higher metabolic risk profile [37] and shortened telomeres are associated with stroke, myocardial infarction, and type 2 diabetes mellitus [38, 39]. The length of type 2 diabetic patients telomeres was demonstrated to be shorter than those of control subjects [40] and the consumption of sugar-sweetened soda was also associated with shorter telomeres [41].

CELLULAR SENESCENCE AND NUTRIENTS

An aged organism differs from a young one by different loss of functions such as resistance to disease, homeostasis, and fertility, as well as an ill-defined "wear and tear." Aging includes, although is not limited to, processes such as cellular senescence and organ senescence. Cellular senescence is a status of irreversible growth arrest. Although senescence contributes to aging and age-related diseases, it must be kept in mind that it also protects cells from cancerous transformation [42]. It is usually mediated by a persistent DNA damage response with an insensibility to mitogenic stimuli and simultaneously an upregulation of tumor suppressor pathways. Senescence may occur from different origins. Replicative senescence is related to telomere attrition. Stress-induced premature senescence occurs after an exposure to stressors such as UV or ionizing radiations. Oncogene-induced senescence occurs after an aberrant activation of oncoproteins or a loss of tumor suppressors. The molecular processes involved in these different forms of senescence largely overlap with the presence of DNA damage, telomere attrition, or persisting DNA damage response. These pathways converge on the inhibition of retinoblastoma protein phosphorylation, which results in the inactivation of the E2F transcription factor and of the target genes involved in cell cycle progression. Thus, senescence implies modifications of gene transcription. Are there experimental procedures using nutrients and or food intake able to reverse cellular senescence? In vitro experiments have indeed partially rejuvenated senescent cells. The activation of cellular growth pathways via the mammalian target of rapamycin (mTOR) pathway is critical to induce senescence. Some strategies involved molecules altering mTOR pathway such as rapamycin (a.k.a. sirolimus) originating from a bacteria of Easter Island. mTOR integrates various extracellular inputs such as nutrients (amino acids and glucose) and growth factors including insulin, IGF-1, and IGF-2. mTOR also "senses" cellular nutrient levels. It is likely that at least part of the beneficial effect of caloric restriction on increase lifespan occurs through a modulation of mTOR signaling hence the research on caloric restriction mimetics [43]. It is also likely that mTOR is a key component of the pathway that is defective in endocrine-mediated insulin signaling that can extend longevity in most model organisms. Other strategies to rejuvenate cells involved individual nutrients that could be available via food intake. For instance, L-carnosine, a natural dipeptide opposed the senescence of fibroblasts [44]. Quercetin reversed cell arrested proliferation [45]. Tocotrienol reversed the senescent morphology and markers and elongated telomeres [46].

EPIGENETICS AND NUTRIENTS

Epigenetics is the study of modifications of DNA and DNA-binding proteins through mechanisms occurring after replication and translation, respectively. These modifications alter gene transcription via DNA compaction and accessibility to transcription proteins. Initial work about the influence of diet on epigenetics has investigated the consequences of maternal diet on in utero development and, more importantly, on subsequent life including aging of offspring. Nutrient-related long-lasting epigenetic changes were suggested to influence the expression of gene somehow involved in the aging process. DNA methylation is the likeliest candidate as post-translational modifications of histones (acetylation, phosphorylation, and methylation) are thought less stable. Aging appears to decrease DNA methylation. However, there are large inter-individual variations of the rate of demethylation; some increase of methylation can even be noted. As there is some familial clustering of the rate of demethylation a genetic component has been suggested. Methylation of some specific genes has been seen during aging (estrogen and retinoic acid receptors, Glutathione S-transferase, etc. [47–49]. Furthermore, it has been shown that the age-related pattern of methylation may change according to the tissue investigated [50]. This is complicated by the fact that cell-type composition may change with age and cell types aging may be heterogeneous. It then appears that as far as aging is concerned DNA methylation has to be investigated in large populations by a tissue-specific approach with a correction of cellular heterogeneity. This is now possible with new available techniques and published results of the methylome in various tissues [51, 52]. When investigating blood tissue as a model of DNA methylome changes with age a recent paper addressed some of these questions showing that age-associated hypomethylated blocks exist associated with local hypermethylation at CpG islands. Interestingly, the differentially methylated regions involved transcription factors [53].

Lastly, some nutrients can be involved in different age-related targets. For instance, the above-mentioned Se also appears to play a role in DNA methylation with different publications reporting a decrease in DNA methylation upon Se supplementation [54]. Se action is not restricted to DNA damage control but also interferes with gene transcription [55], telomere length, and epigenetic modulation in normal cells. Taken together, a score of publications show that DNA methylation is modified during aging and that this is now accessible by specific chips. By comparison, the other epigenetic modifications are less easily investigated. Some posttranslational histone modifications have been related to aging and also appear to be tissue dependent. Epigenetic modifications via interference by RNA can lead to modified gene expression. This has yet to be convincingly linked to the aging processes.

To summarize, epigenetic mechanisms that modulate gene expression are modified during aging. Can these mechanisms be affected by nutrients or more generally by food intake? Considering the abundance of the literature it would appear that the answer is positive although it is not always clear about the mechanisms involved. DNA methylation can be affected by three nutrient related mechanisms: caloric intake, one carbon metabolism, and bioactive nutrients [40].

SUMMARY POINTS

- Molecular and cellular mechanisms of aging may be modulated by nutrients.
- Nutrients provide key proteins involved in aging-limiting processes such as antioxidant capacity, telomere length maintenance.
- Cellular senescence depends, at least in part, on cellular nutrient sensors via the mTOR signaling pathway.
- Many published investigations often indicated strong nutrient-gene interactions with nutrient intake potentially modulating risk factors of many diseases.
- More complex and wider interactions exist between multiple gene variants influencing nutrient metabolism and health outcomes. More robust quantitative assessment of these genetic effects are needed.
- The power of nutrigenetics and nutrigenomics can be further enhanced with the inclusion of high-throughput metabolomic assays. Well controlled dietary intervention studies yielding consistent and valid results are a prerequisite before genetic patterns can be widely applied to individual dietary advice.
- The knowledge from diet-gene interactions will enable more effective and specific interventions for disease prevention based on individualized nutrition.

References

[1] Lovegrove JA, Gitau R. Nutrigenetics and CVD: what does the future hold? Proc Nutr Soc 2008;67(2):206–13.
[2] Lopez-Otin C, Blasco MA, Partridge L, Serrano M, Kroemer G. The hallmarks of aging. Cell 2013;153(6):1194–217.
[3] Corella D, Ordovas JM. Interactions between dietary n-3 fatty acids and genetic variants and risk of disease. Br J Nutr 2012;107(Suppl 2):S271–83.
[4] Merched AJ, Chan L. Nutrigenetics and nutrigenomics of atherosclerosis. Curr Atheroscler Rep 2013;15(6):328.
[5] Ordovas JM. The quest for cardiovascular health in the genomic era: nutrigenetics and plasma lipoproteins. Proc Nutr Soc 2004;63(1):145–52.
[6] Fontaine-Bisson B, Wolever TM, Chiasson JL, Rabasa-Lhoret R, Maheux P, Josse RG, et al. Genetic polymorphisms of tumor necrosis factor-alpha modify the association between dietary polyunsaturated fatty acids and fasting HDL-cholesterol and apo A-I concentrations. Am J Clin Nutr 2007;86(3):768–74.
[7] Lai CQ, Corella D, Demissie S, Cupples LA, Adiconis X, Zhu Y, et al. Dietary intake of n-6 fatty acids modulates effect of apolipoprotein A5 gene on plasma fasting triglycerides, remnant lipoprotein concentrations, and lipoprotein particle size: the Framingham Heart Study. Circulation 2006;113(17):2062–70.
[8] Ferguson JF, Phillips CM, McMonagle J, Perez-Martinez P, Shaw DI, Lovegrove JA, et al. NOS3 gene polymorphisms are associated with risk markers of cardiovascular disease, and interact with omega-3 polyunsaturated fatty acids. Atherosclerosis 2010;211(2):539–44.
[9] Tai ES, Corella D, Demissie S, Cupples LA, Coltell O, Schaefer EJ, et al. Polyunsaturated fatty acids interact with the PPARA-L162V polymorphism to affect plasma triglyceride and apolipoprotein C-III concentrations in the Framingham Heart Study. J Nutr 2005;135(3):397–403.
[10] Merched AJ, Ko K, Gotlinger KH, Serhan CN, Chan L. Atherosclerosis: evidence for impairment of resolution of vascular inflammation governed by specific lipid mediators. Faseb J 2008;22(10):3595–606.
[11] Hersberger M. Potential role of the lipoxygenase derived lipid mediators in atherosclerosis: leukotrienes, lipoxins and resolvins. Clin Chem Lab Med 2010;48(8):1063–73.
[12] Merched AJ, Serhan CN, Chan L. Nutrigenetic disruption of inflammation-resolution homeostasis and atherogenesis. J Nutrigenet Nutrigenomics 2011;4(1):12–24.
[13] Gago-Dominguez M, Castelao JE, Sun CL, Van Den Berg D, Koh WP, Lee HP, et al. Marine n-3 fatty acid intake, glutathione S-transferase polymorphisms and breast cancer risk in postmenopausal Chinese women in Singapore. Carcinogenesis 2004;25(11):2143–7.
[14] Stern MC, Siegmund KD, Corral R, Haile RW. XRCC1 and XRCC3 polymorphisms and their role as effect modifiers of unsaturated fatty acids and antioxidant intake on colorectal adenomas risk. Cancer Epidemiol Biomarkers Prev 2005;14(3):609–15.
[15] Stern MC, Butler LM, Corral R, Joshi AD, Yuan JM, Koh WP, et al. Polyunsaturated fatty acids, DNA repair single nucleotide polymorphisms and colorectal cancer in the Singapore Chinese Health Study. J Nutrigenet Nutrigenomics 2009;2(6):273–9.
[16] Serhan CN, Chiang N, Dalli J. The resolution code of acute inflammation: novel pro-resolving lipid mediators in resolution. Semin Immunol 2015;27(3):200–15.
[17] Hedelin M, Chang ET, Wiklund F, Bellocco R, Klint A, Adolfsson J, et al. Association of frequent consumption of fatty fish with prostate cancer risk is modified by COX-2 polymorphism. Int J Cancer 2007;120(2):398–405.
[18] Fradet V, Cheng I, Casey G, Witte JS. Dietary omega-3 fatty acids, cyclooxygenase-2 genetic variation, and aggressive prostate cancer risk. Clin Cancer Res 2009;15(7):2559–66.
[19] Habermann N, Ulrich CM, Lundgreen A, Makar KW, Poole EM, Caan B, et al. PTGS1, PTGS2, ALOX5, ALOX12, ALOX15, and FLAP SNPs: interaction with fatty acids in colon cancer and rectal cancer. Genes Nutr 2013;8(1):115–26.
[20] Hasty P, Campisi J, Hoeijmakers J, van Steeg H, Vijg J. Aging and genome maintenance: lessons from the mouse? Science 2003;299(5611):1355–9.

REFERENCES

[21] Ferguson LR, Fenech MF. Vitamin and minerals that influence genome integrity, and exposure/intake levels associated with DNA damage prevention. Mutat Res 2012;733(1-2):1–3.

[22] Karunasinghe N, Ryan J, Tuckey J, Masters J, Jamieson M, Clarke LC, et al. DNA stability and serum selenium levels in a high-risk group for prostate cancer. Cancer Epidemiol Biomarkers Prev 2004;13(3):391–7.

[23] Caple F, Williams EA, Spiers A, Tyson J, Burtle B, Daly AK, et al. Inter-individual variation in DNA damage and base excision repair in young, healthy non-smokers: effects of dietary supplementation and genotype. Br J Nutr 2010;103(11):1585–93.

[24] Krajcovicova-Kudlackova M, Valachovicova M, Paukova V, Dusinska M. Effects of diet and age on oxidative damage products in healthy subjects. Physiol Res 2008;57(4):647–51.

[25] Zhao X, Aldini G, Johnson EJ, Rasmussen H, Kraemer K, Woolf H, et al. Modification of lymphocyte DNA damage by carotenoid supplementation in postmenopausal women. Am J Clin Nutr 2006;83(1):163–9.

[26] Gutierrez-Mariscal FM, Perez-Martinez P, Delgado-Lista J, Yubero-Serrano EM, Camargo A, Delgado-Casado N, et al. Mediterranean diet supplemented with coenzyme Q10 induces postprandial changes in p53 in response to oxidative DNA damage in elderly subjects. Age (Dordr) 2012;34(2):389–403.

[27] Halicka HD, Zhao H, Li J, Lee YS, Hsieh TC, Wu JM, et al. Potential anti-aging agents suppress the level of constitutive mTOR- and DNA damage- signaling. Aging (Albany NY) 2012;4(12):952–65.

[28] Valdiglesias V, Giunta S, Fenech M, Neri M, Bonassi S. γH2AX as a marker of DNA double strand breaks and genomic instability in human population studies. Mutat Res 2013;753(1): 24–40.

[29] Tomasello B, Grasso S, Malfa G, Stella S, Favetta M, Renis M. Double-face activity of resveratrol in voluntary runners: assessment of DNA damage by comet assay. J Med Food 2012;15 (5):441–7.

[30] Moores CJ, Fenech M, O'Callaghan NJ. Telomere dynamics: the influence of folate and DNA methylation. Ann NY Acad Sci 2011;1229:76–88.

[31] Panayiotou AG, Nicolaides AN, Griffin M, Tyllis T, Georgiou N, Bond D, et al. Leukocyte telomere length is associated with measures of subclinical atherosclerosis. Atherosclerosis 2010;211(1):176–81.

[32] Bull CF, O'Callaghan NJ, Mayrhofer G, Fenech MF. Telomere length in lymphocytes of older South Australian men may be inversely associated with plasma homocysteine. Rejuvenation Res 2009;12(5):341–9.

[33] Bull C, Fenech M. Genome-health nutrigenomics and nutrigenetics: nutritional requirements or 'nutriomes' for chromosomal stability and telomere maintenance at the individual level. Proc Nutr Soc 2008;67(2):146–56.

[34] Allayee H, Roth N, Hodis HN. Polyunsaturated fatty acids and cardiovascular disease: implications for nutrigenetics. J Nutrigenet Nutrigenomics 2009;2(3):140–8.

[35] Sahin E, Depinho RA. Linking functional decline of telomeres, mitochondria and stem cells during aging. Nature 2010;464 (7288):520–8.

[36] Elks CE, Scott RA. The long and short of telomere length and diabetes. Diabetes 2014;63(1):65–7.

[37] Revesz D, Milaneschi Y, Verhoeven JE, Penninx BW. Telomere length as a marker of cellular aging is associated with prevalence and progression of metabolic syndrome. J Clin Endocrinol Metab 2014;99(12):4607–15.

[38] D'Mello MJ, Ross SA, Briel M, Anand SS, Gerstein H, Pare G. Association between shortened leukocyte telomere length and cardiometabolic outcomes: systematic review and meta-analysis. Circ Cardiovasc Genet 2015;8(1):82–90.

[39] Haycock PC, Heydon EE, Kaptoge S, Butterworth AS, Thompson A, Willeit P. Leucocyte telomere length and risk of cardiovascular disease: systematic review and meta-analysis. Br Med J 2014;349:g4227.

[40] Bacalini MG, Friso S, Olivieri F, Pirazzini C, Giuliani C, Capri M, et al. Present and future of anti-aging epigenetic diets. Mech Aging Dev 2014;136-137:101–15.

[41] Leung CW, Laraia BA, Needham BL, Rehkopf DH, Adler NE, Lin J, et al. Soda and cell aging: associations between sugar-sweetened beverage consumption and leukocyte telomere length in healthy adults from the National Health and Nutrition Examination Surveys. Am J Public Health 2014;104 (12):2425–31.

[42] Rodier F, Campisi J. Four faces of cellular senescence. J Cell Biol 2011;192(4):547–56.

[43] Ingram DK, Roth GS. Calorie restriction mimetics: can you have your cake and eat it, too? Aging Res Rev 2015;20C:46–62.

[44] McFarland GA, Holliday R. Retardation of the senescence of cultured human diploid fibroblasts by carnosine. Exp Cell Res 1994;212(2):167–75.

[45] Chondrogianni N, Kapeta S, Chinou I, Vassilatou K, Papassideri I, Gonos ES. Anti-aging and rejuvenating effects of quercetin. Exp Gerontol 2010;45(10):763–71.

[46] Makpol S, Durani LW, Chua KH, Mohd Yusof YA, Ngah WZ. Tocotrienol-rich fraction prevents cell cycle arrest and elongates telomere length in senescent human diploid fibroblasts. J Biomed Biotechnol 2011;2011:506171.

[47] Wallace K, Grau MV, Levine AJ, Shen L, Hamdan R, Chen X, et al. Association between folate levels and CpG Island hypermethylation in normal colorectal mucosa. Cancer Prev Res (Phila) 2010;3(12):1552–64.

[48] Kwabi-Addo B, Chung W, Shen L, Ittmann M, Wheeler T, Jelinek J, et al. Age-related DNA methylation changes in normal human prostate tissues. Clin Cancer Res 2007;13(13):3796–802.

[49] Ford D, Ions LJ, Alatawi F, Wakeling LA. The potential role of epigenetic responses to diet in aging. Proc Nutr Soc 2011;70 (3):374–84.

[50] Christensen BC, Houseman EA, Marsit CJ, Zheng S, Wrensch MR, Wiemels JL, et al. Aging and environmental exposures alter tissue-specific DNA methylation dependent upon CpG island context. PLoS Genet 2009;5(8):e1000602.

[51] Roadmap Epigenomics C, Kundaje A, Meuleman W, Ernst J, Bilenky M, Yen A, et al. Integrative analysis of 111 reference human epigenomes. Nature 2015;518(7539):317–30.

[52] Ernst J, Kellis M. Large-scale imputation of epigenomic datasets for systematic annotation of diverse human tissues. Nat Biotechnol 2015.

[53] Yuan T, Jiao Y, de Jong S, Ophoff RA, Beck S, Teschendorff AE. An integrative multi-scale analysis of the dynamic DNA methylation landscape in aging. PLoS Genet 2015;11(2):e1004996.

[54] Arai Y, Ohgane J, Yagi S, Ito R, Iwasaki Y, Saito K, et al. Epigenetic assessment of environmental chemicals detected in maternal peripheral and cord blood samples. J Reprod Dev 2011;57(4):507–17.

[55] Dong Y, Ganther HE, Stewart C, Ip C. Identification of molecular targets associated with selenium-induced growth inhibition in human breast cells using cDNA microarrays. Cancer Res 2002;62(3):708–14.

CHAPTER 4

Diet and Longevity Phenotype

Francesco Villa[1], Chiara Carmela Spinelli[1] and Annibale A. Puca[2,3]

[1]Istituto di Tecnologie Biomediche, CNR Segrate, Milan, Italy [2]Dipartimento di Medicina e Chirurgia, Università degli Studi di Salerno, Salerno, Italy [3]IRCCS Multimedica, Milan, Italy

KEY FACTS
- In the last century, human lifespan has obtained the largest increase ever.
- Reaching extreme longevity requires the right genetic make-up, a correct exposition to environmental stimuli, and the occurrence of positive stochastic events.
- The study of model organisms could explain how to ameliorate the interaction with environment.
- Different genetic make-ups produce different responses to the same stimuli and for this reason the study of long-lived individuals' genomes could reveal important functional secrets.
- An immediate and sophisticated link between nutritional functions and genetic predisposition to longevity is represented by epigenetic mechanisms.
- A healthy nutrition style drives for longevity delaying or escaping age-related diseases and activating a complex pathway for improved survival and stress response.

Dictionary of Terms
- *Caloric restriction*: reduced intake of nutritional calories without malnutrition, often called "dietary restriction."
- *Exceptional longevity*: also called "extreme longevity," identified as an age over the average and out of the ordinary.
- *Epigenetic*: describes the study of dynamic alterations in the transcriptional potential of a cell.
- *Unsaturation*: characteristic of a carbon chain that contains carbon—carbon double bonds or triple bonds.
- *Cellular energy sensors*: a group of cellular molecules able to monitor the ratio of AMP/ATP, or other energetic signals, and to manage cellular metabolism.

INTRODUCTION

The number of centenarians in almost all western populations has dramatically increased in the last 100 years. The US 1900 Birth Cohort Study noted that life expectancy was 51.5 years for males and 58.3 years for females and only 1 person in 100,000 reached 100 years of age. Currently about 1 in 10,000 individuals reach 100 years of age and this prevalence will probably soon approach 1 in 5000 [1]. The cause of this increment—quantified as 3 months/year for females—has been attributed to an improvement in diet behaviors and to a reduction in exposure to infection and inflammation [2]. Progress in healthcare science and a healthier lifestyle have reduced or erased many of the avoidable or curable causes of death.

Today, for these reasons, people with the genetic and behavioral characteristics that push life until 100 years are able to become long-lived people. The reaching of this goal is accompanied by a better health condition of people, either physical and/or mental status, as a consequence of demographic selection, that compresses morbidity and mortality toward the end of the lifespan [3]. The cognitive ability, for example, was lost only in the last 3—5 years of life.

Centenarians, despite being exposed to the same environmental conditions as members of the average population, manage to live much longer. The genetic explanation of this compression in morbidity and mortality reflects the correlation of the longevity phenotype with the enrichment of protective alleles and the depletion of detrimental ones. This genetic make-up is heritable, as shown by the familiar clustering of exceptional longevity, and affects at least 25% of the human life-span, and for an even a larger proportion in individuals living to extreme age [4,5]. Thus, long-lived people are a good model for studying the longevity phenotype and age-related pathologies, such as Alzheimer's disease (AD), stroke, cardiovascular disease, diabetes, and Parkinson's disease.

From a nutritional point of view, the improvements in food quality and food-intake behaviors amplify and realize the genetic potential for a good aging. Many studies have demonstrated the ability of calorie restriction (CR), which is defined as a reduced intake of nutritional calories without malnutrition, often called dietary restriction (DR), in maintaining biological systems and increasing lifespan [6]. The molecular signaling pathways that mediate the effects of CR on longevity have been actively studied, and there are evidences that endothelial nitric oxide synthase (eNOS) plays a fundamental role in this process [7].

In addition, a large number of studies are exploring and explaining how nutritional components contribute to a good aging and which foods are favorable for increasing lifespan, and not only in humans [8–13]. Nutrients are necessary for the maintenance of biological functions, including metabolism, growth, and repair. They include organic chemicals, such as carbohydrates, proteins, lipids, and vitamins, and inorganic substances, such as minerals and water [14]. Some foods interact with the organism in a direct, massive, and manifest way, such as fat accumulation in vessels or hepatic intoxication due to chemicals preservatives, while other nutrients regulate gene expression in a fine and complex way with an accurate control of protein system, stress response, DNA damage repair mechanisms, involvement of nervous system, cellular cycle, epigenetic mechanisms, and telomeres length maintenance.

Nutrition and diet are unequivocally the most influential of all the external environmental factors and they are linked to aging and cancer incidence and prognosis [15–17]. The delay of the aging process is linked to epigenetic modifications too, and these are potentiated by nutrition. The importance of the changes within the epigenome also leads to the possibility of using these changes as biomarkers in many age-related diseases [18–20].

The amount of the impact of diet on lifespan depends on the involvement of the particular signaling pathway perturbed by the diet itself in the etiopathology of a disease, but in particular it depends on the incidence of this disease in the population. For example, a very small healthy advantage with respect to a very diffused disease could have an enormous impact on lifespan (ie, cancer).

A promising way to study the correlation of diet and longevity is the analyses of the human genome and the characterization of polymorphisms associated with extreme longevity that has a role in nutrients processing.

For example, polymorphisms in the gene ELOVL6, sited in the longevity-associated locus 4q25, and implicated in the elongation of fatty acid chains, have a protective role against the insulin resistance after a high fat diet [21,22].

Furthermore, a study conducted on a cohort of centenarian's offspring demonstrated that the genetic predisposition to longevity was linked with an improved ability to organize the erythrocyte membrane fatty acids that reflects reduced oxidative stress and increased membrane integrity at the cellular level [23]. This aptitude in managing the polyunsaturated fatty acids (PUFA) is an important example of the close relationship between diet and genetic predispositions.

MAIN TEXT

Nutrient Components and Aging Process

Many studies have illustrated the important role of diet as a modulator of longevity explaining the effects of a single nutrient or complex feeding plan, such as the Mediterranean diet and its variants.

Carbohydrates, organic compounds consisted of carbon, hydrogen, and oxygen, have an important role as signaling molecules, energy sources, and structural components and the abundance of many of them is very harmful for health and accelerates aging. Glucose, for example, is the primary energy source of most living organisms, but its consumption is linked to the inactivation of AMP-activated protein kinase (AMPK) and other proteins belonging to the energy-sensing signaling pathways. These pro-longevity proteins form a complex system of energy management, common to various species, that includes transcription factors (such as FOXO), protein deacetylase (such as Sirtuins), enzymes (such as glyoxalase), and channel proteins (such as aquaporin-1/glycerol channel). In contrast to glucose, other carbohydrates or carbohydrate metabolites have been shown to promote longevity in *Caenorhabditis elegans* [24–27].

Low-carbohydrate diets improve health by reducing several factors which are associated with aging or metabolic defects. Furthermore, reduced carbohydrate intake decreases body weight and consequently reduces the risk factors associated with heart disease. This improvement in general health represents a big step toward a good aging and longevity.

The studies on the function of *proteins* and amino acids in diet have revealed that they have an important role as structural constituents, catalysts for enzymatic reactions, and energy sources. The influence that they have on the aging process is not clear and seems to be linked to the carbohydrates' availability. Many scientists have conducted studies on the P:C ratio and have found that a low-protein/high-carbohydrate diet is associated with health and longevity in different species, such as in mice [28–30]. On the other hand, proteins are not all equivalent. Animal proteins have demonstrated a strong link with the risk of developing tumors, maybe through IGF-1 and TOR signaling pathways, and with excess weight and obesity. Plant proteins, instead, are negatively associated with the same risk factors and this is probably due to the lower amount of methionine content. A limitation of this amino acid intake through diet is associated with an increased lifespan in *Drosophila*, mouse, and rat [31,32]. This phenomenon probably is due to a decrement of production of mitochondrial reactive oxygen species (ROS) and DNA damage [33]. However, the mechanisms of action of methionine and other amino acids are still unclear: several studies demonstrated a positive impact of them on lifespan in many different organisms.

Also *lipids* intake confers a bivalent influence on lifespan. They constitute the main structures in biological membranes, but a high-fat diet is associated with cardiovascular diseases, diabetes, and reduction of lifespan. These negative aspects are due to the inactivation of SIRT1 gene, a cellular sensor for nutrient availability, that affects metabolic dysfunction, insulin resistance, and obesity.

Others important nutritional components are *vitamins* and *minerals*. Humans are not able to synthesize these components and so their correct quantity in diet is fundamental for health. Many diseases and dysfunction are due to a too low provision of these elements and consequently vitamins and minerals are indispensable for a healthy aging. The better studied function of vitamins is in their role as antioxidants. In many different organisms, they are able to moderate levels of ROS acting as cofactors in many diverse biological processes and a deficiency of these components limits the beneficial effects of other components. However, their provision must be balanced, because a high-vitamin diet could provoke detrimental effects on physiological functions, with a negative association with the longevity phenotype. The role of minerals in nutrition is not well studied and the effects of these components are not clear.

There are no evidences that minerals could be associated to longevity, instead it has been demonstrated that a supplementation of diet with selenium, iron, manganese, copper, and zinc is harmful for organisms.

Crucial Role of Fatty Acid Component

One of the most important pathways implicated in the aging process is the management of fatty acids. Many studies have revealed the association of the diverse abundances of particular kinds of fatty acids with different pathologies. For example, omega-3 fatty acid has a protective role for cardiovascular pathology (with respect to the harmful impact of omega-6) but has a detrimental impact on conduction disturbances, like arrhythmias and fibrillations.

This was assessed by the study of Puca et al. [23] on a group of people affected by atrial flutter (AFL) or atrial fibrillation (AF). The phospholipid content of erythrocyte membranes of these subjects was compared with phospholipid content of age-matched controls' erythrocyte membranes. The study correlated the percentage of the main saturated and unsaturated residues of membranes phospholipids together with the sum of saturated (SFA), monounsaturated (MUFA), and polyunsaturated (PUFA n-3 and PUFA n-6) fatty acid residues, and some indicative ratios (SFA/MUFA; Palmitoleic acid-C16:1/Palmitic acid-C16:0; Oleic acid-C18:1/Stearic acid-C18:0). Furthermore, the indexes of peroxidation (PI) and unsaturation (UI) were evaluated. PI value provides information on the amount of substrate for lipid peroxidation based on the contribution of fatty acids according to their degree of unsaturation. UI represents the concentration of the unsaturated product proportional to the sum of the unsaturated product and the saturated substrate. Results indicated values of PI and UI were significantly higher in AF/AFL subjects than in the control groups, as well as the amount of PUFA (omega-3 vs omega-6). Furthermore, the results show that membranes of AF/AFL subjects have significantly lower amounts of MUFA (especially C16:1 and C18:1) and SFA (C18:0). These data mean that membranes of AF/AFL affected subjects are more susceptible to oxidative stress and that there is a direct correlation between PUFA abundance in cells and predisposition to these cardiovascular arrhythmias. This is information that shows the importance of preferring a diet based on MUFA content instead of PUFA.

The same indication was given by another study conducted on a population of nonagenarian and nonagenarian's offspring; offspring of long-living people also show a lower risk of all-cause mortality, cancer-specific mortality, and coronary disease-specific mortality.

The project compared fatty acid percentage of erythrocytes membrane of 41 healthy nonagenarians' offspring with 30 age-matched controls. Analysis of the erythrocyte membranes' fatty acid components was executed by evaluating the fatty acid percentages and correlating the percentages of the main saturated and unsaturated residues of membrane phospholipids, together with some indicative ratios (SFA/MUFA, C16:1/C16:0; C18:1/C18:0). The study also evaluated the presence of some *trans*-fatty acid isomers, which are indicative of an endogenous free radical process. The study revealed that erythrocyte membranes from nonagenarian offspring had significantly higher content of C16:1, *trans* C18:1, and total *trans*-FA than matched controls and significantly reduced content of C18:2 and C20:4 than matched controls. The measure of the PI revealed that it was significantly lower than that in the control groups. These evidences found in the erythrocyte membranes could be realistically expanded to the whole organism and reflect a generalized mechanism of membrane bilayers of cells. This conformational condition of the membrane could alter cell metabolism, the production of ROS, and the activity of membrane-associated proteins [34,35]. Today we can study the composition of centenarian's membranes in order to design a diet that allows the correct intake of the right percentage of every kind of fatty acid. For example, the peroxidizability property must be kept low, maintaining a low intake of PUFA, while levels of *trans* C18:1, C16:1, and C16:1/C16:0 ratio must be kept higher, for a basal condition of free-radical stress, a process that has been shown to induce antioxidant and cytoprotective responses [36].

The importance of evaluating the lipid content of cellular membranes is due to the fact that many of the products of lipid peroxidation are powerful ROS themselves. These ROS are able to attack other PUFA molecules with a domino effect that allows us to consider lipid peroxidation as an autocatalytic, self-propagating process [37]. The products of lipid peroxidation are able to harm surrounding structures, such as protein and nucleic acids, and cells with peroxidation-susceptible membrane fatty acid composition especially. This was named "membrane pacemaker theory" and was implied as a support for the oxidative stress theory of aging. This is one of the various theories of aging that has been proved in more than one species and following its principles lifespan has been experimentally modulated through caloric restriction.

Genomic Studies and Importance of Palmitoleic Acid

Research on nutritional components could give us a direction toward good aging and longevity, but today a new important way to study the aging process consists of the analysis of centenarian people's genome. In this kind of research, the object of the study is not longevity but extreme longevity: centenarian people are perfect models to understand the mechanisms behind aging process and their genetic make-up could reveal how to take advantage from these information for a good aging.

The studies on other species' genome have the same importance, because many of the mechanisms of aging are highly conserved between species and the use of a simpler and shorter-lived organism could give more complete results.

A particular aspect of aging metabolism studied in recent years is the elongation of very long chain fatty acids protein 6 (ELOVL6). This protein is an elongase enzyme that has the role of transforming C16:0 into C18:0, and C16:1 (palmitoleic acid) into C18:1 adding two carbon atoms to the fatty acid chain. The gene that encodes this protein is sited in locus 4q25, an important area of chromosome 4 that presents an high concentration of polymorphisms associated with longevity that interact with lipid pathway management (ie, microsomal triglyceride transfer protein, MTP). Polymorphisms in ELOVL6 gene have been associated with insulin sensitivity and an in vivo experiment in mice demonstrated the importance of palmitoleic acid. KO-mice for ELOVL6 are not able to transform palmitoleic acid into stearinic acid and the abundance of palmitoleic acid avoids the oncoming insulin resistance after a high-fat diet [21,27]. An experiment on genetically modified worms demonstrated that the abundance of palmitoleic acid extended lifespan while several experiments are now trying to explain how these phenotypical effects are associated with this kind of fatty acid.

A study from Clemson University (USA) was performed to investigate the effects of palmitoleic acid infusion in obese sheep [38]. Palmitoleic acid is almost absent from common diets; it is present in little doses in macadamia nuts and in buckthorn oil and the presence in the human organism is due to the desaturation of palmitic acid via stearoyl-CoA desaturase-1. Preliminary data on exogenous administration of palmitoleic acid to bovine adipocytes in vitro shows it downregulates de novo lipogenesis and upregulates fatty acid oxidation to direct fatty acids, forcing cells to expend energy and to avoid lipid storage. Obese sheep, which are animal models with a fat distribution patterns more similar to humans than the other models such as pig or rodents, were infused intravenously with a dose of 10 mg/kg of palmitoleic acid twice daily for 28 days. While carcass traits, visceral adipose depots, and body composition were not altered, levels of palmitoleic acid in serum increased in a linear manner during the treatment. At the same time, levels of arachidonic (C20:4) and eicosapentaneoic (C20:5) acids increased. The total monounsaturated and polyunsaturated n-6 content in serum increased, as well as the ratio of palmitoleic and palmitic acids. Plasma insulin levels decreased and at the end of the treatment the decrement was lower in the obese sheep than in the control sheep infused with the same amount of palmitoleic acid. Plasma glucose levels were unchanged. The analysis of the muscular component of these animals showed that the lipid content of longissimus and subcutaneous muscles was not changed by infusions, but the lipid content in semitendinosus (ST) and mesenteric muscles was decreased. There was no variation in the lipid content of liver. The expression analysis of genes involved in energetic metabolism showed that acetyl-CoA carboxylase and ELOVL6 were upregulated and the phosphorylated form of AMPKa1 was more abundant after treatment. In the subcutaneous adipose and liver tissues the expression of GLUT4 and CPT1B was increased, but not in ST muscle. Taken together, these results suggest a positive association of palmitoleic acid infusions with insulin sensitivity, weight gain, intramuscular adipocyte size, and total lipid content. These effects are probably related to the alterations of AMPK activity in muscles and to the tissue-specific modulation of GLUT4 and CPT1B that regulate glucose uptake and fatty acid oxidation.

Similar results were found by Bolsoni-Lopes [39] who demonstrated that white adipocytes, cells of adipose tissue with an energy storage function, were stimulated by palmitoleic acid toward glucose uptake and expression of the major glucose transporter GLUT4. This enhancement of adipocytes' metabolism is performed by the activity of AMPK that involves the phosphorylation and activation of peroxisome proliferator-activated receptor gamma coactivator 1-alpha (PGC1α) and histone deacetylase 5 (HDAC5).

The Central Position of Energy Sensor Systems

Another important player of these pathways, involved in the management of nutritional component, is cAMP responsive element-binding protein (CREB), a ubiquitous transcription factor that is regulated through phosphorylation and acetylation. Its activation depends on many stimuli of different nature, such as nutritional, hormonal, and growth factors, and causes different effects in various tissues. In neuronal cells, CREB mediated gene expression to promote neuronal survival and growth, differentiation, and plasticity [40]. Conditional deletion of CREB in neurons leads to neurodegeneration and could increase damage inflicted by neurodegenerative diseases. This singular transcription factor is sited between the nutrient sensing and neurotrophic signaling pathways and thus it is able to translate nutritional and dietary healthy behaviors into beneficial neural signals.

Another recent study [41] used a targeted liquid chromatography, mass spectrometry–based metabolomics platform with high analyte specificity, to characterize the metabolic response to exercise. Serial blood samples obtained before and after exercise were analyzed and the magnitude, kinetics, and interrelatedness of plasma metabolic changes in response to acute exercise were defined. In a second step, it was investigated if these metabolites could modulate the expression of transcriptional regulators of metabolism in cell culture and animal models. The results indicated that during strong exercise, such as marathon training, the metabolic pathways were improved for the utilization of fuel substrates: glycolysis, lipolysis, adenine nucleotide catabolism, amino acid catabolism, glycogenolysis, and the presence of small molecules (that indicate oxidative stress and modulation of insulin sensitivity). The metabolites detected have a wide range of health consequences on physiological pathways (especially in skeletal muscle metabolism) and could protect against the onset of pathological conditions (cardiovascular disorders, obesity, and diabetes). The common player of these processes is Nur77, a transcriptional factor that had a threefold increment. Nur77, as well as CREB, is under the control of Ca^{2+}/calmodulin-dependent protein kinase (CAMKIV), which is activated by AMPK and has been associated with exceptional human longevity [42,43]. AMPK is, in turn, responsible for eNOS phosphorylation that causes the beneficial effects of both physical exercise and caloric restriction [44]. Another aspect shared by these two behaviors is the activation of mitochondrial biogenesis, with many metabolic consequences [45].

The Involvement of Sympathetic System

The connection between food intake regulation and modulation of cardiovascular system physiology or tissues lipid content is a relatively well-known mechanism and could appear a more direct and immediate consequence. In recent years more and more research studies have shown the involvement of the nervous system with nutritional lifestyle. To appreciate what kind of influence the diet could have on sympathetic and parasympathetic systems, it has been necessary to perform a long-term experiment on a group of healthy subjects that agreed to undergo a caloric restriction regime for 3–15 years. The CR diet consisted of a variety of nutrient-dense unprocessed foods (eg, vegetables, fruits, nuts egg whites, beans ...) that supplied the recommended daily intake for all essential nutrients. Energy intake was 30% lower in CR diet than in the control diet. Refined foods rich in *trans*-fatty acids and salt were avoided.

An important physiological aspect that evolves in organism life is heart rate variability (HRV), defined as the phenomenon of variation in the time interval between heartbeats, and it is a well-accepted index of autonomic nervous system (ANS) function. The aging process is associated with an alteration of ANS functions such as a progressive decline in HRV and a general increment of homeostatic imbalance. HRV decreases in various diseases such as cardiovascular disease, hypertension, obesity, and inflammatory syndromes. In addition, experiments on mouse models demonstrated that long-term caloric restriction produced a beneficial effect on autonomic function and HRV that is now considered a marker of biological age (ie, it can describe the healthy status of an organism, independently of its chronological age).

Taken together, these observations forced the consideration of HRV values in CR-treated and control subjects to understand if this nutritional style could have an impact on the aging process of ANS. The results of this study show that the control group has higher values of total cholesterol and high-density lipoprotein cholesterol, as expected. Furthermore, systolic and diastolic blood pressures of the CR group were lower than controls of the same age and could be compared with controls 10 years younger. Monitoring of heart rate for 24 hours has shown significant lower values in CR than in control group ($p < 0.001$) and with corresponding virtually higher values of HRV ($p < 0.005$) [46]. Crossing these data with the indications of a study of Mager et al. [47] on caloric restriction and rat models, there is evidence that CR results in decreased sympathetic activity and augmented parasympathetic activity. These results show that caloric restriction not only has the ability to slow aging, prevent or delay several chronic diseases, protect against diabetes, cancer, and cardiovascular disease in animal models, but it is also associated in humans with many metabolic changes. These modulations give protection against heart dysfunctions, hormonal age-related disorders, and ANS decline. This latter discovery has an enormous impact on the regulation of the whole organism and could be implicated in the improvement of aging process of many bodily districts.

The results obtained by this recent study were based on a CR treatment of an average of 7 years of the subjects. This treatment was defined as "caloric restriction" but it is important to highlight that this regime consisted of a very balanced diet, with the correct intake of all necessary nutrients.

Similar results were obtained by Mager et al. [47] with a different form of caloric restriction, called intermittent fasting (IF), where meals are not limited in calories but consumed with decreased frequency. Rats treated with IF showed an improvement in functional and metabolic cardiovascular risk factors, including enhancement of insulin sensitivity, decrease in blood pressure, and improvement of cardiovascular stress adaptation. The IF diet form was shown to be more effective in the improvement of glucose regulation and neuronal resistance to injury. Although it was less clear, the hypothesis that the better condition of cardiovascular function and general vessels resilience was due to an alteration of ANSs had already been formulated. In summary, this study identified a parasympathomimetic effect of caloric restriction.

Another important research study, maybe one of the first that theorized the possibility of nervous system modulation by diet, was performed by Cowen et al. [48]. This study demonstrated that caloric restriction could rescue enteric motor neurons in aged rats. In particular, a very restrictive nutritional regime was shown to increase the age of senescence to 30 months, instead of 24 months, and to prevent significant loss of myenteric neurons up to that age. At the opposite, an *ad libitum* diet was demonstrated to cause a premature loss of myenteric neurons, especially if this diet regime was imposed from a young age. The most involved neuronal cells are the cholinergic group. There is no information about the loss of other populations of neurons due to diet or aging. The loss of cholinergic excitatory motor neurons could explain a long list of age-associated disturbances or disease. First of all the loss of motility in the gut is associated with a direct effect on the increment of constipation, but more important could be the relationship, still unknown, between the alteration in elderly of neuromuscular transmission, with an increment of vulnerability of neurons, and the onset of critical neuronal diseases, such as Parkinson's disease and AD. The fact that NOS neurons are not involved in aging decline could be explained by their function: they use nitric oxide constitutively and for this reason they may have developed a defense system against free radical damage.

Thus, longevity can be linked to the nutritional behavior through another important way: the protection of the ANS and the healthy maintenance of the submitted biological systems.

Epigenetic Linkage of Aging and Nutrition

In recent years, the number of studies around epigenetics has become more and more consistent, and the information that they produce is very innovative and has clarified many aspects of gene expression.

Epigenetic mechanisms, indeed, regulate the expression of genes basing on the external and internal stimuli that the individual receives. Thus, in different phases of life, an organism needs different patterns of protein synthesis because of the age or influence of both the quality and quantity of diet. There are also pathological causes that modulate an epigenetic trait and, more importantly, there are nutrients or drugs that can modulate epigenetic patterns to prevent pathologies, such as the formation and progression of various neoplasms. On the other hand, epigenetic modulations have a genetic foundation and for this reason each individual reacts in a personalized way to different environmental stimuli. It has been demonstrated in the literature that genetic regulation of the epigenome is heritable and this could be an explanation for the rapidity of the establishment of advantageous traits in human. For example, the rapid extension of lifespan in the last century could be linked, in many aspects, with the inheritance of the fine epigenetic mechanisms that allow humans to better profit from environmental signals. Signals, in turn, are becoming more and more characterized and could be moderated. An epigenetic diet, as named recently by Daniel and Tollefsboll [49], can induce epigenetic mechanisms that protect against cancer and aging through the consumption of certain foods, such as soy, grapes, cruciferous vegetables, and green tea [50]. These food groups could be introduced in diet as a helpful therapeutic strategy for medicinal and chemopreventive function. Inside these products, there are sufficient amount of sulforaphane, epigallocatechin-3-gallate, genistein, and resveratrol that act on the epigenome system affecting various mechanisms, such as DNA methyltransferase inhibition, histone modifications via HDAC, histone acetyltransferase inhibition, or noncoding RNA expression. These modulations push the expression of positive genes in physiological cells and proapoptotic genes in cancer cells and the inhibition of detrimental genes in healthy cells and survival genes in bad cells.

In 2012, the Franceschi research group performed an epigenome-wide association study [51] that consisted of the comparison of epigenetic characteristics of a group of mothers (age range 42–83) with those of the offspring group (age range 9–52). This kind of analysis is able to indicate which areas of DNA have a methylation pattern that follows the aging process. The results of this investigation showed an association of aging with three different genes that are more and more methylated with age. One of these is ELOVL2, of which we already know the link with nutritional components: it encodes for a transmembrane protein involved in the synthesis of long (C22 and C24) omega-3 and omega-6 PUFAs [52]. This other piece of the puzzle links the crucial role of PUFAs in biological functions (ie, energy production, modulation of inflammation, and maintenance of cell membrane integrity), and thus the role of ELOVL2, with the biological age of an organism. The ELOVL2 hypermethylation, indeed, continuously increases from the very first stage of life to nonagenarian. In addition to the many advantages that this discovery can give clinically and forensically, it is important because it arises in connection with the fatty acid component in homeostasis and biological age of an individual.

Another point of view was taken by Heyn et al. [53], which compared the epigenome of 20 nonagenarians with the epigenome of 20 newborn controls. This study focused the attention on extreme longevity and tried to discover the epigenetic characteristics that permit reaching this exceptional age. They published a long list of differentially hypermethylated or hypomethylated genes in the long-lived subjects. The results of this study indicate, as expected, that epigenomic conditions are very different at the two extremes of the human lifespan, but a more attractive and interesting evaluation could be the comparison of centenarian epigenome and middle aged/old people epigenome. This matching could highlight the particular characteristic that allows a man to reach extreme longevity and not only old age.

SUMMARY

- The increase of lifespan in the last century is associated with an ameliorated quality of the environment and public health.
- Many nutritional studies explained the effects of nutrients categories on human health and aging.
- Animal models were used to understand how diet influences the aging process.
- Another finer approach in elucidating aging mechanisms is the study of long-lived individuals and genetic characteristics of extreme longevity phenotype.
- Combining information about food content action on organisms and molecular aging mechanisms in extreme longevity models, macronutrients have been classified as positive or detrimental for aging process.
- One of the most important roles is occupied by lipid quantity and quality intake: this component could modulate the homeostasis of many corporal districts.
- Many different pathways are involved in the process regulated simultaneously by diet, physical exercise, and aging: from the management of energy resources to lipid content of muscles.
- Caloric restriction or fasting intermittent diet is the nutritional styles that could bring the most significant improvement of lifespan in all species.

References

[1] Perls TT. The different paths to 100. Am J Clin Nutr 2006;83(2): 484S–7S.

[2] Oeppen J, Vaupel JW. Demography. Broken limits to life expectancy. Science 2002;296(5570):1029–31.

[3] Terry DF, Sebastiani P, Andersen SL, Perls TT. Disentangling the roles of disability and morbidity in survival to exceptional old age. Arch Intern Med 2008;168(3):277–83.

[4] Perls T, Shea-Drinkwater M, Bowen-Flynn J, Ridge SB, Kang S, Joyce E, et al. Exceptional familial clustering for extreme longevity in humans. J Am Geriatr Soc 2000;48(11):1483–5.

[5] Perls TT, Wilmoth J, Levenson R, Drinkwater M, Cohen M, Bogan H, et al. Life-long sustained mortality advantage of siblings of centenarians. Proc Natl Acad Sci U S A 2002;99(12): 8442–7.

[6] Kenyon CJ. The genetics of ageing. Nature 2010;464(7288):504–12.

[7] Nisoli E, Tonello C, Cardile A, Cozzi V, Bracale R, Tedesco L, et al. Calorie restriction promotes mitochondrial biogenesis by inducing the expression of eNOS. Science 2005;310(5746):314–17.

[8] Iwasaki K, Gleiser CA, Masoro EJ, McMahan CA, Seo EJ, Yu BP. The influence of dietary protein source on longevity and age-related disease processes of Fischer rats. J Gerontol 1988;43(1): B5–12.

[9] Iwasaki K, Gleiser CA, Masoro EJ, McMahan CA, Seo EJ, Yu BP. Influence of the restriction of individual dietary components on longevity and age-related disease of Fischer rats: the fat component and the mineral component. J Gerontol 1988;43(1):B13–21.

[10] Maeda H, Gleiser CA, Masoro EJ, Murata I, McMahan CA, Yu BP. Nutritional influences on aging of Fischer 344 rats: II. Pathology. J Gerontol 1985;40(6):671–88.

[11] Masoro EJ. Nutrition and aging—a current assessment. J Nutr 1985;115(7):842–8.

[12] Masoro EJ, Iwasaki K, Gleiser CA, McMahan CA, Seo EJ, Yu BP. Dietary modulation of the progression of nephropathy in aging rats: an evaluation of the importance of protein. Am J Clin Nutr 1989;49(6):1217–27.

[13] Yu BP, Masoro EJ, McMahan CA. Nutritional influences on aging of Fischer 344 rats: I. Physical, metabolic, and longevity characteristics. J Gerontol 1985;40(6):657–70.

[14] Lee D, Hwang W, Artan M, Jeong DE, Lee SJ. Effects of nutritional components on aging. Aging Cell 2015;14(1):8–16.

[15] Li Y, Tollefsbol TO. p16(INK4a) suppression by glucose restriction contributes to human cellular lifespan extension through SIRT1-mediated epigenetic and genetic mechanisms. PLoS One 2011;6(2):e17421.

[16] Meeran SM, Patel SN, Tollefsbol TO. Sulforaphane causes epigenetic repression of hTERT expression in human breast cancer cell lines. PLoS One 2010;5(7):e11457.

[17] Mercken EM, Crosby SD, Lamming DW, JeBailey L, Krzysik-Walker S, Villareal DT, et al. Calorie restriction in humans inhibits the PI3K/AKT pathway and induces a younger transcription profile. Aging Cell 2013;12(4):645–51.

[18] Huffman K. The developing, aging neocortex: how genetics and epigenetics influence early developmental patterning and age-related change. Front Genet 2012;3:212.

[19] Martin SL, Hardy TM, Tollefsbol TO. Medicinal chemistry of the epigenetic diet and caloric restriction. Curr Med Chem 2013;20(32):4050–9.

[20] Ross SA, Dwyer J, Umar A, Kagan J, Verma M, Van Bemmel DM, et al. Introduction: diet, epigenetic events and cancer prevention. Nutr Rev 2008;66(Suppl. 1):S1–6.

[21] Matsuzaka T, Shimano H, Yahagi N, Kato T, Atsumi A, Yamamoto T, et al. Crucial role of a long-chain fatty acid elongase, Elovl6, in obesity-induced insulin resistance. Nat Med 2007;13(10): 1193–202.

[22] Morcillo S, Martin-Nunez GM, Rojo-Martinez G, Almaraz MC, Garcia-Escobar E, Mansego ML, et al. ELOVL6 genetic variation is related to insulin sensitivity: a new candidate gene in energy metabolism. PLoS One 2011;6(6):e21198.

[23] Puca AA, Andrew P, Novelli V, Anselmi CV, Somalvico F, Cirillo NA, et al. Fatty acid profile of erythrocyte membranes as possible biomarker of longevity. Rejuvenation Res. 2008;11(1):63–72.

[24] Denzel MS, Storm NJ, Gutschmidt A, Baddi R, Hinze Y, Jarosch E, et al. Hexosamine pathway metabolites enhance protein quality control and prolong life. Cell 2014;156(6):1167–78.

[25] Edwards CB, Copes N, Brito AG, Canfield J, Bradshaw PC. Malate and fumarate extend lifespan in *Caenorhabditis elegans*. PLoS One 2013;8(3):e58345.

[26] Honda Y, Tanaka M, Honda S. Trehalose extends longevity in the nematode *Caenorhabditis elegans*. Aging Cell 2010;9(4):558–69.

[27] Mouchiroud L, Molin L, Kasturi P, Triba MN, Dumas ME, Wilson MC, et al. Pyruvate imbalance mediates metabolic reprogramming and mimics lifespan extension by dietary restriction in *Caenorhabditis elegans*. Aging Cell 2011;10(1):39–54.

[28] Carey JR, Harshman LG, Liedo P, Muller HG, Wang JL, Zhang Z. Longevity-fertility trade-offs in the tephritid fruit fly, *Anastrepha ludens*, across dietary-restriction gradients. Aging Cell 2008;7(4): 470–7.

[29] Fanson BG, Weldon CW, Perez-Staples D, Simpson SJ, Taylor PW. Nutrients, not caloric restriction, extend lifespan in Queensland fruit flies (*Bactrocera tryoni*). Aging Cell 2009;8(5):514–23.

[30] Min KJ, Tatar M. Restriction of amino acids extends lifespan in *Drosophila melanogaster*. Mech Ageing Dev 2006;127(7):643–6.

[31] Lee BC, Kaya A, Ma S, Kim G, Gerashchenko MV, Yim SH, et al. Methionine restriction extends lifespan of *Drosophila melanogaster* under conditions of low amino-acid status. Nat Commun 2014;5:3592.

[32] Zimmerman JA, Malloy V, Krajcik R, Orentreich N. Nutritional control of aging. Exp Gerontol 2003;38(1–2):47–52.

[33] Sanchez-Roman I, Gomez A, Perez I, Sanchez C, Suarez H, Naudi A, et al. Effects of aging and methionine restriction applied at old age on ROS generation and oxidative damage in rat liver mitochondria. Biogerontology 2012;13(4):399–411.

[34] Carratu L, Franceschelli S, Pardini CL, Kobayashi GS, Horvath I, Vigh L, et al. Membrane lipid perturbation modifies the set point of the temperature of heat shock response in yeast. Proc Natl Acad Sci U S A 1996;93(9):3870–5.

[35] Vigh L, Escriba PV, Sonnleitner A, Sonnleitner M, Piotto S, Maresca B, et al. The significance of lipid composition for membrane activity: new concepts and ways of assessing function. Prog Lipid Res 2005;44(5):303–44.

[36] Mathers J, Fraser JA, McMahon M, Saunders RD, Hayes JD, McLellan LI. Antioxidant and cytoprotective responses to redox stress. Biochem Soc Symp 2004;71:157–76.

[37] Hulbert AJ. Explaining longevity of different animals: is membrane fatty acid composition the missing link? Age (Dordr) 2008;30(2–3):89–97.

[38] Duckett SK, Volpi-Lagreca G, Alende M, Long NM. Palmitoleic acid reduces intramuscular lipid and restores insulin sensitivity in obese sheep. Diabetes Metab Syndr Obes 2014;7:553–63.

[39] Bolsoni-Lopes A, Festuccia WT, Chimin P, Farias TS, Torres-Leal FL, Cruz MM, et al. Palmitoleic acid (n-7) increases white adipocytes GLUT4 content and glucose uptake in association with AMPK activation. Lipids Health Dis 2014;13:199.

[40] Lonze BE, Ginty DD. Function and regulation of CREB family transcription factors in the nervous system. Neuron 2002;35(4): 605–23.

[41] Lewis GD, Farrell L, Wood MJ, Martinovic M, Arany Z, Rowe GC, et al. Metabolic signatures of exercise in human plasma. Sci Transl Med 2010;2(33):33ra7.

REFERENCES

[42] Malovini A, Illario M, Iaccarino G, Villa F, Ferrario A, Roncarati R, et al. Association study on long-living individuals from Southern Italy identifies rs10491334 in the CAMKIV gene that regulates survival proteins. Rejuvenation Res 2011;14(3):283–91.

[43] Racioppi L, Means AR. Calcium/calmodulin-dependent kinase IV in immune and inflammatory responses: novel routes for an ancient traveller. Trends Immunol 2008;29(12):600–7.

[44] Cau SB, Carneiro FS, Tostes RC. Differential modulation of nitric oxide synthases in aging: therapeutic opportunities. Front Physiol 2012;3:218.

[45] Nisoli E, Clementi E, Paolucci C, Cozzi V, Tonello C, Sciorati C, et al. Mitochondrial biogenesis in mammals: the role of endogenous nitric oxide. Science 2003;299(5608):896–9.

[46] Stein PK, Soare A, Meyer TE, Cangemi R, Holloszy JO, Fontana L. Caloric restriction may reverse age-related autonomic decline in humans. Aging Cell 2012;11(4):644–50.

[47] Mager DE, Wan R, Brown M, Cheng A, Wareski P, Abernethy DR, et al. Caloric restriction and intermittent fasting alter spectral measures of heart rate and blood pressure variability in rats. FASEB J 2006;20(6):631–7.

[48] Cowen T, Johnson RJ, Soubeyre V, Santer RM. Restricted diet rescues rat enteric motor neurones from age related cell death. Gut 2000;47(5):653–60.

[49] Daniel M, Tollefsbol TO. Epigenetic linkage of aging, cancer and nutrition. J Exp Biol 2015;218(Pt 1):59–70.

[50] Hardy TM, Tollefsbol TO. Epigenetic diet: impact on the epigenome and cancer. Epigenomics 2011;3(4):503–18.

[51] Garagnani P, Bacalini MG, Pirazzini C, Gori D, Giuliani C, Mari D, et al. Methylation of ELOVL2 gene as a new epigenetic marker of age. Aging Cell 11(6):1132–1134.

[52] Leonard AE, Kelder B, Bobik EG, Chuang LT, Lewis CJ, Kopchick JJ, et al. Identification and expression of mammalian long-chain PUFA elongation enzymes. Lipids 2002;37(8):733–40.

[53] Heyn H, Li N, Ferreira HJ, Moran S, Pisano DG, Gomez A, et al. Distinct DNA methylomes of newborns and centenarians. Proc Natl Acad Sci USA 2012;109(26):10522–7.

CHAPTER 5

Nutrition in the Elderly: General Aspects

Eugenio Mocchegiani

Translational Center Research on Nutrition and Aging, Scientific and Technologic Pole INRCA, Ancona, Italy

KEY FACTS
- Ageing is a challenge for any living organism and human longevity is a complex phenotype.
- Healthy ageing involves the interaction between genes, the environment, and lifestyle factors, particularly diet.
- Ageing and many prevalent conditions in old age are known to affect nutritional needs along lifetime.
- The concept of proper nutrition as the basis for successful aging is a recent development in nutrition research and it is still not completely known how we should adapt nutritional needs to the changes imposed by the aging process, especially in the very old age.
- Good nutrition and appropriate physical exercise are essential for healthy ageing from both a physical and psychological perspective but, it is conceivable that nutritional problems of older people cannot be understood and engaged with the same systems developed for younger adults.
- Current understanding of the biological mechanisms of the ageing process, in particular around the nutrient sensing pathways and gene-nutrient interaction, are rising important challenges and questions around the development of the best dietary strategies and dietary recommendations for adult and elderly.

Dictionary of Terms
- *Malnutrition.* Malnutrition is caused by eating a diet in which nutrients are insufficient or are too much such that it causes health problems. It is a category of diseases that includes undernutrition and overnutrition. Overnutrition can result in obesity and overweight. Undernutrition is sometimes used as a synonym of protein–energy malnutrition (PEM). It differs from calorie restriction because calorie restriction may not result in negative health effects.
- *Caloric restriction.* Caloric restriction (CR) is a dietary regimen that is based on "low" calorie intake (45%) resulting in longer maintenance of youthful health and an increase in both median and maximum lifespan in various species from yeast to nonhuman primates. The mechanism by which CR works is still not well understood. Some explanations include reduced cellular divisions, lower metabolic rates, reduced production of free radicals, reduced DNA damage, and hormesis.
- *Frailty.* Frailty is a geriatric syndrome associated with aging and defined as a clinically recognizable state of increased vulnerability resulting from aging-associated decline in reserve and function across multiple physiologic systems, such that the ability to cope with acute stressors is compromised. Frailty has been defined as meeting three out of five phenotypic criteria indicating compromised energetics: low grip strength, low energy, slowed walking speed, low physical activity, and/or unintentional weight loss. A prefrail stage, in which one or two criteria are present, identifies a subset at high risk of progressing to frailty.
- *Dietary Reference Intake (DRI).* DRI is a system of nutrition recommendations from the US National Academy of Sciences. The DRI provides several different types of reference value:
  - *Estimated Average Requirements (EAR)*, expected to satisfy the needs of 50% of the old people.

- *Recommended Dietary Allowances (RDA)*, the daily dietary intake level of a nutrient considered sufficient to meet the requirements of 97.5% of healthy individuals in each life-stage and sex group. It is calculated based on the EAR and is usually approximately 20% higher than the EAR.
- *Adequate Intake (AI)*, where no RDA has been established, but the amount established is somewhat less firmly believed to be adequate for everyone in the demographic group.
- *Tolerable upper intake levels (UL)*, caution against excessive intake of nutrients that can be harmful in large amounts.
- *Acceptable Macronutrient Distribution Ranges (AMDR)*, a range of intake specified as a percentage of total energy intake. Used for sources of energy, such as fats and carbohydrates.
- *Mini Nutritional Assessment (MNA)*. The MNA is a validated nutrition screening and assessment tool that can identify geriatric patients age 65 and over who are malnourished or at risk of malnutrition. Originally comprised of 18 questions, the current MNA now consists of 6 questions and streamlines the screening process. The revised and current MNA Short Form makes the link to intervention easier and quicker and is now the preferred form of the MNA for clinical use.
- *Body Mass Index (BMI)*. The BMI is a measure of relative size based on the mass and height of an individual. The BMI for a person is their body mass divided by the square of their height—with the value universally being given in units of kg/m^2. BMI was designed to classify average sedentary (physically inactive) populations, with an average body composition. The current value recommendations are as follow: a BMI from 18.5 up to 25 indicates optimal weight, a BMI lower than 18.5 suggests underweight, a number from 25 up to 30 indicates overweight, a number from 30 upwards suggests obesity.
- *Nutrient-sensing pathway*. Nutrient-sensing pathway is a cell's ability to recognize and respond to fuel substrates such as glucose. Each type of fuel used by the cell requires an alternate pathway of utilization and accessory molecules. In order to conserve resources a cell will only produce molecules that it needs at the time. The level and type of fuel that is available to a cell will determine the type of enzymes it needs to express from its genome for utilization. Receptors on the cell membrane's surface, designed to be activated in the presence of specific fuel molecules, communicate to the cell nucleus via a means of cascading interactions. In this way the cell is aware of the available nutrients and is able to produce only the molecules specific to that nutrient type.
- *mammalian Target of Rapamycin (mTOR)*. mTOR is a protein that in humans is encoded by the mTOR gene. mTOR is a serine/threonine protein kinase that regulates cell growth, cell proliferation, cell motility, cell survival, protein synthesis, and transcription. mTOR integrates the input from insulin, growth factors (IGF-1 and IGF-2), and amino acids. mTOR also senses cellular nutrient, oxygen, and energy levels. The mTOR pathway is dysregulated in aging and in human diseases (diabetes, obesity, depression, and cancers). Rapamycin inhibits mTOR by associating with its intracellular receptor FKBP12. The inhibition by rapamycin provides beneficial effects of the drug and in life-span extension.
- *Metallothioneins (MTs)*. MTs are a family of cysteine-rich, low molecular weight proteins. They are localized to the membrane of the Golgi apparatus. MT have the capacity to bind both physiological (zinc, copper, selenium) and xenobiotic (cadmium, mercury, silver, arsenic) heavy metals through the thiol group of its cysteine residues. MT exist in four isoforms and are synthesized primarily in the liver and kidney. Their production is dependent on availability of the dietary minerals (zinc, copper and selenium) and the amino acids histidine and cysteine. Cysteine residues from MT capture harmful oxidant radicals. In this reaction, cysteine is oxidized to cystine, and the metal ions bound to cysteine are liberated to the media. So, free zinc can activate the synthesis of more MT and many other biological functions related to free zinc ion bioavailability. This mechanism has been proposed in the control of oxidative stress and inflammatory/immune response by MT.
- *Nutrigenomics*. Nutrigenomics is a branch of nutritional genomics and is the study of the effects of foods and food constituents on gene expression. This means that nutrigenomics is research focusing on identifying and understanding molecular-level interaction between nutrients and other dietary bioactive compounds with the genome. Nutrigenomics has also been described by the influence of genetic variation on nutrition, by correlating gene expression or Single Nucleodite Polymorphisms (SNPs) with a nutrient's absorption, metabolism, elimination or biological effects. By doing so, nutrigenomics aims to develop rational means to optimize nutrition with respect to the subject's genotype.
- *Single Nucleotide Polymorphism (SNP)*. SNP is a variation at a single position in a DNA sequence among individuals. Recall that the DNA sequence is formed from a chain of four nucleotide bases: A, C, G, and T. If more than 1% of a population does not carry the same nucleotide at a specific position in the DNA sequence, then this variation can be

classified as an SNP. If an SNP occurs within a gene, then the gene is described as having more than one allele. In these cases, SNPs may lead to variations in the amino acid sequence. However SNPs can also occur in noncoding regions of DNA. Although a particular SNP may not cause a disorder, some SNPs are associated with certain diseases. These associations allow the evaluation of an individual's genetic predisposition to develop a disease or the influence of a single nutrient on the DNA (Nutrigenomic approach).

INTRODUCTION

Life expectancy has been increasing in developed countries over the past two centuries at a pace that makes some experts suggest that most babies born in the first decade of the 21st century in developed countries will celebrate their 100th birthdays [1]. Populations of these countries are aging fast for many reasons, and research suggests that aging processes may be amenable to modification, with a postponement of functional limitations. Therefore it is relevant to understand and dispose of precise tools that allow to add years of life in good health in the third age and over. In this context, nutritional knowledge and education are considered useful tools to promote health in elderly patients. The aging process itself is thought to be not a cause of malnutrition in healthy and active elderly people with appropriate lifestyles. However, changes in body composition and organ function, the ability to eat or access food, inadequate dietary intake and the partial loss of taste and smell are associated with aging and may contribute to malnutrition. Over the past decade, the importance of nutritional status has been increasingly recognized in a variety of morbid conditions including cancer, heart disease, and dementia in persons aged 65 and over [2]. Indeed, proper nutrition is an essential part of successful aging, and may delay the onset of diseases [3]. However, the concept of proper nutrition as the basis for successful aging is a recent development in nutrition research and it is still not completely known how we should adapt nutritional needs to the changes imposed by the aging process, especially in the very old age. Although improvement of health and functional trajectories through life depends mostly on improvements in older people, this process should probably start by improving living conditions and lifestyle earlier in life. Progress toward improvement of health is likely to depend on public health to combat many problems, inadequate nutrition being a major area of improvement both in developing and developed countries, the latter because inadequate nutrition may lead to an opposite condition, that is, obesity [4]. Good nutrition and appropriate physical exercise are essential for healthy aging from both a physical and psychological perspective. Therefore, a multidisciplinary life course approach to aging is vital to minimize its complications for quality of life and subsequent public health [5].

Nutrition-related problems in older subjects, formerly ignored, have been gaining prominence in recent years and are now highly relevant, both in research and in usual clinical practice. Having a good nutritional status is not only linked to health and welfare, but is also related to an increased life expectancy with reduced disability, and is an essential component of the therapeutic plan in most chronic diseases. Moreover, food and nutrition is a relevant aspect of most cultures and is related with the individual lifestyle of every person [6].

If good nutrition is key for healthy aging, nutritional assessment and intervention should become part of health care of both healthy and sick older people. Nutritional counseling and intervention should be part of a general care plan that takes into account all aspects of an aged person. The promotion of health in older individuals should incorporate the principles of Gerontology and Geriatric Medicine to public health interventions. At the individual level, strategies of successful aging consist of having the opportunity to make and making healthy lifestyle choices, implementing various self-management techniques. Nutritional programs that aim for high compliance should be individualized, and would have to consider every aspect of old age: beliefs, attitudes, preferences, expectations, and aspirations [7]. However, it is also relevant to understand better the biological alterations of aging and their consequences in order to disentangle malnutrition from other conditions, such as poor nutrition or inadequate nutrition, taking into account that malnutrition is caused by eating a diet in which nutrients are not enough or are too much causing health problems, including obesity and overweight. Among the biological changes naturally occurring in aging, the deregulated nutrient-sensing system seems relevant [8]. Thus, individualized intervention strategies, possibly including the concept of biological age, can be established. On the other hand, it is conceivable that nutritional problems of older people cannot be understood and engaged with using the same systems developed for younger adults. Aging and many prevalent conditions in old age are known to affect nutritional needs through the lifetime. Thus, specific guidelines on nutrition and recommended intakes for older adults are more frequently being considered, but the variability in biological age, morbidity status, pharmacological treatments, and individual genetic background suggest that established general nutritional guidelines could be extremely complicated in such an heterogeneous population [9]. Even laypersons can acknowledge that a healthy and active octogenarian/nonagenarian will not have the same needs and demands as a frail octogenarian/nonagenarian with multiple disabling comorbidities and polypharmacy [10]. Thus, special recommendations for subgroups of individuals are essential, taking into account individual health status and other factors [11]. In this specific chapter, we describe important aspects of nutrition in the elderly. First of all, we address the psychosocio and economic conditions that can lead to malnutrition in the elderly. Then, we discuss the problem of malnutrition with a particular emphasis on the use of supplements in the third age. An additional section is dedicated to the current understanding of the biological mechanisms of the aging process, in particular the nutrient-sensing pathway, and how this knowledge can be related to current dietary strategies and dietary recommendations for the elderly. Finally, we also discuss critical aspects of the nutrigenomic approach for a personalized diet in the elderly that may give an exhaustive picture for a correct diet for an healthy aging.

MALNUTRITION IN THE ELDERLY: DEFINITION AND GENERAL ASPECTS

Malnutrition refers to any nutritional imbalance and in the elderly it is frequently the result of a lack of calories, proteins, and other micronutrients [12]. This condition tends to develop more frequently in the elderly and may significantly affect the quality of life, physical and psychological functions, as well as metabolic and inflammatory status. Therefore, changes in body composition, organ functions, adequate energy intake, and ability to eat or access food are associated with aging and may contribute to malnutrition. Malnutrition is also critical for hospitalized elderly patients as it is strongly related to a high rate of infectious complications and increased mortality rates. In this context, some terms, such as geriatric syndrome, are used to define the nutritional status, particularly the association malnutrition/weight loss, in the elderly. The term "anorexia of aging" is considered a geriatric syndrome and most commonly refers to the loss of appetite and/or reduced food intake beyond that normally expected with physiological aging [13]. This condition is found in up 30% of community-dwelling and institutionalized older adults and is associated with weight loss and higher rates of mortality [14]. Another geriatric syndrome and term associated with weigh loss is the "frailty" in which decreased strength and endurance and increased fatigability are present [15]. Finally, a syndrome completely different from the others, but associated with weight loss, is "cachexia" in which the weight loss is mediated by cytokines released by inflammatory states, often chronic diseases, such as cancer, chronic obstructive pulmonary disease (COPD), congestive heart failure, and rheumatoid arthritis [16]. However, regardless of the term used to define a geriatric syndrome, weight loss and malnutrition are common features of the elderly person that must be taken into serious consideration by the clinical point of view to better address and/or prevent the state of disability due to concomitant onset of diseases associated with age, in which even the persistence of malnutrition can aggravate the clinical picture of the elderly patient. In this respect, the nutritional assessment is useful, necessary, and indispensable to manage in the best way the subject/elderly patient in order to have a decent life and a healthy state. The nutritional assessment (particularly, Mini Nutritional Assessment, MNA), therefore, is a crucial part in the comprehensive evaluation of older adults. The body weight is a critical component of the MNA as weight loss is an early sign of the impaired nutritional status from inadequate caloric intake, as shown by recent meta-analysis of 16 studies on institutionalized older adults, in whom also depression, swallowing issues, and absence of physical activity are strictly associated with weight loss [17]. One advantage of the MNA is that it is applicable to a wide range of elderly patients (ie, from those who are well to hospitalized elderly). The questions are simple and brief. It has a 96% sensitivity and 98% specificity, and a predictive value of 97% which distinguishes patients by their adequate nutritional status (score ≥ 24), risk of malnutrition (score 12–23), and malnourishment (score <17) [18]. A short-form version of the MNA has been developed (MNA-SF) containing six questions and it is strongly correlated with total MNA score ($r = 0.945$), and is applicable for both community dwelling and hospitalized elderly [19,20]. Other screening tools (Nutrition Risk Index and Nutrition Risk Score) have been designed to identify risk of malnutrition in the elderly on admission to hospital [21]. The simplest office measure reflecting undernutrition is a body mass index (BMI = weight in kilograms (kg)/height in meters (m)^2) below 20 kg/m^2 [22]. In the elderly, a convenient approximation of height is arm span [23]. Another easy to remember mnemonic screening tool is the SCALES assessment [24] in order to evaluate older patients regarding: Sadness (depression); Cholesterol levels; Albumin (serum levels <40 g/L); Loss of weight; Eating problems (cognitive and/or physical determinants); and Shopping problems or inability to prepare a meal. A problem with three or more of these areas reflects a high risk for malnutrition [25]. Another tool is the "Determine Your Nutritional Health Checklist" which can be a very useful tool in the community setting [24]. However, it may be of limited use for seniors with cognitive impairment or poor vision because it relies on self-reporting. In the nursing home setting, a validated tool is the amount of food left on a resident's plate. Residents who have more than 25% of their food remaining on their plate are most likely to suffer from protein undernutrition [26]. One estimate suggests that 84% of nursing home patients had intake less than their calculated daily caloric expenditure, and only 5% were receiving nutritional supplements. Those with lower caloric intake had a higher mortality rate than those whose intake more closely matched their caloric expenditure [27]. All these tools are useful in determining the nutritional status of the elderly and therefore useful in order to correct the body weight and condition of malnutrition through specific supplements that are useful and essential in both healthy subjects living at home and in elderly hospitalized. Alongside supplements, proper physical activity has been suggested as physical activity is one of the beneficial components in preventing some geriatric syndrome, such as sarcopenia, and some diseases related with malnutrition, such as diabetes and cardiovascular diseases [28].

CAUSES OF MALNUTRITION IN THE ELDERLY

Screening for malnutrition is recommended in elderly subjects and must be carried out at least once a year in community dwelling elderly and much more frequently in institutionalized and hospitalized elderly. A screening for malnutrition in the old persons is based on the search for risk factors of malnutrition, the estimation of appetite and/or food intake and periodic measurement of body weight that are used to estimate weight loss compared to a previous record. Careful examination of weight loss is particularly important in aging as it is usually associated with onset of frailty, disability, and, finally, mortality. Although lean body mass may decline because of normal physiological changes associated with age [29], a loss of more than 4% per year is an independent predictor of mortality [30]. Rapid weight loss of 5% or more in 1 month is considered significant for the appearance of frailty [31], in which the weight loss and other dimensions (exhaustion, weakness, slowness, and low levels of activity) are the main causes of frailty syndrome [32]. It has been shown that even moderate decline of 5% or more over 3 years is predictive of mortality in older adults [33]. Risk factors for malnutrition can also coincide with causes behind this condition. These causes include oral and dental disorders including loss of dentition, poor fitting dentures, dryness of mouth and other mastication disorders that lead the elderly to eat soft foods, such as cheese and pasta, and to remove meat and other foods that, conversely, are important to sustain the protein and calorie requirements of the organisms [34]. Additional important causes include psychological, social and environmental factors such as isolation from community, the loss of partner or other family member, economic difficulties, medications and hospitalization as well as admission to retirement institutions [35]. These conditions can lead to depressive symptoms, which in turn are known to be strictly related to malnutrition in the elderly. The prevalence of depressive symptoms in older outpatients is around 15% and this prevalence increases from 22% to 34% in hospitalized elderly patients [36]. In this last category, it is estimated that about half of these patients is at risk of undernutrition. However, even if various studies have shown an association between undernutrition and depressive symptoms, it is not easy to define the causality direction between undernutrition and depression. Depression can have a great effect on appetite and eating habits up to anorexia and, at the same time, not eating sufficiently may lead to impaired mood and cognitive performance [37]. Anyway, this information highlights the need for nutritional screening of elderly with depressive symptoms so that an appropriate combination of psychosocial and nutritional interventions can be planned. Last, but not least, possible causes of malnutrition are the diseases associated with aging. Dementia (Alzheimer's or other forms) and other neurological disorders (Parkinson's) can dramatically affect the nutritional status of the elderly [38]. However, this could be a consequence also of other diseases including infectious diseases as a consequence of pain, decompensation mechanisms as well as medications. In industrialized countries the contribution of restrictive diets in the elderly (ie, salt-free, diabetic diets, cholesterol lowering diets, vegan or vegetarian diets, and other diets related to cultural and religious habits) should not be excluded. Even alcohol intake may replace or suppress the consumption of foods with superior nutritional value. Alcohol misuse in the elderly is associated with impaired functional status, poor self-rated health, and depressive symptoms [39]. Table 5.1 lists some of the causes of weight loss and malnutrition in the elderly.

TABLE 5.1 Some Causes of Weight Loss Associated with Malnutrition in the Elderly

• Alcohol and drug abuse	• Dental problems
• Medical problems	• Poor vision
• Dementia	• Medication side effects
• Social isolation	• Functional dependencies
• Depression and other psychiatric disorders	• Environmental factors
• Poverty	• Limited access to or intake of food
• Food attitudes and cultural or religious preferences	• Elder abuse

Therefore, functional, psychological, social, environmental, and economic issues associated with concomitant medical problems may all contribute to malnutrition and weight loss in the elderly [35]. A multidisciplinary geriatric assessment can be helpful to fully address all the complex interacting issues of the frail seniors. This type of comprehensive assessment may include the services of physicians, psychologists, nurses, dieticians, occupational and physical therapists, speech and language pathologists, and social workers, each of which can lend their respective expertise to the effective diagnosis of the functional, psychological, and socioeconomic contributors to malnutrition in the elderly. However, although malnutrition and weight loss are common in elderly populations, they are often underrecognized. One complicating factor is the concomitant presence of a disease tending to present atypically with nonspecific complaints in the elderly individual, making detection and diagnosis of physiological reasons behind malnutrition a greater challenge. On the other hand, the consequence of a malnutrition can be devastating and include loss of strength and function, thereby placing an individual at higher risk for adverse events. Malnutrition is also associated with increased falls and mortality as well as increased hospitalization with thus subsequent negative impact on the quality of life [40]. In this context, some age-related diseases are associated with weight loss and malnutrition.

For example, hyperthyroidism or new onset diabetes are classic examples of medical conditions causing weight loss. Progressive renal or hepatic insufficiency may cause anorexia with highly morbid implications in the elderly. Weight loss related to poor oral intake is also associated with peptic ulcer disease and congestive heart failure, as well as dental or chewing problems. Decreased oral intake may slowly occur in Parkinson's disease, COPD, and Alzheimer's disease [41]. Therefore, the malnutrition state is often associated with the healthy conditions of the old individual. Do not forget that weight loss is also associated with another common aging syndromes related to wasting muscles, such as sarcopenia, even if sarcopenia is not always related to any illness [42]. Anyway, malnutrition and weight loss are often associated and when discussing weight loss and malnutrition there are several commonly used terms that are carefully defined, as reported before.

MALNUTRITION: POSSIBLE CORRECTION WITH SUPPLEMENTS IN THE ELDERLY

In this last decade sound research on nutritional intervention has proved that strategies for individual nutritional care, usually in the form of oral supplements, can prevent and treat malnutrition in old individuals as well as improve the outcome of elderly patients with diseases [43,44]. Malnutrition is one of the first geriatric syndromes and has deleterious effects on the recovery and wellbeing of a wide range of patients and diseases. The goal of nutritional support is to fight malnutrition by supplying all nutrients required for the energy, plastic, and regulatory needs of a given individual, aiming to maintain or restore the functional integrity of the body, including nutritional status, physical and mental function, and quality of life, and to reduce morbidity and mortality.

Following careful nutritional assessment, guidelines have been developed to improve and maintain nutritional status with a focus in preventing weight loss in community-dwelling and hospitalized elderly patients. Clinical studies in hospitalized old patients at risk of undernutrition based on their initial MNA score, an oral supplementation of 200 mL sweet or salty sip feed twice daily (500 kcal, 21 g protein per day) for 2 months maintained body weight and improved MNA score [45]. Recent Canadian guidelines recommend a nutritional intake with foods containing fiber and complex carbohydrates, such as whole grains, vegetables, and fruits. Fat intake should be less than 30% of total caloric intake [41]. The combination among carbohydrates, fat, and protein (casein) should be in a caloric ratio of 3:1:1 respectively [46]. This combination has been proven in different trials in elderly populations at risk of malnutrition with a benefit in increasing body weight by 2.2%, reduction of mortality, and improvement of the healthy status with, however, no significant evidence of improvement in functional benefit or reduction in length of hospital stay with supplements [47]. In a subsequent study, malnourished elders received 600 supplemental calories daily for 3 months had an increment of body weight of 1.5 Kg [48]. The same supplemental in nursing home residents with dementia had the same beneficial effect in body weight but less in memory improvement [49]. Recently, a randomized control trial with 400 calories daily for 3 months in frail community-dwelling older adults showed less decline in short physical performance battery and gait speed than the control group and an improvement in the timed up-and-go test compared with the decline in controls [50]. Formerly known as Recommended Nutritional Intake (RNI), Dietary Reference Intakes (DRI) guidelines have been also revised recently and adjusted for the needs of older adults aged 51–70 years and those over 70. There are increased allowances for the elderly for calcium, magnesium, vitamin D, fluoride, niacin, folate, vitamin B12, vitamin E, and micronutrients (zinc and selenium). A complete list summary of the current DRI for Canada and the United States is listed at http://www.hc-sc.gc.ca/hpfb-dgpsa/onpp-bppn/diet_ref_e.html#3 (this link reports types of nutritional reference values including the recommended dietary allowances (RDA)). Following these guidelines for the elderly, specific nutritional supports have been suggested as preventive dietary strategies for the old populations regarding high cholesterol and fat intake, and their association with ischemic heart disease and diabetes [51,52]. Increasing dietary fiber, such as for example Mediterranean diet, may be useful in the treatment of constipation, glucose intolerance, lipid disorders, and obesity, as well as in preventing diverticular disease and colon cancers [53]. Reduction in sodium has been shown to reduce blood pressure and also reduce the risk of developing hypertension [54]. The recommended adult calcium intake is 1200 mg/day for those over 50 years of age; 400 IU of vitamin D is recommended for ages 50—70, and 600 IU for those over 70 years of age taking into account that a seasonal vitamin D deficiency is recognized that 35—90% of the institutionalized elderly are estimated to be vitamin D deficient [55]. Treatment with vitamin D prevents seasonal variations in vitamin D levels and can reduce hip fractures and the risk of developing osteoporosis [56]. Of interest is also the ecessity to include as nutritional intake in the elderly also some trace elements, such as selenium and zinc, that are useful to reduce oxidative stress and inflammatory status that are usual events in the elderly [57]. Doses of 10—12 mg/day of zinc and 100 mg/day of selenium in old individuals for 2—3 months are useful to reduce the oxidative stress and to maintain the inflammation under control [58]. From all these supplemental trials, it emerges the power that some substances have in preventing the state of malnutrition and in increasing body weight in the individual seniors who are often led to eating foods low in these substances for several reasons. Among them, intrinsic biomolecular reasons present physiologically with advancing age leading the elderly to be a frail person (see the next subchapter). Therefore, adequate nutritional support can be of great help for a satisfactory state of health, especially in preventing the appearance of some age-related diseases with the subsequent disability.

NUTRIENT-SENSING PATHWAYS

Aging is characterized by a progressive loss of physiological integrity, leading to impaired function and increased vulnerability to death. Many molecular mechanisms are involved in inducing the aging process that are defined as hallmarks of aging: genomic instability, telomere attrition, epigenetic alterations, loss of proteostasis, deregulated nutrient-sensing, mitochondrial dysfunction, cellular senescence, stem cell exhaustion, and altered intercellular communication [8]. Of particular interest is the nutrient-sensing pathway because it is related to specific signals from nutrients (also referred to as macronutrients), which are simple organic compounds involved in biochemical reactions that produce energy or are constituents of cellular biomass. Glucose and related sugars, amino acids, and lipids (including cholesterol) are important cellular nutrients, and distinct mechanisms to sense their abundances operate in mammalian cells. In particular, in healthy individuals they control the cellular homeostasis for cellular growth and division with an intracellular de novo synthesis but they must be also obtained from the environment (food). In such a way, circulating nutrient levels within a narrow range are maintained. Since internal nutrient levels do fluctuate, and hence intracellular and extracellular nutrient-sensing mechanisms exist in mammals [59]. In multicellular organisms, nutrients also trigger the release of hormones, which act as long-range signals with noncell autonomous effects, to facilitate the coordination of coherent responses in the organism as a whole [59]. Thus, it is clear the importance that can have a good and proper nutrition in cellular homeostasis through the nutrient-sensitive pathways for the whole life of an organism. Indeed, nutrient sensors modulate lifespan extensions that occur in response to different environmental and physiological signals. Nutrient-sensing pathways are essential to the aging process and longevity because several nutrients can activate different pathways directly or indirectly [60]. Some examples of nutrient-sensing pathways involved in the longevity response are the kinase target of rapamycin (TOR) [61], AMP kinase (AMPK) [62], sirtuins [63] and insulin and insulin/insulin-like growth factor-1 (IGF-1) signaling (IIS) [64].

Many of the genes that act as key regulators of the lifespan also have known functions in nutrient-sensing pathways, and thus are called "nutrient-sensing longevity genes." Genetic polymorphisms or mutations that reduce the functions of these genes (eg, IGF-1, mTOR, and FOXO) have been linked to longevity, both in humans and in model organisms, further illustrating the major impact of trophic and bioenergetic pathways on longevity [65]. Therefore, nutrient-sensing pathways and the related genes are connected to life-span regulation. However, the full understanding of dietary intake and composition to promote human longevity targeting nutrient-sensing pathways require further studies, even if a lot of studies report the beneficial effect of some diets in various cohort of older individuals in the world called "Blue Zones" [66].

Anyway, some of these pathways identified in worms or mice have been recently shown to have human homolog, in particular IIS pathway via IGF-1.

Indeed, it has been observed that natural genetic variants of IGF-1 in nutrient-sensing pathways are associated with increased human life-span [67]. More recently, surprisingly, altered or decreased IGF-1 signaling pathways confer an increased susceptibility to human longevity [68,69]. Paradoxically, IGF-1 levels decline during normal aging, as well as in mouse models of premature aging [70]. Thus, a decreased IIS is a common characteristic of both physiological and accelerated aging, whereas a constitutively decreased IIS extends longevity. These apparently contradictory observations can be accommodated under a unifying model by which IIS downmodulation reflects a defensive response aimed at minimizing cell growth and metabolism in the context of systemic damage [71]. According to this point of view, organisms with a constitutively decreased IIS can survive longer because they have lower rates of cell growth and metabolism and, hence, lower rates of cellular damage. On the other hand, recent studies have demonstrated a significant association between mutations in genes involved in the IIS pathway and extension of human life-span, as shown in some cohorts of centenarians, suggesting that centenarians may harbor rare genetic variations in genes encoding components of the IIS pathway [72]. Moreover, polymorphisms of FOXO transcription factor FOXO3 (part of the IIS pathway) are associated with human longevity in several different Caucasian elderly cohorts [73,74] including Southern Italian centenarians [75] as well as in Chinese centenarians [76]. However, despite the recognition of FOXO3 as a "master gene" in aging, the coding variants of FOXO3 may not be key players for longevity being too rare in multiple ethnic groups [77]. Therefore, the actual functional variant of this gene remains unidentified requiring further studies. Anyway, the IIS is an important part of the nutrient-sensing pathways playing a pivotal role in aging contributing in a large extent to reach the longevity.

In addition to the IIS pathway that participates in glucose-sensing, three other additional related and interconnected nutrient-sensing systems are: mammalian Target Of Rapamycin (mTOR), for the sensing of high amino acid concentrations; AMPK, which senses low-energy states by detecting high AMP levels; and sirtuins, which sense low-energy states by detecting high NAD^+ levels [78]. The mTOR pathway is an evolutionarily conserved nutrient-sensing pathway which has been implicated in the regulation of the life-span and in the response to stress, nutrients, and growth factors [79]. On the other hand, treatment with rapamycin in mice extends longevity [80]. The mTOR kinase is part of two multiprotein complexes, mTORC1 and mTORC2, that regulate essentially all aspects of anabolic metabolism [81]. Genetic downregulation of mTORC1 activity in yeast, worms, and flies extends longevity, suggesting that mTOR inhibition phenocopies Caloric Restriction (CR) [82], which is in turn relevant to reach longevity [83] via possible nutrient-sensing pathways [8]. Indeed, genetically modified mice with low levels of mTORC1 activity but normal levels of mTORC2 have increased life-span [84], and mice deficient in S6K1 (a main mTORC1 substrate) are also long-lived [85]. Therefore, the downregulation of mTORC1/S6K1 appears as the critical mediator of mammalian longevity in relation to mTOR. Moreover, mTOR activity increases during aging in mouse hypothalamic neurons, contributing to age-related obesity, which is reversed by direct infusion of rapamycin to the hypothalamus [86]. These observations, together with those involving the IIS pathway, indicate that intense trophic and anabolic activity, signaled through the IIS or the mTORC1 pathways, are major accelerators of aging whereas inhibition of mTOR activity has beneficial effects during aging. On the other hand, aging may not necessarily be driven by damage, but, in contrast, leads to damage and this process is driven in part by mTOR [87]. However, to date human data are scarce and the details of how mTOR exerts life-span control and antiaging effects are still not fully understood. Notably, it was recently demonstrated that rapamycin reverses the phenotype of cells obtained from patients with Hutchinson-Gilford progeria syndrome, a lethal genetic disorder that mimics rapid aging [88]. Thus, the inhibition of mTOR signaling may have a positive effect on healthy aging and longevity like CR.

A suggestive hypothesis relating to the possible role of how the inhibition of mTOR may affect the healthy status and the subsequent longevity is through the action of specific proteins that are relevant against oxidative stress, named Metallothioneins (MT). In experiments in MT transgenic mice exposed to low temperature, MT protected against cold exposure-induced cardiac anomalies possibly through attenuation of cardiac autophagy via mTOR inhibition [89]. The same role of protection by an upregulation of MT occurs against hypoxia of the kidney through the ERK/mTOR pathway [90]. From these last researches, it is quite impressive to suggest the pivotal role played by MT proteins in order to be included as the core of the action of nutrient-sensing network (mTOR) to reach a satisfactory state of health in the elderly and subsequent longevity. Such an assumption is strongly supported by an increased survival in MT transgenic mice with respect to normal control mice [91]. Another very impressive suggestion for a proper diet in the elderly is that the MT and IGF-1 are closely zinc-dependent [92]. So, in aging a diet that can contain this trace element is essential to have a good health and longevity through a correct nutrient-sensing pathways. This particular assumption is supported by an increased survival in old mice after zinc supplementation [93] and by recovered immune efficiency, antioxidant activity in the elderly after adequate zinc supplementation [94] as well as increased IGF-1 levels after zinc supplementation in conditions of severe zinc deficiency, such as eutrophic children [95]. The other two nutrient sensors, AMPK and sirtuins, act in the opposite direction to IIS and mTOR, meaning that they signal nutrient scarcity and catabolism instead of nutrient abundance and anabolism. Accordingly, their upregulation favors healthy aging. AMPK is a nutrient and energy sensor that might be involved in the regulation of life-span and in the mediation of the beneficial effects of CR. Although this hypothesis is largely unexplored, especially in mammals, it seems likely that the activation of AMPK may have an impact on the activity of FOXO, sirtuin, and the mTOR pathways, which, in turn, have been linked to CR and to the promotion of healthy longevity [8,96]. In particular, AMPK activation has multiple effects on metabolism and, remarkably, shuts off mTORC1 with a subsequent effect on prolonged life-span [97]. With regard to sirtuins, some of seven mammalian sirtuin paralogs can ameliorate various aspects of aging in mice [98]. In particular, transgenic overexpression of mammalian SIRT1 improves aspects of health during aging but does not increase longevity [99]. The beneficial effects of SIRT1 are mainly addressed to improve genomic stability in mouse embryonic stem cells that underwent DNA damage by H_2O_2 [100] and in mutant SIRT1 mice [101] as well as to enhance antioxidant defense in response to fasting signals, via activation of PPARγ coactivator 1a (PGC-1a) [102]. More compelling evidence for a sirtuin-mediated prolongevity role in mammals has been obtained for SIRT6, which regulates genomic stability, NF-kB signaling, and glucose homeostasis through histone H3K9 deacetylation [103,104]. Mutant mice that are deficient in SIRT6 exhibit accelerated aging [105], whereas male transgenic mice overexpressing SIRT6 have a longer life-span than control animals, associated with reduced serum IGF-1 and other indicators of IGF-1 signaling [106]. Interestingly, the mitochondria-located sirtuin SIRT3 has been reported to mediate some of the beneficial effects of CR restriction in longevity [107]. Recently, overexpression of SIRT3 has been reported to improve the regenerative capacity of aged hematopoietic stem cells [108]. Therefore, in mammals, at least three members of the sirtuin family, SIRT1, SIRT3, and SIRT6, contribute to healthy aging and, to some extent, also longevity. In conclusion, current available evidence strongly supports the idea that anabolic signaling accelerates aging and decreased nutrient signaling extends longevity [65]. Further, a pharmacological manipulation that mimics a state of limited nutrient availability, such as rapamycin, can extend longevity in mice [80].

NUTRIENT-GENE INTERACTION IN THE ELDERLY (NUTRIGENOMIC APPROACH)

From the data reported above, it appears that the nutrient-sensing pathways can play a pivotal role in affecting healthy aging. However, this is not enough. The individual genetic background is fundamental for the benefit of the diet and the subsequent healthy aging. It is not certain that a diet with some foods can do well in general for all older people. Each of us has a specific genetic background with specific polymorphisms that can influence in a positive or negative absorption of foods with their subsequent effects. In other words, it is not certain that a food that is of benefit to one person, can be also of benefit to another person. Therefore, it is essential to also consider the individual genetic background for a proper diet and that it is tailored to the presence of specific polymorphisms. Such an assumption is evident if one considers the interaction between genes and nutrients in particular with genes involved in inflammatory/immune response and antioxidant activity interacting with some micronutrients such as zinc, selenium, and vitamin E [58,80]. In this context, the gene expression of MT (a zinc-binding protein) regulates some zinc target inflammatory genes, such as IL-6 [110]. This finding is of great interest in aging and in very old age taking into account that in centenarians the MT gene expression is low like in healthy adults and coupled with good zinc ion bioavailability and satisfactory innate immune response [111]. Therefore, a good MT/zinc—gene interaction is preserved in very old age playing a pivotal role in successful aging and, at the same time, escaping some age-related diseases. Such an assumption is supported by the recent discovery of novel polymorphisms of MT2A and MT1A associated with the inflammatory state, zinc ion bioavailability, and longevity. Old subjects carrying AA genotype for MT2A polymorphism display low zinc ion bioavailability, chronic inflammation, and high risk for atherosclerosis and diabetes type II [112]. Conversely, polymorphism corresponding to A/C (Asparagine/Threonine) transition at +647 nt position in the MT1A coding region is the most involved in the longevity [113]. In addition, +1245 MT1A polymorphism and +647/+1245 MT1A haplotype are implicated in cardiovascular diseases (CVD) [114]. These findings are a clear clue of the relevance of the zinc—MT—gene interaction in inflammation and longevity. Such an interaction is very useful for a possible zinc supplementation in the elderly. Indeed, when the genetic variations of the IL-6 polymorphisms (IL-6-174G/C locus) are associated with also the variations of MT1A +647A/C gene, the plasma zinc deficiency and the altered innate immune response are more evident [115], suggesting that the genetic variations of IL-6 and MT1A are very useful tools for the identification of old people who effectively need zinc supplementation. These results suggest that the daily requirement of zinc might be different in the elderly harboring a different genetic background with thus the possibility of a personalized diet [58]. With regard to selenium, the gene expression of selenoproteins (Glutathione peroxidases (GPxs) and selenoproteins P1 (SEPS1)) are of a great interest because they are involved in the main age-related diseases, such as cancer and CVD [58]. A variety of studies report the role played by selenoproteins GPxs, which protect the cells against oxidative damage. Among them, GPx4 and GPx1 are the more studied because they affect cell growth, apoptosis, and inflammation [116]. A single polymorphism for GPx4 gene (a T/C SNP in position 718) shows that the C variant is related to a higher frequency of colorectal cancer [117]. The recent discovery showing more responsiveness to GPx4 activity in CC subjects than TT ones after selenium supplementation, suggests that a single nucleotide polymorphism (SNP) in the GPx4 gene, a T/C variation at position 718, has also functional consequences against cancer [118]. Such a finding is relevant because it indicates the potential importance of the SNP in the GPx4 gene especially when selenium intake is suboptimal like in cancer. With regard to GPX1 198Pro/Leu variant genotypes, subjects with 198Pro/Leu and 198Leu/Leu genotypes had a significantly higher risk of CVD compared to the 198Pro/Pro carriers, suggesting that Pro/Leu and Leu/Leu genotypes are significantly associated with CVD risk [119]. Of particular interest is the gene encoding the plasma SEPS1 since this protein is an ER membrane protein that participates in the processing and removal of misfolded proteins from ER to cytosol [120]. This system constitutes a SEPS1-dependent regulatory loop in the presence of inflammation. Variation in the SEPS1 gene affects the circulating levels of the inflammatory cytokines IL-1, IL-6 and TNF-α [121]. Given the known association of SEPS1 with the inflammation and CVD, a significant association was found with an increased risk for coronary heart disease in subjects carrying the minor allele of rs8025174 [122], suggesting an involvement of SEPS1 in CVD risk. Despite of all these data showing the relevance of SNPs of selenoproteins in affecting the possible appearance of some age-related diseases, many studies report the role played by selenium deficiency and selenium supplementation in aging and in age-related disease without considering these SNPs, but exclusively based on the selenium content in the plasma with often contradictory data [58]. Thus, there is the need to assess SNPs in human intervention studies using genotyped cohorts in order to show the relationship between selenium and SNPs for the healthy state in aging. With regard to Vitamin E, in vitro experiments and in animal models support the relevance of the Vitamin E–gene interaction in aging and inflammatory age-related diseases [110]. A substantial number of papers reports polymorphisms of genes involved in the uptake, distribution, metabolism, and secretion of the micronutrient. Some genetic polymorphisms and epigenetic modifications may lower the bioavailability and cellular activity of Vitamin E [123,124] influencing a differential susceptibility among old people to specific disorders, such as atherosclerosis, diabetes, CVD, cancers, and neurodegenerative diseases, which could be circumvented by Vitamin E supplementation. Despite these genetic findings, little up-to-date data exist on Vitamin E supplementation on the basis of specific polymorphisms that can be crucial for the effective beneficial effect of Vitamin E. In this context, a paper of Testa et al. [125] shows the relevance of the interaction between Vitamin E and the gene of plasminogen activator inhibitor type 1 (PAI-1), an independent CVD risk factor in diabetic patients closely related to the inflammatory status [126]. The 4G/5G polymorphism of PAI-1 is involved in the incidence of CVD by regulation of PAI-1 levels [127]. A treatment with Vitamin E (500 IU/die for 10 weeks) in old diabetic patients carrying 4G allele provoked a delayed decrease in PAI-1 levels with respect to those carrying 5G/5G genotype [125]. More recently, Belisle et al. [128] proposed that single nucleotide polymorphisms may influence individual response to vitamin E treatment (182 mg/day for 3 years) in terms of pro-inflammatory cytokine production (TNF-α). Old subjects with the A/A and A/G genotypes at TNF-α-308G > A treated with Vitamin E had lower TNF-α production than those with the A allele treated with placebo [128]. Since the A allele at TNF-α-308G > A is associated with higher TNF-α levels [129], these results suggest that the anti-inflammatory effect of Vitamin E may be specific to subjects genetically predisposed to higher inflammation. Anyway, more clinical studies in the gene-Vitamin E supplementation are needed to be carried out in order to better understand the beneficial effects of Vitamin E in aging and age-related diseases.

CONCLUSIONS AND PERSPECTIVES

This chapter summarizes in general some of the healthy aging nutritional secrets to achieve longevity. Presently, it is of great interest to study characteristics and biomolecular/genetic targets in shaping longevity. The realization of healthy aging and longevity is possible taking into account that genetic factors have an incidence of only 30%, whereas the remaining 70% is due to environmental and stochastic factors [130]. Thus, in order to achieve a longer and a healthier life, increased attention must be placed on lifestyle choices, particularly on the diet. There is a huge volume of scientific literature on diet and health but less attention has been paid to dietary patterns. Although it seems unlikely that there is a particular dietary pattern that promotes longevity, the contributing factors that lead to a poor diet up to malnutrition are heterogeneous. Among them, psychosocial conditions are involved but they do not entirely explain the intrinsic reasons that can be the basis of malnutrition resulting in the appearance of age-related diseases. The scientific interest around the nutrient-sensing pathways have greatly increased in recent years as it has been well documented that the modulation of these pathways by diet or pharmaceuticals can have a profound impact on health and thus represent a therapeutic opportunity for the extension of the human life-span and quality of life improvement. The mTOR is considered the most important biomolecular player of the nutrient-sensing pathway. Various experimental nutritional strategies to target mTOR are currently being developed. Among the most innovative strategies in this field, it is relevant to mention the possibility of inhibiting the mTOR protein through the physiological modulation of proteins involved in the regulation of zinc, namely the MT proteins, whose relevance in healthy aging and longevity is well documented. Furthermore, the concept of gene—nutrient interactions could represent an added value for the future development of correct and personalized diet in the elderly. The determination of specific polymorphisms of genes related to inflammatory/immune response and antioxidant activity before nutritional interventions aimed to modulate healthy aging and longevity is thus a desired tool in the current research in nutrition and aging. In conclusion, there is enough background to support the concept that the proper development of this field of research is likely to produce effective nutritional strategies able to promote health and longevity while lowering the risk for the appearance of some age-associated diseases. These strategies might be part of the new paradigm for the biomedical sciences that can be termed "positive biology" [131].

SUMMARY POINTS

- Nutrition-related problems in older subjects have been gaining prominence in recent years. Having a good nutritional status is not only linked to health and welfare, but is also related to an increased life expectancy with reduced disability due to chronic diseases.
- Malnutrition in the elderly is frequently the result of a lack of calories, proteins and other micronutrients. This condition may significantly affect the quality of life, physical and psychological functions as well as metabolic and inflammatory status.
- A screening for malnutrition in the old persons is based on the search for risk factors of malnutrition, the estimation of appetite and/or food intake and periodic measurement of body weight. A moderate decline of 5% or more over three years of the body weight is predictive of mortality in older adults. The causes of malnutrition are various including exhaustion, weakness, slowness, low levels of activity, dental disorders, psychological, social and environmental factors, the loss of partner or other family member, economic difficulties, medications and hospitalization as well as admission to retirement institutions. It is also relevant to underline that some age-related diseases (hyperthyroidism, diabetes, dementia, COPD, sarcopenia) lead to malnutrition.
- Oral supplements can prevent and treat malnutrition in old individuals as well as improve the outcome of elderly patients with diseases. Based on MNA and DRI, supplements with carbohydrates, fat and protein in the ratio 3:1:1 respectively as well as with vitamins (Vit. D, Vit. B12, Vit. E), micronutrients (zinc, selenium), calcium, magnesium, fluoride, niacin, folate, in a combination or alone may prevent the malnutrition and increase the body weight in old individuals.
- It is conceivable that nutritional problems of older people cannot be understood and engaged with the same systems developed for younger adults. While macronutrient supplements are proposed to sustain health of elderly people, caloric restriction is proposed as a health promoting strategy in younger adults. This dichotomy could be in part related to the decay and the role of nutrient sensing pathways in aging.
- The nutrient-sensing pathways are biomolecular processes activated in response to nutrients. These pathways are claimed as hallmarks and determinants of ageing process and longevity. A general deregulation of these pathways is a common characteristic of both physiological and accelerated aging, whereas, paradoxically, their constitutive decrease appears to extend longevity. Some examples of nutrient-sensing pathways involved in the longevity response are the mammalian (or mechanistic) target of rapamycin

(mTOR), AMP kinase (AMPK), sirtuins and insulin and insulin/insulin-like growth factor-1 (IGF-1) signaling (IIS).
- Many of the genes that act as key regulators in nutrient sensing pathways are also called "nutrient-sensing longevity genes". Genetic polymorphisms or mutations that reduce the functions of these genes (IGF-1, mTOR, FOXO) have been associated with longevity, both in humans and in model organisms. Inhibition of TOR activity has been related to longevity and beneficial effects during aging, but may also present important adverse effects. It is thus important to advance our knowledge around the possibility to separate beneficial from adverse effects around the inhibition of mTOR.
- The individual genetic background is another factors that may explain the impact of diet on aging. The concept of gene-nutrient interactions (nutrigenomic approach) could therefore represent an added value for the future development of correct and personalized diet in the elderly.

Acknowledgments

Paper supported by INRCA and by European projects: Zincage (coordinator: Eugenio Mocchegiani, FP6, no FOOD-2004-506850) and Markage (coordinator: Alexander Burkle, FP7, no HEALTH-F4-2008-200880).

References

[1] Christensen K, Doblhammer G, Rau R, Vaupel JW. Ageing populations: the challenges ahead Lancet 2009;374:1196–208.
[2] Coombs JB, Barrocas A, White JV. Nutrition care of older adults with chronic disease: attitudes and practices of physicians and patients. South Med J 2004;97:560–5.
[3] Kiefte-de Jong JC, Mathers JC, Franco OH. Nutrition and healthy aging: the key ingredients. Proc Nutr Soc 2014;73: 249–59.
[4] McKeown RE. The epidemiologic transition: changing patterns of mortality and population dynamics. Am J Lifestyle Med 2009;3:19S–26S.
[5] Shepherd A. Nutrition through the life span. Part 3: adults aged 65 years and over. Br J Nurs 2009;18:301–7.
[6] Woo J. Nutritional strategies for successful aging. Med Clin North Am 2011;95:477–93.
[7] Cruz-Jentoft AJ, Franco A, Sommer P, Baeyens JP, Jankowska E, Maggi A, et al. Silver paper: the future of health promotion and preventive actions, basic research, and clinical aspects of age-related disease—a report of the European Summit on Age-Related Disease. Aging Clin Exp Res 2009;21:376–85.
[8] López-Otín C, Blasco MA, Partridge L, Serrano M, Kroemer G. The hallmarks of aging. Cell 2013;153:1194–217.
[9] Mocchegiani E, Basso A, Giacconi R, Piacenza F, Costarelli L, Pierpaoli S, et al. Diet (zinc)-gene interaction related to inflammatory/immune response in aging: possible link with frailty syndrome? Biogerontology 2010;11:589–95.
[10] Valentini L, Pinto A, Bourdel-Marchasson I, Ostan R, Brigidi P, Turroni S, et al. Impact of personalized diet and probiotic supplementation on inflammation, nutritional parameters and intestinal microbiota—The "RISTOMED project": Randomized controlled trial in healthy older people. Clin Nutr 2014; pii: S0261-5614(14)00251-9
[11] Cannella C, Savina C, Donini LM. Nutrition, longevity and behavior. Arch Gerontol Geriatr 2009;49:19–27.
[12] White JV, Guenter P, Jensen G, Malone A, Schofield M, Academy of Nutrition and Dietetics Malnutrition Work Group, A.S.P.E.N. Malnutrition Task Force, A.S.P.E.N. Board of Directors. Consensus statement of the Academy of Nutrition and Dietetics/American Society for Parenteral and Enteral Nutrition: characteristics recommended for the identification and documentation of adult malnutrition (undernutrition). J Acad Nutr Diet 2012;112:730–8.
[13] Martone AM, Onder G, Vetrano DL, Ortolani E, Tosato M, Marzetti E, et al. Anorexia of aging: a modifiable risk factor for frailty. Nutrients 2013;5:4126–33.
[14] Morley JE. Anorexia of aging: a true geriatric syndrome. J Nutr Health Aging 2012;16:422–5.
[15] Morley JE, Vellas B, van Kan GA, Anker SD, Bauer JM, Bernabei R, et al. Frailty consensus: a call to action. J Am Med Dir Assoc 2013;14:392–7.
[16] Farkas J, von Haehling S, Kalantar-Zadeh K, Morley JE, Anker SD, Lainscak M. Cachexia as a major public health problem: frequent, costly, and deadly. J Cachexia Sarcopenia Muscle 2013;4:173–8.
[17] Tamura BK, Bell CL, Masaki KH, Amella EJ. Factors associated with weight loss, low BMI, and malnutrition among nursing home patients: a systematic review of the literature. J Am Med Dir Assoc 2013;14:649–55.
[18] Vellas B, Guigoz Y, Garry PJ, Nourhashemi F, Bennahum D, Lauque S, et al. The Mini Nutritional Assessment (MNA) and its use in grading the nutritional state of elderly patients. Nutrition 1999;15(2):116–22.
[19] Rubenstein LZ, Harker JO, Salvà A, Guigoz Y, Vellas B. Screening for undernutrition in geriatric practice: developing the short-form mininutritional assessment (MNA-SF). J Gerontol A Bio Sci Med Sci 2001;56A:M366–72.
[20] Ranhoff AH, Gjøen AU, Mowé M. Screening for malnutrition in elderly acute medical patients: the usefulness of MNA-SF. J Nutr Health Aging 2005;9:221–5.
[21] Corish CA, Flood P, Kennedy NP. Comparison of nutritional risk screening tools in patients on admission to hospital. J Hum Nutr Diet 2004;17:133–9.
[22] McWhirter JP, Pennington CR. Incidence and recognition of malnutrition in hospital. BMJ 1994;308:945–8.
[23] Villaverde-Gutiérrez C, Sánchez-López MJ, Ramirez-Rodrigo J, Ocaña-Peinado FM. Should arm span or height be used in calculating the BMI for the older people? Preliminary results. J Clin Nurs 2015;24:817–23.
[24] Nutritional Screening Initiative [online]. URL: <http://www.aafp.org/x16081.xml> [accessed 25.05.05].
[25] Morley JE. Nutritional assessment is a key component of geriatric assessment: the Mini Nutritional Assessment (MNA). Nutrition in the elderly (Suppl 2). Facts and Research in Gerontology. 2nd ed. New York: Serdi Publishing Co.; 1994. p. 1–97.
[26] Beck AM, Oveson L, Schroll M. Validation of the resident assessment instrument triggers in the detection of undernutrition. Age Aging 2001;30:161–5.
[27] Elmstahl S, Persson M, Blabolil V. Malnutrition in geriatric patients: a neglected problem? J Adv Nurs 1997;26:851–5.
[28] Haselwandter EM, Corcoran MP, Folta SC, Hyatt R, Fenton M, Nelson ME. The built environment, physical activity and aging in the united states: a state of the science review. J Aging Phys Act 2015;23:323–9.

[29] Lissner L, Odell PM, D'Agostino RB, Stokes III J, Kreger BE, Belanger AJ, et al. Variability of body weight and healthy outcomes in the Framingham population. N Engl J Med 1991;324:1839–44.

[30] Wallace JI, Schwartz RS, LaCroix AZ, Uhlmann RF, Pearlman RA. Involuntary weight loss in older outpatients: incidence and clinical significance. J Am Geriatr Soc 1995;43:329–37.

[31] Jensen GL, McGee M, Binkley J. Nutrition in the elderly. Gastroenterol Clin North Am 2001;30:313–34.

[32] Fried LP, Tangen CM, Walston J, Newman AB, Hirsch C, Gottdiener J, Cardiovascular Health Study Collaborative Research Group, et al. Frailty in older adults: evidence for a phenotype. J Gerontol A Biol Sci Med Sci 2001;56:M146–56.

[33] Newman AB, Yanez D, Harris T, Duxbury A, Enright PL, Fried LP, Cardiovascular Study Research Group. Weight change in old age and its association with mortality. J Am Geriatr Soc 2001;49:1309–18.

[34] Meydani M. Nutrition interventions in aging and age-associated disease. Ann N Y Acad Sci 2001;928:226–35.

[35] Bartali B, Salvini S, Turrini A, Lauretani F, Russo CR, Corsi AM, et al. Age and disability affect dietary intake. J Nutr 2003;133:2868–73.

[36] Kaiser MJ, Bauer JM, Rämsch C, Uter W, Guigoz Y, Cederholm T, Mini Nutritional Assessment International Group, et al. Frequency of malnutrition in older adults: a multinational perspective using the mini nutritional assessment. J Am Geriatr Soc 2010;58:1734–8.

[37] Hays NP, Roberts SB. The anorexia of aging in humans. Physiol Behav 2006;88:257–66.

[38] Gillette-Guyonnet S, Secher M, Vellas B. Nutrition and neurodegeneration: epidemiological evidence and challenges for future research. Br J Clin Pharmacol 2013;75:738–55.

[39] St John P, Montgomery P, Tyas SL. Alcohol misuse, gender and depressive symptoms in community-dwelling seniors. Geriatr Today J Can Geriatr Soc 2002;5:121–5.

[40] Donini LM, Ricciardi LM, Neri B, Lenzi A, Marchesini G. Risk of malnutrition (over and under-nutrition): validation of the JaNuS screening tool. Clin Nutr 2014;33:1087–94.

[41] Wells JL, Dumbrell AC. Nutrition and aging: assessment and treatment of compromised nutritional status in frail elderly patients. Clin Interv Aging 2006;1:67–79.

[42] Cruz-Jentoft AJ, Landi F, Schneider SM, Zúñiga C, Arai H, Boirie Y, et al. Prevalence of and interventions for sarcopenia in aging adults: a systematic review. Report of the International Sarcopenia Initiative (EWGSOP and IWGS). Age Aging 2014;43:748–59.

[43] Ahmed T, Haboubi N. Assessment and management of nutrition in older people and its importance to health. Clin Interv Aging 2010;5:207–16.

[44] Silver HJ. Oral strategies to supplement older adults' dietary intakes: comparing the evidence. Nutr Rev 2009;67:21–31.

[45] Gazzotti C, Arnaud-Battandier F, Parello M, Farine S, Seidel L, Albert A, et al. Prevention of malnutrition in older people during and after hospitalisation: results from a randomised controlled clinical trial. Age Aging 2003;32:321–5.

[46] Gammack JK, Sanford AM. Caloric supplements for the elderly. Curr Opin Clin Nutr Metab Care 2015;18:32–6.

[47] Milne AC, Potter J, Vivanti A, Avenell A. Protein and energy supplementation in elderly people at risk from malnutrition. Cochrane Database Syst Rev 2009;2:CD003288.

[48] Neelemaat F, Bosmans JE, Thijs A, Seidell JC, van Bokhorst-de van der Schueren MA. Post-discharge nutritional support in malnourished elderly individuals improves functional limitations. J Am Med Dir Assoc 2011;12:295–301.

[49] Stange I, Bartram M, Liao Y, Poeschl K, Kolpatzik S, Uter W, et al. Effects of a low-volume, nutrient- and energy-dense oral nutritional supplement on nutritional and functional status: a randomized, controlled trial in nursing home residents. J Am Med Dir Assoc 2013;14: 628.e1–8.

[50] Kim CO, Lee KR. Preventive effect of protein-energy supplementation on the functional decline of frail older adults with low socioeconomic status: a community-based randomized controlled study. J Gerontol A Biol Sci Med Sci 2013;68:309–16.

[51] Cheng C, Graziani C, Diamond JJ. Cholesterol-lowering effect of the Food for Heart Nutrition Education Program. J Am Diet Assoc 2004;104:1868–72.

[52] Ciardullo AV, Brunetti M, Daghio MM, Bevini M, Feltri G, Novi D, et al. Characteristics of type 2 diabetic patients cared for by general practitioners either with medica nutrition therapy alone or with hypoglycaemic drugs. Diabetes Nutr Metab 2004;17:120–3.

[53] Sofi F, Macchi C, Abbate R, Gensini GF, Casini A. Mediterranean diet and health. Biofactors 2013;39:335–42.

[54] De Santo NG. Reduction of sodium intake is a prerequisite for preventing and curing high blood pressure in hypertensive patients—second part: guidelines. Curr Hypertens Rev 2014;10:77–80.

[55] Compher C, Kim JN, Bader JG. Nutritional requirements of an aging population with emphasis on subacute care patients. AACN Clin Issues 1998;9:441–50.

[56] Holick MF. Vitamin D: a millennium perspective. J Cell Biochem 2003;88:296–307.

[57] Mocchegiani E, Malavolta M, Muti E, Costarelli L, Cipriano C, Piacenza F, et al. Zinc, metallothioneins and longevity: interrelationships with niacin and selenium. Curr Pharm Des 2008;14:2719–32.

[58] Mocchegiani E, Costarelli L, Giacconi R, Malavolta M, Basso A, Piacenza F, et al. Micronutrient-gene interactions related to inflammatory/immune response and antioxidant activity in aging and inflammation. A systematic review. Mech Aging Dev 2014;136–137:29–49.

[59] Rossetti L. Perspective: hexosamines and nutrient-sensing. Endocrinology 2000;141:1922–5.

[60] Efeyan A, Comb WC, Sabatini DM. Nutrient-sensing mechanisms and pathways. Nature 2015;517:302–10.

[61] Hansen M, Taubert S, Crawford D, Libina N, Lee SJ, Kenyon C. Lifespan extension by conditions that inhibit translation in *Caenorhabditis elegans*. Aging Cell 2007;6:95–110.

[62] Greer EL, Dowlatshahi D, Banko MR, Villen J, Hoang K, Blanchard D, et al. An AMPK-FOXO pathway mediates longevity induced by a novel method of dietary restriction in C. elegans. Curr Biol 2007;17:1646–56.

[63] Polito L, Kehoe PG, Forloni G, Albani D. The molecular genetics of sirtuins: association with human longevity and age-related diseases. Int J Mol Epidemiol Genet 2010;1:214–25.

[64] Kaletsky R, Murphy CT. The role of insulin/IGF-like signaling in C. elegans longevity and aging. Dis Model Mech 2010;3:415–19.

[65] Fontana L, Partridge L, Longo VD. Extending healthy life span—from yeast to humans. Science 2010;328:321–6.

[66] Appel LJ. Dietary patterns and longevity: expanding the blue zones. Circulation 2008;118:214–15.

[67] Bonafè M, Barbieri M, Marchegiani F, Olivieri F, Ragno E, Giampieri C, et al. Polymorphic variants of insulin-like growth factor I (IGF-I) receptor and phosphoinositide 3- kinase genes affect IGF-I plasma levels and human longevity: cues for an evolutionarily conserved mechanism of life span control. J Clin Endocrinol Metab 2003;88:3299–304.

[68] Holzenberger M. IGF-I signaling and effects on longevity. Nestle Nutr Workshop Ser Pediatr Program 2011;68:237–45.

[69] Pawlikowska L, Hu D, Huntsman S, Sung A, Chu C, Chen J, et al. Study of osteoporotic fractures. association of common genetic variation in the insulin/IGF1 signaling pathway with human longevity. Aging Cell 2009;8:460–72.

[70] Schumacher B, van der Pluijm I, Moorhouse MJ, Kosteas T, Robinson AR, Suh Y, et al. Delayed and accelerated aging share common longevity assurance mechanisms. PLoS Genet 2008;4:e1000161.

[71] Garinis GA, van der Horst GT, Vijg J, Hoeijmakers JH. DNA damage and aging: new-age ideas for an age-old problem. Nat Cell Biol 2008;10:1241–7.

[72] Suh Y, Atzmon G, Cho MO, Hwang D, Liu B, Leahy DJ, et al. Functionally significant insulin-like growth factor I receptor mutations in centenarians. Proc Natl Acad Sci USA 2008;105:3438–42.

[73] Willcox BJ, Donlon TA, He Q, Chen R, Grove JS, Yano K, et al. FOXO3A genotype is strongly associated with human longevity. Proc Natl Acad Sci USA 2008;105:13987–92.

[74] Flachsbart F, Caliebe A, Kleindorp R, Blanché H, von Eller-Eberstein H, Nikolaus S, et al. Association of FOXO3A variation with human longevity confirmed in German centenarians. Proc Natl Acad Sci USA 2009;106:2700–5.

[75] Anselmi CV, Malovini A, Roncarati R, Novelli V, Villa F, Condorelli G, et al. Association of the FOXO3A locus with extreme longevity in a southern Italian centenarian study. Rejuvenation Res 2009;12:95–104.

[76] Li Y, Wang WJ, Cao H, Lu J, Wu C, Hu FY, et al. Genetic association of FOXO1A and FOXO3A with longevity trait in Han Chinese populations. Hum Mol Genet 2009;18:4897–904.

[77] Donlon TA, Curb JD, He Q, Grove JS, Masaki KH, Rodriguez B, et al. FOXO3 gene variants and human aging: coding variants may not be key players. J Gerontol A Biol Sci Med Sci 2012;67:1132–9.

[78] Houtkooper RH, Williams RW, Auwerx J. Metabolic networks of longevity. Cell 2010;142:9–14.

[79] Sengupta S, Peterson TR, Sabatini DM. Regulation of the mTOR complex 1 pathway by nutrients, growth factors, and stress. Mol Cell 2010;40:310–22.

[80] Harrison DE, Strong R, Sharp ZD, Nelson JF, Astle CM, Flurkey K, et al. Rapamycin fed late in life extends lifespan in genetically heterogeneous mice. Nature 2009;460:392–5.

[81] Laplante M, Sabatini DM. mTOR signaling in growth control and disease. Cell 2012;149:274–93.

[82] Johnson SC, Rabinovitch PS, Kaeberlein M. mTOR is a key modulator of aging and age-related disease. Nature 2013;493:338–45.

[83] Bishop NA, Guarente L. Genetic links between diet and lifespan: shared mechanisms from yeast to humans. Nat Rev Genet 2007;8:835–44.

[84] Lamming DW, Ye L, Katajisto P, Goncalves MD, Saitoh M, Stevens DM, et al. Rapamycin-induced insulin resistance is mediated by mTORC2 loss and uncoupled from longevity. Science 2012;335:1638–43.

[85] Selman C, Tullet JM, Wieser D, Irvine E, Lingard SJ, Choudhury AI, et al. Ribosomal protein S6 kinase 1 signaling regulates mammalian life span. Science 2009;326:140–4.

[86] Yang SB, Tien AC, Boddupalli G, Xu AW, Jan YN, Jan LY. Rapamycin ameliorates age-dependent obesity associated with increased mTOR signaling in hypothalamic POMC neurons. Neuron 2012;75:425–36.

[87] Blagosklonny MV. Revisiting the antagonistic pleiotropy theory of aging: TOR-driven program and quasi-program. Cell Cycle 2010;9:3151–6.

[88] Cao K, Graziotto JJ, Blair CD, Mazzulli JR, Erdos MR, Krainc D, et al. Rapamycin reverses cellular phenotypes and enhances mutant protein clearance in Hutchinson-Gilford progeria syndrome cells. Sci Transl Med 2011;3:89ra58

[89] Jiang S, Guo R, Zhang Y, Zou Y, Ren J. Heavy metal scavenger metallothionein mitigates deep hypothermia-induced myocardial contractile anomalies: role of autophagy. Am J Physiol Endocrinol Metab 2013;304:E74–86.

[90] Kojima I, Tanaka T, Inagi R, Nishi H, Aburatani H, Kato H, et al. Metallothionein is upregulated by hypoxia and stabilizes hypoxia-inducible factor in the kidney. Kidney Int 2009;75:268–77.

[91] Malavolta M, Basso A, Piacenza F, Giacconi R, Costarelli L, Pierpaoli S, et al. Survival study of metallothionein-1 transgenic mice and respective controls (C57BL/6J): influence of a zinc-enriched environment. Rejuvenation Res 2012;15:140–3.

[92] Mocchegiani E, Malavolta M, Costarelli L, Giacconi R, Piacenza F, Lattanzio F, et al. Is there a possible single mediator in modulating neuroendocrine-thymus interaction in aging? Curr Aging Sci 2013;6:99–107.

[93] Mocchegiani E, Romeo J, Malavolta M, Costarelli L, Giacconi R, Diaz LE, et al. Zinc: dietary intake and impact of supplementation on immune function in elderly. Age (Dordr) 2013;35:839–60.

[94] Mocchegiani E, Costarelli L, Giacconi R, Piacenza F, Basso A, Malavolta M. Micronutrient (Zn, Cu, Fe)-gene interactions in aging and inflammatory age-related diseases: implications for treatments. Aging Res Rev 2012;11:297–319.

[95] Alves CX, Vale SH, Dantas MM, Maia AA, Franca MC, Marchini JS, et al. Positive effects of zinc supplementation on growth, GH, IGF1, and IGFBP3 in eutrophic children. J Pediatr Endocrinol Metab 2012;25:881–7.

[96] Salminen A, Kaarniranta K. AMP-activated protein kinase (AMPK) controls the aging process via an integrated signaling network. Aging Res Rev 2012;11:230–41.

[97] Alers S, Löffler AS, Wesselborg S, Stork B. Role of AMPK-mTOR-Ulk1/2 in the regulation of autophagy: cross talk, shortcuts, and feedbacks. Mol Cell Biol 2012;32:2–11.

[98] Houtkooper RH, Pirinen E, Auwerx J. Sirtuins as regulators of metabolism and healthspan. Nat Rev Mol Cell Biol 2012;13:225–38.

[99] Herranz D, Muñoz-Martin M, Cañamero M, Mulero F, Martinez-Pastor B, Fernandez-Capetillo O, et al. SIRT1 improves healthy aging and protects from metabolic syndrome-associated cancer. Nat Commun 2010;1:3–5.

[100] Oberdoerffer P, Michan S, McVay M, Mostoslavsky R, Vann J, Park SK, et al. SIRT1 redistribution on chromatin promotes genomic stability but alters gene expression during aging. Cell 2008;135:907–18.

[101] Wang RH, Sengupta K, Li C, Kim HS, Cao L, Xiao C, et al. Impaired DNA damage response, genome instability, and tumorigenesis in SIRT1 mutant mice. Cancer Cell 2008;14:312–23.

[102] Rodgers JT, Lerin C, Haas W, Gygi SP, Spiegelman BM, Puigserver P. Nutrient control of glucose homeostasis through a complex of PGC-1alpha and SIRT1. Nature 2005;434:113–18.

[103] Kanfi Y, Peshti V, Gil R, Naiman S, Nahum L, Levin, et al. SIRT6 protects against pathological damage caused by diet-induced obesity. Aging Cell 2010;9:162–73.

[104] Kawahara TL, Michishita E, Adler AS, Damian M, Berber E, Lin M, et al. SIRT6 links histone H3 lysine 9 deacetylation to NF-kappaB-dependent gene expression and organismal life span. Cell 2009;136:62–74.

[105] Mostoslavsky R, Chua KF, Lombard DB, Pang WW, Fischer MR, Gellon L, et al. Genomic instability and aging-like phenotype in the absence of mammalian SIRT6. Cell 2006;124:315–29.

[106] Kanfi Y, Naiman S, Amir G, Peshti V, Zinman G, Nahum L, et al. The sirtuin SIRT6 regulates lifespan in male mice. Nature 2012;483:218–21.

[107] Someya S, Yu W, Hallows WC, Xu J, Vann JM, Leeuwenburgh C, et al. SIRT3 mediates reduction of oxidative damage and prevention of age-related hearing loss under caloric restriction. Cell 2010;143:802–12.

[108] Brown K, Xie S, Qiu X, Mohrin M, Shin J, Liu Y, et al. SIRT3 reverses aging-associated degeneration. Cell Rep 2013;3:319–27.

[109] Mocchegiani E, Costarelli L, Giacconi R, Malavolta M, Basso A, Piacenza F, et al. Vitamin E-gene interactions in aging and inflammatory age-related diseases: implications for treatment. A systematic review. Aging Res Rev 2014;14:81–101.

[110] Mazzatti DJ, Malavolta M, White AJ, Costarelli L, Giacconi R, Muti E, et al. Differential effects of in vitro zinc treatment on gene expression in peripheral blood mononuclear cells derived from young and elderly individuals. Rejuvenation Res 2007;10:603–20.

[111] Mocchegiani E, Giacconi R, Cipriano C, Muzzioli M, Gasparini N, Moresi R, et al. MtmRNA gene expression, via IL-6 and glucocorticoids, as potential genetic marker of immunosenescence: lessons from very old mice and humans. Exp Gerontol 2002;37:349–57.

[112] Giacconi R, Cipriano C, Muti E, Costarelli L, Maurizio C, Saba V, et al. Novel −209A/G MT2A polymorphism in old patients with type 2 diabetes and atherosclerosis: relationship with inflammation (IL-6) and zinc. Biogerontology 2005;6:407–13.

[113] Cipriano C, Malavolta M, Costarelli L, Giacconi R, Muti E, Gasparini N, et al. Polymorphisms in MT1a gene coding region are associated with longevity in Italian Central female population. Biogerontology 2006;7:357–65.

[114] Giacconi R, Kanoni S, Mecocci P, Malavolta M, Richter D, Pierpaoli S, et al. Association of MT1A haplotype with cardiovascular disease and antioxidant enzyme defense in elderly Greek population: comparison with an Italian cohort. J Nutr Biochem 2010;21:1008–14.

[115] Mariani E, Neri S, Cattini L, Mocchegiani E, Malavolta M, Dedoussis GV, et al. Effect of zinc supplementation on plasma IL-6 and MCP-1 production and NK cell function in healthy elderly: interactive influence of +647 MT1a and −174 IL-6 polymorphic alleles. Exp Gerontol 2008;43:462–71.

[116] Kim YS, Milner J. Molecular targets for selenium in cancer prevention. Nutr Cancer 2001;40:50–4.

[117] Bermano G, Pagmantidis V, Holloway N, Kadri S, Mowat NA, Shiel RS, et al. Evidence that a polymorphism within the 3′UTR of glutathione peroxidase 4 is functional and is associated with susceptibility to colorectal cancer. Genes Nutr 2007;2:225–32.

[118] Méplan C, Crosley LK, Nicol F, Horgan GW, Mathers JC, Arthur JR, et al. Functional effects of a common single-nucleotide polymorphism (GPX4c718t) in the glutathione peroxidase 4 gene: interaction with sex. Am J Clin Nutr 2008;87:1019–27.

[119] Tang NP, Wang LS, Yang L, Gu HJ, Sun QM, Cong RH, et al. Genetic variant in glutathione peroxidase 1 gene is associated with an increased risk of coronary artery disease in a Chinese population. Clin Chim Acta 2008;395:89–93.

[120] Schomburg L, Schweizer U, Holtmann B, Flohé L, Sendtner M, Köhrle J. Gene disruption discloses role of selenoprotein P in selenium delivery to target tissues. Biochem J 2003;370:397–402.

[121] Curran JE, Jowett JB, Elliott KS, Gao Y, Gluschenko K, Wang J, et al. Genetic variation in selenoprotein S influences inflammatory response. Nat Genet 2005;37:1234–41.

[122] Alanne M, Kristiansson K, Auro K, Silander K, Kuulasmaa K, Peltonen L, et al. Variation in the selenoprotein S gene locus is associated with coronary heart disease and ischemic stroke in two independent Finnish cohorts. Hum Genet 2007;122:355–65.

[123] Rigotti A. Absorption, transport, and tissue delivery of vitamin E. Mol Aspects Med 2007;28:423–36.

[124] Zingg JM, Azzi A, Meydani M. Genetic polymorphisms as determinants for disease-preventive effects of vitamin E. Nutr Rev 2008;66:406–14.

[125] Testa R, Bonfigli AR, Sirolla C, Boemi M, Manfrini S, Mari D, et al. Effect of 4G/5G PAI-1 polymorphism on the response of PAI-1 activity to vitamin E supplementation in type 2 diabetic patients. Diabetes Nutr Metab 2004;17:217–21.

[126] De Taeye B, Smith LH, Vaughan DE. Plasminogen activator inhibitor-1: a common denominator in obesity, diabetes and cardiovascular disease. Curr Opin Pharmacol 2005;5:149–54.

[127] Grubic N, Stegnar M, Peternel P, Kaider A, Binder BR. A novel G/A and the 4G/5G polymorphism within the promoter of the plasminogen activator inhibitor-1 gene in patients with deep vein thrombosis. Thromb Res 1996;84:431–43.

[128] Belisle SE, Leka LS, Delgado-Lista J, Jacques PF, Ordovas JM, Meydani SN. Polymorphisms at cytokine genes may determine the effect of vitamin E on cytokine production in the elderly. J Nutr 2009;139:1855–60.

[129] Cipriano C, Caruso C, Lio D, Giacconi R, Malavolta M, Muti E, et al. The −308G/A polymorphism of TNF-alpha influences immunological parameters in old subjects affected by infectious diseases. Int J Immunogenet 2005;32:13–18.

[130] Schächter F, Cohen D, Kirkwood T. Prospects for the genetics of human longevity. Hum Genet 1993;91:519–26.

[131] Davinelli S, Willcox DC, Scapagnini G. Extending healthy aging: nutrient sensitive pathway and centenarian population. Immun Aging 2012;9:9–12.

CHAPTER 6

Nutrition in the Hospitalized Elderly

Wafaa Mostafa Abd-El-Gawad and Doha Rasheedy

Geriatrics and Gerontology Department, Faculty of Medicine, Ain Shams University, Cairo, Egypt

KEY FACTS
- Nutritional deficiencies are more common among hospitalized patients than community dwelling elderly.
- The hospitalized elderly population is affected by many causes of malnutrition, which can be reversed if it is addressed early.
- Malnutrition is independently associated with poor outcomes and decreased quality of life.
- Management of malnutrition in the elderly population requires a multidisciplinary approach that treats pathology and uses both social and dietary forms of intervention.
- Many simple nutrition rules are clear and easy to be applied in various hospitals setting either regular ward or ICU.
- Ethical consideration should be reviewed cautiously regarding artificial nutrition especial in special group of patients like dementia.

Dictionary of Terms

- *Anorexia*: Anorexia is a true geriatric syndrome defined as a loss of appetite and/or reduced food intake. It is a multifactorial condition associated with multiple negative health outcomes [1]. It generally develops when the accumulated effects of impairments in multiple systems make the older subject more vulnerable to adverse health events [2].
- *Malnutrition*: There are still no universal accepted definitions for the terms of undernutrition or malnutrition. The current guidelines of National Collaborating Centre for Acute Care define malnutrition as state of nutrition in which a deficiency of energy, protein, and/or other nutrients causes measurable adverse effects on tissue/body form, composition, function, or clinical outcome. According to these guidelines, the term "malnutrition" was not used to cover excess nutrient provision [3].
- *Assisted eating*: It is defined as "needing help from another person to be able to eat." Assisted feeding ranges from verbal encouragement and nonverbal prompts (like cutting meat, uncovering the food) up to physical guidance to transfer food from the plate to an individual's mouth [4].
- *Enteral feeding*: It is defined as the process of the delivery of nutritional needs containing protein, carbohydrate, fat, water, minerals, and vitamins, directly into the stomach, duodenum, or jejunum.
- *Parenteral feeding*: It is a way of supplying all the nutritional needs of the body by bypassing the digestive system and dripping nutrient solution directly into a vein. It is used when individuals cannot or should not get their nutrition through eating. It is used when the intestines are obstructed, when the small intestine is not absorbing nutrients properly, or a gastrointestinal fistula (abnormal connection) is present. It is also used when the bowels need to rest and not have any food passing through them like Crohn's disease or pancreatitis, widespread infection (sepsis), and in malnourished individuals to prepare them for major surgery, chemotherapy, or radiation treatment.

INTRODUCTION

Hospitalization is a critical issue for older patients, about 40% of the hospitalized adults are 65 years of age or older, and this percentage is expected to rise as the population continues to age [5].

For many older persons, especially for the frail and the very old, hospitalization is a period of acute stress and increased vulnerability. Elderly patients are susceptible to complications not directly related to the illness for which they are hospitalized. These complications begin immediately upon admission and extend beyond discharge [6]. One of the most debilitating and the highly prevalent complications in the hospital settings is malnutrition [7]. However, it is usually an unrecognized and underestimated consequence of the hospital setting [7]. Malnutrition constitutes an important threat to the independence and quality of life of hospitalized older people [6].

The prevalence of malnutrition among hospitalized elderly varies considerably depending on the population studied and the criteria used for the diagnosis. Nearly 55% of elderly hospitalized patients are undernourished or malnourished on admission [1]. Moreover, the nutritional status of the hospitalized patients tends to deteriorate during their hospital stay in 46–100% of the elderly patients [8], and persists following their discharge into the community [9].

Although the onset of malnutrition depends on the nutritional reserve (independent of the disease status), a connection between malnutrition and disease exists, whereby disease may uncover malnutrition, or malnutrition may have a negative effect on disease. Therefore, malnutrition is often referred to as "a cause and an outcome of hospitalization" [10].

PATHOGENESIS OF MALNUTRITION IN HOSPITALIZED ELDERLY

The Interaction Between Aging, Health Status, Hospital Environment, and Nutritional Status:

The pathogenesis of malnutrition in hospitalized elderly is multifactorial. Aging, disease, and hospital related environmental factors determine the nutritional status and control food intake in the elderly. With aging, there is increased adipose tissue leading to subsequent increase in tumor necrosis factor alpha (TNF-α) and other pro-inflammatory cytokines. These inflammatory cytokines are major contributors of anorexia in the elderly [11].

Many acute and chronic conditions are also accompanied by a pro-inflammatory state mediated by multiple cytokines, such as interleukin-6 (IL-6) and TNF-α. These cytokines alter brain circuitries controlling food intake, have a negative effect on appetite, alter both gustatory and olfactory sensation, delay gastric emptying, increase the resting metabolic rate, enhance skeletal muscle catabolism, and decrease muscle protein synthesis [12].

Hospitalization and acute illness cause marked increase in the cytokines and hormone production that

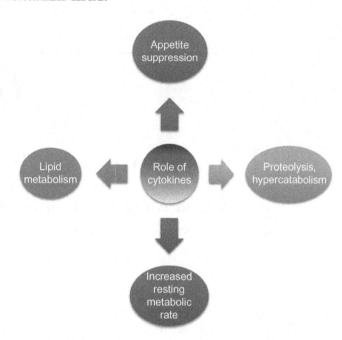

FIGURE 6.1 Role of cytokines in disease related malnutrition.

trigger systemic inflammatory response which increases the resting basal metabolic rate (BMR) in proportionate to the severity of the insult. Moreover, the release of stress related hormones, for example, cortisol and norepinephrine, enhances protein catabolism and aggravates malnutrition [12].

Cytokines such as ILβ-1, IL-2, IFN-γ, and TNF-α may also inhibit feeding by causing nausea and vomiting, decreasing gastric and intestinal motility, modifying gastric acid secretion [13], or by stimulating the release of serotonin, norepinephrine, and dopamine in both the central and peripheral nervous systems which, in turn, suppress food intake. Moreover, some cytokines inhibit muscle protein synthesis by reduction in anabolic hormones (IGF-1 and testosterone), and induction of insulin resistance [14].

Other cytokines induce lipolysis and β-oxidation in fat and liver with subsequent increased very low density lipoprotein (VLDL) synthesis and decreased lipoprotein lipase activity resulting in hypertriglyceridemia. All these processes result in negative energy balance and weight loss [15] (Fig. 6.1).

WHY ARE HOSPITALIZED OLDER ADULTS NUTRITIONALLY VULNERABLE?

Many predisposing factors make older adults susceptible to malnutrition compared to other age groups. Aging itself is a major contributing risk factor for malnutrition. Various physiologic, pathologic, psychologic,

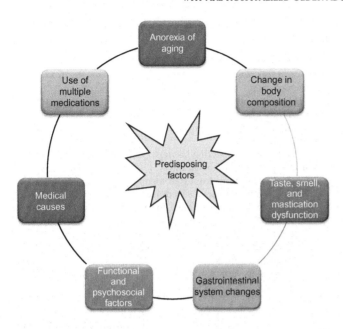

FIGURE 6.2 Predisposing factors for malnutrition in older adults.

and sociologic factors (eg, depression, delirium, social isolation, chronic illness, and medications) increase the risk of anorexia of aging which result in malnutrition [16] (Fig. 6.2).

Age-Related Changes

Anorexia of Aging

Anorexia is a true geriatric syndrome defined as a loss of appetite and/or reduced food intake. It affects over 25% and 30% of elderly men and women, respectively. Anorexia of aging represents one of the major challenges for geriatric medicine given its impact on quality of life, morbidity, and mortality [16].

The mean differences in caloric intake between ages 20–39 and ≥75 years were found to be 897 kcal/day in men and 390 kcal/day in women which indicate loss of appetite and reduced food intake in the elderly group compared to younger adults. This decline is an adaptive response to the reduction in energy expenditure, loss of muscle mass, and reduced physical activities that accompany normal aging [17].

Appetite regulation is a complex process and is regulated by many central and peripheral orexigenic and anorexigenic signals. The pathogenesis of anorexia is thought to involve age-related regression in the activities of particular brain areas, including the hypothalamus, in response to peripheral stimuli (fat cell signals, nutrients, circulating hormones), especially hunger hormones such as ghrelin (orexigenic hormone) and cholecystokinin (anorexigenic or satiety hormone).

These alterations may be responsible for decreased hunger and unintended weight loss in old age [2]. In healthy elderly, anorexigenic signals overcome orexigenic signals leading to prolonged satiety and inhibition of hunger [2,15]. The altered eating pattern in elderly persons with anorexia, is characterized by poorer consumption of nutrient-rich foods (eg, meat, eggs, and fish) [16].

Anorexia itself leads to hypoalbuminemia, increases synthesis of acute-phase proteins such as C-reactive protein, decreases coagulation capacity leading to oxidative stress, and enhances tissue damage, sarcopenia, and decreases quality of life. Moreover, Inflammation, which is linked to the aging process plays a key role in anorexia associated with chronic diseases or cachexia [2,18].

Changes in the Body Composition

Aging is accompanied by a progressive decline in lean body mass and an increase in fat mass with redistribution of fat to central locations within the body. The loss of lean body mass affects mainly skeletal muscle, particularly type II or fast twitch fibers. While, central lean body mass, such as the liver and other splanchnic organs, is relatively preserved [19]. The loss of lean body mass subsequently leads to impaired glucose tolerance, decreased BMR, decreased muscle strength and limited physical activity [19].

Taste, Smell, and Mastication Dysfunction

Alterations in taste and smell in the elderly decreases the enjoyment of food, leads to decreased dietary variety, and promotes dietary use of salt and sugar to compensate for these declines. With aging, olfactory function declines due to reduced mucus secretion, epithelial thinning, and impaired regeneration of olfactory receptor cells [11,16]. However, the alterations in taste sensation with aging are less obvious. There is a slight increase in taste thresholds that occurs due to progressive loss in the number of taste buds per papilla on the tongue sparing those involved in bitter and sour sensations [16]. Weaknesses of the muscles of mastication and decreased chewing efficiency due to dental problems also have a negative effect on food intake in older adults leading to avoidance of foods that are difficult to chew, such as meat, fruit, and vegetables, with subsequent malnutrition [2]. Dysfunction in both taste and smell among the aged can also result from chronic disease and medication use [16].

Gastrointestinal Changes

Multiple physiological and pathological changes in the gastrointestinal tract interfere with food intake. These changes include delayed esophageal emptying and spontaneous gastroesophageal reflux caused by changes in peristaltic activity, and decreased gastric and pancreatic exocrine secretions [2].

With aging, fundal compliance decreases leading to more rapid antral filling and increased central stretch leading to delayed gastric emptying with an increase in satiation [11]. The delayed gastric emptying in the elderly has been also attributed to increased phasic pyloric pressure waves in response to nutrients in the duodenum, age- or disease-related autonomic neuropathy, and a reduction in nitric oxide production by the stomach that reduces relaxation of the fundus [20].

Medical Causes

Multimorbidities are believed to be the main cause of malnutrition in the older persons. With advancing age the burden of chronic and acute diseases increases, which directly impacts the balance of nutritional needs and intake especially with multiple dietary restrictions [21]. Numerous diseases, such as coronary heart diseases, diabetes, and chronic liver disease, are extremely common in the elderly and impose major modifications in their diet. In addition, subclinical catabolic and inflammatory states exist in association with these chronic diseases lead to an increased production of catabolic cytokines, more muscle catabolism, and decreased appetite. The reduced metabolic reserve of muscles due to age and disease related loss of muscle mass results in a reduced ability to cope with any stress. So, even minor stress of short duration can negatively affect the nutritional status of older adults [22].

Disease related malnutrition occurs mainly due to inadequate food intake, increased nutritional requirements, but can also result from complications of the underlying illness such as poor absorption and excessive nutrient losses, drug-related side effects, or a combination of these aforementioned factors [12]. The causes of disease-related malnutrition remain ill-defined but cytokines induced cachexia, hormonal dysregulation, and drug-related side effects are major underlying mechanisms [12].

Use of Multiple Medications

Many medications can cause anorexia or interfere with the ingestion of food. Those most frequently implicated include chemotherapy, morphine derivatives, antibiotics, sedatives, neuroleptics, digoxin, antihistamines, captopril, etc. [23].

Functional and Psychosocial Factors

Functional impairments in basic and instrumental activities of daily living, loneliness, weakness, fatigue,

TABLE 6.1 Effect of Hospitalization on Nutritional State [8,26–28]

1. Exacerbation of already existing risk factors of malnutrition
2. Increased loss of appetite
3. Depressed mood
4. Dysphagia
5. Confusion
6. Isolation and hospital environment
7. Multiple medications
8. Organizational and catering barriers:
   - Interruptions during mealtimes
   - Disruptive behavior from staff members and other patients
   - Missed meals because of scheduled investigatory procedures
   - Inadequate feeding assistance, for example, to open packaging
   - Unpleasant meals or unappealing smells
   - Not given food and fluids between mealtimes
   - Food and fluids were placed out of reach
   - Inflexible mealtimes
   - Lack of menu variability
   - Multiple dietary restrictions

lack of cooking skills, cognitive impairment, and poverty can all predispose to malnutrition [2].

In older humans, depression is the most common cause of anorexia-related weight loss. Depression in older adults can be caused by age-related hormonal dysregulation, for example, diminished testosterone levels and raised leptin levels [2,24].

Hospital Environment-Related Risks

The hospitalization itself regardless the nature or severity of acute illness is a major hazard in the fragile elderly population. It affects every organ in the body with especial impact on the nutritional status. Admission to hospital can aggravate malnutrition due in part to the disease process and in part to the hospital environment that lead to insufficient nutrient intake. The food served in hospitals often lacks the palatable taste and is of poor quality. The choices of food are often limited and the timing of meals may not match the patients' desires. Hospitalized patients frequently experience multiple episodes of fasting prior to undergoing investigations and when ordered nil by mouth without being fed by an alternative route [7,25].

The functional dependency of the patients and the lack of trained staff to assist with feeding contribute to the decreased food intake in hospitals. Moreover, many hospitals lack sufficient training of staff to early recognize and manage the nutritional problems [23]. These risk factors are summarized in Table 6.1.

CLINICAL CONSEQUENCES OF MALNUTRITION IN HOSPITAL

Malnutrition is associated with multiple negative consequences that complicate the course of acute illness and worsen its prognosis. Almost every body system and function is affected by the nutritional status. Malnutrition results in depressed immune response and increased susceptibility to nosocomial infections. It also increases the risk of pressure ulcers, delays wound healing, decreases nutrient intestinal absorption, changes thermoregulation, and deteriorates renal function. On a physical level, malnutrition results in muscle and fat mass loss, reduced respiratory muscle and cardiac function, and atrophy of visceral organs [7].

Moreover, malnutrition is associated with fatigue and apathy, which in turn delays recovery, exacerbates anorexia, and increases convalescence time [7]. Malnutrition was found to be an independent risk factor in the development of delirium [22,29], functional decline, and increased risk of falling [29].

In a systematic review, authors found that in-hospital and 1-year mortality following hospital admission is associated with nutrition and physical function rather than disease diagnosis itself [30]. Malnourished elderly have longer periods of illness, longer hospital stays [7], higher morbidity and mortality rates [31], decreased response and tolerance to treatment, poorer response to regular medical treatment, lower quality of life [7], and significantly higher healthcare costs. On the other hand, nutritional support has been shown to result in a more successful rehabilitation, earlier discharge, fewer infections, and lower mortality rates [32].

MALNUTRITION SCREENING AND ASSESSMENT IN HOSPITALIZED ELDERLY

The early recognition of malnutrition is essential for timely intervention that will prevent subsequent complications. Nutritional assessment should be performed in two stages: first screening, and then a detailed assessment for the malnourished and those at risk for malnutrition.

The optimal nutritional screening for hospitalized elderly should be conducted on admission and weekly thereafter. Malnourished or vulnerable patients should be referred for further assessment to create a nutrition care plan and to monitor the adequacy of nutritional therapy. In the absence of optimal screening programs in hospital, many cases of malnutrition are missed [25]. The need for comprehensive approach for nutrition screening and support is needed to reduce the cost and improve the clinical outcome, such as nutritional status, quality of life, patient satisfaction, morbidity, and mortality [33].

Multidisciplinary comprehensive assessment should be conducted by trained professionals such as physicians, nurses, occupational therapists, and dietitians [2]. The systematic nutrition assessment comprises a history-taking, and physical, clinical, biochemical, anthropometric evaluation, dietary assessment, and functional outcomes [34].

Malnutrition Screening as a Part from Comprehensive Geriatric Assessment (CGA)

The use of malnutrition screening tools is a routine part of CGA in elderly people in different clinical settings. Since it is not easy to perform a complete nutritional assessment for every patient, nutrition screening tools were developed as a method of identifying the vulnerable cases that require a more detailed nutritional assessment [34]. Presumed the multifactorial nature of malnutrition in the elderly and in the absence of a single objective measure or "gold standard," many malnutrition screening tools specific to the older adults have been developed [2].

The clinical characteristics of some of the widely used malnutrition screening tools in hospital setting are shown in Table 6.2.

Assessment of functional status using Katz Activities of Daily Living such as bathing, hygiene, dressing, feeding, using the toilet, and moving around [47] and mental cognitive ability using Mini Mental State Examination [48] and psychological assessment including motivation and mood are integral parts of the CGA and clearly help in screening for the risk factors of malnutrition [34].

The planning for dietary support should be individualized according to the patient's social situation, financial resources, and living arrangements (living independently, alone, in an assisted living facility, or in a skilled nursing facility) [26].

Comprehensive Nutritional Assessment

If the malnutrition screening tool is positive, vulnerable patients (at risk and malnourished) should be subjected for a more detailed nutrition assessment that includes the following (Fig. 6.3).

Nutritional History

The nutritional history is the first step in nutritional assessment. It should address previous medical or surgical problems, history of loss of appetite, dental problems, declining senses of taste and smell, chewing and swallowing abilities, gastrointestinal symptoms, weight loss during the past 3 months and past year, severity of nutritional compromise and the rate of weight decline, dietary habits, and alteration in dietary intake using a dietary dairy. Alcohol use and medications history focusing on gastrointestinal side effects,

TABLE 6.2 Malnutrition Screening Tools [35–46]

Risk assessment tool	Setting	Description	Advantage	Limitation
Subjective global assessment (SGA) [35]	Heavily studied and validated in hospital setting	SGA includes: Five features of patient history: weight change, dietary intake change, GI symptoms, functional capacity, and disease and its relation to nutrition requirements. Five features of physical examination: loss of subcutaneous fat in the triceps region, muscle wasting in the quadriceps and deltoids, ankle edema, sacral edema, and ascites	It incorporates functional capacity as an indicator of malnutrition and also relies heavily on physical signs of malnutrition or malnutrition-inducing conditions. It does not incorporate any laboratory findings	The sensitivity of SGA is dependent on the physical signs of micronutrient deficiency which are usually late in the course of the disease. Thus, SGA is probably not useful as a tool for early detection and is not practical to use for follow up and monitoring during nutritional support
Mini nutrition assessment (MNA) [36]	Most commonly used in elderly patients in different setting. Validated.	Consists of 18 assessment questions including general state, subjective clinical evaluation of nutritional status, anthropometric data, and dietary history	Quick, noninvasive, and inexpensive. Available in different languages	A limiting factor may be accurate assessment of height and weight to obtain BMI in bedridden individuals. Not appropriate for patients who cannot provide reliable information about themselves (ie, patients with Alzheimer's disease, dementia, stroke, etc.) and for patients receiving nutritional support through nasogastric tube feeding [37]
Mini nutrition assessment-short form (MNA-SF) [38]	Used in hospital and geriatric rehabilitation settings	It is the first six questions of MNA. It takes less than 5 minutes to complete, it is used for screening in low risk patients	Quick, noninvasive, and inexpensive	Not appropriate for patients who cannot provide reliable information about themselves (ie, patients with Alzheimer's disease, dementia, stroke, etc.) and for patients receiving nutritional support through nasogastric tube feeding [37]
Malnutrition universal screening tool (MUST) [39]	Developed to detect both undernutrition and obesity in adults, and was designed for use in multiple settings including hospitals and nursing homes	It depends on unintentional weight loss, BMI, disease severity, and problems with food intake	Quick (3–5 minutes), easy, and accurate. May be completed without weighing patients	Need a reliable patient or relative memory/recall. Difficult to use with confused patients
Nutrition risk screening (NRS)—2002 [40]	It is based on an analysis of many controlled clinical trials. Designed for use in hospital setting	Total score is the sum of the score for impaired nutritional status and the score for severity of disease. Impaired nutritional status depends on degree of weight loss or BMI or reduced food intake. Severity of disease depends on the reason for admission (eg, COPD exacerbation mild, moderate or severe, or ICU admission). Additional point given for age 70 years and older	Classifying the patients with respect to their nutritional status and severity of disease. Determining the effect of nutritional intervention on clinical outcome	It requires a subjective evaluation of the severity of the disease which can alter the final result. Its accuracy to detect patients likely to benefit from any means of nutritional support and not as a screening tool for malnutrition per se [37]

(Continued)

TABLE 6.2 (Continued)

Risk assessment tool	Setting	Description	Advantage	Limitation
Geriatric nutritional risk index (GNRI) [41]	It was initially proposed for sub-acute care and long-term setting. There is promising data regarding its suitability for hospital setting [42]	The GNRI was developed as a modification for the nutritional risk index in elderly patients. This index is calculated by using albumin and weight	Suitable tool for patients who cannot provide a reliable self-assessment (advanced dementia, aphasia, or apraxia) or in those cases where patients have parenteral or enteral nutrition	GNRI is an index of risk for development of nutrition-related problems rather than a nutritional screening tool [43]
Malnutrition screening tool (MST) [44]	Validated for use in general medical, surgical and oncology patients	Three questions assess only weight loss, decreased appetite, and recent intake	Quick and easy. It does not require calculation of BMI	Not suitable for monitoring the patients' nutritional status over time [45]
The short nutrition assessment questionnaire (SNAQ) [46]	Validated for hospital inpatient and outpatient use, as well as residential patients	Four items for weight change, appetite, supplements, and tube feeding. Provides an indication for dietetic referrals as well as outlining a nutrition treatment plan	Quick and easy. It does not require calculation of BMI	Not suitable for monitoring the patients' nutritional status over time [45]

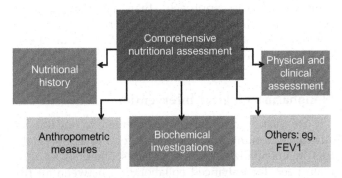

FIGURE 6.3 Components of comprehensive nutritional assessments.

drug–nutrient interaction such as digoxin, antihistaminic, antibiotics, etc. should be obtained [23]. A family member or caregiver is helpful for obtaining an accurate history [26,34].

Physical and Clinical Assessment

The physical examination is the second step in nutritional assessment. The physical examination should focus on the information obtained from medical history and assess the general appearance, musculature, the oral cavity, especially the dentition and ability to swallow, taste, and smell, and gastrointestinal as well as respiratory systems [26,34].

The physical examination should target edema, ascites, and other findings consistent with weight gain or loss. It may not be helpful in detecting early malnutrition in the elderly; the loss of muscle bulk may mimic age-related processes. However, changes in nail, hair, tongue, and angle of the mouth can be seen in case of specific nutrient deficiencies [49].

Anthropometry

The single best measure of malnutrition in elderly is weight loss. The anthropometric indices are simple and inexpensive, but affected by age, gender, and ethnicity [50]. Patient's current weight and body mass index (BMI) are widely used as indicators for malnutrition. BMI is the description of weight in relationship to height. It is unreliable in conditions where muscle mass is replaced with adipose tissue or if there is associated limb edema [50,51]. Individuals are classified as underweight (BMI < 18.5 kg/m^2), normal weight (18.5–24.9 kg/m^2), overweight (25–29.9 kg/m^2), or obese (≥30 kg/m^2) [52]. However, nutrition screening tools such as the mini nutrition assessment-short form (MNA-SF) [38] support the use of higher cut-off points to identify malnutrition in older adults. The rationale for higher BMI cut-off points is that older adults are likely to [53] have a smaller proportion of lean body mass than younger adults (as a result of aging, malnutrition, or inactivity). Research also indicates that moderately higher BMI is protective against mortality [54].

Furthermore, the measurement of height in the elderly is unreliable because of vertebral compression, loss of muscle tone, and postural changes. Hence, using the ulna length is a more suitable alternative [50].

The other classical anthropometric measurements, including skinfolds and circumferences, are useful, but require an appropriate training to achieve acceptable reliability [34].

Advanced technology for assessing body composition including bioelectrical impedance analysis, dual-energy X-ray absorptiometry, computed tomography, and magnetic resonance imaging make anthropometry more reliable in elderly population and hospitalized patients, however, problems regarding access, referral, and reimbursement may restrict their use [55].

The Biochemical Investigations

Laboratory tests can provide an objective measure of the nutritional status; however, no single biochemical marker is considered a satisfactory screening test. Their main value remains for more detailed assessment and for monitoring. Serum proteins synthesized by the liver such as albumin, transferrin, retinol-binding protein, and thyroxine-binding prealbumin are used as markers for nutritional state, however, they are affected by inflammation and infection limiting their usefulness in the acutely ill patients [50].

Other objective measures of malnutrition include low total lymphocyte, low total cholesterol, and the assessment of vitamin and trace element status (including thiamine, riboflavin, pyridoxine, calcium, vitamin D, B12, folate, zinc, and ferritin) [50].

Other Functional Measurements

Measurement of organ function as a determinant of nutritional status, such as FEV1 (forced expiratory volume) or peak expiratory flow rates, are useful in younger adult populations but have limited utility in the very elderly. Assessment of hand muscle strength "hand grip" is an upcoming and promising tool but it is still under investigation [32].

Barriers for Comprehensive Nutritional Assessment

However, barriers in clinical practice are commonly encountered; screening rates in hospitals are only 60–70% at best estimate. The lack of trained staff to assess elderly at risk of malnutrition remains the most encountered barrier for elderly assessment. Barriers for proper nutritional assessment and management are mentioned in Table 6.3 [23,27,56].

MANAGEMENT OF HOSPITALIZATION-ASSOCIATED MALNUTRITION

It is therefore mandatory upon all those who care for the elderly to understand the more limited nature of the compensatory and regulatory mechanisms that maintain normal fluid and food balance in elderly

TABLE 6.3 Barriers for Comprehensive Nutritional Assessment [23,27,56]

1. Lack of staff
2. Lack of staff training, skills, knowledge, and support
3. Patients' short stays
4. Lack of time for what is considered a low priority is commonly cited by nursing staff
5. Lack of prioritization and timely feeding assistance by nursing staff
6. Lack of coordination of shared responsibility between disciplines, including poor interdisciplinary communication

patients, and to incorporate this understanding into the diagnosis and clinical interventions that must be made to provide optimal care for this uniquely susceptible group of patients [2] (Table 6.4). In order to make good use of today's knowledge of nutritional medicine, nutritional support teams are needed in hospitals. These teams must consist of doctors in charge, nutrition, care staff trained in nutrition, dietary assistants, and/or dieticians [25,33,57]. Management of malnutrition in the hospitalized elderly requires a multidisciplinary approach that includes dietary counseling, oral supplements, parenteral nutrition, and enteral nutrition (EN), adapted to patients' nutritional needs [25,57] (Fig. 6.4).

Nonpharmacological Intervention

Care plans should be implemented to ensure an adequate amount (and quality) of food to limit weight loss in hospitalized elderly patients. There is an ultimate need for enhanced collaboration between all the multidisciplinary team involved in providing healthcare [2,27]. A number of interventions are required to enhance the feeding process:

Eliminate All Potential Reversible Risk Factors

It is important to identify and remove all potentially reversible factors that promote weight loss. For example; liberalize the patient's diet and remove or substantially modify dietary restrictions putting in mind the benefit, risk, and life expectancy of the patient. Also replacement of medications that cause anorexia and nausea should be done. Treatment of depression, parkinsonism, and cognitive impairment could help in improving food intake [2,26].

Enhance Food Environment

A good social environment and a pleasant physical atmosphere during meals can have beneficial effects on food intake and increase the time on the table. Many measures can be used to enhance the food environment such as [8,26]:

TABLE 6.4 Estimating Nutritional Requirements of Different Nutrients [32]

Total energy	The total energy intake needed to maintain weight in hospitalized patient is about 1.3 times estimated BMR, and 1.5–1.7 times BMR if weight gain is desired
	The intake of 30–35 kcal/kg/day will therefore meet the needs of most elderly hospital patients' requirements
Protein	The dietary protein intake of 12–15% of total energy is well tolerated in all patients without advanced liver or kidney disease
	The current recommended dietary allowances (RDA) of 0.8 g of protein per kg/day is adequate in healthy elderly. The requirements of the sick elderly are higher, that is, 1–1.5 g/kg/day
	The high protein diet (>15% protein as calories) is debated in elderly patients for fear of precipitating renal problems, but there is no evidence supporting this concern in patients who do not have preexisting renal disease
Fat	It is recommended to increase dietary fat intake during acute illness to comprise 40–60% of energy; however, for long-term treatment this should be reduced to 30%
	The RDA for essential fatty acids (EFA) can be provided by as little as 2–3% of the total calorie intake, that is, only 9–10 g of the EFA, linoleic, and linolenic acid, from animal and vegetable foods
Carbohydrates	Slowly absorbed carbohydrates such as starches should constitute 55–60% of total calories
	Carbohydrates provide the bulk of calories in enteral feedings and in parenteral formulas. However, carbohydrate tolerance diminishes with advancing age. So, carbohydrates should be provided from complex sources and blood sugar should be adequately monitored
Fiber	Dietary fibers whether soluble or nonsoluble are essential for elderly nutritional support. The soluble fibers, such as pectin, break down to short chain fatty acids which are important nutrients for the colonic mucosa. These products may also be absorbed and can meet up to 5% of energy needs. The nonsoluble fibers remain undigested and help to bulk the stool preventing constipation
Fluids	Water is an important nutrient in elderly patients because of their tendency for rapid shifts in fluid compartments. Daily fluid requirements are approximately 1 mL/kcal ingested or 30 mL/kg
Vitamins	Vitamin requirements have not been established for persons over the age of 65 years. However, subclinical vitamin deficiencies are common among elderly persons. The physiologic stress of illness may be sufficient to deplete rapidly any residual stores, and cause deficiencies
Minerals	Minerals (calcium, phosphorus, magnesium, iron, zinc, iodine, chromium, molybdenum, and selenium) requirements do not seem to be altered at old age, but amounts adequate to maintain serum levels must be provided by both enteral and parenteral solutions

- Encouragement of the family to be present at mealtime and to assist in the feeding process.
- Encouragement of eating in a dining room.
- Permitting patient interaction with staff during meals.
- Ensuring that patients are equipped with all necessary sensory aids (glasses, dentures, hearing aids).
- Ensuring that the patient is seated upright at 90°, preferably out of bed and in a chair.
- Ensuring that the food and utensils are removed from wrapped containers and are placed within the patient's reach.
- Removing unpleasant sights, sounds, and smells.
- Allowing for a slower pace of eating; avoiding the removal of the patient's tray too soon.
- Allowing for adequate time for chewing, swallowing, and clearing throat before offering another bite should be allowed.
- Ensuring rapport between patient and feeder.

Up to 70% of elderly hospitalized patients require some feeding assistance [27,28] and protected mealtime is increasingly delivered by patient care assistants as opposed to nurses whose role at mealtimes and responsibilities for patient's nutritional status has diminished in recent years [58,59]. The use of a red tray to identify patients in need of feeding assistance has been encouraged across the United Kingdom and also employed in a South Australian hospital [59,60].

Although some studies have shown that nutritional intake can be improved, the evidence for the effectiveness of feeding assistance in improving patient outcomes is somewhat mixed [60–62].

Enhance the Quality and the Quantity of the Food and Assess the Need for Food Supplements

Apart from the therapeutic benefit of nutritional drinks and supplementary oral nutrition [57], there is nearly no well-known pharmacological treatment identified in medical practice. The use of flavor enhancers,

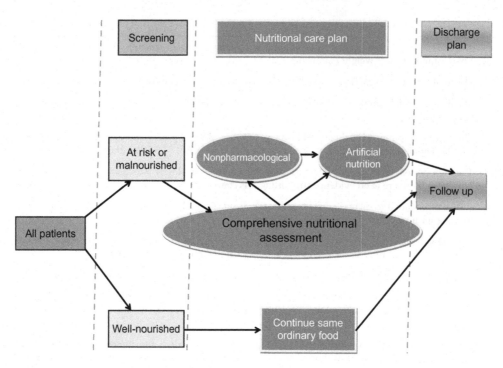

FIGURE 6.4 Scheme of nutritional management for all hospitalized elderly.

adding meat, peanut butter, or protein powder, can enhance food intake and increase the nutritional value of the meals. Considering ethnic food preferences and permitting families to bring specific foods can also increase the appetite of the patient [26].

The use of small, frequent portions of high-energy food with no odor or flavor, such as maltodextrin between meals (fingerfood, snacks, high-energy drinks) and high protein diets and oral nutritional supplements throughout the day have demonstrated improved nutritional, clinical, and functional outcomes and therefore should be provided to all patients not meeting their dietary requirements during hospitalization [10,57,63–65].

Nutritional drinks and oral nutritional supplements should be given between meals rather than mealtimes, or even in the evening as additions to patients' food. In patients with limited food intake, high-calorie drinks with high-calorie content (1.5–2.7 kcal/mL) can be given [57].

There is a debate regarding the benefit of oral supplements. Some studies found minimal or no benefit over dietary counseling alone. Further studies are needed to justify their use especially with high cost on the healthcare system [66,67].

Nutritional Counseling

Special nutritional education for family members providing care and encouraging physical activity between meals could be helpful in enhancing the appetite [57].

Pharmacological Intervention

Pharmacological intervention has a limited role in management of malnutrition. It is mainly indicated to treat specific nutritional deficiencies. The use of some vitamins and/or minerals is sometimes indicated to treat clinical deficiencies [10]. A number of medications have been used to stimulate appetite, but they are not considered as a first line of treatment. Megestrol acetate, dronabinol, and oxandrolone have been used to treat cachexia and anorexia in patients with AIDS and cancer. However, mixed evidence regarding the long-term effectiveness of these agents in the geriatric population has been produced. No drug has received US Food and Drug Administration approval for treating anorexia in geriatric population [26].

Artificial Nutrition

If all the measures stated above have been used with no therapeutic benefit, artificial nutrition must be considered. Many factors should be addressed during providing artificial nutrition. These factors include the health status (underlying disease, comorbidities, expected prognosis, mental/psychological status), patients' wishes, and ethical issues particularly in elderly patients with special problems, for example, advanced dementia, terminal illness [57,68].

The overall nutritional requirements of the older adult do not change [69]. The estimated needs of different nutrients are shown in Table 6.4 [32].

Enteral Nutrition

Enteral feeding is used to meet the nutritional needs of patients with a functional gastrointestinal tract but who are unable to safely swallow [70]. It is usually started through a nasogastric tube, however, in elderly patients in whom EN is anticipated for longer than 4 weeks, placement of a percutaneous endoscopic gastrostomy tube is recommended [71].

Enteral feeding is always favored over parenteral nutrition (PN) because the failure to maintain normal oral nutrition is associated with immunological changes and impairment of the gut associated lymphatic system (GALT), which in turn makes the intestine a source of activated cells and pro-inflammatory stimulants during gut starvation through lymphatic drainage [72].

In addition, EN carried lower risk of infection with increased secretory IgA, minimal or no risk of refeeding syndrome, mechanical and metabolic derangement, and it is more cost effective compared to PN [72]. The data regarding effect of EN on survival, length of hospital stay, and QOL remain conflicting [71].

The EN provided by the gastric route is more physiologic, is easier to administer (ie, bolus feeding with no need for delivery devices for continuous administration), and allows for a larger volume and higher osmotic load than the small intestine. While, postpyloric feeding may be beneficial in patients at high risk of aspiration, severe esophagitis, gastric dysmotility or obstruction, recurrent emesis, and pancreatitis [70].

The complications of EN in elderly are similar to those in other age groups [71]. The most prevalent complications related to enteral feeding are abdominal bloating, vomiting, high gastric residual volume, and increased risk of aspiration especially in mechanically ventilated patients [72].

Parenteral Nutrition

PN is the appropriate route of nutrition support in patients with gastrointestinal tract dysfunction (eg, ileus or other obstruction, severe dysmotility, fistulae, surgical resection, severe malabsorption) that compromise absorption of nutrients. It can be provided by peripheral or central routes [73].

PN support should be instituted in the older person facing a period of starvation of more than 3 days when oral or EN is impossible, and when oral or EN has been or is likely to be insufficient for more than 7–10 days [74].

Standard PN solutions contain carbohydrate (dextrose) and protein (amino acid) concentrates, fat emulsions (soybean or safflower oil with egg phospholipids), micronutrients, and electrolytes. Electrolyte requirements depends on the patient's underlying disease (eg, heart failure, renal dysfunction) and factors such as renal or GI fluid losses [73].

The most common, yet preventable, complications of PN are overfeeding and the refeeding syndrome (in patients with preexisting malnutrition). PN is also associated with many metabolic complications as electrolyte disturbance which can be reduced by early introduction of micronutrients supply [75].

The risk of gastrointestinal immune system failure associated with the lack of EN in many animal studies was reversed by PN fortified by glutamine [72].

Costs and Budgeting of Artificial Nutrition: A Paradigm Shift

Recent data evolved leading to a shift in the attitudes toward supporting enteral feeding. It was believed that resources needed to be spent in the clinical nutrition of patients could not be expected to contribute to cost-efficiency by shortening hospital stays.

Nowadays, however, the available clinical studies and meta-analyses argued against the previous opinion. The early treatment of malnutrition is one of the most effective ways to save money. In both the major Council of Europe resolution and very new programs (Stop Malnutrition), they describe the high number of undernourished patients in European hospitals as totally unacceptable and decisively confirm the medical/clinical consequences and the enormous unnecessary extra costs for healthcare if left untreated [57].

POSTDISCHARGE PLAN

Despite the scarcity of literature and research discussing postdischarge follow up and nutritional support, there is a growing body of evidence to support the use of these services for elderly patients [10,76]. The postdischarge plan consisting of individualized dietary counseling by a dietitian with regular follow up by a general practitioner had a positive effect on both functional and nutritional status outcomes. In a Cochrane study, the structured individualized discharge plans were associated with reductions in LOS and readmission rates, improvements in nutritional status, functional status, and mortality rates [10,77].

NUTRITIONAL ISSUES IN SPECIAL GROUPS OF THE HOSPITALIZED ELDERLY POPULATION

Critically Ill Elderly Patients

Elderly patients in ICUs are considered an exceptionally vulnerable population. This is because most of them have many chronic conditions that impair their oral intake and nutritional status. When coupled with an acute condition, it is likely to be associated with exceptional risk for nutritional decline in ICUs. These critically ill elderly usually consume suboptimal calories and have less than ideal body weight [78]. The catabolic effect on protein stores is one of the most striking features of critically ill patients and it cannot be simply attributed to starvation [72].

However, there is a scarcity of data that has specifically addressed these problems in this particular population. One study demonstrated that malnutrition was evident in 23–34% of elderly patients at the time of admission to either medical or surgical ICU [78].

In an ICU setting, there is a wide variation in patient nutritional state between and within patients on different days [72]. Artificial nutrition in critically ill patients remains an unresolved debate. Many studies revealed that survival in ICU can be improved with earlier introduction and more complete EN or PN delivery [72]. Most of the guidelines recommended early and aggressive nutritional intervention in ICU patients [75]. However, it was reported that the early PN in critically ill patients suppressed the ubiquitin-proteasome pathway. Although this can preserve muscle mass, it also leads to autophagy deficiency in liver and skeletal muscle with subsequent accumulation of toxic protein aggregates that ultimately compromises cell function [79].

Recently, two large trials studied the effect of targeting full-replacement feeding early in critical illness and they found that it did not offer any benefit and may even cause harm in some populations, especially the elderly population. They recommended hypocaloric enteral feeding for up to 7 days in formerly well-nourished patients in the acute phase of critical illness (EPaNIC [80] and EDEN trials [81]) but whether patients with preexisting malnutrition should be otherwise treated differently is still not certain [82].

Based on these observations, it has been speculated that starvation during the early phase of acute illness may be beneficial to the cell recycling system [83]. It is not desirable to interfere with the early catabolic response in critical illness. However, additional research is still needed to explore the exact underlying mechanisms and to identify biomarkers that could classify the patients that are likely to benefit from more aggressive earlier nutrition [82].

Another unresolved controversy is whether to support the use of EN over PN in ICU patients. Although EN is the preferred route preserving GALT, the sicker critically ill patients usually have gastrointestinal intolerance and are more at risk for aspiration, especially in ventilated patients [72].

Patients with Advanced Dementia

Patients with advanced dementia have many swallowing and feeding difficulties that jeopardize their nutritional status and increase their susceptibility to aspiration pneumonia [84]. According to American Society for Parenteral and Enteral Nutrition (ASPEN), EN in patients with advanced dementia is classified as category E, not obligatory in those cases. The decision for EN must be based on effective communication with the caregivers to decide whether it should be used or not [85].

However, the decision whether or not to provide artificial nutrition is often associated with emotional burden. Many surrogates agonize over the withholding and/or withdrawal of artificial nutrition. Frequently caregivers take their decisions without adequate information regarding the clinical course of the patient. Thus providing artificial nutrition to patients with advanced dementia involves many ethical and legal issues and the decision should be based on effective patient and family communication, realistic goals, with respect for patient dignity and autonomy [86].

Behavioral interventions as verbal cues, praise, and mimic techniques can be used to encourage proper meal intake. Verbal cues include reminding the patient to lift the eating utensil, chew, and swallow. The praise for following the cues and acting out lifting the fork to the mouth and then demonstrate a chewing motion and the use of fingerfoods can promote independent feeding in this group of patients [87].

Cancer Patients

Cancer itself represents a state of hypercatabolic state due to the high amount of inflammatory mediators. Patients with cancer represent the commonest vulnerable group candidate to aggressive therapies in the hospital as they may suffer from malnutrition which can decrease the compliance to the oncologic therapies or can be also worsened by such treatments. Supportive nutrition can be given through port systems for patients receiving chemotherapy [4].

Nutrition and End-of-Life Care

Decisions regarding nutrition and hydration during end-of-life are challenging. It is usually difficult for families to accept cessation of nutrition and hydration. The key for optimal nutritional management during the end of life is the good communication between healthcare professionals and the family. The patient's comfort and dignity is the major concern at this point [70]. Legally, the United States Supreme Court ruled in 1990 that artificial nutrition and hydration are not different than other life-sustaining treatments [88].

RECOMMENDATIONS

According to medical consensus of nutritional medicine in order to prevent and treat malnutrition, many practical applications are recommended:

- A qualified nutritional support team must be established in all hospital settings including ICU.
- Immediately upon admission, all patients' nutritional status must be systematically evaluated using established, simple malnutrition screening tool.
- Patients with under-/malnutrition or at risk, based on the screening tool results, should undergo further comprehensive nutritional assessment.
- Those with proved under-/malnutrition or at risk by malnutrition should then receive standardized nutritional intervention.
- Variable and flexible menu with special energy-rich items, including energy-rich snacks (shakes, soups, fingerfood) should be available in all hospitals.
- Based on current evidence, nutritional intervention such as nutritional drinks/oral supplements and, where applicable, supportive supplementary tube feeding must be considered fundamental parts of medical management and prevention in malnourished patients.
- Finally, nutritional medicine must become a crucial part of the training of medical students and specialized physicians.

CONCLUSION

Nutritional deficiencies are more common among the hospitalized elderly. This vulnerable population is subjected to many predisposing factors of malnutrition, which can be reversed if addressed early. Malnutrition is independently associated with poor outcomes and decreased quality of life. Management of malnutrition in the elderly population requires a multidisciplinary approach that treats pathology and uses both social and dietary forms of intervention. Many nutrition rules are simple, clear, and easy to apply in all hospital settings. Ethical considerations should be reviewed cautiously regarding artificial nutrition in special vulnerable population, for example, in patients with late dementia.

SUMMARY

- Hospitalization is a critical issue for older patients as they represent about 40% of the hospitalized adults.
- Malnutrition is common and frequently underrecognized in the hospital setting especially in the elderly.
- Many changes occur in the elderly to make them more susceptible to malnutrition compared to their younger counterparts.
- Malnutrition results primarily due to an increase in nutritional needs or a decrease in food intake or both.
- Predisposing factors for malnutrition in hospitals are anorexia of aging, gastrointestinal system changes, psychosocial, medical factors, and the use of multiple medications.
- Organizational and catering barriers, such as fixed mealtime, play important roles in precipitating or exaggerating malnutrition in hospitalized elderly patients.
- Malnutrition is associated with multiple negative consequences that complicate the course of acute illness and worsen its prognosis.
- The early recognition of malnutrition is essential for timely intervention that will prevent subsequent complications.
- Nutritional assessment should be performed in two stages: initial screening stage, and then comprehensive nutritional assessment for malnourished and those at risk for malnutrition.
- The systematic nutrition assessment comprises a history-taking, and physical, clinical, biochemical, anthropometric evaluation, dietary assessment, and functional outcomes evaluation.
- Nutritional care plans should be implemented to ensure optimal nutritional status in hospitalized elderly patients. They should focus on enhancing food environment, the quality and the quantity of the food, and assess the need for food supplement.
- Behavioral interventions are the first line of management of malnutrition and if they fail to show enough therapeutic benefit, artificial nutrition must be initiated. The pharmacological intervention has a limited role in management of malnutrition; its role is limited to treating specific nutritional deficiencies.
- Choosing the optimal type of artificial feeding depends on considering medical indications (underlying disease, patients' health status,

comorbidities, expected prognosis, and mental/psychological status), patients' preferences, and ethical issues dealing with nutritional intervention in specific situations, for example, palliative care, advanced dementia.
- Postdischarge nutritional plan should include dietary counseling that is tailored for individual patients with a dietitian along with follow up by a general practitioner in order to improve functional and nutritional outcomes.

References

[1] Constans T. Malnutrition in the elderly. Rev Prat 2003;53(3):275–9.

[2] Martone AM, Onder G, Vetrano DL, Ortolani E, Tosato M, Marzetti E, et al. Anorexia of aging: a modifiable risk factor for frailty. Nutrients 2013;5(10):4126–33.

[3] National Collaborating Centre for Acute Care. Nutrition support in adults. oral nutrition support, enteral tube feeding and parenteral nutrition. London: National Collaborating Centre for Acute Care; 2006.

[4] Bozzetti F, Mariani L, Lo Vullo S, SCRINIO Working Group, Amerio ML, Biffi R, et al. The nutritional risk in oncology: a study of 1,453 cancer outpatients. Support Care Cancer 2012;20(8):1919–28.

[5] Healthcare Cost and Utilization Project Facts and Figures. Statistics on hospital-based care in the United States. Agency for Healthcare Research and Quality (AHRQ). <http://www.hcup.us.ahrq.gov/reports/factsandfigures/2008/section1_TOC.jsp>; 2008.

[6] Walsh KA, Bruza JM. Hospitalization of the elderly. Ann Long Term Care 2007;15:18–23.

[7] Barker LA, Gout BS, Crowe TC. Hospital malnutrition: prevalence, identification and impact on patients and the healthcare system. Int J Environ Res Public Health 2011;8:514–27.

[8] Baptiste F, Egan M, Dubouloz-Wilner CJ. Geriatric rehabilitation patients' perceptions of unit dining locations. Can Geriatr J 2014;17(2):38–44.

[9] Gariballa SE. Nutrition and older people: special considerations relating to nutrition and aging. Clin Med 2004;4:411–14.

[10] Agarwala E, Millerb M, Yaxleyb A, Isenringc E. Malnutrition in the elderly: a narrative review. Maturitas 2013;76(4):296–302.

[11] Morley JE. Pathophysiology of the anorexia of aging. Curr Opin Clin Nutr Metab Care 2013;16:27–32.

[12] Schuetz P, Bally M, Stanga Z, Keller U. Loss of appetite in acutely ill medical inpatients: physiological response or therapeutic target? An area of current uncertainty. Swiss Med Wkly 2014;144:w13957.

[13] Jurdana M. Cancer cachexia-anorexia syndrome and skeletal muscle wasting. Radiol Oncol 2009;43(2):65–75.

[14] Schulman RC, Mechanick JI. Metabolic and nutrition support in the chronic critical illness syndrome. Respir Care 2012;57(6):958–77.

[15] Morley JE, Thomas DR, Wilson MMG. Cachexia: pathophysiology and clinical relevance. Am J Clin Nutr 2006;83:735–43.

[16] Champion A. Anorexia of aging. Annals of long-term care. Clin Care Aging 2011;19(10):18–24.

[17] National Center for Health Statistics. Health, United States, 2012: with special feature on emergency care. Hyattsville, MD: National Center for Health Statistics; 2013. <http://www.cdc.gov/nchs/data/hus/hus12.pdf#066>.

[18] Conte C, Cascino A, Bartali B, Donini LM, Rossi-Fanelli F, Laviano A. Anorexia of aging. Curr Nutr Food Sci 2009;5:9–12.

[19] Halter JB, Ouslander JG, Tinetti ME, Studenski S, High KP, Asthana S, editors. Hazzard's geriatric medicine and gerontology. 6th ed. New York: McGraw-Hill Companies, Inc.; 2009.

[20] Hays NP, Roberts SB. The anorexia of aging in humans. Physiol Behav 2006;88:257–66.

[21] van Asselt DZ, van Bokhorst-de van der Schueren MA, van der Cammen TJ, Disselhorst LG, Janse A, Lonterman-Monasch S, et al. Assessment and treatment of malnutrition in Dutch geriatric practice: consensus through a modified Delphi study. Age Aging 2012;41(3):399–404.

[22] Abd-El-Gawad WM, Abou-Hashem RM, El Maraghy MO, Amin GE. Delirium outcomes and its relation to nutritional status in elderly patients admitted to acute geriatric medical ward. J Aging Res Clin Practice 2013;2(2):226–30.

[23] Norman N, Pichard C, Lochs H, Pirlich M. Prognostic impact of disease-related malnutrition. Clin Nutr 2008;27:5–15.

[24] Cabrera MA, Mesas AE, Garcia AR, de Andrade SM. Malnutrition and depression among community-dwelling elderly people. J Am Med Dir Assoc 2007;8:582–4.

[25] Holmes S. Nutrition and eating difficulties in hospitalised older adults. Nurs Standard 2008;22(26):47–57.

[26] Evans C. Malnutrition in the elderly: a multifactorial failure to thrive. Perm J 2005;9(3):38–41.

[27] Kitson AL, Schultz TJ, Long L, Shanks A, Wiechula R, Chapman I, et al. The prevention and reduction of weight loss in an acute tertiary care setting: protocol for a pragmatic stepped wedge randomized cluster trial (the PRoWL project). BMC Health Serv Res 2013;13:299.

[28] Tsang MF. Is there adequate feeding assistance for the hospitalized elderly who are unable to feed themselves? Nutr Diet 2008;65:222–8.

[29] Inouye SK. Delirium in older persons. N Engl J Med 2006;354(11):1157–65.

[30] Thomas JM, Cooney Jr LM, Fried TR. Systematic review: health-related characteristics of elderly hospitalized adults and nursing home residents associated with short-term mortality. J Am Geriatr Soc 2013;61(6):902–11.

[31] Kagansky N, Berner Y, Koren-Morag N, Perelman L, Knobler H, Levy S. Poor nutritional habits are predictors of poor outcome in very old hospitalized patients. Am J Clin Nutr 2005;82(4):784–91.

[32] Stanga Z. Basics in clinical nutrition: nutrition in the elderly. e-SPEN Eur e-J Clin Nutr Metab 2009;4:e289–99.

[33] Bauer J, Capra S. Comparison of a malnutrition screening tool with subjective global assessment in hospitalised patients with cancer—sensitivity and specificity. Asia Pac J Clin Nutr 2003;12(3):257–60.

[34] Jensen GL, Hsiao PY, Wheeler D. Adult nutrition assessment tutorial. J Parenter Enteral Nutr 2012;36(3):267–74.

[35] Detsky AS, McLaughlin JR, Baker JP, Johnston N, Whittaker S, Mendelson RA, et al. What is subjective global assessment of nutritional status? J Parenter Enteral Nutr 1987;11(1):8–13.

[36] Guigoz Y, Vellas B, Garry PJ. Mini nutritional assessment: a practical assessment tool for grading the nutritional state of elderly patients. Facts Res Gerontol 1994;4(2):15–59.

[37] Vanis N, Mesihović R, Saray A, Gornjaković S, Mehmedović A. Comparsion of three nutritional screening tools: to detect nutritional risk in hospitalized patients. Folia Med 2013;48(2):147–53.

[38] Rubenstein LZ, Harker JO, Salva A, Guigoz Y, Vellas B. Screening for undernutrition in geriatric practice: developing the short-form mini nutritional assessment (MNA-SF). J Geront 2001;56A:M366–77.

REFERENCES

[39] Elia M. The "MUST" report. Nutritional screening of adults: a multidisciplinary responsibility. Redditch: British Association for Parenteral and Enteral Nutrition; 2003.

[40] Kondrup J, Rasmussen H, Hamberg O, Stanga Z, An ad hoc ESPEN Working Group. Nutritional risk screening (NRS 2002): a new method based on analysisof controlled clinical trials. Clin Nutr 2003;22(3):321−36.

[41] Bouillanne O, Morineau G, Dupont C, Coulombel I, Vincent JP, Nicolis I, et al. Geriatric nutritional risk index: a new index for evaluating at-risk elderly medical patients. Am J Clin Nutr 2005;82:777e83.

[42] Abd-El-Gawad WM, Abou-Hashem RM, El Maraghy MO, Amin GE. The validity of geriatric nutrition risk index: simple tool for prediction of nutritional-related complication of hospitalized elderly patients. Comparison with mini nutritional assessment. Clin Nutr 2014;33(6):1108−16.

[43] Poulia KA, Yannakoulia M, Karageorgou D, Gamaletsou M, Panagiotakos DB, Sipsas NV, et al. Evaluation of the efficacy of six nutritional screening tools to predict malnutrition in the elderly. Clin Nutr 2012;31:378e385.

[44] Ferguson M, Capra S, Bauer J, Banks M. Development of a valid and reliable malnutrition screening tool for adult acute hospital patients. Nutrition 1999;15:458−64.

[45] Neelemaat F, Meijers J, Kruizenga H, van Ballegooijen H, van Bokhorst-de van der Schueren M. Comparison of five malnutrition screening tools in one hospital inpatient sample. J Clin Nurs 2011;20(15−16):2144−52.

[46] Kruizenga HM, Seidell JC, de Vet HC, Wierdsma NJ, van Bokhorst-de van der Schueren MA. Development and validation of a hospital screening tool for malnutrition: the Short Nutritional Assessment Questionnaire (SNAQ). Clin Nutr 2005;24:75−82.

[47] Katz S, Downs TD, Cash HR, Grotz RC. Progress in development of the index of ADL. Gerontologist 1970;10:20e30.

[48] Folstein MF, Folstein SE, McHugh PR. Mini-mental state. A practical method for grading the cognitive state of patients for the clinician. J Psychiatr Res 1975;12(3):189−98.

[49] Wells JL, Dumbrell AC. Nutrition and aging: assessment and treatment of compromised nutritional status in frail elderly patients. Clin Interv Aging 2006;1(1):67−79.

[50] Harris D, Haboubi N. Malnutrition screening in the elderly population. J R Soc Med 2005;98:411−14.

[51] Morley JE. Update on nutritional assessment strategies from nutrition and health. In: Bales CW, Ritchie CS, editors. Handbook of clinical nutrition and aging. 2nd ed. New York: Humana Press, a part of Springer Science + Business Media, LLC; 2009.

[52] World Health Organization. Global database on body mass index, <http://apps.who.int/bmi/index.jsp?introPage=intro3.html/2010>; 2011 [accessed 13.11.14].

[53] Elia M, Stratton R. An analytical appraisal of nutrition screening tools supported by original data with particular reference to age. Nutrition 2012;28:477−94.

[54] Donini L, Savina C, Gennaro E, et al. A systematic review of the literature concerning the relationship between obesity and mortality in the elderly. J Nutr Health Aging 2012;16(1):89−98.

[55] Herrmann VM. Nutritional assessment. In: Torosian MH, editor. Nutrition for the hospitalized patient: basic science and principles of practice. New York: Marcel Dekker, Inc.; 1995.

[56] Wong S, Gandy J. An audit to evaluate the effect of staff training on the use of Malnutrition Universal Screening Tool. J Hum Nutr Diet 2008;21:405−6.

[57] Löser C. Malnutrition in hospital—the clinical and economic implications. Dtsch Arztebl Int 2010;107(51−52):911−17.

[58] Dickinson A, Welch C, Ager L. No longer hungry in hospital: improving the hospital mealtime experience for older people through action research. J Clin Nurs 2008;17:1492−502.

[59] Jefferies D, Johnson M, Ravens J. Nurturing and nourishing: the nurses' role in nutritional care. J Clin Nurs 2011;20:317−30.

[60] Walton K, Williams P, Bracks J, Zhang QS, Pond L, Smoothy R, et al. A volunteer feeding intakes of assistance program can improve dietary elderly patients—a pilot study. Appetite 2008;51:244−8.

[61] Wright L, Cotter D, Hickson M. The effectiveness of targeted feeding assistance to improve the nutritional intake of elderly dysphagic patients in hospital. J Hum Nutr Diet 2008;21:555−62.

[62] Duncan DG, Beck SJ, Hood K, Johansen A. Using dietetic assistants to improve the outcome of hip fracture: a randomized controlled trial of nutritional support in an acute trauma ward. Age Aging 2006;35:148−53.

[63] Cawood A, Elia M, Stratton R. Systematic review and meta-analysis of the effects of high protein oral nutritional supplements. Aging Res Rev 2012;11:278−96.

[64] Philipson T, Snider J, Lakdawalla D, Stryckman B, Goldman D. Impact of oral nutritional supplementation on hospital outcomes. Am J Manag Care 2013;19(2):121−8.

[65] Somanchi M, Tao X, Mullin GE. The facilitated early enteral and dietary management effectiveness trial in hospitalized patients with malnutrition. J Parenter Enteral Nutr 2011;35(2):209−16.

[66] Milne AC, Potter J, Vivanti A, Avenell A. Protein and energy supplementation in elderly people at risk from malnutrition. Cochrane Database Syst Rev 2009;(2):CD003288.

[67] Silver HJ. Oral strategies to supplement older adults' dietary intakes: comparing the evidence. Nutr Rev 2009;67(1):21−31.

[68] Löser C, Lübbers H, Mahlke R, et al. Der ungewollte Gewichtsverlust des alten Menschen. Dtsch Ärzteblatt 2007;49:3411−20.

[69] Culross B. Gerontol update: Nutrition: meeting the needs of the elderly. ARN Network; 2008.

[70] Kulick D, Deen D. Specialized nutrition support. Am Fam Physician 2011;83(2):173−83.

[71] Volkerta D, Bernerb YN, Berryc E, Cederholmd T, Coti Bertrande P, et al. ESPEN guidelines on enteral nutrition: geriatrics. Clin Nutr 2006;25:330−60.

[72] Griffiths RD, Bongers T. Nutrition support for patients in the intensive care unit. Postgrad Med J 2005;81:629−36.

[73] Halter JB, Ouslander JG, Tinetti ME, Studenski S, High KP, Asthana S, editors. Hazzard's geriatric medicine and gerontology. 6th ed. New York: McGraw-Hill Companies; 2009.

[74] Sobotka L, Schneider SM, Berner YN, Cederholm T, Krznaric Z, et al. ESPEN guidelines on parenteral nutrition: geriatrics. Clin Nutr 2009;28:461−6.

[75] Ziegler TR. Parenteral nutrition in the critically ill patient. N Engl J Med 2009;361(11):1088−97.

[76] Beck A, Kjær S, Hansen B, Storm R, Thal-Jantzen K, Bitz C. Follow-up homevisits with registered dietitians have a positive effect on the functional and nutritional status of geriatric medical patients after discharge: a randomized controlled trial. Clin Rehab 2013;27(6):483−93.

[77] Neelemaat F, Bosmans JE, Thijs A, Seidell JC, van Bokhorst-de van der Schueren MAE. Oral nutritional support in malnourished elderly decreases functional limitations with no extra costs. Clin Nutr 2012;31(2):183−90.

[78] Sheean PM, Peterson SJ, Chen Y, Liu D, Lateef O, Braunschweig CA. Utilizing multiple methods to classify malnutrition among elderly patients admitted to the medical and surgical intensive care units. Clin Nutr 2013;32:752−7.

[79] Vanhorebeek I, Gunst J, Derde S, Derese I, Boussemaere M, Guiza F, et al. Insufficient activation of autophagy allows cellular damage to accumulate in critically ill patients. J Clin Endocrinol Metab 2011;96(4):633–45.

[80] Casaer MP, Mesotten D, Hermans G, et al. Early versus late parenteral nutrition in critically ill adults. N Engl J Med 2011;365:506–17.

[81] The National Heart, Lung, and Blood Institute Acute Respiratory Distress Syndrome (ARDS) Clinical Trials Network, et al. Initial trophic vs full enteral feeding in patients with acute lung injury: the EDEN randomized trial. JAMA 2012;307: 795–803.

[82] Casaer MP, Van den Berghe G. Nutrition in the acute phase of critical illness. N Engl J Med 2014;370:1227–36.

[83] Schuetza P, Ballya M, Stangab Z, Kellerc U. Loss of appetite in acutely ill medical inpatients: physiological response or therapeutic target? An area of current uncertainty. Swiss Med Wkly 2014;144:13957.

[84] Peck G, Dani M, Torrance A, Mir AM. Artificial feeding in patients with advanced dementia. Br J Hosp Med 2014;75(1): C2–4.

[85] McClave SA, Martindale RG, Vanek VW, et al. Guidelines for the provision and assessment of nutrition support therapy in the adult critically ill patient: Society of Critical Care Medicine (SCCM) and American Society for Parenteral and Enteral Nutrition (A.S.P.E.N.). J Parenter Enteral Nutr 2009;33(3):277–316.

[86] Pivi G, Bertolucci P, Schultz R. Nutrition in severe dementia. Curr Gerontol Geriatr Res 2012;2012:983056.

[87] Posthauer ME, Collins N, Dorner B, Sloan C. Nutritional strategies for frail older adults. Adv Skin Wound Care 2013;26(3):128–40.

[88] Cruzan V. Director, Missouri Department of Health, 110 S.Ct. 2841; 1990.

CHAPTER

7

Drug—Nutrient Interactions in the Elderly

Sulaiman Sultan, Zoe Heis and Arshad Jahangir[1]

Sheikh Khalifa bin Hamad Al Thani Center for Integrative Research on Cardiovascular Aging (CIRCA), Aurora University of Wisconsin Medical Group, Aurora Cardiovascular Services, Aurora Research Institute, Aurora Health Care, Milwaukee, WI, USA

KEY FACTS

- Dietary supplement use is common and increasing in the elderly, but not readily disclosed to the healthcare provider.
- Elderly patients with comorbidities using multiple medications and dietary supplements are at increased risk of adverse interactions, which could be clinically serious with drugs that have a narrow therapeutic index.
- Patient-related and drug-nutrient-related factors contribute to drug-nutrient interactions (DNI) and need to be monitored to help reduce adverse DNI.
- Detailed assessment of nutritional status, medication use and potential for DNI should be performed routinely in the elderly with chronic health conditions who take multiple medications and are at greater risk than the general healthy population for adverse DNI.
- Ensuring proper timing of food with drug regimens, and regular monitoring of nutritional status, medication profiles, and laboratory data to evaluate the potential for interactions may help reduce serious adverse DNI in high risk individuals.
- Inclusion of elderly patients in clinical trials and collection of data regarding DNI may help fill knowledge gaps and increase understanding of mechanisms underlying DNI so strategies to prevent adverse DNI in the elderly can be further improved.

Dictionary of Terms

- **Amino acids**: Any one of many acids that occur naturally in living things and that include some that form proteins
- **Antiarrhythmic**: Tending to prevent or relieve cardiac arrhythmia, eg, an antiarrhythmic agent
- **Anticoagulants**: A substance, eg, a drug, that hinders coagulation and especially coagulation of the blood
- **Antihypertensive**: Used or effective against high blood pressure, eg, antihypertensive drugs
- **Atrophy**: Decrease in size or wasting away of a body part or tissue; also: arrested development or loss of a part or organ incidental to the normal development or life of an animal or plant
- **Bioavailability**: The degree and rate at which a substance, eg, a drug, is absorbed into a living system or is made available at the site of physiological activity
- **Ceftriaxone**: A broad-spectrum semisynthetic cephalosporin antibiotic $C_{18}H_{18}N_8O_7S_3$ administered parenterally in the form of its sodium salt
- **Chelate**: To combine with, eg, a metal, so as to form a chelate ring
- **Commensal**: A relation between two kinds of organisms in which one obtains food or other benefits from the other without damaging or benefiting it
- **Comorbidities**: A comorbid condition, eg, diabetes patients tend to present with a long list of comorbidities

[1] AJ is supported by grants from the National Heart, Lung and Blood Institute and the Aurora Health Care Intramural funding.

- Cytochrome: Any of several intracellular hemoprotein respiratory pigments that are enzymes functioning in electron transport as carriers of electrons
- Dextrose: Dextrorotatory glucose — called also grape sugar
- Disposition: A tendency to develop a disease, condition, etc.
- Divalent: Having a chemical valence of two eg, divalent calcium
- DNI: Drug nutrient interactions
- Efficacy: The power to produce a desired result or effect
- Efflux: Something given off in
- Electrolyte: Any of the ions, eg, sodium, potassium, calcium or bicarbonate, that in a biological fluid regulate or affect most metabolic processes, eg, the flow of nutrients into and waste products out of cells
- Endogenously: Growing from or on the inside
- Enteral: Of, relating to, or affecting the intestine
- Enterocytes: Intestinal absorptive cells, are simple columnar epithelial cells found in the small intestine. A glycocalyx surface coat contains digestive enzymes. Microvilli on the apical surface increase surface area for the digestion and transport of molecules from the intestinal lumen
- Emulsion: A mixture of liquids
- Fluoroquinolones: Any of a group of fluorinated derivatives of quinolone that are used as antibacterial drugs
- Fortified: To add material to for strengthening or enriching
- Homeostatic: A relatively stable state of equilibrium or a tendency toward such a state between the different but interdependent elements or groups of elements of an organism, population or group
- Immunosuppression: Suppression, eg, by drugs, of natural immune responses
- Insoluble: Incapable of being dissolved in a liquid and especially water; also: soluble only with difficulty or to a slight degree
- Isoenzyme: Any of two or more chemically distinct but functionally similar enzymes
- Malabsorption: Faulty absorption of nutrient materials from the alimentary canal
- MAOI: Monoamine oxidase inhibitor: any of various antidepressant drugs that increase the concentration of monoamines in the brain by inhibiting the action of monoamine oxidase
- Milieu: The physical or social setting in which people live or in which something happens or develops
- MVM: Multivitamin/multimineral
- Non-prescription: Capable of being bought without a doctor's prescription
- Oxidative: To become combined with oxygen
- Pathophysiologic: The physiology of abnormal states; specifically: the functional changes that accompany a particular syndrome or disease
- Permeability: The quality or state of being permeable
- Pharmacodynamics: A branch of pharmacology dealing with the reactions between drugs and living systems
- Pharmacokinetics: The characteristic interactions of a drug and the body in terms of its absorption, distribution, metabolism and excretion
- Phyto-chemistry: The chemistry of plants, plant processes and plant products
- Physiological: Characteristic of or appropriate to an organism's healthy or normal functioning
- Polymorphisms: The quality or state of existing in or assuming different forms
- Polypharmacy: The practice of administering many different medicines especially concurrently for the treatment of the same disease
- Polyphenols: A polyhydroxy phenol; especially: an antioxidant phytochemical, eg, chlorogenic acid, that tends to prevent or neutralize the damaging effects of free radicals
- Precipitation: The process of forming a precipitate from a solution
- Prescription: A written direction for a therapeutic or corrective agent; *specifically*: one for the preparation and use of a medicine
- Psychotropic: Acting on the mind, eg, psychotropic drugs
- Sequestration: The act of keeping a person or group apart from other people or the state of being kept apart from other people
- Therapeutic Indices: A measure of the relative desirability of a drug for the attaining of a particular medical end that is usually expressed as the ratio of the largest dose producing no toxic symptoms to the smallest dose routinely producing cures
- Thromboembolism: The blocking of a blood vessel by a particle that has broken away from a blood clot at its site of formation
- TPN: Total Parenteral Nutrition
- Trace elements: A chemical element present in minute quantities; especially: one used by organisms and held essential to their physiology
- Transfusions: The process of transfusing fluid into a vein or artery
- Tyramine: A phenolic amine $C_8H_{11}NO$ found in various foods and beverages (as cheese and red wine) that has a sympathomimetic action and is derived from tyrosine
- Vasopressors: Causing a rise in blood pressure by exerting a vasoconstrictor effect

INTRODUCTION

Dietary supplement use has steadily increased in the United States with nearly one-half of all Americans reported to use some form of supplements to improve or maintain overall health, as well as for specific reasons, such as to promote bone or heart health among the elderly [1–4]. According to the National Health and Nutrition Examination Survey (NHANES) dietary supplement usage has increased from 23% in NHANES I (1971–1975) to 35% in NHANES II (1976–1980) and 40% in NHANES III (1988–1994) with another 10% increase after the NANHES III report [5]. The majority used only one supplement; however, a substantial 10% reported taking five or more supplements (Fig. 7.1). Americans spend more than $23 billion out-of-pocket each year on dietary supplements, with multivitamin/multimineral (MVM) products used by nearly one-third [4,6]. The use is particularly prevalent in the elderly, with approximately 70% of adults older than 70 years of age using some form of dietary supplements [6]. This is the same population in which the use of medicinal products is also high, with older adults accounting for approximately one-third of all prescription and nonprescription medication use [7], posing a high risk of drug–nutrient interactions (DNI). Although drug–drug interactions are well recognized, the importance of DNI is slowly getting recognition as a potential factor that may affect outcomes in the elderly with multiple comorbidities [8], polypharmacy [9], reduced functional reserves, altered nutritional status, and aging-associated physiological changes that alter the disposition (absorption, distribution, metabolism, or elimination) of the drug or nutrient and their effect on the body (pharmacodynamics). Up to 45% of patients using dietary supplements with prescription drugs are at risk of interaction, with as many as 29% considered clinically serious [10]. This is particularly the case with drugs with narrow therapeutic indices, such as chemotherapeutics, anticoagulants, and antiarrhythmic agents. Several surveys suggest the percentage of patients who do not disclose their supplement or other complementary and alternative medicine use to healthcare professionals to be as high as more than 60% [11]. Thus, the risk for potential interactions between supplements and medications to go undetected remains substantial. Increasing cognizance of such interaction, including DNI may help reduce these adverse effects. In this chapter, we will summarize different types of DNI, highlight aging-associated alterations in pharmacokinetics and pharmacodynamics that increase predisposition to DNI, and discuss issues in regulating dietary supplements that can potentially improve their safety.

TYPES OF DRUG–NUTRITION INTERACTIONS

DNI may result from a physical, chemical, physiologic, or pathophysiologic interaction between a drug and a nutrient, food in general or nutrition status, and can be classified into four types depending on the site of interaction, as summarized in Table 7.1 [12,13].

Type I or pharmaceutical interactions involve physicochemical reactions that usually occur outside the body, such as in the nutrition or drug delivery device like enteral feeding tube or intravenous (IV) tubing. Total parenteral nutrition (TPN) is a complex formulation of amino acids, fat emulsion, dextrose, vitamins, electrolytes, and trace elements and when administered with IV drugs, can result in physical interactions, such as precipitation of calcium phosphate when given with IV drugs, such as ceftriaxone that in turn can cause serious adverse vascular events that can become life-threatening [14,15]. Enteral nutrition can also chelate drugs such as fluoroquinolones, reducing their bioavailability and effectiveness of TPN, therefore should be administered separate from IV drugs [15].

Type II DNI occurs with oral or enteral administration affecting drug or nutrient bioavailability due to interference with enterocytes metabolizing enzymes (*type IIA interaction*), transporter P glycoprotein protein (*type IIB interaction*) that normally acts as an efflux pump returning the drug to the intestinal lumen, or binding with divalent or trivalent cations (calcium, magnesium, iron, zinc or aluminum) (*type IIC interaction*). A common example of *type IIA* food–drug interaction is between grapefruit juice and drugs metabolized through the intestinal cytochrome-P-450 (Cyp) isoenzyme 3A4 system involved in oxidative metabolism of more than 50% of all prescription medications [16,17]. Active ingredients within grapefruit juice (naringenin, furanokumarin, and bergamottin)

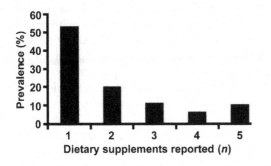

FIGURE 7.1 The number of supplements taken by US adult supplement users, NHANES, 2003–2006, n = 9132. *Source: Adapted from Bailey et al. Dietary Supplement Use in the United States, 2003–2006. J Nutr. 2011;141(2):261–266, with permission from the American Society of Nutrition.*

TABLE 7.1 Classification of Interactions

	Characteristics	Possible effect	Drug	Possible mechanism
Type I	Effects usually occur when substances are in direct physical contact	Physical, biochemical reactions	Ceftriaxone	Binding of drug with TPN or IV calcium containing products
Type II	Limited to substances administered orally or enterally	Increased or decreased bioavailability		
A		Modified enzyme activity	Warfarin	Cranberry juice interferes with the drug
			CYP3A4 substrates (see Table 7.2B)	Grapefruit juice increases bioavailability for several days when taken with *atorvastatin, cyclosporine, diazepam, felodipine, lovastatin, midazolam, nicardipine, nisoldipine, simvastatin, and triazolam*
B		Modified transport mechanism	P-glycoprotein modulators (see Table 7.3) Cyclosporine, digoxin	Water-soluble formulations of vitamin E have increased blood levels of drug
C		Complexing or binding of substances	Isoniazid	Vitamin B6 complexes with the drug in Type IIC interaction: useful in drug overdoses and in minimizing drug side effects
			Norfloxacin, ofloxacin, tetracycline	Large amounts of calcium, iron, magnesium, or zinc may bind *levothyroxine* and prevent complete absorption
Type III	Occur after the substances have reached the systemic circulation	Changed cellular or tissue distribution		Tyramine-containing foods may provoke a hypertensive crisis with *furazolidone, isocarboxazid, isoniazid, linezolid, phenelzine, procarbazine, selegiline* and *tranylcypromine*
Type IV	Affect disposition of substances by liver and kidney	Promoted or impaired clearance or elimination of substances	Lithium	Increased sodium (salt) increases renal excretion of the drug, decreasing its effect
				Potassium-containing salt substitutes increase risk for hyperkalemia when taken with *captopril, enalapril, lisinopril*

inhibit the CYP3A4 enzyme in small-intestine enterocytes, which increases blood levels of several of the common drugs used in the elderly that are CYP3A4 substrates (Tables 7.2A and 7.2B), including statins, midazolam, and terazosin besides others listed in Tables 7.2A and 7.2B. Since the half-life of CYP3A4 is more than 8 hours, its irreversible inhibition by grapefruit juice can increase bioavailability of the affected drugs for 72 hours that increases the risk for adverse effects, such as muscle or liver injury with statins or excessive reduction in blood pressure with calcium channel blockers with repeated juice intake. In postmenopausal women taking estrogen, concern regarding increased risk of breast cancer by inhibiting estrogen metabolism by CYP3A4 has also been raised [18,19]. Individual variability in the expression of intestinal CYP3A4 exists and therefore the response to grapefruit juice could be unpredictable in the elderly, when taken with drugs metabolized by CYP3A4 (Tables 7.2A and 7.2B). This is not an uncommon situation in the elderly with heart disease who may be taking a number of drugs metabolized by this enzyme system, such as calcium channel blocker (felodipine), statin (simvastatin), antihypertensive (terazosin), antiarrhythmic agent (amiodarone), blood thinner (apixaban), or anxiolytic (triazolam). The potential for interaction between grapefruit juice and medications should be discussed with the elderly taking CYP3A metabolized drugs (Table 7.2A) to avoid any potential adverse effect [18]. P-glycoprotein transporter is another protein whose activity is modulated by several drugs (Table 7.3) and nutrients or food products, such as grapefruit juice and vitamin E that can increase the bioavailability of such drugs due to *type IIB DNI*.

TABLE 7.2A Drugs Affecting Cytochrome P450 Enzymes Involved in Anticoagulant Metabolism (Not All-Inclusive)[a]

CYP enzymes (anticoagulant metabolized)	Strong inhibitors ≥ fivefold ↑ in AUC or >80% ↓ in CL	Moderate inhibitors ≥ 2 but <fivefold ↑ in AUC or 50–80% ↓ in CL	Weak inhibitors ≥ 1.25 but <twofold ↑ in AUC or 20–50% ↓ in CL
CYP2C9 (Warfarin)		Amiodarone, fluconazole, miconazole, oxandrolone	Cotrimoxazole, etravirine, fluvastatin, fluvoxamine, metronidazole, sulfinpyrazone, tigecycline, voriconazole, zafirlukast
CYP3A4 (Apixaban, rivoraxaban)	Conivaptan, grapefruit juice (1), azole antifungals (itraconazole, ketoconazole, posaconazole, voriconazole) macrolides (clarithromycin telithromycin), nefazodone, protease inhibitors (boceprevir, indinavir, nelfinavir, ritonavir, saquinavir, telaprevir)	Amprenavir, aprepitant, atazanavir, ciprofloxacin, calcium channel blockers (diltiazem, verapamil), dronedarone, erythromycin, fluconazole, fosamprenavir, grapefruit juice (1), imatinib, valerian	Alprazolam, amiodarone, amlodipine, atorvastatin, bicalutamide, buprenorphin, cilostazol, cimetidine, cyclosporine, fluoxetine, fluvoxamine, ginkgo (2), goldenseal (2), isoniazid, nilotinib, oral contraceptives, ranitidine, ranolazine, tipranavir/ritonavir, zileuton
CYP enzymes	Strong inducers ≥80% ↓ in AUC	Moderate inducers 50–80% ↓ in AUC	Weak inducers 20–50% ↓ in AUC
CYP2C9 (Warfarin)		Carbamazepine, rifampin	Aprepitant, bosentan, phenobarbital, St. John's wort (2,3)
CYP3A4 (Apixaban, rivoraxaban)	Carbamazepine, phenytoin, rifampin, St. John's wort (2,3)	Bosentan, efavirenz, etravirine, modafinil, nafcillin	Amprenavir, aprepitant, armodafinil, echinacea (2), pioglitazone, prednisone, rufinamide
CYP3A4 inhibitors with and without inhibitory effect on P-gp		P-gp inhibitor	Non-P-gp inhibitor
Strong CYP3A inhibitor		Itraconazole, lopinavir/ritonavir, clarithromycin, ritonavir, ketoconazole, conivaptan	Voriconazole, nefazodone
Moderate CYP3A inhibitor		Verapamil, diltiazem, dronedarone, erythromycin	None identified
Weak CYP3A inhibitor		Amiodarone, quinidine, ranolazine, felodipine, azithromycin	Cimetidine

[a] Adapted from: http://www.fda.gov/Drugs/DevelopmentApprovalProcess/DevelopmentResources/DrugInteractionsLabeling/ucm093664.htm#potency [Last accessed 6/20/2015].
(1) The effect of grapefruit juice varies widely among brands and is concentration-, dose-, and preparation-dependent. Studies have shown that it can be classified as a "strong CYP3A inhibitor" when a certain preparation was used (eg, high dose, double strength) or as a "moderate CYP3A inhibitor" when another preparation was used (eg, low dose, single strength).
(2) Herbal product.
(3) The effect of St. John's wort varies widely and is preparation-dependent.

Elderly patients taking P-glycoprotein or CYP3A4 metabolized medications (Tables 7.2A and 7.3), therefore should avoid consumption of grapefruit juice, water-soluble vitamin E, or other food products that can potentially increase bioavailability of these drugs.

Drugs that have the potential for *type IIC DNI* forming insoluble complexes with divalent cations include commonly used drugs in the elderly, such as antimicrobials (tetracycline, ciprofloxacin, norfloxacin, and ofloxacin), anticoagulants (warfarin), and antiepileptic agents (phenytoin) that reduce their absorption. These drugs, therefore, should not be administered within 1–3 hours of taking iron supplements, iron-containing foods, or with milk, yogurt, dairy products, or calcium-fortified juices [20]. Antacids containing magnesium, aluminum, or calcium should also be avoided and enteral feeding withheld for 1 hour before and after the dose of these medications.

Type III DNI occurs after the drug or nutrient has reached the systemic circulation and then undergoes metabolism. Interactions with monoamine oxidase inhibitors (MAOI) or warfarin are examples of such interactions that are of clinical significance. MAO are involved in the breakdown of dopamine and tyramine, neurotransmitters necessary for normal functioning of the nervous system and synthesis of epinephrine and norepinephrine, hormones involved in maintaining blood pressure, heart rate, and day-to-day stress response. Drugs, such as antidepressants, inhibit MAO and elevate levels of epinephrine, norepinephrine, serotonin, and dopamine in the nervous system, which help counteract depression. A number of food products

TABLE 7.2B Substrates for CYP3A4

Analgesics

Acetaminophen, celecoxib, codeine, colchicine, fentanyl

Antiarrhythmics

Diltiazem, disopyramide, quinidine, verapamil

Antimicrobial agents

Dapsone, sulfamethoxazole

Antiplatelets

Cilostazil

Anxiolytics and sedatives

Clonazepam, midazolam, triazolam

Calcium channel blockers

Nifedipine

Chemotherapeutic agents

Cyclophosphamide, docetaxel, etoposide, paclitaxel, vinblastine, vincristine

Erectile dysfunction

Sildenafil, tildenafil

Ergot alkaloids

Ergotamine

(HMG-CoA) reductase inhibitors

Statins

Immunosuppressive agents

Cyclosporine A, sirolimus, tacrolimus

Steroids

Dexamethasone

TABLE 7.3 Drugs Modulating P-glycoprotein (Not All-Inclusive)

Inhibitors

Amiloride, amiodarone, captopril, carvedilol, diltiazem, dipyridamole, doxazosin, dronedarone, felodipine, nifedipine, propranolol, propafenone, quinidine, ranolazine, spironalactone, verapamil, statins

Azithromycin, cefoperazone, ceftriaxone, clarithromycin, conivaptan, cortisol, cyclosporine, erythromycin, haloperidol, itraconazole, ketoconazole, paroxetine, phenothiazines, ritonavir and lopinavir

Tacrolimus, tamoxifen, tricyclic antidepressants

Inducers

Carbamazepine, dexamethasone, phenobarbital, phenytoin, rifampin, St. John's wort, tipranavir/ritonavir

TABLE 7.4 Pressor Agents in Foods and Beverages (Tyramine, Dopamine, Histamine, Phenylethylamine)

Aged cheeses

Aged meats (eg, dry sausage)

Miso

Fava beans or snow pea pods (contain dopamine)

Concentrated yeast extracts (Marmite)

Soy sauce

Sauerkraut, kimchi

Fermented soya beans and paste, teriyaki sauce

Avoid with MAOI medications: phenelzine (Nardil), tranylcypromine (Parnate), isocarboxazid (Marplan), selegiline (Eldepryl) in doses >10 mg/day, and the antibiotic linezolid (Zyvox).

(Table 7.4), such as aged and fermented food contain high levels of tyramine, a metabolic intermediate product in the conversion of the amino acid tyrosine to epinephrine and normally rapidly deaminated by MAO and thus when consumed with MAOI, result in elevated levels of tyramine and other vasopressors, such as epinephrine that can increase blood pressure or heart rate in the elderly to dangerous levels.

Another common type III DNI in the elderly is between foods rich in vitamin K (spinach, leafy green vegetables, broccoli, Brussels sprouts) and vitamin K-antagonist warfarin, whose anticoagulant activity is reduced [21,22], thus increasing the risk for a thromboembolic event in patients with atrial fibrillation or a prosthetic heart valve. Consumption of food high in vitamin K should therefore be avoided or the warfarin dose adjusted to maintain anticoagulation level, while on a consistent intake of Vitamin K rich foods, thus avoiding fluctuation in its efficacy. In addition, vitamin K is produced endogenously by bacterial flora within the intestinal tract and use of antibiotics that can change the flora may result in altered vitamin K levels and anticoagulant efficacy of warfarin [23].

Type IV DNI results from altered elimination of the drug through the kidneys. Dietary salt, rich in sodium, can increase the excretion of lithium, a drug given for manic episodes or manic depressive illness, thus worsening signs and symptoms of mania. On the other hand, reduced sodium and fluid intake may lead to lithium retention and toxicity (drowsiness, muscle weakness, twitching, slurring of speech, and reduced coordination). Similar effects may result from the use of loop diuretics, increasing the risk for lithium toxicity [24]. Therefore, it is important to maintain a steady level of sodium intake in those taking lithium and it may be prudent to monitor concentration or specific gravity of the urine in high-risk patients.

FACTORS AFFECTING DNI IN THE ELDERLY

Factors contributing to DNI include patient-related factors (age, sex, body size/composition, genetics, lifestyle, and comorbidities) and drug–nutrient related factors (amount, route, and time of administration). People with chronic disease who use multiple drugs, particularly those with a narrow therapeutic range, are at particular risk of interactions. Poor nutritional status, impaired organ function, and genetic variants in drug metabolism are common factors that increase predisposition to adverse DNI. There is also a concern about potentially harmful effects of contaminants and other nonnutrient components of some supplements [25–27]. Excess nutrient intake from combinations of dietary supplements and fortified foods, such as breakfast cereals, juices, and milk, is another worry [28].

A variable rate of functional decline of different organ systems, homeostatic mechanisms [29,30], and specific receptor and target organ responses [31,32] with aging may affect drug or nutrient disposition (absorption, distribution, metabolism, and elimination) or action increasing sensitivity toward adverse drug–drug interactions or DNI. The likelihood of adverse DNI in the elderly is increased with the number of comorbidities, medications, and supplements used, overall nutritional status, and presence of critical illnesses [32–34]. Genetic factors also play a role, for example, polymorphism of the methylenetetrahydrofolate reductase gene may affect the amount of pyridoxine, cobalamin, folic acid, and riboflavin required by the body, which is important in determining the threshold intake that prevents certain DNI [35,36].

DNI Affecting Drug Absorption in the Elderly

Practically, there is no significant impact of aging on absorption of most of the drugs, except for those that require special transporter proteins, such as P-glycoprotein or organic anion transporter which could be downregulated with aging or aging-associated diseases increasing sensitivity to DNI [37]. The solubility and intestinal permeability are also important factors affecting the absorption of a drug in the presence of food and thus its clinical impact. Drugs with low solubility but high permeability when administered with food have a greater effect on absorption than those with high solubility and the magnitude of change in the bioavailability with food determines the clinical impact [15]. The bioavailability of certain drugs, such as antibiotics (ampicillin, cloxacillin, ciprofloxacin and penicillin V, azithromycin), retroviral agents (indinavir), antiosteoporotic drugs (alendronate, ibandronate or risedronate), and chemotherapeutic agents (melphalan, mercaptopurine), is affected greatly when taken on a full rather than an empty stomach and are therefore better taken without food [38]. The absorption of antiosteoporotic agents is also significantly reduced when taken with coffee or orange juice. Food can also delay absorption of some of the drugs (such as verapamil) and care should be taken to administer these drugs under similar conditions of food intake to achieve a steady state condition, as fluctuation in drug absorption can result in adverse effect when taken on an empty stomach. Dietary fibers can also decrease absorption of certain drugs taken by the elderly, such as digoxin (for heart failure and/or atrial fibrillation), levothyroxine (for hypothyroidism), tricyclic antidepressants (amitriptyline), antibiotics, or metformin (for type 2 diabetes) [39]. Fatty meals can also affect bioavailability of some drugs (eg, griseofulvin for fungal infection), by changing solubility and interaction time with intestinal mucosa by slowing gastrointestinal (GI) motility.

Phytochemicals in fruits, vegetables, and herbal teas are modulators or substrates of P-glycoproteins and could affect bioavailability of drugs by preventing their efflux through these proteins [40]. Some nutrients, such as vitamin B12, require normal acidity and secretory function of the stomach that produces intrinsic factor for their absorption and any alteration in gastric pH by aging, drugs (antacids, proton-pump inhibitors (PPI), or histamine 2 (H_2) receptor antagonists), or food can alter absorption of drugs requiring acidic medium, such as ketoconazole [41]. Enteric-coated drugs should not be taken with milk, alcohol, or hot beverages that can cause early erosion of the enteric coating making them less acid resistant and likely to be degraded in the acidic medium of the stomach, thus reducing their efficacy [42]. Diets containing large amounts of divalent (calcium, magnesium, iron, or zinc) or trivalent (aluminum) ions or chelative substances may also reduce absorption of drugs such as levothyroxine, antibiotics (ofloxacin, norfloxacin, or tetracycline), or Parkinson's disease drug (entacapone) (Table 7.5) [43].

Aging also causes atrophy of epidermis and dermis with reduction of the barrier function of the skin and thus transdermal absorption may be altered due to changes in blood perfusion [44] affecting bioavailability of drugs, such as transdermal fentanyl patch, in the elderly compared to young subjects [45].

DNI Affecting Drug Distribution in the Elderly

The biodistribution of drugs can also be altered with aging due to changes in body composition resulting in a decrease in total body water and lean body mass, low albumin content, and an increase in total

TABLE 7.5 Drug–Nutrient Interaction

Drug	Food	Effect
ABSORPTION		
Alendronate	Any food intake but water	↓ Absorption
Ciprofloxacin, norfloxacin, ofloxacin	Any food intake	↓ Absorption
Digoxin	Foods high in fiber	↓ Absorption
Metformin	Foods high in fiber	↓ Absorption
Melphalan	Amino acids	↓ Absorption
Indinavir	Any food	↓ Absorption by precipitating the drug
GASTRIC DEGRADATION		
Ampicillin, azithromycin, cloxacillin, isoniazid, penicillin V	Food, acidic juices, carbonated beverages	↑ Gastric degradation of the drug
Erythromycin	Milk, alcohol, hot beverages	↓ Gastric degradation of the drug
SOLUBILITY		
Atovaquone	Fatty meals	↑ Solubility
METABOLISM (BIOAVAILABILITY)[a]		
Warfarin	Large amounts of vitamin K-rich foods and beverages	↓ Metabolism and ↑ Risk of bleeding (Type IIA interaction)
		↓ Anticoagulation effect (Type III interactions)
		Concomitant enteral feeding interferes with warfarin absorption, possibly through protein binding (type II C interaction)
		↓ Anticoagulation effect
Griseofulvin	Fatty meal	Taken with drug stimulates bile secretion and increases bioavailability of the drug
Mercaptopurine	Food	Oxidized by food into inactive metabolites
Phenytoin	Tube feeding	↓ Serum levels, could lead to breakthrough seizures

[a]*Bioavailability (Metabolism)—See Table 7.1 Type IIA Interactions that occur after oral or parenteral administration.*
Cranberry juice interferes with *warfarin*.
Grapefruit juice increases bioavailability for several days when taken with *atorvastatin, cyclosporine, diazepam, felodipine, lovastatin, midazolam, nicardipine, nisoldipine, simvastatin,* and *triazolam*.

body fat. Thus, the volume of distribution of water soluble drugs, like aspirin [44] or lithium [46], decreases in the elderly. Since the plasma concentration of a drug is inversely proportional to the volume of distribution, a decrease in volume of distribution results in greater plasma concentration of the drugs with a potential for adverse effects and toxicity and therefore a dose reduction may be needed in the elderly [46]. With an increase in total body fat content with aging or obesity, there is increased sequestration of lipid-soluble drugs (amiodarone, digoxin, and benzodiazepines) [44] causing an increase in plasma half-life and longer duration to reach a steady-state body concentration or elimination after discontinuation, thus increasing the likelihood of adverse effects [47]. Poor nutritional status with low albumin content in the elderly can further alter distribution of drugs resulting in a larger free fraction or unbound form increasing its pharmacologic effects, with a greater potential for adverse effect with other medications or nutrients, especially with drugs with a narrow therapeutic index (eg, valproate, phenytoin, or warfarin) [48] thus requiring dose reduction and a closer monitoring of patients.

DNI Affecting Drug Metabolism in the Elderly

Most drug metabolism takes place in the liver or intestinal membranes harboring cytochrome P-450 enzyme superfamily or other enzymes that transforms the drug usually from a lipid-soluble to a water-soluble form that can then be eliminated. Activity of this enzyme system is affected by conditions such as starvation, change in diet, disease states, and genetics [49]. For example, the exclusive use of TPN appears to decrease CYP1A and CYP2C activities while increasing CYP2E1 activity [50,51]. Vegetarian diets are associated with decreased CYP1A enzyme activity in Asians, but not in non-Hispanic whites [52]. Induction of CYP enzymes increases the risk of therapeutic failure due to augmentation of metabolic clearance of the drug. For example, charcoal-rich food induces CYP1A enzyme activity in humans. On the contrary, inhibition of CYP enzyme activities increases the risk of drug-related toxicity as bioavailability increases. Aging and aging-associated diseases decrease liver functional capacity, blood flow, and/or synthesis of cytochrome P-450 enzyme system that increases sensitivity to dietary or herbal interactions due to impact on the extent or rate of drug metabolism. Medications that undergo large first pass metabolism are specially affected (eg, fentanyl, propranolol) [53].

Herbal medicines or plant products are commonly used by the elderly along with prescription medicines and could result in serious adverse effects [18]. DNI with use of grapefruit juice and CYP3A4-metabolized drugs have been discussed above in the section on type IIA DNI. In vitro studies and animal studies also indicate that other fruit juices, such as pomegranate or cranberry juice, may also inhibit intestinal CYP3A4 and CYP2C9 [54] resulting in elevated levels; however, human data is limited and the bioavailability of drugs metabolized by these enzymes was not significantly altered in healthy volunteers [55,56]. A clinically important interaction between herbal supplement and medications metabolized through CYP3A4 system (Table 7.2A) is with St. John's Wort (SJW), one of the top selling herbs in the United States, purportedly used for varied conditions and known to induce hepatic cytochrome P450 system (particularly CYP3A4 and the transporter protein P-glycoprotein) [4,18,19,57,58]. Coadministration of SJW with drugs metabolized by CYP3A4 (Table 7.2A) or transported by P-glycoprotein (Table 7.3), reduces their bioavailability, resulting in loss of efficacy that can lead to life-threatening situations [59]. In renal and cardiac transplant recipients, the concentration of cyclosporine, an immunosuppressant agent was significantly reduced by SJW with subsequent transplant organ rejection [19]. Reduced levels of ethinyl estradiol, indinavir, calcium channel blockers, and antiarrhythmic agents have been reported in patients with SJW resulting in unintended pregnancy, recurrence of HIV infection, hypertension, arrhythmia, or other undesirable effects [18,19,59,60]. Concomitant use of SJW with warfarin could lead to subtherapeutic anticoagulation and thromboembolism [61], and the elderly with a history of blood clots, stroke, atrial fibrillation, or prosthetic heart valves should avoid the use of SJW. Hypertensive crisis with foods rich in tyramine also may occur [62]. Serotonin syndrome, a potentially life-threatening adverse drug reaction caused by excess serotonergic activity in the central and peripheral nervous system, has also been reported with concomitant use of antidepressants [63].

DNI Affecting Drug Excretion in the Elderly

Renal excretion (through glomerular filtration or tubular secretion) is a major route of drug removal with lesser extent elimination through feces or other body fluids. Aging leads to a decrease in kidney mass, renal blood flow, and the number of functioning glomeruli with creatinine clearance declining by approximately 1 mL/min/year after age 40 years [64]. Decline in renal function is correlated with the inability of kidneys to effectively eliminate drugs resulting in prolongation of their elimination half-life, increase in steady-state concentrations, and potential for adverse effects. Food and nutrients can alter the resorption of drugs from the renal tubule as described above in type IV DNI between sodium/salt intake and lithium [24]. This is mainly due to the close association of lithium resorption in the tubules with sodium resorption and under conditions of low sodium/salt intake or dehydration, when kidney resorption of sodium is increased, more lithium is absorbed with higher blood levels and increased potential for toxicity. Tubular resorption of drugs that are weak acids or bases occurs in nonionic form of the compound and any change in urinary pH by food (urinary alkalinization by milk, fruits and vegetables) may alter the nonionic versus ionic state of these drugs, such as quinidine, an antiarrhythmic agent (weak base), thus altering its excretion and blood levels.

DNI Affecting Drug Action in the Elderly

Aging-induced changes in cellular function also result in alterations of drug—receptor interactions, receptor—membrane interactions (signal transduction), and postreceptor events in target organs thus affecting the drug's overall effect. There may also be blunting of homeostatic counter regulatory mechanisms making the elderly more vulnerable to the negative effects of individual drugs and potential for adverse DNI. There is paucity of data on age-related pharmacodynamics changes or DNI in the elderly particularly in those aged 80 years and older. The central nervous system is a particularly sensitive drug target in the elderly and adverse interactions with drugs (antidepressants, anxiolytics, antihistamine) after alcohol intake is not uncommon [65]. Elderly are also more sensitive to the sedative effects of benzodiazepines which may cause delirium, ataxia, and frequent falls. The relative risk of hip fractures is doubled in elderly on psychotropic drugs, including benzodiazepines and antidepressants, and use of alcohol or supplements with sedative properties (such as valerian root or leaves and roots of passion flower [*Passiflora incarnata* and *Passiflora edulis*]) should be avoided [66,67]. Caffeine, in beverages or energy drinks, has the potential to increase the adverse effects of drugs, such as methylphenidate, amphetamine, or theophylline, that have a stimulant effect on the nervous system [68]. A concern with regard to a negative impact of caffeine intake on blood glucose control in diabetics has also been raised [69]. The anticoagulant effect of vitamin K-dependent anticoagulant warfarin is reduced by foods rich in vitamin K as discussed above. On the other hand, the bleeding risk is increased when herbal or dietary supplements, such as garlic, ginseng, papaya, mango, or vitamin E, that prolong bleeding or clotting time are used with warfarin in the elderly.

EFFECT OF DRUGS ON NUTRITIONAL STATUS IN ELDERLY

A DNI is considered clinically significant if it alters therapeutic drug response and/or compromises nutrition status [15]. The mechanisms for impairment in nutritional status in the elderly are diverse but commonly result from drug side effects that reduce nutrient bioavailability due to reduction in food intake, digestion, or absorption [15]. Food intake can be influenced by a drug's impact on appetite, taste or smell, or induction of nausea or vomiting [70] (such as with chemotherapeutic agents, aspirin, or nonsteroidal agents) irritating gastrointestinal mucosa or anorexic agents (serotoninergic agents) used specifically for weight loss in obesity that may lead to nutritional deficiency. Several drugs are known to cause metallic or altered taste (dysgeusia), which is especially common among elderly. Chemotherapeutics, including methotrexate and doxorubicin, are a common cause of taste changes but other medicines commonly used in the elderly can alter taste, such as antihypertensives (captopril, enalapril, or diltiazem), antiosteoporotic agents (etidronate), antibiotics (ampicillin, clarithromycin, levofloxacin, tetracyclines), antifungals (amphotericin B, griseofulvin), antihistamines (chlorpheniramine), oral hypoglycemic agents (glipizide), or psychoactive drugs (lithium), which can then affect the nutritional status in the elderly.

Overuse of antibiotics can alter commensal bacteria within the gastrointestinal tract that lead to overgrowth of fungi (*Candida*) or bacteria, resulting in malabsorption and diarrhea [71]. Drugs can also reduce the absorption of specific nutrients by alteration in intestinal absorptive surfaces, metabolizing enzymes, transporter proteins, or gastrointestinal transit time (laxatives, cathartic agents) [72]. Malabsorption syndromes induced by some of these drugs can lead to loss of fat-soluble vitamins (A and E) and minerals (calcium and potassium). On the other hand, drugs such as corticosteroids, anabolic steroids, megesterol, anticonvulsant agents or psychotropic drugs may increase appetite and lead to weight gain. Electrolyte deficiency (potassium, magnesium, sodium, chloride, and calcium) can occur with the prolonged use of loop diuretics (furosemide or bumetanide) in the elderly and therefore require close monitoring and supplementation if needed. Use of thiazide diuretics, on the other hand, can increase blood calcium levels by increasing tubular resorption of calcium, which can be further exaggerated in the elderly on vitamin D supplementation, particularly if there is renal compromise or hyperparathyroidism [73]. Magnesium depletion is associated with the use of cisplatin, a chemotherapeutic agent used in several malignancies, and requires magnesium supplementation [71]. Chelating agents, such as D-pencillamine or ethylenediaminetetraacetic acid increases urinary excretion of zinc that may result in its depletion.

DRUG INTERACTIONS WITH VITAMINS SUPPLEMENTS IN THE ELDERLY

Micronutrients, such as vitamins and minerals, are essential for normal functioning of the body, maintenance of health, and its deficiency or excess can lead to structural or functional changes and disease [12] (Tables 7.6 and 7.7). Most of the vitamins needed for normal functioning (the recommended daily allowance) are covered by regular food intake and a balanced diet [74]. Individuals deficient in vitamins or at risk, such as those with alcoholism, malabsorption, vegan diet, a history of gastric bypass surgery, osteoporosis, as well as those being treated with hemodialysis or parenteral nutrition, could be candidates for MVM supplementation [75]. However, the use of multi-ingredient formulations containing high-dose vitamins and mineral supplements is common in the population with the misconception that vitamin supplementation can prevent cardiovascular or other aging-associated diseases [4]. The use of vitamin supplements is likely safe in healthy individuals but costly and has biological effects with potential for harm with chronic exposure or megadoses, especially when used with prescription medicines in high-risk individuals, such as elderly with cardiovascular or other chronic conditions. Harmful effects may not be evident immediately, but only after a period of time. More than 60,000 vitamin overdose cases were reported to poison control centers just in 2004, with more than 48,000 exposures reported in children younger than 6 years, resulting in 53 major life-threatening outcomes including deaths [76,77], raising concern about indiscriminate use or unnecessary exposure for consumers and their children at home. Overdosage of vitamin supplementation can cause adverse effects directly or indirectly resulting from interactions with medications that can impact intestinal absorption, bioavailability, elimination, or actions at the target organs. Some of these adverse effects are summarized in Table 7.6.

Vitamin B and Folic Acid

Chronic vitamin B6 intake could potentially cause nerve toxicity with symptoms of peripheral neuropathy. Use of isoniazid is associated with vitamin B6

TABLE 7.6 Vitamins–Drug–Nutrient Interactions

Vitamins	Sources (food)	RDA/day (adult)	TUL/day (adult)	Toxicity/side effects	Nutrient–nutrient interactions	Drug–nutrient interactions
Biotin	Yeast, bread, egg, cheese, liver, pork, salmon, avocado, raspberries, cauliflower (raw)	30 mcg	None	Life-threatening eosinophilic pleuro-pericardial effusion in an elderly woman taking combination biotin (10,000 mcg/day) and pantothenic acid (300 mg/day) for 2 months (case report)[a]	Pantothenic acid (large doses) competes with biotin for cellular and intestinal uptake[b]	Reduced biotin blood levels and increase urinary excretion with anticonvulsants[c] Inhibition of biotin small intestine absorption with anticonvulsants, primidone and carbamazepine Decreased bacterial synthesis of biotin with sulfa drugs or other antibiotics
Folate	Lentils, garbanzo beans (chickpeas), asparagus, spinach, lima beans, orange juice, spaghetti, white rice, bread	0.4 mg	None	Can saturate DHFR metabolic capacity with appearance of unmetabolized folic acid in blood[d] causing poor cognition,[e] hematologic abnormalities,[d] and poor immune function[f]	Vitamin C limits folate degradation and improves bioavailability[g] May mask vitamin B12 deficiency Interacts with vitamin B12 and vitamin B6[h] Interacts with riboflavin[i]	NSAIDs (aspirin, ibuprofen) interfere with folate metabolism Phenytoin, phenobarbital, and primidone inhibit intestinal absorption of folate[j] Pancreatin and sulfasalazine decrease folate absorption Cholestyramine and cholestipol decrease absorption of folic acid[k] Methotrexate, trimethoprim, pyrimethamine, triamterene, and sulfasalazine exhibit antifolate activity Pharmacodynamic antagonism of pyrimethamine antiparasitic effect Decreased utilization of dietary folate with triamterene
Niacin	Yeast, meat, poultry, red fish (eg, tuna, salmon), cereals (especially fortified cereals), legumes and seeds. Milk, green leafy vegetables, coffee, and tea	Male 16 mg Female 14 mg	35 mg	Skin flushing, itching, nausea, vomiting, hepatotoxicity,[l,m] impaired glucose tolerance, blurred vision, hyperuricemia, gout[n]	Increases in plasma tryptophan levels[o]	Increased risk of myopathy or rhabdomyolysis with statins[p] Inhibits "uricosuric" effect of sulfinpyrazone[m] Cholestyramine and cholestipol decrease absorption of niacin Flushing and dizziness with nicotine Estrogen and contraceptives (combined) increase niacin synthesis from tryptophan[q] Chemotherapy (long term) can cause symptoms of pellagra Diminished protective effects of the simvastatin-niacin combination with concurrent daily antioxidants therapy (1000 mg/day) of vitamin C (1000 mg), alpha-tocopherol (800 IU), selenium (100 mcg), and beta-carotene (25 mg)[r]
Pantothenic acid (Vitamin B5)	Liver, kidney, yeast, egg yolk, broccoli, fish, shellfish, chicken, milk, yogurt, legumes, mushrooms, avocado, sweet potatoes	5 mg	None	Diarrhea/GI side effects with high intake (10–20 g/day)[m] Life-threatening eosinophilic pleuro-pericardial effusion in an elderly woman taking combination biotin (10,000 mcg/day) and pantothenic acid (300 mg/day) for 2 months (case report)[a]		Oral contraceptive pills may increase requirement of pantothenic acid[u] Additive effects in lowering blood lipids with statins or nicotinic acid[m]

(Continued)

TABLE 7.6 (Continued)

Vitamins	Sources (food)	RDA/day (adult)	TUL/day (adult)	Toxicity/side effects	Nutrient–nutrient interactions	Drug–nutrient interactions
Riboflavin (Vitamin B2)	Milk, egg, almonds, salmon, chicken, beef, broccoli, asparagus, spinach, bread	Male 1.3 mg Female 1.1 mg	None	Excess riboflavin may increase the risk of DNA strand breaks in the presence of chromium (VI), a known carcinogen[s]	Decreases plasma homocysteine levels[i]	Long-term use of the anticonvulsant, phenobarbital, may increase destruction of riboflavin by liver enzymes, increasing the risk of deficiency[t] Chronic alcohol consumption has been associated with riboflavin deficiency[u]
Thiamin	Lentils, green peas, fortified cereals, pork, pecans, spinach, orange, milk, cantaloupe, egg	Male 1.2 mg Female 1.1 mg	None	Life-threatening anaphylactic reactions with large IV doses of thiamin[v]		Diuretics (furosemide) increase urinary excretion of thiamine[w] Long-term phenytoin use reduces blood levels of thiamine[m] Chronic alcohol abuse is associated with thiamine deficiency (low dietary intake, impaired absorption, impaired utilization, and increased excretion)[x]
Vitamin A	Cod liver oil, egg, fortified cereals, butter, milk, sweet potato, carrot, cantaloupe, mango, spinach, broccoli, kale, collards, squash, butternut	Male 900 mcg Female 700 mcg	3000 mcg[y]	Hypervitaminosis A *Acute toxicity*: Nausea, headache, fatigue, anorexia, dizziness, dry skin, desquamation, cerebral edema *Chronic toxicity*: Dry itchy skin, desquamation, anorexia, headache, cerebral edema, and bone and joint pain *Severe cases*: Liver damage, hemorrhage, and coma Increases risk of lung cancer in high risk individuals (smokers)[z] Increased risk of osteoporotic fracture and decreased BMD in older men and women[aa,ab,ac]	Zinc deficiency interferes with vitamin A metabolism[ad, ae] Vitamin A deficiency may exacerbate iron deficiency anemia[af]	Chronic alcohol consumption results in depletion of liver stores of vitamin A and may contribute to alcohol-induced liver damage[ag] Oral contraceptives (combined) increase retinol-binding protein synthesis by the liver and may increase the risk of vitamin A toxicity[m] Increased risk of bleeding with blood thinners (warfarin, fondaparinux, heparin, Plavix, Agatroban, bilvarudin) Decreased vitamin A effectiveness with cholestipol Increased risk of pseudotumor cerebri (BIH) with minocycline
Vitamin B6 (Pyridoxine)	Fish, poultry, nuts, legumes, potatoes, bananas, spinach, hazel nuts, avocado, fortified cereals.	Male 1.7 mg Female 1.5 mg	10 mg[ah]	Sensory neuropathy		Antituberculosis drugs, antiparkinsonians, NSAIDs drugs[ai] and oral contraceptives[aj], interfere with vitamin B6 metabolism Enhances amiodarone-induced photosensitivity reactions Methylxanthines (eg, theophylline) reduce bioavailability of vitamin B6[ak] Decreases the efficacy of phenytoin, phenobarbital, and levodopa[al,am]

Vitamin B12 (Cyanocobalamin)	Clams, mussels, mackerel, crab, salmon, beef, milk, turkey, egg, poultry, brie (cheese) Note: Only bacteria can synthesize vitamin B12[an]	2.4 mcg	None	Previous reports that megadoses of vitamin C destroy vitamin B12 have not been supported[ao] and may have been an artifact of the assay used to measure vitamin B12 levels[ap]	PPI (eg, omeprazole, lansoprazole) and H$_2$ receptor antagonists (eg, cimetidine, famotidine) decrease absorption of vitamin B12 by suppressing gastric acid secretion Drugs decreasing vitamin B12 absorption: Antacids, cholestyramine, cholestipol, chloramphenicol, neomycin, colchicine, nitrous oxide, and metformin Decreased hematological response to cyanocobalamin with chloramphenicol
Vitamin C (Ascorbic Acid)	Orange, grapefruit, kiwi, strawberries, tomatoes, broccoli, potato, spinach, sweet red pepper	Male 90 mg Female 75 mg	2000	Large doses can cause diarrhea and gastrointestinal disturbances Increases urinary oxalate levels which may increase risk of calcium oxalate kidney stones[aq, ar]	Estrogen containing oral contraceptives and aspirin[au] lower vitamin C concentrations Interact with anticoagulant (eg, warfarin) medications Diminishes the protective effects of simvastatin-niacin combination.[av] However, findings are refuted in a much larger RCT
Vitamin D	Salmon, mackerel, sardines, fortified milk, fortified orange juice, fortified cereal, egg yolk	51–70 y 15 mcg (600 IU) ≥71 y 20 mcg (800 IU)	1000 mcg (4000 IU)	Hypervitaminosis D: Hypercalcemia, kidney stones	Drugs increasing metabolism of vitamin D: phenytoin (Dilantin), fosphenytoin, phenobarbital, corticosteroids, carbamazepine, and rifampin[aw] Cholestyramine, cholestipol, and mineral oil decrease intestinal absorption of vitamin D[ax, ay] Ketoconazole reduces serum levels of 1,25-hydroxyvitamin D levels in health men[az] Glucocorticoids and HIV treatment drugs (HAART) increase catabolism of vitamin D[ba] Vitamin D-induced hypercalcemia may precipitate cardiac arrhythmia in patients on digitalis[m,bb]
Vitamin E	Vegetable oils (olive, sunflower and safflower oils), nuts, whole grains, green leafy vegetables, peanuts, carrots, avocado	15 mg	1000 mg (1500 IU)	Increase likelihood of hemorrhage stroke Increase risk of lung cancer in current smokers[bc] 17% higher risk of prostate cancer compared to placebo at doses of 400 IU/day[bd] Supplementation with 400 IU/day of vitamin E has been found to accelerate the progression of retinitis pigmentosa that is not associated with vitamin E deficiency[be]	Increased bleeding risk with warfarin, Plavix, dipyridamole, aspirin, and NSAIDs (eg, ibuprofen) Medications decreasing vitamin E absorption: Cholestyramine, cholestipol, isoniazid, mineral oil, orlistat, sucralfate, olestra Anticonvulsant drugs, such as phenobarbital, phenytoin, or carbamazepine, may decrease plasma levels of vitamin E[m,bf] May diminish the protective effects of simvastatin-niacin combination.[av] However, findings refuted in much larger RCT[bg]

(Continued)

TABLE 7.6 (Continued)

Vitamins	Sources (food)	RDA/day (adult)	TUL/day (adult)	Toxicity/side effects	Nutrient–nutrient interactions	Drug–nutrient interactions
Vitamin K	Phylloquinone (vitamin K₁) major dietary form Vitamin K₂: Menaquinones are primarily of microbial origins and thus commonly found in fermented foods, such as cheese, curds, natto (fermented soybeans and animal livers[bh])	Male 120 mcg Female 90 mcg	None	Allergic reaction Can interfere with the function of glutathione, one of the body's natural antioxidants, resulting in oxidative damage to cell membranes	Large doses of vitamin A and antagonizes vitamin K[bi] Interferes with vitamin K absorption May inhibit vitamin K-dependent carboxylase activity and interfere with the coagulation cascade[bj]	Increases bleeding risk with anticoagulant drugs (eg, warfarin) Decreases anticoagulant effectiveness (eg, warfarin) Prolonged use of broad spectrum antibiotics (eg, cephalosporin, salicylates) can interfere with vitamin K synthesis by intestinal bacteria and lower vitamin K absorption Cholesterol-lowering medications (eg, cholestyramine and cholestipol), orlistat, mineral oil, and the fat substitute, olestra, may affect the absorption of fat-soluble vitamins, including vitamin K[m]

[a]Debourdeau PM, Djezzar S, Estival JL, Zammit CM, Richard RC, Castot AC. Life-threatening eosinophilic pleuropericardial effusion related to vitamins B5 and H. Ann Pharmacother 2001;35(4):424–6.
[b]Zempleni J, Mock DM. Human peripheral blood mononuclear cells: Inhibition of biotin transport by reversible competition with pantothenic acid is quantitatively minor. J Nutr Biochem 1999;10(7):427–32.
[c]Camporeale G, Zempleni J. Biotin. In: Bozman BA, Russell RM, eds. Present knowledge in nutrition. 9th ed. Vol. 1. Washington, DC: ILSI Press; 2006. pp. 314–26.
[d]Kelly P, McPartlin J, Goggins M, Weir DG, Scott JM. Unmetabolized folic acid in serum: acute studies in subjects consuming fortified food and supplements. Am J Clin Nutr 1997;65(6):1790–5.
[e]Morris MS, Jacques PF, Rosenberg IH, Selhub J. Circulating unmetabolized folic acid and 5-methyltetrahydrofolate in relation to anemia, macrocytosis, and cognitive test performance in American seniors. Am J Clin Nutr 2010;91(6):1733–44.
[f]Troen AM, Mitchell B, Sorensen B, Wener MH, Johnston A, Wood B, et al. Unmetabolized folic acid in plasma is associated with reduced natural killer cell cytotoxicity among postmenopausal women. J Nutr 2006;136(1):189–94.
[g]Vertinde PH, Oey I, Hendrickx ME, Van Loey AM, Temme EH. L-ascorbic acid improves the serum folate response to an oral dose of [6S]-5-methyltetrahydrofolic acid in healthy men. Eur J Clin Nutr 2008;62(10):1224–30.
[h]Gerhard GT, Duell PB. Homocysteine and atherosclerosis. Curr Opin Lipidol 1999;10(5):417–28.
[i]Jacques PF, Bostom AG, Wilson PW, Rich S, Rosenberg IH, Selhub J. Determinants of plasma total homocysteine concentration in the Framingham Offspring cohort. Am J Clin Nutr 2001;73(3):613–21.
[j]Apeland T, Mansoor MA, Strandjord RE. Antiepileptic drugs as independent predictors of plasma total homocysteine levels. Epilepsy Res 2001;47(1–2):27–35.
[k]Folate. In: Hendler SS, Rorvik, D.R., ed. PDR for nutritional supplements. 2nd ed. Montvale, NJ: Physicians' Desk Reference Inc.; 2008.
[l]Food and Nutrition Board, Institute of Medicine. Niacin. Dietary Reference Intakes: Thiamin, Riboflavin, Niacin, Vitamin B6, Vitamin B12, Pantothenic Acid, Biotin, and Choline. Washington, DC: National Academy Press; 1998. pp. 123–49.
[m]Hendler SS, Rorvik DR, eds. PDR for nutritional supplements. Montvale, NJ: Medical Economics Company, Inc; 2001.
[n]Vitamins. Drug facts and comparisons. St. Louis, MO: Facts and Comparisons; 2000. pp. 6–33.
[o]Murray MF, Langan M, MacGregor RR. Increased plasma tryptophan in HIV-infected patients treated with pharmacologic doses of nicotinamide. Nutrition 2001;17(7–8):654–6.
[p]Cziraky MJ, Willey VJ, McKenney JM, Kamat SA, Fisher MD, Guyton JR, et al. Risk of hospitalized rhabdomyolysis associated with lipid-lowering drugs in a real-world clinical setting. J Clin Lipidol 2013;7(2):102–8.
[q]Cerranites-Lauraen D, McEltaney NG, Moss J. Niacin. In: Shils M, Olson JA, Shike M, Ross AC, eds. Modern nutrition in health and disease. 9th ed. Baltimore, MD: Williams & Wilkins; 1999:401–11.
[r]Cheung MC, Zhao XQ, Chait A, Albers JJ, Brown BG. Antioxidant supplements block the response of HDL to simvastatin-niacin therapy in patients with coronary artery disease and low HDL. Arterioscler Thromb Vasc Biol 2001;21(8):1320–6.
[s]Sugiyama M. Role of physiological antioxidants in chromium(VI)-induced cellular injury. Free Radic Biol Med 1992;12(5):397–407.
[t]McCormick DB. Riboflavin. In: Shils M, Olson JA, Shike M, Ross AC, eds. Modern nutrition in health and disease. 9th ed. Baltimore, MD: Williams & Wilkins; 1999. pp. 391–9.
[u]Subramanian VS, Subramanya SB, Ghosal A, Said HM. Chronic alcohol feeding inhibits physiological and molecular parameters of intestinal and renal riboflavin transport. Am J Physiol Cell Physiol 2013;305(5):C539–45.
[v]Ruston D, Hoare J, Henderson L, et al. The National Diet & Nutrition Survey: adults aged 19 to 64 years: Nutritional status (anthropometry and blood analytes), blood pressure and physical activity. Vol. 4. London: The Stationary Office; 2004.
[w]Crombleholme WR. Obstetrics. In: Tierney LM, McPhee SJ, Papadakis MA, eds. Current medical treatment and diagnosis. 37th ed. Stamford, CT: Appleton and Lange; 1998. pp. 731–4.
[x]Food and Nutrition Board, Institute of Medicine. Riboflavin. Dietary reference intakes: Thiamin, riboflavin, niacin, vitamin B6, vitamin B12, pantothenic acid, biotin, and choline. Washington, DC: National Academy Press; 1998. pp. 87–122.
[y]Food and Nutrition Board, Institute of Medicine. Vitamin A. Dietary reference intakes for vitamin A, vitamin K, arsenic, boron, chromium, copper, iodine, iron, manganese, molybdenum, nickel, silicon, vanadium, and zinc. Washington, DC: National Academy Press; 2001:82–161.
[z]Omenn GS, Goodman GE, Thornquist MD, Balmes J, Cullen MR, Glass A, et al. Effects of a combination of beta carotene and vitamin A on lung cancer and cardiovascular disease. N Engl J Med 1996;334(18):1150–5.
[aa]Michaelsson K, Lithell H, Vessby B, Melhus H. Serum retinol levels and the risk of fracture. N Engl J Med 2003;348(4):287–94.
[ab]Promislow JH, Goodman-Gruen D, Slymen DJ, Barrett-Connor E. Retinol intake and bone mineral density in the elderly: the Rancho Bernardo Study. J Bone Miner Res 2002;17(8):1349–58.
[ac]Feskanich D, Singh V, Willett WC, Colditz GA. Vitamin A intake and hip fractures among postmenopausal women. JAMA 2002;287(1):47–54.
[ad]Russell RM. The vitamin A spectrum: from deficiency to toxicity. Am J Clin Nutr 2000;71(4):878–84.

[ae]Christian P, West KP, Jr. Interactions between zinc and vitamin A: an update. Am J Clin Nutr 1998;68(Suppl. 2):435S–41S.
[af]Suharno D, West CE, Muhilal, Karyadi D, Hautvast JG. Supplementation with vitamin A and iron for nutritional anaemia in pregnant women in West Java, Indonesia. Lancet 1993;342(8883):1325–8.
[ag]Wang XD. Chronic alcohol intake interferes with retinoid metabolism and signaling. Nutr Rev 1999;57(2):51–9.
[ah]Food and Nutrition Board, Institute of Medicine. Vitamin B6. Dietary reference intakes: Thiamin, riboflavin, niacin, vitamin B6, pantothenic acid, biotin, and choline. Washington, DC: National Academies Press; 1998. pp. 150–195.
[ai]Chang HY, Tang FY, Chen DY, Chih HM, Huang ST, Cheng HD, et al. Clinical use of cyclooxygenase inhibitors impairs vitamin B-6 metabolism. Am J Clin Nutr 2013;98(6):1440–9.
[aj]Morris MS, Picciano MF, Jacques PF, Selhub J. Plasma pyridoxal 5′-phosphate in the US population: the National Health and Nutrition Examination Survey, 2003–2004. Am J Clin Nutr 2008;87(5):1446–54.
[ak]Clayton PT. B6-responsive disorders: a model of vitamin dependency. J Inherit Metab Dis 2006;29(2–3):317–26.
[al]Leklem JE. Vitamin B-6. In: Shils M, Olson JA, Shike M, Ross AC, eds. Nutrition in health and disease. 9th ed. Baltimore, MD: Williams & Wilkins; 1999 pp. 413–22.
[am]Bender DA. Nonnutritional uses of vitamin B6. Br J Nutr 1999;81(1):7–20.
[an]LeBlanc JG, Milani C, de Giori GS, Sesma F, van Sinderen D, Ventura M. Bacteria as vitamin suppliers to their host: a gut microbiota perspective. Curr Opin Biotechnol 2013;24(2):160–8.
[ao]Simon JA, Hudes ES. Relation of serum ascorbic acid to serum vitamin B12, serum ferritin, and kidney stones in US adults. Arch Intern Med 1999;159(6):619–24.
[ap]Food and Nutrition Board, Institute of Medicine. Vitamin B12. Dietary reference intakes: Thiamin, riboflavin, niacin, vitamin B6, pantothenic acid, biotin, and choline. Washington, DC: National Academy Press; 1998. pp. 306–356.
[aq]Traxer O, Huet B, Poindexter J, Pak CY, Pearle MS. Effect of ascorbic acid consumption on urinary stone risk factors. J Urol 2003;170(2 Pt 1):397–401.
[ar]Massey LK, Liebman M, Kynast-Gales SA. Ascorbate increases human oxaluria and kidney stone risk. J Nutr. 2005;135(7):1673–7.
[as]Cheng Y, Willett WC, Schwartz J, Sparrow D, Weiss S, Hu H. Relation of nutrition to bone lead and blood lead levels in middle-aged to elderly men. The Normative Aging Study. Am J Epidemiol 1998;147(12):1162–74.
[at]Simon JA, Hudes ES. Relationship of ascorbic acid to blood lead levels. JAMA 1999;281(24):2289–93.
[au]Basu TK. Vitamin C-aspirin interactions. Int J Vitam Nutr Res Suppl 1982;23:83–90.
[av]Brown BG, Zhao XQ, Chait A, Fisher LD, Cheung MC, Morse JS, et al. Simvastatin and niacin, antioxidant vitamins, or the combination for the prevention of coronary disease. N Engl J Med 2001;345(22):1583–92.
[aw]Grober U, Spitz J, Reichrath J, Kisters K, Holick MF. Vitamin D: Update 2013: From rickets prophylaxis to general preventive healthcare. Dermatoendocrinol. 2013;5(3):331–47.
[ax]Knodel LC, Talbert RL. Adverse effects of hypolipidaemic drugs. Med Toxicol. 1987;2(1):10–32.
[ay]McDuffie JR, Calis KA, Booth SL, Uwaifo GI, Yanovski JA. Effects of orlistat on fat-soluble vitamins in obese adolescents. Pharmacotherapy. 2002;22(7):814–22.
[az]Glass AR, Eil C. Ketoconazole-induced reduction in serum 1,25-dihydroxyvitamin D. J Clin Endocrinol Metab 1986;63(3):766–9.
[ba]Holick MF, Binkley NC, Bischoff-Ferrari HA, Gordon CM, Hanley DA, Heaney RP, et al. Evaluation, treatment, and prevention of vitamin D deficiency: an Endocrine Society clinical practice guideline. J Clin Endocrinol Metab 2011;96(7):1911–30.
[bb]Vitamin D. Natural medicines comprehensive database [Website]. December 3, 2007. Available at: www.naturaldatabase.com. Accessed December 3, 2007.
[bc]Slatore CG, Littman AJ, Au DH, Satia JA, White E. Long-term use of supplemental multivitamins, vitamin C, vitamin E, and folate does not reduce the risk of lung cancer. Am J Respir Crit Care Med 2008;177(5):524–30.
[bd]Klein EA, Thompson IM, Jr., Tangen CM, Crowley JJ, Lucia MS, Goodman PJ, et al. Vitamin E and the risk of prostate cancer: the Selenium and Vitamin E Cancer Prevention Trial (SELECT). JAMA 2011;306(14):1549–56.
[be]Berson EL, Rosner B, Sandberg MA, Hayes KC, Nicholson BW, Weigel-DiFranco C, et al. A randomized trial of vitamin A and vitamin E supplementation for retinitis pigmentosa. Arch Ophthalmol 1993;111(6):761–72.
[bf]Food and Nutrition Board, Institute of Medicine. Vitamin E. Dietary reference intakes for vitamin C, vitamin E, selenium, and carotenoids. Washington, DC: National Academy Press; 2000:186–283
[bg]Collins R, Peto R, Armitage J. The MRC/BHF Heart Protection Study: preliminary results. Int J Clin Pract. 2002;56(1):53–6.
[bh]Holmes MV, Hunt BJ, Shearer MJ. The role of oral vitamin K in the management of oral vitamin K antagonists. Blood Rev. 2012;26(1):1–14.
[bi]Olson RE. Vitamin K. In: Shils M, Olson JA, Shike M, Ross AC, eds. Modern nutrition in health and disease. 9th ed. Baltimore, MD: Lippincott Williams & Wilkins; 1999. pp. 363–80.
[bj]Traber MG. Vitamin E and K interactions—a 50-year-old problem. Nutr Rev 2008;66(11):624–9.
[bk]Institute of Medicine, Food and Nutrition Board. Dietary reference intakes for calcium and vitamin D. Washington, DC: National Academy Press; 2010.

RDA, recommended dietary allowance; TUL, tolerable upper limit; M, males; F, females; CVD, cardiovascular disease; CHD, coronary heart disease; AD, Alzheimer's disease; FNB, Food and Nutrition Board; APL, acute promyelocytic leukemia; OTC, over the counter; ATBC study, alpha-tocopherol, beta-carotene, and vitamin C study; URTI, upper respiratory tract infection; RCT, randomized controlled trial; BIH, benign intracranial hypertension; BMD, bone mineral density; DHFR, dihydrofolate reductase.

DRI is the general term for a set of reference values used to plan and assess nutrient intakes of healthy people. These values, which vary by age and gender, include Hb: hemoglobin synthesis.
RDA: average daily level of intake sufficient to meet the nutrient requirements of nearly all (97–98%) healthy people.
Adequate Intake (AI): established when evidence is insufficient to develop an RDA and is set at a level assumed to ensure nutritional adequacy.
Tolerable Upper Intake Level (UL): maximum daily intake unlikely to cause adverse health effects.[bk]

Adapted from Micromedex Healthcare Series [Database Online]. 2006: Vol 128 (expires 6/2006). Greenwood Village, CO: Thomson Micromedex. For detailed information, see www.micromedex.com.
Adapted from Micronutrient Information Center, Linus Pauling Institute, Oregon State University. For detailed information, see www.lpi.oregonstate.edu/infocenter/vitamins

TABLE 7.7 Mineral–Drug Interactions

Minerals	Sources (food)	RDA/day (adult)	TUL/day (adult)	Toxicity/side effects	Nutrient–nutrient interactions	Drug–nutrient interactions
Calcium	Tofu (prepared with calcium sulfate), yogurt, sardines, cheddar cheese, milk, white beans, orange, kale, pinto beans, broccoli, red beans	51–70 y Male 1200 mg Female 1000 mg ≥ 71 y 1200 mg	2000 mg	*Mild hypercalcemia:* Asymptomatic, loss of appetite, nausea, vomiting, constipation, abdominal pain, fatigue, frequent urination (polyuria), and hypertension[a] *Severe hypercalcemia:* Confusion, delirium, coma, and, if not treated, death[b] Milk-alkali syndrome (hypercalcemia, metabolic alkalosis, renal failure) Concerns for risks of prostate cancer[c] and vascular disease[d] with high intakes of calcium	Decreases absorption of both heme and nonheme iron[e] High sodium intake results in increased loss of calcium in the urine[b] Reduces zinc absorption[f] Increasing dietary protein intake enhances intestinal calcium absorption and urinary calcium excretion[g] Phosphorus increases the excretion of calcium in the urine Caffeine (large amount) increases urinary calcium content for a short time[h] Oxalic acid (oxalate) is the most potent inhibitor of calcium absorption	Hypercalcemia and milk-alkali syndrome with hydrochlorthiazide Increase likelihood of arrhythmias with digoxin[i] Calcium IV decreases efficacy of calcium channel blockers;[j] effect not seen with oral calcium supplementation Decreases absorption of tetracycline, quinolones, bisphosphonates, and levothyroxine Ciprofloxacin, norfloxacin, ofloxacin, and tetracycline will yield insoluble compounds with calcium, decreasing absorption H_2-receptor blockers (eg, cimetidine, famotidine) and PPI (eg, omeprazole) may decrease absorption of calcium carbonate and calcium phosphate[k] Corticosteroids can deplete calcium and increase risk of osteoporosis[l] Reduces plasma concentration of atazanavir Guar gum delays calcium absorption Decreases phosphate absorption Reversal of hypotensive effects of verapamil
Chromium	Broccoli, green beans, potatoes, grape juice, orange, beef, waffle, bagel, banana, turkey breast	Male 30 mcg Female 20 mcg	None	No adverse effects have been convincingly associated with excess intake of trivalent chromium from food or supplements Hexavalent chromium (chromium VI; Cr^{6+}) is a recognized carcinogen. Exposure to hexavalent chromium in dust increases lung cancer incidence and is known to cause dermatitis Chromium picolinate supplementation may increase DNA damage,[m] kidney failure,[n,o] and impair liver function[p]	Chromodulin improves tissue sensitivity to insulin and facilitates glucose transport Iron overload in hereditary hemochromatosis may interfere with chromium transport by competing for transferrin binding contributing to the pathogenesis of diabetes mellitus in patients with hereditary hemochromatosis[q] Diets high in simple sugars (eg, sucrose) result in increased urinary chromium excretion in adults	Little is known about drug interactions with chromium in humans

	Sources	Amount	Toxicity	Nutrient Interactions	Drug Interactions
Copper	Meats, shellfish, nuts, seeds, wheat-bran cereals, liver (beef), crab, mollusks, cashew nuts, hazel nuts, almonds, peanut butter, lentils, mushrooms, chocolate	900 mcg	10 mg	*Acute intoxication:* Abdominal pain, nausea, vomiting, and diarrhea *Severe toxicity:* Liver damage, kidney failure, coma, and death. Immune function and antioxidant status might be affected by higher intakes of copper[r]	Adequate copper nutritional status is necessary for normal iron metabolism and red blood cell formation. Copper deficiency can lead to secondary ceruloplasmin deficiency and hepatic iron overload and/or cirrhosis[s] High supplemental zinc intakes of 50 mg/day may result in copper deficiency	Penicillamine increases the urinary excretion of copper Antacids may interfere with copper absorption when used in very high amounts[t]
Fluoride	Black tea, crab, rice, fish, chicken	Male 4 mg Female 3 mg	10 mg	*Acute toxicity:* Nausea, abdominal pain, vomiting, diarrhea, excessive salivation, tearing, sweating, and generalized weakness[u] Dental fluorosis	Both calcium and magnesium form insoluble complexes with fluoride and are capable of significantly decreasing fluoride absorption when present in the same meal. However, the absorption of fluoride in the form of monofluorophosphate (unlike sodium fluoride) is unaffected by calcium	Calcium supplements, as well as calcium and aluminum-containing antacids, can decrease the absorption of fluoride[v] A diet low in chloride (salt) has been found to increase fluoride retention by reducing urinary excretion of fluoride[w]
Iodine	Salt (iodized), cod, shrimp, fish, milk, egg, seaweed, potato	150 mcg	1100 mcg	*Acute toxicity:* Burning of the mouth, throat, and stomach; fever; nausea; vomiting; diarrhea; a weak pulse; and coma Increased incidence of thyroid papillary cancer with increased iodine intake in observational studies	Selenium deficiency can exacerbate the effects of iodine deficiency[x,y] Deficiencies of vitamin A or iron may exacerbate the effects of iodine deficiency[x-z]	Amiodarone may affect thyroid function Antithyroid medications, propylthiouracil and methimazole, may increase the risk of hypothyroidism Lithium in combination with pharmacologic doses of potassium iodide may result in hypothyroidism Pharmacologic doses of potassium iodide may decrease the anticoagulant effect of warfarin (Coumadin)[x,aa]
Iron	Beef, poultry, oysters, shrimp, tuna, prune juice, potato, kidney beans, lentils, tofu, cashew nuts	8 mg	45 mg	Gastrointestinal irritation, nausea, vomiting, diarrhea, or constipation. Stools will often appear darker in color *Acute toxicity* (doses 20–60 mg/kg) *Stage 1* (1–6 h): Nausea, vomiting, abdominal pain, tarry stools, lethargy, weak and rapid pulse, low blood pressure, fever, difficulty breathing, and coma; *Stage 2:* Symptoms may subside for about 24 h; *Stage 3* (12–48 h): signs of organ failure (cardiovascular, kidney, liver, hematologic, and central	Vitamin A deficiency may exacerbate iron-deficiency anemia[ac] Copper is required for iron transport to the bone marrow for red blood cell formation[t] High doses of iron supplements can decrease zinc absorption[ab] Aluminum reduces iron effectiveness *Enhancers of nonheme iron absorption:* Vitamin C, organic acids (citric, malic, tartaric, lactic acids), meat, fish, poultry[ab,ad] *Inhibitors of nonheme iron absorption:* phytic acid, polyphenols, and soy protein	Medications that decrease stomach acidity, such as antacids, H$_2$ receptor antagonists (eg, cimetidine, ranitidine), and PPI (eg, omeprazole, lansoprazole), may impair iron absorption and reduce iron bioavailability Taking iron supplements at the same time as the following medications may result in decreased absorption and efficacy of the medication: ciprofloxacin, levodopa, levothyroxine, methyldopa, norfloxacin, ofloxacin penicillamine, quinolones, tetracyclines, and bisphosphonate Cholestyramine interferes with iron absorption Allopurinol, a medication used to treat gout, may increase iron storage in the liver and

(Continued)

TABLE 7.7 (Continued)

Minerals	Sources (food)	RDA/day (adult)	TUL/day (adult)	Toxicity/side effects	Nutrient–nutrient interactions	Drug–nutrient interactions
				nervous systems); *Stage 4* (2–6 weeks): long-term damage to the central nervous system, liver (cirrhosis), and stomach may develop[ab]		should not be used in combination with iron supplements[v,ae] Hypothyroidism with levothyroxine
Magnesium	Cereal, brown rice, fish, spinach, almonds, lima beans, peanuts, okra, milk, banana chelates, including magnesium aspartate. Magnesium hydroxide is used as an ingredient in several antacids[ag]	Male 420 mg Female 320 mg	350 mg	Diarrhea, hypotension, lethargy, confusion, renal impairment, dysrhythmias, cardiac arrest	High doses of zinc supplements interfere with the absorption of magnesium[af] Large increases in the intake of dietary fiber have been found to decrease magnesium utilization in experimental studies[ag] Dietary protein may affect magnesium absorption[ah] Vitamin D (calcitriol) slightly increases intestinal absorption of magnesium[ai] Inadequate blood magnesium levels are known to result in low blood calcium levels, resistance to parathyroid hormone (PTH) action and resistance to some of the effects of vitamin D[ao,ag]	Interferes with the absorption of digoxin, nitrofurantoin and certain antimalarial drugs Reduces the efficacy of chlorpromazine, ciprofloxacin, norfloxacin, ofloxacin, penicillamine, oral anticoagulants, quinolone, and tetracycline Bisphosphonates (eg, alendronate) and magnesium should be taken 2 h apart so that the absorption of the bisphosphonate is not inhibited Furosemide (Lasix) and some thiazide diuretics (eg, hydrochlorothiazide) may result in magnesium depletion[v] Long-term use (3 months or longer) of PPI increases the risk of hypomagnesaemia[m] IV magnesium has increased the effects of certain muscle-relaxing medications used during anesthesia Hypermagnesemia with calcitriol, doxercalciferol Decreases drug bioavailability of lansoprazole, delavirdine Decreased plasma concentration of digoxin, atazanavir Hypotension with felodipine Hypoglycemia with nicardipine, felodipine, oral sulfonylureas Bradycardia and decreased cardiac output with labetalol Increased risk of metabolic alkalosis with polystyrene sulfonate Neuromuscular weakness with amikacin, gentamicin, tobramycin
Manganese	Whole grains, nuts, leafy vegetables, teas	Male 2.3 mg Female 1.8 mg	11 mg	Neurological: irritability, tremors, hallucinations, facial muscle spasms, ataxia Pulmonary: cough, acute bronchitis, decreased lung function Individuals with increased risk of manganese toxicity:	Intestinal absorption of manganese is increased during iron deficiency and vice versa[aj] Supplemental calcium (500 mg/day) may decrease bioavailability of manganese Intakes of other minerals, including iron, calcium, and	Magnesium-containing antacids and laxatives and the antibiotic medication, tetracycline, may decrease the absorption of manganese if taken together with manganese-containing foods or supplements[aa] Supplemental magnesium (200 mg/day) has been shown to slightly decrease manganese bioavailability in healthy adults, either by

	Sources	RDA/AI	Deficiency/Toxicity	UL	Drug-Nutrient Interactions
					decreasing manganese absorption or by increasing its excretion[al]
Molybdenum	Legumes (beans, lentils, and peas), grain products, nuts	45 mcg	Hyperuricemia, gout *Acute toxicity*: acute psychosis with hallucinations, seizures	2000 mcg	Study reported that molybdenum intakes to increase urinary copper excretion,[am] however, more recent, well-controlled study indicated that very high dietary molybdenum intakes (up to 1500 mcg/day) did not adversely affect copper nutritional status in eight, healthy young men[an] Little is known about drug interactions with molybdenum in humans
Phosphorus	Dairy food, cereals, meat, fish, beef, lentils, almond, carbonated cola drink, yogurt, milk	700 mg	High serum phosphorus concentrations have been associated with increased rates of CVD and mortality in subjects with or without kidney disease[ao,ap,aq] High serum phosphorus has been shown to impair synthesis of the active form of vitamin D (1,25-dihydroxyvitamin D) in the kidneys, reduce blood calcium and lead to increased PTH release by the parathyroid glands[ar]	19–70 y 4000 mg ≥71 y 3000 mg	Aluminum-containing antacids reduce the absorption of dietary phosphorus by forming aluminum phosphate PPI may limit the efficacy of phosphate-binder therapy in patients with kidney failure[as] Excessively high doses of 1,25-dihydroxyvitamin D or its analogs, may result in hyperphosphatemia[at] Potassium supplements or potassium-sparing diuretics taken together with phosphorus supplements may cause hyperkalemia and life-threatening arrhythmias Hormone replacement therapy causes increased urinary excretion of phosphorous[au]
Potassium	Banana, potato, prune juice, plums, orange, tomato, raisins, artichoke, spinach, sunflower seeds, almonds, molasses	4700 mg	Gastrointestinal symptoms: nausea, vomiting, diarrhea, abdominal discomfort Hyperkalemia: tingling of the hands and feet, muscular weakness, temporary paralysis, cardiac arrhythmia, cardiac arrest[av]	None	*Medications associated with hypokalemia*: potassium-wasting diuretics (eg, thiazide diuretics or furosemide), beta adrenergics, pseudoephedrine, bronchodilators, fludrocortisone, licorice, gossypol, penicillin, caffeine, phenolphthalein, sodium polystyrene sulfonate *Medications associated with hyperkalemia*: potassium sparing diuretics (eg, spironolactone, triamterene, amiloride) ACE inhibitors, NSAIDs, trimethoprim/sulfamethoxazole, pentamidine, heparin, digitalis, beta-blockers, alpha-blockers
Selenium	Brazil nuts, shrimp, crab, salmon, poultry, pork, beef, milk, walnuts	55 mcg	Selenosis → hair and nail brittleness and loss, gastrointestinal disturbances, skin rashes, a garlic-breath	400 mcg	Selenium deficiency may exacerbate the effects of iodine deficiency Selenium supplementation in a small group of elderly individuals Valproic acid has been found to decrease plasma selenium levels

(*Continued*)

TABLE 7.7 (Continued)

Minerals	Sources (food)	RDA/day (adult)	TUL/day (adult)	Toxicity/side effects	Nutrient–nutrient interactions	Drug–nutrient interactions
				odor, fatigue, irritability, nervous system abnormalities	decreased plasma T_4, indicating increased deiodinase activity and thus increased conversion of T_4 to T_3[a]	
Sodium (Chloride)	Most of the sodium and chloride in the diet comes from salt	51–70 y 1.3 g ≥71 y 1.2 g	2.3 g	Hypernatremia → edema (swelling), hypertension, rapid heart rate, difficulty breathing, convulsions, coma, and death		*Medications associated with hyponatremia:* diuretics (hydrochlorthiazide, furosemide), NSAIDs (ibuprofen, naproxen), opiates (codeine, morphine), phenothiazines (promethazine, prochlorperazine), SSRIs (fluoxetine, paroxetine), tricyclic antidepressants (amitriptyline, imipramine) *Medications associated with hyponatremia:* carbamazepine, chlorpropamide, clofibrate, cyclophosphamide, desmopressin, oxytocin, vincristine
Zinc	Oysters, beef, crab, pork, turkey, beans, poultry, cashews, chickpeas, milk, almonds, peanuts, cheese (cheddar)	Male 11 mg Female 8 mg	40 mg	Abdominal pain, diarrhea, nausea, and vomiting Metal fume fever has been reported after the inhalation of zinc oxide fumes. Specifically, profuse sweating, weakness, and rapid breathing may develop within 8 h of zinc oxide inhalation and persist 12–24 h after exposure is terminated[aw] Intranasal zinc is known to cause a loss of the sense of smell (anosmia)[ax]	Long-term consumption of zinc in excess of the tolerable upper intake level (40 mg/day for adults) can result in copper deficiency Iron decreases zinc absorption[ay] Calcium in combination with phytic acid or phytate might affect zinc absorption The bioavailability of dietary folate is increased by the action of a zinc-dependent enzyme, suggesting a possible interaction between zinc and folic acid Zinc deficiency is associated with decreased release of vitamin A from the liver[az] High amounts of zinc decreases copper absorption	Concomitant administration of zinc and certain medications (tetracycline, quinolones, bisphosphonates) may decrease absorption of both zinc and the medication, thus reducing drug efficacy[ae] Penicillamine can result in severe zinc deficiency Valproic acid may precipitate zinc deficiency Prolonged use of diuretics may increase urinary zinc excretion, resulting in zinc deficiency Penicillamine decreases zinc absorption Decreases drug effectiveness of ciprofloxacin, gatifloxacin, ofloxacin, norfloxacin, and tetracycline

[a] Moe SM. Disorders involving calcium, phosphorus, and magnesium. Prim Care 2008;35(2):215–37, v–vi.
[b] Weaver CM. Calcium. In: Erdman JJ, Macdonald I, Zeisel S, eds. Present knowledge in nutrition. 10th ed. John Wiley & Sons, Inc.; 2012.
[c] Gonzalez CA, Riboli E. Diet and cancer prevention: contributions from the European Prospective Investigation into Cancer and Nutrition (EPIC) study. Eur J Cancer 2010;46(14):2555–62.
[d] Bolland MJ, Grey A, Gamble GD, Reid IR. Calcium and vitamin D supplements and health outcomes: a reanalysis of the Women's Health Initiative (WHI) limited-access data set. Am J Clin Nutr 2011;94(4):1144–9.
[e] Scholl TO, Chen X, Stein TP. Maternal calcium metabolic stress and fetal growth. Am J Clin Nutr 2014;99(4):918–25.
[f] Wood RJ, Zheng JJ. High dietary calcium intakes reduce zinc absorption and balance in humans. Am J Clin Nutr 1997;65(6):1803–9.
[g] Food and Nutrition Board, Institute of Medicine. Calcium. Dietary reference intakes for calcium and vitamin D. Washington, DC: The National Academies Press; 2011.
[h] Heaney RP. Bone mass, nutrition, and other lifestyle factors. Nutr Rev 1996;54(4 Pt 2):S3–10.
[i] Vella A, Gerber TC, Hayes DL, Reeder GS. Digoxin, hypercalcaemia, and cardiac conduction. Postgrad Med J 1999;75(887):554–6.
[j] Bania TC, Blaufeux B, Hughes S, Almond GL, Homel P. Calcium alone for severe verapamil toxicity. Acad Emerg Med 2000;7(10):1089–96.
[k] Wright MJ, Proctor DD, Insogna KL, Kerstetter JE. Proton pump-inhibiting drugs, calcium homeostasis, and bone health. Nutr Rev 2008;66(2):103–8.
[l] Gennari C. Differential effect of glucocorticoids on calcium absorption and bone mass. Br J Rheumatol 1993;32 (Suppl. 2):11–4.

[m]Wilhelm SM, Rjater RG, Kale-Pradhan PB. Perils and pitfalls of long-term effects of proton pump inhibitors. Expert Rev Clin Pharmacol 2013;6(4):443–51.

[n]Wasser WG, Feldman NS, D'Agati VD. Chronic renal failure after ingestion of over-the-counter chromium picolinate. Ann Intern Med 1997;126(5):410.

[o]Wani S, Weskamp C, Marple J, Spry L. Acute tubular necrosis associated with chromium picolinate-containing dietary supplement. Ann Pharmacother 2006;40(3):563–6.

[p]Cerulli J, Grabe DW, Gauthier I, Malone M, McGoldrick MD. Chromium picolinate toxicity. Ann Pharmacother 1998;32(4):428–31.

[q]Food and Nutrition Board, Institute of Medicine. Chromium. Dietary reference intakes for vitamin A, vitamin K, boron, chromium, copper, iodine, iron, manganese, molybdenum, nickel, silicon, vanadium, and zinc. Washington, DC: National Academy Press; 2001. pp. 197–223.

[r]Turnlund JR, Keyes WR, Kim SK, Domek JM. Long-term high copper intake: effects on copper absorption, retention, and homeostasis in men. Am J Clin Nutr 2005;81(4):822–8.

[s]Thackeray EW, Sanderson SO, Fox JC, Kumar N. Hepatic iron overload or cirrhosis may occur in acquired copper deficiency and is likely mediated by hypoceruloplasminemia. J Clin Gastroenterol 2011;45(2):153–8.

[t]Turnlund JR. Copper. In: Shils ME, Shike M, Ross AC, Caballero B, Cousins RJ, eds. Modern nutrition in health and disease. 10th ed. Philadelphia, PA: Lippincott Williams & Wilkins; 2006. pp. 286–99.

[u]Whitford GM. Acute toxicity of ingested fluoride. Monogr Oral Sci 2011;22:66–80.

[v]Minerals. Drug facts and comparisons. St. Louis, MO: Facts and Comparisons; 2000. pp. 27–51.

[w]Cerklewski FL. Fluoride bioavailability—nutritional and clinical aspects. Nutr Res 1997;17:907–29.

[x]Food and Nutrition Board, Institute of Medicine. Iodine. Dietary reference intakes for vitamin A, vitamin K, boron, chromium, copper, iodine, iron, manganese, molybdenum, nickel, silicon, vanadium, and zinc. Washington, DC: National Academy Press; 2001. pp. 258–89.

[y]Levander OA, Whanger PD. Deliberations and evaluations of the approaches, endpoints and paradigms for selenium and iodine dietary recommendations. J Nutr 1996;126(Suppl. 9):2427S–34S.

[z]Zimmermann MB, Jooste PL, Pandao CS. Iodine-deficiency disorders. Lancet 2008;372(9645):1251–62.

[aa]Hendler SS, Rorvik DR, eds. PDR for nutritional supplements. Montvale, NJ: Medical Economics Company, Inc; 2001.

[ab]Food and Nutrition Board, Institute of Medicine. Iron. Dietary reference intakes for vitamin A, vitamin K, boron, chromium, copper, iodine, iron, manganese, molybdenum, nickel, silicon, vanadium, and zinc. Washington, DC: National Academy Press; 2001. pp. 290–393.

[ac]Suharno D, West CE, Muhilal, Karyadi D, Hautvast JG. Supplementation with vitamin A and iron for nutritional anaemia in pregnant women in West Java, Indonesia. Lancet 1993;342(8883):1325–8.

[ad]Lynch SR. Interaction of iron with other nutrients. Nutr Rev 1997;55(4):102–10.

[ae]Hendler SS, Rorvik D, eds. PDR for nutritional supplements. 2nd ed. Montvale, NJ: Physicians' Desk Reference Inc.; 2008.

[af]Spencer H, Norris C, Williams D. Inhibitory effects of zinc on magnesium balance and magnesium absorption in man. J Am Coll Nutr 1994;13(5):479–84.

[ag]Rude RK, Shils ME. Magnesium. In: Shils ME, Shike M, Ross AC, Caballero B, Cousins RJ, eds. Modern nutrition in health and disease. 10th ed. Baltimore, MD: Lippincott Williams & Wilkins; 2006. pp. 223–47.

[ah]Schwartz R, Walker G, Linz MD, MacKellar I. Metabolic responses of adolescent boys to two levels of dietary magnesium and protein. I. Magnesium and nitrogen retention. Am J Clin Nutr 1973;26(5):510–18.

[ai]Navarro-Gonzalez JF, Mora-Fernandez C, Garcia-Perez J. Clinical implications of disordered magnesium homeostasis in chronic renal failure and dialysis. Semin Dial. 2009;22(1):37–44.

[aj]Finley JW. Manganese absorption and retention by young women is associated with serum ferritin concentration. Am J Clin Nutr. 1999;70(1):37–43.

[ak]Food and Nutrition Board, Institute of Medicine. Manganese. Dietary reference intakes for vitamin A, vitamin K, boron, chromium, copper, iodine, iron, manganese, molybdenum, nickel, silicon, vanadium, and zinc. Washington, DC: National Academy Press; 2001. pp. 394–419.

[al]Kies C. Bioavailability of manganese. In: Klimis-Tavantzis DL, ed. Manganese in health and disease. Boca Raton, FL: CRC Press, Inc; 1994:39–58.

[am]Food and Nutrition Board, Institute of Medicine. Molybdenum. In: Dietary reference intakes for vitamin A, vitamin K, boron, chromium, copper, iodine, iron, manganese, molybdenum, nickel, silicon, vanadium, and zinc. Washington, DC: National Academy Press; 2001. pp. 420–41.

[an]Turnlund JR, Keyes WR. Dietary molybdenum: effect on copper absorption, excretion, and status in young men. In: Roussel AM, ed. Trace elements in man and animals. Vol. 10. New York, NY: Kluwer Academic Press; 2000. pp. 951–3.

[ao]Dhingra R, Sullivan LM, Fox CS, Wang TJ, D'Agostino RB, Sr., Gaziano JM, et al. Relations of serum phosphorus and calcium levels to the incidence of cardiovascular disease in the community. Arch Intern Med 2007;167 (9):879–85.

[ap]Tonelli M, Sacks F, Pfeffer M, Gao Z, Curhan G. Relation between serum phosphate level and cardiovascular event rate in people with coronary disease. Circulation 2005;112(17):2627–33.

[aq]Palmer SC, Hayen A, Macaskill P, Pellegrini F, Craig JC, Elder GJ, et al. Serum levels of phosphorus, parathyroid hormone, and calcium and risks of death and cardiovascular disease in individuals with chronic kidney disease: a systematic review and meta-analysis. JAMA 2011;305(11):1119–27.

[ar]Calvo MS, Moshfegh AJ, Tucker KL. Assessing the health impact of phosphorus in the food supply: issues and considerations. Adv Nutr 2014;5(1):104–13.

[as]Cerrelli MJ, Shaman A, Meade A, Carroll R, McDonald SP. Effect of gastric acid suppression with pantoprazole on the efficacy of calcium carbonate as a phosphate binder in haemodialysis patients. Nephrology (Carlton) 2012;17 (5):458–65.

[at]Food and Nutrition Board, Institute of Medicine. Phosphorus. Dietary reference intakes: calcium, phosphorus, magnesium, vitamin D, and fluoride. Washington, DC: National Academy Press; 1997. pp. 146–89.

[au]Bansal N, Katz R, de Boer IH, Kestenbaum B, Siscovick DS, Hoofnagle AN, et al. Influence of estrogen therapy on calcium, phosphorus, and other regulatory hormones in postmenopausal women: the MESA study. J Clin Endocrinol Metab 2013;98(12):4890–8.

[av]Mandal AK. Hypokalemia and hyperkalemia. Med Clin North Am 1997;81(3):611–39.

[aw]King JC, Cousins RJ. Zinc. In: Shils ME, Shike M, Ross AC, Caballero B, Cousins RJ, eds. Modern nutrition in health and disease. 10th ed. Baltimore, MD: Lippincott Williams & Wilkins; 2006. pp. 271–85.

[ax]DeCook CA, Hirsch AR. Anosmia due to inhalational zinc: a case report. Chem Senses 2000;25(5):659.

[ay]Sandstrom B. Micronutrient interactions: effects on absorption and bioavailability. Br J Nutr 2001;85(Suppl. 2):S181–5.

[az]Christian P, West KP, Jr. Interactions between zinc and vitamin A: an update. Am J Clin Nutr 1998;68(Suppl. 2):435S–41S.

[ba]Institute of Medicine, Food and Nutrition Board. Dietary reference intakes for calcium and vitamin D. Washington, DC: National Academy Press, 2010.

RDA, recommended dietary allowance; TUL, tolerable upper limit; M, males; F, females; CVD, cardiovascular disease; CHD, coronary heart disease; AD, Alzheimer's disease; FNB, Food and Nutrition Board; APL, acute promyelocytic leukemia; OTC, over the counter; ATBC study, alpha-tocopherol, beta-carotene, and vitamin C study; URTI, upper respiratory tract infection; RCT, randomized controlled trial; SSRI, selective serotonin reuptake inhibitors.

DRI is the general term for a set of reference values used to plan and assess nutrient intakes of healthy people. These values, which vary by age and gender, include Hb: hemoglobin synthesis.
RDA: average daily level of intake sufficient to meet the nutrient requirements of nearly all (97–98%) healthy people.
Adequate Intake (AI): established when evidence is insufficient to develop an RDA and is set at a level assumed to ensure nutritional adequacy.
Tolerable Upper Intake Level (UL): established when evidence is insufficient to develop an RDA and is set at a level assumed to ensure nutritional adequacy.[ba]

Adapted from Micromedex Healthcare Series [Database Online]. 2006: Vol. 128 (expires 6/2006). Greenwood Village, CO: Thomson Micromedex. For detailed information, see www.micromedex.com.
Adapted from Micronutrient Information Center, Linus Pauling Institute, Oregon State University. For detailed information, see www.lpi.oregonstate.edu/mic/minerals

depletion that results in neuropathy, while high doses of vitamin B6 can induce metabolism of anticonvulsants [78] or levodopa [79] with loss of efficacy, which could result in suboptimal control of seizures or Parkinson's disease [80]. Thiamine and folic acid transport can be reduced by polyphenols [81]. Therapeutics targeting cofactors for enzymes involved in cell cycle regulation, such as chemotherapeutics (methotrexate, also used for arthritis) or antimicrobials (trimethoprim or pyrimethamine) may lead to folate deficiency that, in turn, results in megaloblastic anemia and other dysfunction. Use of statins for cholesterol lowering is associated with a reduction in mitochondrial and serum co-enzyme Q, a vitamin-like, fat-soluble quinone present on mitochondrial inner membranes with antioxidant properties and it has been suggested that CoQ10 supplementation might benefit those patients suffering from statin-induced myopathy [8,82]. Coenzyme Q10 is chemically similar to vitamin K, and in animal studies, reduced the anticoagulant effect of warfarin and increased its clearance [83]. Vitamin B12 excess may cause diarrhea, itching and allergic reactions. Vitamin B12 requires an acidic gastric milieu for its removal from dietary proteins and secretion of intrinsic factor for its absorption in the terminal ileum and therefore increase in gastric pH by drugs, such as PPI or H_2-blockers, can reduce Vitamin B12 absorption and blood levels [84,85]. The reduction in B12 blood levels with prolonged PPI use was not prevented by concomitant B12 supplementation [86]. Given the data from majority of clinical studies, PPI should be used with caution in the elderly, known to have a high prevalence of vitamin B12 deficiency [72,85,87]. Prolonged use of metformin in diabetic patients has also been associated with an increase in incidence of vitamin B12 and folic acid deficiency [49]. Monitoring of B12 levels have been recommended in patients on metformin with supplementation if required [8].

Vitamin C

Vitamin C is metabolized to dehydroascorbic acid, which is then oxidized to oxalic acid and its prolonged use, especially in high doses, can increase propensity to form renal stones due to relative hyperoxaluria, crystalluria, and hematuria [88]. In patients with glucose 6 phosphate dehydrogenase deficiency, high doses of vitamin C may cause hemolysis [89]. Vitamin C absorption requires an acidic gastric milieu and any increase in pH, such as with the use of PPI or H_2-blockers, can reduce vitamin C absorption [72]. Vitamin C is depleted by corticosteroids. It enhances iron absorption and should be avoided in high doses in patients with hemochromatosis [90]. Since genetic abnormalities leading to iron overload are common in the population, the potential for iron overload mediated hepatic and cardiac toxicity increases with chronic vitamin C consumption [91]. Similar concerns are present in patients at risk of secondary iron overload, such as those requiring frequent blood transfusions for thalassemia or sickle cell disease [91].

Vitamin E

Studies have raised concerns about the use of large doses of vitamin E demonstrating an increased risk of bleeding, heart failure, cancer, and all-cause mortality [65,66,92–94]. An increased risk of hemorrhagic stroke was demonstrated in participants taking alpha-tocopherol in the alpha-tocopherol, beta-carotene, and vitamin C study (ATBC) study [95] and Physicians Health Study [96]. This raises serious concerns regarding the use of vitamin E in elderly patients with cardiovascular disease (CVD), many of whom are on antiplatelet agents or anticoagulants, and hence are at increased risk for bleeding with concomitant use of vitamin E [97]. Increased risk of bleeding has been reported in patients taking large doses of vitamin E and warfarin concomitantly [97]. A trend toward an increased incidence of prostate cancer was also reported in the Selenium and Vitamin E Cancer Prevention Trial that was terminated prematurely [98].

Vitamin A

Excessive vitamin A intake has been associated with abnormal liver profile, reduced bone mineral density, and increased risk for osteopenia and fractures [99,100]. In the Nurses' Health Study, 72,337 postmenopausal women aged between 34 and 77 years were followed for 18 years. Women taking higher doses of vitamin A (≥3000 μg of retinol equivalents daily) had an increased risk for hip fracture as compared to those taking lower doses (<1250 μg per day) [101]. High levels of vitamin A are also associated with liver abnormalities, neurotoxicity, and birth defects [102,103]. β-carotene as well as long-term retinol has been associated with increased risk of lung cancer, CVD, and death in smokers and those exposed to asbestos [103]. The increased mortality risk with the use of vitamin A, E, and beta-carotene was demonstrated in a Cochrane systematic review [104].

Vitamin D

Excessive vitamin D can cause hypercalcemia resulting in nausea, vomiting, poor appetite, constipation, weakness, and weight loss [105]. An increase in the risk of kidney stones with calcium and phosphate deposition and vascular calcification [106,107] and increased risk of cardiovascular events with calcium supplementation [108] may occur. Concerns regarding induction of hypercalcemia by toxic levels of vitamin D precipitating cardiac arrhythmia in patients on digitalis have also been raised [109]. Vitamin D is metabolized through CYP3A4 and could be affected by drugs that modulate CYP3A4 activity (Table 7.2A). Therefore, several drug–vitamin D interactions have been described, including with cholesterol lowering statins affecting Cyp3A, such as atorvastatin and its metabolites, that were reduced when used concomitantly with vitamin D supplements [73]. CYP3A4 inducers, such as phenobarbital and phenytoin used for seizure control, could increase clearance of vitamin D metabolites and thus lower serum 25 (OH) D levels but the data is mixed and complicated by confounding factors, such as exposure to sunlight and seasonal effects [73]. Concomitant use of thiazide diuretic and vitamin D supplementation should be avoided in the elderly with compromised renal function to avoid hypercalcemia [73].

Vitamin K

Vitamin K antagonizes the effect of warfarin. When vitamin K-containing products are taken together with warfarin, reduction in anticoagulation effect occurs [21,22]. Patients with CVD who are on therapeutic anticoagulation with warfarin to prevent thromboembolism should not take vitamin K. The major microsomal enzyme system involved in warfarin metabolism, CYP2C9, is inhibited by several drugs, including amiodarone and its major metabolite that potentiates the anticoagulant effects of warfarin, increasing the risk of serious bleeding.

DRUG–HERB INTERACTIONS IN THE ELDERLY

The use of herbal supplements is common and continues to increase in the elderly [3,110]. Although perceived as being natural and, therefore, safe, many herbal remedies could potentially have adverse effects when used with prescription medicines that can produce life-threatening consequences [18,19]. These can be easily missed because patients do not routinely disclose the use of these supplements to their healthcare providers [18]. Some of the common herb–drug interactions are summarized in Table 7.8.

DNI IN THE ELDERLY: CHALLENGES AND FUTURE DIRECTIONS

Likelihood of dietary (Fig. 7.2) and herbal supplement interactions with prescription medicine increases in the elderly with an increase in the number of comorbidities and use of multiple medications, however, information on DNI from well-controlled studies are scarce and it is difficult to define the risk for a serious interaction in an individual patient. Most studies, whether in humans, experimental animals or in vitro systems, are conducted not for characterizing adverse effects but for other purposes, such as investigating metabolism, nutrient balance, interactions, or possible benefits [111]. Few high-quality clinical trials have been conducted to determine the efficacy and safety of single-use or paired vitamins/minerals in prevention of chronic diseases that incorporate relevant clinical, genetic, pharmacologic, or physiological characteristics into trial design. Although the general principles of nutrient risk assessment have been established, enormous gaps exist in the information required to conduct a robust risk assessment for many of the nutrients and potential DNI. Dietary or MVM supplements with high nutrient content higher than dietary reference intake (DRI) recommendations contribute substantially to total nutrient intake, raising additional safety concerns with nutritional excess that further increases risk for adverse DNI. Limited regulation of dietary supplements compared to that of prescription medicine pose additional challenges regarding mislabeling, purity, toxicity, and contamination with medications with potential for adverse drug–drug interactions and DNI [112]. Systematic information on the bioavailability and bioequivalence of dietary/herbal supplements in marketed products is scarce and, when provided, not necessarily accurate [113]. Although the Food and Drug Administration (FDA) oversees its use, sufficient regulation of the supplement industry has been difficult to achieve as supplements are regulated as food products under the dietary supplement and education act and not as rigorously assessed as medications despite documented adverse effects and repeated concerns raised [18,19,114,115]. The FDA has insufficient resources and legislative authority to require specific safety data from dietary supplement manufacturers or distributors before their products are made available to the public. Reports of suspected or documented adverse events may be submitted voluntarily to the FDA's MedWatch program [116] or other organizations, such as poison control centers. Yet, adverse events from supplements are typically underreported, a problem also highlighted by the Office of the Inspector General Report [117]. Many adverse events

TABLE 7.8 Herb–Drug Interactions (List Not All-Inclusive)

Herbs	Purported health benefits	Adverse effects	Herb–drug interactions, comments
Alfalfa (*Medicago sativa*)	Hyperlipidemia, asthma, arthritis, diabetes, urinary retention, benign prostatic hypertrophy	Autoimmune hemolytic anemia, systemic lupus erythematosus exacerbation, acute graft rejection due to immune-stimulant properties	Increases bleeding risk with warfarin[a] Decreases effectiveness of oral contraceptives Hypoglycemia when used with antidiabetics Increased photosensitivity with amitriptyline, tetracycline, trimethoprim/sulfamethoxazole, levofloxacin
Aloe vera	Laxative (oral), dermatologic conditions (topical)	Abdominal cramps, diarrhea, hypokalemia, acute hepatitis	Hypokalemia causing digitalis toxicity and arrhythmia[b] Increased blood loss with anesthetic sevoflurane[c] Hypoglycemia with antidiabetic medications[d]
Bitter orange (*Citrus aurantium*)	Indigestion, anorexia, nasal congestion, weight loss	Hypertension, tachycardia, arrhythmias	Decrease effect of antihypertensive medications[e] Tachycardia and hypertension with caffeine use and MAOI Arrhythmogenesis[f,g]
Black cohosh (*Cimicifuga racemosa*)	Premenstrual tension, menopause, osteoporosis	Liver injury?	Conflicting reports with increased incidence of primary breast cancer Potential liver toxicity with statins, amiodarone, acetaminophen, methyldopa, erythromycin, fluconazole Decrease efficacy of cisplatin, CYP2D6 substrates (codeine, amitriptyline, metoprolol, trazodone) Increased lung metastasis of preexisting breast cancer[h]
Butchers broom (*Ruscus aculeatus*)	Circulatory disorders, venous insufficiency, constipation, leg cramps, water retention, hemorrhoids (topical)	Nausea, high blood pressure	Decreases effects of alpha-blockers on blood pressure[i]
Cannabis (*Cannabis sativa* L.)	Nausea, vomiting, anorexia, cachexia, neuropathic pain, asthma, glaucoma	Central nervous system depressant	Potentiates central nervous system depressant effects with alcohol, benzodiazepines, narcotics, and antihistamine use
		Altered patient's response to stress in patients undergoing oral surgery	Increases bleeding risk with warfarin[j] Decreases phenytoin concentration by CYP2C9 expression and increase hydroxylation of liver microsomes[k] Causes manic symptoms with fluoxetine use[l] Reduced blood level of protease inhibitors: indinavir and nelfinavir[m] Alters response to stress and anesthetic medications in oral surgery patients[n]
Capsicum (peppers or bell peppers)	Analgesic, shingles, antidiabetic, trigeminal neuralgia, anti-inflammatory, weight loss	Hypertension, myocardial infarction, coronary vasospasm	Hypertensive crisis with MAOI

Cat's claw (*Uncaria tomentosa*)	Antioxidant, immunomodulation, antiherpetic, anti-inflammatory, immunostimulant	Nausea, vomiting, gastrointestinal discomfort, reversible worsening of Parkinson's disease motor symptoms, allergic interstitial nephritis	Increases bleeding risk with warfarin,[o] ginkgo biloba, garlic, saw palmetto
			Potentiates hypotension with antihypertensive drugs
			Potentiates the effect of benzodiazepines[p]
			Increased plasma concentration of protease inhibitors atazanavir, ritonavir, and saquinavir by inhibition of CYP3A4[q]
			Disulfuram-like reaction when used with metronidazole
Chamomile (*Matricaria recutita*)	Anxiolytic, sedative, anti-inflammatory, spasmolytic, upper respiratory tract infections and premenstrual syndrome		Increases bleeding risk with aspirin or warfarin use[r]
			Increases sedative effects of benzodiazepines and barbiturates[s]
Co-Enzyme Q10 (*Theobroma cacao*)	Antioxidant, used in heart failure, angina, hypertension, and preeclampsia	Increased breast cancer risk in postmenopausal women, insomnia, elevated liver enzymes, rash, nausea, abdominal pain dizziness, photosensitivity, headache, irritability, heartburn, fatigue	Reduces anticoagulant effect of warfarin[t]
Cranberry (*Vaccinium macrocarpon*)	urinary tract infection, diabetes, wounds, *Helicobacter pylori*–induced gastric ulcers, blood and digestive disorders	Potential inhibitory effect on CYP2C9 and CYP3A4 (Tables 7.2A and 7.2B)	Increases bleeding risk with warfarin[u]
Danshen (*Salvia miltiorrhiza*)	Ischemic stroke, angina, myocardial infarction		Increases bleeding risk with warfarin use[v]
			Falsely elevates or lowers serum digoxin levels[w]
			Increases metabolism of CYP substrates (see Table 7.2B)
Devil's claw (*Harpagophytum procumbens*)	Osteoarthritis, myalgia, back pain, atherosclerosis, anorexia, palpitations, difficult childbirth, dyspepsia, bladder and kidney problems	Decreases heart rate and blood pressure	Increases bleeding risk of warfarin[x]
			Inhibits CYP 3A4 and elevates blood levels of drugs summarized in Tables 7.2A and 7.2B
			Inhibits P-glycoprotein activity and can increase blood levels of drugs summarized in Table 7.3[y]
Echinacea purpurea	Immunostimulant, upper respiratory tract infections but efficacy studies have yielded conflicting results	Nausea, dizziness, abdominal discomfort, liver toxicity, thrombotic thrombocytopenic purpura, rash, and anaphylaxis	May induce or inhibit hepatic CYP 3A4 enzymes depending on structure and condition. Inhibition of CYP3A4 enzymes may elevate blood levels of drugs listed in Tables 7.2A and 7.2B[z,aa]
			Inhibits P-glycoprotein activity and may elevate blood levels of drugs listed in Table 7.3[y]
			Increases QT interval with amiodarone or ibutilide
			Increases risk of hepatotoxic effects with statins, fibrates, niacin

(*Continued*)

TABLE 7.8 (Continued)

Herbs	Purported health benefits	Adverse effects	Herb–drug interactions, comments
Ephedra (ma huang)	Weight loss, nasal decongestant, increased alertness boosts athletic performance, common cold/flu, asthma, headache	Increase heart rate, increased blood pressure, fainting, palpitations, insomnia, flushing, headache, motor restlessness, tremors, dry mouth, blurred vision, anorexia, decreased appetite, secondary rhabdomyolysis, renal failure, psychosis, seizure, skin rash, vasculitis, stroke, cardiomyopathy, myocardial infarction, and sudden death	Avoid with caffeinated products. Avoid with drugs causing QT prolongation, antiarrhythmics, and diuretics[ab]
Fenugreek (Trigonella foenum-graecum)	Hyperlipidemia, indigestion, menopausal symptoms, labor induction, appetite stimulant, lactation, diabetes wound and leg ulcers (topical)	Dyspepsia, abdominal distension, hypoglycemia, hypokalemia	Avoid with diuretics; can potentiate hypokalemia[ac]
Garlic (Allium sativum)	Hypertension, hypercholesterolemia, heart disease, hyperglycemia, antimicrobial, immune enhancing effects, HIV, prevention of stomach and colon cancers	Postoperative bleeding, spontaneous spinal epidural hematoma and other bleeding reported	Increases bleeding risk with warfarin[ad] and antiplatelet drugs[ae,af] Hypoglycemia with oral hypoglycemic (chlorpropamide)[ag] Severe GI side effects with ritonavir
Ginger (Zingiber officinale Roscoe)	High cholesterol, motion sickness, chemotherapy-induced nausea, indigestion, arthritis, muscular pain, antispasmodic		Increased risk of bleeding with warfarin[ah]
Ginkgo biloba	Alzheimer's dementia, tinnitus, peripheral arterial disease, and senile macular degeneration	Decrease in blood pressure, potential risk of seizures	Increases risk of bleeding with warfarin[ai], aspirin[aj], and nonsteroidal[ak,al] Increases effects of digoxin[am] Reduces efficacy of antiviral agents (efavirenz)[an] Can cause priapism with risperidone[ao] Worsening of seizure with valproic acid and phenytoin[ap]
Ginseng	Stress reduction, immunomodulation, memory enhancement, attention deficit hyperactivity disorder, sexual dysfunction, alleviation of menopausal symptoms, antidiabetic, antiaging, antifatigue, anticancer	Hepatotoxicity, abnormal vaginal bleeding, gynecomastia, hypertension, mastalgia, manic episode, rapid heartbeat, allergic reactions	Decreases anticoagulation effect of warfarin[aq] Can potentiate arrhythmogenesis with antiarrhythmic agents[ar] Interferes with digoxin assay, leading to falsely elevated levels (Siberian ginseng)[as] Hypoglycemia with oral hypoglycemics, eg, chlorpropamide[at] and herbs lowering blood sugar level (Devil's claw, ginger, fenugreek, gum, guar) Can induce liver enzymes and inhibit CYP3A4.[au] Use with caution in CYP metabolized drugs (Tables 7.2A and 7.2B), hepatitis, or cirrhosis Hepatotoxicity with antiretroviral raltegravir[av] and imattinib mesylate[aw]

Grapefruit juice	Diabetes, CVD, anticancer	Increase risk of bleeding with anticoagulant or antiplatelet agents and can potentiate side effects of drugs metabolized by CYP3A4 (Table 7.2B)
		Increases bleeding risk with warfarin[ax] or antiplatelet (cilostazol) leading to purpura[ay]
		Inhibits CYP3A4 and p-glycoprotein activity elevating serum levels of drugs metabolized in Tables 7.2A–7.3 such as tachycardia, flushing, and hypotension with calcium channel blockers[az], rhabdomyolysis with statins, hypertension and tachycardia with appetite suppressant sibutramine, and hypoglycemia with oral hypoglycemic agent repaglinide
		Potentiates toxicity QT prolongation and Torsades de pointes[ba]
		Causes hypotension with sildanefil, tadalafil or varendafil or nitrates
Hawthorn (*Crataegus* spp.)	Hypertension, coronary heart failure, atherosclerosis, angina pectoris, bradyarrhythmias	Hypotension, tachycardia, arrhythmias
		Increases effects of digoxin[bb]
		Potentiates the effects of vasodilators or drugs used for heart failure, hypertension, angina, and arrhythmias[bc]
Kava (*Piper methysticum*)	Stress, anxiety, sleeping aid	Hepatitis and liver failure
		Potentiates the effects of benzodiazepines causing dystonia
		Interacts with drugs used for Parkinson's disease
		Sedatives (alprazolam, lorazepam, clonazepam) increase central nervous system side effects
		To be avoided with hepatotoxic drugs (acetaminophen, amiodarone, carbamazepine, isoniazid, methotrexate, methyldopa, erythromycin, phenytoin, statins)
Kelp (*Laminaria*)	Hypothyroidism, goiter, cancer, obesity, immune-stimulation, osteoarthritis, indigestion, upper respiratory tract infection, genitourinary infection, promotion of healthy growth of hair, skin, and nails	Uterine stimulant: can dilate the cervix and induce abortion
		Can increase effects of bleeding with warfarin and antiplatelet drugs[bd]
		Due to high iodine content can potentiate action of levothyroxine causing tachycardia, tremors, acne, and hypertension[be]
		Avoid with potassium supplements
Licorice (*Glycyrrhiza glabra*)	Peptic ulcer, catarrhs of upper respiratory tract, hepatitis, cirrhosis, polycystic ovarian syndrome, prostate cancer, eczema	Mineralocorticoid effect, hypertension, hypokalemia, life-threatening ventricular arrhythmias, cardiac arrest, myoglobinuria, tetraparesis, can worsen coronary heart failure (causes sodium retention, edema, and hypertension)
		Decreases anticoagulant effect of warfarin
		Use with caution with drugs that can cause hypokalemia (eg, diuretics)
		Can potentiate digoxin toxicity and arrhythmogenesis[bf]
		Increases plasma prednisolone concentration[bg]
		Increases effects of spironolactone
		Use with caution in hormone-sensitive conditions (breast, uterine, and ovarian cancer, endometriosis and uterine fibroids)

(*Continued*)

TABLE 7.8 (Continued)

Herbs	Purported health benefits	Adverse effects	Herb–drug interactions, comments
Peppermint (*Mentha piperita*)	Aroma, cold, pain (headache, muscle ache, nerve pain), digestive disorders, antitussive, spasmolytic, antioxidant, antimicrobial, anticarcinogenic, radioprotective, analgesic, and anesthetic effects	Heart burn, allergies	Increases the levels of CYP3A4 metabolized drugs (Tables 7.2A and 7.2B) such as felodipine[bh]
			Caution is advised in its co-use with CYP3A4 metabolized drugs
			Caution is also urged for peppermint oil therapy in patients with GI reflux, hiatal hernia, or kidney stones[bi]
Red clover (*Trifolium pratense*)	High cholesterol, menstrual irregularities, mastalgia, osteoporosis, prostatic enlargement, depression, anxiety, favorable influence on markers of cardiovascular function (systemic arterial compliance, endothelial function, and serum HDL cholesterol)	May alter blood clotting and increase risk of surgical bleeding	Increases bleeding risk with warfarin[bj,bk]
			Can induce CYP enzymes. Use with caution with CYP metabolized drugs (Tables 7.2A and 7.2B)[x]
			Due to estrogen-like compounds; avoid red clover use in hormone sensitive conditions (breast cancer, endometrial cancer, ovarian cancer, uterine fibroids, and endometriosis)
Red yeast rice (*Monascus purpureus*)	Hyperlipidemia	Myopathy, rhabdomyolysis, liver damage	Potentiates effects of statins including side effects
Saw palmetto (*Serenoa repens*)	Benign prostatic hyperplasia, genitourinary conditions, azospermia, decreased libido, augmentation of breast size, edema, mild diuretic	Hypertension, fatigue, sexual dysfunction, tachycardia, angina pectoris, extrasystole, intraoperative hemorrhage, bleeding susceptibility	May increase bleeding risk with warfarin or antiplatelet agents[bl]
Soya (*Glycine max* Merr.)	Hyperlipidemia, hot flushes, osteoporosis	Increased risk of endometrial hyperplasia	Should be avoided with contraceptives and estrogens as it decreases effectiveness of estrogen pills
			Decreases anticoagulant effects of warfarin[bm] and other anticoagulants[bm]
			Decreases absorption of levothyroxine[bo]
			Inhibits CYP3A4 and CYP 2C metabolized drugs (Tables 7.2A and 7.2B)[bo,bp]
St. John's wort (*Hypericum perforatum*)	Depression, anxiety, sleep disturbances, mental health conditions	Reversible photosensitivity, induction of mania or hypomania in patients with bipolar disorder, hypertensive crisis (case report)	Increases heart rate and blood pressure with MAOI
			Decreases anticoagulant effects of warfarin[bq]
			Increases activity of clopidogrel with increased bleeding tendency
			Induction of CYP3A4 reduces blood level of drugs metabolized (Tables 7.2A and 7.2B) and therefore decreases effectiveness of statins, antiarrhythmics (precipitating arrhythmias)
			Decreases cyclosporine concentration due to increased clearance (transplant rejection)[br,bs,bt,bu,bv]
			Decreases plasma levels of indinavir (treatment failure in HIV patients)[bw]

| Yohimbine | Aphrodisiac, impotence, erectile dysfunction, orthostatic hypotension | Lightheadedness, confusion, incoordination, submucosal hemorrhage, retrograde amnesia, incoordination, systemic lupus erythematosus-like syndrome, renal failure, exacerbation of panic attacks in post traumatic stress disorder, chest pain, and arrhythmias | Decreases drug levels of ethinyl estradiol (increase breakthrough bleeding and unplanned pregnancies in women taking oral contraceptive pills)[bx] |

Increases bleeding with warfarin, antiplatelet drugs[by] or herbs (angelica, clove, danshen, garlic, ginkgo biloba, ginger)

Elevated blood pressure and tachycardia when used with caffeine, central nervous system stimulants (ephedra), or tyramine-containing products (MAOI)

Decreases blood pressure reduction effect of centrally acting agents (clonidine, guanabenz), ACE inhibitors (hypertension), and beta-blockers (hypertension and increase in heart rate)

[a]Newall CAL, Phillipson J. Herbal medicines: a guide for health-care professionals. London: Pharmaceutical Press; 1996.
[b]Baretta Z, Ghiotto C, Marino D, Jirillo A. Aloe-induced hypokalemia in a patient with breast cancer during chemotherapy. Ann Oncol 2009;20(8):1445–6.
[c]Lee A, Chui PT, Aun CS, Gin T, Lau AS. Possible interaction between sevoflurane and Aloe vera. Ann Pharmacother 2004;38(10):1651–4.
[d]Huseini HF, Kianbakht S, Hajiaghaee R, Dabaghian FH. Antihyperglycemic and antihypercholesterolemic effects of Aloe vera leaf gel in hyperlipidemic type 2 diabetic patients: a randomized double-blind placebo-controlled clinical trial. Planta Med 2012;78(4):311–6.
[e]Bent S, Padula A, Neuhaus J. Safety and efficacy of Citrus aurantium for weight loss. Am J Cardiol 2004;94(10):1359–61.
[f]Nazeri A, Massumi A, Wilson JM, Frank CM, Bensler M, Cheng J, et al. Arrhythmogenicity of weight-loss supplements marketed on the Internet. Heart Rhythm 2009;6(5):658–62.
[g]Jordan S, Murty M, Pilon K. Products containing bitter orange or synephrine: suspected cardiovascular adverse reactions. CMAJ 2004;171(8):993–4.
[h]Davis VL, Jayo MJ, Ho A, Kotlarczyk MP, Hardy ML, Foster WG, et al. Black cohosh increases metastatic mammary cancer in transgenic mice expressing c-erbB2. Cancer Res 2008;68(20):8377–83.
[i]Rubanyi G, Marcelon G, Vanhoutte PM. Effect of temperature on the responsiveness of cutaneous veins to the extract of Ruscus aculeatus. Gen Pharmacol 1984;15(5):431–4.
[j]Yamreudeewong W, Wong HK, Brausch LM, Pulley KR. Probable interaction between warfarin and marijuana smoking. Ann Pharmacother 2009;43(7):1347–53.
[k]Bland TM, Haining RL, Tracy TS, Callery PS. CYP2C-catalyzed delta9-tetrahydrocannabinol metabolism: kinetics, pharmacogenetics and interaction with phenytoin. Biochem Pharmacol 2005;70(7):1096–103.
[l]Stoll AL, Cole JO, Lukas SE. A case of mania as a result of fluoxetine-marijuana interaction. J Clin Psychiatry 1991;52(6):280–1.
[m]Kosel BW, Aweeka FT, Benowitz NL, Shade SB, Hilton JF, Lizak PS, et al. The effects of cannabinoids on the pharmacokinetics of indinavir and nelfinavir. Aids 2002;16(4):543–50.
[n]Gregg JM, Campbell RL, Levin KJ, Ghia J, Elliott RA. Cardiovascular effects of cannabinol during oral surgery. Anesth Analg 1976;55(2):203–13.
[o]Shi JS, Yu JX, Chen XP, Xu RX. Pharmacological actions of Uncaria alkaloids, rhynchophylline and isorhynchophylline. Acta Pharmacol Sin 2003;24(2):97–101.
[p]Quilez AM, Saenz MT, Garcia MD. Uncaria tomentosa (Willd. ex. Roem. & Schult.) DC and Eucalyptus globulus Labill. interactions when administered with diazepam. Phytother Res 2012;26(3):458–61.
[q]Lopez Galera RM, Ribera Pascuet E, Esteban Mur JI, Montoro Ronsano JB, Juarez Gimenez JC. Interaction between cat's claw and protease inhibitors atazanavir, ritonavir and saquinavir. Eur J Clin Pharmacol 2008;64(12):1235–6.
[r]Segal R, Pilote L. Warfarin interaction with Matricaria chamomilla. CMAJ. 2006;174(9):1281–2.
[s]O'Hara M, Kiefer D, Farrell K, Kemper K. A review of 12 commonly used medicinal herbs. Arch Fam Med 1998;7(6):523–36.
[t]Mousa SA. Antithrombotic effects of naturally derived products on coagulation and platelet function. Methods Mol Biol 2010;663:229–40.
[u]Griffiths AP, Beddall A, Pegler S. Fatal haemopericardium and gastrointestinal haemorrhage due to possible interaction of cranberry juice with warfarin. J R Soc Promot Health 2008;128(6):324–6.
[v]Izzat MB, Yim AP, El-Zufari MH. A taste of Chinese medicine! Ann Thorac Surg 1998;66(3):941–2.
[w]Wahed A, Dasgupta A. Positive and negative in vitro interference of Chinese medicine dan shen in serum digoxin measurement. Elimination of interference by monitoring free digoxin concentration. Am J Clin Pathol 2001;116(3):403–8.
[x]Unger M, Frank A. Simultaneous determination of the inhibitory potency of herbal extracts on the activity of six major cytochrome P450 enzymes using liquid chromatography/mass spectrometry and automated online extraction. Rapid Commun Mass Spectrom 2004;18(19):2273–81.
[y]Romiti N, Tramonti G, Corti A, Chieli E. Effects of devil's claw (Harpagophytum procumbens) on the multidrug transporter ABCB1/P-glycoprotein. Phytomedicine 2009;16(12):1095–100.
[z]Gorski JC, Huang SM, Pinto A, Hamman MA, Hilligoss JK, Zaheer NA, et al. The effect of echinacea (Echinacea purpurea root) on cytochrome P450 activity in vivo. Clin Pharmacol Ther 2004;75(1):89–100.
[aa]Tachjian A, Maria V, Jahangir A. Use of herbal products and potential interactions in patients with cardiovascular diseases. J Am Coll Cardiol 2010;55(6):515–25.
[ab]Dresser GK, Spence JD, Bailey DG. Pharmacokinetic-pharmacodynamic consequences and clinical relevance of cytochrome P450 3A4 inhibition. Clin Pharmacokinet 2000;38(1):41–57.
[ac]<https://naturalmedicines.therapeuticresearch.com>
[ad]Evans V. Herbs and the brain: friend or foe? The effects of ginkgo and garlic on warfarin use. J Neurosci Nurs 2000;32(4):229–32.
[ae]Rahman K, Billington D. Dietary supplementation with aged garlic extract inhibits ADP-induced platelet aggregation in humans. J Nutr 2000;130(11):2662–5.
[af]Steiner M, Li W. Aged garlic extract, a modulator of cardiovascular risk factors: a dose-finding study on the effects of AGE on platelet functions. J Nutr 2001;131(3s):980s–4s.
[ag]Aslam M, Stockley IH. Interaction between curry ingredient (karela) and drug (chlorpropamide). Lancet 1979;1(8116):607.
[ah]Samuels N. Herbal remedies and anticoagulant therapy. Thromb Haemost 2005;93(1):3–7.

(Continued)

TABLE 7.8 (Continued)

[al]Matthews MK, Jr. Association of Ginkgo biloba with intracerebral hemorrhage. Neurology 1998;50(6):1933–4.
[aj]Rosenblatt M, Mindel J. Spontaneous hyphema associated with ingestion of Ginkgo biloba extract. N Engl J Med 1997;336(15):1108.
[ak]Meisel C, Johne A, Roots I. Fatal intracerebral mass bleeding associated with Ginkgo biloba and ibuprofen. Atherosclerosis 2003;167(2):367.
[al]Kudolo GB, Wang W, Barrientos J, Elrod R, Blodgett J. The ingestion of Ginkgo biloba extract (EGb 761) inhibits arachidonic acid-mediated platelet aggregation and thromboxane B2 production in healthy volunteers. J Herb Pharmacother 2004;4(4):13–26.
[am]Mauro VF, Mauro LS, Kleshinski JF, Khuder SA, Wang Y, Erhardt PW. Impact of Ginkgo biloba on the pharmacokinetics of digoxin. Am J Ther 2003;10(4):247–51.
[an]Wiegman DJ, Brinkman K, Franssen EJ. Interaction of Ginkgo biloba with efavirenz. Aids 2009;23(9):1184–5.
[ao]Lin YY, Chu SJ, Tsai SH. Association between priapism and concurrent use of risperidone and Ginkgo biloba. Mayo Clin Proc 2007;82(10):1289–90.
[ap]Kupiec T, Raj V. Fatal seizures due to potential herb-drug interactions with Ginkgo biloba. J Anal Toxicol 2005;29(7):755–8.
[aq]Ang-Lee MK, Moss J, Yuan CS. Herbal medicines and perioperative care. JAMA 2001;286(2):208–16.
[ar]Etheridge AS, Black SR, Patel PR, So J, Matheus JM. An in vitro evaluation of cytochrome P450 inhibition and P-glycoprotein interaction with goldenseal, Ginkgo biloba, grape seed, milk thistle, and ginseng extracts and their constituents. Planta Med 2007;73(8):731–41.
[as]McRae S. Elevated serum digoxin levels in a patient taking digoxin and Siberian ginseng. CMAJ. 1996;155(3):293–5.
[at]Izzo AA, Ernst E. Interactions between herbal medicines and prescribed drugs: a systematic review. Drugs 2001;61(15):2163–75.
[au]Christensen LP. Ginsenosides chemistry, biosynthesis, analysis, and potential health effects. Adv Food Nutr Res 2009;55:1–99.
[av]Mateo-Carrasco H, Galvez-Contreras MC, Fernandez-Gines FD, Nguyen TV. Elevated liver enzymes resulting from an interaction between Raltegravir and Panax ginseng: a case report and brief review. Drug Metabol Drug Interact 2012;27(3):171–5.
[aw]Bilgi N, Bell K, Ananthakrishnan AN, Atallah E. Imatinib and Panax ginseng: a potential interaction resulting in liver toxicity. Am Pharmacother 2010;44(5):926–8.
[ax]Brandin H, Myrberg O, Rundlof T, Arvidsson AK, Brenning G. Adverse effects by artificial grapefruit seed extract products in patients on warfarin therapy. Eur J Clin Pharmacol 2007;63(6):565–70.
[ay]Taniguchi K, Ohtani H, Ikemoto T, Miki A, Hori S, Sawada Y. Possible case of potentiation of the antiplatelet effect of cilostazol by grapefruit juice. J Clin Pharm Ther 2007;32(5):457–9.
[az]Tomlinson B, Chow MS. Stereoselective interaction of manidipine and grapefruit juice: a new twist on an old tale. Br J Clin Pharmacol 2006;61(5):529–32.
[ba]Saito M, Hirata-Koizumi M, Matsumoto M, Urano T, Hasegawa R. Undesirable effects of citrus juice on the pharmacokinetics of drugs: focus on recent studies. Drug Saf 2005;28(8):677–94.
[bb]Chang Q, Zuo Z, Harrison F, Chow MS. Hawthorn. J Clin Pharmacol 2002;42(6):605–12.
[bc]Rigelsky JM, Sweet BV. Hawthorn: pharmacology and therapeutic uses. Am J Health Syst Pharm 2002;59(5):417–22.
[bd]Grauffel V, Kloareg B, Mabeau S, Durand P, Jozefonvicz J. New natural polysaccharides with potent antithrombic activity: fucans from brown algae. Biomaterials 1989;10(6):363–8.
[be]Shilo S, Hirsch HJ. Iodine-induced hyperthyroidism in a patient with a normal thyroid gland. Postgrad Med 1986;62(729):661–2.
[bf]Shintani S, Murase H, Tsukagoshi H, Shiigai T. Glycyrrhizin (licorice)-induced hypokalemic myopathy. Report of 2 cases and review of the literature. Eur Neurol 1992;32(1):44–51.
[bg]Homma M, Oka K, Ikeshima K, Takahashi N, Niitsuma T, Fukuda T, et al. Different effects of traditional Chinese medicines containing similar herbal constituents on prednisolone pharmacokinetics. J Pharm Pharmacol 1995;47(6):687–92.
[bh]Dresser GK, Wacher V, Wong S, Wong HT, Bailey DG. Evaluation of peppermint oil and ascorbyl palmitate as inhibitors of cytochrome P4503A4 activity in vitro and in vivo. Clin Pharmacol Ther 2002;72(3):247–55.
[bi]McKay DL, Blumberg JB. A review of the bioactivity and potential health benefits of peppermint tea (Mentha piperita L.). Phytother Res 2006;20(8):619–33.
[bj]Heck AM, DeWitt BA, Lukes AL. Potential interactions between alternative therapies and warfarin. Am J Health Syst Pharm 2000;57(13):1221–7; quiz 8–30.
[bk]Friedman JA, Taylor SA, McDermott W, Alikhani P. Multifocal and recurrent subarachnoid hemorrhage due to an herbal supplement containing natural coumarins. Neurocrit Care 2007;7(1):76–80.
[bl]Yue QY, Jansson K. Herbal drug curbicin and anticoagulant effect with and without warfarin: possibly related to the vitamin E component. J Am Geriatr Soc 2001;49(6):838.
[bm]Schurgers LJ, Shearer MJ, Hamulyak K, Stocklin E, Vermeer C. Effect of vitamin K intake on the stability of oral anticoagulant treatment: dose-response relationships in healthy subjects. Blood 2004;104(9):2682–9.
[bn]Cambria-Kiely JA. Effect of soy milk on warfarin efficacy. Ann Pharmacother 2002;36(12):1893–6.
[bo]Belisle S, Blake J, Basson R, Desindes S, Graves G, Grigoriadis S, et al. Canadian consensus conference on menopause, 2006 update. J Obstet Gynaecol Can 2006;28(2 Suppl. 1):S7–94.
[bp]Foster BC, Vandenhoek S, Hana J, Krantis A, Akhtar MH, Bryan M, et al. In vitro inhibition of human cytochrome P450-mediated metabolism of marker substrates by natural products. Phytomedicine 2003;10(4):334–42.
[bq]Anderson GD, Rosito G, Mohustsy MA, Elmer GW. Drug interaction potential of soy extract and Panax ginseng. J Clin Pharmacol 2003;43(6):643–8.
[br]Jiang X, Williams KM, Liauw WS, Ammit AJ, Roufogalis BD, Duke CC, et al. Effect of St John's wort and ginseng on the pharmacokinetics and pharmacodynamics of warfarin in healthy subjects. Br J Clin Pharmacol 2004;57(5):592–9.
[bs]Breidenbach T, Hoffmann MW, Becker T, Schlitt H, Klempnauer J. Drug interaction of St John's wort with cyclosporin. Lancet 2000;355(9218):1912.
[bt]Ernst E. St John's wort supplements endanger the success of organ transplantation. Arch Surg 2002;137(3):316–9.
[bu]Bauer S, Stormer E, Johne A, Kruger H, Budde K, Neumayer HH, et al. Alterations in cyclosporin A pharmacokinetics and metabolism during treatment with St John's wort in renal transplant patients. Br J Clin Pharmacol 2003;55(2):203–11.
[bv]Mills E, Wu P, Johnston BC, Gallicano K, Clarke M, Guyatt G. Natural health product-drug interactions: a systematic review of clinical trials. Ther Drug Monit 2005;27(5):549–57.
[bw]Piscitelli SC, Burstein AH, Chaitt D, Alfaro RM, Falloon J. Indinavir concentrations and St John's wort. Lancet 2000;355(9203):547–8.
[bx]Yue QY, Bergquist C, Gerden B. Safety of St John's wort (Hypericum perforatum). Lancet 2000;355(9203):576–7.
[by]Berlin I, Crespo-Laumonnier B, Cournot A, Landault C, Aubin F, Legrand JC, et al. The alpha 2-adrenergic receptor antagonist yohimbine inhibits epinephrine-induced platelet aggregation in healthy subjects. Clin Pharmacol Ther 1991;49(4):362–9.

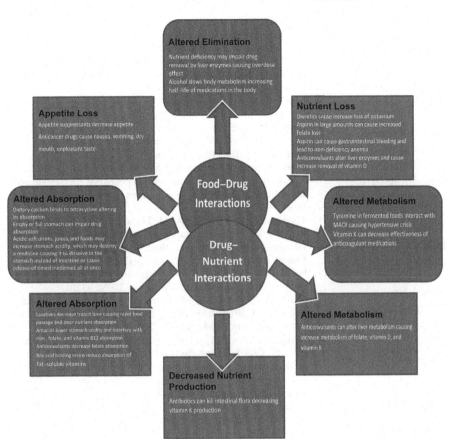

FIGURE 7.2 Dietary nutrient–drug interactions. *Source: Adapted from Oklahoma Cooperative Extension Fact Sheets http://osufacts.okstate.edu and http://pods.dasnr.okstate.edu/docushare/dsweb/Get/Document-2458/T-3120web-2014.pdf.*

related to excess nutrition use, drug–nutrition, or drug–herb interaction go unreported to the FDA because consumers are more likely to associate an adverse event with a drug they are taking rather than a dietary supplement that they consider inherently safe or "natural" and could be embarrassed to tell their healthcare provider that they were self-treating a medical condition using a dietary or herbal supplement.

In order to reduce the risk for adverse DNI, a detailed assessment of nutritional status, medication use and potential for DNI should be performed routinely, even more so in the elderly population with chronic health conditions who take multiple medications and are at greater risk than the general healthy population for adverse DNI. Ensuring proper timing of food with drug regimens and regular monitoring of nutritional status, medication profiles, and laboratory data to evaluate the potential for interactions may help reduce serious adverse DNI [118]. Maintaining a healthy balanced diet rich in vegetables and fruits, whole grains, fat-free and low-fat dairy products, and seafood, and that is low in saturated fats and added sugars as specified by the 2015 dietary guidelines advisory committee may help maintain health in the elderly [119]. Supplements, if needed, should be used sparingly, best decided in consultation with healthcare professionals, who should help patients decide dietary supplement use after considering the risk-benefit ratio, particularly in those on polypharmacy. Strategies to address DNI in long-term care facilities [118], emphasis on education in assessing the need for supplements and concerns associated with their use with medications, and proper coaching of the patient to take an active role in his or her care and communication with healthcare providers may help reduce unnecessary inconvenience and costs associated with preventable adverse DNI in the elderly [120]. Additionally, inclusion of older-elderly patients in clinical trials and nutrition–drug interaction research in geriatric patients will help further improve understanding of mechanisms underlying adverse DNI to design further strategies to prevent adverse DNI in the elderly.

SUMMARY

- Dietary supplement use is common and increasing in the elderly with MVM being the most common supplements consumed.
- Three-fifths of the patients using supplements do not disclose their use to healthcare professionals,

- increasing the risk for potential interactions between supplements and medications to go undetected.
- Elderly patients with comorbidities using multiple medications and dietary supplements are at increased risk of adverse interactions, which could be clinically serious with drugs that have a narrow therapeutic index.
- Factors contributing to DNI include both patient-related and drug–nutrient-related factors and increasing recognition of risk factors particularly in elderly with chronic disease who use multiple drugs require closer monitoring of DNI particularly with drugs with narrow therapeutic index to help reduce adverse DNI.
- Improving regulatory vigilance of dietary supplements manufacturing and distribution and consumers and healthcare providers' education about DNI and its proper reporting may help strategies to reduce adverse events that otherwise go undetected.
- Detailed assessment of nutritional status, medication use and potential for DNI should be performed routinely in the elderly with chronic health conditions who take multiple medications and are at greater risk than the general healthy population for adverse DNI.
- Ensuring proper timing of food with drug regimens, and regular monitoring of nutritional status, medication profiles, and laboratory data to evaluate the potential for interactions may help reduce serious adverse DNI in high risk individuals.
- Inclusion of older-elderly patients in clinical trials and collection of data regarding DNI may help fill gaps in our knowledge and increase understanding of mechanisms underlying adverse DNI so strategies to prevent adverse DNI in the elderly can be further improved.

References

[1] Ervin RB, Wright JD, Reed-Gillette D. Prevalence of leading types of dietary supplements used in the third national health and nutrition examination survey, 1988–94. Adv Data 2004;(349):1–7.

[2] National Institutes of Health State-of-the-Science Panel. National Institutes of Health State-of-the-Science Conference Statement: multivitamin/mineral supplements and chronic disease prevention. Am J Clin Nutr 2007;85(1):257S–64S.

[3] Nahin RL, Barnes PM, Stussman BJ, Bloom B. Costs of complementary and alternative medicine (CAM) and frequency of visits to CAM practitioners: United States, 2007. Natl Health Stat Report 2009;(18):1–14.

[4] Bailey RL, Gahche JJ, Miller PE, Thomas PR, Dwyer JT. Why US adults use dietary supplements. JAMA Intern Med 2013;173(5):355–61.

[5] Rock CL. Multivitamin-multimineral supplements: who uses them? Am J Clin Nutr 2007;85(1):277S–9S.

[6] Bailey RL, Gahche JJ, Lentino CV, Dwyer JT, Engel JS, Thomas PR, et al. Dietary supplement use in the United States, 2003–2006. J Nutr 2011;141(2):261–6.

[7] Chen LH, Liu S, Newell ME, Barnes K. Survey of drug use by the elderly and possible impact of drugs on nutritional status. Drug Nutr Interact 1985;3(2):73–86.

[8] Samaras D, Samaras N, Lang PO, Genton L, Frangos E, Pichard C. Effects of widely used drugs on micronutrients: a story rarely told. Nutrition 2013;29(4):605–10.

[9] Carlson JE. Perils of polypharmacy: 10 steps to prudent prescribing. Geriatrics 1996;51(7): 26–30, 35

[10] Sood A, Sood R, Brinker FJ, Mann R, Loehrer LL, Wahner-Roedler DL. Potential for interactions between dietary supplements and prescription medications. Am J Med 2008;121(3):207–11.

[11] Thomson P, Jones J, Evans JM, Leslie SL. Factors influencing the use of complementary and alternative medicine and whether patients inform their primary care physician. Complement Ther Med 2012;20(1–2):45–53.

[12] Santos CA, Boullata JI. An approach to evaluating drug-nutrient interactions. Pharmacotherapy 2005;25(12):1789–800.

[13] Gardiner P, Phillips R, Shaughnessy AF. Herbal and dietary supplement–drug interactions in patients with chronic illnesses. Am Fam Physician 2008;77(1):73–8.

[14] Mirtallo J, Canada T, Johnson D, Kumpf V, Petersen C, Sacks G, et al. Task force for the revision of safe practices for parenteral nutrition. Safe practices for parenteral nutrition. JPEN J Parenter Enteral Nutr 2004;28(6):S39–70. Erratum in: JPEN J Parenter Enteral Nutr 2006;30(2):177.

[15] Boullata JI, Hudson LM. Drug-nutrient interactions: a broad view with implications for practice. J Acad Nutr Diet 2012;112(4):506–17.

[16] Hanley MJ, Cancalon P, Widmer WW, Greenblatt DJ. The effect of grapefruit juice on drug disposition. Expert Opin Drug Metab Toxicol 2011;7(3):267–86.

[17] Custodio JM, Wu CY, Benet LZ. Predicting drug disposition, absorption/elimination/transporter interplay and the role of food on drug absorption. Adv Drug Deliv Rev 2008;60(6):717–33.

[18] Tachjian A, Maria V, Jahangir A. Use of herbal products and potential interactions in patients with cardiovascular diseases. J Am Coll Cardiol 2010;55(6):515–25.

[19] Jahangir A, Viqar M, Tachjian A. Use of herbal products and potential interactions in patients with cardiovascular diseases. J Am Coll Cardiol 2010;56(11):904–5 author reply 905–9

[20] Neuhofel AL, Wilton JH, Victory JM, Hejmanowsk LG, Amsden GW. Lack of bioequivalence of ciprofloxacin when administered with calcium-fortified orange juice: a new twist on an old interaction. J Clin Pharmacol 2002;42(4):461–6.

[21] Rohde LE, de Assis MC, Rabelo ER. Dietary vitamin K intake and anticoagulation in elderly patients. Curr Opin Clin Nutr Metab Care 2007;10(1):1–5.

[22] Strunets A, Mirza M, Sra J, Jahangir A. Novel anticoagulants for stroke prevention in atrial fibrillation: safety issues in the elderly. Expert Rev Clin Pharmacol 2013;6(6):677–89.

[23] Kurnik D, Lubetsky A, Loebstein R, Almog S, Halkin H. Multivitamin supplements may affect warfarin anticoagulation in susceptible patients. Ann Pharmacother 2003;37(11):1603–6.

[24] Juurlink DN, Mamdani MM, Kopp A, Rochon PA, Shulman KI, Redelmeier DA. Drug-induced lithium toxicity in the elderly: a population-based study. J Am Geriatr Soc 2004;52(5):794–8.

[25] FDA. Steam dietary supplement. <http://www.fda.gov/Safety/MedWatch/SafetyInformation/SafetyAlertsforHumanMedicalProducts/ucm174339.htm>; [accessed 25.06.15].

[26] FDA. FDA warns consumers to avoid red yeast rice products promoted on internet as treatments for high cholesterol products found to contain unauthorized drug. <http://www.fda.gov/NewsEvents/Newsroom/PressAnnouncements/2007/ucm108962.htm>; [accessed 25.06.15].

[27] Navarro M, Wood RJ. Plasma changes in micronutrients following a multivitamin and mineral supplement in healthy adults. J Am Coll Nutr 2003;22(2):124−32.

[28] Neuhouser ML, Patterson RE, Levy L. Motivations for using vitamin and mineral supplements. J Am Diet Assoc 1999;99(7):851−4.

[29] Kenney RA. Physiology of aging. Clin Geriatr Med 1985;1(1):37−59.

[30] Preston CC, Oberlin AS, Holmuhamedov EL, Gupta A, Sagar S, Syed RH, et al. Aging-induced alterations in gene transcripts and functional activity of mitochondrial oxidative phosphorylation complexes in the heart. Mech Ageing Dev 2008;129(6):304−12.

[31] Swift CG. Pharmacodynamics: changes in homeostatic mechanisms, receptor and target organ sensitivity in the elderly. Br Med Bull 1990;46(1):36−52.

[32] Jahangir A, Sagar S, Terzic A. Aging and cardioprotection. J Appl Physiol (1985) 2007;103(6):2120−8.

[33] Salazar JA, Poon I, Nair M. Clinical consequences of polypharmacy in elderly: expect the unexpected, think the unthinkable. Expert Opin Drug Saf 2007;6(6):695−704.

[34] Akamine D, Filho MK, Peres CM. Drug-nutrient interactions in elderly people. Curr Opin Clin Nutr Metab Care 2007;10(3):304−10.

[35] Bailey RL, Dodd KW, Gahche JJ, Dwyer JT, McDowell MA, Yetley EA, et al. Total folate and folic acid intake from foods and dietary supplements in the United States: 2003−2006. Am J Clin Nutr 2010;91(1):231−7.

[36] Saw SM, Yuan JM, Ong CN, Arakawa K, Lee HP, Coetzee GA, et al. Genetic, dietary, and other lifestyle determinants of plasma homocysteine concentrations in middle-aged and older Chinese men and women in Singapore. Am J Clin Nutr 2001;73(2):232−9.

[37] Turnheim K. When drug therapy gets old: pharmacokinetics and pharmacodynamics in the elderly. Exp Gerontol 2003;38(8):843−53.

[38] Chan L-N. Drug-nutrient interactions. In: Shils ME, Shike M, Ross AC, Caballero B, Cousins RJ, editors. Modern nutrition in health and disease. 10th ed. Philadelphia, PA: Lippincott Williams & Wilkins; 2006. p. 1539−53.

[39] González Canga A, Fernández Martínez N, Sahagún Prieto AM, García Vieitez JJ, Díez Liébana MJ, Díez Láiz R, et al. Dietary fiber and its interaction with drugs. Nutr Hosp 2010;25(4):535−9.

[40] Rodríguez-Fragoso L, Reyes-Esparza J. Fruit/vegetable-drug interactions: effects on drug metabolizing enzymes and drug transporters. In: El-Shemy H, editor. Drug discovery. <http://www.intechopen.com/books/drug-discovery/fruit-vegetable-drug-interactions-effects-on-drug-metabolizing-enzymes-and-drug-transporters>; 2013 [accessed 25.06.15].

[41] Thomson AB, Sauve MD, Kassam N, Kamitakahara H. Safety of the long-term use of proton pump inhibitors. World J Gastroenterol 2010;16(19):2323−30.

[42] Schmidt LE, Dalhoff K. Food-drug interactions. Drugs 2002;62(10):1481−502.

[43] Welling PG. Effects of food on drug absorption. Annu Rev Nutr 1996;16:383−415.

[44] Turnheim K. Drug dosage in the elderly. Is it rational? Drugs Aging 1998;13(5):357−79.

[45] Holdsworth MT, Forman WB, Killilea TA, Nystrom KM, Paul R, Brand SC, et al. Transdermal fentanyl disposition in elderly subjects. Gerontology 1994;40(1):32−7.

[46] Sproule BA, Hardy BG, Shulman KI. Differential pharmacokinetics of lithium in elderly patients. Drugs Aging 2000;16(3):165−77.

[47] Spriet I, Meersseman W, de Hoon J, von Winckelmann S, Wilmer A, Willems L. Mini-series: II. Clinical aspects. Clinically relevant CYP450-mediated drug interactions in the ICU. Intensive Care Med 2009;35(4):603−12.

[48] Montamat SC, Cusack BJ, Vestal RE. Management of drug therapy in the elderly. N Engl J Med 1989;321(5):303−9.

[49] Liu Q, Li S, Quan H, Li J. Vitamin B12 status in metformin treated patients: systematic review. PLoS One 2014;9(6):e100379.

[50] Cashman JR, Lattard V, Lin J. Effect of total parenteral nutrition and choline on hepatic flavin-containing and cytochrome P-450 monooxygenase activity in rats. Drug Metab Dispos 2004;32(2):222−9.

[51] Wilkinson GR. The effects of diet, aging and disease-states on presystemic elimination and oral drug bioavailability in humans. Adv Drug Deliv Rev 1997;27(2−3):129−59.

[52] Brodie MJ, Boobis AR, Toverud EL, Ellis W, Murray S, Dollery CT, et al. Drug metabolism in white vegetarians. Br J Clin Pharmacol 1980;9(5):523−5.

[53] Chapron DJ. Drug disposition and response. In: Delafuente JC, Stewart RB, editors. Thearapeutics in the elderly. 3rd ed. Cincinnati, OH: Harvey Whitney Books; 2001. p. 257−88.

[54] Evans WE, McLeod HL. Pharmacogenomics−drug disposition, drug targets, and side effects. N Engl J Med 2003;348(6):538−49.

[55] Srinivas NR. Cranberry juice ingestion and clinical drug-drug interaction potentials; review of case studies and perspectives. J Pharm Pharm Sci 2013;16(2):289−303.

[56] Greenblatt DJ, von Moltke LL, Perloff ES, Luo Y, Harmatz JS, Zinny MA. Interaction of flurbiprofen with cranberry juice, grape juice, tea, and fluconazole: in vitro and clinical studies. Clin Pharmacol Ther 2006;79(1):125−33.

[57] Sparreboom A, Cox MC, Acharya MR, Figg WD. Herbal remedies in the United States: potential adverse interactions with anticancer agents. J Clin Oncol 2004;22(12):2489−503.

[58] Marchetti S, Mazzanti R, Beijnen JH, Schellens JH. Concise review: Clinical relevance of drug-drug and herb-drug interactions mediated by the ABC transporter ABCB1 (MDR1, P-glycoprotein). Oncologist 2007;12(8):927−41.

[59] Piscitelli SC, Burstein AH, Chaitt D, Alfaro RM, Falloon J. Indinavir concentrations and St John's wort. Lancet 2000;355(9203):547−8 Erratum in: Lancet 2001;357(9263):1210

[60] Yue QY, Jansson K. Herbal drug curbicin and anticoagulant effect with and without warfarin: possibly related to the vitamin E component. J Am Geriatr Soc 2001;49(6):838.

[61] Yue QY, Bergquist C, Gerdén B. Safety of St John's wort (Hypericum perforatum). Lancet 2000;355(9203):576−7.

[62] Patel S, Robinson R, Burk M. Hypertensive crisis associated with St. John's wort. Am J Med 2002;112(6):507−8.

[63] Brown RO, Dickerson RN. Drug-nutrient interactions. Am J Manag Care 1999;5(3):345−52 quiz 353−5.

[64] Mühlberg W, Platt D. Age-dependent changes of the kidneys: pharmacological implications. Gerontology 1999;45(5):243−53.

[65] Kompoliti K, Goetz CG. Neuropharmacology in the elderly. Neurol Clin 1998;16(3):599−610.

[66] Klein N, Gazola AC, de Lima TC, Schenkel E, Nieber K, Butterweck V. Assessment of sedative effects of Passiflora edulis f. flavicarpa and Passiflora alata extracts in mice, measured by telemetry. Phytother Res 2014;28(5):706−13.

[67] Fernández-San-Martín MI, Masa-Font R, Palacios-Soler L, Sancho-Gómez P, Calbó-Caldentey C, Flores-Mateo G. Effectiveness of Valerian on insomnia: a meta-analysis of randomized placebo-controlled trials. Sleep Med 2010;11(6): 505–11.

[68] Sorkin BC, Coates PM. Caffeine-containing energy drinks: beginning to address the gaps in what we know. Adv Nutr 2014;5(5):541–3.

[69] Whitehead N, White H. Systematic review of randomised controlled trials of the effects of caffeine or caffeinated drinks on blood glucose concentrations and insulin sensitivity in people with diabetes mellitus. J Hum Nutr Diet 2013;26(2):111–25.

[70] Wunderlich MS. Food and drug interactions. In: Mozayani A, Raymon PL, editors. Hand-book of drug interactions. A clinical and forensic guide. New York, NY: Humana Press; 2004. p. 379–93.

[71] Frankel HE. Drug interactions: basic concepts. In: McCabe JB, Frankel HE, Wolfe JJ, editors. Handbook of food-drug interactions. Boca Raton, FL: CRC Press; 2003. p. 37–45.

[72] Heidelbaugh JJ. Proton pump inhibitors and risk of vitamin and mineral deficiency: evidence and clinical implications. Ther Adv Drug Saf 2013;4(3):125–33.

[73] Robien K, Oppeneer SJ, Kelly JA, Hamilton-Reeves JM. Drug-vitamin D interactions: a systematic review of the literature. Nutr Clin Pract 2013;28(2):194–208.

[74] Marra MV, Boyar AP. Position of the American Dietetic Association: nutrient supplementation. J Am Diet Assoc 2009;109(12):2073–85.

[75] Evatt ML, Terry PD, Ziegler TR, Oakley GP. Association between vitamin B12-containing supplement consumption and prevalence of biochemically defined B12 deficiency in adults in NHANES III (third national health and nutrition examination survey). Public Health Nutr 2010;13(1):25–31.

[76] Watson WA, Litovitz TL, Rodgers Jr GC, Klein-Schwartz W, Reid N, Youniss J, et al. 2004 Annual report of the American association of poison control centers toxic exposure surveillance system. Am J Emerg Med 2005;23(5):589–666.

[77] Bronstein AC, Spyker DA, Cantilena Jr LR, Green JL, Rumack BH, Heard SE. American Association of Poison Control Centers. 2007 Annual Report of the American Association of Poison Control Centers' National Poison Data System (NPDS): 25th annual report. Clin Toxicol (Phila) 2008;46 (10):927–1057.

[78] Hansson O, Sillanpaa M. Letter: pyridoxine and serum concentration of phenytoin and phenobarbitone. Lancet 1976;1 (7953):256.

[79] Leon AS, Spiegel HE, Thomas G, Abrams WB. Pyridoxine antagonism of levodopa in parkinsonism. JAMA 1971;218 (13):1924–7.

[80] MacCosbe PE, Toomey K. Interaction of phenytoin and folic acid. Clin Pharm 1983;2(4):362–9.

[81] Martel F, Monteiro R, Calhau C. Effect of polyphenols on the intestinal and placental transport of some bioactive compounds. Nutr Res Rev 2010;23(1):47–64.

[82] Fedacko J, Pella D, Fedackova P, Hänninen O, Tuomainen P, Jarcuska P, et al. Coenzyme Q(10) and selenium in statin-associated myopathy treatment. Can J Physiol Pharmacol 2013;91(2):165–70.

[83] Mousa SA. Antithrombotic effects of naturally derived products on coagulation and platelet function. Methods Mol Biol 2010;663:229–40.

[84] Shikata T, Sasaki N, Ueda M, Kimura T, Itohara K, Sugahara M, et al. Use of proton pump inhibitors is associated with anemia in cardiovascular outpatients. Circ J 2015;79 (1):193–200.

[85] Lam JR, Schneider JL, Zhao W, Corley DA. Proton pump inhibitor and histamine 2 receptor antagonist use and vitamin B12 deficiency. JAMA 2013;310(22):2435–42.

[86] Dharmarajan TS, Kanagala MR, Murakonda P, Lebelt AS, Norkus EP. Do acid-lowering agents affect vitamin B12 status in older adults? J Am Med Dir Assoc 2008;9(3):162–7.

[87] Solomon LR. Functional cobalamin (vitamin B12) deficiency: role of advanced age and disorders associated with increased oxidative stress. Eur J Clin Nutr 2015;69(6):687–92.

[88] Massey LK, Liebman M, Kynast-Gales SA. Ascorbate increases human oxaluria and kidney stone risk. J Nutr 2005;135 (7):1673–7.

[89] Mehta JB, Singhal SB, Mehta BC. Ascorbic-acid-induced haemolysis in G-6-PD deficiency. Lancet 1990;336(8720):944.

[90] Barton JC, McDonnell SM, Adams PC, Brissot P, Powell LW, Edwards CQ, et al. Management of hemochromatosis. Hemochromatosis Management Working Group. Ann Intern Med 1998;129(11):932–9.

[91] Herbert V, Shaw S, Jayatilleke E. Vitamin C-driven free radical generation from iron. J Nutr 1996;126(Suppl. 4): 1213S-120S. Erratum in: J Nutr 1996;126(7):1902; J Nutr 1996;126(6):1746.

[92] Miller 3rd ER, Pastor-Barriuso R, Dalal D, Riemersma RA, Appel LJ, Guallar E. Meta-analysis: high-dosage vitamin E supplementation may increase all-cause mortality. Ann Intern Med 2005;142(1):37–46.

[93] Lonn E, Bosch J, Yusuf S, Sheridan P, Pogue J, Arnold JM, et al. HOPE and HOPE-TOO Trial Investigators. Effects of long-term vitamin E supplementation on cardiovascular events and cancer: a randomized controlled trial. JAMA 2005;293(11):1338–47.

[94] The Alpha-Tocopherol, Beta Carotene Cancer Prevention Study Group. The effect of vitamin E and beta carotene on the incidence of lung cancer and other cancers in male smokers. N Engl J Med 1994;330(15):1029–35.

[95] Sesso HD, Buring JE, Christen WG, Kurth T, Belanger C, MacFadyen J, et al. Vitamins E and C in the prevention of cardiovascular disease in men: the Physicians' Health Study II randomized controlled trial. JAMA 2008;300(18):2123–33.

[96] Podszun M, Frank J. Vitamin E-drug interactions: molecular basis and clinical relevance. Nutr Res Rev 2014;27(2):215–31.

[97] Traber MG. Vitamin E and K interactions—a 50-year-old problem. Nutr Rev 2008;66(11):624–9.

[98] Lippman SM, Klein EA, Goodman PJ, Lucia MS, Thompson IM, Ford LG, et al. Effect of selenium and vitamin E on risk of prostate cancer and other cancers: the Selenium and Vitamin E Cancer Prevention Trial (SELECT). JAMA 2009;301(1):39–51.

[99] Feskanich D, Singh V, Willett WC, Colditz GA. Vitamin A intake and hip fractures among postmenopausal women. JAMA 2002;287(1):47–54.

[100] Michaëlsson K, Lithell H, Vessby B, Melhus H. Serum retinol levels and the risk of fracture. N Engl J Med 2003;348 (4):287–94.

[101] Institute of Medicine (US) Panel on Micronutrients. Dietary reference intakes for vitamin A, vitamin K, arsenic, boron, chromium, copper, iodine, iron, manganese, molybdenum, nickel, silicon, vanadium, and zinc. Washington, DC: National Academies Press (US); 2001.

[102] Collins MD, Mao GE. Teratology of retinoids. Annu Rev Pharmacol Toxicol 1999;39:399–430.

[103] Goodman GE, Thornquist MD, Balmes J, Cullen MR, Meyskens Jr FL, Omenn GS, et al. The Beta-Carotene and Retinol Efficacy Trial: incidence of lung cancer and cardiovascular disease mortality during 6-year follow-up after stopping beta-carotene and retinol supplements. J Natl Cancer Inst 2004;96(23):1743–50.

[104] Bjelakovic G, Nikolova D, Gluud LL, Simonetti RG, Gluud C. Antioxidant supplements for prevention of mortality in healthy participants and patients with various diseases. Cochrane Database Syst Rev 2008;16(2):CD007176 Update in: Cochrane Database Syst Rev. 2012;3:CD007176.

[105] Hathcock JN, Shao A, Vieth R, Heaney R. Risk assessment for vitamin D. Am J Clin Nutr 2007;85(1):6–18.

[106] Jackson RD, LaCroix AZ, Gass M, Wallace RB, Robbins J, Lewis CE, et al. Women's health initiative investigators. Calcium plus vitamin D supplementation and the risk of fractures. N Engl J Med 2006;354(7):669–83 Erratum in: N Engl J Med 2006;354(10):1102

[107] Rosen CJ, Taylor CL. Common misconceptions about vitamin D—implications for clinicians. Nat Rev Endocrinol 2013;9(7):434–8.

[108] Bolland MJ, Barber PA, Doughty RN, Mason B, Horne A, Ames R, et al. Vascular events in healthy older women receiving calcium supplementation: randomised controlled trial. BMJ 2008;336(7638):262–6.

[109] Zapolski T, Wysokiński A. Safety of pharmacotherapy of osteoporosis in cardiology patients. Cardiol J 2010;17(4):335–43.

[110] Eisenberg DM, Davis RB, Ettner SL, Appel S, Wilkey S, Van Rompay M, et al. Trends in alternative medicine use in the United States, 1990-1997: results of a follow-up national survey. JAMA 1998;280(18):1569–75.

[111] Taylor CL. Highlights of a model for establishing upper levels of intake for nutrients and related substances: report of a Joint FAO/WHO Technical Workshop on Nutrient Risk Assessment, May 2–6, 2005. Nutr Rev 2007;65(1):31–8.

[112] Gershwin ME, Borchers AT, Keen CL, Hendler S, Hagie F, Greenwood MR. Public safety and dietary supplementation. Ann N Y Acad Sci 2010;1190:104–17.

[113] Mills E, Wu P, Johnston BC, Gallicano K, Clarke M, Guyatt G. Natural health product-drug interactions: a systematic review of clinical trials. Ther Drug Monit 2005;27(5):549–57.

[114] Ernst E. How much of CAM is based on research evidence? Evid Based Complement Alternat Med 2011;2011:676490.

[115] Timbo BB, Ross MP, McCarthy PV, Lin CT. Dietary supplements in a national survey: Prevalence of use and reports of adverse events. J Am Diet Assoc 2006;106(12):1966–74.

[116] US Food and Drug Administration. MedWatch. <http://www.fda.gov/medwatch/report/hcp.htm>; [accessed 25.06.15].

[117] Office of Inspector General. Adverse event reporting for dietary supplements. An inadequate safety valve. <https://oig.hhs.gov/oei/reports/oei-01-00-00180.pdf>. Department of Health and Human Services, 2001 [accessed 25.06.15].

[118] Lewis CW, Frongillo Jr EA, Roe DA. Drug-nutrient interactions in three long-term-care facilities. J Am Diet Assoc 1995;95(3):309–15.

[119] Scientific Report of the 2015 Dietary Guidelines Advisory Committee. <http://www.health.gov/dietaryguidelines/2015-scientific-report/> [accessed 25.06.15].

[120] Beyth RJ, Quinn L, Landefeld CS. A multicomponent intervention to prevent major bleeding complications in older patients receiving warfarin. A randomized, controlled trial. Ann Intern Med 2000;133(9):687–95.

CHAPTER

8

Nutritional Biomarkers of Aging

Anne Siepelmeyer[1], Antje Micka[1], Andreas Simm[2] and Jürgen Bernhardt[1]

[1]BioTeSys GmbH, Esslingen, Germany [2]Klinik für Herz- und Thoraxchirurgie, Universitätsklinikum Halle (Saale), Halle (Saale), Germany

KEY FACTS

- Different aging theories based on progressive accumulation of damage caused by reactive species.
- Antioxidant network builds a defense system against reactive species.
- Lower susceptibility for some age-related diseases by consuming a diet rich in antioxidants.
- Aging is affected by endogenous and exogenous influencing factors.
- Biomarker of aging should either alone or in a composite better predict functional capability than chronological age.

Dictionary of Terms

- *Reactive species*: Reactive species is the common term for both free radicals and reactive oxygen species (ROS), which include radicals, such as superoxide radical anion and hydroxyl radical, and nonradicals, such as hydrogen peroxide.
- *Oxidative stress*: Technically, oxidative stress is an imbalance between prooxidative reactive species and antioxidants, in which more prooxidative reactive species than antioxidants exist.
- *Antioxidative network*: Endogenous as well as exogenous antioxidant systems together build up a functional intracellular antioxidative network which is mainly involved in the defense against oxidative stress.
- *Chronological age*: The calendrical age is determined by the date of birth and the number of years a person has lived and is also referred to as chronological age.
- *Biological age*: The discrepancy between subjects with the same chronological age but different biological function, such as health state, is attributed to the biological age. Although little consensus regarding assessment of biological age exists, it is often outlined as a frailty/fitness index [1] or by complex mathematical modeling [2].
- *Biomarker*: A biomarker is an objectively measured characteristic that is evaluated as an indicator of a normal biologic and pathogenic process or of a pharmacologic response to an intervention [3].

INTRODUCTION

For a long time the causal relationship between aging and nutrition has been a subject matter of research. Basically, aging is the progressive accumulation of damage with time, associated with or responsible for the ever-increasing susceptibility to disease and death which accompanies advancing age [4]. Nutrition in terms of a balanced healthy diet has the ability to reduce the risk of malnutrition, deficiency diseases, and moreover can help to prevent chronic diseases and aid recovery from illnesses. But besides these capabilities the main question of the connection between nutrition and healthy longevity still plays a key role in recent research. Hence the major challenge is the characterization of suitable biomarkers to assess these connections. Basically, a biomarker is an objectively measured characteristic that is evaluated as an indicator of a normal biologic and pathogenic process or of a pharmacologic response to an intervention [3]. Moreover, a nutritional biomarker is any parameter that reflects biological consequence of dietary intake or dietary patterns and should indicate the nutritional status with respect to intake or metabolism of dietary constituents [5]. Beyond that, a biomarker of aging is a biological parameter of an organism that either alone or in some multivariate composite will, in the absence of disease, better predict functional capability at some advanced age or remaining lifespan than will the chronological age [6]. All in all, a nutritional biomarker of aging should objectively reflect the influence of dietary intake or the nutritional status of functional performance with regard to age.

Among the different aging theories the class based on damage due to toxic by-products of metabolism or insufficient biological repair mechanisms during aging provides a basis for emphasizing the relationship of nutrition and aging. One of the most noted and complex damage-based theories turned out to be Harman's aging theory based on free radicals which was already provided in the 1950s [7]. Within the free radical theory the major cause of aging was described as a lifelong accumulation of oxidative alterations in proteins, lipids, and DNA induced by free radicals and reactive oxygen species (ROS), which include both radicals, such as superoxide radical anion and hydroxyl radical, and nonradicals, such as hydrogen peroxide. In the following chapter free radicals and ROS will be referred to as reactive species [8]. Although nowadays reactive species are known as being products of normal cellular metabolism and, at low to moderate concentrations, are known to play essential roles in cell signaling and regulation [9], they were mostly seen as toxic by-products within this theory. Possible consequences due to damage of essential cell components by free radicals are impaired function of the cell membrane, enzyme inactivation, effected cell receptors, and decreased protein biosynthesis. Additionally, DNA can be damaged with the result of reduced cell persistence, uncontrolled proliferation, or cellular senescence. The term oxidative stress is used to describe the imbalance between the increased damaging effects of reactive species and insufficient antioxidative capacity. This is in opposition to the antioxidant network which enables controlling of their formation and supports repair mechanisms. The cells' ability to resist oxidative damage is based on a balance between the occurrence of reactive species and the supply with antioxidants.

The antioxidant network is directly and indirectly supported by nutrition. The cells' antioxidant molecules (vitamins, polyphenols, etc.) are mostly derived from dietary fruit and vegetables. Whereas most vitamins themselves have an antioxidative effect, trace elements, such as selenium, copper, and zinc, are crucial for the function of antioxidant enzymes. Therefore this work focuses on the antioxidant network and the contribution of nutrition on healthy aging and longevity.

THE ANTIOXIDANT NETWORK

The defense system against reactive species is built up by enzymatic as well as nonenzymatic antioxidants. The first line of enzymatic antioxidants includes superoxide dismutase (SOD), catalase (CAT), thioredoxin reductase, and glutathione peroxidase (GPx). Nevertheless, a second line of defense is essential to detoxify secondary products which are generated within the interaction of reactive species with biological macromolecules. These include for instance GPx and glutathione S-transferase [10]. Nonenzymatic antioxidants can be divided into hydrophilic as well as lipophilic radical scavengers. Thereby, ascorbate, urate, and glutathione (GSH) belong to the hydrophilic antioxidants; tocopherols, polyphenols as well as carotenoids represent the lipophilic radical scavengers. Most of these parameters interact synergistically and all together build up the antioxidative make-up of a tissue. For an optimal protection against oxidative stress the endogenous antioxidant system (including glutathione) should be intact and additionally an adequate supply with exogenous antioxidants (eg, α-tocopherol, vitamin C, polyphenols) should be guaranteed. Endogenous as well as exogenous systems together build up a functional intracellular antioxidative network. The interaction of α-tocopherol, vitamin C, and glutathione illustrates the network character (see Fig. 8.1): α-tocopherol inhibits lipid peroxidation in cell membranes and lipoproteins. But this function can only be assured if vitamin C is provided in adequate quantity to regenerate the tocopheryl radical. Subsequently, glutathione regenerates the function of vitamin C. Apart from that carotenoids present the cells' most efficient singlet oxygen quencher [12].

In the following sections focus is laid on the individual components of the antioxidative network (enzymatic antioxidants, hydrophilic and lipophilic radical scavengers) as well as their main representatives ((1) SOD, GPx; (2) vitamin C, glutathione; (3) carotenoids, tocopherols, polyphenols) and their causal relationship with aging and age-related diseases.

Enzymatic Antioxidants and Trace Elements

Superoxide Dismutase (Copper and Zinc)

Superoxide dismutases are enzymes which catalyze the reduction of superoxide anions to hydrogen

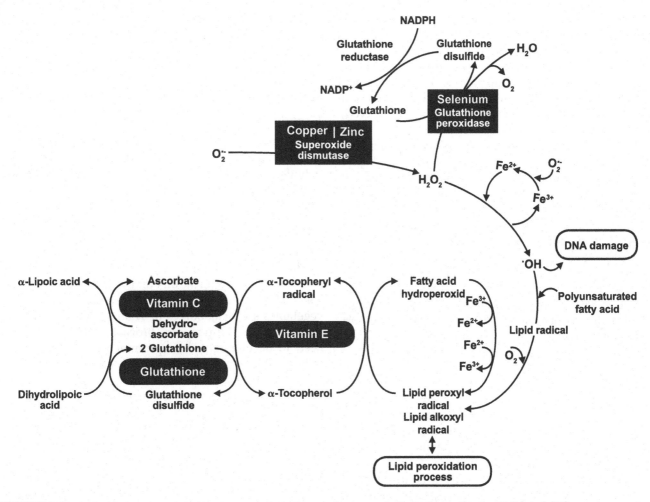

FIGURE 8.1 Antioxidative network and accompanying reactions, pathways, and effects. *Source: Based on Valko et al. [11].*

peroxide and thereby play a critical part in cell defense against the damaging effects of free oxygen radicals. Due to their function SODs are present in all oxygen-metabolizing cells. SODs are divided into different groups according to their capabilities for binding specific metal cofactors. There are SODs containing zinc and copper (Cu/ZnSOD), manganese (MnSOD), or iron (FeSOD) [13]. In the following passage the thematic priority focuses on the zinc and copper containing SOD.

Crucial for the function of Cu/ZnSOD are apparently the trace minerals zinc and copper. The lack of zinc stores in the human body necessitates the requirement for a constant adequate daily supply of dietary zinc [14]. The recommended intake for adults amounts to 10 mg/day for men and 7 mg/day for women in German-speaking countries [15]. Zinc is widely distributed in food, whereby the primary dietary sources of zinc are animal sources, being protein-rich food such as meat, seafood, and eggs [14]. Theoretically, whole grain products are an additional source for zinc. However, plant based food contain high values of zinc-chelating phytates which reduce dietary zinc absorption. Consuming vegetarian diets might therefore require up to 50% more zinc compared to nonvegetarian diets [16].

Estimated values for an adequate copper intake amount to 1.0–1.5 mg/day [15] or an average demand of 11 μg/kg body weight [17]. Dietary sources are mainly cereal products, giblets, and fish.

The intake of zinc decreases during aging due to intestinal malabsorption, inadequate diet, and increased incidence of age-related degenerative diseases [18]. Zinc may play an important role in lowering oxidative stress in cells of elderly subjects [19] and supplementing the elderly's diet with zinc can improve the immune response and reduce oxidative stress markers [18]. Decreasing plasma zinc levels occur with age in both genders. However, the modification in activity of SOD during aging has become a controversial topic of discussion. In one study lower SOD activity was detected in aged subjects compared to a younger control group. Additionally a negative correlation of plasma zinc level and antioxidant enzyme activity to age was observed [20]. In contrast, increasing activity of SOD with increasing age was assessed in another study which might be explained by an adaptive response to an increased level of oxidation products during aging [21]. The conflicting results of these studies might be attributed to methodological differences or individual responses to factors involved, such as nutrition or health state.

Similar to zinc, copper is an integral part of many enzymes (as SOD), which are involved in a number of biological procedures. Different from zinc, no variation in copper levels can be obtained in elderly subjects [20]. Apart from the important function as a metallic cofactor of SOD, free copper, especially during copper-overload, may initiate oxidative damage. In contrast, copper deficiency also leads to higher sensitivity to oxidative damage. As far as this is concerned, vitamin E, β-carotene, and polyphenols provide protection against copper-induced oxidative damage. Additionally, ascorbic acid and zinc indirectly support this process by inhibiting copper uptake indicating the need for a tightly regulated copper concentration [22].

Glutathione Peroxidase (Selenium)

Gluthatione peroxidases are a group of 8 enzymes which are responsible for the reduction of hydrogen peroxide to water and thereby play a crucial role in detoxification and cell protection against oxidative damage [23]. For their antioxidant function GPxs are dependent on selenium [24]. Selenium is an essential trace element and micronutrient for humans and has its biological function mainly as a part of selenoproteins, for example, GPxs. Plasma selenium is not a suitable marker for the functional selenium body pool because it includes not only selenoproteins but also proteins in which selenomethionine nonspecifically substitutes for methionine are included. However, GPx activity can be used as a biomarker for selenium function [25].

The recommended dietary intake for selenium is from 30 to 70 µg/day for adults. Good sources of selenium are meat, seafood, and eggs, but even so nuts or legumes [15]. The tolerable upper intake level for adults is determined to be 300 µg/day. Furthermore, the no adverse observed effect level of selenium is set to 850 µg/day [25].

In older women a lower level of selenium may contribute to inflammation and a higher mortality rate [26]. Similar to SOD, the activity of GPx seems to fade during aging and there is a negative correlation to age [20]. There are several trials dealing with GPx activity in specific disease states. Cardiovascular diseases and diabetes have been shown to be connected with modified activity of GPx [27,28]. On the basis of selenium serum/plasma levels of above 100 µg/L, an inverse relationship to a decreased risk of certain types of cancer (eg, prostate cancer) and total mortality rate is suggested [25].

Hydrophilic Antioxidants

Vitamin C

Vitamin C is referred to as ideal antioxidant due to the fact that ascorbate possesses a low reduction potential which enables the reaction with nearly all relevant reactive species. Vitamin C also acts as a coantioxidant by regenerating α-tocopherol from its radical. Beyond that the ascorbyl radical, which forms when ascorbate scavenges ROS, has a low reactivity and can additionally be converted back to ascorbate.

The recommended intake of vitamin C is established to be 100 mg/day for adults and is increased for smokers (150 mg/day) [15]. In general, fruit, and vegetables are rich sources of vitamin C, whereby sea-buckthorn, sweet peppers, broccoli, and citrus fruit need to be highlighted as prominent specimens with considerable amounts of vitamin C. The vitamin C content in food is decreased by prolonged storage and the cooking process during the preparation of food. In total, five servings of fruit and vegetables are sufficient to cover the requirements. Nevertheless, the optimal intake might be higher and vary with life stage and disease state. A level of supply status is the plasma vitamin C concentration. Thereby a preventive vitamin C concentration of $\geq 50\,\mu mol/L$ is assumed in epidemiologic studies [29].

In comparison to younger people elderly people as well as people suffering from diabetes or hypertension have lowered plasma vitamin C levels [30]. A strong inverse trend for all-cause mortality and blood ascorbate is known [31]. However, ascorbic acid supplementation can substantially reduce oxidative stress, particularly when combined with vitamin E. A vitamin C rich diet associates with protective effects regarding the risk of some types of cancer, for example, upper gastrointestinal tract cancer [32]. Apart from that, the relationship between vitamin C and cognitive function as well as Alzheimer's disease was investigated. Although inconsistencies between study results exist, an optimal supply with vitamin C seems to support cognition and may prevent from Alzheimer's disease [33].

Glutathione

The redox buffer glutathione (GSH) is characterized by the thiol group (-SH) of cysteine. GSH is converted into its dimer glutathione disulfide (GSSG) after reduction of target molecules. In principle, the GSH/GSSG ratio is an indication of the organism's oxidative state [34]. Furthermore its main protective role includes scavenging hydroxyl radicals and singlet oxygen directly or regenerating other antioxidants (vitamin C, E) back to their active forms [10].

Glutathione is synthesized in the cytosol from cysteine, glycine, and glutamate and is abundant in nearly all cell compartments, for example, cytosol, nuclei, and mitochondria [11,35]. However, the oral administration of glutathione supplements is possible but appears problematic. On the one hand it can be assumed that the tripeptide is broken down into the single amino acids in the acidic gastric milieu when using nongastro-resistant preparations. On the other hand even if using gastro-resistant capsules, glutathione might only be slightly membrane-permeable. But besides that, dietary supplementation with the precursors cysteine and glycine seems to restore glutathione concentrations [36].

Due to increased oxidative stress during aging and the fact that aging affects the glutathione redox system, a negative correlation between age and GSH, GSH/GSSG ratio, as well as a positive correlation between age and GPx and GSSG levels appear [37]. Glutathione deficiency during aging occurs due to a pronounced reduction in synthesis. The restoring by dietary supplementation with the amino acids cysteine and glycine suggests an effective approach to decrease oxidative stress in aging [36]. Low levels of glutathione are also found in various diseases. GSH is linked to positive effects with regards to cancer and its therapeutics, alcoholic and nonalcoholic fatty liver disease and Alzheimer's disease [34]. Glutathione loss is strongest in the brain and implicated in Parkinson's disease and neuronal injury following a stroke [38].

Lipophilic Antioxidants

Carotenoids

About 600–700 different carotenoids are known of which α- and β-carotene, lycopene, lutein, and zeaxanthin are the most prominent ones. Thereof, β-carotene is the most known carotenoid and the most often naturally occurring carotene. β-carotene, also known as provitamin A, can be metabolized to vitamin A in different tissues (eg, small intestine, liver). In turn, vitamin A (retinol) can be transformed to retinal, which is essential for vision, or retinoate, which is involved in cell proliferation and cell differentiation via binding to the retinoic acid receptors RAR and RXR. These receptors act as transcriptional regulators. β-carotene itself is an antioxidant as are nearly all carotenoids. Similar to tocopherols, β-carotene possesses chain-breaking abilities. Carotenoids present the cell's most efficient singlet oxygen quencher. But unlike α-tocopherol carotenoids are not depleted by quenching a radical but absorb their energy and reemit this as thermal energy [12].

Carotenoids are secondary plant substances. As naturally occurring color, α- and β-carotene and lycopene appear predominantly in red, orange, and yellow fruit and vegetables, whereas lutein and zeaxanthin occur mainly in green-leaved vegetables. Thereby, the plasma β-carotene level is a marker for fruit and vegetable uptake [15]. Bioavailability of carotenoids is subject to high variability. This might rather be due to individual differences in fat absorption than due to the necessity of presence of fat for carotenoid absorption [15]. A few years ago the average carotenoid supply was estimated to be 5–6 mg/day, but recent insights indicate a higher intake [39]. The estimated value for the intake of β-carotene is extrapolated to be 2–4 mg/day [15].

Carotenoids have many different positive effects on age-related pathological conditions. Lutein and zeaxanthin possess the highest benefit with regard to age-related macular degeneration compared to other carotenoids [40]. Increased levels of plasma carotenoids contribute to lower cardiovascular disease. Additionally, an inverse association between carotenoids and the inflammatory biomarkers C-reactive protein and sICAM-1 could be detected [41]. Carotenoids, particularly lycopene are connected with a decreased risk of different types of cancer, for example, prostate, colon, and cervical cancer [42]. Apart from that, serum carotenoids are positively associated with SOD [41].

Beyond its positive properties, several studies detected toxic effects of β-carotene in smokers. Indeed, the toxicity does not originate from the antioxidative activity or reactive products, but is based on a molecular mechanism. Lung cancer can be caused by chromosomal instability induced by a polycyclic aromatic hydrocarbon which is a component of tobacco smoke. A special isoform of glutathione S-transferase is involved in the protective mechanism against polycyclic aromatic hydrocarbon metabolites, but can also be attenuated by β-carotene [43]. In general, a maximum uptake of 20 mg β-carotene is classified as safe. Starting with the consumption of more than 20 cigarettes/day this limit should not be exceeded [44].

Tocopherols

Vitamin E, the collective term for tocopherols and tocotrienols, is primarily located in the cell and its membranes, where it defends the cell against lipid peroxidation and free radicals. Basically, all natural isoforms of tocopherol show the ability to disrupt radical chain reactions due to inhibition of lipid peroxidation. In this respect, the antioxidative capacity varies between the various derivates. The reestablishment by reduction is supported by ascorbate and glutathione.

α- and γ-tocopherol are the two major forms of vitamin E which occur naturally and which can be found in plasma and red blood cells. Naturally occurring tocopherols are only synthesized by plants, which is why good suppliers are plant oils, for example, wheatgerm oil, sunflower oil, corn oil, rapeseed oil, and soybean oil. Additionally, lipid-rich plant products, for example, nuts, seeds, and grain are a main dietary source of vitamin E. Estimated values for an adequate tocopherol supply amount to 15–12 mg tocopherol equivalents per day for men and 12–11 mg tocopherol equivalents per day for women in Germany (decreasing with age; 1 mg tocopherol equivalent = 1 mg RRR-α-tocopherol = 1,49 IU) [15]. The requirements do not increase with advanced age (>80 years) compared to young age [15]. In order to fulfill the demand plasma tocopherol levels should be within the reference range of 12–46 μmol/L or 0,8 mg/g total lipids for adults [15]. Basic requirements of 4 mg α-tocopherol equivalent are considered necessary to protect from increased lipid peroxidation [45].

According to the available literature a clear statement about vitamin E and aging or age-related diseases cannot be made. Within a meta-analysis, supplementation with vitamin E appears to have no effect on all-cause mortality and therefore should not be recommended as a means of improving longevity [46]. In contrast to this, tocopherol has been associated with disease prevention with regard to cardiovascular health, cancer, and Alzheimer's disease due to its antioxidative activity [42,47]. Furthermore, healthy (Italian) centenarians exhibit higher plasma vitamin E levels and an improved nutritional status compared to younger, but also aged subjects [48]. On the one hand, low dosages of 50 mg/day have beneficial effects on prostate cancer incidences, on the other hand, supplementation of high dosages of 400 IU (268 mg) significantly increase prostate cancer risk. These results are in accordance with a Cochrane systematic review which showed a causal link between vitamin E supplementation and increased mortality rate [49]. A meta-analysis which was conducted later specified the results. Increased mortality after vitamin E supplementation was attributed to the intake of doses above the recommended daily allowance [50]. This points out the importance to focus on recommended intake levels [51].

Polyphenols

Polyphenols are secondary metabolites which are synthesized by plants as a defense mechanism against stressors such as pathogens. They possess antioxidative activity, which is dependent on their chemical structure. In this respect, flavonoids have the most potent antioxidative activity [42]. Many of these substances have radical-scavenging activity accompanied by the ability of quenching chain reactions [10]. Thus, polyphenols can reduce oxidative stress and as a consequence have protective effects against cellular damage.

Several hundred polyphenols are found in fruit, vegetables, and other edible plants and also in their resulting products, such as wine, tea, or chocolate. The most prominent ones are resveratrol, which is for example found in, red grapes and blueberries, and epigallocatechin gallate (EGCG), which is commonly known as the bioactive substance in green tea extract. The estimated uptake of these secondary plant substances amounts to 50–200 mg/day for flavonoids and 200–300 mg/day for phenolic acids in Germany [39].

A great number of nutritional epidemiological studies indicate a lower susceptibility to cardiovascular diseases, cancer, and neurodegenerative disorders in populations generally consuming diets rich in polyphenols [10,42,52]. The Mediterranean diet, which is rich in the olive oil polyphenols hydroxytyrosol and oleuropein, is frequently mentioned. Additionally, the moderate consumption of red wine, occurring with the French diet, is associated with a low risk of coronary heart disease. Another well-known example is the cardioprotective effect of EGCG, which is abundant in green tea, which in turn is related to the far eastern diet. Furthermore, soy products (isoflavones) are positively related to antiartherosclerotic effects [42]. Nevertheless, the effects of soy isoflavones are ambiguous and subject to controversial discussion particularly regarding breast cancer [53].

DISCUSSION

Nutrition seems to play an important role in the aging process. An adequate supply achieved by a balanced diet is prerequisite for the provision with vitamins, essential trace elements, and secondary plant products. The antioxidative network is built by a variety of these substances, each exhibiting special properties. This network is a dynamic and well-matched system involving synergistic effects of several components (see Fig. 8.1). The antioxidative network is affected by exogenous supply and endogenous demand. Consequently nutrition is an influencing key factor. But also conditions consuming antioxidants, for example, diseases influence this network. Antioxidants exhibit in some studies positive properties which include but are not limited to different age-related diseases. In contrast, reactive species positively correlate with most of these diseases such as cardiovascular diseases or cancer [54]. But it is arguable whether the reactive species are really the primary cause of these diseases or just a consequence of the pathological conditions. Using the example of cancer, the answer seems obvious because cancer is evoked by DNA damage. But regarding other diseases like neurologic disorders the relationship seems to be uncertain. DNA damage due to oxidative stress is certainly not the basis for all degenerative conditions [11]. Nevertheless, accumulation of damage is believed to be one relevant cause for aging and therefore reactive species are very probably a crucial cause for aging, too. Basically it can be stated that oxidative stress, health, and aging are closely interconnected, which is also the basis for the general theory of aging published by Liochev: reactive species as well as other causes generate damage. This damage can be of a functional or structural nature. Both interact and result in pathologic conditions and aging. Death is therefore a direct result of pathologic conditions, but not necessarily of aging. People do not die directly due to aging but due to various pathological conditions and aging increases the probability of such pathological conditions to occur [8]. Thus, aging is a complex process including the accumulation of damage in an organism over the course of time (see Fig. 8.2). Therefore, aging should rather be referred to as senescence which is affected by internal factors, such as genes, as well as dynamic external factors, such as lifestyle and diet. Dietary interventions in terms of antioxidant supplementation have been undertaken to verify beneficial effects on longevity. Most of them not merely failed but detected side effects or even an increased mortality rate [56].

Reactive species are not only endogenously generated toxic by-products of the metabolism, but also play an important role in different signal transduction pathways affecting cell growth and differentiation. Beyond that, reactive species are beneficial within the concept of adaptive stress response, commonly known as hormesis [57]. Technically, oxidative stress is an imbalance between prooxidative reactive species and antioxidants, in which more prooxidative reactive species than antioxidants exist. But moreover, controllable low levels of reactive species possess the ability to stimulate the cell's resistance against higher, usually harmful doses of the same species. Nevertheless, this fact does not make reactive species less toxic in all circumstances [8]. The concept of hormesis was further developed to the concept of hormetics, dealing with dietary ingredients which stimulate hormesis [54]. Many antioxidants exhibiting positive effects with regard to degenerative diseases are now examples for hormetics, such as resveratrol. Depending on the concentration and duration of treatment, resveratrol exhibits proliferative as well as proapoptotic properties, increasing levels of reactive species and depleting GSH [54]. Alcohol induces hormesis and when consumed in low doses it is associated with decreased death from all causes [58]. Although alcohol and therefore ethanol abuse refers to the contrary, moderate consumption of alcoholic beverages might be beneficial due to the polyphenol content of for instance red wine and beer. This could be demonstrated in volunteers drinking one or alternatively three drinks of red wine and beer. After consuming one drink, plasma antioxidant activity increased, whereas plasma prooxidant activity was elevated after ingestion of three drinks [59]. During hepatic ethanol metabolism reactive species are generated. Moreover, ethanol induces dyslipidemia and hepatic insulin resistance [60].

A general change in eating habits from basic food toward intensively processed finished products has been observed within the last decades. Additionally, intensive agricultural techniques and infertile soils lead to lower micronutrient concentration of arable crops and food. As a result, cooccurring trace element and vitamin deficiency are also known in western countries although food is abundant [61]. A few years ago metabolic syndrome was an accompanying effect of obesity, but nowadays more and more people with normal weight manifest this syndrome. This might be explained by increased fructose consumption. Within one hundred years fructose uptake was increased

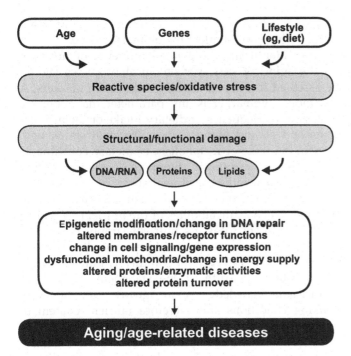

FIGURE 8.2 Scheme of the complex aging process including endogenous and exogenous influencing factors, basic mechanisms involved, and outcomes which manifest aging and age-related diseases. *Source: Based on Jacob et al. [55].*

fivefold to 12% of the total energy uptake in Americans [62]. In general, the consumption of five servings or about 400 g of fruit and vegetables are sufficient to achieve the recommended daily allowance. However, the intake of high levels of fructose by normal diet can lead to the curious situation that excess fruit consumption might have a negative side effect, because the hepatic fructose metabolism generates inflammation and insulin resistance. Evidently, insulin resistance is one manifestation of the metabolic syndrome, cooccurring with adiposity, dyslipidemia, impaired glucose tolerance, and hypertension.

Supplementation of single nutrients or even multivitamin and trace element preparations is the most frequently used method to counter nutrient deficiencies. This might be favorable to rapidly achieve optimum supply but beyond that an added benefit might not be expected. For certain groups with increased requirements such as pregnant and breastfeeding women or smokers and during certain illnesses when requirements may not be covered by nutrition, supplementation of selected nutrients should be mandatory. In general an adequate, balanced nutrition seems to be preferable and should not be replaced by food fortification and nutritional supplements. Another point that supports a balanced nutrition is that phytochemicals of fruit and vegetables might be a prerequisite for a hormetic response and interrelated beneficial effects [54]. The effects of nutrition and dietary supplements are still controversially discussed. There seems to be a fine line between antioxidants acting toxic or healthy. Supplementation with antioxidants needs to be considered very carefully with regard to dosage, which should be oriented toward the tolerable upper intake level, and the duration of treatment [43]. To assess the actual level of supply with micronutrients the establishment of suitable nutritional biomarkers is a focus of current research.

NUTRITIONAL BIOMARKERS OF AGING

Many concepts of biomarkers exist, from basic ones in which a biomarker should originally reflect the consequences of an external chemical, physical, or biological exposure on a biological system [63], through to complex ones including the causal relationship of specific issues such as nutrition or aging. Biomarkers might serve as diagnostic agents, represent the severity or progression of an illness, function as prognostic markers, or illustrate effects of therapeutic interventions. Examples for generally accepted biomarkers which are validated for biological plausibility are the low density lipoprotein (LDL) cholesterol with regards to cardiovascular diseases as well as the glycated hemoglobin (HbA1c) referring to diabetes. The assessment of HbA1c is not only the gold standard for supervision of glycemic control in people with manifest diabetes but also for people with a high risk for diabetes [64]. A significant correlation of LDL cholesterol and cardiovascular diseases was established, proving a benefit toward a better clinical outcome due to a reduction of LDL cholesterol [65]. In contrast to clinical endpoints biomarkers depict a biological process and not necessarily the state of health [66]. In turn, a plausible, powerful, and validated biomarker which exhibits clinical relevance can be used as a surrogate endpoint in clinical research, being a substitute for a clinical endpoint [3]. However, determination of a biomarker's relevance is a crucial point. On the one hand a biomarker should act as an indicator for physiological or pathological processes. This implies the understanding of the complete underlying complex physiology. On the other hand biomarkers need to be measured objectively. This requires essential instructions regarding sample-taking and storage but also other laboratory issues such as reasonable precision and reproducibility of the method. A broad range of suitable specimen for biomarker analysis exist, such as serum, plasma, red and white blood cells, dried blood spot, urine, feces, buccal mucosa cells, adipose tissue, hair, toenails, or mother's milk. Attention should be paid to the fact that with increasing level of specimen complexity the mechanistic linkage to metabolic effects also increases, but the relation to the initial exposure decreases [63]. Additionally, the number of influencing factors which might affect the biomarker rises.

After identifying the general influence of nutrition on the aging process, physiological processes should be assessed in order to identify suitable biomarkers. A nutritional biomarker is any parameter that reflects a biological consequence of dietary intake or dietary patterns and should indicate the nutritional status with respect to intake or metabolism of dietary constituents [5]. Hence, nutritional biomarkers should either directly measure the nutritional status of a nutrient or indirectly reflect dietary intake using a substitutional indicator, such as metabolites of the nutrient. A widely used instrument to assess nutrient supply is the dietary questionnaire which should estimate the quantity of the uptake. However the validity is limited due to the questionable truth of the statements as well as insufficient knowledge about growth, storage, and cooking conditions and therefore nutrient content of the food. Additionally, overestimation and underestimation of serving sizes play a part in contributing to miscalculation of nutrient uptake. In contrast to dietary questionnaires, a biomarker reflects the complex process of absorption, metabolism, tissue distribution, and excretion. This is also referred to as bioavailability.

Interindividual bioavailability of most nutrients varies substantially whereas intraindividual processes remain predominantly consistent. For the most part bioavailability is determined by the absorption of nutrients. In this respect, absorption depends on external influencing factors including matrix effects of different food or interaction with antinutrients, such as phytates which among others obtain zinc-chelating properties and therefore reduce dietary zinc absorption. Beyond that, individual biologic or metabolic characteristics involving feedback control mechanisms and transit time have an impact on nutrient absorption. Therefore, a nutritional biomarker does not necessarily reflect the amount of nutrient ingested with the diet but is related to the nutritional status of the associated nutrient. Total serum carotenoids are a good marker for intake of fruit and vegetables [67] whereas plasma vitamin E poorly predicts the intake and should be related to total plasma cholesterol [68]. Using fasting blood samples and a suitable stabilizer, plasma vitamin C levels should reflect vitamin C uptake of low ranges. In general plasma vitamin C fluctuations are high and after reaching saturation, renal clearance increases [68]. The homeostatic regulation of trace elements like zinc limits the blood zinc level's suitability for biomarker use. Another point which needs to be considered is that nutrient blood concentrations might be subject to circadian variations. Again the example of zinc can be used to illustrate that standardized sampling times are necessary [69]. Nevertheless, plasma zinc is still a widely recommended biomarker of zinc status [70]. A functional biomarker for zinc such as the enzyme activity of Cu/ZnSOD might be a better predictor of nutrient status but is influenced by more than one micronutrient. In contrast, plasma selenium concentrations seem to be a reliable marker of short-term selenium status and whole blood selenium might be suitable as a biomarker for long-term selenium intake [70]. The classic example in which nutrient blood concentrations are unsuitable to detect deficiency is iron. In this case a combination of the biomarkers hemoglobin, ferritin, soluble transferrin receptor, and hepcidin should be used for diagnosing of the varying manifestations of iron deficiency.

Furthermore nutritional concepts such as special diets (eg, the Mediterranean or far eastern diet), caloric restriction or items which are dependent on nutrition like the body mass index (BMI) might influence longevity. The main benefit of the Mediterranean and far eastern diet, which are rich in polyphenols, is the prevention of cardiovascular diseases. A similar effect as the polyphenol consumption was detected after caloric restriction. Caloric restriction means reduced caloric intake while avoiding malnutrition. It decreases formation of reactive species and consequently has a beneficial effect on oxidative damage [71]. For most humans the main problem of caloric restriction is compliance. Alternative approaches with similar benefits might be fasting periods [72]. However, the concept of caloric restriction is matter of controversial discussion and further investigations need to be done to identify the relevant mechanisms. The same applies to the hypothesis, that obesity defined by a BMI $>25 \text{ kg/m}^2$ is a strong predictor of overall mortality [73]. A suitable biomarker for nutrient exposure does not have to be equivalent to a biomarker of nutritional status which in turn is not necessarily synonymous to the nutrient's effect. Therefore the ideal nutritional biomarker reflects exposure, status, and effect of a specific nutrient [74]. Besides the identification of single components a sum parameter representing cumulative antioxidant capacity is worth striving for. Therefore, in absence of the ideal biomarker the combination of different biomarkers might most often be the preferable approach.

A biomarker of aging is a biological parameter of an organism that either alone or in some multivariate composite will, in the absence of disease, better predict functional capability at some advanced age than will the chronological age [6]. Besides the fact that the method for assessing a biomarker of aging should be simple, valid, and universally usable it should depict the aging process itself. Furthermore, a comprehensive biomarker of aging should predict remaining longevity and mortality due to diseases. Therefore it should represent the biological age independently from the chronological age. Subjects with the same chronological age might not necessarily exhibit the same health state. That is why the chronological age seems to be unsuitable to assess a subject's functional decline. This discrepancy in biological function is referred to as biological age. Due to the complexity of human aging, the prediction of biological age by a single marker is implausible. Although little consensus regarding assessment of biological age exists it is often outlined as a frailty/fitness index [1] or by complex mathematical modeling [2]. Since genes, as well as environmental and lifestyle factors, affect aging, biomarkers should presumably be a combination of markers reflecting these influences. A considerable amount of individual lifespan seems to be influenced by lifestyle factors such as nutrition and only one-third can be predicted as owing to genetic factors [75]. A variety of genes associated with longevity have been revealed based on large groups of centenarians. This exceptional longevity might play a key role in the investigation of underlying biological processes [76]. Other factors influencing the aging process are epigenetic mechanisms. Epigenetic mechanisms cause physiological traits which are heritable but not caused by changes in the

DNA sequence but in changes in DNA methylation or histone modifications. These epigenetic mechanisms may play a key role with regard to oxidative stress during aging. Increased oxidative damage leads to a shift in the balance between functional and altered proteins for the benefit of the latter ones. Therefore the quality of proteins rather than the exclusive accumulation of damaged proteins may induce age-related diseases such as diabetes and furthermore aging itself. Additionally, systemic redox regulation with regard to the ability to deal with oxidative stress has been shown to be an important influencing factor for longevity. As a result the generation of reactive species should be prevented to reduce the aging process [77]. Beyond that, epigenetic mechanisms may also result from nutritional interventions during pregnancy and affect physiological equipment of the offspring [78]. Therefore, not only the actual nutritional status affects longevity but early life nutrition should also be included in the development of nutritional intervention strategies. Important issues concerning the plasticity of epigenetics and aging are to determine the critical windows of development, the reversibility of epigenetic imprinting, and the development of suitable biomarkers [78].

CONCLUSION

Sufficient evidence exists that the foundation for the aging process is already laid during the early stages of development in pregnancy. Aging itself is a lifelong process affected by genetic make-up, lifestyle, including nutrition, and the lifespan which was experienced so far. The combination of these exogenous and endogenous influencing factors has an impact on the generation of reactive species as well as on the antioxidant network. An overload of oxidative stress can lead to structural and functional damage of DNA, RNA, proteins, and lipids. However, an oversupply with antioxidants might also entail pathological conditions or increased mortality. Further investigations to improve knowledge of basic physiological and biological processes are necessary and should provide the basis for predicting the optimal supply for each individual. Additionally, biomarkers need to be established to capture hormetic stress response and epigenetics as well as influences of dietary antioxidants on oxidative stress and age-related conditions. Due to the large number of factors involved and the complexity of the aging process the result might be a combination of various biomarkers rather than a single biomarker.

SUMMARY

- Aging is characterized by a progressive accumulation of damage with time (associated with the susceptibility to disease).
- Different aging theories based on damage caused by toxic by-products of the metabolism and/or insufficient biological repair systems during aging exist.
- Reactive species are seen as the major cause of this damage as they lead to oxidative alterations of proteins, lipids, and DNA which is often summarized by the term "oxidative stress."
- The body's ability to resist oxidative damage is based on a balance between the occurrence of reactive species and supply with antioxidants for example by means of nutrition.
- The antioxidant network builds the defense system against reactive oxygen species by controlling their formation and supports repair mechanisms and is built up by enzymatic as well as nonenzymatic antioxidants.
- Copper, zinc, and selenium are important cofactors for enzymes, for example, SOD or GPx which catalyze essential reactions in protection against oxidative damage and detoxification.
- Vitamin C is an ideal antioxidant due to its low reduction potential which enables reactions with nearly all reactive species.
- Glutathione exerts its protective role by scavenging hydroxyl radicals and singlet oxygen.
- Carotenoids such as β-carotene and tocopherols like vitamin E and polyphenols are discussed regarding their relationship to aging and age-related diseases.
- A great number of studies indicate a lower susceptibility for some age-related diseases for subjects consuming a diet rich in antioxidants.
- A biomarker of aging is a biological parameter of an organism that either alone or in composite better predicts the functional capability at some age than will the chronological age.
- Oversupply of antioxidants, particularly β-carotene, might also entail pathologic conditions.
- Foundations for the aging process are already laid during early stages of pregnancy.
- Aging is a lifelong process affected by endogenous influencing factors, such as genes, and exogenous influencing factors, such as lifestyle and nutrition.
- In the absence of the ideal biomarker to reflect the aging process a combination of different biomarkers might be the preferable approach.

References

[1] Mitnitski AB, Graham JE, Mogilner AJ, Rockwood K. Frailty, fitness and late-life mortality in relation to chronological and biological age. BMC Geriatr 2002;2(1):1.

[2] Levine ME. Modeling the rate of senescence: can estimated biological age predict mortality more accurately than chronological age? J Gerontol A Biol Sci Med Sci 2013;68(6):667–74.

[3] Biomarkers Definitions Working Group. Biomarkers and surrogate endpoints: preferred definitions and conceptual framework. Clin Pharmacol Ther 2001;69(3):89–95. Available from: http://dx.doi.org/10.1067/mcp.2001.113989.

[4] Harman D. The aging process. Proc Natl Acad Sci USA 1981;78(11):7124–8.

[5] Potischman N, Freudenheim JL. Biomarkers of nutritional exposure and nutritional status: an overview. J Nutr 2003;133 (Suppl. 3):873S–4S.

[6] Baker III GT, Sprott RL. Biomarkers of aging. Exp Gerontol 1988;23(4–5):223–39.

[7] Harman D. Aging: a theory based on free radical and radiation chemistry. J Gerontol 1956;11(3):298–300.

[8] Liochev SI. Reflections on the theories of aging, of oxidative stress, and of science in general. Is it time to abandon the free radical (oxidative stress) theory of aging? Antioxid Redox Signal.

[9] Thannickal VJ, Fanburg BL. Reactive oxygen species in cell signaling. Am Physiol Lung Cell Mol Physiol 2000;279(6):L1005–28.

[10] Masella R, Di Benedetto R, Vari R, Filesi C, Giovannini C. Novel mechanisms of natural antioxidant compounds in biological systems: involvement of glutathione and glutathione-related enzymes. J Nutr Biochem 2005;16(10):577–86.

[11] Valko M, Leibfritz D, Moncol J, Cronin MT, Mazur M, Telser J. Free radicals and antioxidants in normal physiological functions and human disease. Int J Biochem Cell Biol 2007;39(1):44–84.

[12] Biesalski HK, Köhrle J, Schümann K. Vitamine, spurenelemente und mineralstoffe. 1st ed. Stuttgart: Georg Thieme Verlag; 2002.

[13] Scandalios JG. Oxygen stress and superoxide dismutases. Plant Physiol 1993;101(1):7–12.

[14] Coneyworth LJ, Mathers JC, Ford D. Does promoter methylation of the SLC30A5 (ZnT5) zinc transporter gene contribute to the ageing-related decline in zinc status? Proc Nutr Soc 2009;68(2):142–7.

[15] Referenzwerte für die Nährstoffzufuhr, Deutsche Gesellschaft für Ernährung (DGE), [Konzeption und Entwicklung: Arbeitsgruppe "Referenzwerte für die Nährstoffzufuhr"], 1. Aufl., Frankfurt am Main: Umschau/Braus, 2000; ISBN 3-8295-7114-3.

[16] Hunt JR. Bioavailability of iron, zinc, and other trace minerals from vegetarian diets. Am J Clin Nutr 2003;78(Suppl. 3):633S–9S.

[17] World Health Organization. Trace elements in human nutrition and health. Geneva: WHO; 1996.

[18] Mocchegiani E. Zinc and ageing: third Zincage conference. Immun Ageing 2007;4(5):5.

[19] Jajte JM. Chemical-induced changes in intracellular redox state and in apoptosis. Int J Occup Med Environ Health 1997;10(2):203–12.

[20] Sfar S, Jawed A, Braham H, Amor S, Laporte F, Kerkeni A. Zinc, copper and antioxidant enzyme activities in healthy elderly Tunisian subjects. Exp Gerontol 2009;44(12):812–17.

[21] Mecocci P, Polidori MC, Troiano L, Cherubini A, Cecchetti R, Pini G, et al. Plasma antioxidants and longevity: a study on healthy centenarians. Free Radic Biol Med 2000;28(8):1243–8.

[22] Gaetke LM, Chow CK. Copper toxicity, oxidative stress, and antioxidant nutrients. Toxicology 2003;189(1–2):147–63.

[23] Rotruck JT, Pope AL, Ganther HE, Swanson AB, Hafeman DG, Hoekstra WG. Selenium: biochemical role as a component of glutathione peroxidase. Science 1973;179(4073):588–90.

[24] Fairweather-Tait SJ, Bao Y, Broadley MR, Collings R, Ford D, Hesketh JE, et al. Selenium in human health and disease. Antioxid Redox Signal 2011;14(7):1337–83.

[25] EFSA Journal. Scientific opinion on dietary reference values for selenium. EFSA J 2014;12(10):3846.

[26] Walston J, Xue Q, Semba RD, Ferrucci L, Cappola AR, Ricks M, et al. Serum antioxidants, inflammation, and total mortality in older women. Am J Epidemiol 2006;163(1):18–26.

[27] Kedziora-Kornatowska K, Czuczejko J, Pawluk H, Kornatowski T, Motyl J, Szadujkis-Szadurski L, et al. The markers of oxidative stress and activity of the antioxidant system in the blood of elderly patients with essential arterial hypertension. Cell Mol Biol Lett 2004;9(4A):635–41.

[28] Menon V, Ram M, Dorn J, Armstrong D, Muti P, Freudenheim JL, et al. Oxidative stress and glucose levels in a population-based sample. Diabet Med 2004;21(12):1346–52.

[29] Biesalski HK. Antioxidative vitamine in der prävention. Dtsch Ärztebl 1995;92:B979–83.

[30] Simon JA. Vitamin C and cardiovascular disease: a review. J Am Coll Nutr 1992;11(2):107–25.

[31] Fletcher AE, Breeze E, Shetty PS. Antioxidant vitamins and mortality in older persons: findings from the nutrition add-on study to the medical research council trial of assessment and management of older people in the community. Am J Clin Nutr 2003;78(5):999–1010.

[32] Kaaks R, Tuyns AJ, Haelterman M, Riboli E. Nutrient intake patterns and gastric cancer risk: a case-control study in Belgium. Int J Cancer 1998;78(4):415–20.

[33] Harrison FE, Bowman GL, Polidori MC. Ascorbic acid and the brain: rationale for the use against cognitive decline. Nutrients 2014;6(4):1752–81.

[34] Ribas V, Garcia-Ruiz C, Fernandez-Checa JC. Glutathione and mitochondria. Front Pharmacol 2014;5:151 <http://dx.doi.org/10.3389/fphar.2014.00151>. eCollection@2014:151.

[35] Mari M, Morales A, Colell A, Garcia-Ruiz C, Fernandez-Checa JC. Mitochondrial glutathione, a key survival antioxidant. Antioxid Redox Signal 2009;11(11):2685–700.

[36] Sekhar RV, Patel SG, Guthikonda AP, Reid M, Balasubramanyam A, Taffet GE, et al. Deficient synthesis of glutathione underlies oxidative stress in aging and can be corrected by dietary cysteine and glycine supplementation. Am J Clin Nutr 2011;94(3):847–53.

[37] Erden-Inal M, Sunal E, Kanbak G. Age-related changes in the glutathione redox system. Cell Biochem Funct 2002;20(1):61–6.

[38] Maher P. The effects of stress and aging on glutathione metabolism. Ageing Res Rev 2005;4(2):288–314.

[39] Deutsche Gesellschaft für Ernährung (DGE). Sekundäre Pflanzenstoffe und ihre Wirkung auf die Gesundheit. DGEinfo – Forschung, Klinik, Praxis 2010;01.

[40] Schleicher M, Weikel K, Garber C, Taylor A. Diminishing risk for age-related macular degeneration with nutrition: a current view. Nutrients 2013;5(7):2405–56.

[41] Wang Y, Chun OK, Song WO. Plasma and dietary antioxidant status as cardiovascular disease risk factors: a review of human studies. Nutrients 2013;5(8):2969–3004.

[42] Lee J, Koo N, Min DB. Reactive oxygen species, aging, and antioxidative nutraceuticals. Compr Rev Food Sci Food Saf 2004;3(1):21–33.

[43] Vrolijk MF, Opperhuizen A, Jansen EH, Godschalk RW, Van Schooten FJ, Bast A, et al. The shifting perception on antioxidants: the case of vitamin E and beta-carotene. Redox Biol 2015;4C:272–8 <http://dx.doi.org/10.1016/j.redox.2014.12.017>.

[44] Bund für Lebensmittelrecht und Lebensmittelkunde (BLL). Informationsblatt Beta-Carotin. BLL – Informationsblatt 2006.

[45] Horwitt MK. Status of human requirements for vitamin E. Am J Clin Nutr 1974;27(10):1182–93.

[46] Abner EL, Schmitt FA, Mendiondo MS, Marcum JL, Kryscio RJ. Vitamin E and all-cause mortality: a meta-analysis. Curr Aging Sci 2011;4(2):158–70.

[47] Rizvi S, Raza ST, Ahmed F, Ahmad A, Abbas S, Mahdi F. The role of vitamin E in human health and some diseases. Sultan QaboosUniv Med J 2014;14(2):e157–65.

[48] Junqueira VB, Barros SB, Chan SS, Rodrigues L, Giavarotti L, Abud RL, et al. Aging and oxidative stress. Mol Aspects Med 2004;25(1–2):5–16.

[49] Bjelakovic G, Nikolova D, Gluud LL, Simonetti RG, Gluud C. Antioxidant supplements for prevention of mortality in healthy participants and patients with various diseases. Cochrane Database Syst Rev 2012;3:CD007176. http://dx.doi.org/10.1002/14651858. CD007176.pub2.:CD007176.

[50] Bjelakovic G, Nikolova D, Gluud C. Meta-regression analyses, meta-analyses, and trial sequential analyses of the effects of supplementation with beta-carotene, vitamin A, and vitamin E singly or in different combinations on all-cause mortality: do we have evidence for lack of harm? PLoS ONE 2013;8(9):e74558.

[51] Cardenas E, Ghosh R. Vitamin E: a dark horse at the crossroad of cancer management. Biochem Pharmacol 2013;86(7):845–52.

[52] Khurana S, Venkataraman K, Hollingsworth A, Piche M, Tai TC. Polyphenols: benefits to the cardiovascular system in health and in aging. Nutrients 2013;5(10):3779–827.

[53] Kwon Y. Effect of soy isoflavones on the growth of human breast tumors: findings from preclinical studies. Food Sci Nutr 2014;2(6):613–22.

[54] Birringer M. Hormetics: dietary triggers of an adaptive stress response. Pharm Res 2011;28(11):2680–94.

[55] Jacob KD, Noren HN, Trzeciak AR, Evans MK. Markers of oxidant stress that are clinically relevant in aging and age-related disease. Mech Ageing Dev 2013;134(3–4):139–57.

[56] Bjelakovic G, Nikolova D, Gluud LL, Simonetti RG, Gluud C. Mortality in randomized trials of antioxidant supplements for primary and secondary prevention: systematic review and meta-analysis. J Am Med Assoc 2007;297(8):842–57.

[57] Ristow M, Schmeisser K. Mitohormesis: promoting health and lifespan by increased levels of reactive oxygen species (ROS). Dose Response 2014;12(2):288–341.

[58] Cook R, Calabrese EJ. The importance of hormesis to public health. Environ Health Perspect 2006;114(11):1631–5.

[59] Prickett CD, Lister E, Collins M, Trevithick-Sutton CC, Hirst M, Vinson JA, et al. Alcohol: friend or foe? Alcoholic beverage hormesis for cataract and atherosclerosis is related to plasma antioxidant activity. Nonlinearity Biol Toxicol Med 2004;2(4):353–70.

[60] Lustig RH. Fructose: it's "alcohol without the buzz". Adv Nutr 2013;4(2):226–35.

[61] Fusco D, Colloca G, Lo Monaco MR, Cesari M. Effects of antioxidant supplementation on the aging process. Clin Interv Aging 2007;2(3):377–87.

[62] Vos MB, Kimmons JE, Gillespie C, Welsh J, Blanck HM. Dietary fructose consumption among US children and adults: the third national health and nutrition examination survey. Medscape J Med 2008;10(7):160.

[63] World Health Organization and International Programme on Chemical Safety. Biomarkers and risk assessment: concepts and principles/published under the joint sponsorship of the United Nations environment Programme, the International Labour Organisation, and the World Health Organization. Geneva: World Health Organization; 1993.

[64] World Health Organization. Use of glycated haemoglobin (HbA1c) in the diagnosis of diabetes mellitus. World Health Organization; 2011, WHO/NMH/CHP/CPM/11.1.

[65] Tardif JC, Heinonen T, Orloff D, Libby P. Vascular biomarkers and surrogates in cardiovascular disease. Circulation 2006;113 (25):2936–42.

[66] Strimbu K, Tavel JA. What are biomarkers? Curr Opin HIV AIDS 2010;5(6):463–6.

[67] Campbell DR, Gross MD, Martini MC, Grandits GA, Slavin JL, Potter JD. Plasma carotenoids as biomarkers of vegetable and fruit intake. Cancer Epidemiol Biomarkers Prev 1994;3(6): 493–500.

[68] Mayne ST. Antioxidant nutrients and chronic disease: use of biomarkers of exposure and oxidative stress status in epidemiologic research. J Nutr 2003;133(Suppl. 3):933S–40S.

[69] Markowitz ME, Rosen JF, Mizruchi M. Circadian variations in serum zinc (Zn) concentrations: correlation with blood ionized calcium, serum total calcium and phosphate in humans. Am J Clin Nutr 1985;41(4):689–96.

[70] Hambidge M. Biomarkers of trace mineral intake and status. J Nutr 2003;133(Suppl. 3):948S–55S.

[71] Pallauf K, Giller K, Huebbe P, Rimbach G. Nutrition and healthy ageing: calorie restriction or polyphenol-rich "MediterrAsian" diet? Oxid Med Cell Longev 2013;2013:707421. <http://dx.doi.org/10.1155/2013/707421>. Epub@2013 Aug 28.:707421

[72] Anton S, Leeuwenburgh C. Fasting or caloric restriction for healthy aging. Exp Gerontol 2013;48(10):1003–5.

[73] Whitlock G, Lewington S, Sherliker P, Clarke R, Emberson J, Halsey J, et al. Body-mass index and cause-specific mortality in 900,000 adults: collaborative analyses of 57 prospective studies. Lancet 2009;373(9669):1083–96.

[74] Raiten DJ, Namaste S, Brabin B, Combs Jr G, L'Abbe MR, Wasantwisut E, et al. Executive summary – biomarkers of nutrition for development: building a consensus. Am J Clin Nutr 2011;94(2):633S–50S.

[75] Mathers JC. Nutrition and ageing: knowledge, gaps and research priorities. Proc Nutr Soc 2013;72(2):246–50.

[76] Sebastiani P, Bae H, Sun FX, Andersen SL, Daw EW, Malovini A, et al. Meta-analysis of genetic variants associated with human exceptional longevity. Aging (Albany, NY) 2013;5(9): 653–61.

[77] Ortuño-Sahagún D, Pallàs M, Rojas-Mayorquín AE. Oxidative Stress in Aging: Advances in Proteomic Approaches. Oxid Med Cell longev 2014;2014:18. Available from: http://dx.doi.org/10.1155/2014/573208, Article ID 573208.

[78] Vickers MH. Early life nutrition, epigenetics and programming of later life disease. Nutrients 2014;6(6):2165–78.

CHAPTER 9

Food Preferences in the Elderly: Molecular Basis

Lorenzo M. Donini, Eleonora Poggiogalle and Valeria del Balzo

Food Science and Human Nutrition Research Unit, Medical Pathophysiology, Food Science and Endocrinology Section, Department of Experimental Medicine, Sapienza University of Rome, Rome, Italy

KEY FACTS

- Protein energy malnutrition is associated with functional impairment, comorbidity, delayed recovery from acute events, and increased mortality.
- Physiological modifications together with modifications of clinical/nutritional status and to loss of motivations may represent an important determinant of food choices.

Dictionary of Terms

- *Malnutrition* results from eating a diet in which nutrients are either not enough (undernutrition, protein-calorie malnutrition) or are too much (overnutrition, obesity) such that the modification of nutritional status may predispose to health problems. Malnutrition may refer to energy and/or nutrients (protein, carbohydrates, fat, vitamins, minerals, bioactive molecules, ...).
- *Nutritional status*: the condition of the body in those respects influenced by the diet, the levels of nutrients in the body, and the ability of those levels to maintain normal metabolic integrity. Nutritional status is assessed by measuring body composition, by evaluating food intake, energy and nutrient requirement, and by assessing functional status.
- *Senile anorexia* refers to a loss of appetite and/or reduced food intake. It affects a significant number of elderly people and it is more prevalent in institutionalized and frail individuals. Senile anorexia depends on a multifactorial origin characterized by various combinations of physiological, pathological, environmental, and iatrogenic conditions.
- *Sensory property* of foods depends on the integration of sensory attributes including product taste, texture, and appearance with consumer attitudes and health biases. Both sensory and attitudinal variables determine food preferences, product purchase, and food consumption.
- *Orexigenic and anorexigenic hormones*: the gastrointestinal tract, the central nervous system, and the adipose tissue (AT), referred to as the AT–gut–brain axis, produce a series of hormones with orexigenic and anorexigenic effects. Ghrelin represents a regulatory circuit controlling appetite and energy homeostasis by stimulating the release of other orexigenic peptides and neurotransmitters as well as NPY. Alternatively, anorexigenic CCK, PYY, and leptin have an opposite effect at the hypothalamic level. The differential release of these hormones may act to initiate, maintain, or exacerbate cycles of food restriction.

INTRODUCTION: THE CONCEPT OF NUTRITIONAL FRAILTY

A close relationship exists between nutritional status and health in the elderly. Changes in body composition are physiological during the aging process and are mainly due to the decrease in lean body mass. Furthermore, the presence of chronic diseases, multiple medications, cognitive impairment, depression, and social isolation can act synergistically with the decline in digestive, olfactory, and salivary functions, as well as in hormonal profile, affecting the nutritional status [1,2].

A wealth of studies have shown that elderly people eat less food with an average 30% decline in energy intake between the ages of 20 and 80 years of age (cross-sectional NHANES [III] study from the United States of America; longitudinal study from New Mexico). This anorexia of aging is characterized by reduced appetite, slow eating, smaller meals, and fewer snacks between meals [3].

Senile anorexia may predispose elderly subjects to protein energy malnutrition in particular in the presence of other "pathological" factors associated with aging. Therefore, malnutrition is a very frequent condition in the frailest groups of the population: available data in the literature show that up to 15% of community-dwelling and home-bound elderly, 23—62% of hospitalized subjects, and up to 85% of nursing home residents are malnourished [2,4].

Protein energy malnutrition is associated with functional impairment (due to impaired muscle function and/or decreased bone mass), comorbidity (related to immune dysfunction, anemia, reduced cognitive function, poor wound healing), delayed recovery from acute events, and increased mortality [3,5—7].

Finally elderly subjects may be considered as nutritionally frail: their nutritional status is at an increased vulnerability to stressors. Social, psychological, functional, and clinical difficulties may lead to malnutrition (over- and undernutrition), more easily than in younger subjects. Moreover, in the elderly, the consequences of malnutrition may be more insidious and potentially harmful while the recovery from malnutrition is more difficult [1].

PHYSIOLOGICAL MODIFICATIONS LEADING TO REDUCED ENERGY AND/OR NUTRIENT INTAKE

Physiological changes in gastrointestinal function that occur with aging may have a significant impact on appetite and food intake in older people. When these factors are combined with adverse social and psychological aspects of aging many of this group are at risk of developing sarcopenia and protein-energy malnutrition: leading to an increased risk of acute and chronic illness, hospitalization, and loss of independence, in turn, leading to a greater dependence on community and healthcare facilities [8].

Appetite and energy intake are controlled by a central feeding drive and a peripheral satiety system integrating feedback mechanisms, linking energy intake to usage and storage. These mechanisms include inputs from mechano- and chemoreceptors that respond to the presence of food in the gastrointestinal tract. In particular gastrointestinal satiety signals depends on gastric distension (a reduced ability to extend the gastric fundus is frequent in the elderly and may explain, at least in part, the reduction of food intake) and the interaction of nutrients with gastrointestinal tract receptors that stimulate the release of satiety hormones (cholecystokinin [CCK], glucagon-like peptide-1, gastric inhibitory peptide, and amylin), and inhibit the release of ghrelin, which stimulates feeding. These hormones play a role in both satiation (meal termination) and satiety (time to subsequent meal consumption). In the elderly a relative increase of the effect of satiety hormones (CCK in particular) and a reduction of the action of ghrelin are present [8].

The physiological increased drive to eat, following an acute event leading to a loss of body weight, may be reduced in elderly people and may be explained by the above-mentioned factors. After periods of either underfeeding or overfeeding younger men readjusted their energy intakes, while older men did not, suggesting that long-term energy balance mechanisms may be different in older adults [9].

Molecular circadian clocks might have evolved to synchronize internal metabolic rhythms to predictable environmental cycles to most advantageously time functions such as feeding [10]. In addition, a growing body of evidence suggests reciprocal links between nutrient-sensing pathways and circadian clocks [11]. The mammalian circadian clock is constituted by a coordinator center which integrates cues derived from peripheral tissues with external light:dark cycles. The circadian clock machinery seems to be affected in the elderly and may, in turn, condition nutrient metabolism. During aging there is a decline in circadian rhythms that may affect the overall homeostasis of the organism. The progressive deterioration of neuronal networks in the circadian system may lead to the destabilization of daily circadian oscillations in hormones such as leptin, insulin, and glucagon affecting food intake and metabolic homeostasis [12,13].

Relative satiating effects of different macronutrients may change with age. Oral liquid preloads (either high fat, high carbohydrate, or high protein drinks) result in greater suppression of food intake and fat or carbohydrate consumption in older compared to young adults [8]. In fact among the physiological mechanisms governing dietary selection, postingestive nutritional factors (concentration of the nutrients in the diet) have been shown to influence food intake. Nutrient-sensing mechanisms that play an important role in food intake modulation according to the nutrient content in the diet are affected during the aging process leading to an inappropriate dietary selection and food choice. In self-selection experiments performed in rats, aging induced a significant shift in energetic nutrient preference from carbohydrates to fat while other studies have highlighted the regulation of protein intake as a function of amino acid composition in the diet, through the action of the GCN2 amino acid sensor [14].

A number of diverse signaling proteins (acting as hormones or as neurotransmitters or neuromodulators) seem to affect energy intake. Together with protein receptors on specific target cells and other biochemical factors, signaling proteins relay information on nutritional status. They represent important metabolic satiety signals or signals necessary to increase energy intake. Some of these signaling proteins (neuropeptide Y [NPY], orexins, amylin, CCK, ghrelin, glucagon, pancreatic polypeptide, and peptide YY [PYY]) undergo measurable changes in RNA or protein levels (in animal models or human studies) that are reported to be associated with anorexia or unintentional weight loss [9].

Aging is associated with the general downregulation of hypothalamic peptides that stimulate food intake and unchanged expression of anorexigenic peptides. Experimental studies trying to understand the mechanisms related to aging-related impairments of food intake (performed mainly on rat, and to some extent, on nonhuman primates) found that the expression of NPY, and of its receptors, is highly suppressed in the hypothalamus of old rats. Moreover, the expression of NPY mRNA after fasting was severely depressed in old as compared to young rats. Significant changes are reported for other hypothalamic orexigenic compounds (agouti-related peptide and orexins, in particular) while no effect was found on anorexigens such as cocaine- and amphetamine-regulated transcript and alpha-melanocyte stimulating hormone [15].

Ghrelin and insulin are, respectively, orexigenic and anorexigenic hormones regulating feeding behavior, energy, and glucose metabolism. Moreover they are known to influence aging since they are implicated in learning/memory and cognitive decline in Alzheimer disease. The impaired action of these hormones, which is frequent in the elderly, may therefore affect the eating behavior and therefore nutritional status [16].

MODIFICATIONS OF CHEMOSENSORY FUNCTIONS AND FOOD PREFERENCES IN THE ELDERLY

Individual differences in the ability to detect different tasting compounds (such as bitter taste) has long been recognized as a common genetic trait (eg, TAS2R receptor gene family encode taste receptors on the tongue) that can explain some of the variability in the dietary habits of a population. However, an association between genotype and food preferences has been found among children, but no associations have been reported among older adults, suggesting that environmental factors are more important than genetic influences in food preferences among the elderly [17–19].

In the elderly, tasting and olfactory function are essential in providing healthy nutrition and a balanced diet. Elderly patients are more likely to wear dentures, take medications, and suffer from comorbidities, which, together or alone, make the aging population much more likely to experience sensorial disorders. Elderly patients with chemosensory complaints will most likely use more salt or sugar, and eat less than an adequate amount or quality of food [20].

In particular, taste distortions in human beings have been attributed to various physiological and environmental factors including aging and disease conditions. Aging is associated with a reduced ability to taste foods thereby reducing the possibility to select such foods. In addition, age-related diseases, such as degenerative (eg, stroke, dementia) and cardiovascular diseases, have also been reported to play important roles in the distortion of chemosensory functions as the result of degeneration of regions of the brain that mediate olfactory functions. Moreover oral health, accumulative effects of drug administration, and zinc deficiency (although Zn supplementation in order to restore taste acuity produced conflicting results in different studies) may impair taste perception. It has to be considered that there may be synergistic effects of all or most of the aforementioned factors [17,21–33].

MODIFICATIONS OF CLINICAL AND NUTRITIONAL STATUS LEADING TO CHANGES IN FOOD PREFERENCES

It has been shown that in the elderly, food preferences are greatly affected by health status and functional abilities. The nature of the health condition and its effect on physiological function will drive the dietary change process. The extent to which healthcare providers counsel their patients to incorporate changes into their diet and the mindset and belief system of each individual may also have a strong impact on food choices. Impairment of health and/or functional status may cause a vicious circle. On one hand, the ability to procure and prepare food and eat independently is essential to having an adequate diet. On the other, a poor diet can contribute to frailty and complicate functional limitations. In fact, nutrition surveys of home-dwelling elderly subjects with functional disabilities or poor health have suggested a high prevalence of inadequate food consumption [27,34–37].

Age-associated increases in the production and/or effect of satiating cytokines (secreted in response to acute illness and significant stress) may contribute to the anorexia of aging. In particular interleukin 1 (IL-1), interleukin 6 (IL-6), and tumor necrosis factor alpha (TNF-α) act to decrease food intake and reduce body weight via a number of central and peripheral pathways. Aging itself seems to be associated with stress-like changes in circulating hormonal pattern (increased cortisol and catecholamines and decreased sex hormones and growth hormone and inflammatory pattern (increased production of IL-6, TNF-α, and IL-1). Increased cytokine levels, due to the "stress" of aging per se, or the amplified stressful effects of acute or chronic illness, may be considered an essential contributor to the decline in appetite that occurs in many older people [3].

Cognitive decline and dementia in particular seem to be linked to nutritional status [38]. Nutrition may play an important role in the causation and prevention of age-related cognitive decline and dementia (positive associations have been shown between antioxidants, folate, omega-3 and omega-6 fatty acids, and cognitive performance), while dementia and cognitive decline may cause an important perturbation of eating behavior leading to altered eating patterns and malnutrition [27,33,39,40].

OTHER DETERMINANTS OF FOOD CHOICE AND CONCLUSIONS

Food choices are based upon a complex interaction between the social and environmental context, the individual, and the food. Motivations and perceived barriers may represent an important determinant of food choices made, in particular, by homebound older adults. Key motivations in food choice may be represented by sensory appeal, convenience, and price, while key barriers may embody health, being on a special diet, and being unable to shop. To change eating behaviors, interventions aimed at preventing malnutrition must take into account individuals' self-perceived motivations and barriers to food selection. Considering foods that are tasty, easy to prepare, inexpensive, and that involve caregivers is critical for successful interventions [41].

Behavioral research suggests that amount consumed and rate of eating can be strongly affected by a number of characteristics related to the food consumed. In particular increases in the pleasantness of foods and greater familiarity with foods has been found to result in greater consumption. In different studies the use of condiments and sauces resulted in greater intakes of energy, energy consumed from protein, and energy consumed from fat regardless of premeal hunger and desire to eat, or postmeal pleasantness [42].

The sensory properties of foods and beverages are among the most important determinants of food choice. Hedonic properties of foods are related to flavor components including spices that seem to influence ingestive behavior. These flavor-active compounds are also involved in digestive, absorptive, and metabolic processes through direct activation of signaling pathways or via neurally mediated cephalic phase responses.

However, the efficacy of flavor fortification to increase energy and nutrient intake largely focused on the aging population is not strong, possibly due to methodological issues. Greater short-term energy intake has been reported in a small sample with natural flavors added to lunch and dinner meals of hospitalized elderly and in elderly residents of a nursing home [43–46].

The food industry may have an important role in trying to combine the necessity to ensure nutritional needs and sensory perception to compensate for age-related impairments in odor, flavor, trigeminal mouth feel, and texture perception [32,47,48].

SUMMARY

- A close relationship exists between nutritional status and health in the elderly.
- Changes in body composition (in particular sarcopenia) together with chronic diseases, multiple medications, cognitive impairment, depression, and social isolation can act synergistically with the decline in digestive, olfactory, and salivary functions, as well as in hormonal profile, affecting the nutritional status.
- Senile anorexia may predispose elderly subjects to protein energy malnutrition which is associated with functional impairment (due to impaired muscle function and/or decreased bone mass), comorbidity (related to immune dysfunction, anemia, reduced cognitive function, poor wound healing), delayed recovery from acute events, and increased mortality.
- Physiological modifications (changes in chemosensory function, impairment of gastrointestinal function, modified satiating effects of different macronutrients, and downregulation of signaling proteins and hormones) may lead to reduced energy and/or nutrient intake.
- Modifications of clinical and nutritional status (eg, disability and cognitive decline) together with motivations and perceived barriers may represent an important determinant of food choices.
- The food industry may have an important role in trying to combine the necessity to ensure nutritional needs and sensory perception to compensate for age-related impairments.

References

[1] Donini LM, Scardella P, Piombo L, Neri B, Asprino R, Proietti AR, et al. Malnutrition in elderly: social and economic determinants. J Nutr Health Aging 2013;17(1):9–15.

[2] Chapman IM. Nutritional disorders in the elderly. Med Clin North Am 2006;90:887–907.

[3] Chapman IM. Endocrinology of anorexia of ageing. Best Pract Res Clin Endocrinol Metab 2004;18(3):437–52.

[4] Saffrey MJ. Aging of the enteric nervous system. Mech Ageing Dev 2004;125:266–71.

[5] Donini LM, Savina C, Cannella C. Eating habits and appetite control in the elderly: the anorexia of aging. Int Psychogeriatr 2003;15(1):73–87.

[6] Donini LM, Savina C, Piredda M, Cucinotta D, Fiorito A, Inelmen EM, et al. Senile anorexia in acute-ward and rehabilitation settings. J Nutr Health Aging 2008;12(8):511–17.

[7] Donini LM, Dominguez LJ, Barbagallo M, Savina C, Castellaneta E, Cucinotta D, et al. Senile anorexia in different geriatric settings in Italy. J Nutr Health Aging 2011;15 (9):775–81.

[8] Parker BA, Chapman IM. Food intake and ageing—the role of the gut. Mech Ageing Dev 2004;125:859–66.

[9] Wernette CM, White D, Zizza CA. Signaling proteins that influence energy intake may affect unintentional weight loss in elderly persons. J Am Diet Assoc 2011;111:864–73.

[10] Mazzoccoli G. Clock genes and clock-controlled genes in the regulation of metabolic rhythms. Chronobiol Int 2012;29:227–51.

[11] Peek CB. Nutrient sensing and the circadian clock. Trends Endocrinol Metab 2012;23:312–18.

[12] Kondratova AA, Kondratov RV. The circadian clock and pathology of the ageing brain. Nat Rev Neurosci 2012;13:325–35.

[13] Tevy MF, Giebultowicz J, Pincus Z, Mazzoccoli G, Vinciguerra M. Aging signaling pathways and circadian clock-dependent metabolic derangements. Trends Endocrinol Metab 2013;24 (5):229–37.

[14] Maurin AC, Chaveroux C, Lambert-Langlais S, Carraro V, Jousse C, Bruhat A, et al. The amino acid sensor GCN2 biases macronutrient selection during aging. Eur J Nutr 2012;51:119–26.

[15] Kmiec Z. Central regulation of food intake in ageing. J Physiol Pharmacol 2006;57(Suppl. 6):7–16.

[16] Maejima Y, Kohno D, Iwasaki Y, Yada T. Insulin suppresses ghrelin-induced calcium signaling in neuropeptide Y neurons of the hypothalamic arcuate nucleus. Aging 2011;3(11):1092–7.

[17] Navarro-Allende A, Khataan N, El-Sohemy A. Impact of genetic and environmental determinants of taste with food preferences in older adults. J Nutr Elder 2008;27(3–4):267–76.

[18] Toffanello E, Inelmen E, Imoscopi A, Perissinotto E, Coin A, Miotto F, et al. Taste loss in hospitalized multimorbid elderly subjects. Clin Interv Aging 2013;8:167–74.

[19] Donini LM, Poggiogalle E, Piredda M, Pinto A, Barbagallo M, Cucinotta D, et al. Anorexia and eating patterns in the elderly. PLoS One 2013;8(5):e63539.

[20] Spielman AL. Chemosensory function and dysfunction. Crit Rev Oral Biol Med 1998;9(3):267–91.

[21] Aliani M, Udenigwe CC, Girgih AT, Pownall TL, Bugera JL, Eskin MNA. Zinc deficiency and taste perception in the elderly. Crit Rev Food Sci Nutr 2013;53(3):245–50.

[22] Doty RL, Shah M, Bromley SM. Drug-induced taste disorders. Drug Saf 2008;31:199–215.

[23] Drewnowski A. Sensory control of energy density at different life stages. Proc Nutr Soc 2000;59:239–44.

[24] Heckmann JG, Stossel C, Lang CJG, Neundorfer B, Tomandl B, Hummel T. Taste disorders in acute stroke: a prospective observational study on taste disorders in 102 stroke patients. Stroke 2005;36:1690–4.

[25] Murphy C. The chemical senses and nutrition in older adults. J Nutr Elder 2008;27(3–4):247–65.

[26] Schiffman SS. Critical illness and changes in sensory perception. Proc Nutr Soc 2007;66:331–45.

[27] Shatenstein B. Impact of health conditions on food intakes among older adults. J Nutr Elder 2008;27:333–61.

[28] Lau D. Role of food perception in food selection of the elderly. J Nutr Elder 2008;27:221–46.

[29] Lee JS, Kritchevsky SB, Tylavsky F, Harris TB, Ayonayon HN, Newman AB. Factors associated with impaired appetite in well functioning community-dwelling older adults. J Nutr Elder 2006;26:27–43.

[30] Mak YE, Simmons KB, Gitelman DR, Small DM. Taste and olfactory intensity perception changes following left insular stroke. Behav Neurosci 2005;119:1693–700.

[31] Aliani M, Udenigwe CC, Girgih AT, Pownall TL, Bugera JL, Eskin MNA. Aroma and taste perceptions with Alzheimer disease and stroke. Crit Rev Food Sci Nutr 2013;53(7):760–9.

[32] Koehler J, Leonhaeuser IU. Changes in food preferences during aging. Ann Nutr Metab 2008;52(Suppl. 1):15–19.

[33] Crichton GE, Bryan J, Murphy KJ, Buckley J. Review of dairy consumption and cognitive performance in adults: findings and methodological issues. Dement Geriatr Cogn Disord 2010;30 (4):352–61.

[34] Rahi B, Morais JA, Gaudreau P, Payette H, Shatenstein B. Decline in functional capacity is unaffected by diet quality alone or in combination with physical activity among generally healthy older adults with T2D from the NuAge cohort. Diab Res Clin Pract 2014;105:399–407.

[35] Payette H, Shatenstein B. Determinants of healthy eating in community-dwelling elderly people. Can J Public Health 2005;96:S30–5.

[36] Bernstein M, Munoz N. Position of the academy of nutrition and dietetics: food and nutrition for older adults: promoting health and wellness. J Acad Nutr Diet 2012;112:1255–77.

[37] Shatenstein B, Gauvin L, Keller H, Richard L, Gaudreau P, Giroux F. Baseline determinants of global diet quality in older men and women from the NuAge cohort. J Nutr Health Aging 2013;17:1–7.

[38] Donini LM, De Felice MR, Cannella C. Nutritional status determinants and cognition in the elderly. Arch Gerontol Geriatr 2007;44(Suppl. 1):143–53.

[39] Bryan J. Mechanisms and evidence for the role of nutrition in cognitive ageing. Ageing Int 2004;29:28–45.

[40] Solfrizzi V, Panza F, Capurso A. The role of diet in cognitive decline. J Neural Transm 2003;110:95–110.

[41] Locher JL, Ritchie CS, Roth DL, Sen B, Vickers-Douglas K, Vailas LI. Food choice among homebound older adults: motivation and perceived barriers. J Nutr Health Aging 2009;13 (8):659–64.

[42] Appleton KM. Increases in energy, protein and fat intake following the addition of sauce to an older person's meal. Appetite 2009;52:161–5.

[43] Mattes RD. Spices and energy balance. Physiol Behav 2012;107:584–90.

[44] Henry CJ, Woo J, Lightowler HJ, Yip R, Lee R, Hui E. Use of natural food flavours to increase food and nutrient intakes in hospitalized elderly in Hong Kong. Int J Food Sci Nutr 2003;54:321–7.

[45] Mathey MF, Siebelink E, de Graaf C, Van Staveren WA. Flavor enhancement of food improves dietary intake and nutritional

status of elderly nursing home residents. J Gerontol A Biol Sci Med Sci 2001;56:M200–5.

[46] Dermiki M, Mounayar R, Suwankanit C, Scott J, Kennedy OB, Mottram DS, et al. Maximising umami taste in meat using natural ingredients: effects on chemistry, sensory perception and hedonic liking in young and old consumers. J Sci Food Agric 2013;93:3312–21.

[47] Costa AIA, Jongen WMF. Designing new meals for an ageing population. Crit Rev Food Sci Nutr 2010;50(6):489–502.

[48] Donini LM, Savina C, Cannella C. Nutritional interventions in the anorexia of aging. J Nutr Health Aging 2010;14(6): 494–6.

PART II

MOLECULAR AND CELLULAR TARGETS

CHAPTER

10

Telomeres, Aging, and Nutrition

Varinderpal Dhillon, Caroline Bull and Michael Fenech

CSIRO Food and Nutrition, Adelaide, SA, Australia

KEY FACTS

- Telomeres are nucleoprotein structures at the ends of chromosomes that are essential for maintaining chromosomal stability.
- Telomeric DNA consists of a repeat TTAGGG sequence and together with its complementary sequence, AATCCC, forms a T-loop structure at the terminal ends of the sequence.
- The length of telomeres can vary between 4-11 kilobases in humans. However, telomeres become dysfunctional if they become too short to form the T-loop structure or if they accumulate an excess of DNA strand breaks and base damage.
- Telomere shortening and telomere DNA damage accumulates with age and as a consequence of poor nutrition and/or exposure to endogenous and environmental genotoxins.
- Telomere length in humans has been shown to be influenced beneficially by dietary factors such as the Mediterranean dietary pattern, higher intake of plant foods, adequate folate status, higher intake of 3-fatty acids, and weight loss.
- Telomere shortening in humans has been shown to be associated with increased intake of red meat, processed meat, sweetened carbonated beverages, sodium, white bread and with obesity.

Dictionary of Terms

- Chromosomal instability—a cellular phenotype in which an abnormally high rate of structural chromosome aberrations, including chromosome rearrangements and gene amplifications, are recurringly generated during mitosis by chromosomal breakage-fusion-bridge cycles that are initiated by the formation of dicentric chromosomes caused by telomere end fusions or mis-repair of DNA strand breaks.
- PARPs—polyADPribose polymerases.
- Telomeres—hexamer TTAGGG repeats at the ends of chromosomes.
- Shelterin—Proteins (TRF1, TRF2, POT1, TIN2, TPP1, RAP1 and Tankyrase) associated with telomere DNA to form the shelterin complex which interacts with enzymes such as telomerase and other proteins required for proper telomere maintenance and function.
- Telomeres—hexamer TTAGGG repeats at the ends of chromosomes.
- T/S ratio—number of telomere repeats (T) normalised to the number of copies (S) of a reference gene, such as GAPDH and expressed as a T/S ratio.

INTRODUCTION

What Are Telomeres?

Telomeres are nucleoprotein structures at the end of each chromosome. The nucleic acid sequence of telomeres is a TTAGGG hexamer repeat and its complementary sequence AATCCC [1]. The number of these hexamer repeats can vary greatly from very few repeats to thousands of repeats with most reported telomere lengths varying between 4 and 11 kilobases in humans [2,3]. Several proteins (TRF1, TRF2, POT1, TIN2, TPP1, RAP1, and Tankyrase) are associated with the telomere sequence to form the shelterin complex which interacts with enzymes, such as telomerase, and other proteins required for proper telomere maintenance and function [2,4].

What Are the Mechanisms of Telomere Maintenance?

Telomeres can only function properly if they are long enough to form the T-loop structure (Fig. 10.1) which is essential to prevent the telomeric ends of chromosomes being wrongly identified as double strand breaks (DSBs) in DNA [4,5]. This is important to avoid because DNA repair of DSBs leads to fusion of broken ends; if this were to happen to telomeres it would result in the formation of abnormal chromosomes such as dicentric chromosomes. Dicentric chromosomes are likely to be pulled to opposite poles of the cell forming anaphase bridges which eventually break leading to daughter chromosomes without telomere sequences at one end. This event triggers breakage–fusion–bridge cycles that further amplify chromosomal abnormalities and gene amplification leading to aberrant cells from which cancers and other abnormalities may arise [5–7].

Telomere maintenance is based on at least three functions: (1) ability to restore telomere length to an appropriate size to form the T-loop structure, (2) accurate replication of the telomere sequence, and (3) prevention of damage to the telomere sequence by endogenous or exogenous genotoxins. The ability to restore or lengthen telomeres depends mainly on the action of telomerase and tankyrase [4,8]. Telomerase is the enzyme complex that can add TTAGGG hexamer repeats to the telomere sequence and consists of three major proteins TERT, TERC, and Dyskerin [2,8]. Tankyrase is a member of the shelterin complex and acts as a telomeric poly(ADP-ribose) polymerase to promote telomerase access by inhibiting TRF1, another member of the shelterin complex. Accurate replication of the telomere sequence depends on availability of nutrient cofactors essential for DNA replication which include folate, zinc, and magnesium [9,10]. Surprisingly it has become evident that DNA damage to the telomere sequence (eg, DNA strand breaks, UV-induced thymidine dimers) is not repaired [11,12]. Furthermore such damage could cause replication stress leading to formation of single- or double-stranded DNA breaks in the telomere sequence which could accelerate telomere attrition. Therefore, mitigation of exposure of telomeres to endogenous or exogenous genotoxic agents is important for their integrity to be maintained. A possible approach is to optimize dietary intake of cofactors required for function of antioxidant enzymes (eg, selenium, zinc) to prevent oxidative damage to the telomere sequence [13,14] or intake of nutrients that may prevent UV damage, such as thymidine dimers, to the telomere sequence (eg, folate) [15].

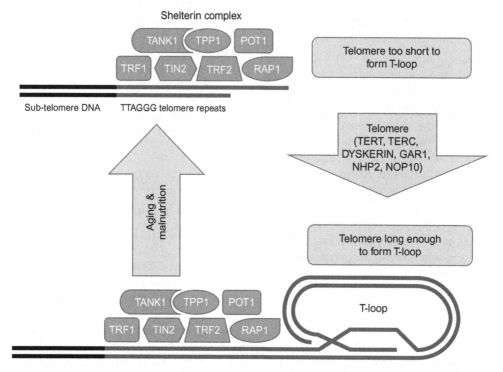

FIGURE 10.1 The telomere and shelterin structure when telomeres are excessively short and when they are long enough to form the T-loop structure. Note: it also possible that the T-loop may not form efficiently due to damage in the telomere sequence.

What Happens When Telomeres Become Too Short?

The actual terminus of mammalian telomeres is not blunt-ended, but consists of a single-stranded 3′-protrusion of the G-strand, known as the 3′ overhang [5,6]. The 3′ overhang of mammalian telomeres varies between 50 and 500 nucleotides, which is considerably longer than the protrusion of most other eukaryotes. T-loops are presumably formed through strand invasion of the duplex telomeric repeat by the 3′ overhang. As indicated above there is a critical length that telomeres must achieve to form the T-loop. PCR mediated analysis of short telomeres in human fibroblasts showed that at least 13 TTAGGG repeats (78 bp) are required to prevent telomere fusions [16,17]. This means a telomere that is less than 78 bp long can be considered to be too short. Other experiments showed that as little as 400 bp of TTAGGG repeats can seed the formation of a fully functional telomere [18,19]. When a telomere becomes so short that T-loops cannot be formed the risk of telomere end-fusions increases and could result in rampant chromosomal instability which may trigger the evolution of cancer cells or senescence [20].

What Happens When Telomeres Are Too Long?

It is not yet clear whether telomeres may become too long for a lengthy and healthy lifespan. However, there are reasons to suspect that excessively long telomeres may not be beneficial. Mice have telomeres that are up to ten times longer than humans, however, their lifespan is 30 times shorter [21]. Therefore, one could argue that it is the quality rather than the length that matters. In addition, telomere length distribution in leukocytes from exceptionally old individuals exhibited overrepresentation of ultra-short telomeres compared with younger groups [22]. In addition both shorter and longer telomeres have been associated prospectively with cognitive decline [23] which suggests that excessively long telomeres may also be detrimental to cell survival or function and that there may be an optimal but restricted telomere length range for longevity. One plausible explanation is that keeping functional telomeres as short as possible has the advantage that it makes them a smaller target for DNA damage [24].

What Happens When Telomeres Are Damaged?

It has been recently shown that damage to telomeric DNA, such as DNA strand breaks, is not efficiently repaired and leads to the accumulation of phosphorylated yH2AX [12]. The persistent DNA damage yH2AX signal induces the senescence associated secretory phenotype (SASP), which is a complex mixture of cytokines that is intended to trigger cell senescence mechanisms to immunologically remove cells with damaged telomeres and trigger cellular regeneration [25,26]. However, with aging the efficiency of immunologically removing the accumulating number of senescent cells is diminished leading to an increasingly pro-inflammatory state [25,27].

TELOMERES AND AGING

How Do Telomeres Change with Age?

Generally telomeres reduce in length with aging. This reduction may be due to telomere attrition that occurs as a result of the telomere end-replication problem [28]. Another explanation is that unrepaired double-stranded DNA breaks lead to the formation of terminal fragments containing telomere sequences that are lost during mitosis as they lack centromeres required to engage with the mitotic spindle. With aging, other insults to the telomere sequence may accumulate, such as point mutations, which may be caused by folate deficiency leading to uracil incorporation instead of thymidine or formation of adducts on DNA bases which may either be small (eg, 8-oxo-guanine) or large (eg, benzo-[a]-pyrene diol epoxide adducts). The accumulation of adducts on the telomere sequence interferes with binding of shelterin proteins, such as TRF1 and TRF2, which play a key role in telomere replication, recruitment of helicases, and T-loop formation [29–32].

Does Telomere Length Predict Healthy Aging?

Several studies have shown that shorter telomere length in white blood cells, buccal cells, or saliva in humans is usually associated cross-sectionally or prospectively with increased risk of degenerative diseases of aging, such as cancer, cardiovascular disease, diabetes, and dementia [33–38]. However, more recently some exceptions have started to appear such as the association of longer telomeres in blood with increased risk of breast cancer, lung cancer, pancreatic cancer, melanoma and cognitive impairment [23,39–44]. This brings to attention the possibility that both critically short telomeres and dysfunctional long telomeres may be hallmarks of unhealthy aging.

Which Factors Apart from Nutrition Affect Telomere Length with Aging?

There are several factors apart from nutrition that can significantly affect telomere length. Mutations in genes coding for proteins required for telomerase activity (hTERT, DKC1) and the shelterin complex (eg, TRF2) lead to abnormal rates of telomere shortening and dysfunction [2]. Furthermore, it is evident from twin studies and parent−child association studies that there is a high heritability of telomere length and paternal age is positively correlated with telomere length [45,46]. There are however some unexpected exceptions such as the recent observation that those with mutations in the breast cancer genes BRCA1 and BRCA2 have longer telomeres than normal and that the increased telomere length occurred in both the carriers who had breast cancer and those who did not [47]. Psychological stress caused by different factors (eg, caring for those who are ill, sexual abuse, harsh parenting, neuroticism, pregnancy-specific stress in the mother, and social disadvantage) and lower socioeconomic status and lower attainment of education by the parents all impact adversely on telomere length in somatic tissues [48]. Among lifestyle factors, smoking is consistently associated with reduced telomere length even when it occurs indirectly via the mother during pregnancy [49]. In addition low physical activity and sedentary behaviors are generally linked with increased telomere attrition [50]. Several studies show that occupational or therapeutic exposure to genotoxins may lead to either lengthening or shortening of telomeres compared to unexposed controls indicating the need to understand both the genetic and environmental factors that may regulate telomere integrity [51−54]. There is growing evidence that inflammation including related conditions such as life-course persistent asthma and obesity contributes to accelerated telomere shortening [55,56]. The mechanisms for these associations are not yet well-defined but there is substantial evidence that increased oxidative stress can damage the telomere sequence by causing oxidation of guanine and inducing DNA strand breaks that either leads to disruption of the binding of shelterin complex proteins to the telomere DNA template or induction of persistent DNA damage response or loss of telomeres due to terminal deletion of the chromosome ends [12,29,57].

NUTRITION AND TELOMERES

Which Dietary Factors Are Associated with Telomere Length in Humans?

Several dietary factors have been associated with telomere length usually measured in blood leukocytes. These associations have been derived from molecular epidemiological cross-sectional or prospective studies examining dietary intake and/or blood micronutrient concentration as well as from dietary intervention studies. A summary of all the studies published to date is provided in Tables 10.1−10.3.

TABLE 10.1 Cross-Sectional Studies—Dietary Intake, Dietary Pattern, or Vitamin Supplement Use and Telomere Length

Participants & country	Longer TL	Shorter TL	Ref.
1958 middle-aged and older Korean adults	Prudent dietary pattern characterized by high intake of whole grains, seafood, legumes, vegetables, and seaweed	Red meat or processed meat and sweetened carbonated beverages	[58]
840 white, black, and Hispanic adults from the Multi-Ethnic Study of Atherosclerosis		After adjustment for age, other demographics, lifestyle factors, and intakes of other foods or beverages, only processed meat intake was associated with telomere length. For every 1 serving/d greater intake of processed meat, the T/S ratio was 0.07 smaller. Categorical analysis showed that participants consuming > or = 1 serving of processed meat each week had 0.017 smaller T/S ratios than did nonconsumers	[59]
4676 disease-free women from nested case-control studies within the Nurses' Health Study (USA)	Greater adherence to the Mediterranean diet was associated with longer telomeres		[60]
217 elderly subjects (in Italy) stratified according Mediterranean diet score (MDS) in low adherence (MDS ≤ 3), medium adherence (MDS 4−5), and high adherence (MDS ≥ 6) groups	High adherence group showed longer leukocyte telomere length ($p = 0.003$) and higher telomerase activity ($p = 0.013$) compared to others		[61]

(Continued)

TABLE 10.1 (Continued)

Participants & country	Longer TL	Shorter TL	Ref.
This study assessed the relationship between telomere length and changes in adiposity indices after a 5-year Mediterranean diet intervention in 521 subjects (55–80 years, 55% women) randomly selected from the PREDIMED-NAVARRA	Higher baseline telomere length significantly predicted a greater decrease in body weight, body mass index, waist circumference and waist to height ratio. The risk of remaining obese after 5 years was lower in those participants who initially had the longest telomeres and increased their TL after intervention (odds ratio = 0.27, 95% CI: 0.03–2.03)		[62]
521 subjects (55–80 years) participating in the Prevención con Dieta Mediterránea randomized trial were assessed over 5 years of a nutritional intervention, which promoted adherence to the Mediterranean diet (MeDiet) in Spain	A higher adherence to the MeDiet pattern strengthens the prevention of telomere shortening among Ala carriers of the variant Pro/Ala (rs1801282) in the PPAR-γ2. This association was modulated by MeDiet because those Ala carriers who reported better conformity to the MeDiet exhibited increased telomere length ($P < 0.001$). Moreover, a reduction in carbohydrate intake (≤ 9.5 g/d) resulted in increased TL among Ala carriers		[63]
1743 multiethnic community residents of New York aged 65 years or older	Higher adherence to a MeDiet was associated with longer leukocyte telomere length (LTL) among whites but not among African Americans and Hispanics. Additionally, a diet high in vegetables but low in cereal, meat, and dairy might be associated with longer LTL among healthy elderly		[64]
5309 USA adults, aged 20 to 65 years, with no history of diabetes or cardiovascular disease, from the 1999 to 2002 National Health and Nutrition Examination Surveys	100% fruit juice was marginally associated with longer telomeres	Sugar-sweetened soda consumption was associated with shorter telomeres	[65]
287 participants (55% males, 6–18 years), who were randomly selected from the GENOI study (Spain)	A positive correlation between dietary total antioxidant capacity and telomere length ($r = 0.157$, $p = 0.007$) was found after adjustment for age and energy intake	Higher white bread consumption was associated with shorter telomeres ($\beta = -0.204$, $p = 0.002$)	[66]
4029 apparently healthy postmenopausal women who participated in the Women's Health Initiative (USA)		Intake of short-to-medium-chain saturated fatty acids (SMSFAs; aliphatic tails of ≤ 12 carbons) was inversely associated with TL. Intakes of nonskim milk, butter, and whole-milk cheese (major sources of SMSFAs) were all inversely associated with TL.	[67]
2284 female participants from the Nurses' Health Study (USA)	Dietary fiber intake was positively associated with telomere length (z score), specifically cereal fiber, with an increase of 0.19 units between the lowest and highest quintiles ($P = 0.007$, P for trend = 0.03)	Polyunsaturated fatty acid intake, specifically linoleic acid intake, was inversely associated with telomere length after multivariate adjustment (-0.32 units; $P = 0.001$, P for trend = 0.05).	[68]
Framingham USA Offspring Study ($n = 1,044$, females = 52.1%, mean age 59 years) using data from samples collected before and after folic acid fortification		Multivitamin use was associated with shorter telomeres in this cohort ($P = 0.015$)	[69]
586 participants (age 35–74 years) in the Sister Study (USA)	After age and other potential confounders were adjusted for, multivitamin use was associated with longer telomeres. Compared with nonusers, the relative telomere length of leukocyte DNA was on average 5.1% longer among daily multivitamin users (P for trend = 0.002). Higher intakes of vitamins C and E from foods were each associated with longer telomeres, even after adjustment for multivitamin use		[70]
766 adolescents aged 14–18 years (50% female, 49% African Americans)		Higher dietary sodium intake was associated with shorter LTL in the overweight/obese group (BMI $\geq$ 85th percentile, $\beta = -0.37$, $P = 0.04$), but not in the normal weight group ($\beta = 0.01$, $P = 0.93$) after adjusting for multiple confounding factors	[71]

TABLE 10.2 Cross-Sectional Studies—Biomarkers of Nutritional Status and Telomere Length

Participants & country	Longer TL	Shorter TL	Ref.
43 younger (18–32 years) and 47 older (65–83 years) adults in South Australia		In older males, there was a significant inverse correlation between telomere length and plasma homocysteine ($r = -0.57$, $p = 0.004$), but this effect was not observed in the younger cohort or in the older female group	[72]
The association of plasma homocysteine and other CVD biomarkers with LTL were assessed in 100 samples drawn from the Singapore Chinese Health Study (SCHS). SCHS, a population-based cohort, recruited Chinese individuals, aged 45–74 years, between 1993 and 1998		After adjustment for age, gender, smoking status, education and dialect, LTL was found to be inversely associated with plasma homocysteine levels (p for trend = 0.014)	[73]
1319 healthy subjects were recruited from a population-based cohort in the UK.		LTL was negatively correlated with plasma homocysteine levels, after adjustment for smoking, obesity, physical activity, menopause, hormone replacement therapy use, and creatinine clearance. The difference in multiply-adjusted LTL between the highest and lowest tertile of homocysteine levels was 111 base pairs ($p = 0.004$), corresponding to 6.0 years of telomeric aging	[74]
195 healthy men in Milan, Italy	When plasma folate concentration was above the median, there was a positive relationship between folate and telomere length	When plasma folate concentration was below the median, there was a negative relationship between folate and telomere length	[75]
Framingham USA Offspring Study ($n = 1,044$, females = 52.1 %, mean age 59 years) using data from samples collected before and after folic acid fortification		The leukocyte TL of the individuals in the fifth quintile of plasma folate was shorter than that of those in the second quintile by 180 bp ($P < 0.01$). There was a linear decrease in leukocyte telomere length with higher plasma folate concentrations in the upper four quintiles of plasma folate (P for trend = 0.001).	[69]
786 participants in the Austrian Stroke Prevention Study 9 mean age of 66 ± 7 58% female)	Multiple linear regression analyses with adjustment for age and sex demonstrated that higher lutein, zeaxanthin, and vitamin C concentrations in plasma were strongly associated with longer telomere length		[76]
2160 women aged 18–79 years (mean age: 49.4) from a large population-based cohort of twins in the UK	Serum vitamin D concentrations were positively associated with LTL ($r = 0.07$, $P = 0.0010$), and this relation persisted after adjustment for age ($r = 0.09$, $P < 0.0001$) and other covariates (age, season of vitamin D measurement, menopausal status, use of hormone replacement therapy, and physical activity; P for trend across tertiles = 0.003)		[77]
43 younger (18–32 years) and 47 older (65–83 years) adults in South Australia	Ca/Mg ratio was positively associated with lymphocyte telomere length ($r = 0.55$, $P = 0.007$)	Negative association between lymphocyte telomere length and both plasma calcium and magnesium levels, ($r = -0.47$, $P = 0.03$ and $r = -0.61$, $P = 0.001$ respectively), in older females; Intriguingly Ca/Mg ratio was positively associated with telomere length ($r = 0.55$, $P = 0.007$)	[78]
437 healthy children ages 3, 6, and 9 years in Western Australia		Lymphocyte TL was inversely associated with plasma zinc ($r = -0.13$, $P < 0.001$)	[79]
125 participants being treated for chronic essential hypertension, 77 old (age range 60–79 years) and 48 very old (age range 80–100 years)	In participants at the age 80–100 years, a significant positive correlation between mean telomere length and intracellular labile zinc ($r = 0.502$; $p < 0.01$) was observed	In participants at the age 80–100 years, a significant inverse correlation existed between % cells with critically short telomeres and intracellular labile zinc ($r = -0.456$; $p < 0.01$)	[80]

TABLE 10.3 Intervention Studies on the Effects of Nutrition on Telomere Length

Participants & country	Intervention and outcome	Ref.
521 subjects (55–80 years, 55% women) from the PREDIMED intervention in Spain	Increase in TL with reduced obesity after a 5-year Mediterranean diet intervention	[62]
54 obese men on calorically restricted weight loss diets in Australia	Telomere length in mid-rectal biopsy tissue increased 4-fold and 10-fold after 12 and 52 weeks on a calorically restricted weight-loss diet respectively. Gain of telomere length was greater if more weight and body fat was lost	[81]
35 men who had biopsy-proven low-risk prostate cancer and had chosen to undergo active surveillance in the USA	Men in the intervention group who followed a program of comprehensive lifestyle changes (physical activity, stress management, and social support), and a diet high in whole foods, plant-based protein, fruits, vegetables, unrefined grains, and legumes, and low in fat (approximately 10% of calories) and refined carbohydrates had increased telomerase and longer LTL after 5 years compared to the men in the control group who underwent active surveillance alone	[82]
Prospective cohort study of 608 ambulatory outpatients in California (USA) with stable coronary artery disease recruited from the Heart and Soul Study between September 2000 and December 2002 and followed up to January 2009 (median, 6.0 years; range, 5.0–8.1 years)	Individuals in the lowest quartile of DHA + EPA experienced the fastest rate of telomere shortening (0.13 telomere-to-single-copy gene ratio [T/S] units over 5 years; 95% confidence interval [CI], 0.09–0.17), whereas those in the highest quartile experienced the slowest rate of telomere shortening (0.05 T/S units over 5 years; 95% CI, 0.02–0.08; $P < .001$ for linear trend across quartiles)	[83]
Double-blind 4-month trial included 106 healthy sedentary overweight middle-aged and older adults in the USA who received (1) 2.5 g/day n-3 PUFAs, (2) 1.25 g/day n-3 PUFAs, or (3) placebo capsules that mirrored the proportions of fatty acids in the typical American diet	The adjusted mean change in telomere length, expressed in base pairs (bp), was an increase of 21 bp for the low-dose group and an increase of 50 bp in the high-dose group compared to a decrease of 43 bp for placebo. Telomere length increased with decreasing n-6:n-3 ratios, $p = 0.02$	[84]
Thirty-three adults ages >65 y with MCI in South Australia were randomized to receive a supplement rich in the long-chain ω-3 PUFAs eicosapentaenoic acid (EPA; 1.67 g EPA + 0.16 g docosahexaenoic acid DHA/d; $n = 12$) or DHA (1.55 g DHA + 0.40 g EPA/d; $n = 12$), versus ω-6 PUFA linoleic acid (LA; 2.2 g/d; $n = 9$) for 6 months	The intervention did not show an increase in telomere length with treatment and there was a trend toward telomere shortening during the intervention period. Telomere shortening was greatest in the LA group ($d = 0.21$) than in the DHA ($d = 0.12$) and EPA groups ($d = 0.06$). Increased erythrocyte DHA levels were associated with reduced telomere shortening ($r = -0.67$; $P = 0.02$) in the DHA group	[85]
A 12-week placebo-controlled intervention trial with zinc carnosine supplement was performed with 84 volunteers from an elderly (65–85 y) South Australian cohort with low plasma Zn levels	Telomere length was not significantly affected but telomere base damage was found to be significantly decreased in the Zn supplemented group ($p < 0.05$)	[14]

The results from dietary pattern studies indicate that a predominantly plant-based diet such as a prudent or Mediterranean diet is consistently associated with longer telomeres [58,60–64]. Some studies revealed associations with specific components of the plant-based diet as being protective, such as total antioxidant capacity of the diet, fruit juice, dietary fiber, and higher intakes of vitamin C and vitamin E [65,66,68,70]. Higher intake of processed meat, red meat, sweetened carbonated beverages, white bread, short- to medium-chain saturated fatty acids and their sources such as whole milk (and products such as butter and cheese), linoleic acid (n6 fatty acid), and higher sodium intake (in those with high BMI) were all associated with shorter telomeres [58,59,65,67,68,70,71]. Multivitamin use was associated with both shorter and longer telomeres [69,70].

Studies on blood nutrient concentrations showed U-shaped relationships between plasma folate and telomere length [69,75] and a reduction of telomere length is associated with higher homocysteine (a metabolic biomarker of folate and/or vitamin B12 deficiency) in older subjects indicating a rather complex relationship with B vitamins [72–74]. Higher plasma zeaxanthin, lutein, vitamin C, vitamin D, and Ca/Mg ratio was associated with longer telomeres but, in contrast, higher plasma concentration of calcium, magnesium, and zinc was associated with shorter telomeres (the latter in children) [76–79]. In the very old (8–100 years), intracellular labile zinc was positively correlated with telomere length and negatively with percentage of cells with critically short telomeres [80]. It is evident that a nutrient

Is There Any Evidence That Telomere Length Can Be Modified by Nutritional Intervention?

Although it is too early to provide definitive answers there are promising results that telomere length may be modifiable by diet and/or lifestyle. The 5-year PREDIMED intervention observed an increase in TL in those assigned to the Mediterranean diet intervention [62]. In men with low-risk prostate cancer a holistic lifestyle program including a plant-based diet significantly improved telomere length and telomerase activity in blood cells [82] whilst a weight loss program on either a high-carbohydrate or high-protein calorie restricted diet increased telomere length in mid-rectal biopsies in obese men [81]. Three studies with ω-3 fatty acid supplementation showed marginal changes in telomere length with trends indicating prevention of telomere shortening with decreasing ω6:ω3 ratios [83–85]. Zinc carnosine supplementation did not affect telomere length but appeared to reduce telomere base damage in those with low plasma zinc; this was the first study to report on telomere base damage in an in vivo human intervention study [14]. Most of these studies have several limitations such as small cohort size and short duration of intervention and therefore can only be considered as providing preliminary evidence.

What Are the Plausible Mechanisms by Which Nutrients Can Affect Telomere Maintenance?

We and others have performed in vitro modeling experiments to obtain some insights on how micronutrient deficiencies or excess may affect DNA and telomere integrity. Deficiencies in micronutrients such as vitamin C, vitamin E, zinc, and selenium may lead to increased susceptibility to oxidative radicals which can either oxidize guanine in the telomere sequence or cause single- or double-stranded breaks in telomeric or subtelomeric sequences [86–92]. Alternatively, deficiencies in vitamins involved in one carbon metabolism could lead to uracil incorporation instead of thymidine and also induce breaks in DNA when excised by one or more of the uracil glycosylases especially when uracils occur within 12 bases from each other on opposite strands of DNA [93]. Our studies showed that folate deficiency initially leads to telomere elongation but subsequently telomeres start to shorten due to terminal deletions of chromosomes; both telomere elongation and shortening occurred concurrently with increased chromosomal instability indicating that telomere lengthening on its own is not an adequate measure of chromosomal stability [94,95]. Another important micronutrient is niacin which is required to generate NAD as a substrate of PARPs, such as tankyrase, in the shelterin complex, which regulates telomere length through its interaction with TRF1 [96,97]. These plausible mechanisms are explained schematically in Fig. 10.2, but it is likely that other important roles for micronutrients in telomere homeostasis will be discovered.

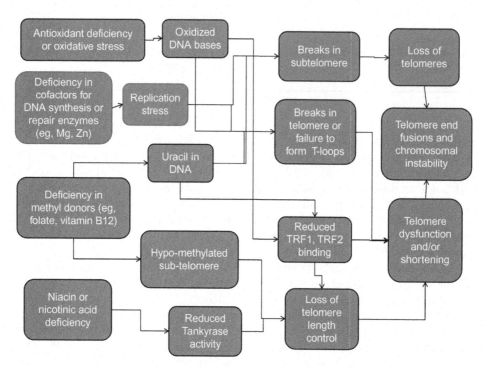

FIGURE 10.2 Plausible mechanisms by which nutritional deficiency can to telomere shortening or telomere dysfunction.

KNOWLEDGE GAPS

Although considerable progress has been made regarding the role of nutrition in maintenance of telomere integrity this progress appears to have been focused entirely on effects relating to telomere length. There is however sufficient reason to argue that prevention of telomere shortening or lengthening is not the only aspect that should be addressed and other measures of telomere dysfunction should also be considered. For example several studies show effects on telomere length of just a few tens or hundreds of bases but this may not be sufficient to make telomeres dysfunctional and might be beneficial if it makes the telomere a smaller target for DNA damage. Furthermore, lengthening could be associated with indirect effects caused by loss of telomere length control due to damages to the telomere sequence or defective activity of shelterin proteins. Therefore, incorporation of measures of telomere base damage, subtelomere methylation, and measures of telomere dysfunction, such as failure to form T-loops and telomere end fusions, should be incorporated to enable a more comprehensive assessment of telomere integrity.

Additionally, polymorphisms in genes coding for proteins required for telomerase activity and for the shelterin complex and other genes coding for polymorphisms in transporters and enzymes involved in micronutrient metabolism may also modify effects exerted by dietary factors on telomere structure and function. Thus a nutrigenetic/nutrigenomic approach may also uncover important new knowledge that could lead to more personalized approaches for maintaining optimal telomere integrity throughout the life-span.

CONCLUSION

It is evident that telomeres are degraded with aging and that nutrition can modify the rate at which these deleterious effects occur. The next challenges for future studies will be to include a more comprehensive assessment of telomere integrity that goes well beyond just measuring telomere length and also incorporates study designs that are sufficiently powered to study nutrient–gene interactive effects on this critical component of the genome.

SUMMARY

Telomeres are nucleoprotein structures at the end of each chromosome. The nucleic acid sequence of telomeres is a TTAGGG hexamer repeat. The number of hexamer repeats can vary greatly from very few to thousands of repeats with most reported telomere lengths varying between 4-11 kilobases in humans. Several proteins (TRF1, TRF2, POT1, TIN2, TPP1, RAP1 and Tankyrase) are associated with telomere DNA to form the shelterin complex which interacts with enzymes such as telomerase and other proteins required for proper telomere maintenance and function. Shortening of telomeres may increase excessively with age and lead to telomere-end-fusions, chromosomal instability and accelerated senescence. Telomeres may become dysfunctional if base damage and/or DNA strand breaks accumulate in the telomere sequence. Several lines of evidence indicate that telomere shortening, telomere base damage and telomere strand breaks may increase due to poor nutrition and that dietary intervention has the potential to improve telomere integrity.

References

[1] Blackburn EH. Telomeres and telomerase: the means to the end (Nobel lecture). Angew Chem Int Ed Engl 2010;49(41):7405–21.

[2] Armanios M, Blackburn EH. The telomere syndromes. Nat Rev Genet 2012;13(10):693–704.

[3] Müezzinler A, Zaineddin AK, Brenner H. A systematic review of leukocyte telomere length and age in adults. Aging Res Rev 2013;12(2):509–19.

[4] Martínez P, Blasco MA. Telomeric and extra-telomeric roles for telomerase and the telomere-binding proteins. Nat Rev Cancer 2011;11(3):161–76.

[5] Aubert G, Lansdorp PM. Telomeres and aging. Physiol Rev 2008;88(2):557–79.

[6] de Lange T. How shelterin solves the telomere end-protection problem. Cold Spring Harb Symp Quant Biol 2010;75:167–77.

[7] Frias C, Pampalona J, Genesca A, Tusell L. Telomere dysfunction and genome instability. Front Biosci (Landmark Ed) 2012;17:2181–96.

[8] Riffell JL, Lord CJ, Ashworth A. Tankyrase-targeted therapeutics: expanding opportunities in the PARP family. Nat Rev Drug Discov 2012;11(12):923–36.

[9] Bernardes de Jesus B, Blasco MA. Telomerase at the intersection of cancer and aging. Trends Genet 2013;29(9):513–20.

[10] Bull C, Fenech M. Genome-health nutrigenomics and nutrigenetics: nutritional requirements or 'nutriomes' for chromosomal stability and telomere maintenance at the individual level. Proc Nutr Soc 2008;67(2):146–56.

[11] Rochette PJ, Brash DE. Human telomeres are hypersensitive to UV-induced DNA damage and refractory to repair. PLoS Genet 2010;6(4):e1000926.

[12] Fumagalli M, Rossiello F, Clerici M, Barozzi S, Cittaro D, Kaplunov JM, et al. Telomeric DNA damage is irreparable and causes persistent DNA-damage-response activation. Nat Cell Biol 2012;14(4):355–65.

[13] Ferguson LR, Karunasinghe N, Zhu S, Wang AH. Selenium and its' role in the maintenance of genomic stability. Mutat Res 2012;733(1–2):100–10.

[14] Sharif R, Thomas P, Zalewski P, Fenech M. Zinc supplementation influences genomic stability biomarkers, antioxidant activity and zinc transporter genes in an elderly Australian population with low zinc status. Mol Nutr Food Res 2015;59(6):1200–12. http://dx.doi.org/10.1002/mnfr.201400784.

[15] Williams JD, Jacobson MK. Photobiological implications of folate depletion and repletion in cultured human keratinocytes. J Photochem Photobiol B 2010;99(1):49−61.

[16] Baird DM, Rowson J, Wynford-Thomas D, Kipling D. Extensive allelic variation and ultrashort telomeres in senescent human cells. Nat Genet 2003;33(2):203−7.

[17] Capper R, Britt-Compton B, Tankimanova M, Rowson J, Letsolo B, Man S, et al. The nature of telomere fusion and a definition of the critical telomere length in human cells. Genes Dev 2007;21(19):2495−508.

[18] Farr C, Fantes J, Goodfellow P, Cooke H. Functional reintroduction of human telomeres into mammalian cells. Proc Natl Acad Sci USA 1991;88(16):7006−10.

[19] Hanish JP, Yanowitz JL, de Lange T. Stringent sequence requirements for the formation of human telomeres. Proc Natl Acad Sci USA 1994;91(19):8861−5.

[20] Rodriguez-Brenes IA, Peskin CS. Quantitative theory of telomere length regulation and cellular senescence. Proc Natl Acad Sci USA 2010;107(12):5387−92.

[21] Calado RT, Dumitriu B. Telomere dynamics in mice and humans. Semin Hematol 2013;50(2):165−74.

[22] Kimura M, Barbieri M, Gardner JP, Skurnick J, Cao X, van Riel N, et al. Leukocytes of exceptionally old persons display ultrashort telomeres. Am J Physiol Regul Integr Comp Physiol 2007;293(6):R2210−17.

[23] Roberts RO, Boardman LA, Cha RH, Pankratz VS, Johnson RA, Druliner BR, et al. Short and long telomeres increase risk of amnestic mild cognitive impairment. Mech Aging Dev 2014;141−142:64−9.

[24] Suram A, Herbig U. The replicometer is broken: telomeres activate cellular senescence in response to genotoxic stresses. Aging Cell 2014;13(5):780−6.

[25] Campisi J. Aging, cellular senescence, and cancer. Annu Rev Physiol 2013;75:685−705.

[26] Rodier F, Muñoz DP, Teachenor R, Chu V, Le O, Bhaumik D, et al. DNA-SCARS: distinct nuclear structures that sustain damage-induced senescence growth arrest and inflammatory cytokine secretion. J Cell Sci 2011;124(Pt 1):68−81.

[27] Ovadya Y, Krizhanovsky V. Senescent cells: SASPected drivers of age-related pathologies. Biogerontology 2014;15(6):627−42.

[28] Wellinger RJ. In the end, what's the problem? Mol Cell 2014;53(6):855−6.

[29] Opresko PL, Fan J, Danzy S, Wilson III DM, Bohr VA. Oxidative damage in telomeric DNA disrupts recognition by TRF1 and TRF2. Nucleic Acids Res 2005;33(4):1230−9.

[30] Opresko PL, Otterlei M, Graakjaer J, Bruheim P, Dawut L, Kølvraa S, et al. The Werner syndrome helicase and exonuclease cooperate to resolve telomeric D loops in a manner regulated by TRF1 and TRF2. Mol Cell 2004;14(6):763−74.

[31] Opresko PL, von Kobbe C, Laine JP, Harrigan J, Hickson ID, Bohr VA. Telomere-binding protein TRF2 binds to and stimulates the Werner and Bloom syndrome helicases. J Biol Chem 2002;277(43):41110−19.

[32] Wood AM, Rendtlew Danielsen JM, Lucas CA, Rice EL, Scalzo D, Shimi T, et al. TRF2 and lamin A/C interact to facilitate the functional organization of chromosome ends. Nat Commun 2014;5:5467. Available from: http://dx.doi.org/10.1038/ncomms6467.

[33] Ma H, Zhou Z, Wei S, Liu Z, Pooley KA, Dunning AM, et al. Shortened telomere length is associated with increased risk of cancer: a meta-analysis. PLoS One 2011;6(6):e2046.

[34] Wentzensen IM, Mirabello L, Pfeiffer RM, Savage SA. The association of telomere length and cancer: a meta-analysis. Cancer Epidemiol Biomarkers Prev 2011;20(6):1238−50.

[35] Haycock PC, Heydon EE, Kaptoge S, Butterworth AS, Thompson A, Willeit P. Leucocyte telomere length and risk of cardiovascular disease: systematic review and meta-analysis. Br Med J 2014;349:g4227. Available from: http://dx.doi.org/10.1136/bmj.g4227.

[36] Willeit P, Raschenberger J, Heydon EE, Tsimikas S, Haun M, Mayr A, et al. Leucocyte telomere length and risk of type 2 diabetes mellitus: new prospective cohort study and literature-based meta-analysis. PLoS One 2014;9(11):e112483.

[37] D'Mello MJ, Ross SA, Briel M, Anand SS, Gerstein H, Paré G. Association between shortened leukocyte telomere length and cardiometabolic outcomes: systematic review and meta-analysis. Circ Cardiovasc Genet 2015;8(1):82−90.

[38] Cai Z, Yan LJ, Ratka A. Telomere shortening and Alzheimer's disease. Neuromolecular Med 2013;15(1):25−48.

[39] Hou L, Zhang X, Gawron AJ, Liu J. Surrogate tissue telomere length and cancer risk: shorter or longer? Cancer Lett 2012;319(2):130−5.

[40] Julin B, Shui I, Heaphy CM, Joshu CE, Meeker AK, Giovannucci E, et al. Circulating leukocyte telomere length and risk of overall and aggressive prostate cancer. Br J Cancer 2015;112(4):769−76.

[41] Machiela MJ, Hsiung CA, Shu X, Seow WJ, Wang Z, Matsuo K, et al. Genetic variants associated with longer telomere length are associated with increased lung cancer risk among never-smoking women in Asia: a report from the female lung cancer consortium in Asia. Int J Cancer 2014;137(2):311−19. http://dx.doi.org/10.1002/ijc.29393.

[42] Campa D, Mergarten B, De Vivo I, Boutron-Ruault MC, Racine A, Severi G, et al. Leukocyte telomere length in relation to pancreatic cancer risk: a prospective study. Cancer Epidemiol Biomarkers Prev 2014;23(11):2447−54.

[43] Pellatt AJ, Wolff RK, Torres-Mejia G, John EM, Herrick JS, Lundgreen A, et al. Telomere length, telomere-related genes, and breast cancer risk: the breast cancer health disparities study. Genes Chromosomes Cancer 2013;52(7):595−609.

[44] Xie H, Wu X, Wang S, Chang D, Pollock RE, Lev D, et al. Long telomeres in peripheral blood leukocytes are associated with an increased risk of soft tissue sarcoma. Cancer 2013;119(10):1885−91.

[45] Hjelmborg JB, Dalgård C, Mangino M, Spector TD, Halekoh U, Möller S, et al. Paternal age and telomere length in twins: the germ stem cell selection paradigm. Aging Cell 2015;14(4):701−3. http://dx.doi.org/10.1111/acel.12334.

[46] Hjelmborg JB, Dalgård C, Möller S, Steenstrup T, Kimura M, Christensen K, et al. The heritability of leucocyte telomere length dynamics. J Med Genet 2015;52(5):297−302.

[47] Pooley KA, McGuffog L, Barrowdale D, Frost D, Ellis SD, Fineberg E, et al. Lymphocyte telomere length is long in BRCA1 and BRCA2 mutation carriers regardless of cancer-affected status. Cancer Epidemiol Biomarkers Prev 2014;23(6):1018−24.

[48] Puterman E, Epel E. An intricate dance: life experience, multisystem resiliency, and rate of telomere decline throughout the lifespan. Soc Personal Psychol Compass 2012;6(11):807−25.

[49] Babizhayev MA, Yegorov YE. Smoking and health: association between telomere length and factors impacting on human disease, quality of life and life span in a large population-based cohort under the effect of smoking duration. Fundam Clin Pharmacol 2011;25(4):425−42.

[50] Ludlow AT, Roth SM. Physical activity and telomere biology: exploring the link with aging-related disease prevention. J Aging Res 2011;2011:790378.

[51] Wong JY, De Vivo I, Lin X, Christiani DC. Cumulative PM(2.5) exposure and telomere length in workers exposed to welding fumes. J Toxicol Environ Health A 2014;77(8):441−55.

REFERENCES

[52] Pavanello S, Pesatori AC, Dioni L, Hoxha M, Bollati V, Siwinska E, et al. Shorter telomere length in peripheral blood lymphocytes of workers exposed to polycyclic aromatic hydrocarbons. Carcinogenesis 2010;31(2):216–21.

[53] Schröder CP, Wisman GB, de Jong S, van der Graaf WT, Ruiters MH, Mulder NH, et al. Telomere length in breast cancer patients before and after chemotherapy with or without stem cell transplantation. Br J Cancer 2001;84(10):1348–53.

[54] Lee JJ, Nam CE, Cho SH, Park KS, Chung IJ, Kim HJ. Telomere length shortening in non-Hodgkin's lymphoma patients undergoing chemotherapy. Ann Hematol 2003;82(8):492–5.

[55] Albrecht E, Sillanpää E, Karrasch S, Alves AC, Codd V, Hovatta I, et al. Telomere length in circulating leukocytes is associated with lung function and disease. Eur Respir J 2014; 43(4):983–92.

[56] Müezzinler A, Zaineddin AK, Brenner H. Body mass index and leukocyte telomere length in adults: a systematic review and meta-analysis. Obes Rev 2014;15(3):192–201.

[57] Passos JF, Saretzki G, von Zglinicki T. DNA damage in telomeres and mitochondria during cellular senescence: is there a connection? Nucleic Acids Res 2007;35(22):7505–13.

[58] Lee JY, Jun NR, Yoon D, Shin C, Baik I. Association between dietary patterns in the remote past and telomere length. Eur J Clin Nutr 2015;69(9):1048–52. http://dx.doi.org/10.1038/ejcn.2015.58.

[59] Nettleton JA, Diez-Roux A, Jenny NS, Fitzpatrick AL, Jacobs Jr. DR. Dietary patterns, food groups, and telomere length in the Multi-Ethnic Study of Atherosclerosis (MESA). Am J Clin Nutr 2008;88(5):1405–12.

[60] Crous-Bou M, Fung TT, Prescott J, Julin B, Du M, Sun Q, et al. Mediterranean diet and telomere length in Nurses' Health Study: population based cohort study. Br Med J 2014;349:g6674. Available from: http://dx.doi.org/10.1136/bmj.g66.

[61] Boccardi V, Esposito A, Rizzo MR, Marfella R, Barbieri M, Paolisso G. Mediterranean diet, telomere maintenance and health status among elderly. PLoS One 2013;8(4):e62781.

[62] García-Calzón S, Gea A, Razquin C, Corella D, Lamuela-Raventós RM, Martínez JA, et al. Longitudinal association of telomere length and obesity indices in an intervention study with a Mediterranean diet: the PREDIMED-NAVARRA trial. Int J Obes (Lond) 2014;38(2):177–82.

[63] García-Calzón S, Martínez-González MA, Razquin C, Corella D, Salas-Salvadó J, Martínez JA, et al. Pro12Ala polymorphism of the PPAR-γ2 gene interacts with a mediterranean diet to prevent telomere shortening in the PREDIMED-NAVARRA randomized trial. Circ Cardiovasc Genet 2015;8(1):91–9.

[64] Gu Y, Honig LS, Schupf N, Lee JH, Luchsinger JA, Stern Y, et al. Mediterranean diet and leukocyte telomere length in a multi-ethnic elderly population. Age (Dordr) 2015;37(2):24 http://dx.doi.org/10.1007/s11357-015-9758-0. [Epub 2015 Mar 8].

[65] Leung CW, Laraia BA, Needham BL, Rehkopf DH, Adler NE, Lin J, et al. Soda and cell aging: associations between sugar-sweetened beverage consumption and leukocyte telomere length in healthy adults from the National Health and Nutrition Examination Surveys. Am J Public Health 2014; 104(12):2425–31.

[66] García-Calzón S, Moleres A, Martínez-González MA, Martínez JA, Zalba G, Marti A, , et al.GENOI Members Dietary total antioxidant capacity is associated with leukocyte telomere length in a children and adolescent population. Clin Nutr 2015;34(4):694–9. pii:S0261-5614(14)00191-5. http://dx.doi.org/10.1016/j.clnu.2014.07.015.

[67] Song Y, You NC, Song Y, Kang MK, Hou L, Wallace R, et al. Intake of small-to-medium-chain saturated fatty acids is associated with peripheral leukocyte telomere length in postmenopausal women. J Nutr 2013;143(6):907–14.

[68] Cassidy A, De Vivo I, Liu Y, Han J, Prescott J, Hunter DJ, et al. Associations between diet, lifestyle factors, and telomere length in women. Am J Clin Nutr 2010;91(5):1273–80.

[69] Paul L, Jacques PF, Aviv A, Vasan RS, D'Agostino RB, Levy D, et al. High plasma folate is negatively associated with leukocyte telomere length in Framingham Offspring cohort. Eur J Nutr 2015;54(2):235–41.

[70] Xu Q, Parks CG, DeRoo LA, Cawthon RM, Sandler DP, Chen H. Multivitamin use and telomere length in women. Am J Clin Nutr 2009;89(6):1857–63.

[71] Zhu H, Bhagatwala J, Pollock NK, Parikh S, Gutin B, Stallmann-Jorgensen I, et al. High sodium intake is associated with short leukocyte telomere length in overweight and obese adolescents. Int J Obes (Lond) 2015;39(8):1249–53. http://dx.doi.org/10.1038/ijo.2015.51.

[72] Bull CF, O'Callaghan NJ, Mayrhofer G, Fenech MF. Telomere length in lymphocytes of older South Australian men may be inversely associated with plasma homocysteine. Rejuvenation Res 2009;12(5):341–9.

[73] Rane G, Koh WP, Kanchi M, Wang R, Yuan JM, Wang X. Association between leukocyte telomere length and plasma homocysteine in Singapore Chinese population. Rejuvenation Res 2015;18(3):203–10.

[74] Richards JB, Valdes AM, Gardner JP, Kato BS, Siva A, Kimura M, et al. Homocysteine levels and leukocyte telomere length. Atherosclerosis 2008;200(2):271–7.

[75] Paul L, Cattaneo M, D'Angelo A, Sampietro F, Fermo I, Razzari C, et al. Telomere length in peripheral blood mononuclear cells is associated with folate status in men. J Nutr 2009;139(7):1273–8.

[76] Sen A, Marsche G, Freudenberger P, Schallert M, Toeglhofer AM, Nagl C, et al. Association between higher plasma lutein, zeaxanthin, and vitamin C concentrations and longer telomere length: results of the Austrian Stroke Prevention Study. J Am Geriatr Soc 2014;62(2):222–9.

[77] Richards JB, Valdes AM, Gardner JP, Paximadas D, Kimura M, Nessa A, et al. Higher serum vitamin D concentrations are associated with longer leukocyte telomere length in women. Am J Clin Nutr 2007;86(5):1420–5.

[78] O'Callaghan NJ, Bull C, Fenech M. Elevated plasma magnesium and calcium may be associated with shorter telomeres in older South Australian women. J Nutr Health Aging 2014;18(2):131–6.

[79] Milne E, O'Callaghan N, Ramankutty P, de Klerk NH, Greenop KR, Armstrong BK, et al. Plasma micronutrient levels and telomere length in children. Nutrition 2015;31(2):331–6.

[80] Cipriano C, Tesei S, Malavolta M, Giacconi R, Muti E, Costarelli L, et al. Accumulation of cells with short telomeres is associated with impaired zinc homeostasis and inflammation in old hypertensive participants. J Gerontol A Biol Sci Med Sci 2009;64(7):745–51.

[81] O'Callaghan NJ, Clifton PM, Noakes M, Fenech M. Weight loss in obese men is associated with increased telomere length and decreased abasic sites in rectal mucosa. Rejuvenation Res 2009;12(3):169–76.

[82] Ornish D, Lin J, Chan JM, Epel E, Kemp C, Weidner G, et al. Effect of comprehensive lifestyle changes on telomerase activity and telomere length in men with biopsy-proven low-risk prostate cancer: 5-year follow-up of a descriptive pilot study. Lancet Oncol 2013;14(11):1112–20.

[83] Farzaneh-Far R, Lin J, Epel ES, Harris WS, Blackburn EH, Whooley MA. Association of marine omega-3 fatty acid levels with telomeric aging in patients with coronary heart disease. J Am Med Assoc 2010;303(3):250–7.

[84] Kiecolt-Glaser JK, Epel ES, Belury MA, Andridge R, Lin J, Glaser R, et al. Omega-3 fatty acids, oxidative stress, and leukocyte telomere length: a randomized controlled trial. Brain Behav Immun 2013;28:16–24.

[85] O'Callaghan N, Parletta N, Milte CM, Benassi-Evans B, Fenech M, Howe PR. Telomere shortening in elderly individuals with mild cognitive impairment may be attenuated with ω-3 fatty acid supplementation: a randomized controlled pilot study. Nutrition 2014;30(4):489–91.

[86] O'Callaghan N, Baack N, Sharif R, Fenech M. A qPCR-based assay to quantify oxidized guanine and other FPG-sensitive base lesions within telomeric DNA. Biotechniques 2011;51(6):403–11.

[87] Sharif R, Thomas P, Zalewski P, Fenech M. The role of zinc in genomic stability. Mutat Res 2012;733(1–2):111–21.

[88] Liu Q, Wang H, Hu D, Ding C, Xu H, Tao D. Effects of trace elements on the telomere lengths of hepatocytes L-02 and hepatoma cells SMMC-7721. Biol Trace Elem Res 2004;100(3):215–27.

[89] Yokoo S, Furumoto K, Hiyama E, Miwa N. Slow-down of age-dependent telomere shortening is executed in human skin keratinocytes by hormesis-like-effects of trace hydrogen peroxide or by anti-oxidative effects of pro-vitamin C in common concurrently with reduction of intracellular oxidative stress. J Cell Biochem 2004;93(3):588–97.

[90] Kashino G, Kodama S, Nakayama Y, Suzuki K, Fukase K, Goto M, et al. Relief of oxidative stress by ascorbic acid delays cellular senescence of normal human and Werner syndrome fibroblast cells. Free Radic Biol Med 2003;35(4):438–43.

[91] Tanaka Y, Moritoh Y, Miwa N. Age-dependent telomere-shortening is repressed by phosphorylated alpha-tocopherol together with cellular longevity and intracellular oxidative-stress reduction in human brain microvascular endotheliocytes. J Cell Biochem 2007;102(3):689–703.

[92] Makpol S, Durani LW, Chua KH, Mohd Yusof YA, Ngah WZ. Tocotrienol-rich fraction prevents cell cycle arrest and elongates telomere length in senescent human diploid fibroblasts. J Biomed Biotechnol 2011;2011:506171.

[93] Fenech M. Folate (vitamin B9) and vitamin B12 and their function in the maintenance of nuclear and mitochondrial genome integrity. Mutat Res 2012;733(1–2):21–33.

[94] Bull CF, Mayrhofer G, Zeegers D, Mun GL, Hande MP, Fenech MF. Folate deficiency is associated with the formation of complex nuclear anomalies in the cytokinesis-block micronucleus cytome assay. Environ Mol Mutagen 2012;53(4):311–23.

[95] Bull CF, Mayrhofer G, O'Callaghan NJ, Au AY, Pickett HA, Low GK, et al. Folate deficiency induces dysfunctional long and short telomeres; both states are associated with hypomethylation and DNA damage in human WIL2-NS cells. Cancer Prev Res (Phila) 2014;7(1):128–38.

[96] Dregalla RC, Zhou J, Idate RR, Battaglia CL, Liber HL, Bailey SM. Regulatory roles of tankyrase 1 at telomeres and in DNA repair: suppression of T-SCE and stabilization of DNA-PKcs. Aging (Albany, NY) 2010;2(10):691–708.

[97] De Boeck G, Forsyth RG, Praet M, Hogendoorn PC. Telomere-associated proteins: cross-talk between telomere maintenance and telomere-lengthening mechanisms. J Pathol 2009;217(3):327–44.

CHAPTER

11

mTOR, Nutrition, and Aging

Giuseppe D'Antona

Department of Public Health, Experimental and Forensic Medicine, CRIAMS Sport Medicine Centre Voghera,
University of Pavia, Pavia, Italy

KEY FACTS
- mTOR sense the nutritional status of the cell.
- mTORC dysregulation plays a fundamental role in human pathology.
- Nutritional regulation of mTORC may be used to counteract diseases.

Dictionary of Terms

- mTOR = mechanistic Target of Rapamycin
- mTORC = mechanistic Target of Rapamycin Complex

Alterations in the target of the rapamycin (TOR) nutrient-sensing pathway are thought to contribute to physiological aging. In normal conditions the activity of the pathway is promoted by mitogens, growth factors, and nutrients (in particular amino acids) thus regulating diverse fundamental functions such as cell growth, proliferation, development, memory, angiogenesis, autophagy, and immune responses. Through life hyperactivation of the mTOR complexes is associated with more rapid aging, whereas hypoactivation leads to reduced translation and increased longevity in several animal species.

MECHANISTIC TARGET OF RAPAMYCIN

mTOR (originally used as an abbreviation of *m*ammalian Target of Rapamycin but now used as an abbreviation of *m*echanistic TOR), an atypical protein kinase essential for organism survival, was first discovered as a target of rapamycin, an FDA-approved bacterial natural product used as a potent immunosuppressant.

To date a single gene encoding for mTOR (*TOR*) has been identified in every eukaryote genome examined except yeast, which may own two *TOR* genes. mTOR is a highly conserved homolog of the yeast protein (2549 amino acid, around 289 kD, NCBI ACCESSION: NP_004949) phosphatidylinositol kinase-related kinase (PIKK), which plays the role of a catalytic subunit of two distinct dimer multiprotein complexes termed mTOR complex 1 (mTORC1) and mTOR complex 2 (mTORC2). In these complexes mTOR acts as a multieffector protein kinase containing a carboxy-terminal serine/threonine protein kinase domain. The C-terminal end of mTOR contains several important elements, including the kinase catalytic domain (KIN). This domain also includes a small region called negative regulatory domain (NRD). Within this region, phosphorylation at specific residues (thr2446, ser2448, and ser2481) leads to higher levels of mTOR activity. Adjacent to the KIN domain is the 11 kDa FKBP12—rapamycin binding domain (FRB) at the N-terminus of the PIKK domain. This domain, which does not exist in the other members of the PIKK family, is the site of the inhibitory interaction between rapamycin and mTOR.

Companions of mTOR, in particular, the regulatory-associated protein of mTOR (RAPTOR) (defining component of mTORC1) and rapamycin-insensitive companion of mTOR (RICTOR) (defining component of mTORC2) are scaffolds for assembling the complexes and for binding to substrates and regulators. RAPTOR is a scaffolding companion to binding TOR signaling (TOS) motif-containing proteins and is essential for all mTORC1 functions [1]. Apart from RAPTOR, mTORC1 consists of mammalian lethal with sec13 protein 8 (mLST8), which is dispensable for mTORC1 activity, proline rich Akt/PKB substrate 40 kDa (PRAs40 also known as AKT1s1), and DEP domain-containing mTOR interacting protein (DEPTOR).

mTORC2 consists of RICTOR, mLST8 protein (also known as Gβl) observed with RICTOR (PROTOR, PRR5), DEPTOR, and mammalian stress-activated map kinase-interacting protein 1 (msIn1; also known as mAPKAP1), which may target mTORC2 to membranes, and SAPK kinase interacting protein 1 (Sin1). Meanwhile mLST8 is necessary for mTORC2 catalytic function. The functional role of PROTOR, which binds RICTOR independent of mTOR and is not fundamental for mTORC2-mediated Akt/PKB activation, is currently unknown. Another unique component of mTORC2 is Sin1 (with at least five alternative splicing isoforms in mammals) which promotes upstream regulator Akt/PKB-Ser473 phosphorylation (see below) required for mTORC2 function in cell survival but is dispensable for mTORC1 function. In fact Sin1 deletion is embryonically lethal and genetic ablation of Sin1 is followed by disruption of RICTOR-mTOR interaction, whereas Akt/PKB-Thr308 phosphorylation is maintained. RICTOR and RAPTOR recruit different substrates to mTOR for phosphorylation and this explains the diverse selection of substrates by the two mTOR complexes. On the other side, mTORC1 and mTORC2 share mLST8 and DEPTOR and these companions may play the role of positive and negative regulators, respectively [2].

Rapamycin ($C_{51}H_{79}NO_{13}$; molecular weight: 914.2) or sirolimus is a macrocyclic lactone (macrolide) produced by the bacterium *Streptomyces hygroscopicus*, first isolated from soil found on Easter Island Rapa Nui and developed as antifungal agent with no antibacterial activity. Based on the discovery of its immunosuppressive properties, its use as an immunosuppressive agent to prevent organ rejection was encouraged and led to FDA approving in 1999. Thereafter, a number of studies demonstrated that rapamycin is also a potent inhibitor of vascular smooth muscle proliferation. These findings led to the development of the rapamycin-eluting stent to treat atherosclerosis in coronary arteries, and its use was FDA-approved in 2003 to prevent in-stent restenosis. Thus, to date, significant experience exists with the clinical application of this drug to prevent organ rejection after kidney transplantation and to prevent occlusion of cardiac stents. It took around 20 years to first identify the molecular target of rapamycin. In 1994 three independent research groups found the protein target of rapamycin [3–5] by selection of spontaneous mutations capable of conferring resistance to the growth inhibitory effect of rapamycin in *Saccharomyces cerevisiae*. These investigations led to the identification of three genes. The first gene identified was RBP1 (homolog of the human peptidyl-prolyl isomerase FKBP12) which was identified in a search for receptors for the immunosuppressant drug with structural homology to rapamycin [6]. Its deletion in yeast was followed by drug resistance while expression of human FKBP12 restored sensitivity to rapamycin thus suggesting that rapamycin forms a toxic complex with FKBP12, inhibiting the function of other cellular proteins. The other two identified genes were TOR1 and TOR2, originally called DRR1 and DRR2, encoding two highly homologous proteins.

To date it is known that in yeast and mammals upon entering the cell rapamycin binds to the FKBP12-rapamycin binding domain of mTORC1 (also known as PPIase FKBP1A). The binding of rapamycin to FKBP12 is followed by dissociation of mTOR from RAPTOR [7] thus preventing the mTORC1 mediated phoshorylation of specific substrates and the translocation of specific cell mRNAs required for G1 to S phase transition. This process ultimately decreases protein synthesis, increases autophagy, and arrests cell growth. Importantly rapamycin is unable to acutely inhibit mTORC2 as the rapamycin–FKBP12 complex cannot interact with the FRB domain of mTORC2. By contrast prolonged rapamycin exposure determines inhibition of mTORC1 and mTORC2 assembly in vitro in cancer cell lines and in vivo in mice thereby inhibiting their activity [8]. In particular, prolonged rapamycin exposure may be responsible for RICTOR (major component of mTORC2) depletion followed by glucose intolerance due to decreased hepatic insulin sensitivity and impaired lifespan in males and not in females, as recently demonstrated in three different mouse models of RICTOR depletion with reduced mTORC2 signaling [9]. This phenomenon may be subsequent to a progressive sequestration of the cellular pool of mTOR due to prolonged rapamycin exposure in a complex with rapamycin–FKBP12, thus making it unavailable for assembly into mTORC2 [10].

NUTRIENTS AND ENERGY STATUS AS UPSTREAM REGULATORS OF mTORC1 AND mTORC2

Four major inputs regulate mTORC1 by cooperating or antagonizing each other and include nutrients (amino acids and glucose), growth factors, energy status, and stress. Therefore different regulatory signaling inputs may converge onto mTORC1 and consist of the PI3K/Akt and Ras/MAPK pathways, as well as AMPK signaling.

PI3K/Akt and Ras/MAPK pathways activation is due to the interaction of growth factors to specific receptor tyrosine kinases [11]. Akt/PKB activation depends on mechanical load as functional overload and hormones as insulin [12].

The fundamental interplay between insulin-mediated PI3K/Akt activation and TORC1 signaling suggests an important role of mTOR in controlling glucose homeostasis and protein synthesis through insulin-dependent and insulin-independent mechanisms [12]. Binding of insulin to the α-subunit of its receptor leads to the autophosphorylation of tyrosine residues, which in turn causes activation (tyr phosphorylation) of the insulin receptor substrate (IRS) proteins, with IRS-1 being the most important of the cells that respond to insulin with glucose transporter (GLUT4) translocation. IRS1 interacts with phosphatidylinositol 3,4,5-triphosphates kinase (PI3K). IRS1 functions as a docking site for many effector proteins bearing an SH2 domain. Phosphorylated IRS1 binds the p85 subunit of PI3-kinase (PI3K), which activates the enzyme leading to generation of phosphatidylinositol 3,4,5-trisphosphate (PIP3) and phosphatidylinositol 3,4-bisphosphate (PIP2) from phosphatidylinositol 4,5-bisphosphate and phosphatidylinositol 4-phosphate, respectively. Modulation of PI3K activity by IRS1 is thus responsible for Akt/PKB activation followed by transmission of the insulin signal and regulates glucose uptake by insulin. PIP3 recruits Akt/PKB to the inner leaflet of the plasma membrane where it binds with high affinity to 3-phosphoinositides and this process is followed by Akt/PKB phosphorylation mediated by two distinct kinases PDK1 on Thr308 and mTOR-rictor, rapamycin insensitive companion of mTOR complex, on Ser473 [13].

PI3K activation is negatively modulated by dephosphorylation of PIP3 via PTEN and SHIP2 3′ phosphatases. Upon activation by PDK1, Akt/PKB regulates a plethora of cellular events, including protein synthesis and degradation, cell cycle progression, glycogen synthesis, and overall cell survival. Akt/PKB is a serine/threonine kinase expressed in three highly homologous isoforms Akt1 (PKBa), Akt2 (PKBb), and Akt3 (PKBg). Insulin activates Akt2 isoform while Insulin Growth Factor 1 (IGF-1) activates Akt1 isoform [14]. Akt/PKB isoforms have an N-terminal pleckstrin homology (PH) domain which mediates binding of Akt/PKB to 3-phosphoinositides, the kinase catalytic domain which displays threonine residue (Thr308 in PKBa/Akt1) whose phosphorylation is necessary for Akt/PKB activation and the C-terminal tail, containing a regulatory phosphorylation site (Ser473 in PKBa/Akt1), activation leads to maximal Akt/PKB activation. Major downstream targets of Akt/PKB include caspase-9, Bad, inhibitory kB kinase, mTOR, tuberous sclerosis complex 1-2 (TSC1-TSC2), and FOXO. Stimulated Akt/PKB inhibits the TSC1-TSC2 complex through phosphorylation of TSC2. The complex acts as a GTPase activating protein (GAP, an heterodimer of TSC1, TSC2, and TBC1D7) for Rheb which is a farnesylated GTPase anchored to the lysosome surface that, once GDP-loaded, is unable to activate mTORC1 [15]. Therefore, the TSC1-TSC2 effectively shuts off the mTORC1 signaling, and GTP loading of Rheb activates the complex (Fig. 11.1). Akt/PKB is also known to promote mTORC1 activation through activation of PRAs40 that binds to inhibitory proteins to mTORC1. If, from one side, Akt/PKB pathway signals mTORC1, from the other mTORC1 is directly involved in the regulation of insulin signaling through IRS1 phosphorylation at different residues. In fact IRS1 phosphorylation on different Ser residues uncouple PI3K from IRS1 which downregulates insulin signaling through the PI3K/Akt pathway, while phosphorylation on different Tyr residues activates the response to insulin. In particular the mechanism of insulin desensitization involves phosphorylation of the Ser/Thr residues on IRS1 and its reduced Tyr phosphorylation.

Prolonged insulin stimulation with nutrient overload promotes Ser phosphorylation of IRS1 through chronic activation of mTORC1 [16] or its downstream effector protein S6 kinase 1 (S6K1). First it has been shown that insulin/amino acid activation of the pathway causes increased IRS1 Ser phosphorylation followed by inhibition of glucose transport in L6 myocytes and 3T3-L1 adipocytes [17]. Next it has been demonstrated that rapamycin administration reduces Ser phosphorylation and prolongs insulin-stimulated PI3K activity at least in skeletal muscle of the mouse. Furthermore, the TSC1-TSC2 complex also appears to be involved as its disruption is followed by S6K1 activation that leads to the decrease of both IRS1 mRNA and protein expression levels. mTORC1 and/or S6K1 phosphorylate diverse Ser residues on IRS1 proteins (rodent/human) including Ser-265/270, Ser-302/307, Ser-307/312, Ser-632/636, and Ser-1097/1101. Importantly rapamycin-insensitive kinases are also involved in the insulin-dependent Ser phosphorylation of IRS1 in the absence of nutrient overload [18] as recently demonstrated for the insulin-mediated activation of 90 ribosomal S6 kinases (RSK) activated by the ERK-MAP kinase pathway, which promotes IRS-1 phosphorylation on Ser-1101, independently of the mTOR/S6K1 pathway [19].

Interestingly the lysosome appears as a major site of interaction between the insulin-IR-PI3K-Akt and the mTOR pathways in response to amino acids exposure. In particular, it has been shown that translocation of mTORC1 to the lysosome is under control of Rag GTPase and Ragulator that is a pentameric complex anchoring Rag to the lysosome and known as LAMTOR1-5 [20]. This process is not inhibited by rapamycin [15]. Once active, Rag recruits mTORC1 probably in the cytoplasm shuttling it back to Rheb on the lysosome surface [20] where it forms a supercomplex

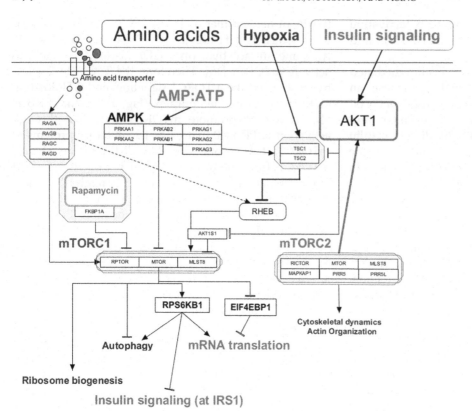

FIGURE 11.1 TOR signaling is responsible for a cellular reaction toward amino acids, energy disposal, and hypoxia, and is highly integrated in AMPK and insulin cascades. mTORC1 regulates mRNA translation, ribosome biogenesis, autophagy, and insulin signaling at IRS1; mTORC2 regulates cytoskeletal response, and, possibly, survival at the Akt/PKB level.

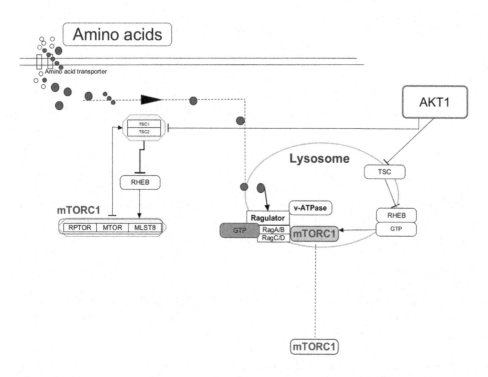

FIGURE 11.2 Amino acids entrance into the cell stimulates the recruitment of mTORC1 in a Rag-dependent manner to the lysosomal surface, whereas Akt inhibits the TSC complex possibly at the lysosome.

including Rag (Fig. 11.2), Ragulator, and vacuolar H^+-adenosine triphosphate ATPase (v-ATPase). The latter is a multisubunit proton pump which works by hydrolyzing ATP at the peripheral cytosolic v1 domain to maintain low pH within the lysosomes, thus contributing to maintaining their function. v-ATPase probably senses the amino acids content within the lysosome leading to the subsequent Rag activation (see below) [21]. In fact depletion of v-ATPase prevents mTORC1 activation and its localization stimulated by amino acids whose depletion strengthens the binding between v-ATPase and Ragulator, whereas amino acids availability weakens this binding.

It is important to underline that cells lacking TSC are unable to respond to growth factors withdrawal but are still responsive to amino acids thus suggesting that mTORC1 activation needs amino acids targeting the lysosome. On the other hand amino acids appear ineffective to activate mTORC1 via Rag when the complex is forcedly localized at the lysosome [22]. Indeed lines of evidence suggest that mTORC1 can be independently activated at different locations within the cell other than the lysosomes, including cytoplasm, nucleus, and mitochondria, thus suggesting that the complex may respond to different cues with phosphorylation of separate substrates.

As chemical inhibitors of glycolysis suppress mTORC1 activity, the complex senses the cellular energy of the cell. This phenomenon suggests that changes in energy disposal may promote convergent upstream regulatory signals on mTORC1. As already known, glycolysis and mitochondrial respiration convert nutrients into energy, which is stored in the form of ATP. Glucose loss, inhibition of glycolysis, or mitochondrial respiration cause a significant reduction of the intracellular ATP levels that determines a change in the intracellular ADP/ATP and AMP/ATP ratios (Fig. 11.1). This change is sensed by heterotrimeric complex AMP dependent protein kinase (AMPK) consisting of two regulatory (β and γ) and one catalytic subunits (α) and activated by binding of ADP or AMP to the regulatory subunit γ or phosphorylation of thr172 by LKB1 kinase. Therefore, ADP and AMP work as allosteric regulators of AMPK. AMPK acts as a key regulator of cellular metabolism and complete loss of AMPK kinase activity is not tolerated at the whole-body level in vivo. In fact Ampk$\alpha 1^{-/-}$ and Ampk$\alpha 2^{-/-}$ mice are embryonic lethal at E10.5. Interestingly deletion of AMPK regulatory subunits is associated with metabolic alterations without significant effects on mouse survival. In particular the Ampk$\gamma 3^{-/-}$ mice show impaired glycogen resynthesis after depletion and muscle-specific Ampk$\beta 1$/Ampk$\beta 2$ DKO mice are physically inactive and display reduced mitochondrial content in muscle cells. This evidence suggests an essential role for AMPK in maintaining mitochondrial content and glucose uptake in skeletal muscle during exercise [23].

Under nutrient deprivation AMPK transmits stress signals to mTORC1 (Fig. 11.1). In fact when the AMP/ATP ratio increases, AMPK phosphorylates TSC2 and RAPTOR [24]. Phosphorylation of TSC2 at ser 1345 inhibits mTORC1 by stimulating the GAP activity of the TSC1-TSC2 complex toward Rheb inactivation. Furthermore AMPK-mediated phosphorylation of RAPTOR leads to it binding to cytosolic chaperone proteins 14-3-3 followed by indirect mTORC1 inhibition allosteric mechanisms [25]. The energy stress may lead to mTORC regulation through diverse mechanisms but changes ATP levels. For example hypoxia may promote the expression of Regulated in development and DNA damage response 1 (*REDD1*; also known as *DDIT4*) thus inhibiting mTORC1 by promoting the subsequent assembly of TSC1-TSC2 complex. It is important to highlight that recent observations suggest further AMPK alternative mechanisms of mTORC1 inhibition, as demonstrated in the presence of energetic dysfunction due to mitochondrial inhibition in the AMPK null cells [26].

The energy status of a cell is tightly linked to amino acids, the building blocks of proteins also used in the synthesis of DNA, glucose, and ATP, which contribute to the overall balance between cell synthesis and catabolism. Therefore, amino acid availability is fundamental for maintenance of the constancy of the cell size over time, strictly linked to ribosomal activity and efficiency. Initial fundamental observations demonstrated that amino acids are required for mTOR activation in the budding yeast [27]. Later the link between amino acids and mTOR was also established in mammals where withdrawal of amino acids from the nutrient medium of Chinese hamster ovary cells (CHO-IR), which overexpress the human insulin receptor when treated with insulin, results in a rapid deactivation of downstream mTOR effectors as p70S6 kinase 1 and eIF-4EBP1, whereas their readdition quickly restore the responsivity of both signals to insulin. It is currently believed that in the postprandial period, when a significant increase in translational efficiency arises despite no change in ribosomal content, translational activation is to be at least partly attributed to Branched Chain Amino Acids (BCAA; leucine, isoleucine, and valine) through the activation of the mTORC1 signaling [28,29], whereas these amino acids are not required to activate the mTORC2 complex. In particular, leucine administration markedly increases protein synthesis concomitant with the hyperactivation of p70S6 kinase 1 and eIF-4EBP1. Hyperphosphorylation of p70S6 kinase 1 and eIF-4EBP1 has been also found following isoleucine exposure but this change is not associated with a concomitant increase of protein synthesis. Rapamycin treatment, responsible for reduced protein synthesis independently of leucine administration, completely prevents leucine-induced eIF-4EBP1 and p70S6 kinase 1 activation. Also rapamycin pretreatment reduces leucine-dependent activation of mTORC1. However rapamycin is unable to completely block the mTORC1 signaling activated by BCAA thus suggesting that this pathway activation is not sufficient to fully explain the activation of BCAA-dependent anabolism.

The mechanisms by which the amino acids are sensed intracellularly is still largely unknown and recent avenues of research identified new potential candidate mediators acting downstream (MAP4K3; RalA; Rab5; GCN2) or upstream (amino acids transporters) of amino acids. Nevertheless, to date the major accepted players of the link between amino acids and mTORC are Rag GTPases and Ragulator. The RRag GTPase family displays four evolutionarily well-conserved members consisting of RagA, RagB, RagC, and RagD. RagA or RagB forms a stable heterodimer with RagC or RagD.

In the absence of amino acids, the Rag GTPases are found in an inactive conformation.

When the levels of intracellular amino acids are high, Rag GTPases recruit mTORC1 to lysosomes and promote its activation [20]. Following amino acid accumulation into the lysosomes, a v-ATPase dependent mechanism, which includes the activation of Ragulator, causes Rag GTPases to switch to the active conformation, in which RagA/RagB is loaded with GTP and RagC/RagD is loaded with GDP. When RagC/RagD is GDP charged RagA/RabB interacts with RAPTOR resulting in mTORC1 recruitment and activation. The importance of Rag GTPases integrity for mTORC1 activation has been confirmed in S. cerevisiae by the expression of the dominant negative mutant of the RacA/RacB that completely abolished mTORC1 activity even in the presence of adequate amino acids stimulation. This evidence suggested that the nucleotide bound state of RacA/RacB, unlike RacC/RacD, is fundamental for amino acids-mediated mTORC1 stimulation [15].

Interestingly it has been found that Rag and amino acids, being unable to directly activate mTORC1 kinase, appear as regulators of spatial disposal of the complex. In fact amino acids depletion determines diffusion of mTORC1 throughout the cytoplasm, whereas in the presence of sufficient amino acids and activated Rag mTORC1 is recruited to the lysosome where it is activated by Rheb in its GTP-bound state [15]. The spatial regulation of mTORC1 activity may explain why growth factors (such as Insulin/IGF1) are unable to efficiently activate the complex in the absence of amino acids.

Recent observations also suggest that Rags interact with the transcription factor EB (TFEB), the master regulator of a gene network that promotes lysosomal biogenesis and autophagy. In basal conditions, TFEB is in the cytoplasm. However, under starvation, TFEB rapidly translocates into the nucleus and activates a transcriptional program that includes genes associated with lysosomal biogenesis and function and genes implicated in the major steps of the autophagy. Under overnutrition, mTORC1 phosphorylates TFEB at several residues (including ser211). This phosphorylation promotes interaction of TFEB with the cytosolic chaperone 14-3-3 and consequent retention of TFEB into the cytosol. In presence of mTORC1 inactivation the TFEB-14-3-3 complex dissociates and TFEB enters the nucleus. This phenomenon is followed by lysosomal biogenesis and autophagy activation. The molecular machinery regulating the association of TFEB to the lysosomal membranes has been recently elucidated and includes an amino acids-dependent mechanism through a direct interaction with active Rags [30]. Therefore, Rags play a critical role in coordinating nutrient availability within the cell and cellular clearance [30].

Apart from the "lysosome model" of amino acids, sensing, a new mechanism involving leucyl-tRNA synthetase (LRS), which normally charges leucine to its tRNA, has been proposed to explain leucine sensing in mammals. In particular LRS has been proposed to act as GAP RagD thus leading to active configuration of the Rag complex [31]. Notwithstanding these initial results, further studies are needed to fully elucidate which is the relative contribution of this mechanism on the overall sensing of amino acids to mTORC1 activation.

Although the new intriguing advancements in understanding the transmission of amino acids signal to mTORC1, it remains to be elucidated how amino acids sensing is actually initiated and which molecules are able to detect their availability and quality.

The scenario arising from these findings open new avenues on the role of endomembranes as key determinants of the cascade activation by nutrients and it is currently unknown whether the lysosome membrane is the only site of Rheb-mediated TORC1 activation.

Upstream regulation of mTORC2. mTORC2 controls growth by regulating lipogenesis, glucose metabolism, actin cytoskeleton, and apoptosis. The regulation of mTORC2 is only beginning to be discovered. Available evidence suggests that growth factors (such as insulin) may directly regulate this complex. In fact given their role in regulating Akt/PKB, it is generally believed that growth factors may control mTORC2, directly or indirectly. Recent studies have strongly suggested that TORC2 may play the role of PDK2 kinase for Akt/PKB ser473. Thus, insulin stimulation of cultured cells promotes ser473 phosphorylation of Akt/PKB by mTORC2. In addition it has been recently demonstrated that Sin1, a unique component of mTORC2, plays a role as a regulator of the Akt/PKB pathway by controlling Akt/PKB ser473 phosphorylation and activation. These studies highlight the role of the SIN1-rictor-mTOR complex in defining the function and specificity of Akt/PKB. Therefore, Akt/PKB acts as an upstream activator of mTOR for TORC1 but since Akt/PKB activation may be dependent on TORC2, Akt/PKB is also a target of mTOR via TORC2.

It is known that upstream signals (Akt/PKB and others) may respond to different growth factors but it is currently unknown how signaling specificity is achieved at the mTORC2 level. The role of splicing variants of mSin1 (mAPKAP1) mTORC2 component as an adaptor between the complex and specific growth factors has been recently proposed but requires further confirmation.

DOWNSTREAM TARGETS OF mTORC1 AND mTORC2

The best characterized function of mTORC1 is the regulation of translation. Protein synthesis proceeds through mRNA translation which includes initiation (the initiator methionyl-tRNA and mRNA bind to 40S ribosomal subunit), elongation (tRNA-bound amino acids are incorporated into growing polypeptide chains according to the mRNA template), and termination (the completed protein is released from the ribosome). The first two steps of mRNA translation are highly regulated at two different levels, that is, the binding of methionyl-tRNA to 40S ribosomal subunit to form 43S preinitiation complex, and recognition, unwinding, and binding of mRNA to the 43S, catalyzed by a multisubunit complex of eukaryotic factors (eIFs).

mTORC1 critical substrates include p70 ribosomal S6 kinase 1 and 2 (S6K1-2) and eIF4 binding proteins (4E-BP1), which associate with mRNAs and regulate translation initiation and progression, thus controlling the rate of protein synthesis. Furthermore, mTORC1 regulates the activity of protein phosphatases (PP2A), which may ultimately control the initiation rate.

4E-BP1, which is now considered a key regulator of cell proliferation, releases the inhibition on eIF4F which mediates translation initiation. When 4E-BP1 is active it blocks the ability of eIF4E to bind to eIF4G (a scaffolding component of eIF4F) by forming an active 4E-BP1-eIF4E complex that precludes mRNA binding to the ribosome. When phosphorylated by mTORC1, inactive 4E-BP1 dissociates from eIF4E which then participates to eIF4F complex formation (eIF4A, eIF4B, eIF4G, eIF4E), thus allowing the complex to interact with the 43S preinitiation and recruits it to the 5′ end of several mRNAs (c-Myc, cyclin D1 and D3, vascular endothelial growth factor VEGF and signal activator and transducer of transcription 3 STAT3). eIF4F complex determines circularization of mRNAs mediated by the association of eIF4G with poli A binding proteins and promote the unwind of secondary structure in the 5′ untranslated regions (UTRs) of diverse mRNAs, which normally impairs the access to ribosome start codon (AUG).

S6K1 is a kinase which requires phosphorylation at two sites as it should be first primed by mTORC1 (at Thr389) and then activated by phosphoinositide-dependent kinase-1 PDK1 on the T-loop by PDK1. Interestingly, a feedback loop involving phosphorylation of mTORC1 at Thr2446 and Ser2448 by S6K1 has been reported but its role is still uncertain.

Once activated by mTORC1, pS6K1 kinase may be targeted to the exon junction complex, where it enhances the translation of newly generated mRNAs through its association with S6K1 Aly/REF-like target (sKAR; known as POlDIP3) and promotes translation initiation and elongation by activating or binding diverse proteins including eEF2K, sKAR, 80 kDa nuclear capbinding protein (CBP80; known as nCBP1), and eIF4B. In particular, pS6K1 kinase regulates the synthesis of ribosomal subunit S6 of the 40S small subunit of ribosome and activates noncatalytic cofactor eIF4B of eIF4A by phosphorylation on Ser422. Once activated eIF4B and S6 can associate with translation initiation complex thus promoting translation. eIF4B enhances the activity of eIF4A which plays the role of a RNA helicase that promotes the unwinding of the UTRs of diverse mRNAs. s6K1 activation also releases the inhibitory effect on eIF4A mediated by phosphorylation-dependent degradation of programmed cell death 4 (PDCD4) which usually blocks the association of eIF4A with the translation preinitiation complex. S6K can also regulate the activity of eEF2K which inactivates eEF2 in a rapamycin/insulin dependent manner. This release of inhibition determines an increase in elongation rate. Recently, the role of mTORC1 in regulating translation was questioned by the finding that rapamycin, that invariantly promotes S6K1 dephosphorylation, does not inhibit 4E-BP1 phosphorylation equally well [32]. On the other hand, Torin1, catalytic site ATP-competitive mTOR inhibitor, restrains both S6K1 and 4E-BP1 phosphorylation and potently dowregulates other mTORC-regulated functions, such as autophagy and proliferation.

Taken together these findings identify mTORC1 as a fundamental regulator of translation and ribosome biogenesis that controls several anabolic and catabolic pathways at the mRNA expression level.

The effects of mTORC1 on other anabolic processes are less well understood and may involve several downstream mTORC1 substrate candidates identified so far that include: sterol regulatory element-binding protein 1 (SREBP1, lipid biosynthesis), peroxisome proliferator-activated receptor γ coactivator α (PGC-1α, mitochondrial biogenesis), STAT3 (cell growth and division, cell movement, apoptosis), PPARa (lipid metabolism in the liver), PPARg (fatty acid storage and glucose metabolism), HIFa (response to hypoxia), yin-yang 1 (YY1, mitochondrial function), RNA polymerase I and III (ribosome biogenesis).

Among them SREBP1 and HIFa are involved in glucose and lipid biosynthesis whereas PGC-1α and YY1 are involved in mitochondrial biogenesis. The SREBP family, belonging to basic helix-loop-helix-leucine zipper (bHLH-Zip), is critical to the regulation of fatty acid and cholesterol biosynthetic gene expression. The SREBP transcription factors family consists of three members: SREBP1a, SREBP1c, and SREBP2. After synthesis SREBPs localize to the endoplasmic reticulum where they bind to the sterol cleavage activating protein (SCAP). The SREBP/SCAP complex translocates to the Golgi where a proteolytic cleavage allows the entry into the nucleus. SREBPs bind to sterol regulatory element (SRE) and E box sequences found in the promoter regions of genes involved in cholesterol and fatty acid biosynthesis. SREBP1 function is positively regulated by mTORC1 through Lipin1 (phosphatidic acid phosphatase) activation [33] and has been associated with advanced glycation endproducts (AGEs) accumulation in skeletal myofibers of animal models of obese diabetes (diabesity) [34]. In particular, the loss of mTORC1-mediated Lipin1 phosphorylation promotes its nuclear entry that is followed by downregulation of nuclear SREBP protein [35]. Interestingly no physical interaction between Lipin1 and SREBP-1 has been found and a possible indirect mechanism has been hypothesized involving the nuclear lamina. Furthermore, a discrepancy between the effects of rapamycin, which does not affect SREBP target gene expression in all cellular contexts, and Torin1, which is more effective in this sense, suggests that the mechanisms through which mTORC1 regulates SREBP are complex and not yet fully understood.

mTORC1 is involved in mitochondrial expression and function [36,37]. It has been established that mTORC1 promotes transcription of PGC-1α, the master regulator of mitochondrial biogenesis and function, and activity of YY1, a mitochondrial zinc finger protein binding to DNA. PGC-1α is also known as a regulator of autophagy and stabilizer of the neuromuscular junction program. Thus PGC-1α links mitochondrial function to tissue (in particular muscle) integrity and its expression is significantly impaired in aging. In cells rapamycin exposure results in reduced gene expression of the mitochondrial transcriptional regulators PGC-1α, estrogen-related receptor α, and nuclear respiratory factors (NRF) and inhibition requires YY1 integrity. In fact mTOR and RAPTOR interacts with YY1 and inhibition of mTOR results in a failure of YY1 to interact with PGC-1α [38]. These changes impair mitochondrial gene expression and oxygen consumption [38]. It has been found that mTORC1 may also exert a direct control of the mitochondrial function in cells. In particular mTORC1 coimmunoprecipitates with the outer mitochondrial membrane proteins VDAC1 (mediators of substrate transport into the mitochondria), Bcl2 and Bcl-xl (mediator of mitochondrial function and cellular apoptosis), probably through involvement of FK506-binding family (as FKBP38). FKBP38 is localized on the mitochondria and has been shown to anchor proteins like Bcl2 and Bcl-xl to this organelle.

Autophagy is the controlled self-degradation of damaged, redundant, or even dangerous cellular components, ranging from individual proteins (microautophagy) to entire organelles (macroautophagy). Autophagy is key in providing substrates for energy production during periods of low extracellular nutrients. mTORC1 actively suppresses autophagy and, conversely, inhibition of mTORC1 (by small molecule or by amino acid withdrawal) strongly induces autophagy [34,35]. In S. cerevisiae, TOR-dependent phosphorylation of autophagy-related 13 (Atg13) disrupts the Atg1—Atg13—Atg17 complex that triggers the formation of the autophagosome [36]. The mammalian homologs of yeast Atg13 and Atg1, ATG13 and UlK1, associate with 200 kDa FAK family kinase-interacting protein (FIP200; a putative ortholog of Atg17) and the mammalian-specific component ATG101 (Fig. 11.2). By phosphorylating ATG13 and UlK1, mTORC1 blocks autophagosome initiation. However, unlike the similar complex in yeast, the formation of the UlK1—ATG13—FIP200—ATG101 complex is not regulated by nutrients [38,39]. mTORC1 also controls the activity of several transcription factors that are implicated in lipid synthesis and mitochondrial metabolism.

Substrates of mTORC2. TORC2, first identified as a mediator of actin cytoskeletal organization and cell polarization, controls members of the AGC kinase family (as Akt/PKB) and cytoskeletal regulators (as Rho1 GDP—GTP exchange protein 2: Rom2 guanine nucleotide exchange factor (GEF) for Rho1 and Rho2). TORC2 phosphorylates Akt/PKB at ser473 in mammals. This activation primes Akt/PKB for further phosphorylation at Thr308 by PDK1 leading to full activation. Importantly the Akt/PKB inhibition of FOXO proteins involved in regulation of metabolism, cell cycle, and apoptosis is regulated by mTORC2. Phosphorylation of FOXO1 and FOXO3 by Akt effectively prevents them from translocating to the nucleus. Thus mTORC2 may favor cell survival through Akt-mediated inhibition of FOXO proteins. Recent findings revealed that mTORC2 plays a variegated role in regulating AGC kinase family members. In particular it has been found that mTORC2 phosphorylates and activates Akt/PKB, serum- and glucocorticoid-regulated kinase (sGK), and protein kinase C (PKC). PKC phosphorylation by mTORC2 determines actin reorganization and immune cell differentiation. sGK is an inhibitor of FOXO proteins and the nTORC2 mediated activation of sGK depends on PROTOR. Considering that the cell cycle progression depends on mTORC1 through sGK, mTORC2 appears as an indirect regulator of mTORC1.

Collectively, these findings place mTORC2 as a key regulator of fundamental cellular processes such as cell-cycle progression, anabolism, and cell survival. Importantly initial results show that downregulation of RICTOR appears to positively regulate protein synthesis as shRNA knockdown of RICTOR stimulates protein synthesis in C2C12 myocytes and the increase was correlated with decreased RICTOR phosphorylation. These results appear to strengthen a possible regulatory link between the action of mTORC1 and mTORC2 but the mechanisms underlying this relationship deserves future rigorous investigations.

mTOR IN SENESCENCE

There is an increasing body of evidence to support a role for mTOR as a major gauge of nutrient signaling and energy balance directly impacting on organismal longevity, particularly in females. Longevity depends on aging that is the failure to maintain homeostasis to physiological and environmental cues. In particular, aging-induced alterations of the balance between energy intake and its consumption leads to related disorders including obesity, sarcopenia (loss of muscle mass and strength), anorexia, glucose intolerance, or type 2 diabetes mellitus. All these conditions may negatively impact on maximal and/or average lifespan, and suppression of growth-promoting cellular processes, which are believed to accelerate the aging process, has been identified as the rational approach to promote well-being and improve survival.

In recent years, dietary intervention and genetic manipulations of nutrient sensing and stress response pathways have been used to extend lifespan of diverse organisms as the budding yeast *S. cerevisiae*, the nematode *Caenorhabditis elegans*, the fruit fly *Drosophila melanogaster*, and the laboratory mouse *Mus musculus*. These interventions consist of dietary restriction (calorie restriction, CR), inhibition of mTOR and insulin/IGF-1-like signaling.

The fundamental role of mTOR signaling as a crossroad between metabolism, nutrition, and longevity was first revealed in *C. elegans* by Vellai and coworkers [39]. In their work they demonstrated that in this organism mutations of RNAi against *let-363*, the worm mTOR gene (as well as against S6 kinase homolog *rsks-1*), extend lifespan. Similar evidences were obtained from mutations in the RAPTOR gene, *daf-15*. In parallel with these results, similar observations emerged from genetic studies in *S. cerevisiae* and *D. melanogaster*. In *S. cerevisiae*, Fabrizio and colleagues first isolated a mutation in SCH9, the yeast ortholog of S6 kinase [40]. Subsequent studies demonstrated that deletion of either SCH9 or TOR1 also extends the yeast lifespan. Lifespan extension was observed in *D. melanogaster* following overexpression of homologs of TSC1 or TSC2 (mTORC1 inhibitors), as well as dominant negative alleles of mTOR or S6 kinase (dS6K), or overexpression of 4E-BP. In mice, genetic manipulation aiming at significantly decreasing mTORC1 expression and function revealed an increase in lifespan as observed in female mice heterozygous for both mTOR and mLST8 or in those carrying two hypomorphic mTOR alleles. Indeed, mutations or RNAi knockdown of several ribosomal protein genes has been shown to extend lifespan in budding yeast and *C. elegans*.

To date consolidated genetic and molecular evidence suggests that under conditions of calorie restriction, (CR where the caloric intake is decreased by 10–50%), the only paradigm that has consistently increased lifespan in a wide variety of organisms, longevity is promoted, at least in part, through a reduction in mTORC1 activity followed by mTOR-dependent changes in metabolism, mRNA translation, and autophagy. In fact, in mice, CR causes changes in gene expression profiles that are similar to those resulting from loss of s6K1, further supporting the view that CR acts through inhibition of mTORC1. Furthermore, treatment with mTORC1 inhibitor rapamycin, extends lifespan in budding yeast, in *C. elegans* and *D. melanogaster* (during adulthood) and in mouse strains. Interestingly, in mice, the same dose of rapamycin equally administered resulted in similar lifespan extension (of about 15% in females and 10% in males) when the delivery regimen was initiated at 9 months of age or at 20 months [41], roughly equivalent to a human age of 60 years, thus suggesting that at least part of the rapamycin- and thus mTORC1-inhibition-mediated beneficial effects on longevity are due to the delay of age-related diseases. Since these initial results, several research groups have replicated the ability of rapamycin and genetic inhibition of mTORC1 signaling to extend lifespan. Other studies have also found a positive effect of rapamycin on lifespan in aged male C57BL/6 mice, in a short-lived, tumor-prone strain of mice (FVB/N HER-2/neu transgenic), and in 129/Sv mice. All this evidence makes this pathway the current best candidate for interventional strategies to slow aging in mammals. Importantly the beneficial effects of rapamycin on lifespan, observed at concentrations about ten times higher than the concentrations currently used in human transplantation medicine, appear to be independent from the drug's antitumor activity as yeast, flies, and worms do not develop cancer and a prolongevity effect of rapamycin is observed in mouse strains with typical lifespan and tumor incidence, such as C57BL/6 and 129/Sv. However, the mechanisms involved in the prolongevity effect of rapamycin/genetic inhibition of mTORC1 signaling are still uncertain.

It has been hypothesized that these interventions may reduce cell stress via a dual-edged mechanism for controlling cellular protein content, that is, the regulation of protein translation via 4E-BP1 and the recycling of proteins via activation of autophagy. The final effect may be a reduction of protein flux through the endoplasmic reticulum. Importantly the lifespan-extending activity of rapamycin may arise independently from S6K1 inhibition. Recent findings demonstrate that that single dose of rapamycin reduces the portion of active ribosomes in liver and muscle tissue, whereas mice chronically treated with multiple doses of rapamycin show no change in ribosome activity. Furthermore, liver and skeletal muscle from $S6K1^{-/-}$ mice have normal ribosomal activity. Thus, as chronic treatment with rapamycin and knockout of S6K1 can extend lifespan in mice they appear to do so without altering ribosome activity [42]. Thus, selective inhibition of the translation of certain transcripts and other unknown mechanisms may play a key role in the antiaging effects of rapamycin.

EFFECTS OF mTORC1 IN AGE-RELATED DISEASES

As healthspan is defined as the period of life free of chronic diseases, the relevance of mTORC1 in longevity should be considered in the context of its correlation with age-related, and potentially deadly, diseases.

Cancer

Deregulation of the mTOR signaling pathway is observed in human cancers. The oncogenic activation of the mTOR signaling contributes to cancer cell growth, proliferation, and survival. In fact mutations or loss-of-function of upstream regulator genes, such as TSC1/2 or LKB1 (liver kinase B1), have been linked to clinical tumor syndromes (Tuberous Sclerosis complex, Peutz–Jeghers syndrome), whereas PI3K and Akt/PKB hyperactivation or genetic loss or mutation of PTEN have been observed in many types of human cancers. Considering that hyperactivation of mTORC1 contributes to cancer initiation and development, targeting the oncogenic mTOR pathway components could potentially be an effective strategy to prevent or treat cancer. Despite the robust anticancer effects in several cancer-prone mouse models (heterozygous p53, heterozygous Rb mice, multiple tumor xenograft models) and cell cultures, rapamycin and derivatives (Temsirolimus and Everolimus, FDA approved) have shown limited potency in clinical trials with the exception of renal cell carcinoma and other rare cancers [43]. The observations that mTORC1 has both rapamycin-sensitive and rapamycin-insensitive substrates, that Akt/PKB activation may follow prolonged mTORC1 inhibition, and that mTORC2 activation is largely unaffected by cellular exposure to rapamycin may contribute to explain the clinical ineffectiveness of the treatments. Promising results arise from combined therapies simultaneously targeting mTORC1, mTORC2 and Akt/PKB, and the clinical impact of these approaches remains to be determined.

Cardiac Dysfunction

Rapamycin's properties against cell growth have been used for cardiovascular benefit. Evidence is accumulating that mTORC1 inhibition may provide, at least in animal models (mouse, rat, zebrafish), protective effects against cardiomyopathy. A beneficial effect of rapamycin (or derivatives) has been observed in hormone-induced cardiomyopathy, cardiac ischemia/reperfusion injury, hypertrophic cardiomyopathy, and dilated cardiomyopathy. In particular rapamycin, through S6K inhibition, significantly attenuates both cardiomyocyte and heart hypertrophy induced by several factors as growth factors, hormones (as angiotensin II-induced increases in protein synthesis in myocytes) or overload and this effect is followed by improvement of cardiac function. Recent evidence also demonstrates that S6K inhibition is followed by beneficial effects on left ventricular function and alleviated pathological cardiac remodeling and cell apoptosis following myocardial infarction in mice due to enhanced Akt/PKB signaling [44].

Neurodegenerative Diseases

In neurodegenerative diseases (only Alzheimer's and Parkinson's diseases are briefly discussed) an accumulation of toxic misfolded proteins may lead to neuronal cell death. This phenomenon suggests that, at least in some cases, defects in autophagy may promote their pathogenesis. Therefore, it was initially hypothesized that stimulation of autophagy by mTOR inhibition would have a beneficial effect via autophagic clearance of these proteins and current research is focused on testing single (mTORC1) or dual (mTORC1 and mTORC2) inhibition as a potential therapeutic strategy in diverse diseases models [45].

Alzheimer's disease (AD) is a progressive neurodegenerative disorder which represents the leading cause of dementia in the aged population. AD symptoms include cognitive impairment with progressive memory deficits and personality changes. The causes of such cognitive decline has been attributed to a progressive synaptic dysfunction and subsequent loss of neurons, which appears to be located in many regions of the brain: mainly neocortex, limbic system, and the subcortical regions. The pathological features of AD are the accumulation of extracellular senile plaques and intracellular neurofibrillary tangles (NFT), leading to chronic inflammatory responses, increase in oxidative stress, mitochondrial dysfunction and neural loss. The two major lesions in AD, plaques and NFT, are caused by distinct proteins, tau in the case of the neurofibrillary tangles and amyloid β-protein in the case of amyloid plaques. The amyloid cascade hypothesis refers to abnormalities in the cleavage of the amyloid precursor protein (APP) resulting in the generation of toxic oligomeric Aβ. Aβ accumulation has been correlated with increased mTOR signaling and increases Akt/PKB levels followed by the appearance of NFT. Importantly inhibition of mTOR signaling appears to be followed by beneficial effects on both behavioral and pathophysiological outcomes in diverse animal models of AD.

Another common debilitating neurodegenerative disease is Parkinson's disease (PD), characterized by bradykinesia, resting tremor, postural and autonomic instability, and rigidity. The pathological signature of PD is the loss of dopaminergic neurons in the substantia nigra and the formation of aggregates (Lewy bodies) in neurons. Alpha-synuclein is the major protein in Lewy bodies and missense mutations in alpha-synuclein (A53T and A30P) cause autosomal dominant, early-onset PD. Lewy bodies are also seen in dementia, the Lewy body variant of Alzheimer's disease, and in other forms of neurodegeneration (brain iron accumulation type I and multiple system atrophy). These diseases, collectively known as alpha-synucleinopathies, suggest that alpha-synuclein is likely to play main role in neurodegeneration. Also in PD case the accumulation of toxic proteins may suggest that defective autophagy may play a role in promoting pathogenesis. As a matter of fact rapamycin-promoted autophagy is followed by alpha-synuclein degradation in alpha-synuclein-overexpressing cells and mice [46], whereas constitutively active form of Akt/PKB (myristoylated Akt/PKB, Myr-Akt), remarkably preserved the dopaminergic nigrostriatal axon projections in a highly destructive neurotoxin model of PD characterized by increased mTOR activity through suppression of macroautophagy by Myr-Akt. On the contrary in the pharmacological PD model induced by the neurotoxin 1,2,3,6-tetrahydro-1-methyl-4-phenylpyridine hydrochloride (MPTP) the mouse mutants display nucleolar disruption restricted to dopaminergic (DA) neurons resulting in increased p53 levels and downregulation of mTOR activity, leading to mitochondrial dysfunction and increased oxidative stress.

Diabesity (Diabetes and Obesity) and Sarcopenia

As known, aging is associated with reduced lean body mass, increased adiposity and insulin resistance. The loss of lean body mass (from the second to the eighth decade of life about 18% in sedentary men and by 27% in sedentary women) is linked to concomitant increase in fat body mass as described by increased body mass index, waist circumference, or waist to-hip ratio. The progressive loss of skeletal muscle mass promotes strength reduction (known as sarcopenia), the change in fiber type composition, including selective loss of fast type II fibers [12], and, considering that the skeletal muscle represents the major glucose acceptor, the appearance of reduced insulin sensitivity. Therefore, late in life, the mechanisms and the reciprocal interactions between loss of lean mass and the arising dysmetabolism are difficult to be distinguished and the relationship between mTOR signaling and age-related diabesity and sarcopenia appears less straightforward than that observed for other age-related phenotypes. Several factors may contribute to sarcopenia, and indirectly, to sarcopenia-related metabolic dysfunction. In particular, a general or selective reduction in muscle protein synthesis (MPS) that is responsible for altered efficiency of the protein turnover and reduced responsiveness to specific stimuli (mechanical stress, anabolic hormones as insulin) and amino acids (essential amino acids) plays a major role. It is currently debated whether mTOR pathway is dysregulated during muscle aging. Recently It has been found that, although the basal levels of activation of mTORC1 and downstream target kinases S6K1 and 4E-BP1 in elderly humans appears almost comparable to the adult, a significant increase in rpS6 235/236 phosphorylation is detectable in 800-day-old mice as compared to the 200-day-old mice, suggesting an increased mTORC1 activity in skeletal muscles at least in aging mice [47]. It has been suggested that decreased AMPK activity may concur to the age-related increase in mTORC1 signaling observed in mice (Fig. 11.3). Importantly a chronic activation of mTOR is observed in adipose tissue in presence of overnutrition (Fig. 11.3) and this change has been linked with cancer, beta cell dysfunction, and fatty liver diseases. It is presently not known if the increased mTORC1 activation correlates with sarcopenia (Fig. 11.3). It has been hypothesized that, apart from the changes in the basal level of mTOR signaling that may justify a plethora of changes due to prolonged inhibitory feedback through

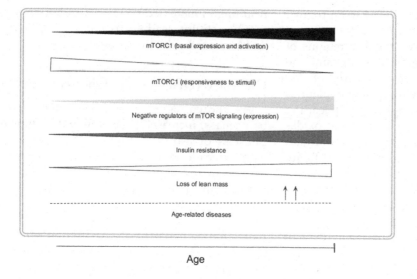

FIGURE 11.3 With age mTOR possibly increases its basal level of expression and activation. This change is paralleled by reduced responsiveness to stimuli (ie, mechanical load), partly ascribable to increased expression of negative regulators. Increased loss of muscle mass and concomitant increased insulin resistance contribute to the arising of age-related diseases. In overnutrition the observed changes in mTOR basal activation, insulin resistance, and loss of mass appear amplified with age. It is currently unknown whether these changes are paralleled by a further decrease of mTOR responsiveness to stimuli and increased expression and function of negative regulators. Magnification of mTOR dysfunction and insulin resistance lead to premature appearance of age-related diseases and/or premature death.

mTORC1/S6K1 cassette to IRS-Akt/PKB signaling, the increase of MPS induced by physiological stimuli as mechanical load and nutrients appears to be attenuated [48], particularly in fast twitch muscles. In agreement the change in MPS to mechanical load is delayed and partially reduced by altered activation response of the mTORC1/S6K1 pathway. In accordance, an increased expression of negative mTORC1 and mTORC2 regulators (DEPTOR, TSC1, and TSC2) as well as translation initiation factor EIF4G2, the cellular differentiation and proliferation factor PRKCA and the transcription factor FOXO1 has been observed with senescence at least in human cell lines. An increased level of complication arises when considering the relationship between sarcopenic obesity and age-related alteration of glucose tolerance. S6K1 knockout mice appear hypoinsulinemic and glucose intolerant due to decreased beta cell size and function [28,29]. Also rapamycin treatment determines glucose intolerance, insulin resistance, and dyslipidemia and these effects has been attributed to inhibition of mTORC2. Despite the apparently detrimental effects of both interventions, S6k1 knockout and mTORC blockage, on metabolism mice are long-lived. Therefore additional studies are needed to determine whether targeted inhibition of mTORC1 can prove useful against diabetes mellitus and/or obesity and whether possible metabolic changes in lipid profile, insulin sensitivity, and glucose homeostasis resulting from chronic mTORC1 inhibition may be harmful.

SUMMARY

- mammalian Target of Rapamycin (mTOR) kinase is a fundamental intersection between cell functions and nutrients and is indispensable for life.

- mTORC1 complex hyperactivation by nutrients may be detrimental in physiological conditions.
- mTORC1 and mTORC2 complexes interactions are only partially known.
- Amino acids and, in particular, essential amino acids are fundamental regulators of mTOR signaling in cells.
- Calorie restriction leads to reduced mTOR signaling activation and pharmacological reduction of mTOR signaling increases lifespan.
- Molecular events associated with calorie restriction and anabolic signaling are increasingly understood and the identification of novel signaling pathways controlling mTORC complexes may open further clinical avenues in human disease.

References

[1] Kim DH, Sarbassov DD, Ali SM, King JE, Latek RR, Erdjument-Bromage H, et al. mTOR interacts with raptor to form a nutrient-sensitive complex that signals to the cell growth machinery. Cell 2002;110(2):163–75.

[2] Peterson TR, Laplante M, Thoreen CC, Sancak Y, Kang SA, Kuehl WM, et al. DEPTOR is an mTOR inhibitor frequently overexpressed in multiple myeloma cells and required for their survival. Cell 2009;137(5):873–86.

[3] Sabatini DM, Erdjument-Bromage H, Lui M, Tempst P, Snyder SH. RAFT1: a mammalian protein that binds to FKBP12 in a rapamycin-dependent fashion and is homologous to yeast TORs. Cell 1994;78(1):35–43.

[4] Brown EJ, Albers MW, Shin TB, Ichikawa K, Keith CT, Lane WS, et al. A mammalian protein targeted by G1-arresting rapamycin-receptor complex. Nature 1994;369(6483):756–8.

[5] Cafferkey R, McLaughlin MM, Young PR, Johnson RK, Livi GP. Yeast TOR (DRR) proteins: amino-acid sequence alignment and identification of structural motifs. Gene 1994;141(1):133–6.

[6] Galat A, Lane WS, Standaert RF, Schreiber SL. A rapamycin-selective 25-kDa immunophilin. Biochemistry 1992;31(8):2427–34.

[7] Sarbassov DD, Ali SM, Kim DH, Guertin DA, Latek RR, Erdjument-Bromage H, et al. Rictor, a novel binding partner of mTOR, defines a rapamycin-insensitive and raptor-independent pathway that regulates the cytoskeleton. Curr Biol 2004;14(14):1296–302.

[8] Lamming DW, Ye L, Katajisto P, Goncalves MD, Saitoh M, Stevens DM, et al. Rapamycin-induced insulin resistance is mediated by mTORC2 loss and uncoupled from longevity. Science 2012;335(6076):1638–43.

[9] Lamming DW, Mihaylova MM, Katajisto P, Baar EL, Yilmaz OH, Hutchins A, et al. Depletion of Rictor, an essential protein component of mTORC2, decreases male lifespan. Aging Cell 2014;13(5):911–17.

[10] Zoncu R, Efeyan A, Sabatini DM. mTOR: from growth signal integration to cancer, diabetes and aging. Nat Rev Mol Cell Biol 2011;12(1):21–35.

[11] Mendoza MC, Er EE, Blenis J. The Ras-ERK and PI3K-mTOR pathways: cross-talk and compensation. Trends Biochem Sci 2011;36:320–8.

[12] D'Antona G, Nisoli E. mTOR signaling as a target of amino acid treatment of the age-related sarcopenia. Interdiscip Top Gerontol 2010;37:115–41.

[13] Sarbassov DD, Guertin DA, Ali SM, Sabatini DM. Phosphorylation and regulation of Akt/PKB by the rictor-mTOR complex. Science 2005;307:1098–101.

[14] Rommel C, Bodine SC, Clarke BA, Rossman R, Nunez L, Stitt TN, et al. Mediation of IGF-1-induced skeletal myotube hypertrophy by PI(3)K/Akt/mTOR and PI(3)K/Akt/GSK3 pathways. Nat Cell Biol 2001;3:1009–13.

[15] Sancak Y, Peterson TR, Shaul YD, Lindquist RA, Thoreen CC, Bar-Peled L, et al. The Rag GTPases bind raptor and mediate amino acid signaling to mTORC1. Science 2008;320:1496–501.

[16] Um SH, Frigerio F, Watanabe M, Picard F, Joaquin M, Sticker M, et al. Absence of S6K1 protects against age- and diet-induced obesity while enhancing insulin sensitivity. Nature 2004;431(7005):200–5.

[17] Nelson BA, Robinson KA, Buse MG. High glucose and glucosamine induce insulin resistance via different mechanisms in 3T3-L1 adipocytes. Diabetes 2000;49(6):981–91.

[18] Tremblay F, Marette A. Amino acid and insulin signaling via the mTOR/p70 S6 kinase pathway. A negative feedback mechanism leading to insulin resistance in skeletal muscle cells. J Biol Chem 2001;276(41):38052–60.

[19] Siebel A, Cubillos-Rojas M, Santos RC, Schneider T, Bonan CD, Bartrons R, et al. Contribution of S6K1/MAPK signaling pathways in the response to oxidative stress: activation of RSK and MSK by hydrogen peroxide. PLoS One 2013;8(9):e75523. Available from: http://dx.doi.org/10.1371/journal.pone.0075523 eCollection 2013

[20] Bar-Peled L, Schweitzer LD, Zoncu R, Sabatini DM. Ragulator is a GEF for the rag GTPases that signal amino acid levels to mTORC1. Cell 2012;150(6):1196–208.

[21] Bar-Peled L, Chantranupong L, Cherniack AD, Chen WW, Ottina KA, Grabiner BC, et al. A tumor suppressor complex with GAP activity for the Rag GTPases that signal amino acid sufficiency to mTORC1. Science 2013;340(6136):1100–6.

[22] Sancak Y, Bar-Peled L, Zoncu R, Markhard AL, Nada S, Sabatini DM. Ragulator-Rag complex targets mTORC1 to the lysosomal surface and is necessary for its activation by amino acids. Cell 2010;141(2):290–303. Available from: http://dx.doi.org/10.1016/j.cell.2010.02.024.

[23] O'Neill HM, Maarbjerg SJ, Crane JD, Jeppesen J, Jørgensen SB, Schertzer JD, et al. AMP-activated protein kinase (AMPK) beta1beta2 muscle null mice reveal an essential role for AMPK in maintaining mitochondrial content and glucose uptake during exercise. Proc Natl Acad Sci USA 2011;108(38):16092–7.

[24] Inoki K, Zhu T, Guan KL. TSC2 mediates cellular energy response to control cell growth and survival. Cell 2003;115(5):577–90.

[25] Gwinn DM, Shackelford DB, Egan DF, Mihaylova MM, Mery A, Vasquez DS, et al. AMPK phosphorylation of raptor mediates a metabolic checkpoint. Mol Cell 2008;30(2):214–26.

[26] Kalender A, Selvaraj A, Kim SY, Gulati P, Brûlé S, Viollet B, et al. Metformin, independent of AMPK, inhibits mTORC1 in a rag GTPase-dependent manner. Cell Metab 2010;11(5):390–401.

[27] Barbet NC, Schneider U, Helliwell SB, Stansfield I, Tuite MF, Hall MN. TOR controls translation initiation and early G1 progression in yeast. Mol Biol Cell 1996;7(1):25–42.

[28] D'Antona G. Essential amino acids supplementation for the prevention and treatment of abdominal obesity. In: Watson RR, editor. Nutrition in the prevention and treatment of abdominal obesity. Academic Press; 2014. p. 447–69.

[29] D'Antona G. Amino acids supplementation as nutritional therapy. In: Watson RR, Dokken B, editors. Glucose intake and utilization in prediabetes and diabetes. Academic Press; 2014. p. 387–95.

[30] Martina JA, Puertollano R. RRAG GTPases link nutrient availability to gene expression, autophagy and lysosomal biogenesis. Autophagy 2013;9(6):928–30.

[31] Han JM, Jeong SJ, Park MC, Kim G, Kwon NH, Kim HK, et al. Leucyl-tRNA synthetase is an intracellular leucine sensor for the mTORC1-signaling pathway. Cell 2012;149(2):410–24.

[32] Thoreen CC, Kang SA, Chang JW, Liu Q, Zhang J, Gao Y, et al. An ATP-competitive mammalian target of rapamycin inhibitor reveals rapamycin-resistant functions of mTORC1. J Biol Chem 2009;284:8023–32.

[33] Porstmann T, Santos CR, Griffiths B, Cully M, Wu M, Leevers S, et al. SREBP activity is regulated by mTORC1 and contributes to Akt-dependent cell growth. Cell Metab 2008;8:224–36.

[34] Mastrocola R, Collino M, Nigro D, Chiazza F, D'Antona G, Aragno M, et al. Accumulation of advanced glycation end-products and activation of the SCAP/SREBP lipogenetic pathway occur in diet-induced obese mouse skeletal muscle. PLoS One 2015;10(3):e0119587.

[35] Peterson TR, Sengupta SS, Harris TE, Carmack AE, Kang SA, Balderas E, et al. mTOR complex 1 regulates lipin 1 localization to control the SREBP pathway. Cell 2011;146(3):408–20.

[36] Valerio A, D'Antona G, Nisoli E. Branched-chain amino acids, mitochondrial biogenesis, and healthspan: an evolutionary perspective. Aging (Albany, NY) 2011;3(5):464–78.

[37] D'Antona G, Ragni M, Cardile A, Tedesco L, Dossena M, Bruttini F, et al. Branched-chain amino acid supplementation promotes survival and supports cardiac and skeletal muscle mitochondrial biogenesis in middle-aged mice. Cell Metab 2010;12(4):362–72.

[38] Cunningham JT, Rodgers JT, Arlow DH, Vazquez F, Mootha VK, Puigserver P. mTOR controls mitochondrial oxidative function through a YY1-PGC-1alpha transcriptional complex. Nature 2007;450(7170):736–40.

[39] Vellai T, Takacs-Vellai K, Zhang Y, Kovacs AL, Orosz L, Müller F. Genetics: influence of TOR kinase on lifespan in *C. elegans*. Nature 2003;426(6967):620.

[40] Fabrizio P, Pozza F, Pletcher SD, Gendron CM, Longo VD. Regulation of longevity and stress resistance by Sch9 in yeast. Science 2001;292(5515):288–90.

[41] Harrison DE, Strong R, Sharp ZD, Nelson JF, Astle CM, Flurkey K, et al. Rapamycin fed late in life extends lifespan in genetically heterogeneous mice. Nature 2009;460(7253):392–5.

[42] Garelick MG, Mackay VL, Yanagida A, Academia EC, Schreiber KH, Ladiges WC, et al. Chronic rapamycin treatment or lack of S6K1 does not reduce ribosome activity in vivo. Cell Cycle 2013;12(15):2493–504.

[43] Xu K, Liu P, Wei W. mTOR signaling in tumorigenesis. Biochim Biophys Acta 2014;1846(2):638–54.

[44] Di R, Wu X, Chang Z, Zhao X, Feng Q, Lu S, et al. S6K inhibition renders cardiac protection against myocardial infarction through PDK1 phosphorylation of Akt. Biochem J 2012;441(1):199–207.

[45] Lipton JO, Sahin M. The neurology of mTOR. Neuron 2014;84(2):275–91.

[46] Webb JL, Ravikumar B, Atkins J, Skepper JN, Rubinsztein DC. Alpha-Synuclein is degraded by both autophagy and the proteasome. J Biol Chem 2003;278(27):25009–13.

[47] Sandri M, Barberi L, Bijlsma AY, Blaauw B, Dyar KA, Milan G, et al. Signalling pathways regulating muscle mass in aging skeletal muscle: the role of the IGF1-Akt-mTOR-FoxO pathway. Biogerontology 2013;14(3):303–23.

[48] Mascaro A, D'Antona G. Acute exposure to essential amino acids activates contraction mediated. mTOR/p70 signaling in soleus muscle of elderly rats. Progr Nutr 2013;15(3):139–45.

CHAPTER

12

Lipid Peroxidation, Diet, and the Genotoxicology of Aging

Petr Grúz

Division of Genetics and Mutagenesis, National Institute of Health Sciences, Tokyo, Japan

KEY FACTS

- Humans can utilize ω-6, ω-3 or ω-9 PUFAs in cellular membranes depending on the dietary intake of the former two which cannot be synthesized *de novo* by vertebrates.
- ω-3 PUFAs counterbalance ω-6 PUFA proinflammatory activities but are highly susceptible to oxidation producing genotoxic lipid peroxides.
- ω-9 PUFAs are the most resilient membrane building blocks selected during evolution of the longevous animal species.
- Seed oils containing mainly ω-6 concentrated PUFAs are excessively used in modern foods of Western nations substituting for traditional fats low in PUFAs.
- Indigenous human populations utilizing traditional saturated fats seem to be protected from the typical chronic degenerative diseases.
- DNA lesions derived from lipid peroxidation of dietary ω-6 and ω-3 PUFAs accumulate in human genome during aging as their repair and detoxification gradually weaken.
- Lipid peroxide derived DNA adducts are mutagenic in several standard mutagenicity assays.
- Seed oil PUFA rich diets fuel the growth of cancers in particular of reproductive organs while diets deficient in ω-6 and ω-3 PUFAs inhibit experimental carcinogenesis.

Dictionary of Terms

- **PUFA**: polyunsaturated fatty acid
- **MUFA**: monounsaturated fatty acid
- **LC-PUFA**: long-chain polyunsaturated fatty acid (20-carbon chain length)
- **VLC-PUFA**: very long-chain polyunsaturated fatty acid (more than 20 carbon chain length)
- **EFA**: essential fatty acid
- **EFAD**: essential fatty acid deficiency, a state characterized by increased levels of ETrA
- **MCFA**: medium chain fatty acids
- **DNA**: deoxyribonucleic acid
- **LA**: 9,12-octadecadienoic acid, linoleic acid (18:2n6)
- **ALA**: 9,12,15-octadecatrienoic acid, α-linolenic acid (18:3n3)
- **AA**: 5,8,11,14-eicosatetraenoic acid, arachidonic acid (20:4n6)
- **ETrA**: 5,8,11-eicosatrienoic acid, mead acid (20:3n9)
- **DHA**: 4,7,10,13,16,19-docosahexaenoic acid, docosahexaenoic acid (22:6n3)
- **ROS**: reactive oxygen species
- **ETC**: electron transport chain (in mitochondrial membrane)
- **TOR**: target of rapamycin (a protein kinase that regulates cell growth)
- **MLSP**: maximum lifespan
- **MDA**: malondialdehyde
- **4-HNE**: 4-hydroxy-2-nonenal
- **4-HPNE**: 4-hydroperoxy-2-nonenal
- **EH**: 2,3-epoxy-4-hydroxynonanal
- **4-ONE**: 4-oxo-2-nonenal
- **4-OHE**: 4-oxo-2-hexenal

- **4-HHE**: 4-hydroxy-2-hexenal
- **13-HODE**: 13-hydroxyoctadecadienoic acid
- **COX-2**: prostaglandin-endoperoxide synthase 2 (cyclooxygenase 2)
- **5-LOX**: arachidonate 5-lipoxygenase
- **PLA2**: phospholipase A2
- **LTA4**: leukotriene A4
- **PGE2**: prostaglandin E2
- **PGH2**: prostaglandin H2
- **PGF2α**: prostaglandin F2α
- **NSAIDs**: nonsteroidal anti-inflammatory drugs
- **LPO**: lipid peroxides
- **AGE**: advanced glycation end products
- **ALE**: advanced lipoxidation end products
- **dG**: deoxyguanosine
- **dA**: deoxyadenosine
- **dC**: deoxycytosine
- **εdG**: 1,N^2-etheno-2'-deoxyguanosine
- **εdA**: 1,N^6-etheno-2'-deoxyadenosine
- **εdC**: 3,N^4-etheno-2'-deoxycytosine
- **7εdG**: heptanone-etheno-2'-deoxyguanosine
- **7εdA**: heptanone-etheno-2'-deoxyadenosine
- **7εdC**: heptanone-etheno-2'-deoxycytosine
- **4εdC**: butanone-etheno-2'-deoxycytosine
- **M_1dG**: 1,N^2-propano-2'-deoxyguanosine
- **γ-HOPdG**: γ-hydroxypropano-2'-deoxyguanosine
- **HNE-dG**: trans-4-hydroxy-2-nonenal-2'-deoxyguanosine (1,N^2-propano-deoxyguanosine type adduct)
- **LTA4-dG**: 5-hydroxy,12-[Guo-N^2-yl]-6,8,11,14-eicosatetraenoic acid
- **8-oxo-dG**: 8-oxo-2'-deoxyguanosine
- **DHT**: dihydrotestosterone
- **AD**: Alzheimer's disease
- **TLS**: translesion DNA synthesis, replication of damaged template
- **PPAR**: peroxisome proliferator-activated receptor
- **NFkB**: nuclear factor kappa-light-chain-enhancer of activated B cells, transcription factor
- **NER**: nucleotide excision repair
- **BER**: base excision repair
- **Polκ**: DNA polymerase κ, DinB homologue

INTRODUCTION

Nearly all aging theories converge at the fact that molecular damage accumulates with advancing age. Since energy sources had been scarce during evolution our ontogenetic program has been optimized to invest into the somatic maintenance and repair only to the extent necessary to ensure successful reproduction. This is best described by the disposable soma theory [1], which is essentially a subset of the antagonistic pleiotropy theory. The cause of biological aging is a failure to repair and replace complex biological molecules which are being continuously damaged by the side-products of metabolism. The downregulation of the repair and maintenance systems also called "determinants of longevity" results in an increase in vulnerability to age-associated diseases and functional decline [2].

The source and perhaps the most important target of the primary damaging molecules designated as Reactive Oxygen Species (ROS) is the energy production machinery in mitochondria. Dysfunctional damaged mitochondria leak more ROS affecting other systems in the body leading for example, to telomere shortening, cellular senescence, and tumorigenesis [3,4]. The whole process has been well described by the mitochondrial theory of aging [5].

Although aging theories describe the process of a gradual breakdown of organism as a whole and in general terms, they don't address the molecular details of the key events occurring at the cellular and tissue levels. The purpose of this chapter is to describe in detail one, and perhaps the most important molecular, aspect of the aging process, which is the accumulation of oxidized lipids. Since all biological membranes contain fatty acids as basic building blocks, their spontaneous oxidation wreaks havoc on the structural integrity of cells and also damages other key molecules such as the DNA. The significance of damage to DNA from lipid peroxidation in the long term cannot be overlooked since it triggers cellular senescence and initiates mutagenesis leading to gradual healthy stem cell exhaustion and the rise of cancer stem cells [6]. Because the fatty acid composition of cellular membranes can be altered through diet some suggestions will be made to render cells and their genetic material more resistant to the damage and deterioration coming along with advancing age. This parallels the evolutionary strategies undertaken by long-living species which minimize the use of oxidation prone lipids in their membranes.

FLEXIBLE FATTY ACID COMPOSITION OF HUMAN CELLS

The PUFAs of the omega-3 and omega-6 series are regarded as the essential fatty acids (EFAs) because vertebrates cannot synthesize them de novo in contrast to the omega-9 series (Fig. 12.1). The long-chain PUFAs (LC-PUFA) with 20 carbons such as AA or EPA used as the precursors of eicosanoids, however, can be synthesized from their shorter counterparts of plant origin, namely linoleic acid (LA) and α-linolenic acid (ALA). The elongation and desaturation processes in the body produce even longer chain VLC-PUFAs with 22 or 24 carbons, such as DHA, which seem to play specific

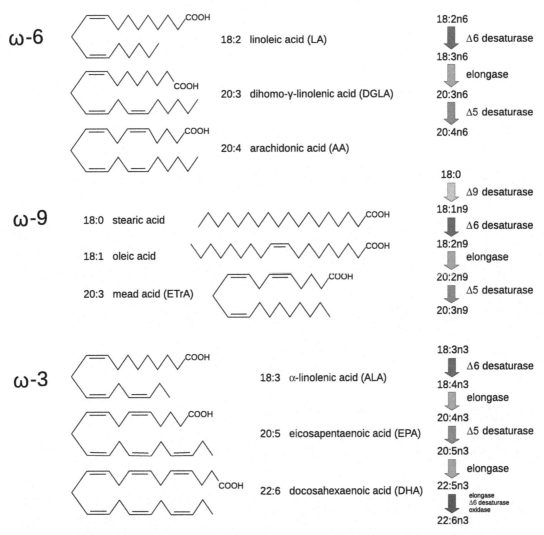

FIGURE 12.1 Schematic representation of three main families of polyunsaturated fatty acids (PUFAs) present in human body and their endogenous synthesis. The structures of the most important members with 18, 20, and 22 carbons chain length referred to in the text are shown. The synthesis of omega-6 and omega-3 start from their PUFA precursors with 18-carbon chain length which are supplied by the diet since vertebrates cannot manufacture them *de novo*.

structural functions in certain tissues such as the retina (Fig. 12.1). Due to the large number of double bonds in the carbon chain of EFAs they remain liquid at low temperatures and are therefore predominantly found in plants and animals, such as the cold water fish, to preserve their biological membranes fluidity.

The proportion of different fatty acids in cellular membranes, adipose tissue stores, and body fluids is strongly influenced by the type of fat in the diet [7,8] and it may take over 2 years to significantly alter the human body fatty acid composition [9].

Durability and Oxidizability of Fatty Acids

Lower double bond content in unsaturated fatty acids leads to a lower sensitivity to in vitro lipid peroxidation, and is associated with a lower concentration of lipid peroxidation products in vivo [10]. But the susceptibility of a membrane to peroxidation is not simply given by its unsaturation index which is just a measure of the density of double bonds in the membrane. By definition, MUFA have a single double-bonded carbon unit ($-C=C-$) per acyl chain while PUFA chains have more than one $-C=C-$ unit per acyl chain. DHA is the most polyunsaturated of the common fatty acids in mammalian membranes and has six such units along its 22-carbon hydrocarbon chain (it is commonly written as 22:6, Fig. 12.1). In naturally occurring polyunsaturates, the $-C=C-$ units are all separated by a single-bonded $-C-$ atom. The hydrogen atoms attached to each of these intermediate $-C-$ atoms are called *bis*-allylic hydrogens, and have the lowest C—H bond-energies of the fatty acid chain.

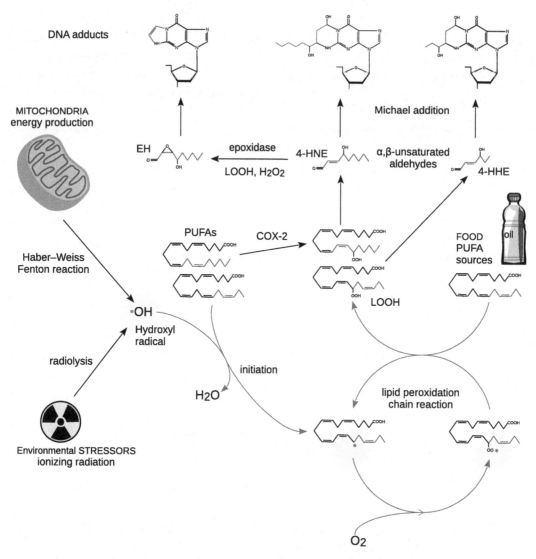

FIGURE 12.2 Pathways leading to the formation of propano- and etheno-DNA adducts from aldehydes and epoxides, respectively. Leaks at the mitochondrial electron transport chain or water radiolysis produces •OH radical that attacks polyunsaturated fatty acids (PUFAs) containing methylene-interrupted double bonds. This triggers a lipid peroxidation chain reaction that provides a continuous supply of free radicals to initiate future peroxidation and thus has potentially devastating effects. The reaction products are lipid hydroperoxides (LOOH) which homolytically decompose to genotoxic α,β-unsaturated aldehydes that directly react with DNA or are transformed to corresponding epoxides which possess even higher reactivity toward DNA. An alternative pathway for the LOOH formation by enzymatic oxidation of PUFAs with cyclooxygenase (COX-2) is also shown.

This makes them the most susceptible to attack by ROS such as the hydroxyl radical produced during aerobic metabolism (Fig. 12.2). For example, DHA (22:6) with its six double bonds and consequently five *bis*-allylic hydrogens per chain is 320 times more susceptible to ROS attack than the common monounsaturated oleic acid (18:1) which has no *bis*-allylic hydrogens in its chain. The peroxidation index takes into account the double bond positioning and is the most accurate measure of unsaturated membrane sensitivity to lipid peroxidation [11].

In vitro experimental evidence suggests that the peroxidation of some major LC-PUFAs, such as AA, could be a more significant source of the Advanced Glycation End Products (AGEs), such as the lysine adducts on proteins, than the glycoxidation reactions [12,13].

Benefits of the Omega-9 PUFAs

In contrast to the cold-blooded worms which synthesize their own omega-6 and omega-3 PUFAs [14], vertebrates can only de novo synthesize omega-9 PUFAs. The omega-9 series PUFA synthesized by humans is termed 5,8,11-eicosatrienoic acid (ETrA) or the Mead acid (Fig. 12.1) and its manufacture has been recently deciphered in detail [15]. Although most tissues prefer to synthesize the LC-PUFAs from their plant omega-6 and omega-3 precursors LA and ALA, these EFAs are not required in large amounts but rather in proper balance to each other.

The omega-9 ETrA is present in normal and young tissues [16] and can be also metabolized to potential signaling molecules derived from the COX [17–19], LOX [20,21], and P450 epoxygenase [22] pathways. Although ETrA cannot be converted into the classical cyclic prostaglandin-type signaling molecules, prostaglandin deficiency is not fatal [23]. Mice deficient in COX-1 live uneventful lives despite a 99% reduction in overall prostaglandin production [23] and a lifetime absence of COX-2 downregulates NFkB and produces resistance to neuroinflammation, excitotoxicity, and carcinogenesis [24–26]. ETrA can be also further elongated and desaturated to VLC-PUFAs [27,28] even in the presence of EFAs in certain vertebrate cells [29]. The metabolites of ETrA have been reported to exert the same biological activities as their counterparts derived from AA acting as a chemoattractant for the immune system [30] or an endogenous agonist for the cannabinoid receptor [31].

LA present in cooking oils is converted into AA which is known to fuel chronic inflammatory processes. Overreactivity of the immune system mediated by the AA metabolites such as PGE2 and LTB4 leads to immunosenescence which is directly linked to aging [32,33]. The common recommendation is to increase the consumption of the omega-3 type fats which interfere with AA metabolization and their LC-PUFA derivatives like EPA are converted into less pro-inflammatory eicosanoids. However, given the higher susceptibility of the omega-3 PUFAs to oxidation it may be wiser to simply reduce the omega-6 intake or rather supplement with ETrA [20,34,35], which is the most stable among the PUFAs and has been found to attenuate toxicity due to lipid peroxidation [36]. The ETrA metabolite LTA3 is also a potent inhibitor of the LTA4 hydrolase, an enzyme responsible for LTB4 synthesis [21,37], which is being targeted by various synthetic inhibitors developed to be used as therapeutic anti-inflammatory agents [38]. The LC-PUFAs of the omega-3 series are known to alter the immune system balance causing a Th1 to Th2 shift which is associated with the development of allergies [39,40]. Indeed the LC-PUFAs in maternal milk are a risk factor for the development of atopy in breastfed infants [41]. EPA has been reported to suppress the activity of NK cells which play an important role in cancer immunity and immune tolerance [42,43].

Although the omega-3 fat consumption is highly advocated by the supplement industry and medical community, there is no evidence that it prevents cancer or cardiovascular diseases [44–46]. Also of concern are reports about some cancer-promoting effects of the omega-3 PUFAs [47–50]. Given the pro-inflammatory nature of the AA metabolites, the omega-6 PUFAs seem to be contributing to carcinogenesis the most. Corn oil rich in LA scores the highest in numerous cancer promotion studies using DMBA-induced adenocarcinomas in rats [51–54]. In the rodent studies the proportion of dietary fat in the fat-rich diets is usually 20% of what approximately corresponds to the amount of fat consumed in the human diet [55–57]. It seems that the dietary intake of corn oil exceeding 3% is sufficient for its cancer promoting effect [58] while less than 1% dietary EFAs is generally considered to lead to EFAD. LA is essential in mammary tumorigenesis with sensitivity increasing proportionally in the range up to 4% [59]. Mice fed only saturated fat diet (lacking appreciable LA amounts) are resistant to the promotion of skin cancer [60]. LA is released from the fat stores upon starvation and its metabolite 13-HODE is a strong cancer growth promoter [61,62]. Omega-6 fatty acids have been also implicated in the activation of chemical mutagens such as the benzo[a]pyrene [63,64] and chronic exposure to them results in direct oxidative DNA damage such as 8-oxo-dG [65] suggesting the generation of superoxide anion and singlet oxygen.

Rats chronically fed specially designed semipurified diets rich in unsaturated fats showed an increased degree of fatty acid unsaturation of postmitotic tissues, such as brain, and had elevated not only lipid but also protein and mtDNA oxidative damage [66]. ROS are generated by the mitochondrial forward electron transport in the presence of PUFAs due to their inhibitory effect on the Complex III of ETC [67,68]. The lower unsaturation of mitochondrial membranes indeed correlates with lower ROS production and increased lifespan [69,70]. Other benefits of dietary EFA restriction relate to calming down an overreactive immune system and include, for example, protection against streptozotocin-induced diabetes [71], kidney injury [72], or suppression of exacerbated immune shock responses to endotoxin [73].

In summary, it seems that the common food-derived PUFAs play important roles in the development of numerous problems associated with aging, such as derailed immunity, susceptibility to tissue peroxidation and inflammation, atherosclerosis, and growth of tumors, and are probably involved in most other health problems even in children. Hence limiting exposure to dietary PUFAs could prove beneficial not only because of decrease in lipid peroxidation [36].

Lessons from Human Populations with Low PUFA Intake

Numerous statistical cohort, clinical, and meta-analysis studies demonstrate the short-term benefits of omega-3 supplements on mortality and surrogate markers of different inflammatory diseases which are explainable by their antagonistic effects on the exaggerated omega-6 AA pro-inflammatory signaling. On the other hand there are examples of positive effects of substituting saturated fat, which is often being demonized as an unhealthy part of diet, for PUFA-rich vegetable oils. Vascular and other chronic diseases are uncommon in the south Asian populations thriving on saturated fat derived from coconuts instead of using the common PUFA-rich seed oils for cooking [74]. Good examples are the Pukapukans and Tokelauans whose diet is reflected in their adipose tissues having eight times less linoleic acid than is present in New Zealand Europeans [75]. Stroke, ischemic heart disease, and the metabolic syndrome are also nonexistent among the inhabitants of Kitava [76,77] eating a similar high saturated fat diet. Even the saturated fat of animal origin containing longer chain fatty acids than present in coconut does not induce deleterious metabolic effects and leads to weight reduction when substituted for dietary carbohydrates [78,79]. High saturated fat diet is associated with diminished coronary artery disease progression in women with the metabolic syndrome [80,81]. France, the country having the oldest recorded person in the world reaching the age of 122, has a low occurrence of coronary heart disease despite a high consumption of saturated fatty acids—what is known as the French paradox [82]. Although this paradox has been attributed to the higher dietary intake of resveratrol from red wine, a recent population study did not confirm any substantial influence of this antioxidant on health status and mortality risk [83]. Cardiovascular disease is also unheard of in the African tribe of Maasai whose potentially atherogenic diet is high in saturated fats and cholesterol [84,85]. High fat ketogenic diets proved also beneficial in the treatment of mental disorders and brain cancers [86,87]. The stable isotope analysis of several Neanderthal remains across Europe suggests that our ancestors were indeed top-level carnivores likely obtaining staple fat from animal sources [88]. It is therefore tempting to speculate that the excessive PUFAs in the current Western diets are directly responsible for the protein and DNA damage underlying molecular pathologies of various degenerative diseases and that a diet containing more saturated fatty acids would be less damaging in the long term. PUFAs have been indeed found to be the causative agent for many AGE/ALE adducts in animal as well as in vitro experiments [13,89,90]. Increased omega-6 consumption has been also linked to higher rates of violence and homicide mortality [91] worldwide and induces obesity epigenetically across generations [92].

LIPID PEROXIDATION AND ITS CONSEQUENCES FOR GENOME STABILITY

The products of lipid peroxidation (LPO) are very reactive and attach to different biomolecules forming bulky adducts responsible for various pathologies [93]. Because of their hydrophobicity and membrane proximity to DNA they readily form DNA lesions and are able to potentiate the activity of other chemical carcinogens [94–96]. Lipid peroxidation dominates the chemistry of DNA adduct formation during inflammation [97,98] and oxidative and nitrative DNA damage occurs at the sites of carcinogenesis [99]. The most studied DNA adducts formed during the peroxidation of PUFAs include the derivatives of acrolein [100–102], malondialdehyde (MDA) [96,103], 4-hydroxy-2-nonenal (4-HNE) [95,104–109], and 4-oxo-2-hexenal (4-OHE) [110–112].

Enzymatic PUFA Oxidation

Specific LPO with signaling functions are formed as a part of the inflammatory response to stressors by enzymatic reactions mediated by the epoxygenases (CYP), eicosanoid cyclooxygenases (COX-1,2), and lipoxygenases (5-LOX) [113,114]. The enzyme phospholipase (PLA2) releases the unsaturated fatty acids from membranes to make them available for this oxidation [115–117]. The mechanisms and significance of the enzymatic oxidation of PUFAs by the lipoxygenase have been covered by Spiteller [118] while the genotoxic properties of the cyclooxygenase metabolites have been extensively studied by Marnett [119].

Eicosanoids are the signaling molecules produced by enzymatic oxidation of PUFAs during the inflammatory response which guide the immune system in tissue repair and warding off infections. The COX-1,2 enzymes produce prostaglandins while 5-LOX gives rise to leukotrienes. The prostaglandins can be further enzymatically activated to MDA [120] or converted to levuglandins [121,122] which generate protein-DNA cross-links potentially leading to strand breaks resulting in clastogenicity. Androgens such as DHT have been reported to promote the MDA guanine adduct formation [123]. The leukotrienes, such as LTA4 [124–126], are able to directly react with DNA bases to form adducts. The series 2 prostaglandins, such as PGE2 and PGF2α, are genotoxic by themselves as is their precursor free AA [127,128].

It is important to understand that both the strength and threshold of the inflammatory response are dependent on the amount of arachidonic acid (AA) in cell membranes because this particular omega-6 PUFA serves as the precursor for the most pro-inflammatory eicosanoids. AA plays an important role in the pro-growth signaling occurring, for example, during development, and is therefore also directly implicated in cancer promotion. It is possible that an excessive TOR signaling leading to hypertrophies associated with aging pathologies is at least partially caused by "overcharging" of tissues with AA. Indeed the phosphatidic acid containing AA chain at its sn2 position is an important natural positive regulator of TOR and competes with rapamycin [129,130].

Various inhibitors of AA metabolization such as the COX inhibitors (NSAIDs) or inhibitors of AA release from phospholipids by PLA2 (glucocorticoids) are used as drugs today to alleviate many chronic inflammatory symptoms and cancer [114]. The inhibition of both COX-2 [131] and 5-LOX [132,133] pathways seems beneficial in suppressing carcinogenesis. Many natural compounds considered as "antioxidants" act the same way. In this respect it is worth mentioning the inhibitors present directly in the traditional cooking oils used as part of the healthy Mediterranean and Japanese diets, that is, hydroxytyrosol and sesamin in olive and sesame oils, respectively [134,135]. In recent years newly discovered anti-inflammatory metabolites of omega-6 and omega-3 fatty acids have also gained attention. In addition to the well-known prostaglandin PGE1 formed from DGLA, the AA derived lipoxins as well as resolvins and protectins synthesized from EPA and DHA, respectively, act during the resolution of the inflammatory response [136].

Nonenzymatic PUFA Oxidation

Although ROS such as the superoxide anion (O_2^-), singlet oxygen, H_2O_2, and hydroxyl radical (OH·) can directly damage DNA and its precursors, this type of damage doesn't form bulky adducts and therefore can be readily repaired or avoided by specific DNA repair systems [137,138]. On the other hand the same ROS can react with the unsaturated fatty acids present in the mitochondrial and nuclear membranes to form reactive LPO (Fig. 12.2). Superoxide also reacts with nitric oxide producing peroxynitrite, which is a powerful oxidant capable of initiating lipid peroxidation [139] similarly to the hydroxyl radical.

Endogenous Oxidation and Reproductive Organ Cancers

From the structural point of view, the most common DNA adduct derived from endogenous lipid peroxidation is a guanine covalently linked to the fatty acid chain at its N_2 position with either open or closed exocyclic conformation. Examples of such guanine DNA adducts with a chain length of 2 is $1,N^2$-etheno-dG, with that of 3 is M_1dG or γ-HOPdG, with that of 9 is HNE-dG, and with that of 20 is LTA4-dG [140].

Breast cancer, the most common cancer in women, could be related to the genotoxic effects of lipid peroxidation products since human breast milk is a rich source of LC-PUFAs [141,142]. It also contains xanthine oxidase, an enzyme that oxidizes xanthine by molecular oxygen and produces superoxide in cells. The extracellular xanthine oxidase is responsible for the sterility of milk because ROS are potent bactericidal agents. Early termination of breastfeeding is thus likely to lead to the accumulation of both bactericidal H_2O_2 and LC-PUFAs susceptible to oxidation in the breast tissue. Such combination would produce large amounts of genotoxic LPO which could even coactivate other procarcinogens such as the aromatic and heterocyclic amines [143,144] or directly initiate and promote mammary carcinogenesis. In fact, genotoxins have been detected in breast lipid and milk samples from different countries [145] but the particular agents have not been characterized or identified yet [146]. The generally inhibitory omega-3 PUFAs have been shown to actually promote development of mammary tumors in a rat model [48]. Perhaps not accidentally, the breast milk from women living in Malawi, one of the countries with the shortest life expectancy, is exceptionally high in both DHA and AA LC-PUFAs [147]. The lipids in milk could have great impact on the progeny given that most of the mutations causing mitochondrial defects at advanced age are actually generated very early in life [148]. Reduction mammoplasty or lactation both reduce breast cancer risk and the reduction is proportional either to the amount of tissue removed or to the total duration of lactation [149].

Evidence is also mounting that the highly unsaturated PUFAs such as ALA may negatively affect the development of prostate cancers [150–154]. Prostate, as its name suggests, is rich in prostaglandins. Jamaica is the country with the highest incidence of prostate cancer in the world and this has been attributed to local diet which is rich in the omega-6 linoleic acid [155,156].

Exogenous Oxidation

The modern Western food is laden with high amounts of vegetable oils containing PUFAs which can be converted to genotoxic LPO even before consumption especially during high temperature cooking [111]. Significant concentrations of oxygenated α,β-unsaturated aldehydes were found in edible oils submitted to frying temperature before they reached 25% of polar compounds [157]. Among the genotoxins detected in these oils and foods were 4-HNE, 4-ONE, 4-HHE, and 4-OHE, as well as 4-hydroxy-2-octenal,

4-oxo-2-decenal, 4,5-epoxy-2-decenal, and 4-oxo-2-undecenal [157,158]. Recently such compounds in foods are receiving a great deal of attention because they are being considered as possible causal agents of numerous diseases, such as chronic inflammation, neurodegenerative diseases, adult respiratory distress syndrome, atherogenesis, diabetes, and different types of cancer [158].

While the oils containing saturated or monounsaturated fatty acids are relatively safe for cooking, the oils with high omega-3 ALA content should be of the greatest concern. The ALA-rich oils as well as other food sources high in omega-3 such as broiled fish have been shown to contain the novel lipid peroxide mutagenic product 4-OHE [111,159]. After oral administration of 4-OHE to mice multiple DNA adducts were detected in esophageal, stomach, and intestinal DNA [110], although it failed to show cancer-initiating activity in a 5-week liver assay [160]. The other oxygenated α,β-unsaturated aldehyde present in common foods is 2-hexenal that has been found to form $1,N^2$-propanodeoxyguanosine-type adducts in liver after gavage to male Fischer rats [161]. Although not promoters, the omega-3 fatty acids could be responsible for the initiation of certain cancers appearing decades later, such as breast cancer [47,48,162]. It is prudent to limit the dietary exposure to LPO particularly in aged individuals since the protective detoxification biotransformations in liver seem to diminish as a consequence of aging [163].

Another negative effect of modern technological manipulation of food at high temperatures is the formation of *trans* fatty acids which are especially abundant in partially hydrogenated seed oils [164]. These, in contrast to the naturally curved *cis*-conformation PUFAs, render the biological membranes stiffer and, more importantly, jam the enzymatic machinery (eg, desaturases) needed to produce LC-PUFAs for both structural and autocrine/paracrine regulatory purposes. Cooking foods containing both PUFAs and cholesterol also produces cholesterol oxidation products such as 7-ketocholesterol found at high levels, for example, in steamed salmon [165]. Both *trans* fatty acids and oxidized cholesterol in diet increase the risk of cardiovascular disease.

Presence of Lipid Peroxidation DNA Adducts in Living Tissues

Short- and long-chain enals and their epoxides derived from oxidized omega-3 and omega-6 PUFAs are endogenous sources of cyclic propano and etheno DNA adducts. Numerous adducts, such as acrolein and crotonaldehyde as well as the longer MDA and 4-HNE derived deoxyguanosine adducts, have been detected in human tissues [166] and implicated in mutagenesis and carcinogenesis [167].

The crotonaldehyde dG adduct has been detected in human DNA extracted from lung and to a lesser extent from liver but not in DNA from blood [168]. The primary propano dG adduct M_1dG derived from the genotoxic bifunctional electrophile MDA was found to be present in leukocytes of healthy human donors at the levels of 5.1 and 6.7 adducts/10^8 bases in females and males, respectively [169]. Other studies report the presence of this adduct at the levels of 1–5 per 10^7 nucleotides in white blood cells and breast tissue of human volunteers and its 2–3 fold rise in the normal tissue of women with breast cancer [119]. The MDA DNA adduct levels rise dramatically 20–45-fold in white cells of females fed LA-rich vegetable oils that has been explained by estrogen redox cycling [170,171]. The acrolein dG adducts, which are preferentially formed in the p53 mutational hotspots in human lung cancer, were detected at concentrations ranging from 16–209 adducts/10^9 nucleotides in human lung DNA collected from both current and ex-smokers [172]. The acrolein adduct has been also found in brain tissues from Alzheimer's disease (AD) subjects and age-matched controls at the levels of 2800–5100 adducts/10^9 nucleosides with significantly higher occurrence in the hippocampal region of the AD subjects [173]. The longer chain AA peroxidation product 4-HNE forms the propano HNE-dG adduct which is a potential promutagenic endogenous lesion found in the DNA of rodent and human tissues and its levels increase with age or under the conditions of increased lipid peroxidation and glutathione depletion [174]. It was present in hippocampus/parahippocampal gyrus and inferior parietal regions of postmortem brains from AD subjects at the frequencies of 556 and 238 per 10^9 nucleosides, respectively [175].

The 4-HNE derived etheno DNA adducts such as εdG and εdA are normally repaired faster than the propano HNE-dG in dividing cells but are present at rather high levels in human sperm [176,177]. Recently characterized novel genotoxin 4-oxo-2-nonenal (4-ONE) derived from LA preferentially forms εdC adduct in dsDNA [178]. This adduct structurally resembles the mutagenic cytosine adduct derived from the potent germ cell mutagen and carcinogen 1,3-butadiene [179,180]. The same compound also produces the εdA adduct [181,182]. The εdA and εdC adducts are present at elevated levels in people having acquired or inherited cancer risk factors such as hemochromatosis, Wilson's disease, inflammatory bowel disease, familial adenomatous polyposis, or being on a high LA diet [171] and are also concentrated in the human atherosclerotic lesions [183]. They have been detected at the levels of 65 and 59 adducts per 10^9 parent nucleotides for εdA and εdC, respectively, in epithelial cell DNA of patients with familial adenomatous polyposis that is 2 to 3 times higher than in unaffected colon tissue and may be the result of upregulated COX-2 [184–186]. In the chronic inflammatory diseases of the digestive tract only εdC was found to be increased in affected colonic mucosa of Crohn's disease (19 times) and of ulcerative colitis patients (4 times) when compared to asymptomatic tissues. In chronic pancreatitis these lesions have been elevated 3 and 28 times for εdA and εdC, respectively, in the inflamed pancreatic tissue [187]. They were also detected in the white blood cells of mother-newborn child pairs with the mothers having 317 and 916 per 10^9 parent nucleotides of εdA and εdC, respectively, and their children having about two times lower levels [188]. Interestingly, the levels of both adducts were inversely related with total plasma cholesterol levels and were lowest on a high saturated fat diet (cocoa butter based) in the apolipoprotein E deficient mouse model. These mice had 1.6 and 4.8 per 10^8 parent nucleotides of εdA and εdC, respectively, in their aortas at 12 weeks of age, which is about half the levels in the wild-type controls. This phenomenon is being attributed to the increased expression of the BER and NER repair enzymes by the high fat diet and hypercholesterolemia [189,190]. But the substitution of PUFAs with the saturated fatty acids and MUFAs could play a role as well since dietary PUFAs such as LA have been shown to elevate the levels of εdA and εdC in the colon and white blood cells of volunteers with lower intake of dietary antioxidants [191,192].

Endogenous DNA adducts derived specifically from the omega-3 fatty acids have not been extensively studied since these fats are not considered cancer promoters. They are accepted rather as chemopreventives since they counteract the pro-inflammatory and growth stimulating AA signaling on multiple fronts (inhibitory eicosanoids, PPARs, NFkB) [193]. Despite their high oxidizability, which makes them even more reactive with DNA, most of the DNA adducts omega-3 LPO produce result in apoptotic cell death under the conditions of suppressed AA eicosanoid signaling. Nevertheless, increased cell death is responsible for different classes of neurodegenerative diseases. Thus DHA, the content of which is 30–50% higher than that of AA in many brain regions, can be oxidized to the potent neurotoxin 4-hydroxy-2-hexenal (4-HHE) as well as to other lipid peroxidation products, such as the F4-neuroprostanes, particularly in patients suffering from the Alzheimer's and Parkinson's diseases [194,195]. It is likely that LPO and subsequently DNA adducts derived from the omega-3 PUFAs can be also formed in vivo [110,111]. Some DNA modifications derived from 4-OHE have been already described in respect to the exogenous oxidation of PUFAs in cooking oils.

Recent analysis of human autopsy tissues by mass spectrometry showed the presence of both omega-6 and omega-3 derived DNA adducts, in some cases at levels exceeding those of the most abundant DNA adduct 8-oxo-dG. The omega-6 adducts 7εdC, 7εdA, and 7εdG (derived from 4-ONE) were present in various human tissues at median values of 10, 15, and 8.6 adducts per 10^8 bases, respectively, with ten times higher levels observed in several cases. The omega-3 DNA adduct 4εdC (derived from 4-OHE) was present at about a seven-fold lower concentration than 7εdC suggesting the endogenous formation of LPO from tissue fatty acids where omega-6 concentration is markedly higher than that of omega-3 [196].

Mutagenicity of Lipid Peroxidation Products

Taken from the carcinogenesis point of view the bacterial Ames test is considered a golden standard among the in vitro assays for cancer initiators and has been widely accepted for genotoxic risk evaluations by governments all over the world. It is based on the reversions of specific mutations responsible for histidine auxotrophy [197]. Interestingly, the longer DNA reactive lipid peroxidation products show only marginal or no mutagenicity in this test. This has been attributed to their high toxicity (particularly the enals with higher molecular weights) due to preferential reactions with the sulfhydryl groups of proteins which are often essential for enzymatic activities. The α,β-unsaturated aldehydes such as acrolein or 4-HNE are soft electrophiles that preferentially form adducts with soft nucleophiles such as the cysteine sulfhydryls of proteins [198] rather than DNA. However, in the presence of a scavenger like glutathione, which suppresses these toxic reactions, normally nonmutagenic compounds, such as the lipid peroxidation product 4-hydroxy-pentenal, become mutagenic [199].

Despite their high toxicity, the mutagenicities of the following LPO have been demonstrated in the Ames test: Acrolein in the TA104 and TA100 strains [199–201], crotonaldehyde using the liquid preincubation assay with the strain TA100 [202], MDA in the frameshift hisD3052 uvrB-proficient strain [103,203], 4-hydroxy-pentenal (4-HNE homolog) in TA104 using the reduced glutathione chase [199], and 4-OHE in the TA100 and TA104 standard strains [112,159]. It appears that the activation increasing preferential reaction with DNA bases is epoxidation since 4-HNE acquires strong mutagenicity in both strains TA100 and TA104 upon conversion to its epoxide (Fig. 12.2) [204]. Moreover this epoxide (EH) is not a substrate for the human epoxide hydrolase which normally detoxifies similar compounds such as B[a]P-epoxide to blunt their in vivo mutagenicity [205]. EH can be produced in vivo in a reaction with the LA hydroperoxide 13-HODE, which is coincidentally known to be a strong cancer growth promoter [62,206], or in decomposition reactions catalyzed by vitamin C [207–210]. It is also possible, that the primary 4-HNE propano DNA adducts are nonmutagenic in nature compared to the smaller etheno EH adducts. That would explain why 4-HNE acts as a potent inducer of the SOS response indicating the presence of DNA damage in bacteria [211]. Indeed the propano-dG DNA modifications formed directly from the enals without epoxidation (Fig. 12.2) are good candidates for the nonmutagenic adducts, because they can be replicated in an error free manner [210,211].

Nonmutagenicity in the bacterial assay does not, however, imply that the aldehydes forming bulky adducts, such as 4-HNE, do not induce mutations in different organisms or in different DNA sequence contexts. This is documented by another widely accepted mutagenicity bioassay utilizing the mouse lymphoma cells heterozygous at the TK locus. In this assay not only acrolein and crotonaldehyde but also 4-HHE and 4-ONE turned clearly mutagenic [212]. It was even possible to demonstrate the mutagenicity of 4-HNE when it was delivered to cells in its protected triacetate form [213]. Moreover, 4-HNE forms DNA adduct preferentially at the third base of codon 249 in the p53 gene which is a mutational hotspot in human cancers [104–106].

Various LPO have been also found to potentiate the mutagenicities of other strong mutagens such as B[a]P and UV light [96]. The levels of the most abundant oxidative stress-induced DNA lesion 8-oxo-dG in fact rise on a diet rich in LA which increases LPO formation [214]. This can be explained by the direct oxidation of DNA or its precursors by lipid peroxidation intermediates, such as the peroxyl radicals of AA [215].

PROTECTIVE MECHANISMS AGAINST LIPID PEROXIDATION

Healthy and young organisms are protected against the negative outcomes of lipid peroxidation on multiple levels. It starts with the prevention of excessive ROS release by regulating their production enzymes and sequestrating free metals. Then the scavenging, trapping, and quenching systems are able to restrain the free radicals leaking, for example, from the mitochondrial respiratory chain. Once lipids get damaged, they can be removed by the detoxification enzymes. Finally if damage to structural or informational molecules occurs, it can be repaired either by the immune system and self-destruction (apoptosis) or by specific DNA repair processes. At last but not least, frequently occurring damage to DNA can be tolerated and "diluted" by an error free chromosomal replication. It is when these protective systems get overwhelmed that the organism starts to age and break down.

Endogenous Antioxidant Systems

The enzymatic antioxidant systems scavenging free radicals leaking from the mitochondrial energy producing machinery restrain the formation of the hydroxyl radical and include superoxide dismutase (SOD), catalase, glutathione peroxidase and the thioredoxin system. The cytosolic glutathione peroxidases also act directly on the lipid peroxides (LOOH) and reduce them. Some of the glutathione-lipid peroxide detoxification products have been recently characterized in great detail [216,217]. Cancer cells often hijack and upregulate the glutathione-utilizing antioxidative defenses as a strategy to protect themselves from apoptosis [218,219].

The nonenzymatic natural LPO quenchers include, for example, carnosine present in tissues like skeletal muscle or the transparent crystalline eye lenses [220–222], bilirubin which protects the neonatal blood plasma from oxidizable PUFAs [223], and the sleep hormone and universal antioxidant melatonin [224]. Biological membranes are then protected internally by scavengers which terminate the lipid peroxidation chain reactions, such as the tocopherols (vitamin E), retinol (vitamin A), or carotenoids (eg, astaxanthin). To counteract the increased susceptibility of omega-3 rich membranes to oxidation animals like fish seem to utilize special antioxidants/radical scavengers termed the furan fatty acids [225]. Other soluble antioxidants such as the uric acid or ascorbic acid (vitamin C) act in synergy with the lipophilic antioxidants and can support them, for example, by recycling of vitamin E by vitamin C.

Exogenous Supplemental Antioxidants

Since humans cannot synthesize some essential antioxidants such as vitamin C or vitamin E they need a steady supply of these compounds in the diet together with other cofactors of the endogenous enzymatic antioxidant systems, such as the minerals selenium, manganese, or zinc. This can be usually achieved by an adequate diet rich in agricultural products.

The question raised in recent years is whether extra supplemental antioxidants have any benefits or rather negative effects on the aging process. It seems that increasing certain protective antioxidants in diet could have negative effects in healthy individuals perhaps by interfering with the intricate endogenous homeostasis control mechanisms. This is demonstrated, for example, in the recent exemplary study by Spindler et al. who found no benefits but rather harmful effects of different combinations of nutraceuticals including antioxidants on murine lifespan [226]. Similarly vitamin E and C supplementation had deleterious consequences on lifespan of wild-derived voles [227]. Beta carotene supplementation increased deaths from lung cancer and ischemic heart disease in a human study [228] and vitamin E markedly increased tumor progression in a lung cancer mouse model [229]. These studies provide laboratory evidence that some of the effects of antioxidants backfire. Particularly adding antioxidant supplements could defeat the body's ability to fight cancer and accelerate tumor growth by disrupting the ROS-p53 axis [230]. In this context it is also important to mention the work of S. Hekimi who showed that upregulating the mitochondrial ROS production could in fact have antiaging effects in worms and mice as a consequence of upregulating the apoptosis pathway and activating the immune system, respectively [231,232]. The adaptive effects of exercise that reverse muscle aging are also blunted by supplementary antioxidants such as vitamins C and E [233,234].

In contrast to the beneficial signaling effects of tuning up the source of oxidative damage (ie, unrestrained output of ROS from mitochondria) on the lifespan there have not been any reports of similar positive effects of increasing the target of this damage (ie, the peroxidation index of biological membranes), for example, by supplementing the omega-3 fatty acids which are sometimes also regarded as "antioxidants." The omega-3 PUFAs upregulate the endogenous antioxidant defenses and calm inflammation but that is also immunosuppressive decreasing T lymphocyte proliferation and natural killer cell activity [42,235] as well as Th1-type cytokine production [39]. In this respect another recent study by Spindler et al. clearly demonstrated that dietary supplementation with fish oils shortened the lifespan of long-lived mice [236]. Omega-3 rich oils also shortened lifespan in the senescence-accelerated mouse model [237], were cataractogenic [238,239], and higher n-3/n-6 ratio increased the risk of glaucoma [240]. Despite their inhibitory effects on cancer promotion the omega-3 fatty acids may be in fact responsible for the initiation of certain reproductive tissue cancers particularly when fed prepubertally [47,48,241]. In an animal model of colon cancer metastasis in rat liver, omega-3s from fish oil actually promoted the colon carcinoma metastasis while omega-6 had no effect compared to the control diet [49,50]. Omega-3-rich diets are reported to have negative effects on murine fecundity and maternal behavior and a detrimental effect on learning ability in old mice [242,243]. So highly unsaturated molecules could also increase production of ROS and perturb mitochondrial metabolism in oocytes and zygotes [244] and reduce endurance in rats [245,246]. Dietary PUFAs directly influence the vasculature and increase aortic plaque formation [247] and some systematic reviews of epidemiological literature conclude that the omega-3 fatty acids are unlikely to prevent cancer or cardiovascular events [44–46]. This is conceivable because an increase in lipid peroxidation after supplementing fish oil cannot be suppressed by extra vitamin E [248].

Tolerance and Repair of Lipid Peroxide DNA Adducts

The major DNA adduct produced by the reaction of various lipid hydroperoxides with DNA is the N^2-dG adduct where the fatty acid-derived chain points into the minor groove of the DNA helix. Although this type of lesion can be repaired by NER [106,249], such a repair is not very efficient. An alternative pathway to deal with this lesion is a replicative bypass by specialized DNA polymerase through the process of translesion DNA synthesis (TLS). The LPO derived minor groove guanine DNA adducts are bypassed in an error free manner by the sequential action of the recently uncovered Y-family DNA polymerases ι and DNA polymerases κ (Polk) [250], thereby suppressing genomic instability and cancer induced by food-derived genotoxins such as 4-HNE [251]. Interestingly in mammals, Polk is strongly expressed in testis [252,253] where both omega-6 and omega-3 LC-PUFAs are concentrated [254] and the content of VLC-PUFAs, such as DHA, greatly increases at the onset of puberty [255,256].

The exocyclic lipid peroxidation-derived DNA adducts affecting bases other than dG have been also detected in tissues and could be considered the main culprit in carcinogenesis because they are not bypassed in an error free manner by the specialized TLS DNA polymerases. If left unrepaired, these lesion are highly mutagenic and that is also why they are specifically targeted by the base excision repair [257,258]. The repair of 1,N^6-ethenoadenine (εdA) is preferentially performed by the MPG and AlkA alkyl purine DNA-glycosylase in mammalian cells and E. coli, respectively, and NER pathway is not involved in its repair [259]. 3,N^4-ethenocytosine (εdC) is then repaired by the thymine or uracil DNA-glycosylases [260]. Moreover, there is also a direct lesion reversal repair for these εdA and εdC adducts carried out by the AlkB protein in Escherichia coli [261]. The bulkier and more complex LPO adducts are then mainly repaired by NER. As an example the functional product of the XPG gene involved in NER suppressed base-substitution mutations induced by AA which was both mutagenic and clastogenic in fibroblasts under peroxidizing conditions [128].

Some of the chemopreventive activities particularly of the omega-3 fatty acids could be mediated by the damage they cause. Certain DNA adducts may prevent the highly processive DNA replication required for the proliferation of symmetrically dividing immortal cancerous cells while allowing asymmetrical DNA replication in normal tissue stem cells to take place. Due to their structural features the telomeres could be especially sensitive to the attack by LPO which may interfere with proper telomere maintenance in cancerous cells. If encountered within the normal replication fork, the PUFA derived DNA adducts are replicated through by the specialized TLS DNA polymerases, but these could not possibly act at the chromosome tips. Thus, telomerase may be a key enzyme inhibited by this sort of DNA damage. The perinuclear localization of telomeres makes them prone to the attacks by LPO produced by the enzymatic oxidation of LC-PUFAs at the nuclear membrane [262].

RESISTANCE TO OXIDATIVE DAMAGE IN LONG-LIVED SPECIES

Unsaturated fatty acids are responsible for the "fluid" nature of functioning biological membranes. It has been shown that it is the introduction of the first double bond to an acyl chain that is primarily responsible for its fluidity at normal physiological temperatures, and introduction of more and more double bonds to the acyl chain has relatively little additional effect on membrane fluidity. Thus substitution of fatty acids with four or six double bonds with those having only two (or sometimes three) double bonds will strongly decrease the susceptibility to lipid peroxidation while maintaining membrane fluidity. This phenomenon may be helpful in longevous animals, and in view of membrane acclimation to low temperature in ectotherms, has been called homeoviscous longevity adaptation [11].

In recent years, it is becoming more apparent that the membrane content of highly unsaturated fatty acids like DHA negatively correlates with the maximal life span (MLSP) of different species and is linked to the accumulation of somatic mutations in mitochondrial DNA [263,264]. Among the bird species significant allometric decline in the percentage of omega-3 PUFAs with increasing body mass has been observed [265]. Because of their high susceptibility to lipid peroxidation, the omega-3 PUFAs are consistently and inversely correlated to MLSP, which gave birth to the membrane pacemaker theory of aging [266,267]. The data collected from different species suggest that membrane phospholipid composition is an important determinant of longevity with the DHA content playing especially important role [268,269] and that the low degree of tissue fatty acid unsaturation of longevous homeothermic animals have been selected during evolution to protect their tissues against oxidative damage [270–272]. The pigeon is a good example having similar metabolic rate and body size as the rat but reaches nearly an order of magnitude longer MLSP (35 vs 4 years) owing to its lower content of the highly unsaturated PUFAs [10,273]. The lower degree of unsaturation of fatty acids in birds renders the avian mitochondria less susceptible to oxidative damage due to much lower rate of lipid peroxidation [274]. Animals such as ruminants naturally feeding on forage high in omega-3 ALA have evolved a special stomach that houses symbiotic microorganisms helping them to reduce PUFA unsaturation by biohydrogenation [275]. A recent lipidomic profiling of 11 mammalian species with MLSP ranging from 3.5 to 120 years confirmed a tight negative correlation of the peroxidizability index and LPO content with longevity [276].

It is interesting that even within the same species (and genome) lowering the fatty acid unsaturation correlates with longer lifespan [267]. An intriguing example is found among the eusocial insects where the queen bees live an order-of-magnitude longer than the workers while having highly monounsaturated membranes with very low content of PUFAs despite originating from the same genome. This difference is contributable to nutrition since the workers increase their PUFA membrane content probably as a result of pollen consumption [277]. Two mouse lines with extended lifespan have been shown to have reduced amount of the highly polyunsaturated omega-3 fatty acid DHA which makes their membranes peroxidation-resistant and extends longevity [278]. Also the increasing content of DHA with age in mice can explain its nearly order of magnitude shorter MLSP than that of the similarly sized naked mole-rat [269,279]. Even the longevity within our human species has been shown to correlate with the membrane building block composition, particularly higher content of the omega-9 MUFAs and reduced content of the omega-6 PUFAs such as LA and AA in nonagenarian offspring [280]. Older Italian subjects consuming diet high in MUFA from olive oil had lower all-cause mortality in contrast to subjects ingesting diet higher in PUFAs and a MUFA-rich diet also protects against cognitive decline [267].

As proposed recently by Zimniak [267] electrophilic stress, which is chemically distinct from oxidative stress, has a causal relationship to longevity. The level of electrophiles such as 4-HNE having deleterious effects on lifespan is dependent mainly on the content of peroxidizable PUFAs in membranes and not on the initiating ROS which are not needed for the propagation of the peroxidation chain reactions. Several interventional studies have been used to demonstrate the causative role of the membrane peroxidizability index in determining the life span. Probably most impressive is the manipulation of membrane unsaturation or the electrophile levels in worms [14] by multiple approaches including single gene mutation, RNAi silencing, transgene expression, or application of 4-HNE scavengers. Silencing the elongases and desaturases resulting in higher MUFA to PUFA ratio increased while silencing the glutathione transferase (GST) or aldehyde dehydrogenases detoxifying 4-HNE decreased the lifespan of Caenorhabditis elegans worms. Likewise overexpressing the murine GST or applying scavengers like carnosine and hydralazine extended the lifespan of worms. Similar manipulations increasing the SFA and MUFA while decreasing the highly peroxidizable omega-3 PUFA in the heart of mice also showed health benefits [90,267].

CONCLUDING REMARKS

Over the last several decades a phobia of animal fats has led food manufacturers and health authorities to replace naturally occurring animal fats with industrially processed vegetable oils in many foods. This in turn has resulted in a massive increase in PUFA intake in humans compared to preindustrial populations. The increased consumption of seed oils containing mainly omega-6 concentrated PUFAs coincides with the epidemic of degenerative diseases characterized by chronic inflammation. Certain PUFAs such as the omega-3 type with the highest level of unsaturation are also being promoted by the supplement industries because of their inhibitory effects on the metabolism of the pro-inflammatory omega-6 PUFA-derived eicosanoids and pro-apoptotic activities serving as a cancer chemoprevention. However, enzymatic and spontaneous oxidation of PUFAs produces reactive electrophiles which attack biomolecules creating DNA and protein adducts. 4-hydroxy-2-alkenals which are generated both in vivo and ingested in the foods are one of the most toxic compound types. For instance it has been estimated that the Korean population is exposed to 16.1 μg of 4-HHE and 4-HNE combined per person daily from their PUFA-rich food sources alone [281]. The risk of these cytotoxic aldehydes for humans could not be quantified due to the lack of a virtually safe dose. Agriculture was developed about 10,000 years ago, and the industrial processing to render vegetable oils only became widely available within the last 50 years. Humans have been evolving for 2.5 million years. This means that high concentrations of grains and vegetable oils in the diet have only been possible for 0.4% and 0.002% of our evolutionary history, respectively. This is certainly not enough to permit sufficient evolutionary adaptation and it is not surprising that introducing massive amounts of these foods never seen before can negatively affect human health.

The evolutionary strategy to increase longevity in animal species teaches us that biological membranes with lower content of the highly unsaturated PUFAs are better fit for the long run due to lower susceptibility to oxidation [282]. ROS mainly attack carbon atoms located between two double bonds on a PUFA chain. Saturated fatty acids and MUFAs lack such configurations of carbon in the acyl chains and to a great extent they are not affected by the ROS, whereas the chains of PUFAs are highly vulnerable to lipid peroxidation. This process is autocatalytic and produces irreversible damage in the membrane and nearby cellular structures. Strengthening the resistance of cellular components to oxidative damage by free radicals has been an attractive antiaging strategy discussed recently [283]. Given the current PUFA-rich human diet, it may be feasible to avert the age-associated degenerative diseases and extend lifespan by simply reducing the content of the highly unsaturated plant-derived PUFA in our diet. Increasing threshold of the inflammatory responses by reducing omega-6 PUFA intake, rather than trying to counteract it with the even more unstable omega-3 PUFAs, could also prove beneficial, for example, for the autoimmune diseases. This is more economical than targeting the mitochondrial membranes with antioxidants designed to protect the crucial phospholipids, such as cardiolipin, from oxidative damage [284–289].

It seems that the least unsaturated omega-9 PUFA naturally synthesized by our body can substitute for the plant-derived omega-6 EFAs in membranes and at least in some enzymatic reactions needed to supply essential signaling molecules. Since the type of PUFAs used in membranes is determined to a large extent by their dietary intake, lowering the dietary omega-6 + 3 fatty acids (EFAs) altogether would allow for the formation of omega-9 ETrA. This could be beneficial particularly past the reproductive age and when physically inactive since the omega-6 derived eicosanoids seem to fuel not only growth but also the aging process afterwards and are a source of damage to the cellular components including DNA. It is generally accepted that the human body requirement for the EFAs is about 1% of calories which is more than met with a normal diet containing meat and vegetables without any added polyunsaturated oils or supplements [290]. Under these conditions, the content of less unsaturated omega-9 MUFA and PUFAs would increase and render the cellular membranes more durable while lessening burden on the protective, repair, and detoxification systems. This may be in fact one of the mechanisms as to how caloric restriction extends the lifespan in mammals since the total intake of the most energy dense nutrients, that is, lipids rich in EFAs, is limiting available space for the utilization of the more stable omega-9 MUFA and PUFAs. When desirable, the EFAs are better supplemented in their shorter 18-carbon length plant precursor form not to bypass the regulatory mechanisms controlling the synthesis of LC-PUFAs [46]. The LC-PUFAs, such as AA, EPA, or DHA, are the "active" substrates for eicosanoid synthesis as well as more prone to lipid peroxidation and their exogenous (over)supply could interfere with the fine tuning of endogenous metabolism.

Due to the advances in modern medicine and healthcare systems the populations in developed countries such as Japan are aging rapidly with increasing numbers of elderly people. Although long-term caloric restriction has been shown to reduce mortality in primates [291], it is considered useful to supplement frail elderly with extra caloric and protein intake. As suggested by Rose [292] the aging may eventually stop at late life but by that time most of the organs including the digestive system already have severely impaired functionality. Thus caloric restriction in aged subjects can easily lead to malnutrition. As demonstrated recently by Khrapko et al. the malfunction of the absorption system is largely due to the clonal expansion of mitochondrial mutations in colonic crypts [148,293]. Since these mutations occur early in life, reducing the source of mitochondrial DNA damage such as the highly oxidizable PUFAs from very young age could have profound effect on the ability to extract nutrients from foods at an old age.

Aging is also associated with an oxidizing shift in the redox state enforced by a sedentary lifestyle and excessive consumption of sugary beverages which further decreases mitochondrial capacity while relying on glycolysis for energy production [294]. This state can be counteracted with exercise or ketogenic diet [79] which increase mitochondrial biogenesis and eliminate damaged mitochondria via the mitophagy [295]. Rather than restricting calories in the elderly, more energy can be supplied from safe fat sources, such as coconut oil, which do not produce LPO during cooking or in vivo due to very low PUFA content. This oil is rich in the saturated medium chain fatty acids (MCFA) which are absorbed more efficiently, bypassing the chylomicrons lymphatic route, to enter the mitochondria independently of the carnitine transport system and undergo preferential oxidation [296]. Since MCFA produce ketone bodies, coconut oil has a similar effect on the mitochondria as fasting or very low-carbohydrate ketogenic diet because it can supply alternative energy to the brain. This has been taken advantage of in alternative treatments of neurodegenerative disorders such as the Alzheimer disease or even autism [87,297,298]. Supplying easily digestible saturated fats in place of PUFAs should also reduce the accumulation of the age pigment lipofuscin and protect the heart [90].

SUMMARY

An unconventional review of lipid peroxidation in its relation to human nutrition and aging is presented. Although emphasis is put on DNA damage as a key molecular target, other aspects of the oxidation of polyunsaturated fatty acids (PUFAs) in the body are covered. This includes the chemical stability of PUFAs, the meaning of their enzymatic oxidation for the immune system, spontaneous oxidation either endogenously in membranes or exogenously during food preparation, the modifications of DNA by lipid peroxidation products and their effects on genome stability. The protective mechanisms against PUFA oxidation are discussed in terms of evolutionary trends and natural cellular defenses, as well as interventions with antioxidant supplements. References to human diseases and population studies are provided together with some detailed experimental data to support the conclusions drawn. It is suggested that minimizing dietary exposure to highly unsaturated PUFAs while not interfering with the internal oxidative stress signaling would increase long term resilience and maximize the utilization of genetic potential for healthy aging.

References

[1] Drenos F, Kirkwood TB. Modelling the disposable soma theory of ageing. Mech Ageing Dev 2005;126(1):99—103.
[2] Hayflick L. The causes of biological ageing are known. In: 19th IAGG world congress of gerontology and geriatrics. International Association of Gerontology and Geriatrics 2009. SA6 044-1.
[3] Passos JF, Saretzki G, Ahmed S, et al. Mitochondrial dysfunction accounts for the stochastic heterogeneity in telomere-dependent senescence. PLoS Biol 2007;5(5):e110.
[4] Lu J, Sharma LK, Bai Y. Implications of mitochondrial DNA mutations and mitochondrial dysfunction in tumorigenesis. Cell Res 2009;19(7):802—15.
[5] Loeb LA, Wallace DC, Martin GM. The mitochondrial theory of aging and its relationship to reactive oxygen species damage and somatic mtDNA mutations. Proc Natl Acad Sci USA 2005;102(52):18769—70.

REFERENCES

[6] Brunet A, Rando TA. Ageing: from stem to stern. Nature 2007;449(7160):288–91.

[7] Actis AB, Perovic NR, Defago D, Beccacece C, Eynard AR. Fatty acid profile of human saliva: a possible indicator of dietary fat intake. Arch Oral Biol 2005;50(1):1–6.

[8] Nakamura T, Takebe K, Tando Y, et al. Serum fatty acid composition in normal Japanese and its relationship with dietary fish and vegetable oil contents and blood lipid levels. Ann Nutr Metab 1995;39(5):261–70.

[9] Beynen AC, Hermus RJ, Hautvast JG. A mathematical relationship between the fatty acid composition of the diet and that of the adipose tissue in man. Am J Clin Nutr 1980;33(1):81–5.

[10] Pamplona R, Portero-Otin M, Requena JR, Thorpe SR, Herrero A, Barja G. A low degree of fatty acid unsaturation leads to lower lipid peroxidation and lipoxidation-derived protein modification in heart mitochondria of the longevous pigeon than in the short-lived rat. Mech Ageing Dev 1999;106(3):283–96.

[11] Hulbert AJ, Pamplona R, Buffenstein R, Buttemer WA. Life and death: metabolic rate, membrane composition, and life span of animals. Physiol Rev 2007;87(4):1175–213.

[12] Refsgaard HH, Tsai L, Stadtman ER. Modifications of proteins by polyunsaturated fatty acid peroxidation products. Proc Natl Acad Sci USA 2000;97(2):611–16.

[13] Fu MX, Requena JR, Jenkins AJ, Lyons TJ, Baynes JW, Thorpe SR. The advanced glycation end product, Nepsilon-(carboxymethyl)lysine, is a product of both lipid peroxidation and glycoxidation reactions. J Biol Chem 1996;271(17):9982–6.

[14] Shmookler Reis RJ, Xu L, Lee H, et al. Modulation of lipid biosynthesis contributes to stress resistance and longevity of C. elegans mutants. Aging (Albany, NY) 2011;3(2):125–47.

[15] Ichi I, Kono N, Arita Y, et al. Identification of genes and pathways involved in the synthesis of Mead acid (20:3n-9), an indicator of essential fatty acid deficiency. Biochim Biophys Acta 2014;1841(1):204–13.

[16] Adkisson HDt, Risener Jr. FS, Zarrinkar PP, Walla MD, Christie WW, Wuthier RE. Unique fatty acid composition of normal cartilage: discovery of high levels of n-9 eicosatrienoic acid and low levels of n-6 polyunsaturated fatty acids. FASEB J 1991;5(3):344–53.

[17] Elliott WJ, Morrison AR, Sprecher H, Needleman P. Calcium-dependent oxidation of 5,8,11-icosatrienoic acid by the cyclooxygenase enzyme system. J Biol Chem 1986;261(15):6719–24.

[18] Lagarde M, Burtin M, Rigaud M, Sprecher H, Dechavanne M, Renaud S. Prostaglandin E2-like activity of 20:3n-9 platelet lipoxygenase end-product. FEBS Lett 1985;181(1):53–6.

[19] Oliw EH, Hornsten L, Sprecher H. Oxygenation of 5,8,11-eicosatrienoic acid by prostaglandin H synthase-2 of ovine placental cotyledons: isolation of 13-hydroxy-5,8,11-eicosatrienoic and 11-hydroxy-5,8,12-eicosatrienoic acids. J Chromatogr B Biomed Sci Appl 1997;690(1–2):332–7.

[20] Stenson WF, Prescott SM, Sprecher H. Leukotriene B formation by neutrophils from essential fatty acid-deficient rats. J Biol Chem 1984;259(19):11784–9.

[21] Jakschik BA, Morrison AR, Sprecher H. Products derived from 5,8,11-eicosatrienoic acid by the 5-lipoxygenase-leukotriene pathway. J Biol Chem 1983;258(21):12797–800.

[22] Oliw EH, Hornsten L, Sprecher H, Hamberg M. Oxygenation of 5,8,11-eicosatrienoic acid by prostaglandin endoperoxide synthase and by cytochrome P450 monooxygenase: structure and mechanism of formation of major metabolites. Arch Biochem Biophys 1993;305(2):288–97.

[23] Langenbach R, Loftin C, Lee C, Tiano H. Cyclooxygenase knockout mice: models for elucidating isoform-specific functions. Biochem Pharmacol 1999;58(8):1237–46.

[24] Fischer SM, Pavone A, Mikulec C, Langenbach R, Rundhaug JE. Cyclooxygenase-2 expression is critical for chronic UV-induced murine skin carcinogenesis. Mol Carcinog 2007;46(5):363–71.

[25] Ma K, Langenbach R, Rapoport SI, Basselin M. Altered brain lipid composition in cyclooxygenase-2 knockout mouse. J Lipid Res 2007;48(4):848–54.

[26] Rao JS, Langenbach R, Bosetti F. Down-regulation of brain nuclear factor-kappa B pathway in the cyclooxygenase-2 knockout mouse. Brain Res Mol Brain Res 2005;139(2):217–24.

[27] Retterstol K, Haugen TB, Woldseth B, Christophersen BO. A comparative study of the metabolism of n-9, n-6 and n-3 fatty acids in testicular cells from immature rat. Biochim Biophys Acta 1998;1392(1):59–72.

[28] Retterstol K, Woldseth B, Christophersen BO. Studies on the metabolism of [1-14C]5.8.11-eicosatrienoic (Mead) acid in rat hepatocytes. Biochim Biophys Acta 1995;1259(1):82–8.

[29] Tocher DR, Dick JR, Sargent JR. Occurrence of 22:3n-9 and 22:4n-9 in the lipids of the topminnow (Poeciliopsis lucida) hepatic tumor cell line, PLHC-1. Lipids 1995;30(6):555–65.

[30] Patel P, Cossette C, Anumolu JR, et al. Structural requirements for activation of the 5-oxo-6E,8Z, 11Z,14Z-eicosatetraenoic acid (5-oxo-ETE) receptor: identification of a mead acid metabolite with potent agonist activity. J Pharmacol Exp Ther 2008;325(2):698–707.

[31] Priller J, Briley EM, Mansouri J, Devane WA, Mackie K, Felder CC. Mead ethanolamide, a novel eicosanoid, is an agonist for the central (CB1) and peripheral (CB2) cannabinoid receptors. Mol Pharmacol 1995;48(2):288–92.

[32] Franceschi C. Inflammaging as a major characteristic of old people: can it be prevented or cured? Nutr Rev 2007;65(12 Pt 2):S173–6.

[33] Vasto S, Candore G, Balistreri CR, et al. Inflammatory networks in ageing, age-related diseases and longevity. Mech Ageing Dev 2007;128(1):83–91.

[34] James MJ, Gibson RA, Neumann MA, Cleland LG. Effect of dietary supplementation with n-9 eicosatrienoic acid on leukotriene B4 synthesis in rats: a novel approach to inhibition of eicosanoid synthesis. J Exp Med 1993;178(6):2261–5.

[35] James MJ, Gibson RA, Cleland LG. Dietary polyunsaturated fatty acids and inflammatory mediator production. Am J Clin Nutr 2000;71(Suppl. 1):343S–8S.

[36] Wey HE, Pyron L, Woolery M. Essential fatty acid deficiency in cultured human keratinocytes attenuates toxicity due to lipid peroxidation. Toxicol Appl Pharmacol 1993;120(1):72–9.

[37] Mancini JA, Waugh RJ, Thompson JA, et al. Structural characterization of the covalent attachment of leukotriene A3 to leukotriene A4 hydrolase. Arch Biochem Biophys 1998;354(1):117–24.

[38] Penning TD. Inhibitors of leukotriene A4 (LTA4) hydrolase as potential anti-inflammatory agents. Curr Pharm Des 2001;7(3):163–79.

[39] Wallace FA, Miles EA, Evans C, Stock TE, Yaqoob P, Calder PC. Dietary fatty acids influence the production of Th1- but not Th2-type cytokines. J Leukoc Biol 2001;69(3):449–57.

[40] Zhang P, Smith R, Chapkin RS, McMurray DN. Dietary (n-3) polyunsaturated fatty acids modulate murine Th1/Th2 balance toward the Th2 pole by suppression of Th1 development. J Nutr 2005;135(7):1745–51.

[41] Stoney RM, Woods RK, Hosking CS, Hill DJ, Abramson MJ, Thien FC. Maternal breast milk long-chain n-3 fatty acids are associated with increased risk of atopy in breastfed infants. Clin Exp Allergy 2004;34(2):194–200.

[42] Thies F, Nebe-von-Caron G, Powell JR, Yaqoob P, Newsholme EA, Calder PC. Dietary supplementation with eicosapentaenoic acid, but not with other long-chain n-3 or n-6 polyunsaturated

[43] Margalit M, Ilan Y. Induction of immune tolerance: a role for natural killer T lymphocytes? Liver Int 2005;25(3):501—4.

[44] Hooper L, Thompson RL, Harrison RA, et al. Risks and benefits of omega 3 fats for mortality, cardiovascular disease, and cancer: systematic review. Br Med J 2006;332(7544):752—60.

[45] MacLean CH, Newberry SJ, Mojica WA, et al. Effects of omega-3 fatty acids on cancer risk: a systematic review. J Am Med Assoc 2006;295(4):403—15.

[46] Peskin BS. Why fish oil fails: a comprehensive 21st century lipids-based physiologic analysis. J Lipids 2014;2014:495761.

[47] Olivo SE, Hilakivi-Clarke L. Opposing effects of prepubertal low- and high-fat n-3 polyunsaturated fatty acid diets on rat mammary tumorigenesis. Carcinogenesis 2005;26(9):1563—72.

[48] Sasaki T, Kobayashi Y, Shimizu J, et al. Effects of dietary n-3-to-n-6 polyunsaturated fatty acid ratio on mammary carcinogenesis in rats. Nutr Cancer 1998;30(2):137—43.

[49] Griffini P, Fehres O, Klieverik L, et al. Dietary omega-3 polyunsaturated fatty acids promote colon carcinoma metastasis in rat liver. Cancer Res 1998;58(15):3312—19.

[50] Klieveri L, Fehres O, Griffini P, Van Noorden CJ, Frederiks WM. Promotion of colon cancer metastases in rat liver by fish oil diet is not due to reduced stroma formation. Clin Exp Metastasis 2000;18(5):371—7.

[51] Abou-el-Ela SH, Prasse KW, Carroll R, Wade AE, Dharwadkar S, Bunce OR. Eicosanoid synthesis in 7,12-dimethylbenz(a) anthracene-induced mammary carcinomas in Sprague-Dawley rats fed primrose oil, menhaden oil or corn oil diet. Lipids 1988;23(10):948—54.

[52] Costa I, Moral R, Solanas M, Escrich E. High-fat corn oil diet promotes the development of high histologic grade rat DMBA-induced mammary adenocarcinomas, while high olive oil diet does not. Breast Cancer Res Treat 2004;86(3):225—35.

[53] Moral R, Solanas M, Garcia G, Colomer R, Escrich E. Modulation of EGFR and neu expression by n-6 and n-9 high-fat diets in experimental mammary adenocarcinomas. Oncol Rep 2003;10(5):1417—24.

[54] Solanas M, Hurtado A, Costa I, et al. Effects of a high olive oil diet on the clinical behavior and histopathological features of rat DMBA-induced mammary tumors compared with a high corn oil diet. Int J Oncol 2002;21(4):745—53.

[55] Astorg P, Arnault N, Czernichow S, Noisette N, Galan P, Hercberg S. Dietary intakes and food sources of n-6 and n-3 PUFA in French adult men and women. Lipids 2004;39(6):527—35.

[56] Flood VM, Webb KL, Rochtchina E, Kelly B, Mitchell P. Fatty acid intakes and food sources in a population of older Australians. Asia Pac J Clin Nutr 2007;16(2):322—30.

[57] Sioen IA, Pynaert I, Matthys C, De Backer G, Van Camp J, De Henauw S. Dietary intakes and food sources of fatty acids for Belgian women, focused on n-6 and n-3 polyunsaturated fatty acids. Lipids 2006;41(5):415—22.

[58] Cave Jr. WT, Jurkowski JJ. Dietary lipid effects on the growth, membrane composition, and prolactin-binding capacity of rat mammary tumors. J Natl Cancer Inst 1984;73(1):185—91.

[59] Ip C, Carter CA, Ip MM. Requirement of essential fatty acid for mammary tumorigenesis in the rat. Cancer Res 1985;45(5):1997—2001.

[60] Reeve VE, Matheson M, Greenoak GE, Canfield PJ, Boehm-Wilcox C, Gallagher CH. Effect of dietary lipid on UV light carcinogenesis in the hairless mouse. Photochem Photobiol 1988;48(5):689—96.

[61] Sauer LA, Dauchy RT. Identification of linoleic and arachidonic acids as the factors in hyperlipemic blood that increase [3H]thymidine incorporation in hepatoma 7288CTC perfused in situ. Cancer Res 1988;48(11):3106—11.

[62] Sauer LA, Dauchy RT, Blask DE, Armstrong BJ, Scalici S. 13-Hydroxyoctadecadienoic acid is the mitogenic signal for linoleic acid-dependent growth in rat hepatoma 7288CTC in vivo. Cancer Res 1999;59(18):4688—92.

[63] Sevanian A, Peterson H. Induction of cytotoxicity and mutagenesis is facilitated by fatty acid hydroperoxidase activity in Chinese hamster lung fibroblasts (V79 cells). Mutat Res 1989;224(2):185—96.

[64] Gower JD. A role for dietary lipids and antioxidants in the activation of carcinogens. Free Radic Biol Med 1988;5(2):95—111.

[65] de Kok TM, ten Vaarwerk F, Zwingman I, van Maanen JM, Kleinjans JC. Peroxidation of linoleic, arachidonic and oleic acid in relation to the induction of oxidative DNA damage and cytogenetic effects. Carcinogenesis 1994;15(7):1399—404.

[66] Pamplona R, Portero-Otin M, Sanz A, Requena J, Barja G. Modification of the longevity-related degree of fatty acid unsaturation modulates oxidative damage to proteins and mitochondrial DNA in liver and brain. Exp Gerontol 2004;39(5):725—33.

[67] Schonfeld P, Wojtczak L. Fatty acids decrease mitochondrial generation of reactive oxygen species at the reverse electron transport but increase it at the forward transport. Biochim Biophys Acta 2007;1767(8):1032—40.

[68] Cocco T, Di Paola M, Papa S, Lorusso M. Arachidonic acid interaction with the mitochondrial electron transport chain promotes reactive oxygen species generation. Free Radic Biol Med 1999;27(1—2):51—9.

[69] Barja G. Mitochondrial oxygen radical generation and leak: sites of production in states 4 and 3, organ specificity, and relation to aging and longevity. J Bioenerg Biomembr 1999;31(4):347—66.

[70] Herrero A, Portero-Otin M, Bellmunt MJ, Pamplona R, Barja G. Effect of the degree of fatty acid unsaturation of rat heart mitochondria on their rates of H2O2 production and lipid and protein oxidative damage. Mech Ageing Dev 2001;122(4):427—43.

[71] Wright Jr. JR, Fraser RB, Kapoor S, Cook HW. Essential fatty acid deficiency prevents multiple low-dose streptozotocin-induced diabetes in naive and cyclosporin-treated low-responder murine strains. Acta Diabetol 1995;32(2):125—30.

[72] Takahashi K, Kato T, Schreiner GF, Ebert J, Badr KF. Essential fatty acid deficiency normalizes function and histology in rat nephrotoxic nephritis. Kidney Int 1992;41(5):1245—53.

[73] Autore G, Cicala C, Cirino G, Maiello FM, Mascolo N, Capasso F. Essential fatty acid-deficient diet modifies PAF levels in stomach and duodenum of endotoxin-treated rats. J Lipid Mediat Cell Signal 1994;9(2):145—53.

[74] Lipoeto NI, Agus Z, Oenzil F, Wahlqvist M, Wattanapenpaiboon N. Dietary intake and the risk of coronary heart disease among the coconut-consuming Minangkabau in West Sumatra, Indonesia. Asia Pac J Clin Nutr 2004;13(4):377—84.

[75] Prior IA, Davidson F, Salmond CE, Czochanska Z. Cholesterol, coconuts, and diet on Polynesian atolls: a natural experiment: the Pukapuka and Tokelau island studies. Am J Clin Nutr 1981;34(8):1552—61.

[76] Lindeberg S, Ahren B, Nilsson A, Cordain L, Nilsson-Ehle P, Vessby B. Determinants of serum triglycerides and high-density lipoprotein cholesterol in traditional Trobriand Islanders: the Kitava Study. Scand J Clin Lab Invest 2003;63(3):175—80.

[77] Lindeberg S, Berntorp E, Nilsson-Ehle P, Terent A, Vessby B. Age relations of cardiovascular risk factors in a traditional Melanesian society: the Kitava Study. Am J Clin Nutr 1997;66(4):845—52.

[78] McClellan WS, Du Bois EF. Clinical calorimetry. XLV. prolonged meat diets with a study of kidney function and ketosis. J Biol Chem 1930;87(3):651–68.

[79] Grieb P, Klapcinska B, Smol E, et al. Long-term consumption of a carbohydrate-restricted diet does not induce deleterious metabolic effects. Nutr Res 2008;28(12):825–33.

[80] Mozaffarian D, Rimm EB, Herrington DM. Dietary fats, carbohydrate, and progression of coronary atherosclerosis in postmenopausal women. Am J Clin Nutr 2004;80(5):1175–84.

[81] Knopp RH, Retzlaff BM. Saturated fat prevents coronary artery disease? An American paradox. Am J Clin Nutr 2004;80(5):1102–3.

[82] Ferrieres J. The French paradox: lessons for other countries. Heart 2004;90(1):107–11.

[83] Semba RD, Ferrucci L, Bartali B, et al. Resveratrol levels and all-cause mortality in older community-dwelling adults. JAMA Intern Med 2014;174(7):1077–84.

[84] Mbalilaki JA, Masesa Z, Stromme SB, et al. Daily energy expenditure and cardiovascular risk in Masai, rural and urban Bantu Tanzanians. Br J Sports Med 2008.

[85] Mann GV, Spoerry A, Gary M, Jarashow D. Atherosclerosis in the Masai. Am J Epidemiol 1972;95(1):26–37.

[86] Zhou W, Mukherjee P, Kiebish MA, Markis WT, Mantis JG, Seyfried TN. The calorically restricted ketogenic diet, an effective alternative therapy for malignant brain cancer. Nutr Metab (Lond) 2007;4:5.

[87] Napoli E, Duenas N, Giulivi C. Potential therapeutic use of the ketogenic diet in autistic spectrum disorders. Front Pediatr 2014;2.

[88] Richards MP, Pettitt PB, Trinkaus E, Smith FH, Paunovic M, Karavanic I. Neanderthal diet at Vindija and Neanderthal predation: the evidence from stable isotopes. Proc Natl Acad Sci USA 2000;97(13):7663–6.

[89] Portero-Otin M, Bellmunt MJ, Requena JR, Pamplona R. Protein modification by advanced Maillard adducts can be modulated by dietary polyunsaturated fatty acids. Biochem Soc Trans 2003;31(Pt 6):1403–5.

[90] Lemieux H, Bulteau AL, Friguet B, Tardif JC, Blier PU. Dietary fatty acids and oxidative stress in the heart mitochondria. Mitochondrion 2011;11(1):97–103.

[91] Hibbeln JR, Nieminen LR, Lands WE. Increasing homicide rates and linoleic acid consumption among five Western countries, 1961–2000. Lipids 2004;39(12):1207–13.

[92] Massiera F, Barbry P, Guesnet P, et al. A Western-like fat diet is sufficient to induce a gradual enhancement in fat mass over generations. J Lipid Res 2010;51(8):2352–61.

[93] Negre-Salvayre A, Coatrieux C, Ingueneau C, Salvayre R. Advanced lipid peroxidation end products in oxidative damage to proteins. Potential role in diseases and therapeutic prospects for the inhibitors. Br J Pharmacol 2008;153(1):6–20.

[94] Baynes JW. The Maillard hypothesis on aging: time to focus on DNA. Ann NY Acad Sci 2002;959:360–7.

[95] Feng Z, Hu W, Tang MS. Trans-4-hydroxy-2-nonenal inhibits nucleotide excision repair in human cells: a possible mechanism for lipid peroxidation-induced carcinogenesis. Proc Natl Acad Sci USA 2004;101(23):8598–602.

[96] Feng Z, Hu W, Marnett LJ, Tang MS. Malondialdehyde, a major endogenous lipid peroxidation product, sensitizes human cells to UV- and BPDE-induced killing and mutagenesis through inhibition of nucleotide excision repair. Mutat Res 2006;601(1–2):125–36.

[97] Pang B, Zhou X, Yu H, et al. Lipid peroxidation dominates the chemistry of DNA adduct formation in a mouse model of inflammation. Carcinogenesis 2007;28(8):1807–13.

[98] Williams MV, Lee SH, Pollack M, Blair IA. Endogenous lipid hydroperoxide-mediated DNA-adduct formation in min mice. J Biol Chem 2006;281(15):10127–33.

[99] Kawanishi S, Hiraku Y, Pinlaor S, Ma N. Oxidative and nitrative DNA damage in animals and patients with inflammatory diseases in relation to inflammation-related carcinogenesis. Biol Chem 2006;387(4):365–72.

[100] Kim SI, Pfeifer GP, Besaratinia A. Lack of mutagenicity of acrolein-induced DNA adducts in mouse and human cells. Cancer Res 2007;67(24):11640–7.

[101] Foiles PG, Akerkar SA, Miglietta LM, Chung FL. Formation of cyclic deoxyguanosine adducts in Chinese hamster ovary cells by acrolein and crotonaldehyde. Carcinogenesis 1990;11(11):2059–61.

[102] Foiles PG, Akerkar SA, Chung FL. Application of an immunoassay for cyclic acrolein deoxyguanosine adducts to assess their formation in DNA of Salmonella typhimurium under conditions of mutation induction by acrolein. Carcinogenesis 1989;10(1):87–90.

[103] Basu AK, Marnett LJ. Molecular requirements for the mutagenicity of malondialdehyde and related acroleins. Cancer Res 1984;44(7):2848–54.

[104] Feng Z, Hu W, Amin S, Tang MS. Mutational spectrum and genotoxicity of the major lipid peroxidation product, trans-4-hydroxy-2-nonenal, induced DNA adducts in nucleotide excision repair-proficient and -deficient human cells. Biochemistry 2003;42(25):7848–54.

[105] Hu W, Feng Z, Eveleigh J, et al. The major lipid peroxidation product, trans-4-hydroxy-2-nonenal, preferentially forms DNA adducts at codon 249 of human p53 gene, a unique mutational hotspot in hepatocellular carcinoma. Carcinogenesis 2002;23(11):1781–9.

[106] Chung FL, Pan J, Choudhury S, Roy R, Hu W, Tang MS. Formation of trans-4-hydroxy-2-nonenal- and other enal-derived cyclic DNA adducts from omega-3 and omega-6 polyunsaturated fatty acids and their roles in DNA repair and human p53 gene mutation. Mutat Res 2003;531(1–2):25–36.

[107] Kowalczyk P, Ciesla JM, Komisarski M, Kusmierek JT, Tudek B. Long-chain adducts of trans-4-hydroxy-2-nonenal to DNA bases cause recombination, base substitutions and frameshift mutations in M13 phage. Mutat Res 2004;550(1–2):33–48.

[108] Fernandes PH, Wang H, Rizzo CJ, Lloyd RS. Site-specific mutagenicity of stereochemically defined 1,N2-deoxyguanosine adducts of trans-4-hydroxynonenal in mammalian cells. Environ Mol Mutagen 2003;42(2):68–74.

[109] Chung FL, Nath RG, Ocando J, Nishikawa A, Zhang L. Deoxyguanosine adducts of t-4-hydroxy-2-nonenal are endogenous DNA lesions in rodents and humans: detection and potential sources. Cancer Res 2000;60(6):1507–11.

[110] Kasai H, Maekawa M, Kawai K, et al. 4-oxo-2-hexenal, a mutagen formed by omega-3 fat peroxidation, causes DNA adduct formation in mouse organs. Ind Health 2005;43(4):699–701.

[111] Kawai K, Matsuno K, Kasai H. Detection of 4-oxo-2-hexenal, a novel mutagenic product of lipid peroxidation, in human diet and cooking vapor. Mutat Res 2006;603(2):186–92.

[112] Maekawa M, Kawai K, Takahashi Y, et al. Identification of 4-oxo-2-hexenal and other direct mutagens formed in model lipid peroxidation reactions as dGuo adducts. Chem Res Toxicol 2006;19(1):130–8.

[113] Phillis JW, Horrocks LA, Farooqui AA. Cyclooxygenases, lipoxygenases, and epoxygenases in CNS: their role and involvement in neurological disorders. Brain Res Rev 2006;52(2):201–43.

[114] Cuendet M, Pezzuto JM. The role of cyclooxygenase and lipoxygenase in cancer chemoprevention. Drug Metabol Drug Interact 2000;17(1−4):109−57.

[115] Nanda BL, Nataraju A, Rajesh R, Rangappa KS, Shekar MA, Vishwanath BS. PLA2 mediated arachidonate free radicals: PLA2 inhibition and neutralization of free radicals by antioxidants − a new role as anti-inflammatory molecule. Curr Top Med Chem 2007;7(8):765−77.

[116] Farooqui AA, Ong WY, Horrocks LA. Inhibitors of brain phospholipase A2 activity: their neuropharmacological effects and therapeutic importance for the treatment of neurologic disorders. Pharmacol Rev 2006;58(3):591−620.

[117] Farooqui AA, Horrocks LA. Phospholipase A2-generated lipid mediators in the brain: the good, the bad, and the ugly. Neuroscientist 2006;12(3):245−60.

[118] Spiteller G. Lipid peroxidation in aging and age-dependent diseases. Exp Gerontol 2001;36(9):1425−57.

[119] Marnett LJ. Oxy radicals, lipid peroxidation and DNA damage. Toxicology 2002;181−182:219−22.

[120] Plastaras JP, Guengerich FP, Nebert DW, Marnett LJ. Xenobiotic-metabolizing cytochromes P450 convert prostaglandin endoperoxide to hydroxyheptadecatrienoic acid and the mutagen, malondialdehyde. J Biol Chem 2000;275(16):11784−90.

[121] Murthi KK, Friedman LR, Oleinick NL, Salomon RG. Formation of DNA-protein cross-links in mammalian cells by levuglandin E2. Biochemistry 1993;32(15):4090−7.

[122] Salomon RG. Levuglandins and isolevuglandins: stealthy toxins of oxidative injury. Antioxid Redox Signal 2005;7(1−2):185−201.

[123] Pathak S, Singh R, Verschoyle RD, et al. Androgen manipulation alters oxidative DNA adduct levels in androgen-sensitive prostate cancer cells grown in vitro and in vivo. Cancer Lett 2007.

[124] Hankin JA, Jones DN, Murphy RC. Covalent binding of leukotriene A4 to DNA and RNA. Chem Res Toxicol 2003;16(4):551−61.

[125] Hankin JA, Murphy RC. Covalent binding of leukotriene A4 to nucleosides, nucleotides, and nucleic acids. Adv Exp Med Biol 2003;525:29−33.

[126] Hankin JA, Murphy RC. Mass spectrometric quantitation of deoxyguanosine and leukotriene A4-deoxyguanosine adducts of DNA. Anal Biochem 2004;333(1):156−64.

[127] Das UN, Ramadevi G, Rao KP, Rao MS. Prostaglandins can modify gamma-radiation and chemical induced cytotoxicity and genetic damage in vitro and in vivo. Prostaglandins 1989;38(6):689−716.

[128] Lim P, Sadre-Bazzaz K, Shurter J, Sarasin A, Termini J. DNA damage and mutations induced by arachidonic acid peroxidation. Biochemistry 2003;42(51):15036−44.

[129] Toschi A, Lee E, Xu L, Garcia A, Gadir N, Foster DA. Regulation of mTORC1 and mTORC2 complex assembly by phosphatidic acid: competition with rapamycin. Mol Cell Biol 2009;29(6):1411−20.

[130] Sun Y, Fang Y, Yoon MS, et al. Phospholipase D1 is an effector of Rheb in the mTOR pathway. Proc Natl Acad Sci USA 2008;105(24):8286−91.

[131] Lee SH, Williams MV, Dubois RN, Blair IA. Cyclooxygenase-2-mediated DNA damage. J Biol Chem 2005;280(31):28337−46.

[132] Chen X, Sood S, Yang CS, Li N, Sun Z. Five-lipoxygenase pathway of arachidonic acid metabolism in carcino-genesis and cancer chemoprevention. Curr Cancer Drug Targets 2006;6(7):613−22.

[133] Chen X, Wang S, Wu N, Yang CS. Leukotriene A4 hydrolase as a target for cancer prevention and therapy. Curr Cancer Drug Targets 2004;4(3):267−83.

[134] Rosignoli P, Fuccelli R, Fabiani R, Servili M, Morozzi G. Effect of olive oil phenols on the production of inflammatory mediators in freshly isolated human monocytes. J Nutr Biochem 2013;24(8):1513−19.

[135] Shimizu S, Akimoto K, Shinmen Y, Kawashima H, Sugano M, Yamada H. Sesamin is a potent and specific inhibitor of delta 5 desaturase in polyunsaturated fatty acid biosynthesis. Lipids 1991;26(7):512−16.

[136] Serhan CN. Pro-resolving lipid mediators are leads for resolution physiology. Nature 2014;510(7503):92−101.

[137] Michaels ML, Miller JH. The GO system protects organisms from the mutagenic effect of the spontaneous lesion 8-hydroxyguanine (7,8-dihydro-8-oxoguanine). J Bacteriol 1992;174(20):6321−5.

[138] Kim JE, Chung MH. 8-Oxo-7,8-dihydro-2′-deoxyguanosine is not salvaged for DNA synthesis in human leukemic U937 cells. Free Radic Res 2006;40(5):461−6.

[139] Hogg N, Kalyanaraman B. Nitric oxide and lipid peroxidation. Biochim Biophys Acta 1999;1411(2−3):378−84.

[140] Gruz P, Shimizu M. Origins of age-related DNA damage and dietary strategies for its reduction. Rejuvenation Res 2010;13(2−3):285−7.

[141] Brenna JT, Varamini B, Jensen RG, Diersen-Schade DA, Boettcher JA, Arterburn LM. Docosahexaenoic and arachidonic acid concentrations in human breast milk worldwide. Am J Clin Nutr 2007;85(6):1457−64.

[142] Das UN. Is metabolic syndrome X a disorder of the brain with the initiation of low-grade systemic inflammatory events during the perinatal period? J Nutr Biochem 2007;18(11):701−13.

[143] Gorlewska-Roberts KM, Teitel CH, Lay Jr. JO, Roberts DW, Kadlubar FF. Lactoperoxidase-catalyzed activation of carcinogenic aromatic and heterocyclic amines. Chem Res Toxicol 2004;17(12):1659−66.

[144] Josephy PD. The role of peroxidase-catalyzed activation of aromatic amines in breast cancer. Mutagenesis 1996;11(1):3−7.

[145] Martin FL, Cole KJ, Weaver G, et al. Genotoxicity of human breast milk from different countries. Mutagenesis 2001;16(5):401−6.

[146] Phillips DH, Martin FL, Williams JA, et al. Mutagens in human breast lipid and milk: the search for environmental agents that initiate breast cancer. Environ Mol Mutagen 2002;39(2−3):143−9.

[147] Yakes Jimenez E, Mangani C, Ashorn P, Harris WS, Maleta K, Dewey KG. Breast milk from women living near Lake Malawi is high in docosahexaenoic acid and arachidonic acid. Prostaglandins Leukot Essent Fatty Acids 2014.

[148] Greaves LC, Nooteboom M, Elson JL, et al. Clonal expansion of early to mid-life mitochondrial DNA point mutations drives mitochondrial dysfunction during human ageing. PLoS Genet 2014;10(9):e1004620.

[149] Grover PL, Martin FL. The initiation of breast and prostate cancer. Carcinogenesis 2002;23(7):1095−102.

[150] Brouwer IA, Katan MB, Zock PL. Dietary alpha-linolenic acid is associated with reduced risk of fatal coronary heart disease, but increased prostate cancer risk: a meta-analysis. J Nutr 2004;134(4):919−22.

[151] Christensen JH, Fabrin K, Borup K, Barber N, Poulsen J. Prostate tissue and leukocyte levels of n-3 polyunsaturated fatty acids in men with benign prostate hyperplasia or prostate cancer. BJU Int 2006;97(2):270−3.

[152] Giovannucci E, Liu Y, Platz EA, Stampfer MJ, Willett WC. Risk factors for prostate cancer incidence and progression in the health professionals follow-up study. Int J Cancer 2007;121(7):1571−8.

REFERENCES

[153] Leitzmann MF, Stampfer MJ, Michaud DS, et al. Dietary intake of n-3 and n-6 fatty acids and the risk of prostate cancer. Am J Clin Nutr 2004;80(1):204—16.

[154] Pandalai PK, Pilat MJ, Yamazaki K, Naik H, Pienta KJ. The effects of omega-3 and omega-6 fatty acids on in vitro prostate cancer growth. Anticancer Res 1996;16(2):815—20.

[155] Ritch CR, Brendler CB, Wan RL, Pickett KE, Sokoloff MH. Relationship of erythrocyte membrane polyunsaturated fatty acids and prostate-specific antigen levels in Jamaican men. BJU Int 2004;93(9):1211—15.

[156] Ritch CR, Wan RL, Stephens LB, et al. Dietary fatty acids correlate with prostate cancer biopsy grade and volume in Jamaican men. J Urol 2007;177(1):97—101 discussion.

[157] Guillén MD, Uriarte PS. Aldehydes contained in edible oils of a very different nature after prolonged heating at frying temperature: presence of toxic oxygenated α,β unsaturated aldehydes. Food Chem 2012;131(3):915—26.

[158] Guillen MD, Goicoechea E. Toxic oxygenated alpha,beta-unsaturated aldehydes and their study in foods: a review. Crit Rev Food Sci Nutr 2008;48(2):119—36.

[159] Kasai H, Kawai K. 4-oxo-2-hexenal, a mutagen formed by omega-3 fat peroxidation: occurrence, detection and adduct formation. Mutat Res 2008;659(1—2):56—9.

[160] Takasu S, Tsukamoto T, Hirata A, et al. Lack of initiation activity of 4-oxo-2-hexenal, a peroxidation product generated from omega-3 polyunsaturated fatty acids, in an in vivo five-week liver assay. Asian Pac J Cancer Prev 2007;8(3):372—4.

[161] Schuler D, Budiawan B, Eder E. Development of a 32P-postlabeling method for the detection of 1, N2-propanodeoxyguanosine adducts of 2-hexenal in vivo. Chem Res Toxicol 1999;12(4):335—40.

[162] Hilakivi-Clarke L, Olivo SE, Shajahan A, et al. Mechanisms mediating the effects of prepubertal (n-3) polyunsaturated fatty acid diet on breast cancer risk in rats. J Nutr 2005;135 (Suppl. 12):2946S—52SS.

[163] Rikans LE, Hornbrook KR. Lipid peroxidation, antioxidant protection and aging. Biochim Biophys Acta 1997;1362 (2—3):116—27.

[164] Mozaffarian D, Katan MB, Ascherio A, Stampfer MJ, Willett WC. Trans fatty acids and cardiovascular disease. N Engl J Med 2006;354(15):1601—13.

[165] Al-Saghir S, Thurner K, Wagner KH, et al. Effects of different cooking procedures on lipid quality and cholesterol oxidation of farmed salmon fish (*Salmo salar*). J Agric Food Chem 2004;52(16):5290—6.

[166] Chung FL, Zhang L, Ocando JE, Nath RG. Role of 1,N2-propanodeoxyguanosine adducts as endogenous DNA lesions in rodents and humans. IARC Sci Publ 1999;150:45—54.

[167] Eder E, Scheckenbach S, Deininger C, Hoffman C. The possible role of alpha, beta-unsaturated carbonyl compounds in mutagenesis and carcinogenesis. Toxicol Lett 1993;67(1—3):87—103.

[168] Zhang S, Villalta PW, Wang M, Hecht SS. Analysis of crotonaldehyde- and acetaldehyde-derived 1,n(2)-propanodeoxyguanosine adducts in DNA from human tissues using liquid chromatography electrospray ionization tandem mass spectrometry. Chem Res Toxicol 2006;19(10):1386—92.

[169] Rouzer CA, Chaudhary AK, Nokubo M, et al. Analysis of the malondialdehyde-2′-deoxyguanosine adduct pyrimidopurinone in human leukocyte DNA by gas chromatography/electron capture/negative chemical ionization/mass spectrometry. Chem Res Toxicol 1997;10(2):181—8.

[170] Fang JL, Vaca CE, Valsta LM, Mutanen M. Determination of DNA adducts of malonaldehyde in humans: effects of dietary fatty acid composition. Carcinogenesis 1996;17(5):1035—40.

[171] Bartsch H, Nair J. Ultrasensitive and specific detection methods for exocyclic DNA adducts: markers for lipid peroxidation and oxidative stress. Toxicology 2000;153(1—3):105—14.

[172] Zhang S, Villalta PW, Wang M, Hecht SS. Detection and quantitation of acrolein-derived 1,N2-propanodeoxyguanosine adducts in human lung by liquid chromatography-electrospray ionization-tandem mass spectrometry. Chem Res Toxicol 2007;20(4):565—71.

[173] Liu X, Lovell MA, Lynn BC. Development of a method for quantification of acrolein-deoxyguanosine adducts in DNA using isotope dilution-capillary LC/MS/MS and its application to human brain tissue. Anal Chem 2005;77(18):5982—9.

[174] Chung FL, Nath RG, Nagao M, Nishikawa A, Zhou GD, Randerath K. Endogenous formation and significance of 1,N2-propanodeoxyguanosine adducts. Mutat Res 1999;424 (1—2):71—81.

[175] Liu X, Lovell MA, Lynn BC. Detection and quantification of endogenous cyclic DNA adducts derived from trans-4-hydroxy-2-nonenal in human brain tissue by isotope dilution capillary liquid chromatography nanoelectrospray tandem mass spectrometry. Chem Res Toxicol 2006;19(5):710—18.

[176] Badouard C, Menezo Y, Panteix G, et al. Determination of new types of DNA lesions in human sperm. Zygote 2008;16 (1):9—13.

[177] Douki T, Odin F, Caillat S, Favier A, Cadet J. Predominance of the 1,N2-propano 2′-deoxyguanosine adduct among 4-hydroxy-2-nonenal-induced DNA lesions. Free Radic Biol Med 2004;37(1):62—70.

[178] Kawai Y, Uchida K, Osawa T. 2′-deoxycytidine in free nucleosides and double-stranded DNA as the major target of lipid peroxidation products. Free Radic Biol Med 2004;36(5):529—41.

[179] Adler ID, Filser J, Gonda H, Schriever-Schwemmer G. Dose response study for 1,3-butadiene-induced dominant lethal mutations and heritable translocations in germs cells of male mice. Mutat Res 1998;397(1):85—92.

[180] Fernandes PH, Lloyd RS. Mutagenic bypass of the butadiene-derived 2′-deoxyuridine adducts by polymerases eta and zeta. Mutat Res 2007;625(1—2):40—9.

[181] Rindgen D, Lee SH, Nakajima M, Blair IA. Formation of a substituted 1,N(6)-etheno-2′-deoxyadenosine adduct by lipid hydroperoxide-mediated generation of 4-oxo-2-nonenal. Chem Res Toxicol 2000;13(9):846—52.

[182] Lee SH, Rindgen D, Bible Jr. RH, Hajdu E, Blair IA. Characterization of 2′-deoxyadenosine adducts derived from 4-oxo-2-nonenal, a novel product of lipid peroxidation. Chem Res Toxicol 2000;13(7):565—74.

[183] Nair J, De Flora S, Izzotti A, Bartsch H. Lipid peroxidation-derived etheno-DNA adducts in human atherosclerotic lesions. Mutat Res 2007;621(1—2):95—105.

[184] Schmid K, Nair J, Winde G, Velic I, Bartsch H. Increased levels of promutagenic etheno-DNA adducts in colonic polyps of FAP patients. Int J Cancer 2000;87(1):1—4.

[185] Bartsch H, Nair J. Potential role of lipid peroxidation derived DNA damage in human colon carcinogenesis: studies on exocyclic base adducts as stable oxidative stress markers. Cancer Detect Prev 2002;26(4):308—12.

[186] Pugh S, Thomas GA. Patients with adenomatous polyps and carcinomas have increased colonic mucosal prostaglandin E2. Gut 1994;35(5):675—8.

[187] Nair J, Gansauge F, Beger H, Dolara P, Winde G, Bartsch H. Increased etheno-DNA adducts in affected tissues of patients suffering from Crohn's disease, ulcerative colitis, and chronic pancreatitis. Antioxid Redox Signal 2006;8(5—6):1003—10.

[188] Arab K, Pedersen M, Nair J, Meerang M, Knudsen LE, Bartsch H. Typical signature of DNA damage in white blood cells: a

[188] pilot study on etheno adducts in Danish mother-newborn child pairs. Carcinogenesis 2009;30(2):282–5.

[189] Godschalk RW, Albrecht C, Curfs DM, et al. Decreased levels of lipid peroxidation-induced DNA damage in the onset of atherogenesis in apolipoprotein E deficient mice. Mutat Res 2007;621(1–2):87–94.

[190] Vogel U, Danesvar B, Autrup H, et al. Effect of increased intake of dietary animal fat and fat energy on oxidative damage, mutation frequency, DNA adduct level and DNA repair in rat colon and liver. Free Radic Res 2003;37(9):947–56.

[191] Fang Q, Nair J, Sun X, Hadjiolov D, Bartsch H. Etheno-DNA adduct formation in rats gavaged with linoleic acid, oleic acid and coconut oil is organ- and gender specific. Mutat Res 2007;624(1–2):71–9.

[192] Hagenlocher T, Nair J, Becker N, Korfmann A, Bartsch H. Influence of dietary fatty acid, vegetable, and vitamin intake on etheno-DNA adducts in white blood cells of healthy female volunteers: a pilot study. Cancer Epidemiol Biomarkers Prev 2001;10(11):1187–91.

[193] Davidson LA, Nguyen DV, Hokanson RM, et al. Chemopreventive n-3 polyunsaturated fatty acids reprogram genetic signatures during colon cancer initiation and progression in the rat. Cancer Res 2004;64(18):6797–804.

[194] Long EK, Murphy TC, Leiphon LJ, et al. Trans-4-hydroxy-2-hexenal is a neurotoxic product of docosahexaenoic (22:6; n-3) acid oxidation. J Neurochem 2008;105(3):714–24.

[195] Fam SS, Murphey LJ, Terry ES, et al. Formation of highly reactive A-ring and J-ring isoprostane-like compounds (A4/J4-neuroprostanes) in vivo from docosahexaenoic acid. J Biol Chem 2002;277(39):36076–84.

[196] Chou PH, Kageyama S, Matsuda S, et al. Detection of lipid peroxidation-induced DNA adducts caused by 4-oxo-2(E)-nonenal and 4-oxo-2(E)-hexenal in human autopsy tissues. Chem Res Toxicol 2010;23(9):1442–8.

[197] Mortelmans K, Zeiger E. The Ames Salmonella/microsome mutagenicity assay. Mutat Res 2000;455(1–2):29–60.

[198] Lopachin RM, Gavin T, Petersen DR, Barber DS. Molecular mechanisms of 4-hydroxy-2-nonenal and acrolein toxicity: nucleophilic targets and adduct formation. Chem Res Toxicol 2009.

[199] Marnett LJ, Hurd HK, Hollstein MC, Levin DE, Esterbauer H, Ames BN. Naturally occurring carbonyl compounds are mutagens in *Salmonella tester* strain TA104. Mutat Res 1985;148(1–2):25–34.

[200] Eder E, Deininger C. Mutagenicity of 2-alkylpropenals in *Salmonella typhimurium* strain TA 100: structural influences. Environ Mol Mutagen 2001;37(4):324–8.

[201] Parent RA, Caravello HE, San RH. Mutagenic activity of acrolein in *S. typhimurium* and *E. coli*. J Appl Toxicol 1996;16(2):103–8.

[202] Neudecker T, Eder E, Deininger C, Henschler D. Crotonaldehyde is mutagenic in *Salmonella typhimurium* TA100. Environ Mol Mutagen 1989;14(3):146–8.

[203] Riggins JN, Marnett LJ. Mutagenicity of the malondialdehyde oligomerization products 2-(3′-oxo-1′-propenyl)-malondialdehyde and 2,4-dihydroxymethylene-3-(2,2-dimethoxyethyl)glutaraldehyde in Salmonella. Mutat Res 2001;497(1–2):153–7.

[204] Chung FL, Chen HJ, Guttenplan JB, Nishikawa A, Hard GC. 2,3-epoxy-4-hydroxynonanal as a potential tumor-initiating agent of lipid peroxidation. Carcinogenesis 1993;14(10):2073–7.

[205] Chen HJ, Gonzalez FJ, Shou M, Chung FL. 2,3-epoxy-4-hydroxynonanal, a potential lipid peroxidation product for etheno adduct formation, is not a substrate of human epoxide hydrolase. Carcinogenesis 1998;19(5):939–43.

[206] Chen HJ, Chung FL. Epoxidation of trans-4-hydroxy-2-nonenal by fatty acid hydroperoxides and hydrogen peroxide. Chem Res Toxicol 1996;9(1):306–12.

[207] Lee SH, Oe T, Blair IA. Vitamin C-induced decomposition of lipid hydroperoxides to endogenous genotoxins. Science 2001;292(5524):2083–6.

[208] Lee SH, Arora JA, Oe T, Blair IA. 4-Hydroperoxy-2-nonenal-induced formation of 1,N2-etheno-2′-deoxyguanosine adducts. Chem Res Toxicol 2005;18(4):780–6.

[209] Blair IA. Lipid hydroperoxide-mediated DNA damage. Exp Gerontol 2001;36(9):1473–81.

[210] Blair IA. DNA adducts with lipid peroxidation products. J Biol Chem 2008;283(23):15545–9.

[211] Benamira M, Marnett LJ. The lipid peroxidation product 4-hydroxynonenal is a potent inducer of the SOS response. Mutat Res 1992;293(1):1–10.

[212] Demir E, Kaya B, Soriano C, Creus A, Marcos R. Genotoxic analysis of four lipid-peroxidation products in the mouse lymphoma assay. Mutat Res 2011;726(2):98–103.

[213] Singh SP, Chen T, Chen L, et al. Mutagenic effects of 4-hydroxynonenal triacetate, a chemically protected form of the lipid peroxidation product 4-hydroxynonenal, as assayed in L5178Y/Tk$^{+/-}$ mouse lymphoma cells. J Pharmacol Exp Ther 2005;313(2):855–61.

[214] Eder E, Wacker M, Lutz U, et al. Oxidative stress related DNA adducts in the liver of female rats fed with sunflower-, rapeseed-, olive- or coconut oil supplemented diets. Chem Biol Interact 2006;159(2):81–9.

[215] Lim P, Wuenschell GE, Holland V, et al. Peroxyl radical mediated oxidative DNA base damage: implications for lipid peroxidation induced mutagenesis. Biochemistry 2004;43(49):15339–48.

[216] Blair IA. Endogenous glutathione adducts. Curr Drug Metab 2006;7(8):853–72.

[217] Jian W, Lee SH, Mesaros C, Oe T, Elipe MV, Blair IA. A novel 4-oxo-2(E)-nonenal-derived endogenous thiadiazabicyclo glutathione adduct formed during cellular oxidative stress. Chem Res Toxicol 2007;20(7):1008–18.

[218] Huang G, Mills L, Worth LL. Expression of human glutathione S-transferase P1 mediates the chemosensitivity of osteosarcoma cells. Mol Cancer Ther 2007;6(5):1610–19.

[219] Wright SC, Wang H, Wei QS, Kinder DH, Larrick JW. Bcl-2-mediated resistance to apoptosis is associated with glutathione-induced inhibition of AP24 activation of nuclear DNA fragmentation. Cancer Res 1998;58(23):5570–6.

[220] Aldini G, Facino RM, Beretta G, Carini M. Carnosine and related dipeptides as quenchers of reactive carbonyl species: from structural studies to therapeutic perspectives. Biofactors 2005;24(1–4):77–87.

[221] Aldini G, Granata P, Carini M. Detoxification of cytotoxic alpha,beta-unsaturated aldehydes by carnosine: characterization of conjugated adducts by electrospray ionization tandem mass spectrometry and detection by liquid chromatography/mass spectrometry in rat skeletal muscle. J Mass Spectrom 2002;37(12):1219–28.

[222] Babizhayev MA. Analysis of lipid peroxidation and electron microscopic survey of maturation stages during human cataractogenesis: pharmacokinetic assay of Can-C N-acetylcarnosine prodrug lubricant eye drops for cataract prevention. Drugs R D 2005;6(6):345–69.

[223] Wiedemann M, Kontush A, Finckh B, Hellwege HH, Kohlschutter A. Neonatal blood plasma is less susceptible to oxidation than adult plasma owing to its higher content of bilirubin and lower content of oxidizable fatty acids. Pediatr Res 2003;53(5):843–9.

[224] Catala A. The ability of melatonin to counteract lipid peroxidation in biological membranes. Curr Mol Med 2007;7(7): 638–49.

[225] Spiteller G. Furan fatty acids: occurrence, synthesis, and reactions. Are furan fatty acids responsible for the cardioprotective effects of a fish diet? Lipids 2005;40(8):755–71.

[226] Spindler SR, Mote PL, Flegal JM. Lifespan effects of simple and complex nutraceutical combinations fed isocalorically to mice. Age (Dordr) 2014;36(2):705–18.

[227] Selman C, McLaren JS, Collins AR, Duthie GG, Speakman JR. Deleterious consequences of antioxidant supplementation on lifespan in a wild-derived mammal. Biol Lett 2013;9(4): 20130432.

[228] Heinonen OP, Albanes D. The effect of vitamin E and beta carotene on the incidence of lung cancer and other cancers in male smokers. N Engl J Med 1994;330(15):1029–35.

[229] Sayin VI, Ibrahim MX, Larsson E, Nilsson JA, Lindahl P, Bergo MO. Antioxidants accelerate lung cancer progression in mice. Sci Transl Med 2014;6(221):221ra15.

[230] Kaiser J. Biomedical research. Antioxidants could spur tumors by acting on cancer gene. Science 2014;343(6170):477.

[231] Hekimi S. Enhanced immunity in slowly aging mutant mice with high mitochondrial oxidative stress. Oncoimmunology 2013;2(4):e23793.

[232] Yee C, Yang W, Hekimi S. The intrinsic apoptosis pathway mediates the pro-longevity response to mitochondrial ROS in C. elegans. Cell 2014;157(4):897–909.

[233] Gomez-Cabrera MC, Domenech E, Romagnoli M, et al. Oral administration of vitamin C decreases muscle mitochondrial biogenesis and hampers training-induced adaptations in endurance performance. Am J Clin Nutr 2008;87(1):142–9.

[234] Ristow M, Zarse K, Oberbach A, et al. Antioxidants prevent health-promoting effects of physical exercise in humans. Proc Natl Acad Sci USA 2009;106(21):8665–70.

[235] Thies F, Nebe-von-Caron G, Powell JR, Yaqoob P, Newsholme EA, Calder PC. Dietary supplementation with gamma-linolenic acid or fish oil decreases T lymphocyte proliferation in healthy older humans. J Nutr 2001;131(7):1918–27.

[236] Spindler SR, Mote PL, Flegal JM. Dietary supplementation with Lovaza and krill oil shortens the life span of long-lived F1 mice. Age (Dordr) 2014.

[237] Umezawa M, Takeda T, Kogishi K, et al. Serum lipid concentrations and mean life span are modulated by dietary polyunsaturated fatty acids in the senescence-accelerated mouse. J Nutr 2000;130(2):221–7.

[238] Lu M, Taylor A, Chylack Jr. LT, et al. Dietary linolenic acid intake is positively associated with five-year change in eye lens nuclear density. J Am Coll Nutr 2007;26(2):133–40.

[239] Mibu H, Nagata M, Hikida M. A study on lipid peroxide-induced lens damage in vitro. Exp Eye Res 1994;58(1):85–90.

[240] Kang JH, Pasquale LR, Willett WC, et al. Dietary fat consumption and primary open-angle glaucoma. Am J Clin Nutr 2004;79(5):755–64.

[241] Brouwer IA. Omega-3 PUFA: good or bad for prostate cancer? Prostaglandins Leukot Essent Fatty Acids 2008;79(3–5):97–9.

[242] Fountain ED, Mao J, Whyte JJ, et al. Effects of diets enriched in omega-3 and omega-6 polyunsaturated fatty acids on offspring sex-ratio and maternal behavior in mice. Biol Reprod 2008;78(2):211–17.

[243] Carrie I, Guesnet P, Bourre JM, Frances H. Diets containing long-chain n-3 polyunsaturated fatty acids affect behaviour differently during development than ageing in mice. Br J Nutr 2000;83(4):439–47.

[244] Wakefield SL, Lane M, Schulz SJ, Hebart ML, Thompson JG, Mitchell M. Maternal supply of omega-3 polyunsaturated fatty acids alter mechanisms involved in oocyte and early embryo development in the mouse. Am J Physiol Endocrinol Metab 2008;294(2):E425–34.

[245] Ayre KJ, Hulbert AJ. Dietary fatty acid profile affects endurance in rats. Lipids 1997;32(12):1265–70.

[246] Turbill C. Mice run faster on a high n-6 polyunsaturated fatty acid diet. In: Society for experimental biology annual meeting. Glasgow; Society for Experimental Biology 2009. p. A5.37. <http://www.sciencedaily.com/releases/2009/06/090629081120.htm>, <http://phys.org/news/2009-06-mice-faster-high-grade-oil.htm>.

[247] Felton CV, Crook D, Davies MJ, Oliver MF. Dietary polyunsaturated fatty acids and composition of human aortic plaques. Lancet 1994;344(8931):1195–6.

[248] Allard JP, Kurian R, Aghdassi E, Muggli R, Royall D. Lipid peroxidation during n-3 fatty acid and vitamin E supplementation in humans. Lipids 1997;32(5):535–41.

[249] Choudhury S, Pan J, Amin S, Chung FL, Roy R. Repair kinetics of trans-4-hydroxynonenal-induced cyclic 1,N2-propano-deoxyguanine DNA adducts by human cell nuclear extracts. Biochemistry 2004;43(23):7514–21.

[250] Wolfle WT, Johnson RE, Minko IG, Lloyd RS, Prakash S, Prakash L. Replication past a trans-4-hydroxynonenal minor-groove adduct by the sequential action of human DNA polymerases iota and kappa. Mol Cell Biol 2006;26(1):381–6.

[251] Temviriyanukul P, Meijers M, van Hees-Stuivenberg S, et al. Different sets of translesion synthesis DNA polymerases protect from genome instability induced by distinct food-derived genotoxins. Toxicol Sci 2012;127(1):130–8.

[252] Velasco-Miguel S, Richardson JA, Gerlach VL, et al. Constitutive and regulated expression of the mouse Dinb (Polkappa) gene encoding DNA polymerase kappa. DNA Repair 2003;2(1):91–106.

[253] Ogi T, Mimura J, Hikida M, Fujimoto H, Fujii-Kuriyama Y, Ohmori H. Expression of human and mouse genes encoding polkappa: testis-specific developmental regulation and AhR-dependent inducible transcription. Genes Cells 2001;6(11): 943–53.

[254] Furland NE, Maldonado EN, Aveldano MI. Very long chain PUFA in murine testicular triglycerides and cholesterol esters. Lipids 2003;38(1):73–80.

[255] Connor WE, Lin DS, Neuringer M. Biochemical markers for puberty in the monkey testis: desmosterol and docosahexaenoic acid. J Clin Endocrinol Metab 1997;82(6):1911–16.

[256] Lin DS, Neuringer M, Connor WE. Selective changes of docosahexaenoic acid-containing phospholipid molecular species in monkey testis during puberty. J Lipid Res 2004;45(3): 529–35.

[257] Gros L, Ishchenko AA, Saparbaev M. Enzymology of repair of etheno adducts. Mutat Res 2003;531(1–2):219–29.

[258] Saparbaev M, Laval J. Enzymology of the repair of etheno adducts in mammalian cells and in Escherichia coli. IARC Sci Publ 1999;150:249–61.

[259] Choudhury S, Adhikari S, Cheema A, Roy R. Evidence of complete cellular repair of 1,N6-ethenoadenine, a mutagenic and potential damage for human cancer, revealed by a novel method. Mol Cell Biochem 2008;313(1–2):19–28.

[260] Saparbaev M, Laval J. 3,N4-ethenocytosine, a highly mutagenic adduct, is a primary substrate for Escherichia coli double-stranded uracil-DNA glycosylase and human mismatch-specific thymine-DNA glycosylase. Proc Natl Acad Sci U S A 1998; 95(15):8508–13.

[261] Delaney JC, Smeester L, Wong C, et al. AlkB reverses etheno DNA lesions caused by lipid oxidation in vitro and in vivo. Nat Struct Mol Biol 2005;12(10):855–60.

[262] Ebrahimi H, Donaldson AD. Release of yeast telomeres from the nuclear periphery is triggered by replication and maintained by suppression of Ku-mediated anchoring. Genes Dev 2008;22(23):3363–74.

[263] Pamplona R, Portero-Otin M, Riba D, et al. Low fatty acid unsaturation: a mechanism for lowered lipoperoxidative modification of tissue proteins in mammalian species with long life spans. J Gerontol A Biol Sci Med Sci 2000;55(6):B286–91.

[264] Barja G. The flux of free radical attack through mitochondrial DNA is related to aging rate. Aging (Milano) 2000;12(5): 342–55.

[265] Hulbert AJ, Faulks S, Buttemer WA, Else PL. Acyl composition of muscle membranes varies with body size in birds. J Exp Biol 2002;205(Pt 22):3561–9.

[266] Hulbert AJ. Life, death and membrane bilayers. J Exp Biol 2003;206(Pt 14):2303–11.

[267] Zimniak P. Relationship of electrophilic stress to aging. Free Radic Biol Med 2011;51(6):1087–105.

[268] Hulbert AJ. The links between membrane composition, metabolic rate and lifespan. Comp Biochem Physiol A Mol Integr Physiol 2006.

[269] Hulbert AJ, Faulks SC, Buffenstein R. Oxidation-resistant membrane phospholipids can explain longevity differences among the longest-living rodents and similarly-sized mice. J Gerontol A Biol Sci Med Sci 2006;61(10):1009–18.

[270] Pamplona R, Portero-Otin M, Ruiz C, Gredilla R, Herrero A, Barja G. Double bond content of phospholipids and lipid peroxidation negatively correlate with maximum longevity in the heart of mammals. Mech Ageing Dev 2000;112(3):169–83.

[271] Portero-Otin M, Bellmunt MJ, Ruiz MC, Barja G, Pamplona R. Correlation of fatty acid unsaturation of the major liver mitochondrial phospholipid classes in mammals to their maximum life span potential. Lipids 2001;36(5):491–8.

[272] Hulbert AJ. Explaining longevity of different animals: is membrane fatty acid composition the missing link? Age (Dordr) 2008;30(2–3):89–97.

[273] Pamplona R, Prat J, Cadenas S, et al. Low fatty acid unsaturation protects against lipid peroxidation in liver mitochondria from long-lived species: the pigeon and human case. Mech Ageing Dev 1996;86(1):53–66.

[274] Skulachev VP. Mitochondria, reactive oxygen species and longevity: some lessons from the Barja group. Aging Cell 2004;3 (1):17–19.

[275] Bauman DE, Perfield JW II, deVeth MJ, Lock AL. New perspectives on lipid digestion and metabolism in ruminants. In: Proc Cornell nutr conf, Cornell University; 2003. p. 175–89. <http://dairynutrition.msu.edu/animal_nutrition/publications>.

[276] Jove M, Naudi A, Aledo JC, et al. Plasma long-chain free fatty acids predict mammalian longevity. Sci Rep 2013;3:3346.

[277] Haddad LS, Kelbert L, Hulbert AJ. Extended longevity of queen honey bees compared to workers is associated with peroxidation-resistant membranes. Exp Gerontol 2007.

[278] Hulbert AJ, Faulks SC, Harper JM, Miller RA, Buffenstein R. Extended longevity of wild-derived mice is associated with peroxidation-resistant membranes. Mech Ageing Dev 2006;127 (8):653–7.

[279] Mitchell TW, Buffenstein R, Hulbert AJ. Membrane phospholipid composition may contribute to exceptional longevity of the naked mole-rat (Heterocephalus glaber): a comparative study using shotgun lipidomics. Exp Gerontol 2007;42(11):1053–62.

[280] Puca AA, Andrew P, Novelli V, et al. Fatty acid profile of erythrocyte membranes as possible biomarker of longevity. Rejuvenation Res 2008;11(1):63–72.

[281] Surh J, Kwon H. Estimation of daily exposure to 4-hydroxy-2-alkenals in Korean foods containing n-3 and n-6 polyunsaturated fatty acids. Food Addit Contam 2005;22(8):701–8.

[282] Catala A. Five decades with polyunsaturated fatty acids: chemical synthesis, enzymatic formation, lipid peroxidation and its biological effects. J Lipids 2013;2013:710290.

[283] Demidov VV. Heavy isotopes to avert ageing? Trends Biotechnol 2007;25(9):371–5.

[284] Kaneko T, Kaji K, Matsuo M. Protective effect of lipophilic derivatives of ascorbic acid on lipid peroxide-induced endothelial injury. Arch Biochem Biophys 1993;304(1):176–80.

[285] Kelso GF, Porteous CM, Hughes G, et al. Prevention of mitochondrial oxidative damage using targeted antioxidants. Ann NY Acad Sci 2002;959:263–74.

[286] Smith RA, Adlam VJ, Blaikie FH, et al. Mitochondria-targeted antioxidants in the treatment of disease. Ann N Y Acad Sci 2008;1147:105–11.

[287] Doughan AK, Dikalov SI. Mitochondrial redox cycling of mitoquinone leads to superoxide production and cellular apoptosis. Antioxid Redox Signal 2007;9(11):1825–36.

[288] Skulachev VP. A biochemical approach to the problem of aging: "megaproject" on membrane-penetrating ions. The first results and prospects. Biochemistry (Mosc) 2007;72(12): 1385–96.

[289] Anisimov VN, Bakeeva LE, Egormin PA, et al. Mitochondria-targeted plastoquinone derivatives as tools to interrupt execution of the aging program. 5. SkQ1 prolongs lifespan and prevents development of traits of senescence. Biochemistry (Mosc) 2008;73(12):1329–42.

[290] Abrams HL. Anthropological research reveals human dietary requirements for optimal health. J Appl Nutr 1982;16(1): 38–45.

[291] Colman RJ, Beasley TM, Kemnitz JW, Johnson SC, Weindruch R, Anderson RM. Caloric restriction reduces age-related and all-cause mortality in rhesus monkeys. Nat Commun 2014; 5:3557.

[292] Shahrestani P, Mueller LD, Rose MR. Does aging stop? Curr Aging Sci 2009;2(1):3–11.

[293] Khrapko K, Turnbull D. Mitochondrial DNA mutations in aging. Prog Mol Biol Transl Sci 2014;127:29–62.

[294] Brewer GJ. Epigenetic oxidative redox shift (EORS) theory of aging unifies the free radical and insulin signaling theories. Exp Gerontol 2010;45(3):173–9.

[295] Santra S, Gilkerson RW, Davidson M, Schon EA. Ketogenic treatment reduces deleted mitochondrial DNAs in cultured human cells. Ann Neurol 2004;56(5):662–9.

[296] Papamandjaris AA, MacDougall DE, Jones PJ. Medium chain fatty acid metabolism and energy expenditure: obesity treatment implications. Life Sci 1998;62(14):1203–15.

[297] Nafar F, Mearow KM. Coconut oil attenuates the effects of amyloid-beta on cortical neurons in vitro. J Alzheimer's Dis 2014;39(2):233–7.

[298] Doty L. Coconut oil for Alzheimer's disease? Clin Pract 2012; 1(2):12–17.

CHAPTER 13

Accumulation of Damage Due to Lifelong Exposure to Environmental Pollution as Dietary Target in Aging

Gabbianelli Rosita[1], Fedeli Donatella[1] and Nasuti Cinzia[2]

[1]Molecular Biology Unit, School of Pharmacy, University of Camerino, Camerino, Italy
[2]Pharmacology Unit, School of Pharmacy, University of Camerino, Camerino, Italy

KEY FACTS
- Toxicology
- Neurodegeneration
- Epigenetics
- Redox system

Dictionary of Terms

- **Toxicology**: is a branch of pharmacology concerned with the study of the adverse effects of chemicals on living organisms. It studies the relationship between dose of chemicals and their effects on the exposed organisms.
- **Neurodegeneration**: a process of progressive loss of function of neurons, including death of neurons.
- **Epigenetics**: study the effect of environmental factors on genome leading to modulation of gene expression. DNA methylation and histone modification represent the two main epigenetic mechanisms that can alter gene expression without altering the nucleotide sequence. Epigenetics modification may or not be hereditable leading to transgenerational effects.
- **Redox system**: System of redox couples of macromolecules such as glutathione, thioredoxin, etc, participating in a reversible chemical reaction in which one reaction is an oxidation and the reverse is a reduction. It participates in cell signaling and modulation of cell function.

INTRODUCTION

Exposure to environmental pollution represents an important risk factor associated with homeostatic unbalance leading to the development of various diseases. Consumption of fruit, vegetables, and cereals, inhalation of suspended particles, and dermal uptake are widely reported to be the means of absorption. Moreover, the identification of xenobiotic metabolites in the urine of the adult and child population represents a useful biomarker of exposure [1]. Several studies also show that low levels of pesticides may negatively affect the brain inducing the loss of neurons that leads to cognitive decline, impaired memory and attention, and motor dysfunction [2,3].

Lifestyle, diet, pesticides, metals, and solvents are potential risk factors of epigenetic changes that affect the development of neurodegeneration and other diseases of adult age. DNA methylation and posttranslational histone modifications are the major epigenetic changes, frequently involved in the modulation of gene expression profiles. The prenatal and postnatal lifetime represents a window of great plasticity for epigenetic modifications because epigenetics is not only the main actor in the process of cellular differentiation but also it can be inherited, having a strong impact on the phenotype of the offspring generations.

In this context, early exposure to environmental pollutants can have long-term effects on health and it can induce transgenerational consequences on the progenies.

The identification of early biomarkers of toxicity and mechanisms associated with the toxicant exposure represents a key approach to manage the effects of pollution on health.

This chapter describes the main environmental pesticides and toxicants and their long-term effects on health. Moreover, the biomarkers of damage that are useful to predict signs of developmental diseases are presented.

XENOBIOTICS

Xenobiotic exposure may take place through food or air, via solids and liquids that accidentally or deliberately come into contact with the skin, and through drinking water.

Apart from nutrients, food contains numerous organic and inorganic toxic compounds which occur naturally in food, but in such minute quantities that they do not cause any adverse effects. Some edible plants contain highly toxic substances which, in the course of normal preparation, undergo degradation or disappear. For example, certain poisonous proteins present in various types of beans, are denatured by cooking. Other substances, such as deoxynivalenol (DON), nivalenol, and T-2 toxin, which may be hazardous to humans, are synthesized by species of fungi, mainly by *Fusarium* mycotoxins that are abundant in cereals (wheat, barley, and maize) and their products. DON is probably the best known and most common contaminant of grains and its occurrence in food represents more than 90% of the total number of samples. Its molecule contains three free hydroxyl groups, which are associated with its toxicity. It affects animal and human health causing acute temporary nausea, vomiting, diarrhea, abdominal pain, and fever. Pathophysiologic effects associated with DON include altered neuroendocrine signaling, pro-inflammatory gene induction, disruption of the growth hormone axis, and altered gut integrity [4].

Foodstuffs also contain a variety of nonnatural and undesirable substances. These may be contaminants such as pesticide residues, heavy metals taken up from polluted soil, residues of veterinary drugs in meat, etc. Pollution of the soil can lead to the occurrence of undesirable substances in food stuffs. In this respect cadmium, mercury, and PCBs are notorious examples.

One route of exposure is the consumption of crops grown in polluted soil. An indirect route of exposure is through eating beef or drinking milk from cows that have grazed on polluted grasslands.

The abundant use of nitrogen fertilizer can lead to excessive nitrate levels in green vegetables. Under certain physiological conditions (ie, low pH condition of saliva), nitrate can be converted to the much more poisonous nitrite and ultimately, in the gut by reacting with proteins, can be converted into more carcinogenic nitrosamine.

Some substances are formed during food processing procedures such as acrylamide during food cooking. The main formation mechanism of acrylamide occurs upon food heating when amino acids interact with sugars in the presence of heat. One noteworthy example of acrylamide formation involves the conventional production of potato chips. During the frying process, fats used for frying can be oxidized and can become converted into acrolein and acrylic acid. Starches in the potato can also be broken down into sugars. This unique mixture of substances can interact in a way that results in unusual amounts of acrylamide formation. Humans metabolize acrylamide to a chemically reactive epoxide, glycidamide, responsible for genotoxic and neurotoxic effects [5].

The use of pesticides worldwide has grown significantly since the 1950s, with approximately 2.3 million tons of industrial pesticides now used annually. With this boom has come larger crop yields and more secure and reliable food sources for growing populations. Many low- and middle-income countries have even been able to produce enough food to sustain a large agricultural export economy, as well as to better feed their own population. These successes have led to an increase of pesticide use. However, the mass application of these agrochemicals has impacted human health and has been linked to a wide range of human health hazards, ranging from short-term impacts, such as headaches and nausea, to chronic impacts like impaired brain development, cancer, reproductive harm, endocrine disruption, cardiovascular problems, reduced immune response, neurodegenerative diseases, etc. [6]. Common pathways for human exposure include inhalation when pesticides are applied through spraying, ingestion of contaminated foods, especially cereals, vegetables, and fruits, and contamination of surface or groundwater and subsequent ingestion.

Children are particularly susceptible to the hazards associated with pesticide use. There is now considerable scientific evidence that the human brain is not fully formed until the age of 12, and childhood exposure to some of the most common pesticides on the market may greatly impact the development of the central nervous system. Children have more skin surface for their size than adults, and absorb proportionally greater amounts of many substances through their lungs and intestinal tracts. Children have not developed their immune system, nervous system, or detoxifying mechanisms completely, and, consequently, they are less capable of fighting the introduction of toxic pesticides into their systems. Many of the activities that children engage in, such as playing in the grass, putting objects into their mouth, and even playing on carpet, increase their exposure to toxic pesticides. Because of the combination of likely increased exposure to pesticides and lack of bodily development to combat the toxic effects of them, children suffer disproportionately from their impacts. For example, a study conducted at the University of California found a six-fold increase in risk factor for autism spectrum disorders for children of women who were exposed to organochlorine pesticides [7].

PYRETHROIDS

Today, pyrethroids are among the most frequently used pesticides because they often replace home and agricultural use of certain restricted or banned insecticides, such as organophosphates [8]. Pyrethroids are selected as replacement pesticides because they are potent insecticides with relatively low mammalian toxicity. Low mammalian toxicity, relative to insects, is attributed to increased levels of detoxifying enzymes in mammals, higher body temperature, and an inherently lower sensitivity of the analogous mammalian ion channel target sites [9]. Additionally, pyrethroids are relatively nonvolatile, thus leading to speculation that inhalation exposure after use for residential pest control would be minimal. The main commercially available pyrethroids include allethrin, bifenthrin, cyfluthrin, lambda cyhalothrin, cypermethrin (CY), deltamethrin, permethrin (PERM), d-phenothrin, resmethrin, and tetramethrin.

Pyrethroids are a class of synthetic insecticides and their chemical structure is based on naturally occurring pyrethrins, which are found in the flowers of *Chrysanthemum cineraraefolium*. The basic pyrethroid structure consists of an acid and an alcohol moiety, with an ester bond. Changes have been progressively introduced to increase their insecticidal potency and decrease their sensitivity to air and light. Most pyrethroids are chiral molecules and exist as mixtures of enantiomers.

In mammals and insects, the primary target of pyrethroids is the nervous system [10]. They are generally divided into two groups depending upon their acute neurobehavioral effects in rodents and the absence (Type I) or presence (Type II) of a cyano group at the carbon of the alcohol moiety. Regardless of the group, they primarily act by disrupting voltage-gated sodium channel (VGSC) function in the axons. Radioligand binding experiments revealed that the pyrethroid binding site is intrinsic to the sodium channel α-subunit [11]. The generally lower sensitivity of the mammalian sodium channel, compared to insect channels, contributes to the favorable selective toxicity of pyrethroids for insects. Nevertheless, binding studies, analyzing the sensitivity of different sodium channel isoforms to pyrethroids, showed a higher pyrethroid affinity for the isoforms Na_v 1.8 and 1.3 expressed respectively in cardiac muscle and brain only during development of mammals, [12]. These data suggest that heart and brain during the developmental phase could be more susceptible to pyrethroid effects in mammals.

With regard to the mechanism of action, pyrethroids slow the rate of VGSC closing, prolong the inward sodium conductance, and then shift the membrane to more polarized potentials [13]. A secondary consequence to cell membrane depolarization is an increased Ca^{+2} influx into the neurons through voltage-gated calcium channel (VGCC) that contributes to impact neuronal synaptic plasticity of neurons [14]. These changes in synaptic transmission may alter neuronal function and may contribute to excitotoxicity and neurodegenerative pathology [15].

In laboratory animals, the most common signs of acute oral poisoning by Type I compounds tend to be hyperexcitability and whole-body tremors, while Type II elicit choreoathetosis and salivation.

They are found in many formulations used in agriculture to control insect pests in crops, forestry, horticulture, and in gardens. They are also widely used as insecticides in indoor environments such as households, warehouses, and farm and public buildings. Pet shampoos, medication used for treating for scabies, and topical louse treatments also contain these compounds. Some pyrethroids, such as permethrin, can be impregnated into clothing and fabric (eg, carpets, blankets, uniforms) as arthropod contact repellents.

Environmental exposure among the general population mostly results from dietary intake and residential application of pyrethroids in gardens and homes for pest-control purposes. Consumption of fresh and cooked fruit and vegetables has been linked to higher levels of exposure, as assessed by biological monitoring and analysis of pesticide residues in raw and processed food products [16]. Low concentrations of pyrethroids have been measured in indoor-air settings, probably due to their low vapor pressure. Nevertheless, additional exposure via ingestion of contaminated household dust may occur after the indoor application of pesticides. Floor dust is one of the major source of exposure in infants and toddlers, with nondietary ingestion (eg, through hand-to-mouth contact) contributing substantially to intake doses [17]. Dermal uptake of pyrethroids can also occur during loading and mixing operations, treatment of pets, and via contact with contaminated work clothes or carpets and other textiles impregnated with pyrethroids for insect protection, such as battle dress uniforms, but the percentage of pyrethroid absorbed dermally is less than oral absorption [18].

The pyrethroids are rapidly metabolized in the liver of mammals by cytochrome P450 oxidation, or esterase hydrolysis followed by the formation of glucuronide and glycine conjugates. Their half-life is short (96 h) and they are mainly excreted in urine as sulfate and glucuronide conjugates. 3-phenoxybenzoic acid (3-PBA) is a metabolite of several commonly used pyrethroids, including permethrin and cypermethrin. This main metabolite has been used as a biomarker to monitor acute or short-term exposure in adults and children from the general population and several studies indicate that children may have higher exposures than adolescents and adults [19]. In addition, pyrethroids are highly lipophilic and by easily crossing the blood—brain barrier they can reach the CNS at concentrations that can be potentially neurotoxic.

The higher hazard of pesticides is not only from acute exposure to high doses but also from long-term exposure to low doses as we will explain in the next section.

ACUTE AND CHRONIC EXPOSURE

In humans, the most frequently reported symptoms after acute exposure are paresthesia of the eyes, face and breast, or symptoms from the respiratory tract, while systemic effects like dizziness, nausea, palpitations, headache, anxiety, hyperactivity, ataxia, salivation, tremor, choreoathetosis, clonic seizure, and confusion are reported after ingestion or inhalation exposures [20]. However, the reports are descriptions of single or a few cases and do not reveal any information about the dose response relationship or the actual frequency of the different health effects.

Less is known about the long-term health effects of repeated exposure to low levels of pyrethroids. Sensory impairments such as irritation and reddening of the skin were reported by soldiers wearing permethrin-impregnated uniforms [21]. Slight changes to a small number of immune response mediators have been observed in workers exposed to pyrethroids [22]. Several epidemiological studies demonstrated the environmental and occupational exposure to pyrethroids by measuring pyrethroid metabolites in urine, and also associated it with altered semen quality, including sperm DNA damage (eg, aneuploidy) and decreased sperm concentration [23].

Several studies investigated the effects of pyrethroid insecticide exposure during pregnancy. These essentially concern growth and neurobehavioral development and hormonal balance in infants. Some of these studies suggest that the use of pyrethroids during the first or second trimester in women with farming jobs was associated with a small, but significant, decrease in birth weight [24].

No clear conclusion can be drawn from the few epidemiological studies that specifically address the potential carcinogenicity of pyrethroids after childhood or adult exposure. A large-scale prospective study of cancer incidence among permethrin-exposed-pesticide applicators found no evidence for increased risk of the different types of cancer that were assessed (melanoma, leukemia, non-Hodgkin lymphoma, or cancer of the colon, rectum, lung, or prostate) [25]. A recent study has suggested that urinary levels of pyrethroid metabolites may be associated with an elevated risk of childhood acute lymphocytic leukemia [26].

EARLY LIFE EXPOSURE AND LONG-TERM EFFECTS

Prenatal and postnatal life represents a key window of plasticity important for the programming of development through the cellular differentiation [27]. During this period exogenous and endogenous factors (ie, nutrition, xenobiotics, stress, hypoxia, infections, hormones, etc.) can induce epigenetic changes leading to the development of diseases in adult age. Transgenerational studies on animal models show that epigenetic changes, such as DNA methylation and histone modifications, may be responsible for transgenerational effects. DNA methylation of CpG islands in the promoter region of genes leads to downregulation of their expression (Fig. 13.1A), while either a down- or upregulation can occur when the methylation is on the regulatory regions of DNA (Fig. 13.1B). Besides, posttranslational modifications, such as acetylation, methylation, phosphorylation, ubiquitination, and sumoyltation of histone proteins (H2A, H2B, H3, and H4), modify their interaction with DNA and finally the chromatin condensation status, which also regulates gene expression. In particular, acetylation of lysine residues by histone acetyltransferase complexes (HCT), increases gene expression decreasing the electrostatic interactions between DNA and lysine. On the other hand, histone deacetylase (HDAC), removing the acetyl group from lysine, shifts the chromatin toward a more condensed status, where gene is off (Fig. 13.2).

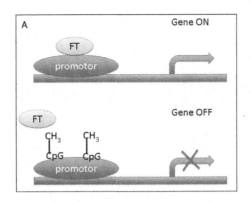
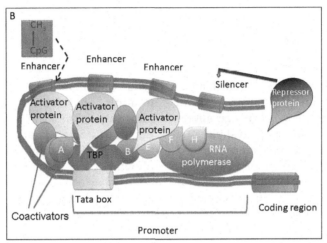

FIGURE 13.1 Effect of DNA methylation, in the promoter (A) and in the regulatory sequences (B) on gene expression.

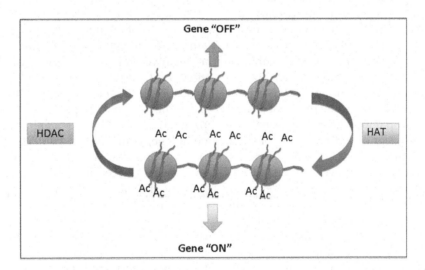

FIGURE 13.2 Histone acetylation by HAT modifying the interactions between DNA and lysine positively modulate gene expression.

These epimutations, are tissue specific and may be reversible according to environmental factor exposure during life (ie, lifestyle, xenobiotics, stress, etc.) [28].

Epidemiological studies support the concept of *developmental programming* or *"Developmental Origins of Health and Diseases"* (DOHaD), suggesting that the more common diseases of adult age (ie, metabolic syndrome, neurodegeneration, cancer, type-2 diabetes) are correlated with early exposure to environmental factors during pregnancy and nursing [29,30]. The role

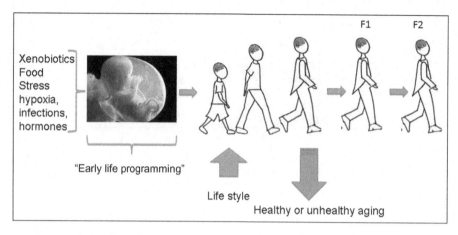

FIGURE 13.3 Impact of environmental factors on developmental programming during prenatal and postnatal life.

of the developmental programming has been demonstrated in both animals and humans. Offspring from undernourished mothers showed in adult age alterations of the glucose metabolism, and dyslipidemia [30,31]. Moreover, offspring (F1) from undernourished women (F0) during the first trimester of pregnancy, developed diabetes and cardiovascular diseases at adult age [32], while the F2 generation developed a less unfavorable metabolic syndrome. Veenendaal et al. report the same transgenerational effects on adult offspring (F2) of prenatally undernourished fathers (F1) [33] (Fig. 13.3).

Lifestyle, diet, pesticides, metals, and solvents are potential risk factors of epigenetic changes that affect the development of neurodegeneration. Several studies describe the neonatal origin of neurodegeneration focusing on the role of pesticide exposure during pre- and postnatal age [34]. Rats treated with low dose of permethrin pesticide for 15 days from the 6th to 21st days of life develop in adult age a Parkinson's like disease, characterized by a decrease of Nurr1 gene expression and dopamine level, an imbalance of the redox system with significant reduction of GSH, and a deficit of spatial working memory [35–37]. As well as an increase of calcium level in the striatum, and increase in the prefrontal cortex and cerebellum was measured, while calcium was decreased in the hippocampus [35]. Furthermore, the early exposure to the permethrin pyrethroid leads to increased free and aggregated alpha synuclein and activation of DNMTs in striatum [38]. Cardiotoxicity following the same early life treatment was also reported at different ages [39]. On the other hand, subchronic permethrin exposure (60 days) in adolescent rats and in early life age induces an important imbalance in the redox system [40]. However, the subchronic treatment seems to promote the oxidative processes more than the treatment in early life. Also early exposure to dieldrin leads to alterations in the dopaminergic system in adult age [41]. Learning deficit was measured following deltametrin exposure due to endoplasmic reticulum stress and apoptotic cell death in the hippocampus [42]. Neuroinflammation was observed in rat following early life paraquat exposure [43]. Permanent loss of olfactory receptor neurons, due to nasal inflammation and respiratory metaplasia, was observed in mice following herbicide 2,6-dichlorobenzonitrile exposure [44].

Studies on children's hair show different level of propoxur and pyrethroids at 2, 4, and 6 years of age, and at 2 years of age, prenatal exposure to propoxur was linked with lesser motor development in children [45,46].

A large-scale study in Norwegian mothers and children (MoBa) shows that prenatal exposure to acrylamide or to polycyclic aromatic hydrocarbons has a negative impact on fetal growth [47,48]. Similarly effect on fetal growth was observed following prenatal exposure to dioxins and PCBs [49].

Studies on two large cohorts of people show that prenatal and early postnatal exposure to tetrachloroethylene, contained in drinking water, improves the risk of epilepsy and certain types of cancer, such as cervical cancer in adult age, and it increases the risk of bipolar disorder and posttraumatic stress disorder [50].

Moreover early exposure to other environmental compounds like metals has been associated with neuronal disorders. Lead exposure has been correlated to central and peripheral damage of the nervous system with myelin loss, cognitive dysfunction, deficit in school performance in children, and DNA damage [51–53]. A large-scale study on Swedish children underlines the neurotoxic effect of lead exposure and its long-term effects: a significant correlation between low blood lead at the age 7–12 and high school performance at the age of 16 and 18–19 in girls and boys was measured [54]. On the other hand, male workers with a history of chronic lead exposure present slowed vegetative nervous system according with individual blood lead level [55]. Lithium exposure via drinking water during early gestation, was associated with impaired fetal size of child [56].

Women living in rural Bangladesh areas having high levels of urinary cadmium produce daughters with decreased head circumference and birth weight [57]. Moreover prenatal exposure to cadmium and arsenic was associated with an impairment in cognitive development and decreased fetal size in females at 5 years old [58,59].

Alterations of cognitive function were measured at 6 months of age in children whose mothers were exposed to lead and cadmium during pregnancy [60].

Changes in the epigenome, in absence of direct exposure, occur through transgenerational inheritance of germline, in which the epigenome undergoes reprogramming during fetal gonadal development. After fertilization, the male germline spread the epigenome changes to somatic cells modulating the health status at adult age. Via this method, epigenetic transgenerational action of various environmental compounds has been studied. Different pesticide mixtures (permethrin, DEET, plastic mixture, dioxin, and hydrocarbons) are able to induce various epigenetic effects in the F1−F3 generations [61]. Early life exposure to the endocrine disruptor vinclozolin, a dicarboximide with fungicide activity, can influence embryonic testis development, improve spermatogenic cell apoptosis at adult age, and this effect on spermatogenesis can be carried on across multiple generations [62]. Although this effect was observed at a dose level higher than NOAL (No Observed Adverse Effect Level) (100 mg/kg/day), it should be considered the unknown role of multiple environmental toxicants [63].

Sperm epimutations were measured in F1, F3, and F4 generations born from female rats exposed to the methoxychlor pesticide during fetal gonadal development (from 8th to 14th gestation days) [64]. Dioxin, which is able to induce liver disease, weight loss, thymic atrophy, and immune suppression in animals, can induce transgenerational effects in F3 generation if animals are exposed to dioxin during gonadal sex determination [65].

A transgenerational effect on fat deposits was measured in the F3 generation of mice following a normal diet and the exposure to the antifouling trybutiltin during pregnancy [66]. Transgenerational effects of heavy metals were measured in rice plants through epigenetic modifications via both maternal and paternal germline [67].

Finally, the early environmental exposure to pesticides and metals can induce long-term effects either following epigenetic modifications or a different detoxifying ability as revealed by the higher metabolites accumulated in the biological fluids of children versus adults. However, it should be underlined that the epigenetic modifications occur easily during early life, but they can be modulated by lifestyle and, in the complexity of healthy and unhealthy status, a key role is related to the type of diet and to the frequency of physical activity during the lifespan.

BIOMARKERS OF DAMAGE

In order to better understand the role of environmental exposures in human disease, accurate and relevant exposure assessment remains a critical and complex component. It is important to note that to accurately measure a toxin (or metabolite) in a biomatrix is not sufficient to correctly assess the exposure to it, but a verified relationship between the intake of the toxin and its "biomeasure" is also required. This aspect is exemplified by the studies on aflatoxin. AFB1 is the most frequently occurring toxin of this family, and it is partially transferred as an unmetabolized molecule to urine; nevertheless, the concentration of AFB1 itself in urine does not reflect the quantity of toxin ingested [68]. On the contrary, a strong dose-response relationship has been demonstrated between aflatoxin exposure and its monohydroxil metabolite AFM1 in breast milk, indicating a higher risk of infant exposures to it. The metabolite aflatoxin-N7-guanine (AF-N7-Gua) in urine has been correlated with aflatoxin intake in chronically exposed individuals and has been defined as an aflatoxin exposure biomarker. Instead, the concentration of aflatoxin-albumin adduct (AF-albumin) in the blood has been strongly correlated with aflatoxin intake [69].

Hemoglobin adducts of acrylamide (AA) and epoxide glycidamide (GA), its main metabolite, quantified in the blood represent good biomarkers, reflecting the average exposure to acrylamide through food intake [70]. In fact, their plasma levels are well correlated to dietary intake of acrylamide, estimated from food frequency questionnaires [71]. Moreover, the ratio of GA-Hb to AA-Hb seems to contribute to the overall risk of individuals for acrylamide-associated disease such as neurotoxicity and cancer. In addition, urine biomarkers, such as mercapturic acids of acrylamide, can be used for validation of dietary intake estimation and ranking of persons with regard to acrylamide exposure.

Levels of arsenic or its methylated metabolites in blood, hair, nails, and urine are used as biomarkers of arsenic exposure in humans and most common laboratory animals. Speciated metabolites in urine, expressed either as inorganic arsenic or as the sum of metabolites, provide the best quantitative estimate of recently absorbed doses of arsenic. Blood arsenic and its metabolites are useful biomarkers only in the case of acute arsenic poisoning or stable chronic high-level exposure. Arsenic is rapidly cleared from blood, and speciation of its chemical forms in blood is difficult. Levels of arsenic in hair and nails can be indicators of past arsenic exposure, provided care is taken to prevent external arsenic contamination of the samples. Arsenic in hair may also be used to estimate relative length of time since an acute exposure [72].

The lifelong exposure to pesticides introduced mainly through the food, has as a consequence the impairment of redox, nervous and immune systems, the damage of proteins, DNA, and other cell compartments. Both the accumulation of damage due to pollutant exposure and aging cause the acceleration of the onset of many pathologies. Considering that the exposure to risk factors occurred years or decades before the diagnosis, the assessment of chronic exposures is difficult to perform in retrospective studies and to associate them with the onset/development of the disease. In this context the identification of early biomarkers associated with the onset of long-term diseases, could represent a useful approach to counterbalance, when possible, the progression of the disease.

Modifications induced by xenobiotics can be assessed by analyzing specific parameters related to redox or immune systems, DNA, neuronal proteins, microbiota, or by identifying analytes present in plasma or urine useful as disease biomarkers.

The environmental contaminants introduced through the food chain, can impair the immune system and activate inflammatory cytokines that are suggested to play a role in human diseases including depression, cardiovascular disease, type-2 diabetes, autoimmune diseases, and neurodegeneration [73–75].

It has been reported [39] that rats treated in early life with permethrin showed in adult phase an increase in the pro-inflammatory cytokines IL-1β, IL-2, IFN-γ, and rat-Rantes that were absent in the oldest rats showing an evident decline in the immune system with the aging. Other studies confirm that pro-inflammatory cytokines increased significantly with exposure to higher level of various pollutants [76] in subjects with coronary artery disease and they are associated with adverse cardiac-related health outcomes [77].

Another important target of xenobiotics introduced through the diet is the redox system. The antioxidant enzyme system represents the primary response of the body to counteract oxidative stress induced by several environmental stimuli (ie, pesticides, metals, and radiations). Redox status has been proven to be an important tool in toxicological evaluation, mostly providing cellular and biochemical mechanism of toxicity for chemicals and drugs.

Oxidative stress reduces the glutathione (GSH) pool and decreases the GSH/GSSG ratio [78]. ROS-mediated toxicity by oxidation of GSH to GSSG in response to pesticide exposure has been demonstrated in many workers [79–82]. Decreased GSH level in plasma [83] of rats treated sub-chronically with permethrin, in genetic modified animal models of Parkinson's disease, and in aged astrocytes cultured from Parkin knockout mice were measured [84].

Many studies report that superoxide dismutase, catalase, and glutathione peroxidase, which represent the main endogenous antioxidant enzymes, can both (1) increase, trying to counterbalance the overproduction of ROS induced by pollutants, or (2) decrease, following prolonged exposure [85]. Higher superoxide anion and hydrogen peroxide production together with an imbalance in the erythrocyte redox system were consequent to polymorphonuclear respiratory burst activation [86–88].

An important biomarker for oxidative stress induced by pollutant exposure is lipid peroxidation, which can be evaluated by measuring malondialdehyde (MDA). The latter is considered as the most significant marker of membrane lipid peroxidation, being one of the major final products of lipid peroxidation, which is produced by free radical-mediated degradative process. It has been reported that MDA increases in the liver and kidney tissue following treatment with glyphosate, an insecticide frequently used in agriculture [89].

Three important transcription factors, Nurr1, NF-κB, and Nrf2, are involved in the regulation of genes modulating the pro-inflammatory response, redox, and immune systems. Nrf2 is an important indicator of cellular oxidative stress because is an oxidant-responsive transcription factor involved in the induction of antioxidant genes and it is a key regulator of the cellular adaptive response to oxidative stress.

Nurr1 plays an essential role in physiological processes (ie, development of dopaminergic neurons, apoptosis in lymphocytes and other cell types) and it is highly inducible in macrophages by inflammatory cytokines and oxidized lipids such as oxidized LDL [90].

Moreover, its gene expression is increased by enhanced NF-κB and CREB-1-binding activity on the gene promoter and also by stimulation of pro-inflammatory cytokines (ie, IL-1β and TNFα) or PGE2 [91]. It represents a specific marker for Parkinson Disease as demonstrated by plasma levels changes measured in lymphocytes of parkinsonian patients.

A recent study on HepG2 cells treated with organic extract contaminants from drinking water reports the increase in Nrf2 protein expression [92] demonstrating the clear response to chemical-induced activation. NF-κB p65, Nurr1, and Nrf2 protein expressions have been studied in leukocytes of rat exposed to permethrin in early life [83]. Leukocytes are important blood components, which represent indicative indices of alterations that have occurred in other tissues, because of their sensitivity to the environmental contaminants able to influence their gene expression during immune differentiation. In the latter study, Nurr1 protein expression increased in leukocytes from treated rats while NF-κB p65, and Nrf2 protein expression increased with the aging but not for the treatment with permethrin; this could be due to the chronic activation of NF-κB that has the capacity to induce the senescent phenotype by increasing inflammatory and homeostatic responses such as apoptosis, autophagy, and tissue atrophy.

The overproduction of reactive oxygen species due to pollutant exposure induces oxidative damage to DNA that is known to be one of the most important mechanisms for the pathogenesis of cancer, cardiovascular, and other chronic diseases [93]. Reactive oxygen species attack DNA, generating intermediates, which can react with DNA and form adducts, including single-strand breaks and the formation of modified bases such as 8-hydroxydeoxyguanosine (8-OHdG) [94]. Thus, the measurement of 8-OHdG may be useful as a marker of oxidative DNA damage. A recent study conducted in Iran has compared the urinary concentration of 8-OHdG between workers in a subway system underground exposed to air pollution and those who work outside the tunnel [95]; the result was that exposed workers presented higher 8-OHdG levels than those measured in the control group. Similar results have been obtained in people exposed to other pollutants such as ethylbenzene [96].

Environmental pollutants represent a risk factors for the onset of neurodegenerative disorders; [97] there are many evidences indicating that the brain responds to diverse inhaled air pollutants, through a common pathway of neuroinflammation [98]. For example, the accumulation of α-synuclein, an highly conserved presynaptic protein, in the brain of PD patients represents a typical response to a inflammatory status [99]. Alpha-synuclein neuronal aggregation and accumulation of 3 nitro-tyrosine and 8-hydroxydeoxyguanosine (8-OHdG), evidence of oxidative stress, were detected in brainstem nuclei of children exposed to environmental pollution in Mexico City [73].

Catecholamines are pivotal signal molecules in the communication between the neuroendocrine system and immune system and serve as neurotransmitters and hormones in the catecholaminergic neuroendocrine system, and immunomodulators in the immune system. Plasma of old rats treated in early life with permethrin presents lower adrenaline and noradrenaline levels with respect to the controls and it could be related to poor activation of the hypothalamic–pituitary–adrenal axis due to the immune system impairment measured in the same animal model [100].

Calcium is another important parameter related to aging processes; its limited intestinal absorption is in fact the cause of osteoporosis. The main hormonal regulator of intestinal calcium absorption is the 1,25-dihydroxyvitamin D3 or calcitriol, whose action is mediated by an intracellular vitamin D receptor protein that regulates the transcription of these vitamin D-responsive genes. Intestinal calcium malabsorption, could be due to a primary hormone deficiency because of lower circulating levels of 1,25-dihydroxyvitamin D3 or to a reduced responsiveness of the intestine to 1,25-dihydroxyvitamin D3. Lower level of calcium in plasma, associated with higher levels of 25-hydroxy-Vit D, were measured in treated old rats. The plasma level of 25-hydroxy-Vit D was higher in treated rats, most probably due to the compensatory stimulation of the parathyroid hormone directed to restore the negative calcium balance through the increased synthesis of vitamin D at the level of kidney [100]. In addition, plasma sodium levels observed in old treated subjects were higher than the controls, and this could be explained by considering the decreased kidney's ability to excrete sodium. Given that the capacity of the kidney to excrete sodium decreases with age, the changes observed for plasma calcium and sodium concentration in exposed rats confirm that permethrin treatment accelerates the aging process.

In conclusion, exposure to xenobiotics impairs various cellular targets that can be used as biomarkers for the screening of correlated diseases.

SUMMARY

- Xenobiotics overviews.
- Acute and chronic toxicity due to xenobiotics.
- Long-term consequences to early-life exposure to toxicants.
- Transgenerational effects of toxicants.
- Biomarkers related to diseases induced by environmental xenobiotics.

References

[1] Saillenfait AM, Ndiaye D, Sabaté JP. Pyrethroids: exposure and health effects – an update. Int J Hyg Environ Health 2015;15: S1438–4639.

[2] Hu R, Huang X, Huang J, Li Y, Zhang C, Yin Y, et al. Long- and short-term health effects of pesticide exposure: a cohort study from China. PLoS One 2015;10(6):e0128766. Available from: http://dx.doi.org/10.1371/journal.pone.0128766.

[3] Parrón T, Requena M, Hernández AF, Alarcón R. Association between environmental exposure to pesticides and neurodegenerative diseases. Toxicol Appl Pharmacol 2011;256:379–85.

[4] Sobrova P, Adam V, Vasatkova A, Beklova M, Zeman L, Kizek R. Deoxynivalenol and its toxicity. Interdiscip Toxicol 2010;3:94–9.

[5] Joint FAO/WHO Expert Committee on Food Additives (JECFA). Health Implications of Acrylamide in Food. World Health Organization; 2002.

[6] Mostafalou S, Abdollahi M. Pesticides and human chronic diseases: evidences, mechanisms, and perspectives. Toxicol Appl Pharmacol 2013;268:157–77.

[7] Shelton JF, Hertz-Picciotto I, Pessah IN. Tipping the balance of autism risk: potential mechanisms linking pesticides and autism. Environ Health Perspect 2012;120:944–51.

[8] Williams MK, Rundle A, Holmes D, Reyes M, Hoepner LA, Barr DB, et al. Changes in pest infestation levels, self-reported pesticide use, and permethrin exposure during pregnancy after the 2000–2001 U.S. Environ Health Perspect 2008;116:1681–8.

[9] Ray DE, Fry JR. A reassessment of the neurotoxicity of pyrethroid insecticides. Pharmacol Ther 2006;111:174–93.

[10] Wolansky MJ, Tornero-Velez R. Critical consideration of the multiplicity of experimental and organismic determinants of pyrethroid neurotoxicity: a proof of concept. J Toxicol Environ Health B Crit Rev 2013;16:453–90.

[11] Smith TJ, Soderlund DM. Action of the pyrethroid insecticide cypermethrin on rat brain IIa sodium channels expressed in Xenopus oocytes. Neurotoxicology 1998;19:823–32.

[12] Meacham CA, Brodfuehrer PD, Watkins JA, Shafer TJ. Developmentally-regulated sodium channel subunits are differentially sensitive to alpha-cyano containing pyrethroids. Toxicol Appl Pharmacol 2008;231:273–81.

[13] Nasuti C, Fattoretti P, Carloni M, Fedeli D, Ubaldi M, Ciccocioppo R, et al. Neonatal exposure to permethrin pesticide causes lifelong fear and spatial learning deficits and alters hippocampal morphology of synapses. J Neurodev Disord 2014;6:7.

[14] Imamura L, Yasuda M, Kuramitsu K, Hara D, Tabuchi A, Tsuda M. Deltamethrin, a pyrethroid insecticide, is a potent inducer for the activity dependent gene expression of brain-derived neurotrophic factor in neurons. J Pharmacol Exp Therm 2006;316:136–43.

[15] Gomez-Villafuertes R, Mellström B, Naranjo JR. Searching for a role of NCX/NCKX exchangers in neurodegeneration. Mol Neurobiol 2007;35:195–202.

[16] Fortes C, Mastroeni S, Pilla MA, Antonelli G, Lunghini L, Aprea C. The relation between dietary habits and urinary levels of 3-phenoxybenzoic acid, a pyrethroid metabolite. Food Chem Toxicol 2013;52:91–6.

[17] Zartarian V, Xue J, Glen G, Smith L, Tulve N, Tornero-Velez R. Quantifying children's aggregate (dietary and residential) exposure and dose to permethrin: application and evaluation of EPA's probabilistic SHEDS-multimedia model. J Expo Sci Environ Epidemiol 2012;22:267–73.

[18] Bradman A, Whitaker D, Quirós L, Castorina R, Claus Henn B, Nishioka M, et al. Pesticides and their metabolites in the homes and urine of farmworker children living in the Salinas Valley, CA. J Expo Sci Environ Epidemiol 2007;17:331–49.

[19] Barr DB, Olsson AO, Wong LY, Udunka S, Baker SE, Whitehead RD, et al. Urinary concentrations of metabolites of pyrethroid insecticides in the general U.S. population: National Health and Nutrition Examination Survey 1999–2002. Environ Health Perspect 2010;118:742–8.

[20] Power LE, Sudakin DL. Pyrethrin and pyrethroid exposures in the United States: a longitudinal analysis of incidents reported to poison centers. J Med Toxicol 2007;3:94–9.

[21] Appel KE, Gundert-Remy U, Fischer H, Faulde M, Mross KG, Letzel S, et al. Risk assessment of Bundeswehr (German Federal Armed Forces) permethrin-impregnated battle dress uniforms (BDU). Int J Hyg Environ Health 2008;211:88–104.

[22] Costa C, Rapisarda V, Catania S, Di Nola C, Ledda C, Fenga C. Cytokine patterns in greenhouse workers occupationally exposed to α-cypermethrin: an observational study. Environ Toxicol Pharmacol 2013;36:796–800.

[23] Ji G, Xia Y, Gu A, Shi X, Long Y, Song L, et al. Effects of non-occupational environmental exposure to pyrethroids on semen quality and sperm DNA integrity in Chinese men. Reprod Toxicol 2011;31:171–6.

[24] Hanke W, Romitti P, Fuortes L, Sobala W, Milkulski M. The use of pesticides in a Polish rural population and its effect on birth weight. Int Arch Occup Environ Health 2003;76:614–20.

[25] Rusiecki J, Patel R, Koutros S, Beane-Freeman L, Landgren O, Bonner MR, et al. Cancer incidence among pesticide applicators exposed to permethrin in the Agricultural Health Study. Environ Health Perspect 2009;117:581–96.

[26] Ding G, Shi R, Gao Y, Zhang Y, Kamijima M, Sakai K, et al. Pyrethroid pesticide exposure and risk of childhood acute lymphocytic leukemia in Shanghai. Environ Sci Technol 2012;46:13480–7.

[27] Hochberg Z, Feil R, Constancia M, Fraga M, Junien C, Carel JC, et al. Child health, developmental plasticity, and epigenetic programming. Endocr Rev 2011;32:159–224.

[28] Gabory A, Attig L, Junien C. Epigenetic mechanisms involved in developmental nutritional programming. World J Diabetes 2011;2:164–75.

[29] McMillen IC, Robinson JS. Developmental origins of the metabolic syndrome: prediction, plasticity, and programming. Physiol Rev 2005;85:571–633.

[30] Reynolds LP, Borowicz PP, Caton JS, Vonnahme KA, Luther JS, Hammer CJ, et al. Developmental programming: the concept, large animal models, and the key role of uteroplacental vascular development. J Anim Sci 2010;88:E61–72.

[31] Barker DJ. Introduction: the window of opportunity. Symposium: novel concepts in the developmental origins of adult health and disease. J Nutr 2007;137:1058–9.

[32] Jiménez-Chillarón JC, Díaz R, Martínez D, Pentinat T, Ramón-Krauel M, Ribó S, et al. The role of nutrition on epigenetic modifications and their implications on health. Biochimie 2012;94:2242–63.

[33] Veenendaal MVE, Painter RC, de Rooij SR, Bossuyt PMM, van der Post JAM, Gluckman PD, et al. Transgenerational effects of prenatal exposure to the 1944–45 Dutch famine. BJOG 2013;120:548–53.

[34] Gapp K, Woldemichael BT, Bohacek J, Mansuy IM. Epigenetic regulation in neurodevelopment and neurodegenerative diseases. Neuroscience 2014;264:99–111.

[35] Carloni M, Nasuti C, Fedeli D, Montani M, Amici A, Vadhana MS, et al. The impact of early life permethrin exposure on development of neurodegeneration in adulthood. Exp Gerontol 2012;47:60–6.

[36] Carloni M, Nasuti C, Fedeli D, Montani M, Vadhana MS, Amici A, et al. Early life permethrin exposure induces long-term brain changes in Nurr1, NF-kB and Nrf-2. Brain Res 2013;1515:19–28.

[37] Nasuti C, Carloni M, Fedeli D, Gabbianelli R, Di Stefano A, Serafina CL, et al. Effects of early life permethrin exposure on spatial working memory and on monoamine levels in different brain areas of pre-senescent rats. Toxicology 2013;303:162–8.

[38] Vadhana MSD, Nasuti C, Carloni M, Fedeli D, Montani M, Amici A, et al. Epigenetic regulation of Nurr1 in striatum of rats exposed to permethrin insecticide. Neurodegenerative Dis 2013;11:287.

[39] Vadhana MSD, Carloni M, Nasuti C, Fedeli D, Gabbianelli R. Early life permethrin insecticide treatment as origin of heart damage in adult rats. Exp Gerontol 2011;46:731–8.

[40] Gabbianelli R, Palan M, Flis DJ, Fedeli D, Nasuti C, Skarydova L, et al. Imbalance in redox system of rat liver following permethrin treatment in adolescence and neonatal age. Xenobiotica 2013;43:1103–10.

[41] Richardson JR, Caudle WM, Wang M, Dean ED, Pennell KD, Miller GW. Developmental exposure to pesticide dieldrin alters the dopamine system and increases neurotoxicity in an animal model of Parkinson's disease. Faseb J 2006;20:1695–7.

[42] Hossain MM, DiCicco-Bloom E, Richardson JR. Hippocampal ER stress and learning deficits following repeated pyrethroid exposure. Toxicol Sci 2015;143:220–8.

[43] Sandström von Tobel J, Zoia D, Althaus J, Antinori P, Mermoud J, Pak HS, et al. Immediate and delayed effects of subchronic Paraquat exposure during an early differentiation stage in 3D-rat brain cell cultures. Toxicol Lett 2014;230:188–97.

[44] Xie F, Fang C, Schnittke N, Schwob JE, Ding X. Mechanisms of permanent loss of olfactory receptor neurons induced by the herbicide 2,6-dichlorobenzonitrile: effects on stem cells and noninvolvement of acute induction of the inflammatory cytokine IL-6. Toxicol Appl Pharmacol 2013;272:598–607.

[45] Ostrea Jr. EM, Reyes A, Villanueva-Uy E, Pacifico R, Benitez B, Ramos E, et al. Fetal exposure to propoxur and abnormal child neurodevelopment at 2 years of age. Neurotoxicology 2012;33:669–75.

[46] Ostrea Jr. EM, Villanueva-Uy E, Bielawski D, Birn S, Janisse JJ. Trends in long term exposure to propoxur and pyrethroids in young children in the Philippines. Environ Res 2014;131:13–16.

[47] Duarte-Salles T, von Stedingk H, Granum B, Gützkow KB, Rydberg P, Törnqvist M, et al. Dietary acrylamide intake during pregnancy and fetal growth—results from the Norwegian mother and child cohort study (MoBa). Environ Health Perspect 2013;121:374–9.

[48] Duarte-Salles T, Mendez MA, Meltzer HM, Alexander J, Haugen M. Dietary benzo(a)pyrene intake during pregnancy and birth weight: associations modified by vitamin C intakes in the Norwegian mother and child cohort study (MoBa). Environ Int 2013;60:217–23.

[49] Papadopoulou E, Caspersen IH, Kvalem HE, Knutsen HK, Duarte-Salles T, Alexander J, et al. Maternal dietary intake of dioxins and polychlorinated biphenyls and birth size in the Norwegian mother and child cohort study (MoBa). Environ Int 2013;60:209–16.

[50] Aschengrau A, Winter MR, Vieira VM, Webster TF, Janulewicz PA, Gallagher LG, et al. Long-term health effects of early life exposure to tetrachloroethylene (PCE)-contaminated drinking water: a retrospective cohort study. Environ Health 2015;12:14–36.

[51] Khanna MM. Boys, not girls, are negatively affected on cognitive tasks by lead exposure: a pilot study. J Environ Health 2015;77:72–7.

[52] Evens A, Hryhorczuk D, Lanphear BP, Rankin KM, Lewis DA, Forst L, et al. The impact of low-level lead toxicity on school performance among children in the Chicago Public Schools: a population-based retrospective cohort study. Environ Health 2015;14:21.

[53] Roy A, Queirolo E, Peregalli F, Mañay N, Martínez G, Kordas K. Association of blood lead levels with urinary 8α isoprostane and 8-hydroxy-2-deoxy-guanosine concentrations in first-grade Uruguayan children. Environ Res 2015;140:127–35.

[54] Skerfving S, Löfmark L, Lundh T, Mikoczy Z, Strömberg U. Late effects of low blood lead concentrations in children on school performance and cognitive. Neurotoxicology 2015;49:114–20. pii: S0161-813X(15)00085-6 http://dx.doi.org/10.1016/j.neuro.2015.05.009.

[55] Böckelmann I, Pfister E, Darius S. Early effects of long-term neurotoxic lead exposure in copper works employees. J Toxicol 2011;2011:11. ID 832519, http://dx.doi.org/10.1155/2011/832519:832519.

[56] Harari F, Langeén M, Casimiro E, Bottai M, Palm B, Nordqvist H, et al. Environmental exposure to lithium during pregnancy and fetal size: a longitudinal study in the Argentinean Andes. Environ Int 2015;77:48–54.

[57] Kippler M, Tofail F, Gardner R, Rahman A, Hamadani JD, Bottai M, et al. Maternal cadmium exposure during pregnancy and size at birth: a prospective cohort study. Environ Health Perspect 2012;120:284–9.

[58] Kippler M, Wagatsuma Y, Rahman A, Nermell B, Persson LÅ, Raqib R, et al. Environmental exposure to arsenic and cadmium during pregnancy and fetal size: a longitudinal study in rural Bangladesh. Reprod Toxicol 2012;34:504–11.

[59] Gardner RM, Kippler M, Tofai F, Bottai M, Hamadani J, Grandér M, et al. Environmental exposure to metals and children's growth to age 5 years: a prospective cohort study. Am J Epidemiol 2013;177:1356–67.

[60] Kim Y, Ha EH, Park H, Ha M, Kim Y, Hong YC, et al. Prenatal lead and cadmium co-exposure and infant neurodevelopment at 6 months of age: the Mothers and Children's Environmental Health (MOCEH) study. Neurotoxicology 2013;35:15–22.

[61] Manikkam M, Guerrero-Bosagna C, Tracey R, Haque MM, Skinner MK. Transgenerational actions of environmental compounds on reproductive disease and identification of epigenetic biomarkers of ancestral exposures. PLoS One 2012;7:e31901.

[62] Guerrero-Bosagna C, Skinner MK. Transgenerational epigenetic actions of environmental compounds. Anim Reprod 2010;7:165–7.

[63] Alyea RA, Gollapudi BB, Rasoulpour RJ. Are we ready to consider transgenerational epigenetic effects in human health risk assessment? Environ Mol Mutagen 2014;55:292–8.

[64] Manikkam M, Haque MM, Guerrero-Bosagna C, Nilsson EE, Skinner MK. Pesticide methoxychlor promotes the epigenetic transgenerational inheritance of adult-onset disease through the female germline. PLoS One 2014;9(7):e102091. Available from: http://dx.doi.org/10.1371/journal.pone.0102091.

[65] Manikkam M, Tracey R, Guerrero-Bosagna C, Skinner MK. Dioxin (TCDD) induces epigenetic transgenerational inheritance of adult onset disease and sperm epimutations. PLoS One 2012;7(9):e46249. http://dx.doi.org/10.1371/journal.pone.0046249 [Epub 2012 Sep 26]

[66] Chamorro-Garcia R, Sahu M, Abbey RJ, Laude J, Pham N, Blumberg B. Transgenerational inheritance of increased fat depot size, stem cell reprogramming, and hepatic steatosis elicited by prenatal exposure to the obesogen tributyltin in mice. Environ Health Perspect 2013;121:359–66.

[67] Ou X, Zhang Y, Xu C, Lin X, Zang Q, Zhuang T, et al. Transgenerational inheritance of modified DNA methylation patterns and enhanced tolerance induced by heavy metal stress in rice (Oryza sativa L.). PLoS One 2012;7(9):e41143 http://dx.doi.org/10.1371/journal.pone.0041143. [Epub 2012 Sep 11]

[68] Groopman JD, Wild CP, Hasler J, Junshi C, Wogan GN, Kensler TW. Molecular epidemiology of aflatoxin exposures: validation of aflatoxin-N7-guanine levels in urine as a biomarker in experimental rat models and humans. Environ Health Perspect 1993;99:107–13.

[69] Turner PC. The molecular epidemiology of chronic aflatoxin driven impaired child growth. Scientifica (Cairo) 2013;2013:152879. Available from: http://dx.doi.org/10.1155/2013/152879.

[70] Duale N, Bjellaas T, Alexander J, Becher G, Haugen M, Paulsen JE, et al. Biomarkers of human exposure to acrylamide and relation to polymorphisms in metabolizing genes. Toxicol Sci 2009;108:90–9.

[71] Bjellaas T, Olesen PT, Frandsen H, Haugen M, Stolen LH, Paulsen JE, et al. Comparison of estimated dietary intake of acrylamide with hemoglobin adducts of acrylamide and glycidamide. Toxicol Sci 2007;98:110−17.

[72] Gomez-Caminero A, Howe P, Hughes M, Kenyon E, Lewis DR, Moore M, et al. Arsenic and arsenic compounds. Environmental Health Criteria, vol. 224. Geneva: WHO; 2001.

[73] Calderón-Garcidueñas L, Mora-Tiscareño A, Gómez-Garza G, et al. Air pollution, cognitive deficits and brain abnormalities: a pilot study with children and dogs. Brain Cogn 2008;68:117−27.

[74] Saxton KB, John-Henderson N, Reid MW, Francis DD. The social environment and IL-6 in rats and humans. Brain Behav Immun 2011;25:1617−25.

[75] Kang HJ, Kim JM, Kim SW, Shin IS, Park SW, Kim YH, et al. Associations of cytokine genes with Alzheimer's disease and depression in an elderly Korean population. J Neurol Neurosurg Psychiatry 2014;86(9):1002−7. pii: jnnp-2014-308469. http://dx.doi.org/10.1136/jnnp-2014-308469.

[76] Delfino RJ, Staimer N, Tjoa T, Polidori A, Arhami M, Gillen DL, et al. Circulating biomarkers of inflammation, antioxidant activity, and platelet activation are associated with primary combustion aerosols in subjects with coronary artery disease. Environ Health Perspect 2008;116:898−906.

[77] Rosenlund M, Berglind N, Pershaden G, Hallquist J, Jonson T, Bellander T. Long-term exposure to urban air pollution and myocardial infarction. Epidemiology 2006;17:383−90.

[78] Owen JB, Butterfield DA. Measurement of oxidized/reduced glutathione ratio. Methods Mol Biol 2010;648:269−77.

[79] Al-Helaly LA, Ahmed TY. Antioxidants and some biochemical parameters in workers exposed to petroleum station pollutants in Mosul City, Iraq. Int Res J Environ Sci 2014;3:31−7.

[80] Possamai FP, Avila Jr. S, Budni P, Backes P, Parisotto EB, Rizelio VM, et al. Occupational airborne contamination in South Brazil: 2. Oxidative stress detected in the blood of workers of incineration of hospital residues. Ecotoxicology 2009;18:1158−64.

[81] Ajiboye TO. Redox status of the liver and kidney of 2,2-dichlorovinyl dimethyl phosphate (DDVP) treated rats. Chem Biol Interact 2010;185:202−7.

[82] Rosin A. The long-term consequences of exposure to lead. Isr Med Assoc J 2009;11:689−94.

[83] Fedeli D, Montani M, Carloni M, Nasuti C, Amici A, Gabbianelli R. Leukocyte Nurr1 as peripheral biomarker of early-life environmental exposure to permethrin insecticide. Biomarkers 2012;1−6 Early Online.

[84] Solano RM, Casarejos MJ, Menéndez-Cuervo J, Rodriguez-Navarro JA, García de Yébenes J, Mena MA. Glial dysfunction in parkin null mice: effects of aging. J Neurosci 2008;28:598−611.

[85] Aycicek A, Erel O, Kocyigit A. Decreased total antioxidant capacity and increased oxidative stress in passive smoker infants and their mothers. Pediatr Int 2005;47:635−9.

[86] Gabbianelli R, Nasuti C, Falcioni G, Cantalamessa F. Lymphocyte DNA damage in rats exposed to pyrethroids: effect of supplementation with Vitamins E and C. Toxicology 2004;203:17−26.

[87] Gabbianelli R, Falcioni ML, Nasuti C, Cantalamessa F, Imada I, Inoue M. Effect of permethrin insecticide on rat polymorphonuclear neutrophils. Chem Biol Interact 2009; 182:245−52.

[88] Nasuti C, Cantalamessa F, Falcioni G, Gabbianelli R. Different effects of Type I and Type II pyrethroids on erythrocyte plasma membrane properties and enzymatic activity in rats. Toxicology 2003;191:233−44.

[89] Waltz E. Glyphosate resistance threaten roundup hegemony. Nat Biotechnol 2010;28:537−8.

[90] Pei L, Castrillo A, Chen M, Hoffmann A, Tontonoz P. Induction of NR4A orphan nuclear receptor expression in macrophages in response to inflammatory stimuli. J Biol Chem 2005;280:29256−62.

[91] Moens U, Kostenko S, Sveinbjørnsson B. The role of mitogen-activated protein kinase-activated protein kinases (MAPKAPKs) in inflammation. Genes (Basel) 2013;4:101−33.

[92] Wang S, Zhang H, Zheng W, Wang X, Andersen ME, Pi J, et al. Organic extract contaminants from drinking water activate Nrf2-mediated antioxidant response in a human cell line. Environ Sci Technol 2013;47:4768−77.

[93] Wu LL, Chiou CC, Chang PY, et al. Urinary 8-OHdG: a marker of oxidative stress to DNA and a risk factor forcancer, atherosclerosis and diabetics. Clin Chim Acta 2004;339: 1−9.

[94] Kawanishi S, Hiraku Y, Oikawa S. Mechanism of guanine-specific DNA damage by oxidative stress and its role in carcinogenesis and aging. Mutat Res 2001;488:65−76.

[95] Mehrdad R, Aghdaei S, Pouryaghoub G. Urinary 8-hydroxy-deoxyguanosine as a biomarker of oxidative DNA damage in employees of subway system. Acta Med Iran 2015;53: 287−92.

[96] Chang FK, Mao IF, Chen ML, Cheng SF. Urinary 8-Hydroxydeoxyguanosine as a biomarker of oxidative DNA damage in workers exposed to ethylbenzene. Ann Occup Hyg 2011;55:519−25.

[97] Moretto A, Colosio C. Biochemical and toxicological evidence of neurological effects of pesticides: the example of Parkinson's disease. Neurotoxicology 2011;32:383−91.

[98] Block ML, Calderon-Garciduenas L. Air pollution: mechanisms of neuroinflammation and CNS disease. Trends Neurosci 2009;32:506−16.

[99] Surguchev A, Surguchov C. Conformational diseases: looking into the eyes. Brain Res Bull 2010;81:12−24.

[100] Fedeli D, Carloni M, Nasuti C, Gambini A, Scocco V, Gabbianelli R. Early life permethrin exposure leads to hypervitaminosis D, nitric oxide and catecholamines impairment. Pestic Biochem Physiol 2013;107(1):93−7.

CHAPTER

14

Nutritional Impact on Anabolic and Catabolic Signaling

Miklós Székely, Szilvia Soós, Erika Pétervári and Márta Balaskó

Department of Pathophysiology and Gerontology, Medical School, University of Pécs, Pécs, Hungary

KEY FACTS

- In the background of middle-aged obesity and aging anorexia/cachexia in humans and mammals, autonomic regulatory changes influencing the balance of anabolic and catabolic systems may be assumed.
- In the regulation of body mass/composition orexigenic/anorexigenic signals from the gastrointestinal tract represent actual feeding state for the brain, while adiposity signals such as leptin and insulin convey information on the nutritional state.
- In addition to aging, nutritional states themselves (eg, obesity or calorie-restriction) may also induce shifts in regulatory processes.
- Obesity induces resistance to catabolic signals, calorie-restriction enhances their efficacy, thus promoting further obesity or maintenance of lean shape.

Dictionary of Terms

- *Catabolic effect*: A combination of appetite reducing and hypermetabolic actions leading to weight loss.
- *Anabolic effect*: A combination of appetite inducing and hypometabolic actions leading to weight gain.
- *Feeding state*: It is an short-term characteristic of energy balance that originates from the gastrointestinal system. Such short-term shifts may be due to a current deficit induced by food deprivation (hunger) or to a recent bout of energy intake (satiety).
- *Nutritional state*: It is a long-term characteristic of energy balance characterized by body weight and body composition (eg, obesity or cachexia).
- *Middle-aged obesity*: In humans and other mammals middle-aged individuals tend to gain weight, especially fat mass. In the background of consequent obesity intrinsic regulatory changes may also be assumed.
- *Aging anorexia*: Food intake of old individuals and old age-groups of mammals decline without any obvious or pathological reason leading to weight loss. In the background intrinsic regulatory changes may also be assumed. Aging sarcopenia (muscle atrophy) may be aggravated by such loss of appetite leading to frailty in the elderly.

INTRODUCTION

In view of the many tons of food an individual consumes throughout his/her life, the relative stability of body mass is most remarkable. In an adult, as little as about 20 g of glucose per day (continuously in excess of the actual need) could result in about 3.5 kg rise in fat stores—plus the bound water—within a year. This could mean an enormous increase in body mass through the lifespan. Although mass, shape, and composition of the body do change with age, these changes are usually gradual and not excessive: some fat tissue often accumulates even normally at middle-age or later on in transition-age, and anorexia/sarcopenia may develop at real old age. Similar changes can also be observed in experimental or domestic animals (less in wildlife). This obviously indicates that the regulation of a balance of uneven food intake versus energy expenditure is under the influence of feedback signaling from the periphery—whether it be the actual short-term feeding state or the more long-term nutritional state, that is, body mass and composition. Although a number of signals originate from the periphery, a balance is reached by activating central anabolic (orexigenic with suppression of energy expenditure) versus catabolic (anorectic with enhancement of energy expenditure) pathways (for reviews see Refs [1–3]). Explanations for the regulation of body energy balance had earlier been attempted by important glucostatic, lipostatic, aminostatic [4], even thermostatic (for reviews see Refs [5,6]), etc. hypotheses, all of which had described a certain facet of the regulatory system, but none of them had been able to give on their own a sufficient answer to this problem. The idea of complex feedback signaling of the feeding and nutritional states appears to be more promising.

Peripheral orexigenic or anorexigenic signals representing the feeding state originate from the gastrointestinal (GI) system: primarily GI-peptides [7,8], endocannabinoids, and also some circulating nutrients, for example, glucose [9]. In young men an exercise-induced energy deficit induced smaller appetite than a similar deficit induced by food deprivation, suggesting the importance of GI signaling (although the subsequent 8-h ad libitum food intakes were similar—this late period represents the nutritional rather than the feeding state) [10]. GI signals are received by central receptors, mainly, but not exclusively in the dorsovagal complex (DVC): (1) the specific sensory inputs carried by afferent vagal fibers plus the sensory structures of the area postrema (AP), (2) the integrating center in the nucleus of the solitary tract (NTS), and (3) the motor/secretory efferent center of the dorsal motor nucleus (DMN) of the vagus [11]. Peripheral humoral signals representing the nutritional state (mainly the circulating leptin, insulin) act primarily in the arcuate nucleus (ARC) (for reviews see Refs [12,13]) using special transport mechanisms [14] to get through the circumventricular organ. Feeding-related signals from the NTS may act also at the ARC. The inputs (Table 14.1) reach hypothalamic nuclei and activate catabolic or anabolic processes—in fact, there is a bidirectional connection between hypothalamic centers and the hindbrain. Abnormalities of signaling may lead to anomalies of body mass/composition.

The opposite problem is how a preexisting abnormality of body mass/composition might modify the expression and efficacy of peripheral signaling mechanisms or their central transmission? Anomalies of body mass and composition increase exponentially.

TABLE 14.1 Neuropeptides in the Control of Food Intake

CNS factors that stimulate food intake	CNS factors that suppress food intake (CNS: central nervous system)
AgRP (agouti-related peptide)	alpha-MSH (POMC, pro-opiomelanocortin derivative alpha-melanocyte-stimulating hormone)
NPY (neuropeptide Y)	CART (cocaine-ampetamine-regulated transcript)
Orexin A and B	CRH (corticotropin-releasing hormone) and urocortins
MCH (melanin-concentrating hormone)	TRH (thyrotropin-releasing hormone)
Galanin	BDNF (brain-derived neurotrophic factor)
GHRH (growth-hormone-releasing hormone)	GALP (galanin-like peptide)
GLP-1 (glucagon-like peptide 1)	Neurotensin and neuromedins
	Nesfatin-1
Peripheral orexigenic signals	Peripheral anorexigenic signals
Ghrelin	Leptin, adiponectin
	Insulin and amylin
	Enterostatin
	Bombesin and GRP (gastrin-releasing peptide)
	Oxyntomodulin
	CCK (cholecystokinin), PYY_{3-36} (peptide YY)
	GLP-1 (glucagon-like peptide-1), GLP-2, GIP (glucose-dependent insulinotropic peptide)

On the one hand, many regions of the world still struggle with the problem of mass-undernutrition. On the other hand, one of the greatest health problems of the 21st century is the increasing number of overweight or even grossly obese persons in the population, starting already at childhood and affecting about 2 billion persons. The abovementioned age-related changes of body mass/composition in humans or animals may be consequences of regulatory changes in energy balance (altered signaling system due to aging or some other, eg, endocrine, factors). Alternatively, primary changes of body adiposity (spontaneous or diet-induced forms) may also excite or inhibit peripheral or central parts of this regulatory system and may lead to further modification of body mass in the course of aging. The questions that arise are: (1) what sort of connection may exist between feeding- or nutritional state-related inputs and alterations of nutritional state?

And (2) can supposedly primary shifts of body mass/composition (including age-related and diet-induced changes) alter the mechanisms of anabolic and catabolic signaling?

CHARACTERISTIC AGE-RELATED AND DIET-INDUCED CHANGES IN BODY MASS/COMPOSITION IN RATS

In mammals, the nutritional state varies with age: the extent is different, but the changes show similar characteristics. Fig. 14.1 demonstrates growth rate, while Table 14.2 shows fat/muscle mass indicators of male Wistar rats (in females the growth rate is different and the endocrine cycles make any evaluation of food intake more difficult). Normally fed (NF) juvenile (6–8 weeks old) male Wistar rats grow quickly and do not accumulate excess fat in their body. Mature adult animals (3–4 months) are already heavier and accumulate both retroperitoneal and epididymal fat. Further, but slower growth (mainly fat accumulation) can be observed until age 6 months, then, from age 12 to 18 months (corresponding to the age of transition) there is again a somewhat accelerated weight gain (obesity), followed by a decline of body mass by the old age of 24 months. At this last stage not only fat but also muscle tissue is lost (Table 14.2).

Both growth rate curves (Fig. 14.1) and the indicators of body composition can be altered by life-long dietary modifications that start at weaning [15]. Calorie-restriction (to 66% of normal calorie intake, CR-rats) suppresses growth rate, that is, young mature body mass remains standard, retroperitoneal fat practically diminishes, and—except for some muscle loss at the latest age—without changes in indicator muscles (Table 14.2) there is no sarcopenia. In contrast, high-fat (60% of calories) diet enhances growth rate and fat mass (HF-rats), although these animals usually have shorter lifespan and start losing fat and muscle already before the age of 18 months. These models allow the investigation of the effects of both age and nutritional state on peripheral versus central signaling mechanisms.

PERIPHERAL SIGNALING, CENTRAL PROCESSING

Gastrointestinal Signals of the Feeding State (Hunger vs Satiety)

One of the first GI peptide hormones, cholecystokinin (CCK), was discovered almost a century ago by its stimulatory effects to duodenal motility and biliary secretion. However, since 1973 it has been known to have neural effects, too: to signal the actual feeding state [16], to cause meal termination (satiation), and also to induce a period of postprandial satiety. CCK primarily forwards signals which represent GI stretch and food composition, and are conveyed by the afferent vagus to the NTS. These signals are basically independent of the nutritional state: satiation can be reached even in severely undernourished persons or

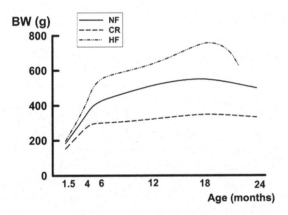

FIGURE 14.1 Mean body mass changes in the course of life in normally fed (NF) male Wistar rats and in similar rats fed from weaning by a calorie-restricted diet (CR) or by a high-fat diet (HF). HF animals had a shorter lifespan. *Source: Soós et al. (unpublished).*

TABLE 14.2 Indicators of Body Composition

	Epididymal fat (g/kg body weight)			Retroperitoneal fat (g/kg body weight)			Tibialis anterior muscle (g/kg body weight)		
Group (age)	HF	NF	CR	HF	NF	CR	HF	NF	CR
3 months		3.7 ± 0.2			3.2 ± 0.2			2.0 ± 0.1	
6 months	11.2 ± 0.8	4.7 ± 0.2	2.6 ± 0.1	19.0 ± 1.0	4.8 ± 0.4	0.7 ± 0.3	1.8 ± 0.1	1.9 ± 0.1	2.0 ± 0.2
12 months		5.3 ± 0.5	2.5 ± 0.5		4.7 ± 0.7	1.2 ± 0.2		1.9 ± 0.1	2.1 ± 0.2
18 months	9.0 ± 1.2	5.5 ± 0.7		15.0 ± 3.6	6.4 ± 1.4		1.5 ± 0.1	1.7 ± 0.1	
24 months		4.1 ± 0.2	2.0 ± 0.9		5.9 ± 0.9	0.2 ± 0.4		1.3 ± 0.1*	1.8 ± 0.2

NF: normally fed, CR: calorie-restricted, HF: high-fat diet-induced obese.
Asterisk (*) indicates significant difference between muscle indicator values of NF24 versus all other groups except for HF18. $P < 0.05$, one-way ANOVA test.

experimental animals, while obese individuals may become hungry upon fasting.

Apart from CCK, such short-term regulation involves many other signals of GI origin. Stretch of the gastric/duodenal/gut wall or altered composition of food enhances also the postprandial secretion of several other peptides produced in the pancreas (glucagon, amylin, pancreatic polypeptide (PP)), together with many lower gut peptides that are produced in mucosal L-cells (glucagon-like peptide (GLP-1), oxyntomodulin (OXM), peptide YY (PYY_{3-36} sequence)) or K-cells (glucose-dependent insulinotropic polypeptide (GIP)). Instant satiation or a period of satiety develops mainly by their action in the brainstem and the hypothalamus: they act via the afferent vagus and, at least some of them, also directly at the ARC [17,18]. The orexigenic ghrelin from the stomach [19] as well as the endocannabinoids of the GI tract [20] possibly also induce orexigenic actions partly via the vagus, at the ARC, and at further hypothalamic nuclei. The brainstem mechanisms may be influenced by tonic neuropeptide and/or monoamine actions from the hypothalamus [21], and conversely, they can influence functions of various hypothalamic nuclei. The hindbrain mechanisms have mostly short-term actions and modify hunger/satiety, although occasional abnormality of some of such actions may result in disorders of body mass, for example, lack of CCK1 receptors is coupled by OLETF (Otsuka Long-Evans Tokushima Fatty)-type obesity in rats, and by lack of anorectic action of peripherally applied CCK [22]. Alterations in the activities of these peptides cannot be excluded in other abnormalities of energy balance, either. Still, such regulation, *per se*, is assumed to explain satiation but not the long-term relative stability or possible abnormalities of weight and/or composition of the body. Apparently, other signals must be fundamental in the regulation of body weight/composition, metabolic rate, and overall energy balance, which signals represent body mass/composition—in interaction with short-term signals [23].

Age-Related Changes of Signaling Feeding State

Intraperitoneal (IP) CCK acts as the endogenous CCK that is produced postprandially in intestinal I-cells: it decreases feeding duration and total food consumption. The effect is exerted via the afferent vagus, NTS, and DVC [12,24,25]. As mentioned previously, lack of CCK1 receptors results in OLETF-type obesity in rats [22,26]. Although centrally (into a cerebral ventricle, ICV) administered CCK has similar anorectic effect—probably as part of the sickness behavior—the peripheral route of administration appears to be more natural, and central CCK may have somewhat different role(s). However, central CCK2 receptor knockout animals are also obese [27].

IP administration of CCK was ineffective in juvenile (aged 6–8 weeks) NF rats (as if "there were almost no physiological ways to suppress their appetite" at juvenile age). The satiating effect of CCK was very pronounced in young 3–4 month-old adult NF animals, still strong at age 6 months, then gradually decreased to nonsignificant levels by age 12 months, before becoming more effective again with further aging and extremely pronounced in old animals [28]. These changes in CCK-sensitivity must be explained mainly by the age and not the nutritional state, since the fat ratios did not change concordantly with efficacy: the fat content was similarly high at age 12 and 18 months and still high at age 24 months (Table 14.2). However, such a sensitivity pattern possibly contributes to the explanation of age-related changes in body weight/composition, that is, to the tendency for decreased satiation and increased adiposity at middle-age, and, together with data on high plasma CCK levels in old persons [29], such a sensitivity pattern might also explain the early satiation, anorexia, and falling body mass at old age. Nevertheless, it will be shown (Sections Obesity and GI Peptides and Signaling Feeding State in Calorie Restriction or Starvation) that nutritional state also has a strong impact on the responsiveness to IP administered CCK.

It needs to be analyzed whether or not similar age-related patterns are valid for other peripheral satiety signals. Unfortunately, only relatively few studies analyzed the efficacy of GI peptides as a function of age, and even then usually only two groups ("young" vs "old") of rodents/subjects were compared. It is a general, although not convincingly proven opinion that with aging the orexigenic signaling becomes weaker and the postprandial anorexigenic one prevails, contributing to the explanation of anorexia of aging [29,30] in old subjects, although this cannot explain the weight gain at earlier stages.

Pancreatic glucagon, amylin, and PP produced by pancreatic islet cells play very similar roles to that of CCK in the short-term inhibition of food intake [24]. They act via the afferent vagus, brainstem, and hypothalamus, with great affinity for Y4 receptors, and lead to anorexia. Glucagon (cleaved from preproglucagon in pancreatic α-cells) suppresses appetite and it can also enhance energy expenditure [31], but its main action is the elevation of blood glucose (in contrast to hypothalamic glucagon, which inhibits hepatic glucose production; [32]). Amylin, co-released with insulin upon food intake, causes anorexia [33] and increases metabolic rate [34,35]. In rats, IP amylin elicited somewhat stronger anorexia at age 13 months than 4 months, and the effect decreased slightly again by 2 years of age [36], although the differences were rather small. In aging humans (as compared with young ones) the food-induced adaptive release of amylin and insulin was attenuated [37], pointing to a delay in satiation (although this may be overcome by enhanced satiating effect of CCK and other GI peptides). Lower gut peptides (GIP, PYY_{3-36} and gut preproglucagon-derived GLP-1, OXM) inhibit feeding, enhance pancreatic β-cell activity and insulin effects, and their plasma levels increase postprandially [38,39]. An anorexigenic role for GIP is supported by the finding that the diet-induced obesity was only moderate in GIP-overexpressing mice [39]. Similarly to CCK, PYY_{3-36} secretion increases with age [29]. IP applied GLP-1 or its agonist exendin-4 induces dose-dependent anorexia, while the antagonist $exendin_{9-39}$ enhances food intake [40]. GIP and GLP-1 also stimulate insulin secretion/sensitivity, and they decrease plasma glucose level (incretin effect; [41]). GLP-1 levels may decrease with age, possibly contributing to type-2 diabetes [42]. OXM exerts a metabolic activity resembling that of GLP-1—it suppresses food intake and enhances energy expenditure [43,44]. The suppression can be blocked by GLP-1R antagonist $exendin_{9-39}$, and the suppression cannot be evoked in GLP-1R null mice [45].

Ghrelin is the only known peripheral orexigenic peptide, produced mainly in mucosal X/A-like cells of oxyntic glands in the empty stomach [46]. However, ghrelin is not strictly a "hunger hormone," it also influences energy homeostasis (suppresses energy expenditure), and moreover it enhances insulin sensitivity [47]. In mice lacking ghrelin receptor the energy expenditure and body temperature are high [48]. The peptide acts at the DVC [49] and at receptors of the ARC by activating neurons expressing orexigenic neuropeptide Y (NPY, which acts at Y1/Y5 receptors) and agouti-related peptide (AgRP), a natural antagonist of anorexigenic melanocortins [50,51]. ICV ghrelin injections for 5 consecutive days caused greater rise of food intake and body weight in young than in middle-aged rats [52]. Independent of nutritional state, the plasma ghrelin levels were significantly lower in elderly than in young persons [29,53]. In old animals fasting failed to enhance ghrelin secretion [54,55], while refeeding could not significantly lower ghrelin levels of old rodents [56], suggesting that the expression of this peptide changes with aging. Furthermore, in old animals exogenous ghrelin was hardly able to stimulate food intake [57], pointing to age-related alteration in ghrelin receptor expression, or to altered postreceptoral function. Ghrelin-receptors were not expressed in white and brown fat tissue of young mice but were detectable in old ones, and old ghrelin-receptor null mice had higher energy expenditure and less white adipose tissue than their young or wild-type old counterparts [58,59]. There is an interaction between ghrelin and CCK at the level of the afferent vagus: CCK pretreatment prevented the ghrelin-induced food intake, except in OLETF rats [60].

Signals of Nutritional State and the Role of Age

Defective feedback signals from body mass/composition, like deficiency or structural abnormality of leptin (eg, *ob/ob* mice), lack or abnormality of its receptors (eg, Koletsky rats, *db/db* mice), as well as leptin resistance (Zucker rats), may block the normal regulation of body mass, resulting in obesity. Resistance to central actions of leptin [61]—what is assumed to often develop by middle-age—may promote weight gain in humans as well as in experimental animals. In contrast, transgenic mice with leptin overexpression lose body fat [62]. Increased leptin sensitivity could contribute to anorexia/sarcopenia in the oldest age groups [15,63], or possibly to other abnormalities coupled with anorexia. Insulin in the CNS has similar roles, in many aspects, to those of leptin [61].

In young men, exposures to 21-day periods of overfeeding or underfeeding (that caused weight gain or weight loss, respectively) were quickly followed by compensation, while in old men the compensatory food intake responses to similar exposures were insufficient for as much as 7 weeks, that is, the original body weight was not restored, presumably due to defective signaling from the increased or decreased body mass [64]. Aging appears to have an impact on responses to changes in nutritional state. The expression of or sensitivity to a number of peripheral and central regulatory peptides conveying the appropriate signals was shown to be altered with age [65].

Leptin

Leptin is produced mainly in white adipose tissue (proportionately to its amount), together with other adiopokines (adiponectin, resistin). From the plasma, leptin reaches the ARC by a special transport system [14] through the blood–brain barrier. Apart from enhancing here the expression of the POMC (proopiomelanocortin)–derived anorexigenic melanocortin (alpha–melanocyte stimulating hormone, α–MSH) and the similarly anorexigenic cocaine–amphetamine-regulated transcript (CART), it also inhibits the expression of orexigenic NPY and AgRP. Such catabolic imbalance of peptides can alter activity at various receptors on second–order neurons of hypothalamic nuclei to finally exert anorectic and hypermetabolic effects [66]. Leptin also has basic (enhancing) influence on hindbrain actions of CCK-like short-term satiety signals [67] and promotes the anorectic effects of GLP-1 [68]. Most investigators described that with age and adiposity some leptin resistance and increase in leptin level can be demonstrated [69–71]. This is often thought to contribute to the fat accumulation or to definite obesity that develops by middle-age or transitional age. However, increased adiposity, *per se*, at any age, may cause elevation of plasma leptin level [69] and leptin resistance [72].

Leptin deficiency in leptin null mice resulted in *ob/ob* type obesity, and in leptin receptor knockout rodents the glucose homeostasis was also impaired [73].

In general, ICV injection or infusion of leptin induces anorexia and hypermetabolism in rats [71,72], although the two effects may not be interconnected. However, the response appears to depend on the adiposity of the animal [61] and on its age [74–76]. There were reports demonstrating that age rather than fat content is responsible for leptin resistance [77]. In contrast, other observations, using 1-week-long ICV leptin infusion came to the conclusion that aging—without obesity (eg, in old rats that were lean due to lifelong calorie restriction)—did not cause leptin resistance, on the contrary, the anorectic/hypermetabolic effects were rather enhanced (Fig. 14.2) and not decreased at old age [15,63].

In mildly obese 18-month-old rats ICV administration of recombinant adeno-associated virus that encoded rat leptin cDNA (rAAV-leptin) inhibited food intake for 25 days, elevated metabolic rate for 83 days, and suppressed body weight for the corresponding period [78]. Plasma leptin levels and hypothalamic NPY mRNA expression exhibited good negative correlation in young rats, but not in more fatty old ones. In old rats high leptin levels were found but the NPY mRNA expression did not change [79]. Lasting elevation of circulating leptin (as in obesity) was shown to reduce hypothalamic leptin receptor expression [80], practically resulting in a form of leptin resistance.

Insulin

According to some data, the cerebrospinal fluid and CNS insulin concentrations are higher than those of the serum, suggesting that—although debated—some of this peptide may be produced within the brain [81]. The circulating insulin can still cross the blood–brain barrier by a saturable transport mechanism [82] and this seems to be the really important source of active insulin in the brain. Similarly to leptin, changes in serum insulin level reflect alterations of body fat mass [83] and the two peptides have many similar steps in their central signal transduction pathways and actions. Accordingly, despite the idea that the great amount of cerebral insulin (or the disorders of its amount/action) may have several functions (eg, in the pathomechanism of Alzheimer disease), the function of regulating

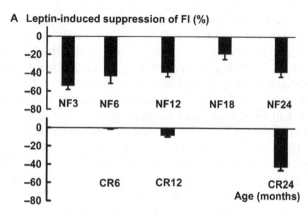

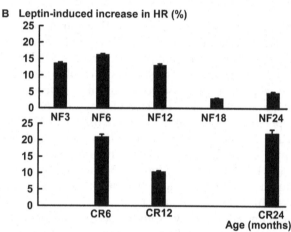

FIGURE 14.2 Part A: Cumulative deviation of food intake (FI) in the course of a 7 day-long ICV leptin infusion as compared with similar infusion of physiological saline in normally fed (NF) versus calorie restricted (CR) rats of various ages. Part B: Deviations of daytime heart rates (HR) in NF and CR rats of various ages on the 5th day of ICV leptin infusion as compared with the corresponding value of saline-infused animals (HR represents metabolic rate). *Source: Pétervári et al. (unpublished).*

metabolic homeostasis [84] is proportional to the serum levels of the peptide. Insulin—just like leptin—acts in the ARC, influences the expression of orexigenic and anorexigenic peptides (NPY/AgRP and melanocortin/CART, respectively) in order to suppress food intake and to enhance metabolic rate [61]. Moreover, the postreceptoral effects of the two peptides share some common pathways [63,84], although they differ in final modulation of the electrical activity of hypothalamic neurons [85]. ICV insulin administration proved that insulin entering the CNS also suppresses hepatic glucose production, independently of serum levels of insulin, glucagon, or other glucoregulatory hormones [86], while ICV applied insulin antibodies acted in the opposite way and induced hyperglycemia.

A recent study [87] analyzed the incidence of insulin resistance in several thousand nondiabetic, nonobese persons aged 30–79 years, and found a positive correlation between resistance and age. Inflammatory processes that are related to aging (or obesity) may also contribute to the development of insulin resistance [88]. Obesity also leads to insulin-resistant diabetes [89]. In animal experiments insulin resistance was demonstrated to develop gradually with aging and obesity [90], including resistance in the CNS [61,71]. Signals of adiposity and body composition (fatty acids, amino acids, leptin, insulin) and fuel sensing mechanisms in the brain are important in the regulation of body weight and glucose homeostasis. Fatty acids can also inhibit insulin transport through the blood–brain barrier, attenuating central insulin activity. By various, not fully clarified mechanisms, central insulin resistance may develop in the brain generally, and specifically in cerebral nuclei involved in the regulation of energy balance. It is often assumed that cerebral insulin resistance is starting point rather than consequence of metabolic syndrome [85]. The consequent obesity further elevates plasma fatty acids, reduces peripheral insulin sensitivity, thereby reinforcing obesity, hyperglycemia, β-cell dysfunction, and diabetes [85,91]. ICV infusion of insulin to 4-month-old rats elicited significant anorexia and also hypermetabolism (Fig. 14.3), but these effects faded gradually with aging and they were below significance level by the age of 24 months.

Central Processing

Collectively, signals representing feeding and/or nutritional states are in functional interaction when transmitted via capsaicin-sensitive fibers of the afferent vagus [3,92] to the brainstem or directly to the ARC—these signals together may influence not only food intake but also metabolic rate and body temperature [93]. This system activates hypothalamic catabolic (eg, melanocortin /α-MSH/, CART) or anabolic (eg, NPY, AgRP) signaling pathways and second-order neuropeptide producing neurons. Such neurons are located partly in the paraventricular nucleus (PVN) which activates anorexigenic mechanisms (eg, thyrotropin releasing hormone (TRH), corticotropin releasing hormone (CRH)), partly in the lateral hypothalamic area (LHA), perifornical area (PFA), and suprachiasmatic nucleus (SCN) which activate orexigenic mechanisms (eg, orexin-A (OXA), melanin concentrating hormone (MCH)). Neural connections exist between PVN and LHA, they also send reverse signals back to the DVC.

Age-related alterations of these central regulatory mechanisms may result in abnormalities of body mass/composition. The anorexigenic responses to ICV α-MSH infusion to NF rats [94,95] were high at ages 4

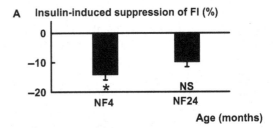
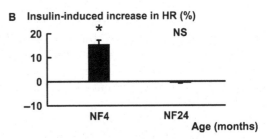
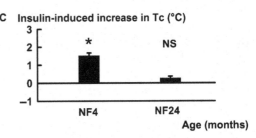

FIGURE 14.3 Part A: Suppression (%) of cumulative food intake (FI) in the course of a 7-day intracerebroventricular insulin infusion as compared with FI during a similar infusion of physiological saline in normally fed (NF) rats aged 4 versus 24 months. Part B: Deviations of daytime heart rates (HR) in NF rats aged 4 versus 24 months on the 4th day of ICV insulin infusion as compared with the corresponding value of saline-infused animals. Part C: Differences in daytime core body temperatures (Tc) on the 4th day of ICV insulin or saline infusion in NF rats aged 4 versus 24 months. Source: Soós et al. (unpublished).

and 6 months, decreased to practically nonsignificant levels by age 12 months, but became very pronounced again at ages 18–24 months. The hypermetabolic effects were highest in middle-aged (12 month old) animals, then they were attenuated with further aging, but the overall catabolic effect was strongest in old animals. Single ICV injections of the peptide caused more pronounced hypermetabolism in the oldest groups than in middle-aged ones, suggesting that old animals were still responsive, but they could not sustain the hypermetabolism for long during infusions. In other studies, age-related monotonous decrease in resposiveness to various centrally applied orexigenic peptides [57,65] has also been described. Some of such changes in central processing might be due to altered receptor expression in connection with altered peripheral signaling (eg, more circulating peptide—less central receptors—smaller effect of ICV applied exogenous peptides). Alternatively, deviations in central processing might also be primary, like the phasic changes of ICV α−MSH effects. In any case, altered central responsiveness to various peptides possibly contributes to age-related differences in the regulation of energy balance.

In contrast to age, data on nutritional state-related changes in responsiveness to central orexigenic/anorexigenic substances or changes in the function of central signaling pathways are rather scarce, however, it cannot be disputed that the central processing of signals is also influenced by the nutritional state [76]. The age-dependent changes in central sensitivity to anorexigenic or orexigenic peptides vary with nutritional state: they are different in NF, CR, and HF animals ([15,63,72,76], Sections Signaling Nutritional State in Obesity and Signaling Nutritional State: Eating Disorders, Calorie Restriction, Chronic Diseases). Additionally, the expression and action of CART were lower in obese than in control rats. The ICV applied NPY evoked stronger feeding response in CR than in NF rats, and practically no response in HF obese ones. In middle-aged NF, but not in CR rats, resistance to the catabolic effects of ICV leptin or insulin injections was found, the resistance was enhanced in HF obese animals [72,76]. For α-MSH, the nutritional state-dependent differences were of similar nature.

DIET-INDUCED CHANGES OF BODY MASS/COMPOSITION

Overfeeding—Obesity

In many studies it is not clarified, whether the altered peptide activities are causes or consequences of concurrent obesity. It is tacitly accepted that abnormal peptide activities may be primarily responsible for the development of increased body mass and adiposity. However, it is remarkable that in diet-induced obese primates (with originally normal GI hormone and leptin, insulin levels) a chronic increase of plasma cytokines is accompanied by elevated levels of circulating leptin, insulin, glucagon, GLP-1, and PYY_{3-36} [96], that is, obesity *per se* may have an impact on peptide-related functions. Moreover, such obesity-induced changes in GI peptides, leptin, and insulin levels/effects may be responsible for sustaining high body mass, once obesity has developed. Obesity (and the concurrent low-grade chronic inflammatory state) may lead to anomalies of insulin production and/or sensitivity, and to peripheral and central insulin resistance [89].

In other studies [97] it was demonstrated that POMC neurons of the ARC of diet-induced obese rodents lost synapses, exhibited increased glial ensheatment of perikarya, decreased stimulation of neighboring NPY cells, and reactive gliosis, making transport through the blood–brain barrier more difficult, that is, in obesity the ARC cytoarchitecture may have changed irreversibly in a way to decrease the efficacy of peripheral signaling.

Obesity and GI Peptides

Both the fasting plasma CCK level and its postprandial rise were significantly lower in morbidly obese women than in lean controls [98]. The age-related changes in sensitivity to the anorectic action of IP CCK seen in NF rats (ie, high sensitivity in young adult and old, with resistance in middle-aged) were also strongly influenced by the nutritional state [28]. In experiments on HF obese rats IP CCK was ineffective already at 6 months of age, when NF rats exhibited high sensitivity to CCK. The HF state not only promoted the development of CCK resistance, but also promoted the reappearance of sensitivity already at age 12 months, when NF rats were least sensitive [28]. Accordingly, HF obese state appeared to speed up the development of age-related changes in CCK-effects. Preexisting obesity seems to induce a vicious circle by decreasing the CCK sensitivity at relatively young age and thereby promoting further obesity.

Plasma PP levels were low in obesity and high in anorexia nervosa, suggesting a possible causal relationship between PP levels and anomalies of body mass [99]. Although infusion of PYY_{3-36} (acting at Y2 receptors) similarly suppressed food intake in lean and obese subjects, both the fasting PYY_{3-36} level and the postprandial rise of the peptide were lower in the obese group [98,100–102]. Obesity-prone rats kept on high-fat diet exhibited reduced GLP-1 levels and GLP-1 receptor activation [103]. GLP-1 did induce anorexia in both lean and obese subjects, however, the obese ones exhibited smaller responses [23,104]. In view of the role of GLP-1 to stimulate insulin release and sensitivity [41], this may also be important in the explanation of type 2 diabetes commonly seen in obesity [105]. Apparently, in diet-induced obesity a suppression of these peptide actions may promote further increase in body mass and adiposity.

Other anorexigenic peptides exhibited changes that were different from the previous ones. In *ob/ob* (but not in lean) mice the plasma amylin (and insulin) levels increased extremely with age, but finally (in old animals) the insulin concentrations exhibited an even greater rise, suggesting pronounced insulin resistance [106]. Plasma amylin [107] and glucagon [108] concentrations were higher in obese elderly patients than in lean ones.

Despite similar gastric ghrelin mRNA levels, the plasma level of the orexigenic ghrelin was higher in undernourished than well-nourished adults [54] and low in obese ones [98,109]. Both acyl ghrelin and desacyl ghrelin levels were low in obese persons, and postprandially mainly the desacyl form decreased slightly, while in controls a suppression of the acyl form was characteristic [110]. Diet-induced obesity in mice caused ghrelin resistance in ARC neurons expressing NPY/AgRP [111]. In obese mice deficient of both CCK1 and CCK2 receptor the fasting ghrelin level was shown to be lower than in controls [112]. These data suggested compensatory changes for ghrelin activity according to body mass.

However, other investigations revealed that in obese adolescents the calorie intake-induced suppression of ghrelin production was defective [113]. Vagal afferent neurons of diet-induced obese rats were also demonstrated to exhibit decreased sensitivity to anorexigenic signals, but enhanced sensitivity to ghrelin [3,114,115], as if obesity elicited a fasting mode-like transformation of the vagal activity, contributing to further maintenance of the obese state.

Signaling Nutritional State in Obesity

Obesity was found to elevate plasma leptin levels, while the levels of adiponectin declined, and resistin was not significantly higher [116]—all these correlated with insulin resistance. Hypothalamic leptin and insulin sensitivities were shown to decrease in diet-induced obesity [72,117]. In rats, after remission of diet-induced obesity (and fall in plasma fatty acids) the insulin transport to the cerebrospinal fluid was found to increase, and the sensitivity to the anorexigenic action of ICV insulin also improved [118]. Diet-induced obesity in mice caused severe but reversible leptin resistance of melanocortin expressing ARC neurons [119]. However, in rats with high-fat diet-induced leptin resistance [120] rAAV-leptin (with lasting high central leptin action) did not promote similar reversibility or normalization of leptin sensitivity. In addition, the diet-induced obesity—as expected in cases of enhanced leptin levels—was associated by suppressed NPY expression in the ARC, however, also by enhanced NPY and CART expression in the dorsomedial hypothalamus [121].

In earlier experiments old calorie-restricted rats still exhibited resistance to the anorectic effects of leptin [77], suggesting that age and not adiposity is responsible for the resistance. However, in recent experiments using lifelong calorie restriction (CR rats) the sensitivity to the anorexigenic as well as the hypermetabolic effects of ICV leptin was maintained, moreover, in the oldest rats it was even enhanced. This was in contrast to HF obese rats in which resistance developed at an earlier phase than in NF animals [15,63]. These data point to the role of fat content rather than age as inducer of leptin resistance.

Obesity and/or aging can also lead to peripheral and central insulin resistance [89]. On the one hand, this might explain the obesity-related type 2 diabetes, and might contribute to maintenance of an obese state. On the other hand, insulin resistance (together with leptin resistance) possibly affects skeletal muscle. In muscle a resistance to insulin decreases the supply of substrates, adipocytokines inhibit muscle cell metabolism, differentiation, and renewal, altogether leading to loss of muscle cells, to decrease of muscle protein anabolism, to ectopic lipid accumulation in muscle [122], and to sarcopenia, often in the presence of still high fat mass, that is, leading to sarcopenic obesity, with lipid infiltration of skeletal muscle and with lipotoxicity.

Roux-en-Y Gastric Bypass (RYGB) Surgery, Other Surgical Methods

RYGB is increasingly used in the surgical treatment of morbid obesity. Fast weight loss after the operation is not a simple consequence of decreased intake/absorption of nutrients, since an increase of weight-adjusted resting metabolic rate [123,124] and higher thermic effect of feeding [125,126] also contribute to the weight loss. This is particularly important, since loss of body weight (malabsorption, malnutrition, starvation), in itself, would evoke a decrease (and not an increase) in resting metabolic rate. Some GI hormones may have a role in this effect of RYGB [127]. After RYGB the production and effect of incretin hormones GLP-1 [41,128], GIP [129] and probably also PYY_{3-36} were enhanced in diet-induced obese mice as well as in humans [100,130], and in humans the meal-induced suppression of ghrelin was more pronounced [131]. The GI signals may have an action via the afferent vagus [132] and the melanocortin system: in MC4R null mice with diet-induced obesity the RYGB-induced weight loss was significantly smaller than in their wild-type counterparts [133]. Both secretion of and sensitivity to insulin are improved after the operation [134], the preoperative hyperglycemia is attenuated, although the glucagon level may remain high [135]. Activation of incretin peptide hormones (GLP-1, GIP) is important not only in the decrease of body weight but also in the improvement of glucose metabolism: some derivatives of exogenous agonist exendin-4 versus structural analogs of GLP-1 (eg, exenatide versus liraglutide) or blockers of its degrading dipeptidyl-peptidase-4 enzyme (eg, sitagliptin) are already used in the medical practice treating type 2 diabetes and obesity.

Some other surgical interventions (eg, sleeve gastrectomy, adjustable gastric banding) cause much smaller changes in incretin functions—these are less effective than RYGB in reducing body weight and hyperglycemia. However, *ob/ob* mice failed to maintain low body mass after RYGB [137].

Apparently, this raises the idea that in the background of massive obesity and diabetes not only gluttony but also some primary disorder of such incretin mechanisms [42] might play a role.

Anorexia—Eating Disorders—Calorie Restriction—Chronic Diseases

Signaling Feeding State in Eating Disorders, Anorexia Nervosa

In anorexia nervosa the baseline PYY_{3-36}, PP, CCK, and insulin plasma levels were high (although with great heterogeneity) [99,136–139], and the postprandial CCK rise started earlier and reached higher peak values than in controls [140]. These seem to suggest that abnormalities of the GI peptides may be responsible for the anorexia. However, since most of these alterations were normalized after special feeding and recovery [141], the changes may also be regarded as consequences rather than causes of the disease. Data gained in cases of simple starvation and refeeding were different from those seen in anorexia nervosa and recovery from the disease.

Similarly to starvation, anorexia nervosa is also a ghrelin hypersecretory state [142], but with decreased responsiveness to ghrelin [143]—obestatin, which is derived from the same prohormone as ghrelin but has an opposite action, increases only moderately [144].

Bulimia nervosa is accompanied by elevated CCK level in the "urge to vomit" stage [145], but not earlier.

Signaling Feeding State in Calorie Restriction or Starvation

In contrast to anorexia nervosa, primary calorie restriction in men resulted in decreased (rather than increased) baseline levels of CCK, with enhanced postprandial rise of the peptide, and also with increased sensitivity to duodenally applied lipids [146]. In undernourished adult persons the plasma ghrelin levels exceeded those in well-nourished ones [54]. In calorie-restricted rodents the baseline levels of CCK, PYY_{3-36} were also low, those of ghrelin were high, in all cases with pronounced changes postprandially [147]. These changes in peptide function appear to be adaptive responses to the calorie-restriction. In contrast to the dietary calorie-deficit, in healthy young men with exercise-induced moderate loss of calorie content the plasma PYY_{3-36} increased and the appetite was suppressed [10].

Plasma ghrelin was higher in undernourished adults than in well-nourished individuals [54], but in elderly subjects malnutrition did not induce a significant rise in ghrelin levels. The role of ghrelin still has not been clarified: while an antighrelin oligonucleotide [148] effectively inhibited the effects of exogenous ghrelin, it had hardly any effect on the calorie-deprivation-induced food intake, although plasma ghrelin level was high in both cases.

In CR rats kept on low energy intake an eventual free food intake was halved by IP CCK even at age 12 months [28] when in NF rats the anorectic effect of IP CCK has already vanished. Accordingly, CR state appeared to delay the development of resistance to CCK (in contrast to the HF state, which promoted it). Lean body composition in CR states maintained CCK sensitivity and this, in turn, tended to sustain the lean state.

Signaling Feeding State in Chronic Wasting Diseases

Chronic wasting diseases often lead to cachexia, that is, to severe fall in body mass, including both fat and lean body mass together with the muscle (sarcopenia). In cancer-induced anorexia–cachexia syndrome, alterations in the feedback loop of food intake regulation involve several neuropeptides [149]. In tumor-bearing states the amount of various cytokines rises. Interleukin-1α may enhance plasma CCK levels, and in mice with various tumors CCK1-receptor antagonists attenuated the anorexia. Furthermore, in such states the ratio of insulin/glucagon decreased, that is, there was a relative glucagon excess which was also promoted by interleukin-6. Glucagon can suppress appetite, and it simultaneously enhances hepatic glucose production, mainly from glucoplastic amino acids of muscle tissue, thus, not only the fat mass, but also the protein (muscle) content decreases, and severe cachexia and sarcopenia may develop.

In other chronic wasting diseases (end-stage renal failure, chronic heart disease, chronic obstructive pulmonary disease, etc.), accumulation of inflammatory cytokines have similar consequences to those in cancer cases [150,151]—ghrelin analogs have been suggested to improve the anorexia [150].

Signaling Nutritional State: Eating Disorders, Calorie Restriction, Chronic Diseases

In anorexia nervosa patients low plasma leptin (and high adiponectin) levels [152] and high levels of cerebrospinal fluid NPY [153] and AgRP [154] were demonstrated, with gradual normalization after recovery—it can be concluded that these peptide changes correlate with the actual fat content. However, other data allow an opposite conclusion: according to these, in the CNS and in the cerebrospinal fluid the expression of both POMC [155] and CRF [156] were high in acute forms of anorexia nervosa, but not in recovered patients. Accordingly, a high anorexigenic tone is coupled with the acute phase of the disease (when the fat content is low) and this tone is normalized upon recovery. The initial nutritional state cannot explain the elevation of anorexigenic factors, but no data support, that a primary increase of the anorexigenic tone would be responsible for the disease either.

Irrespective of whether the energy deficit was due to calorie-deprivation or exercise, in young healthy men the plasma PYY_{3-36} increased (more in case of exercise), and the appetite was suppressed [10]. However, in the brainstem the expression of PYY_{3-36} decreased during acute or chronic calorie-deprivation [157]. Exercise-induced energy deficit resulted in the expected changes in neuropeptide expression in the first-order neurons of rat hypothalamus (suppression of POMC, CART, increase of AgRP, NPY), but in the second-order neurons no changes were observed [158]. In rats calorie restriction between the ages of 6 weeks and 6 months elicited elevation of hypothalamic NPY mRNA with decrease of POMC and growth hormone releasing hormone [159], at the same time low ghrelin levels were also found. In other studies chronic food restriction was followed by decrease in plasma leptin and insulin levels and a corresponding rise of ghrelin [160]. During restoration of body weight, plasma ghrelin and leptin were normalized more quickly than the insulin level. In rats, the hypothalamic mRNA expression for POMC decreased following calorie-restriction but it was soon normalized upon restoration, while the restoration of mRNA for AgRP and NPY was slower, particularly in the ARC [160]. In the LHA versus the ARC/SCN the number of OXA- versus NPY-immunoreactive neurons increased after food deprivation. However, interestingly, these changes were more pronounced in animals that had previously been maintained on a high-fat diet than in those that received a normal diet [161], suggesting that the diet to which an animal had been used prior to restriction and restoration has a strong influence on its regulatory processes.

The functional responses to regulatory peptides are also different in chronically CR and NF rats. The anorectic action of ICV administrations of melanocortins decreased by age 12 months in NF rats, but not in CR animals [76,95,162]. The effects of ICV leptin exhibited a similar pattern: decreased efficacy with age in NF (and HF) rats, maintained efficacy in CR ones [72].

Leptin, as a helical cytokine—together with other cytokines (as in chronic diseases like chronic inflammatory processes, cancer, chronic uremia, chronic heart failure, chronic obstructive pulmonary disease)—particularly if its level is high (as in HF-obese animals), may elicit imbalance of central anabolic versus catabolic neuropeptide mechanisms with catabolic overweight, and may lead to defective metabolic functions in muscle, finally resulting in severe anorexia–cachexia [163,164]. In late phases of HF-obesity the leptin may still suppress NPY expression in ARC [121], causing anorexia and fasting-induced rise in gluconeogenesis, thus, muscle tissue may be lost more quickly than the adipose tissue, and the high body mass may fall only gradually, with altered body composition, and finally may result in sarcopenic obesity.

CONCLUSIONS

The relative stability of body mass depends on a complexity of signals originating from the gastrointestinal system that represent the actual feeding state, and some other signals that represent the more long-term nutritional state. Orexigenic (ghrelin) versus anorexigenic (CCK, PP, PYY_{3-36}, amylin, glucagon, GLP-1, GIP, OXM, etc.) gastrointestinal peptides signalize the feeding state primarily via the afferent vagus and brainstem. Simultaneously, leptin and insulin convey information on the nutritional state (energy content of the body) and inhibit food intake mainly via the arcuate nucleus. Second-order neurons and peptides of hypothalamic nuclei contribute to the regulation of energy balance, establishing anabolic (orexigenic plus hypometabolic) or catabolic (anorexigenic plus hypermetabolic) states. Apart from humans, age-related changes of sensitivity to these factors may contribute to the age-related changes of body mass/composition in rodents: anabolic/catabolic balance in young adults, weakened catabolic tone (resistance to catabolic peptides) with maintained anabolic effects in middle-aged adults of higher fat content (but not in calorie-restricted lean middle-aged animals), and catabolic hypersensitivity in the anorectic old (irrespective of body weight). Conversely, diet-induced variations in nutritional state can alter the peripheral and central sensitivity to signaling mechanisms. Diet-induced or spontaneous obesity leads to resistance to leptin and insulin, and also to anorexigenic gastrointestinal peptides like CCK: both the midlife resistance and the final hypersensitivity appear earlier in obese than in normally fed animals. Such obesity-induced resistance to anorexigenic/catabolic peptides quasireinforces the manifestation of adiposity and diabetes. Incretins (GLP-1, GIP, adipokines, etc.) act in the opposite way: they enhance weight loss and improve glucose homeostasis. In contrast to dietary obese rats, calorie-restricted lean animals exhibit no resistance to leptin signaling at any age (quasipromoting maintenance of a lean shape), however a late hypersensitivity even in these rats contributes to the aging anorexia/sarcopenia. Combined changes of nutritional state and age also influence the ingestive behavior and metabolic responsiveness to hypothalamic peptides.

Acknowledgment

The present review was completed by the help of grants from the Hungarian Society for Scientific Grants (OTKA PD84241) and from the University of Pécs (PTE 34039 KA-OTKA/13-02 and PTE ÁOK KA-2013/13-25).

SUMMARY

- Obesity in the young and middle-aged and frailty of the elderly are severe public health burdens.
- Investigation of age-related regulatory changes aggravating obesity or frailty may promote the development of effective prevention.
- This chapter focuses on the impact of nutritional state on such age-related regulatory changes.
- By suppressing the efficacy of weight reducing regulatory mechanisms, obesity promotes further accumulation of fat.
- Lifelong moderate calorie restriction increases weight-reducing regulatory mechanisms, promoting further weight loss/maintenance of lean shape.
- Prevention of accumulation of fat mass by lifelong calorie restriction helps maintain a healthy regulation of energy balance and avoidance of the vicious circle seen in unmanageable obesity.
- In nonobese aging individuals appropriate energy and nutrient intake need to be maintained (even a slight increase may be advised) to avoid rapid progression of the age-related weight loss.

References

[1] Havel PJ. Peripheral signals conveying metabolic information to the brain: short-term and long-term regulation of food intake and energy homeostasis. Exp Biol Med 2001;226:963−77.
[2] Sobrino Crespo C, Perianes Cachero A, Puebla Jiménez L, et al. Peptides and food intake. Front Endocrinol 2014;5: Article 58. <http://dx.doi.org/10.3389/fendo.2014.00058>.
[3] Dockray GJ. Gastrointestinal hormones and the dialogue between gut and brain. J Physiol 2014;592(14):2927−41.
[4] Lopez M, Tovar S, Vazquez MJ, et al. Peripheral tissue-brain interactions in the regulation of food intake. Proc Nutr Soc 2007;66:131−55.
[5] Brobeck JR. Food intake as a mechanism of temperature regulation. 1948. Obes Res 1997;5:641−5.
[6] Horvath TL, Stachenfeld NS, Diano S. A temperature hypothesis of hypothalamus-driven obesity. Yale J Biol Med 2014;87:149−58.
[7] Smith GP. Satiation: from gut to brain. New York: Oxford University Press; 1998.
[8] Blessing WW. The lower brainstem and bodily homeostasis. New York: Oxford University Press; 1997. p. 323−72.
[9] Sclafani A. Gut-brain nutrient signaling: appetition versus satiation. Appetite 2013;71:454−8.
[10] Deighton K, Batterham RL, Stensel DJ. Appetite and gut peptide responses to exercise and calorie restriction. The effect of modest energy deficits. Appetite 2014;81:52−9.
[11] Young AA. Brainstem sensing of meal-related signals in energy homeostasis. Neuropharmacology 2012;63:31−45.
[12] Konturek SJ, Konturek SW, Pawlik T, et al. Brain-gut axis and its role in the control of food intake. J Physiol Pharmacol 2004;55:137−54.
[13] Valassi E, Scacchi M, Cavagnini F. Neuroendocrine control of food intake. Nutr Metab Cardiovasc Dis 2008;18:158−68.
[14] Banks WA, Kastin AJ, Huang W, et al. Leptin enters the brain by a saturable system independent of insulin. Peptides 1996;17:305−11.
[15] Pétervári E, Rostás I, Soós S, et al. Age versus nutritional state in the development of leptin resistance. Peptides 2014;56:59−67.
[16] Gibbs J, Young RC, Smith GP. Cholecystokinin elicits satiety in rats with open gastric fistulas. Nature 1973;245:323−5.
[17] Duca FA, Covasa M. Current and emerging concepts on the role of peripheral signals in the control of food intake and development of obesity. Br J Nutr 2012;108:778−93.
[18] Yu JH, Kim MS. Molecular mechanisms of appetite regulation. Diabetes Metab J 2012;36:391−8.
[19] Gil-Campos M, Aguilera CM, Cañete R, et al. Ghrelin: a hormone regulating food intake and energy homeostasis. Br J Nutr 2006;96:201−26.
[20] Fride E, Bregman T, Kirkham TC. Endocannabinoids and food intake: newborn suckling and appetite regulation in adulthood. Exp Biol Med (Maywood) 2005;230:225−34.
[21] Bray GA. Afferent signals regulating food intake. Proc Nutr Soc 2000;59:373−84.
[22] Moran TH, Bi S. Hyperphagia and obesity in OLETF rats lacking CCK-1 receptors. Philos Trans R Soc B Biol Sci 2006;361:1211−18.
[23] Suzuki K, Jayasena CN, Bloom SR. Obesity and appetite control. Exp Diab Res 2012; ID 824305, <http://dx.doi.org/10.1155/2012/824305>.
[24] Chaudhri O, Small C, Bloom S. Gastrointestinal hormones regulating appetite. Philos Trans R Soc B Biol Sci 2006;361: 1187−209.
[25] Goebel-Stengel M, Stengel A, Wang L, et al. CCK-8 and CCK-58 differ in their effects on nocturnal solid meal pattern in undisturbed rats. Am J Physiol 2012;303:R850−60.
[26] Tachibana I, Akiyama T, Kanagawa K, et al. Defect in pancreatic exocrine and endocrine response to CCK in genetically diabetic OLETF rats. Am J Physiol 1996;270:G730−7.
[27] Clerc P, Coll Constans MG, Lulka H, et al. Involvement of cholecystokinin 2 receptor in food intake regulation: hyperphagia and increased fat deposition in cholecystokinin 2 receptor-deficient mice. Endocrinology 2007;148:1039−49.
[28] Balaskó M, Rostás I, Füredi N, et al. Age and body composition influence the effects of cholecystokinin on energy balance. Exp Gerontol 2013;48:1180−8.
[29] Moss C, Dhillo WS, Frost G, et al. Gastrointestinal hormones: the regulation of appetite and the anorexia of ageing. J Hum Nutr Diet 2012;25:3−15.

REFERENCES

[30] Di Francesco V, Fantin F, Omizzolo F, et al. The anorexia of aging. Dig Dis 2007;25:129–37.

[31] Kinoshita K, Ozaki N, Takagi Y, et al. Glucagon is essential for adaptive thermogenesis in brown adipose tissue. Endocrinology 2014;155:3484–92.

[32] Mighiu PI, Yue JT, Filippi BM, et al. Hypothalamic glucagon signaling inhibits hepatic glucose production. Nat Med 2013;19:766–72.

[33] Chance WT, Balasubramaniam A, Stallion A, et al. Anorexia following the systemic injection of amylin. Brain Res 1993;607:185–8.

[34] Lutz TA. The role of amylin in the control of energy homeostasis. Am J Physiol 2010;298:R1475–84.

[35] Yang F. Amylin in vasodilation, energy expenditure and inflammation. Front Biosci 2014;19:936–44.

[36] Morley JE, Morley PM, Flood JF. Anorectic effects of amylin in rats over the life span. Pharmacol Biochem Behav 1993;44:577–80.

[37] Dechenes CJ, Verchere CB, Andrikopoulos S, et al. Human aging is associated with parallel reductions in insulin and amylin release. Am J Physiol 1998;275:E785–91.

[38] Small CJ, Bloom SR. Gut hormones and the control of appetite. Trends Endocrinol Metab 2004;15:259–63.

[39] Kim SJ, Nian C, Karunakaran S, et al. GIP-overexpressing mice demonstrate reduced diet-induced obesity and steatosis, and improved glucose homeostasis. PLoS One 2012:e40156 <http://dx.doi.org/10.1371/journal.pone.0040156>.

[40] Williams DL, Baskin DG, Schwartz MW. Evidence that intestinal glucagon-like peptide-1 plays a physiological role in satiety. Endocrinology 2009;150:1680–7.

[41] Holst JJ. The physiology of glucagon-like peptide 1. Physiol Rev 2007;87:1409–39.

[42] Geloneze B, de Oliveira Mda S, Vasques AC, et al. Impaired incretin secretion and pancreatic dysfunction with older age and diabetes. Metabolism 2014;63:922–9.

[43] Pocai A. Action and therapeutic potential of oxyntomodulin. Mol Metab 2014;3:241–51.

[44] Dakin CL, Small CJ, Park AJ, et al. Repeated ICV administration of oxyntomodulin causes a greater reduction in body weight gain than in pair-fed rats. Am J Physiol 2002;283:E1173–7.

[45] Baggio LL, Huang Q, Brown TJ, et al. Oxyntomodulin and glucagon-like peptide-1 differentially regulate murine food intake and energy expenditure. Gastroenterology 2004;127:546–58.

[46] Fry M, Ferguson AV. Ghrelin: central nervous system sites of action in regulation of energy balance. Int J Pept 2010; Article ID 616757. <http://dx.doi.org/10.1155/2010/616757>.

[47] Pradhan G, Samson SL, Sun Y. Ghrelin: much more than a hunger hormone. Curr Opin Clin Nutr Metab Care 2013;16:619–24.

[48] Lin L, Sun Y. Thermogenic characterization of ghrelin receptor null mice. Methods Enzymol 2012;514:355–70.

[49] Gilg S, Lutz TA. The orexigenic effect of peripheral ghrelin differs between rats of different age and with different baseline food intake, and it may in part be mediated by the area postrema. Physiol Behav 2006;87:353–9.

[50] Guan HZ, Li QC, Jiang ZY. Ghrelin acts on dorsal vagal complex to stimulate feeding via arcuate neuropeptide Y/agouti-related peptide neurons activation. Acta Physiol Sin 2010;62:357–64.

[51] Holubová M, Spolcová A, Demianová Z, et al. Ghrelin agonist JMV 1843 increases food intake, body weight and expression of orexigenic neuropeptides in mice. Physiol Res 2013;62:435–44.

[52] Nesic DM, Stevanovic DM, Stankovic SD, et al. Age-dependent modulation of central ghrelin effects on food intake and lipid metabolism in rats. Eur J Pharmacol 2013;710:85–91.

[53] Nass R, Farhy LS, Liu J, et al. Age-dependent decline in acyl-ghrelin concentrations and reduced association of acyl-ghrelin and growth hormone in healthy older adults. J Clin Endocrinol Metab 2014;99:602–8.

[54] Takeda H, Nakagawa K, Okubo N, et al. Pathophysiologic basis of anorexia: focus on the interaction between ghrelin dynamics and the serotonergic system. Biol Pharm Bull 2013;36:1401–5.

[55] Serra-Prat M, Palomera E, Clave P, et al. Effect of age and frailty on ghrelin and cholecystokinin responses to a meal test. Am J Clin Nutr 2009;89:1410–17.

[56] Schneider SM, Al-Jaouni R, Caruba C, et al. Effects of age, malnutrition and refeeding on the expression and secretion of ghrelin. Clin Nutr 2008;27:724–31.

[57] Akimoto Y, Kanai S, Ohta M, et al. Age-associated reduction of stimulatory effect of ghrelin on food intake in mice. Arch Gerontol Geriatr 2012;55:238–43.

[58] Lin L, Saha PK, Ma X, et al. Ablation of ghrelin receptor reduces adiposity and improves insulin sensitivity during aging by regulating fat metabolism in white and brown adipose tissues. Aging Cell 2011;10:996–1010.

[59] Ma X, Lin L, Qin G, et al. Ablations of ghrelin and ghrelin receptor exhibit differential metabolic phenotypes and thermogenic capacity during aging. PLoS One 2011;6:e16391 <http://dx.doi.org/10.1371/journal.pone.0016391>.

[60] Date Y, Toshinai K, Koda S, et al. Peripheral interaction of ghrelin with cholecystokinin on feeding regulation. Endocrinology 2005;146:3518–25.

[61] Schwartz MW, Woods SC, Porte Jr D, et al. Central nervous system control of food intake. Nature 2000;404:661–71.

[62] Ogawa Y, Masuzaki H, Sagawa N, et al. Leptin as an adipocyte- and nonadipocyte-derived hormone. In: Bray GA, Ryan GH, editors. Pennington center nutrition series, vol. 9, Nutrition, genetics and obesity. Baton Rouge, LA: Louisana State University Press; 1999. p. 147–55.

[63] Balaskó M, Soós S, Székely M, et al. Leptin and aging: review and questions with particular emphasis on its role in the central regulation of energy balance. J Chem Neuroanat 2014;61–62:248–55.

[64] Roberts SB, Rosenberg I. Nutrition and aging: changes in the regulation of energy metabolism with aging. Physiol Rev 2006;86:651–67.

[65] Akimoto S, Miyasaka K. Age-associated changes of appetite-regulating peptides. Geriatr Gerontol Int 2010;10:S107–19.

[66] Jéquier E. Leptin signaling, adiposity and energy balance. Ann NY Acad Sci 2002;967:379–88.

[67] Morton GJ, Blevins JE, Williams DL. Leptin action in the forebrain regulates the hindbrain response to satiety signals. J Clin Invest 2005;115:703–10.

[68] Williams DL, Baskin DG, Schwartz MW. Leptin regulation of the anorexic response to glucagon-like peptide-1 receptor stimulation. Diabetes 2006;55:3387–93.

[69] Wolden-Hanson T. Mechanisms of the anorexia of aging in the Brown Norway rat. Physiol Behav 2006;88:267–76.

[70] Scarpace PJ, Tümer N. Peripheral and hypothalamic leptin resistance with age-related obesity. Physiol Behav 2001;74:721–7.

[71] Carrascosa JM, Ros M, Andrés A, et al. Changes in the neuroendocrine control of energy homeostasis by adiposity signals during aging. Exp Gerontol 2009;44:20–5.

[72] Soos S, Balasko M, Jech-Mihalffy A, et al. Anorexic versus metabolic effects of central leptin infusion in rats of various ages and nutritional states. J Mol Neurosci 2010;41:97–104.

[73] D'souza AM, Asadi A, Johnson JD, et al. Leptin deficiency in rats results in hyperinsulinemia and impaired glucose homeostasis. Endocrinology 2014;155:1268–79.

[74] Székely M, Balaskó M, Pétervári E. Peptides and temperature. In: Kastin AJ, editor. Handbook of biologically active peptides. 2nd ed. Amsterdam: Academic Press/Elsevier; 2013. p. 1880–8.

[75] Szekely M, Petervari E, Balasko M. Thermoregulation, energy balance, regulatory peptides: recent developments. Front Biosci (Schol Ed) 2010;2:1009–46.

[76] Székely M, Balaskó M, Soós S, et al. Peptidergic regulation of food intake: changes related to age and body composition ISBN:978-1-61324-183-7 In: Morrison JL, editor. Food intake: regulation, assessing and controlling. Nova Science Publ. Inc.; 2012. p. 83–104.

[77] Gabriely I, Ma XH, Yang XM, et al. Leptin resistance during aging is independent of fat mass. Diabetes 2002;51:1016–21.

[78] Scarpace PJ, Matheny M, Zhang Y, et al. Leptin-induced leptin resistance reveals separate roles for the anorexic and thermogenic responses in weight maintenance. Endocrinology 2002;143:3026–35.

[79] Li H, Matheny M, Tümer N, et al. Aging and fasting regulation of leptin and hypothalamic neuropeptide Y gene expression. Am J Physiol 1998;275:E405–11.

[80] Zhang Y, Scarpace PJ. The role of leptin in leptin resistance and obesity. Physiol Behav 2006;88:249–56.

[81] Molnár G, Faragó N, Kocsis ÁK, et al. GABAergic neurogliaform cells represent local sources of insulin in the cerebral cortex. J Neurosci 2014;34:1133–7.

[82] Banks WA, Jaspan JB, Huang W, et al. Transport of insulin across the blood-brain barrier: saturability at euglycemic doses of insulin. Peptides 1997;18:1423–9.

[83] Porte Jr D, Baskin DG, Schwartz MW. Leptin and insulin action in the central nervous system. Nutr Rev 2002;60: S20–29.

[84] Porte Jr D, Baskin DG, Schwartz MW, et al. Insulin signaling in the central nervous system: a critical role in metabolic homeostasis and disease from C. elegans to humans. Diabetes 2005;54:1264–76.

[85] Pagotto U. Where does insulin resistance start? The brain. Diabetes Care 2009;32:S174–7.

[86] Obici S, Zhang BB, Karkanias G, et al. Hypothalamic insulin signaling is required for inhibition of glucose production. Nat Med 2002;8:1376–82.

[87] Oya J, Nakagami T, Yamamoto Y, et al. Effects of age on insulin resistance and secretion in subjects without diabetes. Intern Med 2014;53:941–7.

[88] Park MH, Kim DH, Lee EK, et al. Age-related inflammation and insulin resistance: a review of their intricate interdependency. Arch Pharm Res 2014;37:1507–14.

[89] Al-Goblan AS, Al-Alfi MA, Khan MZ. Mechanism linking diabetes mellitus and obesity. Diab Metab Syndr Obes:Targ Ther 2014;7:587–91.

[90] De Tata V. Age-related impairment of pancreatic beta-cell function: pathophysiological and cellular mechanisms. Front Endocrinol 2014;5: Article 138. <http://dx.doi.org/10.3389/fendo.2014.00138>.

[91] Könner AC, Brüning JC. Selective insulin and leptin resistance in metabolic disorders. Cell Metab 2012;16:144–52.

[92] Garami A, Balasko M, Szekely M, et al. Fasting hypometabolism and refeeding hyperphagia in rats: effects of capsaicin desensitization of the abdominal vagus. Eur J Pharmacol 2010;644:61–6.

[93] Székely M. The vagus nerve in thermoregulation and energy metabolism. Auton Neurosci Basic Clin 2000;85:26–38.

[94] Pétervári E, Garami A, Soós S, et al. Age-dependence of alpha-MSH-induced anorexia. Neuropeptides 2010;44:315–22.

[95] Pétervári E, Szabad ÁO, Soós S, et al. Central alpha-MSH infusion in rats: disparate anorexic versus metabolic changes with aging. Regul Pept 2011;166:105–11.

[96] Nehete P, Magden ER, Nehete B, et al. Obesity related alterations in plasma cytokines and metabolic hormones in chimpanzees. Int J Inflam 2014;2014:856749. Available from: http://dx.doi.org/10.1155/2014/856749.

[97] Horvath TL, Sarman B, García-Cáceres C, et al. Synaptic input organization of the melanocortin system predicts diet-induced hypothalamic reactive gliosis and obesity. PNAS 2010; 107:14875–80.

[98] Zwirska-Korczala K, Konturek SJ, Sodowski M, et al. Basal and postprandial plasma levels of PYY, ghrelin, cholecystokinin, gastrin and insulin in women with moderate and morbid obesity and metabolic syndrome. J Physiol Pharmacol 2007;58 (Suppl. 1):13–35.

[99] Uhe AM, Szmukler GI, Collier GR, et al. Potential regulators of feeding behavior in anorexia nervosa. Am J Clin Nutr 1992;55:28–32.

[100] Price SL, Bloom SR. Protein PYY and its role in metabolism. Front Horm Res 2014;42:147–54.

[101] Batterham RL, Cohen MA, Ellis SM, et al. Inhibition of food intake in obese subjects by peptide YY_{3-36}. N Engl J Med 2003;349:941–8.

[102] Small CJ, Bloom SR. Gut hormones as peripheral anti obesity targets. Curr Drug Targets CNS Neurol Disord 2004;3:379–88.

[103] Duca FA, Sakar Y, Covasa M. Combination of obesity and high-fat feeding diminishes sensitivity to GLP-1R agonist exendin-4. Diabetes 2013;62:2410–15.

[104] Holst JJ. Enteroendocrine secretion of gut hormones in diabetes, obesity and after bariatric surgery. Curr Opin Pharmacol 2013;13:983–8.

[105] Madsbad S. The role of glucagon-like peptide-1 impairment in obesity and potential therapeutic implications. Diabetes Obes Metab 2014;16:9–21.

[106] Leckström A, Lundquist I, Ma Z. Islet amyloid polypeptide and insulin relationship in a longitudinal study of the genetically obese (ob/ob) mouse. Pancreas 1999;18:266–73.

[107] Reda TK, Geliebter A, Pi-Sunyer FX. Amylin, food intake, and obesity. Obes Res 2002;10:1087–91.

[108] Villareal DT, Banks MR, Patterson BW, et al. Weight loss therapy improves pancreatic endocrine function in obese older adults. Obesity 2008;16:1349–54.

[109] Buss J, Havel PJ, Epel E, et al. Associations of ghrelin with eating behaviors, stress, metabolic factors, and telomere length among overweight and obese women: preliminary evidence of attenuated ghrelin effects in obesity? Appetite 2014;76:84–94.

[110] Dardzińska JA, Małgorzewicz S, Kaska Ł, et al. Fasting and postprandial acyl and desacyl ghrelin levels in obese and non-obese subjects. Endokrynol Pol 2014;65:377–81.

[111] Briggs DI, Enriori PJ, Lemus MB, et al. Diet-induced obesity causes ghrelin resistance in arcuate NPY/AgRP neurons. Endocrinology 2010;151:4745–55.

[112] Sakurai C, Ohta M, Kanai S, et al. Lack of ghrelin secretion in response to fasting in cholecystokinin-A (-1), -B (-2) receptor-deficient mice. J Physiol Sci 2006;56:441–7.

[113] Mittelman SD, Klier K, Braun S, et al. Obese adolescents show impaired meal responses of the appetite-regulating hormones ghrelin and PYY. Obesity (Silver Spring) 2010;18:918–25.

[114] de Lartigue G, Barbier de la Serre C, Espero E, et al. Leptin resistance in vagal afferent neurons inhibits cholecystokinin signaling and satiation in diet induced obese rats. PLoS One 2012;7:e32967. Available from: http://dx.doi.org/10.1371/journal.pone.0032967.

[115] Daly DM, Park SJ, Valinsky WC, et al. Impaired intestinal afferent nerve satiety signalling and vagal afferent excitability in diet induced obesity in the mouse. J Physiol 2011;589:2857–70.

[116] Silha JV, Krsek M, Skrha JV, et al. Plasma resistin, adiponectin and leptin levels in lean and obese subjects: correlations with insulin resistance. Eur J Endocrinol 2003;149:331–5.

[117] Velloso LA, Schwartz MW. Altered hypothalamic function in diet-induced obesity. Int J Obes (Lond) 2011;35:1455–65.

[118] Begg DP, Mul JD, Liu M. Reversal of diet-induced obesity increases insulin transport into the cerebrospinal fluid and restores sensitivity to the anorexic action of central insulin in male rats. Endocrinology 2013;154:1047–54.

[119] Enriori PJ, Evans AE, Sinnayah P, et al. Diet-induced obesity causes severe but reversible leptin resistance in arcuate melanocortin neurons. Cell Metab 2007;5:181–94.

[120] Wilsey J, Zolotukhin S, Prima V, et al. Central leptin gene therapy fails to overcome leptin resistance associated with diet-induced obesity. Am J Physiol 2003;285:R1011–20.

[121] Lee SJ, Verma S, Simonds SE, et al. Leptin stimulates neuropeptide Y and cocaine amphetamine-regulated transcript coexpressing neuronal activity in the dorsomedial hypothalamus in diet-induced obese mice. J Neurosci 2013;33: 15306–17.

[122] Tardif N, Salles J, Guillet C, et al. Muscle ectopic fat deposition contributes to anabolic resistance in obese sarcopenic old rats through eIF2α activation. Aging Cell 2014;13:1001–11.

[123] Faria SL, Faria OP, Buffington C, et al. Energy expenditure before and after Roux-en-Y gastric bypass. Obes Surg 2012;22:1450–5.

[124] Werling M, Olbers T, Fändriks L, et al. Increased postprandial energy expenditure may explain superior long term weight loss after Roux-en-Y gastric bypass compared to vertical banded gastroplasty. PLoS One 2013;8:e60280. Available from: http://dx.doi.org/10.1371/journal.pone.0060280.

[125] Wilms B, Ernst B, Schmid SM, et al. Enhanced thermic effect of food after Roux-en-Y gastric bypass surgery. J Clin Endocrinol Metab 2013;98:3776–84.

[126] Nestoridi E, Kvas S, Kucharczyk J, et al. Resting energy expenditure and energetic cost of feeding are augmented after Roux-en-Y gastric bypass in obese mice. Endocrinology 2012;153:2234–44.

[127] Moran TH, Dailey MJ. Intestinal feedback signaling and satiety. Physiol Behav 2011;105:77–81.

[128] Maning S, Pucci A, Batterham RL. GLP-1: a mediator of the beneficial metabolic effects of bariatric surgery? Physiology 2015;30:50–62.

[129] Moran-Atkin E, Brody F, Fu SW, et al. Changes in GIP gene expression following bariatric surgery. Surg Endosc 2013;27:2492–7.

[130] Reidelberger RD, Haver AC, Apenteng B, et al. Effects of exendin-4 alone and with peptide YY(3-36) on food intake and body weight in diet-induced obese rats. Obesity 2011;19:121–7.

[131] Dirksen C, Jørgensen NB, Bojsen-Møller KN, et al. Gut hormones, early dumping and resting energy expenditure in patients with good and poor weight loss response after Roux-en-Y gastric bypass. Int J Obes 2013;37:1452–9.

[132] Ballsmider LA, Vaughn AC, David M, et al. Sleeve gastrectomy and Roux-en-Y gastric bypass alter the gut-brain communication. Neural Plasticity 2015:1–9. Available from: http://dx.doi.org/10.1155/2015/601985.

[133] Hatoum IJ, Stylopoulos N, Vanhoose AM. Melanocortin-4 receptor signaling is required for weight loss after gastric bypass surgery. J Clin Endocrinol Metab 2012;97:E1023–31.

[134] Jacobsen SH, Olesen SC, Dirksen C, et al. Changes in gastrointestinal hormone responses, insulin sensitivity, and beta-cell function within 2 weeks after gastric bypass in nondiabetic subjects. Obes Surg 2012;22:1084–96.

[135] Eickhoff H, Louro T, Matafome P, et al. Glucagon secretion after metabolic surgery in diabetic rodents. J Endocrinol 2014;223:255–65.

[136] Prince AC, Brooks SJ, Stahl D, et al. Systematic review and meta-analysis of the baseline concentrations and physiologic responses of gut hormones to food in eating disorders. Am J Clin Nutr 2009;89:755–65.

[137] Hao Z, Münzberg H, Rezai-Zadeh K, et al. Leptin deficient ob/ob mice and diet-induced obese mice responded differently to Roux-en-Y bypass surgery. Int J Obes (Lond) 2014. Available from: http://dx.doi.org/10.1038/ijo.2014.189.

[138] Zhang L, Yagi M, Herzog H. The role of NPY and ghrelin in anorexia nervosa. Curr Pharm Des 2012;18:4766–78.

[139] Rigamonti AE, Cella SG, Bonomo SM, et al. Effect of somatostatin infusion on peptide YY secretion: studies in the acute and recovery phase of anorexia nervosa and in obesity. Eur J Endocrinol 2011;165:421–7.

[140] Harty RF, Pearson PH, Solomon TE, et al. Cholecystokinin, vasoactive intestinal peptide and peptide histidine methionine responses to feeding in anorexia nervosa. Regul Pept 1991;36:141–50.

[141] Bailer UF, Kaye WH. A review of neuropeptide and neuroendocrine dysregulation in anorexia and bulimia nervosa. Curr Drug Targets CNS Neurol Disord 2003;2:53–9.

[142] Hotta M, Ohwada R, Akamizu T, et al. Ghrelin increases hunger and food intake in patients with restricting-type anorexia nervosa: a pilot study. Endocr J 2009;56:1119–28.

[143] Broglio F, Gianotti L, Destefanis S, et al. The endocrine response to acute ghrelin administration is blunted in patients with anorexia nervosa, a ghrelin hypersecretory state. Clin Endocr (Oxf) 2004;60:592–9.

[144] Monteleone P, Serritella C, Martiadis V, et al. Plasma obestatin, ghrelin, and ghrelin/obestatin ratio are increased in underweight patients with anorexia nervosa but not in symptomatic patients with bulimia nervosa. J Clin Endocrinol Metab 2008;93:4418–21.

[145] HannonEngel SL, Filin EE, Wolfe BE. CCK response in bulimia nervosa and following remission. Physiol Behav 2013;122: 56–61.

[146] Seimon RV, Taylor P, Little TJ, et al. Effects of acute and longer-term dietary restriction on upper gut motility, hormone, appetite, and energy-intake responses to duodenal lipid in lean and obese men. Am J Clin Nutr 2014;99:24–34.

[147] King JA, Wasse LK, Ewens J, et al. Differential acylated ghrelin, peptide YY$_{3-36}$, appetite, and food intake responses to equivalent energy deficits created by exercise and food restriction. J Clin Endocrinol Metab 2011;96:1114–21.

[148] Teubner BJ, Bartenss TJ. Anti-ghrelin Spiegelmer inhibits exogenous ghrelin-induced increases in food intake, hoarding, and neural activation, but not food deprivation-induced increases. Am J Physiol 2013;305:R323–33.

[149] Inui A. Cancer anorexia-cachexia sydrome: are neuropeptides the key? Cancer Res 1999;49:4493–501.

[150] DeBoer MD. Ghrelin and cachexia: will treatment with GHSR-1a agonists make a difference for patients suffering from chronic wasting syndromes? Mol Cell Endocrinol 2011; 340:97–105.

[151] Mak RH, Cheung W, Cone RD, et al. Orexigenic and anorexigenic mechanisms in the control of nutrition in chronic kidney disease. Pediatr Nephrol 2005;20:427–31.

[152] Brichard SM, Delporte ML, Lambert M. Adiponectiones in anorexia nervosa: a review focusing on leptin and adiponectin. Horm Metab Res 2003;35:337–42.

[153] Kaye WH, Berrettini W, Gwirtsman H, et al. Altered cerebrospinal fluid neuropeptide Y and peptide YY immunoreactivity in anorexia and bulimia nervosa. Arch Gen Psychiatry 1990;47:548–56.

[154] Merle JV, Haas V, Burghardt R, et al. Agouti-related protein in patients with acute and weight-restored anorexia nervosa. Psychol Med 2011;41:2183–92.

[155] Ehrlich S, Weiss D, Burghardt R, et al. Promoter specific DNA methylation and gene expression of POMC in acutely underweight and recovered patients with anorexia nervosa. J Psychiatr Res 2010;44:827–33.

[156] Lawson EA, Holsen LM, Desanti R, et al. Increased hypothalamic-pituitary-adrenal drive is associated with decreased appetite and hypoactivation of food-motivation neurocircuitry in anorexia nervosa. Eur J Endocrinol 2013;169:639–47.

[157] Gelegen C, Chandrana K, Choudhury AI, et al. Regulation of hindbrain PYY expression by acute food deprivation, prolonged caloric restriction, and weight loss surgery in mice. Am J Physiol 2012;303:E659–68.

[158] de Rijke CE, Hillebrand JJG, Verhagen LAW, et al. Hypothalamic neuropeptide expression following chronic food restriction in sedentary and wheel-running rats. J Mol Endocrinol 2005;35:381–90.

[159] Shimokawa I, Fukuyama T, Yanagihara-Outa K, et al. Effects of caloric restriction on gene expression in the arcuate nucleus. Neurobiol Aging 2003;24:117–23.

[160] Kinzig KP, Hargrave SL, Tao EE. Central and peripheral effects of chronic food restriction and weight restoration in the rat. Am J Physiol 2009;296:E282–90.

[161] Park ES, Yi SJ, Kim JS, et al. Changes in orexin-A and neuropeptide Y expression in the hypothalamus of the fasted and high-fat diet fed rats. J Vet Sci 2004;5:295–302.

[162] Pétervári E, Soós S, Székely M, et al. Alterations in the peptidergic regulation of energy balance in the course of aging. Curr Protein Pept Sci 2011;12:316–24.

[163] Dwarkasing JT, van Dijk M, Dijk FJ, et al. Hypothalamic food intake regulation in a cancer-cachectic mouse model. J Cachexia Sarcopenia Muscle 2014;5:159–69.

[164] Grossberg AJ, Scarlett JM, Marks DL. Hypothalamic mechanisms in cachexia. Physiol Behav 2010;100:478–89.

CHAPTER 15

Aging and Its Dietary Modulation by FoxO1 Phosphorylation/Acetylation

Dae Hyun Kim[1], Byung P. Yu[2] and Hae Y. Chung[1]

[1]Molecular Inflammation Research Center for Aging Intervention (MRCA), College of Pharmacy, Pusan National University, Busan, Korea [2]Department of Physiology, The University of Texas Health Science Center at San Antonio, San Antonio, TX, USA

KEY FACTS

- Age-related decline in FoxO activity shortens lifespan through decreased gene expressions of antioxidant enzymes, superoxide dismutase (SOD) and catalase.
- FoxO-targeted enzymes, SOD and catalase scavenge reactive species and decrease oxidative stress.
- Modification of FoxO by acetylation and phosphorylation decreases FoxO activity during aging.
- Insulin level increases with aging, while anti-aging paradigm, calorie restriction (CR) decreases it.
- Upregulation of PI3K/Akt pathway during aging causes the decline of age-related FoxO activity.
- Age-related FoxO1 activity is suppressed through PI3K/Akt pathway during aging, while CR reverses it.

Dictionary of Terms

- *Aging*: Aging is a biological process characterized by progressive decline in physiological function and structural integrity affected by intrinsic and extrinsic factors. The progressive deterioration in maintenance of homeostasis, broadly defined as aging, leads to increased risk of age-related diseases including type 2 diabetes, cardiovascular diseases, dementia, rheumatoid arthritis, osteoporosis, and cancer, and, finally, death. The incidence of such age-related diseases, which cause substantial difficulties to affected individuals and public health issues, is significantly increased since lifespan of human has been extended. Increasing interest on aging research over the past several decades has shed light on a number of profound theories attempting to define underlying mechanisms of aging and extend healthspan.
- *Calorie restriction (CR)*: It has been demonstrated that CR delays age-related biologic changes and decreases risk of age-related diseases across mammalian and nonmammalian species and in both genders. CR is associated with various beneficial physiologic effects including suppression of inflammatory responses in aging, inhibition of protein oxidation in liver and skeletal muscle, and augmentation of immune function. In addition, CR decreases blood insulin and NADH levels, and altered NAD/NADH ratio causes phosphorylation and acetylation of FoxO.
- *FoxO1*: Forkhead box subgroup "O" (FoxO) transcription factors are highly associated with downstream mechanisms of insulin/IGF-1 pathway, and have been suggested to moderately affect lifespan by providing enhanced resistance to oxidative stress, diminishing overproduction of

reactive oxygen species and preventing accumulation of oxidative damage that may accelerate aging. The transcription factor mammalian FoxO family including FoxO1, FoxO3, FoxO4, and FoxO6, is characterized by a highly conserved DNA binding motif, reported as forkhead box O or a winged helix domain. FoxO1 expression is found in brain, liver, pancreas, spleen, heart, lung, thymus, ovary, prostate, and testes. In the nucleus, translocated FoxO1 proteins bind to IRE in the proximal promoter of target genes related to insulin sensitivity, DNA repair, cell cycle, and cell survival, and induce their transcription.

- *Oxidative stress*: The main cause of oxidative stress is imbalance between reduction and oxidation (redox) by systemic overproduction of reactive species (RS), and deteriorated detoxification capability to degrade them. The disruption of balance between RS, such as superoxide anion ($\cdot O_2^-$), hydrogen peroxide (H_2O_2), peroxynitrite ($ONOO^-$), hydroxyl radical ($\cdot OH$), and antioxidant species, such as glutathione (GSH) and thioredoxin (Trx), has been found to be a major factor underlying development of various diseases or aggravation of symptoms in heart failure, myocardial infarction, atherosclerosis, Parkinson's disease, Alzheimer's disease, schizophrenia bipolar disorder, fragile X syndrome, sickle cell disease, and cancer. On the other hand, RS can be beneficial in the immune system as they attack and eliminate pathogens in infection.

- *PI3K/Akt pathway*: PI3K/Akt pathway is one of the main signaling pathways regulated by insulin. PI3K is activated when insulin or growth factors bind to specific receptor tyrosine kinases, then, subsequently, serine/threonine-specific kinase Akt is phosphorylated and activated by PI3K. Akt directly phosphorylates and inhibits members of FoxO subfamily of forkhead transcription factors, enhancing cell survival and proliferation. As insulin binds to its responding receptor, the receptor itself and adaptor molecule, and insulin receptor substrate (IRS) 1 on tyrosine residues are phosphorylated. This leads to PI3K activation which enhances production of phosphatidylinositol 3,4-diphosphate (PIP_2) and phosphatidylinositol 3,4,5-triphosphate (PIP_3). Akt is recruited into the plasma membrane through the series of signaling. Phosphorylation of Thr^{308} and Ser^{473} in Akt is induced by PIP_3-dependent kinase-1 and -2 (PDK-1 and PDK-2), respectively. PDK-1 is activated by PIP2 and PIP3, lipid products of PI3K, whereas the PDK-2 activation mechanism has not been elucidated.

INTRODUCTION

Forkhead box subgroup "O" (FoxO) transcription factors are pivotal downstream targets of inhibiting insulin/IGF-1 signaling and are postulated to affect longevity in part by increasing an organism's oxidative stress resistance, decreasing reactive oxygen species (ROS) production, and slowing the accumulation of oxidative damage that accelerates aging [1]. Some of these antioxidative effects are associated with members of the FoxO family, which, in the absence of insulin/IGF-1 signaling, freely bind to promoters of antioxidant enzymes, thus upregulate their expressions [2]. Although these findings implicate FoxO in the aging process and in age-related diseases, at present, little is known for FoxO's precise role in protective mechanisms.

It has been shown that the ability of FoxO to regulate inflammation reactions plays a major role in many metabolic disorders and diseases, such as obesity, type 2 diabetes mellitus, insulin resistance, hyperlipidemia, and nonalcoholic fatty liver disease (NAFLD) [3]. The notable association between metabolic syndrome and the aging process indicates that inflammation underlines the onset and progression of metabolic syndrome [4]. Furthermore, it has been established that insulin resistance is potentiated by upregulated pro-inflammatory TNF-α and other cytokines during aging [5]. More recent findings on FoxO1 show its important roles in regulating energy homeostasis [6].

Calorie restriction (CR), the most potent dietary antiaging intervention, elicits coordinated and adaptive stress responses at cellular and organism levels; and in *Caenorhabditis elegans* and *Drosophila* models. CR extends lifespan in a FoxO-independent manner [7]. Previous studies on CR clearly showed its efficacy on the aging process, and identified several key regulatory signaling pathways [8]. Epigenetic studies reported that histone modifications and DNA methylation sites on specific gene promoters or in chromosomal regions have direct functional impacts on aging [9]. Importantly, CR has been shown to regulate epigenetically the aging process [10], for instance, by its ability to reverse age-related histone acetylation and DNA methylation changes, thereby, increasing genomic stability [11].

This chapter reviews the important findings on FoxO1 modification in relation to aging and describes its modulation by CR as a possible underlying mechanism influencing the aging process. In addition, we posit that the transcription activities of FoxO1 act as a bridge between aging and age-related changes and suggest a mechanism for the modification of FoxO1 by age-related signaling pathways.

ALTERED FOXO1 TARGET GENES DURING AGING AND THEIR MODULATION BY CR

The evolutionarily conserved FoxO family consists of FoxO1, FoxO3, FoxO4, and FoxO6 in mammals [12]. These FoxO transcription factors are characterized by a highly conserved DNA binding motif, known as forkhead box O or a winged helix domain [13,14]. FoxO are expressed in various tissues, including ovary, prostate, skeletal muscle, brain, heart, lung, liver, pancreas, spleen, thymus, and testes [15].

Recent studies revealed that FoxO regulates various downstream target genes involved in cell cycle, cell death, and the oxidative stress response [12,13]. One of the key mechanisms of FoxO regulation involves its phosphorylation by protein kinase B (also known as Akt) in response to insulin or other growth factors and its subsequent translocation from the nucleus to the cytoplasm [13,14,16]. cAMP-response element-binding protein (CREB)-binding protein (CBP) attenuates the transcriptional activities via acetylation of FoxO1 [17]. However, the acetylation of FoxO1 can be reversed by NAD-dependent deacetylase, silent information regulator 1 (SIRT1), consequently, activating transcription is mediated by FoxO1 [18]. However, when SIRT1 is inactive, FoxO1, which is highly acetylated by CBP, increases the levels of its phosphorylation through the PI3K/Akt pathway [19].

As mentioned earlier, FoxO transcription factors, which are negatively regulated by signals derived from the PI3K/Akt pathways, are involved in the upregulation of a series of target genes that regulate the cell cycle, DNA repair, stress resistance, and cell survival or apoptosis in response to cellular stress [20].

Among members of the FoxO family, FoxO1 was better studied for the regulation of several genes involved in cell cycle arrest (p21), DNA repair (Gadd45a), apoptosis (Bim), and the stress response in response to oxidative stress in liver [21]. It was reported that nuclear FoxO1 activates the transcriptions of stress-inducible genes like manganese superoxide dismutase MnSOD to battle mitochondrial ROS [22]. Moreover, the FoxO1-mediated transcriptions of antioxidant genes such as MnSOD can also be facilitated by the activation of FoxO1 via deacetylation by deacetylase, SIRT1 [23]. As summarized in Table 15.1, many lines of evidence indicate that the function of FoxO affects the expression of various genes in various species and tissues.

CR increases lifespan in various species of experimental animals such as fly, yeast, worm, rat, and monkey. CR has been shown to delay age-related biologic changes and to suppress a number of age-associated pathologic abnormalities in both genders and across mammalian and nonmammalian species [24]. Several CR studies [21,25] illustrated that the functions of FoxO1 were modulated under various conditions such

TABLE 15.1 Functions of FoxO and Target Gene Expressions in Various Tissues

Tissue	Functions	Target genes
Liver	Increase of gluconeogenesis in mice	G6P, PEPCK, PGC-1a
	Reduced triglyceride in pig and mice	ApoCIII, MTP
Pancreatic β-cell	Protection against oxidative stress	MafA, NeuroD
	Repression of β-cell proliferation	PDX-1, NGN3, NKX61, CyclinD1
Adipose tissue	Control of differentiation	P21, PPARγ
Kidney	Protection of lipotoxicity and disease	Bcl-2, Bax, Catalase, MnSOD, Bim
Brain	Protection of neuronal death	Bim, Fas ligand
Skeletal muscle	Repression of differentiation	Atrogin-1, MuRF1
	Induction of muscle atrophy	MAFbx
Lung	Regulation of lung tumor in mice	P21, P27, CyclineD1
	Suppression of lung adenocarcinoma in human	GADD45
Heart	Protection of ischemic heart in mouse	MnSOD, Catalase
	Inhibition of cardiac mass loss in rat	Autophagy genes

as oxidative stress. These findings strongly support the notion that FoxO1 transcription factor is intricately involved in cell death and survival and with the prolonged lifespan as seen in CR [21,26].

EPIGENETIC INFLUENCES ON FOXO1

Aging is predominantly influenced by lifestyle-associated epigenetic modifications on histones and chromatins [27]. Among many extrinsic factors, CR is regarded as the most influential dietary mediator owing to its powerful ability to manipulate epigenetic modifications. FoxO also is involved in a key epigenetic modulation mechanism during aging [28]. In response to insulin or oxidative stress, FoxO proteins are phosphorylated by protein kinase B (PKB, also known as Akt), a downstream kinase of phosphatidylinositol 3-kinase (PI3K) that leads to the translocation of these proteins from the nucleus into cytoplasm [17].

Recently, Li et al. [29] described decreased global DNA methylation in aged animals and the blunting effect of CR. At present, quantitative data on specific modifications of chromatin and histones during aging and CR are sparse; however, increased aberrant hypermethylation of promoter CpG islands and high levels of DNA methylation have been reported in relation to the inflammatory process [30]. In this regard, it is worth noting that age-related hypermethylation was found in pro-inflammatory genes [31], which was suppressed by CR (unpublished data).

The major advantage of epigenetic modification as a prime modifier of underlying gene expression regulation lies in its ability to regulate gene activities without changing DNA sequences. Of the many epigenetic modifications identified, the two most studied are DNA and histone modification. For example, oxidative stress can induce cellular senescence via FoxO transcription factors as well as by deacetylase, SIRT1, but changes in these activities likely lead to cellular malfunction [32]. Structurally, deacetylated histone lysine residues attract negatively charged DNA strands to form a compact chromatin state that would lead to transcriptional repression [28], while methylation of histone lysine residues leads to either activation or repression depending on the location of modified lysine residues [33].

Recent work on CR as a potent epigenetic modifier documented increased deacetylase activity as shown by SIRT1 activation, implying the possibility that global acetylation may underlie the aging process. It is interesting to note that SIRT1 deacetylates FoxO (nonhistone substrate transcription factors) [28,34], and diversely modulates age-related cellular mechanisms.

Furthermore, SIRT1 can interact with several key transcriptional factors and regulatory proteins that may contribute to lifespan extension by CR [35].

A better understanding of the epigenetic interaction between FoxO1 and CR certainly will lead to a clearer view of the modus operandi of FoxO1 in aging and will provide molecular clues for possible interceptive measures against aging and age-associated diseases.

Modifications of FoxO1 by Phosphorylation

One of the key regulatory mechanisms of FoxO factors involves the phosphorylation reaction. Phosphorylation of FoxO by protein kinase B in response to insulin or several growth factors (PKB, also known as Akt), allows translocation from the nucleus to the cytoplasm [13,14,16]. FoxO1 is sensitive to Akt-dependent phosphorylation, and emerging evidence suggests that PI3K/Akt signaling regulates cellular status of oxidative stress [36]. Akt-mediated phosphorylation of FoxO1 inhibits FoxO1 activity through nuclear exclusion; by contrast, oxidative stress-associated phosphorylation of FoxO1 suppresses the translocation of FoxO1 into the nucleus and deactivates the transcription of FoxO1 target genes [25].

Intricate interactions between Akt and FoxO have been reported in the context of the mechanisms of cellular regulation. For example, in yeast, the mutation of Sch 9, which is homologous to Akt, extends lifespan [37] by decreasing Akt signaling and activating FoxO; and an insulin receptor mutation that decreases activity in the insulin/IGF-1-like pathway increases the longevity of fruit flies [38] and mice [39].

Lifespan-extending mutations are often associated with increased resistance to oxidative stress, which is partly mediated by the increased expression of antioxidant genes [40]. It was shown that Akt-regulated phosphorylation of FoxO can reduce levels of cellular oxidative stress by directly increasing the mRNA and protein levels of MnSOD and catalase [41]. Also, Yiu et al. [42] reported that through the Akt/FoxO pathway in mouse, increased renal oxidative stress led to the increased NADPH oxidase activation and suppression of antioxidant enzymes, including MnSOD and catalase in the kidney.

FoxO1 is regulated by signaling molecules, such as cyclin-dependent kinases (CDK)1/2, which catalyze FoxO1 phosphorylation at Ser249. The phosphorylation of FoxO1 favors cell survival in response to DNA damage [43]. However, FoxO1 is a substrate of protein phosphatase 2A (PP2A) that dephosphorylates FoxO1, which in cultured lymphoid FL5.12 cells promotes its nuclear localization [44]. Furthermore, the inhibition of PP2A protects FoxO1 from dephosphorylation and prevents its nuclear localization, leading to a decreased defense mechanism [44].

Modifications of FoxO1: Acetylation and Deacetylation

Four mammalian FoxO transcription factors, all which are negatively regulated by insulin signaling through Akt, are further modified in the nucleus by acetylase CBP/p300, which serves as a transcriptional coactivator [45].

FoxO1 activity is modulated by acetylation via the CBP/p300 factor [46], which enhances its trans-activation activity [47]. In contrast, FoxO1 acetylation at its basic residues, LysK242, Lys245, and Lys262, attenuates FoxO1 activity (corresponding to Lys 242 and Lys 245 in FoxO3) by inhibiting its ability to bind target DNA [19]. Furthermore, the acetylations of FoxO1 and FoxO3 at these three residues by CBP appear to enhance their phosphorylations at Ser253 by Akt/PKB, which inhibits the activity of FoxO1 [19]. These observations indicate the diverse effects of acetylation on the transcriptional activity of FoxO1. In aged rats, it was shown that CBP interacts with FoxO1 and acetylates FoxO1 [25].

Data show that FoxO proteins are substrates of SIRT1, the mammalian ortholog of yeast Sir2 deacetylase [26,48–50]. Furthermore, deacetylation of FoxO proteins by SIRT1 modulates their trans-activation activities, which has been viewed as an important cellular defense mechanism against oxidative stress [20,34,51].

It should be noted that not all phosphorylated FoxO1 proteins are destined for proteasome-mediated degradation in cytoplasm, as there is evidence that cytosolic phosphorylated FoxO1 proteins undergo SIRT1-dependent deacetylation, which results in the reverse translocation of FoxO1 to the nucleus in cultured hepatocytes in response to oxidative stress [51]. SIRT1 activation also promotes the nuclear retention of FoxO1 and enhances its activity in resveratrol-treated cells [51]. The activation of SIRT1 by CR in many experimental models has been proposed as one of the major antiaging mechanisms [52].

Significance of FoxO1 Ubiquitination and Degradation

FoxO1 phosphorylation in response to insulin/IGF-1 results in becoming a target of ubiquitination and proteolytic degradation [53]. In fact, the efficient ubiquitination of FoxO1 depends upon its phosphorylation and cytoplasmic retention. Two studies carried out independently on the molecular basis of ubiquitin-mediated FoxO1 degradation showed that MDM2 serves as an E3 ligase to promote the ubiquitinations of FoxO1 and FoxO4 [54]. The cytosolic ubiquitination and degradation of FoxO1 are also stimulated by the E3 ligase activity of COP1 [55]. Furthermore, FoxO1 ubiquitination serves as a distinct posttranslational mechanism whereby insulin/IGF-1 inhibits FoxO1 activity in cells.

MODULATION OF FOXO1 BY CALORIE RESTRICTION

In aging, oxidatively altered DNA, proteins, lipids, and other cellular components, including antioxidative defense systems, lead to functional deficits, which underlie many age- and inflammatory-related degenerative diseases [56]. FoxO and SIRT1, considered as longevity factors, regulate directly or indirectly the inhibitors of inflammatory factor NF-κB [57], and well maintained FoxO and SIRT1 would guard cellular defenses against continual oxidative stress during aging [58]. The beneficial effects of CR on the aging process are its ability to maintain the integrity of FoxO and SIRT1, as reviewed by Redman and Ravussin [59]. It is evident that the application of CR has preventive and therapeutic potential in age-related disorders, such as, obesity, insulin resistance, type 2 diabetes, atherosclerosis, and cancer [60]. The age-related decrease in FoxO1 transcriptional activity is due to increased phosphorylation, whereas PI3K/Akt and NF-κB activation increase during aging, all which may closely be related to the increased levels of insulin and oxidative stress with age [25].

CR maintains well-regulated immune function and inhibits the inflammatory responses associated with aging [61]. CR also reduces NADH levels, a competitive inhibitor of SIRT1 [62], and decreases the blood insulin level, and the NAD/NADH ratio energy states converge to regulate FoxO by causing phosphorylation and acetylation of FoxO [48].

FoxO1 regulates the aging process in response to dietary cues, and in mammals, the dysregulation of insulin signaling has been associated with obesity and insulin resistance [63]. In addition, FoxOs induce the expression of many genes, including regulators of metabolism, cell cycle, cell death, and the oxidative stress response [64].

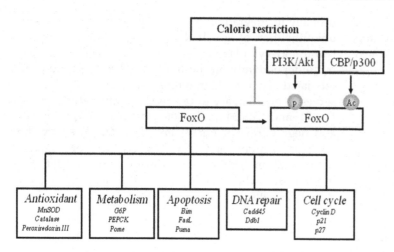

FIGURE 15.1 The functions of FoxO1 targeted genes and their modifications during aging and CR and their involvements during aging and in age-related diseases. *P*, phosphorylation; *Ac*, Acetylation; *MnSOD*, Manganese superoxide dismutase; *G6P*, Glucose-6-phosphatase; *PEPCK*, Phosphoenolpyruvate carboxykinase; *Pomc*, pro-opiomelanocortin; *Bim*, Bcl-2-like protein 11; *Puma*, p53 upregulated modulator of apoptosis; *FasL*, Fas ligand; *Gadd45*, Growth arrest and DNA damage-inducible protein 45; *Ddb1*, Damage-specific DNA-binding protein 1; *p21*, cyclin-dependent kinase inhibitor 1A; *p27*, cyclin-dependent kinase inhibitor.

These newly found effects of CR on the FoxO1, provide a broader molecular basis for the extremely diverse effects of CR, which is the most powerful epigenetic antiaging modifier identified to date.

CONCLUSION

Alterations in the physiology of FoxO family members during the normal aging process and effects of CR on FoxO activities can potentially produce profound changes in the course of aging and related metabolic disorders. A better understanding of FoxO and its alteration should provide intriguing insights into the precise involvement of FoxO in the aging process. Loss of FoxO1 is known to influence age-related insulin resistance and energy metabolism as highlighted under CR conditions, as depicted in Fig. 15.1. FoxO1 is regulated by different growth factors and hormones, and their activities are tightly controlled by posttranslational modifications such as phosphorylation, acetylation, and ubiquitination and transcription factors. Further observations of posttranslational modifications will undoubtedly provide new insights as to how FoxO1 conveys environmental stimuli into specific gene expressions and cellular functions to prevent age-dependent diseases.

SUMMARY

- Aging reflects all biological changes that predominantly show time-dependent, progressive, and physiological declines with increased risk of age-related diseases.
- Age-related redox imbalance is attributed by overproduction of reactive oxygen species or reactive nitrogen species, and deteriorated antioxidative defense mechanism.
- Calorie restriction (CR) retards aging and suppresses age-related deleterious disease, thus improves healthspan and lifespan.
- FoxO1 is a transcription factor which increases a variety of gene expressions related to metabolism, cell cycle, cell death, and oxidative stress responses.
- Phosphorylation and acetylation levels of FoxO1 increase through aging.
- Accumulated data through years of research demonstrate that CR is one of the most powerful epigenetic antiaging modulator. FoxO1 is associated with molecular mechanisms of the anti-inflammatory property of CR among various beneficial effects.

References

[1] Panici JA, Harper JM, Miller RA, Bartke A, Spong A, Masternak MM. Early life growth hormone treatment shortens longevity and decreases cellular stress resistance in long-lived mutant mice. FASEB J 2010;24:5073–9.

[2] Kops GJ, Dansn TB, Polderman PE, Saarloos I, Wirtz KW, Coffer PJ, et al. Forkhead transcription factor FOXO3a protects quiescent cells from oxidative stress. Nature 2002;419:316–21.

[3] Navab M, Gharavi N, Watson AD. Inflammation and metabolic disorders. Curr Opin Clin Nutr Metab Care 2008;11:459–64.

[4] Tereshina EV. Metabolic abnormalities as a basis for age-dependent diseases and aging? State of the art. Adv Gerontol 2009;22:129–38.

[5] Goldberg RB. Cytokine and cytokine-like inflammation markers, endothelial dysfunction, and imbalanced coagulation in development of diabetes and its complications. J Clin Endocrinol Metab 2009;94:3171–82.

[6] Cao Y, Nakata M, Okamoto S, Takano E, Yada T, Minokoshi Y, et al. PDK1-Foxo1 in agouti-related peptide neurons regulates energy homeostasis by modulating food intake and energy expenditure. PLoS One 2011;6:e18324.

[7] Mendelsohn AR, Larrick JW. Dietary restriction: critical cofactors to separate health span from life span bebefitis. Rejuvenation Res 2012;15:523–9.

[8] Fontana L, Partridge L, Longo VD. Extending healthy life span-from yeast to humans. Science 2010;328:321–6.

[9] Calvanese V, Lara E, Kahn A, Fraga MF. The role of epigenetics in aging and age-related diseases. Ageing Res Rev 2009;8: 268–76.

[10] Chouliaras L, van den Hove DL, Kenis G, Keitel S, Hof PR, van Os J, et al. Age-related increase in levels of 5-hydroxymethylcytosine in mouse hippocampus in prevented by caloric restriction. Curr Alzheimer Res 2012;9:536–44.

[11] Vaquero A, Reinberg D. Calorie restriction and the exercise of chromatin. Genes Dev 2009;23:1849–69.

[12] van der Heide LP, Hoekman MFM, Smidt MP. The ins and outs of FoxO shuttling: mechanisms of FoxO translocation and transcriptional regulation. Biochem J 2004;380:297–309.

[13] Accili D, Arden KC. FoxOs at the crossroads of cellular metabolism, differentiation, and transformation. Cell 2004;117:421–6.

[14] Barthel A, Schmoll D, Unterman TG. FoxO proteins in insulin action and metabolism. Trends Endocrinol Metab 2005;16: 183–9.

[15] Hoekman MF, Jacobs FM, Smidt MP, Burbach JP. Spatial and temporal expression of FoxO transcription factors in the developing and adult murine brain. Gene Expr Patterns 2006;6: 134–40.

[16] Biggs WH, Meisenhelder J, Hunter T, Cavenee WK, Arden KC. Protein kinase B/Akt-mediated phosphorylation promotes nuclear exclusion of the winged helix transcription factor FKHR1. Proc Natl Acad Sci USA 1999;96:7421–6.

[17] Brunet A, Bonni A, Zigmond MJ, Lin MZ, Juo P, Hu LS, et al. Akt promotes cell survival by phosphorylating and inhibiting a Forkhead transcription factor. Cell 1999;96:857–68.

[18] Kops GJ, de Ruiter ND, De Vries-Smits AM, Powell DR, Bos JL, Burgering BM. Direct control of the Forkhead transcription factor AFX by protein kinase B. Nature 1999;398:630–4.

[19] Matsuzaki H, Daitoku H, Hatta M, Aoyama H, Yoshimochi K, Fukamizu A. Acetylation of Foxo1 alters its DNA-binding ability and sensitivity to phosphorylation. Proc Natl Acad Sci USA 2005;102:11278–83.

[20] Furukawa-Hibi Y, Kobayashi Y, Chen C, Motoyama N. FOXO transcription factors in cell-cycle regulation and the response to oxidative stress. Antioxid Redox Signal 2005;7:752–60.

[21] Yamaza H, Komatsu T, Wakita S, Kijogi C, Park S, Hayashi H, et al. FoxO1 is involved in the antineoplastic effect of calorie restriction. Aging Cell 2010;9:372–82.

[22] Senapedis WT, Kennedy CJ, Boyle PM, Silver PA. Whole genome siRNA cell-based screen links mitochondria to Akt signaling network through uncoupling of electron transport chain. Mol Biol Cell 2011;22:1791–805.

[23] Jian B, Yang S, Chen D, Chaudry I, Raju R. Influence of aging and hemorrhage injury on Sirt1 expression: possible role of myc-Sirt1 regulation in mitochondrial function. Biochim Biophys Acta 2011;12:1446–51.

[24] Yu BP. Calorie restriction as a potent anti-aging intervention: suppression of oxidative stress. In: Rattan S, editor. Aging intervention and therapies. World Sci. Pub; 2005. p193–217.

[25] Kim DH, Kim JM, Lee EK, Choi YJ, Kim CH, Choi JS, et al. Modulation of FoxO1 phosphorylation/acetylation by baicalin during aging. J Nutr Biochem 2012;23:1277–84.

[26] Motta MC, Divecha N, Lemieux M, Kamel C, Chen D, Gu W, et al. Mammalian SIRT1 represses forkhead transcription factors. Cell 2004;116:551–63.

[27] Fraga ME, Ballestar E, Paz MF, Ropero S, Setien F, Ballestar ML, et al. Epigenetic differences arise during the lifetime of monozygotic twins. Proc Natl Acad Sci USA 2005;102:10604–9.

[28] Ribarič S. Diet and aging. Oxid Med Cell Longev 2012;2012: 741468.

[29] Li Y, Daniel M, Tollefsbol TO. Epigenetic regulation of calorie restriction in aging. BMC Med 2011;9:98.

[30] Stenvinkel P, Karimi M, Johansson S, Axelsson J, Suliman M, Lindholm B, et al. Impact of inflammation on epigenetic DNA methylation – a novel risk factor for cardiovascular disease? J Intern Med 2007;261:488–99.

[31] Issa JP, Ahuja N, Toyota M, Bronner MP, Brentnall TA. Accelerated age-related CpG island methylation in ulcerative colitis. Cancer Res 2001;61:3573–7.

[32] Giannakou ME, Partridge L. The interaction between FOXO and SIRT1: tipping the balance towards survival. Trends Cell Biol 2004;14:408–12.

[33] Kouzarides T. Histone methylation in transcriptional control. Curr Opin Genet Dev 2002;12:198–209.

[34] Guarente L, Picard F. Calorie restriction – the SIR2 connection. Cell 2005;120:473–82.

[35] Duan W. Sirtuins: from metabolic regulation to brain aging. Front Aging Neurosci 2013;5:36.

[36] Erol A. Insulin resistance is an evolutionarily conserved physiological mechanism at the cellular level for protection against increased oxidative stress. BioEssays 2007;29:811–18.

[37] Fabrizio P, Pozza F, Pletcher SD, Gendron CM, Longo VD. Regulation of longevity and stress resistance by Sch9 in yeast. Science 2001;292:288–90.

[38] Tatar M, Kopelman A, Epstein D, Tu MP, Yin CM, Garofalo RS. A mutant Drosophila insulin receptor homolog that extends lifespan and impairs neuroendocrine function. Science 2001;292: 107–10.

[39] Bluher M, Kahn BB, Kahn CR. Extended longevity in mice lacking the insulin receptor in adipose tissue. Science 2003;299: 572–4.

[40] Honda Y, Honda S. The daf-2 gene network for longevity regulates oxidative stress resistance and Mn-superoxide dismutase gene expression in Caenorhabditis elegans. FASEB J 1999;13: 1385–93.

[41] Burgering BM, Medema RH. Decisions on life and death: FOXO Forkhead transcription factors are in command when PKB/Akt is off duty. J Leukoc Biol 2003;73:689–701.

[42] Yiu WH, Mead PA, Jun HS, Mansfield BC, Chou JY. Oxidative stress mediates nephropathy in type Ia glycogen storage disease. Lab Invest 2010;90:620–9.

[43] Huang H, Regan KM, Lou Z, Chen J, Tindall DJ. CDK2-dependent phosphorylation of FOXO1 as an apoptotic response to DNA damage. Science 2006;314:294–7.

[44] Yan L, Lavin VA, Moser LR, Cui Q, Kanies C, Yang E. PP2A regulates the pro- apoptotic activity of FOXO1. J Biol Chem 2008;283:7411–20.

[45] Nasrin N, Ogg S, Cahill CM, Biggs W, Nui S, Dore J, et al. Alexander-Bridges MC. DAF-16 recruits the CREB-binding protein coactivator complex to the insulin-like growth factor binding protein 1 promoter in HepG2 cells. Proc Natl Acad Sci USA 2000;97:10412–17.

[46] van der Heide LP, Smidt MP. Regulation of FoxO activity by CBP/p300-mediated acetylation. Trends Biochem Sci 2005;30: 81–6.

[47] Perrot V, Rechler MM. The coactivator p300 directly acetylates the forkhead transcription factor foxo1 and stimulates foxo1-induced transcription. Mol Endocrinol 2005; 19:2283–98.

[48] Daitoku H, Hatta M, Matsuzaki H, Aratani S, Ohshima T, Miyaqishi M, et al. Silent information regulator 2 potentiates Foxo1-mediated transcription through its deacetylase activity. Proc Natl Acad Sci USA 2004;101:10042–7.

[49] Essers MA, Weijzen S, deVries-Smits AM, Saarloos I, de Ruiter ND, Bos JL, et al. FOXO transcription factor activation by oxidative stress mediated by the small GTPase Ral and JNK. EMBO J 2004;23:4802–12.

[50] Brunet A, Sweeney LB, Sturgill JF, Chua KF, Greer PL, Lin Y, et al. Stress-dependent regulation of FOXO transcription factors by the SIRT1 deacetylase. Science 2004;303:2011–15.

[51] Frescas D, Valenti L, Accili D. Nuclear trapping of the forkhead transcription factor FoxO1 via Sirt-dependent deacetylation promotes expression of glucogenetic genes. J Biol Chem 2005;280:20589–95.

[52] Mercken EM, Hu J, Krzysik-Walker S, Wei M, Li Y, McBurney MW, et al. SIRT1 but not its increased expression is essential for lifespan extension in caloric-restricted mice. Aging Cell 2014;13:193–6.

[53] Aoki M, Jiang H, Vogt PK. Proteasomal degradation of the FoxO1 transcriptional regulator in cells transformed by the P3k and Akt oncoproteins. Proc Natl Acad Sci USA 2004;101:13613–17.

[54] Brenkman AB, de Keizer PL, van den Broek NJ, Jochemsen AG, Burgering BM. Mdm2 induces mono-ubiquitination of FOXO4. PLoS One 2008;3:e2819.

[55] Kato S, Ding J, Pisck E, Jhala US, Du K. COP1 functions as a FoxO1 ubiquitin E3 ligase to regulate FoxO1-mediated gene expression. J Biol Chem 2008;283:35464–73.

[56] Yu BP. Aging and oxidative stress: modulation by dietary restriction. Free Radic Biol Med 1996;21:651–68.

[57] Lee EK, Kim JM, Choi J, Jung KJ, Kim DH, Chung SW, et al. Modulation of NF-κB and FOXOs by baicalein attenuates the radiation-induced inflammatory process in mouse kidney. Free Radic Res 2011;45:507–17.

[58] van der Horst A, Tertoolen LG, de Vries-Smits LM, Frye RA, Medema RH, Burgering BM. FoxO4 is acetylated upon peroxide stress and deacetylated by the longevity protein hSir2 (SIRT1). J Biol Chem 2004;279:28873–9.

[59] Redman LM, Ravussin E. Caloric restriction in humans: impact on physiological, psychological, and behavioral outcomes. Antioxid Redox Signal 2011;14:275–87.

[60] Yu BP, Chung HY. The inflammatory process in aging. Rev Clin Gerontol 2006;16:179–87.

[61] Longo VD, Finch CE. Evolutionary medicine: from dwarf model systems to healthy centenarians? Science 2003;299:1342–6.

[62] Lin SJ, Ford E, Haigis M, Liszt G, Guarente L. Calorie restriction extends yeast life span by lowering the level of NADH. Genes Dev 2004;18:12–16.

[63] Cameron AR, Anton S, Melville L, Houston NP, Dayal S, McDougall GJ, et al. Black tea polyphenols mimic insulin/insulin-like growth factor-1 signaling to the longevity factor FoxO1a. Aging Cell 2008;7:69–77.

[64] Hedrick SM. The cunning little vixen: FoxO and the cycle of life and death. Nat Immunol 2009;10:1057–63.

CHAPTER

16

Epigenetic Responses to Diet in Aging

Dianne Ford

Human Nutrition Research Centre and Institute for Cell and Molecular Biosciences, Newcastle University, Newcastle upon Tyne, UK

KEY FACTS
- DNA exists as part of a complex structure, in association with histone proteins.
- The DNA and histone proteins can undergo a range of specific chemical modifications that can be referred to as "epigenetic marks."
- These epigenetic marks do not change the genetic code as held in the DNA sequence but they do affect how the code is unlocked.
- Epigenetic marks differ between young and old cells; while some of these differences may be simply "bystander" effects of aging, some of them may actually affect how the cell functions and cause it behave as a more "aged" cell.
- Diet, and specific components of food, can affect epigenetic marks so could affect aging through this mechanism.
- Epigenetic marks are most plastic, and are most readily affected by diet, during early development.
- Some age-related changes in epigenetic marks are seen in all tissues and in all individuals. These epigenetic marks may be most useful in terms of providing a "biological readout" of aging.
- Some age-related changes in epigenetic marks may occur at the level of the stem cells from which the mature tissues develop. These epigenetic marks may be the most important with respect to them having functional consequences.
- It is not currently a realistic ambition to reverse or slow down aging specifically through targeting particular epigenetic changes using specific nutritional interventions. However, the current knowledge could be developed to provide a readout of aging that responds to diet and can be used to test if dietary changes are having beneficial actions with respect to aging.

Dictionary of Terms
- *Chromatin*: macromolecular structural "packing" arrangement of DNA in the cell nucleus, including association with histone proteins.
- *CpG site*: DNA sequence comprising cytosine (5′) followed by guanine (3′) within a stretch of DNA, which is a substrate for DNA methylation (on cytosine) in the mammalian genome.
- *DNA methylation*: the addition of a methyl group (chemical group composing one carbon unit) to DNA.
- *Epigenetics*: the chemical modification of the genome (including DNA and histone protein) in a manner that is heritable (passed on through cell division).
- *Epigenetic drift*: the alteration in cell epigenetics that is observed over time/with aging.
- *Epigenome*: epigenetic modifications in a cell in their entirety.
- *Euchromatin*: a configuration of chromatin in which genes are actively expressed (as a result of epigenetic modifications that allow the DNA to be accessible to factors required to drive expression).

- *Heterochromatin*: a configuration of chromatin in which genes are not actively expressed (as a result of epigenetic modifications that "tighten" the association between DNA and histone proteins).
- *Histone covalent modification*: the addition of small chemical groups through covalent bonding to the histone proteins (a form of epigenetic modification).
- *Polycomb group protein gene targets (PCGTs)*: a set of genes that are repressed by the binding of polycomb repressive regulatory protein complexes and that are typically in this inactive state in stem cells.
- *Polycomb repressive complex*: a complex of proteins (polycomb proteins) that bind to and repress particular genes (polycomb group protein gene targets) as a result of effecting epigenetic changes.
- *Sirtuins*: proteins belonging to a family (histone deacetylases) that appear to have roles in aging and/or determining lifespan.
- *Transcription*: the process of "copying" DNA into RNA so that genes are eventually expressed (as proteins).

INTRODUCTION

Features of the epigenetic architecture, including DNA methylation, histone covalent modification, chromatin structure, and small noncoding RNAs, change during aging. It is an established principle that diet, or specific dietary components, can also affect these epigenetic marks. Thus the idea that aging could be slowed through dietary interventions that target epigenetic processes is seductive. However, there are substantial caveats to this suggestion. These include the likelihood that many of these observed epigenetic events may be nonfunctional with respect to affecting the process of aging. Also epigenetic alterations are one component of a plethora of molecular events that contribute to the aging process. While age-related epigenetic drift can be shaped by the environment, there appears to be an underlying genetic influence, and there may be specific windows of nutritional epigenomic plasticity during the life-course when age-protective interventions can be of benefit. Better understanding of the interactions between nutrition, epigenetics, and aging will begin to address these points. Other questions to consider include whether beneficial dietary influences on aging mediated through epigenetic actions directly reverse age-related epigenetic drift or if they have separate but beneficial actions, and if effects of diet on specific cell types—notably stem cells—are particularly pertinent. Better understanding at the fundamental level of how elements of the diet modify the activity of the epigenetic machinery may reveal that specific components have actions that make them particularly attractive targets to develop as diet-based interventions to counteract aging. In this context components that act on targets with key roles in aging, such as sirtuins and polycomb group proteins gene targets (PCGTs), may be particularly relevant. This chapter explores all of these questions, then looks toward to a realistic application of the emerging knowledge over the short to medium term to guide nutritional intervention in the aging process.

EPIGENETIC MODIFICATION OF THE MAMMALIAN GENOME

The concept of epigenetic modification encompasses DNA-sequence-independent genome features that are heritable though cell division. An understanding of epigenetic modification requires knowledge of the structure of chromatin—the macromolecular arrangement of DNA in the nucleus [1]. DNA and histone proteins are the two major constituents of chromatin. Their intimate association forms the nucleosome, comprising approximately 146 base pairs of DNA wound as 1.65 turns of a left-handed helix around a core of basic (positively charged) histone proteins. This histone protein "core" is an octamer, comprising two molecules each of histones H2A, H2B, H3, and H4. The continuous DNA molecule links multiple histone protein cores, in a structure often likened to beads on a string. In mammalian somatic cells the DNA linker region between nucleosomes is approximately 60 bp [2] and the linker histone H1 covers approximately 10 bp as DNA enters and exits the core octamer in an estimated 70% of mammalian nucleosomes. Epigenetic modifications affect gene expression essentially by rendering the DNA in specific regions as "active" euchromatin, in which the interaction between DNA and the histone proteins is such that the DNA is in a "looser" configuration and thus more accessible to regulatory proteins such as transcription factors, or "repressed" heterochromatin, which can be visualized as a structure in which the DNA is more tightly associated with the histone proteins and thus in a configuration less accessible to transcription factors and other regulatory proteins. This chromatin configuration is labile, rather than fixed, and can be influenced by extrinsic factors, including diet, and also by aging.

Epigenetic modifications of the DNA and histone proteins affect chromatin configuration, and are where age and dietary factors have their influence. Epigenetic modification of mammalian DNA comprises mainly methylation on the C5 position of cytosine, which can occur when the cytosine is the 5′ base of a CG pair (CpG dinucleotide). Most CpGs in the genome are methylated, except where they occur at high density in gene promoter regions in clusters often termed CpG islands. CpG islands occur in the promoter regions of approximately 60% of mammalian genes. DNA methylation of CpG islands can occur, and is generally associated with transcriptional repression and with the chromatin being in the inactive heterochromatin configuration. DNA methylation of additional regions, typically of lower CpG density, up to 2 kb upstream of CpG islands, which have been termed CpG island shores, appears also to influence gene expression [3]. Hydroxymethylation of DNA [4] has received less attention than methylation. Thus, current understanding of its functional importance, relationship with aging, and plasticity with respect to dietary remodeling lags well behind that of DNA methylation. However, the potential importance of this epigenetic modification with respect to modifying effects of diet on aging should not be overlooked.

Histone proteins have N-terminal "tails" that are rich in basic amino acid residues and extend out from the more globular histone core. These regions are subject to covalent modifications, which include acetylation at lysine residues on the N-terminal tails of all four core histones and methylation of lysine and arginine residues in the N-terminal tails of histones H3 and H4 [5]. Other covalent modifications to histone proteins include phosphorylation, ubiquitination, glycosylation, ADP-ribosylation, and crotonylation [5]. As a generalization, histone acetylation, though reducing the charge-charge interaction between the positively charged histone tail lysine residues and the negatively charged phosphate residues of the DNA backbone, opens up the nucleosome structure so favors transcriptionally active euchromatin [5]. Histone acetylation may also promote transcription through providing binding sites for transcriptional coactivators. Specific patterns of histone methylation are associated with different chromatin states and thus with transcriptionally active or silent chromatin. For example, methylation of H3K4 and H3K36 is generally associated with gene expression while methylation of H3K3 and H3K27 is generally associated with transcriptional repression (for reviews see Refs [1] and [5]).

Modification of gene expression by small noncoding RNAs is considered an additional form of epigenetic modification and another mechanism susceptible to influences of both diet and aging. Current knowledge about the effects of diet and aging on small noncoding RNAs lags behind understanding of impacts on other forms of epigenetic modification and the likely multiple points of interaction between diet and aging mediated through these molecules must still be uncovered.

EPIGENETIC ALTERATIONS ASSOCIATED WITH AGING: POSSIBLE FUNCTIONAL CONSEQUENCES AND RELATIONSHIP TO A GLOBAL SIGNATURE OF AGING

Differences in epigenetic features between younger and older mammals have been observed with respect to DNA methylation, histone modification, and chromatin architecture. The term "epigenetic drift" has been used to describe these differences. Given the importance of all of these features in the regulation and maintenance of genome function and stability it is highly likely that many of these changes have a causal impact on the aging process. However, showing directly that these observed differences are more than simple bystander effects of cellular aging is challenging. Thus, a direct functional influence of many recorded age-related epigenetic changes has not been demonstrated. Another important point to make that also relates to the need to take a critical and balanced view when considering age-related epigenetic drift is that epigenetic changes observed during aging are only one feature of what could be considered a much more complex signature of the aged phenotype. A recent review identifies 9 interacting elements of this hallmark [6], in which epigenetic alteration features along with genomic instability, telomere attrition, loss of proteostasis, deregulated nutrient sensing, mitochondrial dysfunction, cellular senescence, stem cell exhaustion, and altered intercellular communication. While the authors elegantly rehearse the point that these features are highly interrelated and none is boundaried, the accumulation of advanced glycation end products [7] is added to the representation of these interacting hallmarks shown in Fig. 16.1.

Age-Related Alterations in DNA Methylation

It is often stated that globally DNA methylation decreases with age, reflected, for example, in a reduction in total methylcytosine content of the genome [8,9]. However, this generalization appears not to hold true for all individuals, as longitudinal measurement over 11–16 years in two human cohorts showed that global DNA methylation increased in some individuals [8]. Mapping of changes in DNA methylation to specific sites has revealed both increased and reduced age-related local DNA methylation [10–15], some of which appear to be tissue-specific and others tissue-independent [10,13]. As emphasized already, it is difficult to show direct causal links between these epigenetic changes and features of age-related impaired genome function, such as a breakdown

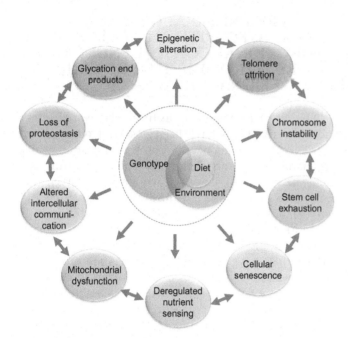

FIGURE 16.1 Interacting hallmarks of aging and extrinsic modifying factors. Nine of the ten hallmarks of aging shown as the periphery of the figure are as previously defined [6]. Accumulation of glycation end products is an added feature. The figure emphasizes that epigenetic alteration is only one of many hallmarks of aging, but that all of these hallmarks are interdependent. The extrinsic modifying factors of genotype and environment are represented in the centre of the figure, and diet is shown as only one component of the environment that can influence the features of aging, including epigenetic alteration.

in transcriptional fidelity, RNA processing errors, impaired DNA repair, and chromosomal instability [6]. However there are arguments to support causality. For example, the fundamental role of epigenetic reprogramming in the process of gamete formation, which must reverse the aging clock to prevent progressively shortened lifespan in each successive generation, provides a compelling argument to support this view; likewise the role of epigenetic reprogramming to restore pluripotency in the success of somatic cell nuclear transfer [16]. Also consistent with the premise that epigenetic changes contribute to aging is that extended lifespan can be inherited transgenerationally in *Caenorhabditis elegans* via genes that are components of a major epigenetic modifier—the histone H3 lysine 4 trimethylation (H3K4me3) complex [17]. Other indications that (at least some specific) changes in DNA methylation have direct causal links with aging include the fact that some changes cluster in particular at the gene targets of polycomb group proteins (PCGTs) [10,14,18], which are important determinants of stem cell differentiation that are repressed in stem cells by mechanisms

involving chromatin modification [19] and tend to be hypermethylated in cancer [20–22]. A further caveat relating to interpreting observed alterations in epigenetic modifications in the context of aging is that differences measured in samples prepared from tissues from younger compared with older animals could reflect a change in the distribution of cell subpopulations that carry different epigenetic marks rather than being a readout of true "epigenetic drift" that occurs on the subcellular level. The results of studies that measured age-related changes in DNA methylation in whole blood exemplify this issue. A number of such studies have shown that age-related changes in DNA methylation in whole blood are linked to a change in the profile of different white blood cell populations [23–25]. However, while some age-related changes in DNA methylation profiles that have been measured are tissue-specific others are tissue-independent [13]. Tissue-independent changes in DNA methylation could be considered robust signatures of epigenetic drift that occurs on the subcellular level. Such tissue-independent signatures have been shown to include 69 CpGs in the promoters of PCGTs [14], 71 CpG sites measured in blood that were highly predictive of age and translated to other tissues [11] and 353 CpG sites proposed to comprise an "epigenetic clock" measured across multiple tissues and cell types and also predictive of age [26]. Importantly, use of both of the latter two signatures to determine the difference between actual chronological age and DNA-methylation-predicted age in a meta-analysis of 4 longitudinal cohorts of older people showed that a difference ("accelerated DNA-methylation aging") of 5 years was associated with a 16–21% higher mortality risk [27]. Interactome nodes of age-related DNA methylation changes discovered through a systems approach further point to age-related DNA methylation drift as disturbing in particular in stem cell differentiation pathways and being tissue-independent [28].

Age-Related Alterations in Histone Modification

Changes in histone modifications observed in older compared with younger mammals include increased histone H4K20 trimethylation and reduced H3K4 trimethylation. Specifically, an increase in H4K20 trimethylation was detected in liver and kidney of older compared with younger rats [29] and was also seen as a feature of the premature aging syndrome Hutchinson–Gilford progeria [30]. H3K4me3 peaks were detected at a larger number of loci in neurons from the prefrontal cortex of infants (<1 year) compared with old adults (>60 years) [31]. Moreover, genetic manipulation in *C. elegans* of components of the protein complexes responsible for maintaining the level of H3K4me3 affect lifespan [32] suggesting a possible causal link between this epigenetic mark and aging.

Age-Related Alterations in Chromatin Architecture

A redistribution of heterochromatin [33] plus net loss [34] appear to be characteristic features of aging cells. This change in chromatin architecture may be the manifestation of observed diminished levels or function in aged cells of proteins that modify chromatin state [35], including chromatin remodeling complexes, such as polycomb group proteins [36], and also components of the NuRD complex [37]. The potential of age-related changes in chromatin architecture to contribute to cellular aging is supported by the observation that loss of function mutations in the *Drosophila* heterochromatin protein HP1α shortened lifespan in a manner rescued by overexpression of HP1α [38].

Age-Related Alterations in Small, Noncoding RNAs

Comparative transcriptional profiling of tissues from younger and older animals has identified among the signatures of age-related change noncoding RNAs, including miRNAs, that target genes that affect stem cell function and/or signaling pathways implicated in determining lifespan [6]. There is thus a strong indication that age-related changes in miRNA expression have an impact on the aging process. Proof of concept that miRNAs can determine lifespan comes from work in model organisms, in particular *Drosophila* and *C. elegans*, that shows changes in lifespan in response to loss or gain of function of specific miRNAs [6].

THE ROLE OF ENVIRONMENTAL VERSUS GENETIC FACTORS IN SHAPING AGING-RELATED EPIGENETIC DRIFT

The fact that the epigenome is plastic invites hypotheses concerning the potential of environmental factors, including diet, to have modifying actions. However, for a balanced view it is important to be aware that genetic factors also may have substantial influence over epigenetic marking (as applies also to other characteristic features of aging; Fig. 16.1). Convincing evidence revealing an impact of genetics on age-related epigenetic drift was the observed clustering of tendency to gain or lose global DNA methylation among family members in a cohort from Utah studied longitudinally over an average of 11 years [8]. Also a difference between chronological age and "biological" age based on DNA methylation profiles was shown to have a heritable component [27].

THE IMPACT OF DIET ON EPIGENETIC ALTERATIONS ASSOCIATED WITH AGING

Arguably, the epigenome may be more susceptible to diet-driven alterations when endogenous remodeling mechanisms are most active, as opposed to under conditions where the epigenome remains relatively static. Such reasoning points toward the period before birth as a likely susceptibility window. In support, a plethora of studies reveal that suboptimal early life nutrition affects epigenetic marks [39]. Also, it is a well-established principle that early life nutrition, and particularly nutrition *in utero*, has consequences for lifelong health. More recent work links suboptimal early nutrition to measures of accelerated cellular aging, including DNA damage and shortening of telomeres (reviewed in Ref. [40]). However, direct mediating effects of these epigenetic changes on the later phenotypic manifestations of aging or age-related diseases have not been demonstrated robustly, and the supportive evidence at present is tenuous.

Recent data indicate that the pattern of accumulation of epigenetic alterations over the life-course *ex utero* is, unsurprisingly, nonlinear and point to early life as a period of more rapid epigenomic change, when dietary influences on aging mediated through epigenetic alterations are thus likely to be most marked. A longitudinal study in twins revealed that methylation of buccal DNA changed markedly between birth and 18 months [41]. At 12 months there was a significant 3% difference in DNA methylation across the genome compared with the profile at birth. Comparison with data on the magnitude of DNA methylation changes observed in people differing in age by several decades—typically in the order of 10–20% [11,14]—highlights the relative rapidity with which DNA methylation is altered during very early life. The findings of a second study that measured DNA methylation in a pediatric cohort, comprising of boys between the ages of 3 and 17 years, were that the profile of DNA methylation in whole blood samples accounted already for most of the variability seen between older adults and mapped to these same loci [42], which suggests that any diet-induced changes that occur during early years are likely to be stable and could thus influence aging.

EPIGENETIC ACTIONS OF SPECIFIC DIETARY COMPONENTS AND PRACTICES

An early indication of the potential of diet to modify gross chromatin structure comprised the observation that the susceptibility of rat liver chromatin to digestion by micrococcal nuclease was altered as a function of diet, with the amount of DNA susceptible to digestion ranging from 71.4% (in rats fed a high-carbohydrate, fat-free diet) to 38.8% (in rats fed a low carbohydrate, protein-free diet) [43].

A large body of work, using predominantly cell line and rodent models but also including research in humans, has shown effects of diet or specific nutrients on epigenetic marking. Other works provide more comprehensive accounts of this body of research (eg, Ref. [39]). An overview of selected dietary components or specific nutritional interventions that have been shown to affect DNA methylation, histone covalent modification, and/or small noncoding RNAs is given below, to provide background and context to evaluating the likelihood that some of these actions translate into effects on aging.

Effects of Diet on DNA Methylation

Methyl Donors

Methyl groups (1-carbon units) for the methylation of DNA are donated by S-adenosylmethionine (SAM) in the reaction catalyzed by the DNA methyltransferases (DNMTs). SAM is generated from 1-carbon donors in the diet, such as folate, methionine, betaine, and vitamin B12, through the folate cycle. The existence of this direct mechanism though which altering the dietary supply of these compounds could alter DNA methylation has been the motivation for much of the research on their potential epigenetic actions. Proof of concept that these agents can have profound effects on DNA methylation is borne out by studies in cultured cells and animal models. The variable yellow *agouti* (A^{vy}) mouse has provided an elegant model in experiments that demonstrate the ability of diets modified with respect to one carbon donors to affect DNA methylation. In this model a genetic insertion at the 5′end of the *agouti* gene provides a cryptic promoter that is suppressed by DNA hypermethylation, resulting in a yellow, rather than brown, coat color (and also associated predisposition to obesity, diabetes, and cancer) [44–46]. Observed directions of change in DNA methylation in response to restricting the supply of 1-carbon units in the diet, in particular at specific genes (as opposed to global DNA methylation), appears variable. As examples, depletion then supplementation of folate induced reversible changes in global and p53 locus-specific DNA methylation in colonic adenocarcinoma cells (SW620 cell line) [47]. Here the direction of change showed a (perhaps more predictable) positive association with folate availability. Similarly, folate deficiency in weanling male rats induced hypomethylation of the p53 tumor suppressor gene after 6 weeks in liver [48] and choline deficiency in pregnant mice induced global and site-specific (*Cdkn3* gene) DNA hypomethylation in fetal brain [49]. In contrast, folate depletion caused hypermethylation of the CpG island of the H-cadherin gene in nasopharyngeal carcinoma cells [50] and a flate-deficient diet induced transient global DNA hypermethylation in weanling male rats [51]. Intervention studies have shown that manipulation of the dietary folate supply can affect DNA methylation also in human subjects. For example, it was found in two separate studies that a low folate diet led to hypomethylation of leukocyte DNA in postmenopausal women, which was revered by subsequent dietary folate supplementation [52,53], and in patients with colorectal adenoma a dietary folate supplement increased DNA methylation in leucocytes and colonic mucosa [54].

Bioactive Phytochemicals

Among the various dietary bioactive phytochemicals shown to affect DNA methylation [39] there is particularly strong evidence that the soyabean isoflavones, in particular genistein, can be active. As examples, genistein, and to a lesser extent the isoflavones biochanin A and daidzein, reversed gene-specific ($p16^{INK4a}$, $RAR\beta$, and *MGMTl*) hypermethylation in both esophageal (KYSE 510) and prostate (PC3 and LNCaP) carcinoma cell lines [55]. In mice, genistein in the diet given postweaning increased DNA methylation at some CpG islands [56] and in the variable yellow *agouti* (A^{vy}) mouse genistein *in utero* affected coat color (a readout of DNA methylation as explained above), and increased DNA methylation at the A^{vy} locus [57]. Significantly, in this latter study these effects persisted into adulthood and were associated with protection against obesity.

Zinc

It seems highly likely that zinc should affect epigenetic modification, given that enzymes involved in epigenetic modification, including the DNA methyltransferases (DNMTs) and histone deacetylases (HDACS) [58], are zinc metalloproteins. Evidence that zinc can have epigenetic effects includes the observation that depressed immune function in mice resulting from a single gestational exposure to zinc deficiency was inherited transgenerationally [59] and a zinc deficient diet in rats lad to that global DNA hypomethylation in the liver [60].

Selenium

Examples of published work showing that DNA methylation can be influenced by the availability of selenium include the observations that removal of selenium from culture medium induced DNA hypomethylation in human intestinal Caco-2 [61] and HT-29 [62] cells and reduced methylation of the *p53* promoter in Caco-2 cells [61]. Also, a selenium-deficient diet induced global DNA hypomethylation in rat liver and colon [61,63].

Vitamin A

Observed effects of vitamin A or its metabolites on DNA methylation include an effect of the active metabolite all-*trans* retinoic acid (ATRA) to demethylate the promoter of the tumor suppressor gene *RARβ2*, possibly only in selected cell types [64,65]. ATRA, but not vitamin A (retinyl palmitate) or 13-*cis*-retinoic acid, in the diet induced global hypomethylation of hepatic DNA in rats [66]. Thus it appears that that only in certain cell types and in certain forms can vitamin A and its derivatives affect DNA methylation.

Protein-Restricted Diet

A low level of protein in the maternal diet during development *in utero* is perhaps the nutritional assault thus far linked most robustly with phenotypic manifestation in older animals through epigenetic action, in particular effects on DNA methylation. In rodents, this intervention is typically associated with low birth weight followed by development of a metabolic phenotype that mimics some of the characteristics of the metabolic syndrome in humans [67,68]. Parallel transgenerationally inherited effects on DNA methylation and expression of genes that are likely mediators of these phenotypic outcomes (eg, *PPARα* (peroxisome proliferator-activated receptor α], *GR1* (glucocorticoid receptor 1), *Acaca* (aceyl-CoA oxidase), and *PEPCK* (phospho-enol-pyruvate carboxykinase) in liver) have been observed [69–71].

Effects of Diet on Histone Modification

Research on the impact of specific dietary agents or interventions to affect histone modification has lagged behind studies on effects on DNA methylation. The body of published data on this topic is thus comparatively small, but this is not necessarily an indication that diet is any less potent a modifier of histone modification than it is of DNA methylation. The short chain fatty acid butyrate, which is generated in the large bowel as a result of bacterial fermentation of dietary fiber, has received particular attention as a dietary-derived compound likely to affect histone modification because of its well-established action as a histone deacetylase (HDAC) inhibitor [72]. A simplistic expectation would be that butyrate should increase histone acetylation and thus exposure generally would be associated with chromatin being in the more active euchromatin configuration. Many observations, largely based on exposure of cell line models in vitro, support this prediction and show effects of butyrate to increase acetylation of histones H3 and H4 (eg, Refs [73–75]). However, local variation in this response occurs, as revealed in HepG2 and colon adenocarcinoma HT29 cells where butyrate led to deacetylated regions close to transcription start sites and concomitant gene repression in spite of an overall increase in the total level of H3 and H4 acetylation [74].

The focus of work exploring the capacity of other dietary agents shown to affect histone modification has been generally (though not exclusively) on histone acetylation and methylation, and currently most of the evidence for effects is from work in cell line models and rodents. From this body of work dietary agents that emerge as having actions include organosulfur compounds, in particular the isothiocyantes, methyl donors, and retinoic acid. As examples: the garlic-derived organosulfur compound diallyl sulfide increased hsitoneacetylation selectively (H3K14, H3K9, H4K12 and H4K16, but not H4K9) and transiently in the Caco-2 human intestinal cell line [76]; trimethylation at H3K9 was reduced in the liver of rats fed a methyl-deficient diet [77]; all-*trans* retinoic acid increased H3 and H4 acetylation at the retinoic acid receptor (RAR) β P2 promoter in the T47D breast cancer cell line accompanied by a concomitant increase in RARβ2 mRNA levels [65]. Protein restriction is highlighted above as an intervention that in utero has been seen to have effects on DNA methylation in mice that are transgenerationally inherited and associated with effects on adult health. Observed effects on histone modification concomitant with these effects on DNA include a shift in the modification state of histones at the glucocorticoid receptor *GR110* promoter in liver toward a more active chromatin configuration in parallel with DNA hypomethylation in this region and increased expression of the corresponding RNA transcript [70].

Effects of Diet on Small Noncoding RNAs

There is good evidence, based on a body of published work, that diet or specific dietary elements can affect the profile of noncoding RNAs, in particular miRNAs, in various cells and tissues, as well as in plasma [78]. For example zinc depletion and repletion in young males affected reversibly a specific serum miRNA signature that included miRNAs associated with inflammation [79], and other work uncovered associations between miRNA profiles and vitamin D status in women during early pregnancy [80]. Also, selenium depletion in human intestinal Caco-2 cells altered the expression of miRNAs that included a group predicted to target selenoprotein transcripts [81]. More provocative is the suggestion that miRNAs in food may survive intact to cross the gut epithelium into the circulation, from where they could potentially have direct and profound functions. While an intriguing notion, the evidence that this process occurs remains equivocal [78].

INTERRELATIONSHIPS BETWEEN AGE-ASSOCIATED AND DIET-INDUCED EPIGENETIC ALTERATIONS

The malleability of the epigenetic architecture and the ability of components of the diet to modify epigenetic features together suggest that modification of food intake is a plausible route to slow the aging process and protect against age-related disease. Factors in the diet could in principle protect against the effects of aging-related epigenetic drift through one of three fundamental mechanisms: they could stabilize the epigenome such that the accumulation of epigenetic "damage" is slowed, they could reverse the age-related accumulation of epigenetic alterations, or they could induce "de novo" epigenetic changes that are in some way protective. These different potential actions of diet are shown in Fig. 16.2. At present, there is an insufficient relevant body of knowledge to evaluate which, if any, of these mechanisms operates to link diet to the aging trajectory. However, the likelihood that the relationship between nutrition, epigenetic alteration, and aging is even predominantly linear and causal is remote. While there may well be instances where a specific nutrient or dietary practice induces an epigenetic effect that then impacts on the aging process, and indeed may directly reverse "pro-aging" epigenetic changes, the interrelationships between aging, epigenetic alteration, and diet are undoubtedly more complex, as represented in Fig. 16.3. Nutrition will undoubtedly affect aging through processes independent of any element of epigenetic action. Even where parallel effects of diet on both epigenetic marks and aging are clear, a causal relationship cannot be assumed, and, as emphasized already, showing the causality of such links is difficult.

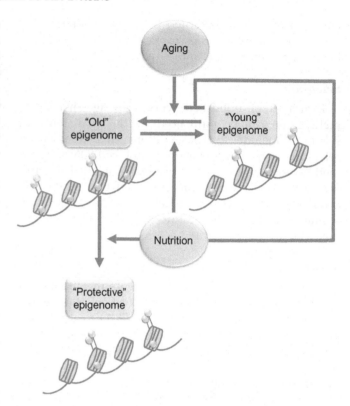

FIGURE 16.2 Possible points of protective actions of nutrition on epigenetic features of aging. Alterations in epigenetic features, shown as a change from a "young" epigenome to an "old" epigenome, are observed during aging. Protective nutritional influences could potentially halt or slow this change or promote its reversal. Alternatively, or additionally, there may be nutritionally stimulated epigenetic features that are distinct from the features of either the "young" or "old" epigenome that are protective. The diagrammatic representations of these three hypothetical epigenomic states show DNA (thick line) wrapped around a series of histone protein cores. Covalent epigenetic modifications to both the DNA (shown in lilac) and the histone proteins (shown in blue (dark gray in print versions), green (light gray in print versions), yellow (white in print versions), and pink (light gray in print versions)) are represented as spheres. Note the different distributions of these covalent modifications in the three different states.

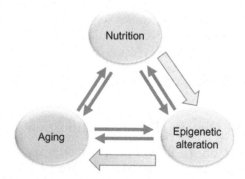

FIGURE 16.3 The multiple interactions between nutrition, epigenetic alteration and aging. The figure emphasizes that a direct causal effect of nutrition on aging mediated through epigenetic alteration (represented as the heavy colored (light gray in print versions) arrows) is only one possible direction of interaction.

A further dimension of this complex interacting network is that epigenetic changes themselves are likely to affect nutrition, as suggested by effects of nutritional interventions associated with altered feeding patterns on epigenetic modification of genes coding for appetite-regulating neuropeptides. For example, altered DNA methylation and histone modification of the gene for pro-opiomelanocortin (POMC) was observed in fetuses of undernourished ewes (which are known to develop into animals predisposed to obesity [82]) and adult rats reared on a high-carbohydrate milk formula, which induced hyperinsulinemia and obesity, had DNA hypomethylation of the neuropeptide Y gene promoter in hypothalamus [83].

Despite the likelihood that age-associated epigenetic alterations associated with aging or age-related diseases are reversible (and thus could, in principle, be reversed by diet), evidence that shows this reversibility directly is currently sparse. A notable example is that in aged mice impaired hippocampal-dependent memory formation (measured by fear-conditioning) was associated with deregulated H4K12 acetylation and an abnormal transcriptional response profile. However, this learning deficit was reversed and the transcriptional response profile reestablished when hippocampal H4K12 acetylation was restored by hippocampal injection of the potent HDAC inhibitor suberoylanilide hydroxamic acid (SAHA) [84].

POTENTIAL MECHANISMS THROUGH WHICH DIET COULD INDUCE EPIGENETIC ALTERATIONS THAT PROTECT AGAINST AGING

As already rehearsed above (and see Fig. 16.2) diet-induced epigenetic alterations that protect against aging could theoretically be de novo epigenetic modifications, layered on top of but counterbalancing damaging age-related epigenetic drift and/or be changes that are direct reversals of this age-related epigenetic drift. However, irrespective of the knowledge of specific target site in the genome and direction of change (be this in DNA methylation or histone modification) the likely fundamental mechanism is common, viz., that dietary components alter availability, expression, and/or activity of the elements of cellular machinery (enzymes and their cofactors) that shape the epigenetic architecture. Potential enzyme targets that could be modified by components of the diet to effect epigenetic change include DNA methyltransferases, DNA demethylases, histone acetyltransferases, histone deacetlyases, histone methyltransferases, and histone demethylases. Dietary effects on cofactor availability have been studied most extensively in the context of methyl group supply. The interrelationship between DNA methylation and histone modification renders it very likely that dietary agents acting primarily at one epigenetic level may produce effects on the other. Some of the evidence for these actions of diet is presented below.

Effects of Diet on Methyl Group Supply

The potential of components of the diet that feed into the 1-carbon cycle to affect the availability of the methyl donor SAM is noted above. In addition to this effect on the availability of substrates to feed into the 1-carbon cycle, dietary agents can influence SAM availability through effects on the expression or activity of specific enzymes in the cycle, as shown in the case of both vitamin A [66] and selenium [85] with respect to expression of the enzyme glycine N-methyltransferase in rat liver.

Effects of Diet on DNMT Activity

Two principal mechanisms through which dietary agents can affect DNMT activity have been identified: (1) direct, competitive inhibition, and (2) in the case of catechol compounds, through catechol-o-methyltransferase (COMT)-catalyzed methylation leading to the generation of SAH, a potent, noncompetitive DNMT inhibitor, from SAM. Several dietary polyphenols, including the tea catechins catechin, epicatechin, and epigallocatechin-3-gallate, and the flavonoids quercetin, fisetin, myricetin, gentisein, daidzein, and biochanin A, all inhibit DNMT activity through one or both of these mechanisms in vitro ([55,86]. A correlation between the efficacy with which the isoflavones genistein, daidzein, and biochanin A inhibited DNMT activity and the ability of these compounds to reverse site-specific DNA hypermethylation in prostate cancer LNCaP and PC3 cells provides evidence that the DNMT-inhibitory effects of the polyphenols are important mediators of their effects on DNA methylation [55]. Selenium (as selenite) has been shown in vitro to be a noncompetitive inhibitor of DNMT activity [87,88] and also to reduce DNMT1 protein expression in HT29 cells [62].

Effects of Diet on Histone Deacetylase Activity

The HDAC-inhibitory action of the short chain fatty acid butyrate is well established in vitro [72], as is the ability of butyrate to induce histone hyperacetylation (see above). Particularly convincing evidence that the

HDAC-inhibitory action is primarily responsible for the gene-regulatory effects of butyrate is that treatment of HT-29 cells with butyrate induced an identical panel of 23 genes (from a pool of 588 measured) as did the well-characterized pharmacological HDAC inhibitor TSA [89]. Other dietary compounds with HDAC inhibitory action include the metabolite of the organosulfate diallyl sulfate S-allylmercaptocysteine, found in garlic, and the cysteine conjugates of the isothiocyanates, derived from glucosinolates found in cruciferous vegetables [58,90], as well as the soybean isoflavone genistein [55].

THE FUNCTIONAL CONSEQUENCES OF STEM CELL AGING AND THE ROLE OF EPIGENETIC ALTERATIONS

Stem cell exhaustion has been cited as a key hallmark of aging [6], which likely results from multiple forms of cellular damage including DNA damage and telomere shortening. However, epigenetic changes are also a probable contributing mechanism. The reversibility of this stem cell functional decline (or stem cell "rejuvenation") points toward there being a (reversible) epigenetic component. The phenomenon has been shown through experiments based on joining the circulatory systems of young and old mice (parabiotic pairing). For example, parabiotic pairing reversed age related loss of the proliferative capacity of muscle satellite cells from mice, coincident with restored activation of NOTCH signaling [91]. Direct evidence that epigenetic changes are a feature of stem cell rejuvenation is lacking, but highly plausible given the central role of epigenetic modification in shaping stem cell function. While attractive, the principle the diet can have an impact on stem cell rejuvenation through affecting the epigenome remains a point to address through studies designed specifically to address the question, which are currently lacking. The fact noted above that the changes in DNA methylation associated with aging cluster at the gene loci of the polycomb repressive complexes, which are key epigenetic regulators of stem cell differentiation, is a further observation consistent with stem cell exhaustion having an epigenetic component. There is also direct evidence that epigenetic changes are a feature of stem cell aging. For example, comparison of the epigenetic profiles of muscle stem cells (quiescent muscle satellite cells) from young and old mice revealed an increase with age in the repressive chromatin mark H3K27me3 [92], low levels of which are associated with pluripotency of ESCs [93,94].

SIRTUINS AS MEDIATORS OF DIET-INDUCED EPIGENETIC ALTERATIONS THAT COUNTERACT AGING

Important roles for epigenetic influences on aging are implicit in the very large body of work that links the actions of sirtuins, which are classed as histone deacetylases, to effects on lifespan and/or other features of aging [6,95]. The mammalian enzyme Sirt1 (an NAD-dependent (class III) histone deacetylase) and its homologs Sir2 in yeast, sir-2.1 in *C. elegans* and dSir2 in *Drosophila*, appear to be mediators of the effect of dietary restriction (DR) to increase lifespan and/or mitigate against features of aging [96,97]. Evidence that supports this view includes the abolition of lifespan extension by DR in mutants that do not express Sir2 in yeast [98], sir-2.1 in *C. elegans* [99], or dSir2 in *Drosophila* [100]. Also consistent with sirtuins having actions that counteract features of aging are effects of transgenic expression of *Sirt1* in mouse models [101]. Such effects include reduced fasting concentrations of insulin, glucose, and cholesterol and reduced adiposity [102], improved glucose tolerance due to increased hepatic insulin sensitivity [103], and reduced production of β-amyloid plaques in brain—the hallmark of Alzheimer's disease [104]. There has been debate concerning the extent to which the sirtuin transgenes contribute to extended lifespan in *C. elegans* and *Drosophila* sir-2.1/dSir2 transgenic strains. Other background genetic factors appear to be alternative explanations for extended lifespan in some cases [105]. However, later work in *Drosophila* that took more careful account of the genetic background of long-lived dSir2 transgenic lines [106] and reassessment of transgenic overexpression of sir2.1 in *C. elegans* [107] substantiated the view that these genes can have positive effects on lifespan.

It is important to note that the deacetylation activity of Sirt1 is not restricted to histones. Deacetylation by Sirt1 of other cellular substrates offers alternative or additional plausible explanations for positive effects on healthspan. These substrates include intermediates in pleiotropic cellular pathways and several key transcription factors linked to aging, for example PGC1α, which controls mitochondrial biogenesis, p53 [96], and many others [101], However, a recent discovery that Sirt1 affects DNA methylation with a bias toward genes that also show altered expression in response to dietary restriction [108] provides sound evidence that epigenetic actions of Sirt1 are a feature of its positive effects on healthspan.

The mammalian sirtuin family comprises seven members, and there is compelling evidence for the involvement of members additional to Sirt1 in influencing aging and/or lifespan. Notably Sirt6 knockout mice aged prematurely [109] and male transgenic mice overexpressing Sirt6 lived longer than control animals in parallel with reduced IGF-1 signaling [110]. Sirt3 is a mitochondrial isoform and appears to contribute to longevity associated with DR due to deacetylation of mitochondrial proteins [111]. Overexpression of Sirt3 in mouse hematopoietic stem cells rescued an age-related decline in colony formation and in ability to regenerate bone marrow in irradiated mice, indicating the potential of Sirt3 to promote longer healthspan [112].

THE UTILITY OF EPIGENETIC ALTERATIONS AS A NUTRITIONALLY MODIFIABLE MARKER OF THE AGING TRAJECTORY

The translational potential of the growing body of knowledge concerning age-related epigenetic drift and effects of diet on epigenetic marking is worth considering. A highly simplistic view would be to posit that aging could be slowed using targeted diet-based interventions to prevent or reverse epigenetic changes that drive aging. This view is unrealistic, at least over the short to medium term, given the many current caveats explored in this chapter. These caveats include the uncertainly that many age-related epigenetic changes play any causal role in aging. Also, any epigenetic changes that do actually have a causal role would be only one component in a multifaceted aging process. Moreover, the full profile of epigenetic alterations the result from any nutritional influence remains difficult both to measure and to interpret functionally. However, there is a strong argument that some signatures of DNA methylation provide a biomarker of the aging trajectory that is more stable, and thus more robust, than other measures such as patterns of gene expression. The finding that discordance between chronological age and a measure of "biological age" based on DNA methylation profiling was associated with increased risk of mortality in later life [27] is a compelling argument to use DNA methylation as a measure of aging. As explained, it is difficult to disentangle the difference in "epigenetic readout" that could result from age-associated changes in the distribution of the heterogeneous cell types (with different epigenetic signatures) that comprise tissues from true epigenetic drift at cellular level. There is thus an argument that robust indicators of likely efficacy of dietary strategies to slow or reverse age-associated epigenetic change are labile loci where consistent changes have been observed in different tissues. Moreover, there is an argument that effects of diet (or indeed other modifying factors) at such loci may be more efficacious than

tissue-specific actions with respect to counteracting the process of aging "holistically." Perhaps, then, a defined aging-linked, tissue-independent, labile DNA methylation signature should be developed as part of a toolkit to indicate the efficacy of dietary changes to slow the aging process that could be useful both for individual health monitoring and also in nutritional research on aging.

SUMMARY

Some epigenetic features are affected by diet, and epigenetic alteration is a hallmark of aging. The extent to which these epigenetic changes are causative, and thus the potential for influences of diet on these features to slow the aging process, requires critical evaluation. Multiple molecular markers of aging interact with epigenetic alteration, and diet is one component of environmental impact that, along with genotype, is likely to influence how epigenetic alterations refine the aging trajectory. Diet could contribute to healthier aging by reversing or slowing age-related causative epigenetic changes or by promoting a different but protective epigenetic state. The plasticity of the epigenome during early development may afford a particular window of opportunity for dietary intervention in aging, and effects on stem cells are likely to be of particular functional importance. Some of the effects of sirtuins to counteract aging may be via epigenetic actions on polycomb group protein gene targets, which have a role in stem cell integrity. Tissue-independent global epigenetic signatures of aging may provide a biological readout of the efficacy of dietary intervention to counteract effects of aging.

References

[1] Schnitzler GR. Chromatic remodelling and histone modifications. In: Choi SW, Friso S, editors. Nutrients and epigenetics. CRC Press (Taylor and Francis Group; 2009. p. 67–103.
[2] Luger K, Mader AW, Richmond RK, Sargent DF, Richmond TJ. Crystal structure of the nucleosome core particle at 2.8 Å resolution. Nature 1997;389(6648):251–60.
[3] Irizarry RA, Ladd-Acosta C, Wen B, Wu Z, Montano C, Onyango P, et al. The human colon cancer methylome shows similar hypo- and hypermethylation at conserved tissue-specific CpG island shores. Nat Genet 2009;41(2):178–86.
[4] Laird A, Thomson JP, Harrison DJ, Meehan RR. 5-hydroxymethylcytosine profiling as an indicator of cellular state. Epigenomics 2013;5(6):655–69.
[5] Zentner GE, Henikoff S. Regulation of nucleosome dynamics by histone modifications. Nat Struct Mol Biol 2013; 20(3):259–66.
[6] Lopez-Otin C, Blasco MA, Partridge L, Serrano M, Kroemer G. The hallmarks of aging. Cell 2013;153(6):1194–217.

[7] Brownlee M. Biochemistry and molecular cell biology of diabetic complications. Nature 2001;414(6865):813–20.

[8] Bjornsson HT, Sigurdsson MI, Fallin MD, Irizarry RA, Aspelund T, Cui H, et al. Intra-individual change over time in DNA methylation with familial clustering. J Am Med Assoc 2008;299(24):2877–83.

[9] Richardson B. Impact of aging on DNA methylation. Ageing Res Rev 2003;2(3):245–61.

[10] Maegawa S, Hinkal G, Kim HS, Shen L, Zhang L, Zhang J, et al. Widespread and tissue specific age-related DNA methylation changes in mice. Genome Res 2010;20(3):332–40.

[11] Hannum G, Guinney J, Zhao L, Zhang L, Hughes G, Sadda S, et al. Genome-wide methylation profiles reveal quantitative views of human aging rates. Molecular cell 2013;49(2):359–67.

[12] Heyn H, Li N, Ferreira HJ, Moran S, Pisano DG, Gomez A, et al. Distinct DNA methylomes of newborns and centenarians. Proc Natl Acad Sci USA 2012;109(26):10522–7.

[13] Teschendorff AE, West J, Beck S. Age-associated epigenetic drift: implications, and a case of epigenetic thrift? Hum Mol Genet 2013;22(R1):R7–15.

[14] Teschendorff AE, Menon U, Gentry-Maharaj A, Ramus SJ, Weisenberger DJ, Shen H, et al. Age-dependent DNA methylation of genes that are suppressed in stem cells is a hallmark of cancer. Genome Res 2010;20(4):440–6.

[15] Rakyan VK, Down TA, Maslau S, Andrew T, Yang TP, Beyan H, et al. Human aging-associated DNA hypermethylation occurs preferentially at bivalent chromatin domains. Genome Res 2010;20(4):434–9.

[16] Rando TA, Chang HY. Aging, rejuvenation, and epigenetic reprogramming: resetting the aging clock. Cell 2012;148(1–2):46–57.

[17] Greer EL, Maures TJ, Ucar D, Hauswirth AG, Mancini E, Lim JP, et al. Transgenerational epigenetic inheritance of longevity in Caenorhabditis elegans. Nature 2011;479(7373):365–71.

[18] Beerman I, Bock C, Garrison BS, Smith ZD, Gu H, Meissner A, et al. Proliferation-dependent alterations of the DNA methylation landscape underlie hematopoietic stem cell aging. Cell Stem Cell 2013;12(4):413–25.

[19] Lee TI, Jenner RG, Boyer LA, Guenther MG, Levine SS, Kumar RM, et al. Control of developmental regulators by polycomb in human embryonic stem cells. Cell 2006;125(2):301–13.

[20] Widschwendter M, Fiegl H, Egle D, Mueller-Holzner E, Spizzo G, Marth C, et al. Epigenetic stem cell signature in cancer. Nat Genet 2007;39(2):157–8.

[21] Ohm JE, McGarvey KM, Yu X, Cheng L, Schuebel KE, Cope L, et al. A stem cell-like chromatin pattern may predispose tumor suppressor genes to DNA hypermethylation and heritable silencing. Nat Genet 2007;39(2):237–42.

[22] Schlesinger Y, Straussman R, Keshet I, Farkash S, Hecht M, Zimmerman J, et al. Polycomb-mediated methylation on Lys27 of histone H3 pre-marks genes for de novo methylation in cancer. Nat Genet 2007;39(2):232–6.

[23] Teschendorff AE, Menon U, Gentry-Maharaj A, Ramus SJ, Gayther SA, Apostolidou S, et al. An epigenetic signature in peripheral blood predicts active ovarian cancer. PLoS One 2009;4(12):e8274.

[24] Houseman EA, Accomando WP, Koestler DC, Christensen BC, Marsit CJ, Nelson HH, et al. DNA methylation arrays as surrogate measures of cell mixture distribution. BMC Bioinformatics 2012;13:86.

[25] Langevin SM, Houseman EA, Christensen BC, Wiencke JK, Nelson HH, Karagas MR, et al. The influence of aging, environmental exposures and local sequence features on the variation of DNA methylation in blood. Epigenetics 2011;6(7):908–19.

[26] Horvath S. DNA methylation age of human tissues and cell types. Genome Biol 2013;14(10):R115.

[27] Marioni RE, Shah S, McRae AF, Chen BH, Colicino E, Harris SE, et al. DNA methylation age of blood predicts all-cause mortality in later life. Genome Biol 2015;16(1):25.

[28] West J, Beck S, Wang X, Teschendorff AE. An integrative network algorithm identifies age-associated differential methylation interactome hotspots targeting stem-cell differentiation pathways. Sci Rep 2013;3:1630.

[29] Sarg B, Koutzamani E, Helliger W, Rundquist I, Lindner HH. Postsynthetic trimethylation of histone H4 at lysine 20 in mammalian tissues is associated with aging. J Biol Chem 2002;277(42):39195–201.

[30] Shumaker DK, Dechat T, Kohlmaier A, Adam SA, Bozovsky MR, Erdos MR, et al. Mutant nuclear lamin A leads to progressive alterations of epigenetic control in premature aging. Proc Natl Acad Sci USA 2006;103(23):8703–8.

[31] Cheung I, Shulha HP, Jiang Y, Matevossian A, Wang J, Weng Z, et al. Developmental regulation and individual differences of neuronal H3K4me3 epigenomes in the prefrontal cortex. Proc Natl Acad Sci USA 2010;107(19):8824–9.

[32] Han S, Brunet A. Histone methylation makes its mark on longevity. Trends Cell Biol 2012;22(1):42–9.

[33] Oberdoerffer P, Sinclair DA. The role of nuclear architecture in genomic instability and ageing. Nat Rev Mol Cell Biol 2007;8(9):692–702.

[34] Tsurumi A, Li WX. Global heterochromatin loss: a unifying theory of aging? Epigenetics 2012;7(7):680–8.

[35] Chambers SM, Shaw CA, Gatza C, Fisk CJ, Donehower LA, Goodell MA. Aging hematopoietic stem cells decline in function and exhibit epigenetic dysregulation. PLoS Biol 2007;5(8):e201.

[36] Pollina EA, Brunet A. Epigenetic regulation of aging stem cells. Oncogene 2011;30(28):3105–26.

[37] Pegoraro G, Kubben N, Wickert U, Gohler H, Hoffmann K, Misteli T. Ageing-related chromatin defects through loss of the NURD complex. Nat cell Biol 2009;11(10):1261–7.

[38] Larson K, Yan SJ, Tsurumi A, Liu J, Zhou J, Gaur K, et al. Heterochromatin formation promotes longevity and represses ribosomal RNA synthesis. PLoS Genet 2012;8(1):e1002473.

[39] Mathers JC, Ford D. Nutrition, epigenetics, and aging. In: Choi SW, Friso S, editors. Nutrients and epigenetics. CRC Press (Taylor and Francis Group); 2009. p. 175–205.

[40] Tarry-Adkins JL, Ozanne SE. The impact of early nutrition on the ageing trajectory. Proc Nutr Soc 2014;73(2):289–301.

[41] Martino D, Loke YJ, Gordon L, Ollikainen M, Cruickshank MN, Saffery R, et al. Longitudinal, genome-scale analysis of DNA methylation in twins from birth to 18 months of age reveals rapid epigenetic change in early life and pair-specific effects of discordance. Genome Biol 2013;14(5):R42.

[42] Alisch RS, Barwick BG, Chopra P, Myrick LK, Satten GA, Conneely KN, et al. Age-associated DNA methylation in pediatric populations. Genome Res 2012;22(4):623–32.

[43] Castro CE, Sevall JS. Alteration of higher order structure of rat liver chromatin by dietary composition. J Nutr 1980;110(1):105–16.

[44] Wolff GL, Kodell RL, Moore SR, Cooney CA. Maternal epigenetics and methyl supplements affect agouti gene expression in A^{vy}/a mice. FASEB J 1998;12(11):949–57.

[45] Waterland RA, Jirtle RL. Transposable elements: targets for early nutritional effects on epigenetic gene regulation. Mol Cell Biol 2003;23(15):5293–300.

[46] Waterland RA, Travisano M, Tahiliani KG. Diet-induced hypermethylation at agouti viable yellow is not inherited transgenerationally through the female. FASEB J 2007;21(12):3380–5.

[47] Wasson GR, McGlynn AP, McNulty H, O'Reilly SL, McKelvey-Martin VJ, McKerr G, et al. Global DNA and p53 region-specific hypomethylation in human colonic cells is induced by folate depletion and reversed by folate supplementation. J Nutr 2006;136(11):2748–53.

[48] Kim YI, Pogribny IP, Basnakian AG, Miller JW, Selhub J, James SJ, et al. Folate deficiency in rats induces DNA strand breaks and hypomethylation within the p53 tumor suppressor gene. Am J Clin Nutr 1997;65(1):46–52.

[49] Niculescu MD, Craciunescu CN, Zeisel SH. Dietary choline deficiency alters global and gene-specific DNA methylation in the developing hippocampus of mouse fetal brains. FASEB J 2006;20(1):43–9.

[50] Jhaveri MS, Wagner C, Trepel JB. Impact of extracellular folate levels on global gene expression. Mol Pharmacol 2001;60(6):1288–95.

[51] Sohn KJ, Stempak JM, Reid S, Shirwadkar S, Mason JB, Kim YI. The effect of dietary folate on genomic and p53-specific DNA methylation in rat colon. Carcinogenesis 2003;24(1):81–90.

[52] Jacob RA, Gretz DM, Taylor PC, James SJ, Pogribny IP, Miller BJ, et al. Moderate folate depletion increases plasma homocysteine and decreases lymphocyte DNA methylation in postmenopausal women. J Nutr 1998;128(7):1204–12.

[53] Rampersaud GC, Kauwell GP, Hutson AD, Cerda JJ, Bailey LB. Genomic DNA methylation decreases in response to moderate folate depletion in elderly women. Am J Clin Nutr 2000;72(4):998–1003.

[54] Pufulete M, Al-Ghnaniem R, Khushal A, Appleby P, Harris N, Gout S, et al. Effect of folic acid supplementation on genomic DNA methylation in patients with colorectal adenoma. Gut 2005;54(5):648–53.

[55] Fang MZ, Chen D, Sun Y, Jin Z, Christman JK, Yang CS. Reversal of hypermethylation and reactivation of p16INK4a, RARbeta, and MGMT genes by genistein and other isoflavones from soy. Clin Cancer Res 2005;11(19 Pt 1):7033–41.

[56] Day JK, Bauer AM, DesBordes C, Zhuang Y, Kim BE, Newton LG, et al. Genistein alters methylation patterns in mice. J Nutr 2002;132(Suppl. 8):2419S–23S.

[57] Dolinoy DC, Weidman JR, Waterland RA, Jirtle RL. Maternal genistein alters coat color and protects A^{vy} mouse offspring from obesity by modifying the fetal epigenome. Environ Health Perspect 2006;114(4):567–72.

[58] Dashwood RH, Myzak MC, Ho E. Dietary HDAC inhibitors: time to rethink weak ligands in cancer chemoprevention? Carcinogenesis 2006;27(2):344–9.

[59] Beach RS, Gershwin ME, Hurley LS. Gestational zinc deprivation in mice: persistence of immunodeficiency for three generations. Science 1982;218(4571):469–71.

[60] Wallwork JC, Duerre JA. Effect of zinc deficiency on methionine metabolism, methylation reactions and protein synthesis in isolated perfused rat liver. J Nutr 1985;115(2):252–62.

[61] Davis CD, Uthus EO, Finley JW. Dietary selenium and arsenic affect DNA methylation in vitro in Caco-2 cells and in vivo in rat liver and colon. J Nutr 2000;130(12):2903–9.

[62] Davis CD, Uthus EO. Dietary selenite and azadeoxycytidine treatments affect dimethylhydrazine-induced aberrant crypt formation in rat colon and DNA methylation in HT-29 cells. J Nutr 2002;132(2):292–7.

[63] Davis CD, Uthus EO. Dietary folate and selenium affect dimethylhydrazine-induced aberrant crypt formation, global DNA methylation and one-carbon metabolism in rats. J Nutr 2003;133(9):2907–14.

[64] Di Croce L, Raker VA, Corsaro M, Fazi F, Fanelli M, Faretta M, et al. Methyltransferase recruitment and DNA hypermethylation of target promoters by an oncogenic transcription factor. Science 2002;295(5557):1079–82.

[65] Sirchia SM, Ren M, Pili R, Sironi E, Somenzi G, Ghidoni R, et al. Endogenous reactivation of the RARbeta2 tumor suppressor gene epigenetically silenced in breast cancer. Cancer Res 2002;62(9):2455–61.

[66] Rowling MJ, McMullen MH, Schalinske KL. Vitamin A and its derivatives induce hepatic glycine N-methyltransferase and hypomethylation of DNA in rats. J Nutr 2002;132(3):365–9.

[67] Kwong WY, Miller DJ, Ursell E, Wild AE, Wilkins AP, Osmond C, et al. Imprinted gene expression in the rat embryo-fetal axis is altered in response to periconceptional maternal low protein diet. Reproduction (Cambridge, England) 2006;132(2):265–77.

[68] Burdge GC, Hanson MA, Slater-Jefferies JL, Lillycrop KA. Epigenetic regulation of transcription: a mechanism for inducing variations in phenotype (fetal programming) by differences in nutrition during early life? Br J Nutr 2007;97(6):1036–46.

[69] Lillycrop KA, Phillips ES, Jackson AA, Hanson MA, Burdge GC. Dietary protein restriction of pregnant rats induces and folic acid supplementation prevents epigenetic modification of hepatic gene expression in the offspring. J Nutr 2005;135(6):1382–6.

[70] Lillycrop KA, Slater-Jefferies JL, Hanson MA, Godfrey KM, Jackson AA, Burdge GC. Induction of altered epigenetic regulation of the hepatic glucocorticoid receptor in the offspring of rats fed a protein-restricted diet during pregnancy suggests that reduced DNA methyltransferase-1 expression is involved in impaired DNA methylation and changes in histone modifications. Br J Nutr 2007;97(6):1064–73.

[71] Burdge GC, Slater-Jefferies J, Torrens C, Phillips ES, Hanson MA, Lillycrop KA. Dietary protein restriction of pregnant rats in the F0 generation induces altered methylation of hepatic gene promoters in the adult male offspring in the F1 and F2 generations. Br J Nutr 2007;97(3):435–9.

[72] Aviram A, Zimrah Y, Shaklai M, Nudelman A, Rephaeli A. Comparison between the effect of butyric acid and its prodrug pivaloyloxymethylbutyrate on histones hyperacetylation in an HL-60 leukemic cell line. Int J Cancer 1994;56(6):906–9.

[73] Demary K, Wong L, Spanjaard RA. Effects of retinoic acid and sodium butyrate on gene expression, histone acetylation and inhibition of proliferation of melanoma cells. Cancer Lett 2001;163(1):103–7.

[74] Rada-Iglesias A, Enroth S, Ameur A, Koch CM, Clelland GK, Respuela-Alonso P, et al. Butyrate mediates decrease of histone acetylation centered on transcription start sites and down-regulation of associated genes. Genome Res 2007;17(6):708–19.

[75] Kida Y, Shimizu T, Kuwano K. Sodium butyrate up-regulates cathelicidin gene expression via activator protein-1 and histone acetylation at the promoter region in a human lung epithelial cell line, EBC-1. Mol Immunol 2006;43(12):1972–81.

[76] Druesne N, Pagniez A, Mayeur C, Thomas M, Cherbuy C, Duee PH, et al. Diallyl disulfide (DADS) increases histone acetylation and p21(waf1/cip1) expression in human colon tumor cell lines. Carcinogenesis 2004;25(7):1227–36.

[77] Pogribny IP, Tryndyak VP, Muskhelishvili L, Rusyn I, Ross SA. Methyl deficiency, alterations in global histone modifications, and carcinogenesis. J Nutr 2007;137(Suppl. 1):216S–22S.

[78] Ross SA, Davis CD. The emerging role of microRNAs and nutrition in modulating health and disease. Ann Rev Nutr 2014;34:305–36.

[79] Ryu MS, Langkamp-Henken B, Chang SM, Shankar MN, Cousins RJ. Genomic analysis, cytokine expression, and microRNA profiling reveal biomarkers of human dietary zinc depletion and homeostasis. Proc Natl Acad Sci USA 108(52):20970–5.

[80] Enquobahrie DA, Williams MA, Qiu C, Siscovick DS, Sorensen TK. Global maternal early pregnancy peripheral blood mRNA and miRNA expression profiles according to plasma 25-hydroxyvitamin D concentrations. J Matern Fetal Neonatal Med 24(8):1002–12.

[81] Maciel-Dominguez A, Swan D, Ford D, Hesketh J. Selenium alters miRNA profile in an intestinal cell line: evidence that miR-185 regulates expression of GPX2 and SEPSH2. Mol Nutr Food Res 57(12):2195–205.

[82] Stevens A, Begum G, Cook A, Connor K, Rumball C, Oliver M, et al. Epigenetic changes in the hypothalamic proopiomelanocortin and glucocorticoid receptor genes in the ovine fetus after periconceptional undernutrition. Endocrinology 151 (8):3652–64.

[83] Mahmood S, Smiraglia DJ, Srinivasan M, Patel MS. Epigenetic changes in hypothalamic appetite regulatory genes may underlie the developmental programming for obesity in rat neonates subjected to a high-carbohydrate dietary modification. J Dev Orig Health Dis 4(6):479–90.

[84] Peleg S, Sananbenesi F, Zovoilis A, Burkhardt S, Bahari-Javan S, Agis-Balboa RC, et al. Altered histone acetylation is associated with age-dependent memory impairment in mice. Science 2010;328(5979):753–6.

[85] Uthus EO, Ross SA, Davis CD. Differential effects of dietary selenium (se) and folate on methyl metabolism in liver and colon of rats. Biol Trace Elem Res 2006;109(3):201–14.

[86] Lee WJ, Shim JY, Zhu BT. Mechanisms for the inhibition of DNA methyltransferases by tea catechins and bioflavonoids. Mol Pharmacol 2005;68(4):1018–30.

[87] Cox R, Goorha S. A study of the mechanism of selenite-induced hypomethylated DNA and differentiation of friend erythroleukemic cells. Carcinogenesis 1986;7(12):2015–18.

[88] Fiala ES, Staretz ME, Pandya GA, El-Bayoumy K, Hamilton SR. Inhibition of DNA cytosine methyltransferase by chemopreventive selenium compounds, determined by an improved assay for DNA cytosine methyltransferase and DNA cytosine methylation. Carcinogenesis 1998;19(4):597–604.

[89] Della Ragione F, Criniti V, Della Pietra V, Borriello A, Oliva A, Indaco S, et al. Genes modulated by histone acetylation as new effectors of butyrate activity. FEBS lett 2001;499(3):199–204.

[90] Myzak MC, Ho E, Dashwood RH. Dietary agents as histone deacetylase inhibitors. Mol Carcinog 2006;45(6):443–6.

[91] Conboy IM, Conboy MJ, Wagers AJ, Girma ER, Weissman IL, Rando TA. Rejuvenation of aged progenitor cells by exposure to a young systemic environment. Nature 2005;433(7027):760–4.

[92] Liu L, Cheung TH, Charville GW, Hurgo BM, Leavitt T, Shih J, et al. Chromatin modifications as determinants of muscle stem cell quiescence and chronological aging. Cell Rep 2013; 4(1):189–204.

[93] Mikkelsen TS, Ku M, Jaffe DB, Issac B, Lieberman E, Giannoukos G, et al. Genome-wide maps of chromatin state in pluripotent and lineage-committed cells. Nature 2007;448 (7153):553–60.

[94] Marks H, Kalkan T, Menafra R, Denissov S, Jones K, Hofemeister H, et al. The transcriptional and epigenomic foundations of ground state pluripotency. Cell 2012;149(3):590–604.

[95] Wakeling LA, Ions LJ, Ford D. Could Sirt1-mediated epigenetic effects contribute to the longevity response to dietary restriction and be mimicked by other dietary interventions? Age (Dordr) 2009;31:327–41.

[96] Guarente L, Picard F. Calorie restriction – the SIR2 connection. Cell 2005;120(4):473–82.

[97] Libert S, Guarente L. Metabolic and neuropsychiatric effects of calorie restriction and sirtuins. Ann Rev Physiol 2013;75:669–84.

[98] Lin SJ, Kaeberlein M, Andalis AA, Sturtz LA, Defossez PA, Culotta VC, et al. Calorie restriction extends Saccharomyces cerevisiae lifespan by increasing respiration. Nature 2002;418 (6895):344–8.

[99] Tissenbaum HA, Guarente L. Increased dosage of a sir-2 gene extends lifespan in Caenorhabditis elegans. Nature 2001;410 (6825):227–30.

[100] Rogina B, Helfand SL. Sir2 mediates longevity in the fly through a pathway related to calorie restriction. Proc Natl Acad Sci USA 2004;101(45):15998–6003.

[101] Donmez G, Guarente L. Aging and disease: connections to sirtuins. Aging Cell 2010;9(2):285–90.

[102] Bordone L, Cohen D, Robinson A, Motta MC, van Veen E, Czopik A, et al. SIRT1 transgenic mice show phenotypes resembling calorie restriction. Aging Cell 2007;6(6):759–67.

[103] Banks AS, Kon N, Knight C, Matsumoto M, Gutierrez-Juarez R, Rossetti L, et al. SirT1 gain of function increases energy efficiency and prevents diabetes in mice. Cell Metab 2008;8 (4):333–41.

[104] Donmez G, Wang D, Cohen DE, Guarente L. SIRT1 suppresses beta-amyloid production by activating the alpha-secretase gene ADAM10. Cell 2010;142(2):320–32.

[105] Burnett C, Valentini S, Cabreiro F, Goss M, Somogyvari M, Piper MD, et al. Absence of effects of Sir2 overexpression on lifespan in C. elegans and Drosophila. Nature 2011;477 (7365):482–5.

[106] Banerjee KK, Ayyub C, Ali SZ, Mandot V, Prasad NG, Kolthur-Seetharam U. dSir2 in the adult fat body, but not in muscles, regulates life span in a diet-dependent manner. Cell Rep 2012;2(6):1485–91.

[107] Viswanathan M, Guarente L. Regulation of Caenorhabditis elegans lifespan by sir-2.1 transgenes. Nature 2011;477(7365): E1–2.

[108] Ions LJ, Wakeling LA, Bosomworth HJ, Hardyman JE, Escolme SM, Swan DC, et al. Effects of Sirt1 on DNA methylation and expression of genes affected by dietary restriction. Age 2013;35:1835–49.

[109] Mostoslavsky R, Chua KF, Lombard DB, Pang WW, Fischer MR, Gellon L, et al. Genomic instability and aging-like phenotype in the absence of mammalian SIRT6. Cell 2006; 124(2):315–29.

[110] Kanfi Y, Naiman S, Amir G, Peshti V, Zinman G, Nahum L, et al. The sirtuin SIRT6 regulates lifespan in male mice. Nature 2012;483(7388):218–21.

[111] Someya S, Yu W, Hallows WC, Xu J, Vann JM, Leeuwenburgh C, et al. Sirt3 mediates reduction of oxidative damage and prevention of age-related hearing loss under caloric restriction. Cell 143(5):802–12.

[112] Brown K, Xie S, Qiu X, Mohrin M, Shin J, Liu Y, et al. SIRT3 reverses aging-associated degeneration. Cell Rep 2013;3(2):319–27.

CHAPTER

17

The Controversy Around Sirtuins and Their Functions in Aging

Yu Sun and Weiwei Dang

Huffington Center on Aging, Baylor College of Medicine, Houston, TX, USA

KEY FACTS
- Sirtuins are a class of evolutionarily conserved protein deacetylases.
- Activation or overexpression of sirtuins provide longevity benefits in many organisms.
- Some sirtuins seem to mediate some effects of CR, though it is only one of the many pathways exploited by CR.
- Whether resveratrol directly activates sirtuins remains a highly contested topic among researchers.
- Some sirtuins act more like proto-oncogenes, some function like tumor suppressors, while others might be considered both depending on the molecular context and the type of cancer.
- Sirtuins are very intriguing targets for various therapeutic purposes, many involves aging.

Dictionary of Terms
- *Lifespan*: The length of time an organism is expected to remain alive.
- *Deacetylase*: The enzyme that removes acetyl groups from lysine amino acid on substrates.
- *HDAC*: Histone deacetylase
- $NAD^+/NADH$: The oxidized and reduced form of nicotinamide adenine dinucleotide. Sirtuins remove acetyl group from substrates to the ADP-ribose moiety of NAD^+ thus performing its function as a deacetylase.
- *Heterochromatin*: The inactive parts of chromatin that consist of tightly packed DNA.
- *rDNA*: The DNA sequences that codes ribosomal RNA.
- *Autophagy*: The intracellular degradation system that delivers unnecessary or dysfunctional cellular components to lysosomes for catabolic actions.
- *Circadian rhythm*: A 24-hour cycle of endogenous change of physical, mental, and behavioral biological process.
- *Senescence*: Senescence means biological aging, referring to both cellular senescence and senescence of the whole organism. Cellular senescence indicates the state of permanent cell-cycle rest of proliferating cells.
- *Hypoxia*: The state of the body or a region of the body that is deprived of oxygen supply.
- *Telomere*: Telomeres are the caps at the end of chromatids that protects chromosomes from deterioration and fusion with adjacent chromosomes. It consists of highly repetitive and tightly packed nucleotides.
- *Calorie restriction*: The dietary strategy that is based on low calorie intake. "Low" can mean either the relative restriction of calorie intake compared to the individual's previous intake, or relative low intake compared to the average intake of the people of similar body type.
- *Resveratrol*: 3,5,4'-trihydroxy-trans-stilbene, a type of natural phenol produced by several plants

including the skin of grapes, blueberries, and raspberries, etc.
- *STAC*: The abbreviation of Sirtuin-activating compounds, the chemical compounds that have effect on Sirtuins.
- *Proto-oncogene*: Oncogenes are the genes that have potential to cause cancer, usually mutated or highly expressed in tumors. Proto-oncogenes are the normal genes that can become oncogenes under certain circumstances, often mutated or overexpressed. They generally code for proteins that help regulate cell growth and differentiation.
- *Tumor suppressor*: The gene that usually control the opposite side of cell growth and proliferation. When this gene is mutated or inhibited, the cell can progress to cancer.

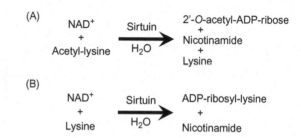

FIGURE 17.1 Chemical reactions catalyzed by sirtuin enzymes. (A) Protein lysine deacetylation by sirtuins requires NAD^+ as a cofactor and releases deacetylated protein, nicotinamide, and 2'-O-acetyl-ADP-ribose. Reactions removing larger acyl groups (malonyl, succinyl, glutaryl, and crotonyl) and lipoyl group are similar to deacetylation. (B) Certain sirtuins are ADP-ribosyltransferases, attaching the ADP-ribose moiety to the ε-amine of lysine, releasing nicotinamide.

INTRODUCTION

Sirtuins are defined as homologs of yeast Sir2, which is encoded by one of the four genes that are required for transcriptional silencing in the budding yeast, *Saccharomyces cerevisiae* [1]. Among these Silent Information Regulators (SIR), Sir2 was found to be an enzyme that removed acetyl groups from acetylated histones. It is unique from other histone deacetylases characterized at the time in that its enzymatic activity requires and consumes NAD^+; therefore, it was categorized as a class III histone deacetylase. Many Sir2 homologs have since been discovered in all model organisms studied, including seven sirtuins, SIRT1 to SIRT7, in mammals [2]. Sirtuins have since been shown to target many nonhistone proteins, and are now known as NAD^+-dependent protein deacetylases.

ENZYMATIC ACTIVITIES OF SIRTUINS

Sirtuins deacetylate acetyl-lysine by coupling the hydrolysis of NAD^+ with transferral of the acetyl group to the ADP-ribose moiety of NAD^+ to form O-acetyl-ADP-ribose and releasing free nicotinamide (Fig. 17.1A). This NAD^+-dependence distinguishes sirtuins from previously identified class I and II histone deacetylases [3]. Because NAD^+ is required for sirtuin activity and the $NAD^+/NADH$ ratio is determined by the nutritional state of the cell, sirtuins are thought to be a direct link between cellular metabolism and protein posttranslational modifications. In addition to the $NAD^+/NADH$ influence, this reaction is also regulated by one of its products, nicotinamide, which is a noncompetitive inhibitor of sirtuins [4]. The other reaction product, O-acetyl-ADP-ribose, has been suggested to facilitate heterochromatin formation and silencing, regulate ion channels and modulate cellular redox state [3].

Some sirtuins carry out reactions that remove larger acetyl moiety-containing groups, such as malonyl, succinyl, and crotonyl groups, as well as the lipoyl group, from their target proteins using a very similar mechanism (see Table 17.1). Some sirtuins do not exhibit deacetylase activity, but rather are ADP-ribosyltransferases, enzymes that add an ADP-ribosyl group to lysines [5] (Fig. 17.1B).

SIRTUINS IN NONMAMMALIAN MODEL ORGANISMS

Yeast Sir2 is a major regulator of transcription silencing at the three heterochromatin-like regions: telomeres, rDNA repeats and the hidden mating type loci *HML* and *HMR*. Sir2 does not have specificity for a particular DNA sequence, but rather is recruited to silencing loci by other DNA binding proteins. At telomeres and the *HML/HMR* loci, it forms a heterotrimeric SIR complex with silencing proteins Sir3 and Sir4; this complex is recruited by Rap1. At the rDNA locus, Sir2 exists in a homotrimeric form in the RENT (REgulator of Nucleolar silencing and Telophase) complex, with Net1 and Cdc14, and is recruited by Fob1 [6]. In a genome-wide analysis, Sir2, together with Hst1 and Sum1, was found to repress genes involved in fermentation, glycolysis, and translation during diauxic shift [7]. In addition to transcriptional repression and silencing, Sir2 also regulates DNA replication at the rDNA locus [8], maintains asymmetric segregation of protein aggregates during cell divisions [9–12], promotes DNA damage repair [13], prevents apoptosis during sustained stress

TABLE 17.1 Summary of Sirtuin Functions

	Localization	Activity	Target	Molecular and cellular function
YEAST				
Sir2	Nucleus, nucleolus	Deacetylation	H4K16, H3K56, H3K4, H3K79, Ifh1	Gene silencing, heterochromatin, transcriptional regulation
Hst1	Nucleus	Deacetylation	H3K4, H4K5, NDT80	Transcription repression, DNA replication, NAD^+ biogenesis, thiamin biosynthesis
Hst2	Cytoplasm	Deacetylation	H4K16	Nucleolar and telomere silencing
Hst3	Nucleus	Deacetylation	H3K56	DNA replication and repair, heterochromatin silencing
Hst4	Nucleus	Deacetylation	H3K56	DNA replication and repair
WORM				
sir-2.1	Nucleus	Deacetylation	H3K9, H4K16, 14-3-3, DAF-16	Transcription silencing, heterochromatin, stress response, calorie restriction
FLY				
dSir2	Nucleus, cytoplasm	Deacetylation	Histone, Dmp53	Transcription silencing, heterochromatin
MAMMAL				
SIRT1	Nucleus, cytoplasm	Deacetylation	H1K26, H4K16, H3K9, H3K14, p53, PGC-1α, NF-κB, FOXO1, FOXO3, FOXO4, Notch, HIF-1α, 14-3-3, PI3K, DNMT1, TORC1, HSF1, Ku70, LC3, BMAL1, PER2, MLL1	Transcription silencing, mitochondria regulation, insulin signaling, tumorigenesis, apoptosis, cell proliferation and survival, tissue regeneration, stem cell differentiation, stress response, autophagy, circadian rhythm, neurodevelopment and memory, cardiovascular protection, delaying replicative senescence
SIRT2	Cytoplasm, nucleus	Deacetylation	H3K18, H4K16, Tubulin, PAR-3, FOXO1, FOXO3, CDH1, CDC20, PGC-1α, p65, RIP, ATP-citrate lyase, HIF-1α, PR-Set7, CDK9, G6PDH, PGAM2, BubR1	Mitosis, nerve myelination and regeneration, brain aging, adipocyte differentiation, genome integrity and stability, oxidative catabolism, oxidative stress response, programmed necrosis, tumor suppression, lifespan regulation
SIRT3	Mitochondria	Deacetylation	LCAD, VLCAD, PDHA1, PDP1, IDH2, GDH, ACS2, GOT2, SOD2, complex I, III, V, OPA1	Fatty acid oxidation, TCA cycle, oxidative phosphorylation, oxidative stress, mitochondria dynamic, mitochondria unfolded protein response
		Decrotonylation	H3K4	
SIRT4	Mitochondria	ADP-ribosylation, lipoamidation	GDH, PDH	TCA cycle, fatty acid oxidation
SIRT5	Mitochondria	Deacetylation, demalonylation, desuccinylation	CPS1, VLCAD	Urea cycle
SIRT6	Nucleus	Deacetylation	H3K9, H3K56, GCN5, p65, BMAL1, SREBP-1, TNF-α	Genome stability, telomere silencing, gluconeogenesis, osteogenic differentiation, autophagy, circadian rhythm, DNA repair, retrotransposition
		ADP-ribosylation	PARP1, KAP1	
SIRT7	Nucleolus	Deacetylation	H3K18, PAF53, GABPβ1	rDNA transcription, oncogenesis, stress response, mitochondria unfolded protein response, mitochondrial function

mediated by stress-activated protein kinase Hog1 [14], and is required for Hsf1 activation induced by heat shock response and ER unfolded protein response [15,16]. Sir2 deacetylates histones and shows activity toward histone H4 lysine 16 (H4K16), histone H3K56, H3K4, and H3K79ac [17]. Recently, nonhistone targets of Sir2 have also been found, such as ribosome protein transcription factor Ifh1 [18].

There are four additional sirtuins in yeast, Hst1—4, all of which are involved in transcription silencing or repression [2]. Hst1 forms a complex with Sum1 and Rfm1, deacetylates H3K4ac and H4K5ac, and regulates the initiation of DNA replication [19,20]. The Hst1/Sum1/Rfm1 complex also functions as a transcriptional repressor for the critical meiotic gene NDT80 [21] and NAD^+ biogenesis genes [22], as well as thiamin biosynthesis genes [23]. Hst2 also regulates transcriptional silencing, but predominantly localizes in the cytoplasm due to a unique nuclear export sequence, regulating its nuclear activity [24,25]. Hst3 and Hst4 deacetylate H3K56ac, a modification found on newly synthesized histones that is critical for DNA replication and repair [26]. Hst3 also contributes to the stability of the silenced heterochromatin state [27].

In the nematode, *Caenorhabditis elegans*, four sirtuins have been identified, with SIR-2.1 showing the highest homology to yeast Sir2. Worm SIR-2.1 deacetylates H3K9ac and H4K16ac and is involved in transcriptional silencing and heterochromatin formation [28]. Its activity on H4K16ac is required for the dosage compensation complex-dependent twofold reduction in transcription on the X chromosome [29]. Several nonhistone targets, including 14-3-3 and DAF-16, have been identified for worm SIR-2.1 and many of them regulate aging and stress responses [30]. In addition, the heat shock response and calorie restriction (CR)-induced resistance to heat shock depend on SIR-2.1 [31]. Deletion of the *sir-2.1* gene moderately shortens lifespan [32]. However, the longevity-promoting effect of *sir-2.1* overexpression has been debated among different groups. The initially observed lifespan extension by either chromosomal duplication or transgene was attributed to secondary mutations unrelated to *sir-2.1* [33,34]. Details of this controversy are discussed below.

In *Drosophila melanogaster*, five sirtuins have been found, with dSir2 being the closest homolog to yeast Sir2. dSir2 shows histone deacetylase activity in vitro and is required for heterochromatin silencing in vivo [35]. Knocking out dSir2 shortens lifespan [36]. Lifespan extension upon dSir2 overexpression has been found using a number of different genetic manipulations, including several different transgenes, different drivers, and tissue-specific and adult-only inductions [37]. However, several of these results have also been challenged and are discussed in detail below [34].

MAMMALIAN SIRTUINS

In mammals, there are seven sirtuins, named SIRT1 to SIRT7. They are involved in a broad range of cellular processes and pathways with distinct cellular localization and molecular targets (Table 17.1).

SIRT1 is the closest mammalian homolog of the yeast Sir2 protein in sequence and is the most studied mammalian sirtuin. SIRT1 predominantly localizes to the nucleus and acts as a deacetylase for histones H1K26 and H4K16 [38], as well as many nonhistone targets (Table 17.1). Deacetylation of p53 by SIRT1 upon DNA damage and oxidative stress results in reduced apoptosis [39,40]. SIRT1 activates PGC-1α (Peroxisome proliferator-activated receptor Gamma Coactivator 1-α) through deacetylation, thus regulating mitochondria biogenesis and activity [41]. Acetylation state of FOXO transcription factors, regulated by SIRT1, is thought to selectively direct them to certain targets, representing another level of regulation of metabolism and stress response mediated by FOXO [5]. Upon inhibition of insulin signaling, SIRT1 is actively shuttled out of the nucleus into the cytoplasm. Many physiological functions of SIRT1 lie in its roles in regulating autophagy: it deacetylates K49 and K51 of LC3, a key component of autophagy pathways, turning on autophagy targeting nuclear proteins under starvation [42,43]. SIRT1 activity is circadian, and it regulates circadian rhythm by deacetylating BMAL1, PER2, and MLL1 in a circadian pattern [44,45]. SIRT1 is also recruited by CLOCK and BMAL1 to clock-controlled genes where it deacetylates histone H3, K9, and K14 [44]. SIRT1 has numerous roles in cancer and apoptosis, which is reviewed in greater detail below. Other newly identified novel functions of SIRT1 include neurodevelopment and memory [46], protection against various neurodegenerative diseases [46,47], cardiovascular protection [48], promoting liver function and regeneration [49], stem cell differentiation and cell fate determination [50—54], and delaying replicative senescence and senescence associated secretory phenotype in primary fibroblasts [55,56].

SIRT2 exists in the cytoplasm, deacetylates tubulin, and regulates skeletal muscle differentiation [5]. It accumulates in neurons in the central nervous system in aging brains and its microtubule deacetylase activity is linked to the pathology of brain aging and neurodegenerative diseases [57]. In the peripheral nervous system, however, SIRT2 appears important for nerve myelination and regeneration by deacetylating Par-3, a critical regulator of cell polarity and myelin assembly in Schwann cells [58]. In adipocytes, SIRT2 regulates metabolism through deacetylating FOXO1 and PCG-1α [5,59]. When FOXO3 is deacetylated by SIRT1 or SIRT2, it is targeted for polyubiquitination and degradation [60]. Tissues from human breast cancers and hepatocellular carcinoma exhibit reduced levels of SIRT2, which regulates the anaphase-promoting complex activity in these tissues by deacetylating CDH1 and CDC20 to preserve genome integrity and antagonize tumorigenesis. The activity of transcription factor NF-κB p65 is regulated by SIRT2 targeting its K310; deacetylation of this residue is required for expression of many NF-κB-dependent genes upon TNF-α stimulation [61]. The RIP1—RIP3 complex is activated by TNF-α to induce programmed necrosis. Deacetylation of RIP lysine 510 by SIRT2 is critical for complex formation and the induction of necrosis [62]. Another tumor suppressor function of SIRT2 lies in its role in deacetylating and destabilizing ATP-citrate lyase, an important enzyme for fatty acid synthesis for membrane expansion [63]. The most recently identified target for SIRT2 in the context of tumor suppression is K709 of hypoxia-inducible factor 1-α (HIF-1α), a critical regulator of hypoxia. Overexpression of SIRT2 destabilizes HIF-1α [64]. Despite its predominant cytoplasmic localization, SIRT2 was recently shown to deacetylate the H4K20 methyltransferase PR-Set7 at K90, which regulates its chromatin localization; thus SIRT2 indirectly influences H4K20 methylation and its associated genome stability [65]. Another role for SIRT2 in maintaining genome stability is to stimulate CDK9 kinase activity by deacetylating K48 in response to replication stress [66]. During bacterial infection, SIRT2 relocates to the nucleus and represses the transcription of a set of genes involved in immune response by deacetylating H3K18ac [67]. Under oxidative stress, SIRT2 targets glucose-6-phosphate dehydrogenase K403 and phosphoglycerate mutase PGAM2 K100 to regulate NADPH homeostasis and promote cell survival [68,69]. Finally, the mitotic checkpoint kinase BubR1, an important longevity regulator, is targeted by SIRT2, which deacetylates K668, thus stabilizing the protein; hyperacetylation of K668 contributes to BubR1 loss during aging [70].

SIRT3 primarily localizes to mitochondria and is the major mitochondrial deacetylase, targeting numerous enzymes playing critical roles in maintaining metabolic homeostasis [71]. It is highly enriched in hematopoietic stem cells and is critical for their maintenance under stress and during aging [72]. In the fatty acid oxidation pathway, SIRT3 deacetylates the long-chain acyl CoA dehydrogenase (LCAD) at conserved K318 and K322, near the active site, as well as the very long-chain acyl-CoA dehydrogenase (VLCAD) K229 [73,74]. The critical link between glycolysis and the citric acid cycle, pyruvate dehydrogenase (PDH) complex, is also regulated by SIRT3, which targets the E1 pyruvate dehydrogenase PDHA1 K321, as well as PDH phosphatase PDP1 K202 [75,76]. Among enzymes involved in the citric acid cycle, both isocitrate dehydrogenase (IDH2) and glutamate dehydrogenase (GDH) are deacetylation targets of SIRT3. Furthermore, SIRT3 deacetylates many components of the electron transport chain, including complex I 39-kDa subunit, complex III core I subunit and complex V ATP synthase F1 α, β, γ subunits [77—79]. The acetyl-coA synthase AceCS2 and glutamate oxaloacetate transaminases are also targets of SIRT3 [71,80]. In addition to these metabolic enzymes, SIRT3 provides protection against oxidative stress by deacetylation and activation of SOD2, an important mitochondrial antioxidant enzyme [71]. In cardiomyocytes, SIRT3 deacetylates and activates OPA1, an important mitochondria fusion protein that is acetylated under pathological stress [81]. SIRT3 was also shown to regulate the mitochondria unfolded protein response; however, its targets in these pathways remain elusive [82]. Although most research on SIRT3 has focused on the processed short isoform of SIRT3 localized in the mitochondria, new evidence suggest that the full-length form of SIRT3 localizes to the nucleus and represses gene expression by deacetylating histones. This nuclear full-length SIRT3, but not the smaller mitochondria isoform, is degraded by the ubiquitin-proteasome pathway upon various stresses, activating its target genes [83]. In addition to functioning as a deacetylase, SIRT3 is also found to decrotonylate H3K4 in vivo [84].

SIRT4 and SIRT5 are also localized to the mitochondria. SIRT4 was initially identified as an enzyme with ADP-ribosyltransferase activity that targeted GDH [5]. Recently, it was found to function as a lipoamidase for the PDH complex E2 component dihydrolipolylysine acetyltransferase, reducing PDH activity [85]. SIRT5 targets many metabolic enzymes to remove large acetyl-containing groups, such as malonyl, succinyl, and glutaryl, in a fashion very similar to deacetylation [86]. Well-characterized SIRT5 targets include carbamoyl phosphate synthetase CPS1, a critical enzyme in the urea cycle, and VLCAD [74,87].

SIRT6 localizes to the nucleus and deacetylates histone H3K9 and H3K56 to maintain genome stability and telomere function [88]. It binds to DNA double-stranded break sites, deacetylates H3K56ac, and recruits chromatin remodeling enzyme SNF2H [89]. SIRT6 deacetylates and activates GCN5, which in turn leads to expression of PGC-1α, a key transcription factor for gluconeogenesis [90]. In bone marrow-derived mesenchymal stem cells, SIRT6 regulates osteogenic differentiation through deacetylation and inactivation of NF-κB p65 [91]. SIRT6's deacetylase activity is required for its role in activating autophagy; however, its direct targets in autophagy regulation remain unknown [92]. SIRT6 also regulates circadian rhythm by recruiting transcription factors BMAL1 and SREBP-1 to their target genes [93]. In addition to deacetylation, SIRT6 removes long-chain acylation from TNF-α, regulating its secretion [94].

Furthermore, SIRT6 functions as an ADP-ribosyltransferase for PARP1 (Poly ADP-ribose polymerase 1) under oxidative stress, stimulating PARP1 function to support DNA repair [95]. It has also been reported that SIRT6 carries out ADP-ribosylation of nuclear corepressor protein KAP1 and represses the activity of LINE1 retrotransposons [96].

SIRT7 is a histone deacetylase that targets H3K18ac at selected gene promoters, repressing their transcription. This activity promotes oncogenesis [97]. Repression of ribosomal genes is mediated through the recruitment of SIRT7 by Myc, leading to hypoacetylation of H3K18, and is important to relieve ER stress [98]. SIRT7 also targets RNA polymerase I subunit PAF53 directly to repress its activity upon stress [99]. Furthermore, SIRT7 collaborates with NRF1 to repress expression of nuclear mitochondrial ribosomal genes and transcription factors to enhance the mitochondria unfolded protein response, a mechanism that regulates hematopoietic stem cell aging [100]. In addition, SIRT7 deacetylates GABPβ1, a master regulator of nuclear-encoded mitochondrial genes, controlling mitochondrial function [101].

In the following sections, we will discuss the major controversial issues around sirtuins and their roles in aging and age-related diseases.

DOES OVEREXPRESSION OF SIRTUINS EXTEND LIFESPAN?

The discovery that sirtuins regulate aging was first made in yeast. In 1995, Kennedy et al. reported that Sir4 regulates yeast lifespan. A mutation in Sir4 prevents it from silencing HML/HMR and telomeres and extends lifespan, despite that *sir4* mutant yeast still have shortened telomeres. The authors proposed that Sir4 localizes to another unknown genomic locus that is critical for yeast lifespan in the absence of normal telomere length [102]. This region was subsequently recognized as the ribosomal gene cluster (rDNA) loci [103]. These findings led to the discovery of one of the causes of aging in yeast: the accumulation of extrachromosomal rDNA circles (ERCs) in old cells due to the inherent instability of the tandemly repeated rDNA locus [104]. Sir2 has previously been shown to repress recombination at rDNA [105]. This protein was quickly shown to be part of the Sir4 complex that regulates the rate of ERC generation, and thus regulates the rate of yeast aging. Furthermore, deletion of Sir2 shortens and overexpression of Sir2 significantly extends yeast lifespan [106]. In this experiment, overexpression was achieved by inserting another copy of the *SIR2* gene into the genome. The strains with twofold expression of Sir2 showed a 30% increase in lifespan. More recently, telomere-associated Sir2 was also shown to be critical to regulating lifespan. Sir2 protein levels are much lower in old cells, leading to increased H4K16 acetylation at telomere regions, which is thought to be another cause of aging in yeast [107]. Sir2 is also required for the asymmetric segregation of damaged and misfolded proteins during cell division, which leads to their accumulation in mother cells and contributes to their aging [9–12].

The longevity effects of Sir2 homologs in other eukaryotic models have also been investigated. However, debates arose from studies in worms and flies due to concerns about inconsistency in experimental results. The first study in *C. elegans* by Tissenbaum et al. showed that three independent extrachromosomal *sir-2.1* transgenic lines had a lifespan extension of about 20–50%. This was also the case with integrated *sir-2.1*-overexpression strains. Furthermore, it was suggested that *sir-2.1* functions in the DAF-2 and DAF-16 insulin-signaling pathway that had been shown to regulate worm longevity [108]. In flies, three GAL4 driver strains that induced greater than fourfold overexpression of dSir2 showed a 29% lifespan increase for females and 18% for males [109]. This effect appeared to be dose-dependent, since a 10% increase in dSir2 expression had no effect on lifespan. Additionally, overexpressing the fly nicotinamadase D-NAAM, either ubiquitously or neuron-specifically, hence activating dSir2 activity by decreasing the levels of nicotinamide, also showed significant increase in lifespan in both males and females [110].

Interestingly, in 2011, a study conducted by several labs across Europe and the United States questioned whether overexpression of Sir2 indeed extends lifespan in worms and flies. In *C. elegans*, Burnett et al. first reexamined the lifespan extension effect of one of the integrated high-copy *sir-2.1* strains. When outcrossed five times to the wild-type strain, the longevity-promoting effect was abrogated [34]. Meanwhile, after outcrossing for six generations, the low-copy *sir-2.1* overexpressing strains failed to show the lifespan extension effect. In *Drosophila*, outcrossing the GAL4-driven dSir2 overexpressing strain still resulted in lifespan extention compared to the wild-type controls, but the GAL4 driver-only control also showed a similar long lifespan, indicating that dSir2 is not the reason for the increased longevity. Another independently created dSir2 overexpression strain also showed no effect on lifespan compared to the transgenic controls in this study [34].

Following up on these reports, several groups reexamined the effect of Sir2 overexpression in worms and flies and reinforced its central role in lifespan regulation. The researchers who originally did the *C. elegans* work revisited their data and found that the original results overestimated the extension of lifespan in a high-copy transgenic *sir-2.1* strain due to an unlinked mutation causing defects in sensory neurons. However, they were still able to observe an approximately 10–14% lifespan extension by *sir-2.1* overexpression with strict background controls [33]. Another group using the same strain also confirmed the lifespan extension effect of *sir-2.1* overexpression [111]. Meanwhile, knocking down *sir-2.1* in the low-copy *sir-2.1* overexpression strain abrogated the longevity promoting effect, demonstrating that *sir-2.1* is indeed responsible for the observed lifespan extension [112]. Moreover, several new studies implicate *sir-2.1* in regulating longevity. Mouchiroud et al. confirmed that *sir-2.1* overexpression extended lifespan [113]. Boosting NAD^+ level in worms promotes longevity in a *sir-2.1*-dependent manner. They found two distinct pathways for NAD^+/*sir-2.1* mediated longevity effect: one involves the insulin-like signaling pathway and the other is related to the mitochondrial unfolded protein response caused by an imbalance of nuclear versus mitochondrial expression of mitochondrial genes. Importantly, this explains a cause of aging: NAD^+ decreases with age due to chronic DNA damage-induced PARP activation.

New evidence from functional studies suggests that SIR-2.1 is indeed intricately involved in aging-regulation in worms. Deletion of *sir-2.1* disrupts the functions of cholinergic neurons in males, resulting in a mating behavior decline similar to older wild-type animals [114]. SIR-2.1 is required for longevity by amino acid supplements [115]. Upregulation of SIR-2.1 has been attributed to the longevity and stress resistance of glutathione treatment [116].

With regards to the fly studies, in Burnett et al.'s paper, they failed to test a previous study that used inducible dSir2 expression to extend lifespan in flies [117]. In fact, another study aplied appropriate genetic controls and still found a lifespan extension that depended on dSir2 levels [118]. Interestingly, they found that increasing dSir2 levels by approximately fivefold showed a 38% increase of median lifespan in females compared to the original 10% increase observed with about threefold overexpression, indicating a dose-dependent effect of lifespan extension by overexpressing dSir2. Another report showed that tissue-specific overexpression of inducible dSir2 in the fat body is sufficient to extend lifespan on a normal diet [119]. These findings were confirmed by another research group [120].

Overall, more studies have confirmed that overexpression of sirtuins, either ubiquitously or in a tissue-specific way, in worms and flies indeed extends lifespan. The conflicting results remind us all about the importance of accurate experimental controls when performing lifespan tests. Better communications are needed to eliminate the differences between strain backgrounds, protocols, and technical operations that might cause the discrepancies between results from different labs. Additionally, it is necessary to test the dose-dependent effect of lifespan extension by sirtuin overexpression with strict controls across a series of strains.

Given the confusing results from these two organisms, it is more critical and informative to look at whether sirtuins also have lifespan extension effects in mammalian models. Although studies from yeast, worms, and flies provided many clues about the function of sirtuins in lifespan extension, these systems are still very different from humans. Therefore, experiments in mammals will better bypass the debates from the lower metazoans while providing valuable knowledge to the aging field. Research initially focused on SIRT1, the closest homolog of Sir2 in mammalian cells. However, in a transgenic mouse model that overexpresses SIRT1 to moderate levels, despite its protecting effects against several aging-related phenotypes such as osteoporosis, glucose intolerance, and metabolic syndrome-related cancers, these mice did not live longer [121]. The explanation could be either that SIRT1 does not function to extend lifespan or moderate overexpression (approximately threefold) was not sufficient to give longevity-promoting phenotype.

SIRT6 is another longevity regulator. SIRT6-deficient mice have severe aging-associated degenerative and premature aging phenotypes [122]. The remarkable evidence came from the Cohen group: overexpression of SIRT6 transgene extended lifespan of male mice [123]. Both median and mean lifespan of males were significantly increased by ~15% in SIRT6 transgenic mice compared to wild-type littermates. There was no significant change in females. SIRT6 was suggested to function in the insulin-like growth factor-1 pathway to mediate longevity effects, which is consistent with the findings in worm studies. With proper controls and two different experimental lines used to counteract any background or site-specific transgenic effects, this study for the first time conclusively showed that sirtuins can extend mammalian lifespan when overexpressed. Another exciting report came from a brain-specific transgenic SIRT1-overexpressing mouse strain (BRASTO). These mice showed significant lifespan extension in both females and males with phenotypes consistent with a delay in aging [124]. However, much more work is needed to address the molecular mechanisms by which SIRT1 extends mouse lifespan and whether other sirtuins may also contribute to longevity in mammals. Nevertheless, these results opened the way for promising aging studies focused on sirtuins.

DO SIRTUINS MEDIATE THE LONGEVITY EFFECTS OF CALORIE RESTRICTION?

Calorie restriction (CR) is the most robust way to slow the process of aging in diverse species, ranging from yeast, worms, and flies to mammals. In yeast, Sir2 was implicated to be a nutrient sensor that was involved in CR-mediated aging regulation. However, in the past decade, this idea has been challenged [125].

CR was shown to extend yeast lifespan in strains with mutated nutrient-sensing pathways or cultured with reduced glucose. This extension was dependent on Sir2 and NAD^+ [126]. Specifically, CR increases the NAD^+/NADH ratio by increasing respiration [127]. Consistent with this, the NAD^+ salvage pathway enzyme Pnc1 was shown to be upregulated by CR, thus facilitating NAD^+ synthesis, which enhances Sir2 activity and ultimately promoting longevity [128]. Furthermore, deletion of Sir2 blocked lifespan extension in different yeast CR models. Similar conclusions were also reached in worms [129] and flies [109].

However, in addition to these reports implicating a role for Sir2 in CR, additional studies found opposite effects. In yeast, CR could still significantly extend lifespan in cells lacking Sir2 as long as the generation of ERCs was inhibited through the *FOB1* deletion [130]. Furthermore, CR was also shown to extend lifespan in strains depleted of Sir2 and two of its paralogs, Hst1 and Hst2 [131]. At the same time, the TOR nutrient sensing kinase signaling pathway emerged as a conserved mechanism behind CR since disruption of TOR1 mimics and is epistatic to CR [132]. Sir2 was also not required for the lifespan extension by *TOR1* deletion, a proposed CR mimetic. Others monitored Sir2 activity by gene silencing and found that the effect of CR on Sir2 function was moderate or undetectable, and much lower than previous suggested [133]. In *C. elegans*, several *sir-2.1*-independent CR mechanisms were proposed. The DAF-2/insulin-like signaling pathway-mediated CR longevity does not require *sir-2.1* [134]. Other mechanisms depending on the mTOR signaling pathway [135], mitochondrial respiration [136], and long-term neuroendocrine control [137] did not require *sir-2.1*. The role of dSir2 in CR-mediated lifespan extension in flies was also challenged in Burnett et al.'s paper. The central controversy focused on the link between nutrient-sensing pathways and lifespan regulation pathways. With different standards for "calorie restriction" in lower organisms, different experimental conditions may trigger different subsets of stress response pathways and metabolic pathways that lead to varying results among different labs. This may explain how various observations were obtained throughout the past decade.

In recent rodent studies, the experimental conditions of CR were better standardized, and thus more consistent conclusions were made on sirtuins' role in CR. The expression of all mammalian sirtuins, except SIRT7, have been reported upregulated in various tissues by CR conditions [138–144]. Many sirtuin targets are critical metabolic enzymes and signaling factors that are essential for adaptation to dietary changes (Table 17.1). Hence, sirtuins are important regulators of cellular response of CR. Furthermore, various sirtuin knockout strains failed to respond to CR retreatment: SIRT1 knockout mice did not show lifespan extension by CR [145,146]; the protective effects of CR on oxidative stress and hearing loss were abrogated in SIRT3 knockout mice [147,148].

These accumulating pieces of evidence suggest the importance of sirtuins in mediating many beneficial effects conferred by CR. Adaptation to metabolic stress is essential for the survival of individuals as well as species. It is not hard to understand that multiple nutrient-sensing and metabolic pathways are affected by CR, including TOR, AMPK, and insulin-signaling. So far, the evidence supports that sirtuins function in these pathways, as part of the metabolic network, regulating physiology, fitness, and lifespan.

IS RESVERATROL A SIRTUIN ACTIVATOR?

Soon after sirtuins emerged as a class of conserved aging regulators, efforts were made to identify chemical inhibitors and potential activators of these enzymes. It was especially intriguing to search for activators since overexpression of sirtuins was shown to have longevity benefits. A screen using an in vitro deacetylation assay featuring a fluorophore-labeled acetylated p53 peptide identified a number of inhibitors and activators for mammalian SIRT1. Among them, resveratrol, a polyphenol enriched in red wine, showed the strongest effect [149]. Strikingly, treating yeast with resveratrol extended their lifespan by 70%. Similar lifespan extension effects by resveratrol treatment were also found in worms and flies, and a sirtuin-dependent CR mimic mechanism was proposed [150]. In *Nothobranchius furzeri*, a very short-lived fish, resveratrol extends lifespan and delayed age-dependent cognitive decline [151]. In mammals, the Sinclair group showed that although resveratrol has no effect on lifespan of mice fed with normal diet, it produces a significant beneficial effect of longevity and health on mice fed on high-fat diet by shifting their metabolic physiology [152]. Consistently, they also reported that resveratrol induced numerous health benefits in these mice, mimicking transcriptional aspects of CR [153].

However, along with these exciting reports were many contradictory findings that lead to controversial conclusions about how resveratrol works. The first question was whether resveratrol indeed extended lifespan. There were conflicting results showing that resveratrol had no or inconsistent effects on lifespan extension in yeast, worms, and flies [154,155]. Second, whether resveratrol actually activates SIRT1 in vitro is heavily argued. It was found that in the original report, the covalently bounded fluorophore was required for SIRT1 enzymatic activation by resveratrol [154]. Along with some other proposed SIRT1 activators, resveratrol in fact interacts with the fluorophore-containing peptides directly, thus showing that the in vitro experimental effects of resveratrol were independent of SIRT1 [156]. Moreover, resveratrol can target other regulators to elicit its physiological functions through SIRT1-dependent or SIRT1-independent pathways. Several studies showed that AMPK is activated by resveratrol in vivo [152,157]. This suggested that SIRT1 might be an indirect target of resveratrol, downstream of AMPK. Indeed, AMPK is known to activate nicotinamide phosphoribosyltransferase (NAMPT), an enzyme required for NAD^+ synthesis, and by activating AMPK, resveratrol may enhance SIRT1 function by this mechanism [158].

So is SIRT1 required for the physiological function of resveratrol in vivo? The answer depends on a number of factors, including the type of tissue, the dose of treatment and the physiological or pathological conditions. It was suggested that, at a low dose, resveratrol functions by activating SIRT1, which deacetylates the AMPK kinase LKB1, activating the AMPK pathway [159,160]. Others suggested that resveratrol activates protein kinase A by inhibiting cAMP phosphodiesterases, which leads to the activation of AMPK and NAMPT, increased levels of NAD^+ and eventually SIRT1 activation [161]. However, in vivo data showed no increase in NAD^+ levels in resveratrol treated animals [159]. Among those who are convinced that resveratrol directly activates sirtuins, there are several different hypotheses for how resveratrol could structurally activate SIRT1. The Steegborn group proposed a model in which resveratrol-like STACs directly interact with the enzyme-substrate complex [162]; while the Sinclair group proposed a direct "assisted allosteric activation" mechanism, in which the binding of substrates with SIRT1 causes the formation of an exosite that facilitates STAC binding, which, in turn, stabilizes the docked substrate [163]. The E320 site at the N-terminus of SIRT1 is especially critical for STAC-mediated activation.

Indeed, these lines of evidence demonstrate the great potential of resveratrol as therapeutics to antagonize aging and age-associated pathologies. Although most of the in vivo studies showed that the therapeutic effects of STACs require sirtuin activation, whether sirtuins are the direct molecular targets of these drugs is unknown. Although some STACs have promising clinical potential, the underlying molecular mechanisms still need to be further clarified.

ARE SIRTUINS PROTO-ONCOGENES OR TUMOR SUPPRESSORS?

Aging is a strong risk factor for most types of adult-onset cancer. As conserved aging regulators, many sirtuins have been shown to promote cellular survival and replicative growth [164]. This is further evidenced by the overexpression of SIRT1 in several types of cancer. However, disruption of SIRT1 was also shown to promote tumorigenesis, while overexpression of SIRT1 and SIRT6 in normal animals and tissue showed tumor suppressor effects [165]. Thus, whether sirtuins are proto-oncogenes or tumor suppressors became a heated discussion.

SIRT1 was first linked to tumorigenesis through its role in deacetylating and inactivating p53, a well-defined tumor suppressor, and promoting cell survival under stress [39,40]. Inhibition of SIRT1 leads to hyperacetylation of p53, thus elevating p53-dependent apoptosis [166], which supports the classification of SIRT1 as a proto-oncogene. SIRT1 also interacts with E2F1, a cell cycle regulator, and interferes with its apoptotic function during DNA damage responses [167]. SIRT1 was found to be expressed in several types of cancer, including primary colon cancer, acute myeloid leukemia, prostate cancers, and nonmelanoma skin cancers [168], supporting that SIRT1 could be a proto-oncogene. Moreover, overexpression of SIRT1 either epigenetically represses the expression or directly inhibits the activity of numerous tumor suppressors, such as FOXO family members, p73, Rb, MLH1, and Ku70 [169]. Consistently, SIRT1 was found to be inhibited by several tumor suppressors: HIC1 (Hypermethylated In Cancer-1) and DIB1 (Deleted in Breast Cancer-1) [170–172]. In a PTEN (Phosphatase and tensin homolog)-deficient thyroid carcinoma model, transgenic overexpression of SIRT1 promoted carcinogenesis by upregulating the c-Myc transcriptional system [173]. Other sirtuins have also been found to be associated with cancer development. SIRT2 is upregulated in acute myeloid leukemia, in which it deacetylates and activates Akt [174]. In colon cancer patients, SIRT3 overexpression is associated with metastasis and poor prognosis [175]. In nonsmall cell lung cancers, SIRT5 overexpression correlates with poor survival; knocking down SIRT5 not only repressed the growth of cancer cells, but also sensitized them to chemotherapy agents [176]. Both SIRT2 and SIRT6 are expressed in retinoblastomas [177]. Finally, SIRT7 is overexpressed in a number of different tumors in which it represses the expression of tumor suppressor genes. Since knocking down SIRT7 could arrest the growth of cancer cells and reverse the transformed phenotype, it has been considered a promising pharmaceutical target [178].

However, in other contexts, it has been reported that sirtuins function as tumor suppressors. Studies in mouse models showed that disruption of SIRT1 in p53 heterozygous mice leads to tumor development in multiple tissues, whereas resveratrol treatment reduced tumorigenesis [179]. In a colon cancer mouse model, activation of SIRT1 by CR significantly reduced tumor formation and morbidity by promoting cytoplasmic localization of oncogenic form of β-catenin [180]. The protection from cancer by overexpression or activation of SIRT1 was also shown in metabolic syndrome-associated liver cancer model [121] and the irradiation-induced p53 heterozygous cancer model, in which SIRT1 functions as a tumor suppressor mainly by promoting genome stability [181]. Another possible tumor suppression mechanism of SIRT1 is through deacetylation and inactivation of HIF-1α, an important factor for cancer growth [182]. In a more recent study, Di Sante et al. showed that loss of SIRT1 promoted prostatic intraepithelial neoplasia, which was supported by clinical data showing that lower SIRT1 expression was associated with poor disease prognosis [183]. Furthermore, SIRT1 deacetylates EZH2 at K348; and elevated EZH2 acetylation has been linked to lung adenocarcinoma [184]. In leukemia cells, SIRT1 deacetylates H3K9ac at the promoters of MLL fusion genes and represses their expression [185]. SIRT3 can also act as a tumor suppressor. Loss of SIRT3 in mice leads to spontaneous mammary gland tumors and SIRT3 suppresses formation of these tumors by modulating ROS in mitochondria [186]. SIRT4 suppresses tumor formation in the model of Myc-induced B-cell lymphoma [187]. SIRT6 was also shown to be a tumor suppressor. Conditional knockout of SIRT6 in mice results in increased number, size, and aggressiveness of tumors; SIRT6 inhibits the growth of these tumors possibly by regulating aerobic glycolysis and ribosomal metabolism [188].

How can these conflicting findings be reconciled? The tumor suppressor function of sirtuins is consistent with the idea that sirtuins, in general, promote longevity and healthspan. However, at the cellular level, many sirtuins promote replicative growth and cell survival through various targets. These activities could certainly be hijacked by cancer cells to promote tumor growth and metastasis. Thus, some sirtuins may show tumor suppressor properties in some cell types and situations, while displaying features of proto-oncogenes in others. Therefore, the molecular and cellular functions of sirtuins should be carefully analyzed for each type of cancer, or even on the basis of individuals, before they can be targeted for therapeutic purposes. In terms of this controversy, some sirtuins, so far, have clearer roles than others. For example, SIRT4 has generally considered a tumor suppressor [187]; while SIRT7 is regarded as a proto-oncogene and a hallmark of certain cancers [178].

THE POTENTIAL OF SIRTUINS AS THERAPEUTIC TARGETS

Despite the many controversies, mammalian sirtuins are clearly implicated as a class of enzymes regulating many cellular processes and playing important functions in diverse tissues and systems. Sirtuin functions have been described in the central/peripheral nerve system, cardiovascular system, immune system, liver, bone, skeletal muscles, stem cells, and tissue regeneration [44,46,48,125,189]. They have also been associated with most major diseases, including cardiovascular diseases, cancer, metabolic disorders, neurodegenerative diseases, arthritis, and osteoporosis, all of which are age-related. For instance, in several types of cancers, knocking down SIRT1 sensitizes cancer cells to radiation and chemotherapies [190,191]. However, the complexity of functions of sirtuins, and their widespread roles and activities, increases the difficulty of determining how best to modulate them therapeutically.

Sirtuins are a class of epigenetic regulators that modulate transcription by deacetylating histones in the promoter region of their target genes. Small molecule regulators targeting sirtuins would provide a robust, rapid, and yet reversible cellular response. Indeed, two histone deacetylase (HDAC) inhibitor drugs have already been approved to treat a certain type of lymphoma, another one is being used for pancreatic cancer, and many more are under clinical trials for more types of cancers [178].

It should be noted that current HDAC inhibitor drugs all target the class I and II HDACs, not sirtuins, which are class III HDACs. Nevertheless, due to the interest in screening for small molecular modulators of sirtuins, many sirtuin-specific inhibitors have been discovered and characterized, several of which have been tested to treat cancer in mouse models [191]. Although resveratrol's function as a sirtuin activator is still under debate, it did show many health benefits in clinical trials [192].

Current controversies and continued debates around sirtuins and their biological functions will inspire more research and eventually lead us to a much better understanding of the molecular mechanisms for this class of enzymes. These mechanistic studies should provide a clearer perspective for their use as targets in future medicine.

CONCLUSIONS

Sirtuins are a class of conserved enzymes that influence aging and longevity through many molecular pathways. Here, we have provided an overview of the most recent findings in sirtuin functions and discussed several major controversial issues around sirtuins. First, more studies, especially those done in mice, suggest that activation or overexpression of sirtuins may provide real longevity benefits. Second, some sirtuins seem to mediate some effects of CR; however, it is only one of the many pathways exploited by CR to realize its longevity effects. Third, despite of years of research and interesting results from clinical tests, whether resveratrol directly activates sirtuins remains a highly contested topic among researchers. Fourth, pertaining to their roles in oncogenesis, some sirtuins act more like proto-oncogenes, some function like tumor suppressors, while others might be considered both depending on the molecular context and the type of cancer. Finally, sirtuins remain intriguing targets for various therapeutic purposes, including many that are related to aging.

SUMMARY

Sirtuins are a class of evolutionarily conserved protein deacetylases and ADP-ribosyltransferases. They are shown to play various and important roles in aging process and generally considered as a pro-longevity molecule. In the past two decades, sirtuins has become a hot target for aging research covering both basic biological mechanisms and therapeutic developments. However, accompanying with their rising fame, the debates on the functional roles of sirtuins in aging regulation and their involvement in age associated diseases, such as cancer, heated up, not only in the scientific community, but also among the general public. In this chapter, we will discuss the controversies that received the most attentions in recent years, hoping to shed light on the potential applications of future sirtuin-based antiaging drug development.

Acknowledgments

We thank Dr Brenna McCauley for critically reading the manuscript and helpful suggestions. This work is supported by NIH grant R00AG037646. WD is a CPRIT scholar R1306.

References

[1] Guarente L. Diverse and dynamic functions of the Sir silencing complex. Nat Genet 1999;23(3):281—5.
[2] Brachmann CB, et al. The SIR2 gene family, conserved from bacteria to humans, functions in silencing, cell cycle progression, and chromosome stability. Genes Dev 1995;9(23):2888—902.
[3] Denu JM. The Sir 2 family of protein deacetylases. Curr Opin Chem Biol 2005;9(5):431—40.
[4] Sanders BD, et al. Structural basis for nicotinamide inhibition and base exchange in Sir2 enzymes. Mol Cell 2007;25(3):463—72.
[5] Houtkooper RH, Pirinen E, Auwerx J. Sirtuins as regulators of metabolism and healthspan. Nat Rev Mol Cell Biol 2012;13(4):225—38.
[6] Huang J, Moazed D. Association of the RENT complex with nontranscribed and coding regions of rDNA and a regional requirement for the replication fork block protein Fob1 in rDNA silencing. Genes Dev 2003;17(17):2162—76.
[7] Li M, et al. Genome-wide analysis of functional sirtuin chromatin targets in yeast. Genome Biol 2012;14(5).
[8] Yoshida K, et al. The histone deacetylases sir2 and rpd3 act on ribosomal DNA to control the replication program in budding yeast. Mol Cell 2014;54(4):691—7.
[9] Erjavec N, Nystrom T. Sir2p-dependent protein segregation gives rise to a superior reactive oxygen species management in the progeny of *Saccharomyces cerevisiae*. Proc Natl Acad Sci USA 2007;104(26):10877—81.
[10] Erjavec N, et al. Accelerated aging and failure to segregate damaged proteins in Sir2 mutants can be suppressed by overproducing the protein aggregation-remodeling factor Hsp104p. Genes Dev 2007;21(19):2410—21.
[11] Orlandi I, et al. Sir2-dependent asymmetric segregation of damaged proteins in ubp10 null mutants is independent of genomic silencing. Biochim Biophys Acta 2010;1803(5):630—8.
[12] Song J, et al. Essential genetic interactors of SIR2 required for spatial sequestration and asymmetrical inheritance of protein aggregates. PLoS Genet 2014;10(7):e1004539.
[13] Mills KD, Sinclair DA, Guarente L. MEC1-dependent redistribution of the Sir3 silencing protein from telomeres to DNA double-strand breaks. Cell 1999;97(5):609—20.
[14] Vendrell A, Posas F. Sir2 plays a key role in cell fate determination upon SAPK activation. Aging 2011;3(12):1163—8.
[15] Nussbaum I, et al. Deteriorated stress response in stationary-phase yeast: Sir2 and Yap1 are essential for Hsf1 activation by heat shock and oxidative stress, respectively. PloS One 2013;9:10.
[16] Weindling E, Shoshana B-N. Sir2 links the unfolded protein response and the heat shock response in a stress response network. Biochem Biophys Res Commun 2015;457(3):473—8.
[17] Bheda P, et al. Biotinylation of lysine method identifies acetylated histone H3 lysine 79 in *Saccharomyces cerevisiae* as a substrate for Sir2. Proc Natl Acad Sci USA 2012;109(16):E916—25.
[18] Downey M, et al. Gcn5 and sirtuins regulate acetylation of the ribosomal protein transcription factor Ifh1. Curr Biol 2013;23(17):1638—48.

[19] Weber JM, Irlbacher H, Ann E-M E. Control of replication initiation by the Sum1/Rfm1/Hst1 histone deacetylase. BMC Mol Biol 2007;9:100.

[20] Guillemette B, et al. H3 lysine 4 is acetylated at active gene promoters and is regulated by H3 lysine 4 methylation. PLoS Genet 2011;7(3):e1001354.

[21] Shin ME, Skokotas A, Winter E. The Cdk1 and Ime2 protein kinases trigger exit from meiotic prophase in *Saccharomyces cerevisiae* by inhibiting the Sum1 transcriptional repressor. Mol Cell Biol 2010;30(12):2996–3003.

[22] Bedalov A, et al. NAD+-dependent deacetylase Hst1p controls biosynthesis and cellular NAD+ levels in *Saccharomyces cerevisiae*. Mol Cell Biol 2003;23(19):7044–54.

[23] Li M, et al. Thiamine biosynthesis in *Saccharomyces cerevisiae* is regulated by the NAD+-dependent histone deacetylase Hst1. Mol Cell Biol 2010;30(13):3329–41.

[24] Perrod S, et al. A cytosolic NAD-dependent deacetylase, Hst2p, can modulate nucleolar and telomeric silencing in yeast. EMBO J 2001;20(1–2):197–209.

[25] Wilson JM, et al. Nuclear export modulates the cytoplasmic Sir2 homologue Hst2. EMBO Rep 2006;7(12):1247–51.

[26] Celic I, et al. The sirtuins hst3 and Hst4p preserve genome integrity by controlling histone h3 lysine 56 deacetylation. Curr Biol 2006;16(13):1280–9.

[27] Dodson AE, Rine J. Heritable capture of heterochromatin dynamics in *Saccharomyces cerevisiae*. Elife 2015;4:e05007.

[28] Wirth M, et al. HIS-24 linker histone and SIR-2.1 deacetylase induce H3K27me3 in the *Caenorhabditis elegans* germ line. Mol Cell Biol 2009;29(13):3700–9.

[29] Wells MB, et al. *Caenorhabditis elegans* dosage compensation regulates histone H4 chromatin state on X chromosomes. Mol Cell Biol 2012;32(9):1710–19.

[30] Berdichevsky A, et al. *C. elegans* SIR-2.1 interacts with 14-3-3 proteins to activate DAF-16 and extend life span. Cell 2006;125(6):1165–77.

[31] Raynes R, et al. Heat shock and caloric restriction have a synergistic effect on the heat shock response in a sir2.1-dependent manner in *Caenorhabditis elegans*. J Biol Chem 2012;287(34):29045–53.

[32] Pan KZ, et al. Inhibition of mRNA translation extends lifespan in *Caenorhabditis elegans*. Aging Cell 2007;6(1):111–19.

[33] Viswanathan M, Guarente L. Regulation of *Caenorhabditis elegans* lifespan by sir-2.1 transgenes. Nature 2011;477(7365):E1–2.

[34] Burnett C, et al. Absence of effects of Sir2 overexpression on lifespan in *C. elegans* and *Drosophila*. Nature 2011;477(7365):482–5.

[35] Blander G, Guarente L. The Sir2 family of protein deacetylases. Annu Rev Biochem 2004;73:417–35.

[36] Astrom SU, Cline TW, Rine J. The *Drosophila melanogaster* sir2+ gene is nonessential and has only minor effects on position-effect variegation. Genetics 2003;163(3):931–7.

[37] Frankel S, Ziafazeli T, Rogina B. dSir2 and longevity in *Drosophila*. Exp Gerontol 2011;46(5):391–6.

[38] Vaquero A, et al. Human SirT1 interacts with histone H1 and promotes formation of facultative heterochromatin. Mol Cell 2004;16(1):93–105.

[39] Vaziri H, et al. hSIR2(SIRT1) functions as an NAD-dependent p53 deacetylase. Cell 2001;107(2):149–59.

[40] Luo J, et al. Negative control of p53 by Sir2alpha promotes cell survival under stress. Cell 2001;107(2):137–48.

[41] Rodgers JT, et al. Nutrient control of glucose homeostasis through a complex of PGC-1alpha and SIRT1. Nature 2005;434(7029):113–18.

[42] Lapierre LR, et al. Transcriptional and epigenetic regulation of autophagy in aging. Autophagy 2015;0.

[43] Huang R, et al. Deacetylation of nuclear LC3 drives autophagy initiation under starvation. Mol Cell 2015;57(3):456–66.

[44] Masri S, Sassone-Corsi P. Sirtuins and the circadian clock: bridging chromatin and metabolism. Sci Signal 2014;7(342):re6.

[45] Aguilar-Arnal L, et al. NAD(+)-SIRT1 control of H3K4 trimethylation through circadian deacetylation of MLL1. Nat Struct Mol Biol 2015;22(4):312–18.

[46] Herskovits AZ, Guarente L. SIRT1 in neurodevelopment and brain senescence. Neuron 2014;81(3):471–83.

[47] Scheibye-Knudsen M, et al. Contribution of defective mitophagy to the neurodegeneration in DNA repair-deficient disorders. Autophagy 2014;10(8):1468–9.

[48] Giblin W, Skinner ME, Lombard DB. Sirtuins: guardians of mammalian healthspan. Trends Genet 2014;30(7):271–86.

[49] Jin J, et al. The reduction of SIRT1 in livers of old mice leads to impaired body homeostasis and to inhibition of liver proliferation. Hepatology 2011;54(3):989–98.

[50] Han MK, et al. SIRT1 regulates apoptosis and Nanog expression in mouse embryonic stem cells by controlling p53 subcellular localization. Cell Stem Cell 2008;2(3):241–51.

[51] Chae HD, Broxmeyer HE. SIRT1 deficiency downregulates PTEN/JNK/FOXO1 pathway to block reactive oxygen species-induced apoptosis in mouse embryonic stem cells. Stem Cells Dev 2011;20(7):1277–85.

[52] Zhang Y, et al. Inhibition of Sirt1 promotes neural progenitors toward motor neuron differentiation from human embryonic stem cells. Biochem Biophys Res Commun 2011;404(2):610–14.

[53] Ma CY, et al. SIRT1 suppresses self-renewal of adult hippocampal neural stem cells. Development 2014;141(24):4697–709.

[54] Diaz-Ruiz A, et al. SIRT1 synchs satellite cell metabolism with stem cell fate. Cell Stem Cell 2015;16(2):103–4.

[55] Yamashita S, et al. SIRT1 prevents replicative senescence of normal human umbilical cord fibroblast through potentiating the transcription of human telomerase reverse transcriptase gene. Biochem Biophys Res Commun 2012;417(1):630–4.

[56] Hayakawa T, et al. SIRT1 suppresses the senescence-associated secretory phenotype through epigenetic gene regulation. PLoS One 2015;10(1):e0116480.

[57] Maxwell MM, et al. The Sirtuin 2 microtubule deacetylase is an abundant neuronal protein that accumulates in the aging CNS. Hum Mol Genet 2011;20(20):3986–96.

[58] Beirowski B, et al. Sir-two-homolog 2 (Sirt2) modulates peripheral myelination through polarity protein Par-3/atypical protein kinase C (aPKC) signaling. Proc Natl Acad Sci USA 2011;108(43):E952–61.

[59] Krishnan J, et al. Dietary obesity-associated Hif1alpha activation in adipocytes restricts fatty acid oxidation and energy expenditure via suppression of the Sirt2-NAD+ system. Genes Dev 2012;26(3):259–70.

[60] Wang F, et al. Deacetylation of FOXO3 by SIRT1 or SIRT2 leads to Skp2-mediated FOXO3 ubiquitination and degradation. Oncogene 2012;31(12):1546–57.

[61] Rothgiesser KM, et al. SIRT2 regulates NF-kappaB dependent gene expression through deacetylation of p65 Lys310. J Cell Sci 2010;123(Pt 24):4251–8.

[62] Narayan N, et al. The NAD-dependent deacetylase SIRT2 is required for programmed necrosis. Nature 2012;492(7428):199–204.

[63] Lin R, et al. Acetylation stabilizes ATP-citrate lyase to promote lipid biosynthesis and tumor growth. Mol Cell 2013;51(4):506–18.

[64] Seo KS, et al. SIRT2 regulates tumour hypoxia response by promoting HIF-1alpha hydroxylation. Oncogene 2015;34(11):1354–62.

[65] Serrano L, et al. The tumor suppressor SirT2 regulates cell cycle progression and genome stability by modulating the mitotic deposition of H4K20 methylation. Genes Dev 2013;27(6):639–53.

[66] Zhang H, et al. SIRT2 directs the replication stress response through CDK9 deacetylation. Proc Natl Acad Sci USA 2013;110(33):13546–51.

[67] Eskandarian HA, et al. A role for SIRT2-dependent histone H3K18 deacetylation in bacterial infection. Science 2013;341(6145):1238858.

[68] Wang YP, et al. Regulation of G6PD acetylation by SIRT2 and KAT9 modulates NADPH homeostasis and cell survival during oxidative stress. EMBO J 2014;33(12):1304–20.

[69] Xu Y, et al. Oxidative stress activates SIRT2 to deacetylate and stimulate phosphoglycerate mutase. Cancer Res 2014;74(13):3630–42.

[70] North BJ, et al. SIRT2 induces the checkpoint kinase BubR1 to increase lifespan. EMBO J 2014;33(13):1438–53.

[71] Bell EL, Guarente L. The SirT3 divining rod points to oxidative stress. Mol Cell 2011;42(5):561–8.

[72] Brown K, et al. SIRT3 reverses aging-associated degeneration. Cell Rep 2013;3(2):319–27.

[73] Bharathi SS, et al. Sirtuin 3 (SIRT3) protein regulates long-chain acyl-CoA dehydrogenase by deacetylating conserved lysines near the active site. J Biol Chem 2013;288(47):33837–47.

[74] Zhang Y, et al. SIRT3 and SIRT5 regulate the enzyme activity and cardiolipin binding of very long-chain acyl-CoA dehydrogenase. PLoS One 2015;10(3):e0122297.

[75] Fan J, et al. Tyr phosphorylation of PDP1 toggles recruitment between ACAT1 and SIRT3 to regulate the pyruvate dehydrogenase complex. Mol Cell 2014;53(4):534–48.

[76] Ozden O, et al. SIRT3 deacetylates and increases pyruvate dehydrogenase activity in cancer cells. Free Radic Biol Med 2014;76:163–72.

[77] Jing E, et al. Sirtuin-3 (Sirt3) regulates skeletal muscle metabolism and insulin signaling via altered mitochondrial oxidation and reactive oxygen species production. Proc Natl Acad Sci USA 2011;108(35):14608–13.

[78] Rahman M, et al. *Drosophila* Sirt2/mammalian SIRT3 deacetylates ATP synthase beta and regulates complex V activity. J Cell Biol 2014;206(2):289–305.

[79] Vassilopoulos A, et al. SIRT3 deacetylates ATP synthase F1 complex proteins in response to nutrient- and exercise-induced stress. Antioxid Redox Signal 2014;21(4):551–64.

[80] Yang H, et al. SIRT3-dependent GOT2 acetylation status affects the malate-aspartate NADH shuttle activity and pancreatic tumor growth. EMBO J 2015.

[81] Samant SA, et al. SIRT3 deacetylates and activates OPA1 to regulate mitochondrial dynamics during stress. Mol Cell Biol 2014;34(5):807–19.

[82] Papa L, Germain D. SirT3 regulates the mitochondrial unfolded protein response. Mol Cell Biol 2014;34(4):699–710.

[83] Iwahara T, et al. SIRT3 functions in the nucleus in the control of stress-related gene expression. Mol Cell Biol 2012;32(24):5022–34.

[84] Bao X, et al. Identification of "erasers" for lysine crotonylated histone marks using a chemical proteomics approach. Elife 2014;3.

[85] Mathias RA, et al. Sirtuin 4 is a lipoamidase regulating pyruvate dehydrogenase complex activity. Cell 2014;159(7):1615–25.

[86] Hirschey MD, Zhao Y. Metabolic regulation by lysine malonylation, succinylation and glutarylation. Mol Cell Proteomics 2015.

[87] Tan M, et al. Lysine glutarylation is a protein posttranslational modification regulated by SIRT5. Cell Metab 2014;19(4):605–17.

[88] Tennen RI, Chua KF. Chromatin regulation and genome maintenance by mammalian SIRT6. Trends Biochem Sci 2011;36(1):39–46.

[89] Toiber D, et al. SIRT6 recruits SNF2H to DNA break sites, preventing genomic instability through chromatin remodeling. Mol Cell 2013;51(4):454–68.

[90] Dominy Jr. JE, et al. The deacetylase Sirt6 activates the acetyltransferase GCN5 and suppresses hepatic gluconeogenesis. Mol Cell 2012;48(6):900–13.

[91] Sun H, et al. SIRT6 regulates osteogenic differentiation of rat bone marrow mesenchymal stem cells partially via suppressing the nuclear factor-kappaB signaling pathway. Stem Cells 2014;32(7):1943–55.

[92] Takasaka N, et al. Autophagy induction by SIRT6 through attenuation of insulin-like growth factor signaling is involved in the regulation of human bronchial epithelial cell senescence. J Immunol 2014;192(3):958–68.

[93] Masri S, et al. Partitioning circadian transcription by SIRT6 leads to segregated control of cellular metabolism. Cell 2014;158(3):659–72.

[94] Jiang H, et al. SIRT6 regulates TNF-alpha secretion through hydrolysis of long-chain fatty acyl lysine. Nature 2013;496(7443):110–13.

[95] Mao Z, et al. SIRT6 promotes DNA repair under stress by activating PARP1. Science 2011;332(6036):1443–6.

[96] Van Meter M, et al. SIRT6 represses LINE1 retrotransposons by ribosylating KAP1 but this repression fails with stress and age. Nat Commun 2014;5:5011.

[97] Barber MF, et al. SIRT7 links H3K18 deacetylation to maintenance of oncogenic transformation. Nature 2012;487(7405):114–18.

[98] Shin J, et al. SIRT7 represses Myc activity to suppress ER stress and prevent fatty liver disease. Cell Rep 2013;5(3):654–65.

[99] Chen S, et al. Repression of RNA polymerase I upon stress is caused by inhibition of RNA-dependent deacetylation of PAF53 by SIRT7. Mol Cell 2013;52(3):303–13.

[100] Mohrin M, et al. Stem cell aging. A mitochondrial UPR-mediated metabolic checkpoint regulates hematopoietic stem cell aging. Science 2015;347(6228):1374–7.

[101] Ryu D, et al. A SIRT7-dependent acetylation switch of GABPbeta1 controls mitochondrial function. Cell Metab 2014;20(5):856–69.

[102] Kennedy BK, et al. Mutation in the silencing gene SIR4 can delay aging in *S. cerevisiae*. Cell 1995;80(3):485–96.

[103] Kennedy BK, et al. Redistribution of silencing proteins from telomeres to the nucleolus is associated with extension of life span in *S. cerevisiae*. Cell 1997;89(3):381–91.

[104] Sinclair DA, Guarente L. Extrachromosomal rDNA circles—a cause of aging in yeast. Cell 1997;91(7):1033–42.

[105] Gottlieb S, Esposito RE. A new role for a yeast transcriptional silencer gene, SIR2, in regulation of recombination in ribosomal DNA. Cell 1989;56(5):771–6.

[106] Kaeberlein M, McVey M, Guarente L. The SIR2/3/4 complex and SIR2 alone promote longevity in *Saccharomyces cerevisiae* by two different mechanisms. Genes Dev 1999;13(19):2570–80.

[107] Dang W, et al. Histone H4 lysine 16 acetylation regulates cellular lifespan. Nature 2009;459(7248):802–7.

[108] Tissenbaum HA, Guarente L. Increased dosage of a sir-2 gene extends lifespan in *Caenorhabditis elegans*. Nature 2001;410(6825):227–30.

[109] Rogina B, Helfand SL. Sir2 mediates longevity in the fly through a pathway related to calorie restriction. Proc Natl Acad Sci USA 2004;101(45):15998–6003.

[110] Balan V, et al. Life span extension and neuronal cell protection by *Drosophila* nicotinamidase. J Biol Chem 2008;283(41):27810–19.

[111] Schmeisser K, et al. Role of sirtuins in lifespan regulation is linked to methylation of nicotinamide. Nat Chem Biol 2013;9(11):693–700.

[112] Rizki G, et al. The evolutionarily conserved longevity determinants HCF-1 and SIR-2.1/SIRT1 collaborate to regulate DAF-16/FOXO. PLoS Genet 2011;7(9):e1002235.

[113] Mouchiroud L, et al. The NAD(+)/Sirtuin pathway modulates longevity through activation of mitochondrial UPR and FOXO signaling. Cell 2013;154(2):430–41.

[114] Guo X, Garcia LR. SIR-2.1 integrates metabolic homeostasis with the reproductive neuromuscular excitability in early aging male *Caenorhabditis elegans*. Elife 2014;3:e01730.

[115] Edwards C, et al. Mechanisms of amino acid-mediated lifespan extension in *Caenorhabditis elegans*. BMC Genet 2015;16(1):8.

[116] Cascella R, et al. S-linolenoyl glutathione intake extends lifespan and stress resistance via Sir-2.1 upregulation in *Caenorhabditis elegans*. Free Radic Biol Med 2014;73:127–35.

[117] Jiang N, et al. Dietary and genetic effects on age-related loss of gene silencing reveal epigenetic plasticity of chromatin repression during aging. Aging 2013;5(11):813–24.

[118] Bauer JH, et al. dSir2 and Dmp53 interact to mediate aspects of CR-dependent lifespan extension in *D. melanogaster*. Aging (Albany NY) 2009;1(1):38–48.

[119] Banerjee KK, et al. dSir2 in the adult fat body, but not in muscles, regulates life span in a diet-dependent manner. Cell Rep 2012;2(6):1485–91.

[120] Hoffmann J, et al. Overexpression of Sir2 in the adult fat body is sufficient to extend lifespan of male and female *Drosophila*. Aging (Albany NY) 2013;5(4):315–27.

[121] Herranz D, et al. Sirt1 improves healthy ageing and protects from metabolic syndrome-associated cancer. Nat Commun 2010;1:3.

[122] Mostoslavsky R, et al. Genomic instability and aging-like phenotype in the absence of mammalian SIRT6. Cell 2006;124(2):315–29.

[123] Kanfi Y, et al. The sirtuin SIRT6 regulates lifespan in male mice. Nature 2012;483(7388):218–21.

[124] Satoh A, et al. Sirt1 extends life span and delays aging in mice through the regulation of Nk2 homeobox 1 in the DMH and LH. Cell Metab 2013;18(3):416–30.

[125] Guarente L. Calorie restriction and sirtuins revisited. Genes Dev 2013;27(19):2072–85.

[126] Lin SJ, Defossez PA, Guarente L. Requirement of NAD and SIR2 for life-span extension by calorie restriction in *Saccharomyces cerevisiae*. Science 2000;289(5487):2126–8.

[127] Lin SJ, et al. Calorie restriction extends *Saccharomyces cerevisiae* lifespan by increasing respiration. Nature 2002;418(6895):344–8.

[128] Anderson RM, et al. Nicotinamide and PNC1 govern lifespan extension by calorie restriction in *Saccharomyces cerevisiae*. Nature 2003;423(6936):181–5.

[129] Wang Y, et al. *C. elegans* 14-3-3 proteins regulate life span and interact with SIR-2.1 and DAF-16/FOXO. Mech Ageing Dev 2006;127(9):741–7.

[130] Kaeberlein M, et al. Sir2-independent life span extension by calorie restriction in yeast. PLoS Biol 2004;2(9):E296.

[131] Kaeberlein M, et al. Comment on "HST2 mediates SIR2-independent life-span extension by calorie restriction". Science 2006;312(5778):1312 author reply 1312

[132] Kaeberlein M, et al. Regulation of yeast replicative life span by TOR and Sch9 in response to nutrients. Science 2005;310(5751):1193–6.

[133] Kaeberlein M. Lessons on longevity from budding yeast. Nature 2010;464(7288):513–19.

[134] Lee GD, et al. Dietary deprivation extends lifespan in *Caenorhabditis elegans*. Aging Cell 2006;5(6):515–24.

[135] Hansen M, et al. Lifespan extension by conditions that inhibit translation in *Caenorhabditis elegans*. Aging Cell 2007;6(1):95–110.

[136] Schulz TJ, et al. Glucose restriction extends *Caenorhabditis elegans* life span by inducing mitochondrial respiration and increasing oxidative stress. Cell Metab 2007;6(4):280–93.

[137] Bishop NA, Guarente L. Two neurons mediate diet-restriction-induced longevity in *C. elegans*. Nature 2007;447(7144):545–9.

[138] Cohen HY, et al. Calorie restriction promotes mammalian cell survival by inducing the SIRT1 deacetylase. Science 2004;305(5682):390–2.

[139] Lombard DB, et al. Mammalian Sir2 homolog SIRT3 regulates global mitochondrial lysine acetylation. Mol Cell Biol 2007;27(24):8807–14.

[140] Nakagawa T, et al. SIRT5 Deacetylates carbamoyl phosphate synthetase 1 and regulates the urea cycle. Cell 2009;137(3):560–70.

[141] Chen YR, et al. Calorie restriction on insulin resistance and expression of SIRT1 and SIRT4 in rats. Biochem Cell Biol 2010;88(4):715–22.

[142] Geng YQ, et al. SIRT1 and SIRT5 activity expression and behavioral responses to calorie restriction. J Cell Biochem 2011;112(12):3755–61.

[143] Luo LL, et al. The effects of caloric restriction and a high-fat diet on ovarian lifespan and the expression of SIRT1 and SIRT6 proteins in rats. Aging Clin Exp Res 2012;24(2):125–33.

[144] Yu W, et al. Short-term calorie restriction activates SIRT14 and 7 in cardiomyocytes in vivo and in vitro. Mol Med Rep 2014;9(4):1218–24.

[145] Boily G, et al. SirT1 regulates energy metabolism and response to caloric restriction in mice. PLoS One 2008;3(3):e1759.

[146] Mercken EM, et al. SIRT1 but not its increased expression is essential for lifespan extension in caloric-restricted mice. Aging Cell 2014;13(1):193–6.

[147] Someya S, et al. Sirt3 mediates reduction of oxidative damage and prevention of age-related hearing loss under caloric restriction. Cell 2010;143(5):802–12.

[148] Qiu X, et al. Calorie restriction reduces oxidative stress by SIRT3-mediated SOD2 activation. Cell Metab 2010;12(6):662–7.

[149] Howitz KT, et al. Small molecule activators of sirtuins extend *Saccharomyces cerevisiae* lifespan. Nature 2003;425(6954):191–6.

[150] Wood JG, et al. Sirtuin activators mimic caloric restriction and delay ageing in metazoans. Nature 2004;430(7000):686–9.

[151] Valenzano DR, et al. Resveratrol prolongs lifespan and retards the onset of age-related markers in a short-lived vertebrate. Curr Biol 2006;16(3):296–300.

[152] Baur JA, et al. Resveratrol improves health and survival of mice on a high-calorie diet. Nature 2006;444(7117):337–42.

[153] Pearson KJ, et al. Resveratrol delays age-related deterioration and mimics transcriptional aspects of dietary restriction without extending life span. Cell Metab 2008;8(2):157–68.

[154] Kaeberlein M, et al. Substrate-specific activation of sirtuins by resveratrol. J Biol Chem 2005;280(17):17038–45.

[155] Bass TM, et al. Effects of resveratrol on lifespan in *Drosophila melanogaster* and *Caenorhabditis elegans*. Mech Ageing Dev 2007;128(10):546–52.

[156] Pacholec M, et al. SRT1720, SRT2183, SRT1460, and resveratrol are not direct activators of SIRT1. J Biol Chem 2010;285(11):8340–51.

REFERENCES

[157] Dasgupta B, Milbrandt J. Resveratrol stimulates AMP kinase activity in neurons. Proc Natl Acad Sci USA 2007;104(17): 7217–22.

[158] Fulco M, et al. Glucose restriction inhibits skeletal myoblast differentiation by activating SIRT1 through AMPK-mediated regulation of NAMPT. Dev Cell 2008;14(5):661–73.

[159] Price NL, et al. SIRT1 is required for AMPK activation and the beneficial effects of resveratrol on mitochondrial function. Cell Metab 2012;15(5):675–90.

[160] Lan F, et al. SIRT1 modulation of the acetylation status, cytosolic localization, and activity of LKB1. Possible role in AMP-activated protein kinase activation. J Biol Chem 2008;283 (41):27628–35.

[161] Park SJ, et al. Resveratrol ameliorates aging-related metabolic phenotypes by inhibiting cAMP phosphodiesterases. Cell 2012;148(3):421–33.

[162] Gertz M, et al. A molecular mechanism for direct sirtuin activation by resveratrol. PLoS One 2012;7(11):e49761.

[163] Hubbard BP, et al. Evidence for a common mechanism of SIRT1 regulation by allosteric activators. Science 2013;339 (6124):1216–19.

[164] Haigis MC, Guarente LP. Mammalian sirtuins—emerging roles in physiology, aging, and calorie restriction. Genes Dev 2006;20(21):2913–21.

[165] Lin Z, Fang D. The roles of SIRT1 in cancer. Genes Cancer 2013;4(3–4):97–104.

[166] Lin Z, et al. USP22 antagonizes p53 transcriptional activation by deubiquitinating Sirt1 to suppress cell apoptosis and is required for mouse embryonic development. Mol Cell 2012;46 (4):484–94.

[167] Wang C, et al. Interactions between E2F1 and SirT1 regulate apoptotic response to DNA damage. Nat Cell Biol 2006;8 (9):1025–31.

[168] Roth M, Chen WY. Sorting out functions of sirtuins in cancer. Oncogene 2014;33(13):1609–20.

[169] Martinez-Pastor B, Mostoslavsky R. Sirtuins, metabolism, and cancer. Front Pharmacol 2012;3:22.

[170] Chen WY, et al. Tumor suppressor HIC1 directly regulates SIRT1 to modulate p53-dependent DNA-damage responses. Cell 2005;123(3):437–48.

[171] Kim JE, Chen J, Lou Z. DBC1 is a negative regulator of SIRT1. Nature 2008;451(7178):583–6.

[172] Zhao W, et al. Negative regulation of the deacetylase SIRT1 by DBC1. Nature 2008;451(7178):587–90.

[173] Herranz D, et al. SIRT1 promotes thyroid carcinogenesis driven by PTEN deficiency. Oncogene 2013;32(34):4052–6.

[174] Dan L, et al. The role of sirtuin 2 activation by nicotinamide phosphoribosyltransferase in the aberrant proliferation and survival of myeloid leukemia cells. Haematologica 2012;97 (4):551–9.

[175] Liu C, et al. The sirtuin 3 expression profile is associated with pathological and clinical outcomes in colon cancer patients. Biomed Res Int 2014;2014:871263.

[176] Lu W, et al. SIRT5 facilitates cancer cell growth and drug resistance in nonsmall cell lung cancer. Tumour Biol 2014;35 (11):10699–705.

[177] Orellana ME, et al. Expression of SIRT2 and SIRT6 in retinoblastoma. Ophthalmic Res 2015;53(2):100–8.

[178] Paredes S, Villanova L, Chua KF. Molecular pathways: emerging roles of mammalian Sirtuin SIRT7 in cancer. Clin Cancer Res 2014;20(7):1741–6.

[179] Wang RH, et al. Impaired DNA damage response, genome instability, and tumorigenesis in SIRT1 mutant mice. Cancer Cell 2008;14(4):312–23.

[180] Firestein R, et al. The SIRT1 deacetylase suppresses intestinal tumorigenesis and colon cancer growth. PLoS One 2008;3(4): e2020.

[181] Oberdoerffer P, et al. SIRT1 redistribution on chromatin promotes genomic stability but alters gene expression during aging. Cell 2008;135(5):907–18.

[182] Lim JH, et al. Sirtuin 1 modulates cellular responses to hypoxia by deacetylating hypoxia-inducible factor 1-alpha. Mol Cell 2010;38(6):864–78.

[183] Di Sante G, et al. Loss of Sirt1 promotes prostatic intraepithelial neoplasia, reduces mitophagy, and delays PARK2 translocation to mitochondria. Am J Pathol 2015;185(1):266–79.

[184] Wan J, et al. PCAF-primed EZH2 acetylation regulates its stability and promotes lung adenocarcinoma progression. Nucleic Acids Res 2015.

[185] Chen CW, et al. DOT1L inhibits SIRT1-mediated epigenetic silencing to maintain leukemic gene expression in MLL-rearranged leukemia. Nat Med 2015;21(4):335–43.

[186] Kim HS, et al. SIRT3 is a mitochondria-localized tumor suppressor required for maintenance of mitochondrial integrity and metabolism during stress. Cancer Cell 2010;17(1):41–52.

[187] Jeong SM, et al. SIRT4 protein suppresses tumor formation in genetic models of Myc-induced B cell lymphoma. J Biol Chem 2014;289(7):4135–44.

[188] Sebastian C, et al. The histone deacetylase SIRT6 is a tumor suppressor that controls cancer metabolism. Cell 2012;151 (6):1185–99.

[189] Ramis MR, et al. Caloric restriction, resveratrol and melatonin: Role of SIRT1 and implications for aging and related-diseases. Mech Ageing Dev 2015;146-148C:28–41.

[190] Tang BL. Sirt1's systemic protective roles and its promise as a target in antiaging medicine. Transl Res 2011;157(5):276–84.

[191] Carafa V, Nebbioso A, Altucci L. Sirtuins and disease: the road ahead. Front Pharmacol 2012;3:4.

[192] Timmers S, Auwerx J, Schrauwen P. The journey of resveratrol from yeast to human. Aging (Albany NY) 2012;4(3):146–58.

C H A P T E R

18

DNA Damage, DNA Repair, and Micronutrients in Aging

María Moreno-Villanueva

Department of Biology, Molecular Toxicology Group, University of Konstanz, Konstanz, Germany

KEY FACTS
- Poor nutrition in elderly and its consequences for genomic stability.
- Nutritional interventions in elderly and their impact on DNA damage protection.

Dictionary of Terms
- *Micronutrient*: A chemical element or substance required in trace amounts indispensable for the normal growth and development of living organisms.
- *DNA lesions*: Sites of damage in the base-pairing or structure of DNA. There are several types of lesions such as mismatch, strand breaks, abasic sites, modified bases, and cross-links.
- *DNA repair*: The cellular mechanisms by which, using specialized enzymes, DNA lesions can be restored to the DNA original state. The relevant DNA repair pathways are base excision repair (BER), nucleotide excision repair (NER), mismatch repair (MMR), homologous recombination (HR), and nonhomologous end joining (NHEJ).
- *Genomic stability*: An increased propensity for genomic alterations when processes involved in maintaining and replicating the DNA are compromised.
- *Enzymatic co-factor*: A substance that usually contains a vitamin or mineral, which forms a complex with specific enzymes and is essential for their activity.
- *Apoptosis*: A genetically programmed sequence of events leading to cell self-destruction and elimination without releasing harmful substances into the surrounding tissue.
- *Micronucleus formation*: Small nuclei originated during cell division when a chromosome or a fragment of a chromosome is not incorporated into one of the daughter nuclei. Several genotoxic compounds induce micronuclei formation.
- *Mutation*: A permanent alteration in the DNA sequence. Mutations can result from unrepaired damage to DNA and significantly increase risk of cancer.

INTRODUCTION

Reactive oxygen species (ROS) are produced continuously in our cells as by-products of various metabolic pathways. On the one hand, ROS play an important role in the modulation of intracellular signaling pathways [1], but on the other hand excessive cellular ROS can lead to oxidative stress and DNA damage. However, cellular antioxidant systems or compounds, including superoxide dismutase (SOD), catalase, glutathione peroxidase (GPx), vitamin E, carotenes, and vitamin C, are responsible for ROS removal ensuring the balance between reactive oxygen species and antioxidant systems. With increasing age, the oxidation products from lipids, nucleic acids, proteins, and sugars are found to increase. Thus, reactive oxygen species might be a determinant in the aging process [2]. Mammalian species with a high metabolic rate, short lifespan and high age-specific cancer rate have a higher rate of oxidative damage than long-lived mammal species with a lower metabolic rate and a lower age-specific cancer rate [3]. Furthermore, there is evidence that both oxidant production and the ability of organisms to respond to oxidative stress are connected to aging and lifespan [4].

Reactive oxygen species generate more than one hundred different types of oxidative DNA lesions, such as base modification, deoxyribose oxidation, single- or double-strand breakage, and DNA-protein cross-links [5]. Excessive or unrepaired DNA damage plays a role not only in carcinogenesis, but is also associated with aging processes. Preventing or eliminating disproportionate ROS production might protect cells against DNA damage and consequently against tumor progression or premature onset of aging-related phenotypes. Therefore, strategies to reduce ROS or to increase antioxidants might be relevant, especially in elderly individuals.

Nutritional intervention strategies might promote healthy aging. Dietary restriction is recognized to extend lifespan and retard aging in several animal models. However, reduced food intake in the elderly increases the risk of malnutrition leading to decline of their body functions and the development of chronic diseases. Recent studies suggest that diet composition, rather than restriction of calorie intake, might affect aging [6]. The effects of dietary restriction on DNA integrity as well as on the cellular ability to repair DNA have been investigated. Dietary restriction was associated with reduced levels of certain types of DNA damage and DNA repair in cells isolated from calorie-restricted rodents compared to cells from ad libitum rodents [7].

Considering the effect of dietary restriction on DNA damage and DNA repair, there is a necessity to investigate which nutrients protect against ROS formation and enhance DNA repair mechanisms as well as their impact on aging processes and tumor progression.

Micronutrient Deficiency and Consequences for Genomic Stability

Several studies report that malnutrition, especially undernutrition, is prevalent among the elderly human population. Older adults experience less of a feeling of hunger and more quickly a feeling of fullness as compared to younger adults. A decrease in both taste and smell in elderly individuals can also cause a decreased interest in food and subsequently a decreased intake of nutrients. Indeed, vitamins and trace elements intake are frequently deficient in the elderly. Micronutrient deficiency induces DNA damage [8]. For instance, deficiency of vitamins such as vitamin C, B12, B6, folic acid, and niacin or trace elements, such as iron or zinc, cause DNA single- and double-strand breaks as well as DNA oxidative lesions [9].

Vitamin B12

Elderly people are especially at risk of developing a vitamin B12 deficiency, which affects 10%−15% of people over the age of 60 years [10]. Vitamin B12 acts as a coenzyme of methionine synthase, which is required for the synthesis of S-adenosyl-methionine (SAM). SAM is the main methyl group donor necessary for DNA methylation reactions. In vitro studies show that vitamin B12 deficiency is significantly correlated with increased micronucleus formation [11].

Folic Acid

The human body needs folic acid to synthesize, repair, and methylate DNA, as well as for cell division and growth. Among persons aged 65−74 and older than 75 years, about 10% and 20%, respectively, are at high risk of folic acid deficiency [12]. Folic acid is required for the synthesis of 2'-desoxythymidine-5'-monophosphate (dTMP) from 2'-desoxyuridine-5'-monophosphat (dUMP). Under conditions of folic acid deficiency dUMP accumulates and uracil is incorporated into the DNA instead of thymine. Excessive misincorporation of uracil into DNA not only leads to point mutation but may also result in chromosome breakage and micronucleus formation [13]. Folic acid deficiency induces chromosomal aberrations, DNA strand breaks, excessive uracil in DNA, micronucleus formation, and DNA hypomethylation in vivo and in vitro [14]. Other data suggest that folic acid deficiency impairs DNA repair in neurons sensitizing them to oxidative damage [15].

Niacin

Niacin, also called vitamin B3, and the nicotinamide adenine dinucleotides, are proposed to play a central role in the aging processes of mammals [16]. Niacin is a precursor of the coenzymes nicotinamide adenine dinucleotide (NAD) and nicotinamide adenine dinucleotide phosphate (NADP). In addition to functions in energy metabolism, niacin, in the form of NAD, is important in cell signaling and DNA repair pathways. It participates in a wide variety of ADP-ribosylation reactions. Poly(ADP-ribose), a negatively charged polymer, is synthesized by Poly(ADP-ribose) polymerase-1 (PARP-1), which plays an important role in DNA repair, maintenance of genomic stability, and cell death signaling, such as apoptosis. A large number of studies have reported that NAD^+ status influences genomic stability. For instance, niacin deficiency increases spontaneous micronucleus formation and sister chromatid exchange (SCE) frequency [17]. Furthermore, studies in animal models as well as epidemiological data from human populations provide evidence that niacin could also have an impact on cancer risk [18].

Trace Elements

Not only vitamins but also trace elements can compromise directly or indirectly DNA integrity. Trace elements are key regulators of metabolic and physiological pathways, which become altered during the aging process. Therefore, trace elements might have the capacity to influence the rate of biological aging [19]. Reduced intake of several trace elements can be particularly challenging for elderly people.

Copper (Cu) is a catalytic cofactor for Cu/Zn superoxide dismutase (SOD) and ceruloplasmin, which are two important antioxidant enzymes. Thus, copper deficiency might promote oxidative stress and DNA damage. Indeed, in vitro oxidatively challenged human lymphoblastoid cells showed higher levels of DNA damage under copper deficiency suggesting a decline in the antioxidant defense system of cells, thereby increasing their susceptibility to oxidative DNA damage [20]. Moreover, copper-deficient rats treated with aflatoxin B1 show a significantly higher induction of the enzymes poly(ADP-ribose) polymerase, DNA polymerase beta (Polβ), and DNA ligase suggesting a higher level of DNA oxidative damage and consequently a stimulation of DNA base excision repair. Rats supplemented with copper were able to bring down the induction of these enzymes to the level observed in normal rats supporting a protective role of copper against the DNA damaging effects of aflatoxin B1 [21]. A more recent study conducted in cattle associated copper deficiency with an increase in the frequency of chromosomal aberrations as well as in DNA damage [22].

Iron (Fe) is a transition-metal ion important for a number of complex processes in the body. Iron deficiency is the most frequent micronutrient deficiency and is particularly prevalent in poor children and women. Moreover, prevalence of anemia increases with advancing age and exceeds 20% in those 85 years and older [23]. Moreover, DNA damage in lymphocytes from patients with iron deficiency anemia was significantly higher than in those from control individuals, while total antioxidant capacity was significantly lower. Adult women with iron deficiency anemia show an increased DNA strand breaks [24], children with iron deficiency anemia present an increase of DNA strand breaks and Fpg-sensitive sites [25]. Iron deficiency has also been associated with imbalances in key physiologic functions, including DNA synthesis and DNA repair [26].

Zinc (Zn) plays a crucial role in several body functions such as growth, development, neuronal functions, and immune response. Zinc deficiency has been observed in elderly individuals [27]. At the cellular level, zinc is an essential component of numerous proteins involved in the defense against oxidative stress, such as Cu/Zn superoxide dismutase (SOD), and is required for maintaining genomic stability, for example, as an important component of DNA repair enzymes such as 8-oxoguanine glycosylase 1 (OGG1), apurinic/apyrimidinic endonuclease (APE) and PARP-1 [28]. OGG1, APE, and PARP-1 are essential enzymes involved in base excision repair (BER), the main pathway responsible for removing damaged DNA bases. OGG1 recognizes and removes the ROS-induced mutagenic base 8-oxoguanine (8-oxoG) creating an apurinic/apyrimidinic (AP) site (baseless site). APE makes an incision at the AP site before a new base can be incorporated and thus DNA damage can be repaired. PARP-1 binds to DNA strand breaks via its Zn-finger motifs and recruits other DNA repair proteins to the damaged site allowing the completion of BER. A positive correlation between cellular poly (ADP-ribosyl)ation capacity and zinc status has been reported before [29]. Thus, decrease of PARP activity in response to Zn depletion can compromise the BER pathway contributing to the accumulation of DNA damage. Indeed, Zn deficiency reduces the ability of cells to repair oxidative DNA damage [30]. Moreover, zinc is not only crucial for DNA repair but also acts as a biological antioxidant therefore zinc deficiency can lead to an increase of ROS and consequently even higher DNA damage accumulation.

Magnesium (Mg) is a cofactor of numerous enzymes, which regulate many important biochemical reactions in the human body. Alterations of Mg metabolism that have been associated to aging include a reduction of Mg intake and intestinal absorption [31]. At the cellular level magnesium not only stabilizes DNA structures, it also functions as a cofactor in DNA synthesis especially with respect to high fidelity. Furthermore, Mg is required for the removal of DNA damage through several repair pathways, such as nucleotide excision repair, base excision repair, and mismatch repair. Animal experiments and epidemiological studies show that magnesium deficiency may decrease membrane integrity and membrane function and increase oxidative stress, cardiovascular heart diseases as well as accelerated aging [32]. Moreover, DNA synthesis is slowed by insufficient Mg and low Mg level affects the efficiency of DNA repair systems [32].

In the face of the importance of micronutrients for the maintenance of the genomic stability and the fact that micronutrients decline with aging, it is not surprising that several cellular pathways involved in antioxidant mechanisms, DNA protection, and DNA repair are impaired in elderly individuals. Indeed, a decrease in the antioxidants glutathione peroxidase and catalase has been found in elderly people [33]. Furthermore, age-associated changes in efficiency of several DNA repair pathways, such as mismatch repair (MMR), base excision repair (BER), nucleotide excision repair (NER), and double-strand break (DSB), has been reported before and are summarized by Gorbunova and collaborators [34].

Taken together, deficiency in key micronutrients, which are not only required as biological antioxidants but also for maintaining genomic stability, may lead to the accumulation of DNA damage. The question arises as to whether nutrition can influence the level of DNA damage and the efficiency of DNA repair mechanisms preventing accelerated aging and age-related diseases, such as cancer.

Nutritional Interventions in the Elderly and their Impact on Genomic Stability

The DNA damage theory of aging proposes that aging is a consequence of accumulation of DNA damage, which occurs during physiological processes when DNA repair mechanisms fail. At a molecular level, there are obvious correlations between premature aging and impaired function of proteins involved in DNA repair pathways. Some progeroid syndromes, a group of rare genetic disorders caused by a single mutation in DNA repair genes and characterized by signs of premature aging phenotype, provide clear evidence of a relationship between defects in DNA repair and accelerated aging. In mammals there are 5 main DNA repair pathways, encoded by approximately 125 genes. Base excision repair (BER) removes the DNA damage that usually occur due to the cellular metabolism itself. Therefore, BER might play a distinguished role in aging and age-related diseases. Indeed, BER is highly compromised in brain cells with increasing age and this could be one of the causes for aging and age-associated neurological abnormalities [35]. Another study suggested that BER capacity may be impaired in aging muscle leading to higher sensitivity to oxidative damage [36].

Both age-related decrease in DNA repair capacity and age-related increase in ROS production, have been identified as the two major causes of age-associated accumulation of DNA damage.

In young healthy organisms an equilibrated nutritional status maintains micronutrients homeostasis. However, micronutrients which are necessary for optimal cellular antioxidant and DNA repair functions decrease with aging, leading to accumulation of DNA damage, higher risk of mutations and hence cancer development. Therefore nutritional intervention strategies to maintain antioxidant capacity as well as to ensure DNA repair efficiency might be relevant in elderly individuals. The most important source of antioxidants is provided by nutrition. Supplemental nutrient administration and addition of selected nutrients to the food (food fortification) has been proposed to have a benefit not only in the elderly but also in healthy young individuals as a strategy to prevent the risk of future age-associated diseases such as cardiovascular diseases, diabetes, and cancer. At the molecular level dietary factors may influence the effectiveness of DNA repair [37]. Calorie restriction (CR) is an experimental nutritional intervention that extends lifespan in laboratory animals. Calorie restriction in rats and mice is associated with reduced ROS production and appears to retard the age-related decline in DNA repair [7], for example, CR can reduce age-dependent decline of nonhomologous end-joining (NHEJ), a pathway that repairs DNA double-strand breaks, in various tissues of rats possibly through upregulation of the DNA repair protein X-ray repair cross-complementing 4 (XRCC4) [38]. Furthermore, the activity of apurinic/apyrimidinic endonuclease (APE), an enzyme involved in BER, declines with increasing age but CR significantly retards this age-dependent process in several regions of the aging rat brain [39]. Regarding dietary factors responsible for reduction of age-dependent decline of DNA repair, there are several studies indicating that intake of certain nutrients might reduce the level of DNA damage or enhance DNA repair. One of the dietary factors responsible for the life extension effect of CR seems to be methionine [40]. Seleno-L-methionine (SeMet) is a naturally occurring amino acid containing selenium and appears to selectively regulate the DNA nucleotide excision repair (NER) pathway. NER is required for the repair of several chemotherapy-induced DNA lesions but cancer cells generally lack a SeMet-inducible DNA repair response. Thus, SeMet might provide a convenient tool in cancer treatment [41]. Another study revealed that Zn supplementation can increase cellular PARP enzymatic capacity in human peripheral blood mononuclear cells (PBMC) from elderly individuals resulting in the maintenance of genome stability and integrity [29]. Also resveratrol, a natural compound found in grapes and red wine, has been shown to extend lifespan and to prevent early mortality. A low dose of dietary resveratrol partially mimics CR and might promote pathways that maintain genomic stability, or prevent epigenetic alterations, and therefore retard some aspects of the aging process [42]. Resveratrol combined with exercise can reverse the effect of aging in liver of old animals, maintaining higher antioxidant activities and decreasing oxidative damage [43]. Coenzyme Q_{10}, also called ubiquinone, is a vitamin-like substance present in the mitochondria. It participates in the generation of energy in the form of adenosine triphosphate (ATP). Rats supplemented on coenzyme Q_{10} are protected from age-related DNA double-strand breaks and have a significantly higher lifespan than nonsupplemented animals [44]. More general daily supplementation with fruit and vegetable extracts reduce the level of DNA damage in peripheral lymphocytes of older people [45].

Genotoxic Effects of Nutritional Interventions

An optimal intake of nutrients might depend on age, health status, and genetic background. This became evident in a study conducted with Down syndrome (DS) patients who experience an accelerated aging. Interestingly, the study showed that the rate of DNA repair in Down syndrome individuals is significantly increased compared to control individuals, but DNA repair rate after Zn supplementation decreases reaching levels similar to the control subjects [46]. Since DS subjects show a higher degree of DNA damage [47], enhanced DNA repair could be explained as result of compensatory mechanism against excessive DNA damage.

An age dependency in protein function can be found in PARP-1, which has been associated with the aging process. PARP-1 plays an essential role not only in genomic maintenance but also in inflammation processes and might act as a longevity assurance factor at younger age (or physiological conditions) or an aging-promoting factor at older age (or pathological conditions) [48]. Therefore stimulation of PARP-1 activity under pathological conditions, such as inflammation, might also be accompanied by side effects.

Several studies have shown that nutritional interventions can extend lifespan in species ranging from yeast to nonhuman primates, but conversely unbalanced intake of micronutrients can also contribute to higher production of ROS. For instance, excessive intake of some trace elements, such as copper and iron, can result in higher DNA damage [49]; in the case of iron, through H_2O_2 lethality generated from iron-mediated Fenton reactions [50]. Furthermore, copper-mediated reactive oxygen species induces DNA double strand breaks in vitro [51]. Regarding genotoxic effects of copper occurring in vivo, there is evidence that toxic levels of copper ions may transiently be reached. However, there are robust mechanisms that maintain copper homeostasis, namely albumin and ceruloplasmin, which, in turn, can bind high quantities of copper with high affinity [52]. Although dietary Zn might counter the Zn deficiency observed in elderly subjects, high levels of Zn could also act as a pro-oxidant inducing DNA damage by triggering a decline in erythrocyte Cu–Zn superoxide dismutase (SOD) [53]. Also, vitamin C-mediated formation of genotoxins has been reported [54].

Calorie restriction can causes beneficial functional changes, but the precise amount of calorie intake necessary for optimal health and maximum longevity in humans is not known. In addition, calorie restriction might not be favorable in specific populations, such as lean persons [55] or elderly people [56]. Excessive calorie restriction can lead to detrimental effects such as anemia, muscle wasting, neurologic deficits, lower extremity edema, weakness, dizziness, lethargy, irritability, and depression [55]. Several studies on whether dietary restriction extends life when initiated at advanced ages show contradictory findings [57].

CONCLUSIONS

Aging is a complex biological process in which nutrient deficiency contributes to accelerated aging and to the development of several age-related diseases. Free radicals produced by metabolic processes lead to DNA damage compromising genomic integrity. Cells from young individuals have an efficient system to ensure a proper balance between free radicals and antioxidants, as well as functional DNA repair pathways, while in old age this balance is disturbed. Nutritional interventions might help to restore this balance protecting against oxidative stress and in turn against progression of degenerative diseases and aging.

Micronutrients regulate many biochemical reactions including metabolic ROS production and the DNA damage response. A balanced micronutrient intake is necessary to ensure cellular functions. Even if several studies show that micronutrient supplementation may decrease the risk of DNA damage and enhance DNA repair, this may depend on several factors like genetic background, age, or health status. Therefore, future studies for identifying an appropriate and balanced micronutrients intake and the role of micronutrients in genomic stability should be considered a fundamental goal in achieving optimal health. Special attention should be dedicated to elderly people since the aging process is considered the major risk factor for disease and death.

SUMMARY

- Micronutrient deficiency induces DNA damage.
- Micronutrients are necessary for the DNA repair machinery.
- Micronutrients deficiency in elderly might have consequences for genomic stability.

- Nutritional strategies to reduce ROS or increase antioxidants might be relevant in elderly individuals.
- Appropriated nutrition in elderly might protect against oxidative DNA damage and enhance DNA repair.

References

[1] Finkel T. Signal transduction by reactive oxygen species. J Cell Biol 2011;194(1):7–15.
[2] Ashok BT, Ali R. The aging paradox: free radical theory of aging. Exp Gerontol 1999;34(3):293–303.
[3] Ames BN, Shigenaga MK. Oxidants are a major contributor to aging. Ann NY Acad Sci 1992;663:85–96.
[4] Finkel T, Holbrook NJ. Oxidants, oxidative stress and the biology of aging. Nature 2000;408(6809):239–47.
[5] Cadet J, Berger M, Douki T, Ravanat JL. Oxidative damage to DNA: formation, measurement, and biological significance. Rev Physiol Biochem Pharmacol 1997;131:1–87.
[6] Maxmen A. Calorie restriction falters in the long run. Nature 2012;488(7413):569.
[7] Haley-Zitlin V, Richardson A. Effect of dietary restriction on DNA repair and DNA damage. Mutation Res 1993;295(4-6):237–45.
[8] Ames BN. Micronutrient deficiencies. A major cause of DNA damage. Ann New York Acad Sci 1999;889:87–106.
[9] Ames BN. Micronutrients prevent cancer and delay aging. Toxicol Lett 1998;102-103:5–18.
[10] Baik HW, Russell RM. Vitamin B12 deficiency in the elderly. Annu Rev Nutr 1999;19:357–77.
[11] Wu X, Cheng J, Lu L. Vitamin B12 and methionine deficiencies induce genome damage measured using the cytokinesis-block micronucleus cytome assay in human B lymphoblastoid cell lines. Nutr Cancer 2013;65(6):866–73.
[12] Clarke R, Refsum H, Birks J, Evans JG, Johnston C, Sherliker P, et al. Screening for vitamin B-12 and folate deficiency in older persons. Am J Clin Nutr 2003;77(5):1241–7.
[13] Blount BC, Mack MM, Wehr CM, MacGregor JT, Hiatt RA, Wang G, et al. Folate deficiency causes uracil misincorporation into human DNA and chromosome breakage: implications for cancer and neuronal damage. Proc Natl Acad Sci USA 1997;94(7):3290–5.
[14] Fenech M. Folate (vitamin B9) and vitamin B12 and their function in the maintenance of nuclear and mitochondrial genome integrity. Mutat Res 2012;733(1-2):21–33.
[15] Kruman II, Kumaravel TS, Lohani A, Pedersen WA, Cutler RG, Kruman Y, et al. Folic acid deficiency and homocysteine impair DNA repair in hippocampal neurons and sensitize them to amyloid toxicity in experimental models of Alzheimer's disease. J Neurosci 2002;22(5):1752–62.
[16] Xu P, Sauve AA. Vitamin B3, the nicotinamide adenine dinucleotides and aging. Mech Aging Dev 2010;131(4):287–98.
[17] Spronck JC, Kirkland JB. Niacin deficiency increases spontaneous and etoposide-induced chromosomal instability in rat bone marrow cells in vivo. Mutat Res 2002;508(1-2):83–97.
[18] Kirkland JB. Niacin and carcinogenesis. Nutr Cancer 2003;46(2):110–18.
[19] Méplan C. Trace elements and aging, a genomic perspective using selenium as an example. J Trace Elem Med Biol 2011;25(Suppl. 1):11–16.
[20] Pan Y, Loo G. Effect of copper deficiency on oxidative DNA damage in Jurkat T-lymphocytes. Free Radic Biol Med 2000;28(5):824–30.
[21] Webster RP, Gawde MD, Bhattacharya RK. Modulation by dietary copper of aflatoxin B1-induced activity of DNA repair enzymes poly (ADP-ribose) polymerase, DNA polymerase beta and DNA ligase. In Vivo 1996;10(5):533–6.
[22] Picco SJ, Abba MC, Mattioli GA, Fazzio LE, Rosa D, De Luca JC, et al. Association between copper deficiency and DNA damage in cattle. Mutagenesis 2004;19(6):453–6.
[23] Patel KV. Epidemiology of anemia in older adults. Semin Hematol 2008;45(4):210–17.
[24] Aslan M, Horoz M, Kocyigit A, Ozgonul S, Celik H, Celik M, et al. Lymphocyte DNA damage and oxidative stress in patients with iron deficiency anemia. Mutat Res 2006;601(1–2):144–9.
[25] Aksu BY, Hasbal C, Himmetoglu S, Dincer Y, Koc EE, Hatipoglu S, et al. Leukocyte DNA damage in children with iron deficiency anemia: effect of iron supplementation. Eur J Pediatr 2010;169(8):951–6.
[26] Prá D, Rech Franke S, Pegas Henriques J, Fenech M. A possible link between iron deficiency and gastrointestinal carcinogenesis. Nutr Cancer 2009;61(4):415–26.
[27] Blumberg J. Nutritional needs of seniors. J Am Coll Nutr 1997;16(6):517–23.
[28] Sharif R, Thomas P, Zalewski P, Fenech M. The role of zinc in genomic stability. Mutat Res 2012;733(1-2):111–21.
[29] Kunzmann A, Dedoussis G, Jajte J, Malavolta M, Mocchegiani E, Burkle A. Effect of zinc on cellular poly(ADP-ribosyl)ation capacity. Exp Gerontol 2008;43(5):409–14.
[30] Song Y, Leonard SW, Traber MG, Ho E. Zinc deficiency affects DNA damage, oxidative stress, antioxidant defenses, and DNA repair in rats. J Nutr 2009;139(9):1626–31.
[31] Barbagallo M, Belvedere M, Dominguez LJ. Magnesium homeostasis and aging. Magnes Res 2009;22(4):235–46.
[32] Hartwig A. Role of magnesium in genomic stability. Mutat Res 2001;475(1-2):113–21.
[33] Akila VP, Harishchandra H, D'Souza V, D'Souza B. Age related changes in lipid peroxidation and antioxidants in elderly people. Indian J Clin Biochem 2007;22(1):131–4.
[34] Gorbunova V, Seluanov A, Mao Z, Hine C. Changes in DNA repair during aging. Nucleic Acids Res 2007;35(22):7466–74.
[35] Rao KS. Free radical induced oxidative damage to DNA: relation to brain aging and neurological disorders. Indian J Biochem Biophys 2009;46(1):9–15.
[36] Joseph A, Tornaletti S, Adhihetty P, Buford T, Wohlgemuth S, Sandesara B, et al. Quantitation of oxidative stress and base excision repair in skeletal muscle of high- and low-functioning elderly individuals. FASEB J 2014;28(1 Suppl. 863.3).
[37] Mathers JC, Coxhead JM, Tyson J. Nutrition and DNA repair — potential molecular mechanisms of action. Curr Cancer Drug Targets 2007;7(5):425–31.
[38] Lee JE, Heo JI, Park SH, Kim JH, Kho YJ, Kang HJ, et al. Calorie restriction (CR) reduces age-dependent decline of nonhomologous end joining (NHEJ) activity in rat tissues. Exp Gerontol 2011;46(11):891–6.
[39] Kisby GE, Kohama SG, Olivas A, Churchwell M, Doerge D, Spangler E, et al. Effect of caloric restriction on base-excision repair (BER) in the aging rat brain. Exp Gerontol 2010;45(3):208–16.
[40] Sanchez-Roman I, Barja G. Regulation of longevity and oxidative stress by nutritional interventions: role of methionine restriction. Exp Gerontol 2013;48(10):1030–42.
[41] Smith ML, Kumar MA. Seleno-L-methionine modulation of nucleotide excision DNA repair relevant to cancer prevention and chemotherapy. Mol Cell Pharmacol 2009;1(4):218–21.
[42] Barger J, Kayo T, Vann J, Arias E, Wang J, Hacker T, et al. A low dose of dietary resveratrol partially mimics caloric restriction and retards aging parameters in mice. PLoS One 2008;4(3(6)):e2264.

REFERENCES

[43] Tung BT, Rodriguez-Bies E, Ballesteros-Simarro M, Motilva V, Navas P, Lopez-Lluch G. Modulation of endogenous antioxidant activity by resveratrol and exercise in mouse liver is age dependent. J Gerontol Series A Biol Sci Med Sci 2014;69(4): 398–409.

[44] Quiles JL, Ochoa JJ, Huertas JR, Mataix J. Coenzyme Q supplementation protects from age-related DNA double-strand breaks and increases lifespan in rats fed on a PUFA-rich diet. Exp Gerontol 2004;39(2):189–94.

[45] Smith MJ, Inserra PF, Watson RR, Wise JA, O'Niell KL. Supplementation with fruit and vegetable extracts may decrease DNA damage in the peripheral lymphocytes of an elderly population. Nutr Res 1999;19(10):1507–18.

[46] Chiricolo M, Musa AR, Monti D, Zannotti M, Franceschi C. Enhanced DNA repair in lymphocytes of Down syndrome patients: the influence of zinc nutritional supplementation. Mutat Res 1993;295(3):105–11.

[47] Tiano L, Littarru GP, Principi F, Orlandi M, Santoro L, Carnevali P, et al. Assessment of DNA damage in Down syndrome patients by means of a new, optimised single cell gel electrophoresis technique. BioFactors 2005;25(1–4):187–95.

[48] Mangerich A, Burkle A. Pleiotropic cellular functions of PARP1 in longevity and aging: genome maintenance meets inflammation. Oxidative Med Cell Longevity 2012;2012:321653.

[49] O'Connor JM. Trace elements and DNA damage. Biochem Soc Trans 2001;29(Pt 2):354–7.

[50] Mello-Filho AC, Meneghini R. Iron is the intracellular metal involved in the production of DNA damage by oxygen radicals. Mutat Res 1991;251(1):109–13.

[51] Bar-Or D, Thomas GW, Rael LT, Lau EP, Winkler JV. Asp-Ala-His-Lys (DAHK) inhibits copper-induced oxidative DNA double strand breaks and telomere shortening. Biochem Biophys Res Commun 2001;282(1):356–60.

[52] Linder MC. The relationship of copper to DNA damage and damage prevention in humans. Mutat Res 2012;733(1-2):83–91.

[53] Abdallah SM, Samman S. The effect of increasing dietary zinc on the activity of superoxide dismutase and zinc concentration in erythrocytes of healthy female subjects. Eur J Clin Nutr 1993; 47(5):327–32.

[54] Lee SH, Oe T, Blair IA. Vitamin C-induced decomposition of lipid hydroperoxides to endogenous genotoxins. Science 2001; 292(5524):2083–6.

[55] Fontana L, Klein S. Aging, adiposity, and calorie restriction. J Am Med Assoc 2007;297(9):986–94.

[56] Kim KI, Kim CH. Calorie restriction in the elderly people. J Korean Med Sci 2013;28(6):797–8.

[57] Masoro EJ. Caloric restriction and aging: controversial issues. J Gerontol Series A Biol Sci Med Sci 2006;61(1):14–19.

19

Neuroprotective Mechanisms of Dietary Phytochemicals: Implications for Successful Brain Aging

Sergio Davinelli[1], Giovanni Scapagnini[1], Guido Koverech[2], Maria Luca[3], Carmela Calandra[3] and Vittorio Calabrese[2]

[1]Department of Medicine and Health Sciences, University of Molise, Campobasso, Italy [2]Department of Biomedical and Biotechnological Sciences, University of Catania, Catania, Italy [3]Department of Medical and Surgical Sciences and Advanced Technologies, University of Catania, Catania, Italy

KEY FACTS

- Several studies indicate that the interconnection between oxidative stress and inflammation is a key determinant of brain aging and cognitive decline.
- Diets rich in phytochemicals and plant-based foods can induce a mild adaptive stress response and modulate hormetic signaling pathways to attenuate pro-inflammatory and oxidative damages.
- The major signaling pathways by which dietary phytochemicals offer neuroprotection include the transcription factors Nrf_2 and NF-κB and the superfamily of serine/threonine kinases MAPKs.
- To date, the challenge is to establish the clinical relevance of dietary phytochemicals and the effective dose range to regulate neuronal function and delay brain aging.

Dictionary of Terms

- **Aging**: Aging is an inherently complex process accompanied by a general dysregulation of different systems and tissues. Stochastic theories suggest that aging is the result of accumulating random changes that negatively affect biological systems. Aging could be the result of the accumulation of toxic byproducts, cellular damage, or other gradual deteriorative process such as such as oxidation, chronic inflammation and telomere shortening.
- **Phytochemicals**: Phytochemicals consist of a large group of nonnutrient compounds that are biologically active in the body. As implied by the name, phytochemicals are found in plants, including fruits, vegetables, legumes, grains, herbs, tea, and spices.
- **Hormesis**: Dose-response phenomenon characterized by a low-dose stimulation and a high-dose inhibition. Hormesis is involved in the biological amplification of adaptive responses and protective mechanisms, leading to the improvement in overall cellular functions and performance.
- **Inflammation**: Inflammation is a crucial component of innate (non-specific) immunity and plays a crucial role in immunosurveillance and host defence. It is characterized by increased blood flow, leucocyte infiltration, and production of chemical mediators, which serves to eliminate toxic agents and initiate the repair of damaged tissue. Chronic low-grade inflammatory state is a pathological feature of a wide range of chronic conditions, including age-related diseases and neurodegeneration.

- **Oxidative stress**: Oxidative stress is described as a condition under which increased production of free radicals, reactive species, and oxidant-related reactions occur and result in cellular damage. However, oxidative stress can be also defined within the context of a changed redox status, where the redox balance between oxidant and antioxidant is disrupted.
- **Neuroprotection**: Neuroprotection is a term generally used to describe the effect of interventions, not necessarily pharmacological, aiming to protect the brain from pathological damage. Moreover, neuroprotection also refers to strategies that interfere with biochemical and molecular events which eventually leads to the neuronal death.

INTRODUCTION

Aging is associated with an inevitable loss of tissue homeostasis and the brain is more vulnerable to the deleterious effects of aging than other organs. Although this vulnerability is not uniform in the brain, a key factor that plays a crucial role in brain aging is the high demand that neurons have for oxidative metabolism in the generation of energy [1]. Therefore, normal brain metabolism is associated with unavoidable and intrinsic mild oxidative stress increasing processes related to senescence and aging. The central nervous system can be affected by more than one process contributing to the occurrence of changes in redox state. For instance, high content of polyunsaturated fatty acids (PUFAs) in neuronal membranes (easily peroxidizable), high consumption of oxygen, limited amount of endogenous antioxidant defenses, low rate of cell turnover, and high production of reactive species signaling make the brain a favored target for oxidative stress [2]. Thus, oxidative stress appears one of the primary physiological mechanism for functional impairment in brain aging and related neurodegenerative disorders, contributing to many types of cellular damage, including DNA damage, mitochondrial damage, telomere attrition, and accumulation of macromolecular waste [3]. The neuropathologic changes associated with brain aging also include deposition of lipofuscin and other insoluble materials, such as amyloid-β (Aβ) plaques and neurofibrillary tangles (NFTs), reduction of the length of neuronal dendritic trees accompanied by a reduction of dendritic spine number in different neuronal populations [1]. All these features characterize "normal brain aging" as well as neurodegenerative disorders and may be attributed to decreased cognitive performances and motor abilities [4,5]. Moreover, it is important to mention that inflammation, being associated with oxidative stress plays an essential role in brain aging. Many chronic neurological disorders associated with advancing age are also characterized by inflammatory reactions. Several components of the inflammatory response produce reactive oxygen species (ROS) in the forms of superoxide, hydrogen peroxide, hydroxyl radical, nitric oxide, hydrochlorous acid, and peroxynitrite [6]. Increased production of ROS by immune system regulates the synthesis of numerous chemokines and pro-inflammatory cytokines, including interleukin (IL)-1, IL-6, and tumor necrosis factor-α (TNF-α). These inflammatory responses activate microglia and astrocytes to generate large amounts of ROS and experimental evidence suggests that neuroinflammation may be considered as a cause and a consequence of chronic oxidative stress. Furthermore, neuroinflammatory processes lead to Aβ and NFTs generation, which are the main factors of cell death and age-associated dementia [7]. Emerging studies show that some food-derived small molecules, also called phytochemicals, may be an effective approach to delay the aging process and age-associated neurodegeneration. In particular, most phytochemicals are secondary plant metabolites and these bioactive compounds are present in a large variety of foods including fruit, vegetables, cereals, nuts, and cocoa/chocolate, as well as in beverages including juice, tea, coffee, and wine [8]. There are several categories of phytochemicals but phenolic compounds, also known as polyphenols, are the most studied. Numerous mechanisms have been proposed for the health benefits of phytochemicals, however, their ability to increase the expression of antioxidant proteins and activate transcription factors that antagonize inflammation have attracted considerable interest because of their role in brain aging [9]. Neuroprotective pathways may be triggered by small concentrations of polyphenols and these "phytonutrients" are not only powerful radical scavengers but they are also important in the chelation of ion metals or in the modulation of signal transduction cascades leading to the expression of neurotrophic factors and antiapoptotic proteins [10]. In this chapter, we summarize the main brain-protective activities of polyphenols that have been studied in cells, animals, and humans. Furthermore, we highlight the importance of oxidative stress and inflammation, the most promising areas in which to explore the role of diet in influencing the biology of brain aging.

OXIDATIVE DAMAGE AND BRAIN AGING

The generation of reactive species in the brain can increase through both spontaneous and enzymatic reactions and the expression of protective proteins are crucial in the preservation of brain homeostasis. Specific neurodegenerative diseases, including Alzheimer's disease (AD), Parkinson's disease (PD), amyotrophic lateral sclerosis (ALS), and stroke, are characterized by increased levels of oxidative markers and damaged cell components [11]. Indeed, oxidative stress is one of the main aging processes that can cause direct damage to cellular architecture within the brain. An age-associated increase in oxidative damage has been shown in neurons of human and rodent brains, and a selective susceptibility of different neuronal populations to oxidative stress has been demonstrated. For example, oligodendrocytes are at a high risk for oxidative damage due to their role in myelin maintenance and production and their limited repair mechanisms, suggesting that white matter may be particularly vulnerable to oxidative activity [12]. Prominent levels of neuronal oxidation can lead to the destruction of cellular components including lipids, proteins, nucleic acids, and ultimately cell death via apoptosis or necrosis [13]. Oxidative damage to cellular components have been described in several neurodegenerative disorders and it has been found also in cellular and animal models of neurodegeneration. To date, accumulating evidence obtained from human studies suggests that oxidative damage to cellular products increases in both central nervous and peripheral systems of many patients with different neurodegenerative diseases. In addition to the impaired function of the mitochondrial electron transport chain, an imbalance between the rate of oxidant generation and the endogenous antioxidant capacity is the primary source of oxidative stress in brain aging [14]. As already mentioned, many intracellular components are vulnerable to oxidative damage and various important classes of macromolecules are particularly liable to the deleterious effects of oxidative modification. For instance, protein carbonyl reflects protein oxidation and can be produced from direct free radical attack on amino acid side chains, glycation, and glycoxidation or from lipid oxidation products [15]. Moreover, there are several studies reporting that oxidative damage to nucleic acids can actively contribute to the background, the onset, and the development of the neurodegenerative disorders [16]. Age-associated accumulation of oxidative DNA damage, as evidenced by the presence of the modified base 8-hydroxydeoxyguanosine (8-OHdG), has been observed in many neuron types, including cerebellar granule cells and retinal ganglion, amacrine, and horizontal cells [17]. Interestingly, it has been recently demonstrated in human neurons that oxidative RNA modification can occur not only in protein-coding RNAs but also in noncoding RNAs, leading to activate inappropriate cell fate pathways [16]. Furthermore, oxidative damage of the brain is characterized by increased lipid peroxidation products in the cerebrospinal fluid (CSF) and plasma, and reduced membrane PUFAs. PUFAs, such as arachidonic acid (AA), are abundant in the aging brain and are highly susceptible to free radical attack. It has been shown that redox changes in membrane fatty acid composition contribute to the deterioration of neuronal functions [18–20]. Besides oxidation of nucleic acids, lipids, and proteins, other modifications of proteins can affect brain performances and contribute to the pathophysiology of neurodegenerative diseases. For instance, synaptic plasticity is known to be dependent on nitric oxide (NO)-mediated signaling pathways. NO is an endogenous gasotransmitter and plays a crucial role in neurotransmission. NO exerts its effects via S-nitrosylation, a redox reaction that occurs on cysteine residues of various proteins. Aging brain processes exacerbate nitrosative stress via excessive production of NO and it has been reported that increased NO levels may produce increased vulnerability to nitrosative stress and contribute to the onset of neurodegenerative diseases [21]. Nitrated α-synuclein (αSyn) protein has been observed in different regions of patients with synucleinopathies including PD, and dementia with Lewy bodies (DLB). Nitration of tau protein has been found in the hippocampus and neocortex of patients with AD, Down syndrome, and in other tau pathologies [22,23].

BRAIN INFLAMMATION AS A FORM OF OXIDATIVE STRESS

Brain aging involves an excess of free radical generation and increasing evidence associates chronic low-grade systemic inflammation with the onset of age-related brain disorders [24,25]. The molecular interactions between the immune and nervous systems are now known to play a pivotal role in oxidative brain injuries and neurodegenerative diseases. Although the immune function of neurons in regulating neuroinflammatory processes remains elusive, all types of glia cells (mainly microglia and astrocytes) are important to maintain the homeostasis and regulate neuroinflammation processes in the central nervous system. The accumulation of reactive microglia in areas of degeneration can sustain a systemic inflammatory response in the brain. Activated microglia and astrocytes can produce ROS, an important defense mechanism against microbial infection, which can, however, contribute to neurodegeneration [26]. Elevated levels of microglial activity were found in aged animals and when chronic inflammation occurs, microglia upregulate a variety of cell surface receptors involved in immune responses and trigger the release of a wide array of neurotoxic products, including pro-inflammatory cytokines. Chronic inflammation produces reactive species and it has been observed that the increase in brain microglial activation may be an early event leading to oxidative damage and depletion of endogenous antioxidants. Interestingly, the relationship between oxidative stress and inflammation in chronic neurodegenerative diseases is exemplified by the cytokine TNF, which is released by activated astrocytes and microglia. TNF can exacerbate inflammation and promote the release of ROS from microglia, thus promoting neurodegeneration. In contrast, it should be noted that TNF not only produce degenerative responses in the brain but can also ameliorate immune responses and promote regeneration and neuroprotection [26]. Chronic pro-inflammatory markers including IL-6, C-reactive protein (CRP), matrix metalloproteinases (MMPs), cytosolic phospholipase A2 (cPLA2), cyclooxygenase-2 (COX-2), and TNF-α are consistently elevated in neurodegenerative diseases, which are largely the result of chronically elevated levels of ROS [27]. Key peripheral blood markers of inflammation have been examined in brain aging to investigate their role in cognitive function. Several studies showed an association between blood levels of inflammatory markers and severity of brain functional impairment. Moreover, it has been hypothesized that IL-6 is a central regulator of the neuroinflammatory responses in aging brain. Animal models and patients with neurodegenerative diseases had higher levels of IL-6 and CRP, providing evidence that peripheral inflammatory mediators can interfere with neurocognitive functions and neurophysiological processes [28,29]. Therefore, reactive species, glial and immune cells form a coordinated network to maintain a proper equilibrium between physiological and pathological process in the central nervous system. The role of free radicals linked to persistent low-basal inflammation and brain aging is schematized in Fig. 19.1.

NEUROPROTECTION OF PHYTOCHEMICALS IN BRAIN AGING

Dietary phytochemicals, at least partly, have a profound effect on many aspects of brain aging, through interactions with the genome which result in altered gene expression. Recent findings suggest that several polyphenolic phytochemicals exhibit biphasic dose responses on cells with low doses activating signaling pathways that result in an increased expression of genes encoding survival proteins [30]. The neuroprotective effects of polyphenols are due to their ability in modulating intracellular signaling cascades involved in the control of neuronal survival, death, and differentiation. Moreover, antioxidant defense enzymes within the brain, such as superoxide dismutase (SOD), catalase (CAT), glutathione peroxidase (GPx), and glutathione S-transferase (GST), are crucial for breaking down the harmful products of oxidative stress. Previous studies have revealed that polyphenol compounds may induce the expression of these enzymes and modulate their activity in the central nervous system. Polyphenols constitute an extremely diverse class of plant secondary metabolites and appear to have many diverse functions in plants, including protective roles. These metabolites are present in a variety of fruits, vegetables, cereals, tea, wine, and fruit juices [31]. A classification of polyphenolic compounds mentioned in this chapter and their food sources is shown in Table 19.1. In the next sections, we will present the main classes of food polyphenols and their emerging role as hormetic inducers of neuroprotective pathways relevant for brain aging.

Resveratrol

The phytoalexin resveratrol (3,5,4′-trihydroxystilbene) is a polyphenolic small molecule present in many plant-derived foods, including grapes, red wine, peanuts, various berries, and cocoa [32]. Although in humans there is still no solid evidence that resveratrol intake can exerts its health benefits on the brain, several short-term supplementation studies showed that resveratrol decreased oxidative stress and inflammation with potential effects on brain performances [33,34]. Whereas many reports have supported the in vitro antioxidant activity of resveratrol, its in vivo efficacy is still controversial, possibly due to the limited knowledge of its bioavailability in humans. However, resveratrol has been shown to protect rat glioma cells from Aβ plaques toxicity by reducing the expression of inducible nitric oxide synthase (iNOS) and COX-2, thus preventing the uncontrolled release of NO and prostaglandin E2 (PGE2) [35]. In astroglial cells, resveratrol prevented ammonia neurotoxicity by modulating oxidative stress and inflammatory responses, thus inhibiting ROS production and cytokine release [36]. Consistent with the hormetic mechanism of action of resveratrol, a preconditioning/treatment was required in many animal studies in which resveratrol was demonstrated to be neuroprotective [37]. Exposure of cats to a high level of arsenic results in oxidative stress and brain damage that can be ameliorated when the cats are pretreated with resveratrol [38]. Furthermore, brain-derived neurotrophic factor (BDNF) has been implicated in regulating neurogenesis and mediates, at least in part, the enhancement of neurogenesis by dietary restriction. Resveratrol elicits gene expression profiles

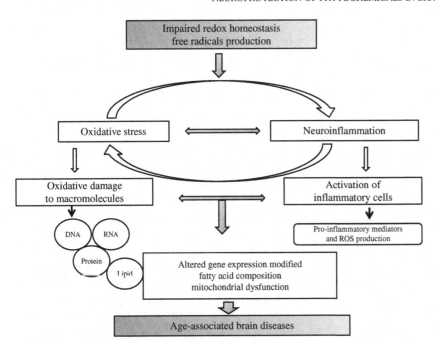

FIGURE 19.1 Oxidative stress, chronic inflammation, and brain aging are closely linked. The figure depicts the central roles of free radicals linking oxidative stress to inflammation and brain aging.

TABLE 19.1 Main Classes of Phytochemicals and Their Food Sources

Category		Name	Food sources
Phenolic compounds			
	Phenolic acids	Curcumin	Turmeric
	Stilbenoids	Resveratrol	Grape, nuts
Flavonoids			
	Flavonols	Quercetin	Onions, tea, apples
	Flavan-3-ols	Catechin	Green and black tea
		Epicatechin	Tea, cocoa, grape
		EGCG	Green tea
	Anthocyanins	Delphinidin	Berry eggplant
Glucosinolates		Sulforaphane	Broccoli, cauliflower

that strongly resemble those induced by calorie restriction and this phytochemical might differentially regulate BDNF expression depending upon the level of stress encountered by neurons.

Curcumin

Curcumin (diferuloylmethane) is present in 31 species of curcuma plants, including *Curcuma longa*, the rhizome of which provides the spice turmeric [39]. Curcumin has a broad spectrum of efficacy in inflammation and oxidative stress-driven diseases, and its beneficial effects have been reviewed previously [37]. A biphasic effect of curcumin on the nervous system have been reported, including a dose-response effect on hippocampal neurogenesis in mice [40]. Therefore, curcumin can stimulate neurogenesis but high concentrations may have an inhibitory effect. Several studies have shown that curcumin has a neuroprotective effect in animal models, and these early studies attributed these effects to the intrinsic antioxidative properties of curcumin. This polyphenolic compound has been proposed in the therapy of neurodegenerative disorders, but translational problems may be due to its pharmacokinetic parameters and bioavailability. However, exposure of neurons to curcumin can protect against oxidative insults, dysfunction, and degeneration in a range of experimental cell culture and animal models [41,42]. For instance, it was shown that curcumin reduced oxidative stress and Aβ-induced inflammatory responses in primary cultured astrocytes attenuating memory deficits in a mouse model of neurodegeneration. Studies in rats and mice demonstrated that curcumin treatment improves oxidative stress, neuroinflammation responses, and cognitive impairment caused by exposure to Aβ plaques. More interestingly, it was observed that curcumin can exert its neuroprotective functions by a mechanism involving BDNF upregulation and TNF-α signaling [37].

Flavan-3-ols: Catechin, (−)-Epicatechin, EGCG

These polyphenols belong to the subgroup named flavan-3-ols, also known as catechins. Flavan-3-ols, similarly to other polyphenols, are effective antioxidants, and this feature contributes to the benefits of their consumption. Their chemical structures vary with respect to the presence or absence of gallate and gallo moieties, and epigallocatechin gallate (EGCG), epigallocatechin, and epicatechin gallate (ECG), are the most abundant food flavan-3-ols, among which EGCG is the most studied [43]. There are numerous examples of the neuroprotective effects of flavan-3-ols in various cell culture and animal models of neurodegenerative disorders. Epicatechin can be found from various foods such as berries, chocolate, grapes, and tea; however, cocoa bean has the highest levels of epicatechin [44]. Epicatechin and its in vivo metabolite 3′-O-methyl epicatechin have potential as protective agents against neuronal degeneration. In particular, preexposure of epicatechin to cultured primary neurons enhanced their resistance to oxidized low-density lipoprotein through selective actions within stress-activated cellular pathways [45]. EGCG also reduced levels of lipid peroxidation and protein oxidation in neurons exposed to advanced glycation end products [46]. In orally treated mice, EGCG reduced myelin-reactive T-cell proliferation and TNF production, and protected neurons against degeneration in a mouse model of multiple sclerosis [47]. However, the neuroprotective effects of EGCG may depend on the dosage, duration of the treatment and animal age at which the intervention begins.

Other Phytochemicals

To date, the neurohormetic protective actions of several phytochemicals have been elucidated. Sulforaphane, which is one of the primary phytochemicals in broccoli, cabbages, and other cruciferous vegetables, can protect brain against oxidative insults. Sulforaphane administration ameliorated cognitive deficits without affecting the aggregation of Aβ in a mouse model of AD. Sulforaphane exposure also protected against neuronal death of nigral dopaminergic neurons by decreasing astrogliosis, microgliosis, and release of pro-inflammatory cytokines in an experimental PD model [48,49]. An example of one of the most studied phytochemicals is quercetin. This flavonoid is abundant in many fruits and vegetables, including onions and berries. Quercetin was shown to inhibit generation of ROS in rat glioma cells [50]. Interestingly, consistent with hormesis concept and dose-response mechanisms, primary neurons treated with low concentrations of quercetin and exposed to Aβ showed reduced oxidative damage (protein carbonyls, and nitrotyrosine). In contrast, higher concentrations of quercetin damaged the neurons [51]. Anthocyanins are one type of flavonoid and anthocyanin-rich fruits (berry, grape) providing beneficial effects against age-related neurodegeneration and cognitive decline. Dietary anthocyanins improved brain functions reducing neuroinflammation mediators, such as IL-1 and TNF-α. Anthocyanins were reported to protect dopaminergic neurons against oxidative stress. The modulation of neural signaling by anthocyanins involves BDNF and its role in neurogenesis [52].

NEUROPROTECTIVE PHYTOCHEMICALS AND MODULATION OF STRESS SIGNALING PATHWAYS

Phytochemicals exert their beneficial effects on the nervous system modulating cellular stress-response signaling pathways [53]. Phenolic compounds such as resveratrol, sulforaphane, curcumin, and catechins are well-established examples of hormesis. At low doses, several phytochemicals are known to enhance neuronal stress resistance by inducing adaptive stress response signaling pathways [54]. In the next sections, we highlight the major hormesis-based mechanisms by which dietary phytochemicals are capable of inducing a mild stress in neuronal cells and enhancing cellular resistance to the mechanisms that determine brain aging.

Nuclear Factor Erythroid 2-Related Factor 2

The transcription factor nuclear factor E2-related factor 2 (Nrf2) has emerged as a key regulator that plays an important role in cellular protection against oxidative stress and inflammation. Nrf2 mediates cellular adaptation to redox stress, regulating the main cellular defense mechanisms evolved to protect cell components from oxidative damage. Nrf2 is ubiquitously expressed in all tissues including the brain, and coordinates the transcription of genes involved in phase II detoxification and antioxidant defense [55]. Some examples include SOD, CAT, GPx, GST, NAD(P)H, quinone oxidoreductase 1 (NQO1), heme oxigenase-1 (HO-1), and the thioredoxin/peroxiredoxin system. Under basal conditions, Nrf2 is kept transcriptionally inactive and sequestered in the cytoplasm by its repressor protein, the Kelch-like ECH-associated protein 1 (Keap1). This binding provides the turnover of Nrf2 through proteasomal degradation and Keap1, a sulfhydryl-rich protein, is a specialized sensor to quantify stress, including ROS and genotoxic chemicals. In response to oxidative and electrophilic stress, Nrf2-Keap1 dissociation is triggered with consequent translocation of Nrf2 to the nucleus, where it heterodimerizes with members of the small musculoaponeruotic fibrosarcoma oncogene family proteins. The formed heterodimer binds the antioxidant response element (ARE) sequence in the promoter regions of genes inducing detoxifying proteins and antioxidant enzymes [56]. Elevated levels of oxidative stress in patients with neurodegenerative diseases have been reported and neurodegeneration may reduce the free radical scavengers activities regulated by Nrf2. Studies have established that Nrf2 nuclear translocation is diminished in hippocampus neurons in AD cases and mRNA and protein levels of Nrf2 are reduced in the motor cortex and spinal cord in amyotrophic lateral sclerosis (ALS) patients [22]. There are several substances involved in the activation of Nrf2 signaling and numerous findings suggest that phytochemicals may exert cytoprotective effects on neurons by Nrf2 activation. Also, pharmacological and genetic studies have been performed in animal models of AD, PD, and ALS to demonstrate the neuroprotective effects of Nrf2-activating phytochemicals [30,37]. A recent study showed that resveratrol promotes the survival of cerebellar granule neurons in rats by enhancing DNA-binding activity of Nrf2 and the expression of its downstream cytoprotective proteins like NQO1 and SOD [57]. Resveratrol induces HO-1 expression via the ARE-mediated transcriptional activation of Nrf2 in PC12 cells and primary neuronal cultures [58,59]. The neuroprotective effects of curcumin have been widely reported and it has been shown to have neuroprotective effects through Nrf2 activation increasing the activity of SOD, GPx, and HO-1 in experimental models of brain aging. In particular, HO-1 appears very active in the brain and its modulation plays an essential role in the pathogenesis of neurodegenerative diseases. Curcumin induces the expression of HO-1 in different brain cells including astrocytes and neurons by activating the Nrf2/ARE pathway, and thereby protecting against damage, inflammation, and oxidative cell death [60]. In addition, dietary intake of epicatechin and EGCG also enhanced Nrf2 protein expression and HO-1 accumulation in the nuclei of primary cortical neurons [61,62]. A low concentration of EGCG induces HO-1 through the ARE/Nrf2 pathway in neurons of the hippocampus, protecting against different types of oxidative damage [63]. In primary neuronal cultures of rat striatum, pretreatment with sulforaphane prevented oxidative stress by raising the intracellular glutathione content via activation of the Nrf2-ARE pathway [64]. Interestingly, the neuroprotection of sulforaphane was also associated with increased expression of the antioxidant enzymes NQO1 and HO1 in cultured mouse hippocampal neurons exposed to oxygen and glucose deprivation [65].

Nuclear Factor-κB

In response to a wide range of biological stimuli, the transcription factor nuclear factor κB (NF-κB) coordinates the expression of inflammatory genes encoding pro-inflammatory cytokines, chemokines, and inducible growth factors. Deregulated NF-κB activity is found in several illnesses, including neurodegenerative diseases, cancer, and chronic inflammation disorders. NF-κB is expressed in all animal tissues and it is located in the cytoplasm in an inactive complex form consisting of two subunits of 50 kDa (p50) and 65 kDa (p65) and an inhibitory subunit called IkB (IkBα or IkBβ). Induction of NF-κB and its translocation to the nucleus depends on the phosphorylation of IkB. In response to activating stimuli, IkB is phosphorylated, ubiquitinated, and degraded by the proteasome, which in turn allows the nuclear translocation of NF-κB to regulate gene expression. Although the exact mechanism is still elusive, it is remarkable to mention the potential cross-talk between the NF-κB and Nrf2 signaling pathways. The role of Nrf2 against inflammation has been related to its ability to antagonize NF-κB. This interconnection affects many signaling pathways to maintain redox homeostasis and coordinate cell fate determination [66]. Data obtained from Nrf2-deficient mice reported decreased expression of phase II detoxifying enzymes in parallel with upregulation of pro-inflammatory cytokines/biomarkers [67]. More interestingly, it has been demonstrated that HO-1 induced by the Nrf2 inhibits the NF-κB transcriptional apparatus, and thereby HO-1 is one of the key mediators for the interplay between Nrf2 and NF-κB [68]. Disturbances in the Nrf2—NF-κB axis can contribute to the onset and progression of neurodegenerative diseases. In neuronal cells, increased NF-κB activity during brain aging is associated with enhanced production of pro-inflammatory cytokines, such as IL-6, TNF-α, and COX-2. Reduced Nrf2 activity is characterized by a decline in HO-1, SOD, and NQO1, thus leading to increased levels of oxidative stress and neuroinflammation. A schematic illustration of relationships between Nrf2 and NF-κB in the brain is depicted in Fig. 19.2. Studies with phytochemicals like resveratrol, curcumin, and sulforaphane have reported beneficial effects on neurons by inhibiting NF-κB and activating Nrf2. In particular, sulforaphane provides neuroprotective mechanisms through activation of Nrf2 and inhibition of NF-κB, which may contribute to its effect in reversing various deficits associated

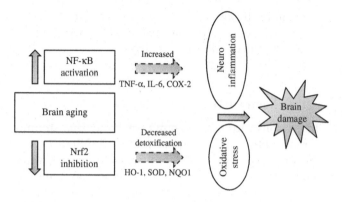

FIGURE 19.2 Schematic illustration on the role of Nrf2–NF-κB axis in bran aging. An imbalance between Nrf2 and NF-κB can lead to increased levels of oxidative stress and neuroinflammation, resulting in structural and functional damage to nervous tissue.

with brain aging [69]. Resveratrol attenuated Aβ-induced microglial inflammation by inhibiting the NF-κB signaling cascade and interfering with IkB phosphorylation [70]. Furthermore, resveratrol exerts anti-inflammatory effects in murine primary microglia and astrocytes by inhibiting NF-κB activation and pro-inflammatory cytokines [71]. Recently, it has been reported that curcumin may improve inflammatory injury, neuronal apoptosis, and activation of microglia/macrophages by inhibiting the NF-κB signaling pathway [72]. Berries rich in anthocyanins can attenuate inflammatory signaling in primary hippocampal neurons and BV-2 mouse microglial cells through a concentration-dependent mechanism involving reduction in NF-κB expression [73]. In human cerebral microvascular cells EGCG significantly suppressed the expression of inflammatory cytokines by a mechanism involving reduction of NF-κB activation [74].

Mitogen-Activated Protein Kinase

Mitogen-activated protein kinases (MAPKs) belong to the superfamily of serine/threonine kinases that mediate a wide range of cellular responses. MAPKs have emerged as critical players that connect various extracellular signals into intracellular response. Based on the degree of sequence homology, MAPK transduction cascades are organized into at least three subfamilies: the extracellular signal regulated kinases (ERKs), the stress activated protein kinase/jun N terminal kinase (JNK), and the p38 MAPK [75]. These kinases are involved in both survival and death pathways in response to different stresses to regulate cellular processes such as cell proliferation, survival, differentiation, and metabolism. The role of ERKs usually is associated with prosurvival signaling, while JNK is involved in the transcriptional regulation of apoptotic pathways. p38 MAPKs are activated by inflammatory cytokines and environmental stresses [76]. Although the mechanism is still unclear, several studies have indicated that phytochemicals and their metabolites may interact within the MAPK signaling pathways, particularly in cancer. However, different phytochemicals may also promote neuroprotection and suppress neuroinflammation by specifically inhibiting JNK and p38. For instance, modified derivatives of resveratrol suppressed the neurotoxicity of the parkinsonian mimetic 6-hydroxydopamine (6-OHDA) in neuroblastoma cells by attenuating phosphorylation of JNK [77]. Resveratrol increased BDNF in rat primary astroglia through a mechanism involving phosphorylation of ERK [78]. Moreover, JNK is involved in dopaminergic neuronal degeneration, and specific inhibitors of JNK may slow the neurodegeneration. Recent in vivo and in vitro studies have shown that curcumin can effectively target the JNK pathway by inhibiting phosphorylation of JNK and preventing dopaminergic neuronal death [79]. Curcumin promoted cell proliferation and hippocampal neurogenesis by activating ERK and p38 kinases [80]. Interestingly, even though ERK is involved in cell survival signaling, quercetin has been shown to stimulate neuronal apoptosis via a mechanism involving the inhibition of ERK, rather than by induction of pro-apoptotic signaling through JNK [81]. Epicatechin, and its metabolites, have been shown to induce phosphorylation of ERK in primary cortical neurons. Epicatechin at nanomolar concentrations acts as a rapid stimulator of ERK and this activation was no longer apparent at higher concentrations (micromolar) [82], suggesting hormetic effects of epicatechin on this pathway. Berry supplementation was demonstrated to partially reverse the deleterious effects of aging on neuronal signaling. Anthocyanin, and its metabolites, improved spatial memory and regulated hippocampal ERK expression in senescence-accelerated mice [83].

CONCLUSIONS

One of the greatest challenges to maintain brain health is to identify how to interrupt the vicious circle between oxidative stress and inflammation in brain aging. Research on bioactive compounds in food plants has rapidly evolved and these phytochemicals may become an important strategy to slow and prevent brain aging. So far, many efforts have been made to achieve a comprehensive understanding of the physiological effects of phytochemicals. Nevertheless, much remains to be elucidated, in particular how phytochemicals might activate neuroprotective pathways. However, current evidence suggests potential therapeutic application of phytochemicals in neuroinflammatory and neurodegenerative disorders. Here, we have described the major classes of phytochemicals and their main targets for dietary interventions. The neurobiological mechanisms of action of certain phytochemicals involve the modulation of specific adaptive stress response pathways.

Many phytochemicals display hormetic biphasic dose responses and this mechanism can enhance cellular resistance to oxidative damage and inflammation. Current evidence points to a crucial role of hormesis to explain the neuroprotective and neuromodulatory effects of phytochemicals at low doses. The challenge now is to establish the exact dose range in which stress response pathways are activated to regulate neuronal function. Future investigations should be carried out in order to determine the actual relevance of phytochemicals for treatment of human brain aging and neurodegeneration.

SUMMARY

Brain energy metabolism is intimately associated with an intrinsic mild oxidative stress contributing to chronic low-grade inflammatory state. There is a substantial amount of evidence suggesting that many dietary phytochemicals modulate signaling pathways involved in the attenuation of inflammation and oxidative stress. Plant-based foods include a number of phytochemicals such as resveratrol, curcumin and flavanols. These compounds act as hormetic inducers of neuroprotective pathways, leading to the expression of cytoprotective and restorative proteins. Future studies need to determine the effective physiological concentrations of these compounds in order preserve cognitive function and delay brain aging.

References

[1] Esiri MM. Ageing and the brain. J Pathol 2007;211:181−7.
[2] Wang X, Michaelis EK. Selective neuronal vulnerability to oxidative stress in the brain. Front Aging Neurosci 2010;22:12.
[3] de Oliveira DM, Ferreira Lima RM, El-Bachá RS. Brain rust: recent discoveries on the role of oxidative stress in neurodegenerative diseases. Nutr Neurosci 2012;15:94−102.
[4] Dickstein DL, Weaver CM, Luebke JI, Hof PR. Dendritic spine changes associated with normal aging. Neuroscience 2013; 251:21−32.
[5] Jellinger KA, Attems J. Neuropathological approaches to cerebral aging and neuroplasticity. Dialogues Clin Neurosci 2013;15:29−43.
[6] Fialkow L, Wang Y, Downey GP. Reactive oxygen and nitrogen species as signaling molecules regulating neutrophil function. Free Radic Biol Med 2007;42:153−64.
[7] Sochocka M, Koutsouraki ES, Gasiorowski K, Leszek J. Vascular oxidative stress and mitochondrial failure in the pathobiology of Alzheimer's disease: a new approach to therapy. CNS Neurol Disord Drug Targets 2013;12:870−81.
[8] Manach C, Scalbert A, Morand C, Rémésy C, Jiménez L. Polyphenols: food sources and bioavailability. Am J Clin Nutr 2004;79:727−47.
[9] Lau FC, Shukitt-Hale B, Joseph JA. Nutritional intervention in brain aging: reducing the effects of inflammation and oxidative stress. Subcell Biochem 2007;42:299−318.
[10] Andrade JP, Assunção M. Protective effects of chronic green tea consumption on age-related neurodegeneration. Curr Pharm Des 2012;18:4−14.
[11] Dasuri K, Zhang L, Keller JN. Oxidative stress, neurodegeneration, and the balance of protein degradation and protein synthesis. Free Radic Biol Med 2013;62:170−85.
[12] Salminen LE, Paul RH. Oxidative stress and genetic markers of suboptimal antioxidant defense in the aging brain: a theoretical review. Rev Neurosci 2014;25:805−19.
[13] Kannan K, Jain SK. Oxidative stress and apoptosis. Pathophysiology 2000;7:153−63.
[14] Calabrese V, Cornelius C, Dinkova-Kostova AT, Calabrese EJ, Mattson MP. Cellular stress responses, the hormesis paradigm, and vitagenes: novel targets for therapeutic intervention in neurodegenerative disorders. Antioxid Redox Signal 2010; 13:1763−811.
[15] Pamplona R, Dalfó E, Ayala V, Bellmunt MJ, Prat J, Ferrer I, et al. Proteins in human brain cortex are modified by oxidation, glycoxidation, and lipoxidation. Effects of Alzheimer's disease and identification of lipoxidation targets. J Biol Chem 2005;280:21522−30.
[16] Nunomura A, Moreira PI, Castellani RJ, Lee HG, Zhu X, Smith MA, et al. Oxidative damage to RNA in aging and neurodegenerative disorders. Neurotox Res 2012;22:231−48.
[17] Klein JA, Ackerman SL. Oxidative stress, cell cycle, and neurodegeneration. J Clin Invest 2003;111:785−93.
[18] Ulmann L, Mimouni V, Roux S, Porsolt R, Poisson JP. Brain and hippocampus fatty acid composition in phospholipid classes of aged-relative cognitive deficit rats. Prostaglandins Leukot Essent Fatty Acids 2001;64:189−95.
[19] Petursdottir AL, Farr SA, Morley JE, Banks WA, Skuladottir GV. Lipid peroxidation in brain during aging in the senescence-accelerated mouse (SAM). Neurobiol Aging 2007;28:1170−8.
[20] Farooqui AA, Horrocks LA. Lipid peroxides in the free radical pathophysiology of brain diseases. Cell Mol Neurobiol 1998;18:599−608.
[21] Nakamura T, Prikhodko OA, Pirie E, Nagar S, Akhtar MW, Oh CK, et al. Aberrant protein S-nitrosylation contributes to the pathophysiology of neurodegenerative diseases. Neurobiol Dis 2015; S0969-9961:00089-3.
[22] Gan L, Johnson JA. Oxidative damage and the Nrf2-ARE pathway in neurodegenerative diseases. Biochim Biophys Acta 2014;1842:1208−18.
[23] Martínez A, Portero-Otin M, Pamplona R, Ferrer I. Protein targets of oxidative damage in human neurodegenerative diseases with abnormal protein aggregates. Brain Pathol 2010;20:281−97.
[24] Rosano C, Marsland AL, Gianaros PJ. Maintaining brain health by monitoring inflammatory processes: a mechanism to promote successful aging. Aging Dis 2012;3:16−33.
[25] Calabrese V, Cornelius C, Mancuso C, Lentile R, Stella AM, Butterfield DA. Redox homeostasis and cellular stress response in aging and neurodegeneration. Methods Mol Biol 2010;610: 285−308.
[26] Fischer R, Maier O. Interrelation of oxidative stress and inflammation in neurodegenerative disease: role of TNF. Oxid Med Cell Longev 2015;2015:610813.
[27] Hsieh HL, Yang CM. Role of redox signaling in neuroinflammation and neurodegenerative diseases. Biomed Res Int 2013; 2013:484613.

[28] Godbout JP, Johnson RW. Interleukin-6 in the aging brain. J Neuroimmunol 2004;147:141–4.

[29] Weaver JD, Huang MH, Albert M, Harris T, Rowe JW, Seeman TE. Interleukin-6 and risk of cognitive decline: MacArthur Studies of Successful Aging. Neurology 2002;59:371–8.

[30] Calabrese V, Cornelius C, Mancuso C, Pennisi G, Calafato S, Bellia F, et al. Cellular stress response: a novel target for chemoprevention and nutritional neuroprotection in aging, neurodegenerative disorders and longevity. Neurochem Res 2008;33:2444–71.

[31] Abuajah CI, Ogbonna AC, Osuji CM. Functional components and medicinal properties of food: a review. J Food Sci Technol 2015;52:2522–9.

[32] Davinelli S, Sapere N, Zella D, Bracale R, Intrieri M, Scapagnini G. Pleiotropic protective effects of phytochemicals in Alzheimer's disease. Oxid Med Cell Longev 2012;2012:386527.

[33] Ghanim H, Sia CL, Abuaysheh S, Korzeniewski K, Patnaik P, Marumganti A, et al. An antiinflammatory and reactive oxygen species suppressive effects of an extract of Polygonum cuspidatum containing resveratrol. J Clin Endocrinol Metab 2010;95:E1–8.

[34] Kennedy DO, Wightman EL, Reay JL, Lietz G, Okello EJ, Wilde A, et al. Effects of resveratrol on cerebral blood flow variables and cognitive performance in humans: a double-blind, placebo-controlled, crossover investigation. Am J Clin Nutr 2010;91:1590–7.

[35] Kim YA, Kim GY, Park KY, Choi YH. Resveratrol inhibits nitric oxide and prostaglandin E2 production by lipopolysaccharide-activated C6 microglia. J Med Food 2007;10:218–24.

[36] Bobermin LD, Quincozes-Santos A, Guerra MC, Leite MC, Souza DO, Gonçalves CA, et al. Resveratrol prevents ammonia toxicity in astroglial cells. PLoS One 2012;7:e52164.

[37] Lee J, Jo DG, Park D, Chung HY, Mattson MP. Adaptive cellular stress pathways as therapeutic targets of dietary phytochemicals: focus on the nervous system. Pharmacol Rev 2014;66:815–68.

[38] Cheng Y, Xue J, Jiang H, Wang M, Gao L, Ma D, et al. Neuroprotective effect of resveratrol on arsenic trioxide-induced oxidative stress in feline brain. Hum Exp Toxicol 2014;33:737–47.

[39] Itokawa H, Shi Q, Akiyama T, Morris-Natschke SL, Lee KH. Recent advances in the investigation of curcuminoids. Chin Med 2008;3:11.

[40] Kim SJ, Son TG, Park HR, Park M, Kim MS, Kim HS, et al. Curcumin stimulates proliferation of embryonic neural progenitor cells and neurogenesis in the adult hippocampus. J Biol Chem 2008;283:14497–505.

[41] Scapagnini G, Colombrita C, Amadio M, D'Agata V, Arcelli E, Sapienza M, et al. Curcumin activates defensive genes and protects neurons against oxidative stress. Antioxid Redox Signal 2006;8:395–403.

[42] Scapagnini G, Caruso C, Calabrese V. Therapeutic potential of dietary polyphenols against brain ageing and neurodegenerative disorders. Adv Exp Med Biol 2010;698:27–35.

[43] Cheynier V. Polyphenols in foods are more complex than often thought. Am J Clin Nutr 2005;81:223S–9S.

[44] Arts ICW, van de Putte B, Hollman PCH. Catechin contents of foods commonly consumed in The Netherlands. 1. Fruits, vegetables, staple foods, and processed foods. J Agricult Food Chem 2000;48:1746–51.

[45] Schroeter H, Spencer JP, Rice-Evans C, Williams RJ. Flavonoids protect neurons from oxidized low-density-lipoprotein-induced apoptosis involving c-Jun N-terminal kinase (JNK), c-Jun and caspase-3. Biochem J 2001;358:547–57.

[46] Lee SJ, Lee KW. Protective effect of (−)-epigallocatechin gallate against advanced glycation endproducts-induced injury in neuronal cells. Biol Pharm Bull 2008;30:1369–73.

[47] Aktas O, Waiczies S, Zipp F. Neurodegeneration in autoimmune demyelination: recent mechanistic insights reveal novel therapeutic targets. J Neuroimmunol 2007;184:17–26.

[48] Kim HV, Kim HY, Ehrlich HY, Choi SY, Kim DJ, Kim Y. Amelioration of Alzheimer's disease by neuroprotective effect of sulforaphane in animal model. Amyloid 2013;20:7–12.

[49] Jazwa A, Rojo AI, Innamorato NG, Hesse M, Fernández-Ruiz J, Cuadrado A. Pharmacological targeting of the transcription factor Nrf2 at the basal ganglia provides disease modifying therapy for experimental parkinsonism. Antioxid Redox Signal 2011;14:2347–60.

[50] Sharma V, Mishra M, Ghosh S, Tewari R, Basu A, Seth P, et al. Modulation of interleukin-1mediated inflammatory response in human astrocytes by flavonoids: implications in neuroprotection. Brain Res Bull 2007;73:55–63.

[51] Ansari MA, Abdul HM, Joshi G, Opii WO, Butterfield DA. Protective effect of quercetin in primary neurons against Abeta (1–42): relevance to Alzheimer's disease. J Nutr Biochem 2009;20:269–75.

[52] Tsuda T. Dietary anthocyanin-rich plants: biochemical basis and recent progress in health benefits studies. Mol Nutr Food Res 2012;56:159–70.

[53] Calabrese V, Cornelius C, Dinkova-Kostova AT, Iavicoli I, Di Paola R, Koverech A, et al. Cellular stress responses, hormetic phytochemicals and vitagenes in aging and longevity. Biochim Biophys Acta 2012;1822:753–83.

[54] Mattson MP, Cheng A. Neurohormetic phytochemicals: low-dose toxins that induce adaptive neuronal stress responses. Trends Neurosci 2006;29:632–9.

[55] Itoh K, Chiba T, Takahashi S, Ishii T, Igarashi K, Katoh Y, et al. An Nrf2/small Maf heterodimer mediates the induction of phase II detoxifying enzyme genes through antioxidant response elements. Biochem Biophys Res Commun 1997;236:313–22.

[56] Kobayashi M, Yamamoto M. Molecular mechanisms activating the Nrf2-Keap1 pathway of antioxidant gene regulation. Antioxid Redox Signal 2005;7:385–94.

[57] Kumar A, Singh CK, Lavoie HA, Dipette DJ, Singh US. Resveratrol restores Nrf2 level and prevents ethanol-induced toxic effects in the cerebellum of a rodent model of fetal alcohol spectrum disorders. Mol Pharmacol 2011;80:446–57.

[58] Chen CY, Jang JH, Li MH, Surh YJ. Resveratrol upregulates heme oxygenase-1 expression via activation of NF-E2-related factor 2 in PC12 cells. Biochem Biophys Res Commun 2005;331:993–1000.

[59] Zhuang H, Kim YS, Koehler RC, Doré S. Potential mechanism by which resveratrol, a red wine constituent, protects neurons. Ann NY Acad Sci 2003;993:276–86.

[60] Cardozo LF, Pedruzzi LM, Stenvinkel P, Stockler-Pinto MB, Daleprane JB, Leite M, et al. Nutritional strategies to modulate inflammation and oxidative stress pathways via activation of the master antioxidant switch Nrf2. Biochimie 2013;95:1525–33.

[61] Shah ZA, Li RC, Ahmad AS, Kensler TW, Yamamoto M, Biswal S, et al. The flavanol (−)-epicatechin prevents stroke damage through the Nrf2/HO1 pathway. J Cereb Blood Flow Metab 2010;30:1951–61.

[62] Davinelli S, Di Marco R, Bracale R, Quattrone A, Zella D, Scapagnini G. Synergistic effect of L-Carnosine and EGCG in the prevention of physiological brain aging. Curr Pharm Des 2013;19:2722–7.

[63] Scapagnini G, Vasto S, Abraham NG, Caruso C, Zella D, Fabio G. Modulation of Nrf2/ARE pathway by food polyphenols: a nutritional neuroprotective strategy for cognitive and neurodegenerative disorders. Mol Neurobiol 2011;44:192–201.

[64] Mizuno K, Kume T, Muto C, Takada-Takatori Y, Izumi Y, Sugimoto H, et al. Glutathione biosynthesis via activation of the nuclear factor E2-related factor 2 (Nrf2) — antioxidant-response element (ARE) pathway is essential for neuroprotective effects of sulforaphane and 6-(methylsulfinyl) hexyl isothiocyanate. J Pharmacol Sci 2011;115:320—8.

[65] Soane L, Li Dai W, Fiskum G, Bambrick LL. Sulforaphane protects immature hippocampal neurons against death caused by exposure to hemin or to oxygen and glucose deprivation. J Neurosci Res 2010;88:1355—63.

[66] Wakabayashi N, Slocum SL, Skoko JJ, Shin S, Kensler TW. When NRF2 talks, who's listening? Antioxid Redox Signal 2010;13:1649—63.

[67] Li W, Khor TO, Xu C, Shen G, Jeong WS, Yu S, et al. Activation of Nrf2-antioxidant signaling attenuates NFkappaB-inflammatory response and elicits apoptosis. Biochem Pharmacol 2008;76:1485—9.

[68] Soares MP, Seldon MP, Gregoire IP, Vassilevskaia T, Berberat PO, Yu J, et al. Heme oxygenase-1 modulates the expression of adhesion molecules associated with endothelial cell activation. J Immunol 2004;172:3553—63.

[69] Negi G, Kumar A, Sharma SS. Nrf2 and NF-κB modulation by sulforaphane counteracts multiple manifestations of diabetic neuropathy in rats and high glucose-induced changes. Curr Neurovasc Res 2011;8:294—304.

[70] Capiralla H, Vingtdeux V, Zhao H, Sankowski R, Al-Abed Y, Davies P, et al. Resveratrol mitigates lipopolysaccharide- and Aβ-mediated microglial inflammation by inhibiting the TLR4/NF-κB/STAT signaling cascade. J Neurochem 120:461—72.

[71] Lu X, Ma L, Ruan L, Kong Y, Mou H, Zhang Z, et al. Resveratrol differentially modulates inflammatory responses of microglia and astrocytes. J Neuroinflammation 2010;7:46.

[72] Zhu HT, Bian C, Yuan JC, Chu WH, Xiang X, Chen F, et al. Curcumin attenuates acute inflammatory injury by inhibiting the TLR4/MyD88/NF-κB signaling pathway in experimental traumatic brain injury. J Neuroinflammation 2014;11:59.

[73] Poulose SM, Fisher DR, Larson J, Bielinski DF, Rimando AM, Carey AN, et al. Anthocyanin-rich açai (*Euterpe oleracea* Mart.) fruit pulp fractions attenuate inflammatory stress signaling in mouse brain BV-2 microglial cells. J Agric Food Chem 2012;60:1084—93.

[74] Li J, Ye L, Wang X, Liu J, Wang Y, Zhou Y, et al. (−)-Epigallocatechin gallate inhibits endotoxin-induced expression of inflammatory cytokines in human cerebral microvascular endothelial cells. J Neuroinflammation 2012;9:161.

[75] Karin M. The regulation of AP-1 activity by mitogen-activated protein kinases. Philos Trans R Soc Lond B Biol Sci 1996;351:127—34.

[76] Johnson GL, Lapadat R. Mitogen-activated protein kinase pathways mediated by ERK, JNK, and p38 protein kinases. Science 2002;298:1911—12.

[77] Chao J, Yu MS, Ho YS, Wang M, Chang RC. Dietary oxyresveratrol prevents parkinsonian mimetic 6-hydroxydopamine neurotoxicity. Free Radic Biol Med 2008;45:1019—26.

[78] Zhang F, Lu YF, Wu Q, Liu J, Shi JS. Resveratrol promotes neurotrophic factor release from astroglia. Exp Biol Med (Maywood) 2012;237:943—8.

[79] Yu S, Zheng W, Xin N, Chi ZH, Wang NQ, Nie YX, et al. Curcumin prevents dopaminergic neuronal death through inhibition of the c-Jun N-terminal kinase pathway. Rejuvenation Res 2010;13:55—64.

[80] Kim SJ, Son TG, Park HR, Park M, Kim MS, Kim HS, et al. Curcumin stimulates proliferation of embryonic neural progenitor cells and neurogenesis in the adult hippocampus. J Biol Chem 2008;283:14497—505.

[81] Spencer JPE, Rice-Evans C, Williams RJ. Modulation of pro-survival Akt/PKB and ERK1/2 signalling cascades by quercetin and its in vivo metabolites underlie their action on neuronal viability. J Biol Chem 2003;278:34783—93.

[82] Schroeter H, Bahia P, Spencer JP, Sheppard O, Rattray M, Cadenas E, et al. (−)Epicatechin stimulates ERK-dependent cyclic AMP response element activity and up-regulates GluR2 in cortical neurons. J Neurochem 2007;101:1596—606.

[83] Tan L, Yang HP, Pang W, Lu H, Hu YD, Li J, et al. Cyanidin-3-O-galactoside and blueberry extracts supplementation improves spatial memory and regulates hippocampal ERK expression in senescence-accelerated mice. Biomed Environ Sci 2014;27:186—9.

CHAPTER

20

Nutritional Modulation of Advanced Glycation End Products

Ma. Eugenia Garay-Sevilla[1], Claudia Luevano-Contreras[1] and Karen Chapman-Novakofski[2]

[1]Department of Medical Science, University of Guanajuato, Leon, Mexico [2]Department of Food Science and Human Nutrition, College of Agricultural, Consumer, and Environmental Sciences, University of Illinois, Urbana-Champaign, IL, USA

KEY FACTS

- AGEs exerts their biological effects by two mechanisms: one independent of receptors by damaging protein structure and one involving receptors which induce pro-inflammatory molecules that could contribute to cellular dysfunction and tissue damage.
- In patients with diabetes and renal failure, intake of AGEs could increase circulating AGEs, affect endothelial function, increase pro-inflammatory cytokine and oxidation markers, and affect insulin resistance.
- Studies with healthy individuals only have shown that intake of AGEs could affect circulating AGEs.
- Safe and effective antiglycation treatments are needed to palliate the potential damage exerted by endogenous and exogenous AGEs. Benfotiamine could be useful, but more evidence is needed.

Dictionary of Terms

- *AGEs:* Advanced glycation end products
- *CML:* Carboxymethyl-lysine
- *CEL:* Carboxyethyl-lysine
- *dAGEs:* Dietary advanced glycation end products
- *DOLD:* 3-deoxyglucosone-derived lysine dimer
- *ELISA:* Enzyme-linked immunosorbent assays
- *FMD:* Flow-mediated dilation of the brachial artery
- *GC/FID:* Gas chromatography with flame ionization detector
- *GC/MS:* Gas chromatography with mass spectrometric detection
- *GOLD:* Glyoxal-derived lysine dimer
- *HMW:* High molecular weight
- *HOMA:* Homeostatic model assessment
- *HPLC:* High-performance liquid chromatography
- *LC-MS/MS:* Liquid chromatography with tandem mass spectrometric detection
- *LMW:* Low molecular weight
- *MOLD:* Methylglyoxal-derived lysine dimer
- *MRP:* Maillard reaction products
- *MALDI-TOF/MS:* Matrix-assisted laser desorption ionization-mass spectrometry with time-of-flight detection

INTRODUCTION

How advanced glycation end products (AGEs) influence and are influenced by the aging process is a growing area of research interest, in part due to their influence on chronic disease development and morbidity. AGEs are formed physiologically (endogenously) and also in food, which is an exogenous source. Because the gradual accumulation of AGEs is the factor that provides AGEs with their substantial influence, the process of aging is also very important in discussing the impact of AGEs on health. This chapter will review AGEs formation, dietary absorption, and metabolism, as well as their impact on aging and chronic disease. The final section addresses interventions with dietary AGEs (dAGEs) and their influence on disease outcomes.

AGEs FORMATION IN VIVO

The AGEs are heterogeneous groups of compound that are formed both outside and inside the body. Physiologically AGEs result from the nonenzymatic reaction of reducing sugar such as glucose, α-oxoaldehydes, and other saccharide derivatives with free amino groups of biological molecules, such as proteins, lipids, and nucleic acids [1].

The Maillard reaction was the first described as responsible for the formation of AGEs; and four types of processes have been identified in the formation of AGEs under physiological conditions: (i) monosaccharide autoxidation (autoxidative glycosylation) or the degradation of saccharides unattached to a protein, (ii) Schiff base fragmentation, (iii) fructosamine degradation, and (iv) α,β-dicarbonyl compounds (methylglyoxal, glyoxal, and 3-deoxyglucosone) formed from lipid peroxidation [2]. The formation of AGEs is catalyzed by transitional metals and is inhibited by reducing compounds such as ascorbate. Many AGEs have been found in physiological systems (Table 20.1).

AGEs have different biological and physiological functions: some are protein cross-links (pentosidine, MOLD, GOLD, and DOLD), some are recognition factors for specific AGE-binding cell-surface receptors (carboxymethyl-lysine (CML), methylglyoxal-derived hydroimidazolone) and some are markers and risk predictors of disease processes [3].

Glycation is one of the most common types of protein modification. This spontaneous damage of proteins affects approximately 0.1–0.2% of the arginine and lysine residues in vivo [4]. Glycation of proteins proceeds with a variable rate and extent during the lifespan of proteins in tissues and body fluid under physiological conditions, but it is more intensive in several disease conditions such as diabetes [5], atherosclerosis [6], neurodegenerative diseases [7], osteoarthritis [8], and renal failure [9].

Glycating agents in vivo include free sugars (glucose), glycolytic intermediates, such as glucose and fructose-6-phosphates, and dicarbonyls and 3-deoxyglucosone. Glycation by sugars proceeds through the early (Schiff bases) and intermediate stages (the Amadori rearrangement to fructosamines) toward the formation of heterogeneous moieties collectively termed AGEs [10]. If the initial glycating agent is glucose the initial product is termed a fructosamine; traditionally considered to be a major source of AGEs in vivo. The best studied fructosamine to date is hemoglobin A1c (HbA1c), which is a widely used and very useful marker of medium- to long-term complications of diabetes [11]. A brief description will be given of the Maillard reaction and alternative pathways.

Maillard Reaction

There are three phases of the Maillard reaction. First, the carbonyl group of a reducing sugar interacts in a nonenzymatic way with an amino acid to form an unstable compound called a Schiff base. Sugars are reactive toward lysine residues while dicarbonyls are mainly reactive toward arginine residues of proteins. During the second phase, the Schiff base may undergo hydrolyzation and generate the original sugar and amino acid; or it could undergo cyclization and further Amadori rearrangements to form more stable compounds (Amadori products). However, under physiological and nonoxidative conditions, 90% of Amadori products could sustain a reversible reaction to the initial sugar and amino acid [12]. Finally, in the last phase, Amadori products can generate AGEs by either oxidative or nonoxidative cleavage. In the oxidative cleavage, the principal AGEs produced is CML, one of the first AGEs characterized in vivo and the major AGEs' biomarker in human tissues [13]. In contrast, the nonoxidative cleavage of Amadori products will produce the α-dicarbonyl derivative 3-deoxyglucosane. This derivative can react with an amino acid and also form CML or other AGEs cross-links, such as pyrraline, pentosidine, or imidazolone, considered the predominant class of AGEs in vivo [14,15].

TABLE 20.1 Types of AGEs

3-deoxyglucosone-derived lysine dimer (DOLD)
Argpyrimidine
Bis(lysyl)imidazolium derivatives
Glyoxal-derived lysine dimers (GOLD)
Hydroimidazolones derived from methylglyoxal, glyoxal, and 3-deoxyglucosone
Methylglyoxal-derived lysine dimers (MOLD)
N^ε-(carboxymethyl)-L-lysine (CML)
N^ε-(1-carboxyethyl)-lysine (CEL)
Pentosidine
Pyrraline

Formation of α-Dicarbonyl: Methylglyoxal, Glyoxal, and 3-Deoxyglucosane

The α-dicarbonyls have been proposed as the primary precursors for AGEs formation, particularly methylglyoxal [16]. There are two pathways that produce α-dicarbonyls. The Namiki pathway occurs when a Schiff base degrades and forms glyoxal; the Wolff pathway involves the autoxidation of monosaccharides, as may be found with the autoxidation of glucose at physiological conditions. However, there are other metabolic intermediates that have been implicated in α-dicarbonyl production and the subsequent AGEs generation. Some of those include glycolytic intermediates (glucose-6-phosphate, glyceraldehyde 3 phosphate, and dihydroacetone phosphate), a polyol pathway intermediate (fructose-6 phosphate), an intermediate from ketone body and threonine metabolism (acetol), and lipid peroxidation also generates methylglyoxal [17,18].

FORMATION OF DIETARY AGEs

The reaction of reducing carbohydrates with amino compounds (Maillard reaction) was described in 1912 by Louis Camille Maillard. It has been used for caramel production, coffee roasting, and bread baking [19]. Protein glycation increases emulsifying activity, improves foaming properties, increases protein solubility, and promotes the formation of compounds with antioxidant activity (extending food shelf life by delaying lipid oxidation) [20]. In the olfactory and visual area, the Maillard reaction leads to the formation of a diverse range of aromatic and color compounds, as well as AGEs [21].

AGEs production in foods (dAGEs) involves the Maillard reaction in a similar manner as in vivo: an amino acid from a protein, amine, or phospholipids reacts in a nonenzymatic fashion with carbonyl groups from reducing sugars as well as with degradation products of carbohydrates, lipids, and ascorbic acid. The resulting products often are referred to in the food science literature as Maillard reaction products (MRP). There are low molecular weight (LMW) Maillard products such as aldehydes, ketones, acryl amides, and AGEs, as well as high molecular weight (HMW) products such as melanoidins [22]. Other products formed during this intricate reaction are furfurals, pyrralines, and α-dicarbonyl compounds, such as methylglyoxal, as well as CML and pentosidine formed in the last stage of the Maillard reaction [23]. Another source of dAGEs could be autoxidation of fatty acids and amino acids [24].

Regardless of the diversity of AGEs, CML has been reported as one of the most abundant in vivo and it was one of the first to be characterized in foods (milk and milk products). For this reason, in most studies, CML is chosen as a marker of AGEs in foods and in vivo [25].

The rate of formation and the diversity of the generated AGEs in food depend on factors such as composition, availability of precursors, presence of transition metals, pH, processing temperature, availability of water, and availability of pro- and antioxidants [26].

METHODS FOR MEASURING AGEs

At present there are several methods to determine AGEs in tissues, biological fluids, and food (Table 20.2); immunochemical and instrumental methods all have their advantages and disadvantages, as do other methodologies. In general, AGEs can be measured by:

1. Enzyme-linked immunosorbent assays (ELISA) using monoclonal or polyclonal antibodies [29,30]. This technique has limitations. For example, the lack of specificity of the antibody, high background responses due to significant glycation adduct content of proteins [31], interference with glycation free adducts [32], and pretreatment techniques, such as heating and alkaline treatment [33]. However, important contributions have been made using this methodology. For example, measurement of CML in plasma and urine [34] and in different foods [26].
2. Fluorescence with excitation and emission wavelengths of 350 and 450 nm, respectively [35], with modification of this method to detect low molecular mass AGE peptides or free adducts [36,37]. These methods have analytical problems, for example, different fluorophores contribute to the global measure of fluorescence each with a different specific fluorescence such that no quantitative calibration can be achieved [4].
3. High-performance liquid chromatography (HPLC) with diode array detector, fluorometric detector, and mass spectrometry [38]. For example, measurement of pentosidine in skin [39] and in meat, such as beef, rabbit, pork, chicken, turkey [40].
4. Gas chromatography with flame ionization detector (GC/FID) or with mass spectrometric detection (GC/MS). This technique requires a precolumn derivatization to convert volatile AGEs in thermosetting compounds. This method is sensitive, reproducible, and accurate [41]. For example, measurement of CML of peritoneal dialysis fluids [42] and infant foods [41].
5. Liquid chromatography with tandem mass spectrometric detection (LC-MS/MS) [4]. It has been reported that this is the best analytical method

TABLE 20.2 Summary of Methods for Measuring Different AGEs in Tissues, Biological Fluids, Food

AGEs	Sample	Method	Reference
N^ε-(carboxymethyl)-L-lysine	Serum	ELISA	[27]
	Plasma	ELISA	[28]
	Urine		
	Plasma	UPLC-MS/MS	[102]
	Serum		
	Peritoneal dialysis fluids	Immunological methods, HPLC, GC/MS	[42]
	Animal-derived foods, carbohydrate-rich foods	ELISA	[26]
	Raw and roasted almonds	HPLC-MS/MS	[103]
	Infant foods	ESI-LC-MS/MS	[41]
		GC-MS	[104]
Pentosidine	Plasma/serum	ELISA, HPLC	[105]
	Plasma	HPLC	[106]
	Skin	HPLC-FLD	[39]
	Skin autofluorescence	AGE reader	[107]
	Urine	HPLC	[108]
	Human cortical bones	UPLC	[109]
	Beef, rabbit, pork, chicken, turkey	HPLC/MS	[40]
	Bread crust extract	RP-HPLC	[110]
Pyrraline	Urine	HPLC	[111, 108]
	Urine	RP HPLC with UV detection	[112]
	Raw and processed cow milk	LC-MS/MS	[43]

ELISA, enzyme-linked immunosorbent assay; UPLC-MS/MS, ultraperformance liquid chromatography-tandem mass spectrometry; HPLC, high-performance liquid chromatography; GC/MS, gas chromatography-mass spectrometry; ESI, electrospray ionization; FLD, postcolumn fluorescence derivatization; UPLC, ultrahigh-pressure liquid chromatography; RP, reverse phase; UV, ultraviolet.

available to quantify glycation adducts. For example, measurement of pyrraline in raw and processed cow milk [43]. Detection in plasma, urine, and dialysate samples has provided a comprehensive and quantitative analysis of glycation adducts [38].

6. Matrix-assisted laser desorption ionization-mass spectrometry with time-of-flight detection (MALDI-TOF/MS) is a promising tool to analyze AGE-modified proteins [44].
7. Measurement of skin autofluorescence is a rapid and noninvasive method, which was validated against biochemical analyses of AGEs in dermal tissue in biopsies taken from the same spot on the arm [45].

Methods of determining dAGEs usually rely on food composition reports and quantitation of dAGEs from dietary records. A more recent food frequency has also been validated to alleviate some of the quantification burden [46–48].

ABSORPTION, METABOLISM, AND ELIMINATION OF dAGEs

It is important to understand the absorption, metabolism, and elimination of dAGEs because of their potential for accumulation and contribution to chronic disease. Evidence from human studies shows dAGEs, such as CML and methylglyoxal, contribute to the levels of circulating AGEs. Cross-sectional and randomized studies have demonstrated the correlation between dAGEs and circulating AGEs. In addition, it has been estimated that 10–50 times more dAGEs are supplied by a conventional diet than those found in plasma or tissues of subjects with uremia [49].

Several researchers have studied the absorption, metabolism, and elimination of dAGEs as well as their precursors. Erbersdobler and Faist [50] reported Amadori products (dAGEs precursors) underwent intestinal absorption by diffusion. ^{14}C-labeled fructose amino acids in rats and humans revealed that urinary excretion of Amadori products after test meals was rapid, while the fecal output was slow, and persisted for 3 days in small quantities. Surprisingly, only 1–3% of the ingested amounts of protein-bound Amadori products were recovered in the urine. A larger percentage of urinary end products have been reported in a review of premelanoidins and melanoidins. These authors' data suggested that 16–30% of absorbed premelanoidins were excreted in the urine and 1–5% for melanoidins. This low recovery of pre- and melanoidins could be due to degradation by digestive microbial enzymes during intestinal transit or retention of premelanoidins by different tissues [51]. To support this former concept, Wiame et al. [52] showed that bacteria normally residing in the large intestine degrades around 80% of dietary Amadori products. However, excretion rates of different Amadori products differ widely, and may depend on the previous intake of dAGEs. In one study, seven healthy subjects ingested a normal diet on days 1 and 5; on days 2, 3, and 4 they ate a diet virtually free of Maillard compounds. Urinary excretion of free pyrraline was directly affected by composition of the diet, decreasing from 4.8 mg/day on day 1 to 0.3 mg/day on days 2, 3, and 4 [53]. More recently, nine healthy volunteers ate either a "no-AGEs" diet or an unrestrictive diet. Urinary excretion of an AGEs precursor, 3-deoxyglucosone, was significantly lower after the "AGEs free diet" [54].

Excretion rates of dAGEs may also depend upon the disease state of the individual. One study of 38 patients with diabetes and with or without renal disease reported that indicators of renal disease were related to the serum rise in AGEs, and that urinary excretion was reduced in those with diabetes and renal disease [55]. However, urinary CML has been reported to increase in diabetes, although these reports are at times conflicting depending on whether type 1 or type 2 diabetes is studied. However, urinary CML is associated with the degree of microalbuminuria present in those with diabetes, and may be a marker for renal disease [28].

MOLECULAR ACTION

AGEs can exert their biological effects by at least two well-identified mechanisms: (i) one independent of receptor which involves modification of intracellular proteins, alteration of extracellular matrix metabolism with damage of cellular function, and modification of circulating proteins with alteration of its function; and (ii) interaction with receptors such as RAGE (receptor for AGEs), and AGEs receptor 1 (AGER1). These receptors serve in the regulation of AGEs uptake and removal [56,57].

The receptor-independent mechanism has been studied in the glycation of lipoproteins, specifically the low-density lipoprotein (LDL) on their apolipoprotein B (ApoB) and phospholipid components. The uptake of the glycated LDL by its receptors will be decreased with its subsequently reduction in clearance, however its uptake by macrophages will be increased with the consequent stimulation of foam cells formation and promotion of atherosclerosis [56].

Whereas the receptor mechanism mainly involves RAGE, which is present constitutively in skin and lungs, but they can be induced by its ligands (AGEs among them) on vascular, renal, hemopoietic, and neuronal/glial cells. RAGE can triggers oxidative stress and inflammation in both acute and chronic diseases [58,59].

The interaction of AGEs and RAGE promotes the activation of the mitogen-activated protein kinases (MAPKs), the phosphatidylinositol-3 kinase (PI3-K), and the nicotinamide adenine dinucleotide phosphate-oxidase (NADPH, a complex of enzymes that produces superoxide) among other signaling pathways. When this complex is upregulated, it increases intracellular oxidative stress. The activation of these signaling pathways will lead to the activation of the transcription factor NF-κB (nuclear factor kappa B), which regulates target genes that are involved in the adaptive and innate immune system as well as the gene for RAGE itself [58]. Thus, activation of NF-κB triggers the transcription of genes for pro-inflammatory cytokines, growth factors, and adhesive molecules, such as tumor necrosis factor α (TNF-α), interleukin 6 (IL-6), well-known inflammation promoters, and vascular cell adhesion molecule 1 (VCAM-1) [60,61]. Because NF-κB also regulates RAGE transcription this will promote the maintenance and amplification of the signal with a sustained response and induction of the pro-inflammatory molecules, which could be responsible for chronic alterations and could contribute to cellular dysfunction and damage target to organs, and ultimately lead to complications as atherosclerosis, cardiovascular disease, nephropathy, and retinopathy among others [62–64] (Fig. 20.1).

Several ligands can interact with RAGE and this interaction seems to depend on the molecule size. It has been found that high molecular weight CML can bind to RAGE, which raised the question if dAGEs could bind to RAGE. Birlouez-Aragon found that after intake of high AGEs diet there was an increase in high-molecular-weight AGEs, therefore dAGEs could be a potential ligand for RAGE [65]. Indeed, it has

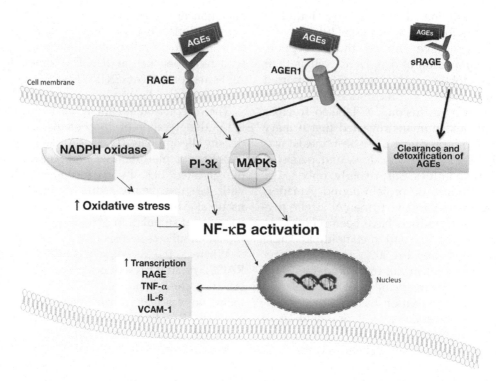

FIGURE 20.1 Molecular action of AGEs: interaction with receptors. Abbreviations: AGE, advanced glycation end product; RAGE, receptor for AGE; AGER1, AGEs receptor 1; sRAGE soluble RAGE; MAPK, mitogen-activated protein kinases; PI-3k, phosphatidylinositol-3 kinase; NAD(P)H oxidase, enzymes complex that produces superoxide; NK-κB, nuclear factor-κB; TNF-α, tumor necrosis factor α; IL-6, interleukin 6; VCAM-1, vascular adhesion molecule 1.

been shown that dAGEs could act as RAGE ligands and activate major signal transduction pathways in vitro [66,67]. dAGEs, together with those made endogenously, could promote a systemic glycoxidant burden, oxidant stress, and cell activation, which increases vulnerability of target tissues to injury [68].

In addition to the transmembrane RAGE, other isoforms have been found. For instance, two forms of circulating RAGE, soluble RAGE (sRAGE) that is produced by proteolytic cleavage from the transmembrane receptor and the endogenous secretory RAGE (esRAGE) that is a natural gene alternative splicing. The function of these isoforms could be removal or neutralization of ligands thus counteracting the damaging effect of AGE–RAGE axis [69].

Additionally, AGER1 suppresses AGEs and their related oxidative stress and inflammation. Overexpression of AGER1 inhibits the epithelial growth factor receptor, suppresses RAGE pro-inflammatory signaling pathways, and is involved in the clearance and possible detoxification of AGEs [70].

AGEs AND NORMAL AGING

Normal aging is associated with a decline in organ function and structural integrity, even in the absence of chronic disease. The theories associated with this decline include oxidative stress, mitochondrial dysfunction, inflammation, and glycation. Most of these have overlapping and synergistic pathways and outcomes from a biochemical and physiological perspective.

Glycation's role in normal aging is related to the accumulation of AGEs and their disruption of normal processes through the aforementioned theories on aging: increased oxidative stress, mitochondrial dysfunction, and increased inflammation. Glycated proteins contribute to oxidative stress because they are reactive compounds and can propagate free radical reactions [71]. In addition, AGEs contribute to oxidative stress through activating NAPH oxidase through the AGE–RAGE interaction [72]. AGEs have proinflammatory consequences through the activation of RAGE, which then leads to an inflammatory cascade, beginning with upregulation of genes such as NF-κB [72]. As proteins are damaged by oxidative stress without adequate repair in normal aging, leading to organ and system decline, proteins damaged by AGEs also can lead to or parallel this aging phenotype [73].

The theoretical frameworks for aging all include uncompensated changes, or poorly compensated changes, that occur over many years. For AGEs to elicit or contribute to this aging process, similarly, longitudinal effects would be needed. For AGEs to accumulate over many years, one of three scenarios is likely: increased production; decreased degradation and excretion; and targeted proteins with a long life, such as collagen or extracellular matrix [74]. However, increased production, degradation, and excretion processes themselves decline with aging. In terms of the third scenario, researchers have found increased AGEs accumulation in proteins and protein complexes with long half-lives [75].

The long accumulation and slow turnover of these glycated proteins further damage organ function and structural integrity. The extent to which this happens may be influenced by genetic predisposition to protein repair as well as the reactivity of the proteins that are glycated. Highly reactive proteins will increase the oxidative stress, which also lead to mitochondrial damage and further diminished function or impaired structure. As with all the theoretical processes of aging, the impact of AGEs is probably influenced by genetic predisposition or resilience, as well as environmental factors.

AGEs AND CHRONIC DISEASES ASSOCIATED WITH OLDER AGE

Just as the AGEs accumulation over time influences the aging process, it also influences chronic disease progression and possibly chronic disease development. Studies of AGEs and chronic disease have included cardiovascular disease; neurodegenerative diseases such as Alzheimer's disease, Parkinson's disease, and vascular dementias; renal disease; rheumatoid arthritis, sarcopenia, cataracts, and other degenerative ophthalmic diseases; and diabetes (Fig. 20.2).

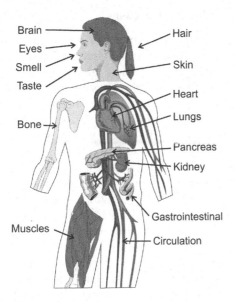

FIGURE 20.2 Organs and body sites where AGEs may accumulate and cause damage.

Cardiovascular Disease

Cardiovascular disease encompasses a variety of conditions affecting the heart, the vascular and arterial systems, and dysfunctional components, such as hypertension. Cardiovascular disease development and progression is influenced by a complex interplay among genetic and environmental factors. AGEs' role in cardiovascular disease appears to be one of promoting inflammation, changing endothelial function, modifying the vascular wall, and amplifying the effects of other mediators of cardiovascular disease, such as lipoproteins.

It has been proposed that collagen-AGEs crosslinking contributes to vascular stiffening, and thus also contributes to atherosclerosis development and progression. In addition, foam cell development is enhanced with increased AGEs through their promotion of adhesion molecules [56]. AGEs may also cause cross-linking of lipoproteins, including LDL, causing LDL to not be recognized by receptors, whereby reducing LDL clearance.

Another way that AGEs could influence cardiovascular functioning is through a decrease of nitric oxide activity. Normally, endothelial cells produce nitric oxide; low levels of nitric oxide indicate endothelial cell dysfunction. This is not the exclusive indicator, with others including increased platelet aggregation and increased reactive oxygen species production, as well as several other indicators. However, nitric oxide is a stimulus for vasodilation, and low nitric oxide diminishes the dilation response and capacity, which in turn affects vascular tone, immune functions, and barrier capacity [76].

With the AGE–RAGE interaction, barrier function is diminished, which results in greater migration of monocytes across the endothelium and thus, foam cell development is promoted [56]. In vitro work has suggested that AGEs accumulation can also decrease the amplitude, width, and duration of calcium-related contractile functions of heart tissue, adding to cardiac dysfunction in diabetes [77].

Neurodegenerative Diseases

The progressive damage or death of neurons is characteristic of neurodegenerative diseases. The most prevalent within this category are Alzheimer's disease and Parkinson's disease. While Alzheimer's disease reflects cognitive decline as a hallmark, characteristics of Parkinson's focus on muscular dysfunction.

Several AGEs have been implicated in cognitive decline and Alzheimer's disease development and progression in particular, primarily through promotion of amyloid tangles and plaques as well as increased beta amyloid and associated toxicity. Albumin has also been found to be transformed from an alpha helical structure to a beta-sheet structure secondary to glycation. A study of 49 nondemented young elderly (mean age 71 years) revealed that dAGEs were associated with a faster decline in memory over a mean follow-up of 35 months [78].

Although AGEs are found in the brain tissue of those without apparent Parkinson's disease, levels have been found to be higher in those with the condition than in age-matched controls. Enhanced expression of RAGE and possibly increased inflammation reflect AGEs influence in neural death and symptoms of Parkinson's disease [7].

Renal Disease

Acute injury can result in resolvable kidney malfunction, but chronic renal disease reflects a progressive loss of filtration and excretion capacity of the kidneys that is not reversible. The etiologies of chronic renal disease include cardiovascular disease and diabetes, as well as chronic inflammation of the kidneys and genetic predisposition.

In chronic kidney disease, the excretion of AGEs is decreased with the decline in glomerular filtration, and the production may be increased secondary to oxidative stress, causing circulating AGEs to be increased. These increased circulating AGEs are not adequately cleared by dialysis, and potentiate the vascular damage associated with chronic kidney disease. The increased plasma AGEs also lead to RAGE activation, which in turn, activates inflammatory cascades and reactive oxygen species production. With this process, renal sclerosis is enhanced by additional cross-linking, and renal function suffers additional decline [79].

Rheumatoid Arthritis

Osteoarthritis reflects deterioration of joints occurring after decades of use, often worse in those who are overweight. Rheumatoid arthritis is an autoimmune disorder that also affects joints. These diseases often share comorbidity with cardiovascular disease.

An increased risk for cardiovascular disease in persons with rheumatoid arthritis is believed to be secondary to chronic inflammation associated with rheumatoid arthritis. This increased inflammation may promote the development of AGEs. Indeed, AGEs have been reported to be elevated in patients with rheumatoid arthritis ($n = 49$) without cardiovascular disease as compared to age- and gender-matched healthy controls. AGEs were measured by skin autofluorescence, and were related to endothelial activation and dysfunction [80]. RAGE activation by increased ligands, such as AGEs, may contribute to the cartilage degradation characteristic of both rheumatoid arthritis and osteoarthritis [58].

Sarcopenia

The loss of muscle tissue and muscle function is represented by the term sarcopenia. Sarcopenia may occur in both under- and overweight, but is more prevalent in older age groups. Sarcopenia can lead to increased physical dependence for activities of daily living as mobility declines, which also contributes to increased morbidity from other chronic diseases.

Collagen cross-linking, inflammation, and endothelial dysfunction are pathways where AGEs have been implicated in sarcopenia. A study of older women ($n = 394$; >65 years) found that those in the upper quartile of CML were more likely to develop walking disability, but other conditions were also associated with the disability, including older age, congestive heart failure, peripheral artery disease, and diabetes [81]. Nevertheless, this study suggests a potential for AGEs in sarcopenia and associated functional disability. An earlier study also reported increased CML and RAGE immunostaining in elderly subjects and in subjects who had gained weight but were not obese. The staining intensities were correlated with each other, as well as with muscle TNF-α ($P \leq .02$) [82]. Although sarcopenia was not diagnosed in these subjects, the authors suggest that sarcopenia could be developing in both the elderly and weight gaining subjects.

Degenerative Eye Diseases

Although there are numerous degenerative eye diseases, those most commonly associated with AGEs are diabetic retinopathy, cataracts, glaucoma, and macular degeneration. The complex function and structure of the eye holds several areas of particular concern relative to the effect of AGEs (Fig. 20.3). In diabetic retinopathy, changes to the blood vessels in the retina of the eye include blood vessel occlusion and proliferation, with vessel leakage possible into the macula. AGEs have been implicated with the breakdown of the inner blood–retinal barrier, which then leads to progressive degeneration. AGEs influence connective tissue growth factor and thus extracellular membrane changes with basal membrane thickening; activation of caspases which leads to pericyte apoptosis; and increased adhesion molecules, with subsequent retinal leucostasis, and increased transjunction permeability. Methylglyoxal AGEs have been particularly implicated in diabetic retinopathy [83].

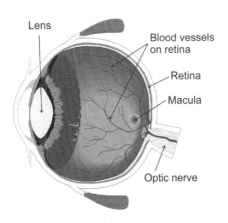

FIGURE 20.3 Structures of the eye where AGEs may accumulate and cause damage.

Myelin, composed primarily of protein and fat, and proteins such as tubulin, and lens crystallins may also be modified to produce AGEs, and thereby initiate processes leading to cataract formation [83]. Cataracts are formed in the eye lens, which is composed mostly of protein and water. Modification of this lens protein can cause clouding, which is a symptom of cataracts. Cataracts are more common in older age, in diabetes, and in those exposed to sunlight for prolonged periods of time.

Glaucoma is a term used to represent a group of optic nerve damage conditions, usually caused by increased eye pressure. One study of age-matched eye donors reported that AGEs accumulation and RAGE upregulation were greater in those with glaucoma as compared to those without glaucoma. Both AGEs and RAGE were greater in older versus younger eyes. An increased rigidity of the structure where the optic nerve exists in the eye is postulated to be a consequence of AGEs accumulation [84].

Age-related macular degeneration (AMD) refers to the conditions where the macula is damaged. The macula is part of the retina where the light-sensing cells contribute to clear, focused vision. As its name implies, AMD is more common in older age, and those with a family history of the condition. The associated oxidative damage, inflammation, changes to vascularization and nervous tissue, and extracellular matrix abnormalities that include cross-linking, all suggest a role for AGEs in AMD [85]. However, in a cross-sectional study of nearly 5000 subjects, higher serum CML levels were not associated with the prevalence of AMD [86].

Diabetes

AGEs have been known to be increased in persons with diabetes, secondary to increased blood glucose [47,48]. Blood proteins can undergo the nonenzymatic reaction of a reducing sugar (glucose) reacting to a reactive amino acid (hemoglobin). HbA1c is a glycated protein that has been used to diagnose and monitor the blood glucose control of those with diabetes. HbA1c represents the average glycemic level over a period of 1–2 months. However, glycation of other proteins occurs during hyperglycemia, such as glycated albumin, which represents a shorter period of time [88].

The increased blood glucose, and possibly increased AGEs themselves, contribute to secondary morbidity with cardiovascular, renal, and eye damage. For instance, a pilot study found increased risk for cardiovascular complications in those with higher dAGEs intake [46–48]. AGEs have also been shown to damage pancreatic beta cells, and thereby may also contribute to the development of diabetes [89].

DIETARY INTERVENTIONS FOR dAGEs

The dAGEs can have an impact on health through a variety of systematic mechanisms. Their effects have been studied in patients with chronic diseases showing that AGEs intake could increase circulating AGEs, accumulate in tissues, affect endothelial function, increase pro-inflammatory cytokine and oxidation markers, and affect insulin resistance. Most of these effects have been studied in clinical trials with dietary manipulation of the amount of AGEs and the first studies were done in patients with renal disease, diabetes or both [90–93]. More recently other studies have also focused on other chronic diseases, healthy subjects and individuals with overweight and obesity, a representative sample of these is presented on the next page (Table 20.3).

A study with 18 nondiabetic patients with peritoneal dialysis, with a reduction of AGEs intake for 4 weeks by exposing meat to different cooking methods found a 34% reduction in serum CML and 35% reduction in methylglyoxal when compared to baseline. In contrast, the subjects in the same study with high AGE diet had elevation of serum CML and methylglyoxal by 29% and 26%, respectively [92]. In a similar study, a decrease in VCAM-1 and TNF-α were also found in the low AGE diet compared with the high AGE diet [91].

Several studies in subjects with diabetes have found similar results. For instance, a study of 13 patients with type 2 diabetes mellitus found that decreasing the intake of dAGEs (meals were provided to participants) for 6 weeks contributed to decreased levels of circulating AGEs and inflammatory markers (VCAM-1 and TNF-α) [93]. In another study, 24 patients with diabetes mellitus were randomized to one of two groups with different diets for 6 weeks (meals were provided), one with a high level of dAGEs and the other with low

TABLE 20.3 Representative Clinical Interventions with dAGEs

	Study	Low AGEs diet [2700–5500 KU AGEs] *[2.2 ± 0.9 mg CML]	High AGEs diet [12,200–16,300 KU AGEs] *[5.4 ± 2.3 mg CML]
[93]	11 DM subjects	CML ↓ 40%	CML ↑ 28%
		CRP ↓ 20%	CRP ↑ 35%
		VCAM-1 ↓ 20%	TNF-α ↑ 80%
[92]	26 RD subjects	CML ↓ 34%	CML ↑ 29%
		MG ↓ 35%	MG ↑ 26%
		CML-LDL ↓ 28%	CML-LDL ↑ 50%
[90]	24 DM subjects	LDL was 50% less	↑ Phosphorylation MAPK
		Glycated and 80% less oxidized	↑ Activity NF-κB
			↑ VCAM-1
[94]	20 DM subjects	CML and MG ↓	TBARS ↑ 21%
			I-CAM and VCAM ↑
			E-selectin ↑ 51%
[65]*	62 healthy subjects	Triglycerides ↓ 9%	Triglycerides, cholesterol, CML plasma, urine ↑
		HOMA ↓ 17%	
[27]	36 DM and healthy subjects	CML, MG ↓	CML ↑
		8-isoprostane ↓	HOMA ↑
		HOMA ↓	

RD, renal disease; DM, diabetes mellitus; AGEs, advanced glycation end products; CML, carboxymethyl-lysine; MG, methylglyoxal; TNF-α, tumor necrosis factor α; CRP, C-reactive protein; VCAM-1, vascular cell adhesion molecule 1; ICAM-1, intracellular adhesion molecule; MDA, malondialdehyde; TBARS, thiobarbituric acid-reactive substance; LDL, low-density lipoprotein; HOMA, homeostatic model assessment; MAPKs, mitogen-activated protein kinases; NF-κB, nuclear factor kappa B.

AGEs. It was found that the LDL in the group with high dAGEs intake was more glycated than in the group with low AGEs intake [90]. In addition, the acute effect of a meal rich in dAGEs has also been measured in type 1 and type 2 diabetes mellitus. It was found that flow-mediated dilation (FMD) of the brachial artery (used as a measure of endothelial function) decreased after a single challenge with dAGEs and inflammatory markers as VCAM-1 increased [94]. Long-term effects of dAGEs on FMD have not been studied, but it is possible that a prolonged exposure could provoke permanent damage of vascular tissue. Indeed, a clinical study carried out over 6 weeks found that LDL in the group with high AGEs intake was more glycated than the low AGEs intake [90]. As mentioned before, it hypothesized that glycated LDL could promote atherosclerosis by decreasing uptake by its receptors and increasing uptake by macrophages [56]. The longest intervention on the effects of dAGEs (4 months) included healthy and subjects with diabetes, and it showed that patients ($n = 18$) assigned to the low AGEs diet had lower levels of serum CML, methylglyoxal, 8-isoprostanes (a lipoxidation marker) and insulin in serum, as well as a lower insulin resistance measured by the homeostatic model assessment (HOMA) when compared with the subjects with the regular AGEs intake (around 20 equivalent of AGEs, measured by 3 days of dietary records) [27]. Lastly, a pilot study showed that subjects with higher intake of AGEs have a higher risk level for cardiovascular disease. dAGEs better explained a high level of risk for cardiovascular disease than other variables from the diet (saturated fat) or any other variable studied [46–48].

Some authors have proposed a relationship between AGEs and cancer, and recently an association was found between dAGEs and pancreatic cancer. A prospective study carried out by the National Health Institute (NIH) and the American Association for Retired Persons (AARP) the NIH-AARP Diet and Health Study explored the association between CML-AGE consumption and pancreatic cancer. The results showed that men in the fifth quartile of CML-AGE consumption had increased pancreatic cancer risk (Hazards Ratio [HR]: 1.43; 95% Confidence Interval [CI]: 1.06, 1.93, $p = 0.003$) but not women (HR: 1.14; 95% CI: 0.76, 1.72, $p = 0.42$). Additionally, men in the highest quartile of red meat consumption had higher risk of pancreatic cancer (HR: 1.35; 95% CI: 1.07, 1.70), which attenuated after adjustment for CML-AGE consumption (HR: 1.20; 95% CI: 0.95, 1.53) [95].

In addition, clinical trials in overweight and obese subjects have been of special interest since they are at higher risk for chronic diseases. Harcourt et al. found that renal function (measured as urinary albumin/creatinine ratio) and the inflammatory profile improved following a low-AGE diet in a group of 11 subjects with overweight and obesity in a crossover design [96]. Whereas Mark et al. found that after 4 weeks with a low AGEs diet the urinary AGEs and the HOMA-IR decreased in a group of 37 overweight women in comparison with their counterpart ($n = 37$) in a high AGE diet [97]. Other clinical trials exploring the effect of a low AGE diet versus a supervised moderate aerobic exercise program (three times per week, 45 min sessions) found that exercise alone did not modify CML-AGEs, however the low AGE diet reduced serum AGE and indices of body fat, and the combined treatment had similar results but also improved lipid profile [98].

However, studies designed to evaluate the impact of dAGEs on healthy individuals have had contradictory results. A recent study in 24 healthy adults did not find changes in endothelial function or inflammatory markers after reducing dAGEs for 6 weeks. In the low-AGE diet group, there was a change in serum CML and urine CML, but there were no significant changes in the group with higher AGEs intake [86,87]. In contrast, Birlouez-Aragon studied 62 healthy adults and found that the group with a low AGE diet had a decrease in total cholesterol, HDL-cholesterol, and triglycerides whereas the group with the high AGE diet had increased CML, fasting insulinemia, and insulin resistance measured by HOMA [65].

Even when some studies have not found changes on inflammatory or oxidation markers, one outcome is similar: there is an association between the consumption of AGEs and their circulatory level. This result supports the earlier concept of Vlassara that dAGEs contribute to the AGEs body pool [93]. Therefore, the search for inhibitors of glycation is of special interest, and the supplementation with thiamine and currently benfotiamine (a derivative of thiamine) suggest a possible inhibition of the formation of AGEs. One of the proposed mechanisms is that benfotiamine increases the endogenous enzyme transketolase, which degrades AGEs. Benfotiamine is a lipophilic compound (allowing it to cross cell membranes easily) and, with a superior absorption, bioavailability, and effectiveness than thiamine [99]. A few clinical trials have been done with benfotiamine. A study by Stracke et al. [100] showed that patients with diabetic neuropathy in the treatment group had fewer symptoms when compared with the placebo group. However, more clinical trials are needed to prove their effectiveness and safety.

SUMMARY

- There are two main sources of AGEs: Physiologically formed AGEs defined as the nonenzymatic reaction of glucose, α-oxoaldehydes, and other saccharide derivates, with proteins, nucleotides, and lipids, in the human body; and exogenous AGEs formed mainly in food.
- There are several methods to determine AGEs in tissues, biological fluids, and food; immunochemical and instrumental methods all have their advantages and disadvantages.
- dAGEs may be important contributors to the accumulation of AGEs and their effect. Therefore, understanding the absorption, metabolism, and excretion is important.
- AGEs exert deleterious effects and have been implicated in aging and several chronic diseases.
- Interventions have shown an association between dAGEs and circulating AGEs, but modifying dAGEs for healthy adults have had ambiguous results.

References

[1] Ames JM. Evidence against dietary advanced glycation endproducts being a risk to human health. Mol Nutr Food Res 2007;51:1085−90.

[2] Ahmed N, Argirov OK, Minhas HS, Cordeiro CA, Thornalley PJ. Assay of advanced glycation endproducts (AGEs): surveying AGEs by chromatographic assay with derivatization by 6-aminoquinolyl-N-hydroxysuccinimidyl-carbamate and application to Nepsilon-carboxymethyl-lysine- and Nepsilon-(1-carboxyethyl)lysine-modified albumin. Biochem J 2002;364:1−14.

[3] Takeuchi M, Yamagishi S. TAGE (toxic AGEs) hypothesis in various chronic diseases. Med Hypotheses 2004;63:449−52.

[4] Thornalley PJ, Battah S, Ahmed N, Karachalias N, Agalou S, Babaei-Jadidi R, et al. Quantitative screening of advanced glycation endproducts in cellular and extracellular proteins by tandem mass spectrometry. Biochem J 2003;375:581−92.

[5] Garay-Sevilla ME, Regalado JC, Malacara JM, Nava LE, Wróbel-Zasada K, Castro-Rivas A, et al. Advanced glycosylation end products in skin, serum, saliva and urine and its association with complications of patients with type 2 diabetes mellitus. J Endocrinol Invest 2005;28:223−30.

[6] Hanssen NM, Stehouwer CD, Schalkwijk CG. Methylglyoxal and glyoxalase I in atherosclerosis. Biochem Soc Trans 2014;42:443−9.

[7] Salahuddin P, Rabbani G, Khan RH. The role of advanced glycation end products in various types of neurodegenerative disease: a therapeutic approach. Cell Mol Biol Lett 2014;19(3):407–37.

[8] Hiraiwa H, Sakai T, Mitsuyama H, Hamada T, Yamamoto R, Omachi T, et al. Inflammatory effect of advanced glycation end products on human meniscal cells from osteoarthritic knees. Inflamm Res 2011;60:1039–48.

[9] Thornalley PJ. Advanced glycation end products in renal failure. J Ren Nutr 2006;16:178–84.

[10] Kanková K. Diabetic threesome (hyperglycaemia, renal function and nutrition) and advanced glycation end products: evidence for the multiple-hit agent? Proc Nutr Soc 2008;67:60–74.

[11] Pickup JC. Diabetic control and its measurement. In: Pickup JC, Williams G, editors. Textbook of diabetes. 3rd ed. Oxford: Blackwell Science; 2003. col 1. 34.31-34. 17.

[12] Cho SJ, Roman G, Yeboah F, Konishi Y. The road to advanced glycation end products: a mechanistic perspective. Curr Med Chem 2007;14:1653–71.

[13] Tessier FJ. The Maillard reaction in the human b1ody. The main discoveries and factors that affect glycation. Pathol Biol 2010;58:214–19.

[14] Zhang Q, Ames JM, Smith RD, Baynes JW, Metz TO. A perspective on the Maillard reaction and the analysis of protein glycation by mass spectrometry: probing the pathogenesis of chronic disease. J Proteome Res 2009;8:754–69.

[15] Robert L, Labat-Robert J, Robert AM. The Maillard reaction. From nutritional problems to preventive medicine. Pathol Biol 2010;58:200–6.

[16] Takeuchi M, Yanase Y, Matsuura N, Yamagishi S, Kameda Y, Bucala R, et al. Immunological detection of a novel advanced glycation end-product. Mol Med 2001;7:783–91.

[17] Beisswenger BG, Delucia EM, Lapoint N, Sanford RJ, Beisswenger PJ. Ketosis leads to increased methylglyoxal production on the Atkins diet. Ann N Y Acad Sci 2005;1043:201–10.

[18] Turk Z. Glycotoxines, carbonyl stress and relevance to diabetes and its complications. Physio Res 2010;59:147–56.

[19] Finot PA. Historical perspective of the Maillard reaction in food science. Ann N Y Acad Sci 2005;1043:1–8.

[20] Oliver CM, Melton LD, Stanley RA. Creating proteins with novel functionality via the Maillard reaction: a review. Crit Rev Food Sci Nutr 2006;46:337–50.

[21] Poulsen MW, Hedegaard RV, Andersen JM, de Courten B, Bügel S, Nielsen J, et al. Advanced glycation endproducts in food and their effects on health. Food Chem Toxicol 2013;60:10–37.

[22] Chuyen NV. Toxicity of the AGEs generated from the Maillard reaction: on the relationship of food-AGEs and biological-AGEs. Mol Nutr Food Res 2006;50:1140–9.

[23] Chao P-C, Hsu C-C, Yin M-C. Analysis of glycative products in sauces and sauce-treated foods. Food Chem 2009;113:262–6.

[24] Zamora R, Hidalgo FJ. Coordinate contribution of lipid oxidation and Maillard reaction to the nonenzymatic food browning. Crit Rev Food Sci Nutr 2005;45:49–59.

[25] Ames JM. Determination of N epsilon-(carboxymethyl)lysine in foods and related systems. Ann N Y Acad Sci 2008;1126:20–4.

[26] Uribarri J, Woodruff S, Goodman S, et al. Advanced glycation end products in foods and a practical guide to their reduction in the diet. J Am Diet Assoc 2010;110:911–16.

[27] Uribarri J, Cai W, Ramdas M, et al. Restriction of advanced glycation end products improves insulin resistance in human type 2 diabetes: potential role of AGER1 and SIRT1. Diab Care 2011;34(7):1610–16.

[28] Coughlan MT, Patel SK, Jerums G, et al. Advanced glycation urinary protein-bound biomarkers and severity of diabetic nephropathy in man. Am J Nephrol 2011;34(4):347–55.

[29] Kizer JR, Benkeser D, Arnold AM, et al. Advanced glycation/glycoxidation endproduct carboxymethyl-lysine and incidence of coronary heart disease and stroke in older adults. Atherosclerosis 2014;235:116–21.

[30] Cai W, Gao QD, Zhu L, Peppa M, He C, Vlassara H. Oxidative stress-inducing carbonyl compounds from common foods: novel mediators of cellular dysfunction. Mol Med 2002;8:337–46.

[31] Koito W, Araki T, Horiuchi S, Nagai R. Conventional antibody against Nepsilon-(carboxymethyl)lysine (CML) shows cross-reaction to Nepsilon-(carboxyethyl)lysine (CEL): immunochemical quantification of CML with a specific antibody. J Biochem 2004;136:831–7.

[32] Agalou S, Ahmed N, Babaei-Jadidi R, Dawnay A, Thornalley PJ. Profound mishandling of protein glycation degradation products in uremia and dialysis. J Am Soc Nephrol 2005;16:1471–85.

[33] Nagai R, Shirakawa J, Fujiwara Y, Ohno R, Moroishi N, Sakata N, et al. Detection of AGEs as markers for carbohydrate metabolism and protein denaturation. Clin Biochem Nutr 2014;55:1–6.

[34] Coughlan MT, Forbes JM. Temporal increases in urinary carboxymethyllysine correlate with albuminuria development in diabetes. Am J Nephrol 2011;34:9–17.

[35] Sebeková K, Podracká L, Blazícek P, Syrová D, Heidland A, Schinzel R. Plasma levels of advanced glycation end products in children with renal disease. Pediatr Nephrol 2001;16:1105–12.

[36] Thomas MC, Tsalamandris C, MacIsaac R, Medley T, Kingwell B, Cooper ME, et al. Low-molecular-weight AGEs are associated with GFR and anemia in patients with type 2 diabetes. Kidney Int 2004;66:1167–72.

[37] Wróbel K, Wróbel K, Garay-Sevilla ME, Nava LE, Malacara JM. Novel analytical approach to monitoring advanced glycosylation end products in human serum with on-line spectrophotometric and spectrofluorometric detection in a flow system. Clin Chem 1997;43:1563–9.

[38] Méndez JD, Xie J, Aguilar-Hernández M, Méndez-Valenzuela V. Trends in advanced glycation end products research in diabetes mellitus and its complications. Mol Cell Biochem 2010;341:33–41.

[39] Vos PA, Welsing PM, deGroot J, et al. Skin pentosidine in very early hip/knee osteoarthritis (CHECK) is not a strong independent predictor of radiographic progression over 5 years follow-up. Osteoarthr Cartil 2013;21:823–30.

[40] Peiretti PG, Medana C, Visentin S, Giancotti V, Zunino V, Meineri G. Determination of carnosine, anserine, homocarnosine, pentosidine and thiobarbituric acid reactive substances contents in meat from different animal species. Food Chem 2011;126:1939–47.

[41] Tareke E, Forslund A, Lindh CH, Fahlgren C, Östman E. Isotope dilution ESI-LC-MS/MS for quantification of free and total Nε-(1-carboxymethyl)-L-lysine and free Nε-(1-carboxyethyl)-L-lysine: comparison of total Nε-(1-carboxymethyl)-L-lysine levels measured with new method to ELISA assay in gruel samples. Food Chem 2013;141:4253–9.

[42] Ruiz MC, Portero-Otín M, Pamplona R, et al. Chemical and immunological characterization of oxidative nonenzymatic protein modifications in dialysis fluids. Perit Dial Int 2003;23:23–32.

[43] Hegele J, Buetler T, Delatour T. Comparative LC-MS/MS profiling of free and protein-bound early and advanced glycation-induced lysine modifications in dairy products. Anal Chim Acta 2008;617:85–96.

REFERENCES

[44] Kislinger T, Humeny A, Peich CC, Becker CM, Pischetsrieder M. Analysis of protein glycation products by MALDI-TOF/MS. Ann N Y Acad Sci 2005;1043:249–59.

[45] Smit AJ, Gerrits EG. Skin autofluorescence as a measure of advanced glycation endproduct deposition: a novel risk marker in chronic kidney disease. Curr Opin Nephrol Hypertens 2010;19:527–33.

[46] Luevano-Contreras C, Durkin T, Pauls M, Chapman-Novakofski K. Development, relative validity, and reliability of a food frequency questionnaire for a case-control study on dietary advanced glycation end products and diabetes complications. Intern J Food Sci Nutr 2013;64(8):1030–5.

[47] Luevano-Contreras C, Garay-Sevilla ME, Chapman-Novakofski K. Role of dietary advanced glycation end products in diabetes mellitus. J Evid Compl Altern Med 2013;18(1):50–66.

[48] Luevano-Contreras C, Garay-Sevilla ME, Preciado-Puga M, Chapman-Novakofski KM. The relationship between dietary advanced glycation end products and indicators of diabetes severity in Mexicans and non-Hispanic whites: a pilot study. Inter J Food Sci Nutr 2013;64(1):16–20.

[49] Henle T. AGEs in foods. Do they play a role in uremia? Kid Inter 2003;63:S145–7.

[50] Erbersdobler HF, Faist V. Metabolic transit of Amadori products. Die Nahrung 2001;45:177–81.

[51] Faist V, Erbersdobler HF. Metabolic transit and in vivo effects of melanoidins and precursor compounds deriving from the Maillard reaction. Ann Nutr Metab 2001;45:1–12.

[52] Wiame E, Delpierre G, Collard F, Van Schaftingen E. Identification of a pathway for the utilization of the Amadori product fructoselysine in Escherichia coli. J Biol Chem 2002;277:42523–9.

[53] Forster A, Kuhne Y, Henle T. Studies on absorption and elimination of dietary Maillard reaction products. Ann N Y Acad Sci 2005;1043:474–81.

[54] Degen J, Beyer H, Heymann B, Hellwig M, Henle T. Dietary influence on urinary excretion of 3-deoxyglucosone and its metabolite 3-deoxyfructose. J Agric Food Chem 2014;62(11):2449–56.

[55] Koschinsky T, He CJ, Mitsuhashi T, et al. Orally absorbed reactive glycation products (glycotoxins): an environmental risk factor in diabetic nephropathy. Proc Nat Acad Sci 1997;94:6474–9.

[56] Stirban A, Gawlowski T, Roden M. Vascular effects of advanced glycation endproducts: clinical effects and molecular mechanisms. Mol Metab 2013;3(2):94–108.

[57] Yamagishi S. Role of advanced glycation end products (AGEs) and receptor for AGEs (RAGE) in vascular damage in diabetes. Exp Geron 2011;46(4):217–24.

[58] Chuah YK, Basir R, Talib H, Tie TH, Nordin N. Receptor for advanced glycation end products and its involvement in inflammatory diseases. Inter J Inflam 2013;2013:403460.

[59] Gao X, Zhang H, Schmidt AM, Zhang C. AGE/RAGE produces endothelial dysfunction in coronary arterioles in type 2 diabetic mice. Am Physio: Heart Circ Physiol 2008;295(2):H491–8.

[60] Fleming TH, Humpert PM, Nawroth PP, Bierhaus A. Reactive metabolites and AGE/RAGE-mediated cellular dysfunction affect the aging process: a mini-review. Gerontol 2011;57(5):435–43.

[61] Bierhaus A, Humpert PM, Morcos M, et al. Understanding RAGE, the receptor for advanced glycation end products. J Mol Med (Berl) 2005;83(11):876–86.

[62] Penfold SA, Coughlan MT, Patel SK, et al. Circulating high-molecular-weight RAGE ligands activate pathways implicated in the development of diabetic nephropathy. Kidney Inter 2010;78(3):287–95.

[63] Ramasamy R, Yan SF, Herold K, Clynes R, Schmidt AM. Receptor for advanced glycation end products: fundamental roles in the inflammatory response: winding the way to the pathogenesis of endothelial dysfunction and atherosclerosis. Ann N Y Acad Sci 2008;1126:7–13.

[64] Yan SF, Ramasamy R, Schmidt AM. The receptor for advanced glycation endproducts (RAGE) and cardiovascular disease. Exp Rev Mole Med 2009;11:e9.

[65] Birlouez-Aragon I, Saavedra G, Tessier FJ, et al. A diet based on high-heat-treated foods promotes risk factors for diabetes mellitus and cardiovascular diseases. Am J Clin Nutr 2010;91(5):1220–6.

[66] Holik AK, Rohm B, Somoza MM, Somoza V. N(epsilon)-Carboxymethyllysine (CML), a Maillard reaction product, stimulates serotonin release and activates the receptor for advanced glycation end products (RAGE) in SH-SY5Y cells. Food Function 2013;4(7):1111–20.

[67] Somoza V, Lindenmeier M, Hofmann T, et al. Dietary bread crust advanced glycation end products bind to the receptor for AGEs in HEK-293 kidney cells but are rapidly excreted after oral administration to healthy and subtotally nephrectomized rats. Ann N Y Acad Sci 2005;1043:492–500.

[68] Vlassara H, Uribarri J. Advanced glycation end products (AGE) and diabetes: cause, effect, or both? Curr Diab Rep 2014;14(1):453.

[69] Sanguineti R, Puddu A, Mach F, Montecucco F, Viviani GL. Advanced glycation end products play adverse proinflammatory activities in osteoporosis. Med Inflam 2014;2014:975872.

[70] Vlassara H, Striker GE. AGE restriction in diabetes mellitus: a paradigm shift. Nature Rev Endo 2011;7(9):526–39.

[71] Grillo MA, Colombatto S. Advanced glycation end-products (AGEs): involvement in aging and in neurodegenerative diseases. Amino Acids 2007;35:29–36.

[72] Van Puyvelde K, Mets T, Njemini R, Beyer I, Bautmans I. Effect of advanced glycation end product intake on inflammation and aging: a systematic review. Nutr Rev 2014;72:638–50.

[73] Baraibar MA, Liu L, Ahmed EK, Friguet B. Protein oxidative damage at the crossroads of cellular senescence, aging, and age-related diseases. Oxid Med Cell Longev 2012;919832.

[74] Contreras CL, Chapman-Novakofski K. Dietary advanced glycation end products and aging. Nutrients 2010;2(12):1247–65.

[75] Nowotny K, Jung T, Grune T, Höhn A. Reprint of "accumulation of modified proteins and aggregate formation in aging". Exp Gerontol 2014;59:3–12.

[76] Kolluru GK, Bir SC, Kevil CG. Endothelial dysfunction and diabetes: effects on angiogenesis, vascular remodeling, and wound healing. Int J Vasc Med 2012;2012:918267.

[77] Yan D, Luo X, Li Y, et al. Effects of advanced glycation end products on calcium handling in cardiomyocytes. Cardiology 2014;129(2):75–83.

[78] West RK, Moshier E, Lubitz I, et al. Dietary advanced glycation end products are associated with decline in memory in young elderly. Mech Ageing Dev 2014;140:10–12.

[79] Gugliucci A, Menini T. The axis AGE-RAGE-soluble RAGE and oxidative stress in chronic kidney disease. Adv Exp Med Biol 2014;824:191–208.

[80] de Groot L, Hinkema H, Westra J, et al. Advanced glycation endproducts are increased in rheumatoid arthritis patients with controlled disease. Arthritis Res Ther 2011;13(6):R205.

[81] Sun K, Semba RD, Fried LP, Schaumberg DA, Ferrucci L, Varadhan R. Elevated serum carboxymethyl-lysine, an advanced glycation end product, predicts severe walking disability in older women: The Women's Health and Aging Study I. J Aging Res 2012;2012:586385.

[82] de la Maza MP, Uribarri J, Olivares D, Hirsch S, Leiva L, Barrera G, et al. Weight increase is associated with skeletal muscle immunostaining for advanced glycation end products, receptor for advanced glycation end products, and oxidation injury. Rejuvenation Res 2008;11(6):1041–8.

[83] Kandarakis SA, Piperi C, Topouzis F, Papavassiliou AG. Emerging role of advanced glycation-end products (AGEs) in the pathobiology of eye diseases. Prog Retin Eye Res 2014;42:85–102.

[84] Tezel G, Luo C, Yang X. Accelerated aging in glaucoma: immunohistochemical assessment of advanced glycation end products in the human retina and optic nerve head. Invest Ophthalmol Vis Sci 2007;48(3):1201–11.

[85] Zarbin MA, Casaroli-Marano RP, Rosenfeld PJ. Age-related macular degeneration: clinical findings, histopathology and imaging techniques. Dev Ophthalmol 2014;53:1–32.

[86] Semba RD, Cotch MF, Gudnason V, et al. Serum carboxymethyllysine, an advanced glycation end product, and age-related macular degeneration: the Age, Gene/Environment Susceptibility-Reykjavik Study. JAMA Ophthalmol 2014;132(4):464–70.

[87] Semba RD, Gebauer SK, Baer DJ, et al. Dietary intake of advanced glycation end products did not affect endothelial function and inflammation in healthy adults in a randomized controlled trial. J Nutr 2014;144(7):1037–42.

[88] Koga M. Glycated albumin; clinical usefulness. Clin Chim Acta 2014;433:96–104.

[89] Vlassara H, Striker GE. Advanced glycation endproducts in diabetes and diabetic complications. Endocrinol Metab Clin North Am 2013;42(4):697–719.

[90] Cai W, He JC, Zhu L, et al. High levels of dietary advanced glycation end products transform low-density lipoprotein into a potent redox-sensitive mitogen-activated protein kinase stimulant in diabetic patients. Circ 2004;110(3):285–91.

[91] Peppa M, Uribarri J, Cai W, Lu M, Vlassara H. Glycoxidation and inflammation in renal failure patients. Am J Kid Dis 2004;43(4):690–5.

[92] Uribarri J. Restriction of dietary glycotoxins reduces excessive advanced glycation end products in renal failure patients. J Am Soc Nephr 2003;14(3):728–31.

[93] Vlassara H, Cai W, Crandall J, et al. Inflammatory mediators are induced by dietary glycotoxins, a major risk factor for diabetic angiopathy. Proc Nat Acad Sci 2002;99(24):15596–601.

[94] Negrean M, Stirban A, Stratmann B, et al. Effects of low- and high-advanced glycation endproduct meals on macro- and microvascular endothelial function and oxidative stress in patients with type 2 diabetes mellitus. Am J Clin Nutr 2007;85(5):1236–43.

[95] Jiao L, Stolzenberg-Solomon R, Zimmerman TP, et al. Dietary consumption of advanced glycation end products and pancreatic cancer in the prospective NIH-AARP Diet and Health Study. Am J Clin Nutr 2015;101(1):126–34.

[96] Harcourt BE, Sourris KC, Coughlan MT, et al. Targeted reduction of advanced glycation improves renal function in obesity. Kidney Intern 2011;80(2):190–8.

[97] Mark AB, Poulsen MW, Andersen S, et al. Consumption of a diet low in advanced glycation end products for 4 weeks improves insulin sensitivity in overweight women. Diab Care 2014;37(1):88–95.

[98] Macias-Cervantes MH, Rodriguez-Soto JM, Uribarri J, Diaz-Cisneros FJ, Cai W, Garay-Sevilla ME. Effect of an advanced glycation end product-restricted diet and exercise on metabolic parameters in adult overweight men. Nutrition 2015;31(3):446–51.

[99] Kousar S, Sheikh MA, Asghar M. Antiglycation activity of thiamin-HCl and benfotiamine in diabetic condition. J Pak Med Assoc 2012;62(10):1033–8.

[100] Stracke H, Gaus W, Achenbach U, Federlin K, Bretzel RG. Benfotiamine in diabetic polyneuropathy (BENDIP): results of a randomised, double blind, placebo-controlled clinical study. Exper Clin Endo Diab 2008;116(10):600–5.

[101] Vlassara H, Palace MR. Glycoxidation: the menace of diabetes and aging. Mount Sinai J Med 2003;70(4):232–41.

[102] Hull GL, Woodside JV, Ames JM, Cuskelly GJ. Hull validation study to compare effects of processing protocols on measured N (ε)-(carboxymethyl)lysine and N (ε)-(carboxyethyl)lysine in blood. J Clin Biochem Nutr. 2013;53(3):129–33.

[103] Zhang G, Huang G, Xiao L, Mitchell AE. Determination of advanced glycation endproducts by LC-MS/MS in raw and roasted almonds (Prunus dulcis). J Agric Food Chem 2011;59(22):12037–46. Available from: http://dx.doi.org/10.1021/jf202515k Epub 2011 Oct 20.

[104] Bosch L, Alegria A, Farre R, Clemente G. Fluorescence and color as markers for the Maillard reaction in milk-cereal based infant foods during storage. Food Chem 2007;105:1135–43.

[105] Nakano M, Kubota M, Owada S, Nagai R. The pentosidine concentration in human blood specimens is affected by heating. Amino Acids 2013;44(6):1451–6. Available from: http://dx.doi.org/10.1007/s00726-011-1180-z Epub 2011 Dec 4.

[106] Slowik-Zylka D, Safranow K, Dziedziejko V, Ciechanowski K, Chlubek D. Association of plasma pentosidine concentrations with renal function in kidney graft recipients. Clin Transplant 2010;24(6):839–47.

[107] Yasuda M, Shimura M, Kunikata H, et al. Relationship of skin autofluorescence to severity of retinopathy in type 2 diabetes. Curr Eye Res 2014;28:1–8.

[108] Aso Y, Takanashi K, Sekine K, Yoshida N, Takebayashi K, Yoshihara K, et al. Dissociation between urinary pyrraline and pentosidine concentrations in diabetic patients with advanced nephropathy. Lab Clin Med 2004;144(2):92–9.

[109] Sroga GE, Karim L, Colón W, Vashishth D. Biochemical characterization of major bone-matrix proteins using nanoscale-size bone samples and proteomics methodology. Mol Cell Proteomics 2011;10(9): M110.006718. http://dx.doi.org/10.1074/mcp.M110.006718.

[110] Pötzsch S, Blankenhorn A, Navarrete Santos A, Silber RE, Somoza V, Simm A. The effect of an AGE-rich dietary extract on the activation of NF-κB depends on the cell model used. Food Funct 2013;4(7):1023–31 http://dx.doi.org/10.1039/c3fo30349g. Epub 2013 Feb 20.

[111] Yoshihara K, Kiyonami R, Shimizu Y, Beppu M. Determination of urinary pyrraline by solid-phase extraction and high performance liquid chromatography. Biol Pharm Bull 2001;24(8):863–6.

[112] Foerster A, Henle T. Glycation in food and metabolic transit of dietary AGEs (advanced glycation end-products): studies on the urinary excretion of pyrraline. Biochem Soc Trans 2003;31(Pt 6):1383–5.

CHAPTER

21

miRNAs as Nutritional Targets in Aging

Robin A. McGregor and Dae Y. Seo

Cardiovascular and Metabolic Disease Center, College of Medicine, Inje University, Busan, Republic of Korea

KEY FACTS
- MicroRNAs (miRNAs) are a class of small noncoding RNAs which are powerful posttranscriptional regulators of gene expression.
- MicroRNAs are altered during ageing in different tissues and also during senescence in different cells.
- MicroRNAs can be modulated by dietary intake and nutraceuticals in animal disease models.
- More studies are necessary in humans to determine whether dietary modulation of miRNAs can be used for the prevention or treatment of age-related diseases.

Dictionary of Terms

- **3'UTR** – The three prime untranslated region is the part of messenger RNA that procedes the translation termination codon. The 3'UTR of mRNA is transcribed from DNA, but is not translated into protein.
- **Exosomes** – are large vesicles between 30–100 nm which are released by cells and found in many biological fluids such as blood, urine, salvia. Exosomes can contain DNA, mRNA, miRNA and proteins.
- **microRNAs (miRNAs)** – are a novel class of small non-coding RNAs which are part of the non-coding genome, which do not code for protein. miRNAs post-transcriptionally regulate protein-coding genes.
- **myomiR** – refers to a miRNA which is highly expressed in skeletal muscle tissue.

- **RISC** – the RISC is a ribosomal induced silencing complex which contains multiple proteins and can bind to different types of double-stranded RNA including small interfering RNA and miRNA.
- **Seed sequence** – the miRNA seed sequence or seed region refers to positions 2–7 from the 5' prime end of the miRNA, which binds the complementary nucleotides in the 3'UTR target mRNAs.
- **Target gene** – refers to a protein-coding gene which harbors a complementary miRNA target site in the 3'UTR of transcribed mRNA.
- **qPCR** – refers to quantitative polymerase chain reaction and is a technique to measure the level of mRNA or miRNA in different cells, tissues or biological fluids. qPCR requires reverse transcription of RNA into cDNA, before amplification and quantification of the template cDNA.

INTRODUCTION

MicroRNAs (miRNAs) are a class of small noncoding RNAs, 19–22 nucleotides in length, which were first discovered over a decade ago [1–3]. Many miRNAs are highly conserved across species. The human genome contains over 1500 miRNA genes, which are encoded in intronic and intergenic regions, or within protein-coding gene regions [4]. miRNAs are expressed in many different cell types and also in a tissue-specific pattern [5]. miRNAs are widely established as powerful posttranscriptional regulators of gene expression, which can fine-tune cellular protein levels within cells and across tissue types [6]. miRNAs can bind to complementary sites with the 3' untranslated region (3'UTR) of target mRNAs, causing translational repression, mRNA deadenylation, or mRNA decay [7], which in turn alters corresponding protein levels. Human protein-coding genes harbor over 45,000 putative miRNA target sites [8]. Each miRNA can regulate several hundred mRNAs [9]. Furthermore, multiple miRNAs can bind to a single mRNA, thus can act combinatorially to posttranscriptionally regulate gene expression [7]. Consequently, miRNAs play a role in almost all biological processes targeting genes such as those involved in cellular development and growth [10], as well as those involved in cellular senescence. Many miRNAs have been identified that may play a role in aging and age-related diseases [11–15]. Antisense targeting methods using cholesterol conjugated chemically modified antisense oligonucleotides or mimics have been successfully used to alter specific miRNAs in liver [16], but this approach is expensive and limited to targeting specific miRNAs. Recently, there has been a growing interest in the potential of a dietary approach to modulate aging and disease-associated miRNAs [11,17–20], because of the association between dietary intake and age-related conditions such as loss of muscle mass, insulin resistance, cardiovascular disease, and type 2 diabetes [21–23]. Furthermore, micronutrients and other dietary compounds including flavonoids and polyphenols are known to modulate transcription and also act via epigenetic mechanisms [18,20,24,25], and therefore are potential candidates for modulating miRNA expression.

miRNA BIOGENESIS AND PROCESSING

miRNAs are initially transcribed by RNA polymerase II into primary miRNA transcripts (pri-miRNAs) that are several hundred nucleotides in length [26]. Pri-miRNAs are cleaved in the nucleus by an RNAase III enzyme called Drosha [27–29], which leaves a ~70–80 nucleotide hairpin precursor miRNA (pre-miRNA), which is exported to the cytoplasm through Exportin-5, a nuclear transport protein [30]. Pre-miRNAs are subsequently cleaved in the cytoplasm by Dicer, leaving an unstable RNA duplex [31]. The RNA duplex rapidly unwinds, releasing the 19–22 nucleotide mature miRNA [31], which is incorporated into a ribosomal induced silencing complex (RISC) and can bind to 6–8 complementary sites in the 3′UTR of protein coding mRNAs [32]. The magnitude of posttranscriptional suppression is related to the degree of complementarity between the miRNA seed sequence and the target mRNA among other factors [8,33]. Importantly, miRNAs have been demonstrated to have a widespread impact on cellular protein levels in vivo [6,34]. The miRNA biogenesis and processing pathway is also regulated at multiple levels by RNA-binding proteins [35]. A recent study revealed that methyltransferase-like 3 (METTL3) binds and methylate's pri-miRNAs, which leaves a mark that is recognized by DGCR8, which subsequently binds to the pri-miRNA for processing by Drosha. Around 14% of all pri-miRNAs have highly conserved stem loops, which have been proposed to act as landing pads for RNA-binding proteins and block miRNA processing [36]. hnRNP A1 has been shown to bind specifically to pri-miR-18a before Drosha processing, and absence of hnRNP A1 blocks miRNA processing [37]. These studies suggest that miRNA biogenesis and processing can be regulated at multiple points, and RNA-binding proteins may regulate specific subgroups of miRNAs, rather than global tissue miRNA expression levels. Currently, we know very little about whether dietary factors can modulate miRNA biogenesis and processing.

CIRCULATING miRNAs

miRNAs are also found in whole blood, plasma, and serum [38]. miRNAs in blood can be either cell-free or contained within platelets, erythrocytes, and nucleated cells. miRNAs found in plasma are remarkably stable even when subject to harsh conditions such as boiling, acidity, alkalinity, and freeze–thaw cycles [39]. Many circulating miRNAs can be found packaged in exosomes, microvesicles, and apoptotic bodies or attached to RNA-binding protein complexes or lipoprotein complexes [40]. Circulating miRNAs are reported to be modulated by aging and by age-related chronic conditions [12]. Therefore, circulating miRNAs are attractive biomarkers of aging and age-related conditions [11]. The origin and function of circulating miRNAs is still not fully understood [41]. Changes in circulating miRNA profiles may be due to active secretion from cells or passive release following acute tissue damage [42]. Exercise-induced muscle damage appears to lead to a shift in circulating miRNAs [43,44]. Senescent cells also represent a possible source of circulating miRNAs with direct relevance to aging [14]. Exosomes and microvesicles are thought to play a role in intercellular communication and provide a vehicle for trafficking proteins, lipids, and RNAs between cells [40]. However, it is difficult to study intracellular trafficking of miRNAs in vivo. Nevertheless, in vitro studies have demonstrated that mouse exosomal RNA can be transferred to human recipient cells and results in expression of mouse proteins not normally present in the recipient cells [45], but it is not known how widespread intracellular trafficking of miRNAs is in vivo.

Changes in circulating miRNAs with aging are likely to occur slowly, due to slow changes in muscle, neural, endothelial, hepatic, and other tissues during aging. It will be important to establish what a healthy circulating miRNA profile looks like, so we can detect when a detrimental shift in circulating miRNAs occurs. A rapidly growing number of studies have measured circulating miRNAs as potential biomarkers in a wide range of clinical conditions [46–48]. Differences in methodologies used for blood sampling, processing, and analysis may contribute to some inconsistencies now emerging between studies [49]. Hematopoietic cells are an abundant source of circulating miRNAs [50] and care needs to be taken during blood sampling and processing to avoid contamination of plasma with hematopoietic-derived miRNAs [51]. Ideally, candidate circulating miRNA biomarkers need to be validated in larger independent cohorts or systematic meta-analysis may help identify candidate miRNA biomarkers that are consistently modulated across multiple cohort studies.

AGING ASSOCIATED miRNAs IN DIFFERENT TISSUES

Age-related differences in miRNAs in various tissues have been reported in a lower organisms, rodents, primates, and humans [11]. In humans, age-related changes occur much more slowly and lifespan is longer compared to lower organisms and rodents. One mouse year is equal to around 30 human years. Therefore, we are reliant on comparing miRNA expression in tissues between younger and older adults. Confounding factors may contribute to any age-associated miRNAs identified when using a cross-sectional design, including factors such as previous diet, physical activity levels, acute injuries, and even environmental exposure [11]. A study of small noncoding RNAs in mononuclear cells of centenarians living near Valencia in Spain revealed that centenarians shared an overlapping small noncoding RNA expression signature with young adults rather than octogenarians [52]. Furthermore, six small noncoding RNAs were more highly expressed in the centenarians compared to the octogenarians and young adults [52]. Independent validation by qPCR confirmed significant overexpression of miR-21, miR-130a, and SCARNA17 in centenarians, but not miR-494 [52]. It was speculated that these small noncoding RNAs play a role in extreme longevity, but conversely these small noncoding RNAs may also be a consequence rather a cause of extreme longevity. It is important to emphasize that small RNA expression in mononuclear cells may not be representative of expression in other tissues, which are also undergoing age-related changes. Mononuclear cells found in the blood, typically include lymphocytes, monocytes or macrophages, which are components of the immune system. Nevertheless, differences in small RNA expression in immune cells may confer protection against infectious diseases and foreign pathogens.

A screen of age-related and senescence-related changes in miRNA expression in several human cells types including endothelial cells, replicated CD8(+) T cells, renal proximal tubular epithelial cells, skin fibroblasts, foreskin, mesenchymal stem cells, and CD8(+) T cells from young and old donors revealed commonly regulated miRNAs [15]. miR-17 was universally downregulated in all cell types tested, miR-19b, miR-20a, and miR-106a were downregulated in 5–6 cell types tested [15]. Several established target transcripts were conversely upregulated including p21/CDKN1A, which is a cyclin-dependent kinase inhibitor and regulates cell cycle progression and cell senescence [15].

Age-related loss of muscle mass and function is termed sarcopenia and can be a precursor to decline in mobility and the onset of frailty [53]. Some research groups have argued that older individuals develop anabolic resistance [54] and therefore have reduced capacity for muscle plasticity in response to nutrients or exercise [55]. A study of changes in miRNA expression in skeletal muscle after an acute bout of resistance exercise found 21 miRNAs were modulated in young men, whereas none were modulated in older men [55]. Furthermore, exercise-induced changes in mRNA expression were blunted in older men [55]. In vitro experiments in myocytes revealed miR-126 as a potential regulator of skeletal muscle growth genes and insulin growth factor 1 (IGF-1) signaling [55]. Other studies have reported higher levels of pri-miRNA transcripts in older men compared to younger men, including pri-miRNA-1-1, -1-2, -133a-1, and -133a [56], which encode the muscle-specific mature miR-1 and miR-133a transcript. miR-1 expression was reported to remain elevated in older men after resistance exercise and an anabolic stimulus, while miR-1 was downregulated in younger men at 3 and 6 h postexercise [56]. A more recent study in young and older men screened changes in 754 miRNAs, 2 h after an acute resistance exercise bout [57]. The study identified 26 miRNAs that were differentially expressed with age and/or exercise, which notably included five miRNAs, which target mRNAs that code for proteins in the Akt-mTOR signaling pathway. The authors suggested the miR-99/100 family as an attractive target for regulating skeletal muscle mass [57]. This miRNA family can bind to the 3′UTR of mTOR, RPTOR, and IGF-1R, which encode major proteins in the Akt-mTOR pathway regulating skeletal muscle growth [57]. Bed rest studies provide a particularly useful model of accelerated aging in humans. Bed rest leads to marked changes in insulin sensitivity, muscle enzyme activity, muscle size, and muscle strength, which can be reversed by a period of resistance exercise [58]. A bed rest study in men over 10 days reported that 13 miRNAs were downregulated and 2 miRNAs were upregulated in skeletal muscle. Notably among the downregulated miRNAs were, miR-206, which is a muscle-specific myomiR involved in skeletal muscle development, and miR-23a, which is associated with the insulin response and atrophy defense [59]. In addition, several of the let-7 family were downregulated, which are involved in cell cycle, differentiation, and glucose homeostasis [59].

DIETARY MODIFICATION OF miRNAs

Evidence is growing that miRNAs are modulated by dietary intake of macronutrients, micronutrients, or natural food-derived compounds, such as flavonoids and polyphenols [18,19,24]. Therefore, specific dietary interventions or nutraceutical supplements may potentially be used to either directly or indirectly target miRNAs altered during aging and in age-related diseases. Most studies to date have focused on the potential of nutraceuticals such as curcumin and resveratrol to modulate miRNAs in a wide variety of cancer cell types. Few studies have assessed dietary or nutraceutical mediated miRNA modulation in in vivo age-related disease models. Caloric restriction is a model often used to study molecular events and interventions to promote longevity. In *Caenorhabditis elegans*, miR-80 expression is high when ab libitum food is available, and conversely low when food is restricted [60]. In addition, a recent study in nonhuman primates reported decreased miR-451, miR-144, miR-18a, and miR-15a, as well as increased miR-181a and miR-181b expression in skeletal muscle during aging [61]. Furthermore, a recent study used deep sequencing to characterize differences in circulating miRNA in young, old, and caloric restricted old mice [62]. Many circulating miRNAs were increased in the old mice, of which 50 were conversely reduced by caloric restriction, the targets of these miRNAs were enriched in genes involved in the regulation of macromolecular biosynthetic processes, regulation of apoptosis, and the Wnt signaling pathway [62]. While prolonged caloric restriction is not recommended in humans, it is plausible that the increased longevity observed in lower organisms may be mediated in part by nutrient sensitive miRNAs and modulation of their target genes. In the following, we will discuss evidence of miRNA modulation by protein, carbohydrate, fat, micronutrients, and other nutraceuticals based on studies to date. Much of the evidence of miRNA modulation by dietary factors is based on findings from cell lines. It will be important in future to establish whether findings from nutraceutical modulation of miRNAs in cell lines can be translated from bench to beside in humans.

miRNAs MODULATED BY PROTEIN

Essential amino acids (EAAs) are well established as potent stimulators of muscle protein synthesis and therefore are particularly useful in medical foods and supplements designed to maintain or increase skeletal muscle mass [63]. EAAs have been shown to modulate miRNA levels in skeletal muscle too [64]. A study in young men found that ingestion of 10 g of EAAs increased the expression of mature miR-499, miR-208b, miR-23a, and miR-1 transcripts, as well as pri-miR-206 after 3 h [64]. These EAA-induced changes were accompanied by alteration of muscle growth related genes [64]. No studies have investigated the time course of changes in miRNA expression levels after feeding EAAs, other amino acids, or different types of proteins. Past studies have shown that whey protein, casein protein, soy protein can all stimulate muscle protein synthesis in humans; leucine is a particularly potent anabolic agent [65]. Future studies are needed to ascertain how and which miRNAs are modulated by whole proteins from different food sources, as well as individual amino acids.

miRNAs MODULATED BY CARBOHYDRATES

Aging is associated with changes in the regulation of glucose homeostasis, related to reduced insulin sensitivity, which can lead to impaired glucose tolerance and increased risk of type 2 diabetes in older adults [66]. Age-related decreases in insulin sensitivity can influence the postprandial response to simple carbohydrates such as glucose, unless sufficient insulin can be produced by the pancreatic beta cells to compensate. However, age-related variability of glucose tolerance is also strongly influenced by fatness and fitness [67]. Impairment in glucose tolerance and the gradual onset of hyperglycemia can remain undetected for several years in older adults, which can lead to a wide range of complications including neuropathy, nephropathy, diabetes retinopathy, susceptibility to infections, osteoporosis, and cardiovascular disease [66,68].

Many cells are sensitive to high-glucose concentrations and also show concomitant modulation of miRNAs. For example, miR-1 expression is increased in rat cardiomyocytes exposed to high glucose, which also promotes apoptosis [69]. miR-1 targets the IGF1 signaling pathway, hence high-glucose induced expression of miR-1 resulted in a posttranscriptional suppression of IGF-1 and increased apoptosis [69]. In pancreatic beta cells, high-glucose levels are reported to lead to an upregulation of miR-29a and increased proliferation [70]. In adipocytes, high-glucose and insulin exposure increases miR-29 expression levels, which leads to an inhibition of insulin signaling [71]. This latter observation has been confirmed in more recent studies, which also showed that miR-22 and miR-27a expressions in adipocytes are sensitive to extracellular glucose.

Diets high in fructose can lead to nonalcoholic fatty liver disease, which is also more prevalent in older adults and can be driven by a variety of factors including high-fat or alcohol intake. A study in Sprague-Dawley rats fed a high-fructose diet or a high-fat and fructose diet compared to a standard diet or high-fat diet for 3 months revealed marked differences in liver miRNA expression dependent on dietary intake [17]. Three miRNAs were consistently downregulated in liver by high fructose intake, including miR-21, miR-122, miR-451, and miR-27, whereas three miRNAs were consistently upregulated in liver by high fructose intake, including miR-200a, miR-200b, and miR-429 [17]. The targets of these miRNAs are involved in the control of lipid and carbohydrate metabolism, signal transduction, cytokine signaling, and apoptosis [17]. The differences in miRNAs were accompanied by histological evidence of liver injury and metabolic dysfunction in the high-fructose fed rats. It was not established whether subsequently reducing fructose intake could reverse high-fructose-induced miRNA expression in liver. Future studies are needed to establish the effect of different types of carbohydrates on miRNA expression, as diets rich in low-glycemic index carbohydrates, which contain a higher proportion of complex sugars rather than simple sugars are recommended to maintain metabolic flexibility and prevent metabolic health-related disorders [72].

miRNAs MODULATED BY DIETARY FAT AND FATTY ACIDS

Aging is associated with increased adipose tissue, which is suggested to be due to reduced metabolically active fat-free tissue mass and hence a decrease in resting metabolic rate, rather than alterations in fat oxidation [73]. Nevertheless, prolonged high dietary fat intake without a corresponding increase in energy expenditure can lead to fat accumulation and related comorbidities, such as insulin resistance and nonalcohol fatty liver disease. Saturated fatty acids have been suggested to play a role in insulin resistance through inhibition of the insulin signaling pathway. Palmitate is a common saturated fatty acid found in Western diets. Palmitate treatment increases miR-29a in skeletal muscle myocytes in vitro, which is in concordance with increases in miR-29a expression induced by a high-fat diet [74]. Whether palmitate directly or indirectly induces miR-29a expression has not been established, nonetheless miR-29a can bind to the 3′UTR of IRS-1 mRNA, which inhibits the translation of IRS-1 into protein, leading to impaired insulin signaling and glucose uptake [74]. Furthermore, saturated fatty acid has also been observed to increase miR-195 expression in hepatocytes [75]. miR-195 can bind to the 3′UTR of INSR mRNA, which reduces insulin receptor protein levels, resulting in impaired insulin signaling and glycogen synthesis [75].

A high-fat diet has been shown to modulate multiple miRNAs in murine and rodent models. In adipose tissue, a prolonged high-fat diet was reported to increase expression of miR-22, miR-342-3p, miR-142-3p, miR-142-5p, miR-21, miR-146a, miR-146b, miR-379 and decrease expression of miR-200b, miR-200c, miR-204, miR-30a*, miR-193, miR-378 and miR-30e*, miR-122, miR-133b, miR-1, miR-30a*, miR-192, and miR-203. Whether changes in these miRNAs play a role in obesity development or are only consequences of excessive fat accumulation is difficult to ascertain from this study, in future a time-course approach may provide new insights in the molecular events underlying diet-induced obesity [76,77]. In skeletal muscle, microarray analysis revealed 30 miRNAs are differentially expressed in mice fed a high-fat diet for 12 weeks compared to normal diet fed mice [78]. Notably several muscle-specific miRNAs were downregulated including miR-1, miR133a, miR-133b, and miR-206, which target muscle development genes [78]. A previous study in individuals with varying insulin resistance and glucose tolerance also found miR-133a and miR-206 were downregulated [79]. In the hypothalamus, high-throughput screening revealed expression of let-7a, miR-9a, miR-30e, miR-132, miR-145, miR-200a, and miR-218 in hypothalamus is sensitive to high dietary fat intake, these groups of miRNAs were conversely regulated by dietary restriction [80]. Finally, in liver more than 50 miRNAs were found to be modulated in diet-induced obese mice. miR-107, which was downregulated, was found to inversely correlated with FASN which encodes fatty acid synthase [81]. A combined high-fat and high-cholesterol diet was reported to induce modulation of miRNAs in liver of baboons, which differed dependent on their lipoprotein cholesterol phenotype [82]. Next generation sequencing revealed 18 miRNAs were differentially expressed in response to a combined high-fat and high-cholesterol diet in baboons with high basal LDL concentration, whereas 10 miRNAs were differentially expressed in response to the same diet in baboons with a low basal LDL concentration [82]. Notably, the miR-29 family members were downregulated in baboons regardless of basal LDL levels [82].

Dysregulated miRNA levels are also seen in subcutaneous adipose tissue of obese humans, however whether these miRNAs represent the cause or consequences of obesity is yet to be established [83]. For example, a recent clinical study of obese individuals found miR-223 and miR-143 levels were significantly lower in their blood compared to normal or overweight individuals [84]. Whether normalization of blood miRNA levels either with a diet or pharmaceutical would reverse obesity is unknown. In addition, many putative targets of high-fat diet sensitive miRNAs remain to be experimentally validated.

Some poly-unsaturated fatty acids (PUFA) are reported to have potential health benefits and modulate miRNA levels. One study showed that omega-3 PUFA treatment can modulate miR-122 and miR-33a levels in diet-induced obese rats fed a cafeteria diet, which was high in both carbohydrate and fat content [85]. The diet-induced increase in miR-33a and miR-122 in the liver was attenuated by omega-3 PUFA treatment [85]. Interestingly, changes in miR-33a levels in peripheral blood mononuclear cells (PBMCs) mirrored changes in the liver, which raises the prospect that miR-33a levels in PBMCs could be used as a biomarker of miR-33a in the liver.

miRNA MODULATION BY DIETARY MICRONUTRIENTS

Dietary intake of micronutrients can be deficient in some elderly individuals [86] and is closely associated with consumption of fruit and vegetables [87]. Micronutrient intake patterns were recently reported to be associated with brain biomarkers of Alzheimer's disease in cognitively normal individuals. Micronutrient intake has also been associated with a variety of age-related conditions such as osteoporosis [88]. Provision of vitamin-D (5000 IU) fortified bread in elderly nursing home residents with low basal vitamin-D levels led to significantly increased bone mineral density in the lumbar spine and hip [89]. Micronutrients are potentially valuable modifiers of miRNA expression levels, because the caloric content of micronutrients is very low compared to macronutrients [90]. In addition, micronutrients can be used to fortify a wide variety of foods. To date studies of micronutrient modulation of miRNA levels have primarily been conducted in cell- or animal-based models.

miRNA MODULATION BY VITAMIN D

The main metabolite of vitamin D is calcitriol (1,25-dihydroxyvitamin D3), which can directly bind to nuclear transcription factors and hence modulate gene expression [91]. Some miRNAs are also regulated by nuclear transcription factors and therefore may be also modulated by vitamin D metabolites. In vitro experiments of calcitriol treatment of colon cancer cells have shown that several miRNAs, including miR-22, miR-146a, and miR-222, are upregulated, while miR-203 is downregulated. These calcitriol induced changes in miRNA expression were accompanied by reduced proliferation. Both miR-222 and miR-22 target genes involved in DNA methylation and histone modification, therefore can modulate many more genes indirectly via epigenetic modification. Calcitriol treatment of other colon cancer cell lines, such as HT-29, modulated additional miRNAs including miR-627, but not miR-22. Notably, miR-627 targets a histone demethylase gene, and blocking miR-627 suppresses the calcitriol-induced antiproliferative effect on HT-29 colon cancer cells [92]. miRNA modulation by the vitamin D metabolite, calcitriol, has been tested in a variety of other cancer cell lines, however studies have typically focused on measuring selected tumor suppressor or oncogenic miRNAs, rather than genome-wide miRNA changes [18]. For example, calcitriol treatment of leukemia cells reduced miR-181a and miR-181b expression and led to cell-cycle arrest, while addition of precursor miR-181a reduced the antiproliferative effects of calcitriol [93]. Calcitriol treatment of several prostate cancer cell lines including RWPE-1, RWPE-2, PrEC, and PrE causes upregulation of miR-100 and miR-125b, which are both tumor-suppressor miRNAs. Similarly, calcitriol treatment of LNCap prostate cancer cells caused the upregulation of another tumor-suppressor miRNA, miR-98. Interestingly, the miR-98 promoter region harbors a VDRE response element, therefore calcitriol treatment can directly induce miR-98 transcription, as well as indirectly through suppression of gene expression [94].

Taken together there is evidence that vitamin D treatment can modulate miRNA expression in cancer cells, which is particularly relevant within the context of aging and prevention of cancer. However, it will be important to establish the effect of vitamin D in normal healthy cells. Typically, supra-physiological concentrations of vitamin D or its metabolites have been used for in vitro experiments on cancer cell lines. Treatment of normal lines or primary human-derived cell lines with similar concentrations of vitamin D is most likely to also suppress proliferation and may result in cell death.

Vitamin D is also associated with modulation of cellular stress. Calcitriol treatment of serum-starved breast cancer cells suppressed the induction of cellular stress related miRNAs, including miR-26b, miR-182, miR-220b/c, and let-7 family members [95]. Cellular stress is a hallmark of aging and it would be useful in future to establish whether calcitriol treatment of normal cells exposed to cellular stressors such as starvation or hypoxia, results in modulation of miRNAs associated with cellular stress pathways.

There have been limited human intervention studies so far on the relationship between vitamin D supplementation and miRNA levels in blood. Cross-sectional analysis revealed a positive correlation between serum 25-hydroxyvitamin D and miR-532-3p levels in the serum of healthy individuals. Twelve months of Vitamin D supplementation compared to placebo revealed a significant reduction of miR-221 only in individuals administrated the placebo treatment [96], but no other serum miRNAs were altered. In another study, in pregnant women with low levels of plasma calcitriol (<25.5 ng/mL) indicative of vitamin D deficiency, 10 miRNAs were lower and 1 miRNA was higher compared to pregnant women with high levels of plasma calcitriol (>31.7 ng/mL) [97]. It is not yet known the functional consequences of diet-induced changes in blood miRNA levels. There have been no studies to date on the effects of vitamin D supplementation in the elderly on blood miRNA levels.

miRNA MODULATION BY FOLATE

Folate is essential for many biological processes and cannot be synthesized de novo, therefore sufficient daily intake of folate is necessary in humans. Folate is used in DNA synthesis, repair, and methylation, in addition to being a cofactor in amino acid metabolism. Folate could potentially modulate miRNA levels by mechanisms involving methylation reactions. Most of the research on folate deficiency-induced miRNA modulation has focused on cancer, due to the role of methyl groups in various cancers. Manipulation of folate availability in various mouse and human cell lines indicates folate deficiency modulates miRNA levels. In mouse embryonic stem cells, folate deficiency leads to reduced growth and apoptosis, which is accompanied by changes in 12 miRNAs [98]. Similar widespread changes in miRNAs have been reported in folate deficient human lymphoblast cells, including upregulation of miR-22 and miR-222, that was reversed by restoring folate availability [99]. Interestingly, miR-22 and miR-222 levels in blood were also observed to be higher in head and neck squamous cell carcinoma patients with low folate intake [99], although this finding remains to be confirmed in a larger cohort.

In vivo studies of rats fed with methyl group deficient diets leads to the development of hepatocellular carcinoma after 12–13 months, which is accompanied by upregulation of hepatic let-7a, miR-21, miR-23, miR-130, miR-190, and miR-17-92, as well as downregulation of miR-122 in comparison to rats fed a methyl sufficient diet [100]. Changes in miR-122 were reversed when methyl group deficient rats were provided with a methyl sufficient diet, which in turn prevented the development of hepatocellular carcinoma [100]. Overlap between aging and cancer pathways suggests findings from manipulation of folate in cancer cells or animal models modulate miRNAs which may be relevant in the context of aging. However, further mechanistic studies are necessary in different tissues which undergo age-related changes that are sensitive to dietary manipulation such as liver, fat, muscle, and endothelium among others.

miRNA MODULATION BY VITAMIN A

The main metabolite of vitamin A is retinoic acid, which plays a role in growth and development. Retinoic acid can bind to nuclear receptors and modulate transcription of genes harboring a retinoic acid response element. Treatment of various cells lines with retinoic acid has been shown to modulate miRNA levels with functional consequences. For example, retinoic acid inhibits proliferation and stimulates differentiation of neuroblastoma cells, with concomitant upregulation of miR-9 and miR-103a, which target the transcriptional factor ID2 that is involved in neuronal related tumorigenesis [101]. Retinoic acid has also been shown to trigger DNA demethylation events during neuroblastoma cell differentiation [102]. The action of retinoic acid on DNA methylation in neuroblastoma cells appears to be partially mediated via increased expression of miR-152, which can bind to DNMT1 [102]. Another study identified the miR-17 family as sensitive to retinoic treatment in neuroblasts, which triggers differentiation. The targets of the miR-17 family include protein coding genes of the mitogen-activated protein kinase (MAPK) signaling pathway, as well as protein coding genes that are involved in proliferation. Retinoic acid has also been demonstrated to modulate multiple miRNAs in leukemia cell lines. Notably, miR-29a and miR-142-3p have been shown to be upregulated by trans-retinoic acid treatment in three different leukemia cell lines, including HL-60,

THP-1, and NB-4 cells. Overexpression of miR-29a or miR-142-3p in hematopoietic stem cells from patients with acute myeloid leukemia or healthy individuals stimulates myeloid differentiation. To date studies of miRNA modulation by retinoic acid have focused on various cancer cell lines. It will be useful in future to establish the effect of retinoic acid on miRNAs in normal cells and aging cells. Retinoic acid treatment has already been tested in a variety of cells and animal disease models, due to the ability of retinoic acid to bind to nuclear receptors, but differences in miRNA expression were not assessed. Expression of the nuclear retinoic acid receptor in peripheral blood mononuclear cells has been reported to decrease with age [103]. Neutrophils appear to be preferentially targeted by retinoic acid in the elderly [104]. Retinoic acid has also been proposed as a topical treatment to prevent age-related changes in skin [105], as well as an oral treatment for preventing age-related cognitive decline [106], but whether these effects are mediated through miRNAs is yet to be established.

miRNA MODULATION BY VITAMIN C

Vitamin C is one of the most widely studied micronutrients and plays a role in many biological processes. Past studies have focused on the potential for vitamin C to increase immunity, as well as for treatment of chronic medical conditions associated with oxidative stress, such as cardiovascular disease and diabetes among others. To date very few studies have investigated whether vitamin C influences miRNA expression in different cells or animal models. Ascorbic acid treatment of multipotent periodontal cells upregulates miR-146 expression and promotes differentiation [107]. In a rat model of mammary carcinogenesis, vitamin C treatment suppressed the 17-beta-estradiol-induced increase in miR-93 and conversely increased nuclear erythroid-related factor 2 (NRF2) target protein levels. Follow-up experiments revealed overexpression of miR-93, posttranscriptionally suppressed both NRF2 and genes regulated by NRF2, as well as promoted carcinogenesis [108]. Conversely, inhibition of miR-93 suppressed carcinogenesis [108]. A recent study, highlighted the role of vitamin C in reprogramming of somatic cells into induced pluripotent stem cells. JHDM1A and JHDM1b have been identified as vitamin-C-dependent H3k36 demethylases, which promote cell cycle progression and suppress senescence. Interestingly, JHDM1B was shown to activate the miR-302/367 cluster, therefore suggesting that vitamin C can modulate histone demethylases partly through miRNA modulation [109]. More systematic studies are required to determine which miRNAs and related targets are modulated by ascorbic acid in different cell lines and the implications for dietary supplementation in humans.

miRNA MODULATION BY VITAMIN E

Vitamin E includes a group of fat soluble compounds that are found naturally in many foods including vegetables, oils, meat, eggs, and fruits, as well as being readily available as a dietary supplement and in fortified foods such as cereals. There are 10 common forms of vitamin E including alpha-, beta-, gamma-, delta- and epsilon-tocopherol, and alpha-, beta-, gamma-, delta- and epsilon-tocotrienol. The alpha-tocopherol form of vitamin E is reported to be the most biologically active and is a potent antioxidant in the glutathione peroxidase pathway, as well as modulating gene expression. Whether the antioxidant effects of alpha-tocopherol are mediated via miRNAs has not been studied. Nevertheless, vitamin E deficiency in rodents has been reported to decrease miR-122 and miR-125b expression in liver, concomitantly with a reduction in plasma cholesterol levels [110]. miR-122 plays a role in the regulation of lipid metabolism and miR-125b is related to inflammation. In addition, a recent study in juvenile Nile tilapia fish reported that diets with higher doses (2500 mg/kg) of alpha-tocopherol decreased superoxide dismutase activity and increased expression of miR-21, miR-223, miR-146a, miR-125b, miR-181a, miR-16, miR-155, and miR-122 in liver. Whereas, diets with no added alpha-tocopherol led to reduced expression of several of these miRNAs, including miR-223, miR-146a, miR-16, and miR-122 [111]. However, there have been no studies to date that have determined whether vitamin E supplementation directly modulates miRNA levels in human cells.

miRNA MODULATION BY DIETARY MINERALS

Healthy diets contain a range of essential dietary minerals, including calcium, phosphorus, potassium, sulfur, sodium, chlorine, and magnesium. In addition, healthy diets contain a range of dietary trace elements including iron, cobalt, copper, zinc, manganese, molybdenum, iodine, bromine, and selenium. These dietary minerals and trace elements are necessary for biochemical reactions. Deficiency in trace elements can occur during caloric restriction and when dietary food variety is limited. Trace element deficiencies have been reported in some elderly groups and the efficacy of trace element supplementation has been suggested to protect against a variety of age-related conditions including cognitive impairment.

To date there have been very few mechanistic studies of miRNA modulation by trace elements. Selenite treatment of human prostate cancer cells has been reported to upregulate miR-34b and miR-34c expression, which target the cell-cycle regulator p53 [112]. A recent study of rats fed selenium deficient diets revealed cardiac dysfunction was associated with upregulation of miR-374, miR-16, miR-199a-5p, miR-195, and miR-30e, and downregulation of miR-3571, miR-675, and miR-450a* [113]. miR-374 targets multiple genes which code for proteins in the Wnt/beta-catenin signaling pathway [113].

miRNA MODULATION BY FLAVONOIDS

Quercetin is a natural flavonol found in a wide variety of foods including fruits, vegetables, and grains. Similar to other flavonoids, quercetin is reported to have potential antioxidant, anti-inflammatory, and anticarcinogenic properties primarily based on in vitro cell experiments. There is in vitro evidence that quercetin can modulate miR-155, which is a pro-inflammatory associated miRNA. Murine macrophages activated with lipopolysaccharide and treated with quercetin downregulated miR-155, as well as mRNA and protein levels of TNF-α [114]. Quercetin also suppressed interleukin 1β, interleukin 6, macrophage inflammatory protein 1α, and inducible nitric oxide synthase genes [114]. There is in vivo evidence that quercetin can modulate miR-125b which is associated with inflammation and miR-122 which is associated with lipid metabolism. Laboratory mice fed a high-fat diet for 6 weeks and treated with quercetin showed higher expression of miR-125b and miR-122, as well as lower expression of inflammatory related genes including interleukin 6, C-reactive protein, monocyte chemoattractant protein 1, and acyloxyacyl hydrolase [115].

Apigenin is a dietary flavone with anti-inflammatory potential, which may be mediated via miRNAs. A high-throughput screen of apigenin treated lipopolysaccharide (LPS)-activated macrophages revealed miR-155 expression was reduced. LPS is an endotoxin that when administered to cells produces a strong immune response. Apigenin treatment of LPS-stimulated macrophages appeared to alter both miR-155a primary and precursor transcripts, which suggests apigenin may regulate miR-155 transcription. Several food types contain a high content of apigenin including celery. An in vivo follow-up study revealed apigenin treatment or a celery-based diet effectively suppressed miR-155 expression in LPS-treated mice [116]. Another study in transgenic mice with glucose intolerance due to overexpression of miR-103 showed that apigenin treatment improved glucose tolerance. Apigenin treatment was hypothesized to inhibit phosphorylation of TRBP, which is a component of the RSIC complex and therefore reduce mature miR-103 levels [117].

miRNA MODULATION BY POLYPHENOLS

Resveratrol is a natural phenol which has been extensively studied for potential health benefits related to cancer, cardiovascular disease, diabetes, aging, and lifespan extension. Resveratrol is found in high concentrations in certain grape varieties, particularly in grape skins, and therefore red wine is a rich source of resveratrol. Dark chocolate and peanuts are also sources of dietary resveratrol. Resveratrol has been observed to directly bind to miR-33 and miR-122 in hepatic cells [118]. Interestingly, resveratrol appears to bind to the mature miR-33 transcript independently or its host-gene SREBP2. This study showed for the first time using ^{1}H NMR spectroscopy that polyphenols can directly bind to miRNAs, revealing a new mechanism through which polyphenols modulate metabolism [118].

A polyphenol mixture containing anthocyanins, flavonols, and phenolic acid derivatives extracted from *Hibiscus sabdariffa*, a variety of plant grown and harvested in Senegal [119]. Atherosclerosis susceptible mice (LDLr$^{-/-}$) were fed high-fat, high-cholesterol diet (22% fat and 0.32% cholesterol, w/w) for 10 weeks and either plain or polyphenol enriched drinking water. Polyphenol treatment reversed changes in miR-103 and miR-107 expression caused by the high fat and cholesterol diet. miR-122 was not altered by the diet, but was markedly reduced in the polyphenol treated mice [119].

Proanthocyanidins are polyphenols that are found in a variety of plants and have a chemical structure of oligomeric flavonoids. Proanthocyanidins are present in high concentrations in grape seeds and also in cocoa beans. An in vitro screening study of HepG2 cells treated with various proanthocyanidins derived from grape seed or cocoa revealed between 6 and 15 miRNAs were differentially modulated out of over 900 miRNAs, including upregulation of miR-1224-3p, miR-197, and miR-532-3p [120]. Another study showed proanthocyanidin treatment can transiently reduce miR-33 and miR-122 levels in hepatocytes, with concomitant increases in the mRNA and protein of target genes, including ATP-binding cassette A1 and fatty acid synthase [121]. Three weeks proanthocyanidin extract supplementation was recently shown to be sufficient to suppress miR-33a and miR-122 expression in liver of diet-induced obese rats [122]. Proanthocyanidin treatment exerted a dose-dependent effect on plasma lipids and liver lipids; however, there was no clear dose-dependent effect on miR-33a or miR-122 expression [122]. A similar study in rats fed a cafeteria diet found proanthocyanidin treatment was able to effectively counteract diet-induced increases in miR-122 and miR-33a expression in liver [85]. Expression of targets of miR-122 and miR-33a was conversely increased including FAS and PPARβ/δ, CPT1A, and ABCA [85].

miRNAs MODULATED BY CURCUMIN

Curcumin treatment has been widely studied in cancer cell lines. Treatment of H460 and A427 lung cancer cells with curcumin increased miR-192-5p and miR-215, which appear to be tumor suppressors of nonsmall cell cancer and promote apoptosis, via targeting of X-linked inhibitor of apoptosis (XIAP) [123]. In pancreatic cancer cells, curcumin treatment increased miR-7 expression, which was accompanied by increased apoptosis. miR-7 targets SETD8, which is a histone-lysine N-methyltransferase that methylates histones and nonhistone proteins that are involved in the cell cycle [124]. In breast cancer cells exposed to bisphenol A, curcumin treatment inhibited proliferation and suppressed miR-19a and miR-19b, and increased expression of miR-19 family target genes [125]. Another study in metastatic breast cancer cells showed that curcumin treatment increased the expression of miR-181b, while also inhibiting proliferation and invasion and promoting apoptosis. Curcumin induced upregulation of miR-181b causes concomitant downregulation of pro-inflammatory cytokines CXCL1 and CXCL2, which are targets of miR-181b [126]. In human prostate cancer cells and murine melanoma cells, a curcumin analog called EF24 was reported to inhibit miR-21 expression, and upregulate several miR-21 target genes including PTEN, in turn promoting apoptosis [127]. In a lung metastasis animal model, EF24 treatment similarly led to decreased miR-21 expression and increased miR-21 target gene expression [127]. Curcumin treatment has also been reported to suppress miR-21 expression in human colon cancer cells, by reducing promoter activity and inhibiting AP-1 promoter binding [128]. In a variety of hepatocellular carcinoma cells, susceptibility to curcumin treatment appears to be determined by the expression of miR-200a/b, as cells expressing high levels of miR-200a/b were more resistance to the pro-apoptotic effects of curcumin [129].

Dietary modulation of miRNAs by curcumin treatment has also been investigated in animal models of liver fibrosis. miR-29b was increased in activated hepatic stellate cells by curcumin treatment, which led to hypomethylation and upregulation of PTEN. miR-29b was confirmed as a target of DNA methyltransferase 3b (DNMT3b) [130]. Thus curcumin action is mediated in part through miR-29b-induced epigenetic regulation. In an in vivo animal model of liver fibrosis induced by carbon tetrachloride (CCL4) injection, miR-199 and miR-200 were confirmed to be upregulated and some of their target genes were downregulated [131]. Curcumin treatment reduced miR-199 and miR-200 expression, which, in turn, normalized the expression of several target genes [131].

Curcumin treatment has also been investigated as a potential treatment for Alzheimer's disease. miR-146a is involved in modulating the innate immune response as well as inflammatory signaling. In stressed primary human neuronal-glial cells, miR-146a expression was upregulated, which was reversed by curcumin treatment and other NF-κB specific inhibitors. Notably miR-146a expression was associated with senile plaque density, as well as synaptic pathology in the Tg2576 transgenic Alzheimer's disease mouse model [132]. miR-146a was also observed to be upregulated in the hippocampus and neocortex of brains from subjects with Alzheimer's disease, while interleukin-1 receptor-associated kinase-1 (IRAK-1) was suppressed. Curcumin treatment reversed the expression of miR-146a and its target IRAK-1 in stressed human astroglial (HAG) cells [133].

miRNAs MODULATED BY MILK

Milk may have beneficial effects on multiple aspects of metabolic health in middle-aged and older adults, primarily related to increased satiety which may aid weight maintenance, as well as increased muscle protein anabolism which may aid maintenance of muscle mass [65]. However, several recent reports have shown miRNAs are present in milk from different sources, including human breast milk and bovine milk [134,135]. Milk miRNAs are predominantly contained inside membrane vesicles such as exosomes, but can also be present in fluid [136]. Isolated milk-derived exosomes containing miRNAs are capable of being taken up by human macrophages [137]. The question still remains whether exogenous diet-derived miRNAs can survive digestion and intestinal absorption to exert functional effects in different tissues in vivo. A study in humans of miRNA changes in peripheral blood mononuclear cells after milk or broccoli consumption indicated miR-29b and miR-200c can be absorbed.

A follow-up study in mice fed with miRNA-depleted milk for 4 weeks indicated that miR-29b concentrations were substantially lower compared to normal milk [138]. However, the horizontal transfer of miRNA between species remains a controversial issue and more evidence is required. Nevertheless, exosomal miRNAs from milk or other foods may exert functional effects even from within the gastrointestinal tract, as exosomal miRNAs have been shown to be resistant to acidic conditions [137].

CONCLUSION

Since the discovery of miRNAs over a decade ago, they have been established as important posttranscriptional regulators of gene expression. miRNAs can target multiple genes causing posttranscription suppression and modulation of cellular protein levels. miRNAs are modulated during cellular development, growth, and senescence. Furthermore, dysregulated miRNA expression occurs during aging and in age-related conditions, therefore miRNAs are attractive as biomarkers and therapeutic targets. In the last five years, evidence has begun emerging that nutrients and diet can modulate miRNA expression in different tissues, which suggests that dietary modulation of miRNAs may explain some of the beneficial effects of specific nutraceuticals on health. However, more in vivo studies are required in animal models of aging and age-related disease to provide causative evidence of the potential for dietary modulation of miRNAs in the context of aging. The challenge in future will be to translate evidence of dietary modulation of miRNAs into useful nutritional recommendations for older adults.

SUMMARY

A growing number of studies now provide evidence that miRNAs can be modulated by dietary factors, including macronutrients (protein, fat, carbohydrate), micronutrients, trace minerals, and nutraceuticals, such as flavonoids and polyphenols.

Many studies to date have focused on nutraceutical modulation of miRNAs in cancer cells, but nutraceutical modulation of miRNAs still needs to be explored further in other cell types.

Current evidence of miRNA modulation by diet or nutraceuticals is heavily reliant on in vitro studies. The challenge in future will be to move towards establishing whether dietary modulation of miRNAs plays an active role in the prevention or treatment of age-related diseases in vivo in animal or human studies.

Acknowledgments

This work was supported by the Ministry of Education, Science and Technology of Korea (2010-0020224).

References

[1] Lagos-Quintana M, Rauhut R, Lendeckel W, Tuschl T. Identification of novel genes coding for small expressed RNAs. Science 2001;294:853–8.

[2] Lau NC, Lim LP, Weinstein EG, Bartel DP. An abundant class of tiny RNAs with probable regulatory roles in *Caenorhabditis elegans*. Science 2001;294:858–62.

[3] Lee RC, Ambros V. An extensive class of small RNAs in *Caenorhabditis elegans*. Science 2001;294:862–4.

[4] Kozomara A, Griffiths-Jones S. miRBase: integrating microRNA annotation and deep-sequencing data. Nucleic Acids Res 2011;39:D152–7.

[5] Sood P, Krek A, Zavolan M, Macino G, Rajewsky N. Cell-type-specific signatures of microRNAs on target mRNA expression. Proc Natl Acad Sci USA 2006;103:2746–51.

[6] Baek D, Villén J, Shin C, Camargo FD, Gygi SP, Bartel DP. The impact of microRNAs on protein output. Nature 2008;455:64–71.

[7] Bartel DP. MicroRNAs: target recognition and regulatory functions. Cell 2009;136:215–33.

[8] Friedman RC, KK-H Farh, Burge CB, Bartel DP. Most mammalian mRNAs are conserved targets of microRNAs. Genome Res 2009;19:92–105.

[9] Lewis BP, Burge CB, Bartel DP. Conserved seed pairing, often flanked by adenosines, indicates that thousands of human genes are microRNA targets. Cell 2005;120:15–20.

[10] Song L, Tuan RS. MicroRNAs and cell differentiation in mammalian development. Birth Defects Res C Embryo Today 2006;78:140–9.

[11] McGregor R, Poppitt S, Cameron-Smith D. Role of microRNAs in the age-related changes in skeletal muscle and diet or exercise interventions to promote healthy aging in humans. Ageing Res Rev 2014;17C:25–33.

[12] Jung HJ, Suh Y. Circulating miRNAs in ageing and ageing-related diseases. J Genet Genomics 2014;41:465–72.

[13] Ibáñez-Ventoso C, Yang M, Guo S, Robins H, Padgett RW, Driscoll M. Modulated microRNA expression during adult lifespan in *Caenorhabditis elegans*. Aging Cell 2006;5:235–46.

[14] Weilner S, Schraml E, Redl H, Grillari-Voglauer R, Grillari J. Secretion of microvesicular miRNAs in cellular and organismal aging. Exp Gerontol 2013;48:626–33.

[15] Hackl M, Brunner S, Fortschegger K, et al. miR-17, miR-19b, miR-20a, and miR-106a are down-regulated in human aging. Aging Cell 2010;9:291–6.

[16] Krützfeldt J, Rajewsky N, Braich R, et al. Silencing of microRNAs in vivo with "antagomirs". Nature 2005;438:685–9.

[17] Alisi A, Da Sacco L, Bruscalupi G, et al. Mirnome analysis reveals novel molecular determinants in the pathogenesis of diet-induced nonalcoholic fatty liver disease. Lab Invest 2011;91:283–93.

[18] Beckett EL, Yates Z, Veysey M, Duesing K, Lucock M. The role of vitamins and minerals in modulating the expression of microRNA. Nutr Res Rev 2014;27:94–106.

[19] Milenkovic D, Jude B, Morand C. miRNA as molecular target of polyphenols underlying their biological effects. Free Radic Biol Med 2013;64:40–51.

[20] Witwer KW. XenomiRs and miRNA homeostasis in health and disease: evidence that diet and dietary miRNAs directly and

[20] indirectly influence circulating miRNA profiles. RNA Biol 2012;9:1147–54.
[21] Michas G, Micha R, Zampelas A. Dietary fats and cardiovascular disease: putting together the pieces of a complicated puzzle. Atherosclerosis 2014;234:320–8.
[22] Kiefte-de Jong JC, Mathers JC, Franco OH. Nutrition and healthy ageing: the key ingredients. Proc Nutr Soc 2014;73:249–59.
[23] Schwab U, Lauritzen L, Tholstrup T, et al. Effect of the amount and type of dietary fat on cardiometabolic risk factors and risk of developing type 2 diabetes, cardiovascular diseases, and cancer: a systematic review. Food Nutr Res 2014;58.
[24] García-Segura L, Pérez-Andrade M, Miranda-Ríos J. The emerging role of MicroRNAs in the regulation of gene expression by nutrients. J Nutrigenet Nutr 2013;6:16–31.
[25] Ross SA, Davis CD. MicroRNA, nutrition, and cancer prevention. Adv Nutr (Bethesda Md) 2011;2:472–85.
[26] Lee Y, Kim M, Han J, et al. MicroRNA genes are transcribed by RNA polymerase II. EMBO J 2004;23:4051–60.
[27] Denli AM, Tops BBJ, Plasterk RHA, Ketting RF, Hannon GJ. Processing of primary microRNAs by the microprocessor complex. Nature 2004;432:231–5.
[28] Gregory RI, Yan K-P, Amuthan G, et al. The microprocessor complex mediates the genesis of microRNAs. Nature 2004;432:235–40.
[29] Lee Y, Ahn C, Han J, et al. The nuclear RNase III Drosha initiates microRNA processing. Nature 2003;425:415–19.
[30] Lund E, Güttinger S, Calado A, Dahlberg JE, Kutay U. Nuclear export of microRNA precursors. Science 2004;303:95–8.
[31] Hutvágner G, Zamore PD. A microRNA in a multiple-turnover RNAi enzyme complex. Science 2002;297:2056–60.
[32] Kim VN, Han J, Siomi MC. Biogenesis of small RNAs in animals. Nat Rev Mol Cell Biol 2009;10:126–39.
[33] Grimson A, KK-H Farh, Johnston WK, Garrett-Engele P, Lim LP, Bartel DP. MicroRNA targeting specificity in mammals: determinants beyond seed pairing. Mol Cell 2007;27:91–105.
[34] Selbach M, Schwanhäusser B, Thierfelder N, Fang Z, Khanin R, Rajewsky N. Widespread changes in protein synthesis induced by microRNAs. Nature 2008;455:58–63.
[35] Ha M, Kim VN. Regulation of microRNA biogenesis. Nat Rev Mol Cell Biol 2014;15:509–24.
[36] Michlewski G, Guil S, Semple CA, Cáceres JF. Posttranscriptional regulation of miRNAs harboring conserved terminal loops. Mol Cell 2008;32:383–93.
[37] Guil S, Cáceres JF. The multifunctional RNA-binding protein hnRNP A1 is required for processing of miR-18a. Nat Struct Mol Biol 2007;14:591–6.
[38] Chen X, Ba Y, Ma L, et al. Characterization of microRNAs in serum: a novel class of biomarkers for diagnosis of cancer and other diseases. Cell Res 2008;18:997–1006.
[39] Mitchell PS, Parkin RK, Kroh EM, et al. Circulating microRNAs as stable blood-based markers for cancer detection. Proc Natl Acad Sci U S A 2008;105:10513–18.
[40] Raposo G, Stoorvogel W. Extracellular vesicles: exosomes, microvesicles, and friends. J Cell Biol 2013;200:373–83.
[41] Ma R, Jiang T, Kang X. Circulating microRNAs in cancer: origin, function and application. J Exp Clin Cancer Res 2012;31:38.
[42] Creemers EE, Tijsen AJ, Pinto YM. Circulating microRNAs novel biomarkers and extracellular communicators in cardiovascular disease? Circ Res 2012;110:483–95.
[43] Uhlemann M, Möbius-Winkler S, Fikenzer S, et al. Circulating microRNA-126 increases after different forms of endurance exercise in healthy adults. Eur J Prev Cardiol 2014;21:484–91.
[44] Roberts TC, Godfrey C, McClorey G, et al. Extracellular microRNAs are dynamic nonvesicular biomarkers of muscle turnover. Nucleic Acids Res 2013;41:9500–13.
[45] Valadi H, Ekström K, Bossios A, Sjöstrand M, Lee JJ, Lötvall JO. Exosome-mediated transfer of mRNAs and microRNAs is a novel mechanism of genetic exchange between cells. Nat Cell Biol 2007;9:654–9.
[46] Khoury S, Tran N. Circulating microRNAs: potential biomarkers for common malignancies. Biomark Med 2015;9:131–51.
[47] Kondkar AA, Abu-Amero KK. Utility of circulating microRNAs as clinical biomarkers for cardiovascular diseases. BioMed Res Int 2015;2015:821823.
[48] Higuchi C, Nakatsuka A, Eguchi J, et al. Identification of circulating miR-101, miR-375 and miR-802 as biomarkers for type 2 diabetes. Metabolism 2015;64:489–97.
[49] Witwer KW. Circulating microRNA biomarker studies: pitfalls and potential solutions. Clin Chem 2015;61:56–63.
[50] Kosaka N, Iguchi H, Ochiya T. Circulating microRNA in body fluid: a new potential biomarker for cancer diagnosis and prognosis. Cancer Sci 2010;101:2087–92.
[51] Pritchard CC, Kroh E, Wood B, et al. Blood cell origin of circulating microRNAs: a cautionary note for cancer biomarker studies. Cancer Prev Res (Phila) 2012;5:492–7.
[52] Serna E, Gambini J, Borras C, et al. Centenarians, but not octogenarians, up-regulate the expression of microRNAs. Sci Rep 2012;2:961.
[53] McGregor RA, Cameron-Smith D, Poppitt SD. It is not just muscle mass: a review of muscle quality, composition and metabolism during ageing as determinants of muscle function and mobility in later life. Longev Healthspan 2014;3:9.
[54] Drummond MJ, Dreyer HC, Pennings B, et al. Skeletal muscle protein anabolic response to resistance exercise and essential amino acids is delayed with aging. J Appl Physiol (Bethesda Md 1985) 2008;104:1452–61.
[55] Rivas DA, Lessard SJ, Rice NP, et al. Diminished skeletal muscle microRNA expression with aging is associated with attenuated muscle plasticity and inhibition of IGF-1 signaling. FASEB J 2014;28:4133–47.
[56] Drummond MJ, McCarthy JJ, Fry CS, Esser KA, Rasmussen BB. Aging differentially affects human skeletal muscle microRNA expression at rest and after an anabolic stimulus of resistance exercise and essential amino acids. Am J Physiol Endocrinol Metab 2008;295:E1333–40.
[57] Zacharewicz E, Della Gatta P, Reynolds J, et al. Identification of microRNAs linked to regulators of muscle protein synthesis and regeneration in young and old skeletal muscle. PLoS One 2014;9:e114009.
[58] English KL, Paddon-Jones D. Protecting muscle mass and function in older adults during bed rest. Curr Opin Clin Nutr Metab Care 2010;13:34–9.
[59] Režen T, Kovanda A, Eiken O, Mekjavic IB, Rogelj B. Expression changes in human skeletal muscle miRNAs following 10 days of bed rest in young healthy males. Acta Physiol (Oxf Engl) 2014;210:655–66.
[60] Vora M, Shah M, Ostafi S, et al. Deletion of microRNA-80 activates dietary restriction to extend C. elegans healthspan and lifespan. PLoS Genet 2013;9:e1003737.
[61] Mercken EM, Majounie E, Ding J, et al. Age-associated miRNA alterations in skeletal muscle from rhesus monkeys reversed by caloric restriction. Aging 2013;5:692–703.
[62] Dhahbi JM, Spindler SR, Atamna H, et al. Deep sequencing identifies circulating mouse miRNAs that are functionally implicated in manifestations of aging and responsive to calorie restriction. Aging 2013;5:130–41.

REFERENCES

[63] Volpi E, Kobayashi H, Sheffield-Moore M, Mittendorfer B, Wolfe RR. Essential amino acids are primarily responsible for the amino acid stimulation of muscle protein anabolism in healthy elderly adults. Am J Clin Nutr 2003;78:250–8.

[64] Drummond MJ, Glynn EL, Fry CS, Dhanani S, Volpi E, Rasmussen BB. Essential amino acids increase microRNA-499, -208b, and -23a and downregulate myostatin and myocyte enhancer factor 2C mRNA expression in human skeletal muscle. J Nutr 2009;139:2279–84.

[65] McGregor RA, Poppitt SD. Milk protein for improved metabolic health: a review of the evidence. Nutr Metab 2013;10:46.

[66] Kalyani RR, Egan JM. Diabetes and altered glucose metabolism with aging. Endocrinol Metab Clin North Am 2013;42:333–47.

[67] Reaven G. Age and glucose intolerance effect of fitness and fatness. Diabetes Care 2003;26:539–40.

[68] Gerich JE. Clinical significance, pathogenesis, and management of postprandial hyperglycemia. Arch Intern Med 2003;163:1306–16.

[69] Yu X-Y, Song Y-H, Geng Y-J, et al. Glucose induces apoptosis of cardiomyocytes via microRNA-1 and IGF-1. Biochem Biophys Res Commun 2008;376:548–52.

[70] Bagge A, Clausen TR, Larsen S, et al. MicroRNA-29a is upregulated in beta-cells by glucose and decreases glucose-stimulated insulin secretion. Biochem Biophys Res Commun 2012;426:266–72.

[71] He A, Zhu L, Gupta N, Chang Y, Fang F. Overexpression of micro ribonucleic acid 29, highly up-regulated in diabetic rats, leads to insulin resistance in 3T3-L1 adipocytes. Mol Endocrinol (Baltim Md) 2007;21:2785–94.

[72] Munsters MJM. Saris WHM. Body weight regulation and obesity: dietary strategies to improve the metabolic profile. Annu Rev Food Sci Technol 2014;5:39–51.

[73] St-Onge M-P, Gallagher D. Body composition changes with aging: the cause or the result of alterations in metabolic rate and macronutrient oxidation? Nutrition (Burbank Los Angel Cty Calif) 2010;26:152–5.

[74] Yang W-M, Jeong H-J, Park S-Y, Lee W. Induction of miR-29a by saturated fatty acids impairs insulin signaling and glucose uptake through translational repression of IRS-1 in myocytes. FEBS Lett 2014;588:2170–6.

[75] Yang W-M, Jeong H-J, Park S-Y, Lee W. Saturated fatty acid-induced miR-195 impairs insulin signaling and glycogen metabolism in HepG2 cells. FEBS Lett 2014;588:3939–46.

[76] McGregor RA, Kwon E-Y, Shin S-K, et al. Time-course microarrays reveal modulation of developmental, lipid metabolism and immune gene networks in intrascapular brown adipose tissue during the development of diet-induced obesity. Int J Obes (Lond) 2013;37:1524–31.

[77] Kwon E-Y, Shin S-K, Cho Y-Y, et al. Time-course microarrays reveal early activation of the immune transcriptome and adipokine dysregulation leads to fibrosis in visceral adipose depots during diet induced obesity. BMC Genomics 2012;13:450.

[78] Chen G-Q, Lian W-J, Wang G-M, Wang S, Yang Y-Q, Zhao Z-W. Altered microRNA expression in skeletal muscle results from high-fat diet-induced insulin resistance in mice. Mol Med Rep 2012;5:1362–8.

[79] Gallagher IJ, Scheele C, Keller P, et al. Integration of microRNA changes in vivo identifies novel molecular features of muscle insulin resistance in type 2 diabetes. Genome Med 2010;2:9.

[80] Sangiao-Alvarellos S, Pena-Bello L, Manfredi-Lozano M, Tena-Sempere M, Cordido F. Perturbation of hypothalamic microRNA expression patterns in male rats after metabolic distress: impact of obesity and conditions of negative energy balance. Endocrinology 2014;155:1838–50.

[81] Park J-H, Ahn J, Kim S, Kwon DY, Ha TY. Murine hepatic miRNAs expression and regulation of gene expression in diet-induced obese mice. Mol Cells 2011;31:33–8.

[82] Karere GM, Glenn JP, VandeBerg JL, Cox LA. Differential microRNA response to a high-cholesterol, high-fat diet in livers of low and high LDL-C baboons. BMC Genomics 2012;13:320.

[83] McGregor RA, Choi MS. microRNAs in the regulation of adipogenesis and obesity. Curr Mol Med 2011;11:304–16.

[84] Kilic ID, Dodurga Y, Uludag B, et al. microRNA-143 and -223 in obesity. Gene 2015;560:140–2.

[85] Baselga-Escudero L, Arola-Arnal A, Pascual-Serrano A, et al. Chronic administration of proanthocyanidins or docosahexaenoic acid reverses the increase of miR-33a and miR-122 in dyslipidemic obese rats. PLoS One 2013;8:e69817.

[86] Chernoff R. Micronutrient requirements in older women. Am J Clin Nutr 2005;81:1240S–5S.

[87] Roberts SB, Hajduk CL, Howarth NC, Russell R, McCrory MA. Dietary variety predicts low body mass index and inadequate macronutrient and micronutrient intakes in community-dwelling older adults. J Gerontol A Biol Sci Med Sci 2005;60:613–21.

[88] Berti V, Murray J, Davies M, et al. Nutrient patterns and brain biomarkers of Alzheimer's disease in cognitively normal individuals. J Nutr Health Aging 2015;19:413–23.

[89] Mocanu V, Stitt PA, Costan AR, et al. Long-term effects of giving nursing home residents bread fortified with 125 microg (5000 IU) vitamin D(3) per daily serving. Am J Clin Nutr 2009;89:1132–7.

[90] Dolara P, Bigagli E, Collins A. Antioxidant vitamins and mineral supplementation, life span expansion and cancer incidence: a critical commentary. Eur J Nutr 2012;51:769–81.

[91] Lamprecht SA, Lipkin M. Chemoprevention of colon cancer by calcium, vitamin D and folate: molecular mechanisms. Nat Rev Cancer 2003;3:601–14.

[92] Padi SKR, Zhang Q, Rustum YM, Morrison C, Guo B. MicroRNA-627 mediates the epigenetic mechanisms of vitamin D to suppress proliferation of human colorectal cancer cells and growth of xenograft tumors in mice. Gastroenterology 2013;145:437–46.

[93] Giangreco AA, Vaishnav A, Wagner D, et al. Tumor suppressor microRNAs, miR-100 and -125b, are regulated by 1,25-dihydroxyvitamin D in primary prostate cells and in patient tissue. Cancer Prev Res (Phila) 2013;6:483–94.

[94] Ting H-J, Messing J, Yasmin-Karim S, Lee Y-F. Identification of microRNA-98 as a therapeutic target inhibiting prostate cancer growth and a biomarker induced by vitamin D. J Biol Chem 2013;288:1–9.

[95] Peng X, Vaishnav A, Murillo G, Alimirah F, Torres KEO, Mehta RG. Protection against cellular stress by 25-hydroxyvitamin D3 in breast epithelial cells. J Cell Biochem 2010;110:1324–33.

[96] Jorde R, Svartberg J, Joakimsen RM, Coucheron DH. Plasma profile of microRNA after supplementation with high doses of vitamin D3 for 12 months. BMC Res Notes 2012;5:245.

[97] Enquobahrie DA, Williams MA, Qiu C, Siscovick DS, Sorensen TK. Global maternal early pregnancy peripheral blood mRNA and miRNA expression profiles according to plasma 25-hydroxyvitamin D concentrations. J Matern Fetal Neonatal Med 2011;24:1002–12.

[98] Liang Y, Li Y, Li Z, et al. Mechanism of folate deficiency-induced apoptosis in mouse embryonic stem cells: cell cycle arrest/apoptosis in G1/G0 mediated by microRNA-302a and tumor suppressor gene Lats2. Int J Biochem Cell Biol 2012;44:1750–60.

[99] Marsit CJ, Eddy K, Kelsey KT. MicroRNA responses to cellular stress. Cancer Res 2006;66:10843–8.

[100] Kutay H, Bai S, Datta J, et al. Downregulation of miR-122 in the rodent and human hepatocellular carcinomas. J Cell Biochem 2006;99:671–8.

[101] Annibali D, Gioia U, Savino M, Laneve P, Caffarelli E, Nasi S. A new module in neural differentiation control: two microRNAs upregulated by retinoic acid, miR-9 and -103, target the differentiation inhibitor ID2. PLoS One 2012;7:e40269.

[102] Das S, Foley N, Bryan K, et al. MicroRNA mediates DNA demethylation events triggered by retinoic acid during neuroblastoma cell differentiation. Cancer Res 2010;70:7874–81.

[103] Brtko J, Rock E, Nezbedova P, et al. Age-related change in the retinoid X receptor beta gene expression in peripheral blood mononuclear cells of healthy volunteers: effect of 13-cis retinoic acid supplementation. Mech Ageing Dev 2007;128:594–600.

[104] Minet-Quinard R, Farges MC, Thivat E, et al. Neutrophils are immune cells preferentially targeted by retinoic acid in elderly subjects. Immun Ageing 2010;7:10.

[105] Quan T, Qin Z, Shao Y, Xu Y, Voorhees JJ, Fisher GJ. Retinoids suppress cysteine-rich protein 61 (CCN1), a negative regulator of collagen homeostasis, in skin equivalent cultures and aged human skin in vivo. Exp Dermatol 2011;20:572–6.

[106] Touyarot K, Bonhomme D, Roux P, et al. A mid-life vitamin A supplementation prevents age-related spatial memory deficits and hippocampal neurogenesis alterations through CRABP-I. PLoS One 2013;8:e72101.

[107] Hung P-S, Chen F-C, Kuang S-H, Kao S-Y, Lin S-C, Chang K-W. miR-146a induces differentiation of periodontal ligament cells. J Dent Res 2010;89:252–7.

[108] Singh B, Ronghe AM, Chatterjee A, Bhat NK, Bhat HK. MicroRNA-93 regulates NRF2 expression and is associated with breast carcinogenesis. Carcinogenesis 2013;34:1165–72.

[109] Wang T, Chen K, Zeng X, et al. The histone demethylases Jhdm1a/1b enhance somatic cell reprogramming in a vitamin-C-dependent manner. Cell Stem Cell 2011;9:575–87.

[110] Gaedicke S, Zhang X, Schmelzer C, et al. Vitamin E dependent microRNA regulation in rat liver. FEBS Lett 2008;582:3542–6.

[111] Tang X-L, Xu M-J, Li Z-H, Pan Q, Fu J-H. Effects of vitamin E on expressions of eight microRNAs in the liver of Nile tilapia (Oreochromis niloticus). Fish Shellfish Immunol 2013;34:1470–5.

[112] Sarveswaran S, Liroff J, Zhou Z, Nikitin AY, Ghosh J. Selenite triggers rapid transcriptional activation of p53, and p53-mediated apoptosis in prostate cancer cells: Implication for the treatment of early-stage prostate cancer. Int J Oncol 2010;36:1419–28.

[113] Xing Y, Liu Z, Yang G, Gao D, Niu X. MicroRNA expression profiles in rats with selenium deficiency and the possible role of the Wnt/β-catenin signaling pathway in cardiac dysfunction. Int J Mol Med 2015;35:143–52.

[114] Boesch-Saadatmandi C, Loboda A, Wagner AE, et al. Effect of quercetin and its metabolites isorhamnetin and quercetin-3-glucuronide on inflammatory gene expression: role of miR-155. J Nutr Biochem 2011;22:293–9.

[115] Boesch-Saadatmandi C, Wagner AE, Wolffram S, Rimbach G. Effect of quercetin on inflammatory gene expression in mice liver in vivo—role of redox factor 1, miRNA-122 and miRNA-125b. Pharmacol Res 2012;65:523–30.

[116] Arango D, Diosa-Toro M, Rojas-Hernandez LS, et al. Dietary apigenin reduces LPS-induced expression of miR-155 restoring immune balance during inflammation. Mol Nutr Food Res 2015;59(4):763–72.

[117] Ohno M, Shibata C, Kishikawa T, et al. The flavonoid apigenin improves glucose tolerance through inhibition of microRNA maturation in miRNA103 transgenic mice. Sci Rep 2013;3:2553.

[118] Baselga-Escudero L, Blade C, Ribas-Latre A, et al. Resveratrol and EGCG bind directly and distinctively to miR-33a and miR-122 and modulate divergently their levels in hepatic cells. Nucleic Acids Res 2014;42:882–92.

[119] Joven J, Espinel E, Rull A, et al. Plant-derived polyphenols regulate expression of miRNA paralogs miR-103/107 and miR-122 and prevent diet-induced fatty liver disease in hyperlipidemic mice. Biochim Biophys Acta 2012;1820:894–9.

[120] Arola-Arnal A, Bladé C. Proanthocyanidins modulate microRNA expression in human HepG2 cells. PLoS One 2011;6:e25982.

[121] Baselga-Escudero L, Bladé C, Ribas-Latre A, et al. Grape seed proanthocyanidins repress the hepatic lipid regulators miR-33 and miR-122 in rats. Mol Nutr Food Res 2012;56:1636–46.

[122] Baselga-Escudero L, Pascual-Serrano A, Ribas-Latre A, et al. Long-term supplementation with a low dose of proanthocyanidins normalized liver miR-33a and miR-122 levels in high-fat diet-induced obese rats. Nutr Res (N Y N) 2015;35:337–45.

[123] Ye M, Zhang J, Zhang J, Miao Q, Yao L, Zhang J. Curcumin promotes apoptosis by activating the p53-miR-192-5p/215-XIAP pathway in nonsmall cell lung cancer. Cancer Lett 2015;357:196–205.

[124] Ma J, Fang B, Zeng F, et al. Curcumin inhibits cell growth and invasion through upregulation of miR-7 in pancreatic cancer cells. Toxicol Lett 2014;231:82–91.

[125] Li X, Xie W, Xie C, et al. Curcumin modulates miR-19/PTEN/AKT/p53 axis to suppress bisphenol A-induced MCF-7 breast cancer cell proliferation. Phytother Res 2014;28:1553–60.

[126] Kronski E, Fiori ME, Barbieri O, et al. miR181b is induced by the chemopreventive polyphenol curcumin and inhibits breast cancer metastasis via down-regulation of the inflammatory cytokines CXCL1 and -2. Mol Oncol 2014;8:581–95.

[127] Yang CH, Yue J, Sims M, Pfeffer LM. The curcumin analog EF24 targets NF-κB and miRNA-21, and has potent anticancer activity in vitro and in vivo. PLoS One 2013;8:e71130.

[128] Mudduluru G, George-William JN, Muppala S, et al. Curcumin regulates miR-21 expression and inhibits invasion and metastasis in colorectal cancer. Biosci Rep 2011;31:185–97.

[129] Liang H-H, Wei P-L, Hung C-S, et al. MicroRNA-200a/b influenced the therapeutic effects of curcumin in hepatocellular carcinoma (HCC) cells. Tumour Biol 2013;34:3209–18.

[130] Zheng J, Wu C, Lin Z, et al. Curcumin upregulates phosphatase and tensin homologue deleted on chromosome 10 through microRNA-mediated control of DNA methylation—a novel mechanism suppressing liver fibrosis. FEBS J 2014;281:88–103.

[131] Hassan ZK, Al-Olayan EM. Curcumin reorganizes miRNA expression in a mouse model of liver fibrosis. Asian Pac J Cancer Prev 2012;13:5405–8.

[132] Li YY, Cui JG, Hill JM, Bhattacharjee S, Zhao Y, Lukiw WJ. Increased expression of miRNA-146a in Alzheimer's disease transgenic mouse models. Neurosci Lett 2011;487:94–8.

[133] Cui JG, Li YY, Zhao Y, Bhattacharjee S, Lukiw WJ. Differential regulation of interleukin-1 receptor-associated kinase-1 (IRAK-1) and IRAK-2 by microRNA-146a and NF-kappaB in stressed human astroglial cells and in Alzheimer disease. J Biol Chem 2010;285:38951–60.

[134] Zhou Q, Li M, Wang X, et al. Immune-related microRNAs are abundant in breast milk exosomes. Int J Biol Sci 2012;8:118–23.

[135] Izumi H, Kosaka N, Shimizu T, Sekine K, Ochiya T, Takase M. Purification of RNA from milk whey. Methods Mol Biol (Clifton NJ) 2013;1024:191–201.

[136] Chen X, Gao C, Li H, et al. Identification and characterization of microRNAs in raw milk during different periods of lactation, commercial fluid, and powdered milk products. Cell Res 2010;20:1128–37.

[137] Izumi H, Tsuda M, Sato Y, et al. Bovine milk exosomes contain microRNA and mRNA and are taken up by human macrophages. J Dairy Sci 2015;98(5):2920–33.

[138] Baier SR, Nguyen C, Xie F, Wood JR, Zempleni J. MicroRNAs are absorbed in biologically meaningful amounts from nutritionally relevant doses of cow milk and affect gene expression in peripheral blood mononuclear cells, HEK-293 kidney cell cultures, and mouse livers. J Nutr 2014; 144:1495–500.

22

Nutritional Modulators of Cellular Senescence In Vitro

Mauro Provinciali[2], Elisa Pierpaoli[2], Francesco Piacenza[1], Robertina Giacconi[1], Laura Costarelli[1], Andrea Basso[1], Rina Recchioni[3], Fiorella Marcheselli[3], Dorothy Bray[4], Khadija Benlhassan[4] and Marco Malavolta[1]

[1]Nutrition and Aging Centre, Scientific and Technological Pole, Italian National Institute of Health and Science on Aging (INRCA), Ancona, Italy [2]Advanced Technology Center for Aging Research, Scientific Technological Area, Italian National Institute of Health and Science on Aging (INRCA), Ancona, Italy [3]Center of Clinical Pathology and Innovative Therapy, Italian National Research Center on Aging (INRCA-IRCCS), Ancona, Italy [4]Immunoclin Corporation, Washington, DC, USA

KEY FACTS

- Cellular senescence is a complex response to stress that contributes to suppress cancer and to initialize mechanisms of repair after tissue injury.
- Senescence cells have to be rapidly removed after their formation to ensure correct functional repair.
- The accumulation of senescent cells is considered a hallmark of aging and is believed to contribute to the aging phenotype and age related disease.
- It is possible to modulate cellular senescence (induce or delay) in vitro with a multitude of compounds including nutritional compounds.
- Modulators of cellular senescence have been identified in vitro from a wide range of nutritional compounds.
- The effects of nutritional compounds on cellular senescence appears to be frequently dependent on cell type, concentration of the compound and condition (ie, stress) that induce senescence.
- Common mechanisms of action of nutritional compounds that can explain their outcome related to the senescent phenotype include activation of NRF2 stress response pathway, interference with metal homeostasis, alteration of metabolic pathways, and epigenetic changes.
- Application of this knowledge in vivo requires further study to clarify the role of senescence in diseases and to identify better the specific cellular target and the modality of action of nutritional compounds.

Dictionary of Terms

- *Cellular senescence*: A complex cellular response to stress and other signals that comprise arrest of cell cycle, upregulation of tumor suppressor pathways, alteration of chromatin, and modification of transcriptional and secretory profile. The concept of senescence is currently rapidly evolving as in the last few years various stimuli and cellular contexts that induce senescence in physiological and pathological processes have been identified.
- *Nutritional compounds or nutritional factors*: this term includes any food component or nutritional supplement that influence physiological or cellular activities.

- *In vitro*: In this chapter, this term refers to studies carried out with cells (primary cells, cancer cells or any other type of cell) outside their normal biological context.
- *Biological Pathway*: is a series of actions among molecules in a cell that leads to a certain product or a change in a cell. Such a pathway can trigger the assembly of new molecules, such as a fat or protein as well as turn genes on and off, or spur a cell to move.
- *Age-related diseases*: are diseases most often seen with increasing frequency with advancing chronological age (ie, cardiovascular diseases including heart disease, stroke and atherosclerosis, cancer, diabetes, chronic obstructive pulmonary disease, etc.). It is still unclear if these diseases are the direct consequence of the aging process itself or if they can be disentangled from the aging process. Age-related diseases do not refer to age-specific diseases, such as childhood diseases, and should also not be confused with accelerated aging diseases, all of which are genetic disorders.

INTRODUCTION

More than 50 years ago, Leonard Hayflick and Paul Moorhead described a phenomenon, that is known today as "replicative senescence," consisting of the irreversible growth arrest of human cells in vitro after a period of apparently normal cell proliferation [1]. Telomere attrition was firstly identified as possibly responsible for this phenomenon [2], but now it is known that other factors can accelerate and/or trigger cellular senescence including DNA damage, oxidative stress, oncogene activation and inactivation, loss of tumor suppressors, nucleolar stress, epigenetic changes and others [3]. This complex response appears to be involved in various processes including tumor suppression, aging, embryogenesis, tissue repair and wound healing. In the last two year there have been at least three excellent comprehensive reviews with a focus on cellular senescence, which highlight the relevance of this field in biology and aging [3–5]. Hence there is growing interest to find modulators of cellular senescence that can be used for a multitude of clinical purposes. Experiments in vitro performed with different models of cellular senescence have provided evidence that it is possible to delay, induce, or accelerate cellular senescence as well as to interfere with the phenotype of senescent cells using bioactive natural compounds [6]. In the first part of this chapter we provide an overview of cellular senescence and its relevance in physiological and pathological process.

In the second part we will focus on the strategies to modulate the process of cellular senescence and will report a collection of the most relevant studies focused on modulation of cellular senescence in vitro by dietary bioactive compounds. Finally we will discuss critical aspects related to their putative mechanisms of action, especially in the context of their possible translation for the development of nutraceuticals.

CELLULAR SENESCENCE

Definition and Triggers of Cellular Senescence

Cellular senescence is a response characterized by a state of metabolic active mitogen insensible growth arrest in which cells show a series of phenotypic alterations including profound chromatin and transcriptional changes. The concept of cellular senescence is continuously evolving and appears to be much more complex than a static endpoint. Perhaps it is still not completely possible to answer the question: what is cellular senescence with a simple definition. Indeed, recent observations support the hypothesis that senescence can be a highly dynamic, multistep process, during which the properties of senescent cells continuously evolve and diversify [3]. A variety of stimuli is able to induce cellular senescence. Telomere erosion, which naturally occurs when cells divide, ultimately leads to "Replicative Senescence." Certain types of DNA damage and reactive oxygen species (ROS) can trigger a premature senescence, which can occur independently by telomere length. Both types of senescence, however, seem to be characterized by the activation of the DNA damage response (DDR) signaling pathway. Well-known triggers of cellular senescence are also activated oncogenes (ie, Ras, BRAF, E2F3) and inactivated tumor suppressor (ie, RB, PTEN, NF1, and VHL), but there are other less investigated inducers of cellular senescence [4]. These include chronic mitogenic signaling (ie, prolonged exposure to interferon-beta), epigenetic, nucleolar, and mitotic spindle stresses. All the types of senescent response induced by different intrinsic and extrinsic stressors (ie, oxidative or other forms of stress, DNA including telomere damage, oncogene activation, and tumor suppressor loss) can be generically termed as "damage induced senescence." However, senescence can be also triggered by signals occurring in the process of embryogenesis where it seems to play a role in tissue remodeling, as well as during processes of polyploidization and cell fusion, thus suggesting that other forms of senescence not strictly related to damage and described as "developmentally programmed senescence" could play a role in physiological processes.

Characterization of Cellular Senescence

Senescent cells differ from other states of cell cycle arrest, such as quiescence or terminal differentiation, by distinctive but not exclusive markers and morphological changes. A flattened and enlarged morphology with increased cellular granularity is generally observed in cultures of senescent cells [7]. However, these morphological changes seems to be not a distinctive feature of senescent cells in vivo, likely as a consequence of the barriers imposed by the 3D tissue architecture [4]. A basic condition of senescent cells is the absence of proliferative markers such as Ki67 and 5-bromodeoxyuridine (BrdU) incorporation but this is clearly insufficient to define the senescent state as it is a common feature of cell cycle arrest. Conversely, the upregulation of tumor suppressor pathways appears a distinctive, yet heterogeneous, feature of senescent cells. These mediators of senescence include various cellular signaling cascades that frequently involve p53 and ultimately activate different cyclin dependent kinase (CDK) inhibitors. The most famous of these CDK inhibitors is p16 (also named INK4A), an inhibitor of CDK4 and CDK6 that is encoded by CDKN2A locus (also known as INK4A/ARF as it encodes also the p53 activator ARF). This locus is known to be epigenetically derepressed with aging [8], likely as a consequence of the loss of Polycomb repressive complexes [9], and to be involved in senescence mediated by ROS as well as oncogenic signaling. Another important piece in the senescence puzzle is the CDK inhibitor p21 (also named WAF1 and encoded by CDKN1A locus) that is pivotal in developmentally regulated senescence [8] and is known to be induced by p53 in most models of damage induced senescence [4]. Oxidative damage and telomeric erosion can activate the DNA damage response (DDR), a signaling pathway in which ATM, ATR, CHK1, and CHK2 kinases activate various cell cycle targets including p53 and its downstream target p21 [10]. Both p16 and p21 as well as other cell cycle inhibitors leads to Rb dephosphorylation/activation thus causing the repression of its target gene, E2F, that is required for the progress of cell cycle. The mechanisms involved in the stable repression of the cell cycle during cellular senescence are still not completely understood but it is likely that chromatin remodeling plays a major role [11]. An important chromatin remodeling of senescent cells seems to consist of an early distension of centric and pericentric satellite heterochromatin [12], termed senescence-associated distension of satellites (SADS). This recent discovery raises the possibility that the formation of SADS could contribute to permanent cell cycle withdrawal as various components of the interphase centromere/kinetochore interacts with Rb and play a role in cell cycle regulation and stabilization of satellite heterochromatin. Regions of highly condensed chromatin called senescence-associated heterochromatin foci (SAHFs) that are characterized by the presence of various chromatin markers (ie, HMGA proteins, HP1g, HIRA, ASF1, macroH2A, and H3K9me3) are also a feature of some senescent cells (mainly p16 and p21 dependent oncogene-induced senescence). These regions have been shown to sequester genes involved in cell-cycle control thus explaining the stable condition of growth arrest during senescence. A key feature of the senescent nucleus related to SAHF formation is the loss of lamin B1 (a component of the nuclear lamina) [13,14]. A decrease in lamin B1, shown in the transition from early to full senescence, is thought to initiate global chromatin changes that are likely to drive the global transcriptome phenotype of senescent cells. These transcriptional changes are responsible for the production of the senescence associated secretory phenotype (SASP), whose function is still not completely understood [15]. The SASP includes cytokines (TGF-beta, IL-6, IL-1, CSFs), chemokines (ie, CXCR2 ligands, CXCL1, CXCL8, GROs), growth factors (ie, IGF-1, IGFBPs, PDGFs, amphiregulin), proteases (ie, MMP-3, MMP-10, collagenase-1), and other molecules (ie, PAI1). The SASP has been shown to be involved in multiple and even pleiotropic phenomena including tissue repair [16], clearance of senescent cells by the immune system [17], propagation of senescence in the neighboring cells [18], as well as driving or exacerbation of age-related pathologies including cancer [19]. The production of the SASP may also vary depending on the cell type and conditions, thus it is conceivable that the respective function in vivo could be dependent upon the systemic- and tissue-specific environment. Anyway, both SASP-related metabolic activity and chromatin remodeling could be responsible for an increase in the lysosomal content of senescent cells, and this enables detection of senescent cells with the histochemical or flow cytometry detection of beta-galactosidase activity at pH 6.0, also known as senescence associated beta-galactosidase (SA-betaGal) activity [20]. However, even if SA-betaGal it is the most widely used senescence biomarkers, it lacks complete specificity as SA-betaGal activity can be increased also in confluent quiescent cells [21]. Senescent cells can also be marked with Sudan Black B, which detects the complex lysosomal aggregate known as lipofuscin, and seems to be less sensitive to quiescence compared to SA-betaGAL [22]. It is still unclear if SA-betaGal activity in senescent cells reflects also increased autophagy [23], the process of digestion of the cell's organelles. Also in this case, the role of autophagy seems to depend on cell type and senescence triggers [24] and appears to be functionally related to the massive SASP protein synthesis [25]. A number of studies have also shown that the mammalian target of rapamycin (mTOR), a master regulator of cell growth and negative regulator of autophagy, positively regulates senescence in different systems [26–28]. However, mTOR seems not to play an essential role in all modes of senescence, because different studies have found that inhibition of mTOR (a process also used to activate autophagy), alternately delays or potentiates aspects of senescence [26–30]. The possibility that mTOR inhibits the initial step of autophagy, whereas the last stage of autophagy (autolysosomes) can facilitate mTOR activation has been recently proposed [31] to explain the observation of simultaneous activation of anabolic (mTOR driven) and catabolic processes (autophagy) in certain models of senescence [32]. Least but not last, it is important to mention the evidence that endogenous retroelements expression and retrotransposition is increased in late senescent cells [33,34]. This appears to be a phenomenon particularly associated with the late senescent stage that can determinate genomic instability and a substantial heterogeneity in the phenotype of senescent cells.

Another open question regards whether the phenomenon of cellular senescence involves also postmitotic cells like neurons and cardiomyocytes. Indeed, these cells are normally irreversibly blocked from reentering the cell cycle but they can accumulate DNA damage and exhibit senescent like phenotypes including heterochromatinization, production of SASP components, and staining with SA-betaGal. A p21-dependent senescence-like phenotype driven by a DDR has been observed in postmitotic neurons from aged mice that was aggravated in brains of late-generation telomerase null mice and ameliorated by caloric restriction [35]. Similarly, cardiomyocytes obtained from aged rats are positively stained for SA-betaGal and protein and RNA levels of CDK inhibitors similarly to young cardiomyocytes treated with DNA damaging agents [36].

In summary, the complexity and heterogeneity of cellular senescence provides evidence of the multifaceted nature of this phenomenon. Indeed, there is substantial evidence that mechanisms that establish senescence are cell type and conditions dependent and that cellular senescence is a dynamic process that involves several phases from the initial trigger to a deep senescence status where phenotypic diversification might be increased.

CELLULAR SENESCENCE IN AGING AND AGE-RELATED DISEASES

Studies of human tissues and cancer-prone mice argue strongly that cellular senescence is one of the most important processes to suppress cancer in vivo [6]. Moreover, the SASP produced by senescent tumor cells developing in premalignant lesions can trigger complementing signaling pathways that mobilize natural killer (NK) cells to eliminate malignant cells [37]. Nevertheless, senescent cells can also drive hyperplastic pathology in the neighboring environment and stimulate malignant phenotypes, likely as a consequence of some SASP components (growth factors, chemokines, and cytokines) [6].

In this scenario, the accumulation of senescent cells in aging together with age-related immune dysfunctions could contribute to explain why cancer incidence is exponentially related to aging. Although it is extremely complex to characterize cellular senescence in vivo, the use of multiple and single biomarkers (CDK inhibitors, SA-betaGal, SASP components, and DNA damage markers) of senescence has provided considerable evidence that senescent cells accumulate in mammalian organs during aging [3]. The presence of senescent cells appears to be particularly marked in tissues affected by age-related pathologies such as atherosclerosis, sarcopenia, and heart failure, osteoporosis, macular degeneration, chronic obstructive pulmonary disease, renal failure, Alzheimer's, and Parkinson's [4]. However, the role of cellular senescence in each disease is still not completely clarified. For example, a detrimental role of cellular senescence has been suggested in type 2 diabetes, cataract, and sarcopenia, where muscle stem cells (satellite cells) undergo a transition from quiescence to irreversible senescence thus impairing the ability to regrowth muscle mass.

Conversely, there are data suggesting a beneficial role of cellular senescence in atherosclerosis, cardiac and liver fibrosis, renal diseases, and wound healing. This is most likely related to a role of cellular senescence in restricting tissues fibrosis and promote tissue repair. The relationship between tissue regeneration and cellular senescence is still unclear. Anyway, an interesting model consisting of the sequence of 3 events has been recently proposed: the first event is the induction of senescence that limits fibrosis and further damage, then senescent cells recruit phagocytic immune cells (via their SASP) that are engaged to promote their clearance, and finally progenitor cells can start to proliferate and regenerate the damaged tissue.

In this context, the accumulation of senescent cells in tissues during aging and age-related diseases is considered a sign of dysfunction in this dynamic mechanism and might play an overt detrimental role. An experimental proof that accumulation of senescent cells is related to the aging phenotype has been provided using a transgenic mouse model in which p16INK4a-expressing cells can be specifically

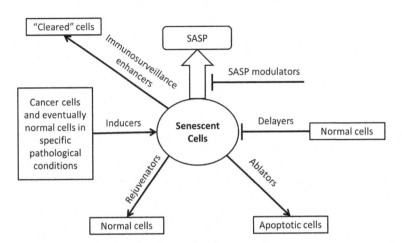

FIGURE 22.1 Modulators of cellular senescence and their potential targets.

eliminated upon drug treatment [38]. In the BubR1 progeroid mouse background, this strategy was shown to delay age-related dysfunction in organs such as adipose tissue (loss of subcutaneous fat), skeletal muscle (sarcopenia) and eye (cataracts) as well as to attenuate progression of already established age-related disorders.

Two major mechanisms are thought to contribute to the accumulation of senescent cells in aging. The first involves processes (ie, epigenetic changes, oxidative damage, and telomere shortening) that are affected by aging and that could increase the rate with which senescent cells are produced in aged tissues. This mechanism may eventually affect also progenitor cells devolved to regenerate the tissues thus limiting the capacity of renewal that is typical of old age. The second involves a diminished efficiency with which senescent cells are removed. Removal of senescent cells is thought to be a competence of the immune system with a process named "senescence immunosurveillance." The major players involved in senescence immunosurveillance are peripheral macrophages, NK cells, and T cells. Notably, immunosenescence-related dysfunctions affect all these players thus suggesting that their efficiency to remove senescent cells could be impaired by the aging process. Another possible problem for the clearance of senescent cells in aging is likely the heterogeneity of the SASP. Indeed, the phenotypic heterogeneity of the SASP produced by late senescent cells could contribute to select those cells that are less able to recall immune cells. An additional factor that could play a role in this phenomenon is the susceptibility to undergo apoptosis of senescent cells. Some senescent cells display resistance to apoptosis in vitro. Although this is not a universal property of senescent cells, it is possible that the accumulation of senescent cells in aged tissues is linked to the survival of cells resistant to apoptosis that escape senescence immunosurveillance.

STRATEGIES TO TARGET CELLULAR SENESCENCE WITH THERAPEUTICAL PERSPECTIVE

Coincident with this increased knowledge, modulators of the dynamics that control senescent-cell formation, fate, and subsequent effect on tissue function have gained critical interest in experimental gerontology and cancer research. There is a growing number of compounds and biomolecular strategies that have been shown to modulate cellular senescence (Fig. 22.1). These approaches can be currently classified into at least 6 categories: (1) rejuvenators of senescent cells; (2) direct ablators of senescent cells (strategies or compounds that induce apoptosis in senescent cells); (3) indirect ablators of senescent cells (strategies or compounds that are able to promote senescence immunosurveillance); (4) SASP modulators; (5) senescence inducers; and (6) senescence delayers.

Strategies to Rejuvenate Senescent Cells

Cellular senescence is considered at a glance an irreversible process of cell cycle arrest. However, there studies that suggest that it is possible to reverse senescence (at least in particular laboratory settings), allowing cells to reenter the cell cycle. One of these ways is to adapt the technology used to generate induced pluripotent stem cells (iPSCs). IPSCs are a particular type of stem cells that can be obtained from adult differentiated cells by genetic reprogramming (ie, delivering a specific set of pluripotency-associated genes). IPSCs, functionally indistinguishable from embryonic stem cells, have been obtained also from in vitro senescing cells by using lentivirus-mediated delivery of six transcription factors (OCT4, SOX2, KLF4, c-MYC, NANOG, and LIN28) [39]. While it is still uncertain to what extent iPSCs can be considered similar to ESCs [40], the potential reversibility of the cell cycle arrest during senescence addresses two important issues. The first is the relevance of epigenetic regulation in the establishment of the senescent state (indeed, it is likely that an irreversibly damaged DNA would have made it not possible to reverse the process). The second is that this mechanism could play a role in normal biological processes, in particular in the development of cancer or in the resistance of cancer cells to therapy induced senescence. Inactivation of p53 pathway in senescent fibroblasts that display low expression of p16 is another technique that is capable to restore cell cycle in senescent cells [41]. Suppression of p53 using shRNA was also shown to induce rapid reentry into the cell cycle of senescent mouse embryonic fibroblasts [42]. However, it should be emphasized that reversibility of senescence appears to be technically feasible in early but not in late senescent stages [43,44]. Inhibitors of mTOR, such as rapamycin (a natural drug isolated from bacteria), have been reported to partially reverse the senescent phenotype in mouse embryonic fibroblasts (MEFs) [27] and primary human fibroblasts (hF) [45]. Interestingly, some bioactive compounds derived from food sources or used as supplements have been shown to display partial rejuvenating effects on senescent cells in vitro. The transfer of fibroblasts approaching senescence from normal medium to a medium supplemented with L-Carnosine (20—50 mM) was reported to partially rejuvenate these cells [46]. Similar results have been also reported after treatment of senescent fibroblasts in vitro with the flavonols quercetin (6—7 μM) [47] and with a tocotrienol rich extract (0.5 mg/mL for 24 h) [48]. Hence, it seems technically possible to overcome the irreversibility of cell cycle arrest of senescent cells with different experimental strategies in vitro. However, since cellular senescence could be a primary defense mechanisms against cancer prone damaged cells, the possibility to overcome cell cycle arrest of senescent cells is currently unlikely to be potentially used for the development of therapies to counteract the exhaustion of stem cells pools in aging.

Direct Ablators of Senescent Cells

Since the finding that p16 positive senescent cells removal in mice counteracts some age-related dysfunction (ie, cataract, sarcopenia, and fat deposits) [38], there has been a strong interest in the identification of compounds or pharmacological strategies that are able to induce apoptosis in senescent cells. This will likely be a target of future development in gene therapy. Indeed, preliminary experiments using ganciclovir combined with the herpes simplex virus thymidine kinase suggest that it is possible to kill senescent cells in vivo with strategies similar to the suicide gene therapies used for cancer [49]. However, the development of strategies to induce apoptosis in senescent cells with pharmacological or natural compounds that interfere with specific metabolic pathways of some senescent cells is already a reality. This new class of compounds, which selectively kill senescent cells, has been recently termed senolytics drugs [50]. The identification of this class of compounds originated from the discovery of an increased expression of specific prosurvival genes (ie, ephrins (EFNB1 or 3), PI3Kδ, p21, BCL-xL, or plasminogen activated inhibitor-2) in senescent cells. Targeting these genes with silencing technology selectively killed senescent cells, but not proliferating or quiescent, differentiated cells. Most importantly, compounds able to target these factors selectively killed senescent cells. Dasatinib and quercetin were effective to kill selectively senescent human fat progenitor cells and endothelial cells, respectively. The combination of dasatinib and quercetin was effective in eliminating various type of senescent cells in vitro and in extending health span in the short living Ercc1-/Δ mice as well as in delaying age-related symptoms and pathology in mice. This is perhaps a breakthrough discovery that opens a new field not only in the treatment of age-related diseases, but also in the improvement of the efficacy of prosenescent therapies for cancer. Anyway, before the publication of these studies, the natural phenol phloretin, an inhibitor of glucose transporters, was found to reduce specifically the viability of therapy-induced senescent lymphoma cells [51]. These cells were also shown to be sensitive to another blocker of glucose transporters, cytochalasin B, and to the pharmacological block of glycolysis by 2-deoxy-D-glucose (2DG), as well as to inhibition of lactate dehydrogenase and of the energy sensor AMPK by sodium oxamate and compound-C, respectively. Suppression of autophagy by 3-MA (3-methyl-adenine, a known inhibitor of phosphatidylinositol 3-kinase class III enzymes) or CQ (chloroquine, a lysosome inhibitor) can also increase apoptosis of human (apoptosis resistant) senescent colorectal cancer cells induced by treatment with low-dosecamptothecin [52]. This is currently one of the most promising fields of research around cellular senescence.

Indirect Ablators of Senescent Cells

Indirect removal of senescent cells might be potentially achieved by potentiating the mechanisms by which the immune system keeps under control senescent cells. Macrophages, neutrophils, CD4 T-cells, and NK cells appear to be the main players of the immune system involved in this process. Immune-mediated clearance of senescent cells is currently largely unexplored. However, this process appears to be mediated by the SASP and seems to involve different players of the immune system in different tissues [53]. Up to now there are only hypotheses of interventions that may act to increase the clearance of senescent cells by acting through the immune system.

A recent senescence inducing strategy, using a combination of the mitotic kinase Aurora A (AURKA) inhibitor with an MDM2 antagonist, was shown to induce p53-mediated senescence and immune clearance of senescent cancer cells by antitumor leukocytes in a manner reliant upon SASP components such as Ccl5, Ccl1, and Cxcl9 [54]. However, a direct "booster" of senescence immune surveillance has still not been found. Possible candidates might include the drug lenalidomide, which was shown to reverse T-cell abnormalities of immunosenescence and to induce hair repigmentation in a clinical case with multiple myeloma [55], and Zn, which was shown to restore multiple parameters of innate immunity in elderly patients [56].

SASP Modulators

Targeting SASP is another attractive aim of research around cellular senescence. Inhibition of SASP could represent a kind of cancer therapy to reduce preneoplastic cell growth, angiogenesis, and invasion. At the same time however, promotion of selected components of SASP could be useful to promote clearance of senescent cells and to enhance repair processes of the tissues in response to damage. There are already various compounds that are potentially able to interfere with the production of pro-inflammatory components of SASP by interfering with NF-κB pathway. Corticosterone and the related glucocorticoid cortisol, well known inhibitors of NF-κB pathway, have been shown to decrease the production and secretion of selected SASP components in human senescent fibroblasts [57]. However, NF-κB is an ubiquitary pathway, thus this approach is likely to be limited by the lack of selectivity toward senescent cells. Interestingly, metformin (an antidiabetic drug) was shown to prevent events required for activation of the NF-κB pathway [58], but at the same time it was shown to promote SASP and reinforce growth arrest in human fibroblasts and in various cancer cell lines [59]. This activity could be related to a possible selective targeting of senescent cells but further research is needed to clarify these aspects. Most importantly, it would be useful to understand these mechanisms for their relevance in type II diabetes therapy, a pathology in which senescence is known to play a crucial but not well-defined role.

Additional compounds that could be able to decrease the inflammatory components of SASP are those used during the treatment of osteoarthritis. These include phycocyanobilin (a tetrapyrrole chromophore commonly found in the blue-green algae spirulina), berberine (an alkaloid usually found in the roots of berberis), and glucosamine (an amino sugar precursor for glycosaminoglycans, which are the major component of joint cartilage), which in turn was shown to inhibit IL-1β-induced activation of NF-κB by specific epigenetic changes [60]. In agreement with this putative role of SASP inhibitor, a decreased risk for cancer and total mortality was observed in regular users of this nutraceutical by prospective epidemiological studies [61]. Inhibition of p38MAPK has been suggested as an effective strategy to reduce the stability of mRNA of SASP factors in senescent cells [62]. Oral administration of a p38MAPK inhibitor (CDD-111, also referred to as SD-0006) in nude mice inhibits SASP production of subcutaneously injected senescent fibroblasts and inhibits SASP-mediated tumor growth driven by senescent fibroblasts. Whether it is important to inhibit the SASP to reduce its negative effects on the microenvironment or to promote its components to enhance senescence immunosurveillance is still intensively debated. Disruption of the NF-κB-mediated SASP has been shown to lead to chemo-resistance in various model of cancer in mice [37,63]. In conclusion, while inhibition of SASP could counteract tissue dysfunction, a possible drawback of this strategy could be related to the inhibition of the inflammatory reaction necessary to activate the clearance of senescent cells by the immune system. Perhaps it would be useful to disentangle the role of single components of the SASP in order to inhibit those who can display the negative effects on the microenvironment while keeping unaltered or boosting those involved in senescence immunosurveillance.

Senescence Inducers

The possibility to take opportunity of senescence response is considered as a key component for therapeutic strategies in the suppression of cancer. However, taking into account the putative beneficial role of cellular senescence in various physio- and pathological processes [4], it is not excluded that induction of cellular senescence might represent in the near future a therapeutic tool for conditions different from cancer (ie, to reduce fibrosis associated to pathological conditions). Anyway, current research is mostly focused on therapy-induced senescence as a functional to improve cancer therapy. Targeting the most common pathways that the tumor uses to escape the cell cycle block is behind the strategy for prosenescence therapy [64]. Moreover, reactivation of the senescence program triggers an innate immune response in vivo, likely mediated by the SASP, which contributes to tumor clearance [17]. Therefore, it is not surprising that compounds potentially able to induce senescence in tumor cells have been widely studied in the last decade. Various anticancer agents, including well-recognized inducers of apoptosis, can act also as inducers of cellular senescence. This indicates that prosenescence strategies may have applications in both early prevention and late stages of cancer development. Also conventional treatments, such as chemo- and radiotherapies, have been shown to preferentially induce premature senescence instead of apoptosis in the appropriate cellular context [65–67]. Moreover, senescence of cancer cells can be achieved with lower doses of anticancer drugs compared to the conventional cytotoxic approach, thus suggesting that prosenescence therapy has the potential to be less severe and with reduced toxic side effects. The recent observation that tumors can be massively infiltrated by a population of (CD11b+Gr-1+) of myeloid cells that protect a fraction of proliferating tumor cells from senescence enhance the availability of targets to engage cellular senescence response in the eradication of cancer [68]. The possibility to target cellular senescence with natural bioactive substances has been also extensively investigated. Examples of natural bioactive substances that can promote senescence of cancer cells in vitro include various phenols such as resveratrol, epigallocatechin-gallate, quercetin, curcumin, and silybin as well as isothiocyanates (ie, the organosulfur compound sulforaphane), methyl-tocols (ie, tocotrienols), and alkaloids (ie, berberine). A possible drawback of prosenescence therapy is the possibility that senescent cancer cells can survive evading immune surveillance and later reenter the cell cycle or promote the growth of new tumors in the microenvironment with their SASP. However, while the reversibility of senescence in vitro is possible by technical artifacts, the concept of reversibility of senescence in vivo is still an unanswered question.

Senescence Delayers

Delay in the onset of cellular senescence is considered nowadays among the most promising strategies to counteract age-related syndromes (ie, cataract, sarcopenia, arthritis) and age-related diseases including cancer. Extension of replicative lifespan of cells in vitro can be achieved by targeting the biomolecules that mediate senescence response. The first breakthrough studies around the possibility of delaying senescence in humans cells were made by overexpression of hTERT, the catalytic subunit of telomerase (ie, the enzyme that elongates telomere) [69]. Repression of p16 expression by antisense technology [70], by stable ectopic expression of the polycomb group proteins (ie, BMI1 CBX7 CBX8) [71–73], as well as by ectopic expression of Id1 (an additional repressor of p16 expression) [74,75] and silencing of E47 (a target of Id1) [76], can also delay cellular senescence in human cells. Inactivation of p53, p21, and Rb is an alternative tool that has been proven to delay or bypass cellular senescence in human fibroblasts [77,78]. However, it is important to consider that direct inhibition of tumor suppressor pathways could not be a useful target for delaying the onset of cells senescence as this is the same mechanism used by many cancer lines to overcome cell cycle blocks. Other studies in vitro suggest that it is possible to delay replicative and stress-induced senescence in primary cells by treatments even with nutritional and pharmacological compounds. Inhibitors of mTOR, such as rapamycin (a natural drug isolated from bacteria), have been reported to delay the onset and partially reverse the senescent phenotype in mouse embryonic (MEFs) [27] and primary human fibroblasts [45]. Lots of different natural compounds known to activate the cytoplasmic oxidative stress system, Nrf2-Keap1 (Kelch-like ECH-associated protein 1), and the downstream antioxidant response elements (AREs) of many cytoprotective genes have been shown to delay cellular senescence in different cells and conditions [79]. In this case, since the effects on cellular senescence is thought to be the consequence of a reduction of damage, there is the possibility to translate these finding into useful clinical tools to prevent age-related pathologies. Similar conclusions can be drawn by the observation that exposure to serum from calorie-restricted animals as well as manipulation of SIRT1 can delay senescence and extend lifespan of normal human fibroblasts in vitro [80].

NUTRITIONAL FACTORS AND CELLULAR SENESCENCE "IN VITRO"

There are two proven facts around the effects of nutritional compound on cellular senescence in vitro. The first is that a multitude of natural compounds can effectively modulate (mainly to induce or delay) cellular senescence in vitro. The second is that most of these compounds have been shown to delay cellular senescence in particular experimental settings and, surprisingly, to induce senescence in others (usually in treatments of cancer cells) (Table 22.1). Here we show that most nutritional compounds claimed to have effects on health and eventually used as supplements modulate cellular senescence in vitro. The compounds

TABLE 22.1 Examples of Divergent Effects of Some Bioactive Dietary Compounds on Cellular Senescence "In Vitro"

Cancer cells					Normal cells	
Concentration (timing)	Model[a] (type of senescence[b])	Effect[c]	Dietary bioactive compound	Effect[c]	Model[a] (type of senescence[b])	Concentration (timing)
50–100 mM (24 h)	hCCC	I	L-Carnosine	D/R	hF (RS)	20–50 mM (chronic)
20–30 mM (chronic)	mTC	I	L-Carnosine			
10–20 µM (chronic)	hGC	I	Resveratrol	I	hMSC	>10 µM (4 d)
30 µM (chronic)	hCCC	I	Resveratrol	D	hMSC (RS)	0.1 µM (30 d)
75–250 µM (72 h)	hCCC	I	Resveratrol	I	hEC	10 µM (chronic)
10–50 µM (chronic)	NSCLC	I	Resveratrol	D	hT (SIPS)	30 µM (72 h during SIPS)
50 µM (96 h, then left recovery for 48 h)	hOC, HACC	I	Resveratrol	I	hMSC	20 µM (chronic)
50 µM (48 h)	mSCC	I	Resveratrol	D	hF, hRPE (SIPS)	50 µM (30 min during H_2O_2 stress or 3 d during other stress)
100 µM (chronic)	hHC	I	Resveratrol	D	pTM (SIPS)	25 µM (chronic during 40% O_2 stress)
50–100 µM (chronic)	mBCC	I	Tocotrienols	D/R	hF (RS)	Gold Tri E 50 0.5 mg/mL (24 h at various PD including senescent hF)
100 µM (chronic)	mBCC	I	Berberine	D	hF (RS)	60 µM (24 h)
10–60 µM (Co-treatment with mitoxantrone stress)	hNSCLC (SIPS)	D	Berberine			
20–40 µM (24–72 h)	hBCC	I	Bisdemethoxycurcumin	D	hF (SIPS)	20 µM (pre-treatment 48 h)
25 µM (24–72 h)	hGC	I	Quercetin	D/R	hF (RS, SIPS)	6–7 µM (chronic)
300 µM (2 h co-treatment with H_2O_2)	mFCL (SIPS)	D	Quercetin			
300 µM (2 h co-treatment during stress)	mFC (SIPS)	D	Quercetin			
0.15–0.3 mM (5 d co-treatment with peroxide)	hBCC (SIPS)	D	Vitamin C			
1 mM (Co-treatment with 4 µM phenylaminonaphthoquinones)	hBCC SIPS	I	Vitamin C	D (RS)	hEmC, hF	0.2 mM (chronic)
1 mM (2 d co-treatment with CKII inhibitor stress)	hCCC	D (SIPS)	Vitamin C	D (RS)	hEC	0.13 mM (chronic)
20 µM (48–72 h)	hLC (SIPS)	I	Ginsenoside Rg1	D (SIPS)	hF	5–20 µM (chronic pre-treatment from 24–30 PD)
			Ginsenoside Rg1	D (RS)	hEPC	1–5 µM (24 h)
			Ginsenoside Rg1	D (SIPS)	hF	5–20 µM (24 h pre-treatment)
20 µM (chronic)	hGC SIPS)	I	Ginsenoside Rg3	D (SIPS)	hOAC	1–2.5 µM (24 h co-treatment with IL-1β stress)
15 µM (chronic)	hCCC, hLC	I	EGCG	D (RS, SIPS)	rVSMC, hF, hAC	50–100 µM (chronic)

[a]Model: hF, human fibroblasts; hCCC, human colon cancer cells; hMSC, Human mesenchymal stem cells; hEC, human endothelial cells; hGC, human glioma cells; hT, human tenocytes; hRPE, Retinal pigment epithelial cells; pTM, Porcine trabecular meshwork; mHER2, mouse breast cancer cells; hKSC, Human keratinocyte stem cells; hNSCLC, human pulmonary nonsmall cell lung carcinoma; mSCC, murine squamous cell carcinoma; mTC, mouse teratocarcinoma cells; mESC, mouse embryonic stem cells; hHC, human hepatoma cells; hOC, human osteosarcoma cells; hACC, human adenocarcinoma cells; hamF, hamster fibroblasts; mFCL, mouse fibroblasts cell line; hBCC, human bladder cancer cells; hLC, human lymphoblastoid cells; hEmC, human embryonic cells; hEPC, human endothelial progenitor cells; hOAC, human osteoarthritic chondrocytes; mFC, mouse fibroblastoma cells; rVSMC, rat vascular smooth muscle cells; hAC, human articular chondrocytes.
[b]Type of senescence: SIPS, stress induced premature senescence; RS, replicative senescence.
[c]Effect: I, inducer of a senescent-like phenotype; R, Rejuvenator of the senescent state; D, delayer of senescence markers onset.
For References see Malavolta et al. [6].

are here grouped and presented according to their structural category. Only documented activity on cellular senescence in vitro (ie, from PubMed or Google search including the name of the compound and the term "cellular senescence") is reported here. Special emphasis is given to the putative mechanisms involved, especially in the cases where the same compound has been shown to display opposite effects in normal versus cancer cells.

Triterpenoid Saponins: Cycloastragenol (also named TA-65), Ginsenosides, Epifriedelanol

Cycloastragenol (TA-65), a small-triterpenoid purified from the root of an Asian medicinal herb, has currently received particular attention for its potential to regulate transcription of telomerase. TA-65 is capable to increase average telomere length and to decrease the percentage of critically short telomeres and DNA damage in MEFs that harbor critically short telomeres [81]. The effects on telomerase appear to be in common with other triterpenoid saponins (ie, ginsenosides RG1 and Rg3). These compounds at concentrations from 1 to 20 μM were reported to protect against IL-1β-, H2O2-, and tert-butylhydroperoxide-induced senescence in human chondrocytes [82], endothelial progenitor cells and fibroblasts [83], respectively. Also another triterpenoid isolated from the root bark of *Ulmus davidiana*, epifriedelanol, was shown to suppress adriamycin-induced cellular senescence as well as replicative senescence in HDFs and HUVECs [84]. Interestingly, in the same range of concentrations that are active in normal cells, ginsenoside Rg3 was reported to induce Akt and p53/p21-dependent senescence in human glioma cells [85] and ginsenoside Rg1 to induce p21-Rb and p16-Rb dependent senescence in human leukemia K562 cells [86]. This likely supports the possibility that ginsenosides interfere with additional mechanisms upstream of telomerase that are differently regulated in cancer cells compared to primary normal cells.

Benzopyranols (or Methyl Tocols): Tocopherols and Tocotrienols (Vitamin E)

Tocopherols and tocotrienols are members of the vitamin E family. Both class of compounds are subdivided into four lipophilic isomers (α-, β-, γ-, and δ-) that are naturally found in different percentages in a number of vegetable oils, wheat germ, barley, and certain types of nuts and grains. In biological systems, their association with lipoproteins, fat deposits, and cellular membranes is thought to protect polyunsaturated fatty acids (PUFA) from peroxidation reactions.

Indeed, this mechanism was indicated as a possible explanation of the antisenescence effects shown by lipophilic antioxidants, including α-tocopherol at 100 μM, in human fibroblasts treated with 8-methoxypsoralen plus ultraviolet-A [87]. In recent years tocotrienols have been shown to display biological properties, including senescence modulatory functions that are much more evident than tocopherols preparations. Treatment with a mixture consisting of ϒ- and/or δ-tocotrienols (0.5 mg/mL) delayed and partially reversed replicative cellular senescence in human fibroblasts [48]. Elongated telomeres and increased telomerase activity as well as alteration in senescence-associated miRNA (mainly miR-34a, miR-24a, miR-20a and miR-449a) [88] were observed in the treated cells compared to controls, thus suggesting the involvement of important posttreatment epigenetic changes.

Conversely, experiments in vitro with cancer cells support a proapoptotic and prosenescence activity of tocotrienols. Tocotrienols were shown to inhibit hTERT in human colorectal adenocarcinoma cell lines at concentration of 10–20 μM [89] as well as to induce senescence-likephenotype in breast cancer cells at 50–100 μM [90]. These effects could be mediated by a possible downregulation of c-Myc and subsequent epigenetic changes leading to derepression of p21 [91]. The selective accumulation of tocotrienols in cancer cells compared to normal cells might at least in part explain these dicothomic results. Another explanation could involve the high affinity of tocotrienols for the estrogen receptor, which in turn may activates the expression of estrogen responsive genes involved in growth arrest in estrogen receptor positive cancer cells [92].

Amino-Acids and Peptides: N-acetylcysteine (NAC), Carnosine

N-acetylcysteine (NAC), a metabolite of the amino acid cysteine that is sold as a dietary supplement, is widely known and used for its antioxidant properties. It was among the compounds that displayed the major protective effects in replicative senescence of endothelial cells [93] and in human fibroblasts induced to senescence by treatment with 8-methoxypsoralen plus ultraviolet-A[87]. NAC pretreatment (1–10 mM range) can attenuate or delay senescence in various models where reactive oxygen species are suspected to participate in the induction of the senescent response. Some examples include 2,3,7,8-tetrachlorodibenzo-p-dioxin (TCDD)-induced senescence in human neuroblastoma and rat pheochromocytoma cells [93], IFN-gamma-induced senescence in Human Umbilical Vascular Endothelial cells (HUVECs) [94], senescence associated with differentiation processes of IPSCs [95], and many more. The inhibition of nuclear export of TERT associated with the age-induced increase in ROS formation has been recently proposed as a mechanism of action.

A natural occurring dipeptide, L-Carnosine, was shown to display similar effects on human fibroblasts [46]. The culture of presenescent fibroblasts in a medium supplemented with L-Carnosine (20–50 mM) leads to a partial cellular rejuvenation with a variable extension of the lifespan. Carnosine's ability to scavenge free radicals is believed to be mainly responsible for these effects. However, this mechanism is not consistent with the inhibitory activity of carnosine on the growth of cultured tumor cells [96]. A possible explanation of these different effects could be related to the Zn binding activity of carnosine. This activity might influence the function of one or more glycolytic enzymes resulting in the decrease of ATP levels and ATP synthesis [97]. Moreover, given the relevance of Zn for p53 transcriptional activity, the involvement of this mechanism in the senescence-modulatory effect of this nutritional compound would deserve appropriate investigation [6]. Other classes of peptides may have a valuable role to counteract senescence. For example, calcitonin-related peptide was reported to inhibit angiotensin II-induced endothelial progenitor cells senescence through upregulation of klotho expression [98].

Organosulfur Compounds: Alpha-Lipoic Acid (Alpha-LA), Sulforaphane

Various organosulfur compounds are claimed to have powerful antioxidant activity. Their capacity to scavenge ROS plays perhaps an important role in their antisenescence activity in vitro. This conclusion can be drawn by various studies on stress-induced senescence. Alpha-LA (5 mM) was shown to delay senescence in human fibroblasts treated with 8-methoxypsoralen plus ultraviolet-Amethoxypsoralen plus ultraviolet-A [87], while Sulforaphane (SFN) at low doses (0.25 and 1 μM) improved mesenchymal stem cells (MSC) proliferation and reduced their senescence related changes [99]. Interestingly, higher doses of SFN (5–20 μM) can induce senescence and even produce cytotoxic effects [99,100]. Besides scavenging ROS, common mechanisms that might explain the effects of organosulfur compounds on cellular senescence include chelation of metals and induction of the antioxidant response pathway mediated by the nuclear factor erythroid 2-related factor 2 (NRF2) [6,79].

Sugars: Glucose

Among nutrients, glucose constitutes a fundamental metabolic fuel, which is catabolized through the glycolytic pathway providing energy in the form of ATP. Following a mechanism that appears to be in line with the modulatory effects of caloric restriction on organism lifespan the energetic stress of glucose restriction can induce an increase of cellular lifespan. In the case of normal human fibroblasts, a glucose depleted medium (below 1 mM) was shown to induce SIRT1-mediated epigenetic changes that repress p16 expression [101] and can maintain hTERT levels [102] with a considerable delay of replicative senescence. Conversely, high glucose conditions (around 25 mM) are commonly used to accelerate senescence in various cellular models [103,104]. Interestingly, glucose restriction in precancerous cells was shown to produce a different epigenetic regulation and cellular fate compared to normal cells [102]. This is likely related to the metabolic shift of cancer cells that in most cases require high rate of glycolysis [105].

Stilbenoids (a Class of Natural Phenols): Resveratrol

The modulatory effects of resveratrol on cellular senescence appear to be related to the same pathways controlled by glucose probably because the activity of this stilbenoid is at least in part mediated by its interaction with the SIRT1 pathway [106]. In this context, it is indicative that a resveratrol derivative was shown to antagonize high-glucose-induced senescence [107]. However, the effects of resveratrol on cellular senescence are still unclear. It appears that very low (0.1 μM) and low doses (5 μM) can delay replicative senescence of human mesenchymal stem cells and human umbilical cord fibroblasts, respectively. Conversely, higher doses (10–50 μM) are required to counteract the effects of stress-induced senescence in various cellular models [6]. These doses and even higher (up to 100 μM) have been also shown to induce senescence-like growth inhibition of cancer cells associated with the increase in the activity and expression of senescence-associated effectors (ie, p53 and p21) as well as with instability of telomeres and DNA damage [108]. Hence, the ability of resveratrol to modulate cellular fate affecting pathways related to cellular senescence depends on the cell types and treatment conditions.

Diarylheptanoids: Curcumin

Apparently in contrast to the anti-inflammatory and antiaging effects claimed to curcumin, its activity in vitro is in line with a clear prosenescence activity. Curcumin was shown to exert antitumor activity by inducing cellular senescence in human colon cancer cells (HCCC) with associated induction of p53 (in p53 responsive lines), p21, and autophagy [109]. Moreover, curcumin was shown to induce a p16-dependent senescence and to suppress the SASP of breast cancer associated fibroblasts [110]. The prosenescence activity of curcumin has been also demonstrated in vasculature related primary human cells. In vascular smooth muscle cells and endothelial cells, curcumin (2.5–7.5 μM) induce senescence by a mechanism which appears to be DNA damage and ATM independent and does not involve increased ROS level [111]. Interestingly, it has been suggested that curcumin can play a beneficial role in the cardiovascular system [112] and there are hypotheses, although controversial, that cellular senescence may also exert beneficial role in cardiovascular diseases [4].

B-Vitamins: Biotin (Vitamin B7)

Biotin is necessary for cell growth, the production of fatty acids, and the metabolism of fats and amino acids. It has been shown that biotin-deficient primary human lung fibroblasts (IMR90 grown in medium with 10% FCS containing 0.29 μg/L biotin, ~0.6% that of normal serum) lose their biotin-dependent carboxylases as well as the biochemical integrity of Krebs cycle and senesced before biotin-sufficient cells [113]. Biotinylation of histones (in particular biotinylation of lysine 12 in Histone H4), an epigenetic mechanism involved in repression of long terminal repeat (LTR) retrotransposons, has been found to be downregulated by telomere attrition and associated to cellular senescence [114]. Hence, covalent modification of histones is likely to play a role in the accelerated senescence caused by biotin deficiency.

Secoiridoids: Oleuropein

Oleuropein is a phenylethanoid, a type of phenolic compound found in the olive leaf. It has been suggested to activate the gerosuppressor AMPK and delay senescence in human primary cells as well as to trigger numerous transcriptomic signatures that can suppress biologically aggressive cancer cells [115]. Oleuropein is rapidly degraded by colonic microflora producing one of its most important metabolites, Hydroxytyrosol (HT). This compound is also found in the olive leaf and olive oil and is claimed to be among the most potent antioxidants found in nature to date. It has been shown to delay cellular senescence of normal human fibroblasts by increasing MnSOD activity and decreasing age-associated mitochondrial reactive oxygen species accumulation [116].

Vitamin C (Ascorbic Acid)

Vitamin C is another intensely studied antioxidants in the context of cellular senescence in vitro. In primary cell culture, physiological concentrations of Vitamin C (0.1–0.2 mM) promote growth, cell division, and cell differentiation while delaying the onset of replicative senescence as well as counteracting the effects of stress induced senescence [6]. Importantly, Vitamin C enhances also primary fibroblast reprogramming to pluripotency, but the underlying mechanisms are unclear [117]. These effects could be related to a possible inhibition of intracellular ROS, prevention of DNA damage, and subsequent repression of cell cycle blocks. This mechanism was proposed in a model of p53-induced senescence of human bladder cancer cells, in a model of RS of human vascular endothelial cells and in a model of CKII inhibition-mediated senescence in hCCC. Moreover, some studies have suggested that there is a close correlation among declines in internal ascorbic acid levels, various disorders, and cellular senescence. Indeed, uptake of ascorbic acid was increased in old cells compared with young cells while the requirement for this vitamin was found to be enhanced by cellular senescence [118].

Trace Elements: Zinc, Copper, Iron, Selenium

The trace elements like zinc, copper, and iron have been shown to display a general prosenescence activity in vitro. Dysregulated zinc has been shown to induce senescence in dermal fibroblasts [6] and the zinc ionophorepyrithione was identified as a major senescence-inducing compound from a screen of 4160 compounds [119]. In addition to these data we have recently established that zinc (50 μM) accelerates senescence in primary human endothelial cells [120]. Interestingly, other trace elements display similar effects. Prosenescence effects mediated by downregulation of Bmi-1 pathway were demonstrated with Cu treatment (250 μM) in human glioblastoma multiforme cells [121]. Conversely, the Se compounds, sodium selenite, methylseleninic acid, and methylselenocysteine (at doses above 0.1 μM), induced an ATM-dependent senescence response likely mediated by ROS in normal cells (MRC-5 and CRL-1790) but not in cancerous cells (PC-3 and HCT 116) [122]. Regarding Fe, increasing intracellular Fe levels through Fe-citrate supplementation or decreasing intracellular iron levels using iron-selective chelators had little effect on cellular lifespan of primary human fibroblasts and markers of cellular senescence when used at subtoxic doses. However, Fe content increased exponentially during cellular senescence in fibroblasts and HUVECs (respectively 10-fold higher and 50-fold higher than young cells) whilst a treatment with low-doses of hydrogen peroxide accelerated Fe accumulation and

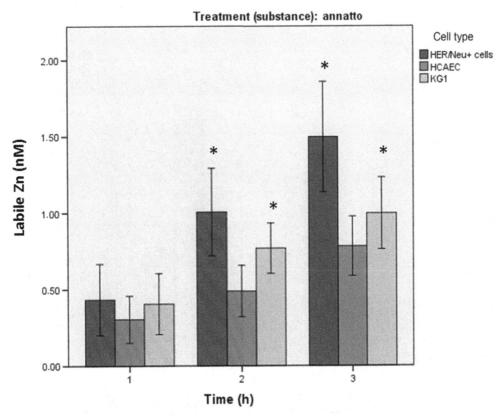

FIGURE 22.2 Evaluation of free Zn by FLuozin-3AM (fluorimetry) after treatment with Annatto tocotrienols (50 μM) in HER/Neu+ breast cancer cells, KG1 cell line, and primary human carotid artery endothelial cells (HCAEC) (all signals significantly increased with time with cell-type differences; *$p<0.05$ compared to HCAEC). Cells were incubated with the intracellular Zn probe FLuozin-3AM 1 μM, added with the nutritional compound, and fluorescence continuously monitored by fluorimetry. Fluorescent signals were converted into an estimation of labile Zn using the formula of Grynkiewicz (Grynkiewicz et al. J Biol Chem. 1985;260:3440-50) after assessment of minimum (with addition of 50 μM TPEN) and maximum (with addition of 100 μM Zn-Pyrithione) fluorescence in separate wells.

senescence-related changes [123]. These results suggest that a common mechanism of action of bioactive dietary factors, especially phenolic compounds, could involve interference with Zn, Cu, and Fe through their oxydryl and chetoester groups or other mechanisms involving intracellular stress response. In this context, there is already evidence that the polyphenols luteolin, apigenin, EGCG, resveratrol, olive oil polyphenols, and caffeic acid can inhibit cell proliferation and induce apoptosis in different cancer cell lines through mobilization of intracellular Cu and generation of ROS [124,125]. We also herein report our evidence of cell-type selective mobilization of Zn after treatment with resveratrol and annatto tocotrienol (Figs. 22.2 and 22.3).

Catechins: Epigallocatechin Gallate (EGCG)

Nontoxic concentrations (15 μM) of epigallocatechin gallate (EGCG) can shorten telomeres, increase SA-betaGal staining, induce chromosomal abnormalities and, most importantly, limit the lifespan of U937 monoblastoidleukemia and HT29 colon adenocarcinoma cell lines [126]. Further experiments in MCF-7 and HL60 cell lines confirmed an inhibitory activity on telomerase activity by EGCG [127]. Alterations in histone modifications, decreased methylation of hTERT promoter and increased binding of the hTERT repressor E2F-1 at the promoter were proposed as mediators of the observed bioactivity. In contrast to the pro-senescent activity of EGCG at 15 μM in U937 and HT29 cells, higher concentration of EGCG (50–100 μM) were shown to the prevent replicative and stress induced senescence in various primary cells, likely as a consequence of its inhibitory activity on p53 [128]. However, these opposite findings are in agreement with the previously observed cancer-selective prooxidant effects induced by EGCG [129].

Alkaloids: Berberine, 1-Deoxynojirimycin

There is a series of scientific evidence about the ability of berberine to exert cell type-specific effects that, in

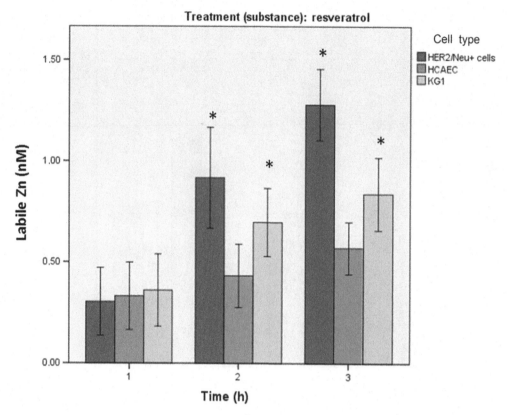

FIGURE 22.3 Evaluation of free Zn by fluozin-3AM (fluorimetry) after treatment with resveratrol (50 μM) in HER/Neu + breast cancer cells, KG1 cell line and primary human carotid artery endothelial cells. (All signals significantly increased with time with cell-type differences; *$p < 0.05$ compared to HCAEC.) For technical details see Fig. 22.3.

certain circumstances, include cell cycle arrest and induction of a senescent-like phenotype. Although several studies show that berberine primarily exerts its anticancer effect by inducing cell cycle arrest, apoptosis, and autophagy, it has been observed that the antitumor effect of berberine on glioblastoma cells is mediated by senescence induction [130]. This effect was reported to be a consequence of the inhibition of RAF-MEK-ERK signaling pathway. Moreover, the antitumor effects of berberine and berberine-derived synthetic derivatives in human HER-2/neu overexpressing breast cancer cells were found to be mediated by increased expression of p53, p21, p16, and PAI-1 mRNAs, thus suggesting that the mechanism of action of berberine may include also the induction of cellular senescence [131]. However, berberine is also reported as an inhibitor of AMPK/mTOR [132] and, consistently with this activity, it was shown (at 5–60 μM) to block stress-induced senescence in lymphoblastoid and cell lung cancers [132] as well as to delay replicative senescence of human fibroblasts [133]. A different alkaloid, 1-deoxynojirimycin, which is obtained from several plants (ie, mulberry) and microorganisms, was also shown to delay cellular senescence that is promoted under high glucose condition in HUVECs [134]. This effect is likely related to its previously recognized activity as an inhibitor of α-glucosidase and suppressor of postprandial hyperglycemia that makes this compound relevant to the pharmaceutical industry.

Flavones and Isoflavones: Apigenin, Genistein

A putative inhibitor of Nrf2 pathway and of protein kinase CKII, apigenin (belonging to flavone class), was shown to induce senescence at 20 μM in human diploid fibroblast IMR-90 cells [135]. Conversely, the isoflavone genistein was shown to display protective activity against cellular senescence in UVB-induced senescence model using human dermal fibroblasts [136]. Downregulation of total and phosphorylated p66Shc on Ser36, as well as FKHRL1 and its phosphorylation on Thr32, were observed after genistein treatment and are likely to be involved in the mechanisms of protection. Since there are still poor data on the effects of flavones and isoflavones on cellular senescence it is difficult at this moment to build a hypothesis on the common mechanism of action.

Flavonols: Quercetin, Morin

Conversely to isoflavone and flavones, various data are available around the modulatory activity of flavonols, especially quercetin, on cellular senescence. One of the most important findings regards the activity of quercetin on terminally senescent human fibroblasts. These cells, treated with 6–7 μM of quercetin for 5 consecutive days were surprisingly shown to restart proliferation compared to the control cultures [47]. Quercetin even at very high concentration (300 μM) was also shown to prevent the premature senescent phenotype and the upregulation of caveolin-1 induced by hydrogen peroxide in different cellular models [137,138]. Similar results were obtained with a relatively lower concentration (50 μM) of quercetin in a model of stress-induced senescence in retinal pigment epithelial cells [139]. An example of positive induction of cellular senescence has been also reported with 25 μM quercetin in C6 rat glioma cells [140]. Senescent modulatory effects of morin were studied with the aim to find a protective agent against UVB irradiation in human keratinocyte stem cells. Treatment of morin in this model suppressed the induction of senescence markers involving the activation of MDM2 and subsequent inhibition of p53 [141].

Cannabinoids: Cannabinol

Cannabinoids include a range of bioactive components of the Cannabis plant. An arrest at all phases of the cell cycle was reported as a main effect of different cannabinoids in leukemic cells. Moreover, gene expression analysis of cannabidiol and Δ(9)-tetrahydrocannabinol (THC) (10 μM) in BV-2 cells revealed an upregulation of senescence effectors including CDKN2A (p16) and CDKN1A (p21), in particular with cannabidiol [142]. Regarding treatments in primary cells there are no data available in literature. However, in our laboratory we have demonstrated that HUVECs cells approaching senescence restart proliferation compared to untreated cells after treatment with cannabinol 0.5 μM (Fig. 22.4). The mechanism involved in the activity of cannabinoids in these models could involve a downstream inhibition of mTOR as a consequence of the activation of the cannabinoid receptors pathway.

COMMENTS AND CONCLUSIONS

This chapter does not encompass the use of nutritional modulators of cellular senescence in vivo as this argument is covered in another large review [6].

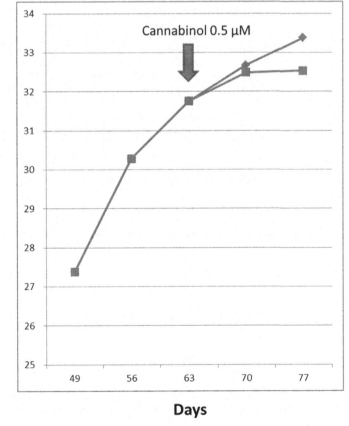

FIGURE 22.4 Effect of cannabinol on growth curve of primary human umbilical endothelial cells (HUVECs) approaching to senescence. The figure display cumulative population doublings (CPDs) versus days in culture. Treatment with cannabinol (in blue (gray in print versions)) slows the growth arrest compared to controls (in red (black in print versions)).

However, the data herein reviewed show that cellular senescence can be modulated by a variety of nutritional compounds. This evidence opens the question of whether there is the possibility to use this knowledge to develop nutraceuticals for therapeutic purposes and if common mechanisms could explain the common cellular outcome in response to different compounds. Regarding this last point it can be observed that all nutritional compounds presented above can interfere with one or more of these biochemical pathways: (1) the nuclear factor erythroid 2-related factor 2 (NRF2) pathway, which is involved in a cytoprotective response to stress; (2) the pathways that regulate homeostasis of trace elements; (3) metabolic pathways that regulate energy consumption and storage. Another possible common mechanism, that might be the direct consequence of an early alteration in one of the pathways above, regards epigenetic modifications in target genes involved in regulation of senescence effectors. The presence of common mechanisms however does not impede that these compounds display a cell-type and condition specific response that makes it difficult to evaluate the real potential for human health. Moreover, we still don't know exactly the role of cellular senescence in aging and in various age-related diseases thus making it complicate to address a specific pathology. Another critical aspect regards the concentrations commonly used in vitro. In most cases concentrations in the micromolar range are used whereas it is more likely that oral consumptions of these polyphenols and other nonnutrient antioxidants can determinate nanomolar or subnanomolar concentrations in blood. This is also complicated by the metabolic rearrangement by gut microbiota and other transformation that occur in our body. Regarding gut microbiota it is noteworthy to mention the recent discovery that colibactin, a genotoxic compound produced by *Escherichia coli*, can induce senescence and promote tumor growth in vivo [143]. The lack of appropriate biomarkers of cellular senescence for human studies and the poor evidence from clinical studies with adjuvants in chemotherapy makes it necessary to investigate this field with further studies.

SUMMARY

- Cellular senescence is an antagonistic hallmark of aging, as it can be a beneficial response that in the long term may turn into dangerous
- Senescent cells accumulate in aged tissues and are currently a major target of developing therapies to prevent or cure age related diseases
- Bioactive dietary compounds (including resveratrol, tocotrienols, berberine, quercetin, curcumin, vitamin C, ginsenosides, polyphenols and cannabinoids) may be used to induce, slow or even partly reverse cellular senescence depending on the experimental settings and on the cellular targets
- A new type of compounds that is able to remove the excessive accumulation of senescent cells in aging (named senolytic compounds) is emerging as potential therapy for a multitude of age related diseases.
- Careful attention in the development of nutritional compounds with senolytic activity should be given (as senescence may exert a beneficial response) and appropriate studies in experimental models are required before planning clinical trials.

References

[1] Hayflick L, Moorhead PS. The serial cultivation of human diploid cell strains. Exp Cell Res 1961;25:585–621.
[2] Counter CM, Meyerson M, Eaton EN, Ellisen LW, Caddle SD, Haber DA, et al. Telomerase activity is restored in human cells by ectopic expression of hTERT (hEST2), the catalytic subunit of telomerase. Oncogene 1998;16:1217–22.
[3] van Deursen JM. The role of senescent cells in ageing. Nature 2014;509:439–46.
[4] (a)Muñoz-Espín D, Serrano M. Cellular senescence: from physiology to pathology. Nat Rev Mol Cell Biol 2014;15:482–96.(b) Bernardes de Jesus B, Blasco MA. Assessing cell and organ senescence biomarkers. Circ Res 2012;111:97–109.
[5] Campisi J. Aging, cellular senescence, and cancer. Annu Rev Physiol 2013;75:685–705.
[6] Malavolta M, Costarelli L, Giacconi R, Piacenza F, Basso A, Pierpaoli E, et al. Modulators of cellular senescence: mechanisms, promises, and challenges from in vitro studies with dietary bioactive compounds. Nutr Res 2014;34:1017–35.
[7] Bayreuther K, Rodemann HP, Hommel R, Dittmann K, Albiez M, Francz PI. Human skin fibroblasts in vitro differentiate along a terminal cell lineage. Proc Natl Acad Sci USA 1988;85:5112–16.
[8] Krishnamurthy J, Torrice C, Ramsey MR, Kovalev GI, Al-Regaiey K, Su L, et al. Ink4a/Arf expression is a biomarker of aging. J Clin Invest 2004;114:1299–307.
[9] Bracken AP, Kleine-Kohlbrecher D, Dietrich N, Pasini D, Gargiulo G, Beekman C, et al. The Polycomb group proteins bind throughout the INK4A-ARF locus and are disassociated in senescent cells. Genes Dev 2007;21:525–30.
[10] von Zglinicki T, Saretzki G, Ladhoff J, d'Adda di Fagagna F, Jackson SP. Human cell senescence as a DNA damage response. Mech Ageing Dev 2005;126:111–17.
[11] Chandra T, Ewels PA, Schoenfelder S, Furlan-Magaril M, Wingett SW, Kirschner K, et al. Global reorganization of the nuclear landscape in senescent cells. Cell Rep Cell Rep 2015; S2211-1247(14)01122-X.
[12] Swanson EC, Manning B, Zhang H, Lawrence JB. Higher-order unfolding of satellite heterochromatin is a consistent and early event in cell senescence. J Cell Biol 2013;203:929–42.
[13] Sadaie M, Salama R, Carroll T, Tomimatsu K, Chandra T, Young AR, et al. Redistribution of the Lamin B1 genomic binding profile affects rearrangement of heterochromatic domains and SAHF formation during senescence. Genes Dev 2013;27:1800–8.

[14] Shah PP, Donahue G, Otte GL, Capell BC, Nelson DM, Cao K, et al. Lamin B1 depletion in senescent cells triggers large-scale changes in gene expression and the chromatin landscape. Genes Dev 2013;27:1787–99.

[15] Coppé JP, Desprez PY, Krtolica A, Campisi J. The senescence-associated secretory phenotype: the dark side of tumor suppression. Annu Rev Pathol 2010;5:99–118.

[16] Demaria M, Ohtani N, Youssef SA, Rodier F, Toussaint W, Mitchell JR, et al. An essential role for senescent cells in optimal wound healing through secretion of PDGF-AA. Dev Cell 2014;31:722–33.

[17] Xue W, Zender L, Miething C, Dickins RA, Hernando E, Krizhanovsky V, et al. Senescence and tumour clearance is triggered by p53 restoration in murine liver carcinomas. Nature 2007;445:656–60.

[18] Acosta JC, Banito A, Wuestefeld T, Georgilis A, Janich P, Morton JP, et al. A complex secretory program orchestrated by the inflammasome controls paracrine senescence. Nat Cell Biol 2013;15:978–90.

[19] Krtolica A, Parrinello S, Lockett S, Desprez PY, Campisi J. Senescent fibroblasts promote epithelial cell growth and tumorigenesis: a link between cancer and aging. Proc Natl Acad Sci USA 2001;98:12072–7.

[20] Kurz DJ, Decary S, Hong Y, Erusalimsky JD. Senescence-associated (beta)-galactosidase reflects an increase in lysosomal mass during replicative ageing of human endothelial cells. J Cell Sci 2000;113:3613–22.

[21] Yang NC, Hu ML. The limitations and validities of senescence associated-beta-galactosidase activity as an aging marker for human foreskin fibroblast Hs68 cells. Exp Gerontol 2005;40:813–19.

[22] Georgakopoulou EA, Tsimaratou K, Evangelou K, Fernandez Marcos PJ, Zoumpourlis V, Trougakos IP, et al. Specific lipofuscin staining as a novel biomarker to detect replicative and stress-induced senescence. A method applicable in cryo-preserved and archival tissues. Aging (Albany NY) 2013;5:37–50.

[23] Gerland LM, Peyrol S, Lallemand C, Branche R, Magaud JP, Ffrench M. Association of increased autophagic inclusions labeled for beta-galactosidase with fibroblastic aging. Exp Gerontol 2003;38:887–95.

[24] Deschênes-Simard X, Lessard F, Gaumont-Leclerc MF, Bardeesy N, Ferbeyre G. Cellular senescence and protein degradation: breaking down cancer. Cell Cycle 2014;13:1840–58.

[25] Young ARJ, Narita M, Ferreira M, Kirschner K, Sadaie M, Darot JFJ, et al. Autophagy mediates the mitotic senescence transition. Genes Dev 2009;23:798–803.

[26] Demidenko ZN, Blagosklonny MV. At concentrations that inhibit mTOR, resveratrol suppresses cellular senescence. Cell Cycle 2009;8:1901–4.

[27] Pospelova TV, Leontieva OV, Bykova TV, Zubova SG, Pospelov VA, Blagosklonny MV. Suppression of replicative senescence by rapamycin in rodent embryonic cells. Cell Cycle 2012;11:2402–7.

[28] Garbers C, Kuck F, Aparicio-Siegmund S, Konzak K, Kessenbrock M, Sommerfeld A, et al. Cellular senescence or EGFR signaling induces Interleukin 6 (IL-6) receptor expression controlled by mammalian target of rapamycin (mTOR). Cell Cycle 2013;12:3421–32.

[29] Wall M, Poortinga G, Stanley KL, Lindemann RK, Bots M, Chan CJ, et al. The mTORC1 inhibitor everolimus prevents and treats Eμ-Myc lymphoma by restoring oncogene-induced senescence. Cancer Discov 2013;3:82–95.

[30] Kennedy AL, Adams PD, Morton JP. Ras, PI3K/Akt and senescence: paradoxes provide clues for pancreatic cancer therapy. Small GTPases 2011;2:264–7.

[31] Salama R, Sadaie M, Hoare M, Narita M. Cellular senescence and its effector programs. Genes Dev 2014;28:99–114.

[32] Narita M, Young ARJ, Arakawa S, Samarajiwa SA, Nakashima T, Yoshida S, et al. Spatial coupling of mTOR and autophagy augments secretory phenotypes. Science 2011;332:966–70.

[33] Cardelli M, Giacconi R, Malavolta M, Provinciali M. Endogenous retroelements as key players in pathogenic processes related to cellular and organismal senescence: new therapeutic targets in human diseases. Curr Drug Targets 2015; [in press].

[34] Sedivy JM, Kreiling JA, Neretti N, De Cecco M, Criscione SW, Hofmann JW, et al. Death by transposition - the enemy within? Bioessays 2013;35:1035–43.

[35] Jurk D, Wang C, Miwa S, Maddick M, Korolchuk V, Tsolou A, et al. Postmitotic neurons develop a p21-dependent senescence-like phenotype driven by a DNA damage response. Aging Cell 2012;11:996–1004.

[36] Maejima Y, Adachi S, Ito H, Hirao K, Isobe M. Induction of premature senescence in cardiomyocytes by doxorubicin as a novel mechanism of myocardial damage. Aging Cell 2008;7:125–36.

[37] Chien Y, Scuoppo C, Wang X, Fang X, Balgley B, Bolden JE, et al. Control of the senescence-associated secretory phenotype by NF-κB promotes senescence and enhances chemosensitivity. Genes Dev 2011;25:2125–36.

[38] Baker DJ, Wijshake T, Tchkonia T, LeBrasseur NK, Childs BG, van de Sluis B, et al. Clearance of p16Ink4a-positive senescent cells delays ageing-associated disorders. Nature 2011;479:232–6.

[39] Lapasset L, Milhavet O, Prieur A, Besnard E, Babled A, Aït-Hamou N, et al. Rejuvenating senescent and centenarian human cells by reprogramming through the pluripotent state. Genes Dev 2011;25:2248–53.

[40] Feng Q, Lu SJ, Klimanskaya I, Gomes I, Kim D, Chung Y, et al. Hemangioblastic derivatives from human induced pluripotent stem cells exhibit limited expansion and early senescence. Stem Cells 2010;28:704–12.

[41] Beauséjour CM, Krtolica A, Galimi F, Narita M, Lowe SW, Yaswen P, et al. Reversal of human cellular senescence: roles of the p53 and p16 pathways. EMBO J 2003;22:4212–22.

[42] Dirac AM, Bernards R. Reversal of senescence in mouse fibroblasts through lentiviral suppression of p53. J Biol Chem 2003;278:11731–4.

[43] Kang HT, Lee CJ, Seo EJ, Bahn YJ, Kim HJ, Hwang ES. Transition to an irreversible state of senescence in HeLa cells arrested by repression of HPV E6 and E7 genes. Mech Ageing Dev 2004;125:31–40.

[44] Takahashi A, Ohtani N, Yamakoshi K, Iida S, Tahara H, Nakayama K, et al. Mitogenic signalling and the p16INK4a-Rb pathway cooperate to enforce irreversible cellular senescence. Nat Cell Biol 2006;8:1291–7.

[45] Kolesnichenko M, Hong L, Liao R, Vogt PK, Sun P. Attenuation of TORC1 signaling delays replicative and oncogenic RAS-induced senescence. Cell Cycle 2012;11:2391–401.

[46] McFarland GA, Holliday R. Retardation of the senescence of cultured human diploid fibroblasts by carnosine. Exp Cell Res 1994;212:167–75.

[47] Chondrogianni N, Kapeta S, Chinou I, Vassilatou K, Papassideri I, Gonos ES. Anti-ageing and rejuvenating effects of quercetin. Exp Gerontol 2010;45:763–71.

[48] Makpol S, Durani LW, Chua KH, Mohd Yusof YA, Ngah WZ. Tocotrienol-rich fraction prevents cell cycle arrest and elongates telomere length in senescent human diploid fibroblasts. J Biomed Biotechnol 2011;2011:506171.

[49] Laberge RM, Adler D, DeMaria M, Mechtouf N, Teachenor R, Cardin GB, et al. Mitochondrial DNA damage induces apoptosis in senescent cells. Cell Death Dis 2013;4:e727.

[50] Zhu Y, Tchkonia T, Pirtskhalava T, et al. The Achilles' heel of senescent cells: from transcriptome to senolytic drugs. Aging Cell 2015; [in press].

[51] Dörr JR, Yu Y, Milanovic M, Beuster G, Zasada C, Däbritz JH, et al. Synthetic lethal metabolic targeting of cellular senescence in cancer therapy. Nature 2013;501:421−5.

[52] Zhang JW, Zhang SS, Song JR, Sun K, Zong C, Zhao QD, et al. Autophagy inhibition switches low-dosecamptothecin-induced premature senescence to apoptosis in human colorectal cancer cells. Biochem Pharmacol 2014;90:265−75.

[53] Hoenicke L, Zender L. Immune surveillance of senescent cells—biological significance in cancer- and non-cancer pathologies. Carcinogenesis 2012;33:1123−6.

[54] Vilgelm AE, Pawlikowski JS, Liu Y, Hawkins OE, Davis TA, Smith J, et al. Mdm2 and aurora kinase a inhibitors synergize to block melanoma growth by driving apoptosis and immune clearance of tumor cells. Cancer Res 2015;75:181−93.

[55] Dasanu CA, Mitsis D, Alexandrescu DT. Hair repigmentation associated with the use of lenalidomide: graying may not be an irreversible process!. J Oncol Pharm Pract 2013;19:165−9.

[56] Mocchegiani E, Romeo J, Malavolta M, Costarelli L, Giacconi R, Diaz LE, et al. Zinc: dietary intake and impact of supplementation on immune function in elderly. Age (Dordr) 2013; 35:839−60.

[57] Laberge RM, Zhou L, Sarantos MR, Rodier F, Freund A, de Keizer PL, et al. Glucocorticoids suppress selected components of the senescence-associated secretory phenotype. Aging Cell 2012;11:569−78.

[58] Moiseeva O, Deschênes-Simard X, Pollak M, Ferbeyre G. Metformin, aging and cancer. Aging (Albany NY) 2013;5: 330−1.

[59] Menendez JA, Cufí S, Oliveras-Ferraros C, Martin-Castillo B, Joven J, Vellon L, et al. Metformin and the ATM DNA damage response (DDR): accelerating the onset of stress-induced senescence to boost protection against cancer. Aging (Albany NY) 2011;3:1063−77.

[60] Imagawa K, de Andrés MC, Hashimoto K, Pitt D, Itoi E, Goldring MB, et al. The epigenetic effect of glucosamine and a nuclear factor-kappa B (NF-kB) inhibitor on primary human chondrocytes—implications for osteoarthritis. Biochem Biophys Res Commun 2011;405:362−7.

[61] Pocobelli G, Kristal AR, Patterson RE, Potter JD, Lampe JW, Kolar A, et al. Total mortality risk in relation to use of less-common dietary supplements. Am J Clin Nutr 2010; 91:1791−800.

[62] Alspach E, Flanagan KC, Luo X, Ruhland MK, Huang H, Pazolli E, et al. p38MAPK plays a crucial role in stromal-mediated tumorigenesis. Cancer Discov 2014;4:716−29.

[63] Liu Y, Hawkins OE, Su Y, Vilgelm AE, Sobolik T, Thu YM, et al. Targeting aurora kinases limits tumour growth through DNA damage-mediated senescence and blockade of NF-κB impairs this drug-induced senescence. EMBO Mol Med 2013;5:149−66.

[64] Nardella C, Clohessy JG, Alimonti A, Pandolfi PP. Pro-senescence therapy for cancer treatment. Nat Rev Cancer 2011;11:503−11.

[65] Gewirtz DA, Holt SE, Elmore LW. Accelerated senescence: an emerging role in tumor cell response to chemotherapy and radiation. Biochem Pharmacol 2008;76:947−57.

[66] Suzuki M, Boothman DA. Stress-induced premature senescence (SIPS)–influence of SIPS on radiotherapy. J Radiat Res 2008;49:105−12.

[67] Lee M, Lee JS. Exploiting tumor cell senescence in anticancer therapy. BMB Rep 2014;47:51−9.

[68] Di Mitri D, Toso A, Chen JJ, Sarti M, Pinton S, Jost TR, et al. Tumour-infiltrating Gr-1 + myeloid cells antagonize senescence in cancer. Nature 2014;515:134−7.

[69] Bodnar AG, Ouellette M, Frolkis M, Holt SE, Chiu CP, Morin GB, et al. Extension of life-span by introduction of telomerase into normal human cells. Science 1998;279:349−52.

[70] Duan J, Zhang Z, Tong T. Senescence delay of human diploid fibroblast induced by anti-sense p16INK4a expression. J Biol Chem 2001;276:48325−31.

[71] Itahana K, Zou Y, Itahana Y, Martinez JL, Beausejour C, Jacobs JJ, et al. Control of the replicative life span of human fibroblasts by p16 and the polycomb protein Bmi-1. Mol Cell Biol 2003;23:389−401.

[72] Gil J, Bernard D, Martinez D, Beach D. Polycomb CBX7 has a unifying role in cellular lifespan. Nat Cell Biol 2004;6:67−72.

[73] Dietrich N, Bracken AP, Trinh E, Schjerling CK, Koseki H, Rappsilber J, et al. Bypass of senescence by the polycomb group protein CBX8 through direct binding to the INK4A-ARF locus. EMBO J 2007;26:1637−48.

[74] Nickoloff BJ, Chaturvedi V, Bacon P, Qin JZ, Denning MF, Diaz MO. Id-1 delays senescence but does not immortalize keratinocytes. J Biol Chem 2000;275:27501−4.

[75] Tang J, Gordon GM, Nickoloff BJ, Foreman KE. The helix-loop-helix protein id-1 delays onset of replicative senescence in human endothelial cells. Lab Invest 2002;82:1073−9.

[76] Zheng W, Wang H, Xue L, Zhang Z, Tong T. Regulation of cellular senescence and p16(INK4a) expression by Id1 and E47 proteins in human diploid fibroblast. J Biol Chem 2004; 279:31524−32.

[77] Wei W, Herbig U, Wei S, Dutriaux A, Sedivy JM. Loss of retinoblastoma but not p16 function allows bypass of replicative senescence in human fibroblasts. EMBO Rep 2003;4:1061−6.

[78] Brown JP, Wei W, Sedivy JM. Bypass of senescence after disruption of p21CIP1/WAF1 gene in normal diploid human fibroblasts. Science 1997;277:831−4.

[79] Malavolta M, Pierpaoli E, Giacconi R, Costarelli L, Piacenza F, Basso A, et al. Senescence of cancer cells induced by natural substances: concerns related to NRF2 signaling and senescence immunosurveillance. Curr Drug Targets 2015; [in press].

[80] de Cabo R, Liu L, Ali A, Price N, Zhang J, Wang M, et al. Serum from calorie-restricted animals delays senescence and extends the lifespan of normal human fibroblasts in vitro. Aging 2015; [in press].

[81] Bernardes de Jesus B, Schneeberger K, Vera E, Tejera A, Harley CB, Blasco MA. The telomerase activator TA-65 elongates short telomeres and increases health span of adult/old mice without increasing cancer incidence. Aging Cell 2011;10:604−21.

[82] So MW, Lee EJ, Lee HS, Koo BS, Kim YG, Lee CK, et al. Protective effects of ginsenoside Rg3 on human osteoarthritic chondrocytes. Mod Rheumatol 2013;23:104−11.

[83] Shi AW, Gu N, Liu XM, Wang X, Peng YZ. Ginsenoside Rg1 enhances endothelial progenitor cell angiogenic potency and prevents senescence in vitro. J Int Med Res 2011;39: 1306−18.

[84] Yang HH, Son JK, Jung B, Zheng M, Kim JR. Epifriedelanol from the root bark of Ulmus davidiana inhibits cellular senescence in human primary cells. Planta Med 2011;77:441−9.

[85] Sin S, Kim SY, Kim SS. Chronic treatment with ginsenoside Rg3 induces Akt-dependent senescence in human glioma cells. Int J Oncol 2012;4:1669−74.

[86] Liu J, Cai SZ, Zhou Y, Zhang XP, Liu DF, Jiang R, et al. Senescence as a consequence of ginsenoside rg1 response on k562 human leukemia cell line. Asian Pac J Cancer Prev 2012;13:6191−6.

[87] Briganti S, Wlaschek M, Hinrichs C, Bellei B, Flori E, Treiber N, et al. Small molecular antioxidants effectively protect from PUVA-induced oxidative stress responses underlying fibroblast senescence and photoaging. Free Radic Biol Med 2008; 45:636–44.

[88] Khee SG, Yusof YA, Makpol S. Expression of senescence-associated microRNAs and target genes in cellular aging and modulation by tocotrienol-rich fraction. Oxid Med Cell Longev 2014;2014:725929.

[89] Eitsuka T, Nakagawa K, Miyazawa T. Down-regulation of telomerase activity in DLD-1 human colorectal adenocarcinoma cells by tocotrienol. Biochem Biophys Res Commun 2006;348:170–5.

[90] Pierpaoli E, Viola V, Pilolli F, Piroddi M, Galli F, Provinciali M. Gamma- and delta-tocotrienols exert a more potent anti-cancer effect than alpha-tocopheryl succinate on breast cancer cell lines irrespective of HER-2/neu expression. Life Sci 2010;86:668–75.

[91] Gui CY, Ngo L, Xu WS, Richon VM, Marks PA. Histone deacetylase (HDAC) inhibitor activation of p21WAF1 involves changes in promoter-associated proteins, including HDAC1. Proc Natl Acad Sci USA 2004;101:1241–6.

[92] Nesaretnam K, Meganathan P, Veerasenan SD, Selvaduray KR. Tocotrienols and breast cancer: the evidence to date. Genes Nutr 2012;7:3–9.

[93] Wan C, Liu J, Nie X, Zhao J, Zhou S, Duan Z, et al. 2, 3, 7, 8-Tetrachlorodibenzo-P-dioxin (TCDD) induces premature senescence in human and rodent neuronal cells via ROS-dependent mechanisms. PLoS One 2014;9:e89811.

[94] Kim KS, Kang KW, Seu YB, Baek SH, Kim JR. Interferon-gamma induces cellular senescence through p53-dependent DNA damage signaling in human endothelial cells. Mech Ageing Dev 2009;130:179–88.

[95] Berniakovich I, Laricchia-Robbio L, Izpisua Belmonte JC. N-acetylcysteine protects induced pluripotent stem cells from in vitro stress: impact on differentiation outcome. Int J Dev Biol 2012;56:729–35.

[96] Holliday R, McFarland GA. Inhibition of the growth of transformed and neoplastic cells by the dipeptide carnosine. Br J Cancer 1996;73:966–71.

[97] Hipkiss AR, Cartwright SP, Bromley C, Gross SR, Bill RM. Carnosine: can understanding its actions on energy metabolism and protein homeostasis inform its therapeutic potential? Chem Cent J 2013;7:38.

[98] Zhou Z, Hu CP, Wang CJ, Li TT, Peng J, Li YJ. Calcitonin gene-related peptide inhibits angiotensin II-induced endothelial progenitor cells senescence through up-regulation of klotho expression. Atherosclerosis 2010;213:92–101.

[99] Zanichelli F, Capasso S, Cipollaro M, Pagnotta E, Cartenì M, Casale F, et al. Dose-dependent effects of R-sulforaphane isothiocyanate on the biology of human mesenchymal stem cells, at dietary amounts, it promotes cell proliferation and reduces senescence and apoptosis, while at anti-cancer drug doses, it has a cytotoxic effect. Age (Dordr) 2012;34:281–93.

[100] Meeran SM, Patel SN, Tollefsbol TO, Zanichelli, et al. Sulforaphane causes epigenetic repression of hTERT expression in human breast cancer cell lines. PLoS One 2010;5: e11457.

[101] Li Y, Tollefsbol TO. p16(INK4a) suppression by glucose restriction contributes to human cellular lifespan extension through SIRT1-mediated epigenetic and genetic mechanisms. PLoS One 2011;6:e17421.

[102] Li Y, Liu L, Tollefsbol TO. Glucose restriction can extend normal cell lifespan and impair precancerous cell growth through epigenetic control of hTERT and p16 expression. FASEB J 2010;24:1442–53.

[103] Zhang B, Cui S, Bai X, Zhuo L, Sun X, Hong Q, et al. SIRT3 overexpression antagonizes high glucose accelerated cellular senescence in human diploid fibroblasts via the SIRT3-FOXO1 signaling pathway. Age (Dordr) 2013;35:2237–53.

[104] Mortuza R, Chen S, Feng B, Sen S, Chakrabarti S. High glucose induced alteration of SIRTs in endothelial cells causes rapid aging in a p300 and FOXO regulated pathway. PLoS One 2013;8:e54514.

[105] Warburg O. On the origin of cancer cells. Science 1956;123:309–14.

[106] Hubbard BP, Gomes AP, Dai H, Li J, Case AW, Considine T, et al. Evidence for a common mechanism of SIRT1 regulation by allosteric activators. Science 2013;339:1216–19.

[107] Yuan Q, Peng J, Liu SY, Wang CJ, Xiang DX, Xiong XM, et al. Inhibitory effect of resveratrol derivative BTM-0512 on high glucose-induced cell senescence involves dimethylaminohydrolase/asymmetric dimethylarginine pathway. Clin Exp Pharmacol Physiol 2010;37:630–5.

[108] (a)Bishayee A, Politis T, Darvesh AS. Resveratrol in the chemoprevention and treatment of hepatocellular carcinoma. Cancer Treat Rev 2010;36:43–53. (b)Luo H, Yang A, Schulte BA, Wargovich MJ, Wang GY. Resveratrol induces premature senescence in lung cancer cells via ROS-mediated DNA damage. PLoS One 2013;8:e60065.

[109] Mosieniak G, Adamowicz M, Alster O, Jaskowiak H, Szczepankiewicz AA, Wilczynski GM, et al. Curcumin induces permanent growth arrest of human colon cancer cells: link between senescence and autophagy. Mech Ageing Dev 2012;133:444–55.

[110] Hendrayani SF, Al-Khalaf HH, Aboussekhra A. Curcumin triggers p16-dependent senescence in active breast cancer-associated fibroblasts and suppresses their paracrine procarcinogenic effects. Neoplasia 2013;15:631–40.

[111] Grabowska W, Kucharewicz K, Wnuk M, Lewinska A, Suszek M, Przybylska D, et al. Curcumin induces senescence of primary human cells building the vasculature in a DNA damage and ATM-independent manner. Age (Dordr) 2015;37:9744.

[112] Shen LR, Parnell LD, Ordovas JM, Lai CQ. Curcumin and aging. Biofactors 2013;39:133–40.

[113] Atamna H, Newberry J, Erlitzki R, Schultz CS, Ames BN. Biotin deficiency inhibits heme synthesis and impairs mitochondria in human lung fibroblasts. J Nutr 2007;137:25–30.

[114] Wijeratne SS, Camporeale G, Zempleni J. K12-biotinylated histone H4 is enriched in telomeric repeats from human lung IMR-90 fibroblasts. J Nutr Biochem 2010;21:310–16.

[115] Corominas-Faja B, Santangelo E, Cuyàs E, Micol V, Joven J, Ariza X, et al. Computer-aided discovery of biological activity spectra for anti-aging and anti-cancer olive oil oleuropeins. Aging (Albany NY) 2014;6:731–41.

[116] Sarsour EH, Kumar MG, Kalen AL, Goswami M, Buettner GR, Goswami PC. MnSOD activity regulates hydroxytyrosol-induced extension of chronological lifespan. Age (Dordr) 2012;34:95–109.

[117] Esteban MA, Wang T, Qin B, Yang J, Qin D, Cai J, et al. Vitamin C enhances the generation of mouse and human induced pluripotent stem cells. Cell Stem Cell 2010;6:71–9.

[118] Saitoh Y, Morishita A, Mito S, Tsujiya T, Miwa N. Senescence-induced increases in intracellular oxidative stress and enhancement of the need for ascorbic acid in human fibroblasts. Mol Cell Biochem 2013;380:129–41.

[119] Ewald JA, Peters N, Desotelle JA, Hoffmann FM, Jarrard DF. A high-throughput method to identify novel senescence-inducing compounds. J Biomol Screen 2009;14:853–8.

[120] Malavolta M., Ford D. Accelerated senescence of human coronary endothelial cells by a moderately excessive zinc environment. Zinc-Net conference; London, 3 Nov 2014.

[121] Li Y, Hu J, Guan F, Song L, Fan R, Zhu H, et al. Copper induces cellular senescence in human glioblastoma multiforme cells through downregulation of Bmi-1. Oncol Rep 2013; 29:1805—10.

[122] Wu M, Kang MM, Schoene NW, Cheng WH. Selenium compounds activate early barriers of tumorigenesis. J Biol Chem 2010;285:12055—62.

[123] Killilea DW, Wong SL, Cahaya HS, Atamna H, Ames BN. Iron accumulation during cellular senescence. Ann N Y Acad Sci 2004;1019:365—7.

[124] Khan HY, Zubair H, Faisal M, Ullah MF, Farhan M, Sarkar FH, et al. Plant polyphenol induced cell death in human cancer cells involves mobilization of intracellular copper ions and reactive oxygen species generation: a mechanism for cancer chemopreventive action. Mol Nutr Food Res 2014;58: 437—46.

[125] Bhat SH, Azmi AS, Hadi SM. Prooxidant DNA breakage induced by caffeic acid in human peripheral lymphocytes: involvement of endogenous copper and a putative mechanism for anticancer properties. Toxicol Appl Pharmacol 2007;218:249—55.

[126] Naasani I, Seimiya H, Tsuruo T. Telomerase inhibition, telomere shortening, and senescence of cancer cells by tea catechins. Biochem Biophys Res Commun 1998;249:391—6.

[127] Berletch JB, Liu C, Love WK, Andrews LG, Katiyar SK, Tollefsbol TO. Epigenetic and genetic mechanisms contribute to telomerase inhibition by EGCG. J Cell Biochem 2008;103:509—19.

[128] Han DW, Lee MH, Kim B, Lee JJ, Hyon SH, Park JC. Preventive effects of epigallocatechin-3-O-gallate against replicative senescence associated with p53 acetylation in human dermal fibroblasts. Oxid Med Cell Longev 2012;2012: 850684.

[129] Yamamoto T, Hsu S, Lewis J, Wataha J, Dickinson D, Singh B, et al. Green tea polyphenol causes differential oxidative environments in tumor versus normal epithelial cells. J Pharmacol Exp Ther 2003;307:230—6.

[130] Liu Q, Xu X, Zhao M, Wei Z, Li X, Zhang X, et al. Berberine induces senescence of human glioblastoma cells by downregulating the EGFR-MEK-ERK signaling pathway. Mol Cancer Ther 2015;14:355—63.

[131] Pierpaoli E, Arcamone AG, Buzzetti F, Lombardi P, Salvatore C, Provinciali M. Antitumor effect of novel berberine derivatives in breast cancer cells. Biofactors 2013;39:672—9.

[132] Zhao H, Halicka HD, Li J, Darzynkiewicz Z. Berberine suppresses gero-conversion from cell cycle arrest to senescence. Aging (Albany NY) 2013;5:623—36.

[133] Halicka HD, Zhao H, Li J, Lee YS, Hsieh TC, Wu JM, et al. Potential anti-aging agents suppress the level of constitutive mTOR- and DNA damage- signaling. Aging (Albany NY) 2012;4:952—65.

[134] Shuang E, Kijima R, Honma T, Yamamoto K, Hatakeyama Y, Kitano Y, et al. 1-Deoxynojirimycin attenuates high glucose-accelerated senescence in human umbilical vein endothelial cells. Exp Gerontol 2014;55:63—9.

[135] Ryu SW, Woo JH, Kim YH, Lee YS, Park JW, Bae YS. Downregulation of protein kinase CKII is associated with cellular senescence. FEBS Lett 2006;580:988—94.

[136] Wang YN, Wu W, Chen HC, Fang H. Genistein protects against UVB-induced senescence-like characteristics in human dermal fibroblast by p66Shc down-regulation. Dermatol Sci 2010;58:19—27.

[137] Volonte D, Zhang K, Lisanti MP, Galbiati F. Expression of caveolin-1 induces premature cellular senescence in primary cultures of murine fibroblasts. Mol Biol Cell 2002;13: 2502—17.

[138] Dasari A, Bartholomew JN, Volonte D, Galbiati F. Oxidative stress induces premature senescence by stimulating caveolin-1 gene transcription through p38 mitogen-activated protein kinase/Sp1-mediated activation of two GC-rich promoter elements. Cancer Res 2006;66:10805—14.

[139] Kook D, Wolf AH, Yu AL, Neubauer AS, Priglinger SG, Kampik A, et al. The protective effect of quercetin against oxidative stress in the human RPE in vitro. Invest Ophthalmol Vis Sci 2008;49:1712—20.

[140] Zamin LL, Filippi-Chiela EC, Dillenburg-Pilla P, Horn F, Salbego C, Lenz G. Resveratrol and quercetin cooperate to induce senescence-like growth arrest in C6 rat glioma cells. Cancer Sci 2009;100:1655—62.

[141] Lee J, Shin YK, Song JY, Lee KW. Protective mechanism of morin against ultraviolet B-induced cellular senescence in human keratinocyte stem cells. Int J Radiat Biol 2014; 90:20—8.

[142] Juknat A, Pietr M, Kozela E, Rimmerman N, Levy R, Coppola G, et al. Differential transcriptional profiles mediated by exposure to the cannabinoids cannabidiol and Δ9-tetrahydrocannabinol in BV-2 microglial cells. Br J Pharmacol 2012;165:2512—28.

[143] Dalmasso G, Cougnoux A, Delmas J, Darfeuille-Michaud A, Bonnet R. The bacterial genotoxin colibactin promotes colon tumor growth by modifying the tumor microenvironment. Gut Microbes 2014;5:675—80.

PART III

SYSTEM AND ORGAN TARGETS

CHAPTER 23

Nutrition, Diet Quality, and Cardiovascular Health

Li Wang, Geeta Sikand and Nathan D. Wong

Division of Cardiology, Medical Sciences, University of California, Irvine, CA, USA

KEY FACTS
- Cardiovascular diseases are aging-related.
- Low saturated fat diet should achieve an optimal macronutrient composition.
- A cardioprotective dietary pattern contains foods high in bioactive compounds.
- Long-term dietary intervention is the key for successful primary and secondary prevention of cardiovascular diseases.

Dictionary of Terms
- *ASCVD*: Atherosclerotic cardiovascular disease
- *CHO*: Carbohydrates
- *CVD*: Cardiovascular disease
- *DBP*: Diastolic blood pressure
- *HDL-C*: High-density lipoprotein cholesterol
- *LDL-C*: Low-density lipoprotein cholesterol
- *MetS*: Metabolic Syndrome
- *MNT*: Medical Nutrition Therapy
- *MUFA*: Monounsaturated fatty acids
- *PUFA*: Polyunsaturated fatty acids
- *RCT*: Randomized controlled trial
- *RDN*: Registered Dietitian Nutritionist
- *ROS*: Reactive oxygen species
- *SBP*: Systolic blood pressure
- *SFA*: Saturated fatty acids
- *TC*: Total cholesterol
- *TG*: Triglycerides

INTRODUCTION

Cardiovascular disease (CVD) includes coronary heart disease (CHD), cerebrovascular disease (including transient ischemic attack and stroke), hypertension, peripheral artery disease, rheumatic heart disease, arrhythmias, congenital heart disease, and heart failure. CVD is ranked as the leading cause of morbidity and mortality in both developed and developing countries; the global burden of CVD is still rising [1]. The 2015 update on heart disease statistics from the American Heart Association (AHA) reported that the prevalence of CVD in American adults in 2012 was 35%, an estimated 85.6 million adults (>1 in 3) have at least one type of CVD. Of these, 43.7 million are estimated to be ≥60 years of age. CVD as the listed underlying cause of death accounted for 31.3% all deaths, or ≈1 of every 3 deaths in the United States. By 2030, 43.9% of the US population is projected to have some form of CVD [2].

The initiation of atherosclerosis results from a combination of abnormalities in lipoprotein metabolism, oxidative stress, thrombosis, and chronic inflammation [3]. Multiple factors are involved in the development of CVD, including nonmodifiable risk factors (age, gender, and family history), and modifiable risk factors (including dyslipidemia, hypertension, hyperglycemia, poor diet, smoking, overweight/obesity, physical inactivity) [4–6]. The role of nutrition in the initiation, progression, and reversal of CVD is well established. The purpose of this chapter is to review our current understanding of the role that nutrition plays in the prevention and treatment of CVD associated with aging.

MOLECULAR BASIS AND MODIFIABLE RISK FACTORS OF CARDIOVASCULAR DISEASE

The current model for the pathogenesis of atherosclerotic cardiovascular disease (ASCVD) is based on several hypotheses, including vascular response to injury, vascular wall retention of low-density lipoprotein (LDL) particles, and oxidative modification of LDL. Overall, atherosclerosis is proposed to be an inflammatory disease of arteries. Modified forms of LDL, such as acetylated LDL and oxidized LDL, can be taken up by macrophages by the scavenger receptor, leading to substantial cholesterol accumulation and foam cell formation. The activated macrophage produces inflammatory cytokines and leads to inflammatory processes and the evolution of a more advanced atherosclerotic lesion [7].

Aging is associated with a redistribution of both fat and lean tissue within the body. Along with aging, physical inactivity, menopause, hormonal changes, reduced fatty acid utilization, resistance to leptin, and increased oxidative stress all expedite intra-abdominal fat accumulation and the loss of lean body mass (sarcopenia) [8]. Increases in visceral fat play a major role in the pathogenesis of insulin resistance that induces metabolic syndrome. Metabolic syndrome is a multicomponent risk factor for CVD and type 2 diabetes that reflects the clustering of cardiometabolic risk factors (excess abdominal adiposity, dyslipidemia, elevated blood pressure, and hyperglycemia) and is also related to other cardiovascular risk factors, such as insulin resistance, nonalcoholic fatty liver disease, and prothrombotic and pro-inflammatory states [9]. It increases the likelihood of developing CVD within 10 years, even in the absence of diabetes [10]. The criteria for defining metabolic syndrome (MetS) include three or more of the following: (1) elevated waist circumference (according to population and country-specific definitions); (2) triglyceride (TG) ≥150 mg/dL; (3) high density lipoprotein (HDL)-cholesterol <40 mg/dL in men and <50 mg/dL in women; (4) blood pressure of 130/85 mmHg or greater or on treatment; and (5) fasting glucose 100 mg/dL or greater or on hypoglycemic treatment. Most criteria of MetS are associated with prediabetes or diabetes; however, many investigators prefer to separate out those who have frank diabetes from those with metabolic syndrome [10]. At the same time, vessel aging leads to intimal and medial thickening (vascular remodeling) as well as gradual loss of arterial elasticity, resulting in vascular stiffness and subsequently hypertension [11]. Vascular aging and atherosclerosis are also associated with cellular senescence. Cellular senescence impairs cell proliferation resulting in irreversible growth arrest and impairs survival, due to an accumulation of nuclear and mitochondrial DNA damage, increased reactive oxygen species (ROS), and a pro-inflammatory state [11]. A summary of aging and atherosclerosis is presented in Fig. 23.1.

Recommended nutrition and medical interventions target modifiable risk factors and can significantly reduce the progression of atherosclerosis and CVD. These modifiable risk factors are presented in Table 23.1 [4–6]. The 2013 American College of Cardiology (ACC) /AHA Guideline on the Assessment of Cardiovascular Risk estimates 10-year and lifetime ASCVD risk from age, sex, race, total cholesterol, HDL cholesterol, systolic blood pressure, blood pressure lowering medication use, diabetes status, and smoking status [13].

NUTRIENTS, FOODS AND ASCVD RISK FACTORS

Lipids and Lipoproteins

The primary focus of prevention of ASCVD is on lowering LDL-C. Non-HDL-C has been treated as a secondary therapeutic target for CVD since the residual risk is mainly due to low HDL-C and TG-rich lipoprotein levels. The atherogenic lipoprotein phenotype associated with insulin resistance includes high TG, increased very low-density lipoprotein (VLDL), low HDL-C, and increased small, dense LDL particles (sdLDL), the latter of which are more susceptible to in vivo oxidation in the artery wall [14]. The following are general dietary recommendations for achieving optimal lipids and lipoproteins:

- *Reducing dietary saturated fat and achieving an optimal macronutrient intake*
  The 2013 AHA/ACC Guideline on Lifestyle Management to Reduce CVD Risk recommends a dietary pattern aimed on LDL-lowering that emphasizes intake of vegetables, fruits, and whole grains; includes low-fat dairy products, poultry, fish, legumes, nontropical vegetable oils and nuts; and limits intake of sweets, sugar-sweetened beverages, and red meats [15]. Specifically, a focus of dietary recommendations is reducing saturated fat and *trans* fat intake, primarily as a means of lowering LDL-cholesterol concentrations. The 2010 Dietary Guidelines and the 2015 Dietary Guidelines Advisory Committee (DGAC) [16,17] recommend

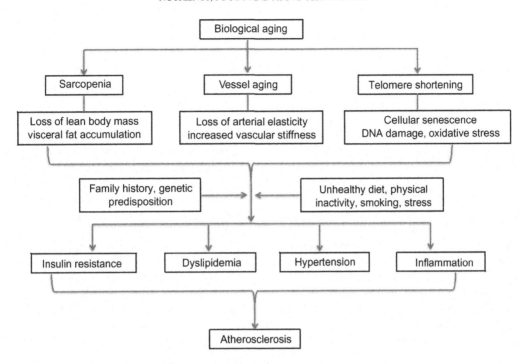

FIGURE 23.1 A summary of aging and other risk factors in the development of atherosclerosis.

TABLE 23.1 Modifiable Risk Factors and Ideal Levels for CVD Health [4–6]

Modifiable risk factors	Goals for modifiable risk factors
Weight	*BMI*: 18.5–24.9
Lipid profile[a]	*Total cholesterol*: <200 mg/dL
	LDL-C: <100 mg/dL; <70 mg/dL if with clinical CVD or diabetes
	Non-HDL-C: <130 mg/dL; <100 mg/dL if with clinical CVD or diabetes
	HDL-C: Men: ≥40 mg/dL; Women: ≥50 mg/dL
	Triglycerides: <150 mg/dL
Blood pressure	*Systolic*: <120 mmHg; *Diastolic*: <80 mmHg
Glycemic control	*Blood glucose*: <100 mg/dL
Physical activity	150 min moderate activity or 75 min vigorous activity per week
Smoking	Stop smoking
Diet	Follow a cardioprotective dietary pattern
Stress	Keep stress as low as possible

[a]*The LDL-C and nonHDL-C targets are endorsed by International Atherosclerosis Society [5] and National Lipid Association [6]. Instead, 2013 ACC/AHA Guideline suggests initiating statin therapy and lifestyle change to reduce risk of ASCVD in four statin eligible groups of individuals: (1) those with clinical ASCVD, (2) primary elevations of LDL-C ≥90 mg/dL, (3) aged 40–75 years old with diabetes and LDL-C 70–189 mg/dL without clinical ASCVD, and (4) those aged 40–75 years old with LDL-C 70–189 mg/dL and 10-year ASCVD risk ≥7.5% [12].*

healthy dietary patterns that contain less than 10% calories from saturated fat (SFA) for lowering LDL-C. The 2013 AHA/ACC Lifestyle Guidelines recommend 5–6% of calories from SFA [15].

Multiple clinical trials and epidemiological data have demonstrated that dietary SFAs raise LDL-C level by decreasing LDL receptor activity, protein, and mRNA abundance, while unsaturated fatty acids increase these variables [18]. Intake of *trans* fatty acids induces an increase in cholesterol synthesis and LDL-C level. *Trans* fatty acids decrease HDL-C too. In 2013, the FDA made a tentative determination on the ban of any use of partially hydrogenated oils that contain *trans* fatty acids [19]. SFAs exist broadly in the common western diet, such as animal fats and dairy desserts. Nonhydrogenated vegetable oils that are high in unsaturated fats and relatively low in SFA (eg, soybean, corn, olive, and canola oils) instead of animal fats (eg, butter, cream, beef tallow, and lard) or tropical oils (eg, palm, palm kernel, and coconut oils) should be recommended as the primary source of dietary fat [17].

A meta-analysis of 60 controlled trials shows all SFAs except stearic acid increase LDL-C and HDL-C [20]. A recent Cochrane review concluded that reducing SFAs lowered the risk of CVD events by 14% [21]. In contrast, two recent meta-analyses including prospective cohort studies showed poor association of dietary SFA intake with CHD [22,23]. However, many of the studies that were included (more than half) in these meta-analyses used 24-hour recalls or some other dietary assessment method with known limitations to assess long-term dietary habits raises questions about the reliability of the results [24]. Also, a reduction in SFA intake must be evaluated in the context of food source and the replacement by other macronutrients. Mensink and Katan [20,25] found that replacing 1% of SFA with an equal amount of CHO, MUFA, or PUFA led to comparable LDL-C reductions: 1.2, 1.3, and 1.8 mg/dL, respectively. Replacing 1% of SFA with carbohydrate (CHO), monounsaturated fatty acid (MUFA), or polyunsaturated fatty acid (PUFA) also lowered HDL-C by 0.4, 1.2, and 0.2 mg/dL, respectively. Replacing 1% of CHO by an equal amount of MUFA or PUFA raised LDL-C by 0.3 and 0.7 mg/dL, raised HDL-C by 0.3 and 0.2 mg/dL, and lowered TG by 1.7 and 2.3 mg/dL, respectively [20,25]. The ratio of total cholesterol to HDL-C (TC/HDL-C), a stronger predictor of CVD risk than total or LDL cholesterol alone, did not change when SFA was replaced by CHO, but the ratio significantly decreased when SFA was replaced by unsaturated fats, especially PUFA [20]. Current macronutrient intake status in US adults and the recommended intake range of each macronutrient is shown in Table 23.2.

- *Carbohydrates*

  Replacing SFA with CHO also reduces total and LDL cholesterol, but significantly increases TG and reduces HDL-C. Large prospective studies showed replacing total fat with CHO does not lower CVD risk [17]. However, a diet high in complex CHO, whole grain foods and dietary fiber, and low in SFA, has benefits on CVD risk [27]. It was estimated that every serving (28 g/day) of whole grain consumption was associated with a 5% (95% confidence interval [CI]: 2–7%) lower total morality or a 9% (95% CI: 4–13%) lower CVD mortality [27]. In contrast, total grain intake and refined grain intake were not associated with CVD risk [28]. Overall, there is insufficient evidence exists to conclude the effect on CHD risk of replacing SFAs with CHO although there might be a benefit if the CHO composition includes more whole grain products and has a low glycemic index.

- *Polyunsaturated fat*

  Clinical, epidemiological, and mechanistic studies consistently show that the LDL-C and risk of CHD are reduced when SFAs are replaced with PUFAs. For every 1% of energy intake from SFA replaced with PUFA, incidence of CHD is reduced by 2–3% [29]. However, intakes of PUFAs have been limited to ≤10% of energy due to their potential adverse effects, including a reduction of HDL-C level and increased susceptibility of LDL to oxidation [30]. Omega-3 (n-3) and omega-6 (n-6) PUFAs are essential nutrients that cannot be synthesized by humans [26]. The parent fatty acid for the n-3 series is α-linolenic acid (ALA; 18:3n-3), and the parent fatty acid for the n-6 series is linoleic acid (LA; 18:2n-6). Bioconversion of these PUFAs leads to the production of several LC fatty acids including eicosapentaenoic acid (EPA; 20:5n-3) and docosahexaenoic acid (DHA; 22:4n-6) in the n-3 series, and arachidonic acid (AA; 20:4n-6) in the n-6 series. However, the bioconversion of the n-3 series to these very long chain FAs is low [26]. Research have confirmed the protective effect of both marine-derived and plant derived n-3 FAs against CVD [31]. The evidence for plant-based n-3 FAs (rich in ALA) is less conclusive than that for marine-based, longer chain n-3 FAs, EPA and DHA [32]. Flaxseeds, walnuts, and canola oil are sources of ALA, while seafood, such as salmon, mackerel, herring, and sardines, are the richest sources of EPA and DHA [33,34]. ALA can reduce total and LDL cholesterol and TG; EPA and DHA can increase HDL-C and reduce TG [35]. LA is the primary n-6 FA in the diet, found in vegetable oils, such as corn, soybean, and safflower oil, as well as many seeds and nuts. Using LA to replace SFA can decrease LDL-C and improve LDL/HDL-C. The possible mechanism by which n-6 PUFAs decrease cholesterol include upregulation of the LDL receptor and increased cholesterol 7 α-hydrolase (CYP7) acitivity through the upregulation on liver X receptor (LXR), whereas n-3 PUFAs decrease TG by decreasing lipogenesis and VLDL secretion, increasing LPL activity, and increasing reverse cholesterol transport.

TABLE 23.2 Current Macronutrient Intake Status in US Adults and the Recommended Intake Range of Macronutrients

Macronutrients	NHANES 2010–2012[a]	Recommend intake[b]
Total fat (% kcal)	33%	20–35%
SFA (% kcal)	11%	≤7%
PUFA (% kcal)	7%	Up to 10%
MUFA (% kcal)	12%	Up to 20%
CHO (% kcal)	49%	45–65%
Protein (% kcal)	15%	10–35%

[a]Data source: NAHNES 2010–2012.
[b]Based on Institute of Medicine's Acceptable Macronutrient Distribution Range [26], Dietary Guidelines 2010 [16], and AHA/ACC diet and lifestyle guidelines [15].

- *Monounsaturated fatty acids*

  Clinical studies have demonstrated convincingly that dietary MUFA beneficially affect MetS risk versus dietary CHO since the potential cardioprotective, high-MUFA Mediterranean diet was examined in the Seven Countries Study [36]. Substituting SFA with MUFA lowers LDL-C, increases HDL-C, and lowers TC/HDL-C ratio. Replacing 5% calories from SFA with MUFA could potentially reduce CHD risk by 7.5% via reduction in the TC/HDL-C ratio [37]. However, there is a lack of randomized controlled trials (RCTs) directly confirm the casual link. Studies suggest using dietary MUFA to replace SFA and refined CHO promotes healthy blood lipid profiles, mediates blood pressure, improves insulin sensitivity and regulates glucose levels. Moreover, metabolism of dietary MUFA may affect body composition and ameliorating the risk of obesity [38]. Epidemiologic studies conducted with western populations have reported a neutral or positive association of dietary MUFA and CHD risk [39,40], but high-MUFA foods in the Mediterranean diet (olive oil and nuts) have demonstrated cardioprotective benefits in epidemiologic studies [41–43] and a long-term RCT [44]. Insufficient evidence exists to judge the effect on CHD risk of replacing SFAs with MUFAs, mainly because the available data on MUFAs are limited and confounded by the major food sources of MUFAs (eg, dairy and meats) in Western dietary patterns. It is evident that we need to identify the favorable food sources of MUFA in a healthy diet, such as extra virgin olive oil, nuts and avocados. Unlike SFA and PUFA with a recommend limit, MUFA intakes are determined by calculating the difference, that is, MUFA (% energy) = total fat (% of energy)—SFA (% of energy)—PUFA (% of energy)—*trans* fat (% energy), thus, MUFA intakes will range with respect to the CHO and total fat composition of the diet. Current dietary guidelines recommend the intake range of total fat is 20–35% of total energy, including MUFA up to 20% (Table 23.2).

- *Dietary cholesterol*

  Previously, the Dietary Guidelines for Americans recommended that cholesterol intake be limited to no more than 300 mg/day. The 2015 DGAC discontinued this recommendation because available evidence shows no strong relationship between consumption of dietary cholesterol and serum cholesterol, consistent with the conclusions of the 2013 AHA/ACC guideline [15,17]. Typically, dietary cholesterol suppresses the activity of LDL receptors, thereby inhibiting cholesterol clearance from plasma and increasing LDL-C levels [45]. However, the causal link is not established because there are hyperresponders and hyporesponders on LDL-C to dietary cholesterol intake [46]. An overview of epidemiological studies reported no association between eating one egg a day ($\approx$200 mg cholesterol) and CVD risk in nondiabetic men and women [47]. In contrast, several cohort studies have shown that persons with type 2 diabetes who consume one egg a day have up to a two-fold increase in CVD risk over type 2 diabetes that consume less than one egg per week [48–50]. Furthermore, consumption of $\geq$2 eggs/day has been associated with a greater risk of heart failure among male physicians (1.64 [95 % CI: 1.08–2.49], $P < 0.006$ for trend) [51]. Since LDL receptor protein expression is decreased with aging [52], older adults tend to have more hyperresponders to dietary cholesterol impacting on LDL–C level. Thus, limiting dietary cholesterol intake <300 mg/day may be reasonable for people with or at high risk of CVD or type 2 diabetes, especially in the aged population.

  Because of the recent controversy surrounding recommendations for dietary cholesterol intake, most recently the National Lipid Association (NLA) [52a] completed a thorough review of the effects of dietary cholesterol on atherogenic lipid levels, including well-controlled randomized controlled trials (RCTs) and meta-analyses of RCTs. This review indicated that, for each 100 mg/day of dietary cholesterol, blood levels of LDL-C increase by an average of ~1.9 mg/dL and that within the population, there are hyper- and hypo-responders where some individuals experience little or no increase in LDL-C in response to a greater intake of dietary cholesterol, whereas others experience greater increases. As there is no method available to determine hyper- vs. hypo-responders, the NLA recently recommended that dietary cholesterol be limited to < 200 mg per day as part of a cardioprotective dietary pattern to reduce LDL-C and non-HDL-C levels by limiting egg yolks to 2–4 a week and avoiding organ meats like liver and gizzards.

- *Diet portfolio to achieve LDL-C lowering*

  A moderate body of evidence suggest that soy protein decreases total cholesterol (TC) and LDL-C in both normo- or hyperlipidemic individuals [53]. When plant sterols (1 g/1000 kcal), viscous fiber (8.2 g/1000 kcal), and plant protein ($\approx$50 g/1000 kcal) were incorporated into a diet already low in dietary cholesterol (99 mg/1000 kcal) and saturated fat (7.7% of energy), both normal and hyperlipidemic subjects had a significant decrease in TC (−22.3%), LDL-C (−29.0%), the TC/HDL-C

TABLE 23.3 Change in Dietary Factors and Estimated LDL Reduction

Dietary component	Recommendation	Approximate LDL reduction
Major:		
Saturated fat	<7% calories	8–10%
Weight reduction	Lose 10 lb	5–8%
Other:		
Soy protein	25 g/day	5%
Viscous fiber	5–10 g/day	3–5%
Plant sterol/stanol esters	2 g/day	6–15%
Cumulative estimate		25–35%

Adapted from the Third Report of the National Cholesterol Education Program (NCEP) Expert Panel on Detection, Evaluation and Treatment of High Blood Cholesterol in Adults-Adult Treatment Panel III [3] and the clinical trials on diet portfolio [55] and soy protein [53].

ratio (−19.8%), and the LDL-C/HDL-C ratio (−26.5%) compared to baseline [54]. Based on these blood lipid results, the calculated CHD risk reduction of this Portfolio Diet is 30% [54]. A summary of change in dietary factors and estimated LDL reduction is shown in Table 23.3.

Blood Pressure

Both blood lipids and blood pressure could be significantly affected by dietary factors and have been treated as the primary therapeutic targets in CVD prevention and therapy [15]. Both a healthy dietary pattern and a reduced sodium intake lower BP independently. A combined DASH (Dietary Approaches to Stop Hypertension) dietary pattern with lower sodium intake could lead to greater BP reduction outcomes. The 2013 AHA/ACC Lifestyle Management Guideline recommended lowering sodium intake to less than 2400 mg/day. A further reduction to 1500 mg/day in sodium intake is recommended for those at higher risk or for those who seek further BP lowering. A reduction of sodium intake by 1000 mg/day at any intake level will further reduce BP [15]. A reduction in sodium intake by approximately 1000 mg/day reduces CVD events by about 30% [56]. Changes in sodium intake influence blood volume; increased sodium levels cause an increase in water retention resulting in an increase in blood volume and blood pressure [57,58]. A meta-analysis of 13 cohort studies by Strazzullo and colleagues reported that an increased sodium intake increased risk for stroke and cardiovascular disease; 2000 mg/day increase in intake resulting in a 23% higher risk for stroke [59]. Other evidence, however, has shown that a restricted sodium intake (1840 mg/day) in persons with heart failure significantly increased hospitalization and mortality when compared to persons on an intake of 2760 mg/day [60], indicating that restriction may be harmful in certain populations [61]. The 2015 DGAC report concurs with the 2013 AHA/ACC Lifestyle Management Guideline report that evidence is not sufficient to determine the association between sodium intake and the development of heart failure [15,17].

There is a moderate body of evidence for an association between potassium intake and blood pressure levels [62]. A higher potassium intake has been associated with lower blood pressure in adults [63–67] and several observational studies have shown that an increase in potassium intake decreases risk of stroke and coronary heart disease [34]. Meta-analyses of clinical and cohort studies showed that an increase in potassium intake, for an average supplemental level between 44 and 86 mmol/day, lowered blood pressure [65,66]. However, evidence is not sufficient to determine whether increasing dietary potassium intake lowers blood pressure and CHD risk. Foods high in potassium includes beans, dark leafy greens, potatoes, squash, banana, mushrooms, and avocados. The 2015 DGAC report stated that "in observational studies with appropriate adjustments (eg, blood pressure, sodium intake), higher dietary potassium intake is associated with lower risk for stroke" [17].

Other minerals like calcium and magnesium also have potential benefits on blood pressure, but the evidence is not adequate to determine their effects on CVD risk. Also, consumption of calcium and magnesium is limited to relatively few specific food groups, it was unlikely to make a recommendation on the food groups to increase or decrease consumption of the mineral [15].

Insulin Resistance and Metabolic Syndrome

Aging increases the risk of MetS, mainly due to the visceral fat accumulation induced insulin resistance. The treatment of MetS aims to improve insulin sensitivity and prevent the associated metabolic and cardiovascular abnormalities. Since many individuals with MetS are overweight, dietary treatment should be primarily focused on weight reduction, especially on reduction of visceral fat. An optimal diet for decreasing risk of MetS includes three basics: (1) reduced SFA; (2) increased vegetables, fruit, legumes, nuts and low-glycemic index CHO; and (3) energy intake balance for weight loss. In traditional low-fat diets, fats are often replaced with refined CHO and this is of particular concern because such diets are generally associated

TABLE 23.4 Summary of Responses of ALA, EPA, and DHA on Metabolic Syndrome Risk Factors [35]

Risk factor	Physiological responses	
	ALA	EPA and DHA
Adiposity	None	Reduces visceral adiposity
		Increases mitochondrial biogenesis and oxidative metabolism
Dyslipidemia	Reduces total cholesterol, LDL-C and TG	Increases HDL-C
		Reduces TG
		Increases lipoprotein lipase activity
		Reduces TG lipase activity
Insulin resistance	Animal studies show improved glucose tolerance and insulin resistance index	Unknown
Hypertension	Reduces systolic blood pressure	Inhibits renin secretion
		Reduces the formation of thromboxane A2
Oxidative stress	Data inconclusive	Does not reduce oxidative damage but improves serum antioxidant enzymes levels. Animal studies show reasonable evidence of reducing oxidative damage but also show prooxidant responses to EPA and DHA at higher doses
Inflammation	Reduces pro-inflammatory cytokines and LOX/COX metabolites of AA	

Adapted from Poudyal et al. [35].

with atherogenic dyslipidemia (hypertriglyceridemia and low HDL-C concentrations) [68]. Therefore, dietary advice should put more emphasis on optimizing types of dietary fat than reducing total fat. Dietary CHO increases blood glucose levels, particularly in the postprandial period, and consequently increases insulin levels and plasma TG. The detrimental effects of a high-CHO diet on plasma glucose/insulin, TG/HDL or fibrinolysis occur only when CHO foods with a high glycemic index are consumed, while the side effects are attenuated if the diet is based largely on fiber-rich, low-glycemic-index foods [69].

Evidence has shown that a high MUFA diet significantly improves insulin sensitivity compared to a high SFA diet. However, this beneficial effect of monounsaturated fat disappears when total fat intake exceeds 38% of total energy [69]. Recent data suggest a role for preferential oxidation and metabolism of dietary MUFA, influencing body composition and ameliorating central obesity [38].

Protein intake is also important to maintain body lean mass for older adults. The AHA Science Advisory takes a more modest stance, and recommends that diets contain 15–20% of calories from protein [70]. Higher protein diets must be planned carefully to incorporate important nutrients found in fruits, vegetables, and whole grains and to avoid protein-rich foods excessively high in total fat, saturated fat, and dietary cholesterol [70].

N-3 PUFAs have multiple effects on MetS risk factors including dyslipidemia, insulin resistance, visceral adiposity, hypertension and oxidative stress [35]. Summary of effects of ALA, EPA, and DHA on MetS is shown in Table 23.4 [35].

Obesity

Approximately 70% of adults who are overweight and 75% of those who are obese have one or more cardiometabolic risk factors [15]. Overweight and obese adults can achieve weight loss through a variety of dietary patterns when calorie intake is controlled. Clinically meaningful weight losses that were achieved ranged from 4 to 12 kg at 6-month follow-up [17]. Thereafter, slow weight regain is observed, with total weight loss at 1 year of 4–10 kg and at 2 years of 3–4 kg. However, some dietary patterns may be more beneficial in the long term for cardiometabolic health [17]. A moderate amount of evidence demonstrates that intake of dietary patterns with less than 45% calories from CHO or more than 35% calories from protein are not more effective than other diets for weight loss or weight maintenance, are difficult to maintain over the long term, and may be less safe [17]. The 2013 AHA/ACC/TOS Guideline for the Management of Overweight and Obesity in Adults [71] recommends reducing dietary energy intake via:

- Specification of an energy intake target that is less than that required for energy balance, usually 1200–1500 kcal/day for women and

1500–1800 kcal/day for men (kcal levels are usually adjusted for the individual's body weight and physical activity levels);
- Estimation of individual energy requirements according to expert guidelines and prescription of an energy deficit of 500 kcal/day or 750 kcal/day or 30% energy deficit;
- Dietary patterns should be individualized as a variety of dietary approaches can produce weight loss in overweight and obese adults if reduction in dietary energy intake is achieved. At least 14 visits over 6 months with a Registered Dietitian Nutritionist (RDN) for behavior modification and personalized meal planning.

Oxidative Stress and Telomere Length

Telomere length is an indicator of biological aging and linked to pathology of CVD. Telomere length has been shown to be positively associated with nutritional status in human and animal studies. Various nutrients influence telomere length potentially through mechanisms that reflect their role in cellular functions including inflammation, oxidative stress, DNA integrity, DNA methylation, and activity of telomerase, the enzyme that adds the telomeric repeats to the ends of the newly synthesized DNA [72]. Folate, vitamin B12, nicotinamide, vitamin A and D, vitamin C and E, zinc, magnesium, omega-3 fatty acids, tea, and grape seed polyphenols are associated with longer telomere length via the enhancement of telomerase activity and DNA methylation, or the decrease of oxidative stress [72]. Studies on food groups showed that dietary fiber intake from whole grains, vegetable, and fruits consumption were associated with longer telomere length, but the processed meat consumption showed an inverse association with telomere length [72]. An analysis on 4676 women from the Nurses' Health Study concluded that greater adherence to the Mediterranean diet was associated with longer leukocyte telomeres [73].

The benefit of antioxidant bioactive compounds on CVD risk is to reduce oxidant stress and inflammation. Mechanistic studies have shown that individual antioxidants can successfully reduce antioxidant stress in vitro. However, clinical trials have failed to justify the routine use of antioxidant supplements (vitamins C and E) for the prevention and treatment of CVD [74]. In contrast, current evidence supports recommending consumption of a diet high in food sources of antioxidants instead of antioxidant supplements to reduce risk of CVD [75]. Recent studies also suggest that the antioxidant potential of plant foods is a function of the synergistic effects of numerous antioxidant compounds [76], which may explain the "antioxidant paradox" of antioxidant supplements.

In conclusion, a cardioprotective diet including foods high in antioxidants, vitamins, minerals, polyphenols, n-3 fatty acids, and dietary fiber can reduce oxidative stress, improve telomere length, and reduce biological aging linked risk for CVD.

FUNCTIONAL FOODS

Conventional Functional Foods in Cardioprotective Dietary Patterns

Under current US food regulations, functional foods have no definition despite providing a health benefit beyond the traditional nutrients [77]. Functional foods include whole foods along with fortified, enriched, or enhanced foods that could be beneficial to health within the context of a consistent well balanced diet [78]. All cardioprotective dietary patterns recommend consumption of conventional functional foods including whole grains, legumes, vegetables, fruits, nuts, seeds, and fish. These foods contain more than 8000 flavonoids that are associated with prevention of CVD [79,80]. Strong evidence shows that cardioprotective dietary patterns including the DASH dietary pattern (Dietary Approaches to Stop Hypertension), Mediterranean Style Dietary Pattern, HEI (Healthy Eating Index) dietary pattern by the United States Department of Agriculture (USDA), and the Alternate Healthy Eating Index (AHEI) by the American Heart Association have had a significant beneficial impact on all cause and cardiovascular mortality among older adults [81]. A summary of the effects of major recommended foods on CVD risk factors are presented in Table 23.5.

Whole Grains

Whole grains such as whole wheat, rye, oats, and barley have protective effects against obesity, type 2 diabetes (T2DM), cardiovascular disease, hypertension, metabolic syndrome, and various cancers [82]. Functional components in whole grains include nondigestible complex polysaccharides, for example, soluble and insoluble fibers, inulin, beta-glucan and resistant starches. Additional bioactive components in whole grains include carotenoids, phytates, phytoestrogens, phenolic acids (ferulic acids, vanilic acid, caffeic acid, syringic acid, P-cumaric acid) and tocopherols [83,84]. Moderate evidence has shown that consumption of foods rich in cereal fiber or mixtures of whole grains and bran is associated with a reduced risk of obesity (level of evidence: B/C), T2DM (level of evidence: B), or CVD (level of evidence: B) [85].

TABLE 23.5 Effects of Recommended Foods by Lifestyle Guidelines on CVD Risk Factors

Foods	Effects on CVD risk factors[a]
Fruits and vegetables	↓LDL-C, ↓BP, ↑glycemic control, ↓ oxidative stress
Whole grains versus refined CHO	↓LDL-C, ↓BP, ↑glycemic control
Soy, legumes	↓LDL-C, ↓BP
Nuts, seeds	↓LDL-C, ↓BP, ↑glycemic control, ↑HDL-C, ↓ oxidative stress
Vegetable oils versus solid fat	↓LDL-C
Low fat/skim versus full-fat dairy products	↓LDL-C, ↓BP
Lean meat, poultry versus high-fat meat	↓LDL-C, ↓BP
Seafood	↓TG, ↓BP, ↓ arrhythmia, ↓inflammation

[a]Source: original studies that have been reviewed by scientific reports of DGAC [17] and AHA/ACC Task Force on Practice Guidelines [15].

Legumes

Pinto, dark red kidney, and black beans were reported to be associated with improved dyslipidemia, postprandial glycemic responses, and weight management in T2DM patients [86]. Soybeans have antioxidant, anti-inflammatory and hypotensive effects [87]. A meta-analysis of 30 studies based on 2013 adults with normal or mild hypercholesterolemia showed that soy protein (25 g/day) lowered LDL-C 6% [53].

Fruits and Vegetables

Evidence from clinical trials and meta-analyses show that fruits and vegetables are associated with reduced risk of CVD, diabetes, obesity, and metabolic syndrome. Intake of fruits and vegetables showed improvements in TG and hemoglobin A1C levels, antioxidants, oxidative stress, and inflammatory markers and risk of carotid atherosclerosis [88].

Of the several dietary factors implicated in CHD prevention, the evidence is most consistent for fruits and vegetables [89,90], which may be due to their rich content of dietary fiber, micronutrients, and beneficial bioactive compounds. Both antioxidant (polyphenols, flavonoids, and carotenoids) and nonantioxidant (phytosterols) bioactive compounds have been intensively studied, inspired by many epidemiologic studies that have shown protective effects of plant-based diets on CVD. Recently, low fruit and vegetable consumption has been identified as one of the top risk factors for causing the greatest "loss of health" in the Global Burden of Disease 2010 Study [91].

Nuts

Moderate evidence has shown that consumption of unsalted peanuts and tree nuts, specifically walnuts, almonds, and pistachios, in the context of an isocaloric diet have a favorable impact on CVD risk factors, particularly serum lipid levels [16]. A recent meta-analysis [36] reported that nut consumption was associated with a reduction in risk of incident ischemic heart disease (IHD). Consumption of nuts was inversely associated with fatal IHD (six studies; 6749 events; RR per four servings weekly, 28.4 g; 0.76; 95% CI: 0.69, 0.84) and inversely associated with nonfatal IHD (four studies; 2101 events; RR per four servings weekly, 28.4 g; 0.78; 95% CI: 0.67, 0.92; I2 = 0%) [36]. Nuts lower inflammation by reducing C-reactive protein (hs CRP), interleukin 6, fibrinogen and by increasing adiponectin [92]. Nuts could have a beneficial effect on the endothelium due to high L-arginine content that leads to an increase in nitric oxide production [92].

Fish and Seafood

Consumption of eicosapentanoic acid (EPA) + docosahexaenoic acid (DHA) (250–500 mg/day) was associated with a 25% lower risk of coronary heart disease (CHD) death [93]. Higher plasma levels of phospholipid n-3 fatty acids were associated with lower CV mortality and incident CVD. From the first quintile to the fifth, as plasma phospholipid n-3 FA levels increased a reduction was noted in total CVD mortality (−35%), CHD deaths (−40%), arrhythmic deaths (−48%), fatal and nonfatal CHD (−28%), and ischemic stroke (−37%), $P < 0.05$ for all [94]. Blood levels of total long-chain n-3 FA and DHA were associated with a lower risk of incident atrial fibrillation (−29% for EPA + DPA + DHA and −23% for DHA; $P < 0.01$ for both) [95]. Based on seven prospective studies, a meta-analysis showed a decrease in relative risk for heart failure comparing the highest to the lowest fish intake (−15%) along with marine n-3 FA intake (−14%; $P < 0.05$ for both) [96]. Current US dietary guidelines recommend consumption of two servings (6–8 oz) per week to consume 250–500 mg/day of EPA and DHA for prevention of coronary heart disease [34].

Green Tea, Dark Chocolate, and Red Wine

In addition to the above functional foods found conventionally in currently recommended cardioprotective dietary patterns, benefits have also been noted from green tea, dark chocolate, and red wine in the prevention of CVD [97–102].

Epidemiological, clinical, and experimental evidence supports the role of green tea in preventing CVD [97].

A meta-analysis of 5 studies on green tea showed a significant association between highest green tea consumption and a decrease in risk of CAD, specifically one cup per day was associated with a 10% reduction in CAD risk and another meta-analysis of 13 studies on black tea was not associated with a reduction in CAD risk [98].

Cocoa polyphenols contain catechins, epicatechins, and procyanidins. These flavonols scavenge ROS and upregulate antioxidant defenses to improve endothelial hyper permeability by acting as antioxidants, antihypertensives, anti-inflammatory, antiatherogenic and antithrombotic effect, and also improve insulin sensitivity, vascular endothelial function, and activation of nitric oxide [99]. A meta-analysis based on seven studies reported a positive association between with the highest levels of chocolate consumption and an adjusted lower risk for CVD (RR = 0.63 (95% CI 0.44–0.90) and a 29% reduced risk of stroke compared with the lowest levels [100]. However, there is a lack of well-designed clinical studies demonstrating a cardioprotective benefit of chocolate. Uncontrolled consumption of chocolate candies could lead to high caloric and high sugar consumption [101].

High polyphenol content in red wine (but not alcohol) is associated with antioxidant, anti-inflammatory, and vasodilator effects. Evidence from small RCTs reported that red wine showed greater benefit in insulin resistance, lipid profiles, and endothelial function versus other alcoholic beverages [78]. A daily consumption of 275 mL/day of dealcoholized red wine decreased systolic and diastolic blood pressure by increasing nitric oxide levels [103].

DIETARY PATTERNS AND CLINICAL OUTCOMES

Dietary recommendations on nutrients for reducing CVD risk should be emphasizing foods that characterize healthy dietary patterns. 2015 DGAC identifies that a healthy dietary pattern is higher in vegetables, fruits, whole grains, low- or nonfat dairy, seafood, legumes, and nuts; moderate in alcohol; lower in red and processed meats; and low in sugar-sweetened foods and drinks and refined grains [17]. Cardioprotective dietary patterns that consistently associated with prevention of ASCVD are low in saturated fat, salt, and *trans*-fatty acids and encourage consumption of seafood and plant-based foods such as vegetables, fruits, legumes, lentils, nuts, seeds, and whole grains [104]. Strong evidence shows that cardioprotective dietary patterns have greater impact on all cause and cardiovascular mortality include the DASH dietary pattern (Dietary Approaches to Stop Hypertension), the Mediterranean Style Dietary Pattern, the HEI (Healthy Eating Index) dietary pattern by the USDA, and the AHEI (Alternate Healthy Eating Index) food pattern by the AHA among older adults [81].

The 2013 AHA/ACC Lifestyle Guidelines focus on four dietary patterns designed for the prevention of CVD: Dietary Approaches to Stop Hypertension (DASH) diet, a Mediterranean-style diet, a vegetarian diet, and the National Cholesterol Education Program Adult Treatment Panel III (NCEP ATP III) recommended Therapeutic Lifestyle Changes (TLC) diet. With these cardioprotective dietary patterns there are many options to accommodate individual preferences so as to enhance adherence to current guidelines for reducing CVD risk.

Dietary Approaches to Stop Hypertension (DASH) Dietary Pattern

The DASH dietary pattern emphasizes fruits, vegetables, and low-fat dairy products; incorporates whole grains, fish, nuts, and poultry; and reduces intake of red meats, sweets, and sugar-sweetened beverages [105]. Consequently, nutrient targets are higher for potassium (4700 mg/day), magnesium (500 mg/day), and calcium (1240 mg/day) and lower for total fat (27% total energy), saturated fat (6% total energy), and cholesterol (150 mg/day) in a diet providing 2100 kcal/day; CHO) and protein intake comprise 55% and 18% of total energy, respectively [105]. Other DASH-style dietary patterns have been evaluated with higher unsaturated fat content and higher protein content [68]. The DASH diet has been shown to decrease CVD risk factors, including: systolic and diastolic blood pressure (−5.5 and −3.0 mmHg, respectively), total cholesterol (−13 mg/dL), and LDL-C (−10.7 mg/dL) compared to a typical American diet [105,106]. The blood pressure-lowering effect of the DASH dietary pattern compared to typical US dietary intake was significant in all populations studied, with this effect being greater in individuals with hypertension than in those who were prehypertensive. The total and LDL-cholesterol lowering effects of the diet did not differ by race or baseline lipid concentration [106]. Overall, the DASH diet decreased 10-year CHD risk by 18% compared to the control diet, that is, the typical American diet [107]. In a study that evaluated the original DASH diet, a higher unsaturated fat- DASH diet (37% FAT, 15% PRO, 48% CHO), and a higher protein-DASH diet (27% FAT, 25% PRO, 48% CHO), all three diets reduced systolic (−8.2 to −9.5 mmHg) and diastolic (−4.1 to −5.2 mmHg) blood pressure and total (−12.4 to −19.9 mg/dL) and LDL-cholesterol (−11.6 to −14.2 mg/dL) concentrations in prehypertensive and stage 1 hypertensive individuals [68]. HDL-C significantly decreased in both the original DASH diet

(−1.4 mg/dL) and the higher protein DASH diet (−2.6 mg/dL), while HDL-C remained unchanged during the higher fat DASH diet [68]. The higher protein and higher fat DASH diets both significantly decreased TG [68]. Overall, the original DASH dietary pattern, the higher protein-DASH diet, the higher unsaturated fat-DASH diet, and the DASH diet with reduced sodium all favorably affect blood pressure, blood lipids, and CVD risk [34].

Mediterranean-Style Dietary Patterns

Although there is no uniform definition of a Mediterranean diet, some common attributes of the Mediterranean diet include: increased consumption of fruits, vegetables, whole grains, legumes, nuts, seeds, and olive oil; low to moderate consumption of red wine (nonIslamic countries), fish, poultry, and dairy products; and decreased consumption of red meat [108]. The major food sources of dietary MUFA in Mediterranean diet are nuts and olive oil, what are also high in other bioactive compounds that may have additional benefits on lowering the risk of CVD and metabolic syndrome. Epidemiological and clinical evidence both show benefits of a Mediterranean-style diet on CVD risk factors. In a prospective cohort study of a Greek population, greater adherence to a Mediterranean diet was associated with decreased total mortality, CHD death, and cancer death [109]. In the Nurses' Health Study, women who followed a Mediterranean diet most closely were at a lower relative risk (RR) for CHD (0.71) and stroke (0.87) incidence and total CVD mortality (0.61) compared to women with the lowest adherence [110]. A recent meta-analysis confirmed these results, reporting decreased total mortality (0.92) and CVD-related incidence or mortality (0.90) with a 2-point increase in the Mediterranean diet scoring index [111]. In 2013, the PREDIMED (Prevención con Dieta Mediterránea) trial reported that a Mediterranean diet (supplemented with either extra-virgin olive oil or nuts) reduced the incidence of major CVD events by approximately 30% after 5 years intervention in men and women (50–80 years of age) who were at high risk for CVD [44]. A meta-analysis of 50 studies on Mediterranean diet concluded the protective role of the Mediterranean diet on components of metabolic syndrome, including waist circumference (−0.42 cm, 95% CI: −0.82 to −0.02), HDL-C (1.17 mg/dL, 95% CI: 0.38–1.96), TG (−6.14 mg/dL, 95% CI: −10.35 to −1.93), systolic BP (−2.35 mmHg, 95% CI: −3.51 to −1.18) and diastolic BP (−1.58 mmHg, 95% CI: −2.02 to −1.13), and glucose (−3.89 mg/dL, 95% CI: −5.84 to −1.95) [112]. However, due to the lack of uniform definition of the Mediterranean diet in the RCTs and cohort studies, the strength of evidence for Mediterranean diet lowering CVD is still estimated as low in the 2013 AHA/ACC Lifestyle Management Guidelines. Since a higher Mediterranean diet scoring index tends to have reduced total mortality, CVD risk factors, and CVD incidence, the 2015 DGAC report provided a Mediterranean-style food pattern based on the healthy U.S.-Style food pattern due to the recent data supporting the health-related benefits of a Mediterranean-style diet [17].

Vegetarian Dietary Pattern

Vegetarian is a broadly encompassing term used for a variety of different categories: ovo-lactovegetarians do not consume meat or fish; ovo-vegetarians do not consume meat, fish, or dairy products; lacto-vegetarians do not consume meat, fish, or eggs; vegans do not consume any animal products; while raw vegans, Su vegetarians, and fruitarians do not consume any animal products or vegetables in the Allium family (eg, onions, garlic) [113]. Vegetarian diets emphasize fruits, vegetables, whole grains, legumes, nuts, seeds, and soy foods and include little or no animal products. Vegetarians typically consume increased fiber, soy, CHO, potassium, magnesium, folic acid, n-6 PUFAs, nonheme iron (the less biologically absorbable form of iron found predominantly in plants), and vitamin C, and have decreased total calories, total fat, SFA, cholesterol, and sodium compared to nonvegetarians [34,114].

Vegetarians tend to have lower mortality from CHD, specifically, a 24% decrease in mortality from ischemic heart disease (IHD) compared to nonvegetarians [115,116]. Epidemiological data indicate that vegetarians have a decreased prevalence of diabetes and hypertension [115,117] and a significantly lower BMI than nonvegetarians [115,117–119]. Clinical interventions show decreased total (−7.6 to −26.6%) and LDL-cholesterol (−9.2 to −37.4%) in participants randomized to a plant-based diet versus a typical Western diet [120]. A vegan combination diet (plant sterols 1.2 g/1000 kcal, soy protein 16.2 g/1000 kcal, viscous fiber 8.3 g/1000 kcal, almonds 16.6 g/1000 kcal) has been shown to decrease total cholesterol (−26.6 vs −9.9%), LDL-C (−35.0 vs −12.1%), the total cholesterol: HDL-C ratio (−20.8 vs −2.6), and the LDL: HDL-C ratio (−30.0 vs −5.1) compared to a low-fat control diet and similarly to first generation statin drug therapy [121]. It was estimated that lipid profile improvements may be responsible for the overall 31.7% decrease in CHD mortality witnessed in individuals who follow plant-based, vegetarian, and vegan diets [121].

A concern for vegetarians, especially vegans, is that they typically have lower levels of circulating serum ferritin and vitamin B12 and decreased phospholipid-incorporated n-3 PUFAs [118,119,122]. Vegetarians also show elevated blood homocysteine level [122] that has been associated with increased risk of atherosclerosis [123]. Individuals consuming vegetarian diets should therefore monitor their B-12 levels, consume B-12 fortified foods, and consider taking a B-12 supplement. In addition, decreased dietary n-3 PUFA (especially longer-chain n-3 fatty acids) intake in vegetarians may contribute to platelet aggregation and platelet structural changes, which may be associated with an increased tendency for thrombosis [113].

MEDICAL NUTRITION THERAPY FOR MANAGEMENT OF CARDIOVASCULAR RISK FACTORS

In the United States, Registered Dietitian Nutritionists (RDN) are qualified nutrition professionals who provide personalized nutrition counseling referred to as medical nutrition therapy (MNT) for the management of dyslipidemia and related cardiovascular risk factors, including hypertension, overweight/obesity, prediabetes, diabetes, and MetS [6,15,104,124]. The ACC/AHA/TOS guidelines for overweight and obesity management recommend at least 14 MNT visits over 6 months with an RDN [124]. MNT includes nutrition assessment, nutrition diagnosis, nutrition intervention, nutrition monitoring, and evaluation [104,125,126]. In a recent review of 34 studies by Lin et al. [127], the following benefits were noted:

- MNT reduced LDL-C by 3.43 mg/dL (95% CI, 5.37−1.49) at 12−24 months (25 studies); improvement in CVD risk factors include: hypertension, dyslipidemia, MetS, and impaired fasting glucose or glucose tolerance (34 studies).
- Systolic blood pressure was reduced by 2.03 mmHg (95% CI, 2.91−1.15) (31 studies) and diastolic blood pressure by 1.38 mmHg (95% CI, 1.92−0.84) at 12−24 months (24 studies). Fasting glucose was reduced by 2.08 mg/dL (95% CI, 3.29−0.88) at 12−24 months (22 studies).
- Improved weight outcomes by a pooled mean difference of 0.26, using standardized units (95% CI, 0.35−0.16) (34 studies).
- Diabetes incidence was reduced by a relative risk of 0.58 (95% CI, 0.37−0.89) at 12−24 months (8 studies).
- Improvements in intakes of fruits, vegetables, and fiber and reduction in saturated fat intake without an overall increase in sugar or total calories consumed (8 studies) [127].

A Cochrane review of nine studies [128] concluded that MNT by a RDN is recommended for treating patients with dyslipidemia and other coronary heart disease risk factors. The Academy of Nutrition and Dietetics Expert Panel on Disorders of Lipid Metabolism recommends that the RDN should provide multiple MNT visits (3−6 visits) over 8−2 weeks to further improve a patient's lipid profile. Two studies reported that the magnitude of LDL-C reduction increased with additional visits or time spent with the RDN. One study reported further reduction in LDL-C (↓21% with four RDN visits [180 min] vs ↓12% with two RDN visits [120 min] ($P < 0.027$) [129,130] and three studies in this Cochrane review also noted a reduction in the use of lipid lowering medications when MNT was provided by a RDN [129−131]. The Academy of Nutrition and Dietetics recommends that if a patient is on lipid-lowering medications, the RDN should provide three or four MNT visits averaging 45 min per session over a 6−8 week period to improve the patient's lipid profile [132].

In conclusion, Multiple MNT visits with a RDN is important for helping patients adopt cardioprotective dietary patterns and to manage dyslipidemia and related CVD risk factors such as diabetes, prediabetes, hypertension, and MetS.

SUMMARY

- Cardiovascular disease represents the leading cause of morbidity and mortality globally. Cardiovascular risk increases significantly with age due to factors such as aging-related visceral fat accumulation, arterial stiffness, inflammation, cellular senescence, and oxidative stress.
- Key lifestyle-related risk factors are modifiable and include dyslipidemia, hypertension, hyperglycemia, obesity, and inflammation and have important nutritional determinants.
- Nutritional guidelines recommend to reduce intake of saturated and *trans*-fats, sodium, and sugar and other simple CHO can make a significant impact on improving lipids, blood pressure, insulin sensitivity, inflammation, oxidative stress, and other key cardiovascular risk factors.
- Functional foods include whole foods along with fortified, enriched or enhanced foods that could be beneficial to health within the context of a consistent well balanced diet. Cardioprotective dietary patterns recommend consumption of conventional functional foods including whole grains, legumes, vegetables, fruits, nuts, seeds, and fish.
- Recommended dietary patterns, such as those rich in fresh vegetables, fruits, whole grain products,

fiber, and low in sodium and sugar rich foods and specifically a DASH, Mediterranean, or vegetarian dietary pattern can reduce levels of atherogenic lipids, blood pressure, and help to prevent cardiovascular disease and/or diabetes.
- Medical nutrition therapy provided by a registered dietitian/nutritionist over multiple sessions is a key for achieving required lifestyle behavioral change over the long term to provide the required cardiovascular benefits.

References

[1] World Health Organization. Global status report on noncommunicable diseases 2010. Description of the global burden of NCDs, their risk factors and determinants. Geneva, Switzerland: World Health Organization; 2011.

[2] Mozaffarian D, Benjamin EJ, Go AS, Arnett DK, Blaha MJ, Cushman M, et al. Executive summary: heart disease and stroke statistics-2015 update a report from the american heart association. Circulation 2015;131(4):434–41.

[3] Hansson GK. Inflammation, atherosclerosis, and coronary artery disease. N Engl J Med 2005;352(16):1685–95.

[4] Sacco RL. The New American Heart Association 2020 goal: achieving ideal cardiovascular health. J Cardiovasc Med 2011;12(4):255–7.

[5] Grundy SM, Arai H, Barter P, Bersot TP, Betteridge DJ, Carmena R, et al. An International Atherosclerosis Society Position Paper: global recommendations for the management of dyslipidemia. J Clin Lipidol 2013;7:561–5.

[6] Jacobson TA, Ito MK, Maki KC, Orringer CE, Bays HE, Jones PH, et al. National Lipid Association recommendations for patient-centered management of dyslipidemia: part 1–executive summary. J Clin Lipidol 2014;8(5):473–88.

[7] Steinberg D, Parthasarathy S, Carew TE, Khoo JC, Witztum JL. Beyond cholesterol. Modifications of low-density lipoprotein that increase its atherogenicity. N Engl J Med 1989;320(9):15–924.

[8] Beaufrere B, Morio B. Fat and protein redistribution with aging: metabolic considerations. Eur J Clin Nutr 2000;54:S48–53.

[9] Grundy SM. Pre-diabetes, metabolic syndrome, and cardiovascular risk. J Am Coll Cardiol 2012;59(7):635–43.

[10] Alberti K, Eckel RH, Grundy SM, Zimmet PZ, Cleeman JI, Donato KA, et al. Harmonizing the metabolic syndrome a joint interim statement of the International Diabetes Federation Task Force on Epidemiology and Prevention; National Heart, Lung, and Blood Institute; American Heart Association; World Heart Federation; International Atherosclerosis Society; and International Association for the Study of Obesity. Circulation 2009;120(16):1640–5.

[11] Wang JC, Bennett M. Aging and atherosclerosis mechanisms, functional consequences, and potential therapeutics for cellular senescence. Circ Res 2012;111(2):245–59.

[12] Stone NJ, Robinson JG, Lichtenstein AH, Goff DC, Lloyd-Jones DM, Smith SC, et al. Treatment of blood cholesterol to reduce atherosclerotic cardiovascular disease risk in adults: synopsis of the 2013 American College of Cardiology/American Heart Association cholesterol guideline. Ann Intern Med 2014;160 (5):339–43.

[13] Goff DC, Lloyd-Jones DM, Bennett G, Coady S, D'Agostino RB, Gibbons R, et al. 2013 ACC/AHA guideline on the assessment of cardiovascular risk, a report of the American College of Cardiology/American Heart Association Task Force on Practice Guidelines. J Am Coll Cardiol 2014;63(25 Pt A):2886.

[14] Austin MA, King MC, Vranizan KM, Krauss RM. Atherogenic lipoprotein phenotype. A proposed genetic marker for coronary heart disease risk. Circulation 1990;82(2):495–506.

[15] Eckel RH, Jakicic JM, Ard JD, Miller NH, Hubbard VS, Nonas CA, et al. 2013 AHA/ACC guideline on lifestyle management to reduce cardiovascular risk—a report of the American College of Cardiology/American Heart Association task force on practice guidelines. Circulation 2014;129(25 Suppl. 2):S76–99.

[16] Scientific Report of the 2010 Dietary Guidelines Advisory Committee. USDA 2010.

[17] Scientific Report of the 2015 Dietary Guidelines Advisory Committee. USDA 2015.

[18] Fernandez ML, West KL. Mechanisms by which dietary fatty acids modulate plasma lipids1. J Nutr 2005;135(9):2075–8.

[19] Brownell KD, Pomeranz JL. The trans-fat ban—food regulation and long-term health. N Engl J Med 2014;370(19):1773–5.

[20] Mensink RP, Zock PL, Kester AD, Katan MB. Effects of dietary fatty acids and carbohydrates on the ratio of serum total to HDL cholesterol and on serum lipids and apolipoproteins: a meta-analysis of 60 controlled trials. Am J Clin Nutr 2003;77 (5):1146–55.

[21] Hooper L, Summerbell CD, Thompson R, Sills D, Roberts FG, Moore HJ, et al. Reduced or modified dietary fat for preventing cardiovascular disease. Cochrane Database Syst Rev 2012;(5): CD002137.

[22] Siri-Tarino PW, Sun Q, Hu FB, Krauss RM. Meta-analysis of prospective cohort studies evaluating the association of saturated fat with cardiovascular disease. Am J Clin Nutr 2010;91 (3):535–46.

[23] Chowdhury R, Warnakula S, Kunutsor S, Crowe F, Ward HA, Johnson L, et al. Association of dietary, circulating, and supplement fatty acids with coronary risk: a systematic review and meta-analysis. Ann Intern Med 2014;160(6):398–406.

[24] Katan MB, Brouwer IA, Clarke R, Geleijnse JM, Mensink RP. Saturated fat and heart disease. Am J Clin Nutr 2010;92 (2):459–60.

[25] Mensink RP, Katan MB. Effect of dietary fatty acids on serum lipids and lipoproteins. A meta-analysis of 27 trials. Arterioscler Thromb Vasc Biol 1992;12(8):911–19.

[26] Trumbo P, Schlicker S, Yates AA, Poos M, Food and Nutrition Board of the Institute of Medicine, The National Academies. Dietary reference intakes for energy, carbohydrate, fiber, fat, fatty acids, cholesterol, protein, and amino acids. J Am Diet Assoc 2002;102(11):1621–30.

[27] Wu H, Flint AJ, Qi Q, van Dam RM, Sampson LA, Rimm EB, et al. Association between dietary whole grain intake and risk of mortality: two large prospective studies in US men and women. J Am Med Assoc Intern Med 2015;175(3):373–84.

[28] Mellen PB, Walsh TF, Herrington DM. Whole grain intake and cardiovascular disease: a meta-analysis. Nutr Metab Cardiovasc Dis 2008;18(4):283–90.

[29] Mozaffarian D, Micha R, Wallace S. Effects on coronary heart disease of increasing polyunsaturated fat in place of saturated fat: a systematic review and meta-analysis of randomized controlled trials. PLoS Med 2010;7(3):e1000252.

[30] Moreno JJ, Teresa Mitjavila M. The degree of unsaturation of dietary fatty acids and the development of atherosclerosis (review). J Nutr Biochem 2003;14(4):182–95.

[31] Balk E, Chung M, Lichtenstein A, Chew P, Kupelnick B, Lawrence A, et al. Effects of Omega-3 fatty acids on cardiovascular risk factors and intermediate markers of cardiovascular disease. Agency for Healthcare Research and Quality, U.S.

[31] Department of Health and Human Services, Evidence Report/Technology Assessment (summery), 2004;(93):1–6.

[32] Wang C, Harris WS, Chung M, Lichtenstein AH, Balk EM, Kupelnick B, et al. n-3 Fatty acids from fish or fish-oil supplements, but not alpha-linolenic acid, benefit cardiovascular disease outcomes in primary- and secondary-prevention studies: a systematic review. Am J Clin Nutr 2006;84(1):5–17.

[33] Harris WS, Miller M, Tighe AP, Davidson MH, Schaefer EJ. Omega-3 fatty acids and coronary heart disease risk: clinical and mechanistic perspectives. Atherosclerosis 2008;197(1):12–24.

[34] Dietary Guidelines for Americans, 2010. USDA 2010.

[35] Poudyal H, Panchal SK, Diwan V, Brown L. Omega-3 fatty acids and metabolic syndrome: effects and emerging mechanisms of action. Prog Lipid Res 2011;50(4):372–87.

[36] Afshin A, Micha R, Khatibzadeh S, Mozaffarian D. Consumption of nuts and legumes and risk of incident ischemic heart disease, stroke, and diabetes: a systematic review and meta-analysis. Am J Clin Nutr 2014;100(1):278–88.

[37] Kuipers R, De Graaf D, Luxwolda M, Muskiet M, Dijck-Brouwer D, Muskiet F. Saturated fat, carbohydrates and cardiovascular disease. Neth J Med 2011;69(9):372–8.

[38] Gillingham LG, Harris-Janz S, Jones PJ. Dietary monounsaturated fatty acids are protective against metabolic syndrome and cardiovascular disease risk factors. Lipids 2011;46(3):209–28.

[39] Oh K, Hu FB, Manson JE, Stampfer MJ, Willett WC. Dietary fat intake and risk of coronary heart disease in women: 20 years of follow-up of the nurses' health study. Am J Epidemiol 2005;161(7):672–9.

[40] Jakobsen MU, O'Reilly EJ, Heitmann BL, Pereira MA, Bälter K, Fraser GE, et al. Major types of dietary fat and risk of coronary heart disease: a pooled analysis of 11 cohort studies. Am J Clin Nutr 2009;89(5):1425–32.

[41] Knoops KT, de Groot LC, Kromhout D, Perrin A-E, Moreiras-Varela O, Menotti A, et al. Mediterranean diet, lifestyle factors, and 10-year mortality in elderly European men and women. J Am Med Assoc 2004;292(12):1433–9.

[42] Renaud S, De Lorgeril M, Delaye J, Guidollet J, Jacquard F, Mamelle N, et al. Cretan Mediterranean diet for prevention of coronary heart disease. Am J Clin Nutr 1995;61(6):1360S–7S.

[43] Luo C, Zhang Y, Ding Y, Shan Z, Chen S, Yu M, et al. Nut consumption and risk of type 2 diabetes, cardiovascular disease, and all-cause mortality: a systematic review and meta-analysis. Am J Clin Nutr 2014;100(1):256–69.

[44] Estruch R, Ros E, Salas-Salvadó J, Covas M-I, Corella D, Arós F, et al. Primary prevention of cardiovascular disease with a Mediterranean diet. N Engl J Med 2013;368(14):1279–90.

[45] Katan MB. The response of lipoproteins to dietary fat and cholesterol in lean and obese persons. Curr Cardiol Rep 2006;8(6):446–51.

[46] Katan MB, Beynen AC, de Vries JH, Nobels A. Existence of consistent hypo- and hyper-responders to dietary cholesterol in man. Am J Epidemiol 1986;123(2):221–34.

[47] Kritchevsky SB, Kritchevsky D. Egg consumption and coronary heart disease: an epidemiologic overview. J Am Coll Nutr 2000;19(Suppl. 5):549S–55S.

[48] Djousse L, Gaziano JM. Egg consumption in relation to cardiovascular disease and mortality: the Physicians' Health Study. Am J Clin Nutr 2008;87(4):964–9.

[49] Hu FB, Stampfer MJ, Rimm EB, Manson JE, Ascherio A, Colditz GA, et al. A prospective study of egg consumption and risk of cardiovascular disease in men and women. J Am Med Assoc 1999;281(15):1387–94.

[50] Tanasescu M, Cho E, Manson JE, Hu FB. Dietary fat and cholesterol and the risk of cardiovascular disease among women with type 2 diabetes. Am J Clin Nutr 2004;79(6):999–1005.

[51] Djousse L, Gaziano JM. Egg consumption and risk of heart failure in the Physicians' Health Study. Circulation 2008;117(4):512–16.

[52] Mamo JC, Watts GF, Barrett PHR, Smith D, James AP, Pal S. Postprandial dyslipidemia in men with visceral obesity: an effect of reduced LDL receptor expression? Am J Physiol Endocrinol Metab 2001;281(3):E626–32.

[52a] Jacobson TA, Maki KC, Orringer CE, Jones PH, Kris-Etherton P, Sikand G, et al. NLA expert panel. National Lipid Association recommendations for patient-centered management of dyslipidemia: part 2. J Clin Lipidol 2015;9(6S):S1–122.

[53] Anderson JW, Bush HM. Soy protein effects on serum lipoproteins: a quality assessment and meta-analysis of randomized, controlled studies. J Am Coll Nutr 2011;30(2):79–91.

[54] Jenkins DJA, Kendall CWC, Faulkner D, Vidgen E, Trautwein EA, Parker TL, et al. A dietary portfolio approach to cholesterol reduction: combined effects of plant sterols, vegetable proteins, and viscous fibers in hypercholesterolemia. Metabolism 2002;51(12):1596–604.

[55] Kendall CW, Jenkins DJ. A dietary portfolio: maximal reduction of low-density lipoprotein cholesterol with diet. Curr Atheroscler Rep 2004;6(6):492–8.

[56] National Heart Lung and Blood Institute. Lifestyle interventions to reduce cardiovascular risk: systematic evidence review from the lifestyle work group. US Department of Health and Human Services, National Institutes of Health; 2013. https://www.nhlbi.nih.gov/sites/www.nhlbi.nih.gov/files/lifestyle.pdf.

[57] Sheng H-W. Sodium, chloride and potassium. In: Stipanuk M, editor. Biochemical and physiological aspects of human nutrition. Philadelphia, PA: W.B. Saunders Company; 2000. p. 686–710.

[58] Food and Nutrition Board IoM. Sodium and chloride. Dietary reference intakes for water, potassium, sodium, chloride and sulfate. Washington, DC: National Academies Press; 2004. p. 247–392.

[59] Strazzullo P, D'Elia L, Kandala NB, Cappuccio FP. Salt intake, stroke, and cardiovascular disease: meta-analysis of prospective studies. Br Med J 2009;339:b4567.

[60] Cohen HW, Hailpern SM, Alderman MH. Sodium intake and mortality follow-up in the Third National Health and Nutrition Examination Survey (NHANES III). J General Intern Med 2008;23(9):1297–302.

[61] Jessup M, Abraham WT, Casey DE, Feldman AM, Francis GS, Ganiats TG, et al. 2009 focused update: ACCF/AHA guidelines for the diagnosis and management of heart failure in adults: a report of the American College of Cardiology Foundation/American Heart Association Task Force on Practice Guidelines: developed in collaboration with the International Society for Heart and Lung Transplantation. Circulation 2009;119(14):1977–2016.

[62] James PA, Oparil S, Carter BL, Cushman WC, Dennison-Himmelfarb C, Handler J, et al. 2014 Evidence-based guideline for the management of high blood pressure in adults: report from the panel members appointed to the Eighth Joint National Committee (JNC 8). J Am Med Assoc 2014;311:507–20.

[63] Houston MC, Harper KJ. Potassium, magnesium, and calcium: their role in both the cause and treatment of hypertension. J Clin Hypertens (Greenwich) 2008;10(7 Suppl. 2):3–11.

[64] Whelton PK, He J. Potassium in preventing and treating high blood pressure. Semin Nephrol 1999;19(5):494–9.

REFERENCES

[65] Cappuccio FP, MacGregor GA. Does potassium supplementation lower blood pressure? A meta-analysis of published trials. J Hypertens 1991;9(5):465–73.

[66] Geleijnse JM, Kok FJ, Grobbee DE. Blood pressure response to changes in sodium and potassium intake: a metaregression analysis of randomised trials. J Hum Hypertens 2003;17(7):471–80.

[67] Whelton PK, He J, Cutler JA, Brancati FL, Appel LJ, Follmann D, et al. Effects of oral potassium on blood pressure. Meta-analysis of randomized controlled clinical trials. J Am Med Assoc 1997;277(20):1624–32.

[68] Appel LJ, Sacks FM, Carey VJ, Obarzanek E, Swain JF, Miller III ER, et al. Effects of protein, monounsaturated fat, and carbohydrate intake on blood pressure and serum lipids. J Am Med Assoc 2005;294(19):2455–64.

[69] Riccardi G, Rivellese A. Dietary treatment of the metabolic syndrome—the optimal diet. Br J Nutr 2000;83(S1):S143–8.

[70] St. Jeor ST, Howard BV, Prewitt TE, Bovee V, Bazzarre T, Eckel RH. Dietary protein and weight reduction: a statement for healthcare professionals from the Nutrition Committee of the Council on Nutrition, Physical Activity, and Metabolism of the American Heart Association. Circulation 2001;104(15):1869–74.

[71] Jensen MD, Ryan DH, Apovian CM, Ard JD, Comuzzie AG, Donato KA, American College of Cardiology/American Heart Association Task Force on Practice Guidelines; Obesity Society, et al. 2013 AHA/ACC/TOS guideline for the management of overweight and obesity in adults: a report of the American College of Cardiology/American Heart Association Task Force on Practice Guidelines and The Obesity Society. Circulation 2014;129(25 Suppl. 2):S102–38.

[72] Paul L. Diet, nutrition and telomere length. J Nutr Biochem 2011;22(10):895–901.

[73] Crous-Bou M, Fung TT, Prescott J, Julin B, Du M, Sun Q, et al. Mediterranean diet and telomere length in Nurses' Health Study: population based cohort study. Br Med J 2014;349:g6674.

[74] Vivekananthan DP, Penn MS, Sapp SK, Hsu A, Topol EJ. Use of antioxidant vitamins for the prevention of cardiovascular disease: meta-analysis of randomised trials. Lancet 2003;361(9374):2017–23.

[75] Joshipura KJ, Hu FB, Manson JE, Stampfer MJ, Rimm EB, Speizer FE, et al. The effect of fruit and vegetable intake on risk for coronary heart disease. Ann Intern Med 2001;134(12):1106–14.

[76] Liu RH. Health benefits of fruit and vegetables are from additive and synergistic combinations of phytochemicals. Am J Clin Nutr 2003;78(3):517S–20S.

[77] Mirmiran P, Bahadoran Z, Azizi F. Functional foods-based diet as a novel dietary approach for management of type 2 diabetes and its complications: a review. World J Diabetes 2014;5(3):267.

[78] Crowe KM, Francis C. Position of the academy of nutrition and dietetics: functional foods. J Acad Nutr Diet 2013;113(8):1096–103.

[79] Vauzour D, Rodriguez-Mateos A, Corona G, Oruna-Concha MJ, Spencer JPE. Polyphenols and human Health: prevention of disease and nechanisms of action. Nutrients 2010;2:1106–31.

[80] Pandey KB, Rizvi SI. Plant polyphenols as dietary antioxidants in human health and disease. Oxid Med Cell Longev 2009;2(5):270–8.

[81] Reedy J, Krebs-Smith SM, Miller PE, Liese AD, Kahle LL, Park Y, et al. Higher diet quality is associated with decreased risk of all-cause, cardiovascular disease, and cancer mortality among older adults. J Nutr 2014;144(6):881–9.

[82] Borneo R, León AE. Whole grain cereals: functional components and health benefits. Food Funct 2012;3(2):110–19.

[83] Ye EQ, Chacko SA, Chou EL, Kugizaki M, Liu S. Greater whole-grain intake is associated with lower risk of type 2 diabetes, cardiovascular disease, and weight gain. J Nutr 2012;142(7):1304–13.

[84] Okarter N, Liu RH. Health benefits of whole grain phytochemicals. Crit Rev Food Sci Nutr 2010;50(3):193–208.

[85] Cho SS, Qi L, Fahey GC, Klurfeld DM. Consumption of cereal fiber, mixtures of whole grains and bran, and whole grains and risk reduction in type 2 diabetes, obesity, and cardiovascular disease. Am J Clin Nutr 2013;98(2):594–619.

[86] Helmstädter A. Beans and diabetes: phaseolus vulgaris preparations as antihyperglycemic agents. J Med Food 2010;13(2):251–4.

[87] Liu CF, Pan TM. Beneficial effects of bioactive peptides derived from soybean on human health and their production by genetic engineering. In: Hany El-Shemy, editor. Soybean and health, agricultural and biological sciences; September 12, 2011, ISBN 978-953-307-535-8.

[88] Toh J, Tan VM, Lim PC, Lim S, Chong MF. Flavonoids from fruit and vegetables: a focus on cardiovascular risk factors. Curr Atheroscler Rep 2013;15(12):1–7.

[89] Dauchet L, Amouyel P, Hercberg S, Dallongeville J. Fruit and vegetable consumption and risk of coronary heart disease: a meta-analysis of cohort studies. J Nutr 2006;136(10):2588–93.

[90] He FJ, Nowson CA, MacGregor GA. Fruit and vegetable consumption and stroke: meta-analysis of cohort studies. Lancet 2006;367(9507):320–6.

[91] Lim SS, Vos T, Flaxman AD, Danaei G, Shibuya K, Adair-Rohani H, et al. A comparative risk assessment of burden of disease and injury attributable to 67 risk factors and risk factor clusters in 21 regions, 1990–2010: a systematic analysis for the Global Burden of Disease Study 2010. Lancet 2013;380(9859):2224–60.

[92] Ros E. Nuts and novel biomarkers of cardiovascular disease. Am J Clin Nutr 2009;89(5):1649S–56S.

[93] Hoekstra J, Hart A, Owen H, Zeilmaker M, Bokkers B, Thorgilsson B, et al. Fish, contaminants and human health: quantifying and weighing benefits and risks. Food Chem Toxicol 2013;54:18–29.

[94] Mozaffarian D, Lemaitre RN, King IB, Song X, Huang H, Sacks FM, et al. Plasma phospholipid long-chain ω-3 fatty acids and total and cause-specific mortality in older adults: a cohort study. Ann Intern Med 2013;158(7):515–25.

[95] Wu JH, Lemaitre RN, King IB, Song X, Sacks FM, Rimm EB, et al. Association of plasma phospholipid long-chain omega-3 fatty acids with incident atrial fibrillation in older adults the cardiovascular health study. Circulation 2012;125(9):1084–93.

[96] Djoussé L, Akinkuolie AO, Wu JH, Ding EL, Gaziano JM. Fish consumption, omega-3 fatty acids and risk of heart failure: a meta-analysis. Clin Nutr 2012;31(6):846–53.

[97] Zuo X, Tian C, Zhao N, Ren W, Meng Y, Jin X, et al. Tea polyphenols alleviate high fat and high glucose-induced endothelial hyperpermeability by attenuating ROS production via NADPH oxidase pathway. BMC Res Notes 2014;7(1):120.

[98] Wang Z-M, Zhou B, Wang Y-S, Gong Q-Y, Wang Q-M, Yan J-J, et al. Black and green tea consumption and the risk of coronary artery disease: a meta-analysis. Am J Clin Nutr 2011;93(3):506–15.

[99] Grassi D, Desideri G, Ferri C. Protective effects of dark chocolate on endothelial function and diabetes. Curr Opin Clin Nutr Metab Care 2013;16(6):662–8.

[100] Farham B. Chocolate consumption and cardiometabolic disorders. Contin Med Educ 2011;29(10):431.

[101] Fernández-Murga L, Tarín J, García-Perez M, Cano A. The impact of chocolate on cardiovascular health. Maturitas 2011;69(4):312–21.

[102] Li H, Förstermann U. Red wine and cardiovascular health. Circ Res 2012;111(8):959–61.

[103] Chiva-Blanch G, Urpi-Sarda M, Ros E, Arranz S, Valderas-Martinez P, Casas R, et al. Dealcoholized red wine decreases systolic and diastolic blood pressure and increases plasma nitric oxide short communication. Circ Res 2012;111(8):1065–8.

[104] Sikand G. Cardio-protective dietary patterns and preventions of athersclerotic cardiovascular disease. Am Soc Prev Cardiol (ASPC) Man Prev Cardiol 2014;15:142–55.

[105] Appel LJ, Moore TJ, Obarzanek E, Vollmer WM, Svetkey LP, Sacks FM, et al. A clinical trial of the effects of dietary patterns on blood pressure. DASH Collaborative Research Group. N Engl J Med 1997;336(16):1117–24.

[106] Obarzanek E, Sacks FM, Vollmer WM, Bray GA, Miller III ER, Lin PH, et al. Effects on blood lipids of a blood pressure-lowering diet: the Dietary Approaches to Stop Hypertension (DASH) Trial. Am J Clin Nutr 2001;74(1):80–9.

[107] Chen ST, Maruthur NM, Appel LJ. The effect of dietary patterns on estimated coronary heart disease risk. Circ Cardiovasc Qual Outcomes 2010;3(5):484–9.

[108] Kris-Etherton P, Eckel RH, Howard BV, St. Jeor S, Bazzarre TL. Lyon diet heart study: benefits of a Mediterranean-style, national cholesterol education program/american heart association Step I dietary pattern on cardiovascular disease. Circulation 2001;103(13):1823–5.

[109] Trichopoulou A, Costacou T, Bamia C, Trichopoulos D. Adherence to a Mediterranean diet and survival in a Greek population. N Engl J Med 2003;348(26):2599–608.

[110] Fung TT, Rexrode KM, Mantzoros CS, Manson JE, Willett WC, Hu FB. Mediterranean diet and incidence of and mortality from coronary heart disease and stroke in women. Circulation 2009;119(8):1093–100. Available from: http://dx.doi.org/10.1161/circulationaha.108.816736.

[111] Sofi F, Abbate R, Gensini GF, Casini A. Accruing evidence on benefits of adherence to the Mediterranean diet on health: an updated systematic review and meta-analysis. Am J Clin Nutr 2010;92(5):1189–96.

[112] Kastorini C-M, Milionis HJ, Esposito K, Giugliano D, Goudevenos JA, Panagiotakos DB. The effect of Mediterranean diet on metabolic syndrome and its componentsa meta-analysis of 50 studies and 534,906 individuals. J Am Coll Cardiol 2011;57(11):1299–313.

[113] Li D. Chemistry behind vegetarianism. J Agric Food Chem 2011;59(3):777–84.

[114] Duo L, Sinclair AJ, Mann NJ, Turner A, Ball MJ. Selected micronutrient intake and status in men with differing meat intakes, vegetarians and vegans. Asia Pac J Clin Nutr 2000;9(1):18–23.

[115] Fraser GE. Vegetarian diets: what do we know of their effects on common chronic diseases? Am J Clin Nutr 2009;89(5):1607S–12S.

[116] Key TJ, Fraser GE, Thorogood M, Appleby PN, Beral V, Reeves G, et al. Mortality in vegetarians and nonvegetarians: detailed findings from a collaborative analysis of 5 prospective studies. Am J Clin Nutr 1999;70(3):516S–24S.

[117] Fraser GE. Associations between diet and cancer, ischemic heart disease, and all-cause mortality in non-hispanic white California seventh-day adventists. Am J Clin Nutr 1999;70(3):532S–8S.

[118] Li D, Sinclair AJ, Mann NJ, Turner A, Ball MJ. Selected micronutrient intake and status in men with differing meat intakes, vegetarians and vegans. Asia Pac J Clin Nutr 2000;9(1):18–23.

[119] Haddad EH, Berk LS, Kettering JD, Hubbard RW, Peters WR. Dietary intake and biochemical, hematologic, and immune status of vegans compared with nonvegetarians. Am J Clin Nutr 1999;70(3):586S–93S.

[120] Ferdowsian HR, Barnard ND. Effects of plant-based diets on plasma lipids. Am J Cardiol 2009;104(7):947–56.

[121] Jenkins DJA, Kendall CWC, Marchie A, Faulkner DA, Wong JMW, de Souza R, et al. Effects of a dietary portfolio of cholesterol-lowering foods vs lovastatin on serum lipids and C-reactive protein. J Am Med Assoc 2003;290(4):502–10. Available from: http://dx.doi.org/10.1001/jama.290.4.502.

[122] Elmadfa I, Singer I. Vitamin B-12 and homocysteine status among vegetarians: a global perspective. Am J Clin Nutr 2009;89(5):1693S–8S.

[123] Gerhard GT, Duell PB. Homocysteine and atherosclerosis. Curr Opin Lipidol 1999;10(5):417–28.

[124] Jensen MD, Ryan DH, Apovian CM, Ard JD, Comuzzie AG, Donato KA, et al. 2013 AHA/ACC/TOS guideline for the management of overweight and obesity in adults: a report of the American College of Cardiology/American Heart Association Task Force on Practice Guidelines and The Obesity Society. J Am Coll Cardiol 2014;63(25–PA):2985–3023.

[125] Franz MJBJL, Pereira RF. ADA pocket guide to lipid disorders, hypertension, diabetes and weight management. Acad Nutr Diet 2012.

[126] Hark L, Deen D, Pruzansky A. Overview of Nutrition Assessment in Clinical Care in Hark L & Morrison G. Medical Nutrition & Disease: A Case Based Approach. Wiley Blackwell; 2009;3–57.

[127] Lin JS, O'Connor EA, Evans CV, Senger CA, Rowland MG, Groom HC. Behavioral Counseling to Promote a Healthy Life. style for Cardiovascular Disease Prevention in Persons With Cardiovascular Risk Factors: An Updated Systematic Evidence Review for the U.S. Preventive Services Task Force [Internet]. Agency for Healthcare Research and Quality (US); U.S. Preventive Services Task Force Evidence Syntheses 2014; Report No.:13-05179-EF-1.

[128] McCoin M, Sikand G, Johnson EQ, Kris-Etherton PM, Burke F, Carson JAS, et al. The effectiveness of medical nutrition therapy delivered by registered dietitians for disorders of lipid metabolism: a call for further research. J Am Diet Assoc 2008;108(2):233–9.

[129] Sikand G, Kashyap ML, Yang I. Medical nutrition therapy lowers serum cholesterol and saves medication costs in men with hypercholesterolemia. J Am Diet Assoc 1998;98(8):889–94.

[130] Sikand G, Kashyap ML, Wong ND, Hsu JC. Dietitian intervention improves lipid values and saves medication costs in men with combined hyperlipidemia and a history of niacin noncompliance. J Am Diet Assoc 2000;100(2):218–24.

[131] Pritchard DA, Hyndman J, Taba F. Nutritional counselling in general practice: a cost effective analysis. J Epidemiol Community Health 1999;53(5):311–16.

[132] Academy of Nutrition and Dietetics, Evidence Analysis Library. <http://www.andeal.org/topic.cfm?menu=5300&cat=4533>; [accessed 20.02.15].

[133] National Cholesterol Education Program Expert Panel (NCEP). Third Report of the National Cholesterol Education Program (NCEP) Expert Panel on Detection, Evaluation, and Treatment of High Blood Cholesterol in Adults (Adult Treatment Panel III). Washington (DC): National Cholesterol Education Program, National Heart Lung and Blood Institute, National Institute of Health, 2002:V19–V22.

CHAPTER 24

The Influences of Dietary Sugar and Related Metabolic Disorders on Cognitive Aging and Dementia

Shyam Seetharaman

Department of Psychology, St. Ambrose University, Davenport, IA, USA

KEY FACTS

- Modern dietary lifestyles have diverged from our ancestors to include substantial amounts of refined and artificial sugars, especially after technological advancements in food processing techniques and the industrial revolution
- The increased prevalence of metabolic abnormalities seen in cardiovascular disease and type II diabetes may be the result of increased refined sugar consumption
- There are common insulin and blood glucose-related mechanisms which underlie the development of both cardiovascular and neurological disturbances
- There are relationships between sugar consumption, type II diabetes, insulin resistance, cardiovascular disease and the onset of cognitive impairment, abnormal brain aging, and dementia
- Insulin resistance and insulin-related mechanisms seem to play a crucial role in mediating cognitive and brain changes observed in Alzheimer's Disease
- Low sugar diets and normalizing insulin are effective in ameliorating symptoms of Alzheimer's Disease
- There may be too much emphasis placed on the importance of dietary fat and cholesterol on health, and more focus is needed on investigating the influences of dietary sugar and blood sugar on both physical and brain health.

Dictionary of Terms

- *CVD*: Cardiovascular Disease
- *T2D*: Type II Diabetes
- *GI*: Glycemic Index
- *GL*: Glycemic Load
- *BDNF*: Brain-Derived Neurotropic Factor
- *CREB*: Cyclic AMP Response Binding Element Protein
- *ROS*: Reactive Oxygen Species
- *HFCS*: High Fructose Corn Syrup
- *LTP*: Long-Term Potentiation
- *STZ*: Streptozotocin
- *AD*: Alzheimer's Disease
- *SMC*: Smooth Muscle Cell
- *IGF*: Insulin-like Growth Factor
- *Ang-II*: Angiotensin-II
- *AGE*: Advanced Glycation End Product
- *LDL*: Low Density Lipoprotein
- *MI*: Myocardial Infarction
- *VaD*: Vascular Dementia
- *SCE*: Spontaneous Cerebral Emboli
- *RAGE*: Receptors for Advanced Glycation End Products
- *Apo*: Apolipoprotein

INTRODUCTION

The influence of diet on health has become of major concern in recent years. According to the American Heart Association, cardiovascular disease (CVD) is the single largest killer of Americans [1]. In addition to contributing to poor cardiovascular health, accumulating evidence suggests that diet has a detrimental effect on brain and cognitive aging. Much of the research in this area has identified dietary saturated fat and cholesterol as the main contributors to poor cognitive health and accelerated brain aging. However, an accumulating body of evidence suggests that refined sugar consumption may not only be a substantial factor in the etiology of physical disorders, such as CVD [2], but also cognitive deficits and neurological impairment underlying the development of dementia [3].

As indicated by the United States Department of Agriculture, there has been an increase in refined sugar consumption over the past several decades [4]. In turn, this increased consumption has been related to the development of development of insulin resistance syndrome, type II diabetes (T2D), and CVD [5]. With this information, this chapter will focus on: (1) Research assessing the effects of dietary sugar on brain and cognitive impairment; (2) how diet influences the development of metabolic disorders such as CVD and T2D; (3) the influence of metabolic disorders on abnormal cognitive and brain aging; (4) dementia and glucose dysregulation; and (5) a discussion of dietary treatment strategies for poor cognitive aging and dementia. In summary, this chapter provides unique insight into common mechanistic pathways shared in the etiology of both cardiovascular and neurological abnormalities that may lead to accelerated brain aging. Gaining knowledge in this area is important in the development of strategies that may not only help to prevent cardiovascular/metabolic abnormalities, but also in the identification of possible risk factors and treatments for cognitive deficits in an increasingly aging population.

THE INFLUENCE OF DIETARY SUGAR ON BRAIN AND COGNITIVE IMPAIRMENT

Carbohydrates/Sugar

Carbohydrates have different physical and chemical structures. They can be classified according to three principal groups: sugars, oligosaccharides, and polysaccharides. In turn, each of these subclassifications can be further divided according to their monosaccharide chemical composition. The term *sugars* refers to monosaccharides (glucose, galactose, fructose), disaccharides (sucrose, lactose, trehalose), and polyols (sorbitol, mannitol), which are considered to be *simple or refined*. The effects of dietary carbohydrates on physiological responses have been studied for decades and seem to differ depending on, in part, the type of carbohydrate and rates of digestibility [6].

A major source of carbohydrates in the human diet is starch, a complex energy source found in many plants, and can be beneficial for health [6]. The consumption of simple sugars has been related to the development of various diseases, such as CVD [5], a major source of global health concern [7]. A commonly utilized measure to predict the physiological effects of various forms of carbohydrates is the glycemix index (GI). The GI is used to illustrate the relative rate of glucose absorption from foods and, hence, the rate at which glucose appears in the bloodstream. For this reason, plasma glucose is often referred to as "blood sugar." Simple sugars are indexed by a higher GI value compared to more complex carbohydrates, although glycemic load (GL) is increasing in usage since it takes into account the amount of carbohydrates ingested in addition to its potency in elevating plasma glucose [6]. Evidence indicates that people replacing low GI/GL foods (ie, simple carbohydrates/refined sugars) for high GI/GL foods (ie, complex carbohydrates/less refined sugars) demonstrated decreases in glucose and insulin elevations, which have been used as dietary treatments for glucose intolerance [8].

Diet and Brain Impairment

An increasing body of evidence has shown that animals maintained on high refined sugar diets exhibit neurobiological and behavioral changes indicative of cognitive impairment. Such studies are useful in that dietary factors can be experimentally manipulated. This allows for the causal assessment of the specific effects of dietary factors on brain changes and behavioral measures of cognitive abilities. Thus, understanding findings from animal research provides a basis for understanding sugar-induced cognitive impairments observed in humans.

A majority of animal studies in this field have implicated high saturated fat consumption in the development of neurobiological and behavioral impairments related to, for instance, learning and memory. Specifically, investigators have found that rats maintained on a high saturated fat diet for 2, 6, or 24 months demonstrated significantly lower expression of hippocampal brain-derived neurotropic factor (BDNF), critical for synaptic plasticity and learning, relative to controls maintained on control diets for the same duration. They also found an inhibition of cyclic AMP-response element-binding protein (CREB) in the hippocampus, a protein required for various forms of memory regulated by BDNF [9]. This is corroborated by results showing that high fat diet-induced inhibitions in BDNF [10] and synapsin I [9], a molecule which modulates neurotransmitter release by BDNF, critical in hippocampus synaptic functioning. Similarly, Baran and colleagues found that rats fed diets high in saturated fat for 3 weeks exhibited significant retractions in hippocampal dendrites relative to controls, again suggesting inhibitions of this brain structure critical to learning and memory processes [11]. In other measures of hippocampus functioning, one study showed that rats maintained on a high fat–high sugar diet for 8 months demonstrated significant inhibitions of spine density, and long-term potentiation measured in the hippocampus, indicative of impaired neuronal communication [12].

Although studies such as these seem to suggest that high fat feeding impairs hippocampus functioning, it is important to note that in a majority of studies, simultaneous feeding of saturated fat and sugar are employed. It is unclear, therefore, whether the fat or sugar components of these diets contribute to the observed changes. It is possible that high sugar content itself facilitated deleterious hippocampal functioning. The literature in this area provides insight into underlying biological processes, such as oxidative stress, blood sugar, and insulin resistance, which may explain the source of these impairments.

Oxidative Stress. Oxidative stress refers to an imbalance between normal free radical production and an inability of cells to buffer against them, which can contribute to cellular damage [13]. A molecular sign of oxidative stress is the presence of reactive oxygen species (ROS) [14]. In addition to impairments in synaptic plasticity, findings showed that high sugar diets produced significant increases in protein oxidation, suggestive of protein damage [15–16], as well as hippocampal levels of ROS [15]. These results suggest that high sugar diet-induced impairments in plasticity may be modulated by oxidative stress. In turn, research suggests that elevations in blood sugar may play a critical role in the oxidative stress response.

Blood Sugar. Elevations in plasma glucose are a main source of free radical production, a biomarker of oxidative stress [17]. Elevations in blood sugar, which can be the result of high sugar diets, have been related to impaired memory. Sucrose supplemented animals, along with elevations in plasma glucose, showed impairments in spatial learning [18] and declarative memory [19] performance compared with control diet rats. Other work found that rats fed high fructose corn syrup (HFCS), exhibited significant reductions in hippocampal dendritic spines, BDNF, and long-term potentiation (LTP) compared to controls. Furthermore, HFCS fed animals had impaired spatial memory performance which was accompanied by significant elevations in blood glucose [12]. Chronic elevations in plasma glucose produced by high sugar diets have been posited to produce disturbances in intracellular secondary messenger systems and elevations in ROS, all implicated in contributing substantially to neuronal loss and cognitive impairment [3]. Here, elevations in blood sugar may have contributed to increased oxidative damage and, as a result, produced cellular damage.

Insulin Resistance. Insulin stimulates the uptake of glucose into cells for energy. Insulin resistance arises when cells become resistant to the actions of insulin. The pancreas, in an attempt to overcome this resistance, secretes more insulin and, therefore, it can result in an overproduction of insulin (ie, hyperinsulinemia). Hyperinsulinemia and insulin resistance can arise as a result of chronic elevations in plasma glucose, which may be produced by the ingestion of foods high in refined sugar [6].

Insulin resistance may be a crucial mediator in the relationship between high sugar diet and cognitive/brain impairments by affecting synaptic plasticity in the hippocampus. To this point, animals genetically depleted of insulin receptors exhibited significant reductions in phosphorylated Akt and GSK3β, two of the main downstream targets of growth and neurotropic factors [20]. Some findings have suggested that normalizing insulin signaling may protect against neurobiological impairments. For instance, animals injected with streptozotocin (STZ; a pharmacological agent which depletes insulin, insulin receptor and insulin-like growth factor activity) showed significant reductions in insulin binding, as well as insulin receptor and insulin-like growth factor receptor expression in the hippocampus, thereby mimicking an insulin resistant state. Additionally, STZ animals were impaired on spatial memory performance relative to controls. Importantly, STZ-induced impairments were prevented with administration of an insulin sensitizing agent. Here, normalizing insulin signaling protected against impairments in brain and behavioral assessments of memory impairment and elevations in ROS [21]. These findings suggest that insulin signaling may modulate free radical production and, in turn, contribute to oxidative stress-induced brain impairments governing brain functioning.

Diet and Cognitive Impairment

Research, in addition to examining specific brain changes, has also focused on studying the effects of high saturated fat feeding on behavioral measures of cognitive abilities in experimental models. Specifically, findings have illustrated that rats maintained on high saturated fat diets demonstrate impaired performance on spatial memory tasks, where rats are trained to locate a hidden platform in a water maze and subsequently tested for their memory of the platform location [9,22,23]. Wu and colleagues found that rats fed a high fat diet for 2 months demonstrated impaired performance on their ability to learn a hidden platform location, and spent significantly less time in the platform target zone at testing compared to controls [16]. These studies indicating that high fat diets impair behavioral measures of cognition in experimental models are, again, confounded by the high carbohydrate content present in the high fat diets given to animals.

To illustrate the need to rigorously examine the specific influence of dietary carbohydrates on cognition, some findings have indicated that high sugar intake alone produced cognitive impairments. For instance, rats given a high fructose diet were impaired on a spatial learning task [24]. Additionally, a high sucrose diet was shown to impair performance on a novel object recognition task, often utilized in rat models to measure declarative memory [19]. Other measures of cognition showed that rats fed a diet high in carbohydrate content demonstrated heightened anxiety by exploring a novel environment for significantly less time than controls [17]. Further, compared to animals maintained on a sugar free (complex vegetable starch) or refined sucrose diet, rats supplemented with honey showed less behavioral signs of anxiety and spatial memory impairment [25]. Therefore, it is possible that honey, a less refined sugar rich in antioxidant properties, alleviated anxiety and memory deficits compared to the ingestion of a more refined sugar. Animal research provides insight into the potential detrimental effects of dietary sugar on the brain and behavior.

Although the human literature is limited, existing evidence indicates that individuals consuming high levels of sugar demonstrate impaired cognitive performance across various measures. Other studies indicate that individuals ingesting high refined sugar foods demonstrated impaired verbal recall [26,27], recognition memory [28], and reaction time [29] compared to those eating complex carbohydrates. In other findings, dementia-free older adults with higher levels of refined sugar consumption showed poorer overall performance on tests of perceptual speed and spatial performance. Further, older adults showed greater rates of cognitive decline over time on assessments of general cognitive ability, perceptual speed, verbal ability, and spatial performance [30]. Other findings show that sugar consumption may also be particularly harmful to those with T2D [31]. In this fashion, sugar consumption, in combination with T2D, may exacerbate cognitive deficits, possibly through oxidative stress-related mechanisms.

Accumulating evidence provides support for the theory that CVD may mediate the relationship between sugar consumption and cognitive impairment. Since T2D is a risk factor for CVD [32], it is important to understand research assessing the impact of dietary sugars on the development of CVD. Analysis of the literature provides support for the notion that a high refined sugar dietary lifestyle is a risk factor for CVD, which, in turn, may underlie the development of cognitive dysfunction and neurodegenerative disorders, such as Alzheimer's disease (AD) [33]. In order to gain insight into the mechanistic link between CVD and accelerated brain aging, it is important to understand the etiology of CVD, its physiological precursors, and biological consequences. This provides the foundation for discussing how dietary sugar may facilitate CVD and, in turn, increase the risk for cognitive abnormalities. A key biological process in the progression of CVD is the development of atherosclerotic plaques.

METABOLIC DISORDERS

Atherosclerosis

Atherosclerosis, a main feature of CVD, is an inflammatory process [2,34], characterized by the development of arterial plaques and damage to the endothelium (ie, the inner lining of blood vessels and arteries) functioning [35]. Atherosclerotic plaques are the result of, in part, the proliferation and migration of monocytes, lymphocytes and smooth muscle cells into the arterial wall after injury to the endothelium on the surface of the arterial wall. Accumulating substances in the arterial wall can produce an atheroma (ie, "bulge") which can eventually rupture resulting in a cardiac event. Under normal metabolic conditions, the endothelium (surface lining of arteries) maintains a nonadhesive, smooth surface, which acts to inhibit abnormal growth of SMCs and damage may compromise this protective state [2]. Research suggests that abnormalities in insulin signaling contribute to the damage observed on the endothelial surface, abnormal SMC growth, and, in turn, atherosclerosis.

The atherosclerotic process involves the transition of SMCs from a static to a dynamic state, where they can proliferate, migrate, and accumulate inside the arterial wall, contributing to the atherosclerotic process. This process seems to be governed, in part, by insulin and insulin-like growth factors. Monocytes and macrophages which proliferate in the arterial wall have been shown to secrete insulin-like growth factor-I (IGF-I). These elevations, in turn, interact with surrounding SMCs thereby modulating their accumulation and, in turn, contribute to the atherosclerotic plaque [36]. Underlying this biological process is IGF-I, a potent chemo-attractant to SMCs which facilitates SMC hypertrophy [37]. Studies also show that elevated insulin levels are associated with the proliferation of SMCs [5].

Hyperinsulinemia promotes the growth and migration of SMCs, in part, by activating the renin-angiotensin system. Specifically, activation of angiotensin- II (ang-II [5]) through increased ang-II enzymes [38] seems to underlie this insulin-related progression. Ang-II, in turn, facilitates the production of ROS, an inflammatory biomarker key in the atherosclerotic process [5,39]. ROS further inhibits the synthesis of NO, a potent vasodilator [5]. Therefore, activation of ang-II under hyperinsulinemia conditions may act as a vasoconstrictor, thereby restricting blood flow and increasing the risk of CVD through inflammatory-mediated processes, as well as by increasing indices of coagulation [32]. The relationship between insulin resistance and CVD has also been illustrated in epidemiological studies. One study, which directly measured insulin resistance and atherosclerosis, directly found significant negative relationship between insulin sensitivity and intima-media thickness [40], suggesting that this specific marker of atherosclerosis was directly related to insulin resistance.

Glycation and Atherosclerosis

Glycation refers to the irreversible binding of glucose with proteins and lipids and is an important factor in the pathogenesis of atherosclerosis. Understanding the process of glycation leads to greater insight into mechanistic link between glycation, CVD, and cognitive deficits.

Increased glycation may lead to the production of advanced glycation end products (AGEs). Some work indicates that AGEs are more likely to be engulfed by macrophages and taken into the arterial wall, facilitating atherosclerosis and the pathogenesis of CVD. Additionally, glycated products, such as low density lipoproteins (LDL) in the bloodstream are at increased susceptibility to be oxidized [41]. Oxidized LDL, in turn, has been shown to facilitate the atherosclerotic process [42–44], and is a strong predictor of CVD, independent of total LDL levels [45]. The glycation of LDL, therefore, has been a major focus of study in relation to atherosclerosis and its pathogenesis.

Early in vitro studies showed that LDL glycation varies as a function of glucose concentration. Studies also revealed, interestingly, that both T2D and elevated blood sugar is significantly related to increased protein glycation [46]. This suggests that dietary measures to control blood sugar levels may be important in mitigating glycation. In addition, macrophages, which accumulate in the walls of arteries contributing to the development of atheromas, contain a surface receptor for AGEs which bind the products. In the vessel wall, macrophages may engulf the AGEs and enhance foam cell formation, the release of pro-inflammatory cytokines, and generally contribute to oxidative damage to arterial walls. AGEs may also inhibit the activity of nitric oxide, which plays a role in the elasticity of vessel walls. In this fashion, rigidity is promoted, which may contribute to increased sheer stress and the propensity for a cardiac event [47]. Therefore, reductions in dietary sugar, which may lead to elevations in blood sugar and insulin abnormalities, may be an important strategy in minimizing cardiovascular risk and, as will be discussed, poor cognitive aging.

Type II Diabetes, Insulin Resistance, and Atherosclerosis

T2D is a physiological disease characterized by an elevation on blood glucose levels and abnormal insulin metabolism. Individuals with the disease may exhibit a state of insulin resistance, arising from an increase in insulin secretion to counteract the decrease in insulin sensitivity in bodily tissues [48]. Over time, a reduced sensitivity to insulin may develop known as insulin resistance, a primary deficit characteristic of T2D [49]. Substantial evidence has indicated T2D to be a significant risk factor for CVD [32,40,50]. Additionally, insulin seems to mediate the development of atherosclerosis via several mechanisms, such as the accumulation of oxLDL [51] and changes in IGF [36]. Other work reveals that PPAR (an insulin sensitizing agent) agonist administration blocked the differentiation of monocytes into activated macrophages suggesting an amelioration of the atherosclerotic process [52]. As discussed in the following section, evidence indicates that metabolic abnormalities, such as T2D and CVD are strongly related to accelerated brain aging.

METABOLIC DISORDERS AND BRAIN AGING

Type II Diabetes and Cognition

Experimental studies indicate that T2D patients assigned to high GI diets scored significantly worse on immediate [31] and delayed [53] word recall compared to control and low GI diet groups. Interestingly, findings also showed significant correlations between increased levels of plasma glucose and impaired memory performance [31,53]. T2D patients also show impairments in performance on executive functioning, immediate verbal recall, information processing speed, and reaction time tests [54,55].

In case control and large population-based studies, there are robust findings indicating a relationship between T2D and impaired immediate and delayed verbal/visuospatial recall, verbal fluency, mental flexibility, psychomotor speed and semantic memory performance [56]. It is important to note, however, that conclusions based on case control studies should be treated with caution since, typically, T2D participants are part of outpatient groups and, therefore, findings may not generalize to other populations. Additionally, inconsistencies across studies may arise due to the lack of control of comorbidities, such as hypertension and depression, which are independently related to impaired cognition [56,57]. There is a lack of research examining the influence of refined sugar diets on long-term cognitive functioning. Existing evidence does indicate, however, a relationship between T2D and cognitive decline with age [57]. This body of literature, in turn, provides some insight into the potential harmful effects of dietary sugars, which may lead to T2D [4], on cognitive aging.

Type II Diabetes, Cognitive Aging, and Dementia

Findings suggest that T2D is related to cognitive declines in verbal fluency [58], executive functioning, verbal recall memory, mental speed, motor speed [55], information processing, and executive functioning [59]. Additionally, T2D patients were impaired on organization, attention [60], as well as planning and sequencing [61], which may also target executive processes. Importantly, these decrements were present after controlling for hypertension, which is a significant risk factor for cerebrovascular injury [62] and stroke [63], both which can potentiate cognitive decline. Additional findings have shown that T2D diagnosis at mid-life increases the risk for dementia, and related cognitive declines, assessed later in life [64]. Interestingly, those that reported a 15-year or more duration of T2D demonstrated accelerated cognitive decline compared to those having the disease for a shorter amount of time [61], suggesting that high sugar intake contributing to the development of T2D may exert harmful effects on cognition in a compounding fashion over time.

Dementia is characterized by a rapid decline in cognitive abilities, such as memory loss, and can eventually lead to more progressive, debilitating neurological disorders. A growing body of literature indicates a strong link between T2D, insulin resistance and dementia [54,65,66]. In a meta-analysis study, 14 longitudinal large population-based studies showed that the incidences of total dementia was increased in those diagnosed with T2D [67]. Strachan and colleagues analyzed controlled studies which examined the relationship between T2D and cognitive functioning. They found that, in studies with adequate statistical power to detect between-group differences, diabetics demonstrated impaired performance on cognitive tasks, such as verbal memory, psychomotor function, and concentration relative to healthy counterparts [68].

Evidence for T2D-mediated changes in brain structures governing cognitive functioning was provided in a study where investigators found that, through MRI examinations, elderly subjects with T2D exhibited significant reduction in the volumes of medial temporal lobe structures, such as the hippocampus and amygdala [69]. These findings suggest that T2D may mediate the development of cognitive functioning associated with these structures, such as learning, memory, and emotion. In support of this notion, neuropsychological scores on memory and general cognition were significantly worse in patients with T2D compared to healthy controls [55]. These results corroborate with evidence indicating that AD patients exhibited smaller hippocampal volumes relative to healthy controls [70]. Chronic ingestion of refined sugars, therefore, may increase the risk of T2D and progressive cognitive decline potentially characteristic of dementia.

Cardiovascular Disease, Cognitive Decline, and Dementia

Findings based on large prospective studies have found that previous myocardial infarction (MI) served to shift population scores of cognitive functioning lower compared to those without MI or CVD [71]. A more recent study indicated a significant relationship between CVD and cognitive decline in studies with an average follow-up period of 8 years [72]. In a related study, CVD, independent of stroke, increased the risk of AD, vascular dementia (VaD), as well as "mixed" dementia, where both cerebrovascular and neurodegenerative processes were present [73]. Further, AD and VaD were found to be significantly associated atherosclerosis severity [74]. These findings suggest that CVD may potentiate the onset of dementia and related neurobiological and cognitive impairments.

CVD may facilitate cognitive decline through the formation of spontaneous cerebral emboli (SCEs). CVD can increase the chances of SCE as a result of blood vessels breaking loose, traveling through the bloodstream and blocking vessels [75]. SCE production, in turn, increases the risk of stroke or transient ischemic attacks, and leukoaroasis (changes in cerebral white matter resulting from multiple microvessel infarcts), contributing to neuronal loss and cellular injury, which are common in dementia, especially VaD [75]. The development of age-related cognitive decline, are the results of complex cascades of interacting mechanisms. Chronic elevations in blood sugar increases the chances of glycation and AGEs, which have been found in both plaques and tangles in AD brains, and found to facilitate plaque aggregation in vitro [33].

Studies showing a relationship between CVD and dementia enable us to clearly assess the intimate nature of the brain-heart connection. For instance, in a longitudinal cohort study, researchers studied a group of healthy individuals for 5 years to investigate relationships between various markers of CVD and cognitive impairment. Neuropsychiatric testing assessed dementia prevalence using the Mini Mental State Examination, which measures general intelligence, language, visuoperceptual abilities and executive control. They found that, at the follow-up time point, rates of dementia, independent of stroke prevalence, were significantly higher in those with CVD compared to those without Additionally, to assess specific CVD markers, the investigators found a significant relationship between arterial thickening-related atherosclerotic changes and rates of total dementia [73]. Importantly, their findings indicated that statistical adjustment for hypertension, cholesterol, and smoking did not attenuate the observed associations. These results may indicate that the prevalence of CVD, and specific markers of atherosclerosis, may reduce the threshold for dementia, and common mechanisms may underlie the development of the two disorders.

The Rotterdam study is a single center prospective follow-up study assessing individuals 55–94 years of age, and has been the focus of researchers examining risk factors and prevalence of CVD prevalence. Breteler and colleagues analyzed the distribution of cognitive function in this population and its relationship with atherosclerosis and CVD. Findings revealed that, independent of age and education level, the presence of plaques in the carotid arteries, and the presence of arterial atherosclerotic disease, assessed by ultrasound, were significantly related to poorer performance on cognitive testing, measured utilizing the mini mental state examination [71]. In another Rotterdam analysis, researchers examined 284 patients diagnosed with dementia and found that the frequencies of total dementia were positively related to the prevalence of atherosclerosis. Additionally, findings revealed that rates of dementia increased with the degree of atheroscelerosis, measured by carotid arterial wall thickness and widening [74]. Interestingly, these investigators also showed that the relationship between atherosclerosis and dementia prevalence was unchanged after adjusting for total cholesterol suggesting that factors, other than total cholesterol, may be of greater importance. In fact, high total cholesterol levels have been associated with reduced risk of dementia [76–78].

In recent years, there has been increased focus on the relationship between dietary carbohydrates, CVD, and related risk factors, such as insulin resistance and T2D on dementia development. Examining the literature in this area suggests that dementia may be the result of an insulin resistant brain state. Some researchers, in fact, have classified AD as type III diabetes [21]. This is suggestive of common biological processes which underlie both metabolic disturbances and dementia.

DEMENTIA AND METABOLIC DISTURBANCES

Glucose Hypometabolism

One of the main characteristics of the AD brain is the dramatic reduction in glucose metabolism. Since the brain relies heavily on glucose utilization for energy, the integrity and functionality of neuronal communication is highly dependent on glucose usage. Some work has indicated that impaired glucose metabolism occurs early on in the progression of AD and correlates with clinical measures of cognitive abilities in these patients [79]. Research suggests that a dysregulation of glucose metabolism can ultimately lead to the production of ROS, thereby heightening oxidative stress-induced cell death. Furthermore, oxidative stress is associated with the aggregation of plaques and tangles in AD brains [80].

Increased metabolism of glucose in the AD brain may indicate an insulin-resistant type of state which could be exacerbated by a high carbohydrate dietary lifestyle. To this point, investigators have shown a relationship between insulin resistance and AD and, in fact, some have referred to AD as being type III diabetes. The mechanisms by which insulin resistance may impact AD are not clearly understood. Growing evidence, however, suggests that insulin and the development of insulin resistance may facilitate the accumulation of plaques [81], which could facilitate cognitive impairment associated with AD.

Alzheimer's Disease and Insulin Resistance

Consistent research has indicated that insulin resistance and T2D are significant risk factors for AD [82,83]. Additionally, AD may be closely related to insulin resistance stemming from an inhibition of insulin-like growth factor activity. One study, for example, found that, relative to control samples, IGF-1 receptors, as well as IGF-II expression were downregulated in the hippocampi of AD postmortem tissues [84]. Reduced levels of growth factor receptor expression may exert inhibitory effects on insulin signaling and, therefore, be indicative of indicate insulin resistance. Furthermore, there are indications that IGF-I promotes neuronal survival [85,86]. Insulin-induced exacerbation of AD-like brain changes were also indicating by significant increases in Aβ deposition in the brains of individuals who were offspring of AD patients after insulin infusions [87], indicating that genetic factors may also play a role in insulin-mediated AD responses.

Experimental nonhuman animal work has contributed most substantially to the literature regarding the effects of insulin resistance on the formation of AD-related changes. For instance, one study showed that rats administered STZ, which depletes brain insulin, showed decreases in IGF-I and IGF-II expression in the temporal lobe and cerebellar regions. Additionally, treatment with PPAR (an insulin sensitizing agent) recovered these deficits. Other results showed that STZ increased the phosphorylation of tau proteins which, again, was recovered with PPAR treatment. Indications of insulin-mediated cytotoxicity were found in results revealing that STZ produced elevations in p53 (a proapoptosis gene), which was blocked with PPAR administration. In spatial learning testing, STZ impaired working memory performance, which was prevented with PPAR treatment [21]. To further illustrate the possible mediation of AD-related changes by insulin, one group of investigators found a significant increase in the phosphorylation of tau proteins in an insulin receptor genetic knockout mouse model [20]. Others found that transgenic mice lacking the gene for insulin degrading enzyme, which has been found to breakdown Aβ (the main constituent of plaques observed in AD brains), exhibited significant increases in cerebral Aβ accumulation relative to controls. These investigators also showed that mice in the experimental group exhibited hyperinsulinemia and glucose intolerance, supporting the notion that insulin could mediate AD-like brain responses [88]. In one human study, investigators showed that healthy adults receiving infusions of insulin, to hyperinsulinemic levels, showed significant increases in plasma Aβ levels relative to the placebo group [89]. These findings suggest that insulin resistance and insulin-related mechanisms could play a crucial role in mediating AD-related symptoms.

High refined sugar consumption may result in insulin resistance which, in turn, could exacerbate AD and related symptoms. Cao and colleagues, for instance, showed that transgenic AD mice given free access to sucrose, developed hyperinsulinemia after 5 weeks, indicative of insulin resistance onset. Additionally, they found a significant increase in cortical Aβ expression, accompanied by impaired spatial learning. Finally, they also showed elevated blood glucose responses during testing relative to controls [81], suggesting that elevated blood sugar may produce insulin resistance and, in turn, exacerbate AD-like responses. These findings are supported by human literature showing significant correlations between high blood glucose and impaired memory performance in AD patients [90]. In addition to the potential for insulin resistance, chronic blood sugar elevations increases the potentially for the glycation of proteins and lipids which, as mentioned previously, may facilitate the atherosclerotic process. Increasing evidence also suggests that glycation may play a critical role in the progression of AD.

Alzheimer's Disease and Glycation

The formation and accumulation of AGEs in tissue may play a critical role in the pathogenesis of AD and may be an important biomarker for the disease [91]. Studies have indicated that AD brains contain high levels of AGEs relative to healthy age-matched controls [92]. Some in vitro work indicates that Aβ peptides modified by AGEs accelerate the formation of plaque aggregation [93]. In addition, tau proteins extracted from AD tissue show an increased propensity to be glycated relative to healthy tissue [94]. Also, examinations of AD patients have revealed elevated AGE levels in Aβ deposits of these individuals [91,95]. Investigations have also indicated that AGE receptors (RAGE) may provide an indication of the effects of AGE influences on cognitive impairments associated with AD. Some animal work, for instance, showed that transgenic mice overexpressing both RAGE and amyloid precursor protein were significantly impaired on spatial learning tasks and exhibited impairments in hippocampal synaptic activity relative to controls [96]. These results suggest that glycation and the formation of AGEs may facilitate neurobiological changes accompanied by AD as well as cognitive dysfunction.

There is strong evidence that certain genetic factors increase the risk of individuals to developing AD. Established research indicates that the apolipoprotein (apo) E4 allele is the most important genetic risk factor for AD [97] with AGE-binding activity greater in E4 positive individuals. This suggests that glycation may increase the likelihood of AD in those genetically predisposed to developing the disease [92]. This is supported by research indicating that apoE4 positive patients exhibited larger plaque depositions than those homozygous for the apoE3 isoform of the gene [98].

Existing evidence in this area suggests that a dietary lifestyle consisting of high levels of carbohydrate intake could result in chronic elevations in blood sugar, thereby increasing the production of AGEs, and facilitating an insulin resistant brain state, which could all mediate both CVD and neurotoxicity associated with AD. To this point, dietary treatment strategies for AD have become of increased interest in recent decades. New lines of evidence suggest that diets low in sugar may be effective in treating AD.

DIET AND BRAIN HEALTH

Ketogenic Diets

The ketogenic diet, consisting of high-fat and very low carbohydrate content, was developed in the 1920s, and was primarily found to be effective in the treatment of epilepsy. The traditional ketogenic diet consists of 80–90% fat, with carbohydrate and protein comprising the remainder of the diet composition. In recent years, there has been increased interest in the potential of ketogenic diets to treat neurological disorders, such as AD. Accumulating evidence indicates that ketone bodies, a by-product of strict adherence to the diet, may exert protective influences on neurodegeneration accompanied by neurological disorders, including AD. Research examining the effects of strictly limiting carbohydrate intake in AD patients may be of great importance in treating the disease, as there is some evidence indicating that high carbohydrate intake is associated with poor cognitive performance on neuropsychiatric scales measuring memory, motor abilities, and social engagement in patients with the disease [99].

The underlying metabolic mechanism utilized by the ketogenic diet is the conversion of fat to ketone bodies, as opposed to glucose metabolism as a main energy source. There are some indications that ketone bodies may exert their protective efficacy against neurodegeneration by increasing mitochondrial respiration [100] and decreasing the production of ROS [101], which may, in turn, protect against cell death [102]. One of the main ketone bodies generated by fat conversion under strict adherence to the diet is β-hydroxybutyrate. This substance represents an alternative energy source to glucose and has been the focus of work aimed at treating cognitive problems related to neurological diseases.

In one study, AD patients were given daily administration of a ketogenic agent for 90 days, thereby significantly increasing the levels of the ketone body β-hydroxybutyrate from baseline. AD patients were then given a battery of tests to measure cognitive abilities. They found that, relative to baseline pretreatment time points, patients significantly improved on cognitive measures of, for instance, memory, language, and orientation [103]. It is important, and perhaps crucial, to note that these cognitive improvements seemed to be more effective in those without the apoE4 genetic variant. In light of this evidence, further testing with this ketogenic compound is needed to examine whether this may be a potential treatment even for those genetically at risk for developing AD. Nonetheless, administration of ketogenic agents, and perhaps prescribing ketogenic diets, may be effective in treating cognitive dysfunction associated with AD, especially in nonpredisposed individuals. Other findings supporting the protective influence of ketone bodies revealed a positive correlation between plasma levels of β-hydroxybutyrate with memory performance in memory-impaired adults [104].

Experimental work has also served to illustrate the protective efficacy of ketone bodies against deleterious neurobiological changes accompanying AD. One study, for example, showed that transgenic AD mice maintained on a ketogenic diet for 43 days exhibited significant elevations in β-hydroxybutyrate levels and significant reductions in Aβ depositions relative to controls [105]. Another study showed that direct incubation of β-hydroxybutyrate protected hippocampal neurons in culture against Aβ toxicity [106], also illustrating the potentially for by-products of ketogenic diets to be neuroprotective. Findings such as these are in contrast to findings indicating that ketogenic diets actually produce increases in Aβ deposition and, as a result, were said to contribute to the progression of AD and related symptoms [107,108]. This conflict can be addressed by noting that the diets in those studies were also high in sugar content. It is, therefore, not possible to conclude that fat itself contributed to AD in these studies, as carbohydrates may be the most detrimental factor.

Adherence to a low carbohydrate diets may elevate ketone body levels, which may be an ideal treatment for patients with AD to ameliorate cognitive symptoms associated with the disease. Some researchers have posited that it may be difficult to implement a ketogenic diet for substantial lengths of time in AD patients because of their seemingly stronger preference for sugar [109]. Here, it can be seen how treatment with nondietary ketogenic compounds may be of use.

Concerns over the use of ketogenic diets to treat cognitive deficits revolve around its potential to exert detrimental effects on cardiovascular health. There have been, however, no substantiated associations between ketosis and abnormal physiological profiles [110]. In fact, some evidence indicates that ketogenic diets produce improvements in cardiovascular health, such as decreased triglyceride levels [111]. It may be of great clinical relevance, therefore, to consider ketogenic diets as part of treatments strategies for CVD and AD.

CONCLUSIONS

Research suggests that dietary carbohydrates, through their influence on CVD and risk factors for the disease, may also play a critical role in the development of dementia. Existing literature suggests that there are mechanisms by which the etiologies of CVD and dementia are linked. Specifically, a high carbohydrate dietary lifestyle increases the chances of elevated blood sugar [49]. This, in turn, could lead to an insulin resistant state and the development of T2D, both which are suggested to be risk factors for CVD [32] and dementia [66]. Additionally, chronic blood glucose elevation may result in increased glycation of proteins, which also seem to play a critical role in the inflammatory process of atherosclerosis [2] a main feature of CVD, as well as the progression of AD [92]. The apparent effectiveness of ketogenic diets on ameliorating AD and related symptoms brings to light the importance of continuing research examining the specific influence of dietary carbohydrates on affecting cognitive functioning. Reviewing existing research also suggests that there may be too much emphasis placed on the importance of dietary cholesterol on health [112] and, in fact, limiting cholesterol intake may be detrimental to mental health [76]. This also falls in line with evidence suggesting that dietary fat and total cholesterol may not produce CVD [112–114].

In developing treatment strategies for CVD a multitude of factors need be taken into consideration. There is an urgent need, however, for further research in order to elucidate the specific mechanisms of the association between dietary carbohydrates, CVD and the development of debilitating neurodegenerative disorders causing severe impairments in cognition.

FINAL REMARKS

The evolution of humans began approximately 2.6 million to 10,000 years ago during the Paleolithic period [115]. Historical and archaeological evidence portrays Paleolithic man as being lean, fit, and largely free of signs indicating chronic disease [116]. In terms of dietary lifestyle, the Paleolithic time period consisted of hunter-gatherer societies surviving off food foraged or hunted from plants and animals available in the natural surroundings. There are indications that animal sources comprised more than half of the hunter-gatherer diet, and that they consumed very low amounts of carbohydrates [117]. The dietary lifestyle in contemporary Western societies has drastically diverged from that of Paleolithic man, with the onset of the agricultural revolution. Technological advancements in food processing techniques, beginning with the industrial revolution in the nineteenth century approximately 200 years ago, and more so in recent decades, reflects this drastic shift in diets. Dietary lifestyles have changed from our ancestors, who derived the majority of their nutrition from high protein and animal fat, to a lifestyle consisting of refined and artificial sugars [5].

The increased prevalence of CVD could be the result of, in part, sugar-induced metabolic disturbances produced by a lack of adaptation to starch and sugars. Increased refined sugar intake could also be a major factor in facilitating cognitive dysfunction and the development of neurodegenerative diseases such as AD. In light of the available evidence, there is an urgent need for further research examining specific dietary factors, such as refined sugar, on the etiology of cognitive dysfunction and dementia, especially prevalent in aging populations.

It should be emphasized that dementias, such as AD, are complex, multifaceted neurodegenerative disorders, consisting of various stages. Therefore, dietary considerations may be only one factor which may influence the etiology of such disease. Better insight into the direct link between high sugar dietary lifestyles and dementia is essential for public health policy in order to identify risk factors for, and develop treatment strategies for such cognitive disorders.

SUMMARY

1. Diets high in refined sugar produce neurobiological, cognitive, and behavioral impairments, especially related to memory.
2. High sugar diets may result in brain impairment through oxidative damage produced by elevations in blood sugar and insulin resistance.
3. Hyperinsulinemia contributes to the inflammatory process leading to the development of atherosclerosis in cardiovascular disease.
4. High blood sugar increases the risk for glycation (irreversible binding of glucose with proteins) which, in turn, facilitates the atherosclerotic process.
5. Type II Diabetes is related to cardiovascular abnormalities through dysregulations in insulin signaling and oxidative stress.
6. Type II Diabetes and cardiovascular disease are related to accelerated brain and cognitive aging, including the development of dementia.
7. Alzheimer's disease is characterized by a diabetic brain state, including disruptions in insulin signaling and glucose hypometabolism.

8. Glycation facilitates abnormal brain changes found in Alzheimer's disease (eg, the development of plaques).
9. Ketogenic (very low sugar) diets may be effective in the treatment of dementia.
10. There is an urgent need for public health policy to further address the detrimental effects of refined sugar on poor cognitive aging.

References

[1] Lloyd-Jones D, Adams RJ, Brown TM, et al. Heart disease and stroke statistics — 2010 update: a report from the American Heart Association. Circulation; 121(7): e46—e215.
[2] Ross R. Atherosclerosis — an inflammatory disease. N Engl J Med 1999;340(2):115—26.
[3] Stephan BC, Wells JC, Brayne C, Albanese F, Siervo M. Increased fructose intake as a risk factor for dementia. J Gerontol A Biol Sci Med Sci; 65(8): 809—814.
[4] Gross LS, Li L, Ford ES, Liu S. Increased consumption of refined carbohydrates and the epidemic of type 2 diabetes in the United States: an ecologic assessment. Am J Clin Nutr 2004;79(5):774—9.
[5] Kopp W. The atherogenic potential of dietary carbohydrate. Prev Med 2006;42(5):336—42.
[6] Leonard WR, Snodgrass JJ, Robertson ML. Evolutionary perspectives on fat ingestion and metabolism in humans. In: Montmayeur JP, le Coutre J, editors. Fat detection: taste, texture, and post ingestive effects. Boca Raton (FL) USA: CRC Press; 2010.
[7] Popkin BM, Nielsen SJ. The sweetening of the world's diet. Obes Res 2003;11(11):1325—32.
[8] Liu S, Willett WC, Stampfer MJ, et al. A prospective study of dietary glycemic load, carbohydrate intake, and risk of coronary heart disease in US women. Am J Clin Nutr 2000;71(6):1455—61.
[9] Molteni R, Barnard RJ, Ying Z, Roberts CK, Gomez-Pinilla F. A high-fat, refined sugar diet reduces hippocampal brain-derived neurotrophic factor, neuronal plasticity, and learning. Neuroscience 2002;112(4):803—14.
[10] Pistell PJ, Morrison CD, Gupta S, et al. Cognitive impairment following high fat diet consumption is associated with brain inflammation. J Neuroimmunol; 219(1—2): 25—32.
[11] Baran SE, Campbell AM, Kleen JK, et al. Combination of high fat diet and chronic stress retracts hippocampal dendrites. Neuroreport 2005;16(1):39—43.
[12] Stranahan AM, Norman ED, Lee K, et al. Diet-induced insulin resistance impairs hippocampal synaptic plasticity and cognition in middle-aged rats. Hippocampus 2008;18(11):1085—8.
[13] Baynes JW. Role of oxidative stress in development of complications in diabetes. Diabetes 1991;40(4):405—12.
[14] Touyz RM. Reactive oxygen species, vascular oxidative stress, and redox signaling in hypertension: what is the clinical significance? Hypertension 2004;44(3):248—52.
[15] Molteni R, Wu A, Vaynman S, Ying Z, Barnard RJ, Gomez-Pinilla F. Exercise reverses the harmful effects of consumption of a high-fat diet on synaptic and behavioral plasticity associated to the action of brain-derived neurotrophic factor. Neuroscience 2004;123(2):429—40.
[16] Wu A, Ying Z, Gomez-Pinilla F. The interplay between oxidative stress and brain-derived neurotrophic factor modulates the outcome of a saturated fat diet on synaptic plasticity and cognition. Eur J Neurosci 2004;19(7):1699—707.
[17] Souza CG, Moreira JD, Siqueira IR, et al. Highly palatable diet consumption increases protein oxidation in rat frontal cortex and anxiety-like behavior. Life Sci 2007;81(3):198—203.
[18] Jurdak N, Lichtenstein AH, Kanarek RB. Diet-induced obesity and spatial cognition in young male rats. Nutr Neurosci 2008;11(2):48—54.
[19] Jurdak N, Kanarek RB. Sucrose-induced obesity impairs novel object recognition learning in young rats. Physiol Behav 2009;96(1):1—5.
[20] Schubert M, Gautam D, Surjo D, et al. Role for neuronal insulin resistance in neurodegenerative diseases. Proc Natl Acad Sci USA 2004;101(9):3100—5.
[21] de la Monte SM, Tong M, Lester-Coll N, Plater Jr. M, Wands JR. Therapeutic rescue of neurodegeneration in experimental type 3 diabetes: relevance to Alzheimer's disease. J Alzheimers Dis 2006;10(1):89—109.
[22] Goldbart AD, Row BW, Kheirandish-Gozal L, Cheng Y, Brittian KR, Gozal D. High fat/refined carbohydrate diet enhances the susceptibility to spatial learning deficits in rats exposed to intermittent hypoxia. Brain Res 2006;1090(1):190—6.
[23] Pathan AR, Gaikwad AB, Viswanad B, Ramarao P. Rosiglitazone attenuates the cognitive deficits induced by high fat diet feeding in rats. Eur J Pharmacol 2008;589(1—3):176—9.
[24] Ross AP, Bartness TJ, Mielke JG, Parent MB. A high fructose diet impairs spatial memory in male rats. Neurobiol Learn Mem 2009;92(3):410—16.
[25] Chepulis LM, Starkey NJ, Waas JR, Molan PC. The effects of long-term honey, sucrose or sugar-free diets on memory and anxiety in rats. Physiol Behav 2009;97(3—4):359—68.
[26] Benton D, Ruffin MP, Lassel T, et al. The delivery rate of dietary carbohydrates affects cognitive performance in both rats and humans. Psychopharmacology (Berl) 2003;166(1):86—90.
[27] Nabb SL, Benton D. The effect of the interaction between glucose tolerance and breakfasts varying in carbohydrate and fibre on mood and cognition. Nutr Neurosci 2006;9(3—4):161—8.
[28] Ingwersen J, Defeyter MA, Kennedy DO, Wesnes KA, Scholey AB. A low glycaemic index breakfast cereal preferentially prevents children's cognitive performance from declining throughout the morning. Appetite 2007;49(1):240—4.
[29] Lloyd HM, Green MW, Rogers PJ. Mood and cognitive performance effects of isocaloric lunches differing in fat and carbohydrate content. Physiol Behav 1994;56(1):51—7.
[30] Seetharaman S, Andel R, McEvoy C, Dahl Aslan AK, Finkel D, Pedersen NL. Blood glucose, diet-based glycemic load and cognitive aging among dementia-free older adults. J Gerontol A Biol Sci Med Sci 2014.
[31] Greenwood CE, Kaplan RJ, Hebblethwaite S, Jenkins DJ. Carbohydrate-induced memory impairment in adults with type 2 diabetes. Diabetes Care 2003;26(7):1961—6.
[32] Ginsberg HN. Insulin resistance and cardiovascular disease. J Clin Invest 2000;106(4):453—8.
[33] Martins IJ, Hone E, Foster JK, et al. Apolipoprotein E, cholesterol metabolism, diabetes, and the convergence of risk factors for Alzheimer's disease and cardiovascular disease. Mol Psychiatry 2006;11(8):721—36.
[34] Fan J, Watanabe T. Inflammatory reactions in the pathogenesis of atherosclerosis. J Atheroscler Thromb 2003;10(2):63—71.
[35] Brunner H, Cockcroft JR, Deanfield J, et al. Endothelial function and dysfunction. Part II: association with cardiovascular risk factors and diseases. A statement by the Working Group on Endothelins and Endothelial Factors of the European Society of Hypertension. J Hypertens 2005;23(2):233—46.

[36] Arnqvist HJ, Bornfeldt KE, Chen Y, Lindstrom T. The insulin-like growth factor system in vascular smooth muscle: interaction with insulin and growth factors. Metabolism 1995;44(10 Suppl. 4):58–66.

[37] Zhu B, Zhao G, Witte DP, Hui DY, Fagin JA. Targeted overexpression of IGF-I in smooth muscle cells of transgenic mice enhances neointimal formation through increased proliferation and cell migration after intraarterial injury. Endocrinology 2001;142(8):3598–606.

[38] Kamide K, Rakugi H, Nagai M, et al. Insulin-mediated regulation of the endothelial renin-angiotensin system and vascular cell growth. J Hypertens 2004;22(1):121–7.

[39] Brasier AR, Recinos 3rd A, Eledrisi MS. Vascular inflammation and the renin-angiotensin system. Arterioscler Thromb Vasc Biol 2002;22(8):1257–66.

[40] Howard G, O'Leary DH, Zaccaro D, et al. Insulin sensitivity and atherosclerosis. The Insulin Resistance Atherosclerosis Study (IRAS) Investigators. Circulation 1996;93(10):1809–17.

[41] Sobal G, Menzel J, Sinzinger H. Why is glycated LDL more sensitive to oxidation than native LDL? A comparative study. Prostaglandins Leukot Essent Fatty Acids 2000;63(4):177–86.

[42] Boullier A, Bird DA, Chang MK, et al. Scavenger receptors, oxidized LDL, and atherosclerosis. Ann NY Acad Sci 2001;947:214–22 discussion 22–3

[43] Bjorkhem I, Henriksson-Freyschuss A, Breuer O, Diczfalusy U, Berglund L, Henriksson P. The antioxidant butylated hydroxytoluene protects against atherosclerosis. Arterioscler Thromb 1991;11(1):15–22.

[44] Ishigaki Y, Katagiri H, Gao J, et al. Impact of plasma oxidized low-density lipoprotein removal on atherosclerosis. Circulation 2008;118(1):75–83.

[45] Holvoet P, Kritchevsky SB, Tracy RP, et al. The metabolic syndrome, circulating oxidized LDL, and risk of myocardial infarction in well-functioning elderly people in the health, aging, and body composition cohort. Diabetes 2004;53(4):1068–73.

[46] Schleicher E, Deufel T, Wieland OH. Non-enzymatic glycosylation of human serum lipoproteins. Elevated epsilon-lysine glycosylated low density lipoprotein in diabetic patients. FEBS Lett 1981;129(1):1–4.

[47] Bucala R, Tracey KJ, Cerami A. Advanced glycosylation products quench nitric oxide and mediate defective endothelium-dependent vasodilatation in experimental diabetes. J Clin Invest 1991;87(2):432–8.

[48] Bergman RN, Finegood DT, Kahn SE. The evolution of beta-cell dysfunction and insulin resistance in type 2 diabetes. Eur J Clin Invest 2002;32(Suppl. 3):35–45.

[49] Whitney E, DeBruyne LK, Pinna K, Rolfes SR. Nutrition for health and health and healthcare. Belmont (CA): Thompson; 2007.

[50] Juhan-Vague I, Alessi MC, Vague P. Increased plasma plasminogen activator inhibitor 1 levels. A possible link between insulin resistance and atherothrombosis. Diabetologia 1991;34(7):457–62.

[51] Singh BM, Mehta JL. Interactions between the renin-angiotensin system and dyslipidemia: relevance in the therapy of hypertension and coronary heart disease. Arch Intern Med 2003;163(11):1296–304.

[52] Combs CK, Johnson DE, Karlo JC, Cannady SB, Landreth GE. Inflammatory mechanisms in Alzheimer's disease: inhibition of beta-amyloid-stimulated proinflammatory responses and neurotoxicity by PPARgamma agonists. J Neurosci 2000;20(2):558–67.

[53] Papanikolaou Y, Palmer H, Binns MA, Jenkins DJ, Greenwood CE. Better cognitive performance following a low-glycaemic-index compared with a high-glycaemic-index carbohydrate meal in adults with type 2 diabetes. Diabetologia 2006;49(5):855–62.

[54] Awad N, Gagnon M, Messier C. The relationship between impaired glucose tolerance, type 2 diabetes, and cognitive function. J Clin Exp Neuropsychol 2004;26(8):1044–80.

[55] van Harten B, Oosterman J, Muslimovic D, van Loon BJ, Scheltens P, Weinstein HC. Cognitive impairment and MRI correlates in the elderly patients with type 2 diabetes mellitus. Age Ageing 2007;36(2):164–70.

[56] Stewart R, Liolitsa D. Type 2 diabetes mellitus, cognitive impairment and dementia. Diabet Med 1999;16(2):93–112.

[57] Messier C. Impact of impaired glucose tolerance and type 2 diabetes on cognitive aging. Neurobiol Aging 2005;26(Suppl. 1):26–30.

[58] Kanaya AM, Barrett-Connor E, Gildengorin G, Yaffe K. Change in cognitive function by glucose tolerance status in older adults: a 4-year prospective study of the Rancho Bernardo study cohort. Arch Intern Med 2004;164(12):1327–33.

[59] Manschot SM, Brands AM, van der Grond J, et al. Brain magnetic resonance imaging correlates of impaired cognition in patients with type 2 diabetes. Diabetes 2006;55(4):1106–13.

[60] Knopman D, Boland LL, Mosley T, et al. Cardiovascular risk factors and cognitive decline in middle-aged adults. Neurology 2001;56(1):42–8.

[61] Gregg EW, Yaffe K, Cauley JA, et al. Is diabetes associated with cognitive impairment and cognitive decline among older women? Study of Osteoporotic Fractures Research Group. Arch Intern Med 2000;160(2):174–80.

[62] Johnston SC, O'Meara ES, Manolio TA, et al. Cognitive impairment and decline are associated with carotid artery disease in patients without clinically evident cerebrovascular disease. Ann Intern Med 2004;140(4):237–47.

[63] Elias MF, Sullivan LM, D'Agostino RB, et al. Framingham stroke risk profile and lowered cognitive performance. Stroke 2004;35(2):404–9.

[64] Xu W, Qiu C, Gatz M, Pedersen NL, Johansson B, Fratiglioni L. Mid- and late-life diabetes in relation to the risk of dementia: a population-based twin study. Diabetes 2009;58(1):71–7.

[65] Allen KV, Frier BM, Strachan MW. The relationship between type 2 diabetes and cognitive dysfunction: longitudinal studies and their methodological limitations. Eur J Pharmacol 2004;490(1-3):169–75.

[66] Convit A. Links between cognitive impairment in insulin resistance: an explanatory model. Neurobiol Aging 2005;26(Suppl. 1):31–5.

[67] Biessels GJ, Staekenborg S, Brunner E, Brayne C, Scheltens P. Risk of dementia in diabetes mellitus: a systematic review. Lancet Neurol 2006;5(1):64–74.

[68] Strachan MW, Deary IJ, Ewing FM, Frier BM. Is type II diabetes associated with an increased risk of cognitive dysfunction? A critical review of published studies. Diabetes Care 1997;20(3):438–45.

[69] den Heijer T, Vermeer SE, van Dijk EJ, et al. Type 2 diabetes and atrophy of medial temporal lobe structures on brain MRI. Diabetologia 2003;46(12):1604–10.

[70] Convit A, De Leon MJ, Tarshish C, et al. Specific hippocampal volume reductions in individuals at risk for Alzheimer's disease. Neurobiol Aging 1997;18(2):131–8.

[71] Breteler MM, Claus JJ, Grobbee DE, Hofman A. Cardiovascular disease and distribution of cognitive function in elderly people: the Rotterdam Study. BMJ 1994;308(6944):1604–8.

[72] Anstey K, Christensen H. Education, activity, health, blood pressure and apolipoprotein E as predictors of cognitive change in old age: a review. Gerontology 2000;46(3):163–77.

[73] Newman AB, Fitzpatrick AL, Lopez O, et al. Dementia and Alzheimer's disease incidence in relationship to cardiovascular disease in the Cardiovascular Health Study cohort. J Am Geriatr Soc 2005;53(7):1101–7.

[74] Hofman A, Ott A, Breteler MM, et al. Atherosclerosis, apolipoprotein E, and prevalence of dementia and Alzheimer's disease in the Rotterdam Study. Lancet 1997;349(9046):151–4.

[75] Stampfer MJ. Cardiovascular disease and Alzheimer's disease: common links. J Intern Med 2006;260(3):211–23.

[76] Mielke MM, Zandi PP, Sjogren M, et al. High total cholesterol levels in late life associated with a reduced risk of dementia. Neurology 2005;64(10):1689–95.

[77] Reitz C, Tang MX, Luchsinger J, Mayeux R. Relation of plasma lipids to Alzheimer's disease and vascular dementia. Arch Neurol 2004;61(5):705–14.

[78] Romas SN, Tang MX, Berglund L, Mayeux R. APOE genotype, plasma lipids, lipoproteins, and AD in community elderly. Neurology 1999;53(3):517–21.

[79] Mosconi L, Pupi A, De Leon MJ. Brain glucose hypometabolism and oxidative stress in preclinical Alzheimer's disease. Ann NY Acad Sci 2008;1147:180–95.

[80] Lambert MP, Barlow AK, Chromy BA, et al. Diffusible, nonfibrillar ligands derived from Abeta1-42 are potent central nervous system neurotoxins. Proc Natl Acad Sci USA 1998;95(11):6448–53.

[81] Cao D, Lu H, Lewis TL, Li L. Intake of sucrose-sweetened water induces insulin resistance and exacerbates memory deficits and amyloidosis in a transgenic mouse model of Alzheimer's disease. J Biol Chem 2007;282(50):36275–82.

[82] Ott A, Stolk RP, Hofman A, van Harskamp F, Grobbee DE, Breteler MM. Association of diabetes mellitus and dementia: the Rotterdam Study. Diabetologia 1996;39(11):1392–7.

[83] Kuusisto J, Koivisto K, Mykkanen L, et al. Association between features of the insulin resistance syndrome and Alzheimer's disease independently of apolipoprotein E4 phenotype: cross sectional population based study. Br Med J 1997;315(7115):1045–9.

[84] Steen E, Terry BM, Rivera EJ, et al. Impaired insulin and insulin-like growth factor expression and signaling mechanisms in Alzheimer's disease – is this type 3 diabetes? J Alzheimers Dis 2005;7(1):63–80.

[85] Bondy CA, Cheng CM. Signaling by insulin-like growth factor 1 in brain. Eur J Pharmacol 2004;490(1–3):25–31.

[86] Dore S, Kar S, Quirion R. Insulin-like growth factor I protects and rescues hippocampal neurons against beta-amyloid- and human amylin-induced toxicity. Proc Natl Acad Sci USA 1997;94(9):4772–7.

[87] Haan MN, Wallace R. Can dementia be prevented? Brain aging in a population-based context. Annu Rev Public Health 2004;25:1–24.

[88] Farris W, Mansourian S, Chang Y, et al. Insulin-degrading enzyme regulates the levels of insulin, amyloid beta-protein, and the beta-amyloid precursor protein intracellular domain in vivo. Proc Natl Acad Sci USA 2003;100(7):4162–7.

[89] Fishel MA, Watson GS, Montine TJ, et al. Hyperinsulinemia provokes synchronous increases in central inflammation and beta-amyloid in normal adults. Arch Neurol 2005;62(10):1539–44.

[90] Craft S, Zallen G, Baker LD. Glucose and memory in mild senile dementia of the Alzheimer type. J Clin Exp Neuropsychol 1992;14(2):253–67.

[91] Yamagishi S, Nakamura K, Inoue H, Kikuchi S, Takeuchi M. Serum or cerebrospinal fluid levels of glyceraldehyde-derived advanced glycation end products (AGEs) may be a promising biomarker for early detection of Alzheimer's disease. Med Hypotheses 2005;64(6):1205–7.

[92] Li YM, Dickson DW. Enhanced binding of advanced glycation endproducts (AGE) by the ApoE4 isoform links the mechanism of plaque deposition in Alzheimer's disease. Neurosci Lett 1997;226(3):155–8.

[93] Vitek MP, Bhattacharya K, Glendening JM, et al. Advanced glycation end products contribute to amyloidosis in Alzheimer's disease. Proc Natl Acad Sci USA 1994;91(11):4766–70.

[94] Colaco CA, Ledesma MD, Harrington CR, Avila J. The role of the Maillard reaction in other pathologies: Alzheimer's disease. Nephrol Dial Transplant 1996;11(Suppl. 5):7–12.

[95] Ledesma MD, Bonay P, Avila J. Tau protein from Alzheimer's disease patients is glycated at its tubulin-binding domain. J Neurochem 1995;65(4):1658–64.

[96] Arancio O, Zhang HP, Chen X, et al. RAGE potentiates Abeta-induced perturbation of neuronal function in transgenic mice. EMBO J 2004;23(20):4096–105.

[97] Petot GJ, Friedland RP. Lipids, diet and Alzheimer's disease: an extended summary. J Neurol Sci 2004;226(1–2):31–3.

[98] Strittmatter WJ, Roses AD. Apolipoprotein E and Alzheimer's disease. Proc Natl Acad Sci USA 1995;92(11):4725–7.

[99] Young KW, Greenwood CE, van Reekum R, Binns MA. A randomized, crossover trial of high-carbohydrate foods in nursing home residents with Alzheimer's disease: associations among intervention response, body mass index, and behavioral and cognitive function. J Gerontol A Biol Sci Med Sci 2005;60(8):1039–45.

[100] Maalouf M, Rho JM, Mattson MP. The neuroprotective properties of calorie restriction, the ketogenic diet, and ketone bodies. Brain Res Rev 2009;59(2):293–315.

[101] Maalouf M, Sullivan PG, Davis L, Kim DY, Rho JM. Ketones inhibit mitochondrial production of reactive oxygen species production following glutamate excitotoxicity by increasing NADH oxidation. Neuroscience 2007;145(1):256–64.

[102] Gasior M, Rogawski MA, Hartman AL. Neuroprotective and disease-modifying effects of the ketogenic diet. Behav Pharmacol 2006;17(5–6):431–9.

[103] Henderson ST, Vogel JL, Barr LJ, Garvin F, Jones JJ, Costantini LC. Study of the ketogenic agent AC-1202 in mild to moderate Alzheimer's disease: a randomized, double-blind, placebo-controlled, multicenter trial. Nutr Metab (Lond) 2009;6:31.

[104] Reger MA, Henderson ST, Hale C, et al. Effects of beta-hydroxybutyrate on cognition in memory-impaired adults. Neurobiol Aging 2004;25(3):311–14.

[105] Van der Auwera I, Wera S, Van Leuven F, Henderson ST. A ketogenic diet reduces amyloid beta 40 and 42 in a mouse model of Alzheimer's disease. Nutr Metab (Lond) 2005;2:28.

[106] Kashiwaya Y, Takeshima T, Mori N, Nakashima K, Clarke K, Veech RL. D-beta-hydroxybutyrate protects neurons in models of Alzheimer's and Parkinson's disease. Proc Natl Acad Sci USA 2000;97(10):5440–4.

[107] Refolo LM, Malester B, LaFrancois J, et al. Hypercholesterolemia accelerates the Alzheimer's amyloid pathology in a transgenic mouse model. Neurobiol Dis 2000;7(4):321–31.

[108] George AJ, Holsinger RM, McLean CA, et al. APP intracellular domain is increased and soluble Abeta is reduced with diet-induced hypercholesterolemia in a transgenic mouse model of Alzheimer's disease. Neurobiol Dis 2004;16(1):124–32.

[109] Wolf-Klein GP, Silverstone FA, Levy AP. Sweet cravings and Alzheimer's disease. J Am Geriatr Soc 1991;39(5):535–6.

[110] Perez-Guisado J. Arguments in favor of ketogenic diets. Internet J Nutr Wellness 2007;4:1–24.

[111] Dashti HM, Bo-Abbas YY, Asfar SK, et al. Ketogenic diet modifies the risk factors of heart disease in obese patients. Nutrition 2003;19(10):901–2.

[112] Colpo A. LDL cholesterol: "Bad" cholesterol, or bad science? J Am Phys Surg 2005;10:83–9.

[113] Colpo A. The Great Cholesterol Con. Raleigh, NC, USA: Lulu Press Inc.; 2006.

[114] Yerushalmy J, Hilleboe HE. Fat in the diet and mortality from heart disease; a methodologic note. NY State J Med 1957;57(14):2343–54.

[115] O'Keefe Jr. JH, Cordain L. Cardiovascular disease resulting from a diet and lifestyle at odds with our Paleolithic genome: how to become a 21st-century hunter-gatherer. Mayo Clin Proc 2004;79(1):101–8.

[116] Eaton SB, Konner M, Shostak M. Stone agers in the fast lane: chronic degenerative diseases in evolutionary perspective. Am J Med 1988;84(4):739–49.

[117] Cordain L, Miller JB, Eaton SB, Mann N. Macronutrient estimations in hunter-gatherer diets. Am J Clin Nutr 2000;72(6):1589–92.

CHAPTER

25

Dietary Factors Affecting Osteoporosis and Bone Health in the Elderly

Pawel Glibowski

Department of Milk Technology and Hydrocolloids, University of Life Science in Lublin, Lublin, Poland

KEY FACTS

- More than 200 million people suffer from osteoporosis worldwide.
- Insufficient calcium intake can raised risk of bone fractures.
- The most abundant sources of calcium are rennet-coagulated ripening cheeses among dairy products and sesame seeds among other foods.
- Although 90% of vitamin D is synthesized in the skin, adequate vitamin D intake is associated with a lower risk of osteoporotic hip fractures.
- Calcium supplementation is not necessary when appropriate quantities of dairy products are consumed.
- Increased intake of dairy foods and fruits may decrease the risk of osteoporosis in postmenopausal women.
- Low vegetable protein intake is associated with lower bone mineral density.
- Abstainers have a higher risk of hip fracture compared with those who consume from 7 to 14 g of ethanol per day.

Dictionary of Terms

- *AI*: adequate intake
- *PRI*: population reference intake
- *RDA*: reference daily allowance
- *RDI*: reference daily intake
- *EAR*: estimated average requirement
- *UL*: tolerable upper intake level
- *25OHD*: 25-hydroxyvitamin D
- *PTH*: serum parathyroid hormone
- *BMD*: bone mineral density

INTRODUCTION

Osteoporosis is a growing problem in the world. Due to aging societies, there is considerable increase in the number of osteoporotic fractures worldwide. To prevent them, there is a need to promote good nutritional and lifestyle habits. Many excellent reviews and reports precisely explain the role of nutrients in osteoporosis and bone health management. Two of the most important nutrients are calcium and vitamin D and their influence on the human body have been very well discussed, for example, by the European Food Safety Authority Panel [1], the WHO/FAO report [2], or recommendations published by the American Institute of Medicine [3]. However, these documents focus mainly on the nutrients per se, not the role of foods. Although it is not possible to discuss factors affecting osteoporosis and bone health in the elderly without referring to recommended Dietary Reference Values, in this chapter however, diet ingredients are discussed mostly.

OSTEOPOROSIS

Osteoporosis is a systemic skeletal disease, characterized by low bone mineral density and abnormal bone microarchitecture, which in consequence leads to

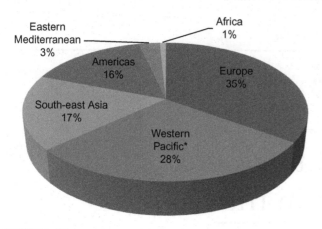
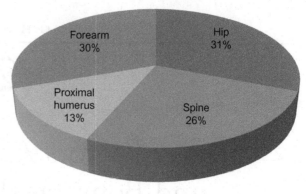

FIGURE 25.1 Share (%) of osteoporotic fractures in men and women aged 50 years or more in 2000, by WHO regions [7]. *Includes Australia, China, Japan, New Zealand, and the Republic of Korea.

FIGURE 25.2 Estimated number of osteoporotic fractures in the world, in men and women aged 50 years or more in 2000 [7].

fragility and increased susceptibility to fractures [4]. In Europe, USA, and Japan, osteoporosis has been diagnosed in 75 million individuals [5]. More than 200 million people suffer from osteoporosis worldwide [6] and due to the aging of society the incidence of osteoporosis will increase. Fig. 25.1 shows the share of osteoporotic fractures in men and women aged 50 years or more in different WHO regions. Osteoporosis increases the risk of fractures, it is associated with increased mortality, increased morbidity, limitations in physical function, pain, and losses in health-related quality of life [8]. Osteoporosis is a disease affecting both sexes, however due to the late start of losing bone mass and a milder course with no sudden changes in hormonal activity, osteoporosis develops less often in men than in women [4]. Seventy to eighty percent of people suffering from osteoporosis are women. Women have a 40—50% risk of having a fracture during their lifetime, while men have a 13—22% risk [9]. Ciesielczuk et al. [4] in their study registered that due to osteoporosis the wrist, forearm, and hip were broken most frequently. As well as those, osteoporotic fractures can affect the spine, shoulder, or femoral neck. According to the WHO [7], more than 50% of fractures concern the spine and hip (Fig. 25.2).

There are two critical periods in a lifetime affecting bone mass. One of these is childhood and adolescence. Because adolescence is a time of rapid skeletal growth, during this period, adequate intake of calcium leads to high peak bone mass which decreases the risk of osteoporosis in the elderly. About 85—90% of adult bone mass is acquired by the age of 18 in girls and the age of 20 in boys [10]. Ninety-nine percent bone mass is achieved by age 26 [11]. Adequate nutrition and regular participation in physical activity are important factors in achieving and maintaining optimal bone mass [10].

The second crucial period is the elderly, generally, and in case of woman, postmenopausal period. Women generally lose about 1—2% of their bone per year during and after menopause [12]. In the case of older man this loss may reach 0.5—1% per year [13]. During this period, adequate nutrition and the use of supplements help to reduce and sometimes even stop these losses.

DIETARY REFERENCE VALUES FOR CALCIUM

Calcium is a macromineral that is very important in nerve transmission, constriction and dilation of blood vessels, and muscle contraction [10]. 99% of calcium in a human body is in bones and teeth and for this reason adequate calcium intake is important for optimal bone health. Insufficient calcium intake can cause low bone mass, a risk factor for osteoporosis, and consequently raise the risk of bone fractures.

Many institutions worldwide recommend different dietary reference values for calcium. Table 25.1 shows these values for adults with special emphasis on older adults and postmenopausal women.

In Dietary Guidelines for Americans [10], the recommended Reference Daily Intake for adults 51+ are 1200 mg calcium per day for men and women. The newest recommendations for Americans and Canadians published Institute of Medicine [3]. IOM recommends a lower Reference Daily Allowance (RDA) for calcium to be 1000 mg per day for males aged 51—70. The other recommendations are in agreement with Dietary Guidelines for Americans 2010. The abovementioned recommendations are based, inter alia, on calcium balance data.

The newest European Food Safety Authority Panel recommendations are based mainly on the calcium excretion data [1]. Calcium balance data were collected from a number of carefully controlled metabolic

TABLE 25.1 Overview of Dietary Reference Values for Calcium (mg/day) for Older Adults

	Dietary guidelines for Americans (2010)	WHO/FAO (2004)	EFSA (2015)	EFSA (2010)	Dietary reference intakes for Japanese (2010)
Female 51+ (RDA)	1200				
Male 51+ (RDA)	1200				
Male ≥65 (PRI)		1300			
Postmenopausal female (PRI)		1300			
Adults ≥25 years (PRI)			950		
Female 50+ (RDA)				1200	
General population (RDA)				800	
Female ≥15 (RDA)					650
Male ≥50 (RDA)					700

studies and analyzed to determine the value where calcium intake equals calcium losses via urine and feces. In addition, dermal losses of calcium were added to derive an average requirement of 750 mg/day. Finally, the Population Reference Intake (PRI) was established on 950 mg calcium per day. That value represents the 97.5th percentile of the distribution of the individual predictions for calcium intake. Final PRI is lower than former recommendations for females aged 50 and older, but higher than general recommendation.

The highest recommendations are suggested by the WHO [2]. Their Population Reference Intake for males aged 65 and older, as well as postmenopausal women is 1300 mg calcium per day. These recommendations result from an analysis of many prospective trials and a meta-analysis suggesting that calcium at this level, mainly achieved by supplementation, helps in bone mineralization.

The lowest recommendations for calcium intakes are for Asians. In the Dietary Reference Intakes for Japanese [14], the Estimated Average Requirement (EAR) was calculated by considering the calcium accumulated in the body, excreted by urine, lost via skin, and assuming apparent absorption rate. The final RDA is 650 mg calcium for females aged 15 plus and 700 mg for males aged 50 and older [15]. Very similar recommendations can be found in Dietary Reference Intakes for Koreans [16]. Harinarayan et al. [17] indicate that the recommended dietary allowances of calcium in India for men and women equal 400 mg calcium per day. Some explanation of these recommendations was given by Barrett-Connor et al. [18]. According to these authors, Asian women have a lower bone mineral density (BMD) than white or black women due to their relatively small body size, genetics, lifestyle, and culture.

FOOD SOURCES OF CALCIUM

Milk and dairy products appeared in the human diet during the agricultural revolution, approximately 10,000 year ago [11]. According to different studies milk is a major food source of calcium in Western countries, contributing 36–70% of the dietary calcium generally [11,19]. Practically, it is really hard to achieve dietary calcium recommendations without consuming dairy products. Besides calcium and vitamin D, dairy foods provide substantial amounts of other essential nutrients including potassium, phosphorus, riboflavin, vitamin B_{12}, protein, zinc, magnesium, and vitamin [11].

Table 25.2 summarizes the content of calcium in different foods. The most abundant sources of calcium are rennet-coagulated ripening cheeses among dairy products and sesame seeds among other foods. Since cheese usually contains a lot of saturated fatty acids, it is not recommended by many nutritionists, especially for older adults. For example, Dietary Guidelines for Americans, 2010 [10] recommends consuming fat-free or low-fat milk and milk products. Milk and fermented milk beverages, like yoghurt are easily consumed in large volume, while other foods rich in calcium are either consumed occasionally (spinach) or not in big quantities (white bean). Quite interesting sources of calcium are nuts and some seeds and dry fruits. They can be an alternative calcium source for those who have cows' milk protein allergy. In contrast to western countries, most calcium ingested in eastern and black African countries comes from vegetables and fish [22].

One should remember that the availability of calcium present in the food is different. Calcium bioavailability from milk and dairy products is approximately 30% [23]. However, calcium in plant-based

TABLE 25.2 Content of Calcium in Different Foods [10,20,21]

	Product	Serving size	Calcium in serving size (mg)	Calcium (mg/100 g)
Dairy products	Milk (skim, low fat, whole)	1 cup/250 mL	293–305	117–122
	Yogurt (nonfat, low-fat, whole milk, fruit)	8 ounces/227 g	275–452	121–199
	Kefir	1 cup/250 mL	260	103
	Cream (9–30%)	1 ounce/28 g	24–31	86–109
	Buttermilk	1 cup/250 mL	275–300	110–120
	Ice cream	0.5 cup/125 mL/80 g	100–125	125–155
	Cottage cheese	0.5 cup/125 mL	65	52
	Quark (skim, low fat, whole)	1½ ounces/42 g	37–40	88–96
	Rennet-coagulated ripening cheeses (eg, brie, cheddar, gouda, emmentaler)	1½ ounces/42 g	252–351	600–835
Other products	White bean, dry seeds	1 cup/250 mL	400	163
	Broccoli, cooked	1 cup/250 mL	180	48
	Spinach, cooked	1 cup/250 mL	240	93
	Figs, dried, uncooked	1 cup /250 mL	300	203
	Kiwi, raw	1 cup/250 mL	50	25
	Almonds, toasted unblanched	1 ounce/28 g	67–80	239–286
	Sesame seeds, whole roasted	1 ounce/28 g	280	1000
	Sardines, canned	3 ounces/84 g	326–370	388–440
	Molasses, blackstrap	1 tablespoon	135	600

foods is less available due to the presence of fiber or phytin or calcium's chemical form [17]. However, there are several studies suggesting that soluble fiber (inulin, fructooligosaccharose, oligofructose) enhances calcium absorption. These fructans are fermented by prohealthy intestinal bacteria and products of this fermentation decrease pH in the colon increasing macrominerals absorption [24].

On the other hand, high calcium intake is linked with higher risk of cardiovascular events and mortality. The Tolerable Upper Intake Level (UL) for calcium is 2000–2500 mg [1,3]. Although it is not practically possible to reach this level in a diet, it is possible to reach it and exceed by taking supplements or by consuming an excess of supplemented foods.

DIETARY REFERENCE VALUES FOR VITAMIN D

Vitamin D is very important in calcium regulation and bone metabolism. The most important consequences of vitamin D deficiency include secondary hyperparathyroidism, accelerated bone loss, increased bone turnover, proximal muscle weakness, increased body sway, falls, osteoporosis, and fractures [25]. Vitamin D is synthesized in the skin from ultraviolet irradiation, which mainly comes from the sun. Although 90 % of vitamin D is endogenic, there are some food sources of this vitamin. The two major forms are vitamin D_2 (ergocalciferol) and vitamin D_3 (cholecalciferol). Vitamin D from the diet has to be activated in the human body. There are several intermediate compounds before forming the final biologically active vitamin D. One of this is a serum 25-hydroxyvitamin D (25OHD), a very often used marker allowing the estimation of a person's nutrition. A 25OHD level of less than 27–30 nmol/L is associated with an increased risk for developing rickets. Practically all persons are sufficient with serum 25OHD levels of at least 50 nmol/L (20 ng/mL) [3], however according to Verhaar [26] the optimal level for intestinal calcium resorption is about 80 nmol/L.

The Institute of Medicine [3] estimated recommendations for vitamin D, based on measures of serum 25OHD level as a reflection of total vitamin D

TABLE 25.3 Overview of Dietary Reference Values for Vitamin D (μg/day) for Older Adults

	IOM (2011)	WHO/FAO (2004)	EFSA (2010)	Dietary reference intakes for Japanese (2010)
Female, male 1–70 (RDA)	15			
Female, male >70 (RDA)	20			
Adults 51–65 (RDI)		10		
Adults 65+ (RDI)		15		
Female 50+ (RDA)			20	
General population (RDA)			5	
Female ≥18 (AI)				5.5
Male ≥18 (AI)				5.5

exposure. 50 nmol 25OHD/L of serum was estimated as a level reasonable for nearly all the population. Although increasing intake of vitamin D results in not quite linear higher blood levels of 25OHD, the final Reference Daily Allowance (15 μg vitamin D/day) should cover the needs of 97.5 % of USA and Canadian society. The RDA value for persons aged 70 years and more was increased to 20 μg vitamin D/day, due to changes in bone density and fracture risk (Table 25.3).

FAO/WHO recommendations were established after analysis of several studies that found that modest increases in vitamin D intake reduces the rate of bone loss and the incidence of hip fractures. These findings have led the FAO/WHO to recommend an intake of 10–15 μg/day vitamin D for the elderly [2].

According to Japanese recommendations concerning vitamin D, only the adequate intake was established due to limitations on the available data. 5.5 μg vitamin D/day was set, because this level allowed serum parathyroid hormone (PTH) level to keep low, and prevent bone demineralization [14,27].

EFSA, on the basis of available scientific data, estimated that 800 I.U. (20 μg) of vitamin D from all sources should be consumed daily by women 50 years and older in order to obtain a reduction of bone loss. Five micrograms of vitamin D is recommended for the general population. According to EFSA's opinion, such an intake allows a serum concentration of 25OHD to be maintained above 50 nmol/L during winter [28].

The differences in Dietary Reference Values are quite striking, especially between Japan and Western countries. Although vitamin D deficiency is related to osteoporosis in observational studies in Western populations, completely different conclusions were drawn from analysis taken in the Lanzhou, which is located on the northwestern inland of China. Zhen et al. [25], in a cross-sectional study, analyzed serum 25OHD levels and BMD in 2942 men and 7158 women aged 40–75 years. They concluded that vitamin D deficiency is prevalent in the middle-aged and elderly northwestern Chinese population but reduced 25OHD levels were not associated with an increased osteoporosis risk.

FOOD SOURCES OF VITAMIN D

The main sources of vitamin D are fish (Table 25.4), especially those with red meat, like salmon or tuna. In addition, vitamin D can be found in eggs, meat, and dairy products; however these products are not very rich sources of this vitamin. Mushrooms deserve special attention. The very vast range of vitamin D content in this product depends on the method of cultivation. Vitamin D_2 is almost totally absent in cultivated mushrooms, while some wild mushrooms contained very high concentrations of this vitamin. The level of this compound depends on exposition to UV light of wavelengths 280–320 nm [29]. For this reason, wild mushrooms which grow exposed to the sun can contain almost two-fold higher levels of vitamin D in comparison to those which grow in the shade. Nowadays, it is possible to buy cultivated mushrooms with higher levels of vitamin D, because they were additionally treated with UV light during cultivation.

Since eggs and dairy products do not contain very high levels of vitamin D, mushrooms can be an interesting source of this vitamin, especially for vegetarians. The only problem is the availability of this product for consumers. Although mushrooms are cultivated in many countries worldwide, the choice is rather limited to a few species. Edible wild mushrooms naturally grown in forests, parks, and meadows give much better choice. However, because of many cases of people eating poisonous mushrooms, foraging for wild mushrooms is limited only to those who are very experienced.

TABLE 25.4 Content of Vitamin D in Different Foods [10]

Product	Serving size	Vitamin D in serving size (µg)	Vitamin D (µg /100 g)
Salmon (cooked, smoked, canned)	3 ounces/84 g	11–20	13–24
Tuna (canned in oil or water, drained)	3 ounces/84 g	3.8–5.7	4.5–6.8
Sardines, canned	3 ounces/84 g	3.1	3.7
Herring, pickled	3 ounces/84 g	2.4	2.9
Cod, cooked	3 ounces/84 g	1	1.2
Pork	3 ounces/84 g	0.6–2.2	0.7–2.6
Milk (skim, low fat, whole)	1 cup/250 mL	2.9–3.2	1.2–1.3
Egg, hard-boiled	1 large	0.7	1.4
Mushrooms	½ cup/125 mL	0.6–24	0.3–15

THE EFFECT OF DAIRY PRODUCTS ON OSTEOPOROSIS

Clinical and epidemiological studies published since 1939 indicate that consuming adequate amounts of calcium or calcium-rich foods, such as milk and other dairy foods, throughout life helps to optimize peak bone mass development by age 30 or earlier, slow age-related bone loss, and reduce the risk of osteoporotic fracture in later adult years [1,11].

McCabe et al. [30] analyzed the relationship between baseline dairy intake and mean percentage change in femoral neck bone mineral density in 121 white men and women supplemented with 750 mg Ca/day or placebo over 4 years. Subjects were classified into two groups on the basis of whether they consumed less than 1.5 or more than 1.5 servings of dairy products/day (450 mg Ca). A mean percentage change in femoral neck bone mineral density and total hip BMD in subjects consuming dairy products (more than 450 mg Ca/day at baseline) was not evident when compared to those who were provided with more than 450 mg Ca/day in dairy products and additionally supplemented their diet with calcium. From this study it can be concluded that supplementation is not necessary when appropriate quantities of dairy products are consumed. However, those who had deficient diets had more than 2% mean percentage change in femoral neck bone mineral density.

Shin and Joung [31] in a study of 3735 Korean postmenopausal women identified four dietary patterns using factor analysis. These patterns were as follows: "meat, alcohol, and sugar," "vegetables and soya sauce," "white rice, kimchi, and seaweed," and "dairy and fruit." Statistical analysis showed that a diet rich in dairy and fruit products is positively associated with the BMD of five regions of the femur (femoral neck, trochanter, intertrochanter, Ward's triangle and total) and the lumbar spine. Shin and Joung [31] in this study analyzed different nutrient intakes, and as expected the "dairy and fruit" dietary pattern score was positively associated with the most nutrients, including Ca, except carbohydrates and Na. Even though the dietary intakes of the subjects were assessed using a single 24 h recall, which might not represent the individual's usual intake, the results showed that subjects consuming more dairy and fruit products had a 53% lower risk of osteoporosis in the lumbar spine, while those with a diet based on white rice, kimchi, and seaweed had a 40% higher risk of osteoporosis in the lumbar spine. Since the traditional Korean diet is based on white rice, vegetables, and fermented foods, an increased intake of dairy foods and fruits may decrease the risk of osteoporosis in postmenopausal women [31].

Storm et al. [32] analyzed sixty older postmenopausal women without osteoporosis. The group consuming 4 glasses of milk/day averaged a calcium intake of 1028 mg/day and the placebo-treated women consumed a mean of 683 mg/day. After two years, the placebo group lost 3.0% of their greater trochanteric (GT) bone mineral density (BMD) ($p < 0.03$ vs baseline) and dietary supplemented women sustained minimal loss from the GT (-1.5%; $p = 0.30$). What is more interesting, femoral bone loss occurred exclusively during the two winters of the study and there was no change in GT BMD during summer. Serum 25-OH vitamin D level decreased more than 20% in both groups during the winter months but returned to baseline in the summer and parathyroid hormone levels increased about 20% during winter but did not return to baseline during the summers. Statistical analysis showed that total calcium intake was the strongest predictor of bone loss from the hip. They concluded that calcium supplementation prevents bone loss in elderly women by suppressing bone turnover during the winter when serum 25-OH vitamin D declines and serum PTH increases [32].

On the other hand, there are studies questioning the role of dietary calcium. Feskanich et al. [19] examined calcium and vitamin D intakes, milk consumption, and use of calcium supplements during 18 years of follow-up in 72,337 postmenopausal women with repeated measures of dietary intake and supplement use. During the 18 years of follow-up, 7466 women (10%) reported a diagnosis of osteoporosis. Six hundred three incident hip fractures caused by low or moderate trauma (eg, slipping on ice, falling from the height of a chair) were identified. Fractures caused by high trauma (eg, skiing, falling down a flight of stairs) were excluded from analysis (15% of the reported fractures). After statistical analysis, they concluded that neither milk nor a high-calcium diet appears to reduce the risk of osteoporotic hip fractures in postmenopausal women. Only an adequate vitamin D intake is associated with a lower risk of such incidents. Although the authors indicate that higher milk consumption conferred a weak, insignificant reduction in fracture risk—women consuming more than 1.5 glasses of milk/day had a relative risk of 0.83 compared with women who consumed less than 1 glass/week—they did not observe a dose response relation (P for trend = 0.21), and with higher daily intakes of more than 2.5 glasses of milk, there was still no evidence of a protective effect [19].

Beside calcium, milk contains 19 minerals that are considered to be nutritionally essential for health. An excellent review concerning the role of calcium, phosphorus, magnesium, sodium, potassium, and zinc in bone health is presented by Cashman [23]. Cashman [23] analyzed the role of phosphorous which is critical in bone formation. There is no phosphorus deficiency in an average diet. The ratio of phosphorus:calcium seems to be quite controversial. Although the excess phosphorus increases parathyroid hormone levels, some studies suggest that there are no links between high levels of phosphorus intake and lower bone mass or higher rates of bone loss in humans. What is more important, the P:Ca ratio in milk is well balanced [23].

Another macromineral present in milk and milk beverages is magnesium, although dairy products are not the main food sources of this nutrient. While magnesium deficiency has been identified as a possible, risk factor for osteoporosis in humans, the effect of magnesium on bone health requires further investigation [23]. In particular, the excess of magnesium in a diet seems controversial because it is able to decrease the serum levels of calcium [33].

VITAMIN K AND BONE HEALTH

Vitamin K is important in bone metabolism and probably has a positive influence on bone health. Osteocalcin, one of the bone matrix proteins is vitamin K dependent. This protein may play an important role in maintaining bone mineral density (BMD) and reducing the incidence of fractures through incremental bone formation [34].

There are two major forms of vitamin K—K_1 and K_2. The main sources of vitamin K_1 are green and leafy vegetables. Vitamin K_2 is formed by some fermentation bacteria. Low vitamin K intake is associated with an increased risk of hip fracture, but different studies have been inconsistent regarding its effect on BMD [35]. According to Asakura et al. [36] Vitamin K was one of the most frequently administered drugs for the treatment of patients with osteoporosis in Japan.

SEMINUTRITIONAL FACTORS AFFECTING OSTEOPOROSIS AND RISK OF FRACTURES

Besides nutrition, bone density and strength depend on many factors like genetics, age, physical activity, gender, ethnicity, endocrine changes, lifestyles, general health condition, smoking, and alcohol consumption as well as taking drugs [4,23]. One of the most important factors affecting osteoporosis is physical activity. Exercise is necessary to stimulate bone modeling and remodeling while calcium, vitamin D, and other nutrients are important substrates for bone mineralization [37]. According to Daly et al. [37], calcium in the diet may improve the effects of exercise on bone mass both in children and adults, particularly in those with inadequate intakes. Interestingly, combining exercise and calcium appears to have little effect on bone structure or strength, which is predominantly regulated by mechanical loading. However, insufficient intake can compromise the skeletal response to loading.

There are alternative studies trying to find new dependences. According to Feskanich et al. [38], greater milk consumption during teenage years was not associated with a lower risk of hip fracture in older adults. The positive association observed in men was partially mediated through attained height, which may suggest that height can be an independent risk factor for fracture.

Barrera et al. [39] confirmed the protective effect of a high BMI on femoral neck bone mineral density among elderly subjects. They studied 615 women and 230 men with mean age, 75 ± 4.4 years. One quarter of women and 11% of men had osteoporosis.

The age-adjusted odds ratios for femoral osteoporosis were 0.34 and 0.13 for women and men with a BMI between 25 and 30 kg/m^2, respectively. The odds ratios for women and men with a BMI between 30 and 35 kg/m^2 were 0.21 and 0.09, respectively. The risk for osteoporosis among men and women with a BMI above 30 kg/m^2 was approximately 33% compared with subjects with a normal BMI.

Kouda et al. [40] in a large-scale study of elderly Japanese men revealed that an alcohol intake of below 55 g/day was positively correlated to BMD, whilst alcohol intake of ≥55 g/day was inversely correlated to BMD. The studied group counted 1665 subjects aged ≥65 years with no diseases or drug therapy affecting bone mineral density. Adjusted total hip BMD of men with alcohol intake above 39 g/day was 0.90 g/cm^2 and that of abstainers was 0.85 g/cm^2. These results are in agreement with many other studies. Berg et al. [41] in their meta-analysis concluded that abstainers had a higher risk of hip fracture compared with those who consume from 7 to 14 g of ethanol per day. The mechanisms hiding behind these observations remain unknown as yet.

Muraki et al. [42] studied lifestyle factors associated with BMD. After a cross-sectional study with 632 women age ≥60 years they concluded that patients with the habit of green tea drinking had significantly higher BMD than patients without such a habit. Multiple regression analysis showed that the effect of green tea on BMD was independent of age and BMI. The authors suggest that the connection between green tee drinking and BMD can be related to flavonoids present in the infusions having a weak estrogenic effect, which may increase BMD.

A quite interesting study was published by Varenna et al. [43] who analyzed 3301 postmenopausal women to find a link between hypertension and osteoporosis. Osteoporosis was diagnosed by lumbar dual-energy X-ray absorptiometry and hypertension was defined by blood pressure data and/or the use of antihypertensive medication. The subjects were divided into four groups on the basis of dairy food consumption which was evaluated using a weekly food-frequency questionnaire. The lowest quartile corresponded to an intake less than 400 mg Ca per day and the higher quartile more than 800 mg per day. After statistical analysis Varenna et al. [43] concluded that osteoporosis and hypertension are associated in postmenopausal women, and a low dairy intake may increase the risk of both diseases. 25.7% women were found to be affected by osteoporosis and 25% by hypertension. The proportion of subjects in the lowest quartile of dairy intake was significantly higher among women with osteoporosis compared to nonosteoporotic women (32.0 vs 24.1%). Additionally, women affected by hypertension were more often found in the osteoporosis group (32.2 vs 22.5%). A link between calcium intake and blood pressure can be the result of numerous biological mechanisms, ie, association between reduced calcium intake and increasing vascular tone and blood pressure. High sodium intake, typical for subjects with hypertension, could affect a negative Ca balance [44] or dietary calcium may influence the activity of the renin–angiotensin system critical in blood pressure regulation [45,46].

Link between osteoporosis and the cardiovascular system were also found by Abou-Raya and Abou-Raya [47]. 126 consecutive patients aged 65 years and above, with moderate to severe congestive heart failure (CHF) were screened for osteoporosis. The results suggest that the increased bone loss in conjunction with CHF is likely to increase fracture risk.

PROTEIN CONSUMPTION AND BONE HEALTH

According to Hampson et al. [48], elderly men and women with a lower protein intake have been shown to have increased bone loss. A few studies showed that hip fracture patients have diets particularly deficient in protein and energy and protein supplementation reduced femoral bone loss and a shortened rehabilitation time [48]. On the other hand, excess of protein in a diet can be problematic as well.

Sellmeyer at al. [49] and Weikert et al. [50] showed that a high ratio of dietary animal to vegetable protein increased the rate of bone loss and the risk of fracture in postmenopausal women. Animal protein is supposed to provide a higher dietary acid load and consequently affect increased calcium excretion and negative calcium balance [51]. However, Fentron et al. [52] in a meta-analysis showed that increasing the diet acid load does not promote skeletal bone mineral loss or osteoporosis. Promotion of the "alkaline diet" to prevent calcium loss is not justified. This conclusion supports other later studies. Beasley at al. [53] examined cross-sectional and longitudinal associations between baseline dietary protein and bone mineral density among 560 females aged 14–40 years. The role of protein source (animal or vegetable) and participant characteristics were considered. Data from this longitudinal study suggest that a higher protein intake does not have an adverse effect on bone in premenopausal women. However cross-sectional analyses suggested that low vegetable protein intake is associated with lower BMD.

FUTURE TRENDS

Literature data suggests that calcium intake may need to be adjusted for dietary factors (eg, animal/plant protein consumed ratio, sodium and vitamin D intake) and nondietary factors, for example, sun exposure (regarding latitude and geographic location). Although high calcium intakes promote bone health, maybe in the future recommended daily intakes will depend on additional factors. For example, physical activity enhances BMD. Even longer sleepers—sleeping more than 8 h—have higher odds of osteoporosis compared to those who sleep shorter [54]. Bearing in mind that recommended dietary intake is very hard to achieve, because even three dairy servings per day give calcium intake well below the recommended dietary intake [55], maybe in future studies it will be possible to estimate the level of main factors affecting osteoporosis development and in this way to recommend more precise calcium as well as vitamin D intakes. Another conclusion which can be drawn from the above review is that considering osteoporosis separately as a health problem can be insufficient. Since there are links between osteoporosis and diseases of the cardiovascular system [43,47], perhaps dietary approaches should be broader, and not only focus on a few food ingredients, like calcium or vitamin D. This conclusion is supported by positive correlations between the decreased risk of osteoporosis and moderate alcohol or green tea drinking [39,42].

SUMMARY

- Milk and dairy products are the most abundant sources of calcium and positively affect bone health in the elderly, preventing from bone fractures.
- Vitamin D is very important in bone metabolism and the main sources of vitamin D are fish, especially those with red meat.
- Besides dairy products, increased intake of fruits and vegetables may decrease the risk of osteoporosis.
- Moderate alcohol intake decrease the risk of osteoporotic fractures in older adults.
- Calcium intake may need to be adjusted for dietary factors (eg, animal/plant protein consumed ratio, sodium and vitamin D intake) and nondietary factors as sun exposure, smoking, or physical activity.

References

[1] EFSA NDA Panel (EFSA Panel on Dietetic Products, Nutrition and Allergies), 20YY; Draft Scientific opinion on Dietary Reference Values for calcium. EFSA Journal 20YY; volume (issue): NNNN, 82 pp. <http://dx.doi.org/10.2903/j.efsa.20YY.NNNN> Available online: <www.efsa.europa.eu/efsajournal>.

[2] WHO/FAO (World Health Organization/Food and Agriculture Organization of the United Nations), 2287 2004. Vitamin and mineral requirements in human nutrition: report of a joint FAO/WHO expert 2288 consultation, Bangkok, Thailand, September 21–30, 1998.

[3] IOM (Institute of Medicine). Dietary reference intakes for calcium and Vitamin D. Washington, DC: The National Academies Press; 2011.

[4] Ciesielczuk N, Glibowski P, Szczepanik J. Awareness of factors affecting osteoporosis obtained from a survey on retired polish subjects. Rocz Państw Zakl Hig 2014;65(2):147–53.

[5] Wadolowska L, Sobas K, Szczepanska JW, Slowinska MA, Czlapka-Matyasik M, Niedzwiedzka E. Dairy products, dietary calcium and bone health: possibility of prevention of osteoporosis in women: the polish experience. Nutrients 2013;5:2684–707.

[6] Reginster JY, Burlet N. Osteoporosis: a still increasing prevalence. Bone 2006;38(2 Suppl. 1):S4–9.

[7] WHO. WHO scientific group on the assessment of osteoporosis at primary health care level Summary Meeting Report Brussels, Belgium; 2007 May 5–7, 2004, WHO Press, Geneva, Switzerland.

[8] Lötters FJB, Lenoir-Wijnkoop I, Fardellone P, Rizzoli R, Rocher E, Poley MJ. Dairy foods and osteoporosis: an example of assessing the health-economic impact of food products. Osteoporos Int 2013;24:139–50.

[9] Johnell O, Kanis J. Epidemiology of osteoporotic fractures. Osteoporis Int 2005;16(Suppl. 2):S3–7.

[10] Dietary Guidelines for Americans. 7th ed. U.S. Department of agriculture and U.S. Department of health and human services, 2014. Washington, DC: U.S. Government Printing Office; 2010.

[11] Huth PJ, DiRienzo DB, Miller GD. Major scientific advances with dairy foods in nutrition and health. J Dairy Sci 2006; 89:1207–21.

[12] Tenta R, Moschonis G, Koutsilieris M, Manios Y. Calcium and vitamin D supplementation through fortified dairy products counterbalances seasonal variations of bone metabolism indices: the postmenopausal health study. Eur J Nutr 2011;50:341–9.

[13] Scholtissen S, Guillemin F, Bruyère O, Collette J, Dousset B, Kemmer C, et al. Assessment of determinants for osteoporosis in elderly men. Osteoporos Int 2009;20:1157–66.

[14] Dietary Reference Intakes for Japanese (2010) National Institute of Health and Nutrition, Japan 2011–2012.

[15] Uenishi K, Ishimi Y, Nakamura K, Kodama H, Esashi T. Dietary reference intakes for Japanese 2010: macrominerals. J Nutr Sci Vitaminol 2013;59(Suppl.):S83–90.

[16] Dietary Reference Intake for Korean. The Korean Nutrition Society. South Korea: Hanareum Press, Seoul; 2010.

[17] Harinarayan CV, Ramalakshmi T, Prasad UV, Sudhakar D, Srinivasarao PVLN, Sarma KVS, et al. High prevalence of low dietary calcium, high phytate consumption, and vitamin D deficiency in healthy south Indians. Am J Clin Nutr 2007;85:1062–7.

[18] Barrett-Connor E, Siris ES, Wehren LE, et al. Osteoporosis and fracture risk in women of different ethnic groups. J Bone Miner Res 2005;20:185–94.

[19] Feskanich D, Willett WC, Colditz GA. Calcium, vitamin D, milk consumption, and hip fractures: a prospective study among postmenopausal women. Am J Clin Nutr 2003;77:504–11.

[20] Calcium Content of Foods, University of California San Francisco, <http://www.ucsfhealth.org/education/calcium_content_of_selected_foods/>.

[21] Kunachowicz H, Nadolna I, Iwanow K. Wartość odżywcza wybranych produktów spożywczych i typowych potraw (The nutritional value of selected foods and typical dishes). Wydawnictwo Lekarskie PZWL Warszawa 2012.

[22] Menkes CJ. Prevention and treatment of deficiency diseases with milk and dairy products [Lait et produits laitiers dans la prévention et le traitement des maladies par carence]. Bull Acad Natl Med 2008;192(4):739–47.

[23] Cashman KD. Milk minerals (including trace elements) and bone health. Int Dairy J 2006;16:1389–98.

[24] Coudray C, Bellanger J, Castiglia-Delavaud C, Remesy C, Vermorel M, Rayssignuier Y. Effect of soluble or partly soluble dietary fibres supplementation on absorption and balance of calcium, magnesium, iron and zinc in healthy young men. Eur J Clin Nutr 1997;51:375–80.

[25] Zhen D, Liu L, Guan C, Zhao N, Tang X. High prevalence of vitamin D deficiency among middle-aged and elderly individuals in northwestern China: its relationship to osteoporosis and lifestyle factors. Bone 2015;71:1–6.

[26] Verhaar HJJ. Medical treatment of osteoporosis in the elderly. Aging Clin Exp Res 2009;21(6):407–13.

[27] Tanaka K, Terao J, Shidoji Y, Tamai H, Imai E, Okano T. Dietary reference intakes for Japanese 2010: fat-soluble vitamins. J Nutr Sci Vitaminol 2013;59(Suppl.):S57–66.

[28] Scientific Opinion in relation to the authorisation procedure for health claims on calcium and vitamin D and the reduction of the risk of osteoporotic fractures by reducing bone loss pursuant to Article 14 of Regulation (EC) No 1924/2006, EFSA Panel on Dietetic Products, Nutrition and Allergies, EFSA J 2010;8(5):1609.

[29] Mattila P, Lampi A-M, Ronkainen R, Toivo J, Piironen V. Sterol and vitamin D_2 contents in some wild and cultivated mushrooms. Food Chem 2002;76:293–8 (020).

[30] McCabe LD, Martin BR, McCabe GP, Johnston CC, Weaver CM, Peacock M. Dairy intakes affect bone density in the elderly. Am J Clin Nutr 2004;80:1066–74.

[31] Shin S, Joung H. A dairy and fruit dietary pattern is associated with a reduced likelihood of osteoporosis in Korean postmenopausal women. Br J Nutr 2013;110:1926–33.

[32] Storm D, Eslin R, Porter ES, Musgrave K, Vereault D, Patton C, et al. Calcium supplementation prevents seasonal bone loss and changes in biochemical markers of bone turnover in elderly New England women: a randomized placebo-controlled trial. J Clin Endocrinol Metab 1998;83(11):3817–25.

[33] Wang S, Lin S, Zhou Y. Changes of total content of serum magnesium in elderly Chinese women with osteoporosis. Biol Trace Elem Res 2006;110:223–31.

[34] Miki T, Nakatsuka K, Naka H, Kitatani K, Saito S, Masaki H, et al. Vitamin K2 (menaquinone 4) reduces serum undercarboxylated osteocalcin level as early as 2 weeks in elderly women with established osteoporosis. J Bone Miner Metab 2003;21:161–5.

[35] Fujita Y, Iki M, Tamaki J, Kouda K, Yura A, Kadowaki E, et al. Association between vitamin K intake from fermented soybeans, natto, and bone mineral density in elderly Japanese men: the Fujiwara-kyo osteoporosis risk in men (FORMEN) study. Osteoporos Int 2012;23:705–14.

[36] Asakura H, Myou S, Ontachi Y, Mizutani T, Kato M, Saito M, et al. Vitamin K administration to elderly patients with osteoporosis induces no hemostatic activation, even in those with suspected vitamin K deficiency. Osteoporos Int 2001;12:996–1000.

[37] Daly RM, Duckham RL, Gianoudis J. Evidence for an interaction between exercise and nutrition for improving bone and muscle health. Curr Osteoporos Rep 2014;12:219–26.

[38] Feskanich D, Bischoff-Ferrari HA, Frazier AL, Willett WC. Milk consumption during teenage years and risk of hip fractures in older adults. JAMA Pediatr 2014;168(1):54–60.

[39] Barrera G, Bunout D, Gattás V, de la Maza MP, Leiva L, Hirsch, et al. A high body mass index protects against femoral neck osteoporosis in healthy elderly subjects. Nutrition 2004;20:769–71.

[40] Kouda K, Iki M, Fujita Y, Tamaki J, Yura A, Kadowaki E, et al. Alcohol intake and bone status in elderly Japanese men: baseline data from the Fujiwara-kyo osteoporosis risk in men (FORMEN) study. Bone 2011;49:275–80.

[41] Berg KM, Kunins HV, Jackson JL, Nahvi S, Chaudhry A, Harris Jr KA, et al. Association between alcohol consumption and both osteoporotic fracture and bone density. Am J Med 2008;121:406–18.

[42] Muraki S, Yamamoto S, Ishibashi H, Oka H, Yoshimura N, Kawaguchi H, et al. Diet and lifestyle associated with increased bone mineral density: cross-sectional study of Japanese elderly women at an osteoporosis outpatient clinic. J Orthop Sci 2007;12:317–20.

[43] Varenna M, Manara M, Galli L, Binelli L, Zucchi F, Sinigaglia L. The association between osteoporosis and hypertension: the role of a low dairy intake. Calcif Tissue Int 2013;93:86–92.

[44] Teucher B, Dainty JR, Spinks CA, Majsak-Newman G, Berry DJ, Hoogewerff JA, et al. Sodium and bone health: impact of moderately high and low salt intakes on calcium metabolism in postmenopausal women. J Bone Miner Res 2008;23:1477–85.

[45] Resnick LM. The role of dietary calcium in hypertension. A hierarchal overview. Am J Hyperten 1999;12:99–112.

[46] Guan XX, Zhou Y, Li JY. Reciprocal roles of angiotensin II and angiotensin II receptors blockade (ARB) in regulating Cbfa1/RANKL via cAMP signaling pathway: possible mechanism for hypertension-related osteoporosis and antagonistic effect of ARB on hypertension-related osteoporosis. Int J Mol Sci 2011;12:4206–13.

[47] Abou-Raya S, Abou-Raya A. Osteoporosis and congestive heart failure (CHF) in the elderly patient: double disease burden. Arch Gerontol Geriatr 2009;49:250–4.

[48] Hampson G, Martin FC, Moffat K, Vaja S, Sankaralingam S, Cheung J, et al. Effects of dietary improvement on bone metabolism in elderly underweight women with osteoporosis: a randomised controlled trial. Osteoporos Int 2003;14:750–6.

[49] Sellmeyer DE, Stone KL, Sebastian A, Cummings SR. A high ratio of dietary animal to vegetable protein increases the rate of bone loss and the risk of fracture in postmenopausal women. Study of osteoporotic fractures research group. Am J Clin Nutr 2001;73:118–22.

[50] Weikert C, Walter D, Hoffmann K, Kroke A, Bergmann MM, Boeing H. The relation between dietary protein, calcium and bone health in women: results from the EPIC-Potsdam cohort. Ann Nutr Metab 2005;49:312–18.

[51] Langsetmo L, Hanley DA, Prior JC, Barr SI, Anastassiades T, Towheed T, et al. Dietary patterns and incident low-trauma fractures in postmenopausal women and men aged >50 y: a population-based cohort study. Am J Clin Nutr 2011;93:192–1929.

[52] Fenton TR, Lyon AW, Eliasziw M, Tough SC, Hanley DA. Metaanalysis of the effect of the acid-ash hypothesis of osteoporosis on calcium balance. J Bone Miner Res 2009;24:1835–40.

[53] Beasley JM, Ichikawa LE, Ange BA, Spangler L, LaCroix AZ, Ott SM, et al. Is protein intake associated with bone mineral density in young women? Am J Clin Nutr 2010;91:1311–16.

[54] Kobayashi D, Takahashi O, Deshpande GA, Shimbo T, Fukui T. Association between osteoporosis and sleep duration in healthy middle-aged and elderly adults: a large-scale, cross-sectional study in Japan. Sleep Breath 2012;16:579–83.

[55] van den Berg P, van Haard PM, van den Bergh JP, Niesten DD, van der Elst M, Schweitzer DH. First quantification of calcium intake from calcium-dense dairy products in Dutch fracture patients (the Delft cohort study). Nutrients 2014;6(6):2404–18.

C H A P T E R

26

The Aging Muscle and Sarcopenia: Interaction with Diet and Nutrition

Emanuele Marzetti, Riccardo Calvani, Matteo Tosato and Francesco Landi

Department of Geriatrics, Neurosciences and Orthopedics, Catholic University of the Sacred Heart School of Medicine, Rome, Italy

KEY FACTS

- The age-related loss of muscle mass and function (sarcopenia) is associated with multiple adverse health outcomes.
- Sarcopenia is a major determinant of physical frailty, a condition of increased vulnerability to developing negative health-related events.
- The age-associated "physiologic" decline in appetite may be aggravated by health, social and financial conditions, leading to overt malnutrition.
- Inadequate quantitative or qualitative food intake contributes to the development of sarcopenia and frailty.
- Supplementation with specific nutrients (eg, protein, amino acid metabolites, vitamin D, creatine, polyunsaturated fatty acids) amplifies the beneficial effects of physical activity on muscle health and functional status.

Dictionary of Terms

- *Anorexia of aging*: age-related decline in appetite and/or food intake.
- *Frailty*: state of increased vulnerability to internal and/or external stressors with greater risk of incurring adverse events.

- *Multidomain intervention*: therapeutic approach employing complementary and synergistic treatments to address complex health conditions (eg, sarcopenia and frailty).
- *Recommended dietary allowance (RDA)*: average daily level of intake sufficient to meet the nutrient requirements of nearly all (97–98%) healthy people.
- *Sarcopenia*: pathologic reduction of muscle mass and function during aging.

INTRODUCTION

Nutrition is a major determinant of the health status throughout the life course. Indeed, inadequate nutrition, either excessive or insufficient, is associated with the development and progression of virtually all major health conditions, including cancer, cardiovascular disease, chronic kidney disease, diabetes mellitus, obesity, immunodepression, neurodegeneration, etc. [1]. In advanced age, inadequate quantitative and/or qualitative food intake has also emerged as a main contributing factor to the development of sarcopenia and frailty [2].

The age-related loss of muscle mass and function (sarcopenia) is one of the most pervasive changes that accompany aging. As acknowledged by Rosenberg [3], "There may be no single feature of age-related decline that could more dramatically affect ambulation, mobility, calorie intake, and overall nutrient intake and status, independence, breathing, etc."

Sarcopenia is indeed associated with a multitude of adverse health outcomes, among which falls, disability, institutionalization, and mortality are certainly the most worrisome.

The relationship between inadequate nutrition and sarcopenia is bidirectional. In fact, the ingestion of insufficient amounts of calories and specific nutrients promotes the development of sarcopenia. Once sarcopenia has developed, the ability to shop for and prepare adequate meals may be impaired. This vicious circle may lead to severe malnutrition, weight loss, and progression of sarcopenia into disability and, eventually, death [4].

The maintenance of an adequate nutritional status, through dietary optimization and eventual targeted supplementation, is therefore instrumental for preserving muscle health and functional ability into advanced age [2].

AGE-RELATED CHANGES IN DIETARY INTAKE AND EATING HABITS

Food intake declines gradually throughout adulthood. The third National Health and Nutrition Examination Survey (NHANES III) [5] has shown that that energy intake peaks during late adolescence and young adulthood and declines thereafter. Remarkably, relative to the 20- to 29-year age group, persons 80+ year−old ingest approximately 1000 fewer calories per day, regardless of gender [5].

The origin of declining food intake in old age is multifactorial. Advancing age per se is associated with a "physiologic" reduction in appetite, known as anorexia of aging, that can eventually evolve into pathologic anorexia and malnutrition [6]. Common conditions in old age, such as poor dentition, acute and chronic medical conditions, medications and polypharmacy, functional limitations, depression, cognitive decline, social isolation, and financial constraints, all contribute to decreasing food intake in late life [7]. As a consequence, malnutrition is highly prevalent among older adults, with rates of 5−20% in the community and exceeding 60% in institutionalized elderly [8].

Although it is virtually impossible to discern changes in food preferences attributable to cohort differences from pure age-related modifications, older persons seem to develop predilection for energy-dilute foods, such as grains, vegetables, and fruits, in place of energy-dense sweets and protein-rich nutrients [6]. This phenomenon, coupled with the overall decrease in food intake, is thought to play a major role in the development of sarcopenia [2].

NUTRITIONAL INTERVENTIONS AGAINST SARCOPENIA AND FRAILTY: PRELIMINARY CONSIDERATIONS

The changes in dietary behaviors outlined above suggest that, when designing a nutritional intervention against sarcopenia, several factors must be considered, in order to (1) provide an adequate calorie intake; (2) ensure the provision of appropriate nutrients, taking into account age, sex, physical activity level, eventual comorbidities, and medications; and (3) provide the adequate quality and quantity of nutrients at the right time, that is, when there is a physiological need.

A major, unresolved issue involves the duration of the intervention that maximizes the benefits and reduces the risk of adverse effects. Indeed, while 6 months are typically considered the minimum time frame to expect sizable changes in muscle mass [9], it is unclear whether nutritional supplementations induce linear gains in muscle mass over time or a ceiling effect occurs before substantial improvements in muscle health have been attained. In addition, it is presently unclear to what extent nutritional interventions impact muscle function. For instance, nitrogen balance does not appear to directly relate to functional outcomes in older adults [10]. To further complicate the matter, there is a lack of consensus on clinically meaningful thresholds that distinguish normal from abnormal values of muscle mass and function [11].

It should also be considered that achieving a true nutritional supplementation in older people is often challenging. Indeed, older adults prescribed with nutritional supplements tend to proportionally decrease their dietary intake. As a result, the total daily energy intake remains often unchanged in spite of the supplementation [12]. On the other hand, no drug is currently recommended to stimulate appetite in old age. For instance, metoclopramide has shown to improve symptoms of early satiety; however, its long-term use is associated with several side effects, mainly extrapyramidal symptoms. Other orexigenic drugs (eg, testosterone, megesterol, meclobemide, tetrahydrocannabinol, cyproheptadine, cholecystokinin antagonists, such as loxiglumide) have been associated with numerous adverse effects, including cardiovascular events, delirium, and abdominal symptoms [13]. Finally, while nutritional supplementation may be per se sufficient at improving muscle health in old age [14], compelling evidence indicates that the combination of nutrition and exercise is required to prevent or reverse sarcopenia [15].

Based on the current state of the science, the nutritional factors whose deficiency is linked to sarcopenia and frailty include protein, several amino acids, and vitamin D. Growing evidence indicates that supplementation with the above-listed nutrients as well as with other dietary components may be harnessed as a strategy to restore muscle health in old age.

PROTEIN SUPPLEMENTATION: A MATTER OF QUANTITY AND QUALITY

Adequate protein intake is advocated as the most effective remedy against sarcopenia [16]. Dietary protein is the major source of amino acids needed for muscle protein synthesis. Absorbed amino acids also stimulate muscle anabolism through direct and indirect actions [17]. In particular, leucine (especially abundant in whey) is recognized as the master dietary regulator of muscle protein anabolism, owing to its ability to activate the mammalian target of rapamycin (mTOR) pathway [18] and inhibit protein breakdown by the ubiquitin-proteasome system (UPS) [19].

Data from the Health, Aging, and Body Composition Study, a prospective cohort study conducted in over 3075 community-living older persons, showed that participants in the highest quintile of protein intake lost approximately 40% less appendicular lean mass over 3 years of follow-up than did those in the lowest quintile [20]. The association remained significant after adjustments for several potential confounders, including total energy intake. The relevance of dietary protein to physical function is further supported by findings from the Invecchiare in Chianti (InCHIANTI) study showing that low protein intake is independently associated with physical frailty [21]. Further to this point, a low intake of protein and leucine has recently been associated with reduced muscle mass in older hip fractured patients, regardless of gender and several confounding factors [22]. Similarly, an inverse correlation was reported between protein intake and muscle mass in healthy, sedentary older women [23]. Interestingly, animal protein intake emerged as the only independent predictor of muscle mass. This observation echoes previous findings by Pannemans et al. [24], who reported lower net protein synthesis in older women consuming a high vegetable-protein diet relative to those on a high animal-protein regimen. The different muscle anabolic potency evoked by the two protein sources may be linked to the higher content in branched-chain amino acids, especially leucine, of animal protein relative to their plant-derived counterparts.

In spite of the beneficial effects of dietary protein on muscle trophism, more than one-third of older persons ingest less than the recommended dietary allowance (RDA) for protein (0.8 g/kg/day), and virtually no older adult introduces the highest acceptable macronutrient distribution range (AMDR) for protein (35% of total energy intake) [25]. In addition, the extraction of dietary amino acids by the splanchnic bed is higher in advanced age, which can lead to lower peripheral amino acid availability [26]. The ability to upregulate protein synthesis in response to anabolic stimuli, such as protein intake and physical exercise, is also blunted in the aged muscle [27]. However, such an "anabolic resistance" may be overcome by the ingestion of protein-rich meals [28] and high-quality protein food [29].

An increase in protein intake above 0.8 g/kg/day is deemed as necessary to maintain muscle protein homeostasis in advanced age [16]. Indeed, a daily protein intake of 1.0 g/kg is presently considered the minimum amount required to preserve muscle mass in old age [30]. Higher intakes are advisable for older persons engaged in physical exercise ($\geq$1.2 g/kg/day) or suffering from acute or chronic diseases (1.2–1.5 g/kg/day), unless presenting with severe kidney disease (ie, estimated GFR <30 mL/min/1.73 m^2) [31].

The protein ingestion pattern is another important aspect to be taken into account. For instance, a pulse-feeding pattern, in which 80% of the daily protein intake is provided in one meal, has been shown to improve whole-body protein retention to a greater extent than the same amount of protein distributed evenly across four meals [30]. However, the ingestion of large protein quantities in a single occasion may be difficult to maintain over the long term. Furthermore, the consumption of more than 30 g of protein per meal does not appear to further stimulate muscle protein anabolism in either young or old individuals [29]. Hence, it is agreed that a moderate amount of high-quality protein spread equally throughout the day (ie, 25–30 g per meal) may elicit an optimal 24-hour anabolic response [29,32]. With regard to the timing of protein ingestion, the maximal muscle protein synthesis occurs approximately 60 min after the end of physical exercise. Hence, older subjects engaged in physical activity should be advised to ingest a protein meal 1 hour after the completion of the exercise session [33].

In summary, the provision of adequate amounts of protein and amino acids, especially essential amino acids and leucine, is necessary to support muscle anabolism. The current RDA for protein seems to be insufficient to maintain muscle health in old age. The ingestion of 1.0–1.2 g of protein/kg/day may represent a reasonable and safe nutritional target to pursue in older adults. Indeed, there is no evidence for a detrimental effect of such an intake in elderly with normal renal function. Protein restriction should only be considered in subjects with advanced chronic kidney disease.

VITAMIN D

Vitamin D is a secosteroid prohormone generated in the skin through photolysis of 7-dehydrocholesterol by UV light or ingested with food. The biologically active form is achieved through a two-step hydroxylation. The first reaction occurs largely in the liver to produce 25-hydroxy vitamin D (calcidiol). This molecule circulates bound to a specific transport protein and is indicative of vitamin D status. The second hydroxylation takes place mainly in the kidney to generate 1α,25-dihydroxy vitamin D (calcitriol), the biologically active metabolite of vitamin D. Vitamin D receptors (VDRs) have been found in many tissues and organs. With regard to the skeletal muscle, the activation of nuclear VDR initiates a genomic pathway that modulates the transcription of genes involved in calcium uptake, phosphate transport, phospholipid metabolism, and satellite cell proliferation and terminal differentiation [34]. Binding of vitamin D to cell surface VDR triggers a nongenomic pathway involved in calcium transport and protein synthesis [35].

Low serum levels of vitamin D are commonly observed in advanced age. Indeed, up to 60% of persons older than 60 years are vitamin D deficient (<30 ng/mL) [36]. Prevalence rates are higher in subjects 80+ year old and in institutionalized elderly, as a consequence of poorer dietary intake, decreased sunshine exposure, and more severe multimorbidity [37].

Vitamin D deficiency has been associated with a multitude of negative health outcomes, including osteoporosis, diabetes mellitus, cancer, cardiovascular disease, cognitive decline, multiple sclerosis, and greater susceptibility to infections [38]. In addition, studies have shown that older adults with vitamin D deficiency are at higher risk of postural instability, falls, and sarcopenia [39,40]. On the other hand, supplementation with vitamin D, either alone or combined with physical exercise, improves muscle mass and function [41,42] and decreases the risk of falls [43]. Although a recent meta-analysis has questioned the fall-preventive properties of vitamin D supplementation [44], there is sufficiently robust evidence to obtain routine measurement of serum 25-hydroxy vitamin D in older persons diagnosed with sarcopenia and recommend supplementation (800 IU/day) to those with values lower than 100 nmol/L [45].

CREATINE

Creatine (Cr) is a nonprotein tripeptide composed of glycine, arginine, and methionine. The compound is synthesized endogenously in the liver and kidney and introduced with foods (eg, lean red meat, tuna, and salmon). De novo synthesized and dietary-derived Cr is taken up by muscle cells and neurons and converted into the high-energy metabolite phosphocreatine (PCr) by creatine kinase. During the initial phase of an intense muscular effort, PCr donates a phosphate group to ADP to form ATP. At rest or during periods of low activity, excess ATP is used to regenerate PCr from Cr. Apart from its function as an energy buffer, Cr promotes muscle anabolism by stimulating muscle protein kinetics, satellite cell activity, and anabolic hormone secretion, and reducing reactive oxygen species generation [2].

Advancing age has been associated with reduced muscle content of Cr/PCr in skeletal muscle [46]. Cr depletion is thought to have a greater impact on type II muscle fibers, which rely on anaerobic metabolism to generate fast contractions [47] and are those mostly affected during aging [48]. Cr supplementation has shown to increase intramuscular Cr concentration and improve muscle mass and strength in older subjects engaged in resistance exercise training [49–53]. Notably, the combination of Cr and whey protein may potentiate the muscle anabolic effects of the two nutrients during resistance exercise [52], suggesting that Cr and high-quality protein sources could synergistically stimulate muscle anabolism.

In summary, Cr has emerged as an effective nutritional aid able to improve muscle mass and function in older adults practicing resistance exercise. In these subjects, short-term Cr supplementation (eg, 5–20 g/day for 2 weeks) may be advisable [2]. Further research is however necessary to establish the optimal dosing and timing of Cr supplementation in sarcopenic elderly and determine the risk of possible adverse events associated with this intervention (eg, renal damage).

POLYUNSATURATED FATTY ACIDS

Recent research suggests that supplementation with polyunsaturated fatty acids, especially long-chain omega-3 fatty acids, may be harnessed as a nutritional countermeasure against muscle wasting associated with aging and several disease conditions [53]. The consumption of fatty fish (the richest dietary source of omega-3 fatty acids) has indeed been correlated with grip strength in older community dwellers [54]. With regard to omega-3 fatty acid supplementation, small clinical studies have shown that these compounds elicit a muscle anabolic response, sustained by mTOR activation [55] and blunting of inflammation [56].

These observations suggest that an adequate intake of omega-3 fatty acids could offer therapeutic gain against sarcopenia. Available data indicate that 1 g/day of omega-3 fatty acids may help support muscle anabolism in advanced age. Such an intake is ensured by the consumption of two servings of fatty fish a week (eg, salmon, mackerel, herring, lake trout, sardines, and albacore tuna) or can be obtained via pharmacologic supplementation [2].

AMINO ACID METABOLITES AND PRECURSORS: β-HYDROXY β-METHYLBUTYRATE AND ORNITHINE α-KETOGLUTARATE

β-Hydroxy β-methylbutyrate (HMB) is a metabolite of leucine that possesses anticatabolic, anabolic, and lipolytic properties. At the muscle level, HMB inhibits protein breakdown via UPS downregulation, stimulates protein synthesis through mTOR activation, promotes sarcolemmal integrity, and increases fatty acid oxidative capacity [57]. Preclinical data also suggest that HMB stimulates satellite cell proliferation [58] and inhibits myonuclear apoptosis [59].

HMB supplementation (3 g/day) for 8 weeks has shown to improve muscle strength and body composition in older persons participating in a strength training program [60]. Furthermore, administration of a nutritional mixture containing HMB (2 g), arginine (5 g), and lysine (1.5 g) for 12 weeks improved measurements of physical performance, muscle strength, body composition, and whole-body protein synthesis in sedentary older women [61]. Similar benefits were observed in sedentary elderly men and women supplemented for 1 year with a nutrition cocktail of HMB (2–3 g), arginine (5–7.5 g), and lysine (1.5–2.25 g) [62]. Notably, daily supplementation with HMB (3 g) has been shown to preserve muscle mass in older adults during 10 days of bed rest [63].

Ornithine α-ketoglutarate (OKG), an ionic salt formed by two molecules of ornithine and one of α-ketoglutarate, is the precursor of several amino acids, such as glutamate, glutamine, arginine, and proline, and of other bioactive compounds, including polyamines, citrulline, α-ketoisocaproate, and nitric oxide [64]. These molecules are important modulators of protein metabolism and hemodynamics in the skeletal muscle. OKG also acts as a potent secretagogue of anabolic hormones, including insulin and growth hormone [65]. OKG administration, either orally or intravenously, has shown to stimulate muscle protein anabolism in experimental model of cancer cachexia, endotoxemia, burn injury, and traumata [66]. Small clinical trials also reported improvements in muscle protein anabolism during the postoperative period [67,68]. Interestingly, administration of OKG (10 g/day for 2 months) ameliorated appetite, nutritional status, and body composition in ambulatory elderly patients recovering from acute illnesses [69].

In summary, HMB and OKG have emerged as candidate nutritional aids against sarcopenia. Larger scale clinical trials are necessary to determine their optimal dosage as well as timing and duration of the supplementation.

CONCLUSION

The nutritional status is a major determinant of the person's well-being. In old age, inadequate nutrition is an important, modifiable factor contributing to sarcopenia and physical frailty [2]. As such, nutrition plays a role in the definition of the interventions aimed at restoring robustness and contrasting sarcopenia [70]. A careful assessment of dietary habits, with special attention to nutritional patterns potentially affecting muscle trophism, should therefore be routinely carried out in older adults [6]. Given the capacity of nutrition to provide beneficial effects in multiple systems, acting at biological, clinical, and social levels, dietary optimization nutrition may be considered a true "multidomain" intervention [71]. In the context of sarcopenia, the provision of adequate amounts of high-quality protein, vitamin D, and omega-3 fatty acids may offer substantial benefits both in prevention and treatment interventions. Supplementation with specific nutrients (eg, branched-chain amino acids, Cr, HMB, and OKG) is advisable in older persons engaged in physical exercise programs.

SUMMARY

- Inadequate food intake is commonly observed in older adults and is associated with a vast array of negative health outcomes, including disability, loss of independence, and mortality.
- Malnutrition is causatively linked to the geriatric syndromes of sarcopenia and physical frailty.
- The maintenance of an adequate nutritional status, through dietary optimization and eventual targeted supplementation, is recognized as a fundamental factor for preserving muscle health and functional ability in advanced age.
- The provision of adequate amounts of protein, certain amino acids, and vitamin D may amplify the beneficial effects of physical activity on muscle mass and function in older persons.
- Specific nutrient supplementation regimens may offer therapeutic gain against sarcopenia and physical frailty in the context of multimodal treatment strategies.

References

[1] Diet and Disease—Food and Nutrition Information Center United States Department of Agriculture, National Agricultural Library. Available at http://fnic.nal.usda.gov/diet-and-disease [accessed on 20.03.15].

[2] Calvani R, Miccheli A, Landi F, Bossola M, Cesari M, Leeuwenburgh C, et al. Current nutritional recommendations and novel dietary strategies to manage sarcopenia. J Frailty Aging 2013;2:38–53.

[3] Rosenberg I. Summary comments. Am J Clin Nutr 1989;50:1231–3.

[4] Volkert D. The role of nutrition in the prevention of sarcopenia. Wien Med Wochenschr 2011;161:409–15.

[5] Briefel RR, McDowell MA, Alaimo K, Caughman CR, Bischof AL, Carroll MD, et al. Total energy intake of the US population: the third National Health and Nutrition Examination Survey, 1988–1991. Am J Clin Nutr 1995;62(5 Suppl.):1072S–80S.

[6] Martone AM, Onder G, Vetrano DL, Ortolani E, Tosato M, Marzetti E, et al. Anorexia of aging: a modifiable risk factor for frailty. Nutrients 2013;5:4126–33.

[7] Vandewoude MF, Alish CJ, Sauer AC, Hegazi RA. Malnutrition-sarcopenia syndrome: is this the future of nutrition screening and assessment for older adults? J Aging Res 2012;2012:651570.

[8] Sieber CC. Nutritional screening tools—How does the MNA compare? Proceedings of the session held in Chicago May 2–3, 2006 (15 Years of Mini Nutritional Assessment). J Nutr Health Aging 2006;10:488–92; discussion 492–4.

[9] Cesari M, Fielding RA, Pahor M, Goodpaster B, Hellerstein M, van Kan GA, et al. Biomarkers of sarcopenia in clinical trials-recommendations from the International Working Group on Sarcopenia. J Cachexia Sarcopenia Muscle 2012;3:181–90.

[10] Wolfe RR. Protein summit: consensus areas and future research. Am J Clin Nutr 2008;87:1582S–3S.

[11] Marzetti E. Imaging, functional and biological markers for sarcopenia: the pursuit of the golden ratio. J Frailty Aging 2012;1:97–8.

[12] Milne AC, Avenell A, Potter J. Meta-analysis: protein and energy supplementation in older people. Ann Intern Med 2006;144:37–48.

[13] Morley JE. Orexigenic and anabolic agents. Clin Geriatr Med 2002;18:853–66.

[14] Tieland M, van de Rest O, Dirks ML, van der Zwaluw N, Mensink M, van Loon LJ, et al. Protein supplementation improves physical performance in frail elderly people: a randomized, double-blind, placebo-controlled trial. J Am Med Dir Assoc 2012;13:720–6.

[15] Dickinson JM, Volpi E, Rasmussen BB. Exercise and nutrition to target protein synthesis impairments in aging skeletal muscle. Exerc Sport Sci Rev 2013;41:216–23.

[16] Landi F, Marzetti E, Bernabei R. Perspective: Protein: what kind, how much, when? J Am Med Dir Assoc 2013;14:66–7.

[17] Luiking YC, Deutz NE, Memelink RG, Verlaan S, Wolfe RR. Postprandial muscle protein synthesis is higher after a high whey protein, leucine-enriched supplement than after a dairy-like product in healthy older people: a randomized controlled trial. Nutr J 2014;13:9.

[18] Anthony JC, Yoshizawa F, Anthony TG, Vary TC, Jefferson LS, Kimball SR. Leucine stimulates translation initiation in skeletal muscle of postabsorptive rats via a rapamycin-sensitive pathway. J Nutr 2000;130:2413–19.

[19] Nakashima K, Ishida A, Yamazaki M, Abe H. Leucine suppresses myofibrillar proteolysis by down-regulating ubiquitin-proteasome pathway in chick skeletal muscles. Biochem Biophys Res Commun 2005;336:660–6.

[20] Houston DK, Nicklas BJ, Ding J, Harris TB, Tylavsky FA, Newman AB, et al. Dietary protein intake is associated with lean mass change in older, community-dwelling adults: the Health, Aging, and Body Composition (Health ABC) Study. Am J Clin Nutr 2008;87:150–5.

[21] Bartali B, Frongillo EA, Bandinelli S, Lauretani F, Semba RD, Fried LP, et al. Low nutrient intake is an essential component of frailty in older persons. J Gerontol A Biol Sci Med Sci 2006;61:589–93.

[22] Calvani R, Martone AM, Marzetti E, Onder G, Savera G, Lorenzi M, et al. Pre-hospital dietary intake correlates with muscle mass at the time of fracture in older hip-fractured patients. Front Aging Neurosci 2014;6:269.

[23] Lord C, Chaput JP, Aubertin-Leheudre M, Labonté M, Dionne IJ. Dietary animal protein intake: association with muscle mass index in older women. J Nutr Health Aging 2007;11:383–7.

[24] Pannemans DL, Wagenmakers AJ, Westerterp KR, Schaafsma G, Halliday D. Effect of protein source and quantity on protein metabolism in elderly women. Am J Clin Nutr 1998;68:1228–35.

[25] Kerstetter JE, O'Brien KO, Insogna KL. Low protein intake: the impact on calcium and bone homeostasis in humans. J Nutr 2003;133:855S–61S.

[26] Boirie Y, Gachon P, Beaufrère B. Splanchnic and whole-body leucine kinetics in young and elderly men. Am J Clin Nutr 1997;65:489–95.

[27] Koopman R. Dietary protein and exercise training in ageing. Proc Nutr Soc 2011;70:104–13.

[28] Symons TB, Schutzler SE, Cocke TL, Chinkes DL, Wolfe RR, Paddon-Jones D. Aging does not impair the anabolic response to a protein-rich meal. Am J Clin Nutr 2007;86:451–6.

[29] Symons TB, Sheffield-Moore M, Wolfe RR, Paddon-Jones D. A moderate serving of high-quality protein maximally stimulates skeletal muscle protein synthesis in young and elderly subjects. J Am Diet Assoc 2009;109:1582–6.

[30] Arnal MA, Mosoni L, Boirie Y, Houlier ML, Morin L, Verdier E, et al. Protein pulse feeding improves protein retention in elderly women. Am J Clin Nutr 1999;69:1202–8.

[31] Bauer J, Biolo G, Cederholm T, Cesari M, Cruz-Jentoft AJ, Morley JE, et al. Evidence-based recommendations for optimal dietary protein intake in older people: a position paper from the PROT-AGE Study Group. J Am Med Dir Assoc 2013;14:542–59.

[32] Paddon-Jones D, Rasmussen BB. Dietary protein recommendations and the prevention of sarcopenia. Curr Opin Clin Nutr Metab Care 2009;12:86–90.

[33] Paddon-Jones D, Short KR, Campbell WW, Volpi E, Wolfe RR. Role of dietary protein in the sarcopenia of aging. Am J Clin Nutr 2008;87:1562S–6S.

[34] Ceglia L, Harris SS. Vitamin D and its role in skeletal muscle. Calcif Tissue Int 2013;92:151–62.

[35] Montero-Odasso M, Duque G. Vitamin D in the aging musculoskeletal system: an authentic strength preserving hormone. Mol Aspects Med 2005;26:203–19.

[36] Souberbielle JC, Cormier C, Kindermans C, Gao P, Cantor T, Forette F, et al. Vitamin D status and redefining serum parathyroid hormone reference range in the elderly. J Clin Endocrinol Metab 2001;86:3086–90.

[37] Braddy KK, Imam SN, Palla KR, Lee TA. Vitamin D deficiency/insufficiency practice patterns in a veterans health administration long-term care population: a retrospective analysis. J Am Med Dir Assoc 2009;10:653–7.

[38] Holick MF. Vitamin D deficiency. N Engl J Med 2007;357:266–81.

[39] Dhesi JK, Bearne LM, Moniz C, Hurley MV, Jackson SH, Swift CG, et al. Neuromuscular and psychomotor function in elderly subjects who fall and the relationship with vitamin D status. J Bone Miner Res 2002;17:891–7.

[40] Genaro Pde S, Pinheiro Mde M, Szejnfeld VL, Martini LA. Secondary hyperparathyroidism and its relationship with sarcopenia in elderly women. Arch Gerontol Geriatr 2015;60: 349–53.

[41] Drey M, Zech A, Freiberger E, Bertsch T, Uter W, Sieber CC, et al. Effects of strength training versus power training on physical performance in prefrail community-dwelling older adults. Gerontology 2012;58:197–204.

[42] Ceglia L, Niramitmahapanya S, da Silva Morais M, Rivas DA, Harris SS, Bischoff-Ferrari H, et al. A randomized study on the effect of vitamin D_3 supplementation on skeletal muscle morphology and vitamin D receptor concentration in older women. J Clin Endocrinol Metab 2013;98:E1927–35.

[43] Bischoff HA, Stähelin HB, Dick W, Akos R, Knecht M, Salis C, et al. Effects of vitamin D and calcium supplementation on falls: a randomized controlled trial. J Bone Miner Res 2003; 18:343–51.

[44] Bolland MJ, Grey A, Gamble GD, Reid IR. Vitamin D supplementation and falls: a trial sequential meta-analysis. Lancet Diabetes Endocrinol 2014;2(7):573–80.

[45] Morley JE, Argiles JM, Evans WJ, Bhasin S, Cella D, Deutz NE, et al. Nutritional recommendations for the management of sarcopenia. J Am Med Dir Assoc 2010;11:391–6.

[46] Smith SA, Montain SJ, Matott RP, Zientara GP, Jolesz FA, Fielding RA. Creatine supplementation and age influence muscle metabolism during exercise. J Appl Physiol (1985) 1998; 85:1349–56.

[47] Rondanelli M, Faliva M, Monteferrario F, Peroni G, Repaci E, Allieri F, et al. Novel insights on nutrient management of sarcopenia in elderly. Biomed Res Int 2015;2015:524948.

[48] Marzetti E, Lees HA, Wohlgemuth SE, Leeuwenburgh C. Sarcopenia of aging: underlying cellular mechanisms and protection by calorie restriction. Biofactors 2009;35:28–35.

[49] Gotshalk LA, Volek JS, Staron RS, Denegar CR, Hagerman FC, Kraemer WJ. Creatine supplementation improves muscular performance in older men. Med Sci Sports Exerc 2002;34:537–43.

[50] Brose A, Parise G, Tarnopolsky MA. Creatine supplementation enhances isometric strength and body composition improvements following strength exercise training in older adults. J Gerontol A Biol Sci Med Sci 2003;58:11–19.

[51] Tarnopolsky M, Zimmer A, Paikin J, Safdar A, Aboud A, Pearce E, et al. Creatine monohydrate and conjugated linoleic acid improve strength and body composition following resistance exercise in older adults. PLoS One 2007;2:e991.

[52] Candow DG, Little JP, Chilibeck PD, Abeysekara S, Zello GA, Kazachkov M, et al. Low-dose creatine combined with protein during resistance training in older men. Med Sci Sports Exerc 2008;40:1645–52.

[53] Gualano B, Macedo AR, Alves CR, Roschel H, Benatti FB, Takayama L, et al. Creatine supplementation and resistance training in vulnerable older women: a randomized double-blind placebo-controlled clinical trial. Exp Gerontol 2014; 53:7–15.

[54] Di Girolamo FG, Situlin R, Mazzucco S, Valentini R, Toigo G, Biolo G. Omega-3 fatty acids and protein metabolism: enhancement of anabolic interventions for sarcopenia. Curr Opin Clin Nutr Metab Care 2014;17:145–50.

[55] Robinson SM, Jameson KA, Batelaan SF, Martin HJ, Syddall HE, Dennison EM, et al. Diet and its relationship with grip strength in community-dwelling older men and women: the Hertfordshire cohort study. J Am Geriatr Soc 2008;56:84–90.

[56] Smith GI, Atherton P, Reeds DN, Mohammed BS, Rankin D, Rennie MJ, et al. Dietary omega-3 fatty acid supplementation increases the rate of muscle protein synthesis in older adults: a randomized controlled trial. Am J Clin Nutr 2011;93:402–12.

[57] Cornish SM, Chilibeck PD. Alpha-linolenic acid supplementation and resistance training in older adults. Appl Physiol Nutr Metab 2009;34:49–59.

[58] Wilson GJ, Wilson JM, Manninen AH. Effects of beta-hydroxy-beta-methylbutyrate (HMB) on exercise performance and body composition across varying levels of age, sex, and training experience: A review. Nutr Metab (Lond) 2008;5:1.

[59] Alway SE, Pereira SL, Edens NK, Hao Y, Bennett BT. β-Hydroxy-β-methylbutyrate (HMB) enhances the proliferation of satellite cells in fast muscles of aged rats during recovery from disuse atrophy. Exp Gerontol 2013;48:973–84.

[60] Hao Y, Jackson JR, Wang Y, Edens N, Pereira SL, Alway SE. β-Hydroxy-β-methylbutyrate reduces myonuclear apoptosis during recovery from hind limb suspension-induced muscle fiber atrophy in aged rats. Am J Physiol Regul Integr Comp Physiol 2011;301:R701–15.

[61] Vukovich MD, Stubbs NB, Bohlken RM. Body composition in 70-year-old adults responds to dietary beta-hydroxy-beta-methylbutyrate similarly to that of young adults. J Nutr 2001;131:2049–52.

[62] Flakoll P, Sharp R, Baier S, Levenhagen D, Carr C, Nissen S. Effect of beta-hydroxy-beta-methylbutyrate, arginine, and lysine supplementation on strength, functionality, body composition, and protein metabolism in elderly women. Nutrition 2004;20:445–51.

[63] Baier S, Johannsen D, Abumrad N, Rathmacher JA, Nissen S, Flakoll P. Year-long changes in protein metabolism in elderly men and women supplemented with a nutrition cocktail of beta-hydroxy-beta-methylbutyrate (HMB), L-arginine, and L-lysine. JPEN J Parenter Enteral Nutr 2009;33:71–82.

[64] Deutz NE, Pereira SL, Hays NP, Oliver JS, Edens NK, Evans CM, et al. Effect of β-hydroxy-β-methylbutyrate (HMB) on lean body mass during 10 days of bed rest in older adults. Clin Nutr 2013;32:704–12.

[65] Cynober L. Ornithine alpha-ketoglutarate as a potent precursor of arginine and nitric oxide: a new job for an old friend. J Nutr 2004;134(10 Suppl.):2858S–62S; discussion 2895S.

[66] Vaubourdolle M, Cynober L, Lioret N, Coudray-Lucas C, Aussel C, Saizy R, et al. Influence of enterally administered ornithine alpha-ketoglutarate on hormonal patterns in burn patients. Burns Incl Therm Inj 1987;13:349–56.

[67] Walrand S. Ornithine alpha-ketoglutarate: could it be a new therapeutic option for sarcopenia? J Nutr Health Aging 2010;14:570–7.

[68] Wernerman J, Hammarkvist F, Ali MR, Vinnars E. Glutamine and ornithine-alpha-ketoglutarate but not branched-chain amino acids reduce the loss of muscle glutamine after surgical trauma. Metabolism 1989;38(8 Suppl. 1):63–6.

[69] Wernerman J, Hammarqvist F, von der Decken A, Vinnars E. Ornithine-alpha-ketoglutarate improves skeletal muscle protein synthesis as assessed by ribosome analysis and nitrogen use after surgery. Ann Surg 1987;206:674–8.

[70] Brocker P, Vellas B, Albarede JL, Poynard T. A two-centre, randomized, double-blind trial of ornithine oxoglutarate in 194 elderly, ambulatory, convalescent subjects. Age Ageing 1994; 23:303–6.

[71] Martone AM, Lattanzio F, Abbatecola AM, Carpia DL, Tosato M, Marzetti E, et al. Treating sarcopenia in older and oldest old. Curr Pharm Des 2015;21:1715–22.

[72] Kelaiditi E, Guyonnet S, Cesari M. Is nutrition important to postpone frailty? Curr Opin Clin Nutr Metab Care 2015;18:37–42.

CHAPTER

27

Nutritional Status and Gastrointestinal Health in the Elderly

Luzia Valentini, Amelie Kahl and Ann-Christin Lindenau

Department of Agriculture and Food Sciences, University of Applied Sciences Neubrandenburg,
Neubrandenburg, Germany

KEY FACTS

- Physiological aging of the intestinal tract may cause increased intestinal symptoms but is unlikely to affect nutritional status.
- In disease, however, aging of the gastrointestinal tract predisposes to increased and faster susceptibility to malnutrition, especially when combined with polypharmacy, care-dependence, and altered diet.

Dictionary of Terms

- *CCK*: cholecystokinin
- *CrP*: C-reactive protein
- *GERD*: gastroesophageal reflux disease
- *IL*: interleukin
- *JAM*: junctional adhesion molecule
- *NSAID*: nonsteroidal anti-inflammatory drug
- *PYY*: peptide YY

INTRODUCTION

Malnutrition is common in older adults [1] and a functioning gastrointestinal tract is important to maintain a good nutritional status. Therefore, it is reasonable to question if aging itself leads to compromised function of the gastrointestinal tract and predisposes older humans to nutritional risk independent of disease.

Intestinal dysfunction can lead to symptoms, impaired quality of life, and can contribute to malnutrition in older people. In an Italian trial Pilotto and colleagues [2] investigated UPPER gastrointestinal symptoms in 3100 community-living older people (60–96 years, mean 72) who had neither significant cognitive impairment nor advanced cancer. As many as 43% had upper gastrointestinal symptoms. Overall, 30% reported indigestion, 22% reflux syndrome, and 14% abdominal pain [2]. In a trial in the United States, Ferch and colleagues [3] investigated the prevalence of LOWER gastrointestinal symptoms in 242 community-living adults (60–84 years, mean 76). In total, 38% demonstrated lower gastrointestinal symptoms, with 32% reporting chronic constipation, 42% fecal incontinence, and 1% diarrhea. Overall, 11% fulfilled the Rome III criteria for irritable bowel syndrome (IBS), which parallels the general population [3]. Fecal incontinence together with chronic constipation or IBS resulted in significantly lower quality of life [3].

Thus, the prevalence of gastrointestinal symptoms is high in free-living older people, but it is unclear if these symptoms alone already contribute to compromised nutritional status. This chapter highlights the physiological and pathophysiological changes associated with the aging of the human gastrointestinal tract, including intestinal barrier function and microbiota composition. It will mainly deal with the question to which extent these physiological changes affect the nutritional status in older persons. Special emphasis is laid on the physiological aspects of healthy aging. We report on human research rather than animal studies whenever possible.

ENTERIC NERVOUS SYSTEM

The enteric nervous system plays a central role in the regulation of the gastrointestinal function [4]. The motor activity of the intestine is controlled by neural plexuses between the circular and longitudinal muscle layers (myenteric plexus) and in the submucosa [5]. Aging is associated with loss of neurons in both the myenteric and submucosal plexus, beginning in adulthood and progressing thereafter [6]. Predominantly cholinergic rather than nitrergic neurons are involved in the neuronal loss in human colon specimens over an age range of 33–99 years [7] with parallel loss of enteric glia examined in aging rodents [6]. The losses are greater in the distal than proximal gut [8].

In addition, specialized pacemaker cells, the interstitial cells of Cajal (ICC), decline in numbers in human aging tissue [9]. ICC set the underlying rhythm of depolarization necessary for the motor activity [5]. ICC decrease in both stomach and colon in a linear fashion (approximately 13% every 10 years) between the third and tenth decades [9].

Upper gastrointestinal sensory function was recently investigated in older individuals (>60 years) compared to younger persons (<40 years), by means of a nutrient drink test [10]. The older group reported substantially lower scores for abdominal pain and nausea [10].

Suggested mechanisms for neuronal loss with aging include oxidative stress and mitochondrial dysfunction [6]. Despite being extensive, the changes in the enteric nervous system do not seem to result in major physiological changes in the intestine, indicating enormous redundancy of these neurons and a large adaptive activity [4,5]. Still, possible effects on prolonged gastrointestinal transit time, particularly in the colon, and contribution to constipation cannot be excluded [5].

TASTE AND SMELL (CHEMOSENSATION)

The enjoyment of food depends on olfaction and taste sensation.

The major component involved in food tasting is *olfaction* [4], owing to volatile compounds released from food in the mouth that stimulate odor receptors in the nasal epithelium. All these sensations are integrated in the orbitofrontal cortex. The human genome encodes about 800 olfactory receptors [11], which by type are G-coupled receptors [11]. The complex background of a multitude of odor receptors and receptor interactions must be known to interpret the relative crude testing done in humans to understand the effects of aging on oral sensation [4].

About 50% of the population over 65 years of age and 75% of those aged 80 years and more have impaired olfaction [12]. Over the lifespan, the loss of olfactory receptors is at the rate of about 10% per decade [13]. In persons with Alzheimer's disease [14] the alterations in odor perception are even greater with a reduction in the olfactory tracts and their myelination [15].

In regard to *taste* receptors it is still unclear whether they and their intracellular signal transduction are altered with aging [4]. The recognition of bitter, sweet, and umami tastes are dependent on the activation of G-protein-coupled receptors in the taste buds [16]. Each taste bud has a single receptor that recognizes a single flavor. Interestingly, similar-protein coupled taste receptors are present throughout the mucosa of the gastrointestinal tract [17]. Those intestinal taste receptors transmit signals that regulate nutrient transporter expression, nutrient uptake, and possibly gastrointestinal hormone and neurotransmitter release [17].

In general, the effect of changes in taste sensation on total food intake with aging is felt to be minor [4]. Alterations in saliva due to xerostomia (increased salt content), mouth hygiene and periodontal disease, medications, and cigarette smoking are all thought to be more important for altered taste perception than aging [4].

Nevertheless, taste changes may play a role in food choice [4]. There are preliminary data to suggest that altering meal composition [18] and using taste enhancers [19] can encourage increased food intake in older persons.

In summary, olfaction seems to be more important to enjoy food than do basic taste sensations [4]. It has been estimated that changes in taste and smell associated with aging account for approximately a 100 calories decrease per day in food intake between the ages of 20 and 80 years [4]. It does not appear to be a major factor contributing to malnutrition in healthy older people.

THE ANOREXIA OF AGING

The concept physiological anorexia of aging was introduced by Morley and Silver [20] in 1988. They suggested that the major effects were due to altered release of anorexogenic (satiety) hormones and neurotransmitters in the gut and subsequent changes in the activation of the ascending vagal fibers. In one study, 10 community-dwelling healthy older persons (mean 77 years of age) were compared to 9 younger adults (mean 32 years of age) and showed higher and more persistent elevation of cholecystokinin (CCK) and peptide YY (PYY) concentrations after a meal, coupled with longer gastric emptying time and reduced cholecystic emptying [21]. In the older group postprandial satiety lasted significantly longer [21]. This study confirmed previous results of increased CCK and PYY release in duodenal enteroendocrine cells with age [22]. Older people also seem to have earlier satiation compared to younger adults when eating a meal [23]. This early satiation appears to be due to decreased stomach fundal compliance leading to more rapid antral stretch [24] (see also section "Stomach, Gastric Emptying, Postprandial Hypotension, and Acid Secretion").

These changes combined can help to explain why older humans face more difficulties to upregulate their food intake after a period of food deprivation [25]. Overall, in free-living older people the effects of physiological anorexia of aging seem not to impact the body mass index on the population level [26]. The reasons for anorexia of aging are multifactorial and predominantly due to altered fundal compliance, delayed gastric emptying, and increased activity of CCK [4,21]. As energy requirements decrease with age, they might also be part of physiological adaption. There is consensus that the much stronger contributor to malnutrition in older age is the pathophysiological anorexia of aging, including poverty, loneliness, social isolation, depression, and medical conditions [27].

OROPHARYNGEAL CAPACITIES AND SWALLOWING

With aging, there is a decrease in saliva production and possibly alterations in dentition, which affect the formation of the food bolus [4,6]. In the oropharyngeal phase, there is a reduced driving force of the tongue and reduced pharyngeal swallowing, all of which favors increased retention of food in the valleculae and piriform sinuses [6]. This is probably in part compensated for by relaxation of the upper esophageal sphincter and well-preserved pharyngeal peristalsis leading to minimal changes in the swallowing function in healthy aging [28,29]. Nevertheless, the gag reflex is apparently absent in 40% of the healthy older people [6].

In independent-living facilities, up to 15% of older persons have problems with swallowing and half of these feel that dysphagia substantially reduced their quality of life [30]. The existence of dysphagia and oropharyngeal or gastric aspiration in older patients are important factors in the occurrence of aspiration pneumonia, but are not sufficient to cause aspiration pneumonia in the absence of other risk factors [31]. In healthy older people the glottal closure protecting against aspiration is well-preserved [32]. Salivary flow and swallowing can eliminate Gram-negative bacilli from the oropharynx and normal immunity and pulmonary clearance will avoid the development of aspiration pneumonia in absence of disease [31].

Among frail older people with dysphagia, however, those with unsafe swallowing have an approximately 50% 1-year mortality compared to 13% in those whose swallowing is deemed safe [33]. The pathomechanisms contributing to dysphagia and aspiration in frailty are abnormally slow laryngeal closure, upper sphincter opening and vertical hyoid motion, and weaker tongue propulsion compared to healthy young people [34]. Parkinson's disease, stroke, and cerebrovascular disease are main risk factors for oropharyngeal dysphagia in older patients [5,31]. The severity of dysphagia increases with disease progression [35]. Silent aspiration of oropharyngeal bacteria is often and mostly due to unrecognized swallowing disorders [5,36]. Diminished production of saliva due to medications and oral/dental disease, leading to poor oral hygiene resulting in oropharyngeal colonization with pathogenic organism, impaired immunity or pulmonary clearance, and a decreased cough reflex is a combination of risk factors that will increase the development of aspiration pneumonia in disease.

In conclusion, there are several changes in the oropharyngeal capacities with aging, but in the absence of disease, they seem to have little clinical significance. According to Britton and McLaughlin [37], all patients with dysphagia or aspiration are recommended to be thoroughly investigated to ascertain an underlying cause rather than attributing the change solely to aging.

ESOPHAGUS, ACHALASIA, AND GASTROESOPHAGEAL REFLUX DISEASE

In the esophagus with age, there is reduction in the amplitude of peristaltic contractions, increased nonpropulsive contractions, decreased esophageal all compliance, and a reduction in the lower esophageal sphincter pressure [6]. These changes are of little clinical significance in physiological aging and the absence of major disease.

A pathological condition arising from these changes is *esophageal achalasia*. This is an esophageal motility disorder involving the smooth muscle layer of the esophagus, the lower esophageal sphincter [38], and the loss of esophageal ganglia. *Achalasia* is characterized by swallowing difficulties, regurgitation, and sometimes chest pain. The incidence peaks in the third and seventh decades of life [39]. Older patients tend to have a longer duration, but less severe symptoms than younger patients [39].

In the older population, the prevalence of symptomatic *gastroesophageal reflux disease (GERD)* seems high with 40% having occasional and 20% weekly reflux symptoms [40,41]. Interestingly, this prevalence is not higher than in younger adults. In a recent report on 1107 US adults with a mean age of 46 years, heartburn was shown in 59% and regurgitation in 39% of cases [42]. In this study, higher visceral anxiety and being divorced, but not age, was associated with GERD [42]. Multiple factors contribute to GERD, several of which might be influenced by age, including hiatus hernia, lower esophageal sphincter pressure, acid production and clearance, salivary and bicarbonate secretion, and delayed gastric emptying [41,43]. Interestingly, the severity of reflux esophagitis increases with age, but that of symptoms decreases [41,44]. A recent study of GERD patients confirmed that those of at least 60 years have reduced perception of esophageal acid infusion and respond with a lower swallow frequency [45].

STOMACH, GASTRIC EMPTYING, POSTPRANDIAL HYPOTENSION, AND ACID SECRETION

As mentioned above (see section "The Anorexia of Aging"), the aging process seems to reduce fundus compliance leading to earlier arrival of food at the antrum with earlier antral stretch in smaller antral volume as well as early and prolonged satiety [21].

Delayed gastric emptying might predispose to reflux. Motor function of the stomach is nevertheless relatively well preserved with healthy aging [6]. The slowing of gastric emptying concerns mainly large solid meals [46], whereas small meals fail to demonstrate any changes [47]. Other changes in gastric motility are usually secondary to disease process such as diabetes mellitus, neurological or connective tissue disease, rather than a consequence of aging per se. For example, 70–100% of Parkinson's patients attending neurology clinics have abnormally delayed gastric emptying [5,48]. Those patients also often report nausea, vomiting, and bloating. In addition to L-dopa, delayed gastric emptying also might contribute to these symptoms [48].

Another phenomenon common in old age is the mild, meal-induced decrease in blood pressure. *Postprandial hypotension* can be found at all meal times [49] and can result in falls and cardiovascular events [50]. The blood pressure nadir varies but generally occurs within 30–90 minutes postmeal [49]. Postprandial hypertension is inversely correlated to the rate of gastric emptying [51] meaning that slowing gastric emptying improves its clinical presentation. Size and nutrient content of meals affect the magnitude of the decrease in postprandial blood pressure, with sugars (mono- and disaccharides) being particularly implicated [49,52]. The underlying mechanisms for postprandial hypotension are unclear, but adenosine-induced splanchnic vasodilatation and increase in superior mesenteric artery blood flow may play a role [49].

Hypochlorhydria is particularly common in older people and gastric acid secretion was long thought to decline with age, This was probably confounded by atrophic gastritis caused by *Helicobacter pylori* infection [41], because hypochlorhydria has its highest prevalence in those previously or currently affected by *H. pylori* [41]. Hypochlorhydria can predispose to iron malabsorption, small bowel bacterial overgrowth, and vitamin B12 deficiency if associated with atrophic gastritis [37].

Overall, reduced fundal compliance may contribute to early satiety, postprandial hypotension is ameliorated by low-glycemic food and hypochlorhydria seems to be mostly the result of *H. pylori* infection rather than aging.

SMALL BOWEL, NUTRIENT ABSORPTION, AND SMALL INTESTINAL PERMEABILITY

There is little evidence to show any structural or functional change in the small bowel mucosa attributable to the normal healthy aging process. Lipski and colleagues investigated distal duodenal biopsy specimen from healthy adults and found no correlation between age and duodenal surface epithelium, crypts and lamina propria, height of villi and surface epithelium, depths of crypts, crypt to villus ratio, the number of intraepithelial lymphocytes, duodenal architecture, enterocytes, or brush borders [53]. Trbojevic-Stankovic and colleagues performed morphometric studies in 24 jejunal and 25 ileal biopsies from healthy people younger than 60 years of age and older people of at least 60 years of age [54]. Older people had narrower jejunal villi and wider ileal villi but mucosa epithelium height, crypt numerical density, villous height, crypts and villous perimeter, diameter, and epithelium height were similar [54]. Most studies have shown no effect on small bowel transit [6,55].

Changes in satiety hormones are discussed in the section "The Anorexia of Aging." The release of the "incretin" hormones glucagon-like peptide-1 (GLP-1) and glucose-dependent insulinotropic polypeptide (GIP) in response to small intestinal nutrient seems not to be deficient in healthy aging [56].

Most probably the changes with healthy aging in *nutrient absorption* are small and clinically insignificant

[37,57,58]. Concerning macronutrients, there is no malabsorption of fat or protein in older persons into their nineties [59]. Despite reduced postprandial serum bile acids concentrations, no correlation between age and 72-hour fecal fat excretion was found [60]. Absorption of fat may take longer in older people [61] and delayed absorption of fat may induce postprandial satiety and reduce nutrient intake [62]. Amino acid absorption, for example, tyrosine, arginine, and aspartic acid, decline in aging rodents [63], but systematic human studies are still missing [58]. Furthermore, there is no evidence to support significant changes in the brush border enzymes without the presence of concurrent disease [64,65].

Regarding carbohydrates, a decreased urinary excretion of orally ingested D-xylose was observed in older humans indicating a decreased absorption of sugars in the intestine [66]. However, the difference got lost when normalizing the results for kidney function [66]. The amount of glucose transporters such as sodium-glucose cotransporter 1 (SGLT1), glucose transporter 2 (GLUT2), and NA^+K^+-ATPase is similar in young and old rats [67]. The prevalence of lactose intolerance is similar in younger and older humans [68] and suggests that lactose malabsorption is not more common with aging.

Also in regard to micronutrient absorption there is also little change with aging (Table 27.1). Increased prevalence of vitamin B12 deficiency is attributed mainly to atrophic gastritis, which occurs in approximately 10–30% of older adults and impairs splitting vitamin B12 from its protein binding [69]. Another potential cause for vitamin B12 deficiency is long-term use of drugs that block gastric acid secretion (histamine H2 receptor antagonists or proton pump inhibitors), which are commonly used for treating GERD and peptic ulcer disease [70].

There is reduced fractional absorption rate of calcium with aging [71]. Furthermore, younger persons increase calcium absorption with a low-calcium diet, but older persons fail to do this [72]. In the presence of atrophic gastritis, calcium carbonate is not transformed into soluble calcium chloride [73]. A decline in vitamin D receptors in the intestinal epithelium contribute to decreased calcium and vitamin D absorption with age [74].

Small intestinal permeability: The function of an intact intestinal barrier is to prevent the permeation of antigens, endotoxins, pathogens, and other pro-inflammatory substances into the human body. Intestinal dysintegrity might trigger inflammation locally or systemically, and cause disease [37,75].

Small intestinal permeability does not appear to change with healthy aging up to about 80 years of age [76,77] (see Fig. 27.1). In healthy older people,

TABLE 27.1 Effects of Physiological Aging on Absorption of Nutrients

Nutrient	Effect
Protein	None
Fat	None
Cholesterol	Increase
Carbohydrate	None
Thiamin (vitamin B1)	None
Riboflavin (vitamin B2)	None
Niacin (vitamin B3)	None
Pyridoxalphosphat (vitamin B6)	None
Folate (vitamin B9)	Decrease
Vitamin B12	Decreased with atrophic gastritis
Vitamin A	Increase
Vitamin C	Increase
Vitamin D	Decrease
Vitamin K	None
Zinc	None
Magnesium	None
Iron	None
Calcium	Decrease

Combined from Refs. [37] and [57].

low-grade inflammation alone, defined as C-reactive protein (CrP) concentrations between 1 and 10 mg/L, does not seem to impair small intestinal permeability [76]. However, presence of low-grade inflammation together with even minor comorbidities, such as diabetes type 2, is associated with increased intestinal permeability in older adults [76]. These results suggest that low-grade inflammation makes the intestinal barrier more vulnerable to insults from minor disease challenges. Intestinal permeability is increased in the presence of nonalcoholic fatty liver disease [78], chronic heart failure [79], and malnutrition [80]. Higher use of drugs (eg, NSAID) plays an important role in the development of higher permeability in older people [58].

Two recent studies investigated the intestinal permeability of patients with Parkinson's disease. One study shows changes pointing toward alterations in the enterocyte brush border membrane [81], the second demonstrates an increased permeability in about 25–30% of Parkinson's patients [82]. In these patients, increased intestinal permeability is not associated with increased gastrointestinal symptoms [82].

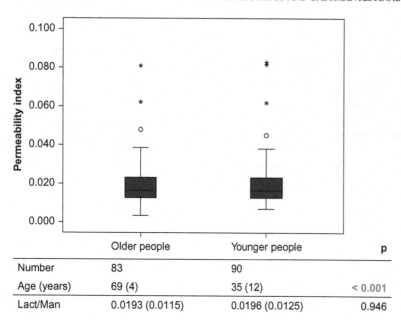

FIGURE 27.1 **Intestinal permeability in older adults.** The figure displays the lactulose/mannitol ratio (=permeability index) of 83 overall healthy older people compared to 90 healthy younger people showing no difference in intestinal permeability with higher age (Valentini L, unpublished data).

The production of cytokines by intestinal epithelial cells in the gastrointestinal tract significantly increases with aging [83]. Some of these cytokines, such as interleukin (IL)-1b, increase the leakiness in intestinal epithelial tight junction, via activation of both canonical and noncanonical NF-kB pathways [84]. The age-associated remodeling of the epithelial barrier includes reduced expression of the tight junction proteins such as zonula occludens (ZO)-1, occludin, and junctional adhesion molecule (JAM)-A [85].

In summary, in the absence of disease, there is currently no evidence within the literature to support any significant alteration in small bowel function associated with aging. Therefore, in patients presenting with clinical features suggestive of small bowel malabsorption, active investigation for underlying pathology is indicated [37]. Polypharmacy and diseases of old age can lead to impaired barrier function in older people.

COLON, CONSTIPATION, AND GUT MICROBIOTA

Colonic transit time appears to increase with age due to a decline in propulsive activity in the colon [86] (see section "Enteric Nervous System"). Saad and colleagues [55] investigated 243 humans with chronic constipation aged 18–79 years (mean 44 years) with wireless motility capsules and found an increase of colonic transit time by 29 minutes for every 1 year increase in age. Increased colonic transit time allows increased water absorption and makes older individuals more prone to constipation.

As many as 50% of community-dwelling older people report *constipation* and over 70% of nursing homes residents [87]. Factors additionally contributing to constipation include poor mobility, comorbidities (eg, stroke, dementia, depression, diabetes, or Parkinson's disease), medications (opiates and anticholinergics), and impaired anorectal sensation. Often more than one contributing cause can be identified [87]. In healthy older persons, decreased fiber and fluid intake and lower physical activity are the most prominent factors [88].

Dysphagia, impaired gastric emptying, and constipation occur frequently in patients with Parkinson's disease. Recent evidence indicates that these gastrointestinal features are evident in the course of the disease, even before formal neurological diagnosis, probably reflecting the involvement of the dorsal motor nucleus of the vagus and the enteric nervous system [89].

Intestinal microbiota: The human microbiota is an individual and adaptable collection of bacteria and its balanced composition is important for the maintenance of health [90–93]. Some main functions of the microbiota are the regulation of the mucosal immune system, releasing nutritional contents from the diet, synthesizing vitamins and cofactors [93,94], and providing energy as short-chain fatty acids (SCFAs) [91,94].

Claesson et al. [95] compared 161 Irish older people (>65 years) with 9 younger (28–46 years) adults. The microbiota of the older group was dominated by 57% Bacteroidetes and 40% Firmicutes, whereas in the younger group, the dominant phylum was Firmicutes with 51% and Bacteroidetes followed with 41% [95].

Claesson concluded older adults show an atypical Firmicutes/Bacteroidetes ratio compared to younger adults [95], which was in line with results of Mariat et al. [92]. However, there were big interindividual differences in the microbiota composition. The variations were extreme with the proportion of Bacteroidetes, ranging from 3% to 92% and that of Firmicutes varying from 7% to 94% in the older individuals [95]. Moreover, the results concerning changes in Firmicutes and Bacteroidetes in older adults vary due to nationality and age of the subjects [96].

Variations in the microbiota composition of Irish older adults were also found in the other phyla like Proteobacteria (11–23%) and Actinobacteria (0–8%), the latter including Bifidobacteria [95].

Mueller et al. [97,98] compared 145 healthy older people (61–100 years) from four European countries with 85 young adults with a mean age of 34 years and showed no age-related microbiota changes in the French and Swedish study arms [97,98]. In the German and Italian groups some changes were present, however, they were inconsistent between the two countries [97,98].

Biagi et al. [99] showed that the structure and diversity of gut microbiota were largely comparable between younger adults (20–40 years) and healthy older people up to the age of about 80 years. Centenarians, however, demonstrated interesting differences [99]. The microbiota of centenarians was still dominated by Bacteroidetes and Firmicutes but in comparison to younger adults, it was less diverse [99]. In centenarians the *Clostridium* cluster XIVa was decreased, especially the SCFA-producers like *Ruminococcus*, *Roseburia*, *Eubacterium rectale*, and *E. hallii* [99]. Low fiber diets are associated with decreased *Clostridium* cluster XIVa [100,101], and might have contributed to those results. The Bifidobacteria species were significantly lower and the gut microbiota were enriched with facultative anaerobes (eg, *Fusobacterium*, *Bacillus*, *Staphylococcus*, *Escherichia coli*) [99].

In the Elderment study [95], the microbiota of 26 of 161 older adults was again analyzed after 3 months. The microbiota seems to be temporal stable [93] and the observed interindividual difference in the microbiota composition were greater than the differences within individuals [95].

Older adults who lived in long-stay residential care differed from the microbiota of free-living older people, even if they lived in the same ethnogeographic region [96,102]. The microbiota of institutionalized older adults contained higher proportions of phylum Bacteroidetes and lower proportions of *Bifidobacterium* and *Clostridium* cluster IV. A 2.5-fold decrease of the *Bifidobacterium* species was described by Bartosch et al. for hospitalized and antibiotic receiving older adults [103,104]. In addition, an increase of facultative anaerobes, especially after antibiotic treatment was observed [91,104–106].

In the Elderment study [107] not only the influence of the living situation but also the influence of the diet on the microbiota was analyzed. The older adults who consumed a diet enriched in animal products and with high glycemic index had a greater abundance of *Bacteroides* and *Alistipes* species as those who consumed a diet rich in plant-based foods. Those harbored an abundance of bacteria from the genus *Prevotella* [102,107].

A change from community-dwelling to long-term residential care apparently correlates with changes in the diet and subsequent reduced microbiota diversity [107]. Higher levels of butyrate was found in fecal water of community-dwelling subjects compared to residents in long-term care [100,102].

The microbiota composition is characterized by the hosts' dietary habits [96,108]. Thus modification of the diet and dietary habits, due to different reasons, is one of the main factors for microbiota changes in older adults [91].

These age-related alterations lead to a less variable diet with a decreased intake of high-fiber and protein-rich food and an increased intake of high sugar and high fat food with a low-nutrient density [107]. Malnutrition is often a consequence of these age-related diets and can consequently lead to changes in the gut microbiome [109]. Furthermore, the frequent use of antibiotics and other medications [95,103] and changes in the lifestyle [103] and living situation of the elderly (community-dwelling or long-term care) cause a microbiota change [109].

In summary, the main change within the colon is an increased transit time that is related to both reduced neurotransmitters and receptors [37]. Although this can cause symptoms of constipation and flatulence, it is unlikely to affect an individual's nutritional status [37]. During adulthood the microbiota composition seems to be relatively stable up to about 80 years of life but diseases of old age can reduce biodiversity and stability of the microbiota [96,99,103,110]. An overview of microbial changes in disease is shown in Fig. 27.2.

CONCLUSION

"Old age" spans over nearly 40 years of life that are characterized by health or advancing morbidity, multimorbidity, and increasing intake of medical drugs, bodily limitations, and possibly care dependence, all to a very different degree. Thus, finding consistent answers is challenging in many circumstances. Many

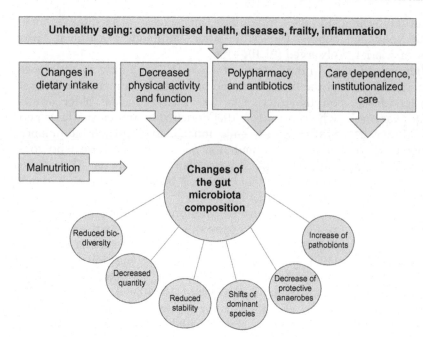

FIGURE 27.2 **The unhealthy aging–microbiota interaction.** Unhealthy aging is characterized by compromised health, diseases, frailty, and inflammation. Unhealthy aging and diseases often result in a dietary intake that is less variable, with a decreased content of food rich in fiber, proteins, and vegetables. The dietary changes can also contribute to malnutrition. Another effect is decreased physical activity due to compromised bodily functions, which predisposes to constipation and reduced bacteria excretion. Moreover, diseases in older adults often lead to polypharmacy and to frequent use of antibiotics. Due to care dependence, it is often necessary to live in institutionalized care. These consequences of unhealthy aging lead to an intestinal microbiota with a reduced biodiversity, a decreased quantity of bacteria, and a reduced stability of the microbiota. Moreover, shifts of the dominant species with a decrease of protective anaerobes and increased pathobionts are observed.

TABLE 27.2 Changes in Physiological Aging

	Reduced	Minor or no change	Increased
Enteric nervous system	• Cholinergic neurons • Interstitial cells of Cajal • Enteric glia	• Nitrergic neurons	
Oropharynx	• Olfaction • Saliva production	• Taste sensation • (Swallowing)	
Esophagus	• Sensation		Achalasia
Stomach	• Sensory functions • (Mucus) • (Bicarbonate secretion)	• Motor function • Gastric emptying • GERD • (hypochlorhydria)	Postprandial hypotension
Small intestine		• Intestinal motor function • Small bowel transit • Incretin hormone secretion (GLP-1, GIP) • Duodenal, ileal, and jejunal architecture, enterocytes or brush borders • Nutrient absorption • Intestinal permeability	Postprandial hypotension
Colon	• Transit time • Rectal compliance • (Ano)rectal sensation	• Colon transit time • Gut microbiota composition	Constipation Fecal incontinence

scientific investigations do not precisely characterize their populations or use heterogenic study populations, which sometimes hamper the interpretation of results. Still, physiological aging of the intestinal tract in absence of disease may cause increased gastrointestinal symptoms in older people, but it is unlikely that relevant changes in nutritional status occur. A summary of changes can be found in Table 27.2. However, in disease with polypharmacy, care-dependence, and altered diet, aging-associated changes in the intestinal tract predisposes for increased and faster susceptibility for malnutrition.

SUMMARY

- In healthy aging changes in the enteric nervous system include continuous loss of neurons and enteric glia with minor physiological consequences.

- Changes in smell and olfaction do not seem to have major effects on caloric intake in physiological aging.
- Poverty, loneliness, social isolation, depression, and medical conditions are much stronger predictors of anorexia than aging itself.
- There are several changes in the oropharyngeal capacities with aging, but they have little significance in the absence of disease.
- In free-living older people the prevalence of gastroesophageal reflux disease is similar to younger people.
- Changes in the stomach can cause earlier satiety even in healthy aging.
- In the absence of disease, there is currently no evidence to support any significant changes in small bowel function, in small intestinal permeability and only minor changes in nutrient absorption.
- In healthy aging the increased colonic transit time can cause symptoms of constipation but not changes in nutritional status.
- The intestinal microbiota is stable up to at least 80 years of age in healthy older people.

References

[1] Agarwal E, Miller M, Yaxley A, Isenring E. Malnutrition in the elderly: a narrative review. Maturitas 2013;76(4):296–302.

[2] Pilotto A, Maggi S, Noale M, Franceschi M, Parisi G, Crepaldi G. Association of upper gastrointestinal symptoms with functional and clinical characteristics in elderly. World J Gastroenterol 2011;17(25):3020–6.

[3] Ferch CC, Chey WD. Irritable bowel syndrome and gluten sensitivity without celiac disease: separating the wheat from the chaff. Gastroenterology 2012;142(3):664–6.

[4] Bhutto A, Morley JE. The clinical significance of gastrointestinal changes with aging. Curr Opin Clin Nutr Metab Care 2008;11(5):651–60.

[5] Rayner CK, Horowitz M. Physiology of the ageing gut. Curr Opin Clin Nutr Metab Care 2013;16(1):33–8.

[6] Bitar K, Greenwood-Van Meerveld B, Saad R, Wiley JW. Aging and gastrointestinal neuromuscular function: insights from within and outside the gut. Neurogastroenterol Motil 2011;23(6):490–501.

[7] Bernard CE, Gibbons SJ, Gomez-Pinilla PJ, et al. Effect of age on the enteric nervous system of the human colon. Neurogastroenterol Motil 2009;21(7): 746–e46.

[8] Wiskur B, Greenwood-Van Meerveld B. The aging colon: the role of enteric neurodegeneration in constipation. Curr Gastroenterol Rep 2010;12(6):507–12.

[9] Gomez-Pinilla PJ, Gibbons SJ, Sarr MG, et al. Changes in interstitial cells of cajal with age in the human stomach and colon. Neurogastroenterol Motil 2011;23(1):36–44.

[10] Gururatsakul M, Holloway RH, Adam B, Liebregts T, Talley NJ, Holtmann GJ. The ageing gut: diminished symptom response to a standardized nutrient stimulus. Neurogastroenterol Motil 2010;22(3): 246–e77.

[11] Kambere MB, Lane RP. Co-regulation of a large and rapidly evolving repertoire of odorant receptor genes. BMC Neurosci 2007;8(Suppl. 3):S2.

[12] Duffy VB. Variation in oral sensation: implications for diet and health. Curr Opin Gastroenterol 2007;23(2):171–7.

[13] Loo AT, Youngentob SL, Kent PF, Schwob JE. The aging olfactory epithelium: neurogenesis, response to damage, and odorant-induced activity. Int J Dev Neurosci 1996;14(7–8):881–900.

[14] Djordjevic J, Jones-Gotman M, De Sousa K, Chertkow H. Olfaction in patients with mild cognitive impairment and Alzheimer's disease. Neurobiol Aging 2008;29(5):693–706.

[15] Christen-Zaech S, Kraftsik R, Pillevuit O, et al. Early olfactory involvement in Alzheimer's disease. Can J Neurol Sci 2003;30(1):20–5.

[16] He W, Yasumatsu K, Varadarajan V, et al. Umami taste responses are mediated by alpha-transducin and alpha-gustducin. J Neurosci 2004;24(35):7674–80.

[17] Depoortere I. Taste receptors of the gut: emerging roles in health and disease. Gut 2014;63(1):179–90.

[18] Hollis JH, Henry CJ. Dietary variety and its effect on food intake of elderly adults. J Hum Nutr Diet 2007;20(4):345–51.

[19] Mathey MF, Siebelink E, de Graaf C, Van Staveren WA. Flavor enhancement of food improves dietary intake and nutritional status of elderly nursing home residents. J Gerontol A Biol Sci Med Sci 2001;56(4):M200–5.

[20] Morley JE, Silver AJ. Anorexia in the elderly. Neurobiol Aging 1988;9(1):9–16.

[21] Di Francesco V, Zamboni M, Dioli A, et al. Delayed postprandial gastric emptying and impaired gallbladder contraction together with elevated cholecystokinin and peptide YY serum levels sustain satiety and inhibit hunger in healthy elderly persons. J Gerontol A Biol Sci Med Sci 2005;60(12):1581–5.

[22] MacIntosh CG, Andrews JM, Jones KL, et al. Effects of age on concentrations of plasma cholecystokinin, glucagon-like peptide 1, and peptide YY and their relation to appetite and pyloric motility. Am J Clin Nutr 1999;69(5):999–1006.

[23] Cook CG, Andrews JM, Jones KL, et al. Effects of small intestinal nutrient infusion on appetite and pyloric motility are modified by age. Am J Physiol 1997;273(2 Pt 2):R755–61.

[24] Sturm K, Parker B, Wishart J, et al. Energy intake and appetite are related to antral area in healthy young and older subjects. Am J Clin Nutr 2004;80(3):656–67.

[25] Roberts SB, Fuss P, Heyman MB, et al. Control of food intake in older men. JAMA 1994;272(20):1601–6.

[26] Destatis (DSB). Körpermaße nach Altersgruppen—Ergebnisse des Mikrozensus 2013 [Body measures according to age group—results of the microcensus 2013], <https://www.destatis.de/DE/ZahlenFakten/GesellschaftStaat/Gesundheit/GesundheitszustandRelevantesVerhalten/Tabellen/KoerpermasseInsgesamt.html>; 2015 [accessed 02.05.15].

[27] Donini LM, Savina C, Cannella C. Eating habits and appetite control in the elderly: the anorexia of aging. Int Psychogeriatr 2003;15(1):73–87.

[28] Shaker R, Ren J, Podvrsan B, et al. Effect of aging and bolus variables on pharyngeal and upper esophageal sphincter motor function. Am J Physiol 1993;264(3 Pt 1):G427–32.

[29] Grande L, Lacima G, Ros E, et al. Deterioration of esophageal motility with age: a manometric study of 79 healthy subjects. Am J Gastroenterol 1999;94(7):1795–801.

[30] Chen PH, Golub JS, Hapner ER, Johns 3rd MM. Prevalence of perceived dysphagia and quality-of-life impairment in a geriatric population. Dysphagia 2009;24(1):1–6.

[31] Kikawada M, Iwamoto T, Takasaki M. Aspiration and infection in the elderly : epidemiology, diagnosis and management. Drugs Aging 2005;22(2):115–30.

[32] Ren J, Shaker R, Zamir Z, Dodds WJ, Hogan WJ, Hoffmann RG. Effect of age and bolus variables on the coordination of the

glottis and upper esophageal sphincter during swallowing. Am J Gastroenterol 1993;88(5):665—9.

[33] Australian and New Zealand Society for Geriatric Medicine. Australian and New Zealand Society for Geriatric Medicine. Position statement—dysphagia and aspiration in older people*. Australas J Ageing 2011;30(2):98—103.

[34] Rofes L, Arreola V, Romea M, et al. Pathophysiology of oropharyngeal dysphagia in the frail elderly. Neurogastroenterol Motil 2010;22(8):851—8 e230

[35] Leow LP, Huckabee ML, Anderson T, Beckert L. The impact of dysphagia on quality of life in ageing and Parkinson's disease as measured by the swallowing quality of life (SWAL-QOL) questionnaire. Dysphagia 2010;25(3):216—20.

[36] Ramsey D, Smithard D, Kalra L. Silent aspiration: what do we know? Dysphagia 2005;20(3):218—25.

[37] Britton E, McLaughlin JT. Ageing and the gut. Proc Nutr Soc 2013;72(1):173—7.

[38] Park W, Vaezi MF. Etiology and pathogenesis of achalasia: the current understanding. Am J Gastroenterol 2005;100(6):1404—14.

[39] Schechter RB, Lemme EM, Novais P, Biccas B. Achalasia in the elderly patient: a comparative study. Arq Gastroenterol 2011;48(1):19—23.

[40] Scholl S, Dellon ES, Shaheen NJ. Treatment of GERD and proton pump inhibitor use in the elderly: practical approaches and frequently asked questions. Am J Gastroenterol 2011;106(3):386—92.

[41] Poh CH, Navarro-Rodriguez T, Fass R. Review: treatment of gastroesophageal reflux disease in the elderly. Am J Med 2010;123(6):496—501.

[42] Cohen E, Bolus R, Khanna D, et al. GERD symptoms in the general population: prevalence and severity versus care-seeking patients. Dig Dis Sci 2014;59(10):2488—96.

[43] Zuchelli T, Myers SE. Gastrointestinal issues in the older female patient. Gastroenterol Clin North Am 2011;40(2):449—66 x

[44] Becher A, Dent J. Systematic review: ageing and gastro-oesophageal reflux disease symptoms, oesophageal function and reflux oesophagitis. Aliment Pharmacol Ther 2011;33(4):442—54.

[45] Chen CL, Yi CH, Liu TT, Orr WC. Altered sensorimotor responses to esophageal acidification in older adults with GERD. Scand J Gastroenterol 2010;45(10):1150—5.

[46] Brogna A, Loreno M, Catalano F, et al. Radioisotopic assessment of gastric emptying of solids in elderly subjects. Aging Clin Exp Res 2006;18(6):493—6.

[47] Madsen JL, Graff J. Effects of ageing on gastrointestinal motor function. Age Ageing 2004;33(2):154—9.

[48] Heetun ZS, Quigley EM. Gastroparesis and Parkinson's disease: a systematic review. Parkinsonism Relat Disord 2012;18(5):433—40.

[49] Carey BJ, Potter JF. Cardiovascular causes of falls. Age Ageing 2001;30(Suppl. 4):19—24.

[50] Luciano GL, Brennan MJ, Rothberg MB. Postprandial hypotension. Am J Med 2010;123(3):281.e1—6.

[51] Jones KL, Tonkin A, Horowitz M, et al. Rate of gastric emptying is a determinant of postprandial hypotension in non-insulin-dependent diabetes mellitus. Clin Sci 1998;94(1):65—70.

[52] Jansen RW, Connelly CM, Kelley-Gagnon MM, Parker JA, Lipsitz LA. Postprandial hypotension in elderly patients with unexplained syncope. Arch Intern Med 1995;155(9):945—52.

[53] Lipski PS, Bennett MK, Kelly PJ, James OF. Ageing and duodenal morphometry. J Clin Pathol 1992;45(5):450—2.

[54] Trbojevic-Stankovic JB, Milicevic NM, Milosevic DP, et al. Morphometric study of healthy jejunal and ileal mucosa in adult and aged subjects. Histol Histopathol 2010;25(2):153—8.

[55] Saad R, Semler JR, Eilding GE, Chey WD. Advancing age is associated with progressive delays in colon transit in patients with chronic constipation. (Abstract presentation). Gastroenterology 2012;142(5,S1):S710—11.

[56] Trahair LG, Horowitz M, Rayner CK, et al. Comparative effects of variations in duodenal glucose load on glycemic, insulinemic, and incretin responses in healthy young and older subjects. J Clin Endocrinol Metab 2012;97(3):844—51.

[57] Morley JE. The aging gut: physiology. Clin Geriatr Med 2007;23(4):757—67, v—vi.

[58] Meier J, Sturm A. The intestinal epithelial barrier: does it become impaired with age? Dig Dis 2009;27(3):240—5.

[59] Russell RM. Factors in aging that effect the bioavailability of nutrients. J Nutr 2001;131(Suppl. 4):1359S—1361SS.

[60] Arora S, Kassarjian Z, Krasinski SD, Croffey B, Kaplan MM, Russell RM. Effect of age on tests of intestinal and hepatic function in healthy humans. Gastroenterology 1989;96(6):1560—5.

[61] Salemans JM, Nagengast FM, Tangerman A, et al. Effect of ageing on postprandial conjugated and unconjugated serum bile acid levels in healthy subjects. Eur J Clin Invest 1993;23(3):192—8.

[62] Wisen O, Hellstrom PM. Gastrointestinal motility in obesity. J Intern Med 1995;237(4):411—18.

[63] Chen TS, Currier GJ, Wabner CL. Intestinal transport during the life span of the mouse. J Gerontol 1990;45(4):B129—33.

[64] Holt PR. Intestinal malabsorption in the elderly. Dig Dis 2007;25(2):144—50.

[65] O'Mahony D, O'Leary P, Quigley EM. Aging and intestinal motility: a review of factors that affect intestinal motility in the aged. Drugs Aging 2002;19(7):515—27.

[66] Lindi C, Marciani P, Faelli A, Esposito G. Intestinal sugar transport during ageing. Biochim Biophys Acta 1985;816(2):411—14.

[67] Drozdowski L, Woudstra T, Wild G, Clandinin MT, Thomson AB. The age-associated decline in the intestinal uptake of glucose is not accompanied by changes in the mRNA or protein abundance of SGLT1. Mech Ageing Dev 2003;124(10—12):1035—45.

[68] Wilt TJ, Shaukat A, Shamliyan T, et al. Lactose intolerance and health. Evid Rep Technol Assess (Full Rep) 2010;192:1—410.

[69] Allen LH. How common is vitamin B-12 deficiency? Am J Clin Nutr 2009;89(2):693S—696SS.

[70] Jensen GL, Mirtallo J, Compher C, et al. Adult starvation and disease-related malnutrition: a proposal for etiology-based diagnosis in the clinical practice setting from the International Consensus Guideline Committee. Clin Nutr 2010;29(2):151—3.

[71] Shapses SA, Sukumar D, Schneider SH, Schlussel Y, Brolin RE, Taich L. Hormonal and dietary influences on true fractional calcium absorption in women: role of obesity. Osteoporos Int 2012;23(11):2607—14.

[72] Ireland P, Fordtran JS. Effect of dietary calcium and age on jejunal calcium absorption in humans studied by intestinal perfusion. J Clin Invest 1973;52(11):2672—81.

[73] Wood RJ, Serfaty-Lacrosniere C. Gastric acidity, atrophic gastritis, and calcium absorption. Nutr Rev 1992;50(2):33—40.

[74] Perry 3rd HM, Horowitz M, Morley JE, et al. Longitudinal changes in serum 25-hydroxyvitamin D in older people. Metabolism 1999;48(8):1028—32.

[75] Soeters PB, Luyer MD, Greve JW, Buurman WA. The significance of bowel permeability. Curr Opin Clin Nutr Metab Care 2007;10(5):632—8.

[76] Valentini L, Ramminger S, Haas V, et al. Small intestinal permeability in older adults. Physiol Rep 2014;2(4):e00281.

[77] Saltzman JR, Kowdley KV, Perrone G, Russell RM. Changes in small-intestine permeability with aging. J Am Geriatr Soc 1995;43(2):160—4.

[78] Miele L, Valenza V, La Torre G, et al. Increased intestinal permeability and tight junction alterations in nonalcoholic fatty liver disease. Hepatology 2009;49(6):1877–87.

[79] Sandek A, Rauchhaus M, Anker SD, von Haehling S. The emerging role of the gut in chronic heart failure. Curr Opin Clin Nutr Metab Care 2008;11(5):632–9.

[80] Norman K, Pirlich M, Schulzke JD, et al. Increased intestinal permeability in malnourished patients with liver cirrhosis. Eur J Clin Nutr 2012;66(10):1116–19.

[81] Davies KN, King D, Billington D, Barrett JA. Intestinal permeability and orocaecal transit time in elderly patients with Parkinson's disease. Postgrad Med J 1996;72(845):164–7.

[82] Salat-Foix D, Tran K, Ranawaya R, Meddings J, Suchowersky O. Increased intestinal permeability and Parkinson disease patients: chicken or egg? Can J Neurol Sci 2012;39(2):185–8.

[83] Man AL, Gicheva N, Nicoletti C. The impact of ageing on the intestinal epithelial barrier and immune system. Cell Immunol 2014;289(1–2):112–18.

[84] Al-Sadi R, Ye D, Said HM, Ma TY. IL-1beta-induced increase in intestinal epithelial tight junction permeability is mediated by MEKK-1 activation of canonical NF-kappaB pathway. Am J Pathol 2010;177(5):2310–22.

[85] Tran L, Greenwood-Van Meerveld B. Age-associated remodeling of the intestinal epithelial barrier. J Gerontol A Biol Sci Med Sci 2013;68(9):1045–56.

[86] Salles N. Basic mechanisms of the aging gastrointestinal tract. Dig Dis 2007;25(2):112–17.

[87] Rao SS, Go JT. Update on the management of constipation in the elderly: new treatment options. Clin Interv Aging 2010;5:163–71.

[88] Morley JE. Constipation and irritable bowel syndrome in the elderly. Clin Geriatr Med 2007;23(4):823–32 vi–vii.

[89] Jost WH. Gastrointestinal dysfunction in Parkinson's disease. J Neurol Sci 2010;289(1–2):69–73.

[90] Garagnani P, Pirazzini C, Giuliani C, et al. The three genetics (nuclear DNA, mitochondrial DNA, and gut microbiome) of longevity in humans considered as metaorganisms. Biomed Res Int 2014;2014:560340.

[91] Woodmansey EJ. Intestinal bacteria and ageing. J Appl Microbiol 2007;102(5):1178–86.

[92] Mariat D, Firmesse O, Levenez F, et al. The Firmicutes/Bacteroidetes ratio of the human microbiota changes with age. BMC Microbiol 2009;9:123.

[93] O'Toole PW. Changes in the intestinal microbiota from adulthood through to old age. Clin Microbiol Infect 2012;18(Suppl. 4):44–6.

[94] Sekirov I, Russell SL, Antunes LC, Finlay BB. Gut microbiota in health and disease. Physiol Rev 2010;90(3):859–904.

[95] Claesson MJ, Cusack S, O'Sullivan O, et al. Composition, variability, and temporal stability of the intestinal microbiota of the elderly. Proc Natl Acad Sci USA 2011;108(Suppl. 1): 4586–91.

[96] Biagi E, Candela M, Turroni S, Garagnani P, Franceschi C, Brigidi P. Ageing and gut microbes: perspectives for health maintenance and longevity. Pharmacol Res 2013;69(1):11–20.

[97] Brussow H. Microbiota and healthy ageing: observational and nutritional intervention studies. Microb Biotechnol 2013;6 (4):326–34.

[98] Mueller S, Saunier K, Hanisch C, et al. Differences in fecal microbiota in different European study populations in relation to age, gender, and country: a cross-sectional study. Appl Environ Microbiol 2006;72(2):1027–33.

[99] Biagi E, Nylund L, Candela M, et al. Through ageing, and beyond: gut microbiota and inflammatory status in seniors and centenarians. PloS One 2010;5(5):e10667.

[100] Candela M, Biagi E, Brigidi P, O'Toole PW, De Vos WM. Maintenance of a healthy trajectory of the intestinal microbiome during aging: a dietary approach. Mech Ageing Dev 2014;136–137:70–5.

[101] Walker AW, Ince J, Duncan SH, et al. Dominant and diet-responsive groups of bacteria within the human colonic microbiota. ISME J 2011;5(2):220–30.

[102] Claesson MJ, Jeffery IB, Conde S, et al. Gut microbiota composition correlates with diet and health in the elderly. Nature 2012;488(7410):178–84.

[103] Lakshminarayanan B, Stanton C, O'Toole PW, Ross RP. Compositional dynamics of the human intestinal microbiota with aging: implications for health. J Nutr Health Aging 2014;18(9):773–86.

[104] Bartosch S, Fite A, Macfarlane GT, McMurdo ME. Characterization of bacterial communities in feces from healthy elderly volunteers and hospitalized elderly patients by using real-time PCR and effects of antibiotic treatment on the fecal microbiota. Appl Environ Microbiol 2004;70(6):3575–81.

[105] Hopkins MJ, Macfarlane GT. Changes in predominant bacterial populations in human faeces with age and with Clostridium difficile infection. J Med Microbiol 2002;51 (5):448–54.

[106] Hopkins MJ, Sharp R, Macfarlane GT. Age and disease related changes in intestinal bacterial populations assessed by cell culture, 16S rRNA abundance, and community cellular fatty acid profiles. Gut 2001;48(2):198–205.

[107] O'Connor EM, O'Herlihy EA, O'Toole PW. Gut microbiota in older subjects: variation, health consequences and dietary intervention prospects. Proc Nutr Soc 2014;73(4):441–51.

[108] Thomas F, Hehemann JH, Rebuffet E, Czjzek M, Michel G. Environmental and gut bacteroidetes: the food connection. Front Microbiol 2011;2:93.

[109] Voreades N, Kozil A, Weir TL. Diet and the development of the human intestinal microbiome. Front Microbiol 2014; 5:494.

[110] Biagi E, Candela M, Fairweather-Tait S, Franceschi C, Brigidi P. Aging of the human metaorganism: the microbial counterpart. Age (Dordrecht, Netherlands) 2012;34(1):247–67.

CHAPTER

28

Can Nutritional Intervention Counteract Immunosenescence in the Elderly?

Sarah J. Clements[1] and Simon R. Carding[1,2]

[1]Gut Health & Food Safety Research Programme, Institute of Food Research, Norwich, UK
[2]Norwich Medical School, University of East Anglia, Norwich, UK

KEY FACTS
- With increasing age the immune system declines which is known as immunosenescence.
- Immunosenescence appears, at least in part, to be effected by dietary intake, giving potential for small but significant dietary alterations to improve aging immune function; such as response to vaccinations.
- More work is required but some findings show that increasing polyphenol and fiber intake or changing which types of fat we consume can influence certain immune cell types, such as dendritic cells and T cells.
- Many parts of the immune system have shown changes as a result of nutritional intervention, showing potential for targeted approaches.
- Traditional diets, such as the Mediterranean diet, might be of crucial importance in developing specific recommendations for an elderly population.

Dictionary of Terms

- *Immunosenescence*: the progressive decline in immune function associated with age.
- *Antigen presenting cells (APCs)*: specific innate immune cells responsible for linking the innate and adaptive immune system; DCs are professional APCs.
- *Thymic atrophy*: damage to the thymus that occurs with age resulting in reduced production of "new" naïve T cells.
- *VDJ-recombination*: the random organization of V_H, D_H, and J_H gene segments that occurs during B cell development to create a vast array of antigen receptors.
- *B cell repertoire*: the variety of immunoglobulins specific to different antigens, which reduced with age.
- *Immunonutrition*: changes to the functionality of the immune system using nutritional intervention.

INTRODUCTION

The immune system comprises the innate immune system, which is the first line of defense against pathogens and is fast acting, nonspecific and short lived [1], and the adaptive immune system which is slower to respond, is antigen specific, and longer lasting due the generation of long-lived memory T and B cells [2]. During aging there is a progressive decline in immune function which in the elderly adversely affects response to vaccines and resistance to infection [3,4] and has been termed immunosenescence [5].

This chapter will summarize the characteristic changes to the immune system with age that impacts the innate and adaptive immune responses, highlighting the key areas for any intervention. The evidence that dietary intervention can beneficially impact on the aged immune system will be summarized, identifying how small but significant nutritional changes might result in improved quality of life of the elderly.

AGING AND THE INNATE IMMUNE SYSTEM

There are reports of age-associated alterations in the innate immune system with alterations in numbers being restricted to specific cell populations, while the functionality of most cell types is affected. In particular, antigen presenting cells (APCs) and dendritic cells (DCs) are affected by aging, which impacts on adaptive immune responses.

Dendritic Cells

DCs circulate in the blood and can be divided into two subsets; myeloid (mDC) and plasmacytoid (pDC) DCs. DCs are professional APCs and crucially link the innate and adaptive immune response. Numerous observations of reductions in numbers of pDCs with age have been made [6–9]. Additionally, both subsets are substantially reduced in frail elderly as compared to healthy elderly and young subjects [9]. These findings are not however uniform with some [10] describing no changes in either DC population with age. Among mDCs no changes in their distribution have been reported [11]. There is therefore controversy surrounding the impact of aging on DCs, which may be in part due to differences in experimental methodologies, the age range of subjects representing the elderly, and whether or not the strict SENIEUR protocol [12,13] was used for recruiting study participants.

DC functionality is affected by aging as evidenced by [14] where older DCs are shown to be causal in defective $CD4^+$ and $CD8^+$ T-cell responses [15] while the "age" of the T cells themselves appeared less important, highlighting a reduction in the functional ability of DCs to present antigen to T cells. Functional defects of aged DCs also include reduced IFN-γ production by pDCs in mice and elderly human subjects compared to young controls [7,8,16–18]. Alterations in DC cytokine secretion may account for the age-related impairment of T-cell responses [19]. The age-related impairment in IFN-γ expression has been associated with reduction in phosphorylation of interferon regulatory factor-7 (IFR-7), which is required to initiate transcription of genes encoding IFN-I and -III [17]. Surface expression of CD40, CD80, CD86, and MHC II [20] and CD54, CD86, CD80, and HLD-DR [10,21] was similar among young and old DCs (mDCs and pDCs), suggesting that DC maturation is not impaired with age. Indeed, expression of maturation markers in response to influenza virus vaccine was comparable among DCs from young and elderly subjects with no differences seen in IL-12 and TNF-α production [22]. Other groups, however, have observed reduced expression of HLA-DR on DCs from elderly subjects compared to young individuals [23], and mature monocyte-derived DCs (Mo-DCs) from elderly subjects had reduced expression of CD25 with expression of ICAM-1 also being lower among LPS-stimulated Mo-DCs [24].

In considering the central importance of DCs in linking innate and adaptive responses, the age-associated defects in their function is likely to be a major contributing factor in declining T-cell responses with age and immunosenescence in general.

Monocytes/Macrophages

Monocytes and macrophages contribute to host defense and innate immunity by contributing to antigen presentation, phagocytosis, and production of numerous factors and chemicals toxic to various bacterial pathogens. Age-associated increases in intermediate and nonclassical monocyte populations have been observed with an alteration in the proportion of monocytes subsets with age, as well as reduced expression of CD38, CD62L, and CD115, and increased expression of CD11b [25]. The expression of Toll-like receptor (TLR)-4 [25] was also reduced by age, in addition to other TLRs and their associated downstream signaling pathways leading to induction of cytokine production [26]. Expression of the costimulatory molecules CD80 and CD86 by activated monocytes after stimulation of TLRs was defective in elderly subjects which impacted on post-influenza virus vaccine cell expansion [27]. No effect of age on monocyte antigen presenting ability has been described, although altered expression of MHC class II molecules has been observed in elderly versus young subjects [28,29]. While bone marrow–derived macrophages from aged mice have reduced expression of MHC class II molecules, in addition to other age-associated alterations this is not seen in humans [30–32]. Additionally, reduced TNF-α, macrophage-inhibitory protein (MIP)-1α, MIP-1β, and MIP-2, RANTES and eotaxin secretion by macrophages was observed in the skin of aged mice [33], while MCP-1 was increased [34] in parallel with an increased numbers of macrophages at the wound site but with reduced phagocytic ability.

Neutrophils

Neutrophils contribute to innate immunity via their ability to produce reactive oxygen species (ROS) and by their phagocytic ability. Neutrophils are the most numerous phagocytic cells within the blood and are the first cells found at the site of infection or tissue damage [1,35]. Their numbers appear to remain constant with advancing age [1]. The impact of aging on neutrophil function is however less clear but of considerable significance for host defense from infection. Impairments in phagocytosis, in terms of the number of neutrophils available for phagocytosis and the amount of bacteria ingested [36], have been observed in the elderly. The causes of this are unclear with no obvious change in opsonizing complement and immunoglobulin (Ig) that enhance and promote phagocytosis being described. Neutrophil extracellular trap (NET) formation has been shown to be defective with age in elderly human subjects when neutrophils were primed with TNF-α and stimulated with IL-8 or LPS, compared to cells from young subjects [37]. Of note, the respiratory burst and spontaneous generation of ROS is significantly elevated among neutrophils from old compared to young subjects [36,38]. However, others have found no difference in ROS generation [38,39]. The short lifespan of polymorphonuclear neutrophils (PMNs) can be extended in vitro using the growth factor GM-CSF in cells from young subjects, but this is not replicated in cells from elderly subjects, which may be related to age-associated alterations in the Jak/STAT intracellular signaling pathway [40,41].

Natural Killer Cells

Natural killer (NK) cells play a central role in immune surveillance and killing of virus-infected cells and tumor cells [42,43]. However, the effectiveness of this cytotoxicity declines with age and is therefore directly implicated in immunosenescence; the incidence and mortality of viral infections increases in the elderly [44,45]. NK cells are divided into two main subsets: $CD56^{DIM}$ and $CD56^{BRIGHT}$, along with a third subset $CD56^-CD16^+$, all of which have differing functions. With advanced age the total number of NK cells increases [42,46,47], although this may not occur within all NK cell populations. Whereas the $CD56^-CD16^+$ cell population [48] and the $CD56^{DIM}$ population, considered to be mature cytotoxic effector NK cells, appear to increase with age [12], the number of $CD56^{BRIGHT}$ cells, which are the more immature and developing NK cells, may decrease [42,46]. Interestingly, these age-associated changes in NK cells have been investigated in the context of obesity-associated inflammation, demonstrating that $CD56^{BRIGHT}CD16^+$ cells are significantly negatively associated with inflammation, while $CD56^{DIM}CD16^+$ and $CD56^-CD16^+$ NK cells were significantly positively associated with body mass index (BMI) [46]. Other phenotypic changes observed among NK cells include increased expression of HLA-DR and CD95, while CD69 expression decreased among NK cells from elderly compared to young subjects [12,49]. Conversely, other studies have seen no effect of age on NK cell function [42,49], however, while cytotoxic potential was preserved in the oldest old (80–100 years), some subjects aged 60–80 years did not demonstrate the same response; preservation of cytotoxic potential of NK cells may be a factor associated with longevity. Similarly, middle-aged subjects displayed a significantly lower NK cell activity compared to young controls and centenarians [50], suggesting that the immune system of centenarians is affected differently by age. Functional decline and reduced response of NK cells has been observed in other studies [12,48] where a reduction in proliferation and CD69 induction was observed; this study implemented the SENIEUR protocol for recruitment of elderly participants [12], and selected only the healthiest elderly subjects, ruling out the effect of other diseases on this response.

NKT cells have been less extensively studied although their numbers may be augmented with increasing age [51]. Other studies, however, have identified a reduction in the number of NKT cells. It is possible that these contrasting findings may be due to differing effects on different subsets of NKT cells, the invariant NKT (iNKT) and the NKT-like lymphocytes [52–54]. iNKT cells are thought to decrease with age, while the NKT-like lymphocyte may increase [52].

AGING AND THE ADAPTIVE IMMUNE SYSTEM

Immunosenescence is mainly associated with detrimental changes to the adaptive immune system, which is partially attributed to thymic atrophy and involution and the loss of functional thymic tissue [55], resulting in a reduction in T-cell production. However, the total number of peripheral (blood) T cells appears to remain constant throughout life [42]. Additionally, production of B cells by the bone marrow has been suggested to be impaired with age [56], resulting in an imbalance in the lymphocyte pool consisting of progressively more memory cells and fewer naïve cells. With age, numbers of activated B cells ($CD69^+$), $CD8^+$ T cells, and naïve ($CD45RA^+CCR7^+$) T cells decrease significantly while the CD4:CD8 T cell ratio and the number of effector memory T cells ($CD45RA^-CCR7^-$) increase [57], with males being more susceptible to immunosenescence than females.

Cell-Mediated Adaptive Immune Responses

The production of "new" T cells with a diverse repertoire of T cell receptors (TCRs) from the thymus is an inefficient and energy intensive process but is required in order to protect the host from infection during reproductive age. Once humans approach 60 years old this process deteriorates [58] with accumulating numbers of memory and effector $CD8^+$ T cells, while the numbers of naïve $CD8^+$ T cells decreases [59–61]. By comparison, the distribution of naïve and effector $CD4^+$ T cells appears to be less affected by age [59,61,62]. The profile of cytokine production by T cells changes with aging with increasing production of pro-inflammatory cytokines and other factors which has led to the use of the term inflammaging [63]. This phenomenon has been investigated in the context of in vitro studies using non antigen-specific stimuli (eg, PMA/ionomycin) which has shown that naïve, memory, and cytotoxic $CD8^+$ T cells from elderly subjects produced higher levels of IL-2, IFN-γ, and TNF-α than equivalent cells from young subjects [60]. Naïve $CD4^+$ T cells from elderly subjects produced less IFN-γ, with memory $CD4^+$ T cells producing more IL-4 compared to cells from young subjects [62]. IFN-γ and IL-4 are important in influencing Ig production and in particular which Ig isotypes are produced by mature B cells (Ig switch factors) with IL-4 inducing class switching for IgG1 and IgE producing B cells [64] and type I interferons inducing increased antibody titers for all IgG subclasses [65,66]. Reversing these age-associated alterations in the T cell–derived class switching cytokines may prove to be important in alleviating problems associated with immunosenescence and declining antibody (IgG) production. $CD8^+$ T cells produce significantly less IFN-γ and display a reduced ability to mount responses against influenza when antigen is presented by older DCs that have impaired TNF-α production [15].

Age-related decline in thymic output and the generation of new naïve populations of T cells alters the makeup and balance of peripheral naïve and memory T cells populations. The half-life of memory T cells is much shorter than that of naïve T cells [67] which contribute to a decline in TCR repertoire with age. As noted above, with increasing age the effector functions of $CD4^+$ and $CD8^+$ T cells decline and compounds the ability of the immune system to effectively combat infections and tumor growth. The extent to which $CD4^+$ and $CD8^+$ T cells are affected by the aging process may not be uniform, with differences noted in the oligoclonal expansion within each T cell compartment [61] and in accessory molecule (KIR, CD28, CD57) expression with differences in the extent of CD28 expression [61]. The age-related changes in $CD8^+$ T cells have been attributed to lifelong, chronic, virus infection and in particular, cytomegalovirus infection (CMV) which causes progressive and chronic antigenic stimulation and oligoclonal expansion of activated $CD8^+$ T cells [68–71]. These observations provide an explanation for the observed clonal expansions of the $CD8^+$ T cell pool seen with increasing age, though others have refuted this explanation [61].

Humoral Adaptive Immune Responses

Aging of the B cell compartment has not been extensively investigated and findings to date have demonstrated varying age-associated alterations. Although similar concentrations of antibody are produced in adults of all ages in response to vaccinations, the functionality of antibodies and ability to opsonize bacteria is reduced in elderly subjects (>77 years) [72,73]. Additionally, the elderly demonstrate progressive reductions in their response to vaccinations in a year on year fashion [74,75]. It is not surprising therefore that the elderly have greater susceptibility to infections [3,76], and in particular to viruses. In terms of the composition of the B cell pool it appears the overall number of B cells remains the same but fewer IgM-expressing cells were observed within lymph nodes of elderly subjects [77], suggesting that the ratio of B cell subsets within the B cell pool may be subject to change with age. Numbers of memory B cells increased in elderly subjects [78], although the size of the study population used meant it was underpowered. The absolute total number of B cells in old subjects (60–80 years) has been observed to be significantly lower than that of younger adults (18–60 years) [42]. Other groups have observed increases in IgG and IgA, while IgM and IgD decreased with age [79] suggesting an increase in memory B cells in the periphery with developing or naïve B cells expressing membrane IgM or IgD, which after antigen driven selection and differentiation leads to memory B cells expressing IgG, IgA, or IgE; naïve IgD^+CD27^- B cells decreased and IgD^-CD27^-, exhausted memory B cells, increased with age [80,81].

The underlying defects contributing to these functional defects with aging are not understood but they may relate to recombination events that occur during B cell development within the variable region of the Ig heavy chain. This involves the random reassortment of V_H, D_H, and J_H gene segments [82] to create a vast repertoire of antigen receptors enabling a wide array of antigens to be recognized. The effect of aging on this process is evidenced by *Streptococcus pneumoniae* infection in mice, which identified differences in antibody structures and the protective ability of antibodies between young and old mice [83,84]. More recently similar findings have been observed in elderly humans after vaccination with the pneumococcal vaccine in which the V_H3-07 and V_H3-74 region encoding genes were more frequently expressed in young than elderly with the opposite seen for V_H3-21 and V_H3-30 genes, accompanied by a loss of oligoclonality with aging in response to vaccines [85]. This demonstrates a significant alteration in the Ig-V_H gene repertoire with age. Analysis of the Ig response to influenza virus and pneumococcal vaccines using high-throughput sequencing revealed a potential delay in clonal expansion with age as well as significant age-related changes in the size of the CDR3 region, which comprise the junctions between the V, D, and J gene segments [86], of IgA [87] and IgM clonotypes but not IgG [88]. Recently, increased B cell clonality scores, indicative of a persistence of one or more large clonal B cell populations, in the oldest subjects (72–89 years) have been observed [89] suggesting a diminishing B cell repertoire with age. Along with evidence of reduced B cell diversity between CDR3 regions of the Ig heavy chain from elderly subjects [90], which appeared to correlate with health status and could indicate a reduction in bone marrow output of naïve B cell and thus an accumulation of antigen experienced cells as a result. There have, however, also been observations of a preserved diversity of VDJ-recombination in aged subjects [89,91], while CDR3 was observed to be shorter in length when compared to that of young subjects [91].

Studies examining the ability of B cells to act as APCs in older subjects have revealed no differences between young and old subjects although there was diversity between subjects, with the B cells from some older subjects showing inferior antigen presentation compared to B cells from younger subjects [29]. This may be associated with the age-associated alterations in V, D, J gene rearrangements and diminishing B cell repertoire since the IgM and MHC class II molecules were still able to take up and present antigen to T cells.

These findings may be significant should it prove possible to counteract the observed reductions in B cell diversity, as it may then be possible to restore immune competency in the elderly.

TARGETS OF NUTRITIONAL INTERVENTION

It is clear that the immune system experiences a number of changes as a result of aging, and that some components are affected to greater extents than others. There is growing interest in the interaction of nutritional intake on the immune system (immunonutrition) with mounting research into the effects different nutrients can have on immune function and their mechanism of action. These effects are not all beneficial, as will be detailed below, and have been observed for polyphenols, fatty acids, particularly polyunsaturated fatty acids (PUFAs), and fiber.

Polyphenols are compounds within many of the foods and beverages commonly consumed, which confer health benefits [92,93]. Typical sources are red wine, fruits, vegetables, cereals, tea, and cocoa. Recent research has highlighted the need to investigate metabolites of polyphenols since they undergo extensive changes in the body, particularly in the large intestine [93]. A key product of anthocyanin metabolism is protocatechuic acid (PCA), which, along with the isothiocyanate sulforaphane, can influence DCs by impairing production of pro-inflammatory cytokines including IL-6, IL-8, IL-12, and IL-23 in response to LPS stimulation [94,95], while neither DC maturation nor T-cell activation were affected by sulforaphane, suggesting that the NF-κB signaling pathway is somehow affected.

Since there have been reports of PUFA intake influencing antigen presentation [96–98] including altering the MHC II complex on B cells [99] and also lipid raft size within plasma membranes [100–103], it is likely that B cells and DCs are key targets for nutritional intervention. Activation of B cells by n-3 PUFAs in vitro resulted in reduced IL-6 secretion with docosahexanoic acid (DHA) not inhibiting MHC II or CD69 expression nor promoting apoptosis, unlike palmitic acid [104]. While feeding with a high fat diet enriched with n-3 PUFA increased CD69 surface expression and secretion of IL-6 and IFN-γ by B cells, this was also observed when using physiologically relevant doses of n-3 PUFA but the ability of B cells to stimulate $CD4^+$ T cells was suppressed [100]. When considering an elderly population the detrimental effects of high intakes of n-3 PUFAs in this population must be taken into account since high intakes of PUFA by the elderly may increase susceptibility to infection [105].

Vitamin E has been shown to reduce the production of 2-series prostaglandins (PGE_2) in mice [106,107]. This is an important observation since PGE_2 is known to increase with age and is associated with increased inflammation and dysregulated immune function [106]. The production of PGE_2 has been shown to be regulated by the short-chain fatty acid, butyrate [108].

Additionally, the consumption of dietary fiber, which leads to the production of short-chain fatty acids as a result of bacterial fermentation in the colon, has been shown to have effects on effector and regulatory T-cells [109]. Both acetate and propionate increased naïve T cell differentiation into Th17 cells, as well as increasing expression of IL-10. Short-chain fatty acids have been implicated in affecting the balance of anti-inflammatory and pro-inflammatory cells and their products [110] though additional effects of fiber alone have shown that some of the anti-inflammatory effects were a result of direct contact between fiber and intestinal epithelial cells (IECs) and DCs [111] with the effect being dependent on the type of fiber. However, as this work relied on an in vitro approach it may not be directly applicable to the in vivo situation since not all relevant cells, including the mucous layer, are represented.

Hence, the use of nutritional intervention in terms of immunomodulation is important and should be further investigated in aging subjects. Evidence of immunomodulatory effects in aging subjects is limited to focusing on single nutrients or individual food groups.

CAN NUTRITION IMPACT ON IMMUNE COMPONENTS AFFECTED BY AGE?

Since cells of the immune system contain large quantities of PUFAs, within their membranes, antioxidants have a crucial role in protecting them from oxidative damage due to constant exposure to ROS that are by-products of normal processes such as energy generation [87,112,113]. This is important as oxidative damage appears to accumulate with age, which brings with it an increased incidence of infection [113]. A number of studies have investigated the effects of single vitamins (E, C, A, D) or minerals (zinc) in the elderly and have demonstrated modulation of cytokine production, reductions in bactericidal activity and directed migration of neutrophils, increasing antibody titers for vaccine antigens, increasing numbers of naïve T cells, and reducing levels of PGE_2 [87,107,114–117]. While some studies have shown no changes in the immune parameters tested, this does include prospective cohort studies which rely on self-report data [112,118,119]; see Table 28.1 for details. One study demonstrated the importance of dietary carotenoids through a depletion–repletion study, with one of the main outcomes showing that carotenoid supplementation decreased $CD4-CD45RO^+$ memory T cell numbers in all age groups, which is of importance in the elderly since naïve T-cell production is reduced with age [87]. However, vitamins and minerals are not usually consumed in isolation but within foods, which means that studying them alone is of less value. Many of these micronutrients are found in large quantities within fruits and vegetables, the consumption of which by elderly subjects has been shown to improve the antibody response to the Pneumovax II vaccine in those consuming five or more portions per day of fruits and vegetables, with each additional portion suggested to increase total IgG response by 18% [120]. Polyphenols have similar effects with the consumption of polyphenol-rich biscuits (naturally present in cereals) increasing proliferative response upon LPS stimulation, as well as increased NK cell activity and chemotaxis indexes in aged mice [121]. Cocoa consumption by elderly subjects has shown significant reductions in the monocyte surface markers VLA-4, CD40, and CD36, along with reduced circulatory levels of ICAM-1 and P-selectin [122]. Unfortunately, relatively few similar studies have been carried out in elderly humans.

The gastrointestinal tract is one of the physical barriers protecting the host from the external environment, with the intestinal epithelial barrier playing a major role, not just as a physical barrier but via sophisticated structures involving M cells and Peyer's patches to sample luminal antigens and to orchestrate appropriate immune responses to them [131]. Numerous studies have investigated the use of probiotic bacteria to target parameters of aging with one suggested mechanism of protective effects being adherence of probiotic bacteria to the epithelial cells to form a protective barrier [132–134]. However, such protective effects do not appear to be limited to this mechanism since increased levels of Escherichia coli specific IgG1 and IgA antibodies and pro-inflammatory cytokines IFN-γ, MCP-1, and TNF-α were observed after pathogen challenge in an aged mouse model compared to young control animals [135]. The authors claimed evidence of an impact of probiotics on redressing the Th2 to Th1 imbalance seen in immunosenescence. It is important to note however that it is difficult to extrapolate findings from animal models to humans. A study using human low density cells, which are typically 98–100% HLA-DR$^+$ and strongly induce T-cell proliferations with very few allogenic T cells, as a source of human blood–derived DCS (LDCs) investigated the effect of four different strains of probiotics on young and aged LDCs. Old LDCs failed to respond to probiotics in terms of integrin-β7 expression and induction of T-cell activation with no effect of probiotics on DC-T cell priming seen with either young or old DCs, though the type of DCs used may impact on the observed results [136]; for further studies and more details refer to Table 28.2. It is difficult to summarize the beneficial effects of probiotics as a whole because different strains were used in various studies and showed varying effects indicating that probiotic effects are very likely strain and possibly subject specific.

Dietary fatty acids include saturated fatty acids (SFA), monounsaturated fatty acids (MUFAs), and PUFAs with the effect of each on immunosenescence varying. It is crucial that the most efficacious amounts of each type are incorporated in the diet to have

TABLE 28.1 Studies Investigating the Immunomodulatory Effects of Single Nutrients

Nutrient/food	Dosage	Study design and duration	N =	Main outcomes	Reference
MICRONUTRIENTS					
Lycopene and β-carotene	13.3 mg lycopene 8.2 mg β-carotene	RCT, PC, DB 12 weeks	52 free-living elderly subjects	• No change in cell counts, lymphocyte proliferation, or cytokine production	[112]
β-carotene	15, 30, 45, or 60 mg/day	RCT, PC, single-blinded 7 months	20 "elderly" subjects (mean age 56 years)	• Significantly increased $CD4^+$ T cells, NK cells, and markers of expression with supplementation ≥30 mg • Increased ratio helper: suppressor T cells at highest dosage	[116]
Carotenoids	30 mg β-carotene, 15 mg lycopene, 9 mg lutein	Depletion (3 weeks)-repletion (5 weeks) of carotenoids	98 males, 26 of which aged 60–75 years	• Significantly decreased bactericidal activity and directed migration of neutrophils with diet • Significantly higher number of positive reactions for DTH skin test response in the oldest subjects (70–75 years), though carotenoids had no effect	[87]
Cocoa polyphenols	40 g	RCT, CO, 8 week duration	42 subjects (69.7 ± 11.5) years	• Significant reduction in expression of VLA-4, CD40, and CD36 expression on monocytes after cocoa consumption • Significantly lower circulatory levels of P-selectin and ICAM-1 after cocoa consumption	[122]
Vitamin E	60, 200, and 800 mg/day	RCT, DB, PC 4 months	88 elderly subjects (≥65 years)	• Increased DTH and a sixfold increase in antibody titer for hepatitis B for 200 mg/day, compared to placebo and other dosages	[117]
Vitamin E	Control diet 30 ppm, supplemented diet 500 ppm DL-α-tocopherol acetate	Animal feeding study	140 specific pathogen-free C57BL/6NCrlBR mice (70 young, 70 old)	• Significant reduction in lung viral titers postinfection in vitamin E supplemented old mice • Increased IL-2 and IFN-γ and decreased TNF-α production postinfection with vitamin E • Reduced prostaglandin E_2 production by vitamin E supplemented old mice	[107]
Vitamin E	200 IU DL-α-tocopherol/day	RCT, PC, DB 1 year	451 elderly subjects (68–100 years)	• Significant interaction with IFN-γ, TNF-α, IL-1β, and IL-6 compared to baseline	[115]
Vitamin E	Control, 30 ppm, supplemented diet 500 ppm D-α-tocopheryl acetate	Animal feeding study/ex vivo analysis of human neutrophils	Male C57BL/6 mice. 12 mice/group	• Increased resistance to pneumococcal infection in old mice receiving the supplemented diet • Reductions in TNF-α and IL-6 productions by old supplemented mice • Reduced human PMN transepithelial migration postinfection with vitamin E	[123]
Zinc	Average intakes 3.3–52.2 mg/day	Observational study, one visit	201 elderly subjects (≥80 years)	• Significant association between plasma zinc levels and CRP and leukocyte count	[124]

(Continued)

TABLE 28.1 (Continued)

Nutrient/food	Dosage	Study design and duration	N =	Main outcomes	Reference
Vitamin D	Vitamin D serostatus 31 ± 11 ng/mL	Prospective cohort study	1103 community-dwelling adults (64 ± 10 years)	• No consistent effect of vitamin D on seroprotection or seroconversion for any vaccine strains tested	[119]
Vitamins E, A, and zinc	Serostatus: Vitamin A 72.9 ± 5.4, vitamin E 1750.7 ± 809.24, Zinc 80.3 ± 11.6 (all in µg/dL)	Observational prospective cohort study	205 Elderly subjects (65–92 years)	• No association of vitamin E, A, or zinc with prevaccination or postvaccination seroprotection or seroconversion for any vaccine antigens	[118]
MACRONUTRIENTS					
n-3 PUFA: EPA	Low (3 g) Moderate (6 g) High (9 g EPA)	RCT, DB, PC. 12 weeks	93 young (18–42 years) and 62 elderly (53–70 years) subjects	• Significant effect on percentage of neutrophils engaging in respiratory burst in response to E. coli with EPA, inversely affected by dose • Significant reduction in PGE_2 across all EPA doses upon LPS stimulation	[125]
n-3 PUFA: EPA	2 g ALA/day or, 700 mg DHA or, 700 mg AA or, 700 mg GLA or, 720 mg EPA, and 280 mg DHA in FO	RCT, PC, DB, parallel 12 weeks	46 healthy subjects (55–75 years)	• FO significantly decreased NK cell activity at 12 weeks compared to 0 and 4 weeks	[126]
n-3 PUFA: EPA	2 g ALNA/day, or ~700 mg GLA, or ~700 mg AA, or ~700 mg DHA, or 720 mg EPA, and 280 mg DHA in FO	RCT, PC, DB, parallel 12 weeks	46 subjects (55–75 years)	• GLA and FO significantly decreased lymphocyte stimulation index in response to all but the highest dose of Con A	[127]
n-3 and n-6 fatty acids	0.22–5.29 g/day ALNA 1.46–53.16 g/day linoleic acid	Prospective cohort study, 10-year follow-up	38,378 males (44–79 years)	• Increased intake of ALNA or linoleic acid were associated with reduced risk of pneumonia, with 31% reduction with each 1 g/day ALNA increase in intake	[128]
Fish oil	30 mg EPA and 150 mg DHA/day	RCT, DB, PC 6 weeks	20 elderly subjects (70–83 years)	• Significantly reduced lymphocyte proliferative responses in FO group	[129]
Fish oil	Low dose (0.4 g/day) High dose (1.8 g/day)	RCT, PC, single-blinded 26 weeks	302 elderly subjects (≥65 years)	• 900 uniquely changed genes in the high dose FO group compared to control • Significantly reduced expression of genes involved in inflammatory pathways with high dose	[130]

RCT, randomized controlled trial; DB, double-blind; PC, placebo-controlled; CO, cross-over; ALA, α-linoleic acid; ALNA, α-linolenic acid; GLA, γ-linolenic acid; DHA, docasahexanoic acid; EPA, eicosapentanoic acid; AA, arachidonic acid; FO, fish oil.

beneficial immunomodulatory effects. The PUFA, eicosapentanoic acid (EPA), has been tested in healthy young and older men and showed reduced neutrophil respiratory burst in older men in a dose-dependent fashion as well as elevated incorporation of EPA into plasma and mononuclear cell (MNC) phospholipids of MNCs of older men [125]. High intakes of n-3 PUFAs, especially in women, can impair immune responses by increasing lipid peroxidation and suppressing production of IL-2 and T–cell proliferation [142] with consumption of very low doses of fish oils by elderly subjects demonstrating diminishing effects on lymphocyte proliferation in response to all mitogens tested [129]. It is, however, imperative to consider the doses tested in terms of actual fish consumption. Since a dose of 0.4 g EPA + DHA corresponds to 2 portions of fish per week, the aforementioned studies used dosages greater than this, with one using doses equivalent to 10 portions per week, which are much greater than the current recommendations of 2 portions of fish per week [143], and is considerably greater than the 85 g of fish per week that UK elderly currently consume [144].

There have, however, been observations from prospective investigations showing that men who

TABLE 28.2 Studies Investigating the Immunomodulatory Effects of Functional Foods

Nutrient/food	Dosage	Study design and duration	N =	Main outcomes	Reference
PROBIOTICS					
Lactobacillus rhamnosus (MTCC 5897)	1×10^9 CFU/mL in fermented milk	In vivo, aging mice dietary supplementation and pathogen challenge	6 animals/group 8 animals/group for pathogen challenge	• Interleukin levels altered to levels near that of young counterparts with the probiotic • Increased production of IFN-γ, MCP-1, and TNF-α upon stimulation and reduced IL-4 and IL-10 • Highly significant decrease in pathogen colonization in all areas tested in probiotic treated group • Increase in E. coli specific IgG1 and IgE antibody production in probiotic treated group	[135]
Lactobacillus delbrueckii subsp. Bulgaricus 8481	At least 3×10^7 bacteria per capsule	Human randomized intervention DB, PC, MC; 6 months	47 elderly subjects (>65 years)	• Modulated percentages of CD8$^+$ T cells and NK cells • Increase in naïve T cells and decrease in memory T cell subsets associated with increased generation of immature T cells evaluated via TREC content in peripheral T cells • Decreased IL-8 secretion • Effects lost once probiotics stopped	[137]
Lactobacillus casei Shirota	1.3×10^{10} CFUs/day	RCT, PC, single-blinded, CO study; 12 week duration	30 healthy subjects (55–74 years)	• NK cell activity was significantly increased with probiotic consumption compared to baseline but not placebo • The ratio of IL-10:IL-12 was significantly increased with probiotic compared to placebo	[133]
L. rhamnosus HN001 or B. lactis HN019	5×10^{10} CFUs/day (L. rhamnosus) 5×10^9 CFUs/day (B. lactis)	Base diet run in controlled, 3-stage dietary intervention, blinded; 9 week duration	27 healthy elderly subjects (60–84 years)	• Significant increases in CD56$^+$ NK cell numbers with consumption of probiotics • Significant increases in tumoricidal activity of PBMCs upon probiotic consumption; greater responses observed in subjects over 70 years	[138]
B. lactis HN019	5×10^{10}/day or 5×10^9/day	3 stage dietary supplementation trial (run in, supplementation, washout), 3 week duration	30 healthy elderly subjects (63–84 years)	• Increased total, helper and activated T cells, as well as increased NK cells after probiotic consumption • Typical dose showing significant increase in phagocytosis and tumoricidal activity by polymorphonuclear and mononuclear cells, compared to pretreatment	[134]
B. infantus 52486, B. longum SP 07/3, L. rhamnosus GG, and L. casei Shirota	1×10^7 CFUs/mL	Ex vivo assays using human blood	Pooled samples from young (20–30 years) and old (65–75 years)	• DCs were more responsive to probiotics with age (increased CD80 expression) • TGF-β, TNF-α, IFN-γ influenced by probiotic exposure and age,	[136]

(Continued)

TABLE 28.2 (Continued)

Nutrient/food	Dosage	Study design and duration	N =	Main outcomes	Reference
				with most efficient production of TGF-β in the old DCs • Reduced ability of young or old DCs exposed to probiotics to prime T cells as assessed by levels of T-cell proliferation	
PREBIOTICS					
β-galactooligosaccharides (B-GOS)	5.5 g/day	RCT, DB, CO human intervention with elderly subjects	41 elderly subjects (64–79 years)	• Increased IL-10 production and decreased IL-6, IL-1β, and TNF-α production	[139]
Oligosaccharides	1.3 g/250 mL	Prospective RCT, DB, 12 week duration	74 community-dwelling elderly and/or nursing home residents (84 ± 7 years)	• Reduced TNF-α and IL-6 mRNA in the oligosaccharide group compared to the controls	[140]
70% raftilose and 30% raftiline mixture	6 g/day	PC, RCT, DB human intervention; 28 week duration	43 elderly subjects (≥70 years)	• No differences in antibody titers against influenza (A or B) and *S. pneumoniae* between groups • No difference in stimulation induced lymphocyte proliferation between groups • No differences in IL-4 or IFN-γ secretion between groups	[141]
SYNBIOTICS					
Lactitol and *L. acidophilus* NCFM	2×10^9 CFUs/g (5–5.5 g/sachet)	RCT, PC, DB, parallel, 2 week run in period followed by 2 week intervention	47 healthy elderly subjects (≥65 years)	• Significant differences in PGE_2 concentrations with symbiotic treatment	[132]

RCT, randomized controlled trial; DB, double-blind; PC, placebo-controlled; CO, cross-over; MC; multicentered; CFU, colony forming units.

consumed more α-linolenic acid, linoleic acid, and n-3 and n-6 PUFAs from fish had a reduced risk of pneumonia [128], though this data is based on food frequency questionnaires which includes an element of reporter bias. A comparative study in women found opposite results [105], suggesting the effects of these dietary lipids may be gender specific. Additionally, an investigation of high daily doses (1.8 g) and low daily doses (0.4 g) DHA + EPA in elderly subjects showed reductions in plasma free fatty acids (FFAs) with the high dose showing 900 unique changes in gene expression with gene profiling, including reduced levels of expression of a number of pro-inflammatory gene products, including NF-κB, and Ig-like receptors [130]. The overall picture is that any beneficial effects are specific and restricted to particular immune functions. Fish oil supplementation (EPA + DHA) significantly reduced lymphocyte proliferation in a time- and dose-dependent manner which is similar to the results using γ-linolenic acid (GLA) although secretion of IL-2 and IFN-γ was unaffected by GLA [127]. It is important to note that α-linolenic acid (from flax seed oil) had no impact on any immune parameters tested. Further investigation into various fatty acids, α-linolenic acid (ALNA), γ-linolenic, arachidonic acid (AA), DHA, and fish oil (EPA + DHA), showed that GLA, AA, and fish oil caused changes to plasma phospholipids but only fish oil caused a reduction in NK cell activity [126], suggesting that the effect seen by fish oil may be due to the EPA content, not DHA.

In order to achieve consistent amounts of the fatty acids under investigation the participants in these studies consumed capsules, which does not take into account the food matrix in which the fatty acids are normally consumed. Soybean oils with varying fatty acid compositions have been developed to investigate the effects of changing the ratio of n-6 to n-3 PUFAs (LA:ALA). Oils high in oleic acid and low in SFAs increased lymphocyte proliferation in response to phytohemagglutin (PHA), while oils with low α-linolenic acid reduced lymphocyte proliferation [145]. This highlights that, although the most studied fatty acids are

TABLE 28.3 The Nutritional Effects on Aged Immune Cells

Cell type	Age-associated effects	Evidence of potential dietary impact on aged cells
DCs	• Reduction in pDC numbers • Reduction in IFN-γ production by pDCs	• Increased costimulatory expression and production of TNF-α, TGF-β, and IFN-γ with probiotics
Monocytes/macrophages	• Increase in intermediate and nonclassical monocytes • Defective costimulatory receptor expression • Reduced phagocytic ability	
Neutrophils	• Defective NET formation • Impaired phagocytosis • Potential impact on ROS generation	• Increased numbers of cells involved in respiratory burst with EPA
NK cells	• Increase in total numbers • Decrease in CD56^{BRIGHT} cells • Increase in CD56$^-$CD16$^+$ and CD56^{DIM} cells • Increased expression of HLA-DR and CD95	• Modulated percentages and activity with probiotics
T cells	• Reduced production of naïve T cells • Increase in memory T cells • Increased production of pro-inflammatory cytokines	• Increased numbers of naïve T cells with β-carotene • Reduced CD4-CD45RO$^+$ memory T cells with carotenoids • Increased lymphocyte proliferation with high oleic acid containing oils (MUFAs)
B cells	• Increased production of memory B cells • Defects in B cell Ig repertoire diversity	• Increased antibody titers for vaccinations with increased fruit + vegetable intake • Reduced risk of pneumonia with n-3 PUFA (controversial; opposing results found) • Increased IgG1 and IgE antibody production specific for E. coli with probiotics

n-3 and n-6 PUFAs, MUFAs may be more beneficial for improving immune function during aging. Therefore, while studies using capsules is useful they do not reflect the dietary response, and consequently whole foods and the whole diet situation needs to be studied to give more relevant data; a summary of the immunomodulatory effects of nutrition is given in Table 28.3. It is therefore, more relevant to refer to studies that have focused on the entire diet consumed by elderly subjects. A key example is the Mediterranean diet (Med Diet) which has been considered to have beneficial health effects, in particular, related to cardiovascular health and longevity [146–149], as well as cognitive health [150,151]. This dietary pattern incorporates many of the beneficial dietary components already discussed within one diet, which includes for example, increased amounts of fruits and vegetables, oily fish and wholegrain cereals, reduced consumption of red and processed meats and moderate consumption of red wine with meals. Since these polyphenols, n-3 PUFAs and wholegrains have been shown to improve health alone, the impact of them combined is of interest.

Investigation of the polyphenols present in virgin olive oil (VOO) within a Med Diet demonstrated that compared to consumption of "washed" olive oil (WOO) the diet containing VOO significantly decreased plasma IFN-γ and s-P-selectin, as well as downregulating the expression of a number of genes in PBMCs involved in the inflammatory process, oxidative stress, and DNA damage [152]. Recently, the effects of the Med Diet enriched with MUFA, a SFA enriched diet or a low fat and high carbohydrate diet enriched with n-3 PUFA in elderly subjects demonstrated that the Med Diet was associated with lower levels of NF-κB expression compared to the SFA rich diet (in the fasting state) and lower expression of the p65 subunit within 2 h of consumption. Additionally, consumption of the SFA rich diet caused increased gene and plasma expression of MCP-1, while a carbohydrate (CHO) rich diet induced increased expression of TNF-α mRNA compared to that for the Med Diet [153]. Similarly, consumption of Med Diets supplemented with VOO, or nuts caused significantly reduced monocyte expression of CD49d and CD40 when comparing the VOO Med Diet to the SFA diet, as well as plasma levels of VCAM-1, ICAM-1, IL-6, and CRP [154]. These studies indicate an antiinflammatory effect of the Med Diet, and in particular olive oil. Additional evidence of different fats impacting on expression of pro-inflammatory cytokines in the postprandial state in young, healthy men emphasizes this point and that these effects may be translatable to an elderly population [153–155]. Adherence to the Med Diet was shown to be negatively associated with plasma levels of IL-8, while assessment of the individual components showed that olive oil was negatively associated with plasma levels of IL-6, IL-8, MCP-1, and

TNF-α, with all of these inflammatory markers being positively associated with red meat consumption, and MCP-1 and TNF-α were positively associated with nonrefined cereal consumption [156].

Additional dietary patterns include the Okinawan diet and the New Nordic diet (NND), which have been associated with health benefits and longevity [157–163]. The Okinawan diet combines dietary patterns from China, Southeast Asia, and Japan with the majority of energy provision from sweet potatoes, which are high in fiber, vitamins, and minerals as well as polyphenols. The Okinawan diet also consists of large amounts of seaweed, which is high in EPA and DHA, leafy vegetables, and soy as the principle source of protein [159,161]. This diet is also associated with increased longevity, like the Med Diet; however, the effects are thought to be due to caloric restriction [161]. However, to date immune parameters have not been investigated so further research is required to define the potential effects of this dietary pattern in relation to immunosenescence. The NND was developed to produce a diet which not only addresses health but also is palatable, ecologically sustainable, and that includes regional foods. The dietary pattern of the NND is very similar to that of the Med Diet with increased intake of fruits and vegetables but with emphasis on berries, cabbage, root vegetables, and legumes, and increased intake of wholegrains and fish is also encouraged [164,165]. There have been few studies investigating the impact of this dietary pattern on health outcomes in the elderly as of yet, however, the latest reports show that in adults aged on average 55 years with metabolic syndrome (MetS) the NND reduced levels of IL-1Ra, which is a marker of inflammation in obesity and MetS, and this increased with length of adherence; all other inflammatory markers remained unaffected [160].

CONCLUSIONS

So far reductions in expression of pro-inflammatory markers have been shown when investigating the effects of the Med Diet on elderly subjects [152–154]; with other dietary patterns showing some potential. However, as the adaptive immune system is substantially altered with age and T-cell proliferation has been shown to be effected by fat intake [127,142] there is potential for fatty acids to impact on B cells. Thus, this raises the potential for restoration of immune competencies in elderly people and counteracting the reduction in diversity observed with increased age. With further research it may be possible to improve the defects in aged immune function providing future dietary guidelines, specific to the elderly.

SUMMARY

- While the entire immune system is affected by the aging process, the adaptive immune system and the professional antigen presenting cells, DCs, are particularly affected.
- With age T and B lymphocyte pools accumulate increasing numbers of memory cells, while naïve cells decrease.
- The B cell Ig repertoire has reduced diversity with increasing age.
- Nutritional intervention of immune parameters has been effective with the use of polyphenols, fatty acids, and fiber.
- Key targets of nutritional intervention include T cells and DCs, with potential for effects on B cells.
- Single nutrient studies have provided useful information; however, it is more appropriate to investigate effects of immunonutrition using the whole diet.
- Diets such as the Med Diet have demonstrated some promising findings relating to ameliorating some deleterious effects of immunosenescence such as reducing production of inflammatory markers.
- The Okinawan Diet and the New Nordic Diet may also have potential in improving immune function, but further research is required.
- There may be potential to restore immune competencies in elderly people if the reduction in diversity can be counteracted.
- This research could result in future dietary guidelines specific to the elderly.

References

[1] Solana R, Tarazona R, Gayoso I, Lesur O, Dupuis G, Fulop T. Innate immunosenescence: effect of aging on cells and receptors of the innate immune system in humans. Semin Immunol 2012;24(5):331–41.

[2] Bonilla FA, Oettgen HC. Adaptive immunity. J Allergy Clin Immunol 2010;125:S33–40.

[3] Wordsworth D, Dunn-Walters D. The ageing immune system and its clinical implications. Rev Clin Gerontol 2011;21:110–24.

[4] Castle SC. Impact of age-related immune dysfunction on risk of infections. Z Gerontol Geriatr 2000;33:341–9.

[5] Ostan R, Bucci L, Capri M, Salvioli S, Scurti M, Pini E, et al. Immunosenescence and immunogenetics of human longevity. Neuroimmunomodulation 2008;15:224–40.

[6] Pérez-Cabezas B, Naranjo-Gómez M, Fernández MA, Grífols JR, Pujol-Borrell R, Borràs FE. Reduced numbers of plasmacytoid dendritic cells in aged blood donors. Exp Gerontol 2007;42:1033–8.

[7] Shodell M, Siegal FP. Circulating, interferon-producing plasmacytoid dendritic cells decline during human ageing. Scand J Immunol 2002;56:518–21.

[8] Canaday D, Amponsah N, Jones L, Tisch D, Hornick T, Ramachandra L. Influenza-induced production of interferon-alpha is defective in geriatric individuals. J Clin Immunol 2010;30:373–83.

[9] Jing Y, Shaheen E, Drake RR, Chen N, Gravenstein S, Deng Y. Aging is associated with a numerical and functional decline in plasmacytoid dendritic cells, whereas myeloid dendritic cells are relatively unaltered in human peripheral blood. Hum Immunol 2009;70:777–84.

[10] Agrawal A, Agrawal S, Cao J-N, Su H, Osann K, Gupta S. Altered innate immune functioning of dendritic cells in elderly humans: a role of phosphoinositide 3-kinase-signaling pathway. J Immunol 2007;178:6912–22.

[11] Della Bella S, Bierti L, Presicce P, Arienti R, Valenti M, Saresella M, et al. Peripheral blood dendritic cells and monocytes are differently regulated in the elderly. Clin Immunol 2007;122:220–8.

[12] Borrego F, Alonso M, Galiani M, Carracedo J, Ramirez R, Ostos B, et al. NK phenotypic markers and IL2 response in NK cells from elderly people. Exp Gerontol 1999;34:253–65.

[13] Ligthart GJ, Corberand JX, Fournier C, Galanaud P, Hijmans W, Kennes B, et al. Admission criteria for immunogerontological studies in man: the SENIEUR protocol. Mech Ageing Dev 1984;28:47–55.

[14] Pereira LF, Duarte de Souza AP, Borges TJ, Bonorino C. Impaired in vivo CD4+ T cell expansion and differentiation in aged mice is not solely due to T cell defects: decreased stimulation by aged dendritic cells. Mech Ageing Dev 2011;132:187–94.

[15] Liu WM, Nahar TE, Jacobi RH, Gijzen K, van Beek J, Hak E, et al. Impaired production of TNF-alpha by dendritic cells of older adults leads to a lower CD8+ T cell response against influenza. Vaccine 2012;30:1659–66.

[16] Prakash S, Agrawal S, Cao J-n, Gupta S, Agrawal A. Impaired secretion of interferons by dendritic cells from aged subjects to influenza. Age 2013;35(5):1785–97.

[17] Sridharan A, Esposo M, Kaushal K, Tay J, Osann K, Agrawal S, et al. Age-associated impaired plasmacytoid dendritic cell functions lead to decreased CD4 and CD8 T cell immunity. Age 2011;33:363–76.

[18] Stout-Delgado HW, Yang X, Walker WE, Tesar BM, Goldstein DR. Aging impairs IFN regulatory factor 7 up-regulation in plasmacytoid dendritic cells during TLR9 activation. J Immunol 2008;181:6747–56.

[19] Agrawal A, Agrawal S, Gupta S. Dendritic cells in human aging. Exp Gerontol 2007;42:421–6.

[20] Wong CP, Magnusson KR, Ho E. Aging is associated with altered dendritic cells subset distribution and impaired proinflammatory cytokine production. Exp Gerontol 2010;45:163–9.

[21] Lung TL, Saurwein-Teissl M, Parson W, Schönitzer D, Grubeck-Loebenstein B. Unimpaired dendritic cells can be derived from monocytes in old age and can mobilize residual function in senescent T cells. Vaccine 2000;18:1606–12.

[22] Saurwein-Teissl M, Schonitzer D, Grubeck-Lobenstein B. Dendritic cell responsiveness to stimulation with influenza vaccine is unimpaired in old age. Exp Gerontol 1998;33:625–31.

[23] Pietschmann P, Hahn P, Kudlacek S, Thomas R, Peterlik M. Surface markers and transendothelial migration of dendritic cells from elderly subjects. Exp Gerontol 2000;35:213–24.

[24] Ciaramella A, Spalletta G, Bizzoni F, Salani F, Caltagirone C, Bossù P. Effect of age on surface molecules and cytokine expression in human dendritic cells. Cell Immunol 2011;269:82–9.

[25] Hearps AC, Martin GE, Angelovich TA, Cheng WJ, Maisa A, Landay AL, et al. Aging is associated with chronic innate immune activation and dysregulation of monocyte phenotype and function. Aging Cell 2012;11:867–75.

[26] Alvarez-Rodriguez L, Lopez-Hoyos M, Garcia-Unzueta M, Amado JA, Cacho PM, Martinez-Taboada VM. Age and low levels of circulating vitamin D are associated with impaired innate immune function. J Leukoc Biol 2012;91:829–38.

[27] Van Duin D, Allore HG, Mohanty S, Ginter S, Newman FK, Belshe RB, et al. Prevaccine determination of the expression of costimulatory B7 molecules in activated monocytes predicts influenza vaccine responses in young and older adults. J Infect Dis 2007;195:1590–7.

[28] Villanueva J, Solana R, Alonso M, Pena J. Changes in the expression of HLA-class II antigens on peripheral blood monocytes from aged humans. Dis Markers 1989;8:85–91.

[29] Clark HL, Banks R, Jones L, Hornick TR, Higgins PA, Burant CJ, et al. Characterization of MHC-II antigen presentation by B cells and monocytes from older individuals. Clin Immunol 2012;144:172–7.

[30] Varas A, Sacedón R, Hernandez-López C, Jiménez E, García-Ceca J, Arias-Díaz J, et al. Age-dependent changes in thymic macrophages and dendritic cells. Microsc Res Tech 2003;62:501–7.

[31] Vetvicka V, Tlaskalovar-Hogenova H, Pospisil M. Impaired antigen presenting function of macrophages from aged mice. Immunol Invest 1985;14:105–14.

[32] Herrero C, Marqués L, Lloberas J, Celada A. IFN-γ−dependent transcription of MHC class II IA is impaired in macrophages from aged mice. J Clin Invest 2001;107:485–93.

[33] Agius E, Lacy KE, Vukmanovic-Stejic M, Jagger AL, Papageorgiou A-P, Hall S, et al. Decreased TNF-α synthesis by macrophages restricts cutaneous immunosurveillance by memory CD4+ T cells during aging. J Exp Med 2009;206:1929–40.

[34] Swift ME, Burns AL, Gray KL, DiPietro LA. Age-related alterations in the inflammatory response to dermal injury. J Invest Dermatol 2001;117:1027–35.

[35] Uciechowski P, Rink L. Basophil, eosinophil, and neutrophil functions in the elderly. Immunology of aging. Springer; 2014. p. 47–63.

[36] Butcher S, Chahal H, Nayak L, Sinclair A, Henriquez N, Sapey E, et al. Senescence in innate immune responses: reduced neutrophil phagocytic capacity and CD16 expression in elderly humans. J Leukoc Biol 2001;70:881–6.

[37] Hazeldine J, Harris P, Chapple IL, Grant M, Greenwood H, Livesey A, et al. Impaired neutrophil extracellular trap formation: a novel defect in the innate immune system of aged individuals. Aging Cell 2014;13:690–8.

[38] Ogawa K, Suzuki K, Okutsu M, Yamazaki K, Shinkai S. The association of elevated reactive oxygen species levels from neutrophils with low-grade inflammation in the elderly. Immun Ageing 2008;5:13.

[39] Tortorella C, Ottolenghi A, Pugliese P, Jirillo E, Antonaci S. Relationship between respiratory burst and adhesiveness capacity in elderly polymorphonuclear cells. Mech Ageing Dev 1993;69:53–63.

[40] Fortin CF, Larbi A, Lesur O, Douziech N, Fulop T. Impairment of SHP-1 down-regulation in the lipid rafts of human neutrophils under GM-CSF stimulation contributes to their age-related, altered functions. J Leukoc Biol 2006;79:1061–72.

[41] Larbi A, Douziech N, Fortin C, Linteau A, Dupuis G, Fulop Jr T. The role of the MAPK pathway alterations in GM-CSF modulated human neutrophil apoptosis with aging. Immun Ageing 2005;2:6.

[42] Garff-Tavernier L, Béziat V, Decocq J, Siguret V, Gandjbakhch F, Pautas E, et al. Human NK cells display major phenotypic and functional changes over the life span. Aging Cell 2010;9:527–35.

[43] Hazeldine J, Lord JM. The impact of ageing on natural killer cell function and potential consequences for health in older adults. Ageing Res Rev 2013;12:1069–78.

[44] Public Health England. Surveillance of influenza and other respiratory viruses, including novel respiratory viruses, in the United Kingdom: Winter 2012−2013; 2013.

[45] Public Health England. PHE Weekly National Influenza Report. Summary of UK surveillance of influenza and other seasonal respiratory illnesses; 2015. p. 1−11.

[46] Campos C, Pera A, Lopez-Fernandez I, Alonso C, Tarazona R, Solana R. Proinflammatory status influences NK cells subsets in the elderly. Immunol Lett 2014;162:298−302.

[47] Almeida-Oliveira A, Smith-Carvalho M, Porto LC, Cardoso-Oliveira J, dos Santos Ribeiro A, Falcão RR, et al. Age-related changes in natural killer cell receptors from childhood through old age. Hum Immunol 2011;72:319−29.

[48] Di Lorenzo G, Balistreri CR, Candore G, Cigna D, Colombo A, Romano GC, et al. Granulocyte and natural killer activity in the elderly. Mech Ageing Dev 1999;108:25−38.

[49] Solana R, Mariani E. NK and NK/T cells in human senescence. Vaccine 2000;18:1613−20.

[50] Sansoni P, Cossarizza A, Brianti V, Fagnoni F, Snelli G, Monti D, et al. Lymphocyte subsets and natural killer cell activity in healthy old people and centenarians [see comments]. Blood 1993;82:2767−73.

[51] Faunce DE, Palmer JL, Paskowicz KK, Witte PL, Kovacs EJ. CD1d-restricted NKT cells contribute to the age-associated decline of T cell immunity. J Immunol 2005;175:3102−9.

[52] Peralbo E, Alonso C, Solana R. Invariant NKT and NKT-like lymphocytes: Two different T cell subsets that are differentially affected by ageing. Exp Gerontol 2007;42:703−8.

[53] Peralbo E, DelaRosa O, Gayoso I, Pita ML, Tarazona R, Solana R. Decreased frequency and proliferative response of invariant Vα24Vβ11 natural killer T (iNKT) cells in healthy elderly. Biogerontology 2006;7:483−92.

[54] DelaRosa O, Tarazona R, Casado JG, Alonso C, Ostos B, Peña J, et al. Vα24+ NKT cells are decreased in elderly humans. Exp Gerontol 2002;37:213−17.

[55] Aspinall R, Andrew D. Thymic involution in aging. J Clin Immunol 2000;20:250−6.

[56] Lescale C, Dias S, Maës J, Cumano A, Szabo P, Charron D, et al. Reduced EBF expression underlies loss of B-cell potential of hematopoietic progenitors with age. Aging Cell 2010;9:410−19.

[57] Yan J, Greer JM, Hull R, O'Sullivan JD, Henderson RD, Read SJ, et al. The effect of ageing on human lymphocyte subsets: comparison of males and females. Immun Ageing 2010;7:4.

[58] Buchholz VR, Neuenhahn M, Busch DH. CD8+ T cell differentiation in the aging immune system: until the last clone standing. Curr Opin Immunol 2011;23:549−54.

[59] Saule P, Trauet J, Dutriez V, Lekeux V, Dessaint J-P, Labalette M. Accumulation of memory T cells from childhood to old age: central and effector memory cells in CD4+ versus effector memory and terminally differentiated memory cells in CD8+ compartment. Mech Ageing Dev 2006;127:274−81.

[60] Zanni F, Vescovini R, Biasini C, Fagnoni F, Zanlari L, Telera A, et al. Marked increase with age of type 1 cytokines within memory and effector/cytotoxic CD8+ T cells in humans: a contribution to understand the relationship between inflammation and immunosenescence. Exp Gerontol 2003;38:981−7.

[61] Czesnikiewicz-Guzik M, Lee W-W, Cui D, Hiruma Y, Lamar DL, Yang Z-Z, et al. T cell subset-specific susceptibility to aging. Clin Immunol 2008;127:107−18.

[62] Alberti S, Cevenini E, Ostan R, Capri M, Salvioli S, Bucci L, et al. Age-dependent modifications of Type 1 and Type 2 cytokines within virgin and memory CD4+ T cells in humans. Mech Ageing Dev 2006;127:560−6.

[63] Franceschi C, Capri M, Monti D, Giunta S, Olivieri F, Sevini F, et al. Inflammaging and anti-inflammaging: a systemic perspective on aging and longevity emerged from studies in humans. Mech Ageing Dev 2007;128:92−105.

[64] Snapper CM, Finkelman FD, Paul WE. Differential regulation of IgG1 and IgE synthesis by interleukin 4. J Exp Med 1988;167:183−96.

[65] Le Bon A, Schiavoni G, D'Agostino G, Gresser I, Belardelli F, Tough DF. Type I interferons potently enhance humoral immunity and can promote isotype switching by stimulating dendritic cells in vivo. Immunity 2001;14:461−70.

[66] Le Bon A, Thompson C, Kamphuis E, Durand V, Rossmann C, Kalinke U, et al. Cutting edge: enhancement of antibody responses through direct stimulation of B and T cells by type I IFN. J Immunol 2006;176:2074−8.

[67] Macallan DC, Wallace D, Zhang Y, de Lara C, Worth AT, Ghattas H, et al. Rapid turnover of effector−memory CD4+ T cells in healthy humans. J Exp Med 2004;200:255−60.

[68] Vasto S, Colonna-Romano G, Larbi A, Wikby A, Caruso C, Pawelec G. Role of persistent CMV infection in configuring T cell immunity in the elderly. Immun Ageing 2007;4:1510.

[69] Connor LM, Kohlmeier JE, Ryan L, Roberts AD, Cookenham T, Blackman MA, et al. Early dysregulation of the memory CD8(+) T cell repertoire leads to compromised immune responses to secondary viral infection in the aged. Immun Ageing 2012;9:28.

[70] Akbar AN, Fletcher JM. Memory T cell homeostasis and senescence during aging. Curr Opin Immunol 2005;17:480−5.

[71] Khan N, Shariff N, Cobbold M, Bruton R, Ainsworth JA, Sinclair AJ, et al. Cytomegalovirus seropositivity drives the CD8 T cell repertoire toward greater clonality in healthy elderly individuals. J Immunol 2002;169:1984−92.

[72] Schenkein JG, Park S, Nahm MH. Pneumococcal vaccination in older adults induces antibodies with low opsonic capacity and reduced antibody potency. Vaccine 2008;26:5521−6.

[73] Kolibab K, Smithson SL, Shriner AK, Khuder S, Romero-Steiner S, Carlone GM, et al. Immune response to pneumococcal polysaccharides 4 and 14 in elderly and young adults. I. Antibody concentrations, avidity and functional activity. Immun Ageing 2005;2:10.

[74] Goronzy JJ, Fulbright JW, Crowson CS, Poland GA, O'Fallon WM, Weyand CM. Value of immunological markers in predicting responsiveness to influenza vaccination in elderly individuals. J Virol 2001;75:12182−7.

[75] Murasko DM, Bernstein ED, Gardner EM, Gross P, Munk G, Dran S, et al. Role of humoral and cell-mediated immunity in protection from influenza disease after immunization of healthy elderly. Exp Gerontol 2002;37:427−39.

[76] Dorshkind K, Montecino-Rodriguez E, Signer RA. The ageing immune system: is it ever too old to become young again? Nat Rev Immunol 2009;9:57−62.

[77] Lazuardi L, Jenewein B, Wolf AM, Pfister G, Tzankov A, Grubeck-Loebenstein B. Age-related loss of naïve T cells and dysregulation of T-cell/B-cell interactions in human lymph nodes. Immunology 2005;114:37−43.

[78] Macallan DC, Wallace DL, Zhang Y, Ghattas H, Asquith B, de Lara C, et al. B-cell kinetics in humans: rapid turnover of peripheral blood memory cells. Blood 2005;105:3633−40.

[79] Listì F, Candore G, Modica MA, Russo M, Lorenzo GD, Esposito-Pellitteri M, et al. A study of serum immunoglobulin levels in elderly persons that provides new insights into B cell immunosenescence. Ann N Y Acad Sci 2006;1089:487−95.

[80] Colonna-Romano G, Buffa S, Bulati M, Candore G, Lio D, Pellicanò M, et al. B cells compartment in centenarian offspring and old people. Curr Pharm Des 2010;16:604−8.

[81] Colonna-Romano G, Bulati M, Aquino A, Pellicanò M, Vitello S, Lio D, et al. A double-negative (IgD − CD27 −) B cell population is increased in the peripheral blood of elderly people. Mech Ageing Dev 2009;130:681−90.

[82] Huang C, Stewart AK, Schwartz RS, Stollar BD. Immunoglobulin heavy chain gene expression in peripheral blood B lymphocytes. J Clin Invest 1992;89:1331−43.

[83] Nicoletti C, Yang X, Cerny J. Repertoire diversity of antibody response to bacterial antigens in aged mice. III. Phosphorylcholine antibody from young and aged mice differ in structure and protective activity against infection with Streptococcus pneumoniae. J Immunol 1993;150:543−9.

[84] Nicoletti C, Borghesi-Nicoletti C, Yang XH, Schulze DH, Cerny J. Repertoire diversity of antibody response to bacterial antigens in aged mice. II. Phosphorylcholine-antibody in young and aged mice differ in both VH/VL gene repertoire and in specificity. J Immunol 1991;147:2750−5.

[85] Kolibab K, Smithson SL, Rabquer B, Khuder S, Westerink MAJ. Immune response to pneumococcal polysaccharides 4 and 14 in elderly and young adults: analysis of the variable heavy chain repertoire. Infect Immun 2005;73:7465−76.

[86] Benichou J, Glanville J, Prak ETL, Azran R, Kuo TC, Pons J, et al. The restricted DH gene reading frame usage in the expressed human antibody repertoire is selected based upon its amino acid content. J Immunol 2013;190:5567−77.

[87] Farges M-C, Minet-Quinard R, Walrand S, Thivat E, Ribalta J, Winklhofer-Roob B, et al. Immune status is more affected by age than by carotenoid depletion-repletion in healthy human subjects. Br J Nutr 2012;108:2054−65.

[88] Wu Y-CB, Kipling D, Dunn-Walters DK. Age-related changes in human peripheral Blood IGH Repertoire Following Vaccination. Front Immunol 2012;3:193.

[89] Wang C, Liu Y, Xu LT, Jackson KJL, Roskin KM, Pham TD, et al. Effects of aging, cytomegalovirus infection, and EBV infection on human B cell repertoires. J Immunol 2014;192:603−11.

[90] Gibson KL, Wu Y-C, Barnett Y, Duggan O, Vaughan R, Kondeatis E, et al. B-cell diversity decreases in old age and is correlated with poor health status. Aging Cell 2009;8:18−25.

[91] Chong Y, Ikematsu H, Yamaji K, Nishimura M, Kashiwagi S, Hayashi J. Age-related accumulation of Ig VH gene somatic mutations in peripheral B cells from aged humans. Clin Exp Immunol 2003;133:59−66.

[92] Bravo L. Polyphenols: chemistry, dietary sources, metabolism, and nutritional significance. Nutr Rev 1998;56:317−33.

[93] Del Rio D, Costa L, Lean M, Crozier A. Polyphenols and health: what compounds are involved? Nutr Metab Cardiovasc Dis 2010;20:1−6.

[94] Del Cornò M, Varano B, Scazzocchio B, Filesi C, Masella R, Gessani S. Protocatechuic acid inhibits human dendritic cell functional activation: role of PPARγ up-modulation. Immunobiology 2014;219:416−24.

[95] Geisel J, Bruck J, Glocova I, Dengler K, Sinnberg T, Rothfuss O, et al. Sulforaphane protects from T cell-mediated autoimmune disease by inhibition of IL-23 and IL-12 in dendritic cells. J Immunol 2014;192:3530−9.

[96] Shaikh SR, Edidin M. Polyunsaturated fatty acids, membrane organization, T cells, and antigen presentation. Am J Clin Nutr 2006;84:1277−89.

[97] Shaikh SR, Edidin M. Immunosuppressive effects of polyunsaturated fatty acids on antigen presentation by human leukocyte antigen class I molecules. J Lipid Res 2007;48:127−38.

[98] Teague H, Rockett BD, Harris M, Brown DA, Shaikh SR. Dendritic cell activation, phagocytosis and CD69 expression on cognate T cells are suppressed by n-3 long-chain polyunsaturated fatty acids. Immunology 2013;139:386−94.

[99] Rockett BD, Melton M, Harris M, Bridges LC, Shaikh SR. Fish oil disrupts MHC class II lateral organization on the B-cell side of the immunological synapse independent of B-T cell adhesion. J Nutr Biochem 2013;24(11):1810−16.

[100] Rockett BD, Teague H, Harris M, Melton M, Williams J, Wassall SR, et al. Fish oil increases raft size and membrane order of B cells accompanied by differential effects on function. J Lipid Res 2012;53:674−85.

[101] Shaikh SR. Biophysical and biochemical mechanisms by which dietary N-3 polyunsaturated fatty acids from fish oil disrupt membrane lipid rafts. J Nutr Biochem 2012;23:101−5.

[102] Shaikh SR, Teague H. N-3 fatty acids and membrane microdomains: from model membranes to lymphocyte function. Prostaglandins Leukot Essent Fatty Acids 2012;87:205−8.

[103] Gurzell EA, Teague H, Harris M, Clinthorne J, Shaikh SR, Fenton JI. DHA-enriched fish oil targets B cell lipid microdomains and enhances ex vivo and in vivo B cell function. J Leukoc Biol 2012;93(4):463−70.

[104] Rockett BD, Salameh M, Carraway K, Morrison K, Shaikh SR. n-3 PUFA improves fatty acid composition, prevents palmitate-induced apoptosis, and differentially modifies B cell cytokine secretion in vitro and ex vivo. J Lipid Res 2010;51:1284−97.

[105] Alperovich M, Neuman MI, Willett WC, Curhan GC. Fatty acid intake and the risk of community-acquired pneumonia in U.S. women. Nutrition 2007;23:196−202.

[106] Wu D, Mura C, Beharka AA, Han SN, Paulson KE, Hwang D, et al. Age-associated increase in PGE2 synthesis and COX activity in murine macrophages is reversed by vitamin E. Am J Physiol 1998;275:C661−8.

[107] Han S, Wu D, Ha W, Beharka A, Smith D, Bender B, et al. Vitamin E supplementation increases T helper 1 cytokine production in old mice infected with influenza virus. Immunology 2000;100:487−93.

[108] Cox MA, Jackson J, Stanton M, Rojas-Triana A, Bober L, Laverty M, et al. Short-chain fatty acids act as antiinflammatory mediators by regulating prostaglandin E2 and cytokines. World J Gastroenterol 2009;15:5549.

[109] Park J, Kim M, Kang SG, Jannasch AH, Cooper B, Patterson J, et al. Short-chain fatty acids induce both effector and regulatory T cells by suppression of histone deacetylases and regulation of the mTOR-S6K pathway. Mucosal Immunol 2015;8:80−93.

[110] Meijer K, de Vos P, Priebe MG. Butyrate and other short-chain fatty acids as modulators of immunity: what relevance for health? Curr Opin Clin Nutr Metab Care 2010;13:715−21.

[111] Bermudez-Brito M, Sahasrabudhe NM, Rösch C, Schols HA, Faas MM, Vos P. The impact of dietary fibers on dendritic cell responses in vitro is dependent on the differential effects of the fibers on intestinal epithelial cells. Mol Nutr Food Res 2015;59:698−710.

[112] Corridan BM, O'Donoghue M, Hughes DA, Morrissey PA. Low-dose supplementation with lycopene or beta-carotene does not enhance cell-mediated immunity in healthy free-living elderly humans. Eur J Clin Nutr 2001;55:627−35.

[113] Martin I, Grotewiel MS. Oxidative damage and age-related functional declines. Mech Ageing Dev 2006;127:411−23.

[114] Schmoranzer F, Fuchs N, Markolin G, Carlin E, Sakr L, Sommeregger U. Influence of a complex micronutrient supplement on the immune status of elderly individuals. Int J Vitam Nutr Res 2009;79:308−18.

[115] Belisle SE, Leka LS, Dallal GE, Jacques PF, Delgado-Lista J, Ordovas JM, et al. Cytokine response to vitamin E

[116] Watson RR, Prabhala RH, Plezia PM, Alberts DS. Effect of beta-carotene on lymphocyte subpopulations in elderly humans: evidence for a dose-response relationship. Am J Clin Nutr 1991;53:90–4.

[117] Meydani S, Meydani M, Blumberg JB, et al. Vitamin E supplementation and in vivo immune response in healthy elderly subjects: A randomized controlled trial. JAMA 1997;277:1380–6.

[118] Sundaram ME, Meydani SN, Vandermause M, Shay DK, Coleman LA. Vitamin E, vitamin A, and zinc status are not related to serologic response to influenza vaccine in older adults: an observational prospective cohort study. Nutr Res 2014;34:149–54.

[119] Sundaram ME, Talbot HK, Zhu Y, Griffin MR, Spencer S, Shay DK, et al. Vitamin D is not associated with serologic response to influenza vaccine in adults over 50 years old. Vaccine 2013;31:2057–61.

[120] Gibson A, Edgar JD, Neville CE, Gilchrist SE, McKinley MC, Patterson CC, et al. Effect of fruit and vegetable consumption on immune function in older people: a randomized controlled trial. Am J Clin Nutr 2012;96(6):1429–36.

[121] De la Fuente M. Murine models of premature ageing for the study of diet-induced immune changes: improvement of leucocyte functions in two strains of old prematurely ageing mice by dietary supplementation with sulphur-containing antioxidants. Proc Nutr Soc 2010;69:651–9.

[122] Monagas M, Khan N, Andres-Lacueva C, Casas R, Urpí-Sardà M, Llorach R, et al. Effect of cocoa powder on the modulation of inflammatory biomarkers in patients at high risk of cardiovascular disease. Am J Clin Nutr 2009;90:1144–50.

[123] Ghanem ENB, Clark S, Du X, Wu D, Camilli A, Leong JM, et al. The α-tocopherol form of vitamin E reverses age-associated susceptibility to *Streptococcus pneumoniae* lung infection by modulating pulmonary neutrophil recruitment. J Immunol 2015;194:1090–9.

[124] De Paula RC, Aneni EC, Costa APR, Figueiredo VN, Moura FA, Freitas WM, et al. Low zinc levels is associated with increased inflammatory activity but not with atherosclerosis, arteriosclerosis or endothelial dysfunction among the very elderly. BBA Clin 2014;2:1–6.

[125] Rees D, Miles EA, Banerjee T, Wells SJ, Roynette CE, Wahle KW, et al. Dose-related effects of eicosapentaenoic acid on innate immune function in healthy humans: a comparison of young and older men. Am J Clin Nutr 2006;83:331–42.

[126] Thies F, Nebe-von-Caron G, Powell JR, Yaqoob P, Newsholme EA, Calder PC. Dietary supplementation with eicosapentaenoic acid, but not with other long-chain n−3 or n−6 polyunsaturated fatty acids, decreases natural killer cell activity in healthy subjects aged > 55 years. Am J Clin Nutr 2001;73:539–48.

[127] Thies F, Nebe-von-Caron G, Powell JR, Yaqoob P, Newsholme EA, Calder PC. Dietary supplementation with gamma-linolenic acid or fish oil decreases T lymphocyte proliferation in healthy older humans. J Nutr 2001;131:1918–27.

[128] Merchant AT, Curhan GC, Rimm EB, Willett WC, Fawzi WW. Intake of n−6 and n−3 fatty acids and fish and risk of community-acquired pneumonia in US men. Am J Clin Nutr 2005;82:668–74.

[129] Bechoua S, Dubois M, Véricel E, Chapuy P, Lagarde M, Prigent A-F. Influence of very low dietary intake of marine oil on some functional aspects of immune cells in healthy elderly people. Br J Nutr 2003;89:523–31.

[130] Bouwens M, van de Rest O, Dellschaft N, Bromhaar MG, de Groot LC, Geleijnse JM, et al. Fish-oil supplementation induces antiinflammatory gene expression profiles in human blood mononuclear cells. Am J Clin Nutr 2009;90:415–24.

[131] Man AL, Gicheva N, Nicoletti C. The impact of ageing on the intestinal epithelial barrier and immune system. Cell Immunol 2014;289:112–18.

[132] Ouwehand AC, Tiihonen K, Saarinen M, Putaala H, Rautonen N. Influence of a combination of Lactobacillus acidophilus NCFM and lactitol on healthy elderly: intestinal and immune parameters. Br J Nutr 2009;101:367–75.

[133] Dong H, Rowland I, Thomas L, Yaqoob P. Immunomodulatory effects of a probiotic drink containing *Lactobacillus casei* Shirota in healthy older volunteers. Eur J Nutr 2013;1–11.

[134] Gill HS, Rutherfurd KJ, Cross ML, Gopal PK. Enhancement of immunity in the elderly by dietary supplementation with the probiotic *Bifidobacterium lactis* HN019. Am J Clin Nutr 2001;74:833–9.

[135] Sharma R, Kapila R, Dass G, Kapila S. Improvement in Th1/Th2 immune homeostasis, antioxidative status and resistance to pathogenic *E. coli* on consumption of probiotic *Lactobacillus rhamnosus* fermented milk in aging mice. Age 2014;36(4):9686.

[136] You J, Dong H, Mann ER, Knight SC, Yaqoob P. Probiotic modulation of dendritic cell function is influenced by ageing. Immunobiology 2014;219:138–48.

[137] Antonio Moro-Garcia M, Alonso-Arias R, Baltadjieva M, Fernandez Benitez C, Fernandez Barrial MA, Diaz Ruisanchez E, et al. Oral supplementation with *Lactobacillus delbrueckii* subsp *Bulgaricus* 8481 enhances systemic immunity in elderly subjects. Age 2013;35:1311–26.

[138] Gill H, Rutherfurd K, Cross M. Dietary probiotic supplementation enhances natural killer cell activity in the elderly: an investigation of age-related immunological changes. J Clin Immunol 2001;21:264–71.

[139] Vulevic J, Drakoularakou A, Yaqoob P, Tzortzis G, Gibson GR. Modulation of the fecal microflora profile and immune function by a novel trans-galactooligosaccharide mixture (B-GOS) in healthy elderly volunteers. Am J Clin Nutr 2008;88:1438–46.

[140] Schiffrin EJ, Thomas DR, Kumar VB, Brown C, Hager C, Van't Hof MA, et al. Systemic inflammatory markers in older persons: the effect of oral nutritional supplementation with prebiotics. J Nutr Health Aging 2007;11:475–9.

[141] Bunout D, Hirsch S, de la Maza M, Munoz C, Haschke F, Steenhout P, et al. Effects of prebiotics on the immune response to vaccination in the elderly. JPEN J Parenter Enteral Nutr 2002;26:372–6.

[142] Meydani SN, Endres S, Woods MM, Goldin BR, Soo C, Morrill-Labrode A, et al. Oral (n-3) fatty acid supplementation suppresses cytokine production and lymphocyte proliferation: comparison between young and older women. J Nutr 1991;121:547–55.

[143] Weichselbaum E, Coe S, Buttriss J, Stanner S. Fish in the diet: a review. Nutr Bull 2013;38:128–77.

[144] Bates B, Lennox A, Prentice A, Bates C, Swan G. National Diet and Nutrition Survey: Headline results from years 1, 2 and 3 (combined) of the Rolling Programme (2008/2009–2010/11); 2012. p. 1–79.

[145] Han SN, Lichtenstein AH, Ausman LM, Meydani SN. Novel soybean oils differing in fatty acid composition alter immune functions of moderately hypercholesterolemic older adults. J Nutr 2012;142:2182–7.

[146] Perez-Lopez FR, Chedraui P, Haya J, Cuadros JL. Effects of the Mediterranean diet on longevity and age-related morbid conditions. Maturitas 2009;64:67–79.

[147] Trichopoulou A, Costacou T, Bamia C, Trichopoulos D. Adherence to a Mediterranean diet and survival in a Greek population. N Engl J Med 2003;348:2599–608.

[148] Knoops KT, de Groot LC, Kromhout D, Perrin AE, Moreiras-Varela O, Menotti A, et al. Mediterranean diet, lifestyle factors, and 10-year mortality in elderly European men and women: the HALE project. JAMA 2004;292:1433–9.

[149] Lasheras C, Fernandez S, Patterson AM. Mediterranean diet and age with respect to overall survival in institutionalized, nonsmoking elderly people. Am J Clin Nutr 2000;71:987–92.

[150] Psaltopoulou T, Kyrozis A, Stathopoulos P, Trichopoulos D, Vassilopoulos D, Trichopoulou A. Diet, physical activity and cognitive impairment among elders: the EPIC-Greece cohort (European Prospective Investigation into Cancer and Nutrition). Public Health Nutr 2008;11:1054–62.

[151] Valls-Pedret C, Lamuela-Raventos RM, Medina-Remon A, Quintana M, Corella D, Pinto X, et al. Polyphenol-rich foods in the Mediterranean diet are associated with better cognitive function in elderly subjects at high cardiovascular risk. J Alzheimers Dis 2012;29:773–82.

[152] Konstantinidou V, Covas M-I, Muñoz-Aguayo D, Khymenets O, de la Torre R, Saez G, et al. In vivo nutrigenomic effects of virgin olive oil polyphenols within the frame of the Mediterranean diet: a randomized controlled trial. FASEB J 2010;24:2546–57.

[153] Camargo A, Delgado-Lista J, Garcia-Rios A, Cruz-Teno C, Yubero-Serrano EM, Perez-Martinez P, et al. Expression of proinflammatory, proatherogenic genes is reduced by the Mediterranean diet in elderly people. Br J Nutr 2012;108:500–8.

[154] Mena M-P, Sacanella E, Vazquez-Agell M, Morales M, Fitó M, Escoda R, et al. Inhibition of circulating immune cell activation: a molecular antiinflammatory effect of the Mediterranean diet. Am J Clin Nutr 2009;89:248–56.

[155] Bellido C, López-Miranda J, Blanco-Colio LM, Pérez-Martínez P, Muriana FJ, Martín-Ventura JL, et al. Butter and walnuts, but not olive oil, elicit postprandial activation of nuclear transcription factor κB in peripheral blood mononuclear cells from healthy men. Am J Clin Nutr 2004;80:1487–91.

[156] Dedoussis GV, Kanoni S, Mariani E, Cattini L, Herbein G, Fulop T, et al. Mediterranean diet and plasma concentration of inflammatory markers in old and very old subjects in the ZINCAGE population study. Clin Chem Lab Med 2008;46:990–6.

[157] Adamsson V, Reumark A, Fredriksson IB, Hammarström E, Vessby B, Johansson G, et al. Effects of a healthy Nordic diet on cardiovascular risk factors in hypercholesterolaemic subjects: a randomized controlled trial (NORDIET). J Intern Med 2011;269:150–9.

[158] Poulsen SK, Due A, Jordy AB, Kiens B, Stark KD, Stender S, et al. Health effect of the New Nordic Diet in adults with increased waist circumference: a 6-mo randomized controlled trial. Am J Clin Nutr 2014;99:35–45.

[159] Sho H. History and characteristics of Okinawan longevity food. Asia Pac J Clin Nutr 2001;10:159–64.

[160] Uusitupa M, Hermansen K, Savolainen M, Schwab U, Kolehmainen M, Brader L, et al. Effects of an isocaloric healthy Nordic diet on insulin sensitivity, lipid profile and inflammation markers in metabolic syndrome—a randomized study (SYSDIET). J Intern Med 2013;274:52–66.

[161] Willcox BJ, Willcox DC, Todoriki H, Fujiyoshi A, Yano K, He Q, et al. Caloric restriction, the traditional Okinawan diet, and healthy aging. Ann N Y Acad Sci 2007;1114:434–55.

[162] Gavrilova NS, Gavrilov LA. Comments on dietary restriction, Okinawa diet and longevity. Gerontology 2012;58:221–3.

[163] Åkesson A, Andersen LF, Kristjánsdottir AG, Roos E, Trolle E, Voutilainen E, et al. Health effects associated with foods characteristic of the Nordic diet: a systematic literature review. Food Nutr Res 2013;57:. Available from: http://dx.doi.org/10.3402/fnr.v57i0.22790.

[164] Mithril C, Dragsted LO, Meyer C, Blauert E, Holt MK, Astrup A. Guidelines for the New Nordic Diet. Public Health Nutr 2012;15:1941–7.

[165] Mithril C, Dragsted LO, Meyer C, Tetens I, Biltoft-Jensen A, Astrup A. Dietary composition and nutrient content of the New Nordic Diet. Public Health Nutr 2013;16:777–85.

CHAPTER
29

Glucose Metabolism, Insulin, and Aging: Role of Nutrition

Massimo Boemi, Giorgio Furlan and Maria P. Luconi
UOC Malattie Metaboliche e Diabetologia, INRCA-IRCCS, Ancona, Italy

KEY FACTS
- Aging is associated to alterations in glucose metabolism.
- Impairment of insulin secretion and action have been demonstrated in aging individuals.
- Glucose homeostasis is linked to cognitive decline.
- Inflammation and oxidative stress alter glucose tolerance.
- Change in body composition partially account for altered glucose metabolism in the elderly.
- Diet influences aging processes.

Dictionary of Terms
- *IVGTT (intravenous glucose tolerance test)*: a method to estimate insulin sensitivity based on intravenous glucose infusion.
- *SIR Sirtuin or Sir2*: silent mating-type information regulation 2 proteins with deacylase activity, including deacetylase, desuccinylase, activity. Sirtuins have been implicated in a wide range of cellular processes, such as transcription, apoptosis, inflammation, and stress resistance, as well as energy efficiency and alertness during low-calorie diets.
- *GLUTs (Glucose transporter)*: a class of membrane proteins that allow the transport of glucose over plasma membranes.
- *RNS (reactive nitrogen species)*: family of antimicrobial molecules derived from nitric oxide and superoxide. Reactive nitrogen species cause nitrosative stress.
- *ROS (Reactive oxygen species)*: reactive molecules containing oxygen formed as a natural byproduct of the normal metabolism of oxygen. ROS have important roles in cell signaling and homeostasis. Increase of ROS levels as a consequence of environmental stress results in damage to cell structures oxidative stress).
- *mTOR (mammalian target of rapamycin)*: protein encoded by the MTOR gene with enzymatic activity. mTOR is serine/threonine protein kinase that regulates cell growth, cell proliferation, cell motility, cell survival, protein synthesis, and transcription.
- *AGEs (Advanced glycation end-products)*: compounds derived from the glycation reaction, which refers to the irreversible addition of a carbohydrate to a protein without the involvement of an enzyme. AGEs overproduction is involved in aging, diabetes vascular complications, and oxidative stress

Aging is characterized by a reduction in whole-body carbohydrate metabolism. Glucose tolerance declines with age and, currently, 30% of older adults in the United States meet the criteria for diabetes diagnosis, and a 4.5-fold increase in those aged 65 years and older with diabetes has been projected from 2005 to 2050 [1]. At the same time, hyperglycemia, as the result of the reduced glucose tolerance, exacerbates and accelerates aging processes such as those involved in oxidative damage, DNA repair, collagen cross-linking, and capillary basement membrane thickening [2–5].

From a clinical perspective, impaired glucose tolerance and diabetes are emerging risk factors for cognitive decline and dementia [6,7] and in a more general way hyperglycemia worsens geriatric syndromes leading to frailty [8].

In elder populations the first expression of glucose intolerance is the increase in the postprandial blood glucose levels; after an oral glucose tolerance test (OGTT), fasting blood glucose level rises by 0.06 mmol/L per decade and the 2-h level by 0.5 mmol/L.

Several factors predispose the elder individual to diabetes. The genetic component plays a pivotal role in glucose metabolism alterations as demonstrated by studies in elderly identical twins [9] and in specific ethnic groups; some candidate genes have been identified and particularly one of them, the TCF7L2 gene [10,11], is associated with a nearly 40% increase in the risk of developing type 2 diabetes (T2DM). Other genes are associated both with unsuccessful aging and diabetes [12].

On this genetic background other environmental factors and age-associated conditions, such as central and sarcopenic obesity [13], inactivity, comorbidities, and polypharmacology, would act leading to impaired glucose tolerance and diabetes. Nevertheless, age results as an independent determinant of glucose intolerance [14] and age-dependent changes in carbohydrate metabolism have been identified.

The possible mechanisms underlying these changes are discussed in this chapter.

ALPHA AND BETA CELL MASS

A decline in insulin secretion is crucial for the development of T2DM. The loss of beta cell mass and function may be ascribed to several mechanisms, and among those glucose and lipid overload have been indicated as the most important [15,16]. In an in vitro model of human islets, age correlates with glucose induced beta cell apoptosis [17] and beta cell function declines at a rate of about 1% per year with age in glucose tolerant Caucasian individuals [18]. Obese individuals have an increase in relative beta cell volume of about 50% compared to nondiabetic lean controls; obese individuals with impaired fasting glucose (IGT) and T2DM have a 40—63% beta cell volume deficit compared to obese nondiabetic controls indicating that these subject are unable to adaptively increase their beta cell volume. Beta cell apoptosis is increased by three- to ten-fold in obese and lean diabetic individuals compared to obese and lean nondiabetic individuals, respectively [19,20]. Nevertheless, in normal aging, both alpha and beta cell mass appear to be quite stable and, contrary to rodents, human islet cells have shown a low rate of regeneration and proliferation; beta cell replication decreases age-dependently while apoptosis remains stable [21]. The long lifespan of human islet cells and the reduced plasticity of this cell population contrast with the current opinion on the expansion of cell mass when hormone demand is increased, such as in aging. Expansion of beta cell mass during obesity seems to be mainly attributable to neogenesis [22]. Even if the possibility that neogenetic processes from ductal and hematopoietic cells in aging individuals are a balancing effect of increased apoptosis, as seen in rodents [23], cannot be excluded, these phenomena seem both to be marginal in glucose homeostasis maintenance during aging. Thus it is more likely that a compensatory increase in insulin secretion may account for a functional rather than an anatomical process [24—26]. Whereas in diabetic subjects a relative and absolute increase in alpha cell mass has been documented [27], there is no evidence of an expansion of this cellular line during aging.

INSULIN SECRETION, METABOLISM, AND CLEARANCE

Aging is associated with insulin secretion dysfunction leading to glucose metabolism disturbances and accounting for the increased prevalence of diabetes in the elderly population. Beta cell secretory ability can be evaluated by several techniques, the most relevant being Homa beta cell, IVGTT, Disposition Index using Minimal Model, Oral Glucose Tolerance Test (OGTT), and Hyperglycemic Clamp Technique; the latter is considered the golden standard. Nevertheless each method shows advantages but also limitations and disadvantages [28]. All these techniques have been used for estimating beta cell function in aging, giving often contrasting results. Apart from the techniques applied in the different studies, multiple factors can account for conflicting results such as a small magnitude of the aging effect, coexisting confounding factors, such as obesity or decreased physical activity, and the degree of insulin resistance. Taken as a whole, the literature shows that almost all the indices of insulin secretion are potentially affected in aging, indicating a decreased beta cell secretory reserve [29] but this defect would not entirely explain the reduced glucose tolerance in the elderly population.

The hypothesis that beta cell sensitivity to glucose stimuli is reduced in elderly people is still under debate. Elahi and colleagues [30] using hyperglycemic clamp did not find any differences in the first and second phase of insulin response in elderly volunteers compared to middle-aged and young individuals. In contrast, when insulin secretion is matched to sustained continuous glucose stimuli, elderly people show a decrease in mass and amplitude of rapid insulin pulse accounting for a reduced responsiveness of the beta cell to the glucose [31]. The aging beta cell appears also less able to properly adapt insulin secretion to the oscillating blood glucose. In addition insulin secretion in response to arginine stimulus is reduced by 48% in older compared to younger people [32].

When beta cell function is assessed by an OGTT-based disposition index, a measure of the capacity for insulin secretion adjusted for insulin sensitivity, an age-dependent decline is evident [33].

Applying a sophisticated research design, Bazu and colleagues demonstrated an impairment of insulin secretion in elderly individuals [34]. The study was conducted using both a mixed meal and an intravenous glucose tolerance test and insulin secretion was considered in light of the degree of insulin resistance. The defect concerned the first and the second phase of insulin secretion as well as glucose-sensitivity.

Aging is also associated to disruption in the rhythmicity of insulin release. Insulin is normally secreted in a pulsatile way consisting of rapid low amplitude pulses occurring every 8–15 min and ultradian pulses with a periodicity of 60–140 min having a larger amplitude. The rapid pulses inhibit hepatic glucose production and the ultradian ones stimulate peripheral glucose disposal. Both patterns appear to be dysregulated in elderly individuals as the rapid pulses are reduced in amplitude and the ultradian in frequency [35,36].

The glucose-induced insulin secretion has as the first step the translocation of the glucose transporters (GLUTs) and especially the GLUT2 on the cell surface. While demonstrated in rodent and primate models, at present time there is no evidence of defect of beta cell GLUTs in nondiabetic elderly people; on the contrary some evidence exists regarding a lower glucose oxidation and an higher lipid oxidation in the elderly compared to young. The enhancement of the Randle cycle may lead to a reduced ATP generation and consequently to a reduced insulin secretion [37]. Data obtained in rats demonstrated an impairment of insulin granule exocytosis due to a diminished calcium uptake in response to an age-dependent decline of potassium efflux [38]. More recently evidence on the role of the NAD-dependent acetylase proteins belonging to the SIR2 family in both senescence and metabolism has raised interest for a possible implication of at least two of them, namely SIRT1 and SIRT4, in beta cell survival, insulin secretion, and peripheral glucose utilization but data on humans are lacking.

Another putative mechanism for reduced insulin action is linked to an age-dependent impairment of the conversion of proinsulin to insulin. Even though proinsulin concentration and the proinsulin/insulin ratio increase with advancing age, this could be ascribed to the age-related decline in kidney function given that the renal metabolic clearance is normally greater for proinsulin than for insulin [39].

Incretins are gut hormones that potentiate insulin release and diminish glucagon secretion in response to meals. A reduced incretin action has been recognized as one of the key factors leading to hyperglycemia in diabetes. Two incretins, the glucagon-like peptide 1 (GLP-1) and the glucose-dependent insulinotropic polypeptide (GIP), have been extensively studied in diabetic individuals. In elder people, basal GIP levels do not differ from those of young and middle-aged people while the response of both GIP and GLP-1 to oral glucose can be normal or increased. This latter condition is most likely attributable to a reduced activity of the dipeptidyl peptidase IV (DPPIV), the enzyme that cleaves and inactivate the hormones, in old people [40,41]. Using clamp technique, the potentiation of insulin secretion in response to GIP infusion is reduced by 48% at low glycemic plateau but does not differ between young and old individuals at high glycemic plateaux accounting for a partial reduction of beta cell sensitivity to this hormone, a defect that disappears at elevated glucose concentration.

Total body insulin clearance is lower in elderly people despite a higher hepatic insulin extraction [42,34]; this evidence can partially account for the unchanged basal and early postmeal insulin levels, reported by many authors in elderly people, masking the reduction in insulin production. The reduction of insulin clearance could be attributable to the age-related decline in kidney function but other mechanisms involving vascular insulin proteases or insulin receptor binding can be invoked. However, it has been proposed that the increased hepatic extraction may represent a balancing mechanism for the reduced insulin clearance. More recently, in healthy 70-year-old individuals with nonalcoholic fatty liver disease, which is one of the components in the insulin-resistant state, a reduction in hepatic insulin extraction has been reported [43].

Alpha cell activity is preserved during aging; neither basal nor stimulate glucagon level vary and alpha cell sensitivity to alanine is normal; furthermore, glucagon suppression by both hyperglycemia and hyperinsulinemia, as well as its metabolic clearance rate, do not differ in young and older individuals [42].

INSULIN ACTION

Insulin is the main regulator of glucose metabolism. Insulin action begins with the binding of insulin to a heterotetrameric receptor on the cell membrane of the target cells. Insulin receptors are membrane glycoproteins composed of two separate insulin-binding (alpha-subunits) and two signal transduction (beta-subunits) domains. The insulin signal is further propagated through a phosphorylation network involving other intracellular substances.

Insulin exerts its metabolic action in the so-called insulin-dependent tissue, mainly muscle and adipose tissue. The liver is not such a tissue but insulin action on the hepatocyte is central in glucose metabolism.

Insulin resistance can be defined as a subnormal biological response to normal insulin concentrations. Thus it reflects the inability of insulin to increase glucose uptake and utilization in an individual as much as it does in a normal population; in clinical practice, insulin resistance refers to a state in which a given concentration of insulin is associated with a subnormal glucose response.

It is well established that aging is a major risk factor for the development and progression of insulin resistance, central obesity, type 2 diabetes, and cardiovascular disease, but what links aging and metabolic dysfunction at the molecular and cellular level is less known. A cluster of age-related alterations has been recognized responsible for insulin resistance: (1) oxidative stress and acquisition of mitochondrial dysfunctions; (2) systemic inflammation; (3) changes in body composition, in term of reduction of lean muscle mass (sarcopenia) and redistribution of fat mass with a greater increase of intraabdominal fat; (4) hormonal changes, specifically reduction in growth hormone secretion (somatopenia), late onset hypogonadism, impairment of thyroid and adrenal functions, altered secretion of appetite-related peptides (leptin, ghrelin, and adiponectin); and (5) influence of energy balance [44].

Resting on the results of the hyperinsulinemic-euglicemic clamp studies, insulin sensitivity appears to be reduced in older versus younger adults. The reduction in insulin effectiveness can be potentially explained by some of the above-cited conditions, such as obesity and inactivity, as well as by mitochondrial dysfunction, increased oxidative stress, and inflammation, nevertheless it appears to be decreased even after adjustment for these variables.

Hepatic glucose production (HGP) plays a pivotal role in glucose homeostasis both in the fasting and in the postprandial state. The ability of insulin in suppressing hepatic glucose production is not reduced in elderly individuals. On the contrary, at a physiological level HGP suppression occurs more rapidly in older than in younger subjects probably by virtue of a delayed suppression of endogenous insulin release. Thus HGP seems not to play a significant role in decreased glucose tolerance of elderly people.

On the contrary, there is a general consensus on the fact that aging is accompanied by an increased peripheral tissue insulin resistance, mainly at the level of skeletal muscles. Skeletal muscle is responsible for 70–80% of whole body insulin-stimulated glucose uptake and is therefore generally considered the most important site of insulin resistance. Lean body mass decreases with age with a loss of approximately 30% from the third to the seventh decade and 40–45% from the eighth decade [45]. Muscle mass, a change in the proportion of different fiber types within the muscle and a lower density of capillary supply to the tissue, could be virtually evoked as causes for a reduced insulin action but it is unlikely that this mechanism per se could explain the development of insulin resistance in aging.

Muscular lipid infiltration may play an important role in the adverse metabolic profile associated with muscle loss in aging. Intermuscular fat has been correlated with insulin resistance in numerous studies independently of body weight, fat mass, or percentage body fat. Muscular lipid accumulation is higher in insulin-resistant state [46].

Aging is also characterized by the accumulation of visceral fat and hepatic lipid, both of which are associated with insulin resistance and metabolic markers of cardiovascular disease and systemic inflammation [47].

It has been demonstrated that glucose can stimulate its own uptake in the absence of insulin, an effect which is known as "glucose effectiveness" or noninsulin-mediated glucose uptake.

In healthy elderly subjects glucose effectiveness is impaired during fasting, but is normal during hyperglycemia. It has also been demonstrated that elderly patients with diabetes have an even greater impairment in glucose effectiveness than healthy elderly subjects although the cause of this abnormality is uncertain, it may relate to a decreased ability of glucose to recruit glucose transporters to the cell surface in these patients [48].

In the past decade much effort has been applied to understanding the role of oxidative stress and low-grade systemic inflammation in the process of aging and to studying the key pathway involved in the regulation of glucose metabolism, the mammalian target of rapamycin (mTOR) and the insulin/IGF1-like signaling pathway (IIS).

Oxidative stress is recognized as a key participant in the development of insulin resistance in the elderly.

The reactive oxygen species (ROS), generated via incomplete reduction of oxygen, and the reactive nitrogen species (RNS), produced via the reaction of the nitric oxide with superoxide, induce a cumulative damage to cellular macromolecules during aging [49,50].

The oxidative phosphorylation represents the main mitochondrial process by which the electrons are removed and transferred from organic molecules to oxygen and the energy released is used to synthesize ATP. Thus mitochondria are the main source of superoxide anion and the principal target of oxidative damage (known as the mitochondrial free radical theory of aging) [51]. Cumulative oxidative alterations of proteins, lipids, and DNA (mtDNA) during the lifetime impairs the respiratory chain components, leading to an insufficient supply of energy to cells and finally apoptosis [52].

Evidence suggests a link between oxidative stress and insulin resistance. In pancreatic beta cells, the dysfunction of mitochondrial energy metabolism impairs glucose-induced insulin secretion [53,54]. Hydrogen peroxide is able to activate the same IIS signaling pathway, which thus induces the GLUT4 translocation from intracellular vesicles to the plasma membrane with glucose uptake in adipocytes and muscles and lipogenesis in adipose tissue [55,56]. In a state of augmented oxidative stress, the effectors for tyrosine kinase activity of the insulin receptor (IR), IRS molecules, have been found hyperphosphorylated, released from the internal membrane, and subjected to increased protein degradation. From this perspective, mitochondrial impairment may explain the failure of insulin signaling and insulin resistance [57]. Some authors have hypothesized that the decreases in number and mitochondrial function constitute the mechanism by which triglycerides have been found accumulated in skeletal muscle and liver, thus interfering with insulin-stimulated glucose metabolism [58,59]. Petersen et al. using magnetic resonance spectroscopy have shown in elderly participants an increased intramuscular and intrahepatic fat accumulation and an impaired synthesis of ATP; these patients were markedly insulin resistant as compared with young controls [53]. Furthermore, enhanced lipid peroxidation markers have been reported in animal models of diabetes and obesity and in plasma and urine of diabetic patients [60]. A sustained hyperglycemia may contribute to oxidative stress: the urinary excretion rates of 8-iso-PDG2α, an isoprostane isomer derived from free radical-mediated oxidation of arachidonic acid, has been found to be significantly higher in patients with type 2 diabetes as compared with nondiabetic healthy subjects [61].

Finally an elevated ROS production has been proposed as a marker of accelerated senescence; Passos et al. observed in senescent cells higher ROS levels, mtDNA damage, and shorter telomere [62].

Several large epidemiologic studies have described in older adults a chronic low-grade inflammation, revealed by increased levels of cytokines, especially CRP, IL-6, and TNF-α. The proposed mechanisms include: (1) the shift to more visceral adiposity, which exerts the role of active endocrine organ by secreting cytokines and adipokines; (2) declining levels of sex hormones after andropause and menopause, both testosterone and estrogen are suggested to be able to inhibit the secretion of IL-6; and (3) cumulative oxidative damage, which further invokes an inflammatory response. Pro-inflammatory cytokines play a role in developing insulin resistance. In human studies, CRP and IL-6 have been found to predict diabetes development, even though in some of them the association became nonsignificant after adjustment for possible confounders, such as body mass index, smoking, and systolic blood pressure [63].

Mammalian TOR is an evolutionarily conserved cytoplasmic protein which belongs to the phosphoinositide 3-kinase (PI3K)-related protein kinases (PIKK) family. It is the target of rapamycin, a macrolide with immunomodulatory properties produced by the bacterium *Streptomyces hygroscopicus* [64,65]. mTOR is the catalytic subunit of two distinct complexes, mTORC1 and mTORC2. They share some components and differentiate for a unique accessory protein: mTORC1 contains the regulatory-associated protein of mTOR (RAPTOR), and mTORC2 the rapamycin-insensitive companion of mTOR (RICTOR) [66,67]. mTORC1 integrates signals from growth factors (insulin and insulin-like growth factors (IGFs)), nutrients (especially amino acids), energy status, and various stressors (hypoxia and DNA damage) to inhibit autophagy and promote mRNA translation, cell mass increase (especially in skeletal muscle), lipogenesis, and mitochondrial biogenesis (proliferation and function). Little is known about the upstream activators of mTORC2, probably its interaction with growth factors determines glucose uptake and glycogen synthesis in peripheral tissues and inhibition of hepatic gluconeogenesis [68].

The inhibition of mTORC1-regulated processes (mRNA translation and autophagy) through dietary restriction or rapamycin has been documented to extend life in yeasts, worms, flies, and rodents and to delay the incidence of age-related disease, including cancer, neurodegeneration, cardiovascular disease, and diabetes in rodents and rhesus monkeys [69,70]. An inhibited mRNA translation reduces the demands on the protein folding systems and maintains protein homeostasis. The evidence come from different species, even if the specific mRNAs seem to differ [71]. Furthermore, studies in yeast and invertebrates support that activation of autophagy through mTORC1 inhibition, is required to extend lifespan [72,73]. Autophagy declines with age; the accumulation of damage (unfolded proteins, endoplasmic reticulum stress, and degenerate mitochondria) may contribute to the development and progression of age-related disease [74]. mTORC1 also regulates mitochondrial function; its inhibition in yeast and its lack in adipose tissue of mice resulted in a shift toward greater respiration [75,76].

mTOR increases pancreatic beta cell mass and function in response to signals from glucose, amino acids, and fatty acids; this mechanism helps to compensate the age-related development of insulin resistance. However, hyperactivation of mTORC1 during aging or in the condition of chronic overabundance of nutrients, contributes to alteration of energetic homeostasis [77]. Interestingly, mTORC1 and insulin/IGF1-like signaling pathways are linked: mTOR is activated by IIS through AKT and negatively regulates IIS through S6 kinase 1(S6K1)-mediated phosphorylation of IRS-1 at the cellular membrane with consequent insulin desensitization [78,79].

The upregulation of mTORC1 triggers an S6K1-IRS-1 negative feedback loop which translates into reduced cellular glucose uptake and glycogen synthesis in liver and muscle, increased hepatic gluconeogenesis, excess fat deposition in white adipose tissue, and ectopic lipogenesis in liver and muscle. This deregulation of mTOR signaling contributes to the development of hyperinsulinemia, insulin resistance, hyperglycemic condition, and, from a clinical perspective, obesity [80]. Metformin has been shown to be an inhibitor of mTORC1: it decreases the phosphorylation of its substrates S6K1 and 4E-BP1 and increases longevity in some species [81,82].

A LINK BETWEEN INSULIN AND CENTRAL NERVOUS SYSTEM AGING: BRAIN INSULIN RESISTANCE

Insulin plays a key role in the central nervous system (CNS). This is demonstrated by the fact that insulin receptors are very abundant in the brain [83]. Insulin receptors are located in the synapses of both astrocytes and neurons. Although insulin and insulin receptors are abundant in the brain, they are selectively distributed, with high concentrations in the olfactory bulb, cerebral cortex, hippocampus, hypothalamus, amygdala, and septum [84].

It is controversial as to if insulin is synthesized in the adult brain or if it is readily transported into the central nervous system across the blood—brain barrier by a saturable, receptor-mediated process.

Raising peripheral insulin levels acutely elevates brain and cerebrospinal fluid insulin levels, whereas prolonged peripheral hyperinsulinemia downregulates blood—brain barrier insulin receptors and reduces insulin transport into the brain.

Due to structural and functional homology, both insulin and IGF-1 can bind and activate both their receptors: insulin receptor (IR) and IGF-1 receptor (IGF-1R) [85]. Following the binding of the receptor with its ligand, two main pathways can be activated: the PI3K/Akt/GSK-3beta and the Ras/Raf-1/ERK [86].

The activation of the first signaling cascade (PI3K/Akt) results in inhibition of cellular apoptosis [87] and increased synthesis of proteins involved in neuronal antioxidant defense [88] as, for example, increased expression of the Cu/Zn-superoxide dismutase [89], stimulation of NFR-2/antioxidant responsive element ARE [90], and glucose metabolism; another target of the signaling cascade is CREB with an improvement of mitochondrial membrane potential, intracellular levels of ATP, and hexokinase activity: this improves glucose metabolism in adult neurons and axonal outgrowth [91].

The parallel signaling cascade (ERK) has traditionally been a mitogenic role [92]. Some authors suggest it is involved in synaptic plasticity and cell death after oxidative stress [93] and excitotoxicity mediated by N-methyl D-aspartate (NMDA) receptor [94].

Insulin performs multiple functions within the CNS: regulation of glucose metabolism, involvement in synaptic transmission and memory/learning, and a neuroprotective role.

As for glucose metabolism, it is known that a group of neurons called glucosensing neurons, are located in the hypothalamic arcuate nucleus and regulate energy homeostasis of the organism [95]. Two antagonistic neuronal populations are present: orexigenic neurons coexpressing neuropeptide Y (NPY) and agouti-related peptide (AgRP) and the anorexigenic neurons that produce proopiomelanocortin (POMC) and cocaine and amphetamine related-transcript (CART [96]. Both of the above physiological levels of blood sugar and intracerebroventricular administration of insulin reduce the level of expression of NPY/AgRP and increase those of POMC/CART, leading to a reduced ratio of orexigenic/anorexigenic signals and therefore reduced body weight. On the contrary, an alteration of brain insulin signal (as is the case of insulin resistance state), determines activation of the neurons containing NPY, AgRP, and GABA, resulting in an orexigenic effect and increase of body weight [97]. In addition, insulin at the hypothalamic level regulates hepatic glucose production: the central insulin infusion suppresses hepatic glucose production, while the alteration of hypothalamic insulin signaling stimulates gluconeogenesis [98].

Another important role that insulin plays in the CNS is its involvement in memory/learning and in synaptic transmission. Insulin has been shown to improve memory/learning in rats [97] and in healthy humans after nasal administration, without changes in peripheral blood glucose levels [99]. These effects may be mediated by the known actions of insulin on synaptic transmission (promotion of the release of epinephrine and norepinephrine from adrenergic terminals [100], inhibition of synaptic reuptake of norepinephrine, stimulation of neuronal uptake of serotonin [101]), and the most recent discoveries about the actions of insulin confirm this hormonal activity (modulation of the NMDA receptor and the increased influx of intracellular calcium and reinforcement of synaptic communication between neurons) [102].

Neuroprotection is an additional role that insulin plays in the brain. Both insulin and IGF-1 attenuate apoptosis of neurons subjected to oxidative stress or other harmful conditions [88]. Indeed the signal IR/IGF-1R improves neuronal metabolism of glucose and increases the antioxidant defenses [103].

The harmful effect of insulin deficiency in the brain in the case of type 2 diabetes mellitus is an important evidence of the key role of insulin in the CNS. During type 2 diabetes mellitus, hyperinsulinemia determines a downregulation of transport of insulin within the brain resulting in a reduction of the insulin signaling pathways and glycolytic enzymes [104]. This alters the mitochondrial electron transport chain (which can be further altered by aging and by amyloid-beta peptide) and leads to oxidative stress and metabolic dysfunctions [105].

During diabetes and chronic hyperglycemia, an increased activity of mitochondrial nitric oxide synthase (NOS) may occur, leading to increased levels of nitric oxide (NO) and inhibition of ATP synthase; this can lead to an alteration in the production of ATP and cell death [106]. Moreover, chronic hyperglycemia can aggravate oxidative stress through advanced glycation endproducts (AGEs) formation, autooxidation of glucose, endoplasmic reticulum stress, and alteration of antioxidant defenses [107].

Diabetes can cause brain damage through another fundamental mechanism that is hypoglycemia. Mainly in elderly patients, recurrent hypoglycemia alters glucose metabolism and protein synthesis, accelerates lipolysis, and alters the ion homeostasis as well as mitochondrial function; ultimately it causes neuronal dysfunction especially in areas of the brain involved in memory and learning [108].

Insulin appears to play a key role in the brain even in the etiology of age-related neurodegenerative disorders: the most important of which is Alzheimer's disease (AD). AD is a degenerative disease characterized by a progressive deterioration of cognition [109]. Hallmarks histological of the disease are senile plaques (formed by deposition of amyloid beta) and intracellular neurofibrillary tangles (composed of hyperphosphorylated tau protein) [110]. Indeed the accumulation of amyloid beta is a physiological process that occurs during aging, but in the case of AD this process is expanded, probably due both to overproduction of the amyloid beta and a reduction of its catabolism [111]. Abnormal metabolism of amyloid beta seems to be secondary to oxidative stress and metabolic and mitochondrial disorders [112].

Impaired insulin signaling and reduced glucose transport in the brain seem to be central in the development of the disease. Not surprisingly AD and type 2 diabetes are epidemiologically strongly related. Type 2 diabetes increases the risk of developing AD [113], and patients with AD have a higher incidence of diabetes [114]. Furthermore, the two pathologies, as age-related degenerative processes, share many common features: CNS insulin resistance, abnormal IR-mediated signal, reduced transport into the cells, and altered intracellular glucose metabolism [115]. In both diseases there is a reduction of levels of insulin mRNA expression, IR and IGF-1R, insulin receptor substrate (IRS-1 and IRS-2) [109]. Moreover amyloid beta, and in particular the soluble amyloid beta oligomers, have been shown to alter the function of the IR resulting in metabolic and neurotrophic damage of neurons [116]. The abovementioned evidence led to defining AD as a state of cerebral insulin resistance or type 3 diabetes [117].

In addition, insulin modulates the metabolism of amyloid beta, and the activation of IR and IGF-1R via the signaling cascade of PI3K/Akt pathway inhibits the production of amyloid beta and its abnormal intracellular accumulation [118]. Insulin also competes with amyloid beta for the same catabolic enzyme (insulin-degrading enzyme IDE), a metal-protease that degrades both the substances. In insulin resistance states, this enzyme can be inhibited with subsequent impairment of the degradation of amyloid beta and this process enhances its neurotoxicity [119]. Insulin and IGF-1 may also modulate the phosphorylation of tau protein, but not all authors reached the same conclusions. Some have shown that treatment with insulin can reduce the phosphorylation of this protein [120], whereas other authors reported that exposure to insulin by some human brain cells can lead to the hyperphosphorylation of tau protein [121].

The decline of brain insulin action is not found only in pathological conditions (eg, diabetes and AD) but also in a physiological condition and it can be demonstrated in aging.

Aging is associated with changes in the metabolism of insulin, reduced expression of IRs and IR mRNA, reduced accessibility of insulin to the brain, lower affinity of insulin to its binding sites, and, finally, changes in intracellular signaling pathways [122,123].

A growing body of evidence suggests that aging is associated with a state of chronic low-grade inflammation, in the same way as occurs in obesity and during high fat diet (HFD) [124].

Following a diet rich in fatty acids or as a result of obesity-mediated release of pro-inflammatory cytokines, activation of the inflammatory cascade occurs in the hypothalamus [125].

During normal aging, the basal hypothalamic level of activated NF-kB increases gradually leading to a pro-inflammatory state [126].

Hypothalamic inflammation is expressed mainly through the activation of inflammatory kinases: c-Jun N-terminal kinase (JNK) and nuclear factor-kB (NF-kB) [127]. Furthermore, obesity, HFD and the activation of NF-kB pathways, can lead to endoplasmic reticulum (ER) stress that further activates NF-kB so that a vicious circle occurs [128]. The inflammatory cascade directly determines insulin resistance, mainly by interfering with the phosphorylation of second messengers after IR activation [96]. For example, the activation of JNK inhibits the phosphorylation of IRS and thus ultimately reduces the action of the insulin signal [129].

The inflammatory condition is further exacerbated in the brains of individuals with AD; a vicious circle is created as inflammation facilitates the formation and deposition of amyloid beta [130], and, on the other hand, amyloid beta stimulates and perpetuates the inflammation of the brain [131].

PREVENTING GLUCOSE METABOLISM ALTERATION

Over recent years, improvements in standards of living and health care have resulted in a significant increase in life expectancy and of frailty, defined as an increased vulnerability to stress in old age. The Cardiovascular Health Study Collaborative Research Group has provided the criteria for defining frailty, including a decline in strength, unintentional weight loss, fatigue, low physical activity level, and slow walking speed. Frail older adults have high risk for frequent falls, hospitalizations, disability, and death [132,133].

As previous described, the age-related changes in body composition include a body fat redistribution with increase of intraabdominal and intrahepatic fat, and a progressive loss of skeletal muscle mass and muscle strength, known as sarcopenia. The prevalence of sarcopenia is estimated to range from 6% to 26% depending on age, sex, and measurement of muscle mass; in persons aged 80 and older the prevalence increases to over 50% [134,135]. Sarcopenia explains the functional decline in older adults and recognizes a multifactorial etiology: lifestyle factors (smoking, alcohol consumption, limited physical activity, and sun exposure), socioeconomic status (income, education, and occupation), nutritional habits (reduction of protein intake), altered muscle structure (increased fatty infiltration in skeletal muscle and myostatin expression, impaired muscle metabolism, mitochondrial dysfunction with impaired sensitivity to insulin), inflammatory status, hormonal changes (lower secretion of growth hormone and testosterone), and chronic diseases [136–143].

On the other hand, longevity is the result of numerous factors, including genetic, environmental, and medical factors (preventive and curative medicine), but it is also recognized that there is a stochastic component [144]. These determinants are still under investigation, but it is clear that protective behaviors may result in exceptional longevity: abstaining from smoking and alcohol consumption, constant and moderate physical activity, social engagement, and a diet rich in vegetables, fruits, and whole grains.

Older adults should engage in regular physical activity and avoid an inactive lifestyle in order to increase energy expenditure and attenuate the risk for developing and progression of most chronic diseases. Unfortunately, the arthritic disorders, and heart and pulmonary failure contribute to a sedentary life with advancing age [145]. A study of approximately 13,000 elderly people (median age at baseline of 74 years) followed up for 28 years, has showed in those who have spent any amount of time in physical activity a 15–35% lower risk of mortality than those who have spent no time in physical activity [146]. Similar results have been shown by the longitudinal Survey in Europe on Nutrition and the Elderly: a Concerted Action (SENECA) Study and by the Italian Silver Network Home Care project [147,148]. Physical activity is associated with beneficial metabolic changes: reduced body weight and fat mass, decrease in blood concentrations of insulin, proinsulin, insulin-like growth factors, glucose and lipids, increased energy expenditure, changes in sex hormone levels, and immune function. The improvement in metabolic profile reduces consequently the prevalence of cardiovascular disease and cancer [149].

Regular exercise is the only strategy found to prevent frailty, improve sarcopenia and physical function in older adults by guaranteeing the maintenance of muscle strength, postural balance, functional independence, and health-related quality of life. Resistance exercise training is more effective in increasing muscle mass and strength, whereas endurance exercise is superior for maintaining and improving maximum aerobic power. Recommendations should include a balanced program of both endurance and strength exercises, performed regularly, at least 3 days a week [150].

In the last years, the sarcopenic obesity, born from the confluence of longevity and augmented incidence of obesity, has been recognized as an emerging problem. It has a prevalence of 20% in older adults [151,152]. The combination of both obesity and sarcopenia augments the risk for poor health-related outcomes, disability (including dyskinesia and difficulties of endurance), metabolic comorbidities, and mortality [153–156]. Diabetes, as well as obesity, has been found to accelerate the progression of sarcopenia and physical disability in older adults.

The effects of weight loss in older adults remains controversial because of the concomitant loss in lean body mass and exacerbation of sarcopenia [157]. Several studies have focused on the association between body mass index (BMI) in the elderly and mortality, with contrasting results. A collaborative analysis of 57 prospective studies which included almost 900,000 adults, has shown that each 5-point increase in BMI over 25 kg/m² is associated with an increase in mortality of 30% in patients above 70 years and of 15% in those above 80 years [158]. Among 13 prospective studies conducted in people aged ≥65 years during a follow-up of 3–23 years, none of them has recognized in mild-to-moderate overweight (BMI range 20–27) a risk factor for all-cause mortality. Few of them have described an association between BMI over 27 kg/m² and mortality, but the same association seems to disappear in adults above 75 years [159]. The reason for this contrasting result is not clear, the high BMI may reflect an increase of total (and abdominal) fat mass with adverse effects or, on the contrary, may reflect a higher lean mass, a nutritional reserve in periods of illness and disease [160].

It has also been recognized that a low BMI, especially below 20 kg/m², is associated with an increased mortality among elderly people, probably due to loss of peripheral and respiratory muscle mass and to coexisting illness [158,160].

Some studies have examined the role of physical activity in obese older adults. Davidson et al. have randomized abdominally obese older patients in 4 groups: resistance exercise, aerobic exercise, resistance and aerobic exercise (combined exercise), or nonexercise control. After 6 months, significant improvements in functional limitations have been found in all groups compared with controls and amelioration of insulin resistance in the aerobic exercise and the combined exercise group but not in the resistance exercise group. Thus the combination of resistance and aerobic exercise appears to be the optimal exercise strategy for simultaneous reduction in insulin resistance and functional limitation in this kind of patients [161]. However, Manini et al. have found that one year of moderate intensity physical activity intervention involving aerobic, strength, balance, and flexibility exercise (150 min per week) may improve physical function in older adults, but the positive benefits are attenuated with obesity [162].

Other authors have illustrated the superior benefits of a combined treatment program, including both diet and exercise. Villareal et al. have studied 107 obese adults, 65 years of age or older, randomly assigned to a control group, a weight-management (diet) group, an exercise group, or a weight-management-plus-exercise (diet-exercise) group. Body weight decreased by 10% in the diet group and by 9% in the diet-exercise group, but did not decrease in the exercise group or the control group. Lean body mass and bone mineral density at the hip decreased less in the diet-exercise group than in the diet group; furthermore, strength, balance, and gait improved consistently in the same group [163]. Similarly, Santanasto and colleagues have examined 36 overweight to moderately obese sedentary older adults for 6 months. They were randomized into either a physical activity plus weight loss (both aerobic and resistance exercise training and a healthy-eating weight loss plan based on the Diabetes Prevention Program), or physical activity plus successful aging health education programs. In the weight loss group, the loss of muscle fat infiltration, thigh fat, muscle area, and strength were greater, but thigh fat area decreased 6-fold in comparison to lean area. Significant improvements on the SPPB score (Short Physical Performance Battery) were achieved in this group probably due to an inverse correlation between the score and change in fat [164].

Regarding older adults with diabetes, only a few studies have specifically analyzed the effects of exercise on physical function. Allet et al. have shown that in diabetic participants in physiotherapeutic training group after 12 weeks there was a significant improvement in gait speed, balance, muscle strength, and joint mobility compared to participants in the no-treatment control group [165]. Geirsdottir et al. have investigated muscle mass and physical function before and after a resistance exercise program of 12 weeks in participants with prediabetes or type 2 diabetes mellitus compared with healthy controls. All groups experienced significant improvement in muscle strength and function. Glucose and triacylglycerol improved significantly in the healthy group, glucose level alone improved significantly in the prediabetic group, whereas in patients with T2DM no metabolic parameters changed significantly [166]. Furthermore, a balance exercise program has been found able to improve balance and trunk proprioception in individuals with diabetic neuropathy [167].

Physical activity seems also able to reduce the levels of age-related inflammatory markers, probably due to a greater decrease in visceral fat. In this respect, the Cardiovascular Health Study (CHS), the MacArthur Studies of Successful Aging and the InCHIANTI study have shown a beneficial association between inflammatory markers and physical activity [168–170].

NUTRITION AND AGING

In recent decades, great advances have been made in research on aging processes and molecular pathways involved in aging and age-related diseases: cardiovascular disease, type 2 diabetes, and Alzheimer's disease. These findings can be applied to the field of nutrition: some foods seem to accelerate and facilitate the processes of aging, while other foods can have the opposite effect, slowing malicious processes and reducing the incidence of age-related chronic diseases.

The two pathways most studied and most involved in the aging process are: mTOR (mammalian target of rapamycin) [71] and ISS (insulin/insulin-like growth factor signaling) [171]. Both are nutrient-sensing pathways: they can be triggered by nutrient in the diet; carbohydrates trigger especially the ISS pathways [171], while amino acids trigger primarily the mTOR pathways [172]. ISS is activated by the binding of the transmembrane receptor for insulin and IGF-1 with its ligand, and promotes the growth and proliferation of cells; mTOR is an intracellular protein and is a potent protein translation.

A large amount of data on experimental animals suggests that the loss or reduction of the signal ISS and mTOR may increase lifespan. This also seems to be true in humans: polymorphisms of IGF1 pathways give a benefit in terms of health and lifespan [173], centenarian Ashkenazi Jewish women have mutations in the receptor for IGF1 [174], in Laron dwarfism there is a reduction in the incidence of cancer and diabetes, in spite of the frequent incidence of overweight [175].

Therefore, some foods may continuously stimulate the ISS and mTOR systems; this would reduce the lifespan through several mechanisms: reduced clearance of protein aggregates, increased protein aggregation and proteotoxicity, inflammation, reduced expression of antioxidant proteins [176], atherosclerosis, neurodegenerative diseases, cancer, osteoporosis, and other age-related diseases. Aging could be interpreted as a process in which cells are bombarded by stimuli to growth, hyperactivity, and replication [177]; these pulses can also be obtained from food and excesses of certain nutrients: a diet low in these harmful foods could slow the aging process.

Amino acids activate the mTOR pathways. Therefore, a diet with avoidance of an overload of amino acids can be beneficial [178]. Animal data support this thesis. In humans, it was shown that a diet rich in red meat can have deleterious effects on health and it may increase mortality [179]. A diet rich in animal protein increases the likelihood of cancer, type 2 diabetes, and other age-related diseases [180]. Conversely the proteins of plant origin do not have the same effect [180]. The explanation could be that the plant protein does not contain methionine and sulfur-rich amino acids. All this supports the recommendation to replace red meat with vegetable proteins such as beans, nuts, and tofu [181]. However, it is important to prevent sarcopenia in the elderly; many authors suggest a good protein intake in old age. In fact, the reduction in protein intake leads to a reduction of muscle mass and body weight, but it does not seem to reduce the duration of life. Moreover, the substances that naturally inhibit mTOR pathways can be present in foods: ethylxantine in coffee [182], epigallocatechin gallate (EGCG) in green tea [183], quercetin in fruit, vegetables, and spices [184], and resveratrol in red grape skin [185]. Other substances such as curcumin and extracts of cinnamon may reduce the effects of the activation of mTOR: they can reduce the formation of beta amyloid and tau protein aggregation (two main features of Alzheimer's disease) [186].

The other way related to the aging process is ISS pathways: this way is activated in particular from carbohydrates. In fact, studies show that a diet with a high glycemic index and high glycemic load, is correlated with increased risk of type 2 diabetes, stroke, cardiovascular disease, and other age-related diseases [187]. A diet high in carbohydrates with a high glycemic index may have harmful effects through other mechanisms, such as the formation of advanced glycation end products (AGEs) [188].

On the contrary, some micronutrients may have a protective effect: anthocyanidis content in blueberries increase insulin sensitivity [189], extracts of cinnamon reduce glycated hemoglobin and fasting glucose in subjects with type 2 diabetes mellitus [190], the cacao increases insulin sensitivity and reduces the risk of cardiovascular disease [191].

Caloric restriction also appears to be useful, regardless of which macronutrients are recruited. Both caloric restriction and fasting reduces drastically the ISS and mTOR signaling in many animal species [69]. In humans, long-term caloric restriction leads to an improvement of metabolic parameters and reduction of atherosclerotic processes [192]. Furthermore, intermittent fasting (fasting every other day or fasting for several days per week) seems able to reduce LDL cholesterol levels and increase insulin sensitivity [193]. Fasting could also be useful in treating cancer [194].

Among the many diets available, there are some that meet the characteristics required from the anti-aging diet. In particular the Mediterranean diet. Characteristics of this diet are: abundance of plant foods (fruits, vegetables, bread, cereals, legumes, nuts, and seeds), fresh, and seasonal; olive oil is the principal source of fat; dairy products consumed in low/moderate amount; eggs consumed with moderation, red meat consumed rarely, red wine in moderate amounts.

The Mediterranean diet has been shown to be able to reduce major cardiovascular risk factors: it reduces the levels of blood pressure, serum cholesterol, and blood sugar; it also appeared to reduce the levels of inflammation in diabetics [195] and in patients with metabolic syndrome [196].

The Mediterranean diet is also effective in the elderly: in the PREDIMED study, in just three months on the diet, the elderly participants showed an improvement in the lipid profile [197] and a reduction of the blood concentration of pro-inflammatory molecules [198].

Finally, the Mediterranean diet seems to be also associated with a lower incidence of depression [199], Alzheimer's disease [200], and other types of cognitive decline in the elderly [201].

CONCLUSIONS

Glucose metabolism modifications in the elderly are caused by a combination of genetic and environmental factors acting together with age-related changes in carbohydrate handling. At the pancreatic level, there are functional rather than anatomical changes regarding beta cell's ability to produce and secrete insulin that characterize the aging process. On the other side, insulin action is impaired at muscular level while its hepatic action is not diminished. The increase in peripheral insulin resistance can be attributable to environmental-driven processes, such as those leading to modification in body mass composition. The increase in intraabdominal fat and in intramuscular fat and the contemporary reduction in muscle mass decrease both insulin and noninsulin-mediated glucose disposal. The accumulation of fat accounts for the development of a low-grade chronic inflammation in aging people and subsequent insulin resistance; different pathways link inflammation and glucose metabolism impairment. Among them, those leading to an increase of oxidative stress can affect energy handling and cause mitochondrial damage. Change in lifestyle via combined treatment programs addressing both dietary habits and physical activity levels can contrast and slow the development of visceral obesity and sarcopenia thus ameliorating carbohydrate metabolism. Altered glucose metabolism and impaired insulin action are linked to brain aging. There is evidence that in the brain insulin is a key hormone for both cell survival and normal function which opens the door to new scenarios for the prevention of cognitive decline and dementia in aging. Finally, dietary components can influence aging processes acting on glucose-related pathways so that a healthy diet can favor a healthy aging.

SUMMARY

1. Abnormal glucose metabolism is a frequent but not a necessary component of aging and genetically predisposed individuals develop reduced glucose tolerance and diabetes.
2. Changes in beta and alpha-cell mass do not explain the reduction in glucose tolerance, thus functional rather than anatomic changes account for glucose metabolism disturbances in the elderly.
3. The most relevant modification in glucose metabolism are attributable to a decline in insulin secretion, disturbances in insulin release, and a reduction of peripheral insulin action.
4. Hepatic insulin sensitivity seems not to be reduced in aging while it is impaired at muscle and fat tissues level.
5. Insulin resistance is possibly driven by the increased low grade inflammation and oxidative damage, central to these processes are the mTOR and the insulin/IGF1-like signaling pathways.
6. Insulin plays a central role in CNS function and its action is involved at different levels in brain function and in neuroprotection, both in diabetic and nondiabetic individuals.
7. Impaired insulin signaling and reduced glucose transport in the brain seem to play a key role in the development of Alzheimer's disease.
8. Age-related changes in body composition, such as increase in intraabdominal, intrahepatic, and intramuscular fat, are common features of aging as well as a reduction in muscle mass and strength (sarcopenia).
9. Age-related modification of body composition are mainly due to incorrect lifestyle and represent main drivers for insulin resistance development.
10. Combined treatment programs addressing both dietary habits and sedentarity are effective in reversing glucose metabolism alterations due to body mass composition modifications.
11. Dietary components can modulate mTOR and the insulin/IGF1-like signaling pathways.

References

[1] Centers for Disease Control and Prevention. National diabetes statistics report: estimates of diabetes and its burden in the United States, 2014. Atlanta (GA): U.S. Department of Health and Human Services; 2014.

[2] Jacob KD, Noren Hooten N, Trzeciak AR, Evans MK. Markers of oxidant stress that are clinically relevant in aging and age-related disease. Mech Ageing Dev 2013;134:139–57.

[3] Lombard DB, Chua KF, Mostoslavsky R, Franco S, Gostissa M, Alt FW. DNA repair, genome stability, and aging. Cell 2005;120(4):497–512.

[4] Haus JM, Carrithers JA, Trappe SW, Trappe TA. Collagen, cross-linking, and advanced glycation end products in aging human skeletal muscle. J Appl Physiol 2007;103:2068–76.

[5] Paneni F, Costantino S, Cosentino F. Molecular pathways of arterial aging. Clin Sci (Lond) 2015;128(2):69–79.

[6] Umegaki H, Hayashi T, Nomura H, Yanagawa M, Nonogaki Z, Nakshima H, et al. Cognitive dysfunction: an emerging concept of a new diabetic complication in the elderly. Geriatr Gerontol Int 2013;13(1):28–34.

[7] Dash SK. Cognitive impairment and diabetes. Recent Pat Endocr Metab Immune Drug Discov 2013;7(2):155–65.

[8] Araki A, Ito H. Diabetes mellitus and geriatric syndromes. Geriatr Gerontol Int 2009;9(2):105–14.

[9] Vaag A, Henriksen JE, Madsbad S, Holm N, Beck-Nielsen H. Insulin secretion, insulin action, and hepatic glucose production in identical twins discordant for non-insulin-dependent diabetes mellitus. J Clin Invest 1995;95(2):690–8.

[10] Helgason A, Pálsson S, Thorleifsson G, Grant SF, Emilsson V, Gunnarsdottir S, et al. Refining the impact of TCF7L2 gene variants on type 2 diabetes and adaptive evolution. Nat Genet 2007;39:218–25.

[11] Kahn SE, Suvag S, Wright LA, Utzschneider KM. Interactions between genetic background, insulin resistance and β-cell function. Diabetes Obes Metab 2012;14(Suppl. 3):46–56.

[12] Capri M, Salvioli S, Sevini F, Valensin S, Celani L, Monti D, et al. The genetics of human longevity. Ann NY Acad Sci 2006;1067:252–63.

[13] Addison O, Marcus RL, Lastayo PC, Ryan AS. Intermuscular fat: a review of the consequences and causes. Int J Endocrinol 2014;2014:309570.

[14] Shimokata H, Muller DC, Fleg JL, Sorkin J, Ziemba AW, Andres R. Age as independent determinant of glucose tolerance. Diabetes 1991;40(1):44–51.

[15] Campos C. Chronic hyperglycemia and glucose toxicity: pathology and clinical sequelae. Postgrad Med 2012;124(6):90–7.

[16] Sharma RB, Alonso LC. Lipotoxicity in the pancreatic beta cell: not just survival and function, but proliferation as well? Curr Diab Rep 2014;14(6):492.

[17] Maedler K, Schumann DM, Schulthess F, Oberholzer J, Bosco D, Berney T, et al. Aging correlates with decreased beta-cell proliferative capacity and enhanced sensitivity to apoptosis: a potential role for Fas and pancreatic duodenalhomeobox-1. Diabetes 2006;55(9):2455–62.

[18] Chiu KC, Lee NP, Cohan P, Chuang LM. Beta cell function declines with age inglucose tolerant Caucasians. Clin Endocrinol (Oxf) 2000;53(5):569–75.

[19] Butler AE, Janson J, Bonner-Weir S, Ritzel R, Rizza RA, Butler PC. Beta-cell deficit and increased beta-cell apoptosis in humans with type 2 diabetes. Diabetes 2003;52:102–10.

[20] Gunasekaran U, Gannon M. Type 2 diabetes and the aging pancreatic beta cell. Aging (Albany, NY) 2011;3(6):565–75.

[21] Reers C, Erbel S, Esposito I, Schmied B, Büchler MW, Nawroth PP, et al. Impaired islet turnover in human donor pancreata with aging. Eur J Endocrinol 2009;160(2):185–91.

[22] Saisho Y, Butler AE, Manesso E, Elashoff D, Rizza RA, Butler PC. β-cell mass and turnover in humans: effects of obesity and aging. Diabetes Care 2013;36(1):111–17.

[23] Manesso E, Toffolo GM, Saisho Y, Butler AE, Matveyenko AV, Cobelli C, et al. Dynamics of beta-cell turnover: evidence for beta-cell turnover and regeneration from sources of beta-cells other than beta-cell replication in the HIP rat. Am J Physiol Endocrinol Metab 2009;297(2):E323–30.

[24] Kushner JA. The role of aging upon β cell turnover. J Clin Invest 2013;123(3):990–5.

[25] Cnop M, Igoillo-Esteve M, Hughes SJ, Walker JN, Cnop I, Clark A. Longevity of human islet α- and β-cells. Diabetes Obes Metab 2011;13(Suppl. 1):39–46.

[26] Gong Z, Muzumdar RH. Pancreatic function, type 2 diabetes, and metabolism in aging. Int J Endocrinol 2012;2012:320482.

[27] Clark A, Wells CA, Buley ID, Cruickshank JK, Vanhegan RI, Matthews DR, et al. Islet amyloid, increased A-cells, reduced B-cells and exocrine fibrosis: quantitative changes in the pancreas in type 2 diabetes. Diabetes Res 1988;9(4):151–9.

[28] Choi CS, Kim MY, Han K, Lee MS. Assessment of β-cell function in human patients. Islets 2012;4(2):79–83.

[29] De Tata V. Age-related impairment of pancreatic beta-cell function: pathophysiological and cellular mechanisms. Front Endocrinol (Lausanne) 2014;5:138.

[30] Elahi D, Muller DC, McAloon-Dyke M, Tobin JD, Andres R. The effect of age on insulin response and glucose utilization during four hyperglycemic plateaux. Exp Gerontol 1993;28(4–5):393–409.

[31] Meneilly GS, Veldhuis JD, Elahi D. Disruption of the pulsatile and entropic modes of insulin release during an unvarying glucose stimulus in elderly individuals. J Clin Endocrinol Metab 1999;84(6):1938–43.

[32] Chen M, Bergman RN, Pacini G, Porte Jr. D. Pathogenesis of age-related glucose intolerance in man: insulin resistance and decreased beta-cell function. J Clin Endocrinol Metab 1985;60(1):13–20.

[33] Komada H, Sakaguchi K, Takeda K, Hirota Y, Hashimoto N, Okuno Y, et al. Age-dependent decline in β-cell function assessed by an oral glucose tolerance test-based disposition index. J Diabetes Investig 2011;2(4):293–6.

[34] Basu R, Breda E, Oberg AL, Powell CC, Dalla Man C, Basu A, et al. Mechanisms of the age-associated deterioration in glucose tolerance: contribution of alterations in insulin secretion, action, and clearance. Diabetes 2003;52(7):1738–48.

[35] Scheen AJ, Sturis J, Polonsky KS, Van Cauter E. Alterations in the ultradian oscillations of insulin secretion and plasma glucose in aging. Diabetologia 1996;39(5):564–72.

[36] Meneilly GS, Ryan AS, Veldhuis JD, Elahi D. Increased disorderliness of basal insulin release, attenuated insulin secretory burst mass, and reduced ultradian rhythmicity of insulin secretion in older individuals. J Clin Endocrinol Metab 1997;82(12):4088–93.

[37] Bonadonna RC, Groop LC, Simonson DC, DeFronzo RA. Free fatty acid and glucose metabolism in human aging: evidence for operation of the Randle cycle. Am J Physiol 1994;266(3 Pt 1):E501–9.

[38] Ammon HP, Fahmy A, Mark M, Wahl MA, Youssif N. The effect of glucose on insulin release and ion movements in isolated pancreatic islets of rats in old age. J Physiol 1987;384:347–54.

[39] Røder ME, Schwartz RS, Prigeon RL, Kahn SE. Reduced pancreatic B cell compensation to the insulin resistance of aging: impact on proinsulin and insulin levels. J Clin Endocrinol Metab 2000;85(6):2275–80.

[40] Korosi J, McIntosh CH, Pederson RA, Demuth HU, Habener JF, Gingerich R, et al. Effect of aging and diabetes on the enteroinsular axis. J Gerontol A Biol Sci Med Sci 2001;56(9):M575–9.

[41] Meneilly GS, Demuth HU, McIntosh CH, Pederson RA. Effect of ageing and diabetes on glucose-dependent insulinotropic polypeptide and dipeptidyl peptidase IV responses to oral glucose. Diabet Med 2000;17(5):346–50.

[42] Ahrén B, Pacini G. Age-related reduction in glucose elimination is accompanied by reduced glucose effectiveness and increased hepatic insulin extraction in man. J Clin Endocrinol Metab 1998;83(9):3350–6.

[43] Finucane FM, Sharp SJ, Hatunic M, Sleigh A, De Lucia Rolfe E, Aihie Sayer A, et al. Liver fat accumulation is associated with reduced hepatic insulin extraction and beta cell dysfunction in healthy older individuals. Diabetol Metab Syndr 2014;6(1):43.

[44] Michalakis K, Goulis DG, Vazaiou A, Mintziori G, Polymeris A, Abrahamian-Michalakis A. Obesity in the ageing man. Metabolism 2013;62(10):1341–9.

[45] Simonson DC, DeFronzo RA. Glucagon physiology and aging: evidence for enhanced hepatic sensitivity. Diabetologia 1983;25(1):1–7.

[46] Johannsen DL, Ravussin E. Obesity in the elderly: is faulty metabolism to blame? Aging health 2010;6(2):159–67.

[47] Kuk JL, Saunders TJ, Davidson LE, Ross R. Age-related changes in total and regional fat distribution. Ageing Res Rev 2009;8(4):339–48.

[48] Meneilly GS. In: Sinclair AJ, editor. Diabetes in old age. 3rd ed. John Wiley & Sons, Ltd; 2009.

[49] Dela F, Helge JW. Insulin resistance and mitochondrial function in skeletal muscle. Int J Biochem Cell Biol 2013;45(1):11–15.

[50] Harman D. Aging: a theory based on free radical and radiation chemistry. J Gerontol 1956;11(3):298–300.

[51] Harman D. The biologic clock: the mitochondria? J Am Geriatr Soc 1972;20(4):145–7.

[52] Pak JW, Herbst A, Bua E, Gokey N, McKenzie D, Aiken JM. Mitochondrial DNA mutations as a fundamental mechanism in physiological declines associated with aging. Aging Cell 2003;2(1):1–7.

[53] Petersen KF, Befroy D, Dufour S, Dziura J, Ariyan C, Rothman DL, et al. Mitochondrial dysfunction in the elderly: possible role in insulin resistance. Science 2003;300(5622):1140–2.

[54] Liesa M, Shirihai OS. Mitochondrial dynamics in the regulation of nutrient utilization and energy expenditure. Cell Metab 2013;17(4):491–506.

[55] Higaki Y, Mikami T, Fujii N, Hirshman MF, Koyama K, Seino T, et al. Oxidative stress stimulates skeletal muscle glucose uptake through a phosphatidylinositol 3-kinase-dependent pathway. Am J Physiol Endocrinol Metab 2008;294(5):E889–97.

[56] May JM, de Haën C. The insulin-like effect of hydrogen peroxide on pathways of lipid synthesis in rat adipocytes. J Biol Chem 1979;254(18):9017–21.

[57] Powell DJ, Hajduch E, Kular G, Hundal HS. Ceramide disables 3-phosphoinositide binding to the pleckstrin homology domain of protein kinase B (PKB)/Akt by a PKCzeta-dependent mechanism. Mol Cell Biol 2003;23(21):7794–808.

[58] Kelley DE, He J, Menshikova EV, Ritov VB. Dysfunction of mitochondria in human skeletal muscle in type 2 diabetes. Diabetes 2002;51(10):2944–50.

[59] Shulman GI. Cellular mechanisms of insulin resistance. J Clin Invest 2000;106(2):171–6.

[60] Davì G, Ciabattoni G, Consoli A, Mezzetti A, Falco A, Santarone S, et al. In vivo formation of 8-iso-prostaglandin f2alpha and platelet activation in diabetes mellitus: effects of improved metabolic control and vitamin E supplementation. Circulation 1999;99(2):224–9.

[61] Monnier L, Mas E, Ginet C, Michel F, Villon L, Cristol JP, et al. Activation of oxidative stress by acute glucose fluctuations compared with sustained chronic hyperglycemia in patients with type 2 diabetes. J Am Med Assoc 2006;295(14):1681–7.

[62] Passos JF, Saretzki G, Ahmed S, Nelson G, Richter T, Peters H, et al. Mitochondrial dysfunction accounts for the stochastic heterogeneity in telomere-dependent senescence. PLoS Biol 2007;5(5):e110.

[63] Singh T, Newman AB. Inflammatory markers in population studies of aging. Ageing Res Rev 2011;10(3):319–29.

[64] Kapahi P, Chen D, Rogers AN, Katewa SD, Li PW, Thomas EL, et al. With TOR, less is more: a key role for the conserved nutrient-sensing TOR pathway in aging. Cell Metab 2010;11(6):453–65.

[65] Blagosklonny MV. Calorie restriction: decelerating mTOR-driven aging from cells to organisms (including humans). Cell Cycle 2010;9(4):683–8.

[66] Hara K, Maruki Y, Long X, Yoshino K, Oshiro N, Hidayat S, et al. Raptor, a binding partner of target of rapamycin (TOR), mediates TOR action. Cell 2002;110(2):177–89.

[67] Sarbassov DD, Ali SM, Kim DH, Guertin DA, Latek RR, Erdjument-Bromage H, et al. Rictor, a novel binding partner of mTOR, defines a rapamycin-insensitive and raptor-independent pathway that regulates the cytoskeleton. Curr Biol 2004;14(14):1296–302.

[68] Shimobayashi M, Hall MN. Making new contacts: the mTOR network in metabolism and signalling crosstalk. Nat Rev Mol Cell Biol 2014;15(3):155–62.

[69] Fontana L, Partridge L, Longo VD. Extending healthy life span – from yeast to humans. Science 2010;328(5976):321–6.

[70] Colman RJ, Anderson RM, Johnson SC, Kastman EK, Kosmatka KJ, Beasley TM, et al. Caloric restriction delays disease onset and mortality in rhesus monkeys. Science 2009;325(5937):201–4.

[71] Johnson SC, Rabinovitch PS, Kaeberlein M. mTOR is a key modulator of ageing and age-related disease. Nature 2013;493(7432):338–45.

[72] Hansen M, Chandra A, Mitic LL, Onken B, Driscoll M, Kenyon C. A role for autophagy in the extension of lifespan by dietary restriction in C. elegans. PLoS Genet 2008;4(2):e24.

[73] Alvers AL, Wood MS, Hu D, Kaywell AC, Dunn Jr WA, Aris JP. Autophagy is required for extension of yeast chronological life span by rapamycin. Autophagy 2009;5(6):847–9.

[74] Mizushima N, Levine B, Cuervo AM, Klionsky DJ. Autophagy fights disease through cellular self-digestion. Nature 2008;451(7182):1069–75.

[75] Pan Y, Schroeder EA, Ocampo A, Barrientos A, Shadel GS. Regulation of yeast chronological life span by TORC1 via adaptive mitochondrial ROS signaling. Cell Metab 2011;13(6):668–78.

[76] Polak P, Cybulski N, Feige JN, Auwerx J, Rüegg MA, Hall MN. Adipose-specific knockout of raptor results in lean mice with enhanced mitochondrial respiration. Cell Metab 2008;8(5):399–410.

[77] Blagosklonny MV. TOR-centric view on insulin resistance and diabetic complications: perspective for endocrinologists and gerontologists. Cell Death Dis 2013;4:e964.

[78] Um SH, Frigerio F, Watanabe M, Picard F, Joaquin M, Sticker M, et al. Absence of S6K1 protects against age-and diet-induced obesity while enhancing insulin sensitivity. Nature 2004;431(7005):200–5 Erratum in: Nature. 2004;431(7007):485.

[79] Takano A, Usui I, Haruta T, Kawahara J, Uno T, Iwata M, et al. Mammalian target of rapamycin pathway regulates insulin signaling via subcellular redistribution of insulin receptor substrate 1 and integrates nutritional signals and metabolic signals of insulin. Mol Cell Biol 2001;21(15):5050–62.

[80] Newgard CB, An J, Bain JR, Muehlbauer MJ, Stevens RD, Lien LF, et al. A branched-chain amino acid-related metabolic signature that differentiates obese and lean humans and contributes to insulin resistance. Cell Metab 2009;9(4):311–26 Erratum in: Cell Metab. 2009;9(6):565–66.

[81] Dowling RJ, Zakikhani M, Fantus IG, Pollak M, Sonenberg N. Metformin inhibits mammalian target of rapamycin-dependent translation initiation in breast cancer cells. Cancer Res 2007;67(22):10804–12.

[82] Lamming DW, Ye L, Sabatini DM, Baur JA. Rapalogs and mTOR inhibitors as anti-aging therapeutics. J Clin Invest 2013;123(3):980–9.

[83] Zhao W, Chen H, Xu H, Moore E, Meiri N, Quon MJ, et al. Brain insulin receptors and spatial memory. Correlated changes in gene expression, tyrosine phosphorylation, and signaling molecules in the hippocampus of water maze trained rats. J Biol Chem 1999;274(49):34893–902.

[84] Salkovic-Petrisic M, Hoyer S. Central insulin resistance as a trigger for sporadic Alzheimer-like pathology: an experimental approach. J Neural Transm Suppl 2007;72:217–33.

[85] Conejo R, Lorenzo M. Insulin signaling leading to proliferation, survival, and membrane ruffling in C2C12 myoblasts. J Cell Physiol 2001;187(1):96–108.

[86] van der Heide LP, Ramakers GM, Smidt MP. Insulin signaling in the central nervous system: learning to survive. Prog Neurobiol 2006;79(4):205–21.

[87] Erol A. An integrated and unifying hypothesis for the metabolic basis of sporadic Alzheimer's disease. J Alzheimers Dis 2008;13(3):241–53.

[88] Duarte AI, Santos P, Oliveira CR, Santos MS, Rego AC. Insulin neuroprotection against oxidative stress is mediated by Akt and GSK-3beta signaling pathways and changes in protein expression. Biochim Biophys Acta 2008;1783(6):994–1002.

[89] Rojo AI, Salinas M, Martín D, Perona R, Cuadrado A. Regulation of Cu/Zn-superoxide dismutase expression via the phosphatidylinositol 3 kinase/Akt pathway and nuclear factor-kappaB. J Neurosci 2004;24(33):7324–34.

[90] Li F, Szabó C, Pacher P, Southan GJ, Abatan OI, Charniauskaya T, et al. Evaluation of orally active poly(ADP-ribose) polymerase inhibitor in streptozotocin-diabetic rat model of early peripheral neuropathy. Diabetologia 2004;47(4):710–17.

[91] Huang TJ, Verkhratsky A, Fernyhough P. Insulin enhances mitochondrial inner membrane potential and increases ATP levels through phosphoinositide 3-kinase in adult sensory neurons. Mol Cell Neurosci 2005;28(1):42–54.

[92] Groop PH, Forsblom C, Thomas MC. Mechanisms of disease: Pathway-selective insulin resistance and microvascular complications of diabetes. Nat Clin Pract Endocrinol Metab 2005;1(2):100–10.

[93] Purves T, Middlemas A, Agthong S, Jude EB, Boulton AJ, Fernyhough P, et al. A role for mitogen-activated protein kinases in the etiology of diabetic neuropathy. FASEB J 2001;15(13):2508–14.

[94] Haddad JJ. N-methyl-D-aspartate (NMDA) and the regulation of mitogen-activated protein kinase (MAPK) signaling pathways: a revolving neurochemical axis for therapeutic intervention? Prog Neurobiol 2005;77(4):252–82.

[95] Porte Jr D, Baskin DG, Schwartz MW. Insulin signaling in the central nervous system: a critical role in metabolic homeostasis and disease from C. elegans to humans. Diabetes 2005;54(5):1264–76.

[96] Vogt MC, Brüning JC. CNS insulin signaling in the control of energy homeostasis and glucose metabolism – from embryo to old age. Trends Endocrinol Metab 2013;24(2):76–84.

[97] Wada A, Yokoo H, Yanagita T, Kobayashi H. New twist on neuronal insulin receptor signaling in health, disease, and therapeutics. J Pharmacol Sci 2005;99(2):128–43.

[98] Girard J. The inhibitory effects of insulin on hepatic glucose production are both direct and indirect. Diabetes 2006;55(Suppl. 2):S65–9.

[99] Benedict C, Hallschmid M, Hatke A, Schultes B, Fehm HL, Born J, et al. Intranasal insulin improves memory in humans. Psychoneuroendocrinology 2004;29(10):1326–34.

[100] Sauter A, Goldstein M, Engel J, Ueta K. Effect of insulin on central catecholamines. Brain Res 1983;260(2):330–3.

[101] Raizada MK, Shemer J, Judkins JH, Clarke DW, Masters BA, LeRoith D. Insulin receptors in the brain: structural and physiological characterization. Neurochem Res 1988;13(4):297–303.

[102] Skeberdis VA, Lan J, Zheng X, Zukin RS, Bennett MV. Insulin promotes rapid delivery of N-methyl-D-aspartate receptors to the cell surface by exocytosis. Proc Natl Acad Sci USA 2001;98(6):3561–6.

[103] Duarte AI, Santos MS, Oliveira CR, Rego AC. Insulin neuroprotection against oxidative stress in cortical neurons – involvement of uric acid and glutathione antioxidant defenses. Free Radic Biol Med 2005;39(7):876–89.

[104] Piroli GG, Grillo CA, Charron MJ, McEwen BS, Reagan LP. Biphasic effects of stress upon GLUT8 glucose transporter expression and trafficking in the diabetic rat hippocampus. Brain Res 2004;1006(1):28–35.

[105] Moreira PI, Santos MS, Moreno AM, Seiça R, Oliveira CR. Increased vulnerability of brain mitochondria in diabetic (Goto-Kakizaki) rats with aging and amyloid-beta exposure. Diabetes 2003;52(6):1449–56.

[106] Mastrocola R, Restivo F, Vercellinatto I, Danni O, Brignardello E, Aragno M, et al. Oxidative and nitrosative stress in brain mitochondria of diabetic rats. J Endocrinol 2005;187(1):37–44.

[107] Li ZG, Zhang W, Sima AA. The role of impaired insulin/IGF action in primary diabetic encephalopathy. Brain Res 2005;1037(1–2):12–24.

[108] Singh P, Jain A, Kaur G. Impact of hypoglycemia and diabetes on CNS: correlation of mitochondrial oxidative stress with DNA damage. Mol Cell Biochem 2004;260(1–2):153–9.

[109] Moreira PI, Duarte AI, Santos MS, Rego AC, Oliveira CR. An integrative view of the role of oxidative stress, mitochondria and insulin in Alzheimer's disease. J Alzheimers Dis 2009;16(4):741–61.

[110] Selkoe DJ. Translating cell biology into therapeutic advances in Alzheimer's disease. Nature 1999;399(6738 Suppl.):A23–31.

[111] Moreira PI, Santos MS, Sena C, Nunes E, Seiça R, Oliveira CR. CoQ10 therapy attenuates amyloid beta-peptide toxicity in brain mitochondria isolated from aged diabetic rats. Exp Neurol 2005;196(1):112–19.

[112] Eckert A, Keil U, Marques CA, Bonert A, Frey C, Schüssel K, et al. Mitochondrial dysfunction, apoptotic cell death, and Alzheimer's disease. Biochem Pharmacol 2003;66(8):1627–34.

[113] Ott A, Stolk RP, van Harskamp F, Pols HA, Hofman A, Breteler MM. Diabetes mellitus and the risk of dementia: the Rotterdam Study. Neurology 1999;53(9):1937–42.

[114] Janson J, Laedtke T, Parisi JE, O'Brien P, Petersen RC, Butler PC. Increased risk of type 2 diabetes in Alzheimer's disease. Diabetes 2004;53(2):474–81.

[115] Rasgon N, Jarvik L. Insulin resistance, affective disorders, and Alzheimer's disease: review and hypothesis. J Gerontol A Biol Sci Med Sci 2004;59(2):178–83 discussion 184–92.

[116] Townsend M, Mehta T, Selkoe DJ. Soluble Abeta inhibits specific signal transduction cascades common to the insulin receptor pathway. J Biol Chem 2007;282(46):33305–12.

[117] Lester-Coll N, Rivera EJ, Soscia SJ, Doiron K, Wands JR, de la Monte SM. Intracerebral streptozotocin model of type 3 diabetes: relevance to sporadic Alzheimer's disease. J Alzheimers Dis 2006;9(1):13–33.

[118] Phiel CJ, Wilson CA, Lee VM, Klein PS. GSK-3alpha regulates production of Alzheimer's disease amyloid-beta peptides. Nature 2003;423(6938):435–9.

[119] Rhein V, Eckert A. Effects of Alzheimer's amyloid-beta and tau protein on mitochondrial function – role of glucose metabolism and insulin signalling. Arch Physiol Biochem 2007;113(3):131–41.

[120] Hong M, Lee VM. Insulin and insulin-like growth factor-1 regulate tau phosphorylation in cultured human neurons. J Biol Chem 1997;272(31):19547–53.

[121] Lesort M, Johnson GV. Insulin-like growth factor-1 and insulin mediate transient site-selective increases in tau phosphorylation in primary cortical neurons. Neuroscience 2000;99(2):305–16.

[122] Fernandes ML, Saad MJ, Velloso LA. Effects of age on elements of insulin-signaling pathway in central nervous system of rats. Endocrine 2001;16(3):227–34.

[123] De Felice FG. Alzheimer's disease and insulin resistance: translating basic science into clinical applications. J Clin Invest 2013;123(2):531–9.

[124] Thaler JP, Guyenet SJ, Dorfman MD, Wisse BE, Schwartz MW. Hypothalamic inflammation: marker or mechanism of obesity pathogenesis? Diabetes 2013;62(8):2629–34. Available from: http://dx.doi.org/10.2337/db12-1605.

[125] Velloso LA, Schwartz MW. Altered hypothalamic function in diet-induced obesity. Int J Obes (Lond) 2011;35(12):1455–65.

[126] Zhang G, Li J, Purkayastha S, Tang Y, Zhang H, Yin Y, et al. Hypothalamic programming of systemic ageing involving IKK-β, NF-κB and GnRH. Nature 2013;497(7448):211–16.

[127] De Souza CT, Araujo EP, Bordin S, Ashimine R, Zollner RL, Boschero AC, et al. Consumption of a fat-rich diet activates a proinflammatory response and induces insulin resistance in the hypothalamus. Endocrinology 2005;146(10):4192–9.

[128] Zhang X, Zhang G, Zhang H, Karin M, Bai H, Cai D. Hypothalamic IKKbeta/NF-kappaB and ER stress link overnutrition to energy imbalance and obesity. Cell 2008;135(1):61–73.

[129] Aguirre V, Uchida T, Yenush L, Davis R, White MF. The c-Jun NH(2)-terminal kinase promotes insulin resistance during association with insulin receptor substrate-1 and phosphorylation of Ser(307). J Biol Chem 2000;275(12):9047–54.

[130] Yamamoto M, Kiyota T, Horiba M, Buescher JL, Walsh SM, Gendelman HE, et al. Interferon-gamma and tumor necrosis factor-alpha regulate amyloid-beta plaque deposition and beta-secretase expression in Swedish mutant APP transgenic mice. Am J Pathol 2007;170(2):680–92.

[131] Akiyama H, Barger S, Barnum S, Bradt B, Bauer J, Cole GM, et al. Inflammation and Alzheimer's disease. Neurobiol Aging 2000;21(3):383–421.

[132] Fried LP, Tangen CM, Walston J, Newman AB, Hirsch C, Gottdiener J, et al. Frailty in older adults: evidence for a phenotype. J Gerontol A Biol Sci Med Sci 2001;56(3):M146–56.

[133] Fried LP, Xue QL, Cappola AR, Ferrucci L, Chaves P, Varadhan R, et al. Nonlinear multisystem physiological dysregulation associated with frailty in older women: implications for etiology and treatment. J Gerontol A Biol Sci Med Sci 2009;64(10):1049–57.

[134] Baumgartner RN, Koehler KM, Gallagher D, Romero L, Heymsfield SB, Ross RR, et al. Epidemiology of sarcopenia among the elderly in New Mexico. Am J Epidemiol 1998;147(8):755–63 Erratum in: Am J Epidemiol 1999;149(12):1161.

[135] Iannuzzi-Sucich M, Prestwood KM, Kenny AM. Prevalence of sarcopenia and predictors of skeletal muscle mass in healthy, older men and women. J Gerontol A Biol Sci Med Sci 2002;57(12):M772–7.

[136] Sarti S, Ruggiero E, Coin A, Toffanello ED, Perissinotto E, Miotto F, et al. Dietary intake and physical performance in healthy elderly women: a 3-year follow-up. Exp Gerontol 2013;48(2):250–4.

[137] Milanović Z, Pantelić S, Trajković N, Sporiš G, Kostić R, James N. Age-related decrease in physical activity and functional fitness among elderly men and women. Clin Interv Aging 2013;8:549–56.

[138] Marcus RL, Addison O, Dibble LE, Foreman KB, Morrell G, Lastayo P. Intramuscular adipose tissue, sarcopenia, and mobility function in older individuals. J Aging Res 2012;2012:629637.

[139] Léger B, Derave W, De Bock K, Hespel P, Russell AP. Human sarcopenia reveals an increase in SOCS-3 and myostatin and a reduced efficiency of Akt phosphorylation. Rejuvenation Res 2008;11(1):163–175B.

[140] Lanza IR, Nair KS. Muscle mitochondrial changes with aging and exercise. Am J Clin Nutr 2009;89(1):467S–71S.

[141] Stenholm S, Rantanen T, Heliövaara M, Koskinen S. The mediating role of C-reactive protein and handgrip strength between obesity and walking limitation. J Am Geriatr Soc 2008;56(3):462–9.

[142] Yeap BB, Paul Chubb SA, Lopez D, Ho KK, Hankey GJ, Flicker L. Associations of insulin-like growth factor-I and its binding proteins and testosterone with frailty in older men. Clin Endocrinol (Oxf) 2013;78(5):752–9.

[143] Eichholzer M, Barbir A, Basaria S, Dobs AS, Feinleib M, Guallar E, et al. Serum sex steroid hormones and frailty in older American men of the Third National Health and Nutrition Examination Survey (NHANES III). Aging Male 2012;15(4):208–15.

[144] Christensen K, Doblhammer G, Rau R, Vaupel JW. Ageing populations: the challenges ahead. Lancet 2009;374(9696):1196–208.

[145] American College of Sports Medicine, Chodzko-Zajko WJ, Proctor DN, Fiatarone Singh MA, Minson CT, Nigg CR, et al. American College of Sport Medicine position stand. Exercise and physical activity for older adults. Med Sci Sports Exerc 2009;41(7):1510–30.

[146] Paganini-Hill A, Kawas CH, Corrada MM. Activities and mortality in the elderly: the Leisure World cohort study. J Gerontol A Biol Sci Med Sci 2011;66(5):559–67.

[147] de Groot LC, Verheijden MW, de Henauw S, Schroll M, van Staveren WA, SENECA Investigators. Lifestyle, nutritional status, health, and mortality in elderly people across Europe: a review of the longitudinal results of the SENECA study. J Gerontol A Biol Sci Med Sci 2004;59(12):1277–84.

[148] Landi F, Cesari M, Onder G, Lattanzio F, Gravina EM, Bernabei R. Physical activity and mortality in frail, community-living elderly patients. J Gerontol A Biol Sci Med Sci 2004;59(8):833–7.

[149] Samitz G, Egger M, Zwahlen M. Domains of physical activity and all-cause mortality: systematic review and dose-response meta-analysis of cohort studies. Int J Epidemiol 2011;40(5):1382–400.

[150] Landi F, Marzetti E, Martone AM, Bernabei R, Onder G. Exercise as a remedy for sarcopenia. Curr Opin Clin Nutr Metab Care 2014;17(1):25–31.

[151] Arterburn DE, Crane PK, Sullivan SD. The coming epidemic of obesity in elderly Americans. J Am Geriatr Soc 2004;52(11):1907–12.

[152] Kim YS, Lee Y, Chung YS, Lee DJ, Joo NS, Hong D, et al. Prevalence of sarcopenia and sarcopenic obesity in the Korean population based on the Fourth Korean National Health and Nutritional Examination Surveys. J Gerontol A Biol Sci Med Sci 2012;67(10):1107–13.

[153] Rolland Y, Lauwers-Cances V, Cristini C, Abellan van Kan G, Janssen I, Morley JE, et al. Difficulties with physical function associated with obesity, sarcopenia, and sarcopenic-obesity in community-dwelling elderly women: the EPIDOS (EPIDemiologie de l' OSteoporose) Study. Am J Clin Nutr 2009;89(6):1895–900.

[154] Baumgartner RN, Wayne SJ, Waters DL, Janssen I, Gallagher D, Morley JE. Sarcopenic obesity predicts instrumental activities of daily living disability in the elderly. Obes Res 2004;12(12):1995–2004.

[155] Vincent HK, Vincent KR, Lamb KM. Obesity and mobility disability in the older adult. Obes Rev 2010;11(8):568–79.

[156] Dominguez LJ, Barbagallo M. The cardiometabolic syndrome and sarcopenic obesity in older persons. J Cardiometab Syndr 2007;2(3):183–9.

[157] Waters DL, Ward AL, Villareal DT. Weight loss in obese adults 65 years and older: a review of the controversy. Exp Gerontol 2013;48(10):1054–61.

[158] Prospective Studies Collaboration, Whitlock G, Lewington S, Sherliker P, Clarke R, Emberson J, et al. Body-mass index and cause-specific mortality in 900 000 adults: collaborative analyses of 57 prospective studies. Lancet 2009;373(9669):1083–96.

[159] Heiat A, Vaccarino V, Krumholz HM. An evidence-based assessment of federal guidelines for overweight and obesity as they apply to elderly persons. Arch Intern Med 2001;161(9):1194–203.

[160] Rizzuto D, Fratiglioni L. Lifestyle factors related to mortality and survival: a mini-review. Gerontology 2014;60(4):327–35.

[161] Davidson LE, Hudson R, Kilpatrick K, Kuk JL, McMillan K, Janiszewski PM, et al. Effects of exercise modality on insulin resistance and functional limitation in older adults: a randomized controlled trial. Arch Intern Med 2009;169(2):122–31.

[162] Manini TM, Newman AB, Fielding R, Blair SN, Perri MG, Anton SD, et al. Effects of exercise on mobility in obese and nonobese older adults. Obesity (Silver Spring) 2010;18(6):1168–75.

[163] Villareal DT, Chode S, Parimi N, Sinacore DR, Hilton T, Armamento-Villareal R, et al. Weight loss, exercise, or both and physical function in obese older adults. N Engl J Med 2011;364(13):1218–29.

[164] Santanasto AJ, Glynn NW, Newman MA, Taylor CA, Brooks MM, Goodpaster BH, et al. Impact of weight loss on physical function with changes in strength, muscle mass, and muscle fat infiltration in overweight to moderately obese older adults: a randomized clinical trial. J Obes 2011;2011.

[165] Allet L, Armand S, de Bie RA, Golay A, Monnin D, Aminian K, et al. The gait and balance of patients with diabetes can be improved: a randomised controlled trial. Diabetologia 2010;53(3):458–66.

[166] Geirsdottir OG, Arnarson A, Briem K, Ramel A, Jonsson PV, Thorsdottir I. Effect of 12-week resistance exercise program on body composition, muscle strength, physical function, and glucose metabolism in healthy, insulin-resistant, and diabetic elderly Icelanders. J Gerontol A Biol Sci Med Sci 2012;67(11):1259–65.

[167] Song CH, Petrofsky JS, Lee SW, Lee KJ, Yim JE. Effects of an exercise program on balance and trunk proprioception in older adults with diabetic neuropathies. Diabetes Technol Ther 2011;13(8):803–11.

[168] Geffken DF, Cushman M, Burke GL, Polak JF, Sakkinen PA, Tracy RP. Association between physical activity and markers of inflammation in a healthy elderly population. Am J Epidemiol 2001;153(3):242–50.

[169] Reuben DB, Judd-Hamilton L, Harris TB, Seeman TE, MacArthur Studies of Successful Aging. The associations between physical activity and inflammatory markers in high-functioning older persons: MacArthur Studies of Successful Aging. J Am Geriatr Soc 2003;51(8):1125–30.

[170] Elosua R, Bartali B, Ordovas JM, Corsi AM, Lauretani F, Ferrucci L, et al. Association between physical activity, physical performance, and inflammatory biomarkers in an elderly population: the InCHIANTI study. J Gerontol A Biol Sci Med Sci 2005;60(6):760–7.

[171] Bartke A, Sun LY, Longo V. Somatotropic signaling: trade-offs between growth, reproductive development, and longevity. Physiol Rev 2013;93(2):571–98.

[172] Wullschleger S, Loewith R, Hall MN. TOR signaling in growth and metabolism. Cell 2006;124(3):471–84.

[173] Pawlikowska L, Hu D, Huntsman S, Sung A, Chu C, Chen J, et al. Association of common genetic variation in the insulin/IGF1 signaling pathway with human longevity. Aging Cell 2009;8(4):460–72.

[174] Suh Y, Atzmon G, Cho MO, Hwang D, Liu B, Leahy DJ, et al. Functionally significant insulin-like growth factor I receptor mutations in centenarians. Proc Natl Acad Sci USA 2008;105(9):3438–42.

[175] Guevara-Aguirre J, Balasubramanian P, Guevara-Aguirre M, Wei M, Madia F, Cheng CW, et al. Growth hormone receptor deficiency is associated with a major reduction in pro-aging signaling, cancer, and diabetes in humans. Sci Transl Med 2011;3(70):70ra13.

[176] Kenyon CJ. The genetics of ageing. Nature 2010;464(7288):504–12 Erratum in: Nature. 2010;467(7315):622.

[177] Gems D, de la Guardia Y. Alternative perspectives on aging in Caenorhabditis elegans: reactive oxygen species or hyperfunction? Antioxid Redox Signal 2013;19(3):321–9.

[178] Efeyan A, Zoncu R, Sabatini DM. Amino acids and mTORC1: from lysosomes to disease. Trends Mol Med 2012;18(9):524–33.

[179] Rohrmann S, Overvad K, Bueno-de-Mesquita HB, Jakobsen MU, Egeberg R, Tjønneland A, et al. Meat consumption and mortality – results from the European Prospective Investigation into Cancer and Nutrition. BMC Med 2013;11:63.

[180] Levine ME, Suarez JA, Brandhorst S, Balasubramanian P, Cheng CW, Madia F, et al. Low protein intake is associated with a major reduction in IGF-1, cancer, and overall mortality in the 65 and younger but not older population. Cell Metab 2014;19(3):407–17.

[181] Pan A, Sun Q, Bernstein AM, Schulze MB, Manson JE, Stampfer MJ, et al. Red meat consumption and mortality: results from 2 prospective cohort studies. Arch Intern Med 2012;172(7):555–63.

[182] Wanke V, Cameroni E, Uotila A, Piccolis M, Urban J, Loewith R, et al. Caffeine extends yeast lifespan by targeting TORC1. Mol Microbiol 2008;69(1):277–85.

[183] Van Aller GS, Carson JD, Tang W, Peng H, Zhao L, Copeland RA, et al. Epigallocatechin gallate (EGCG), a major component of green tea, is a dual phosphoinositide-3-kinase/mTOR inhibitor. Biochem Biophys Res Commun 2011;406(2):194–9.

[184] Bruning A. Inhibition of mTOR signaling by quercetin in cancer treatment and prevention. Anticancer Agents Med Chem 2013;13(7):1025–31.

[185] Liu M, Wilk SA, Wang A, Zhou L, Wang RH, Ogawa W, et al. Resveratrol inhibits mTOR signaling by promoting the interaction between mTOR and DEPTOR. J Biol Chem 2010;285(47):36387–94.

[186] Yang F, Lim GP, Begum AN, Ubeda OJ, Simmons MR, Ambegaokar SS, et al. Curcumin inhibits formation of amyloid beta oligomers and fibrils, binds plaques, and reduces amyloid in vivo. J Biol Chem 2005;280(7):5892–901.

[187] Sieri S, Brighenti F, Agnoli C, Grioni S, Masala G, Bendinelli B, et al. Dietary glycemic load and glycemic index and risk of cerebrovascular disease in the EPICOR cohort. PLoS One 2013;8(5):e62625.

[188] Singh R, Barden A, Mori T, Beilin L. Advanced glycation end-products: a review. Diabetologia 2001;44(2):129–46.

[189] Muraki I, Imamura F, Manson JE, Hu FB, Willett WC, van Dam RM, et al. Fruit consumption and risk of type 2 diabetes: results from three prospective longitudinal cohort studies. Br Med J 2013;347:f5001.

[190] Lu T, Sheng H, Wu J, Cheng Y, Zhu J, Chen Y. Cinnamon extract improves fasting blood glucose and glycosylated hemoglobin level in Chinese patients with type 2 diabetes. Nutr Res 2012;32(6):408–12.

[191] Buitrago-Lopez A, Sanderson J, Johnson L, Warnakula S, Wood A, Di Angelantonio E, et al. Chocolate consumption and cardiometabolic disorders: systematic review and meta-analysis. Br Med J 2011;343:d4488.

[192] Heilbronn LK, de Jonge L, Frisard MI, DeLany JP, Larson-Meyer DE, Rood J, et al. Effect of 6-month calorie restriction on biomarkers of longevity, metabolic adaptation, and oxidative stress in overweight individuals: a randomized controlled trial. J Am Med Assoc 2006;295(13):1539–48 Erratum in: J Am Med Assoc. 2006;295(21):2482.

[193] Varady KA, Bhutani S, Church EC, Klempel MC. Short-term modified alternate-day fasting: a novel dietary strategy for weight loss and cardioprotection in obese adults. Am J Clin Nutr 2009;90(5):1138–43.

[194] Cheng CW, Adams GB, Perin L, Wei M, Zhou X, Lam BS, et al. Prolonged fasting reduces IGF-1/PKA to promote hematopoietic-stem-cell-based regeneration and reverse immunosuppression. Cell Stem Cell 2014;14(6):810–23.

[195] Ciccarone E, Di Castelnuovo A, Salcuni M, Siani A, Giacco A, Donati MB, et al. A high-score Mediterranean dietary pattern is associated with a reduced risk of peripheral arterial disease in Italian patients with type 2 diabetes. J Thromb Haemost 2003;1(8):1744–52.

[196] Alvarez León E, Henríquez P, Serra-Majem L. Mediterranean diet and metabolic syndrome: a cross-sectional study in the Canary Islands. Public Health Nutr 2006;9(8A):1089–98.

[197] Estruch R, Martínez-González MA, Corella D, Salas-Salvadó J, Ruiz-Gutiérrez V, Covas MI, et al. Effects of a Mediterranean-style diet on cardiovascular risk factors: a randomized trial. Ann Intern Med 2006;145(1):1–11.

[198] Fitó M, Guxens M, Corella D, Sáez G, Estruch R, de la Torre R, et al. PREDIMED Study Investigators. Effect of a traditional Mediterranean diet on lipoprotein oxidation: a randomized controlled trial. Arch Intern Med 2007;167(11):1195–203.

[199] Sánchez-Villegas A, Bes-Rastrollo M, Martínez-González MA, Serra-Majem L. Adherence to a Mediterranean dietary pattern and weight gain in a follow-up study: the SUN cohort. Int J Obes (Lond) 2006;30(2):350–8.

[200] Scarmeas N, Stern Y, Tang MX, Mayeux R, Luchsinger JA. Mediterranean diet and risk for Alzheimer's disease. Ann Neurol 2006;59(6):912–21.

[201] Solfrizzi V, Colacicco AM, D'Introno A, Capurso C, Torres F, Rizzo C, et al. Dietary intake of unsaturated fatty acids and age-related cognitive decline: a 8.5-year follow-up of the Italian Longitudinal Study on Aging. Neurobiol Aging 2006;27(11):1694–704.

CHAPTER
30

Nutritional Status in Aging and Lung Disease

R. Antonelli Incalzi[1], N. Scichilone[2], S. Fusco[3] and A. Corsonello[3]

[1]Department of Geriatric Medicine, University Campus Bio-Medico, Rome, Italy [2]Biomedical Department of Internal and Specialist Medicine, Section of Pulmonology, University of Palermo, Palermo, Italy [3]Unit of Geriatric Pharmacoepidemiology, Italian National Research Center on Aging (INRCA), Cosenza, Italy

KEY FACTS

- Lung diseases, though to a variable extent and through different mechanisms, impact the nutritional status.
- Undernutrition may impact health-related quality of life; the association between underweight and increased mortality risk has been well established in patients with lung diseases.
- Pulmonary cachexia is caused by a combination of several physiological and pathophysiological alterations, such as local and systemic inflammation, increased respiratory work, hypoxia, inadequate energy intake, and/or increased energy expenditure.
- For these reasons, nutritional screening is considered as an essential component of integrated lung disease management.
- Nutritional assessment and support seems to be a key component of this approach, but the inherent scientific evidence is too limited to reach firm conclusions and, then, to suggest clear cut dietary interventions.

Dictionary of Terms

- *COPD*: is characterized by airflow limitation that is usually progressive and associated with an enhanced chronic inflammatory response to noxious particles or gas in the airways and the lungs. Exacerbations and comorbidities contribute to the overall severity in individual patients.
- *Asthma*: is a chronic inflammatory disorder of the airways in which many cells and biochemical pathways play a role. The chronic inflammation is associated with airway responsiveness that leads to recurrent episodes of wheezing, breathlessness, chest tightness, and coughing, particularly at night or in the early morning. These episodes are usually associated with widespread, but variable, airflow obstruction within the lung that is often reversible either spontaneously or with treatment.
- *Emphysema*: is a chronic respiratory disease where there is overinflation of the air sacs (alveoli) in the lungs, causing a decrease in lung function, and often, breathlessness. It is a phenotypic variant of COPD.
- *Bronchiectasis*: is a condition in which an area of the bronchial tubes is permanently and abnormally widened, with accompanying infection.
- *Lung cancer*: is a cancer that forms in tissues of the lung, usually in the cells lining air passages. The two main types are small cell lung cancer and nonsmall cell lung cancer.
- *Sarcopenia*: is a syndrome characterized by progressive and generalized loss of skeletal muscle mass and strength with a risk of adverse outcomes such as physical disability, poor quality of life and death.
- *Pulmonary cachexia*: (including anorexia) has been defined as disproportionate cytokine-driven loss of skeletal muscle, which is reflected in the loss of free-fat mass associated with an accelerated decline in functional status and can affect patients with any type of advanced lung disease.

- *Mini nutritional assessment*: is structured into 18 questions grouped in four rubrics (anthropometry, general status, dietary habits, and self-perceived health and nutrition states), the MNA provides a multidimensional assessment of the patients.
- *Bode index*: is a multidimensional scoring system and capacity index used to test patients who have been diagnosed with chronic obstructive pulmonary disease (COPD) and to predict long-term outcomes for them.

INTRODUCTION

Pulmonary diseases are heterogeneous in nature. COPD is the most common chronic respiratory condition in the elderly, but the prevalence of asthma, interstitial lung disease, bronchiectasis, and other conditions in the elderly population is worthy of consideration. All these chronic respiratory disorders, though to a variable extent and through different mechanisms, impact the nutritional status and are variably amenable to nutritional interventions. Most of them are typical age-related diseases: the mean age of COPD patients is 64 ± 9 year [1], but that of patients on long-term oxygen therapy (LTOT) is over 72 years [2], whereas the mean age at presentation of primary lung fibrosis is 66 years [3]. Thus, these disorders are expected to add to the effect of age itself on the nutritional status. Furthermore, COPD may be considered to potentiate the aging process: telomere shortening has been observed in the alveolar, bronchial, and endothelial cells of the COPD lung, which is older than expected based on the biological age [4]. On the other hand, being an umbrella definition, COPD encompasses a variety of conditions variably impacting the nutritional status. Indeed, a hypercatabolic form of COPD seems to account for loss of free fat mass and ensuing sarcopenia in a minority of patients [5], but deconditioning per se frequently accounts for decreased muscle mass. Selected nutritional deficits are also common in COPD patients and have the potential for affecting the health status and the course of the disease. For instance, vitamin D serum level has been positively associated with lung volumes in the third National Health and Nutrition Examination Survey [6] and is inversely related to lung function decline in smokers [7], while low serum levels are common in the elderly with respiratory diseases [8]. Furthermore, important regional variations in nutritional status and nutrient intake have been reported, with Mediterranean countries experiencing a lower prevalence of sarcopenia and a better nutrient intake [9]. Thus, nutritional status eventually results form a complex interplay among the disease status, social conditions and aging itself. Furthermore, the onset of the respiratory insufficiency, that is, of hypoxemia, dramatically impacts the nutritional status, whichever is the baseline respiratory condition, but LTOT, if correctly tailored to the individual needs, can to some extent smooth this effect [10]. Unfortunately, complications of chronic respiratory diseases such as depression and cognitive impairment, may further impair the nutrient intake, and the same is true of comorbidity, mobility limitations, and lack of support. Accordingly, the nutritional status of older respiratory patients variably reflects nonrespiratory and primarily respiratory factors (Fig. 30.1).

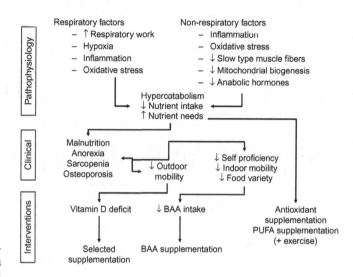

FIGURE 30.1 Respiratory and non-respiratory factors affecting nutritional status and related clinical characteristics and therapeutic approach.

NUTRITION AND LUNG DISEASE: PATHOLOGICAL PATHWAYS

Aging per se is associated with important changes in body composition. During adulthood, the body undergoes a progressive loss of fat-free mass and an increase in fat mass. Nonmuscle lean tissue is preserved and the most significant reductions in fat-free mass occur in the muscle compartment [11]; static and dynamic muscle strength also decreases and type II muscle fibers tend to atrophy. Changes in muscle mass have profound effects, reducing basal metabolic rate and limiting maximal exercise capacity. Additionally, in the presence of chronic diseases, nutritional status is tightly linked to underlying diseases, and indices of nutritional status are usually considered as markers of disease severity.

Among purely respiratory factors affecting nutritional status in patients with lung diseases, increased respiratory work and hypoxia deserve mention. Increased respiratory work is frequently observed in COPD patients where it is mediated by increased inspiratory resistance (mainly related to mucosa inflammation and thickening in COPD, or to destruction of small airways in emphysema), elastance of lung and chest wall (especially during exercise or when breathing rate increases), expiratory resistance, and hyperinflation (reducing the strength of inspiratory muscle contraction, decreasing the ventilatory reserve capacity, and increasing dyspnea). All the above mechanisms contribute to increase energy output among respiratory patients [12]. Hypoxia activates the sympathetic nervous system, which in turn increases systemic inflammation, ROS, and TNF-α secretion. TNF-α upregulates the transcription factor NF-κB, which promotes protein breakdown in cells, providing a molecular mechanism for muscle loss [13]. Thus, chronic inflammation significantly contributes to the link between lung diseases and malnutrition not only by locally increasing respiratory work, but also by affecting muscle structure and function [14].

Muscle mass reduction is the result of an imbalance between protein breakdown and synthesis, and it represents an important nonrespiratory factor. Sedentary habits, systemic inflammation, and use of steroids are involved in COPD-related muscle loss [5]. A central actor in the control of protein degradation is proteinkinase B (AKT). In its phosphorylated form, AKT contrasts protein degradation by downregulating two muscle-specific E3 ligases, muscle ring finger-1 (MuRF-1) and muscle atrophy F-box (atrogin-1), which are involved in the transfer of activated ubiquitin to the proteins targeted for degradation [15]. Recently, an increase of mRNA of MuRF-1, atrogin-1, and FoxO-1 was found in the quadriceps muscle of patients with COPD compared with controls matched for age and level of activity [16]. Fiber type shift characterizes limb muscle in COPD. Biopsies from limb muscle in patients with COPD showed a selective loss of type I and IIa fibers, while glycolytic type IIb fibers were relatively spared [17]. Mitochondria density is decreased in the limb muscle of patients with pulmonary diseases, especially in COPD [18]. One obvious reason is that in COPD muscle there is an increased proportion of type IIb fibers, which are poor in mitochondria. However, other important elements are involved. Hypoxia-inducible factors (HIF-1) are known to induce mitochondrial autophagy and to inhibit mitochondrial biogenesis [19]. A smaller number of mitochondria allows sparing of oxygen consumption, protecting the muscle from oxidative stress. Nevertheless, biopsies obtained from limb (vastus lateralis) and respiratory (external intercostalis) muscles in patients with COPD showed an increased reactive oxygen species production [20].

Changes in nutrient intake also play a relevant role (Fig. 30.2). Appetite is known to be regulated by a fine balance between orexigenic and anorexigenic neuropeptides in the hypothalamus and this balance frequently bends toward anorexia in chronic lung diseases

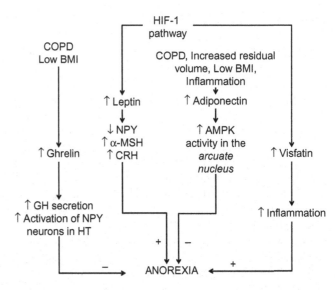

FIGURE 30.2 Effects of COPD on pathway regulating appetite.

[21]. Ghrelin, an endogenous ligand of the growth hormone (GH) secretagogue receptor, promotes anabolism by stimulating GH secretion and activating neuropeptide Y neurons in the hypothalamus. The role of ghrelin in cachexia is not completely understood. However, circulating ghrelin is inversely correlated to the body mass index (BMI) [11], and has been reported to increase among COPD patients [22], which could be interpreted as a compensatory mechanism to anorexia. Leptin, a hormone mainly secreted by adipocytes, has an anorectic effect by negatively regulating neuropeptides involved in food intake such as neuropeptide Y or positively regulating others involved in anorexia such the α-melanocyte stimulating hormone or corticotrophin-releasing hormone. It is known that the human leptin gene is induced by HIF-1, which suggest a potential role in lung disease-related anorexia [23].

Visfatin, an adipocytokine involved in the maturation of B-cell precursors, has pro-inflammatory activity and is increased by hypoxia via the HIF-1 pathway [24]. Adiponectin is a protein secreted from adipocytes, just like leptin, but with opposite actions: adiponectin enhances appetite and acts to reduce the fatty acids in muscle tissues. The blood adiponectin level is elevated in COPD patients, with the levels reported to be positively correlated with plasma TNF-α levels as well as with the residual volume, and to be negatively correlated with the body weight [25]. Thus, adiponectin seems to mark and to contrast a hypercatabolic status in the emphysematous variant of COPD.

Reduced intake of selected nutrients is frequently observed among older patients with lung diseases. This is more relevant for vitamin D because reduced outdoor mobility may affect its synthesis in the skin mediated by the UV radiation. Vitamin D deficiency is a risk factor for respiratory infection because of the important regulatory activities of vitamin D on genes expression in leukocytes, synthesis of antimicrobial peptides, lymphocyte response to antigens, and production of inflammatory cytokines [26]. Additionally, vitamin D is crucial for maintaining lower extremity function and reducing risk of falls. Proximal muscle weakness, diffuse muscle pain, and gait impairments are well-known clinical symptoms of vitamin D deficiency [27]. Moreover vitamin D is essential for bone growth and bone health preservation, and its deficit, together with systemic inflammation and steroids provide substantial contribution to the well known association between COPD and osteoporosis [28].

Bronchiectasis also contribute to increase the risk of malnutrition [29]. In patients with bronchiectasis, high plasma levels of inflammatory cytokines have been described in association with a reduction in fat-free mass (FFM), increased muscle proteolysis, respiratory exacerbations, severe phenotypes, and worse pulmonary function, even in patients who are clinically stable [30].

Patients with lung cancer exhibit relevant changes in nutritional status that often lead to cancer cachexia. Although the etiology of cancer-induced cachexia remains to be fully understood, several cellular and molecular mechanisms have been proposed such as systemic inflammation, oxidative stress, metabolic disturbances, and nutritional abnormalities [31]. The myostatin and activin IIB system is also a relevant proposed mechanism contributing to muscle protein catabolism. Indeed, it was clearly demonstrated that the soluble receptor antagonist of myostatin (sActRIIB) improved muscle mass loss, survival, and physical activity in cancer cachectic rodents [32]. Furthermore, mitogen-activated protein kinases (MAPK) and nuclear factor (NF)-kB, which are central regulators of gene expression, redox balance, and metabolism, have also been shown to play a major role in adaptive or maladaptive responses to cellular stress within skeletal muscles. MAPK activation also seems to mediate oxidative stress-induced muscle atrophy. Interestingly, MAPK signaling was also shown to be involved in enhanced expression of proteasome via proteolysis-inducing factor in C2C12 myotubes [33]. NF-kB was also demonstrated to participate in the process of muscle wasting under several conditions such as sepsis, cancer cachexia [34], and COPD [35].

NUTRITIONAL ASSESSMENT FOR PULMONARY DISEASE

As previously discussed, changes in the nutritional status often parallel, although to a different extent, the progression of respiratory diseases, and may impact health-related quality of life [36]. In this scenario, the association between underweight and increased mortality risk has been well established in retrospective studies on COPD patients [37]. For these reasons, nutritional screening is considered as an essential component of integrated COPD management.

Though more frequent in subjects at the advanced stages of the respiratory disease, malnutrition may occur earlier and insidiously, and serum biomarkers of malnutrition may not be detected at earlier stages. The risk of malnutrition is estimated in individuals who involuntary lose >10% of their weight in the last three months, or when body weight is <90% of the ideal weight; a condition of <85% of the ideal body weight is suggestive of malnutrition. A large number of clinical signs indicate nutritional deficiencies, including wasting, dry skin, thin hair, depigmented nails, bone and joint pain, and edema. As mentioned earlier, malnutrition can also occur for dramatic depletion of lean body mass in the context of normal body weight, resulting in decreased muscle function, which may also affect respiratory muscles [38], and health status [39].

Methods for nutritional assessment are summarized in Table 30.1. Twenty-four-hour dietary recall represent the most simple screening method [40]. BMI measurement is also very simple to apply, but it should be considered that a variable loss of skeletal muscle can manifest in up to 25% of COPD patients with normal BMI [41].

The Malnutrition Universal Screening Tool (MUST) consists of a five-step screening tool (based on BMI, history of unexplained weight loss, and acute illness effect) that allows to derive a malnutrition risk score on which to develop a care plan both in hospitals and in the community [45]. Studies have shown that it has a high predictive validity in the hospital environment (length of stay, mortality in older people, and discharge destination in orthopedic patients) [46], as well as in respiratory patients admitted to a pulmonology hospital department [42].

One of the most common and sensitive methods to assess the risk of (or the presence of) malnutrition is the Mini Nutritional Assessment (MNA) [47]. The MNA test is a two-step procedure (screening for risk of malnutrition followed by global assessment of the nutritional conditions), which evaluates the overall health status of elderly subjects by rating cognitive function, functional status, walking, balance, and socioeconomic status. It includes anthropometric variables (BMI, upper arm and calf circumference, recent weight loss), dietary habit (food and liquid intake, number of meals, and feeding autonomy), general status (global assessment of lifestyle, medication, acute stress, and mood changes), and self-perception of quality of life

TABLE 30.1 Methods for Nutritional Assessment for Patients with Lung Diseases

Method	Description	Evidence
24-h dietary recall [40]	Interview during which the patient recalls all food consumed in the previous 24 h	Simplest method to assess dietary intake, although it may not represent a patient's typical intake.
Body mass index (BMI) [41]	Weight/(height)^2 (kg/m^2) World Health Organization categories: underweight BMI <18.5 normal 18.5 to 24.9 overweight 25 to 29.9 obese 30 to 39.9 extreme obesity >40	Loss of height caused by vertebral collapse or loss of muscle tone (elderly patients) may reduce accuracy Ascites and edema may reduce accuracy. Inability of BMI to identify unintentional weight loss as a single assessment. BMI may not reflect all tissue components, and therefore normal values do not rule out the occurrence of lean mass depletion.
Malnutrition Universal Screening Tool (MUST) [42]	BMI History of unexplained weight loss Acute illnesses A malnutrition risk score is calculated	Validated in hospital and community settings Predicts mortality in respiratory patients admitted to a pulmonology hospital department.
Mini Nutritional Assessment (MNA) [43]	Anthropometric variables (BMI, upper arm and calf circumference, recent weight loss) Dietary habit (food and liquid intake, number of meals and feeding autonomy) General status (global assessment of lifestyle, medication, acute stress, and mood changes) Self-perception of quality of life and nutrition	MNA predicts the perception of dyspnea in elderly individuals with COPD and normal BMI. MNA is an independent correlate of COPD severity and age.
Body composition [44]	Bioimpedance analysis (BIA) Dual energy X-ray absorptiometry (DEXA)	BIA is easy to apply and may provide clinical benefit in the assessment of COPD patients DEXA is more expensive than BIA, and less easy to apply for screening or clinical purposes.

and nutrition. The MNA is divided into two sections: the screening phase is followed by the assessment phase if the minimum score of 12 is not attained. The MNA provides a total score ranging 0 to 30 points, which allows to differentiate among undernourished (≤17 points), at risk of malnutrition (17.5–23.5), and well-nourished (>23.5). MNA was found to predict the perception of dyspnea in elderly individuals with COPD and normal BMI [48]. When the MNA questionnaire was applied in COPD subjects of different severity of disease to assess the nutritional status [43], it was independently correlated with disease severity and age, suggesting that it may best explain the complexity of the different components of the nutritional status.

The assessment of nutritional status can be incorporated in a multidimensional grading system, which also takes into account respiratory symptoms, the exercise capacity, and the spirometric measure of airflow. The BODE index requires BMI, degree of airflow obstruction measured by means of spirometry (percentage of predicted FEV$_1$), dyspnea as assessed by the modified Medical Research Council dyspnea scale, and exercise capacity as determined by a 6-min walk test. The BODE index was shown to be better than lung function in predicting the risk of death in patients with COPD [49] and hospitalization for COPD [50]. The discriminative capacity of the BODE index in predicting mortality was recently confirmed in a 15-year prospective cohort study [51], conducted in elderly individuals suffering from COPD. The BODE index is, however, impractical for debilitated persons. In one prospective study of survival, 26% of the original cohort of 327 COPD participants either could not complete the 6-minute walk test or could not perform adequate spirometry [52]. In the effort to overcome this limitation, the "quasi-BODE" index was constructed by using proxy measures of lung function (PEF) and 6-min WT (questionnaires), and validated in elderly populations with and without COPD [53]. The quasi-BODE score confirmed to predict mortality in elderly population, allowing a test that can be more practically administered to older adults, especially those with severe physical limitations. For the same reasons, the 6-min WT can be replaced by the incremental shuttle walking test, an alternative measure of exercise capacity to derive a simpler version of the BODE index (named the i-BODE), which was demonstrated to predict mortality in COPD [54].

As regards body composition, deuterium dilution is currently considered the gold standard for assessment, but its use is restricted to highly specialized hospitals or research. Two reasonable compromises are bioimpedance analysis (BIA) and dual energy X-ray absorptiometry (DEXA) [44]. The use of bioelectrical impedance analysis (BIA) in body composition assessment has been validated in COPD patients [55]. The dual energy X-ray absorptiometry (DEXA) analysis allows measurements of body composition and skeletal muscle, as well as mineral bone density, which are compromised to a large extent in elderly patients suffering from COPD [56]. The amount of FFM assessed by BIA was shown to be lower than with DEXA in patients with COPD, especially in men [57]. However, DEXA is not without potential limitation, because it provides systematically higher values for FFM compared with deuterium dilution [58]. Furthermore, DEXA is more expensive [59] and results differ between the different commercial devices [44].

Accurate measures of body composition are obtained with two imaging techniques, such as computerized tomography (CT) [60] and magnetic resonance imaging (MRI) [61]. Typically, these methods provide regional estimates of skeletal muscle by means of cross-sectional images, which allow also to detect muscle infiltration from adipose tissue and to quantify fat-free skeletal muscle. Total muscle area and fat-free skeletal muscle area, calculated from cross-sectional images, can be integrated from head to toe, to calculate total muscle and fat-free skeletal muscle volumes. Actually, only few studies of CT have considered the assessment of body composition in respiratory patients [62,63]. Anyway, these methods are very expensive, are not easily accessible, and are not routinely indicated to study muscle mass, but have been used mainly for research purposes. They require a highly specialized staff, specific software, and a relatively large amount of time. A further limitation of CT includes radiation exposure.

Finally, a screening by serum 25-hydroxy vitamin D levels measurement is currently indicated for COPD patients having low BMI and/or taking corticosteroids. However, in the context of chronic lung disease the potential association with vitamin D is further confounded by the strong association with reduced physical activity, which is directly linked to an individual's level of sun exposure and disease severity; thus, it is unclear whether vitamin D is simply an indirect marker of reduced physical activity and, consequently, is an innocent bystander in disease pathogenesis [64]. One further issue still to be addressed is the direct effect of inflammation on circulating 25(OH)D. Given that there is tissue upregulation of the conversion of 25(OH)D to 1,25(OH)2D during infections, one might predict that immediate circulating store of 25(OH)D would be depleted during infectious exacerbations of lung diseases [65].

NUTRITIONAL INTERVENTION IN LUNG DISEASE

Chronic respiratory diseases in the elderly are optimally managed in a multidimensional and multispecialistic perspective [66]. Nutritional assessment and support seems to be a key component of this approach, but the inherent scientific evidence is too limited to reach firm conclusions and, then, to suggest clear-cut dietary interventions. Indeed, the vast majority of the available studies are on small series of mainly adult subjects and many suffer from some methodological weakness. However, the scientific evidence pertaining to the frail and multimorbid elderly is likely true also for the COPD patient. Indeed, chronic noncommunicable diseases share selected long term effects on nutritional status and benefit from the same nutritional interventions. Selected disease-specific interventions will also be described.

COPD

It is since the classic study by Fiatarone et al. that the role of dietary supplementation in the context of a multidimensional approach having physical rehabilitation as central has been proven to benefit even disabled elderly [67]. Also, in elderly and disabled COPD patients combining rehabilitation and dietary supplementation can improve muscle mass and function [68]. Strategies reducing the systemic inflammation and the respiratory work are theoretically able to make nutritional interventions more effective. Indeed, even the alveolar overstretching has been proved to promote local inflammation and ensuing spill over [69]. Thus, nutritional interventions should never been considered out of an integrated and comprehensive therapeutic approach. Central to this approach is counseling: smoking is associated with comparable nutritional deficits in COPD and non-COPD elderly, likely reflecting the anorexigen and pro-inflammatory effects of smoke [70]. Correcting hypoxia is similarly important because hypoxia promotes the synthesis of anorexigen peptides such as ghrelin and leptin, induces oxidative stress and inflammation, and promotes a shift from oxidative metabolism to glycolytic fibers in nonrespiratory [71]. Thus, hypoxia has an important net catabolic effect which is especially dreadful in older patients who also experience the catabolic effect of aging itself.

Correcting the daily intake to provide the patient with the daily estimated average requirement represents the first step of nutritional approach given that the available knowledge allows the recommendation of only a few COPD-specific dietary interventions. Indeed, the high lipid diet proposed to decrease the CO_2 production might benefit only COPD patients with hypercarbia and severely reduced ventilatory reserve [72]. On the other hand, a meta-analysis of studies done in COPD patients concludes that nutrient supplementation can increase body weight and, to a lesser extent, free-fat mass in COPD patients, mainly if undernourished [73]. However, the majority of COPD patients studied in these meta-analyzed trials were adult and not elderly. This limits the generalizability of results to the elderly population.

Correcting selected and well-proved nutrient deficits through supplements is also mandatory. Some specific nutritional interventions look promising and need special consideration:

1. Branched-chain amino acids (leucine, isoleucine, valine) are key components of muscle myofibrillar proteins and, if needed, an important source of energy as an alternative or addition to glucose and fatty acids. Their plasmatic concentrations and muscle content have repeatedly been reported to be decreased in COPD mainly due to higher protein turnover [74]. The supplementation of leucine has the potential for improving muscle metabolism and structure due to the reduced first pass splanchnic extraction of leucine by the liver and, then, the increased availability of leucine to the muscle, as if leucine were selectively addressed to meet the need of the skeletal muscle in COPD patients [75]. There is also evidence that it is preferable to supplement all the three BAAs in order to obtain the desired metabolic effect [76]. However, this evidence pertains to a COPD population with mean age of 68 years; thus, it is unclear to what extent older COPD patients can benefit from BAA supplementation. It should also be considered that the BAA amount to be supplemented, should be tailored to the glomerular filtration rate, which is frequently depressed despite normal serum creatinine in COPD [77].
2. Both mitochondrial biogenesis and phenotypic expression of slow contracting and highly resistant muscle fibers, which depend upon several mechanisms, are reduced in the peripheral muscle of COPD patients [78]. PUFA could reverse one of these mechanisms, that is, they could promote peroxisome proliferator-activated receptors (PPARs) synthesis and activity, which is implicated in mitochondrial biogenesis. This likely explains the positive effect of PUFA supplementation on exercise capacity in COPD [79].
3. The highly common serum vitamin D deficit (see previous sections) and the ensuing rise in PTH drives important negative effects, such as insulin resistance, increased renin activity and systemic inflammation, and decreased muscle strength [80]. Furthermore, vitamin D deficit is associated with depressed immunity, the effects of which are especially evident in the context of tubercular infection [81], and vitamin D supplementation can reduce the frequency of COPD exacerbation in severely deficient patients [82]. Finally, an inverse association between vitamin D serum level and both FEV1 and FVC has been found in the NHANES survey [6], but not in the Hertfordshire Cohort Study [83]. Thus, there is evidence to supplement vitamin D to deficient COPD patients, though formal trials on the effect of supplementations are not available. Guidelines for supplementation will not differ from those designed for the overall elderly population [84]. In the event of coexistent chronic renal failure, a common finding in elderly COPD people [77], 1-alpha-25-OH vitamin D must be prescribed in order to compensate for the decreased 1-alpha hydroxylase renal activity.
4. Iron deficit is a common cause of anemia in COPD patients. However, it should be appreciated that, given that myoglobin has lower affinity to iron than hemoglobin, muscle weakness may be the presenting complaint of iron deficit prior to the onset of anemia [85]. Thus, a high suspicion is needed to timely diagnose this deficit. Supplementation will conform to general rules [86].
5. Deficits of antioxidants and folate have been reported in COPD patients. However, there is no robust evidence to either screen for them systematically or provide precautionary or on demand supplements [87].

A summary of available nutritional interventions in elderly patients with COPD is provided in Table 30.2.

Asthma

A link between obesity, mainly visceral obesity, and asthma has been proved and ascribed to several mechanisms [88]. However, there is no formal trial testing the effect of nutritional intervention on asthma severity in obese patients. On the other hand, vitamin D deficit has been repeatedly associated with asthma and the most likely pathogenetic link has been hypothesized to be the immune dysfunction and the decreased steroid responsiveness secondary to this deficit [89]. However, most of the evidence pertains to children and pregnant women, whereas there is paucity of data on elderly

TABLE 30.2 Summary of Nutritional Interventions for Elderly Patients with COPD

Intervention	Rationale	Evidence
Correction of Hypoxia [71]	Increase the generation of ROS and TNF-α which in turn may give rise to inflammatory changes leading to cachexia.	Reduction of catabolic effect which is especially dreadful in older patients who also experience the catabolic effect of aging itself.
PUFA [79]	PUFA are the natural ligands of PPARs, and they may imply a stimulating effect on muscle fat oxidative metabolism by inhibiting classic NF-kB signaling or stimulating PGC-1 a/PPAR signaling.	RCTs showed that polyunsaturated fatty acid supplementation as an adjunct to exercise training indeed significantly enhanced improvement in endurance exercise capacity in COPD.
Branched-chain amino acids (leucine, isoleucine, valine) [74]	Their plasmatic concentrations and muscle content have repeatedly been reported to be decreased in COPD mainly due to higher protein turn over.	Significant improvement of body weight, handgrip strength, decrease airflow limitation and increase quality of life.
Vitamin D [82,83]	Serum vitamin D deficit drives important negative effects such as insulin resistance, increased renin activity and systemic inflammation, decreased muscle strength.	Vitamin D supplementation can reduce the frequency of COPD exacerbation in severely deficient patients and can improve respiratory function.
Iron [86]	Iron deficit is a common cause of anemia in COPD patients.	Correction of the iron deficiency in COPD patients can improve the anemia and may improve the dyspnea.
Antioxidant [87]	Deficits of antioxidants have been reported in COPD patients.	There is no robust evidence to either screen for them systematically or provide precautionary or on demand supplements.

asthmatics. Thus, it seems reasonable to screen for vitamin D deficit and, if needed, to supplement elderly asthmatics not differently from how it seems reasonable to do so for elderly people with chronic diseases.

The finding of a better control of asthma in people consuming a diet rich in linolenic acid also refers to a young-adult population [90]. However, supplementation with omega 3 fatty acids as well as antioxidants could not improve asthma control [91].

In conclusion, there is no evidence that nutritional intervention can benefit asthmatic patients. Accordingly, there are no recommendations for older patients.

Lung Cancer

Nutritional interventions, mainly based on omega 3 fatty acid supplementation or dietary advice have been proved able to improve nutritional status and quality of life [92]. They also qualify as a primary component of multidimensional rehabilitation programs [93]. However, there is no evidence of any benefit on survival [94]. Thus, no recommendation can be provided for elderly lung cancer patients except for general rules in managing or preventing malnutrition, mainly for elderly people undergoing chemotherapy or radiation therapy.

CONCLUSIONS

Nutritional depletion in elderly patients with LD is common and has a negative impact on respiratory and peripheral muscle function. Additionally, in the presence of chronic diseases, nutritional status is tightly linked to underlying diseases, and indices of nutritional status are usually considered as markers of disease severity. Cachexia in LD is a multifactorial disease process characterized by loss of appetite, growth hormone resistance, and pro-inflammatory immune activation. Reduced appetite is part of the catabolic/anabolic imbalance in the process of wasting in LD, especially in COPD. Hypoxia and chronic inflammation significantly contributes to the link between lung diseases and malnutrition not only by locally increasing respiratory work, but also affecting muscle structure and function. Chronic respiratory diseases in the elderly are optimally managed in a multidimensional and multispecialistic perspective. Nutritional assessment and support seems to be a key component of this approach, but the inherent scientific evidence is too limited to reach firm conclusions and, then, to suggest clear-cut dietary interventions. However, careful nutritional support and correction of poor nutritional intake should be considered crucial to enhancing physical well-being and function.

SUMMARY POINTS

- Aging per se is associated with relevant changes in nutrient intake, metabolism, and body composition.
- Lung diseases have a negative impact on nutritional status, especially among older patients.

- Respiratory and nonrespiratory factors contribute to the association between lung diseases and malnutrition.
- Nutritional assessment, for example, Mini Nutritional Assessment, is of paramount importance in older patients with lung diseases.
- Bioimpedance analysis is an useful screening method to identify sarcopenic patients.
- Nutritional interventions that include correction of hypoxia, nutritional support, and correction of poor nutritional intake are crucial to enhancing physical well-being and function.

References

[1] Schols AM, Broekhuizen R, Weling-Scheepers CA, Wouters EF. Body composition and mortality in chronic obstructive pulmonary disease. Am J Clin Nutr 2005;82(1):53—9.

[2] Neri M, Melani AS, Miorelli AM, Zanchetta D, Bertocco E, Cinti C, et al. Long-term oxygen therapy in chronic respiratory failure: a Multicenter Italian Study on Oxygen Therapy Adherence (MISOTA). Respir Med 2006;100(5):795—806.

[3] Meltzer EB, Noble PW. Idiopathic pulmonary fibrosis. Orphanet J Rare Dis 2008;3:8.

[4] Savale L, Chaouat A, Bastuji-Garin S, Marcos E, Boyer L, Maitre B, et al. Shortened telomeres in circulating leukocytes of patients with chronic obstructive pulmonary disease. Am J Respir Crit Care Med 2009;179(7):566—71.

[5] Ceelen JJ, Langen RC, Schols AM. Systemic inflammation in chronic obstructive pulmonary disease and lung cancer: common driver of pulmonary cachexia? Curr Opin Support Palliat Care 2014;8(4):339—45.

[6] Black PN, Scragg R. Relationship between serum 25-hydroxyvitamin d and pulmonary function in the third national health and nutrition examination survey. Chest 2005;128(6):3792—8.

[7] Lange NE, Sparrow D, Vokonas P, Litonjua AA. Vitamin D deficiency, smoking, and lung function in the Normative Aging Study. Am J Respir Crit Care Med 2012;186(7):616—21.

[8] Hirani V. Associations between vitamin D and self-reported respiratory disease in older people from a nationally representative population survey. J Am Geriatr Soc 2013;61(6):969—73.

[9] Coronell C, Orozco-Levi M, Gea J. COPD and body weight in a Mediterranean population. Clin Nutr 2002;21(5):437 author reply -8.

[10] Chambellan A, Chailleux E, Similowski T. Prognostic value of the hematocrit in patients with severe COPD receiving long-term oxygen therapy. Chest 2005;128(3):1201—8.

[11] Broekhuizen R, Grimble RF, Howell WM, Shale DJ, Creutzberg EC, Wouters EF, et al. Pulmonary cachexia, systemic inflammatory profile, and the interleukin 1beta -511 single nucleotide polymorphism. Am J Clin Nutr 2005;82(5):1059—64.

[12] Loring SH, Garcia-Jacques M, Malhotra A. Pulmonary characteristics in COPD and mechanisms of increased work of breathing. J Appl Physiol 1985;107(1):309—14.

[13] Langen RC, Schols AM. Inflammation: friend or foe of muscle remodelling in COPD? Eur Respir J 2007;30(4):605—7.

[14] Piehl-Aulin K, Jones I, Lindvall B, Magnuson A, Abdel-Halim SM. Increased serum inflammatory markers in the absence of clinical and skeletal muscle inflammation in patients with chronic obstructive pulmonary disease. Respiration 2009;78(2):191—6.

[15] Stitt TN, Drujan D, Clarke BA, Panaro F, Timofeyva Y, Kline WO, et al. The IGF-1/PI3K/Akt pathway prevents expression of muscle atrophy-induced ubiquitin ligases by inhibiting FOXO transcription factors. Mol Cell 2004;14(3):395—403.

[16] Doucet M, Russell AP, Leger B, Debigare R, Joanisse DR, Caron MA, et al. Muscle atrophy and hypertrophy signaling in patients with chronic obstructive pulmonary disease. Am J Respir Crit Care Med 2007;176(3):261—9.

[17] Sandri M, Sandri C, Gilbert A, Skurk C, Calabria E, Picard A, et al. Foxo transcription factors induce the atrophy-related ubiquitin ligase atrogin-1 and cause skeletal muscle atrophy. Cell 2004;117(3):399—412.

[18] Gosker HR, Hesselink MK, Duimel H, Ward KA, Schols AM. Reduced mitochondrial density in the vastus lateralis muscle of patients with COPD. Eur Respir J 2007;30(1):73—9.

[19] Zhang H, Gao P, Fukuda R, Kumar G, Krishnamachary B, Zeller KI, et al. HIF-1 inhibits mitochondrial biogenesis and cellular respiration in VHL-deficient renal cell carcinoma by repression of C-MYC activity. Cancer Cell 2007;11(5):407—20.

[20] Puente-Maestu L, Lazaro A, Humanes B. Metabolic derangements in COPD muscle dysfunction. J Appl Physiol 1985;114(9):1282—90.

[21] Schols AM, Ferreira IM, Franssen FM, Gosker HR, Janssens W, Muscaritoli M, et al. Nutritional assessment and therapy in COPD: a European Respiratory Society statement. Eur Respir J 2014;44(6):1504—20.

[22] Uzum AK, Aydin MM, Tutuncu Y, Omer B, Kiyan E, Alagol F. Serum ghrelin and adiponectin levels are increased but serum leptin level is unchanged in low weight Chronic Obstructive Pulmonary Disease patients. Eur J Intern Med 2014;25(4):364—9.

[23] Simler N, Grosfeld A, Peinnequin A, Guerre-Millo M, Bigard AX. Leptin receptor-deficient obese Zucker rats reduce their food intake in response to hypobaric hypoxia. Am J Physiol Endocrinol Metab 2006;290(3):E591—7.

[24] Segawa K, Fukuhara A, Hosogai N, Morita K, Okuno Y, Tanaka M, et al. Visfatin in adipocytes is upregulated by hypoxia through HIF1alpha-dependent mechanism. Biochem Biophys Res Commun 2006;349(3):875—82.

[25] Tomoda K, Yoshikawa M, Itoh T, Tamaki S, Fukuoka A, Komeda K, et al. Elevated circulating plasma adiponectin in underweight patients with COPD. Chest 2007;132(1):135—40.

[26] Pfeffer PE, Hawrylowicz CM. Vitamin D and lung disease. Thorax 2012;67(11):1018—20.

[27] Wicherts IS, van Schoor NM, Boeke AJ, Visser M, Deeg DJ, Smit J, et al. Vitamin D status predicts physical performance and its decline in older persons. J Clin Endocrinol Metab 2007;92(6):2058—65.

[28] Romme EA, Smeenk FW, Rutten EP, Wouters EF. Osteoporosis in chronic obstructive pulmonary disease. Expert Rev Respir Med 2013;7(4):397—410.

[29] Ionescu AA, Nixon LS, Luzio S, Lewis-Jenkins V, Evans WD, Stone MD, et al. Pulmonary function, body composition, and protein catabolism in adults with cystic fibrosis. Am J Respir Crit Care Med 2002;165(4):495—500.

[30] Martinez-Garcia MA, Soler-Cataluna JJ, Perpina-Tordera M, Roman-Sanchez P, Soriano J. Factors associated with lung function decline in adult patients with stable noncystic fibrosis bronchiectasis. Chest 2007;132(5):1565—72.

[31] Argiles JM, Busquets S, Lopez-Soriano FJ. Anti-inflammatory therapies in cancer cachexia. Eur J Pharmacol 2011;668(Suppl. 1):S81—6.

[32] Busquets S, Toledo M, Orpi M, Massa D, Porta M, Capdevila E, et al. Myostatin blockage using actRIIB antagonism in mice bearing the Lewis lung carcinoma results in the improvement

[33] of muscle wasting and physical performance. J Cachexia Sarcopenia Muscle 2012;3:37−43.

[33] Smith HJ, Tisdale MJ. Signal transduction pathways involved in proteolysis-inducing factor induced proteasome expression in murine myotubes. Br J Cancer 2003;89(9):1783−8.

[34] Moore-Carrasco R, Busquets S, Almendro V, Palanki M, Lopez-Soriano FJ, Argiles JM. The AP-1/NF-kappaB double inhibitor SP100030 can revert muscle wasting during experimental cancer cachexia. Int J Oncol 2007;30(5):1239−45.

[35] Agusti A, Morla M, Sauleda J, Saus C, Busquets X. NF-kappaB activation and iNOS upregulation in skeletal muscle of patients with COPD and low body weight. Thorax 2004;59(6):483−7.

[36] Potter J, Klipstein K, Reilly JJ, Roberts M. The nutritional status and clinical course of acute admissions to a geriatric unit. Age Ageing 1995;24(2):131−6.

[37] Schols AM, Slangen J, Volovics L, Wouters EF. Weight loss is a reversible factor in the prognosis of chronic obstructive pulmonary disease. Am J Respir Crit Care Med 1998;157(6 Pt 1):1791−7.

[38] Engelen MP, Schols AM, Does JD, Wouters EF. Skeletal muscle weakness is associated with wasting of extremity fat-free mass but not with airflow obstruction in patients with chronic obstructive pulmonary disease. Am J Clin Nutr 2000;71(3):733−8.

[39] Shoup R, Dalsky G, Warner S, Davies M, Connors M, Khan M, et al. Body composition and health-related quality of life in patients with obstructive airways disease. Eur Respir J 1997;10(7):1576−80.

[40] Lee H, Kim S, Lim Y, Gwon H, Kim Y, Ahn JJ, et al. Nutritional status and disease severity in patients with chronic obstructive pulmonary disease (COPD). Arch Gerontol Geriatr 2013;56(3):518−23.

[41] Schols AM, Soeters PB, Dingemans AM, Mostert R, Frantzen PJ, Wouters EF. Prevalence and characteristics of nutritional depletion in patients with stable COPD eligible for pulmonary rehabilitation. Am Rev Respir Dis 1993;147(5):1151−6.

[42] Maia I, Xara S, Dias I, Parente B, Amaral TF. Nutritional screening of pulmonology department inpatients. Rev Port Pneumol 2014;20(6):293−8.

[43] Battaglia S, Spatafora M, Paglino G, Pedone C, Corsonello A, Scichilone N, et al. Ageing and COPD affect different domains of nutritional status: the ECCE study. Eur Respir J 2011;37(6):1340−5.

[44] Maltais F, Decramer M, Casaburi R, Barreiro E, Burelle Y, Debigare R, et al. An official American Thoracic Society/European Respiratory Society statement: update on limb muscle dysfunction in chronic obstructive pulmonary disease. Am J Respir Crit Care Med 2014;189(9):e15−62.

[45] Ellia M. Development and use of the 'Malnutrition Universal Screening Tool' ('MUST') for Adults. British Association of Parenteral and Enteral Nutrition; 2003.

[46] Kondrup J, Allison SP, Elia M, Vellas B, Plauth M. ESPEN guidelines for nutrition screening 2002. Clin Nutr 2003;22(4):415−21.

[47] Guigoz Y, Lauque S, Vellas BJ. Identifying the elderly at risk for malnutrition. The Mini Nutritional Assessment. Clin Geriatr Med 2002;18(4):737−57.

[48] Scichilone N, Paglino G, Battaglia S, Martino L, Interrante A, Bellia V. The mini nutritional assessment is associated with the perception of dyspnoea in older subjects with advanced COPD. Age Ageing 2008;37(2):214−17.

[49] Celli BR, Cote CG, Marin JM, Casanova C, Montes de Oca M, Mendez RA, et al. The body-mass index, airflow obstruction, dyspnea, and exercise capacity index in chronic obstructive pulmonary disease. N Engl J Med 2004;350(10):1005−12.

[50] Moberg M, Vestbo J, Martinez G, Williams JE, Ladelund S, Lange P, et al. Validation of the i-BODE index as a predictor of hospitalization and mortality in patients with COPD participating in pulmonary rehabilitation. COPD 2014;11(4):381−7.

[51] Pedone C, Scarlata S, Forastiere F, Bellia V, Antonelli Incalzi R. BODE index or geriatric multidimensional assessment for the prediction of very-long-term mortality in elderly patients with chronic obstructive pulmonary disease? A prospective cohort study. Age Ageing 2014;43(4):553−8.

[52] Ko FW, Tam W, Tung AH, Ngai J, Ng SS, Lai K, et al. A longitudinal study of serial BODE indices in predicting mortality and readmissions for COPD. Respir Med 2011;105(2):266−73.

[53] Roberts MH, Mapel DW, Bruse S, Petersen H, Nyunoya T. Development of a modified BODE index as a mortality risk measure among older adults with and without chronic obstructive pulmonary disease. Am J Epidemiol 2013;178(7):1150−60.

[54] Williams JE, Green RH, Warrington V, Steiner MC, Morgan MD, Singh SJ. Development of the i-BODE: validation of the incremental shuttle walking test within the BODE index. Respir Med 2012;106(3):390−6.

[55] Schols AM, Wouters EF, Soeters PB, Westerterp KR. Body composition by bioelectrical-impedance analysis compared with deuterium dilution and skinfold anthropometry in patients with chronic obstructive pulmonary disease. Am J Clin Nutr 1991;53(2):421−4.

[56] Barker BL, McKenna S, Mistry V, Pancholi M, Patel H, Haldar K, et al. Systemic and pulmonary inflammation is independent of skeletal muscle changes in patients with chronic obstructive pulmonary disease. Int J Chron Obstruct Pulmon Dis 2014;9:975−81.

[57] Rutten EP, Spruit MA, Wouters EF. Critical view on diagnosing muscle wasting by single-frequency bio-electrical impedance in COPD. Respir Med 2010;104(1):91−8.

[58] Engelen MP, Schols AM, Heidendal GA, Wouters EF. Dual-energy X-ray absorptiometry in the clinical evaluation of body composition and bone mineral density in patients with chronic obstructive pulmonary disease. Am J Clin Nutr 1998;68(6):1298−303.

[59] Walter-Kroker A, Kroker A, Mattiucci-Guehlke M, Glaab T. A practical guide to bioelectrical impedance analysis using the example of chronic obstructive pulmonary disease. Nutr J 2011;10:35.

[60] Sjostrom L. A computer-tomography based multicompartment body composition technique and anthropometric predictions of lean body mass, total and subcutaneous adipose tissue. Int J Obes 1991;15(Suppl. 2):19−30.

[61] Selberg O, Burchert W, Graubner G, Wenner C, Ehrenheim C, Muller MJ. Determination of anatomical skeletal muscle mass by whole body nuclear magnetic resonance. Basic Life Sci 1993;60:95−7.

[62] Braunschweig CA, Sheean PM, Peterson SJ, Gomez Perez S, Freels S, Troy KL, et al. Exploitation of diagnostic computed tomography scans to assess the impact of nutrition support on body composition changes in respiratory failure patients. J Parenter Enteral Nutr 2014;38(7):880−5.

[63] Marquis K, Debigare R, Lacasse Y, LeBlanc P, Jobin J, Carrier G, et al. Midthigh muscle cross-sectional area is a better predictor of mortality than body mass index in patients with chronic obstructive pulmonary disease. Am J Respir Crit Care Med 2002;166(6):809−13.

[64] Strine TW, Balluz LS, Ford ES. The associations between smoking, physical inactivity, obesity, and asthma severity in the general US population. J Asthma 2007;44(8):651−8.

REFERENCES

[65] Finklea JD, Grossmann RE, Tangpricha V. Vitamin D and chronic lung disease: a review of molecular mechanisms and clinical studies. Adv Nutr 2011;2(3):244–53.

[66] Antonelli Incalzi R, Pedone C, Pahor M. Multidimensional assessment and treatment of the elderly with COPD. Eur Respir Mon 2009;35–55.

[67] Fiatarone MA, O'Neill EF, Ryan ND, Clements KM, Solares GR, Nelson ME, et al. Exercise training and nutritional supplementation for physical frailty in very elderly people. N Engl J Med 1994;330(25):1769–75.

[68] van de Bool C, Mattijssen-Verdonschot C, van Melick PP, Spruit MA, Franssen FM, Wouters EF, et al. Quality of dietary intake in relation to body composition in patients with chronic obstructive pulmonary disease eligible for pulmonary rehabilitation. Eur J Clin Nutr 2014;68(2):159–65.

[69] Vlahakis NE, Schroeder MA, Limper AH, Hubmayr RD. Stretch induces cytokine release by alveolar epithelial cells in vitro. Am J Physiol 1999;277(1 Pt 1):L167–73.

[70] Obase Y, Mouri K, Shimizu H, Ohue Y, Kobashi Y, Kawahara K, et al. Nutritional deficits in elderly smokers with respiratory symptoms that do not fulfill the criteria for COPD. Int J Chron Obstruct Pulmon Dis 2011;6:679–83.

[71] Raguso CA, Luthy C. Nutritional status in chronic obstructive pulmonary disease: role of hypoxia. Nutrition 2011;27(2):138–43.

[72] Ferreira I, Brooks D, Lacasse Y, Goldstein R. Nutritional intervention in COPD: a systematic overview. Chest 2001;119(2):353–63.

[73] Ferreira IM, Brooks D, White J, Goldstein R. Nutritional supplementation for stable chronic obstructive pulmonary disease. Cochrane Database Syst Rev 2012;12:CD000998.

[74] Engelen MP, Wouters EF, Deutz NE, Menheere PP, Schols AM. Factors contributing to alterations in skeletal muscle and plasma amino acid profiles in patients with chronic obstructive pulmonary disease. Am J Clin Nutr 2000;72(6):1480–7.

[75] Engelen MP, De Castro CL, Rutten EP, Wouters EF, Schols AM, Deutz NE. Enhanced anabolic response to milk protein sip feeding in elderly subjects with COPD is associated with a reduced splanchnic extraction of multiple amino acids. Clin Nutr 2012;31(5):616–24.

[76] Engelen MP, Rutten EP, De Castro CL, Wouters EF, Schols AM, Deutz NE. Supplementation of soy protein with branched-chain amino acids alters protein metabolism in healthy elderly and even more in patients with chronic obstructive pulmonary disease. Am J Clin Nutr 2007;85(2):431–9.

[77] Incalzi RA, Corsonello A, Pedone C, Battaglia S, Paglino G, Bellia V. Chronic renal failure: a neglected comorbidity of COPD. Chest 2010;137(4):831–7.

[78] Remels AH, Schrauwen P, Broekhuizen R, Willems J, Kersten S, Gosker HR, et al. Peroxisome proliferator-activated receptor expression is reduced in skeletal muscle in COPD. Eur Respir J 2007;30(2):245–52.

[79] Broekhuizen R, Wouters EF, Creutzberg EC, Weling-Scheepers CA, Schols AM. Polyunsaturated fatty acids improve exercise capacity in chronic obstructive pulmonary disease. Thorax 2005;60(5):376–82.

[80] Lee JH, O'Keefe JH, Bell D, Hensrud DD, Holick MF. Vitamin D deficiency an important, common, and easily treatable cardiovascular risk factor? J Am Coll Cardiol 2008;52(24):1949–56.

[81] Korthals Altes H, Kremer K, Erkens C, van Soolingen D, Wallinga J. Tuberculosis seasonality in the Netherlands differs between natives and nonnatives: a role for vitamin D deficiency? Int J Tuberc Lung Dis 2012;16(5):639–44.

[82] Lehouck A, Mathieu C, Carremans C, Baeke F, Verhaegen J, Van Eldere J, et al. High doses of vitamin D to reduce exacerbations in chronic obstructive pulmonary disease: a randomized trial. Ann Intern Med 2012;156(2):105–14.

[83] Shaheen SO, Jameson KA, Robinson SM, Boucher BJ, Syddall HE, Sayer AA, et al. Relationship of vitamin D status to adult lung function and COPD. Thorax 2011;66(8):692–8.

[84] Holick MF, Binkley NC, Bischoff-Ferrari HA, Gordon CM, Hanley DA, Heaney RP, et al. Evaluation, treatment, and prevention of vitamin D deficiency: an Endocrine Society clinical practice guideline. J Clin Endocrinol Metab 2011;96(7):1911–30.

[85] DeLoughery TG. Microcytic anemia. N Engl J Med 2014;371(14):1324–31.

[86] Aspuru K, Villa C, Bermejo F, Herrero P, Lopez SG. Optimal management of iron deficiency anemia due to poor dietary intake. Int J Gen Med 2011;4:741–50.

[87] Rahman I, MacNee W. Antioxidant pharmacological therapies for COPD. Curr Opin Pharmacol 2012;12(3):256–65.

[88] Lv N, Xiao L, Camargo Jr. CA, Wilson SR, Buist AS, Strub P, et al. Abdominal and general adiposity and level of asthma control in adults with uncontrolled asthma. Ann Am Thorac Soc 2014;11(8):1218–24.

[89] Paul G, Brehm JM, Alcorn JF, Holguin F, Aujla SJ, Celedon JC. Vitamin D and asthma. Am J Respir Crit Care Med 2012;185(2):124–32.

[90] Barros R, Moreira A, Fonseca J, Delgado L, Castel-Branco MG, Haahtela T, et al. Dietary intake of alpha-linolenic acid and low ratio of n-6:n-3 PUFA are associated with decreased exhaled NO and improved asthma control. Br J Nutr 2011;106(3):441–50.

[91] Allan K, Devereux G. Diet and asthma: nutrition implications from prevention to treatment. J Am Diet Assoc 2011;111(2):258–68.

[92] van der Meij BS, Langius JA, Spreeuwenberg MD, Slootmaker SM, Paul MA, Smit EF, et al. Oral nutritional supplements containing n-3 polyunsaturated fatty acids affect quality of life and functional status in lung cancer patients during multimodality treatment: an RCT. Eur J Clin Nutr 2012;66(3):399–404.

[93] Harada H, Yamashita Y, Misumi K, Tsubokawa N, Nakao J, Matsutani J, et al. Multidisciplinary team-based approach for comprehensive preoperative pulmonary rehabilitation including intensive nutritional support for lung cancer patients. PLoS One 2013;8(3):e59566.

[94] Bourdel-Marchasson I, Blanc-Bisson C, Doussau A, Germain C, Blanc JF, Dauba J, et al. Nutritional advice in older patients at risk of malnutrition during treatment for chemotherapy: a two-year randomized controlled trial. PLoS One 2014;9(9):e108687.

CHAPTER 31

How Nutrition Affects Kidney Function in Aging

Christina Chrysohoou[1], Georgios A. Georgiopoulos[1] and Ekavi N. Georgousopoulou[2]

[1]1st Cardiology Clinic, University of Athens, Athens, Greece [2]Department of Nutrition-Dietetics, School of Health and Education, Harokopio University, Athens, Greece

KEY FACTS

- Regular fish consumption, independent of overall dietary and other measured lifestyle habits, seems to be a non-pharmacological means in the maintenance of preserved kidney function, especially among elders.
- Green tea polyphenols have shown renal protection.
- Low PUFA consumption has been associated with kidney dysfunction.
- Excessive intake of protein, sodium, potassium, and alcohol has been associated with the progression of kidney failure.
- Omega-3 fatty acids supplementation has shown beneficial impact in chronic kidney disease.
- Mediterranean type of diet has been linked with improved rates of renal function, due to its anti-inflammatory and antioxidative properties.
- A skilled dietitian will incorporate a patient's food preferences, adequate calories, and a proper distribution of foods while encouraging compliance and avoiding inappropriate dietary fats.

Dictionary of Terms

- *Polyunsaturated fatty acids (PUFAs)*: are fatty acids that contain more than one double bond in their backbone. This class includes many important compounds, such as essential fatty acids and those that give drying oils their characteristic property.
- *Monounsaturated fats (MUFAs)*: are fatty acids that have one double bond in the fatty acid chain with all of the remainder carbon atoms being single-bonded.
- *Omega-3 fatty acids*: (also called ω-3 fatty acids or n-3 fatty acids) are polyunsaturated fatty acids (PUFAs) with a double bond (C=C) at the third carbon atom from the end of the carbon chain.
- *Glomerular filtration rate (GFR)*: is the volume of fluid filtered from the renal (kidney) glomerular capillaries into the Bowman's capsule per unit time.
- *Creatinine clearance rate*: is the volume of blood plasma that is cleared of creatinine per unit time and is a useful measure for approximating the GFR.
- *Mediterranean diet*: the traditional dietary patterns of Greece, Southern Italy, and Spain. The principal aspects of this diet include proportionally high consumption of olive oil, legumes, unrefined cereals, fruits, and vegetables, moderate to high consumption of fish, moderate consumption of dairy products (mostly as cheese and yogurt), moderate wine consumption, and low consumption of meat and meat products.
- *Chronic kidney disease (CKD)*: also known as chronic renal disease, is a progressive loss of renal function over a period of months or years.
- *Cardiovascular disease (CVD)*: is a class of diseases that involve the heart or blood vessels. Common CVDs include: ischemic heart disease (IHD),

stroke, hypertensive heart disease, rheumatic heart disease (RHD), aortic aneurysms, cardiomyopathy, atrial fibrillation, congenital heart disease, endocarditis, and peripheral artery disease (PAD), among others.
- *Albuminuria*: is a pathological condition wherein albumin is present in the urine.
- *KDIGO*: the acronym stands for Kidney Disease: Improving Global Outcomes and is the global organization developing and implementing evidence based clinical practice guidelines in kidney disease. It is an independent volunteer-led self-managed charity incorporated in Belgium accountable to the public and the patients it serves.
- *Uric acid (UA)*: is a heterocyclic compound of carbon, nitrogen, oxygen, and hydrogen with the formula $C_5H_4N_4O_3$. It forms ions and salts known as urates and acid urates, such as ammonium acid urate. Uric acid is a product of the metabolic breakdown of purine nucleotides.
- *Gout*: (also known as *podagra* when it involves the big toe) is a medical condition usually characterized by recurrent attacks of acute inflammatory arthritis—a red, tender, hot, swollen joint. The metatarsal-phalangeal joint at the base of the big toe is the most commonly affected (approximately 50% of cases).
- *Polyphenols*: are a structural class of mainly natural, but also synthetic or semi synthetic, organic chemicals characterized by the presence of large multiples of phenol structural units.

INTRODUCTION

The kidney participates in body homeostasis by regulating acid—base balance, electrolyte concentrations, extracellular fluid volume, and arterial blood pressure. Many of the kidney's functions are accomplished by relatively simple mechanisms of filtration, reabsorption, and secretion, which take place in the nephron, which is the structural unit of the kidneys. The kidney generates 180 liters of filtrate a day, while reabsorbing a large percentage, allowing for the generation of only approximately 2 liters of urine. The level of renal function is usually described by the Glomerular filtration rate (GFR), which is defined as the volume of fluid filtered from the renal glomerular capillaries into the Bowman's capsule per unit time. The decreased level of the kidney function can gradually cause chronic kidney disease. Chronic kidney disease is defined according to the presence or absence of kidney damage and level of kidney function. Defining stages of chronic kidney disease requires "categorization" of continuous measures of kidney function, and the "cut-off levels" between stages are inherently arbitrary, but it has been widely accepted that $GFR < 60 \text{ mL/min}/1.73 \text{ m}^2$ for more than three months, with or without kidney damage should be considered chronic kidney disease.

Recently, the Australian healthcare system ran a survey in order to estimate the financial cost of chronic kidney disease in comparison to cardiovascular disease (CVD). Between 2012 and 2020, prevalence, per-patient expenditure, and total disease expenditure associated with CKD are estimated to increase significantly more rapidly than CVD. Total CKD-related expenditure is estimated to increase by 37%, compared to 14% in CVD. Recent studies in high-risk cardiovascular populations and patients with cardiovascular disease have consistently revealed that the decreased level of renal function is an independent risk factor for cardiovascular mortality. Especially in patients with severe kidney injury, defined as an estimated glomerular filtration rate less than $60 \text{ mL/min}/1.73 \text{ m}^2$, an increased prevalence of several cardiovascular risk factors may occur, including volume expansion secondary to sodium retention, arterial hypertension, and insulin resistance with impaired glucose tolerance. Chronic kidney disease has also been associated with oxidative stress, and increased circulating levels of inflammatory markers, phosphate retention with medial vascular calcification, increased parathyroid hormone concentrations causing valvular calcification and dysfunction, anemia, and left ventricular hypertrophy, all of which can increase cardiovascular risk. In an analysis of four community-based studies that enrolled 22,634 subjects without a history of cardiovascular disease, chronic kidney disease was shown to be a risk factor for the primary composite outcome of myocardial infarction, fatal coronary artery disease, stroke, and diabetes; while in a recent case-controlled study of 6432 men, chronic kidney disease was related with the existence of several cardiovascular risk factors, like advanced age, diabetes mellitus, and smoking, and was independently associated with cardiovascular mortality The role of kidney function with mortality has long been investigated in the context of heart failure; where in patients hospitalized with acute decompensation, even small changes in renal function can influence prognosis. Furthermore, in patients undergoing cardiac surgery, acute kidney injury has been identified as the strongest risk factor for death; while even a small increase in serum creatinine concentration after cardiothoracic surgery is related with increased complications and mortality. During aging there is a decline in renal function, due to a decrease in the structure and function ability of the nephrons, accumulation of fibrosis, and reduction in blood flow, which is more evident in the case of systolic heart failure. The weight of each kidney is reduced from 200 to 350 g at the age of 40 years old to 200 g at the age of 80 years old. The glomerular filtration rate is even reduced by 8 mL/min/1.73 m^2 each decade of aging. Thus the renal clearance of medications during advanced aging is reduced by 7.8 times.

RENAL FUNCTION IN AGING

The main function of the kidney is the regulation of extracellular fluid; thus even mild abnormalities can impair the balance of extracellular volume. According to this pathophysiologic sequence, one of the initial events in the pathogenesis of essential arterial hypertension is renal sodium retention. Additionally, volume overload increases left ventricular preload and transmural myocardial pressure, leading according to Laplace's law to an increase in left ventricular mass and eventually left ventricular hypertrophy, which has been recognized as an independent risk factor of cardiovascular mortality. In a recent study, people with impaired renal function, compared to those with normal renal function, had increased prevalence of hypertension and increased levels of systolic and diastolic blood pressure, although the impact of even mild to moderate renal dysfunction on cardiovascular morbidity was independent of hypertension status. It seems that the chronic exposure of cardiovascular disease-free individuals to even mild increases in left ventricular preload and left ventricular mass can cause structural changes in the myocardium due to the deposition of excessive interstitial fibrillar collagen and can lead to cardiovascular events independently of the level of arterial hypertension.

Between lifestyle factors, diet seems to play an important role in the prevention of renal failure. Although, there is no specific dietary pattern for preventing renal failure, excessive intake of protein, sodium, potassium, and alcohol have shown to be related with the progression of renal failure even among elderly individuals; while the well-balanced Mediterranean type of diet has been related with preserved renal function. The role of dietary habits may be related with the observation that chronic kidney disease is also associated with increased inflammation process, including plasma homocysteine, fibrinogen, and uric acid levels. Recent studies have illustrated that elderly subjects with even mild kidney disease had increased fast serum levels of homocysteine, the latter could be also attributed to a higher consumption of meat and carbohydrates, instead of the closer adherence to the Mediterranean type of diet that was evident in those with normal renal function. Additionally uric acid has been recognized as an independent risk factor for cardiovascular events in patients affected by hypertension, diabetes, and coronary heart disease and is related to total mortality; while it has been also related with deterioration of left ventricular function. Mediterranean type of diet has been related with lower incidence of cardiovascular diseases, metabolic disorders, and several types of cancer mainly due to the benefits on lipids metabolism, arterial blood pressure levels, on the control of body mass index, as well as on the reduction of inflammation and coagulation process. Although excessive alcohol consumption is a risk factor for hypertension and stroke, evidence for an association with chronic kidney disease is conflicting. White et al., in a prospective study of 6259 adults without a history of alcohol dependence, revealed that alcohol intake of ≥ 30 g/day was associated with an increased risk of albuminuria after adjustment for age, sex, and baseline kidney function, and with a reduced risk of glomerular filtration rate <60 mL/min/1.73 m^2, compared with consumption of <10 g/day. Physical activity has been related with reduced mortality in the general population and also in patients with chronic kidney disease. On the other hand, exercise-associated hyponatremia, due to inappropriate retention of body water in combination with the nonosmotic stimuli of the antidiuretic hormone and the higher than normal fluid intake, has emerged as an important complication of prolonged endurance physical activities. In the Third National Health and Nutrition Examination Survey, glomerular filtration rate was related with activity status and duration, but negatively with activity frequency in those without metabolic syndrome who performed larger numbers of activity varieties after an adjustment for confounders. It seems that higher level of physical exercise can lead to even mild reductions in renal function.

Especially in the elderly, chronic kidney disease is also associated with oxidative stress and increased circulating levels of inflammatory markers, phosphate retention with medial vascular calcification, increased parathyroid hormone concentrations causing valvular calcification and dysfunction, anemia, and left ventricular hypertrophy. Therefore, as in elderly individuals the deterioration of kidney function increases mortality, the study of modifiable lifestyle factors, like nutrition, that may be related to kidney disease, deserves special attention. During aging, glomerular filtration is reduced and this reduction is not in accordance with the reduction of creatinine clearance. Thus is due to the significant reduction in skeletal muscle mass that accompanies aging, creating a false higher creatinine clearance than inulin clearance. Urine formation consists of three basic processes: glomerular filtration, tubular secretion, and tubular reabsorption. Several disease conditions can interfere with these functions. Inflammatory and degenerative diseases can involve the small blood vessels and membranes in the nephrons. Urinary tract infections and kidney stones can interfere with normal drainage, causing further infection and tissue damage. Circulatory disorders, such as hypertension, can damage the small renal arteries. Other diseases, such as diabetes, gout, and urinary tract abnormalities can lead to impaired function, infection, or obstruction. Toxic agents such as insecticides, solvents, and certain drugs may also harm renal tissue. There was a notion that a reduction in protein consumption can reduce the symptoms of renal failure, but this idea has been abandoned as it became clear that hypoalbuminuria in patients under renal filtration is a main factor for increased mortality. Thus, a daily consumption of protein less that 1 g/kg is not recommended for patients. On this basis a high-protein diet with fish, poultry, pork, or eggs at every meal may be recommended. By contrast, in patients with chronic kidney disease, before the stage of hemodialysis, protein intake is recommended to vary between 0.6 and 0.8 g/kg, including a high concentration of adequate amino acids. In patients with moderately severe CKI, average baseline values of serum bicarbonate (23 mM), phosphorus (3.6 mg/dL), and urea nitrogen (30 mg/dL) are significantly lower in those who successfully reduce their protein intake by 0.2 g of protein/kg per day for one year. Diets with a higher concentration of protein can cause uremic symptoms with vomiting, acidosis, and edema (in combination with increased uptake of salt) due to the inability to absorb higher protein intake in parallel with impaired water waste. Water dilutes the urine and keeps calcium, oxalates, and uric acid in solution. In research studies, in those subjects whose total fluid intake (from all sources) over 24 h was roughly 2.5 L, the risk of a stone was about one-third less than that of subjects drinking only half that much. In particular, elderly people have a reduced sense of thirst; while their kidneys cannot respond to the excretion of vasopressin hormone to reduce the waste of fluids. This can lead to further dehydration and limitation of glomerular filtration rate. In chronic kidney disease the approach for managing elevated serum phosphate, through the use of phosphate binders as an adjunct to the restriction of dietary intake, has also been recognized. The KDIGO guidelines continue to recommend restricting dietary phosphate in combination with other treatments, however the evidence is poor. The effect of diet is also important even for elderly subjects with chronic kidney disease under hemodialysis, as presented in a Japanese study that used data from 3080 general-population participants in the Hisayama study (year 2007), and data from 1355 hemodialysis patients in the Japan Dialysis Outcomes and Practice Patterns Study (JDOPPS: years 2005–2007). The researchers identified three food groups (meat, fish, and vegetables) within the Hisayama population data and used principal components analysis in order to reveal dietary patterns. They identified three dietary patterns: well-balanced, unbalanced, and other and after adjusting for potential confounders, adherence to the unbalanced diet was associated with a two-fold higher risk of developing important clinical events. Thus, hemodialysis patients might not benefit from a so-called well-balanced diet with regard to the food groups that were highlighted as beneficial, that is, meat, fish, and vegetables.

POLYUNSATURATED FATTY ACIDS

Among the various nutritional factors, fish and polyunsaturated fatty acids (PUFA) consumption has been favorably related to the prognosis and progression of kidney function, mainly due to the antiinflammatory properties of PUFA, as well as to its negative correlation with blood pressure levels, a risk factor for chronic kidney disease. A large cross-sectional study that recruited 5316 participants without diabetes that examined the role of several macronutrients intake including total-, animal-, and plant-protein, carbohydrates, simple sugar, fructose, total fat, saturated fatty acids, poly- and monounsaturated-fatty acids (PUFA and MUFA), and n-3 and n-6 fatty acids with kidney function, reported that kidney function was better in the highest quartile of plant protein intake, PUFA, and n-6 fatty acids. In contrast, the risk of CKD increased in the highest quartile of animal protein intake as compared to the lowest quartile, independently of hypertension status. However, it could be supported that the later associations may be confounded by overall healthy dietary habits and other favorable lifestyle behaviors, like physical activity.

In another prospective study with 2600 aged ≥50 years, the association between dietary intakes of PUFA (n-3, n-6, and α-linolenic acid), fish, and the prevalence of chronic kidney disease was investigated. Participants in the highest quartile of long-chain n-3 PUFA intake had a significantly reduced likelihood of having CKD as compared to subjects in the lowest quartile of intake. Moreover, α-linolenic acid intake was positively associated with CKD, but total n-3 PUFA or total n-6 PUFA were not independently associated with CKD. The highest compared with the lowest quartile of fish consumption was associated with a significantly reduced likelihood of CKD, independently of several potential confounders. The authors suggested that adherence to a diet rich in n-3 PUFA and fish could have a beneficial role in maintaining healthy kidney function for middle-aged subjects. This study was not focused on elderly subjects that are the age group with the highest kidney function failure.

As data from observational studies concerning fish consumption in elderly population were limited, a published work in the IKARIA study aimed to investigate the association of fish consumption with kidney function, under the context of a healthy dietary pattern, the Mediterranean diet, in a long-lived population of elderly people that participated. The IKARIA study, which was a cross-sectional health survey carried out in the Province of Ikaria Island from June 2009 to October 2009, evaluated, among several other factors, nutritional habits and renal function. During this period, 673 elderly males and females (ie, above the age of 65 years), all long-term residents of the island, were voluntarily enrolled in the study; 328 of the participants were males (48.7%, mean age 75 ± 7 years, 3% were >90 years) and 339 were females (50.3%, mean age 75 ± 6 years, 2.3% were >90 years). The main interest of this cohort of subjects was that Ikaria island inhabitants have been recognized as having one of the highest longevity rates universally with a high percentage of healthy aging. While in Europe only 0.1% of population lives long (over 90 years old), in Ikaria Island the percentage of longevity rises 10-fold, and approaches 1%. This study revealed that after adjusting for physical activity status, MedDietScore, and history of hypertension and diabetes fish consumption (g/day) was positively associated with higher CCr rate in both males and females. Furthermore, for every 100 g/day increase of fish consumption, a significant increase of the likelihood of having good CCr levels, that is, CCr >60 mL, by 121% is anticipated. Additionally, it was evident that although the beneficial effects of fish intake on kidney function was apparent irrespective of the level of adherence to the traditional Mediterranean dietary pattern, the effect was much higher among females who had higher adherence rate to this healthy dietary pattern. This finding was more prominent among females who closely followed the Mediterranean diet. Moreover, the beneficial effects of long-term fish intake on kidney function were observed in the elderly suffering from several other comorbidities, extending by this way the current scientific knowledge about the benefits of PUFA through fish consumption on kidney function in middle-aged or diseased-only populations.

MEDITERRANEAN TYPE OF DIET

Lifestyle modification has been proposed as a simple, relatively inexpensive approach for the improvement of kidney function. Although, there is no specific dietary pattern for preventing kidney failure, excessive intake of protein, sodium, potassium, and alcohol have been shown to be correlated with the progression of kidney failure. Especially regarding protein intake, not only the amount but also the composition of proteins has an impact on kidney function, as proteins coming from animal sources (with saturated fatty acids) seem to have a harmful impact. On the contrary, the Mediterranean type of diet has been linked with improved rates of CCr. Such a diet was reported to be high in anti-inflammatory and antioxidative properties, which are attributed to the high content in fruits, vegetables, legumes, cereals, and olive oil as the main added lipid, with moderate to low animal protein intake and alcohol drinking. Over the past 25 years, several clinical studies have revealed the beneficial role of omega-3 (n-3) polyunsaturated fatty acids (PUFA) on normal growth and development of human's cell. Nowadays, n-3 fatty acids intake seems to play an important role in the primary and secondary prevention of atherosclerosis, arterial hypertension, diabetes mellitus, inflammatory and autoimmune disorders, kidney failure, and cancer. Modern agriculture technology has diminished n-3 fatty acids content in many foods, giving an emphasis on the production of vegetable oils from corn, sunflower seeds, safflower seeds, cottonseed, and soybeans. The essential n-3 fatty acids, mostly derived from fish consumption, replace n-6 fatty acids in platelets' erythrocytes and liver cells, resulting in decreased production of prostaglandin E2 and thromboxane A2, and thus in the inhibition of platelet aggregation and leukocytes chemotaxis and adherence. In the CHIANTI study, a population-based epidemiological study in elderly individuals, low plasma PUFA was associated with kidney dysfunction over the 3 years of follow-up; while the relative concentration of n-3 fatty acid was independently and significantly associated with low circulating levels of IL-6, IL-1ra, and TNF-alpha, and high circulating levels of sIL-6r and TGF-beta, as well as downregulating platelet activating factor (PAF) biosynthesis and upregulating PAF catabolism. Omega-3 fatty acids supplementation has shown a beneficial impact in chronic kidney disease, although the utility and optimal dosing needs to be more clearly defined. Even in the case of diabetes mellitus, n-3 supplementation has shown to reduce the probability of presenting diabetic nephropathy, preventing albuminuria, glomerulosclerosis, tubulointerstitial fibrosis, inflammation, and arterial hypertension associated with long-term diabetes.

As regard to the Mediterranean dietary pattern, the Northern Manhattan Study, a prospective, multiethnic, observational cohort collected data from 900 participants between 1993 and 2008, took serum creatinine measurements a mean 6.9 years apart. Mean baseline age of the participants was 64 years, 59% of them were women and the mean baseline estimated GFR was 83.1 mL/min/1.73 m^2. The primary outcome was incident estimated GFR at <60 mL/min/1.73 m^2 using the Modification of Diet in Renal Disease formula. The incident estimated GFR of <60 mL/min/1.73 m^2 developed in 14% of the participants within the follow-up period. In adjusted models, every 1-point increase in the MeDi score, indicating increasing adherence to a Mediterranean diet, was associated with decreased odds of incident estimated GFR of <60 mL/min/1.73 m^2 (odds ratio, 0.83; 95% confidence interval, 0.71 to 0.96). Thus it can be presumed that adherence to a Mediterranean diet was associated with a reduced incidence of estimated GFR of <60 mL/min/1.73 m^2 in a multiethnic cohort, with a relatively high mean age of 64 years at baseline, suggesting that this association is also important for elderly subjects.

The definition of the Mediterranean diet is difficult, as for all dietary patterns. The most common definition is the one that was firstly introduced by UNESCO that suggests more a lifestyle pattern than a dietary pattern. Specifically, UNESCO define Mediterranean diet as "a set of skills, knowledge, rituals, symbols and traditions concerning crops, harvesting, fishing, animal husbandry, conservation, processing, cooking, and particularly the sharing and consumption of food. Eating together is the foundation of the cultural identity and continuity of communities throughout the Mediterranean basin. It is a moment of social exchange and communication, an affirmation and renewal of family, group or community identity. The Mediterranean diet emphasizes values of hospitality, neighbourliness, intercultural dialogue and creativity, and a way of life guided by respect for diversity. It plays a vital role in cultural spaces, festivals and celebrations, bringing together people of all ages, conditions and social classes. It includes the craftsmanship and production of traditional receptacles for the transport, preservation and consumption of food, including ceramic plates and glasses. Markets also play a key role as spaces for cultivating and transmitting the Mediterranean diet during the daily practice of exchange, agreement and mutual respect."

HYPERURICEMIA AND DIET

Hyperuricemia (ie, serum uric acid (UA) concentration above 450 μmol/L or 7.0 mg/dL in males and 360 μmol/L or 6.0 mg/dL in females), has been linked with increased oxidative stress, and, in some studies, with increased cardiovascular morbidity and mortality. Serum UA concentration reflects the interactions of four major processes: dietary purine intake, endogenous purine metabolism, urinary urate excretion, and intestinal uricolysis. Among several factors, like high arterial blood pressure levels, diuretic and aspirin use, excess body weight, male sex, alcohol abuse, aging, renal insufficiency and family history, diet has been involved in the development of hyperuricemia and gout. Experimental metabolic studies, mainly in animals, have illustrated a link between purine-rich diets and increased serum UA levels. Furthermore, a positive association between protein intake from animal sources and prevalence of hyperuricemia, as well as an inverse association with protein from plant sources has been revealed. Among other therapeutic modalities, diet has been associated with the prevention of hyperuricemia, as consumption with purified purines increase serum UA levels in animal and humans. It has been estimated that about two-thirds of the daily purine load is generated endogenously from turnover of cells, while one-third is derived from the diet. Purine-rich foods include animal meats (ie, beef, pork, lamb, organ meats, and meat extracts), seafood (ie, fish fillets, tuna, shrimp, lobster, clams, etc.) and plants (ie, yeast extracts, peas, beans, lentils, asparagus, and mushrooms). By contrast, dairy products (ie, milk, cheese, yogurt, ice cream), grains and their products (ie, bread, pasta, cereals), vegetables, fruits, nuts, sugars, and sweets are low in purines. In this study those individuals in the highest tertile of UA levels showed lower protein consumption from milk, whereas they had higher total protein intake and higher protein intake from meat sources, compared with those in the lowest UA tertile. The Third National Health and Nutrition Examination Survey (NHANES-III) for the years 1988—1994, examined the relationship between the intake of purine-rich foods, proteins, and dairy products and serum UA levels. Increased meat and seafood consumption were associated with higher serum uric acid levels, but, interestingly the total protein intake was not. By contrast, consumption of dairy products was inversely associated with serum UA levels, while consumption of purine-rich foods, like meat and products, was associated with repeated gout attacks. The Mediterranean type of diet is a widely studied dietary pattern that has been described in the early 1960s in the northern Mediterranean basin and emphasizes the consumption of fat, primarily from foods high in monounsaturated fatty acids, high consumption of fruits, vegetables, legumes, nuts, whole grain cereals, and moderate consumption of fish, poultry, and wine, with relatively low consumption of red meat and meat products. Due to the antioxidants and anti-inflammatory properties of the aforementioned dietary pattern, it has also been recognized as a nonpharmaceutical means for the prevention of renal failure, CVD, and some types of cancer, even in elderly populations. In recent studies, it seems that, even a slight increase in the MedDietScore (which reflects higher adherence to this healthy dietary pattern) was associated with a lower likelihood of having hyperuricemia. Although the effects of a Mediterranean type of diet on the incidence of hyperuricemia have not been studied thoroughly up to now, there is experimental evidence to indicate that diets enriched in both gamma-linolenic acid in plant seed oil and eicosapentaenoic acid in fish oil, significantly suppress urate crystal-induced inflammation in animal models. The antioxidative effects of various food groups that are included in the Mediterranean diet, like fruits and vegetables, as well as red wine and olive oil, have already been reported. Moreover, Mediterranean diet has been associated with increased total antioxidant capacity and decreased oxidized LDL levels, which may partially explain the beneficial role of this diet on the cardiovascular system. In particular, vegetables, fruit and olive oil that have a central position in the Mediterranean diet, contain a variety of compounds with an antioxidant capacity, such as vitamin C and E, carotenoids and polyphenols, and have been linked with low oxidative stress in the general population.

High UA levels have been associated with high alcohol intake. In an observational study that studied only males for a follow-up period of 12 years, alcohol drinking was strongly and independently associated with an increased risk of gout. Specifically, the risk of gout was 2.5 times higher among those who consumed 50 g or more alcohol per day compared with those who abstained from alcohol. Additionally, moderate wine consumption (2 glasses/day) showed no significant effect on the aforementioned risk. Several mechanisms have been proposed linking alcohol consumption with hyperuricemia. It seems that during excess consumption, alcohol is converted to lactic acid, which in turn inhibits UA secretion by the renal proximal tubules, leading to reduced renal UA excretion. Also, chronic alcohol consumption increases purine and UA production by accelerating the degradation of adenosine triphosphate to adenosine monophosphate. In the NHANES-III study, alcohol drinking was associated with higher serum UA levels in both genders. Mediterranean type of diet is characterized by moderate alcohol consumption; this inherent characteristic of the traditional Mediterranean diet could be associated with lower UA levels, and consequently gout risk. In the present work a positive relationship was observed between UA levels and alcohol intake, confirming the previously mentioned knowledge; however, the majority of the participants reported moderate alcohol drinking (ie, about 1–3 wine glasses per day) conforming with this dietary prototype, a fact that could partially explain the inverse relationship observed between adherence to the Mediterranean diet and UA levels. In another cross-sectional study of middle-aged Chinese people, where the usual dietary habits are plant-based, hyperuricemia was positively associated with animal source protein and inversely associated with plant source protein consumption; although those associations were insignificant in multivariate analysis. Furthermore, soy food consumption was inversely associated with hyperuricemia.

POLYPHENOLS AND RENAL FUNCTION

The green tea extract contains several polyphenols. The major polyphenol in the extract was epigallocatechin gallate (~50%). Previous studies showed that epicatechin and epicatechin gallate had similar protective effects on liver ischemia/reperfusion injury and liver transplantation as green tea extracts containing multiple polyphenol components. In some other studies, the relative activities of the various polyphenolic components to inhibit oxidation and injury were variable. It was also shown that a combination of epigallocatechin gallate, epicatechin gallate, epigallocatechin, and epicatechin in the molar ratio 5:2:2:1 provided optimal protective effects against lipid peroxidation. Studies should be performed in the future to evaluate the efficacies of each polyphenolic component and various combinations of polyphenols on mitochondrial biogenesis (MB) in cultured renal cells and in vivo. Green tea polyphenols are free radical and singlet oxygen scavengers. Beneficial effects of green tea polyphenols in the prevention/treatment of cardiovascular, hepatic, renal, neural, pulmonary and intestinal diseases, cancer, diabetes, arthritis, shock, and decreases in ischemia/reperfusion injury and drug/chemical toxicity in various organs/tissues have been widely reported and many of these effects are presumably due to their antioxidant and anti-inflammatory properties.

It is important to state that the food that contains the phenolic agents seem to play a very important role in this association. In another clinical trial, the phenolic contents in olive oil were investigated for their effect on urinary proteomic biomarkers. The authors evaluated the impact of supplementation with two types of olive oil, either low or high in phenolics, Overall 69 apparently healthy participants were randomly allocated to supplementation with a daily 20-mL dose of olive oil either low or high in phenolics

(18 compared with 286 mg caffeic acid equivalents per kg, respectively) for 6 weeks. Urinary proteomic biomarkers were measured at baseline and 3 and 6 weeks. The authors claimed that consumption of both olive oil types improved the proteomic CVD score at endpoint compared with baseline, but not for CKD or diabetes proteomic biomarkers. It should be mentioned that the study sample were middle-aged and apparently healthy subjects and that the study sample was relatively small. The phenolic content of several foods could be beneficial for kidney function, but the specific food composition might play a moderating role in this association.

It is a fact that renal failure represents a major public health issue related with increased total mortality, especially among elders. Thus, there is an emerging need to implement preventive strategies in order to fight the upcoming burden on this condition. Based on the present work, it was observed that regular fish consumption, independent of overall dietary and other measured lifestyle habits, seems to be a nonpharmacological means of the maintenance of preserved kidney function among the elderly. This is in accordance with previous investigations on the role of omega-3 PUFA on kidney failure in terms of secondary prevention. Therefore, public health strategies could give further attention on the role of fish, and omega-3 fatty acids consumption in the fight against kidney failure, in terms of primary prevention among the elders.

SUMMARY

The kidney participates in body homeostasis by regulating acid-base balance, electrolyte concentrations, extracellular fluid volume, and arterial blood pressure. During aging there is a decline in renal function, due to a decrease in the structure and function ability of the nephrons, accumulation of fibrosis, and reduction in blood flow, which is more evident in the case of systolic heart failure. Between lifestyle factors, diet seems to play an important role in the prevention of renal failure. Although, there is no specific dietary pattern for preventing renal failure, excessive intake of protein, sodium, potassium, and alcohol have shown to be related with the progression of renal failure even among elderly individuals; while the well-balanced Mediterranean type of diet has been related with preserved renal function. The role of dietary habits may be related with the observation that chronic kidney disease is also associated with increased inflammation process, including plasma homocysteine, fibrinogen, and uric acid levels. This is especially evident in the elderly, where chronic kidney disease is also associated with oxidative stress and increased circulating levels of inflammatory markers, phosphate retention with medial vascular calcification, increased parathyroid hormone concentrations causing valvular calcification and dysfunction, anemia, and left ventricular hypertrophy. public health strategies could give further attention on the role of fish, and omega-3 fatty acids consumption in the fight against kidney failure, in terms of primary prevention among the elders.

References

[1] Clapp WL. Renal anatomy. In: Zhou XJ, Laszik Z, Nadasdy T, D'Agati VD, Silva FG, editors. Silva's diagnostic renal pathology. New York: Cambridge University Press; 2009.

[2] Guyton A, Hall J. Chapter 26: urine formation by the kidneys: I. Glomerular filtration, renal blood flow, and their control. In: Gruliow R, editor. Textbook of medical physiology (book). 11th ed. Philadelphia (PA): Elsevier Inc.; 2006. p. 308–25.

[3] Essue BM, Wong G, Chapman J, Li Q, Jan S. How are patients managing with the costs of care for chronic kidney disease in Australia? A cross-sectional study. BMC Nephrol 2013;14:5.

[4] Coresh J, Astor B, Sarnak M. Evidence for increased cardiovascular disease risk in patients with chronic kidney disease. Curr Opin Nephrol Hypertens 2004;13:73–81.

[5] Weiner DE, Tighiouart H, Amin MG, Stark PC, MacLeod B, Griffith JL, et al. Chronic kidney disease as a risk factor for cardiovascular disease and all-cause mortality: a pooled analysis of community-based studies. J Am Soc Nephrol 2004;15:1307–15.

[6] Shlipak MG, Fried LF, Crump C, Bleyer AJ, Manolio TA, Tracy RP, et al. Elevations of inflammatory and procoagulant biomarkers in elderly persons with renal insufficiency. Circulation 2003;107:87–92.

[7] Weiner DE, Tighiouart H, Stark PC, Amin MG, MacLeod B, Griffith JL, et al. Kidney disease as a risk factor for recurrent cardiovascular disease and mortality. Am J Kidney Dis 2004;44:198–206.

[8] Schrier RW. Role of diminished renal function in cardiovascular mortality. J Am Coll Cardiol 2006;47:1–8.

[9] Panagiotakos DB, Pitsavos C, Stefanadis C. Dietary patterns: a Mediterranean diet score and its relation to clinical and biological markers of cardiovascular disease risk. Nutr Metab Cardiovasc Dis 2006;16:559–68.

[10] (a)Cockcroft DW, Gault MH. Prediction of creatinine clearance from serum creatinine. Nephron 1976;16:31–41.(b)National Kidney Foundation. K/DOQI clinical practice guidelines for chronic kidney disease: evaluation, classification, and stratification. Am J Kidney Dis 2002;39:S1–266.

[11] Schrier RW. Body fluid volume regulation in health and disease: a unifying hypothesis. Ann Intern Med 1990;113:155–9.

[12] Go A, Chertow G, Fan D, McCulloch C, Hsu CY. Chronic kidney disease and the risks of death, cardiovascular events, and hospitalization. N Engl J Med 2004;351:1296–305.

[13] Murphy S, Rigatto C, Parfrey P. Cardiac disease in chronic renal disease. In: Schrier RW, editor. Diseases of the kidney and urinary tract. 7th ed. Philadelphia (PA): Lippincott Williams Wilkins; 2001. p. 2795–814.

[14] Fang J, Alderman MH. Serum UA and cardiovascular mortality the NHANES I epidemiologic follow-up study, 1971–1992. National Health and Nutrition Examination Survey. J Am Med Assoc 2000;283:2404–10.

[15] Chrysohoou C, Panagiotakos DB, Pitsavos C, Das UN, Stefanadis C. Adherence to the Mediterranean diet attenuates inflammation and coagulation process in healthy adults: the ATTICA study. J Am Coll Cardiol 2004;44:152–8.

[16] Panagiotakos DB, Matalas AL. Back to the ancient diet: a matter of urgency for Southern Mediterranean countries. Nutr Metab Cardiovasc Dis 2009.

[17] Trichopoulou A, Costacou T, Bamia C, Trichopoulos D. Adherence to a Mediterranean diet and survival in a Greek population. N Engl J Med 2003;348:2599–608.

[18] O'Meara E, Chong KS, Gardner RS, Jardine AG, Neilly JB, McDonagh TA. The Modification of Diet in Renal Disease (MDRD) equations provide valid estimations of glomerular filtration rates in patients with advanced heart failure. Eur J Heart Fail 2006;8:63–7.

[19] White SL, Polkinghorne KR, Cass A, Shaw JE, Atkins RC, Chadban SJ. Alcohol consumption and 5-year onset of chronic kidney disease: the AusDiab study. Nephrol Dial Transplant 2009;24:2464–72.

[20] Verbalis JG. Renal function and vasopressin during marathon running. Sports Med 2007;37:455–8.

[21] Finkelstein J, Joshi A, Hise MK. Association of physical activity and renal function in subjects with and without metabolic syndrome: a review of the Third National Health and Nutrition Examination Survey (NHANES III). Am J Kidney Dis 2006;48:372–82.

[22] Ruf JC. Overview of epidemiological studies on wine, health and mortality. Drugs Exp Clin Res 2003;29:173–9.

[23] Johnson RJ, Rideout BA. Uric acid and diet – insights into the epidemic of cardiovascular disease. N Engl J Med 2004;350:1071–3.

[24] Jee SH, Lee SY, Kim MT. Serum UA and risk of death from cancer, cardiovascular disease or all causes in men. Eur J Cardiovasc Prev Rehabil 2004;11:185–91.

[25] Niskanen LK, Laaksonen DE, Nyyssonen K, Alfthan G, Lakka HM, Lakka TA, et al. UA level as a risk factor for cardiovascular and all-cause mortality in middle-aged men: a prospective cohort study. Arch Intern Med 2004;164:1546–51.

[26] Roubenoff R, Klag MJ, Mead LA, Liang KY, Seidler AJ, Hochberg MC. Incidence and risk factors for gout in white men. J Am Med Assoc 1991;266:3004–7.

[27] Choi HK, Curhan G. Coffee consumption and risk of incident gout in women: the Nurses' Health Study. Am J Clin Nutr 2010;92:922–7.

[28] Villegas R, Xiang YB, Elasy T, Xu WH, Cai H, Cai Q, et al. Purine-rich foods, protein intake, and the prevalence of hyperuricemia: the Shanghai Men's Health Study. Nutr Metab Cardiovasc Dis 2012;22:409–16.

[29] Choi HK, Liu S, Curhan G. Intake of purine-rich foods, protein and dairy products and relationship to serum levels of uric acid. The Third National Health and Nutrition Examination Survey. Arthritis Rheum 2005;52:283–9.

[30] Matalas A, Zampelas A, Stavrinos V. The Mediterranean Diet. 1st ed. New York: CRC Press; 2001.

[31] Chrysohoou C, Panagiotakos DB, Pitsavos C, Skoumas J, Zeimbekis A, Kastorini CM, et al. Adherence to the Mediterranean diet is associated with renal function among healthy adults: the ATTICA study. J Ren Nutr 2010;20:176–84.

[32] Chrysohoou C, Tsitsinakis G, Siassos G, Psaltopoulou T, Galiatsatos N, Metaxa V, et al. Fish consumption moderates depressive symptomatology in elderly men and women from the IKARIA study. Cardiol Res Pract 2010;2011:219578.

[33] Wen CP, David Cheng TY, Chan HT, Tsai MK, Chung WS, Tsai SP, et al. Is high serum uric acid a risk marker or a target for treatment? Examination of its independent effect in a large cohort with low cardiovascular risk. Am J Kidney Dis 2010;56:273–88.

[34] Strazzullo P, Puig JG. Uric acid and oxidative stress: relative impact on cardiovascular risk? Nutr Metab Cardiovasc Dis 2007;17:409–14.

[35] Filippatos GS, Ahmed MI, Gladden JD, Mujib M, Aban IB, Love TE, et al. Hyperuricaemia, chronic kidney disease, and outcomes in heart failure: potential mechanistic insights from epidemiological data. Eur Heart J 2011;32:712–20.

[36] Tate GA, Mandell BF, Karmali RA, et al. Suppression of monosodium urate crystal-induced acute inflammation by diets enriched with gamma-linolenic acid and eicosapentaenoic acid. Arthritis Rheum 1988;31:1543–51.

[37] Carluccio MA, Siculella L, Ancora MA, Massaro M, Scoditti E, Storelli C, et al. Olive oil and red wine antioxidant polyphenols inhibit endothelial activation: antiatherogenic properties of Mediterranean diet phytochemicals. Arterioscler Thromb Vasc Biol 2003;23:622–9.

[38] Pitsavos C, Panagiotakos DB, Tzima N, Chrysohoou C, Economou M, Zampelas A, et al. Adherence to the Mediterranean diet is associated with total antioxidant capacity in healthy adults: the ATTICA study. Am J Clin Nutr 2005;82:694–9.

[39] Briante R, Febbraio F, Nucci R. Antioxidant properties of low molecular weight phenols present in the Mediterranean diet. J Agric Food Chem 2003;51:6975–81.

[40] Fam AG. Gout: excess calories, purines, and alcohol intake and beyond. Response to a urate-lowering diet. J Rheumatol 2005;32:773–7.

[41] Lippi G, Montagnana M, Luca Salvagno G, Targher G, Cesare Guidi G. Epidemiological association between uric acid concentration in plasma, lipoprotein(a), and the traditional lipid profile. Clin Cardiol 2010;33:E76–80.

[42] Williams PT. Effects of diet, physical activity and performance, and body weight on incident gout in ostensibly healthy, vigorously active men. Am J Clin Nutr 2008;87:1480–7.

[43] Yuzbashian E, Asghari G, Mirmiran P, Hosseini FS, Azizi F. Associations of dietary macronutrients with glomerular filtration rate and kidney dysfunction: Tehran lipid and glucose study. J Nephrol 2015;28:173–80.

[44] Tsuruya K, Fukuma S, Wakita T, Ninomiya T, Nagata M, Yoshida H, et al. Dietary patterns and clinical outcomes in hemodialysis patients in Japan: a cohort study. PLoS One 2015;10:e0116677.

[45] Silva S, Bronze MR, Figueira ME, Siwy J, Mischak H, Combet E, et al. Impact of a 6-wk olive oil supplementation in healthy adults on urinary proteomic biomarkers of coronary artery disease, chronic kidney disease, and diabetes (types 1 and 2): a randomized, parallel, controlled, double-blind study. Am J Clin Nutr 2015;101:44–54.

[46] UNESCO. Intangible heritage lists [internet]. 2010 [cited at 14.02.12]. Available from: <www.unesco.org/culture/ich/en/RL/00394>.

[47] Khatri M, Moon YP, Scarmeas N, Gu Y, Gardener H, Cheung K, et al. The association between a Mediterranean-style diet and kidney function in the Northern Manhattan Study cohort. Clin J Am Soc Nephrol 2014;9:1868–75.

[48] Ash S, Campbell KL, Bogard J, Millichamp A. Nutrition prescription to achieve positive outcomes in chronic kidney disease: a systematic review. Nutrients 2014;6(1):416–51.

CHAPTER

32

The Role of Nutrition in Age-Related Eye Diseases

Bamini Gopinath

Centre for Vision Research, The Westmead Institute, The University of Sydney, Sydney, NSW, Australia

KEY FACTS

- Incidence of age-related eye disease is expected to rise with the aging population.
- Oxidative damage and inflammation are thought to be involved in the etiology of these age-related eye diseases.
- Recent research has focused on the hypothesis that nutritional factors could protect against these degenerative eye diseases, because they are amenable to medication, by acting on food habits or by supplementation with specific nutrients.
- Recommendation of food choices rich in vitamins C and E, beta-carotene, zinc, lutein and zeaxanthin, and omega-3 fatty acids to older adults might offer a cost-effective strategy to potentially prevent the onset and progression of disabling eye diseases.

Dictionary of Terms

- **Age-related macular degeneration (AMD):** AMD is an abnormality of the retinal pigment epithelium that leads to photoreceptor degeneration of the overlying central retina, or macula, and loss of central vision.
- **Cataract:** Cataract is when there is loss of lens clarity and media opacities secondary to denaturation of lens proteins and oxidative damage. This condition can range from being asymptomatic to causing moderate visual impairment to almost complete blindness.
- **Diabetic Macular Edema (DME):** DME is characterized by fluid accumulation within the macular layers of the retina, leading to distortion and blur.
- **Diabetic Retinopathy (DR):** Chronically high blood sugar from diabetes is associated with damage to the tiny blood vessels in the retina; leading to DR. DR can cause blood vessels in the retina to leak fluid or hemorrhage, distorting vision. In its most advanced stage, new abnormal blood vessels proliferate on the surface of the retina, which can lead to scarring and cell loss in the retina.
- **Glaucoma:** Glaucoma affects the optic nerve connecting the eye to the brain. Glaucoma is often caused by high intraocular pressure, a result of a blockage in the eye's drainage system.
- **Antioxidants:** Antioxidants are found in certain foods and may prevent some of the damage caused by free radicals by neutralising them. These include the nutrient antioxidants, vitamins A, C and E, and the minerals copper, zinc and selenium.

AGE-RELATED CHANGES IN THE EYE

The primary function of the eye is to receive and focus light and then transform that photic energy to chemical and electrical signals that are sent to the brain. There, they are recoded into the images that we describe when we talk about our sense of sight and seeing [1]. As the eye ages, it undergoes a number of physiologic changes that may increase susceptibility to disease [2]. Consequently, the incidence and prevalence of diseases such as age-related macular degeneration (AMD), glaucoma, and vascular occlusive diseases increases significantly with age [3].

Ocular structure, function, and blood flow undergo demonstrable age-related changes [3]. Specifically, these changes include loss of cells in the ganglion layer, loss of retinal pigment epithelial cells and photoreceptors, changes to the optic nerve, reduced blood flow, condensation of the vitreous gel, loss of endothelial cells, and meibomian gland dysfunction [2–5]. Moreover, visual function changes include decreased visual acuity, diminished visual field sensitivity, reduced contrast sensitivity, and increased dark adaptation threshold [3,6].

Age-related vascular changes that occur systemically also affect ocular vascular beds. Studies show that ocular blood flow generally diminishes with age, which may result from an atherosclerotic process and narrowing of the retinal vessels [2,3]. Vascular changes in the eye are thought to impede regulation of blood pressure and flow, limiting the exchange of nutrients and removal of metabolic waste and creating conditions of ischemia [2,3]. Dysfunction of the vascular endothelium, a monolayer of cells covering the inner surface of blood vessels, is a common pathological feature of a number of age-related diseases and is thought to be a factor in some ocular diseases, such as glaucoma [2,3]. In the aging eye, endothelial dysfunction leads to decreased production of nitric oxide, thereby increasing vascular tone and vasoconstriction, and restricting blood flow [2,3]. Reduced nitric oxide levels have been reported in the aqueous humor of patients with glaucoma, suggesting that vascular endothelial dysfunction plays a role in the pathophysiology of this disease [2,3].

With advanced age, changes also affect the ocular surface. Structural and functional changes to the cornea that occur with age can affect its ability to refract light and repair itself, and can leave it more vulnerable to infection [5]. As the cornea ages, there is an increase in epithelial permeability, which may represent a breakdown of epithelial barrier function [2,5]. Further, there is a gradual decrease in the number of corneal endothelial cells with age that may adversely affect endothelial function and leave the cornea more vulnerable to hypoxic stress [2,5].

DEGENERATIVE EYE DISEASES

The prevalence of visual impairment and blindness increases exponentially with age [7]. When decreased visual acuity and visual field loss become significant, they begin to affect patients' ability to perform activities of daily living, such as reading, writing, driving, and ambulating [8]. Therefore, visual impairment contributes significantly to disability among older adults, which represents a major public health challenge to our aging countries [9,10]. Worldwide, the main causes of blindness and visual impairment are degenerative eye diseases, which affect the different structures of the eye: the retina (AMD, diabetic retinopathy (DR), and macular edema), the lens (cataract), and the optic nerve (glaucoma) [11,12]. Their etiologies are multifactorial, involving genetic and environmental factors, and will be discussed in further detail below [12,13].

Cataract

Cataract is the most common cause of visual impairment and blindness worldwide, with approximately 17.7 million people blinded by this disease [11]. Cataract formation is a slow, progressive, and irreversible process that occurs at the site of the native crystalline lens. This lens, along with the cornea, serves to bring objects into focus by modulating refractive power according to object distance [8]. Aging brings about loss of lens clarity and media opacities secondary to denaturation of lens proteins and oxidative damage [14]. Symptoms include blurred vision, glare, halos around lights, poor vision at night and in dim settings, diminished contrast sensitivity, and monocular double vision [8]. The presence of cataract does not necessitate cataract surgery, as this condition can range from being asymptomatic to causing moderate visual impairment to almost complete blindness [8]. Patients are monitored for deteriorating visual acuity and/or symptoms that interfere with normal activities of daily living. Timing for cataract surgery is considered by weighing specific individual needs against the risks of surgery [8].

Risk factors for cataract aside from aging in otherwise healthy eyes include gender, illiteracy, rural residence, ultraviolet light exposure, smoking, certain medication intake, and an unhealthy lifestyle [15,16]. Other local ocular risk factors for cataract include trauma, inflammation, vitreoretinal surgery, and topical steroid use [8].

Age-Related Macular Degeneration

AMD is the leading cause of blindness among people of European descent who are over 65 years of age [7]. AMD thus poses a major public health problem with significant economic and social impact [17]. Advanced AMD can be nonneovascular (dry, atrophic, or nonexudative) or neovascular (wet or exudative).

Advanced AMD can result in loss of central visual acuity and lead to severe and permanent visual impairment and blindness. Whereas dry AMD accounts for 80–90% of all cases of advanced disease, more than 90% of AMD patients with severe loss of central vision manifest wet or exudative AMD [17,18]. AMD is an abnormality of the retinal pigment epithelium (RPE) that leads to photoreceptor degeneration of the overlying central retina, or macula, and loss of central vision [17,19,20]. AMD is characterized by subretinal deposits, known as *drusen*, that measure greater than 60 μm and hyper- or hypopigmentation of the RPE [17].

Although advancing age is the greatest risk factor associated with the development of AMD, environmental and lifestyle factors may significantly affect individual risk [17]. Smoking is an important, modifiable factor that has been consistently associated with a twofold increased risk for developing AMD (odds ratios [OR] ranging from 1.8 to 3) [21]. Oxidative stress and antioxidant depletion have been implicated in retinal damage from smoking, although the precise mechanism in AMD remains unclear [22,23]. Other factors that have been reported to influence risk for AMD include sunlight exposure, alcohol consumption, increased plasma fibrinogen levels, diet, hypertension, body mass index, and iris color [17,21,24]. Dietary parameters, antioxidants in particular, such as carotenoids, zinc, and vitamins A and E, may provide a protective benefit against AMD. These will be discussed in detail later in this chapter.

Epidemiological studies have demonstrated differences in the prevalence of AMD based on ethnicity, with prevalence among Caucasians being greater than that among nonwhite groups [25,26]. Such ethnic differences may reflect genetic as well as environmental risk factors [17]. AMD has a significant genetic component [27,28], as several twin studies have shown significantly higher concordance rates between monozygotic twins as compared to concordance between dizygotic twins [29,30]. Familial aggregation studies from general populations [31,32] and tertiary eye care centers [27] reveal that family members of individuals with AMD are at increased risk (2.4- to 19.8-fold) for developing the disease relative to individuals with no family history [17].

AMD is a common, complex disease with numerous associated AMD genetic loci [13]. Genetic linkage studies found several peaks in many chromosomes, including 1 and 10 where the initial AMD genes were located. Candidate gene studies based on genes associated with macular and retinal Mendelian diseases were done, but did not yield strong or consistent results [13]. Since 2005, at least 20 known genes have been confirmed for AMD [33]. Many are in the complement pathway [13]. Some genes play a role in pathways related to high-density lipoprotein cholesterol, collagen, and extracellular matrix and angiogenesis [34,35], and the pathway is not yet known for several genes [13]. Rare, highly penetrant mutations also contribute to AMD risk, including *CFH* R1210C, which is one of the first instances in which a common complex disease variant led to the discovery of a rare penetrant mutation and the first one reported for AMD [36]. Rare variants in the genes *C3*, *CFI*, and *C9* also confer high risk of AMD [37,38].

Several interactions between genetic and environmental factors for AMD have been reported. There is an interaction between docosahexaenoic acid (DHA) intake and *ARMS2/HTRA1* on risk of developing geographic atrophy [39]. Smoking increases risk for all *CFH* and *ARMS2/HTRA1* genotypes [40–42]. In monozygotic twins discordant for signs of AMD, smoking was heavier for the twin with the more advanced stage of AMD and dietary intakes of betaine and methionine were higher for twins with less advanced AMD [43]. These effects of smoking and diet suggest that epigenetic changes can modify gene expression and lead to different phenotypes in genetically identical individuals. Methylation changes were also evaluated in a small study of AMD-discordant twin pairs [13].

Glaucoma

Glaucoma is the leading cause of irreversible blindness worldwide, and the second-leading cause of blindness in the United States [44,45]. Of the various identified forms of glaucoma (eg, neovascular, pigmentary, mixed mechanism), the two major ones are open-angle, the most common form, and closed angle (a much less common form in the United States). The terminology is derived from the appearance of the anterior chamber angle (open or closed), which is critical to the diagnosis [2,44,45]. Glaucoma is further distinguished between primary (not having an identifiable cause) and secondary forms, and also by intraocular pressure (IOP), which may be elevated or normal (normal tension glaucoma) in patients with primary open-angle glaucoma (OAG) [2]. Vision loss associated with OAG is generally progressive and irreversible, affecting the peripheral visual field in the early disease stage and central visual acuity in the late disease stage [2,46].

The mechanisms underlying glaucoma are not well understood, although years of studies have shown that the largest risk factors are elevated IOP, age, and genetics [8,47]. Smoking, glucocorticoid use, and diabetes are potentially involved in glaucoma pathogenesis, but their effects are most likely mediated by genetic susceptibility [47]. Eleven genes and multiple loci have been identified as contributing factors [47]. These genes act by a number of mechanisms, including mechanical stress, ischemic/oxidative stress, and neurodegeneration [47].

Diabetic Eye Diseases

DR represents the most prevalent retinal vascular disease and is the leading cause of blindness among the working population in developed countries [8]. A recent pooled analysis of population-based studies on DR estimated that 35% of all patients with diabetes manifested some form of DR, 7% exhibited proliferative diabetic retinopathy (PDR), and 7% exhibited diabetic macular edema (DME) [8,48]. DME can manifest itself at any stage of DR and results from vascular permeability after endothelial damage by the chronically hyperglycemic state [8,49]. DME is characterized by fluid accumulation within the macular layers of the retina, leading to distortion and blur [8]. Diabetes also markedly increases the risk for cataract and diabetic cataracts develop at earlier ages than in nondiabetics. People with diabetes also have twice the risk of developing glaucoma, relative to nondiabetic persons [1].

Contributing risk factors for DR are duration of diabetes, hemoglobin A1C level, and blood pressure [8,48]. Other established risk factors for DR include dyslipidemia, ethnicity (African American, Hispanic, and South Asian), type 1 diabetes, pregnancy, puberty, and cataract surgery [8,49]. Among patients with DR, vision loss can result from DME and PDR manifested as macular ischemia, vitreous hemorrhage, and tractional retinal detachment [8]. Heritability has been estimated as high as 52% for the advanced form of proliferative retinopathy with retinal neovascularization [13]. However, genetic studies have not yet identified large or consistent genetic susceptibility loci for this disease [50].

THE RELATIONSHIP BETWEEN NUTRITION AND HEALTHY OCULAR STRUCTURE AND FUNCTION

Recent research has been increasingly focused on efforts to stop the progression of age-related eye diseases or to prevent the damage leading to these conditions. Hence, nutritional intervention is becoming recognized as a part of these efforts [51]. Compared to most other organs, the eye is particularly susceptible to oxidative damage due to its exposure to light and high metabolism. Recent literature indicates that nutrients important in vision health include vitamins and minerals with antioxidant functions (eg, vitamins C and E, carotenoids [lutein, zeaxanthin, β-carotene], zinc) [52], and compounds with anti-inflammatory properties (omega-3 fatty acids, eicosapentaenoic acid [EPA], DHA) [53] may ameliorate the risk for age-related eye diseases [51].

ANTIOXIDANTS AND AGE-RELATED EYE DISEASES

Oxidative mechanisms may play an important role in the pathogenesis of age-related eye disease, in particular cataract and AMD [54]. As the retina is particularly susceptible to oxidative stress, the use of antioxidants may prevent its damage due to reactive oxygen species. The latter plays a role in modulating the nuclear factor erythroid 2-related factor 2, involved in the regulation of expression of genes encoding antioxidant proteins [23]. Much effort has been engaged in the research of modulating antioxidant balance by appropriate diet or supplementation with specific micronutrients to prevent or delay the development or progression of age-related eye diseases [52]. The most prominent study being the Age-Related Eye Disease Study (AREDS), which was an 11-center, double-masked clinical trial designed to evaluate the effect of high-dose vitamins and zinc on AMD progression and visual acuity [55]. In 2001, after an average follow-up time of 6.3 years, the study reported that treatment with a combination of antioxidants and zinc reduced the risk of progression to advanced AMD (OR = 0.72; 95% CI = 0.52–0.98; $P = 0.007$). The risk reduction for those taking the formulation was about 25%. Analyses were also done for those most likely to benefit from an effective treatment, for whom the 5-year event rates were between 18% and 43%. Comparisons with placebo found a significant risk reduction for antioxidants plus zinc (OR = 0.66; 95% CI = 0.47–0.91; $P = 0.001$). At the conclusion of the trial, the study group recommended that persons with at least moderate risk of progression to advanced AMD should consider taking a supplement similar to the AREDS antioxidant plus zinc formulation [55].

Many studies have addressed the importance of antioxidants in the control of abnormalities in diabetic retinas [56,57]; however, many of these studies have indicated the inability of these antioxidants to lower blood hexose levels [56,58,81]. Other studies have indicated the inability of some antioxidants to inhibit lipid peroxidation in diabetic eyes [81]. In addition, significant oxidative damage has been demonstrated in human trabecular meshwork cells of patients with glaucoma [59], causing elevated IOP and visual field damage [60,61]. It is evident from the literature that oxidative stress mechanisms play a critical role in the pathogenesis of glaucoma [61] and that antioxidants could play a protective role against the development of glaucoma. Finally, laboratory and epidemiologic evidence linking oxidative stress to cataract formation have led investigators to assess the role of antioxidant intake in the development of age-related cataract. Supplementation with vitamins with antioxidant properties such as beta-carotene and vitamins C and E has been proposed as candidate interventions to prevent or slow progression of cataract [62]. Several observational studies have noted protective associations for various antioxidants. However, in totality, the evidence from the large number of observational studies that have examined this association has been inconsistent [62].

Vitamin C

Vitamin C, also known as ascorbic acid, is an important water-soluble vitamin. Vitamin C is available in many forms, but there is little scientific evidence that any one form is absorbed better or has more activity than another. Most experimental and clinical research uses ascorbic acid or sodium ascorbate [51]. Vitamin C is a highly effective antioxidant, protecting essential molecules in the body, such as proteins, lipids, carbohydrates, DNA, and RNA, from damage by free radicals and reactive oxygen species that can be generated during normal metabolism as well as through exposure to toxins and such pollutants as cigarette smoke [51]. The eye has a particularly high metabolic rate, and thus, requires antioxidant protection. Plasma concentrations of vitamin C, an indicator of intake, are related to levels in the eye tissue [63]. In the eye, vitamin C may also be able to regenerate other antioxidants, such as vitamin E [51,64].

Vitamin C is present in high levels in normal human lens, however, in experimental cataract, the levels of vitamin C have been shown to be reduced [65]. Other studies have also shown the anticataract action of vitamin C, whereby the mechanism is primarily as an antioxidant. Studies on animals have shown the importance of this vitamin against human age-related cataracts [66,67]. A robust analysis of observational studies indicates that vitamin C intake is also likely to be most effective in reducing the risk of nuclear cataract [68]. Decreases in risk of approximately 40% have been reported in a majority of studies for intakes above approximately 135 mg/day or blood concentrations of 6 μM of vitamin C [68]. However, anticataract studies using vitamin C also suggest that vitamin C can prevent only nuclear cataract and not cortical or posterior subcapsular cataracts in humans [69]. Hence, the efficacy of vitamin C in preventing all types of cataract in humans is unclear, and this has been shown to be the case in certain long-term studies in humans who used vitamin C as a supplement [65,70,71].

The protective effect of vitamin C on risk of developing or worsening of AMD remains equivocal, despite 14 studies having assessed the association between vitamin C intake or supplement use and AMD risk [52]. In an Italian study [72], serum vitamin C levels were significantly lower ($P < 0.05$) in subjects with late versus early AMD patients. In contrast, a prospective Australian study [73] found that higher vitamin C intake from diet and supplements was associated with increased risk for early AMD (intake in the 5th quintile, OR = 2.3; 95% CI = 1.3–4.0; P for trend = 0.002). While a prospective Dutch study [74] did not find that subjects with a higher dietary intake of vitamin C alone had reduced risk of AMD (per 1 − SD increase, relative risk [RR] = 1.02; 95% CI = 0.94–1.10), but they found that an above-median intake of all four nutrients, β-carotene, vitamin C, vitamin E, and zinc, was associated with a 35% reduced risk (HR = 0.65; 95% CI = 0.46–0.92) of AMD.

A single nucleotide polymorphism in the gene related to this vitamin C transporter has been significantly associated with a higher risk of primary OAG [75]. One study showed that vitamin C levels were lower in the glaucoma patients than in the control subjects, uric acid levels were higher, and levels of the other substances examined were not significantly different [76,77]. Higher intake of certain fruits and green leafy vegetables high in vitamins A and C and carotenoids were associated with a decreased likelihood of glaucoma in older African-American women [78,79]. Further, a large US prospective study showed that supplement consumption of vitamin C was associated with decreased odds of glaucoma (OR = 0.47; 95% CI = 0.23–0.97), but serum levels were not associated with glaucoma prevalence (OR = 0.94; 95% CI = 0.42–2.11) [80]. These incongruous results highlight the limitation of self-reported diagnosis and nutrient intake compared with objective measurements [79].

Vitamin C has been shown to act as an aldose reductase inhibitor of the hyperglycemia-induced polyol pathway, affecting retinal blood flow [81] in animal models and human supplementation trials [82]. However, the results of a clinical trial of an aldose reductase inhibitor did not prevent DR [83,87]. A large epidemiological study observed a nonsignificant relationship between retinopathy and the intake of vitamin C from food or from food and supplements combined in a sample of participants with type 2 diabetes [87]. Dietary antioxidant therapy did not appear to be an effective treatment for the prevention of diabetic complications in this sample and is associated with a higher risk of retinopathy in some subsamples of this population. However, there was a protective effect against retinopathy with the intake of vitamin C in supplements [87].

Vitamin E

Vitamin E can be described as a family of eight fat-soluble antioxidants: four tocopherols (α-, β-, γ-, and δ-) and four tocotrienols (α-, β-, γ-, and δ-). α-Tocopherol is the form of vitamin E that is actively maintained in the human body and also the major form in blood and tissues [51,84]. It is also the chemical form that meets the recommended daily allowance (RDA) for vitamin E. The main function of α-tocopherol in humans appears to be that of an antioxidant. Fats, which are an integral part of all cell membranes, are vulnerable to destruction through oxidation by free radicals [51]. α-Tocopherol attacks free radicals to prevent a chain reaction of lipid oxidation. This is important, given that the retina is highly concentrated in fatty acids [53]. When a molecule of α-tocopherol neutralizes a free radical, it is altered in such a way that its antioxidant capacity is lost. However, other antioxidants, such as vitamin C, are capable of regenerating the antioxidant ability of α-tocopherol [84]. Other functions of α-tocopherol that would be of benefit to ocular health include effects on the expression and activities of molecules and enzymes in immune and inflammatory cells. Further, α-tocopherol has been shown to inhibit platelet aggregation and to improve vasodilation [51,85].

Vitamin E supplements often provide significantly higher levels of this nutrient than can be obtained in the diet [52]. A variety of studies examined relationships between vitamin E supplement use, dietary intake and plasma levels, and risk for AMD. The results from the observational studies were mixed [52]. Delcourt et al. [86] found that subjects with a higher fasting blood α-tocopherol–lipid ratio had reduced risk of late AMD (neovascular AMD or geographic atrophy) (the highest vs lowest quintile of the ratio, OR = 0.18; 95% CI = 0.05–0.67; P for trend = 0.004), but they did not find any significant association with fasting blood α-tocopherol (not the ratio) levels. An Italian study [72] found that serum vitamin E levels and vitamin E/cholesterol ratios were significantly lower ($P <$ 0.05) in subjects with late AMD ($N = 29$) than in early ARM ($N = 19$) and in older control subjects ($N = 24$). A Dutch study [74] found that subjects with a higher dietary intake of vitamin E alone had reduced risk of AMD (per 1 − SD increase, RR = 0.92; 95% CI = 0.84–1.00), and an above-median intake of all four nutrients, β-carotene, vitamin C, vitamin E, and zinc, was associated with a 35% reduced risk (HR = 0.65; 95% CI = 0.46–0.92).

Similar to vitamin C, vitamin E has been shown to be primarily effective against age-related cataract [87], and earlier reports have also looked at the possibility of clinical application of this vitamin [88] against cataract [57]. Nevertheless, mounting experimental evidence and randomized human trials have clearly shown that vitamin E affords no protection against cataract in humans [57,89]. Specifically, in studies completed since 2007 (approximately 56,000 subjects enrolled), there also appeared to be no effect of vitamin E on risk of cataract in any part of the lens. This includes retrospective, cross-sectional studies as well as prospective studies regarding risk of cataract associated with vitamin E intake, plasma levels, or supplementation [68].

Vitamin E has been shown to improve insulin sensitivity in short-term supplementation trials [90,91]. Vitamin E has been shown to inhibit the hyperglycemia-induced diacylglycerol protein kinase C pathway in both human retinal tissue and diabetic rat retinal tissue [87,92]. A large US study observed different associations between vitamin E intake and the odds of DR among blacks and whites [87]. Associations were direct in whites and inverse, although not statistically significant, in blacks [87].

In the anterior chamber, the aqueous humor vitamin E content is low, but an increasing intake of vitamin E may affect the antioxidant defenses in the aqueous humor by increasing the concentration of glutathione [93]. In a large US study, the largest contributor to vitamin E intake was supplements, and no clear associations with duration or dose of vitamin E and primary OAG was observed [94]. Hence, the overall inverse associations with dietary vitamin E observed in this study were speculated to be due to chance or to foods with high vitamin E content containing other nutrients that may protect against primary OAG risk [94]. A US national population-based sample of adults aged 40 years and older also did not find conclusive evidence that supplementary consumption of vitamin E, as determined by self-report or by measurement of serum levels, is related to the self-reported prevalence of glaucomatous disease [80].

Zinc

Zinc is important in maintaining the health of the retina, given that zinc is an essential constituent of many enzymes [95] and needed for optimal metabolism of the eye [51]. Zinc ions are present in the enzyme superoxide dismutase, which plays an important role in scavenging superoxide radicals. As related to the eye, zinc plays important roles in antioxidant and immune function [51].

Zinc has been proposed to have a role in AMD prevention because of its structural role in antioxidant enzymes [96,97]. Zinc is found in high concentrations in regions of the retina that are affected by AMD [98]. Zinc in the retina and RPE is also believed to interact with taurine and vitamin A, modify photoreceptor plasma membranes, regulate the light-rhodopsin reaction, and modulate synaptic transmission [99]. Retinal zinc content has been shown to decline with age [97]. However, the benefit of zinc is still debated and a recent meta-analysis confirmed that available data about its role in prevention of AMD are inconclusive [97]. Specifically, this meta-analysis showed that based on the six cohort studies no conclusions can be made on the associations between dietary zinc intake (from foods and supplements) and the incidence of early, late, and any AMD since the results were inconsistent [97]. The strongest evidence for zinc in the treatment of AMD comes from AREDS [41]. AREDS showed that zinc supplementation alone reduced the risk of progression to advanced AMD in those with intermediate AMD and those with advanced AMD in one eye [55]. Treatment with zinc alone did not reduce the risk of loss of visual acuity in AREDS subjects. Based on observation alone, data suggest that zinc supplementation may improve visual acuity in early AMD patients, but not in patients with advanced AMD [55,97].

Although some epidemiological studies have shown zinc involvement in the development of cataracts, the lowest concentration of zinc in crystalline lenses has been detected in patients with mature age-related cataract, while the highest concentrations have been detected in patients with traumatic cataracts [100,101]. However, the results from AREDS done in the United States showed no beneficial effect of supplement zinc with cupric oxide on the development or progression of cataracts [100,102].

Zinc has been shown to protect the retina from diabetes-induced increased lipid peroxidation and decreased glutathione levels in rats either by stabilizing the membrane structure or by inducing metallothionein synthesis [81,103]. The precise mechanism by which zinc exerts its protective effects against retinal damage remains unclear, but there is a strong possibility that it could help decrease oxidative damage [81].

Carotenoids

The major carotenoids which are found in the lens and have been related to lens health are lutein, zeaxanthin, and riboflavin [52]. Levels of β-carotene, the best known carotenoid because of its importance as a vitamin A precursor, are vanishingly low in the lens [52]. β-Carotene is an orange pigment commonly found in fruits and vegetables and belongs to a class of compounds called carotenoids [51].

Data from over 17,000 subjects, including the Alpha Tocopherol Beta Carotene intervention, indicate that this carotenoid does not affect risk of cortical, nuclear, PSC, or "any" cataract or risk of cataract extraction [52,69,89,104]. Intakes or blood levels of α-carotene, lycopene, cryptoxanthin, and total carotenoids were also evaluated for possible relations with risk of onset or risk of progression of various forms of cataract, but records from as many as 190,000 subjects indicate little effect of these nutrients [89]. Prospective data from a US study showed that those who supplemented with vitamin A had a reduced risk of cortical cataract (OR = 0.42; 95% CI = 0.24−0.73) [15]. Further, Dherani et al. [104] found that those with the highest blood levels of retinol had a reduced risk of nuclear cataract, (OR = 0.56; 95% CI = 0.33−0.96) and mixed cataract (OR = 0.58; 95% CI = 0.37−0.91) [89]. In a large European study, high intakes of vitamin A were also associated with increased risk of any cataract (incidence risk ratio = 1.28; 95% CI = 1.07−1.54) [89,105].

An increased risk for the development of AMD in individuals with a high dietary intake of β-carotene has been reported [106,107]. Moreover, several adverse effects of supplementation with β-carotene have been described, in particular an increase in the likelihood of developing lung cancer [108]. Although some studies have reported no association [109], others have investigated the effect of dietary supplementation of β-carotene on the risk of lung cancer and found a direct relationship [110]. Taken together, this evidence discourages the use of β-carotene in the prevention of AMD due to the potential side effects and poor efficacy in reducing the risk of AMD [107,108]. Despite proscription against β-carotene supplementation in current smokers in AREDS2, β-carotene was still associated with a greater risk of lung cancer in AREDS2 participants (2% vs 0.9%), especially in those who had previously been smokers. This finding is clinically relevant, as 50% of participants in AREDS and AREDS2 with AMD were former smokers, and 91% of those who developed lung cancer in AREDS2 were former smokers [41,111].

There have been very few studies that have looked at β-carotene and glaucoma. However, there has been some research done around vitamin A intake (of which β-carotene is a precursor of) and glaucoma. A US national population-based sample of adults aged 40 years and older did not find conclusive evidence that supplementary consumption of vitamins A, as determined by self-report or by measurement of serum levels, is related to the self-reported prevalence of glaucomatous disease [80]. Coleman et al. [112] have investigated the effect of diet on glaucoma incidence among women, and found that the consumption of certain foods, such as collards, kale, and carrots, appeared to be protective against glaucoma, however, analysis of the constituent nutrients did not reveal a significant protective effect related to the intake of vitamin A. In an analysis examining the relationship between the consumption of various nutritional components and OAG confirmed by chart review Kang et al. [94] also found no appreciable effect of vitamin A intake on the incidence of glaucoma [80].

When β-carotene was administered with vitamins C, E, and other possible antioxidants, numerous diabetes-related changes to the retina were shown to be prevented [113,114]. Dene et al. showed that oxidative stress markers were reduced in the retina of diabetic rats after receiving a combination of antioxidant therapy which primarily included β-carotene [113]. One of the few epidemiological studies to assess the link between β-carotene and DR showed that among those taking insulin, increased intake of β-carotene was associated with a risk for severity of DR (OR = 3.31, $P = 0.003$, and 2.99, $P = 0.002$, respectively, for the ninth and tenth deciles compared to the first quintile) [115].

Lutein and Zeaxanthin

Lutein and zeaxanthin are carotenoids found in high quantities in green leafy vegetables. Unlike β-carotene, these two carotenoids do not have vitamin A activity [116]. Of the 20–30 carotenoids found in human blood and tissues [117], only lutein and zeaxanthin are found in the lens and retina [118,119]. Lutein and zeaxanthin are concentrated in the macula or central region of the retina, and are referred to as macular pigment. In addition to their role as antioxidants, lutein and zeaxanthin are believed to limit retinal oxidative damage by absorbing incoming blue light and/or quenching reactive oxygen species [51,120].

Most retrospective and prospective studies indicate that intake or blood levels of lutein and zeaxanthin do not modulate risk of cortical, nuclear, or posterior subcapsular cataract [68], although a Finnish study showed that those with the highest plasma levels of lutein (RR = 0.58; 95% CI = 0.35–0.98) and zeaxanthin (RR = 0.59; 95% CI = 0.35–0.99) had a reduced risk of nuclear cataract [68]. Prospective data from epidemiological studies [52] also suggest that elevated lutein and zeaxanthin status is associated with diminished risk of nuclear cataract [68]. A cross-sectional analysis of 1443 Indian subjects supports this data and revealed that high zeaxanthin blood levels were protective against nuclear cataract ($P < 0.03$) [104], but this was not observed in carotenoids in AREDS analysis [68,121]. Similarly, AREDS2 showed that supplementation with 10 mg of lutein and 2 mg of zeaxanthin had no effect on risk of any type of cataract, nor did it improve visual acuity. However, a subgroup analysis did show a beneficial effect of lutein and zeaxanthin on cataract risk (HR = 0.70; 95% CI = 0.53–0.94) in those patients with the lowest baseline intake of these carotenoids [89,122].

Some of the dietary sources of lutein and zeaxanthin have been largely investigated as a protective factor in AMD [108]. Primary studies have focused on the dietary intake of carotenoids, which is obtained from vegetables (eg, kale, spinach, and Brussels sprouts) [123], as well as supplements. The first evidence of an association between the consumption of fruits and vegetables and the risk of AMD was reported in 1988, with the publication of data obtained from the first National Health and Nutrition Examination Survey [124]. Some studies reported a direct association between the higher intake of lutein/zeaxanthin from foods and a reduced likelihood of AMD [125,126]. AREDS2 showed that adding lutein and zeaxanthin to the AREDS formula resulted in an additional beneficial effect of about 10% beyond the effects of the original AREDS formulation in reducing the risk of progressing to advanced AMD, and when β-carotene was removed, the incremental benefit increased to 18%, possibly due to amelioration of competitive absorption effects [41,111]. Those who derived the most benefit from the addition of lutein and zeaxanthin were those in the lowest quintile of dietary lutein and zeaxanthin intake [41,111]. Ma et al. reported that the intake of lutein and zeaxanthin supplements in patients with early AMD could improve the macular pigment optical density and visual function [127]. Also, in a large meta-analysis published by Ma et al. [128], it was revealed that a high dietary intake of lutein and zeaxanthin was useful in reducing the risk of late AMD, with no effect on early AMD [108].

In a recent US study of African-American women, higher intake of lutein/zeaxanthin showed a near significant trend toward reduced odds of glaucoma diagnosis [78]. A large US cohort study observed an inverse association observed between lutein/zeaxanthin and primary OAG risk in the 4-year lagged analyses, which the authors speculated may be a chance finding, and suggested that it deserves further evaluation, particularly in relation to high-tension glaucoma [94]. Further, an animal study showed that supplementation with lutein/zeaxanthin could protect against glaucoma optic neuropathy [129]. However, more prospective and intervention studies are warranted to conclusively establish the mechanisms by which lutein/zeaxanthin supplementation could reduce risk of glaucoma.

Lutein has been shown to attenuate oxidative stress in experimental models of early DR [130,131]. Animal studies have shown that zeaxanthin administration in diabetic rats prevents an increase in retinal oxidative stress and pro-inflammatory cytokines (eg, VEGF, ICAM-1) [132,133]. A community-based study of persons with type 2 diabetes showed a protective role for a higher combined lutein/zeaxanthin and lycopene concentration against DR, after adjustment for potential confounders [131]. However, epidemiological studies assessing the association between lutein/zeaxanthin and risk of diabetic eye diseases remain relatively scarce.

Polyunsaturated Fatty Acids

In addition to the antioxidants cited above, the omega-3 fatty acids DHA and EPA are thought to be important in AMD prevention. The omega-3 fatty acids have a number of actions that provide neuroprotective effects in the retina [51]. This includes modulation of metabolic processes affecting oxidative stress, inflammation, and vascularization [53]. DHA is a key fatty acid found in the retina, and is present in large amounts in this tissue [134]. Tissue DHA status affects retinal cell-signaling mechanisms involved in photo-transduction [53]. It has been suggested that atherosclerosis of the blood vessels that supply the retina contributes to the risk of AMD, similar to the mechanism involved in coronary heart disease [51], suggesting that the same dietary fats related to coronary heart disease may also be related to AMD [51,135]. In addition, long-chain omega-3 fatty acids may have another role in the function of the retina. Biophysical and biochemical properties of DHA may affect photoreceptor-membrane function by altering permeability, fluidity, thickness, lipid-phase properties, and the activation of membrane-bound proteins [51,53].

AREDS study has confirmed a protective effect of dietary total n-3 PUFA intake in neovascular AMD and central geographic atrophy [108,136,137]. Furthermore, Sangiovanni et al. confirmed the inverse association between dietary DHA intake and fish consumption and the risk of neovascular disease [136]. Conversely, dietary arachidonic acid intake was directly associated with neovascular AMD prevalence [136]. The results from the Blue Mountain Eye Study confirmed these associations, providing further evidence of the protective role of fish, n-3 PUFA, and a low intake of foods rich in linoleic acid [106]. Taken together, these supporting data motivated further studies to investigate the effects of specific fat supplements on AMD [107]. Johnson et al. reported a positive relationship between DHA supplementation and an increase in macular pigment optical density, after 4 months of treatment [138]. Omega-6 PUFA intake was reported as being directly associated with intermediate AMD signs [107]. Parekh and colleagues found no association between omega-3 PUFA and AMD. Moreover, in the sample analyzed by Parekh et al., a high correlation between the intake of omega-3 PUFA and omega-6 PUFA was found ($r = 0.8$). Nevertheless, several hypotheses have been made, such as: the polyunsaturated fatty acids may modulate the gastrointestinal uptake of lutein and zeaxanthin, their transport by lipoproteins, or their concentration in the macular area [107,139]. Similarly, higher fish intake was associated with a lower risk of AMD progression among patients with lower linoleic acid intake [140]. A subsequent study performed on the AREDS population showed a protective effect of DHA and EPA in the progression to advanced AMD. Moreover, consuming a diet rich in DHA protects against the progression of early AMD, particularly among subjects not taking AREDS supplements [107,141].

In contrast, a recent review of randomized controlled trials concluded that there was no sufficient evidence to support the role of increasing levels of dietary omega-3 PUFA to prevent or slow the progression of AMD [142]. Given these results, both omega-3 PUFA and omega-6 PUFA need further studies to determine the impact that these PUFA have on AMD [107].

A 42% lower risk of nuclear cataract was found in consumers of 0.5–1.42 g/day of omega-3 fatty acids (found in flaxseed, walnuts, salmon, shrimp, and other seafood) [143]. A 17% or 12% decreased risk of nuclear or any cataract extraction, respectively, was found in women with elevated intake of omega-3 fatty acids (specifically EPA and DHA) [144]. A large 6-year Spanish study [145] found that subjects with 0.05% and 0.2% of total energy intake as omega-6 fatty acids (found predominantly in oils such as sunflower, corn, and soybean oils) had a reduced risk of any cataract (OR = 0.54; 95% CI = 0.29–0.99) compared with those with the lowest intake. There was a trend of increased risk of extraction of any type of cataract with increased consumption of linoleic acid in a US women's study ($P = 0.04$) [144]. The increased consumption of linoleic acid (but not arachidonic acid) may explain the increased risk of nuclear cataract observed in those with high polyunsaturated fatty acid intake [68].

Risk factors such as dietary fat are believed to potentially influence the IOP [146]. In fact, the omega-6 derived eicosanoids include prostaglandin F2α, which has been demonstrated to lower IOP [147]. Indeed, an analog of prostaglandin F2α called latanoprost is a widespread antiglaucomatous drug that decreases IOP by 25–35% [148] by increasing uveoscleral outflow [146,149]. Some studies have reported an inverse association between omega-3 intake and glaucoma [150,151]. Accordingly, Ren et al. [151] found decreased omega-3 PUFA levels in glaucoma patients compared with their healthy siblings. A review article [152] postulated that cod liver oil, as a combination of vitamin A and omega 3 fatty acids, should be beneficial for the treatment of glaucoma. A French nationwide case-control study [153] showed that primary OAG was significantly associated with low consumption of fatty fish (OR = 2.14; 95% CI = 1.10–4.17; $P = 0.02$) and walnuts (OR = 2.02; 95% CI = 1.18–3.47; $P = 0.01$). These results suggested a protective effect of omega-3 fatty acids against primary OAG. Conversely, a recent Spanish cohort study [146] found no association between omega-3 intake and the risk of glaucoma but found a direct association between omega 3:6 ratio intake and risk of glaucoma.

Alterations in the levels and metabolism of PUFAs could trigger initiation and progression of DR and other related retinal diseases and associated angiogenic processes [154]. Studies have noted that increased dietary intake of omega-3 PUFAs prevents retinopathy in both type 1 diabetes [155] and a model of retinopathy of prematurity [156]. A recent study showed that DHA-rich dietary supplementation prevented retinal vessel loss as evidenced by the decreased number of acellular capillaries in retinas of type 2 diabetic animals [157].

CONCLUSIONS AND RECOMMENDATIONS

The incidence of age-related eye diseases is expected to rise with the aging of the population [51]. Attention has been increasingly focused on efforts to stop the progression of eye diseases or to prevent the damage leading to these conditions [51]. The hypothesis that antioxidant and anti-inflammatory nutrients may be of benefit in age-related eye health is plausible, given the role of oxidative damage and inflammation in the etiology of age-related eye diseases [51]. Great strides have been made for AMD in terms of prevention, slowing progression, and treatment of the neovascular form of the disease. However, treatments can be improved for the neovascular forms of AMD and need to be developed for the dry stages [13]. Cataract has a successful surgical intervention, but preventive measures and other types of therapies could provide enormous economic benefit [13]. DR can be reduced with cultural and lifestyle changes and behavioral modification, as well as improved therapeutic targets [13]. Glaucoma requires elucidation of more definitive environmental and genetic factors to enable earlier diagnosis, prevention, and intervention [13].

There is a growing interest in the role of nutritional factors in these diseases, because they are amenable to modification, by acting on food habits or by supplementation with specific nutrients [12]. To date, the evaluation of a single nutrient in the prevention of age-related eye diseases has not been entirely consistent. The inconsistencies among studies in terms of which nutrients and the amount of nutrients required for protection make it difficult to make specific recommendations for dietary intakes [51]. It is likely that nutrients are acting synergistically to provide protection. Therefore, it may be more practical to recommend food choices rich in vitamins C and E, β-carotene, zinc, lutein and zeaxanthin, and omega-3 fatty acids [51]. A healthy diet including a variety of fresh fruit and vegetables, legumes, lean meats, dairy, fish, and nuts will have many benefits and will be a good source of the antioxidant vitamins and minerals implicated in the etiology of age-related ocular health. Although long-term dietary intake should provide a more cost-effective strategy than supplementation, due to the unavailability of good nutrition and sufficient resources for lifestyles, which are associated with diminished risk for degenerative eye diseases, among the impoverished some supplementation might be considered [52]. Moreover, creating an awareness of nutrients and their dietary sources, perhaps through an educational tool [52] and public health campaigns, may be valuable in assisting older adults in the community to incorporate key foods/nutrients and/or supplements that could potentially prevent the onset or progression of disabling eye diseases.

SUMMARY

Worldwide the main causes of visual impairment and blindness are degenerative eye diseases, which affect the different structures of the eye: the retina (age-related macular degeneration, AMD; diabetic retinopathy, DR; and macular edema); the lens (cataract); and the optic nerve (glaucoma). Nutritional interventions are being increasingly recognized as important in stopping the progression of these age-related eye diseases or preventing the damage leading to these conditions. Recent research indicates that nutrients important in vision health include vitamins and minerals with antioxidant functions (eg, vitamins C and E, carotenoids, and zinc) and compounds with anti-inflammatory properties (omega-3 fatty acids), which may minimize the risk for age-related eye diseases. However, the evaluation of a single nutrient in the prevention of age-related eye diseases has not entirely been consistent. Future research should likely focus on how these nutrients are acting synergistically to provide protection.

References

[1] Chader GJ, Taylor A. Preface: the aging eye: normal changes, age-related diseases, and sight-saving approaches. Invest Ophthalmol Vis Sci 2013;54(14): ORSF1-ORSF4.

[2] Akpek EK, Smith RA. Overview of age-related ocular conditions. Am J Manag Care 2013;19(Suppl. 5):S67–75.

[3] Ehrlich R, Kheradiya NS, Winston DM, Moore DB, Wirostko B, Harris A. Age-related ocular vascular changes. Graefes Arch Clin Exp Ophthalmol 2009;247(5):583–91.

[4] Ding J, Sullivan DA. Aging and dry eye disease. Exp Gerontol 2012;47(7):483–90.

[5] Faragher RG, Mulholland B, Tuft SJ, Sandeman S, Khaw PT. Aging and the cornea. Br J Ophthalmol 1997;81(10):814–17.

[6] Salvi SM, Akhtar S, Currie Z. Ageing changes in the eye. Postgrad Med J 2006;82(971):581–7.

[7] Congdon N, O'Colmain B, Klaver CC, et al. Causes and prevalence of visual impairment among adults in the United States. Arch Ophthalmol (Chicago, IL: 1960) 2004;122(4):477–85.

[8] Voleti VB, Hubschman JP. Age-related eye disease. Maturitas 2013;75(1):29–33.

[9] Cigolle CT, Langa KM, Kabeto MU, Tian Z, Blaum CS. Geriatric conditions and disability: the health and retirement study. Ann Intern Med 2007;147(3):156–64.

[10] Laitinen A, Sainio P, Koskinen S, Rudanko SL, Laatikainen L, Aromaa A. The association between visual acuity and functional limitations: findings from a nationally representative population survey. Ophthalmic Epidemiol 2007;14(6):333–42.

[11] Resnikoff S, Pascolini D, Etya'ale D, et al. Global data on visual impairment in the year 2002. Bull World Health Organ 2004;82 (11):844–51.

[12] Delcourt C, Korobelnik J-F, Barberger-Gateau P, et al. Nutrition and age-related eye diseases: the ALIENOR (Antioxydants, LIpides Essentiels, Nutrition et Maladies OculaiRes) Study. J Nutr Health Aging 2010;14(10):854–61.

[13] Seddon JM. Genetic and environmental underpinnings to age-related ocular diseases. Invest Ophthalmol Vis Sci 2013;54(14): ORSF28-ORSF30.

[14] Michael R, Bron AJ. The ageing lens and cataract: a model of normal and pathological ageing. Philos Trans R Soc B Biol Sci 2011;366(1568):1278–92.

[15] Klein BEK, Knudtson MD, Lee KE, et al. Supplements and age-related eye conditions. The beaver dam eye study. Ophthalmology 2008;115(7):1203–8.

[16] Rao GN, Khanna R, Payal A. The global burden of cataract. Curr Opin Ophthalmol 2011;22(1):4–9.

[17] Chen Y, Bedell M, Zhang K. Age-related macular degeneration: genetic and environmental factors of disease. Mol Interv 2010;10(5):271–81.

[18] Bressler NM. Early detection and treatment of neovascular age-related macular degeneration. J Am Board Fam Pract 2002;15 (2):142–52.

[19] Haddad S, Chen CA, Santangelo SL, Seddon JM. The genetics of age-related macular degeneration: a review of progress to date. Surv Ophthalmol 2006;51(4):316–63.

[20] Rattner A, Nathans J. Macular degeneration: recent advances and therapeutic opportunities. Nat Rev Neurosci 2006;7(11):860–72.

[21] Smith W, Assink J, Klein R, et al. Risk factors for age-related macular degeneration: pooled findings from three continents. Ophthalmology 2001;108(4):697–704.

[22] Ni Dhubhghaill SS, Cahill MT, Campbell M, Cassidy L, Humphries MM, Humphries P. The pathophysiology of cigarette smoking and age-related macular degeneration. Adv Exp Med Biol 2010;664:437–46.

[23] Cano M, Thimmalappula R, Fujihara M, et al. Cigarette smoking, oxidative stress, the anti-oxidant response through Nrf2 signaling, and age-related macular degeneration. Vision Res 2010;50(7):652–64.

[24] Hirvela H, Luukinen H, Laara E, Sc L, Laatikainen L. Risk factors of age-related maculopathy in a population 70 years of age or older. Ophthalmology 1996;103(6):871–7.

[25] Klein R, Klein BE, Knudtson MD, et al. Prevalence of age-related macular degeneration in 4 racial/ethnic groups in the multi-ethnic study of atherosclerosis. Ophthalmology 2006;113 (3):373–80.

[26] Bressler SB, Munoz B, Solomon SD, West SK. Racial differences in the prevalence of age-related macular degeneration: the Salisbury Eye Evaluation (SEE) Project. Arch Ophthalmol (Chicago, IL: 1960) 2008;126(2):241–5.

[27] Seddon JM, Ajani UA, Mitchell BD. Familial aggregation of age-related maculopathy. Am J Ophthalmol 1997;123(2):199–206.

[28] Klaver CC, Wolfs RC, Assink JJ, van Duijn CM, Hofman A, de Jong PT. Genetic risk of age-related maculopathy. Population-based familial aggregation study. Arch Ophthalmol (Chicago, IL: 1960) 1998;116(12):1646–51.

[29] Hammond CJ, Webster AR, Snieder H, Bird AC, Gilbert CE, Spector TD. Genetic influence on early age-related maculopathy: a twin study. Ophthalmology 2002;109(4):730–6.

[30] Grizzard SW, Arnett D, Haag SL. Twin study of age-related macular degeneration. Ophthalmic Epidemiol 2003;10 (5):315–22.

[31] Smith W, Mitchell P. Family history and age-related maculopathy: the Blue Mountains Eye Study. Aust N Z J Ophthalmol 1998;26(3):203–6.

[32] Klein BE, Klein R, Lee KE, Moore EL, Danforth L. Risk of incident age-related eye diseases in people with an affected sibling: the beaver dam eye study. Am J Epidemiol 2001;154(3):207–11.

[33] Fritsche LG, Chen W, Schu M, et al. Seven new loci associated with age-related macular degeneration. Nat Genet 2013;45 (4):433–9 9e1–2.

[34] Neale BM, Fagerness J, Reynolds R, et al. Genome-wide association study of advanced age-related macular degeneration identifies a role of the hepatic lipase gene (LIPC). Proc Natl Acad Sci U S A 2010;107(16):7395–400.

[35] Yu Y, Bhangale TR, Fagerness J, et al. Common variants near FRK/COL10A1 and VEGFA are associated with advanced age-related macular degeneration. Hum Mol Genet 2011;20 (18):3699–709.

[36] Raychaudhuri S, Iartchouk O, Chin K, et al. A rare penetrant mutation in CFH confers high risk of age-related macular degeneration. Nat Genet 2011;43(12):1232–6.

[37] Seddon JM, Yu Y, Miller EC, et al. Rare variants in CFI, C3 and C9 are associated with high risk of advanced age-related macular degeneration. Nat Genet 2013;45(11):1366–70.

[38] van de Ven JP, Nilsson SC, Tan PL, et al. A functional variant in the CFI gene confers a high risk of age-related macular degeneration. Nat Genet 2013;45(7):813–17.

[39] Reynolds R, Rosner B, Seddon JM. Dietary omega-3 fatty acids, other fat intake, genetic susceptibility, and progression to incident geographic atrophy. Ophthalmology 2013;120(5):1020–8.

[40] Lim LS, Mitchell P, Seddon JM, Holz FG, Wong TY. Age-related macular degeneration. Lancet 2012;379(9827):1728–38.

[41] Age-Related Eye Disease Study 2 Research Group. Lutein + zeaxanthin and omega-3 fatty acids for age-related macular degeneration: the Age-Related Eye Disease Study 2 (AREDS2) randomized clinical trial. JAMA 2013;309(19):2005–15.

[42] Seddon JM, Francis PJ, George S, Schultz DW, Rosner B, Klein ML. Association of CFH Y402H and LOC387715 A69S with progression of age-related macular degeneration. JAMA 2007;297(16):1793–800.

[43] Seddon JM, Reynolds R, Shah HR, Rosner B. Smoking, dietary betaine, methionine, and vitamin D in monozygotic twins with discordant macular degeneration: epigenetic implications. Ophthalmology 2011;118(7):1386–94.

[44] Chang EE, Goldberg JL. Glaucoma 2.0: neuroprotection, neuroregeneration, neuroenhancement. Ophthalmology 2012;119(5):979–86.

[45] Kwon YH, Fingert JH, Kuehn MH, Alward WL. Primary open-angle glaucoma. N Engl J Med 2009;360(11):1113–24.

[46] Quigley HA. Glaucoma. Lancet 2011;377(9774):1367–77.

[47] Doucette LP, Rasnitsyn A, Seifi M, Walter MA. The interactions of genes, age, and environment in glaucoma pathogenesis. Surv Ophthalmol 2015;60(4):310–26.

[48] Yau JW, Rogers SL, Kawasaki R, et al. Global prevalence and major risk factors of diabetic retinopathy. Diabetes care 2012;35(3):556–64.

[49] Cheung N, Mitchell P, Wong TY. Diabetic retinopathy. Lancet 2010;376(9735):124–36.

[50] Sobrin L, Green T, Sim X, et al. Candidate gene association study for diabetic retinopathy in persons with type 2 diabetes: the Candidate gene Association Resource (CARe). Invest Ophthalmol Vis Sci 2011;52(10):7593–602.

[51] Rasmussen HM, Johnson EJ. Nutrients for the aging eye. Clin Interv Aging 2013;8:741–8.

[52] Chiu CJ, Taylor A. Nutritional antioxidants and age-related cataract and maculopathy. Exp Eye Res 2007;84(2):229–45.

[53] SanGiovanni JP, Chew EY. The role of omega-3 long-chain polyunsaturated fatty acids in health and disease of the retina. Prog Retin Eye Res 2005;24(1):87–138.

[54] Christen WG, Glynn RJ, Hennekens CH. Antioxidants and age-related eye disease current and future perspectives. Ann Epidemiol 1996;6(1):60–6.

[55] Chew EY, Clemons TE, Agrón E, et al. Long-term effects of vitamins C and E, β-carotene, and zinc on age-related macular degeneration: AREDS Report No. 35. Ophthalmology 2013;120(8):1604–11 e4.

[56] Kowluru RA, Engerman RL, Case GL, Kern TS. Retinal glutamate in diabetes and effect of antioxidants. Neurochem Int 2001;38(5):385–90.

[57] Obrosova IG, Minchenko AG, Marinescu V, et al. Antioxidants attenuate early up regulation of retinal vascular endothelial growth factor in streptozotocin-diabetic rats. Diabetologia 2001;44(9):1102–10.

[58] Kowluru RA, Engerman RL, Kern TS. Abnormalities of retinal metabolism in diabetes or experimental galactosemia VIII. Prevention by aminoguanidine. Curr Eye Res 2000;21(4):814–19.

[59] Izzotti A, Sacca SC, Cartiglia C, De Flora S. Oxidative deoxyribonucleic acid damage in the eyes of glaucoma patients. Am J Med 2003;114(8):638–46.

[60] Sacca SC, Pascotto A, Camicione P, Capris P, Izzotti A. Oxidative DNA damage in the human trabecular meshwork: clinical correlation in patients with primary open-angle glaucoma. Arch Ophthalmol (Chicago, IL: 1960) 2005;123(4):458–63.

[61] Mousa A, Kondkar AA, Al-Obeidan SA, et al. Association of total antioxidants level with glaucoma type and severity. Saudi Med J 2015;36(6):671–7.

[62] Mathew MC, Ervin AM, Tao J, Davis RM. Antioxidant vitamin supplementation for preventing and slowing the progression of age-related cataract. Cochrane Database Syst Rev 2012;6: Cd004567.

[63] Taylor A, Jacques PF, Nowell T, et al. Vitamin C in human and guinea pig aqueous, lens and plasma in relation to intake. Curr Eye Res 1997;16(9):857–64.

[64] Carr AC, Frei B. Toward a new recommended dietary allowance for vitamin C based on antioxidant and health effects in humans. Am J Clin Nutr 1999;69(6):1086–107.

[65] Thiagarajan R, Manikandan R. Antioxidants and cataract. Free Radic Res 2013;47(5):337–45.

[66] Ravindran RD, Vashist P, Gupta SK, et al. Inverse association of vitamin C with cataract in older people in India. Ophthalmology 2011;118(10):1958–1965.e2.

[67] Yoshida M, Takashima Y, Inoue M, et al. Prospective study showing that dietary vitamin C reduced the risk of age-related cataracts in a middle-aged Japanese population. Eur J Nutr 2007;46(2):118–24.

[68] Weikel KA, Garber C, Baburins A, Taylor A. Nutritional modulation of cataract. Nutr Rev 2014;72(1):30–47.

[69] Tan AG, Mitchell P, Flood VM, et al. Antioxidant nutrient intake and the long-term incidence of age-related cataract: the Blue Mountains Eye Study. Am J Clin Nutr 2008;87(6):1899–905.

[70] Gritz DC, Srinivasan M, Smith SD, et al. The antioxidants in prevention of cataracts study: effects of antioxidant supplements on cataract progression in South India. Br J Ophthalmol 2006;90(7):847–51.

[71] Christen WG, Glynn RJ, Sesso HD, et al. Age-related cataract in a randomized trial of vitamins E and C in men. Arch Ophthalmol 2010;128(11):1397–405.

[72] Simonelli F, Zarrilli F, Mazzeo S, et al. Serum oxidative and antioxidant parameters in a group of Italian patients with age-related maculopathy. Clin Chim Acta 2002;320(1–2):111–15.

[73] Flood V, Smith W, Wang JJ, Manzi F, Webb K, Mitchell P. Dietary antioxidant intake and incidence of early age-related maculopathy: the Blue Mountains Eye Study. Ophthalmology 2002;109(12):2272–8.

[74] Van Leeuwen R, Boekhoorn S, Vingerling JR, et al. Dietary intake of antioxidants and risk of age-related macular degeneration. J Am Med Assoc 2005;294(24):3101–7.

[75] Zanon-Moreno V, Ciancotti-Olivares L, Asencio J, et al. Association between a SLC23A2 gene variation, plasma vitamin C levels, and risk of glaucoma in a Mediterranean population. Mol Vis 2011;17:2997–3004.

[76] Yuki K, Murat D, Kimura I, Ohtake Y, Tsubota K. Reduced-serum vitamin C and increased uric acid levels in normal-tension glaucoma. Graefes Arch Clin Exp Ophthalmol 2010;248(2):243–8.

[77] Grover AK, Samson SE. Antioxidants and vision health: facts and fiction. Mol Cell Biochem 2014;388(1–2):173–83.

[78] Giaconi JA, Yu F, Stone KL, et al. The association of consumption of fruits/vegetables with decreased risk of glaucoma among older African-American women in the study of osteoporotic fractures. Am J Ophthalmol 2012;154(4):635–44.

[79] Bussel II, Aref AA. Dietary factors and the risk of glaucoma: a review. Ther Adv Chronic Dis 2014;5(4):188–94.

[80] Wang SY, Singh K, Lin SC. Glaucoma and vitamins A, C, and E supplement intake and serum levels in a population-based sample of the United States. Eye (Lond, Engl) 2013;27(4):487–94.

[81] Van den Enden MK, Nyengaard JR, Ostrow E, Burgan JH, Williamson JR. Elevated glucose levels increase retinal glycolysis and sorbitol pathway metabolism. Implications for diabetic retinopathy. Invest Ophthalmol Vis Sci 1995;36(8):1675–85.

[82] Cunningham JJ, Mearkle PL, Brown RG. Vitamin C: an aldose reductase inhibitor that normalizes erythrocyte sorbitol in insulin-dependent diabetes mellitus. J Am Coll Nutr 1994;13(4):344–50.

[83] A randomized trial of sorbinil, an aldose reductase inhibitor, in diabetic retinopathy. Arch Ophthalmol 1990;108(9):1234–44.

[84] Traber MG. Utilization of vitamin E. BioFactors (Oxford, Engl) 1999;10(2–3):115–20.

[85] Traber MG. Does vitamin E decrease heart attack risk? summary and implications with respect to dietary recommendations. J Nutr 2001;131(2):395s–7s.

[86] Delcourt C, Cristol JP, Tessier F, Léger CL, Descomps B, Papoz L. Age-related macular degeneration and antioxidant status in the POLA study. Arch Ophthalmol 1999;117(10):1384–90.

[87] Sen CK, Khanna S, Roy S. Tocotrienols: vitamin E beyond tocopherols. Life Sci 2006;78(18):2088–98.

[88] Ayala MN, Soderberg PG. Vitamin E can protect against ultraviolet radiation-induced cataract in albino rats. Ophthalmic Res 2004;36(5):264–9.

[89] Christen WG, Glynn RJ, Gaziano JM, et al. Age-related cataract in men in the selenium and vitamin e cancer prevention trial eye endpoints study: a randomized clinical trial. JAMA Ophthalmol 2015;133(1):17–24.

[90] Millen AE, Klein R, Folsom AR, Stevens J, Palta M, Mares JA. Relation between intake of vitamins C and E and risk of diabetic retinopathy in the Atherosclerosis Risk in Communities Study. Am J Clin Nutr 2004;79(5):865–73.

[91] Paolisso G, Di Maro G, Galzerano D, et al. Pharmacological doses of vitamin E and insulin action in elderly subjects. Am J Clin Nutr 1994;59(6):1291–6.

[92] Bursell SE, Clermont AC, Aiello LP, et al. High-dose vitamin E supplementation normalizes retinal blood flow and creatinine clearance in patients with type 1 diabetes. Diabetes care 1999;22(8):1245–51.

[93] Costagliola C, Iuliano G, Menzione M, Rinaldi E, Vito P, Auricchio G. Effect of vitamin E on glutathione content in red blood cells, aqueous humor and lens of humans and other species. Exp Eye Res 1986;43(6):905–14.

[94] Kang JH, Pasquale LR, Willett W, et al. Antioxidant intake and primary open-angle glaucoma: a prospective study. Am J Epidemiol 2003;158(4):337–46.

[95] Trumbo P, Yates AA, Schlicker S, Poos M. Dietary reference intakes: vitamin A, vitamin K, arsenic, boron, chromium, copper, iodine, iron, manganese, molybdenum, nickel, silicon, vanadium, and zinc. J Am Diet Assoc 2001;101(3):294–301.

[96] O'Dell BL. Role of zinc in plasma membrane function. J Nutr 2000;130(Suppl. 5S):1432s–6s.

[97] Vishwanathan R, Chung M, Johnson EJ. A systematic review on zinc for the prevention and treatment of age-related macular degeneration. Invest Ophthalmol Vis Sci 2013;54(6):3985–98.

[98] Prasad AS. Zinc deficiency in humans: a neglected problem. J Am Coll Nutr 1998;17(6):542–3.

[99] Grahn BH, Paterson PG, Gottschall-Pass KT, Zhang Z. Zinc and the eye. J Am Coll Nutr 2001;20(Suppl. 2):106–18.

[100] Miao X, Sun W, Miao L, et al. Zinc and diabetic retinopathy. J Diabetes Res 2013;2013:425854.

[101] Jeru I. The role of zinc in the appearance of cataract. Oftalmologia (Bucharest, Romania: 1990) 1997;41(4):329–32.

[102] Age-Related Eye Disease Study Research Group. A randomized, placebo-controlled, clinical trial of high-dose supplementation with vitamins C and E and beta carotene for age-related cataract and vision loss: AREDS report no. 9. Arch Ophthalmol (Chicago, IL: 1960) 2001;119(10):1439–52.

[103] Moustafa SA. Zinc might protect oxidative changes in the retina and pancreas at the early stage of diabetic rats. Toxicol Appl Pharmacol 2004;201(2):149–55.

[104] Dherani M, Murthy GV, Gupta SK, et al. Blood levels of vitamin C, carotenoids and retinol are inversely associated with cataract in a North Indian population. Invest Ophthalmol Vis Sci 2008;49(8):3328–35.

[105] Appleby PN, Allen NE, Key TJ. Diet, vegetarianism, and cataract risk. Am J Clin Nutr 2011;93(5):1128–35.

[106] Tan JSL, Wang JJ, Flood V, Rochtchina E, Smith W, Mitchell P. Dietary antioxidants and the long-term incidence of age-related macular degeneration: the Blue Mountains Eye Study. Ophthalmology 2008;115(2):334–41.

[107] Evans JR, Lawrenson JG. Antioxidant vitamin and mineral supplements for preventing age-related macular degeneration. Cochrane Database Syst Rev (Online) 2012;6.

[108] Zampatti S, Ricci F, Cusumano A, Marsella LT, Novelli G, Giardina E. Review of nutrient actions on age-related macular degeneration. Nutr Res (New York, NY) 2014;34(2):95–105.

[109] Hennekens CH, Buring JE, Manson JE, et al. Lack of effect of long-term supplementation with beta carotene on the incidence of malignant neoplasms and cardiovascular disease. N Engl J Med 1996;334(18):1145–9.

[110] Omenn GS, Goodman GE, Thornquist MD, et al. Risk factors for lung cancer and for intervention effects in CARET, the Beta-Carotene and Retinol Efficacy Trial. J Natl Cancer Inst 1996;88(21):1550–9.

[111] Hobbs RP, Bernstein PS. Nutrient supplementation for age-related macular degeneration, cataract, and dry eye. J Ophthalmic Vis Res 2014;9(4):487–93.

[112] Coleman AL, Stone KL, Kodjebacheva G, et al. Glaucoma risk and the consumption of fruits and vegetables among older women in the study of osteoporotic fractures. Am J Ophthalmol 2008;145(6):1081–9.

[113] Kowluru RA, Engerman RL, Kern TS. Abnormalities of retinal metabolism in diabetes or experimental galactosemia. VI. Comparison of retinal and cerebral cortex metabolism, and effects of antioxidant therapy. Free Radic Biol Med 1999;26(3–4):371–8.

[114] Dene BA, Maritim AC, Sanders RA, Watkins III JB. Effects of antioxidant treatment on normal and diabetic rat retinal enzyme activities. J Ocul Pharmacol Ther 2005;21(1):28–35.

[115] Mayer-Davis EJ, Bell RA, Reboussin BA, Rushing J, Marshall JA, Hamman RF. Antioxidant nutrient intake and diabetic retinopathy: the San Luis valley diabetes study. Ophthalmology 1998;105(12):2264–70.

[116] Johnson EJ. The role of carotenoids in human health. Nutr Clin Care 2002;5(2):56–65.

[117] Parker RS. Carotenoids in human blood and tissues. J Nutr 1989;119(1):101–4.

[118] Yeum KJ, Shang FM, Schalch WM, Russell RM, Taylor A. Fat-soluble nutrient concentrations in different layers of human cataractous lens. Curr Eye Res 1999;19(6):502–5.

[119] Bone RA, Landrum JT, Tarsis SL. Preliminary identification of the human macular pigment. Vision Res 1985;25(11):1531–5.

[120] Krinsky NI. Possible biologic mechanisms for a protective role of xanthophylls. J Nutr 2002;132(3):540s–2s.

[121] Moeller SM, Voland R, Tinker L, et al. Associations between age-related nuclear cataract and lutein and zeaxanthin in the diet and serum in the carotenoids in the Age-Related Eye Disease Study, an Ancillary Study of the Women's Health

Initiative. Arch Ophthalmol (Chicago, IL: 1960) 2008;126(3):354–64.

[122] Chew EY, SanGiovanni JP, Ferris FL, et al. Lutein/zeaxanthin for the treatment of age-related cataract: AREDS2 randomized trial report no. 4. JAMA Ophthalmol 2013;131(7):843–50.

[123] Sommerburg O, Keunen JEE, Bird AC, Van Kuijk FJGM. Fruits and vegetables that are sources for lutein and zeaxanthin: the macular pigment in human eyes. Br J Ophthalmol 1998;82(8):907–10.

[124] Goldberg J, Flowerdew G, Smith E, Brody JA, Tso MO. Factors associated with age-related macular degeneration: an analysis of data from The first National Health and Nutrition Examination Survey. Am J Epidemiol 1988;128(4):700–10.

[125] Sin HPY, Liu DTL, Lam DSC. Lifestyle modification, nutritional and vitamins supplements for age-related macular degeneration. Acta Ophthalmol (Copenh) 2013;91(1):6–11.

[126] Cho E, Seddon JM, Rosner B, Willett WC, Hankinson SE. Prospective study of intake of fruits, vegetables, vitamins, and carotenoids and risk of age-related maculopathy. Arch Ophthalmol 2004;122(6):883–92.

[127] Ma L, Yan SF, Huang YM, et al. Effect of lutein and zeaxanthin on macular pigment and visual function in patients with early age-related macular degeneration. Ophthalmology 2012;119(11):2290–7.

[128] Ma L, Dou HL, Wu YQ, et al. Lutein and zeaxanthin intake and the risk of age-related macular degeneration: a systematic review and meta-analysis. Br J Nutr 2012;107(3):350–9.

[129] Neacsu A, Oprean C, Curea M, Tuchila G, Trifu M. Neuroprotection with carotenoids in glaucoma. Oftalmologia (Bucharest, Romania: 1990) 2003;59(4):70–5.

[130] Miranda M, Muriach M, Roma J, et al. Oxidative stress in a model of experimental diabetic retinopathy: the utility of peroxynitrite scavengers. Arch Soc Esp Oftalmol 2006;81(1):27–32.

[131] Brazionis L, Rowley K, Itsiopoulos C, O'Dea K. Plasma carotenoids and diabetic retinopathy. Br J Nutr 2009;101(2):270–7.

[132] Kowluru RA, Menon B, Gierhart DL. Beneficial effect of zeaxanthin on retinal metabolic abnormalities in diabetic rats. Invest Ophthalmol Vis Sci 2008;49(4):1645–51.

[133] Kowluru RA, Zhong Q, Santos JM, Thandampallayam M, Putt D, Gierhart DL. Beneficial effects of the nutritional supplements on the development of diabetic retinopathy. Nutr Metab (Lond) 2014;11(1):8.

[134] Fliesler SJ, Anderson RE. Chemistry and metabolism of lipids in the vertebrate retina. Prog Lipid Res 1983;22(2):79–131.

[135] Snow KK, Seddon JM. Do age-related macular degeneration and cardiovascular disease share common antecedents? Ophthalmic Epidemiol 1999;6(2):125–43.

[136] SanGiovanni JP, Chew EY, Clemons TE, et al. The relationship of dietary lipid intake and age-related macular degeneration in a case-control study: AREDS report no. 20. Arch Ophthalmol 2007;125(5):671–9.

[137] SanGiovanni JP, Agrón E, Meleth AD, et al. ω-3 long-chain polyunsaturated fatty acid intake and 12-y incidence of neovascular age-related macular degeneration and central geographic atrophy: AREDS report 30, a prospective cohort study from the Age-Related Eye Disease Study. Am J Clin Nutr 2009;90(6):1601–7.

[138] Johnson EJ, Chung HY, Caldarella SM, Max Snodderly D. The influence of supplemental lutein and docosahexaenoic acid on serum, lipoproteins, and macular pigmentation. Am J Clin Nutr 2008;87(5):1521–9.

[139] Delyfer MN, Buaud B, Korobelnik JF, et al. Association of macular pigment density with plasma omega-3 fatty acids: the PIMAVOSA study. Invest Ophthalmol Vis Sci 2012;53(3):1204–10.

[140] Seddon JM, Cote J, Rosner B. Progression of age-related macular degeneration: association with dietary fat, transunsaturated fat, nuts, and fish intake. Arch Ophthalmol 2003;121(12):1728–37.

[141] Chiu CJ, Klein R, Milton RC, Gensler G, Taylor A. Does eating particular diets alter the risk of age-related macular degeneration in users of the Age-Related Eye Disease Study supplements? Br J Ophthalmol 2009;93(9):1241–6.

[142] Lawrenson JG, Evans JR. Omega 3 fatty acids for preventing or slowing the progression of age-related macular degeneration. Cochrane Database Syst Rev (Online) 2012;11.

[143] Townend BS, Townend ME, Flood V, et al. Dietary macronutrient intake and five-year incident cataract: the Blue Mountains Eye Study. Am J Ophthalmol 2007;143(6):932–9.

[144] Lu M, Cho E, Taylor A, Hankinson SE, Willett WC, Jacques PF. Prospective study of dietary fat and risk of cataract extraction among US women. Am J Epidemiol 2005;161(10):948–59.

[145] Martinez-Lapiscina EH, Martinez-Gonzalez MA, Guillen Grima F, Olmo Jimenez N, Zarranz-Ventura J, Moreno-Montanes J. Dietary fat intake and incidence of cataracts: the SUN Prospective study in the cohort of Navarra, Spain. Med Clin (Barc) 2010;134(5):194–201.

[146] Perez de Arcelus M, Toledo E, Martinez-Gonzalez MA, Sayon-Orea C, Gea A, Moreno-Montanes J. Omega 3:6 ratio intake and incidence of glaucoma: the SUN cohort. Clin Nutr (Edinburgh, Scotland) 2014;33(6):1041–5.

[147] Camras CB, Alm A. Initial clinical studies with prostaglandins and their analogues. Surv Ophthalmol 1997;41(Suppl. 2):S61–8.

[148] Stjernschantz JW. From PGF2α-isopropyl ester to latanoprost: a review of the development of xalatan: the proctor lecture. Invest Ophthalmol Vis Sci 2001;42(6):1134–45.

[149] Schachtschabel U, Lindsey JD, Weinreb RN. The mechanism of action of prostaglandins on uveoscleral outflow. Curr Opin Ophthalmol 2000;11(2):112–15.

[150] Nguyen CTO, Bui BV, Sinclair AJ, Vingrys AJ. Dietary omega 3 fatty acids decrease intraocular pressure with age by increasing aqueous outflow. Invest Ophthalmol Vis Sci 2007;48(2):756–62.

[151] Ren H, Magulike N, Ghebremeskel K, Crawford M. Primary open-angle glaucoma patients have reduced levels of blood docosahexaenoic and eicosapentaenoic acids. Prostaglandins Leukot Essent Fatty Acids 2006;74(3):157–63.

[152] Huang WB, Fan Q, Zhang XL. Cod liver oil: a potential protective supplement for human glaucoma. Int J Ophthalmol 2011;4(6):648–51.

[153] Renard JP, Rouland JF, Bron A, et al. Nutritional, lifestyle and environmental factors in ocular hypertension and primary open-angle glaucoma: an exploratory case-control study. Acta Ophthalmol (Copenh) 2013;91(6):505–13.

[154] Shen J, Bi Y-L, Das UN. Potential role of polyunsaturated fatty acids in diabetic retinopathy. Arch Med Sci 2014;10(6):1167–74.

[155] Opreanu M, Tikhonenko M, Bozack S, et al. The unconventional role of acid sphingomyelinase in regulation of retinal microangiopathy in diabetic human and animal models. Diabetes 2011;60(9):2370–8.

[156] Connor KM, SanGiovanni JP, Lofqvist C, et al. Increased dietary intake of omega-3-polyunsaturated fatty acids reduces pathological retinal angiogenesis. Nat Med 2007;13(7):868–73.

[157] Tikhonenko M, Lydic TA, Opreanu M, et al. N-3 polyunsaturated fatty acids prevent diabetic retinopathy by inhibition of retinal vascular damage and enhanced endothelial progenitor cell reparative function. PLoS One 2013;8(1):e55177.

PART IV

HEALTH EFFECTS OF DIETARY COMPOUNDS AND DIETARY INTERVENTIONS

CHAPTER

33

Vitamin D Nutrient-Gene Interactions and Healthful Aging

Mark R. Haussler[1], Rimpi K. Saini[1], Marya S. Sabir[2], Christopher M. Dussik[2], Zainab Khan[2], G. Kerr Whitfield[1], Kristin P. Griffin[1], Ichiro Kaneko[1,2] and Peter W. Jurutka[1,2]

[1]Department of Basic Medical Sciences, College of Medicine, University of Arizona, Phoenix, AZ, USA
[2]School of Mathematical and Natural Sciences, Arizona State University, Glendale, AZ, USA

KEY FACTS

- The renal vitamin D metabolite/hormone is 1,25D.
- VDR binds 1,25D and mediates its actions on gene transcription.
- 1,25D/VDR signals intestinal absorption of calcium and phosphate for bone mineralization.
- 1,25D also induces FGF23, klotho, and CYP24A1 to curtail age-related pathologies caused by phosphate and 1,25D excess.
- Beneficial nutrients that are low-affinity VDR ligands include curcumin, docosahexenoic acid, and delphinidin.
- Resveratrol and SIRT1 potentiate VDR signaling to promote antiaging.
- Optimum levels of vitamin D facilitate longevity by delaying the chronic diseases of aging such as cardiovascular failure, cancer, osteoporosis, ectopic calcification, as well as loss of hearing and cognition.

Dictionary of Terms

- 1,25-Dihydroxyvitamin D_3 (1,25D): the vitamin D_3 sterol hormone.

- Vitamin D receptor (VDR): the protein that binds 1,25D and mediates its actions.
- Osteoporosis: a bone disease characterized by thinning of the skeletal mineral, resulting in fractures.
- Ectopic calcification: bony deposits in soft tissues
- Fibroblast growth factor-23 (FGF23): a peptide hormone secreted by bone cells (osteocytes) that reduces 1,25D and phosphate in the bloodstream.
- Klotho: a peptide hormone produced in the kidneys that promotes longevity by acting as a co-receptor for FGF23.
- Neuroprotection: preservation of the central nervous system from degeneration and disease.
- Anti-inflammatory: acting to prevent or curb inflammation.
- Anticancer nutrient: a dietary substance that prevents cancer.

VITAMIN D: FROM NUTRIENT TO TIGHTLY REGULATED HORMONE

The hormone precursor, vitamin D_3, either can be obtained in the diet or synthesized from 7-dehydrocholesterol in skin in a nonenzymatic, UV light-dependent reaction (Fig. 33.1). Vitamin D_3 is then transported to the liver, where it is hydroxylated to generate 25-hydroxyvitamin D_3 (25D), the major circulating form of vitamin D_3 that is assayed to quantitate

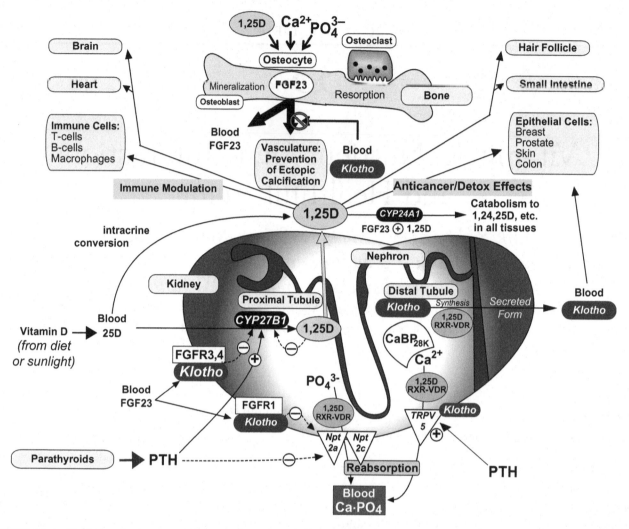

FIGURE 33.1 The kidney is the nexus of healthful aging. The kidney responds to 1,25D, FGF23, and PTH to regulate vitamin D bioactivation and calcium/phosphate reabsorption, and serves as an endocrine source of 1,25D and klotho. Thus, the kidney is the endocrine nexus of health by conserving calcium, eliminating phosphate, and producing 1,25D and klotho "fountain of youth" hormones. Renal hormones 1,25D (shaded in light blue (light gray in print versions)) and klotho (shaded in dark blue (white type on black background in print versions)) reach beyond bone mineral homeostasis to delay other chronic disorders of aging besides osteoporosis, such as cardiovascular disease, epithelial cell cancers, autoimmune disease, hair loss, and neuropsychiatric conditions.

clinical vitamin D status. The final step in the production of the hormonal form occurs mainly, but not exclusively, in the kidney, via a closely regulated 1α-hydroxylation reaction (Fig. 33.1). The cytochrome P450-containing (CYP) enzyme that catalyzes 1α-hydroxylation is mitochondrial CYP27B1. 1,25-dihydroxyvitamin D_3 (1,25D) circulates to various target tissues to exert its endocrine actions that are mediated by the vitamin D receptor (VDR). Many of the long-recognized functions of 1,25D involve the regulation of calcium and phosphate absorption, raising the blood levels of these ions to facilitate bone mineralization, as well as activating bone resorption as part of the skeletal remodeling cycle [1].

In addition to effecting bone mineral homeostasis by functioning at the small intestine and bone, 1,25D also acts through its VDR mediator to influence a number of other cell types. These extraosseous actions of 1,25D-VDR include differentiation of certain cells in skin [2] and in the immune system [3] (Fig. 33.1). Notably, the skin and the immune system are recognized as extrarenal sites of CYP27B1 action to produce 1,25D locally for autocrine and paracrine effects [4,5], creating important intracrine systems (Fig. 33.1) for extraosseous 1,25D-VDR functions amplifying the renal endocrine actions of 1,25D-VDR [6]. Higher circulating 25D levels are likely required for optimal intracrine actions of 1,25D (Fig. 33.1). This insight stems from a multitude of epidemiologic studies reporting the associations between low 25D levels and chronic disease, coupled with statistically significant protection against a host of pathologies by much higher concentrations of circulating 25D [7]. Thus, locally produced 1,25D appears to be capable of benefiting the vasculature to reduce the risk of heart attack and stroke, controlling the adaptive immune system to lower the prevalence of autoimmune disease while boosting the innate immune system to fight infection, effecting xenobiotic detoxification, and exerting anti-inflammatory and anticancer pressure on epithelial cells prone to fatal malignancies. Finally, recent evidence indicates that 1,25D also affects behavior, perhaps emerging as a new strategy in the prevention and/or treatment of various neuropsychiatric disorders [8]. In conclusion, many of the biological responses to the 1,25D hormone are both endocrine and intracrine in nature, rendering this nutrient metabolite a potent regulator of all cells in the body that express VDR, thereby explaining the panoply of 1,25D/VDR-gene interactions that promulgate healthful aging.

The major inducer of CYP27B1 in kidney is parathyroid hormone (PTH), the calcemic peptide hormone secreted during hypocalcemia [9]. When VDR expressed in the parathyroid glands is liganded with 1,25D, PTH synthesis is suppressed by a direct action on gene transcription [10]. This negative feedback loop, which limits the stimulation of CYP27B1 by PTH under low calcium conditions, serves to curtail the bone-resorbing effects of PTH in anticipation of 1,25D-mediated increases in both intestinal calcium absorption and bone resorption, thus preventing hypercalcemia. Understanding of the homeostatic control of phosphate emanates from characterization of unsolved familial hypo- or hyperphosphatemic disorders which we now know are caused by deranged levels of bone-derived fibroblast growth factor 23 (FGF23) [11]. The major repressor of CYP27B1 in kidney is FGF23, the phosphaturic peptide hormone secreted during hyperphosphatemia. In brief, FGF23 has emerged as a dramatic new phosphate regulator, and a second phosphaturic hormone after PTH (Fig. 33.1). We [12] and others [13] proved that 1,25D induces the release of FGF23 from bone, specifically from osteocytes of the osteoblastic lineage (Fig. 33.1), a process that is independently stimulated by high circulating phosphate levels. Thus, in a striking and elegant example of biological symmetry, PTH is repressed by 1,25D and calcium, whereas FGF23 is induced by 1,25D and phosphate, protecting mammals against hypercalcemia and hyperphosphatemia, respectively, either of which can elicit ectopic calcification that is often associated with aging.

Also illustrated in Fig. 33.1, using the kidney as an example, is an important feedback mechanism by which the 1,25D/VDR-mediated endocrine or intracrine signal is terminated in all target cells, namely the action of CYP24A1, an enzyme that initiates the process of 1,25D catabolism [14]. The *CYP24A1* gene is transcriptionally activated by 1,25D [15,16], as well as by FGF23 (Fig. 33.1). Conversely, the *CYP27B1* gene is repressed by FGF23 and 1,25D, with the latter regulation effected by epigenetic demethylation [17] in a short feedback loop to limit the production of 1,25D [18]. Therefore, the vitamin D endocrine system is elegantly governed by feedback controls of vitamin D bioactivation that interpret bone mineral ion status, and via feedforward induction of 1,25D catabolism, to prevent hypervitaminosis D. The vitamin D intracrine system, in contrast, appears to be dependent more on the availability of ample 25D substrate to generate local 1,25D to lower the risk of chronic diseases of the epithelial (skin, colon, etc.), immune, cardiovascular, and nervous systems.

THE KIDNEY IS THE NEXUS OF HEALTHFUL AGING

Fig. 33.1 also illustrates our thesis that the kidney is the nexus of healthful aging by virtue of its ability to produce the 1,25D and klotho hormones, as well as to conserve calcium and eliminate phosphate, all within the context of a healthspan circuit between bone and kidney that prevents ectopic calcification. Excessive circulating levels of 1,25D, phosphate, or calcium are sensed by the bone osteocyte to induce FGF23 expression and secretion (Fig. 33.1, top center). Skeletally-derived FGF23 impacts the kidney to elicit reduced 1,25D synthesis and to stimulate the elimination of phosphate, thereby effecting healthful control of active vitamin D and phosphate. We assert that maintaining phosphate and active vitamin D at optimum levels is a vital factor in maintaining a healthy lifespan. As shown in Fig. 33.1, the kidney responds to 1,25D, FGF23, and PTH to regulate vitamin D bioactivation/catabolism and calcium/phosphate reabsorption, while serving as an endocrine source of 1,25D and α-klotho (hereafter referred to as "klotho"). Thus, the kidney is the endocrine nexus of health by conserving calcium, eliminating phosphate, and producing 1,25D and klotho "fountain of youth" hormones. In other words, the skeleton and kidneys orchestrate bone mineral and vitamin D metabolism to promote healthful aging and the quality of life. This conclusion is supported by the hyperphosphatemic phenotype of klotho- and FGF23-null mice, which includes high levels of 1,25D, short lifespan/premature aging, ectopic calcification, arteriosclerosis, osteoporosis, muscle atrophy, skin atrophy, and hearing loss [19,20]. Finally, the hyperphosphatemic/aging phenotype of klotho- and FGF23-null mice is rescued by knocking out either VDR [21] or CYP27B1

[22], establishing that excess 1,25D, acting through VDR, is eliciting the observed pathologies.

Given the apparent pro-aging potency of excess 1,25D, there was some surprise when we first proposed that 1,25D, either alone, or in combination with bona fide antiaging factors like klotho, is a mediator of healthful aging [23]. This hypothesis could explain the epidemiologic/association studies which suggest that 25D, in the newly recognized optimal high-normal range in blood, confers a lower risk of virtually all of the fatal diseases of aging such as heart attack, stroke, and cancers. Thus, as depicted schematically in Fig. 33.1, the endocrine/intracrine actions of vitamin D/klotho protect the vascular system, as well as epithelial cells (breast, prostate, colon, skin, etc.) that are subject to fatal cancers [24], the immune system [3,25,26] (Fig. 33.1), and likely the central nervous system [8,27]. With respect to the chronic diseases of aging, it is now becoming clear that the kidney represents the nexus of control, and we contend that klotho is a third renal hormone after 1,25D and erythropoietin. Therefore, in this chapter we emphasize the importance of renal health during aging, and unveil the kidney as a focal point for the prevention of chronic diseases. The role of renal-directed healthspan control is supported clinically by the numerous deleterious aging-related outcomes that are consistently observed in patients suffering from chronic kidney disease (CKD).

Moreover, we assert that 1,25D is an antiaging/wellness hormone through its ability to induce klotho, a bona fide antisenescence principle, and argue that optimum levels of 25D manifest as antiaging. However, *both* VDR-null and vitamin D toxic animals display similar pro-aging phenotypes [27]. This dichotomy with respect to 1,25D action is analogous to the actions of the only other known toxic fat-soluble vitamin, namely vitamin A and its active retinoic acid metabolite. In physiologic quantities, retinoids mediate optimal epithelial cell differentiation and barrier formation, embryonic development, etc., yet pharmacologic excesses of vitamin A and retinoic acid yield epithelial pathologies such as gastroenteritis and exfoliation, and embryopathy, respectively. Whereas no feedback control exists in the vitamin A system, the FGF23/klotho endocrine system allows the body to keep vitamin D in check by repressing CYP27B1 and inducing CYP24A1 [28]. Thus, we hypothesize that CYP24A1 represents a second antiaging gene, along with klotho, that is expressed in response to the vitamin D hormone and elicits feedback control to prevent pro-aging vitamin D excess. Additionally, we assert that the 1,25D-VDR/FGF23/klotho/CYP24A1 system is analogous to the "sister" paradigms [29] of bile acids-FXR/FGF15,19 and polyunsaturated fatty acids-PPARα/FGF21 that protect the vasculature via cholesterol catabolism (ie, CYP7A1) and fatty acid clearance (ie, β-oxidation), respectively. The 1,25D-VDR/FGF23/klotho/CYP24A1 antiaging axis limits 1,25D, phosphate, ectopic calcification, inflammation and fibrosis. A greater fundamental understanding of vitamin D and phosphate homeostasis as well as the pathophysiology of aging disorders (eg, osteoporotic fractures, muscle weakness, atherosclerosis, ectopic calcification, hearing loss, skin atrophy, myocardial infarction, and ischemic stroke) is therefore emerging. We envision that prevention and treatment of diseases of aging could not only include statins, a low saturated fatty acid/glucose diet, and exercise, to reverse the pathology of atheromas, obesity, and Type II diabetes, but would also feature enhanced ("high normal") levels of circulating vitamin D and klotho to prevent hypertension, improve the underlying cellular infrastructure of the vasculature, preclude ectopic calcification, and maintain a fracture-free skeleton through remodeling. These clinical benefits will be realized through the actions of the FGF23/klotho/1,25D/phosphate axis, which are analogous to the healthful cardiovascular effects of interrupting the FGF19/FXR/bile acid axis to enhance cholesterol clearance, and the activation of the FGF21/PPARα/polyunsaturated fatty acids axis to stimulate degradation and impede synthesis of saturated fatty acids and cholesterol. Strikingly, there is evidence [30] that the FGF19, FGF21, and FGF23 sibling axes form a network to achieve optimal lipid, carbohydrate, and mineral homeostasis. Finally, α-klotho, which acts as a coreceptor/mediator of FGF23 effects, appears to function similarly to mediate the effects of FGF19 and FGF21, perhaps explaining the broad chronic disease spectrum of the α-klotho null mouse [19,30].

MECHANISM OF GENE REGULATION BY LIGANDED VDR

The hormonal metabolite of vitamin D_3, 1,25D, acts as a classic nuclear receptor ligand that binds specifically to the vitamin D receptor to control the transcription of a multitude of genes [1]. Liganded VDR attracts one of the retinoid X receptors (RXR) into a heterodimer that then recognizes vitamin D responsive elements (VDREs) in the vicinity of target genes, regulating their expression via the recruitment of comodulator complexes that modify chromatin to effect either induction or repression of the cognate mRNA [31–33]. Various domains of the 427 amino acid human VDR are highlighted on a linear schematic of the protein presented in Fig. 33.2, with the two major functional regions being the N-terminal zinc finger DNA binding domain (DBD), and the

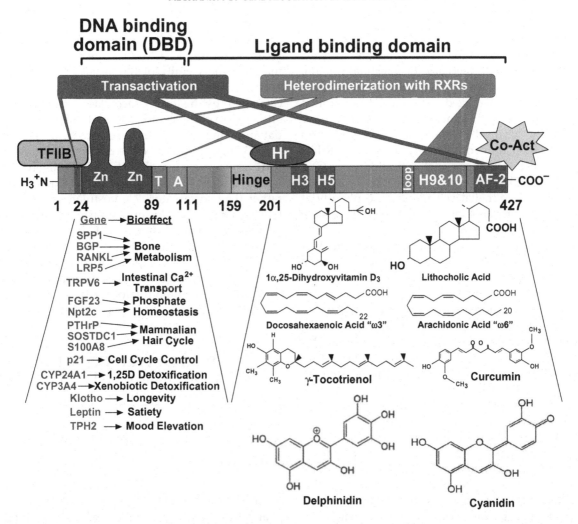

FIGURE 33.2 Functional domains in human VDR. Highlighted at the left is the human VDR zinc finger DNA binding domain that, in cooperation with the corresponding domain in the RXR heteropartner, mediates direct association with the target genes listed at the lower left, leading to the indicated physiological effects. The official gene symbol for bone Gla protein (BGP) is BGLAP, for RANKL is TNFSF11, for Npt2c is SLC34A3, for PTHrP is PTHLH, and for klotho is KL. Below the ligand-binding domain (at the right) are illustrated selected VDR ligands, including several novel ligands discussed in the text. Also shown above the schematic are the proposed interactions with TFIIB, hairless (Hr) and coactivators (Co-Act), the latter of which associate with the activation function-2 (AF-2) domain.

C-terminal ligand binding (LBD)/heterodimerization domain. The original X-ray crystallographic structure of the VDR LBD consisting of 12 α-helices [34] has been updated (PDB 3A78). The pregnane X receptor (PXR) X-ray crystal structure [35] allows for comparison between the ligand binding/heterodimerization domain of VDR and that of its closest relative in the nuclear receptor superfamily. The PXR LBD has a particularly large ligand binding pocket (1150 Å^3) [36]; the ligand binding pocket of VDR, is also large (approx. 700 Å^3) compared with other nuclear receptors that have been crystallized. Thus, like PXR, VDR has a sizable binding pocket to accommodate a diverse variety of lipophilic ligands beyond 1,25D as discussed below. Finally, VDR and PXR each heterodimerize with RXR to signal detoxification of xenobiotics and overlap somewhat in their target gene repertoires, which are laden with *CYPs*.

The diversity of ligands for PXR is especially broad and includes not only endogenous steroids, but also the secondary bile acid lithocholic acid (LCA), the antibiotic rifampicin, as well as xenobiotics, such as hyperforin, the active ingredient of St. John's wort [37]. We have identified several nutritional lipids as low affinity VDR ligands that apparently function locally in high concentrations. Fig. 33.2 reveals that these novel VDR ligands include ω3- and ω6-essential polyunsaturated fatty acids (PUFAs), docosahexaenoic acid (DHA) and arachidonic acid, respectively, the vitamin E derivative γ-tocotrienol, and curcumin [38], which is a turmeric-derived polyphenol found in curry, as well as the anthocyanidins delphinidin and cyanidin found in pigmented fruits and vegetables [39,40]. Thus, it is now recognized that VDR binds many nutrient ligands beyond the 1,25D hormone. VDR appears to have evolved as a "specialty" regulator of intestinal calcium absorption and hair growth in terrestrial animals, providing both a mineralized skeleton for locomotion in a calcium-scarce environment, and physical protection against the harmful UV radiation of the sun. Yet VDR has retained its evolutionarily ancient, PXR-like ability to effect xenobiotic detoxification via CYP induction [41]. VDR may complement PXR by serving as a guardian of epithelial cell integrity, especially at environmentally or xenobiotically exposed sites, such as skin, intestine, and kidney.

The structure of 1,25D-occupied human VDR, heterodimerized with full-length RXRα, docked on a VDRE, and bound with a single coactivator, has been determined in solution via Small Angle X-ray Scattering and Fluorescence Resonance Energy Transfer techniques [42]. The allosteric communication between the interaction surfaces of the VDR-RXR complex has been solved by hydrogen-deuterium exchange [43]. These advances render it possible to visualize how the DBD and the ligand binding/heterodimerization domains are arranged relative to one another, and how their binding to ligand, DNA, and coactivators influence one another. Fig. 33.3A illustrates in schematic fashion how the hormonal ligand could be influencing VDR to interact more efficaciously with its heterodimeric partner, VDREs, and coactivators. The key event in the allosteric model presented in Fig. 33.3 is the binding of a ligand, depicted as 1,25D; however, the model likely applies to any of the alternative low-affinity ligands pictured in Fig. 33.2, provided their concentration is sufficiently high. The presence of ligand in the VDR binding pocket results in a dramatic conformational change in the position of helix 12 at the C-terminus of VDR, bringing it to the "closed" position to serve in its AF2 role as part of a platform for coactivator binding [44–46]. The attraction of a coactivator to the helix-3, -5, and -12 platform of liganded VDR likely allosterically stabilizes the VDR-RXR heterodimer on the VDRE, and may even assist in triggering strong heterodimerization by inducing the VDR LBD to migrate to the 5' side of the RXR LBD, and in so doing rotate the RXR LBD 180 degrees employing the driving force of the ionic and hydrophobic interactions between helices 9 and 10 in hVDR and the corresponding helices in RXR (Fig. 33.3A). There is also evidence that DNA binding, likely to a positive VDRE, influences the stability of helix 12 for coactivator binding [43]. Alternative ligands may serve as selective VDR modulators that drive VDRE- or comodulator-specific target gene regulation.

Ligand-dependent repression of gene transcription by VDR-RXR involves the recruitment of nuclear receptor repressor(s) to alter the architecture of chromatin in the vicinity of the target gene to that of heterochromatin. This restructuring of chromatin is catalyzed by histone deacetylases and demethylases attracted to the receptor-tethered corepressor. The initial targeting of the repressed gene, as illustrated in Fig. 33.3B, appears to be the docking of liganded VDR-RXR on a negative VDRE, which likely conforms liganded VDR in such a way that it binds corepressor rather than coactivator. We postulate that the information driving this allosteric transformation of VDR is intrinsic to the negative VDRE DNA sequence [47]. With both induction (Fig. 33.3A) and repression (Fig. 33.3B) of gene expression by VDR ligands, the process is regulated by deacetylation/acetylation of VDR and/or its comodulators as described later in this chapter.

VDR-MEDIATED CONTROL OF NETWORKS OF GENES VITAL FOR HEALTHFUL AGING

Vitamin D and Phosphate Homeostasis Attenuate Senescence (CYP24A1, FGF23, and *klotho*)

As discussed above, whereas the predominant action of 1,25D-VDR is promoting intestinal calcium and phosphate absorption to prevent osteopenia, the initial signal for this function is PTH reacting to low calcium, whereas the hormonal agent that feedback controls these events to preclude ectopic calcification is FGF23. In this fashion, bone resorption and mineralization remain coupled to protect the integrity of the mineralized skeleton. FGF23 functions acutely in concert with PTH, and chronically when PTH is suppressed by calcium and 1,25D. In fact, FGF23 directly represses PTH [48] to abolish the activation of CYP27B1 by PTH, while at the same time appropriating from PTH the role of phosphate elimination. Like PTH, FGF23 inhibits renal Npt2a and Npt2c to elicit phosphaturia [28] (Fig. 33.1). In contrast to PTH that is downregulated by 1,25D in parathyroid glands, FGF23 is upregulated by 1,25D in osteocytes [11,12,49], a major source of endocrine FGF23 production by bone. Hyperphosphatemia enhances osteocytic FGF23 production independently of 1,25D, rendering FGF23 the perfect phosphaturic counter-1,25D hormone because it inhibits renal phosphate reabsorption, and 1,25D biosynthesis via repression of CYP27B1, while enhancing 1,25D degradation by inducing CYP24A1 in all tissues (Fig. 33.1). In this fashion, FGF23 allows osteocytes to communicate with

(A) Ligand-dependent activation

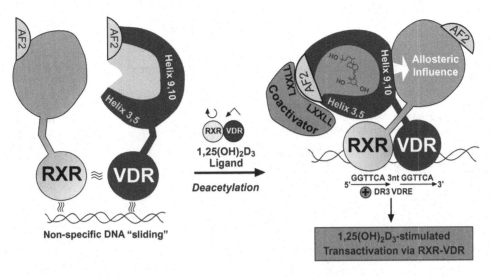

(B) Ligand-dependent repression

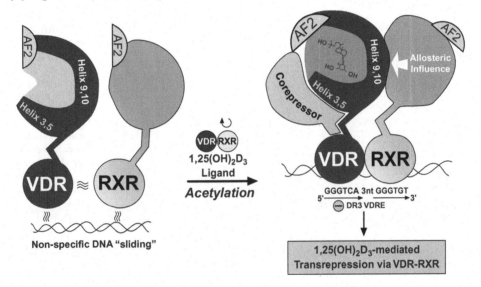

FIGURE 33.3 Proposed mechanisms of gene induction and repression by VDR. (A) Allosteric model of RXR-VDR activation after binding 1,25D and coactivator, deacetylation, and docking on a high affinity positive VDRE (mouse osteopontin). See text for explanation. (B) Allosteric model for VDR-RXR inactivation after binding 1,25D and corepressor, acetylation, and docking in reverse polarity on a high affinity negative VDRE (chicken PTH). See text for explanation.

the kidney to govern circulating 1,25D as well as phosphate levels, thereby preventing excess 1,25D function and hyperphosphatemia. FGF23 signals via renal FGFR/klotho coreceptors to promulgate phosphaturia [50], repress CYP27B1 [51] and induce CYP24A1 [28,50] (Fig. 33.1).

CYP24A1

Upregulation of CYP24A1 by 1,25D and by FGF23 may be a key feedback link whereby circulating and local 1,25D levels are maintained optimally to prevent the pro-ectopic calcification and possibly even pro-aging properties of the potent 1,25D hormone. We contend that this CYP24A1 induction [28] maintains intracrine as well as endocrine 1,25D homeostasis. Thus, CYP24A1 constitutes, beyond klotho, a second antiaging gene controlling phosphate, calcium and active vitamin D. Indeed, mice with ablation of the *CYP24A1* gene die early as a result of 1,25D toxicity [52].

Stimulation of *CYP24A1* transcription is well understood and, although active VDREs have been detected over 50 kb downstream of the transcription start site [53], regulation is primarily executed at the level of the proximal promoter, the sequence of which is listed for

```
-648
AAATTCTACAAACTCCCCTTCTTGCTCAAGTTAAGAAAGTCTCCTCTTCTGGTGCATTTCAGTAAGACTCA
AATCCTCCCCACCCTGGGAGGCGCAGAAAGCCAAACTTCCTCCAAAAAAAAAAAGGCAAAAAAAAAAAAAA
AATCACTTCAGTCCAGGCTGGGGTATCTGGCTCCCCGGGAGGCGCCCGGGCTCCCCGGGGCCCTGGCAGA
CGCCGGCAGCTTTTCTGGGCCCGCACTCGGGGACCTCGCCCGCCCGGCATCGCGATTGTGCAAGCGCCGGG
CGGCAACCACGGCCGCCGCTGCCGGCTCCTGCCCGCCGGGGAGGGCGGGGAGGCGCGTTCGAAGCACACC
CGGTGAACTCCGGGCTTCGCATGACTTCCTGGGGGTTATCTCCGGGGTGGAGTCTGCCGCCCCCACCCAC
CTCCCGCGCCCAGCGAACATAGCCCCGGTCACCCCAGGCCCGGACGCCCTCGCTCACCTCGCTGACTCCAT
CCTCCTTCCACCCCCCTCCCCTGGGTCCCGCGTCCCTCGGAGTCTGGCCAGCCGGGGGCCACTCCGCCCT
CCTCTGCGTGCTCATTGGCCACCCAGGGCATGCTCTGTCTCCATAAATGCATGGTCCCTGGGCATAGGAAC
ATGGAGAGG[TSS]gacaggaggaaacgcagcgccagcagcatctcatctaccctccttgacacctcccg
tggctccagccagacccctagaggtcagccttgcggaccaacaggaggactcccagctttccttttcaaga
ggtccccagacaccggccaccctcttccagcccctgcggccagtgcaaggaggcaccaatg

Functional DR3 VDREs  C/EBP site   Bold Italics: CRE/AP-1 sites  EGR1 site
Candidate DR3 VDRE spanning a putative vitamin D stimulatory element
ETS1 site  RUNX2 site   CCAAT box   atg start codon
```

FIGURE 33.4 Nucleotide sequence of the human *CYP24A1* promoter. The transcription start site [TSS] is shown, as is the ATG start codon (underlined) for translation. The color-coded legend for cis-elements is depicted below the sequence. In the print version, the legend for shades of gray that correspond to color highlights is as follows: lightest gray, ETS1 site (TCCATCCTCC); next lightest gray, CCAAT box (GCTCATTGGC); next lightest gray, candidate DR3 VDRE (TAGCCCCGGTCACCC); next lightest gray, EGR1 site (CGCCCCCAC); medium gray, CRE/AP-1 sites (TGACTTCC and TGACTCCA); slightly darker than medium gray, functional DR3 VDREs (CACACCCGGTGAACT and CGCCCTCGCTCACCT); darker yet than medium gray, C/EBP site (TTGTGCAAG); darkest gray, RUNX2 site (CACCCCC).

```
-828
TCCTCTGATCAGCCAGCAGTGCCGTTCCAGTCCTCCAAATGAGTCTCCTTCCTATTGGCAAAGCCATAATTGCC
AGTTTAGTTCCCTGCCTCATCCAGACGAGGGAAACTGAGAAACCAGATCTTGCCATTTTTGCTGACCTCAAGAC
CTTGTTTTCTTTCTTCTTGCCTTGAGGTCTTTGGGAATGGTGATGGGAGGTGTTGAGCTCATGCTCCCACCTAA
CACCCTTCCTGGGGTTTGTGTATGGGGGTAGGGGGGAGTCTCATTTGCCTGATAGCATCACTTATGACCATATA
TCAAGACACTTGCCAGATGCAACAGCCAGGAGTAAGCTCCAAGAACACACTTGGCAGCTGGAGGAAAAGGGCTT
AAGCAAACCAAAACAAGGACACTGGAGGGAGATGAGTTAGCGAGGAGGCGGCTTTCTGGTTTTCTGGGGTTTTT
TTTGTTTGTTTGTTTCAGTACTGCTGGCTGCCTTCACACTTCCTGATGGAAGTGGGgacAGGTCAACAAATGAC
CCAGGGTCACAGATAACTTTTGCCCACACATCATTCACTTATGGGAGCACTGGCTTGAAATTGAGGGGTGTGTG
CGTGCATGTATGTGTGTGCCTGGAACTGACGCGCCTTCCGCAAGCCTAAGAAGTCTGGGCTTTTTCTTTGAATG
GATGATTACAACACAGAGGATGTGGCGGCATTGTTTTTCCTGCTTGATGTCACACCACCACCCTTTAAAGTCC
CGGGGAAAAAAGGAGGGAATCTAGCCCAGGATCCCCACCTCAGTTCTCAGCTTCTTCCTAGGAAGAAGAGAAA
GCCAGCAAGGGCCCAGCCTGTCTGGGAGTGTCAGATTTCAAACTCAGCATTAGCCACTCAGTGCTGTGCAATG

VDRE@-334:  AGTGGGgacAGGTCA   Ets1-site   GATA-Site   Lef-1-site   CRE/AP1-site
```

FIGURE 33.5 Nucleotide sequence of the mouse *FGF23* promoter. The color-coded legend for cis-elements is depicted above the sequence. The transcription start site begins at the 5' end of the underlined sequence. The TATA-box and the ATG start codon for translation are highlighted in light green (Medium gray in print versions). In the print version, the legend for shades of gray that correspond to color highlights is as follows: lightest gray, Ets-1 sites (CTTCCT, CTTCCG, and AGGATG); next lightest gray, Lef-1 site (TTTGAATG); next lightest gray, CRE/AP-1 site (TGATGTCA); darker than medium gray, functional DR3 VDRE (AGTGGGgacAGGTGA); darkest gray, GATA sites (GATAA and GATTA).

the human gene in Fig. 33.4. Key *cis*-elements, mostly occurring antisense, are two functional DR3 VDREs (light purple highlight), a candidate DR3 VDRE (gray highlight) overlapping a putative vitamin D stimulatory element, a C/EBP site (teal highlight), an EGR1 site (light green highlight), two AP1 sites (bold italics, cyan highlight), an ETS1 site (yellow highlight), and a RUNX2 site (red highlight). In the print version, see legend for gray shade correspondence to color highlights. The C/EBP and RUNX2 sites are consistent with the composition of VDRE-containing *cis*-regulatory modules (CRMs) in other vitamin D-controlled genes [54], and the ETS1 site is reminiscent of the control of the FGF23 proximal promoter by 1,25D as discussed below (Fig. 33.5), although the ETS1 site in CYP24A1 overlaps an AP1 *cis*-element. Of potential mechanistic significance is the EGR1 site, as FGF23 signals via EGR1 [55] to induce CYP24A1. Therefore, all of the elements with the potential for control of *CYP24A1* transcription by 1,25D and FGF23 are present in the proximal promoter (Fig. 33.4), possibly explaining the striking inducibility of CYP24A1 to catalyze the catabolism of the potent 1,25D hormone.

FGF23

The significance of FGF23 in regulating bone mineral homeostasis is confirmed by the two dominant characteristics of the FGF23 knockout mouse, namely hyperphosphatemia and ectopic calcification [20]. FGF23 null mice also possess markedly elevated 1,25D in blood, generating the additional phenotypes of skin atrophy, osteoporosis, vascular disease, and emphysema. Many of these pathologies are also the consequence of hypervitaminosis D [27], indicating that 1,25D must be "detoxified" and sustained in an optimal range to maintain healthful aging. As discussed above, the biological effects of 1,25D are curtailed by CYP24A1-catalyzed catabolism of 1,25D, providing an "off" signal once the hormone has executed its physiologic modulation of gene expression. With respect to FGF23, we [12] originally reported that, in osteocyte-like UMR-106 cells, FGF23 mRNA levels are dramatically upregulated by 1,25D; this FGF23 induction is potentiated by leptin, and inhibited by IL-6 [56]. Functional VDREs were identified in the human FGF23 gene at −35.7, −16.2, and +8.6 kb in relation to the transcription start site, and each of the three candidate VDREs is located in a cluster of binding sites for C/EBP and RUNX2 [56], consistent with the concept of cis-regulatory modules for control of osteoblast expressed genes by the vitamin D hormone [57]. Based upon the observation that FGF23 induction by 1,25D is partially cycloheximide sensitive [23], and the fact that 1,25D upregulates ETS1, a transcription factor that cooperates with VDR [58], we also concluded that 1,25D induces FGF23 production directly (primarily) via multiple VDREs, and indirectly (secondarily) via stimulation of ETS1 expression, with VDR and ETS1 cooperating in the induction of FGF23 through DNA looping and generation of euchromatin architecture [56]. However, it has been reported that the FGF23 gene region in mouse osteocytes is not marked by detectable, 1,25D-dependent VDR/RXR binding sites when ChIP-seq analysis is carried out [59]. Consequently, we reexamined the mouse FGF23 proximal promoter (DNA sequence listed in Fig. 33.5) for transactivator binding sites that might mediate control of FGF23 mRNA expression. As depicted in Fig. 33.5, we observed a collection of consensus cis-elements between −110 and −347 bp in relation to the start of transcription. Most prominent in this collection are three ETS1, two GATA, and single AP1 and LEF-1 sites, with the latter AP1 and LEF-1 cis-elements often associated with VDREs in 1,25D-regulated genes [60,61]. Importantly, ETS1, which was implicated as a partner with VDR in controlling FGF23 in a previous study [56], possesses a cis-docking site in the mouse FGF23 promoter that is only six bp 5′ of a newly discovered candidate VDRE (AGTGGGgacAGGTCA), highlighted in light purple in Fig. 33.5. We therefore hypothesized that the −110 to −347 bp region of the FGF23 promoter constitutes a cis-regulatory module (CRM) anchored by the adjacent ETS1 and VDRE sites. To test this hypothesis, we utilized the mouse FGF23 promoter (1.0 kb) and progressively truncated promoter fragments kindly supplied by Drs. Ito and Miyamoto of the Tokushima University [62], to generate luciferase gene–reporter constructs. These constructs were transfected into K562 cells and evaluated for 1,25D-responsiveness. The results (Fig. 33.6A and 33.6B) implied the existence of a VDRE between −200 and −400 bp in the mouse FGF23 promoter. To determine if this VDRE corresponds to the newly revealed candidate VDRE shown in Fig. 33.5, this VDRE and/or its adjacent ETS1 site were inactivated by point mutation within the context of a −0.6 kb promoter fragment-luciferase construct, and the mutated plasmids were examined for transcriptional stimulation by 1,25D. The observation that inactivation by site-directed mutagenesis of either the VDRE or ETS1 elements abolishes transcriptional stimulation by 1,25D (Fig. 33.6C) provides evidence for a novel, composite ETS1-VDRE cis-element that may be central to the CRM in the mouse FGF23 promoter that mediates at least part of the response of this gene to induction by 1,25D. Our results further suggest that regulation by phosphate does not reside in the proximal 1.0 kb promoter, because no significant difference in reporter gene expression is observed when the truncation series is exposed to low (0.9 mM, Fig. 33.6A) versus high (3.0 mM, Fig. 33.6B) phosphate concentration. Finally, we probed the responsiveness of the FGF23 promoter to calcium, as David et al. [63] recently called attention to calcium as a stimulator of FGF23 secretion by osteoblasts. Indeed, we provide evidence for a regulatory site that responds to high (6 mM) calcium within the proximal FGF23 promoter. As illustrated in Fig. 33.6D, high calcium significantly (1.7- to 2.0-fold) induces the −1.0 kb mouse FGF23 promoter–reporter construct. Therefore, as depicted schematically at the top (center) of Fig. 33.1, we conclude that 1,25D and calcium comprise FGF23 inducers that function by stimulating the proximal promoter of the mouse gene (and likely other species as the CRM is conserved across species). Phosphate, another FGF23 secretagogue (Fig. 33.1, top center), apparently functions by a mechanism independent of the proximal promoter.

Klotho

Phosphate is abundant in a normal diet and is a fundamental biological component of not only mineralized bone, but also essential biomolecules such as DNA, RNA, phospholipids, phosphoproteins, ATP, and metabolic intermediates. Yet phosphate excess may act as a pro-senescence factor independently of hypervitaminosis D by contributing to ectopic calcification and arteriosclerosis, COPD, chronic kidney disease, and loss of hearing. Fortunately, FGF23 and klotho are designed to eliminate excess phosphate and therefore promote increased healthspan. Klotho appears to have systemic antiaging properties independent of its

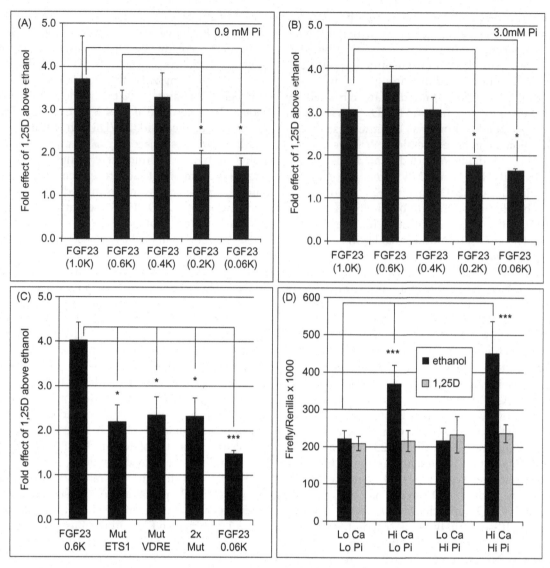

FIGURE 33.6 Dissection of the mouse FGF23 proximal promoter. (A) A −1.0 kb promoter fragment linked to a luciferase reporter and transfected into K562 cells in the presence of 0.9 mM (low) phosphate stimulates transcription in response to 1,25D by 3 to 4-fold. This transcriptional effect is significantly reduced to less than 2-fold between −0.4 kb and −0.2 kb when progressively truncated promoter fragments are tested. (B) Profile of transcriptional stimulation by 1,25D of FGF23 promoter fragments is unaffected by 3.0 mM (high) phosphate concentration. (C) A −0.6 kb promoter fragment-luciferase construct of the mouse FGF23 promoter yields a 4-fold response to 1,25D. This effect is significantly diminished to 2.2- to 2.4-fold by mutation of either the VDRE or ETS1 site, or both simultaneously, with all responses not significantly different from the promoter-less (−0.06 kb) control construct. (D) High (6.0 mM) calcium, but not high (3.0 mM) phosphate, significantly (1.7- to 2.0-fold) induces the −1.0 kb mouse FGF23 promoter-reporter construct.

phosphaturic actions, perhaps through its glycosyl hydrolase enzymatic activity [64]. Conversely, although FGF23 is antiaging at the kidney by eliciting phosphate elimination and detoxifying 1,25D, its "off-target" actions could actually be pro-aging in terms of coronary artery disease, as well as possible neoplastic actions in the colon [65], and it is possible that these off-target FGF23 pathologies are opposed by secreted klotho [11]. Upregulation of klotho by 1,25D [66] is consistent with potentiation of FGF23 signaling in the kidney and perhaps protection of other cell types (eg, vascular and colon) where a secreted form of klotho is considered a potentially beneficial renal hormone [67].

Klotho is the only reported single gene mutation that leads to a premature aging phenotype in the mouse [19], and a recessive inactivating mutation in the human klotho gene elicits a phenotype of severe tumoral calcinosis [68]. Klotho- and FGF23-null mice have identical hyperphosphatemic phenotypes of short lifespan/premature aging, ectopic calcification, arteriosclerosis, osteoporosis, muscle atrophy, skin atrophy, and hearing loss [19,20]. Klotho is a known coreceptor for FGF23, whereas β-klotho, the product of a separate gene, is a coreceptor for other FGF hormones. (α-) Klotho exists in multiple forms [69]: a full-length (130 kDa) transmembrane coreceptor with a minimal cytoplasmic region of 11 amino acids and two extracellular domains with homology to glycosyl hydrolases, at least three proteolyzed forms that are shed into the circulation [70], and finally a hypothetical 80 kDa secreted form (s-klotho) produced by alternative splicing in exon 3 that generates a protein species possessing a portion of the extracellular domain containing one of the glycosyl hydrolase domains [71]. Because of the potential significance of these forms in klotho biology, much remains to be learned about how their expression is regulated. Thurston et al. [72] have demonstrated that TNF-α and γ-interferon are suppressors of renal klotho expression; the FGF23 ligand is also a putative repressor of klotho expression [13]. With the exception of a preliminary report by Tsujikawa et al. [73] who treated mice with various dietary and pharmacologic regimens, including vitamin D, inducers of klotho are poorly characterized. However, as detailed below, it has been reported recently [66] that 1,25D significantly induces klotho mRNA expression at the cellular and molecular level in human and mouse renal cell lines.

Analysis by Forster et al. [66] of RNA isolated from mouse distal convoluted tubule (mpkDCT) cells, the primary expression site for klotho in kidney, with primers designed to capture both alternatively spliced mRNAs for the membrane and secreted forms of klotho, demonstrated that 1,25D treatment induces mRNA expression of either the membrane or secreted splice forms of klotho mRNA, suggesting that 1,25D may be capable of both amplifying FGF23 responsiveness and eliciting secretion of circulating klotho hormone, with the s-klotho hormonal isoform being more responsive to 1,25D induction. Interestingly, curcumin, an alternative VDR ligand [38], selectively upregulates membrane klotho mRNA in mpkDCT cells [66], indicating that distinct VDR ligands can differentially modulate the membrane and secreted forms of klotho. These data lead to the hypothesis that designer vitamin D analogs could promote the healthful aging benefits of systemic klotho without accentuating FGF23 action to perhaps elicit hypophosphatemia. Bioinformatic analysis [66] of both the human and mouse klotho genes revealed 17 candidate VDREs in the mouse gene and 11 putative VDREs in the human gene [66]. When assessed for functionality by cotransfection of reporter constructs into human kidney (HK-2) cells (Fig. 33.7), only the mouse VDRE at −35 kb (mKL-12) and the human

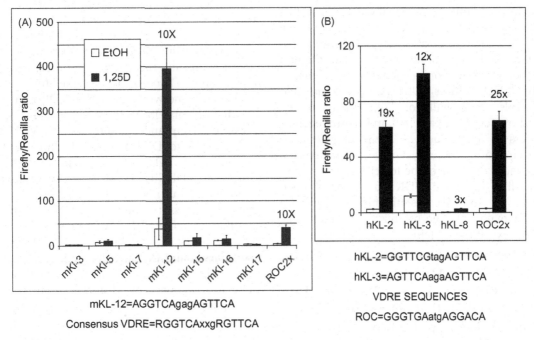

FIGURE 33.7 Functional activity of candidate mouse and human klotho VDREs. Candidate VDREs were cloned into a pLUC-MCS reporter vector, cotransfected into HK-2 human kidney cells along with a pSG5-VDR cDNA expression plasmid and treated with 1,25D (10^{-8} M) for 24 h. Firefly luciferase values were normalized to expression of Renilla luciferase. Data are depicted as the Firefly to Renilla luciferase ratio, with the fold effect of 1,25D on top of key black bars. (A) Analysis of seven mouse klotho candidate VDREs, revealing that only the mouse klotho VDRE located at −35 kb (mKl-12) displays transactivation ability. (B) Transfection of candidate human klotho VDREs demonstrates a striking (>10-fold) 1,25D responsiveness of VDREs corresponding to sequences at −46 kb (hKL-2) and −31 kb (hKL-3), but not +3.2 kb (hKL-8). The sequences of the active VDREs in the mouse and human klotho genes are listed; they are very similar to proven VDREs, with the mouse VDRE at −35 kb (mKL-12) conforming exactly to the consensus VDRE.

VDREs at −46 kb and −31 kb (hKL-2 and hKL-3) display a potency similar to the established rat osteocalcin (ROC) VDRE [66]. Thus, it appears that 1,25D-liganded VDR-RXR induces klotho expression by binding to functional VDREs in the range of 31 to 46 kb 5′ of the transcriptional start site of both the human and mouse klotho genes. In combination with the data of Tsujikawa et al. [73] that 1,25D increases steady-state klotho mRNA levels in mouse kidney, in vivo, the results shown in Fig. 33.7 indicate that 1,25D is the first discovered inducer of the longevity gene, klotho.

Detoxification of Endo- and Xenobiotics (CYP3A4, SULT, and CBS)

A common theme for VDR and PXR is the induction of CYP enzymes that participate in xenobiotic detoxification. A major target for VDR and PXR in humans is CYP3A4 [74–76], for which the detoxification substrates include lithocholic acid (LCA) [77]. The precursor to LCA, chenodeoxycholic acid, is produced in the liver and converted to LCA via 7-dehydroxylation by gut bacteria. LCA is not a good substrate for the enterohepatic bile acid reuptake system, and thus remains in the enteric tract and passes to the colon, where it can exert carcinogenic effects [78]. Thus, natural ligands for VDR, including the high affinity 1,25D hormonal metabolite and the lower affinity, nutritionally-modulated bile acids, seem to possess the important potential to serve as agents for promoting detoxification of LCA and possibly other intestinal endo- or xenobiotics, with the end result likely being a reduction in colon cancer incidence. Additionally, 1,25D induces SULT2A, an enzyme that detoxifies sterols via 3α-sulfation [79]. Finally, in the realm of cardiovascular disease [80] and neurodegenerative disorders of aging such as Alzheimer's disease [81], excess circulating homocysteine is considered a negative risk factor because of the toxicity of this methionine metabolite. Liganded VDR has recently been shown [82] to induce cystathionine β-synthase (Table 33.1), a major enzyme catalyzing the metabolic elimination of toxic homocysteine.

VDR Ligands Promote Healthspan via Regulation of Antiaging Genes (SPP1, TRPV6, LRP5, RANKL, OPG, BGP, Defensins/Cathelicidins, IL-17/23, p21/53, NF-κB, FOXOs, and TPH2)

1,25D-VDR regulates the expression of numerous genes that encode bone and mineral homeostasis effectors for which appropriate control can be considered to facilitate healthful aging, and several of these are discussed below. The first, osteopontin or SPP1 (Table 33.1) is induced by 1,25D in osteoblasts, where it triggers ossification; SPP1 also serves as an inducible inhibitor of vascular calcification and associated disease, an important antiaging action [118]. Intestinal calcium uptake is mediated, in part, by 1,25D-VDR induction of TRPV6 [88,97] (Table 33.1). TRPV6 is a key calcium channel gene product that supplies dietary calcium via transport to build the mineralized skeleton and thereby not only prevents rickets, but also delays the inevitable calcium leaching from bone in senile osteoporosis. 1,25D significantly induces LRP5 [88,89] (Table 33.1), a gene product that promotes osteoblastogenesis via enhanced canonical Wnt signaling, and is thereby anabolic to bone [119]. The expression of RANKL (Table 33.1), which is catabolic to bone, is enhanced by 1,25D-VDR [23] and mediates bone resorption through osteoclastogenesis. OPG, the soluble decoy receptor for RANKL that tempers its activity, is simultaneously repressed [23] to amplify the bioeffect of RANKL. Therefore, a well-mineralized bone, in response to SPP1, TRPV6, LRP5, and OPG, as well as an actively remodeled bone as a result of RANKL action, is a healthy bone that is less susceptible to fractures associated with aging and senile osteoporosis. Osteocalcin (BGP, Table 33.1) is another gene classically induced by 1,25D in osteoblasts. Utilizing BGP null animals, it has been shown that normal osteocalcin expression is important for robust, fracture-resistant bones [120]. Moreover, osteocalcin has been identified as a bone-secreted hormone that both improves insulin release from pancreatic β-cells and increases insulin metabolic responsiveness [121], perhaps delaying Type II diabetes mellitus.

The final network of 1,25D-VDR-regulated genes encodes factors impacting cell survival/cancer, the immune system and metabolism. The VDR null mouse is sensitized to DMBA-induced skin cancer [122] as well as UV light-induced skin malignancy [123]. Moreover, VDR likely reduces risk for many cancers by inducing the p53 [124] and p21 [124] (Table 33.1) tumor suppressors, as well as DNA mismatch repair enzymes in colon [125]. VDR knockout mice exhibit enhanced colonic proliferation [126] plus amplified mammary gland ductal extension, end buds and density [127], indicating that the fundamental actions of VDR to promote cell differentiation and apoptosis [128] play an important role in reducing the risk of age-related epithelial cell cancers such as those of the breast and colon.

In the case of immune function, 1,25D-VDR induces cathelicidin [26] to activate the innate immune system to fight infection (Table 33.1), and represses IL-17 [3] to temper the adaptive immune system and lower the risk of autoimmune disorders such as type I diabetes mellitus, multiple sclerosis,

TABLE 33.1 VDREs in Genes Directly Modulated in Their Expression by 1,25D and Possibly Other VDR Ligands

Gene	Bioeffect	Type	Location	5′-Half	Spacer	3′-Half	Ref.	Group
rBGP	Bone metabolism	Positive	−456	GGGTGA	atg	AGGACA	[83]	Bone
mBGP	Bone metabolism	Negative	−444	GGGCAA	atg	AGGACA	[84]	Bone
hBGP	Bone metabolism	Positive	−485	GGGTGA	acg	GGGGCA	[85]	Bone
mSPP1	Bone metabolism	Positive	−757	GGTTCA	cga	GGTTCA	[86]	Bone
mSPP1	Bone metabolism	Positive	−2000	GGGTCA	tat	GGTTCA	[87]	Bone
mLRP5	Bone anabolism	Positive	+656	GGGTCA	ctg	GGGTCA	[88]	Bone
mLRP5	Bone anabolism	Positive	+19 kb	GGGTCA	tgc	AGGTTC	[89]	Bone
rRUNX2	Bone anabolism	Negative	−78	AGTACT	gtg	AGGTCA	[90]	Bone
mRANKL	Bone resorption	Positive	−22.7 kb	TGACCT	cctttg	GGGTCA	[91]	Bone
mRANKL	Bone resorption	Positive	−76 kb	GAGTCA	ccg	AGTTGT	[92]	Bone
mRANKL	Bone resorption	Positive	−76 kb	GGTTGC	ctg	AGTTCA	[92]	Bone
cIntegrin-beta3	Bone resorption, platelet aggregation	Positive	−756	GAGGCA	gaa	GGGAGA	[93]	Bone
cCarbonic anhydrase II	Bone resorption, brain function	Positive	−39	AGGGCA	tgg	AGTTCG	[94]	Bone
cPTH	Mineral homeostasis	Negative	−60	GGGTCA	gga	GGGTGT	[95]	Mineral
mVDR	Auto-regulation of VDR	Positive	+8467	GGGTTA	gag	AGGACA	[96]	Mineral
hTRPV6	Intestinal Ca^{2+} transport	Positive	−1270	AGGTCA	ttt	AGTTCA	[97]	Mineral
hTRPV6	Intestinal Ca^{2+} transport	Positive	−2100	GGGTCA	gtg	GGTTCG	[97]	Mineral
hTRPV6	Intestinal Ca^{2+} transport	Positive	−2155	AGGTCT	tgg	GGTTCA	[97]	Mineral
hTRPV6	Intestinal Ca^{2+} transport	Positive	−4287	GGGGTA	gtg	AGGTCA	[97]	Mineral
hTRPV6	Intestinal Ca^{2+} transport	Positive	−4337	CAGTCA	ctg	GGTTCA	[97]	Mineral
hNPT2a	Renal phosphate reabsorption	Positive	−1963	GGGGCA	gca	AGGGCA	[98]	Mineral
hNpt2c	Renal phosphate reabsorption	Positive	−556	AGGTCA	gag	GGTTCA	[88]	Mineral
hFGF23	Renal phosphate elimination	Positive	−35.7 kb	GGGAGA	atg	AGGGCA	[31,56]	Mineral
hFGF23	Renal phosphate elimination	Positive	−32.9 kb	TGAACT	caaggg	AGGGCA	[31,56]	Mineral
hFGF23	Renal phosphate elimination	Positive	−16.2 kb	TAACCC	tgcttt	AGTTCA	[31,56]	Mineral
hFGF23	Renal phosphate elimination	Positive	+8.6 kb	AGGGCA	gga	AGGACA	[31,56]	Mineral
mFGF23	Renal phosphate elimination	Positive	−334	AGTGGG	gac	AGGTCA	Figs. 33.5 and 33.6 herein	Mineral
hklotho	Renal phosphate elimination	Positive	−31 kb	AGTTCA	aga	AGTTCA	[66]	Mineral
hklotho	Renal phosphate elimination	Positive	−46 kb	GGTTCG	tag	AGTTCA	[66]	Mineral
mklotho	Renal phosphate elimination	Positive	−35 kb	AGGTCA	gag	AGTTCA	[66]	Mineral

(Continued)

TABLE 33.1 (Continued)

Gene	Bioeffect	Type	Location	5′-Half	Spacer	3′-Half	Ref.	Group
rCYP24A1	1,25D detoxification	Positive	−151	AGGTGA	gtg	AGGGCG	[15]	Detox
rCYP24A1	1,25D detoxification	Positive	−238	GGTTCA	gcg	GGTGCG	[16]	Detox
hCYP24A1	1,25D detoxification	Positive	−164	AGGTGA	gcg	AGGGCG	[99]	Detox
hCYP24A1	1,25D detoxification	Positive	−285	AGTTCA	ccg	GGTGTG	[99]	Detox
hCYP3A4	Xenobiotic detoxification	Positive	−169	TGAACT	caaagg	AGGTCA	[75,76]	Detox
hCYP3A4	Xenobiotic detoxification	Positive	−7.7 kb	GGGTCA	gca	AGTTCA	[74]	Detox
rCYP3A23	Xenobiotic detoxification	Positive	−120	AGTTCA	tga	AGTTCA	[76,100]	Detox
hMDR1	P-glycoprotein, drug resistance	Positive	−7863	AGTTCA	atg	AGGTAA	[101]	Detox
hMDR1	P-glycoprotein, drug resistance	Positive	−7853	AGGTCA	agtt	AGTTCA	[101]	Detox
hp21	Cell cycle control	Positive	−765	AGGGAG	att	GGTTCA	[102]	Cell life
hFOXO1	Cell cycle control	Positive	−2856	GGGTCA	cca	AGGTGA	[103]	Cell life
hIGFBP-3	Cell proliferation/apoptosis	Positive	−3282	GGTTCA	ccg	GGTGCA	[104]	Cell life
hInvolucrin	Skin barrier function	Positive	−2083	GGCAGA	tct	GGCAGA	[105]	Cell life
hPLD1	Keratinocyte differentiation	Positive	−246	GGGTGA	tgc	GGTCGA	[106]	Cell life
hCCR10	Homing of T-cells to skin	Positive	−110	GGGTCT	acg	GGGTCA	[107]	Cell life
rPTHrP	Mammalian hair cycle	Negative	−805	AGGTTA	ctc	AGTGAA	[108]	Cell life
hSOSTDC1	Mammalian hair cycle	Negative	−6215	AGGACA	gca	GGGACA	[109]	Cell life
hHairless	Mammalian hair cycle	Positive	−7269	TGGTGA	gtg	AGGTCA	[109]	Cell life
rVEGF	Angiogenesis	Positive	−2730	AGGTGA	ctc	AGGGCA	[110]	Cell life
hMIS	Müllerian-inhibiting substance	Positive	−381	GGGTGA	gca	GGGACA	[111]	Cell life
hHLA-DRB1	Major histocompatibility complex	Positive	−1	GGGTGG	agg	GGTTCA	[112]	Immune
hCAMP	Antimicrobial peptide	Positive	−615	GGTTCA	atg	GGTTCA	[113]	Immune
hKSR-1	Monocytic differentiation	Positive	−8156	GGTGCA	tat	AGGTCA	[114]	Immune
hKSR-2	Monocytic differentiation	Positive	−2501	AGTTCA	gca	TGGTCA	[115]	Immune
hKSR-2	Monocytic differentiation	Positive	+3185	GGTTCA	aac	AGTTCT	[115]	Immune
mInsig-2	Regulation of lipid synthesis	Positive	−2470	AGGGTA	acg	AGGGCA	[116]	Metabolism
hPCFT	Intestinal folate transporter	Positive	−1680	AGGTTA	ttc	AGTTCA	[117]	Metabolism
hCystathionine β synthase	Homocysteine clearance	Positive	+5983	GGGTTG	atg	AGTTCA	[82]	Metabolism
hTryptophan hydroxylase 2	Serotonin synthesis	Positive	−9971	TGGTCA	att	AGTTCA	[8]	Metabolism
			−7059	AGGTCA	att	TGGTCA		CNS

lupus, and rheumatoid arthritis. 1,25D-VDR is antiinflammatory by blunting NFκB [129] and COX2 [130], and inflammation is considered a common denominator in maladies such as heart disease and stroke, as well as cancer. In addition, 1,25D-VDR induces FOXO3 [131], a significant molecular player in preventing oxidative damage, the leading candidate for the cause of aging [132]. Finally, VDREs have been identified in the *tryptophan hydroxylase 2 (TPH2)* gene that encodes the enzyme catalyzing the rate-limiting step in serotonin synthesis in the central nervous system [8,133]. Fig. 33.8A lists these VDREs in the human *TPH2* gene, and the data in Fig. 33.8B reveal that these VDREs are functionally responsive to 1,25D when inserted upstream of reporter genes and transfected into intact cells. Their activity is statistically significant but modest in magnitude (somewhat less active than the CYP3A4 VDRE), yet *TPH2* mRNA concentrations are upregulated by 1,25D treatment in human U87 glioblastoma (Fig. 33.8C, 2.4 to 2.9-fold) and embryonic kidney cells (Fig. 33.8D, 2.4-fold). The data in Fig. 33.8 indicate that 1,25D may be capable of raising serotonin levels in the brain to elevate mood and serve as a potential defense against depression, as well as possibly supporting cognitive function and prosocial behavior, while curbing impulsive behavior. In summary, through control of vital genes, VDR allows one to age well by delaying fractures, ectopic calcification, malignancy, oxidative damage, infections, autoimmunity, inflammation/pain, cardiovascular and neuropsychiatric diseases.

Another nutritionally derived lipid, resveratrol, is a potent antioxidant found in the skin of red grapes and as a component in red wine that may serve to augment 1,25D-VDR antiaging actions [134]. As shown in Fig. 33.9A (top panel), 1,25D activates (20-fold over vehicle control) VDR-mediated transcription of CYP24A1 in human embryonic kidney cells. Significantly, the combined presence of 1,25D and resveratrol results in synergism (394%) in VDR

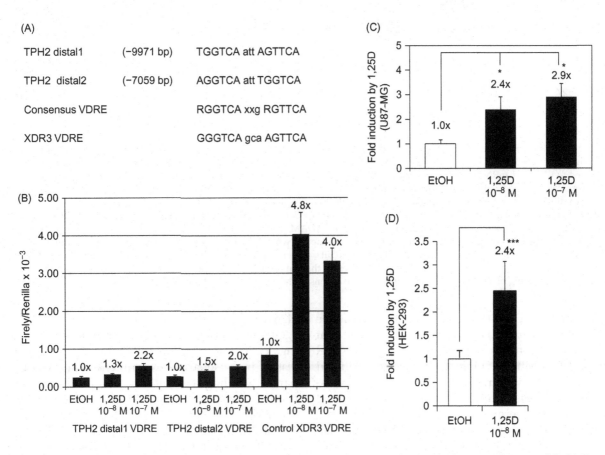

FIGURE 33.8 1,25D induces *TPH2* mRNA in human brain cells. (A) Candidate VDREs in the human *TPH2* gene [8]. (B) Functional activity of putative VDREs in the human *TPH2* gene when linked into a luciferase reporter construct and transfected into U87 human glioblastoma cells, and treated in culture with 1,25D (10 or 100 nM) for 24 hours. Data are depicted as a ratio of firefly (test) to Renilla (control) luciferase. XDR3 represents the positive control, a distal VDRE from the human CYP3A4 gene [76]. (C) Induction of *TPH2* mRNA by 1,25D in U87-MG human glioblastoma cells treated for 24 hours with the indicated concentrations of hormone. Data are depicted as a fold effect of 1,25D. (D) Induction of *TPH2* mRNA by 1,25D in HEK-293 human embryonic kidney cells treated for 24 hours with the indicated concentration of hormone.

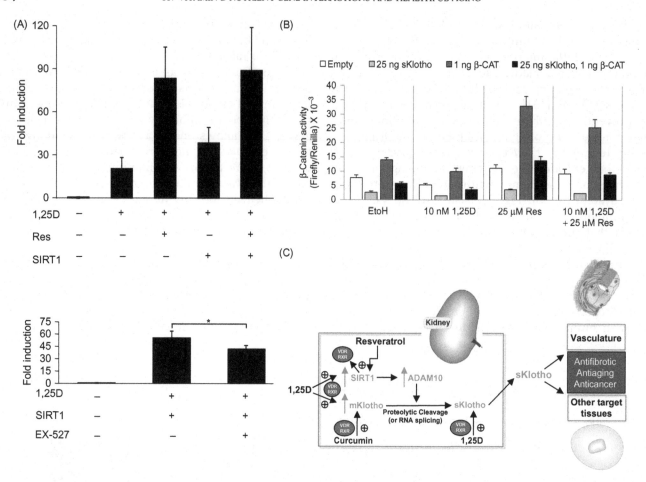

FIGURE 33.9 Resveratrol and SIRT1 cooperate with 1,25D to enhance VDR signaling. (A) *Upper panel*: HEK-293 cells were treated with the indicated ligands or SIRT1, and endogenous *CYP24A1* in human embryonic kidney cells was measured by real-time PCR using human primers directed to *CYP24A1* mRNA. *Lower panel*: As in the upper panel, but treatments included the selective SIRT1 inhibitor, EX-527. (B) HEK-293 cells were cotransfected with a firefly luciferase plasmid containing a β-catenin responsive element along with the indicated expression plasmids encoding soluble (s)-klotho or β-catenin (β-CAT), with firefly luciferase results normalized to Renilla luciferase. (C) A hypothetical model for resveratrol activation of VDR via stimulation of SIRT1 [136]. SIRT1 catalyzes deacetylation of VDR (to increase the capacity of 1,25D binding), RXR, or comodulators. SIRT1 activation also leads to ADAM10 stimulation [137] to produce soluble (s)-klotho via ADAM10-mediated cleavage of membrane (m)-klotho. Curcumin induces m-klotho, while 1,25D stimulates SIRT1 activity [138] as well as expression of both m- and s-klotho. The integration of these regulatory circuits, which are controlled by the levels of nutritionally derived bioactive "healthy" lipids (1,25D, curcumin, and resveratrol), culminates in the elaboration of s-klotho from the kidney to exert proposed endocrine antiaging effects consisting of antifibrogenic and antineoplastic actions in the vasculature and other target tissues.

transactivation of CYP24A1 compared to 1,25D alone (Fig. 33.9A, top panel). This result indicates that resveratrol functions as a potentiator of 1,25D-VDR signaling. Competitive binding assays of tritiated 1,25D-bound VDR with resveratrol reveal that resveratrol, rather than displacing, actually *enhances* tritiated 1,25D binding to overexpressed hVDR in a COS-7 cell extract [135]. Thus, resveratrol confers VDR with an increased capacity for 1,25D binding, likely by activating SIRT1 to deacetylate VDR and/or one of its comodulators. To test this hypothesis, we cotransfected a SIRT expression plasmid and observed that SIRT1, like resveratrol, boosts (185%) 1,25D-induced transcription of CYP24A1 (Fig. 33.9A, upper panel), and the combination of SIRT1 and resveratrol results in the greatest (426%) transcriptional activation of CYP24A1 by 1,25D-liganded VDR. To confirm that SIRT1 is activating VDR, the selective SIRT1 inhibitor, EX-527, was tested in a separate experiment (Fig. 33.9A, lower panel), yielding a significant diminution of transcriptional activation of CYP24A1 by 1,25D-liganded VDR. Therefore, resveratrol affects VDR indirectly, likely via the ability of resveratrol to potentiate 1,25D binding to VDR by stimulating SIRT1, an enzyme known to deacetylate nuclear receptors. These data illuminate a pathway for cross-talk between two nutritionally derived lipids, vitamin D and resveratrol, both of which converge on VDR signaling.

To determine the generality of resveratrol modulation of VDR, we also tested 1,25D ± resveratrol in the context of klotho regulation of β-catenin action. We noted that klotho is not only antiaging, but has also been recognized as an anticancer peptide [139]. One mechanism for the antineoplastic action of s-klotho is inhibition of Wnt/β-catenin signaling, especially in the colon. The data in Fig. 33.9B show that s-klotho is a potent suppressor of both endogenous and exogenous β-catenin activity, a phenomenon potentiated by 1,25D-VDR but not by resveratrol, suggesting that resveratrol/SIRT action in the context of klotho/VDR signaling may represent a more complex, multilayered array of signal transduction cross-talk. In kidney, s-klotho is documented to serve as a tumor suppressor and inhibit tumor proliferation through the regulation of the IGF-1 signaling pathway. Klotho also binds directly to type-II TGF-β receptors and prevents TGF-β1 binding, thereby blocking TGF-β1 action; accordingly, s-klotho abolishes the fibrogenic effects of TGF-β1 [139]. Since 1,25D and curcumin induce klotho (discussed above), while resveratrol activates SIRT1 [136], we propose the existence of a molecular mechanism (Fig. 33.9C) that orchestrates collaborative cross-talk in 1,25D/klotho/resveratrol/SIRT signaling to achieve the antifibrogenic, antiaging, and anticancer activities of klotho, in vivo, that is thought to occur in the vasculature and other tissues such as kidney and colon [67].

Compilation of VDREs

Table 33.1 provides a list of VDR-RXR target genes recognized by the combined DNA binding domain zinc fingers of the two receptors, and their C-terminal extensions. These VDR-RXR controlled genes encode proteins that determine bone growth and remodeling, bone mineral homeostasis, detoxification, the mammalian hair cycle, cell proliferation/differentiation, apoptosis, lipid metabolism, immune and CNS function, and longevity. In general, VDREs possess either a direct repeat of two hexanucleotide half-elements with a spacer of three nucleotides (DR3) or an everted repeat of two half-elements with a spacer of six nucleotides (ER6) motif, with DR3s being the most common. Many VDREs occur as a single copy in the proximal promoter of vitamin D-regulated genes, however, CYP24A1 VDREs are at least bipartite, as is the human CYP3A4 VDRE, with the 5′ DR3 located some 7.7 kb upstream of the proximal ER6 VDRE in the latter case. Studies of 1,25D-controlled CYP genes introduced the concepts of multiplicity and remoteness to VDREs, concepts reinforced by ChIP or ChIP scanning [88,89,97,140] of genomic DNA surrounding the transient receptor potential vanilloid type 6 (TRPV6), LRP5, and receptor activator of nuclear factor κB ligand (RANKL) genes, uncovering novel VDREs at some distance from the transcription start site (Table 33.1). Indeed, the recent genome-wide study of the VDR/RXR cistrome cited above found that the vast majority (98%) of VDR/RXR binding sites in LS180 cells were located >500 bp upstream or downstream from the transcriptional start site of the nearest gene [141]. Strikingly, in the human genome it is thought that via epigenetic modification to "open-up" enhancers for transcription factor binding in chromatin, functional enhancers often occur as much as 250 kb 5′ or 3′ of the transcription start-site in a regulated gene. The most attractive model is that remote VDREs are juxtapositioned with more proximal VDREs via DNA looping in chromatin, creating a single platform that supports the transcriptional machinery [56].

CONCLUSION AND PERSPECTIVES

The healthful aging facets of vitamin D and novel VDR ligands described in the present chapter reveal new roles for vitamin D and its nutritional surrogates that go far beyond vitamin D as a simple promoter of dietary calcium and phosphate absorption to ensure adequate bone mineralization. We now understand that vitamin D hormonal ligands and their nuclear receptor also mediate both the sculptured delimiting and remodeling of the skeleton as well as the prevention of ectopic calcification through novel peptide mediators, such as FGF23 and klotho. These latter roles of 1,25D-VDR can be considered as protective against osteoporotic fractures on the one hand, and on the other hand in reducing the ravages of ectopic calcification which occur with aging, especially with respect to diminishing cardiovascular calcification and mortality [142]. It is striking indeed that the calcemic and phosphatemic hormone, 1,25D, and its receptor coevolved mechanisms for countermanding the potential deleterious effects of calcification. These mechanisms include feedback repression of calcemic PTH in one endocrine loop, and feed-forward induction of FGF23 in yet a second endocrine loop. FGF23 acts as a check and balance on bone mineral by delimiting skeletal calcification, retarding ectopic calcification, mostly via its phosphaturic action, and feedback repressing 1,25D production by the kidney. Although both 1,25D/PTH/Ca and 1,25D/FGF23/PO$_4$ are intricate and essential axes for mineral homeostasis, they represent only "the tip of the iceberg" in vitamin D and VDR functions significant to health (Fig. 33.10). Thus, the traditional bone and mineral (antirachitic/antiosteoporotic) effects, as well as newly recognized bone anabolic and counter-ectopic calcification functions, comprise the fraction of the iceberg above the water line. As illustrated, the bulk of the actions mediated by VDR are novel extraosseous effects

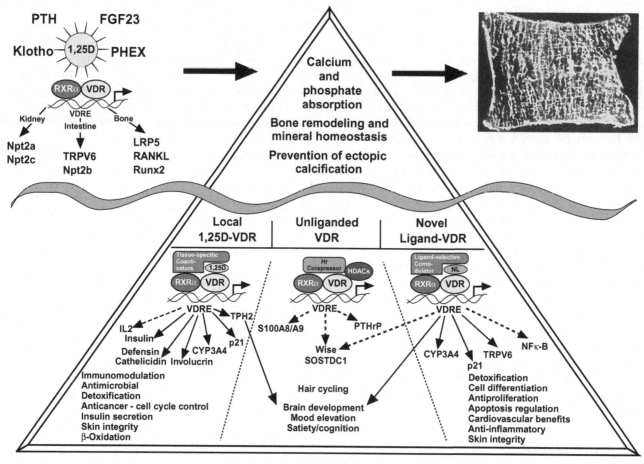

FIGURE 33.10 Osseous and extraosseous effects mediated by VDR. Shown at the upper left are the calcemic and phosphaturic hormones that participate in feedback loops to maintain bone mineral homeostasis as discussed in the text; a normally mineralized human vertebral body with its trabeculations is illustrated at the upper right. The upper portion depicts actions of 1,25D-liganded VDR to maintain bone health, including interactions with other hormones (PTH, FGF23 and klotho). The lower portion summarizes the many extraosseous effects of VDR in the 1,25D-bound, unliganded, and novel ligand-bound states. Repressive actions of VDR are depicted as dashed arrows. We hypothesize that VDR occupied by locally generated 1,25D uses cell context specific coactivators, and that VDR occupied by a novel ligand (NL, lower right of figure) may utilize ligand selective comodulators. Numerous tissues besides the kidney express the 1α-OHase enzyme, including cells of the immune and central nervous systems, the pancreas, skin, etc. This locally produced 1,25D does not contribute significantly to circulating 1,25D, but it retains the capacity to be active in a cell- and tissue-specific manner. Examples of local 1,25D-VDR actions include repression of IL-2 in T-cells [143], along with induction of defensin and cathelicidin as local antimicrobial effectors [26]. By locally stimulating the expression of genes, the vitamin D/VDR system emerges, likely redundantly with other regulators, as an immunomodulator that stimulates the innate and suppresses the adaptive immune system to effect both antimicrobial and antiautoimmune actions, as a detoxifier of xenobiotics to be chemoprotective, as a controller of cell proliferation and regulator of apoptosis to reduce cancer, and as a moderator of type II diabetes by promoting insulin release as well as possibly enhancing fatty acid β-oxidation via induction of FOXO1 (Table 33.1). Besides the utilization of novel ligands, another possibility obviating the need to locally generate 1,25D would be for VDR to function unliganded. VDR, but not vitamin D, is required to sustain the mammalian hair cycle [144]. Thus, as depicted in the lower center, the Hr corepressor could function as a surrogate VDR "ligand" to suppress SOSTDC1, S100A8/A9 [23], or other genes that normally keep the hair cycle in check. Also, unlike the case of intestine, kidney and bone, calbindin induction by VDR does not require vitamin D in brain [145]. VDR is widely expressed in the central nervous system, as is Hr, raising the possibility that unliganded VDR, perhaps along with Hr, acts in select neurons. Notably, it has been reported that VDR-null mice exhibit behavioral abnormalities including anxiety [146]. Furthermore, 1,25D/VDR exerts neuroprotective actions against excitotoxicity, and induces serotonin mood elevator to support cognitive function and prosocial behavior.

that are diagrammed in the submerged portion of the iceberg.

We propose that VDR binds one or more naturally occurring nonvitamin D ligands to effect many of its extraosseous actions. Indeed, we have identified several potential examples of nonvitamin D related VDR ligands, including LCA, curcumin, γ-tocotrienol, PUFAs, and anthocyanidins. Because VDR is capable of binding alternative lipid ligands, albeit with low affinity, the receptor may have retained its promiscuity for ligand binding that presumably originated with its primitive detoxification function, similar to that of its evolutionary cousin PXR. The question remains whether, in the course of its evolution, VDR function was further refined and endowed with higher affinity for local bioactive ligands that would explain the broad health benefits of vitamin D and other lipid nutrients beyond bone. In summary, through control of key genes via a mechanism that is potentiated by longevity agents such as resveratrol and SIRT1, we profess that VDR feeds the "fountain of youth" and allows one to age well by delaying fractures, ectopic calcification, oxidative damage, infections, autoimmunity, inflammation, pain, cardiovascular disease, neuropsychiatric disorders, and malignancy.

SUMMARY

1,25-Dihydroxyvitamin D_3 (1,25D) is the endocrine metabolite of vitamin D that signals through binding to the vitamin D receptor (VDR). The ligand—receptor complex transcriptionally regulates genes that encode factors promoting intestinal calcium and phosphate absorption plus bone remodeling, maintaining a skeleton with reduced risk of age-related osteoporotic fractures. 1,25D/VDR signaling further exerts feedback control of mineral ions via regulation of FGF23, klotho, and CYP24A1 to prevent age-related, ectopic calcification, and associated pathologies. Vitamin D also elicits xenobiotic detoxification, oxidative stress reduction, antimicrobial defense, immunoregulation, anti-inflammatory/anticancer actions, and cardiovascular benefits. 1,25D exerts neuroprotective actions against excitotoxicity, and induces serotonin mood elevator to support cognitive function and prosocial behavior. Nutrient, low-affinity VDR ligands including curcumin, polyunsaturated fatty acids, and delphinidin/anthocyanidins initiate VDR signaling, whereas longevity agents such as resveratrol and SIRT1 potentiate VDR signaling. Therefore, liganded VDR modulates the expression of a network of genes that facilitates health span by delaying the chronic diseases of aging.

References

[1] Haussler MR, Whitfield GK, Kaneko I, et al. Molecular mechanisms of vitamin D action. Calcif Tissue Int 2013;92(2):77–98.
[2] Bikle DD, Pillai S. Vitamin D, calcium and epidermal differentiation. Endocr Rev 1993;14:3–19.
[3] Mora JR, Iwata M, von Andrian UH. Vitamin effects on the immune system: vitamins A and D take centre stage. Nat Rev Immunol 2008;8(9):685–98.
[4] Omdahl JL, Morris HA, May BK. Hydroxylase enzymes of the vitamin D pathway: expression, function, and regulation. Annu Rev Nutr 2002;22:139–66.
[5] Adams JS, Singer FR, Gacad MA, et al. Isolation and structural identification of 1,25-dihydroxyvitamin D_3 produced by cultured alveolar macrophages in sarcoidosis. J Clin Endocrinol Metab 1985;60(5):960–6.
[6] Jacobs ET, Martinez ME, Jurutka PW. Vitamin D: marker or mechanism of action? Cancer Epidemiol Biomarkers Prev 2011;20(4):585–90.
[7] Bikle D. Extrarenal synthesis of 1,25-dihydroxyvitamin D and its health implications. Clin Rev Bone Miner Metab 2009;7(2):114–25.
[8] Patrick RP, Ames BN. Vitamin D hormone regulates serotonin synthesis. Part 1: relevance for autism. FASEB J 2014;28(6):2398–413.
[9] Hughes MR, Brumbaugh PF, Haussler MR, Wergedal JE, Baylink DJ. Regulation of serum 1a,25-dihydroxyvitamin D_3 by calcium and phosphate in the rat. Science 1975;190:578–80.
[10] DeMay MB, Kiernan MS, DeLuca HF, Kronenberg HM. Sequences in the human parathyroid hormone gene that bind the 1,25-dihydroxyvitamin D_3 receptor and mediate transcriptional repression in response to 1,25-dihydroxyvitamin D_3. Proc Natl Acad Sci USA 1992;89:8097–101.
[11] Bergwitz C, Juppner H. Regulation of phosphate homeostasis by PTH, vitamin D, and FGF23. Annu Rev Med 2010;61:91–104.
[12] Kolek OI, Hines ER, Jones MD, et al. 1α,25-dihydroxyvitamin D_3 upregulates FGF23 gene expression in bone: the final link in a renal-gastrointestinal-skeletal axis that controls phosphate transport. Am J Physiol Gastrointest Liver Physiol 2005;289(6): G1036–42.
[13] Quarles LD. Endocrine functions of bone in mineral metabolism regulation. J Clin Invest 2008;118(12):3820–8.
[14] St-Arnaud R. CYP24A1-deficient mice as a tool to uncover a biological activity for vitamin D metabolites hydroxylated at position 24. J Steroid Biochem Mol Biol 2010;121(1–2):254–6.
[15] Ohyama Y, Ozono K, Uchida M, et al. Identification of a vitamin D-responsive element in the 5′ flanking region of the rat 25-hydroxyvitamin D_3 24-hydroxylase gene. J Biol Chem 1994;269(14):10545–50.
[16] Zierold C, Darwish HM, DeLuca HF. Identification of a vitamin D-responsive element in the rat calcidiol (25-hydroxyvitamin D_3) 24-hydroxylase gene. Proc Natl Acad Sci USA 1994;91:900–2.
[17] Kouzmenko A, Ohtake F, Fujiki R, Kato S. Hormonal gene regulation through DNA methylation and demethylation. Epigenomics 2010;2(6):765–74.
[18] Murayama A, Takeyama K, Kitanaka S, et al. Positive and negative regulations of the renal 25-hydroxyvitamin D_3 1α-hydroxylase gene by parathyroid hormone, calcitonin, and 1α,25(OH)$_2D_3$ in intact animals. Endocrinology 1999;140(5):2224–31.
[19] Kuro-o M, Matsumura Y, Aizawa H, et al. Mutation of the mouse klotho gene leads to a syndrome resembling ageing. Nature 1997;390(6655):45–51.
[20] Shimada T, Kakitani M, Yamazaki Y, et al. Targeted ablation of Fgf23 demonstrates an essential physiological role of FGF23 in phosphate and vitamin D metabolism. J Clin Invest 2004;113(4):561–8.
[21] Hesse M, Frohlich LF, Zeitz U, Lanske B, Erben RG. Ablation of vitamin D signaling rescues bone, mineral, and glucose homeostasis in Fgf-23 deficient mice. Matrix Biol 2007;26(2):75–84.
[22] Sitara D, Razzaque MS, St-Arnaud R, et al. Genetic ablation of vitamin D activation pathway reverses biochemical and skeletal anomalies in Fgf-23-null animals. Am J Pathol 2006;169(6):2161–70.
[23] Haussler MR, Haussler CA, Whitfield GK, et al. The nuclear vitamin D receptor controls the expression of genes encoding factors which feed the "fountain of youth" to mediate healthful aging. J Steroid Biochem Mol Biol 2010;121:88–97.

[24] Mordan-McCombs S, Valrance M, Zinser G, Tenniswood M, Welsh J. Calcium, vitamin D and the vitamin D receptor: impact on prostate and breast cancer in preclinical models. Nutr Rev 2007;65(8 Pt 2):S131–3.

[25] Raghuwanshi A, Joshi SS, Christakos S. Vitamin D and multiple sclerosis. J Cell Biochem 2008;105(2):338–43.

[26] Liu PT, Stenger S, Li H, et al. Toll-like receptor triggering of a vitamin D-mediated human antimicrobial response. Science 2006;311(5768):1770–3.

[27] Keisala T, Minasyan A, Lou YR, et al. Premature aging in vitamin D receptor mutant mice. J Steroid Biochem Mol Biol 2009;115(3–5):91–7.

[28] Shimada T, Hasegawa H, Yamazaki Y, et al. FGF-23 is a potent regulator of vitamin D metabolism and phosphate homeostasis. J Bone Miner Res 2004;19(3):429–35.

[29] Kurosu H, Kuro-o M. The Klotho gene family and the endocrine fibroblast growth factors. Curr Opin Nephrol Hypertens 2008;17(4):368–72.

[30] Wu X, Li Y. Role of FGF19 induced FGFR4 activation in the regulation of glucose homeostasis. Aging (Albany, NY) 2009;1(12):1023–7.

[31] Haussler MR, Jurutka PW, Mizwicki M, Norman AW. Vitamin D receptor (VDR)-mediated actions of $1\alpha,25(OH)$vitamin D: genomic and non-genomic mechanisms. Best Pract Res Clin Endocrinol Metab 2011;25(4):543–59.

[32] Pike JW, Meyer MB. Fundamentals of vitamin D hormone-regulated gene expression. J Steroid Biochem Mol Biol 2014;144(Pt A):5–11.

[33] Pike JW, Lee SM, Meyer MB. Regulation of gene expression by 1,25-dihydroxyvitamin D_3 in bone cells: exploiting new approaches and defining new mechanisms. Bonekey Rep 2014;3:482.

[34] Rochel N, Wurtz JM, Mitschler A, Klaholz B, Moras D. The crystal structure of the nuclear receptor for vitamin D bound to its natural ligand. Mol Cell 2000;5(1):173–9.

[35] Watkins RE, Wisely GB, Moore LB, et al. The human nuclear xenobiotic receptor PXR: structural determinants of directed promiscuity. Science 2001;292(5525):2329–33.

[36] Watkins RE, Maglich JM, Moore LB, et al. 2.1 A crystal structure of human PXR in complex with the St. John's wort compound hyperforin. Biochemistry (Mosc) 2003;42(6):1430–8.

[37] Moore LB, Maglich JM, McKee DD, et al. Pregnane X receptor (PXR), constitutive androstane receptor (CAR), and benzoate X receptor (BXR) define three pharmacologically distinct classes of nuclear receptors. Mol Endocrinol 2002;16(5):977–86.

[38] Bartik L, Whitfield GK, Kaczmarska M, et al. Curcumin: a novel nutritionally derived ligand of the vitamin D receptor with implications for colon cancer chemoprevention. J Nutr Biochem 2010;21:1153–61.

[39] Hoss E, Austin HR, Batie SF, Jurutka PW, Haussler MR, Whitfield GK. Control of late cornified envelope genes relevant to psoriasis risk: upregulation by 1,25-dihydroxyvitamin D and plant-derived delphinidin. Arch Dermatol Res 2013;305(10):867–78.

[40] Austin HR, Hoss E, Batie SF, et al. Regulation of late cornified envelope genes relevant to psoriasis risk by plant-derived cyanidin. Biochem Biophys Res Commun 2014;443(4):1275–9.

[41] Whitfield GK, Dang HTL, Schluter SF, et al. Cloning of a functional vitamin D receptor from the lamprey (Petromyzon marinus), an ancient vertebrate lacking a calcified skeleton and teeth. Endocrinology 2003;144(6):2704–16.

[42] Rochel N, Ciesielski F, Godet J, et al. Common architecture of nuclear receptor heterodimers on DNA direct repeat elements with different spacings. Nat Struct Mol Biol 2011;18(5):564–70.

[43] Zhang J, Chalmers MJ, Stayrook KR, et al. DNA binding alters coactivator interaction surfaces of the intact VDR-RXR complex. Nat Struct Mol Biol 2011;18(5):556–63.

[44] Jurutka PW, Hsieh J-C, Remus LS, et al. Mutations in the 1,25-dihydroxyvitamin D_3 receptor identifying C-terminal amino acids required for transcriptional activation that are functionally dissociated from hormone binding, heterodimeric DNA binding and interaction with basal transcription factor IIB, in vitro. J Biol Chem 1997;272:14592–9.

[45] Masuyama H, Brownfield CM, St-Arnaud R, MacDonald PN. Evidence for ligand-dependent intramolecular folding of the AF-2 domain in vitamin D receptor-activated transcription and coactivator interaction. Mol Endocrinol 1997;11(10):1507–17.

[46] Rachez C, Gamble M, Chang CP, Atkins GB, Lazar MA, Freedman LP. The DRIP complex and SRC-1/p160 coactivators share similar nuclear receptor binding determinants but constitute functionally distinct complexes. Mol Cell Biol 2000;20(8):2718–26.

[47] Whitfield GK, Jurutka PW, Haussler CA, et al. Nuclear vitamin D receptor: structure-function, molecular control of gene transcription, and novel bioactions. In: Feldman D, Pike JW, Glorieux FH, editors. Vitamin D. 2nd ed Oxford (UK): Elsevier Academic Press; 2005. p. 219–61.

[48] Ben-Dov IZ, Galitzer H, Lavi-Moshayoff V, et al. The parathyroid is a target organ for FGF23 in rats. J Clin Invest 2007;117(12):4003–8.

[49] Liu S, Tang W, Zhou J, et al. Fibroblast growth factor 23 is a counter-regulatory phosphaturic hormone for vitamin D. J Am Soc Nephrol 2006;17(5):1305–15.

[50] Razzaque MS. The FGF23-Klotho axis: endocrine regulation of phosphate homeostasis. Nat Rev Endocrinol 2009;5(11):611–19.

[51] Perwad F, Zhang MY, Tenenhouse HS, Portale AA. Fibroblast growth factor 23 impairs phosphorus and vitamin D metabolism in vivo and suppresses 25-hydroxyvitamin D-1α-hydroxylase expression in vitro. Am J Physiol Renal Physiol 2007;293(5):F1577–83.

[52] Masuda S, Byford V, Arabian A, et al. Altered pharmacokinetics of $1\alpha,25$-dihydroxyvitamin D_3 and 25-hydroxyvitamin D_3 in the blood and tissues of the 25-hydroxyvitamin D-24-hydroxylase (Cyp24a1) null mouse. Endocrinology 2005;146(2):825–34.

[53] Meyer MB, Goetsch PD, Pike JW. A downstream intergenic cluster of regulatory enhancers contributes to the induction of CYP24A1 expression by $1\alpha,25$-dihydroxyvitamin D_3. J Biol Chem 2010;285(20):15599–610.

[54] Pike JW, Meyer MB, Martowicz ML, et al. Emerging regulatory paradigms for control of gene expression by 1,25-dihydroxyvitamin D(3). J Steroid Biochem Mol Biol 2010;121:130–5.

[55] Urakawa I, Yamazaki Y, Shimada T, et al. Klotho converts canonical FGF receptor into a specific receptor for FGF23. Nature 2006;444(7120):770–4.

[56] Saini RK, Kaneko I, Jurutka PW, et al. 1,25-dihydroxyvitamin D (3) regulation of fibroblast growth factor-23 expression in bone cells: evidence for primary and secondary mechanisms modulated by leptin and interleukin-6. Calcif Tissue Int 2013;92(4):339–53.

[57] Meyer MB, Goetsch PD, Pike JW. Genome-wide analysis of the VDR/RXR cistrome in osteoblast cells provides new mechanistic insight into the actions of the vitamin D hormone. J Steroid Biochem Mol Biol 2010;121(1–2):136–41.

[58] Dittmer J. The biology of the Ets1 proto-oncogene. Mol Cancer 2003;2:29.

[59] St John HC, Bishop KA, Meyer MB, et al. The osteoblast to osteocyte transition: epigenetic changes and response to the vitamin D_3 hormone. Mol Endocrinol 2014;28(7):1150–65.

[60] Ozono K, Liao J, Kerner SA, Scott RA, Pike JW. The vitamin D-responsive element in the human osteocalcin gene: association with a nuclear proto-oncogene enhancer. J Biol Chem 1990;265:21881–8.

[61] Luderer HF, Gori F, Demay MB. Lymphoid enhancer-binding factor-1 (LEF1) interacts with the DNA-binding domain of the vitamin D receptor. J Biol Chem 2011;286(21):18444–51.

[62] Ito M, Sakai Y, Furumoto M, et al. Vitamin D and phosphate regulate fibroblast growth factor-23 in K-562 cells. Am J Physiol Endocrinol Metab 2005;288(6):E1101–9.

[63] David V, Dai B, Martin A, Huang J, Han X, Quarles LD. Calcium regulates FGF-23 expression in bone. Endocrinology 2013;154(12):4469–82.

[64] Cha SK, Hu MC, Kurosu H, Kuro-o M, Moe O, Huang CL. Regulation of renal outer medullary potassium channel and renal K(+) excretion by Klotho. Mol Pharmacol 2009;76(1):38–46.

[65] Jacobs E, Martinez ME, Buckmeier J, Lance P, May M, Jurutka P. Circulating fibroblast growth factor-23 is associated with increased risk for metachronous colorectal adenoma. J Carcinog 2011;10:3.

[66] Forster RE, Jurutka PW, Hsieh JC, et al. Vitamin D receptor controls expression of the anti-aging klotho gene in mouse and human renal cells. Biochem Biophys Res Commun 2011;414(3):557–62.

[67] Wang Y, Sun Z. Klotho gene delivery prevents the progression of spontaneous hypertension and renal damage. Hypertension 2009;54(4):810–17.

[68] Ichikawa S, Imel EA, Kreiter ML, et al. A homozygous missense mutation in human KLOTHO causes severe tumoral calcinosis. J Clin Invest 2007;117(9):2684–91.

[69] Kuro-o M. Klotho. Pflugers Arch 2010;459(2):333–43.

[70] Chen CD, Podvin S, Gillespie E, Leeman SE, Abraham CR. Insulin stimulates the cleavage and release of the extracellular domain of Klotho by ADAM10 and ADAM17. Proc Natl Acad Sci USA 2007;104(50):19796–801.

[71] Matsumura Y, Aizawa H, Shiraki-Iida T, Nagai R, Kuro-o M, Nabeshima Y. Identification of the human klotho gene and its two transcripts encoding membrane and secreted klotho protein. Biochem Biophys Res Commun 1998;242(3):626–30.

[72] Thurston RD, Larmonier CB, Majewski PM, et al. Tumor necrosis factor and interferon-gamma down-regulate klotho in mice with colitis. Gastroenterology 2010;138(4):1384–94 1394 e1–2.

[73] Tsujikawa H, Kurotaki Y, Fujimori T, Fukuda K, Nabeshima Y. Klotho, a gene related to a syndrome resembling human premature aging, functions in a negative regulatory circuit of vitamin D endocrine system. Mol Endocrinol 2003;17(12):2393–403.

[74] Makishima M, Lu TT, Xie W, et al. Vitamin D receptor as an intestinal bile acid sensor. Science 2002;296(5571):1313–16.

[75] Thummel KE, Brimer C, Yasuda K, et al. Transcriptional control of intestinal cytochrome P-4503A by 1α,25-dihydroxy vitamin D_3. Mol Pharmacol 2001;60(6):1399–406.

[76] Thompson PD, Jurutka PW, Whitfield GK, et al. Liganded VDR induces CYP3A4 in small intestinal and colon cancer cells via DR3 and ER6 vitamin D responsive elements. Biochem Biophys Res Commun 2002;299(5):730–8.

[77] Araya Z, Wikvall K. 6α-hydroxylation of taurochenodeoxycholic acid and lithocholic acid by CYP3A4 in human liver microsomes. Biochim Biophys Acta 1999;1438(1):47–54.

[78] Kozoni V, Tsioulias G, Shiff S, Rigas B. The effect of lithocholic acid on proliferation and apoptosis during the early stages of colon carcinogenesis: differential effect on apoptosis in the presence of a colon carcinogen. Carcinogenesis 2000;21(5):999–1005.

[79] Echchgadda I, Song CS, Roy AK, Chatterjee B. Dehydroepiandrosterone sulfotransferase is a target for transcriptional induction by the vitamin D receptor. Mol Pharmacol 2004;65(3):720–9.

[80] Guilliams TG. Homocysteine – a risk factor for vascular diseases: guidelines for the clinical practice. JANA 2004;7:11–24.

[81] Seshadri S, Beiser A, Selhub J, et al. Plasma homocysteine as a risk factor for dementia and Alzheimer's disease. N Engl J Med 2002;346(7):476–83.

[82] Kriebitzsch C, Verlinden L, Eelen G, et al. 1,25-dihydroxyvitamin D3 influences cellular homocysteine levels in murine preosteoblastic MC3T3-E1 cells by direct regulation of cystathionine β-synthase. J Bone Miner Res 2011;26(12):2991–3000.

[83] Terpening CM, Haussler CA, Jurutka PW, Galligan MA, Komm BS, Haussler MR. The vitamin D-responsive element in the rat bone Gla protein gene is an imperfect direct repeat that cooperates with other cis-elements in 1,25-dihydroxyvitamin D_3-mediated transcriptional activation. Mol Endocrinol 1991;5(3):373–85.

[84] Lian JB, Shalhoub V, Aslam F, et al. Species-specific glucocorticoid and 1,25-dihydroxyvitamin D responsiveness in mouse MC3T3-E1 osteoblasts: dexamethasone inhibits osteoblast differentiation and vitamin D down-regulates osteocalcin gene expression. Endocrinology 1997;138(5):2117–27.

[85] Kerner SA, Scott RA, Pike JW. Sequence elements in the human osteocalcin gene confer basal activation and inducible response to hormonal vitamin D_3. Proc Natl Acad Sci USA 1989;86:4455–9.

[86] Noda M, Vogel RL, Craig AM, Prahl J, DeLuca HF, Denhardt DT. Identification of a DNA sequence responsible for binding of the 1,25-dihydroxyvitamin D_3 receptor and 1,25-dihydroxyvitamin D_3 enhancement of mouse secreted phosphoprotein 1 (Spp-1 or osteopontin) gene expression. Proc Natl Acad Sci USA 1990;87:9995–9.

[87] Pike JW, Meyer MB, Watanuki M, et al. Perspectives on mechanisms of gene regulation by 1,25-dihydroxyvitamin D_3 and its receptor. J Steroid Biochem Mol Biol 2007;103(3–5):389–95.

[88] Barthel TK, Mathern DR, Whitfield GK, et al. 1,25-dihydroxyvitamin D_3/VDR-mediated induction of FGF23 as well as transcriptional control of other bone anabolic and catabolic genes that orchestrate the regulation of phosphate and calcium mineral metabolism. J Steroid Biochem Mol Biol 2007;103(3–5):381–8.

[89] Fretz JA, Zella LA, Kim S, Shevde NK, Pike JW. 1,25-dihydroxyvitamin D_3 regulates the expression of low-density lipoprotein receptor-related protein 5 via deoxyribonucleic acid sequence elements located downstream of the start site of transcription. Mol Endocrinol 2006;20(9):2215–30.

[90] Drissi H, Pouliot A, Koolloos C, et al. 1,25-$(OH)_2$-vitamin D_3 suppresses the bone-related Runx2/Cbfa1 gene promoter. Exp Cell Res 2002;274(2):323–33.

[91] Haussler MR, Haussler CA, Bartik L, et al. Vitamin D receptor: molecular signaling and actions of nutritional ligands in disease prevention. Nutr Rev 2008;66(10 Suppl. 2):S98–112.

[92] Kim S, Yamazaki M, Zella LA, Shevde NK, Pike JW. Activation of receptor activator of NF-κB ligand gene expression by 1,25-dihydroxyvitamin D_3 is mediated through multiple long-range enhancers. Mol Cell Biol 2006;26(17):6469–86.

[93] Cao X, Ross FP, Zhang L, MacDonald PN, Chappel J, Teitelbaum SL. Cloning of the promoter for the avian integrin b_3 subunit gene and its regulation by 1,25-dihydroxyvitamin D_3. J Biol Chem 1993;268(36):27371–80.

[94] Quelo I, Machuca I, Jurdic P. Identification of a vitamin D response element in the proximal promoter of the chicken carbonic anhydrase II gene. J Biol Chem 1998;273(17): 10638−46.

[95] Liu SM, Koszewski N, Lupez M, Malluche HH, Olivera A, Russell J. Characterization of a response element in the 5′-flanking region of the avian (chicken) PTH gene that mediates negative regulation of gene transcription by 1,25-dihydroxyvitamin D_3 and binds the vitamin D_3 receptor. Mol Endocrinol 1996;10:206−15.

[96] Zella LA, Kim S, Shevde NK, Pike JW. Enhancers located in the vitamin D receptor gene mediate transcriptional autoregulation by 1,25-dihydroxyvitamin D_3. J Steroid Biochem Mol Biol 2007;103(3−5):435−9.

[97] Meyer MB, Watanuki M, Kim S, Shevde NK, Pike JW. The human transient receptor potential vanilloid type 6 distal promoter contains multiple vitamin D receptor binding sites that mediate activation by 1,25-dihydroxyvitamin D_3 in intestinal cells. Mol Endocrinol 2006;20(6):1447−61.

[98] Taketani Y, Segawa H, Chikamori M, et al. Regulation of type II renal Na^+-dependent inorganic phosphate transporters by 1,25-dihydroxyvitamin D_3. Identification of a vitamin D-responsive element in the human NAPi-3 gene. J Biol Chem 1998;273(23):14575−81.

[99] Zou A, Elgort MG, Allegretto EA. Retinoid X receptor (RXR) ligands activate the human 25-hydroxyvitamin D_3-24-hydroxylase promoter via RXR heterodimer binding to two vitamin D- responsive elements and elicit additive effects with 1,25- dihydroxyvitamin D_3. J Biol Chem 1997;272(30):19027−34.

[100] Barwick JL, Quattrochi LC, Mills AS, Potenza C, Tukey RH, Guzelian PS. Trans-species gene transfer for analysis of glucocorticoid-inducible transcriptional activation of transiently expressed human CYP3A4 and rabbit CYP3A6 in primary cultures of adult rat and rabbit hepatocytes. Mol Pharmacol 1996;50(1):10−16.

[101] Saeki M, Kurose K, Tohkin M, Hasegawa R. Identification of the functional vitamin D response elements in the human MDR1 gene. Biochem Pharmacol 2008;76(4):531−42.

[102] Liu M, Lee MH, Cohen M, Bommakanti M, Freedman LP. Transcriptional activation of the Cdk inhibitor p21 by vitamin D_3 leads to the induced differentiation of the myelomonocytic cell line U937. Genes Dev 1996;10(2):142−53.

[103] Wang TT, Tavera-Mendoza LE, Laperriere D, et al. Large-scale in silico and microarray-based identification of direct 1,25-dihydroxyvitamin D_3 target genes. Mol Endocrinol 2005;19 (11):2685−95.

[104] Peng L, Malloy PJ, Feldman D. Identification of a functional vitamin D response element in the human insulin-like growth factor binding protein-3 promoter. Mol Endocrinol 2004;18 (5):1109−19.

[105] Bikle DD, Ng D, Oda Y, Hanley K, Feingold K, Xie Z. The vitamin D response element of the involucrin gene mediates its regulation by 1,25-dihydroxyvitamin D_3. J Invest Dermatol 2002;119(5):1109−13.

[106] Kikuchi R, Sobue S, Murakami M, et al. Mechanism of vitamin D_3-induced transcription of phospholipase D1 in HaCat human keratinocytes. FEBS Lett 2007;581(9):1800−4.

[107] Shirakawa AK, Nagakubo D, Hieshima K, Nakayama T, Jin Z, Yoshie O. 1,25-dihydroxyvitamin D_3 induces CCR10 expression in terminally differentiating human B cells. J Immunol 2008;180(5):2786−95.

[108] Falzon M. DNA sequences in the rat parathyroid hormone-related peptide gene responsible for 1,25-dihydroxyvitamin D_3-mediated transcriptional repression. Mol Endocrinol 1996;10:672−81.

[109] Hsieh JC, Estess RC, Kaneko I, Whitfield GK, Jurutka PW, Haussler MR. Vitamin D receptor-mediated control of Soggy, Wise, and Hairless gene expression in keratinocytes. J Endocrinol 2014;220(2):165−78.

[110] Cardus A, Panizo S, Encinas M, et al. 1,25-dihydroxyvitamin D_3 regulates VEGF production through a vitamin D response element in the VEGF promoter. Atherosclerosis 2009;204(1):85−9.

[111] Malloy PJ, Peng L, Wang J, Feldman D. Interaction of the vitamin D receptor with a vitamin D response element in the Mullerian-inhibiting substance (MIS) promoter: regulation of MIS expression by calcitriol in prostate cancer cells. Endocrinology 2009;150(4):1580−7.

[112] Ramagopalan SV, Maugeri NJ, Handunnetthi L, et al. Expression of the multiple sclerosis-associated MHC class II Allele HLA-DRB1*1501 is regulated by vitamin D. PLoS Genet 2009;5(2):e1000369.

[113] Gombart AF, Borregaard N, Koeffler HP. Human cathelicidin antimicrobial peptide (CAMP) gene is a direct target of the vitamin D receptor and is strongly up-regulated in myeloid cells by 1,25-dihydroxyvitamin D_3. FASEB J 2005;19(9):1067−77.

[114] Wang X, Wang TT, White JH, Studzinski GP. Induction of kinase suppressor of RAS-1(KSR-1) gene by 1α,25-dihydroxyvitamin D_3 in human leukemia HL60 cells through a vitamin D response element in the 5′-flanking region. Oncogene 2006;25(53):7078−85.

[115] Wang X, Wang TT, White JH, Studzinski GP. Expression of human kinase suppressor of Ras 2 (hKSR-2) gene in HL60 leukemia cells is directly upregulated by 1,25-dihydroxyvitamin D(3) and is required for optimal cell differentiation. Exp Cell Res 2007;313(14):3034−45.

[116] Lee S, Lee DK, Choi E, Lee JW. Identification of a functional vitamin D response element in the murine Insig-2 promoter and its potential role in the differentiation of 3T3-L1 preadipocytes. Mol Endocrinol 2005;19(2):399−408.

[117] Eloranta JJ, Zair ZM, Hiller C, Hausler S, Stieger B, Kullak-Ublick GA. Vitamin D_3 and its nuclear receptor increase the expression and activity of the human proton-coupled folate transporter. Mol Pharmacol 2009;76(5):1062−71.

[118] Weissen-Plenz G, Nitschke Y, Rutsch F. Mechanisms of arterial calcification: spotlight on the inhibitors. Adv Clin Chem 2008;46:263−93.

[119] Milat F, Ng KW. Is Wnt signalling the final common pathway leading to bone formation? Mol Cell Endocrinol 2009;310 (1−2):52−62.

[120] Poundarik AA, Diab T, Sroga GE, Ural A, Boskey AL, Gundberg CM, et al. Dilatational band formation in bone. Proc Natl Acad Sci USA 2012;109(47):19178−83.

[121] Ferron M, Hinoi E, Karsenty G, Ducy P. Osteocalcin differentially regulates beta cell and adipocyte gene expression and affects the development of metabolic diseases in wild-type mice. Proc Natl Acad Sci USA 2008;105(13):5266−70.

[122] Zinser GM, Sundberg JP, Welsh J. Vitamin D_3 receptor ablation sensitizes skin to chemically induced tumorigenesis. Carcinogenesis 2002;23(12):2103−9.

[123] Ellison TI, Smith MK, Gilliam AC, MacDonald PN. Inactivation of the vitamin D receptor enhances susceptibility of murine skin to UV-induced tumorigenesis. J Invest Dermatol 2008;128(10):2508−17.

[124] Audo I, Darjatmoko SR, Schlamp CL, et al. Vitamin D analogues increase p53, p21, and apoptosis in a xenograft model of human retinoblastoma. Invest Ophthalmol Vis Sci 2003;44(10):4192−9.

[125] Sidelnikov E, Bostick RM, Flanders WD, et al. Effects of calcium and vitamin D on MLH1 and MSH2 expression in rectal mucosa of sporadic colorectal adenoma patients. Cancer Epidemiol Biomarkers Prev 2010;19(4):1022−32.

[126] Kallay E, Pietschmann P, Toyokuni S, et al. Characterization of a vitamin D receptor knockout mouse as a model of colorectal hyperproliferation and DNA damage. Carcinogenesis 2001;22(9):1429–35.

[127] Zinser G, Packman K, Welsh J. Vitamin D_3 receptor ablation alters mammary gland morphogenesis. Development 2002;129(13):3067–76.

[128] Egan JB, Thompson PA, Vitanov MV, et al. Vitamin D receptor ligands, adenomatous polyposis coli, and the vitamin D receptor FokI polymorphism collectively modulate beta-catenin activity in colon cancer cells. Mol Carcinog 2010;49(4):337–52.

[129] Cohen-Lahav M, Shany S, Tobvin D, Chaimovitz C, Douvdevani A. Vitamin D decreases NFκB activity by increasing IκBα levels. Nephrol Dial Transplant 2006;21(4):889–97.

[130] Moreno J, Krishnan AV, Swami S, Nonn L, Peehl DM, Feldman D. Regulation of prostaglandin metabolism by calcitriol attenuates growth stimulation in prostate cancer cells. Cancer Res 2005;65(17):7917–25.

[131] Eelen G, Verlinden L, Meyer MB, et al. 1,25-Dihydroxyvitamin D3 and the aging-related forkhead box O and sestrin proteins in osteoblasts. J Steroid Biochem Mol Biol. 2013;136:112–19.

[132] Lin MT, Beal MF. The oxidative damage theory of aging. Clin Neurosci Res 2003;2(5):305–15.

[133] Patrick RP, Ames BN. Vitamin D and the omega-3 fatty acids control serotonin synthesis and action, part 2: relevance for ADHD, bipolar, schizophrenia, and impulsive behavior. FASEB J 2015;29(6):2207–22.

[134] Hayes DP. Resveratrol and vitamin D: significant potential interpretative problems arising from their mutual processes, interactions and effects. Med Hypotheses 2011;77(5):765–72.

[135] Dampf-Stone A, Batie S, Sabir M, et al. Resveratrol potentiates vitamin D and nuclear receptor signaling. J Cell Biochem 2015;116(6):1130–43.

[136] Baur JA. Resveratrol, sirtuins, and the promise of a DR mimetic. Mech Ageing Dev 2010;131(4):261–9.

[137] Donmez G, Wang D, Cohen DE, Guarente L. SIRT1 suppresses beta-amyloid production by activating the α-secretase gene ADAM10. Cell 2010;142(2):320–32.

[138] An BS, Tavera-Mendoza LE, Dimitrov V, et al. Stimulation of Sirt1-regulated FoxO protein function by the ligand-bound vitamin D receptor. Mol Cell Biol 2010;30(20):4890–900.

[139] Xu Y, Sun Z. Molecular basis of klotho: from gene to function in aging. Endocr Rev 2015;36(2):174–93.

[140] Kim S, Yamazaki M, Shevde NK, Pike JW. Transcriptional control of receptor activator of nuclear factor-κB ligand by the protein kinase A activator forskolin and the transmembrane glycoprotein 130-activating cytokine, oncostatin M, is exerted through multiple distal enhancers. Mol Endocrinol 2007;21(1):197–214.

[141] Meyer MB, Goetsch PD, Pike JW. VDR/RXR and TCF4/beta-catenin cistromes in colonic cells of colorectal tumor origin: impact on c-FOS and c-MYC gene expression. Mol Endocrinol 2012;26(1):37–51.

[142] Stubbs JR, Liu S, Tang W, et al. Role of hyperphosphatemia and 1,25-dihydroxyvitamin D in vascular calcification and mortality in fibroblastic growth factor 23 null mice. J Am Soc Nephrol 2007;18(7):2116–24.

[143] Haussler MR, Whitfield GK, Haussler CA, et al. The nuclear vitamin D receptor: biological and molecular regulatory properties revealed. J Bone Miner Res 1998;13(3):325–49.

[144] Sakai Y, Kishimoto J, Demay MB. Metabolic and cellular analysis of alopecia in vitamin D receptor knockout mice. J Clin Invest 2001;107(8):961–6.

[145] Clemens TL, Zhou XY, Pike JW, Haussler MR, Sloviter RS. 1,25-dihydroxyvitamin D receptor and vitamin D-dependent calcium binding protein in rat brain: Comparative immunocytochemical localization. In: Norman AW, Schaefer K, Grigoleit H-G, Herrath DV, editors. Vitamin D: chemical, biochemical and clinical update. Berlin: Walter de Gruyter; 1985. p. 95–6.

[146] Kalueff AV, Keisala T, Minasyan A, Kuuslahti M, Miettinen S, Tuohimaa P. Behavioural anomalies in mice evoked by "Tokyo" disruption of the vitamin D receptor gene. Neurosci Res 2006;54(4):254–60.

CHAPTER

34

Carotenoid Supplements and Consumption: Implications for Healthy Aging

Karin Linnewiel-Hermoni[1], Esther Paran[2] and Talya Wolak[2,3]

[1]Clinical Biochemistry and Pharmacology, Ben-Gurion University of the Negev and Soroka Medical Center, Beer Sheva, Israel [2]Hypertension and Vascular Research Laboratory, Ben-Gurion University of the Negev and Soroka Medical Center, Beer Sheva, Israel [3]Hypertension Unit, Faculty of Health Sciences, Ben-Gurion University of the Negev and Soroka Medical Center, Beer Sheva, Israel

KEY FACTS
- Carotenoids have been credited with the prevention of many age-related conditions.
- Carotenoids function as antioxidants and anti-inflammatory agents.
- Carotenoids are highly lipophilic molecules located inside cell membranes, protecting the membrane from oxidative stress damage.
- Of several hundred carotenoids identified in nature, only 40 are present in a typical human diet, and about half have been identified in human blood and tissues.
- Regular carotenoid consumption improves vascular health, protects the skin from aging and cancer, and has an anticarcinogenic effect in prostate and breast cancer.

Dictionary of Terms

- *Carotenoids*: a group of red, orange, or yellow polyisoprenoid hydrocarbons with symmetrical tetraterpene skeleton formed by the tail-to-tail linkage of the two C_{20} moieties synthesized by photosynthesis.
- *Reactive oxygen species*: chemically reactive molecules containing oxygen that have an unpaired electron, rendering them extremely reactive.

- *Electrophile/antioxidant response element (EpRE/ARE)*: a transcription system that induces the expression of phase II detoxifying enzymes (antioxidant enzymes) which eliminates harmful substances.
- *Nuclear factor kappa B (NF-κB)*: nuclear factor kappa-light-chain-enhancer of activated B cells is a protein complex that enhances inflammatory response by induction of DNA transcription. It is involved in cellular responses to stimuli such as stress, cytokines, free radicals, ultraviolet irradiation, oxidized LDL, and infection.
- *Vascular cell adhesion molecule 1*: VCAM-1 mediates the adhesion of lymphocytes, monocytes, eosinophils, and basophils to the vascular endothelium.
- *Intercellular adhesion molecule 1*: ICAM-1 leukocytes bind to endothelial cells via ICAM-1.
- *Matrix metalloproteinase*: MMPs are zinc-dependent endopeptidases; other family members are adamalysins, serralysins, and astacins. The MMPs belong to a larger family of proteases.

INTRODUCTION

As life expectancy increases, so does the proportion of older people in the population, and, accordingly, the incidence of chronic disease. In light of this situation, new approaches to promoting or maintaining health in this age group are of great interest. Today, it is widely accepted that diet can provide desirable health benefits beyond basic nutrition. The dietary approach to promoting healthy aging and preventing age-related diseases is currently a popular research topic, largely because its absence of toxicity makes it suitable for the general aging population.

Carotenoids function as antioxidant antiinflammatory agents. Carotenoids and their derivatives also modulate various cellular signaling pathways, as will be discussed in this chapter. They have been associated with the prevention of many age-related conditions and chronic diseases, including cancer, cardiovascular disease (CVD), age-related macular degeneration (AMD) and loss of visual function, skin aging, and more. New evidence also suggests a beneficial effect of carotenoids on cognitive function and muscle strength. This chapter will focus on the role of carotenoids in CVD, cancer, and skin aging.

STRUCTURE AND FUNCTION

While carotenoids are necessary to maintain the normal health and behavior of animals, including humans, nearly all animals are unable to synthesize carotenoids and therefore rely on their diet to obtain these compounds [1]. Several hundred carotenoids have been identified in nature, but only 40 are present in a typical human diet, and approximately 20 have been identified in human blood and tissues [2].

The carotenoids are a group of red, orange, or yellow pigmented polyisoprenoid hydrocarbons synthesized in all organisms capable of photosynthesis, such as prokaryotes and higher plants as well as most bacteria [3,4]. The main sources of carotenoids in human plasma are yellow-orange-red fruits and green leafy vegetables [5]. Lycopene is the main pigment in tomatoes, while lutein is found in a number of vegetables (eg, cabbage, corn, and broccoli). The carotenoids can be divided into those with provitamin A activity and those that have no role in the formation of vitamin A [6]. The provitamin A carotenoids are metabolized by humans into retinol. These carotenoids include alpha-, beta-, and gamma-carotene and alpha- and beta-cryptoxanthin. Among the nonprovitamin A carotenoids, the most abundant are lutein, zeaxanthin, and lycopene [7]. Carotenoids are transported in the plasma of humans and animals exclusively by lipoproteins. Chylomicrons are responsible for the transport of carotenoids from the intestinal mucosa to the bloodstream via the lymphatics [8]. After ingestion, the carotenoids incorporate into mixed lipid micelles in the lumen and are taken up by the intestinal mucosa and incorporated into chylomicrons. These chylomicrons are released into the lymph, where they are digested by lipoprotein lipase, resulting in the release of carotenoids. Next, they are further distributed in the plasma by the use of very low-density lipoproteins (VLDL), low-density lipoproteins (LDL), and high-density lipoproteins (HDL) [8]. The most abundant carotenoids in plasma include lycopene, beta-carotene, and lutein [9]. Their plasma half-life is relatively long (days to weeks) due to their fat solubility, limited phase II metabolism, and impaired renal clearance [10]. Their plasma concentration is $2\,\mu M$ [9]. In the plasma, the carotenoids are categorized into carotenes (beta-carotene, lycopene) and xanthophylls (lutein). (Xanthophylls contain oxygen atoms, while carotenes are purely hydrocarbons with no oxygen.) Carotenes are fat-soluble and thus tend to be localized in LDL in circulation. Xanthophylls, which contain at least one hydroxyl group, are more polar than carotenes and are evenly distributed between LDL and HDL [11]. After leaving the blood stream, carotenoids are mainly accumulated in the liver and adipose tissues; a relatively high amount was also reported in the adrenal gland, corpus luteum, testes, skin, retina (macula), kidney, and ovary, while in brain stem tissue their concentration was below the detection limit [12,13].

Most carotenoids exhibit a characteristic, symmetrical tetraterpene skeleton formed by the tail-to-tail linkage of the two C_{20} moieties [14]. Carotenoids, as highly lipophilic molecules, are typically located inside cell membranes. Strict hydrocarbons, such as beta-carotene or lycopene, are arranged exclusively within the inner part of the lipid bilayer. More polar pigment molecules, containing attached oxygen atoms (eg, lutein, zeaxanthin) are oriented roughly perpendicular to the membrane surface, exhibiting their hydrophilic parts to the aqueous environment [15]. This membrane incorporation may result in a prominent enhancement of the membranes' resistance to oxidative stress.

Due to their antioxidative effect, carotenoids may attenuate age-related pathological processes, including atherosclerosis, macular degeneration, skin aging, and cancer. However, their plasma level decreases with aging, probably due to reduced intestinal absorption [16].

CAROTENOIDS AND OXIDATIVE STRESS

Reactive oxygen species (ROS) and reactive nitrogen species (RNS) are generated in a variety of pathological as well as physiological process. They are products of UV light, X-ray and gamma-ray irradiation, are generated in metal-catalyzed reactions, and are air pollutants. They also produced in the cell by neutrophils and macrophages during inflammation and as byproducts of mitochondria-catalyzed electron transport reactions [17]. ROS/RNS have a dual effect in the biological system; they can be both harmful and beneficial to living organisms [18]. Some of their beneficial effects, such as cellular defense against infectious agents, are present during acute inflammation. Another beneficial effect is that at low concentrations they induce a mitogenic response. However, at high concentrations, ROS are responsible for damage of the cell structures and its organelles, namely: lipids, membranes, proteins, and nucleic acids. [19]. This cellular damage is reduced by the antioxidant action of nonenzymatic antioxidants and antioxidant enzymes [20]. Despite the antioxidant defense system, oxidative damage accumulates during the lifetime, and ROS/RNS cause irreversible damage to DNA, proteins, and lipids. This type of damage is particularly harmful when it is continuous, as it is, for example in chronic inflammatory states. Thus, ROS/RNS play a major role in the development of age-dependent diseases, among them cancer, atherosclerosis, skin aging, and macular degeneration [21].

It is well known that carotenoids exhibit free radical scavenging properties [22]. They are known to be very efficient physical quenchers of singlet oxygen ($^1O_2$), as well as potent scavengers of other ROS [14,23]. The interaction of carotenoids with $^1O_2$ depends largely on physical quenching, which involves direct energy transfer between both molecules. The energy of singlet molecular oxygen is transferred to the carotenoid molecule to yield ground state oxygen and a triplet excited carotene. Instead of further chemical reactions, the carotenoid returns to ground state, dissipating its energy by interaction with the surrounding solvent, and it can be reused several fold in such quenching cycles [24]. The most efficient carotenoid is the open ring carotenoid lycopene, which contributes up to 30% to total carotenoids in humans [25]. Among the various radicals formed under oxidative conditions in the organism, carotenoids most efficiently react with peroxyl radicals. They are generated in the process of lipid peroxidation, and scavenging of this species interrupts the reaction sequence, finally leading to damage of lipophilic compartments. Due to their lipophilicity and specific property of scavenging peroxyl radicals, carotenoids are thought to play an important role in the protection of cellular membranes and lipoproteins against oxidative damage [26].

In addition to their antioxidant properties, carotenoids possess anti-inflammatory characteristics [27]. Chronic inflammation, in contrast to acute inflammation, is an ongoing damaging process that causes constant cellular damage and results in the development of illness such as cancer and CVD [28,29]. As antioxidant, anti-inflammatory substances, carotenoids play a role in the attenuation of these morbidities. The combined antioxidative and anti-inflammatory effect of carotenoids was demonstrated by our group in vitro, showing that lycopene and lutein significantly improved endothelial function as measured by increased nitric oxide (NO) and decreased endothelin (ET-1) release as well as attenuating the expression of leukocytes adhesion molecules (intercellular adhesion molecule-1 [ICAM-1] and vascular cell adhesion molecule-1 [VCAM-1]) and inhibiting nuclear factor kappa B (NF-κB) activation in transfected endothelial cells [30]. The anti-inflammatory effect of carotenoids was also demonstrated in clinical trials. Riso et al. evaluated the effects of 10-day lutein-rich broccoli (250 g/day) intake on plasma markers of inflammation in young healthy male smokers. After 10 days of consumption there was a significant reduction in the inflammatory marker CRP [31]. Another study conducted on a healthy young population showed comparable results: high vegetable and fruit consumption resulted in an inverse association between beta-carotene and inflammatory markers such as interleukin-6 and tumor necrosis factor-alpha (TNF-α) [32]. These studies suggest that in young subjects in a disease-free state, carotenoid consumption has a beneficial effect. In the following sections, we will try to evaluate whether this effect might influence disease and health in old age.

CAROTENOIDS AND VASCULAR HEALTH AND ATHEROSCLEROSIS

Atherosclerosis can be regarded as vascular aging. This process is characterized by excessive formation of ROS in the vascular bed [33]. The accumulation of ROS in aging atherosclerotic arteries results in endothelial injury. The damaged and aged endothelium has decreased NO bioavailability [34]. NO is the main antioxidant mediator. Endothelium-derived NO halts atherosclerotic vascular aging by maintaining normal organ blood flow via flow/shear stress-mediated vasodilatation. This preservation of flow-mediated vasodilatation is one of the hallmarks of the healthy endothelium [35,36]. Another antiatherosclerotic mechanism induced by NO is decreased vascular inflammation and thrombosis [34].

The antioxidant vascular effect of carotenoids has been demonstrated in both experimental and clinical studies. An interventional study conducted by our group demonstrated that the reduction in blood pressure induced by tomato extract (lycopene and beta-carotene) was accompanied by the elevation of plasma nitrate levels [37]. The antioxidant ability was further demonstrated by the carotenoid lycopene in human umbilical vein endothelial cells (HUVECs). Pretreatment of HUVECs with lycopene decreased the formation of TNF-α–induced ROS. The lycopene pretreated cells also had reduced expression of monocyte adhesion molecules [1]. Another study examined the protective effect of lycopene on injured endothelial cells induced by ROS and showed that lycopene-pretreated endothelial cells had increased viability, decreased apoptosis rate, and downregulation of the expressions of p53 and caspase-3 mRNA [38]. Further confirmation of lycopene's ability to reduce the effect of the inflammatory atherosclerotic process by serving as an ROS and inflammation inhibitor comes from a study performed on healthy male volunteers that demonstrated that lycopene supplementation improved their endothelial function, significantly decreased the inflammatory markers hsCRP, sICAM-1, and sVCAM-1, and improved atherosclerotic risk factors (lipid profile and systolic blood pressure level) [39].

The ability of carotenoids to diminish ROS levels was also found in vascular smooth muscle cells (VSMCs). Lo et al. showed that lutein reduced PDGF-induced intracellular ROS production and attenuated ROS-(H_2O_2-)-induced ERK1/2 and p38 MAPK activation [40].

In contrast to the above studies, other research has shown more modest effects of carotenoids. A study in which New Zealand White (NZW) rabbits were fed a high-fat diet with or without lycopene demonstrated that the lycopene group showed a significant reduction in their total cholesterol and LDL cholesterol serum levels, but failed to demonstrate a difference in either plaque size or endothelial function [41]. Comparable results were also found in a human study. Biomarkers of vascular oxidative stress and inflammation were unaffected after short-term (1 week) lycopene supplementation [42]. However, in a longer study (30 days) that examined the effect of the lycopene-rich Mediterranean diet, it was found that avoiding this lycopene-rich diet resulted in a significant increase of plasma ICAM-1 [43].

One of the initial events in atherosclerosis is the formation of the highly atherogenic oxidation of LDL [44,45]. The oxidized LDL particles present in the subintimal space chemo-attract inflammatory cells [46]. As mentioned above, the fat-soluble lycopene in the circulation is carried by LDL. An in vitro study using extracted human LDL particles demonstrated that the lycopene in LDL was resistant to myeloperoxidase activity. This resistance might be one of the mechanisms that protect LDL from oxidation and thus reduce atherosclerosis [47].

This ability of carotenoids to decrease the formation of oxidized LDL received further confirmation from a clinical study performed on young, healthy, lean men and women (mean age = 27 ± 8 years; mean body mass index = 22 ± 2). Study participants consumed high-fat meals on two occasions. On one occasion, the high-fat meal also included a processed tomato product (containing a high concentration of lycopene), and on the second occasion tomato products were not included in their meal. Although both meals increased postprandial lipid concentrations, tomato consumption significantly attenuated the levels of high-fat meal–induced LDL oxidation and interleukin-6 [48].

It might be said that carotenoids have proved to have antioxidative and anti-inflammatory abilities. Increased NO bioavailability improves vascular health and delays vascular aging. Still, it is unknown whether lycopene and other carotenoids have a protective rather than a therapeutic role, and whether their beneficial effect is restricted to healthy subjects in whom they serve as primary prevention substances, or whether can also change the history of the atherosclerotic process.

Can Carotenoids Improve Atherosclerotic Risk Factors and Early Atherosclerosis?

Diabetes mellitus, arterial hypertension, hyperlipidemia, metabolic syndrome, and abdominal obesity are known risk factors for atherosclerosis and its target organ damage [43].

In the previous section, the favorable effect of lycopene on plasma oxidized LDL was mentioned. Another study showed that lycopene can also reduce plasma LDL via reduction in the expression of the 3-hydroxy-3-methylglutaryl coenzyme A (HMG-CoA) reductase, an effect that resembles the mechanism of statins [49]. This interplay between statins and carotenoids was further demonstrated in subjects with mild and moderate hypercholesterolemia. Statin treatment resulted in plasma LDL reduction, together with carotenoid elevation [50]. Regarding obesity, a small but rather definitive study showed that obese subjects have significantly lower levels of plasma carotenoids than nonobese subjects [51].

Another well-known atherosclerotic risk factor is hypertension. In two studies, our group demonstrated a beneficial effect of carotenoids on blood pressure. We showed that 8 weeks treatment with tomato extract in Grade 1 hypertensive patients resulted in a significant decrease in systolic blood pressure 144 (SE ± 1.1) to 134 mmHg (SE ± 2, $P < 0.001$), and a decrease in diastolic blood pressure from 87.4 mmHg (SE ± 1.2) to 83.4 mmHg (SE ± 1.2, $P < 0.05$) [52]. Furthermore, the blood pressure lowering effect of tomato extract was demonstrated in hypertensive patients with Grade 2 hypertension. Adding tomato extract to regular antihypertensive treatment resulted in significant blood pressure reduction [37]. Further evidence that supports carotenoids' antihypertensive properties comes from a large follow-up study in which 4412 young adults (age 18–30 years) were recruited from four US cities (as part of the CARDIA study). At recruitment, four serum carotenoids (alpha-carotene, beta-carotene, lutein/zeaxanthin, and cryptoxanthin) and lycopene were measured. After 20 years of follow-up there was a significant inverse association between serum carotenoid at time 0- and 20-year hypertension incidence (relative hazard per SD increase of sum of four carotenoids: 0.91; 95% confidence interval = 0.84–0.99). It is worth mentioning that lycopene was unrelated to hypertension in any model [53]. However, a randomized case control study failed to demonstrate any antihypertensive effect of 6 months of tomato extract treatment on blood pressure reduction among middle-aged (51.2 ± 12.1 years) subjects with prehypertension (systolic BP 120–139 mmHg or diastolic BP 80–89 mmHg) [5].

The same protective effect of carotenoid consumption for long periods was found regarding the development of glucose intolerance. In a 9-year longitudinal study that included 1389 volunteers aged 59–71 years, the relationship between plasma carotenoid at baseline and incidence of dysglycemia was examined. It was demonstrated that the risk of dysglycemia remained significantly lower in participants who were in the highest quartile of total plasma carotenoid at recruitment compared with participants in the lowest quartile [40]. Furthermore, in the CARDIA study, both insulin resistance and diabetes incidence (after 15 years) were inversely associated with the baseline level of serum carotenoid concentrations in nonsmokers, but not in current smokers [40]. A recent study that examined the association between serum carotenoids and metabolic syndrome found that higher serum carotenoid levels were associated with a lower prevalence of metabolic syndrome and fewer abnormal metabolic syndrome components in middle-aged and elderly Chinese adults [54].

Subclinical atherosclerosis may be regarded as an early sign of atherosclerosis and vascular aging. One of the surrogate markers for this early phase of atherosclerosis is microalbuminuria. A cross-sectional study showed that increased levels of serum beta-carotene were independently associated with lower risk of albuminuria among Japanese women [55]. Other measurements of early atherosclerosis include arterial stiffness and carotid intima-media thickness (CIMT). Increase in arterial stiffness results in left ventricular hypertrophy and augmentation in systolic hypertension [56]. The intima-media thickness was proved to be a good surrogate marker of atherosclerosis and predictor of coronary heart disease (CHD) [57] and ischemic stroke [58].

A study performed on healthy women (aged 35–75 years) showed an independent inverse relationship between circulating lycopene and brachial-ankle pulse wave velocity (increased arterial stiffness results in increased brachial-ankle pulse wave velocity) [39]. Another study from the same group showed that high plasma lycopene levels were associated with reduction in aortic stiffness in patients with metabolic syndrome [59]. However, in contrast to these observational studies, a 12-week interventional study conducted on overweight, healthy, middle-aged individuals demonstrated that lycopene consumption did not result is a reduction of blood pressure, arterial stiffness, or waist circumference [60].

A prospective, cross-sectional study was conducted on a population of 640 participants (men and women 35–78 years old) who were asymptomatic with respect to carotid artery disease and were seen at the Cardiology Unit (part of the ACADIM Study). Among participants with CIMT ≥ 0.8 mm, lycopene and beta-carotene were all significantly lower when compared with participants with CIMT < 0.8 mm [61]. Another case-control study performed on a middle-aged (45–68 years) Asian population found that lutein levels were significantly lower in subjects with early atherosclerosis determined as an increase in CIMT) compared to subjects with normal CIMT. However, there was no change regarding aortic stiffness [62–65]. A study was conducted on 40 patients with early atherosclerosis not diagnosed as suffering from CVD, but with increased aortic stiffness and CIMT. These patients had lower serum concentrations of the carotenoids lutein and zeaxanthin than healthy controls [64].

The results of the above-mentioned observational studies suggest that carotenoid consumption for long periods (9–20 years) probably has a beneficial effect on cardiovascular risk profile and early atherosclerosis. However, the interventional study in which carotenoids were consumed for several weeks to several months demonstrated only controversial results.

The Effect of Carotenoid Consumption on Major Cardiovascular Events

According to our current knowledge, no major clinical trials have proven that food supplements change the outcomes of patients with known CVD. However, guidelines recently published by the American College of Cardiology/American Heart Association emphasized the importance of fruit and vegetable consumption for cardiovascular health [66]. As mentioned, carotenoids are found in a large variety of fruits and vegetables. In this section we will attempt to elucidate whether carotenoid consumption can alter CVD.

In a rat myocardial infarction (MI) model it was demonstrated that lycopene consumption for 28 days improved cardiac function and ventricular remodeling [67].

Patients with coronary artery disease (CAD) had significantly lower plasma concentrations of HDL cholesterol and beta-carotene and increased levels of all inflammatory markers [68]. Comparable results were demonstrated in patients with documented CAD. These subjects had decreased plasma carotenoid levels in comparison to subjects free of CAD [69]. In a case-control study conducted on Chinese men and women (45–74 years old), lutein was found to have a cardioprotective effect. High levels of plasma beta-cryptoxanthin and lutein measured at baseline were associated with decreased risk of developing acute MI after adjustment for multiple risk factors for CHD [70,71]. However, other studies did not demonstrate the favorable effect of carotenoids on the development of ischemic heart disease. A study was conducted in France and Northern Ireland on 9758 men aged 50–59 years who were free of CHD at baseline. After 5 years' follow-up, 150 incident cases of CHD (nonfatal MI and fatal CHD) were compared with 285 controls matched for age. The baseline serum carotenoid level was not different between cases and controls [72]. Another observational study, including 1224 female participants in the Women's Health Initiative-Observational Study (WHI-OS), also failed to demonstrate a protective cardiovascular effect of diet rich in carotenoids [73].

An animal model using the atherosclerotic apo $E^{-/-}$ mouse demonstrated the protective role of carotenoids on the development of cerebrovascular disease. In this model, Angiotensin II (AgII) induced cerebral aneurysm and ischemia/infarction of the brain. Combined treatment of AgII and beta-carotene resulted in reduction of the size of brain infarction and a complete abolishment in the formation of aneurysm. These favorable effects of beta-carotene were accompanied by a significant reduction of local cerebral inflammation [74]. A large intervention study was performed with 29,133 middle-aged male smokers who received either vitamin E 50 mg/day or beta-carotene 20 mg/day, or both, or placebo, for a median of 6.1 years. The study was a double-blind, randomized trial with a 2×2 factorial design. At baseline, 1700 men had type 2 diabetes. Of these men, 662 were diagnosed with first-ever macrovascular complication. Neither supplementation affected the risk of macrovascular complication or total mortality during the intervention period [75].

According to the known studies, the beneficial effect of carotenoids on major cardiovascular and cerebrovascular events is unclear. Cross-sectional studies demonstrated a positive correlation between low carotenoid levels and adverse cardiovascular outcome. However, to the best of our knowledge, no interventional randomized control double blind study has shown a decrease in major cardiovascular and cerebrovascular events due to utilization of carotenoids. Although carotenoids have beneficial antioxidant properties, their ability to protect the individual is limited to the period preceding the development of CVD. Their consumption should be regarded as a primary prevention strategy.

EFFECT OF CAROTENOIDS ON SKIN AGING

Skin aging is a continuous process affected by endogenous and environmental factors. Topical skin products that attempt to prolong skin's youthful appearance are widely available. Nonetheless, the effects of nutrition on skin parameters such as texture, color, moisture, and other physiological properties, such as defense from UV radiation, have long been appreciated. The understanding of the direct role of nutrition in skin aging triggered not only nutritional awareness, but also a growing interest in the development of nutritional supplements and functional food products to benefit human skin.

The primary preventable cause of skin aging is exposure to UV radiation. It has been estimated that approximately 80% of facial skin aging is attributed to UV-exposure [76], which generates ROS in the skin. These can harm DNA, lipids, and proteins by causing local oxidative stress and initiating an inflammatory response [57,62]. As a result, prolonged exposure to sunlight causes not only erythema (redness of the skin), but also premature skin aging, inflammation, and increased risk of cancer [57,71,77]. Skin aging includes loss of elasticity, drying, and wrinkling. The skin photoaging process is to a great extent influenced by the activity of matrix-degrading enzymes called matrix metalloproteinases (MMPs). MMP family members are induced in response to UV exposure, and trigger degradation of collagen, which is important for maintaining skin strength and elasticity [71,78]. Degradation of collagen can lead to wrinkles, which characterize the skin photoaging process.

Dietary antioxidants and specifically carotenoids accumulate in human skin [62,79] and provide an important line of defense against UV-induced skin damage. Carotenoids, as plant pigments, function in the protection of the plant against excess light. Therefore, it is not surprising that these molecules have been researched for their ability to confer endogenous photoprotection to human skin. The most prominent carotenoids in human skin are lycopene and beta-carotene [79], possibly reflecting a specific function of these carotenoids in skin photoprotection and the defense against oxidative damage. Carotenoid concentration in the skin is dependent on regular consumption of dietary carotenoids from fruits and vegetables. Notably, the antioxidant reservoir of the skin, and specifically the carotenoid reservoir, is depleted as a result of natural aging as well as stress factors such as illness, UV and infrared radiation of the sun, smoking, and alcohol consumption. Upon exposure to UV light, more skin lycopene than beta-carotene is destroyed, suggesting a role of lycopene in mitigating oxidative damage in tissues [80]. Notably, in the healthy population, lycopene skin level has been found to be influenced by age, with older people having lower lycopene levels compared to younger people under 40 years old [81]. Diminution of the carotenoid reservoir leaves the aged skin unprotected.

The mechanism of action by which carotenoids contribute to the prevention of skin aging and photoaging has been explored in several in vitro, in vivo, and human studies [77,82–85].

Carotenoids (particularly lycopene) are powerful dietary antioxidants. Nonetheless, they exert their function not only by scavenging ROS generated in the skin in response to UV or natural aging, but also by activating the antioxidant defense mechanism of the cell, namely the electrophile antioxidant transcription system (EpRE/ARE) and its key transcription factor, nuclear factor erythroid 2-related factor 2 (Nrf2) [49,86]. Induction of the EpRE/ARE transcription system is of great importance in cellular protection against oxidative and carcinogenic stress, as will be further discussed in the next section, which deals with cancer.

Lycopene was also shown to attenuate inflammatory processes [87,88] and diminish DNA damage in vitro in skin cells [77,89]. The anticancer effect of carotenoids also includes cell cycle inhibition, cellular differentiation, apoptosis, and gap-junction communication [90]. The molecular mechanism underlying the antiaging, antiwrinkling effect of carotenoids includes a reduction in the expression of MMP family proteins in skin cells [78,91–93]. For example, beta-carotene suppressed UVA-induction of MMP-1, MMP-3, and MMP-10, three major matrix metalloproteases involved in photoaging in skin keratinocytes (the predominant cell type in the epidermis) [93]. However, in studies performed in skin fibroblasts, while astaxanthin attenuated UV-induced MMP-1 mRNA level [92], lycopene and beta-carotene had a positive effect only when combined with vitamin E, which improved their stability and enhanced their uptake by the cells [78].

Reduction in MMPs was also evident in human studies. In the work of Rizwan et al. [91], healthy volunteers were supplemented with lycopene-rich tomato paste for 12 weeks. Following supplementation, the UV-induced MMP-1 mRNA level was reduced in the tomato paste group compared to the control group, supporting a possible contribution to wrinkle prevention by carotenoid consumption. Indeed, a significant correlation was found between forehead skin roughness and lycopene concentration in the skin of 40- to 50-year-old men and women [94]. Importantly, 12-week supplementation with a combination of carotenoids (including lycopene, lutein, beta-carotene, alpha-tocopherol, and selenium) improved parameters related to skin structure, such as increased skin density and thickness, and improved roughness in healthy volunteers. Altogether, these studies suggest a beneficial effect of carotenoids on skin appearance related to the aging process.

Evidence for the protective effect of carotenoids on skin health (eg, prevention of UV-induced damage as well as prevention of skin cancer) has been found in numerous human studies as well [77,91,95–97]. The efficacy of lycopene- and beta-carotene-rich products or supplements in preventing UV-induced erythema has been shown in several intervention studies [91,95,97]. For example, in healthy subjects, erythema on dorsal skin (back) was significantly diminished after 8 weeks of carotenoid supplementation (mainly beta-carotene). Erythema suppression was greater when the carotenoid supplement was combined with vitamin E [98]. The photoprotective effect of lycopene containing tomato extract was shown in the work of Aust et al. [95], in which UV-induced erythema was evaluated. Twelve week supplementation with tomato extract or a tomato extract-based drink fortified with phytoene and phytofluene caused a greater reduction in UV-induced erythema than that obtained with synthetic lycopene. While the amounts of lycopene ingested from these three different sources was similar (about 10 mg/day) and so were the serum lycopene and total skin carotenoids, the protective effect of the tomato-derived supplements was more pronounced than that of synthetic lycopene and reached a 48% reduction in UV-induced erythema in the tomato drink group. The difference in the efficacy of the tomato-based products compared to synthetic lycopene was attributed to the presence of additional tomato phytonutrients, such as the carotenoids phytoene and phytofluene, which absorb light in the UVA and UVB range [95].

As mentioned above, UV exposure accompanied by oxidative and inflammatory processes enhances the risk of precancerous lesions and skin cancer. Moreover, it is well known that skin cancer risk correlates with age. For example, melanoma is a malignancy diagnosed mainly in the fifth and sixth decades of life [57]. The effect of carotenoids on skin cancer has been examined in several in vitro studies. For instance, UV-induced DNA damage was lessened by astaxanthin in skin fibroblast cells in vitro [99]. Furthermore, hydrogen peroxide-induced DNA damage was diminished by lycopene treatment in squamous cell carcinoma cells [89]. Notably, in humans supplemented with lycopene-rich tomato paste, a mitochondrial DNA deletion of 3895 base pairs following UV radiation was significantly reduced [91].

Another means by which carotenoids contribute to skin cancer prevention is regulation of apoptosis. It was shown that carotenoids such as beta-carotene are able to induce apoptosis in melanoma cells in vitro [84,100]. The molecular mechanism of this effect involves Bcl-2 suppression and p53 activation [84] eventually leading to activation of a caspase cascade [84,100]. Beta-carotene was also shown to inhibit tumor-specific angiogenesis in vitro and reduce inflammatory cytokines, such as TNF-α, that play a role in carcinogenesis [84]. The preventive effect of lycopene on skin cancer was also studied in vitro. It was found to inhibit platelet-derived growth factor-BB (PDGF-BB)-induced dermal fibroblast migration and reduce PDGF-BB-induced signaling [101] and thus may contribute to arresting the progression of melanoma [102].

A protective effect of carotenoids against UV-induced skin cancer was shown in animal studies as well. For example, oral administration of beta-carotene had a protective effect against skin tumor development in animals subjected to UV irradiation. [103,104]. In addition, oral administration of lycopene protected against 7,12-imethylbenz(a)anthracene-induced and 12-O-tetradecanoylphorbol-13-acetate-promoted (DMBA/TPA)-induced cutaneous carcinoma by delaying tumor formation and reducing tumor incidence and volume. Moreover, in this animal model, lycopene activated the antioxidant defense mechanism of the cells, reduced ROS formation, and protected against loss of glutathione [105].

Nonetheless, in human studies, inconsistent results have been observed regarding the effect of carotenoids on skin carcinogenesis. While some prospective studies found an association between a carotenoid rich diet and a reduction in risk for skin cancer (eg, melanoma) [106], other studies found no association between dietary intake or serum level of specific carotenoids such as lycopene or beta-carotene and the risk of several types of skin cancer [107,108].

Notably, expression of the carotenoid central cleavage enzyme BCO1 [109] but not BCO2 [110] was found in keratinocytes of the squamous epithelium of skin. This result is in line with the known effect of vitamin A and retinol, which can be formed by the cleavage of beta-carotene, on skin health and appearance. Today, it is possible to measure the carotenoids in human skin in vivo, online, and noninvasively by resonance Raman spectroscopy. The level of carotenoids in the skin is indicative of the complete antioxidative network of human skin and the nutritional status of the individual [85].

Consumption of carotenoid rich foods may reduce skin photoaging, UV-induced erythema, and stress-induced signaling such as inflammatory processes in the skin. Moreover, the absence of these compounds leaves the aged skin unprotected. These findings, which support the effect of nutrition on skin health, suggest the existence of a scientific basis for the concept of "beauty from within."

CAROTENOIDS AND CANCER PREVENTION

Cancer is an age-related disease. For example, according to the American Cancer Society, in 2015, over half the estimated new cases of cancer (all types) will be diagnosed in people over 65 years of age [63]. The second leading causes of cancer deaths in the United States today are prostate cancer for men and breast cancer for women [63].

Epidemiological studies suggest an association between the consumption of fruits and vegetables and reduced incidence of cancer [111,112]. A protective effect has been attributed to carotenoids, which are abundant in fruits and vegetables [11,113–117].

Anticancer activity is perhaps the most researched and best characterized biological activity of carotenoids. Beta-carotene has received much attention due to its presence in many foods [116]. However, other carotenoids such as lycopene, the main tomato carotenoid, have become the subject of intense investigation in recent years. For example, in a meta-analysis of the epidemiologic literature related to tomato consumption and cancer prevention, Giovannucci found that most of the reviewed studies reported an inverse association between tomato intake or lycopene concentration in blood and the risk of numerous types of cancer [114,118,119].

The anticancer effect of carotenoids was also observed in vivo [120–122] and in vitro [123–127]. Several carotenoids inhibit the proliferation of different human cancer cell lines. For example, lycopene was shown to inhibit mammary, endometrial, lung, and leukemic cancer cell growth in a dose-dependent manner [125–127]. The molecular mechanism of the chemo-preventive effect of carotenoids has been intensively explored. Nonetheless, a complete and comprehensive understanding has yet to be elucidated (Table 34.1).

TABLE 34.1 Carotenoids Characteristics: Molecular Action and Their Beneficial Clinical Effect

Characteristics	Molecular action	Clinical outcome
Free radical scavenging: − Physical quenchers of singlet oxygen ($^1O_2$) − Scavenging peroxyl radicals	− Protection of cellular membranes against oxidative damage − Protects LDL from oxidation Defense from UV radiation	− Reduces plasma oxidized LDL − Reduces skin photoaging and cancer development
Induction of the EpRE/ARE transcription system	Detoxify many harmful substances	Anticarcinogenic
Increased NO and decreased ET-1	Improves endothelial function	Attenuates the development of early atherosclerosis
Anti-inflammatory	− Decreases expression of serum and vascular vCAM and iCAM − TNF-α reduction	Attenuates atherosclerosis and cancer development
Reduction in the expression of MMP family proteins	Reduced degradation of collagen	Improves skin strength and elasticity
Bcl-2 suppression and p53 activation	Regulation of apoptosis	Cancer prevention: melanoma, breast prostate, lung, endometrium, and leukemia
Attenuation of NF-κB activation	Reduction in inflammatory cytokine production and cancer cell proliferation	− Antiatherosclerotic − Anticarcinogenic
Inhibition of steroid hormone signaling	− Downregulating the expression of the 5-alpha-reductase enzyme − Attenuates the estrogen-induced proliferation of mammary and endometrial cancer cells	Prevention of sex hormone-dependent cancers: endometrium, breast, and prostate

Carotenoids are lipophilic plant pigments typically containing a series of conjugated double bonds. This chemical structure confers their ability to function as excellent singlet oxygen quenchers and free radical scavengers [25]. Increased ROS level results in oxidative stress that can damage DNA, proteins, and lipids, all involved in cancer initiation and progression. Therefore, the antioxidant ability of carotenoids is thought to contribute to their chemopreventive properties. Importantly, the structure of carotenoids makes them susceptible to oxidative cleavage, resulting in the formation of different oxidation products found to contribute to their anticancer effect. These oxidation products, for example, apo-lycopenals, the products of lycopene oxidative cleavage, were found in tomatoes, tomato products, and mammalian tissues [128], implying their significance and possible influence on human health. It is worth noting that the anticancer effect of carotenoids and their derivatives relies not solely on their chemical antioxidant properties, but also on the modulation of key signaling pathways in the cells. Some of these important pathways will be discussed in this chapter.

Modulation of cell proliferation and apoptosis was found to result from regulation of strategic proteins involved in these processes. For example, lycopene was found to regulate the expression of tumor suppressor protein p53 as well as its target proteins, such as the cell cycle protein cyclin D1, the cell cycle inhibitor p21, and the apoptotic proteins Bax-1 and cleaved caspase 3 in vitro and in vivo [38,129,130]. A similar trend was observed for other carotenoids, such as beta-carotene [131] and astaxanthin [132]. Moreover, lycopene was found to interfere with growth factor signaling in vitro (eg, insulin-like growth factor I) [133]. Notably, in colon cancer patients, tomato lycopene extract supplementation decreased insulin-like growth factor-I levels [134]. Another signal crucial for the regulation of precancerous cell proliferation is gap junction communication. Several carotenoids were found to upregulate the expression of connexin43, a major protein in the assembly of the communication channel. Interestingly, different oxidation products of carotenoids can induce gap junction communication [135–137]. Modulation of a vast array of proteins by carotenoids and their derivatives implies their ability to modulate transcription. Indeed, these compounds regulate key transcription systems in the cell, contributing to their chemopreventive effect (Table 34.2).

Electrophile/Antioxidant Response Element

Induction of the EpRE/ARE transcription system gives rise to the activation of phase 2 enzymes, a major cellular strategy for reducing cancer risk. These enzymes detoxify many harmful substances by converting them to hydrophilic metabolites that can be

TABLE 34.2 Carotenoids and Antiatherosclerotic Effect

Antiatherosclerotic characteristics	Study category (reference number)
Improves endothelial function	In vitro and interventional clinical study [37]
Reduced vascular inflammation	Observational clinical study [39,43,48]
Reduction in plasma LDL level	Animal model [41], clinical study [48,50], and in vitro study [49]
Blood pressure lowering	Observational and interventional clinical study [37,52,53]
Glucose tolerance improvement	Observational clinical study [40]
Metabolic syndrome improvement	Observational clinical study [54]
Reduction in vascular stiffness	Observational clinical study [39,59]
Reduction in carotid intima-media	Observational clinical study [61–65]
Cardiac protection	Animal model [69], Observational clinical study [70,71]
Cerebrovascular protection	Animal model [74]

readily excreted from the body. Thus, their induction is associated with enhanced susceptibility to chemical carcinogenesis [27]. Carotenoids were found to activate this cancer preventive system in numerous cancer cell lines [27,49,138]. For example, lycopene was found to activate EpRE/ARE-dependent transcription and its major activating transcription factor, Nrf2 in hepatoma, mammary, and prostate cancer cells. This activation was accompanied by elevated expression of phase II detoxifying enzymes such as NAD(P)H: quinone oxidoreductase and gamma-glutamylcysteine synthetase [138]. A similar effect was observed in vivo [132,139]. Other carotenoids, such as beta-carotene, phytoene, and astaxanthin had somewhat lower effect [138]. Notably, we have recently shown that carotenoid oxidized derivatives mediate this activation [140].

Nuclear Factor Kappa B and Inflammation

The NF-κB transcription system plays a key role in inflammatory processes and chronic inflammatory diseases such as cancer. NF-κB activation increases pro-inflammatory cytokine production and cancer cell proliferation, decreases apoptosis, and promotes tumor metastasis, all of which lead to the progression of cancer. A prevalent view is that NF-κB is a pivotal link between inflammation and cancer. Indeed, constitutive NF-κB activation has been observed in many human cancers (Fig. 34.1).

Carotenoids can attenuate NF-κB activation in various types of cancer cells [27,141,142]. For example, beta-carotene attenuates NF-κB binding activity in human leukemia and colon adenocarcinoma cells [141]. Moreover, dietary supplementation of lycopene or tomato extract inhibited nonalcoholic steatohepatitis-promoted hepatocarcinogenesis in rats. The molecular mechanism underpinning this anticancer effect included inhibition of NF-κB dependent transcription, accompanied by a reduced expression of NF-κB target genes such as pro-inflammatory cytokines [139]. Importantly, we recently showed that oxidized carotenoid derivatives are the active mediators in NF-κB inhibition, and that the molecular mechanism of this inhibition includes direct interaction with key thiol groups of different proteins in the NF-κB signaling pathway. Inhibition of NF-KB by carotenoids or their derivatives is considered a promising therapeutic approach for blocking tumor growth (Fig. 34.2).

Nuclear Receptors: Steroid Hormones (Estrogen/Androgen Receptors)

It has been suggested that carotenoids, particularly lycopene, play a significant role in the prevention of sex hormone-dependent cancers. Among these are the major human malignancies: estrogen-dependent breast and endometrial cancers in women and androgen-dependent prostate cancer in men. Estrogens and androgens play a pivotal role in the initiation and progression of these malignancies [143]. Therefore, inhibition of steroid hormone signaling was suggested as a molecular mechanism underpinning the protective effect of carotenoids on these cancers.

The strongest association of carotenoids with the prevention of prostate cancer has been documented for lycopene [118]. Prostate cancer is the second leading cause of cancer deaths in men in the United States [63]. Notably, age is the most important nonmodifiable risk factor for this malignancy. In fact, prostate cancer has the steepest age-incidence curve of all cancers, with a rapid increase in the seventh decade [144]. Due to the role of androgens in prostate cancer pathology, the lowering of testosterone levels or the reduction of

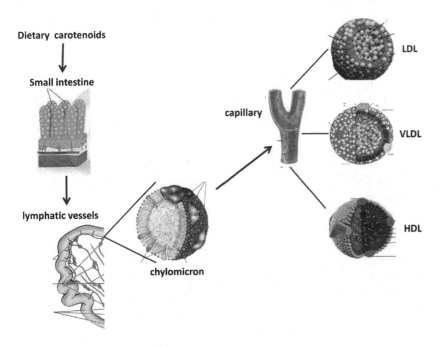

FIGURE 34.1 Carotenoids circulation. After ingestion, the carotenoids incorporate into mixed lipid micelles in the lumen and are taken up by the intestinal mucosa and incorporated into chylomicrons. These chylomicrons are released into the lymph. Next, they are further distributed in the plasma by the use of VLDL, LDL, and HDL.

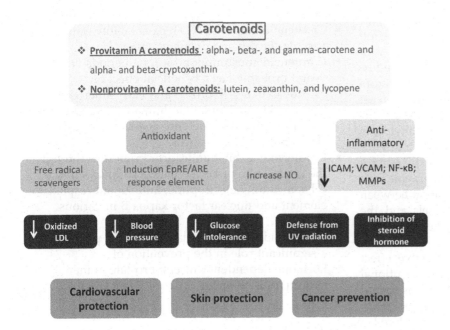

FIGURE 34.2 Carotenoids molecular and clinical properties. Dietary carotenoids serve as antioxidative and anti-inflammatory elements. They are direct oxidative stress scavengers and also activate the antioxidative EpRE/ARE response. Their anti-inflammatory properties include reduction of expression of ICAM-1, VCAM-1, NF-κB, and the MMPs proteins. As such, carotenoids improve cardio-metabolic profile, protect the skin from UV-related damage, and decrease cancer susceptibility (Electrophile/antioxidant response element, EpRE/ARE; nuclear factor kappa B, NF-κB; vascular cell adhesion molecule 1, VCAM-1; intercellular adhesion molecule 1, ICAM-1; matrix metalloproteinase, MMPs).

androgen-induced signaling are considered desired therapeutic strategies for dietary chemoprevention. Indeed, the protective mechanism of lycopene includes these two strategies [145–147]. For instance, in a rat model of hormone-dependent prostate cancer, lycopene supplementation reduced local testosterone activation by downregulating the expression of the 5-alpha-reductase enzyme that converts testosterone to its active, more potent form, dihydrotestosterone [145]. In addition, we have recently shown that carotenoids can also interfere with the androgen signaling pathway. Astaxanthin, phytoene+phytofluene, and, predominantly, lycopene attenuated dihydrotestosterone-induced prostate cancer cell growth. This attenuation resulted at least partly from the inhibition of androgen-induced transcription via the androgen response element (ARE), accompanied by a reduction in the level of prostate-specific antigen, an important marker for androgenic activity [148]. Significantly, low concentrations of different active nutritional ingredients, including carotenoids, act in combination to produce additive or synergistic effects on prostate cell proliferation as well as inhibition of androgen signaling, suggesting that the beneficial effects of fruits, vegetables, and other dietary ingredients reside in the combined effect of diversified phytonutrients [148]. Interestingly, a similar combined effect was observed for the induction of the EpRE/ARE transcription system discussed above. In line with these results, in prostate cancer patients, a lycopene-rich tomato extract was found to modulate these same two regulatory pathways: androgen/estrogen metabolism and the EpRE/ARE transcription system [149]. It should not go unnoticed that a pivotal role of androgens has been identified not only in the development of prostate cancer but also in the risk of benign prostate hyperplasia (BPH), a condition common in older men that causes prostate enlargement, interferes with urine flow, and causes great discomfort. Fifty percent of men over the age of 50 will suffer from BPH, and that probability increases to 90% for men over 80 [150]. Importantly, lycopene was also found to be effective in the prevention and treatment of benign prostatic hyperplasia in men [150].

Breast cancer is the second leading cause of cancer deaths in women today [63]. Estrogens play an important role in both the development and the progression of the malignant process of this cancer [151]. As a result, inhibitors of estrogenic activity such as antiestrogens or specific estrogen receptor modulators (SERMs) are clinically used for the prevention of breast cancer or its recurrence in women at high risk for this malignancy. Carotenoids such as lycopene, phytoene, phytofluene, and beta-carotene were found to attenuate the estrogen-induced proliferation of mammary and endometrial cancer cells [152]. Moreover, epidemiological studies suggest an inverse association between carotenoid intakes or their circulating levels and the risk of breast cancer in postmenopausal women [115,117,153].

The molecular mechanism of this protective effect includes inhibition of ER-dependent transcriptional activity via the estrogen response element as found in mammary and endometrial cancer cells [152]. Importantly, while estrogenic activity is considered a risk factor in cancer cells, this activity is crucial for maintaining bone health, particularly in postmenopausal women. Therefore, inhibition of estrogenic activity in bone cells could put postmenopausal women at risk of developing osteoporosis. Interestingly, we found that in osteoblast bone forming cells, not only do carotenoids not inhibit the estrogenic signal, but they also induced estrogen-dependent transcription and the expression of beneficial target genes of this pathway. Indeed, animal [154,155] and human [156,157] studies support the beneficial effect of lycopene and other carotenoids on bone health.

CONCLUSIONS

Carotenoids cannot be produced by mammals; their dietary source is yellow−orange−red fruits and green leafy vegetables. They function as antioxidants and as such have a role in attenuation of skin aging and prevention of skin cancer. They also reduce the risk of other cancers (such as prostate and breast). Due their antioxidant ability carotenoids improve vascular health and can attenuate atherosclerosis, especially, if they are consumed for a period of at least 10 years.

SUMMARY

- The main sources of carotenoids in human plasma are yellow−orange−red fruits and green leafy vegetables.
- Carotenoids are transported in the plasma of humans and animals exclusively by lipoproteins.
- Carotenoids exhibit free radical scavenging properties.
- Carotenoids increase NO bioavailability improving vascular health and delaying aging.
- Carotenoid consumption for long periods (9−20 years) probably has a beneficial effect on cardiovascular risk profile and early atherosclerosis.
- Carotenoids accumulate in human skin and provide an important line of defense against UV-induced skin aging and skin cancer.
- Carotenoids exhibit anticancer properties through the modulation of key signaling pathways in the cells, including electrophile/antioxidant response element and nuclear factor kappa B in various cancers.
- Carotenoids and particularly lycopene play a significant role in the prevention of sex hormone-dependent cancers: prostate in men and breast in women.

Acknowledgment

I thank Prof. Yossi Levy and Prof. Yoav Sharoni (Ben-Gurion University of the Negev) for their input during the preparation of this manuscript.

References

[1] Nisar N, Li L, Lu S, Khin NC, Pogson BJ. Carotenoid metabolism in plants. Mol Plant 2015;8:68−82.
[2] Rao AV, Rao LG. Carotenoids and human health. Pharmacol Res 2007;55:207−16.
[3] Najm W, Lie D. Dietary supplements commonly used for prevention. Prim Care 2008;35:749−67.

REFERENCES

[4] Penuelas J, Munne-Bosch S. Isoprenoids: an evolutionary pool for photoprotection. Trends Plant Sci 2005;10:166–9.

[5] Maiani G, Caston MJ, Catasta G, Toti E, Cambrodon IG, Bysted A, et al. Carotenoids: actual knowledge on food sources, intakes, stability and bioavailability and their protective role in humans. Mol Nutr Food Res 2009;53(Suppl. 2):S194–218.

[6] Heber D, Lu QY. Overview of mechanisms of action of lycopene. Exp Biol Med (Maywood) 2002;227:920–3.

[7] Bohn T. Bioavailability of non-provitamin a carotenoids. Curr Nutr Food Sci 2008;4:240–58.

[8] Parker RS. Absorption, metabolism, and transport of carotenoids. FASEB J 1996;10:542–51.

[9] Olmedilla B, Granado F, Southon S, Wright AJ, Blanco I, Gil-Martinez E, et al. Serum concentrations of carotenoids and vitamins A, E, and C in control subjects from five European countries. Br J Nutr 2001;85:227–38.

[10] Rock CL, Swendseid ME, Jacob RA, McKee RW. Plasma carotenoid levels in human subjects fed a low carotenoid diet. J Nutr 1992;122:96–100.

[11] Krinsky NI, Johnson EJ. Carotenoid actions and their relation to health and disease. Mol Aspects Med 2005;26:459–516.

[12] Stahl W, Schwarz W, Sundquist AR, Sies H. cis-trans isomers of lycopene and beta-carotene in human serum and tissues. Arch Biochem Biophys 1992;294:173–7.

[13] Darvin ME, Fluhr JW, Meinke MC, Zastrow L, Sterry W, Lademann J. Topical beta-carotene protects against infra-red-light-induced free radicals. Exp Dermatol 2011;20:125–9.

[14] Fiedor J, Burda K. Potential role of carotenoids as antioxidants in human health and disease. Nutrients 2014;6:466–88.

[15] Wisniewska A, Subczynski WK. Effects of polar carotenoids on the shape of the hydrophobic barrier of phospholipid bilayers. Biochim Biophys Acta 1998;1368:235–46.

[16] Yeum KJ, Russell RM. Carotenoid bioavailability and bioconversion. Annu Rev Nutr 2002;22:483–504.

[17] Cadenas E. Biochemistry of oxygen toxicity. Annu Rev Biochem 1989;58:79–110.

[18] Valko M, Izakovic M, Mazur M, Rhodes CJ, Telser J. Role of oxygen radicals in DNA damage and cancer incidence. Mol Cell Biochem 2004;266:37–56.

[19] Poli G, Leonarduzzi G, Biasi F, Chiarpotto E. Oxidative stress and cell signalling. Curr Med Chem 2004;11:1163–82.

[20] Halliwell B. Antioxidants in human health and disease. Annu Rev Nutr 1996;16:33–50.

[21] Valko M, Rhodes CJ, Moncol J, Izakovic M, Mazur M. Free radicals, metals and antioxidants in oxidative stress-induced cancer. Chem Biol Interact 2006;160:1–40.

[22] Burton GW. Antioxidant action of carotenoids. J Nutr 1989;119:109–11.

[23] Fiedor J, Fiedor L, Haessner R, Scheer H. Cyclic endoperoxides of beta-carotene, potential pro-oxidants, as products of chemical quenching of singlet oxygen. Biochim Biophys Acta 2005;1709:1–4.

[24] Conn PF, Schalch W, Truscott TG. The singlet oxygen and carotenoid interaction. J Photochem Photobiol B 1991;11:41–7.

[25] Di Mascio P, Kaiser S, Sies H. Lycopene as the most efficient biological carotenoid singlet oxygen quencher. Arch Biochem Biophys 1989;274:532–8.

[26] Sies H, Stahl W. Vitamins E and C, beta-carotene, and other carotenoids as antioxidants. Am J Clin Nutr 1995;62:1315S–21S.

[27] Kaulmann A, Bohn T. Carotenoids, inflammation, and oxidative stress—implications of cellular signaling pathways and relation to chronic disease prevention. Nutr Res 2014;34:907–29.

[28] Reuter S, Gupta SC, Chaturvedi MM, Aggarwal BB. Oxidative stress, inflammation, and cancer: how are they linked? Free Radic Biol Med 2010;49:1603–16.

[29] Libby P, Ridker PM, Maseri A. Inflammation and atherosclerosis. Circulation 2002;105:1135–43.

[30] Armoza A, Haim Y, Bashiri A, Wolak T, Paran E. Tomato extract and the carotenoids lycopene and lutein improve endothelial function and attenuate inflammatory NF-kappaB signaling in endothelial cells. J Hypertens 2013;31:521–9 discussion 529.

[31] Riso P, Vendrame S, Del Bo C, Martini D, Martinetti A, Seregni E, et al. Effect of 10-day broccoli consumption on inflammatory status of young healthy smokers. Int J Food Sci Nutr 2014;65:106–11.

[32] Holt EM, Steffen LM, Moran A, Basu S, Steinberger J, Ross JA, et al. Fruit and vegetable consumption and its relation to markers of inflammation and oxidative stress in adolescents. J Am Diet Assoc 2009;109:414–21.

[33] Taddei S, Virdis A, Ghiadoni L, Salvetti G, Bernini G, Magagna A, et al. Age-related reduction of NO availability and oxidative stress in humans. Hypertension 2001;38:274–9.

[34] Dai DF, Rabinovitch PS, Ungvari Z. Mitochondria and cardiovascular aging. Circ Res 2012;110:1109–24.

[35] Celermajer DS, Sorensen KE, Spiegelhalter DJ, Georgakopoulos D, Robinson J, Deanfield JE. Aging is associated with endothelial dysfunction in healthy men years before the age-related decline in women. J Am Coll Cardiol 1994;24:471–6.

[36] Selvaraju V, Joshi M, Suresh S, Sanchez JA, Maulik N, Maulik G. Diabetes, oxidative stress, molecular mechanism, and cardiovascular disease—an overview. Toxicol Mech Methods 2012;22:330–5.

[37] Paran E, Novack V, Engelhard YN, Hazan-Halevy I. The effects of natural antioxidants from tomato extract in treated but uncontrolled hypertensive patients. Cardiovasc Drugs Ther 2009;23:145–51.

[38] Tang X, Yang X, Peng Y, Lin J. Protective effects of lycopene against H2O2-induced oxidative injury and apoptosis in human endothelial cells. Cardiovasc Drugs Ther 2009;23:439–48.

[39] Akimoto Y, Horinouchi T, Shibano M, Matsushita M, Yamashita Y, Okamoto T, et al. Nitric oxide (NO) primarily accounts for endothelium-dependent component of beta-adrenoceptor-activated smooth muscle relaxation of mouse aorta in response to isoprenaline. J Smooth Muscle Res 2002;38:87–99.

[40] Akbaraly TN, Fontbonne A, Favier A, Berr C. Plasma carotenoids and onset of dysglycemia in an elderly population: results of the Epidemiology of Vascular Ageing Study. Diabetes Care 2008;31:1355–9.

[41] Lorenz M, Fechner M, Kalkowski J, Frohlich K, Trautmann A, Bohm V, et al. Effects of lycopene on the initial state of atherosclerosis in New Zealand White (NZW) rabbits. PLoS One 2012;7:e30808.

[42] Denniss SG, Haffner TD, Kroetsch JT, Davidson SR, Rush JW, Hughson RL. Effect of short-term lycopene supplementation and postprandial dyslipidemia on plasma antioxidants and biomarkers of endothelial health in young, healthy individuals. Vasc Health Risk Manag 2008;4:213–22.

[43] Blum A, Monir M, Khazim K, Peleg A, Blum N. Tomato-rich (Mediterranean) diet does not modify inflammatory markers. Clin Invest Med 2007;30:E70–4.

[44] Rizzo M, Berneis K, Koulouris S, Pastromas S, Rini GB, Sakellariou D, et al. Should we measure routinely oxidised and atherogenic dense low-density lipoproteins in subjects with type 2 diabetes? Int J Clin Pract 2010;64:1632–42.

[45] Rizzo M, Kotur-Stevuljevic J, Berneis K, Spinas G, Rini GB, Jelic-Ivanovic Z, et al. Atherogenic dyslipidemia and oxidative stress: a new look. Transl Res 2009;153:217–23.

[46] Stocker R, Keaney Jr JF. Role of oxidative modifications in atherosclerosis. Physiol Rev 2004;84:1381–478.

[47] Chew PY, Riley L, Graham DL, Rahman K, Lowe GM. Does lycopene offer human LDL any protection against myeloperoxidase activity? Mol Cell Biochem 2012;361:181–7.

[48] Burton-Freeman B, Talbot J, Park E, Krishnankutty S, Edirisinghe I. Protective activity of processed tomato products on postprandial oxidation and inflammation: a clinical trial in healthy weight men and women. Mol Nutr Food Res 2012;56:622–31.

[49] Palozza P, Catalano A, Simone R, Cittadini A. Lycopene as a guardian of redox signalling. Acta Biochim Pol 2012;59:21–5.

[50] Ryden M, Leanderson P, Kastbom KO, Jonasson L. Effects of simvastatin on carotenoid status in plasma. Nutr Metab Cardiovasc Dis 2012;22:66–71.

[51] Markovits N, Ben Amotz A, Levy Y. The effect of tomato-derived lycopene on low carotenoids and enhanced systemic inflammation and oxidation in severe obesity. Isr Med Assoc J 2009;11:598–601.

[52] Engelhard YN, Gazer B, Paran E. Natural antioxidants from tomato extract reduce blood pressure in patients with grade-1 hypertension: a double-blind, placebo-controlled pilot study. Am Heart J 2006;151:100.

[53] Hozawa A, Jacobs Jr. DR, Steffes MW, Gross MD, Steffen LM, Lee DH. Circulating carotenoid concentrations and incident hypertension: the Coronary Artery Risk Development in Young Adults (CARDIA) study. J Hypertens 2009;27:237–42.

[54] Liu J, Shi WQ, Cao Y, He LP, Guan K, Ling WH, et al. Higher serum carotenoid concentrations associated with a lower prevalence of the metabolic syndrome in middle-aged and elderly Chinese adults. Br J Nutr 2014;112:2041–8.

[55] Suzuki K, Honjo H, Ichino N, Osakabe K, Sugimoto K, Yamada H, et al. Association of serum carotenoid levels with urinary albumin excretion in a general Japanese population: the Yakumo study. J Epidemiol 2013;23:451–6.

[56] Vlachopoulos C, Aznaouridis K, Stefanadis C. Prediction of cardiovascular events and all-cause mortality with arterial stiffness: a systematic review and meta-analysis. J Am Coll Cardiol 2010;55:1318–27.

[57] Amaro-Ortiz A, Yan B, D'Orazio JA. Ultraviolet radiation, aging and the skin: prevention of damage by topical cAMP manipulation. Molecules 2014;19:6202–19.

[58] Freitas D, Alves A, Pereira A, Pereira T. Increased intima-media thickness is independently associated with ischemic stroke. Arq Bras Cardiol 2012;98(6):497–504.

[59] Yeo HY, Kim OY, Lim HH, Kim JY, Lee JH. Association of serum lycopene and brachial-ankle pulse wave velocity with metabolic syndrome. Metabolism 2011;60:537–43.

[60] Thies F, Masson LF, Rudd A, Vaughan N, Tsang C, Brittenden J, et al. Effect of a tomato-rich diet on markers of cardiovascular disease risk in moderately overweight, disease-free, middle-aged adults: a randomized controlled trial. Am J Clin Nutr 2012;95:1013–22.

[61] Riccioni G, D'Orazio N, Palumbo N, Bucciarelli V, Ilio E, Bazzano LA, et al. Relationship between plasma antioxidant concentrations and carotid intima-media thickness: the Asymptomatic Carotid Atherosclerotic Disease in Manfredonia Study. Eur J Cardiovasc Prev Rehabil 2009;16:351–7.

[62] Schagen SK, Zampeli VA, Makrantonaki E, Zouboulis CC. Discovering the link between nutrition and skin aging. Dermato Endocrinol 2012;4:298–307.

[63] Siegel R, Ma J, Zou Z, Jemal A. Cancer statistics, 2014. CA Cancer J Clin 2014;64:9–29.

[64] Xu XR, Zou ZY, Huang YM, Xiao X, Ma L, Lin XM. Serum carotenoids in relation to risk factors for development of atherosclerosis. Clin Biochem 2012;45:1357–61.

[65] Zou Z, Xu X, Huang Y, Xiao X, Ma L, Sun T, et al. High serum level of lutein may be protective against early atherosclerosis: the Beijing atherosclerosis study. Atherosclerosis 2011;219: 789–93.

[66] Eckel RH, Jakicic JM, Ard JD, de Jesus JM, Houston Miller N, Hubbard VS, American College of Cardiology/American Heart Association Task Force on Practice G, et al. 2013 AHA/ACC guideline on lifestyle management to reduce cardiovascular risk: a report of the American College of Cardiology/American Heart Association Task Force on Practice Guidelines. J Am Coll Cardiol 2014;63:2960–84.

[67] Wang X, Lv H, Gu Y, Wang X, Cao H, Tang Y, et al. Protective effect of lycopene on cardiac function and myocardial fibrosis after acute myocardial infarction in rats via the modulation of p38 and MMP-9. J Mol Histol 2014;45:113–20.

[68] Muzakova V, Kand'ar R, Meloun M, Skalicky J, Kralovec K, Zakova P, et al. Inverse correlation between plasma beta-carotene and interleukin-6 in patients with advanced coronary artery disease. Int J Vitam Nutr Res 2010;80:369–77.

[69] Lidebjer C, Leanderson P, Ernerudh J, Jonasson L. Low plasma levels of oxygenated carotenoids in patients with coronary artery disease. Nutr Metab Cardiovasc Dis 2007;17:448–56.

[70] Koh WP, Yuan JM, Wang R, Lee YP, Lee BL, Yu MC, et al. Plasma carotenoids and risk of acute myocardial infarction in the Singapore Chinese Health Study. Nutr Metab Cardiovasc Dis 2011;21:685–90.

[71] Kohl E, Landthaler M, Szeimies RM. Skin aging. Hautarzt 2009;60(quiz 934):917–33.

[72] Gey KF, Ducimetiere P, Evans A, Amouyel P, Arveiler D, Ferrieres J, et al. Low plasma retinol predicts coronary events in healthy middle-aged men: the PRIME Study. Atherosclerosis 2010;208:270–4.

[73] Horn LV, Tian L, Neuhouser ML, Howard BV, Eaton CB, Snetselaar L, et al. Dietary patterns are associated with disease risk among participants in the Women's Health Initiative Observational Study. J Nutr 2012;142:284–91.

[74] Gopal K, Nagarajan P, Raj TA, Jahan P, Ganapathy HS, Mahesh Kumar MJ. Effect of dietary beta carotene on cerebral aneurysm and subarachnoid haemorrhage in the brain apo $E^{-/-}$ mice. J Thromb Thrombolysis 2011;32:343–55.

[75] Kataja-Tuomola MK, Kontto JP, Mannisto S, Albanes D, Virtamo JR. Effect of alpha-tocopherol and beta-carotene supplementation on macrovascular complications and total mortality from diabetes: results of the ATBC Study. Ann Med 2010;42:178–86.

[76] Uitto J. Understanding premature skin aging. N Engl J Med 1997;337:1463–5.

[77] Ascenso A, Ribeiro H, Marques HC, Oliveira H, Santos C, Simoes S. Chemoprevention of photocarcinogenesis by lycopene. Exp Dermatol 2014;23:874–8.

[78] Offord EA, Gautier JC, Avanti O, Scaletta C, Runge F, Kramer K, et al. Photoprotective potential of lycopene, beta-carotene, vitamin E, vitamin C and carnosic acid in UVA-irradiated human skin fibroblasts. Free Radic Biol Med 2002;32:1293–303.

[79] Scarmo S, Cartmel B, Lin H, Leffell DJ, Welch E, Bhosale P, et al. Significant correlations of dermal total carotenoids and dermal lycopene with their respective plasma levels in healthy adults. Arch Biochem Biophys 2010;504:34–9.

[80] Ribaya-Mercado JD, Garmyn M, Gilchrest BA, Russell RM. Skin lycopene is destroyed preferentially over beta-carotene during ultraviolet irradiation in humans. J Nutr 1995;125:1854–9.

[81] Meinke MC, Lauer A, Taskoparan B, Gersonde I, Lademann J, Darvin MF. Influence on the carotenoid levels of skin arising from age, gender, body mass index in smoking/non-smoking individuals. Free Rad Antiox 2011;1:15–20.

[82] Chinembiri TN, du Plessis LH, Gerber M, Hamman JH, du Plessis J. Review of natural compounds for potential skin cancer treatment. Molecules 2014;19:11679–721.

[83] Guruvayoorappan C, Kuttan G. Beta-carotene inhibits tumor-specific angiogenesis by altering the cytokine profile and inhibits the nuclear translocation of transcription factors in B16F-10 melanoma cells. Integr Cancer Ther 2007;6:258–70.

[84] Guruvayoorappan C, Kuttan G. β-Carotene down-regulates inducible nitric oxide synthase gene expression and induces apoptosis by suppressing bcl-2 expression and activating caspase-3 and p53 genes in B16F-10 melanoma cells. Nutr Res 2007;27:336–42.

[85] Lademann J, Meinke MC, Sterry W, Darvin ME. Carotenoids in human skin. Exp Dermatol 2011;20:377–82.

[86] Sharoni Y, Linnewiel-Hermoni K, Khanin M, Salman H, Veprik A, Danilenko M, et al. Carotenoids and apocarotenoids in cellular signaling related to cancer: a review. Mol Nutr Food Res 2012;56:259–69.

[87] Palozza P, Parrone N, Catalano A, Simone R. Tomato lycopene and inflammatory cascade: basic interactions and clinical implications. Curr Med Chem 2010;17:2547–63.

[88] Yang TH, Lai YH, Lin TP, Liu WS, Kuan LC, Liu CC. Chronic exposure to Rhodobacter sphaeroides extract Lycogen prevents UVA-induced malondialdehyde accumulation and procollagen I down-regulation in human dermal fibroblasts. Int J Mol Sci 2014;15:1686–99.

[89] Kowalczyk MC, Walaszek Z, Kowalczyk P, Kinjo T, Hanausek M, Slaga TJ. Differential effects of several phytochemicals and their derivatives on murine keratinocytes in vitro and in vivo: implications for skin cancer prevention. Carcinogenesis 2009;30:1008–15.

[90] Leone A, Zefferino R, Longo C, Leo L, Zacheo G. Supercritical CO(2)-extracted tomato Oleoresins enhance gap junction intercellular communications and recover from mercury chloride inhibition in keratinocytes. J Agric Food Chem 2010;58:4769–78.

[91] Rizwan M, Rodriguez-Blanco I, Harbottle A, Birch-Machin MA, Watson RE, Rhodes LE. Tomato paste rich in lycopene protects against cutaneous photodamage in humans in vivo: a randomized controlled trial. Br J Dermatol 2011;164:154–62.

[92] Suganuma K, Nakajima H, Ohtsuki M, Imokawa G. Astaxanthin attenuates the UVA-induced up-regulation of matrix-metalloproteinase-1 and skin fibroblast elastase in human dermal fibroblasts. J Dermatol Sci 2010;58:136–42.

[93] Wertz K, Seifert N, Hunziker PB, Riss G, Wyss A, Lankin C, et al. Beta-carotene inhibits UVA-induced matrix metalloprotease 1 and 10 expression in keratinocytes by a singlet oxygen-dependent mechanism. Free Radic Biol Med 2004;37:654–70.

[94] Darvin M, Patzelt A, Gehse S, Schanzer S, Benderoth C, Sterry W, et al. Cutaneous concentration of lycopene correlates significantly with the roughness of the skin. Eur J Pharm Biopharm 2008;69:943–7.

[95] Aust O, Stahl W, Sies H, Tronnier H, Heinrich U. Supplementation with tomato-based products increases lycopene, phytofluene, and phytoene levels in human serum and protects against UV-light-induced erythema. Int J Vitam Nutr Res 2005;75:54–60.

[96] Kopcke W, Krutmann J. Protection from sunburn with beta-Carotene—a meta-analysis. Photochem Photobiol 2008;84:284–8.

[97] Stahl W, Heinrich U, Wiseman S, Eichler O, Sies H, Tronnier H. Dietary tomato paste protects against ultraviolet light-induced erythema in humans. J Nutr 2001;131:1449–51.

[98] Stahl W, Heinrich U, Jungmann H, Sies H, Tronnier H. Carotenoids and carotenoids plus vitamin E protect against ultraviolet light-induced erythema in humans. Am J Clin Nutr 2000;71:795–8.

[99] Lyons NM, O'Brien NM. Modulatory effects of an algal extract containing astaxanthin on UVA-irradiated cells in culture. J Dermatol Sci 2002;30:73–84.

[100] Palozza P, Serini S, Torsello A, Di Nicuolo F, Maggiano N, Ranelletti FO, et al. Mechanism of activation of caspase cascade during beta-carotene-induced apoptosis in human tumor cells. Nutr Cancer 2003;47:76–87.

[101] Chiang HS, Wu WB, Fang JY, Chen DF, Chen BH, Huang CC, et al. Lycopene inhibits PDGF-BB-induced signaling and migration in human dermal fibroblasts through interaction with PDGF-BB. Life Sci 2007;81:1509–17.

[102] Wu WB, Chiang HS, Fang JY, Hung CF. Inhibitory effect of lycopene on PDGF-BB-induced signalling and migration in human dermal fibroblasts: a possible target for cancer. Biochem Soc Trans 2007;35:1377–8.

[103] Lambert LA, Wamer WG, Wei RR, Lavu S, Chirtel SJ, Kornhauser A. The protective but nonsynergistic effect of dietary beta-carotene and vitamin E on skin tumorigenesis in Skh mice. Nutr Cancer 1994;21:1–12.

[104] Mathews-Roth MM, Krinsky NI. Carotenoid dose level and protection against UV-B induced skin tumors. Photochem Photobiol 1985;42:35–8.

[105] Shen C, Wang S, Shan Y, Liu Z, Fan F, Tao L, et al. Chemomodulatory efficacy of lycopene on antioxidant enzymes and carcinogen-induced cutaneum carcinoma in mice. Food Funct 2014;5:1422–31.

[106] Millen AE, Tucker MA, Hartge P, Halpern A, Elder DE, Guerry D, et al. Diet and melanoma in a case-control study. Cancer Epidemiol Biomarkers Prev 2004;13:1042–51.

[107] Fung TT, Hunter DJ, Spiegelman D, Colditz GA, Speizer FE, Willett WC. Vitamins and carotenoids intake and the risk of basal cell carcinoma of the skin in women (United States). Cancer Causes Control 2002;13:221–30.

[108] Fung TT, Spiegelman D, Egan KM, Giovannucci E, Hunter DJ, Willett WC. Vitamin and carotenoid intake and risk of squamous cell carcinoma of the skin. Int J Cancer 2003;103:110–15.

[109] Lindqvist A, Andersson S. Cell type-specific expression of beta-carotene 15,15′-mono-oxygenase in human tissues. J Histochem Cytochem 2004;52:491–9.

[110] Lindqvist A, He YG, Andersson S. Cell type-specific expression of beta-carotene 9′,10′-monooxygenase in human tissues. J Histochem Cytochem 2005;53:1403–12.

[111] Riboli E, Norat T. Epidemiologic evidence of the protective effect of fruit and vegetables on cancer risk. Am J Clin Nutr 2003;78:559S–69S.

[112] Aune D, Chan DS, Vieira AR, Rosenblatt DA, Vieira R, Greenwood DC, et al. Fruits, vegetables and breast cancer risk: a systematic review and meta-analysis of prospective studies. Breast Cancer Res Treat 2012;134:479–93.

[113] Etminan M, Takkouche B, Caamano-Isorna F. The role of tomato products and lycopene in the prevention of prostate cancer: a meta-analysis of observational studies. Cancer Epidemiol Biomarkers Prev 2004;13:340–5.

[114] Giovannucci E. Tomatoes, tomato-based products, lycopene, and cancer: review of the epidemiologic literature. J Natl Cancer Inst 1999;91:317–31.

[115] Eliassen AH, Hendrickson SJ, Brinton LA, Buring JE, Campos H, Dai Q, et al. Circulating carotenoids and risk of breast cancer: pooled analysis of eight prospective studies. J Natl Cancer Inst 2012;104:1905–16.

[116] van Poppel G, Goldbohm RA. Epidemiologic evidence for beta-carotene and cancer prevention. Am J Clin Nutr 1995;62:1393S–402S.

[117] Aune D, Chan DS, Vieira AR, Navarro Rosenblatt DA, Vieira R, Greenwood DC, et al. Dietary compared with blood concentrations of carotenoids and breast cancer risk: a systematic review and meta-analysis of prospective studies. Am J Clin Nutr 2012;96:356–73.

[118] Giovannucci E. Tomato products, lycopene, and prostate cancer: a review of the epidemiological literature. J Nutr 2005;135:2030S–1S.

[119] Wei MY, Giovannucci EL. Lycopene, tomato products, and prostate cancer incidence: a review and reassessment in the PSA screening era. J Oncol 2012;2012:271063.

[120] Sharoni Y, Giron E, Rise M, Levy J. Effects of lycopene-enriched tomato oleoresin on 7,12-dimethyl-benz[a]anthracene-induced rat mammary tumors. Cancer Detect Prev 1997;21:118–23.

[121] Narisawa T, Fukaura Y, Hasebe M, Ito M, Aizawa R, Murakoshi M, et al. Inhibitory effects of natural carotenoids, alpha-carotene, beta-carotene, lycopene and lutein, on colonic aberrant crypt foci formation in rats. Cancer Lett 1996;107:137–42.

[122] Kim JM, Araki S, Kim DJ, Park CB, Takasuka N, Baba-Toriyama H, et al. Chemopreventive effects of carotenoids and curcumins on mouse colon carcinogenesis after 1,2-dimethylhydrazine initiation. Carcinogenesis 1998;19:81–5.

[123] Prakash P, Russell RM, Krinsky NI. In vitro inhibition of proliferation of estrogen-dependent and estrogen-independent human breast cancer cells treated with carotenoids or retinoids. J Nutr 2001;131:1574–80.

[124] Pastori M, Pfander H, Boscoboinik D, Azzi A. Lycopene in association with alpha-tocopherol inhibits at physiological concentrations proliferation of prostate carcinoma cells. Biochem Biophys Res Commun 1998;250:582–5.

[125] Nahum A, Hirsch K, Danilenko M, Watts CK, Prall OW, Levy J, et al. Lycopene inhibition of cell cycle progression in breast and endometrial cancer cells is associated with reduction in cyclin D levels and retention of p27(Kip1) in the cyclin E-cdk2 complexes. Oncogene 2001;20:3428–36.

[126] Levy J, Bosin E, Feldman B, Giat Y, Miinster A, Danilenko M, et al. Lycopene is a more potent inhibitor of human cancer cell proliferation than either alpha-carotene or beta-carotene. Nutr Cancer 1995;24:257–66.

[127] Amir H, Karas M, Giat J, Danilenko M, Levy R, Yermiahu T, et al. Lycopene and 1,25-dihydroxyvitamin D3 cooperate in the inhibition of cell cycle progression and induction of differentiation in HL-60 leukemic cells. Nutr Cancer 1999;33:105–12.

[128] Kopec RE, Riedl KM, Harrison EH, Curley Jr. RW, Hruszkewycz DP, Clinton SK, et al. Identification and quantification of apo-lycopenals in fruits, vegetables, and human plasma. J Agric Food Chem 2010;58:3290–6.

[129] Palozza P, Sheriff A, Serini S, Boninsegna A, Maggiano N, Ranelletti FO, et al. Lycopene induces apoptosis in immortalized fibroblasts exposed to tobacco smoke condensate through arresting cell cycle and down-regulating cyclin D1, pAKT and pBad. Apoptosis 2005;10:1445–56.

[130] Liu C, Russell RM, Wang XD. Lycopene supplementation prevents smoke-induced changes in p53, p53 phosphorylation, cell proliferation, and apoptosis in the gastric mucosa of ferrets. J Nutr 2006;136:106–11.

[131] Stivala LA, Savio M, Quarta S, Scotti C, Cazzalini O, Rossi L, et al. The antiproliferative effect of beta-carotene requires p21waf1/cip1 in normal human fibroblasts. Eur J Biochem 2000;267:2290–6.

[132] Tripathi DN, Jena GB. Astaxanthin intervention ameliorates cyclophosphamide-induced oxidative stress, DNA damage and early hepatocarcinogenesis in rat: role of Nrf2, p53, p38 and phase-II enzymes. Mutat Res 2010;696:69–80.

[133] Karas M, Amir H, Fishman D, Danilenko M, Segal S, Nahum A, et al. Lycopene interferes with cell cycle progression and insulin-like growth factor I signaling in mammary cancer cells. Nutr Cancer 2000;36:101–11.

[134] Walfisch S, Walfisch Y, Kirilov E, Linde N, Mnitentag H, Agbaria R, et al. Tomato lycopene extract supplementation decreases insulin-like growth factor-I levels in colon cancer patients. Eur J Cancer Prev 2007;16:298–303.

[135] Hix LM, Lockwood SF, Bertram JS. Upregulation of connexin 43 protein expression and increased gap junctional communication by water soluble disodium disuccinate astaxanthin derivatives. Cancer Lett 2004;211:25–37.

[136] Hanusch M, Stahl W, Schulz WA, Sies H. Induction of gap junctional communication by 4-oxoretinoic acid generated from its precursor canthaxanthin. Arch Biochem Biophys 1995;317:423–8.

[137] Aust O, Ale-Agha N, Zhang L, Wollersen H, Sies H, Stahl W. Lycopene oxidation product enhances gap junctional communication. Food Chem Toxicol 2003;41:1399–407.

[138] Ben-Dor A, Steiner M, Gheber L, Danilenko M, Dubi N, Linnewiel K, et al. Carotenoids activate the antioxidant response element transcription system. Mol Cancer Ther 2005;4:177–86.

[139] Wang Y, Ausman LM, Greenberg AS, Russell RM, Wang XD. Dietary lycopene and tomato extract supplementations inhibit nonalcoholic steatohepatitis-promoted hepatocarcinogenesis in rats. Int J Cancer 2010;126:1788–96.

[140] Linnewiel K, Ernst H, Caris-Veyrat C, Ben-Dor A, Kampf A, Salman H, et al. Structure activity relationship of carotenoid derivatives in activation of the electrophile/antioxidant response element transcription system. Free Radic Biol Med 2009;47:659–67.

[141] Palozza P, Serini S, Torsello A, Di Nicuolo F, Piccioni E, Ubaldi V, et al. Beta-carotene regulates NF-kappaB DNA-binding activity by a redox mechanism in human leukemia and colon adenocarcinoma cells. J Nutr 2003;133:381–8.

[142] Huang CS, Fan YE, Lin CY, Hu ML. Lycopene inhibits matrix metalloproteinase-9 expression and down-regulates the binding activity of nuclear factor-kappa B and stimulatory protein-1. J Nutr Biochem 2007;18:449–56.

[143] Kaarbo M, Klokk TI, Saatcioglu F. Androgen signaling and its interactions with other signaling pathways in prostate cancer. Bioessays 2007;29:1227–38.

[144] Cuzick J, Thorat MA, Andriole G, Brawley OW, Brown PH, Culig Z, et al. Prevention and early detection of prostate cancer. Lancet Oncol 2014;15:e484–92.

[145] Siler U, Barella L, Spitzer V, Schnorr J, Lein M, Goralczyk R, et al. Lycopene and vitamin E interfere with autocrine/paracrine loops in the Dunning prostate cancer model. FASEB J 2004;18:1019–21.

[146] Anderson ML. A preliminary investigation of the enzymatic inhibition of 5alpha-reduction and growth of prostatic carcinoma cell line LNCap-FGC by natural astaxanthin and Saw Palmetto lipid extract in vitro. J Herb Pharmacother 2005;5:17–26.

[147] Wan L, Tan HL, Thomas-Ahner JM, Pearl DK, Erdman Jr. JW, Moran NE, et al. Dietary tomato and lycopene impact androgen signaling- and carcinogenesis-related gene expression during early TRAMP prostate carcinogenesis. Cancer Prev Res (Phila) 2014;7:1228–39.

[148] Linnewiel-Hermoni K, Khanin M, Danilenko M, Zango G, Amosi Y, Levy J, et al. The anti-cancer effects of carotenoids and other phytonutrients resides in their combined activity. Arch Biochem Biophys 2015;572:28–35.

[149] Magbanua MJ, Roy R, Sosa EV, Weinberg V, Federman S, Mattie MD, et al. Gene expression and biological pathways in tissue of men with prostate cancer in a randomized clinical trial of lycopene and fish oil supplementation. PLoS One 2011;6:e24004.

[150] Schwarz S, Obermuller-Jevic UC, Hellmis E, Koch W, Jacobi G, Biesalski HK. Lycopene inhibits disease progression in patients with benign prostate hyperplasia. J Nutr 2008;138:49−53.

[151] Henderson BE, Ross R, Bernstein L. Estrogens as a cause of human cancer: the Richard and Hinda Rosenthal Foundation award lecture. Cancer Res 1988;48:246−53.

[152] Hirsch K, Atzmon A, Danilenko M, Levy J, Sharoni Y. Lycopene and other carotenoids inhibit estrogenic activity of 17beta-estradiol and genistein in cancer cells. Breast Cancer Res Treat 2007;104:221−30.

[153] Cui Y, Shikany JM, Liu S, Shagufta Y, Rohan TE. Selected antioxidants and risk of hormone receptor-defined invasive breast cancers among postmenopausal women in the Women's Health Initiative Observational Study. Am J Clin Nutr 2008;87:1009−18.

[154] Iimura Y, Agata U, Takeda S, Kobayashi Y, Yoshida S, Ezawa I, et al. Lycopene intake facilitates the increase of bone mineral density in growing female rats. J Nutr Sci Vitaminol (Tokyo) 2014;60:101−7.

[155] Iimura Y, Agata U, Takeda S, Kobayashi Y, Yoshida S, Ezawa I, et al. The protective effect of lycopene intake on bone loss in ovariectomized rats. J Bone Miner Metab 2015;33:270−8.

[156] Mackinnon ES, Rao AV, Josse RG, Rao LG. Supplementation with the antioxidant lycopene significantly decreases oxidative stress parameters and the bone resorption marker N-telopeptide of type I collagen in postmenopausal women. Osteoporos Int 2011;22:1091−101.

[157] Sahni S, Hannan MT, Blumberg J, Cupples LA, Kiel DP, Tucker KL. Protective effect of total carotenoid and lycopene intake on the risk of hip fracture: a 17-year follow-up from the Framingham Osteoporosis Study. J Bone Miner Res 2009;24:1086−94.

CHAPTER

35

Mechanisms of Action of Curcumin on Aging: Nutritional and Pharmacological Applications

Ana C. Carvalho[1,2], Andreia C. Gomes[2], Cristina Pereira-Wilson[1] and Cristovao F. Lima[1]

[1]Department of Biology, CITAB - Centre for the Research and Technology of Agro-Environmental and Biological Sciences, University of Minho, Braga [2]Department of Biology, CBMA - Centre of Molecular and Environmental Biology, University of Minho, Braga, Portugal

KEY FACTS

- The ancestral use of turmeric in traditional medicine and the extensive research over the last decades suggest that curcumin can be exploited for future nutritional and pharmacological interventions.
- The peculiar chemical structure of curcumin is behind its diverse biological activities and ability to interact and regulate multiple molecular targets.
- The effects of curcumin in aging and longevity are related with its multitargeting capacity.
- Intense research is underway to envisage the future application of curcumin in the improvement of human healthspan and longevity.

Dictionary of Terms

- *Curry powder*: Curry powder is a blend of up to 30 (average 10) different spices and herbs; turmeric is one of the basic components, which gives curry its characteristic golden color.
- *Functional food*: A natural or processed food that contains biologically active components that provide health benefits beyond that of basic nutrients it contains.
- *Apoptosis*: A type of programmed cell death in which an active sequence of events regulated by different groups of executioner and regulatory molecules leads to the self-destruction of the cell without releasing intracellular constituents to intercellular space.
- *Glutathione*: A tripeptide of glycine, cysteine, and glutamic acid, existing in reduced (GSH) and oxidized (GSSG; glutathione disulfide) forms, which is an important intracellular antioxidant and cofactor of different enzymes that has numerous roles in protecting cells from oxidants and in maintaining cellular thiol-disulfide redox state.
- *Hormesis*: In the biology/medicine field, hormesis is a process in which exposure to a low dose or moderate stressor (that is damaging at higher concentrations) elicits adaptive beneficial responses, which may result in health promotion effects.
- *Epigenome*: Potentially heritable chemical modifications to the DNA and associated structures (eg, histone proteins) that can result in changes of genetic expression independent of the DNA sequence of a gene. Unlike the genome, the

epigenome can be modified by environmental factors including dietary constituents.
- *Proteasome*: Is a multicomponent enzymatic system incorporating different regulators and a catalytic core with different proteases responsible for the degradation of a large portion of soluble intracellular proteins. The proteasome is an important component of the intracellular system for the turnover of proteins.
- *Autophagy*: A catabolic process of cellular self-digestion in which proteins and organelles are engulfed by double-membrane autophagosomes and degraded in lysosomes by proteases, leading to recycling of macromolecular constituents.
- *Mammalian target of rapamycin (mTOR)*: Evolutionarily conserved serine/threonine protein kinase that regulates protein synthesis and degradation processes important for cell growth, proliferation, motility and survival. mTOR is a key autophagic regulator, whose inhibition activates autophagy.
- *Tetrahydrocurcumin (THC)*: A metabolite of curcumin that retains many effects of curcumin, such as the antioxidant and anti-inflammatory activities. Curcumin is metabolized to THC after oral ingestion by reductases found in the intestinal epithelium, and the structures vary only by the lack of the double bonds in the seven carbon linker of curcumin.
- *Nanoparticle*: A small particle generally between 1 and 100 nanometers in size. In biomedical research, several types of nanoparticles are being developed to carry and deliver compounds to particular biological targets.

INTRODUCTION

The polyphenol curcumin (diferuloylmethane) is an active compound found in the rhizome of *Curcuma longa*, which has been used since ancient times as a spice, coloring ingredient, and component in traditional medicine. Native to tropical Asia, *C. longa* is used in these countries, from India to China, to treat several ailments including several inflammatory symptoms, respiratory conditions, sinusitis, and abdominal pain. Extensive research over the last decades has increased remarkably our knowledge about curcumin. Indeed, more than 7000 articles are listed in the US National Institutes of Health PubMed database (consulted in March 2015) [1], most of them being published in the last 10 years (Fig. 35.1A). The data gathered in these studies confirmed many of the attributed biological effects of *C. longa*, and of curcumin in particular, such as their antiseptic, anti-inflammatory, and wound healing ability. The exponential growth of biomedical research with curcumin was also observed in the area of aging (Fig. 35.1B). In fact, in recent years, the effects of curcumin on the aging process and on several age-related diseases, including cancer, diabetes, cardiovascular, and neurodegenerative diseases, have been investigated with the aim for their treatment and/or prevention.

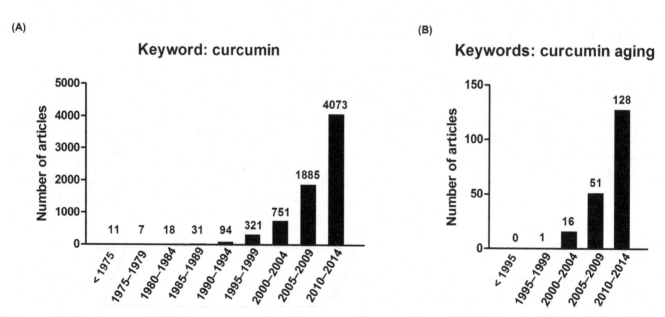

FIGURE 35.1 Number of publications found in the US National Institutes of Health PubMed database [1] using the keyword "curcumin" (A) or the keywords "curcumin, aging" (B).

In this chapter we will review the current knowledge on the effects of curcumin in the biology of aging, mainly the effects in senescence, the lifespan of experimental organisms, and the potential therapeutic benefits in age-related diseases. The impact of curcumin in several molecular targets related with hallmarks of aging will also be reviewed. The research in progress, intended to increase the bioavailability of curcumin and its impact in the current and future nutritional and pharmacological applications for improving healthspan and longevity, will be extensively discussed.

CURCUMIN AND ITS TRADITIONAL USES

Curcumin is the main curcuminoid found in the rhizome of the perennial plant *C. longa* Linn, a member of the ginger family (Zingiberaceae). *Curcuma longa* is indigenous to South and Southeast Asia but is nowadays widely cultivated in tropical areas of Asia and Central America. Curcuminoids, including curcumin, demethoxycurcumin, and bisdemethoxycurcumin, have also been isolated from *Curcuma mangga*, *Curcuma zedoaria*, *Costus speciosus*, *Curcuma xanthorrhiza*, *Curcuma aromatica*, *Curcuma phaeocaulis*, *Etlingera elatior*, and *Zingiber cassumunar* [2]. Together they are responsible for the yellow color of turmeric (the powder that results from the ground dried rhizome of *C. longa*), which is commonly used as a spice and in phytomedicine [2,3]. Curcumin (1,7-bis-(4-hydroxy-3-methoxyphenyl)-1,6-heptadiene-3,5-dione) corresponds to about 2–5% of turmeric and was first isolated in 1815 by Vogel and Pelletier, with its chemical structure determined by Milobedzka and colleagues in 1910 (Fig. 35.2) [4]. Curcumin is a yellow-orange hydrophobic compound insoluble in water and ether but soluble in dimethylsulfoxide, acetone, ethanol, and oils. It generically consists of two ferulic acid residues joined by a methylene bridge, and exhibit two main tautomeric forms (Fig. 35.2): the keto-enol form (also known as enol form) and the diketo form (also known as keto form). The enol form of curcumin is the predominant in most solvents and has important implications in the activity

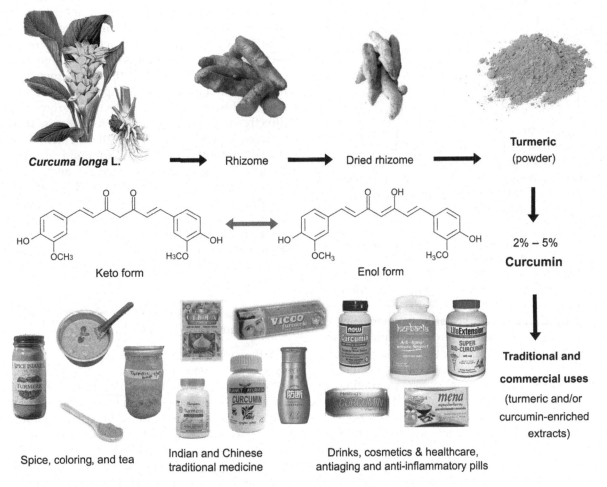

FIGURE 35.2 From *Curcuma longa* to turmeric to curcumin: its traditional and commercial uses.

of this polyphenol, such as metal quelation and interaction with proteins [5,6]. Recently it was shown that in water containing mixtures, the keto tautomeric form of curcumin increases and dominates with a high percentage of water [6].

Turmeric has been used for thousand years as a dietary spice, food preservative, and as a coloring agent of food and other materials, such as cosmetics, textile fibers, and paper [2,3,7]. Turmeric is frequently used in Asian cooking, particularly in India, Pakistan, and Thailand, to improve the palatability and presentation of food preparations. It is one of the ingredients of curry powder, and gives it its distinctive yellowish color and flavor [2,3,8]. In fact, curcumin is an approved natural food coloring (E100) additive that is used at low concentrations in butter, canned fish, cheese, mustard, pastries, and other foods [9]. Turmeric is also used in tea preparations, particularly by the Japanese Okinawa population [8]. Turmeric also has been included for centuries in preparations of traditional Indian and Chinese medicine to treat various conditions such as allergy, anorexia, asthma, biliary disorders, bronchial hyperactivity, cough, diabetic wounds, liver disorders, rheumatism, rhinitis, sinusitis, sprains, and swelling [10]. In the old Hindu texts of traditional medicine of Ayurveda (meaning knowledge of long life), turmeric is recommended for its aromatic, stimulant, and carminative properties [11]. In traditional Chinese medicine, it is used to treat diseases associated with abdominal pain [2], and as a major constituent of Jiawei-Xiaoyao-san medicinal formula, *C. longa* has been used to manage dyspepsia, stress, and depression/mood-related ailments [12]. In addition, in the Indian subcontinent, turmeric mixed with slaked lime has been used topically as a household remedy for the treatment of wounds, inflammation, burns, and skin diseases [11,13].

The importance of this medicinal plant as a spice and in folk medicine, not only in Asian countries but also in the western world, prompted the growing research with turmeric extracts and their main constituents over the past 30 years (see Fig. 35.1). Many of the biological activities attributed to turmeric were confirmed by experimental scientific studies, such as its antimicrobial, antioxidant, and anti-inflammatory properties [2,8,14,15]. Curcumin has been identified as one of the main active compounds responsible for turmeric's effects. For example, the promotion of wound healing by turmeric was confirmed for curcumin from preclinical to intervention studies in humans [2,16–18]. The pleiotropic and multitargeting capability of curcumin combined with its apparently safe profile identified it as a potential therapeutic phytochemical against cancer, lung and liver diseases, aging, neurological diseases, metabolic diseases, and cardiovascular diseases [7,12]. The public interest in the wide range of effects of curcumin paved the way to its commercial use worldwide in several products (Fig. 35.2) including antiaging and immunomodulatory pills, *functional foods*, soaps, and cosmetics [12,15].

BIOCHEMICAL AND MOLECULAR TARGETS OF CURCUMIN: AN OVERVIEW

Extensive research over the last decades has revealed that curcumin has antioxidant, antibacterial, antifungal, antiviral, anti-inflammatory, antiproliferative and proapoptotic effects [2]. How a single compound can exhibit all these effects is the subject of intense investigation [19]. Accumulating evidence suggests that the pleiotropic effects of curcumin are dependent on its ability to interact and regulate multiple biological molecular targets. Mechanistic investigations attribute the diverse biological activities to the peculiar chemical structure of curcumin, in particular the two aromatic O-methoxy phenol groups, the α, β-unsaturated β-diketo moiety and the seven carbon linker [5]. These chemical structural features allow curcumin to possess antioxidant activity and to interact directly with different biomolecules through noncovalent and covalent binding [5,20]. Apart from the direct binding of curcumin to key components of cellular signaling pathways, various molecular targets are also indirectly modulated resulting in either increased or decreased expression/activity [20]. The main biochemical and molecular targets of curcumin include transcription factors, growth factors and receptors, inflammatory cytokines, enzymes and protein kinases, adhesion molecules, cell cycle and *apoptosis*-related proteins, among others (Table 35.1). Effects of curcumin on these targets were recently reviewed by different authors, for example, see Gupta et al. [20], Zhou et al. [21], Shishodia [7], Prasad et al. [4], and Shanmugam et al. [22].

Aging-Related Molecular Targets

At the cellular level, aging is characterized by the progressive accumulation of molecular damage, leading to impaired function and increased vulnerability to cell death [23–25] (see also chapter: "Molecular and Cellular Basis of Aging" for details). The main causes of aging are attributed to the deterioration and failure of genetic pathways and biochemical processes conserved in evolution involved in the cellular maintenance and repair mechanisms [23]. The loss of physiological integrity during aging is the primary risk

TABLE 35.1 Different Known Molecular Targets of Curcumin

Molecular targets of curcumin
Transcription factors
AHR, AP-1, ATF3, β-catenin, C/EBP, CHOP, CTCF, EGR1, EpRE, HIF1, HSF1, NF-κB, NOTCH1, Nrf2, PPAR-γ, STAT1, 3, 4, 5, WT1
Growth factors and receptors
AR, CTGF, CXCR1, 2, 4, EGF, EGFR, ESR, FasR, FGF, HGF, HRH2, INSR, ITPR, LDLR, NGF, PDGF, TF, TGF-β1, VEGF
Inflammatory cytokines
IFN-γ, IL-1, 2, 5, 6, 8, 12, 18, MIP-1, 2, MCP-1, TNF-α
Enzymes and protein kinases
5-LOX, ATPase, CDPK, COX-2, FAK, GCL, GST, HO-1, INOS, IRAK, JAK, MAPK, MMP, MTOR, NQO1, ODC, PHK, PKA, PKB/AKT, PKC, SIRT1, SRC, SYK
Adhesion molecules
ELAM-1, ICAM-1, VCAM-1
Cell cycle and apoptosis-related proteins
Cyclin D1, p53, BCL2, BCL-XL, IAP, BAX, BAK1, CASP3, 7, 8, 9, DR4, 5

Adapted from Zhou et al. [21], Shishodia [7], and Prasad et al. [4].

factor for major human pathologies, including cancer, diabetes, cardiovascular disorders, and neurodegenerative diseases [25]. Recently, Lopez-Otin et al. proposed nine cellular and molecular hallmarks of aging, which are genomic instability, telomere attrition, epigenetic alterations, proteostasis imbalance, deregulated nutrient sensing, mitochondrial dysfunction, cellular senescence, stem cell exhaustion, and altered intercellular communication [25]. These hallmarks that characterize the aging process aim to help design further studies on the molecular mechanisms of aging and on the development of interventions that will improve human healthspan [25]. In the following subsections we review the effects of curcumin on biochemical and molecular targets that impact on some of the hallmarks of aging (Fig. 35.3).

Genomic Instability

The genome is constantly being challenged by endogenous and exogenous threats that result in DNA damage and genomic instability [26]. A highly conserved and complex DNA repair machinery is usually able to maintain the integrity of the genome [25]. However, excessive DNA damage or insufficient DNA repair favors the aging process. The accumulation of genetic damage and alterations during aging inevitably also associate aging and cancer [27].

In several experimental settings, curcumin showed a higher antioxidant activity that can be useful to protect DNA and other biomolecules from oxidative damage. The direct antioxidant activity of curcumin is dependent on its ability to work as a free radical scavenger and to chelate transition metal ions. The first activity is mainly dependent on the electron donation of the phenolic OH groups, whereas the latter is dependent on the β-diketo group [5]. Curcumin also possesses indirect antioxidant activities by upregulating several antioxidant and cytoprotective enzymes [28–30]. This effect is known to be mediated by Nrf2, a basic leucine zipper (bZIP) transcription factor, that upon activation accumulates in the nucleus and, in heterodimeric combination with small Maf transcription factors, binds to ARE and recruits the basal transcriptional machinery to activate the transcription of genes encoding stress-responsive and cytoprotective enzymes and related proteins [31,32]. Under normal conditions, the Kelch-like ECH-associated protein (Keap1) forms a complex with cullin3 (Cul3) and represses Nrf2 by presenting it for ubiquitination and proteasomal degradation. Upon stimulation with curcumin, the cysteine residues of Keap1 are modified resulting in conformational changes that eliminate the capacity of Keap1 to repress Nrf2 [31,32]. Curcumin contains electrophilic α, β-unsaturated carbonyl groups (β-diketo moiety) that can react selectively with nucleophiles such as thiols, leading to formation of Michael adducts [29]. This Michael addition ability of curcumin allows its covalent binding to nucleophilic cysteine sulfhydryls and the selenocysteine moiety, affecting the activity of several proteins/enzymes [5,20], including the inhibition of Keap1. Curcumin was also shown to interact with glutathione (GSH) forming glutathionylated products [33]. Moreover, the induction of antioxidant defenses by curcumin was associated with a

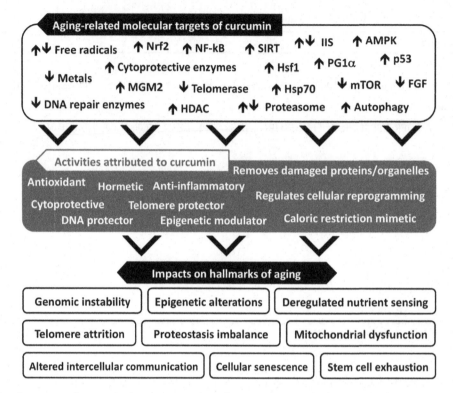

FIGURE 35.3 Molecular targets of curcumin that impact on the hallmarks of aging, with the associated attributed activities. Arrows in the first box indicate increase/activation (↑) or decrease/inhibition (↓) of the target by curcumin.

transient decrease of GSH levels that impact on cellular redox state [30,34]. This transient mild impairment of thiol-disulfide redox state by curcumin may also influence indirectly redox signaling through activation of Nrf2, resulting in a compensatory antioxidant response in cells, including the increase in GSH synthesis to restore cellular redox state [30]. Both the direct and indirect antioxidant effects of curcumin are thus regarded as useful in the slow down of aging and prevention of age-related diseases that have been associated with the deleterious effects of free radicals and oxidative stress, such as cancer, neurodegenerative disorders, and cardiovascular diseases [35].

The O^6-methylguanine-DNA methyltransferase (MGMT), a DNA repair protein that protects the cellular genome and critical oncogenic genes from the mutagenic action of endogenous and exogenous alkylating agents, was also shown to be induced by curcumin, probably by the ability of this polyphenol to increase the availability of cysteine [36]. Higher content of cysteine drives the synthesis of cysteine-rich and cysteine-sufficient proteins including MGMT [36].

Curcumin at high concentrations, however, is also known to induce reactive oxygen species (ROS), DNA damage, and cytotoxicity, which are linked to one of its anticancer mechanisms [37]. Curcumin is also known to inhibit several DNA repair enzymes and mechanisms [37–39]. However, when curcumin induces mild stress and DNA damage to normal human skin fibroblasts, cells recover and repair DNA in a few hours [30]. It is possible that in this case *hormesis* is involved [40]. According to this concept of intervention, mild toxic treatments may generate a beneficial compensatory response by increasing the maintenance and repair mechanisms of the cells [41,42]. Thus, although higher concentrations of curcumin are deleterious, which may be useful for cancer treatment, lower doses may slow down aging and be chemopreventive by inducing a hormetic response. Interestingly, in a human intervention study, the ability of curcumin to prevent DNA damage and to enhance the repair potential was shown in a human population in West Bengal (India) chronically exposed to arsenic [43].

Telomere Attrition

Telomeres are specialized nucleoprotein structures on the extremities of eukaryotic chromosomes that protect them and are particularly implicated in the aging process [44]. Gradual telomere shortening (attrition) in normal somatic cells during consecutive rounds of replication leads to critically short telomeres that induce replicative senescence, the irreversible loss of division potential of somatic cells [45,46]. In addition, it was shown that telomeres are a preferential target of genotoxic stress and ROS-induced DNA damage, which has important consequences for the aging process [44]. The antioxidant actions of curcumin may therefore have important implications in healthy aging by preventing oxidative damage to telomeres.

Telomerase, a reverse transcriptase that maintains telomere length, is practically absent in normal somatic cells and highly activated in most tumor cells enabling their replicative immortality [47,48]. In recent years, telomerase has been proposed as a potential target for cancer therapy and thus extensive investigations have been carried out in the search for compounds capable of inhibiting telomerase [49]. Several studies have demonstrated that curcumin inhibits telomerase activity in a dose- and time-dependent manner by suppressing the translocation of human telomerase reverse transcriptase (hTERT) from the cytosol to the nucleus [49] and by decreasing hTERT expression [50,51].

Both telomeric and nontelomeric DNA damage has been shown to induce and stabilize senescence contributing to the aging phenotype [52]. Recently, it was shown that systemic chronic inflammation in mice accelerates aging via ROS-mediated exacerbation of telomere dysfunction and cell senescence [53]. The preferential accumulation of telomere-dysfunctional senescent cells in this mouse model of chronic inflammation was blocked by anti-inflammatory or antioxidant treatments [53]. Therefore, the known anti-inflammatory action of curcumin (see Ref. [21] for review) may help to prevent telomere attrition contributing to the potential promotion of healthspan by this phytochemical. Many age-related diseases as well as normal and pathological aging have been for some time associated with chronic low-grade inflammation [54,55]. In fact, when the expression of an NF-κB inhibitor was activated in the aged skin of transgenic mice this tissue was rejuvenated and that was accompanied by the restoration of the transcriptional signature associated to young age [56]. In addition, genetic and pharmacological inhibition of NF-κB signaling was shown to prevent age-associated parameters in different mouse models of accelerated aging [57,58]. Interestingly, Zhang and colleagues showed that mice with blocked NF-κB activation live longer than untreated mice [59]. Therefore, the global imbalance between the lifelong inflammatory processes and the anti-inflammatory networks is proposed to be a major driving force for frailty and common age-related pathologies [54]. Thus, curcumin's action as an anti-inflammatory agent and an efficient inhibitor of NF-κB suggests its potential contribution to the promotion of health aging [60]. In particular, the inhibition of the NF-κB transcriptional activity by curcumin suppresses expression of various cell survival and proliferative genes and consequently leads to cell cycle arrest, inhibition of proliferation, and induction of apoptosis, all of which may have positive implications in the combat against cancer [21].

The anti-inflammatory action of curcumin may also influence another hallmark of aging—altered intercellular communication. Such alterations during aging have been frequently associated with pro-inflammatory signals that result, for example, from the secretome of senescent cells, dysfunctional immune system, tissue damage, and enhanced NF-κB activation [25,52]. The anti-inflammatory activity of curcumin has been associated to its ability to inhibit various pro-inflammatory cytokines, including interferon-γ (IFN-γ), interleukin (IL)-1, 2, 5, 6, 8, 12 and 18, macrophage inflammatory protein (MIP)-1 and 2, monocyte chemoattractant protein (MCP)-1, and tumor necrosis factor alpha (TNF-α) (reviewed in Refs [21] and [7]).

Epigenetic Alterations

Epigenetic alterations, including changes in DNA methylation, histone modifications and chromatin remodeling, are associated with aging and several age-related diseases [61]. Epigenetic modulation constitutes an important mechanism by which dietary components can selectively activate or inactivate gene expression and exert their biological activities [62,63]. Recent evidence has shown that curcumin possesses different epigenetic effects including the modulation of histone deacetylases (HDAC 1, 3, 8 and SIRT1, 7), histone acetyltransferases (p300/CBP), DNA methyltransferase (DNMT1, 2), and several miRNAs [63–66]. Therefore, the exploitation of curcumin as a regulator of the epigenome holds great promise for promotion of healthy aging and prevention of diseases. Deserving special attention are the effects of curcumin on sirtuins—NAD^+-dependent HDACs. Sirtuins and their functions in aging have been the subject of intense research because increased expression of sirtuins has been shown to have beneficial effects on aging or even increase the lifespan in different model organisms [25,67]. In view of the increased longevity afforded by caloric restriction in several laboratory organisms, in particular through activation of sirtuins and other epigenetic effects, mounting evidence is being gathered to show that many phytochemicals might have health benefits by mimicking caloric restriction [66]. This can also be the case for curcumin.

Proteostasis Imbalance

Recent studies show that aging and diverse age-related pathologies, such as diabetes, and Alzheimer's and Parkinson's diseases, are associated with impaired protein homeostasis—proteostasis [68,69]. Proteostasis involves mechanisms of protein stabilization by molecular chaperones, including the heat shock family of proteins (Hsp), and mechanisms of protein degradation by the lysosome and the proteasome [68,70]. These mechanisms function in a coordinated manner to restore the structure or to remove and degrade misfolded proteins, thus preventing the accumulation of damaged components and assuring the continuous renewal of intracellular proteins [25]. Several studies demonstrated that proteostasis is altered with aging [68] and that its perturbation is detrimental to aging whereas genetic manipulations that improve proteostasis have been shown to delay aging in mammals [25].

Curcumin has been shown to modulate several effectors of proteostasis that may have a profound impact in improving healthy aging. In particular, curcumin was able to stimulate the heat stress-induced expression of Hsp70 [71]. Curcumin alone was also reported to induce the heat shock response, in particular causing the nuclear translocation of the heat shock transcription factor 1 and increasing the expression of Hsp70 at transcriptional and translational levels [72–75]. Several studies indicate that curcumin inhibits the proteasome, usually at higher concentrations and linked with induction of cell death in cancer cells [74,76,77]. On the other hand, others have shown the opposite effect with lower curcumin concentrations [78]. This is consistent with the principle of hormesis and may have positive implications for longevity. Contrarily to reports identifying the anticancer effects of curcumin through proteasome inhibition, other studies demonstrated its anticancer potential by inducing the proteasome to promote the degradation of some oncogenic and angiogenic proteins [74,79]. Curcumin has also been shown to efficiently induce autophagy in numerous studies [74,80,81], and also because of that, curcumin has been classified as a caloric restriction mimetic [82]. The induction of autophagy by curcumin can, however, be a double-edged sword in cancer treatment. Whereas chronic induction may lead to autophagic cancer cell death, moderate induction may promote autophagic survival in nutrient-limiting and low-oxygen conditions characteristic of internal regions of tumors [83,84].

Deregulated Nutrient Sensing

Lifespan is regulated by highly conserved nutrient sensing pathways that are controlled by insulin/insulin-like growth factor 1 (IGF1) signaling (IIS; that participates in glucose sensing), mammalian target of rapamycin (mTOR; related to amino acid sensing), and AMP-activated protein kinase (AMPK) and sirtuins that sense low-energy states (by detecting high levels of AMP and NAD^+, respectively) [25,85,86]. Several studies have shown that increasing or restricting dietary intake affects the aging process and the onset of several age-related diseases. High nutrient intake shortens lifespan and accelerates age-associated disorders, while moderate nutrient intake extends lifespan and delays (or attenuates) age-related diseases [86,87]. Moreover, dietary restriction increases lifespan in all investigated model organisms, supporting the idea that deregulated nutrient sensing is a relevant characteristic of aging [88].

The effects of curcumin on IIS have been controversial, since some studies reported that curcumin activates this pathway when stimulated by mitogenic factors [30,89–91], whereas others report the opposite [92–94]. Some of these later studies link Akt inhibition with decreased mTOR signaling and eventually induction of autophagy by curcumin [92,93]. Other authors observed that mTOR inhibition by curcumin was independent of Akt [89,95]. The increased sensitivity to insulin conferred by curcumin has been linked with its beneficial effects on diabetes [90,96]. AMPK activation is also being described as a mechanism for the health beneficial effects of curcumin [90,97,98]. In fact, this particular effect of curcumin was associated with neuroprotection and improvement of aging-related cerebrovascular dysfunction in mice [99,100]. Overall, the regulation of nutrient sensing pathways by curcumin mimics a state of limited nutrient availability that is known to extend longevity, and therefore the connection between this polyphenol and the potential promotion of human healthspan is not surprising.

Mitochondrial Dysfunction

Mitochondrial dysfunction has a profound impact on the aging process and contributes to multiple aging-associated pathologies [25]. The progressive mitochondrial dysfunction that occurs with aging results in decreased ATP generation and increased ROS production, which in turn causes further mitochondrial and global cellular damage [101]. Some studies demonstrated that mitochondrial dysfunction could in fact accelerate aging in mammals [25]. Curcumin, through the described antioxidant effects, and induction of Nrf2, autophagy and sirtuins, might in fact control mitochondria function conferring protection against aging and age-associated diseases. For example, activation of SIRT1 by curcumin [102,103] may impact on mitochondrial biogenesis through the involvement of the peroxisome proliferator-activated receptor-γ (PPAR-γ), coactivator-1α (PGC-1α), and the removal of damaged mitochondria by autophagy [25]. In fact, curcumin was shown to induce the transcriptional coactivator PGC-1α [104–106] and to induce autophagy (see above).

Cellular Senescence

Cellular senescence refers to the state of irreversible growth arrest that occurs due to DNA damage, mitogenic and oncogentic stress, and/or epigenomic perturbations that derepress the INK4/ARF locus [107,108]. Although senescence has an important underlying tumor suppressive role, it is also involved in aging and related pathologies by contributing to the age-related loss of tissue renewal and function [52]. Senescence can, however, in the long run, have adverse effects promoting cancer growth by virtue of the senescence-associated secretory phenotype (SASP), characterized by the release of ROS and an array of pro-inflammatory molecules [107]. Senescence is usually associated with the activation of the $p16^{INK4a}$/pRb and p53/p21 pathways that drive the irreversible growth arrest [107,108]. Due to its antioxidant, anti-inflammatory, and DNA repair promoting activity, curcumin may help to restrain the increase in senescent cells during aging and thereby help control the impact of the senescence phenotype in the surrounding tissue microenvironment.

Curcumin has also been shown to induce p53 [109–111]. This transcriptional factor plays a central role in maintaining genome stability and integrates and responds to a multitude of stresses to exert its function in tumor suppression, such as by inducing cell cycle arrest, apoptosis, and/or senescence [112]. P53 is also an important player in the regulation of aging and longevity: it may accelerate aging in case of severe and chronic activation of p53, but its low grade and regulated activation may extend lifespan [25,113]. Therefore, curcumin may confer hormetic health benefits by moderately inducing p53 in a transient and regulated manner, thus preventing cancer and promoting healthy aging. At higher doses and in continuous administration, curcumin is reported to be cytostatic and to induce senescence in normal cells [114] (and our own unpublished data). The senescence induction by curcumin at higher concentrations may be useful in cancer treatment [110,111,115].

Stem Cell Exhaustion

Although the beneficial compensatory response of activation of p53 and INK4a/ARF aims to avoid the propagation of damaged and potentially oncogenic cells and its consequences on aging and cancer, under persistent activation of these senescence pathways the regenerative capacity of progenitor cells can be exhausted or saturated [25]. Under these extreme conditions, p53 and INK4a/ARF responses can become deleterious and accelerate aging which, in conjugation with the perturbations of the immunological system during aging, eventually results in the accumulation of senescent cells and consequent loss of the tissue's normal functions [25,107]. Like rapamycin that enhances the generation of mouse induced pluripotent stem cells and restoration of self-renewal in hematopoietic stem cells of aged mice by mTORC1 inhibition [116,117], curcumin was also shown to have some capacity to regulate cellular reprogramming [117], which may counteract stem cell exhaustion that is associated with aging.

Recently, loss of quiescence in the aged muscle stem cell niche of mice was also shown to be due to increased fibroblast growth factor 2 (FGF2) signaling [118]. This eventually results in stem cell depletion and diminished regenerative capacity [25,118]. Therefore, the recognized inhibition of FGF signaling by curcumin, which is associated with angiogenesis and tumorigenesis [7,25,119,120], may also be a promising approach to reduce stem cell exhaustion during aging.

CURCUMIN AND AGING

Curcumin has been viewed as a promising phytochemical in future interventions for improving human healthspan and longevity. Although there are no epidemiologic studies with curcumin intake being the sole variable, several investigations with animal models have associated it with the prevention and treatment of age-related diseases and the potential to delay the aging process [121].

Curcumin in Aging and Longevity

Over the past years, experiments in model organisms demonstrated that dietary curcumin can prolong lifespan by inducing cellular stress response pathways that are controlled by highly conserved signaling mechanisms, including AMPK, IIS, mTOR, and sirtuin pathways [85,86]. Until recently, different studies demonstrated that curcumin and its metabolite, tetrahydrocurcumin (THC), increase mean lifespan of three model organisms useful in identifying compounds that can prolong the lifespan of humans—*Caenorhabditis elegans*, *Drosophila melanogaster*, and *Mus musculus* [122–128]. These effects on increased lifespan were, however, not observed in yeast [129].

The roundworm, *C. elegans*, and the fruit fly, *D. melanogaster*, are popular models for studying aging and longevity because of their short lifespan, rapid generation time and well-defined genetics [130]. In 2011, curcumin was shown to prolong in about 45% the mean lifespan of the nematode *C. elegans* through a mechanism that involves the regulation of protein homeostasis [131]. In the same year, Liao and colleagues reported an increase in mean and maximal lifespan of *C. elegans* when the synchronized L1 larvae were allowed to develop to adulthood in the presence of curcumin [123]. This lifespan extension was associated with decreased levels of ROS and age-related lipofuscin content, and increased resistance to heat stress. Several mutant strains in selected stress- and lifespan-relevant genes did not show prolonged lifespan suggesting their gene products were required for curcumin-mediated increased longevity in C. elegans, unraveling the involvement of diverse modes of action and signaling pathways [123].

Several studies have also investigated the effects of curcumin on the lifespan and aging of D. melanogaster. Suckow and Suckow reported an increase of 12 days (18% increase) in the mean lifespan of wild-type fruit flies maintained on media containing curcumin 1 mg/mL [122]. The authors suggested that the effect of curcumin was mediated by induction of superoxide dismutase (SOD) activity. Lee et al. reported gender- and genotype-specific lifespan extension in D. melanogaster: by 19% in females of Canton-S strain exposed to 100 mM curcumin, but not in males; by 16% in males of Ives strain given 250 mM curcumin, but not in females [124]. The increase in lifespan of Canton-S flies by curcumin was associated with increased expression of several aging and stress-associated genes [124]. Other authors also confirmed the ability of curcumin to increase the lifespan in fruit flies [125,128,132]. Shen and colleagues confirmed that curcumin increases SOD and that was associated with decreased lipid peroxidation [125]. Two independent studies reported no negative effects on flies' fecundity by curcumin [128,132]. Turmeric was also shown to increase mean and maximal lifespan of D. melanogaster [133]. The THC metabolite of curcumin was also shown to extend mean lifespan in both male and female flies and to inhibit the oxidative stress response by regulating the evolutionarily conserved signaling pathways foxo and Sir2 [127]. Although curcumin did not cause lifespan extension in the mammalian mouse model [134], THC supplementation did in male C57BL/6 mice as reported by Kitani and colleagues [126].

Curcumin has also been studied in normal human cells, a model used to study the molecular basis of cellular aging by serial cell subcultivation, which undergo progressive aging and culminate in irreversible growth arrest (replicative senescence) [135]. Our group demonstrated that curcumin induces stress responses in normal human skin fibroblasts through induction of redox stress and associated with activation of Nrf2 signaling [30]. This effect resulted in the induction of several antioxidant and cytoprotective enzymes as well as in the increase of GSH content, effects that were substantially impaired in late passage senescent cells. This hormetic response by curcumin conferred significant protection against a further oxidant challenge [30]. The ability of curcumin to induce Nrf2 might have important implications in slowing down the aging process. Several studies demonstrated that the expression of Nrf2 and its target genes decline during aging and disease [136–139]. In fact, the decline of Nrf2 transcriptional activity causes the age-related loss of glutathione synthesis, which adversely affects cellular thiol redox balance, leaving cells highly susceptible to different stresses [136]. In addition, the ortholog of mammalian Nrf2 has been associated with oxidative stress tolerance and aging modulation in Drosophila [139–141]. Therefore, modulation of Nrf2 signaling pathway by curcumin may positively regulate healthspan by hormesis. In fact, stimulation of maintenance and repair mechanisms by repeated exposure to mild stressors has been recognized as a promising strategy to prevent the age-related accumulation of molecular damage [23,24]. Many lines of evidence, including our own studies and those described in the previous section on curcumin's molecular targets, indicate that it acts as a hormetin. From the point of view of aging prevention and healthspan, curcumin has positive effects at lower concentrations but is detrimental at higher concentrations. This hormetic response elicited by curcumin was also observed in most studies with model organisms cited earlier. The higher tested doses did not cause lifespan extension, and only mild and appropriate doses of curcumin were effective as compared with untreated controls (see for example Refs [123] and [128]).

Considering the beneficial and hormetic effects of curcumin in human normal cells, we recently investigated whether curcumin would protect human skin fibroblasts from undergoing replicative senescence in vitro. For that, middle passage cells were grown either continuously (except in the day for cell attachment after subcultivation) or intermittently (two times a week for 3 h with 5 μM of curcumin) in the presence of curcumin. No positive effects were observed and, in fact, curcumin was even detrimental inducing the stoppage of cell division and the appearance of morphological and physiological features typical of replicative senescence. In the continuous treatment with curcumin, among the concentrations tested, only the lower one (1 μM) did not affect the growth curve (the cumulative population doublings). Therefore, contrary to the positive effects with carnosol that we recently reported [142], curcumin at the tested conditions was not capable of ameliorating the physiological state of cells during replicative senescence. The fact that curcumin decreases GSH levels in the initial hours of incubation [30] as part of its hormetic response may explain the unfavorable impact in the growth of human fibroblasts, an hypothesis raised by Satoh and collaborators [143]. The ability of curcumin to induce senescence was also recently shown in primary human cells (vascular smooth muscle and endothelial cells derived from aorta) [114]. We also suggested previously the possible induction of premature senescence in normal cells by high concentrations of curcumin due to chronic induction of oxidative stress and Akt signaling [30]. It may happen that the beneficial effects of curcumin in healthspan are not attained by itself but when consumed mixed with different bioactive phytochemicals that will antagonize the drawbacks of curcumin. These possibilities should be explored in the design of future studies from preclinical to animal models and to human intervention trials.

Curcumin in Aging and Disease

By virtue of its multitarget ability demonstrated in numerous studies, curcumin has been advocated as effective against a wide range of diseases including cancer, cardiovascular, inflammatory, liver, lung, metabolic, neurological, renal, and other diseases [2,4], which are summarized in Table 35.2. The molecular and biochemical targets of curcumin on aging hallmarks have been linked with the protection against several age-related diseases using animal models, specifically on cancer (reviewed in Refs [22] and [144]), diabetes (reviewed in Refs [145], [146], and [147]), cardiovascular disorders (reviewed in Refs [145] and [148]), and neurodegenerative diseases (reviewed in Refs [9], [12], and [149]).

Curcumin and cancer is a major area of research, and data largely agree that this polyphenol inhibits carcinogenesis at multiple levels [22]. In a systematic review of published results, Ghorbani et al. concluded that curcumin also has antihyperglycemic and insulin sensitizer effects [96]. In addition, curcumin regulates the expression of AMPK, PPARγ, and NF-κB in livers of diabetic db/db mice, suggesting its beneficial effect for treatment of diabetes complications [98]. Recently, pretreatment with curcumin was shown to confer cardioprotection by attenuating mitochondrial oxidative damage in ischemic rat hearts through activation of SIRT1 signaling [102].

More recently, a strong effort has been channeled to exploiting these effects of curcumin for tackling neurodegenerative diseases. Using different transgenic *Drosophila* as models of Alzheimer's disease, curcumin led to up to 75% extended lifespan associated with improvement of amyloid fibril conversion, locomotor

TABLE 35.2 Diseases Potentially Targeted by Curcumin According to the Literature Review Done by Aggarwal et al. [2] and Prasad et al. [4]

Diseases and symptoms targeted by curcumin
Cancer
Bone, brain, breast, gastrointestinal cancers (esophagus, stomach, liver, pancreas, intestine, rectum), genitourinary cancers (bladder, kidney, prostate), gynecologic cancers (cervix, ovary, uterus), hematologic cancers (leukemia, lymphoma, myeloma), lung, oral, skin
Cardiovascular diseases
Atherosclerosis, cardiomyopathy, myocardial infarction, stroke
Inflammatory diseases
Allergy, colitis, Crohn's disease, eczema, gallstone, inflammatory bowel disease, multiple sclerosis, pancreatitis, psoriasis, rheumatoid arthritis, sinusitis, systemic sclerosis, ulcer
Kidney diseases
Chronic kidney disease, diabetic nephropathy, renal failure, ischemia and reperfusion
Liver diseases
Alcoholic liver disease, cirrhosis, fibrosis
Lung diseases
Asthma, bronchitis, chronic obstructive pulmonary disease, cystic fibrosis
Metabolic diseases
Diabetes, hyperlipidemia, hypoglycemia, hypothyroidism, obesity
Neurological diseases
Alzheimer's disease, depression, epilepsy, Parkinson's disease
Other diseases and symptoms
Cataract, fatigue, fever, hemorrhage, osteoporosis, septic shock, wound healing

activity, and reduced neurotoxicity [150]. Wang et al. also showed the ability of curcumin to protect both morphological and behavioral defects in these models [151]. In *Drosophila* expressing human wild-type α-synuclein in neurons as a model of Parkinson's disease, curcumin significantly delayed the loss of activity pattern, reduced oxidative stress and apoptosis, and increased the lifespan of transgenic flies [152]. Dietary curcumin supplementation also regulated molecules involved in energy homeostasis, such as AMPK, in rat brain tissue after induced trauma, which may be important for brain functional recovery [153]. Using rats and mice, Pu et al. demonstrated that curcumin improves aging-related cerebrovascular dysfunction via the AMPK/mitochondrial uncoupling protein 2 (UCP2) pathway [99]. Recently, curcuminoids intake by female rats also demonstrated several improvements on brain age-related mitochondrial dysfunction parameters [154]. Curcumin given intragastrically attenuated cognitive deficits of senescence-accelerated mouse prone 8 (SAMP8 mice) [155]. Overall, these reports validate curcumin as an interesting drug candidate for the prevention of age-associated neurodegenerative disorders.

NUTRITIONAL AND PHARMACOLOGICAL APPLICATIONS OF CURCUMIN IN AGING

Although there is intense research regarding different effects associated with curcumin, epidemiological and clinical studies showing health beneficial effects from its consumption or use as a phytopharmaceutic are scarce. The pharmacological application of curcumin is constrained by its poor solubility and low bioavailability. New delivery systems are being developed to overcome this limitation and may be a significant step forward toward its applicability in the treatment and prevention of several age-related diseases.

Epidemiological Data and Clinical Trials

Therapeutic and preventive effects against several age-related diseases have been attributed to curcumin and, therefore, it has acquired great importance as a possible antiaging therapeutic. Although many preclinical studies support these health promoting effects, there are no consistent epidemiological data and clinical trials that unequivocally link curcumin with healthy aging. Although Indians and Chinese have used turmeric since ancient times, systematic collection of epidemiological data is influenced by other factors that affect the drawing of reliable conclusions. These include the presence of other ingredients commonly found in the diet of these ethnic groups, such as other spices and food additives, high percentages of vegetarianism, and use of alternative medicines [156].

Relevant prospective epidemiological research studies that link curcumin (turmeric) consumption and aging are being performed in one of world's longest-lived populations—the Okinawans [157]. This Japanese population from the Ryukyu Islands has the world's longest life and health expectancy with the lowest mortality rates from a multitude of chronic diseases of aging [157]. The traditional Okinawan diet is characterized by low fat intake, particularly saturated, and high carbohydrate intake, together with a very abundant ingestion of calorie-poor and antioxidant-rich orange-yellow root vegetables (turmeric), sweet potatoes, and green leafy vegetables [158,159]. These nutritional data support the view that mild caloric restriction (10−15%) and high consumption of foods with antioxidant and caloric restriction-mimetic properties play a role in the extended healthspan and lifespan of the Okinawans [159]. More research is nevertheless needed to clearly understand the function of curcumin in this equation. Present research data support that curcumin has in fact some activities that mimic the biological effects of caloric restriction [66,82,132].

Epidemiological data have been collected supporting a positive relationship between turmeric consumption and cognitive function in the elderly. It has been reported that the prevalence of Alzheimer's disease in India among people of ages between 70 and 79 years is 4.4-fold lower than that of the United States [160]. In addition, a meta-analysis showed that dementia incidence in East Asian countries (that have the highest consumption of curry) is lower than in Europe, where a higher incidence of Alzheimer's disease tends to exist [161]. In another study with the multiethnic population of Singapore, the association between curry consumption and cognitive function in elderly was investigated in more than 1000 older adults over 60 years of age. The study found a significant beneficial effect on cognitive functioning associated with low-to-moderate levels of curry consumption in elderly Asian subjects [162]. The authors stated however that these findings should be interpreted with caution since they come from a cross-sectional data analysis and do not establish a clear and direct causal effect of curry consumption and improvement of cognitive function [162].

The overall lower cancer rates in the Indian population, especially colorectal, prostate, pancreatic, and lung cancers, as compared with western countries, have also been used as an argument for the possible correlation between curry consumption and decreased cancer risk [156,163]. However, to better explore the relationship between curcumin consumption and healthspan well-designed epidemiological studies are needed, in particular longitudinal follow up cohorts of elderly persons and that specifically investigate parameters and biomarkers relevant for aging and age-related diseases. Though, several confounding factors such as lifestyle, genetics, and other dietary components, will make it difficult to have a clear picture.

Supported by the promising activities and multitargets identified for curcumin in in vitro and animal studies, several clinical trials and human intervention studies have already been performed or are underway. On the website www.clinicaltrials.gov are documented more than 50 phase I and/or phase II clinical trials with turmeric and more than 100 using curcumin. To date, human intervention studies with curcumin have focused mainly in the treatment of an existing health problem or disease, such as cancer and neurodegeneration, demonstrating in some cases its potential as a promising drug. Considering the cardioprotection and lipid lowering ability of curcumin established in preclinical studies, a meta-analysis was recently performed with data from randomized controlled trials that measured parameters related with blood lipids [164]. No effects could be associated with curcumin supplementation on serum total cholesterol, LDL-C, triglycerides, and HDL-C levels.

The clinical use of curcumin is greatly limited by its poor oral bioavailability (see details below). In fact, Sahebkar made some suggestions for a more robust assessment of the lipid-modulating properties of curcumin, such as the use of bioavailability-improved formulations of curcumin with longer supplementation duration in randomized controlled trials conducted in dyslipidemic subjects [164]. Recently, DiSilvestro et al. reported the use of lipidated curcumin in a low dose supplementation study on healthy middle aged people [165]. In this form curcumin is expected to have better absorption performance and the authors found a variety of potentially health promoting effects. Among them, curcumin decreased triglyceride levels but did not affect cholesterol parameters; decreased plasma beta amyloid protein concentrations, which may impact on Alzheimer's disease development; and, increased salivary radical scavenging capacity (direct antioxidant activity) and plasma catalase enzyme activity (indirect antioxidant activity) [164]. With better designed and longer intervention studies as well as the development of different approaches to increase curcumin bioavailability we may attain very soon promising results demonstrating the beneficial effects of this polyphenol for the treatment/prevention of age-related diseases and the promotion of healthy aging.

Curcumin Bioavailability and Pharmacokinetics

From the studies conducted in humans, curcumin has a good safety profile with no toxicities reported in phase I clinical trials at dosages as high as 8 g/day [166,167]. Together with the safety profile documented for curcumin present in the diet and in phytomedicinal extracts, its use at high dosages that far exceed the ones in ancient tradition seems also to be well tolerated by humans. Contributing to that may be the reduced water solubility of curcumin, its limited intestinal absorption, and rapid metabolism and excretion, which results in low bioavailability [168]. After oral ingestion of 8 g of curcumin the peak serum concentration was observed after 1 to 2 h with an average value of $1.8 \pm 1.9\ \mu M$ [166]. At dosages below 3.6 g/day no detectable amount of curcumin and metabolites were observed [167]. Consistent with the findings in animal models, curcumin was also shown to be efficiently metabolized, particularly in the intestine, resulting in, for example, glucuronide and sulfate conjugates [13]. The metabolite tetrahydrocurcumin has been reported to retain some of the activities of the parent compound, and may therefore contribute to the pharmacological effects of curcumin taken orally [169]. Although some pilot studies with humans suggest poor systemic availability of curcumin when administered orally, probably not enough to exert pharmacologic activity in tissues such as liver and brain, such effects may be possible in gastrointestinal tissues and oral mucosa [13,167]. Patients with colorectal cancer receiving 3.6 g of curcumin daily in capsules resulted in detectable levels of curcumin (together with its sulfate and glucuronide metabolites) in normal and malignant colorectal tissue at concentrations compatible to exert pharmacologic activity [170]. In view of its lipophilicity, topical application of curcumin might also be effective. Topical treatment of burned and photodamaged skin with curcumin in a gel base was shown to possess healing and repair effects [18].

Maximal dietary intake has been estimated at 1.5 g of turmeric per person per day in certain South East Asian communities [171], and therefore the consumption of curcumin is much lower. Thus, based on the prospective human studies reported above, exposure to curcumin or its active metabolites (apart from gastrointestinal tract) from normal diet consumption is expected to be insufficient to reach any biological activity. However, these phase I and phase II clinical trials do not consider the possible involvement of other factors present in the diet that may boost curcumin bioavailability. Compounds and other polyphenols present in the food matrix may interact positively for enhancement of bioavailability of phenolic compounds. Upon epithelial uptake, certain flavonoids may reduce and/or inhibit phase II metabolic enzymes and influence efflux transporters such as p-glycoprotein [172]. Dietary lipids may also increase polyphenol bioaccessibility, specially for hydrophobic molecules such as curcumin and other polyphenol aglycones [172]. These notions are already being explored to increase curcumin bioavailability and might be used in the future in functional foods for prevention and pharmacological strategies in aging interventions. Rather than a high dose of curcumin, a low dose supplement of lipidated curcumin showed diverse health promoting effects in healthy middle-aged people [164]. Zou et al. are also developing an excipient corn oil-in-water emulsion as a food matrix to increase oral bioavailability of curcumin [173]. Although excipient foods have no bioactivity themselves, they may promote that of coingested bioactives [174]. These excipients may increase the release of lipophilic curcumin from food matrices, improve its solubility in gastrointestinal fluids, and enhance epithelium cell permeability. On the other hand, the piperine compound from black pepper has been shown to synergize with curcumin to increase significantly its bioavailability. Piperine, which inhibits the intestinal and hepatic metabolic enzymes involved in the glucuronidation, was successfully used to increase serum concentration of curcumin in rats and humans [175]. This principle was already tested in clinical trials in patients with multiple myeloma and against mild cognitive impairment in the United States (see www.clinicaltrials.gov; search for curcumin), but without significant or available results. The use of adjuvants to enhance curcumin bioavailability by increasing absorption and decreasing metabolic clearance are also being explored in new drug delivery systems, such as nanoparticles and self-microemulsions [176,177].

New Delivery Systems

Different approaches have been developed to overcome the poor oral bioavailability of curcumin due to various physicochemical and physiological processes. Apart from the use of adjuvants to decrease curcumin metabolization (see above), synthetic analogs and conjugates are being developed to be metabolically stable [12,168,178,179]. Nevertheless, much of the effort has been focused on the development of new and improved drug delivery systems for curcumin by means of nanotechnology. Several types of biocompatible and biodegradable nanoparticles are being developed to suitably encapsulate and deliver curcumin at higher concentrations to target organs or tissues in order to improve curcumin's therapeutic efficacy. These include liposomes, polymeric nanoparticles, nanoemulsions, phospholipid formulations, cyclodextrins, and nanogels (for review see Refs [15] and [180]).

Nanoparticles can increase solubility and stability of polyphenols in biological fluids, enhance their absorption in several epithelia, prevent them from premature metabolization and excretion from the body, and allow specific targeting, ultimately resulting in improved bioactivity with minor side effects [181]. Most of the nanodelivery studies are related with the effects of curcumin in the treatment of several ailments, mainly cancer [182–184], brain damage [185–188], and inflammation [189–191]. Some of these new bioavailable curcumin formulations are being already tested in phase I clinical trials [192].

These nanotechnological innovative approaches might also be useful in nutraceutical applications of curcumin to prevent age-related diseases and to improve healthspan. However, as a very recent field, nanoparticles still present many difficulties and limitations to their use. Major challenges are the potential toxicity of nanoparticles, their efficient absorption through the different biological barriers until the target tissue is reached, and cost [181]. Another possible limitation of the use of these drug delivery systems is the potential side-effects of highly effective curcumin delivery to target organs. The improvement of curcumin bioavailability by these new formulations might induce toxic effects that have not been addressed yet in the interventions done with free curcumin [9]. For example, some adverse events experienced by volunteers that ingested very high doses of curcumin, such as nausea, diarrhea, and rash [193], might be exponentially exacerbated. In addition, toxic effects and strong prooxidant effects observed for high concentrations of curcumin in in vitro studies may also happen in vivo with these efficient delivery vehicles [9]. Instead, other alternative concepts of intervention, such as exploiting the hormetic effects of curcumin and possible involvement of the gut microbiome are worthy of future investigation regarding the promotion of healthy aging, and should be considered in the design of novel clinical trials.

Applications

As mentioned before, due to the age-old use of turmeric as a food condiment and in phytomedicine, and its attributed beneficial effects against several ailments, curcumin and curcumin-containing plant extracts have been already applied in several commercial products sold worldwide, but especially in Southeast Asian countries and the United States [15]. Although the efficacy of some of these products has never been scientifically proven, the pleiotropic biological activities of curcumin have been supported in recent years by hundreds of in vitro and in vivo preclinical studies as well as some human intervention studies. Although its chemical proprieties (lipophilicity and intense yellow color) and low bioavailability are hampering the development of curcumin for clinical applications, novel formulations and drug delivery systems may lead the way to a promising future for the use of this compound. Its multitargeting faculty enables curcumin to modulate the activity of several proteins and signaling pathways, which together with its apparent safe profile and low cost, make curcumin an ideal candidate for nutritional interventions to improve healthspan and longevity. Data collected in animal models and in humans are also encouraging for the use of curcumin in the treatment of cancer, stroke, Alzheimer's disease, and Parkinson's disease [12].

An interesting application of curcumin is its use against skin aging and photoaging. Its cosmetic use for skin care and aging and wound healing is already a reality [15]. These applications are supported by several studies, from animals to humans. Curcumin has shown in vitro the ability to hormetically stimulate wound healing of human fibroblasts [16], and to be effective in the treatment of burn wounds in rats and humans [17,18]. Evidence from the literature indicates the ability of curcumin to enhance collagen deposition, tissue remodeling, and wound contraction as important processes for proper wound healing [194]. Application of curcumin in new formulations has been identified as essential for optimizing its therapeutic effect in wound healing [194]. Turmeric extracts and curcumin also significantly inhibited the ultraviolet-induced photodamage in mice and humans [18,195–198]. Anti-inflammatory and antioxidant actions of curcumin are proposed to mediate not only the burn wound healing capacity of this polyphenol but also its antiphotoaging effects [18,194,199]. Therefore, all these data support the use of curcumin in cosmeceuticals, which combined with the development of new formulations may help to maximize its beneficial skin antiaging effects.

The hormetic effects of curcumin discussed above can also have tremendous relevance on the preventive potential of this polyphenol against aging and age-related diseases. The hormetic modulation of aging by nutritional factors, by inducing mild stress and increasing antioxidant defenses and other cellular maintenance and repair pathways, has been increasingly recognized as potentially applicable to aging intervention strategies [200]. However, in order to fully explore the mild stress-induced hormesis attained by curcumin, the problem of ensuring that tissues and organs are exposed to the correct dose to attain positive effects needs to be overcome. Further research in new formulations and nanotechnologies to improve bioavailability of curcumin and to ensure the correct dosage to obtain beneficial effects from mild stress and avoid toxicity are needed to envisage the use of curcumin as a hormetin.

Fig. 35.4 schematically shows the possible future applications of curcumin for stimulation of a healthy

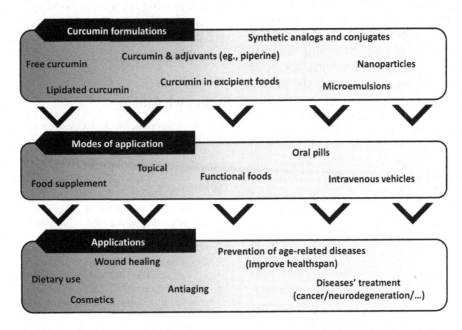

FIGURE 35.4 Actual and possible future applications of curcumin for stimulation of healthy aging: formulations involved and modes of application.

aging, which may depend on forthcoming research in relevant formulations, nanoparticles, food matrices, and routes of applications. The different approaches, whether with the aim of treating age-related diseases, preventing them, or intervening in the basic process of aging, will need to be tested through adequate experimental studies with animal models and with human subjects. This will be essential to validate the ancient and novel attributed nutritional and pharmacological actions of curcumin, ultimately seeking the achievement of healthy aging and healthspan extension.

CONCLUDING REMARKS

The ancient use of turmeric for treatment of several ailments and the discovery that the polyphenolic curcumin is one of its most active compounds have guided in the last decades thousands of studies regarding this natural compound. Several biological activities, putative cellular targets, and potential therapeutic effects were identified in numerous preclinical studies confirming many folk uses of turmeric and attributed new effects to curcumin. The potential beneficial effects of curcumin on aging arise from the modulation of several molecular and biochemical targets relevant in the context of the recent proposed aging hallmarks, as well as by its lifespan extension capabilities in different model organisms. There is a lack, however, of epidemiological and clinical evidence with human subjects of the health-promoting effects and increased longevity provided by curcumin. In view of its low bioavailability, which is hampering its clinical application, different curcumin formulations, including drug nanodelivery systems, are being developed to target specific tissues at therapeutic concentrations.

Considering the hormetic effects of curcumin and potential to induce senescence at high doses, appropriate tissue exposure needs to be established with proper experimental design in order to foresee the use of curcumin in nutritional interventions for improving healthspan and longevity. The potential interactions with other phytochemicals, the effect of food matrices, and the use of new curcumin formulations should be explored in the context of functional foods. Considering the aging of global population and increased incidence of age-related diseases and morbidity, both in developed and developing countries, new and cost-effective strategies need to be employed. The multitargeting ability of curcumin may constitute a challenging opportunity as a golden nutraceutical for a healthy aging.

SUMMARY

- Turmeric has been used in traditional medicine for thousands of years to treat several medical conditions.
- Curcumin is the main active constituent of turmeric and it has powerful antioxidant and anti-inflammatory properties.
- Numerous biochemical and molecular targets relevant to the hallmarks of aging are regulated by curcumin.
- Curcumin extends the lifespan of model organisms, including worms and flies.
- Many lines of evidence demonstrate the potential of curcumin for the prevention and/or treatment of age-related diseases, such as cancer and neurodegenerative diseases.
- Human intervention studies and clinical trials are being conducted to demonstrate that curcumin increases healthspan.
- The nutritional and pharmacological applications of curcumin are restricted by its poor solubility and low bioavailability.
- Different formulations are being developed to increase curcumin bioavailability, including new delivery systems like nanoparticles.
- Applications of curcumin are already a reality in wound healing and cosmetics against skin aging.
- The caloric restriction mimetic and hormetic effects of curcumin may be explored for future aging nutraceutical applications.
- Considering the global population aging, curcumin may constitute an interesting low cost and effective natural compound for interventions to achieve healthy aging.

Acknowledgments

ACC is supported by a doctoral grant (SFRH/BD/86953/2012) awarded by Fundação para a Ciência e a Tecnologia (FCT).

References

[1] US National Institutes of Health, PubMed database. <www.ncbi.nlm.nih.gov/pubmed>. [accessed 16.03.15].
[2] Aggarwal BB, Sundaram C, Malani N, Ichikawa H. Curcumin: the Indian solid gold. Adv Exp Med Biol 2007;595:1–75.
[3] Govindarajan VS. Turmeric – chemistry, technology, and quality. Crit Rev Food Sci Nutr 1980;12:199–301.
[4] Prasad S, Gupta SC, Tyagi AK, Aggarwal BB. Curcumin, a component of golden spice: from bedside to bench and back. Biotechnol Adv 2014;32:1053–64.
[5] Priyadarsini KI. Chemical and structural features influencing the biological activity of curcumin. Curr Pharm Des 2013;19:2093–100.

[6] Manolova Y, Deneva V, Antonov L, Drakalska E, Momekova D, Lambov N. The effect of the water on the curcumin tautomerism: a quantitative approach. Spectrochim Acta A Mol Biomol Spectrosc 2014;132:815–20.

[7] Shishodia S. Molecular mechanisms of curcumin action: gene expression. Biofactors 2013;39:37–55.

[8] Gupta SC, Sung B, Kim JH, Prasad S, Li S, Aggarwal BB. Multitargeting by turmeric, the golden spice: from kitchen to clinic. Mol Nutr Food Res 2013;57:1510–28.

[9] Chin D, Huebbe P, Pallauf K, Rimbach G. Neuroprotective properties of curcumin in Alzheimer's disease – merits and limitations. Curr Med Chem 2013;20:3955–85.

[10] Shishodia S, Singh T, Chaturvedi MM. Modulation of transcription factors by curcumin. Adv Exp Med Biol 2007;595: 127–48.

[11] Ammon HP, Wahl MA. Pharmacology of *Curcuma longa*. Planta Med 1991;57:1–7.

[12] Witkin JM, Li X. Curcumin, an active constiuent of the ancient medicinal herb *Curcuma longa* L.: some uses and the establishment and biological basis of medical efficacy. CNS Neurol Disord Drug Targets 2013;12:487–97.

[13] Sharma RA, Steward WP, Gescher AJ. Pharmacokinetics and pharmacodynamics of curcumin. Adv Exp Med Biol 2007;595: 453–70.

[14] Goel A, Kunnumakkara AB, Aggarwal BB. Curcumin as 'curecumin': from kitchen to clinic. Biochem Pharmacol 2008;75: 787–809.

[15] Prasad S, Tyagi AK, Aggarwal BB. Recent developments in delivery, bioavailability, absorption and metabolism of curcumin: the golden pigment from golden spice. Cancer Res Treat 2014;46:2–18.

[16] Demirovic D, Rattan SIS. Curcumin induces stress response and hormetically modulates wound healing ability of human skin fibroblasts undergoing ageing in vitro. Biogerontology 2011;12:437–44.

[17] Kulac M, Aktas C, Tulubas F, et al. The effects of topical treatment with curcumin on burn wound healing in rats. J Mol Histol 2013;44:83–90.

[18] Heng MCY. Signaling pathways targeted by curcumin in acute and chronic injury: burns and photo-damaged skin. Int J Dermatol 2013;52:531–43.

[19] Alpers DH. The potential use of curcumin in management of chronic disease: too good to be true? Curr Opin Gastroenterol 2008;24:173–5.

[20] Gupta SC, Prasad S, Kim JH, et al. Multitargeting by curcumin as revealed by molecular interaction studies. Nat Prod Rep 2011;28:1937–55.

[21] Zhou H, Beevers CS, Huang S. The targets of curcumin. Curr Drug Targets 2011;12:332–47.

[22] Shanmugam MK, Rane G, Kanchi MM, et al. The multifaceted role of curcumin in cancer prevention and treatment. Molecules 2015;20:2728–69.

[23] Rattan SIS. Increased molecular damage and heterogeneity as the basis of aging. Biol Chem 2008;389:267–72.

[24] Rattan SIS. Molecular gerontology: from homeodynamics to hormesis. Curr Pharm Des 2014;20:3036–9.

[25] López-Otín C, Blasco MA, Partridge L, Serrano M, Kroemer G. The hallmarks of aging. Cell 2013;153:1194–217.

[26] Aguilera A, García-Muse T. Causes of genome instability. Annu Rev Genet 2013;47:1–32.

[27] Belancio VP, Blask DE, Deininger P, Hill SM, Jazwinski SM. The aging clock and circadian control of metabolism and genome stability. Front Genet 2014;5:455.

[28] Motterlini R, Foresti R, Bassi R, Green CJ. Curcumin, an antioxidant and anti-inflammatory agent, induces heme oxygenase-1 and protects endothelial cells against oxidative stress. Free Radic Biol Med 2000;28:1303–12.

[29] Balogun E, Hoque M, Gong P, et al. Curcumin activates the haem oxygenase-1 gene via regulation of Nrf2 and the antioxidant-responsive element. Biochem J 2003;371:887–95.

[30] Lima CF, Pereira-Wilson C, Rattan SIS. Curcumin induces heme oxygenase-1 in normal human skin fibroblasts through redox signaling: relevance for anti-aging intervention. Mol Nutr Food Res 2011;55:430–42.

[31] Son TG, Camandola S, Mattson MP. Hormetic dietary phytochemicals. Neuromolecular Med 2008;10:236–46.

[32] Surh Y-J, Kundu JK, Na H-K. Nrf2 as a master redox switch in turning on the cellular signaling involved in the induction of cytoprotective genes by some chemopreventive phytochemicals. Planta Med 2008;74:1526–39.

[33] Awasthi S, Pandya U, Singhal SS, et al. Curcumin-glutathione interactions and the role of human glutathione S-transferase P1-1. Chem Biol Interact 2000;128:19–38.

[34] Kunwar A, Sandur SK, Krishna M, Priyadarsini KI. Curcumin mediates time and concentration dependent regulation of redox homeostasis leading to cytotoxicity in macrophage cells. Eur J Pharmacol 2009;611:8–16.

[35] Dinkova-Kostova AT, Talalay P. Direct and indirect antioxidant properties of inducers of cytoprotective proteins. Mol Nutr Food Res 2008;52(Suppl. 1):S128–38.

[36] Niture SK, Velu CS, Smith QR, Bhat GJ, Srivenugopal KS. Increased expression of the MGMT repair protein mediated by cysteine prodrugs and chemopreventative natural products in human lymphocytes and tumor cell lines. Carcinogenesis 2007;28:378–89.

[37] Sun B, Ross SM, Joseph Trask O, et al. Assessing dose-dependent differences in DNA-damage, p53 response and genotoxicity for quercetin and curcumin. Toxicol Vitr 2013;27: 1877–87.

[38] Lu H-F, Yang J-S, Lai K-C, et al. Curcumin-induced DNA damage and inhibited DNA repair genes expressions in mouse-rat hybrid retina ganglion cells (N18). Neurochem Res 2009;34: 1491–7.

[39] Charles C, Nachtergael A, Ouedraogo M, Belayew A, Duez P. Effects of chemopreventive natural products on non-homologous end-joining DNA double-strand break repair. Mutat Res Genet Toxicol Environ Mutagen 2014;768:33–41.

[40] Mattson MP. Hormesis defined. Ageing Res Rev 2008;7:1–7.

[41] Rattan SIS. Targeting the age-related occurrence, removal, and accumulation of molecular damage by hormesis. Ann NY Acad Sci 2010;1197:28–32.

[42] Calabrese V, Cornelius C, Cuzzocrea S, Iavicoli I, Rizzarelli E, Calabrese EJ. Hormesis, cellular stress response and vitagenes as critical determinants in aging and longevity. Mol Aspects Med 2011;32:279–304.

[43] Roy M, Sinha D, Mukherjee S, Biswas J. Curcumin prevents DNA damage and enhances the repair potential in a chronically arsenic-exposed human population in West Bengal, India. Eur J Cancer Prev 2011;20:123–31.

[44] Hewitt G, Jurk D, Marques FDM, et al. Telomeres are favoured targets of a persistent DNA damage response in ageing and stress-induced senescence. Nat Commun 2012;3:708.

[45] Harley CB. Telomere loss: mitotic clock or genetic time bomb? Mutat Res 1991;256:271–82.

[46] Blackburn EH, Greider CW, Szostak JW. Telomeres and telomerase: the path from maize, Tetrahymena and yeast to human cancer and aging. Nat Med 2006;12:1133–8.

[47] Kim NW, Piatyszek MA, Prowse KR, et al. Specific association of human telomerase activity with immortal cells and cancer. Science 1994;266:2011–15.

[48] Hanahan D, Weinberg RA. Hallmarks of cancer: the next generation. Cell 2011;144:646–74.
[49] Chakraborty S, Ghosh U, Bhattacharyya NP, Bhattacharya RK, Roy M. Inhibition of telomerase activity and induction of apoptosis by curcumin in K-562 cells. Mutat Res 2006;596: 81–90.
[50] Lee JH, Chung IK. Curcumin inhibits nuclear localization of telomerase by dissociating the Hsp90 co-chaperone p23 from hTERT. Cancer Lett 2010;290:76–86.
[51] Khaw AK, Hande MP, Kalthur G, Hande MP. Curcumin inhibits telomerase and induces telomere shortening and apoptosis in brain tumour cells. J Cell Biochem 2013;114:1257–70.
[52] Correia-Melo C, Hewitt G, Passos JF. Telomeres, oxidative stress and inflammatory factors: partners in cellular senescence? Longev Heal 2014;3:1.
[53] Jurk D, Wilson C, Passos JF, et al. Chronic inflammation induces telomere dysfunction and accelerates ageing in mice. Nat Commun 2014;2:4172.
[54] Franceschi C, Capri M, Monti D, et al. Inflammaging and anti-inflammaging: a systemic perspective on aging and longevity emerged from studies in humans. Mech Ageing Dev 2007;128:92–105.
[55] Chung HY, Cesari M, Anton S, et al. Molecular inflammation: underpinnings of aging and age-related diseases. Ageing Res Rev 2009;8:18–30.
[56] Adler AS, Sinha S, Kawahara TLA, Zhang JY, Segal E, Chang HY. Motif module map reveals enforcement of aging by continual NF-κB activity. Genes Dev 2007;21:3244–57.
[57] Osorio FG, Bárcena C, Soria-Valles C, et al. Nuclear lamina defects cause ATM-dependent NF-κB activation and link accelerated aging to a systemic inflammatory response. Genes Dev 2012;26:2311–24.
[58] Tilstra JS, Robinson AR, Wang J, et al. NF-κB inhibition delays DNA damage-induced senescence and aging in mice. J Clin Invest 2012;122:2601–12.
[59] Zhang G, Li J, Purkayastha S, et al. Hypothalamic programming of systemic ageing involving IKK-β, NF-κB and GnRH. Nature 2013;497:211–16.
[60] Chung S, Yao H, Caito S, Hwang J-W, Arunachalam G, Rahman I. Regulation of SIRT1 in cellular functions: role of polyphenols. Arch Biochem Biophys 2010;501:79–90.
[61] Talens RP, Christensen K, Putter H, et al. Epigenetic variation during the adult lifespan: cross-sectional and longitudinal data on monozygotic twin pairs. Aging Cell 2012;11: 694–703.
[62] Meeran SM, Ahmed A, Tollefsbol TO. Epigenetic targets of bioactive dietary components for cancer prevention and therapy. Clin Epigenetics 2010;1:101–16.
[63] Reuter S, Gupta SC, Park B, Goel A, Aggarwal BB. Epigenetic changes induced by curcumin and other natural compounds. Genes Nutr 2011;6:93–108.
[64] Lewinska A, Wnuk M, Grabowska W, et al. Curcumin induces oxidation-dependent cell cycle arrest mediated by SIRT7 inhibition of rDNA transcription in human aortic smooth muscle cells. Toxicol Lett 2015;233:227–38.
[65] Vahid F, Zand H, Nosrat-Mirshekarlou E, Najafi R, Hekmatdoost A. The role dietary of bioactive compounds on the regulation of histone acetylases and deacetylases: a review. Gene 2015;562:8–15.
[66] Martin SL, Hardy TM, Tollefsbol TO. Medicinal chemistry of the epigenetic diet and caloric restriction. Curr Med Chem 2013;20:4050–9.
[67] Herranz D, Muñoz-Martin M, Cañamero M, et al. Sirt1 improves healthy ageing and protects from metabolic syndrome-associated cancer. Nat Commun 2010;1:3.

[68] Koga H, Kaushik S, Cuervo AM. Protein homeostasis and aging: the importance of exquisite quality control. Ageing Res Rev 2011;10:205–15.
[69] Powers ET, Morimoto RI, Dillin A, Kelly JW, Balch WE. Biological and chemical approaches to diseases of proteostasis deficiency. Annu Rev Biochem 2009;78:959–91.
[70] Hartl FU, Bracher A, Hayer-Hartl M. Molecular chaperones in protein folding and proteostasis. Nature 2011;475:324–32.
[71] Kato K, Ito H, Kamei K, Iwamoto I. Stimulation of the stress-induced expression of stress proteins by curcumin in cultured cells and in rat tissues in vivo. Cell Stress Chaperones 1998;3:152–60.
[72] Teiten M-H, Reuter S, Schmucker S, Dicato M, Diederich M. Induction of heat shock response by curcumin in human leukemia cells. Cancer Lett 2009;279:145–54.
[73] Berge U, Kristensen P, Rattan SIS. Hormetic modulation of differentiation of normal human epidermal keratinocytes undergoing replicative senescence in vitro. Exp Gerontol 2008;43:658–62.
[74] Murakami A. Modulation of protein quality control systems by food phytochemicals. J Clin Biochem Nutr 2013;52:215–27.
[75] Maiti P, Manna J, Veleri S, Frautschy S. Molecular chaperone dysfunction in neurodegenerative diseases and effects of curcumin. Biomed Res Int 2014;2014:495091.
[76] Hasima N, Aggarwal BB. Targeting proteasomal pathways by dietary curcumin for cancer prevention and treatment. Curr Med Chem 2014;21:1583–94.
[77] Yoon MJ, Kang YJ, Lee JA, et al. Stronger proteasomal inhibition and higher CHOP induction are responsible for more effective induction of paraptosis by dimethoxycurcumin than curcumin. Cell Death Dis 2014;5:e1112.
[78] Ali RE, Rattan SIS. Curcumin's biphasic hormetic response on proteasome activity and heat-shock protein synthesis in human keratinocytes. Ann NY Acad Sci 2006;1067:394–9.
[79] Gao Y, Shi Q, Xu S, et al. Curcumin promotes KLF5 proteasome degradation through downregulating YAP/TAZ in bladder cancer cells. Int J Mol Sci 2014;15:15173–87.
[80] Pietrocola F, Lachkar S, Enot DP, et al. Spermidine induces autophagy by inhibiting the acetyltransferase EP300. Cell Death Differ 2015;22:509–16.
[81] Hasima N, Ozpolat B. Regulation of autophagy by polyphenolic compounds as a potential therapeutic strategy for cancer. Cell Death Dis 2014;5:e1509.
[82] Mariño G, Pietrocola F, Madeo F, Kroemer G. Caloric restriction mimetics: natural/physiological pharmacological autophagy inducers. Autophagy 2014;10:1879–82.
[83] Shintani T, Klionsky DJ. Autophagy in health and disease: a double-edged sword. Science 2004;306:990–5.
[84] Kantara C, O'Connell M, Sarkar S, Moya S, Ullrich R, Singh P. Curcumin promotes autophagic survival of a subset of colon cancer stem cells, which are ablated by DCLK1-siRNA. Cancer Res 2014;74:2487–98.
[85] Kenyon CJ. The genetics of ageing. Nature 2010;464:504–12.
[86] Haigis MC, Yankner BA. The aging stress response. Mol Cell 2010;40:333–44.
[87] Haigis MC, Sinclair DA. Mammalian sirtuins: biological insights and disease relevance. Annu Rev Pathol 2010;5:253–95.
[88] Fontana L, Partridge L, Longo VD. Extending healthy life span — from yeast to humans. Science 2010;328:321–6.
[89] Johnson SM, Gulhati P, Arrieta I, et al. Curcumin inhibits proliferation of colorectal carcinoma by modulating Akt/mTOR signaling. Anticancer Res 2009;29:3185–90.
[90] Kang C, Kim E. Synergistic effect of curcumin and insulin on muscle cell glucose metabolism. Food Chem Toxicol 2010;48: 2366–73.

[91] Song Z, Wang H, Zhu L, et al. Curcumin improves high glucose-induced INS-1 cell insulin resistance via activation of insulin signaling. Food Funct 2015;6:461–9.

[92] Aoki H, Takada Y, Kondo S, Sawaya R, Aggarwal BB, Kondo Y. Evidence that curcumin suppresses the growth of malignant gliomas in vitro and in vivo through induction of autophagy: role of Akt and extracellular signal-regulated kinase signaling pathways. Mol Pharmacol 2007;72:29–39.

[93] Beevers CS, Li F, Liu L, Huang S. Curcumin inhibits the mammalian target of rapamycin-mediated signaling pathways in cancer cells. Int J Cancer 2006;119:757–64.

[94] Youreva V, Kapakos G, Srivastava AK. Insulin-like growth-factor-1-induced PKB signaling and Egr-1 expression is inhibited by curcumin in A-10 vascular smooth muscle cells. Can J Physiol Pharmacol 2013;91:241–7.

[95] Beevers CS, Chen L, Liu L, Luo Y, Webster NJG, Huang S. Curcumin disrupts the Mammalian target of rapamycin-raptor complex. Cancer Res 2009;69:1000–8.

[96] Ghorbani Z, Hekmatdoost A, Mirmiran P. Anti-hyperglycemic and insulin sensitizer effects of turmeric and its principle constituent curcumin. Int J Endocrinol Metab 2014;12: e18081.

[97] Ejaz A, Wu D, Kwan P, Meydani M. Curcumin inhibits adipogenesis in 3T3-L1 adipocytes and angiogenesis and obesity in C57/BL mice. J Nutr 2009;139:919–25.

[98] Jiménez-Flores LM, López-Briones S, Macías-Cervantes MH, Ramírez-Emiliano J, Pérez-Vázquez V. A PPARγ, NF-κB and AMPK-dependent mechanism may be involved in the beneficial effects of curcumin in the diabetic db/db mice liver. Molecules 2014;19:8289–302.

[99] Pu Y, Zhang H, Wang P, et al. Dietary curcumin ameliorates aging-related cerebrovascular dysfunction through the AMPK/uncoupling protein 2 pathway. Cell Physiol Biochem 2013;32:1167–77.

[100] Li Y, Li J, Li S, et al. Curcumin attenuates glutamate neurotoxicity in the hippocampus by suppression of ER stress-associated TXNIP/NLRP3 inflammasome activation in a manner dependent on AMPK. Toxicol Appl Pharmacol 2015;286:53–63.

[101] Green DR, Galluzzi L, Kroemer G. Mitochondria and the autophagy-inflammation-cell death axis in organismal aging. Science 2011;333:1109–12.

[102] Yang Y, Duan W, Lin Y, et al. SIRT1 activation by curcumin pretreatment attenuates mitochondrial oxidative damage induced by myocardial ischemia reperfusion injury. Free Radic Biol Med 2013;65:667–79.

[103] Sun Q, Jia N, Wang W, Jin H, Xu J, Hu H. Activation of SIRT1 by curcumin blocks the neurotoxicity of amyloid-β25-35 in rat cortical neurons. Biochem Biophys Res Commun 2014;448: 89–94.

[104] Hann SS, Chen J, Wang Z, Wu J, Zheng F, Zhao S. Targeting EP4 by curcumin through cross talks of AMP-dependent kinase alpha and p38 mitogen-activated protein kinase signaling: the role of PGC-1α and Sp1. Cell Signal 2013;25: 2566–74.

[105] Chin D, Hagl S, Hoehn A, et al. Adenosine triphosphate concentrations are higher in the brain of APOE3- compared to APOE4-targeted replacement mice and can be modulated by curcumin. Genes Nutr 2014;9:397.

[106] Zhai X, Qiao H, Guan W, et al. Curcumin regulates peroxisome proliferator-activated receptor-γ coactivator-1α expression by AMPK pathway in hepatic stellate cells in vitro. Eur J Pharmacol 2015;746:56–62.

[107] Campisi J. Aging, cellular senescence, and cancer. Annu Rev Physiol 2013;75:685–705.

[108] Tchkonia T, Zhu Y, van Deursen J, Campisi J, Kirkland JL. Cellular senescence and the senescent secretory phenotype: therapeutic opportunities. J Clin Invest 2013;123:966–72.

[109] Jiang MC, Yang-Yen HF, Lin JK, Yen JJ. Differential regulation of p53, c-Myc, Bcl-2 and Bax protein expression during apoptosis induced by widely divergent stimuli in human hepatoblastoma cells. Oncogene 1996;13:609–16.

[110] Su C-C, Wang M-J, Chiu T-L. The anti-cancer efficacy of curcumin scrutinized through core signaling pathways in glioblastoma. Int J Mol Med 2010;26:217–24.

[111] Mosieniak G, Adamowicz M, Alster O, et al. Curcumin induces permanent growth arrest of human colon cancer cells: link between senescence and autophagy. Mech Ageing Dev 2012;133:444–55.

[112] Feng Z, Lin M, Wu R. The regulation of aging and longevity: a new and complex role of p53. Genes Cancer 2011;2:443–52.

[113] Matheu A, Maraver A, Klatt P, et al. Delayed ageing through damage protection by the Arf/p53 pathway. Nature 2007;448:375–9.

[114] Grabowska W, Kucharewicz K, Wnuk M, et al. Curcumin induces senescence of primary human cells building the vasculature in a DNA damage and ATM-independent manner. Age (Dordr) 2015;37:9744.

[115] Hendrayani S-F, Al-Khalaf HH, Aboussekhra A. Curcumin triggers p16-dependent senescence in active breast cancer-associated fibroblasts and suppresses their paracrine procarcinogenic effects. Neoplasia 2013;15:631–40.

[116] Chen C, Liu Y, Liu Y, Zheng P. mTOR regulation and therapeutic rejuvenation of aging hematopoietic stem cells. Sci Signal 2009;2: ra75.

[117] Chen T, Shen L, Yu J, et al. Rapamycin and other longevity-promoting compounds enhance the generation of mouse induced pluripotent stem cells. Aging Cell 2011;10:908–11.

[118] Chakkalakal JV, Jones KM, Basson MA, Brack AS. The aged niche disrupts muscle stem cell quiescence. Nature 2012;490:355–60.

[119] Mohan R, Sivak J, Ashton P, et al. Curcuminoids inhibit the angiogenic response stimulated by fibroblast growth factor-2, including expression of matrix metalloproteinase gelatinase B. J Biol Chem 2000;275:10405–12.

[120] Shao Z-M, Shen Z-Z, Liu C-H, et al. Curcumin exerts multiple suppressive effects on human breast carcinoma cells. Int J Cancer 2002;98:234–40.

[121] Sikora E, Scapagnini G, Barbagall o M. Curcumin, inflammation, ageing and age-related diseases. Immun Ageing 2010;7:1.

[122] Suckow BK, Suckow MA. Lifespan extension by the antioxidant curcumin in *Drosophila melanogaster*. Int J Biomed Sci 2006;2:402–5.

[123] Liao VH-C Yu C-W, Chu Y-J, Li W-H, Hsieh Y-C, Wang T-T. Curcumin-mediated lifespan extension in *Caenorhabditis elegans*. Mech Ageing Dev 2011;132:480–7.

[124] Lee K-S, Lee B-S, Semnani S, et al. Curcumin extends life span, improves health span, and modulates the expression of age-associated aging genes in *Drosophila melanogaster*. Rejuvenation Res 2010;13:561–70.

[125] Shen L-R, Xiao F, Yuan P, et al. Curcumin-supplemented diets increase superoxide dismutase activity and mean lifespan in *Drosophila*. Age (Dordr) 2013;35:1133–42.

[126] Kitani K, Osawa T, Yokozawa T. The effects of tetrahydrocurcumin and green tea polyphenol on the survival of male C57BL/6 mice. Biogerontology 2007;8:567–73.

[127] Xiang L, Nakamura Y, Lim Y-M, et al. Tetrahydrocurcumin extends life span and inhibits the oxidative stress response by regulating the FOXO forkhead transcription factor. Aging (Albany, NY) 2011;3:1098–109.

[128] Chandrashekara KT, Popli S, Shakarad MN. Curcumin enhances parental reproductive lifespan and progeny viability in *Drosophila melanogaster*. Age (Omaha) 2014;36:9702.

[129] Choi K-M, Lee H-L, Kwon Y-Y, Kang M-S, Lee S-K, Lee C-K. Enhancement of mitochondrial function correlates with the extension of lifespan by caloric restriction and caloric restriction mimetics in yeast. Biochem Biophys Res Commun 2013;441:236–42.

[130] Shen L-R, Parnell LD, Ordovas JM, Lai C-Q. Curcumin and aging. Biofactors 2013;39:133–40.

[131] Alavez S, Vantipalli MC, Zucker DJS, Klang IM, Lithgow GJ. Amyloid-binding compounds maintain protein homeostasis during ageing and extend lifespan. Nature 2011;472:226–9.

[132] Soh J-W, Marowsky N, Nichols TJ, et al. Curcumin is an early-acting stage-specific inducer of extended functional longevity in *Drosophila*. Exp Gerontol 2013;48:229–39.

[133] Rawal S, Singh P, Gupta A, Mohanty S. Dietary intake of *Curcuma longa* and *Emblica officinalis* increases life span in *Drosophila melanogaster*. Biomed Res Int 2014;2014:910290.

[134] Strong R, Miller RA, Astle CM, et al. Evaluation of resveratrol, green tea extract, curcumin, oxaloacetic acid, and medium-chain triglyceride oil on life span of genetically heterogeneous mice. J Gerontol A Biol Sci Med Sci 2013;68:6–16.

[135] Rattan S.I.S. Cell senescence in vitro. eLS. <http://dx.doi.org/10.1002/9780470015902.a0002567.pub3>; 2012.

[136] Suh JH, Shenvi SV, Dixon BM, et al. Decline in transcriptional activity of Nrf2 causes age-related loss of glutathione synthesis, which is reversible with lipoic acid. Proc Natl Acad Sci USA 2004;101:3381–6.

[137] Suzuki M, Betsuyaku T, Ito Y, et al. Down-regulated NF-E2-related factor 2 in pulmonary macrophages of aged smokers and patients with chronic obstructive pulmonary disease. Am J Respir Cell Mol Biol 2008;39:673–82.

[138] Przybysz AJ, Choe KP, Roberts LJ, Strange K. Increased age reduces DAF-16 and SKN-1 signaling and the hormetic response of *Caenorhabditis elegans* to the xenobiotic juglone. Mech Ageing Dev 2009;130:357–69.

[139] Rahman MM, Sykiotis GP, Nishimura M, Bodmer R, Bohmann D. Declining signal dependence of Nrf2-MafS-regulated gene expression correlates with aging phenotypes. Aging Cell 2013;12:554–62.

[140] Sykiotis GP, Bohmann D. Keap1/Nrf2 signaling regulates oxidative stress tolerance and lifespan in *Drosophila*. Dev Cell 2008;14:76–85.

[141] Lewis KN, Mele J, Hayes JD, Buffenstein R. Nrf2, a guardian of healthspan and gatekeeper of species longevity. Integr Comp Biol 2010;50:829–43.

[142] Carvalho AC, Gomes AC, Pereira-Wilson C, Lima CF. Redox-dependent induction of antioxidant defenses by phenolic diterpenes confers stress tolerance in normal human skin fibroblasts: Insights on replicative senescence. Free Radic Biol Med 2015;83:262–72.

[143] Satoh T, McKercher SR, Lipton SA. Nrf2/ARE-mediated antioxidant actions of pro-electrophilic drugs. Free Radic Biol Med 2013;65:645–57.

[144] Park W, Amin ARMR, Chen ZG, Shin DM. New perspectives of curcumin in cancer prevention. Cancer Prev Res (Phila) 2013;6:387–400.

[145] Shehzad A, Rehman G, Lee YS. Curcumin in inflammatory diseases. Biofactors 2013;39:69–77.

[146] Zhang D-W, Fu M, Gao S-H, Liu J-L. Curcumin and diabetes: a systematic review. Evid Based Complement Alternat Med 2013;2013:636053.

[147] Rivera-Mancía S, Concepción Lozada-García M, Pedraza-Chaverri J. Experimental evidence for curcumin and its analogues for management of diabetes mellitus and its associated complications. Eur J Pharmacol 2015;756:30–7.

[148] Khurana S, Venkataraman K, Hollingsworth A, Piche M, Tai TC. Polyphenols: benefits to the cardiovascular system in health and in aging. Nutrients 2013;5:3779–827.

[149] Monroy A, Lithgow GJ, Alavez S. Curcumin and neurodegenerative diseases. Biofactors 2013;39:122–32.

[150] Caesar I, Jonson M, Nilsson KPR, Thor S, Hammarström P. Curcumin promotes A-beta fibrillation and reduces neurotoxicity in transgenic *Drosophila*. PLoS One 2012;7:e31424.

[151] Wang X, Kim J-R, Lee S-B, et al. Effects of curcuminoids identified in rhizomes of *Curcuma longa* on BACE-1 inhibitory and behavioral activity and lifespan of Alzheimer's disease *Drosophila* models. BMC Complement Altern Med 2014;14:88.

[152] Siddique YH, Naz F, Jyoti S. Effect of curcumin on lifespan, activity pattern, oxidative stress, and apoptosis in the brains of transgenic *Drosophila* model of Parkinson's disease. Biomed Res Int 2014;2014:606928.

[153] Sharma S, Zhuang Y, Ying Z, Wu A, Gomez-Pinilla F. Dietary curcumin supplementation counteracts reduction in levels of molecules involved in energy homeostasis after brain trauma. Neuroscience 2009;161:1037–44.

[154] Rastogi M, Ojha RP, Sagar C, Agrawal A, Dubey GP. Protective effect of curcuminoids on age-related mitochondrial impairment in female Wistar rat brain. Biogerontology 2014;15:21–31.

[155] Sun CY, Qi SS, Zhou P, et al. Neurobiological and pharmacological validity of curcumin in ameliorating memory performance of senescence-accelerated mice. Pharmacol Biochem Behav 2013;105:76–82.

[156] Sinha R, Anderson DE, McDonald SS, Greenwald P. Cancer risk and diet in India. J Postgrad Med 2003;49:222–8.

[157] Okinawa Centenarian Study. <www.okicent.org>. [accessed 12.02.15].

[158] Willcox DC, Willcox BJ, Todoriki H, Suzuki M. The Okinawan diet: health implications of a low-calorie, nutrient-dense, antioxidant-rich dietary pattern low in glycemic load. J Am Coll Nutr 2009;28(Suppl.):500S–16S.

[159] Willcox BJ, Willcox DC. Caloric restriction, caloric restriction mimetics, and healthy aging in Okinawa: controversies and clinical implications. Curr Opin Clin Nutr Metab Care 2014;17:51–8.

[160] Ganguli M, Chandra V, Kamboh MI, et al. Apolipoprotein E polymorphism and Alzheimer's disease: the Indo-US cross-national dementia study. Arch Neurol 2000;57:824–30.

[161] Jorm AF, Jolley D. The incidence of dementia: a meta-analysis. Neurology 1998;51:728–33.

[162] Ng T-P, Chiam P-C, Lee T, Chua H-C, Lim L, Kua E-H. Curry consumption and cognitive function in the elderly. Am J Epidemiol 2006;164:898–906.

[163] Hutchins-Wolfbrandt A, Mistry AM. Dietary turmeric potentially reduces the risk of cancer. Asian Pac J Cancer Prev 2011;12:3169–73.

[164] Sahebkar A. A systematic review and meta-analysis of randomized controlled trials investigating the effects of curcumin on blood lipid levels. Clin Nutr 2014;33:406–14.

[165] DiSilvestro RA, Joseph E, Zhao S, Bomser J. Diverse effects of a low dose supplement of lipidated curcumin in healthy middle aged people. Nutr J 2012;11:79.

[166] Cheng AL, Hsu CH, Lin JK, et al. Phase I clinical trial of curcumin, a chemopreventive agent, in patients with high-risk or pre-malignant lesions. Anticancer Res 2001;21:2895–900.

[167] Hsu C-H, Cheng A-L. Clinical studies with curcumin. Adv Exp Med Biol 2007;595:471–80.

[168] Anand P, Kunnumakkara AB, Newman RA, Aggarwal BB. Bioavailability of curcumin: problems and promises. Mol Pharm 2007;4:807–18.

[169] Wang K, Qiu F. Curcuminoid metabolism and its contribution to the pharmacological effects. Curr Drug Metab 2013;14:791–806.

[170] Garcea G, Berry DP, Jones DJL, et al. Consumption of the putative chemopreventive agent curcumin by cancer patients: assessment of curcumin levels in the colorectum and their pharmacodynamic consequences. Cancer Epidemiol Biomarkers Prev 2005;14:120–5.

[171] Eigner D, Scholz D. Ferula asa-foetida and *Curcuma longa* in traditional medical treatment and diet in Nepal. J Ethnopharmacol 1999;67:1–6.

[172] Bohn T. Dietary factors affecting polyphenol bioavailability. Nutr Rev 2014;72:429–52.

[173] Zou L, Liu W, Liu C, Xiao H, McClements DJ. Utilizing food matrix effects to enhance nutraceutical bioavailability: increase of curcumin bioaccessibility using excipient emulsions. J Agric Food Chem 2015;63:2052–62.

[174] McClements DJ, Xiao H. Excipient foods: designing food matrices that improve the oral bioavailability of pharmaceuticals and nutraceuticals. Food Funct 2014;5:1320–33.

[175] Shoba G, Joy D, Joseph T, Majeed M, Rajendran R, Srinivas PS. Influence of piperine on the pharmacokinetics of curcumin in animals and human volunteers. Planta Med 1998;64:353–6.

[176] Moorthi C, Krishnan K, Manavalan R, Kathiresan K. Preparation and characterization of curcumin-piperine dual drug loaded nanoparticles. Asian Pac J Trop Biomed 2012;2:841–8.

[177] Grill AE, Koniar B, Panyam J. Co-delivery of natural metabolic inhibitors in a self-microemulsifying drug delivery system for improved oral bioavailability of curcumin. Drug Deliv Transl Res 2014;4:344–52.

[178] Anand P, Thomas SG, Kunnumakkara AB, et al. Biological activities of curcumin and its analogues (Congeners) made by man and Mother Nature. Biochem Pharmacol 2008;76:1590–611.

[179] Yanagisawa D, Ibrahim NF, Taguchi H, et al. Curcumin derivative with the substitution at C-4 position, but not curcumin, is effective against amyloid pathology in APP/PS1 mice. Neurobiol Aging 2015;36:201–10.

[180] Ghalandarlaki N, Alizadeh AM, Ashkani-Esfahani S. Nanotechnology-applied curcumin for different diseases therapy. Biomed Res Int 2014;2014:394264.

[181] Wang S, Su R, Nie S, et al. Application of nanotechnology in improving bioavailability and bioactivity of diet-derived phytochemicals. J Nutr Biochem 2014;25:363–76.

[182] Lee W-H, Loo C-Y, Young PM, Traini D, Mason RS, Rohanizadeh R. Recent advances in curcumin nanoformulation for cancer therapy. Expert Opin Drug Deliv 2014;11:1183–201.

[183] Tabatabaei Mirakabad FS, Akbarzadeh A, Milani M, et al. A Comparison between the cytotoxic effects of pure curcumin and curcumin-loaded PLGA-PEG nanoparticles on the MCF-7 human breast cancer cell line. Artif Cells Nanomed Biotechnol. Published online September 17, 2014. <http://dx.doi.org/10.3109/21691401.2014.955108>.

[184] Zhang L, Qi Z, Huang Q, et al. Imprinted-like biopolymeric micelles as efficient nanovehicles for curcumin delivery. Colloids Surf B Biointerfaces 2014;123:15–22.

[185] Kakkar V, Muppu SK, Chopra K, Kaur IP. Curcumin loaded solid lipid nanoparticles: an efficient formulation approach for cerebral ischemic reperfusion injury in rats. Eur J Pharm Biopharm 2013;85:339–45.

[186] Ray B, Bisht S, Maitra A, Maitra A, Lahiri DK. Neuroprotective and neurorescue effects of a novel polymeric nanoparticle formulation of curcumin (NanoCurcTM) in the neuronal cell culture and animal model: implications for Alzheimer's disease. J Alzheimers Dis 2011;23:61–77.

[187] Doggui S, Sahni JK, Arseneault M, Dao L, Ramassamy C. Neuronal uptake and neuroprotective effect of curcumin-loaded PLGA nanoparticles on the human SK-N-SH cell line. J Alzheimers Dis 2012;30:377–92.

[188] Gregory M, Sarmento B, Duarte S, et al. Curcumin loaded MPEG-PCL di-block copolymer nanoparticles protect glioma cells from oxidative damage. Planta Med 2014;80:P2N13.

[189] Yadav VR, Suresh S, Devi K, Yadav S. Novel formulation of solid lipid microparticles of curcumin for anti-angiogenic and anti-inflammatory activity for optimization of therapy of inflammatory bowel disease. J Pharm Pharmacol 2009;61:311–21.

[190] Yadav VR, Prasad S, Kannappan R, et al. Cyclodextrin-complexed curcumin exhibits anti-inflammatory and antiproliferative activities superior to those of curcumin through higher cellular uptake. Biochem Pharmacol 2010;80:1021–32.

[191] Young NA, Bruss MS, Gardner M, et al. Oral administration of nano-Emulsion curcumin in mice suppresses inflammatory-induced NFκB signaling and macrophage migration. PLoS One 2014;9:e111559.

[192] US National Institutes of Health, ClinicalTrials.gov. <www.clinicaltrials.gov>. [accessed 02.03.15].

[193] Gupta SC, Patchva S, Aggarwal BB. Therapeutic roles of curcumin: lessons learned from clinical trials. AAPS J 2013;15:195–218.

[194] Akbik D, Ghadiri M, Chrzanowski W, Rohanizadeh R. Curcumin as a wound healing agent. Life Sci 2014;116:1–7.

[195] Sumiyoshi M, Kimura Y. Effects of a turmeric extract (*Curcuma longa*) on chronic ultraviolet B irradiation-induced skin damage in melanin-possessing hairless mice. Phytomedicine 2009;16:1137–43.

[196] Agrawal R, Kaur IP. Inhibitory effect of encapsulated curcumin on ultraviolet-induced photoaging in mice. Rejuvenation Res 2010;13:397–410.

[197] Heng MCY. Curcumin targeted signaling pathways: basis for anti-photoaging and anti-carcinogenic therapy. Int J Dermatol 2010;49:608–22.

[198] Kaur CD, Saraf S. Topical vesicular formulations of *Curcuma longa* extract on recuperating the ultraviolet radiation-damaged skin. J Cosmet Dermatol 2011;10:260–5.

[199] Cheppudira B, Fowler M, McGhee L, et al. Curcumin: a novel therapeutic for burn pain and wound healing. Expert Opin Investig Drugs 2013;22:1295–303.

[200] Rattan SIS. Rationale and methods of discovering hormetins as drugs for healthy ageing. Expert Opin Drug Discov 2012;7:439–48.

CHAPTER

36

One-Carbon Metabolism: An Unsung Hero for Healthy Aging

Eunkyung Suh[1], Sang-Woon Choi[1] and Simonetta Friso[2]

[1]Chaum Life Center, School of Medicine, CHA University, Seoul, Korea [2]Department of Medicine, School of Medicine, University of Verona, Verona, Italy

KEY FACT

- One-carbon nutrients, such as water-soluble B vitamins, methionine, and choline, may delay the aging process and development of age-associated diseases through one-carbon metabolism.

Dictionary of Terms

- *Epigenetics*: an inheritable but reversible phenomenon that affects gene expression without altering DNA base sequence.
- *One-carbon metabolism*: a biochemical network that ultimately delivers one-carbon units (methyl moiety) to many biological methylation reactions that are critical to maintain homeostasis in our body.
- *DNA methylation*: a biological methylation that methylates cytosine base at the CpG residues in DNA by DNA methyltransferases using S-adenosylmethionine, the universal methyl donor.
- *Nutrient and gene interaction*: a nutrient conveys a health effect according to the genotype of a certain gene.
- *Transsulfuration pathway*: homocysteine from methionine, an essential amino acid, is converted to cystathionine and further to cysteine, a nonessential amino acid, which ultimately synthesizes taurine and glutathione.

INTRODUCTION

One-carbon metabolism is a sort of biochemical network that has several critical pathways including nucleotide synthesis, biological methylation, and transsulfuration pathways [1]. Many nutrients, such as B-vitamins, methionine, choline, betaine, serine, and glycine, are directly involved in one-carbon metabolism, while other nutrients, such as retinoic acid and selenium, indirectly influence one-carbon metabolism. S-adenosylmethionine, a metabolite of one-carbon metabolism synthesized from methionine, is the universal methyl donor to numerous biological methylation reactions including DNA methylation and histone methylation, both of which are so-called epigenetic phenomena. Epigenetic phenomena are important mechanisms by which nutrients can affect gene expression without altering DNA base pairs and change phenotypes including disease susceptibility. Epigenetic phenomena are fundamentally influenced by aging. S-adenosylmethionine is also a precursor of polyamines, such as spermidine and spermine, which have recently been highlighted to have inhibitory effects on the aging process through their effects on cell death [2].

One-carbon metabolism also transfers a one-carbon unit (methyl group) to the nucleotide synthesis pathway which is essential for DNA replication and repair. Inadequate nucleotide synthesis due to impaired one-carbon metabolism can increase the risk of mutagenesis and induce aberrant DNA repair. Through the transsulfuration pathway, one-carbon metabolism synthesizes cysteine from methionine and ultimately synthesizes taurine and glutathione, which are important antioxidants. One-carbon metabolism is also linked to other energy metabolisms, such as lipid and carbohydrate metabolisms. Therefore, one-carbon metabolism can influence and be influenced by those metabolisms.

One-carbon metabolism can affect many health conditions including aging and age-associated diseases. Nevertheless, compared to individual one-carbon nutrients, such as B-vitamins, methionine, and choline, the health effects of one-carbon metabolism are not well-understood, even though one-carbon metabolism regulates numerous important metabolisms and pathways. In the present chapter we address the importance of one-carbon metabolism in delaying both the aging process and the development of age-associated diseases.

ONE-CARBON METABOLISM

In one-carbon metabolism, one-carbon units (methyl moiety) from methyl donor nutrients are transferred to homocysteine remethylation, the synthesis of purine and thymidine, and the biological methylation of DNA, RNA, protein, and many other methyl recipients, such as small molecules and lipids. Water-soluble B vitamins such as folate, vitamin B6, vitamin B12, and vitamin B2 play key roles as coenzymes, while methionine, choline, betaine, and serine are methyl-donors (Fig. 36.1).

The sole function of folate is as a coenzyme of one-carbon metabolism with several different types of reduced forms. Tetrahydrofolate, a reduced form of folate, receives a methyl group from serine catalyzed by serine hydroxymethyltransferase (SHMT), which is a reversible reaction. The conversion of serine to glycine produces a carbon unit to tetrahydrofolate and converts to 5,10-methylenetetrahydrofolate which can be utilized to synthesize nucleotides. Methylation of deoxyuridine monophosphate by 5,10-methylenetetrahydrofolate and thymidylate synthase (TS) produces deoxythymidine monophosphate, which is one of the four nucleotides used in building DNA, and dihydrofolate, which will turn into tetrahydrofolate by dihydrofolate reductase (DHFR). DHFR is the target of methotrexate, a cancer chemotherapeutic agent that inhibits cancer cell DNA synthesis. Purines are generated through the intermediate 10-formyltetrahydrofolate, which is derived from 5,10-methylenetetrahydrofolate (Fig. 36.1).

5,10-methylenetetrahydrofolate can also be converted to 5-methyltetrahydrofolate when catalyzed by methylenetetrahydrofolate reductase (MTHFR), which is an irreversible reaction. The folate cycle is then

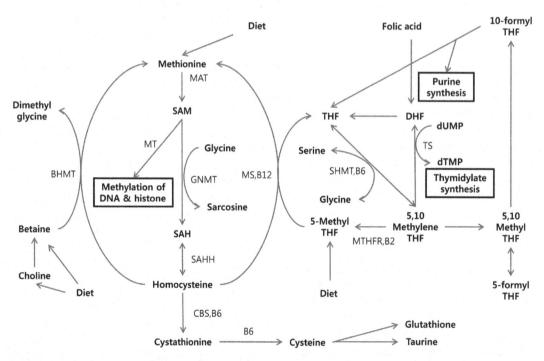

FIGURE 36.1 One-carbon metabolism. *BHMT*, betaine homocysteine methyltransferase; *CBS*, cystathionine-β-synthase; *dTMP*, deoxythymidine monophosphate; *dUMP*, deoxyuridine monophosphate, *GNMT*, glycine N-methyltransferase; *MAT*, methionine adenosyltransferase; *MS*, methionine synthase; *MT*, methyltransferase; *MTHFR*, methylenetetrahydrofolate reductase; *SHMT*, serine hydroxymethyltransferase; *SAHH*, S-adenosylhomocysteine hydrolase; *THF*, tetrahydrofolate; *TS*, thymidylate synthase.

coupled with homocysteine remethylation to regenerate methionine. 5-methyltetrahydrofolate donates a methyl group to homocysteine, methylating and converting it into methionine through methionine synthase (MS) and vitamin B12. The adenylation of methionine produces S-adenosylmethionine catalyzed by methionine adenosyltransferase (MAT). S-adenosylmethionine functions as a methyl donor for many biological methylation reactions, including DNA, RNA, histone, protein, lipid, and small molecules, as well as polyamine synthesis [3]. However, S-adenosylmethionine is most frequently utilized for the synthesis of phosphatidylcholine, which contributes to the lipid membrane content [4], the formation of creatine, which is a nitrogenous compound that helps to supply energy to muscle, and a substrate for glycine N-methyltransferase (GNMT) that converts S-adenosylmethionine to S-adenosylhomocysteine in the liver.

After donating a methyl group to biological reactions, S-adenosylmethionine turns into S-adenosylhomocysteine and then later on into homocysteine. The reaction between S-adenosylhomocysteine and homocysteine is reversible. Once homocysteine is formed, it can be immediately exported to the bloodstream, remethylated into methionine or undergo the transsulfuration pathway because homocysteine is toxic to cells. Blood homocysteine is excreted through the kidney and it is the reason why reduced renal function elevates plasma homocysteine levels. In the transsulfuration pathway, homocysteine reacts with serine through vitamin B6-dependent enzyme cystathionine β-synthase (CBS) to form cystathionine and then into cysteine, which is utilized to form taurine and glutathione (Fig. 36.1).

In summary, water-soluble B vitamins and methyl donor nutrients are involved in one-carbon metabolism which ultimately regulates methyl transfer to the nucleotide synthesis pathway and the biological methylation pathway as well as the transsulfuration pathway, all of which are dedicated to maintaining the homeostasis of various metabolisms and body functions.

AGING, AGE-ASSOCIATED DISEASE, AND ONE-CARBON METABOLISM

Influence of Aging on One-Carbon Metabolism

When renal function is working normally, hyperhomocysteinemia indicates a derangement of one-carbon metabolism because it suggests an inadequacy in the remethylation pathway or the transsulfuration pathway. Interestingly, plasma homocysteine elevations have been shown to be associated with aging, which suggests that aging is also associated with the alteration of one-carbon metabolism. The prevalence of hyperhomocysteinemia (over 15 μmol/L) in the elderly population older than 65 years may be as high as 30% according to the Framingham Study [5].

The Hordaland Homocysteine Study, a population-based study to show lifestyle factors determining homocysteine levels, showed that older age is associated with elevated homocysteine levels, along with male sex, smoking status, high coffee consumption, and lack of exercise [6]. In addition to these lifestyle factors, the Framingham Offspring Study showed chronic alcohol consumption and serum creatinine levels to be determinants of hyperhomocysteinemia [7]. Alcohol deranges one-carbon metabolism by wasting methionine and choline to replenish taurine and glutathione to reduce alcoholic liver injury as well as by methyl folate trapping through inhibiting MS, which induces functional folate deficiency and thereby reduces methyl transfer to homocysteine remethylation.

Ames dwarf mice, which are deficient in the growth hormone, prolactin and thyroid-stimulating hormone, live significantly longer than their litter mates and have significantly higher activity of liver MAT, CBS, cystathionase, and GNMT, all of which are involved in one-carbon metabolism. In this model, S-adenosylmethionine is decreased and S-adenosylhomocysteine is increased, which suggests that the increment of MAT activity produces more methionine but the increment of GNMT activity pushes S-adenosylmethionine toward the transsulfuration pathway, which probably reduces age-associated oxidative damage through increased production of glutathione and taurine [8].

Numerous studies suggest that aging is associated with altered one-carbon metabolism. But we still do not know whether there is a definite cause and effect relationship between altered one-carbon metabolism induced by aging and the development of age-associated diseases.

Methionine Restriction Diet

Over the years, many studies were done to find clues to longevity. Based on the results of the rodent studies, low calorie intake was found to be associated with an increase in lifespan. Study data show that a 40% caloric restriction lowers the generation rate of mitochondrial reactive oxygen species and decreases the oxidative damage of mitochondrial DNA and proteins in rodent organs [9]. Several studies have also suggested that decreased intake of particular components of the diet, specifically methionine, an essential sulfur containing amino acid, can be an important factor in extending the lifespan [10]. In rodent studies, isocaloric 80% methionine restriction was shown to increase longevity in F344 rats [11] and mice [12]. 80% methionine restriction was also shown to decrease the incidence of age-related degenerative diseases and disease-associated markers, such as decreases in serum glucose, insulin, cholesterol, triglycerides and leptin [13]. Interestingly, it has been demonstrated that methionine restriction reduces tumor development in several animal models. A recent paper demonstrated that dietary methionine restriction inhibits prostatic intraepithelial neoplasia in TRAMP mice [14], which is an animal model of prostate cancer. It appears that methionine restriction is also important in preventing cancer through inhibiting tumor growth. In fact, some tumor cells have a greater dependence on methionine for growth compared to normal cells. Specifically, these are tumor cells of the bladder, breast, colon, glioma, kidney, melanoma, and prostate. Growth of these tumor cells can be limited or inhibited in the absence of methionine [15]. Collectively, methionine restriction may extend animal life span as well as reduce the development and progress of cancer.

One-Carbon Metabolism and Neurocognitive Disease

Cognitive decline and dementia are common in old age. Many cross-sectional and longitudinal studies have reported significant associations between dementia and low blood levels of folate and vitamin B12 or high levels of homocysteine [16]. A few studies did not report a significant association between plasma levels of vitamin B12 or homocysteine and dementia [17]. Among the intervention studies of folate supplementation, Fioravanti et al. [18] observed a significant improvement of memory test scores in the folate-treated group in a randomized-controlled study. Another study, the Folate After Coronary Intervention Trial (FACIT), a 3-year randomized controlled trial, showed a significant improvement in memory, information processing speed, and sensori-motor speed in the folic acid group [19]. Despite the many studies done, there is no definite evidence that supplementation with vitamin B12 and folate improves cognitive decline or dementia, even though it may normalize homocysteine levels [20].

Choline, a folate-independent methyl donor in one-carbon metabolism, is essential for neurodevelopment and brain function. The Hordaland Health Study demonstrated that low plasma free choline concentrations are associated with poor cognitive performance and that there is an interaction between low choline and low vitamin B12 on cognitive performance.

Many population-based cross-sectional studies have shown that high plasma homocysteine levels are associated with low cognitive function [21] and dementia [22]. In 1992, it was suggested that elevated homocysteine levels may be a marker of abnormal one-carbon metabolism and that this may contribute to Alzheimer's disease [23]. Since then, it has been reported that elevated homocysteine levels are an independent risk factor for Alzheimer's disease [24]. Proposed mechanisms include the atherosclerotic tendency of homocysteine, and studies have shown that homocysteine can be directly toxic to cultured neuronal cells [25]. Various studies over the years have been conducted to see the effect of homocysteine-lowering treatments through vitamin B supplementation. A few studies have demonstrated improved vascular endothelial function and reduced risk of stroke through folic acid and vitamin B12 supplementations [26]. Another common disease in the elderly is depression, which can affect their quality of life. A population based cross-sectional study showed that homocysteine was significantly associated with depression [27], but the results were not consistent for depression.

One-carbon metabolism is important for brain function and maintenance. It highly influences age-associated neurocognitive decline, but the exact mechanism is not yet clear. The results of intervention through one-carbon metabolism are still contradictory.

One-Carbon Metabolism and Cancer

Although epidemiologic studies strongly indicate that the alteration of one-carbon metabolism is highly associated with the development of several cancers, the mechanism has not yet been clearly elucidated. Over the past two decades, the most probable mechanism has been the derangement of the biological methylation pathway and nucleotide synthesis pathways as we described in the introduction section. Low B-vitamins, such as folate, vitamin B12, and vitamin B6, low methyl donor nutrients, such as methionine and choline, or derangement of one-carbon metabolism induced by alcohol consumption have been studied based on these two mechanisms. However the results were not consistent and reversal of deranged one-carbon nutrients by supplementation with one-carbon nutrients could not reduce the risk of cancer. Rather, too much supplementation raised concern that it may increase the risk of cancer. Those inconsistent observations caused us to change our scope regarding the relationship between one-carbon metabolism and cancer.

In one-carbon metabolism, glycine and serine conduct a kind of methyl shuttle. Glycine receives a methyl group from serine through the SHMT reaction, which is a reversible reaction. Serine can be synthesized from glucose through glycolysis so that carbon derived from glucose can be utilized in one-carbon metabolism. Interestingly, it has been shown that glycine uptake and catabolism can promote tumorigenesis as well as serine *de novo* synthesis through glycolysis, which is also associated with tumorigenesis [28]. Glycine also receives a methyl group from S-adenosylmethionine catalyzed by GNMT. Interestingly, *GNMT*-deficient mice can spontaneously develop primary liver cancer and the polymorphism of the *SHMT* gene is also associated with the risk of cancer. The upregulation of serine/glycine metabolism correlates with cell proliferation and poor prognosis in several tumors. Glycine conversion was shown to increase the proliferation rate in cancer cells [29]. Thus, glycine/serine metabolism could be a new target for the investigation of the relationship between one-carbon metabolism and cancer.

In fact, one-carbon metabolism is known to play an important role in the cellular redox balance. The reduction of 5,10-methylenetetrahydrofolate by MTHFR produces one molecule of $NADP^+$ for each turn of the folate cycle. Modulation of the $NADP^+/NADPH$ ratio contributes to the maintenance of redox status. The transsulfuration pathway produces glutathione, a tripeptide with components of cysteine, glycine, and glutamate, which is also important for the maintenance of the $NADP^+/NADPH$ ratio and redox balance. Therefore, alterations in these metabolic pathways can affect cancer cell growth.

An accumulated body of research strongly indicates that the alteration of one-carbon metabolism is highly associated with carcinogenesis. However, we still do not know the exact mechanism. Further studies with better designs are needed to clarify it.

One-Carbon Metabolism and Cardiovascular Disease

Epidemiologic studies have strongly suggested that the status of B vitamins, such as folate, vitamin B12, and vitamin B6, is associated with the risk of cardiovascular diseases. Many intervention studies were conducted to determine whether dietary supplementation with B vitamins can actually reduce the risk of cardiovascular diseases, but the results were not satisfactory.

Homocysteine is the most extensively studied one-carbon metabolite in the field of one-carbon metabolism and cardiovascular diseases. Epidemiologic studies clearly demonstrated the association of total plasma homocysteine levels and the risk of cardiovascular diseases supplemented by cultured cell and animal studies that elucidated the mechanisms by which homocysteine increases the development of cardiovascular diseases. It was suggested that homocysteine can increase atherosclerosis by endothelial injury, platelet activation, and thrombus formation [30]. However, two big clinical trials, the Heart Outcomes Prevention Evaluation (HOPE) study and the Norwegian Vitamin Trial (NORVIT) study, reported the negative results that decrement of blood homocysteine by B-vitamin supplementation could not protect from the recurrence of major cardiac events. Rather, it increased the recurrence and development of major complications. Thereafter, only a few meta-analyses have suggested the beneficial effects of one-carbon nutrients on the prevention of cardiovascular diseases, without suggesting any significant mechanisms.

Recently a genome-based study, which was conducted as a part of a case-control study nested in the prospective cohort, European prospective Investigation into Cancer and Nutrition (EPIC) cohort, evaluated the DNA methylation status of candidate genes selected from the whole genome analysis. Five increased methylated DNA regions were found that included transcobalamine II (*TCN2*) gene promoter, *CBS* gene 5′UTR, and aminomethyltransferase (*AMT*) gene body from male myocardial infarction subjects as well as paraoxonase 1 (*PON1*) gene body/1st exon and *CBS* gene 5′UTR from female myocardial infarction. An inverse association between B vitamins intake and DNA methylation of these genes was observed [2]. These findings indicate that DNA methylation patterns in specific regions of one-carbon metabolism genes may affect the cardiovascular disease risk by low B vitamins intake, suggesting an epigenetic connection between one-carbon metabolism and cardiovascular diseases.

INFLAMMATION AND ONE-CARBON METABOLISM

Vitamin B6 and Inflammation

In a cross-sectional study of 1976 women selected from the Women's Health Initiative Observational Study [31], plasma vitamin B6 as pyridoxal-5′-phosphate, plasma vitamin B12, plasma folate, and RBC folate were measured as nutritional biomarkers. Serum C-reactive protein and serum amyloid A were measured as inflammation biomarkers. Homocysteine and cysteine were measured as biomarkers that reflect the global one-carbon metabolism status. Plasma pyridoxal-5′-phosphate, RBC folate, homocysteine, and cysteine were identified as independent predictors of C-reactive protein, while plasma pyridoxal-5′-phosphate, vitamin B12, RBC folate, and homocysteine were identified as predictors of serum amyloid A. This study suggests that both individual B-vitamins and one-carbon metabolism are associated with inflammation. Among them, the relationship between vitamin B6 and inflammation has been most highlighted, especially after the mandatory folate fortification era.

Vitamin B6 exists in several forms including pyridoxine, pyridoxine 5′-phosphate, pyridoxal, pyridoxal 5′-phosphate, pyridoxamine, and pyridoxamine 5′phosphate. Pyridoxal 5′-phosphate is the biologically active form which acts as a cofactor for over 140 enzyme reactions that are involved in the metabolism of proteins, lipids, and carbohydrates, and the synthesis or metabolism of hemoglobin, neurotransmitters, nucleic acids, and one carbon units [32]. The relationship between vitamin B6 and inflammation has been demonstrated—plasma concentrations of pyridoxal 5′-phosphate, in particular, were reduced during inflammation. The underlying mechanisms may include altered tissue distribution or increased catabolism via pyridoxal to pyridoxic acid, but the exact mechanism remains unclear.

Investigators observed that patients with rheumatoid arthritis had reduced plasma pyridoxal-5′-phosphate concentrations compared to healthy controls. Furthermore, among the rheumatoid arthritis patients, pyridoxal-5′-phosphate was lower in patients with severe types and with higher levels of inflammation markers [33]. In another study it was found that plasma pyridoxal-5′-phosphate was approximately 35% lower in individuals with elevated C-reactive protein levels (≥ 6 mg/L) than individuals with normal C-reactive protein levels (<6 mg/L) [34]. An analysis of the National Health and Nutrition Examination Survey data also showed an inverse association between vitamin B6 blood levels and inflammation [35].

The association between vitamin B6 and inflammation was also observed in patients with inflammatory bowel disease. In a human study, plasma pyridoxal-5′-phosphate concentrations were significantly lower in the inflammatory bowel disease patients than in healthy controls. Among the patient group, the prevalence of low plasma PLP was significantly higher in patients with an active disease compared to those with a quiescent disease (26.9% vs 2.9%; $p \leq .001$) [36]. The other study showed that vitamin B6 deficiency in healthy elderly adults impairs interleukin-2 production and peripheral blood lymphocyte proliferation to both T- and B-cell mitogens, which were reversed after repletion of vitamin B6 [37].

It is not yet clear whether low vitamin B6 in inflammation is just a result or a cause of inflammation. The research investigating the cause and effect relationship between vitamin B6 and inflammation is important, because we may find a simple strategy to reduce inflammation using vitamin B6. Of note is that inflammation is one of the most important risk factors for the aging process as well as the development of age-associated diseases and metabolic diseases.

S-Adenosylmethionine and Inflammation

S-adenosylmethionine is not found in food, even though its precursor, methionine, is plentiful in many protein foods. However, S-adenosylmethionine is found in almost every tissue and fluid in the body. Other than the universal methyl donor and a precursor of polyamine, S-adenosylmethionine has been known to have an anti-inflammatory effect. S-adenosylmethionine has also been recommended for the treatment of chronic liver diseases, including alcoholic liver injury and nonalcoholic fatty liver diseases.

S-adenosylmethionine has been shown to lower lipopolysaccharide (LPS)-induced expression of the pro-inflammatory cytokine TNF-α and increase the expression of the anti-inflammatory cytokine IL-10 in macrophages. In a cell culture study human monocytic THP1 cells were differentiated into macrophages and treated with S-adenosylmethionine for 24 h, followed by stimulation with LPS. Compared to nontreated cells, S-adenosylmethionine treatment significantly increased global DNA methylation. A DNA methylation microarray found 765 differentially methylated regions associated with 918 genes, which are associated with cardiovascular disease. A subset of genes that were differentially hypomethylated also showed altered expression levels of those genes by S-adenosylmethionine. This study indicates that S-adenosylmethionine, a metabolite of one-carbon metabolism, may convey its anti-inflammatory effect through DNA methylation [38].

INTERACTIONS BETWEEN ONE-CARBON METABOLISM GENES AND NUTRIENTS IN AGING

MTHFR and Folate

Changes in folate status, quantitatively by decreased intake or qualitatively by polymorphic variants of *MTHFR*, may influence DNA integrity and DNA methylation patterns. The enzyme MTHFR catalyzes the conversion of 5,10-methylenetetrahydrofolate, which is essential for DNA synthesis, into 5-methyltetrahydrofolate, which is the primary circulating form of folate that provides methyl groups for homocysteine remethylation to reform methionine [38]. The most studied common polymorphism is *C* to *T* substitution at 677 base (C677T) that results in an alanine to valine conversion in the protein (A222V) [39]. Another common polymorphism is *A* to *C* substitution at 1298 base (*A1298C*), resulting in an alanine to glutamate conversion in the protein. Studies have shown that in *C* to *T* polymorphisms, the heterozygotes (*677CT*) and homozygous (*677TT*) variants have reduced enzyme activity of 65% and 30%, respectively, compared with the wild types (*677CC*) [40]. Further, individuals with the homozygous (*677TT*) variant have been shown to have lowered plasma folate and higher homozygous levels [41]. Similarly, in the variant of *A1298C*, enzyme activity in vitro has been shown to be lowered in homozygous variants (*1298CC*) and heterozygotes (*1298AC*), compared with the wild types (*1298AA*) [42].

A body of studies has suggested the association of *MTHFR* polymorphisms and the risk of colorectal cancer. A modest reduction of colorectal cancer risk was reported in *677TT* individuals compared to those with the *677CC* genotype [43]. Furthermore, it was also found that *677TT* individuals are at reduced colorectal cancer risk when folate is replete, while at increased risk when folate is deplete [44]. Thereafter, the interaction between folate and MTHFR became a paradigm of nutrient and gene interaction and has been extensively studied in other cancers including gastric cancer, breast cancer, and cervical cancer but the results were not always consistent.

There is evidence that interaction between folate and *MTHFR* could also be associated with neurocognitive disease. In an animal study using young and old mice, *Mthfr* deficiency induced brain-region specific impairment of the methylation of Ser/Thr protein phosphatase 2A (*pp2a*) and the expression levels of *Pp2a* and leucine carboxyl methyltransferase (*Lcmt1*) were decreased in the hippocampus and cerebellum. Dietary folate deficiency significantly decreased *Lcmt1* and methylated *Pp2a* levels in all brain regions of aged wild-type mice, and further exacerbated the regional effects of *Mthfr* deficiency in aged heterozygote mice. It appears that *Mthfr* deficiency and the *Mthfr* gene–diet interaction could lead to disruption of neuronal homeostasis, and increase the risk for a variety of neuropsychiatric disorders, including age-related diseases like sporadic Alzheimer's disease [38].

Telomeres, which consist of tandem repeats of DNA, cap the ends of chromosomes. The telomere length is reduced with increasing cell divisions except when the enzyme telomerase is active, such as in stem cells or cancer cells. Thus, the telomere length has been regarded as a marker of cell senescence and aging. Telomere length is epigenetically regulated by DNA methylation that could be modulated by folate. In a human study that measured telomere length in peripheral blood mononuclear cells, plasma concentration of folate was associated with telomere length of peripheral blood mononuclear cells. When plasma folate concentration was above the median, there was a positive relationship between folate and telomere length. However, when plasma folate concentration was below the median, there was an inverse relationship between folate and telomere length. The *MTHFR* 677 *C* to *T* polymorphism was weakly associated ($P = 0.065$) with increased telomere length at below-median folate status. This study suggests that folate and *MTHFR* interaction may influence the aging process [41].

MS and Vitamin B12

The enzyme MS catalyzes the remethylation of homocysteine to methionine through transfer of a methyl group from 5-methyltetrahydrofolate to homocysteine. Vitamin B12 acts as a cofactor in this reaction. Polymorphism of the *MS* gene of *A* to *G* transition at position 2756, which replaces aspartic acid with glycine, has been studied [45].

In a Chinese case-control study the *MS2756G* allele was associated with a higher risk of breast cancer in individuals with low folate intake, vitamin B6, and vitamin B12, but the association disappeared among subjects with moderate and high intake of folate, vitamin B6, and vitamin B12. This result indicates that the *MS 2756AG* polymorphisms are associated with the risk of breast cancer and intakes of folate, vitamin B6, and vitamin B12 influence the association [42]. In another human study, individuals with the variant *MS2756 GG* genotype showed a 5-fold decreased risk of acute lymphoblastic leukemia compared to individuals with wild-type (OR 0.20; 95% CI 0.02-1.45) [46].

MAT and S-Adenosylmethionine

Studies demonstrated that all mammal cells express the enzyme MAT, but activity is highest in the liver [47]. Those with a deficiency in hepatic MAT activity show hypermethioninemia [48]. Two genes, *MAT1A* and *MAT2A*, encode the two homologous catalytic subunits MATα1 and MATα2, respectively [49]. The α1 subunits organizes into two MAT isoenzymes, MAT I and MAT III, while the α2 subunit makes the MAT isoenzyme MAT II. *MAT1A* is expressed mostly in the liver and *MAT2A* is widely distributed in other tissues [49]. Fetal liver expresses only *MAT2A* but not *MAT1A*. With age, *MAT1A* expression exceeds the *MAT2A* expression and adult liver expresses mainly *MAT1A* [50]. Thus, *MAT1A* expression can be considered as differentiated liver phenotype [51]. In a study using a cell line model that expresses different *MAT* forms, cells that express *MAT1A* was shown to have higher levels of S-adenosylmethionine and DNA methylation and was also shown to grow more slowly compared to the cells that expressed *MAT2A* [52]. Thus, the type of *MAT* expressed by the cells resulted in different S-adenosylmethionine levels and also, in return, showed that the S-adenosylmethionine level can affect *MAT* expression. In a cultured cell study with human hepatocytes, when the S-adenosylmethionine level decreased, the *MAT2A* gene expression was rapidly expressed but the effect was diminished when S-adenosylmethionine was added to the medium [53], indicating that a decrease of the hepatic S-adenosylmethionine level causes a switch of *MAT* expression, resulting in dedifferentiation of liver cells [54].

The expression of hepatic *MAT1A* gene is downregulated in chronic liver diseases such as hepatocellular carcinoma, liver cirrhosis and alcoholic hepatitis patients [55]. The effects of chronic hepatic S-adenosylmethionine deficiency have been well described by the *MAT1A* deficient mouse model [56], in which the liver is more susceptible to injury. Thus, in various animal models, S-adenosylmethionine treatment has improved liver injury effectively [57]. However, evidence in human liver disease is still limited.

MAT1A and Vitamin B6 and Folate

In a study using subjects enrolled in the Boston Puerto Rican Health Study and the Nutrition, Aging, and Memory in Elders Study, *MAT1A* variants were strongly associated with hypertension and stroke. Homozygotes of the *MAT1A d18777A* (rs3851059) allele had a significantly higher risk of stroke (OR: 4.30; 95% CI: 1.34, 12.19; $P = 0.006$), whereas *3U1510A* (rs7087728) homozygotes had a significantly lower risk of hypertension (OR: 0.67; 95% CI: 0.48, 0.95; $P = 0.022$) and stroke (OR: 0.35; 95% CI: 0.15, 0.82; $P = 0.015$). Improving folate and vitamin B6 status decreased cardiovascular disease risk of only a subset of the population, depending on genotype. These findings suggest that this gene and nutrient interaction may have an effect on cardiovascular disease risk [40].

CONCLUSION AND FUTURE PERSPECTIVES

One-carbon metabolism is critical to body functions including many biological and chemical processes, especially DNA and cellular metabolisms as well as energy metabolism. Many nutrients directly involved in one-carbon metabolism and other nutrients indirectly influence it. It seems that many nutrients convey their health effects through one-carbon metabolism. However, one-carbon metabolism is not well known compared to individual nutrients that are involved in one-carbon metabolism.

Obviously aging fundamentally alters one-carbon metabolism and the imbalance of one-carbon metabolism can cause unhealthy conditions including age-associated diseases. However, we still do not know the exact mechanism by which altered one-carbon metabolism in the aging process increases the development of age-associated diseases. In the future we may need to change our paradigm to investigate those mechanisms from conventional research that gave us inconsistent data. However, inflammation, epigenetics and gene-nutrient interaction still could be candidate mechanisms that are worthwhile to investigate. One-carbon metabolism could be a good target for the prevention and treatment of age-associated diseases and can provide good clinical and research biomarkers if we delineate the exact mechanism.

Abbreviations

SHMT	serine hydroxymethyltransferase,
TS	thymidylate synthase,
DHFR	dihydrofolate reductase,
MTHFR	methylenetetrahydrofolate reductase,
MS	methionine synthase,
MAT	methionine adenosyltransferase,
CBS	cystathionine β-synthase,
TCN2	transcobalamine II,
AMT	aminomethyltransferase,
PON1	paraoxonase 1,
pp2a	protein phosphatase 2A,
Lcmt1	leucine carboxyl methyltransferase.

SUMMARY

- One-carbon metabolism is a sort of biochemical network that has several critical pathways including nucleotide synthesis, biological methylation, and transsulfuration pathways.
- Water-soluble B vitamins such as folate, vitamin B6, vitamin B12 and vitamin B2 play key roles as coenzymes, while methionine, choline, betaine, and serine are methyl-donors in one-carbon metabolism.
- S-adenosylmethionine, a metabolite of one-carbon metabolism synthesized from methionine, is the universal methyl donor to numerous biological methylation reactions including DNA methylation and histone methylation.
- Aging is associated with altered one-carbon metabolism but we still do not know whether there is a certain cause and effect relationship between altered one-carbon metabolism by aging and the development of age-associated diseases.
- One-carbon metabolism could be a good target for the prevention and treatment of age-associated diseases.

References

[1] Choi SW, Mason JB. Folate and carcinogenesis: an integrated scheme. J Nutr 2000;130(2):129–32.

[2] Moschou PN, Roubelakis-Angelakis KA. Polyamines and programmed cell death. J Exp Bot 2014;65(5):1285–96.

[3] Heby O, Persson L. Molecular genetics of polyamine synthesis in eukaryotic cells. Trends Biochem Sci 1990;15(4):153–8.

[4] Spector AA, Yorek MA. Membrane lipid composition and cellular function. J Lipid Res 1985;26(9):1015–35.

[5] Selhub J, Jacques PF, Wilson PW, Rush D, Rosenberg IH. Vitamin status and intake as primary determinants of homocysteinemia in an elderly population. J Am Med Assoc 1993;270(22):2693–8.

[6] Nygard O, Refsum H, Ueland PM, Vollset SE. Major lifestyle determinants of plasma total homocysteine distribution: the Hordaland Homocysteine Study. Am J Clin Nutr 1998;67(2):263–70.

[7] Jacques PF, Bostom AG, Wilson PW, Rich S, Rosenberg IH, Selhub J. Determinants of plasma total homocysteine concentration in the Framingham Offspring cohort. Am J Clin Nutr 2001;73(3):613–21.

[8] Soda K, Kano Y, Chiba F, Koizumi K, Miyaki Y. Increased polyamine intake inhibits age-associated alteration in global DNA methylation and 1,2-dimethylhydrazine-induced tumorigenesis. PLoS One 2013;8(5):e64357.

[9] Gredilla R, Barja G. Minireview: the role of oxidative stress in relation to caloric restriction and longevity. Endocrinology 2005;146(9):3713–17.

[10] Sanchez-Roman I, Gomez A, Perez I, et al. Effects of aging and methionine restriction applied at old age on ROS generation and oxidative damage in rat liver mitochondria. Biogerontology 2012;13(4):399–411.

[11] Richie Jr. JP, Leutzinger Y, Parthasarathy S, Malloy V, Orentreich N, Zimmerman JA. Methionine restriction increases blood glutathione and longevity in F344 rats. FASEB J 1994;8(15):1302–7.

[12] Sun L, Sadighi Akha AA, Miller RA, Harper JM. Life-span extension in mice by preweaning food restriction and by methionine restriction in middle age. J Gerontol A Biol Sci Med Sci 2009;64(7):711–22.

[13] Perrone CE, Mattocks DA, Jarvis-Morar M, Plummer JD, Orentreich N. Methionine restriction effects on mitochondrial biogenesis and aerobic capacity in white adipose tissue, liver, and skeletal muscle of F344 rats. Metabolism 2010;59(7):1000–11.

[14] Sinha R, Cooper TK, Rogers CJ, et al. Dietary methionine restriction inhibits prostatic intraepithelial neoplasia in TRAMP mice. Prostate 2014;74(16):1663–73.

[15] Cavuoto P, Fenech MF. A review of methionine dependency and the role of methionine restriction in cancer growth control and life-span extension. Cancer Treat Rev 2012;38(6):726–36.

[16] Seshadri S, Beiser A, Selhub J, et al. Plasma homocysteine as a risk factor for dementia and Alzheimer's disease. N Engl J Med 2002;346(7):476–83.

[17] Ravaglia G, Forti P, Maioli F, et al. Elevated plasma homocysteine levels in centenarians are not associated with cognitive impairment. Mech Ageing Dev 2000;121(1-3):251–61.

[18] Fioravanti M, Ferrario E, Massaia M, et al. Low folate levels in the cognitive decline of elderly patients and the efficacy of folate as a treatment for improving memory deficits. Arch Gerontol Geriatr 1998;26(1):1–13.

[19] Durga J, van Boxtel MP, Schouten EG, et al. Effect of 3-year folic acid supplementation on cognitive function in older adults in the FACIT trial: a randomised, double blind, controlled trial. Lancet 2007;369(9557):208–16.

[20] Dali-Youcef N, Andres E. An update on cobalamin deficiency in adults. Q J Med 2009;102(1):17–28.

[21] Miller JW, Green R, Ramos MI, et al. Homocysteine and cognitive function in the Sacramento Area Latino Study on Aging. Am J Clin Nutr 2003;78(3):441–7.

[22] Tu MC, Huang CW, Chen NC, et al. Hyperhomocysteinemia in Alzheimer dementia patients and cognitive decline after 6 months follow-up period. Acta Neurol Taiwan 2010;19(3):168–77.

[23] McCaddon A, Kelly CL. Alzheimer's disease: a 'cobalaminergic' hypothesis. Med Hypotheses 1992;37(3):161–5.

[24] Smith AD. The worldwide challenge of the dementias: a role for B vitamins and homocysteine? Food Nutr Bull 2008;29(2 Suppl.):S143–72.

[25] Kruman II, Culmsee C, Chan SL, et al. Homocysteine elicits a DNA damage response in neurons that promotes apoptosis and hypersensitivity to excitotoxicity. J Neurosci 2000;20(18):6920–6.

[26] Bazzano LA, He J, Ogden LG, et al. Dietary intake of folate and risk of stroke in US men and women: NHANES I Epidemiologic Follow-up Study. National Health and Nutrition Examination Survey. Stroke 2002;33(5):1183–8.

[27] Bjelland I, Tell GS, Vollset SE, Refsum H, Ueland PM. Folate, vitamin B12, homocysteine, and the MTHFR 677C->T polymorphism in anxiety and depression: the Hordaland Homocysteine Study. Arch Gen Psychiatry 2003;60(6):618–26.

[28] Snell K, Natsumeda Y, Weber G. The modulation of serine metabolism in hepatoma 3924A during different phases of cellular proliferation in culture. Biochem J 1987;245(2):609–12.

[29] Tedeschi PM, Markert EK, Gounder M, et al. Contribution of serine, folate and glycine metabolism to the ATP, NADPH and purine requirements of cancer cells. Cell Death Dis 2013;4:e877.

[30] Harker LA, Slichter SJ, Scott CR, Ross R. Homocystinemia. Vascular injury and arterial thrombosis. N Engl J Med 1974;291(11):537–43.

[31] Brown-Borg HM, Rakoczy S, Wonderlich JA, Armstrong V, Rojanathammanee L. Altered dietary methionine differentially impacts glutathione and methionine metabolism in long-living growth hormone-deficient Ames dwarf and wild-type mice. Longev Healthspan 2014;3(1):10.

[32] Mooney S, Leuendorf JE, Hendrickson C, Hellmann H. Vitamin B6: a long known compound of surprising complexity. Molecules 2009;14(1):329–51.

[33] Roubenoff R, Roubenoff RA, Selhub J, et al. Abnormal vitamin B6 status in rheumatoid cachexia. Association with spontaneous tumor necrosis factor alpha production and markers of inflammation. Arthritis Rheum 1995;38(1):105–9.

[34] Friso S, Jacques PF, Wilson PW, Rosenberg IH, Selhub J. Low circulating vitamin B(6) is associated with elevation of the inflammation marker C-reactive protein independently of plasma homocysteine levels. Circulation 2001;103(23):2788–91.

[35] Morris MS, Sakakeeny L, Jacques PF, Picciano MF, Selhub J. Vitamin B-6 intake is inversely related to, and the requirement is affected by, inflammation status. J Nutr 2010;140(1):103–10.

[36] Saibeni S, Cattaneo M, Vecchi M, et al. Low vitamin B(6) plasma levels, a risk factor for thrombosis, in inflammatory bowel disease: role of inflammation and correlation with acute phase reactants. Am J Gastroenterol 2003;98(1):112–17.

[37] Meydani SN, Ribaya-Mercado JD, Russell RM, Sahyoun N, Morrow FD, Gershoff SN. Vitamin B-6 deficiency impairs interleukin 2 production and lymphocyte proliferation in elderly adults. Am J Clin Nutr 1991;53(5):1275–80.

[38] Sontag JM, Wasek B, Taleski G, et al. Altered protein phosphatase 2A methylation and Tau phosphorylation in the young and

aged brain of methylenetetrahydrofolate reductase (MTHFR) deficient mice. Front Aging Neurosci 2014;6:214.

[39] Kang SS, Zhou J, Wong PW, Kowalisyn J, Strokosch G. Intermediate homocysteinemia: a thermolabile variant of methylenetetrahydrofolate reductase. Am J Hum Genet 1988;43(4):414—21.

[40] Lai CQ, Parnell LD, Troen AM, et al. MAT1A variants are associated with hypertension, stroke, and markers of DNA damage and are modulated by plasma vitamin B-6 and folate. Am J Clin Nutr 2010;91(5):1377—86.

[41] Paul L, Cattaneo M, D'Angelo A, et al. Telomere length in peripheral blood mononuclear cells is associated with folate status in men. J Nutr 2009;139(7):1273—8.

[42] Jiang-Hua Q, De-Chuang J, Zhen-Duo L, Shu-de C, Zhenzhen L. Association of methylenetetrahydrofolate reductase and methionine synthase polymorphisms with breast cancer risk and interaction with folate, vitamin B6, and vitamin B 12 intakes. Tumour Biol 2014;35(12):11895—901.

[43] Hubner RA, Houlston RS. MTHFR C677T and colorectal cancer risk: a meta-analysis of 25 populations. Int J Cancer 2007;120(5):1027—35.

[44] Chen J, Giovannucci E, Kelsey K, et al. A methylenetetrahydrofolate reductase polymorphism and the risk of colorectal cancer. Cancer Res 1996;56(21):4862—4.

[45] van der Put NM, van der Molen EF, Kluijtmans LA, et al. Sequence analysis of the coding region of human methionine synthase: relevance to hyperhomocysteinaemia in neural-tube defects and vascular disease. Q J Med 1997;90(8):511—17.

[46] Gemmati D, Ongaro A, Scapoli GL, et al. Common gene polymorphisms in the metabolic folate and methylation pathway and the risk of acute lymphoblastic leukemia and non-Hodgkin's lymphoma in adults. Cancer Epidemiol Biomarkers Prev 2004;13(5):787—94.

[47] Finkelstein JD. Methionine metabolism in mammals. J Nutr Biochem 1990;1(5):228—37.

[48] Ubagai T, Lei KJ, Huang S, Mudd SH, Levy HL, Chou JY. Molecular mechanisms of an inborn error of methionine pathway. Methionine adenosyltransferase deficiency. J Clin Invest 1995;96(4):1943—7.

[49] Kotb M, Mudd SH, Mato JM, et al. Consensus nomenclature for the mammalian methionine adenosyltransferase genes and gene products. Trends Genet 1997;13(2):51—2.

[50] Gil B, Casado M, Pajares MA, et al. Differential expression pattern of S-adenosylmethionine synthetase isoenzymes during rat liver development. Hepatology 1996;24(4):876—81.

[51] Mato JM, Lu SC. Role of S-adenosyl-L-methionine in liver health and injury. Hepatology 2007;45(5):1306—12.

[52] Cai J, Mao Z, Hwang JJ, Lu SC. Differential expression of methionine adenosyltransferase genes influences the rate of growth of human hepatocellular carcinoma cells. Cancer Res 1998;58(7):1444—50.

[53] Martinez-Chantar ML, Latasa MU, Varela-Rey M, et al. L-methionine availability regulates expression of the methionine adenosyltransferase 2A gene in human hepatocarcinoma cells: role of S-adenosylmethionine. J Biol Chem 2003;278(22):19885—90.

[54] Garcia-Trevijano ER, Latasa MU, Carretero MV, Berasain C, Mato JM, Avila MA. S-adenosylmethionine regulates MAT1A and MAT2A gene expression in cultured rat hepatocytes: a new role for S-adenosylmethionine in the maintenance of the differentiated status of the liver. FASEB J 2000;14(15):2511—18.

[55] Cai J, Sun WM, Hwang JJ, Stain SC, Lu SC. Changes in S-adenosylmethionine synthetase in human liver cancer: molecular characterization and significance. Hepatology 1996;24(5):1090—7.

[56] Lu SC, Alvarez L, Huang ZZ, et al. Methionine adenosyltransferase 1A knockout mice are predisposed to liver injury and exhibit increased expression of genes involved in proliferation. Proc Natl Acad Sci USA 2001;98(10):5560—5.

[57] Mato JM, Alvarez L, Ortiz P, Pajares MA. S-adenosylmethionine synthesis: molecular mechanisms and clinical implications. Pharmacol Ther 1997;73(3):265—80.

CHAPTER 37

Iron Metabolism in Aging

Laura Silvestri[1,2]

[1]Division of Genetics and Cell Biology, IRCCS San Raffaele Scientific Institute, Milan, Italy
[2]Vita Salute University, Milan, Italy

KEY FACTS

- Iron is essential for life due to its chemical properties as electron acceptor and donor.
- There are no physiologic mechanisms for iron excretion except cell desquamation.
- Alteration of the systemic iron control by the hepcidin/ferroportin axis may cause disorders like genetic iron refractory iron deficiency anemia and hemochromatosis.
- Manipulation of the hepcidin activating pathway could be beneficial for anemia of aging when due to inappropriately high hepcidin because of a mild inflammatory status.

Dictionary of Terms

- *IRP* (Iron Regulatory Proteins): IRP1 and IRP2 are cytosolic proteins that sense iron concentration and posttranscriptionally regulate the expression/translation of iron-related genes to maintain cellular iron homeostasis. IRP1 is an RNA binding protein in its apo form when iron is low. When iron increases, IRP1 binds the iron-sulfur cluster, synthesized mainly by mitochondria, and becomes a cytosolic aconitase. IRP2, highly homologous to IRP1, does not bind the iron-sulfur cluster and its level is regulated by an iron-dependent ubiquitin-proteasome degradation.
- *IRE* (Iron Responsive Elements): IREs are highly conserved IRP binding sites located in the 5′ or 3′ UTR of target genes. Binding of IRPs to the 5′ IRE inhibits protein translation through steric hindrance, whereas binding to the 3′ UTR increases mRNA stability of the corresponding gene.
- *Hepcidin (HAMP)*: It is a liver peptide hormone that belongs to the defensin family. It is produced mainly by the hepatocytes and by inflammatory macrophages. It is the main regulator of iron homeostasis since it binds and degrades the unique iron exporter ferroportin (FPN1), thus controlling the amount of iron released into the bloodstream. Hepcidin levels are regulated mainly at transcriptional levels by the BMP-SMAD and STAT3 pathways and by CREBH in response to endoplasmic reticulum stress and gluconeogenesis.
- *Ferroportin* (FPN1): Ferroportin is a plasma membrane protein with 12 transmembrane domains well conserved among species. It is ubiquitously expressed but it has a central role as an iron exporter in enterocytes, macrophages, and hepatocytes. In enterocytes, FPN1 is localized in the basolateral membrane, facing the bloodstream. FPN1 levels are controlled through several mechanisms: (i) by an IRP-dependent posttranscriptional mechanism through the IRE in the 5′ UTR; (ii) by a posttranslational mechanism through hepcidin-mediated degradation; and (iii) by a transcriptional regulation mediated by heme and inflammation. Mutations in FPN1 cause type 4 hemochromatosis.
- *DMT1*: DMT1 is a transmembrane protein responsible for proton-coupled divalent metal transport, such as ferrous iron, manganese, and cobalt. Protons provide the driving force for metal transport. It is ubiquitously expressed on the plasma membrane and in the endosomal

compartment of cells. In duodenum, DMT1 is localized in the apical membrane of enterocytes, facing the intestinal lumen to import metals from the diet.
- *Hephaestin* (HEPH): Hephaestin is a membrane-anchored, multicopper ferroxidase, expressed in the enterocytes, that oxidizes ferrous to ferric iron, a form suitable for binding to TF. HEPH is highly homologous to CP.
- *Ceruloplasmin* (CP): It is a multicopper ferroxidase, highly similar to hephaestin, that is synthesized by an alternative splicing event as a soluble form, secreted mainly by the hepatocytes and a GPI-anchored isoform, expressed mainly by macrophages, astrocytes, and immune cells. It cooperates with FPN1 to export iron from all cells except enterocytes. Mutations in CP gene cause aceruloplasminemia, a genetic recessive syndrome characterized by low serum iron, increased liver iron, diabetes, and brain iron deposition.
- *Hemojuvelin* (HJV): It is a GPI-anchored protein that belongs to the Repulsive Guidance Molecule (RGM) family. It is expressed in hepatocytes, skeletal muscle, and heart. In the hepatocytes it functions as a BMP-coreceptor that positively modulates hepcidin. Inactivation of HJV in human and mice causes juvenile hemochromatosis (type 2) and severe iron overload. HJV exists as a membrane isoform, the BMP-coreceptor, and as a soluble protein, released in the extracellular space by a furin-mediated cleavage. The soluble counterpart acts as a BMP-SMAD decoy molecule.
- *Transferrin Receptor 2* (TFR2): It is a type II transmembrane protein highly expressed in hepatocytes and erythroid cells. TFR2 is mutated in type 3 hemochromatosis. In the hepatocytes TFR2 regulates hepcidin expression through a mechanism not yet fully characterized. TFR2 is shed from the plasma membrane in the presence of apo-transferrin (apo-TF) and it is stabilized on the cell surface in the presence of holo-TF. In erythroid cells TFR2 binds EPO receptor and modulates EPO receptor sensitivity to EPO.
- *HFE (HLA-H)*: It is ubiquitously expressed and highly homologous to the major histocompatibility complex class I molecules. In the liver HFE activates hepcidin through a still unknown mechanism. Although HFE is ubiquitously expressed, mice with conditional inactivation of Hfe in the liver develop a hemochromatosis-like phenotype, suggesting that the major function of Hfe is the regulation of hepcidin. Indeed mutations in HFE genes cause type 1 hemochromatosis.
- *Autophagy*: It is an adaptive process that promotes cell survival during starvation by maintaining energy production. It requires sequestration of cytoplasmic components in double-membrane structures called autophagosomes and delivers to lysosomes for degradation.

INTRODUCTION

Iron is an essential element for all living organisms and because of its ability to coordinate with proteins and to donate or accept electrons, it is required for the activity of enzymes involved in oxidation-reduction reactions essential for energy production, metabolites, and DNA synthesis as well as reactive oxygen species generation and oxygen transport [1,2]. However, due to its high chemical reactivity, iron can be toxic when accumulated due to its propensity to produce reactive oxygen species through the Fenton's reaction. This extremely reactive species can damage DNA, lipids, and proteins. To avoid this effect, iron concentration in the body should be tightly regulated and iron should be always bound to proteins. Iron circulates in the bloodstream bound to transferrin (TF), an iron carrier glycoprotein that contains two specific high-affinity binding sites for iron. Iron is taken up by cells through the TF–TF receptor endosomal cycle. In the cell, iron is utilized mainly by mitochondria for iron-sulfur cluster and heme synthesis. Excess cytoplasmic iron is bound to ferritin, an iron-storage protein conserved among species, that forms a globular structure composed by 24 subunits and that can accommodate 4500 iron atoms in a soluble and nontoxic form. Alternatively it is released in the extracellular environment through the iron exporter ferroportin (FPN1).

The human body contains about 3–4 g of iron: 60–70% is contained in hemoglobin in circulating erythrocytes and their bone marrow precursors; 10% in myoglobin, cytochromes, iron-containing enzymes; 20–30% is stored in ferritin and hemosiderin (mainly in the liver and macrophages); and 1% is bound to TF in the bloodstream. Due to the poor iron bioavailability, organisms have developed very efficient mechanisms for iron absorption and conservation.

Under steady state conditions, 1–2 mg of iron is absorbed through the diet every day and the same amount is lost through physiologic duodenal and skin cell desquamation. Due to its high reactivity, iron is delivered to organs and tissues mainly bound to the carrier protein TF. This represents the exclusive source of iron for erythropoiesis, where it is used for hemoglobin synthesis. Other cells and tissues, such as liver, pancreas, and heart, require less iron and can also utilize nontransferrin bound iron (NTBI). Recycling of iron occurs through the process of erythrophagocytosis operated by spleen macrophages which provide most of the usable iron for erythropoiesis through degradation of senescent red blood cells, thus recycling about 25 mg iron per day.

Since there are no active mechanisms for iron excretion and due to its ability to produce reactive oxygen species through the Fenton's reaction, iron concentration in the body should be tightly regulated both at cellular and systemic levels. Cellular iron homeostasis is regulated by the Iron Responsive Element/Iron Regulatory Proteins (IRE/IRP) system. Systemic iron homeostasis is regulated by the hepcidin-ferroportin axis. Perturbation of the mechanisms that control cellular and systemic iron metabolism causes pathologic conditions characterized by accumulation of iron or iron deficiency and anemia [2].

DIETARY IRON AND INTESTINAL IRON ABSORPTION

Iron absorption is a complex process that occurs mainly at the brush border villi of duodenal enterocytes; it reflects not only the iron content in the diet but also the bioavailability of iron. The iron content of a normal diet is about 12–15 mg per day. Dietary iron is composed by nonheme ferric (Fe^{3+}) iron, present in plant-derived and animal-derived foods, and heme iron, contained mainly in hemoglobin and myoglobin from animal-derived foods. Although nonheme iron is less bioavailable, heme iron has a higher rate of absorption and contributes about 40% of total absorbed iron [3].

Ferric iron (Fe^{3+}), reduced to ferrous species (Fe^{2+}) by low pH of the gastric juice or by the reductase DCytB, is taken up by the mucosal cells through Divalent Metal Transporter (DMT1), and only about 1 mg (or 5–10%) of dietary iron is transferred into the systemic circulation.

Nonheme-iron present in vegetables and cereals is less bioavailable to absorption than heme iron contained in red meat, since iron is present in complex with phytates, polyphenols, and tannates in tea. This explains why iron deficiency is especially common in developing countries where the diet is predominantly cereal-based. The duodenum is the site of maximal iron absorption due to the low pH. The luminal transporter DMT1 is upregulated in iron deficiency and by hypoxia inducible factor 2 (HIF2-α) activated by systemic and local hypoxia. In the duodenal cells iron may be incorporated into ferritin, when iron is not needed and the iron exporter FPN1 is not on the basolateral cell surface. In this case it is lost when the dying cells are exfoliated into the lumen. When the body needs iron, FPN1 is stabilized on the plasma membrane of enterocytes and transports iron through the basolateral membrane into the circulation in cooperation with hephaestin, a copper-containing oxidase that converts Fe^{2+} to Fe^{3+} to allow iron transfer to circulating TF.

Absorbed heme-iron, which enters the enterocytes through a mechanism not yet clarified, is released by heme oxygenase 1 (HO-1), then is either used, stored, or exported through FPN1.

REGULATION OF CELLULAR IRON HOMEOSTASIS

The IRE/IRP system is involved in the posttranscriptional regulation of iron-related genes and controls intracellular iron levels through regulation of its uptake, utilization, storage, and release. IRPs are cytosolic proteins that coordinate the posttranslational regulation of iron-related genes whereas IREs are highly conserved regions of 25–30 nucleotides that form a hairpin structure in the 5′ or 3′ untranslated region (UTR) of target genes [4] (Fig. 37.1A). The hairpin is characterized by a CAGUGN (N can be A, C, or U) apical loop sequence conserved in all IREs. To efficiently bind IRPs, the C at position 1 forms a base pair with the G at position 5 [5]. IREs have been identified several years ago in a single copy in the 5′ UTR of H and L ferritin (FT), and in multiple copies in the 3′ UTR of TFR1 [6] (Fig. 37.1A). In conditions of low iron, IRP1 and 2 bind to the IRE of FTs and TFR1, thus decreasing iron storage and increasing iron uptake: they block translation of FTs since the binding to the 5′ UTR IRE impedes ribosome binding and translation of the protein by steric hindrance, and stabilize TFR1 transcript, thus increasing its protein levels, by binding of IRPs to the 3′ UTRs of TFR1 that masks the mRNA to endonuclease digestion. On the contrary, in the presence of iron the activity of IRPs as RNA binding proteins is reduced, thus increasing FTs translation, decreasing TFR1 mRNA levels and allowing iron storage and reduced iron uptake. Thanks to computational and biochemical studies other iron-related genes have been characterized for the presence of IREs in the 5′ or 3′ UTR. They include genes involved in iron entry and export, as DMT1 and FPN1; the erythroid form of 5-aminolevulinic acid synthase 2 (ALAS2), involved in heme biosynthesis, mitochondrial aconitase (ACO2); the cell cycle regulator CDC14A and one of the genes important for the regulation of signaling pathway in hypoxia; and the HIF2-α gene. DMT1 and CDC14A mRNA are stabilized in conditions of low iron, as TFR1, since they have a single IRE in the 3′ UTR of the gene. Translation of FPN1, HIF2-α, mitochondrial aconitase, and ALAS2 is repressed as that of ferritin in conditions of low iron due to IRPs binding to IREs in the 5′ UTR of

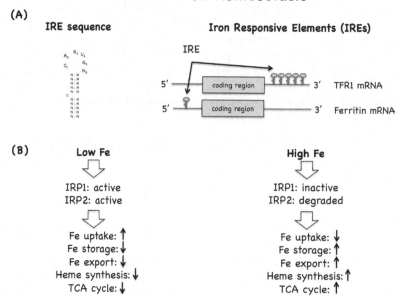

FIGURE 37.1 Regulation of cellular iron homeostasis. (A) *Left panel*: schematic representation of an IRE motif. *Right panel*: the IREs are present in the 5′ or 3′ UTR of target genes. Binding of IRPs in the 5′ UTR blocks translation through steric hindrance (eg, Ferritin H and L, FPN1, ALAS2, ACO2), whereas binding of IRPs in the 3′ UTR stabilizes the mRNA (eg, TFR1 and DMT1).
(B) *Left panel*: in conditions of low iron, both IRP1 and IRP2 act as RNA binding proteins thus increasing iron uptake (TFR1) and decreasing iron storage (ferritin H and L), iron export (FPN1), heme synthesis (ALAS2), and TCA cycle (ACO2). *Right panel*: in conditions of high iron, IRP1 binds the iron-sulfur cluster and becomes a cytosolic aconitase, whereas IRP2 is degraded via proteasome. In this condition, iron uptake is reduced whereas iron storage and export are both increased. In addition, heme synthesis and TCA cycle are upregulated.

these genes (Fig. 37.1B). To further increase the level of complexity, DMT1, FPN1, and CDC14A mRNA exist also as non-IRE isoforms that originate from cell specific alternative splicing. For example, duodenal enterocytes and erythroid cells express an FPN1 transcript which lacks the IRE and is not repressed in iron-deficiency to skip the IRP-dependent repression of intestinal iron uptake, even when duodenal cells are iron poor, and to facilitate restriction of erythropoiesis in response to low systemic iron [7]. In addition to alternative splicing, other posttranslational mechanisms participate in the regulation of iron-related genes. FPN1, the only known iron exporter, is regulated also by hepcidin, the peptide hormone upregulated in conditions of high iron that triggers FPN1 internalization and degradation. In addition, HIF2-α is regulated by oxygen levels, through a mechanism that involves oxygen-dependent prolyl hydroxylases (PHDs) and the E3 ubiquitin ligase von Hippel Lindau (VHL). In conditions of high oxygen/high iron, HIF2-α is degraded via proteasome (Fig. 37.1B).

IRP1 and IRP2 are cytosolic proteins that coordinate the regulation of IRE-containing genes through iron-dependent binding to IRE in the 5′ or 3′ UTR regions of target genes. They share 56% of sequence identity and are homologous to mitochondrial aconitase. IRPs sense intracellular iron with two mechanisms: IRP1 can accommodate a [4Fe-4S] cluster in its active site, thus becoming a cytosolic aconitase that converts citrate in isocitrate. When iron is low, IRP1 functions as a RNA binding protein; in iron replete conditions, IRP1 has a cytosolic enzymatic activity. Iron regulates IRP2 through iron-dependent ubiquitination and proteasomal degradation: FBXL5, a member of the E3 ubiquitin ligase complex, has a hemerythrin domain that stabilizes the protein through the binding to oxygen and iron, thus increasing the ubiquitin ligase activity. When iron is low or in hypoxia, FBXL5 is destabilized thus reducing the ubiquitin ligase activity and increasing the stability of IRP2 [8] (Fig. 37.1B).

IRPs are ubiquitously expressed, with IRP1 highly expressed in kidney, liver, and brown fat, and IRP2 more present in the central nervous system [9]. The function of IRP1 and 2 is not redundant. $Irp2^{-/-}$ mice display microcytic hypochromic anemia due to erythroid reduction of Tfr1 and low iron availability for hemoglobin synthesis [10]. In addition they show significant iron accumulation in the duodenum, liver, and neurons. Pharmacological upregulation of $Irp1$ reverted the neurodegenerative phenotype of $Irp2^{-/-}$ mice but not microcytic anemia, suggesting that Irp2 has a fundamental role in the regulation of iron homeostasis in erythroid cells [11]. $Irp1^{-/-}$ animals maintained an iron-deficient (ID) diet showing an increased number of red blood cells and higher erythropoietin (EPO) levels than wild-type littermates, due to constitutive activation of Hif2α in the kidney, the major activator of Epo in response to hypoxia. Since Hif2-α has an IRE in its 5′ UTR, genetic inactivation of *Irp1* derepresses Hif2-α translation in conditions of low iron, leading to uncontrolled EPO production and erythrocytosis [12].

REGULATION OF SYSTEMIC IRON HOMEOSTASIS

Upon dietary absorption, iron is safely transported to organs and tissues by TF, a glycoprotein that can bind two ferric iron atoms. Before TF binding, ferrous iron is oxidized by hephaestin in duodenal cells or by ceruloplasmin (CP) in macrophages and hepatocytes, basolateral membrane oxidases that mediate iron efflux from the cell in cooperation with FPN1. Iron circulates mainly bound to TF and is taken up by cells through transferrin receptor 1 (TFR1). Although TFR1 is ubiquitously expressed, it is particularly relevant as an iron source in erythroid, hepatic, muscular, and nervous system cells. Indeed genetic inactivation of *Tfr1* in mice is embryonically lethal and is characterized by disrupted erythropoiesis and neurologic development defects [13]. At physiologic pH, TFR1 binds 2Fe-TF and endocytosis is initiated through clathrin-coated pits that invaginate the plasma membrane to form endosomes containing the ligand-receptor complexes. Following acidification (pH 5–5.5), iron is released from TF, reduced to ferrous iron and exported to cytosol through DMT1. TFR1-TF complexes are recycled back to the cell surface where TF dissociates from TFR1 at neutral pH. In the cell a little amount of iron can form the so-called labile iron pool (LIP), a pool of bioactive iron that can be imported into mitochondria for iron-sulfur cluster and heme biosynthesis, or can be incorporated into enzymes and proteins. Excess iron can be stored into ferritin and/or released in the circulation through the basolateral transmembrane protein FPN1, the only known iron exporter in mammals that releases iron into the bloodstream. Circulating iron is stored in the liver, mainly hepatocytes, and in macrophages, both expressing high levels of FPN1. When circulating iron is low, the metal is released from stores and dietary absorption is increased. The opposite occurs in conditions of high iron. This homeostatic mechanism has been conserved in evolution to avoid iron excess and iron-dependent damage of DNA, lipids, and proteins.

The organ that senses iron levels and that coordinates iron entry into the bloodstream is the liver.

The liver is a heterogeneous organ composed of several cell types: hepatocytes constitute about 60–70% of the cells, the others are resident macrophages (Kuppfer cells), liver sinusoidal endothelial cells, and hepatic stellate cells. Excess iron is accumulated mainly in hepatocytes. These cells regulate dietary iron absorption and iron release from the stores in a paracrine manner through the production of a soluble hormone named hepcidin. Hepcidin binds to FPN1 and triggers FPN1 internalization and degradation, thus reducing iron export from enterocytes, macrophages, and the liver (Fig. 37.2).

Iron-Dependent and Erythropoiesis-Dependent Regulation of Hepcidin

The mature form of hepcidin is a 25-amino acid cationic peptide that contains four disulfide bonds. It is synthesized as an 84 amino acid propeptide that is maturated before secretion by a proconvertase cleavage operated by the proconvertase furin [14]. Hepcidin is expressed at high level by hepatocytes and to a lesser extent by pro-inflammatory macrophages and it is regulated by iron and inflammation at transcriptional levels. In conditions of high iron, hepcidin mRNA is upregulated very rapidly and the mature protein is released in the circulation to bind and induce FPN1 degradation [14], thus reducing iron entry from duodenal enterocytes and iron exit from the stores (liver and spleen macrophages). Iron in enterocytes is lost due to physiologic cell desquamation, whereas iron in the liver and spleen is stored in ferritin and could be released (likely through the mechanism of ferritin degradation named ferritinophagy, see below) when hepcidin decreases and FPN1 is on the cell surface.

Since hepcidin is the key orchestrator of systemic iron homeostasis, its level is tightly regulated and closely related to body iron concentration.

The main signaling pathway activating hepcidin is the BMP-SMAD pathway. On the cell surface, the activation of BMP type I [15] and type II receptors [16], together with the coreceptor hemojuvelin (HJV) [16], leads to SMAD1/5/8 phosphorylation and interaction with the cargo protein SMAD4. This multiprotein complex translocates to the nucleus and activates BMP target genes, as hepcidin. In the liver, BMP receptors (BMPRs) are activated mainly by BMP6 (Fig. 37.3). The BMP-SMAD pathway is central in the regulation of hepcidin in vivo since inactivation of *Bmp6* and liver conditional deletion of *Bmprs* and *Smad4* in mice cause iron overload due to impaired hepcidin production. In humans, mutations in *HFE2* gene, coding for HJV, cause a severe form of hemochromatosis [17].

Systemic Iron Homeostasis

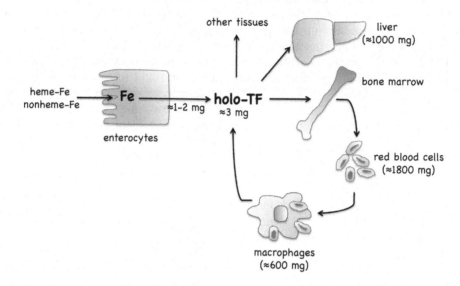

FIGURE 37.2 **Regulation of systemic iron homeostasis.** Dietary iron is absorbed by duodenal enterocytes and entered into the bloodstream through FPN1. Iron is delivered to organs and tissues bound to TF. Most of the circulating iron is used by erythropoiesis for hemoglobin synthesis. Excess iron is stored mainly in the liver. Iron is recycled by macrophages that degrade senescent red blood cells.

Hepcidin Regulation

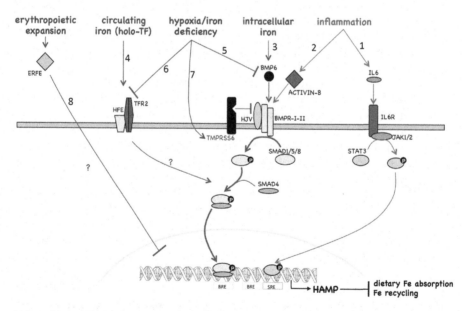

FIGURE 37.3 **Regulation of hepcidin in the hepatocytes.** Hepcidin is increased in inflammation mainly by IL6 (1) through activation of the JAK2-STAT3 pathway and the positive modulation of the BMP-SMAD pathway through activin B (2). Intracellular iron activates hepcidin through transcriptional upregulation of BMP6 and the BMP-SMAD pathway (3); circulating iron acts on HFE-TFR2 to positively modulate hepcidin, through a mechanism not yet fully clarified (4). Iron deficiency inhibits hepcidin through different mechanisms: (i) it reduces BMP6 expression (5); (ii) it reduces TFR2 plasma membrane levels and it favors the HFE-TFR2 dissociation (6); (iii) it stabilizes the hepcidin inhibitor TMPRSS6 (7). Increased erythropoiesis downregulates hepcidin through the release of ERFE (8).

The BMP-SMAD pathway is positively modulated by HFE, the ubiquitously expressed atypical MHC class I molecule that interact with TFR1, and by TFR2, a protein highly homologous to TFR1 but with a different function and expression profile. Inactivation of HFE and TFR2 in humans and mice causes iron overload of different severity due to inappropriate hepcidin levels. The mechanism of action of these two genes and their role in hepcidin regulation are not yet fully clarified (Fig. 37.3). It has been proposed that in a condition of iron deficiency TFR2 is reduced on the plasma membrane and HFE is bound to TFR1, while in a condition of iron overload TFR2 is stabilized on the cell surface, HFE is displaced by holo-TF to the binding to TFR1 and becomes able to interact with TFR2. The interaction of the two proteins activates a signaling pathway that positively modulates hepcidin (Fig. 37.3) [18]. However, this model has been recently criticized for the following reasons: (i) mice with inactivation of both Hfe and Tfr2 develop a more severe phenotype than animals with single inactivation of the single genes [19]; this has been confirmed also in humans [19]; (ii) patients with mutations in HFE maintain the ability to activate hepcidin in response to increase iron, although the fold chance is impaired compared to normal individuals, whereas patients with inactivation of TFR2 loose the iron-mediated hepcidin response [20]; and (iii) in cells expressing physiologic levels of HFE and TFR2, the two proteins do not interact in a proximity ligation assay [21]. Overall these findings suggest that HFE and TFR2 have nonredundant function in the regulation of hepcidin.

Hepcidin Regulation by Iron

Iron activates hepcidin with different mechanisms: (i) increased liver iron transcriptionally activates BMP6 [22]; (ii) increased iron bound to TF stabilizes TFR2 on the cell surface [23]; and (iii) binding of iron loaded TF to its receptor TFR1 displaces it from the binding to HFE that likely becomes able to interact with TFR2 and activates hepcidin [18].

The sole hepcidin inhibitor whose role has been clearly demonstrated in vivo is TMPRSS6, encoding matriptase-2, a type II transmembrane serine protease expressed exclusively in the liver. Genetic inactivation of TMPRSS6 in humans [24] and mice [25,26] causes IRIDA, a rare genetic disorder characterized by iron deficiency anemia due to high hepcidin levels. It was demonstrated that TMPRSS6 downregulates hepcidin by cleaving HJV [27–30] (Fig. 37.3). Although the formal proof on the role of TMPRSS6 in HJV cleavage in vivo is still lacking, inactivation of Hjv [31] and Bmp6 [32] in Tmprss6 KO mice reverts the IRIDA phenotype, indicating that Tmprss6 is functionally upstream of Hjv and Bmp6 and negatively modulates the BMP-SMAD pathway, in accordance with HJV being the physiologic TMPRSS6 substrate.

TMPRSS6 is supposed to be active in iron deficiency, since inactivation of the protease causes the inability to downregulate hepcidin even if the patients are severely ID, and inactive in iron overload (Fig. 37.3). However, genetic inactivation of the protease in the hemochromatosis Hfe [33] and Tfr2 [34] KO animals reverts the iron overload phenotype because of low hepcidin levels, suggesting a function role of Tmprss6 even in iron overload. Hypoxia responsive elements have been identified in the TMPRSS6 promoter region and in vitro chemical hypoxia transcriptionally activates TMPRSS6 [35,36]. However, in vivo this transcriptional regulation seems not to be relevant [37], suggesting that TMPRSS6 is regulated mainly with posttranslational mechanisms. Indeed acute iron deficiency stabilizes Tmprss6 in rats [38] through a mechanism than reduces the iron-mediated decrease of the protease [39].

Hepcidin is efficiently suppressed in iron deficiency through several mechanisms: (i) BMP6 downregulation [39]; (ii) destabilization of the HFE-TFR2 complex [18]; (iii) increased stability of TMPRSS6 [38]; and (iv) increased EPO-mediated erythropoiesis [40] (Fig. 37.3).

Hepcidin Regulation by Erythropoiesis

About 70% of iron in the body is present in the form of heme in hemoglobin in red blood cells. Erythropoiesis consumes 25 mg of iron daily that derives mainly from degradation of senescent red blood cells. In conditions of increased erythropoiesis, as during growth, hypoxia, iron deficiency anemia, bleeding, and in genetic diseases of red blood cells, especially of the globin synthesis, hepcidin is strongly downregulated through a mechanism not fully clarified, to favor iron supply to the bone marrow for hemoglobin synthesis. In all these conditions EPO levels are increased. However, the inhibition of hepcidin is not directly caused by EPO, but it is mediated by the expanded erythropoietic activity. In fact inhibition of erythropoiesis in mice by using cytotoxic drugs as carboplatin, doxorubicin or irradiation blocks hepcidin downregulation by EPO [41,42], suggesting that erythropoiesis controls hepcidin levels by releasing soluble inhibitory factor(s) (Fig. 37.3).

Several molecules have been proposed to act as hepcidin inhibitors: Growth Differentiation Factor 15 (GDF15) [43], Twisted Gastrulation BMP Signaling Modulator 1 (TWSG1) [44], Platelet-Derived Growth Factor B chain dimer (PDGF-BB) [45], and erythroferrone (ERFE) [46]. GDF15 and TWSG1 are bone morphogenetic protein family members that have been found increased in the sera of β-thalassemia patients characterized by ineffective erythropoiesis and low hepcidin. In vitro they inhibit hepcidin at high concentration [43,44]. However, *Gdf15* KO mice are still able to reduce hepcidin as wild type animals during bleeding-induced erythropoiesis, suggesting that in vivo Gdf15 does not function as an erythroid hepcidin inhibitor [47]. PDGF-BB, associated with vascular remodeling and tumor angiogenesis, has been recently proposed to be the hypoxia-mediated hepcidin inhibitor [45]. In the liver PDGF-BB inhibits the transcription factor CREBH that has been shown to regulate hepcidin in response to endoplasmic reticulum stress and gluconeogenesis. Circulating PDGF-BB promotes also extramedullary erythropoiesis through activation of EPO production [48] and could thus inhibit hepcidin indirectly through erythropoietic expansion. ERFE, encoded by *FAM132B*, is a member of the orphan TNF-alpha superfamily that has been recently identified as the erythroid hepcidin inhibitor in vivo. It has been discovered searching for genes encoding for secreted proteins transcriptionally upregulated in the bone marrow before the suppression of hepcidin in the liver [46]. *Erfe* is highly expressed in the bone marrow and fetal liver and its expression is induced in bone marrow and spleen by stress erythropoiesis, such as phlebotomy and EPO injection, and in ineffective erythropoiesis, such as in β-thalassemia mice, but it is not directly affected by hypoxia and inflammation. Mice with genetic inactivation of *Erfe* fail to suppress hepcidin after erythropoietic stimulation by phlebotomy and exhibit a delay in the recovery from anemia compared with wild-type animals [46]. How Erfe inhibits hepcidin is still unknown, although the BMP-SMAD pathway does not seem to be affected.

TMPRSS6 is one of the liver-expressed proteins that could potentially be involved in the erythroid-mediated hepcidin inhibition. In β-thalassemia mice inactivation of *Tmprss6* increases hepcidin and ameliorates both iron overload and erythropoiesis, although EPO levels remain similar to β-thalassemia mice with *Tmprss6* [49] and *Erfe* is high [46,50]. This suggests that a functional Tmprss6 is required for Erfe-mediated hepcidin suppression.

Inflammation-Dependent Regulation of Hepcidin

Of all the other micronutrients, iron is fundamental in the regulation of host-pathogen interactions because of its role as a microbial growth factor. Pathogens developed multiple strategies to acquire iron and an adequate supply of this metal is linked to pathogen proliferation and virulence. Indeed several studies both in animal models and humans clearly established that the host's iron status influences infection and that iron supplementation worsens infectious diseases. It is well known that an iron supplementation trial in anemic young children in Pemba island, a region endemic for malaria infection, was prematurely stopped due to severe adverse effects and death in iron treated children [51]. During infection, the host activates a physiologic response that parallels activation of the immune system, that is, the reduction of circulating iron thus causing the so-called anemia of infection/inflammation, a process that now is known to be mediated mainly, but not exclusively, by hepcidin.

Hepcidin, like many other proteins expressed in the liver, is a type II acute phase protein transcriptionally activated in conditions of inflammation and/or infection. Activation is mediated mainly by the TLR4-IL6-STAT3 signaling pathway (Fig. 37.3): in infection, hepcidin activation by IL6 degrades FPN1 thus reducing dietary iron absorption and sequestering iron in hepatocytes and macrophages. Degradation of FPN1 in macrophages, the cells that recycle iron from senescent red blood cells, rapidly induces hypoferremia due to high iron consumption in erythropoiesis, thus controlling the proliferation of extracellular pathogens. This process is of relevance also in chronic inflammatory conditions, such as autoimmunity and cancer, when a chronic disease triggers a low but persistent activation of pro-inflammatory cytokines thus causing the so-called "anemia of chronic disease" (ACD). In this case restriction of iron for erythropoiesis and impairment of erythroid differentiation are caused by pro-inflammatory cytokines. Hepcidin upregulation is not the only mechanism activated by the host for pathogen growth restriction. Recently a hepcidin-independent hypoferremia has been identified [52]. Stimulation of TLR2/6 by FSL-1, a synthetic lipoprotein derived from *Mycoplasma salivarium*, or PAM3CSK4, a synthetic triacylated lipopeptide (LP) that mimics the acylated amino terminus of bacterial LPs, causes a strong downregulation of FPN1 mRNA and protein in bone marrow-derived macrophages in the liver and spleen without changing hepcidin expression. As a consequence, serum iron and TF saturation are both reduced and spleen and liver iron content increased, due to FPN1-mediated iron retention. Overall, these results strengthen the primary role of the hepcidin-ferroportin axis in hypoferremia due to infection and/or inflammation.

DEREGULATION OF THE HEPCIDIN-FERROPORTIN AXIS

In physiologic conditions hepcidin levels are strongly controlled and closely related to body iron levels to avoid excess iron accumulation or iron deficiency.

TABLE 37.1 Genetic Conditions Associated with Hepcidin Deregulation

Disease	Gene	Interactors	Hepcidin levels	Phenotype
HH type 1 OMIM 235200	HFE	β2M	Decreased/inappropriate	Hepatocyte iron overload
		TFR1		
		TFR2		
		HJV		
HH type 2 OMIM 602390	HAMP	FPN1	Low/undetectable	Severe systemic iron overload
	HJV	TMPRSS6	Low/undetectable	Severe systemic iron overload
		NEO1		
		TFR2		
		HFE		
HH type 3 OMIM 604250	TFR2	HFE	Low	Hepatocyte iron overload
		HJV		
HH type 4 OMIM 606069	FPN	HEPCIDIN	High (gain of function mutations)	Systemic iron overload
			Normal/low (loss of function mutations)	Iron-restricted anemia and liver and spleen iron overload (macrophages)
β-Thalassemia OMIM 613985	β-GLOBIN	α-GLOBIN	Low/inappropriate	Severe iron overload
IRIDA OMIM 206200	TMPRSS6	HJV	High/inappropriate	Iron deficiency anemia, refractory to oral iron

HH, hereditary hemochromatosis; IRIDA, iron refractory iron deficiency anemia.

An imbalance between iron levels and hepcidin production can cause two pathological conditions (Table 37.1). The first, characterized by low hepcidin and high iron accumulation, is typical of hereditary hemochromatosis and β-thalassemia. The second, characterized by high hepcidin production and low iron, is typical of iron refractory iron deficiency anemia (IRIDA) and of acquired conditions, such as anemia of inflammation (AI) or ACD. The identification of the genes mutated in hemochromatosis and IRIDA shed light on the signaling pathways that control hepcidin expression [53].

Disorders with Hepcidin Deficiency

Hemochromatosis was described for the first time in the 19th century as "bronze diabetes" because of the skin pigmentation due to excessive melanin deposition stimulated by excess iron. It is caused by mutations in five different genes (Table 37.1) that have central roles in the regulation of hepcidin expression and function. The most common form of Hemochromatosis is due to mutations in the *HFE* gene (Type 1), which encodes an MHC class I-like protein, ubiquitously expressed. HFE in the hepatocytes regulates hepcidin through a mechanism not well clarified (Fig. 37.2). The most common variant is the Cys282Tyr mutation that impairs HFE binding to the shuttle protein β2 microglobulin for plasma membrane localization. As a consequence, in vitro HFE^{C282Y} is retained into the endoplasmic reticulum [54]. Mutations in HFE are characterized by low penetrance and a late onset phenotype with mild accumulation of iron especially in the liver and pancreas. In particular the C282Y variant is characterized by a founder effect and has a high frequency in individuals of North European origin [55].

Mutations in *hepcidin* or *HJV* genes cause a severe form of juvenile hemochromatosis (type 2; Table 37.1). HJV is a GPI-anchored protein that belongs to the Repulsive Guidance Molecule family, highly expressed in the liver, heart, and skeletal muscle. In the liver, HJV acts as a BMP-coreceptor and is one of the main activators of hepcidin expression. HJV pathogenic variants are impaired in cell surface localization [56] or have impaired BMPs binding [17]. Patients with *HJV* mutations have low/undetectable hepcidin levels that cause a rapid rate of iron accumulation and toxicity, especially in the heart.

Mutations in *TFR2* cause hemochromatosis type 3 (Table 37.1), characterized by an early onset and a more severe phenotype compared to type 1 hemochromatosis. TFR2 is a type II transmembrane protein highly homologous to TFR1, highly expressed in the liver and erythroid precursors. In the liver TFR2 activates hepcidin through a mechanism not fully clarified, whereas in erythroid cells TFR2 binds to EPO receptor [57] and regulates EPO receptor sensitivity to EPO [40]. Like its homolog TFR1, TFR2 exists as a membrane and soluble isoform that is shed from the cell surface in conditions of low holo-TF [58]. In contrast, TFR2 is stabilized on the cell surface when holo-TF is increased, thus coordinating erythropoiesis and hepcidin production according to circulating iron availability [59].

Mutations in the *ferroportin* gene cause hemochromatosis type 4 (Table 37.1). It is an autosomal dominant disease responsible for two distinct phenotypes. Mutations that impaired FPN1 plasma membrane localization are responsible for a loss of function of the protein that causes macrophage iron retention and iron-restricted erythropoiesis [60], a phenotype different from pure hemochromatosis. Mutations that impair hepcidin binding to FPN1, rendering FPN1 resistant to hepcidin degradation, cause a hemochromatosis phenotype [61] with high hepcidin levels [62].

Animal models of the hemochromatosis diseases are available and recapitulate the human phenotype.

Inappropriate hepcidin regulation is a feature also of iron loading anemias, such as β-thalassemia (Table 37.1), characterized by ineffective erythropoiesis and low hepcidin levels that cause deregulated iron absorption [63]. Notwithstanding iron overload, hepcidin cannot be upregulated because the expanded erythropoiesis inhibits its transcription.

Disorders of Hepcidin Excess

In transgenic mice, hepcidin overexpression causes iron deficiency anemia due to FPN1 degradation and the inability to absorb iron from the diet and to release iron from the stores [64], thus confirming that hepcidin levels should be tightly regulated. Genetic inactivation of the hepcidin inhibitor *Tmprss6* in mice causes a phenotype highly similar to what is observed in transgenic mice with hepcidin overexpression. *Tmprss6* KO animals are characterized by microcytic hypochromic anemia refractory to oral and partially refractory to parenteral iron. The phenotype is due to inappropriately high levels of hepcidin and increased degradation of FPN1. Interestingly, mutant mice also have high EPO levels. In humans, *TMPRSS6* mutations cause a pediatric disorder named IRIDA (Table 37.1), a rare recessive anemia characterized by small hypochromic red blood cells and normal/high hepcidin, that results in inappropriately considering the iron deficiency conditions in which hepcidin is usually strongly downregulated. Interestingly, *TMPRSS6* common polymorphic variants have been associated with hemoglobin levels, erythrocytes indices, and serum iron in genome-wide association studies and also with hepcidin levels [65–67].

Hepcidin has been found to be increased also in acquired conditions, such as in chronic infections or inflammations, in cancer, and in chronic renal failure [68]. Hepcidin excess, due to increased pro-inflammatory cytokines, is responsible for iron sequestration in macrophages and other cells, thus causing restriction of iron for erythropoiesis that ultimately leads to anemia of chronic disorders or ACD. In chronic renal failure, anemia-reduced hepcidin excretion and impaired EPO production contribute to anemia.

The continuous progress in our understanding of iron metabolism has stimulated translational research to manipulate the hepcidin/FPN1 pathway [69] for the correction of diseases characterized by hepcidin deregulation. In particular hepcidin agonists (such as mini-hepcidins, TMPRSS6 inhibitors) have been proposed to treat iron overload in thalassemia, while hepcidin antagonists (antibodies, ASO) are under development to block excessive hepcidin in inflammation. Although prevalently tested in preclinical models, some of these compounds have entered clinical trials [70].

IRON HOMEOSTASIS AND HEPCIDIN IN THE ELDERLY

Anemia is common in the elderly. The prevalence of anemia in the elderly increases with age and about 20% of people over 85 years have anemia [71]. Population studies have shown that it is associated mainly with iron deficiency and chronic diseases [72]. Anemia in this population causes an increased risk of hospitalization and death, although the specific mechanisms by which anemia affect health-related outcomes in the elderly are unknown [73]. One-fifth of older people with anemia have AI that can be caused by chronic disease, such as rheumatoid arthritis, inflammatory bowel disease, cancer, and chronic kidney disease [74]. Iron deficiency, the most common disorder of nutrition worldwide, is frequent in the elderly because of inappropriate diet. Iron deficiency can be also caused by chronic disorders or secondarily to reduced dietary absorption of the metal or gastrointestinal blood loss, which are especially frequent with aging.

Of special interest is the understanding of whether iron homeostasis is maintained in the elderly. In conditions of normal iron homeostasis, serum hepcidin levels show a strong difference according to gender and ages. While in males hepcidin concentration is stable through all decades of life, in females it changes strikingly, being lower in premenopausal than in postmenopausal women [66,75]. Studies on hepcidin in the elderly are limited. In both males and females, hepcidin shows a strong correlation with serum ferritin that is maintained at all ages in males and decreases in female [66,75]. Homozygosity for HFE C282Y, the causes of HH type 1, has been found with the same frequency in old people than in the general population in a single study, suggesting that the iron overload associated with HFE mutation is not detrimental to survival [76].

In aged subjects, the mechanisms that mediate iron and inflammation-dependent hepcidin regulation are maintained and intact, since old subjects with iron deficiency anemia have low hepcidin, and aged people with AI have high hepcidin levels, due to a combination of factors including the inflammatory status, decreased hepcidin excretion, or low EPO [74].

Indeed hepcidin is elevated in aged people and strongly correlated with the inflammatory marker C-reactive protein (CRP) and EPO [74]. Interestingly, high hepcidin levels are found also in aged people with the so-called unexplained anemia, that may be caused by physiologic changes, such as stem cell aging, low grade chronic inflammation, and subclinical kidney impairment. The studies have been limited by the availability of hepcidin tests and by problems in comparing results obtained with different methods to doses of hepcidin [75].

Even studies in animal models are limited. However, in agreement with the theory of inflammation in aging, naturally aged rats present a pro-inflammatory status, as assessed by upregulation of the inflammatory cytokine IFN-γ, kidney function impairment and iron metabolism changes with low serum iron and increased soluble TFR and serum ferritin [77]. According to impaired renal function, EPO levels were decreased and hepcidin increased suggesting that dysregulation of iron metabolism contributes to age-related anemias [77].

Whether iron deregulation contributes to the development of age-related disorders is a controversial issue. Epidemiological studies suggest that elevated iron stores are a risk factor for aging complications, such as cardiovascular and metabolic diseases. Metabolic syndrome is a collection of risk factors, such as high levels of triglycerides, low HDL-cholesterol, high blood pressure, and high glucose levels, and predisposes to heart disease, stroke, and diabetes [78]. In metabolic syndrome and diabetes, iron deposition in liver and pancreas can increase oxidative damage and contribute to insulin deficiency and resistance. In addition, increased iron deposition in adipocytes due to increased hepcidin could cause insulin resistance through impaired adiponectin production [79]. In cardiovascular disease, iron withheld in macrophages and foam cells due to high hepcidin may increase the susceptibility to the formation of atherosclerotic plaques [78]. Aging is characterized also by accumulation of functional impairment in cells, tissues, and organs. In this context, iron homeostasis could influence the metabolic pathways and stress response of the organism due to the crucial role of this metal in ROS production and mitochondrial activities. However, the molecular mechanisms remain to be investigated.

Autophagy and Iron Metabolism

Recently, the role of autophagy in determining the lifespan of model organisms has become increasingly important to help in understanding the mechanism of aging. Autophagy is a process that is used by the cell to generate energy-rich compounds when needed, for example, in conditions of restriction of nutrients. Inhibition of autophagy has been associated with degenerative changes such as those observed in aging. On the contrary, autophagy stimulation can extend the lifespan in animal models: for example caloric restriction, that activates autophagy, is known to increase the lifespan of several animal models including nonhuman primates, such as rhesus monkeys, and also to reduce the incidence of aging-related diseases, such as diabetes, cardiovascular disease, cancer, and brain atrophy [80]. The positive effect of caloric restriction has been confirmed also in humans in the control of lifestyle in patients with metabolic disorders [81].

Autophagy is essential in iron metabolism, since ferritin, the protein responsible for iron storage, is selectively degraded by lysosomes in iron deficiency through a process that has been named "ferritinophagy." Among the autophagosome-enriched proteins, NCOA4 was identified as the cargo receptor that specifically binds heavy and light ferritin chains [82]. Ferritin cages are degraded via autophagy to release iron in conditions of iron deprivation through a mechanism not yet clarified. Interestingly, cells with inactivation of NCOA4 are unable to degrade ferritin and have low levels of bioavailable iron, as shown by increased IRP2 levels and TFR1 stabilization. The mechanism seems to be conserved in vivo since mice with genetic inactivation of Ncoa4 show accumulation of iron in the spleen suggesting an imbalance between iron storage and recycling [83].

We speculate that in aging the impairment of autophagy decreases ferritin degradation and limits bioavailable iron, thus providing the cells with a false signal of iron deficiency leading to an increase in iron absorption. In agreement, aging is associated with increased hepatic iron accumulation in rats [84] and stimulation of autophagy through caloric restriction ameliorates this phenotype [85]. Further advances in understanding iron metabolism will offer new tools to explore iron metabolism in the process of aging.

FUTURE PERSPECTIVE

With the continuous increase of life expectancy and decrease in birth rate, the health of the old population is becoming a socially important problem at least in developed countries. The contribution of iron both in terms of iron deficiency, AI, or iron overload is well defined in pilot studies that should be extended. In addition, mechanisms of iron absorption and recycling are still clarified only partially. We need to understand how heme absorption occurs, how autophagy is important in iron recycling and in iron deficiency, and whether and how iron accumulated in inflammation in macrophages may cause damage. A full understanding of how iron impacts on disorders of aging will allow prevention or targeted treatment with novel compounds that are at present under development to modulate systemic iron homeostasis [86].

SUMMARY

- Iron is an important component of the diet, a nutritional element essential for many cell functions, the most important being oxygen transport.
- Iron homeostasis is tightly regulated both at cell and systemic level in order to avoid both deficiency and overload.
- Hepcidin, the key regulator of iron metabolism, is produced mainly by the liver.
- Hepcidin binds and degrades the unique iron exporter ferroportin, thus controlling iron entry into the bloodstream.
- Hepcidin is controlled mainly by iron, inflammatory cytokines, and erythroid needs.
- Both iron deficiency and iron maldistribution are frequent in anemia of the elderly.
- Mild inflammation characterizing aging-related disorders may explain the observed iron dysregulation.

Acknowledgments

I am indebted to Clara Camaschella for suggestions and helpful discussion. I apologize to all colleagues whose work was not cited due to space limitations.

References

[1] Andrews NC, Schmidt PJ. Iron homeostasis. Annu Rev Physiol 2007;69:69−85.
[2] Ganz T, Nemeth E. Hepcidin and iron homeostasis. Biochim Biophys Acta 2012;1823(9):1434−43.
[3] Hurrell R, Egli I. Iron bioavailability and dietary reference values. Am J Clin Nutr 2010;91(5):1461S−1467SS.
[4] Wilkinson N, Pantopoulos K. The IRP/IRE system in vivo: insights from mouse models. Front Pharmacol 2014;5:176.
[5] Piccinelli P, Samuelsson T. Evolution of the iron-responsive element. RNA 2007;13(7):952−66.
[6] Hentze MW, Caughman SW, Casey JL, Koeller DM, Rouault TA, Harford JB, et al. A model for the structure and functions of iron-responsive elements. Gene 1988;72(1−2):201−8.
[7] Zhang DL, Hughes RM, Ollivierre-Wilson H, Ghosh MC, Rouault TA. A ferroportin transcript that lacks an iron-responsive element enables duodenal and erythroid precursor cells to evade translational repression. Cell Metab 2009;9(5):461−73.
[8] Salahudeen AA, Thompson JW, Ruiz JC, Ma HW, Kinch LN, Li Q, et al. An E3 ligase possessing an iron-responsive hemerythrin domain is a regulator of iron homeostasis. Science 2009;326(5953):722−6.
[9] Zhang DL, Ghosh MC, Rouault TA. The physiological functions of iron regulatory proteins in iron homeostasis—an update. Front Pharmacol 2014;5:124.
[10] Galy B, Ferring D, Minana B, Bell O, Janser HG, Muckenthaler M, et al. Altered body iron distribution and microcytosis in mice deficient in iron regulatory protein 2 (IRP2). Blood 2005;106(7):2580−9.
[11] Ghosh MC, Tong WH, Zhang D, Ollivierre-Wilson H, Singh A, Krishna MC, et al. Tempol-mediated activation of latent iron regulatory protein activity prevents symptoms of neurodegenerative disease in IRP2 knockout mice. Proc Natl Acad Sci USA 2008;105(33):12028−33.
[12] Anderson SA, Nizzi CP, Chang YI, Deck KM, Schmidt PJ, Galy B, et al. The IRP1-HIF-2alpha axis coordinates iron and oxygen sensing with erythropoiesis and iron absorption. Cell Metab 2013;17(2):282−90.
[13] Levy JE, Jin O, Fujiwara Y, Kuo F, Andrews NC. Transferrin receptor is necessary for development of erythrocytes and the nervous system. Nat Genet 1999;21(4):396−9.
[14] Valore EV, Ganz T. Posttranslational processing of hepcidin in human hepatocytes is mediated by the prohormone convertase furin. Blood Cells Mol Dis 2008;40(1):132−8.
[15] Nemeth E, Tuttle MS, Powelson J, Vaughn MB, Donovan A, Ward DM, et al. Hepcidin regulates cellular iron efflux by binding to ferroportin and inducing its internalization. Science 2004;306(5704):2090−3.
[16] Mayeur C, Lohmeyer LK, Leyton P, Kao SM, Pappas AE, Kolodziej SA, et al. The type I BMP receptor Alk3 is required for the induction of hepatic hepcidin gene expression by interleukin-6. Blood 2014;123(14):2261−8.
[17] Babitt JL, Huang FW, Wrighting DM, Xia Y, Sidis Y, Samad TA, et al. Bone morphogenetic protein signaling by hemojuvelin regulates hepcidin expression. Nat Genet 2006;38(5):531−9.
[18] Goswami T, Andrews NC. Hereditary hemochromatosis protein, HFE, interaction with transferrin receptor 2 suggests a molecular mechanism for mammalian iron sensing. J Biol Chem 2006;281(39):28494−8.
[19] Delima RD, Chua AC, Tirnitz-Parker JE, Gan EK, Croft KD, Graham RM, et al. Disruption of hemochromatosis protein and transferrin receptor 2 causes iron induced liver injury in mice. Hepatology 2012;56(2):585−93.
[20] Girelli D, Trombini P, Busti F, Campostrini N, Sandri M, Pelucchi S, et al. A time course of hepcidin response to iron

challenge in patients with HFE and TFR2 hemochromatosis. Haematologica 2011;96(4):500—6.
[21] Rishi G, Crampton EM, Wallace DF, Subramaniam VN. In situ proximity ligation assays indicate that hemochromatosis proteins Hfe and transferrin receptor 2 (Tfr2) do not interact. PloS One 2013;8(10):e77267.
[22] Pietrangelo A, Caleffi A, Henrion J, Ferrara F, Corradini E, Kulaksiz H, et al. Juvenile hemochromatosis associated with pathogenic mutations of adult hemochromatosis genes. Gastroenterology 2005;128(2):470—9.
[23] Kautz L, Meynard D, Monnier A, Darnaud V, Bouvet R, Wang RH, et al. Iron regulates phosphorylation of Smad1/5/8 and gene expression of Bmp6, Smad7, Id1, and Atoh8 in the mouse liver. Blood 2008;112(4):1503—9.
[24] Finberg KE, Heeney MM, Campagna DR, Aydinok Y, Pearson HA, Hartman KR, et al. Mutations in TMPRSS6 cause iron-refractory iron deficiency anemia (IRIDA). Nat Genet 2008;40(5):569—71.
[25] Du X, She E, Gelbart T, Truksa J, Lee P, Xia Y, et al. The serine protease TMPRSS6 is required to sense iron deficiency. Science 2008;320(5879):1088—92.
[26] Folgueras AR, de Lara FM, Pendas AM, Garabaya C, Rodriguez F, Astudillo A, et al. Membrane-bound serine protease matriptase-2 (Tmprss6) is an essential regulator of iron homeostasis. Blood 2008;112(6):2539—45.
[27] Silvestri L, Pagani A, Nai A, De Domenico I, Kaplan J, Camaschella C. The serine protease matriptase-2 (TMPRSS6) inhibits hepcidin activation by cleaving membrane hemojuvelin. Cell Metab 2008;8(6):502—11.
[28] Silvestri L, Guillem F, Pagani A, Nai A, Oudin C, Silva M, et al. Molecular mechanisms of the defective hepcidin inhibition in TMPRSS6 mutations associated with iron-refractory iron deficiency anemia. Blood 2009;113(22):5605—8.
[29] Silvestri L, Rausa M, Pagani A, Nai A, Camaschella C. How to assess causality of TMPRSS6 mutations? Hum Mutat 2013;34(7):1043—5.
[30] De Falco L, Silvestri L, Kannengiesser C, Moran E, Oudin C, Rausa M, et al. Functional and clinical impact of novel TMPRSS6 variants in iron-refractory iron-deficiency anemia patients and genotype-phenotype studies. Hum Mutat 2014;35(11):1321—9.
[31] Finberg KE, Whittlesey RL, Fleming MD, Andrews NC. Downregulation of Bmp/Smad signaling by Tmprss6 is required for maintenance of systemic iron homeostasis. Blood 2010;115(18):3817—26.
[32] Lenoir A, Deschemin JC, Kautz L, Ramsay AJ, Roth MP, Lopez-Otin C, et al. Iron-deficiency anemia from matriptase-2 inactivation is dependent on the presence of functional Bmp6. Blood 2011;117(2):647—50.
[33] Finberg KE, Whittlesey RL, Andrews NC. Tmprss6 is a genetic modifier of the Hfe-hemochromatosis phenotype in mice. Blood 2011;117(17):4590—9.
[34] Nai A, Pellegrino RM, Rausa M, Pagani A, Boero M, Silvestri L, et al. The erythroid function of transferrin receptor 2 revealed by Tmprss6 inactivation in different models of transferrin receptor 2 knockout mice. Haematologica 2014;99(6):1016—21.
[35] Lakhal S, Schodel J, Townsend AR, Pugh CW, Ratcliffe PJ, Mole DR. Regulation of type II transmembrane serine proteinase TMPRSS6 by hypoxia-inducible factors: new link between hypoxia signaling and iron homeostasis. J Biol Chem 2011;286(6):4090—7.
[36] Maurer E, Gutschow M, Stirnberg M. Matriptase-2 (TMPRSS6) is directly up-regulated by hypoxia inducible factor-1: identification of a hypoxia-responsive element in the TMPRSS6 promoter region. Biol Chem 2012;393(6):535—40.
[37] Rausa M, Pagani A, Nai A, Campanella A, Gilberti ME, Apostoli P, et al. Bmp6 expression in murine liver non parenchymal cells: a mechanism to control their high iron exporter activity and protect hepatocytes from iron overload? PloS One 2015;10(4):e0122696.
[38] Zhang AS, Anderson SA, Wang J, Yang F, DeMaster K, Ahmed R, et al. Suppression of hepatic hepcidin expression in response to acute iron deprivation is associated with an increase of matriptase-2 protein. Blood 2011;117(5):1687—99.
[39] Zhao N, Nizzi CP, Anderson SA, Wang J, Ueno A, Tsukamoto H, et al. Low intracellular iron increases the stability of matriptase-2. J Biol Chem 2015;290(7):4432—46.
[40] Nai A, Lidonnici MR, Rausa M, Mandelli G, Pagani A, Silvestri L, et al. The second transferrin receptor regulates red blood cell production in mice. Blood 2015;125(7):1170—9.
[41] Pak M, Lopez MA, Gabayan V, Ganz T, Rivera S. Suppression of hepcidin during anemia requires erythropoietic activity. Blood 2006;108(12):3730—5.
[42] Vokurka M, Krijt J, Sulc K, Necas E. Hepcidin mRNA levels in mouse liver respond to inhibition of erythropoiesis. Physiol Res 2006;55(6):667—74.
[43] Tanno T, Bhanu NV, Oneal PA, Goh SH, Staker P, Lee YT, et al. High levels of GDF15 in thalassemia suppress expression of the iron regulatory protein hepcidin. Nat Med 2007;13(9):1096—101.
[44] Tanno T, Porayette P, Sripichai O, Noh SJ, Byrnes C, Bhupatiraju A, et al. Identification of TWSG1 as a second novel erythroid regulator of hepcidin expression in murine and human cells. Blood 2009;114(1):181—6.
[45] Sonnweber T, Nachbaur D, Schroll A, Nairz M, Seifert M, Demetz E, et al. Hypoxia induced downregulation of hepcidin is mediated by platelet derived growth factor BB. Gut 2014;63(12):1951—9.
[46] Kautz L, Jung G, Valore EV, Rivella S, Nemeth E, Ganz T. Identification of erythroferrone as an erythroid regulator of iron metabolism. Nat Genet 2014;46(7):678—84.
[47] Casanovas G, Vujic Spasic M, Casu C, Rivella S, Strelau J, Unsicker K, et al. The murine growth differentiation factor 15 is not essential for systemic iron homeostasis in phlebotomized mice. Haematologica 2013;98(3):444—7.
[48] Xue Y, Lim S, Yang Y, Wang Z, Jensen LD, Hedlund EM, et al. PDGF-BB modulates hematopoiesis and tumor angiogenesis by inducing erythropoietin production in stromal cells. Nat Med 2012;18(1):100—10.
[49] Nai A, Pagani A, Mandelli G, Lidonnici MR, Silvestri L, Ferrari G, et al. Deletion of TMPRSS6 attenuates the phenotype in a mouse model of beta-thalassemia. Blood 2012;119(21):5021—9.
[50] Silvestri L, Gelsomino GR, Nai A, Rausa M, Pagani A, Camaschella C. Is Tmprss6 required for hepcidin inhibition by erythroferrone? Blood-56th ASH Annual Meeting and Exposition 2014; 124(21).
[51] Sazawal S, Black RE, Ramsan M, Chwaya HM, Stoltzfus RJ, Dutta A, et al. Effects of routine prophylactic supplementation with iron and folic acid on admission to hospital and mortality in preschool children in a high malaria transmission setting: community-based, randomised, placebo-controlled trial. Lancet 2006;367(9505):133—43.
[52] Guida C, Altamura S, Klein FA, Galy B, Boutros M, Ulmer AJ, et al. A novel inflammatory pathway mediating rapid hepcidin-independent hypoferremia. Blood 2015.
[53] Hentze MW, Muckenthaler MU, Galy B, Camaschella C. Two to tango: regulation of mammalian iron metabolism. Cell. 2010;142(1):24—38.
[54] Waheed A, Parkkila S, Zhou XY, Tomatsu S, Tsuchihashi Z, Feder JN, et al. Hereditary hemochromatosis: effects of C282Y and H63D mutations on association with beta2-microglobulin,

intracellular processing, and cell surface expression of the HFE protein in COS-7 cells. Proc Natl Acad Sci USA 1997;94(23): 12384–9.

[55] Milman N, Pedersen P. Evidence that the Cys282Tyr mutation of the HFE gene originated from a population in Southern Scandinavia and spread with the Vikings. Clin Genet 2003;64 (1):36–47.

[56] Silvestri L, Pagani A, Fazi C, Gerardi G, Levi S, Arosio P, et al. Defective targeting of hemojuvelin to plasma membrane is a common pathogenetic mechanism in juvenile hemochromatosis. Blood 2007;109(10):4503–10.

[57] Forejtnikova H, Vieillevoye M, Zermati Y, Lambert M, Pellegrino RM, Guihard S, et al. Transferrin receptor 2 is a component of the erythropoietin receptor complex and is required for efficient erythropoiesis. Blood 2010;116(24):5357–67.

[58] Pagani A, Vieillevoye M, Nai A, Rausa M, Ladli M, Lacombe C, et al. Regulation of cell surface transferrin receptor-2 by iron-dependent cleavage and release of a soluble form. Haematologica 2015.

[59] Silvestri L, Nai A, Pagani A, Camaschella C. The extrahepatic role of TFR2 in iron homeostasis. Front Pharmacol 2014;5:93.

[60] Schimanski LM, Drakesmith H, Merryweather-Clarke AT, Viprakasit V, Edwards JP, Sweetland E, et al. In vitro functional analysis of human ferroportin (FPN) and hemochromatosis-associated FPN mutations. Blood 2005;105(10):4096–102.

[61] Drakesmith H, Schimanski LM, Ormerod E, Merryweather-Clarke AT, Viprakasit V, Edwards JP, et al. Resistance to hepcidin is conferred by hemochromatosis-associated mutations of ferroportin. Blood 2005;106(3):1092–7.

[62] Sham RL, Phatak PD, Nemeth E, Ganz T. Hereditary hemochromatosis due to resistance to hepcidin: high hepcidin concentrations in a family with C326S ferroportin mutation. Blood 2009;114(2):493–4.

[63] Origa R, Galanello R, Ganz T, Giagu N, Maccioni L, Faa G, et al. Liver iron concentrations and urinary hepcidin in beta-thalassemia. Haematologica 2007;92(5):583–8.

[64] Nicolas G, Bennoun M, Porteu A, Mativet S, Beaumont C, Grandchamp B, et al. Severe iron deficiency anemia in transgenic mice expressing liver hepcidin. Proc Natl Acad Sci USA 2002;99(7):4596–601.

[65] Benyamin B, Ferreira MA, Willemsen G, Gordon S, Middelberg RP, McEvoy BP, et al. Common variants in TMPRSS6 are associated with iron status and erythrocyte volume. Nat Genet 2009;41(11):1173–5.

[66] Traglia M, Girelli D, Biino G, Campostrini N, Corbella M, Sala C, et al. Association of HFE and TMPRSS6 genetic variants with iron and erythrocyte parameters is only in part dependent on serum hepcidin concentrations. J Med Genet 2011;48(9):629–34.

[67] Nai A, Pagani A, Silvestri L, Campostrini N, Corbella M, Girelli D, et al. TMPRSS6 rs855791 modulates hepcidin transcription in vitro and serum hepcidin levels in normal individuals. Blood 2011;118(16):4459–62.

[68] Weiss G, Goodnough LT. Anemia of chronic disease. N Engl J Med 2005;352(10):1011–23.

[69] Ganz T, Nemeth E. The hepcidin-ferroportin system as a therapeutic target in anemias and iron overload disorders. Hematology Am Soc Hematol Educ Program 2011;2011:538–42.

[70] van Eijk LT, John AS, Schwoebel F, Summo L, Vauleon S, Zollner S, et al. Effect of the antihepcidin Spiegelmer lexaptepid on inflammation-induced decrease in serum iron in humans. Blood 2014;124(17):2643–6.

[71] den Elzen WP, Willems JM, Westendorp RG, de Craen AJ, Assendelft WJ, Gussekloo J. Effect of anemia and comorbidity on functional status and mortality in old age: results from the Leiden 85-plus Study. CMAJ 2009;181(3–4):151–7.

[72] Tettamanti M, Lucca U, Gandini F, Recchia A, Mosconi P, Apolone G, et al. Prevalence, incidence and types of mild anemia in the elderly: the "Health and Anemia" population-based study. Haematologica 2010;95(11):1849–56.

[73] Riva E, Tettamanti M, Mosconi P, Apolone G, Gandini F, Nobili A, et al. Association of mild anemia with hospitalization and mortality in the elderly: the Health and Anemia population-based study. Haematologica 2009;94(1):22–8.

[74] den Elzen WP, de Craen AJ, Wiegerinck ET, Westendorp RG, Swinkels DW, Gussekloo J. Plasma hepcidin levels and anemia in old age. The Leiden 85-Plus Study. Haematologica 2013;98 (3):448–54.

[75] Galesloot TE, Vermeulen SH, Geurts-Moespot AJ, Klaver SM, Kroot JJ, van Tienoven D, et al. Serum hepcidin: reference ranges and biochemical correlates in the general population. Blood 2011;117(25):e218–25.

[76] Coppin H, Bensaid M, Fruchon S, Borot N, Blanche H, Roth MP. Longevity and carrying the C282Y mutation for haemochromatosis on the HFE gene: case control study of 492 French centenarians. BMJ 2003;327(7407):132–3.

[77] Costa E, Fernandes J, Ribeiro S, Sereno J, Garrido P, Rocha-Pereira P, et al. Aging is associated with impaired renal function, INF-gamma induced inflammation and with alterations in iron regulatory proteins gene expression. Aging Dis 2014;5 (6):356–65.

[78] Basuli D, Stevens RG, Torti FM, Torti SV. Epidemiological associations between iron and cardiovascular disease and diabetes. Front Pharmacol 2014;5:117.

[79] Gabrielsen JS, Gao Y, Simcox JA, Huang J, Thorup D, Jones D, et al. Adipocyte iron regulates adiponectin and insulin sensitivity. J Clin Invest 2012;122(10):3529–40.

[80] Rubinsztein DC, Marino G, Kroemer G. Autophagy and aging. Cell 2011;146(5):682–95.

[81] Levine B, Kroemer G. Autophagy in the pathogenesis of disease. Cell 2008;132(1):27–42.

[82] Mancias JD, Wang X, Gygi SP, Harper JW, Kimmelman AC. Quantitative proteomics identifies NCOA4 as the cargo receptor mediating ferritinophagy. Nature 2014;509(7498):105–9.

[83] Dowdle WE, Nyfeler B, Nagel J, Elling RA, Liu S, Triantafellow E, et al. Selective VPS34 inhibitor blocks autophagy and uncovers a role for NCOA4 in ferritin degradation and iron homeostasis in vivo. Nat Cell Biol 2014;16(11):1069–79.

[84] Bloomer SA, Han O, Kregel KC, Brown KE. Altered expression of iron regulatory proteins with aging is associated with transient hepatic iron accumulation after environmental heat stress. Blood Cells Mol Dis 2014;52(1):19–26.

[85] Xu J, Jia Z, Knutson MD, Leeuwenburgh C. Impaired iron status in aging research. Int J Mol Sci 2012;13(2):2368–86.

[86] Fung E, Nemeth E. Manipulation of the hepcidin pathway for therapeutic purposes. Haematologica 2013;98(11):1667–76.

CHAPTER

38

Dietary Mineral Intake (Magnesium, Calcium, and Potassium) and the Biological Processes of Aging

Nicolas Cherbuin

Centre for Research on Ageing, Health and Wellbeing, The Australian National University, Canberra, ACT, Australia

KEY FACTS

- Dietary Mg, Ca, and K and their concentration in the body contribute to the modulation of biological processes known to be implicated in the aging process.
- Increased dietary intake of Mg and K is associated with better cardiovascular health but the evidence is less clear for dietary Ca.
- Ca supplementation appears to be associated with a modestly increased cardiovascular risk.
- Increased dietary intake of Mg, Ca, and K is associated with lower blood pressure
- Higher K intake is associated with a decreased risk of stroke. Similar but weaker evidence is also available for Mg and Ca.
- Higher dietary intake of Ca, and to a lesser extent K and Mg, is associated with a decreased risk of type 2 diabetes.
- While the evidence demonstrating a protective effect of dietary intake of Mg, Ca, and K and cardiovascular health as well as T2D strongly suggests a similar effect on brain aging and cognitive decline, since CVD and T2D are important and known risk factors for brain and cognitive health, research in this area is lacking.

Dictionary of Terms

- *Apoptosis*: Sequence of programmed biological events as part of developmental/aging processes which lead to cell death but do not result in the release of harmful substances in the surrounding tissue.
- *Insulin resistance*: Decreased capacity of cells to respond to the action of insulin and which leads to impaired ability to store glucose in cells.

INTRODUCTION

Magnesium (Mg), Calcium (Ca), and Potassium (K) are some of the most abundant cations in the body and are major micronutrients whose daily intake is essential to maintaining good health. While their involvement in biological processes is complex and multifaceted this chapter will only review and discuss their involvement in some major mechanisms implicated in biological aging and specifically those relevant to cardiovascular, brain, and cognitive health.

In a first section, the involvement of these minerals in biological processes particularly relevant to the aging process such as inflammation, oxidative stress, DNA damage, telomere shortening, and apoptosis will be briefly summarized. Next evidence showing an association between dietary intake, cardiometabolic health, and the development of cardiovascular disease and diabetes will be presented. The third section will discuss the evidence linking dietary intake, neurodegeneration, and brain aging. And finally, the last section will review the evidence indicating an association between dietary intake, cognitive decline, and dementia.

DIETARY INTAKE AND BIOLOGICAL MECHANISMS IMPLICATED IN THE AGING PROCESS

Magnesium

Mg is involved in over 300 chemical reactions contributing to biological processes essential to life, including glycolysis, phosphorylation, growth factor messaging, cell proliferation, mitochondrial energy production, cellular glucose uptake, and regulation of smooth muscle contractility [1–3]. The recommended daily Mg intake is 4–6 mg/kg/day (~300–400 mg/day) [4] as replacement for the amount of Mg lost mostly through normal glomerular filtration. However, epidemiological evidence suggests that Mg dietary intake is inadequate in many countries and particularly in those following a western diet [5]. Importantly, because Mg is mostly stored in tissues with only 1% remaining in extracellular space, serum Mg is not a very accurate measure of depletion across the body. Moreover, a number of conditions, including type 2 diabetes (T2D), the metabolic syndrome (MetS), and renal failure, all known to be strongly associated with increasing age, are thought to both lead to and be partly caused by magnesium deficiency or hypomagnesemia [1].

The mechanisms linking low Mg levels and pathophysiology are not fully understood but have been studied extensively. Intracellular Mg is necessary in most enzymatic systems involved in DNA processing, DNA damage repair, and replication, as well as in cell cycle control and apoptosis [6]. Mg also contributes to counteracting oxidative stress and in protecting telomeres against shortening. Consistent with a critical role in these mechanisms, Mg deficiency has been shown to be associated with activation of proinflammatory processes and increased oxidative stress [7–9] which are known to lead to DNA damage, decreased telomerase activity and telomere shortening [8,10], and apoptosis [11]. In addition, Mg is a Ca antagonist. Thus, it competes for calcium cellular membrane binding sites and it stimulates Ca sequestration which contributes to the maintenance of lower free Ca concentration in cells which is important for many cellular functions and critically in the contractility of cardiac muscle cells [1].

An implication of this interaction between Mg and Ca is that their effect cannot easily be considered independently of each other and particularly when considering dietary effects as the intake of Mg and Ca is highly correlated. A good example of this problem is illustrated by a study which demonstrated that in older women higher plasma Mg and Ca levels were associated with shorter telomere length [12], which contradicts the evidence reviewed above suggesting higher Mg levels being associated with longer telomeres. However, an additional finding of this study was that the ratio between Ca and Mg levels was associated with greater telomere length. Thus, in light of the literature a more likely interpretation of these findings is that a negative association between Mg and telomere length might have been an artifact of dietary intake where those who consumed greater quantities of Ca also consumed greater quantities of Mg. However, when this association was controlled for proportionally greater Mg levels relative to Ca levels, it was associated with longer, healthier telomeres.

Mg is also implicated in glucose metabolism. It seems to be involved in insulin-mediated cellular uptake and intracellular glucose utilization such that decrease in free cellular Mg concentration appears to contribute to insulin resistance [2,13].

Evidence from epidemiological studies in humans is also consistent with a role of dietary Mg intake in mechanisms known to be implicated in the aging process and briefly discussed above. For example, low Mg intake has been found in a recent meta-analysis to be associated with increased C-reactive protein (CRP) levels, a general marker of systemic inflammation [14].

Calcium

Calcium plays a central role in a large number of critical biological mechanisms at the cellular and organism level. This includes cellular physiology, maintenance of cell membrane electric potential, muscle fibers contractility, synaptic signaling through neurotransmitter release, enzymatic function, bone formation, and many others [15]. The recommended daily Ca intake is 1000–1200 mg/day in adults [4]. However, recent research suggests that minimum intake is not met, particularly in countries where the western diet is prevalent [16]. Dietary Ca intake has been shown to directly and promptly increase serum levels and particularly when supplements are used [17].

As Mg, Ca is likely to contribute to the aging process through the modulation of inflammatory, oxidative stress, and apoptosis processes, although the pathways involved are different. Laboratory studies have shown that intracellular Ca levels are directly implicated in apoptosis, and excessive Ca flux into cells triggers apoptotic processes [18,19]. Indeed, Ca has been shown to suppress the action of 1α,25-dihydroxycholecalciferol and thereby increases apoptosis in some tissues [20]. While this effect could contribute negatively to the aging process if applied to certain organs, such as the brain, heart, or vasculature, recent findings suggest that this is not the case. Indeed, this action has been mostly demonstrated in adipose tissue where higher Ca levels have been found to produce increased fat cell apoptosis, decreased fatty acid synthesis, and increased lipolysis and therefore reduced adipose cell counts and tissue volume [21]. Because 1α,25-dihydroxycholecalciferol appears to promote abdominal fat deposition through the promotion of glucocorticoid production, Ca-mediated suppression of its activity is also likely to decrease the accumulation of abdominal fat [22]. The involvement of Ca in apoptosis has also been demonstrated in relation to epithelial cells and is thought to underpin the anticancerous effect detected in colon cancer [23,24]. In addition, cellular Ca signaling is implicated in the modulation of reactive oxygen species in human adipocytes and thereby is thought to downregulate oxidative stress and inflammation processes [25].

With these in vitro findings in mind a critical question is whether such effects can also be demonstrated in vivo. To address this question, Zemel and Sun [26] investigated the effect of a basal diet with suboptimal but not deficient Ca content, a high Ca diet (no dairy), and a high Ca-dairy diet on inflammatory markers in transgenic mice over a 3-week period. They found that the high Ca and high-dairy diets led to lower measures of oxidative stress and inflammatory cytokines (IL6). The same authors also assessed inflammatory measures in blood samples of obese men and women who had participated in two previous clinical trials of high-dairy eucaloric and hypocaloric diets over 24 weeks. Consistent with their findings in mice they found that compared to the low-dairy controls, where no change was observed pre- and postintervention, CRP levels were 11% lower in the high-dairy eucaloric diet and 29% lower in the high-dairy hypocaloric diet after treatment. Furthermore, in another study, obese mice fed a diet high in Ca supplemented with vitamin D, had higher levels of apoptosis in adipose tissue which led to a decrease in adiposity in these animals compared to obese controls [27]. This effect was not observed when vitamin D alone was administered. Surprisingly, to date an effect on body weight has not been confirmed in humans [28,29].

Another action of dietary Ca on body fat occurs through the digestive process. It has been shown in rats that dietary Ca precipitates fatty acids and bile acids thus reducing the absorption of fatty acids in the body and increasing their fecal concentration [30]. Consistent results were also demonstrated in a randomized controlled trial (RCT). Postmenopausal women ($n = 223$) receiving treatment for hyperlipidemia or osteoporosis and who were randomly allocated to receive 1 g/day of Ca over 1 year were found to have a greater change in HDL to LDL ratio (0.05 more; 95% CI 0.02–0.08) than those on placebo [26]. This effect was driven by both an increase in HDL (7%) and a decrease in LDL (6%) levels. A similar effect was not replicated in a sample of older men [29]. Together these findings suggest that high dietary Ca intake is associated with lower oxidative stress, lower levels of inflammation, and greater reductions in adipose tissue although not necessarily a decrease in body weight. Furthermore, effects may vary in men and women.

In contrast, and somewhat contradictory to the evidence reported above, Callaghan et al. [12] reported that higher plasma Ca levels were associated with shorter telomere length. Since increased Ca levels are associated with lower inflammation and oxidative stress, and since inflammation and oxidative stress have been clearly shown to be associated with longer telomeres in the literature, an opposite association would have been expected. Since this effect was observed only in older women but not in younger women or men it is possible that hormonal or other sex-related factors, and not Ca levels per se, drive this association. However, future research needs to address this question in more detail.

In addition to the effects discussed above, chronic inflammatory states have been shown to produce dysfunctions in cellular Ca signaling. It is specifically the case in the central nervous system where Ca signaling dysregulation is associated with changes in neuronal structure and function which is likely to contribute to the progression of neurodegenerative disorders [30]. Consequently, since higher Ca dietary intake contributes to the downregulation of pro-inflammatory processes it follows that lower intake should be associated with higher inflammatory levels and therefore greater impairment in cellular Ca signaling. Consequently the interaction between inflammatory processes and Ca signaling pathways contributes to a feedback loop that is likely to produce further neurodegeneration [30].

Finally, it should be noted that while Ca is necessarily involved in the calcification processes particularly relevant to the stiffening and occlusion of blood vessels that occurs in aging and which is known to be associated with coronary heart disease [31], Ca dietary intake or plasma levels are not necessarily or obviously to blame for these pathological processes. Evidence investigating the link between higher dietary intake or supplementation and higher levels of vascular calcification

TABLE 38.1 Deleterious Mechanisms Through Which Lower Levels of Mg, Ca, and K Have Been Shown to Influence Some Processes Implicated in Biological Aging

	Inflammation	Oxidative stress	Apoptosis	Impaired glucose metabolism	Intracellular signaling
Mg	↑ Deficiency → increased CRP supplementation → decreased CRP	↑ Deficiency → increased oxidative stress / DNA damage	↑ Lower levels → increased apoptosis → risk to vascular and brain cells? telomere shortening	↑ Lower levels → increased insulin resistance	↓ Lower Ca antagonist activity → calcification
Ca	↑ Lower levels → increased CRP high levels → decreased pro-inflammatory cytokines	↑ Lower level → increased oxidative stress	↓ Lower levels → decrease in apoptosis in fat cells → increased adiposity, more abdominal fat? telomere shortening	↑ Lower levels → increased fat absorption	? Lower levels → decreased cellular calcification
K	↑ Low levels → increased sodium retention → inflammation	↑ Low levels → less oxidase inhibition	?	?	↓ Lower levels → sodium retention

is unclear. One recent study (*n* = 23,652) in asymptomatic middle-aged Koreans (40.8 years; 83.5% male) found that dietary calcium was not associated with coronary calcification [32]. However, higher Ca serum levels were associated with a two-fold increased risk of calcification. Other studies have also reported similar associations between serum Ca and coronary calcification. In addition, other factors such as oxidative stress, inflammatory processes, and high phosphate levels also make substantial contributions to the calcification process [33]. Given these somewhat inconsistent findings more research needs to be conducted in this area.

Potassium

Potassium is the most abundant cation in living cells and plays a major role in maintaining an electrical potential between the inside and outside of cells, and as such, is critical to cellular excitability of muscle cells and neurons with particular relevance to motor, cardiovascular, and nervous systems' function. The recommended daily K intake is 3800–4700 mg/day [34]. However, recent research suggests that, as for Mg and Ca, minimum intake is frequently not met [35].

K also contributes to oxidative stress and inflammation processes although perhaps somewhat less so than Mg and Ca. In contrast, whereas cellular K contributes to apoptotic processes, there is no clear reason or evidence suggesting that this role is significantly modulated by either serum or dietary K. In addition, apart for some anti-inflammatory action it appears that much of K's contribution to mechanisms involved in aging processes occurs through the pathophysiology of hypertension and cardiovascular disease [36] on the one hand and through the modulation of glucose metabolism and insulin resistance on the other.

K is known to be implicated in the regulation of sodium levels in the body such that low K plasma levels lead to greater retention and reabsorption of sodium in the kidneys. As chronically high sodium levels are strong contributors to the development of hypertension, low K levels also contribute to this cascade and in the development of CVD more broadly. As high sodium levels lead to the overproduction of reactive oxygen species and therefore oxidative stress and inflammation, increased K levels contribute indirectly to reducing this action and also more directly through inhibition of oxidase activity which has been demonstrated to have a protective effect on cardiac function [37]. In addition, K contributes to the vasodilation of blood vessels and consequently low K levels further contribute to raising blood pressure. Other mechanisms including the modulation of aldosterone production, sympathetic activation, arterial stiffening, and regulation of central blood pressure parameters are also thought to be positively influenced by K levels [36] (Table 38.1).

DIETARY INTAKE AND CARDIOVASCULAR AND METABOLIC EFFECTS

As discussed above, Mg, Ca, and, possibly to a lesser extent, K body concentration and intake are involved in the modulation of biological processes (inflammation, oxidative stress, apoptosis, DNA maintenance, glucose metabolism) essential to health albeit to varying degrees and through different pathways. Since these processes are also linked to the development of cardiovascular disease [38,39], T2D [40,41], and the metabolic syndrome [42], it follows logically that intake of these minerals should be linked to the development of these pathologies. Indeed, substantial and generally consistent evidence demonstrating such associations is available. In addition, the link between K and cardiometabolic health appears to be more strongly mediated through glucose metabolism modulation and through tuning of muscle fibers tone and contractility. In this section, the evidence demonstrating an association between dietary intake of Mg, Ca, and K and cardiovascular disease and type 2 diabetes will be discussed in detail.

Magnesium

First, it has been shown that dietary Mg intake affects the mechanical properties of the vasculature with rats on an Mg deficient diet developing thicker intima-media than controls and animals on an Mg supplemented diet [43]. Moreover, a linear negative association between plasma Mg concentration and wall thickness was detected suggesting that increased intake may counteract wall thickening. In bovine cell models increased Mg concentration led to decreased vascular calcification [44]. Similar findings were also observed in humans. For example, in patients on hemodialysis Mg supplementation helped reduce the intima-media thickness in the carotid [45]. And in 2695 individuals (53 years, SD 11) free of cardiovascular disease and participating in the Framingham Heart Study, dietary MG was inversely associated with coronary calcification [46]. This was the case despite controlling for a large number of sociodemographic and cardiometabolic variables (SBP, cholesterol, BMI, smoking, fasting insulin). However, negative findings have also been reported [47].

The prediction that thickening and calcification of blood vessels as well as increased atherosclerotic plaque deposition associated with chronic inflammation would contribute to higher blood pressure, coronary heart disease, and ultimately, premature death attributable to cardiovascular disease is also generally supported by the literature. Spontaneously hypertensive rats have been found to have lower serum and tissue Mg levels [2,48]. Particularly noteworthy is that Mg supplementation in this animal model led to lower blood pressure levels but only in young animals in the prehypertensive stage and not in older animals [48,49]. This may suggest greater scope for intervention at younger ages. Although Mg supplementation has also been found to reduce blood pressure in humans and not only in younger samples. In an RCT of 48 individuals suffering from mild hypertension, supplementation with Mg (600 mg daily for 12 weeks) led to a 5.6 mmHg ($p = 0.001$) decrease in systolic and 2.8 mmHg ($p = 0.002$) in diastolic 24-h blood pressure [50]. This was also the case in middle-aged women ($n = 90$) with mild to moderate hypertension in a Mg supplementation (782 mg daily for 6 months) intervention which was associated with a decrease of 2.7 mmHg in SBP ($p = 0.18$) and 3.4 mmHg in DBP ($p = 0.003$) compared to a placebo group [51]. A third RCT of Mg supplementation over an 8-week period in 60 individuals suffering from essential hypertension showed that Mg supplementation was also associated with a small (2–3.7 mmHg) but significant decrease in systolic blood pressure [52].

Importantly, these effects appear to be clinically significant at the population level. A recent systematic review has shown that higher dietary magnesium intake and plasma concentration are associated with lower incidence of cardiovascular events [53]. And higher magnesium intake has been shown to be associated with lower mortality in a large study ($n = 7216$) of men and women aged 55 to 80 years [54]. While in a clinical environment, Mg supplementation in heart failure patients led to significant decreases in systemic inflammation (CRP) [55]. In addition in a large epidemiological study in Swedish women ($n = 34{,}670$, 49–83 years) dietary Mg intake was associated with a decreased risk of ischemic stroke but only in women with a history of hypertension (RR 0.63, 95% CI 0.42–0.93) [56].

A robust link has also been demonstrated between Mg intake and the development of T2D and MetS. Rats fed a high-fructose low-magnesium diet were shown to have higher levels of oxidative stress and lower antioxidant levels [57] which suggests that low Mg intake may at least in part contribute to the development of T2D through increased systemic inflammation and oxidative stress [58]. In addition, CVD is also a risk factor for T2D and conversely, T2D is a risk factor for CVD [59]. Therefore, any increased risk of CVD attributable to dietary Mg intake may further contribute to the risk of developing T2D and vice versa. Irrespective of the mechanisms involved, individuals with T2D or MetS have been found to have lower Mg serum levels [60], and two recent systematic reviews have confirmed that higher intake of Mg and increase in MG intake were associated with decreased risk (RR 0.77–0.86) of T2D [61,62]. Interestingly, the association between Mg levels and T2D risk appears to be partly genetically determined [63].

Calcium

Research investigating associations between dietary Ca and cardiometabolic health appears somewhat inconsistent. A line of evidence suggests that dietary Ca is beneficial, particularly in relation to blood pressure regulation while another seems to suggest negative effects associated with vascular calcification and greater incidence of coronary heart disease and possibly stroke.

Strong evidence demonstrating a blood pressure lowering effect of dietary Ca and Ca supplementation exists. A meta-analysis of 40 RCT studies ($n = 2492$) investigating the effect of Ca treatment on blood pressure found that supplementation (mean 1.2 g/day) was associated with a 1.86 mmHg (95%CI 0.81–2.91) and a 0.99 mmHg (95%CI 0.37–1.61) decrease in systolic and diastolic blood pressure [64]. Effects were even greater when restricted to people with low Ca intake.

The evidence is less clear in relation to risk of myocardial infarction, stroke and CVD mortality. The fact that Ca intake is associated with somewhat lower blood pressure would suggest that risk of myocardial infarction and stroke should also be reduced. However, this is not always the case. For example, in a large epidemiological study in Swedish women ($n = 34,670$, 49–83 years) while dietary Ca intake was overall not associated with an increased risk of stroke, in women with a history of hypertension higher dietary Ca was associated with an almost two-fold increased risk (highest vs lowest quartile, RR 0.63, 95% CI 0.42–0.93) [56]. Although, at a later follow-up (median 19 years) these effects were not detectable in relation to all-cause and cardiovascular mortality except in those using supplements and with a Ca dietary intake above 1.4 g/day for whom all-cause mortality risk was increased (HR 2.57, 95% CI 1.19–5.55) [65]. Moreover, a large RCT (1 g Ca and 400 UI vitamin D) in 36,282 postmenopausal women living in the community demonstrated a modest increased risk of myocardial infarction or stroke associated with supplementation [66]. However, this was only the case for those women who did not already take personal Ca supplementation (RR 1.16, 95% CI 1.02–1.32) while no additional risk was detected in those who did. While in a smaller RCT of Ca supplementation in postmenopausal women ($n = 1471$, 74.3 years) treated (1 g/day) for 5 years, there was a trend of upward risk of developing cardiovascular events (myocardial infarction, stroke or sudden death; RR 1.47, 95% CI 0.97–2.23) in the treatment compared to placebo group [67]. Although these results were weakened when unreported events identified through the New Zealand database of hospital admissions were considered. Moreover, in a longitudinal study involving 388,229 individuals aged 50 to 71 years over a 12-year follow-up, supplemental calcium intake in men (RR 1.20, 95% CI 1.05–1.36) but not in women (RR 1.06, 95% CI 0.96–1.18) was associated with an increased risk of CVD death. In men, this risk was specific to heart disease (RR 1.19, 95% CI 1.03–1.37) but was not significant in relation to cerebrovascular disease (RR 1.14, 95% CI 0.81–1.61). However, CVD mortality was unrelated to dietary intake in men or women.

In contrast, in a very large Japanese epidemiological study in women ($n = 85,764$; 34–59 years) and free of diagnosed CVD or cancer, Ca intake was associated with a decreased risk of ischemic stroke (lowest vs highest quartile RR 0.69, 95% CI 0.50–0.95) [68]. However, this effect was not linear and intakes higher than 600 mg/day did not appear to decrease risk further. Moreover, the effect was stronger for dairy than nondairy Ca. Similarly, in another large European epidemiological study with an 11-year follow-up ($n = 23,980$, 35–64 years), dietary Ca in the third quartile was associated with a decreased risk of myocardial infarction (HR 0.68, 95% CI 0.50–0.93) and of stroke (HR 0.69, 95% CI 0.50–0.94) compared to the lowest quartile [69]. However, users of Ca supplements were at higher risk of myocardial infarction (RR 1.86, 95% CI 1.17–2.96).

When considered together these findings suggest that dietary Ca and supplementation are associated with lower blood pressure but that only dietary Ca is consistently associated with a lower risk of myocardial infarction, stroke and other CVD events while supplementation seems to be associated with a modest increased risk of CVD events and particularly when supplementation occurs in the context of a higher dietary Ca intake.

In relation to glucose metabolism the evidence, although limited, is very consistent. Both large studies (lowest vs third quintile, OR 0.72, 95% CI 0.62–0.84) [70] and meta-analyses (OR 0.82, 95% CI 0.72–0.93) [71] indicate that higher dietary Ca intake is associated with a decreased risk of incident T2D. Consistent association have also been demonstrated in relation to insulin resistance [72].

Potassium

Evidence supporting a blood pressure lowering action of K intake is strong [73]. Apart from the evidence discussed above, a systematic review and meta-analysis of 27 trials with a median intake of 1720 mg/day concluded that K supplementation was associated with a decrease of 2.42 mmHg (95% CI 1.08–3.75) in systolic and 1.57 mmHg (95% CI 0.50–2.65) in diastolic blood pressure [73]. Also, the response was larger in hypertensive than normotensive individuals. A more recent meta-analysis covering fewer trials confirmed these results and suggested effects may be even stronger [74].

Support for an association between K intake and incident stroke is equally strong. In a very large

TABLE 38.2 Adverse Cardiometabolic Outcomes Associated with Lower Dietary Intake and Plasma/Serum Levels of Mg, Ca, and K

	Blood pressure	Stroke	CVD	T2D
Mg	Lower intake → higher systolic and diastolic blood pressure	Lower intake → higher risk of ischemic stroke	Lower intake → greater vascular calcification/thicker intima-media lower dietary intake → higher incidence of cardiovascular events	Lower intake → higher incidence of T2D
Ca	Lower intake → higher systolic and diastolic blood pressure	Lower dietary intake → higher risk of stroke	Lower dietary intake → higher risk of myocardial infarction supplementation → higher risk of heart disease and cardiovascular events	Lower dietary intake → higher incidence of T2D and insulin resistance
K	Supplementation → lower systolic and diastolic blood pressure	Lower dietary intake and supplementation → higher risk of stroke	?	? lower intake → higher incidence of T2D

Japanese epidemiological study in women ($n = 85,764$; 34–59 years) and free of diagnosed CVD or cancer, it was found that K intake was associated with a trend in decreased risk of ischemic stroke (lowest vs highest quartile RR 0.72, 95% CI 0.51–1.01) [68]. In addition in a large epidemiological study in Swedish women ($n = 34,670$, 49–83 years) dietary K intake was associated with a decreased risk of hemorrhagic stroke (RR 0.64, 95% CI 0.45–0.92) and of ischemic stroke (RR 0.56, 95% CI 0.38–0.84) but only in women with a history of hypertension [56]. More recently, in 90,137 postmenopausal women (50–79 years) followed up for 11 years dietary K (mean 2611 mg/day) was inversely associated with all-cause mortality (OR 0.90, 95% CI 0.85–0.95), all stroke (OR 0.88, 95% CI 0.79–0.98), and ischemic stroke (OR 0.84, 95% CI 0.74–0.96) [75]. With the risk of ischemic stroke being more apparent in women who were not hypertensive. Findings of research including both men and women is also consistent with the results reported above [76]. And a systematic review and meta-analyses of such investigations (11 studies, $n = 127,038$) confirmed that K intake was associated with a lower risk of stroke (RR 0.76, 0.66–0.89) but not of cardiovascular disease or coronary heart disease [74].

Finally, available evidence investigating the association between K intake and T2D is also suggestive of a protective effect. In a study following 12,209 participants over a 9-year follow-up lower K serum levels (<4 mEq/L compared to 5.0–5.5 mEq/L) were associated with an increased risk of incident T2D (HR 1.64, 95% CI 1.29–2.08) [77]. Consistent results were also found based on K urinary excretion measurements [78]. However, in a smaller study while higher K serum levels were associated with a better insulin resistance index, neither serum nor dietary K intake was associated with long-term diabetes risk [78]. Converging evidence showing that diuretics (Thiazide) use which lowers K serum levels is associated with an increased risk of T2D [79] while use of antihypertensive medication (renin–angiotensin–aldosterone system inhibitors) which increase K serum levels is associated with a decreased risk [80]. Thus, available data, while limited, point to a protective effect of K intake against T2D.

For completeness it should be noted that K intake can lead to hyperkalemia in a number of conditions and diseases such as chronic renal insufficiency. Hyperkalemia is a very serious condition if not promptly treated and can lead to cardiac arrhythmias including ventricular fibrillation and cardiac arrest as well as other deleterious impacts on other systems. However, this topic is beyond the scope of this review (Table 38.2).

DIETARY INTAKE AND BRAIN AGING

As Mg, Ca, and, to a lesser extent, K contribute to similar mechanisms (systemic inflammation, oxidative stress, apoptosis, cellular signaling, glucose metabolism, etc.) which are known to impact cerebral health through cell death, decreased neurogenesis, demyelination, neuronal shrinkage, and abnormal cellular metabolism, it is expected that dietary intake of these minerals will also contribute to and modulate brain aging. In addition, both CVD and T2D are known risk factors for neurodegeneration [81,82] and therefore are likely to indirectly mediate the effects of Mg, Ca, and K on brain health through the effects reviewed in the previous section. In this section, the evidence demonstrating more direct associations between dietary intake of these minerals and brain aging will be reviewed.

Magnesium

Although this is an emerging field of inquiry, available evidence seems to confirm direct and/or indirect links between Mg dietary intake and brain aging. For example, in an animal experiment contrasting three groups with induced cerebral ischemia, induced ischemia with Mg-sulfate treatment, and sham controls, Mg treatment resulted in lower extracellular glutamate release and lower cell death in the hippocampus [83]. The exact mechanisms underlying these findings are not completely clear but are thought to involve increased vasodilation and decreased vasospasm. Indeed, in a study comparing patients hospitalized for subarachnoid hemorrhage who were or were not treated with intravenous Mg supplementation, a greater proportion (55% more) of those who did not receive Mg supplementation developed vasospasms [84]. Thus, these and other findings [85] suggest that Mg supplementation may improve blood flow and decrease neuronal deaths in acute ischemic conditions. Consistent findings were also reported in relation to Mg protective effects for traumatic brain injury [86–89]. How these findings relate to the effect Mg dietary intake may have on typical brain aging is not clear but since a significant contributor to cerebral senescence is vascular health it is not unreasonable to hypothesize that it includes decreasing low grade brain tissue inflammation, reducing apoptosis, and improving or better maintaining vascularization.

Another area where evidence linking Mg intake and brain aging is accumulating relates to Alzheimer's pathology. A number of studies have reported lower Mg plasma and brain levels in Alzheimer' disease (AD) patients [90–92]. Moreover, in animal models of AD it was found that Mg supplementation (intraperitoneal injection) increased rats' Mg brain levels and led to a decrease in Tau hyperphosphorylation, a reversal of dendritic abnormalities, and maintenance of cognitive function [93]. Also, in an in vitro study, low dosage of Mg promoted the production of amyloid beta precursor protein implicated in the pathological cascade leading to amyloid plaque formation in AD [94]. Consistent with these findings, Li and colleagues showed in mice that Mg treatment led to reduced production of amyloid beta precursor protein and decreased synaptic loss [95]. They also noted that long-term magnesium supplementation increased CSF concentration by only 15% while total brain concentration increased by 30%.

In aggregate, these findings suggest specific pathological pathways through which Mg deficient diet may contribute to brain aging and how supplementation may help preserve cerebral health and function. It should be noted, however, that many of these findings relied on laboratory or animal models and that supplementation was often conducted through channels other than diet. Consequently, further research relying on dietary Mg supplementation is required.

Despite this limitation the available literature provides generally convincing links between Mg intake and brain aging.

Calcium

As discussed above, dietary Ca appears to decrease blood pressure and to be associated with a decreased risk of stroke, particularly when the source is dairy. Limited evidence is also available showing that higher Ca serum levels at admission for acute ischemic stroke are associated with smaller infarct volumes [96]. In contrast, Ca supplementation seems to be associated with a modest increased risk of stroke and vascular disease. It is consequently difficult to predict whether higher Ca intake should be associated with better or worse cerebral health and whether it is likely to slow down or speed up brain aging. However, given the other biological mechanisms to which dietary Ca contributes (oxidative stress, inflammation, apoptosis, glucose metabolism) it might be expected that it would exert a generally more positive influence particularly when dietary Ca and not supplementation is considered.

Unfortunately the evidence on this topic is not so clear. Vascular (coronary, aortic, carotid) calcification, which, as reviewed above, may be exacerbated by increased Ca serum levels, has been found to be associated with greater incidence of cerebral infarcts and white matter lesions [7]. This is consistent with results from other studies which found that higher dietary Ca intake and serum levels were associated with larger brain lesion volumes [97] and particularly in men [98]. Although recent findings suggest that this effect is mostly driven by intake of Ca supplements and not dietary Ca which would correspond well with the evidence reviewed in the previous section.

In addition, findings from a postmortem study of brains from older individuals (83.3 years, SD = 7.5) suggest that calcification does not only occur in the vasculature but also in some brain regions (putamen) [99]. Other studies indicate that calcium levels also increase in other parts of the brain with aging [99,100]. Importantly, the level of a protein (calbindin), which buffers Ca in the brain to maintain optimal Ca signaling, is known to decrease with aging [101,102]. And, at least in animal models, lack of this protein leads to accelerated brain and cognitive aging [103]. Thus, while the extent to which dietary Ca might exacerbate these processes is not established, such mechanism may explain possible links between higher Ca intake, particularly through supplementation, and impaired brain function in aging. Overall insufficient evidence is available to conclusively assess whether Ca intake is protective or deleterious to brain aging and further research needs to be conducted to resolve this question.

TABLE 38.3 Brain Health Associations with Dietary Intake and Plasma/Serum Levels of K^+, Mg^{2+}, and Ca^{2+}

	Neurodegeneration	Structure	Function
Mg	Supplementation → improved blood flow after stroke and traumatic brain injury? lower intake → higher Alzheimer's pathology some mediated through CVD and T2D	? but some mediated through CVD and T2D	? but some mediated through CVD and T2D
Ca	Cellular Ca signaling dysregulation → neuronal shrinkage and remodeling → neurodegeneration?some mediated through CVD and T2D	Increased Ca concentration in the brain with aging → calcification	? but some mediated through CVD and T2D
K	? but some mediated through CVD and T2D	? but some mediated through CVD and T2D	? but some mediated through CVD and T2D

Potassium

Since higher K intake within the normal range and outside pathological conditions such as renal disease has an overwhelmingly beneficial effect on mechanisms implicated in aging and is associated with lower blood pressure, better cardiovascular health, and lower risk of stroke and T2D, it is expected it would also be associated with better cerebral health and slower brain aging. Unfortunately, very little data on the association between K intake and brain health are available. It appears that K ion homeostasis in the brain is impaired in AD [104] and that this dysregulation may be partly caused by amyloid beta 40 [105]. However, at this stage there is no indication that dietary intake of K or supplementation play any significant role in this pathological process although a protective effect of K concentration on rat hippocampal neurons has been demonstrated in vitro [106]. Much more research is required in this field before any effect of K intake on brain aging can be assessed (Table 38.3).

DIETARY INTAKE AND COGNITIVE AGING

Magnesium

A number of animal studies have investigated the effect of Mg supplementation on cognitive outcomes after traumatic brain injury and reported a protective effect at short- and long-term follow-ups and particularly in relation to memory and learning function [107,108]. Interestingly, not only supplementation posttrauma but also before the traumatic event was found to be protective [88]. In humans Mg supplementation is not a recommended treatment for traumatic brain injury but it is being considered [87]. While not conclusive this evidence might suggest that higher dietary Mg intake, if reflected by higher brain levels, might have a preventative effect, at least in the context of major trauma. This is important because traumatic brain injury is a known risk factor for cognitive decline [109] and has been shown to be associated with a two- to three-fold increased risk of developing dementia [110,111]. It is therefore noteworthy that individuals with Alzheimer's disease and other types of dementia have been found to have significantly different serum, cerebrospinal fluid, and brain tissue Mg levels in some studies [112–114], although not all [115]. Moreover, lower serum levels appear to be associated with greater levels of cognitive impairment among AD patients [114].

While it is not clear whether body Mg levels are causally related to disease development or whether they are a consequence of the pathological processes involved, recent studies investigating associations between Mg dietary intake and incidence of mild cognitive impairment (MCI, a preclinical stage to dementia) and dementia provide support to the former. Indeed, in an epidemiological study with a 17-year follow-up surveying 1081 participants Ozawa et al. [116] showed that higher Mg intake was associated with a 50–100% lower risk of developing all-cause dementia, and particularly vascular dementia (lowest vs highest quartile of intake, HR 0.26, 95% CI 0.11–0.61). Given the relatively young mean age at baseline (69 years), the long follow-up, and the fact that participants with prevalent dementia at the start of the study were excluded from analyses and that important covariates were controlled for, these findings provide convincing evidence supporting a link between dietary Mg and cognitive decline. However, as these effects were detected in a Japanese cohort it is possible they were driven by culturally specific factors associated with Mg dietary intake. Moreover, as intakes of Mg, Ca, and K are highly correlated and since Ozawa et al. found significant associations with these three minerals and a decreased risk of dementia, the possibility that intake of Ca or K is associated with decreased risk cannot be discounted.

In our own research [117] we have also investigated the associations between Mg intake and cognitive decline. We followed 1406 individuals (62.5 years, 52% female) participating in the PATH Through Life

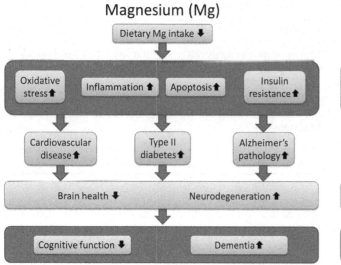

FIGURE 38.1 Pathways linking dietary intake of magnesium, biological mechanisms, cardiometabolic health, and brain and cognitive aging.

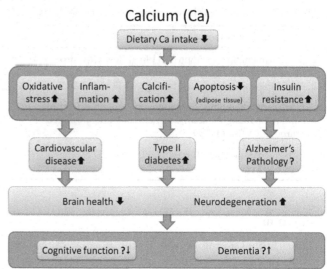

FIGURE 38.2 Pathways linking dietary intake of calcium, biological mechanisms, cardiometabolic health, and brain and cognitive aging.

project, a large longitudinal study of aging conducted in Australia, over 8 years and we investigated associations between Mg dietary intake and incidence of MCI and other mild cognitive disorders. As per Ozawa et al. [116] we excluded from analyses those who were cognitively impaired at baseline and controlled for a large number of important covariates including age, sex, education, body mass index, activity level, diabetes, hypertension, depression, smoking, alcohol, and caloric intake. In fully adjusted models we found that higher Mg intake was associated with a 14-fold decreased risk of developing MCI (HR 0.07, 95% CI 0.01−0.56) and a more than two-fold decreased risk of developing a broader category of mild cognitive disorders (HR 0.04, 95% CI 0.22−0.99). Because in addition to other covariates these analyses also controlled for intake of Ca and K it seems less likely that they were driven solely by intake of these other minerals. However, as for previous studies, the possibility that these effects were driven by other factors (dietary or not) correlated with Mg intake cannot be excluded. Interestingly, this apparent protective effect of Mg was mostly present in men who on average had a lower intake. Thus, as suggested in other studies investigating risk of stroke and hypertension, it is possible that Mg deficiency might be more critical than Mg intake *per se* (Fig. 38.1).

Calcium

As in relation to brain aging, little evidence linking dietary Ca and cognitive aging is available. Ozawa et al. [116] found in 1,081 older Japanese individuals followed over 17 years that higher Ca intake was associated with a circa 50% lower risk of developing all-cause dementia (lowest vs highest quartile of intake, HR 0.64, 95% CI 0.41−1.00). However, in our own research [117] surveying 1406 individuals (62.5 years, 52% female) over 8 years we failed to find any significant association between Ca dietary intake and incidence of MCI and other mild cognitive disorders. A recent systematic review found only four studies contrasting plasma Ca levels in AD patients vs control subjects and of those only one found that Ca levels were significantly lower in the AD group [118]. Other studies also found that CSF levels were lower in AD [119,120]. Thus, although most of the available evidence points to lower Ca levels being associated with poorer cognitive outcomes, it is too limited to be conclusive (Fig. 38.2).

Potassium

The paucity of findings reporting on associations between K intake and brain aging are largely mirrored in relation to cognitive aging. As for Mg and Ca, Ozawa et al. [116] found that higher K dietary intake was associated with a lower risk of developing all-cause dementia (lowest vs highest quartile of intake, HR 0.52, 95% CI 0.30−0.91) and vascular dementia (HR 0.20, 95% CI 0.07−0.56). However, we were not able to detect such effects in relation to MCI [117] (Fig. 38.3).

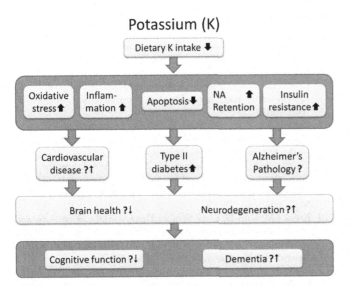

FIGURE 38.3 Pathways linking dietary intake of potassium, biological mechanisms, cardiometabolic health, and brain and cognitive aging.

DISCUSSION

The evidence reviewed in this chapter supports the view that dietary intake of Mg, Ca, and K is associated with important processes implicated in biological aging. Similarly, it is reasonably clear that intake of these minerals is associated with a protective effect on cardiometabolic health. What is less clear is whether this effect is mostly present in those who have a diet deficient in these elements or whether it applies more broadly across the normal intake range. Some research is suggestive of a sexual dimorphisms of these effects and possibly of nonlinear effects across the lifespan.

The relative consistency of findings showing a beneficial effect of Ca dietary intake but a deleterious impact of supplementation on cardiovascular events is quite striking. Even if the latter effect is relatively modest, this finding emphasizes the benefit of managing micronutrients' intake through diet rather than through supplements where feasible.

The evidence is a lot less clear in relation to the influence of dietary intake of Mg, Ca, and K on brain and cognitive aging. The sparse available evidence is generally suggestive of a protective effect congruent to that seen with respect to cardiovascular health and T2D. In addition, as noted above cardiovascular disease contributes to the development of T2D and it is also the case that T2D contributes to the development of CVD [121,122]. Moreover, CVD and T2D are known risk factors for brain aging and cognitive decline. Consequently, any contribution of dietary intake on pathological aging and cognitive decline is likely to be effected through a number of direct and indirect mechanisms. Therefore, although evidence linking dietary intake of Mg, Ca, and K with cognitive health is lacking, the case for an indirect effect through cardiometabolic factors is so strong that it is probably reasonable to assume that these minerals have an important impact on cognitive aging. However, we will need to await definitive research in this area to reach a conclusion.

SUMMARY

- Mg, Ca, and P play essential roles in biological processes and are implicated in biological aging.
- Dietary intake and body concentrations of these minerals modulate inflammatory, oxidative stress, apoptosis, and cellular signalling processes as well as glucose metabolism.
- Action of Magnesium, Calcium and Potassium on physiological processes tends to be protective.
- Overall dietary intake of these minerals is linked to better cardiovascular health, lower risk of stroke and myocardial infarction, and decreased risk of type 2 diabetes.
- Some evidence, particularly for CA suggests that dietary intake is preferable to supplementation.
- Limited evidence suggests that Mg, Ca, and P intake exerts a positive influence on brain health.
- Emerging evidence is also suggestive of a protective effect of these minerals against cognitive decline.

Acknowledgments

Nicolas Cherbuin is funded by Australian Research Council Future Fellowship No. 120100227.

References

[1] Swaminathan R. Magnesium metabolism and its disorders. Clin Biochem Rev 2003;24:47–66.
[2] Cunha AR, Umbelino B, Correia ML, Neves MF. Magnesium and vascular changes in hypertension. Int J Hypertens 2012;2012:754250.
[3] Laurant P, Touyz RM. Physiological and pathophysiological role of magnesium in the cardiovascular system: implications in hypertension. J Hypertens 2000;18:1177–91.
[4] Standing Committee on the Scientific Evaluation of Dietary Reference Intakes FaNB, Institute of Medicine. Dietary reference intakes for calcium, phosphorus, magnesium, vitamin D, and fluoride. Washington: National Academies Press; 1997.
[5] Ford ES, Mokdad AH. Dietary magnesium intake in a national sample of US adults. J Nutr 2003;133:2879–82.
[6] Hartwig A. Role of magnesium in genomic stability. Mutat Res 2001;475:113–21.
[7] Bo S, Pisu E. Role of dietary magnesium in cardiovascular disease prevention, insulin sensitivity and diabetes. Curr Opin Lipidol 2008;19:50–6.

[8] Shah NC, Shah GJ, Li Z, Jiang XC, Altura BT, Altura BM. Short-term magnesium deficiency downregulates telomerase, upregulates neutral sphingomyelinase and induces oxidative DNA damage in cardiovascular tissues: relevance to atherogenesis, cardiovascular diseases and aging. Int J Clin Exp Med 2014; 7:497–514.

[9] Martin H, Uring-Lambert B, Adrian M, et al. Effects of long-term dietary intake of magnesium on oxidative stress, apoptosis and ageing in rat liver. Magnes Res 2008;21:124–30.

[10] Killilea DW, Ames BN. Magnesium deficiency accelerates cellular senescence in cultured human fibroblasts. Proc Natl Acad Sci USA 2008;105:5768–73.

[11] Altura BM, Shah NC, Jiang XC, et al. Short-term magnesium deficiency results in decreased levels of serum sphingomyelin, lipid peroxidation, and apoptosis in cardiovascular tissues. Am J Physiol Heart Circ Physiol 2009;297:H86–92.

[12] O'Callaghan NJ, Bull C, Fenech M. Elevated plasma magnesium and calcium may be associated with shorter telomeres in older South Australian women. J Nutr Health Aging 2014; 18:131–6.

[13] Kolterman OG, Gray RS, Griffin J, et al. Receptor and postreceptor defects contribute to the insulin resistance in noninsulin-dependent diabetes mellitus. J Clin Invest 1981;68:957–69.

[14] Dibaba DT, Xun P, He K. Dietary magnesium intake is inversely associated with serum C-reactive protein levels: meta-analysis and systematic review. Eur J Clin Nutr 2014;68:510–16.

[15] Wang L, Manson JE, Sesso HD. Calcium intake and risk of cardiovascular disease: a review of prospective studies and randomized clinical trials. Am J Cardiovasc Drugs 2012;12:105–16.

[16] Cordain L, Eaton SB, Sebastian A, et al. Origins and evolution of the Western diet: health implications for the 21st century. Am J Clin Nutr 2005;81:341–54.

[17] Reid IR, Schooler BA, Hannan SF, Ibbertson HK. The acute biochemical effects of four proprietary calcium preparations. Aust NZJ Med 1986;16:193–7.

[18] Mattson MP, Chan SL. Calcium orchestrates apoptosis. Nat Cell Biol 2003;5:1041–3.

[19] Pinton P, Giorgi C, Siviero R, Zecchini E, Rizzuto R. Calcium and apoptosis: ER-mitochondria Ca(2+) transfer in the control of apoptosis. Oncogene 2008;27:6407–18.

[20] Zemel MB, Shi H, Greer B, Dirienzo D, Zemel PC. Regulation of adiposity by dietary calcium. FASEB J 2000;14:1132–8.

[21] Sun X, Zemel MB. Calcium and 1,25-dihydroxyvitamin D3 regulation of adipokine expression. Obesity (Silver Spring) 2007;15:340–8.

[22] Sun X, Morris KL, Zemel MB. Role of calcitriol and cortisol on human adipocyte proliferation and oxidative and inflammatory stress: a microarray study. J Nutrigenet Nutrigenomics 2008;1: 30–48.

[23] Penman I, Liang Q, Bode J, Eastwood M, Arends M. Dietary calcium supplementation increases apoptosis in the distal murine colonic epithelium. J Clin Pathol 2000;53:302–7.

[24] Huncharek M, Muscat J, Kupelnick B. Colorectal cancer risk and dietary intake of calcium, vitamin D, and dairy products: a meta-analysis of 26,335 cases from 60 observational studies. Nutr Cancer 2009;61:47–69.

[25] Sun X, Zemel MB. 1α,25-dihydroxyvitamin D3 modulation of adipocyte reactive oxygen species production. Obesity (Silver Spring) 2007;15:1944–53.

[26] Reid IR, Mason B, Horne A, et al. Effects of calcium supplementation on serum lipid concentrations in normal older women: a randomized controlled trial. Am J Med 2002;112:343–7.

[27] Sergeev IN, Song Q. High vitamin D and calcium intakes reduce diet-induced obesity in mice by increasing adipose tissue apoptosis. Mol Nutr Food Res 2014;58:1342–8.

[28] Reid IR, Horne A, Mason B, Ames R, Bava U, Gamble GD. Effects of calcium supplementation on body weight and blood pressure in normal older women: a randomized controlled trial. J Clin Endocrinol Metab 2005;90:3824–9.

[29] Reid IR, Ames R, Mason B, et al. Effects of calcium supplementation on lipids, blood pressure, and body composition in healthy older men: a randomized controlled trial. Am J Clin Nutr 2010;91:131–9.

[30] Govers MJ, Van der Meet R. Effects of dietary calcium and phosphate on the intestinal interactions between calcium, phosphate, fatty acids, and bile acids. Gut 1993;34:365–70.

[31] Pletcher MJ, Tice JA, Pignone M, Browner WS. Using the coronary artery calcium score to predict coronary heart disease events: a systematic review and meta-analysis. Arch Intern Med 2004;164:1285–92.

[32] Kwak SM, Kim JS, Choi Y, et al. Dietary intake of calcium and phosphorus and serum concentration in relation to the risk of coronary artery calcification in asymptomatic adults. Arterioscler Thromb Vasc Biol 2014;34:1763–9.

[33] Reid IR. Effects of calcium supplementation on circulating lipids: potential pharmacoeconomic implications. Drugs Aging 2004;21:7–17.

[34] Standing Committee on the Scientific Evaluation of Dietary Reference Intakes FaNB, Institute of Medicine. Dietary reference intakes for water, potassium, sodium, chloride, and sulfate. Washington: National Academies Press; 2005.

[35] Drewnowski A, Maillot M, Rehm C. Reducing the sodium-potassium ratio in the US diet: a challenge for public health. Am J Clin Nutr 2012;96:439–44.

[36] Castro H, Raij L. Potassium in hypertension and cardiovascular disease. Semin Nephrol 2013;33:277–89.

[37] Matsui H, Shimosawa T, Uetake Y, et al. Protective effect of potassium against the hypertensive cardiac dysfunction: association with reactive oxygen species reduction. Hypertension 2006;48:225–31.

[38] Golia E, Limongelli G, Natale F, et al. Inflammation and cardiovascular disease: from pathogenesis to therapeutic target. Curr Atheroscler Rep 2014;16:435.

[39] Mangge H, Becker K, Fuchs D, Gostner JM. Antioxidants, inflammation and cardiovascular disease. World J Cardiol 2014;6:462–77.

[40] Devaraj S, Dasu MR, Jialal I. Diabetes is a proinflammatory state: a translational perspective. Expert Rev Endocrinol Metab 2010;5:19–28.

[41] Alexandraki K, Piperi C, Kalofoutis C, Singh J, Alaveras A, Kalofoutis A. Inflammatory process in type 2 diabetes: the role of cytokines. Ann NY Acad Sci 2006;1084:89–117.

[42] Misiak B, Leszek J, Kiejna A. Metabolic syndrome, mild cognitive impairment and Alzheimer's disease — the emerging role of systemic low-grade inflammation and adiposity. Brain Res Bull 2012;89:144–9.

[43] Laurant P, Hayoz D, Brunner H, Berthelot A. Dietary magnesium intake can affect mechanical properties of rat carotid artery. Br J Nutr 2000;84:757–64.

[44] Kircelli F, Peter ME, Sevinc Ok E, et al. Magnesium reduces calcification in bovine vascular smooth muscle cells in a dose-dependent manner. Nephrol Dial Transplant 2012;27:514–21.

[45] Turgut F, Kanbay M, Metin MR, Uz E, Akcay A, Covic A. Magnesium supplementation helps to improve carotid intima media thickness in patients on hemodialysis. Int Urol Nephrol 2008;40:1075–82.

[46] Hruby A, O'Donnell CJ, Jacques PF, Meigs JB, Hoffmann U, McKeown NM. Magnesium intake is inversely associated with coronary artery calcification: the Framingham Heart Study. JACC Cardiovasc Imaging 2014;7:59–69.

[47] Cosaro E, Bonafini S, Montagnana M, et al. Effects of magnesium supplements on blood pressure, endothelial function and metabolic parameters in healthy young men with a family history of metabolic syndrome. Nutr Metab Cardiovasc Dis 2014; 24:1213–20.

[48] Touyz RM, Milne FJ. Magnesium supplementation attenuates, but does not prevent, development of hypertension in spontaneously hypertensive rats. Am J Hyperten 1999;12:757–65.

[49] Sado T, Oi H, Sakata M, et al. Aging impairs the protective effect of magnesium supplementation on hypertension in spontaneously hypertensive rats. Magnes Res 2007;20:196–9.

[50] Hatzistavri LS, Sarafidis PA, Georgianos PI, et al. Oral magnesium supplementation reduces ambulatory blood pressure in patients with mild hypertension. Am J Hypertens 2009;22:1070–5.

[51] Witteman JC, Grobbee DE, Derkx FH, Bouillon R, de Bruijn AM, Hofman A. Reduction of blood pressure with oral magnesium supplementation in women with mild to moderate hypertension. Am J Clin Nutr 1994;60:129–35.

[52] Kawano Y, Matsuoka H, Takishita S, Omae T. Effects of magnesium supplementation in hypertensive patients: assessment by office, home, and ambulatory blood pressures. Hypertension 1998;32:260–5.

[53] Qu X, Jin F, Hao Y, et al. Magnesium and the risk of cardiovascular events: a meta-analysis of prospective cohort studies. PLoS One 2013;8:e57720.

[54] Guasch-Ferre M, Bullo M, Estruch R, et al. Dietary magnesium intake is inversely associated with mortality in adults at high cardiovascular disease risk. J Nutr 2014;144:55–60.

[55] Almoznino-Sarafian D, Berman S, Mor A, et al. Magnesium and C-reactive protein in heart failure: an anti-inflammatory effect of magnesium administration? Eur J Nutr 2007;46:230–7.

[56] Larsson SC, Virtamo J, Wolk A. Potassium, calcium, and magnesium intakes and risk of stroke in women. Am J Epidemiol 2011;174:35–43.

[57] Chaudhary DP, Boparai RK, Bansal DD. Implications of oxidative stress in high sucrose low magnesium diet fed rats. Eur J Nutr 2007;46:383–90.

[58] Rayssiguier Y, Gueux E, Nowacki W, Rock E, Mazur A. High fructose consumption combined with low dietary magnesium intake may increase the incidence of the metabolic syndrome by inducing inflammation. Magnes Res 2006;19:237–43.

[59] Gress TW, Nieto FJ, Shahar E, Wofford MR, Brancati FL. Hypertension and antihypertensive therapy as risk factors for type 2 diabetes mellitus. N Engl J Med 2000;342:905–12.

[60] Corica F, Corsonello A, Ientile R, et al. Serum ionized magnesium levels in relation to metabolic syndrome in type 2 diabetic patients. J Am Coll Nutr 2006;25:210–15.

[61] Schulze MB, Schulz M, Heidemann C, Schienkiewitz A, Hoffmann K, Boeing H. Fiber and magnesium intake and incidence of type 2 diabetes: a prospective study and meta-analysis. Arch Intern Med 2007;167:956–65.

[62] Larsson SC, Wolk A. Magnesium intake and risk of type 2 diabetes: a meta-analysis. J Intern Med 2007;262:208–14.

[63] Chan KHK, Chacko SA, Song Y, et al. Genetic variations in magnesium-related ion channels may affect diabetes risk among african american and hispanic american women. J Nutr 2015.

[64] van Mierlo LA, Arends LR, Streppel MT, et al. Blood pressure response to calcium supplementation: a meta-analysis of randomized controlled trials. J Hum Hypertens 2006;20:571–80.

[65] Michaëlsson K, Melhus H, Warensjö Lemming E, Wolk A, Byberg L. Long term calcium intake and rates of all cause and cardiovascular mortality: community based prospective longitudinal cohort study. Br Med J 2013;346.

[66] Bolland MJ, Grey A, Avenell A, Gamble GD, Reid IR. Calcium supplements with or without vitamin D and risk of cardiovascular events: reanalysis of the Women's Health Initiative limited access dataset and meta-analysis. Br Med J 2011;342:d2040.

[67] Bolland MJ, Barber PA, Doughty RN, et al. Vascular events in healthy older women receiving calcium supplementation: randomised controlled trial. Br Med J 2008;336:262–6.

[68] Iso H, Stampfer MJ, Manson JE, et al. Prospective study of calcium, potassium, and magnesium intake and risk of stroke in women. Stroke 1999;30:1772–9.

[69] Li K, Kaaks R, Linseisen J, Rohrmann S. Associations of dietary calcium intake and calcium supplementation with myocardial infarction and stroke risk and overall cardiovascular mortality in the Heidelberg cohort of the European Prospective Investigation into Cancer and Nutrition study (EPIC-Heidelberg). Heart (British Cardiac Society) 2012;98:920–5.

[70] Villegas R, Gao YT, Dai Q, et al. Dietary calcium and magnesium intakes and the risk of type 2 diabetes: the Shanghai women's health study. Am J Clin Nutr 2009;89:1059–67.

[71] Pittas AG, Lau J, Hu FB, Dawson-Hughes B. The role of vitamin D and calcium in type 2 diabetes. A systematic review and meta-analysis. J Clin Endocrinol Metab 2007;92:2017–29.

[72] Wu T, Willett WC, Giovannucci E. Plasma C-peptide is inversely associated with calcium intake in women and with plasma 25-hydroxy vitamin D in men. J Nutr 2009;139:547–54.

[73] Geleijnse JM, Kok FJ, Grobbee DE. Blood pressure response to changes in sodium and potassium intake: a metaregression analysis of randomised trials. J Hum Hypertens 2003;17:471–80.

[74] Aburto NJ, Hanson S, Gutierrez H, Hooper L, Elliott P, Cappuccio FP. Effect of increased potassium intake on cardiovascular risk factors and disease: systematic review and meta-analyses. Br Med J 2013;346:f1378.

[75] Seth A, Mossavar-Rahmani Y, Kamensky V, et al. Potassium intake and risk of stroke in women with hypertension and non-hypertension in the Women's Health Initiative. Stroke 2014;45:2874–80.

[76] Bazzano LA, He J, Ogden LG, et al. Dietary potassium intake and risk of stroke in US men and women: national health and nutrition examination survey I epidemiologic follow-up study. Stroke 2001;32:1473–80.

[77] Chatterjee R, Yeh HC, Shafi T, et al. Serum and dietary potassium and risk of incident type 2 diabetes mellitus: the Atherosclerosis Risk in Communities (ARIC) study. Arch Intern Med 2010;170:1745–51.

[78] Chatterjee R, Colangelo LA, Yeh HC, et al. Potassium intake and risk of incident type 2 diabetes mellitus: the Coronary Artery Risk Development in Young Adults (CARDIA) Study. Diabetologia 2012;55:1295–303.

[79] Elliott WJ, Meyer PM. Incident diabetes in clinical trials of antihypertensive drugs: a network meta-analysis. Lancet 2007;369:201–7.

[80] Tocci G, Paneni F, Palano F, et al. Angiotensin-converting enzyme inhibitors, angiotensin II receptor blockers and diabetes: a meta-analysis of placebo-controlled clinical trials. Am J Hypertens 2011;24:582–90.

[81] Moreira PI. Alzheimer's disease and diabetes: an integrative view of the role of mitochondria, oxidative stress, and insulin. J Alzheimers Dis 2012;30(Suppl. 2):S199–215.

[82] Qiu C, Fratiglioni L. A major role for cardiovascular burden in age-related cognitive decline. Nat Rev Cardiol 2015.

[83] Kang SW, Choi SK, Park E, et al. Neuroprotective effects of magnesium-sulfate on ischemic injury mediated by modulating the release of glutamate and reduced of hyperreperfusion. Brain Res 2011;1371:121–8.

[84] Chia RY, Hughes RS, Morgan MK. Magnesium: a useful adjunct in the prevention of cerebral vasospasm following aneurysmal subarachnoid haemorrhage. J Clin Neurosci 2002;9:279–81.

[85] Meloni BP, Campbell K, Zhu H, Knuckey NW. In search of clinical neuroprotection after brain ischemia: the case for mild hypothermia (35 degrees C) and magnesium. Stroke 2009;40:2236–40.

[86] Lee JS, Han YM, Yoo DS, et al. A molecular basis for the efficacy of magnesium treatment following traumatic brain injury in rats. J Neurotrauma 2004;21:549–61.

[87] Sen AP, Gulati A. Use of magnesium in traumatic brain injury. Neurotherapeutics 2010;7:91–9.

[88] Enomoto T, Osugi T, Satoh H, McIntosh TK, Nabeshima T. Pre-injury magnesium treatment prevents traumatic brain injury-induced hippocampal ERK activation, neuronal loss, and cognitive dysfunction in the radial-arm maze test. J Neurotrauma 2005;22:783–92.

[89] Browne KD, Leoni MJ, Iwata A, Chen XH, Smith DH. Acute treatment with MgSO$_4$ attenuates long-term hippocampal tissue loss after brain trauma in the rat. J Neurosci Res 2004; 77:878–83.

[90] Durlach J. Magnesium depletion and pathogenesis of Alzheimer's disease. Magnes Res 1990;3:217–18.

[91] Vural H, Demirin H, Kara Y, Eren I, Delibas N. Alterations of plasma magnesium, copper, zinc, iron and selenium concentrations and some related erythrocyte antioxidant enzyme activities in patients with Alzheimer's disease. J Trace Elem Med Biol 2010;24:169–73.

[92] Andrasi E, Pali N, Molnar Z, Kosel S. Brain aluminum, magnesium and phosphorus contents of control and Alzheimer-diseased patients. J Alzheimers Dis 2005;7:273–84.

[93] Xu Z-P, Li L, Bao J, et al. Magnesium protects cognitive functions and synaptic plasticity in streptozotocin-induced sporadic Alzheimer's model. PLoS One 2014;9:e108645.

[94] Yu J, Sun M, Chen Z, et al. Magnesium modulates amyloid-beta protein precursor trafficking and processing. J Alzheimers Dis 2010;20:1091–106.

[95] Li W, Yu J, Liu Y, et al. Elevation of brain magnesium prevents and reverses cognitive deficits and synaptic loss in Alzheimer's disease mouse model. J Neurosci 2013;33:8423–41.

[96] Buck BH, Liebeskind DS, Saver JL, et al. Association of higher serum calcium levels with smaller infarct volumes in acute ischemic stroke. Arch Neurol 2007;64:1287–91.

[97] Payne ME, Anderson JJ, Steffens DC. Calcium and vitamin D intakes may be positively associated with brain lesions in depressed and nondepressed elders. Nutr Res (New York, NY) 2008;28:285–92.

[98] Payne ME, Pierce CW, McQuoid DR, Steffens DC, Anderson JJ. Serum ionized calcium may be related to white matter lesion volumes in older adults: a pilot study. Nutrients 2013;5:2192–205.

[99] Tohno Y, Tohno S, Azuma C, et al. Mineral composition of and the relationships between them of human basal ganglia in very old age. Biol Trace Elem Res 2013;151:18–29.

[100] Tohno Y, Tohno S, Ongkana N, et al. Age-related changes of elements and relationships among elements in human hippocampus, dentate gyrus, and fornix. Biol Trace Elem Res 2010;138:42–52.

[101] Iacopino AM, Christakos S. Specific reduction of calcium-binding protein (28-kilodalton calbindin-D) gene expression in aging and neurodegenerative diseases. Proc Natl Acad Sci USA 1990;87:4078–82.

[102] de Jong GI, Naber PA, Van der Zee EA, Thompson LT, Disterhoft JF, Luiten PG. Age-related loss of calcium binding proteins in rabbit hippocampus. Neurobiol Aging 1996;17:459–65.

[103] Moreno H, Burghardt NS, Vela-Duarte D, et al. The absence of the calcium-buffering protein calbindin is associated with faster age-related decline in hippocampal metabolism. Hippocampus 2012;22:1107–20.

[104] Shieh CC, Coghlan M, Sullivan JP, Gopalakrishnan M. Potassium channels: molecular defects, diseases, and therapeutic opportunities. Pharmacol Rev 2000;52:557–94.

[105] Vitvitsky VM, Garg SK, Keep RF, Albin RL, Banerjee R. Na$^+$ and K$^+$ ion imbalances in Alzheimer's disease. Biochim Biophys Acta 2012;1822:1671–81.

[106] Pike CJ, Balazs R, Cotman CW. Attenuation of beta-amyloid neurotoxicity in vitro by potassium-induced depolarization. J Neurochem 1996;67:1774–7.

[107] Vink R, O'Connor CA, Nimmo AJ, Heath DL. Magnesium attenuates persistent functional deficits following diffuse traumatic brain injury in rats. Neurosci Lett 2003;336:41–4.

[108] Hoane MR. Assessment of cognitive function following magnesium therapy in the traumatically injured brain. Magnes Res 2007;20:229–36.

[109] Moretti L, Cristofori I, Weaver SM, Chau A, Portelli JN, Grafman J. Cognitive decline in older adults with a history of traumatic brain injury. Lancet Neurol 2012;11:1103–12.

[110] Lee Y-K, Hou S-W, Lee C-C, Hsu C-Y, Huang Y-S, Su Y-C. Increased risk of dementia in patients with mild traumatic brain injury: a nationwide cohort study. PLoS One 2013;8: e62422.

[111] Fleminger S, Oliver DL, Lovestone S, Rabe-Hesketh S, Giora A. Head injury as a risk factor for Alzheimer's disease: the evidence 10 years on; a partial replication. J Neurol Neurosurg Psychiatry 2003;74:857–62.

[112] Barbagallo M, Belvedere M, Di Bella G, Dominguez LJ. Altered ionized magnesium levels in mild-to-moderate Alzheimer's disease. Magnes Res 2011;24:S115–21.

[113] Bostrom F, Hansson O, Gerhardsson L, et al. CSF Mg and Ca as diagnostic markers for dementia with Lewy bodies. Neurobiol Aging 2009;30:1265–71.

[114] Cililer AE, Ozturk S, Ozbakir S. Serum magnesium level and clinical deterioration in Alzheimer's disease. Gerontology 2007;53:419–22.

[115] Gerhardsson L, Lundh T, Minthon L, Londos E. Metal concentrations in plasma and cerebrospinal fluid in patients with Alzheimer's disease. Dement Geriatr Cogn Disord 2008;25:508–15.

[116] Ozawa M, Ninomiya T, Ohara T, et al. Self-reported dietary intake of potassium, calcium, and magnesium and risk of dementia in the Japanese: the Hisayama Study. J Am Geriatr Soc 2012;60:1515–20.

[117] Cherbuin N, Kumar R, Sachdev PS, Anstey KJ. Dietary mineral intake and risk of mild cognitive impairment: the PATH through Life Project. Front Aging Neurosci 2014;6:4.

[118] Lopes da Silva S, Vellas B, Elemans S, et al. Plasma nutrient status of patients with Alzheimer's disease: systematic review and meta-analysis. Alzheimers Dement 2014;10:485–502.

[119] Basun H, Forssell LG, Wetterberg L, Winblad B. Metals and trace elements in plasma and cerebrospinal fluid in normal aging and Alzheimer's disease. J Neural Transm Park Dis Dement Sect 1991;3:231–58.

[120] Subhash MN, Padmashree TS, Srinivas KN, Subbakrishna DK, Shankar SK. Calcium and phosphorus levels in serum and CSF in dementia. Neurobiol Aging 1991;12:267–9.

[121] Ergul A, Abdelsaid M, Fouda AY, Fagan SC. Cerebral neovascularization in diabetes: implications for stroke recovery and beyond. J Cereb Blood Flow Metab 2014;34:553–63.

[122] Hamilton SJ, Watts GF. Endothelial dysfunction in diabetes: pathogenesis, significance, and treatment. Rev Diabet Stud 2013;10:133–56.

CHAPTER

39

Zinc: An Essential Trace Element for the Elderly

Peter Uciechowski and Lothar Rink

Institute of Immunology, Medical Faculty, RWTH Aachen University, Aachen, Germany

KEY FACTS

- Zinc is essential for the elderly because of its effect on biochemical, biological, and immune properties, especially the T cell system.
- Zinc deficiency as well as mild zinc-deficiency is documented within the aged population.
- Zinc deficiency and immunosenescence in the elderly is involved in the development of certain diseases influencing and causing morbidity.
- Zinc deficiency and immunosenescence may contribute to a reduced vaccination.
- Zinc is important for cell-mediated immune responses and is able to correct defects in the immune response.
- Zinc supplementation restore cytokine synthesis in old subjects with high pro-inflammatory cytokine levels (inflammaging) when compared to younger donors.
- Zinc supplementation affects the reduction of markers of oxidative stress.
- Zinc supplementation will be an effective therapeutic agent for old individuals with zinc deficiency, and for the treatment of chronic diseases, and some malignancies.
- Methods are needed to define zinc deficiency, and (genetic) biomarkers to define dosage, time of treatment, and application form of zinc.

Dictionary of Terms

- **Immunosenescence:** a term for the condition that elderly people are more susceptible to microbial infections because of impaired immune functions.
- **Inflammaging:** a term for a pro-inflammatory condition defined by increased serum levels of pro-inflammatory and reduced anti-inflammatory cytokines in the elderly.
- **Recommended daily allowance (RDA) for zinc:** individuals >19 years old: 11 mg/day for men and 8 mg/day for women (USA). European Community and WHO: 9.4 to 10 mg for men and 6.5 to 7.1 mg for women.
- **Solute-linked carrier (SLC):** zinc transporter families; SLC39A (Zrt, Irt-like protein (ZIP)1 to ZIP14) carry zinc into the cytosol, SLC30A (ZnT1 to ZnT10) transport zinc out of the cytosol.
- **Zinc deficiency:** Zinc concentrations measured in serum or plasma defining zinc deficiency <10.7 μM, corresponding <70 μg/dl.

INTRODUCTION

It is now more than 50 years ago that Prasad and coworkers showed that zinc is essential for humans [1,2]. The human body comprises 2–3 g zinc, mostly bound to proteins. Zinc is a component of more than 300 enzymes, necessary for catalytic activity and structural stabilization, and as a cofactor. Many transcription factors contain zinc in zinc fingers and similar structural motifs [3–5]. Consequently, zinc is essentially involved in cell activation and division, transcription, DNA synthesis, and apoptosis [6–8].

Although zinc is a heavy metal, zinc ions have only a very restricted acute toxicity, but zinc deficiency is accompanied by severe defects. It is noteworthy to mention that the prevalence of zinc deficiency is calculated to affect over two billion people, especially in developing countries [9,10]. Severe zinc deficiency is involved in retardation of growth and development in children, retarded genital development and hypogonadism, dermatitis and impaired wound healing, alopecia, poor pregnancy outcomes and teratology, and impaired immune function [9]. But also milder forms of zinc deficiency have an effect on immunity [11].

It has been documented that zinc deficiency modifies cell numbers, immune responses, and functions of the innate and adaptive immune system [4,5,7,11,12]. Hence, the immune system with its high proliferation and differentiation rates is dependent on a constant and sufficient zinc supply [11,13,14], but zinc homeostasis is also important for signal transduction and control of apoptosis [5,15,16]. To ensure a consistent supply of zinc, many proteins are involved in homeostatic regulation, such as zinc transporters and zinc-binding proteins [15]. Most zinc is stored within the cells and further zinc is taken up from the plasma, but this zinc pool is only one percent of total body content [3]. It is therefore obvious that changes in zinc uptake, retention, sequestration, or secretion rapidly result in zinc deficiency and impair zinc-dependent functions, particularly in the immune system.

Aging also affects the development and properties of the immune system with changes more precisely documented for the adaptive part [7,13,17–24]. Elderly people are more susceptible to microbial infections, with rising rates of morbidity and mortality because of impaired immune functions, termed immunosenescence [21,25]. Additionally, these changes result in higher incidences of inflammatory and autoimmune disorders, and cancer [5,26]. Aged people also reveal increased serum levels of pro-inflammatory and reduced anti-inflammatory cytokines, a condition named inflammaging [7,27].

There are notable similarities between age-dependent consequences and the effects of zinc deprivation on modifications of the immune system. Atrophy of the thymus, reduced cellular activity and functions of immune cells, and increased levels of pro-inflammatory cytokines are examples for shared outcomes of both conditions [3,7,11,13].

Zinc is essential for the elderly because of its effect on biochemical, biological, and immune properties [3,7,28]. Hence, zinc deficiency in the elderly has severe consequences for maintaining immune competence and is involved in the development of certain diseases influencing and causing morbidity [3,5,7,13,14,28–30]. Many studies exist in which negative effects of zinc deficiency on immune responses or properties could be reversed or normalized by zinc supplementation [7,12,14,31]. That is the reason why zinc supplementation of the elderly to prevent or treat diseases accompanied with zinc deficiency is suggested by many groups [3,7,10,11,13,14,29,32,33]. In recent approaches the role of epigenetic mechanisms and genetic variations in influencing an old individual's response to zinc supplementation was investigated [34,35].

In the following, the zinc status within aging, the relationship between zinc and immunosenescence, and the effect of zinc supplementation on the elderly, as well as nutrient−gene interactions, are discussed.

ZINC INTAKE, ZINC DEFICIENCY, AND ZINC STATUS OF THE ELDERLY

A sufficient uptake of zinc is not reached in many old individuals [36], and there are multifactorial causes, such as physiological, social, psychological, and economic factors, which are indicated to be responsible for zinc malnutrition. Additionally, other factors during aging promote the inadequate intake of zinc and other micronutrients. These are impaired mastication and intestinal absorption, particularly depending on the composition of the food, changes in zinc transporters, restricted mobility, drug interactions and those with other bivalent ions, as well as medication, such as diuretics [7,13].

The regular supply of cells with zinc is homeostatically regulated by specialized proteins. At least three dozen proteins homeostatically control the vesicular storage and subcellular distribution of zinc [37]. But no zinc-specific storage compartment within the human body exists. The intracellular level of zinc is regulated via the transport through the plasma membrane [15], storage in and secretion from vesicular structures termed zincosomes [24,38], and by binding to metallothioneins (MT), and other zinc-binding proteins such as S100A8 and S100A9 [4,37]. Zinc homeostasis is also regulated by two different solute-linked carrier (SLC) families, consisting of 10 human SLC30A (ZnT1 to ZnT10) and 14 human SLC39A (Zrt, Irt-like protein (ZIP1 to ZIP14)) members. While ZIP proteins carry zinc into the cytosol, all ZnT proteins transport zinc out of the cytosol (into cellular compartments or the extracellular space) [4,15,39]. These transporters show a tissue-specific expression and differential location in cellular compartments. Their stability and subcellular distribution is dependent on zinc availability. The expression of zinc transporters is regulated by cytokines and hormones, ensuring a proper adjustment to changes in the environment or physiological demands [15,40].

The decrease in the cellular uptake of zinc is also due to modified cell membrane expressions of zinc transporters [7]. In addition, promoter hypermethylation of SLC30A5 (Znt5) is suggested to lead to impaired zinc adsorption in the intestinal lumen but only with the assumption that this transporter has bidirectional functions [41]. Finally, diseases occurring with enhanced frequencies in the elderly like diabetes are accompanied by zinc deficiency [9,13,42,43]. By joining all these factors together, an inadequate nutritional intake of zinc has clinical consequences in young and old individuals and promotes the risk to develop degenerative diseases.

In the United States, the recommended daily allowance (RDA) for zinc in individuals ≥ 19 years old is 11 mg/day for men and 8 mg/day for women [3,9,13,44]. In the European Community and by the WHO RDAs between 9.4 to 10 mg for men and 6.5 to 7.1 mg for women are recommended [9]. Curiously, no specific RDA for old people has been defined. However, a zinc uptake below these values may predict a mild zinc deficiency which may also depend on other factors including a potential adaption of the metabolism to impaired zinc intake.

Zinc in the elderly is often found beneath the plasma/serum concentration of younger individuals. The reference serum zinc values are 70–110 µg/dL, respectively 10.7–16.8 µM [13]. The sufficient zinc intake has been defined as >67% of the RDA, but only 42.9% of free-living old individuals show an adequate intake of zinc (mild zinc deficiency) [7,13,45]. Many studies, for example, the ZENITH project, the ZINCAGE project, the Japan study, a German study from the Max Rubner-Institute (MRI), and a Federal Research Center for Nutrition and Food: National Nutrition Survey II, part 2, 2008, investigating old individuals came to the same results [34,46,47]. The German MRI study (participants aged 65–80 years) showed that 44% of men and 27% of women did not fulfill the RDA. The Third National Health and Nutrition Examination Survey (NHANES) from the United States reported that only 42.5% of the participants with an age ≥ 71 years had a sufficient (≥ 77% of the RDA) zinc intake, demonstrating an impairment of zinc intake with aging [48]. In summary, within the aged population a zinc deficiency as well as mild zinc-deficiency is documented, but hard to define [29].

In this context, it is necessary to analyze and define zinc deficiency of individuals by specific parameters. In most of the studies, the zinc concentrations were measured in serum or plasma defining zinc deficiency <10.7 µM, corresponding <70 µg/dL. Although many studies described suboptimal zinc intake and decreased zinc plasma and serum concentrations with age (NHANES), a clear prevalence of zinc deficiency in healthy elderly could not demonstrated. This was confirmed by data from the second NHANES in which over 13,400 serum samples were analyzed. The serum zinc levels rose into the third decade of life, and then declined with age [49]. In an early study from 1971 where 284 participants from age 20 to 84 years had been included, a significant linear decrease of plasma zinc with age was found [50]. There are also findings showing significantly reduced zinc levels when comparing old and young individuals [18,51], and when serum zinc of the oldest old (≥ 90 years) were measured [52]. In contrast, other studies did not detect a high prevalence of zinc deficiency in the elderly [46,53]. But the degree of severity of zinc deficiency can vary between different countries due to different foods, eating habits, or lifestyles [7,34]. Taken together, there is a tendency of undersupply of zinc intake and a zinc decrease in serum/plasma with age but most old people have values located within the reference ranges.

An obvious explanation is that in most of the studies, healthy old people fulfilling the criteria of the SENIEUR protocol [54] have been examined. This group, with regard to the importance of zinc for immune capacity, probably displays a normal zinc status in contrast to normal old, ill, or institutionalized old people.

To confirm this hypothesis, studies comparing healthy, free-living, ill, or institutionalized old people in fact found zinc deficiency in institutionalized subjects [55,56]. Further studies also revealed zinc deficiency in hospitalized elderly (<10.7 µM plasma/serum levels) or zinc concentrations <70 µg/dL in serum [57–59]. No differences between plasma and whole blood zinc levels were found when healthy and chronically ill old patients were compared, but a significant decrease in leukocyte zinc was measured [51,60].

Whether zinc deficiency in sick or hospitalized old people is caused by insufficient intake of zinc through hospital diets remains to be clarified. One study showed only an intake of 39% of RDA in homebound ill patients, while hospitalized patients with lower serum and leukocyte levels of zinc do not vary in their mean dietary zinc intake [61].

The relationship of zinc status and immunity has also been reported. These studies showed that only a small difference (1.5 µM) in zinc levels can affect skin testing energy or responsiveness to Diphtheria vaccination [62,63].

Determining the zinc status by measurement of zinc concentrations in serum or plasma is not optimal. In experiments inducing zinc deficiency in young subjects, significant effects on the secretion of interferon (IFN)-γ, interleukin (IL)-2, and tumor necrosis factor (TNF)-α, together with an imbalance in the Th1/Th2 system were detected, notably, plasma zinc was not significantly changed [64]. Moreover, old healthy people displayed plasma zinc levels within reference ranges [65], but had enhanced IL-6 secretion and impaired NK cell cytotoxicity. Therefore, reduced plasma or serum zinc concentrations are indicators for zinc deficiency, but zinc deficiency can take place even when serum/plasma levels are within the reference ranges [9].

In recent years, the determination of the labile intracellular zinc using specific zinc probes in leukocytes became obvious to be a more accurate parameter for present and future studies [3,7,24]. Within the ZINCAGE project, the zinc score (based on the determination of the zinc content in food and the individual quantity of the food intake) may define the zinc status correlated with zinc plasma levels dependent on age [7,66].

There is one outstanding subgroup of old people who have achieved "successful aging," without suffering from age-related diseases, so-called centenarians [67]. Consequently, the analysis of the zinc status was undertaken to find a possible contribution of zinc and successful aging. Nonagenarian and centenarian individuals are reported to have low zinc dietary intake and zinc deficiency [52,65,68]. A significant decrease of serum zinc was detected in oldest subjects >90 years compared to subjects younger than 65 years and ones aged 65–89 years suggesting a continuing decline [52]. Although zinc deficiency correlates with chronic inflammation [69,70], it is assumed that centenarians possess a lower inflammatory status and are able to release sufficient amounts of zinc to maintain an adequate status in immune and antioxidant activity [65]. Thus, the intact immune capacity of centenarians appears to be independent of the measured zinc status.

Interestingly, there are only few studies reporting a high frequency of zinc deficiency in the elderly. One study found zinc-deficient subjects of 90 years and older compared to reference data of the same laboratory from younger subjects [52,68]. In a different study, mean serum levels of 61.8 μg/dL were observed in old subjects from South Africa (mean age 71.7 years) that implies that 76.3% of the study population were zinc-deficient (<70 μg/dL) [71]. Others described that 42.9% of the old subjects investigated had a sufficient intake of zinc (>67% RDA) [45]. Notably, this study and other studies used higher RDAs of 15 mg (male) and 12 mg (female). However, when using the current, lower RDAs these participants would show a higher zinc deficiency. Additionally, by measuring their granulocyte and lymphocyte zinc content, 30% were also defined to be zinc-deficient. It is conspicuous that the zinc plasma concentrations did not indicate zinc deficiency in these subjects, confirming the weakness of this parameter [45].

ZINC AND IMMUNOSENESCENCE

The consequences of zinc deficiency on immunity could also be demonstrated in animal studies [72]. Gestational zinc deficiency reduced immune function in mice observed until the third filial generation [73]. In addition, zinc deficiency in rats appears to affect postthymic T-cell maturation with fewer new T cells in the periphery contributing to immunodeficiency [74]. The striking example of severe zinc deficiency in humans is the syndrome *acrodermatitis enteropathica*, a rare autosomal recessive disorder caused by a mutation in the ZIP4 gene. The intestinal zinc transporter ZIP4 takes up zinc from food and the exocrine pancreas. This disorder displays severe immune defects plus thymic atrophy leading to death within few years without treatment, but zinc supplementation of 1 mg/kg body weight is able to adjust all symptoms [75–77]. Thus, zinc malabsorption and its effects on the immune function and health can be corrected by zinc treatment [14].

Old people have increased susceptibility to infections, autoimmune diseases, and cancer [21]. But one has to address that a lot of other factors including lifestyle and underlying diseases significantly influence the onset in each individual. The immune capacity to build an effective immune response declines with age, starting around 60–65 years. Since the changes of the immune system during zinc deprivation and aging are largely similar, a potential relationship between immunosenescence and zinc deficiency is indicated.

The restricted immune capacity of the elderly is associated with increased incidence and mortality of infectious diseases such as pneumonia [78] and tuberculosis [79], or reactivation of herpes zoster [80]. A lot of studies could demonstrate the beneficial effects of zinc supplementation with respect to infectious diseases in humans [28]. As an example, when zinc was given (as sulfate, 7 mg per day) in a combination of a mixture of vitamins and minerals for one year, individuals with low serum zinc (<70 μg/dL, corresponding to 30% of the study group) showed a higher risk to develop pneumonia than those with adequate zinc levels [81].

The frequency of autoimmune diseases is enhanced during aging, along with higher autoantibody levels [13,82]. The latter could not be detected in centenarians [83]. However, the reported reduction in IgE production with age may dampen the risk to develop allergies [84].

Finally, cancer occurs more frequently in the old population [85]. There is an 11-fold higher incidence of cancer and a 15-fold higher mortality in old subjects over 65 years compared to younger ones [86].

Since minimal changes in the zinc status are able to influence immune responses, disturbed bioavailability or defects in leukocyte activity can be affected by zinc. This is supported by the observations that zinc interferes with signal transduction, and inflammatory signaling in immune cells, including protein kinase C, casein kinase (CK)2, lymphocyte protein-tyrosine kinase (lck), cyclic nucleotide phosphodiesterases, mitogen-activated protein kinase (MAPK) phosphatases, and protein-tyrosine kinases as targets [13,87–91]. Additionally, inhibiting effects of zinc on activating [92] and signaling of the transcription factor nuclear factor (NF)-κB pathway, responsible for the expression of pro-inflammatory cytokines (eg, IL-1β, TNF-α), have been described [4,93].

Innate Immunity

Neutrophil granulocytes (polymorphonuclear neutrophils, PMN) represent the largest group of leukocytes. They build the first line of defense against pathogenic microorganisms, fighting them by phagocytosis, via release of antimicrobial molecules, and production of reactive oxygen species (ROS). They are also inflammatory and regulatory cells with the ability to secrete chemokines and cytokines, and to recruit other immune cells to the site of infection [94,95]. The higher frequency of microbial infections in the elderly is often thought to be caused by an impaired T cell function, but may also be a result from reduced PMN functions. Although the total number of PMN is not different between old and young subjects, phagocytosis, chemotaxis, oxidative burst, and intracellular killing are modified [19,20,96]. PMN from the elderly display an impairment in chemotaxis and a lower resistance to apoptosis, based on a reduced antiapoptotic effect after stimulation with lipopolysaccharide (LPS), granulocyte colony-stimulating factor (G-CSF), and granulocyte-monocyte (GM)-CSF [19,94,95]. Their differentiation within and their recruitment from the bone marrow is diminished [95]. Zinc deprivation leads to reduced phagocytosis, influences the production of reactive oxygen species, and PMN chemotaxis, and decreased formation of neutrophil extracellular traps (NETs) [4,5,12,97]. Thus, zinc deficiency affects the major defense mechanisms of PMN leading to reduced capability to efficiently fight infections which in turn may shorten the healthy life span of the elderly. Moroni et al. found in nonagenarians that a sufficient intracellular zinc ion bioavailability could preserve the oxidative burst by PMN which is caused by downregulating IL-6 signaling [98]. Zinc administration restores some functions of PMN, and reduce infections (Fig. 39.1).

In contrast to other immune cells, several functions of monocytes and macrophages are even increased in old people. The number of monocytes in the blood of old subjects remain equal compared to younger ones, and no changes in chemotaxis, phagocytosis, and oxidative burst are observed. Although no alterations in the expression of many cytokines, adhesion molecules, and HLA-DR could be detected, the monocyte accessory function for T cells is reduced [99]. The synthesis of pro-inflammatory cytokines (IL-1, IL-6, IL-8, and TNF-α) is significantly increased after LPS stimulation, as well as IL-6, IL-8, monocyte chemoattractant protein (MCP)-1, macrophage inflammatory protein-1α, and TNF-α levels in plasma are positively correlated with age [13,100–102]. Zinc signals are required for the synthesis of pro-inflammatory cytokines (IL-1, IL-6 and TNF-α) in monocytes [92]. Zinc depletion promotes monocyte adhesion and leads to enhanced phagocytosis and oxidative burst, but to a reduction of IL-6 and TNF-α in monocytes, suggesting that monocytes may change to basic innate defensive functions in response to zinc deficiency [103]. Calcitriol-dependent monocyte differentiation is increased under zinc depletion, suggesting a blocking function of zinc on maturation of monocytes [104]. However, high-dose zinc supplementation is able to inhibit the secretion of pro-inflammatory cytokines from LPS-stimulated monocytes [90], and can also restore IFN-α production from monocytes of elderly subjects (Fig. 39.1) [105].

NK and NKT cells belong to the innate immune system. NK cells are the first line of defense to delete cancer- and virus-infected cells. Since there is a higher incidence of malignancies and viral infections in the old population, the integrity of their function, especially cytotoxicity, is of main importance. In the elderly, a rise of total numbers and higher percentages of NK cells in the circulation are found [7,106–110]. In contrast, NK cells of the elderly display impaired cytotoxicity and proliferation after IL-2 stimulation, along with reduced calcium signaling and CD69 expression [110–112]. Zinc is also needed for the binding of HLA-C on target cells to killer inhibitory receptors on NK cells; therefore, zinc deficiency may lead to more nonspecific killing [113].

Muzzioli et al. reported that in vitro incubation with zinc (10 μM) enhanced the generation of NK cells from CD34$^+$ progenitor cells in young (CD56$^+$ CD16$^-$ cells) and old (CD56$^-$ CD16$^+$ or CD56$^+$CD16$^+$ cells) subjects caused by amplified GATA-3 transcription factor

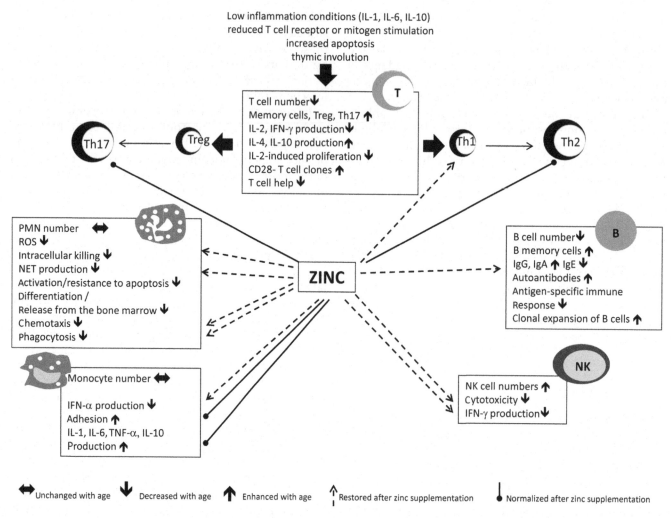

FIGURE 39.1 Age-dependent changes in the immune system: effect of zinc. During zinc deficiency and with aging the number, function, and responses of leukocytes are changed. Zinc counteracts these effects by adjusting the Th1–Th2 and Th17-Treg imbalance leading to a restored antibody production, and normalized Th1 and Treg functions. This improves the success of vaccination, Th1 cell-mediated defense, and lowers the incidence of autoimmune disorders. The effect of zinc on pro-inflammatory cytokines decreases the development of chronic inflammatory diseases. In addition, zinc supplementation ameliorates the innate defense against viruses and cancer by restoration of IFN-γ, IFN-α production, and NK cell cytolytic activity. And finally, the enhanced phagocytic and chemotactic activity, and impaired ROS and NET generation by PMN after zinc administration supports the destruction of invading microorganisms and, therefore, reduces infections.

expression [114]. Studies with old animals and in humans showed that zinc deficiency is involved in reduced NK cell cytotoxicity [107]. The beneficial influence of zinc was supported by observations that in vivo (12 mg zinc/day) and in vitro (1 μM zinc) supplementation for 1 month completely restored NK cell cytotoxicity in old subjects and mice [7,107,115]. In vitro zinc supplementation (30 or 60 μM, 8 d) revealed a higher IFN-γ and decreased IL-10 release and a higher proportion of NK cells [116]. Although data from centenarians and very old mice support the importance of zinc supply to correct NK cell properties, it should be noted that centenarians preserved NK cell cytotoxicity, zinc ion bioavailability, and sufficient IFN-γ synthesis [65].

NKT cells expressing either αβ or γδ T-cell receptors (TCR) produce Th1 (IFN-γ) and Th2 (IL-4) cytokines and are strongly involved in defense against viruses, tumors, and intracellular bacteria. Zinc enhanced liver cell NKT cytotoxicity in old and very old mice indicating that improved capacities of NKT cells and good zinc ion bioavailability may be important for successful aging [106].

Adaptive Immunity

The main changes induced by aging or zinc deficiency affect the T-cell system.

Within aging, delayed proliferation after exposure to mitogens or stimulation via the CD3/T-cell receptor

(TCR), and an altered CD4/CD8 ratio have been observed [25]. Higher expression of the death receptor CD95 (Fas), and the increase of proapoptotic BAX in combination with a decrease of antiapoptotic BCL-2 and p53 result in enhanced apoptosis of T cells from old subjects [117].

Thymic involution leads to a lower generation of naïve T cells (CD45RA$^+$), and to increased numbers of memory (CD45R0$^+$) T cells. Additionally, the increased production of pro-inflammatory cytokines and impaired IFN-α generation observed with age influence the regulation of T helper cell subpopulations. During aging, a decrease of Th1 (IFN-γ and IL-2) cytokines is observed, while Th2 (IL-4 and IL-10) cytokines increase [18]. Changes in the levels of Th1 and Th2 cytokines result in an imbalance of the Th1/Th2 paradigm (Fig. 39.1). The shift to higher Th2 cytokine production affects a chronic low grade of inflammation (inflammaging) [27]. This condition is also accompanied by increased serum levels of pro-inflammatory cytokines IL-1β, IL-6, TNF-α, and the acute phase protein C-reactive protein (CRP) as well as reduced anti-inflammatory IL-10 [7,27].

Regulatory T cells (Tregs) play an important role in the regulation of T-cell-mediated immune responses through suppression of T-cell proliferation and secretion of inhibitory cytokines, such as IL-10 and transforming growth factor-beta [118]. Gregg and coworkers found an increase in peripheral blood CD4$^+$CD25^{high} Tregs associated with aging [119]. Pro-inflammatory Th17 cells are involved in the defense against extracellular bacteria and in autoimmune disorders [120]. Healthy elderly have increased differentiation of IL-17-producing effector cells but decreased IL-17 production from memory CD4$^+$T cells compared to the healthy young [121]. Even the balance of Th17 and regulatory T cells (Treg) is disturbed during aging [122]. Another feature of T cell senescence is the lack of the costimulation molecule CD28 on T cells; CD28 is important for T cell activation, proliferation, and survival. This CD28 negative subpopulation is present in large numbers, particularly within the CD8 subset, leading to deficient Th cell function for B cells and to reduced cytotoxic CD8$^+$ T-cell function [123].

Notably, an imbalance of Th1/Th2 cytokines is also detected in zinc deficiency characterized by impaired IFN-γ, IL-2 and TNF-α release (Th1 cytokine response), and amplified or less affected Th2 response (IL-4, IL-6, IL-10) [124,125]. Zinc supplementation can restore the Th1/Th2 balance [124]. High doses of zinc can inhibit the development of Th17 cells in mice [126]. Experiments using zinc-sufficient HUT-78 cells showed that mRNA levels of IFN-γ, IL-12Rβ2, and T-bet in PMA/PHA-stimulated cells were enhanced when compared to zinc-deficient cells. Although the intracellular zinc levels were not increased, zinc supplementation appears to favor Th1 differentiation [127]. Prasad and coworkers described that decreased IL-2 synthesis and NF-κB activation in peripheral blood mononuclear cells (PBMC) from old subjects with zinc deficiency could be corrected by 45 mg/day zinc gluconate administration for 6 months [128]. The beneficial zinc effects on T cells are displayed in Fig. 39.1.

One of the hallmarks of immunosenescence is thymic involution [129]. Reduced numbers of naïve T cells are also the result of the loss of thymic hormone activity necessary for T-cell maturation and differentiation [130].

Thymic involution regardless of age also occurs in zinc deficiency [131,132]. A reduction of zinc availability leads to increased thymocyte apoptosis, either by elevating glucocorticoid production or by the negative regulatory function of zinc in immune cell apoptosis [6,133]. Oral zinc supplementation in old mice (zinc sulfate) restores serum thymic hormone (thymulin) activity [134], lower zinc concentrations may also contribute to thymic involution by elevating apoptosis during T-cell maturation and selection [135,136]. Additionally, the thymic output measured by T-cell receptor rearrangement excision circles analysis is clearly diminished during aging and in zinc deficiency leading to lower numbers of naïve T cells in the circulation [129]. The situation is different in centenarians where the thymic output can sufficiently maintained by IL-7 [137]. Interestingly, IL-7 and the IL-7 receptor mediate their effects through zinc finger proteins which in turn are regulated by different zinc finger proteins [7,138].

B Cells and Humoral Immunity

During aging, changes are also detected in B cells where a reduction in B cell number is found (Fig. 39.1). Opposite to the reduced B cell numbers an increased production of immunglobulin (Ig) A and several IgG subclasses has been observed (Fig. 39.1) [139]. The impairment of humoral immunity also results from a defect in the help from T cells in combination with modified cytokine production, since several cytokines involved in B cell regulation are altered during aging [23]. This impaired interaction with T helper cells may contribute to the declined response to vaccination [18,19], but also a loss of antibody affinity was detected. Zinc deprivation resulted in a disturbed antibody production by murine B cells [140], and a higher level of apoptosis in pre-B cells was detected in zinc-deficient mice [6]. Mature cells were less affected by zinc induced apoptosis due to a higher expression of

BCL-2 [6]. Therefore, zinc deficiency and immunosenescence may contribute to a reduced response to neoantigens, which may also affect vaccination.

An enhancement of organ-specific and nonorgan-specific autoantibodies in the elderly was found at the same time. The level of nonorgan-specific autoantibodies increase with age while 90 years old individuals have lower organ-specific autoantibody levels compared to younger elderly [141]. Another observation with age is the occurrence of increased clonal expansion of B cells; a possible relationship with increased incidence of lymphocyte malignancies is assumed [141]. The role of zinc in regulating humoral immunity is not clearly defined but there are several indications that zinc supplementation may influence humoral immunity by blocking IL-6 release and amplify IFN-γ production [90,142]. A direct influence of zinc in antibody production has not been finally clarified, but zinc supplementation administered before vaccination promoted antibody generation [5,136,143]. In addition, restored T cell help after zinc supplementation may also contribute to normalized antibody production and improved responses to vaccination (Fig. 39.1).

Additionally, the in vitro supply of zinc promotes the antiviral activity of IFN-α, but not IFN-β and -γ [144].

The age-dependent alterations in the immune system and the effect of zinc are summarized in Fig. 39.1.

ZINC SUPPLEMENTATION IN ELDERLY SUBJECTS

Several studies found a beneficial effect of zinc supplementation on human health. But there are inconsistent data on its beneficial effect upon both, innate and adaptive, immune functions which are due to the usage of different doses, the period of time supplying zinc, as well as the given salt of zinc. In recent reviews, it was reported that zinc supplementation at the doses recommended by RDA when administered as zinc gluconate, aspartate, or acetate, and less using zinc sulfate, delivered the best results [3,7,13].

When zinc was given at high doses (400 mg zinc as sulfate; 90 mg elemental zinc) 15 days before influenza vaccination for 60 days, it had no influence on $CD3^+$, $CD4^+$, and CD8 T cell numbers, and humoral immunity since antibody titer against influenza viral antigens were unchanged to controls [136]. Treatment with 45 mg elemental zinc as gluconate during one year significantly decreased the incidence of infections of elderly in comparison to the placebo group [8]. To examine the influence of zinc supplementation on the immune system, especially on T cells, healthy subjects over 70 years of age received 220 mg zinc sulfate (corresponding to 50 mg of elemental zinc) twice a day for one month in an earlier study. When compared to controls not supplemented with zinc, the proportion of T cells was significantly increased. However, the numbers of circulating T cells as well as the response of T cells to in vitro mitogenic stimulation were not modified [143], and the zinc status was not evaluated. The authors found enhanced delayed type hypersensitivity (DTH) reaction and increased response to tetanus toxoid [143].

Although zinc is a nearly nontoxic essential trace element, excess of zinc intake above the Tolerable Upper Intake Level (UL) (40 mg) can lead to acute adverse effects, including nausea, vomiting, appetite loss, abdominal cramps, headache, and diarrhea [44]. High zinc intake for longer periods results in reduced copper absorption, changed iron metabolism, enhanced apoptosis, and decreased immune function including heavy neutropenia and anemia [5,44,145]. Secondly, zinc supplementation, even in pharmacological concentrations, can affect the adaptive immune response by inhibiting T-cell-dependent allogenic reactions and by reducing B cell numbers momentarily [13,146,147]. Therefore, results from studies that used zinc doses above the UL, especially for a longer period, should be critically viewed.

Additional studies supported the increased effect after zinc supplementation on DTH, but a low number of participants (5–13), different dosages of zinc (55, 60, 30 mg/d) and the lack of control groups attenuate the conclusion of these studies [45,148,149]. Although Cossack and Prasad detected an increase of plasma and cellular zinc as well as increased zinc content in granulocytes and lymphocytes after supplementation, the observed DTH reactions might be a result from repeated testing [13,45,148].

By examining other parameters, beneficial effects of zinc supplementation on T-cell function have been described. First, zinc supplemented (25 mg/d) residents of a retirement home showed significant enhanced numbers of cytotoxic T cells and activated (HLA-DR positive) T helper cells [150]. The increase in HLA-DR positive cells was accompanied by enhanced total T cell numbers, whereas the percentage of activated cells in T cells was unchanged [146,150].

Other studies investigated thymulin activity. In a first study where age-related decreased plasma zinc and decreased plasma thymulin activity have been detected, thymulin activity could be restored by in vitro supply of zinc. This was based on inactivity of thymulin by declined plasma zinc, not by thymic involution [151]. In another study, a partial recovery of thymulin activity after in vitro zinc supplementation was observed in 44 institutionalized old subjects. A 16 week crossover, 8 weeks application of 20 mg zinc/d, then 8 weeks placebo, in the same study led to an significantly enhancement of active thymulin in serum of lean subjects with a body mass index of 21 only [152]. Additionally, others detected significantly increased serum thymulin activity in zinc-deficient elderly subjects after zinc supplementation [45].

Although many studies support a beneficial effect of zinc supplementation on T-cell properties, this seems not to be generally the case. Bogden et al. treated old subjects with placebo, 15 mg, or 100 mg zinc per day; all three groups were supplied with multivitamins and mineral supplements to exclude possible deficiencies in other micronutrients. The authors collected data at the beginning of the study, after three months, and after one year [62,153,154]. They did not find any responses related to dosage of zinc, DTH, and lymphocyte proliferation to different antigens after three months. The beneficial effect of zinc supplementation in subjects with adequate zinc levels may be temporary, because the body may adapt to higher zinc intake [153]. This is supported by findings that increased NK cell activity was observed in the group receiving 100 mg zinc after 3 months, but not after 6 or 12 months. However, an increase in DTH was detected in all three groups after one year.

There might be several explanations why zinc supplementation had no effect on T-cell activity in this study. First, the participants had no zinc deficiency (mean 13 μM; 85 μg/dL) plasma zinc), and it is the first study with a placebo group [154]. Other causes might be, again, repeated testing, an enhancing effect of the administered multivitamin and mineral supplements, or with them and zinc, as well as adaptation to zinc supplementation during the longer supplementation period.

Hodkinson et al. started a six month, placebo-controlled supplementation study with 15 and 30 mg zinc per day (as gluconate) examining the immune status of 93 healthy Irish subjects (55 and 70 years) [155]. Positive correlations between erythrocyte zinc and numbers of T lymphocytes ($CD3^+$), naïve T cells ($CD3^+/CD45RA^+$), NKT cells ($CD3^+/CD16^+/CD56^+$), and activated T cells ($CD25^+HLA-DR^+$), were found. Additionally, the erythrocyte zinc content was inversely correlated with granulocyte phagocytosis and serum zinc with CRP levels [146]. The participants supplemented with 15 mg zinc/day showed an increased helper to cytotoxic T cell ratio. After 3 months, reduced B cell amounts were detected in the 30 mg group, but not in the two other groups. By testing further parameters, zinc supplementation had no effect on, for example, granulocyte phagocytosis, cytokine production by monocytes or inflammation markers. Like in the study performed by Bogden and coworkers, the subjects of this study had mean serum zinc concentrations of 13 μM (85 μg/dL) indicating that there are no long-term effects of zinc supplementation on most immune parameters in old people without zinc deficiency [146].

Zinc supplementation affects the reduction of markers of oxidative stress. Therefore, zinc has also antioxidant effects but in a more indirect manner: via stabilization of cell membranes, as a component of antioxidant enzymes (Cu/Zn superoxide dismutase, SOD1, eSOD), by replacing redox active metal ions (copper, iron), blocking of NAPDH oxidase (inhibition of superoxide-production), and induction and maintenance of radical scavenger MT [4,13,70].

Two studies only found an increase of SOD1 after short- and long-term zinc supplementation [156,157]. In a recent study, Venneria and coworkers evaluate the effect of long-term zinc supplementation (placebo, 15 mg/d and 30 mg/d) on plasma and cellular redox status markers in 108 healthy old subjects from Italy without zinc deficiency for 6 months. The authors did not find any changes in values of carotenoids, vitamin A and E in plasma, glutathione, thiol groups, malondialdehyde, percentage of hemolysis, and methemoglobin in erythrocytes [158]. This study supports the results of former studies [156,157] that zinc supplementation of healthy elderly with an adequate zinc nutritive status and normal macro- and micronutrients intakes seems to be an inefficient way to increase antioxidant defense [158].

There are many publications reporting that zinc supplementation is not only able to improve but also restore immune function. In a study by Prasad, 35% of the old individuals were considered basically zinc-deficient defined by plasma zinc concentrations, and showed significantly higher percentage of cells producing IL-1β and TNF-α [8]. The authors described a significant higher expression of IL-2 mRNA in zinc-deficient old people after zinc supplementation (45 mg/d, as elemental zinc). These observations were consistent with other reports including cell culture experiments with HUT-78 cells from this group [64,127]. Moreover, after zinc supply enhanced T-cell function by decreasing unspecific activated T cells and increased T-cell response to mitogenic stimulation have been observed [8,159,160]. In the 2007 study, the mean incidence of infections was lower in the zinc-supplemented group, as well as a decreased production of TNF-α and IL-10 compared to the placebo group. In this context, zinc supplementation restored cytokine synthesis in old subjects with high pro-inflammatory cytokine levels when compared to younger donors [160].

Otherwise, the influence of zinc supplementation on cytokine expression is not restricted to the elderly population. A study of young individuals aged 19–31 years receiving 15 mg zinc sulfate per day showed amplified IFN-γ mRNA in stimulated T cells as well as enhanced IL-1β and TNF-α mRNA expression in LPS-stimulated granulocytes and monocytes [39]. The zinc supplementation studies of the elderly are summarized in Table 39.1.

TABLE 39.1 Zinc Supplementation Studies in the Elderly

Age	Zinc administration	Number of participants	Effect	References
60–89 years	15 or 100 mg zinc as acetate 3 months	36 P 36 S, 15 31 S,100	No effect on DTH and proliferation of lymphocytes (in vitro)	[153]
60–89 years	15 or 100 mg zinc as acetate 12 months	24 P 20 S,15 9 S,100	Negative influence on DTH, enhanced NK cell activation (only after 3 months)	[154]
65–78 years	60 mg zinc as acetate 4.5 months	8 S	Increase in DTH, plasma, and cellular zinc	[148]
Over 70 years	100 mg zinc as sulfate 1 month	15 C 15 S	Enhanced DTH, T cells, response to tetanus toxoid	[143]
64–76 years	55 mg zinc as sulfate four weeks	5 S	Enhanced DTH	[149]
50–80 years	30 mg zinc as gluconate 6 months	13 S	Improvement of plasma thymulin activity, DTH and IL-1	[45]
73–106 years	20 mg zinc as gluconate 8 weeks	44 P/S	Improved thymulin activity	[152]
55–70 years	15/30 mg zinc as gluconate 6 months	31 P 28/34 S	No influence on markers of inflammation or immunity	[146]
64–100 years	90 mg zinc as sulfate 60 days	190 C 160 S	No influence of zinc on vaccination (response to influenza)	[136]
Over 65 years	25 mg zinc as sulfate 3 months	30 P 28 S	Enhancement of CD4$^+$DR$^+$ T cells and cytotoxic T cells in contrast to placebo	[150]
65–82 years	10 mg zinc as aspartate 7 weeks	19 S	Decrease of reduced levels of activated Th cells and basal IL-6 release from PBMC, T-cell response enhanced	[159,160]
55–87 years	45 mg zinc as gluconate 12 months 45 mg zinc as gluconate 6 months	25 P 24 S 6 P 6 S	Impaired incidence of infections enhanced IL-2 mRNA after PHA stimulation (ex vivo)	[8,128]
60–92 years	10 mg zinc as aspartate 7 weeks	108 C/S	SOD1, SOD3 activity increased	[157]
55–87 years	15 mg and 30 mg zinc as gluconate 3 and 6 months	387	No effect on antioxidant markers, SOD1 activity increased	[156]
70–80 years	15 mg 30 mg 6 months	108 C/S	No effect on antioxidant markers	[158]

The concentrations of zinc are shown as elemental zinc. C, groups without zinc supplementation, control; S, groups with zinc supplementation; P, groups with placebo; DTH, delayed type of hypersensitivity; SOD1, Cu/Zn superoxide dismutase; SOD3, extracellular superoxide dismutase. Table was modified according to Ref. [13].

The main question remains whether zinc supplementation is useful to correct the immune competence, dampen oxidative stress and inflammation, and therefore has beneficial effects on the elderly population. As mentioned before, there are contradictory results about the success of zinc treatment on improvement of innate and adaptive immunity, since many studies used different doses, times of supplementation, and zinc in different forms [7,13]. But even short-term zinc supplementation corrects thymic endocrine activity, $CD4^+$ cell amounts, proliferation capacity of lymphocytes, NK cell cytotoxicity, pro-inflammatory cytokines, and reduces relapsing infections in elderly and old infected patients [7,8,29,65,159]. These observations strongly underline the usefulness of zinc supplementation in the therapy of diseases, and the elderly population with zinc deficiency. In this context, future challenges will be (1) to define methods that clearly prove zinc deficiency, (2) to define (genetic) biomarkers which enable the choice of elderly people that need zinc supplementation, and (3) to define dosage, time of treatment, and application form. Nevertheless, the long-term objective is the determination of an individually tailored treatment with zinc.

EPIGENETIC EVENTS, POLYMORPHISM AND GENETIC MARKERS: RELATIONSHIP TO ZINC DEFICIENCY IN THE ELDERLY

In the last years, many studies were conducted to address the question of whether and how genetic modifications regulate zinc distribution and homeostasis in the elderly. On the other hand, identifying genetic motifs or polymorphisms that predict the need for zinc supply is another aim to improve healthy aging [66] (reviewed in Ref. [35]). Additionally, the restoration of altered genes or polymorphisms of genes regulating zinc is another goal to ameliorate, for example, the immune function of the elderly.

On example of gene–nutrient interaction is that zinc supply enhances SLC30A1,7,8 (Znt1,7,8) zinc transporter expression in vitro and in vivo, even if the underlying mechanisms have to be uncovered [7]. As already outlined, correlations between methylation of specific CpG sites in the SLC30A5 (ZnT5) promoter and age has been reported potentially resulting in declined zinc absorption [41]. Recently, the same group also identified a specific sequence, the zinc transcriptional regulatory element (ZTRE), in the ZnT5 promoter. ZTRE is also found in several genes of the SLC30 family, including SLC30A10 (ZnT10), which is repressed by zinc [161]. The authors concluded that zinc-induced signals may be repressed via this regulatory motif.

Reduced SLC39A6 (ZIP6) expression and dysregulation in aging was related to enhanced pro-inflammatory response, and an increase in ZIP6 promoter methylation using an in vitro cell culture system and an aged mouse model. Dietary supplementation could reconstitute zinc status and led to reduced aged-associated inflammation. These experiments indicate that age-dependent epigenetic dysregulation of zinc transporter expression affects cellular zinc levels and promotes inflammation that can be adjusted by zinc supplementation [162]. According to this, zinc is also involved in the epigenetic regulation of human IL-1β and TNF-α genes. Long-term zinc deprivation leads to enhanced accessibilities of the human IL-1β and TNF-α promoters and, consequently, to significantly higher expression of these cytokines in promyeloid cells [163]. In addition, elevated ROS production was detected, suggesting that zinc deficiency may promote the low inflammatory and oxidant status in the elderly by epigenetic modulation.

The promoters of MT and some zinc transporters are regulated by the metal-response element binding transcription factor (MTF)-1; its binding is controlled through the stabilization of its zinc-finger motifs by soluble cellular zinc [40,164,165]. But future studies will show if MTF-1 is modulated by zinc supplementation and age.

Single nucleotide polymorphisms in pro-inflammatory genes, MT, and zinc transporter genes may also play a role in gene–nutrient interaction in the elderly. Within the ZINCAGE study, genotype data for the -174G/C IL-6 SNP were analyzed in 819 healthy old Europeans. A significant interaction of zinc diet score and GG (-174G/C) genotype on higher plasma IL-6 level was found indicating gene–nutrient interaction [66]. Elderly carriers with the GG genotype of the IL-6 -174 locus have enhanced IL-6 production, low intracellular zinc ion availability, impaired innate immune response and enhanced MT, whereas those with GC and CC genotypes display adequate intracellular zinc and innate immune response, and may reach centenarian age [166]. A former study of this group revealed that a zinc-aspartate supplementation of a healthy low grade inflamed elderly population resulted in alterations in IL-6 and MCP-1 concentrations and NK lytic activity, suggesting an interactive effect of polymorphic alleles of MT1A and IL-6 genes on zinc [115]. The effects of zinc significantly increased when the MT1A + 647 C/A transition was present, indicating that the interleukin-6 and MT1A genes control zinc-regulated gene expression [167]. In this context, also SNP -209 A/G in the promoter of the MT2A gene is associated with higher IL-6 plasma levels, hyperglycemia, and zinc deficiency in old diabetic/atherosclerotic patients carrying the AA genotype [168]. Hence, genetic variations of IL-6 and MT1/2A may be useful tools for the identification of old people who effectively need zinc treatment to restore NK cell cytotoxicity and improve zinc status. A screening of the SNP +1245 MT1A in 110 healthy old participants (72 ± 6 years) from the ZINCAGE study, supplied with 10 mg/day zinc aspartate for 7 weeks, was performed [169]. Although +1245 MT1A G+ carriers revealed increased advanced glycation end-products and ROS production in PBMCs after zinc supplementation basically, no differences of these parameters including MT activity were observed between different genotypes. Only higher intracellular labile zinc and higher NO-induced release of zinc in PBMC of G+ individuals were detected [169]. The heat shock protein (Hsp)70 1267 A/G SNP was found to be related to altered TNF-α, IL-6 plasma levels and to zinc plasma levels [170]. The allele and genotype distribution of Hsp70 1267 A/G were similar among various European countries, in contrast to TNF-α-308 G/A SNP. Thus, Hsp70 1267 A/G SNP is indicated as a putative marker for individual susceptibility to inflammatory diseases [170].

Taken together, several genetic markers and motifs in zinc transporters, zinc binding proteins, and pro-inflammatory cytokine genes are candidates to identify elderly people for zinc supplementation or to define the optimal intake of micronutrients at the molecular level. But one has to take into account that the frequencies of genetic variations (SNPs) are differently distributed in various countries and continents [170,171].

CONCLUSIONS

Zinc is not only needed for cell-mediated immune response, it is able to correct defects in the immune response. Therefore, zinc supplementation will be an effective therapeutic agent for old individuals with zinc deficiency, and for the treatment of chronic diseases, and some malignancies [14,29]. But methods to clearly define zinc deficiency as well as equal conditions for zinc application should be basic prerequisites before starting treatment. Genetic biomarkers to identify old people who need zinc supplementation will be important tools for individual therapies in the future.

SUMMARY POINTS

- Zinc deficiency affects both parts of the immune system but mainly the adaptive part.
- There are striking similarities in modified immune capabilities observed with age and zinc deficiency.
- The elderly show an inadequate intake of zinc, but not a clear prevalence of zinc deficiency.
- Minor changes in zinc can modify the immune response of elderly with mild zinc deficiency.
- Most studies show an improvement of the immune status of the elderly after zinc supply.
- Zinc supplementation can decrease chronic inflammation and enhance the resistance to infection in old subjects.

References

[1] Prasad AS, Farid Z, Sandstead HH, Miale A, Schulert AR. Zinc metabolism in patients with syndrome of iron deficiency anemia hepatosplenomegaly dwarfism and hypogonadism. J Lab Clin Med 1963;61(4):537–49.
[2] Prasad AS. Discovery of human zinc deficiency: 50 years later. J Trace Elem Med Biol 2012;26(2–3):66–9.
[3] Haase H, Mocchegiani E, Rink L. Correlation between zinc status and immune function in the elderly. Biogerontology 2006;7(5–6):421–8.
[4] Haase H, Rink L. Zinc signals and immune function. Biofactors 2014;40(1):27–40.
[5] Maywald M, Rink L. Zinc homeostasis and immunosenescence. J Trace Elem Med Biol 2015;29:24–30.
[6] Fraker PJ, King LE. Reprogramming of the immune system during zinc deficiency. Annu Rev Nutr 2004;24:277–98.
[7] Mocchegiani E, Romeo J, Malavolta M, Costarelli L, Giacconi R, Diaz LE, et al. Zinc: dietary intake and impact of supplementation on immune function in elderly. Age 2013;35(3):839–60.
[8] Prasad AS, Beck FWJ, Bao B, Fitzgerald JT, Snell DC, Steinberg JD, et al. Zinc supplementation decreases incidence of infections in the elderly: effect of zinc on generation of cytokines and oxidative stress. Am J Clin Nutr 2007;85(3):837–44.
[9] Maret W, Sandstead HH. Zinc requirements and the risks and benefits of zinc supplementation. J Trace Elem Med Biol 2006;20(1):3–18.
[10] Prasad AS. Zinc in human health: effect of zinc on immune cells. Mol Med 2008;14(5–6):353–7.
[11] Ibs KH, Rink L. Zinc-altered immune function. J Nutr 2003;133(5):1452S–6S.
[12] Haase H, Rink L. Multiple impacts of zinc on immune function. Metallomics 2014;6(7):1175–80.
[13] Haase H, Rink L. The immune system and the impact of zinc during aging. Immun Ageing 2009;6:9.
[14] Rink L. Zinc in human health. IOS Press; 2011.
[15] Cousins RJ, Liuzzi JP, Lichten LA. Mammalian zinc transport, trafficking, and signals. J Biol Chem 2006;281(34):24085–9.
[16] Truong-Tran AQ, Carter J, Ruffin RE, Zalewski PD. The role of zinc in caspase activation and apoptotic cell death. Biometals 2001;14(3–4):315–30.
[17] Alvarez-Rodriguez L, Lopez-Hoyos M, Munoz-Cacho P, Martinez-Taboada VM. Aging is associated with circulating cytokine dysregulation. Cell Immunol 2012;273(2):124–32.
[18] Cakman I, Rohwer J, Schutz RM, Kirchner H, Rink L. Dysregulation between TH1 and TH2 T cell subpopulations in the elderly. Mech Ageing Dev 1996;87(3):197–209.
[19] Ibs KH, Rink L. The immune system of elderly. Z Gerontol Geriatr 2001;34(6):480–5.

[20] Panda A, Arjona A, Sapey E, Bai FW, Fikrig E, Montgomery RR, et al. Human innate immunosenescence: causes and consequences for immunity in old age. Trends Immunol 2009;30(7):325–33.

[21] Pawelec G, Larbi A, Derhovanessian E. Senescence of the human immune system. J Comp Pathol 2010;141:S39–44.

[22] Rink L, Seyfarth M. Features of immunological investigations in elderly. Z Gerontol Geriatr 1997;30(3):220–5.

[23] Rink L, Cakman I, Kirchner H. Altered cytokine production in the elderly. Mech Ageing Dev 1998;102(2–3):199–209.

[24] Haase H, Rink L. Functional significance of zinc-related signaling pathways in immune cells. Annu Rev Nutr 2009;29:133–52.

[25] Pawelec G, Adibzadeh M, Pohla H, Schaudt K. Immunosenescence – aging of the immune-system. Immunol Today 1995;16(9):420–2.

[26] Grubeck-Loebenstein B, Della Bella S, Iorio AM, Michel JP, Pawelec G, Solana R. Immunosenescence and vaccine failure in the elderly. Aging Clin Exp Res 2009;21(3):201–9.

[27] Franceschi C, Capri M, Monti D, Giunta S, Olivieri F, Sevini F, et al. Inflammaging and anti-inflammaging: a systemic perspective on aging and longevity emerged from studies in humans. Mech Ageing Dev 2007;128(1):92–105.

[28] Shankar AH, Prasad AS. Zinc and immune function: the biological basis of altered resistance to infection. Am J Clin Nutr 1998;68(2):447S–63S.

[29] Prasad AS. Discovery of human zinc deficiency: its impact on human health and disease. Adv Nutr 2013;4(2):176–90.

[30] Prasad AS. Impact of the discovery of human zinc deficiency on health. J Trace Elem Med Biol 2014;28(4):357–63.

[31] Rink L, Kirchner H. Zinc-altered immune function and cytokine production. J Nutr 2000;130(5):1407S–11S.

[32] Prasad AS. Impact of the discovery of human zinc deficiency on health. J Am Coll Nutr 2009;28(3):257–65.

[33] Prasad AS. Zinc: an antioxidant and anti-inflammatory agent: role of zinc in degenerative disorders of aging. J Trace Elem Med Biol 2014;28(4):364–71.

[34] Mocchegiani E, Giacconi R, Costarelli L, Muti E, Cipriano C, Tesei S, et al. Zinc deficiency and IL-6-174G/C polymorphism in old people from different European countries: effect of zinc supplementation. ZINCAGE study. Exp Gerontol 2008;43(5):433–44.

[35] Mocchegiani E, Costarelli L, Giacconi R, Malavolta M, Basso A, Piacenza F, et al. Micronutrient-gene interactions related to inflammatory/immune response and antioxidant activity in ageing and inflammation. A systematic review. Mech Ageing Dev 2014;136:29–49.

[36] Sandstead HH, Henriksen LK, Greger JL, Prasad AS, Good RA. Zinc nutriture in the elderly in relation to taste acuity, immune-response, and wound-healing. Am J Clin Nutr 1982;36(5):1046–59.

[37] Maret W. Zinc biochemistry: from a single zinc enzyme to a key element of life. Adv Nutr 2013;4(1):82–91.

[38] Wellenreuther G, Cianci M, Tucoulou R, Meyer-Klaucke W, Haase H. The ligand environment of zinc stored in vesicles. Biochem Biophys Res Commun 2009;380(1):198–203.

[39] Aydemir TB, Blanchard RK, Cousins RJ. Zinc supplementation of young men alters metallothionein, zinc transporter, and cytokine gene expression in leukocyte populations. Proc Natl Acad Sci USA 2006;103(6):1699–704.

[40] Lichten LA, Cousins RJ. Mammalian zinc transporters: nutritional and physiologic regulation. Annu Rev Nutr 2009;29:153–76.

[41] Coneyworth LJ, Mathers JC, Ford D. Does promoter methylation of the SLC30A5 (ZnT5) zinc transporter gene contribute to the ageing-related decline in zinc status? Proc Nutr Soc 2009;68(2):142–7.

[42] Jansen J, Karges W, Rink L. Zinc and diabetes – clinical links and molecular mechanisms. J Nutr Biochem 2009;20(6):399–417.

[43] Jansen J, Rosenkranz E, Overbeck S, Warmuth S, Mocchegiani E, Giacconi R, et al. Disturbed zinc homeostasis in diabetic patients by in vitro and in vivo analysis of insulinomimetic activity of zinc. J Nutr Biochem 2012;23(11):1458–66.

[44] Institute of Medicine, Food and Nutrition Board. Dietary reference intakes for Vitamin A, Vitamin K, arsenic, boron, chromium, copper, iodine, iron, manganese, molybdenum, nickel, silicon, vanadium and zinc. Washington (DC): National Academy Press; 2001.

[45] Prasad AS, Fitzgerald JT, Hess JW, Kaplan J, Pelen F, Dardenne M. Zinc-deficiency in elderly patients. Nutrition 1993;9(3):218–24.

[46] Andriollo-Sanchez M, Hininger-Favier I, Meunier N, Toti E, Zaccaria M, Brandolini-Bunlon M, et al. Zinc intake and status in middle-aged and older European subjects: the ZENITH study. Eur J Clin Nutr 2005;59:S37–41.

[47] Kogirima M, Kurasawa R, Kubori S, Sarukura N, Nakamori M, Okada S, et al. Ratio of low serum zinc levels in elderly Japanese people living in the central part of Japan. Eur J Clin Nutr 2007;61(3):375–81.

[48] Briefel RR, Bialostosky K, Kennedy-Stephenson J, McDowell MA, Ervin RB, Wright JD. Zinc intake of the US population: findings from the third National Health and Nutrition Examination Survey, 1988–1994. J Nutr 2000;130(5):1367S–73S.

[49] Hotz C, Peerson JM, Brown KH. Suggested lower cutoffs of serum zinc concentrations for assessing zinc status: reanalysis of the second National Health and Nutrition Examination Survey data (1976–1980). Am J Clin Nutr 2003;78(4):756–64.

[50] Lindeman RD, Clark ML, Colmore JP. Influence of age and sex on plasma and red-cell zinc concentrations. J Gerontol 1971;26(3):358–63.

[51] Bunker VW, Hinks LJ, Lawson MS, Clayton BE. Assessment of zinc and copper status of healthy elderly people using metabolic balance studies and measurement of leukocyte concentrations. Am J Clin Nutr 1984;40(5):1096–102.

[52] Ravaglia G, Forti P, Maioli F, Bastagli L, Facchini A, Mariani E, et al. Effect of micronutrient status on natural killer cell immune function in healthy free-living subjects aged >=90 y. Am J Clin Nutr 2000;71(2):590–8.

[53] Vir SC, Love AHG. Zinc and copper status of the elderly. Am J Clin Nutr 1979;32(7):1473–6.

[54] Ligthart GJ, Corberand JX, Fournier C, Galanaud P, Hijmans W, Kennes B, et al. Admission criteria for immunogerontological studies in man – the senieur protocol. Mech Ageing Dev 1984;28(1):47–55.

[55] Goode HF, Penn ND, Kelleher J, Walker BE. Evidence of cellular zinc depletion in hospitalized but not in healthy elderly subjects. Age Ageing 1991;20(5):345–8.

[56] Worwag M, Classen HG, Schumacher E. Prevalence of magnesium and zinc deficiencies in nursing home residents in Germany. Magnes Res 1999;12(3):181–9.

[57] Belbraouet S, Biaudet H, Tebi A, Chan N, Gray-Donald K, Debry G. Serum zinc and copper status in hospitalized vs. healthy elderly subjects. J Am Coll Nutr 2007;26(6):650–4.

[58] Girodon F, Blache D, Monget AL, Lombart M, Brunet-Lecompte P, Arnaud J, et al. Effect of a two-year supplementation with low doses of antioxidant vitamins and/or minerals in elderly subjects on levels of nutrients and antioxidant defense parameters. J Am Coll Nutr 1997;16(4):357–65.

[59] Pepersack T, Rotsaert P, Benoit F, Willems D, Fuss M, Bourdoux P, et al. Prevalence of zinc deficiency and its clinical relevance among hospitalised elderly. Arch Gerontol Geriatr 2001;33(3):243–53.

[60] Bunker VW, Delves HT. Accurate determination of selenium in biological-materials without perchloric-acid for digestion. Anal Chim Acta 1987;201:331–4.

[61] Stafford W, Smith RG, Lewis SJ, Henery E, Stephen PJ, Rafferty J, et al. A study of zinc status of elderly institutionalized patients. Age Ageing 1988;17(1):42–8.

[62] Bogden JD, Oleske JM, Munves EM, Lavenhar MA, Bruening KS, Kemp FW, et al. Zinc and immunocompetence in the elderly — base-line data on zinc nutriture and immunity in unsupplemented subjects. Am J Clin Nutr 1987;46(1): 101–9.

[63] Kreft B, Wohlrab J, Fischer M, Uhlig H, Skolziger R, Marsch WC. Analysis of the serum zinc level in patients with atopic dermatitis, psoriasis vulgaris and in subjects with healthy skin. Hautarzt 2000;51(12):931–4.

[64] Beck FWJ, Prasad AS, Kaplan J, Fitzgerald JT, Brewer GJ. Changes in cytokine production and T cell subpopulations in experimentally induced zinc-deficient humans. Am J Physiol Endocrinol Metab 1997;272(6):E1002–7.

[65] Mocchegiani E, Muzzioli M, Giacconi R, Cipriano C, Gasparini N, Franceschi C, et al. Metallothioneins/PARP-1/IL-6 interplay on natural killer cell activity in elderly: parallelism with nonagenarians and old infected humans. Effect of zinc supply. Mech Ageing Dev 2003;124(4):459–68.

[66] Kanoni S, Dedoussis GV, Herbein G, Fulop T, Varin A, Jajte J, et al. Assessment of gene-nutrient interactions on inflammatory status of the elderly with the use of a zinc diet score — ZINCAGE study. J Nutr Biochem 2010;21(6):526–31.

[67] Franceschi C, Monti D, Sansoni P, Cossarizza A. The immunology of exceptional individuals — the lesson of centenarians. Immunol Today 1995;16(1):12–16.

[68] Ravaglia G, Forti P, Maioli F, Nesi B, Pratelli L, Savarino L, et al. Blood micronutrient and thyroid hormone concentrations in the oldest-old. J Clin Endocrinol Metab 2000;85(6):2260–5.

[69] Mocchegiani E, Costarelli L, Giacconi R, Cipriano C, Muti E, Malavolta M. Zinc-binding proteins (metallothionein and alpha-2 macroglobulin) and immunosenescence. Exp Gerontol 2006;41(11):1094–107.

[70] Prasad AS. Zinc: role in immunity, oxidative stress and chronic inflammation. Curr Opin Clin Nutr Metab Care 2009;12(6):646–52.

[71] Oldewage-Theron WH, Samuel FO, Venter CS. Zinc deficiency among the elderly attending a care centre in Sharpeville, South Africa. J Hum Nutr Diet 2008;21(6):566–74.

[72] Beisel WR. Single nutrients and immunity. Am J Clin Nutr 1982;35(2):417–68.

[73] Beach RS, Gershwin ME, Hurley LS. Gestational zinc deprivation in mice — persistence of immunodeficiency for 3 generations. Science 1982;218(4571):469–71.

[74] Blewett HJ, Taylor CG. Dietary zinc deficiency in rodents: effects on T-cell development, maturation and phenotypes. Nutrients 2012;4(6):449–66.

[75] Kury S, Dreno B, Bezieau S, Giraudet S, Kharfi M, Kamoun R, et al. Identification of SLC39A4, a gene involved in acrodermatitis enteropathica. Nat Genet 2002;31(3):239–40.

[76] Neldner KH, Hambidge KM. Zinc therapy of acrodermatitis enteropathica. N Engl J Med 1975;292(17):879–82.

[77] Wang K, Zhou B, Kuo YM, Zemansky J, Gitschier J. A novel member of a zinc transporter family is defective in acrodermatitis enteropathica. Am J Hum Genet 2002;71(1):66–73.

[78] Plouffe JF, Breiman RF, Facklam RR. Bacteremia with Streptococcus pneumoniae — Implications for therapy and prevention. J Am Med Assoc 1996;275(3):194–8.

[79] Davies PDO. Tuberculosis in the elderly. J Antimicrob Chemother 1994;34:93–100.

[80] Donahue JG, Choo PW, Manson JE, Platt R. The incidence of Herpes-Zoster. Arch Intern Med 1995;155(15):1605–9.

[81] Meydani SN, Barnett JB, Dallal GE, Fine BC, Jacques PF, Leka LS, et al. Serum zinc and pneumonia in nursing home elderly. Am J Clin Nutr 2007;86(4):1167–73.

[82] Mariotti S, Sansoni P, Barbesino G, Caturegli P, Monti D, Cossarizza A, et al. Thyroid and other organ-specific autoantibodies in healthy centenarians. Lancet 1992;339(8808):1506–8.

[83] Steinmann G, Hartwig M. Immunology of Centenarians. Immunol Today 1995;16(11):549.

[84] Schwarzenbach HR, Nakagawa T, Conroy MC, Deweck AL. Skin reactivity, basophil de-granulation and ige levels in aging. Clin Allergy 1982;12(5):465–73.

[85] Fulop T, Kotb R, Fortin CF, Pawelec G, de Angelis F, Larbi A. Potential role of immunosenescence in cancer development aging, cancer, and age-related diseases: common mechanism? Ann N Y Acad Sci 2010; 1197:158–165.

[86] Yancik R. Cancer burden in the aged — an epidemiologic and demographic overview. Cancer 1997;80(7):1273–83.

[87] Honscheid A, Dubben S, Rink L, Haase H. Zinc differentially regulates mitogen-activated protein kinases in human T cells. J Nutr Biochem 2012;23(1):18–26.

[88] Huse M, Eck MJ, Harrison SC. A Zn^{2+} ion links the cytoplasmic tail of CD4 and the N-terminal region of Lck. J Biol Chem 1998;273(30):18729–33.

[89] Kim PW, Sun ZYJ, Blacklow SC, Wagner G, Eck MJ. A zinc clasp structure tethers Lck to T cell coreceptors CD4 and CD8. Science 2003;301(5640):1725–8.

[90] von Bulow V, Rink L, Haase H. Zinc-mediated inhibition of cyclic nucleotide phosphodiesterase activity and expression suppresses TNF-α and IL-1 beta production in monocytes by elevation of guanosine 3′,5′-cyclic monophosphate. J Immunol 2005;175(7):4697–705.

[91] Wilson M, Hogstrand C, Maret W. Picomolar concentrations of free zinc(II) ions regulate receptor protein-tyrosine phosphatase beta activity. J Biol Chem 2012;287(12):9322–6.

[92] Haase H, Ober-Bloebaum JL, Engelhardt G, Hebel S, Heit A, Heine H, et al. Zinc signals are essential for lipopolysaccharide-induced signal transduction in monocytes. J Immunol 2008;181(9):6491–502.

[93] von Bulow V, Dubben S, Engelhardt G, Hebel S, Plumakers B, Heine H, et al. Zinc-dependent suppression of TNF-α production is mediated by protein kinase A-induced inhibition of Raf-1, IκB kinase beta, and NF-κB. J Immunol 2007;179(6):4180–6.

[94] Schroder AK, Rink L. Neutrophil immunity of the elderly. Mech Ageing Dev 2003;124(4):419–25.

[95] Wessels I, Jansen J, Rink L, Uciechowski P. Immunosenescence of polymorphonuclear neutrophils. ScientificWorldJournal 2010;10:145–60.

[96] Shaw AC, Joshi S, Greenwood H, Panda A, Lord JM. Aging of the innate immune system. Curr Opin Immunol 2010;22(4):507–13.

[97] Hasan R, Rink L, Haase H. Zinc signals in neutrophil granulocytes are required for the formation of neutrophil extracellular traps. Innate Immun 2013;19(3):253–64.

[98] Moroni F, Paolo M, Rigo A, Cipriano C, Giacconi R, Recchioni R, et al. Interrelationship among neutrophil efficiency, inflammation, antioxidant activity and zinc pool in very old age. Biogerontology 2005;6(4):271–81.

[99] Rich EA, Mincek MA, Armitage KB, Duffy EG, Owen DC, Fayen JD, et al. Accessory function and properties of monocytes from healthy elderly humans for T-lymphocyte responses to mitogen and antigen. Gerontology 1993;39(2):93–108.

[100] Gabriel P, Cakman I, Rink L. Overproduction of monokines by leukocytes after stimulation with lipopolysaccharide in the elderly. Exp Gerontol 2002;37(2−3):235−47.

[101] Mariani E, Cattini L, Neri S, Malavolta M, Mocchegiani E, Ravaglia G, et al. Simultaneous evaluation of circulating chemokine and cytokine profiles in elderly subjects by multiplex technology: relationship with zinc status. Biogerontology 2006;7(5−6):449−59.

[102] Maylor EA, Simpson EEA, Secker DL, Meunier N, Andriollo-Sanchez M, Polito A, et al. Effects of zinc supplementation on cognitive function in healthy middle-aged and older adults: the ZENITH study. Br J Nutr 2006;96(4):752−60.

[103] Mayer LS, Uciechowski P, Meyer S, Schwerdtle T, Rink L, Haase H. Differential impact of zinc deficiency on phagocytosis, oxidative burst, and production of pro-inflammatory cytokines by human monocytes. Metallomics 2014;6(7):1288−95.

[104] Dubben S, Honscheid A, Winkler K, Rink L, Haase H. Cellular zinc homeostasis is a regulator in monocyte differentiation of HL-60 cells by 1α,25-dihydroxyvitamin D-3. J Leukoc Biol 2010;87(5):833−44.

[105] Cakman I, Kirchner H, Rink L. Zinc supplementation reconstitutes the production of interferon-alpha by leukocytes from elderly persons. J Interferon Cytokine Res 1997;17(8):469−72.

[106] Mocchegiani E, Malavolta M. NK and NKT cell functions in immunosenescence. Aging Cell 2004;3(4):177−84.

[107] Mocchegiani E, Giacconi R, Cipriano C, Malavolta M. NK and NKT cells in aging and longevity: role of zinc and metallothioneins. J Clin Immunol 2009;29(4):416−25.

[108] Solana R, Alonso MC, Pena J. Natural killer cells in healthy aging. Exp Gerontol 1999;34(3):435−43.

[109] Solana R, Mariani E. NK and NK/T cells in human senescence. Vaccine 2000;18(16):1613−20.

[110] Solana R, Campos C, Pera A, Tarazona R. Shaping of NK cell subsets by aging. Curr Opin Immunol 2014;29:56−61.

[111] Borrego F, Alonso MC, Galiani MD, Carracedo J, Ramirez R, Ostos B, et al. NK phenotypic markers and IL2 response in NK cells from elderly people. Exp Gerontol 1999;34(2):253−65.

[112] Solana R, Tarazona R, Gayoso I, Lesur O, Dupuis G, Fulop T. Innate immunosenescence: effect of aging on cells and receptors of the innate immune system in humans. Semin Immunol 2012;24(5):331−41.

[113] Rajagopalan S, Winter CC, Wagtmann N, Long EO. The Ig-related killer-cell inhibitory receptor binds zinc and requires zinc for recognition of Hla-C on target-cells. J Immunol 1995;155(9):4143−6.

[114] Muzzioli M, Stecconi R, Moresi R, Provinciali M. Zinc improves the development of human CD34 + cell progenitors towards NK cells and increases the expression of GATA-3 transcription factor in young and old ages. Biogerontology 2009;10(5):593−604.

[115] Mariani E, Neri S, Cattini L, Mocchegiani E, Malavolta M, Dedoussis GV, et al. Effect of zinc supplementation on plasma IL-6 and MCP-1 production and NK cell function in healthy elderly: interactive influence of + 647 MT1a and -174 IL-6 polymorphic alleles. Exp Gerontol 2008;43(5):462−71.

[116] Metz CHD, Schroder AK, Overbeck S, Kahmann L, Plumakers B, Rink L. T-helper type 1 cytokine release is enhanced by in vitro zinc supplementation due to increased natural killer cells. Nutrition 2007;23(2):157−63.

[117] Mcleod JD. Apoptotic capability in ageing T cells. Mech Ageing Dev 2000;121(1−3):151−9.

[118] Sakaguchi S, Wing K, Yamaguchi T. Dynamics of peripheral tolerance and immune regulation mediated by Treg. Eur J Immunol 2009;39(9):2331−6.

[119] Gregg R, Smith CM, Clark FJ, Dunnion D, Khan N, Chakraverty R, et al. The number of human peripheral blood CD4(+) CD25(high) regulatory T cells increases with age. Clin Exp Immunol 2005;140(3):540−6.

[120] Noack M, Miossec P. Th17 and regulatory T cell balance in autoimmune and inflammatory diseases. Autoimmun Rev 2014;13(6):668−77.

[121] Lee JS, Lee WW, Kim SH, Kang Y, Lee N, Shin MS, et al. Age-associated alteration in naive and memory Th17 cell response in humans. Clin Immunol 2011;140(1):84−91.

[122] Schmitt V, Rink L, Uciechowski P. The Th17/Treg balance is disturbed during aging. Exp Gerontol 2013;48(12):1379−86.

[123] Weng NP, Akbar AN, Goronzy J. CD28(−) T cells: their role in the age-associated decline of immune function. Trends Immunol 2009;30(7):306−12.

[124] Prasad AS. Effects of zinc deficiency on Th1 and Th2 cytokine shifts. J Infect Dis 2000;182:S62−8.

[125] Uciechowski P, Kahmann L, Plumakers B, Malavolta M, Mocchegiani E, Dedoussis G, et al. TH1 and TH2 cell polarization increases with aging and is modulated by zinc supplementation. Exp Gerontol 2008;43(5):493−8.

[126] Kitabayashi C, Fukada T, Kanamoto M, Ohashi W, Hojyo S, Atsumi T, et al. Zinc suppresses T(h)17 development via inhibition of STAT3 activation. Int Immunol 2010;22(5):375−86.

[127] Bao B, Prasad AS, Beck FWJ, Bao GW, Singh T, Ali S, et al. Intracellular free zinc up-regulates IFN-gamma and T-bet essential for Th-1 differentiation in Con-A stimulated HUT-78 cells. Biochem Biophys Res Commun 2011;407(4):703−7.

[128] Prasad AS, Bao B, Beck FWJ, Sarkar FH. Correction of interleukin-2 gene expression by in vitro zinc addition to mononuclear cells from zinc-deficient human subjects: a specific test for zinc deficiency in humans. Transl Res 2006;148(6):325−33.

[129] Mitchell WA, Meng I, Nicholson SA, Aspinall R. Thymic output, ageing and zinc. Biogerontology 2006;7(5-6):461−70.

[130] Arnold CR, Wolf J, Brunner S, Herndler-Brandstetter D, Grubeck-Loebenstein B. Gain and loss of T cell subsets in old age-age-related reshaping of the T cell repertoire. J Clin Immunol 2011;31(2):137−46.

[131] Dardenne M. Zinc and immune function. Eur J Clin Nutr 2002;56:S20−3.

[132] Mocchegiani E, Malavolta M, Muti E, Costarelli L, Cipriano C, Piacenza F, et al. Zinc, metallothioneins and longevity: interrelationships with niacin and selenium. Curr Pharm Des 2008;14(26):2719−32.

[133] Taub DD, Longo DL. Insights into thymic aging and regeneration. Immunol Rev 2005;205:72−93.

[134] Sbarbati A, Mocchegiani E, Marzola P, Tibaldi A, Mannucci R, Nicolato E, et al. Effect of dietary supplementation with zinc sulphate on the aging process: a study using high field intensity MRI and chemical shift imaging. Biomed Pharmacother 1998;52(10):454−8.

[135] King LE, Osati-Ashtiani F, Fraker PJ. Apoptosis plays a distinct role in the loss of precursor lymphocytes during zinc deficiency in mice. J Nutr 2002;132(5):974−9.

[136] Provinciali M, Montenovo A, Di Stefano G, Colombo M, Daghetta L, Cairati M, et al. Effect of zinc or zinc plus arginine supplementation on antibody titre and lymphocyte subsets after influenza vaccination in elderly subjects: a randomized controlled trial. Age Ageing 1998;27(6):715−22.

[137] Nasi M, Troiano L, Lugli E, Pinti M, Ferraresi R, Monterastelli E, et al. Thymic output and functionality of the IL-7/IL-7 receptor system in centenarians: implications for the neolymphogenesis at the limit of human life. Aging Cell 2006;5(2):167−75.

[138] Saba I, Kosan C, Vassen L, Moroy T. IL-7R-dependent survival and differentiation of early T-lineage progenitors is regulated by the BTB/POZ domain transcription factor Miz-1. Blood 2011;117(12):3370−81.

[139] Paganelli R, Quinti I, Fagiolo U, Cossarizza A, Ortolani C, Guerra E, et al. Changes in circulating B-cells and immunoglobulin classes and subclasses in a healthy aged population. Clin Exp Immunol 1992;90(2):351–4.

[140] Depasqualejardieu P, Fraker PJ. Interference in the development of a secondary immune-response in mice by zinc deprivation — persistence of effects. J Nutr 1984;114(10):1762–9.

[141] Weksler ME, Szabo P. The effect of age on the B-cell repertoire. J Clin Immunol 2000;20(4):240–9.

[142] Driessen C, Hirv K, Rink L, Kirchner H. Induction of cytokines by zinc ions in human peripheral-blood mononuclear-cells and separated monocytes. Lymphokine Cytokine Res 1994;13(1):15–20.

[143] Duchateau J, Delepesse G, Vrijens R, Collet H. Beneficial-effects of oral zinc supplementation on the immune-response of old-people. Am J Med 1981;70(5):1001–4.

[144] Berg K, Bolt G, Andersen H, Owen TC. Zinc potentiates the antiviral action of human IFN-α tenfold. J Interferon Cytokine Res 2001;21(7):471–4.

[145] Prasad AS, Brewer GJ, Schoomaker EB, Rabbani P. Hypocupremia induced by zinc therapy in adults. J Am Med Assoc 1978;240(20):2166–8.

[146] Hodkinson CF, Kelly M, Alexander HD, Bradbury I, Robson PJ, Bonham MP, et al. Effect of zinc supplementation on the immune status of healthy older individuals aged 55–70 years: the ZENITH study. J Gerontol Ser A Biol Sci Med Sci 2007;62(6):598–608.

[147] Faber C, Gabriel P, Ibs KH, Rink L. Zinc in pharmacological doses suppresses allogeneic reaction without affecting the antigenic response. Bone Marrow Transplant 2004;33(12):1241–6.

[148] Cossack ZT. T-Lymphocyte dysfunction in the elderly associated with zinc-deficiency and subnormal nucleoside phosphorylase-activity — effect of zinc supplementation. Eur J Cancer Clin Oncol 1989;25(6):973–6.

[149] Wagner PA, Jernigan JA, Bailey LB, Nickens C, Brazzi GA. Zinc nutriture and cell-mediated-immunity in the aged. Int J Vitam Nutr Res 1983;53(1):94–101.

[150] Fortes C, Agabiti N, Fano V, Pacifici R, Forastiere F, Virgili F, et al. Zinc supplementation and plasma lipid peroxides in an elderly population. Eur J Clin Nutr 1997;51(2):97–101.

[151] Fabris N, Amadio L, Licastro F, Mocchegiani E, Zannotti M, Franceschi C. Thymic hormone deficiency in normal aging and downs-syndrome — is there a primary failure of the thymus. Lancet 1984;1(8384):983–6.

[152] Boukaiba N, Flament C, Acher S, Chappuis P, Piau A, Fusselier M, et al. A physiological amount of zinc supplementation — effects on nutritional, lipid, and thymic status in an elderly population. Am J Clin Nutr 1993;57(4):566–72.

[153] Bogden JD, Oleske JM, Lavenhar MA, Munves EM, Kemp FW, Bruening KS, et al. Zinc and immunocompetence in elderly people — effects of zinc supplementation for 3 months. Am J Clin Nutr 1988;48(3):655–63.

[154] Bogden JD, Oleske JM, Lavenhar MA, Munves EM, Kemp FW, Bruening KS, et al. Effects of one year of supplementation with zinc and other micronutrients on cellular-immunity in the elderly. J Am Coll Nutr 1990;9(3):214–25.

[155] Hodkinson CF, Kelly M, Coudray C, Gilmore WS, Hannigan BM, O'Connor JM, et al. Zinc status and age-related changes in peripheral blood leukocyte subpopulations in healthy men and women aged 55–70 y: the ZENITH study. Eur J Clin Nutr 2005;59:S63–7.

[156] Andriollo-Sanchez M, Hininger-Favier I, Meunier N, Venneria E, O'Connor JM, Maiani G, et al. No antioxidant beneficial effect of zinc supplementation on oxidative stress markers and antioxidant defenses in middle-aged and elderly subjects: the zenith study. J Am Coll Nutr 2008;27(4):463–9.

[157] Mariani E, Mangialasche F, Feliziani FT, Ceechetti R, Malavolta M, Bastiani P, et al. Effects of zinc supplementation on antioxidant enzyme activities in healthy old subjects. Exp Gerontol 2008;43(5):445–51.

[158] Venneria E, Intorre F, Foddai MS, Azzini E, Palomba L, Raguzzini A, et al. Antioxidant effect of zinc supplementation on both plasma and cellular red-ox status markers in a group of elderly Italian population. J Nutr Health Aging 2014;18(4):345–50.

[159] Kahmann L, Uciechowski P, Warmuth S, Malavolta M, Mocchegiani E, Rink L. Effect of improved zinc status on T helper cell activation and TH1/TH2 ratio in healthy elderly individuals. Biogerontology 2006;7(5–6):429–35.

[160] Kahmann L, Uciechowski P, Warmuth S, Pliimdkers B, Gressner AM, Malavolta M, et al. Zinc supplementation in the elderly reduces spontaneous inflammatory cytokine release and restores T cell functions. Rejuvenation Res 2008;11(1):227–37.

[161] Coneyworth LJ, Jackson KA, Tyson J, Bosomworth HJ, van der Hagen E, Hann GM, et al. Identification of the human Zinc Transcriptional Regulatory Element (ZTRE) a palindromic protein-binding DNA sequence responsible for zinc-induced transcriptional repression. J Biol Chem 2012;287(43):36567–81.

[162] Wong CP, Ho E. Zinc and its role in age-related inflammation and immune dysfunction. Mol Nutr Food Res 2012;56(1):77–87.

[163] Wessels I, Haase H, Engelhardt G, Rink L, Uciechowski P. Zinc deficiency induces production of the proinflammatory cytokines IL-1 beta and TNF-α in promyeloid cells via epigenetic and redox-dependent mechanisms. J Nutr Biochem 2013;24(1):289–97.

[164] Lichten LA, Liuzzi JP, Cousins RJ. Zinc suppresses hepatic Zip10 expression through activation of MTF-1. Faseb J 2007;21(5):A170.

[165] Lichten LA, Ryu MS, Guo L, Embury J, Cousins RJ. MTF-1-mediated repression of the Zinc transporter Zip10 Is alleviated by zinc restriction. Plos One 2011;6(6).

[166] Mocchegiani E, Costarelli L, Giacconi R, Piacenza F, Basso A, Malavolta M. Zinc, metallothioneins and immunosenescence: effect of zinc supply as nutrigenomic approach. Biogerontology 2011;12(5):455–65.

[167] Mazzatti DJ, Malavolta M, White AJ, Costarelli L, Giacconi R, Muti E, et al. Effects of interleukin-6-174C/G and metallothionein 1A+647A/C single-nucleotide polymorphisms on zinc-regulated gene expression in ageing. Exp Gerontol 2008;43(5):423–32.

[168] Giacconi R, Cipriano C, Muti E, Costarelli L, Maurizio C, Saba V, et al. Novel-209A/G MT2A polymorphism in old patients with type 2 diabetes and atherosclerosis: relationship with inflammation (IL-6) and zinc. Biogerontology 2005;6(6):407–13.

[169] Giacconi R, Simm A, Santos AN, Costarelli L, Malavolta M, Mecocci P, et al. Influence of +1245 A/G MT1A polymorphism on advanced glycation end-products (AGEs) in elderly: effect of zinc supplementation. Genes Nutr 2014;9(5).

[170] Giacconi R, Costarelli L, Malavolta M, Piacenza F, Galeazzi R, Gasparini N, et al. Association among 1267 A/G HSP70-2,-308 G/A TNF-α polymorphisms and pro-inflammatory plasma mediators in old ZincAge population. Biogerontology 2014;15(1):65–79.

[171] Uciechowski P, Oellig EM, Mariani E, Malavolta M, Mocchegiani E, Rink L. Effects of human Toll-like receptor 1 polymorphisms on ageing. Immun Ageing 2013;10.

C H A P T E R

40

Testing the Ability of Selenium and Vitamin E to Prevent Prostate Cancer in a Large Randomized Phase III Clinical Trial: The Selenium and Vitamin E Cancer Prevention Trial

Barbara K. Dunn[1], Ellen Richmond[2], Darrell E. Anderson[3] and Peter Greenwald[4]

[1]Chemopreventive Agent Development Research Group, Division of Cancer Prevention, National Cancer Institute/National Institutes of Health, Bethesda, MD, USA
[2]Gastrointestinal and Other Cancers Research Group, Division of Cancer Prevention, National Cancer Institute/National Institutes of Health, Bethesda, MD, USA [3]Gray Sourcing, Inc., La Mesa, CA, USA [4]Office of the Director, National Cancer Institute/National Institutes of Health, Bethesda, MD, USA

KEY FACTS
- Randomized: participants were randomly assigned to one of four possible treatment groups
- Prospective: participants free from prostate cancer were followed over time
- Double-blind: participants did not know to which treatment group they were assigned
- Placebo-controlled: effects of the different treatments were compared to the effects of inert treatments
- 2 × 2 factorial design: two agents were tested alone and in combination
- Doses: 200 µg/day L-selenomethoinine and 400 mg/day racemic mix of α-tocopherol
- Sample size: 35,533 men
- Primary outcome: incident prostate cancer rates
- Intervention length: 7–12 years

- Predetermined comparisons: vitamin E versus placebo, selenium versus. placebo, vitamin E plus selenium versus placebo, vitamin E plus selenium versus vitamin E alone, vitamin E plus selenium versus selenium alone

Dictionary of Terms

List of Abbreviations
- *ATBC*—The Alpha-Tocopherol Beta-Carotene Cancer Prevention Trial
- *NPC*—Nutritional Prevention of Cancer trial
- *OR*—odds ratio
- *PSA*—prostate-specific antigen
- *RR*—relative risk
- *SELECT*—Selenium and Vitamin E Cancer Prevention Trial
- *SU.VI.MAX*—Supplementation en Vitamins et Mineraux Antioxidants Trial

- *SWOG*—Southwest Oncology Group
- *USPSTF*—United States Preventive Services Task Force

Definitions of Words and Terms

- *Cancer prevention* refers to the prevention of initiation of the carcinogenic process as well as the halting, reversal, or slowing down of ongoing carcinogenesis so that the pre-malignant state never transforms into invasive cancer.
- *Chemoprevention* is the prevention of disease before it becomes clinically evident by intervening with pharmaceutical agents/drugs. Some authors include intervention with nutritional agents in chemoprevention. In this paper the term chemoprevention applies specifically to the prevention, or reduction in risk, of cancer.
- The *primary endpoint* of a randomized clinical trial is the most important outcome that we are interested in. We want to see if the intervention, whether a drug or a nutritional agent, leads to an improvement in the primary disease endpoint. In chemoprevention of cancer trials, this endpoint is usually the occurrence, or incidence, of cancer. The statistical design of a trial is based on its primary endpoint.
- The *randomized clinical trial* is often considered the "gold standard" of study design for trials of drug interventions in humans. The randomization, whereby participants are randomly assigned to each of the two or more drug groups or placebo (the control) group, ensures that there is no bias in the assignments. This is the only way that any benefit derived from the drug intervention can be trusted as being real and not the result of a bias in the assignment of participants to the different groups.
- *SELECT* is an acronym for the Selenium and Vitamin E Cancer Prevention Trial. This clinical trial, which was run by the Division of Cancer Prevention in the National Cancer Institute together with the Southwest Oncology Group (SWOG), tested the two nutrients in the title for their ability to reduce the risk of prostate cancer in a population of men at moderately increased risk by virtue of age.
- *Selenium* is a trace element that is essential in the diet.
- *Vitamin E* is a vitamin that is essential in the diet.

INTRODUCTION

Prostate Cancer

Prostate cancer is the most commonly diagnosed noncutaneous cancer, and is the second leading cause of cancer death among men in the United States (US). Nearly 3 million men in the US are living with prostate cancer [1], with one in six expected to be diagnosed with this disease in his lifetime. In 2014, an estimated 233,000 new cases will be diagnosed in the US and approximately 29,480 men will die of prostate cancer [2]. Effective treatment with surgery or radiation, largely palliative, results in a five-year survival rate approaching 100% when diagnosed in the localized or regional stages, as are 81% and 12% of prostate cancers, respectively; when diagnosed in the distant stage (4% of all prostate cancers), the five-year survival rate drops to 28%. Although most known risk factors for prostate cancer, including age, race, and genetic factors are nonmodifiable, a number of risk factors—obesity, physical activity, and possibly dietary factors—are potentially modifiable.

Early detection of prostate cancer, specifically by screening clinically healthy men with prostate-specific antigen (PSA), has been a standard part of clinical practice for a number of years. Since PSA screening was instituted, the incidence of prostate cancer has risen rapidly, far exceeding prostate cancer mortality which has remained essentially stable. This incidence-mortality relationship is highly suggestive of overdiagnosis, which has led to overtreatment in men with non-fatal disease [3], in some cases resulting in decreased quality-of-life due to side effects like declining urinary, bowel, or sexual functioning [1,4]. In 2012, the United States Preventive Services Task Force changed its stance to strongly recommend against PSA screening for prostate cancer, a screening modality that it once endorsed [5]. Because prostate cancer has a long natural history, mainly nonmodifiable risk factors, and an incidence rate that far exceeds the mortality rate, a focus on prevention over screening or early detection offered a more appealing way to decrease the burden of this disease.

Early efforts to decrease prostate cancer incidence through chemoprevention included the use of the antiandrogens finasteride and dutasteride. However, two large multicenter randomized controlled clinical trials found that both agents slightly increased the rates of high grade prostate cancers [6,7]. Due to concerns about increasing risks of high grade prostate cancer, a US Food and Drug Administration (FDA) advisory panel voted overwhelmingly not to approve finasteride or dutasteride for prostate cancer prevention [8]. Concomitant with the interest and subsequent waning interest in antiandrogens, an independent approach to prostate cancer chemoprevention evolved using nutritional agents, specifically vitamin E and selenium. Secondary analyses in two large-scale chemoprevention trials designed to study reduction of other cancers, as well as further controlled intervention trials, human observational studies, and preclinical studies, pointed to selenium and vitamin E as having potential to decrease the risk of prostate cancer [9,10]. Importantly, however, some preclinical studies suggested no benefit, but few of these were published [4,11]. Based on these background observations, the Selenium and Vitamin E Cancer Prevention Trial (SELECT) was conceived and undertaken to test the chemopreventive efficacy of these micronutrients. In this chapter, we will describe the rationale, results, and implications of SELECT.

Selenium

Sources, Metabolism, and Biological Activities of Selenium

Selenium is a nutritionally essential trace mineral found in soil in the inorganic forms of selenate (SeO_4^{2-}) and selenite (SeO_3^{2-}). Selenium is converted by plants from these inorganic to organic forms, including the major form L-selenomethionine and, in lesser amounts, L-selenocysteine. Therefore, selenium concentrations in foods are dependent on the selenium content of the soil in which the foods or ingredients were grown. Increasing fortification of the food system with selenium and increasing nationwide shipping of food products, however, have lessened concerns that low selenium soil content might affect the selenium status of individuals [12].

The most abundant amounts of selenium are found in Brazil nuts, meats, fish, eggs, and cereals, with lesser amounts in cruciferous vegetables, garlic, and mushrooms. Selenium is also found in dietary supplements, including multivitamin/multimineral supplements, as selenomethionine or as selenized yeast (baker's yeast grown in a selenium-rich medium) can provide up to 2000 µg/g selenium, over 90% of which is selenomethionine.

A central role for dietary selenium, including the major dietary forms selenomethionine, selenocysteine, selenate, or selenite, is to provide the starting material for synthesis of selenoproteins (Fig. 40.1) [13]. All dietary forms of selenium can be used for selenoprotein synthesis via their conversion to hydrogen selenide. Clinical efficacy of selenium, including its chemopreventive effects, derives from the activities of intracellular selenium metabolites, including methyl selenol (reviewed in Ref. [4]).

Nutritional Requirements

The nutritional requirements for selenium are not clearly defined [4]. The recommended dietary allowance (RDA) for adult men and women for selenium is 55 µg/day [12]. This is based on the daily selenium intake necessary to achieve maximal activity of GPx-3, a selenoprotein that maintains redox balance by detoxification of hydrogen peroxide. Because not all selenoproteins' activity levels would be maximal with the RDA intake level, the suggestion has been made that an RDA of 80 µg/day may be more appropriate for men to achieve selenium balance, although the physiological implications for maximal activity levels of all selenoproteins at higher intakes of selenium remain unknown. Another approach is to examine the estimated average requirement (EAR) of selenium, which represents the average daily intake level estimated to meet the requirements of half of the healthy individuals in a group. The EAR for adults is 45 µg/day.

Based on studies from China on the health effects of various levels of selenium, deficiency symptoms (eg, immunoincompetency, progression of viral infections, and reproductive symptoms) became apparent with intakes <11 µg/day and toxicity symptoms (eg, hair and nail brittleness/loss, gastrointestinal disturbances, skin rash, fatigue, or irritability) appeared with 800 µg/day of selenium intake. Based on this observation, the tolerable upper limit (UL) for selenium intake is estimated to be 400 µg/day. In the US most adults have adequate selenium intake. Based on data from the National Health and Nutrition Examination Survey (NHANES) 2003–06, less than 1% of adults had intakes below the estimated average requirements (EAR) and even fewer had intakes above the tolerable UL [14], with approximately 19% of males reporting taking a dietary supplement that contained selenium.

Vitamin E

Sources, Metabolism, and Biological Activities of Vitamin E

Tocopherols, collectively known as "vitamin E," are the major lipid-soluble antioxidants in cell membranes and act as peroxyl and alkoxyl free radical scavengers (reviewed in Ref. [15]). Naturally occurring vitamin E has eight different forms: the α-, β-, γ-, and δ-isomers of tocopherol and tocotrienol, with α-tocopherol the most active, the only isomer that is maintained in the plasma, and the only isomer that is recognized to meet human requirements.

The most abundant dietary sources of α-tocopherol are nuts, seeds, and oils, with lesser amounts in green leafy vegetables or fortified cereal products. The primary function of vitamin E is as a free radical scavenger/antioxidant in the lipid phase of cell membranes. α-tocopherol does, however, exhibit other cellular activities, such as regulation of cell growth, proliferation, apoptosis, adhesion, angiogenesis, and inflammation (reviewed in Ref. [16]).

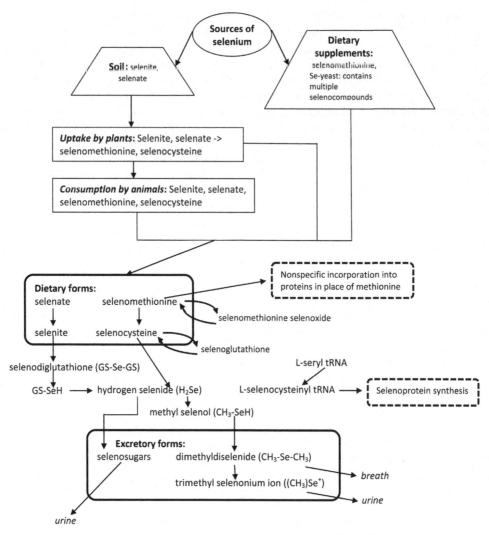

FIGURE 40.1 Selenium metabolism and distribution. Selenium, in various forms, may enter the body through consumption of foods or through the use of dietary supplements. Selenium in the soil exists as selenite, selenate, selenomethionine, or selenocysteine, and after uptake by plants and consumption by animals, selenomethionine may be nonspecifically incorporated into proteins in place of methionine. Selenium not incorporated in selenomethionine may be converted to hydrogen selenide, a key intermediate for selenoprotein synthesis; hydrogen selenide also may be converted to excretory forms. *Adapted from Ref. [4].*

Nutritional Requirements

The RDA for vitamin E for adult men and women is 15 mg; the EAR is 12 mg. Vitamin E deficiency manifests as peripheral neuropathy, is exceedingly rare, and occurs almost exclusively as a result of genetic abnormalities in alpha-tocopherol transfer protein (α-TTP) or as a result of protein-energy malnutrition. Rather, The RDA is not based on vitamin E deficiency resulting from low dietary intake, which has never been clinically described [12], but instead is based on amounts of vitamin E sufficient to prevent hydrogen peroxide-induced hemolysis. Little is known about possible harmful effects of extremely high levels of vitamin E, although even prior to the release of SELECT results, adverse events in recent clinical trials had begun to emerge, highlighting the need for more research into this area, particularly concerning long-term use of vitamin E supplements [17]. The tolerable upper limit for vitamin E intake is 1000 mg/day, based on the lowest observed intake level that resulted in hemorrhagic events [12].

Because of the multiple forms of vitamin E and α-tocopherol, a useful approach to expressing vitamin E intake and requirements is in terms of biological activity. One international unit (IU) of vitamin E activity is defined as having the activity of 1 mg of all-racemic α-tocopherol acetate, 0.67 mg D-α-tocopherol, or 0.74 mg D-α-tocopherol acetate [12]. The RDA of 15 mg corresponds to an RDA of 22.4 IU. NHANES analyses in 2003–06 show that, in stark contrast to selenium, $93.3 \pm 0.4\%$ of people >2 years of age did not meet the vitamin E requirement from food [14]. Among users of vitamin E-containing dietary supplements, 60.3 ± 0.8% did not meet the RDA requirement. Similar to selenium supplements, vitamin E supplements were more likely to be used by older individuals.

RATIONALE FOR SELECT

Prostate Cancer Prevention by Selenium

Potential antitumorigenic mechanisms of selenium have been examined in both in vitro and in vivo studies. Supplemental selenium enhances the activity of selenoenzymes, increases carcinogen metabolism, augments natural killer cell-dependent cytotoxicity, and perturbs tumor cell growth and cell cycle progression. In a review of over 100 animal studies that investigated selenium's chemopreventive efficacy and mechanisms of action of selenium, two-thirds pointed to a significant reduction in tumor incidence from selenium; half of these studies showed tumor reductions of 50% or more at a variety of cancer sites [18].

The first observational evidence suggesting that selenium had cancer preventive activity came from ecological analyses, which pointed to an association between a higher incidence of certain cancers and mortality with selenium-deficient regions of the US compared to selenium-replete areas (reviewed in Ref. [4]). Additional case-control studies have revealed a general trend between higher selenium levels (assessed by prediagnostic blood levels, serum levels, toenail selenium levels, or dietary selenium intake) and decreased total cancer incidence and mortality (reviewed in Ref. [18]). For prostate cancer, some selenium studies concur with this general association, while others do not. A matched case-control study nested within the prospective US Cohort Health Professionals Follow-Up Study showed that the highest quintile of toenail selenium level was associated with a decreased risk of advanced prostate cancer compared to the lowest quintile (reviewed in Ref. [4]). In contrast, no such association was observed in two nested case-control studies among men in the European Prospective Investigation into Cancer and Nutrition (EPIC) cohort. EPIC investigators noted that plasma selenium was not associated with prostate cancer risk [15]. The basis for the discrepancies remains unclear, although the EPIC authors noted that the European cohort had substantially lower plasma selenium concentrations than those found in men in the US, suggesting that even the highest selenium levels of men in the EPIC cohort may have been below levels necessary for cancer prevention. The World Cancer Research Fund/American Institute for Cancer Research (WCRF/AICR) reviewed 17 cohort studies, three ecological studies, 14 case-control studies, and one clinical trial (NPC, see below) and concluded that selenium and foods containing selenium probably protect against prostate cancer. This WCRF/AICR conclusion was based on observed dose-response relationships together with evidence for plausible mechanisms (reviewed in Ref. [15]). A meta-analysis contained within this report, which included many studies published before the start of SELECT, indicated a 5% decrease in the risk of prostate cancer and a 13% decrease in the risk of advanced or aggressive prostate cancer for every 10 ng/mL increase in plasma selenium or a 20% decrease in the risk of advanced or aggressive prostate cancer for every 100 ng/g increase in toenail selenium.

Trials in Qidong and Linxian, China, were among the first large randomized trials to demonstrate cancer preventive activities of selenium in humans. These trials focused on liver, gastric, and esophageal, that is, not prostate, cancers. Qidong is a selenium-deficient region where a community intervention trial showed that areas receiving salt supplemented with sodium selenite (providing 50–80 μg of selenium per day) experienced a 35.1% decrease in incidence of primary liver cancer over 8 years compared to control areas (reviewed in Ref. [4]). After the intervention was discontinued and the selenized salt was no longer administered, primary liver cancer increased in incidence in the intervention areas.

Selenium supplementation in multivitamin/multimineral formulations also has been studied for chemoprevention of esophageal and gastric cardia cancers in individuals with esophageal dysplasia. The Linxian Nutritional Intervention Trials found that the group receiving 50 μg selenium, 30 mg vitamin E, and 15 mg β-carotene daily (Factor D) showed a decrease in cancer mortality of 13% and a decrease in stomach cancer mortality of 21% over 5 years compared to those not receiving Factor D [19]. A second Linxian trial found that individuals with esophageal cancer randomized to a multivitamin/multimineral supplement containing 12 micronutrients (including 50 μg selenium) and a separate 15 mg β-carotene supplement daily showed a 4% decrease in total cancer mortality, 8% decrease in gastric/esophageal cancer mortality, and 16% decrease in esophageal cancer mortality, although none of these decreases was statistically significant [20].

The Nutritional Prevention of Cancer (NPC) Trial was the main study underlying the hypothesis that selenium prevents prostate cancer and prompting the use of selenium in SELECT [10]. This randomized, double-blinded, placebo-controlled trial enrolled 1312 participants who were from the eastern US and had a history of skin cancer. Participants were randomized to receive either 200 μg of selenium in the form of selenized yeast or a placebo daily. The primary aim was to test the efficacy of selenium supplementation in preventing nonmelanoma skin cancer, while secondary endpoints included total cancer incidence, total cancer mortality, and mortality due to lung, prostate, and colorectal cancers. The trial outcomes were reported with 8271 person-years of follow-up. Although the selenium intervention did not favorably affect the primary endpoint of skin cancer, the incidence of prostate cancer was reduced in the selenium group by 64% after 4.5 years and by 49% after 10 years. This decrease was most pronounced in former smokers. A stratified analysis showed that selenium supplementation reduced the incidence of prostate cancer in men in the lowest two tertiles of baseline selenium status, but not in those in the highest tertile of baseline selenium status. In addition to baseline selenium status, baseline PSA levels also exhibited a significant interaction with treatment group. Prostate cancer risk was reduced only in men whose baseline PSA levels were ≤4.0 ng/mL, although no significant interaction was observed between treatment and baseline PSA [4]. This trial examined prostate cancer incidence and mortality as a secondary endpoint, greatly stimulating interest in the use of selenium to prevent prostate cancer and highlighting the point that only specific subgroups of men may benefit.

Prostate Cancer Prevention by Vitamin E

The cellular functions of vitamin E are more extensive than the usual described antioxidant activities (reviewed in Ref. [15]). Animal studies have shown that vitamin E prevents various chemically induced tumors, some of which are hormonally mediated. In rats receiving various doses of chemotherapeutic agents, vitamin E has been shown to slow the growth of prostate cancer.

Many foods contain vitamin E but generally at low concentrations, which creates difficulty in assessing dietary intake. Therefore, serum or plasma α-tocopherol concentrations are often used to assess vitamin E status. Observational studies have only inconsistently shown a possible beneficial association between vitamin E status (α-tocopherol circulating levels) or intake and prostate cancer risk (reviewed in Ref. [15]). Prediagnostic serum or plasma vitamin E concentrations have been shown to be lower years prior to prostate cancer diagnosis in cases compared to noncases in a number of positive studies. In contrast, other case-control and cohort studies have reported no such association [15]. The Alpha-Tocopherol Beta-Carotene (ATBC) Cancer Prevention Trial (see below) contains one nested cohort analysis that showed no association between any baseline measures of vitamin E (serum α-tocopherol, dietary vitamin E) and prostate cancer, except when the analysis was limited to the intervention group. A subsequent case-control study, also nested within the ATBC, contained 100 randomly selected incident prostate cancer cases and 200 matched controls. In this study high baseline serum levels for both α- and γ-tocopherol were associated with decreased prostate cancer risk between the highest and lowest tertile for each; this association was stronger in the α-tocopherol-supplemented arm. In contrast, such associations have not been consistently supported by other observational analyses.

A number of randomized clinical trials have tested the effects of high-dose vitamin E supplements (above the 15 mg, or 33 IU, all rac α-tocopherol dietary level). The results of these trials suggested benefits for noncancer endpoints, such as a decrease in coronary heart disease risk. Regarding cancer, the ATBC trial generated the most convincing evidence that vitamin E is associated with a decrease in prostate cancer risk. In this randomized, double-blind, placebo-controlled trial, 29,133 male smokers 50–69 years old were randomized to α-tocopherol (50 mg synthetic dl-α-tocopherol acetate) daily and β-carotene (20 mg) daily alone or in combination [9]. Although the ATBC trial had as its primary endpoint lung cancer incidence, the risk of lung cancer was not affected by α-tocopherol acetate after 5–8 years of follow-up. Paradoxically, the incidence of lung cancer increased among men receiving β-carotene. In contrast, prostate cancer incidence, which was a prespecified secondary endpoint, showed a statistically significant 32% decrease with α-tocopherol. This preventive effect appeared to be stronger for clinically evident cases (stages B–D disease), with participants who received α-tocopherol showing a 40% decrease in disease. Prostate cancer mortality, though based on fewer events, also showed a statistically significant decrease of 41% among the 14,564 men taking vitamin E compared to the men not receiving vitamin E [9]. These findings for prespecified secondary endpoints offered strong support for testing vitamin E in a prospective clinical trial of prostate cancer prevention.

Prostate Cancer Prevention by Selenium and Vitamin E Combined

Although selenium and vitamin E both have antiprostate cancer effects, they are mechanistically distinct, modulating different cellular processes (reviewed in Ref. [15]). These differing anticancer mechanisms are borne out in observations indicating a substantially more pronounced effect of selenium and vitamin E together on molecular markers of anticancer cellular activity than that observed with either nutrient alone. In one study, vitamin E inhibited LNCaP prostate adenocarcinoma cell growth by 47%, selenium by 37%, and the combination by 78%. The combination also induced significantly more apoptosis (37–43%) in prostate cancer cells; this was accompanied by greater increases in the protein Bax/Bcl-2 ratio. In addition, levels of PCNA protein, a marker of cell proliferation, were reduced to a greater extent with the combination than with either nutrient alone. These additive anticancer effects supported the inclusion of a combined treatment arm in SELECT.

Few prostate cancer prevention studies in animal models with tumor endpoints have used selenium or vitamin E (Table 40.1). Only two studies incorporated designs relevant to SELECT. These studies, which tested selenium and vitamin E alone or in combination on prostate cancer incidence and multiplicity, reported null findings [11,21]. Other in vivo selenium and prostate cancer prevention studies either only used selenium in combination with other agents and not alone or used other forms of selenium such as methylseleninic acid or selenite [24].

SELECT

Rationale and Objectives

The Selenium and Vitamin E Cancer Prevention Trial (SELECT) was funded by the US National Cancer Institute (NCI) and implemented by the Southwest Oncology Group (SWOG). The primary objective of SELECT was to assess the efficacy of selenium and vitamin E alone and in combination on the incidence of prostate cancer. Additional prespecified secondary endpoints were included in the study design: prostate cancer-free survival, all-cause mortality, incidence and mortality of other cancers including lung and colorectal cancer, overall cancer incidence and survival, and disease potentially impacted by chronic administration of selenium and/or vitamin E. Other planned studies included monitoring serious cardiovascular events, assessment of quality of life, investigation of the relationship between serum micronutrient levels and prostate cancer risk, and the evaluation of biological and

TABLE 40.1 In Vivo Animal Studies That Assessed Prostate Cancer Prevention by Selenium

	Trial design in studies with direct relevance to SELECT	Agent selection in studies with selenium and vitamin E or other agents	Other forms of selenium studied in animal models
Animal model	1. N-Nitroso-N-methylurea (MNU) + testosterone-treated Wistar-Unilever rats [11] 2. Testosterone + estradiol-treated NBL rat [21]	1. Lady transgenic mice [22] 2. Lady transgenic mice [23]	1. Transgenic adenocarcinoma mouse prostate (TRAMP) model [24] 2. MNU + testosterone-treated Wistar rats [25]
Agent and dose	1. a. L-selenomethionine (1.5 or 3 mg/kg diet) b. DL-α-tocopherol (4000 or 2000 mg/kg diet) c. L-selenomethionine (3 mg/kg diet) + DL-α-tocopherol (2000 or 5000 mg/kg diet) d. Selenized yeast (target Se levels of 9 or 3 mg/kg diet) e. Control 1. a. L-selenomethionine (1.5 or 3.0 mg/kg diet) b. DL-α-tocopherol (4000 or 2000 mg/kg diet) c. Control	1. a. α-tocopherol succinate (800 IU) + L-selenomethionine (200 μg) + lycopene (50 mg) b. α-tocopherol succinate (800 IU) + L-selenomethionine (200 μg) c. Control 1. a. α-tocopherol succinate (800 IU) + L-selenomethionine (200 μg) + lycopene (50 mg) b. Control	1. a. Methylseleninic acid (3 mg selenium/kg body weight) for 10 weeks b. Methylseleninic acid (3 mg selenium/kg body weight) for 16 weeks c. Control 1. a. Sodium selenite (4 mg/L in drinking water/day) b. Control
Results	1. No effect of selenium on prostate cancer incidence in any group 2. No effect on prostate tumor incidence, multiplicity, or mortality in any group	1. No effect on prostate cancer incidence in either group; increased survival ($p < 0.0001$) in both treatment groups compared to controls 2. Four-fold decrease in prostate cancer incidence in treatment group ($p < 0.0001$) compared to controls	1. Decreased cancer-specific mortality ($p_{10weeks} = 0.0078$, $p_{16weeks} = 0.0385$) in methylseleninic acid group compared to control group 2. Decreased prostate cancer multiplicity by 44.6% in the sodium selenite group compared to control group; no effect of sodium selenite on prostate intraepithelial neoplasia

Animal model chemoprevention studies relevant to the SELECT study design. Two studies published after the SELECT was underway investigated rodent prostate cancer models for the effects of L-selenomethionine and DL-α-tocopherol; each reported null findings. The effects of L-selenomethionine and/or vitamin E in combination with lycopene, but not alone, or of the effects of other forms of selenium have been investigated in other studies.
Adapted from Ref. [4].

genetic markers associated with the risk of prostate cancer [26].

Agents: Formulations and Doses

L-selenomethionine was chosen as the selenium formulation for SELECT despite the use of selenized yeast in the hypothesis-generating NPC trial [27]. This recommendation was made by an NCI-sponsored panel of experts and was based on large batch-to-batch variability and lack of commercial availability of selenized yeast. Since laboratory analyses showed L-selenomethionine to be the predominant selenium species in selenized yeast available at the time, there was reason to believe that the daily dose of 200 µg was similar to the dose of 200 µg selenized yeast used in the NPC trial [26]. While 200 µg L-selenomethionine might be expected to deliver more selenium than 200 µg selenized yeast, direct comparisons between the two doses were difficult to make.

The racemic mix of α-tocopherol, containing the D- and L-isomers was the form used in the ATBC trial which showed its association with reduced prostate cancer incidence. Therefore, this formulation of vitamin E was chosen. The dose of 400 mg/day was selected based on its use in vitamin supplements, which suggested safety, and on its potential benefits for noncancer diseases, including cardiovascular disease and Alzheimer's disease. However, this chosen dose was eight times higher than the 50 mg/day used in the ATBC study [28].

Trial Design and Outcome Ascertainment

SELECT was a prospective, randomized, double-blind, placebo-controlled clinical trial using a 2 × 2 factorial design. Eligible healthy men who were at elevated risk for prostate cancer were randomized to the following four groups: selenium alone (daily oral doses of 200 µg selenium plus placebo); vitamin E alone (daily oral doses of 400 mg α-tocopherol plus placebo); 200 µg selenium plus 400 mg α-tocopherol; or two placebos. The original plan was for the study to last for 12 years including 5 years for accrual and 7–12 years of intervention and follow-up. The following five predetermined comparisons were addressed: vitamin E versus placebo; selenium versus placebo; vitamin E plus selenium versus placebo; vitamin E plus selenium versus vitamin E alone; and vitamin E plus selenium versus selenium alone. A sample size of 32,400 men was required in order to provide adequate power to detect ≥25% decreases in the incidence of prostate cancer for selenium or vitamin E alone and an additional 25% decrease for selenium and vitamin E combined compared to either agent alone.

Prostate cancer was assessed on routine clinical diagnostic evaluations, which included yearly digital rectal exams (DRE) and serum PSA measurements. Biopsies were performed only at the discretion of study physicians or "for cause," that is, for participants with either suspicious DREs or elevated PSA levels. Unlike the previous NCI-sponsored prostate cancer chemoprevention trial comparing finasteride to placebo, no end-of-study biopsies were implemented.

Cohort

Elevated risk of prostate cancer was the basis for eligibility for SELECT and was determined by age and/or African ancestry. Caucasian men were required to be ≥55 years and African American men ≥50 years old in order to be eligible. The rationale for this age difference is that African American men aged 50–55 have an incidence of prostate cancer that is comparable to that of Caucasian men aged 55–60. Participating men had to be healthy, have a total PSA ≤4.0 ng/mL, a DRE not suspicious for cancer, no previous prostate cancer or high grade prostate intraepithelial neoplasia, normal blood pressure, no current use of anticoagulation therapy, and be willing to stop taking off-study supplements.

Recruitment Strategies, Accrual, and Adherence

Recruitment Strategies

The SELECT recruitment strategies were formulated as a robust study-specific plan well in advance of protocol finalization to maximize timely participant accrual [29–31]. The trial's plan included a recruitment- and adherence-focused coordinating center with designated staff, a clinical design with recruitment feasibility, selection of a large number of sites with documented accrual capabilities, and a comprehensive media kickoff. The Recruitment and Adherence Committee (RAC) was composed of an advisory committee of experienced research clinicians and a Minority and Medically Underserved (MMUS) subcommittee. Prior to study initiation the central recruitment and adherence staff was an operational unit who refined the recruitment plan, developing specific recruitment and retention strategies and materials, and working closely with the RAC [32].

The SELECT media campaign was a highly coordinated effort that was deployed through a distribution of materials to 800 national and regional print and electronic media outlets that targeted minority, health professional, and advocacy groups, as well as the public at large. The direct effect on the number of randomizations is difficult to quantify, but data from the National Cancer Institute's Communication Information Service, which received 6400 calls during the first week of the media launch suggested that the target audience was reached.

The SELECT trial design required baseline blood and toenail samples, a blood sample at 5 years, clinic visits with a limited physical examination and assessment for adherence and adverse events every 6 months (annually for those with prostate cancer) and a commitment to refrain from over-the-counter selenium and vitamin E. Testing for PSA and DRE per local site standard of care was recommended, but not required [33]. This relatively nondemanding protocol, testing two fairly nontoxic agents, more than likely contributed to participants' willingness to participate [33–36].

Accrual

Accrual to SELECT began on August 22, 2001. In a 3-year period, 35,533 eligible men from the US, Canada, and Puerto Rico were enrolled; this enrollment rate exceeded the goals for both number of participants and for anticipated length of the accrual period. The cohort included 22% minorities (15% African American, 6% Hispanic, and 1% other) [33]. Successful randomization was evident in the fact that known prostate cancer risk factors, including age, race, baseline PSA, and smoking status were equally balanced among the four treatment groups [30,37]. Not only was SELECT the largest randomized chemoprevention trial ever conducted, but it had the largest percentage of African American participants ever randomized to this type of study [32]. In addition, because comorbidities generally exclude a higher percentage of African American men from clinical trials [35,38,39], SELECT recruitment allowed participation of men with stable comorbidities.

Adherence

Participant adherence can be particularly challenging in long-lasting cancer prevention clinical trials. To give study candidates an opportunity to decide if they were willing to commit to participation and protocol adherence (most significantly, abstaining from vitamins not provided by the study) for the planned 7–12 year study, there was a 28–90 day prerandomization period, after which they would return to the clinic if they chose to enroll. To foster continued adherence, staff considered the participants' convenience and comfort by sending visit reminders, assisting with transportation problems, having flexible clinic hours and maintaining a pleasant clinic experience. Efforts were made to acknowledge the participants' time and dedication to the study by offering materials from certificates of appreciation to items with the study logo such as key chains or post-it notes. In short, the relationship between the study site staff and the participants is arguably the most important factor in bonding the participant to the study.

Study agent adherence was assessed via participant diary and pill count. Bioadherence, as measured by nutrient serum levels in a subset of patients, was also used to assess adherence and is described in detail elsewhere [31].

Primary and Secondary Endpoint Results

Two separate publications, one in 2009 and one in 2011, reported results of the SELECT trial [30,37]. Before publication of the results, in September 2008 an independent Data and Safety Monitoring Committee (DSMC) unanimously voted to discontinue the use of study supplements because neither agent showed evidence of benefit. The first SELECT results publication included data current as of October 23, 2008, the date on which study sites were advised to discontinue supplement administration. The median follow-up time was 5.46 years (range was 4.17–7.33 years). No significant differences in rates of prostate cancer were observed among the four intervention arms (Table 40.2). The hazard ratio (HR) of prostate cancer for the vitamin E group was 1.13, for selenium 1.04, and for the combination group 1.05, relative to the placebo group. The 13% increase in prostate cancer risk in the vitamin E group was statistically nonsignificant ($p = 0.06$), but troubling nonetheless.

Participants continued to be followed for additional events following the October 2008 study supplement discontinuation. On May 20, 2011, the DSMC reviewed the data and recommended reporting the follow-up findings [37]. Data included in the second SELECT report were collected through July 2011 and contained an additional 54,464 person-years of follow-up. An additional 521 prostate cancers were diagnosed: 113 in the placebo group, 147 in the vitamin E group, 143 in the selenium group, and 118 in the selenium + vitamin E group. No statistically significant differences in the rates of prostate cancer detection in the selenium group versus placebo or in the selenium + vitamin E group versus placebo (Table 40.2); however, risk of prostate cancer in the vitamin E group was increased 17% compared to the placebo group, and in contrast to the 13% increase in the first report, now appeared to be significant (HR = 1.17, 99% CI: 1.004–1.36, $p = 0.008$). The increased risk in the vitamin E group began approximately 3–4 years after randomization based on cumulative incidence curves of prostate

TABLE 40.2 Primary Endpoint Results from SELECT Reported in the First and Second Reports

	Prostate cancer		Method of diagnosis, n (%)		Gleason score, n (%)				
	No. events	HR (99% CI)	Prostate biopsy	Other/ unknown	2–6	4–6	7	8–10	Not graded
First Reports, October 2008 [30]									
Placebo (*n* = 8696)	416	1 (ref)	404 (97)	12 (3)	240 (66)		101 (28)	24 (7)	51
Vitamin E (*n* = 8737)	473	1.13 (0.95–1.35)	458 (97)	15 (3)	249 (63)		124 (31)	23 (6)	77
Selenium (*n* = 8752)	432	1.04 (0.87–1.24)	419 (97)	13 (3)	217 (60)		124 (34)	20 (6)	71
Selenium + Vitamin E (*n* = 8703)	437	1.05 (0.88–1.25)	420 (96)	17 (4)	220 (60)		115 (32)	30 (8)	72
Second Report, July 2011 [37]									
Placebo (*n* = 8696)	529	1 (ref)	n.r.	n.r.		286 (69)	102 (24)	31 (7)	110
Vitamin E (*n* = 8737)	620	1.17 (1.004–1.36)[a]	n.r.	n.r.		310 (67)	118 (25)	37 (8)	155
Selenium (*n* = 8752)	575	1.09 (0.93–1.27)	n.r.	n.r.		281 (64)	135 (31)	26 (6)	133
Selenium + Vitamin E (*n* = 8702)	555	1.05 (0.89–1.22)	n.r.	n.r.		281 (63)	124 (28)	40 (9)	110

[a]$p = 0.008$. n.r.: not reported.
This table shows that there were no reported significant differences in prostate cancer incident in the first SELECT report, but a 17% increase in prostate cancer incidence in the group receiving vitamin E alone in the second report.
ref: reference value.

cancer by supplement group [4]. Although increased risk of prostate cancer was evident in the vitamin E only group, no such increase appeared in the vitamin E+ selenium group.

Overall, in both reports, most prostate cancers were diagnosed by prostate biopsies administered due to abnormal PSA or DRE results, most were early stage and low Gleason grade, and stage, grade, and PSA levels did not differ by treatment group.

Several a priori secondary endpoints were prespecified, including other cancers (including colorectal and lung cancer), total cancer incidence, cardiovascular events, diabetes, and deaths. No significant differences among supplement groups were observed for any of these endpoints (Table 40. 3). Adherence to study supplements was assessed by using pill counting and participant diaries for all participants and by measuring serum levels of selenium and cholesterol-adjusted α-tocopherol and γ-tocopherol levels in a bioadherence subcohort. Mean adherence by pill count was 83% at year 1 and 65% at year 5. Serum selenium and α-tocopherol levels rose only in participants randomized to receive those agents and not in the other groups, demonstrating good adherence and minimal drop-ins. Additional assessment of drop-in rates was done by asking participants whether they took either supplement; rates were 3.1% or less for vitamin E and 1.8% or less for selenium [30].

DISCUSSION: Explanations for the Results of SELECT

The most direct explanation of the results of SELECT is that neither agent has a potent preventive effect on prostate cancer and that the prior hypothesis-generating data from randomized trials in humans merely reflected secondary endpoints that did not carry statistical validity. In fact, laboratory data mimicking the design of SELECT showed no statistically significant reductions in prostate cancer incidence with either selenium as L-selenomethionine or vitamin E alone or together in rodents [11]. Nevertheless, interest remained (and remains) in pursuing selenium as a potential cancer preventive agent for prostate as well as other cancers and alternative hypotheses to explain the null findings center on the agent formulation and dose that were chosen, the cohort, or the study design. The findings of significantly increased prostate cancer incidence in association with vitamin E have quelled interest in this agent for this purpose.

Agents

Selenium

Selenium Formulation

One possible explanation for the null selenium results is that the wrong form of selenium was used in

TABLE 40.3 Secondary Endpoint Results and Adverse Outcomes from the SELECT First and Second Reports

	Trial arm												
	Any cancer (including prostate cancer)		Lung cancer		Colorectal cancer		Other primary cancer		Diabetes		Cardiovascular (CV) events[a]		Deaths, all cause	
	No. events	HR (99% CI)	No. events	HR (99% CI)	No. events	HR (99% CI)	No. events	HR (99% CI)	No. events	HR (99% CI)	No. events	HR (99% CI)	No. events	HR (99% CI)
First Report, October 2008 [26]														
Placebo ($n = 8696$)	824	1 (ref)	67	1 (ref)	60	1 (ref)	306	1 (ref)	669	1 (ref)	1050	1 (ref)	382	1 (ref)
Vitamin E ($n = 8737$)	856	1.03 (0.91–1.17)	67	1.00 (0.64–1.55)	66	1.09 (0.69–1.73)	274	0.89 (0.72–1.10)	700	1.04 (0.91–1.18)	1034	0.98 (0.88–1.09)	358	0.93 (0.77–1.13)
Selenium ($n = 8752$)	837	1.01 (0.89–1.15)	75	1.12 (0.73–1.72)	63	1.05 (0.66–1.67)	292	0.95 (0.77–1.17)	724	1.07 (0.94–1.22)	1080	1.02 (0.92–1.13)	378	0.99 (0.82–1.19)
Selenium + Vitamin E ($n = 8703$)	846	1.02 (0.90–1.16)	78	1.16 (0.76–1.78)	77	1.28 (0.82–2.00)	290	0.94 (0.76–1.16)	660	0.97 (0.85–1.11)	1041	0.99 (0.89–1.10)	359	0.94 (0.77–1.13)
Second Report, July 2011 [37]											*CV events, grade ≥ 4*[b]			
Placebo ($n = 8696$)	1108	1 (ref)	92	1 (ref)	75	1 (ref)	579	1 (ref)	869	1 (ref)	969	1 (ref)	564	1 (ref)
Vitamin E ($n = 8737$)	1190	1.07 (0.96–1.19)	104	1.11 (0.76–1.61)	85	1.09 (0.72–1.64)	570	0.97 (0.83–1.14)	918	1.05 (0.93–1.17)	909	0.93 (0.83–1.05)	571	1.01 (0.86–1.17)
Selenium ($n = 8752$)	1132	1.02 (0.92–1.14)	94	1.02 (0.70–1.50)	74	0.96 (0.63–1.46)	557	0.96 (0.83–1.13)	913	1.04 (0.93–1.17)	939	0.97 (0.86–1.09)	551	0.98 (0.84–1.14)
Selenium + Vitamin E ($n = 8702$)	1149	1.02 (0.92–1.12)	104	1.11 (0.76–1.62)	93	1.21 (0.81–1.81)	594	1.02 (0.92–1.14)	875	0.99 (0.89–1.12)	943	0.97 (0.86–1.09)	542	0.96 (0.82–1.12)

[a] No second report.
[b] No first report.

This table shows results from the first and second SELECT reports for results of secondary endpoints and adverse events. No significant differences were reported in any treatment group for endpoints/outcomes. Due to multiple cancers in some SELECT participants, the numbers for specific types of cancers (lung, colorectal, other primary cancers, or prostate cancers) in this table may not sum to the number of individuals (categorized as "any cancer" as listed in columns 2 and 3).

SELECT. Organic selenium was chosen although inorganic selenium had been shown to have better in vitro anticancer activities. The problem with these inorganic forms is that they have been linked to DNA single strand breaks [40]. As discussed, the selection of L-selenomethionine over selenized yeast was based on the batch-to-batch variability and lack of availability of the yeast form of selenium. L-selenomethionine is the predominant selenium species in the yeast, although selenized yeast contains numerous other selenocompounds with varying and possibly uncharacterized chemopreventive efficacy (discussed in Ref. [4]). Another problem with selenomethionine is its nonspecific incorporation into proteins in place of methionine, which diverts the selenium away from its potential active chemopreventive form(s) (Fig. 40.1). Other selenocompounds like selenocysteine, selenite, or selenate, are not used nonspecifically in proteins and therefore are more likely to be converted to the potentially antitumorigenic metabolites via methyl selenol [13]. Despite the suspected differences in L-selenomethionine content between pure L-selenomethionine and selenized yeast, recent findings have indicated that selenomethionine and selenized yeast have similar biological activities in prostatic tissue of dogs (reviewed in Ref. [4]). Many investigators have held the opinion that the formulation of selenium used in SELECT was probably not a major reason for the null results. Caution must be exercised in drawing this conclusion, however, since laboratory evidence suggests that specific forms of selenium may differ with respect to their biological effects [41]. The relative effects of these different selenium formulations in humans are beginning to be investigated. In a clinical trial randomizing 69 healthy men to selenomethionine versus two doses of selenized yeast, reduction in biomarkers of oxidative stress were observed following supplementation with selenized yeast but not selenomethionine [42].

Selenium Dose

Beyond selenium formulation, the dose of selenium may not have been optimal. A daily dose of L-selenomethionine 200 μg/day was used in SELECT, whereas the NPC trial used a dose of 200 μg/day of selenized yeast. An optimal dose of selenium for cancer prevention had not been established prior to SELECT, nor has one been established to date. It is likely that a narrow range of optimal doses exist, and these doses may depend on the baseline selenium status of the individual [43].

Baseline Selenium Status

A secondary analysis of the NPC found an accentuated inverse relationship between baseline plasma selenium status and prostate cancer incidence [44]. In this analysis, selenium supplementation in men with moderate and low plasma selenium at baseline decreased the risk of prostate cancer by more than 75%, but had no effect among men with high plasma selenium at baseline. Based on the NPC reports, a case–cohort study nested within SELECT investigated the effects of selenium and vitamin E supplementation in 1739 cases (489 high-grade cases) by baseline selenium status, which was determined by toenail selenium concentrations [45]. The optimal range of toenail selenium levels for prostate cancer risk reduction in humans has been estimated to be 119–137 ng/mL; any additional selenium above this would either offer no benefit or cause harm [43]. Baseline selenium in the absence of supplementation showed no association with prostate cancer risk. However, supplementation with selenium alone or in combination with vitamin E increased the risk of high-grade prostate cancer in men with high baseline toenail selenium levels. These data are consistent with studies in dogs, which suggest a U-shaped dose response [43] and is consistent with NPC results where protection was only seen in men with baseline serum selenium concentrations <123.2 ng/mL. Furthermore, in the small clinical trial of Richie et al. significant reductions of the relevant biomarkers of oxidative stress were observed with the selenized yeast only in men belonging to the lowest tertile of baseline selenium [42].

Environmental selenium may serve as a surrogate for baseline selenium levels. The NPC trial recruited men from low environmental selenium areas in the eastern coastal US, in contrast to SELECT, which recruited across the US and Canada, leading to large discrepancies in baseline selenium status between men in the two trials; 114 ng/mL compared to 135 ng/mL, respectively (reviewed in Ref. [15]). It is possible that the baseline serum selenium concentrations of men in even the lowest quartile in SELECT may be above concentrations at which supplemental selenium could offer protection.

The importance of baseline selenium status, emphasized in the observations of Richie et al. [42], was likely not adequately recognized during planning of SELECT. A specified baseline selenium concentration was not established as an enrollment criterion, nor was a soil selenium-deficient region considered for the study location. Because SELECT was a very large trial, it was necessary to be inclusive with regards to geography in order to accrue an adequate number of participants. Additional complexity derives from increasing fortification of the food system with selenium in the years since NPC. This makes it prohibitively difficult to ensure low baseline selenium status by limiting the study's geographical area to one with low soil selenium.

Vitamin E Formulation

Regarding the lack of benefit from vitamin E supplementation, one possible explanation is the choice of α-tocopherol alone as the intervention. High doses of α-tocopherol decrease blood and tissue levels of δ-tocopherols [46]. Furthermore, the adherence cohort, which showed good adherence to the intervention according to increased serum α-tocopherol levels following initiation of the intervention, also showed concomitant decreases in γ-tocopherol levels in the vitamin E-assigned groups. Although at the nutritional level, all forms of vitamin E are presumed to have cancer preventive properties, at supranutritional levels α-tocopherol is not preventive [47]. In contrast, γ-tocopherol has strong anti-inflammatory activity and may be the most effective form of vitamin E for cancer prevention.

The Effect of Other Nutrients

The effect of other nutrients associated with prostate cancer has been investigated using the SELECT biorepository. Of particular interest has been investigation of plasma 25-hydroxy vitamin D (25(OH)D), which reflects endogenous and exogenous vitamin D sources. Evidence from laboratory and animal studies show that high doses of 25(OH)D inhibit proliferation and differentiation in prostate cancer cell lines and rodent models [48]. Inconclusive and conflicting results from epidemiological studies (prospective as well as nested case-control) have suggested some association between vitamin D levels and prostate cancer risk (reviewed in Ref. [48]). Results from large, population-based studies have identified a U-shaped curve of risk with increasing levels of vitamin D intake, although the direction of the U-shape and consistency in linear risk models make a definitive understanding of vitamin D levels and prostate cancer risk equivocal (reviewed in Ref. [49]). A nested case-cohort study of 1731 cases and 3203 controls from SELECT was conducted to investigate whether baseline plasma concentration, adjusted for season of collection, was associated with the risk of total prostate cancer and Gleason score 2–6, 7–10, and 8–10 prostate cancer [50]. Analysis indicated a U-shaped association of vitamin D with total cancer risk across quintiles. Results for African-Americans showed that vitamin D was associated with a reduced risk of Gleason 7–10 cancer only, with no evidence of dose-response or a U-shaped association. Overall conclusions from this study suggest that both low and high vitamin D concentrations were associated with increased risk of prostate cancer, with stronger associations for high-grade disease [50].

Genetics

Genetic factors were not a focus of NPC or SELECT, but they may nevertheless affect selenium status or an individual's response to selenium supplementation. Polymorphisms in genes that encode either selenoproteins or proteins that function in selenium metabolism may influence outcomes (reviewed in Ref. [4]). For example, the AA genotype polymorphism of codon 16 (rs4880) of SOD2, which encodes the mitochondrial antioxidant enzyme manganese superoxide dismutase, has been shown to confer a lower risk of total prostate cancer and of clinically aggressive prostate cancer among men with higher selenium levels compared to those with lower selenium levels. This protection was much weaker in men with VV or VA genotypes. Depending on the polymorphism and the genotype, the survival time was either increased or decreased. Additional interactions have been shown between selenium and the rs561104 polymorphism, where high levels of selenium were associated with decreased prostate cancer mortality only in those with the increased risk homozygous variant genotype and not in those with the wild-type genotype. Finally, the genotype of GPX1, which encodes the selenoprotein GPx1, recently was shown to be a determinant of selenium requirements. Among 161 men and women, those with the GPX1 679 (rs1050450) T/T genotype had significantly lower plasma selenium levels than those with the C/C genotype. The mean plasma selenium level in this study was 142.0 ng/mL, slightly above that of men in SELECT.

Polymorphisms in the prostate cancer tumor suppressor gene NKX3.1 have been identified from animal and in vitro studies that show exposure to antioxidant supplements increased prostate epithelial proliferation through dysregulation of genetic pathways responsible for the regulation of ROS and elevated oxidative stress [51]. Two cancer-related polymorphisms in NKX3.1 have been identified; the variant rs11781886 has been shown to alter the binding of the SP1 transcription factor through the reduction of NKX3.1 mRNA expression [52] and the variant rs2228013 has been shown to alter NKX3.1 phosphorylation and DNA binding activity in vitro by coding for a variant NKX3.1 protein [53]. These variants were investigated in a study conducted using the SELECT biorepository to determine if the NKX3.1 polymorphisms were associated with overall prostate cancer risk and the risk of low-grade and high-grade prostate cancer among men randomized to take vitamin E and/or selenium supplements [54]. Results indicated that in the selenium arm the CC genotype at rs11781886 was significantly associated with an increased risk for total and low grade prostate cancer; this same CC genotype was associated with an increased risk of high-grade prostate cancer in the vitamin E arm. Having one C allele (CT genotype) was associated with a significantly increased risk of total and high grade prostate cancers in the vitamin E arm, but a marginally significant increase in the selenium arm. There was no effect on prostate cancer risk in any of the intervention arms with the rs2228013 variant [54].

Based on these studies, analyses of SELECT data according to participants' genotypes may elucidate relationships between supplemental selenium and prostate cancer risk that were not evident in the SELECT population as a whole.

Age

Intervention in men older than 50 years of age may be too late, given the long natural history of prostate cancer (reviewed in Ref. [4]). The critical window, if any, for prostate cancer chemoprevention, has not been clearly defined. In an analogous situation in breast cancer, also a hormonally driven cancer, an inverse association was observed between adolescent soy food intake and adult breast cancer, but not with adult soy intake and adult breast cancer. A clearer understanding of the timing by which selenium influences prostate carcinogenesis might help researchers to better predict an optimal age range at which selenium supplementation should take place. Regardless, results from NPC suggest that intervention in men over the age of 50 still has the potential to yield some anticancer benefit.

Study Design

In NPC, prostate cancer incidence was a secondary outcome measure, and in SELECT, prostate cancer incidence was a primary outcome measure. This important distinction may help to explain the different results. Both trials were adequately powered to detect differences of predetermined magnitudes in their respective primary outcomes only. In clinical trials with multiple outcomes, primary endpoints, unlike secondary outcomes, must be designated a priori in order to protect that endpoint from concerns about the observed results being due to chance due to multiple testing. In NPC, results on skin cancer were protected while results for prostate cancer were not. NPC results were particularly open to chance findings due to the small sample size of 64 prostate cancer cases in 1312 participants.

Lag time between intervention and effect in SELECT also may have influenced the primary outcome. SELECT was designed to test the effects on prostate cancer after 7–12 years of vitamin E or selenium supplementation. Possibly a much longer period of time is required to see such an impact.

IMPLICATIONS AND FUTURE DIRECTIONS

Following the discontinuation of supplement administration and publication of the primary data, SELECT transitioned into an observational cohort study, the SELECT Centralized Follow-up Study (SELECT CFU). As of December 2011, 17,761 participants, 58% of the 32,569 SELECT participants who were still alive and not refusing further contact, had enrolled in SELECT CFU [55]. The SELECT biorepository, maintained by SWOG, contains over 100,000 banked tissue samples, including blood, tissue, and toenail samples. To date, at least 15 projects utilizing these samples have been approved, with continuing plans for additional projects.

Regarding vitamin E, the United States Preventive Services Task Force (USPSTF) issued a grade D recommendation against the use of vitamin E supplements for the prevention of cardiovascular disease or cancer. This recommendation is based on moderate certainty of lack of a net benefit of vitamin E supplementation.

Selenium, however, remains an area of interest for prostate cancer chemoprevention, with caveats. One of the most important findings of SELECT was that there is a need to better understand selenium biology to aid researchers in choosing the appropriate doses and formulations of selenium, as well as the appropriate cohorts and study design [13]. One promising area of selenium research is the ongoing characterization of selenium's antiDNA damage activities [43]. Furthermore, the tissue specificity, function, regulation, and enzyme kinetics of many of the 25 selenoproteins identified in humans remain uncharacterized [56]. Ongoing characterization of these proteins will contribute to a better understanding of selenium's mechanisms of action, which is critical, along with a better understanding of proper doses, selenium formulations, and subpopulations that may benefit, before any new large phase III clinical trials are undertaken. Conflicting or inconclusive results from clinical and epidemiological studies underscore the difficulty in making nutritional recommendations for cancer prevention to the general population and bring to light the need for studies using subgroups at greater risk or who may benefit more than others [57]. To date, the best evidence indicates that men with low serum selenium concentrations or those that live in selenium-deficient regions may represent the optimal cohort for studying prostate cancer prevention by selenium, and more research is necessary to determine whether further refinement of eligibility criteria by age or selenoprotein genotype may also be useful.

SUMMARY

- Prostate cancer is the most commonly diagnosed non-cutaneous cancer in men in the United States.
- The goal of this trial was to study selenium and vitamin E as prostate cancer prevention strategies, as they are naturally occurring essential micronutrients with promising preclinical, observational, and clinical data suggesting risk-reducing efficacy.
- The Selenium and Vitamin E Cancer Prevention Trial (SELECT) was conducted to test the efficacy of selenium and vitamin E alone and in combination on the incidence of prostate cancer.
- Supplemental selenium is available as L-selenomethionine or selenized yeast. The major component of selenized yeast is L-selenomethionine, but selenized yeast contains other components.
- The majority of men in the United States have adequate selenium status but inadequate vitamin E intake.
- SELECT was a prospective, randomized, double-blinded, placebo-controlled, 2x2 factorial design clinical trial of selenium and vitamin E alone or in combination in eligible healthy mean who were old enough to be considered at elevated risk for prostate cancer.
- No significant differences in prostate cancer incidence were reported in the first report of SELECT, but in the 2011 report, the investigators showed that there was a 17% increase in prostate cancer incidence in the group receiving vitamin E alone.
- The null findings regarding selenium in SELECT may be partially attributable to the formulation of selenium chosen (L-selenomethionine as opposed to selenized yeast), the relatively high baseline selenium levels in the trial cohort, age or genetic factors in the cohort, or the study design of SELECT.
- The null findings regarding vitamin E may in part be due to the form of vitamin E used in the trial, namely α-tocopherol.
- There is a need for a better understanding of the biology of selenium and vitamin E before additional clinical trials that test the cancer preventive efficacy of either agent are undertaken.

References

[1] DeSantis C, Lin C, Mariotto A, et al. Cancer treatment and survivorship statistics, 2014. CA Cancer J Clin 2014;64:252–71.
[2] Siegel R, Naishadham D, Jemal A. Cancer statistics, 2013. CA Cancer J Clin 2013;63:11–30.
[3] Welch H, Black W. Overdiagnosis in cancer. J Natl Cancer Inst 2010;102:605–13.
[4] Nicastro H, Dunn B. Selenium and prostate cancer prevention: insights from the Selenium and Vitamin E Cancer Prevention Trial (SELECT. Nutrients 2013;5:1122–48.
[5] Moyer V. Screening for prostate cancer: U.S. Preventive Services Task Force recommendation statement. Ann Intern Med 2012;157:120–34.
[6] Thompson Jr I, Goodman P, Tangen C, et al. Long-term survival of participants in the prostate cancer prevention trial. N Engl J Med 2003;369:603–10.
[7] Andriole G, Bostwick D, Brawley O, REDUCE Study Group, et al. Effect of dutasteride on the risk of prostate cancer. N Eng J Med 2010;362:1192–202.
[8] Theoret M, Ning Y, Zhang J, Justice R, Keegan P, Pazdur R. The risks and benefits of 5α-reductase inhibitors for prostate-cancer prevention. N Eng J Med 2011;365:97–9.
[9] Heinonen O, Albanes D, Virtamo J, et al. Prostate cancer and supplementation with alpha-tocopherol and beta-carotene: incidence and mortality in a controlled trial. J Natl Cancer Inst 1998;90:440–6.
[10] Clark L, Combs G, Turnbull B, et al. Effects of selenium supplementation for cancer prevention in patients with carcinoma of the skin. A randomized controlled trial. Nutritional Prevention of Cancer Study Group. J Am Med Assoc 1996;276:1957–63.
[11] McCormick D, Rao K, Johnson W, Bosland M, Lubet R, Steele V. Null activity of selenium and vitamin E as cancer chemopreventive agents in the rat prostate. Cancer Prev Res (Phila) 2010;3:381–92.
[12] Institute of Medicine (U.S.) Panel on Dietary Antioxidants and Related Compounds. Dietary reference intakes for vitamin C, vitamin E, selenium, and carotenoids : a report of the Panel on Dietary Antioxidants and Related Compounds, Subcommittees on Upper Reference Levels of Nutrients and of Interpretation and Use of Dietary Reference Intakes, and the Standing Committee on the Scientific Evaluation of Dietary Reference Intakes, Food and Nutrition Board, Institute of Medicine. Washington (DC): National Academy Press; 2000. p. 506.
[13] Hatfield D, Gladyshev V. The outcome of Selenium and Vitamin E Cancer Prevention Trial (SELECT) reveals the need for better understanding of selenium biology. Mol Interv 2009;9:18–21.
[14] Fulgoni 3rd V, Keast D, Bailey R, Dwyer J. Foods, fortificants, and supplements: where do Americans get their nutrients? J Nutr 2011;141:1847–54.
[15] Dunn B, Richmond E, Minasian L, Ryan A, Ford L. A nutrient approach to prostate cancer prevention: the Selenium and Vitamin E Cancer Prevention Trial (SELECT). Nutr Cancer 2010;62:896–918.
[16] Azzi A. Molecular mechanism of alpha-tocopherol action. Free Radic Biol Med 2007;43:16–21.
[17] The Alpha-Tocopherol Beta Carotene Cancer Prevention Study Group. The effect of vitamin E and beta carotene on the incidence of lung cancer and other cancers in male smokers. N Eng J Med 1994;330:1029–35.
[18] Combs Jr G, Gray W. Chemopreventive agents: selenium. Pharmaco Ther 1998;79:179–92.
[19] Blot W, Li J, Taylor P, et al. Nutrition intervention trials in Linxian, China: supplementation with specific vitamin/mineral combinations, cancer incidence, and disease-specific

[20] Li J, Taylor P, Li B, et al. Nutrition intervention trials in Linxian, China: multiple vitamin/mineral supplementation, cancer incidence, and disease-specific mortality among adults with esophageal dysplasia. J Natl Cancer Inst 1993;85:1492–8.

mortality in the general population. J Natl Cancer Inst 1993;85:1483–92.

[21] Ozten N, Horton L, Lasano S, Bosland M. Selenomethionine and alpha-tocopherol do not inhibit prostate carcinogenesis in the testosterone plus estradiol-treated NBL rat model. Cancer Prev Res (Phila) 2010;3:371–80.

[22] Venkateswaran V, Fleshner N, Sugar L, Klotz L. Antioxidants block prostate cancer in lady transgenic mice. Cancer Res 2004;64:5891–6.

[23] Venkateswaran V, Klotz L, Ramani M, et al. A combination of micronutrients is beneficial in reducing the incidence of prostate cancer and increasing survival in the lady transgenic model. Cancer Prev Res (Phila) 2009;2:473–83.

[24] Wang L, Bonorden M, Li G, et al. Methyl-selenium compounds inhibit prostate carcinogenesis in the transgenic adenocarcinoma of mouse prostate model with survival benefit. Cancer Prev Res (Phila) 2009;2:484–95.

[25] Bespalov V, Panchenko A, Murazov I, Chepik O. Influence of sodium selenite on carcinogenesis of the prostate and other organs induced by methylnitrosourea and testosterone in rats. Vopr Onkol 2011;57:486–92.

[26] Lippman S, Goodman P, Klein E, et al. Designing the Selenium and Vitamin E Cancer Prevention Trial (SELECT). J Natl Cancer Inst 2005;97:94–102.

[27] Clark L, Dalkin B, Krongrad A, Combs G, Turnbull B, Slate E. Decreased incidence of prostate cancer with selenium supplementation: results of a double-blind cancer prevention trial. Br J Urol 1998;81:730–4.

[28] The Alpha-Tocopherol Beta Carotene Cancer Prevention Study Group. The alpha-tocopherol, beta-carotene lung cancer prevention study: design, methods, participant characteristics, and compliance. Ann Epidemiol 1994;4:1–10.

[29] Thompson I, Pauler D, Goodman P, et al. Prevalence of prostate cancer among men with a prostate-specific antigen level < or = 4.0 ng per milliliter. N Eng J Med 2004;350:2239–46.

[30] Lippman S, Klein E, Goodman P, et al. Effect of selenium and vitamin E on risk of prostate cancer and other cancers: the Selenium and Vitamin E Cancer Prevention Trial (SELECT). J Am Med Assoc 2009;301:39–51.

[31] Hughes G, Cutter G, Donahue R, et al. Recruitment in the Coronary Artery Disease Risk Development in Young Adults (Cardia) Study. Control Clin Trials 1987;8(Suppl. 4):68S–73S.

[32] Age-Related Eye Disease Study Research Group. A randomized, placebo-controlled, clinical trial of high-dose supplementation with vitamins C and E, beta carotene, and zinc for age-related macular degeneration and vision loss: AREDS report no. 8. Arch Ophthalmol 2001;119:1417–36.

[33] Cook E, Moody-Thomas S, Anderson K, et al. Minority recruitment to the Selenium and Vitamin E Cancer Prevention Trial (SELECT). Clin Trials 2005;2:436–42.

[34] Blanton S, Morris D, Prettyman M, et al. Lessons learned in participant recruitment and retention: the EXCITE trial. Phys Ther 2006;86:1520–33.

[35] Mills K, Stewart A, King A, et al. Factors associated with enrollment of older adults into a physical activity promotion program. J Aging Health 1996;8:96–113.

[36] Brandt C, Argraves S, Money R, Ananth G, Trocky N, Nadkarni P. Informatics tools to improve clinical research study implementation. Contemp Clin Trials 2006;27:112–22.

[37] Klein E, Thompson Jr I, Tangen C, et al. Vitamin E and the risk of prostate cancer: the Selenium and Vitamin E Cancer Prevention Trial (SELECT). J Am Med Assoc 2011;306:1549–56.

[38] Vickers A. How to improve accrual to clinical trials of symptom control 2: design issues. J Soc Integr Oncol 2007;5:61–4.

[39] Getz K, Wenger J, Campo R, Seguine E, Kaitin K. Assessing the impact of protocol design changes on clinical trial performance. Am J Ther 2008;15:450–7.

[40] Lu J, Jiang C, Kaeck M, et al. Dissociation of the genotoxic and growth inhibitory effects of selenium. Biochem Pharmacol 1995;50:213–19.

[41] Christensen MJ. Selenium and prostate cancer prevention: what next--if anything? Cancer Prev Res (Phila) 2014;7:781–5.

[42] Richie Jr J, Das A, Calcagnotto A, et al. Comparative effects of two different forms of selenium on oxidative stress biomarkers in healthy men: a randomized clinical trial. Cancer Prev Res (Phila) 2014;7:796–804.

[43] Chiang E, Shen S, Kengeri S, et al. Defining the optimal selenium dose for prostate cancer risk reduction: insights from the U-shaped relationship between selenium status, DNA damage, and apoptosis. Dose Response 2009;8:285–300.

[44] Duffield-Lillico A, Dalkin B, Reid M, Nutritional Prevention of Cancer Study Group, et al. Selenium supplementation, baseline plasma selenium status and incidence of prostate cancer: an analysis of the complete treatment period of the Nutritional Prevention of Cancer Trial. BJU Int 2003;91:608–12.

[45] Kristal A, Darke A, Morris J, et al. Baseline selenium status and effects of selenium and vitamin E supplementation on prostate cancer risk. J Natl Cancer Inst 2014;106:djt456.

[46] Ju J, Picinich S, Yang Z, et al. Cancer-preventive activities of tocopherols and tocotrienols. Carcinogenesis 2010;31:533–42.

[47] Yang C, Suh N, Kong A. Does vitamin E prevent or promote cancer? Cancer Prev Res (Phila) 2012;5:701–5.

[48] Ahn J, Peters U, Albanes D, Prostate, Lung, Colorectal, and Ovarian (PLCO) Trial Project Team, et al. Serum vitamin D concentration and prostate cancer risk: a nested case–control study. J Natl Cancer Inst 2008;100:796–804.

[49] Schenk J, Till C, Tangen C, et al. Serum 25-hydroxyvitamin D concentrations and risk of prostate cancer: results from the prostate cancer prevention trial. Cancer Epidemiol Biomarkers Prev 2014;23:1484–93.

[50] Kristal A, Till C, Song X, et al. Plasma vitamin D and prostate cancer risk: results from the Selenium and Vitamin E Cancer Prevention Trial. Cancer Epidemiol Biomarkers Prev 2014;23:1494–504.

[51] Martinez E, Anderson P, Abdulkadir S. Antioxidant treatment promotes prostate epithelial proliferation in Nkx3.1 mutant mice. PLoS One 2012;7:e46792.

[52] Akamatsu S, Takata R, Ashikawa K, et al. A functional variant in NKX3.1 associated with prostate cancer susceptibility down-regulates NKX3.1 expression. Hum Mol Genet 2010;19:4265–72.

[53] Gelmann E, Steadman D, Ma J, et al. Occurrence of NKX3.1 C154T polymorphism in men with and without prostate cancer and studies of its effect on protein function. Cancer Res 2002;62:2654–9.

[54] Martinez E, Darke A, Tangen C, et al. A functional variant in NKX3.1 associated with prostate cancer risk in the Selenium and Vitamin E Cancer Prevention Trial (SELECT). Cancer Prev Res (Phila) 2014;7:950–7.

[55] Goodman P, Hartline J, Tangen C, et al. Moving a randomized clinical trial into an observational cohort. Clin Trials 2012;10:131–42.

[56] Davis C, Tsuji P, Milner J. Selenoproteins and cancer prevention. Ann Rev Nutr 2012;32:73–95.

[57] Nicastro H, Trujillo E, Milner J. Nutrigenomics and cancer prevention. Curr Nutr Rep 2012;1:37–43.

CHAPTER

41

Iodine Intake and Healthy Aging

Leyda Callejas, Shwetha Mallesara and Philip R. Orlander

Division of Endocrinology, Diabetes and Metabolism, University of Texas Health Science Center, Houston, TX, USA

KEY FACTS

- The most effective way to supplement iodine in areas of deficiency is salt iodization.
- Recommended dosages of daily iodine supplementation according to the World Health Organization (WHO) are as follows: pregnant and lactating women 250 μg/d, children 12 years or older and adults 150 μg/d, children between 6–12 years 120 μg/d, and children less than 6 years 90 μg/d.
- The iodine status of populations can be assessed by using the prevalence of goiter, biomarkers of iodine exposure such as urinary iodine concentration (UIC), biomarkers of function (serum thyroglobulin levels), and thyroid function tests.
- The relationship between the iodine intake of a population and the occurrence of thyroid disease is U-shaped, with an increased risk from both low and high iodine intakes.
- Maternal dietary iodine and thyroid hormones are critical over the entire course of gestation, despite the fetal gland starting T4 production at approximately 12 weeks.
- Thyroid hormones play a critical role in the central nervous system of the fetus and newborn, affecting neuronal migration, differentiation, myelination, and synaptogenesis. Iodine deficiency during fetal development can result in cretinism.
- Children from chronically iodine deficient areas show impaired intellectual and motor skills and delayed growth. Iodine excess in children is associated with goiter and thyroid dysfunction.

Dictionary of Terms

- *Goiter*: Abnormal enlargement of the thyroid gland, it can be diffuse, affecting the whole gland or focal, characterized by the appearance of nodules.
- *Cretinism*: Consequence of iodine deficiency during pregnancy characterized by: cognitive impairment associated to either a predominant neurological syndrome or hypothyroidism and motor deficits.
- *Hypothyroidism*: Medical condition caused by deficient thyroid hormone production.
- *Hyperthyroidism*: Medical condition caused by excessive thyroid hormone production.
- *Thyroiditis*: Inflammation of thyroid gland which can have many etiologies including infection or autoimmunity.

INTRODUCTION

Iodine (I) is a nonmetallic element, which is acquired only through the diet. Iodine is a crucial component of thyroid hormones. The development and function of the human body, which is regulated by thyroid hormones, relies entirely on the availability and adequate dietary intake of I. Both iodine excess and deficiency can have adverse effects on health [1–5].

Bernard Courtois first described the element in 1811, when he saw a violet vapor arising from seaweed ash during the manufacture of gunpowder for Napoleon's army. Joseph Louis Gay-Lussac subsequently suggested the term "iodine," derived from the Greek for "violet" [3].

In the mid-1890s, Eugen Baumann identified iodine in thyroid glands. Around 1917, it was understood that thyroid disorders were related to iodine deficiency and could be prevented by iodine supplementation. During the 1920s, oral iodine supplementation for goiter prophylaxis was introduced in Switzerland and the United States [3,4].

At present, the only physiologic role known for iodine in the human body is in the synthesis of thyroid hormones. Thyroid hormones, and therefore iodine, are essential for life as they regulate many key enzymatic reactions. Major target organs include the developing brain, muscle, heart, pituitary, and kidney [1−8].

SOURCES OF IODINE

Iodine is usually found in rock; it gets dissolved in water and washed away by erosion. It then reaches the ocean where it is taken up by marine organisms. These organisms metabolize it into compounds, which can be volatilized to the atmosphere and deposited in the clouds. Rain re-deposits iodine onto the land, where is taken up by soil bacteria and other terrestrial organisms and released back into the soil and into ground water [3,9] (see Fig. 41.1).

Most of the earth's iodine is found in oceans, and iodine content in the soil varies according to regions. The iodine content of food depends on the iodine content of the soil in which it is grown. The more exposed a soil surface, the more iodine has been removed by erosion. Mountainous regions and flooded river valleys are among the most severely iodine−deficient areas in the world [1−5,7].

The native iodine content of most foods and beverages is low. Commonly consumed foods provide 3 to 80 μg per serving. Seawater, seaweed, and fish are a rich source of iodine. Marine life can concentrate the iodine from seawater. A wide variety of food contains iodine, including eggs, meat, milk and milk products, cereal grains, dried legumes, dried vegetables, and dried fruits, but in much lower quantity than fishes and shellfish. Iodine can also be found in varying amounts in drinking water [2,3,6,7,10]. Processed foods may contain slightly higher levels of iodine due to the addition of iodized salt or food additives, such as calcium iodate and potassium iodate. Dairy products are relatively good sources of iodine because iodine is commonly added to animal feed. Please see Table 41.1 for common sources of dietary iodine [2,3,7].

Vegetarianism and low seafood consumption can result in low iodine intakes. Many substances can affect iodine metabolism and thyroid hormone production, collectively known as goitrogens. Goitrogenic substances promote goiter formation if consumed chronically. Some anions, such as perchlorate, thiocyanate, bromated nitrates, and chlorate, impair the rate of iodine trapping by NIS (Sodium iodide symporter). Some of these inhibitors of iodine absorption can be found in foods. Perchlorate has been found in foods such as cow's milk. Cruciferous vegetables, such as cabbage, kale, cauliflower, broccoli, turnips, and rapeseed, contain glucosinolates, whose metabolites compete with iodine for thyroidal uptake. Cassava, lima beans, linseed, sorghum, and sweet potato contain thiocyanates or their precursors, which react with iodine reducing its bioavailability. Thiocyanates have also been associated to smoking. Soybean isoflavones have been found to inhibit thyroid hormone synthesis [1−3,5−7,10,11]. Other micronutrients and vitamins have been linked to thyroid function, for example, selenium, vitamin A, and iron. Selenium deficiency impairs the conversion from T4 to T3, which is done via deiodinases that use it as a cofactor. There are also selenium dependent glutathione peroxidases, which protect the thyroid gland against oxidative stress [1−3,5,12,13].

FIGURE 41.1 Schematic representation of iodine's cycle in the environment. *Source: modified from Ref. [9].*

TABLE 41.1 Dietary Iodine Sources

Common sources of dietary iodine
Seaweed- including kelp, dulce, and nori
Shellfish
Saltwater fish
Cheese
Cow's milk
Eggs
Dairy products: yogurt, ice cream
Iodized table salt
Iodine containing multivitamins

The medium iodine intake in the United States is 240–300 μg/day in men and 190–210 μg/day in women.

Recommended dosages of daily iodine supplementation according to the World Health Organization (WHO) are as follows:

- Pregnant and lactating women 250 μg/d
- Children over 12 years and adults 150 μg/d
- Children between 6–12 years of age 120 μg/d
- Children less than 6 years 90 μg/d

For children 0–6 months of age, iodine supplementation should be given through breast milk. This implies that the child is exclusively breastfed and that the lactating mother received iodine supplementation. In areas where food fortified with iodine is not available, iodine supplementation is required for children of 7–24 months of age [1,3,5,6,14].

The most effective way to supplement iodine intake is through salt iodization. Ideally all salts for human (food industry and household) and livestock consumption should be iodinized. Unfortunately this is rarely achieved as food industries are often reluctant to use iodized salt, and many countries do not iodize salt for livestock [14].

WHO recommends 20–40 mg iodine/kg salt, depending on local salt intake. Iodine can be added to salt in the form of potassium iodide (KI) or potassium iodate (KIO3). As KIO3 has higher stability in the presence of salt impurities, humidity, and porous packaging, it is the recommended form in tropical countries and those with low-grade salt. Iodine is usually added after the salt has been dried using one of two techniques. The "wet method" where a solution of KIO3 is dripped or sprayed on to salt passing by on a conveyor belt or the "dry method" where KI or KIO3 powder is sprinkled over the dry salt [3,14].

Iodization of salt may not be practical in certain regions—for example in areas where communications are poor or where there are numerous small-scale salt producers. Iodized oil supplements can be used in these areas as an alternative. Iodized oil is prepared by esterification of the unsaturated fatty acids in seed or vegetable oils, and addition of iodine to the double bonds. It can be given orally or by intramuscular injection. The intramuscular route has a longer duration of action, but oral administration is more common because of its simplicity. Usual doses are 200–400 mg iodine/year and it is often targeted to women of childbearing age, pregnant women, and children (see Table 41.2). Iodine can also be given as potassium iodide or iodate as drops or tablets. Single oral doses of potassium iodide monthly (30 mg) or biweekly (8 mg) can provide adequate iodine for school-age children. Lugol's iodine, containing approximately 6 mg iodine per drop offers another simple way to deliver iodine [3,14].

Most people are very tolerant of excess iodine intake from food. It is rare for diets of natural foods to supply more than 2000 μg of iodine/day, and most diets supply less than 1000 μg of iodine/day. However, some people living in coastal Asia whose diets contain large amounts of seaweed and fish have been found to have iodine intakes ranging from 50,000 to 80,000 μg (50–80 mg) of iodine/day [2,5,14–16].

Acute iodine poisoning is rare and usually caused by ingestion of several grams and causes gastrointestinal symptoms, such as abdominal pain, nausea, vomiting, and diarrhea, as well as cardiovascular symptoms, coma, and cyanosis. Excess iodine intake can cause a very rare skin disorder called iodermia, which is characterized by acneiform lesions, pruritic rash, and urticaria [2,5,14–16].

IODINE AND ITS ROLE IN THE HUMAN BODY

Iodine is crucial to the synthesis of thyroid hormones. Thyroid hormones (THs) are essential for normal growth and development, and during all stages of life. Through direct and permissive effects on biochemical cell functions, regulatory genes, and other hormones, THs influence virtually all biological systems from the point of conception onward. THs are involved in many processes including the differentiation of embryonic and adult brain stem cells; regulation of early embryonic cell migration, differentiation, and maturation; regulation of embryonic and postnatal somatic growth; regulation of embryonic and postnatal development of the brain and eyes; generation of energy in mitochondria; and regulation of brain function and neurogenesis [2,4–6,8,14,17].

During evolution, a concentrating mechanism developed that permitted iodine found in the environment to be stored and available for hormone production, even in times of scarcity. Thus, vertebrates developed a thyroid gland [4,9].

The human body contains approximately 10–20 mg of iodine; about 70–80% of it is concentrated in the thyroid gland. It is primarily obtained through the diet but it is also a component of some medications. Iodine is ingested in a range of forms, including iodide (I^-), molecular iodine (I_2), and iodate (IO_3^-), which are reduced to iodide in the gut prior to absorption. The amount of iodine absorbed is largely dependent on the level of dietary iodine intake, rather than on its chemical form or the composition of the diet [1,3–5,7,9,10,14].

TABLE 41.2 Recommendations for Iodine Supplementation in Pregnancy and Infancy in Areas Where <90% of Households Are Using Iodized Salt and the Median UIC Is <100 μg/L in Schoolchildren

Population	Iodine supplementation
Women of child-bearing age	• A single annual oral dose of 400 mg of iodine as iodized oil • A daily oral dose of iodine as potassium iodide should be given so that the total iodine intake meets the recommendation of 150 μg/day of iodine
Women who are pregnant or lactating	• A single annual oral dose of 400 mg of iodine as iodized oil • A daily oral dose of iodine as potassium iodide should be given so that the total iodine intake meets the recommendation of 150 μg/day of iodine • Iodine supplements should not be given to a woman who has already been given iodized oil during her current pregnancy or up to 3 months before her current pregnancy started
Children aged 0–6 months	• A single oral dose of 100 mg of iodine as iodized oil • A daily oral dose of iodine as potassium iodide should be given so that the total iodine intake meets the recommendation of 90 μg/day of iodine • Should be given iodine supplements only if the mother was not supplemented during pregnancy or if the child is not being breast-fed
Children aged 7–24 months old	• A single annual oral dose of 200 mg of iodine as iodized oil as soon as possible after reaching seven months of age • A daily oral dose of iodine as potassium iodide should be given so that the total iodine intake meets the recommendation of 90 μg/day of iodine

Reproduced from Ref. [14].

Absorption and Metabolism

Iodide absorption takes place predominantly in the stomach and upper small intestine. The different forms of iodine are reduced to iodide in the gut before absorption [1,3]. In healthy adults, the absorption of iodide is greater than 90%. It is absorbed via an active transport protein on the apical surfaces of enterocytes called the sodium−iodide symporter (NIS). NIS expression is downregulated when the concentration of iodide from food increases. Once in the circulation, the thyroid gland and the kidney quickly take up the iodide. The thyroid accumulates iodide depending on iodine and thyroid hormone homeostasis [1,3–5].

During lactation, the mammary glands concentrate iodine and secrete it into milk for the newborn. Other tissues take up small amounts of iodine, including the salivary glands, gastric mucosa, and choroid plexus. Iodine has no proven function in these tissues; however, there are suggestions of generalized immune system support [1,3–5].

NIS is located on the basal membrane in thyroid follicular cells and is the key enzyme responsible for iodide accumulation. The activity of NIS is three to four times greater in the thyroid than in any other tissue in the body. This allows the gland to accumulate and sequester iodide from the blood. Iodide concentration in the cytoplasm of follicular cells is more than 40 times greater than in the plasma [1,3,4].

If there is adequate or surplus dietary iodine supply, less than 10% of absorbed iodide is taken up by the thyroid. When dietary iodide is less abundant, the fraction of iodide taken up by the thyroid from the blood increases up to about 80%. Thyroid-stimulating hormone (TSH) and plasma iodide regulate NIS expression in thyroid follicular cells, determining iodide uptake by the thyroid gland. In hypothyroidism there is increased pituitary secretion of TSH, which results in upregulation of NIS expression. During the euthyroid or hyperthyroid states, TSH secretion is decreased and does not stimulate NIS expression. High concentrations of iodide in plasma directly decrease NIS expression [1,3,4].

NIS utilizes the energy released by the inward translocation of sodium down its electrochemical gradient to be able to translocate iodide against its electrochemical gradient. The sodium gradient as the driving force for the iodide uptake is generated by Na^+/K^+-ATPase. This active transport mediated by NIS can be competitively inhibited by thiocyanate, perchlorate, pertechnetate, and perrhenate [1,3,4,18].

Within the thyroid, iodide is moved to the colloid space via pendrin, a sodium-independent iodide/chloride transport protein. The thyroperoxidase (TPO) enzyme then organifies iodide by binding it to tyrosine residues on thyroglobulin forming monoiodotyrosine (MIT) or diiodotyrosine (DIT). Thyroxine (T4) is formed from the combination of two DIT molecules, whereas Triiodothyroinine (T3) is formed from one

DIT and one MIT molecule. Increased TSH stimulates thyroglobulin proteolysis and release of thyroid hormones into circulation. Depending on overall iodine status, the colloid space stores enough covalently bound iodine in thyroglobulin to account for several weeks to months worth of hormone secretion [1,3–5].

Distribution and Elimination

Absorbed iodide is distributed through the extracellular space with a half-life of approximately 10 h in plasma. The half-life may vary, as there is more rapid thyroid uptake and increased GFR in states of iodine deficiency or hyperthyroidism. When dietary iodide intake is abundant, approximately 90% of ingested iodide is excreted in urine and the remainder in feces. Urine iodine clearance is approximately 40 mL/min. Fecal losses vary but are generally low (10–20 μg/day), and small amounts are also lost via the skin through sweat [3,5,10].

NIS protein is also expressed in the renal tubular system. Renal iodide clearance remains constant as a percentage of filtered iodide in plasma, even if there is variable iodine intake. This means there is decreased urinary iodide when iodine intake is low and increased urinary iodide when intake is high. Thus, the measurement of urinary iodine can be a useful clinical and epidemiologic tool. Iodide clearance in the kidney also varies with thyroid status, being lower in hypothyroidism and increased in hyperthyroidism, which can be explained by changes of glomerular filtration [3,10,18].

Recycling

Intrathyroidal and extrathyroidal recycling of iodide occurs after GI absorption and may be up- or downregulated depending on the level of dietary iodine. The inactive MIT and DIT released during proteolysis of thyroglobulin accounts for approximately 80% of recaptured intrathyroidal iodide. Extrathyroidal iodine recycling may occur when absorbed iodide is excreted by salivary or gastric glands into the upper alimentary tract to be reabsorbed as previously described. Free iodide is released from the conversion of T4 to T3 in peripheral tissues and enters the general circulation, where it may be reused or eliminated [3,5].

Deiodinases are the primary iodine-recycling enzyme. There are three different types of deiodinases. Type I is present in the liver, kidney, and thyroid. It is responsible for the clearance of plasma reverse T3 (rT3) and a major source of circulating T3. Type II is found in the human brain, anterior pituitary, and thyroid and allows for local T3 production in these tissues. Type III is expressed in the human brain, placenta, and fetal tissues. It helps to regulate intracellular T3 levels as it inactivates T3 more effectively than T4. These deiodinases have been demonstrated to be selenoproteins. rT3 and T3 can be further deiodinated in the liver and are conjugated to either sulfur or glucuronide before excretion in the bile. An enterohepatic circulation of TH as intestinal flora deconjugates some of these compounds and promotes the reuptake of TH [3,8].

This can be summarized in Fig. 41.2, which shows the metabolism of iodine in the human body.

EPIDEMIOLOGY

Iodine deficiency and impaired thyroid hormone production have many adverse effects throughout the human lifespan from the intrauterine stage to old age, as shown in Table 41.3. Pregnant and lactating women, women of reproductive age, and children younger than 3 years are considered to be at high risk of iodine deficiency disorders [1–3,5,6,14,19].

Thyroid enlargement (goiter) is the earliest and most visible effect of iodine deficiency; however cognitive impairment is its most deleterious consequence. Neuronal migration, glial differentiation, and myelination of the central nervous system are dependent on normal thyroid hormone concentrations. Iodine deficiency (ID) is the most preventable cause of mental retardation in the world. Inadequate iodine intake results in hypothyroidism and goiter at all ages. Mild ID also has a significant impact on reproductive health, increasing miscarriages, stillbirths, and perinatal mortality, and is the most common cause of reproductive failure worldwide [1,3–7,10,14,19–22].

Control of ID is thus a critical and achievable development goal for governments. Salt iodization has been the central strategy to achieve this goal. Through salt iodization, ID is among the simplest and least expensive of nutritional deficiencies to prevent [20,22].

There has been remarkable progress in the global effort to eliminate ID over the past two decades. In 1993, the WHO estimated that goiter and ID affected 110 countries. From 2003 to 2011, the number of iodine-deficient countries decreased from 54 to 32 and the number of countries with adequate iodine intake increased from 67 to 105. Currently, 71% of the global population has access to iodized salt, up from 20% in 1990 [20,22].

Systematic reviews have confirmed the benefits of correcting iodine deficiency. Of two recent publications, one looked at 89 studies that provided iodized salt to populations and recorded a significant 72–76% reduction in risk for low intelligence (defined as IQ

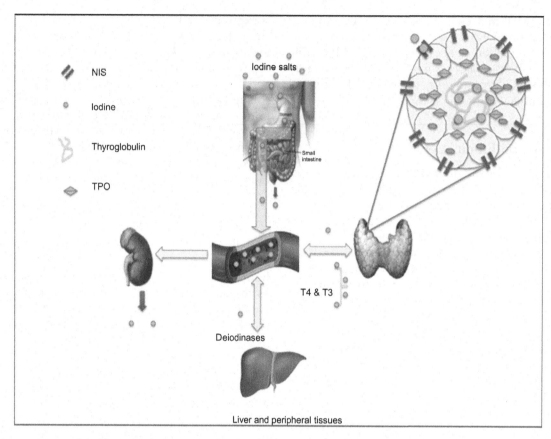

FIGURE 41.2 Iodine metabolism in the human body. *Source: modified from Ref. [6].*

TABLE 41.3 The Iodine Deficiency Disorders and Their Health Consequences, by Age Group

Age group	Consequences
All ages	Goiter
	Increased susceptibility of the thyroid gland to nuclear radiation
	In severe iodine deficiency, hypothyroidism
Fetus	Abortion
	Stillbirth
	Congenital anomalies
	Perinatal mortality
Neonate	Infant mortality
	Endemic cretinism
Child and adolescent	Impaired mental function
	Delayed physical development
Adults	Impaired mental function
	Reduced work productivity
	Toxic nodular goiter
	Hyperthyroidism

Reproduced from Ref. [19].

<70) and an 8.2–10.5 point overall increase in IQ. The second systematic review similarly concluded that iodine-sufficient children have a 6.9–10.2 point higher IQ than iodine-deficient children [4,19,21].

Currently approximately 1.9 billion of the world population is estimated to have inadequate iodine intake. Of these 285 million are school aged children ($^1/_3$ of school-aged children). The prevalence of iodine deficiency is lowest in the Americas (13.7%) and highest in Europe (44.2%). Southeast Asia represents about 31% of the global population with insufficient iodine intake. See Table 41.4 [1,23,24].

The iodine status of populations can be assessed by using the prevalence of goiter, biomarkers of iodine exposure, such as urinary iodine concentration (UIC), biomarkers of function (serum thyroglobulin levels), and thyroid function tests [1,3,5,6,14].

In the past, iodine nutrition was evaluated according to thyroid size and goiter rate (GR). In the 1990s, a simplified grading system was released, to try to reduce interobserver variation. Grade 0 was defined as a thyroid that is not palpable or visible; grade 1 was defined as an enlarged gland that is palpable but not visible when the neck is in the normal position; grade 2 was defined as a thyroid that is clearly visible when

TABLE 41.4 Number of Countries, Proportion and Number of School Aged Children (SAC) and the General Population with Insufficient Iodine Intake by WHO Region 2011

WHO region	Insufficient iodine intake <100 mcg/L				
	SAC			General population	
	Countries (n)	Proportion (%)	Total (millions)	Proportion (%)	Total (millions)
Africa	10	39.3	57.9	40.0	321.1
Americas	2	13.7	14.6	13.7	125.7
South-East Asia	0	31.8	76.0	31.6	541.3
Europe	11	43.9	30.5	44.2	393.3
Eastern Mediterranean	4	38.6	30.7	37.4	199.2
Western Pacific	5	18.6	31.2	17.3	300.8
Global Total	32	29.8	240.9	28.5	1881.2

Reproduced from Ref. [24].

TABLE 41.5 Indicators of Iodine Status in Populations

Measure	Age group	Application
Median urinary iodine concentration (ng/mL or µg/L)	School-aged children (6–12 years) and pregnant women	See Table 41.6
Rate of goiter measured by palpation (%)	School-aged children	Extent of iodine deficiency by frequency of goiter:
		None: 0–4·9%
		Mild: 5–19·9%
		Moderate: 20–29·9%
		Severe: ≥30%
Rate of goiter measured by ultrasound (%)	School-aged children	Extent of iodine deficiency by frequency of goiter:
		None: 0–4·9%
		Mild: 5–19·9%
		Moderate: 20–29·9%
		Severe: ≥30%
TSH concentrations (mU/L)	Neonates	A frequency of <3% of TSH concentration values >5 mU/L shows iodine sufficiency in a population (when samples collected >48 h after birth)
Serum or dried blood spot thyroglobulin values	School-aged children	Reference interval in iodine-sufficient children is 4–40 µg/L

Modified from Ref. [19].

the neck is in a normal position. It was recommended that the GR be used to define severity ID in populations (see Table 41.5) [3,14,19,24].

After the introduction of iodized salt programs in areas of endemic goiter, it was reported that although thyroid size decreased as iodine intake increased, thyroid size did not return to normal for months or years after ID correction, and the GR remained elevated (>5%), particularly among older children and adults. Because of this long lag-time in the resolution of goiter, the GR is difficult to interpret for several years after iodized salt introduction, since it reflects both a population's history of iodine nutrition as well as its present status [24].

Due to the limitations of using goiter rate to determine iodine status, new methods for assessing this parameter were sought. Urinary iodine concentration (UIC) is an excellent indicator of recent iodine intake because ≥92% of dietary iodine is absorbed and, in healthy, iodine-replete adults >90% is excreted in the

TABLE 41.6 Epidemiological Criteria for Assessing Iodine Nutrition Based on Median Urinary Iodine Concentrations of School-Age Children (≥6 Years)

Median urinary Iodine (ng/mL or µg/L)	Iodine intake	Iodine status
<20	Insufficient	Severe iodine deficiency
20–49	Insufficient	Moderate iodine deficiency
50–99	Insufficient	Mild iodine deficiency
100–199	Adequate[a]	Adequate iodine nutrition
200–299	Above requirements	Likely to provide adequate intake for pregnant/lactating women but may pose a slight risk of more than adequate intake in the overall population
>300	Excessive	Risk for adverse health consequences (eg, iodine-induced hyperthyroidism, autoimmune thyroid diseases)

[a]Applies to adults but not to pregnant and lactating women.
For pregnant women, median urinary iodine concentrations of 150–249 ng/mL and for lactating women, median urinary iodine concentrations of above 100 ng/mL represent adequate iodine intake WHO categories for median urinary iodine concentrations in school-age children and adults.
Adapted from Ref. [20].

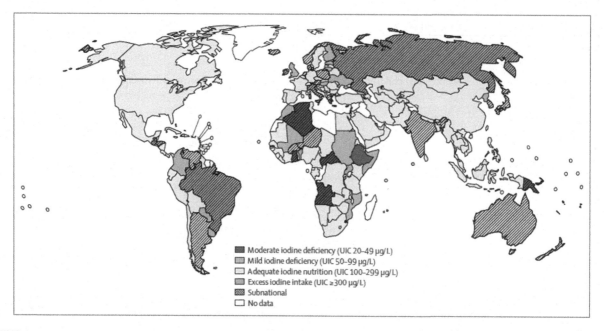

FIGURE 41.3 National iodine status as per urinary iodine concentration in school children. *Source: reproduced from Ref. [19].*

urine within 24–48 h. UICs are usually measured in spot urine collections. Based in the UIC we can classify populations as iodine deficient, iodine sufficient, or with excessive iodine intake. See Tables 41.5 and 41.6 and Fig. 41.3 [1,3,5,6,14,19,22,24].

TSH is considered as a sensitive indicator of iodine status for the newborn but is less predictive in adults. For this reason it can be used in addition to UIC when assessing iodine status for the newborns but not for other age groups. In older children and adults, serum TSH may be slightly elevated in ID, however, levels often remain within the normal range. Serum thyroglobulin (Tg) is considered as a sensitive indicator for school-aged children and adults and appears to be a better alternative to UIC for estimating the longer term iodine intake. An elevated Tg level is well correlated with iodine deficiency [1,3,19].

Iodine Status and Thyroid Disorders

The thyroid gland is able to autoregulate the use of iodine for thyroid hormone production according to its availability. In healthy adults, the mean daily uptake and turnover of iodine by the thyroid is approximately 95 µg. In order to remain euthyroid, the daily iodine

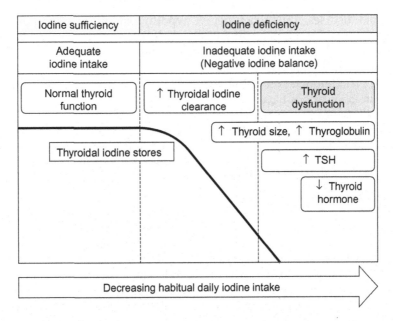

FIGURE 41.4 The physiological stages of iodine status. *Source: reproduced from Ref. [20].*

intake must be sufficient to enable the thyroid to turn over 95 μg iodine per day. If dietary iodine intake decreases, thyroid function is maintained by increasing thyroidal clearance of circulating iodine. Deficient iodine intake triggers secretion of TSH and increases the expression of NIS to increase the uptake of iodine into thyroid cells. The thyroid accumulates a larger proportion of the ingested iodine and reuses the iodine from the degradation of TH, which reduces renal clearance of iodine. Short-term deficits in iodine intake are buffered by intrathyroidal stores (up to 20 mg in iodine-sufficient areas) [3,5,19,20,25,26].

However, in chronically low iodine intake, thyroidal reserves will be depleted and iodine turnover will need to be increasingly replaced by dietary iodine supply. Eventually, low dietary intake will limit thyroid hormone synthesis. Thus ID can be broadly defined in two phases: inadequate iodine intake, leading to chronic iodine deficiency and thyroid dysfunction. Fig. 41.4 illustrates a simplified model of human iodine and thyroid status at different stages (left to right) of iodine intake: sufficient iodine intake, low iodine intake without thyroid dysfunction, and chronically low iodine intake with hypothyroidism. The scientific evidence is limited with regard to the absolute levels of habitual daily iodine intake at which thyroid stores decrease and thyroid dysfunction occurs [19,20,25,26].

The prolonged thyroid hyperactivity associated with such adaptations can lead to thyroid growth. There is an increased tendency to mutations during follicular cell proliferation, which can lead to multifocal autonomous growth and function (autonomous nodules) [26].

Excessive iodine intake can also cause thyroid dysfunction. Following transient exposure to high iodine levels, the synthesis of thyroid hormone is normally inhibited via the acute Wolff-Chaikoff effect. It is thought this is mediated by the generation of inhibitory substances that affect thyroid peroxidase activity, and increased intrathyroidal iodine concentration that suppresses deiodinase activity. If excessive iodine exposure persists, the thyroid is able to "escape" from the acute Wolff-Chaikoff effect within approximately 2 weeks. This is accomplished, in part, by downregulating NIS on the basolateral membrane. This occurs within 24 h after exposure to excess iodine. The result is decreased intrathyroidal iodine concentrations and a decrease of iodinated substances that inhibit thyroid hormone production leading to a subsequent increase in thyroid hormone synthesis. In individuals with dysregulation of the thyroid follicular cell (such as Hashimoto's thyroiditis or inborn errors of metabolism), excess iodine exposure can induce thyroid dysfunction, which might be transient or permanent [2,4–6,16,21,25,27].

The Jod Basedow phenomenon, or iodine-induced hyperthyroidism, occurs most commonly in individuals with a history of nontoxic diffuse or nodular goiters, living in areas of iodine deficiency that are exposed to an increase in iodine in the environment [15,16,19,27].

In summary, the relationship between the iodine intake of a population and the occurrence of thyroid disease is U-shaped, with an increased risk from both low and high iodine intakes. Even small increases in iodine intake in previously iodine-deficient populations can change the pattern of thyroid diseases. Thus when performing studies one needs to consider not only present intake, but also the history of iodine intake of the population as well as varying environmental and genetic factors that modify the effects of iodine intake on thyroid disorders. For example, white populations have a higher risk of autoimmune thyroiditis than Africans, which makes the former more susceptible to hypothyroidism if iodine intakes are excessive [19,25,26].

IODINE AND THE HUMAN LIFESPAN

Pregnancy

In early gestation, maternal TH production increases by approximately 25—50% in response to increase of serum binding proteins and increased stimulation of TSH receptor by human chorionic gonadotropin. The placenta is a source of deiodinase that increases the degradation of thyroxine thus increasing the TH demand [1,28,29].

The placenta regulates the transfer of maternal THs via placental thyroid hormone transporters, thyroid hormone binding proteins, and enzymes such as deiodinases, sulfotransferases, and sulfatases. Iodine transport is regulated as well. Placental NIS protein levels correlate with gestational age during early pregnancy and increase with increased placental vascularization. This allows for the fetus to obtain an adequate iodide supply to meet thyroid hormone synthesis [29].

The placenta expresses deiodinases type II and III. Various thyroid hormone transporters have been located at the apical and basolateral membranes of the syncytiotrophoblasts in the placenta. Transthyretin, a serum thyroid hormone binding protein, appears to play an important role in the delivery of maternal thyroid hormone to the developing fetus. The placenta secretes transthyretin into maternal and fetal circulations. This placental protein is secreted into the maternal placental circulation where it can be taken up by trophoblasts and be translocated to the fetal circulation, forming a "shuttle system" [29].

Pregnant women are at a higher risk of I insufficiency, as their daily requirement is higher than that of the general population. The increased requirement is due to several factors: an increased maternal TH production to both maintain maternal euthyroidism and to transfer TH to the fetus in the early first trimester, the need for iodine transfer to the fetus particularly in later gestation, and the increase in renal iodine clearance. If a chronically iodine-deficient woman becomes pregnant she has no thyroid iodine stores to meet these increased needs and subsequently goiter and hypothyroidism can occur. Maternal I insufficiency can result in a wide spectrum of growth and mental impairment in the newborn [1,3,4,17].

Fetal Development

Severe endemic goiter and cretinism are observed in iodine-deficient regions. The link between ID and cretinism was confirmed in 1966, after a trial in Papua New Guinea. Iodized oil was administered to a population, with reduction of the incidence of cretinism and goiter in the treated group when the supplementation was provided before the onset of pregnancy [3,4,28].

Maternal dietary I and T4 are critical over the entire course of gestation. Even after the fetus's thyroid starts producing THs, the mother's T4 contribution is still crucial. At birth, maternal T4 accounts for 20% to 50% of the T4 measured in cord blood. After birth, maternal I, which is transferred by NIS into the milk, is critical for the newborn's development [4]. the recommendation for iodine intake during lactation is based on the woman's daily requirement plus the amount of iodine that is loss in human milk [3,5].

Thyroid hormone receptors are present in the fetal brain by nine weeks, making the central nervous system's (CNS) development sensitive to TH deficiency as early as the first trimester. THs play a critical role in the CNS of the fetus and newborn, affecting neuronal migration, differentiation, myelination, and synaptogenesis. Development of the cerebral cortex, cochlea, and basal ganglia during the first trimester of gestation is dependent on thyroid hormones. They are also necessary for brain growth and differentiation in the third trimester (see Fig. 41.5) [3,4,17,23].

Prior to the onset of fetal thyroid function, maternal TH is critical for fetal development and is provided via placental vessels. The early effects of iodine deficiency in the fetus are due to reduced maternal T4 transfer prior to onset of fetal TH production [4,17,23].

In humans, thyroid morphogenesis is completed by gestation week 7. NIS expression is the limiting step in the terminal differentiation and onset of thyroid function. The thyroid gland is considered fully differentiated at gestation week 11, once follicular cells have fully polarized, formed follicles, and fetal thyroid begins to take up I and synthesize TH [4,17].

The consequences of ID during gestation depend upon the timing and severity of the hypothyroidism. TH insufficiency can cause severe neurodevelopmental defects, cretinism being the gravest. However, even a slight reduction in maternal T4 production during gestation can impair cognitive development in the newborn and result in decreased IQ [3,4,17,28].

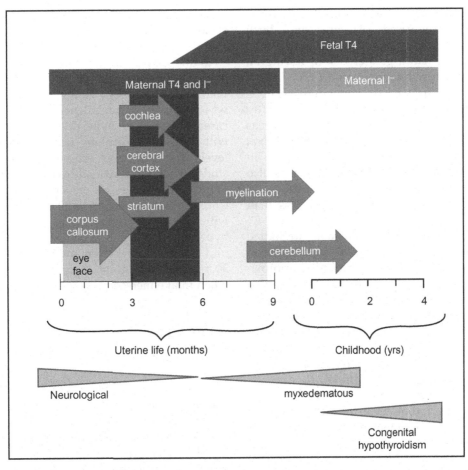

FIGURE 41.5 THs in human development and effects of ID and maternal hypothyroidism on fetal development. Blue arrows (gray in print versions) indicate the time of development of specific brain areas and body features that depend on adequate T4 levels and can be affected by low T4 at the indicated time of gestation. As evidenced by the red bars (dark gray in print versions), the timing of the deficiency will affect clinical outcome. *Source: reproduced from Ref. [4].*

Severe ID in utero causes cretinism, of which there are two forms depending on the timing of the iodine deficiency during fetal life, neurological (earlier) and myxedematous (later). Worldwide neurological cretinism is the most common subtype. The clinical features for neurological cretinism include mental retardation, defects of hearing and speech, squint, impaired voluntary motor activity involving spastic diplegia or paresis of the lower limbs, and disorders of stance, with spastic gait and ataxia. The children are usually euthyroid, but can develop goiter and hypothyroidism in some cases [3–5,17].

In myxedematous cretinism, there is hypothyroidism associated with dwarfism, myxedema, dry, thickened skin, sparseness of hair and nails, deep hoarse voice, sexual retardation, retarded maturation of body parts, skeletal retardation, weak abdominal muscles, poor bowel function, and delayed tendon reflexes. There is a typical facies with wide-set eyes, saddle-nose deformity with retarded maturation of nasoorbital configurations, mandibular atrophy, and thickened lips [3,4,17].

Iodine treatment during pregnancy in areas of severe iodine deficiency result in better psychomotor development scores and higher IQ scores in offspring when compared to placebo [3,23].

Both thyroid and growth hormone (GH) are essential for growth and development even during fetal life. Thyroid hormone is required for normal GH expression in vitro and in vivo, and in animal studies TH promotes GH secretion and modulates its effect at the receptor. Thyroid hormone also directly affects epiphyseal growth, bone maturation, and stature [3,17].

Supplementation of iodine has been found to have a positive impact on the anthropometric measures at birth and decreases infant mortality. Repletion of severely iodine-deficient pregnant women in China improved head circumference and reduced microcephaly. In an area of Algeria with endemic goiter, administration of iodine prior to conception and during first trimester increased placental and birth weights. In Spain, it was observed that women, who were iodine deficient, had a greater risk of having an infant that was small for gestational age [3,17].

Infant survival is improved to women whose ID is corrected before or during pregnancy, as seen after adding potassium iodate to irrigation water in ID areas of China. Studies in Indonesia show infant survival may also be improved by iodine supplementation in the newborn period. The use of iodized salt was also associated with lower prevalence of malnutrition and mortality in neonates, infants, and children less than 5 years [3,17].

The fetus and newborn are vulnerable to iodine excess. The fetal thyroid does not acquire the capacity to suppress the acute Wolff-Chaikoff effect until approximately 36 weeks of gestation. Excessive maternal iodine intake can cause neonatal goiter and hypothyroidism. It has been suggested that the fetal thyroid is more sensitive to the inhibitory effect of iodine. The upper limit of acceptable iodine intake in pregnancy is controversial. Excessive iodine supplementation during pregnancy had a negative effect on birth weight [15,17,22].

Thyroidal iodine turnover rate is more rapid in infants, thus adequate iodine levels in breast milk are important for neurodevelopment of infants. Breastfed infants are reliant on maternal dietary iodine intake. After delivery, maternal TH production and urinary iodine excretion return to normal, however iodine is being concentrated by the mammary gland. Breast milk provides approximately 100 μg/d of iodine. The iodine content in cow milk is dependent on the amount of iodine that is consumed by the animal [5,25,29].

Childhood

Many cross-sectional studies comparing cognitive and/or motor function in children from chronically iodine-deficient and sufficient areas show impaired intellectual and motor skills in children from these areas. They were unable to make a distinction between persistent effects from deficiency in utero versus the effects of current iodine status [3,17].

Differences in psychomotor development in ID children become apparent after the age of 2. The problems range from small neurologic changes to impaired learning ability and poor performance in school/formal testing of psychomotor function [3,23].

Cross sectional studies on ID and child growth have yielded mixed results. There may be confounding based on anthropometric measurements reflecting an earlier iodine status. Hypothyroidism is a well-recognized cause of short stature in children. Iodine repletion's effect on growth is likely due to improved thyroid function [3,17].

Iodine excess in children is associated with goiter and thyroid dysfunction. In Japanese children with intakes of more than 20 mg of iodine per day, there was an increased prevalence of goiter (3–9%), but no cases of hypo- or hyperthyroidism. Chinese children with iodine rich drinking water had elevations of TSH levels and increased rates of goiter. A multicenter study of school aged children in 12 countries found a higher prevalence of elevated thyroglobulin levels. Chronic dietary intakes >50 mg a day are associated with increased thyroid size. Onset of mild thyroid hyperstimulation as evidenced by increased thyroglobulin levels occurs when iodine intake increases over 30 mg/day and goiter begins to appear in children when intake increases above 40–50 mg/day [3,15,22].

Adults/Elderly

The recommended daily requirement of iodine for adults was estimated based on the uptake and turnover of iodine by the thyroid of euthyroid adults. Current recommendations do not have specific guidelines for dietary intake in the elderly population. There is lack of data on this specific population. Studies in elderly Europeans suggest that this subgroup has the same level of iodine intake as the rest of the population. However due to an age-related decreased energy and nutrient intake, elderly subjects have an increased risk of insufficient iodine intake compared to the rest of the adult population [3,30–32].

Iodine Deficiency

In populations with mild-to-moderate iodine deficiency, serum thyroglobulin concentrations and thyroid size usually increase in the population, whereas serum TSH, T3, and T4 are often still in the normal range. In regions of mild iodine deficiency, many people develop simple diffuse goiters and some develop nodules. The cause of diffuse goiter in the setting of mild iodine deficiency is unclear as mean TSH levels are not usually increased. On the contrary, populations with mild to moderate iodine deficiency might have lower mean TSH concentrations than do sufficient populations. This difference is explained by an increase in the prevalence of thyroid nodularity and multinodular toxic goiter in populations with mild-to-moderate iodine deficiency, particularly in adults older than 60 years. If multinodular toxic goiters are prevalent in a population, overall mean TSH concentration will be lowered [3,19,33].

In regions with moderate to severe iodine deficiency, mean serum TSH concentration often increases slightly whereas T4 remains normal. Many affected individuals develop subclinical hypothyroidism. As the severity of the deficiency worsens TSH continues to rise. T3 production increases slightly or remains unchanged and T4 decreases because of preferential secretion of T3 by the thyroid in the setting of iodine deficiency. This preferential secretion of T3 occurs because the activity of T3 is roughly four times that of T4, but T3 needs only 75% as much iodine for its synthesis. In moderate to severe iodine deficiency, the concentration of serum TSH is usually inversely correlated with that of T4, but not with that of T3, which suggests a closer feedback control of TSH secretion by T4 than by T3. In chronic, severe iodine deficiency many individuals have high TSH concentrations and most develop goiter. If thyroidal iodine is exhausted, mean concentrations of T4 and T3 decrease, TSH concentration increases, and there is an increase in overt hypothyroidism in the population [3,19,33].

Populations with mild to moderate iodine deficiency have a higher prevalence of hyperthyroidism and lower mean serum TSH concentrations than populations with adequate or excessive iodine intakes [3,19].

In populations with severe iodine deficiency, the prevalence of hypothyroidism is higher than in areas of optimum iodine intake. However, in the setting of mild-to-moderate iodine deficiency, the prevalence of subclinical and overt hypothyroidism is generally lower than in areas of optimum or excessive iodine intake [3,19].

The relation between iodine intake and risk for diffuse goiter also demonstrated a U-shaped curve, with increased risk at deficient and excess intakes, whereas risk for nodular goiter seems to be increased only in deficient populations [19].

Correction of iodine deficiency in adult populations, irrespective of severity, reduces mean thyroid size and the prevalence of diffuse goiter within a few years. Although iodine repletion usually does not reduce the prevalence of thyroid nodularity in adults older than 50 years because of fibrotic changes in nodules, it does reduce the risk for development of nodular disease in younger adults [19].

Iodine Excess

Iodine-induced hypothyroidism can occur if there is a failure of adaptive mechanisms.

The common denominator of the states predisposing to iodine-induced hypothyroidism probably is a slightly elevated TSH or persistent thyroid-stimulating antibodies, which keep the NIS activated and intrathyroidal iodide concentration high, thereby preventing an escape from the Wolff-Chaikoff effect. The other proposed mechanism is induction of thyroid hypofunction due to iodine-induced autoimmune thyroiditis although this has not been proven in all studies [3,16,27].

An increment in iodine intake in deficient populations usually increases the incidence of hyperthyroidism. The increase tends to be greater and more severe if iodine fortification is excessive and the preexisting iodine deficiency was severe. The mechanism by which this occurs seems to vary in the different age groups. Observations after introduction of salt supplementation in Denmark demonstrated that in the population with ages between 20 and 39 years the occurrence of hyperthyroidism seemed to be autoimmune in origin versus in the elderly where it was associated to thyroid nodularity. Individuals at highest risk for development of hyperthyroidism are adults older than 60 years with nodular thyroid disease. Follicular cells in these nodules can be insensitive to TSH control, and if the iodine supply is suddenly increased, these cells can overproduce thyroid hormone [3,16,19].

The increase in incidence of hyperthyroidism after a properly monitored introduction of iodine into populations with mild-to-moderate iodine deficiency is transient (up to 10 years) because the resulting iodine sufficiency in the population reduces the future risk for the development of autonomous thyroid nodules [3,16,19].

Higher mean TSH concentrations in populations with excessive iodine intake could be explained by a more general phenomenon of iodine downregulation of thyroid function. Increasing iodine intakes in a population leads to a small increase in the incidence of mild subclinical hypothyroidism; this increase seems to occur more often in individuals positive for thyroid antibodies [19].

Thyroid Autoimmunity

It is difficult to ascertain the relationship between iodine intake and thyroid autoimmunity as there is conflicting data. In animal studies it has been observed in mice that if they are first fed an iodine-deficient diet, and then switched to one with iodine excess, they dose-dependently develop thyroid cell damage suggesting autoimmune disease [15,16].

Individuals with multinodular goiter often have circulating antibodies, likely from antigen release by the abnormal gland. Thus antibodies are common in populations with iodine deficiency and multinodular goiter. An increase in iodine intake in a previously deficient population can induce thyroid autoimmunity. Epidemiologic studies performed in China, Turkey, and Denmark suggest that supplementation with iodized salt increases the prevalence of autoimmune thyroid disease, be it overt or subclinical hypothyroidism, hyperthyroidism, or both. This phenomenon seemed to be dose dependent at three urinary iodine excretion levels (marginally low/more than adequate/excessive); the prevalence of subclinical hypothyroidism was 0.9%, 2.9%, and 6.1% [16,26].

Iodine intake does not seem to influence the occurrence of thyroid autoantibodies in those under the age of 45. The prevalence of thyroid autoantibodies increases with age, and are more frequently seen in people between 60 and 65 years. Thyroid autoantibodies are seen more commonly in the moderately iodine-deficient populations. The presence thyroid autoimmunity, characterized by lymphocytic infiltration, rises as iodine intake in the population increases. Interestingly, an increase in circulating antibodies is not seen in populations with chronic stable high iodine intake. This discrepancy may be explained in part by the observation that not everyone with autoimmune thyroid disease has circulating antibodies as observed in autopsy series where the frequency of finding lymphocytic infiltration is higher than that of thyroid antibodies [15,16,19,26,34].

Thyroid Cancer

The overall incidence of thyroid cancer in populations does not seem to be affected by the usual range of iodine intakes from dietary sources. A systematic review reported no association of thyroid cancer risk with dietary iodine intake. Although several ecological studies have suggested an increase in papillary thyroid cancer after the introduction of iodized salt to populations, this could be confounded by a change in diagnostic surveillance. For example, in the USA, the prevalence of thyroid cancer has been steadily increasing over the last decades; however, dietary iodine intake over the same period has decreased [5,14,19].

Differences in iodine intake between regions might affect the distribution of thyroid cancer subtypes: populations in areas of optimum iodine intake seem to have fewer follicular thyroid cancers, but more papillary thyroid cancers. Chronic iodine deficiency seems to be a risk factor for follicular thyroid cancer and, possibly, anaplastic thyroid cancer. Correction of iodine deficiency might be beneficial in that it reduces the goiter in populations, which is a major risk factor for thyroid cancer. Also in the event of nuclear fallout, if iodine sufficient, the thyroid would be exposed to lower doses of radioactive iodine and the risk for development of thyroid cancer would therefore also be lower [3,5,14,19].

SUMMARY

- Iodine is a nonmetallic trace element, which is an essential component of thyroid hormones.
- Iodine is obtained from diet, sources include seafood, seaweed, grains, and dairy products. The iodine content of food depends on the content of the soil in which it is grown.
- Certain geographic areas like mountainous regions and flooded river valleys are endemic for iodine deficiency.
- Iodine deficiency (ID) is the most preventable cause of mental retardation in the world. Inadequate iodine intake results in hypothyroidism and goiter at all ages.
- Pregnant and lactating women, women of reproductive age, and children younger than 3 years are considered to be at high risk of iodine deficiency disorders.
- Iodine deficiency and excess can cause thyroid dysfunction in adults and the elderly including hypothyroidism, hyperthyroidism, and goiter.

References

[1] Doggui R, El Atia J. Review: iodine deficiency: physiological, clinical and epidemiological features, and pre-analytical considerations. Ann Endocrinol 2015;76:59–66.

[2] Higdon J. Iodine. Micronutrient Information Center. Linus Pauling institute. Oregon State University. Retrieved from: < http://lpi.oregonstate.edu/infocenter/minerals/iodine >; 2001. Last updated March 2010 [accessed 12/2014].

[3] Zimmermann MB. Iodine deficiency. Endocr Rev 2009;30: 376–408.

[4] Portulano C, Paroder-Belenitsky M, Carrasco N. The Na^+/I^- (NIS): mechanism and medical impact. Endocr Rev 2014;35:106–49.

[5] Institute of Medicine (U.S.). Panel on Micronutrients. 8 Iodine in Dietary reference intakes for vitamin A, vitamin K, arsenic, boron, chromium, copper, iodine, iron, manganese, molybdenum, nickel, silicon, vanadium, and zinc: a report of the Panel on Micronutrients. National Academy Press; 2001. p. 258–89.

[6] Report of a joint FAO/WHO expert consultation Bangkok, Thailand. Iodine in Human vitamin and mineral requirements. FAO; 2001 [Chapter 12].

[7] Iodine in Recommended Nutrient Intakes for Malaysia. A report of the technical working group on nutritional guidelines. Published by: National Coordinating Committee on Food and Nutrition (NCCFN) Ministry of Health Malaysia. <http://www2.moh.gov.my/images/gallery/rni/16_chat.pdf>; 2005.

[8] Yen P. Physiological and molecular basis of thyroid hormone action. Physiol Rev 2001;81(3):1097–126.

[9] Crockford, S. Evolutionary roots of iodine and thyroid hormones in cell–cell signaling from the symposium "cell–cell signaling drives the evolution of complex traits" presented at the annual meeting of the Society for Integrative and Comparative Biology. January 3–7, 2009, at Boston, Massachusetts. Integr Comp Biol 49(20):155–166. <http://dx.doi.org/10.1093/icb/icp053>.

[10] Ristic-Medic D, Novakovic R, Glibetic M, Gurinovic M. EUURECA — Estimating iodine requirements for deriving dietary reference values. Crit Rev Food Sci Nutr 2013;53:1051−63.

[11] Laurberg P, Pedersen IB, Carlé A, Andersen S. Chapter 28 — the relationship between thiocyanate and iodine. In: Comprehensive handbook of Iodine: nutritional, biochemical, pathologic and therapeutic aspects, Elsevier; 2009. p. 275−81.

[12] Thomson CD, Campbell JM, Miller J, Seaff SA, Livingstone V. Selenium and iodine supplementation: effect on thyroid function of older New Zealanders. Am J Clin Nutr 2009;90:1038−46.

[13] Hess SY. The impact of common micronutrient deficiencies on iodine and thyroid metabolism: the evidence from human studies. Best Pract Res Clin Endocrinol Metab 2010;24(1):117−32.

[14] Zimmermann MB. Iodine requirements and the risks and benefits of correcting iodine deficiency in populations. J Trace Elem Med Biol 2008;22:81−92.

[15] Sun X, Shan Z, Teng W. Effects of increased iodine intake on thyroid disorders. Endocrinol Metab 2014;29:240−7.

[16] Bürgi H. Iodine excess. Best Practice Res Clin Endocrinol Metab 2010;24(1):107−15.

[17] Zimmerman M. The role of iodine in human growth and development. Semin Cell Dev Biol 2011;22:625−52.

[18] Spitzwieg C, Dutton CM, Castro MR, Bergert ER, Goellner JR, Heufelder AE, et al. Expression of the sodium iodide symporter in human kidney. Kidney Int 2001;59:1013−23. Available from: http://dx.doi.org/10.1046/j.1523-1755.2001.0590031013.x.

[19] Zimmermann MB, Boelaert K. Iodine deficiency and thyroid disorders. Lancet Diabetes Endocrinol 2015;3(4):286−95 doi: 10.1016/S2213-8587(14)70225-6

[20] Zimmermann MB, Andersson M. Assessment of iodine nutrition in populations: past, present and future. Nutr Rev 2012;70(10):553−70.

[21] Pearce E, Andersson M, Zimmermann MB. Global iodine nutrition: where do we stand in 2013? Thyroid 2013;23(5):523−6.

[22] Zimmerman M. Iodine deficiency and excess in children: worldwide status in 2013. Endocr Pract 2013;19(2):839−46.

[23] Melse-Boonstra A, Jaiswal N. Iodine deficiency in pregnancy, infancy and childhood and its consequences on brain development. Best Pract Res Clin Endocrinol Metab 2010;24(1):29−38.

[24] Zimmermann MB, Andersson M. Global iodine nutrition a remarkable leap forward in the past decade. International Council for Control of Iodine Deficiency Disorders IDD Newsletter 2012;40(1):1−5.

[25] Lauberg P, Pedersen B, Knudsen N, Ovesen L, Andersen S. Environmental iodine intake affects the type of nonmalignant thyroid disease. Thyroid 2001;11(5):457−66.

[26] Laurberg P, Cerqueira C, Ovesen L, Rasmussen LB, Perrild H, Andersen S, et al. Iodine intake as a determinant of thyroid disorders in populations. Best Pract Res Clin Endocrinol Metab 2010;24(1):13−27.

[27] Leung AM, Braverman LE. Consequences of iodine excess. Nat Rev Endocrinol 2014;10:136−42 Published online 17 December 2013; <http://dx.doi,org/10.1038/nrendo.2013.251>.

[28] Leung A, Pearce A. Iodine nutrition in pregnancy and lactation. Endocrinol Metabol Clin N Am 2011;40:765−77.

[29] Lazarus JH. Thyroid regulation and dysfunction in the pregnant patient. South Dartmouth, MA: Endocrine Education Inc.; 2015 [Chapter 14]. Available from: www.thyroidmanager.org1/2015.

[30] Wood RJ, Suter PM, Russel RM. Mineral requirements of elderly people. Am J Clin Nutr 1995;62:493−505.

[31] Destefani SA, Corrente JE, Paiva SAR, Mazeto GMFS. Prevalence of iodine intake inadequacy in elderly Brazilian women. A cross sectional study. J Nutr Health Aging 2015;19(2):137−40.

[32] Pedersen AN, Rasmussen LB. Chapter 116 — iodine intake in the European elderly. Comprehensive handbook of iodine: nutritional, biochemical, pathologic and therapeutic aspects. Elsevier; 2009. p. 1139−46.

[33] Dayan CM, Panicker V. Chapter 5 — interpretation of thyroid function tests and their relationship to iodine nutrition: changes in TSH, free T4, and free T3 resulting from iodine deficiency and iodine excess. Comprehensive handbook of iodine: nutritional, biochemical, pathologic and therapeutic aspects. Elsevier; 2009. p. 47−54.

[34] Andersen S, et al. Iodine deficiency influences thyroid autoimmunity in old age — a comparative population-based study. Maturitas 2012;71:39−43.

CHAPTER

42

Vitamin B12 Requirements in Older Adults

Esmée L. Doets[1] and Lisette CPGM de Groot[2]

[1]FBR, Fresh Food and Chains, Wageningen University and Research Center, Wageningen, The Netherlands
[2]Division of Human Nutrition, Wageningen University and Research Center, Wageningen, The Netherlands

INTRODUCTION

Since their discovery in the early part of the twentieth century, the importance of the water-soluble B complex vitamins to the maintenance of healthy tissue and prevention of disease has been recognized. Today, research on the actions of these vitamins is leading to the understanding that beyond the prevention of deficiency diseases, these vitamins function in the body to maintain optimal health, and to protect against chronic disease. This is particularly important for the growing aging population, where changes in energy requirement, and decline in metabolic efficiency make it difficult to ensure adequate vitamin intake. The B vitamins include thiamin (B1), riboflavin (B2), niacin (B3), pantothenic acid (B5), pyridoxine (B6), biotin (B7), folate (B9), and vitamin B12. Except for vitamin B12, which is stored in the liver, the water-soluble vitamins are not stored and, therefore, regular intakes are needed. In the elderly, malabsorption of vitamin B12 from food is the main cause of deficiency and explains why depletion occurs with aging, and why vitamin B12 deficiency is becoming one of the dominant nutritional concerns later in life. Therefore, this paper focuses on vitamin B12. It addresses associations between dietary intake, vitamin B12 status, and health outcomes, including vitamin B12 body stores, cognitive function, bone health, and biomarkers of vitamin B12 status. In this way it provides insight in vitamin B12 requirements later in life, considering different approaches to derive them.

VITAMIN B12

An adequate dietary supply of vitamin B12 is essential for normal blood formation and neurological function [1–3]. Although only small quantities of the vitamin are needed, deficiency may have serious consequences for health and development of populations across the world [4,5]. Elderly people in particular are vulnerable to vitamin B12 deficiency resulting from inadequate dietary intake or from malabsorption of the vitamin due to age-related changes in the gastrointestinal tract [6,7]. Estimates of the prevalence of vitamin B12 deficiency among elderly people range from 5% to 60% depending on the diagnostic criteria used [4,8,9].

Most countries provide recommendations on the amount of dietary vitamin B12 needed to maintain health for nearly all apparently healthy people within a defined population [10]. These recommendations are estimated from a statistical distribution of vitamin B12 requirements for meeting a specific health criterion [11]. For adults and the elderly, current recommendations on vitamin B12 intake range from 1.4 to 3.0 μg/day with requirements defined as the intake needed to maintain adequate hematological status or body stores by compensating daily losses [12]. Data that have been used to estimate vitamin B12 requirements in adults and elderly people so far, originate from a limited number of depletion–repletion and balance studies published between 1958 and 1991. These studies merely focused on the prevention of severe deficiency [12]. The prevention of subclinical deficiency, chronic diseases, or long-term outcomes were not taken into account.

Recently, four systematic reviews summarized available dose-response data on vitamin B12 requirements of healthy adults and elderly people focusing on different indicators of health: vitamin B12 body stores [13], cognitive function [14], bone health [15], and biomarkers of vitamin B12 status [16]. The aim of this chapter is to give an overview of the evidence and to provide estimates of vitamin B12 requirements based on different health indicators. Furthermore, pros and cons of different approaches that may be used for estimating vitamin B12 requirements will be discussed.

APPROACHES FOR ESTIMATING VITAMIN B12 REQUIREMENTS

Basically there are two approaches for estimating the distribution of vitamin B12 requirements: (1) based on physiological needs estimated by basal losses assuming a specified level of bioavailability from the usual diet (the factorial approach); and (2) based on associations between intake and physiological or clinical health outcomes using biomarkers that correlate both with intake and a disease or physiological state (for simplicity referred to as "status" throughout this paper) as an intermediary between intake and health (dose-response approach). The methodology [13–16] to systematically review, summarize, and evaluate evidence on vitamin B12 requirements has recently been developed by the EURRECA Network of Excellence [17,18].

VITAMIN B12 REQUIREMENTS—A FACTORIAL APPROACH

Daily vitamin B12 losses are most commonly measured using the whole body counting method (WBC). With this method, subjects are given labeled vitamin B12 and the rate of loss of vitamin B12 (μg/day) is calculated based on the decrease in rate of radioactivity during follow-up. This rate indicates the mean decay in radioactivity per day as a percentage of the radioactivity measured after redistribution of the labeled vitamin B12 through different body compartments (ie, at the time that a constant excretion rate was established). To derive estimates of daily losses, the rate of loss must be combined with estimates of vitamin B12 body stores.

As an approximation to the WBC method, daily vitamin B12 losses can be estimated by determining vitamin B12 excretion in bile, accounting for the partial reabsorption of this excreted vitamin B12 in the small intestine. However, such data are limited, largely inconsistent—with estimates varying from 0.8 to 35 μg/day [13]—and therefore not usable.

To estimate vitamin B12 requirements with the factorial approach, estimates of daily vitamin B12 losses need to be corrected for bioavailability of the vitamin from the usual diet:

- Daily loss = rate of loss (%) × body stores
- Requirement = daily losses/bioavailability [13].

According to the current body of evidence, *the rate of loss* averages 0.13% of total body stores per day (95% confidence interval = 0.10; 0.15, $I^2 = 91.5\%$, $p < 0.0001$) as estimated from a pooled analysis of five studies including a total of 52 subjects. Estimates of mean body stores in apparently healthy people range between 1.1 and 3.9 mg (4 studies) [13]. This means that among apparently healthy individuals, absolute amounts of vitamin B12 lost per day will most likely range between 1.4 and 5.1 μg/day [13].

Absorption of vitamin B12 depends on the ingested dose and food source and varies substantially between individuals. Data from eight studies including 79 subjects, jointly demonstrate that the overall relation between intake of vitamin B12 and the absorbed amount is linear: $\log_e$(the amount absorbed) = 0.7694* $\log_e$(intake) − 0.9614 ($R^2 = 0.78$). Based on this equation and assuming that three meals per day equally contribute to compensating the losses of vitamin B12 (1.4–5.1 μg), each meal should provide between 1.3 and 6.9 μg; in turn, this would maintain body stores between 1.1 and 3.9 mg.

When evaluating the total body of evidence for the factorial approach, it should be noted that the evidence could only be derived from relatively old studies published between 1958 and 1991 and can hardly be updated because this requires invasive methods or the use of isotopes that do not comply with current ethical standards.

VITAMIN B12 REQUIREMENTS— ASSOCIATIONS BETWEEN VITAMIN B12 INTAKE, HEALTH OUTCOMES, AND BIOMARKERS

The dose-response approach for estimating nutrient requirements is based on associations between intake and physiological or clinical health outcomes [19–21]. For vitamin B12, traditionally, maintenance of an adequate hematological status, as measured by stable hemoglobin levels, normal mean cell volume, and normal reticulocyte response, was used as a marker for setting recommendations. Requirements for the preservation of cognitive functioning or bone health or for maintaining vitamin B12 status (as measured by selected biomarkers) may well differ. Estimates for these can be be derived using statistical methods combining datalinks between vitamin B12 intake and status and/or health indicators as recently proposed by van der Voet et al. [22], DerSimonian et al. [23], and Higgins et al. [24]. For this purpose data from well-defined observational and intervention studies (Table 42.1) are integrated, in order to estimate mean intakes needed to achieve serum/plasma serum/plasma vitamin B12

TABLE 42.1 Inclusion Criteria for Exposure and Outcome Measures per Study Design

	Randomized controlled trials	Observational studies
Vitamin B12 intake	• Oral supplements • Fortified foods • Natural food sources	• Validated food frequency questionnaire • Dietary history method • 24-hour recall method for at least 3 days • Food record/diary for at least 3 days
Vitamin B12 status	• Serum/plasma vitamin B12 • Serum/plasma methylmalonic acid • Serum/plasma holotranscobalamin	
Health outcomes	• Cognitive function including incidence of dementia or Alzheimer's disease, global cognition test scores, domain-specific test scores • Bone health parameters including fracture risk, bone mineral density, or bone turnover markers	

concentrations of 150, 200, 258, or 300 pmol/L. These concentrations are generally accepted as cut-off values for clinical vitamin B12 deficiency (150 pmol/L and 200 pmol/L), or suggested as cut-off for subnormal vitamin B12 status (258 pmol/L and 300 pmol/L) [25–27].

Vitamin B12 and Cognitive Function

A recent systematic review on dose-response evidence from two RCTs and 23 prospective cohort studies on the relation of vitamin B12 intake and/or status with cognitive function in adults and elderly people is summarized in Table 42.2 [14]. Clearly the association between *vitamin B12 intake or status* and cognitive function was consistently absent for Alzheimer's disease (3 studies), inconsistent for global cognition scores (3 studies), and limited or inconsistent for specific cognitive domains. More data are needed to support dose-response evidence on sensitive markers of vitamin B12 status (methylmalonic acid (MMA) and holotranscobalamin (holoTC)): four out of five cohort studies report that low vitamin B12 status was significantly associated with incidence of either dementia, Alzheimer's disease, or global cognition. These findings are in line with those presented in the World Alzheimer Report 2014: there is no evidence strong enough at this time to include cognitive outcomes in the setting of vitamin B12 recommendations.

Vitamin B12 and Bone Health

The evidence-base summarizing the relation of vitamin B12 intake and status with indicators of bone health includes dose response evidence from 22 observational studies Table 42.3 [15]. Evidence on associations between intake and fractures and between intake and bone mineral density (BMD) was limited and showed no or inconsistent associations. A meta-analysis on prospective cohort data showed a modest decrease in risk of fractures of 4% per 50 pmol/L increase in serum/plasma vitamin B12 concentrations. Findings for BMD were largely inconsistent. Meta-analyses including cross-sectional studies showed that serum/plasma vitamin B12 was not associated with BMD at femoral neck, lumbar spine or total hip. However, out of nine remaining studies that could not be combined in a meta-analyses, six studies reported that low serum/plasma vitamin B12 was significantly associated with low BMD and three studies observed no association between serum/plasma vitamin B12 and BMD. Overall, the evidence-base is suggestive of an association between serum/plasma vitamin B12 concentrations and fractures. However, data from intervention studies are needed to strengthen the evidence-base before considering vitamin B12 and bone health links for estimating vitamin B12 requirements.

Vitamin B12 Intake and Markers for Vitamin B12 Status

Dose-response evidence on the relation between vitamin B12 intake via diet, enriched foods, or supplements, and different markers of vitamin B12 status (serum/plasma concentrations of vitamin B12, MMA and holoTC) in adults and the elderly is summarized in Table 42.4 [16] in which the heterogeneous dose-response relation between vitamin B12 intake and serum/plasma vitamin B12 concentrations is estimated as: $\log_e(\text{serum/plasma vitamin B12}) = 0.17 \cdot \log_e(\text{vitamin B12 intake}) + 5.47$ [16]. This relation indicates that a doubling of vitamin B12 intake, for example, from 1.5 to 3 µg, increases serum/plasma vitamin B12 concentrations by $2^{0.17}$-fold, which corresponds to 12.5% (95% confidence interval: 11.0%, 14.9%), that is, from 254 to 286 pmol/L. For Methylmalonic acid (MMA) a doubling in vitamin B12 intake resulted in a 7% decrease in MMA concentrations (95% confidence interval: −10%, −4%). For holoTC the number of RCTs was small.

Based on the dose-response relationship between vitamin B12 intake and serum/plasma vitamin B12 concentrations as described above, estimates of the vitamin B12 intakes needed to achieve concentrations of 150, 200, 258 or 300 pmol/L are 0.07, 0.36, 1.6, or 4.0 µg/day, respectively.

TABLE 42.2 Main Findings of a Systematic Literature Review on Vitamin B12 Intake or Status in Relation to Cognitive Function

Outcome	Exposure	Study design (publication year)	N total (range) no of cases	Main results
Dementia	Vitamin B12 intake	1 Cohort (2009)	3634; 352 cases	Risk of dementia does not significantly change with increasing intake
	Serum/plasma vitamin B12	5 Cohorts (1994–2008)	3446 (370–1332); 375 cases	Risk of dementia does not significantly change with increasing serum/plasma vitamin B12[a]
	HoloTC	1 Cohort (2009)	213; 83 cases	No significant continuous association, moderate holoTC is significantly associated with reduced risk of dementia
Alzheimer's disease	Vitamin B12 intake	3 Cohorts (2005–2009)	5254 (579–3634); 431 cases	Risk of Alzheimer's disease does not significantly change with increasing intake. Pooled analysis: RR per μg dietary vitamin B12 = 0.99; 95% CI = 0.99, 1.00, $I^2 = 0\%$, p = 0.92
	Serum/plasma vitamin B12	3 Cohorts (1994–2005)	1596 (370–816); 160 cases	Risk of Alzheimer's disease does not significantly change with increasing serum/plasma vitamin B12
	HoloTC	2 Cohorts (2009–2010)	484; 78 cases	1 study (n = 271) showed a reduced risk of Alzheimer's disease with increasing holoTC, 1 study (n = 213) showed no significant continuous association, but moderate holoTC is significantly associated with a reduced risk of Alzheimer's disease
Global cognition	Vitamin B12 intake	1 RCT, 2 cohorts (2002–2005)	3070 (31–3718)	No significant effect of/association between intake and cognitive scores
	Serum/plasma vitamin B12	10 Cohorts (2002–2009)	4807	8 studies (n = 3752) showed no significant association between serum/plasma vitamin B12 and cognitive scores, 2 studies (n = 1055) showed a positive association[b]
	MMA	3 Cohorts (2007–2009)	2039	Inconsistent findings
	HoloTC	2 Cohorts (2007–2009)	1523	Inconsistent findings
Domain-specific cognition	Vitamin B12 intake	1 RCT, 2 cohorts (1997–2006)	554 (111–321)	Inconsistent findings for executive function (3 studies) and memory (3 studies). No significant association between intake and scores for language (1 study) or speed (1 study)
	Serum/plasma vitamin B12	5 Cohorts (1997–2006)	3082	4 studies (n = 1404) showed no significant association between serum/plasma vitamin B12 and memory, 1 study (n = 1678) showed an increased risk of memory deficit at lower serum/plasma vitamin B12[c]. Inconsistent findings for executive function (3 studies). No association between serum/plasma vitamin B12 and scores for language (1 study) or speed (1 study)

[a] Pooled analysis of 4 studies (n = 2630): RR per 50 pmol/L serum/plasma vitamin B12 = 1.00; 95% CI = 0.98, 1.02, $I^2 = 9.1\%$, p = 0.35.
[b] Pooled analysis of 4 studies (n = 1579): β for annual change in global cognition z-score per 50 pmol/L serum/plasma vitamin B12 additional to change as a result of aging = 0.00, 95% CI = −0.00, 0.01, $I^2 = 42.6\%$, p = 0.16.
[c] Pooled analysis of 4 studies (n = 3460): β for associations between memory z-scores follow-up per 50 pmol/L serum/plasma vitamin B12 concentrations at baseline = 0.01, 95% CI: −0.01, 0.03, $I^2 = 0.0\%$, p = 0.99.
RR, Relative risk; 95% CI, 95% confidence interval; HoloTC, holotranscobalamin; RCT, Randomized Controlled Trial; MMA, methylmalonic acid.

TABLE 42.3 Main Findings of a Systematic Literature Review on Vitamin B12 Intake or Status in Relation to Indicators of Bone Health

Outcome	Exposure	Study design (publication year)	N total (range) no of cases sex	Main results
Fractures	Vitamin B12 intake	1 Cohort (2008)	1800 360 cases Women	Risk for fractures does not change with increasing quartiles of intake
	Serum/plasma vitamin B12	4 Cohorts (2005–2008)	7539 (702–4761) 458 cases Mixed	Risk of fractures significantly changes with increasing serum/plasma vitamin B12 Pooled analysis: RR per 50 pmol/L serum/plasma vitamin B12 = 0.96, 95% CI = 0.92, 1.00, I^2 = 0.0%, p = 0.84
BMD	Vitamin B12 intake	4 Cross-sectional (2004–2008)	10,114 (1241–5304) Women (3 studies) Mixed (1 study)	2 studies showed significant correlations between intake and BMD in a specific subpopulation (MTHFR CC/TT polymorphism, n = 1700) or at a specific time point (n = 1241), 2 studies (n = 7173) showed no significant association
	Serum/plasma vitamin B12	12 cross-sectional, 2 cohorts (2003–2012)	11,719 (83–5329) Women (9 studies) Men (1 study) Mixed (4 studies)	7 studies (n = 5033) showed significant associations between low serum/plasma vitamin B12 and BMD at least at one site[a], 7 studies (n = 6686) showed no significant association
	MMA	1 cross-sectional (2005)	1550 Mixed	Increasing serum MMA significantly associated with lower BMD

[a]Pooled analyses of cross-sectional studies showed no significant association between serum/plasma vitamin B12 and BMD at different sites in women (Femoral neck (3 studies), n = 674): β per 50 pmol/L = 0.00, 95% CI = −0.13, 0.14, I^2 = 0.0%, p = 0.40; lumbar spine (4 studies, n = 862): β per 50 pmol/L = −2.25, 95% CI = −7.98, 3.49, I^2 = 99.5%, p < 0.0001; total hip (4 studies, n = 1037): β per 50 pmol/L = −2.23, 95% CI = −10.38, 5.92, I^2 = 97.7, p = 0.0001).
RR, Relative risk; 95% CI, 95% confidence interval; BMD, bone mineral density; MMA, methylmalonic acid.

TABLE 42.4 Main Findings of a Systematic Literature Review on the Relation between Vitamin B12 Intake and Markers of Vitamin B12 Status

Intake (μg/d)[a]	Status marker	Study design (publication year)	N total (range)	Main results
2.1–1000	Serum/plasma vitamin B12	37 RCTs (1985–2009)	3398 (21–217)	12.5% increase in status for every doubling of intake $β^b$ = 0.17 (95% CI: 0.15, 0.20; I2 = 88%)
0.7–10.5	Serum/plasma vitamin B12	19 observational (cross-sectional data) (1984–2009)	12,570 (64–2156)	7% increase in status for every doubling of intake $β^{b,c}$ = 0.10 (95% CI: 0.06, 0.14; I2 = 94%)
2.5–987	Serum/plasma MMA	9 RCT's (2001–2009)	850 (38–178)	7% decrease in status for every doubling of intake $β^b$ = 20.11 (95% CI: 20.15, 20.06; I2 = 76%)
9.6–987	Serum/plasma Holo-TC	3 RCTs (2006–2008)	350 (103–142)	Inadequate data for meta-analysis. All three studies showed that vitamin B12 supplementation increased holoTC

[a]Intake via supplements, diet or enriched foods.
[b]The regression coefficient represents the difference in the log_e-transformed predicted value of serum or plasma vitamin B-12 status for each one-unit difference in the log_e-transformed value in vitamin B-12 intake.
[c]No regression coefficients from observational studies were included in the dose-response meta-analysis as these are subject to attenuation due to measurements errors [20–22].
RCT, randomized controlled trial; 95% CI, 95% confidence interval; MMA, methylmalonic acid; HoloTC, holotranscobalamin.

IV. HEALTH EFFECTS OF DIETARY COMPOUNDS AND DIETARY INTERVENTIONS

Statistical modeling of intake and status data from observational studies provides estimates of the estimated requirements for vitamin B12 of 1.9, 2.6, 3.5, or 4.1 μg/day, respectively.

FROM ESTIMATES OF REQUIREMENTS TO VITAMIN B12 RECOMMENDATIONS

The distribution of vitamin B12 requirements for a specific health outcome is basic to the setting of recommendations. Using the common assumption of a 10% coefficient of variation in nutrient requirements it can be derived as the average requirement (ANR) ± 2 × its SD. Such intake suffices for 97.5% of the population to reach the health criterion. The classical criterion—maintenance of an adequate hematological status—is debatable for several reasons: (1) the use of single-study evidence, for example, the required intake needed to maintain hematological status, as proposed by the Institute of Medicine [28], was based on one single study including 7 subjects that were not able to absorb vitamin B12 from foods due to pernicious anemia [29]. (2) Although the biological role of vitamin B12 in blood cell formation is well-defined [30–32] it has been estimated that 19–28% of patients with pernicious anemia do not have anemia, and 17–33% have a normal mean cell volume [33,34]. (3) Potentially irreversible neurologic disorders due to vitamin B12 deficiency may occur in the absence of anemia or an elevated mean cell volume [3,35,36]. It is even suggested that the occurrence of neurological complications is inversely correlated with the degree of anemia; patients who are less anemic show more prominent neurological complications and vice versa [37,38]. An overview of vitamin B12 requirements and recommendations based on more recent approaches is presented in Table 42.5.

According to the factorial approach vitamin B12 recommendations for apparently healthy people range between 3.8 and 20.7 μg/day depending on the underlying assumptions. Strictly speaking this range—based on stores from patients suffering from malabsorption with normal hematological status and low serum vitamin B12—is reflecting minimum requirements for intake to prevent clinical symptoms of severe deficiency [41]. What stores suffice for the general population to maintain health remains unclear.

So far, the evidence-base on the relation between vitamin B12 intake or status and the long-term health outcomes cognitive function and bone health is too limited for being used to estimate vitamin B12 requirements and subsequently set recommendations. Numerous factors may contribute to this where in the cognitive research domain the variability in cognitive outcomes is substantial, populations are diverse, determinants may interact, and nutritional impacts are small [42–47], and where the inclusion of novel technologies (brain imaging) and biomarkers (holoTC and MMA) [48–50] may well advance the research field. Such advances are also envisaged in studying bone health [51], when involving stronger study design and markers of bone metabolism [52–55].

The systematic review on intake and status of vitamin B12 provided estimates of the dose-response relation between vitamin B12 intake and serum/plasma vitamin B12 concentrations. Based on these dose-response relations we were able to estimate mean intakes needed to achieve serum/plasma concentrations of 150, 200, 258, and 300 pmol/L. Recommendations based on these ANRs are 0.08, 0.44, 2.0, and 4.7 μg/day, respectively (Table 42.5). Recommended intakes of vitamin B12 for the prevention of subclinical deficiency (cut off of 300 pmol/L) based on RCT's or derived with the stochastic model are similar to intakes needed for the maintenance of normal body stores: 2.3, 3.2, 4.2, and 5.0 μg/day, respectively (Table 42.5). These results are also supported by observations from five studies that showed that markers for vitamin B12 status (serum/plasma vitamin B12, MMA, and holoTC) leveled off at daily intakes between 4 and 10 μg/day [56–60].

CONCLUDING REMARKS

This chapter illustrates that evidence underlying current ANRs of vitamin B12 (maintenance of minimal stores or hematological status) is old and has large uncertainties, whereas evidence on long-term health outcomes in relation to vitamin B12 intake or status is yet very limited and not suitable for setting recommendations. The relation between vitamin B12 intake and markers of vitamin B12 status seem the best alternative, and stochastic methods may provide a good opportunity to integrate dose-response data from different study types.

Intakes needed to maintain normal body stores or to prevent subclinical deficiency may be twice as high as current recommendations, but a missing piece of information is the amount of vitamin B12 intake associated with observable health benefits beyond the prevention of signs and symptoms of deficiency. Further intervention research on biomarkers intermediating B12 intake and the health effects, as well as epidemiological research on these biomarkers in relation to health outcomes is indicated. To improve estimates of requirements, advanced models for integrating the evidence are needed with assumptions consistent with the strengths and limitations of epidemiological and physiological studies.

TABLE 42.5 Overview of Vitamin B12 Requirements and Recommendations Based on Different Health Outcomes and Derived with Different Approaches

Health outcome	Approach	Evidence-base (number of studies and subjects)	ANR	Recommendation type	Recommendation (µg/day)
CURRENT RECOMMENDATIONS					
Minimal body stores (0.5 µg)[a]	Factorial	7 experimental, $n = 3-16$	2.0	ANR + 20% or 40%	2.4 or 2.8
Hematological status [a]	Dose-response	1 experimental, $n = 7$	2.0	ANR + 20%	2.4
RECOMMENDATIONS BASED ON REVIEWS DESCRIBED IN THIS PAPER					
Normal body stores (1.1–3.9 µg)	Factorial	13 experimental, $n = 135$	n.a.	Adequate intake[b]	3.8–20.7
Cognitive performance	Dose-response	2 RCT's, 19 observational, $n = 13915$	Insufficient data	–	–
Bone health	Dose-response	22 observational, $n = 29372$	Insufficient data	–	–
Serum/plasma vitamin B12 cut-off 150 pmol/L	Dose-response[c]	37 RCT's, $n = 3398$	0.07	ANR + 20%	0.08
cut-off 200 pmol/L			0.36	ANR + 20%	0.44
cut-off 258 pmol/L			1.6	ANR + 20%	2.0
cut-off 300 pmol/L			4.0	ANR + 20%	4.0
Serum/plasma vitamin B12 cut-off 150 pmol/L	Dose-response[d]	19 observational, $n = 12570$[e]	1.9	ANR + 20%	2.3
cut-off 200 pmol/L			2.6	ANR + 20%	3.2
cut-off 258 pmol/L			3.5	ANR + 20%	4.2
cut-off 300 pmol/L			4.1	ANR + 20%	5.0

[a] Adapted from Doets et al. [12].
[b] Adequate intake is used when the average nutrient requirement and consequently the recommended intake cannot be estimated. The Adequate Intake is based on observed or experimentally determined approximations of nutrient intake by a group (or groups) of healthy people.
[c] Model A as described by van der Voet et al. [22].
[d] Model C as described by van der Voet et al. [22].
[e] Additional data on status measurement errors [21] and variation in usual vitamin B12 intake [39,40] were used to estimate vitamin B12 requirements with this model.
ANR, Average Nutrient Requirement; n.a., not applicable.

SUMMARY

- Evidence underlying current ANRs of vitamin B12 (maintenance of minimal stores or hematological status) is old and has large uncertainties.
- Intakes needed to maintain normal body stores or to prevent subclinical deficiency may be twice as high as current recommendations.
- A missing piece of information is the amount of vitamin B12 intake associated with observable health benefits beyond the prevention of signs and symptoms of deficiency.
- To improve estimates of requirements, advanced models for integrating the evidence are needed with assumptions consistent with the strengths and limitations of epidemiological and physiological studies.

Acknowledgments

This research was undertaken as an activity of the European Micronutrient Recommendations Aligned (EURRECA) Network of Excellence (www.eurreca.org), funded by the European Commission Contract Number FP6 036196-2 (FOOD).

None of the authors has a conflict of interest. The following authors contributed to this chapter and we would like to express our gratitude for their contributions; Carla Dullemeijer, Janneke P van Wijngaarden, Olga W Souverein, Rosalie AM Dhonukshe-Rutten, Adriënne EJM Cavelaars, Hilko van der Voet and Pieter van't Veer.

References

[1] Carmel R. Current concepts in cobalamin deficiency. Ann Rev Med 2000;51:357–75.
[2] Reynolds EH. Neurological aspects of folate and vitamin-B12 metabolism. Clin Haematol 1976;5(3):661–96.
[3] Stabler SP, Allen RH, Savage DG, Lindenbaum J. Clinical spectrum and diagnosis of cobalamin deficiency. Blood 1990;76(5):871–81.

[4] Allen LH. How common is vitamin B-12 deficiency? Am J Clin Nutr 2009;89(2):693S–6S.

[5] de Benoist B. Conclusions of a WHO technical consultation on folate and vitamin B12 deficiencies. Food Nutr Bull 2008;29 (Suppl. 2):S238–44.

[6] Baik HW, Russell RM. Vitamin B12 deficiency in the elderly. Ann Rev Nutr 1999;19:357–77.

[7] Carmel R. Discussion: causes of vitamin B-12 and folate deficiencies. Food Nutr Bull 2008;29(2):S35–7.

[8] Andres E, Vidal-Alaball J, Federici L, Loukili NH, Zimmer J, Kaltenbach G. Clinical aspects of cobalamin deficiency in elderly patients. Epidemiology, causes, clinical manifestations, and treatment with special focus on oral cobalamin therapy. Eur J Intern Med 2007;18(6):456–62.

[9] McLean E, de Benoist B, Allen LH. Review of the magnitude of folate and vitamin B-12 deficiencies worldwide. Food Nutr Bull 2008;29(2):S38–51.

[10] Doets EL, de Wit LS, Dhonukshe-Rutten RAM, Cavelaars AEJM, Raats MM, Timotijevic L, et al. Current micronutrient recommendations in Europe: towards understanding their differences and similarities. Eur J Nutr 2008;47(Suppl. 1):17–40.

[11] Yates AA. Using criteria to establish nutrient intake values (NIVs). Food Nutr Bull 2007;28(1):S38–50.

[12] Doets EL, Cavelaars A, Dhonukshe-Rutten RAM, van't Veer P, de Groot L. Explaining the variability in recommended intakes of folate, vitamin B-12, iron and zinc for adults and elderly people. Public Health Nutr 2012;15(5):906–15.

[13] Doets EL, in't Veld PH, Szczecinska A, Dhonukshe-Rutten RAM, Cavelaars AEJM, van't Veer P, et al. Systematic review on daily vitamin B12 losses and bioavailability for deriving recommendations on vitamin B12 intake with the factorial approach. Ann Nutr Metab 2012;62(4):311–22.

[14] Doets EL, van Wijngaarden JP, Szczecińska A, Dullemeijer C, Souverein OW, Dhonukshe-Rutten RAM, et al. Vitamin B12 intake and status and cognitive function in elderly people. Epidemiol Rev 2013;35(1):2–21.

[15] van Wijngaarden JP, Doets EL, Szczecińska A, Souverein OW, Duffy ME, Dullemeijer C, et al. Vitamin B12, folate, homocysteine and bone health in adults and elderly people: a systematic review with meta-analyses. J Nutr Metab 2013;2013. Available from: http://dx.doi.org/10.1155/2013/486186.

[16] Dullemeijer C, Souverein OW, Doets EL, van der Voet H, van Wijngaarden JP, de Boer WJ, et al. Systematic review with dose-response meta-analyses between vitamin B12 intake and EURRECA's prioritized biomarkers of vitamin B12 including randomized controlled trials and observational studies in adults and elderly. Am J Clin Nutr 2013;97(2):390–402.

[17] Matthys C, van TVP, de Groot L, Hooper L, Cavelaars AE, Collings R, et al. EURRECAs approach for estimating micronutrient requirements. Int J Vitam Nutr Res 2011;81(4):256–63.

[18] Higgins JPT, Green S, editors. Cochrane handbook for systematic reviews of interventions version 5.1.0 [updated March 2011]. The Cochrane Collaboration, 2011. Available from: <http://www.cochrane-handbook.org>.

[19] Souverein OW, Dullemeijer C, van't Veer P, van der Voet H. Transformations of summary statistics as input in meta-analysis for linear dose-response models on a logarithmic scale: a methodology developed within EURRECA. BMC Med Res Methodol 2012;12:57. Available from: http://dx.doi.org/10.1186/1471-2288-12-57.

[20] Kipnis V, Subar AF, Midthune D, Freedman LS, Ballard-Barbash R, Troiano RP, et al. Structure of dietary measurement error: results of the OPEN biomarker study. Am J Epidemiol 2003;158(1):14–21.

[21] McKinley MC, Strain JJ, McPartlin J, Scott JM, McNulty H. Plasma homocysteine is not subject to seasonal variation. Clin Chem 2001;47(8):1430–6.

[22] van der Voet H, de Boer WJ, Souverein OW, Doets EL, van't Veer P. A statistical method to base nutrient recommendations on meta-analysis of intake and health-related status biomarkers. PLoS One 2014;9(3):e93171. Available from: http://dx.doi.org/10.1371/journal.pone.0093171. eCollection 2014.

[23] DerSimonian R, Laird N. Meta-analysis in clinical trials. Control Clin Trials 1986;7(3):177–88.

[24] Higgins JPT, Thompson SG. Quantifying heterogeneity in a meta-analysis. Stat Med 2002;21(11):1539–58.

[25] Carmel R, Sarrai M. Diagnosis and management of clinical and subclinical cobalamin deficiency: advances and controversies. Curr Hematol Rep 2006;5(1):23–33.

[26] Smith DA, Refsum H. Do we need to reconsider the desirable blood level of vitamin B12? J Intern Med 2012;271(2):179–82.

[27] Yetley EA, Pfeiffer CM, Phinney KW, Bailey RL, Blackmore S, Bock JL, et al. Biomarkers of vitamin B-12 status in NHANES: a roundtable summary. Am J Clin Nutr 2011;94(1):313S–21S.

[28] Food and Nutrition Board, Institute of Medicine (FNB/IOM). Dietary reference intakes for thiamin, riboflavin, niacin, vitamin B6, folate, vitamin B12, pantothenic acid, biotin, and choline. Washington (DC): National Academy Press; 1998.

[29] Darby WJ, Bridgforth EB, Brocquy JL, Clark SL, Dutradeoliveira J, Kevany J, et al. Vitamin-B12 requirement of adult man. Am J Med 1958;25(5):726–32.

[30] Chanarin I. The megaloblastic anaemias. Blackwell Scientific; 1969.

[31] Myhre E. Studies on megaloblasts in vitro. II. Maturation of nucleated red cells in pernicious anemia before and during treatment with vitamin B12. Scand J Clin Lab Invest 1964;16:320–31.

[32] Samson D, Halliday D, Chanarin I. Reversal of ineffective erythropoiesis in pernicious anaemia following vitamin B12 therapy. Br J Haematol 1977;35(2):217–24.

[33] Carmel R. Pernicious anemia. The expected findings of very low serum cobalamin levels, anemia, and macrocytosis are often lacking. Arch Intern Med 1988;148(8):1712–14.

[34] Savage DG, Lindenbaum J, Stabler SP, Allen RH. Sensitivity of serum methylmalonic acid and total homocysteine determinations for diagnosing cobalamin and folate deficiencies. Am J Med 1994;96(3):239–46.

[35] Carmel R, Lau KH, Baylink DJ, Saxena S, Singer FR. Cobalamin and osteoblast-specific proteins. N Engl J Med 1988;319(2):70–5.

[36] Lindenbaum J, Savage DG, Stabler SP, Allen RH. Diagnosis of cobalamin deficiency: II. Relative sensitivities of serum cobalamin, methylmalonic acid, and total homocysteine concentrations. Am J Hematol 1990;34(2):99–107.

[37] Healton EB, Savage DG, Brust JC, Garrett TJ, Lindenbaum J. Neurologic aspects of cobalamin deficiency. Medicine 1991;70(4):229–45.

[38] Savage DG, Lindenbaum J. Neurological complications of acquired cobalamin deficiency: clinical aspects. Baillieres Clin Haematol 1995;8(3):657–78.

[39] van Rossum CTM, Fransen HP, Verkaik-Kloosterman J, Buurma-Rethans EJM, Ocké MC. Dutch National Food Consumption Survey 2007–2010: diet of children and adults aged 7 to 69 years. Report 350050006, RIVM, 2011. Available from: <http://www.rivm.nl/bibliotheek/rapporten/350050006.pdf>.

[40] Roodenburg AJ, van Ballegooijen AJ, Dötsch-Klerk M, van der Voet H, Seidell JC. Modelling of usual nutrient intakes: potential impact of the choices programme on nutrient intakes in young dutch adults. PLoS One 2013;8(8):e72378 doi:10.1371/journal.pone.0072378. eCollection 2013.

[41] Anderson BB. Investigations into the Euglena methods of assay of vitamin B12; the results obtained on human serum and liver

using a improved method of assay. Ph.D. thesis, University of London, 1965.

[42] Aisen PS, Andrieu S, Sampaio C, Carrillo M, Khachaturian ZS, Dubois B, et al. Report of the task force on designing clinical trials in early (predementia) AD. Neurology 2011;76(3):280–6.

[43] De Jager CA, Kovatcheva A. Summary and discussion: methodologies to assess long-term effects of nutrition on brain function. Nutr Rev 2010;68:S53–8.

[44] McCracken C. Challenges of long-term nutrition intervention studies on cognition: discordance between observational and intervention studies of vitamin B12 and cognition. Nutr Rev 2010;68:S11–15.

[45] Smith AD, Refsum H, Smith AD, Refsum H. Vitamin B-12 and cognition in the elderly. Am J Clin Nutr 2009;89(2):707S–11S.

[46] Vellas B, Andrieu S, Sampaio C, Coley N, Wilcock G, European Task Force G. Endpoints for trials in Alzheimer's disease: a European task force consensus. Lancet Neurol 2008;7(5):436–50.

[47] Vellas B, Andrieu S, Sampaio C, Wilcock G, European Task Force G. Disease-modifying trials in Alzheimer's disease: a European task force consensus. Lancet Neurol 2007;6(1):56–62.

[48] de Lau LML, Smith AD, Refsum H, Johnston C, Breteler MMB. Plasma vitamin B12 status and cerebral white-matter lesions. J Neurol Neurosurg Psychiatry 2009;80(2):149–57.

[49] Vogiatzoglou A, Refsum H, Johnston C, Smith SM, Bradley KM, de Jager C, et al. Vitamin B12 status and rate of brain volume loss in community-dwelling elderly. Neurology 2008;71(11):826–32.

[50] Smith AD, Smith SM, de Jager CA, Whitbread P, Johnston C, Agacinski G, et al. Homocysteine-lowering by B vitamins slows the rate of accelerated brain atrophy in mild cognitive impairment: a randomized controlled trial. Plos One 2010;5(9):e12244.

[51] Herrmann M, Umanskaya N, Traber L, Schmidt-Gayk H, Menke W, Lanzer G, et al. The effect of B-vitamins on biochemical bone turnover markers and bone mineral density in osteoporotic patients: a 1-year double blind placebo controlled trial. Clinl Chem Lab Med 2007;45(12):1785–92.

[52] Herrmann M, Peter Schmidt J, Umanskaya N, Wagner A, Taban-Shomal O, Widmann T, et al. The role of hyperhomocysteinemia as well as folate, vitamin B(6) and B(12) deficiencies in osteoporosis: a systematic review. Clin Chem Lab Med 2007;45(12):1621–32.

[53] Dhonukshe-Rutten RAM, Pluijm SMF, de Groot LCPGM, Lips P, Smit JH, van Staveren WA. Homocysteine and vitamin B12 status relate to bone turnover markers, broadband ultrasound attenuation, and fractures in healthy elderly people. J Bone Miner Res 2005;20(6):921–9.

[54] Green TJ, McMahon JA, Skeaff CM, Williams SM, Whiting SJ. Lowering homocysteine with B vitamins has no effect on biomarkers of bone turnover in older persons: a 2-y randomized controlled trial. Am J Clin Nutr 2007;85(2):460–4.

[55] van Wijngaarden JP, Dhonukshe-Rutten RAM, van Schoor NM, van der Velde N, Swart KMA, Enneman AW, et al. Rationale and design of the B-PROOF study, a randomized controlled trial on the effect of supplemental intake of vitamin B12 and folic acid on fracture incidence. BMC Geriatr 2011;11:80. Available from: http://dx.doi.org/10.1186/1471-2318-11-80.

[56] Bor MV, Lydeking-Olsen E, Moller J, Nexo E. A daily intake of approximately 6 microg vitamin B-12 appears to saturate all the vitamin B-12-related variables in Danish postmenopausal women. Am J Clin Nutr 2006;83(1):52–8.

[57] Bor MV, von Castel-Roberts KM, Kauwell GP, Stabler SP, Allen RH, Maneval DR, et al. Daily intake of 4 to 7 microg dietary vitamin B-12 is associated with steady concentrations of vitamin B-12-related biomarkers in a healthy young population. Am J Clin Nutr 2010;91(3):571–7.

[58] Kwan LL, Bermudez OI, Tucker KL. Low vitamin B-12 intake and status are more prevalent in Hispanic older adults of Caribbean origin than in neighborhood-matched non-Hispanic whites. J Nutr 2002;132(7):2059–64.

[59] Tucker KL, Rich S, Rosenberg I, Jacques P, Dallal G, Wilson PW, et al. Plasma vitamin B-12 concentrations relate to intake source in the Framingham Offspring study. Am J Clin Nutr 2000;71(2):514–22.

[60] Vogiatzoglou A, Smith AD, Nurk E, Berstad P, Drevon CA, Ueland PM, et al. Dietary sources of vitamin B-12 and their association with plasma vitamin B-12 concentrations in the general population: the Hordaland Homocysteine study. Am J Clin Nutr 2009;89(4):1078–87.

CHAPTER 43

Vitamin C, Antioxidant Status, and Cardiovascular Aging

Ammar W. Ashor[1,2], Mario Siervo[1] and John C. Mathers[1]

[1]Human Nutrition Research Centre, Institute of Cellular Medicine, Newcastle University, Newcastle on Tyne, UK
[2]College of Medicine, University of Al-Mustansiriyah, Baghdad, Iraq

KEY FACTS

- The aging phenotype is caused by the accumulation of damage to the cell's macromolecules which limit, or derange, function throughout the body.
- Cardiovascular disease (CVD) is the most common cause of death among the elderly in Western countries.
- Damage to the endothelium increases CVD risk.
- Oxidative stress and inflammation damage the endothelium and, together with senescent cells, contribute to reduced endothelial function with age.
- Nitric oxide (NO) is secreted from endothelial cells and regulates several aspects of cardiovascular function including vascular tone, platelet aggregation, thrombosis, monocyte adhesion and smooth muscle proliferation.
- Vitamin C is important for vascular health through its roles as an antioxidant nutrient, as a co-factor of collagen synthesis and in augmenting NO bioavailability.
- Older people with lower vitamin C status and at higher CVD risk may benefit from additional vitamin C intake.

Dictionary of Terms

- *Aging*: The processes involved from conception to death; an evolution of metabolic, endocrinological, and physiological change that occurs in the individual from inception through birth, childhood, adolescence, adulthood, and senescence.
- *Antioxidants*: Agents that prevent or inhibit oxidation reactions; in particular they are agents that prevent the oxidation of unsaturated fatty acids. Vitamins A, E, and C act as antioxidants. Some food additives prolong product freshness by acting as antioxidants.
- *Atherosclerosis*: The process in which fatty and fibrous deposits cause thickening and hardening of the arterial walls.
- *Cardiovascular disease*: A disease of the heart or circulation. This broad term encompasses coronary heart disease, peripheral vascular disease, and stroke.
- *Coronary arteries*: The arteries supplying blood to the muscle of the heart.
- *Coronary heart disease (CHD) (or ischemic heart disease)*: Heart disease resulting from the build-up of fatty deposits in the lining of the coronary arteries. It may cause angina, a heart attack, or sudden death.
- *Cytokines*: Small, hormone-like proteins released by leukocytes, endothelial cells, and other cells to promote an inflammatory immune response to an injury.
- *Dyslipidemia*: An abnormal concentration in the blood of one or more lipids, such as an elevated low-density lipoprotein (LDL) cholesterol level or a depressed high-density lipoprotein (HDL) cholesterol level.
- *Endothelin*: A substance produced by the body that plays an important role in regulating blood flow.
- *Endothelium*: The membrane lining various vessels and cavities of the body, including the heart and blood vessels. It consists of a fibrous layer covered

- *with thin flat cells, which render the surface perfectly smooth and secrete the fluid for its lubrication.*
- *Epigenetic*: Describes a factor or mechanism that changes the expression of a gene or genes without changing their DNA sequence. In more general terms, an epigenetic factor is something that changes the phenotype without changing the genotype.
- *Free radical*: Molecules that contain an unpaired electron and are therefore highly reactive.
- *Heterogeneous (heterogeneity)*: Denotes dissimilarity. This term can be used to denote the discrepant results obtained within or between epidemiological studies or randomized trials that can be either statistical (meaning that studies used different statistical methods) or clinical (meaning that studies evaluated different types of subjects, treatments or outcomes).
- *Homogeneous (homogeneity)*: Denotes similarity. For example, if results are similar and consistent from one study to another, then the results are said to be homogeneous.
- *Inflammation*: The reaction of the body to any injury, which may be the result of trauma, infection or chemicals. In response, local blood vessels dilate, increasing blood flow to the injured site and white blood cells invade the affected tissue engulfing bacteria or other foreign bodies.
- *Interleukins*: Molecules made by leukocytes that are involved in signaling between cells of the immune system.
- *Leukotrienes*: A type of eicosanoid. These are a group of hormones that are derived from arachidonic acid and have a role in allergic or inflammatory reactions in the body.
- *Meta-analysis*: A discipline that reviews critically and combines statistically the results of previous research in an attempt to summarize the totality of the evidence relating to a particular medical issue.
- *Mutation*: A heritable change in genetic material (ie, a change which can potentially be passed from parent to child). This change may occur in a gene or in a chromosome, and may take the form of a chemical rearrangement, or a partial loss or gain of genetic material.
- *Oxidation*: A chemical reaction that involves the loss of electrons; it usually, but not always, involves direct participation of oxygen.
- *Oxidative stress*: A condition in which the production of oxidants and free radicals exceeds the body's ability to inactivate them.
- *Peroxynitrite*: A nitrogen reactive species (ONOOH) formed by the reaction between nitric oxide and superoxide under inflammatory conditions.
- *Reactive oxygen species (ROS)*: A collective term that includes free radicals of oxygen and nonradical derivatives of oxygen, such as hydrogen peroxide and singlet oxygen.
- *Single gene polymorphism (SNP)*: Changes in genetic structure between individuals differ by a single nucleotide.
- *Vasoconstriction*: Narrowing of the blood vessels resulting from contracting of the muscular wall of the vessels.
- *Vasodilation*: Widening of blood vessels resulting from relaxation of the muscular wall of the vessels.

INTRODUCTION

Overview of the Biology of Aging

The aging process is characterized by a progressive decline of cellular integrity and function resulting from the structural modification of macromolecules including formation of oxidized lipid species, advanced glycated products, nitrosylated proteins, and DNA mutations and epimutations. The accumulation of modified molecules and their incorporation into cellular components are responsible for the structural and functional deterioration of tissues and organs with time.

Whilst the complexity of the biological mechanisms contributing to the aging process is still poorly understood, a comprehensive summary of some of these mechanisms has been proposed recently by Lopez-Otin et al. [1]. These authors proposed the following set of hallmarks of aging, that is, genomic instability, telomere attrition, epigenetic alterations, loss of proteostasis, deregulated nutrient-sensing, mitochondrial dysfunction, cellular senescence, stem cell exhaustion, and altered intercellular communication, which are seen across diverse species [1]. Many factors contribute to age-related molecular damage but it seems likely that much damage is due to three common stressors, namely, oxidative stress/redox changes, inflammation, and metabolic stress [2]. The accumulation of molecular damage in stem cells which leads to cell death, senescence, or loss of regenerative function is likely to have a particularly profound effect on aging because of the role of stem cells in maintaining tissue integrity. The cumulative load of (macro)molecular damage, together with alterations in tissue cellularity, causes diminished functional capacity and increases the risk of age-related frailty, disability, and disease, including cardiovascular disease (CVD). In humans, factors such as nutrition which modulate the aging trajectory and the risk of age-related diseases must do so by affecting the acquisition of molecular damage with age or by altering the body's ability to defend itself against that damage [2]. For example, nutrition influences DNA repair—a key cellular defense mechanism which ensures the integrity of the genome [3]. Although the mechanisms through which dietary factors affect DNA repair systems are poorly understood, our recent studies suggest that this may include altered expression of the genes encoding key repair proteins such as the base excision repair (BER) enzyme apurinic/apyrimidinic endonuclease 1 (APE1) and that this may involve epigenetic mechanisms [4].

Aging and Cardiovascular Disease Risk

Cardiovascular diseases are the most common cause of death among elderly populations in the Western World [5]. Age-specific mortality rates from heart disease and stroke increase exponentially with age and account for more than 40% of all deaths among individuals aged 65–74 years and almost 60% at age 85 years and older [6]. In the UK, although death rates from CVD have been falling for four decades, ischemic heart disease is ranked number one for years of life lost due to premature mortality (YLLs). Importantly, lifestyle factors, including smoking, poor diet, and lack of physical activity, are the major causes of morbidity measured by disability-adjusted life years (DALYs) [7].

As noted above, aging is associated with complex structural and functional changes in all tissues including the vascular system and these changes increase CVD risk independent of other risk factors such as hypertension, diabetes, or hypercholesterolemia [8]. These functional changes include widespread endothelial dysfunction, dilation of the central arteries and increased arterial stiffness [9,10]. Development of strategies to attenuate aging of the vascular system could make a substantial contribution to lowering CVD risk and improving the quality of life of older people [11].

The endothelium is a monolayer of cells separating the vascular lumen from the rest of the blood vessel. In totality, the human endothelium weighs approximately 1.5 kg and has a surface area of >700 m² making it one of the major organs of the human body [12]. In addition to its role as a physical barrier between the circulating blood and the underlying vascular tissue, the endothelium has vital paracrine, endocrine, and autocrine functions [13]. Arterial endothelial dysfunction refers to functional changes in the normal endothelial phenotype of the arteries that may contribute to the development, and clinical expression, of atherosclerosis and other vascular disorders. These alterations include a shift toward a vasoconstrictor, procoagulant, proliferative, and pro-inflammatory state [14].

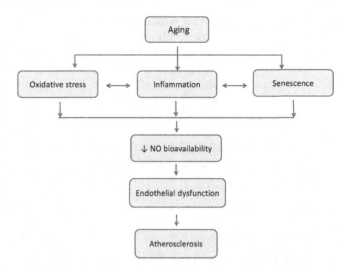

FIGURE 43.1 Mechanisms involved in the development of vascular aging.

Therefore, factors which influence endothelial integrity and endocrine function will be important targets for interventions to reduce age-related CVD risk (Fig. 43.1). Nitric oxide (NO) is the key molecule secreted from endothelial cells which regulates vascular tone, platelet aggregation, thrombosis, monocyte adhesion, and smooth muscle proliferation and reduced NO bioavailability is a hallmark of endothelial dysfunction [15].

FACTORS THAT IMPAIR ENDOTHELIAL FUNCTION WITH AGING

Oxidative Stress

The accumulation of endogenous oxygen radicals and the consequent oxidative modification of cellular macromolecules (lipids, proteins and nucleic acids) have been suggested to contribute to aging in all organisms [16]. Increased production of free radicals, secondary to mitochondrial dysfunction, causes oxidative damage to cells including vascular cells [17].

Excessive production of reactive oxygen species (ROS) leads to NO inadequacy and impaired endothelial function [15]. The ROS superoxide, in excessive amounts, reacts rapidly with NO leading to its conversion into a highly reactive intermediate, peroxynitrite (ONOO$^-$), which reduces substantially the biological half-life of NO [18]. ONOO$^-$ damages lipids, proteins, and nucleic acids [19]. In addition, ONOO$^-$ may deteriorate endothelial function by impairing the production of the vasodilator prostacyclin (PGI$_2$) through tyrosine nitration of the enzyme prostacyclin synthase [18]. Further reduction in the bioavailability of NO occurs secondary to ONOO$^-$-induced inactivation of tetrahydrobiopterin (BH4) which is an essential cofactor for endothelial nitric oxide synthase (eNOS) function. In the absence of this coenzyme, eNOS reduces molecular oxygen rather than L-arginine, resulting in the production of superoxide rather than NO, a phenomenon known as "NOS uncoupling" [20]. Finally, oxidative stress may enhance the production of asymmetric dimethylarginine (ADMA), an endogenous inhibitor of eNOS [21], which causes endothelial dysfunction in normal healthy volunteers [22].

Inflammation

Chronic inflammation is a driver of aging and contributes to the pathology of many age-related diseases including atherosclerosis [23]. Observational and experimental studies have demonstrated the importance of inflammation as a determinant of an unhealthy aging phenotype. For example, the Whitehall II study reported that a high level of interleukin 6 almost halved the odds of successful aging after 10 years (OR = 0.53) and increased the risk of cardiovascular events and noncardiovascular mortality [24]. In addition, there appears to be a synergistic interaction between DNA damage and inflammatory signals and, in mice, chronic inflammation aggravates telomere dysfunction and cellular senescence independent of genetic and environmental factors [25]. In this recent study, administration of anti-inflammatory agents rescued telomere dysfunction, prevented cellular senescence, and supported tissue regenerative potential [25].

Growing evidence suggests important cross-talk between oxidative stress, inflammatory processes, and the onset of endothelial dysfunction prior to atherosclerosis [6]. Reactive oxygen species induce pro-inflammatory changes in vascular endothelium—described as "endothelial activation"—which involves secretion of autocrine/paracrine factors, leukocyte-endothelial interaction and the upregulation of expression of cellular adhesion molecules [26].

Oxidative stress activates redox-sensitive transcription factors including the activator protein (AP-1) and nuclear factor-κB (NF-κB), increasing the expression of cytokines (tumor necrosis factor-α [TNF-α], interleukin-1 [IL-1], and IL-6), adhesion molecules (intracellular adhesion molecule (ICAM), vascular cell adhesion molecule [VSAM]), and pro-inflammatory enzymes (inducible nitric oxide synthase (iNOS) and cyclooxygenase-2 (COX-2)) [11].

Aging is associated with higher levels of cytokines, especially TNF-α, IL-1β, and IL-6, which mediate the acute phase protein C-reactive protein (CRP). These factors contribute significantly to the pro-inflammatory microenvironment and facilitate the development of vascular dysfunction [11]. Among middle-aged and older adults, the Framingham Heart Study showed that brachial FMD is inversely related to CRP, IL-6, and ICAM inflammatory markers [27]. Furthermore, inhibition of NF-κB signaling improved endothelial function significantly in middle-aged and older adults [28].

Senescence

Cellular senescence is characterized by permanent loss of mitotic capability associated with morphological and functional changes and impaired cellular homeostasis [29]. Senescent cells are characterized by the absence of proliferative markers, by the presence of senescent-associated β-galactosidase activity, by increased expression of tumor suppressor genes and cell cycle inhibitors, by increased DNA damage markers, more nuclear foci of constitutive heterochromatin and prominent secretion of signaling molecules [30]. Without functional telomeres, DNA at the ends of chromosomes could erode each time a cell divides, that is, during mitosis [31]. In addition to telomere length shortening, alterations in several signaling pathways contribute to the occurrence of cellular senescence [32]. Most of these signaling pathways activate p53 and they converge in activation of the cyclin-dependent kinase (CDK) inhibitors p16, p15, p21, and p27 [30].

All the risk factors for atherosclerosis including oxidative stress, inflammation, smoking, diabetes, and hypertension have been associated with accelerated telomere shortening. Furthermore, clinical trials showed that the presence of atherosclerosis is independently associated with increased telomere shortening [33]. The induction of senescence may be related to the enhanced susceptibility of telomeres to oxidative damage [34]. Telomere length in endothelial cells is inversely proportional to the age of the patients [35] and this shortening is exacerbated in elderly patients with coronary artery disease [36]. Cross-sectional studies demonstrate that those with increased arterial stiffness, an indicator of vascular aging, have shorter telomeres [37]. Hypertensive patients have shorter telomeres than their normotensive peers and hypertensives with shorter telomeres are more likely to develop atherosclerosis over 5 years follow-up [11]. The above observations suggest that telomerase length might be a good marker for cardiovascular aging but several issues limit the interpretation of these studies. Characterization and quantification of telomere length is difficult and not all methods are reliable. Secondly, the small sample size and heterogeneity of the populations included in these studies limits confidence. Lastly, the cross-sectional design of many of these studies precludes definitive conclusions particularly in the context of CVD which has multiple risk factors, which influence and confound each other [33].

Vascular endothelial cell senescence in vivo has been observed [38]. The development of more senescent endothelial cells was associated with a shift from an antiatherosclerotic phenotype (characterized by decreased levels of NO, eNOS activity and shear stress-induced NO production) to a proatherosclerotic phenotype (indicated by increased reactive oxygen species, thromboxane A_2 and endothelin-1). These observations implicate endothelial cell senescence in the initiation and progression of atherosclerosis [39]. Nitric oxide increases telomerase activity and promotes mobilization of endothelial progenitors cells (EPCs) which have the potential to delay endothelial cell aging by replacing damaged endothelial cells to maintain physical and functional integrity of the endothelium [31].

ANTIOXIDANT VITAMINS AS A STRATEGY TO DELAY VASCULAR AGING

Antioxidants are endogenous or exogenous elements which inhibit or delay the oxidation of biological macromolecules (proteins, lipids, and nucleic acids). Endogenous antioxidant defenses include both enzymatic (superoxide dismutase, glutathione peroxidases and catalases) and nonenzymatic (uric acid, glutathione, thiols and albumin) components. Exogenous antioxidant defenses include some vitamins (vitamins C and E and the pro-vitamin A molecule beta carotene) and polyphenols which are present in relatively high concentrations in plant-based foods including vegetables, fruits and wholegrain products [17].

Furthermore, the causal role played by oxidative stress in the pathogenesis of cardiovascular diseases and the, overall, positive associations between high antioxidant intake and lower incidence of cardiovascular diseases and mortality found in epidemiological studies have stimulated the design of large clinical trials (eg, the Women's Health Study and the Physicians' Health Study II) testing the antioxidant hypothesis. The results of these seminal studies were not encouraging since they found no evidence for beneficial effects of supplementation with large doses of antioxidant vitamins on cardiovascular outcomes [40,41]. However, follow-up, secondary analyses of the same datasets and subsequent meta-analyses have reevaluated some of the results and have suggested that antioxidant vitamin supplementation might be beneficial only among subjects who experience a prooxidant state due to high oxidative damage and/or low antioxidant status and not for the whole population [42]. In other words, the subgroup of participants experiencing a prooxidant state might be those with the potential to benefit from enhanced antioxidant status. In recently published meta-analyses of the effects of vitamin C and vitamin E on endothelial function, we showed positive effects in participants with lower baseline plasma vitamin E and C concentrations [43].

In the United States, total energy intake decreases substantially with age (likely due to age-related decrease in energy expenditure in physical activity) and this is usually associated with concomitant decline in intakes of several nutrients including calcium, iron, zinc, B vitamins, and vitamins C and E [44]. In the following sections, we will focus on vitamin C as a prototype of antioxidant vitamins, discussing in more detail the molecular mechanisms which may modulate the effects of vitamin C on cardiovascular aging.

Vitamin C (Ascorbic Acid)

Ascorbic acid is a six-carbon compound which was isolated first in 1928. Its structure was determined in 1933 and the L-isomer is the active form. The human being is one of the few mammals unable to synthesize vitamin C (ascorbic acid). The inability to synthesize vitamin C results from the lack of the last enzyme in the pathway of vitamin C synthesis (gulonolactone oxidase). A classical set of symptoms called scurvy, now known to be due to inadequate vitamin C intake, has been known for centuries [45].

Daily Requirement, Dietary Sources, Dietary Intakes, and Population Status of Vitamin C

Vitamin C is found mainly in fruits and vegetables with the highest concentrations in cantaloupe, grapefruit, kiwi, mango, orange, tangerine, and strawberries. As a supplement, vitamin C is available as a tablet or in powder form and it is included in many multivitamin formulations [46].

The US recommended daily allowance (RDA) for vitamin C is 75 mg for adult women and 90 mg for men with an upper tolerable limit of 2000 mg/day for both. However, recommendations for Vitamin C intake varies between countries with lower requirements set by the UK (40 mg/day) whereas the Japan's National Institute of Health and Nutrition recommends an intake of 100 mg/day. The latter level of intake is associated with plasma concentrations of vitamin C of ~70 μmol/L and with greater intakes circulating concentrations of vitamin C will reach, or exceed, the renal threshold of 80 μmol/L and result in increasing urinary excretion [47].

Many epidemiological studies have demonstrated that a significant proportion of the western population suffer from hypovitaminosis C (estimated as a plasma concentration less than 28 μmol/L) [47,48]. In the third National Health and Nutrition Examination Survey (NHANES III), about 10% of the included American population suffered from severe vitamin C deficiency (plasma concentration less than 11 μmol/L) [47]. Additionally, 20% of males and 15% of females had marginal vitamin C deficiency (determined as plasma concentrations between 11 and 28 μmol/L) [49]. The UK National Dietary Nutrition Survey showed that the mean plasma vitamin C concentration for men aged 19 to 64 years was 49.9 μmol/L and for women aged 19 to 64 years 55.2 μmol/L and both were above the level indicative of biochemical depletion. Prevalence of vitamin C deficiency (<11 μmol/L) was approximately 10% in men and women aged 65 to 74 years but increased progressively with age (~16% in people 75 to 84 years and ~19% in those 85 years and over) and approached 40% in older people living in care homes. In addition, poor socioeconomic status is a risk factor for vitamin C deficiency. The Low Income Diet and Nutrition Survey (2003–2005) of a representative sample of the low-income/materially deprived UK population estimated that 25% of men and 16% of women had deficient plasma vitamin C concentrations.

The age-related increase in vitamin C deficiency may be due to a decline among older people in the capacity to absorb vitamin C as a consequence of reduced activity of sodium-dependent vitamin C transporter proteins in the intestine. This reduction in vitamin C absorptive capacity is associated with chronic reduction in vitamin C concentration in elderly [50]. A meta-analysis of the relationship between vitamin C intake and plasma concentration of vitamin C found that, at the same intake (60 mg/d), circulating concentrations were 25% less among older adults (>65 years) than in younger adults. The authors concluded that the elderly would require double the oral intake of vitamin C, as compared with younger subjects, to maintain plasma concentration of 50 μmol/L [51,52].

CARDIOVASCULAR EFFECTS OF VITAMIN C

An overview of the pathways linking vitamin C and cardiovascular aging is provided in Fig. 43.2.

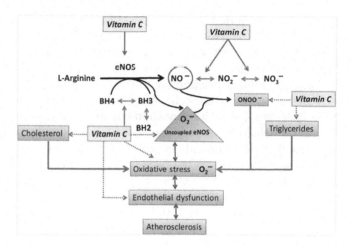

FIGURE 43.2 Pathways by which vitamin C may reverse or halt the process of atherosclerosis.

Antioxidant Effect

Vitamin C can be oxidized by free radicals, reactive oxygen species, and compounds that react with free radicals and which are changed to radicals themselves such as tocopheroxyl radicals [53]. Vitamin C counteracts the potential damaging effects of oxidative stress by its ability to quench aqueous reactive oxygen and nitrogen species and by helping to regenerate the antioxidant vitamin E [54]. In addition, vitamin C may mitigate the earliest stages of atherosclerosis through the following mechanisms: the prevention of the oxidation of LDL and reduced uptake and modification by activated macrophages, stimulation of endothelial cell proliferation and prevention of apoptosis, and decreased recruitment and proliferation of VSMCs and increased type IV collagen synthesis [54].

Nitric Oxide-Sparing Effect

Vitamin C contributes to maintenance of the total pool of NO through several different mechanisms. (1) Superoxide free radicals are highly reactive molecules which, in excess, may react with NO leading the formation of the harmful free radical peroxynitrite. Vitamin C inactivates the superoxide free radical leading to subsequent increase in the bioavailability of the protective NO [55]. (2) Vitamin C stabilizes and increases the synthesis of BH4, a cofactor which is essential for the function of eNOS. BH4 deficiency leads to eNOS uncoupling and the production of superoxide instead of NO [20]. (3) Inhibition of the enzyme arginase and consequent increase in L-arginine bioavailability [55] (4) Vitamin C may enhance the activity of eNOS by acting as a cofactor in hydroxylation reactions that lead to NO release. In addition, vitamin C may protect the eNOS enzyme from S-nitrosylation and, consequently, inactivation [56]. (5) Vitamin C may enhance the functional activity of NO by preserving guanylate cyclase activity and, hence, the formation of cGMP, the second messenger for NO production [57]. (6) Vitamin C-induced decomposition of nitrosothiol into NO via two main pathways: (a) low concentrations of ascorbic acid acting as a reducing agent for copper which reacts with the nitrosothiol, giving NO and disulphide, and (b) high concentrations of ascorbic acid inducing a nucleophilic attack at the nitroso nitrogen, followed by decomposition of the nitroso ascorbate formed to NO and dehydroascorbic acid [55]. (7) Enhancing the nitrate-nitrite-NO pathway. In vitro and animal studies have shown that vitamin C may convert nitrate and nitrite to NO in the gastrointestinal tract and plasma leading to increased availability of NO in the tissues [58,59].

A recently conducted systematic review and meta-analysis showed that vitamin C supplementation significantly enhances EF in patients with cardiometabolic disorders, an effect which may be attributed to the antioxidant and NO-sparing roles of vitamin C [60]. Despite the significant enhancement of EF with vitamin C supplementation, there was no significant effect on indices of arterial stiffness [61]. When used in high doses, as in most supplementation studies, vitamin C may have a prooxidant effect which may deteriorate, rather than ameliorate, arterial stiffness.

Lipid-Lowering Effect

Vitamin C facilitates the conversion of cholesterol into bile acids by modifying the activity of the rate-limiting enzyme cholesterol 7a-hydroxylase involved in bile acid synthesis and, therefore, may help lower blood cholesterol concentrations. Animals deficit in vitamin C exhibit hypercholesterolemia and the subsequent administration of vitamin C restores cholesterol concentrations to normal [62]. Furthermore, human observational studies showed that high plasma concentrations of vitamin C were associated with more favorable lipid profiles [63,64] but whether this is a causal relationship is uncertain. A meta-analysis of 13 trials which recruited participants with hypercholesterolemia showed that vitamin C supplementation reduced concentrations of LDL-C and triglycerides significantly with no significant effect on HDL-C concentration [65].

Blood Pressure-Lowering Effect

A recently published systematic review and meta-analysis has shown that vitamin C can lower the systolic and diastolic blood pressure significantly. The pooled changes in SBP and DBP were -3.84 mmHg and -1.48 mmHg, respectively. The proposed mechanism is the ability of vitamin C to reduce oxidative stress and to increase NO availability [66]. The blood pressure lowering effects could also be related to increased conversion of nitrite into NO in the gastric environment. The effects of enhanced NO supply via the nitrate-nitrite pathway in improving blood pressure control and endothelial function have been supported by two recent meta-analyses [67,68].

GENETIC INFLUENCES ON VITAMIN C STATUS AND IMPACT ON CARDIOVASCULAR RISK

Vitamin C homeostasis is influenced by several common genetic variants [69]. These genetic variations fall into two major categories. The first category are variants in genes which encode proteins responsible for vitamin C absorption and transport, the sodium-dependent vitamin C transporters 1 and 2 (SVCT1, SVCT2) which are encoded by the *SLC23A1* and *SLC23A2* genes respectively. SVCT1 is expressed in epithelial cells, notably the enterocytes of the small bowel where it transports vitamin C into the body. SVCT2 is expressed widely throughout the body especially in metabolically active cells. The second category of genes are those related to the antioxidant and redox activities of vitamin C and include variants in genes affecting iron homeostasis and oxidative stress such as haptoglobin (*HP*) and the glutathione S-transferases (*GSTs*) [69]. Transcriptional regulation of the SLC23 genes controls the tissue distribution of vitamin C transporters (SVCTs) and is responsible for the maintenance of vitamin C concentrations in cells, tissues and extracellular fluids [70].

Polymorphisms in Genes Encoding Vitamin C Transporters

Sodium-Dependent Vitamin C Transporter 1 (SVCT1)

The SVCT1 is located in the apical membrane of epithelial cells in liver, kidney, lung, small intestine, and pancreas and is encoded by the gene *SLC23A1*. SVCT1 expression in enterocytes appears to be the primary mechanism for vitamin C uptake from the gut although there may also be a role for SVCT2 which is found in the basolateral membrane [71]. Recently, following expression of human SCVT1 in *Xenopus laevis* oocytes, the purification and initial structural analysis of SCVT1 has been reported which will open the way to greater insights into structure-function relationships for this vitamin C transporter [72]. SVCT1 expression in the proximal tubules of the kidney also plays a major role in whole-body vitamin C homeostasis by controlling vitamin C loss into the urine. There are many genetic variants in *SLC23A1*, some of which are associated with reduced, and others with increased, plasma vitamin concentration (for more details, see review Ref. [69]). Human studies showed that lower serum vitamin C concentrations were associated with the rs6886922, rs34521685, rs35817838, and rs33972313 variants in *SLC23A1* [69,73]. In contrast, rs10063949 was associated with increased plasma concentration of vitamin C [73,74]. Despite the evidence for associations between *SLC23A1* genotype and plasma vitamin C concentration, there is only weak evidence for association between these genetic variants and risk of cardiovascular diseases [69,75]. Use of the Mendelian Randomization approach with *SLC23A1* genotype as a surrogate for lifelong vitamin C exposure has been used recently to question whether there is a causal relationship between vitamin C and cardiometabolic health [75].

Sodium-Dependent Vitamin C Transporter 2 (SVCT2)

SVCT2 is a high affinity vitamin C transporter that regulates the tissue accumulation of ascorbate from plasma; SVCT2 is expressed in nearly all cells and is encoded by the gene *SLC23A2* [70]. Two previous studies found contradictory results for relationships between *SLC23A2* genotype and vitamin C status [74,76]. The first study found no correlation between serum ascorbate concentrations and two common SNPs (rs6139591 and rs268116) in *SLC23A2* [74], while the second study observed lower concentration of vitamin C in individuals with the rs1279683 variant [76]. Since, SVCT2 regulates the concentration of vitamin C within cells, future studies are required to investigate possible effects of *SLC23A2* genotype on intracellular vitamin C status [69].

Since vitamin C is a cofactor in the hydroxylation of proline and lysine residues within procollagen during collagen synthesis, vitamin C deficiency may reduce the collagen content of the atherosclerotic plaques leading to vulnerable plaques that are more likely to rupture [52,77]. A recently conducted study showed that the rs6139591 polymorphism in *SLC23A2* was associated with acute coronary syndrome and that this risk was augmented in participants with low dietary vitamin C intake [78].

Polymorphisms Related to Antioxidant Enzymes and Redox Activity

Haptoglobin

Haptoglobin (Hp) is a hemoglobin-binding protein synthesized in the liver and released into the circulation. The main function of Hp is to bind free hemoglobin released from lysed erythrocytes and thus prevent the iron within hemoglobin from reacting with molecular oxygen to produce the free radical superoxide [79]. The human *HP* gene is polymorphic, with two alleles found in the population at relatively high frequency (HP^1 and HP^2). Individuals homozygous for the HP^1 are termed Hp1-1 and heterozygous individuals are termed Hp2-1 [69]. Compared with Hp1-1, homozygous individuals with the Hp2-2 phenotype have increased circulating concentrations of free iron or of iron bound to the porphyrin ring of hemoglobin. Such high concentrations of redox-active iron enhance the generation of superoxide and hydroxyl radicals and, consequently, oxidative stress. Moreover, individuals with the Hp2-2 phenotype have high rates of ascorbate oxidation which is mirrored by chronically low plasma vitamin C concentrations [80,81].

Glutathione S-Transferases (GSTs)

GSTs are enzymes involved in the detoxification of harmful electrophilic endogenous and exogenous compounds. GSTs action involves the conjugation of glutathione to target molecules. There are multiple variants in the genes encoding the GSTs families, that is, *GSTM1-5* and *GSTT1-2*. Both *GSTM1* and *GSTT1* have deletion variants which occur at relatively high frequencies in human populations from diverse ethnic backgrounds [82]. A common homozygotic deletion of the *GSTM1* gene nullifies its activity completely leading to a nonfunctional phenotype. Similarly, a deletion polymorphism in *GSTT1* leads to lack of enzyme activity [81]. These genetic variants of GSTs are associated with vitamin C status [83]. The proposed mechanism is that glutathione induces the reduction of dihydroascorbic acid to ascorbic acid and, therefore, prevents its degradation [84]. A systematic review of a small number of studies found no relationship between copy number variants in the GSTs M1 and T1 and risk of ischemic vascular disease [85]. In contrast, a more recent systematic review reported significantly increased risk of hypertension in those carrying the null genotypes of both GSTs M1 and T1 [86].

CONCLUSIONS

Aging is associated with complex structural and functional changes in all tissues including the vascular system and these changes increase CVD risk independent of other risk factors, such as hypertension, diabetes, or hypercholesterolemia. The accumulation of macromolecular damage with time is responsible for the molecular and cellular dysfunction which characterizes aging tissues and so dietary factors such as vitamin C which appear to modulate vascular aging must do so by reducing macromolecular damage, increasing repair processes or both. Through its roles (1) as an antioxidant nutrient, (2) as a cofactor in collagen synthesis, and (3) in augmenting NO bioavailability, a strong case can be made for the importance of vitamin C in maintaining good cardiovascular function throughout the life-course. However, the null outcomes from large-scale clinical trials of supplementation with (large doses of) vitamin C, often in combination with other so-called antioxidant nutrients, provide compelling evidence that there are unlikely to be population-wide benefits on cardiovascular health of vitamin C supplementation in older people. It seems more likely that those population groups with low vitamin C status, because of inadequate vitamin C intake or for genotypic reasons, or with high levels of oxidative damage may benefit from extra vitamin C intake. This opens the way for more effective public health interventions using a personalized (or stratified) approach [87].

SUMMARY

The accumulation of damage to the cell's molecules is responsible for the insidious, age-related decline in function and for the increased risk of frailty, disability and disease among older people. These changes occur in almost all of the body's cells including those of the cardiovascular system so that, at a population level, increasing age is the biggest risk factor for cardiovascular disease (CVD). Age-dependent damage to the endothelium occurs through oxidative stress and inflammation and leads to increasing prevalence of senescent cells. Nitric oxide (NO) which is secreted from endothelial cells, regulates several aspects of cardiovascular function including vascular tone, platelet aggregation, thrombosis, monocyte adhesion and smooth muscle proliferation. Vitamin C is important for vascular health through its roles as an antioxidant nutrient, as a co-factor of collagen synthesis and in augmenting NO bioavailability. However, a number of large randomised controlled trials have found no evidence for beneficial effects of supplementation of middle-aged and older people with large doses of antioxidant vitamins, including vitamin C. Substantial proportions of adults in the UK and USA have inadequate vitamin C status and vitamin C deficiency increases with ageing. In addition, vitamin C requirements are modulated by genotype, including variants in SLC23A1 the gene which encodes sodium-dependent vitamin C transporter 1. Older individuals with lower vitamin C status because of inadequate vitamin C intake or for genotypic reasons, or those with high levels of oxidative damage, may benefit from extra vitamin C intake.

References

[1] Lopez-Otin C, Blasco MA, Partridge L, Serrano M, Kroemer G. The hallmarks of aging. Cell 2013;153(6):1194—217.
[2] Mathers JC. Impact of nutrition on the ageing process. Br J Nutr 2015;113(Suppl.):S18—22.
[3] Tyson J, Mathers JC. Dietary and genetic modulation of DNA repair in healthy human adults. Proc Nutr Soc 2007;66(1):42—51.
[4] Langie SA, Kowalczyk P, Tomaszewski B, Vasilaki A, Maas LM, Moonen EJ, et al. Redox and epigenetic regulation of the APE1 gene in the hippocampus of piglets: the effect of early life exposures. DNA Repair (Amst) 2014;18:52—62.
[5] Virdis A, Ghiadoni L, Giannarelli C, Taddei S. Endothelial dysfunction and vascular disease in later life. Maturitas 2010;67(1):20—4.
[6] Ungvari Z, Kaley G, de Cabo R, Sonntag WE, Csiszar A. Mechanisms of vascular aging: new perspectives. J Gerontol A Biol Sci Med Sci 2010;65(10):1028—41.
[7] Murray CJ, Richards MA, Newton JN, Fenton KA, Anderson HR, Atkinson C, et al. UK health performance: findings of the Global Burden of Disease Study 2010. Lancet 2013;381(9871):997—1020.
[8] Jani B, Rajkumar C. Ageing and vascular ageing. Postgrad Med J 2006;82(968):357—62.
[9] Mitchell GF, Parise H, Benjamin EJ, Larson MG, Keyes MJ, Vita JA, et al. Changes in arterial stiffness and wave reflection with advancing age in healthy men and women: the Framingham Heart Study. Hypertension 2004;43(6):1239—45.
[10] Taddei S, Virdis A, Ghiadoni L, Salvetti G, Bernini G, Magagna A, et al. Age-related reduction of NO availability and oxidative stress in humans. Hypertension 2001;38(2):274—9.
[11] El Assar M, Angulo J, Vallejo S, Peiro C, Sanchez-Ferrer CF, Rodriguez-Manas L. Mechanisms involved in the aging-induced vascular dysfunction. Front Physiol 2012;3:132.
[12] Lehr HA, Germann G, McGregor GP, Migeod F, Roesen P, Tanaka H, et al. Consensus meeting on "Relevance of parenteral vitamin C in acute endothelial dependent pathophysiological conditions (EDPC)". Eur J Med Res 2006;11(12):516—26.
[13] Sena CM, Pereira AM, Seica R. Endothelial dysfunction — a major mediator of diabetic vascular disease. Biochim Biophys Acta 2013;1832(12):2216—31.
[14] Seals DR, Jablonski KL, Donato AJ. Aging and vascular endothelial function in humans. Clin Sci (Lond) 2011;120(9):357—75.
[15] Forstermann U. Nitric oxide and oxidative stress in vascular disease. Pflugers Arch 2010;459(6):923—39.
[16] Gil Del Valle L. WITHDRAWN: oxidative stress in aging: theoretical outcomes and clinical evidences in humans. Biomed Pharmacother 2010.
[17] Fusco D, Colloca G, Lo Monaco MR, Cesari M. Effects of antioxidant supplementation on the aging process. Clin Interv Aging 2007;2(3):377—87.
[18] Munzel T, Gori T, Bruno RM, Taddei S. Is oxidative stress a therapeutic target in cardiovascular disease? Eur Heart J 2010;31(22):2741—8.
[19] Gori T, Munzel T. Oxidative stress and endothelial dysfunction: therapeutic implications. Ann Med 2011;43(4):259—72.
[20] Schmidt TS, Alp NJ. Mechanisms for the role of tetrahydrobiopterin in endothelial function and vascular disease. Clin Sci (Lond) 2007;113(2):47—63.
[21] Sibal L, Agarwal SC, Home PD, Boger RH. The Role of asymmetric dimethylarginine (ADMA) in endothelial dysfunction and cardiovascular disease. Curr Cardiol Rev 2010;6(2):82—90.
[22] Vallance P, Leone A, Calver A, Collier J, Moncada S. Accumulation of an endogenous inhibitor of nitric oxide synthesis in chronic renal failure. Lancet 1992;339(8793):572—5.

[23] Chung HY, Cesari M, Anton S, Marzetti E, Giovannini S, Seo AY, et al. Molecular inflammation: underpinnings of aging and age-related diseases. Ageing Res Rev 2009;8(1):18–30.

[24] Akbaraly TN, Hamer M, Ferrie JE, Lowe G, Batty GD, Hagger-Johnson G, et al. Chronic inflammation as a determinant of future aging phenotypes. Can Med Assoc J 2013;185(16): E763–70.

[25] Jurk D, Wilson C, Passos JF, Oakley F, Correia-Melo C, Greaves L, et al. Chronic inflammation induces telomere dysfunction and accelerates ageing in mice. Nat Commun 2014;2:4172.

[26] Herrera MD, Mingorance C, Rodriguez-Rodriguez R, Alvarez de Sotomayor M. Endothelial dysfunction and aging: an update. Ageing Res Rev 2010;9(2):142–52.

[27] Vita JA, Keaney Jr. JF, Larson MG, Keyes MJ, Massaro JM, Lipinska I, et al. Brachial artery vasodilator function and systemic inflammation in the Framingham Offspring Study. Circulation 2004;110(23):3604–9.

[28] Pierce GL, Lesniewski LA, Lawson BR, Beske SD, Seals DR. Nuclear factor-κB activation contributes to vascular endothelial dysfunction via oxidative stress in overweight/obese middle-aged and older humans. Circulation 2009;119(9):1284–92.

[29] Erusalimsky JD. Vascular endothelial senescence: from mechanisms to pathophysiology. J Appl Physiol (1985) 2009;106(1):326–32.

[30] Munoz-Espin D, Serrano M. Cellular senescence: from physiology to pathology. Nat Rev Mol Cell Biol 2014;15(7):482–96.

[31] Farsetti A, Grasselli A, Bacchetti S, Gaetano C, Capogrossi MC. The telomerase tale in vascular aging: regulation by estrogens and nitric oxide signaling. J Appl Physiol (1985) 2009;106(1):333–7.

[32] van Deursen JM. The role of senescent cells in ageing. Nature 2014;509(7501):439–46.

[33] Zietzer A, Hillmeister P. Leucocyte telomere length as marker for cardiovascular ageing. Acta Physiol (Oxf) 2014;211(2):251–6.

[34] Kurz DJ, Decary S, Hong Y, Trivier E, Akhmedov A, Erusalimsky JD. Chronic oxidative stress compromises telomere integrity and accelerates the onset of senescence in human endothelial cells. J Cell Sci 2004;117(Pt 11):2417–26.

[35] Aviv H, Khan MY, Skurnick J, Okuda K, Kimura M, Gardner J, et al. Age dependent aneuploidy and telomere length of the human vascular endothelium. Atherosclerosis 2001;159(2):281–7.

[36] Ogami M, Ikura Y, Ohsawa M, Matsuo T, Kayo S, Yoshimi N, et al. Telomere shortening in human coronary artery diseases. Arterioscler Thromb Vasc Biol 2004;24(3):546–50.

[37] Nawrot TS, Staessen JA, Holvoet P, Struijker-Boudier HA, Schiffers P, Van Bortel LM, et al. Telomere length and its associations with oxidized-LDL, carotid artery distensibility and smoking. Front Biosci (Elite Ed) 2010;2:1164–8.

[38] Minamino T, Miyauchi H, Yoshida T, Ishida Y, Yoshida H, Komuro I. Endothelial cell senescence in human atherosclerosis: role of telomere in endothelial dysfunction. Circulation 2002;105(13):1541–4.

[39] Minamino T, Komuro I. Vascular cell senescence: contribution to atherosclerosis. Circ Res 2007;100(1):15–26.

[40] Ridker PM, Cook NR, Lee IM, Gordon D, Gaziano JM, Manson JE, et al. A randomized trial of low-dose aspirin in the primary prevention of cardiovascular disease in women. N Engl J Med 2005;352(13):1293–304.

[41] Sesso HD, Buring JE, Christen WG, Kurth T, Belanger C, MacFadyen J, et al. Vitamins E and C in the prevention of cardiovascular disease in men: the Physicians' Health Study II randomized controlled trial. J Am Med Assoc 2008;300(18):2123–33.

[42] Cesari M, Cerullo F, Demougeot L, Zamboni V, Gambassi G, Vellas B. Antioxidant supplementation in older persons. In: Laher I, editor. Systems biology of free radicals and antioxidants. Berlin Heidelberg: Springer; 2014. p. 3899–927.

[43] Ashor AW, Siervo M, Lara J, Oggioni C, Afshar S, Mathers JC. Effect of vitamin C and vitamin E supplementation on endothelial function: a systematic review and meta-analysis of randomised controlled trials. Br J Nutr 2015;113:1182–94.

[44] Thomas DR. Vitamins in aging, health, and longevity. Clin Interv Aging 2006;1(1):81–91.

[45] Gropper SS, Smith JL, Groff JL. The water-soluble vitamins. Advanced nutrition and human metabolism. 5th ed. Wadsworth: Belmont; 2009.

[46] Padayatty SJ, Katz A, Wang Y, Eck P, Kwon O, Lee JH, et al. Vitamin C as an antioxidant: evaluation of its role in disease prevention. J Am Coll Nutr 2003;22(1):18–35.

[47] Frei B, Birlouez-Aragon I, Lykkesfeldt J. Authors' perspective: what is the optimum intake of vitamin C in humans? Crit Rev Food Sci Nutr 2012;52(9):815–29.

[48] Lykkesfeldt J, Poulsen HE. Is vitamin C supplementation beneficial? Lessons learned from randomised controlled trials. Br J Nutr 2010;103(9):1251–9.

[49] Schleicher RL, Carroll MD, Ford ES, Lacher DA. Serum vitamin C and the prevalence of vitamin C deficiency in the United States: 2003–2004 National Health and Nutrition Examination Survey (NHANES). Am J Clin Nutr 2009;90(5):1252–63.

[50] Visioli F, Hagen TM. Nutritional strategies for healthy cardiovascular aging: focus on micronutrients. Pharmacol Res 2007;55(3):199–206.

[51] Brubacher D, Moser U, Jordan P. Vitamin C concentrations in plasma as a function of intake: a meta-analysis. Int J Vitam Nutr Res 2000;70(5):226–37.

[52] Li Y, Schellhorn HE. New developments and novel therapeutic perspectives for vitamin C. J Nutr 2007;137(10):2171–84.

[53] Duarte TL, Lunec J. Review: when is an antioxidant not an antioxidant? A review of novel actions and reactions of vitamin C. Free Radic Res 2005;39(7):671–86.

[54] Aguirre R, May JM. Inflammation in the vascular bed: importance of vitamin C. Pharmacol Ther 2008;119(1):96–103.

[55] May JM. How does ascorbic acid prevent endothelial dysfunction? Free Radic Biol Med 2000;28(9):1421–9.

[56] Holowatz LA. Ascorbic acid: what do we really no? J Appl Physiol (1985) 2011;111(6):1542–3.

[57] May JM, Harrison FE. Role of Vitamin C in the function of the vascular endothelium. Antioxid Redox Signal 2013.

[58] Lundberg JO, Weitzberg E, Gladwin MT. The nitrate-nitrite-nitric oxide pathway in physiology and therapeutics. Nat Rev Drug Discov 2008;7(2):156–67.

[59] Sibmooh N, Piknova B, Rizzatti F, Schechter AN. Oxidation of iron-nitrosyl-hemoglobin by dehydroascorbic acid releases nitric oxide to form nitrite in human erythrocytes. Biochemistry 2008;47(9):2989–96.

[60] Ashor AW, Lara J, Mathers JC, Siervo M. Effect of vitamin C on endothelial function in health and disease: a systematic review and meta-analysis of randomised controlled trials. Atherosclerosis 2014;235(1):9–20.

[61] Ashor AW, Siervo M, Lara J, Oggioni C, Mathers JC. Antioxidant vitamin supplementation reduces arterial stiffness in adults: a systematic review and meta-analysis of randomized controlled trials. J Nutr 2014;144(10):1594–602.

[62] Chambial S, Dwivedi S, Shukla KK, John PJ, Sharma P. Vitamin C in disease prevention and cure: an overview. Indian J Clin Biochem 2013;28(4):314–28.

[63] Dobson HM, Muir MM, Hume R. The effect of ascorbic acid on the seasonal variations in serum cholesterol levels. Scott Med J 1984;29(3):176–82.

[64] Khaw KT, Bingham S, Welch A, Luben R, Wareham N, Oakes S, et al. Relation between plasma ascorbic acid and mortality in men and women in EPIC-Norfolk prospective study: a prospective population study. European Prospective Investigation into Cancer and Nutrition. Lancet 2001;357(9257):657–63.

[65] McRae MP. Vitamin C supplementation lowers serum low-density lipoprotein cholesterol and triglycerides: a meta-analysis of 13 randomized controlled trials. J Chiropr Med 2008;7(2):48–58.

[66] Juraschek SP, Guallar E, Appel LJ, Miller 3rd ER. Effects of vitamin C supplementation on blood pressure: a meta-analysis of randomized controlled trials. Am J Clin Nutr 2012;95(5):1079–88.

[67] Siervo M, Lara J, Ogbonmwan I, Mathers JC. Inorganic nitrate and beetroot juice supplementation reduces blood pressure in adults: a systematic review and meta-analysis. J Nutr 2013;143(6):818–26.

[68] Lara J, Ashor AW, Oggioni C, Ahluwalia A, Mathers JC, Siervo M. Effects of inorganic nitrate and beetroot supplementation on endothelial function: a systematic review and meta-analysis. Eur J Nutr 2015.

[69] Michels AJ, Hagen TM, Frei B. Human genetic variation influences vitamin C homeostasis by altering vitamin C transport and antioxidant enzyme function. Annu Rev Nutr 2013;33:45–70.

[70] Burzle M, Hediger MA. Functional and physiological role of vitamin C transporters. Curr Top Membr 2012;70:357–75.

[71] Boyer JC, Campbell CE, Sigurdson WJ, Kuo SM. Polarized localization of vitamin C transporters, SVCT1 and SVCT2, in epithelial cells. Biochem Biophys Res Commun 2005;334(1):150–6.

[72] Boggavarapu R, Jeckelmann JM, Harder D, Schneider P, Ucurum Z, Hediger M, et al. Expression, purification and low-resolution structure of human vitamin C transporter SVCT1 (SLC23A1). PLoS One 2013;8(10):e76427.

[73] Timpson NJ, Forouhi NG, Brion MJ, Harbord RM, Cook DG, Johnson P, et al. Genetic variation at the SLC23A1 locus is associated with circulating concentrations of L-ascorbic acid (vitamin C): evidence from 5 independent studies with >15,000 participants. Am J Clin Nutr 2010;92(2):375–82.

[74] Cahill LE, El-Sohemy A. Vitamin C transporter gene polymorphisms, dietary vitamin C and serum ascorbic acid. J Nutrigenet Nutrigenomics 2009;2(6):292–301.

[75] Wade KH, Forouhi NG, Cook DG, Johnson P, McConnachie A, Morris RW, et al. Variation in the SLC23A1 gene does not influence cardiometabolic outcomes to the extent expected given its association with L-ascorbic acid. Am J Clin Nutr 2015;101(1):202–9.

[76] ZanonMoreno V, Ciancotti-Olivares L, Asencio J, Sanz P, Ortega-Azorin C, Pinazo-Duran MD, et al. Association between a SLC23A2 gene variation, plasma vitamin C levels, and risk of glaucoma in a Mediterranean population. Mol Vis 2011;17:2997–3004.

[77] Orbe J, Rodriguez JA, Arias R, Belzunce M, Nespereira B, Perez-Ilzarbe M, et al. Antioxidant vitamins increase the collagen content and reduce MMP-1 in a porcine model of atherosclerosis: implications for plaque stabilization. Atherosclerosis 2003;167(1):45–53.

[78] Dalgard C, Christiansen L, Vogel U, Dethlefsen C, Tjonneland A, Overvad K. Variation in the sodium-dependent vitamin C transporter 2 gene is associated with risk of acute coronary syndrome among women. PLoS One 2013;8(8):e70421.

[79] Schaer DJ, Vinchi F, Ingoglia G, Tolosano E, Buehler PW. Haptoglobin, hemopexin, and related defense pathways-basic science, clinical perspectives, and drug development. Front Physiol 2014;5:415.

[80] Cahill LE, El-Sohemy A. Haptoglobin genotype modifies the association between dietary vitamin C and serum ascorbic acid deficiency. Am J Clin Nutr 2010;92(6):1494–500.

[81] Schwartz B. New criteria for supplementation of selected micronutrients in the era of nutrigenetics and nutrigenomics. Int J Food Sci Nutr 2014;65(5):529–38.

[82] Block G, Shaikh N, Jensen CD, Volberg V, Holland N. Serum vitamin C and other biomarkers differ by genotype of phase 2 enzyme genes GSTM1 and GSTT1. Am J Clin Nutr 2011;94(3):929–37.

[83] Cahill LE, Fontaine-Bisson B, El-Sohemy A. Functional genetic variants of glutathione S-transferase protect against serum ascorbic acid deficiency. Am J Clin Nutr 2009;90(5):1411–17.

[84] Horska A, Mislanova C, Bonassi S, Ceppi M, Volkovova K, Dusinska M. Vitamin C levels in blood are influenced by polymorphisms in glutathione S-transferases. Eur J Nutr 2011;50(6):437–46.

[85] Norskov MS, Frikke-Schmidt R, Loft S, Sillesen H, Grande P, Nordestgaard BG, et al. Copy number variation in glutathione S-transferases M1 and T1 and ischemic vascular disease: four studies and meta-analyses. Circ Cardiovasc Genet 2011;4(4):418–28.

[86] Eslami S, Sahebkar A. Glutathione-S-transferase M1 and T1 null genotypes are associated with hypertension risk: a systematic review and meta-analysis of 12 studies. Curr Hypertens Rep 2014;16(6):432.

[87] Celis-Morales C, Lara J, Mathers JC. Personalising nutritional guidance for more effective behaviour change. Proc Nutr Soc 2015;74:130–8.

44

Omega-3 Fatty Acids in Aging

Natalia Úbeda, María Achón and Gregorio Varela-Moreiras

Departamento de Ciencias Farmacéuticas y de la Salud, Facultad de Farmacia, Universidad CEU San Pablo, Boadilla del Monte, Madrid, Spain

KEY FACTS

- Results about the relationship of omega-3 fatty acids and neurodegenerative disorders are promising and encourage further research in order to establish a safe and effective treatment solution for cognitive impairment and subsequent disability in elderly population.
- In individuals with mild cognitive impairment or age-related cognitive impairment, higher intakes of n-3 PUFA appear to be positive, whereas when Alzheimer's disease was already established no clear benefit is achieved.
- Controversial results have been published in recent reviews and meta-analyses reporting both positive and negative findings on the effectiveness of omega-3 fatty acids in cardiovascular health in the elderly. In consequence, some authors propose that the beneficial effects of omega-3 FA are not as large as previously implied and current recommendations for widespread use should be tempered.
- Even very low doses of omega-3 fatty acids may be sufficient to affect the immune responses of the elderly subjects. In rheumatoid arthritis, it has been reported a fairly consistent, but modest, benefit of marine omega-3 PUFAs on joint swelling and pain, duration of morning stiffness, global assessments of pain and disease activity, and use of nonsteroidal anti-inflammatory drugs.
- Omega-3 fatty acids may be a useful tool for the prevention and treatment of sarcopenia.
- The association of a good omega-3 fatty acids status and better maintenance of muscle mass is promising but clearly needs further specific trials.
- There could be a possible relation between higher omega-3 FA and better functional mobility in older adults.
- Fish oil promotes weight maintenance or gain during active oncology treatment. Certainly, larger and specific cancer localization studies in older adults are needed to confirm some encouraging results relating to colorectal and endometrial cancer.
- The potential adverse effects of omega-3 supplementation appear mild-moderate at worst and seem unlikely to be of clinical significance.
- Dietary omega-3 PUFA in later life may be of benefit in reducing total mortality and, therefore, increasing the quality of life.

Dictionary of Terms

- *Aging*: a persistent decline in the age-specific fitness components of an organism due to internal physiological degeneration. It is a process that is genetically determined and environmentally modulated.
- *Bone mineral density (BMD)*: Bone mineral density is a measure of bone density, reflecting the strength of bones as represented by calcium and other type of minerals content. The BMD tests are used to detect

- osteopenia (mild bone loss) and osteoporosis (more severe bone loss).
- *Cardiovascular diseases*: Cardiovascular diseases are a group of disorders of the heart and blood vessels and include: coronary heart disease, cerebrovascular disease, peripheral arterial disease, rheumatic heart disease, congenital heart disease, deep vein thrombosis, and pulmonary embolism—blood clots in the leg veins, which can dislodge and move to the heart and lungs.
- *Cognitive impairment*: Cognitive deficit or cognitive impairment is an inclusive term to describe any characteristic that acts as a barrier to the cognition process. The term may describe deficits in global intellectual performance, such as mental retardation, it may describe specific deficits in cognitive abilities (learning disorders, dyslexia), or it may describe drug-induced cognitive/memory impairment. Cognitive deficits may be congenital or caused by environmental factors such as brain injuries, neurological disorders, or mental illness. Cognitive impairment ranges from mild to severe.
- *Docosahexaenoic acid (DHA)*: Docosahexaenoic acid is an omega-3 fatty acid whose structure is a carboxylic acid with a 22-carbon chain and six cis double bonds, with the first double bond located at the third carbon from the omega end.
- *Eicosapentaenoic acid (EPA)*: Eicosapentaenoic acid is an omega-3 fatty acid whose structure is a carboxylic acid with a 20-carbon chain and five cis double bonds, with the first double bond located at the third carbon from the omega end.
- *Polyunsaturated fatty acids (PUFA)*: Polyunsaturated fatty acids are classified as omega-3 (n-3) or omega-6 (n-6) depending on whether their first double bond is located on the third or sixth carbon from the terminal methyl group.
- *Quality of life*: Quality of life could be described as an optimal patient functioning and well-being. At its most essential level, quality of life may be understood to be both subjective and multidimensional. Subjective, because it is best measured from the patient's perspective. Multidimensional, because its measurement requires an investigator to inquire about a variety of areas of the patient's life, including physical well-being, functional ability, emotional well-being, and social well-being.

INTRODUCTION

Nutrition is one of the major determinants of successful aging, defined as the ability to maintain three key elements: low risk of disease and disease-related disability, high mental and physical function, and active engagement of life [1–2]. Physiological and functional changes that occur with aging can result in changes in nutrient needs [3–4]. Knowledge of nutrient requirements of older adults is growing yet is still inadequately documented [5–6]. Older adults have unique nutrient needs [7]. Strategies to prevent and/or reduce morbidity in the elderly are therefore required as the population ages worldwide. Increasing the intake of omega-3, particularly the long-chain n-3 PUFA may be one such strategy [8–10]. Moreover, since in most populations, current intakes of PUFA and especially n-3 PUFA are insufficient for optimal health [11–12].

Fat is an important source of energy and facilitates the absorption of fat-soluble dietary components, such as vitamins. Additionally, fats and oils are important sources of essential fatty acids (EFA). Fats also play an important role by enhancing the taste and acceptability of foods and fat components contribute to the texture, flavor, and aroma of foods, a key factor to maintain appropriate appetite during the aging process [11].

The role of fats in nutrition has moved fast and serves to underscore the recent realization in human nutrition that the quality of the fat supply in terms of the parent n-6 and n-3 essential fatty acids as well as their longer chain, more unsaturated derivatives—arachidonic, eicosapentaenoic and docosahexaenoic acids—plays a vital role in human health from conception through every stage of human development, maturation, and aging [12]. In terms of health and disease the essential fatty acids and their derivatives interact at multiple levels, including cell membrane composition, metabolism, signal transduction and amplification, and gene expression. Furthermore, they influence cell growth and differentiation, tissue repair, apoptosis and cell death and many physiological and pathological processes including immunity and inflammation, which are of special potential interest in the aging process [11]. Omega-3 fatty acids are now generally recognized as potentially beneficial for optimal function of the cardiovascular system in adults [13–14]. Because of the increasing risk of deteriorating health of the cardiovascular system and brain with age, it is important to establish whether healthy aging is associated with changes in plasma omega-3 fatty acid content or response to fortification/supplementation [15]. It is becoming increasingly evident that long-chain PUFAs from the n-3 family appear to be neuroprotective and may also have unique properties in affecting neurobiology, both of critical interest during the aging process [16–23]. Intake of n-3 PUFA has also been associated with potential benefits in other age-related morbidities, including rheumatoid arthritis, depression, and macular degeneration. In fact, surprisingly, in both plasma and red blood cells, several studies have reported that the content of eicosapentanoic acid (EPA) and docosahexaenoic acid (DHA) rises significantly from the second to the seventh decade of life [24–29]. Normally, higher plasma (and red cell) plasma omega-3 fatty acid status in the elderly would seem to be due to higher fish/seafood intake but may also be due to aging-related changes in omega-3 metabolism. However, these findings are not universal, since other studies have found that the intake of polyunsaturated fatty acids decreases with age [30] and most older people eat less that the recommended amounts. In contrast to their proposed actions in childhood, where n-3 long-chain polyunsaturated fatty acids are required for healthy development of brain tissue, in older age they are more likely to act in a protective and health-maintaining manner. For example, n-3 are known to inhibit hepatic triglyceride synthesis and, by modifying eicosanoid function, cause vascular relaxation, a diminished inflammatory process, and decrease platelet aggregation [31].

Nevertheless, these studies in the elderly as discussed later in this chapter, show important limitations, including a small number of subjects. In addition, although 65 years of age is frequently used as reference, there is no official or widely accepted definition of elderly, so the cut-off used in the studies is often as low as 50 years of age. Furthermore, information about the health status, blood chemistry, cognitive function or physical activity of elderly research subjects is rarely given in sufficient detail to establish whether or not the data reported are for the so-called healthy elderly.

The physiological aging process, its association with quality of life, and the impact of omega-3 fatty acids intake and/or status is the focus of this chapter.

LIPIDS, AGING, AND BRAIN FUNCTION

The cost of brain and mental disorders has increased significantly and the cost now exceeds all other related health problems [32]. It has been reported that omega-3 fatty acids are strongly linked to brain function during aging [13]. Lipids can potentially play an important role in preventing the progression to mild cognitive impairment and associated diseases, either preventing the onset of cardiovascular risk factors, by closely related vascular changes that occur normally in many cases, and also by direct mechanisms, since the brain contains a high proportion of PUFA lipids n-3 and n-6 [8].

DHA is the only n-3 fatty acid used as an important structural and functional component of photoreceptor neurons and their synapses signaling over 600 million years of evolution. This is one of the critical reasons for the absolute necessity of DHA for human brain. Mammals cannot synthesize n-3 polyunsaturated long-chain fatty acid (LC-PUFA) from anything other than n-3 fatty acids; therefore, direct consumption of DHA is required to target the tissue and plasma levels that have been associated with cognitive protection [16]. DHA is concentrated in the gray matter, and very small amounts are found in purified myelin. DHA is the predominant LC-PUFA in the brain, constituting from 30% to 40% of total LCPUFA in the cerebral cortex. DHA, as an integral component of neuronal membranes, is involved in many brain functions such as the fluidity of the cell membrane, the receptor affinity, and modulating transduction molecules/signaling [19,33]. DHA can also modulate apoptosis, neuronal differentiation, and ion channels [34].

EPA, although is not stored in the cell membranes of the brain, may reduce various neurological disorders, since it has a role in brain function, possibly by interacting with the signals mediated by arachidonic acid [35,36]. In addition, some authors reported that both DHA and EPA can reduce oxidative stress, have an anti-inflammatory action and neurotransmission and participate in gene expression through cytosolic and nuclear interaction with various peroxisome proliferator-activated receptors (PPARs) [37]. In addition, omega-3 fatty acids, by improving the composition of cell membranes, can stimulate the development and regeneration of nerve cells [38]. The mechanisms involved are complex and multiple, reflecting the extraordinary diversity of the functions performed by polyunsaturated fatty acids, from the modulation of the dynamic properties of the membranes, for the production of active mediators, and the regulation of the expression gene. Brain lipids, which may influence each of these pathways in each life stage, are, therefore, essential elements of brain development [35,39].

The neural system is widely developed during the prenatal period and the first years of life, and is influenced by several factors. There is convincing evidence that neural development determines the functional capacity of the brain in the long term, including the aging process. The concentration of LCPUFA in neuronal tissue generally increases with age over the first 2 decades and then levels decrease, and it has been proposed that changes in the lipid composition are associated with changes in central nervous system function [15,40]. Although the underlying cause of this change in the concentration of LCPUFA is not fully known, it is observed that dietary intake throughout life has an important role in determining the lipid composition of the brain in old age [23].

OMEGA-3 FATTY ACIDS AND BRAIN HEALTH IN AGING

Effects of Omega-3 Fatty Acids on Cognitive Function in Normal Aging

The potential role for omega-3 fatty acids in the prevention of cognitive decline has become of high interest during the past two decades. Scientists have carried out different types of studies (epidemiological, interventional, and experimental) in order to prove any potential positive relation with different results, which are briefly reviewed further.

In 2010, the NIH State of the Science Conference Panel concluded that insufficient evidence is available to recommended the use of any primary prevention therapy for Alzheimer's disease (AD) of age-related cognitive decline [41], although "the most consistent evidence is available for longer-chain omega-3 fatty acids with several studies showing an association with reduced risk of cognitive decline." The Panel also acknowledged that promising research is under way. Since then, several studies and meta-analyses have been published. In a recent systematic review we evaluated the effects of omega-3 fatty acids on cognitive function in normal aging and healthy older people [42]. We included a total of 26 articles. Thirteen of them were longitudinal studies, ten were cross-sectional (three of them included longitudinal and cross-sectional analysis), and six were randomized, double-blind controlled trials. All of these studies used the Mini Mental State Examination (MMSE) for evaluate global cognitive function, which includes questions about orientation to time and place, registration, attention and calculation, recall, language, etc. and/or other specific tests (eg, verbal fluency, speed, visuospatial skills). Most of them considered cognitive impairment as a MMSE score <25 and cognitive decline was defined as a drop of more than two points in the MMSE over the period studied; however cognitive decline is defined uniquely according to each study, and thus comparing outcomes across studies must be interpreted with caution. The content of omega-3 fatty acid was evaluated through dietary information (intake of fish or DHA/EPA) with appropriate questionnaires or concentration in plasma and/or erythrocyte membrane.

Most of the studies showed an inverse relationship between an adequate fish consumption or n-3 fatty acids intake (diet or supplements) or total content in erythrocyte membrane and cognitive status or reduced cognitive decline during aging. Two of the articles showed some improvements but other measures remained unchanged. Vercambre et al. [43] observed an inverse association between fish consumption and n-3 fatty acids intake and cognitive decline but not significant for functional impairment. Dullemeijer et al. [44] concluded that higher plasma proportions of n-3 fatty acids are associated with less decline of cognitive performance in sensorimotor speed and complex speed but not in memory, information-processing speed, and word fluency.

Five of the studies showed no positive effects of n-3 fatty acids on cognitive function. Two RCTs [21,45] used high doses or EPA plus DHA and olive oil or oleic acid as placebo but did not find any significant changes in cognitive domains over 6 and 24 months, respectively. Both of them recommended that further longer trials should be conducted. Three cross-sectional studies [46–48] showed no significant differences in n-3 PUFA intake between controls and cases of cognitive decline/impairment, and two of them [47,48] also did not show any significant associations between fatty fish or n-3 PUFA intake and cognitive change over 5 and 6 years, respectively. Furthermore, another two additional studies [49,50], showed no association between omega-3 FA supplementation or levels and cognitive decline in an older population.

Some RCTs of the negative studies used olive oil or oleic acid as a placebo. Rosales [51] published a letter in which he explained the possible weakness for these studies. For example, as indicated by the authors, the study population might already consume a sufficient amount of PUFA in their diets and thus not be sensitive to the dose of DHA/EPA provided, but they have also rejected that the control group might result in maintaining cognitive function in later life, comparable to the effects of DHA/EPA, because they consumed an extra amount of oleic acid (olive oil) which could provide benefits, albeit by different mechanisms. Actually, prospective studies have shown that the Mediterranean diet is associated with slower cognitive decline and a reduced risk of progression [52,53]. Oleic acid, the major component of olive oil, has recently been shown to provide a satiety factor, oleylethanolamide, which enhances memory consolidation without crossing the blood–brain barrier [54]. This is important in verbal learning, organization, and memory. Oleoythanolamide is a mediator in maintaining cognitive function that it is not related to vascular or other nonvascular biological mechanisms (ie, metabolic, oxidative, and inflammatory). Thus, this evidence suggests that these RCTs should consider the benefits of oleic acid as well as omega-3 PUFA in protecting against age-associated cognitive decline and maintaining cognitive function in later life.

Also in 2012, a meta-analysis of 10 RCTs was published [55]. Positive effects could be concluded for omega-3 FA supplementation in participants with mild cognitive impairments (MCI). This conclusion was especially true for the domains of immediate recall, attention, and speed. However, no effects could be observed in either patients with AD or healthy individuals.

In general, and as also concluded by other authors [34,56], longitudinal studies are mainly positive, indicating that high fish intake and high DHA blood levels are protective against subsequent cognitive decline. In intervention studies there are some discrepancies, but it seems that when DHA is given to individuals with MCI or age-related cognitive impairment the data appear to be positive, whereas when AD was already established not clear benefit is achieved (see later section).

It is important, however, taking into account the limitations of these studies. For example, according to Cederholm et al. [34] longitudinal studies with negative observation might never have been published in the international literature. On the other hand, the competing risk of death is a potential peril leading to an underestimation of the protective effects of EPA and DHA because it is plausible that low fish intake increases cardiovascular risk burden and that death occurs before reaching the age at which one is likely to develop cognitive decline. Also, the sample size, heterogeneity in the population studied, different methods for estimating the intake of omega-3 fatty acids and/or blood concentration, the wide variety of questionnaires to determine cognitive development, etc., could affected the observed results [42]. With regard to intervention studies, Cederholm et al. [34] explained that the duration has not been long enough. The DHA dose varied extensively and so far the basal DHA status has not been taken into account. Furthermore, it is doubtful as to whether beneficial effects of DHA supplementation can be observed in individuals who already have sufficient DHA intake and tissue content.

Dacks et al. [56] described some variables that may impact the effect of omega-3 LCPUFA intake including omega-3 FA baseline levels, Apolipoprotein ε4 (APOE) genotype, and vegetarian and vegan diets. The cognitive benefits of high omega-3 FA intake may be specific to APOE noncarriers because in carriers blood levels are reported to be higher and less responsive to supplementation. On the contrary, vegetarians and vegans have low plasma and tissue levels of omega-3 LCPUFA, but no evidence for increased risk of dementia or decreased cognitive function, probably due to decreased n-6 PUFA content in the diet and subsequently decreased n-6 to n-3 ratio and inflammatory signaling through arachidonic acid.

Effects of Omega-3 Fatty Acids in Alzheimer's Disease and Related Dementia

To study the relationship between omega-3 FA consumption and/or blood levels and the risk of dementia or AD, some epidemiological and trials were also carried out in elderly people. The levels of EPA, DHA, and total omega-3 PUFA are significantly decreased in peripheral blood tissues in patients affected by dementia [57] supporting the important role of n-3 PUFAs in the pathophysiology of this disease. Most, but not all, epidemiologic studies have reported that fish and seafood intake associates with decreased risk of dementia. In contrast, there are other studies that did not observe any beneficial effect (revised in Ref. [22]). Therefore, more studies are needed to better understand the relationship between omega-3 FA status and neurodegeneration.

A list of reviews also reported results of RCT of omega-3 FA treatment in AD patients with negative data and no effects on brain volume decline or biomarkers of inflammation or AD in the CSF and plasma. Some posthoc analyses and small RCTs suggest benefit in early stages of cognitive decline or mild cognitive impairment (revised in Refs. [56,58]). Dangour et al. [59] included in a systematic review three trials that enrolled older people with cognitive function impairments ranging from defined age-related cognitive decline to moderate AD [60–62]. One of these three trials identified some evidence of a benefit from DHA supplementation, whereas the other two trials of omega-3 FA supplementation among cognitively healthy and cognitively impaired older people did not support the use of these PUFA for the prevention of cognitive decline.

In conclusion, results are promising and further research in order to establish a safe and effective treatment solution for cognitive impairment and subsequent disability in elderly population is urgently needed. Promoting higher intakes of n-3 PUFA in the diet or specific supplements may have substantial benefits in reducing the risk of cognitive decline and maintaining a healthy brain during aging. However, dose-effect has not been yet well established. It is also plausible that cognitive protection is limited to APOE ε4 noncarriers and that omega-3 FA supplementation must be applied prophylactically at or before early stages of decline [56]. Therefore, the potential benefits for health outcomes of such personalized recommendations for fish and EPA/DHA intake still have to be evaluated [63]. More prospective studies, as well as interventions studies investigating the association between n-3 PUFA and domain-specific measures are needed to clarify the current conflicting results observed in the literature.

OMEGA-3 FATTY ACIDS AND CARDIOVASCULAR DISEASE

In 2012, cardiovascular diseases (CVD) were the leading cause of noncommunicable disease deaths worldwide (17.5 million deaths) [64]. In 2011, 34% of deaths attributable to CVD in the United States occurred before the age of 75 years, which is younger than the current average life expectancy of 78.7 years. The prevalence of coronary heart disease, myocardial infarction and angina pectoris is higher, however, in the group over 80 years versus 60–79 years in both genders [65]. Each year CVD causes over 1.9 million deaths in the European Union (EU). The percentage of people who die from diseases of the circulatory system increases with age and the majority of people dying from stroke and ischemic heart disease are 70 years and over [66].

In the early 2000s, evidence from large-scale epidemiological studies suggested that people at risk of cardiovascular heart disease (CHD) benefit from consuming omega-3 fatty acids from plants and marine sources. Also, in randomized clinical trials that enrolled patients with coronary heart disease, omega-3 fatty acids supplement significantly reduced cardiovascular (CV) events. On this basis, the American Heart Association (AHA) recommended that all adults should eat fish (particularly fatty fish) at least two times a week and patients with documented CHD should consume ~1 g of EPA and DHA (combined) per day [67].

The exact mechanisms by which omega-3 performs its functions are still a matter of debate, but the main mechanisms proposed are plaque stabilization, lipid profile, anti-inflammatory, blood pressure, heart failure, or antiarrhythmic properties (reviewed in Ref. [68]).

However, controversial results have been published in recent reviews and meta-analyses reporting both positive and negative findings on the effectiveness of omega-3 fatty acids. This is a very alive and current field and, actually, the number of articles has duplicated or triplicated in the last five years. While most observational studies have found inverse associations between dietary omega-3 FA and death from CHD [69–71] or total and cause-specific (mainly cardiovascular) mortality [72], randomized trials of omega-3 supplementation have shown contradictory results: reduction of cardiovascular events, cardiac death, and coronary events by approximately 8–18% reported by some authors [68,73,74] or null effects in all-cause mortality, cardiac death, sudden death, arrhythmia, myocardial infarction, or stroke reported by others [75–77]. Some authors propose that the beneficial effects of omega-3 FA are not as large as previously implied and recommendations for widespread use should be tempered.

Different meta-analyses included a variable number of RCTs and assigned several inclusion criteria but according to Delgado-Lista et al. [68] these have to be accepted with caution, because the number of studies included are hardly ever designed with the same type of patients (primary or secondary prevention, age, gender), under the same conditions (time of intervention), or with the same intervention (dose or type of fatty acid). Another fact to be noted is the difference in the medication given to the patients that could minimize the positive effects of omega-3, and the background diet, since the effects of omega-3 in populations where its mean consumption is low may be higher than in those populations where the dietary habits include oil fish as a habitual food.

It is possible that the general population, and especially those that suffer from cardiovascular disease, were concerned about the importance of fish consumption and that even "control" subjects in the randomized studies have a higher intake of marine products and are receiving the benefits of their intake of fish. In this sense, Mozaffarian et al. [72] demonstrated a threshold effect and a nonlinear relationship of dietary versus circulating omega-3 FA with a steepest dose-response up to ~400 mg/d consumption. Previously, the same authors [78] observed, in a meta-analysis of cohort studies and randomized trials, a significant, nonlinear threshold relationship between dietary omega-3 FA and CHD mortality with greatest benefits up to ~250 mg/d. Therefore, all these findings support and average target dietary range of 250–400 mg/d EPA + DHA, and highlight the potential benefits of modest omega-3 FA intake for primary prevention in older adults. Hence, dietary or supplemental omega-3 FA may be most beneficial for people with little to no consumption.

Another issue to take into account when evaluating the apparent disparity of results between studies is population genetics, especially regarding the modulation of omega-3 intake on the lipid profile. Variation in the composition of the placebo compound may also contribute to the neutrality of the findings, as we previously explained in the preceding section.

Some authors evaluated the effect of subgroups in their studies. Wang et al. [73] observed that the evidence for the benefits of fish oil is stronger in secondary than in primary-prevention settings. Marik and Varon [74] explained that the benefit appeared to depend on the patient's risk stratification with a reduction in deaths in high risk patients and a reduction in nonfatal cardiovascular events in moderate risk patients. This observation likely reflects the fact that the number of deaths in the moderate risk group was very low, while these patients are at a higher risk of developing nonfatal cardiovascular events related to progressive atherosclerotic disease.

De Oliveira et al. [79] examined the race effect in a large prospective cohort of multiethnic Americans, and found that higher circulating EPA and DHA were each inversely associated with markers of inflammation and prospectively associated with lower CVD incidence. They observed also an inverse association of seafood-derived long-chain n-3 PUFAs, but not plant-derived n-3 or n-6 PUFAs, with CVD incidence in all races in the multiethnic cohort, including White, Chinese, African American, and Hispanic.

On the other hand, a remarkable effect of gender was shown in some studies. For example, The Risk and Prevention Study Collaborative Group [80] concluded that there was no significant benefit of n-3 FA in reducing the risk of death from CV causes or hospital admission for CV causes, but in the prespecified subgroup analyses, the only significant interaction was between the efficacy of n-3 FA and sex with protective effect in women. Larsson et al. [81] conducted a meta-analysis of prospective studies on the relation between n-3 FA intake and stroke. They showed no overall association between omega-3 PUFA intake and stroke, but suggested that women might benefit from a higher intake of these PUFAs (RR 0.80, 95% CI 0.65–0.99). Kromhout et al. [82] showed that a low dose of EPA-DHA had no effect on the rate of major CV events in patients who had had a myocardial infarction but a prespecified analysis according to sex showed that there were significantly fewer major CV events among women who received alpha linolenic acid (ALA) than among women who received placebo. However, Lemaitre et al. [83] found little evidence of an association between ALA and congestive heart failure (CHF) in subgroups based on age, sex, diabetes, fish consumption, BMI, or *FADS2* genotype.

Another consideration could be the method of culinary food processing. Belin et al. [84] and Mozaffarian et al. [85] found that consumption of ≥5 servings of baked/broiled fish per week was independently associated with lower risk for incident heart failure (HF) by 30% and 32% respectively, whereas consumption of >1 serving of fried fish per week was independently associated with increased HF risk by 48% and 35% respectively. The first study was carried out in postmenopausal women and the second in elderly [83].

In summary, better evidence is required to support current recommendations for the intake of fish oil or omega-3 FA supplements in relation to the prevention or treatment of CVD. Generalizability of the overview findings may be questioned and more studies are needed to evaluate specific outcomes and populations stratified by sex, age, CV risk, medication use, race, etc.

EFFECTS OF OMEGA-3 FATTY ACIDS ON IMMUNE FUNCTION IN NORMAL AGING

Aging is a multifactorial process involving decreased immune functions [86]. Omega-3 polyunsaturated fatty acids have been associated with human health benefits in immune function in several pathologies, mainly against inflammatory and autoinmune diseases. However, the effects of omega-3 fatty acids on functional variables of the immune system in healthy aging remain controversial.

In the already mentioned systematic review [42], we found five studies (four RCT and one cross-sectional and longitudinal) examining the effect of omega-3 fatty acids on immune function. All four RCT studies assessed the effect of moderate dietary supplementation with omega-3 fatty acids (EPA or EPA + DHA) on lymphocyte proliferation and immune biomarkers known to be altered in normal healthy aging. Even very low doses of omega-3 fatty acids resulted in significantly decreased proliferative responses of lymphocytes in two of the studies [86,87]. In the study by Bechoua et al. [86], this was accompanied by a marked significant ($P < 0.05$) increase of their cytosolic cyclic nucleotide phosphodiesterase (PDE) activity (+56–57%) and an increase ($P < 0.05$) in cyclic nucleotide intracellular levels. At the same time, the glutathione peroxidase activity was markedly depressed ($P < 0.01$). In contrast, the cross sectional study [88] reported significant positive correlations ($P < 0.05$) between PHA-induced proliferation and intake of DHA and EPA. However, these authors stated that intakes of DHA plus EPA were inadequate for the studied population when compared to recommended intakes. They also suggested that dietary EPA in vivo might interact differently compared to in vitro studies where EPA is added to cell cultures. On the other hand, a moderate amount of extra supplementary EPA (720 mg/d) resulted in a decrease of NK cell activity in the same elderly population [89]. This decline (48%) was, however, fully reversed by 4 weeks after supplementation had ceased.

The effects of different amounts of EPA on innate immune outcomes in older males compared to young were also evaluated [27]. EPA was incorporated in a linear dose-response fashion into plasma and mononuclear cell (MNC) phospholipids; incorporation was greater in the older men. This increased incorporation was associated with decreased production of prostaglandin E2 (PGE2) by MNCs. Also, EPA treatment caused a dose-dependent decrease in neutrophil respiratory burst only in the older men. These five above mentioned studies therefore suggest that even very low doses of omega-3 fatty acids may be sufficient to affect the immune responses of the elderly subjects.

It is important to remark, however, that evidence on the influence of omega-3 polyunsaturated fatty acids on immune function is mainly described in rheumatoid arthritis. In this sense, a recent systematic review evaluating 23 randomized controlled trials, reports a fairly consistent, but modest, benefit of marine omega-3 PUFAs on joint swelling and pain, duration of morning stiffness, global assessments of pain and disease activity, and use of nonsteroidal anti-inflammatory drugs [90].

EFFECTS OF OMEGA-3 FATTY ACIDS ON MUSCLE MASS AND FUNCTION IN NORMAL NONPATHOLOGICAL AGING

It is well recognized that a progressive loss of muscle mass, strength, and function, that is, sarcopenia, occurs with aging. Approximately 1–2% of muscle mass per year is lost after the age of 50 [11]. Aging is associated with a reduced ability to capture blood-borne amino acids as protein, a decreased protein synthesis, a decrease in signaling proteins capacity to indicate the presence of amino acids, and a resistance to the effects of insulin in reducing muscle proteolysis and improving muscle anabolic response. In addition, age-related immobility negatively influences the maintenance of muscle protein synthesis. Therefore, it has been hypothesized that the stimulation of protein synthesis induced by omega-3 fatty acids supplementation might be useful for the prevention and treatment of sarcopenia of aging [11]. In fact, there are studies suggesting the usefulness of omega-3 fatty acid supplementation in patients at high risk of sarcopenia, such as cancer patients [91,92]. However, there is a lack of concluding evidence in healthy aging people.

The study of Smith et al. [93] is one of the very few controlled, clinical trials that shows evidence supporting dietary omega-3 intervention promoting muscle protein synthesis in older adults, and elucidates a possible mechanism of action, namely a reduction of the muscle anabolic resistance to plasma amino acids in this compromised physiological situation. The objective of the study was to evaluate the effect of omega-3 fatty acid supplementation for 8 weeks on the rate of muscle protein synthesis in the elderly population (69–73 years). Omega-3 fatty acid supplementation (1.86 g EPA and 1.50 g DHA) augmented the hyperaminoacidemia-hyperinsulinemia-induced increase in the rate of muscle protein synthesis (from $0.009 \pm 0.005\%/h$ above basal values to $0.031 \pm 0.003\%/h$ above basal values; $P < 0.01$), which was accompanied by larger increases in muscle mTORSer2448 ($P = 0.08$) and p70s6kThr389 ($P < 0.01$) phosphorylation, two key elements of intramuscular signal transduction proteins involved in the regulation of muscle protein synthesis.

A cohort study [94] examined the relationship between diet (fatty fish consumption) and grip strength in older men and women living in their own homes. Of the dietary factors considered in relation to grip strength, the most important was fatty fish consumption. An increase in grip strength of 0.43 kg (95% confidence interval CI = 0.13–0.74) in men ($P = 0.005$) and 0.48 kg (95% CI = 0.24–0.72) in women ($P < 0.001$) was observed for each additional portion of fatty fish consumed per week. The authors of these two studies suggest that these important influences on muscle function in older men and women raise the possibility that omega-3 fatty acids may be a useful tool for the prevention and treatment of sarcopenia.

The outcomes of other human studies are inconclusive. In a cross-sectional analysis involving 247 older adults (>60 years), no correlation between total omega-3 fatty acids self-reported intake and lower extremity muscular function was found [95]. To our knowledge, there is only one up-to-date randomized double-blind study on this topic, which assigned 126 postmenopausal women to either 1.2 g EPA + DHA or placebo supplements. After 6 months of the intervention, the supplemented women significantly improved their walking speed, compared to controls. However, no differences were found between groups when studying other markers of frailty, such as grip strength and body composition [96].

In conclusion, the association of a good omega-3 fatty acids status and better maintenance of muscle mass is still promising and clearly needs further specific trials.

EFFECTS OF OMEGA-3 FATTY ACIDS ON BONE HEALTH

Bone remodeling is a process that replaces old bone with new. Excess resorption over formation leads to loss of bone mass, osteoporosis, and increased fracture risk, with possible additional functional losses and increased morbidity and mortality [97]. Recently highlighted has been the potential role of PUFA in regulation of bone remodeling through different pathways, including opposing effects on inflammatory cytokines, modulation of prostaglandin E2 (PGE2) production, enhancement of calcium transport, and reduction of urinary calcium excretion [98]. These authors also reviewed plausible biological mechanisms and included gene transcription as an important one. In this sense, peroxisome proliferator activator receptor gamma (PPARγ) is a transcription factor with negative effects on bone homeostasis, that may be activated by the PUFA. Treatment with omega-6 fatty acids inhibits proliferation of osteoblasts via increased expression of PPARγ; however, DHA favors osteoblastogenesis due to its binding affinity for PPARγ. PUFA might also affect bone via other pathways, by increasing nitric oxide production and by promoting osteoblastic differentiation via increased production of insulin-like growth factor-1 (IGG-1) and parathyroid hormone. It is important to note that most of these proposed pathways are reported in animal studies. The same is also true for the suggested effect of PUFA leading to a decreased osteoclast maturation, as reviewed by Mangano et al. [98].

Most studies on PUFA and bone health in aging have reported inconsistent findings. In the above mentioned study by Rousseau et al. [95], a positive relation between total omega-3 intake and hip bone mineral density (BMD) was suggested, but the study had some limitations since it did not control for confounding variables such as age, sex, and energy intake. Moreover, individual omega-3 fatty acids were not examined. Even RCT studies are not concluding, or do not allow to conclude a clear effect only related to PUFA, since they are intervention studies including supplemental vitamin D, vitamin K, or multisupplemented foods [99,100]. Another promising study was undertaken with postmenopausal osteoporotic women [101]; 25 patients received 900 mg omega-3 fatty acid capsules or placebo per day for 6 months. Urine level of pyridinoline (Pyd), a bone resorption marker, decreased significantly ($P < 0.05$) in the treatment group, thus indicating that omega-3 fatty acids can decrease bone resorption, but no effect on bone formation was shown after 6 months of treatment.

The most convincing evidence in favor of PUFA and bone comes from the Framingham Osteoporosis Study (FOS) [102]. A cohort study evaluated the effects of omega-3 fatty acids on bone health in aging. This study investigated the associations between dietary polyunsaturated fatty acid and fish intakes and hip bone mineral density at baseline (1988–1989, $n = 854$) and the changes four years later in the same individuals ($n = 623$) with a mean age of 75 years. High intakes (>3 servings/wk) of fish relative to lower intakes were associated with maintenance of femoral neck BMD (FN-BMD) in men and in women ($P < 0.05$), thus suggesting that fish consumption may protect against bone loss. This study was adjusted for confounders, used valid methods to measure outcomes, and described withdrawals and dropouts.

Farina and her group [103] also conducted the first longitudinal study to examine the association among individual types of short and long-chain PUFA with hip fracture risk in men and women. Higher α-linoleic acid (ALA) intake was associated with lower hip fracture risk (P-trend = 0.02). Participants in the highest quartile of ALA intake had a 54% lower risk of hip fracture than those in the lowest quartile (Q4 vs Q1: HR = 0.46, 95% CI = 0.26–0.83). Higher arachidonic acid (AA) intake was associated with lower hip fracture risk in men (P-trend = 0.05) but not in women. Men in the highest quartile of AA intake had an 80% lower risk of hip fracture than those in the lowest quartile (Q4 vs Q1: HR = 0.20, 95% CI = 0.04–0.96). No significant associations were observed among intakes of total fish, dark fish, tuna, or dark fish + tuna and hip fracture in either simple or multivariable-adjusted models in the combined sample of men and women. Their findings suggested that ALA intake may reduce the risk of hip fracture in women and men, and that AA may reduce the risk of hip fracture in men. The authors also stated that the protective effect of ALA on hip fracture risk in humans may be independent of BMD. Thus, ALA might reduce hip fracture via alternative mechanisms, such as protective effects on bone quality not measured by BMD, including a reduction in bone turnover.

Among nested case-control studies, the latest outcomes from The Women's Health Initiative Study (WHI) also reported stimulating results [104]. This study examined red blood cells (RBC) PUFAs as predictors of hip fracture risk in postmenopausal women. Lower hip fracture risk was associated with higher RBC α-linolenic acid, EPA and total omega-3 PUFAs. Conversely, hip fracture nearly doubled with the highest RBC omega-6/omega-3 ratio, suggesting that high omega-6 intake may overcome the benefits of the omega-3 FA.

Taken together, all these sparse data suggest a possible relation between higher omega-3 FA and better functional mobility in older adults. And even though results are limited and in their initial stages, they could be considered as encouraging. At present, it is generally recognized that future clinical trials and prospective studies are required to determine the long-term benefits of dietary or supplemental PUFA upon bone outcomes.

OMEGA-3 AND CANCER IN AGING

EPA and DHA, used along with anticancer drugs, have improved cancer treatment outcome, in vitro and in vivo studies. Clinical studies have reported some positive results with omega-3 supplements in oncologic young and adult patients. However, there are no systematic reviews assessing the effectiveness of EPA and or DHA omega-3 fatty acids supplementation during cancer therapies in aging on outcome improvement. The only systematic review addressing this question is related to the adult population [105]. This recent review reported that there are beneficial effects of omega-3 fatty acids supplements in patients undergoing chemotherapy and/or radiotherapy on different outcomes, such as body weight, body composition, immune and inflammatory markers, oxidative status, and quality of life. Among them, the most evident beneficial effect described is that fish oil promotes weight maintenance or gain during active oncology treatment. Other important outcomes analyzed such as decreased tumor size and prolonging patient survival were not shown [105].

Among the few studies conducted in older adults, the remarkable one came from the Vitamins and Lifestyle (VITAL) cohort, where EPA/DHA intake, and its primary sources, fish oil supplement use and dark fish consumption, were evaluated in relation to colorectal cancer (CRC) risk. Participants were 50–76 years old. The population taking fish oil supplements for at least three years experienced 49% lower CRC risk than nonusers. The association between fish oil use and decreased CRC risk was primarily observed for men and for colon cancer. The authors concluded that associations between long-chain PUFA intake and CRC may vary by gender, subsite, and genetic risk, providing additional insight into the potential role of long-chain PUFAs in cancer prevention [106]. Conversely, in a population of 22,494 women from the same VITAL study, those in the highest compared with the lowest quintile of dietary EPA + DHA intake had a 79% increased risk of endometrial cancer (95% CI: 16%, 175%; P-trend = 0.026). Furthermore, results were similar for EPA and DHA measured individually and for fish intake. It is important to note that when data were stratified by body mass index (in kg/m^2; <25 or ≥25), increases in risk of long-chain omega-3 PUFAs were restricted to overweight and obese women, and statistically significant reductions in risk were observed for normal-weight women. The authors conclude that randomized trials are needed to confirm this overall increased risk of endometrial cancer reported [107]. To our knowledge, this study is the first to prospectively examine the intake of specific omega-3 PUFAs in association with endometrial cancer incidence. Also in women, the recent study by Hutchins-Wiese et al. [108] reported a positive effect in bone resorption, since it was inhibited in fish oil supplemented postmenopausal breast cancer survivors, in a randomized double-blind placebo controlled pilot study that included 38 women. However, inflammatory markers were not altered. Another promising study in women evaluated the associations between endometrial cancer risk and intake of fatty acids and fish, in a population-based sample of 556 incident cancer cases and 533 age-matched controls. The authors concluded that dietary intake of long-chain PUFAs EPA and DHA in foods and supplements may have protective associations against the development of endometrial cancer. However, the latter population was 35–81 years old [109].

Certainly, larger and specific cancer localization studies in older adults are needed to confirm all the encouraging above mentioned results.

EFFECTS OF OMEGA-3 FATTY ACIDS ON QUALITY OF LIFE AND MORTALITY IN NORMAL AGING

Two single studies (both of them RCT) assessed the influence of omega-3 fatty acids in the quality of life of normal aging [110,111]. Moreover, the baseline characteristics of the population of the two studies were the same, 302 independently living older individuals. The studies differed in the outcomes measured. The first one investigated the effect of EPA and DHA (1800 mg/d EPA + DHA, 400 mg/d EPA or placebo, for 26 wk) on mental well-being. The second one evaluated the effect on physical health, psychological health, social relationships, and satisfaction with environment through the World Health Organization Quality of Life questionnaire (WHOQOL). Plasma concentrations of EPA + DHA increased by 238% in the high-dose group, and 51% in the low dose fish-oil group, compared with the placebo group, reflecting an excellent compliance. However, treatment with neither 1800 mg nor 400 mg differentially affected any of the measures and geriatric scales of mental well-being. Following the same pattern, median baseline total WHOQOL scores ranged from 107 to 110 in the three groups and were not significantly different from each other. Treatment with 1800 mg to 400 mg EPA-DHA did not affect total Quality of Life (QOL) or any of its separate domains after 26 weeks of intervention. For the interpretation of these results, however, it is quite interesting to note that the placebo capsules contained mainly oleic acid, which might somehow exert an effect.

Another interesting concern, in terms of quality of life, is the possibility of potential adverse effects associated with omega-3 supplementation in older adults. Moreover, since omega-3 fatty acid supplementation is becoming increasingly popular. However, given its antithrombotic properties, the potential for severe adverse effects (SAE), such as bleeding, has safety implications particularly in older adults. In this sense, a systematic review of randomized control trials explored the potential for SAE and nonsevere adverse effects (non-SAE) associated with omega-3 supplementation in older adults. A total of ten studies involving 994 initially healthy older adults aged >60 years were included. No SAE were reported, and there were no significant differences in the total adverse effects rate between intervention and placebo groups. The authors concluded the potential for adverse effects (AE) appear mild to moderate at worst and are unlikely to be of clinical significance [112].

Two studies estimating the effect of omega-3 fatty acid on mortality in healthy aging were retrieved. Folsom and Demissie [113] investigated the diet of a group of 720 postmenopausal women with low cancer and coronary heart disease risk. There was an inverse age- and energy-adjusted association between total mortality and fish intake, with a relative risk of 0.82 (95% confidence interval: 0.74, 0.91) for the highest versus lowest quintile. Estimated marine omega-3 fatty acid intake was not associated with total or cause-specific mortality. A more recent intervention study [114] showed results similar to those above mentioned. A group of elderly men ($n = 282$) received a total of 2.4 g omega-3 PUFA in two capsules twice daily (49% EPA and 35% DHA) for 36 months. The authors observed in this supplemented population a tendency toward reduction in all-cause mortality that, despite the low number of participants, reached almost statistical significance ($P < 0.063$).

Most randomized trials have actually tested the effects of adding supplements to the diet and evaluated a possible secondary prevention, limiting inference for dietary omega-3-PUFA or primary prevention. Furthermore, observational studies have assessed self-reported dietary FA intakes, rather than objective biomarkers. In order to clarify the relation of PUFA and mortality in older adults, the compelling study by Mozzafarian et al. [72] investigated associations of plasma phospholipid EPA, DPA, DHA, and total omega-3 PUFA with total and cause-specific mortality, among generally healthy older adults not taking fish or oil supplements. Participants were 2692 US older adults (average age 75 years), free of coronary heart disease (CHD), stroke, or heart failure. In this prospective study, circulating individual and total omega-3-PUFA were associated with lower total mortality, with 27% lower risk across total omega-3-PUFA quintiles. Associations appeared strongest for cardiovascular deaths. The observed mortality differences corresponded to approximately 2.2 more years of remaining life after age 65 in people with higher versus lower omega-3-PUFA levels. Because these biomarkers were measured specifically among older adults, the authors suggest that dietary omega-3 PUFA in later life may be of benefit in reducing total mortality and, therefore, increasing the quality of life.

FINAL CONCLUSIONS

Nutrition is one of the major determinants of successful aging. It is becoming increasingly evident that omega-3 fatty acids appear to have a protective effect on some prevalent pathologies during aging such as neurodegenerative disorders, cardiovascular disease, immune function, bone health, muscle tonus, cancer, and general quality of life.

Omega-3 fatty acids are involved in many brain functions, such as the fluidity of the cell membrane, the receptor affinity, modulating transduction molecules/signaling, modulating apoptosis, modulating neuronal differentiation and ion channels, reducing oxidative stress, having an anti-inflammatory action, neurotransmission, and participating in gene expression. While most of the observational studies in older people showed an inverse relationship between an adequate fish consumption or omega-3 fatty acids intake (diet or supplements) or total content in erythrocyte membrane and cognitive status or less cognitive decline during aging, in intervention studies there are some discrepancies, but it seems that when DHA is given to individuals with mild cognitive impairment or age-related cognitive impairment the data appear to be positive, whereas when Alzheimer's disease was already established no clear benefit is achieved. Results are promising and encourage further research in order to establish a safe and effective treatment solution.

In the early 2000s, evidence from large-scale epidemiological studies suggested that people at risk of cardiovascular heart disease (CHD) benefit from consuming omega-3 fatty acids. The exact mechanisms by which omega-3 performs its functions are still under debate, but the main mechanisms proposed are plaque stabilization, lipid profile, antiinflamatory effect, blood pressure, heart failure, or anti-arrhythmic properties. However, controversial results have been published in recent reviews and meta-analyses reporting both positive and negative findings on the effectiveness of omega-3 fatty acids. In consequence, some authors propose that the beneficial effects of omega-3 FA are not as large as previously implied and recommendations for widespread use should be tempered. Generalizability of the overview finding may be questioned and more studies are needed to evaluate specific outcomes and populations stratified by sex, age, CV risk, medication use, race, etc.

This chapter also deals with the effects of omega-3 fatty acids on immune function, bone health, muscle tonus, cancer, and general quality of life in aging. Results are limited and in their initial stages—even though the elderly population is the largest population studied—but they could be considered as promising. It may be concluded that the number of studies and the methodology employed clearly lack sufficient evidence, and that further specific and larger trials, as well as prospective studies, are required to definitely determine the effects of omega-3 fatty acids on aging and on quality of life.

SUMMARY

- Increasing the intake of omega-3, particularly the long-chain n-3 PUFA may be one strategy to prevent and/or reduce morbidity in the elderly.
- It is becoming increasingly evident that these fatty acids appear to be neuroprotective. In general, longitudinal studies are mainly positive, indicating a protective effect against subsequent cognitive decline. In intervention studies there are some discrepancies, but it seems that when DHA is given to individuals with mild cognitive impairment or age-related cognitive impairment the data appear to be positive, whereas when Alzheimer's disease was already established no clear benefit is achieved. Results are promising and encourage further research in order to establish a safe and effective treatment solution.
- In the early 2000s, evidence from large-scale epidemiological studies suggested that people at risk of cardiovascular heart disease (CHD) benefit from consuming omega-3 fatty acids. Nowadays, some authors propose that the beneficial effects of omega-3 FA are not as large as previously implied and recommendations for widespread use should be tempered. Generalizability of the overview finding may be questioned and more studies are needed to evaluate specific outcomes and populations stratified by sex, age, CV risk, medication use, race, etc.
- This chapter also deals with the effects of omega-3 fatty acids on immune function, bone health, muscle tonus, cancer, and general quality of life in aging. Results are limited and in their initial stages—even though the elderly population is the largest group—but they could be considered as promising. It may be concluded that the number of studies and the methodology employed clearly lacks sufficient evidence, and that further specific and larger trials, as well as prospective studies, are required to definitely determine the effects of omega-3 fatty acids on aging and on quality of life.

References

[1] Knoops KTB, de Groot L, Kromhout D, et al. Mediterranean diet, lifestyle factors, and 10-year mortality in elderly European men and women, the HALE project. J Am Med Assoc 2004;292:1433–9.

[2] Kozlowska K, Szczecinska A, Roszkowski W, et al. Patterns of healthy lifestyle and positive health attitudes in older Europeans. J Nutr Health Aging 2008;12:728–34.

[3] Vincent D, Lauque S, Lanzmann D, et al. Changes in dietary intakes with age. J Nutr Health Aging 1998;2:45–8.

[4] Wakimoto P, Block G. Dietary intake, dietary patterns, and changes with age: an epidemiological perspective. J Gerontol A Biol Sci Med Sci 2001;56A:65–80.

[5] Blanc S, Schoeller DA, Bauer D, et al. Energy requirements in the eight decade of life. Am J Clin Nutr 2004;79:303–10.

[6] Foote JA, Giuliano AR, Harris RB. Older adults need guidance to meet nutritional recommendations. J Am Coll Nutr 2000;19:628–40.

[7] Dean M, Raats MM, Grunert KG, Lumbers M. Factors influencing eating a varied diet in old age. Public Health Nutr 2009;12:2421–7.

[8] Buhr G, Bales CW. Nutritional supplements for older adults: review and recommendations-Part I. J Nutr Elder 2009;28:5–29.

[9] Everitt AV, Hilmer SN, Brand-Miller JC, Jamieson HA, Truswell AS, Sharma AP, et al. Dietary approaches that delay age-related diseases. Clin Interv Aging 2006;1:11–31.

[10] Lichtenstein A, Rasmussen H, Yu W, et al. Modified my pyramid for older adults. J Nutr 2008;138:5–11.

[11] Molfino A, Gioia G, Fanelli FR, Muscaritoli M. The role for dietary omega-3 fatty acids supplementation in older adults. Nutrients 2014;6:4058–72. Available from: http://dx.doi.org/10.3390/nu6104058.

[12] Swanson D, Block R, Mousa SA. Omega-3 fatty acids EPA and DHA: health benefits throughout life. Adv. Nutr 2012;3:1–7.

[13] Riediger ND, Othman RA, Miyoung S, et al. A systemic review of the roles of n-3 fatty acids in health and disease. J Am Diet Assoc 2009;109:668–79.

[14] Kamphuis MH, Geerlings MI, Tijhuis MAR, et al. Depression and cardiovascular mortality: a role for n-3 fatty acids? Am J Clin Nutr 2006;84:1513–17.

[15] Whelan J. (n-6) and (n-3) Polyunsaturated fatty acids and the aging brain: food for thought. J Nutr 2008;138:2521–2.

[16] Bazan NG. Cell survival matters: docosahexanoic acid signaling, neuroprotection and photoreceptors. Trends Neurosci 2006;29:263–71.

[17] Beydoun MA, Kaufman JS, Satia JA, et al. Plasma n-3 fatty acids and the risk of cognitive decline in older adults: the Atherosclerosis Risk in Communities Study. Am J Clin Nutr 2007;85(4):1103–11.

[18] Bourre JM, Paquotte P. Seafood (wild and farmed) for the elderly: contribution to the dietary intakes of iodine, selenium, DHA and vitamins B12 and D. J Nutr Health Aging 2008;12:186–92.

[19] Cole GM, Frautschy SA. DHA may prevent age-related dementia. J Nutr 2010;140:869–74.

[20] Dangour A, Uauy R. N-3 long-chain polyunsaturated fatty acids for optimal function during brain development and ageing. Asia Pac J Clin Nutr 2008;17(S1):185–8.

[21] Dangour AD, Allen E, Elbourne D, et al. Effect of 2-y n-3 long-chain polyunsaturated fatty acid supplementation on cognitive

[22] Huang TL. Omega-3 fatty acids, cognitive decline, and Alzheimer's disease: a critical review and evaluation of the literature. J Alzheimers Dis 2010;21:673–90.

[23] Lukiw WJ, Bazan NG. Docosahexanoic acid and the aging brain. J Nutr 2008;138:2510–14.

[24] Caprari P, Scuteri A, Salvati AM, et al. Aging and red blood cell membrane: a study of centenarians. Exp Gerontol 1999;34:47–57.

[25] Sands SA, Reid KJ, Windsor SL, Harris WS. The impact of age, body mass index, and fish intake on the EPA and DHA content of human erythrocytes. Lipids 2005;40:343–7.

[26] Itomura M, Fujioka S, Hamazaki K, et al. Factors influencing EPA + DHA levels in red blood cells in Japan. In Vivo 2008;22:131–5.

[27] Rees D, Miles EA, Banerjee T, et al. Dose-related effects of eicosapentaenoic acid on innate immune function in healthy humans: a comparison of young and older men. Am J Clin Nutr 2006;83(2):331–42.

[28] Plourde M, Tremblay-Mercier J, Fortier M, Pifferi F, Cunnane SC. Eicosapentaenoic acid decreases postprandial beta-hydroxybutyrate and free fatty acid responses in healthy young and elderly. Nutrition 2009;25:289–94.

[29] Vandal M, Freemantle E, Tremblay-Mercier J, et al. Plasma omega-3 fatty acid response to a fish oil supplement in the healthy elderly. Lipids 2008;43:1085–9.

[30] Garry PJ, Hunt WC, Koehler KM, VanderJagt DJ, Vellas BJ. Longitudinal study of dietary intakes and plasma lipids in healthy elderly men and women. Am J Clin Nutr 1992;55:682–8.

[31] Uauy R, Valenzuela A. Marine oils: the health benefits of n-3 fatty acids. Nutrition 2000;16:680–4.

[32] Andlin-Sobocki P, Jonsson B, Wittchen HU, Olesen J. Cost of disorders of the brain in Europe. Eur J Neurol 2005;12(Suppl. 1):1–27.

[33] Salem Jr N, Litman B, Kim H-Y, Gawrisch K. Mechanisms of action of docosahexaenoic acid in the nervous system. Lipids 2001;36:945–59.

[34] Cederholm T, Salem Jr. N, Palmblad J. ω-3 fatty acids in the prevention of cognitive decline in humans. Adv Nutr 2013;4(6):672–6.

[35] Alessandri JM, Guesnet P, Vancassel S, Astorg P, Denis I, Langelier B, et al. Polyunsaturated fatty acids in the central nervous system: evolution of concepts and nutritional implications throughout life. Reprod Nutr Dev 2004;44:509–38.

[36] Crawford MA, Casperd NM, Sinclair AJ. The long chain metabolites of linoleic and linolenic acids in liver and brain in herbivores and carnivores. Comp Biochem Physiol 1976;54B:395–401.

[37] Young G, Conquer J. Omega-3 fatty acids and neuropsychiatric disorders. Reprod Nutr Dev 2005;45:1–28.

[38] Newman PE. Alzheimer's disease revisited. Med Hypotheses 2000;54:774–6.

[39] Freeman MP, Hibbeln JR, Wisner KL, Davis JM, Mischoulon D, Peet M, et al. Omega-3 fatty acids: evidence basis for treatment and future research in psychiatry. J Clin Psychiatry 2006;67(12):1954–67.

[40] Söderberg M, Edlund C, Kristensson K, Dallner G. Lipid compositions of different regions of the human brain during aging. J Neurochem 1990;54:415–23.

[41] Daviglus ML, Bell CC, Berrettini W, Bowen PE, Connolly Jr ES, Cox NJ, et al. NIH state-of-the-science conference statement: preventing Alzheimer's disease and cognitive decline. NIH Consens State Sci Statements 2010;27(4):1–30.

[42] Úbeda N, Achón M, Varela-Moreiras G. Omega 3 fatty acids in the elderly. Br J Nutr 2012;107(Suppl. 2):S137–51.

[43] Vercambre MN, Boutron-Ruault MC, Ritchie K, et al. Long-term association of food and nutrient intakes with cognitive and functional decline: a 13-year follow-up study of elderly French women. Br J Nutr 2009;102(3):419–27.

[44] Dullemeijer C, Durga J, Brouwer IA, et al. n-3 fatty acid proportions in plasma and cognitive performance in older adults. Am J Clin Nutr 2007;86(5):1479–85.

[45] Van de Rest O, Geleijnse JM, Kok FJ, et al. Effect of fish oil on cognitive performance in older subjects: a randomized, controlled trial. Neurology 2008;71(6):430–8.

[46] Ortega RM, Requejo AM, Andrés P, et al. Dietary intake and cognitive function in a group of elderly people. Am J Clin Nutr 1997;66(4):803–9.

[47] Laurin D, Verreault R, Lindsay J, et al. Omega-3 fatty acids and risk of cognitive impairment and dementia. J Alzheimers Dis 2003;5(4):315–22.

[48] Van de Rest O, Spiro 3rd A, Krall-Kaye E, et al. Intakes of (n-3) fatty acids and fatty fish are not associated with cognitive performance and 6-year cognitive change in men participating in the Veterans Affairs Normative Aging Study. J Nutr 2009;139(12):2329–36.

[49] Andreeva VA, Kesse-Guyot E, Barberger-Gateau P, Fezeu L, Hercberg S, Galan P. Cognitive function after supplementation with B vitamins and long-chain omega-3 fatty acids: ancillary findings from the SU.FOL.OM3 randomized trial. Am J Clin Nutr 2011;94(1):278–86.

[50] Ammann EM, Pottala JV, Harris WS, Espeland MA, Wallace R, Denburg NL, et al. ω-3 fatty acids and domain-specific cognitive aging: secondary analyses of data from WHISCA. Neurology 2013;81(17):1484–91.

[51] Rosales FJ. No differential effect between docosahexaenoic acid and oleic acid in preventing cognitive decline. Am J Clin Nutr 2011;93(2):476–7.

[52] Panza F, Frisardi V, Seripa D, et al. Dietary unsaturated fatty acids and risk of mild cognitive impairment. J Alzheimers Dis 2010;21:867–70.

[53] Valls-Pedret C, Lamuela-Raventós RM, Medina-Remón A, Quintana M, Corella D, Pintó X, et al. Polyphenol-rich foods in the Mediterranean diet are associated with better cognitive function in elderly subjects at high cardiovascular risk. J Alzheimers Dis 2012;29(4):773–82.

[54] Campolongo P, Roozedndaal B, Trezza V, et al. Fat-induced satiety factor oleoylethanolamide enhances memory consolidation. Proc Natl Acad Sci USA 2009;106:8027–31.

[55] Mazereeuw G, Lanctôt KL, Chau SA, Swardfager W, Herrmann N. Effects of ω-3 fatty acids on cognitive performance: a meta-analysis. Neurobiol Aging 2012;33(7):1482.

[56] Dacks PA, Shineman DW, Fillit HM. Current evidence for the clinical use of long-chain polyunsaturated n-3 fatty acids to prevent age-related cognitive decline and Alzheimer's disease. J Nutr Health Aging 2013;17(3):240–51.

[57] Lin PY, Chiu CC, Huang SY, Su KP. A meta-analytic review of polyunsaturated fatty acid compositions in dementia. J Clin Psychiatry 2012;73(9):1245–54.

[58] Sinn N, Milte CM, Street SJ, Buckley JD, Coates AM, Petkov J, et al. Effects of n-3 fatty acids, EPA v. DHA, on depressive symptoms, quality of life, memory and executive function in older adults with mild cognitive impairment: a 6-month randomised controlled trial. Br J Nutr 2012;107(11):1682–93.

[59] Dangour AD, Andreeva VA, Sydenham E, Uauy R. Omega 3 fatty acids and cognitive health in older people. Br J Nutr 2012;107(Suppl. 2):S152–8.

[60] Yurko-Mauro K, McCarthy D, Rom D, Nelson EB, Ryan AS, Blackwell A, et al. Beneficial effects of docosahexaenoic acid on cognition in age-related cognitive decline. Alzheimers Dement 2010;6(6):456–64.

[61] Freund-Levi Y, Eriksdotter-Jönhagen M, Cederholm T, Basun H, Faxén-Irving G, Garlind A, et al. Omega-3 fatty acid treatment in 174 patients with mild to moderate Alzheimer's disease: OmegAD study: a randomized double-blind trial. Arch Neurol 2006;63(10):1402–8.

[62] Quinn JF, Raman R, Thomas RG, Yurko-Mauro K, Nelson EB, Van Dyck C, et al. Docosahexaenoic acid supplementation and cognitive decline in Alzheimer's disease: a randomized trial. J Am Med Assoc 2010;304(17):1903–11.

[63] Samieri C, Lorrain S, Buaud B, Vaysse C, Berr C, Peuchant E, et al. Relationship between diet and plasma long-chain n-3 PUFAs in older people: impact of apolipoprotein E genotype. J Lipid Res 2013;54(9):2559–67.

[64] World Health Organization. Cardiovascular disease. Available in: <http://www.who.int/cardiovascular_diseases/en/>. [visited 28.02.15].

[65] Mozaffarian D, Benjamin EJ, Go AS, Arnett DK, Blaha MJ, Cushman M, et al. Executive summary: heart disease and stroke statistics-2015 update: a report from the American Heart Association. Circulation 2015;131(4):434–41.

[66] Nichols M, Townsend N, Scarborough P, Rayner M. European cardiovascular disease statistics. 2012 ed. European Heart Network and European Society of Cardiology; 2012.

[67] Kris-Etherton PM, Harris WS, Appel LJ, AHA Nutrition Committee. American Heart Association. Omega-3 fatty acids and cardiovascular disease: new recommendations from the American Heart Association. Arterioscler Thromb Vasc Biol 2003;23(2):151–2.

[68] Delgado-Lista J, Perez-Martinez P, Lopez-Miranda J, Perez-Jimenez F. Long chain omega-3 fatty acids and cardiovascular disease: a systematic review. Br J Nutr 2012;Suppl. 2:S201–13.

[69] Mente A, de Koning L, Shannon HS, Anand SS. A systematic review of the evidence supporting a causal link between dietary factors and coronary heart disease. Arch Intern Med 2009;169(7):659–69.

[70] Mozaffarian D, Wu JH. Omega-3 fatty acids and cardiovascular disease: effects on risk factors, molecular pathways, and clinical events. J Am Coll Cardiol 2011;58(20):2047–67.

[71] Zheng J, Huang T, Yu Y, Hu X, Yang B, Li D. Fish consumption and CHD mortality: an updated meta-analysis of seventeen cohort studies. Public Health Nutr 2012;15(4):725–37.

[72] Mozaffarian D, Lemaitre RN, King IB, Song X, Huang H, Sacks FM, et al. Plasma phospholipid long-chain ω-3 fatty acids and total and cause-specific mortality in older adults: a cohort study. Ann Intern Med 2013;158(7):515–25.

[73] Wang C, Harris WS, Chung M, Lichtenstein AH, Balk EM, Kupelnick B, et al. n-3 Fatty acids from fish or fish-oil supplements, but not alpha-linolenic acid, benefit cardiovascular disease outcomes in primary- and secondary-prevention studies: a systematic review. Am J Clin Nutr 2006;84(1):5–17.

[74] Marik PE, Varon J. Omega-3 dietary supplements and the risk of cardiovascular events: a systematic review. Clin Cardiol 2009;32(7):365–72.

[75] Kwak SM, Myung SK, Lee YJ, Seo HG, Korean Meta-analysis Study Group. Efficacy of omega-3 fatty acid supplements (eicosapentaenoic acid and docosahexaenoic acid) in the secondary prevention of cardiovascular disease: a meta-analysis of randomized, double-blind, placebo-controlled trials. Arch Intern Med 2012;172(9):686–94.

[76] Kotwal S, Jun M, Sullivan D, Perkovic V, Neal B. Omega 3 fatty acids and cardiovascular outcomes: systematic review and meta-analysis. Circ Cardiovasc Qual Outcomes 2012;5(6):808–18.

[77] Rizos EC, Ntzani EE, Bika E, Kostapanos MS, Elisaf MS. Association between omega-3 fatty acid supplementation and risk of major cardiovascular disease events: a systematic review and meta-analysis. J Am Med Assoc 2012;308(10):1024–33.

[78] Mozaffarian D, Rimm EB. Fish intake, contaminants, and human health: evaluating the risks and the benefits. J Am Med Assoc 2006;296(15):1885–99.

[79] de Oliveira Otto MC, Wu JH, Baylin A, Vaidya D, Rich SS, Tsai MY, et al. Circulating and dietary omega-3 and omega-6 polyunsaturated fatty acids and incidence of CVD in the Multi-Ethnic Study of Atherosclerosis. J Am Heart Assoc 2013;2(6):e000506.

[80] Risk and Prevention Study Collaborative Group, Roncaglioni MC, Tombesi M, Avanzini F, Barlera S, Caimi V, et al. n-3 fatty acids in patients with multiple cardiovascular risk factors. N Engl J Med 2013;368(19):1800–8.

[81] Larsson SC, Orsini N, Wolk A. Long-chain omega-3 polyunsaturated fatty acids and risk of stroke: a meta-analysis. Eur J Epidemiol 2012;27(12):895–901.

[82] Kromhout D, Giltay EJ, Geleijnse JM, Alpha Omega Trial Group. n-3 fatty acids and cardiovascular events after myocardial infarction. N Engl J Med 2010;363(21):2015–26.

[83] Lemaitre RN, Sitlani C, Song X, King IB, McKnight B, Spiegelman D, et al. Circulating and dietary α-linolenic acid and incidence of congestive heart failure in older adults: the Cardiovascular Health Study. Am J Clin Nutr 2012;96(2):269–74.

[84] Belin RJ, Greenland P, Martin L, Oberman A, Tinker L, Robinson J, et al. Fish intake and the risk of incident heart failure: the Women's Health Initiative. Circ Heart Fail 2011;4(4):404–13.

[85] Mozaffarian D, Bryson CL, Lemaitre RN, Burke GL, Siscovick DS. Fish intake and risk of incident heart failure. J Am Coll Cardiol 2005;45(12):2015–21.

[86] Bechoua S, Dubois M, Véricel E, et al. Influence of very low dietary intake of marine oil on some functional aspects of immune cells in healthy elderly people. Br J Nutr 2003;89(4):523–31.

[87] Thies F, Nebe-von-Caron G, Powell JR, et al. Dietary supplementation with gamma-linolenic acid or fish oil decreases T lymphocyte proliferation in healthy older humans. J Nutr 2001;131(7):1918–27.

[88] Wardwell L, Chapman-Novakofski K, Herrel S, et al. Nutrient intake and immune function of elderly subjects. J Am Diet Assoc 2008;108(12):2005–12.

[89] Thies F, Nebe-von-Caron G, Powell JR, et al. Dietary supplementation with eicosapentaenoic acid, but not with other long-chain n-3 or n-6 polyunsaturated fatty acids, decreases natural killer cell activity in healthy subjects aged >55 y. Am J Clin Nutr 2001;73(3):539–48.

[90] Miles EA, Calder PC. Influence of marine n-3 polyunsaturated fatty acids on immune function and a systematic review of their effects on clinical outcomes in rheumatoid arthritis. Br J Nutr 2012;107(Suppl. 2):S171–84.

[91] Murphy RA, Mourtzakis M, Chu QS, Reiman T, Mazurak VC. Skeletal muscle depletion is associated with reduced plasma (n-3) fatty acids in non-small cell lung cancer patients. J Nutr 2010;140(9):1602–6.

[92] Ryan AM, Reynolds JV, Healy L, Byrne M, Moore J, Brannelly N, et al. Enteral nutrition enriched with eicosapentaenoic acid (EPA) preserves lean body mass following esophageal cancer surgery: results of a double-blinded randomized controlled trial. Ann Surg 2009;249(3):355–63.

[93] Smith GI, Atherton P, Reeds DN, et al. Dietary omega-3 fatty acid supplementation increases the rate of muscle protein synthesis in older adults: a randomized controlled trial. Am J Clin Nutr 2011;93(2):402—12.

[94] Robinson SM, Jameson KA, Batelaan SF, Hertfordshire Cohort Study Group, et al. Diet and its relationship with grip strength in community-dwelling older men and women: the Hertfordshire cohort study. J Am Geriatr Soc 2008;56(1):84—90.

[95] Rousseau JH, Kleppinger A, Kenny AM. Self-reported dietary intake of omega-3 fatty acids and association with bone and lower extremity function. J Am Geriatr Soc 2009;57(10):1781—8.

[96] Hutchins-Wiese HL, Kleppinger A, Annis K, Liva E, Lammi-Keefe CJ, Durham HA, et al. The impact of supplemental n-3 long chain polyunsaturated fatty acids and dietary antioxidants on physical performance in postmenopausal women. J Nutr Health Aging 2013;17(1):76—80.

[97] Leibson CL, Tosteson AN, Gabriel SE, et al. Mortality, disability, and nursing home use for persons with and without hip fracture: a population-based study. J Am Geriatr Soc 2002;50(10):1644—50.

[98] Mangano KM, Sahni S, Kerstetter JE, Kenny AM, Hannan MT. Polyunsaturated fatty acids and their relation with bone and muscle health in adults. Curr Osteoporos Rep 2013;11(3):203—12.

[99] Lappe J, Kunz I, Bendik I, Prudence K, Weber P, Recker R, et al. Effect of a combination of genistein, polyunsaturated fatty acids and vitamins D3 and K1 on bone mineral density in postmenopausal women: a randomized, placebo-controlled, double-blind pilot study. Eur J Nutr 2013;52(1):203—15.

[100] Martin-Bautista E, Muñoz-Torres M, Fonolla J, Quesada M, Poyatos A, Lopez-Huertas E. Improvement of bone formation biomarkers after 1-year consumption with milk fortified with eicosapentaenoic acid, docosahexaenoic acid, oleic acid, and selected vitamins. Nutr Res 2010;30(5):320—6.

[101] Salari Sharif P, Asalforoush M, Ameri F, Larijani B, Abdollahi M. The effect of n-3 fatty acids on bone biomarkers in Iranian postmenopausal osteoporotic women: a randomized clinical trial. Age 2010;32:179—86.

[102] Farina EK, Kiel DP, Roubenoff R, Schaefer EJ, Cupples LA, Tucker KL. Protective effects of fish intake and interactive effects of long-chain polyunsaturated fatty acid intakes on hip bone mineral density in older adults: the Framingham Osteoporosis Study. Am J Clin Nutr 2011;93(5):1142—51.

[103] Farina EK, Kiel DP, Roubenoff R, Schaefer EJ, Cupples LA, Tucker KL. Dietary intakes of arachidonic acid and alpha-linolenic acid are associated with reduced risk of hip fracture in older adults. J Nutr 2011;141(6):1146—53.

[104] Orchard TS, Ing SW, Lu B, Belury MA, Johnson K, Wactawski-Wende J, et al. The association of red blood cell n-3 and n-6 fatty acids with bone mineral density and hip fracture risk in the women's health initiative. J Bone Miner Res 2013;28(3):505—15.

[105] de Aguiar Pastore Silva J, de Souza Fabre ME, Waitzberg DL. Omega-3 supplements for patients in chemotherapy and/or radiotherapy: A systematic review. Clin Nutr 2015;34(3):359—66. Available from: http://dx.doi.org/10.1016/j.clnu.2014.11.005.

[106] Kantor ED, Lampe JW, Peters U, Vaughan TL, White E. Long-chain omega-3 polyunsaturated fatty acid intake and risk of colorectal cancer. Nutr Cancer 2014;66(4):716—27.

[107] Brasky TM, Neuhouser ML, Cohn DE, White E. Associations of long-chain n-3 fatty acids and fish intake with endometrial cancer risk in the VITamins and lifestyle cohort. Am J Clin Nutr 2014;99:599—608.

[108] Hutchins-Wiese HL, Picho K, Watkins BA, Li Y, Tannenbaum S, Claffey K, et al. High-dose eicosapentaenoic acid and docosahexaenoic acid supplementation reduces bone resorption in postmenopausal breast cancer survivors on aromatase inhibitors: a pilot study. Nutr Cancer 2014;66(1):68—76.

[109] Arem H, Neuhouser ML, Irwin ML, Cartmel B, Lu L, Risch H, et al. Omega-3 and omega-6 fatty acid intakes and endometrial cancer risk in a population-based case-control study. Eur J Nutr 2013;52(3):1251—60.

[110] Van de Rest O, Geleijnse JM, Kok FJ, et al. Effect of fish-oil supplementation on mental well-being in older subjects: a randomized, double-blind, placebo-controlled trial. Am J Clin Nutr 2008;88(3):706—13.

[111] Van de Rest O, Geleijnse JM, Kok FJ, et al. Effect of fish oil supplementation on quality of life in a general population of older Dutch subjects: a randomized, double-blind, placebo-controlled trial. J Am Geriatr Soc 2009;57(8):1481—6.

[112] Villani AM, Crotty M, Cleland LG, James MJ, Fraser RJ, Cobiac L, et al. Fish oil administration in older adults: is there potential for adverse events? A systematic review of the literature. BMC Geriatr 2013;13:41.

[113] Folsom AR, Demissie Z. Fish intake, marine omega-3 fatty acids, and mortality in a cohort of postmenopausal women. Am J Epidemiol 2004;160(10):1005—10.

[114] Einvik G, Klemsdal TO, Sandvik L, et al. A randomized clinical trial on n-3 polyunsaturated fatty acids supplementation and all-cause mortality in elderly men at high cardiovascular risk. Eur J Cardiovasc Prev Rehabil 2010;17(5):588—92.

CHAPTER 45

Vitamin E, Inflammatory/Immune Response, and the Elderly

Eugenio Mocchegiani and Marco Malavolta

Translational Center Research on Nutrition and Ageing, Scientific and Technologic Pole, INRCA, Ancona, Italy

KEY FACTS

- Vitamin E is an antioxidant and anti-inflammatory compound for the whole life of an organism with a special emphasis in aging and in some inflammatory age-related diseases.
- Vitamin E deficiency impairs both humoral and cell-mediated immune functions.
- Vitamin E supplementation not exceeding 400 IU/day is useful for a correct inflammatory/immune response in aging.
- The more known isoform of vitamin E (α-tocopherol) seems to have the major properties either as an antioxidant or anti-inflammatory agent. However, another isoform of vitamin E (γ-tocopherol) and tocotrienols (δ-tocotrienol) seems to have more precise antioxidant properties in affecting the inflammatory/immune response in aging and in age-related diseases.
- The interaction of vitamin E with genes related to its bioactivity is fundamental for the success of the clinical trials with vitamin E supplementation in aging and in restoring the inflammatory/immune response.

Dictionary of Terms

- *Vitamin E*: Vitamin E is a lipid-soluble vitamin found in cell membranes and circulating lipoproteins that functions as a nonenzymatic antioxidant scavenging toxic free radicals. It refers to a group of eight compounds that possesses a similar chemical structure comprising a chromanol ring with a 16-carbon side chain and includes all isoforms of tocopherols (α, β, γ, δ) and tocotrienols (α, β, γ, δ).
- *Inflammation*: Inflammation is part of the complex biological response of body tissues to harmful stimuli, such as pathogens, damaged cells, or irritants. Inflammation is a protective response that involves immune cells and molecular mediators. The purpose of inflammation is to eliminate the initial cause of cell injury, clear out necrotic cells and tissues damaged from the original insult and the inflammatory process, and to initiate tissue repair. Inflammation is considered as a mechanism of innate immunity, as compared to adaptive immunity, which is specific for each pathogen. Inflammation can be classified as either acute or chronic. Acute inflammation is the initial response of the body to harmful stimuli and is achieved by the increased movement of plasma and leukocytes (especially granulocytes) from the blood into the injured tissues. A series of biochemical events propagates and matures the inflammatory response, involving in particular the immune system. Chronic inflammation, such as in aging, leads to a progressive shift in the type of cells present at the site of inflammation and is characterized by simultaneous destruction and healing of the tissue from the inflammatory process.
- *Free radical theory*: The free-radical theory of aging was formally proposed by Denham Harman in 1956 and postulates that the inborn process of aging is caused by cumulative oxidative damage to cells

by free radicals produced during aerobic respiration. Free radicals are atoms or molecules with single unpaired electrons. They are unstable and highly reactive, as they attack nearby molecules in order to steal their electrons and gain stability, causing radical chain reactions to occur. Free radicals are generated in vivo primarily within mitochondria during mitochondrial electron transport as well as by other physiological processes. Harman later extended the free-radical theory of aging to incorporate the role of mitochondria in the generation of free radicals and other reactive oxygen species (ROS). The theory proposes that the rate of oxidative damage to mitochondrial DNA primarily determines life span.

INTRODUCTION

Aging is a complex biological phenomenon often accompanied by various socioeconomic changes having a great impact on the nutritional status, needs of the elderly individual, and on the increased incidence of disability due to the commonly onset of some chronic diseases. Among the latter, cardiovascular and neurodegenerative diseases, diabetes, cancer, infections, are closely related to a deficiency in the nutritional status and to the presence of a chronic inflammatory condition [1]. Various factors contribute to the nutritional deficiency in aging with subsequent chronic inflammation and oxidative stress. Under this profile, free radicals and oxidative stress have been recognized as important factors in the biology of aging and in many age-associated degenerative diseases. A time-dependent shift in the antioxidant/prooxidant balance, which leads to higher free radical generation, increased oxidative stress, and dysregulation of cellular function, is the basis for the free radical theory of aging [2,3]. This theory is commonly manifested with phenotypic changes and functional deterioration in later life [4]. The changes are mainly due to reactive oxygen species (ROS) production owing to oxidative stress leading to a damage to DNA, lipid, and proteins with subsequent altered cellular homeostasis and integrity [5]. As a result, the cell has an elaborate system to maintain a proper balance between the levels of free radicals and antioxidants to ensure the integrity of cellular components [6]. This balance is absent in old age due to the presence of high ROS production and antioxidant deficiencies [7]. It has long been postulated that supplementation with dietary antioxidant can alleviate the redox imbalance and thereby protect against the deteriorating effects of oxidative stress, inflammation, progression of degenerative diseases, and aging. In this context, many micronutrients in the diet may fight oxidative stress and delay aging. Among them, vitamin E is considered one of the most potent liposoluble antioxidant to delay aging and to prevent some age-related degenerative diseases [8,9]. In this chapter, we report the main role of vitamin E as a powerful antioxidant especially in maintaining the efficiency of the immune system in aging and the possible personalized supplementation considering the vitamin E—gene interaction.

BIOLOGY AND INTAKE OF VITAMIN E

Vitamin E is a lipid-soluble vitamin found in cell membranes and circulating lipoproteins that functions as a nonenzymatic antioxidant scavenging toxic free radicals. It refers to a group of eight compounds that possess a similar chemical structure comprising a chromanol ring with a 16-carbon side chain and includes all isoforms of tocopherols ($\alpha, \beta, \gamma, \delta$) and tocotrienols ($\alpha, \beta, \gamma, \delta$) [10]. Its most active and abundant form is α-tocopherol, which is considered the major chain-breaking antioxidant in plasma, in cell membranes, and in tissues [11], capable of reacting directly with chain-carrying radicals and consequently interrupting the oxidative chain reactions [12]. α-Tocopherol serves as a peroxyl radical scavenger that protects polyunsaturated fatty acids in membranes and lipoproteins [13]. Apart from its antioxidant property, vitamin E has been reported to also enhance immune response [14] and to modulate DNA repair systems [15] and signal transduction pathways [16]. Advances in gene chip and array technology have led to the discovery of novel vitamin E-sensitive genes that in turn regulate signal transduction pathways. Therefore, polymorphisms in genes involved in vitamin E tissue uptake, export and metabolism may be important determinants for the biological activity of vitamin E itself. Therefore, genetic determinants, environmental and lifestyle factors play important roles in the effective biological activity of vitamin E in aging and in the development of age-associated diseases. In this context, it is relevant to consider the different forms of vitamin E for its possible beneficial effect. The current formulation of vitamin E consists primarily of α-tocopherol, but recent research has suggested that tocotrienol, the lesser known form of vitamin E, appears superior regarding its antioxidant properties [17] and possesses unique biological functions unrelated to antioxidant activity not shared by tocopherol [18]. Even among the tocopherols, particular importance is placed on the other isomers because supplementation with large doses of α-tocopherol alone has been reported to deplete the availability of γ-tocopherol, thus denying the benefits of γ-tocopherol that are not shared by α-tocopherol [19]. Therefore, it has been suggested that the full benefits of vitamin E are better achieved by supplementation with the full spectrum of vitamin E isomers (α-,β-,γ-,δ-tocopherol) and the corresponding tocotrienols [20,21]. Following this suggestion, the Recommended Dietary Allowance (RDA) has established that the vitamin E intake has to be from 7.0 to 11.1 mg/day from conventional diets. In particular, 10 mg/day (median 7.53 mg/day) in men and 7.57 mg/day (median 5.90 mg/day) in women [22]. Despite that, few studies report the Vitamin E deficiency in humans. Lipid malabsorption, deterioration of lipoprotein metabolism, and genetic factors in α-tocopherol transfer protein (α-TTP) result in vitamin E deficiency [9], which is mainly associated to peripheral neuropathy and ataxia [23,24]. In humans, vitamin E is taken up in the jejunum, the proximal part of the small intestine, where the first phase of the uptake is dependent on the amount of lipids, bile, and pancreatic esterases. Unspecific absorption occurs at the intestinal brush membrane by passive diffusion, where, together with triglycerides, cholesterol and apolipoproteins, vitamin E (all its isomers) is reassembled into chylomicrons by the Golgi of the mucosa cells. The chylomicrons are then stored as secretory granula and excreted by exocytosis into the lymphatic system from where they in turn reach the bloodstream. Intravascular degradation of the chylomicrons proceeds via endothelial lipoprotein lipase, a prerequisite for the hepatic uptake of tocopherols [10] and subsequent storage in the liver, in which α-TTP governs the hepatic uptake of vitamin E [25]. α-TTP in the liver specifically sorts out RRR-α-tocopherol (a natural derivate of vitamin E) from all incoming tocopherols for incorporation into plasma lipoproteins in exerting their antioxidant functions [26]. Following its systemic delivery in plasma, tissue-specific distribution and specific regulation of α-tocopherol occur [27]. From all these studies, it emerges a pivotal role played by α-TTP in the economy of vitamin E intake, uptake, and distribution within the body. A deficiency in α-TTP gene expression and also mutation lead to the development of a variety of diseases, such as neurodegeneration, cardiovascular diseases, diabetes, and compromised immune response, which are, in turn, associated to the aging process. In the cases of α-TTP mutation, it is relevant to note that α-tocopherol absorption is normal, but the clearance in the removal of vitamin E results more rapid than its supply. This phenomenon can be explained by the chylomicron form in which vitamin E finds itself, which is more susceptible to degradation and elimination rather than the stable lipoprotein-associated form. This fact implies that in presence of α-TTP mutation (for example in Ataxia with vitamin E deficiency, AVED) a continuous depletion of vitamin E both at cellular and subcellular level occurs with no antioxidant defense [28]. In the absence of a sufficient α-tocopherol content, cell signaling becomes altered and a plethora of deleterious phenomena emerge [29], including a deficiency in immune cell signaling such as CD4 T-cell function, via sphingolipid metabolism [30]. This fact becomes relevant in aging because of the presence of an impaired inflammatory/immune response [31] and an altered sphingolipid composition in CD4 + T cells [30] associated with a possible diet Vitamin E deficiency due to the presence of the intestinal malabsorption [9].

VITAMIN E, INFLAMMATORY/IMMUNE RESPONSE, AND AGING

As reported above, vitamin E is the most effective chain-breaking, lipid-soluble antioxidant in biologic membranes of all cells. Immune cells are particularly enriched in vitamin E because their high polyunsaturated fatty acid content puts them at especially high risk for oxidative damage [32]. Free-radical damage to immune cell membrane lipids may ultimately impair their ability to respond normally to pathogenic challenge with subsequent impaired inflammatory/immune response and development of inflammatory diseases [33]. Available evidence suggests beneficial effects of supplemental vitamin E on immune function and related diseases. Results from animal and human studies indicate that vitamin E deficiency impairs both humoral and cell-mediated immune functions [21]. Taking into account the efficiency of vitamin E in restoring cell-mediated immunity of T cells in the aged [34], several double-blind, placebo-controlled clinical trials tested the effect of vitamin E on immune system in elderly as well as in old animals (see review in Ref. [14]). In the 1990s, Meydani et al. [34] suggested that a short-term vitamin E supplementation could improve immune responsiveness and some clinically relevant indexes of T cell-mediated immunity in healthy elderly. In particular, vitamin E supplementation (800 mg/day of α-tocopheryl acetate) for 30 days significantly improves DTH response, ex vivo T cells proliferation, and IL-2 production concomitantly with a reduction of PGE_2 synthesis by PBMCs and plasma lipid peroxides [34]. The same group tested the effect of lower doses of vitamin E on free-living elderly (≥65 years) indicating that subjects consuming 200 mg/day of vitamin E had a significant increase in DTH and in antibody titer to hepatitis B and to tetanus vaccine compared with placebo group and with subjects supplemented with 60 mg/day and 800 mg/day of vitamin E [35]. It was shown that a longer (6 months) supplementation of vitamin E in healthy elderly subjects (65–80 years) affected the production of IL-2, IFN-γ (typical Th1 cytokines) and IL-4 (typical Th2 cytokine) by PBMCs after stimulation with mitogens. In particular, IL-2 and IL-4 production increased while IFN-γ production decreased in the groups receiving vitamin E [36]. Moreover, healthy elderly subjects receiving a diet supplementation with vitamin E (200 mg/daily) for 3 months showed an improvement of mitogen-induced lymphocytes proliferative response and IL-2 production, NK cell activity, chemotaxis and phagocytosis of neutrophils, and a decrease in neutrophil adherence and superoxide anion production. From all these findings, it emerges that vitamin E is an immunoregulator nutrient in elderly with an effect especially in cell-mediated and innate immunity with thus a possible role in preventing some inflammatory diseases. Such an assumption is also supported by the findings in old animals and "in vitro" models explaining also the mechanisms by how vitamin E works. In particular, vitamin E can enhance T cell-mediated function by directly influencing membrane integrity and signal transduction in T cells mainly affecting CD3/TCR complex as well as CD36 gene expression [37] and the subsequent cascade of key activators in the signal transduction. Among them, PKC, ICAM-1, ZAP-70, LAT, Vav, and nuclear factor-κB (NF-κB) (the latter at nuclear level) play a key role in activating IL-2 gene into the T cells (CD4+) [21,29]. A preincubation with vitamin E in purified spleen T cells from young and old mice increased both cell-dividing and IL-2-producing capacity of naive T cells from old mice, with no effect on memory T cells. These results were of particular interest because they indicated, on one side an effect of vitamin E on genes involved in cell cycle (Ccnb2, Cdc2, and Cdc6) and therefore in cell proliferation; on the other side, they pin-point that vitamin E has a direct immune-enhancing effect via increased IL-2 production. This fact is relevant because it suggests that vitamin E can reverse the age-associated reduction in activation-induced division on naïve T cells, which represent a T-cell subset exhibiting the greatest age-related defects [38]. Moreover, vitamin E is able to reverse the age-associated increase of macrophages synthesis of PGE$_2$, a well-known potent T-cell suppressor and inflammatory mediator [39]. It was also reported that PGE$_2$, apart from being immunosuppressive, regulates the balance of activity between Th1 and Th2 subsets in favor of the latter [40]. Thus, it was speculated that, through its action on PGE$_2$ synthesis, vitamin E stimulates Th1-like immune responses [34]. Alternatively, vitamin E exerts its immune-enhancing effect through inhibiting COX activity without altering COX-1 or COX-2 expression at either protein or mRNA level [41], via a possible reduction of peroxynitrite production, which is a molecule able to upregulate COX-2 activity without changing its expression [42]. In particular, vitamin E reduces COX activity in old macrophages by decreasing nitric oxide (NO) production, which leads in turn to lower production of ONOO with subsequent antioxidant effect [42]. However, the mechanism of vitamin E in affecting T cells is more complex because it also involves lipid rafts on the cell membrane [43] together with another key signaling transducer SHP-1 [44]. In old age, SHP-1 increases due to its NO phosphorylation because of reduced actions of lipid rafts by ROS [45]. The enhancing of SHP-1 in old age blocks Zap-70 and LAT with subsequent negative effect in IL-2 production by IL-2 gene [29]. Vitamin E supplementation in old age, through its double action on CD3/TCR complex (consequently on ZAP-70 and LAT activation) and on lipid rafts by reducing ROS with subsequent low expression (20%) of SHP-1, is able to induce a correct signaling cascade for a satisfactory IL-2 production by naïve CD4+ T cells from old mice [29]. From these findings, it emerges that the action of vitamin E upon the immune system is very complex involving a wide range of signaling transducers that are the subject of continuing investigations. There is a general agreement that ROS contribute to the age-related decline in T-cell function, probably by damaging the lipid moieties of membranes, as well as enzymatic and structural proteins [46]. Thus, the best known function of vitamin E, as a highly lipophilic antioxidant element, may provide an important mechanistic basis, by neutralizing ROS-mediated damage of membrane lipids or associated adapter proteins/kinases [29], and CD4+ naïve T cells have an enhanced susceptibility to oxidative damage [47]. The antioxidant effect of vitamin E may not be only restricted to modulating CD4+ T-cell function, but also to its influence on the activities of several enzymes involved in signal transduction pathways especially those ones related to the inflammation and, consequently, to a correct inflammatory/immune response. For example, vitamin E (α-tocopherol isoform) inhibits PKC [48]. This aspect is relevant taking into account that PKC is involved both in cell-mediated immune response and in cell proliferation [49]. While, on one side, the action of PKC is fundamental in young-adult age during a possible transient inflammatory state; on the other side, in chronic inflammation, such as in aging, an overexpression of PKC may lead to the recruitment of an abnormal number of inflammatory cells in the inflammatory sites through the adhesion molecules (ICAM-1) worsening the just precarious inflammatory picture of aging [50]. Vitamin E (especially the isomer α-tocopherol) is able to reduce the abnormal inflammatory/immune response by monocytes decreasing significantly the superoxide anion release [51] and downregulating the gene expression of extracellular MAP-kinase (ERK 1/2), p38 and NF-κB [52]. As a consequence, the production of pro-inflammatory cytokines (IL-1β) and the expression of adhesion molecules (ICAM-1), via an inhibition of the 5 lipoxygenase pathway, is reduced [53]. Therefore, vitamin E (especially α-tocopherol) has a direct role, via CD3/TCR complex and lipid rafts, on the immune cells acting as an antioxidant agent, whereas it has an indirect role acting on the inflammatory state, via MAPK-kinases and NF-κB inactivation with thus anti-inflammatory properties. Therefore, a right intake and cellular content of vitamin E is pivotal in aging because of impaired T-cell function, altered inflammatory/immune response, increased oxidative stress and chronic inflammation with the risk to develop age-related inflammatory diseases [31]. Such an assumption is strongly supported by the recent findings in centenarian subjects, who show a satisfactory vitamin E content [54] coupled with a satisfactory degree of antioxidant activity, reduced inflammation [31], as well as good performances in inflammatory/immune response [55]. As such, many age-related diseases can be escaped with the achievement of an healthy state and longevity. However, high vitamin E intake may be harmful affecting also mortality [56]. Thus, strong caution has to be used in vitamin E supplementation in aging and in age-related diseases. The vitamin E–gene interactions may be a useful tool for a personalized supplementation avoiding its possible toxic effect because an interaction with other micronutrients might occur leading to an unbalance among micronutrients, as it occurs for other micronutrients [1].

VITAMIN E–GENE INTERACTIONS

Vitamin E family (tocopherols and tocotrienols) contains various isoforms with potent antioxidant and anti-inflammatory properties. For this reason, a lot of clinical trials in humans have been carried out but, unfortunately, with contradictory and inconsistent results (see review in Ref. [57]). Since vitamin E interacts with cell receptors (eg, LDL receptor) and transcription factors (eg, pregnane X receptor) thereby driving (redox-regulated) gene expression (eg, scavenger receptor CD36) and it modulates protein levels (eg, glutathione) and changes enzyme activity levels (eg, protein kinase C), the interaction of vitamin E and the genes codifying these proteins is crucial for the effects of vitamin E supplementation. Modulation of enzyme transcription and/or activity by vitamin E has been shown in genes involved in oxidative stress, proliferation, inflammation, and apoptosis. Such genes include, SOD, NO synthase, cyclooxygenase-2, NAPDH oxidase, NF-κB, phospholipase A2, protein phosphatase 2A, 5-lipooxygenase, activator protein-1, cytochrome P450, BCL2-like 1, and a lot of other genes [58–61]. To obtain a comprehensive understanding of the genes affected by vitamin E, preliminary global gene expression profile experiments using DNA arrays in rat liver and hepatocellular liver carcinoma cells (HepG2) have been conducted over the short-term (49 days) and long-term (290 days) of vitamin E deficiency and then supplemented with vitamin E (RRR-α-tocopheryl acetate) [62,63]. Differential gene expression by DNA arrays comprising up to 7000 genes were measured before and after vitamin E supplementation showing that vitamin E has more long-term rather than short-term effects thereby suggesting that the interaction of vitamin E with genes can have a long term effect. It is noteworthy that the more significant results "in vitro" (HepG2 cells) and "in vivo" experimental rats were obtained using natural vitamin E (RRR-α-tocopheryl acetate) rather than synthetic vitamin E (all-rac-tocopheryl acetate) suggesting that the benefit of vitamin E–gene interactions comes more from the diet rather than from a supplementation [63]. Subsequently, array technology showed a wide range of genes affected by vitamin E, including genes related to inflammation and cell adhesion, cell cycle, and extracellular matrix [63] (Table 45.1). Many of these genes play an important role in many inflammatory age-related diseases especially atherosclerosis and CVD, in particular genes related to the cellular adhesion molecules induced by cytokines inside the human vascular endothelia, such as VCAM-1 expressed at the macrophage surfaces [64], L-selectin from pulmonary macrophages [65], and Mac-1 (CD11/CD18) induced by oxLDL within monocytes [66]. In this context, experiments conducted in vitro and in experimental animal models (for example in APO E null mice) have shown that the interaction between vitamin E and genes related to oxLDL is fundamental for the benefit of vitamin E supplementation inducing a decrement in the amount of cellular lipid peroxides with subsequent inhibition of macrophage uptake of oxLDL [38] suggesting the relevance of the vitamin E–gene interaction in aging and inflammatory age-related diseases. In this regard, a substantial number of papers reports polymorphisms of genes involved in the uptake, distribution, metabolism, and secretion of the micronutrient. A number of genetic polymorphisms and epigenetic modifications (that can occur in the homozygote or heterozygote state) may lower the bioavailability and cellular activity of vitamin E [27,67] (Table 45.2) influencing a differential susceptibility among the people to specific disorders, such as atherosclerosis, diabetes, CVD, cancers, and neurodegenerative diseases, which could be circumvented by vitamin E supplementation. Despite of these genetic findings, few data exist up to date in vitamin E

TABLE 45.1 Some Target Genes Regulated at Transcriptional Level by Vitamin E

Gene class	Gene	Function	Effect of α-tocopherol
Scavenger receptors	CD36, SR-BI, SR-AI/II	Uptake of oxLDL	Inhibition
Extracellular matrix	E-Selectin, L-Selectin, ICAM-1, Integrins, Mac-1	Rolling and adhesion of monocytes/macrophages, platelet adhesion	Inhibition
	Collagen α1, glycoprotein IIb, VCAM-1		
Inflammatory cytokines	TGF-β, IL-4, IL-1β, TNF-α	Inflammation and chemotaxis of inflammatory cells	Inhibition
Cell cycle regulation	P27	Inhibition of smooth muscle cells proliferation and aortic thickening, Induction of proliferation	Induction
	Cyclin D1, Cyclin E, Cyclin B2,		Inhibition
	Cyclin-dependent kinase5, Cdc6-related protein		
Apoptosis	CD95L (CD95 APO-1/Fas ligand), Bcl2-L1, Birc5	Induction of apoptosis	Inhibition
Regulation of transcription	NF-κB, AP-1, PKC	Induction of inflammatory genes	Inhibition
	Kruppel-like factor3, Ikaros	Induction of immune response	Induction
Chemotaxis	Ccl2, MCP-1	Migration and infiltration of monocytes/macrophages	Inhibition
Antioxidant defense	Gamma-glutamyl cisteinyl synthetase (GCS)	Involved in glutathione biosynthesis pathway	Induction
Detoxification	P450-Cytochromes (Cyp3A, Cyp4F2), Pregnane	Detoxification of exogenous and endogenous compounds	Induction
	X receptor (PXR)		
Cell proliferation	MMP-1, MMP-19	Tissue remodeling and inflammatory/immune response	Inhibition
Lipid metabolism	ApoE, PPAR-γ, LDL-R	Lipid uptake, delivery transport	Inhibition
Vascular defence	Haptoglobin (Hp)	Formation of haptoglobin-hemoglobin	Induction
	PAI-1	(Hp–Hb) complex	Induction
		Inhibition of fibrinolysis and degradation of blood clots	

For single references related to the specific gene class see Refs [16,59,64].

supplementation on the basis of specific polymorphisms that can be crucial for the beneficial effect of vitamin E (Table 45.3). In this context, an interesting paper of Testa et al. [70] shows the relevance of the interaction between vitamin E and the gene of plasminogen activator inhibitor type 1 (PAI-1), an independent CVD risk factor, which increases in patients with DM and is closely related to the inflammatory state [75]. The 4G/5G polymorphism of PAI-1 is involved in the incidence of cardiovascular disease by regulation of PAI-1 levels [76]. A treatment with vitamin E (500 IU/die for 10 weeks) in old diabetic patients carrying 4G allele provoked a delayed decrease in PAI-1 levels with respect to those carrying 5G/5G genotype [73]. This finding demonstrates that 4G/5G polymorphism mainly influences the rate of decrease of PAI-1 after supplementation with vitamin E in diabetes. More recently, Belisle et al. [72] proposed that single nucleotide polymorphisms may influence individual response to vitamin E treatment (182 mg/day for 3 years) in terms of pro-inflammatory cytokine production (TNF-α). Old subjects with the A/A and A/G genotypes at TNF-α-308G > A treated with vitamin E had lower TNF-α production than those with the A allele treated with placebo. Since the A allele at TNF-α-308G > A is associated with higher TNF-α levels [77], these results suggest that the anti-inflammatory effect of vitamin E may be specific to subjects genetically predisposed to higher inflammation. Moreover, the interactions between vitamin E and Hp gene or ApoE gene are intriguing. In particular Hp2-2 genotype is involved in diabetic retinopathy with CVD

TABLE 45.2 Some Relevant Genes Possibly Affecting Vitamin E Bioactivity in Relation to Their Polymorphisms

Candidate genes	Function in relation to vitamin E	Effects on vitamin E bioactivity by polymorphisms	References
Haptoglobin (Hp)	Increased free radicals in vitamin E deficiency	Increased free radicals in Hp-2-2 genotype	[68]
Apolipoprotein E (ApoE)	Increased free radicals in vitamin E deficiency; plasma lipoprotein on vitamin E turnover	ApoE4 genotype is associated with increased levels of vitamin E	[69]
SR-BI scavenger receptor	Vitamin E uptake and transport	Influence of vitamin E levels in cell and tissue	[69]
CD36 scavenger receptor	Reduced gene expression of CD36 by vitamin E with no formation of foam cells	Influence on the responsiveness to vitamin E	[70]
LDL-receptor	Removal of LDL from plasma	Influence on plasma lipid profile	[71]
α-Tocopherol transfer protein (α-TTP)	Vitamin E retention in plasma	Influence of plasma and tissue level of vitamin E	[71]
Pregnane X receptor (PXR)	Vitamin E-mediated gene expression	Influence on PXR target genes for detoxification	[71]
P450-cytochromes (Cyp3A, Cyp4F2)	Vitamin E metabolism	Influence on metabolites deriving from detoxification	[71]
TNF-α	Decreased inflammation	Influence on better inflammatory/immune response by vitamin E	[72]
Plasminogen activator inhibitor type 1 (PAI-1)	Control of fibrinolysis by Vitamin E	Delayed and low production of PAI-1	[73]
Tocopherol associated protein (TAP1, TAP2, TAP3)	Vitamin E binding, uptake, signal transduction, gene expression	Influence on vitamin E on cellular activity	[71]
Afamin	Vitamin E transport into the brain	Influence of vitamin E in the nervous system	[74]
Lipoprotein lipase (LPL)	Transfer of vitamin E from lipoprotein into peripheral tissues	Influence on vitamin E content in plasma, tissues and cells	[69]

TABLE 45.3 Vitamin E Supplementation on the Basis of Some Polymorphisms Affecting Vitamin E Bioactivity

Dose of vitamin E	Condition	Gene target	Genotype	Effect	References
500 IU/day for 10 weeks	Type 2 diabetes n.93 ≥60 years	4G/5G Polymorphism of PAI-1	4G/4G 4G/5G 5G/5G	Faster decrement in PAI-1 Low decrement in PAI-1	[73]
182 mg/day for 3 years	Healthy aging n.617 ≥65 years	TNF-α-308 G/A	A/A A/G G/G	Low TNFα-production No effect	[72]
400 IU/day for 18 months	Type 2 diabetes n.726 ≥65 years	Haptoglobin (Hp)	Hp 1-1 Hp 2-1 Hp 2-2	No effect on cardiovascular events (MI, stroke, mortality) Reduction of cardiovascular events (MI, stroke, mortality)	[68]

complications. Supplementation with vitamin E in Hp-2-2 genotype shows potent preventive effects [68,78]. With regard to ApoE, ApoE4 genotype is associated with increased morbidity and mortality, and represents a significant risk factor for CVD and late-onset Alzheimer's disease (AD) [79]. ApoE is an important modulator of many stages of the lipoprotein metabolism as well as possesses immunomodulatory/anti-inflammatory properties. An increasing number of studies in cell lines [80], transgenic rodents [81] and AD [82] indicate higher oxidative stress and pro-inflammatory state associated with the ε4 allele

[82]. AD carrying E4 allele better counteract to the adverse effect of oxidative stress and chronic inflammation than do non-E4 carriers [83]. Therefore, the polymorphisms of Hp and ApoE may be crucial points for the benefit of vitamin E supplementation in diabetes, AD, in inflammation and neurodegeneration. In addition, a significant number of genes was found to be regulated by vitamin E, such as nerve growth factor, dopaminergic neurotransmitters, and clearance of amyloid-β in the rat brain [84]. Of interest is also the influence of vitamin E in miRNA, in particular in miRNA-122a and miRNA-125b that are related with increased inflammation (TNF-α) [63]. Thus, the reduced miRNA-125b levels observed in vitamin E-deficient rats may be associated with an enhanced inflammatory response, as previously described [85]. These findings indicate that vitamin E regulates cell signaling not only at the mRNA level but also at the miRNA level. From all the data relating to vitamin E–gene interactions emerges the pivotal role played by the specific genetic background for a positive effect of vitamin E as an antioxidant and anti-inflammatory agent. However, the study in this field of miRNA is still in its infancy and future research is required for definitive guidelines addressed to a more correct and personalized vitamin E supplementation in relation to miRNA.

CONCLUSIONS AND PERSPECTIVES

While there is no doubt as to the relevance of vitamin E as an antioxidant and anti-inflammatory compound for the whole life of an organism with a special emphasis in aging and in some inflammatory age related diseases, a critical point is the translation of the benefit of vitamin E in human clinical trials. Experiments in various cell cultures and in different animal models have clearly shown that vitamin E is an essential dietary compound for the efficiency of many body homeostatic mechanisms with a particular focus on the immune system. In particular, the cell-mediated immunity and the inflammatory/immune response are preserved by the lipid peroxide formation on CD4+ cells both in aging and inflammatory age-related diseases. As a consequence, the production of IL-2 is satisfactory with a good immune response to external noxae. On the other hand, the presence of good circulating levels in centenarians of vitamin E coupled with satisfactory antioxidant activity and immune response [54], clearly testify the relevance of vitamin E in the economy of the immune and antioxidant performances required to achieve healthy aging and longevity. However, the various isoforms of the vitamin E family do not have similar beneficial effects. The more known isoform of vitamin E (α-tocopherol) seems to have the major properties either as an antioxidant or anti-inflammatory agent. However recently, another isoform of vitamin E (γ-tocopherol) and tocotrienols (δ-tocotrienol) seem to have more precise antioxidant properties in affecting the inflammatory/immune response in aging and in age-related diseases. These effects have been obtained mainly in animal models. When transferred in humans, contradictory data exist on the benefit of various isoforms of vitamin E. It is possible that inadequate subject selection (by sex, vitamin E status, genetic polymorphisms) and the dosage and chemical form of vitamin E administered may partly explain the contradictory data. It is also relevant to note that the effect of vitamin E is greater over the long-term than short-term and the dosage of vitamin E should not exceed 400 IU/day. In this context, it is relevant to note that high doses of vitamin E might induce mortality in old people with atherosclerosis [86] as well as in old frail people [87] with still undefined and unclear mechanisms explaining the mortality by high dose of vitamin E. However, the major incongruence in human clinical trials may be related to the specific genetic background from each individual. Such an assumption is supported by two different approaches with vitamin E supplementation in restoring the inflammatory/immune response in aging [72] and in reducing the insulin resistance in DM [73] on the basis of TNF-α and PAI-1 polymorphisms, respectively. Moreover, polymorphisms of ApoE may be useful for vitamin E supplementation against oxidative stress and inflammation in late Alzheimer's disease [83]. An intriguing point is that vitamin E supplementation in diabetic patients carrying Hp 2-2 genotype leads to a low risk of developing CVD [68]. Therefore, the interaction of vitamin E with genes related to its bioactivity is fundamental for the success of the clinical trials with vitamin E supplementation in aging and in restoring the inflammatory/immune response.

SUMMARY POINTS

- Aging is a complex biological phenomenon in which the deficiency of the nutritional state combined with the presence of chronic inflammation and oxidative stress contribute to the development of many age-related diseases.
- Supplementation with vitamin E can protect against the deteriorating effects of oxidative stress, progression of degenerative diseases, altered inflammatory/immune response, and aging.

- Vitamin E influences many genes related to the inflammatory/immune response supporting a lot of clinical trials in old humans and in inflammatory age-related diseases, which however, have given contradictory and inconsistent results and even indicated a dangerous role of vitamin E, including mortality.
- The more plausible gap is the poor consideration of the vitamin E-gene interactions, especially polymorphisms affecting Vitamin E bioactivity, that may open new roadmaps for a correct and personalized vitamin E supplementation in aging and age-related diseases with satisfactory results in order to reach healthy aging and longevity.

References

[1] Mocchegiani E, Costarelli L, Giacconi R, Piacenza F, Basso A, Malavolta M. Micronutrient (Zn, Cu, Fe)-gene interactions in ageing and inflammatory age-related diseases: implications for treatments. Ageing Res Rev 2012;11:297–319.

[2] Harman D. Free radical theory of aging: dietary implications. Am J Clin Nutr 1972;25:839–43.

[3] Liochev SI. Reactive oxygen species and the free radical theory of aging. Free Radic Biol Med 2013;60:1–4.

[4] Harman D. Origin and evolution of the free radical theory of aging: a brief personal history, 1954–2009. Biogerontology 2009;10:773–81.

[5] Vina J, Borras C, Mohamed K, Garcia-Valles R, Gomez-Cabrera MC. The free radical theory of ageing revisited. The cell signaling disruption theory of aging. Antioxid Redox Signal 2013;19:779–87.

[6] Villanueva C, Kross RD. Antioxidant-induced stress. Int J Mol Sci 2012;13:2091–109.

[7] Poljsak B, Milisav I. The neglected significance of "antioxidative stress". Oxid Med Cell Longev 2012;2012:480895–907.

[8] Meydani M. Vitamin E. Lancet 1995;345:170–5.

[9] Niki E, Traber MG. A history of vitamin E. Ann Nutr Metab 2012;61:207–12.

[10] Brigelius-Flohe R, Traber MG. Vitamin E: function and metabolism. FASEB J 1999;13:1145–55.

[11] Burton GW, Cheeseman KH, Doba T, Ingold KU, Slater TF. Vitamin E as an antioxidant in vitro and in vivo. Ciba Found Symp 1983;101:4–18.

[12] Palace VP, Hill MF, Farahmand F, Singal PK. Mobilization of antioxidant vitamin pools and hemodynamic function after myocardial infarction. Circulation 1999;99:121–6.

[13] Burton GW, Cheng SC, Webb A, Ingold KU. Vitamin E in young and old human red blood cells. Biochem Biophys Acta 1986;860:84–90.

[14] Pae M, Meydani SN, Wu D. The role of nutrition in enhancing immunity in aging. Aging Dis 2012;3:91–129.

[15] Claycombe KJ, Meydani SN. Vitamin E and genome stability. Mutat Res 2001;475:37–44.

[16] Azzi A, Gysin R, Kempna P, Munteanu A, Negis Y, Villacorta L, et al. Vitamin E mediates cell signaling and regulation of gene expression. Ann NY Acad Sci 2004;1031:86–95.

[17] Yoshida Y, Niki E, Noguchi N. Comparative study on the action of tocopherols and tocotrienols as antioxidant: chemical and physical effects. Chem Phys Lipids 2003;123:63–75.

[18] Aggarwal BB, Sundaram C, Prasad S, Kannappan R. Tocotrienols, the vitamin E of the 21st century: its potential against cancer and other chronic diseases. Biochem Pharmacol 2010;80:1613–31.

[19] Jiang Q, Christen S, Shigenaga MK, Ames BN. Gamma-tocopherol, the major form of vitamin E in the US diet, deserves more attention. Am J Clin Nutr 2001;74:714–22.

[20] Brigelius-Flohe R, Kelly FJ, Salonen JT, Neuzil J, Zingg JM, Azzi A. The European perspective on vitamin E: current knowledge and future research. Am J Clin Nutr 2002;76:703–16.

[21] Wu D, Meydani SN. Age-associated changes in immune and inflammatory responses: impact of vitamin E intervention. J Leukoc Biol 2008;84:900–14.

[22] Weber P, Bendich A, Machlin LJ. Vitamin E and human health: rationale for determining recommended intake levels. Nutrition 1997;13:450–60.

[23] Bromley D, Anderson PC, Daggett V. Structural consequences of mutations to the alpha-tocopherol transfer protein associated with the neurodegenerative disease ataxia with vitamin E deficiency. Biochemistry 2013;52:4264–73.

[24] Di Donato I, Bianchi S, Federico A. Ataxia with vitamin E deficiency: update of molecular diagnosis. Neurol Sci 2010;31:511–15.

[25] Stocker A. Molecular mechanisms of vitamin E transport. Ann NY Acad Sci 2004;1031:44–59.

[26] Traber MG, Kayden HJ. Preferential incorporation of alpha-tocopherol vs gamma-tocopherol in human lipoproteins. Am J Clin Nutr 1989;49:517–26.

[27] Rigotti A. Absorption, transport, and tissue delivery of vitamin E. Mol Aspects Med 2007;28:423–36.

[28] Azzi A, Ricciarelli R, Zingg JM. Non-antioxidant molecular functions of alpha-tocopherol (vitamin E). FEBS Lett 2002;519:8–10.

[29] Azzi A. Molecular mechanism of alpha-tocopherol action. Free Radic Biol Med 2007;43:16–21.

[30] Molano A, Huang Z, Marko MG, Azzi A, Wu D, Wang E, et al. Age-dependent changes in the sphingolipid composition of mouse CD4+ T cell membranes and immune synapses implicate glucosylceramides in age-related T cell dysfunction. PLoS One 2012;7:e47650.

[31] Franceschi C, Capri M, Monti D, Giunta S, Olivieri F, Sevini F, et al. Inflammaging and anti-inflammaging: a systemic perspective on aging and longevity emerged from studies in humans. Mech Ageing Dev 2007;128:92–105.

[32] Coquette A, Vray B, Vanderpas J. Role of vitamin E in the protection of the resident macrophage membrane against oxidative damage. Arch Int Physiol Biochim 1986;94:S29–34.

[33] Meydani SN, Han SN, Wu D. Vitamin E and immune response in the aged: molecular mechanisms and clinical implications. Immunol Rev 2005;205:269–84.

[34] Meydani SN, Barklund MP, Liu S, Meydani M, Miller RA, Cannon JG, et al. Vitamin E supplementation enhances cell-mediated immunity in healthy elderly subjects. Am J Clin Nutr 1990;52:557–63.

[35] Meydani SN, Meydani M, Blumberg JB, Leka LS, Siber G, Loszewski R, et al. Vitamin E supplementation and in vivo immune response in healthy elderly subjects. A randomized controlled trial. J Am Med Assoc 1997;277:1380–6.

[36] Pallast EG, Schouten EG, de Waart FG, Fonk HC, Doekes G, von Blomberg BM, et al. Effect of 50- and 100-mg vitamin E supplements on cellular immune function in noninstitutionalized elderly persons. Am J Clin Nutr 1999;69:1273–81.

[37] Ozer NK, Negis Y, Aytan N, Villacorta L, Ricciarelli R, Zingg JM, et al. Vitamin E inhibits CD36 scavenger receptor expression in hypercholesterolemic rabbits. Atherosclerosis 2006;184:15–20.

[38] Adolfsson O, Huber BT, Meydani SN. Vitamin E-enhanced IL-2 production in old mice: naive but not memory T cells show increased cell division cycling and IL-2-producing capacity. J Immunol 2001;167:3809–17.

[39] Wu D, Mura C, Beharka AA, Han SN, Paulson KE, Hwang D, et al. Age-associated increase in PGE2 synthesis and COX activity in murine macrophages is reversed by vitamin E. Am J Physiol 1998;275:C661–8.

[40] Phipps RP, Stein SH, Roper RL. A new view of prostaglandin E regulation of the immune response. Immunol Today 1991;12:349–52.

[41] O'Leary KA, de Pascual-Tereasa S, Needs PW, Bao YP, O'Brien NM, Williamson G. Effect of flavonoids and vitamin E on cyclooxygenase-2 (COX-2) transcription. Mutation Res 2004;551:245–54.

[42] Beharka AA, Wu D, Serafini M, Meydani SN. Mechanism of vitamin E inhibition of cyclooxygenase activity in macrophages from old mice: role of peroxynitrite. Free Radic Biol Med 2002;32:503–11.

[43] Catalgol B, Kartal-Ozer N. Lipid rafts and redox regulation of cellular signaling in cholesterol induced atherosclerosis. Curr Cardiol Rev 2010;6:309–24.

[44] Fulop Jr. T, Douziech N, Larbi A, Dupuis G. The role of lipid rafts in T lymphocyte signal transduction with aging. Ann NY Acad Sci 2002;973:302–4.

[45] Fortin CF, Larbi A, Lesur O, Douziech N, Fulop Jr. T. Impairment of SHP-1 down-regulation in the lipid rafts of human neutrophils under GM-CSF stimulation contributes to their age-related, altered functions. J Leukoc Biol 2006;79:1061–72.

[46] Larbi A, Kempf J, Pawelec G. Oxidative stress modulation and T cell activation. Exp Gerontol 2007;42:852–8.

[47] Lohmiller JJ, Roellich KM, Toledano A, Rabinovitch PS, Wolf NS, Grossmann A. Aged murine T-lymphocytes are more resistant to oxidative damage due to the predominance of the cells possessing the memory phenotype. J Gerontol A Biol Sci Med Sci 1996;51:B132–40.

[48] Boscoboinik D, Szewczyk A, Hensey C, Azzi A. Inhibition of cell proliferation by alpha-tocopherol. Role of protein kinase C. J Biol Chem 1991;266:6188–94.

[49] Baier G. The PKC gene module: molecular biosystematics to resolve its T cell functions. Immunol Rev 2003;192:64–79.

[50] Battaini F, Pascale A. Protein kinase C signal transduction regulation in physiological and pathological aging. Ann NY Acad Sci 2005;1057:177–92.

[51] Wigg SJ, Tare M, Forbes J, Cooper ME, Thomas MC, Coleman HA, et al. Early vitamin E supplementation attenuates diabetes-associated vascular dysfunction and the rise in protein kinase C-beta in mesenteric artery and ameliorates wall stiffness in femoral artery of Wistar rats. Diabetologia 2004;47:1038–46.

[52] Ekstrand-Hammarstrom B, Osterlund C, Lilliehook B, Bucht A. Vitamin E down-modulates mitogen-activated protein kinases, nuclear factor-kappaB and inflammatory responses in lung epithelial cells. Clin Exp Immunol 2007;147:359–69.

[53] Kato E, Sasaki Y, Takahashi N. Sodium dl-α-tocopheryl-6-O-phosphate inhibits PGE2 production in keratinocytes induced by UVB, IL-1β and peroxidants. Bioorg Med Chem 2011;19:6348–55.

[54] Mecocci P, Polidori MC, Troiano L, Cherubini A, Cecchetti R, Pini G, et al. Plasma antioxidants and longevity: a study on healthy centenarians. Free Radic Biol Med 2000;28:1243–8.

[55] Mocchegiani E, Giacconi R, Cipriano C, Muzzioli M, Gasparini N, Moresi R, et al. MtmRNA gene expression, via IL-6 and glucocorticoids, as potential genetic marker of immunosenescence: lessons from very old mice and humans. Exp Gerontol 2002;37:349–57.

[56] Miller III ER, Pastor-Barriuso R, Dalal D, Riemersma RA, Appel LJ, Guallar E. Meta-analysis: high-dosage vitamin E supplementation may increase all-cause mortality. Ann Intern Med 2005;142:37–46.

[57] Mocchegiani E, Costarelli L, Giacconi R, Malavolta M, Basso A, Piacenza F, et al. Vitamin E-gene interactions in aging and inflammatory age-related diseases: implications for treatment. A systematic review. Ageing Res Rev 2014;14:81–101.

[58] Munteanu A, Zingg JM, Ogru E, Libinaki R, Gianello R, West S, et al. Modulation of cell proliferation and gene expression by alpha-tocopheryl phosphates: relevance to atherosclerosis and inflammation. Biochem Biophys Res Commun 2004;318:311–16.

[59] Lirangi M, Meydani M, Zingg JM, Azzi A. α-Tocopheryl-phosphate regulation of gene expression in preadipocytes and adipocytes. Biofactors 2012;38:450–7.

[60] Kaga E, Karademir B, Baykal AT, Ozer NK. Identification of differentially expressed proteins in atherosclerotic aorta and effect of vitamin E. J Proteomics 2013;92:260–73.

[61] Zingg JM, Han SN, Pang E, Meydani M, Meydani SN, Azzi A. In vivo regulation of gene transcription by alpha- and gamma-tocopherol in murine T lymphocytes. Arch Biochem Biophys 2013;538:111–19.

[62] Barella L, Muller PY, Schlachter M, Hunziker W, Stocklin E, Spitzer V, et al. Identification of hepatic molecular mechanisms of action of alpha-tocopherol using global gene expression profile analysis in rats. Biochem Biophys Acta 2004;1689:66–74.

[63] Rimbach G, Moehring J, Huebbe P, Lodge JK. Gene-regulatory activity of alpha-tocopherol. Molecules 2010;15:1746–61.

[64] Zapolska-Downar D, Zapolski-Downar A, Markiewski M, Ciechanowicz A, Kaczmarczyk M, Naruszewicz M. Selective inhibition by alpha-tocopherol of vascular cell adhesion molecule-1 expression in human vascular endothelial cells. Biochem Biophys Res Commun 2000;274:609–15.

[65] Sabat R, Kolleck I, Witt W, Volk H, Sinha P, Rustow B. Immunological dysregulation of lung cells in response to vitamin E deficiency. Free Radic Biol Med 2001;30:1145–53.

[66] Terasawa Y, Manabe H, Yoshida N, Uemura M, Sugimoto N, Naito Y, et al. Alpha-tocopherol protects against monocyte Mac-1 (CD11b/CD18) expression and Mac-1-dependent adhesion to endothelial cells induced by oxidized low-density lipoprotein. Biofactors 2000;11:221–33.

[67] Zingg JM, Azzi A, Meydani M. Genetic polymorphisms as determinants for disease-preventive effects of vitamin E. Nutr Rev 2008;66:406–14.

[68] Milman U, Blum S, Shapira C, Aronson D, Miller-Lotan R, Anbinder Y, et al. Vitamin E supplementation reduces cardiovascular events in a subgroup of middle-aged individuals with both type 2 diabetes mellitus and the haptoglobin 2-2 genotype: a prospective double-blinded clinical trial. Arterioscler Thromb Vasc Biol 2008;28:341–7.

[69] Borel P, Moussa M, Reboul E, Lyan B, Defoort C, Vincent-Baudry S, et al. Human plasma levels of vitamin E and carotenoids are associated with genetic polymorphisms in genes involved in lipid metabolism. J Nutr 2007;137:2653–9.

[70] Zingg JM, Ricciarelli R, Andorno E, Azzi A. Novel 5′ exon of scavenger receptor CD36 is expressed in cultured human vascular smooth muscle cells and atherosclerotic plaques. Arterioscler Thromb Vasc Biol 2002;22:412–17.

[71] Döring F, Rimbach G, Lodge JK. In silico search for single nucleotide polymorphisms in genes important in vitamin E homeostasis. IUBMB Life 2004;56:615–20.

[72] Belisle SE, Leka LS, Delgado-Lista J, Jacques PF, Ordovas JM, Meydani SN. Polymorphisms at cytokine genes may determine the effect of vitamin E on cytokine production in the elderly. J Nutr 2009;139:1855–60.

[73] Testa R, Bonfigli AR, Sirolla C, Boemi M, Manfrini S, Mari D, et al. Effect of 4G/5G PAI-1 polymorphism on the response of PAI-1 activity to vitamin E supplementation in Type 2 diabetic patients. Diabetes Nutr Metab 2004;17:217–21.

REFERENCES

[74] Voegele AF, Jerkovic L, Wellenzohn B, Eller P, Kronenberg F, Liedl KR, et al. Characterization of the vitamin E-binding properties of human plasma afamin. Biochemistry 2002;41:14532–8.

[75] De Taeye B, Smith LH, Vaughan DE. Plasminogen activator inhibitor-1: a common denominator in obesity, diabetes and cardiovascular disease. Curr Opin Pharmacol 2005;5:149–54.

[76] Grubic N, Stegnar M, Peternel P, Kaider A, Binder BR. A novel G/A and the 4G/5G polymorphism within the promoter of the plasminogen activator inhibitor-1 gene in patients with deep vein thrombosis. Thromb Res 1996;84:431–43.

[77] Cipriano C, Caruso C, Lio D, Giacconi R, Malavolta M, Muti E, et al. The −308G/A polymorphism of TNF-alpha influences immunological parameters in old subjects affected by infectious diseases. Int J Immunogenet 2005;32:13–18.

[78] Vardi M, Blum S, Levy AP. Haptoglobin genotype and cardiovascular outcomes in diabetes mellitus – natural history of the disease and the effect of vitamin E treatment. Meta-analysis of the medical literature. Eur J Int Med 2012;23:628–32.

[79] Liu CC, Kanekiyo T, Xu H, Bu G. Apolipoprotein E and Alzheimer disease: risk, mechanisms and therapy. Nat Rev Neurol 2013;9:106–18.

[80] Huebbe P, Jofre-Monseny L, Boesch-Saadatmandi C, Minihane AM, Rimbach G. Effect of apoE genotype and vitamin E on biomarkers of oxidative stress in cultured neuronal cells and the brain of targeted replacement mice. J Physiol Pharmacol 2007;58:683–98.

[81] Jofre-Monseny L, de Pascual-Teresa S, Plonka E, Huebbe P, Boesch-Saadatmandi C, Minihane AM, et al. Differential effects of apolipoprotein E3 and E4 on markers of oxidative status in macrophages. Br J Nutr 2007;97:864–71.

[82] Jofre-Monseny L, Minihane AM, Rimbach G. Impact of apoE genotype on oxidative stress, inflammation and disease risk. Mol Nutr Food Res 2008;52:131–45.

[83] Tanzi RE, Bertram L. New frontiers in Alzheimer's disease genetics. Neuron 2001;32:181–4.

[84] Rota C, Rimbach G, Minihane AM, Stoecklin E, Barella L. Dietary vitamin E modulates differential gene expression in the rat hippocampus: potential implications for its neuroprotective properties. Nutr Neurosci 2005;8:21–9.

[85] Yamaoka S, Kim HS, Ogihara T, Oue S, Takitani K, Yoshida Y, et al. Severe vitamin E deficiency exacerbates acute hyperoxic lung injury associated with increased oxidative stress and inflammation. Free Radic Res 2008;42:602–12.

[86] Saremi A, Arora R. Vitamin E and cardiovascular disease. Am J Ther 2010;17:e56–65.

[87] Bjelakovic G, Nikolova D, Gluud LL, Simonetti RG, Gluud C. Antioxidant supplements for prevention of mortality in healthy participants and patients with various diseases. Cochrane Database Syst. Rev. 2012;3: Art. No: CD007176.

CHAPTER 46

Polyphenols and Aging

E. Paul Cherniack

Division of Geriatrics and Palliative Medicine, Miller School of Medicine, University of Miami, Bruce W. Carter Miami VA Medical Center, Miami, FL, USA

KEY FACTS

- Several polyphenols increase the lifespan of invertebrates or fish: resveratrol, quercetin, curcumin, cathecin, tannic acid, gallic acid, chicoric acid, caffeic acid, rosmarinic acid, mycetin, and kaempferol.
- None has yet to do so in other vertebrates.
- Polyphenols act on important age and metabolic regulatory genes, such as mTOR and AMPK.

Dictionary of Terms

- *Polyphenol*: A molecule containing two or more phenol rings.
- *Reactive oxygen species*: Oxygen bearing molecules that can chemically react with intracellular structures which cells use to send a signal but which sometimes cause damage if the local environments cause too high a concentration to occur.
- *Autophagy*: A process by which cells recycle molecules, most commonly through organelles called lysosomes.
- *Klotho gene*: A gene that produces an insulin-regulatory protein found in cell membranes.
- *mTOR (mammalian target of rapamycin)*: a gene coding for an enzyme important in energy regulation and cell growth.

INTRODUCTION

Despite the ubiquity of aging, and the effort and interest spent trying to fathom the mechanisms behind the process, our understanding is constantly evolving. This understanding is shaped by the fortuitous discovery of compounds that influence lifespan and age-related morbidities, such as plant polyphenols. In order to better comprehend the potential interaction between plant polyphenols and animal aging, it may be helpful to review the current theoretical of the aging process.

OVERVIEW OF THE PROCESS OF AGING

Presently, two theories receive the most attention in the published scientific literature, both of which have implications for efforts to alter the course of aging. The first of these suggests that aging is the result of the continued activation of intracellular genetic energy regulatory pathways, which had been useful earlier in life, inappropriately at a later age due to the failure of the function of the counterregulatory suppressor genes [1]. Particularly important are the energy regulatory gene AMPK and the mTOR gene complex, which promote cell growth and hypertrophy [1,2]. The increase in lifespan in several invertebrate species and mice by the mTOR inhibitor rapamycin, when the Mtorc1 gene is activated, and when Rictor, a component of mTORC1, is deleted transgenically provide experimental evidence for this theory [3–6]. AMPK and mTORc1 also regulate autophagy, the process by which cells recycle organelles and molecules. According to this theory, aging might be modified by supplying substances that interrupt signaling along the appropriate genetic pathways.

An alternative hypothesis suggests that aging is the result of accumulated mitochondrial injury caused by putative toxins, such as reactive oxygen species (ROS). In this model, loss of NAD+ and an increase in hypoxic-inducible factor (HIF-1α) blocks interaction between the nucleus and mitochondria [7,8]. Supplementation of NAD+ to old mice improves biomarkers of mitochondrial health in myocytes, but the mice do not regain strength [8]. The theory implies that aging might be reversed by substances that control these toxins or preserve mitochondrial functioning.

In fact, both of these theories might not be mutually exclusive. One of the actions of mTOR is to stimulate HIF-1α production [9]. While a comprehensive review of the theories delineating the process of cellular aging is far beyond the scope of this manuscript, and other theories exist, taken together both theories imply targets for modification of the aging process, and some agents or classes of molecules exist that might act to retard aging according to both theories.

PLANT POLYPHENOL FUNCTION AND XENOBIOTIC EFFECTS ON THE AGING PROCESS

Plant polyphenols are molecules containing multiple phenolic rings that provide multiple intracellular functions. In plants, they may induce cells to strengthen their cell walls in response to the stress of UV light energy, and scavenge excess (ROS) [10]. As cells may also utilize ROS for intracellular signaling, polyphenols may regulate message transduction within plants [10]. Furthermore, they may act defensively to deter consumption by animals [11].

Animals commonly consume plants, ingesting and absorbing their polyphenols. The pharmacokinetics of a small proportion of polyphenols found as food have been studied, largely in rodents and humans. Polyphenols are extensively conjugated in the intestine before absorption, which may alter their activity [12]. Absorbed metabolites are additionally modified in the liver, and the low bioavailability of ingested polyphenols has been a long-standing concern of researchers [13].

Interest in the consumption of polyphenols as anti-aging substances has been spurred by a phenomenon termed the "French Paradox," characterized by a below expected death rate from heart disease considering a large saturated fat consumption by the population [14]. Further analysis led to the possibility that plant polyphenol intake might be a protective factor, especially the red wine polyphenol resveratrol [14].

Subsequently, much research on the cellular level, and on an organismal level in invertebrates, fish, and rodents has been conducted to ascertain its effects on the aging process [15]. Many polyphenols at some concentration increase lifespan including resveratrol, quercetin, and curcumin [16].

In most, but not all of these species thus far, resveratrol extends lifespan [15]. This, the most extensively studied polyphenol, augments the lifespan of the yeast *Saccharomyces cerevisiae* by 70% [17]. The polyphenol extended the lifespans of the fruit fly *Drosophila melanogaster* by 29% and the flatworm *Caenorhabditis elegans* by 20% [18]. A fish *N. fuzuri* lived 50% longer, but mice fed normally did not gain longevity [19–21]. However, mice that consumed a high-fat lifespan-curtailing diet regained a normal lifespan when resveratrol was added [21].

The polyphenol quercetin boosted longevity in the yeast *S. cerevisiae* by 20% [22,23]. In one study, yeast grown in a concentration of between 100 μM and 200 μM augmented lifespan, but a 600 μM, concentration reduced survival, suggesting determination of optimal dose may be an important concept in applying polyphenols to enhance longevity [24].

Curcumin, a polyphenol found in the common spice turmeric increases the lifespan of yeast, flies, and mice. *Caenorhabditis elegans* exposed to 20 μM of the polyphenol curcumin lived 39% longer, but those given 200 μM failed to gain any survival benefit [25]. Interestingly, curcumin also increases lifespan in *D. melanogster*, but male flies optimized longevity at a higher concentration than female [26].

Numerous other polyphenols augment longevity in *C. elegans*. Cathecin, a polyphenol derived from an East Asian tree, enhanced the lifespan at a 200 μM dose [27]. Tannic acid, found in numerous plants and parts, including oak, sumac, and gallnuts, raised mean lifespan by 17% at a concentration of 50 μM, but the increase tapered off at higher doses [28]. Gallic acid, derived again from sumac, oak, and gallnuts, raised flatworm length of life at 300 μM by a mean 10% [28]. Chicoric acid, from the chicory plan increased *C. elegans* lifespan by 11% [29]. The flatworms lived 8% longer grown in 100–200 μM caffeic acid, obtained from mushrooms, ferns, and the Eucalyptus tree [16,24]. The fruit polyphenol ellagic acid in a 50 μM concentration raised longevity in worms by 9% [28]. An herbal polyphenol, rosmarinic acid enhanced *C. elegans* lifespan by 10% at 200 μM [24]. Mycertin, a polyphenol contained in numerous plants, augmented flatworm longevity a mean 18%. Finally, kaempferol, derived from a wide variety of fruits, vegetables and tea augment work lifespan by a mean 5.6% [30].

INTRACELLULAR POLYPHENOL EFFECTS

Polyphenols have multiple actions that might alter the process of cellular aging. Firstly, polyphenols act on genes that regulate mTor and AMPK activity [31]. Several polyphenols, including the most well studied polyphenol, resveratrol, binds to protein deacetylase genes called sirtuins, which, in both invertebrates and vertebrates, inhibit mTor [32]. In addition, polyphenols induce molecular pathways to increase AMPK, which also shares common molecular targets with sirtuins [31]. Polyphenols block phosphodiesterases that recycle cyclic AMP, augmenting sirtuin levels, and inducing intracellular calcium to raise AMPK concentrations [33,34]. Polyphenols also can act directly upon on sirtuins, fitting into the peptide binding pocket of the sirtuin [35]. In addition to resveratrol, phloridzin [36], a polyphenol found in apples, activates the sirtuin gene in yeast, augmenting lifespan.

The result of polyphenol action on mTOR and AMPK pathways may be to maintain autophagy and mitochondrial functioning [37]. Autophagy can be an important strategy for the cell to maintain the function of its organelles, such as mitochondria, and recycle defective proteins, in situations of nutrient scarcity and stress [38]. Autophagy is also observed in dying cells, including programmed cell death [38].

To complicate matters, some polyphenols, such as genistein, actually inhibit autophagy, while others, such as naringin, protect cells against agents that suppress autophagy [39]. Structural differences between polyphenols may account for these differences [39]. Interestingly, several polyphenols, including resveratrol and quercetin, may have different effects on autophagy, which can be an intracellular strategy to remove or replace defective proteins or organelles, depending on intracellular conditions [33,40]. When the intracellular nutrient supply is ample, autophagy is promoted, while when nutrients are scarce, autophagy is downregulated [41–43]. Polyphenols, such as quercetin, ECGC, and curcumin, in addition to acting on other targets in the mTOR or AMPK pathways, activate different PI3K molecules (in the case of quercetin), ERK and Akt (curcumin), or Beclin1 (ECGC), which may have different effects on autophagy [38]. Resveratrol promotes autophagy in vitro in ovarian and breast cancer cells by multiple mechanisms [44,45]. Old rats that consumed water with a 20 kg/d concentration of a red grape polyphenol extract for a month experienced improved myocyte mitochondrial autophagy and biogenesis, while measurements of oxidative stress were unaltered [46].

The effects of polyphenols on autophagy and other processes important to aging may also be mediated by their effects on free radicals and cellular oxidative stress. To generalize about the actions of polyphenols on intracellular activity is difficult, as not all polyphenols have similar actions. Some chelate, bind ROS, while in addition to, or alternatively, bind receptors, and participate in multiple signaling pathways [47]. Some of the effect of polyphenols on ROS and autophagy may be "hermetic," implying a beneficial effect at lower polyphenol concentrations that induce stress responses, and a toxic effect at concentrations that are too great [38]. When young (3 months old) and old (20 months old) rats received 0.5–1 g/kg/d of an olive leave extract (a plentiful source of polyphenols) for 60 days, tissue concentrations of several ROS species, including glutathione, diene conjugate, and malondialdehyde, were reduced although the concentrations of several enzymes metabolizing ROS, superoxide dismutase and glutathione transferase, remained unchanged [48]. *Drosophila melanogaster* exposed to an experimental oxidative stress by the consumption of paraquat and iron in their drinking water but that consumed 0.1 mM ECGC for a day lived longer and had greater locomotor activity than flies unprotected by ECGC [49].

An additional mechanism by which polyphenols exert an intracellular effect on ROS may involve its influence on apoptosis [50]. Apoptosis is complex process triggered by both internal mechanisms, involving the mitochondria, and external, in which a specific cell membrane receptor is bound. ROS transduce apoptotic signals to the mitochondria. Apoptosis may be desirable in some situations, for example, inducing neoplastic cell death, but in other situations, such as neurodegenerative diseases, be unwanted. The polyphenols quercetin and epigallocathecin-3–gallate (ECGC) can amplify apoptosis modulation through its action on ROS or by modification of downstream genetic targets, such as caspase proteins. In fact, the relationship between polyphenols and apoptosis is a finely nuanced one, because polyphenols can also inhibit apoptosis. Furthermore, the same polyphenols might promote or retard apoptosis depending on the intracellular redox state. Investigations of the influence of polyphenols on apoptosis, in combination with others, imply the exact relationship between apoptosis and the process of aging awaits further clarification, although autophagy can result in apoptosis. Silibinin, for example, reduces ROS concentrations, augments autophagy and apoptosis in neoplastic cells, but preserves neurons and hepatocytes against toxins [38,51–53].

The effects of polyphenols on oxidative stress may be mediated intracellularly through mitochondria.

When rats consumed 7 mmg/kg/d of red wine polyphenols for 6 months, antioxidant processing enzymes in myocyte mitochondrial were enhanced [54]. An extract derived from strawberries with antioxidant properties protected fibroblasts grown in the presence of hydrogen peroxide to create oxidative stress from mitochondrial damage [55]. Several polyphenols, rosmarinic acid, quercetin, and caffeic acid both activated stress response genes, reduced oxidative stress markers, and augmented the antioxidative capacity and the lifespan of C. elegans [24].

In fact, mitochondria may be important mediators of the effect of polyphenols on aging. Aside from, or, perhaps, in addition to, their effect on ROS, polyphenols stimulate mitochondrial biogenesis [56]. This enhancement of mitochondrial biogenesis may involve sirtuins, although this has not been proven yet by direct experimentation [56].

Polyphenols also can impact intracellular functioning through their effects on the Klotho gene. Activation of the Klotho gene expresses a peptide found in the cell membrane that induces the transcription factors insulin-like growth factor 1 and transforming growth factor(TGF)-1β. Animals that lack the genes are plagued with shortened lifespans and higher rates of age-related diseases [57]. Resveratrol alters the expression of Klotho by modulating transcriptional factor activator protein-1 and activating transcription factor-3(ATF-3) [57]. Renal murine cells grown in resveratrol augmented Klotho mRNA expression and ATF-3 and c-jun gene concentrations [57].

Other transcription factors, such a NF-κβ, may also mediate polyphenol activity [38]. These may include enhancement or downregulation of autophagy and programmed cell death.

Epigenetic effects may be additional mechanisms by which polyphenols alter cellular aging. At least 24 different polyphenols are known to have epigenetic effects including changes to histone structure, addition of methyl groups to DNA, retardation of histone deacetylase function, and blockage of RNA activity [58]. The well-known interaction between polyphenols and sirtuins is an important example. Polyphenol induced epigenetic modification creates multiple intracellular and organismic actions, such as reduction of proinflammatory gene activation via the MAPK and NK-κβ signal transduction pathways, fat synthesis, and an increase in lipid metabolism [58]. Mitochondria derived from old *Podospora anserine* display increased concentrations of a protein methylating enzyme, S-adenosyl-methionine dependent methyl-transferase (PaMTH1), that binds with flavonoids, one group of polyphenols than mitochondria from younger organisms [59], Increasing PaMTH1 concentrations improved organism health and lowering levels of proteins revealing evidence of ROS injury [60]. Organisms genetically engineered without the gene to encode for the enzyme exhibit a reduced lifespan, but when they are transfected with a plasmid containing a PaMTH1 gene, their lifespans are restored [61,62].

An intriguing possibility is that polyphenols may exert activity through their effect on microRNAs (miRNAs) [63]. MiRNAs have multiple effects on intracellular activity, including modulation of DNA and RNA processing enzymes. The activation of miRNAS may promote the differentiation of stem cells, and promote inflammation. Numerous polyphenols alter the expression of microRNAs including resveratrol, curcumin, and isoflavones.

ORGAN SYSTEM POLYPHENOL EFFECTS

As a manifestation of its intracellular effects, polyphenols have the potential to alter organ system aging in multicellular organisms. While it may distract from the purpose of this chapter to outline the relationship of polyphenols to the aging process to provide a comprehensive consideration of research on the mechanism of action of polyphenols in specific disease processes, it may be helpful to exemplify the mechanisms by which these substances might act to retard organ system aging. For example, polyphenols can promote cellular differentiation. Polyphenols can increase stem cell numbers necessary to maintain an organ system. Mice that received resveratrol supplementation of 5 mg/kg resveratrol for 21 days augmented their production of multipotential progenitor hemapoietic cells [64]. In C. elegans, a series of mulberry leaf polyphenols were found to increase lifespan [65]. Their activity was noted to require the activation of multiple germline signaling genes [65].

Polyphenols can restore organ system function through enhancement of cellular functions that improve intercellular interaction. Mice given 30 mg/kg/d of catechin for 3–12 months after treatments with acetylcholine to induce endothelial dysfunction experienced improved endothelial cell adhesion and altered ROS processing enzyme levels [66].

Another important vascular consequence of polyphenol action is its enhancement of endothelial nitrous oxide (NO) activity that induces vasodilation, an important response mechanism in vascular homeostasis. Blood vessels in aging rodents exhibit increased concentrations of ROS and lower dilatory responsiveness to NO [67]. S6kI, a molecule that mTOR regulates determines the vasodilatory response. Pretreatment of middle age rats with 100 mg/kg of red wine polyphenols for one month restored the age-related loss vascular relaxation response to NO in mesenteric artery rings [67]. In another experiment, rats receiving the same dose of grape-derived polyphenols improved vasodilation in mesenteric artery rings, corrected cellular deficits in angiotensin converting enzymes receptors, and reduced concentrations of angiotensin converting enzyme [68]. Resveratrol inhibited activation of S6KI, a molecule induced by mTOR, that reduces the ability of endothelial cells to respond to NO and ROS [69]. ECGC augments mRNA creation by endothelial nitric oxide synthetase, and quercetin augments NO production in bovine endothelial tissue [70]. Resveratrol and other polyphenols also enhance vasodilation through its antagonism of several molecules that result in vascular constriction and platelet adhesion including COX1, p38 MAP kinase, PKC, TXA2, and TXB2 [70].

Polyphenols may also protect the heart by preserving autophagy. When elderly (26 months old) rats were treated with the ROS stimulating-drug doxorubicin, the concentrations of several proteins associated with autophagy in cardiac myocytes declined. However, when the rats exposed to doxorubicin consumed 50 mg/kg/d day resveratrol combined with a 20% calorie restriction (40% calorie restriction lengthens lifespan and augments autophagy) for one and a half months, the expression of autophagic proteins was preserved, which did not occur when rats were diet-restricted in the absence of resveratrol [71].

An additional mechanism by which polyphenols can regulate organ systems, such as the vasculature, is through their anti-inflammatory effects [72]. Polyphenols both in vitro and in vivo stimulate the production of antinflammatory cytokines. Polyphenols attenuate signaling through the NF-κβ pathway, an important signal transducer for inflammation involving COX-2, vascular endothelial growth factor, matrix metalloproteinases (MMPs), TNF-α, and interleukins 1, 2, and 6 [73]. The polyphenol luteolin attenuated transmission along the NF-κβ signaling pathway and activation of genes coding for inflammatory cytokines in bone marrow dendritic, pancreatic, and colon epithelial cells [74–76].

FUTURE DIRECTIONS

Although the potential impact of polyphenols on the aging process has generated much interest and research, many important questions remain unanswered that will hopefully be resolved in future research. To generalize, these questions might be grouped into two categories: what should be the target endpoints of research, and what form of polyphenols should be used to study these.

Proper selection of endpoints is critical in determination of whether polyphenols might in fact be efficacious. Reviews on the effect of polyphenols on aging have generally considered two different, but possibly related types of endpoints: modification of lifespan, and modification of diseases that occur more commonly with advanced that frequently impact lifespan (eg, cardiovascular diseases, neurodegenerative diseases, and cancer) [16,77]. While, this chapter has concentrated primarily on description of the former rather than the latter to maintain its focus on the aging process rather than the complex individual pathophysiologic of discreet diseases, polyphenols may enhance both. Research conducted in vitro and in vivo on unicellular organisms, invertebrates, and some rodent species suggest this [16]. However, the form of polyphenol and dose used to augment lifespan optimally might not be the same as to protect against or treat diseases more commonly found in old age, such as cancer. Further study ought to establish if this is the case.

In addition, new technologies are available that might help researchers to predict which of the plethora of polyphenols and their metabolites might exhibit the greatest promise in retarding the aging process. One, called network analysis, analyzes the spectrum of genes a molecule activates, allowing someone with knowledge of which genes are activated in a given cellular signal transduction pathway to select the molecule that induces the appropriate genes [78]. In a recent investigation, the authors applied this technology to the polyphenol icariin derived from *Epimedium* plants to determine its potential effect on sirtuins, nitrous oxidase synthetase, and pro-inflammatory cytokines, among others [78]. Another technique applies a software program, known as Prediction of Activity Spectra for Substances (PASS) to predict, based on a molecule's structure and known effects from prior research, what effects that molecule, or molecules with a similar structure, ought to have on an organism [11]. In one study, investigators used PASS to analyze multiple related polyphenol compounds derived from extra-virgin olive oil on expected effects related to aging [11].

Another important issue, and one that has engendered much discussion among scientists, is the bioavailability of polyphenols [72]. Most polyphenols are almost completely metabolized in the colon and liver [79,80]. In the quantities present in natural foodstuffs, after ingestion they appear in only trace quantities in the serum. The metabolites might also possess activity, but the pharmacokinetics of few polyphenols or their metabolites have been studied. Many researchers have questioned if they can be delivered in a form that would significantly alter the course of aging or age-related diseases.

A related issue is the appropriate form of the polyphenol that would best alter the process of aging. Most polyphenols are found in plants, and are ingested as plant products, consumed as a processed food, extract, compound, or purified molecule, or injected as a molecule or compound [16]. In plants or extracts, the polyphenol is found in combination with other polyphenols and molecules that might augment or retard activity. Resveratrol, the best-studied polyphenol, has often been evaluated as a purified molecule, but also in grape seed or red wine extracts. The ideal forms of antiaging substances have yet to be studied or determined.

Several investigations have examined the influence of plant extracts on the process of aging. A combination of polyphenols from a green tea, 45% of which was composed of the polyphenol epigallocatechin gallate (ECGC), was tested in *D. melanogaster* [81]. Interestingly, male fruit flies consuming the extract boosted their mean lifespan, but not females. In addition, there are numerous plants, fruits, and extracts that increase invertebrate and rodent longevity. These include blueberry, cocoa, apple, tea, pomegranate, *Ludwigia octovalvis* (a tropical plant), and betony (a wildflower extract) [16,82–85]. The increase in lifespan observed from ingestion of these products has been attributed to polyphenols, although plant products contain a plethora of molecular substances besides polyphenols that might augment lifespan, and thus, one cannot, at this point, attribute benefit to one or more specific polyphenols.

In nature, polyphenols are sometimes found conjugated to other compounds, such as polymers, that alter their properties [86]. Polyphenols may be synthetically conjugated to other substances such as nanoparticles and proteins [86]. Theoretically the activity of polyphenols on the aging process might be enhanced by conjugation, perhaps by increasing lipophilicity, improving their absorption and bioavailability [86]. ECGC and catechin have been synthetically polymerized [86].

An additional consideration for further research is whether or not polyphenol supplementation might have any harm. The published scientific literature provides little evidence about toxicity. In one investigation of rats, doses of 3 g/kg/d for 30 days of resveratrol, the most-well studied polyphenol, produced renal injury and leukocytosis [87], and in another study reduced concentrations of hepatic antioxidant enzymes [88].

Finally, although individual polyphenols lengthen lifespan in a number of invertebrate, or vertebrate species, they might be combined with other strategies to retard aging in higher organisms. In the multiyear RESTRIKAL trial, resveratrol is used together with a 30% calorie restriction in lemurs that live for a decade [89]. Such trials will help determine whether higher organisms will benefit from the promise polyphenols confer on improving the aging process in lower organisms.

SUMMARY

- Several polyphenols augment the lifespan of multiple invertebrate and vertebrate species, in higher organisms in combination with dietary modification.
- Polyphenols may influence aging by acting on energy-regulatory genes.
- Polyphenols may alter the aging process through effects on concentrations of reactive oxygen species.
- Polyphenols may modulate the aging process via a gene called Klotho or processes during or after transcription of nuclear DNA.
- Polyphenols may influence aging through its effect on an intracellular recycling process called autophagy.
- Polyphenols may impact aging through its effect on mitochondria.
- Further research will be necessary to decide what forms of polyphenols are the best candidates, perhaps using more sophisticated computerized genetic and molecular analyses.

References

[1] Blagosklonny MV. Aging and immortality: quasi-programmed senescence and its pharmacologic inhibition. Cell Cycle 2006;5:2087–102.

[2] Xu J, Ji J, Yan XH. Cross-talk between AMPK and mTOR in regulating energy balance. Crit Rev Food Sci Nutr 2012;52:373–81.

[3] Lamming DW, Mihaylova MM, Katajisto P, Baar EL, Yilmaz OH, Hutchins A, et al. Depletion of Rictor, an essential protein component of mTORC2, decreases male lifespan. Aging Cell 2014;13:911–17.

[4] Robida-Stubbs S, Glover-Cutter K, Lamming DW, Mizunuma M, Narasimhan SD, Neumann-Haefelin E, et al. TOR signaling and rapamycin influence longevity by regulating SKN-1/Nrf and DAF-16/FoxO. Cell Metab 2012;15:713–24.

[5] Medvedik O, Lamming DW, Kim KD, Sinclair DA. MSN2 and MSN4 link calorie restriction and TOR to sirtuin-mediated lifespan extension in *Saccharomyces cerevisiae*. PLoS Biol 2007;5:e261.

[6] Bjedov I, Toivonen JM, Kerr F, Slack C, Jacobson J, Foley A, et al. Mechanisms of life span extension by rapamycin in the fruit fly *Drosophila melanogaster*. Cell Metab 2010;11:35–46.

[7] Gomes AP, Price NL, Ling AJ, Moslehi JJ, Montgomery MK, Rajman L, et al. Declining NAD(+) induces a pseudohypoxic state disrupting nuclear-mitochondrial communication during aging. Cell 2013;155:1624–38.

[8] Mendelsohn AR, Larrick JW. Partial reversal of skeletal muscle aging by restoration of normal NAD(+) levels. Rejuvenation Res 2014;17:62–9.

[9] Leontieva OV, Blagosklonny MV. M(o)TOR of pseudo-hypoxic state in aging: rapamycin to the rescue. Cell Cycle 2014;13:509–15.

[10] Brunetti C, Di Ferndinando M, Agati G, Tattini M. Multiple functions of polyphenols in plants inhabiting unfavorable Mediterranean areas. Environ Exp Botany 2014;103:106–17.

[11] Corominas-Faja B, Santangelo E, Cuyas E, Micol V, Joven J, Ariza X, et al. Computer-aided discovery of biological activity spectra for anti-aging and anti-cancer olive oil oleuropeins. Aging (Albany, NY) 2014;6:731–41.

[12] Manach C, Williamson G, Morand C, Scalbert A, Remesy C. Bioavailability and bioefficacy of polyphenols in humans. I. Review of 97 bioavailability studies. Am J Clin Nutr 2005;81:230S–42S.

[13] Hu M. Commentary: bioavailability of flavonoids and polyphenols: call to arms. Mol Pharm 2007;4:803–6.

[14] St Leger AS, Cochrane AL, Moore F. Factors associated with cardiac mortality in developed countries with particular reference to the consumption of wine. Lancet 1979;1:1017–20.

[15] Cherniack EP. The potential influence of plant polyphenols on the aging process. Forsch Komplementmed 2010;17:181–7.

[16] Uysal U, Seremet S, Lamping JW, Adams JM, Liu DY, Swerdlow RH, et al. Consumption of polyphenol plants may slow aging and associated diseases. Curr Pharm Des 2013;19:6094–111.

[17] Howitz KT, Bitterman KJ, Cohen HY, Lamming DW, Lavu S, Wood JG, et al. Small molecule activators of sirtuins extend *Saccharomyces cerevisiae* lifespan. Nature 2003;425:191–6.

[18] Wood JG, Rogina B, Lavu S, Howitz K, Helfand SL, Tatar M, et al. Sirtuin activators mimic caloric restriction and delay ageing in metazoans. Nature 2004;430:686–9.

[19] Valenzano DR, Cellerino A. Resveratrol and the pharmacology of aging: a new vertebrate model to validate an old molecule. Cell Cycle 2006;5:1027–32.

[20] Valenzano DR, Terzibasi E, Genade T, Cattaneo A, Domenici L, Cellerino A. Resveratrol prolongs lifespan and retards the onset of age-related markers in a short-lived vertebrate. Curr Biol 2006;16:296–300.

[21] Baur JA, Pearson KJ, Price NL, Jamieson HA, Lerin C, Kalra A, et al. Resveratrol improves health and survival of mice on a high-calorie diet. Nature 2006;444:337–42.

[22] Pietsch K, Saul N, Menzel R, Sturzenbaum SR, Steinberg CE. Quercetin mediated lifespan extension in *Caenorhabditis elegans* is modulated by age-1, daf-2, sek-1 and unc-43. Biogerontology 2008.

[23] Saul N, Pietsch K, Menzel R, Steinberg CE. Quercetin-mediated longevity in *Caenorhabditis elegans*: Is DAF-16 involved? Mech Ageing Dev 2008;129:611–13.

[24] Pietsch K, Saul N, Chakrabarti S, Sturzenbaum SR, Menzel R, Steinberg CE. Hormetins, antioxidants and prooxidants: defining quercetin-, caffeic acid- and rosmarinic acid-mediated life extension in *C. elegans*. Biogerontology 2011;12:329–47.

[25] Liao VH, Yu CW, Chu YJ, Li WH, Hsieh YC, Wang TT. Curcumin-mediated lifespan extension in *Caenorhabditis elegans*. Mech Ageing Dev 2011;132:480–7.

[26] Lee KS, Lee BS, Semnani S, Avanesian A, Um CY, Jeon HJ, et al. Curcumin extends life span, improves health span, and modulates the expression of age-associated aging genes in *Drosophila melanogaster*. Rejuvenation Res 2010;13:561–70.

[27] Saul N, Pietsch K, Menzel R, Sturzenbaum SR, Steinberg CE. Catechin induced longevity in *C. elegans*: from key regulator genes to disposable soma. Mech Ageing Dev 2009;130:477–86.

[28] Saul N, Pietsch K, Sturzenbaum SR, Menzel R, Steinberg CE. Diversity of polyphenol action in *Caenorhabditis elegans*: between toxicity and longevity. J Nat Prod 2011;74:1713–20.

[29] Schlernitzauer A, Oiry C, Hamad R, Galas S, Cortade F, Chabi B, et al. Chicoric acid is an antioxidant molecule that stimulates AMP kinase pathway in L6 myotubes and extends lifespan in *Caenorhabditis elegans*. PLoS One 2013;8:e78788.

[30] Grunz G, Haas K, Soukup S, Klingenspor M, Kulling SE, Daniel H, et al. Structural features and bioavailability of four flavonoids and their implications for lifespan-extending and antioxidant actions in *C. elegans*. Mech Ageing Dev 2012;133:1–10.

[31] Agarwal B, Baur JA. Resveratrol and life extension. Ann NY Acad Sci 2011;1215:138–43.

[32] de Boer VC, de Goffau MC, Arts IC, Hollman PC, Keijer J. SIRT1 stimulation by polyphenols is affected by their stability and metabolism. Mech Ageing Dev 2006;127:618–27.

[33] Quideau S, Deffieux D, Pouysegu L. Resveratrol still has something to say about aging!. Angew Chem Int Ed Engl 2012;51:6824–6.

[34] Park SJ, Ahmad F, Philp A, Baar K, Williams T, Luo H, et al. Resveratrol ameliorates aging-related metabolic phenotypes by inhibiting cAMP phosphodiesterases. Cell 2012;148:421–33.

[35] Gertz M, Nguyen GT, Fischer F, Suenkel B, Schlicker C, Franzel B, et al. A molecular mechanism for direct sirtuin activation by resveratrol. PLoS One 2012;7:e49761.

[36] Xiang L, Sun K, Lu J, Weng Y, Taoka A, Sakagami Y, et al. Anti-aging effects of phloridzin, an apple polyphenol, on yeast via the SOD and Sir2 genes. Biosci Biotechnol Biochem 2011;75:854–8.

[37] Menendez JA, Joven J. Energy metabolism and metabolic sensors in stem cells: the metabostem crossroads of aging and cancer. Adv Exp Med Biol 2014;824:117–40.

[38] Pallauf K, Rimbach G. Autophagy, polyphenols and healthy ageing. Ageing Res Rev 2013;12:237–52.

[39] Gordon PB, Holen I, Seglen PO. Protection by naringin and some other flavonoids of hepatocytic autophagy and endocytosis against inhibition by okadaic acid. J Biol Chem 1995;270:5830–8.

[40] Armour SM, Baur JA, Hsieh SN, Land-Bracha A, Thomas SM, Sinclair DA. Inhibition of mammalian S6 kinase by resveratrol suppresses autophagy. Aging (Albany, NY) 2009;1:515–28.

[41] Alayev A, Doubleday PF, Berger SM, Ballif BA, Holz MK. Phosphoproteomics reveals resveratrol-dependent inhibition of Akt/mTORC1/S6K1 signaling. J Proteome Res 2014.

[42] Alayev A, Berger SM, Kramer MY, Schwartz NS, Holz MK. The combination of rapamycin and resveratrol blocks autophagy and induces apoptosis in breast cancer cells. J Cell Biochem 2014.

[43] Alayev A, Sun Y, Snyder RB, Berger SM, Yu JJ, Holz MK. Resveratrol prevents rapamycin-induced upregulation of autophagy and selectively induces apoptosis in TSC2-deficient cells. Cell Cycle 2014;13:371–82.

[44] Scarlatti F, Maffei R, Beau I, Codogno P, Ghidoni R. Role of non-canonical Beclin 1-independent autophagy in cell death induced by resveratrol in human breast cancer cells. Cell Death Differ 2008;15:1318–29.

[45] Opipari Jr. AW, Tan L, Boitano AE, Sorenson DR, Aurora A, Liu JR. Resveratrol-induced autophagocytosis in ovarian cancer cells. Cancer Res 2004;64:696–703.

[46] Laurent C, Chabi B, Fouret G, Py G, Sairafi B, Elong C, et al. Polyphenols decreased liver NADPH oxidase activity, increased muscle mitochondrial biogenesis and decreased gastrocnemius age-dependent autophagy in aged rats. Free Radic Res 2012;46:1140–9.

[47] Obrenovich ME, Nair NG, Beyaz A, Aliev G, Reddy VP. The role of polyphenolic antioxidants in health, disease, and aging. Rejuvenation Res 2010;13:631–43.

[48] Coban J, Oztezcan S, Dogru-Abbasoglu S, Bingul I, Yesil-Mizrak K, Uysal M. Olive leaf extract decreases age-induced

oxidative stress in major organs of aged rats. Geriatr Gerontol Int 2014;14:996–1002.

[49] Ortega-Arellano HF, Jimenez-Del-Rio M, Velez-Pardo C. Life span and locomotor activity modification by glucose and polyphenols in *Drosophila melanogaster* chronically exposed to oxidative stress-stimuli: implications in Parkinson's disease. Neurochem Res 2011;36:1073–86.

[50] Giovannini C, Masella R. Role of polyphenols in cell death control. Nutr Neurosci 2012;15:134–49.

[51] Saller R, Brignoli R, Melzer J, Meier R. An updated systematic review with meta-analysis for the clinical evidence of silymarin. Forsch Komplementmed 2008;15:9–20.

[52] Duan W, Jin X, Li Q, Tashiro S, Onodera S, Ikejima T. Silibinin induced autophagic and apoptotic cell death in HT1080 cells through a reactive oxygen species pathway. J Pharmacol Sci 2010;113:48–56.

[53] Marrazzo G, Bosco P, La Delia F, Scapagnini G, Di Giacomo C, Malaguarnera M, et al. Neuroprotective effect of silibinin in diabetic mice. Neurosci Lett 2011;504:252–6.

[54] Charles AL, Meyer A, Dal-Ros S, Auger C, Keller N, Ramamoorthy TG, et al. Polyphenols prevent ageing-related impairment in skeletal muscle mitochondrial function through decreased reactive oxygen species production. Exp Physiol 2013;98:536–45.

[55] Giampieri F, Alvarez-Suarez JM, Mazzoni L, Forbes-Hernandez TY, Gasparrini M, Gonzalez-Paramas AM, et al. Polyphenol-rich strawberry extract protects human dermal fibroblasts against hydrogen peroxide oxidative damage and improves mitochondrial functionality. Molecules 2014;19: 7798–816.

[56] Ungvari Z, Sonntag WE, de Cabo R, Baur JA, Csiszar A. Mitochondrial protection by resveratrol. Exerc Sport Sci Rev 2011;39:128–32.

[57] Hsu SC, Huang SM, Chen A, Sun CY, Lin SH, Chen JS, et al. Resveratrol increases anti-aging Klotho gene expression via the activating transcription factor 3/c-Jun complex-mediated signaling pathway. Int J Biochem Cell Biol 2014;53:361–71.

[58] Ayissi VB, Ebrahimi A, Schluesenner H. Epigenetic effects of natural polyphenols: a focus on SIRT1-mediated mechanisms. Mol Nutr Food Res 2014;58:22–32.

[59] Averbeck NB, Jensen ON, Mann M, Schagger H, Osiewacz HD. Identification and characterization of PaMTH1, a putative O-methyltransferase accumulating during senescence of *Podospora anserina* cultures. Curr Genet 2000;37:200–8.

[60] Kunstmann B, Osiewacz HD. Over-expression of an S-adenosylmethionine-dependent methyltransferase leads to an extended lifespan of Podospora anserina without impairments in vital functions. Aging Cell 2008;7:651–62.

[61] Kunstmann B, Osiewacz HD. The S-adenosylmethionine dependent O-methyltransferase PaMTH1: a longevity assurance factor protecting Podospora anserina against oxidative stress. Aging (Albany, NY) 2009;1:328–34.

[62] Knab B, Osiewacz HD. Methylation of polyphenols with vicinal hydroxyl groups: a protection pathway increasing organismal lifespan. Cell Cycle 2010;9:3387–8.

[63] Lancon A, Michaille JJ, Latruffe N. Effects of dietary phytophenols on the expression of microRNAs involved in mammalian cell homeostasis. J Sci Food Agric 2013;93:3155–64.

[64] Rimmele P, Lofek-Czubek S, Ghaffari S. Resveratrol increases the bone marrow hematopoietic stem and progenitor cell capacity. Am J Hematol 2014;89:E235–8.

[65] Zheng S, Liao S, Zou Y, Qu Z, Shen W, Shi Y. Mulberry leaf polyphenols delay aging and regulate fat metabolism via the germline signaling pathway in *Caenorhabditis elegans*. Age (Dordr) 2014;36:9719.

[66] Gendron ME, Thorin-Trescases N, Mamarbachi AM, Villeneuve L, Theoret JF, Mehri Y, et al. Time-dependent beneficial effect of chronic polyphenol treatment with catechin on endothelial dysfunction in aging mice. Dose Response 2012;10:108–19.

[67] Dal-Ros S, Bronner C, Auger C, Schini-Kerth VB. Red wine polyphenols improve an established aging-related endothelial dysfunction in the mesenteric artery of middle-aged rats: role of oxidative stress. Biochem Biophys Res Commun 2012;419: 381–7.

[68] Idris Khodja N, Chataigneau T, Auger C, Schini-Kerth VB. Grape-derived polyphenols improve aging-related endothelial dysfunction in rat mesenteric artery: role of oxidative stress and the angiotensin system. PLoS One 2012;7:e32039.

[69] Rajapakse AG, Yepuri G, Carvas JM, Stein S, Matter CM, Scerri I, et al. Hyperactive S6K1 mediates oxidative stress and endothelial dysfunction in aging: inhibition by resveratrol. PLoS One 2011;6:e19237.

[70] Khurana S, Venkataraman K, Hollingsworth A, Piche M, Tai TC. Polyphenols: benefits to the cardiovascular system in health and in aging. Nutrients 2013;5:3779–827.

[71] Dutta D, Xu J, Dirain ML, Leeuwenburgh C. Calorie restriction combined with resveratrol induces autophagy and protects 26-month-old rat hearts from doxorubicin-induced toxicity. Free Radic Biol Med 2014;74:252–62.

[72] Queen BL, Tollefsbol TO. Polyphenols and aging. Curr Aging Sci 2010;3:34–42.

[73] Magrone T, Jirillo E. Potential application of dietary polyphenols from red wine to attaining healthy ageing. Curr Top Med Chem 2011;11:1780–96.

[74] Kim EK, Kwon KB, Song MY, Han MJ, Lee JH, Lee YR, et al. Flavonoids protect against cytokine-induced pancreatic beta-cell damage through suppression of nuclear factor kappaB activation. Pancreas 2007;35:e1–9.

[75] Kim JS, Jobin C. The flavonoid luteolin prevents lipopolysaccharide-induced NF-κB signalling and gene expression by blocking IκB kinase activity in intestinal epithelial cells and bone-marrow derived dendritic cells. Immunology 2005;115:375–87.

[76] Kim JA, Kim DK, Kang OH, Choi YA, Park HJ, Choi SC, et al. Inhibitory effect of luteolin on TNF-α-induced IL-8 production in human colon epithelial cells. Int Immunopharmacol 2005;5:209–17.

[77] Sadowska-Bartosz I, Bartosz G. Effect of antioxidants supplementation on aging and longevity. Biomed Res Int 2014;2014:404680.

[78] Schluesener JK, Schluesener H. Plant polyphenols in the treatment of age-associated diseases: revealing the pleiotropic effects of icariin by network analysis. Mol Nutr Food Res 2014;58:49–60.

[79] Smoliga JM, Blanchard O. Enhancing the delivery of resveratrol in humans: if low bioavailability is the problem, what is the solution? Molecules 2014;19:17154–72.

[80] Lewandowska U, Szewczyk K, Hrabec E, Janecka A, Gorlach S. Overview of metabolism and bioavailability enhancement of polyphenols. J Agric Food Chem 2013;61:12183–99.

[81] Lopez T, Schriner SE, Okoro M, Lu D, Chiang BT, Huey J, et al. Green tea polyphenols extend the lifespan of male *Drosophila melanogaster* while impairing reproductive fitness. J Med Food 2014.

[82] Lin WS, Chen JY, Wang JC, Chen LY, Lin CH, Hsieh TR, et al. The anti-aging effects of *Ludwigia octovalvis* on Drosophila melanogaster and SAMP8 mice. Age (Dordr) 2014;36:689–703.

[83] Peng C, Chan HY, Huang Y, Yu H, Chen ZY. Apple polyphenols extend the mean lifespan of Drosophila melanogaster. J Agric Food Chem 2011;59:2097–106.

[84] Palermo V, Mattivi F, Silvestri R, La Regina G, Falcone C, Mazzoni C. Apple can act as anti-aging on yeast cells. Oxid Med Cell Longev 2012;2012:491759.

[85] Sunagawa T, Shimizu T, Kanda T, Tagashira M, Sami M, Shirasawa T. Procyanidins from apples (*Malus pumila* Mill.) extend the lifespan of *Caenorhabditis elegans*. Planta Med 2011;77:122−7.

[86] Cirillo G, Curcio M, Vittorio O, Iemma F, Restuccia D, Spizzirri UG, et al. Polyphenol conjugates and human health: a perspective review. Crit Rev Food Sci Nutr 2014;56:326−37.

[87] Crowell JA, Korytko PJ, Morrissey RL, Booth TD, Levine BS. Resveratrol-associated renal toxicity. Toxicol Sci 2004;82:614−19.

[88] Rocha KK, Souza GA, Ebaid GX, Seiva FR, Cataneo AC, Novelli EL. Resveratrol toxicity: effects on risk factors for atherosclerosis and hepatic oxidative stress in standard and high-fat diets. Food Chem Toxicol 2009;47:1362−7.

[89] Dal-Pan A, Blanc S, Aujard F. Resveratrol suppresses body mass gain in a seasonal non-human primate model of obesity. BMC Physiol 2010;10:11.

CHAPTER 47

Potential of Asian Natural Products for Health in Aging

Bernice Cheung[1,2], Macy Kwan[1], Ruth Chan[1,2], Mandy Sea[1,2] and Jean Woo[1,2]

[1]Department of Medicine and Therapeutics, The Chinese University of Hong Kong, Shatin, Hong Kong
[2]Center for Nutritional Studies, The Chinese University of Hong Kong, Shatin, Hong Kong

KEY FACTS

- Delay the onset of aging can reduce the physical, mental, and financial burdens of individuals and society.
- Some Asian natural products show potential beneficial effects on disease preventing based on evidence from in vitro and in vivo studies.
- These natural products are rich in phytochemicals and generally provide antioxidant, cardioprotective, and immunomodulation effects.
- The absence of consistent and reproducible data from human studies leads to the difficulties to draw conclusions on the protective effects on aging among these foods.
- Large-scale well-designed randomized controlled trials are required in dietary and clinical settings to fully examine the putative effects of these Asian natural products and the underlying mechanisms involved, and to allow more definitive conclusions to be drawn.

Dictionary of Terms

- *β-amyloid*: Crucially involved in Alzheimer's disease as the main component of the amyloid plaques found in the brains of Alzheimer patients.
- *Alzheimer's disease*: The most common form of dementia. It is a complex neurodegenerative disease with a progressive cognitive and functional impairment. It is characterized by the presence of amyloid plaques and neurofibrillary tangles.
- *Antioxidant*: Inhibits oxidation reaction by removing free radical. Insufficient antioxidant causes oxidative stress and may damage or kill the cells.
- *Neuroprotection*: To preserve the structure and function of neuron cells. It aims to prevent, slow down, or provide potential treatment options on the progress of many central nervous system disorders including neurodegenerative diseases.
- *Phytochemical*: Biologically active, naturally existing substances in plants show potential health benefits in human.

INTRODUCTION

Aging is a global epidemic. Our bodies undergo progressive deterioration in physical functions, loss of homeostasis, and increased susceptibility to degenerative diseases. It causes increase in physical, mental, and financial burdens. Delaying the onset of aging is an important topic in public health. Free radical theory has been proposed to explain aging [1]. Reactive oxygen species (ROS) which are produced during intracellular mechanisms, cause DNA damage, lipid peroxidation, and protein oxidation and lead to aging. The brain, consuming high levels of oxygen and being rich in polyunsaturated fatty acids, is particular vulnerable to oxidative damage [2]. Antioxidant systems, for example, superoxide dismutase (SOD), catalase, glutathione peroxidase (GSH-Px), glutathione reductase, vitamin C, and vitamin E are the main defense systems against free radicals [1]. Eliminating the formation of free radicals and reducing oxidative stress are crucial in reducing the rate of aging and the risk of chronic diseases.

It is postulated that any food with a great antioxidant capacity can be a potential candidate for delaying aging. Food may offer a polypharmacology approach in promoting healthy aging. The antioxidant rich foods and the antiaging capabilities of food-derived bioactive compounds have been widely investigated based on Western culture [1,3]. In Asia, there has been a long history of using natural food in delaying the progression of aging based on the traditional medicine theories. However, there are limited reviews regarding this area. In addition, a large percentage of the aging population has at least one chronic disease and neurodegenerative disease is high on the list of the aged. Since, neurodegenerative diseases prevalence are lower in East Asian countries than in Western populations [4], the studies of Asians food on neuroprotective effects are of special interest. This review aims to discuss the health benefits and the antioxidant effects of the natural products that are commonly consumed in Asia and their potential neuroprotective abilities. The inclusions of these products are based on their popularity and current availability of evidence.

ASIAN NATURAL PRODUCTS

Ginseng

Ginseng refers to the root of several species in the plant genus *Panax* (C.A. Meyer *Araliaceae*). Among them, *Panax ginseng* is the most widely used. For traditional medicine, ginseng has been reported to enhance stamina, physical performance, and general vitality in healthy individuals, to resist against diseases and to cure diseases [5]. Ginsenosides or ginseng saponins are the major active ingredients in ginseng [6]. The ginsenosides content varies depending on the *Panax* species, the plant age, the part of plant, the preservation method, the season of harvest, and the extraction method [7]. Ginsenosides exhibit anti-inflammatory, antioxidant, and antiapoptotic mechanisms and exert various effects on the immune system and the nervous system [5–8] (Table 47.1). Systematic review shows that *P. ginseng* has promising results for improving glucose metabolism and moderating the immune response [8].

Neuroprotective effects have been shown in *P. ginseng* and ginsenosides in vitro and in vivo models [5,6]. Significant improvement in learning and memory has been observed in aged and brain damaged rats after administration of ginseng powder [6]. It is suggested that the neuroprotective effects may attribute to the direct effect of ginsenosides on central nervous system and the indirect effects by increasing oxygen and glucose supply to the brain thus facilitating the cognitive function [9]. However, the overall evidence of ginseng in human is inconclusive. An observational study reported that regular use of ginseng product during long periods of time (up to 2 years) by healthy participants did not provide any quantifiable beneficial effects on memory performance [10]. Both positive and negative effects of ginseng on memory are reported in randomized controlled trials (Table 47.2). Most studies had a treatment period of 12 weeks with one study investigating the acute effect after two days of treatment among the younger population. In general, ginseng appeared to have beneficial effects for improvement of some aspects of cognitive behavior. In a study among

TABLE 47.1 Potential Health Benefits of Ginseng

Health benefits	Possible actions
Moderate immune response	• Inhibit interleukin-1β, interleukin-6 gene expression. • Enhance interferon induction, phagocytosis, natural killer cells, B and T cells. • More resistant to infection by *Staphylococcus aureus*, *Escherichia coli*, and *Salmonella typhi*.
Cardioprotective effects	• Decrease systemic blood pressure. • Enhance vasodilation. • Reduce plasma cholesterol. • Reduce arterial stiffness.
Glucose tolerance	• Decrease circulating glucose in patients with diabetes. • Modulate insulin secretion.
Antiulcer effects	• Against gastritis through increasing mucus secretion.
Anticancer effects	• Suppress proliferation in human cancer cells. • Inhibit tumor angiogenesis and metastasis.
Antioxidant effects	• Prevent overproduction of nitric oxide and free radical medicated lipid peroxidation.

TABLE 47.2 Summary of Clinical Trials of Ginseng in Cognitive Function Tests

Reference	Type of study	Participants	Intervention	Results
[9]	Randomized, double-blind, placebo-controlled, two period crossover design	30 subjects (15 male), mean age 20	G115 (200 mg of *Panax ginseng* extract) for 2 days with a 7-day washout period	Treatment group had significant effect on speed of attention. No significant differences were found between quality of memory, speed of memory, continuity of attention, working memory, and secondary memory.
[11]	Randomized, double-blind, placebo-controlled trial	112 subjects (38 male), age 40 and above	400 mg of Gerimax ginseng extract for 8 to 9 weeks	No statistically significant differences on visual simple reaction time tests, 5-min letter and symbol cancellation test, verbal fluency test, Logical Memory and Reproduction Test, the Rey-Ostreith Complex Figure test. Significant performance improvements for treatment group on auditory simple reaction time tests, computerized Wisconsin Card Sort test.
[12]	Randomized, open-label pilot study	61 subjects (24 males) with Alzheimer's disease (AD), age 60–80	Low dose Korean red ginseng (KRG) (4.5 g/day, $n = 15$) High dose KRG (9 g/day $n = 15$), for 12 weeks, control group n = 31	High dose group showed significant improvement on ADAS-cog and CDR. MMSE did not significantly improve
[13]	Randomized, open-label prospective study	97 consecutive AD patients, age 47–83	*P. ginseng* powder 4.5 g/day or 9 g/day for 12 weeks; after discontinued of administration, monitored for another 12 weeks	Treatment group showed significant improvement in MMSE; the effect no longer exited after ginseng discontinuation. No significant difference was found between dosages.
[14]	Randomized, double-blind, placebo-controlled trial	60 subjects (18 male), age 51–65 with age-associated memory impairment (AAMI)	Ginseng-containing vitamin complex for 9 months	Treatment group had significantly higher result in Randt Memory test.

healthy adults, administration of 400 mg ginseng extract for 8 to 9 weeks showed improvement on auditory simple reaction time tests and computerized Wisconsin Card Sort test [11]. In the other study among AD subjects, those receiving 9 g/day Korean red ginseng showed improvement on ADAS-cog (Alzheimer's Disease Assessment Scale) and CDR (Clinical Dementia rating) but not MMSE (Mini Mental State Examination) [12]. While the other group, administered 4.5 g *P. ginseng* powder to subjects with AD resulted in improvement of MMSE [13]. Acute administration of ginseng to healthy young volunteers resulted in improvement of speed of attention, indicating a beneficial effect on participant's ability to allocate attentional process to a particular task [9]. Based on current evidence, ginseng is generally considered as safe and appears to have some beneficial effects on certain cognition tests but higher quality evidence is needed to further confirm the effect.

Tea

Tea is defined as a beverage brewed from the *Camellia sinensis* plant. It is the most popular beverage consumed after water. Flavonoids are the most abundant components in tea and comprise of catechins, theaflavins (TF), and Thearubigins (TR). Tea can be classified to black (fermented), green (nonfermented), or oolong (semifermented), depending on the degree of oxidation. Green tea has the highest amount of catechins followed by oolong tea and black tea [15]. Catechins act as powerful hydrogen-donating antioxidants and free radical scavengers of reactive oxygen and nitrogen species [16]. Table 47.3 summarizes the potential health benefits of tea [17,18].

In vitro and animal model studies suggest catechins have neuroprotective actions via their modulation of antioxidant capacities, iron chelating activities, and signal transduction pathways [15]. Accumulated β-amyloid might impair neuronal synapses and dendrites and cause local oxidative stress reaction and increase risk of AD [19]. While (−)-epigallocatechin-3-gallate (EGCG), a major form of catechins, has the ability to inhibit the formation, extension, and stabilization of β-amyloid in vivo and in vitro [20]. Both black tea and green tea inhibited human acetylocholinesterase activity and displayed protective action against β-amyloid induced toxicity [20]. Several observational

TABLE 47.3 Potential Health Benefits of Tea

- Prevent various types of cancer including skin cancer, prostate cancer, lung cancer, breast cancer
- Decrease risk of cardiovascular disease and stroke
- Decrease absorption of triglycerides and cholesterol
- Antioxidant
- Promote weight loss. Increase energy expenditure and promote fat oxidation
- Affect glucose metabolism and decrease risk of diabetes
- Decrease risk of rheumatoid arthritis
- Improve bone health
- Increase alertness, jitteriness, and positively affect mood

TABLE 47.4 Observational Studies on the Association Between Tea Consumption and Cognitive Health

Origin of study	Participants	Results
Singapore, cross-sectional study [20]	716 Chinese residents in Singapore, age ≥55	Total tea consumption was independently associated with better performances on global cognition ($p = 0.03$), memory ($p = 0.01$), executive function ($p = 0.009$), and information processing speed ($p = 0.001$). Both black/oolong tea and green tea consumption were associated with better cognitive performance.
Japan, cross-sectional study [21]	1003 Japanese, age ≥70	High consumption of green tea was associated with lower prevalence of cognitive impairment. The ORs for cognitive impairment associated with different frequencies of green tea consumption were 1.00 for ≤3 cups/week, 0.62 (95% CI 0.33, 1.19) for 4–6 cups/week or 1 cup/day, 0.46 (95% CI 0.30, 0.72) for ≥2 cups/day (p for trend <0.0006).
Singapore, cohort, cross-sectional longitudinal study [23]	2501 Chinese resided in Singapore age ≥55 at baseline, 1438 subjects were reassessed after median 16 months	Higher level of total tea consumption was significantly associated with a lower risk of cognitive impairment and cognitive decline in the adjusted models. For cognitive impairments, low, medium, and high levels of tea intake (with reference to no tea intake), the ORs = 0.56, 0.45, 0.37 respectively. For cognitive decline, low, medium, and high levels of tea intake (with reference to no tea intake), the ORs = 0.74, 0.78, 0.57, respectively. Only with the consumption of black or oolong tea was significantly associated with much lower odds of association with cognitive impairment in cross-sectional analysis and with cognitive decline in longitudinal analysis.
China, population-based longitudinal cohort study [22]	7139 Chinese, age 80–115 at baseline, ongoing follow-up for 7 years	Tea drinkers had higher verbal fluency scores throughout the follow-up period. Daily tea drinkers had a steeper slope of cognitive decline as compared with nondrinkers.
Japan, population-based prospective study [24]	723 Japanese, age >69 at baseline, 490 in mean follow-up for 4.9 years	More frequent consumption of green tea was associated with lower incidence of dementia and mild cognitive impairment. For overall cognitive decline, those consuming green tea every day ORs = 0.32 (95% CI 0.16, 0.64); those consuming green tea 1–6 days per week ORs = 0.47 (95% CI 0.25, 0.86) compared with those did not consume green tea at all. For the incidence of dementia, those consumed green tea daily ORs = 0.26 (95% CI 0.06, 1.06).
China, cross-sectional study [25]	681 Chinese, age ≥90.	In men, those with cognitive impairment had significantly lower prevalence of habits of tea consumption ($p = 0.041$ for former tea drinker, $p = 0.044$ for current tea drinker). No significant differences were found in women.

studies support the protective effect of tea on cognitive health (Table 47.4). A cross sectional study based on Comprehensive Geriatric Assessment (CGA) was the first study to examine the association between consumption of green tea and cognitive function in Asian. After adjustment for potential confounders, higher consumption of green tea was associated with a lower prevalence of cognitive impairment [21]. It is suggested that tea drinking helps to build up cognitive reserve which can compress the expression of brain damage. As a result of extra reserve, tea drinkers would have delayed manifestation of pathological processes in brain but once it begins, the progression rate will be faster [22]. However, it is inconclusive to show green tea has better neuroprotective effects over black tea. Either black, oolong tea, or green tea, but not coffee consumption, among Singapore Chinese were associated with better cognitive performance [20]. Most evidence for the potential effect of black and oolong tea on lowering prevalence of cognitive impairment is concluded in Singapore Longitudinal Aging Studies [23]. In a recent study in Japan, consumption of green tea, but not black tea or coffee, was associated with reduced risk of cognitive decline [24]. There are different tea drinking habits geographically. Since green tea is common in Japan, while black or oolong tea is common in Singapore and China, it results in research difficulties to show one is better than the other in a single country study [23,24]. Studies involving different ethnic groups are important.

Soy

Consumption of soy and soy products are estimated to be 10 to 40-fold higher in Asian compared to Western populations [26]. In Japan, soy foods contributed 6.5% to 12.8% of total protein intake [27]. Phytoestrogens are the most bioactive components in soy. It can interact with estrogen receptors and mediate estrogenic response. Isoflavones is one of the subclasses of phytoestrogens and genistein is the most potent isoflavones and daidzein and glycitein are the other active forms [26]. Studies have suggested various health benefits of soy (Table 47.5) [27–29]. High soy food consumption was associated with lower breast cancer risk in Singaporean women [30]. Soy food consumption improved vaginal cytology and bone mineral content in postmenopausal women [30]. Soy has been demonstrated to have strong antioxidant effects in hypercholesterolemic subjects after supplementation for 42 days regardless of dietary protein source [31]. In a randomized cross-over design study, 42 postmenopausal women replaced red meat from the Dietary Approach to Stop Hypertension diet (DASH) by soy protein or soy nuts showed significant increase in total antioxidant capacity through an 8-week study period [32].

Phytoestrogens have demonstrated neuroprotective effects in neuron cultures [26]. However, the results of soy on human cognitive function are contradictive (Table 47.6). Positive results are usually found in Caucasian studies. In the UK, subjects receiving 60 mg isoflavone/day for 12 weeks showed significant improvement in delayed recall or pictures, immediate story recall, and sustained attention [33]. In the US, subjects treated with various dosages of isoflavones for 6 months resulted in improvement of some cognitive function tests [34,35]. In the longest intervention study in the US which lasted for 2.5 years, subjects receiving 25 g of isoflavone rich soy protein showed improvement in visual memory [36]. On the other hand, several Asian studies showed no beneficial effects for soy or even harmful effects in cognitive health. Observational studies in Shanghai and Indonesia reported high tofu intake was associated with cognitive impairment. 517 adults (age 50–95) in Shanghai showed a trend of weekly tofu intake and increase risk of cognitive impairment by 20% [37]. In Indonesia, a study revealed that high tofu intake had a negative association with HVLT (Hopkins Verbal Learning test) after controlling for age, sex, education, and other dietary intakes [38]. Another clinical trial in Hong Kong also failed to support the significant effect of isoflavones on various domains of cognitive function. 176 women (age 55–76) were randomized and the treatment group received a daily oral intake of 80 mg soy-derived isoflavones for 6 months. The treatment group did not show significant improvement in the cognitive function tests [39].

Though health benefits have been extensively evaluated, conflicting results are shown with respect to cognitive function. Several possible reasons have been discussed to explain the inconclusive findings. First, the variations of participant characteristics may influence the effect of isoflavones on cognitive function. It is suggested that effects of isoflavones differ between ethnic groups. The capacity to convert isoflavones into their bioactive metabolites has been shown to vary cross-culturally based on their habitual intake [26]. It is estimated that approximately 25–30% of Western adults have the ability to break down diadzein into its bioactive metabolite equol, whereas in populations where soy is more regularly consumed, the proportion of equol producers is 50–60% [45]. Second, different intervention studies used different sources and types

TABLE 47.5 Potential Health Benefits of Soy or Soy Extracts

- Relief of menopausal symptoms
- Protection against breast cancer and prostate cancer
- Protection against coronary heart disease
- Reduction of adipose tissue mass and improvement of blood cholesterol level
- Prevention of osteoporosis
- Alleviation of insulin resistance state and lower the risk of type 2 diabetes onset
- Act as antioxidants by lowering the production of ROS

TABLE 47.6 Evidence of Soy on Cognition in Adults

Randomized controlled trials showing positive effects

Origin of study	Participants	Intervention	Results
UK Randomized, double-blind, placebo-controlled study [33]	33 postmenopausal women, age 55–65	18 subjects took soya supplement (Solgen 40) (60 mg isoflavone/day) for 12 weeks	Treatment group showed significant improvement in recall of pictures ($p < 0.03$), immediate story recall ($p < 0.06$), learning the first compound discrimination ($p < 0.01$), and learning the reversal of compound discrimination ($p < 0.05$).
USA Randomized, double-blind, placebo-controlled study [34]	56 women, age 55–74, postmenopausal at least 2 years	27 women took two pills per day (110 mg total isoflavones) for 6 months	Treatment group showed significant improvement in performance in category fluency ($p = 0.02$). Treatment group showed a nonsignificant improvement in verbal memory and Trails B ($p = 0.08$). Treatment group with younger women (age 50–59) showed significant improvement in Trails B ($p = 0.007$).
USA Randomized, double-blind, placebo-controlled, pilot study [35]	30 postmenopausal women, age 62–89	15 subjects, ingested 100 mg/day soy isoflavones for 6 months	Treatment group showed improvement on visual-spatial memory ($p < 0.01$) and construction ($p = 0.01$), verbal fluency ($p < 0.01$), and speeded dexterity ($p = 0.04$).
USA Randomized, double-blind, placebo-controlled trial [36]	313 menopausal women, age 45–92	154 subject received daily 25 g of isoflavone rich soy protein (52 mg of genistein, 36 mg of daidzein and 3 mg glycitein) for 2.5 years. Milk-protein matched placebo	Treatment group showed greater improvement on visual memory factor ($p = 0.018$), but no significant difference in executive/expressive/visuospatial factor, verbal episodic memory factor as compared to the placebo group.

Randomized controlled trials showing no or negative results

Study	Participants	Intervention	Results
USA Randomized, double-blind, placebo-controlled trial [40]	79 postmenopausal women, age 48–65	4 experimental groups: placebo, cow's milk and placebo, soy milk (72 mg isoflavones/day) and placebo, cow's milk and isoflavone supplement (70 mg isoflavones/day) for 16 weeks	Soy isoflavones did not improve selective attention, visual long-term memory, short-term visuospatial memory, or visuospatial working memory. Soy milk group showed a decline in verbal working memory compared to soy supplement and control group.
Netherlands Randomized, double-blind, placebo-controlled trial [41]	202 postmenopausal women, age 60–75	100 subjects received 25.6 g of isoflavone-rich soy protein containing 99 mg of isoflavones for 12 months 25.6 g of milk protein matched placebo	Cognitive function did not differ significantly between groups.

Randomized controlled trials showing no or negative results

Study	Participants	Intervention	Results
Hong Kong Randomized, double-blind, placebo-controlled, parallel group trial [39]	168 postmenopausal women, age 55–76	80 women took 80 mg soy-derived isoflavones for 6 months	No significant difference between groups on standardized neuropsychological tests of memory, executive, function, attention, motor control, language and visual perception, and global cognitive function.

Observational studies showing no or negative results

Origin of study	Participants	Isoflavone intake measure	Results
China Observational cross-sectional study [37]	517 subjects, age 50–95	Food Frequency Questionnaire (FFQ)	Weekly higher intake of tofu was associated with worse memory performance using Hopkins Verbal Learning test ($p = 0.01$) after controlling for variates. Among older elderly (68 or above), high tofu intake increased risk of cognitive impairment indicative of dementia ($p = 0.04$).
Indonesia Cross-sectional study [38]	719 subjects, age 52–98	FFQ	High tofu consumption was associated with worse memory.
USA Longitudinal study [42]	3734 men, 502 women/spouses, age 71–93	FFQ and information provided by the husband as proxy for wife's diet, baseline conducted in 1965–1967, follow up in 1971–1974. Cognitive assessment on 1991–1993	High midlife tofu consumption was significantly associated with poor cognitive test performance, enlargement of ventricles, and low brain weight independently in both sexes.
Indonesia Cross-sectional study [43]	142 subjects, age 56–97	FFQ	No association with soy consumption and delayed recall performance. For the younger age group (mean age 67), weekly tofu consumption was positively linear associated to immediate recall. In older age group (mean age 80), no association was reported.
Netherlands Cross-sectional study [44]	301 women age 60–75	Dietary isoflavones and lignan intake was assessed with FFQ in the year preceding enrolment	No association between dietary isoflavones intake with memory, processing capacity, and speed and executive function.

of soy isoflavones and such differences may lead to different bioavailability and possibly affect the overall effects of the studies [26]. Third, there is an "age-dependent" hypothesis suggested. Cells may undergo pathological changes responding differently to estrogenic compounds. The ability to produce equol in our body declines with age [26]. Research targeted at different age groups is crucial. Additional well controlled large-scale studies among different ethnic groups are needed to confirm the beneficial effects of soy.

Ginkgo

Ginko biloba has been identified as a valuable plant in China for more than 2000 years. It is used as a medicine for memory and age-related deterioration in China. A well-defined extract, EGb 761 is prepared as a dry powder and contains two main groups of active compounds, flavonoids (w24%) and terpenoids (w6%) [46]. The terpenoid fraction is composed of ginkgolides and bilobalides, which are exclusively in the *G. biloba* tree [47]. Ginkgo exerts a combination of effects, including antioxidant, anti-inflammatory, and antiapoptosis, increasing blood supply by dilating blood vessels, reducing blood viscosity, and modifying neurotransmitter system [46,48,49]. Ginkgo can scavenge many ROS, such as hydroxyl radical, peroxyl, and oxoferryl radicals, and superoxide anions, and nitric oxide in vitro [47]. Animals treated with EGb 761 showed a decrease in peroxide generation and an increase in SOD and catalase activities in the hippocampus, striatum, and substantia nigra [50]. Posttreatment of EGb 761 reversed brain damage in animal models [47].

It is believed that *G. biloba* has neuroprotective and cognitive enhancing properties. However, one observational study and two large randomized controlled trials and the Cochrane review showed negative results. It is reported that regular use of *G. biloba* as supplements during long periods of time (up to 2 years) by healthy participants did not enhance memory performance [10]. The Ginkgo Evaluation of Memory (GEM) study was a randomized, double-blinded, placebo-controlled clinical trial conducted in the US among community-dwelling subjects. In the GEM study, 1545 subjects received 120 mg extract of *G. biloba* and were followed up through 6.1 years. Annual rates of decline in z scores did not differ between *G. biloba* and placebo groups in any domains, including memory, attention, visuospatial, and executive functions. For the 3MSE (Modified Mini-Mental State Examination) and ADAS-Cog, there were no differences in rates of change between treatment groups (for 3MSE, $p = 0.71$; for ADAS-Cog, $p = 0.97$). The overall dementia rate was 3.3 per 100 person-years in *G. biloba* group and 2.9 per 100 person-years in the placebo group. The hazard ratio for *G. biloba* compared with placebo for all-cause dementia was 1.12 (95% CI 0.94, −1.33, $p = 0.21$) [51,52]. In another study in France under the GuidAge clinical trial, 1406 adults with reported memory complaints (aged 70 years or above) received a twice per day dose of 120 mg *G. biloba* extract and had median follow up of 5.0 years. There were no between-group differences for the development of AD or mixed dementia [53].

The negative results from the above studies may be due to an overoptimistic view of the preventive effect of *G. biloba*. There is a particularly long predementia phase, studies between 3 to 6 years may not be enough to show the effects [49]. It is reported that there is a major limitation in the Cochrane review as it combined evaluation of patients of self-reported cognitive complaints without validated diagnostic criteria, less rigorous randomization and allocation schemes and, therefore, may lead to a high risk of bias [49,54]. Two recent meta-analyses pooled 9 randomized controlled trials respectively and included studies with validated diagnosis criteria. They evaluated the clinically therapeutic effect of *G. biloba* and showed effective results of standardized *G. biloba* extract for cognitive impairment and dementia. In patients with AD, the standardized changes scores on ADAS-cog or SKT (Syndrom-Kurz test) as a clinical outcome were greater for the ginkgo group than for the placebo group (95% CI −1.16, −0.10; $z = 2.35$, $N = 6$, $p = 0.02$) [54]. The other study concluded that EGb 761 at 240 mg/day is able to stabilize or slow decline in cognition, function, behavior, and global change at 22−26 weeks in cognitive impairment and dementia subjects [49]. Patients with AD showed a significant effect of ginkgo on the change scores of ADAS-cog (95% CI −2.94, −2.13; $z = 12.26$, $N = 9$; $p < 0.001$) [49].

Herbs and Spices

Curcumin

Curcumin, a yellow pigment present in the rhizome of turmeric (*Curcuma longa*), has been widely used in Southeast Asia since ancient days. Curcuminoids are the active components in curcumin and responsible for the majority medicinal properties [55]. In vivo and in vitro researches have shown that curcumin is a potent antioxidant and is capable of decreasing oxidative stress. It can bind to iron, manganese, and copper and is reported to modulate antioxidant properties [56]. Curcumin inhibits nitric oxide and ROS production in macrophages and inhibits lipooxygenase and cyclooxygenase in fibroblast cells of rats [56]. Curcumin induces cell apoptosis and inhibits cell proliferation. It possesses anticancer activity in various cancer cell lines, such as colorectal cancer, prostate cancer, bladder cancer, cervical cancer, and lung cancer [55]. Curcumin also exhibits antimicrobial, antiprotozoal, antiviral, and antifungal properties [55]. Irritable bowel syndrome prevalence was significantly reduced after administration of curcumin extract [55]. Clinical research shows that curcumin can regulate lipid metabolism and lower levels of cholesterol. Curcumin has the potential ability to improve endothelial function and prevent diabetic vascular complications [55].

Curcumin is able to inhibit the formation of β-amyloid fibrils in vitro and protect against β-amyloid-induced cell death. Low doses of curcumin attenuated β-amyloid overexpression and decreased oxidative damage in brain injured mice [57]. An epidemiologic study for neuroprotective effect of curcumin was first illustrated in Asia. Curry is the predominant dietary source of curcumin intake. In a population-based cohort of 1010 nondemented Asians aged 60–93 years, subjects who consumed curry "occasionally" and "often or very often" had significantly better MMSE score than subjects who "never or rarely" consumed curry in the adjusted model ($p = 0.023$) [58]. However, two clinical trials failed to demonstrate the efficacy of curcumin. A 6-months randomized, double-blind, placebo-controlled trial in Hong Kong was conducted in patients with AD. Patients were given 1 or 4 g curcumin per day orally. The neuroprotective effect of curcumin cannot be demonstrated due to the lack of cognitive decline in the placebo group [59]. In another study in the US, thirty subjects with mild-to-moderate problem AD were randomized to receive placebo, 2 g, or 4 g of curcumin C3 complex per day for 24 weeks. There were no differences between treatment groups in the change in ADAS-Cog, NPI (neuropsychiatric inventory), ADCS-ADL (Alzheimer's disease cooperative Study Activities of Daily Living), or MMSE scores [60]. The authors suggested that the bioavailability of oral curcumin might have limited the results of these studies. Although curcumin readily penetrates the blood–brain barrier, it is relatively low in intestinal absorption and is rapidly metabolize in liver followed by elimination through the gall bladder. Curcumin is eliminated unchanged. Curcumin would have appeared as a slower decline rather than an improvement in cognition [57,59,60]. A longer duration of study with a more sensitive test is required in the future to reveal the role of curcumin in neuroprotection.

Saffron

Saffron, the most expensive spice, is the dried elongated stigmas from the flower *Crocus sativus*. This spice is widely used in India. The four major bioactive compounds in saffron are crocin, crocetin, picrocrocin, and safranal. These compounds contribute to the sensory profile and health promoting properties of saffron [61].

Crocin acts as an antioxidant by quenching free radicals, protecting cells and tissues against oxidation [62,63]. Pretreatment cells with crocin inhibited ROS generation exposed to acrylamide [62]. Administration of 50 mg of saffron dissolved in 100 mL milk for 6 weeks in 10 healthy volunteers and 10 patients showed a significant decrease in lipoprotein oxidation susceptibility [64]. Research has demonstrated several health benefits of saffron: improving digestion, preventing gastric disorder and ulcer, reducing insulin resistance, lowering serum triglyceride and cholesterol level, and anticarcinogenic, anti-inflammatory, antidepressive, and antianxiety effects [61].

Neuroprotective effects of saffron have been revealed in animal and human studies. Administering crocins at 30 mg/kg to rats with AD induced by intracerebroventricular streptozocin showed a significant attenuation in learning and memory performance and spatial cognition evaluation [65]. In a human study, 46 patients with mild-to-moderate severity of AD were randomized into saffron group (30 mg/day) or placebo for 16 weeks. ADAS-cog differed significantly between the two groups after 16 weeks ($p < 0.0001$). The changes at week 16 compared to baseline were: -3.69 ± 1.69 (mean ± SD) and 4.08 ± 1.34 for saffron and placebo, respectively. The Clinical Dementia Rating Scale also showed a significant difference between the two groups ($p < 0.0001$) [66]. In addition, two studies provided preliminary evidence of a possible therapeutic effect of saffron extract in the treatment of patients with AD. Donepezil, an acetylcholinesterase inhibitor (ACEI) and memantine, an N-methyl-D-aspartate (NMDA) receptor antagonist are the most widely accepted drugs in attenuating some of the AD-related symptoms in severe stages. Clinical trials revealed comparable results of saffron extract and these medications on AD patients. Fifty-four Persian-speaking adults with mild-to-moderate AD aged 55 years or older were recruited. Participants were randomly assigned to receive saffron 30 mg/day (15 mg capsule twice per day) or donepezil 10 mg/day (5 mg twice per day). Saffron was found to be as effective as donepezil in the treatment of mild-to-moderate AD after 22 weeks of intervention. The frequency of adverse effects was similar between groups with the exception of vomiting, which occurred significantly more frequently in the donepezil group [67]. The other study showed saffron extract was comparable with memantine in reducing cognitive decline in patients with AD. In a randomized double-blind parallel-group study, 68 patients aged 60 above with moderate-to-severe AD received memantine (20 mg/day) or saffron extracts (30 mg/day) capsules. After 12 months intervention, there was no significant difference between the two groups in the score changes from baseline to the endpoint on Severe Cognitive Impairment Rating Scale (SCIRS), Functional Assessment Staging (FAST) and adverse events [68].

Garlic

Garlic (*Allium sativum*) is commonly consumed in Asia as a flavoring agent. Garlic contains a unique organosulfur compound, allicin. Extract of fresh garlic aged over a prolonged period of time produces Aged Garlic Extract (AGE). It is an odorless product and highly bioavailable. It contains antioxidant phytochemicals that prevent oxidative stress and inhibits lipid peroxidation [69]. Garlic has been reported to have hepatoprotective, immune-enhancing, anticancer, and chemopreventive activities [70]. It also has the ability to scavenge oxidants and increase antioxidant levels [70]. Ingested garlic at the daily dose of 0.1/kg body weight for one month in human showed a significantly lower malondialdehyde (MDA) levels and significantly higher erythrocyte GSH-Px and SOD activities [71]. Black garlic, which has been recently introduced to the Korean and Japanese markets as a health product by aging whole garlic at high temperature and in high humidity, showed even stronger antioxidant activities when compared to raw garlic [72]. Various beneficial effects of garlic and garlic extract are summarized in Table 47.7 [70,73,74].

Garlic protects the brain from loss of intellectual capacity and memory and has a potentially positive and preventive therapeutic effect in the treatment of AD in animal studies. It improved spatial memory deficits which were associated with aging [75]. AGE improved learning abilities and memory retention and increased longevity in the senescence-accelerated mouse [76]. Garlic has been found to possess antioxidant and neuroprotective activities in animal studies. There is a need to continue more investigations on the type, the dosage, and the mechanisms regarding the protective activities in human.

Ginger

Ginger, the rhizome of *Zingiber officinale*, is a common spice that has culinary use in Asian countries. Numerous active compounds are present in ginger including gingerol and shogaol [77]. These compounds have been reported to exhibit many pharmacological and physiological functions including immunomodulatory, antitumorigenic, anti-inflammatory, antiapoptotic, antihyperglycemic, antilipidemic and antiemetic actions (Table 47.8) [77–79]. Ginger juice has a protective effect by decreasing lipid peroxidation and increasing GSH-PX, SOD, catalase in rats [80]. It has been reported that ginger is capable of inhibiting chemotherapy-induced adverse reactions and improving the quality of life of cancer patients [81]. A recent systematic review suggested that ginger could be considered as a harmless and possible effective alternative option for women suffering from pregnancy associated nausea and vomiting [82].

The neuroprotective effect of ginger is suggested to be due to the presence of phenolic and flavonoids compounds [77]. In an animal study, ginger root extract had the ability to prevent behavioral dysfunction in the β-amyloid-induced AD model in rats [83]. Currently, there have been limited trials on human. With the strong antioxidant effects of ginger and the positive neuroprotective effect in animal studies, it is suggested that ginger is a promising product in fighting the ravages of aging and neurodegenerative diseases.

TABLE 47.7 Various Health Benefits of Garlic Based on Current Studies

Health benefits	Possible actions
Cardioprotective effects	Enhance vasodilatory response. Lower total cholesterol and low density lipoprotein (LDL). Prevent development of ventricle hypertrophy. Reduce glucose level.
Anticancer effects	Induce cell apoptosis. Suppress breast, blood, bladder, gastric, oral cavity, colorectal, skin, uterus, esophagus and lung cancers.
Anti-infection effects	Prevent cold and flu symptoms. Antimicrobial activity against *Staphylococcus aureus*, *Pseudomonas aeruginosa* and *Escherichia coli*. Antiviral and antiparasite activities.
Antioxidant effects	Scavenge oxidants. Increase antioxidant enzymes, for example, GSH-Px, SOD. Inhibit lipid peroxidation.

TABLE 47.8 Summary of the Physiological Functions of Ginger

Physiological functions	Protective effects
Anti-inflammatory effects	Suppress the synthesis of pro-inflammatory cytokines. Downregulate the induction of inflammatory genes.
Hepato-protective effects	Decrease fasting blood glucose. Decrease liver enzymes levels. Increase activities of antioxidant enzymes in liver.
Antiemetic effects	Relief the severity in nausea and vomiting.
Antilipidemic effects	Reduce fructose-induced elevation of lipid levels. Reduce body weight. Lower serum cholesterol and triglyceride level.
Gastroprotective effects	Prevent gastric ulcer via increasing mucin secretion.
Antitumor effects	Control tumor development through upregulation of tumor suppressor gene. Induce apoptosis.

Black Rice

Rice is the main staple food in many Asian countries. Pigmented rice is commonly consumed as a health-promoting food. The pigments are due to the presence of variety of flavones, tannin, phenolics, sterols, tocols, amino acids, and essential oils [84]. Black rice (*Oryza sativa L. indica*) derives its name because of the rich natural anthocyanin compounds, a group of reddish to purple water soluble flavonoids. It is mainly located in the aleurone layer [84].

Anthocyanins are responsible for the major functional components in black rice. There is a mixture of anthocyanins in which cyanidin-3-glucoside and peonidin-3-glucoside are not found in white rice [84]. Black rice pigmented fractions exhibit antioxidant activities and free radical scavenging capacities in cells [85]. When the outer layer of black rice was fed to rabbits, it significantly decreased MDA level when compared with white rice outer layer fraction [86]. Increased hepatic SOD and CAT activities were observed in black rice extract treated mice [84]. The antioxidant effect of black rice has also been demonstrated in human. When black rice pigment fractions were supplemented for 6 months in patients with coronary heart disease in China in a randomized double-blind placebo-controlled trial, greatly enhanced plasma total antioxidant capacity and significantly reduced plasma level of high sensitive C-reactive protein were observed in the treatment group in comparison to the control group [87]. Black rice also exhibits cardioprotective effects. Dietary supplementation of black rice in animals showed decreasing in serum triglyceride, total cholesterol, and nonhigh density lipoprotein cholesterol level and improving insulin sensitivity [84]. Anthocyanins extracted from black rice protected mouse brain neuron against death at low concentration [88]. Though there is limited study on human, the strong antioxidant effects of black rice postulate its neuroprotective effect.

Marine Algae

Marine algae or seaweeds are traditional Asian foods particularly common in Japan and Korea. It is consumed as a snack or as ingredients in dishes or soups. Marine algae have two major bioactive compounds: sulfated polysaccharides (SPs) and polyphenol. The major SPs include fucoidan and laminaran [89].

Marine algae have been shown to scavenge ROS production and protect against oxidative stress in in vivo and in vitro studies [89–91]. Consumption of seaweed increased the endogenous antioxidant enzymes SOD, GSH-Px and catalase activities in vivo [90,92]. The antioxidant effect can be further supported by the antitumor effects. Human and monkey cancer cell lines were inhibited by seaweed extract [90]. Oral administration of seaweeds can cause a significant decrease in incidence of carcinogenesis in vivo. SPs have antiproliferative activity in human leukemic monocyte lymphoma and inhibit tumor growth in mice [90]. In addition, seaweed is a potential hypolipidemic agent. It reduced serum triglyceride, total cholesterol, and LDL and elevated HDL in mice [92]. Seaweed extracts have blood pressure lowering dipeptides and angiotensin converting inhibitory properties in vitro. Seaweed extracts are also capable of prolonging partial thromboplastin time, prothrombin time, and thrombin time and thus reducing the risk of cardiovascular disease [89,92]. In addition, antiobesity effects of seaweed extract, including reducing body weight, increasing resting energy expenditure, and improving insulin sensitivity, have been demonstrated in obese patients with nonalcoholic fatty liver disease [92].

Several studies have provided insight into the neuroprotective effects of marine algae including antioxidant, antineuroinflammatory, and cholinesterase inhibitory activity and the inhibition of neuronal death. Fucoidan inhibited neurotoxic effects of β-amyloid in rat neurons via blocking the generation of ROS [93]. Fucoidan was also strongly neuroprotective against AD in vivo and in vitro [93]. Seaweed extracts altered the levels of neurotransmitters modified by the ethanol treatment in mice. They increased the level of acetylcholine, and exerted anticholinesterase activities and resulted in memory-enhancing abilities [94].

Mushrooms

Mushrooms have long been regarded as health-promoting foods worldwide. L-ergothioneine is a unique sulfur containing amino acid that cannot be synthesized by human and is only available from certain dietary sources, especially in fungi. It is a stable antioxidant in that it does not autooxidize at physiologic pH, does not promote generation of hydroxyl radicals, and may serve as a final defense against oxidation in cells [94]. Among cultivated mushrooms, Lingzhi has been used in folk medicine in China and many Asian countries. It is believed that Lingzhi (Ganoderma lucidum) is attributed to a range of beneficial effects including enhancing longevity, increasing youthful vigor and vitality, and as a remedy for illness [94]. Polysaccharides and triterpenes are two major active compounds in Lingzhi.

Researches have shown various health benefits of Lingzhi and its anticancer abilities are widely investigated. The growth of different tumors cells were inhibited in vitro, possibly through its effect on enhancing the body immune system with a broad spectrum of immune-modulation activities [95]. A Cochrane review also concluded that Lingzhi could be administered as an alternative adjunct to the conventional treatment of cancer in consideration of its potential of enhancing tumor response and stimulating host immunity [96]. Lingzhi has the ability to promote free radical scavenging. Ten healthy Chinese subjects (age 22–56) ingested a single dose of 1.1 g Lingzhi powder and showed an increase in plasma FRAP (Ferric Reducing/Antioxidant Power) over the 3 hours monitoring period. Ingestion of 3.3 g Lingzhi caused a significant postingestion increase in plasma antioxidant capacity with a peak at 90 min [94]. Lingzhi and its extracts exerted hepatoprotective effects by preventing liver damage induced by alcohol and protecting against liver damage in rats [95]. One-third of chronic hepatitis B patients returned to normal aminotransferase level after taking Lingzhi extracts for 6 months [97]. Glucans from Lingzhi fruiting bodies could lower postprandial glucose level in diabetes patients [95]. Lingzhi also showed an ability to regulate lipid metabolism. In a double-blind, placebo-controlled, cross-over study, healthy individuals following an administration of 1.44 g Lingzhi/day for 4 weeks showed a slight trend in the reduction of cholesterol, LDL, and triglycerides [98].

Several studies have suggested the neuroprotective effects of Lingzhi against oxidative stress in vitro and has the ability to induce neuronal differentiation. Lingzhi attenuated β-amyloid induced synaptotoxicity by preserving a synaptic density protein in cultured neurons and preserved neurons in AD [94]. It is suggested that Lingzhi is a potential candidate for the treatment of aging-related neurodegenerative diseases. More clinical studies have to be done to confirm this effect.

Another commonly used mushroom in East Asian countries is Hericium erinaceus. It is an edible mushroom with medicinal value, which is also known as Lion's Mane Mushroom or Hou Tou Gu in Chinese or Yamabushitake in Japanese. Polysaccharides extract of H. erinaceus has significant anticancer and immunomodulation activities. It enhanced T cells and macrophages secretion and enhanced the expression of cytokines [99]. In animal studies, H. erinaceus exerted hypolipidemic effects by reducing plasma total cholesterol, LDL, and triglyceride and hepatic HMG-CoA reductase activity [99]. The antioxidant index, free radical scavenging activity, and lipid peroxidation inhibitory activities have been reported in H. erinaceus [99]. The mycelium extract of H. erinaceus is rich in phenolic content and has potential ferric reducing antioxidant power. The fresh fruit body extract was found to have the potent, 1-diphenyl-2-picrylhydrazyl radical scavenging activity. And, oven-dried fruit body extract was excellent in reducing the extent of carotene bleaching [99].

Hericium erinaceus reports to have activities related to nerve and brain health. The polysaccharides of H. erinaceus can induce neuronal differentiation and promote neuronal survival. It has the ability to prevent the impairments of spatial short-term and visual recognition memory induced by β-amyloid peptide in animals [99]. Daily oral administration of H. erinaceus could promote nerve regeneration after injury in rats [100]. The neuroprotective effect on human was demonstrated. A double-blind, parallel-group, placebo-controlled trial was performed on 30 Japanese (age 50–80 years old) with mild cognitive impairment. The subjects took four Yamabushitake-containing tablets (each contained 96% of H. erinaceus) or placebo tablets

TABLE 47.9 Potential Health Benefits of Wolfberry

General health benefits	Possible actions
Hepatoprotective effects	Prevent alcohol fatty liver. Reduce serum liver enzymes level. Increase antioxidants abilities of liver.
Weight management	Increase postprandial energy expenditure. Reduce waist circumference. Reduce risk of metabolic syndrome.
Cardiovascular benefits	Reduce serum total cholesterol and TG and increase HDL.
Anticancer effects	Inhibit the growth of transplantable sarcoma. Reduce the lipid peroxidation.
Immune modulation	Increase macrophage phagocytosis. Increase activity of T cells, cytotoxic T cells, and natural killer cells.
Antioxidant effects	Increase serum SOD and GSH-Px levels and decrease MDA.
Eye health effects	Protects against light damage to the retinal pyramid, rod cell layer, outer nuclear layer, and retinal pigmented epithelium.

three times a day for 16 weeks. The intervention group showed significantly increased scores on the cognitive function scale (Revised Hasegawa Dementia Scale (HDS-R)) compared with the placebo group ($p < 0.001$). The scores decreased at week 4 after the termination. This study suggested the effectiveness of H. erinaceus in the prevention or the treatment of dementia and cognitive dysfunction [101].

Wolfberry

Goji or wolfberry (Lycium barbarum) is used as a food and medicinal plant in China. It is generally consumed in soups, with rice or added to a meat and vegetables dish. Polysaccharides, carotenoids, and flavonoids are the major metabolites in wolfberry [102]. A meta-analysis which pooled four randomized controlled trials has confirmed various health effects of L. barbarum polysaccharides. Intake of goji among 18–72 years old resulted in statistically significant improvements in neurological/psychological performance and overall feelings of health and well-being [103]. Polysaccharides and flavonoids exhibit radical scavenging activity toward superoxide anion and reducing capacity which are similar to the synthetic antioxidant [102]. Fifty healthy Chinese adults (age 55–72) who were administered 120 mL/day of standardized L. barbarum fruit juice (equivalent to at least 150 g of fresh fruit) showed a significant increase in serum level of SOD, GSH-Px, and a significant decrease in lipid peroxidation as indicated by a decrease level of MDA [104]. Other health benefits of wolfberry are summarized in Table 47.9 [105,106].

Polysaccharides of L. barbarum have been reported to prevent neuronal death in both necrosis and apoptosis. Goji has been shown to protect rat cortical neurons against β-amyloid induced toxicity in vitro [102]. Feeding animals with daily L. barbarum polysaccharides extract can restore motor activity, restore memory index, and increase SOD levels [105]. A randomized, double-blind, placebo-controlled clinical study administered standardize L. barbarum fruit juice at 120 mg/day (equivalent to at least 150 g fresh fruit) for 30 days in 60 older healthy adults (age 55–72). The treatment group showed a statistically significant increase in the number of lymphocytes, better general feelings of well-being, and a tendency for increased short-term memory and focus between pre- and postintervention [104]. L. barbarum has been used as a traditional herbal medicine in Asian countries and no known toxicity have been reported. Further studies are needed to clarify the mechanisms of L. barbarum promoting healthy aging.

Dansen

Dansen is a dried root of Salvia miltiorrhiza commonly used in traditional Chinese herbal medicine either alone or mixed with other herbs [107]. Lipophilic and hydrophilic compounds are the two major classes of active components in Dansen. Tanshinone I, IIA, IIB, and cryptotanshinone are the most abundant lipophilic compounds, whereas phenolic acids are the major hydrophilic compounds [108]. Various in vitro and in vivo studies suggested that Dansen has a potent antioxidant activity. Sodium tanshinone IIA sulfonate (STS) has free radical scavenger properties which inhibited LDL oxidation resulting in the attenuation of cardiac cell hypertrophy [108]. The hydrophilic components also have antioxidant capacity by increasing activity of SOD and GPx and decreasing the level of MDA and ROS production significantly in vitro [109]. Dansen has been demonstrated as a promising natural medicine to prevent cardiovascular diseases. It promotes microcirculation, dilates coronary arteries, enhances blood flow, prevents uptake and oxidation of LDL, and protects from ischemia-reperfusion injury [108]. Therapeutic effects of STS were examined in five hospitalized pulmonary arterial hypertension patients. After 8 weeks of STS infusion, the average pulmonary arterial systolic pressure and the right ventricle size were lowered as compared between pre- and postintervention. All patients performed better exercise capacity as indicated by the improvement of a 6-min walking distance [110].

Dansen is postulated as containing neuroprotective agent. Salvianolic B, a Dansen extract was found to promote neuronal stem progenitor cells proliferation in vitro and in vivo [111]. Another extract Salvianolic acid A is a neuroprotective agent against β-amyloid induced toxicity in vitro [111]. Tanshinone IIA and IIB readily penetrated the blood—brain barrier and reached a peak concentration 60 min after intraperitoneal injection [112]. Twenty-four hours after middle cerebral artery occlusion, brain infarct volume was reduced following treatment with Tanshinone in mice. The preventive effect was accompanied by a significant decrease in the observed neurological deficit [112].

Gegen

Gegen is the root of *Pueraia lobata*, also known as the root of Kudzu, which has been widely used in traditional Chinese medicine for promoting circulation and increasing blood flow. Puerarin, daidzin and daidzein are the major active components in Gegen and responsible for many profound pharmacological actions. Daidzein combines with daidzin to form 7-0-glycoside daidzein, which is another major active component in Gegen, contains antioxidant and phytoestrogenic properties [113].

Puerarin, being the most abundant component in Gegen, has been identified as performing a variety of functions, including antihypertension, antiarrhythmic, antioxidant, antiischemic, antiapoptotic, antidiabetic, and neuroprotective properties [113,114]. Gegen has been used for the treatment of cardiovascular disease and type 2 diabetes [113]. Puerarin plays an important role in cell regulation, involving cell expression and receptor functions. Puerarin suppressed blood pressure by upregulating the hepatic Angiotensin II type I receptor (AT1) and angiotensin-converting enzyme 2 (ACE2) mRNA expression. It suppressed rennin activity in rats with the intake of 100, 200 mg/kg/day for three weeks [115]. Puerarin is capable of lowering glucose levels in diabetic patients. In a dose-dependent high glucose treatment, puerarin enhanced glucose uptake of insulin resistant adipocytes and upregulated the protein expression of GLUT-4 in skeletal muscle to cause the effects of lowering blood glucose and insulin levels [114]. Puerarin also protects against cerebral ischemia and reperfusion. An in vivo study showed that rats with middle cerebral artery occlusion received puerarin (100 mg/kg, i.p.) resulting in improving neurological functions and diminishing infarct and edema volume [116]. Moreover, puerarin protected primary hippocampal neurons against apoptosis and necrosis caused by glutamate and oxygen/glucose deprivation [114]. Puerarin significantly increased the spontaneous behavior and explorative response in the aging mice in the open field test and improved their learning memory ability [117]. Though there have been limited clinical trials of the single use of Gegen, it can be a potential candidate to suppress aging-related neuronal cell apoptosis and dysfunction of the memory system.

LIMITATIONS OF RESEARCH

There has been a long history of using natural products for health in aging based on unique traditional medicine theories in Asia. However, clinical evidence of certain food on healthy aging remains inconclusive. This is due to several reasons. First, the holistic approach of traditional medicine results in research difficulties. In Asian countries, aging is regarded as a process of progressive decline of "vital energy" in the body [2]. Balancing different components in the body is considered essential for antiaging and disease prevention [2]. Food is considered to be multifunctional and to correct the overall imbalance of vital energy components. Since it does not target a single organ specifically, it is difficult in identifying the appropriate biomarkers to be measured in epidemiological studies [2].

Secondly, these natural products are consumed as food in Asian countries with habitual daily intake for health maintenance and disease prevention rather than disease treatment. These food acts as disease-modifying agents for presymptomatic individuals or those with earliest onset. There are insufficient biomarkers or methods to evaluate the effectiveness in the presymptomatic stages [118]. Also, a lifetime habitual intake of food could have had some overall protective effects in the study population, even among the placebo group [39]. Such smaller between-group differences may lessen the association to be explored and thus may lead to inconclusive results in different studies.

Thirdly, the investigation of single component may not be enough to show the sufficient effect of the food [26]. Herbal crude extracts are usually used to standardize the chemical and biological components in order to provide quality control of natural food. However, food contains vitamins, minerals, phytochemicals, and antioxidants that act synergistically rather than working in isolation. It is possible that the complex mixture of bioactive compounds in food cannot be substituted by a single purified component.

SUMMARY

- Due to the irreversible deterioration with the aging process, older people are susceptible to various chronic diseases.
- Ginseng has promising benefits on cardioprotection, moderate immune response, antiulcer, anticancer and antioxidants effects. Observation and clinical studies however cannot confirm the cognitive protection effects of ginseng.
- Various observational studies have revealed the beneficial effects of tea on cognitive health. However, there is no evidence to show that green tea is more potent than black/oolong tea on cognitive reserve.
- Isoflavones, one of the bioactive compounds in soy can relieve menopausal symptoms, protect against breast and prostate cancer, cardiovascular disease and osteoporosis. Difference in the bioavailability of isoflavones between ethnic groups and age lead to inconclusive results in its neuroprotective effect.
- Ginkgo is used as a medicine for memory and age-related deterioration in China. Two meta-analyses show effective results for ginkgo for cognitive impairment and dementia.
- Curcumin, saffron, garlic, and ginger are commonly used spices in Asia. They exhibit various biological functions including antioxidant, anticarcinogenic, immunomodulation, and cardioprotective effects. More studies are needed regarding the dosage and the mechanism on their protective activities in human.
- Anthocyanin gives the dark purple color of black rice and is responsible for its biological functions. It is a strong antioxidant and provides cardioprotective and neuroprotective effects in vitro and in vivo.
- Lingzhi and Lion's Mane have shown anticancer, immunomodulation, and antioxidant activities. Both exert activities related to nerve and brain health. Lingzhi preserved neurons in AD and Lion's Mane showed increase in cognitive function score in an intervention study.
- Wolfberry is commonly consumed as food in soups or dishes. A meta-analysis showed statistically significant improvements in neurological and psychological performance and overall feelings of health after taking wolfberry in the treatment group when compared with the placebo group. Further studies focusing on its antiaging effects are required.
- Dansen and Gegen have been widely used in traditional Chinese medicine to prevent cardiovascular disease and promote circulation. Tanshinone, the active component in dansen, and puerarin, the active component in gegen, have shown antioxidant effects and are beneficial to improve neurological function in injured or aged animals.

References

[1] Peng C, Wang X, Chen J, Jiao R, Wang L, Li YM, et al. Biology of ageing and role of dietary antioxidants. BioMed Res Int 2014;2014:831–41.
[2] Ho YS, So KF, Chang RC. Anti-aging herbal medicine – how and why can they be used in aging-associated neurodegenerative diseases? Ageing Res Rev 2010;9(3):354–62.
[3] Ferrari CK. Functional foods, herbs and nutraceuticals: towards biochemical mechanisms of healthy aging. Biogerontology 2004;5(5):275–89.
[4] Prince M, Bryce R, Albanese E, Wimo A, Ribeiro W, Ferri CP. The global prevalence of dementia: a systematic review and metaanalysis. Alzheimers Dement 2013;9(1):63–75 e2.
[5] Cho IH. Effects of *Panax ginseng* in neurodegenerative diseases. J Ginseng Res 2012;36(4):342–53.
[6] Radad K, Gille G, Liu L, Rausch WD. Use of ginseng in medicine with emphasis on neurodegenerative disorders. J Pharmacol Sci 2006;100(3):175–86.
[7] Kennedy DO, Scholey AB. Ginseng: potential for the enhancement of cognitive performance and mood. Pharmacol Biochem Behav 2003;75(3):687–700.
[8] Shergis JL, Zhang AL, Zhou W, Xue CC. *Panax ginseng* in randomised controlled trials: a systematic review. Phytother Res 2013;27(7):949–65.
[9] Sünram-Lea SI, Birchall RJ, Wesnes KA, Petrini O. The effect of acute administration of 400 mg of *Panax ginseng* on cognitive performance and mood in healthy young volunteers. Curr Top Nutraceutical Res 2005;3(1):10.
[10] Persson J, Bringlov E, Nilsson LG, Nyberg L. The memory-enhancing effects of ginseng and *Ginkgo biloba* in healthy volunteers. Psychopharmacology 2004;172(4):430–4.
[11] Sorensen H, Sonne J. A double-masked study of the effects of ginseng on cognitive functions. Aging Clin Exp Res 1996;8(6):4.
[12] Heo JH, Lee ST, Chu K, Oh MJ, Park HJ, Shim JY, et al. An open-label trial of Korean red ginseng as an adjuvant treatment for cognitive impairment in patients with Alzheimer's disease. Eur J Neurol 2008;15(8):865–8.
[13] Lee ST, Chu K, Sim JY, Heo JH, Kim M. *Panax ginseng* enhances cognitive performance in Alzheimer's disease. Alzheimer Dis Assoc Disord 2008;22(3):222–6.

[14] Neri M, Andermarcher E, Pradelli JM, Salvioli G. Influence of a double blind pharmacological trial on two domains of well-being in subjects with age associated memory impairment. Arch Gerontol Geriatr 1995;21(3):241–52.

[15] Song J, Xu H, Liu F, Feng L. Tea and cognitive health in late life: current evidence and future directions. J Nutr Health Aging 2012;16(1):31–4.

[16] Mandel S, Youdim MB. Catechin polyphenols: neurodegeneration and neuroprotection in neurodegenerative diseases. Free Radic Biol Med 2004;37(3):304–17.

[17] Khan N, Mukhtar H. Tea and health: studies in humans. Curr Pharm Des 2013;19(34):6141–7.

[18] Vuong QV. Epidemiological evidence linking tea consumption to human health: a review. Crit Rev Food Sci Nutr 2014;54(4):523–36.

[19] Kim HG, Oh MS. Herbal medicines for the prevention and treatment of Alzheimer's disease. Curr Pharm Des 2012;18(1):57–75.

[20] Feng L, Gwee X, Kua EH, Ng TP. Cognitive function and tea consumption in community dwelling older Chinese in singapore. J Nutr Health Aging 2010;14(6):433–8.

[21] Kuriyama S, Hozawa A, Ohmori K, Shimazu T, Matsui T, Ebihara S, et al. Green tea consumption and cognitive function: a cross-sectional study from the Tsurugaya Project 1. Am J Clin Nutr 2006;83(2):355–61.

[22] Feng L, Li J, Ng TP, Lee TS, Kua EH, Zeng Y. Tea drinking and cognitive function in oldest-old Chinese. J Nutr Health Aging 2012;16(9):754–8.

[23] Ng TP, Feng L, Niti M, Kua EH, Yap KB. Tea consumption and cognitive impairment and decline in older Chinese adults. Am J Clin Nutr 2008;88(1):224–31.

[24] Noguchi-Shinohara M, Yuki S, Dohmoto C, Ikeda Y, Samuraki M, Iwasa K, et al. Consumption of green tea, but not black tea or coffee, is associated with reduced risk of cognitive decline. PloS One 2014;9(5):e96013.

[25] Huang CQ, Dong BR, Zhang YL, Wu HM, Liu QX. Association of cognitive impairment with smoking, alcohol consumption, tea consumption, and exercise among Chinese nonagenarians/centenarians. Cogn Behav Neurol 2009;22(3):190–6.

[26] Soni M, Rahardjo TB, Soekardi R, Sulistyowati Y, Lestariningsih, Yesufu-Udechuku A, Irsan A, et al. Phytoestrogens and cognitive function: a review. Maturitas 2014;77(3):209–20.

[27] Messina M, Messina V. The role of soy in vegetarian diets. Nutrients 2010;2(8):855–88.

[28] D'Adamo CR, Sahin A. Soy foods and supplementation: a review of commonly perceived health benefits and risks. Altern Ther Health Med 2014;20(Suppl. 1):39–51.

[29] Behloul N, Wu G. Genistein: a promising therapeutic agent for obesity and diabetes treatment. Eur J Pharmacol 2013;698(1–3):31–8.

[30] Karyadi D, Lukito W. Functional food and contemporary nutrition-health paradigm: tempeh and its potential beneficial effects in disease prevention and treatment. Nutrition 2000;16(7–8):697.

[31] Vega-Lopez S, Yeum KJ, Lecker JL, Ausman LM, Johnson EJ, Devaraj S, et al. Plasma antioxidant capacity in response to diets high in soy or animal protein with or without isoflavones. Am J Clin Nutr 2005;81(1):43–9.

[32] Azadbakht L, Kimiagar M, Mehrabi Y, Esmaillzadeh A, Hu FB, Willett WC. Dietary soya intake alters plasma antioxidant status and lipid peroxidation in postmenopausal women with the metabolic syndrome. Br J Nutr 2007;98(4):807–13.

[33] Duffy R, Wiseman H, File SE. Improved cognitive function in postmenopausal women after 12 weeks of consumption of a soya extract containing isoflavones. Pharmacol Biochem Behav 2003;75(3):721–9.

[34] Kritz-Silverstein D, Von Muhlen D, Barrett-Connor E, Bressel MA. Isoflavones and cognitive function in older women: the SOy and Postmenopausal Health In Aging (SOPHIA) Study. Menopause 2003;10(3):196–202.

[35] Gleason CE, Carlsson CM, Barnet JH, Meade SA, Setchell KD, Atwood CS, et al. A preliminary study of the safety, feasibility and cognitive efficacy of soy isoflavone supplements in older men and women. Age and Aging 2009;38(1):86–93.

[36] Henderson VW, St John JA, Hodis HN, Kono N, McCleary CA, Franke AA, et al. Long-term soy isoflavone supplementation and cognition in women: a randomized, controlled trial. Neurology 2012;78(23):1841–8.

[37] Xu X, Xiao S, Rahardjo TB, Hogervorst E. Tofu intake is associated with poor cognitive performance among community-dwelling elderly in China. J Alzheimer's Dis 2015;43(2):669–75.

[38] Hogervorst E, Sadjimim T, Yesufu A, Kreager P, Rahardjo TB. High tofu intake is associated with worse memory in elderly Indonesian men and women. Dement Geriatr Cogn Disord 2008;26(1):50–7.

[39] Ho SC, Chan AS, Ho YP, So EK, Sham A, Zee B, et al. Effects of soy isoflavone supplementation on cognitive function in Chinese postmenopausal women: a double-blind, randomized, controlled trial. Menopause 2007;14(3 Pt 1):489–99.

[40] Fournier LR, Ryan Borchers TA, Robison LM, Wiediger M, Park JS, Chew BP, et al. The effects of soy milk and isoflavone supplements on cognitive performance in healthy, postmenopausal women. J Nutr Health Aging 2007;11(2):155–64.

[41] Kreijkamp-Kaspers S, Kok L, Grobbee DE, de Haan EH, Aleman A, Lampe JW, et al. Effect of soy protein containing isoflavones on cognitive function, bone mineral density, and plasma lipids in postmenopausal women: a randomized controlled trial. J Am Med Assoc 2004;292(1):65–74.

[42] White LR, Petrovitch H, Ross GW, Masaki K, Hardman J, Nelson J, et al. Brain aging and midlife tofu consumption. J Am Coll Nutr 2000;19(2):242–55.

[43] Hogervorst E, Mursjid F, Priandini D, Setyawan H, Ismael RI, Bandelow S, et al. Borobudur revisited: soy consumption may be associated with better recall in younger, but not in older, rural Indonesian elderly. Brain Res 2011;1379:206–12.

[44] Kreijkamp-Kaspers S, Kok L, Grobbee DE, de Haan EH, Aleman A, van der Schouw YT. Dietary phytoestrogen intake and cognitive function in older women. J Gerontol A Biol Sci Med Sci 2007;62(5):556–62.

[45] Setchell KD, Clerici C. Equol: pharmacokinetics and biological actions. J Nutr 2010;140(7):1363S–8S.

[46] Diamond BJ, Bailey MR. Ginkgo biloba: indications, mechanisms, and safety. Psychiatr Clin North Am 2013;36(1):73–83.

[47] Rojas P, Montes P, Rojas C, Serrano-Garcia N, Rojas-Castaneda JC. Effect of a phytopharmaceutical medicine, Ginko biloba extract 761, in an animal model of Parkinson's disease: therapeutic perspectives. Nutrition 2012;28(11–12):1081–8.

[48] Birks J, Grimley Evans J. Ginkgo biloba for cognitive impairment and dementia. Cochrane Database Syst Rev 2009;(1):CD003120.

[49] Tan MS, Yu JT, Tan CC, Wang HF, Meng XF, Wang C, et al. Efficacy and adverse effects of Ginkgo biloba for cognitive impairment and dementia: a systematic review and meta-analysis. J Alzheimer's Dis 2015;43(2):589–603.

[50] Bridi R, Crossetti FP, Steffen VM, Henriques AT. The antioxidant activity of standardized extract of Ginkgo biloba (EGb 761) in rats. Phytother Res 2001;15(5):449–51.

[51] DeKosky ST, Williamson JD, Fitzpatrick AL, Kronmal RA, Ives DG, Saxton JA, et al. Ginkgo biloba for prevention of dementia: a randomized controlled trial. J Am Med Assoc 2008;300(19):2253–62.

[52] Snitz BE, O'Meara ES, Carlson MC, Arnold AM, Ives DG, Rapp SR. Ginkgo biloba for preventing cognitive decline in older adults: a randomized trial. J Am Med Assoc 2009;302 (24):2663–70.

[53] Vellas B, Coley N, Ousset PJ, Berrut G, Dartigues JF, Dubois B, et al. Long-term use of standardised Ginkgo biloba extract for the prevention of Alzheimer's disease (GuidAge): a randomised placebo-controlled trial. Lancet Neurol 2012;11(10):851–9.

[54] Weinmann S, Roll S, Schwarzbach C, Vauth C, Willich SN. Effects of Ginkgo biloba in dementia: systematic review and meta-analysis. BMC Geriatr 2010;10:14.

[55] Fan X, Zhang C, Liu DB, Yan J, Liang HP. The clinical applications of curcumin: current state and the future. Curr Pharm Des 2013;19(11):2011–31.

[56] Noorafshan A, Ashkani-Esfahani S. A review of therapeutic effects of curcumin. Curr Pharm Des 2013;19(11):2032–46.

[57] Monroy A, Lithgow GJ, Alavez S. Curcumin and neurodegenerative diseases. Biofactors 2013;39(1):122–32.

[58] Ng TP, Chiam PC, Lee T, Chua HC, Lim L, Kua EH. Curry consumption and cognitive function in the elderly. Am J Epidemiol 2006;164(9):898–906.

[59] Baum L, Lam CW, Cheung SK, Kwok T, Lui V, Tsoh J, et al. Six-month randomized, placebo-controlled, double-blind, pilot clinical trial of curcumin in patients with Alzheimer's disease. J Clin Psychopharmacol 2008;28(1):110–13.

[60] Ringman JM, Frautschy SA, Teng E, Begum AN, Bardens J, Beigi M, et al. Oral curcumin for Alzheimer's disease: tolerability and efficacy in a 24-week randomized, double blind, placebo-controlled study. Alzheimer's Res Ther 2012;4(5):43.

[61] Wang S, Melnyk JP, Marcone MF. Chemical and biological properties of the world's most expensive spice: saffron. Food Res Int 2010;43:9.

[62] Mehri S, Abnous K, Mousavi SH, Shariaty VM, Hosseinzadeh H. Neuroprotective effect of crocin on acrylamide-induced cytotoxicity in PC12 cells. Cell Mol Neurobiol 2012;32(2):227–35.

[63] Mousavi SH, Tayarani NZ, Parsaee H. Protective effect of saffron extract and crocin on reactive oxygen species-mediated high glucose-induced toxicity in PC12 cells. Cell Mol Neurobiol 2010;30(2):185–91.

[64] Verma SK, Bordia A. Antioxidant property of Saffron in man. Indian J Med Sci 1998;52(5):205–7.

[65] Khalili M, Hamzeh F. Effects of active constituents of Crocus sativus L., crocin on streptozotocin-induced model of sporadic Alzheimer's disease in male rats. Iran Biomed J 2010;14 (1–2):59–65.

[66] Akhondzadeh S, Sabet MS, Harirchian MH, Togha M, Cheraghmakani H, Razeghi S, et al. Saffron in the treatment of patients with mild to moderate Alzheimer's disease: a 16-week, randomized and placebo-controlled trial. J Clin Pharm Ther 2010;35(5):581–8.

[67] Akhondzadeh S, Shafiee Sabet M, Harirchian MH, Togha M, Cheraghmakani H, Razeghi S, et al. A 22-week, multicenter, randomized, double-blind controlled trial of Crocus sativus in the treatment of mild-to-moderate Alzheimer's disease. Psychopharmacology 2010;207(4):637–43.

[68] Farokhnia M, Shafiee Sabet M, Iranpour N, Gougol A, Yekehtaz H, Alimardani R, et al. Comparing the efficacy and safety of Crocus sativus L. with memantine in patients with moderate to severe Alzheimer's disease: a double-blind randomized clinical trial. Hum Psychopharmacol 2014;29(4):351–9.

[69] Borek C. Antioxidant health effects of aged garlic extract. J Nutr 2001;131(3s):1010S–5S.

[70] Amagase H, Petesch BL, Matsuura H, Kasuga S, Itakura Y. Intake of garlic and its bioactive components. J Nutr 2001;131 (3s):955S–62S.

[71] Avci A, Atli T, Erguder IB, Varli M, Devrim E, Aras S, et al. Effects of garlic consumption on plasma and erythrocyte antioxidant parameters in elderly subjects. Gerontology 2008;54 (3):173–6.

[72] Choi IS, Cha HS, Lee YS. Physicochemical and antioxidant properties of black garlic. Molecules 2014;19(10):16811–23.

[73] Chan JY, Yuen AC, Chan RY, Chan SW. A review of the cardiovascular benefits and antioxidant properties of allicin. Phytother Res 2013;27(5):637–46.

[74] Amagase H. Clarifying the real bioactive constituents of garlic. J Nutr 2006;136(Suppl. 3):716S–25S.

[75] Majewski M. Allium sativum: facts and myths regarding human health. Rocz Panstw Zakl Hig 2014;65(1):1–8.

[76] Moriguchi T, Saito H, Nishiyama N. Anti-ageing effect of aged garlic extract in the inbred brain atrophy mouse model. Clin Exp Pharmacol Physiol 1997;24(3–4):235–42.

[77] Rahmani AH, Shabrmi FM, Aly SM. Active ingredients of ginger as potential candidates in the prevention and treatment of diseases via modulation of biological activities. Int J Physiol, Pathophysiol Pharmacol 2014;6(2):125–36.

[78] Azam F, Amer AM, Abulifa AR, Elzwawi MM. Ginger components as new leads for the design and development of novel multi-targeted anti-Alzheimer's drugs: a computational investigation. Drug Des, Dev Ther 2014;8:2045–59.

[79] Ali BH, Blunden G, Tanira MO, Nemmar A. Some phytochemical, pharmacological and toxicological properties of ginger (Zingiber officinale Roscoe): a review of recent research. Food Chem Toxicol 2008;46(2):409–20.

[80] Sharma P, Singh R. Dichlorvos and lindane induced oxidative stress in rat brain: protective effects of ginger. Pharmacogn Res 2012;4(1):27–32.

[81] Kundu JK, Na HK, Surh YJ. Ginger-derived phenolic substances with cancer preventive and therapeutic potential. Forum Nutr 2009;61:182–92.

[82] Viljoen E, Visser J, Koen N, Musekiwa A. A systematic review and meta-analysis of the effect and safety of ginger in the treatment of pregnancy-associated nausea and vomiting. Nutr J 2014;13:20.

[83] Zeng GF, Zhang ZY, Lu L, Xiao DQ, Zong SH, He JM. Protective effects of ginger root extract on Alzheimer's disease-induced behavioral dysfunction in rats. Rejuvenation Res 2013;16(2):124–33.

[84] Deng GF, Xu XR, Zhang Y, Li D, Gan RY, Li HB. Phenolic compounds and bioactivities of pigmented rice. Crit Rev Food Sci Nutr 2013;53(3):296–306.

[85] Hu C, Zawistowski J, Ling W, Kitts DD. Black rice (Oryza sativa L. indica) pigmented fraction suppresses both reactive oxygen species and nitric oxide in chemical and biological model systems. J Agric Food Chem 2003;51(18):5271–7.

[86] Ling WH, Wang LL, Ma J. Supplementation of the black rice outer layer fraction to rabbits decreases atherosclerotic plaque formation and increases antioxidant status. J Nutr 2002;132 (1):20–6.

[87] Wang QHP, Zhang M, Xia M, Zhu H, Ma J, Hou M, et al. Supplementation of black rice pigment fraction improves antioxidant and anti-inflammatory status in patients with coronary heart disease. Asia Pac J Clin Nutr 2007;16(Suppl. 1):7.

[88] Dong BXH, Zheng YX, Xu X, Wan FJ, Zhou XM. Effect of anthocyanin-rich extract from black rice on the survival of mouse brain neuron. Cell Biol Int 2008;2008(32):1.

[89] Ngo DH, Kim SK. Sulfated polysaccharides as bioactive agents from marine algae. Int J Biol Macromol 2013;62:70–5.

[90] Wijesekara I, Yoon NY, Kim SK. Phlorotannins from Ecklonia cava (Phaeophyceae): biological activities and potential health benefits. Biofactors 2010;36(6):408–14.

[91] Cornish ML, Garbary DJ. Antioxidants from macroalgae: potential applications in human health and nutrition. Algae 2010;25(4):17.

[92] Wijesekara IPR, Kim SK. Biological activities and potential health benefits of sulfated polysaccharides derived from marine algae. Carbohydr Polym 2011;84:8.

[93] Pangestuti R, Kim SK. Neuroprotective effects of marine algae. Mar Drugs 2011;9(5):803–18.

[94] Myung CS, Shin HC, Bao HY, Yeo SJ, Lee BH, Kang JS. Ethanol-treated mice: possible involvement of the inhibition of acetylcholinesterase. Arch Pharm Res 2005;28(6):691–8.

[95] Zhou X, Lin J, Yin Y, Zhao J, Sun X, Tang K. Ganodermataceae: natural products and their related pharmacological functions. Am J Chin Med (Gard City, NY) 2007;35(4):559–74.

[96] Jin X, Ruiz Beguerie J, Sze DM, Chan GC. *Ganoderma lucidum* (Reishi mushroom) for cancer treatment. Cochrane Database Syst Rev 2012;6:CD007731.

[97] Gao J, Inagaki Y, Li X, Kokudo N, Tang W. Research progress on natural products from traditional Chinese medicine in treatment of Alzheimer's disease. Drug Discoveries Ther 2013;7(2):46–57.

[98] Wachtel-Galor S, Tomlinson B, Benzie IF. *Ganoderma lucidum* ("Lingzhi"), a Chinese medicinal mushroom: biomarker responses in a controlled human supplementation study. Br J Nutr 2004;91(2):263–9.

[99] Khan MA, Tania M, Liu R, Rahman MM. *Hericium erinaceus*: an edible mushroom with medicinal values. J Complementary Integr Med 2013;10.

[100] Wong KH, Kanagasabapathy G, Naidu M, David P, Sabaratnam V. *Hericium erinaceus* (Bull.: Fr.) Pers., a medicinal mushroom, activates peripheral nerve regeneration. Chin J Integr Med 2014:26.

[101] Mori K, Inatomi S, Ouchi K, Azumi Y, Tuchida T. Improving effects of the mushroom Yamabushitake (*Hericium erinaceus*) on mild cognitive impairment: a double-blind placebo-controlled clinical trial. Phytother Res: PTR 2009;23(3):367–72.

[102] Potterat O. Goji (*Lycium barbarum* and *L. Chinese*): phytochemistry, pharmacology and safety in the perspective of traditional uses and recent popularity. Planta Med 2010;76(1):7–19.

[103] Paul Hsu CH, Nance DM, Amagase H. A meta-analysis of clinical improvements of general well-being by a standardized *Lycium barbarum*. J Med Food 2012;15(11):1006–14.

[104] Amagase H, Sun B, Borek C. *Lycium barbarum* (Goji) juice improves in vivo antioxidant biomarkers in serum of healthy adults. Nutr Res 2009;29(1):19–25.

[105] Amagase H, Farnsworth R. A review of botanical characteristics, phytochemistry, clinical relevance in efficacy and safety of *Lycium barbarum* fruit (Goji). Food Res Int 2011;44:16.

[106] Amagase H, Nance DM. *Lycium barbarum* increases caloric expenditure and decreases waist circumference in healthy overweight men and women: pilot study. J Am Coll Nutr 2011;30(5):304–9.

[107] Cheng TO. Cardiovascular effects of Danshen. Int J Cardiol 2007;121(1):9–22.

[108] Zhou L, Zuo Z, Chow MS. Danshen: an overview of its chemistry, pharmacology, pharmacokinetics, and clinical use. J Clin Pharmacol 2005;45(12):1345–59.

[109] Seetapun S, Yaoling J, Wang Y, Zhu YZ. Neuroprotective effect of Danshensu derivatives as anti-ischaemia agents on SH-SY5Y cells and rat brain. Biosci Rep 2013;33(4):.

[110] Wang J, Lu W, Wang W, Zhang N, Wu H, Liu C, et al. Promising therapeutic effects of sodium tanshinone IIA sulfonate towards pulmonary arterial hypertension in patients. J Thorac Dis 2013;5(2):169–72.

[111] Hugel HM, Jackson N. Danshen diversity defeating dementia. Bioorg Med Chem Lett 2014;24(3):708–16.

[112] Lam BY, Lo AC, Sun X, Luo HW, Chung SK, Sucher NJ. Neuroprotective effects of tanshinones in transient focal cerebral ischemia in mice. Phytomedicine 2003;10(4):286–91.

[113] Zhang Z, Lam TN, Zuo Z. Radix Puerariae: an overview of its chemistry, pharmacology, pharmacokinetics, and clinical use. J Clin Pharmacol 2013;53(8):787–811.

[114] Wong KH, Li GQ, Li KM, Razmovski-Naumovski V, Chan K. Kudzu root: traditional uses and potential medicinal benefits in diabetes and cardiovascular diseases. J Ethnopharmacol 2011;134(3):584–607.

[115] Ye XY, Song H, Lu CZ. Effect of puerarin injection on the mRNA expressions of AT1 and ACE2 in spontaneous hypertension rats. Zhongguo Zhong Xi Yi Jie He Za Zhi 2008;28(9):824–7.

[116] Xu XH, Zheng XX, Zhou Q, Li H. Inhibition of excitatory amino acid efflux contributes to protective effects of puerarin against cerebral ischemia in rats. Biomed Environ Sci 2007;20(4):336–42.

[117] Xu XH, Zhao TQ. Effects of puerarin on D-galactose-induced memory deficits in mice. Acta Pharmacol Sin 2002;23(7):587–90.

[118] Fusco D, Colloca G, Lo Monaco MR, Cesari M. Effects of antioxidant supplementation on the aging process. Clin Interv Aging 2007;2(3):377–87.

CHAPTER 48

Calorie Restriction in Humans: Impact on Human Health

Eric Ravussin, L. Anne Gilmore and Leanne M. Redman

Pennington Biomedical Research Center, Baton Rouge, LA, USA

KEY FACTS
- Prolonged CR has been shown to extend both the median and maximal lifespan in a variety of animal species.
- CALERIE is a randomized controlled trial that is testing the effects of prolonged CR in humans on biomarkers of aging and the rate of living theory hypothesis.
- Metabolic adaptation—total energy expenditure is reduced beyond the expected level for the reduction in the metabolizing mass (fat-free and fat mass: FFM and FM) following caloric restriction.
- Somatotropic axis—one of the major hormonal systems regulating postnatal growth in mammals. It interacts with the central nervous system on several levels including growth hormone (GH) and insulin-like growth factor-I (IGF-I) signaling pathways.
- Calorie restriction mimetics—Compounds that mimic the biochemical and functional effects of caloric restriction.

Dictionary of Terms

- Caloric restriction (CR)—consuming fewer calories than needed for weight maintenance.
- ad libitum—"at one's pleasure." In the case of dietary intake, the individual eats as much as desired without external restriction.
- Intermittent fasting—periods of abstinence from food and drink.
- CALERIE trial—Comprehensive Assessment of the Long-term Effect of Reducing Intake of Energy sponsored by the National Institutes of Aging. CALERIE was a multicenter randomized controlled trial designed to determine feasibility, safety, and effects of CR on predictors of longevity, disease risk factors, and quality of life in nonobese humans.
- Rate of living theory—A hypothesis stating that lowering of the metabolic rate reduces the flux of energy with a consequential lowering of reactive oxygen species and rate of oxidative damage to vital tissues.

WHY CALORIE RESTRICTION?

Since the first report of prolonged lifespan in rodents more than 70 years ago [1], calorie restriction (CR) has been gaining momentum as an intervention with the potential to ward off age-associated diseases and delay death. While the first observations were reported in rodents, similar observations have been reported across a wide range of species including yeast, worms, spiders, flies, fish, mice, and rats [2], but the effects of CR in longer-lived species remains unknown. The results reported thus far from three nonhuman primate colonies, however, suggest that CR might have a similar effect in these animals. Two of the largest longitudinal studies in Rhesus monkeys agreed that CR was beneficial for several markers of metabolic health, including weight and body composition, blood lipids, and cancer incidence but disagreed that young-onset CR reduced all-cause mortality and age-related death [3,4]. Differences in diet composition and supplementation are two possible reasons for the conflicting results as well as a slight CR in the control animals, genetic diversity, and age at time of CR initiation.

MECHANISMS TO ACHIEVE CALORIC RESTRICTION

Traditional Caloric Restriction

Traditional caloric restriction is long-term daily restriction of dietary intake, typically defined by a 20–50% lowering of energy intake below habitual levels. This traditional form of caloric restriction is proven to be effective in facilitating weight loss and improving metabolic biomarkers, but adherence to these regimens long-term is difficult which often leads to weight regain as individuals become less adherent over time [5]. While traditional caloric restriction is likely the most common and well known dietary program, additional programs of caloric restriction include (1) increasing physical activity to produce a calorie deficit, (2) fasting on alternate days, and (3) a modified fast on alternate days and we speculate in the future, caloric restriction achieved through dietary supplements or mimetics.

Physical Activity to Produce Caloric Deficit

The primary aim of caloric restriction is to achieve a prolonged state of negative energy balance (more kilocalories expended than consumed). In addition to the traditional dietary restriction, the calorie deficit can also be achieved through increasing daily energy expenditure through physical activity. Increased physical activity can in theory be the solitary contribution to produce a daily caloric deficit, but achieving a caloric deficit of 30% below weight maintenance (500–1000 kcal/d) through physical activity alone can prove to be challenging for most people. For this reason, physical activity interventions are often paired with dietary caloric restriction as a means to produce CR.

Intermittent Fasting

Intermittent fasting or periods of abstinence from food and drink has been a common religious practice since ancient times. Intermittent fasting, as a means of caloric restriction is gaining in popularity. In research, intermittent fasting has encompassed various regimens including alternate day fasting (AFD), modified AFD, and the 5:2 diet [6]. AFD, which has achieved much scientific focus, consists of a day of ad libitum eating often referred to as the "feed day," followed by a day with no caloric consumption called the "fast day." This alternating pattern of food intake is continued for the duration of the dietary intervention. Modifications to this strict regime of feeding and fasting were developed with the mindset to facilitate a higher level of individual adherence over longer durations. Modified AFD, typically referred to in the literature as alternate day modified fast or ADMF, allows for some caloric intake on the fast day, though severely restricted (~75% caloric restriction). The 5:2 diet, or the Fast Diet, prescribes only 2 days of severe caloric restriction per week. Through these varying methods of intermittent fasting, the level of overall caloric restriction achieved can be equivalent to the traditional caloric restriction programs, but ease of implementation (less calorie counting) and improved long-term compliance is evident with intermittent fasting. Overeating on the "feed day" due to elevated hunger followed on from the "fast day" is obviously a concern with these approaches, however, studies on intermittent fasting have concluded that even after fasting every other day, participants report no compensatory eating and high levels of satiety throughout the duration of the study. This observation probably reflects an adaptation to the intermittent fasting regimen achieved within a few weeks [7]. Overall, intermittent fasting is novel and a potentially more efficacious intervention for weight loss, preservation of lean mass, and improved metabolic health in humans. To our knowledge there is only one study that was designed to test the effects of alternate day feeding on nutrition in the aged. Termed the Hunger Study, 60 men with alternating days of fasting and feeding, received an average of 1500 kcal per day for 3 years which amounted to approximately 35% CR. The 60 other men were fed ad libitum. The initial report from this study was brief, however analyses conducted several years later [8] indicated that the death rate tended to be lowered in the intermittent fasting group and hospital admissions were reduced in these individuals by approximately 50% (123 days for CR vs 219 days for Control). Studies that compare intermittent fasting with traditional CR for the ability of intermittent fasting to attenuate age-related disease are underway (NCT02420054, NCT01964118, NCT02148458, NCT02169778).

EVIDENCE OF THE BENEFITS FOR CALORIC RESTRICTION IN HUMANS

In humans, data from controlled trials is lacking and, no long-term prospective trials of CR have been conducted with survival being the primary end-point [9]. There is, however, a lot that can be learned from controlled trials of CR where biomarkers of aging are measured and a handful of epidemiological and cross-sectional observations in longer-lived humans, centenarians, and individuals who self-impose CR.

Centenarians from Okinawa Have Life-Long Exposure to CR

Probably the most intriguing epidemiological evidence supporting the role of CR in lifespan extension in humans comes from the Okinawans [10]. Compared to most industrialized countries, Okinawa, Japan has 4–5 times the average number of centenarians, with an estimated 50 in every 100,000 people [11]. Reports from the Japanese Ministry of Health, Labor, and Welfare show that both the average (50th percentile) and maximum (99th percentile) lifespan are increased in Okinawans. From age 65, the expected lifespan in Okinawa is 24.1 years for women and 18.5 years for men compared to 19.3 years for women and 16.2 years for men in the United States [12]. What is interesting about the Okinawan population is that a low calorie intake was reported in school children on the island more than 40 years ago and later studies confirmed a 20% CR in adults residing on Okinawa compared to mainland Japan [13]. A recent estimate of the energy balance in a cohort of Okinawa septuagenarians during youth to middle age suggested a 10–15% energy deficit [14]. This energy deficit can be attributed to laborious occupations and daily activities as farmers and a diet that was rich in nutrients yet low in energy density [14]. The nutrient-dense diet of the Okinawans allowed for a negative energy balance while providing an abundance of vitamins, minerals, antioxidants, and flavonoids [12]. Unfortunately, with the increase in American presence on Okinawa in recent years (location of US military base and increased availability of fast food chains), unpublished reports suggest that the longevity-promoting lifestyle (ie, CR and high levels of physical activity) is threatened and may be reversed.

Unexpected CR in Biosphere 2

Biosphere 2 was an enclosed 3.15-acre ecological laboratory that housed seven ecosystems or biomes resembling the earth: rainforest, savannah, ocean, marsh, desert, and agriculture and human/animal habitats [15]. For 2 years, eight individuals, including Dr Roy Walford, were completely isolated within this "mini-world," where 100% of the air and water was recycled and all the food grown inside. Due to unforeseen problems with agriculture early on, food supply became quickly insufficient. Food intake for the eight individuals was projected at ~2500 kcal/d and estimates from food records maintained by one of the biospherians suggested diets were restricted by ~750 kcal/d in each person during the first 6 months. The resulting ~15% weight loss in the Biospherians was associated with many physiological, hematological, biochemical, and metabolic alterations [16,17] consistent with calorie-restricted rodents and primates, including reductions in insulin, core temperature, and metabolic rate (Fig. 48.1) [4].

The CALERIE Trials

For the past 15 years, The National Institute on Aging (NIA) sponsored the CALERIE (Comprehensive Assessment of the Long-term Effect of Reducing Intake of Energy) trial. Three clinical sites were involved in the CALERIE trials: Washington University in St Louis, MO; Tufts University in Boston, MA; and the Pennington Biomedical Research Center in Baton Rouge, LA. CALERIE was conducted in two phases. The goal of Phase 1 was to determine the feasibility as well as efficacy of different CR modalities (diet only, exercise only, or diet and exercise) and to inform the design of a larger trial to be conducted across the three study centers. Thereby in Phase 1, each study center was testing a slightly different primary outcome and as a result carried out independent study protocols over 6–12 months. At Tufts University the aim of the Phase 1 study was to investigate the role of CR as well as the glycemic index on changes in body composition and cardiometabolic risk factors [18]. At Washington University, the Phase I study was conducted in middle aged individuals and was designed to compare the effect of CR achieved by diet alone versus exercise alone [19]. The Phase 1 study at Pennington Biomedical Research Center was a 6-month study where 48 men and women were randomized to one of four treatment groups for 6 months [20–28]. For the CR group, overweight individuals were restricted to 75% (a 25% CR) of their weight maintenance energy requirements assessed by doubly labeled water [29]. The other groups were: (1) CR plus exercise group, for which the calorie deficit was also 25% from weight maintenance but half (12.5%) was achieved by CR and half (12.5%) by increasing energy expenditure with structured aerobic exercise; (2) a low calorie diet group in which participants consumed 890 kcal/d to achieve a 15% weight loss and thereafter followed a weight maintenance diet; and (3) a healthy diet control group that followed a weight-maintaining diet based on the American Heart Association Step 1 diet. A sample menu can be found in Table 48.1. The effects of the CR interventions were determined from changes in various physiological and psychological endpoints after 3 and 6 months.

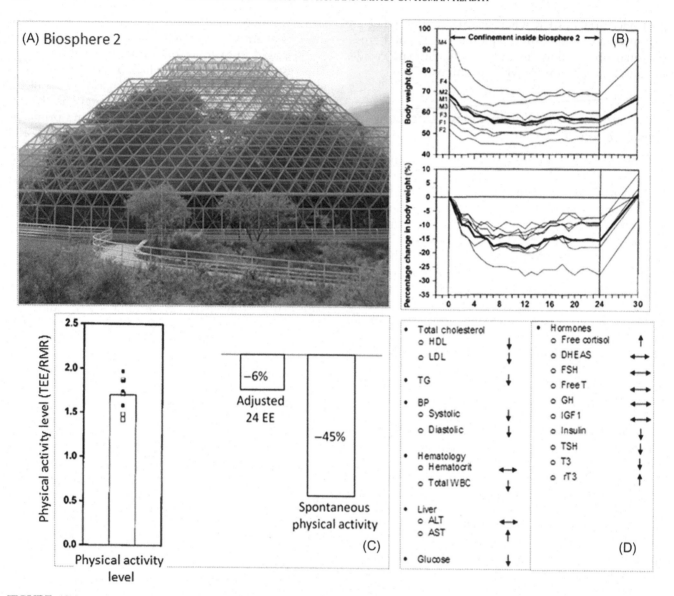

FIGURE 48.1 Biosphere 2 (A), a 3.15-acre ecological enclosure provided an unexpectedly low availability of food for eight individuals who were housed inside for 2 years in the early 1990s. This study of nature of CR resulted in ~15% weight loss. (B) Changes in energy expenditure (24 EE) and physical activity measured in a metabolic chamber (SPA) or as the ratio of total daily energy expenditure to resting metabolic rate (TDEE/RMR) (C) and many hematological, biochemical, and metabolic alterations (D) consistent with calorie-restricted rodents and primates including reductions in insulin, core temperature, and metabolic rate [15,17].

On the basis of the findings of the Phase 1 studies, the CALERIE Phase 2 study was a multicenter randomized controlled trial designed to determine feasibility, safety, and effects of CR on predictors of longevity, disease risk factors, and quality of life (QOL) in nonobese humans ($22.0 \leq BMI < 28$ kg/m^2) aged 21–51 years [30]. The intervention was designed to achieve 25% CR, defined as a 25% reduction from ad libitum baseline energy intake as determined by doubly labeled water studies. The 25% CR was imposed from Day 1 and individuals were provided with all meals for the first 27 days (three different 9 day menus) to help facilitate early adherence to the intervention as well as to teach behavioral strategies such as portion control. Throughout the intervention, CR subjects were able to track adherence using projection of weight change which was derived from our Phase 1 studies that predicted weekly changes in body weight for 1 year of 25% CR [31]. The weight projection provided the weekly expected weight change reaching 15.5% weight loss by 1 year, with an acceptable range of 11.9–22.1%, followed by weight maintenance. The primary outcomes were change from baseline resting metabolic rate (RMR) adjusted for

TABLE 48.1 Sample Menus of Food Provided in the CALERIE Phase I Study at 1500, 1800, and 2100 kcal/day

Day 1

Breakfast				Lunch				Dinner			
1500	1800	2100		1500	1800	2100		1500	1800	2100	
½ c	⅔ c	¾ c	Low-fat granola with raisins				Greek wrap	1 c	1¼ c	1⅓ c	Lentils with olives and feta
1 c	1 c	1 c	Skim milk	1	1	1½	10 inch tortilla	½ c	⅔ c	¾ c	Couscous
1	1	1	Banana	2 T	2½ T	3 T	Hummus	½ c	¾ c	¾ c	Zucchini
				½ c	½ c	½ c	Cucumber	¾ c	1 c	1 c	Strawberries
				⅓ c	⅓ c	⅓ c	Tomato				
				1½ T	1½ T	1½ T	Onion				
				1¼ T	1¼ T	1¾ T	Olives				
				1½ T	2 T	3 T	Feta cheese				
				⅓ c	½ c	½ c	Cheddar cheese				
				1⅓ c	1⅔ c	1⅔ c	Red grapes				

Day 2

Breakfast				Lunch				Dinner			
1500	1800	2100		1500	1800	2100		1500	1800	2100	
¼ c	½ c	½ c	Oatmeal	1⅔ c	2 c	2¼ c	Pesto pasta	1 c	1 c	1¼ c	Greek-style potatoes
⅔ c	⅔ c	¾ c	Peaches	¾ oz	¾ oz	1¼ oz	Chicken breast	3 oz	3 oz	3 oz	Salmon steak
1 T	1 T	1½ T	Almonds	1	1	1	Apple	¾ c	¾ c	¾ c	Green beans
½ c	1 c	1 c	Skim milk	1	1	2	Dinner roll	¾ c	1 c	1 c	Mandarin orange

weight change (RMR residual) and core temperature. It was hypothesized that CR would induce metabolic adaptations, specifically (1) decrease in RMR adjusted for changes in body composition and (2) decrease in core body temperature. Change in RMR was defined as "RMR residual," that is, the difference between an individual's RMR measured by indirect calorimetry during the intervention and RMR predicted from a regression of RMR as a function of fat mass (FM) and fat-free mass (FFM) in participants at baseline. Such metabolic adaptations to CR in laboratory animals have been proposed to slow aging by reducing metabolic production of reactive oxygen species (ROS) and/or lowering core temperature [32,33]. Lower core temperature has also been found to predict human longevity in longitudinal studies [34]. Secondary outcomes included changes in plasma triiodothyronine (T3) and TNF-α based on evidence suggesting relationships of the thyroid axis and inflammatory mediators to longevity and health span and effects of CR on these factors [9,35–37]. Exploratory outcomes included risk factors for age-related conditions and psychological responses. Study outcomes were evaluated at baseline, 6, 12, 18, and 24 months with a primary focus on baseline, 12 and 24 months.

A total of 218 individuals started the 25% CR intervention, with 82% of CR ($N = 117$) and 95% of AL ($N = 71$) completing the 2-year protocol [30]. The Phase 2 cohort was predominantly female (69.7%) and Caucasian (77.1%) aged between 20 and 50. At baseline, these individuals had normal blood pressures, fasting blood glucose, insulin, and lipids. The daily energy intake in the CR group declined by 480 ± 20 kcal/d during the first 6 months of intervention but then stabilized at approximately 234 ± 19 kcal/d below baseline for the remainder of the trial. This resulted in approximately 12% over the 2-year trial ($19.5 \pm 0.8\%$ during the first 6 months and $9.1 \pm 0.7\%$ on average for the remainder of the study). There was no evidence of CR in the AL control group. CALERIE therefore achieved significant CR over the 2-year trial in nonobese persons and resulted in sustained weight loss which was maintained at 7.6 ± 0.3 kg (~10% from baseline) throughout the 24 months. Importantly the CALERIE studies also indicated that the degree of CR achieved in the studies is tolerable and safe, and with no adverse effects on QOL. Long-term follow-up on individuals in the CALERIE trials will further help to understand the prolonged benefits of CR, assuming CR was continued on biomarkers of aging. More detailed findings from the CALERIE trials are later in the chapter.

POTENTIAL MECHANISMS FOR CR AND MORE HEALTHFUL AGING

Caloric Restriction May Alter the "Rate of Living" and "Oxidative Stress"

The aging process may be influenced by CR through a reduction in the "rate of living," [38] leading ultimately to reduced oxidative damage. The rate of living theory suggests that increased metabolism and thus increased production of ROS leads to a shorter lifespan. An ongoing controversy among investigators appears to be whether chronic CR actually leads to metabolic slowing or rate of living. This phenomenon is termed "metabolic adaptation," and reflects a reduction in the metabolic rate that is larger than expected for the loss of metabolic mass of the organism that occurs as a result of the caloric deficit [2]. Unfortunately results from rats and monkeys does not shed light on this hypothesis because most of the collected data needs to be reevaluated using appropriate statistical methods of normalizing the observed changes in metabolic rate for changes in metabolic size [39]. For example, Blanc et al. [40] recently calculated a 13% reduction in resting energy expenditure after adjusting for FFM in an 11-year-long study of energy restricted monkeys. Yet Selman et al. [8], using doubly labeled water to measure total energy expenditure, reported that calorie-restricted rats expended 30–50% more energy than expected. Yamada et al. [41] explain that while there is a decrease in resting energy expenditure, calorie-restricted monkeys also have an increased physical activity level and intensity, which might explain the greater than expected energy expenditure.

The "free radical theory of aging" or "oxidative stress" hypothesis is one of the well-supported theories of aging. It is widely accepted that the metabolic rate of an organism is a major factor in the rate of aging and is inversely related to its lifespan [42]. Additionally, since 1–3% of consumed oxygen is associated with the production of ROS, namely superoxide ($O_2^{\bullet-}$), hydrogen peroxide ($H_2O_2^{\bullet-}$), and the hydroxyl ion ($OH^{\bullet-}$) [43], the production of these highly reactive molecules from normal aerobic metabolism is also in direct proportion to an organism's metabolic rate. Many investigators have shown that modulation of the oxidative stress of an organism through prolonged CR is able to retard the aging process in various species, including mammals [44,45]. As a result of increased oxygen consumption, aerobic exercise is associated with increased production of ROS in muscle tissues [46]. However, exercise training boosts the antioxidant capacity of skeletal muscle, probably resulting in decreased overall oxidative stress [47]. Mitochondria consume the majority of cellular oxygen resulting in the production of ROS [48]. In humans and monkeys, long-term energy restriction has been shown to induce robust increases in PGC-1 and mitochondrial biogenesis, which in turn is hypothesized to delay the onsets of sarcopenia, as well as loss of muscle function [49,50]. Despite an increase in mitochondrial biogenesis, CR results in improved mitochondrial function, decreased total body oxygen consumption, and therefore decreased production of ROS [20].

Biomarkers of Longevity

A "biomarker of aging or longevity" is considered to be any parameter that reflects physiological or functional age; it must undergo significant age-related changes, be slowed or reversed by treatments that increase longevity (eg, CR), and must be reliably measured. Numerous biomarkers have been identified in rodents and primates, including body temperature and hormones such as DHEA-S and insulin (Fig. 48.2) [34]. The most comprehensive evaluation of biomarkers of health and aging was investigated in the Phase 1 CALERIE study at Pennington Biomedical and the physiological and behavioral effects has been published in over 20 independent manuscripts. From the CALERIE Phase 1 study we learned that two out of the three biomarkers of longevity [34] were improved with a 6-month program of 25% CR [9]. Significant reductions were observed in both fasting insulin concentrations ($-29 \pm 6\%$) and core body temperature ($-0.20 \pm 0.05°C$), whereas DHEA-S was unchanged by the intervention. Despite being only a 6-month program of CR, these findings echo the results previously reported in nonhuman primates and rodents on CR and data from long-lived men in The Baltimore Longitudinal Study of Aging [34]. In CALERIE Phase 2, however, only small declines in 24-hour core temperature at 12 and 24 months was observed with the same 25% CR intervention, but the small declines did not differ significantly from the change in the control group who did not receive a dietary intervention [30]. These findings are obviously at odds, however, perhaps a greater degree of CR is needed to induce changes in core temperature. The degree of CR in the Phase 1 study in overweight persons was ~19% over the course of the program [9] whereas the degree of CR in the Phase 2 study was only 12% at 24 months. Of note, a lower core temperature has also been observed in individuals who are self-selected practitioners of CR [51].

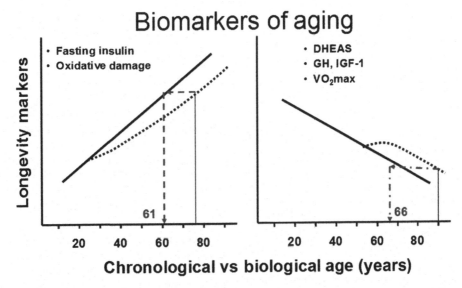

FIGURE 48.2 Can CR improve biological age and extend chronological age? This figure summarizes some of the potential biomarkers of aging. It is hypothesized that CR will change the biological trajectory of these biomarkers and therefore improve biological age and extend chronological age. For example, the left panel shows an individual aged 75 years. With prolonged CR it is hypothesized that fasting insulin and oxidative damage will be reduced in this individual. The *dotted line* represents the theoretical effects of CR. Therefore, an individual although 75 will have a biological age 17 years younger. Similarly the individual on the right at 90 years with prolonged CR will be biologically similar to an individual aged 66 years.

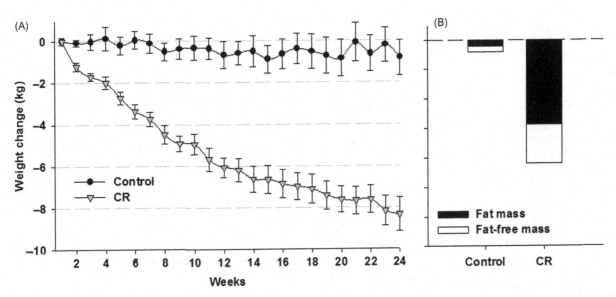

FIGURE 48.3 Our 6-month study of 25% CR resulted in a progressive decline in body weight that reached ~10% at the completion of the study [45]. Body composition analysis by dual X-ray absorptiometry showed that the loss of tissue mass was attributable to significant reductions in both FM (CR: −24 ± 3%) and FFM (CR: −4 ± 1%).

Body Composition

Throughout the 6-month intervention there was a progressive decline in body weight that reached ~10% for the CR group at the completion of the study (Fig. 48.3) [21]. Body composition analysis by dual X-ray absorptiometry and multislice computed tomography showed that the loss of tissue mass was attributable to significant reductions in both FM (CR: −24 ± 3%) and FFM (CR: −4 ± 1%). There was a 27% decrease in both visceral and subcutaneous fat depots, but it was interesting to note that the fat distribution within the abdomen was not altered by CR [21]. We also observed a reduction in subcutaneous abdominal mean fat cell size by ~20%, a lowering of hepatic lipid by ~37% but no change in skeletal muscle lipid content [26].

Metabolic Adaptation and Oxidative Stress

One of the most popular proposed theories by which CR promotes lifespan extension is the "rate of living theory" [52]. It is hypothesized that a lowering of the metabolic rate reduces the flux of energy with a consequential lowering of ROS and rate of oxidative damage to vital tissues [42]. Indeed, CR is associated with a robust decrease in energy metabolism, including an absolute lowering of RMR (or sleeping metabolic rate, SMR), and thermic effect of meals and a decrease in the energy cost of physical activity. However, as mentioned earlier, whether total energy expenditure is reduced beyond the expected level (ie, metabolic adaptation) for the reduction in the metabolizing mass (FFM and FM) following CR is a matter of debate.

In CALERIE Phase 1, as expected, absolute 24-hour energy expenditure and SMR (both measured in a respiratory chamber) were significantly reduced from baseline with CR ($p < 0.001$). Importantly, however, both 24-hour sedentary and sleeping energy expenditures were reduced ~6% beyond what was expected for the loss of metabolic mass, that is, FFM and FM [9]. This metabolic adaptation was also observed for RMR measured by a ventilated hood indirect calorimeter [22]. These physiological responses were associated with a reduced amount of oxidative stress as measured by DNA damage. DNA damage was reduced from baseline after 6 months in CR ($p = 0.0005$), but not in controls [9]. In addition, 8-oxo7,8-dihidro-2'deoxyguanosine was also significantly reduced from baseline in CR subjects ($p < 0.0001$). These data confirm findings in animals that CR reduces energy metabolism, oxidative stress to DNA, both potentially attenuating the aging process.

Similarly, in CALERIE Phase 2, the absolute change in energy expenditure, measured as RMR at 12 and 24 months for the 25% CR group was $5.9 \pm 0.7\%$ and $5.0 \pm 0.9\%$ at 12 and 24 months, respectively, and this decline significantly exceeded those individuals in the control group. As mentioned earlier in the chapter, the residuals of RMR which essentially separates the change in RMR from the loss in FFM and FM, shows that RMR was significantly reduced by the CR intervention when compared to the control group at 12 months (48 ± 9 kcal/d vs 14 ± 12 in AL, $p = 0.04$) however the effect was no longer evident after 24 months [30].

Physical Activity

Daily energy expenditure has three major components: RMR, the thermic effect of food (TEF), and the energy cost of physical activity. Investigation of changes in physical activity are important in studies of CR not only because the contribution of physical activity to daily energy expenditure is variable, but also because it is not known if individuals volitionally or nonvolitionally decrease their level of physical activity in an attempt to conserve energy [53]. In the CALERIE Phase 1 we observed no change in spontaneous physical activity (SPA) when measured in a respiratory chamber [22] which is consistent with earlier reports of no alterations in SPA or posture allocation in obese individuals following weight loss [54,55]. These findings are not surprising if the current hypothesis that SPA is biologically determined is true [55,56]. Interestingly with a measure of energy metabolism in free-living conditions (doubly labeled water) which also includes physical activity, we found that a metabolic adaptation was evident after 3 months (-386 ± 69 kcal/d) but not after 6 months of CR (Fig. 48.4) [29]. Interestingly, this adaptation was evident even after total daily energy expenditure (TDEE) was adjusted for the changes in sedentary energy metabolism (24-hour or sleeping energy expenditure) which indicates that changes in other components of daily energy expenditure, mostly as physical activity and less via diet-induced thermogenesis, are also involved in the metabolic slowing with CR. In support of this, physical activity level calculated by either the ratios of TDEE to RMR or SMR [22], or TDEE adjusted for the change in SMR, was significantly reduced at month 3 by 12% and returned toward baseline values after 6 months of intervention. Interestingly, despite lower physical activity levels, participants reported an improvement in physical functioning, a primary component of QOL. All the effects of CR on physiological outcomes are summarized in Table 48.2. We believe this effect of CR is likely robust as the same observations were made in the larger CR cohort in the Phase 2 studies. The energy expenditures associated with TDEE was significantly reduced in the CR group after 12 and 24 months of the CR intervention and independently from the change in body composition. Whether this reflects a reduction in physical activity levels or improved metabolic efficiency (reduced EE for a given amount of work), we cannot determine from the CALERIE studies.

Endocrine Responses

Thyroid Function

Short-term studies of CR in humans have reported alterations in thyroid function. Four weeks of complete fasting resulted in a decrease in T3 and an increase in reverse triiodothyronine (rT3), which was associated with a reduction in metabolic rate [57]. The CRONIES

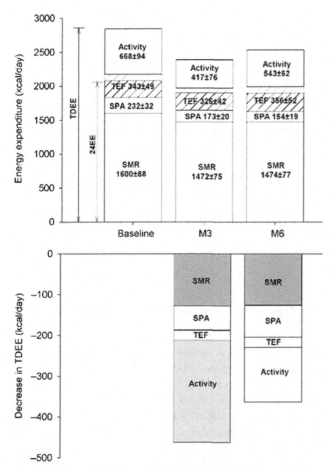

FIGURE 48.4 The effect of CR on all components of daily energy expenditure (*top panel*). The components of energy expenditure were determined by combining a measure of sedentary energy expenditure in the metabolic chamber (SMR, sleeping metabolic rate; SPA, spontaneous physical activity; TEF, thermic effect of food) and free-living energy expenditure by doubly labeled water (physical activity). The changes in TDEE after 3 and 6 months of CR (*bottom panel*) are shown and those representing a metabolic adaptation (larger than due to weight loss) are highlighted in *gray* [29]. Combining two state-of-the-art methods (indirect calorimetry in the metabolic chamber and doubly labeled water) for quantifying precisely the complete energy expenditure response to CR in nonobese individuals, we identified a reduction in sedentary energy expenditure that was 6% larger than what could be accounted for by the loss in metabolic size, that is, a "metabolic adaptation" [9] and a metabolic adaptation in the free-living situation as well. This adaptation comprised not only a reduction in cellular respiration (energy cost of maintaining cells, organs and tissue alive) but also a decrease in free-living activity thermogenesis, highlighted in *blue* (behavioral adaptation).

TABLE 48.2 Summary of the Psychological and Behavioral Responses to 6 Months of CR in Humans

Psychological/behavioral responses
Development of eating disorder symptoms
↓ Disinhibition
↓ Binge eating
↓ Concern about body size and shape
↔ Fear of fatness
↔ Purgative behavior
Depressed mood
↓ MAEDS Depression scale
↔ Beck Depression Inventory II
Subjective feelings of hunger
↓ Eating Inventory, Perceived Hunger Scale
Quality of life
↑ Physical functioning
↔ Vitality
Cognitive performance
↔ Verbal memory
↔ Short-term memory and retention
↔ Visual perception and memory
↔ Attention/concentration

(a self-selected group engaging in long-term CR) have significantly lower T3, but not thyroxine (T4) or thyroid-stimulating hormone, concentrations compared to age-, sex-, and weight-matched controls [58].

In the CALERIE Phase 1 study, plasma T3 concentrations were reduced from baseline in the CR group after 3 ($p<0.01$) and 6 months ($p<0.02$) of intervention [9]. Similar results were found for the change in plasma T4 in response to the treatment. When the data of the subjects in the three CR groups were combined into one intervention sample, we observed significant linear relationships between the change in plasma thyroid hormones and the degree of 24-hour metabolic adaptation after 3 months of intervention (T3; $r = 0.40$, $p = 0.006$ and T4; $r = 0.29$, $p = 0.05$) [9]. Similarly in CALERIE Phase 2, there were substantial decreases from baseline within the normal range in circulating T3 in CR ($16 \pm 1.5\%$ at month 12, $22 \pm 1.4\%$ at 24 months) significantly exceeded changes in AL [30].

Leptin

Leptin, which influences body composition and energy balance by regulating energy intake and expenditure, also decreases with CR [59–61]. Metabolic adaptation or a drop in energy expenditure as a result of CR that is greater than what is expected on the basis of changes in weight and energy stores. We found that the CR-mediated change in leptin is a significant factor for the metabolic adaptation but it is independent of the change in body composition. This suggests that leptin response to CR diets could be a possible biomarker for aging [59].

The Somatotropic Axis

Aging is marked by a reduction in both growth hormone (GH) and insulin-like growth factor-1 (IGF-1) concentrations in healthy adults, resulting from a reduced amount of GH secreted at each burst without alterations of burst frequency or GH half-life [62]. Unlike for rodents, weight loss via CR in humans increases GH [63]. After 6 months of CR in the Phase 1 study, 11-hour mean GH concentrations were not changed with CR nor was the secretory dynamics in terms of the number of secretion events, secretion amplitude, and secretion mass [64]. The fasting plasma concentration of ghrelin, a GH secretagogue was significantly increased from baseline, but IGF-1 was unaffected. Despite a significant reduction in weight and visceral fat and an improvement in insulin sensitivity, mean GH concentrations were not altered by the 6-month intervention. In agreement with this observation was the finding that both GH and IGF-1 were not affected by the chronic food shortage experienced by the individuals in Biosphere 2 [16].

DHEA-S

Given the evidence from cross-sectional [65] and longitudinal studies [66] that DHEA-S declines with age. DHEA-S, which is a metabolite of DHEA, an abundant steroid hormone in the body, is considered to be a reliable endocrine marker of human aging and longevity [67]. It was hypothesized that CR will delay or attenuate the age-associated decline in DHEA-S. In our 6-month study in young individuals (37 ± 2 years), we observed no alteration in DHEA-S [9]. Similarly, DHEA-S was not changed with 2 years of energy restriction in the individuals within Biosphere 2 [16]. To our knowledge, there has been no report of DHEA-S levels in those individuals from the Calorie Restriction Society (CRONIES) who are self-imposing CR. The lack of agreement between the human and nonhuman primate data is believed to be due to first, the chronological age of the subjects at the onset of CR, and second, to the duration of CR. Young adult monkeys undergoing CR for 3–6 years had an age-related decline in DHEA-S of 3% compared to 30% in monkeys fed ad libitum [68]. In contrast, CR initiated in older animals (~ 22 years) did not attenuate the age-associated decline in DHEA-S [69]. These explanations remain to be tested in longer term studies of CR in humans.

PSYCHOLOGICAL AND BEHAVIORAL EFFECTS OF CR

CR in humans might prove to have positive effects on physical health and longevity, resulting in the practice of CR or the identification of CR mimetics. Very little is known about the effect of CR on the QOL and, furthermore, if people attempt to follow CR for health promotion, important questions must be answered about possible negative effects of CR on psychological well-being, cognitive functioning, mood, and subjective feelings of appetite. Determining the effect of CR on these parameters is critical to learn if adhering to a CR regimen is feasible and if CR has unintended negative consequences that would offset the potential of its health benefits. Again the Phase 1 study of CALERIE provided a unique opportunity to examine the effect of 6 months of CR on psychological and behavioral endpoints. We here summarize the effects of 6 months of CR on the development of eating disorder symptoms, QOL, mood (symptoms of depression), subjective ratings of appetite, and cognitive function.

Development of Eating Disorder Symptoms

One of the most pressing concerns about CR is that the adherence to a sustained reduction in food intake will potentiate the development of symptoms of eating disorders. This concern is based in part on the Keys [53] study, which found that 50% CR for 6 months among healthy men was associated with the development of eating disorder symptoms, for example, binge eating [70]. Additionally, CR or the intent to restrict intake has been associated with the onset of eating disorders, including anorexia [71], bulimia nervosa [72], and binge-eating disorder [62]. Hence, there is a need to examine both the benefit and potential harm of CR in humans, particularly for people who are not obese, and to answer important safety questions before CR is recommended [73,74].

In our study, participants completed an assessment battery that included: (1) the Multifactorial Assessment of Eating Disorder Symptoms (MAEDS), which measures six symptom domains associated with eating disorders (binge eating, purgative behavior, depression, fear of fatness, avoidance of forbidden foods, restrictive eating) [75], (2) the Eating Inventory, which measures dietary restraint, disinhibition, and perceived hunger [76], and (3) the Body Shape Questionnaire (BSQ) [77], which measures concern about body size and shape.

As reported by Williamson et al. [27], the three "dieting" groups in CALERIE, including the CR group, reported higher dietary restraint scores in comparison to the Control group at months 3 and 6, but measures of eating disorder symptoms did not increase and some decreased. All groups, except the control group, reported a significant reduction in disinhibition at month 6, whereas binge eating decreased in all groups at months 3 and 6. Concern about body size/shape decreased at 3 and 6 months among the three dieting groups but did not change in the control group. The Fear of Fatness and Purgative Behavior subscales of the MAEDS did not change during CR.

Subjective Feelings of Hunger

The ability of people to adhere to a strict CR diet could be limited by feelings of increased hunger. We evaluated change in appetite ratings during CR using the perceived hunger scale of the Eating Inventory [76] and the Visual Analogue Scales (VAS), which have been found to be reliable and valid measures of appetite: hunger, fullness, desire to eat, satisfaction, and prospective food consumption [59]. During the 6-month Phase 1 study, appetite ratings changed, but the changes among the dieting groups were not different from those in the control group. Moreover, based on the perceived hunger scale of the Eating Inventory, hunger was reduced in the CR group at month 6 [27].

QOL and Mood

The Minnesota Semi Starvation study [53] indicated that CR can negatively affect mood; therefore, the effect of CR on mood and QOL becomes an important factor when considering the feasibility of CR in humans. During CALERIE Phase 1, the Medical Outcomes Study Short-Form 36 Health Survey (SF-36) [60,61] was used to measure QOL, and the Beck Depression Inventory II [78] and depression scale of the MAEDS were used to measure mood. Our results indicate that depressed mood, measured by the BDI-II, did not change during the trial. Additionally, in the CR group, scores on the MAEDS depression subscale decreased at 3 and 6 months in comparison to baseline [27]. Together, the results indicate that CR had no negative effect on mood during this trial and, in fact, symptoms of depressed mood, measured with the MAEDS, decreased in the CR group.

The SF-36 was used to test the effects of CR on two components of QOL—physical functioning and vitality. All dieting groups, but not the control group, had improved physical functioning during the trial. For the CR group, physical functioning was significantly improved at both months 3 and 6 but CR had no significant effect on vitality.

Cognitive Function and Performance

Self-reported dieting or CR has been associated with deficits in cognitive performance (eg, memory and concentration deficits) [79,80]. Nevertheless, cognitive impairment is frequently mediated by preoccupation with food and body weight [81], suggesting that obsessive thoughts about food and weight, rather than CR, negatively affect cognitive performance. If CR has negative effects on cognitive performance, the feasibility of CR in humans would be in doubt.

In our trial, cognitive performance was evaluated empirically at baseline and months 3 and 6 with a comprehensive neuropsychological battery [23]. Verbal memory was measured with the Rey Auditory and Verbal Learning Test (RAVLT) [82], short-term memory and retention with the Auditory Consonant Trigram (ACT) [83,84], visual perception and memory with the Benton Visual Retention Test (BVRT) [85], and attention/concentration with the Conners' Continuous Performance Test-II (CPT-II) [86]. During CR, no pattern of memory or attention/concentration deficits emerged. The degree of daily energy deficit also was not correlated with change in cognitive performance; hence, these data indicate that CR did not have a negative effect on cognitive performance [23]. All the effects of CR on psychological and behavioral outcomes are summarized in Table 48.2.

The psychological and behavioral findings from CALERIE provide important information about the feasibility and safety of CR in humans. CR was not associated with the development of eating disorder symptoms, decreased QOL, depressed mood, or cognitive impairment. In fact, many of these endpoints improved, and changes in subjective ratings of appetite were similar in the CR group to those of the control group. These results suggest that CR might be feasible and have few unintended consequences, at least among overweight individuals.

CR AND THE DEVELOPMENT OF CHRONIC DISEASE

Elevated levels of LDL, excessive ROS generation, hypertension, and diabetes are all potential causes for the development of endothelial dysfunction, a precipitating event in the progression of atherosclerosis. These factors are believed to initiate an inflammatory response in the injured endothelial tissue. Long-term CR is associated with sustained reductions in factors related to endothelial dysfunction in humans, such as decreased blood pressure [87], reduced levels of total plasma cholesterol and triglycerides [16], and reduced markers of inflammation such as C-reactive protein, NF-κB, and plasminogen activator inhibitor type-1 [88–91]. A long-term CR study in humans supports the feasibility of using CR to protect against atherosclerosis by showing a 40% reduction in carotid artery intima-media thickness in CR participants relative to a control group [92]. Additionally, long-term CR counters expected age-related cardiac autonomic changes resulting in function equal to that of an average human 20 years younger [49].

Strong evidence shows that long-term CR in lean and obese subjects improves insulin sensitivity, a mechanism by which CR may act to extend lifespan [16,93]. Improved insulin sensitivity is due in part to the downregulation of IGF-1/insulin pathway that results in decreased P13K and AKT transcripts in skeletal muscle [91]. Additionally, prolonged CR reduces fasting glucose and insulin concentration, two factors believed to contribute to the aging process due to protein glycation [94] and mitogenic action [95], respectively. This compelling evidence suggests that weight loss due to CR may be the most effective means of improving insulin sensitivity, thereby decreasing the risk for the development of diabetes mellitus.

Cardiovascular and Diabetes Risk Factors

With heart disease and stroke ranked numbers 1 and 3 in the causes of death in the United States [96], delaying the progression of atherosclerotic cardiovascular disease (CVD) may be one potential mechanism by which CR promotes longevity. The risk factors for CVD, including abnormalities in blood lipids, blood pressure, hemostatic factors, inflammatory markers, and endothelial function, are worsened with aging [97,98]. At least a portion of these age-related changes appear to be secondary to increases in adiposity and/or reductions in physical activity [99,100] and, therefore, may be amenable to improvements through prolonged CR. Six months CR significantly reduced triacylglycerol and factor VIIc by 18% and 11%, respectively [101]. HDL-cholesterol was increased and fibrinogen, homocysteine, and endothelial function were not changed. According to total and HDL cholesterol (expressed as their ratio), systolic blood pressure, age, and gender, estimated 10-year CVD risk was 28% lower after only 6 months of CR.

Insulin resistance is an early metabolic abnormality that precedes the development of hyperglycemia, hyperlipidemia, and overt type 2 diabetes. Both insulin resistance and β-cell dysfunction are associated with obesity [102–104]. CR reduces FM and delays the development of age-associated diseases such as type 2 diabetes. In humans with obesity it is well established that CR and weight loss improve insulin sensitivity [105,106]. The effects of CR on insulin sensitivity and, diabetes risk however are not well understood in individuals who are overweight or normal weight. In our study of 6 months CR in overweight individuals we observed a 40% improvement in insulin sensitivity in the CR group, although this did not reach significance ($p = 0.08$; p-values assess level of statistical significance. In most cases to be statistically significant a p-value of ≤ 0.05 must be reached) [50]. The acute insulin response to glucose, however, was significantly decreased from baseline (CR: $29 \pm 7\%$, $p < 0.01$), indicating an improvement in β-cell responsiveness to glucose. Furthermore the Phase 2 CALERIE study in overweight and normal weight individuals also showed that CR results in improvements in glucose and insulin action as demonstrated by a significant decrease in HOMA-IR and also glucose tolerance from an oral glucose tolerance test in the CR group compared to the control subjects. Interestingly we note that these effects while evident after 12 months when most of the weight loss occurred, were also evident after 24 months after 12 months of weight maintenance. This probably suggests that CR-mediated improvements in glucose metabolism are not only tied to weight loss [30].

COULD CR INCREASE LONGEVITY IN HUMANS?

The wealth of CR literature in rodents allows us to address some important questions relating to the practicality and feasibility of CR in humans. Relevant and practical questions are (1) How much CR do we need to improve age-related health and possibly longevity? (2) How long do we need to sustain CR in order to obtain these benefits? Analysis of 24 published studies of CR in rodents (CR up to 55%) indicated a strong negative relationship between survival and energy intake [107] and a positive relationship between the duration of CR and longevity. Using the prediction equations derived from the rodent data above [107], we and others estimated that a 5-year life extension could be induced by 20% CR starting at age 25 and sustained for 52 years, that is, the life expectancy of a male in the United States. However, if a 30% CR was initiated at age 55 for the next 22 years, the gain would only be 2 months (Fig. 48.5).

Certainly there are individuals who self-impose CR with the CRON (Calorie Restriction with Optimal Nutrition) diet for health and longevity. A group of 18 CRONIES (only three women) have recently been studied after 3–15 years of CR [92,109]. Dietary analysis indicated an energy intake ~50% less than age-matched controls. In terms of body composition, the mean BMI of the males was 19.6 ± 1.9 kg/m^2 with an extremely low percent body fat of ~7%. Atherosclerosis risk factors including total cholesterol, LDL-c, HDL-c, and triglycerides fell within the 50th percentile of values for people in their age group. This report provides further evidence that longer term CR

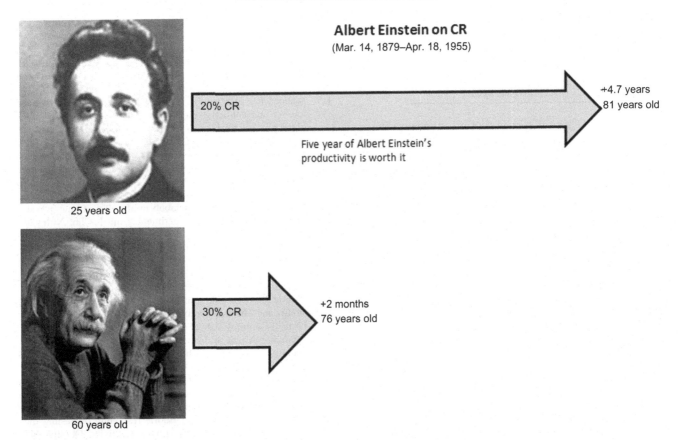

FIGURE 48.5 How can CR impact lifespan in humans? By extrapolating the data from rodents to humans [107], one can predict the potential effect of CR in humans [108]. As an example, if Albert Einstein started a 20% CR diet at 25 years of age, he could have increased his life by approximately 5 years. On the other hand, undertaking a 30% CR diet 45 years later (age 60) would have extended his life by only 2 months. Therefore, CR needs to be initiated early in adult life to significantly increase life expectancy.

is highly effective in lowering the risk of developing coronary heart disease and other age-related comorbidities. It remains to be seen if the CRONIES live longer than their age and sex matched counterparts.

CONCLUSION

While the rodent and primate data indicate that lifespan extension is possible with CR, collective analysis of the rodent data suggest that intensity and onset of CR required to induce these effects is probably not suitable for many individuals [108]. Epidemiological studies certainly support the notion that a reduced energy intake that is nutritionally sound improves age-associated health. While results of the first randomized trials of CR, albeit short in duration suggest a reduction in risk of age-related disease and improvements in some biomarkers of longevity, the ultimate effect of this intervention on lifespan in humans will probably never be determined in the scientific setting. The CALERIE studies provide the first evidence from an RCT that CR up to 2 years is feasible in humans and can be sustained without adverse effects on QOL. The degree of CR sufficient to affect some potential modulators of longevity that have been induced by CR in laboratory animal studies is probably in excess of 15% restriction of energy intake from the usual diet. However a lesser degree of CR is needed to influence factors associated with longevity in human observational studies, and to diminish risk factors for age-related cardiovascular and metabolic diseases. The potential impact for a CR regimen to influence human lifespan and health span has not been determined from clinical trials. Future studies can potentially clarify the effect of CR on human health and aging by assessing effects of differing degrees and durations of CR in humans.

However, it is a challenge for most individuals to practice CR in an "obesogenic" environment so conducive to overfeeding. Only a very few number of people will be able to practice a lifestyle of CR and probably benefit from it. There is therefore a need to search for organic or inorganic compounds that mimic the biological effects of CR. If such compounds often called "CR mimetics" (such as resveratrol [110,111]) prove viable in humans, individuals for the most part will opt to enjoy the effects of antiaging via a "pill" rather than CR.

SUMMARY

- Calorie restriction (CR) is a dietary intervention hypothesized to improve quality of life and extend lifespan.
- Mechanisms of this CR-mediated lifespan extension possibly involve alterations in energy metabolism, oxidative damage, insulin sensitivity, and functional changes in both the neuroendocrine and sympathetic nervous systems.

Acknowledgments

Studies of CR in nonobese individuals were supported by funding from the National Institutes of Health (grant number U01AG20478, R01AG029914, T32DK064584).

References

[1] McCay CM, Crowel MF, Maynard LA. The effect of retarded growth upon the length of the life span and upon th eultimate body size. J Nutr 1935;(10):63—79.
[2] Heilbronn LK, Ravussin E. Calorie restriction and aging: review of the literature and implications for studies in humans. Am J Clin Nutr 2003;78(3):361—9.
[3] Bodkin NL, et al. Mortality and morbidity in laboratory-maintained Rhesus monkeys and effects of long-term dietary restriction. J Gerontol A Biol Sci Med Sci 2003;58(3):212—19.
[4] Lane MA, et al. Energy balance in rhesus monkeys (*Macaca mulatta*) subjected to long-term dietary restriction. J Gerontol A Biol Sci Med Sci 1995;50(5):B295—302.
[5] Thomas DM, et al. Effect of dietary adherence on the body weight plateau: a mathematical model incorporating intermittent compliance with energy intake prescription. Am J Clin Nutr 2014;100(3):787—95.
[6] Johnstone A. Fasting for weight loss: an effective strategy or latest dieting trend? Int J Obes (Lond) 2015;39(5):727—33.
[7] Klempel MC, et al. Dietary and physical activity adaptations to alternate day modified fasting: implications for optimal weight loss. Nutr J 2010;9:35.
[8] Selman C, et al. Energy expenditure of calorically restricted rats is higher than predicted from their altered body composition. Mech Ageing Dev 2005;126(6—7):783—93.
[9] Heilbronn LK, et al. Effect of 6-month calorie restriction on biomarkers of longevity, metabolic adaptation, and oxidative stress in overweight individuals: a randomized controlled trial. JAMA 2006;295(13):1539—48.
[10] Kagawa Y. Impact of Westernization on the nutrition of Japanese: changes in physique, cancer, longevity and centenarians. Prev Med 1978;7(2):205—17.
[11] Japan Ministry of Health Law. Journal of health and welfare statistics. Tokyo: Health and Welfare Statistics Association; 2005.
[12] Willcox DC, et al. Caloric restriction and human longevity: what can we learn from the Okinawans? Biogerontology 2006;7(3):173—7.
[13] Suzuki M, Wilcox BJ, Wilcox CD. Implications from and for food cultures for cardiovascular disease: longevity. Asia Pac J Clin Nutr 2001;10(2):165—71.
[14] Willcox BJ, et al. Caloric restriction, enerhy balance and healthy aging in Okinawans and Americans: biomarker differences in Septuagenarians. Okinawan J Am Stud 2007;4:62—74.
[15] Walford RL, Harris SB, Gunion MW. The calorically restricted low-fat nutrient-dense diet in Biosphere 2 significantly lowers blood glucose, total leukocyte count, cholesterol, and blood pressure in humans. Proc Natl Acad Sci USA 1992;(2389):11533—7.
[16] Walford RL, et al. Calorie restriction in biosphere 2: alterations in physiologic, hematologic, hormonal, and biochemical parameters in humans restricted for a 2-year period. J Gerontol A Biol Sci Med Sci 2002;57(6):B211—24.
[17] Weyer C, et al. Energy metabolism after 2 y of energy restriction: the biosphere 2 experiment. Am J Clin Nutr 2000;(472):946—53.
[18] Das SK, et al. Long-term effects of 2 energy-restricted diets differing in glycemic load on dietary adherence, body composition, and metabolism in CALERIE: a 1-y randomized controlled trial. Am J Clin Nutr 2007;85(4):1023—30.
[19] Racette SB, et al. One year of caloric restriction in humans: feasibility and effects on body composition and abdominal adipose tissue. J Gerontol A Biol Sci Med Sci 2006;61(9):943—50.
[20] Civitarese AE, et al. Calorie restriction increases muscle mitochondrial biogenesis in healthy humans. PLoS Med 2007;4(3):e76.
[21] Redman LM, et al. Effect of calorie restriction with or without exercise on body composition and fat distribution. J Clin Endocrinol Metab 2007;92(3):865—72.
[22] Martin CK, et al. Effect of calorie restriction on resting metabolic rate and spontaneous physical activity. Obesity (Silver Spring) 2007;15(12):2964—73.
[23] Martin CK, et al. Examination of cognitive function during six months of calorie restriction: results of a randomized controlled trial. Rejuvenation Res 2007;10(2):179—90.
[24] Martin CK, et al. Empirical evaluation of the ability to learn a calorie counting system and estimate portion size and food intake. Br J Nutr 2007;98(2):439—44.
[25] Williamson DA, et al. Measurement of dietary restraint: validity tests of four questionnaires. Appetite 2007;48(2):183—92.
[26] Larson-Meyer DE, et al. Effect of calorie restriction with or without exercise on insulin sensitivity, beta-cell function, fat cell size, and ectopic lipid in overweight subjects. Diabetes Care 2006;29(6):1337—44.
[27] Williamson DA, et al. Is caloric restriction associated with development of eating disorder syndromes? Results from the CALERIE trial. Health Psychol 2008;27(1):S32—42.
[28] Anton SD, et al. Psychosocial and behavioral pre-treatment predictors of weight loss outcomes. Eat Weight Disord 2008;13(1):30—7.
[29] Redman LM, et al. Metabolic and behavioral compensations in response to caloric restriction: implications for the maintenance of weight loss. PLoS One 2009;4(2):e4377.
[30] Ravussin E, et al. A two-year randomized controlled trial of human caloric restriction: feasibility and effects on predictors of health span and longevity. J Gerontol A Biol Sci Med Sci 2015; In press.
[31] Pieper C, et al. Development of adherence metrics for caloric restriction interventions. Clin Trials 2011;8(2):155—64.
[32] Speakman JR, Mitchell SE. Caloric restriction. Mol Aspects Med 2011;32(3):159—221.
[33] Conti B. Considerations on temperature, longevity and aging. Cell Mol Life Sci 2008;65(11):1626—30.

REFERENCES

[34] Roth GS, et al. Biomarkers of caloric restriction may predict longevity in humans. Science 2002;(5582297):811.

[35] Holloszy JO, Fontana L. Caloric restriction in humans. Exp Gerontol 2007;42(8):709–12.

[36] Bowers J, et al. Thyroid hormone signaling and homeostasis during aging. Endocr Rev 2013;34(4):556–89.

[37] Chung HY, et al. Molecular inflammation: underpinnings of aging and age-related diseases. Ageing Res Rev 2009;8(1):18–30.

[38] Sacher GA, Duffy PH. Genetic relation of life span to metabolic rate for inbred mouse strains and their hybrids. Fed Proc 1979;(238):184–8.

[39] Ravussin E, Bogardus C. Relationship of genetics, age, and physical-fitness to daily energy-expenditure and fuel utilization. Am J Clin Nutr 1989;(549):968–75.

[40] Blanc S, et al. Energy expenditure of rhesus monkeys subjected to 11 years of dietary restriction. J Clin Endocrinol Metab 2003;(188):16–23.

[41] Yamada Y, et al. Long-term calorie restriction decreases metabolic cost of movement and prevents decrease of physical activity during aging in rhesus monkeys. Exp Gerontol 2013;48(11):1226–35.

[42] Sohal RS, Allen RG. Relationship between metabolic rate, free radicals, differentiation and aging: a unified theory. Basic Life Sci 1985;35:75–104.

[43] Alexeyev MF, Ledoux SP, Wilson GL. Mitochondrial DNA and aging. Clin Sci (Lond) 2004;107(4):355–64.

[44] Sohal RS, Weindruch R. Oxidative stress, caloric restriction, and aging. Science 1996;273(5271):59–63.

[45] Weindruch R, et al. The retardation of aging in mice by dietary restriction: longevity, cancer, immunity and lifetime energy intake. J Nutr 1986;116(4):641–54.

[46] Fulle S, et al. The contribution of reactive oxygen species to sarcopenia and muscle ageing. Exp Gerontol 2004;39(1):17–24.

[47] Sachdev S, Davies KJ. Production, detection, and adaptive responses to free radicals in exercise. Free Radic Biol Med 2008;44(2):215–23.

[48] Ames BN, Shigenaga MK, Hagen TM. Mitochondrial decay in aging. Biochim Biophys Acta 1995;1271(1):165–70.

[49] Stein PK, et al. Caloric restriction may reverse age-related autonomic decline in humans. Aging Cell 2012;11(4):644–50.

[50] McKiernan SH, et al. Cellular adaptation contributes to calorie restriction-induced preservation of skeletal muscle in aged rhesus monkeys. Exp Gerontol 2012;47(3):229–36.

[51] Soare A, et al. Long-term calorie restriction, but not endurance exercise, lowers core body temperature in humans. Aging (Albany NY) 2011;3(4):374–9.

[52] Sacher GA. Life table modifications and life prolongation. In: Finch CE, Hayflick L, editors. Handbook of the biology of aging. New York: van Nostrand Reinold; 1977. p. 582–638.

[53] Keys A, et al. The biology of human tarvation. Minneapolis, MN: University of Minnesota Press; 1950.

[54] Ravussin E, et al. Energy expenditure before and during energy restriction in obese patients. Am J Clin Nutr 1985;41(4):753–9.

[55] Levine JA, et al. Interindividual variation in posture allocation: possible role in human obesity. Science 2005;307(5709):584–6.

[56] Zurlo F, et al. Spontaneous physical activity and obesity: cross-sectional and longitudinal studies in Pima Indians. Am J Physiol 1992;263(2 Pt 1):E296–300.

[57] Vagenakis AG, et al. Diversion of peripheral thyroxine metabolism from activating to inactivating pathways during complete fasting. J Clin Endocrinol Metab 1975;41(1):191–4.

[58] Fontana L, et al. Effect of long-term calorie restriction with adequate protein and micronutrients on thyroid hormones. J Clin Endocrinol Metab 2006;91(8):3232–5.

[59] Flint A, et al. Reproducibility, power and validity of visual analogue scales in assessment of appetite sensations in single test meal studies. Int J Obes Relat Metab Disord 2000;24(1):38–48.

[60] Ware Jr. JE, Sherbourne CD. The MOS 36-item short-form health survey (SF-36). I. Conceptual framework and item selection. Med Care 1992;30(6):473–83.

[61] Ware Jr. JE, Kosinski M, Gandek B. SF-36 Health Survey: manual & interpretation guide 1993. Lincoln, RI: QualityMetric, Inc; 2002.

[62] Williamson DA, Martin CK. Binge eating disorder: a review of the literature after publication of DSM-IV. Eat Weight Disord 1999;4(3):103–14.

[63] Smith SR. The endocrinology of obesity. In: Bray G, editor. Endocrinology and metabolism clinics of North America. Philadelphia, PA: W.B Saunders; 1996. p. 921–42.

[64] Redman LM, et al. The effect of caloric restriction interventions on growth hormone secretion in nonobese men and women. Aging Cell 2010;9(1):32–9.

[65] Orentreich N, et al. Age changes and sex differences in serum dehydroepiandrosterone sulfate concentrations throughout adulthood. J Clin Endocrinol Metab 1984;59(3):551–5.

[66] Orentreich N, et al. Long-term longitudinal measurements of plasma dehydroepiandrosterone sulfate in normal men. J Clin Endocrinol Metab 1992;75(4):1002–4.

[67] Roth GS, et al. Aging in rhesus monkeys: relevance to human health interventions. Science 2004;305(5689):1423–6.

[68] Lane MA, et al. Dehydroepiandrosterone sulfate: a biomarker of primate aging slowed by calorie restriction. J Clin Endocrinol Metab 1997;82(7):2093–6.

[69] Urbanski HF, et al. Effect of caloric restriction on the 24-hour plasma DHEAS and cortisol profiles of young and old male rhesus macaques. Ann NY Acad Sci 2004;1019:443–7.

[70] Garner DM. Psychoeducational principles in treatment. In: Garner DM, Garfinkel PE, editors. Handbook of treatment for eating disorders. New York: Guilford Press; 1997. p. 145–77.

[71] Williamson DA. Assessment of eating disorders: obesity, anorexia, and bulimia nervosa. Elmsford, NY: Pergamon Press; 1990.

[72] Polivy J, Herman CP. Dieting and binging. A causal analysis. Am Psychol 1985;40(2):193–201.

[73] Vitousek KM, et al. Calorie restriction for longevity: II—The systematic neglect of behavioural and psychological outcomes in animal research. Eur Eat Disord Rev 2004;12:338–60.

[74] Vitousek KM, Gray JA, Grubbs KM. Caloric restriction for longevity: I—Paradigm, protocols and physiological findings in animal research. Eur Eat Disord Rev 2004;12:279–99.

[75] Anderson DA, et al. Development and validation of a multifactorial treatment outcome measure for eating disorders. Assessment 1999;6(1):7–20.

[76] Stunkard AJ, Messick S. Eating inventory manual (The Psychological Corporation). San Antonio, TX: Harcourt Brace & Company; 1988.

[77] Cooper PJ, Taylor MJ, Cooper Z, Fairburn CG. The development and validation of the Body Shape Questionnaire. Int J Eat Disord 1987;6:485–94.

[78] Beck AT, Brown GK, Steer RA. Beck depression inventory-II. San Antonio, TX: Psychological Corporation; 1996.

[79] Green MW, Rogers PJ. Impairments in working memory associated with spontaneous dieting behaviour. Psychol Med 1998;28(5):1063–70.

[80] Kemps E, Tiggemann M, Marshall K. Relationship between dieting to lose weight and the functioning of the central executive. Appetite 2005;45(3):287–94.

[81] Kemps E, Tiggemann M. Working memory performance and preoccupying thoughts in female dieters: evidence for a selective central executive impairment. Br J Clin Psychol 2005;44 (Pt 3):357–66.

[82] Schmidt M. Rey Auditory and Verbal Learning Test: a handbook. Los Angeles, CA: Western Psychological Services; 1996.

[83] Peterson LR. Short-term memory. Sci Am 1966;215(1):90–5.

[84] Peterson LR, Peterson MJ. Short-term retention of individual verbal items. J Exp Psychol 1959;58:193–8.

[85] Sivan AB. Benton Visual Retention Test. 5th ed. San Antonio, TX: The Psychological Corporation, Harcourt Brace & Company; 1992.

[86] Conners CK. Conners' Continuous Performance Test (CPT II). Toronto: Multi-Health Systems, Inc; 2000.

[87] Velthuis-te Wierik EJ, et al. Energy restriction, a useful intervention to retard human ageing? Results of a feasibility study. Eur J Clin Nutr 1994;(248):138–48.

[88] Heilbronn LK, Noakes M, Clifton PM. Energy restriction and weight loss on very-low-fat diets reduce C-reactive protein concentrations in obese, healthy women. Arterioscler Thromb Vasc Biol 2001;21(6):968–70.

[89] Mavri A, et al. Subcutaneous abdominal, but not femoral fat expression of plasminogen activator inhibitor-1 (PAI-1) is related to plasma PAI-1 levels and insulin resistance and decreases after weight loss. Diabetologia 2001;44(11):2025–31.

[90] Bastard JP, et al. Elevated levels of interleukin 6 are reduced in serum and subcutaneous adipose tissue of obese women after weight loss. J Clin Endocrinol Metab 2000;85(9):3338–42.

[91] Mercken EM, et al. Calorie restriction in humans inhibits the PI3K/AKT pathway and induces a younger transcription profile. Aging Cell 2013;12(4):645–51.

[92] Fontana L, et al. Long-term calorie restriction is highly effective in reducing the risk for atherosclerosis in humans. Proc Natl Acad Sci USA 2004;101(17):6659–63.

[93] Walford RL, et al. Physiologic changes in humans subjected to severe, selective calorie restriction for two years in biosphere 2: health, aging, and toxicological perspectives. Toxicol Sci 1999;(2 Suppl. 52):61–5.

[94] Robertson RP. Chronic oxidative stress as a central mechanism for glucose toxicity in pancreatic islet beta cells in diabetes. J Biol Chem 2004;279(41):42351–4.

[95] Stenkula KG, et al. Expression of a mutant IRS inhibits metabolic and mitogenic signalling of insulin in human adipocytes. Mol Cell Endocrinol 2004;221(1–2):1–8.

[96] Rosamond W, et al. Heart disease and stroke statistics—2007 update: a report from the American Heart Association Statistics Committee and Stroke Statistics Subcommittee. Circulation 2007;115(5):e69–171.

[97] Celermajer DS, et al. Aging is associated with endothelial dysfunction in healthy men years before the age-related decline in women. J Am Coll Cardiol 1994;24(2):471–6.

[98] Mendall MA, et al. C reactive protein and its relation to cardiovascular risk factors: a population based cross sectional study. Br Med J 1996;312(7038):1061–5.

[99] DeSouza CA, et al. Regular aerobic exercise prevents and restores age-related declines in endothelium-dependent vasodilation in healthy men. Circulation 2000;102(12):1351–7.

[100] Mora S, et al. Association of physical activity and body mass index with novel and traditional cardiovascular biomarkers in women. JAMA 2006;295(12):1412–19.

[101] Lefevre M, et al. Caloric restriction alone and with exercise improves CVD risk in healthy non-obese individuals. Atherosclerosis 2009;203(1):206–13.

[102] Forsey RJ, et al. Plasma cytokine profiles in elderly humans. Mech Ageing Dev 2003;124(4):487–93.

[103] Matsumoto K, et al. Inflammation and insulin resistance are independently related to all-cause of death and cardiovascular events in Japanese patients with type 2 diabetes mellitus. Atherosclerosis 2003;169(2):317–21.

[104] Utzschneider KM, et al. Impact of intra-abdominal fat and age on insulin sensitivity and beta-cell function. Diabetes 2004;53(11):2867–72.

[105] Dengel DR, et al. Distinct effects of aerobic exercise training and weight loss on glucose homeostasis in obese sedentary men. J Appl Physiol 1996;81(1):318–25.

[106] Niskanen L, et al. The effects of weight loss on insulin sensitivity, skeletal muscle composition and capillary density in obese non-diabetic subjects. Int J Obes Relat Metab Disord 1996;20(2):154–60.

[107] Merry BJ. Calorie restriction and age-related oxidative stress. Ann NY Acad Sci 2000;908:180–98.

[108] Redman LM, Ravussin E. Could calorie restriction increase longevity in humans? Aging Health 2007;3(1):1–4.

[109] Fontana L, et al. Long-term effects of calorie or protein restriction on serum IGF-1 and IGFBP-3 concentration in humans. Aging Cell 2008;7(5):681–7.

[110] Baur JA, et al. Resveratrol improves health and survival of mice on a high-calorie diet. Nature 2006;444(7117):337–42.

[111] Pearson KJ, et al. Resveratrol delays age-related deterioration and mimics transcriptional aspects of dietary restriction without extending life span. Cell Metab 2008;8(2):157–68.

CHAPTER 49

Prebiotics and Probiotics in Aging Population: Effects on the Immune-Gut Microbiota Axis

Thea Magrone and Emilio Jirillo

Department of Basic Medical Sciences Neuroscience and Sensory Organs, University of Bari, Bari, Italy

KEY FACTS

- In healthy, gut microbiota and immune system interact each other for maintaining immune homeostasis.
- In elderly, alteration of microbiota may contribute to inflammaging.
- Food nutrients (*eg*, vitamins) induce an anti-inflammatory pathway in the gut.
- Prebiotics and probiotics act on the immune-gut microbiota axis, thus inducing a conditions of immune homeostasis

Dictionary of Terms

- **Microbiota:** The mass of microbes present in our body.
- **Microbiome:** The expression of microbe genomes.
- **Prebiotics:** Food ingredients which stimulate both growth and activity of one or more bacteria in the colon.
- **Probiotics:** Live microbial cells whose administration in adequate amounts contributes to host health.

INTRODUCTION

Nowadays, DNA sequencing and bioinformatics techniques have greatly contributed to a better understanding of human microbiota and microbiome, respectively [1]. Even if these two terms are indiscriminately used, they have different meanings since microbiota represents the mass of microbes which our body harbors, while microbiome is rather the expression of microbe genomes [2]. Each of us possesses a unique microbiome which differentiates us from one another [3]. Furthermore, microbiota contributes to our health generating beneficial food metabolites [eg, short chain fatty acids (SCFAs), vitamins, and amino acids], participating in the development of the immune system, maintaining immune homeostasis, repairing gut epithelium after infectious events, and regulating enteric nerve functions [2,4].

Ontogenetically, microorganisms and immune cells have coevolved through fetal life to senescence. According to a recent report [5], microbiota was identified in human placenta specimens and compared to other districts such as oral, nasal, gut, vaginal, and respiratory mucosa. The placenta microbiota was mostly composed of Firmicutes, Tenericutes, Proteobacteria, Bacteroidetes, and Fusobacteria phyla and its profile is quite similar to that of the oral mucosa, being associated with a remote history of antinatal infection (urinary tract infection in the first trimester of pregnancy) and preterm birth. Bearing in mind the above concepts, it is conceivable that placenta microbiota may be able to activate T regulatory (Treg) cells (see also in the next paragraphs for further details) which, in turn, are involved in dampening maternal immune response, thus preventing fetus rejection. In support of the above hypothesis, there is recent evidence that human fetal placenta can induce the antiinflammatory subset of M2 macrophages as well as Treg cells which both promote fetal tolerance [6].

The microbiome undergoes major changes during the first 3 years of life. However, within the first 3 weeks of life skin microbiota has already become personalized [7]. The method of delivery seems to affect the precocious composition of microbiota since cesarean section favors the development of an infantile microbiota quite similar to that of adult human skin [7]. Conversely, in infants born through vaginal delivery, there is evidence for skin colonization with a microbiota close to that of human vagina.

The strict interaction between gut microbiota and immune cells is supported by the demonstration that in germ-free mice the development of lymphoid tissue is very poor, thus attributing a role to these microorganisms in the lymphopoiesis [8,9]. In humans, the link between gut microbiota and immune response may account for susceptibility or resistance to disease. In fact, eradication of gut microbiota following antibiotic therapy increases the frequency of infections such as oral candidiasis or *Clostridium difficile* colitis (which is epidemic in hospitals) likely via reduced activation of natural and adaptive immune responses [10]. In this context, fecal transplants from healthy individuals could reduce symptoms in patients with refractory *C. difficile* infections, normalizing gut microbiota composition [11]. After birth, diet seems to modulate the immune-gut microbiota axis, and, for instance, it is well known that hypercaloric regimen may account for disease outcome. Obesity as a result of a wrong dietary lifestyle is associated to disruption of gut microbiota and immune alteration [12]. In particular, the intestinal balance between pro-inflammatory and antiinflammatory cytokines has been shown to play an important role in keeping healthy conditions or generating disease. Actually, as recently demonstrated by us [13], since childhood the inflammatory interleukin (IL)-17/antiinflammatory (IL)-10 balance is strictly connected to the type of diet. In fact, in children who practiced healthy dietary recommendations (Mediterranean type diet) levels of salivary IL-10 were higher than those observed in children who did not practice a healthy eating regimen. Vice versa, in the latter IL-17 salivary levels were more elevated than in the former (Fig. 49.1).

The above cited pattern of dietary behavior becomes more dramatic in adulthood with a large predominance of obese people and/or patients with metabolic syndrome [14]. Especially in the elderly, the impact of an unbalanced diet is quite heavy in terms of senile frailty characterized by obesity, diabetes, cardiovascular disease (CVD), and neurodegeneration [10]. Oxidative stress and inflammation represent a common pathogenetic denominator in frail elderly patients, thus

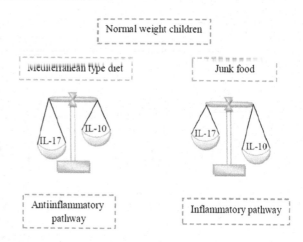

FIGURE 49.1 Different patterns of salivary cytokines in normal weight children who practiced or did not practice healthy food recommendations.

aggravating the progressive impairment of the immune system and gut microbiota composition [15].

On these grounds, in the present review, at first, the immune-gut microbiota axis will be illustrated, and, then, special focus will be concentrated on the ability of prebiotics and probiotics to correct the immune microbial dysfunctions in elderly.

THE RELATIONSHIP BETWEEN GUT MICROBIOTA AND IMMUNE SYSTEM IN ELDERLY

With special reference to the composition of gut microbiota, it is maximally represented in the colon with minimal differences between the transcending colon and sigmoid colon from the same person than it is with the transcending colon from another subject [1]. These taxonomic differences in microbiota composition may have pathogenic implications in terms of outcome of disease, such as inflammatory bowel disease (IBD) (Crohn's disease and ulcerative colitis), obesity, diabetes, and autism.

Age-dependent variations of microbiota have been observed with age both at infancy and old age, while in adulthood intestinal microbial composition is quite stable [10]. However, interindividual diversity has been reported in the microbiota composition and efforts have been made to characterize a conserved microbiota among human beings.

In the adult, microbiota is mainly composed by Bacteroidetes and Firmicutes whose equilibrium is altered by diet [3,16]. In fact, the western diet, rich in fat and sugar, leads to a prevalence of Bacteroidetes in the microbiota, while a dietary regimen rich in fibers increases the number of Firmicutes [17]. In elderly, the number of Bacteroidetes significantly augments with a decrease of Bifidobacteria, *Bacteroides*, and *Clostridium* cluster IV. Evidence has been provided that the increase of certain bacterial strains such as *Ruminococcus*, *Atopobium*, and Enterobacteriaceae seems to be associated with frailty in elderly [18]. Especially, in long-stay resident cohorts certain bacterial species such as *Alistipes* and *Oscillibacter* seem to be predominant in comparison to healthy community-dwelling subjects [18] (Fig. 49.2).

Methagenomic archaea seem to be more predominant in elderly and their production of methane from hydrogen and CO_2 can slow bowel transit with an increase in pH values in the colon [19,20]. In turn, pH shifts may favor growth of harmful bacterial strains, even including *Bacteroides* species [21].

From a metabolic point of view, many colonic anaerobes are able to ferment dietary carbohydrates in order to form SCFAs, such as acetate, butyrate and propionate, thus contributing to gut health [22]. Butyrate levels are lower in elderly than in healthy adults with an increase in branched chain fatty acids, ammonia, and phenols, and this may be due to low fiber intake [23]. Of note, butyrate along with propionate and acetate contributes to the host's energy, thus signaling via free fatty acid intestinal receptors and regulating anorexogenic receptors [24]. In particular, SCFAs are able to activate intestinal neoglucogenesis via a gut-brain neural circuit involving the fatty acid receptor FFAR3.

Three major pathways are used by gut microbes to produce SCFAs as indicated below [25,26]. Acetate is either directly generated from acetyl CoA or via the Wood-Ljungdahl pathway using formate. Propionate can be generated from phosphoenolpyruvate through the succinate decarboxylation pathway or the acrylate pathway in which lactate is reduced to propionate. Condensation of two molecules of acetyl CoA results

FIGURE 49.2 Microbiota composition in the various ages of life.

Fetal placenta:
Firmicutes,
Tenericutes,
Proteobacteria,
Bacteroidetes,
Fusobacteria

Infancy:
1. Vaginal delivery is responsible for the development of a human vaginal microbiota
2. Cesarian section is responsible for the development of a microbiota equivalent to that of human adult skin microbiota

Adulthood: Prevalence of Bacteroidetes or Firmicutes depending on diet:
1 Western diet rich in fat and sugar promotes growth of Bacteroidetes
2 A diet rich in fiber is associated to a predominance of Firmicutes

Elderly:
Increase in Bacteroidetes.
Decrease in Bifidobacteria, Bacteroidetes, and Clostridium cluster IV.
Increase in *Ruminococcus*, *Atopobium*, and Enterobacteriaceae which may be involved in frailty.

in butyrate by the enzyme butyrate-kinase or utilizing exogenously derived acetate by means of the enzyme butyryl-CoA acetate-CoA-transferase.

The major part of SCFAs is transported across the apical membrane of colonocytes under dissociated form by an HCO_3^- exchanger of unknown identity or one of the known symporters, monocarboxylate transporter 1 or sodium-dependent monocarboxylate transporter. A small part may be transported via passive diffusion. The part of SCFAs that is not oxidized by colonocytes is instead transported through the basolateral membrane. This mode of transport can be mediated by an unknown HCO_3^- exchanger, monocarboxylate transporter 4 or monocarboxylate transporter 5. Finally, the proposed mechanisms by which SCFAs increase fatty acid oxidation in liver, muscle, and brown adipose tissue are illustrated below. In muscle and liver, SCFAs phosphorylate and activate 5' adenosine monophosphate-activated protein kinase directly by increasing the AMP/ATP ratio and indirectly via the Ffar2-leptin pathway in white adipose tissue. In the same tissue, SCFAs decrease insulin sensitivity via Ffar2, thereby decreasing fat storage. In addition, binding of SCFAs to Ffar2 leads to release of the G (i/o) protein, the subsequent inhibition of adenylate cyclase and an increase of the ATP/cAMP ratio. This, in turn, leads to the inhibition of protein kinase A and the subsequent inhibition of hormone-sensitive lipase, leading to a decreased lipolysis and reduced plasma free fatty acids.

In relation to SCFA altered generation in elderly, it is well known that butyrate may derive from lactate formed by *Bifidobacterium* species [27] but accumulation of fecal lactate has been observed in aged people as a consequence of a reduced number of gut bacteria which utilize lactate and produce butyrate [28].

Quite interestingly, SCFAs in the gut are able to regulate T-cell differentiation into effector and Treg cells depending on cytokine milieu [29]. For instance, acetate administration to *Citrobacterium rodentium* infected mice augmented the induction of T helper 1 (Th1) and Th17 cells but diminished the anti-CD3 dependent inflammation via release of IL-10. On the other hand, orally administered butyrate at high levels increased IL-23 secretion by dendritic cells (DCs) which resulted in enhanced activation of Th17 cells, thus aggravating dextran sulfate sodium-induced colitis. It is likely that the altered microbiota in the elderly may lead to a predominant SCFA-induced inflammation rather than immune tolerance.

Many factors are involved in the senile alteration of microbiota. For instance, diet strongly contributes to the modulation of gut microbiota which is often unbalanced in the elderly because of an inappropriate intake of calcium, vitamins (especially vitamin D), and fiber (fruit and vegetables) [18]. Also use of a broad spectrum of antibiotics in the elderly can favor growth of bacteria such as *C. difficile*, which possesses an elevated carriage rate in care home residents [30]. Finally, administration of proton pump inhibitors, nonsteroidal antiinflammatory drugs, and opioids has been associated to modification of gut microbiota in elderly [31–33].

Under normal circumstances, bacterial components of the gut microbiota and immune cells actively interact for the maintenance of the immune homeostasis [34]. *Bacteroides fragilis* renders gut DCs more tolerogenic through activation of Treg cells and release of IL-10 [35,36]. By contrast, segmented filamentous bacteria stimulate DCs to release of IL-6 and IL-23 with activation of Th17 cells [37]. In turn, Th17 cells release IL-17 (A–F) which perpetuates a condition of chronic inflammation [38] (Fig. 49.3).

Evidence has been provided that epithelial cells produce less antimicrobial peptides (α-defensins), reactive oxygen species, and secretory immunoglobulin A (sIgA), thus permitting a growth of pathogens in the gut of elderly people. In turn, bacterial products via stimulation of toll-like receptors (TLRs) and nucleotide-binding oligomerization domain (NOD)-

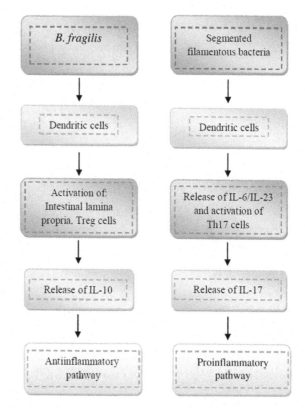

FIGURE 49.3 Intestinal microbiota and induction of Treg cells and Th17 cells.

like receptors trigger release of pro-inflammatory cytokines and chemokines from epithelial cells [15]. On these grounds, the altered microbiota in the elderly along with a condition of immunosenescence seems to aggravate the low grade inflammatory status, the so-called inflammaging. [39,40]. In this respect, aging is characterized by a reduction of CD34+ hematopoietic stem cells with scarce formation of T- and B-common lymphoid progenitors and reduced net proliferation of both lymphoid and myeloid precursors [41]. Functionally, aged phagocytes (polymorphonuclear cells, monocytes/macrophages) exhibit multiple deficits of chemotaxis, ingestion, and digestion, as well as reduced expression of TLR-2 and TLR-4 [41]. At the same time, in vitro studies have demonstrated exaggerated release of inflammatory cytokines and chemokines from lipopolysaccharide-stimulated monocytes and natural killer (NK) cells [41]. With special reference to NK cells, at the single cell level, their cytotoxicity is decreased in elderly, while their percentage and expression of CD56 marker are increased in cytomegalovirus positive old persons [41]. Aged DCs, the major type of antigen presenting cells (APCs), display an impaired ability to present antigens to T cells owing to a deficit of migration, chemotaxis, pinocytosis, and ingestion [41]. This impaired antigen presentation seems to affect processing of self-antigen by T and B cells, thus leading to a breakdown of peripheral tolerance. In addition, various alterations of T-cell receptor signaling, cognate helper function, and cytokine production have been reported as well [41]. Furthermore, in aged humans reduced frequency of naïve CD4+ and CD8+ lymphocytes seems to be compensated by the expansion of T memory cells [42]. In general terms, data concerning T-cell function in elderly are quite conflicting. In fact, release of interferon (IFN)-γ and IL-4 by aged T cells as well as Treg cell frequency have been reported as within normal ranges, decreased, or increased [42]. All these discrepancies may arise from different criteria of selection of old subjects and healthy/disease status of enrolled individuals.

With special reference to B cells, quantitative and qualitative reductions of antibodies have been demonstrated in the elderly [43]. Aged mature follicular B cells have been shown to produce less quantity of antibodies in view of a reduced T cell help as well as an intrinsic B-cell deficit [44]. In aged mice, Ig isotype class switching and somatic hypermutation are diminished, thus leading to a low affinity maturation of antibodies [45,46]. Finally, according to a recent view, inflammaging may account for the diminished B-cell lymphopoiesis with a functional impairment of the pre-B cell receptor and alteration of the diversity of the B-cell repertoire [47]. Taken together, these dysfunctions may promote autoantibody formation and scarce humoral immune response against pathogens in elderly (Table 49.1).

TABLE 49.1 Dysfunctions of Immune Cells in Elderly

	Innate immunity	Adaptive immunity
Phagocytes (granulocytes, monocytes, macrophages)	Reduction of chemotaxis, ingestion, digestion, and expression of TLRs; increased release of pro-inflammatory cytokines	Increased release of pro-inflammatory cytokines
NK cells	Diminished cytotoxicity; increase in percentage and expression of CD57 marker	
Dendritic cells	Decreased migration, chemotaxis, pinocytosis, ingestion, and antigen presentation	
T lymphocytes		Reduction of CD34+ hematopoietic stem cells with poor formation of T- and B-common lymphoid progenitors; reduced frequency of naïve CD4+ and CD8+ cells and Treg cells; increased expansion of T memory cells; conflicting data on IFN-γ and IL-4 secretion
B lymphocytes		Quantitative and qualitative reduction of antibodies; Ig isotype switching and somatic hypermutation reductions and low affinity maturation of antibodies; mature follicular B cells produce less quantity of antibodies in view of a reduced T-cell help and/or intrinsic defects of B lymphocytes; effects of inflammaging on pre-B cell receptor function and alteration of the B-cell repertoire diversity

FOOD INTAKE AND MODULATION OF THE IMMUNE-GUT MICROBIOTA AXIS

Gut is primarily involved in the digestion and absorption of food, but the presence of a robust immune system allows protection against the invasion of pathogens [48]. Conversely, the gut immune system is unresponsive to food antigens via mechanisms of tolerance [49]. In fact, interruption of tolerance seems to account for inflammation and food allergy development [50].

Food nutrients, which have been absorbed, metabolically produced, or newly generated by microbial components of the gut microbiota, are essential for the development of the immune system and maintenance of immune homeostasis [51,52]. Even if vitamins do not represent the actual topic of the present chapter, one cannot help to mention them as essential nutrients. Briefly, retinoic acid, a metabolite of vitamin A, in the gut environment is involved in the differentiation of Treg cells, and enhancement of IL-22 release from $\gamma\delta$ T cells and innate lymphoid cells for the maintenance of immune protection [53]. Vitamin D and its metabolite 1,25-dihydroxyvitamin D possess specific receptors on intestinal macrophages and DCs, the so-called vitamin D receptors [54]. Therefore, vitamin D induces differentiation of Treg cells, as well as production of cathelicidin and α-defensin 2 from macrophages, epithelial cells, and Paneth cells [55,56]. Finally, vitamin B9, derived from spinach and broccoli, binds to the folate receptor 4 expressed on intestinal Treg cells, thus maintaining their survival [57]. Taken together, the example of vitamins as fine modulators of the gut immunity, suggests that their introduction with diet or, in alternative, supplementation should be performed in the course of senile malnutrition.

EFFECTS OF PREBIOTICS AND PROBIOTICS ON THE IMMUNE-GUT MICROBIOTA AXIS

Prebiotics

Prebiotics are usually defined as nondigestible food ingredients which stimulate both growth and activity of one or more bacteria in the human colon, thus contributing to host health [58]. For instance, fructo-oligosaccharides (FOS) have been shown to increase colonic bifidobacteria when human volunteers ingested oligofructose or inulin [59]. Galacto-oligosaccharides (GOS) have also been shown to increase bacteria in healthy old subjects [60]. In addition, GOS were able to enhance phagocytic and NK cell activity while decreasing the production of pro-inflammatory cytokines [60]. Quite interestingly, biopsy samples taken from old subjects following 2 week treatment with FOS and inulin were cultivated for evaluation of bacteria [61]. Data showed an increase in *Bifidobacterium* counts as well as in number of Eubacteria, a specific butyrate producing group.

There is evidence that intake of whole grains and vegetables may prevent alteration of gut microbiota in aging with less risk of chronic disease [62]. In addition, a whole grain cereal-rich diet represents a source of fiber, also replacing servings of refined grains [63]. In addition, SCFAs derived from the microbial metabolism of fibers seem to be essential for maintaining a regular intestinal immune response [64,65]. In a recent report, the relationship between consumption of fiber and microbiota fecal concentrations were evaluated in institutionalized aged people [66]. Quite interestingly, in this study for the first time it was demonstrated that potato consumption increased fecal concentration of acetic, propionic, and butyric acids. It is likely that cellulose may in part explain the increase in SCFA levels. On the other hand, apple intake led to a slight increase of SCFAs. This may be due to the fact that apples contain low amounts of fiber and the elderly usually eat pealed apples, thus reducing the quality of insoluble fiber [66]. Evidence has been provided that consumption of whole grain barley, brown rice, and a mixture of the two exert a beneficial effect on the human microbiota, increasing microbial diversity [62]. In addition, whole grain barley and brown rice in combination reduced plasma levels of IL-6 as a consequence of gut microbiota modification.

In vitro studies have demonstrated that four flours (whole grain rye, whole grain wheat, chickpeas and lentils 50:50, and barley milled grains) were able to modulate microbiota composition and its metabolic activity using a three-stage continuous fermentative system, thus simulating the human colon [67]. In particular, a significant increase in *Bifidobacterium* and *Desulfuvibrionales* spp. were observed with a decrease in *Roseburia/Escherichia* rectal.

Experimentally, the effects on leukocyte function of diet supplementation with polyphenol-rich cereals in prematurely aging mice (PAM) have been reported [68]. In fact, cereals are enriched in polyphenols, such as ferulic and diferulic acid, endowed with antioxidant activities. Polyphenols are mostly concentrated in the outermost aleuron layers, bran and germ of grains. In particular, PAM, which underwent the above dietary regimen exhibited increased phagocytic activity, NK cell function, lymphocyte proliferation, and IL-2 release. In the light of these experimental results, addition of polyphenols to dietary cereals and their consumption may represent an important therapeutic strategy to reduce oxidative stress and inflammation in aging. In this framework, our own group has documented the beneficial effects of administration of Leucoselect Phytosome (a compound enriched in polyphenols isolated from red grape seeds) to frail elderly people [42]. Following treatment we could detect a condition of immune recovery represented by increased serum levels of IFN-γ and IL-2, thus indicating restoration of the Th1 response in frail aged people.

Despite the beneficial effects exerted by prebiotics on the immune-gut microbiota axis, other studies have reported interindividual variations in response to fibers. For instance, in certain volunteers no effects have been detected following consumption of prebiotics and these individuals were considered as nonresponders [69,70]. Many predominant bacteria which possess more nutritional flexibility were not affected by dietary interventions [71]. Furthermore, polysaccharide utilization by gut bacteria may vary and this should be taken into account in future studies [72]. Finally, among confounding factors, whole grains may differ in their composition and magnitude of inducible healthy effects [62].

Also milk oligosaccharides may represent new prebiotics, even including L-fucose, D-glucose, and D-galactose which are the third largest contingent of human milk [73] (see below for further details).

Probiotics

Probiotics are defined as live microbial cells whose administration in adequate amounts contributes to host health [74]. Among the major beneficial mechanisms exerted by probiotics, induction of Treg cells [75] has been included. Probiotic-induced tolerance seems to be mediated by tolerogenic APCs with generation of Treg cells and release of IL-10 [76]. In fact, oral application of *Bifidobacterium infantis* could reduce NF-κB activation, thus increasing numbers of mucosal and splenic Treg cells, thus contrasting *Salmonella thyphimurium*-dependent activation of NF-κB [77]. Induction of Treg cells and their survival by probiotics depends on the recognition receptors TLR-2 and NOD2/CARD15 [78]. For instance, muramyldipeptide (MDP) from the cell wall of Gram-positive bacteria is able to reduce Fas-induced Treg cell apoptosis [79]. In addition, binding of MDP to NOD2 may promote NF-κB activation in FoxP3+ cells [79]. Of note, NOD2 deficiency aggravates graft versus host disease and experimental colitis in mice with subsequent reduction of FoxP3+ cells [80]. In mice *Lactobacillus salivarius* administration protected from colitis via generation of Treg cells, but this effect was abrogated in NOD2 deficient mice [81]. Probiotics can also generate antiinflammatory effects via regulation of exogenous bacterial ATP [82]. In fact, ATP enhances FoxP3 transcription following A2 adrenergic receptor engagement. Also an increase in transforming growth factor-β and decrease of IL-6 levels were detected in response to ATP. Finally, Treg cells convert ATP to adenosine which exerts an immunosuppressive role [83]. Conversely, evidence has been provided that ATP accumulation tends to induce Th17 cells which are inflammatory [84]. This discrepancy may depend on both probiotic microbial composition and subsets of gut DCs involved.

Both Gram-positive Lactobacilli and Bifidobacteria and Gram-negative organisms such as *Escherichia coli* have been used in human trials [85]. Consumption of Bifidobacteria by elderly people seems to be beneficial for their capacity to synthesize vitamins, even including folate, and generate acetate and lactate, which, in turn, arrests growth of pathogens [86]. Administration of *Bifidobacterium* species to aged people led to their increase in the stool as well as reduction of inflammatory status [87]. In another study with healthy elderly subjects, intake of milk containing *Bifidobacterium lactis* increased fecal *Bifidobacterium* counts [88]. Fecal Lactobacilli and Enterococci increase with a reduction of Enterobacteria was also observed when Bifidobacteria were administered to elderly people [89]. In another group of elderly nursing home residents administered with *Bifidobacterium longum* an increase in *Bifidobacterium bifidum* and *Bifidobacterium breve* was reported [90].

With special reference to the effects of probiotics on aged immune system a few reports have been published. *Bifidobacterium lactis* HNO19 administration to elderly people was very effective in the recovery of phagocytic, NK, and T cell functions [41]. Fermented milk containing *Lactobacillus casei* shirota strain when administered to aged people with *Norovirus* gastroenteritis could reduce fever duration [41]. Moreover, probiotic administration has been shown to augment vaccine efficacy increasing antibody titers and NK and polymorphonuclear activities, thus leading to decreased incidence of influenza and fever [41].

Yoghurt is a food derived from milk fermentation by several strains of lactic acid bacteria. Experimentally, yoghurt administered colitis mice underwent a reduction of inflammatory recurrences and kept an antiinflammatory profile of gut cytokines [91]. This effect has been related to changes of intestinal microbiota with an increase in bifidobacteria population. This antiinflammatory effect was also related to the ability of yoghurt feeding to inhibit tumor growth in a model of murine colon cancer. In this model, the mechanisms of action of yoghurt were dependent on release of IL-10 and decrease of procarcinogenic enzymes in the gut. In the light of these results, administration of yoghurt to elderly people may exert beneficial effects in terms of antiinflammation and antineoplastic activities.

Synbiotics

Synbiotics result from the association of pre- and probiotics. The effect of a synbiotic on the elderly fecal microbiota has been assessed in vitro, using fecal batch cultures and three-stage continuous culture system [92]. *Bifidobacterium longum* and isomaltooligosaccharides significantly increased *Bifidobacterium* counts, while decreasing *Bacteroides* counts. The same was true for the combination of *L. fermentum* and FOS. In addition, the various combinations of probiotics and prebiotics were also able to increase the concentration of acetic acid which is the main end-products of *Bacteroides* and *Bifidobacterium* species.

In healthy old women, administration of *B. bifidum* and *B. lactis* with inulin increased *Bifidobacterium* counts up to the third-post feeding week [93]. In another study, 43 older volunteers were administered with *B. longum* and inulin-based prebiotic Synergy 1 in a randomized, double-blind placebo controlled, 4-week crossover trial [94]. A significant increase in *Bifidobacterium* counts as well as in Actinobacteria and Firmicutes was observed. Conversely, counts of Proteobacteria were significantly reduced. An increase in butyrate was also detected. Quite interestingly, serum tumor necrosis factor (TNF)-α levels were reduced by synbiotic feeding after 2 and 4 weeks. Also decrease of IL-6, IL-8, and monocyte chemotactic protein-1 was observed but only for a period of 2 weeks. Increase in Enterobacteria in older people has been associated to elevated release of IL-6 and IL-8. In the present study reduction of Proteobacteria has been documented after synbiotic treatment.

In our own research, a synbiotic composed by *Lactobacillus rhamnosus* Gorbach and Goldin and oligofructose was administered to 10 healthy free-living elderly people twice a day for 1 month [95]. Serum cytokine levels were evaluated before treatment (T0) and after treatment (T1). The reduced levels of IL-1β, IL-12, and IL-10, at T0 remained unchanged at T1. Serum concentrations of IL-6 and IL-8 increased at T1 when compared to those detected at T0 as well as to placebo group and the younger counterpart. It is well known that IL-6 increases with age, thus contributing to inflammaging, even if its association to senile frailty has not yet been clarified [96]. However, no side effects have been demonstrated in our group of elderly people which may be related to the elevation of this cytokine. Also in the case of IL-8 evidence has been provided for increased concentrations of this mediator in aging and, in this context, others found a correlation between nutritional deficiencies and elevation of IL-8 serum levels [41]. There is evidence for a protective effect exerted by the axis C3a/C5a/IL-8 against *Cryptococcus neoformans* [41]. Therefore, it is postulated that IL-8 may also afford protection in our group of elderly people treated with the synbiotic. To support our hypothesis, it is important mentioning that administration of *B. lactis* and *B. bifidum* along with inulin could lead to a dramatic reduction of winter infections in another group of elderly people [41].

Among synbiotics, human milk should be included for its content in Bifidobacteria and oligosaccharides. Human milk oligosaccharides (HMO) have been shown to favor growth of probiotics, thus contributing to the gut microbiota development in infants [97]. In addition, HMO represent decoy receptors for bacteria that possess high affinity binding to oligosaccharide receptors present on gut epithelial cells in infants. Moreover, human milk is enriched in immune substances, such as sIgA, IFN-γ, lactoferrin, fibronectin, mucins, and lysozyme, which protect neonates against invading pathogens [98]. Over the past few years, niche milks, even including donkey's and goat's milk, have been reappraised as substitutes of human and bovine milks. Both niche milks are also endowed with antiinflammatory and immunomodulating activities according to our own in vitro studies [99,100]. Also in vivo, administration of donkey's and goat's milk to free-living elderly (200 mL/day for 1 month) was able to influence peripheral cytokine levels [101]. Actually, donkey's milk administration increased serum levels of IL-8 and IL-6, while goat's milk intake led to a decrease of both cytokines. According to a recent report [102], fermented goat's milk in the presence of *Lactobacillus plantarum* was able to inhibit melanogenesis via production of radical scavenging peptides and antioxidants. These experiments offer new opportunities to create probiotics for treatment of skin pigment disorders or melanoma.

FUTURE TRENDS

According to recent evidence, the use of probiotics can also be beneficial to other organs as well as just to the intestines. For instance, *Lactobacillus reuteri* when administered to mice with osteoporosis could reverse bone resorption by stopping osteoclast growth [103]. In human menopause, a decrease in estrogen may account for the increased release of TNF-α which promotes osteoclast formation. Since *L. reuteri* converts L-histidine into histamine which inhibits TNF-α production, this probiotic may be used in women to limit the postmenopausal osteoporotic process. *Lactobacillus reuteri* can also attenuate the inflammatory pathway in colitis mice, reducing TNF-α production [104]. In this framework, it has been demonstrated that soy milk fermented in the presence of *L. plantarum* could diminish systolic and diastolic blood pressure in rats [102]. These hypotensive effects have been demonstrated to depend on suppression of angiotensin converting enzyme activity in kidney and liver and increased nitric oxide and superoxide dismutase activities.

Both osteoporosis and hypertension are serious complications in elderly patients and, therefore, introduction of dietary probiotics since adulthood may prevent bone deterioration and hypertension.

It is well known that dysbiosis might be implicated in the pathogenesis of IBD with an alteration of microbiota. Chronic intestinal inflammation is associated to the development of colorectal cancer (CRC), which increases in the elderly [11]. In this context, germ-free mice injected with *Bacteroides thetaiotamicron* undergo mucosal gene expression modifications in the colon, angiogenesis, and immune responses, which lead to colon cancer development [11]. In addition, the enterotoxigenic *B. fragilis* secretes the toxin BFT in close proximity of the CRC mucosa which may act as human matrix metalloproteinases, thus promoting cancer growth [105]. Therefore, normalization of the altered microbiota in the elderly by probiotics may prevent mucosal inflammation and CRC development. Also in veterinary medicine, feeding probiotics to livestock could replace the use of antibiotics. For instance, in piglets administration of probiotics could reduce pathogenic *E. coli* strains which adhere to intestinal mucosa [106]. In conclusion, probiotic administration could reduce the abundant use of antibiotics in livestock kept in crowded conditions, which are more susceptible to undergo infectious disease.

In the light of the above considerations, probiotics for their therapeutic effects have lately been termed "bugs as drugs." Some bacteria can be effective in their near-native form, others can be genetically modified to make them safer and also enhance their therapeutic potential. However, synbiotics taking advantage of the beneficial effects of both prebiotics and probiotics may represent novel more efficacious substances to fight specific disease. Finally, prebiotics and probiotics mostly interact with the host immune system and, therefore, immunomodulation may represent a therapeutic solution when traditional drugs have failed.

Another novel issue related to the interaction between microbiota and its host is the development of the indoor microbiota [107]. In this respect, the home microbiome project [108] has clearly demonstrated that the indoor microbiota is mostly composed by our own microbiome. In our homes and in other indoor urban environments, we have created a personalized microbiota and if its composition is in dysbiosis with our body, this may be detrimental in the long run. For instance, in hospitals the negative consequences of a pathobiome on patient health may be prevented by a course of selected probiotics and an environment which promotes the development of microbes more appropriate for patients [109].

In conclusion, the indoor microbiota can affect our immune responses and, therefore, normalization of a noxious microbiota is needed for the reassessment of the microbe-immune axis.

SUMMARY POINTS

- The decay of the immune responsiveness in elderly leads to a condition of inflammaging.
- Under normal conditions, in the gut the dendritic cell-T regulatory cell axis when activated triggers the anti-inflammatory pathway.
- Besides vitamins A and D which act on the above cited axis, prebiotics and probiotics may be effective in the induction of the anti-inflammatory pathway in elderly.
- Probiotics for their immune modulating effects have been termed as "bugs as drugs" for the correction of the intestinal dysbiosis.
- The new concept of the indoor microbiota indicates that our personalized microbiota should be corrected when altered as in the case of a pathobiome in hospitals.

Abbreviations

APCs	antigen presenting cells
CVD	cardiovascular disease
CRC	colorectal cancer
DCs	dendritic cells
FOS	fructo-oligosaccharides
GOS	galacto-oligosaccharides
HMO	human milk oligosaccharide
IBD	inflammatory bowel disease
IFN	interferon
IL	interleukin
MDP	muramyldipeptide
NK	natural killer
NOD	nucleotide-binding oligomerization domain
PAM	prematurely aging mice
SCFAs	short chain fatty acids
sIgA	secretory immunoglobulin A
Th	T helper
TLRs	toll-like receptors
TNF	tumor necrosis factor
Treg	T regulatory cells

Acknowledgment

Thea Magrone was a recipient of a contract in the context of the project "Bioscience and Health" (B&H) (PONa3_00395).

References

[1] Califf K, Gonzalez A, Knight R, Caporaso GJ. The human microbiome: getting personal. Microbe 2014;9:410–15.

[2] Casadevall A, Pirofski LA. What is a host? Incorporating the microbiota into the damage-response framework. Infect Immun 2015;83:2–7. Available from: http://dx.doi.org/10.1128/IAI.02627-14.

[3] Eckburg PB, Bik EM, Bernstein CN, et al. Diversity of the human intestinal microbial flora. Science 2005;308:1635–8.

[4] Flint HJ, Scott KP, Louis P, Duncan SH. The role of the gut microbiota in nutrition and health. Nat Rev Gastroenterol Hepatol 2012;9:577–89. Available from: http://dx.doi.org/10.1038/nrgastro.2012.156.

[5] Aagaard K, Ma J, Antony KM, Ganu R, Petrosino J, Versalovic J. The placenta harbors a unique microbiome. Sci Transl Med 2014;6: 237ra65. http://dx.doi.org/10.1126/scitranslmed.3008599.

[6] Svensson-Arvelund J, Mehta RB, Lindau R, et al. The human fetal placenta promotes tolerance against the semiallogeneic fetus by inducing regulatory T cells and homeostatic M2 macrophages. J Immunol 2015;194:1534–44. Available from: http://dx.doi.org/10.4049/jimmunol.1401536.

[7] Costello EK, Lauber CL, Hamady M, Fierer N, Gordon JI, Knight R. Bacterial community variation in human body habitats across space and time. Science 2009;326:1694–7. Available from: http://dx.doi.org/10.1126/science.1177486.

[8] Cash HL, Whitham CV, Behrendt CL, Hooper LV. Symbiotic bacteria direct expression of an intestinal bactericidal lectin. Science 2006;313:1126–30.

[9] Bry L, Falk PG, Midtvedt T, Gordon JI. A model of host-microbial interactions in an open mammalian ecosystem. Science 1996;273:1380–1383.

[10] Tojo R, Suárez A, Clemente MG, et al. Intestinal microbiota in health and disease: role of bifidobacteria in gut homeostasis. World J Gastroenterol 2014;20:15163–76. Available from: http://dx.doi.org/10.3748/wjg.v20.i41.15163.

[11] Robinson CJ, Lee EL, Eribo BE, Ashktorab H, Brim H. Colon cancer and IBD: potential link to race, microbiota. Microbe 2013;8:243–8.

[12] Magrone T, Jirillo E. Childhood obesity: immune response and nutritional approaches. Front Immunol 2015;24:76. Available from: http://dx.doi.org/10.3389/fimmu.2015.00076.

[13] Vitale E, Jirillo E, Magrone T. Determination of body mass index and physical activity in normal weight children and evaluation of salivary levels of interleukin 10 and interleukin 17. Clin Immunol Endocr Metab Drugs 2014;1:81–8.

[14] Weiman S. Studies link gut inflammation, obesity, diabetes to microbiome. Microbe 2014;9:394.

[15] Magrone T, Jirillo E. The interaction between gut microbiota and age-related changes in immune function and inflammation. Immun Ageing 2013;10:31. Available from: http://dx.doi.org/10.1186/1742-4933-10-31.

[16] Kim BS, Jeon YS, Chun J. Current status and future promise of the human microbiome. Pediatr Gastroenterol Hepatol Nutr 2013;16:71–9. Available from: http://dx.doi.org/10.5223/pghn.2013.16.2.71.

[17] Wu GD, Chen J, Hoffmann C, et al. Linking long-term dietary patterns with gut microbial enterotypes. Science 2011;334:105–8. Available from: http://dx.doi.org/10.1126/science.1208344.

[18] Claesson MJ, Jeffery IB, Conde S, et al. Gut microbiota composition correlates with diet and health in the elderly. Nature 2012;488:178–84. Available from: http://dx.doi.org/10.1038/nature11319.

[19] Mihajlovski A, Dore J, Alric M, Brugere J-F. Molecular evaluation of the humangut methanogenic archaeal micorbiota reveals an age-associated increase of diversity. Environ Microbiol Rep 2012;2:272–80.

[20] Pimentel M, Lin HC, Enayati P, van den Burg B, et al. Methane, a gas produced by enteric bacteria slows intestinal transit and augments small intestinal contractile activity. Am J Physiol 2006;290:G1089–95.

[21] Lewis SJ, Heaton KW. Increasing butyrate concentration in the distal colon by accelerating intestinal transit. Gut 1997;41:245–51.

[22] Pryde SE, Duncan SH, Hold GL, Stewart CS, Flint HJ. The microbiology of butyrate formation in the human colon. FEMS Microbiol Lett 2002;217:133–9.

[23] Laurin D, Brodeur JM, Bourdages J, Vallée R, Lachapelle D. Fibre intake in elderly individuals with poor masticatory performance. J Can Dent Assoc 1994;60:443–6 449.

[24] Sleeth ML, Thompson EL, Ford HE, Zac-Varghese SE, Frost G. Free fatty acid receptor 2 and nutrient sensing: a proposed role for fibre, fermentable carbohydrates and short-chain fatty acids in appetite regulation. Nutr Res Rev 2010;23:135–45. Available from: http://dx.doi.org/10.1017/S095442241000000.

[25] den Besten G, van Eunen K, Groen AK, Venema K, Reijngoud DJ, Bakker BM. The role of short-chain fatty acids in the interplay between diet, gut microbiota, and host energy metabolism. J Lipid Res 2013;54:2325–40. Available from: http://dx.doi.org/10.1194/jlr.R036012.

[26] Duncan SH, Barcenilla A, Stewart CS, Pryde SE, Flint HJ. Acetate utilization and butyryl coenzyme A (CoA):acetate-CoA transferase in butyrate-producing bacteria from the human large intestine. Appl Environ Microbiol 2002;68:5186–90.

[27] Vernia P, Caprilli R, Latella G, Barbetti F, Magliocca FM, Cittadini M. Fecal lactate and ulcerative colitis. Gastroenterology 1988;95:1564–8.

[28] Bartosch S, Fite A, Macfarlane GT, McMurdo ME. Characterization of bacterial communities in feces from healthy elderly volunteers and hospitalized elderly patients by using real-time PCR and effects of antibiotic treatment on the fecal microbiota. Appl Environ Microbiol 2004;70:3575–81.

[29] Park J, Kim M, Kang SG, et al. Short-chain fatty acids induce both effector and regulatory T cells by suppression of histone deacetylases and regulation of the mTOR-S6K pathway. Mucosal Immunol 2015;8:80–93. Available from: http://dx.doi.org/10.1038/mi.2014.44.

[30] Pérez-Cobas AE, Artacho A, Knecht H, et al. Differential effects of antibiotic therapy on the structure and function of human gut microbiota. PLoS One 2013;8:e80201. Available from: http://dx.doi.org/10.1371/journal.pone.0080201.

[31] Williams C, McColl KE. Review article: proton pump inhibitors and bacterial overgrowth. Aliment Pharmacol Ther 2006;23:3–10.

[32] Bjarnason I, Takeuchi K. Intestinal permeability in the pathogenesis of NSAID-induced enteropathy. J Gastroenterol 2009;44:23–9. Available from: http://dx.doi.org/10.1007/s00535-008-2266-6.

[33] Pappagallo M. Incidence, prevalence, and management of opioid bowel dysfunction. Am J Surg 2001;182:11S–8S.

[34] Magrone T, Jirillo E. The interplay between the gut immune system and microbiota in health and disease: nutraceutical intervention for restoring intestinal homeostasis. Curr Pharm Des 2013;19:1329–42.

[35] Round JL, Mazmanian SK. Inducible Foxp3+ regulatory T-cell development by a commensal bacterium of the intestinal microbiota. Proc Natl Acad Sci USA 2010;107:12204–9.

[36] Denning TL, Wang YC, Patel SR, et al. Lamina propria macrophages and dendritic cells differentially induce regulatory and

interleukin 17-producing T cell responses. Nat Immunol 2007;8:1086—94.
[37] Ivanov II, Atarashi K, Manel N, et al. Induction of intestinal Th17 cells by segmented filamentous bacteria. Cell 2009;139:485—98.
[38] Chewning JH, Weaver CT. Development and survival of Th17 cells within the intestines: the influence of microbiome- and diet-derived signals. J Immunol 2014;193:4769—77. Available from: http://dx.doi.org/10.4049/jimmunol.1401835.
[39] Candore G, Caruso C, Jirillo E, Magrone T, Vasto S. Low grade inflammation as a common pathogenetic denominator in age-related diseases: novel drug targets for anti-ageing strategies and successful ageing achievement. Curr Pharm Des 2010;16:584—96.
[40] Franceschi C, Campisi J. Chronic inflammation (inflammaging) and its potential contribution to age-associated diseases. J Gerontol A Biol Sci Med Sci 2014;69:S4—9. Available from: http://dx.doi.org/10.1093/gerona/glu057.
[41] Magrone T, Jirillo E. Disorders of innate immunity in human ageing and effects of nutraceutical administration. Endocr Metab Immune Disord Drug Targets 2014;14:272—82.
[42] Magrone T, Pugliese V, Fontana S, Jirillo E. Human use of Leucoselect® Phytosome® with special reference to inflammatory-allergic pathologies in frail elderly patients. Curr Pharm Des 2014;20:1011—19.
[43] Cancro MP, Hao Y, Scholz JL, et al. B cells and aging: molecules and mechanisms. Trends Immunol 2009;30:313—18. Available from: http://dx.doi.org/10.1016/j.it.2009.04.005.
[44] Scholz JL, Diaz A, Riley RL, Cancro MP, Frasca D. A comparative review of aging and B cell function in mice and humans. Curr Opin Immunol 2013;25:504—10. Available from: http://dx.doi.org/10.1016/j.coi.2013.07.006.
[45] Frasca D, Van der Put E, Riley RL, Blomberg BB. Reduced Ig class switch in aged mice correlates with decreased E47 and activation-induced cytidine deaminase. J Immunol 2004;172:2155—62.
[46] Yang X, Stedra J, Cerny J. Relative contribution of T and B cells to hypermutation and selection of the antibody repertoire in germinal centers of aged mice. J Exp Med 1996;183:959—70.
[47] Riley RL. Impaired B lymphopoiesis in old age: a role for inflammatory B cells? Immunol Res 2013;57:361—9. Available from: http://dx.doi.org/10.1007/s12026-013-8444-5.
[48] Sassone-Corsi M, Raffatellu M. No vacancy: how beneficial microbes cooperate with immunity to provide colonization resistance to pathogens. J Immunol 2015;194:4081—7. Available from: http://dx.doi.org/10.4049/jimmunol.1403169.
[49] Pabst O, Mowat AM. Oral tolerance to food protein. Mucosal Immunol 2012;5:232—9. Available from: http://dx.doi.org/10.1038/mi.2012.4.
[50] Scurlock AM, Vickery BP, Hourihane JO, Burks AW. Pediatric food allergy and mucosal tolerance. Mucosal Immunol 2010;3:345—54. Available from: http://dx.doi.org/10.1038/mi.2010.21.
[51] Veldhoen M, Brucklacher-Waldert V. Dietary influences on intestinal immunity. Nat Rev Immunol 2012;12:696—708. Available from: http://dx.doi.org/10.1038/nri3299.
[52] Spencer SP, Belkaid Y. Dietary and commensal derived nutrients: shaping mucosal and systemic immunity. Curr Opin Immunol 2012;24:379—84. Available from: http://dx.doi.org/10.1016/j.coi.2012.07.006.
[53] Kunisawa J, Kiyono H. Vitamins mediate immunological homeostasis and diseases at the surface of the body. Endocr Metab Immune Disord Drug Targets 2015;15:25—30.
[54] Liu PT, Stenger S, Li H, et al. Toll-like receptor triggering of a vitamin D-mediated human antimicrobial response. Science 2006;311:1770—3.
[55] Wang TT, Nestel FP, Bourdeau V, et al. Cutting edge: 1,25-dihydroxyvitamin D3 is a direct inducer of antimicrobial peptide gene expression. J Immunol 2004;173:2909—12.
[56] Yim S, Dhawan P, Ragunath C, Christakos S, Diamond G. Induction of cathelicidin in normal and CF bronchial epithelial cells by 1,25-dihydroxyvitamin D(3). J Cyst Fibros 2007;6:403—10.
[57] Kinoshita M, Kayama H, Kusu T, et al. Dietary folic acid promotes survival of Foxp3+ regulatory T cells in the colon. J Immunol 2012;189:2869—78. Available from: http://dx.doi.org/10.4049/jimmunol.1200420.
[58] Gibson GR, Roberfroid MB. Dietary modulation of the human colonic microbiota: introducing the concept of prebiotics. J Nutr 1995;125:1401—12.
[59] Gibson GR, Beatty ER, Wang X, Cummings JH. Selective stimulation of bifidobacteria in the human colon by oligofructose and inulin. Gastroenterology 1995;108:975—82.
[60] Vulevic J, Drakoularakou A, Yaqoob P, Tzortzis G, Gibson GR. Modulation of the fecal microflora profile and immune function by a novel trans-galactooligosaccharide mixture (B-GOS) in healthy elderly volunteers. Am J Clin Nutr 2008;88:1438—46.
[61] Langlands SJ, Hopkins MJ, Coleman N, Cummings JH. Prebiotic carbohydrates modify the mucosa associated microflora of the human large bowel. Gut 2004;53:1610—16.
[62] Walter J, Martínez I, Rose DJ. Holobiont nutrition: considering the role of the gastrointestinal microbiota in the health benefits of whole grains. Gut Microbes 2013;4:340—6. Available from: http://dx.doi.org/10.4161/gmic.24707.
[63] US Department of Health and Human Services & Department of Agriculture. Dietary guidelines for Americans 2005. 6th ed. Washington, DC: US Government Printing Office; 2005.
[64] Schwiertz A, Taras D, Schäfer K, et al. Microbiota and SCFA in lean and overweight healthy subjects. Obesity (Silver Spring) 2010;18:190—5. Available from: http://dx.doi.org/10.1038/oby.2009.167.
[65] Maslowski KM, Mackay CR. Diet, gut microbiota and immune responses. Nat Immunol 2011;12:5—9. Available from: http://dx.doi.org/10.1038/ni0111-5.
[66] Cuervo A, Salazar N, Ruas-Madiedo P, Gueimonde M, González S. Fiber from a regular diet is directly associated with fecal short-chain fatty acid concentrations in the elderly. Nutr Res 2013;33:811—16. Available from: http://dx.doi.org/10.1016/j.nutres.2013.05.016.
[67] Schwiertz A, Taras D, Schäfer K, et al. In vitro fermentation of potential prebiotic flours from natural sources: impact on the human colonic microbiota and metabolome. Mol Nutr Food Res 2012;56:134252. Available from: http://dx.doi.org/10.1002/mnfr.201200046.
[68] Alvarez P, Alvarado C, Puerto M, et al. Improvement of leukocyte functions in prematurely aging mice after five weeks of diet supplementation with polyphenol-rich cereals. Nutrition 2006;22:913—21.
[69] Lefevre M, Jonnalagadda S. Effect of whole grains on markers of subclinical inflammation. Nutr Rev 2012;70:387—96. Available from: http://dx.doi.org/10.1111/j.1753-4887.2012.00487.x.
[70] Jonnalagadda SS, Harnack L, Liu RH, et al. Putting the whole grain puzzle together: health benefits associated with whole grains—summary of American Society for Nutrition 2010 Satellite Symposium. J Nutr 2011;141:1011S—22S. Available from: http://dx.doi.org/10.3945/jn.110.132944.

[71] Tilg H, Kaser A. Gut microbiome, obesity, and metabolic dysfunction. J Clin Invest 2011;121:2126−32. Available from: http://dx.doi.org/10.1172/JCI58109.

[72] Ze X, Duncan SH, Louis P, Flint HJ. Ruminococcus bromii is a keystone species for the degradation of resistant starch in the human colon. ISME J 2012;6:1535−43. Available from: http://dx.doi.org/10.1038/ismej.2012.4.

[73] Host A, Halken S. Cow's milk allergy: where have we come from and where are we going? Endocr Metab Disord Drug Targets 2014;14:2−8.

[74] Hume ME. Historic perspective: prebiotics, probiotics, and other alternatives to antibiotics. Poult Sci 2011;90:2663−9. Available from: http://dx.doi.org/10.3382/ps.2010-01030.

[75] Magrone T, Jirillo E. Intestinal regulatory T cells: their function and modulation by dietary nutrients. Nutr Ther Metab 2014;32:157−65.

[76] Issazadeh-Navikas S, Teimer R, Bockermann R. Influence of dietary components on regulatory T cells. Mol Med 2012;18:95−110. Available from: http://dx.doi.org/10.2119/molmed.2011.00311.

[77] O'Mahony C, Scully P, O'Mahony D, et al. Commensal-induced regulatory T cells mediate protection against pathogen-stimulated NF-kappaB activation. PLoS Pathog 2008;4:e1000112. Available from: http://dx.doi.org/10.1371/journal.ppat.1000112.

[78] Foligne B, Zoumpopoulou G, Dewulf J, et al. A key role of dendritic cells in probiotic functionality. PLoS One 2007;2: e313.

[79] Rahman MK, Midtling EH, Svingen PA, et al. The pathogen recognition receptor NOD2 regulates human FOXP3+ T cell survival. J Immunol 2010;184:7247−56. Available from: http://dx.doi.org/10.4049/jimmunol.0901479.

[80] Penack O, Smith OM, Cunningham-Bussel A, et al. NOD2 regulates hematopoietic cell function during graft-versus-host disease. J Exp Med 2009;206:2101−10. Available from: http://dx.doi.org/10.1084/jem.20090623.

[81] Macho Fernandez E, Valenti V, Rockel C, et al. Anti-inflammatory capacity of selected Lactobacilli in experimental colitis is driven by NOD2-mediated recognition of a specific peptidoglycan-derived muropeptide. Gut 2011;60:1050−9. Available from: http://dx.doi.org/10.1136/gut.2010.232918.

[82] Zarek PE, Huang CT, Lutz ER, et al. A2A receptor signaling promotes peripheral tolerance by inducing T-cell anergy and the generation of adaptive regulatory T cells. Blood 2008;111:251−9.

[83] Mandapathil M, Hilldorfer B, Szczepanski MJ, et al. Generation and accumulation of immunosuppressive adenosine by human CD4+CD25highFOXP3+ regulatory T cells. J Biol Chem 2010;285:7176−86. Available from: http://dx.doi.org/10.1074/jbc.M109.047423.

[84] Atarashi K, Nishimura J, Shima T, et al. ATP drives lamina propria T(H)17 cell differentiation. Nature 2008;455:808−12. Available from: http://dx.doi.org/10.1038/nature07240.

[85] Kleerebezem M, Vaughan EE. Probiotic and gut Lactobacilli and Bifidobacteria: molecular approaches to study diversity and activity. Annu Rev Microbiol 2009;63:269−90. Available from: http://dx.doi.org/10.1146/annurev.micro.091208.073341.

[86] Toward R, Montandon S, Walton G, Gibson GR. Effect of prebiotics on the human gut microbiota of elderly persons. Gut Microbes 2012;3:57−60. Available from: http://dx.doi.org/10.4161/gmic.19411.

[87] Ouwehand AC, Bergsma N, Parhiala R, et al. Bifidobacterium microbiota and parameters of immune function in elderly subjects. FEMS Immunol Med Microbiol 2008;53:18−25. Available from: http://dx.doi.org/10.1111/j.1574-695X.2008.00392.x.

[88] Gill HS, Rutherfurd KJ, Cross ML, Gopal PK. Enhancement of immunity in the elderly by dietary supplementation with the probiotic Bifidobacterium lactis HN019. Am J Clin Nutr 2001;74:833−9.

[89] Ahmed M, Prasad J, Gill H, Stevenson L, Gopal P. Impact of consumption of different levels of Bifidobacterium lactis HN019 on the intestinal microflora of elderly human subjects. J Nutr Health Aging 2007;11:26−31.

[90] Lahtinen SJ, Tammela L, Korpela J, et al. Probiotics modulate the Bifidobacterium microbiota of elderly nursing home residents. Age (Dordr) 2009;31:59−66. Available from: http://dx.doi.org/10.1007/s11357-008-9081-0.

[91] Galdeano CM, Nunez IN, Carmuega E, de Moreno de LeBlanc A, Perdigon G. Role of probiotics and functional foods in health: gut immune stimulation by two prebiotico strains and a potential probiotic yoghurt. Endocr Metab Immune Disord Drug Targets 2015;15:37−45. Available from: http://dx.doi.org/10.2174/1871530314666141216121349.

[92] Likotrafiti E, Tuohy KM, Gibson GR, Rastall RA. An in vitro study of the effect of probiotics, prebiotics and synbiotics on the elderly faecal microbiota. Anaerobe 2014;27:50−5. Available from: http://dx.doi.org/10.1016/j.anaerobe.2014.03.009.

[93] Bartosch S, Woodmansey EJ, Paterson JC, McMurdo ME, Macfarlane GT. Microbiological effects of consuming a synbiotic containing Bifidobacterium bifidum, Bifidobacterium lactis, and oligofructose in elderly persons, determined by real-time polymerase chain reaction and counting of viable bacteria. Clin Infect Dis 2005;40:28−37.

[94] Macfarlane S, Cleary S, Bahrami B, Reynolds N, Macfarlane GT. Synbiotic consumption changes the metabolism and composition of the gut microbiota in older people and modifies inflammatory processes: a randomised, double-blind, placebo-controlled crossover study. Aliment Pharmacol Ther 2013;38:804−16. Available from: http://dx.doi.org/10.1111/apt.12453.

[95] Amati L, Marzulli G, Martulli M, et al. Administration of a synbiotic to free-living elderly and evaluation of serum cytokines. A pilot study. Curr Pharm Des 2010;16:854−8.

[96] Franceschi C, Valensin S, Bonafè M, et al. The network and the remodeling theories of aging: historical background and new perspectives. Exp Gerontol 2000;35:879−96.

[97] Musilova S, Rada V, Vlkova E, Bunesova V. Beneficial effects of human milk oligosaccharides on gut microbiota. Benefic Microbes 2014;5:273−83. Available from: http://dx.doi.org/10.3920/BM2013.0080.

[98] Jirillo F, Magrone T. Anti-inflammatory and anti-allergic properties of donkey's and goat's milk. Endocr Metab Immune Disord Drug Targets 2014;14:27−37.

[99] Jirillo F, Martemucci G, D'Alessandro AG, et al. Ability of goat milk to modulate healthy human peripheral blood lymphomonocyte and polymorphonuclear cell function: in vitro effects and clinical implications. Curr Pharm Des 2010;16:870−6.

[100] Jirillo F, Jirillo E, Magrone T. Donkey's and goat's milk consumption and benefits to human health with special reference to the inflammatory status. Curr Pharm Des 2010;16:859−63.

[101] Amati L, Marzulli G, Martulli M, et al. Donkey and goat milk intake and modulation of the human aged immune response. Curr Pharm Des 2010;16:864−9.

[102] Weiman S. Bugs as drugs: bacteria as therapeutic against diseases. Microbe 2014;9:437−41.

[103] Britton RA, Irwin R, Quach D, et al. Probiotic *L. reuteri* treatment prevents bone loss in a menopausal ovariectomized mouse model. J Cell Physiol 2014;229:1822–30. Available from: http://dx.doi.org/10.1002/jcp.24636.

[104] McCabe LR, Irwin R, Schaefer L, Britton RA. Probiotic use decreases intestinal inflammation and increases bone density in healthy male but not female mice. J Cell Physiol 2013;228:1793–8. Available from: http://dx.doi.org/10.1002/jcp.24340.

[105] Sears CL, Pardoll DM. Perspective: alpha-bugs, their microbial partners, and the link to colon cancer. J Infect Dis 2011;203:306–11. Available from: http://dx.doi.org/10.1093/jinfdis/jiq061.

[106] Bednorz C, Guenther S, Oelgeschläger K, et al. Feeding the probiotic *Enterococcus faecium* strain NCIMB 10415 to piglets specifically reduces the number of *Escherichia coli* pathotypes that adhere to the gut mucosa. Appl Environ Microbiol 2013;79:7896–904. Available from: http://dx.doi.org/10.1128/AEM.03138-13.

[107] Lax S, Nagler CR, Gilbert JA. Our interface with the built environment: immunity and the indoor microbiota. Trends Immunol 2015;36:121–3. Available from: http://dx.doi.org/10.1016/j.it.2015.01.001.

[108] Lax S, Smith DP, Hampton-Marcell J, et al. Longitudinal analysis of microbial interaction between humans and the indoor environment. Science 2014;345:1048–52.

[109] Zaborin A, Smith D, Garfield K, et al. Membership and behavior of ultra-low-diversity pathogen communities present in the gut of humans during prolonged critical illness. MBio 2014;5: e01361–14. http://dx.doi.org/10.1128/mBio.01361-14.

50

Vegetables and Fruit in the Prevention of Chronic Age-Related Diseases

Kirsten Brandt

Food Quality and Health Research Group, Human Nutrition Research Centre, School of Agriculture, Food and Rural Development, Newcastle University, Newcastle upon Tyne, UK

KEY FACTS

- Eating a lot of vegetables and fruit makes you live longer than if you don't eat these foods.
- Some types of vegetables and fruits improve health more than others.
- We want to know which natural chemicals are responsible for the health benefits. Since then we can choose the fruits and vegetables with highest content of these beneficial substances, and obtain additional health benefits.
- Previously scientists believed that ageing was caused by too many free radicals, and that the benefits of fruits and vegetables were caused by antioxidant effects, "mopping up the dangerous free radicals."
- However we now know that antioxidant effects of food have no benefits for human health, unless the diet is seriously deficient in vitamin C and vitamin E.
- The only other diet that makes you live longer is to eat less. "Caloric restriction" slows down the ageing process so it takes more years before you start getting the diseases of old age, such as heart disease, stroke or cancer.
- Vegetables and fruit contain substances called phytochemicals, and some of the phytochemicals slow down growth and other processes in the cell just like caloric restriction. So these phytochemicals may be causing the health benefits of eating fruits and vegetables.
- If this is the case, in the future it will become possible to choose vegetables and fruits with specific contents of anti-aging phytochemicals, and thus enhance the benefits of vegetables and fruits in the diet.

Dictionary of Terms

- *Antioxidant*: A compound that can neutralize free radicals.
- *Bioactive compound*: A compound that can enhance human health in the concentrations found in food, even though it is not an essential nutrient.
- *Free radical*: A very reactive type of chemical compound, usually formed by reactions that involve oxygen. One free radical molecule can react with many other molecules and damage them, until it is neutralized by an antioxidant.
- *Intervention (experimental) studies*: Using changes to people's normal lives to study factors affecting human health. For example, in a nutritional intervention study participants receive pills with nutrients or placebo (dummy pills), in order to determine whether the nutrients can improve the health of the participants.
- *Observational (epidemiological) studies*: Using data about people's normal lives to study factors affecting human health. For example, nutritional epidemiology compares what people eat with

information about their health in order to identify healthy foods.
- *Phytochemical*: A bioactive compound produced by a plant.

EFFECTS OF INTAKE OF VEGETABLES AND FRUIT ON MORTALITY AND MORBIDITY

The importance of fruits and vegetable consumption as part of a healthy diet is well recognized both in the scientific community and the general population. "An apple a day" is supposed to "keep the doctor away," and promotion of increased intake of fruit and vegetables is an important responsibility of health authorities across the World [1]. However, fruits and vegetables are not particularly nutrient dense foods, and our understanding of their role in the prevention of age-related diseases is surprisingly patchy, particularly if compared with other relevant factors such as smoking or menopause.

Epidemiology

Already in 1898, Williams asserted [2] that increased mortality from cancer may be caused by "deficiency in fresh vegetable food." Since then, prospective epidemiological studies often show protective effects (negative correlations) of intake of fruits and vegetables with incidence of subsequent age-related chronic diseases or conditions. Data from large numbers of studies and with thousands of participants are available for major diseases, such as cardiovascular disease [3,4], type 2 diabetes [5], and cancer [6], and in all these cases relatively high fruit and vegetable intakes seem to result in small but significant associations with reduced risks. Typically the risk of each of these diseases are reduced by 5—10% in the study populations, when comparing the 20% of participants with the highest total intake of fruits and vegetables with the 20% with the lowest intake (highest versus lowest quintile). In these major studies and meta-analyses, the effect of fruit and vegetable intakes on the disease incidences are adjusted for confounding factors, such as body mass index and food energy intake [6] and in some cases even plasma lipids [7]. This means that any indirect effects of the consumption of vegetables and fruit on BMI or other aspects of energy balance (which on its own is a known risk factor for cancer) would be excluded from the results, indicating that these estimates may be considered conservative.

For other age-related conditions, less data are available, although associations with beneficial effects of fruits and vegetables are published for hypertension [8], skin aging [9], cognitive performance [10], dental caries [11], and telomere shortening [12].

The most definitive prospective assessments are associations with overall mortality. In their 2013 paper Bellavia et al. [13] found very strong and significant nonlinear effects of fruit and vegetable intake on mortality among participants with low intakes, such as 1 serving per day corresponding to 17 months shorter life expectancy than 5 servings per day, while increases beyond 5 a day did not affect mortality. Previous prospective studies tended to report much smaller effects, often not significant or only marginally so [14], however, this might be due to development in statistical analysis methods, in particular improved methods to detect nonlinear associations [13].

However, it is often observed that strong unadjusted associations become progressively less significant when adjusted for relevant confounders such as education and income, which are well known independent predictors of longevity and healthy aging [15].

There could be several possible reasons for this, each of which may be more or less important for an individual study, since they are not mutually exclusive: One possibility is that some of the associations of health outcomes with fruit and vegetable intake represent "false positives" caused by residual confounding with other factors such as smoking [16], health awareness [17] or social rank [18], which are often not accurately assessed in epidemiological studies and themselves correlated with both disease risk and fruit and vegetable intake. Another possibility is related to the use of questionnaires such as Food Frequency Questionnaires, 24-hour Dietary Recall, or Food Diaries to assess the intakes of fruits and vegetables. While questionnaires are cost-effective tools for collection of dietary data from large cohorts, these methods are still relatively imprecise [19,20], making it difficult to accurately assess correlations with relatively rare events such as serious disease or death [16]. This well-recognized problem may cause "false negative" effects, where correlations are obscured by high variability [21].

One way to address this question is to compare with epidemiological studies, where measured plasma concentrations of carotenoids or of vitamin C are used as proxy measures of vegetable and fruit intake (as "biomarkers of intake"), rather than relying on the subjective self-reported intakes [22]. Studies using biomarkers of intake often show stronger correlations than those based on questionnaires, even with lower numbers of participants. For example, a meta-analysis of questionnaire-based prospective studies of

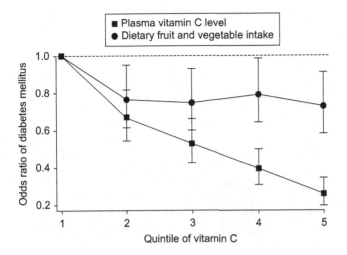

FIGURE 50.1 Odds ratio of diabetes mellitus by quintiles of plasma vitamin C level and fruit and vegetable intake, adjusted for age and sex: European Prospective Investigation of Cancer–Norfolk study. For plasma vitamin C analysis, the sample size was 19 246. For fruit and vegetable analysis, the sample size was 21 831. Error bars indicate 95% confidence intervals. *Source: from Ref. [23].*

associations with type 2 diabetes found odds ratios of high versus low intake of fruits and vegetables of around 0.9 [5], while the corresponding values in studies using plasma concentrations of vitamin C or beta-carotene as biomarkers of intake were 0.38 [23] and 0.42 [24] respectively (Fig. 50.1).

Using a combination of plasma concentrations of vitamin C, beta-carotene, and lutein, the odds ratio for the highest versus the lowest quartile came down to 0.35 [25]. Plasma concentrations of plant-derived compounds are still affected by many different confounders, including genetic variants affecting nutrient metabolism, which modulate the effect of intake on plasma concentrations [26,27]. However, for morbidities where the use of biomarkers improves the strength of correlations with intake of fruits and vegetables, this clearly indicates that inaccurate intake estimates are a major source of error and may mask even quite strong correlations. The problem is that analysis of biological markers in biological fluids such as plasma is more expensive per participant than administering a questionnaire, so improvements in precision may be offset by reduced numbers of participants.

Even when intake is accurately measured and assessed, some types of potential confounders remain. Effects may be related to only certain fruits or vegetables. For example, some studies indicate that benefits may be linked with specific plant species, such as carrots for coronary heart disease [28] or cruciferous vegetables for colon cancer [29], while specific processing methods such as tinning of fruit may even increase mortality [30]. A large study of the risk of pancreatic cancer using data from across Europe showed that associations with specific biomarkers of fruit and vegetable intake differed significantly among three regions [31] corresponding to differences in the types of vegetables and fruits consumed in each region [32]. The aggregation of all species and forms of vegetables and fruit will "dilute" those factors and reduce significances of associations with health outcomes.

High intakes of vegetables and fruit often relate to specific dietary patterns, in which cases they are correlated with certain other dietary components, for example, olive oil and wine (Mediterranean pattern) [21]. This further complicates efforts to study the contributions from the vegetables and fruits.

Overall, epidemiological studies are essential as background observations to generate hypotheses about which environmental factors including diet may affect the age-related diseases. Epidemiological data may also provide opportunities to test (disprove) mechanistic hypotheses that predict specific associations. However, it is important to keep in mind that no matter how significant, epidemiological associations between a biomarker of fruit and vegetable intake and morbidity (such as plasma vitamin C and type 2 diabetes [23]) does not provide any information about the role of this specific compound in terms of cause and effect. The epidemiological association would be similar if any other compound in the fruits and vegetables caused the risk reduction, or even if the consumption of the fruits and vegetables replaced another component in the diet that increases the risk of contracting diabetes.

Intervention Studies

It is not realistically feasible to carry out randomized placebo-controlled studies in humans of the effects of high versus low intake of fruits and vegetables during the multiyear timescales relevant for testing effects on age-related diseases. For ethical and practical reasons this would be very difficult to do in humans, and even if it could be done, such a trial would probably not be cost-effective for what information it would provide.

In contrast it is possible to carry out shorter-term randomized interventions (weeks or months) that focus on assessment of changes in biomarkers of effect and are not fully placebo-controlled. A few intervention trials assessed the effects of controlling the intake of a mixture of vegetables and fruits, within ranges directly comparable to the variations assessed in epidemiological surveys. To the extent that those studies show an effect, they tend to support the results from prospective epidemiology. For example, an intervention to increase fruit and vegetable intake from 2 to 5 portions per day for 16 weeks achieved a marginally significant improvement in grip strength [33] and significantly increased response to vaccinations [34]. In another trial mildly hypertensive participants consumed 1, 3, or 6 portions per day for 12 weeks, resulting in dose-dependent significantly improved blood flow [35]. Both trials showed reductions in biomarkers of inflammation [36], as well as changes in plasma biomarkers of intake confirming the compliance. In contrast several studies have shown that interventions aiming only to increase intake of fruits and vegetables do not reduce body weight [37]. However, the actual achieved changes in diets tend to be smaller than intended when study participants are provided with foods to consume in addition to their habitual diet, since they often compensate by reducing the intake of nonintervention vegetables and fruit [34]. So also for interventions the effects should be considered minimum estimates.

As observed in the epidemiological studies, all fruits and vegetables are probably not equal, and quite a lot of intervention studies have focused on one plant species or even a specific product. A study with 60 volunteers consuming 85 g watercress per day for 8 weeks showed significantly improved resistance to DNA damage in lymphocytes [38], indicating a reduced risk of cancer; consumption of 3 kiwifruits per day for 8 weeks reduced blood pressure significantly in mildly hypertensive participants [39], while 2 kiwifruits per day for 4 weeks had no effect on blood pressure [40] but did improve plasma cholesterol composition as a biomarker of cardiovascular disease risk [41]. Other fruits that have been tested in similar studies include strawberries [42], where 500 g per day for 30 days improved blood lipid profiles and other biomarkers of cardiovascular disease risk; similar effects as well as improvement in platelet function were obtained from a combination of different berries, 2 portions per day for 8 weeks [43].

Animal studies offer potential to overcome many of the difficulties of human studies, in particular for age-related diseases due to their short lifespans, and the possibility to access tissues such as brain and heart as part of the experiment.

For example, supplementation with strawberry or spinach, blueberries, or grapes have shown improved cognitive function in aging rodents [44–47]. Glycemic control was improved in diabetic rodents fed *Satsuma mandarin* or grape [48,49], and grape powder enhanced cardiovascular function in salt-stressed rats [50]. A study with *Drosophila* fruit flies showed that addition

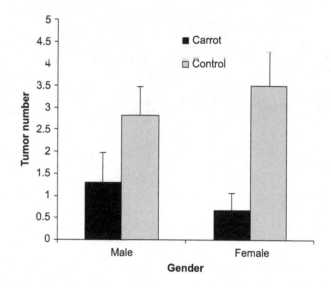

FIGURE 50.2 Mean tumor number at 12 weeks in APC^{Min} mice fed control diet ($N = 28$) or carrot-supplemented diet ($N = 19$). Error bars represent Standard Error of the Mean. *Source: from Ref. [53].*

of broccoli extract to the feed provided a dose-dependent enhancement of survival when the flies were exposed to toxins [51], and an apple extract increased lifespan in *Caenorhabditis elegans* nematode worms [52]. Saleh et al. found that addition of freeze-dried carrot to the diet of mice with the APC^{Min} mutation, which causes the development of multiple spontaneous intestinal tumors, reduced the number of visible (macroscopic) tumors substantially, indicating a protective effect in relation to cancer [53]—mice do eat carrots, at least in moderation, so the result of this experiment is more likely to be relevant for humans (Fig. 50.2).

Still, animal studies suffer from their own suite of methodological problems, in particular for studies of entire foods rather than single compounds.

This includes obvious problems related to the differences among species in digestive processes and relevance of vegetables and fruit as components of a normal diet. However, in some cases results can be completely skewed by issues that are much more difficult to recognize. For example, a 1998 study using the Min mouse model tested the effect of a vegetable–fruit mixture using heat-treated pellets, and found that tumor numbers significantly increased rather than the expected reduction [54]. This was a conundrum until 2012, when another study demonstrated the generation of carcinogens, such as furan, from vegetables when exposed to high-temperature processing methods at low humidity, exactly the condition used when food materials are incorporated into pellets to use for rodent studies [55].

CONSTITUENTS OF VEGETABLES AND FRUIT THAT MAY AFFECT AGE-RELATED MORBIDITY

Almost any dietary component for which fruits and vegetables constitute the primary dietary source has been mentioned as a potential cause of the health-promoting effects of consumption of vegetable and fruit, and for some of these the corresponding hypotheses have been tested. However, in most cases those experiments have disproved the hypothesis, so this is still an area of active research where new discoveries are likely to change the state of the art within a few years.

Constituents with potential effects on human health can be grouped according to their presumed roles in humans.

Nutrients Serving Key Functions in Human Metabolism

This group includes potassium, folate, vitamins C and E, and vitamin A obtained from digestion of beta-carotene and other provitamin A carotenoids. Lutein and zeaxanthin are also in this category, due to their role in vision [56].

Compounds like these are commercially available and affordable in substantial quantities and considered safe to consume in relevant quantities. So it is relatively easy to carry out randomized placebo-controlled intervention studies to investigate the effect of increasing the intake of each of these nutrients, and therefore plenty of data are available. However, the results are generally inconclusive, in particular when tested in the general population (as general use supplements) rather than to correct specific malnutrition problems in selected populations. While some nutrients can be shown to exert short-term effects under specific circumstances, the clear conclusion is that no single plant-derived nutrient or combination of nutrients consistently benefit age-related chronic disease to an extent that could explain the benefits of a comparable consumption of vegetables and fruit [57,58]. For example, supplementation with folate and other B-vitamins can reduce hyperhomocysteinemia to a similar extent as a corresponding intake of vegetables and fruit, but it has no detectable benefit on the chronic diseases usually associated with hyperhomocysteinemia such as CVD [59].

Antioxidant Effects of Plant-Derived Compounds After Consumption

For many of these compounds (vitamins C and E, lutein, and zeaxanthin), their key role as nutrients in the human body is as antioxidants: these compounds are critical members of a network of substances, which serve to keep the oxidative stress within a healthy range and prevent excessive formation or persistence of free radicals which can damage cells and other structures through their exceptionally high chemical reactivity.

When the substantial (unadjusted) associations between lower risk of age-related diseases such as cancer and higher plasma concentrations of carotenoids, in particular beta-carotene [60] were discovered, this was then combined with the "free radical theory of aging," first published in 1956 [61]. This resulted in a hypothesis that could be described as "dietary antioxidants as antiaging drugs" [62]. Numerous research groups initiated studies to test this presumed beneficial effect, including large-scale randomized placebo-controlled trials in populations at high risk of cancer and cardiovascular disease, ensuring sufficient power to reliably assess the expected reductions of 20–30% [63,64]. However, to everyone's surprise and horror, the supplementation caused an increase in cancer risk rather than the expected reduction [65]. The researchers involved strongly discouraged the use of beta-carotene supplements [66], and the European Food Safety Authority banned the use of health claims based on the antioxidant hypothesis [67]. However, despite the clear evidence and strongly worded advice from key researchers in the field, both the hypothesis and the supplements still seem to be popular, in particular the notion that reactive oxygen species (ROS) are detrimental and that keeping the levels as low as possible at all times will improve health in general. Many of the researchers studying the effects of fruits and vegetables on age-related chronic diseases still appear to use this authoritatively disproven hypothesis as the basis for interpretation of their results [31,68]. This topic has even been selected as a prime example of how claims continue to be made in the literature even after a hypothesis has been definitively disproved [69]. Recently other more balanced hypotheses are being published [70], which focus on the need to understand and optimize the role of oxidative stress rather than single-mindedly attempt to eliminate it.

Whether or not the antioxidative effect of plant constituents eventually turns out to have any relevance for human health or not, this chronicle of hype and despair has clearly put off many promising scientists from working in this field. It is probably a major reason why since 1996 research on antiaging substances in general and cancer chemoprevention in particular has been very limited, providing a possible explanation for the relative lack of progress in this field [71].

More detailed analyses of the nutritional roles of these fruit- and vegetable-derived nutrients are provided in the corresponding previous chapters.

Nonnutrient Components Implicated in Digestive Tract Function or Bioavailability of Nutrients

This group comprises components usually considered beneficial: fiber, including prebiotic oligo- and polysaccharides; and components traditionally considered as having negative effects on human health, but where recent studies indicate possible benefits, at least in populations where the diet tends to provide more nutrients than optimal for health: phytic acid and nitrate.

Several of the presumed beneficial components have been tested in randomized placebo-controlled intervention trials, but with very modest outcomes compared with the effects corresponding to equivalent intakes of vegetables and fruit. For example, relatively long-term trials of insoluble fibers [72] or resistant starch [73] showed no effect on colorectal cancer. While the intake of certain seed-derived fibers such as beta-glucans can reduce plasma cholesterol content, this may still be a transient effect reflecting adaptation to a high-fiber diet, rather than indicating an actual reduction in risk of cardiovascular disease [74]. The association is stronger with cardiovascular disease [75], in accordance with improvement of blood lipid profiles and the primary mechanism of the benefits [76]. However, since most dietary fiber is derived from cereals and other seeds rather than from vegetables and fruit [77], even strong evidence for protection against age-related chronic diseases would not explain the benefits of vegetables and fruit in this context.

Antinutrients, such as phytate, are best known for reducing the bioavailability of iron and other minerals, and due to this have detrimental effects on the health of malnourished or at-risk populations [78]. However, in developed countries and other well-nourished populations, where mineral intake generally is higher than the nutritional requirement, phytate has been clearly shown to reduce the morbidity of chronic diseases in humans as well as in animal models [79]. On the other hand, as for fiber, cereals are the dominant source of dietary phytate, with vegetables and fruit providing only around 30% of typical intake of phytate in the UK [80].

Dietary nitrate has recently received quite a lot of attention, with several studies indicating a range of health benefits, while other studies are interpreted as showing harmful effects, as recently extensively reviewed by Habermeyer et al. [81]. Nitrate from plants are converted into nitrite and then NO, which has been shown to reduce blood pressure and improve vascular function in young healthy volunteers, for example, after consumption of beetroot juice [82]. In contrast, studies of older people, many of whom are prehypertensive or hypertensive, tend to show smaller or no effects of dietary nitrate interventions [83]. Nitrate/nitrite may alternatively be converted into carcinogenic nitrosamines, in particular in the absence of vitamin C, which inhibits this reaction [81]. Certain vegetables, particularly leafy salad vegetables, but not fruit, are the dominant source of dietary nitrate, so any beneficial or harmful effects of nitrate would modulate the effects of overall vegetable intake on human health. Some of the inconsistencies are caused by confounding of dietary nitrate from three main sources: vegetables (containing high concentrations of nitrate together with high concentration of vitamin C), cured foods such as meat and preserved vegetables (containing nitrite as well as nitrate, no or little vitamin C), or water (containing nitrate in low concentrations, correlated with contaminations with human or animal waste). This may be the reason why studies in different populations with different ratios of these sources tend to show different correlations between nitrate intake and morbidity [84]. Nitrate is actively excreted into the salivary gland, providing a minimal nitrite concentration in the stomach even if the diet is free from nitrate (since 40–100% of plasma nitrate is generated endogenously from arginine), indicating that the presence of nitrate in saliva directly benefits health, probably by enhancing the stomach acid's ability to kill bacteria [85]. Overall, the evidence is still insufficient to conclude to what extent nitrate may be responsible for some of the beneficial effects of vegetable intake, and more studies are clearly required [81].

Phytochemicals Without Known Direct Nutritional Requirement

Most biologically active compounds in fruits and vegetables (phytochemicals) are categorized as secondary metabolites. These are compounds with no clear role in primary metabolism, which means that none of these compounds are specifically required for human development or health. However, each phytochemical may have other effects on human consumers, which could be beneficial or harmful or, probably most often, dose-dependent (beneficial in low doses, harmful if too much is consumed).

A well-known example of a typical dose-dependent phytochemical is caffeine. While the sources of caffeine (coffee, tea, cocoa, and a few other plants) are not used in quite the same way as typical fruits or vegetables, caffeine's properties in other respects are indeed typical of many different phytochemicals found in these foods: a strong bitter taste, which helps us to avoid too high doses; inhibitory effect on pests and diseases threatening the plants it occurs in; the occurrence of several similar compounds (analogs) with similar effects on humans (theophylline and theobromine in cocoa and tea); relatively rapid excretion from the human body via the urine, eliminating most of it within hours of its consumption.

Until the end of the twentieth century, the dominant perception of phytochemicals in general was that they should be considered "contaminants," all potentially harmful and mostly without any benefits for humans. So at that time the main purpose of studying phytochemicals was to predict and prevent intoxications by banning or replacing foods with too high contents. For example, a project to produce an early overview of phytochemicals in European food plants was called "Nettox," and the catalogue entitled "List of Food Plant Toxicants" [86].

Flavonoids

However, in 1993 a key paper was published, where dietary intake of flavonols, a subgroup of flavonoids, which is a type of polyphenols, was linked with reduced risk of cardiovascular disease and mortality [87]. This was hypothesized to be caused by antioxidant effects, even though this type of compound was known to not be a vitamin or provitamin. In the paper it was emphasized that the correlation with flavonol intake was stronger than correlations with the intakes of each individual food that were major sources of flavonols in this cohort: tea, onions, and apples.

This happened at the height of the beta-carotene antioxidant hype, and spurred a drive to measure and ideally increase the "antioxidant power" of fruits, vegetables, and foods derived from them, such as wine [88]. The publicity about these two types of antioxidants mutually reinforced each other, creating an impression among both scientists and nonscientists that modulation of antioxidant capacity was an actual biological effect rather than an untested hypothesis. However, in contrast to beta-carotene, flavonoids are much more complicated to synthesize or isolate, so it was and is difficult to obtain sufficient amounts of flavonols or related compounds to be able to carry out full-scale dietary intervention studies. After the negative results about beta-carotene were published in 1996 [65,89], it became much more difficult to obtain ethical approval for large-scale supplementation studies. However, this did not eliminate the "antioxidant theory of aging," instead the flavonoids and similar polyphenols were written about and sold as supplements for many years, without having to face the burden of data contradicting the expectations. When actual mechanisms of genuine biological effects were eventually discovered, such as the pathways by which dietary phytochemicals induce enzymes that enhance their metabolism and excretion, the researchers chose to name it the "antioxidant responsive element" [90]. This resulted in phytochemicals with no antioxidant capacity in the chemical sense being described as antioxidants [91] because they activated the same genetic pathways as the real antioxidants. "Antioxidant" became such a popular buzzword that several of the scientists who originally developed the hypothesis have published strongly worded statements criticizing the indiscriminate and often uninformed use of the term [92,93].

A few studies do unequivocally demonstrate beneficial health effects of flavonoids, at least in animal models. For example, a rat study measuring the extent of damage after a simulated heart attack [94], compared diets made from different genotypes of corn chosen to be identical except for their color (yellow or black), reflecting the absence or presence of the purple pigments anthocyanins, another flavonoid subgroup. However, there is still a strong tendency to claim that antioxidant effects of flavonoid phytochemicals are responsible for any benefit of any flavonoid-containing food. For example, in a study demonstrating reduced tumor formation in a rat model of intestinal cancer after feeding with blackcurrant extract [95], the discussion focused on the anthocyanin content of the food [96], even though a previous study on APC^{Min} mice fed white currants (same species of fruit, but a variety with no anthocyanins) had shown similar effect as the anthocyanin-containing fruit [97].

Flavonoids are only one of many types of polyphenols. Together with phenolic acids, stilbenes, quinones, gallotannins, and others, they comprise a wide range of groups and subgroups with many different types of structures (see Chapter "Mechanisms and Effects of Polyphenols in Aging").

Some of these subgroups are known to have specific bioactivities not shared with all phenolics or even all flavonoids, and which may help to explain some of the effects of fruits and vegetables on age-related diseases. For example, the polymeric condensed tannins, also known as proanthocyanidins, bind to specific proteins in saliva, leading to the perception of astringency [98], which is an important aspect of the taste of wine, green tea, cranberries, and other fruits and drinks containing these compounds. High content of condensed tannins in animal feed is well known to reduce their growth rate [99], and is quite possible that these salivary proteins arose during evolution of humans as a protective mechanism to counteract harmful effects of condensed tannins in fruits and vegetables. In populations where obesity is a major risk factor for chronic diseases, such growth-inhibiting effects may indirectly protect health by reducing the prevalence of obesity, and explain why adjustment for BMI could weaken the associations between fruit and vegetable intake and, for example, cancer incidence, as previously mentioned.

Nonvitamin A Carotenoids

Fruits and vegetables contain a large number of yellow, orange, and red compounds, which are fat-soluble carotenoids. This group of compounds are used as pigments by plants, they are very stable inside live plant tissue and are biosynthesized via the terpene pathway, which is very efficient in terms of energy and nutrients. While a few types of carotenoids can be metabolized to retinol (vitamin A) in the human body, most of them have no known nutritional function for us. If vegetables or fruit are consumed together with fat, or even fried in it, the carotenoids are absorbed with the food and accumulate in fat tissues, which take on a yellow color—this can also be observed with animals that consume grass or other fresh plant foods (Fig. 50.3).

Due to the epidemiological associations as described above, a lot of interest has focused on some of these carotenoids. Lycopene is the major pigment in tomatoes and some other red fruits like watermelon, and several studies have found significant correlations between high intake of lycopene and low risk of prostate cancer [101,102]. One animal study aimed to assess what proportion of the effect of tomato consumption is due to lycopene [103]. So vitamin E-deficient rats were fed diets supplemented with either red tomatoes, isolated lycopene, or yellow tomatoes (without lycopene), and oxidative stress measured in their heart tissue. As shown in Fig. 50.4, tomatoes had substantial and significant effect on the stress level, irrespective of color, while the lycopene-fed rats did not differ from the controls, despite just as high levels of lycopene in the plasma as in those fed red tomatoes.

The author of the present chapter is not aware of any appropriately powered study of lycopene, which unequivocally demonstrates that any relevant health effect of lycopene consumption could not be primarily caused by some other at present unknown constituent of tomatoes (or other lycopene-containing plant food).

Diet Patterns

In recent years, the recurrent disappointments in the search for "cause and effect" of the benefits of fruit and vegetable intake have forced researchers of health benefits of fruit and vegetable intake to take a further step away from the historic focus on antioxidants and other nutrients. This is generally developing in one of two directions: either focusing on the food as part of a diet pattern rather than as a source of specific beneficial compounds, or searching for bioactive compounds that are neither nutrients nor antioxidants, which usually means natural pesticides.

Regarding the diet pattern concept [104], a common argument for it is that since intervention studies have failed to confirm the benefits of individual nutrients, any effect of diet must either be due to synergies among different foods, or that intake of a food (eg, vegetables) by simple substitution reduces the intake of another, harmful food (eg, cured meat), resulting in a relative benefit which is really just a reduction in a risk. The diet pattern concept allows the scientists to retain the classical division of foods and their components into "essential nutrients" and "harmful contaminants." It provides useful tools to link epidemiological research with public health interventions, such as the use of diets like the "Mediterranean diet" as an educational tool for people who want to improve their diet [105]. All commonly used diet pattern systems give high scores to diets with a high intake of vegetables and fruits [106], and all show significant correlations with reduced mortality [106]. In other words, they confirm the beneficial effect of consumption of fruits and vegetables on age-related disease, without adding much new information in relation to the present chapter.

Natural Pesticides

The other direction differs from the previously described approaches by not presupposing any particular nutritional role of any particular compounds, but simply assessing the compounds found in a plant with a particular health benefit to try to find a match between the observed properties of the food and the properties of compounds characteristic for this plant. This approach has scientifically been very successful with *Brassica* vegetables [107], where a large body of research has since confirmed the beneficial effects of glucosinolates (or rather their degradation products isothiocyanates) on human health [108,109]. However, the glucosinolates also clearly illustrate the main drawback of this approach, which is that these chemically reactive compounds were made by the plant as natural pesticides, to protect it against diseases or pests (including herbivores). So while these compounds may well be responsible for the nutritional benefits of the fruits and vegetables, the same compounds also have various harmful effects if consumed in high amounts. For glucosinolates, these harmful effects include interference with iodine uptake, resulting in goiter in humans and livestock [110]. While it is recognized that such toxicant effects do not normally have any negative health consequences when each vegetable is consumed in moderate amounts and as part of a balanced diet [86], there are nevertheless serious concerns about toxicity, in particular in cases where a particular phytochemical becomes famous for some health benefit or other. If a bioactive compound is taken out of its normal food context and consumed in pure form and

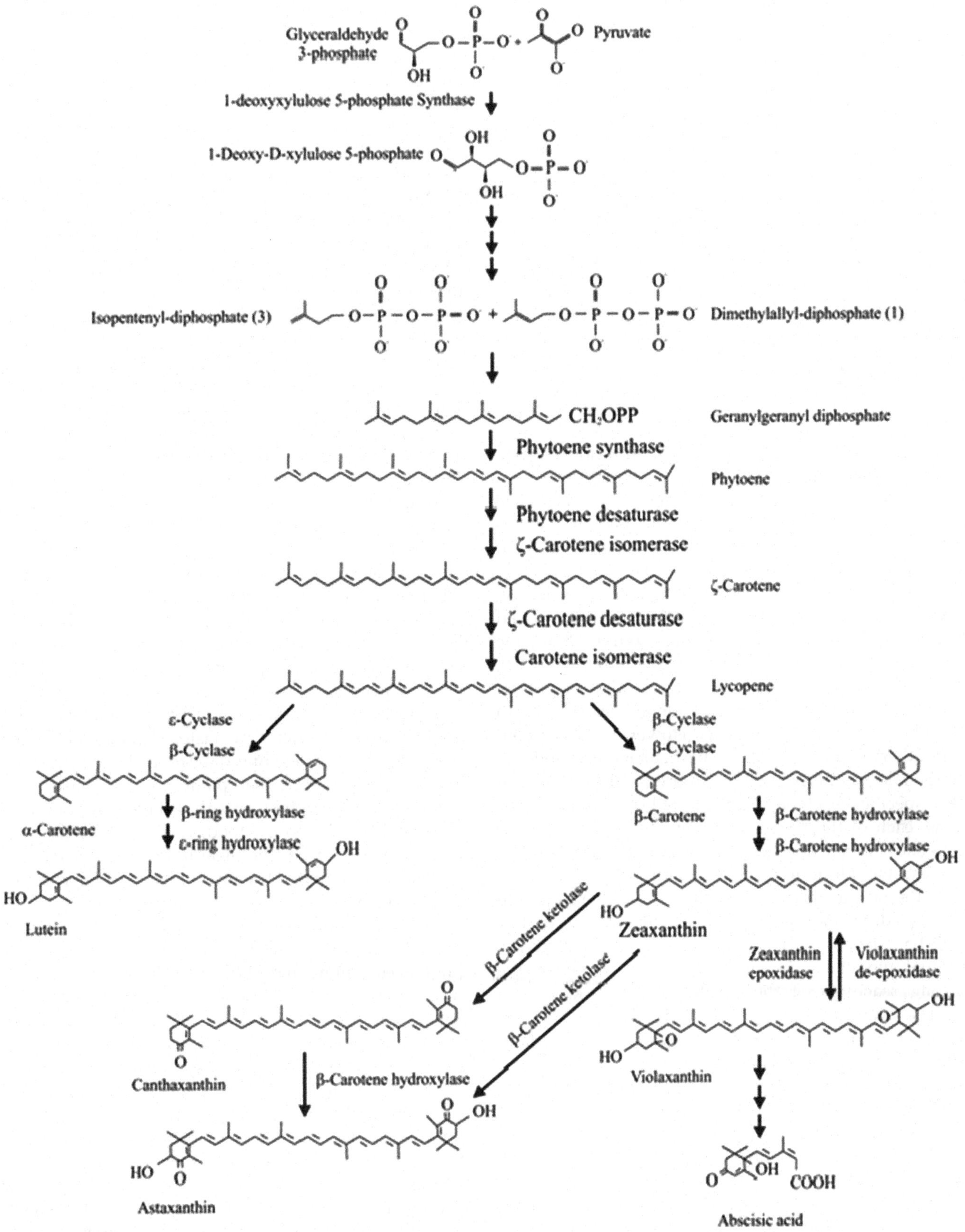

FIGURE 50.3 Carotenoid biosynthesis pathways and structures of common carotenoids found in vegetables and fruits. *Source: from Ref. [100] with permission.*

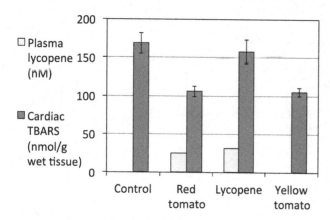

FIGURE 50.4 Effects of tomatoes and lycopene on oxidative stress in vitamin E-deficient rats, and concentration of lycopene in plasma. Error bars represent Standard Error of the Mean. *Source: drawn using data from Ref. [103].*

much higher doses than what occur in a normal diet, any risk of toxic effects is multiplied several times.

Another example is carrots, where epidemiological associations show that alpha-carotene tends to have as good or better association with cancer risk as beta-carotene [111,112]. While beta-carotene occurs widely in many different vegetables and fruits, including tomato, alpha-carotene is much less widespread among different plant species, and around 90% of the intake can come from only one species, the carrot [32]. It is clear from Fig. 50.3 that there is nothing special about the structure of alpha-carotene compared with the other common carotenoids. However, carrots contain other phytochemicals, in particular polyacetylenes of the falcarinol-type, that are (more or less) unique for this specific vegetable. When tested against leukemia cells, each of the three main polyacetylenes from carrots show much stronger anticancer effect in vitro than the carotenoids found in the same food [113]. Due to this, the polyacetylenes have been implicated as potential health-promoting phytochemicals [114], and they will be isolated and tested in the same APC^{Min} mouse animal model as shown in Fig. 50.2. However, the strong reactivity of the polyacetylenes is known to induce skin allergy among plant nursery workers occupationally exposed to ornamental plants containing high concentrations of these compounds [115], and large doses are neurotoxic to mice [116].

Table 50.1 and Fig. 50.5 show a (nonexhaustive) list of examples of such natural pesticide phytochemicals, which occur in common vegetables or fruit, have no known nutritional role (in the sense of a compound being essential for human nutrition), may have beneficial effects on the risk or the progression of one or more age-related diseases, and are known to exert toxic effects when consumed in excess.

Several other properties are common for many, if not all, of this type of phytochemicals. Firstly their effects tend to be surprisingly unspecific, other than simply enhancing lifespan. While the evidence published for each compound/vegetable tends to be stronger for a particular disease such as cancer, similar effects are often found for other noncommunicable diseases such as cardiovascular disease or Type 2 diabetes. Many of these compounds seem to affect the same types of pathways in the human body. As xenobiotics (compounds foreign to the human body) they tend to activate the corresponding pathways that metabolize and eliminate xenobiotics from the body [129,130]. Many of these phytochemicals have also been shown to interact with the nutrient-sensing protein mTOR, which then increases the ability of the human cell to protect itself from, for example, cancer-inducing conditions [121,131], although at the expense of growth and nutrient accumulation. If a wide range of bitter, slightly toxic phytochemicals all "nudge" the body toward a more defensive state of metabolism, similar to what can be induced by caloric restriction [132,133], then this can explain why they all seem to have some benefit for a range of age-related diseases. In this context it is notable that the hormesis of these compounds, that they have beneficial effects at certain low doses despite toxic effects at higher doses, seems to be a fundamental characteristic of their mode of action [114], and has now been recognized as such in relation to the effect on mTOR [134].

Another aspect is that they tend to be bitter or have another strong characteristic taste. This may be related to their function as feeding deterrents. However, it has had the consequence that plant breeders have done their best to reduce the content of these compounds to obtain milder taste in the food, and thus may inadvertently have reduced the health benefits of modern vegetables compared with older stronger-tasting varieties [135].

Another common property is that where it has been possible to assess, there appears to be very substantial differences among individuals in how they react to these phytochemicals. The same dose tends to have very different effects on different people. For glucosinolates, this has been shown to be related to specific genetic polymorphisms in enzymes involved in the uptake and metabolism of these phytochemicals [29].

As long as these phytochemicals are consumed only in specific foods, the real risk of serious harm is negligible, even though the highest doses occurring in food probably quite regularly exceed the "Lowest Observed Adverse Effect Level" (LOAEL) for the most susceptible individuals. This is not due to some magic "naturalness" making these natural pesticides any less toxic than other compounds with similar effects. However,

TABLE 50.1 Examples of Natural Pesticides, Bioactive Compounds from Vegetables and Fruits, Which May Contribute to Prevention of Chronic Age-Related Diseases

Type of compound	Main food source	Examples of disease(s) benefited	Toxic effects (in humans unless other specified)	Bitter/spicy taste
Glucosinolates (eg, progoitrin)	*Brassica* vegetables	Cancer [107,108]	Goiter, from food intake [117,118]	Yes
Linamarin (cyanogenic glucoside)	Cassava	Cancer [119]	Paralysis (konzo) from food intake [120]	Yes
Gingerols (eg, 6-gingerol)	Ginger	Cancer [121]	?	Yes
Capsaicin	Chili pepper	Idiopathic pain [122]	Breathing difficulties, from aerosol [123]	Yes
Glycoalkaloids (eg, solanine)	Potato, tomato	Cardio-vascular diseases, cancer, bacterial infection [124]	Diarrhea, from food intake [124]	Yes
Polyacetylenes (eg, falcarinol)	Carrots	Cancer [113,125,126]	Skin allergy, from external exposure [115]	Yes
Organosulfur compounds (eg, S-allyl-cysteine)	Garlic, onion	Cancer [127]	Liver injury (in animals), from supplement intake [127]	Yes

These compounds have no known role as essential nutrients, however they are all known or suspected to contribute to reduction of morbidity from age-related chronic diseases as well as having undesirable (toxic) effects when consumed in excess. The author's hypothesis is that these effects are exerted primarily through general pathways activated by a range of moderately toxic compounds, rather than through specific actions of each compound. Which would imply that moderate toxicity could be used as a selection criterion when searching for natural compounds responsible for the health benefits of consumption of vegetables and fruits [114].

humans (and other animals) possess a mechanism specifically adapted to protect us against such toxins, this mechanism is called "Conditioned Taste Aversion" [136]. The general principle is that an encounter with a food that is followed by nausea or similar gastrointestinal symptoms immediately induces a long-lasting aversion to the taste of this food [136]. So as long as a compound has a strong, characteristic taste, and is consumed in a food or drink, not a pill or a capsule, the Conditioned Taste Aversion will control the intake to ensure that an individual's personal susceptibility is not repeatedly exceeded. A very few plant phytochemicals are so toxic that one portion may be harmful or even fatal (eg, high-linamarin cassava or high-glycoalkaloid potatoes), such products are not allowed to be sold to the general public [86].

OPPORTUNITIES TO IMPROVE INTAKE AND EFFECT

Once it becomes well established which components of the vegetables and fruit are responsible for the beneficial effects of their intake on age-related chronic diseases, then it will become possible to select certain products with particular benefits. On the one hand this relates to the choice of species and types of fruits and vegetables, which almost certainly have very different effects on human health and should not all be lumped together in too broad categories. Examples of compounds with very variable contents include nitrate, carotenoids, phenolic compounds, polyacetylenes, etc. On the other hand, since as mentioned above the mechanisms of different bioactive compounds may be more similar than what has until now been recognized, it is likely that it will be relatively easy to substitute one type of vegetable for another to suit individual preferences, no need for all of us to eat only the one vegetable that is better than all the others, since it probably does not exist.

The other important aspect where the benefits of vegetable and fruit intake on age-related diseases can be enhanced is the variety of processing and postharvest treatments that many vegetables and fruit undergo before consumption. In cases where the active compound is known or suspected, such as for the *Brassica* vegetables, it is already known that much or most of the phytochemicals are lost during processing [137], in particular for highly processed foods that undergo several processing steps. However, often this is not an inevitable consequence of cooking, very small changes in the process can make a very large difference. For example, if a carrot is diced, blanched by immersion in hot water, frozen, and then reheated by immersion in hot water, then it will lose not only most of each bioactive water-soluble phytochemical and vitamin, but also most of the compounds responsible for its taste. If a similar carrot is boiled or steamed whole, then diced, frozen, and reheated using microwave or steam, the concentration of nutrients, phytochemicals and taste compounds will easily be twice as high as with the standard process. So by rearranging

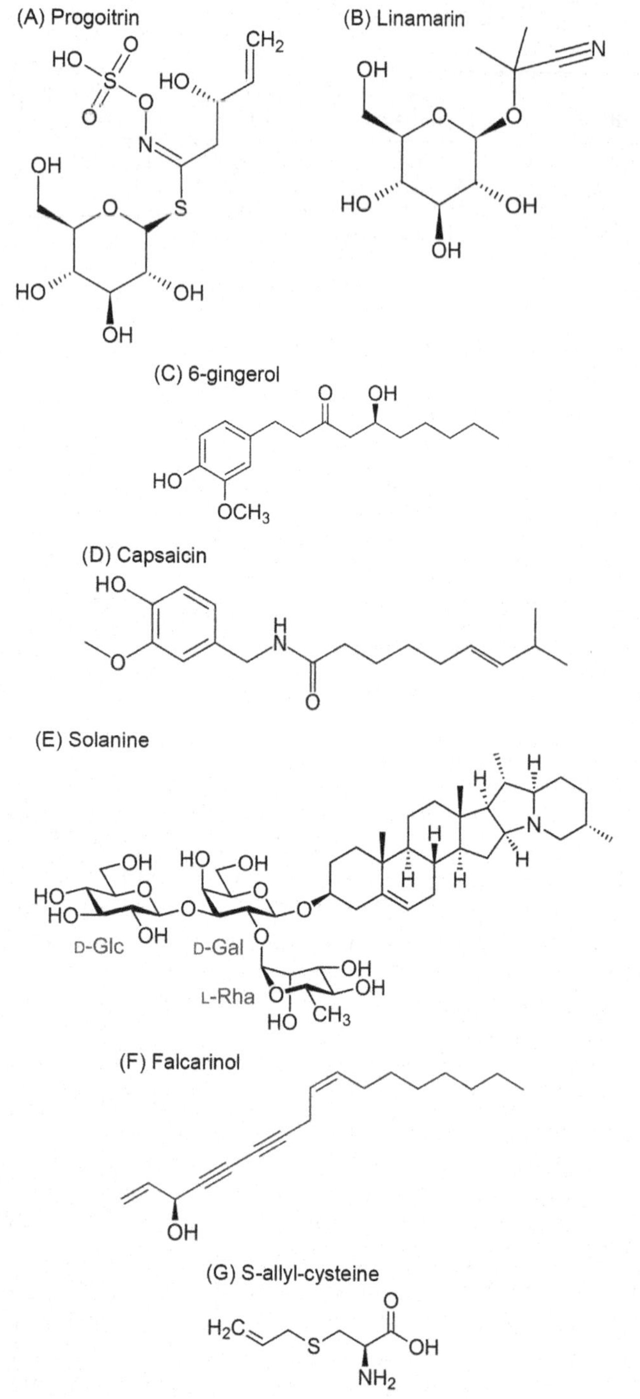

FIGURE 50.5 Structures of the bioactive compounds described in Table 5. (A) Progoitrin; [128] (B) Linamarin; (C) 6-gingerol; (D) Capsaicin; (E) Solanine; (F) Falcarinol; (G) S-allyl-cysteine. Figures in public domain unless otherwise indicated.

the sequence of processes taking into account the potential diffusion gradients at each step, both nutritional and sensory quality can be improved.

At present the public health messages in this area focus on persuading people to consume more fruits and vegetables, without any particular guidance on which ones may be better than others. If the studies mentioned above are fully completed, our understanding of how relevant components can be retained or even enhanced in the fruits and vegetables would be substantially improved. Implementation of the results into practical food production could lead to very substantial increases in intake of these components, even without necessarily increasing the overall intake of this food group.

SUMMARY

- Consumption of vegetables and fruits is clearly associated with a reduced risk of dying from cancer, cardiovascular diseases and diabetes.
- The 20% of people with highest intake of fruit and vegetables typically experience 10% lower risk of premature death. This is highly significant, although much less than some other factors such as not smoking.
- Not all vegetables and fruits are equal, some types appear to benefit human health more than others. Most likely this means that the substances that are responsible for the health benefits occur in higher concentrations in the most beneficial types of produce.
- If we can identify the beneficial substances, it would be feasible to choose the most 'concentrated' fruits and vegetables. This would substantially increase the content of these substances in the diet, and could thus provide major health improvements without radically changing the diet.
- An old theory (from 1956) suggested that the health benefits of vegetables and fruits could be primarily caused by antioxidant effects, via reduction of the level of free radicals in the body, since this was assumed to protect tissues against damage and thus reduce disease risk. We now know that this is incorrect, since excess free radicals in human tissues are not a general cause of ageing or disease (although can be a consequence of these). Numerous studies have shown that antioxidant effects of food have no benefits for human health, unless the diet is seriously deficient in vitamin C and vitamin E.
- In contrast, recent research indicates that a variety of moderately toxic phytochemicals (natural pesticides) may inhibit pathways controlling the ageing process, in a similar way as "caloric restriction," which is known to extend lifespan.
- This knowledge is helping to define a way forward for this research, to design experiments to identify the real beneficial substances among all the different compounds present in the plant foods.
- Some of the beneficial phytochemicals have antioxidant effects, but most of them do not. This demonstrates how other properties of these same

compounds, such as slowing down the growth and ageing at cell level, are more important for the human health benefits.

References

[1] WHO. Increasing fruit and vegetable consumption to reduce the risk of noncommunicable diseases. <http://www.who.int/elena/titles/bbc/fruit_vegetables_ncds/en/>; 2015 [accessed 29.05.15].

[2] Williams WR. Remarks on the mortality from cancer. Lancet 1898;152(3912):481–2.

[3] He FJ, Nowson CA, Lucas M, MacGregor GA. Increased consumption of fruit and vegetables is related to a reduced risk of coronary heart disease: meta-analysis of cohort studies. J Hum Hypertens 2007;21(9):717–28.

[4] Dauchet L, Amouyel P, Dallongeville J. Fruits, vegetables and coronary heart disease. Nat Rev Cardiol 2009;6(9):599–608.

[5] Wu Y, Zhang D, Jiang X, Jiang W. Fruit and vegetable consumption and risk of type 2 diabetes mellitus: a dose-response meta-analysis of prospective cohort studies. Nutr Metab Cardiovasc Dis 2015;25(2):140–7.

[6] Boffetta P, Couto E, Wichmann J, et al. Fruit and vegetable intake and overall cancer risk in the European prospective investigation into cancer and nutrition (EPIC). J Natl Cancer Inst 2010;102(8):529–37.

[7] Dauchet L, Ferrieres J, Arveiler D, et al. Frequency of fruit and vegetable consumption and coronary heart disease in France and Northern Ireland: the PRIME study. Br J Nutr 2004;92(6):963–72.

[8] Singh R, Fedacko J, Pella D, et al. Prevalence and risk factors for prehypertension and hypertension in five Indian cities. Acta Cardiol 2011;66(1):29–37.

[9] Nagata C, Nakamura K, Wada K, et al. Association of dietary fat, vegetables and antioxidant micronutrients with skin ageing in Japanese women. Br J Nutr 2010;103(10):1493–8.

[10] Hosking DE, Nettelbeck T, Wilson C, Danthiir V. Retrospective lifetime dietary patterns predict cognitive performance in community-dwelling older Australians. Br J Nutr 2014;112(2):228–37.

[11] Yoshihara A, Watanabe R, Hanada N, Miyazaki H. A longitudinal study of the relationship between diet intake and dental caries and periodontal disease in elderly Japanese subjects. Gerodontology 2009;26(2):130–6.

[12] Marcon F, Siniscalchi E, Crebelli R, et al. Diet-related telomere shortening and chromosome stability. Mutagenesis 2012;27(1):49–57.

[13] Bellavia A, Larsson SC, Bottai M, Wolk A, Orsini N. Fruit and vegetable consumption and all-cause mortality: a dose-response analysis. Am J Clin Nutr 2013;98(2):454–9.

[14] Hung HC, Joshipura KJ, Jiang R, et al. Fruit and vegetable intake and risk of major chronic disease. J Natl Cancer Inst 2004;96(21):1577–84.

[15] Jatrana S, Blakely T. Socio-economic inequalities in mortality persist into old age in New Zealand: study of all 65 years plus, 2001–04. Ageing Society 2014;34(6):911–29.

[16] Key T. Fruit and vegetables and cancer risk. Br J Cancer 2011;104(1):6–11.

[17] Mathe N, Van der Meer L, Agborsangaya CB, et al. Prompted awareness and use of Eating Well with Canada's Food Guide: a population-based study. J Hum Nutr Diet 2015;28(1):64–71.

[18] Daly M, Boyce C, Wood A. A social rank explanation of how money influences health. Health Psychol 2015;34(3):222–30.

[19] Henriquez-Sanchez P, Sanchez-Villegas A, Doreste-Alonso J, Ortiz-Andrellucchi A, Pfrimer K, Serra-Majem L. Dietary assessment methods for micronutrient intake: a systematic review on vitamins. Br J Nutr 2009;102:S10–37.

[20] Ortiz-Andrellucchi A, Sanchez-Villegas A, Doreste-Alonso J, de Vries J, de Groot L, Serra-Majem L. Dietary assessment methods for micronutrient intake in elderly people: a systematic review. Br J Nutr 2009;102:S118–49.

[21] Kiefte-de Jong JC, Mathers JC, Franco OH. Nutrition and healthy ageing: the key ingredients. Proc Nutr Soc 2014;73(2):249–59.

[22] Michels KB, Welch AA, Luben R, Bingham SA, Day NE. Measurement of fruit and vegetable consumption with diet questionnaires and implications for analyses and interpretation. Am J Epidemiol 2005;161(10):987–94.

[23] Harding A-H, Wareham NJ, Bingham SA, et al. Plasma vitamin C level, fruit and vegetable consumption, and the risk of new-onset type 2 diabetes mellitus – the European Prospective Investigation of Cancer-Norfolk prospective study. Arch Intern Med 2008;168(14):1493–9.

[24] Akbaraly T, Fontbonne A, Favier A, Berr C. Plasma carotenoids and onset of dysglycemia in an elderly population – results of the Epidemiology of Vascular Ageing study. Diabetes Care 2008;31(7):1355–9.

[25] Cooper AJM, Sharp SJ, Luben RN, Khaw KT, Wareham NJ, Forouhi NG. The association between a biomarker score for fruit and vegetable intake and incident type 2 diabetes: the EPIC-Norfolk study. Eur J Clin Nutr 2015;69(4):449–54.

[26] Lietz G, Oxley A, Leung W, Hesketh J. Single nucleotide polymorphisms upstream from the beta-carotene 15,15′-monoxygenase gene influence provitamin a conversion efficiency in female volunteers. J Nutr 2012;142(1):161S–5S.

[27] Block G, Shaikh N, Jensen CD, Volberg V, Holland N. Serum vitamin C and other biomarkers differ by genotype of phase 2 enzyme genes GSTM1 and GSTT1. Am J Clin Nutr 2011;94(3):929–37.

[28] Oude Griep LM, Verschuren WMM, Kromhout D, Ocke MC, Geleijnse JM. Colours of fruit and vegetables and 10-year incidence of CHD. Br J Nutr 2011;106(10):1562–9.

[29] Tse G, Eslick GD. Cruciferous vegetables and risk of colorectal Neoplasms: a systematic review and meta-analysis. Nutr Cancer 2014;66(1):128–39.

[30] Aasheim ET, Sharp SJ, Appleby PN, et al. Tinned fruit consumption and mortality in three prospective cohorts. Plos One 2015;10(2).

[31] Jeurnink SM, Ros MM, Leenders M, et al. Plasma carotenoids, vitamin C, retinol and tocopherols levels and pancreatic cancer risk within the European Prospective Investigation into Cancer and Nutrition: a nested case-control study plasma micronutrients and pancreatic cancer risk. Int J Cancer 2015;136(6):E665–76.

[32] O'Neill ME, Carroll Y, Corridan B, et al. A European carotenoid database to assess carotenoid intakes and its use in a five-country comparative study. Br J Nutr 2001;85(4):499–507.

[33] Neville CE, Young IS, Gilchrist SECM, et al. Effect of increased fruit and vegetable consumption on physical function and muscle strength in older adults. Age 2013;35(6):2409–22.

[34] Gibson A, Edgar J, Neville C, et al. Effect of fruit and vegetable consumption on immune function in older people: a randomized controlled trial. Am J Clin Nutr 2012;96(6):1429–36.

[35] McCall DO, McGartland CP, McKinley MC, et al. Dietary intake of fruits and vegetables improves microvascular function in hypertensive subjects in a dose-dependent manner. Circulation 2009;119(16):2153–60.

[36] Nadeem N, Woodside JV, Neville CE, et al. Serum amyloid A-related inflammation is lowered by increased fruit and vegetable intake, while high-sensitive C-reactive protein, IL-6 and E-selectin remain unresponsive. Br J Nutr 2014;112(7):1129–36.

[37] Kaiser KA, Brown AW, Brown MMB, Shikany JM, Mattes RD, Allison DB. Increased fruit and vegetable intake has no discernible effect on weight loss: a systematic review and meta-analysis. Am J Clin Nutr 2014;100(2):567–76.

[38] Gill CIR, Haldar S, Boyd LA, et al. Watercress supplementation in diet reduces lymphocyte DNA damage and alters blood antioxidant status in healthy adults. Am J Clin Nutr 2007;85(2):504–10.

[39] Svendsen M, Tonstad S, Heggen E, et al. The effect of kiwifruit consumption on blood pressure in subjects with moderately elevated blood pressure: a randomized, controlled study. Blood Press 2015;24(1):48–54.

[40] Gammon CS, Kruger R, Brown SJ, Conlon CA, von Hurst PR, Stonehouse W. Daily kiwifruit consumption did not improve blood pressure and markers of cardiovascular function in men with hypercholesterolemia. Nutr Res 2014;34(3):235–40.

[41] Gammon CS, Kruger R, Minihane AM, Conlon CA, von Hurst PR, Stonehouse W. Kiwifruit consumption favourably affects plasma lipids in a randomised controlled trial in hypercholesterolaemic men. Br J Nutr 2013;109(12):2208–18.

[42] Alvarez-Suarez JM, Giampieri F, Tulipani S, et al. One-month strawberry-rich anthocyanin supplementation ameliorates cardiovascular risk, oxidative stress markers and platelet activation in humans. J Nutr Biochem 2014;25(3):289–94.

[43] Erlund I, Koli R, Alfthan G, et al. Favorable effects of berry consumption on platelet function, blood pressure, and HDL cholesterol. Am J Clin Nutr 2008;87(2):323–31.

[44] Joseph JA, Shukitt-Hale B, Denisova NA, et al. Long-term dietary strawberry, spinach, or vitamin E supplementation retards the onset of age-related neuronal signal-transduction and cognitive behavioral deficits. J Neurosci 1998;18(19):8047–55.

[45] Shukitt-Hale B, Carey A, Simon L, Mark DA, Joseph JA. Effects of Concord grape juice on cognitive and motor deficits in aging. Nutrition 2006;22(3):295–302.

[46] Joseph JA, Shukitt-Hale B, Denisova NA, et al. Reversals of age-related declines in neuronal signal transduction, cognitive, and motor behavioral deficits with blueberry, spinach, or strawberry dietary supplementation. J Neurosci 1999;19(18):8114–21.

[47] Williams CM, El Mohsen MA, Vauzour D, et al. Blueberry-induced changes in spatial working memory correlate with changes in hippocampal CREB phosphorylation and brain-derived neurotrophic factor (BDNF) levels. Free Radic Biol Med 2008;45(3):295–305.

[48] Sugiura M, Ohshima M, Ogawa K, Yano M. Chronic administration of *Satsuma mandarin* fruit (*Citrus unshiu* Marc.) improves oxidative stress in streptozotocin-induced diabetic rat liver. Biol Pharm Bull 2006;29(3):588–91.

[49] Cheng DM, Pogrebnyak N, Kuhn P, Krueger CG, Johnson WD, Raskin I. Development and phytochemical characterization of high polyphenol red lettuce with anti-diabetic properties. Plos One 2014;9(3).

[50] Seymour EM, Singer AAM, Bennink MR, et al. Chronic intake of a phytochemical-enriched diet reduces cardiac fibrosis and diastolic dysfunction caused by prolonged salt-sensitive hypertension. J Gerontol Ser A Biol Sci Med Sci 2008;63(10):1034–42.

[51] Li Y, Chan H, Huang Y, Chen Z. Broccoli (*Brassica oleracea* var. botrytis L.) improves the survival and up-regulates endogenous antioxidant enzymes in *Drosophila melanogaster* challenged with reactive oxygen species. J Sci Food Agric 2008;88(3):499–506.

[52] Vayndorf EM, Lee SS, Liu RH. Whole apple extracts increase lifespan, healthspan and resistance to stress in *Caenorhabditis elegans*. Journal of Functional Foods 2013;5(3):1235–43.

[53] Saleh H, Garti H, Carroll M, Brandt K. Effect of carrot feeding to APC(Min) mouse on intestinal tumours. Proc Nutr Soc 2013;72(OCE4):E183.

[54] van Kranen H, van Iersel P, Rijnkels J, Beems D, Alink G, van Kreijl C. Effects of dietary fat and a vegetable-fruit mixture on the development of intestinal neoplasia in the Apc(Min) mouse. Carcinogenesis 1998;19(9):1597–601.

[55] Duan HY, Barringer SA. Changes in furan and other volatile compounds in sliced carrot during air-drying. J Food Process Preserv 2012;36(1):46–54.

[56] Johnson EJ. Role of lutein and zeaxanthin in visual and cognitive function throughout the lifespan. Nutr Rev 2014;72(9):605–12.

[57] Dolara P, Bigagli E, Collins A. Antioxidant vitamins and mineral supplementation, life span expansion and cancer incidence: a critical commentary. Eur J Nutr 2012;51(7):769–81.

[58] Mayne S, Ferrucci L, Cartmel B, Cousins R. Lessons learned from randomized clinical trials of micronutrient supplementation for cancer prevention. Annu Rev Nutr 2012;32:369–70.

[59] Lippi G, Plebani M. Hyperhomocysteinemia in health and disease: where we are now, and where do we go from here? Clin Chem Lab Med 2012;50(12):2075–80.

[60] Peto R, Doll R, Buckley JD, Sporn MB. Can dietary beta-carotene materially reduce human cancer rates? Nature 1981;290(5803):201–8.

[61] Harman D. Aging: a theory based on free radical and radiation chemistry. J Gerontol 1956;11(3):298–300.

[62] Machlin LJ, Bendich A. Free-radical tissue-damage – protective role of antioxidant nutrients. FASEB J 1987;1(6):441–5.

[63] ATBC Cancer Prevention Study Group T. The alpha-tocopherol, beta-carotene lung cancer prevention study: design, methods, participant characteristics, and compliance. Ann Epidemiol 1994;4(1):1–10.

[64] Omenn GS, Goodman G, Thornquist M, et al. The beta-carotene and retinol efficacy trial (CARET) for chemoprevention of lung cancer in high risk populations: smokers and asbestos-exposed workers. Cancer Res 1994;54(7 Suppl.):2038s–43s.

[65] Albanes D, Heinonen OP, Taylor PR, et al. Alpha-tocopherol and beta-carotene supplements and lung cancer incidence in the Alpha-Tocopherol, Beta-Carotene Cancer Prevention study: effects of base-line characteristics and study compliance. J Natl Cancer Inst 1996;88(21):1560–70.

[66] Omenn GS. Chemoprevention of lung cancer: the rise and demise of beta-carotene. Annu Rev Public Health 1998;19:73–99.

[67] EFSA Panel on Dietetic Products, Nutrition and Allergies (NDA). Guidance on the scientific requirements for health claims related to antioxidants, oxidative damage and cardiovascular health. EFSA J 2011;9(12):2474.

[68] Pantavos A, Ruiter R, Feskens EF, et al. Total dietary antioxidant capacity, individual antioxidant intake and breast cancer risk: the Rotterdam study. Int J Cancer 2015;136(9):2178–86.

[69] Tatsioni A, Bonitsis NG, Ioannidis JPA. Persistence of contradicted claims in the literature. J Am Med Assoc 2007;298(21):2517–26.

[70] Ristow M, Schmeisser S. Extending life span by increasing oxidative stress. Free Radic Biol Med 2011;51(2):327–36.

[71] Potter JD. The failure of cancer chemoprevention. Carcinogenesis 2014;35(5):974–82.

[72] Parkin D, Boyd L. Cancers attributable to dietary factors in the UK in 2010 III. Low consumption of fibre. Br J Cancer 2011;105:S27–30.

[73] Mathers JC, Movahedi M, Macrae F, et al. Long-term effect of resistant starch on cancer risk in carriers of hereditary colorectal cancer: an analysis from the CAPP2 randomised controlled trial. Lancet Oncol 2012;13(12):1242–9.

[74] Brownlee I. The physiological roles of dietary fibre. Food Hydrocolloids 2011;25(2):238–50.

REFERENCES

[75] Liu L, Wang S, Liu J. Fiber consumption and all-cause, cardiovascular, and cancer mortalities: a systematic review and meta-analysis of cohort studies. Mol Nutr Food Res 2015;59(1):139–46.

[76] Anderson JW. Dietary fiber prevents carbohydrate-induced hypertriglyceridemia. Curr Atheroscler Rep 2000;2(6):536–41.

[77] Bradbury K, Appleby P, Key T. Fruit, vegetable, and fiber intake in relation to cancer risk: findings from the European Prospective Investigation into Cancer and Nutrition (EPIC). Am J Clin Nutr 2014;100(1):394S–398SS.

[78] Joy EJM, Ander EL, Young SD, et al. Dietary mineral supplies in Africa. Physiol Plant 2014;151(3):208–29.

[79] Kumar V, Sinha AK, Makkar HPS, Becker K. Dietary roles of phytate and phytase in human nutrition: a review. Food Chem 2010;120(4):945–59.

[80] Amirabdollahian F, Ash R. An estimate of phytate intake and molar ratio of phytate to zinc in the diet of the people in the United Kingdom. Public Health Nutr 2010;13(9):1380–8.

[81] Habermeyer M, Roth A, Guth S, et al. Nitrate and nitrite in the diet: how to assess their benefit and risk for human health. Mol Nutr Food Res 2015;59(1):106–28.

[82] Kapil V, Weitzberg E, Lundberg JO, Ahluwalia A. Clinical evidence demonstrating the utility of inorganic nitrate in cardiovascular health. Nitric Oxide Biol Chem 2014;38:45–57.

[83] Siervo M, Lara J, Jajja A, et al. Ageing modifies the effects of beetroot juice supplementation on 24-hour blood pressure variability: an individual participant meta-analysis. Nitric Oxide 2015;47:97–105.

[84] Bottex B, Dorne J-L, Carlander D, et al. Risk-benefit health assessment of food. Food fortification and nitrate in vegetables. Trends Food Sci Tech 2008;19:S113–19.

[85] Lundberg JO, Weitzberg E, Cole JA, Benjamin N. Opinion – Nitrate, bacteria and human health. Nat Rev Microbiol 2004;2(7):593–602.

[86] Holm S, Alexander J, Andersson C, et al. NETTOX list of food plant toxicants. Søborg, Denmark: Danish Veterinary and Food Administration; 1998.

[87] Hertog MGL, Feskens EJM, Hollman PCH, Katan MB, Kromhout D. Dietary antioxidant flavonoids and risk of coronary heart-disease – the Zutphen Elderly study. Lancet 1993;342(8878):1007–11.

[88] Frankel EN, Kanner J, German JB, Parks E, Kinsella JE. Inhibition of oxidation of human low-density-lipoprotein by phenolic substances in red wine. Lancet 1993;341(8843):454–7.

[89] Omenn GS, Goodman GE, Thornquist MD, et al. Risk factors for lung cancer and for intervention effects in CARET, the beta-carotene and retinol efficacy trial. J Natl Cancer Inst 1996;88(21):1550–9.

[90] Xie T, Belinsky M, Xu YH, Jaiswal AK. Are-mediated and Tre-mediated regulation of gene-expression – response to xenobiotics and antioxidants. J Biol Chem 1995;270(12):6894–900.

[91] Enrique Guerrero-Beltran C, Calderon-Oliver M, Pedraza-Chaverri J, Irasema Chirino Y. Protective effect of sulforaphane against oxidative stress: recent advances. Exp Toxicol Pathol 2012;64(5):503–8.

[92] Frankel EN, German JB. Antioxidants in foods and health: problems and fallacies in the field. J Sci Food Agric 2006;86(13):1999–2001.

[93] Hollman PCH, Cassidy A, Comte B, et al. The biological relevance of direct antioxidant effects of polyphenols for cardiovascular health in humans is not established. J Nutr 2011;141(5):989S–1009S.

[94] Toufektsian M-C, de Lorgeril M, Nagy N, et al. Chronic dietary intake of plant-derived anthocyanins protects the rat heart against ischemia-reperfusion injury. J Nutr 2008;138(4):747–52.

[95] Bishayee A, Mbimba T, Thoppil RJ, et al. Anthocyanin-rich black currant (*Ribes nigrum* L.) extract affords chemoprevention against diethylnitrosamine-induced hepatocellular carcinogenesis in rats. J Nutr Biochem 2011;22(11):1035–46.

[96] Thoppil RJ, Bhatia D, Barnes KF, et al. Black currant anthocyanins abrogate oxidative stress through Nrf2-mediated antioxidant mechanisms in a rat model of hepatocellular carcinoma. Curr Cancer Drug Targets 2012;12(9):1244–57.

[97] Rajakangas J, Misikangas M, Paivarinta E, Mutanen M. Chemoprevention by white currant is mediated by the reduction of nuclear beta-catenin and NF-kappaB levels in Min mice adenomas. Eur J Nutr 2008;47(3):115–22.

[98] Brandao E, Soares S, Mateus N, de Freitas V. In vivo interactions between procyanidins and human saliva proteins: effect of repeated exposures to procyanidins solution. J Agric Food Chem 2014;62(39):9562–8.

[99] Elkin RG, Rogler JC, Sullivan TW. Comparative effects of dietary tannins in ducks, chicks, and rats. Poult Sci 1990;69(10):1685–93.

[100] Hannoufa A, Hossein Z. Carotenoid biosynthesis. <http://lipidlibrary.aocs.org/plantbio/carotenoids/index.htm>; 2015 [accessed 03.06.15].

[101] Zu K, Mucci L, Rosner BA, et al. Dietary lycopene, angiogenesis, and prostate cancer: a prospective study in the prostate-specific antigen era. J Natl Cancer Institute 2014;106(2).

[102] Chen J, Song Y, Zhang L. Lycopene/tomato consumption and the risk of prostate cancer: a systematic review and meta-analysis of prospective studies. J Nutr Sci Vitaminol (Tokyo) 2013;59(3):213–23.

[103] Gitenay D, Lyan B, Rambeau M, Mazur A, Rock E. Comparison of lycopene and tomato effects on biomarkers of oxidative stress in vitamin E deficient rats. Eur J Nutr 2007;46(8):468–75.

[104] Hu FB. Dietary pattern analysis: a new direction in nutritional epidemiology. Curr Opin Lipidol 2002;13(1):3–9.

[105] Lara J, McCrum L-A, Mathers JC. Association of Mediterranean diet and other health behaviours with barriers to healthy eating and perceived health among British adults of retirement age. Maturitas 2014;79(3):292–8.

[106] Schwingshackl L, Hoffmann G. Diet quality as assessed by the healthy eating index, the alternate healthy eating index, the dietary approaches to stop hypertension score, and health outcomes: a systematic review and meta-analysis of cohort studies. J Acad Nutr Diet 2015;115(5):780–U268.

[107] Verhoeven DTH, Verhagen H, Goldbohm RA, vandenBrandt PA, vanPoppel G. A review of mechanisms underlying anticarcinogenicity by brassica vegetables. Chem Biol Interact 1997;103(2):79–129.

[108] Fimognari C, Turrini E, Ferruzzi L, Lenzi M, Hrelia P. Natural isothiocyanates: genotoxic potential versus chemoprevention. Mutat Res Rev Mutat Res 2012;750(2):107–31.

[109] Latte KP, Appel K-E, Lampen A. Health benefits and possible risks of broccoli – an overview. Food Chem Toxicol 2011;49(12):3287–309.

[110] Stoewsand GS. Bioactive organosulfur phytochemicals in Brassica-oleracea vegetables – a review. Food Chem Toxicol 1995;33(6):537–43.

[111] Min K-b, Min J-y. Serum carotenoid levels and risk of lung cancer death in US adults. Cancer Sci 2014;105(6):736–43.

[112] Bates CJ, Hamer M, Mishra GD. Redox-modulatory vitamins and minerals that prospectively predict mortality in older British people: the National Diet and Nutrition Survey of people aged 65 years and over. Br J Nutr 2011;105(1):123–32.

[113] Zaini RG, Brandt K, Clench MR, Le Maitre CL. Effects of bioactive compounds from carrots (*Daucus carota* L.),

polyacetylenes, beta-carotene and lutein on human lymphoid leukaemia cells. Anticancer Agents Med Chem 2012;12(6):640−52.

[114] Brandt K, Christensen L, Hansen-Moller J, et al. Health promoting compounds in vegetables and fruits: a systematic approach for identifying plant components with impact on human health. Trends Food Sci Technol 2004;15(7−8):384−93.

[115] Hansen L, Boll PM. The polyacetylenic falcarinol as the major allergen in *Schefflera arboricola*. Phytochemistry 1986;25(2):529−30.

[116] Bernart MW, Cardellina JH, Balaschak MS, Alexander MR, Shoemaker RH, Boyd MR. Cytotoxic falcarinol oxylipins from *Dendropanax arboreus*. J Nat Prod 1996;59(8):748−53.

[117] Erdogan MF. Thiocyanate overload and thyroid disease. Biofactors 2003;19(3−4):107−11.

[118] Mezgebu Y, Mossie A, Rajesh P, Beyene G. Prevalence and severity of iodine deficiency disorder among children 6−12 years of age in Shebe Senbo district, Jimma Zone, Southwest Ethiopia. Ethiop J Health Sci 2012;22(3):196−204.

[119] Clero E, Doyon F, Chungue V, et al. Dietary patterns, goitrogenic food, and thyroid cancer: a case-control study in French Polynesia. Nutr Cancer 2012;64(7):929−36.

[120] Banea-Mayambu JP, Tylleskar T, Gitebo N, Matadi N, Gebre-Medhin M, Rosling H. Geographical and seasonal association between linamarin and cyanide exposure from cassava and the upper motor neurone disease konzo in former Zaire. Trop Med Int Health 1997;2(12):1143−51.

[121] Singh BN, Singh HB, Singh A, Naqvi AH, Singh BR. Dietary phytochemicals alter epigenetic events and signaling pathways for inhibition of metastasis cascade. Cancer Metastasis Rev 2014;33(1):41−85.

[122] Peppin JF, Pappagallo M. Capsaicinoids in the treatment of neuropathic pain: a review. Ther Adv Neurol Dis 2014;7(1):22−32.

[123] Johnson W. Final report on the safety assessment of capsicum annuum extract, capsicum annuum fruit extract, capsicum annuum resin, capsicum annuum fruit powder, capsicum frutescens fruit, capsicum frutescens fruit extract, capsicum frutescens resin, and capsaicin. Int J Toxicol 2007;26:3−106.

[124] Friedman M. Potato glycoalkaloids and metabolites: roles in the plant and in the diet. J Agric Food Chem 2006;54(23):8655−81.

[125] Kobaek-Larsen M, Christensen LP, Vach W, Ritskes-Hoitinga J, Brandt K. Inhibitory effects of feeding with carrots or (−)-falcarinol on development of azoxymethane-Induced preneoplastic lesions in the rat colon. J Agric Food Chem 2005;53(5):1823−7.

[126] Tan K, Killeen D, Li Y, Paxton J, Birch N, Scheepens A. Dietary polyacetylenes of the falcarinol type are inhibitors of breast cancer resistance protein (BCRP/ABCG2). Eur J Pharmacol 2014;723:346−52.

[127] Iciek M, Kwiecien I, Wlodek L. Biological properties of garlic and garlic-derived organosulfur compounds. Environ Mol Mutagen 2009;50(3):247−65.

[128] Chebi G. Structure of progoitrin. In: progoitrin.png So, editor. <http://creativecommons.org/licenses/by-sa/3.0/>; 2014.

[129] Aggarwal BB, Shishodia S. Molecular targets of dietary agents for prevention and therapy of cancer. Biochem Pharmacol 2006;71(10):1397−421.

[130] Itoh K, Chiba T, Takahashi S, et al. An Nrf2 small Maf heterodimer mediates the induction of phase II detoxifying enzyme genes through antioxidant response elements. Biochem Biophys Res Commun 1997;236(2):313−22.

[131] Cerella C, Gaigneaux A, Dicato M, Diederich M. Antagonistic role of natural compounds in mTOR-mediated metabolic reprogramming. Cancer Lett 2015;356(2):251−62.

[132] Blagosklonny MV. Prospective treatment of age-related diseases by slowing down aging. Am J Pathol 2012;181(4):1142−6.

[133] Blagosklonny MV. Molecular damage in cancer: an argument for mTOR-driven aging. Aging Us 2011;3(12):1130−41.

[134] Blagosklonny MV. Hormesis does not make sense except in the light of TOR-driven aging. Aging Us 2011;3(11):1051−62.

[135] Drewnowski A, Gomez-Carneros C. Bitter taste, phytonutrients, and the consumer: a review. Am J Clin Nutr 2000;72(6):1424−35.

[136] Scalera G. Effects of conditioned food aversions on nutritional behavior in humans. Nutr Neurosci 2002;5(3):159−88.

[137] McNaughton SA, Marks GC. Development of a food composition database for the estimation of dietary intakes of glucosinolates, the biologically active constituents of cruciferous vegetables. Br J Nutr 2003;90(3):687−97.

CHAPTER

51

Current Nutritional Recommendations: Elderly Versus Earlier Stage of Life

Carol Wham[1] and Michelle Miller[2]

[1]School of Food and Nutrition, Massey University, Auckland, New Zealand
[2]Nutrition and Dietetics, Flinders University, Adelaide, Australia

KEY FACTS

- Older people who live a long life tend not to develop a chronic disease and have delays in functional decline.
- Good nutrition and physical activity helps to maintain muscle mass, strength and function.
- In older people a body mass index in the overweight range is associated with optimal survival.
- The challenge for older people is to meet the same nutrient needs as when they were younger by consuming less calorific but more nutrient dense foods and beverages (for example low fat milk versus regular milk).
- Protein is an important determinant of muscle mass and function and foods such as fish, poultry, meat, eggs, dairy products, legumes and nuts need to be incorporated into meals.
- Evidence suggests that spreading protein intake evenly over meals may be beneficial.
- Regular consumption of a variety of fruit, vegetables and wholegrains is important to maintain a healthy bowel function.
- Micronutrient needs can easily be achieved with a healthy eating pattern consisting of three meals and healthy snacks.

Dictionary of Terms

- *BMR (Basal metabolic rate)*: The amount of energy required for cellular metabolic processes and functions of organs.
  Definitions adapted from the FNB:IOM DRI process—Food and Nutrition Board: Institute of Medicine [1–8].
- *EAR (Estimated Average Requirement)*: A daily nutrient level estimated to meet the requirements of half the healthy individuals in a particular life stage and gender group.
- *RDI (Recommended Dietary Intake)*: The average daily dietary intake level that is sufficient to meet the nutrient requirements of nearly all (97–98%) healthy individuals in a particular life stage and gender group.
- *AI (Adequate Intake)* (used when an RDI cannot be determined): The average daily nutrient intake level based on observed or experimentally determined approximations or estimates of nutrient intake by a group (or groups) of apparently healthy people that are assumed to be adequate.
- *EER (Estimated Energy Requirement)*: The average dietary energy intake that is predicted to maintain energy balance in a healthy adult of defined age, gender, weight, height, and level of physical activity, consistent with good health.
- *UL (Upper Level of Intake)*: The highest average daily nutrient intake level likely to pose no adverse health effects to almost all individuals in the general population. As intake increases above the UL, the potential risk of adverse effects increases.

INTRODUCTION

Optimal nutrition is a key determinant of successful aging as food is not only critical to physiological well-being but also contributes to social, cultural, and psychological quality of life [9]. The process of aging affects nutrient needs; requirements for some nutrients may be reduced while requirements for others may be increased. Mounting evidence indicates nutrient requirements may differ alongside the myriad of changes associated with advancing years.

Older adults are a heterogeneous group and have unique nutrient needs and the process of aging occurs at different rates in different people. In general, the main health risk for younger population groups is weight gain and being overweight whereas in advanced age older adults are vulnerable to eating too little energy with associated weight loss.

Physical function changes with aging. Characteristically both lean body mass and basal metabolic rate (BMR) tend to decline with age, concurrent with body composition changes slowly over time. There is a decrease in skeletal muscle, smooth muscle, and muscle that affects vital organ function. A loss of cardiac muscle for example, may reduce cardiac capacity. Gastric atrophy has an increased prevalence in older people [10]; levels of hydrochloric acid secretion are reduced and may contribute to impaired absorption of nutrients such as calcium, iron, and vitamin B12. Use of multiple medications (polypharmacy) may also cause diminish nutrient absorption. In addition, there is a reduction of antibodies, hormones, and enzymes, as well as a decline in bone density. Along with the reduction in skeletal muscle, total body water tends to decrease and body fat may increase proportionally. These changes in body composition affect the body's metabolism, nutrient intake, absorption, storage, utilization and excretion of nutrients, and overall nutrient requirements. As BMR declines proportionately with the decline in muscle tissue an older person's energy requirement per kilogram of body weight tends to be reduced. An overall decline in food intake may compromise dietary variety which is positively associated with nutritional quality and positive health outcomes [11].

Pathways to nutritional health in aging populations are complex and multifactorial and a variety of physical, social, and psychological factors may contribute to increased risk of malnutrition. Older people have a higher prevalence of chronic disease, tend to be sedentary, and higher patterns of morbidity occur in malnourished older people [12]. Increased functional difficulties, cognitive decline, and increased comorbidities may all lead to malnutrition in advanced age [13].

Differences in gender may be evident. Older women are more likely than men to report poorer health and have multiple chronic diseases [14], which can escalate age-related muscle loss and result in poor function [15]. They are more likely to be widowed and live alone compared to men, both of which are known nutrition risk factors. In living alone people have higher levels of weight loss [9], decreased energy intake, and poorer dietary variety perhaps resulting from decreased interaction at meal times [16]. Women who lose their spouse report higher levels of food insecurity and difficulty accessing food due to transport difficulties [17], and companionship is an important preventative measure especially after the loss of a spouse. Meal sharing increases food intake which is positively correlated with nutritional quality as well as health outcomes [18]. Although not a normal part of aging, depression is a common problem in older adults and depressive symptoms are a risk factor for impaired nutrition status [19–21]. Depression is likely to increase the risk of malnutrition through reduced appetite, food intake, and physical capacity [19,22]. Several studies of older people also show a relationship between nutrition risk and health-related quality of life (HRQOL) [12,23,24], which underpins the importance of optimizing the nutritional health of older people.

Most countries throughout the world have established their own nutrient recommendations to assess the dietary adequacy of individuals. However the nutritional requirements of older adults, especially the oldest old, are not well defined and there is a scarcity of information regarding their specific nutrient needs. Nutrient recommendations are used by dietitians, nutritionists, food legislators, and the food industry and are pivotal for dietary assessment, dietary modeling, as well as food formulation and labeling. Further to this, nutrient recommendations are integral to menu planning, with residential aged care facilities and ambulatory community food services, such as Meals on Wheels, using these recommendations as a basis for menu planning. As such, there is a pressing need for recommendations to be relevant and thus allow the nutritional needs of growing older populations to be addressed.

While nutritional recommendations for older people are frequently stratified by age, chronological age does not depict an older person's functional ability or quality of life. In the development of the 2010 Nutritional Guidelines for Older People in Finland the heterogeneity of older people was recognized [25]. To identify the nutritional needs of older people in relation to functional ability, older individuals were divided into four groups according to functional ability and illnesses.

Accordingly the recommendations for home dwelling healthy older people were similar to middle aged people; those for home dwelling with multiple diseases and risk of frailty emphasized unintentional weight loss; those for home care with multiple diseases and functional disabilities assessed for risk of malnutrition and acknowledged a high risk of malnutrition in older individual in residential care.

The unique nutritional needs of older people is a challenge for the future but with life expectancy at its highest throughout many parts of the world there is a pressing need for nutritional recommendations to be consistent with the best available evidence and feasible at the practice level to help maximize healthy aging in later life.

MEETING THE NUTRITIONAL NEEDS OF OLDER PERSONS: CURRENT RECOMMENDATIONS

Throughout the world it is widely recognized that a wide variety of food cultures provide a nutritious range of foods that can promote healthy aging. Macronutrient and micronutrient requirements tend not to be decreased with age, and some (ie, protein and calcium) increase with age. By consuming a variety of foods from the main food groups, older individuals can meet recommended macronutrient and micronutrient intakes and achieve energy balance.

For older adults who are overweight or obese a lower fat intake may be an appropriate strategy to reduce weight and negate chronic disease. However, weight loss at any weight can be detrimental to the health of an older adult [26] and should be attempted only under the supervision or guidance of a qualified dietitian. Overweight and obesity can mask the presence of sarcopenia which has been demonstrated to increase the risk of frailty, reduce physical function, and diminish capacity for exercise [27,28]. Thus, reducing body weight without compounding an already diminishing muscle mass is complex. About 60% of total energy should come from carbohydrates, with emphasis on complex carbohydrates. Glucose tolerance may decrease with advancing years. Complex carbohydrates put less stress on the circulating blood glucose than do refined carbohydrates. Such a regime also enhances dietary fiber intake. Adequate fiber, together with adequate fluid, helps maintain normal bowel function. An overview of the nutrient requirements for older people is provided in this chapter alongside nutrient intake recommendations from the WHO, EU, United Kingdom, and United States.

ENERGY

Energy requirements for older adults vary widely according to gender, body size, and physical activity and may be similar to younger adults whilst good health and physical function are maintained. Reduction in lean body mass, BMR, and overall physical activity may all contribute to an overall reduction in the energy needs in older adults compared with younger people.

Energy balance is achieved when the energy intake from food and drinks equals the energy expended for metabolic processes (BMR and thermic effect of food, TEF) and physical activity. A change in energy intake or output leads to a positive or negative energy balance. A positive energy balance results in body tissue being deposited as fat and an increase in body weight. A negative energy balance results in body tissue being mobilized and a loss of body weight. A change in body weight, particularly in body fatness, may have important implications for the health and functional status of older adults. In older people physical activity therefore has an important role in maintaining energy balance, and consequently a healthy body weight. Planned and incidental physical activity is also valuable for its role in maintaining muscle mass and strength, stimulating appetite, and maintaining social connections.

Based on data from the US Institute of Medicine of the National Academies database for individuals aged between 20 and 100 years, there is evidence of a progressive decline in total energy expenditure and physical activity level with advancing age [29]. This has important implications for defining dietary energy requirements for older people. BMR accounts for 45–70% of daily energy expenditure. This includes energy for cell metabolism, synthesis and metabolism of enzymes and hormones, transport of substances around the body, maintenance of body temperature, ongoing functioning of muscles, and brain function. BMR declines with age at an estimated 1–2% a decade due in part to the change in body composition. The TEF accounts for a further 10% of daily energy expenditure. Energy expenditure for physical activity is defined as the increase in metabolic rate above BMR and TEF, and is the most variable component of energy expenditure. Energy is expended through both planned and incidental physical activity, and older people generally have lower physical activity levels than younger adults although exceptions do apply. An inadequate energy intake may lead to nutrient deficiencies and can augment functional decline. This may contribute to further deterioration of health in older vulnerable individuals. Meeting energy requirements

TABLE 51.1 Equations for the Prediction of Resting Energy Expenditure in Older Adults

Source	Equation (kJ/day)	
	Male	Female
Harris & Benedict [b]	$(57.5 \times wt) + (20.93 \times ht) - (28.35 \times A) + 278$	$(40.0 \times wt) + (7.74 \times ht) - (19.56 \times A) + 2741$
Mifflin et al.[c]	$[(10 \times wt) + (6.25 \times ht) - (5 \times A) + 5] \times 4.2$	$[(10 \times wt) + (6.25 \times ht) - (5 \times A) - 161] \times 4.2$
Schofield (over 60 years)[d]	$[(0.049 \times wt) + 2.459] \times 1000$	$[(0.038 \times wt) + 2.755] \times 1000$
WHO (over 60 years)[e]	$(36.8 \times wt) + (4719.5 \times ht^a) - 4481$	$(38.5 \times wt) + (2665.2 \times ht^a) - 1264$
Fredrix et al.[f]	$[1641 + (10.7 \times wt) - (9.0 \times A) - 203] \times 4.2$	$[1641 + (10.7 \times wt) - (9.0 \times A) - 203(2)] \times 4.2$
Luhrmann et al.[g]	$3169 + (50 \times wt) - (15.3 \times A) + 746$	$3169 + (50 \times wt) - (15.3 \times A)$

[a]Height in meters.
[b]Harris JA, Benedict FG. A biometric study of basal metabolism in man. Carnegie Institution of Washington (1919).
[c]Mifflin MD, St Jeor ST, Hill LA, Scott BJ, Daugherty SA, Koh YO. A new predictive equation for resting energy expenditure in healthy individuals. Am J Clin Nutr 1990;51(2):241–7.
[d]Schofield WN. Predicting basal metabolic rate, new standards and review of previous work. Hum Nutr Clin Nutr 1985;39:5–41.
[e]FAO/WHO/UNU. Energy and protein requirements. Report of a Joint FAO/WHO/UNU expert consultation. Technical report series No. 724. World Health Organization.
[f]Fredrix EWHM, Soeters PB, Deerenberg IM, Kester ADM, Von Meyenfeldt MF, Saris WHM. Resting and sleeping energy expenditure in the elderly. Eur J Clin Nutr 1990;44(10):741–7.
[g]Luhrmann PM, Herbert BM, Krems C, Neuhauser-Berthold M. A new equation especially developed for predicting resting metabolic rate in the elderly for easy use in practice. Eur J Nutr 2002;41(3):108–13.
[h]Yaxley A, Cibich C, Miller M. Energy expenditure in healthy, community-dwelling older adults; are predictive equations valid? 16th International Congress of Dietetics. Sydney, New South Wales September 5–8; 2012.
wt = weight (kilograms); ht = height (centimetres); A = age (years); WHO = World Health Organization.
Miller equation (unpublished results; Yaxley ICD abstract)[h]
Male REE (kJ) = $282.630 + (-15.124 \times Age) + (24.481 \times Ht) + (31.870 \times Wt)$
Female REE (kJ) = $282.630 + (-15.124 \times Age) + (24.481 \times Ht) + (31.870 \times Wt) + (-243.226)$

is particularly difficult for those adults showing characteristics of frailty. Eating at least three meals a day and where possible, suitable snacks that are energy- and nutrient-dense, is especially important for the frail old.

With advancing age, maintaining an adequate energy intake and weight can be a challenge. Older people are less able than younger adults to make compensatory increases in their energy intake and are less able to regulate weight and therefore regain any lost weight. Food intake in older people may be compromised due to sensory (taste and smell) deficits and impaired sensory specific society which leads to less variety seeking behavior. Physiological changes in gastrointestinal function that occur with aging may have an adverse impact on appetite and lead to a decrease in food intake. Combined with poor dentition, chronic illness, and adverse social and psychological factors, such as bereavement and depression, older adults tend to be less hungry than younger adults and may be at increased risk of malnutrition.

Estimates of total energy requirements for older adults are complicated given the evidence to suggest that the desirable healthy weight range should be set higher for improved health outcomes [30] and the potential that the current estimates are based on predictive equations that have not been validated in this age group and may therefore overestimate requirements as a result of a decline in muscle mass with age [31].

There have been some recently developed predictive equations for use with older people; however, there is much work still to be done. In practice it is recommended to use an established equation and monitor weight status closely as a multitude of factors can influence energy expenditure, particularly in those with multiple medical conditions or recovering from surgery (Table 51.1). Observing a trend in change in body weight, beyond what might be expected from usual day to day fluid shifts, is the best clinical indicator of energy balance (Table 51.2).

The challenge for older people is to meet the same nutrient needs as when they were younger, yet consume fewer calories. As such the consumption of foods high in nutrients in relation to their energy content is a prudent choice. Such foods are considered "nutrient-dense." For example, low-fat milk is more nutrient dense than regular milk. Its nutrient content is the same, but it has fewer calories because it has less fat.

PROTEIN

Older adults have a higher requirement for protein compared to younger adults; older adults usually eat less, including less protein [33] but have higher protein needs to offset the resistance to the positive effects of dietary protein on protein synthesis (anabolic resistance) as well as the elevated metabolism of

TABLE 51.2 Prescribed Energy Requirements

Source	Recommendation
WHO (2002)[a]	Propose energy requirements of old age are 1.4−1.8 multiples of the BMR to promote body weight at different levels of activity
EU EFSA (1993)[b]	Based on actual body weights without desirable physical activity
	Men (years)
	30−59: 11.3 MJ
	60−74: 9.2 MJ
	≥75: 8.0 MJ
	Women (years)
	30−59: 8.5 MJ
	60−74: 7.8 MJ
	≥75: 7.3 MJ
US IOM (2006)[c]	Prescribed according to age and gender with the goal of maintaining a BMI of 22 kg/m^2, consistent with the midpoint of the healthy weight range
UK SACN (2011)[d]	EAR values for all adults
	Men (175 cm) 10.9 MJ/day or 2605 kcal/day
	Women (162 cm) 8.7 MJ/day or 2079 kcal/day
	Adults aged 75 +
	Men (170 cm) 9.6 MJ/day or 2294 kcal/day
	Women (155 cm) 7.7 MJ/day or 1840 kcal/day

[a]WHO (2002). Keep fit for life: meeting the nutritional needs of older persons. Geneva, World Health Organization, Tufts University School of Nutrition and Policy: 83.
[b]Reports of the scientific committee for food (1993). 31st series. Nutrient and energy intakes for the European community. European Commission. Luxembourg. Available at: http://ec.europa.eu/food/fs/sc/scf/out89.pdf
[c]IOM (Institute of Medicine) Dietary Reference Intakes: The Essential Guide to Nutrient Requirement. Washington, DC: The National Academies Press; 2006.
[d]Scientific Advisory Committee on Nutrition. Dietary reference values for energy. SACN Reports and Position Statements. London: P. H. England; 2011.

inflammatory conditions such as chronic obstructive pulmonary disease (COPD) [34]. There is mounting evidence that the existing recommended dietary intakes (RDIs) for protein are too low for older people [35] and do not take into consideration age-related changes in metabolism and immunity [36]. Findings suggest that protein intake greater than the RDI can help older people to improve immune status and wound healing as well as muscle mass, strength, and function [37].

The most recent recommendation on protein requirements published by the joint World Health Organization/Food and Agriculture Organization of the United Nations/United Nations University (WHO/FAO/UNU) expert consultation suggest a recommended dietary allowance (RDA) of 0.83 g/kg for adults [38]. This is considered a "safe level of intake" identified as the 97.5th percentile of the population distribution requirement. The report concluded that the protein requirement of elderly people did not differ from younger adults. This conclusion was partly based on nitrogen balance data which showed the mean protein requirement did not differ between younger (21−46 years) and older (64−81 years) healthy adults: 0.61 (SD 0.14) versus 0.58 (SD 0.12) g protein/kg body weight per day [39]. Other national and international authorities concur with this recommendation. The European Food Safety Authority (EFSA) 2012 Dietary References Values for protein also propose a population reference value of 0.83 g/kg for all adults including older adults [40]. Similarly in the United States the RDA for protein is 0.8 g/kg/day, set at two standard deviations above the Estimated Average Requirement (EAR, 0.66 g/kg/day) where the needs of most (97−98%) of the adult population should be met [41]. These recommendations were again based on a meta-analysis of nitrogen-balance studies by Rand et al. [42] given recommendations for an optimum intake of protein for healthy older people based on health outcomes has not been established. In France however a higher protein intake of 1.0 /kg per day for adults ≥75 years is recommended which is established on considerations about protein metabolism regulation in older adults [43].

National Health and Nutrition Examination Survey data indicate that mean protein intakes of Americans, including those more than 70 years of age, meet or exceed the RDA. However, variability is large. A significant proportion of older adults (approximately 10–25%) eat less protein than the RDA and approximately 5–9% of older persons, particularly women, consume less than the EAR of protein [44]. Considering the EAR is insufficient for about half of the population, the proportion of older adults at risk of inadequate protein intake is potentially large.

Rigorously controlled, nitrogen-balance studies are necessarily small and over recent years only a modest number of short-term nitrogen-balance experiments have been conducted to estimate the protein needs of older adults. The results of these studies are mixed and inconclusive. Some suggest that the requirement for total dietary protein is not different for healthy older adults than for younger adults and that the allowance estimate does not differ statistically from the RDA [39]. Others indicate that higher intakes are needed [45,46] to meet the dietary needs of virtually all healthy older adults. Notably, nitrogen-balance studies have not addressed the possibility that protein intake above the RDA could prove beneficial in healthy individuals. Prevention of nitrogen loss may be an inadequate outcome for older adults, especially for those with a significant loss of lean body mass.

Although there is insufficient longer term research with defined health outcomes to specify an optimal intake for protein there is mounting evidence that increasing protein intake beyond 0.8 g/kg may enhance protein anabolism and help reduce the progressive loss of lean mass with aging. Protein intake has been demonstrated to be an important determinant of muscle mass and function. Among a group of healthy older women with a protein intake of 0.45 g/kg body weight/day muscle mass and strength decreased over a period of 9 weeks. By contrast in women who consumed twice the amount of protein intake (0.92 g/kg body weight/day), muscle mass remained stable and muscle strength improved [47]. It has also been demonstrated that chronic ingestion of the RDA for protein results in reduced skeletal muscle size in weight-stable older adults although there was no change in muscle function detected [48]. Findings from the Health, Aging and Body Composition (Health ABC) cohort indicate that a lower energy-adjusted protein intake in healthy older adults is associated with a larger loss of lean body mass over a period of 3 years of observation [49]. Among the 2066 older adults aged 70–79 years in the Health ABC study the median protein intake ranged between 0.7 g/kg in the lowest quintile and 1.1 g/kg in the highest quintile with a reported loss in lean mass of 0.85 kg in the quintile with lowest protein intake versus a loss of 0.45 kg in the quintile with the highest protein intake. This translated into a 40% less decrease in lean mass over three years in participants in the highest quintile of protein intake compared with the lowest quintile, and suggests a clear linkage between protein intake and muscle change in older adults. Furthermore data from the InChianti and the Women's Health Initiative cohort studies indicate a higher protein intake is associated with reduced risk of muscle strength loss and incident frailty [50,51].

Not only do older adults usually eat less protein compared to younger adults but they often consume less high biological value animal protein [52] due to factors such as chewing difficulty, fear of increasing cholesterol, perceived intolerances [49], as well as cost and access to such sources.

Although muscle mass decreases in older people, the formation of muscle protein can be stimulated by higher availability of protein; so it is imperative that an adequate protein intake is maintained. An adequate protein intake is especially important to maintain a healthy functional status and decrease the risk of prolonged infections that lead to hospitalization [53]. The pattern of protein intake may also be important to stimulate protein synthesis in older adults and there is some evidence that spreading protein intake evenly over meals may be beneficial [54]. However, further studies are needed to determine the optimal pattern of intake to improve muscle strength and function.

In summary, the optimal protein intake for older adults to meet the requirements of maintaining nitrogen balance and preservation of muscle mass and function remains to be ascertained. Recent recommendations from the European Society for Clinical Nutrition and Metabolism (ESPEN) suggest higher dietary protein intakes for older adults (65 years plus) compared to younger adults [34]. It is suggested the diet should provide between 1.0 and 1.2 g protein/kg body weight/day for healthy older people [27] and between 1.2 and 1.5 g protein/kg body weight/day for older people who are malnourished or at nutrition risk [34]. To limit age-related decline in muscle mass, strength and function resistance exercise training is also recommended [55].

Notably, higher protein intakes are now recommended by the PROT-AGE study group appointed by the European Geriatric Medicine Society (EUGMS) (1.0–1.2 g protein/kg body weight/day for healthy older people; ≥1.2 g protein/kg body weight/day for active and exercising older adults; and 1.2–1.5 g protein/kg body weight/day for older adults who have acute or chronic disease) [35]. Similarly, in

TABLE 51.3 Protein Requirements

Source	Recommendation
WHO/FAO [38]	0.83 g/kg/day for all adults
EU [40]	0.83 g/kg/day for all adults
US [41]	0.8 g/kg/day for all adults
UK [57]	Reference nutrient intakes gram/person/day
	Men 19–50 years 55.5
	Men 51+ years 53.3
	Women 19–50 years 45
	Women 51+ years 46.5

acids) and glycerol. Total fatty acid intake should therefore average 30% of total energy and total fat including glycerol 33% of energy including alcohol or 35% of energy excluding alcohol.

The EFSA provides proposals for specific fatty acids [59]. Recommended saturated fatty acid and *trans* fatty acid intake should be as low as possible. An adequate intake (AI) of 4 E% for linolenic acid, an AI for alpha-linoleic acid of 0.5 E% and an AI of 250 mg for eicosapentaenoic acid plus docosahexaenoic acid was set for adults. Recommendations concur that saturated fats should be kept as low as possible within the context of a nutritionally adequate diet as well as *trans* fats, which are not required by the body.

Norway, The Nordic Nutrition Recommendations suggest a safe intake of 1.2–1.5 g protein/kg body weight/day for healthy older people or approximately 15–20% of total energy intake [56] (Table 51.3).

FAT

The recommendations for dietary fat are not provided in the form of an EAR or the RDA for adults as insufficient data are available to identify a defined intake level for fat based on maintaining fat balance or on the prevention of chronic diseases. Rather, recommendations are based on an acceptable macronutrient distribution range (AMDR), reported as a percentage of energy from fat. Broadly the FAO report of an expert consultation on fat and fatty acids [58] proposed a minimum of 15% energy from fat to ensure adequate consumption of total energy, essential fatty acids, and fat soluble vitamins and a maximum of 30–35% of total energy for most individuals.

The Dietary Guidelines for Americans (HHS/USDA, 2005) and the EFSA recommend keeping fat intake between 20% and 35% of energy. The rationale is that within this range, based on observations of dietary intakes, no overt nutritional deficiencies or adverse effects on blood fats or body weight have been observed. It has been noted that higher fat intakes can still be compatible with both good health and normal body weight, depending on the type of foods eaten and physical activity levels.

The UK Committee on Medical Aspects of Food Policy (COMA) [57] concludes that dietary reference values for fat should be calculated from the summation of reference values for individual classes of fatty acids (saturated fatty acids, monounsaturated fatty acids, polyunsaturated fatty acids, *trans* fatty

CARBOHYDRATE

While there are no dietary recommendations in the form of EAR and RDI for carbohydrate as there is insufficient evidence to support these, the acceptable intake of carbohydrate is implicit for reducing chronic disease risk. EFSA has given an acceptable range of carbohydrate intake (sugars and starchy carbohydrates combined), known as a reference intake range. Diets containing between 45% and 60% of daily energy from carbohydrates, combined with reduced fat and saturated fat intake, improve metabolic risk factors for chronic disease. No specific intake or upper limit for intake of total sugars or added sugars is set [59,60].

DIETARY FIBER

Recommendations for dietary fiber and its components (cellulose, hemicellulose, lignin, pectin, resistant starch) have been established with consideration given to gastrointestinal function and adequate laxation. Based on the available evidence on bowel function, the EUFIC Panel considers dietary fiber intakes of 25 g per day to be adequate for normal laxation [59].

The US Food and Nutrition Board [41] set an AI for total dietary fiber of 3.4 g per MJ (14 g per 1000 kcal) based on the energy adjusted median intake associated with the lowest risk of coronary heart disease (CHD) in observational studies. The AI corresponds to 25 g per day for women and 38 g per day for men aged 14–50 years, respectively.

It is recognized that dietary fiber has a major role in bowel function and gastrointestinal symptoms, such as constipation, have been linked to low fiber intakes [41].

WATER

Water is particularly important in older age due to the decline in kidney function, use of medications, such as diuretics, and subsequent consequences of dehydration including constipation, confusion, bladder infections, functional decline, falls, or stroke. Inadequate fluid intake can also affect saliva production which is essential for the maintenance of food oral health. However, it may be challenging for older adults to achieve adequate fluid requirements as a result of a reduction in the thirst mechanism that occurs with age. It is important for older people to drink regularly even when they are not thirsty because of their potentially low water reserves and to moderately increase their salt intake when they sweat. Foods such as fruit and vegetables contribute an important source of water to total intake and need to be encouraged.

Observed water intakes in older people appear to be lower than in younger adults. This is likely to be because older people are less thirsty and drink less fluid when they are fluid deprived compared to younger individuals [61]. Even when offered a highly palatable selection of drinks following fluid deprivation older adults appear to fail to ingest sufficient fluid to replenish their body water deficit [62]. Older adults may also be more sensitive to heat stress and subsequent water depletion leading to heat exhaustion [63] and ultimately loss of consciousness and heat stroke.

The EFSA Panel on Dietetic Products, Nutrition, and Allergies (NDA) has set the same AIs for water for older adults and younger adults as 2.0 L/day (P95 3.1 L) for women and 2.5 L/day (P95 4.0 L) for men [64]. AIs were derived from a combination of observed intakes in population groups with desirable osmolarity values of urine and desirable water volumes per energy unit consumed.

In the United States the AI for total water (drinking water, beverages, and foods) is 3.7 L/day of total water for men and 2.7 L/day of total water for women. [65]. The AI for older adults (>70 years) is also the same as for younger adults (51–70 years). This AI is based on the median total water intake of young adults rather than the older age group, in order to ensure that total water intake is not limited as a result of a potential declining ability to consume adequate amounts in response to thirst.

MICRONUTRIENTS

An older adult who is eating well in terms of quality and quantity and is not experiencing or recovering from an acute illness will likely achieve an AI of all micronutrients. While some micronutrients are required in larger amounts in older age, these amounts are easily achievable within a healthy, well-balanced diet which meets energy and macronutrient recommendations. In some circumstances there may be risk of inadequacy (eg, poor appetite or limited access to food, gastrointestinal disease, recovery from illness or surgery, drug–nutrient interactions) and there are key micronutrients that are generally affected whereby dietary education and counseling or even micronutrient supplementation may be necessary. Some of the micronutrient deficiencies that more commonly occur in aging include vitamin B12, folate, calcium, vitamin D, potassium, magnesium, and iron.

Vitamin B12

While recommendations for dietary vitamin B12 intake are generally not thought to be any different to younger adults there are physiological changes that occur with aging that can be responsible for increasing risk of deficiency. Vitamin B12 is primarily found in animal products such as fish, poultry, meat, eggs, and dairy but is increasingly also found in fortified foods, particularly breakfast cereals. A vitamin B12 deficiency commonly manifests as fatigue, anemia, and depression and can be corrected with relative ease through diet or supplementation.

Folate

Much like vitamin B12, the recommendation for dietary folate intake is not considered to be dissimilar to recommendations for younger adults. Folate is found across a wide variety of foods including fruit and vegetables, legumes and nuts, dairy, poultry, meat, and eggs. Despite being commonly found in the food supply and older adults not generally having any greater requirement than younger adults, folate can become a nutrient at risk if access and availability of fresh food is limited or there is a significant decline in appetite. The primary clinical sign of deficiency is megaloblastic anemia, commonly manifesting as fatigue, weakness, and headache.

Calcium and Vitamin D

These nutrients are vital for bone health and both are consistently recommended in higher amounts compared to younger adults. Food sources high in calcium include milk, cheese, yoghurt, fortified foods (eg, juice, breakfast cereal, breads) and some fish (eg, sardines). Unfortunately food sources of vitamin D are limited (eg, some fish, fortified foods, liver) and while exposure to sunlight provides an endogenous source, disease or disability can prevent adequate exposure. Supplementation of both calcium and vitamin D is not uncommon with increasing age in addition to prescribing regular exercise for prevention and treatment of osteoporosis, falls, and fractures.

Potassium

This nutrient is increasingly being suggested to be integral to good cardiovascular health hence is very relevant with increasing age where risk of cardiovascular disease is more common. Despite this, the dietary recommendations for older adults are routinely the same as for younger adults. Good food sources of potassium include fruits and vegetables, milk, and unprocessed meats and while potassium deficiency is not common, it can manifest in older adults as a result of poor appetite and lack of regular access to fresh foods. Symptoms of deficiency include muscle cramp, nausea, fatigue, and weakness.

Magnesium

Much like potassium, the dietary recommendations for magnesium in older adults are the same as for younger adults. Magnesium is found in many foods with those considered rich sources including nuts, legumes, whole grains, and most green vegetables. Magnesium has many functions including energy production and in bone health hence a deficiency can be debilitating with symptoms akin to potassium deficiency but also including confusion, delirium and it its most severe form, heart failure.

Iron and Zinc

Iron and zinc dietary recommendations for older adults are largely the same as for younger adults with the only exception being recommendations for older women being reduced for younger women secondary to menopause. Iron plays an important role in the body with signs and symptoms of deficiency being fatigue and decreased immunity. The most probable cause of iron deficiency anemia among older adults would be inadequate dietary intake or blood loss from conditions such as ulcers, polyps, or intestinal cancer. Good sources of dietary iron are red meat, offal, and fortified breakfast cereals. Similar to iron, zinc plays an important role in immunity but also has a role to play in wound healing and maintaining the senses of taste and smell, important for optimal health of an older adult. Good sources of dietary iron are also routinely also good sources of dietary zinc although zinc can also be found in reasonable amounts in foods including pulses, nuts and legumes, wholegrain cereals, and dairy products.

SUMMARY

- Older adults are a diverse population and are living longer, healthier lives than ever before. Indeed centenarians are no longer unique. They comprise the fastest growing segment of older adults in countries such as Japan and characteristically have delays in any kind of physical decline or onset of chronic disease. Nevertheless the rates of change with aging differ among individuals and older adults can be vulnerable to nutritional inadequacies.
- Food intake is typically decreased in older adults and energy requirements are lower. The challenge for older people is to meet the same nutrient needs as when they were younger, yet consume fewer calories.
- Many older adults have special nutrient requirements because aging affects absorption, use and excretion of nutrients.
- Protein intake may be lower in older adults and this may contribute to the age-related loss of muscle mass which can lead to a decrease in muscle strength and greater risk of functional impairment and disability. Reduced muscle strength and function can affect an older adult's quality of life.
- Nutritional recommendations vary by country. In the United States the dietary reference intakes separate the cohort of people aged over 50 years into two groups, those aged 50–70 years and those aged 71 years and older. In the future determining and implementing nutrient recommendations for older adults is likely to become more complex as the population grows and there is increasing prevalence of chronic disease secondary to obesity during adulthood.
- Recommendations need to encourage a more liberalized diet for older adults as emerging evidence suggests this can be beneficial in reducing the risk of frailty [66].
- Consumer shifts in attitudes toward food and expectations of food services with advancing age places a need for continuous review of the recommendations for this complex group. Recommendations need to be consistent with the best available evidence and need to be feasible at the practice level.

References

[1] Food and Nutrition Board: Institute of Medicine. Dietary Reference Intakes for calcium, phosphorus, magnesium, vitamin D and fluoride. Washington, DC: National Academy Press; 1997.

[2] Food and Nutrition Board: Institute of Medicine. Dietary reference intakes for thiamin, riboflavin, niacin, vitamin B6, folate, vitamin B12, pantothenic acid, biotin, and choline. Washington, DC: National Academy Press; 1998.

[3] Food and Nutrition Board: Institute of Medicine. Dietary reference intakes. A risk assessment model for establishing upper intake level for nutrients. Washington, DC: National Academy Press; 1998.

[4] Food and Nutrition Board: Institute of Medicine. Dietary reference intakes for vitamin C, vitamin E, selenium and carotenoids. Washington, DC: National Academy Press; 2000.

[5] Food and Nutrition Board: Institute of Medicine. Dietary reference intakes. Applications in dietary assessment. Washington, DC: National Academy Press; 2000.

[6] Food and Nutrition Board: Institute of Medicine. Dietary reference intakes for vitamin A, vitamin K, arsenic, boron, chromium, copper, iodine, iron, manganese, molybdenum, nickel, silicon, vanadium and zinc. Washington, DC: National Academy Press; 2001.

[7] Food and Nutrition Board: Institute of Medicine. Dietary reference intakes for energy, carbohydrate, fiber, fat, fatty acids, cholesterol, protein and amino acids (Macronutrients). Washington, DC: National Academy Press; 2002.

[8] Food and Nutrition Board: Institute of Medicine. Dietary Reference Intakes for water, potassium, sodium, chloride and sulfate. Panel on the dietary reference intakes for electrolytes and water. Washington, DC: National Academy Press; 2004.

[9] American Dietetic Association. Position paper of the American dietetic association: nutrition across the spectrum of aging. J Am Diet Assoc 2005;105:616–33.

[10] The Eurohepygast Study Group. Risk factors for atrophic chronic gastritis in a European population: results of the Eurohepygast study. Gut 2002;50(6):779–85.

[11] Donini LM, Savina C, Cannella C. Eating habits and appetite control in the elderly: the anorexia of aging. Int Psychogeriatr 2003;15(1):73–87.

[12] Johansson L, Sidenvall B, Malmberg B, Christensson L. Who will become malnourished? A prospective study of factors associated with malnutrition in older persons living at home. J Nutr Health Aging 2009;13(10):855–61.

[13] Cereda E, Pedrolli C, Zagami A, Vanotti A, Piffer S, Faliva M, et al. Nutritional risk, functional status and mortality in newly institutionalised elderly. Br J Nutr 2013;110(10):1903–9.

[14] Castel H, Shahar D, Harman-Boehm I. Gender differences in factors associated with nutritional status of older medical patients. J Am Coll Nutr 2006;25(2):128–34.

[15] Payette H. Nutrition as a determinant of functional autonomy and quality of life in aging: a research program. Can J Physiol Pharmacol 2005;88(11):1061–70.

[16] Locher J, Robinson C, Roth D, Ritchie C, Burgio K. The effect of the presence of others on caloric intake in homebound older adults. J Gerontol Med Sci 2005;60A(11):1475–8.

[17] Locher J, Ritchie C, Roth D, Sawyer Baker P, Bodner E, Allman R. Social isolation, support, and capital and nutritional risk in an older sample: ethnic and gender differences. Soc Sci Med 2005;60:747–61.

[18] Bernstein MA, Tucker KL, Ryan ND, et al. Higher dietary variety is associated with better nutritional status in frail elderly people. J Am Diet Assoc 2002;102:1096–104.

[19] Ávila-Funes JA, Gray-Donald K, Payette H. Association of nutritional risk and depressive symptoms with physical performance in the elderly: the Quebec longitudinal study of nutrition as a determinant of successful aging (NuAge). J Am Coll Nutr 2008;27(4):492–8.

[20] Payette H, Gueye NDR, Gaudreau P, Morais JA, Shatenstein B, Gray-Donald K. Trajectories of physical function decline and psychological functioning: The Québec longitudinal study on nutrition and successful aging (NuAge). J Gerontol B Psychol Sci Soc Sci 2011;66B(Suppl. 1):i82–90.

[21] Wham CA, McLean C, Teh R, Moyes S, Peri K, Kerse N. The BRIGHT trial: what are the factors associated with nutrition risk? J Nutr Health Aging 2014;18(7):692–7.

[22] Cabrera MAS, Mesas AE, Garcia ARL, de Andrade SM. Malnutrition and depression among community-dwelling elderly people. J Am Med Dir Assoc 2007;8(9):582–4.

[23] Keller HH, Ostbye T, Goy R. Nutritional risk predicts quality of life in elderly community-living Canadians. J Gerontol Med Sci 2004;59A(1):68–74.

[24] Kvamme J-M, Olsen J, Florholmen J, Jacobsen B. Risk of malnutrition and health-related quality of life in community-living elderly men and women: the Tromsø study. Qual Life Res 2011;20(4):575–82.

[25] Suominen MH, Jyvakorpi SK, Pitkala KH, Finne-Soveri H, Hakala P, Mannisto S, et al. Nutritional guidelines for older people in Finland. J Nutr Health Aging 2014;18(10):861–7.

[26] Bannerman E, Miller MD, Daniels LA, Cobiac L, Giles LC, Whitehead C, et al. Anthropometric indices predict physical function and mobility in older Australians: the Australian longitudinal study of ageing. Public Health Nutr 2002;5(5):655–62.

[27] Cruz-Jentoft AJ, Baeyens JP, Bauer J r M, Boirie Y, Cederholm T, Landi F, et al. Sarcopenia: European consensus on definition and diagnosis. Age Ageing 2010;39(4):412–23.

[28] Thomas DR. Loss of skeletal muscle mass in aging: examining the relationship of starvation, sarcopenia and cachexia. Clin Nutr 2007;26(4):389–99.

[29] Roberts SB, Dallal GE. Energy requirements and aging. Public Health Nutr 2005;8(7A):1028–36.

[30] Rejeski WJ, Marsh AP, Chmelo E, Rejeski JJ. Obesity, intentional weight loss and physical disability in older adults. Obes Rev 2010;11(9):671–85.

[31] NHMRC. Nutrient reference values for Australia and New Zealand including recommended dietary intakes. Canberra: National Health and Medical Research Council; 2006.

[32] Food and Agriculture Organization, World Health Organization and United Nations University. Energy and protein requirements. Geneva: WHO technical report series; 1985.

[33] Volpi E, Campbell WW, Dwyer JT, Johnson MA, Jensen GL, Morley JE, et al. Is the optimal level of protein intake for older adults greater than the recommended dietary allowance? J Gerontol A Biol Sci Med Sci 2012;68(6):677–81.

[34] Deutz NEP, Bauer JM, Barazzoni R, Biolo G, Boirie Y, Bosy-Westphal A, et al. Protein intake and exercise for optimal muscle function with aging: Recommendations from the ESPEN Expert Group. Clin Nutr 2014;33(6):929–36.

[35] Bauer J, Biolo G, Cederholm T, Cesari M, Cruz-Jentoft AJ, Morley JE, et al. Evidence-based recommendations for optimal dietary protein intake in older people: a position paper from the PROT-Age study group. J Am Med Dir Assoc 2013;14(8):542–59.

[36] Clegg A, Young J, Iliffe S, Rikkert MO, Rockwood K. Frailty in elderly people. Lancet 2013;381(9868):752–62.

[37] Wolfe R, Miller S, Miller K. Optimal protein intake in the elderly. Clin Nutr 2008;27(5):675–84.

[38] WHO/FAO/UNU. Protein and amino acid requirements in human nutrition: Report of a joint FAO/WHO/UNU expert consultation. WHO Technical Series. Geneva: World Health Organization; 2007.

[39] Campbell WW, Johnson CA, McCabe GP, Carnell NS. Dietary protein requirements of younger and older adults. Am J Clin Nutr 2008;88(5):1322–9.

[40] EFSA Panel on Dietetic Products. Nutrition and allergies (NDA). Scientific opinion on dietary reference values for protein. EFSA J 2012;10(2):66.

[41] IoM (Institute of Medicine). Dietary reference intakes for energy, carbohydrate, fiber, fat, fatty acids, cholesterol, protein, and amino acids. Washington, DC: The National Academies Press; 2005. p. 1357.

[42] Rand WM, Pellett PL, Young VR. Meta-analysis of nitrogen balance studies for estimating protein requirements in healthy adults. Am J Clin Nutr 2003;77(1):109–27.

[43] AFSSA. (2007). Apport en protéines: consommation, qualité, besoins et recommandations. p. 461.

[44] Fulgoni VL. Current protein intake in America: analysis of the National Health and Nutrition Examination Survey, 2003–2004. Am J Clin Nutr 2008;87(5):1554S–7S.

[45] Kurpad AV, Vaz M. Protein and amino acid requirements in the elderly. Eur J Clin Nutr 2000;54(6):S131–42.

[46] Morse MH, Haub MD, Evans WJ, Campbell WW. Protein requirement of elderly women: nitrogen balance responses to three levels of protein intake. J Gerontol A Biol Sci Med Sci 2001;56(11):M724–30.

[47] Castaneda C, Charnley JM, Evans WJ, Crim MC. Elderly women accommodate to a low-protein diet with losses of body cell mass, muscle function, and immune response. Am J Clin Nutr 1995;62(1):30–9.

[48] Campbell WW, Trappe TA, Wolfe RR, Evans WJ. The recommended dietary allowance for protein may not be adequate for older people to maintain skeletal muscle. J Gerontol A Biol Sci Med Sci 2001;56(6):M373–80.

[49] Houston DK, Nicklas BJ, Ding J, Harris TB, Tylavsky FA, Newman AB, et al. Dietary protein intake is associated with lean mass change in older, community-dwelling adults: the Health, Aging, and Body Composition (Health ABC) Study. Am J Clin Nutr 2008;87(1):150–5.

[50] Bartali B, Frongillo EA, Stipanuk MH, Bandinelli S, Salvini S, Palli D, et al. Protein intake and muscle strength in older persons: does inflammation matter? J Am Geriatr Soc 2012;60(3):480–4.

[51] Beasley JM, LaCroix AZ, Neuhouser ML, Huang Y, Tinker L, Woods N, et al. Protein intake and incident frailty in the women's health initiative observational study. J Am Geriatr Soc 2010;58(6):1063–71.

[52] Gaffney-Stomberg E, Insogna K, Rodriguez N, Kerstetter J. Increasing dietary protein requirements in elderly people for optimal muscle and bone health. J Am Geriatr Soc 2009;57(6):1073–9.

[53] Volpi E, Campbell WW, Dwyer JT, Johnson MA, Jensen GL, Morley JE, et al. Is the optimal level of protein intake for older adults greater than the recommended dietary allowance? J Gerontol A Biol Sci Med Sci 2013;68(6):677–81.

[54] Bouillanne O, Curis E, Hamon-Vilcot B, Nicolis I, Chretien P, Schauer N, et al. Impact of protein pulse feeding on lean mass in malnourished and at-risk hospitalized elderly patients: a randomized controlled trial. Clin Nutr 2013;32(2):186–92.

[55] Lanza IR, Short DK, Short KR, Raghavakaimal S, Basu R, Joyner MJ, et al. Endurance exercise as a countermeasure for aging. Diabetes 2008;57(11):2933–42.

[56] Pedersen AN, Cederholm T. Health effects of protein intake in healthy elderly populations: a systematic literature review. Food Nutr Res 2014;58:. Available from: http://dx.doi.org/10.3402/fnr.v58.23364.

[57] Department of Health [London]. Dietary reference values for food energy and nutrients for the United Kingdom. Report of the panel on dietary reference values of the committee on medical aspects of food policy. London: HMSO; 1991. p. 41.

[58] FAO. Fats and fatty acids in human nutrition. Report of an expert consultation. FAO Food and Nutrition Paper. Geneva: FAO; 2008.

[59] EFSA Panel on Dietetic Products. Nutrition and allergies (NDA). Scientific opinion on dietary reference values for fats, including saturated fatty acids, polyunsaturated fatty acids, monounsaturated fatty acids, trans fatty acids, and cholesterol. EFSA J 2010;8(3):1461 [p. 107].

[60] EFSA panel on dietetic products, NDA. Scientific opinion on dietary reference values for carbohydrates and dietary fibre. EFSA J 2010;8(3):1462.

[61] Phillips PA, Rolls BJ, Ledingham JGG, Forsling ML, Morton JJ, Crowe MJ, et al. Reduced thirst after water deprivation in healthy elderly men. N Engl J Med 1984;311(12):753–9.

[62] Phillips PA, Johnston CI, Gray L. Disturbed fluid and electrolyte homoeostasis following dehydration in elderly people. Age Ageing 1993;22(1):S26–33.

[63] Davidhizar R, Dunn CL, Hart AN. A review of the literature on how important water is to the world's elderly population. Int Nurs Rev 2004;51:159–66.

[64] EFSA Panel on Dietetic Products Nutrition and Allergies (NDA). Scientific opinion on dietary reference values for water. EFSA J 2010;8(3):1459–507.

[65] Institute of Medicine. Dietary reference intakes for water, potassium, sodium, chloride, and sulfate. Washington, DC: The National Academies Press; 2005.

[66] Baulderstone L, Yaxley A, Luszcz M, Miller M. Diet liberalization in older Australians decreases frailty without increasing risk of developing chronic disease. J Frailty Aging 2012;1(4):174–82.

Index

Note: Page numbers followed by "*f*" and "*t*" refer to figures and tables, respectively.

A

AA. *See* Arachidonic acid (AA)
AARP. *See* American Association for Retired Persons (AARP)
Absorbed metabolites, 650
Absorption, 586–587
ACC. *See* American College of Cardiology (ACC)
Acceptable macronutrient distribution range (AMDR), 42, 357, 729
ACD. *See* Anemia of chronic disease (ACD)
ACE2. *See* Angiotensin-converting enzyme 2 (ACE2)
ACEI. *See* Acetylcholinesterase inhibitor (ACEI)
Acetate, 695–696
O-Acetyl-ADP-ribose, 228
Acetylation, 211
Acetylcholinesterase inhibitor (ACEI), 667–668
Achalasia, 365–366
Acid
  secretion, 366
  urates, 424
Acrolein dG adducts, 162
Acrylamide, 178
ACT. *See* Auditory Consonant Trigram (ACT)
Activating transcription factor-3 (ATF-3), 652
Activator protein-1 (AP-1), 612
Acute exposure, 180
Acute inflammation, 637
Acute iodine poisoning, 585
AD. *See* Alzheimer's disease (AD)
Adaptive immune system, 377, 556–557. *See also* Innate immune system
  cell-mediated adaptive immune responses, 378
  humoral adaptive immune responses, 378–379
ADAS-cog. *See* Alzheimer's Disease Assessment Scale-cog (ADAS-cog)
ADCSADL. *See* Alzheimer's disease cooperative Study Activities of Daily Living (ADCSADL)
Adenosine triphosphate (ATP), 246–247
Adequate intake (AI), 42, 345, 723, 729
Adequate protein intake, 357
Adiponectin, 413
Adipose tissue (AT), 121
ADMA. *See* Asymmetric dimethylarginine (ADMA)

Adults/elderly, 594
Advanced dementia, patients with, 68
Advanced glycation end product (AGE), 147–148, 158, 263–264, 335, 393, 399, 402. *See also* Dietary advanced glycation end product (dAGE)
  and chronic diseases associated with older age, 269–271
    cardiovascular disease, 269
    degenerative eye diseases, 270–271
    diabetes, 271
    neurodegenerative diseases, 269–270
    renal disease, 270
    Rheumatoid Arthritis, 270
    Sarcopenia, 270
  dietary AGEs formation, 265
  formation in vivo, 264
    α-dicarbonyl formation, 265
    Maillard reaction, 264
  methods for measuring, 265–266, 266*t*
  molecular action, 267–268
  and normal aging, 268–269
  types, 264*t*
Adverse effects (AE), 630–631
AE. *See* Adverse effects (AE)
AF. *See* Atrial fibrillation (AF)
AF-N7-Gua. *See* Aflatoxin-N7-guanine (AF-N7-Gua)
AFD. *See* Alternate day fasting (AFD)
AFL. *See* Atrial flutter (AFL)
Aflatoxin-N7-guanine (AF-N7-Gua), 183
AGE. *See* Advanced glycation end product (AGE); Aged garlic extract (AGE)
AGE receptors (RAGE), 338
Age-related alterations
  chromatin architecture, 217
  in DNA methylation, 215–216
  histone modification, 216–217
  interacting hallmarks of aging and extrinsic modifying factors, 216*f*
  small, noncoding RNAs, 217
Age-related changes, 191
  anorexia of aging, 59
  changes in body composition, 59
  in dietary intake and eating habits, 356
  in eye, 433–434
  gastrointestinal changes, 59–60
  of signaling feeding state, 192–193
  taste, smell, and mastication dysfunction, 59

Age-related diseases, 110, 114, 294. *See also* Chronic age-related diseases prevention
Age-Related Eye Disease Study (AREDS), 436, 441
Age-related eye diseases
  age-related changes in eye, 433–434
  antioxidants and, 436
    carotenoids, 439–440
    lutein, 440–441
    polyunsaturated fatty acids, 441–442
    vitamin C, 437–438
    vitamin E, 438
    zeaxanthin, 440–441
    zinc, 438–439
  degenerative eye diseases, 434–436
  relationship between nutrition and healthy ocular structure and function, 436
Age-related macular degeneration (AMD), 271, 433–435, 474
Age-related morbidity
  constituents of vegetables and fruit, 711
    antioxidant effects of plant-derived compounds, 711
    nonnutrient components, 712
    nutrients in human metabolism, 711
    phytochemicals, 712–717
Age-related phenotypes, 12
Aged garlic extract (AGE), 668
AGEs receptor 1 (AGER1), 267
AgII. *See* Angiotensin II (AgII)
Aging, 22, 109–110, 207, 235, 393, 412, 533, 552, 609, 621, 627, 638–641, 649, 659–660. *See also* Omega-3 (ω-3) fatty acids
  anorexia, 189
  biology, 610–611
    principles, 4, 4*t*
  and curcumin, 499–502
  and CVD risk, 611
  dietary intake and biological mechanisms in
    calcium, 538–540
    magnesium, 538
    mechanisms through lower levels of Mg, Ca, and K, 540*t*
    potassium, 540
  epigenetic linkage of, 36–37
  factors impairing endothelial function with inflammation, 612
    oxidative stress, 611–612
    senescence, 612–613

Aging (*Continued*)
  homeodynamics and homeodynamic space, 5–6
  lipids and brain function, 623
  nutrition and food for aging interventions, 6, 402–403
  nutritional hormetins, 6–7
  occurrence, accumulation, and consequences of molecular damage, 4–5
  process, 32–33, 156, 649–650, 682
  renal function in, 425–426
  sarcopenia, 189
  telomeres and, 131–132
  theories, 156
  xenobiotic effects on, 650
Aging-related molecular targets, 494–495
  cellular senescence, 498–499
  deregulated nutrient sensing, 498
  epigenetic alterations, 497
  genomic instability, 495–496
  mitochondrial dysfunction, 498
  proteostasis imbalance, 497–498
  stem cell exhaustion, 499
  telomere attrition, 496–497
Agouti-related peptide (AgRP), 193, 398–399
AgRP. *See* Agouti-related peptide (AgRP)
AHA. *See* American Heart Association (AHA)
AHEI. *See* Alternate Healthy Eating Index (AHEI)
AI. *See* Adequate intake (AI); Anemia of inflammation (AI)
Akt/PKB, 143
AKT1s1. *See* Proline rich Akt/PKB substrate 40 kDa (PRAs40)
ALA. *See* α-Linolenic acid (ALA)
ALAS2. *See* 5-Aminolevulinic acid synthase 2 (ALAS2)
Albumin, 247
Albuminuria, 424
Alcohol, 350
Alkaloids, 305–306
All-*trans* retinoic acid (ATRA), 219
Allium sativum. *See* Garlic (*Allium sativum*)
ALNA. *See* α-Linolenic acid (ALA)
ALOX, 24
ALOX15, 24
Alpha cell mass, 394
α-Dicarbonyl formation, 265
α-klotho, 451–452
Alpha-LA. *See* Alpha-lipoic acid (Alpha-LA)
α-Linolenic acid (ALA), 156–157, 318, 382–384, 627, 629
Alpha-lipoic acid (Alpha-LA), 303
α−MSH. *See* Alpha−melanocyte stimulating hormone (α−MSH)
Alpha-synuclein (αSyn), 151, 252–253
α-tocopherol, 114, 438, 569–572, 638–639
α-tocopherol transfer protein (α-TTP), 638–639
α-TTP. *See* α-Tocopherol transfer protein (α-TTP)
Alpha-Tocopherol Beta-Carotene (ATBC) Cancer Prevention Trial, 572

Alpha−melanocyte stimulating hormone (α−MSH), 193–194
ALS. *See* Amyotrophic lateral sclerosis (ALS); Average lifespan (ALS)
Alternate day fasting (AFD), 678
Alternate Healthy Eating Index (AHEI), 322
Alzheimer's disease (AD), 32, 150–151, 162, 252–253, 334, 399, 544, 624, 641–644, 659
  and glycation, 338–339
  and insulin resistance, 337–338
  and related dementia, 625
Alzheimer's Disease Assessment Scale-cog (ADAS-cog), 660–661
Alzheimer's disease cooperative Study Activities of Daily Living (ADCSADL), 667
AMD. *See* Age-related macular degeneration (AMD)
AMDR. *See* Acceptable macronutrient distribution range (AMDR)
American Association for Retired Persons (AARP), 272–273
American College of Cardiology (ACC), 316
American Heart Association (AHA), 315, 626
American Society for Parenteral and Enteral Nutrition (ASPEN), 68
Ames dwarf mice, 515
Amino-acids, 302–303, 402
  metabolites and precursors, 359
5-Aminolevulinic acid synthase 2 (ALAS2), 525–526
Aminomethyltransferase (AMT), 517
AMP-activated protein kinase (AMPK), 32, 47, 145, 498
AMPK. *See* AMP-activated protein kinase (AMPK)
AMT. *See* Aminomethyltransferase (AMT)
Amyloid precursor protein (APP), 150–151
Amyloid-β (Aβ) plaques, 252
Amyotrophic lateral sclerosis (ALS), 252–253, 256–257
Anabolic effect, 189
Anabolic resistance, 357
Androgen response element (ARE), 482–484
Androgen(s), 160
  receptors, 482–484
Anemia, 532
Anemia of chronic disease (ACD), 530
Anemia of inflammation (AI), 530–531
Angiotensin II (AgII), 478
Angiotensin II type I receptor (AT1), 672
Angiotensin-converting enzyme 2 (ACE2), 672
Animal studies, 572, 710
Anorexia, 57
  of aging, 44–45, 59, 356, 364–365
  nervosa, 198
Anorexigenic hormones, 121
Anorexigenic peptides, 196
ANS. *See* Autonomic nervous system (ANS)
Anthocyanins, 256, 669
Anthropometry, 63–64
Antiaging genes regulation, VDR ligands promoting healthspan via, 460–465

Anticancer activity, 480
Antigen presenting cell (APC), 375–376, 696–697
Antinutrients, 712
Antioxidant network, 109–110, 111f
  enzymatic antioxidants and trace elements, 110–112
    glutathione peroxidase, 112
    SOD, 110–112
  hydrophilic antioxidants
    GSH, 112–113
    Vitamin C, 112
  lipophilic antioxidants
    carotenoids, 113
    polyphenols, 114
    tocopherols, 113–114
Antioxidant response element (ARE), 256–257, 300, 473
Antioxidant(s), 609, 659, 707, 713
  and age-related eye diseases, 436
    carotenoids, 439–440
    lutein, 440–441
    PUFAs, 441–442
    vitamin C, 437–438
    vitamin E, 438
    zeaxanthin, 440–441
    zinc, 438–439
  antioxidant enzymes, polymorphisms to, 616
  depletion, 435
  effect, 614
  plant-derived compounds effects, 711
  systems, 659–660
  vitamins
    as strategy to delaying vascular aging, 613
    vitamin C, 613–614
AP. *See* Area postrema (AP)
AP site. *See* Apurinic/apyrimidinic site (AP site)
AP-1. *See* Activator protein-1 (AP-1)
APC. *See* Antigen presenting cell (APC)
APE. *See* Apurinic/apyrimidinic endonuclease (APE)
Apigenin, 285, 306
apo. *See* Apolipoprotein (apo)
Apo E4 allele. *See* Apolipoprotein E4 (apo E4) allele
ApoA-V. *See* Apolipoprotein A-V (ApoA-V)
ApoB. *See* Apolipoprotein B (ApoB)
Apolipoprotein (apo), 23
Apolipoprotein A-V (ApoA-V), 23
Apolipoprotein B (ApoB), 267
Apolipoprotein E4 (apo E4) allele, 338, 625, 641–644
Apoptosis, 243, 491, 537
APP. *See* Amyloid precursor protein (APP)
Appetite, 122
  regulation, 59
Apurinic/apyrimidinic endonuclease (APE), 245–247
  APE1, 610–611
Apurinic/apyrimidinic site (AP site), 245
Arachidonic acid (AA), 160–161, 252–253, 382–384, 629

Arcuate nucleus (ARC), 190
ARE. *See* Androgen response element (ARE); Antioxidant response element (ARE)
Area postrema (AP), 190
AREDS. *See* Age-Related Eye Disease Study (AREDS)
Arterial endothelial dysfunction, 611
Artificial nutrition, 66–67
  costs and budgeting of, 67
  EN, 67
  PN, 67
Ascorbic acid, 304
Ascorbyl radical, 112
ASCVD. *See* Atherosclerotic cardiovascular disease (ASCVD)
Asian natural products
  black rice, 669
  dansen, 671–672
  gegen, 672
  Ginkgo, 666
  ginseng, 660–661
  herbs and spices
    curcumin, 666–667
    garlic, 668
    ginger, 668
    saffron, 667–668
  limitations of research, 672–673
  marine algae, 669–670
  mushrooms, 670–671
  soy, 663–666
  tea, 661–663
  wolfberry, 671
ASPEN. *See* American Society for Parenteral and Enteral Nutrition (ASPEN)
Assisted eating, 57
Asthma, 411, 417–418
Asymmetric dimethylarginine (ADMA), 611–612
AT. *See* Adipose tissue (AT)
AT1. *See* Angiotensin II type I receptor (AT1)
Ataxia with vitamin E deficiency (AVED), 638–639
ATBC Cancer Prevention Trial. *See* Alpha-Tocopherol Beta-Carotene (ATBC) Cancer Prevention Trial
ATF-3. *See* Activating transcription factor-3 (ATF-3)
Atherosclerosis, 315–316, 317f, 334–335, 609
  carotenoids and, 475
    effect of carotenoid consumption, 477–478
    early atherosclerosis, 476–477
    improving atherosclerotic risk factors, 476–477
  glycation and, 335
  type II diabetes, insulin resistance, and, 335
Atherosclerotic cardiovascular disease (ASCVD), 316. *See also* Cardiovascular disease (CVD)
  risk factors, 316–322
ATP. *See* Adenosine triphosphate (ATP)
ATRA. *See* All-*trans* retinoic acid (ATRA)
Atrial fibrillation (AF), 33
Atrial flutter (AFL), 33

Auditory Consonant Trigram (ACT), 687
AURKA. *See* Mitotic kinase Aurora A (AURKA)
Autonomic nervous system (ANS), 36. *See also* Central nervous system (CNS)
Autophagy, 148, 227, 492, 524, 533–534, 649, 651
AVED. *See* Ataxia with vitamin E deficiency (AVED)
Average lifespan (ALS), 3
Aβ plaques. *See* Amyloid-β (Aβ) plaques

B

B cell(s), 557–558, 697
  repertoire, 375
B-vitamins, 304, 599
Bacteroides fragilis, 696
Basal metabolic rate (BMR), 58, 723–724
Base excision repair (BER), 243, 245–246, 610–611
Basic helix-loop-helix-leucine zipper (bHLH-Zip), 147–148
Basic leucine zipper (bZIP), 495–496
BCAA. *See* Branched chain amino acids (BCAA)
BDNF. *See* Brain-derived neurotrophic factor (BDNF)
Benfotiamine, 273
Benign prostate hyperplasia (BPH), 482–484
Benton Visual Retention Test (BVRT), 687
Benzopyranols, 302
BER. *See* Base excision repair (BER)
Berberine, 305–306
Beta cell mass, 394
β-amyloid, 659
β-carotene, 113, 439–440, 480
BH4. *See* Tetrahydrobiopterin (BH4)
bHLH-Zip. *See* Basic helix-loop-helix-leucine zipper (bHLH-Zip)
BIA. *See* Bioimpedance analysis (BIA)
Bifidobacterium species, 699
Bioactive compounds, 707, 718f
Bioactive phytochemicals, 218
Bioavailability, 116–117
Biochemical investigations, 64
Bioimpedance analysis (BIA), 416
Biological age, 11–12, 109, 117–118
Biological pathway, 294
Biomarker(s), 109
  of damage, 183
    antioxidant enzyme system, 184
    arsenic in hair, 183–184
    calcium, 185
    catecholamines, 185
    environmental contaminants, 184
    environmental pollutants, 185
    exposure to xenobiotics, 185
    HepG2 cells, 184–185
    lifelong exposure to pesticides, 184
    MDA, 184
    modifications, 184
    Nrf2, 184
    Nurr1, 184

    overproduction of reactive oxygen species, 185
    oxidative stress, 184
    ratio of GA-Hb to AA-Hb, 183
  of longevity, 682
Biosphere 2, 679, 680f
Biotin, 304, 599
Black rice (*Oryza sativa L. indica*), 669
Blood
  pressure, 320
  pressure-lowering effect, 615
  sugar, 332–333
Blue Zones, 47
BMD. *See* Bone mineral density (BMD)
BMI. *See* Body mass index (BMI)
BMP receptor (BMPR), 527
BMP-SMAD pathway, 529
BMPR. *See* BMP receptor (BMPR)
BMR. *See* Basal metabolic rate (BMR)
Bode index, 412, 415
Body composition, 683
Body mass index (BMI), 42, 44–45, 63, 116–117, 401, 413
Body mass/composition
  characteristic age-related and diet-induced changes in, 191
  diet-induced changes, 196–199
Body shape questionnaire (BSQ), 686
Body weight, 44–45
Bone
  health, 352
  mineral homeostasis, 450–451
  remodeling, 628–629
  omega-3 fatty acids on, 628–629
  systematic literature review on, 603t
  vitamin B12 and, 601
Bone mineral density (BMD), 345, 347, 350–351, 601, 621–622, 629
BPH. *See* Benign prostate hyperplasia (BPH)
Brain
  function, 623
  health in aging, 339
  impairment, 332–333
  inflammation as form of oxidative stress, 253–254, 255f
  insulin resistance, 398
    aging, 400
    brain insulin action, 399
    diabetes, 399
    inflammatory condition, 400
    insulin and IGF-1, 398
    insulin in CNS, 399
    NMDA, 398
  omega-3 fatty acids effects
    in AD and related dementia, 625
    on cognitive function in normal aging, 624–625
Brain aging, 252–253. *See also* Cognitive aging
  dietary intake and, 543
  calcium, 544
  magnesium, 543–544
  potassium, 545
  metabolic disorders and, 335–337

Brain aging (Continued)
  neuroprotection of phytochemicals in, 254–256
  oxidative damage and, 252–253
Brain-derived neurotrophic factor (BDNF), 254–255, 332–333
Branched chain amino acids (BCAA), 145, 417
BrdU. See 5-Bromodeoxyuridine (BrdU)
Breakage–fusion–bridge cycles, 130
Breast cancer, 161, 484
5-Bromodeoxyuridine (BrdU), 295–296
Bronchiectasis, 411, 414
Bronze diabetes, 531
BSQ. See Body shape questionnaire (BSQ)
Bulimia nervosa, 198
BVRT. See Benton Visual Retention Test (BVRT)
bZIP. See Basic leucine zipper (bZIP)

C

C to T substitution at 677 base (C677T), 518–519
c-Jun N-terminal kinase (JNK), 258, 400
C-reactive protein (CRP), 253–254, 323, 363, 367, 538, 557, 612
C677T. See C to T substitution at 677 base (C677T)
CAD. See Coronary artery disease (CAD)
Calcitriol, 282
Calcitriol-dependent monocyte differentiation, 555
Calcium (Ca), 537, 730–731
  dietary intake
    biological mechanisms in aging process, 538–540
    and brain aging, 544
    and cognitive aging, 546
    dietary intake and cardiovascular and metabolic effects, 541–542
  dietary reference values for, 346–347, 347t
  food sources of, 347–348, 348t
  pathways linking, 546f
CALERIE trial. See Comprehensive Assessment of Long-term Effect of Reducing Intake of Energy (CALERIE) trial
Calmodulin-dependent protein kinase (CAMKIV), 35
Caloric restriction (CR), 31–32, 36, 41, 48, 149, 198, 207–208, 227, 230, 246–247, 402–403, 677–678
  altered FOXO1 target genes during aging and modulation by, 209–210
  and chronic disease development, 687–688
  FoxO1 modulation, 211–212
  in humans, 678–679
    Biosphere 2, 679, 680f
    CALERIE trials, 679–682
    centenarians from Okinawa, 679
    food in CALERIE Phase I Study, 681t
  increasing longevity in humans, 688–689
  intermittent fasting, 678
  mechanisms and healthful aging
    biomarkers of longevity, 682
    body composition, 683
    endocrine responses, 684–686
    metabolic adaptation, 684
    oxidative stress, 682, 684
    physical activity, 684
    rate of living, 682
  physical activity to producing caloric deficit, 678
  psychological and behavioral effects, 686
    cognitive function and performance, 687
    eating disorder symptoms development, 686–687
    QOL and mood, 687
    subjective feelings of hunger, 687
  psychological and behavioral responses, 685t
  traditional CR, 678
Calorie restriction. See Caloric restriction (CR)
Calorie Restriction Society (CRONIES), 686
Calorie Restriction with Optimal Nutrition (CRON), 688–689
Camellia sinensis, 661
CAMKIV. See Calmodulin-dependent protein kinase (CAMKIV)
cAMP responsive element-binding protein (CREB), 35, 209
Cancer, 150, 235, 296–298, 332–333, 708–709, 711, 716
  in aging, 629–630
  and curcumin, 501
  incidence, 568–569, 571–572
    prostate, 574, 576, 580
  mortality, 568, 571
  nutrigenetics of omega-3 PUHA in, 24–25
  one-carbon metabolism and, 516–517
  patients, 68
  prevention, 568
    carotenoids and, 480
    EpRE/ARE transcription system, 481–482
    inflammation, 482
    NF-κB, 482
    nuclear receptors, 482–484
Candidate gene studies, 435
Cannabinoids, 307
Cannabinol, 307
Carbohydrates, 32, 318, 332, 385–386, 729
  miRNAs modulation by, 280–281
Carbon tetrachloride (CCL4), 286
Carboxymethyl-lysine (CML), 264
Cardiac dysfunction, 150
Cardioprotective diet, 322
Cardioprotective dietary patterns, 315
  conventional functional foods in, 322–323
  fish, 323
  fruits and vegetables, 323
  legumes, 323
  nuts, 323
  seafood, 323
  whole grains, 322
Cardiovascular disease (CVD), 22, 49–50, 94, 269, 315, 331–332, 336–337, 423–425, 474, 540–541, 609–611, 622, 626, 688, 694, 708, 710, 713
  calcium, 541–542
  magnesium, 541
  modifiable risk factors and ideal levels for, 317t
  molecular basis and modifiable risk factors of, 316
  nutrients, foods and ASCVD risk factors
    blood pressure, 320
    insulin resistance and metabolic syndrome, 320–321
    lipids and lipoproteins, 316–320
    obesity, 321–322
    oxidative stress and telomere length, 322
  nutrigenetics of omega-3 PUFA in, 22–24
  omega-3 fatty acids and, 626–627
  one-carbon metabolism and, 517
  potassium, 542–543
Cardiovascular effects of vitamin C, 614–615
Cardiovascular (CV) events, 626
  effect of carotenoid consumption, 477–478
Cardiovascular Health Study (CHS), 402
Cardiovascular heart disease (CHD), 626
Cardiovascular risk factors, 688
Carnosine, 303
Carotenoid(s), 113, 439–440, 473–474
  and antiatherosclerotic effect, 482t
  biosynthesis pathways, 715f
  and cancer prevention, 480
    EpRE/ARE transcription system, 481–482
    inflammation, 482
    NF-κB, 482
    nuclear receptors, 482–484
  characteristics, 481t
  circulation, 483f
  molecular and clinical properties, 483f
  and oxidative stress, 474–475
  on skin aging, 478–480
  structure and function, 474
  and vascular health and atherosclerosis, 475
    effect of carotenoid consumption, 477–478
    early atherosclerosis, 476–477
    improving atherosclerotic risk factors, 476–477
Carotid intima-media thickness (CIMT), 477
CART. See Cocaine–amphetamine-regulated transcript (CART)
Casein kinase (CK), 555
CAT. See Catalase (CAT)
Catabolic effect, 189
Catalase (CAT), 110, 254
Cataract, 434
Catechins, 305
Catecholamines, 185
Catecholo-methyltransferase (COMT), 221
Cathecin, 650
CBP. See CREB-binding protein (CBP)
CBP/p300 factor, 211
CBP80. See 80 kDa nuclear capbinding protein (CBP80)
CBS. See Cystathionine β-synthase (CBS)

CCK. *See* Cholecystokinin (CCK)
CCL4. *See* Carbon tetrachloride (CCL4)
CDD-111, 299
CDK. *See* Cyclin-dependent kinases (CDK)
CDKN2A locus, 295–296
CDR. *See* Clinical dementia rating (CDR)
Cell-mediated adaptive immune responses, 378
Cellular energy sensors, 31
Cellular iron homeostasis regulation, 525–527, 526f
Cellular senescence, 27, 293, 498–499, 612
  in aging and age-related diseases, 296–297
  characterization of, 295–296
  definition and triggers of, 294
  modulators and potential targets, 297f
  nutritional factors and, 300–302
    alkaloids, 305–306
    amino-acids and peptides, 302–303
    B-vitamins, 304
    benzopyranols, 302
    cannabinoids, 307
    catechins, 305
    diarylheptanoids, 303–304
    divergent effects of bioactive dietary compounds, 301t
    flavones, 306
    flavonols, 307
    isoflavones, 306
    organosulphur compounds, 303
    secoiridoids, 304
    stilbenoids, 303
    sugars, 303
    trace elements, 304–305
    triterpenoid saponins, 302
    vitamin C, 304
  strategies to target cellular senescence with therapeutical perspective, 297
    direct ablators of senescent cells, 298
    indirect ablators of senescent cells, 298–299
    SASP modulators, 299
    senescence delayers, 300
    senescence inducers, 299–300
  strategies to rejuvenate senescent cells, 297–298
Centenarians, 554
  from Okinawa, 679
Central nervous system (CNS), 398, 592. *See also* Autonomic nervous system (ANS)
Central processing, 195–196
Cerebrospinal fluid (CSF), 252–253
Ceruloplasmin (CP), 245, 247, 524, 527
CGA. *See* Comprehensive geriatric assessment (CGA)
CHD. *See* Cardiovascular heart disease (CHD); Coronary heart disease (CHD)
Chemoprevention, 568
Chemosensation, 364
Chemotherapeutics, 82
CHF. *See* Congestive heart failure (CHF)
CHIANTI study, 427
Chicoric acid, 650
Childhood, 594

Chinese hamster ovary cells (CHO-IR), 145
Chloroquine (CQ), 298
CHO-IR. *See* Chinese hamster ovary cells (CHO-IR)
Cholecystokinin (CCK), 122, 191–192, 363–365
Cholesterol 7 α-hydrolase (CYP7), 318
Choline, 516
Chromatin, 213
  chromatin architecture, age-related alterations in, 217
Chronic age-related diseases prevention
  bioactive compounds, 718f
  constituents of vegetables and fruit, 711–717
  effects of intake of vegetables and fruit on mortality and morbidity, 708
    epidemiology, 708–709
    intervention studies, 709–710
    odds ratio of diabetes mellitus, 709f
  improving intake and effect, 717–718
Chronic disease development, 687–688
Chronic exposure, 180
Chronic inflammation, 475, 612
Chronic kidney disease (CKD), 423–425, 452
Chronic obstructive pulmonary disease (COPD), 44–45, 411–412, 416–417, 726–727
Chronic renal disease. *See* Chronic kidney disease (CKD)
Chronic wasting diseases, 198
Chronological age, 11–13, 109
Chrysanthemum cineraraefolium, 179
CHS. *See* Cardiovascular Health Study (CHS)
Chylomicrons, 474
CIMT. *See* Carotid intima-media thickness (CIMT)
Circadian rhythm, 227
cis-regulatory module (CRM), 455–456
CK. *See* Casein kinase (CK)
CKD. *See* Chronic kidney disease (CKD)
Clinical dementia rating (CDR), 660–661, 667–668
Clinical trials, 502–503, 568–570, 572, 580
CML. *See* Carboxymethyl-lysine (CML)
CMV infection. *See* Cytomegalovirus (CMV) infection
CNS. *See* Central nervous system (CNS)
Cocaine–amphetamine-regulated transcript (CART), 193–194
Cocoa polyphenols, 324
Coenzyme Q_{10}, 82–94, 246–247
Cognition, 335–336
Cognitive aging, 336. *See also* Brain aging
  dietary intake and
    calcium, 546
    magnesium, 545–546
    potassium, 546
Cognitive decline, 336–337, 624–625
Cognitive function
  systematic literature review on, 602t
  vitamin B12 and, 601
Cognitive impairment, 333–334, 622
Colon, 368–369
Colorectal cancer (CRC), 701

COMA. *See* UK Committee on Medical Aspects of Food Policy (COMA)
Comprehensive Assessment of Long-term Effect of Reducing Intake of Energy (CALERIE) trial, 677, 679–682
Comprehensive geriatric assessment (CGA), 61, 661–663
Comprehensive nutritional assessment, 61–64, 63f
  barriers for, 64, 64t
Computerized tomography (CT), 416
COMT. *See* Catecholo-methyltransferase (COMT)
Conditioned Taste Aversion, 716–717
Confounding factors, 279
Congestive heart failure (CHF), 352, 627
Constipation, 368–369
Continuous Performance Test-II (CPT-II), 687
COPD. *See* Chronic obstructive pulmonary disease (COPD)
Copper (Cu), 245, 304–305
Coronary arteries, 609
Coronary artery disease (CAD), 478
Coronary heart disease (CHD), 315, 323, 325, 423–424, 477, 609, 631
Corticotropin releasing hormone (CRH), 195
COX-2. *See* Cyclooxygenase-2 (COX-2)
CP. *See* Ceruloplasmin (CP)
CpG site, 213
cPLA2. *See* Cytosolic phospholipase A2 (cPLA2)
CPT-II. *See* Continuous Performance Test-II (CPT-II)
CPT1B, 34–35
CQ. *See* Chloroquine (CQ)
CR. *See* Caloric restriction (CR)
CRC. *See* Colorectal cancer (CRC)
Creatine (Cr), 358
Creatinine clearance rate, 423
CREB. *See* cAMP responsive element-binding protein (CREB)
CREB-binding protein (CBP), 209
Cretinism, 583
CRH. *See* Corticotropin releasing hormone (CRH)
Critically Ill elderly patients, 68
CRM. *See* *cis*-regulatory module (CRM)
Crocin, 667
CRON. *See* Calorie Restriction with Optimal Nutrition (CRON)
CRONIES. *See* Calorie Restriction Society (CRONIES)
Crotonaldehyde dG adduct, 162
CRP. *See* C-reactive protein (CRP)
CSF. *See* Cerebrospinal fluid (CSF)
CT. *See* Computerized tomography (CT)
Cu/ZnSOD. *See* Zinc and copper superoxide dismutase (Cu/ZnSOD)
Cullin3 (Cul3), 495–496
Curcuma longa, 493–494
Curcumin (diferuloylmethane), 255, 303–304, 492, 650, 666–667
  and aging, 499
  and disease, 501–502, 501t
  and longevity, 499–501

Curcumin (diferuloylmethane) (Continued)
  biochemical and molecular targets, 494, 495t
    aging-related molecular targets, 494–499
    impact on hallmarks of aging, 496f
  miRNAs modulation by, 286
  nutritional and pharmacological applications in aging, 502
    actual and possible future applications, 505f
    applications, 504–506
    clinical trials, 502–503
    curcumin bioavailability and pharmacokinetics, 503–504
    epidemiological data, 502–503
    new delivery systems, 504
  publications in US, 492f
  traditional uses, 493–494
Curcuminoids, 666–667
Curry, 667
  powder, 491
CV events. See Cardiovascular (CV) events
CVD. See Cardiovascular disease (CVD)
Cyclin-dependent kinases (CDK), 210–211, 295–296, 612
Cycloastragenol, 302
Cyclooxygenase-2 (COX-2), 24–25, 253–254, 612
Cyp isoenzyme 3A4 system. See Cytochrome-P-450 (Cyp) isoenzyme 3A4 system
CYP-containing enzyme. See Cytochrome P450 (CYP)-containing enzyme
CYP24A1 gene, 451, 455–456
  nucleotide sequence of human, 456f
CYP7. See Cholesterol 7 α-hydrolase (CYP7)
Cystathionine β-synthase (CBS), 515
Cysteine, 517–518
Cytochrome P450 (CYP)-containing enzyme, 449–450
Cytochrome-P-450 (Cyp) isoenzyme 3A4 system, 75–77, 78t
Cytokines, 58, 609
Cytomegalovirus (CMV) infection, 378
Cytosolic phospholipase A2 (cPLA2), 253–254

D

1,25D. See 1,25-Dihydroxyvitamin D_3 (1,25D)
25D. See 25-Hydroxyvitamin D_3 (25D)
DA neurons. See Dopaminergic (DA) neurons
dAGE. See Dietary advanced glycation end product (dAGE)
Dairy products, 347, 584
DALYs. See Disability-adjusted life years (DALYs)
Damage induced senescence, 294
Dansen, 671–672
Dark chocolate, 323–324
DASH. See Dietary approaches to stop hypertension (DASH)
Data and Safety Monitoring Committee (DSMC), 575

DBD. See DNA binding domain (DBD)
DCs. See Dendritic cells (DCs)
DDIT4. See Regulated in development and DNA damage response 1 (REDD1)
DDR. See DNA damage response (DDR)
Deacetylase, 227
Deacetylation, 211
Degenerative eye diseases, 270–271, 434
  AMD, 434–435
  cataract, 434
  diabetic eye diseases, 436
  glaucoma, 435–436
Deiodinases, 587
Delayed gastric emptying, 366
Delayed type hypersensitivity (DTH), 558
Dementia, 336–337, 516
  Alzheimer's disease, 337–338
    and glycation, 338–339
    and insulin resistance, 337–338
  glucose hypometabolism, 337
  and metabolic disturbances, 337–339
Dementia with Lewy bodies (DLB), 252–253
Dendritic cells (DCs), 376, 696
Dentition, 365
2-Deoxy-D-glucose (2DG), 298
3-Deoxyglucosane, 265
Deoxynivalenol (DON), 178
1-Deoxynojirimycin, 305–306
Deregulated nutrient sensing, 498
2′-Desoxythymidine-5′-monophosphate (dTMP), 244
2′-Desoxyuridine-5′-monophosphat (dUMP), 244
Determinants of longevity, 156
Detoxification of endo-and xenobiotics, 460
Developmental Origins of Health and Diseases (DOHaD), 181–182
Developmentally programmed senescence, 294
DEXA. See Dual energy X-ray absorptiometry (DEXA)
2DG. See 2-Deoxy-D-glucose (2DG)
DGAC. See Dietary Guidelines Advisory Committee (DGAC)
DGCR8, 278
DHA. See Docosahexaenoic acid (DHA)
DHEA-S, 686
DHFR. See Dihydrofolate reductase (DHFR)
Diabesity, 151–152
Diabetes, 271, 399
  diabetic eye diseases, 436
  risk factors, 688
Diabetic macular edema (DME), 436
Diabetic retinopathy (DR), 434
Diarylheptanoids, 303–304
Dicentric chromosomes, 130
Diet, 32
  and brain impairment, 332–333
  and cognitive impairment, 333–334
  correlation, 32
  function of proteins and amino acids in, 33
  hyperuricemia and, 428–429
  ketogenic diets, 339
  Mediterranean type, 427–428
  patterns, 714

Diet-induced changes, 191
  body mass/composition, 196–199
    overfeeding—obesity, 196
      obesity and GI peptides, 196–197
      RYGB Surgery, 197–198
      signaling nutritional state in obesity, 197
  signaling feeding state
    calorie restriction, 198
    chronic wasting diseases, 198
    in eating disorders, anorexia nervosa, 198
Dietary advanced glycation end product (dAGE), 263–264. See also Advanced glycation end product (AGE)
  absorption, metabolism, and elimination of, 266–267
  dietary interventions for, 271–273
  formation, 265
  representative clinical interventions with, 272t
Dietary approaches to stop hypertension (DASH), 320, 324, 663
  dietary pattern, 324–325
Dietary cholesterol, 319
Dietary fat, miRNAs modulation by, 281–282
Dietary fatty acids, 23, 380–382
  miRNAs modulation by, 281–282
Dietary fiber, 729
Dietary Guidelines Advisory Committee (DGAC), 316–318
Dietary intake, 540–541
  aging process in
    calcium, 538–540
    deleterious mechanisms through lower levels of Mg, Ca, and K, 540t
    magnesium, 538
    potassium, 540
  and brain aging, 543
    calcium, 544
    magnesium, 543–544
    potassium, 545
  and cardiovascular and metabolic effects
    calcium, 541–542
    magnesium, 541
    potassium, 542–543
  and cognitive aging
    calcium, 546
    magnesium, 545–546
    potassium, 546
Dietary micronutrients, miRNAs modulation by, 282
Dietary minerals, miRNAs modulation by, 284–285
Dietary nitrate, 712
Dietary patterns and clinical outcomes, 324
  DASH dietary pattern, 324–325
  Mediterranean-style dietary patterns, 325
  2013 AHA/ACC Lifestyle Guidelines, 324
  Vegetarian dietary pattern, 325–326
Dietary phytochemicals, neuroprotective mechanisms of
  aging, 252

INDEX

brain
  aging, 252–253
  inflammation as form of oxidative stress, 253–254, 255f
  neuroprotection of phytochemicals in brain aging, 254, 255t
    curcumin, 255
    Flavan-3-ols, 256
    other phytochemicals, 256
    resveratrol, 254–255
  oxidative damage, 252–253
  stress signaling pathways, neuroprotective phytochemicals and modulation of, 256–258
Dietary protein, 357
Dietary reference intake (DRI), 41–42, 46–47, 95–103
Dietary reference values
  for calcium, 346–347, 347t
  for vitamin D, 348–349, 349t
Dietary restriction (DR), 31–32, 222. See also Caloric restriction (CR)
Dietary sugar influence on brain and cognitive impairment
  carbohydrates/sugar, 332
  diet and brain impairment, 332–333
  diet and cognitive impairment, 333–334
Dietary supplements, 569
Dietetic Products, Nutrition, and Allergies (NDA), 730
Diferuloylmethane. See Curcumin (diferuloylmethane)
Digestive tract function, nonnutrient components in, 712
Digital rectal exams (DRE), 574
Dihydrofolate reductase (DHFR), 514
1,25-Dihydroxyvitamin D_3 (1,25D), 449–450, 452
Diiodotyrosine (DIT), 586–587
Diketo form, 493–494
Dioxin, 183
Dipeptidyl peptidase IV (DPPIV), 395
Direct ablators of senescent cells, 298
Disability-adjusted life years (DALYs), 611
Distribution and elimination, 587
DIT. See Diiodotyrosine (DIT)
Divalent Metal Transporter (DMT1), 523–525
DLB. See Dementia with Lewy bodies (DLB)
DME. See Diabetic macular edema (DME)
DMN. See Dorsal motor nucleus (DMN)
DMT1. See Divalent Metal Transporter (DMT1)
DNA
  damage and nutrients, 25–26
  lesions, 243
  repair, 243
DNA binding domain (DBD), 452–453
DNA damage response (DDR), 294–296
DNA methylation, 27, 213, 513
  age-related alterations in, 215–216
  effects of diet on
    bioactive phytochemicals, 218
    methyl donors, 218
    protein-restricted diet, 219

Selenium, 219
Vitamin A, 219
Zinc, 218–219
DNA methyltransferase (DNMT), 218–219, 497
  DNMT3b, 286
DNI. See Drug–nutrient interaction (DNI)
DNMT. See DNA methyltransferase (DNMT)
Docosahexaenoic acid (DHA), 23, 318, 323, 379, 435, 441, 453–454, 622–623
DOHaD. See Developmental Origins of Health and Diseases (DOHaD)
DON. See Deoxynivalenol (DON)
Dopaminergic (DA) neurons, 151
Dorsal motor nucleus (DMN), 190
Dorsovagal complex (DVC), 190
Dose-response approach, 600–601
Double strand break (DSB), 130, 245–246
Down syndrome (DS), 247
DPPIV. See Dipeptidyl peptidase IV (DPPIV)
DR. See Diabetic retinopathy (DR); Dietary restriction (DR)
DRE. See Digital rectal exams (DRE)
DRI. See Dietary reference intake (DRI)
Drosha, 278
Drosophila, 501–502
Drug absorption in elderly, 79
Drug action in elderly, 81
Drug distribution in elderly, 79–80
Drug excretion in elderly, 81
Drug metabolism in elderly, 80–81
Drug–herb interactions in elderly, 95, 96t
Drug–nutrient interaction (DNI), 75, 80t
  challenges and future directions, 95–103
  classification, 76t
  dietary, 103f
    supplement use, 75
  drugs
    affecting cytochrome P450 enzymes, 77t
    drug–herb interactions in elderly, 95, 96t
    effect on nutritional status in elderly, 82
    modulating P-glycoprotein, 78t
  factors affecting DNI in elderly, 79
    drug absorption in elderly, 79
    drug action in elderly, 81
    drug distribution in elderly, 79–80
    drug excretion in elderly, 81
    drug metabolism in elderly, 80–81
  number of supplements, 75f
  pressor agents in foods and beverages, 78t
  type I, 75
  type II, 75–77
  type IIC, 77
  type III, 77–78
  type IV, 78
  with vitamins supplements in elderly, 82
    mineral–drug interactions, 88t
    vitamin A, 94
    vitamin B and folic acid, 82–94
    vitamin C, 94
    vitamin D, 94–95
    vitamin E, 94
    vitamin K, 95
    vitamins–drug–nutrient interactions, 83t

Drusen, 434–435
Dry method, 585
DS. See Down syndrome (DS)
DSB. See Double strand break (DSB)
dSir2, 233
DSMC. See Data and Safety Monitoring Committee (DSMC)
DTH. See Delayed type hypersensitivity (DTH)
dTMP. See 2'-Desoxythymidine-5'-monophosphate (dTMP)
Dual energy X-ray absorptiometry (DEXA), 416
dUMP. See 2'-Desoxyuridine-5'-monophosphat (dUMP)
Duodenum, 525
DVC. See Dorsovagal complex (DVC)
Dyslipidemia, 609
Dysregulated miRNA levels, 282

E

EAA. See Essential amino acid (EAA)
EAR. See Estimated average requirement (EAR)
Early atherosclerosis, 476–477
Early life exposure and long-term effects, 180
  alterations of cognitive function, 183
  effect of DNA methylation, 181f
  DOHaD, 181–182
  effect on fetal growth, 182
  impact of environmental factors, 182f
  epigenetic modifications, 183
  epimutations, 181
  histone acetylation, 181f
  lead exposure, 182–183
  lifestyle, diet, pesticides, metals, and solvents, 182
  sperm epimutations, 183
  studies on children's hair, 182
  transgenerational effect on fat deposits, 183
Eating disorders, 198
  symptoms development, 686–687
Eating Inventory, 686
4E-BP1. See eIF4 binding proteins (4E-BP1)
ECG. See Epicatechin gallate (ECG)
Ectopic calcification, 451–452, 454–455, 457–458
EER. See Estimated Energy Requirement (EER)
EFA. See Essential fatty acid (EFA)
EFSA. See European Food Safety Authority (EFSA)
EGCG. See Epigallocatechin gallate (EGCG); (−)-Epigallocatechin-3-gallate (EGCG)
Eicosanoids, 160
Eicosapentaenoic acid (EPA), 23, 318, 436, 622
Eicosapentanoic acid (EPA), 323, 380–382, 622–623
5,8,11-Eicosatrienoic acid (ETrA), 158–159
eIF4 binding proteins (4E-BP1), 147
Elderly, 425–426
  hepcidin in, 532

Elderly (*Continued*)
    autophagy, 533–534
    iron metabolism, 533–534
    immune cells dysfunctions in, 697t
    relationship between gut microbiota and immune system in elderly, 694
        B cells, 697
        Bacteroides fragilis, 696
        methagenomic archaea, 695
        microbiota composition in ages of life, 695f
        pathways, 695–696
        Treg cells and Th17 cells intestinal microbiota and induction, 696f
    zinc supplementation in elderly person, 558, 560t
        antioxidant effects, 559
        cytokine expression, 559
        T cell activity, 559
        T cell response, 559
        zinc intake, 558
        zinc treatment, 561
Electrophile response element (EpRE), 473, 479
Electrophilic stress, 166–167
ELISA. *See* Enzyme-linked immunosorbent assays (ELISA)
Elongation of very long chain fatty acids protein 2 (ELOVL2), 37
Elongation of very long chain fatty acids protein 6 (ELOVL6), 32, 34
ELS. *See* Essential lifespan (ELS)
Emphysema, 411
EN. *See* Enteral nutrition (EN)
End-of-life care, 68–69
Endocrine responses
    DHEA-S, 686
    leptin, 685
    somatotropic axis, 686
    thyroid function, 684–685
Endogenous antioxidant systems, 164
Endogenous DNA adducts, 163
Endogenous oxidation, 161
Endogenous secretory RAGE (es RAGE), 268
Endoplasmic reticulum (ER), 400
Endothelial activation, 612
Endothelial nitric oxide synthase (eNOS), 32, 611–612
Endothelin (ET-1), 475, 609
Endothelium, 609–611
Energy, 725–726, 727t
    balance, 189–190, 192, 725
    energy sensor systems, central position of, 35
    intake, 122
Enol form. *See* Keto-enol form
eNOS. *See* Endothelial nitric oxide synthase (eNOS)
Enteral feeding, 57
Enteral nutrition (EN), 64, 67
Enteric nervous system, 364
Environmental pollution, exposure to, 177
Environmental selenium, 578
Enzymatic antioxidants, 110–112
Enzymatic co-factor, 243

Enzymatic PUFA oxidation, 160–161
Enzyme-linked immunosorbent assays (ELISA), 265
EPA. *See* Eicosapentaenoic acid (EPA); Eicosapentanoic acid (EPA)
EPIC cohort. *See* European prospective Investigation into Cancer and Nutrition cohort (EPIC cohort)
Epicatechin, 256
Epicatechin gallate (ECG), 256
Epidemiological studies, 435, 502–503, 516–517, 709–710
Epifriedelanol, 302
Epigallocatechin, 256
Epigallocatechin gallate (EGCG), 114, 256, 305, 402
(−)-Epigallocatechin-3-gallate (EGCG), 661–663
Epigenetic alterations, 497
    role of, 222
Epigenetic responses to diet in aging
    age-associated and diet-induced epigenetic alterations interrelationships, 220–221
    age-related alterations
        chromatin architecture, 217
        in DNA methylation, 215–216
        histone modification, 216–217
        interacting hallmarks of aging and extrinsic modifying factors, 216f
        small, noncoding RNAs, 217
    diet effect, 221
        on DNMT activity, 221
        HDAC, 221–222
        on methyl group supply, 221
    epigenetic actions of specific dietary components, 218–220
        diet effects on DNA methylation, 218–219
        diet effects on histone modification, 219
        diet effects on small noncoding RNAs, 219–220
    epigenetic alterations, 222
        impact of diet, 217
        of mammalian genome, 214–215
        as nutritionally-modifiable marker of aging trajectory, 223
    epigenetic drift, 215
        role of environmental *vs.* genetic factors, 217
    sirtuins, 222–223
    stem cell aging, functional consequences of, 222
Epigenetic(s), 27–28, 31, 213, 513, 610
    clock, 215–216
    drift, 213, 215–216
    linkage of aging and nutrition, 36–37
    marks, 213
    modification of mammalian genome, 214–215
    phenomena, 513
Epigenome, 213, 491–492
Epigenome-wide association study, 37
EPO. *See* Erythropoietin (EPO)
EpRE. *See* Electrophile response element (EpRE)

EpRE/ARE transcription system, 481–482
ER. *See* Endoplasmic reticulum (ER)
ERCs. *See* Extrachromosomal rDNA circles (ERCs)
ERFE. *See* Erythroferrone (ERFE)
ERK. *See* Extracellular signal regulated kinase (ERK)
Erythroferrone (ERFE), 529–530
Erythropoiesis, hepcidin regulation by, 527, 529–530
Erythropoiesis-dependent regulation of hepcidin, 527–530
Erythropoietin (EPO), 526–527
es RAGE. *See* Endogenous secretory RAGE (es RAGE)
Esophageal achalasia, 365
Esophagus, 365–366
ESPEN. *See* European Society for Clinical Nutrition and Metabolism (ESPEN)
Essential amino acid (EAA), 280
Essential fatty acid (EFA), 156–157, 622
Essential lifespan (ELS), 3, 6
Estimated average requirement (EAR), 41, 345, 347, 569, 723, 727
Estimated Energy Requirement (EER), 723
Estrogen receptors, 482–484
ET-1. *See* Endothelin (ET-1)
1,N^6-Ethenoadenine (ϵdA), 165–166
3,N^4-Ethenocytosine (ϵdC), 165–166
ETrA. *See* 5,8,11-Eicosatrienoic acid (ETrA)
EU. *See* European Union (EU)
Euchromatin, 213
EUGMS. *See* European Geriatric Medicine Society (EUGMS)
European Food Safety Authority (EFSA), 727
European Food Safety Authority Panel recommendations, 346–347
European Geriatric Medicine Society (EUGMS), 728–729
European prospective Investigation into Cancer and Nutrition cohort (EPIC cohort), 517, 571
European Society for Clinical Nutrition and Metabolism (ESPEN), 728
European Union (EU), 626
EURRECA Network of Excellence, 600
Exceptional longevity, 31
Exogenous oxidation, 161–162
Exogenous supplemental antioxidants, 164–165
Experimental studies. *See* Intervention studies
Extracellular signal regulated kinase (ERK), 258
Extrachromosomal rDNA circles (ERCs), 232
Extreme longevity. *See* Exceptional longevity

F

FACIT. *See* Folate After Coronary Intervention Trial (FACIT)
"False negative" effects, 708
"False positives" effects, 708
FAO. *See* Food and Agriculture Organization (FAO)

FAST. See Functional Assessment Staging (FAST)
"Fast day", 678
Fat, 622, 729
Fat mass (FM), 677, 680–681
Fat-free mass (FFM), 414, 677, 680–681
Fatty acid(s)
  component, 33–34
  durability and oxidizability of, 157–158
  ω-3. See Omega-3 fatty acids (ω-3 fatty acids)
  ω-6. See Omega-6 fatty acids (ω-6 fatty acids)
FDA. See Food and Drug Administration (FDA)
"Feed day", 678
Feeding state, 189
  age-related changes of signaling, 192–193
  gastrointestinal signals of, 191–192
Femoral neck BMD (FN-BMD), 629
Ferric iron (Fe^{3+}), 525
Ferric Reducing/Antioxidant Power (FRAP), 670
Ferritinophagy, 533
Ferroportin (FPN1), 523–524, 532
FeSOD. See Iron superoxide dismutase (FeSOD)
Fetal development, 592–594
FEV1. See Forced expiratory volume (FEV1)
FFA. See Free fatty acid (FFA)
FFM. See Fat-free mass (FFM)
Fibroblast growth factor 2 (FGF2), 499
Fibroblast growth factor 23 (FGF23), 451, 456–457
  dissection of the mouse, 458f
  nucleotide sequence of human, 456f
FIP200. See 200 kDa FAK family kinase-interacting protein (FIP200)
Fish, 323
FKBP12–rapamycin binding domain (FRB), 141
FKBP38, 148
Flavan-3-ols, 256
Flavones, 306
Flavonoids, 713
  miRNAs modulation by, 285
Flavonols, 307
Flexible fatty acid composition of human cells, 156–157
  benefits of Omega-9 PUFAs, 158–159
  durability and oxidizability of fatty acids, 157–158
  families of polyunsaturated fatty acids, 157f
  lessons from human populations with low PUFA intake, 159–160
  pathways, 158f
Floor dust, 179–180
Flow-mediated dilation (FMD), 271–272
FM. See Fat mass (FM)
FMD. See Flow-mediated dilation (FMD)
FN-BMD. See Femoral neck BMD (FN-BMD)
Folate, 518–520, 599, 730
  cycle, 514–515
  miRNAs modulation by, 283
  sole function, 514

Folate After Coronary Intervention Trial (FACIT), 516
Folic acid, 82–94, 244
Food. See also Diet
  for aging interventions, 6
  bioactive components, 21
  choices, 124
  intake, 189, 192–193, 196
    declines, 356
  nutrients, 698
  preferences in elderly
    chemosensory function modifications and, 123
    clinical and nutritional status modifications, 123–124
    determinants of food choices, 124
    nutritional frailty, 121–122
    physiological modifications, 122–123
  sources
    of calcium, 347–348, 348t
    vitamin D, 349, 350t
Food and Agriculture Organization (FAO), 727
Food and Drug Administration (FDA), 95–103, 568–569
Food-derived small molecules. See Phytochemical(s)
Foodstuffs, 178
Forced expiratory volume (FEV1), 64
Forkhead box subgroup "O" (FoxO), 207–208, 209t
  FOXO3, 47–48, 365
  proteins, 211
FOS. See Framingham Osteoporosis Study (FOS); Fructooligosaccharides (FOS)
FoxO. See Forkhead box subgroup "O" (FoxO)
FoxO1, 207–208
  epigenetic influences on, 210–211
  modifications
    acetylation, 211
    deacetylation, 211
    by phosphorylation, 210–211
  modulation by calorie restriction, 211–212
  significance ubiquitination and degradation, 211
  target genes during aging and modulation by CR, 209–210
FPN1. See Ferroportin (FPN1)
FR. See Free radicals (FR)
Frailty, 41
  nutritional interventions, 356–357
Framingham Osteoporosis Study (FOS), 629
FRAP. See Ferric Reducing/Antioxidant Power (FRAP)
FRB. See FKBP12–rapamycin binding domain (FRB)
Free fatty acid (FFA), 382–384
Free radical theory of aging (FRTA), 4, 682
Free radicals (FR), 4, 610, 707
  theory, 637–638
French Paradox, 650
FRTA. See Free radical theory of aging (FRTA)
Fructooligosaccharides (FOS), 698

Fruits, 323
  constituents affecting age-related morbidity, 711
    antioxidant effects of plant-derived compounds, 711
    nonnutrient components, 712
    nutrients in human metabolism, 711
    phytochemicals, 712–717
  effects of intake on mortality and morbidity, 708
    epidemiology, 708–709
    intervention studies, 709–710
    odds ratio of diabetes mellitus, 709f
  natural pesticides, bioactive compounds, 717t
Functional annotation analysis, 16–18
Functional Assessment Staging (FAST), 667–668
Functional food(s), 491
  conventional functional foods in cardioprotective dietary patterns, 322–323
  green tea, dark chocolate, and red wine, 323–324

G

G-CSF. See Granulocyte colony-stimulating factor (G-CSF)
Galacto-oligosaccharides (GOS), 698
GALT. See Gut associated lymphatic system (GALT)
γ-linolenic acid (GLA), 382–384
γ-tocopherol, 114, 572
Ganoderma lucidum. See Lingzhi (*Ganoderma lucidum*)
GAP. See GTPase activating protein (GAP)
Garlic (*Allium sativum*), 668
Gas chromatography with flame ionization detector (GC/FID), 265
Gas chromatography with mass spectrometric detection (GC/MS), 265
Gastric atrophy, 724
Gastric emptying, 366
Gastroesophageal reflux disease (GERD), 363, 365–366
Gastrointestinal signals of feeding state, 191–192
Gastrointestinal system (GI system), 190
Gastrointestinal tract, 380
GC/FID. See Gas chromatography with flame ionization detector (GC/FID)
GC/MS. See Gas chromatography with mass spectrometric detection (GC/MS)
GDF15. See Growth Differentiation Factor 15 (GDF15)
GDH. See Glutamate dehydrogenase (GDH)
GEF. See Guanine nucleotide exchange factor (GEF)
Gegen, 672
GEM study. See Ginkgo Evaluation of Memory (GEM) study
Gene expression profiles/whole transcriptome analysis, 11–12, 16–18

Genes and nutrients interactions, One-carbon metabolism
  folate, 518–519
  MAT and S-adenosylmethionine, 519–520
  MAT1A and vitamin B-6 and folate, 520
  MS and vitamin B-12, 519
  MTHFR, 518–519
Genistein, 306
Genome-based study, 517
Genomic instability, 495–496
Genomic stability, 243
  micronutrient deficiency and consequences for, 244
    folic acid, 244
    niacin, 244–245
    trace elements, 245–246
    vitamin B12, 244
  nutritional interventions in elderly and impact on, 246–247
Genotoxic effects of nutritional interventions, 247
GERD. *See* Gastroesophageal reflux disease (GERD)
GFR. *See* Glomerular filtration rate (GFR)
GH. *See* Growth hormone (GH)
Ghrelin, 123, 193
GI. *See* Glycemix index (GI)
GI system. *See* Gastrointestinal system (GI system)
Ginger, 668
Ginkgo, 666
Ginkgo Evaluation of Memory (GEM) study, 666
Ginko biloba, 666
Ginseng, 660–661
  clinical trials in cognitive function tests, 661t
  health benefits, 660t
  saponins, 660
Ginsenosides, 302, 660
GIP. *See* Glucose-dependent insulinotropic polypeptide (GIP)
GL. *See* Glycemic load (GL)
GLA. *See* γ-linolenic acid (GLA)
Glaucoma, 271, 435–436
Glomerular filtration rate (GFR), 423–424
Glucagon-like peptide (GLP), 192
  GLP-1, 366, 395
Glucose, 32, 303
  effectiveness, 396
  hypometabolism, 337
  metabolism, 398–399
    preventing alteration, 400–402
  tolerance, 725
Glucose transporter (GLUT), 393, 395
  GLUT2, 367
  GLUT4, 34–35
Glucose-dependent insulinotropic polypeptide (GIP), 192, 366, 395
GLUT. *See* Glucose transporter (GLUT)
Glutamate dehydrogenase (GDH), 231
Glutathione (GSH), 110, 112–113, 208, 491, 495–496
Glutathione disulfide (GSSG), 491
Glutathione peroxidase (GPx), 49–50, 110, 112, 243–244, 254, 659–660

Glutathione peroxidase (GSH-Px). *See* Glutathione peroxidase (GPx)
Glutathione S-transferase (GST), 24, 166–167, 254, 615–616
Glycation, 264, 335, 338–339
Glycemic control, 710
Glycemic load (GL), 332
Glycemix index (GI), 332
  peptides, 196–197
Glycine, 516–517
Glycine N-methyltransferase (GNMT), 514–515
Glyoxal, 265
GM. *See* Granulocyte-monocyte (GM)
GNMT. *See* Glycine N-methyltransferase (GNMT)
Goiter, 583
Goiter rate (GR), 588–589
Goitrogens, 584
Goji. *See* Wolfberry (*Lycium barbarum*)
GOS. *See* Galacto-oligosaccharides (GOS)
Gout, 424
GPx. *See* Glutathione peroxidase (GPx)
GR. *See* Goiter rate (GR)
Grade 0, 588–589
Granulocyte colony-stimulating factor (G-CSF), 555
Granulocyte-monocyte (GM), 555
Greater trochanteric (GT), 350–351
Green tea, 323–324, 429, 661
Growth Differentiation Factor 15 (GDF15), 529–530
Growth hormone (GH), 413, 593, 677, 686
GSH. *See* Glutathione (GSH)
GSSG. *See* Glutathione disulfide (GSSG)
GST. *See* Glutathione S-transferase (GST)
GT. *See* Greater trochanteric (GT)
GTPase activating protein (GAP), 143
Guanine nucleotide exchange factor (GEF), 148
Gut associated lymphatic system (GALT), 67
Gut microbiota, 368–369
  and immune system relationship in elderly, 694
    B cells, 697
    Bacteroides fragilis, 696
    methagenomic archaea, 695
    microbiota composition in ages of life, 695f
    pathways, 695–696
    Treg cells and Th17 cells intestinal microbiota and induction, 696f
Gβl. *See* Mammalian lethal with sec13 protein 8 (mLST8)

H

H3K4me3. *See* Histone H3 lysine 4 trimethylation (H3K4me3)
H4K16. *See* Histone H4 lysine 16 (H4K16)
HAG cells. *See* Human astroglial (HAG) cells
Hand grip, 64
Haptoglobin (HP), 615–616
Hazard ratio (HR), 575
HbA1c. *See* Hemoglobin A1c (HbA1c)

HCCC. *See* Human colon cancer cells (HCCC)
HCT. *See* Histone acetyltransferase complexes (HCT)
HDAC. *See* Histone deacetylase (HDAC)
HDL. *See* High density lipoprotein (HDL)
HDS-R. *See* Revised Hasegawa Dementia Scale (HDS-R)
Health, Aging and Body Composition (Health ABC), 728
Health-related quality of life (HRQOL), 724
Healthful aging
  kidney as nexus, 450f, 451–452
  VDR-mediated control of genes networks vital for, 454–465
Healthspan, 150
Healthy aging, 43–44, 48–49, 363, 366–367
Healthy diets, 284–285
Healthy Eating Index (HEI), 322
Heart failure (HF), 627
Heart Outcomes Prevention Evaluation (HOPE) study, 517
Heart rate variability (HRV), 36
Heat shock proteins (HSP), 7, 497–498, 561–562
Heat shock response (HSR), 6–7
HEI. *See* Healthy Eating Index (HEI)
Heme oxigenase-1 (HO-1), 256–257, 525
Heme-iron, 525
Hemochromatosis, 531
  animal models, 532
Hemoglobin A1c (HbA1c), 264, 271
Hemojuvelin (HJV), 524, 527
Hepatic glucose production (HGP), 396
Hepcidin, 523
  in elderly, 532
    autophagy, 533–534
    iron metabolism, 533–534
  excess disorders, 532
  hepcidin deficiency, disorders with, 531–532
  regulation, 527
    by erythropoiesis, 529–530
    by Iron, 529
  regulation in hepatocytes, 528f
Hepcidin-ferroportin axis deregulation, 530
  disorders of hepcidin excess, 532
  disorders with hepcidin deficiency, 531–532
  genetic conditions with, 531t
HepG2 cells, 184–185
Hephaestin (HEPH), 524
Hericium erinaceus, 670
Heterochromatin, 213–214, 227
Heterogeneity. *See* Heterogeneous
Heterogeneous, 610
HF. *See* Heart failure (HF)
hF. *See* Human fibroblasts (hF)
HFCS. *See* High fructose corn syrup (HFCS)
HFD. *See* High fat diet (HFD)
HGP. *See* Hepatic glucose production (HGP)
4-HHE. *See* 4-Hydroxy-2-hexenal (4-HHE)
HIF-1. *See* Hypoxia-inducible factor (HIF-1)
High density lipoprotein (HDL), 316, 474, 609

High fat diet (HFD), 400
High fructose corn syrup (HFCS), 333
High molecular weight (HMW), 265
High-fat diet, 281–282
High-performance liquid chromatography (HPLC), 265
Histone
  covalent modification, 214
  modification
    age-related alterations in, 216–217
    effects of diet on, 219
  proteins, 215
Histone acetyltransferase complexes (HCT), 180
Histone deacetylase (HDAC), 180, 218–219, 227, 236, 497
  HDAC5, 35
Histone H3 lysine 4 trimethylation (H3K4me3), 215–216
Histone H4 lysine 16 (H4K16), 228–230
HJV. See Hemojuvelin (HJV)
HK-2 cells. See Human kidney-2 (HK-2) cells
HMB. See β-Hydroxy β-methylbutyrate (HMB)
HMG-CoA. See 3-Hydroxy-3-methylglutaryl coenzyme A (HMG-CoA)
HMO. See Human milk oligosaccharides (HMO)
HMW. See High molecular weight (HMW)
4-HNE. See 4-Hydroxy-2-nonenal (4-HNE)
HO-1. See Heme oxigenase-1 (HO-1)
13-HODE, 159
Holotranscobalamin (holoTC), 601
HOMA. See Homeostatic model assessment (HOMA)
Homeodynamic(s), 3, 5–6
  space, 3, 5–6
Homeostatic model assessment (HOMA), 271–272
Homocysteine, 517–518
Homogeneity. See Homogeneous
Homogeneous, 610
Homologous recombination (HR), 243
HOPE study. See Heart Outcomes Prevention Evaluation (HOPE) study
Hopkins Verbal Learning test (HVLT), 663
Hordaland Health Study, 516
Hordaland Homocysteine Study, 515
Hormesis, 3, 6, 256, 491, 716
Hormetin, 3
Hospital environment-related risks, 60
Hospitalization, 57
Hospitalization-associated malnutrition management, 64–67
  estimating nutritional requirements, 65t
  nonpharmacological intervention, 64–66
  pharmacological intervention, 66
Hospitalized elderly, nutrition in, 57
  clinical consequences of malnutrition in hospital, 60–61
  hospitalized older adults nutritionally vulnerable, 58–60
    age-related changes, 59–60
    functional and psychosocial factors, 60
    hospital environment-related risks, 60

effect of hospitalization on nutritional state, 60t
medical causes, 60
use of multiple medications, 60
malnutrition screening and assessment, 61–64, 62t
  anthropometry, 63
  barriers for comprehensive nutritional assessment, 64, 64t
  biochemical investigations, 64
  nutritional history, 61–63
  other functional measurements, 64
  physical and clinical assessment, 63
management of hospitalization-associated malnutrition, 64–67
  estimating nutritional requirements, 65t
  nonpharmacological intervention, 64–66
  pharmacological intervention, 66
nutritional issues in special groups
  cancer patients, 68
  critically Ill elderly patients, 68
  nutrition and end-of-life care, 68–69
  patients with advanced dementia, 68
pathogenesis of malnutrition, 58
postdischarge plan, 67
predisposing factors for malnutrition, 59f
recommendations, 69
role of cytokines, 58f
Hou Tou Gu in Chinese. See Hericium erinaceus
HP. See Haptoglobin (HP)
HPLC. See High-performance liquid chromatography (HPLC)
HR. See Hazard ratio (HR); Homologous recombination (HR)
HRQOL. See Health-related quality of life (HRQOL)
HRV. See Heart rate variability (HRV)
HSP. See Heat shock proteins (HSP)
HSR. See Heat shock response (HSR)
HT. See Hydroxytyrosol (HT)
hTERT. See Human telomerase reverse transcriptase (hTERT)
Human astroglial (HAG) cells, 286
Human colon cancer cells (HCCC), 303–304
Human fibroblasts (hF), 297–298
Human kidney-2 (HK-2) cells, 459–460
Human milk oligosaccharides (HMO), 700
Human telomerase reverse transcriptase (hTERT), 497
Human umbilical vascular endothelial cell (HUVEC), 302–303, 475–476
Human(s)
  breast milk, 161
  CR in, 678–679
    Biosphere 2, 679, 680f
    CALERIE trials, 679–682
    centenarians from Okinawa, 679
    food in CALERIE Phase I Study, 681t
  CR increasing longevity in, 688–689
  human body, iodine in, 585
    absorption, 586–587
    distribution and elimination, 587
    metabolism, 586–587, 588f
    recycling, 587

human metabolism, nutrients in, 711
microbiota, 368
Humoral adaptive immune responses, 378–379
Humoral immunity, 557–558
Hunger subjective feelings, 687
HUVEC. See Human umbilical vascular endothelial cell (HUVEC)
HVLT. See Hopkins Verbal Learning test (HVLT)
Hydrogen peroxide, 397
Hydrophilic antioxidants. See Lipophilic antioxidants
  GSH, 112–113
  vitamin C, 112
β-Hydroxy β-methylbutyrate (HMB), 359
4-Hydroxy-2-hexenal (4-HHE), 163
4-Hydroxy-2-nonenal (4-HNE), 160, 163–164
3-Hydroxy-3-methylglutaryl coenzyme A (HMG-CoA), 476
4-Hydroxy-pentenal. See 4-Hydroxy-2-nonenal (4-HNE)
8-Hydroxydeoxyguanosine (8-OHdG), 185, 252–253
6-Hydroxydopamine (6-OHDA), 258
Hydroxytyrosol (HT), 304
25-Hydroxyvitamin D (25OHD), 345, 348
25-Hydroxyvitamin D_3 (25D), 449–450
Hyperinsulinemia, 335
Hyperthyroidism, 583
Hyperuricemia and diet, 428–429
Hypoalbuminemia, 59
Hypochlorhydria, 366
Hypothalamic inflammation, 400
Hypothyroidism, 583, 594
Hypoxia, 227
Hypoxia-inducible factor (HIF-1), 413
  HIF-1α, 650
  HIF2-α, 525

I

IBD. See Inflammatory bowel disease (IBD)
ICAM. See Intracellular adhesion molecule (ICAM)
ICC. See Interstitial cells of Cajal (ICC)
ID. See Iodine deficiency (ID); Iron deficient (ID) diet
IDH2. See Isocitrate dehydrogenase (IDH2)
IECs. See Intestinal epithelial cells (IECs)
IF. See Intermittent fasting (IF)
IFN. See Interferon (IFN)
IFR-7. See Interferon regulatory factor-7 (IFR-7)
Ig. See Immunglobulin (Ig)
IGFs. See Insulin-like growth factors (IGFs)
IIS. See Insulin-like growth factor-1 signaling (IIS)
IKARIA study, 427
IL. See Interleukin (IL)
Immune cells, 693–694
Immune function, omega-3 fatty acids on, 627–628
Immune system, 12, 18, 375, 450–451
  aging and adaptive, 377–379

Immune system (Continued)
　aging and innate, 376–377
　and gut microbiota relationship in elderly, 694
　　B cells, 697
　　Bacteroides fragilis, 696
　　methagenomic archaea, 695
　　microbiota composition in ages of life, 695f
　　pathways, 695–696
　　Treg cells and Th17 cells intestinal microbiota and induction, 696f
　immunomodulatory effects
　　of functional foods, 383t
　　of single nutrients, 381t
　nutrition impact on immune components, 380–386
　nutritional effects on aged immune cells, 385t
　nutritional intervention targets, 379–380
Immune-gut microbiota axis, 694
　DNA sequencing, 693
　food intake and modulation, 698
　future trends, 700–701
　immune cells dysfunctions in elderly, 697t
　microbiota, 693–694
　patterns of salivary cytokines, 694f
　prebiotics on, 698–699
　probiotics on, 699
　relationship between gut microbiota and immune system in elderly, 694
　　B cells, 697
　　Bacteroides fragilis, 696
　　methagenomic archaea, 695
　　microbiota composition in ages of life, 695f
　　pathways, 695–696
　　Treg cells and Th17 cells intestinal microbiota and induction, 696f
　synbiotics on, 700
Immunglobulin (Ig), 557–558
Immunomodulatory effects
　functional foods, 383t
　single nutrients, 381t
Immunonutrition, 375
Immunosenescence, 375, 377
　zinc and, 554
　　adaptive immunity, 556–557
　　age-dependent changes in immune system, 556f
　　B cells, 557–558
　　humoral immunity, 557–558
　　innate immunity, 555–556
In vitro studies, 294, 698
InCHIANTI. See Invecchiare in Chianti (InCHIANTI)
Incretins, 395
Indirect ablators of senescent cells, 298–299
Induced pluripotent stem cells (iPSCs), 297–298
Inducible nitric oxide synthase (iNOS), 254–255, 612
Inflammaging, 18, 696–697
Inflammation, 482, 610, 637, 694
Inflammatory bowel disease (IBD), 694

Inflammatory/immune response, 639–641
Influence of diet on health, 331–332
INK4A/ARF. See CDKN2A locus
iNKT cells. See Invariant NKT (iNKT) cells
Innate immune system. See also Adaptive immune system
　aging and, 376
　　DCs, 376
　　macrophages, 376
　　monocytes, 376
　　neutrophils, 376–377
　　NK cells, 377
Innate immunity, 555–556
iNOS. See Inducible nitric oxide synthase (iNOS)
INSR mRNA, 281
Insulin, 123, 194–195, 396. See also Brain insulin resistance
　epidemiologic studies, 397
　insulin-dependent tissue, 396
　mTORC1, 398
　　mTORC1-regulated processes, 397–398
　muscular lipid infiltration, 396
　oxidative stress, 396–397
　resistance, 320–321, 333, 335, 337–338, 396, 537, 688
　secretion, metabolism, and clearance, 394–395
Insulin degrading enzyme, 338
Insulin receptor (IR), 397–398
Insulin receptor substrate (IRS), 143, 208, 399
　IRS1, 143
Insulin-like growth factor-1 signaling (IIS), 47
Insulin-like growth factors (IGFs), 397
　IGF-I, 47, 143, 279–280, 334, 498, 628–629, 677, 686
Intercellular adhesion molecule 1 (ICAM-1), 473, 475
Interferon (IFN), 553–554, 696–697
　IFN-γ, 497
Interferon regulatory factor-7 (IFR-7), 376
Interleukin (IL), 252, 363, 368, 497, 553–554, 610
　IL-1, 123–124
　IL-6, 58, 123–124, 267, 612
　IL-17, 694
Interleukin-1 receptor-associated kinase-1 (IRAK-1), 286
Intermittent fasting (IF), 36, 677
International unit (IU), 570–571
Interstitial cells of Cajal (ICC), 364
Intervention studies, 707
Intestinal dysfunction, 363
Intestinal epithelial cells (IECs), 379–380
Intestinal microbiota, 368
Intracellular adhesion molecule (ICAM), 612
Intraocular pressure (IOP), 435
Intraorgan specific biological phenotypes, 15–16
Intraperitoneal CCK (IP CCK), 192
Intravenous glucose tolerance test (IVGTT), 393
Invariant NKT (iNKT) cells, 377
Invecchiare in Chianti (InCHIANTI), 357
Iodermia, 585

Iodine (I), 583
　cycle in environment, 584f
　epidemiology, 587
　　criteria for assessing iodine nutrition, 590t
　　ID disorders, 588t
　　indicators of iodine status in populations, 589t
　　iodine nutrition, 588–589
　　iodine status, 590–592
　　national iodine status, 590f
　　physiological stages of iodine status, 591f
　　SAC, 589t
　　systematic reviews, 587–588
　　thyroid disorders, 590–592
　　UIC, 589–590
　excess, 595
　in human body, 585
　　absorption, 586–587
　　distribution and elimination, 587
　　metabolism, 586–587
　　recycling, 587
　and human lifespan
　　adults/elderly, 594
　　childhood, 594
　　fetal development, 592–594
　　iodine deficiency, 594–595
　　iodine excess, 595
　　pregnancy, 592
　　THs in human development, 593f
　　thyroid autoimmunity, 595–596
　　thyroid cancer, 596
　sources, 584–585, 584t
　supplementation in pregnancy and infancy, 586t
Iodine deficiency (ID), 587, 594–595
Iodine-induced hypothyroidism, 595
IOP. See Intraocular pressure (IOP)
IP CCK. See Intraperitoneal CCK (IP CCK)
iPSCs. See Induced pluripotent stem cells (iPSCs)
IR. See Insulin receptor (IR)
IRAK-1. See Interleukin-1 receptor-associated kinase-1 (IRAK-1)
IRE. See Iron Responsive Elements (IRE)
IRIDA. See Iron refractory iron deficiency anemia (IRIDA)
Iron (Fe), 245, 304–305, 524, 731
　absorption, 525
　cellular iron homeostasis regulation, 525–527, 526f
　deficit, 417
　deregulation of hepcidin-ferroportin axis, 530–532
　dietary, 525
　hepcidin in elderly, 532
　　autophagy, 533–534
　　iron metabolism, 533–534
　homeostasis, 532
　　autophagy, 533–534
　　iron metabolism, 533–534
　homeostasis, 532–534
　intestinal iron absorption, 525
　iron-dependent regulation of hepcidin, 527–530

metabolism, 533–534
  systemic iron homeostasis regulation, 527–530, 528f
Iron deficient (ID) diet, 526–527
Iron refractory iron deficiency anemia (IRIDA), 531
Iron Regulatory Proteins (IRP), 523, 525–526
Iron Responsive Elements (IRE), 523, 525–526
Iron superoxide dismutase (FeSOD), 110–111
IRP. See Iron Regulatory Proteins (IRP)
IRS. See Insulin receptor substrate (IRS)
Ischemic heart disease (IHD). See Coronary heart disease (CHD)
Isocitrate dehydrogenase (IDH2), 231
Isoflavones, 306
IU. See International unit (IU)
IVGTT. See Intravenous glucose tolerance test (IVGTT)

J

Japan Dialysis Outcomes and Practice Patterns Study (JDOPPS), 425–426
JNK. See c-Jun N-terminal kinase (JNK)
Junctional adhesion molecule (JAM), 363

K

Karyomegaly, 12–13
200 kDa FAK family kinase-interacting protein (FIP200), 148
80 kDa nuclear capbinding protein (CBP80), 147
KDIGO. See Kidney Disease Improving Global Outcomes (KDIGO)
Kelch-like ECH-associated protein 1 (Keap1), 7, 256–257, 495–496
Keto form. See Diketo form
Keto-enol form, 493–494
Ketogenic diets, 339
Ketone body levels, 339
Ki67, 295–296
Kidney, 451
  nexus of healthful aging, 450f, 451–452
Kidney Disease Improving Global Outcomes (KDIGO), 424
  guidelines, 425–426
Kidney, functions of, 424
  in aging, 425–426
    Australian healthcare system, 424–425
    hyperuricemia and diet, 428–429
    Mediterranean type of diet, 427–428
    polyphenols and, 429–430
    PUFA, 426–427
Kinase catalytic domain, 141
Klotho gene, 457–460, 649, 652
Knowledge gaps, 137
Kuppfer cells, 527

L

L-Carnosine, 303
L-ergothioneine, 670
L-selenomethionine, 574
LA. See Linoleic acid (LA)

Labile iron pool (LIP), 527
Lactation, 586
Latanoprost, 442
Lateral hypothalamic area (LHA), 195
LBD. See Ligand binding domain (LBD)
LC. See Long chain (LC)
LC-MS/MS. See Liquid chromatography with tandem mass spectrometric detection (LC-MS/MS)
LC-PUFA. See Long chain PUFAs (LC-PUFA)
LCA. See Lithocholic acid (LCA)
LCAD. See Long-chain acyl CoA dehydrogenase (LCAD)
lck. See Lymphocyte protein-tyrosine kinase (lck)
Lcmt1. See Leucine carboxyl methyltransferase (Lcmt1)
LDL. See Low-density lipoprotein (LDL)
Legumes, 323
Lenalidomide, 299
Leptin, 193–194, 685
Leucine, 357
Leucine carboxyl methyltransferase (Lcmt1), 519
Leucyl-tRNA synthetase (LRS), 146
Leukocytes, 184–185
Leukotrienes, 160, 610
LHA. See Lateral hypothalamic area (LHA)
Life expectancy, 43
Lifespan, 227, 498
Lifespan-extending mutations, 210
Ligand binding domain (LBD), 452–453
Ligands, 267–268
Lingzhi (Ganoderma lucidum), 670
Linoleic acid (LA), 156–157
Lion's Mane Mushroom. See Hericium erinaceus
LIP. See Labile iron pool (LIP)
Lipid peroxidation (LPO), 160
  and consequences for genome stability, 160
  enzymatic PUFA oxidation, 160–161
  nonenzymatic PUFA oxidation, 161
    endogenous oxidation, 161
    exogenous oxidation, 161–162
    reproductive organ cancers, 161
  presence of lipid peroxidation DNA adducts in living tissues, 162
  mutagenicity of lipid peroxidation products, 163–164
  protective mechanisms against, 164
    endogenous antioxidant systems, 164
    exogenous supplemental antioxidants, 164–165
    tolerance and repair of lipid peroxide DNA adducts, 165–166
Lipid peroxide DNA adducts, tolerance and repair of, 165–166
Lipid-lowering effect, 615
Lipids, 33, 316–320, 623
Lipin1, 147–148
Lipofuscin, 12–13, 295–296
  index, 15–16
Lipofuscinosis, 13
Lipopeptide (LP), 530

Lipophilic antioxidants. See Hydrophilic antioxidants
  carotenoids, 113
  polyphenols, 114
  tocopherols, 113–114
Lipopolysaccharide (LPS), 518, 555
Lipoprotein lipase (LPL), 23
Lipoproteins, 316–320
Liquid chromatography with tandem mass spectrometric detection (LC-MS/MS), 265–266
Lithocholic acid (LCA), 453–454, 460
Liver, 527
  vacuolization, 12
Liver kinase B1 (LKB1), 150
Liver X receptor (LXR), 318
LKB1. See Liver kinase B1 (LKB1)
LMW. See Low molecular weight (LMW)
LOAEL. See Lowest Observed Adverse Effect Level (LOAEL)
Long chain (LC), 25
Long chain PUFAs (LC-PUFA), 156–157, 623
Long terminal repeat (LTR), 304
Long-chain acyl CoA dehydrogenase (LCAD), 231
Long-lived species, resistance to oxidative damage in, 166–167
Long-term dietary intervention, 315
Long-term oxygen therapy (LTOT), 412
Long-term potentiation (LTP), 333
Longevity, 3–4, 47–50, 109–110, 114, 117–118
  correlation, 32
  principles, 4t
Low molecular weight (LMW), 265
Low-carbohydrate diets, 32–33
Low-density lipoprotein (LDL), 116, 267, 316, 335, 474, 609
Lowest Observed Adverse Effect Level (LOAEL), 716–717
LP. See Lipopeptide (LP)
LPL. See Lipoprotein lipase (LPL)
LPO. See Lipid peroxidation (LPO)
LPS. See Lipopolysaccharide (LPS)
LRS. See Leucyl-tRNA synthetase (LRS)
LTOT. See Long-term oxygen therapy (LTOT)
LTP. See Long-term potentiation (LTP)
LTR. See Long terminal repeat (LTR)
Lung cancer, 411, 418
Lung disease
  nutrition and, 412–414
  nutritional intervention, 416
    asthma, 417–418
    COPD, 416–417
    elderly patients with COPD, 418t
    lung cancer, 418
Lutein, 440–441
LXR. See Liver X receptor (LXR)
Lycium barbarum. See Wolfberry (Lycium barbarum)
Lycopene, 474, 479
Lymphocyte protein-tyrosine kinase (lck), 555
Lysosomes, 143–145, 649

M

3-MA. *See* 3-Methyl-adenine (3-MA)
Macronutrients, 47, 725, 729
Macrophage inflammatory protein (MIP).
    See Macrophage-inhibitory protein
    (MIP)
Macrophage-inhibitory protein (MIP), 376,
    497
Macrophages, 335, 376, 555
Macular pigment, 440
MAEDS. *See* Multifactorial Assessment of
    Eating Disorder Symptoms (MAEDS)
Magnesium (Mg), 245, 537, 731
    deficiency, 351
    depletion, 82
    dietary intake
        biological mechanisms in aging process,
            538
        and brain aging, 543–544
        and cardiovascular and metabolic
            effects, 541
        and cognitive aging, 545–546
    pathways linking, 546f
Magnetic resonance imaging (MRI), 416
Maillard reaction, 264
Maillard reaction products (MRP), 265
Maintenance and repair systems (MARS), 5,
    5t
Malnutrition, 41, 44–45, 57, 121, 363, 414
    in elderly, 44–45
        causes of, 45–46, 45t
        possible correction with supplements in
            elderly, 46–47
Malnutrition Universal Screening Tool
    (MUST), 414
Malondialdehyde (MDA), 160, 184, 668
Mammalian genome, epigenetic modification
    of, 214–215
Mammalian lethal with sec13 protein 8
    (mLST8), 141–142
Mammalian sirtuins, 230–232
Mammalian stress-activated map kinase-
    interacting protein 1 (msIn1), 142
Mammalian target of rapamycin (mTOR), 27,
    42, 48, 141, 295–296, 357, 393,
    396–397, 492, 498, 649, 716. *See also*
    mTOR complex 1 (mTORC1)
    functional role of PROTOR, 142
    mTORC1, 141
        downstream targets, 147–149
        effects in age-related diseases, 150–152
        nutrients and energy status as upstream
            regulators of, 142–147
    mTORC2, 141
        downstream targets, 147–149
        nutrients and energy status as upstream
            regulators of, 142–147
    PPIase FKBP1A, 142
    RAPTOR, 141–142
    in senescence, 149–150
    single gene encoding, 141
Manganese superoxide dismutase (MnSOD),
    110–111
MAOI. *See* Monoamine oxidase inhibitors
    (MAOI)

mAPKAP1. *See* Mammalian stress-activated
    map kinase-interacting protein 1
    (msIn1)
MAPKs. *See* Mitogen-activated protein
    kinases (MAPKs)
Marine life, 584
Markers of frailty, 12
Marine algae, 669–670
MARS. *See* Maintenance and repair systems
    (MARS)
MAT. *See* Methionine adenosyltransferase
    (MAT)
Matrix metalloproteinase, 473
Matrix metalloproteinases (MMPs), 253–254,
    478, 653
Matrix-assisted laser desorption ionization-
    mass spectrometry with time-of-flight
    detection (MALDI-TOF/MS), 266
Max Rubner-Institute (MRI), 553
Maximal life span (MLSP), 166
Maximum lifespan (MLS), 3
MB. *See* Mitochondrial biogenesis (MB)
MCH. *See* Melanin concentrating hormone
    (MCH)
MCI. *See* Mild cognitive impairment (MCI)
MCP. *See* Monocyte chemoattractant protein
    (MCP)
MDA. *See* Malondialdehyde (MDA)
mDC. *See* Myeloid DC (mDC)
MDP. *See* Muramyldipeptide (MDP)
Mead acid, 158–159
Mechanistic target of rapamycin (mTOR).
    See Mammalian target of rapamycin
    (mTOR)
Med Diet. *See* Mediterranean diet (Med Diet)
Medical nutrition therapy (MNT), 326
Medical nutrition therapy for management
    of cardiovascular risk factors, 326
Mediterranean diet (Med Diet), 114,
    384–385, 403, 423, 427–428, 714
Mediterranean-style dietary patterns, 325
MEFs. *See* Mouse embryonic fibroblasts
    (MEFs)
Melanin concentrating hormone (MCH),
    195
Membrane pacemaker theory, 34
Mesenchymal stem cells (MSC), 303
Meta-analysis, 610
Metabolic adaptation, 677, 682, 684
Metabolic disorders
    atherosclerosis, 334–335
    and brain aging
        cardiovascular disease, cognitive
            decline, and dementia, 336–337
        type II diabetes, cognitive aging, and
            dementia, 336
        type II diabetes and cognition, 335–336
    glycation and atherosclerosis, 335
    type II diabetes, insulin resistance, and
        atherosclerosis, 335
Metabolic effects, 540–541
    calcium, 541–542
    magnesium, 541
    potassium, 542–543
Metabolic rate, 192, 195, 197

Metabolic syndrome (MetS), 316, 320–321,
    386, 533, 538
Metabolism, 586–587
Metal-response element binding
    transcription factor (MTF), 561
Metallothioneins (MT), 42, 48–49, 552–553
Methionine adenosyltransferase (MAT),
    514–515, 519–520
    MAT1A, 520
Methionine restriction diet, 515–516
Methionine synthase (MS), 514–515, 519
Methyl donors, 218
Methyl tocols. *See* Benzopyranols
3-Methyl-adenine (3-MA), 298
Methylate's pri-miRNAs, 278
Methylenetetrahydrofolate reductase
    (MTHFR), 514–515, 518–519
5,10-Methylenetetrahydrofolate, 514–515
Methylglyoxal, 265
    AGEs, 270
O^6-Methylguanine-DNA methyltransferase
    (MGMT), 496
Methylmalonic acid (MMA), 601
Methyltransferase-like 3 (METTL3), 278
MetS. *See* Metabolic syndrome (MetS)
MGMT. *See* O^6-Methylguanine-DNA
    methyltransferase (MGMT)
MI. *See* Myocardial infarction (MI)
Microbiome, 369, 693
Micronucleus formation, 243
Micronutrients, 82, 136, 243, 571, 730
    calcium, 730–731
    folate, 730
    iron, 731
    magnesium, 731
    potassium, 731
    vitamin B12, 730
    vitamin D, 730–731
    zinc, 731
Microorganisms, 693–694
MicroRNAs (miRNAs), 277–278, 652
    aging associated, 279–280
    biogenesis and processing, 278
    circulating, 278–279
    dietary modification of, 280
    modulation
        by carbohydrates, 280–281
        by curcumin, 286
        by dietary fat and fatty acids, 281–282
        by dietary micronutrients, 282
        by dietary minerals, 284–285
        by flavonoids, 285
        by folate, 283
        by milk, 286–287
        by polyphenols, 285–286
        by protein, 280
        by vitamin A, 283–284
        by vitamin C, 284
        by vitamin D, 282–283
        by vitamin E, 284
Microsomal triglyceride transfer protein
    (MTP), 34
Middle-aged obesity, 189
Mild cognitive impairment (MCI), 545,
    624–625

Milk, 347
    miRNAs modulation by, 286–287
    oligosaccharides, 699
Min mouse model, 710
Minerals, 33, 537–538
Mini Mental State Examination (MMSE), 624, 660–661
Mini nutrition assessment-short form (MNA-SF), 63
Mini nutritional assessment (MNA), 42, 44–45, 412, 414–415
Minnesota Semi Starvation study, 687
Minority and Medically Underserved (MMUS), 574
MIP. See Macrophage-inhibitory protein (MIP)
miR-1 expression, 279–280
miR-7 expression, 286
miR-7 targets SETD8, 286
miR-9, 283–284
miR-23a, 279–280
miR-34b, 285
miR-34c, 285
miR-103a, 283–284
miR-107, 281–282
miR-122, 285
miR-125b, 285
miR-152, 283–284
miR-155, 285
miR-192–5p, 286
miR-206, 279–280
mIR-215, 286
miR-374, 285
miRNAs. See MicroRNAs (miRNAs)
Mismatch repair (MMR), 243, 245–246
MIT. See Monoiodotyrosine (MIT)
Mitochondria, 652
    density, 413
Mitochondrial biogenesis (MB), 429
Mitochondrial dysfunction, 16–18, 498
Mitochondrial free radical theory of aging
Mitogen-activated protein kinases (MAPKs), 258, 267, 283–284, 397, 414, 555
Mitotic kinase Aurora A (AURKA), 299
MLS. See Maximum lifespan (MLS)
MLSP. See Maximal life span (MLSP)
mLST8. See Mammalian lethal with sec13 protein 8 (mLST8)
MMA. See Methylmalonic acid (MMA)
MMPs. See Matrix metalloproteinases (MMPs)
MMR. See Mismatch repair (MMR)
MMSE. See Mini Mental State Examination (MMSE)
MMUS. See Minority and Medically Underserved (MMUS)
MNA. See Mini nutritional assessment (MNA)
MNA-SF. See Mini nutrition assessment-short form (MNA-SF)
MNC. See Mononuclear cell (MNC)
MnSOD. See Manganese superoxide dismutase (MnSOD)
MNT. See Medical nutrition therapy (MNT)

Mo-DCs. See Monocyte-derived DCs (Mo-DCs)
Modified Mini-Mental State Examination (3MSE), 666
Molecular circadian clocks, 122
Molecular nutrition
    aging, 22
    cellular senescence and nutrients, 27
    DNA damage and nutrients, 25–26
    epigenetics and nutrients, 27–28
    food bioactive components, 21
    nutrigenetics
        of Omega-3 PUFA in CVD, 22–24
        of omega-3 PUHA in cancer, 24–25
    nutrigenomic exploration, 22f
Monoamine oxidase inhibitors (MAOI), 77–78
Monocyte chemoattractant protein (MCP), 497, 555
Monocyte-derived DCs (Mo-DCs), 376
Monocytes, 376, 555
Monoiodotyrosine (MIT), 586–587
Mononuclear cell (MNC), 380–382, 627–628
Monounsaturated fatty acid (MUFA), 33, 316–319, 380–382, 423, 426
Mood, 687
Morin, 307
Mouse distal convoluted tubule (mpkDCT) cells, 459–460
Mouse embryonic fibroblasts (MEFs), 297–298
mpkDCT cells. See Mouse distal convoluted tubule (mpkDCT) cells
MPS. See Muscle protein synthesis (MPS)
MPTP. See 1,2,3,6-Tetrahydro-1-methyl-4-phenylpyridine hydrochloride (MPTP)
MRI. See Magnetic resonance imaging (MRI); Max Rubner-Institute (MRI)
MRP. See Maillard reaction products (MRP)
MS. See Methionine synthase (MS)
MSC. See Mesenchymal stem cells (MSC)
3MSE. See Modified Mini-Mental State Examination (3MSE)
msIn1. See Mammalian stress-activated map kinase-interacting protein 1 (msIn1)
MT. See Metallothioneins (MT)
MTF. See Metal-response element binding transcription factor (MTF)
MTHFR. See Methylenetetrahydrofolate reductase (MTHFR)
mTOR. See Mammalian target of rapamycin (mTOR)
mTOR complex 1 (mTORC1), 141. See also Mammalian target of rapamycin (mTOR)
    downstream targets, 147
        anabolic processes, 147
        autophagy, 148
        4E-BP1, 147
        p70 ribosomal S6 kinase, 147
        PGC-1α, 148
        PolDIP3, 147
        S6K1, 147
        SREBP1 and HIFa, 147–148
        substrates of mTORC2, 148

    effects in age-related diseases, 150
        cancer, 150
        cardiac dysfunction, 150
        diabesity, 151–152
        neurodegenerative diseases, 150–151
        sarcopenia, 151–152
    nutrients and energy status as upstream regulators of, 142–147
        Akt/PKB, 143
        amino acids entrance, 144f, 146
        AMPK, 145
        cells lacking TSC, 145
        CHO-IR, 145
        IRS, 143
        levels of intracellular amino acids, 146
        LRS, 146
        lysosome, 143–145
        PI3K/Akt and Ras/MAPK pathways activation, 142–143
        Rag and amino acids, 146
        REDD1, 145
        S6K1, 143
        TFEB, 146
        TOR signaling, 144f
        upstream signals, 146–147
mTOR complex 2 (mTORC2), 141
mTORC1. See mTOR complex 1 (mTORC1)
MTP. See Microsomal triglyceride transfer protein (MTP)
MUFA. See Monounsaturated fatty acid (MUFA)
Multifactorial Assessment of Eating Disorder Symptoms (MAEDS), 686
Multimorbidities, 60
Multivitamin/multimineral (MVM), 75
Muramyldipeptide (MDP), 699
MuRF-1. See Muscle ring finger-1 (MuRF-1)
Muscle mass and function
    omega-3 fatty acids on, 628
Muscle mass reduction, 413
Muscle protein synthesis (MPS), 151–152
Muscle ring finger-1 (MuRF-1), 413
Muscular lipid infiltration, 396
Mushrooms, 670–671
MUST. See Malnutrition Universal Screening Tool (MUST)
Mutagenicity of lipid peroxidation products, 163–164
Mutation, 243, 610
MVM. See Multivitamin/multimineral (MVM)
Myelin, 271
Myeloid DC (mDC), 376
Myocardial infarction (MI), 336
Myxedematous cretinism, 593

N

n-3 fatty acids. See Omega-3 (ω-3) fatty acids
N-3 PUFAs, 321
n-6 fatty acids. See Omega-6 (ω-6) fatty acids
N-acetylcysteine (NAC), 302–303
N-methyl D-aspartate (NMDA), 398, 667–668
NAC. See N-acetylcysteine (NAC)

NAD. See Nicotinamide adenine dinucleotide (NAD)
NAD(P)H, quinone oxidoreductase 1 (NQO1), 256–257
NAD⁺-dependent protein deacetylases, 228
NAD⁺/NADH, 227
NADP. See Nicotinamide adenine dinucleotide phosphate (NADP)
NADPH. See Nicotinamide adenine dinucleotide phosphateoxidase (NADPH)
NAFLD. See Nonalcoholic fatty liver disease (NAFLD)
Namiki pathway, 265
NAMPT. See Nicotinamide phosphoribosyltransferase (NAMPT)
Nanoparticles, 492, 504
National Cancer Institute, 568
National Cholesterol Education Program Adult Treatment Panel III (NCEP ATP III), 324
National Health and Nutrition Examination Survey (NHANES), 75, 553, 569
NHANES III, 356
National Health Institute (NIH), 272–273
National Institute on Aging (NIA), 679
Natural killer (NK) cells, 296, 377, 696–697
Natural pesticides, 714–717
NCEP ATP III. See National Cholesterol Education Program Adult Treatment Panel III (NCEP ATP III)
NCOA4, 533
NDA. See Dietetic Products, Nutrition, and Allergies (NDA)
Negative regulatory domain (NRD), 141
NER. See Nucleotide excision repair (NER)
NET. See Neutrophil extracellular trap (NET)
Nettox, 713
Network analysis, 653
Neural system, 623
Neurocognitive disease, 516
Neurodegenerative diseases, 150–151, 269–270
Neurofibrillary tangles (NFT), 150–151, 252
Neuropeptide Y (NPY), 123, 193, 398–399
Neuropeptides in control of food intake, 190t
Neuroprotection, 399, 659
  neuroprotective effects, 660–661
Neuropsychiatric inventory (NPI), 667
Neutrophil extracellular trap (NET), 376–377, 555
Neutrophils, 376–377
  granulocytes, 555
New Nordic diet (NND), 386
New Zealand White (NZW) rabbits, 476
NF. See Nuclear factor (NF)
NF juvenile male Wistar rats. See Normally fed (NF) juvenile male Wistar rats
NF-κB. See Nuclear factor kappa B (NF-κB)
NFT. See Neurofibrillary tangles (NFT)
NHANES. See National Health and Nutrition Examination Survey (NHANES)
NHANES-III. See Third National Health and Nutrition Examination Survey (NHANES-III)

NHEJ. See Nonhomologous end joining (NHEJ)
NIA. See National Institute on Aging (NIA)
Niacin, 244–245, 599
Nicotinamide, 228
Nicotinamide adenine dinucleotide (NAD), 244–245
Nicotinamide adenine dinucleotide phosphate (NADP), 244–245
Nicotinamide adenine dinucleotide phosphateoxidase (NADPH), 267
Nicotinamide phosphoribosyltransferase (NAMPT), 235
NIH. See National Health Institute (NIH)
NIS. See Sodium–iodide symporter (NIS)
Nitrate, 712
Nitric oxide (NO), 23, 252–253, 399, 475, 611, 639–641
Nitric oxide synthase (NOS), 23, 399
  uncoupling, 611–612
Nitric oxide-sparing effect, 614–615
Nitrite, 712
Nitrogen-balance studies, 728
Nitrous oxide (NO), 652–653
NK cells. See Natural killer (NK) cells
NMDA. See N-methyl D-aspartate (NMDA)
NND. See New Nordic diet (NND)
NNFC. See Nonnutritional food components (NNFC)
NO. See Nitric oxide (NO); Nitrous oxide (NO)
NOD. See Nucleotide-binding oligomerization domain (NOD)
Non-HDL-C, 316–320
Non-SAE. See Nonsevere adverse effects (non-SAE)
Nonalcoholic fatty liver disease (NAFLD), 208
Noncommunicable diseases, 716
Nonenzymatic antioxidants, 110
Nonheme-iron, 525
Nonhomologous end joining (NHEJ), 243, 246–247
Noninsulin-mediated glucose uptake, 396
Nonnutrient components, 712
Nonnutritional food components (NNFC), 6
Nonpharmacological intervention, 64–66. See also Pharmacological intervention
  assessing need for food supplements, 65–66
  eliminating all potential reversible risk factors, 64
  enhancing food environment, 64–65
  enhancing quality and quantity of food, 65–66
  nutritional counseling, 66
  scheme of nutritional assessment of all patients, 66f
Nonsevere adverse effects (non-SAE), 630–631
Nonsteroidal anti-inflammatory drug (NSAID), 363
Nontransferrin bound iron (NTBI), 524–525
Nonvitamin A carotenoids, 714

Nordic Nutrition Recommendations, 728–729
Normal aging, 268–269
Normal nonpathological aging, omega-3 fatty acids effect, 628
Normally fed (NF) juvenile male Wistar rats, 191
Norwegian Vitamin Trial (NORVIT) study, 517
NOS. See Nitric oxide synthase (NOS)
Nothobranchius furzeri, 234–235
NPC. See Nutritional Prevention of Cancer (NPC)
NPI. See Neuropsychiatric inventory (NPI)
NPY. See Neuropeptide Y (NPY)
NRD. See Negative regulatory domain (NRD)
NRF. See Nuclear respiratory factors (NRF)
Nrf2. See Nuclear factor erythroid 2-related factor 2 (Nrf2)
NSAID. See Nonsteroidal anti-inflammatory drug (NSAID)
NTBI. See Nontransferrin bound iron (NTBI)
NTS. See Nucleus of the solitary tract (NTS)
Nuclear factor (NF), 555
Nuclear factor erythroid 2-related factor 2 (Nrf2), 7, 184, 256–257, 284, 303
Nuclear factor kappa B (NF-κB), 257–258, 267, 400, 414, 473, 475, 482, 612, 639–641
Nuclear receptors, 482–484
Nuclear respiratory factors (NRF), 148
Nucleotide excision repair (NER), 243, 245–247
Nucleotide-binding oligomerization domain (NOD), 696–697
Nucleus of the solitary tract (NTS), 190
Nur77, 35
Nurr1, 184
Nutrient(s), 32, 513, 579
  absorption, 366–368
  bioavailability, nonnutrient components in, 712
  cellular senescence and, 27
  components, 32–33
  DNA damage and, 25–26
  epigenetics and, 27–28
  and gene interaction, 513
  in human metabolism, 711
  immunomodulatory effects of single, 381t
  nutrient-gene interaction in elderly, 49–50
  nutrient-sensing
    longevity genes, 47
    mechanisms, 122–123
    pathway, 42, 47–49
Nutrigenetics, 21, 22f
  of omega-3 PUFA in CVD, 22–24
  of omega-3 PUHA in cancer, 24–25
Nutrigenomics, 21, 42, 49–50
  exploration, 22f
Nutritional Prevention of Cancer (NPC), 571–572, 580
Nutritional factors. See Nutritional/nutrition—compounds

Nutritional/nutrition, 32, 68–69, 109–110, 355, 622
  and aging, 402–403
  for aging interventions, 6
  assessment for pulmonary disease, 414–416
  biomarkers of aging, 109–110, 116–118
    antioxidant network, 110–114
    complex aging process, 115f
    ROS, 110
  compounds, 293
  counseling, 66
  in elderly, 43
    individualized intervention strategies, 43–44
    nutrient-gene interaction in elderly, 49–50
    nutrient-sensing pathway, 47–49
    nutrition-related problems, 43
  epigenetic linkage, 36–37
  frailty, 121–122
  history, 61–63
  hormetins, 6–7
  impact on immune components, 380–386
  intervention, 68
    in elderly and impact on genomic stability, 246–247
    genotoxic effects, 247
    strategies, 244
    targets, 379–380
  and lung disease, 412–414
    nutritional intervention in, 416–418
  methods for nutritional assessment, 415t
  new delivery systems, 504
    actual and possible future applications, 505f
    applications, 504–506
  nutritional assessment for pulmonary disease, 414–416
  and pharmacological applications of curcumin, 502
    clinical trials, 502–503
    curcumin bioavailability and pharmacokinetics, 503–504
    epidemiological data, 502–503
  recommendations
    carbohydrate, 729
    dietary fiber, 729
    energy, 725–726, 727t
    equations for prediction of resting energy expenditure, 726t
    fat, 729
    micronutrients, 730–731
    for older people, 724–725
    older persons, 725
    optimal nutrition, 724
    pathways to nutritional health, 724
    protein, 726–729
    water, 730
  state, 189
  state signals and age role, 193–195
    insulin, 194–195
    leptin, 193–194
  status, 121
  status and gastrointestinal health in elderly, 363
    achalasia, 365–366
    anorexia of aging, 364–365
    colon, 368–369
    constipation, 368–369
    enteric nervous system, 364
    esophagus, 365–366
    GERD, 365–366
    gut microbiota, 368–369
    intestinal dysfunction, 363
    oropharyngeal capacities, 365
    small bowel, nutrient absorption, and small intestinal permeability, 366–368
    stomach, gastric emptying, postprandial hypotension, and acid secretion, 366
    swallowing, 365
    taste and smell, 364
  and telomeres, 132–136
Nuts, 323
NZW rabbits. See New Zealand White (NZW) rabbits

O

OAG. See Open-angle glaucoma (OAG)
Obesity, 196–197, 321–322
  and GI peptides, 196–197
  signaling nutritional state in, 197
Observational studies, 707–708
Octamer, 214
Odds ratios (OR), 435
OGG1. See 8-Oxoguanine glycosylase 1 (OGG1)
OGTT. See Oral glucose tolerance test (OGTT)
25OHD. See 25-hydroxyvitamin D (25OHD)
6-OHDA. See 6-Hydroxydopamine (6-OHDA)
8-OHdG. See 8-Hydroxydeoxyguanosine (8-OHdG)
4-OHE. See 4-Oxo-2-hexenal (4-OHE)
OKG. See Ornithine α-ketoglutarate (OKG)
Okinawan diet, 386
Older persons, nutritional needs for, 725
Oleoythanolamide, 624
Oleuropein, 304
Olfaction, 364
Oligodendrocytes, 252–253
Omega-3 (ω-3) fatty acids, 33, 358–359, 423, 622–623. See also Polyunsaturated fatty acid (PUFA)
  on bone health, 628–629
  and brain health in aging
    in AD and related dementia, 625
    on cognitive function in normal aging, 624–625
  and cancer in aging, 629–630
  and CVD, 626–627
  on immune function in normal aging, 627–628
  on muscle mass and function, 628
  on quality of life and mortality in aging, 630–631
Omega-3 rich diets, 165
Omega-6 (ω-6) fatty acids, 622
4-ONE. See 4-Oxo-2-nonenal (4-ONE)
One-carbon metabolism, 513, 514f
  aging, age-associated disease and cancer, 516–517
    and cardiovascular disease, 517
    influence, 515
    methionine restriction diet, 515–516
    and neurocognitive disease, 516
  genes and nutrients interactions in aging
    folate, 518–519
    MAT and S-adenosylmethionine, 519–520
    MAT1A and vitamin B-6 and folate, 520
    MS and vitamin B-12, 519
    MTHFR, 518–519
  inflammation and
    S-adenosymethionine and, 518
    vitamin B-6 and, 517–518
  one-carbon units, 514
Open-angle glaucoma (OAG), 435
OR. See Odds ratios (OR)
Oral glucose tolerance test (OGTT), 394
Orexigenic hormones, 121
Orexin-A (OXA), 195
Organic selenium, 576–578
Organosulphur compounds, 303
Ornithine α-ketoglutarate (OKG), 359
Oropharyngeal capacities, 365
Oryza sativa L. indica. See Black rice (*Oryza sativa L. indica*)
Osteoarthritis, 270
Osteocalcin, 460
Osteoporosis, 345–346
  calcium
    dietary reference values for, 346–347, 347t
    food sources of, 347–348, 348t
  dairy products effect on, 350–351
  estimated number of osteoporotic fractures, 346f
  risk of fractures, 351–352
  seminutritional factors affecting, 351–352
  share of osteoporotic fractures, 346f
  vitamin D
    dietary reference values for, 348–349, 349t
    food sources, 349, 350t
3′ Overhang of mammalian telomeres, 131
OXA. See Orexin-A (OXA)
Oxidation, 610
Oxidative mechanisms, 436
Oxidative phosphorylation, 397
Oxidative stress, 109, 208, 322, 333, 435, 610, 694
  caloric restriction altering, 682
  carotenoids and, 474–475
  hypothesis, 682
  in insulin resistance development, 396–397
  metabolic adaptation and, 684
OXM. See Oxyntomodulin (OXM)
8-OxoG. See 8-oxoguanine (8-oxoG)
4-Oxo-2-hexenal (4-OHE), 160
4-Oxo-2-nonenal (4-ONE), 162–163
8-Oxoguanine glycosylase 1 (OGG1), 245
8-Oxoguanine (8-oxoG), 245
Oxyntomodulin (OXM), 192

P

p53, 235–236, 338
p70 ribosomal S6 kinase, 147
PAD. See Peripheral artery disease (PAD)
PAF. See Platelet activating factor (PAF)
PAI-1. See Plasminogen activator inhibitor type 1 (PAI-1)
Palmitoleic acid, genomic studies and importance, 34–35
PAM. See Prematurely aging mice (PAM)
Pancreatic polypeptide (PP), 192
Pantothenic acid, 599
Parathyroid hormone (PTH), 345, 349, 451
Paraventricular nucleus (PVN), 195
Parenteral feeding, 57
Parenteral nutrition (PN), 67
Parkinson's disease (PD), 151, 252–253
PARP. See Poly(ADP-ribose) polymerase (PARP)
Participant adherence, 575
PASS. See Prediction of Activity Spectra for Substances (PASS)
3-PBA. See 3-Phenoxybenzoic acid (3-PBA)
PBMC. See Peripheral blood mononuclear cells (PBMC)
PCA. See Protocatechuic acid (PCA)
PCGTs. See Polycomb group protein gene targets (PCGTs)
PCr. See Phosphocreatine (PCr)
PD. See Parkinson's disease (PD)
pDC. See Plasmacytoid DC (pDC)
PDCD4. See Programmed cell death 4 (PDCD4)
PDE. See Phosphodiesterase (PDE)
PDGF-BB. See Platelet-derived growth factor B chain dimer (PDGF-BB)
PDH. See Pyruvate dehydrogenase (PDH)
PDK-1, 208
PDR. See Proliferative diabetic retinopathy (PDR)
PEM. See Protein–energy malnutrition (PEM)
Peptide YY (PYY), 123, 363–365
Peptides, 302–303
Perifornical area (PFA), 195
Peripheral artery disease (PAD), 423–424
Peripheral blood mononuclear cells (PBMC), 246–247, 557
Peripheral orexigenic or anorexigenic signals, 190
PERM. See Permethrin (PERM)
Permethrin (PERM), 179
Peroxisome proliferator activated receptors (PPARs), 23, 417, 623
 PPAR-a, 23–24
 PPAR-γ, 498, 628–629
Peroxisome proliferator-activated receptor γ coactivator α (PGC-1α), 35, 48–49, 147–148, 498
Peroxynitrite, 161, 610
Personalized diet, 43–44, 49–50
Personalized nutrition, 22f
Pesticides, 178
PETA. See Protein error theory of aging (PETA)

PFA. See Perifornical area (PFA)
PGC-1α. See Peroxisome proliferator-activated receptor γ coactivator α (PGC-1α)
PGE2. See Prostaglandin E2 (PGE2)
PGI$_2$. See Prostacyclin (PGI$_2$)
pH 6.0, 295–296
PH domain. See Pleckstrin homology (PH) domain
PHA. See Phytohemagglutin (PHA)
Pharmacological intervention, 66. See also Nonpharmacological intervention
 artificial nutrition, 66–67
PHDs. See Prolyl hydroxylases (PHDs)
Phenolic compounds, 256
3-Phenoxybenzoic acid (3-PBA), 180
Phosphate, 457–458
 homeostasis attenuates senescence, 454–460
Phosphatidylinositol 3-kinase (PI3K), 143, 210, 267, 397
 PI3K/Akt pathway, 142–143, 208
Phosphatidylinositol 3,4-bisphosphate (PIP2), 143, 208
Phosphatidylinositol 3,4,5-triphosphates kinase. See Phosphatidylinositol 3-kinase (PI3K)
Phosphatidylinositol 3,4,5-trisphosphate (PIP3), 143, 208
Phosphocreatine (PCr), 358
Phosphodiesterase (PDE), 627
Phosphoinositide 3-kinase. See Phosphatidylinositol 3-kinase (PI3K)
Phospholipase (PLA2), 160
Phosphorylated FoxO1 proteins, 211
Physical activity, 684
Phytoalexin resveratrol, 254–255
Phytochemical(s), 252, 659, 708, 712
 carotenoid biosynthesis pathways, 715f
 diet patterns, 714
 flavonoids, 713
 natural pesticides, 714–717
 nonvitamin A carotenoids, 714
Phytoestrogens, 663
Phytohemagglutin (PHA), 384–385
PI3K. See Phosphatidylinositol 3-kinase (PI3K)
PI3K-related protein kinases (PIKK), 141, 397
PIP2. See Phosphatidylinositol 3,4-bisphosphate (PIP2)
PIP3. See Phosphatidylinositol 3,4,5-trisphosphate (PIP3)
PKC. See Protein kinase C (PKC)
PLA2. See Phospholipase (PLA2)
Plant polyphenol function, 650
Plasma ghrelin, 198
Plasmacytoid DC (pDC), 376
Plasminogen activator inhibitor type 1 (PAI-1), 49–50, 641–644
Platelet activating factor (PAF), 427
Platelet-derived growth factor B chain dimer (PDGF-BB), 480, 529–530
Pleckstrin homology (PH) domain, 143

PMNs. See Polymorphonuclear neutrophils (PMNs)
PN. See Parenteral nutrition (PN)
Podagra. See Gout
POlDIP3, 147
Polk, 165
Poly(ADP-ribose) polymerase (PARP), 24
 PARP-1, 244–245, 247
Poly(ADP-ribose), 244–245
Polycomb group protein gene targets (PCGTs), 214–216
Polycomb repressive complex, 214
Polymeric condensed tannins, 285–286, 713
Polymorphisms
 to antioxidant enzymes and redox activity, 616
 in genes encoding vitamin C transporters, 615–616
 SVCT1, 615–616
 SVCT2, 616
Polymorphonuclear neutrophils (PMNs), 376–377, 555
Polyphenol(s), 114, 252, 254, 379, 424, 649
 future directions, 653–654
 intracellular effects, 651–652
 miRNAs modulation by, 285–286
 organ system effects, 652–653
 plant polyphenol function, 650
 and renal function, 429–430
Polyunsaturated fat, 318
Polyunsaturated fatty acid (PUFA), 22, 32–33, 302, 316–318, 358–359, 379, 423, 426–427, 441–442, 622. See also Saturated fatty acid (SFA)
 to health benefits, 282
 in neuronal membranes, 252
 nutrigenetics of omega-3, 22–24
 omega-3 PUFAs, 442
 omega-9 PUFAs benefits, 158–159
 VDR ligands, 453–454
POMC. See Proopiomelanocortin (POMC)
Population reference intake (PRI), 345
Population-based cross-sectional studies, 516
Postdischarge plan, 67
Postprandial hypotension, 366
Potassium (K), 537, 731
 dietary intake
  biological mechanisms in aging process, 540
  and brain aging, 545
  and cognitive aging, 546
  dietary intake and cardiovascular and metabolic effects, 542–543
 pathways linking, 547f
Potassium iodate (KIO$_3$), 585
Potassium iodide (KI), 585
PP. See Pancreatic polypeptide (PP)
PP2A. See Protein phosphatases 2A (PP2A)
PPARs. See Peroxisome proliferator activated receptors (PPARs)
PPI. See Proton-pump inhibitors (PPI)
PPIase FKBP1A, 142
PRAs40. See Proline rich Akt/PKB substrate 40 kDa (PRAs40)
Prebiotics, 698–699

Prediction of Activity Spectra for Substances (PASS), 653
PREDIMED trial. *See* Prevención con Dieta Mediterránea (PREDIMED) trial
Pregnancy, 592
Pregnane X receptor (PXR), 452–453
Prematurely aging mice (PAM), 698–699
Prevención con Dieta Mediterránea (PREDIMED) trial, 325
PRI. *See* Population reference intake (PRI)
Primary endpoint, 568
Primary miRNA (pri-miRNAs) transcripts, 278
Proanthocyanidins. *See* Polymeric condensed tannins
Probiotics, 699
Procarcinogens, 161
Programmed cell death 4 (PDCD4), 147
Proliferative diabetic retinopathy (PDR), 436
Proline rich Akt/PKB substrate 40 kDa (PRAs40), 141–142
Prolyl hydroxylases (PHDs), 525–526
Proopiomelanocortin (POMC), 193–194, 220–221, 398–399
Prostacyclin (PGI$_2$), 611–612
Prostaglandin E2 (PGE2), 254–255, 379, 627–629
Prostaglandin synthase (PTGS), 24
Prostaglandins, 160
Prostate, 161
Prostate cancer, 568–569, 574
  prevention by selenium, 571–572
  prevention by selenium and vitamin E combination, 572–573
  prevention by vitamin E, 572
Prostate-specific antigen (PSA), 568
Prostatic intraepithelial neoplasia, 236
Proteasome, 492
Protein error theory of aging (PETA), 5
Protein kinase C (PKC), 148
Protein phosphatases 2A (PP2A), 147, 210–211, 519
Protein(s), 33, 356–357
  consumption, 352
  energy malnutrition, 122
  ingestion pattern, 357
  intake, 321
  kinase B, 209–210
  miRNAs modulation by, 280
  National Health and Nutrition Examination Survey, 728
  nitrogen-balance studies, 728
  older adults, 726–727
  optimal protein intake, 728
  PROT-AGE study group, 728–729
  protein-restricted diet, 219
  recommendation on, 727
  supplementation, 357
Protein–energy malnutrition (PEM), 41
Proteolytic degradation, 211
Proteostasis imbalance, 497–498
Proto-oncogene, 228
Protocatechuic acid (PCA), 379
Proton-pump inhibitors (PPI), 79
Provitamin A. *See* β-carotene

PSA. *See* Prostate-specific antigen (PSA)
PTGS. *See* Prostaglandin synthase (PTGS)
PTH. *See* Parathyroid hormone (PTH)
Pueraia lobata, 672
PUFA. *See* Polyunsaturated fatty acid (PUFA)
Pulmonary cachexia, 411
Pulmonary diseases, 412
  nutritional assessment for, 414–416
PVN. *See* Paraventricular nucleus (PVN)
PXR. *See* Pregnane X receptor (PXR)
Pyrethroids, 179
  C. cineraraefolum, 179
  environmental exposure, 179–180
  formulations, 179
  higher hazard of pesticides, 180
  3-PBA, 180
  primary target, 179
  VGCC, 179
Pyridoxal 50-phosphate, 518
Pyridoxine, 599
Pyruvate dehydrogenase (PDH), 231
PYY. *See* Peptide YY (PYY)

Q

Qidong, 571
Quality of Life (QOL), 622, 630, 680–681, 687
  omega-3 fatty acids in aging, 630–631
Quercetin, 256, 285, 307

R

RAC. *See* Recruitment and Adherence Committee (RAC)
RAF-MEK-ERK signaling pathway, 305–306
RAGE. *See* AGE receptors (RAGE)
Randomized clinical trial, 568
Randomized controlled trial (RCT), 319, 539
RANKL. *See* Receptor activator of nuclear factor κB ligand (RANKL)
Rapamycin, 142
Rapamycin-insensitive companion of mTOR (RICTOR), 141–142, 397
RAPTOR. *See* Regulatory-associated protein of mTOR (RAPTOR)
Ras/MAPK pathways, 142–143
Rat osteocalcin (ROC), 459–460
Rate of living, 682
  theory, 677
RAVLT. *See* Rey Auditory and Verbal Learning Test (RAVLT)
RBC. *See* Red blood cells (RBC)
RCT. *See* Randomized controlled trial (RCT)
RDA. *See* Recommended daily allowance (RDA); Recommended dietary allowance (RDA); Reference daily allowance (RDA)
RDI. *See* Recommended dietary intake (RDI); Reference daily intake (RDI)
RDN. *See* Registered Dietitian Nutritionist (RDN)
rDNA. *See* Ribosomal gene cluster DNA (rDNA)
Reactive nitrogen species (RNS), 393, 396–397, 474–475

Reactive oxygen species (ROS), 4, 25–26, 33, 109, 156, 244, 294, 316, 333, 393, 473–475, 496, 555, 610, 637–638, 649, 659–660, 680–681, 711
  DNA inducing by, 110
  via incomplete reduction of oxygen, 396–397
  inflammatory response producing, 252
  in intracellular signalling pathways modulation, 243–244
  neutrophils, 376–377
  putative toxins, 650
  superoxide, 611–612
Reactive species (RS), 109, 208
Receptor activator of nuclear factor κB ligand (RANKL), 465
Receptor-independent mechanism, 267
Recommended daily allowance (RDA), 438, 553, 613–614
Recommended dietary allowance (RDA), 42, 46–47, 357, 569, 638–639, 727
Recommended dietary intake (RDI), 723, 726–727
Recommended Nutritional Intake (RNI), 46–47
Recruitment and Adherence Committee (RAC), 574
Recycling, 587
Red blood cells (RBC), 629
Red wine, 323–324
REDD1. *See* Regulated in development and DNA damage response 1 (REDD1)
Reduction and oxidation (Redox), 208
  polymorphisms to activity, 616
  system, 184
Reference daily allowance (RDA), 345–346
Reference daily intake (RDI), 345
Registered Dietitian Nutritionist (RDN), 322, 326
Regulated in development and DNA damage response 1 (REDD1), 145
Regulatory T cells (Tregs cells), 557, 693–694
  intestinal microbiota and induction, 696f
Regulatory-associated protein of mTOR (RAPTOR), 141–142, 397
Renal disease, 270
Renal function, 515
  in aging, 425–426
  polyphenols and, 429–430
Repair, 4, 6
Replicative senescence, 294
Reproductive organ cancers, 161
Repulsive Guidance Molecule (RGM), 524
Resting metabolic rate (RMR), 680–681
  residual, 680–681
Resveratrol, 114, 227–228, 234–235, 246–247, 254–255, 285, 303, 463–464
Retinal pigment epithelium (RPE), 434–435
Retinoic acid, 283–284
Retinoid X receptors (RXR), 452–453
Reverse T3 (rT3), 587
Revised Hasegawa Dementia Scale (HDS-R), 670–671

Rey Auditory and Verbal Learning Test (RAVLT), 687
RGM. *See* Repulsive Guidance Molecule (RGM)
RHD. *See* Rheumatic heart disease (RHD)
Rheumatic heart disease (RHD), 423–424
Rheumatoid Arthritis, 270
Riboflavin, 599
Ribosomal gene cluster DNA (rDNA), 227, 232
Ribosomal induced silencing complex (RISC), 278
Ribosomal S6 kinases (RSK), 143
RICTOR. *See* Rapamycin-insensitive companion of mTOR (RICTOR)
RISC. *See* Ribosomal induced silencing complex (RISC)
RMR. *See* Resting metabolic rate (RMR)
RNI. *See* Recommended Nutritional Intake (RNI)
RNS. *See* Reactive nitrogen species (RNS)
ROC. *See* Rat osteocalcin (ROC)
Rodent studies, 515–516
Root of Kudzu. *See* Pueraia lobata
ROS. *See* Reactive oxygen species (ROS)
Rotterdam study, 337
Roux-en-Y Gastric Bypass (RYGB) Surgery, 197–198
RPE. *See* Retinal pigment epithelium (RPE)
RRag GTPase, 146
RS. *See* Reactive species (RS)
RSK. *See* Ribosomal S6 kinases (RSK)
rT3. *See* Reverse T3 (rT3)
RXR. *See* Retinoid X receptors (RXR)
RYGB Surgery. *See* Roux-en-Y Gastric Bypass (RYGB) Surgery

S

S-Adenoslymethionine, 513
S-Adenosylhomocysteine, 515
S-Adenosylmethionine (SAM), 218, 244, 515, 519–520
  and inflammation, 518
S6 kinase 1 (S6K1), 143, 147
SAC. *See* School Aged Children (SAC)
sActRIIB. *See* Soluble receptor antagonist of myostatin (sActRIIB)
SADS. *See* Senescence-associated distension of satellites (SADS)
SAE. *See* Severe adverse effects (SAE)
Saffron, 667–668
SAHA. *See* Suberoylanilide hydroxamic acid (SAHA)
SAHFs. *See* Senescence-associated heterochromatin foci (SAHFs)
SAM. *See* S-adenoslymethionine (SAM)
SAMP8 mice. *See* Senescence-accelerated mouse prone 8 (SAMP8 mice)
SAPK kinase interacting protein 1 (Sin1), 142
Sarcopenia, 151–152, 270, 279–280, 355–356, 400, 411
  nutritional interventions, 356–357
SASP. *See* Senescence associated secretory phenotype (SASP)

Saturated fatty acid (SFA), 33, 380–382. *See also* Polyunsaturated fatty acid (PUFA)
  intake, 729
SCAP. *See* Sterol cleavage activating protein (SCAP)
SCE. *See* Sister chromatid exchange (SCE)
SCEs. *See* Spontaneous cerebral emboli (SCEs)
SCFAs. *See* Short-chain fatty acids (SCFAs)
Schiff base, 264
School Aged Children (SAC), 589t
SCIRS. *See* Severe Cognitive Impairment Rating Scale (SCIRS)
SCN. *See* Suprachiasmatic nucleus (SCN)
SD-0006. *See* CDD-111
sdLDL. *See* Small, dense LDL particles (sdLDL)
Se-Cys. *See* Selenocysteine (Se-Cys)
Se-Met. *See* Selenomethionine (Se-Met)
Seafood, 323
Seaweeds. *See* Marine algae
Secoiridoids, 304
Secondary metabolites, 712
Secretory immunoglobulin A (sIgA), 696–697
SELECT. *See* Selenium and Vitamin E Cancer Prevention Trial (SELECT)
SELECT Centralized Follow-up Study (SELECT CFU), 580
Selenate (SeO_4^{2-}), 569
Selenite (SeO_3^{2-}), 569
Selenium (Se), 26, 112, 219, 304–305, 568. *See also* Zinc (Zn)
  baseline selenium status, 578
  dose, 578
  formulation, 576–578
  metabolism and distribution, 570f
  nutritional requirements, 569
  prostate cancer
    prevention by, 571–572
    prevention by selenium and vitamin E combination, 572–573
  sources, metabolism, and biological activities, 569
Selenium and Vitamin E Cancer Prevention Trial (SELECT), 568–569
  accrual, 575
  adherence, 575
  agents, 574
  cohort, 574
  implications and future directions, 580–581
  primary endpoint results, 575–576, 576t
  prostate cancer
    prevention by selenium, 571–572
    prevention by selenium and vitamin E combination, 572–573
    prevention by vitamin E, 572
  rationale and objectives, 573–574
  recruitment strategies, 574–575
  for results, 576
    age, 580
    agents, 576–579
    genetics, 579–580

  effect of other nutrients, 579
  study design, 580
  secondary endpoint results, 575–576, 577t
  trial design and outcome ascertainment, 574
  in vivo animal studies, 573t
Seleno-L-methionine (SeMet), 246–247
Selenocysteine (Se-Cys), 26
Selenomethionine (Se-Met), 26
Selenoproteins P1 (SEPS1), 49–50
SeMet. *See* Seleno-L-methionine (SeMet)
Semitendinosus (ST), 34–35
SENECA Study. *See* Survey in Europe on Nutrition and Elderly Concerted Action (SENECA) Study
Senescence, 227
  delayers, 300
  immunosurveillance, 297
  inducers, 299–300
Senescence associated secretory phenotype (SASP), 131, 295–296, 498–499
  modulators, 299
Senescence-accelerated mouse prone 8 (SAMP8 mice), 501–502
Senescence-associated distension of satellites (SADS), 295–296
Senescence-associated heterochromatin foci (SAHFs), 295–296
Senescent cells
  direct ablators of, 298
  indirect ablators of, 298–299
  strategies to rejuvenate, 297–298
Senile anorexia, 121–122
Sensory property of foods, 121
SEPS1. *See* Selenoproteins P1 (SEPS1)
2-Series prostaglandins (PGE2). *See* Prostaglandin E2 (PGE2)
Serine, 516–517
Serine hydroxymethyltransferase (SHMT), 514
SERMs. *See* Specific estrogen receptor modulators (SERMs)
Serum-and glucocorticoid-regulated kinase (sGK), 148
Severe adverse effects (SAE), 630–631
Severe Cognitive Impairment Rating Scale (SCIRS), 667–668
SF-36 Health Survey. *See* Short-Form 36 (SF-36) Health Survey
SFA. *See* Saturated fatty acid (SFA)
SFN. *See* Sulforaphane (SFN)
sGK. *See* Serum-and glucocorticoid-regulated kinase (sGK)
SGLT1. *See* Sodium-glucose cotransporter 1 (SGLT1)
SHMT. *See* Serine hydroxymethyltransferase (SHMT)
Short-chain fatty acids (SCFAs), 368, 693
Short-Form 36 (SF-36) Health Survey, 687
sIgA. *See* Secretory immunoglobulin A (sIgA)
Signaling feeding state
  calorie restriction, 198
  chronic wasting diseases, 198
  in eating disorders, anorexia nervosa, 198

Signaling nutritional state
    eating disorders, calorie restriction, chronic diseases, 198–199
    in obesity, 197
Silent Information Regulators (SIR), 228
    SIR2, 222, 228
    SIRT1-dependent deacetylation, 211
    SIRT1, 33, 209, 222, 230
    SIRT2, 230–231
    SIRT3, 222–223, 231
    SIRT4, 231
    SIRT5, 231
    SIRT6, 48–49, 231–232
    SIRT7, 232
Sin1. See SAPK kinase interacting protein 1 (Sin1)
Single Nucleodite Polymorphisms (SNPs), 42–43, 49–50, 610
Single nucleotide polymorphism, 437
Single nucleotide polymorphisms, 561–562
Singlet oxygen ($^1O_2$), 475
SIR. See Silent Information Regulators (SIR)
Sirolimus, 142
Sirtuin(s), 48, 214, 222–223, 228, 393, 651
    enzymatic activities, 228
    functions, 229t
    longevity effects of calorie restriction, 234
    mammalian sirtuins, 230–232
    in nonmammalian model organisms, 228–230
    overexpression, 232–233
    proto-oncogenes or tumor suppressors, 235–236
    resveratrol sirtuin activator, 234–235
    as therapeutic targets, 236–237
Sister chromatid exchange (SCE), 244–245
SJW system. See St. John's Wort (SJW) system
Skeletal muscle, 358, 396
Skin aging, carotenoids and, 478
    carotenoid rich foods, 480
    dietary antioxidants, 478–479
    mechanism of action, 479
    photoprotective effect of lycopene, 479
    skin cancer prevention, 480
    UV exposure, 479–480
Skin autofluorescence measurement, 266
SKT. See Syndrom-Kurz test (SKT)
SLC. See Solute-linked carrier (SLC)
Small, dense LDL particles (sdLDL), 316–320
Small bowel mucosa, 366–368
Small intestinal permeability, 366–368, 368f
Small noncoding RNAs, 279
    age-related alterations, 217
    effects of diet on, 219–220
Smell, 364
Smoking, 435
SNPs. See Single Nucleodite Polymorphisms (SNPs)
SOD. See Superoxide dismutase (SOD)
Sodium tanshinone IIA sulfonate (STS), 671–672
Sodium-dependent vitamin C transporters 1 (SVCT1), 615

Sodium-dependent vitamin C transporters 2 (SVCT2), 615
Sodium-glucose cotransporter 1 (SGLT1), 367
Sodium–iodide symporter (NIS), 586
    protein, 587
Soluble RAGE (sRAGE), 268
Soluble receptor antagonist of myostatin (sActRIIB), 414
Solute-linked carrier (SLC), 552–553
Somatotropic axis, 677, 686
Southwest Oncology Group (SWOG), 568, 573–574
Soy, 663–666
    on cognition in adults, 664t
    health benefits, 663t
SPA. See Spontaneous physical activity (SPA)
Spearman rank correlation analysis, 16
Specific estrogen receptor modulators (SERMs), 484
Sperm epimutations, 183
Spontaneous cerebral emboli (SCEs), 336–337
Spontaneous physical activity (SPA), 684
SPs. See Sulfated polysaccharides (SPs)
sRAGE. See Soluble RAGE (sRAGE)
SRE. See Sterol regulatory element (SRE)
SREBP1. See Sterol regulatory element-binding protein 1 (SREBP1)
ST. See Semitendinosus (ST)
St. John's Wort (SJW) system, 80–81
Stem cell
    exhaustion, 222, 499
    functional consequences of aging, 222
Steroid hormones, 482–484
Sterol cleavage activating protein (SCAP), 147–148
Sterol regulatory element (SRE), 147–148
Sterol regulatory element-binding protein 1 (SREBP1), 147–148
Stilbenoids, 303
Streptozotocin (STZ), 333
Stress, 4–6
    MAPKs, 258
    neuroprotective phytochemicals and modulation of signaling pathway, 256
    NF-κB, 257–258
    Nrf2, 256–257
    response pathway, 7
Stroke, 542–545, 543t
STS. See Sodium tanshinone IIA sulfonate (STS)
STZ. See Streptozotocin (STZ)
Suberoylanilide hydroxamic acid (SAHA), 221
Subjective feelings of hunger, 687
Sudan Black B, 295–296
Sugar(s), 303, 332
Sulfated polysaccharides (SPs), 669
Sulforaphane (SFN), 256, 303
Superoxide dismutase (SOD), 110–112, 164, 243–245
    activity, 500
    antioxidant defense enzymes, 254
    antioxidant systems, 659–660
    Cu/Zn, 245, 247

Suprachiasmatic nucleus (SCN), 195
Survey in Europe on Nutrition and Elderly Concerted Action (SENECA) Study, 400
SVCT1. See Sodium-dependent vitamin C transporters 1 (SVCT1)
Swallowing, 365
SWOG. See Southwest Oncology Group (SWOG)
Sympathetic system, 35–36
Synbiotics, 700
Syndrom-Kurz test (SKT), 666
Systematic review, 599–601
Systemic iron homeostasis regulation, 527, 528f
    hepcidin regulation, 527
        by erythropoiesis, 529–530
        by Iron, 529

T

T cell receptors (TCRs), 378, 556
T regulatory cells. See Regulatory T cells (Tregs cells)
T2D. See Type II diabetes (T2D)
T3. See Triiodothyroinine (T3)
TA-65. See Cycloastragenol
Tankyrase, 130
Target of rapamycin (TOR), 47, 141–142
Taste, 364
TC. See Total cholesterol (TC)
TC/HDL-C. See Total cholesterol to HDL-C (TC/HDL-C)
TCDD. See 2,3,7,8-Tetrachlorodibenzo-p-dioxin (TCDD)
TCN2. See Transcobalamine II (TCN2)
TCRs. See T cell receptors (TCRs)
TDEE. See Total daily energy expenditure (TDEE)
Tea, 661–663
    health benefits, 662t
    observational studies, 662t
TEF. See Thermic effect of food (TEF)
Telomerase, 497
Telomere(s), 129–130, 130f, 227, 519
    and aging
        factors, 132
        telomere length predict healthy aging, 131
        telomeres change with age, 131
    attrition, 294, 496–497
    damaged, 131
    erosion, 294
    knowledge gaps, 137
    length, 322
    maintenance mechanisms, 130
    nutrition and
        cross-sectional studies, 132t, 134t
        dietary factors, 132–136
        intervention studies, 135t
        plausible mechanisms, 136, 136f
        telomere length, 136
    too long, 131
    too short, 131
2,3,7,8-Tetrachlorodibenzo-p-dioxin (TCDD), 302–303

1,2,3,6-Tetrahydro-1-methyl-4-phenylpyridine hydrochloride (MPTP), 151
Tetrahydrobiopterin (BH4), 611–612
Δ(9)-Tetrahydrocannabinol (THC), 307
Tetrahydrocurcumin (THC), 492, 499
TF. See Theaflavins (TF); Transferrin (TF)
TF receptor 2 (TFR2), 524
TFEB. See Transcription factor EB (TFEB)
TFR1. See Transferrin receptor 1 (TFR1)
TFR2. See TF receptor 2 (TFR2)
Tg. See Thyroglobulin (Tg)
TG. See Triglyceride (TG)
TGF. See Transforming growth factor (TGF)
Th1/Th2 balance, 557
Th17 cells, 557
   intestinal microbiota and induction, 696f
THC. See Tetrahydrocurcumin (THC); Δ(9)-Tetrahydrocannabinol (THC)
Theaflavins (TF), 661
Thearubigins (TR), 661
Therapeutic Lifestyle Changes (TLC), 324
Thermic effect of food (TEF), 684, 725
Thiamin, 599
Thioredoxin (Trx), 208
Third National Health and Nutrition Examination Survey (NHANES-III), 428–429, 614
Threonine residue, 143
THs. See Thyroid hormones (THs)
Thymic atrophy, 375
Thymidylate synthase (TS), 514
Thyroglobulin (Tg), 590
Thyroid
   autoimmunity, 595–596
   cancer, 596
   enlargement, 587
   function, 684–685
   gland, 590–591
Thyroid hormones (THs), 585
   receptors, 592
Thyroid-stimulating hormone (TSH), 586
Thyroiditis, 583
Thyroperoxidase (TPO), 586–587
Thyrotropin releasing hormone (TRH), 195
Thyroxine (T4), 586–587
Tissue-specific biological phenotypes, 16
TLC. See Therapeutic Lifestyle Changes (TLC)
TLR. See Toll-like receptor (TLR)
TLS. See Translesion DNA synthesis (TLS)
TMPRSS6 proteins, 529–530
TNF. See Tumor necrosis factor (TNF)
Tocopherols, 113–114, 302, 569
Tocotrienols, 302
Tolerable upper intake levels (Tolerable UL), 42
Toll-like receptor (TLR), 376, 696–697
TOR. See Target of rapamycin (TOR)
TOR signaling (TOS), 141–142, 144f
Total cholesterol (TC), 319–320
Total cholesterol to HDL-C (TC/HDL-C), 316–318
Total daily energy expenditure (TDEE), 684
Toxic agents, 425–426

TPO. See Thyroperoxidase (TPO)
TR. See Thearubigins (TR)
Trace elements, 110–112, 245–246, 284–285, 304–305
Trans fatty acid intake, 729
Trans-fatty acid isomers, 33–34
Transcobalamine II (TCN2), 517
Transcription, 214
Transcription factor EB (TFEB), 146
Transferrin (TF), 524
Transferrin receptor 1 (TFR1), 527
Transforming growth factor (TGF), 652
Transient receptor potential vanilloid type 6 (TRPV6), 465
Translesion DNA synthesis (TLS), 165
Transsulfuration pathway, 513
Tregs cells. See Regulatory T cells (Tregs cells)
TRH. See Thyrotropin releasing hormone (TRH)
Triglyceride (TG), 316
3,5,4′-Trihydroxystilbene. See Phytoalexin resveratrol
Triiodothyroinine (T3), 586–587
Triterpenoid saponins, 302
TRPV6. See Transient receptor potential vanilloid type 6 (TRPV6)
Trx. See Thioredoxin (Trx)
TS. See Thymidylate synthase (TS)
TSC1. See Tuberous sclerosis complex 1 (TSC1)
TSH. See Thyroid-stimulating hormone (TSH)
Tuberous sclerosis complex 1 (TSC1), 143
Tuberous sclerosis complex 2 (TSC2), 143
Tumor necrosis factor (TNF), 553–554, 700
   TNF-α, 23, 58, 123–124, 252, 267, 475, 497, 612
Tumor suppressor, 228
Turmeric, 494
Twisted Gastrulation BMP Signaling Modulator 1 (TWSG1), 529–530
2013 AHA/ACC Lifestyle Guidelines, 324
Two-dimensional gel electrophoresis, 5
TWSG1. See Twisted Gastrulation BMP Signaling Modulator 1 (TWSG1)
Type II diabetes (T2D), 332, 334–336, 538
   T2DM, 322, 394
Type II muscle fibers, 358
Type IIA interaction, 75–77
Type IIB interaction, 75–77
Type IIC interaction, 75–77

U

UA. See Uric acid (UA)
Ubiquionone. See Coenzyme Q_{10}
Ubiquitin-proteasome system (UPS), 357
Ubiquitination, 211
UCP2. See Uncoupling protein 2 (UCP2)
UI level. See Upper intake (UI) level
UIC. See Urinary iodine concentration (UIC)
UK Committee on Medical Aspects of Food Policy (COMA), 729
UL. See Upper limit (UL)

Uncoupling protein 2 (UCP2), 501–502
Unexplained anemia, 533
United Nations University (UNU), 727
United States (US), 568
United States Department of Agriculture (USDA), 322
United States Preventive Services Task Force (USPSTF), 580
Unraveling stochastic aging processes in mouse liver, 12
   correlation matrix of chronological age and pathology parameters, 14t
   future perspectives, 18–19
   gene expression profiles related to pathological aging parameters, 16–18
   heat maps of genes, 17f
   intraorgan specific biological phenotypes, 15–16
   pathological parameters, 13–15
   pathology ranking
      for lipofuscin accumulation in liver and brain, 16f
      for multiple endpoints in liver, 15f
   ranking of mice, 14f
   set of six genes, 13
   tissue-specific biological phenotypes, 16
   Venn diagram of genes, 18f
Unsaturated fatty acids, 166. See also Polyunsaturated fatty acid (PUFA); Saturated fatty acid (SFA)
Unsaturation, 31
Untranslated region (UTR), 525–526
   3′ UTR, 277–278
   5′ UTR, 147
UNU. See United Nations University (UNU)
Upper gastrointestinal sensory function, 364
Upper intake (UI) level, 345, 348, 558, 723
Upper limit (UL), 569
UPS. See Ubiquitin-proteasome system (UPS)
Urates, 424
Uric acid (UA), 424–425
Urinary iodine concentration (UIC), 588–590
US Food and Nutrition Board, 729
USDA. See United States Department of Agriculture (USDA)
USPSTF. See United States Preventive Services Task Force (USPSTF)
UTR. See Untranslated region (UTR)

V

Vacuolar H^+-adenoside triphosphate ATPase (v-ATPase), 143–145
VaD. See Vascular dementia (VaD)
VAS. See Visual Analogue Scales (VAS)
Vascular
   aging, 316
   carotenoids and health, 475
      effect of carotenoid consumption, 477–478
      early atherosclerosis, 476–477
      improving atherosclerotic risk factors, 476–477
Vascular cell adhesion molecule (VSAM), 612